MW00358727

SURVEY OF APPLICABLE MATHEMATICS

Prof. RNDr. KAREL REKTORYS, Dr.Sc.

Contributory chapters by

Asst Prof. KAREL DRÁBEK,
Prof. RNDr. MIROSLAV FIEDLER, Dr.Sc.,
RNDr. OTTO FISCHER,
JAROSLAV FUKA, C.Sc.,
Prof. Dipl. Eng. Dr. JAROSLAV HÁJEK, Dr.Sc.,
Asst Prof. FRANTIŠEK KEJLA,
Asst Prof. BOŘIVOJ KEPR,
Dipl. Eng. KVĚTA KORVASOVÁ,
Prof. JINDŘICH NEČAS, Dr.Sc.,
Prof. RNDr. FRANTIŠEK NOŽIČKA,
Prof. Dr. VÁCLAV PLESKOT,
Asst Prof. OLGA POKORNÁ, C.Sc.,
RNDr. MILAN PRAGER, C.Sc.,
RNDr. OTTO VEJVODA, C.Sc.,
Asst Prof. VÁCLAV VILHELM, C.Sc.,
RNDr. EMIL VITÁSEK, C.Sc.,
RNDr. MILOSLAV ZELENKA, C.Sc.

SURVEY OF APPLICABLE MATHEMATICS

KAREL REKTORYS, Editor

English translation edited by
Staff of the Department of Mathematics,
University of Surrey

THE M.I.T. PRESS
Massachusetts Institute of Technology
Cambridge, Massachusetts

Translated by Dr. Rudolf Výborný *et al.* 1968

English edition first published in 1969 by Iliffe Books Ltd,
Dorset House, Stamford Street,
London, S. E. 1. in co-edition with SNTL - Publishers of Technical Literature, Prague

Published in the U.S.A. by the M.I.T. Press,
Massachusetts Institute of Technology,
Cambridge, Massachusetts

Library of Congress Catalog Card Number 68-25586

Printed in Czechoslovakia

2. Trigonometric and Inverse Trigonometric Functions. Hyperbolic and Inverse Hyperbolic Functions

By Václav Vilhelm

3. Some Formulae (Areas, Circumferences, Volumes, Surfaces, Centroids, Moments of Inertia)

By Václav Vilhelm

CONTENTS

4. Plane Curves and Constructions

By Karel Drábek

5. Plane Analytic Geometry

By Miloslav Zelenka

9. Differential Geometry

By Bořivoj Kepr

10. Sequences and Series of Constant Terms. Infinite Products

By Karel Rektorys

11. Differential Calculus of Functions of a Real Variable

By Karel Rektorys

12. Functions of Two or More Variables

By Karel Rektorys

13. Integral Calculus of Functions of One Variable

By Karel Rektorys

14. Integral Calculus of Functions of Two or More Variables

By Karel Rektorys

15. Sequences and Series with Variable Terms (Sequences and Series of Functions)

By Karel Rektorys

16. Orthogonal Systems. Fourier Series. Some Special Functions (Bessel Functions, etc.)

By Karel Rektorys

17. Ordinary Differential Equations

By Karel Rektorys

18. Partial Differential Equations

By Karel Rektorys

19. Integral Equations

By Karel Rektorys

20. Functions of a Complex Variable

By Karel Rektorys

21. Conformal Mapping

By Jaroslav Fuka

22. Some Fundamental Concepts from the Theory of Sets and Functional Analysis

By Karel Rektorys

23. Calculus of Variations

By František Nožička

A. *Problems of the First Category (Elementary Problems of the Calculus of Variations)*

B. *Problems of the Second Category*
(Extrema of Functionals of the Form $\int_a^b F(x_1, y_1, \ldots, y_n, y_1', \ldots, y_n')\,dx$)

C. *Problems of the Third Category*
(Extrema of Functionals of the Type $\int_a^b F(x, y, y', \ldots, y^{(n)})\,dx$)

D. *Problems of the Fourth Category (Functionals Depending on a Function of n Variables)*

E. *Problems of the Fifth Category (Variational Problems with "Movable (Free) Ends of Admissible Curves")*

F. *Problems of the Sixth Category (The Isoperimetric Problem in the Simplest Case)*

24. Variational (Direct) Methods for Solving Boundary Value Problems of Differential Equations

By Milan Prager

25. Approximate Solution of Ordinary Differential Equations

By Otto Vejvoda and Karel Rektorys

A. *Initial-Value Problems*

B. *Boundary-Value Problems*

C. *Eigenvalue Problems*

D. *Periodic Solutions*

26. Solution of Partial Differential Equations by Infinite Series (by the Fourier Method)

By Karel Rektorys

27. Solution of Partial Differential Equations by the Finite-Difference Method

By Emil Vitásek

28. Integral Transforms (Operational Calculus)

By Jindřich Nečas

29. Approximate Solution of Fredholm Integral Equations

By Karel Rektorys

30. Numerical Methods in Linear Algebra

A. *Solution of Systems of Linear Algebraic Equations*

By Olga Pokorná

B. *Numerical Calculation of Eigenvalues of a Matrix*

By Květa Korvasová

31. Numerical Solution of Algebraic and Transcendental Equations

By Miroslav Fiedler

32. Nomography and Graphical Analysis. Interpolation. Differences

By Václav Pleskot

A. *Nomography and Graphical Analysis*

B. *Interpolation. Differences*

33. Probability Theory

By JAROSLAV HÁJEK

34. Mathematical Statistics

By JAROSLAV HÁJEK

35. Method of Least Squares. Fitting Curves to Empirical Data. Elements of the Calculus of Observations

By OTTO FISCHER

A. *Fitting Curves to Empirical Data. Regression*

B. *General Problems of the Least-Squares Method*

C. *Elements of the Calculus of Observations*

FOREWORD

By Professor F. M. Arscott, M.Sc., Ph.D., F. I. M. A.,
Department of Mathematics, University of Surrey

A mathematician, pausing two-thirds of the way through the twentieth century to look back, might feel a justifiable pride in the progress of his subject. Many old problems have been solved and others absorbed into wider questions, while new branches of the subject have appeared at frequent intervals and blossomed rapidly. Meanwhile, other scientific disciplines grow more mathematical; it is said that really good work in physics, chemistry or engineering requires a first degree in mathematics, and even disciplines which were never regarded as scientific are proving susceptible to mathematical analysis.

This coincidence of an explosion of mathematical activity with greatly enlarged scope for its application is, unhappily, overshadowed by a communication barrier. Between those who have mathematical knowledge and those who wish to use it there lies a great gulf. One can try to bridge this by bringing to the notice of abstract mathematicians the intriguing and challenging problems waiting for them in other fields – but mathematicians are not easily tempted from their ivory towers. This book starts, instead, from the other side, putting into the hands of the users of mathematics an array of powerful tools, of whose existence they may be unaware, with precise directions for their use. To achieve this in a reasonable compass something had to be sacrificed and the authors took the bold step of omitting virtually all proofs – an unorthodox but highly sensible procedure, since otherwise the book might have been ten times its present size. As it is, these covers contain the equivalent of a small library of standard texts on the uses of mathematics.

It is no coincidence that such a book should come from the Continent, for it is especially in Germany and eastern Europe that there flourishes the subject of "Angewandte Mathematik" – better described as "useful", "utilisable" or "applicable" mathematics rather than by the literal translation of "applied mathematics", which in Britain means something very different. For the English translation, therefore, the title "Survey of Applicable Mathematics" has been chosen.

The task of editing the translation has been interesting and congenial. We have sought to produce a text in good mathematical English while preserving all the distinctive features of the original. Notation has been left practically unchanged; only where Czech and English usages differ significantly have changes been made.

Terminology has sometimes proved more difficult, such as when direct translation produced a term which, though clear and acceptable, was not generally used. In such cases we have usually retained the equivalent of the Czech original, with a note giving the common English alternatives; thus matrices are described as "regular" rather than "non-singular", though the latter is given as an equivalent term. When, however, serious confusion might result, or a different English term has become completely standard, we have made the necessary changes.

An extensive revision of the bibliography has also been made, giving a fuller guide to current British and American literature. Translations of Russian literature have been referenced whenever they could be traced, names of Russian authors being transliterated according to the practice of the London Mathematical Society.

My colleagues and I have found the editing of this book an exciting and stimulating experience; throughout we have had the inestimable benefit of Professor Rektorys's advice and help and in commending this book to the English-speaking mathematical world we would pay our own tribute to the scholarship and imagination of Professor Rektorys and his co-authors.

PREFACE TO THE ENGLISH EDITION

The task of translating this book into English has been pleasant but rather difficult. I must express appreciation to my colleagues Vl. Dlab, K. Komínek and R. Výborný, who translated the greater part of the text, and also acknowledge the generous assistance rendered by A. Žaludová, who revised the whole translation.

In the preparation and editing of this English translation I have received invaluable help from Professor F. M. Arscott and his colleagues at the University of Surrey. Without their co-operation it would be difficult to imagine a successful production of this English edition.

To all these individuals, and also to Iliffe Books Ltd, I want to express once more my sincere thanks.

Prague *K. R.*

PREFACE TO THE CZECH EDITION

In recent years several books dealing with special fields of mathematics (for example, Angot's *Applied Mathematics for Electronic Engineering,* and others) have been published in Czechoslovakia. They have supplied readers with information, in a condensed form, about those mathematical disciplines which find employment in these particular fields.

This volume has been published as a result of the initiative of the Česká matice technická (Czech Scientific Institution for Propagation of Technical Literature). In particular, the late Professor Vyčichlo devoted much of his time and organisational powers to make clear questions concerning fundamental features and conception of this volume. The authors have attempted to produce a comprehensive work for the use of a very wide circle of readers, and the book comprises the great majority of mathematical disciplines applied in technology, yet the contributions have been prepared in such a way that a reader with only limited theoretical knowledge of mathematics can easily follow them. This volume contains, therefore, a survey of results in applicable mathematics needed by engineering graduates or other research workers, or by undergraduates and teachers of technological subjects. It is also intended to be of service to theoretical research workers in such related disciplines as physics, geodesy, etc., and to mathematicians themselves.

It was not easy to select the subject matter and to present it in a form acceptable to such a varied body of readers. During Professor Vyčichlo's lifetime an extensive survey was made in order to ascertain the views of a number of outstanding technologists; some of the opinions expressed regarding selection and presentation of subject matter showed extensive disagreement, but it was possible to formulate an outline plan for the selection of subject matter and its mode of presentation — even though some questions remained unanswered.

It was not possible to include any specialised disciplines used only in narrow fields of technology. Thus, electrical engineers may miss the theory of transmission, while readers particularly interested in solving systems of linear algebraic equations may regret the absence of reference to cracovians. On the other hand it was clearly necessary not only to include current mathematical topics but also to pay considerable attention to approximate methods. Prominent among the latter are approximate methods in algebra, including solution of systems of linear equations, of transcendental equations and algebraic equations of higher degree, and the determination of eigenvalues of matrices, while in the field of analysis we have included approximate methods for the solution of differential equations (especially partial differential equations) and of integral equations; these are not yet adequately treated in technological literature. The book also includes comprehensive tables of integrals, of sums of series, and of solved differential equations, while in the chapter on statistics (Chapter 34) attention is devoted to the subject of quality control. The book does not, however, include the theory of computers or the technique of linear programming; these disciplines are developing so rapidly at the present time that any description would be out of date before it appeared in print.

The leading experts in our country were invited to write the contributions on individual subjects. While each author was allowed a certain degree of freedom, the editor-in-chief (of the Czech edition) has ensured the maintenance of a consistent style of treatment throughout the book. I should like to thank all the authors for their great patience and for incorporating my suggestions into their work.

The book omits proofs of the theorems and derivation of the results, but theorems and formulae are complemented with explanatory remarks and appropriate examples; in choosing these examples we have sought to include those which not only provide suitable illustration but also have practical importance. In stating the results we have borne in mind the varying standards of mathematical education and skill of the readers for whom the book is intended. In algebra, for instance, we write: "If a, b are real or complex numbers, then ...", instead of: "If a, b are complex numbers, then ..."; a mathematician may legitimately object that the second statement is sufficient because real numbers are a special category of complex numbers, but by using the first form of statement we leave the mathematically less advanced reader in no doubt that the result is valid for real numbers as well as complex.

Some sections of this volume are not, and by their nature cannot be, truly original — for instance, the tables of integrals and of solved differential equations. These tables were abstracted from different books, mainly from the Czechoslovak edition of [48] and from [211], and have been carefully checked.

Although the authors and annotators of the various chapters worked with extreme care, the possibility cannot be excluded that some errors remain undetected, and we shall be very grateful to any readers who inform us of such errors. The authors of individual chapters are responsible for their accuracy, while the editor-in-chief takes overall responsibility for the general outline of the book; he will be grateful for any criticism relating to the selection of the subject matter and its presentation.

The book is divided into chapters and sections, which are numbered according to the decimal system, so that 5.3 for example, means Chapter 5, section 3. In each section the theorems, examples, etc., are numbered in order and are quoted by means of that number; if, for instance, in a certain section Example 1 is quoted, this refers to Example 1 of the current section. If, however, we refer to an example from another section, then the number of that section is given before the number of the example. Similarly a reference such as (5.3.2) relates to equation (2) of section 5.3; generally the page is also quoted for the reader's convenience.

The bibliography is to be found at the end of the book; in the text we refer to a work merely by quoting (in square brackets) its number in the list of references.

Grateful acknowledgement is due to the Česká matice technická and the Publishers of Technological Literature (SNTL) in Prague. I am indebted also to many friends and colleagues, particularly to Professor V. Dašek, who has read the greater part of the manuscript, to I. Babuška for his great work in the organisation of the project and for his many valuable suggestions, and to K. Drábek for preparing the diagrams. Thanks are also due to E. Jokl, M. Josífko, M. Pišl, Č. Vitner, J. Výborná and R. Výborný for their most careful revision of the manuscript and their many contributions to the improvement of the whole work. I have also to thank the Prometheus printing house for their extremely competent work.

K.R.

Prague

LIST OF SYMBOLS AND NOTATION

Symbols and notation are arranged according to their logical connections with various parts of mathematics.

The reader should note that it is often difficult to put a symbol or notation precisely in its appropriate place; it may happen therefore that he will have to look up a notation in a different place from that which he anticipated.

Symbol or Notation	Meaning
	Algebra
$=$	(is) equal to
$\equiv$	(is) identically equal to
$\neq$	(is) not equal to
$\not\equiv$	(is) not identically equal to
$\approx$	is approximately equal to
$\doteq$	is equal, after rounding off, to
$<$	(is) smaller than, is less than
$>$	is greater than
$\leqq$	is less than or equal to ...
$\geqq$	is greater than or equal to ...
$+$	plus; positive sign
$-$	minus; negative sign
$.\,,\ \times$	multiplied by; this sign is often omitted, e.g. instead of $a\,.\,b$ we often write ab
$:\,;\ -\,;\ /$	divided by; over; in the text we often write, for example, $1/(2n+1)$ instead of $\dfrac{1}{2n+1}$; obviously, $1/2n+1$ stands for $\dfrac{1}{2n}+1$

Symbol or Notation	Meaning
( ), [ ], { }	parentheses or round brackets, square brackets, curly brackets respectively
$\sqrt[n]{a}$	the n-th root of a (the sign $\sqrt{\ }$ is called the radicle or the radical or the radical sign), instead of $\sqrt[2]{a}$ we write simply $\sqrt{a}$
$\|a\|$	the absolute value of the number a
a^n	$\underbrace{a \,.\, a \,.\, \ldots \,.\, a}_{n\text{-times}}$; the n-th power
a^{-n}	$\dfrac{1}{a^n}$
$n!$	$1 \,.\, 2 \,.\, \ldots \,.\, n$ (n-factorial or factorial n); e.g. $3! = = 1 \,.\, 2 \,.\, 3 = 6$
$(2n)!!$	$2 \,.\, 4 \,.\, 6 \,.\, \ldots \,.\, 2n$; e.g. $6!! = 2 \,.\, 4 \,.\, 6 = 48$
$\dbinom{r}{k}$	$\dfrac{r(r-1)\ldots(r-k+1)}{k!}$; r any real number
$\dbinom{n}{k}$, C_n^k	$\dfrac{n(n-1)\ldots(n-k+1)}{k!} = \dfrac{n!}{k!\,(n-k)!}$ the binomial coefficient, n a positive integer
$\sum$	the sum of, the summation sign; e.g. $\sum_{k=1}^{3} a_k = a_1 + a_2 + a_3$; $\sum_k a_k$ means: we sum over all values of k considered
$\prod$	the product; $\prod_{k=1}^{3} a_k = a_1 a_2 a_3$
$\log_b a$	the logarithm of a to the base b
$\log a$	the common or Briggs logarithm (to the base 10)
$\ln a$	the natural or Napierian logarithm (to the base e)
$\in$	is an element of; e.g. $x \in [a, b]$ means: x is (or lies) in the interval $[a, b]$;
$\notin$	is not an element of
$\subset$	the sign of inclusion; e.g. $M \subset N$ (see § 1.23)
$\cup$	the union (the sum); e.g. $M \cup N$; often written $M + N$ (see § 1.23)
$\cap$	the intersection (or the product); e.g. $M \cap N$ (see § 1.23)

Symbol or Notation	Meaning
$\max(a_1, a_2, \ldots, a_n)$	the greatest of the numbers $a_1, a_2, \ldots, a_n$
$\min(a_1, a_2, \ldots, a_n)$	the least of the numbers $a_1, a_2, \ldots, a_n$
$\boldsymbol{a} = (a_1, a_2, \ldots, a_n)$	n-component vector (or vector of order n) with components (coordinates) $a_1, a_2, \ldots, a_n$
$\mathbf{A} = \left\Vert \begin{matrix} a_{11}, a_{12}, a_{13} \\ a_{21}, a_{22}, a_{23} \end{matrix} \right\Vert$	the 2 by 3 matrix (see § 1.16)
$\mathbf{A}'$	the transpose $\mathbf{A}'$ of a matrix $\mathbf{A}$
$\mathbf{A}^{-1}$	the inverse of a matrix $\mathbf{A}$
$\mathbf{E}$	the identity matrix
$\mathbf{0}$	the zero-matrix
$\mathbf{A} \sim \mathbf{B}$	the matrices $\mathbf{A}$, $\mathbf{B}$ are equivalent
$A = \left\vert \begin{matrix} a_{11}, a_{12} \\ a_{21}, a_{22} \end{matrix} \right\vert$	the determinant of order 2 or of the second order (see § 1.17)
$[a_1, b_2, c_3]$	the abbreviated notation for a determinant $\left\vert \begin{matrix} a_1, a_2, a_3 \\ b_1, b_2, b_3 \\ c_1, c_2, c_3 \end{matrix} \right\vert$
A_{ik}	the minor belonging to the element a_{ik}
A_{ik}	the cofactor belonging to the element a_{ik}

Geometry

Symbol or Notation	Meaning
$\parallel$	(is) parallel to
$\uparrow\uparrow$	(is) parallel to ... and of the same orientation
$\uparrow\downarrow$	(is) parallel to ... and of the opposite orientation
$\perp$	(is) perpendicular to
$\triangle$	the triangle; e.g. $\triangle ABC$ stands for a triangle with the vertices A, B, C
$^\circ$ degree, $'$ minute, $''$ second	in the sexagesimal measure of angles
arc α	the arc, the radian (circular) measure of an angle α; if the magnitude of an angle α is given in degrees then $$\text{arc}\,\alpha = \frac{\pi\alpha}{180};$$ e.g. for $\alpha = 90^\circ$ $$\text{arc}\,\alpha = \frac{\pi \,.\, 90}{180} = \frac{\pi}{2}$$

Symbol or Notation	Meaning
rad	the radian, the unit angle in circular measure; 1 rad $\doteq 57°17'44{\cdot}8''$
AB	the segment with the end points A, B
$\overline{AB}$	the length of a segment AB
$C \prec D$	on an oriented straight line, a point C lies before a point D
(x, y)	the rectangular coordinates of a point in the plane
(x, y, z)	the rectangular coordinates of a point in space
(ϱ, φ)	the plane-polar coordinates
(r, φ, z)	the cylindrical coordinates of a point in space
(r, φ, ϑ)	the spherical coordinates of a point in space

Vectors in Geometry, Vector Calculus, Vector Analysis

Symbol or Notation	Meaning
$\boldsymbol{a}$	a vector
$\overrightarrow{\boldsymbol{AB}}$	the vector with the initial (starting) point A and the end (terminal) point B
a, $\lvert\boldsymbol{a}\rvert$	the length (module) of a vector $\boldsymbol{a}$
$\boldsymbol{i}, \boldsymbol{j}, \boldsymbol{k}$	the principal (unit or coordinate) vectors in the axes x, y, z of a cartesian coordinates system
$\boldsymbol{r}$	the radius vector of a point (x, y, z) (a vector with the initial point $(0, 0, 0)$ and the end point (x, y, z))
$k\boldsymbol{a}$	k-multiple of the vector $\boldsymbol{a}$ (k being a scalar)
$\boldsymbol{a} . \boldsymbol{b}$, $\boldsymbol{ab}$	the scalar (inner) product of vectors $\boldsymbol{a}$, $\boldsymbol{b}$ (§ 7.1)
$\boldsymbol{a} \times \boldsymbol{b}$, $\boldsymbol{a} \wedge \boldsymbol{b}$, $[\boldsymbol{ab}]$	the vector (cross, outer) product of vectors $\boldsymbol{a}$, $\boldsymbol{b}$ (§ 7.1)
$[\boldsymbol{abc}]$, $\boldsymbol{abc}$	the mixed product (or the trivector) of vectors $\boldsymbol{a}, \boldsymbol{b}, \boldsymbol{c}$; $\boldsymbol{abc} = (\boldsymbol{a} \times \boldsymbol{b}) . \boldsymbol{c}$ (§ 7.1)
$\boldsymbol{a}'(t)$, $\boldsymbol{a}^{(n)}(t)$ respectively	the first, the n-th derivative, respectively, of a vector $\boldsymbol{a}$ with respect to the scalar variable t, i.e. $\dfrac{\mathrm{d}\boldsymbol{a}(t)}{\mathrm{d}t}$ or $\dfrac{\mathrm{d}^n\boldsymbol{a}(t)}{\mathrm{d}t^n}$ (§ 7.2)

Symbol or Notation	Meaning
grad u, ∇u	the gradient of u (§ 7.2)
div $\boldsymbol{a}$, $\nabla \boldsymbol{a}$	the divergence of a vector $\boldsymbol{a}$ (§ 7.2)
curl $\boldsymbol{a}$, rot $\boldsymbol{a}$, $\nabla \times \boldsymbol{a}$	the curl of a vector $\boldsymbol{a}$
∇	the Hamilton nabla operator
a^{ijkl}_{mn}	the four-times contravariant and two-times covariant tensor

Analysis (Differential and Integral Calculus)

Symbol or Notation	Meaning
(a, b) or $[a, b]$	an open or closed interval respectively (for details see § 11.1)
$x \in [a, b]$	x belongs to the interval $[a, b]$, x is in the interval $[a, b]$
$[a, b] \times [c, d]$	the cartesian product of the intervals $[a, b]$, $[c, d]$ {in the cartesian coordinate system in a plane the product is a rectangle with the vertices (a, c), (b, c), (b, d), (a, d)}
$\{a_n\}$	a sequence with general term a_n
$\lim\limits_{n\to\infty} a_n = a$	the sequence $\{a_n\}$ possesses a limit a
$\lim\limits_{n\to\infty} a_n = +\infty$	the sequence $\{a_n\}$ diverges to $+\infty$
$\limsup\limits_{n\to\infty} a_n$, $\overline{\lim\limits_{n\to\infty}}\, a_n$	the greatest limiting point of a sequence $\{a_n\}$ (§ 10.1)
$\liminf\limits_{n\to\infty} a_n$, $\underline{\lim\limits_{n\to\infty}}\, a_n$	the least limiting point of a sequence $\{a_n\}$ (§ 10.1)
$\sum\limits_{n=1}^{\infty} a_n$	the infinite series with general term a_n
$\prod\limits_{n=1}^{\infty} a_n$	the infinite product with general term a_n
$f(x)$, $g(x)$, ...	functions of a single variable x
$f(x, y)$, $g(x, y)$, ...	functions of two variables x, y
$f(g(x))$, $f(g(x, y), h(x, y))$	composite functions
$O(f(x))$, $o(f(x))$	(see § 11.4, p. 414)
$\max\limits_{a\leqq x\leqq b} f(x)$ or $\min\limits_{a\leqq x\leqq b} f(x)$	the maximum or minimum value of a function $f(x)$ on an interval $[a, b]$
$\sup\limits_{a\leqq x\leqq b} f(x)$ or $\inf\limits_{a\leqq x\leqq b} f(x)$	the least upper bound (the supremum) or the greatest lower bound (the infimum) of a function $f(x)$ on the interval $[a, b]$ (on the supremum and infimum, see § 1.3)

Symbol or Notation	Meaning
$\lim\limits_{x\to a} f(x) = A$	the function $f(x)$ has the limit A at the point a
$\lim\limits_{x\to a+} f(x) = B,\ f(a+0) = B$	the function $f(x)$ possesses the right-hand limit B at the point a
$\lim\limits_{x\to a-} f(x) = -\infty,$ $f(a-0) = -\infty$	the function $f(x)$ has the infinite left-hand limit $-\infty$ at the point a
$\lim\limits_{x\to+\infty} f(x) = C$	the function $f(x)$ has the limit C at the point $+\infty$
$y',\ f'(x),\ \dfrac{\mathrm{d}y}{\mathrm{d}x},\ \dfrac{\mathrm{d}f}{\mathrm{d}x}$	the (first) derivative of the function $y = f(x)$
$y^{(n)},\ f^{(n)}(x),\ \dfrac{\mathrm{d}^n y}{\mathrm{d}x^n},\ \dfrac{\mathrm{d}^n f}{\mathrm{d}x^n}$	the n-th derivative of the function $y = f(x)$; we write y'', y''', $f''(x)$, $f'''(x)$ instead of $y^{(2)}$, $y^{(3)}$, $f^{(2)}(x)$, $f^{(3)}(x)$
$\mathrm{d}y,\ \mathrm{d}f(x)$	the differential of the function $y = f(x)$
$\delta y,\ \delta f(x)$	the variation of the function $y = f(x)$
$\dfrac{\partial f}{\partial x},\ f'_x,\ f_x$	the partial derivative of the function f (of several variables) with respect to x
$\dfrac{\partial f}{\partial n},\ \dfrac{\partial f}{\partial \nu}$	the derivative in the direction of the outward normal
$\dfrac{\partial^2 f}{\partial x^2},\ f''_{xx},\ f_{xx}$	the second partial derivative of the function f with respect to x
$\dfrac{\partial^2 f}{\partial x\,\partial y},\ f''_{xy},\ f_{xy}$	the second mixed derivative of the function f; $\dfrac{\partial^2 f}{\partial x\,\partial y} = \dfrac{\partial}{\partial x}\left(\dfrac{\partial f}{\partial y}\right)$
$\mathrm{d}_x f$	the partial differential of the function f (of several variables)
$\mathrm{d}f$	the total differential of the function f
$\dfrac{D(y_1, y_2, \ldots, y_n)}{D(x_1, x_2, \ldots, x_n)}$	the functional determinant (the Jacobian) of the system of functions $y_1, y_2, \ldots, y_n$ with respect to the variables $x_1, x_2, \ldots, x_n$; cf. § 12.7
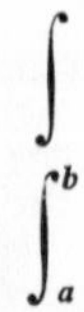 $\int$	the indefinite integral (the primitive)
$\int_a^b$	the definite integral between the limits a, b

Symbol or Notation	Meaning
$\int_a^{\infty}$	the improper integral (§ 13.8)
$\int_a^b f\,\mathrm{d}g$	the Stieltjes integral
$[f(x)]_a^b$	$f(b) - f(a)$
$\iint_O$	the double integral over (or in) a region O
$\iiint_O$	the triple integral over (or in) a region O
$\int_k$	the line integral over (or along) a curve k
$\iint_S$	the surface integral over a surface S
$f \in L_2(a, b)$	a function f is square integrable in the interval $[a, b]$
(f, g)	the scalar (inner) product of functions (§ 16·2)
$\lVert f \rVert$	the norm of the function (§ 16·2)
a region of type A	(see § 14·1)
a solid of type A	(see § 14·1)
a function of type B	(see § 14·1)
π	the number π; $\pi \doteq 3{\cdot}141\ 592\ 654$
e	the base of natural logarithms; $\mathrm{e} \doteq 2{\cdot}718\ 281\ 828$
C	Euler's constant; $C \doteq 0{\cdot}577\ 215\ 665$
M	the modulus of common logarithms; $M = \log \mathrm{e} \doteq 0{\cdot}434\ 294\ 48$
m	the modulus of natural logarithms; $m = \ln 10 \doteq 2{\cdot}302\ 585\ 09$
$\sin x$	the sine
$\cos x$	the cosine
$\tan x$	the tangent
$\cot x$	the cotangent
$\sec x$	the secant
$\operatorname{cosec} x$	the cosecant
$\arcsin x$	the arc sine

Symbol or Notation	Meaning
$\arccos x$	the arc cosine
$\arctan x$	the arc tangent
$\operatorname{arccot} x$	the arc cotangent
$\sinh x$	the hyperbolic sine
$\cosh x$	the hyperbolic cosine
$\tanh x$	the hyperbolic tangent
$\coth x$	the hyperbolic cotangent
$\operatorname{arsinh} x$ $\operatorname{arcosh} x$ $\operatorname{artanh} x$ $\operatorname{arcoth} x$	the inverse of the hyperbolic $\left\{\begin{array}{l}\text{sine}\\ \text{cosine}\\ \text{tangent}\\ \text{cotangent}\end{array}\right\}$
a^x	the exponential function with the base a, or the general exponential function
e^x	the exponential function (we often write $\exp x$, particularly when the argument is rather cumbersome; e.g. $\exp(x/at) = e^{x/at}$)
$\log_a x$	the logarithm of x to the base a
$\ln x$	the natural logarithm of x
$\Gamma(x)$	the Gamma function
$B(p, q)$	the Beta function
$J_r(x)$	the Bessel function of the first kind of order r
$Y_r(x)$	the Bessel function of the second kind of order r
$P_n(x)$	the Legendre polynomial of degree n
$Y_n(\vartheta, \varphi)$	a spherical function of the n-th order
$T_n(x)$	the Chebyshev polynomial of degree n
$L_n(x)$	the Laguerre polynomial of degree n
$H_n(x)$	the Hermite polynomial of degree n
$F(\alpha, \beta, \gamma, x)$	the hypergeometric series (function)
$\operatorname{Si}(x)$	the sine integral (§ 13·1)
$\operatorname{Ci}(x)$	the cosine integral (§ 13·1)
$\operatorname{li}(x)$	the logarithmic integral (§ 13.1)
$\operatorname{Ei}(x)$	(see § 13.1)
$\operatorname{erf}(x)$	the error function; $\operatorname{erf}(x) = \frac{2}{\sqrt{\pi}} \int_0^x e^{-t^2}\,dt$
$\operatorname{erfc}(x)$	$\operatorname{erfc}(x) = 1 - \operatorname{erf}(x) = \frac{2}{\sqrt{\pi}} \int_x^\infty e^{-t^2}\,dt$

Symbol or Notation	Meaning
$F(k, \varphi)$	the Legendre elliptic integral of the first kind in the normal form
$E(k, \varphi)$	the Legendre elliptic integral of the second kind in the normal form
K	the complete elliptic integral of the first kind; $\mathrm{K} = F(k, \frac{1}{2}\pi)$
E	the complete elliptic integral of the second kind; $\mathrm{E} = E(k, \frac{1}{2}\pi)$
sn u, cn u, dn u	Jacobian elliptic functions; see p. 591

Functions of a Complex Variable

Symbol or Notation	Meaning
i, j	the imaginary unit, $\mathrm{i}^2 = -1$, $\mathrm{i}^3 = -\mathrm{i}$ (in electrical engineering j is used instead of i)
Re α, R[α], R(α)	the real part of the complex number α
Im α, I[α], I(α)	the imaginary part of the complex number α
$\|\alpha\|$	the absolute value (modulus) of the complex number α
arg α, ampl α, amp α	the argument (amplitude) of the complex number α
$\bar{\alpha}$	the (complex) conjugate of the complex number α
$\ln(z)$	the natural logarithm z (a multi-valued function)
$\ln_0 z$	the principal branch of the function $\ln z$ (a single-valued function)
res $f(z)$, res $[f(z)]_{z=z_k}$	the residue of a function $f(z)$ at the point z_k (§ 20·5)
$\int_{a-i\infty}^{a+i\infty} f(z)\,\mathrm{d}z$	the integral of the function $f(z)$ along the straight line $x = a$ which is parallel to the imaginary axis

Probability Theory, Statistics, Method of Least Squares, Calculus of Observations

Symbol or Notation	Meaning
n_A	the number of occurrences, the frequency
p	the probability of an event
$P\{x_a \leqq x \leqq x_b\}$	the probability that the random variable x lies in the interval $[x_a, x_b]$
$F(x)$	the distribution function of a random variable

Symbol or Notation	Meaning
$E(x)$	the expectation of a random variable
$D(x),\ \sigma_x^2$	the variance of a random variable
σ_x	the standard deviation
$C(x)$	the coefficient of variation
$\tilde{x}$	the median
$\tilde{x}_P$	the P-quantile
$\Phi\left(\dfrac{x-\mu}{\sigma}\right)$	the distribution function of a normal distribution with the expectation μ and the variance σ^2
$\operatorname{Cov}(x, y),\ \sigma_{xy}$	the covariance of a pair of random variables x, y
ϱ_{xy}	the correlation coefficient of a pair of random variables
$\bar{x}$	the sample average (mean)
s_x^2	the sample variance
s_x	the sample standard deviation
s_{xy}	the sample covariance
c_x	the sample coefficient of variation
r_{xy}	the sample correlation coefficient
b	the sample regression coefficient of y on x
$\bar{x}_{\mathrm{G}}$	the geometric mean
$\bar{x}_{\mathrm{H}}$	the harmonic mean
n	the size of a sample
T_{l}	the lower tolerance limit
T_{u}	the upper tolerance limit
R	the sample range
$(y_{1\mathrm{l}})$ or $(y_{1\mathrm{u}})$	the lower or upper outer control limit
$(y_{2\mathrm{l}})$ or $(y_{2\mathrm{u}})$	the lower or upper inner control limit
S_0	the sum of squares of the residuals
e_i	the measurement errors

Some other Notations

Symbol or Notation	Meaning
H	the Hilbert space (Chap. 22)
B	the Banach space (Chap. 22)
$D(A),\ D_A$	the domain of definition of the operator A (Chap. 22)
$R(A),\ R_A$	the range of the operator A (Chap. 22)
$\lVert A\rVert$	the norm of the operator A (Chap. 22)
A^*	the adjoint operator to A (Chap. 22)

Symbol or Notation	Meaning
$\alpha,\ \beta$	the scale moduli (Chap. 32)
$\Delta^n f$	the n-th forward difference (Chap. 32)
$\nabla^n f$	the n-th backward difference (Chap. 32)
$F(p)$	Laplace transform of the function $f(t)$ (Chap. 28)

1. ARITHMETIC AND ALGEBRA

By Václav Vilhelm

References: [5], [13], [22], [24], [28], [36], [42], [50], [51], [58], [75], [79], [84], [104], [105], [108], [111], [140], [143], [170], [172], [178], [186], [228], [260], [284], [293], [311], [356], [390], [406].

1.1. Some Concepts of Logic

By a *sentence* is to be understood any statement concerning which it is meaningful to say that its content is true (it holds), or false (it does not hold).

The *opposite* or *contradictory of a sentence A* (denoted by not-A or A') is a sentence defined in the following way: The sentence not-A is true if the sentence A is false, and vice versa.

Example 1. "All chairs in the room are occupied" is an example of a sentence. Its opposite is the sentence "Not all chairs in the room are occupied", i.e. "There is at least one unoccupied chair in the room".

If A, B are two sentences, then one can construct from them new sentences in various ways. First, let us introduce the concept of *implication*.

We say that "the sentence A implies the sentence B" or "B follows from A" or "if A is true, then B is true" or "B is a necessary condition for A" or "A is a sufficient condition for B" (in notation $A \Rightarrow B$), if the truth of the sentence B follows from the truth of the sentence A. (If the sentence A is false, then the sentence B can be either true or false.) In an implication $A \Rightarrow B$, A is called the *premise* (*cause*) and B the *conclusion* (*effect*) of the implication.

Example 2. The implications "If a is an integer divisible by four, then a is even" and "If the sum of the angles of a triangle is 120°, then every triangle is a right-angled triangle" are true. (The premise of the second implication is false and thus the implication is true.)

Another sentence combined from the sentences A, B is *equivalence*:

We say, that "the sentence A is equivalent to the sentence B" or "A is true if and only if B is true" or "A is a necessary and sufficient condition for B" (in notation $A \Leftrightarrow B$), if the sentences A and B are either both true or both false.

Example 3. A typical example of an equivalence is the sentence "A triangle is equilateral if and only if all its angles are equal".

REMARK 1. The equivalence $A \Leftrightarrow B$ is true if and only if both $A \Rightarrow B$ and $B \Rightarrow A$ are true.

REMARK 2. The sentence $A \Rightarrow B$ is equivalent to the sentence not-$B \Rightarrow$ not-A.

REMARK 3. Mathematical theorems usually have the form of an implication or an equivalence; e.g. "If a function $f(x)$ possesses a finite derivative at a point x_0, then it is continuous at x_0", "A quadratic equation with real coefficients has two distinct real roots if and only if its discriminant is positive."

1.2. Natural, Integral and Rational Numbers

Natural numbers are the numbers 1, 2, 3, 4, 5, ...

Natural numbers satisfy the *principle of complete* (or *mathematical*) *induction* or *finite induction*, namely:

If M is any set of natural numbers which contains the number 1 and which has the further property that if it contains the number n it also contains the number $n + 1$, then M contains *all* natural numbers.

REMARK 1. This principle is "intuitively evident": If a set M has the properties assumed in the above principle, then it contains the number 1. Hence, the property $n \in M \Rightarrow n + 1 \in M$ implies that the set M contains the numbers $1 + 1 = 2, 2 + 1 = 3$ etc.

The principle of complete induction is the basis of "*proofs by complete induction*". To make the principle of such proofs clear, let us consider an example.

Example 1. Let $q \neq 1$. Then, for any natural number k, the formula

$$1 + q + q^2 + q^3 + \ldots + q^k = \frac{q^{k+1} - 1}{q - 1}$$

holds.

We shall prove this statement by complete induction. Let M be the set of those natural numbers k, for which the statement is valid. Evidently, the statement is true for $k = 1$ and thus $1 \in M$. Let us assume that the statement is true for $k = n$, i.e.

$n \in M$. Then

$$1 + q + \ldots + q^n + q^{n+1} = \frac{q^{n+1} - 1}{q - 1} + q^{n+1} = \frac{q^{n+1} - 1 + q^{n+2} - q^{n+1}}{q - 1} =$$

$$= \frac{q^{n+2} - 1}{q - 1};$$

hence the statement is true also for $n + 1$, i.e. $n + 1 \in M$. Since the statement holds for $k = 1$, the set M contains, in accordance with the principle of complete induction, all natural numbers and therefore the statements holds for any natural number k.

Integers are obtained by extending the set of all natural numbers by the numbers 0 (zero) and $-1, -2, -3, \ldots$.

The numbers 1, 2, 3, ... are called *positive*, the numbers $-1, -2, -3, \ldots$ *negative.*

Definition 1. The fact that an integer a is positive, or negative, is denoted by $a > 0$, or $a < 0$, respectively.

We say that a number a is *less* than or *greater* than a number b if the difference $b - a > 0$, or $b - a < 0$, and in that case we write $a < b$, or $a > b$, respectively.

REMARK 2. The notation $a \leqq b$ means that either $a < b$ or $a = b$; similarly for $a \geqq b$.

Theorem 1. *By the relation $<$ the integers are ordered. This ordering has the following properties (a, b, c, d stand for integers):*

A. *For any two integers a, b one and only one of the following relations holds:*

$$a < b, \quad a > b, \quad a = b.$$

B. $$a < b, \quad b < c \Rightarrow a < c.$$

C. $$a < b, \quad c \leqq d \Rightarrow a + c < b + d.$$

D. $$a < b, \quad c > 0 \Rightarrow ac < bc.$$

E. $$a < b, \quad c < 0 \Rightarrow ac > bc.$$

REMARK 3. The properties A–E express the *basic rules of inequalities.* D and E imply that $bc > 0 \Leftrightarrow b > 0, c > 0$ or $b < 0, c < 0$.

Rational numbers are obtained by extending the set of all integers by fractions, i.e. numbers of the form p/q with integers p and q, $q \neq 0$. The equality $p/q = p'/q'$ holds if and only if $pq' = p'q$.

Theorem 2. *Any rational number can be written in the form a/b, where a is an integer and b a natural number.*

Theorem 3. *Rational numbers can be added, subtracted, multiplied and divided; these operations satisfy the following rules (a, b, c stand for rational numbers):*

1. $(a + b) + c = a + (b + c)$ (*associative law for addition*).
2. $a + b = b + a$ (*commutative law for addition*).
3. *For every* a, $a + 0 = a$.
4. *For every* a, *there exists a number* $-a$ *such that* $a + (-a) = 0$.
5. $(ab)c = a(bc)$ (*associative law for multiplication*).
6. $ab = ba$ (*commutative law for multiplication*).
7. *For every* a, $a \,.\, 1 = a$.
8. *For every* $a \neq 0$, *there exists a number* a' *such that* $aa' = 1$. (We write $a' = a^{-1}$ or $a' = 1/a$.)
9. $(a + b)c = ac + bc$ (*distributive law*).

Remark 4. Addition, multiplication and division of fractions (rational numbers) are performed according to the following rules:

$$\frac{a_1}{b_1} + \frac{a_2}{b_2} = \frac{a_1 b_2 + a_2 b_1}{b_1 b_2}, \quad \frac{a_1}{b_1}\frac{a_2}{b_2} = \frac{a_1 a_2}{b_1 b_2},$$

$$\frac{\dfrac{a_1}{b_1}}{\dfrac{a_2}{b_2}} = \frac{a_1}{b_1}\frac{b_2}{a_2} = \frac{a_1 b_2}{a_2 b_1}.$$

In the last rule we assume, of course, that $a_2/b_2 \neq 0$, i.e. $a_2 \neq 0$.

Theorem 4. *The rational numbers can be ordered in the following way: If* $a = p/q$, $b = p'/q'$, *where* p, p' *are integers and* q, q' *natural numbers, then* $a \lesseqgtr b$ *according as* $pq' \lesseqgtr p'q$. *This order agrees with that of the integers and satisfies the rules* A–E *of Theorem* 1.

1.3. Real Numbers

The ordered set of the rational numbers is *dense* (i.e. between any two different rational numbers there is an infinity of rational numbers), but it has *gaps*; this means that there exist partitions of the set of the rational numbers into two non-empty classes A, B such that

1° $A \cup B$ (see Definition 1.23.2, p. 83) is the set of all rational numbers;

2° for every number $a \in A$ and every number $b \in B$, the relation $a < b$ holds;

3° the set A has no greatest number and the set B has no least number. (One can get such a partition by defining e.g. the class B to contain all positive rational numbers x satisfying $x^2 > 2$ and the class A all the other rational numbers.)

Filling up these gaps by new, so-called *irrational numbers*, we extend the set of rational numbers and so get the *real numbers* (for the detailed theory see e.g. [16]).

Theorem 1. *The rules* 1–9 *of Theorem* 1.2.3 *also hold for addition and multiplication of real numbers.*

Theorem 2. *The real numbers can be ordered in such a way that this order corresponds to that of the rational numbers and the rules* A–E *of Theorem* 1.2.1 *hold.*

Theorem 3. *Every irrational number can be expressed in the form of an infinite non-periodic decimal fraction. Rational numbers are expressed by finite or infinite periodic decimal fractions.*

Definition 1. A real number α is said to be *algebraic* if it is a root of some algebraic equation $x^n + a_1x^{n-1} + \ldots + a_n = 0$ with rational coefficients $a_1, a_2, \ldots, a_n$. If α is not algebraic, it is called *transcendental.* For example, the numbers e, π are transcendental.

Definition 2. A set M of real numbers is said to be *bounded above* (*or bounded below*), if there exists a real number a which is greater (or less) than any number belonging to M, respectively. The set M is said to be *bounded if* it is bounded above as well as below.

Definition 3. Let M be a set of real numbers. A real number ξ is called the *least* (*exact*) *upper bound of M* (l.u.b., briefly; we shall write $\xi = \sup M$), if 1° $a \leqq \xi$ for every $a \in M$, 2° ξ is the least number having the property 1°.

Similarly: A number η is called *the greatest* (*exact*) *lower bound of M* (g.l.b.; $\eta = \inf M$) if 1° $a \geqq \eta$ for every $a \in M$, 2° η is the greatest number having the property 1°.

Example 1. Let M be the set of all numbers $0, \frac{1}{2}, \frac{2}{3}, \frac{3}{4}, \ldots$ [i.e. the numbers of the form $(n-1)/n$, where $n = 1, 2, 3, \ldots$]. The set M is bounded, since every number of M is greater than, say, -1 and less than, say, 5. The least upper bound of this set is the number 1, for every $x \in M$ satisfies $x \leqq 1$ (in fact, $x < 1$) and for every (fixed) number $a < 1$ there exists a number of the form $(k-1)/k$ (k being a natural number), in the set M, such that $(k-1)/k > a$. [The choice of $k > 1/(1-a)$ is sufficient]. The greatest lower bound of the set M is evidently 0.

The following theorem states the fundamental property of the ordering of real numbers:

Theorem 4. *Every non-empty set of real numbers bounded above or bounded below possesses a least upper bound or a greatest lower bound respectively. Therefore, there are no gaps in the ordering of real numbers.*

Theorem 5. *If a set M of real numbers possesses a greatest element (maximum, denoted by* $\max M$*), then* $\sup M = \max M$*. Similarly if there exists a least element (minimum, denoted by* $\min M$*) in M, then* $\inf M = \min M$.

Theorem 6. *Between any two different real numbers there is an infinity of rational as well as an infinity of irrational numbers.*

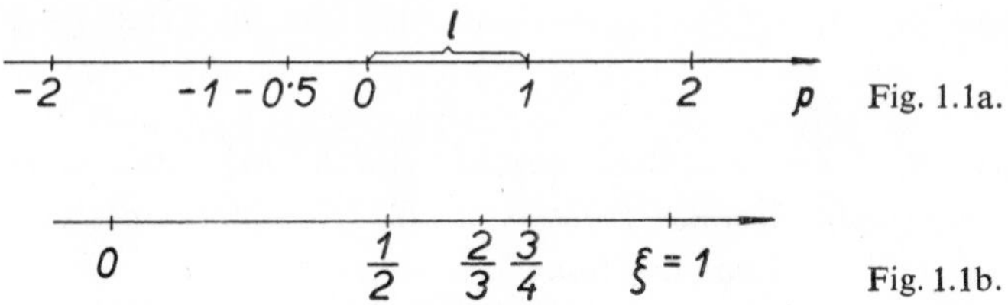

Fig. 1.1a.

Fig. 1.1b.

REMARK 1 (The *numbered scale* or *continuum*, or *axis of real numbers*). Real numbers can be represented by points on a straight line. If we choose, on the straight line p, the origin O, a certain orientation of the straight line (Fig. 1.1a), and a unit of length l, then, to every real number a, there corresponds one and only one point A on the line p, whose coordinate is a; conversely, every point on the line p has a certain coordinate. The straight line p is then called the *numbered scale* or *continuum*. The points of the numbered scale are often identified with real numbers. A number a is less than a number b if and only if the point representing a is to the left of the point representing b on the numbered scale. Fig. 1.1b illustrates several numbers of the set M of Example 1 and the least upper bound ξ of this set.

REMARK 2. On so-called rounded off numbers and operations on them (abbreviated multiplication etc.), see § 32.15.

1.4. Inequalities between Real Numbers. Absolute Value

Theorem 1. *Inequalities between real numbers satisfy the rules* A–E *of Theorem* 1.2.1.

Theorem 2.

$$0 < a < b \Rightarrow 0 < \frac{1}{b} < \frac{1}{a},$$

$$a < b < 0 \Rightarrow \frac{1}{b} < \frac{1}{a} < 0.$$

Theorem 3. *The inequality*

$$ax + b > 0 \tag{1}$$

with $\begin{Bmatrix} a > 0 \\ a < 0 \\ a = 0,\ b > 0 \end{Bmatrix}$ *holds if and only if* $\begin{Bmatrix} x > -b/a \\ x < -b/a \\ x \text{ is arbitrary} \end{Bmatrix}$.

For $a = 0$, $b \leqq 0$, *there is no* x *satisfying* (1).

Theorem 4. *The solution of the inequality*

$$ax^2 + bx + c > 0\,, \quad a \neq 0\,, \tag{2}$$

is as follows:
If the polynomial $f(x) = ax^2 + bx + c$ *has real zeros, then*

$$f(x) = a(x - \alpha_1)(x - \alpha_2)\,, \quad \alpha_1 \leqq \alpha_2\,;$$

so

(a) *if* $a > 0$, *then* $f(x) > 0$ *for all* $x < \alpha_1$ *and for all* $x > \alpha_2$;

(b) *if* $a < 0$, *then* $f(x) > 0$ *for all* x *satisfying* $\alpha_1 < x < \alpha_2$. *If the polynomial* $f(x)$ *has no real roots, then it can be expressed in the form*

$$f(x) = a[(x + c)^2 + d] \quad \text{with} \quad d > 0\,,$$

and thus,

(a) *if* $a > 0$, *then* $f(x) > 0$ *for every real number* x;

(b) *if* $a < 0$, *then* $f(x) < 0$ *always and the inequality* $f(x) > 0$ *has no solution.*

REMARK 1. On inequalities between powers see Theorem 1.9.2, p. 52.

REMARK 2. Simple inequalities which one meets in practice are frequently reducible to inequalities of the type (1) or (2). When solving an inequality of the form $P(x)/Q(x) > 0$, where $P(x)$ and $Q(x) \neq 0$ are polynomials, without common real zeros, the following theorem is useful: The function $P(x)/Q(x)$ changes its sign only in the neighbourhood of the zeros of odd multiplicity of the polynomials $P(x)$ and $Q(x)$. Thus, knowing the zeros of these polynomials and the sign of the function P/Q at one point where this function is non-zero, we can solve the given inequality quite easily. The procedure is illustrated in the following example.

Example 1. Let us solve the inequality

$$\frac{2x - 5}{x - 1} > 3$$

(i.e. find all real x for which this inequality holds). First, we transform the inequality to the form

$$\frac{2x - 5}{x - 1} - 3 > 0\,, \quad \text{i.e.} \quad \frac{-x - 2}{x - 1} > 0\,.$$

The polynomials $P(x) = -x - 2$ and $Q(x) = x - 1$ have the single zeros -2 and 1. Since $P(0)/Q(0) = 2 > 0$, the function P/Q is positive in the interval $(-2, 1)$ and negative in the intervals $(-\infty, -2)$, $(1, +\infty)$. Hence the given inequality is satisfied for all x of the *open* interval $(-2, 1)$ and only for them (Fig. 1.2); these values of x represent the solution of the given inequality.

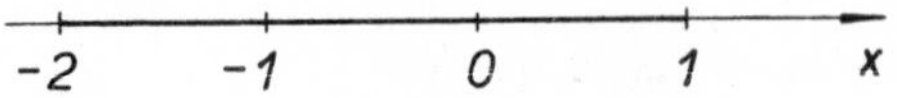

Fig. 1.2. −2 −1 0 1 x

Definition 1. The *absolute value of a real number a* (denoted by $|a|$) is defined as follows:

$$|a| = a \quad \text{for} \quad a \geqq 0\,, \quad |a| = -a \quad \text{for} \quad a < 0\,.$$

Theorem 5. $|a| > 0$ *for* $a \neq 0$; $|0| = 0$; $|a| = \sqrt{(a^2)}$.

Theorem 6. $|a + b| \leqq |a| + |b|$ (*triangle inequality*).

Theorem 7. $||a| - |b|| \leqq |a + b|$.

Theorem 8.
$$|ab| = |a|\,|b|; \quad \left|\frac{a}{c}\right| = \frac{|a|}{|c|} \quad \textit{for} \quad c \neq 0.$$

Theorem 9. *Let* $k > 0$. *Then the inequality* $|a - b| < k$ *is equivalent to the inequalities* $b - k < a < b + k$. (The number $|a - b|$ is equal to the distance between the points a and b on the numbered scale.)

1.5. Further Inequalities. Means

Theorem 1 (*Hölder's Inequality*). *Let* $a_1, \ldots, a_n$, $b_1, \ldots, b_n$ *be real or complex numbers; let* $q > 1$, $q' = q/(q - 1)$. *Then*

$$\left|\sum_{k=1}^{n} a_k b_k\right| \leqq \sum_{k=1}^{n} |a_k b_k| \leqq \left(\sum_{k=1}^{n} |a_k|^q\right)^{1/q} \left(\sum_{k=1}^{n} |b_k|^{q'}\right)^{1/q'}.$$

Theorem 2 (*Cauchy's Inequality*). *Let* $a_1, \ldots, a_n$, $b_1, \ldots, b_n$ *be real or complex numbers. Then*

$$\left|\sum_{k=1}^{n} a_k b_k\right|^2 \leqq \left(\sum_{k=1}^{n} |a_k|^2\right)\left(\sum_{k=1}^{n} |b_k|^2\right)$$

(see Theorem 1 for $q = 2$).

Theorem 3 (*Minkowski's Inequality*). *Let $a_1, \ldots, a_n, b_1, \ldots, b_n$ be real or complex numbers, $q \geqq 1$. Then*

$$\left(\sum_{k=1}^{n} |a_k + b_k|^q\right)^{1/q} \leqq \left(\sum_{k=1}^{n} |a_k|^q\right)^{1/q} + \left(\sum_{k=1}^{n} |b_k|^q\right)^{1/q}.$$

Definition 1. The number $\frac{1}{n}(a_1 + \ldots + a_n)$ is called the *arithmetic mean* of the numbers $a_1, \ldots, a_n$. If these numbers are non-negative, then the number $\sqrt[n]{(a_1 a_2 \ldots a_n)}$ is said to be the geometric mean and the number $\sqrt{\left[\frac{1}{n}(a_1^2 + \ldots + a_n^2)\right]}$ the *quadratic mean* or *root-mean-square* (r.m.s.) of the numbers $a_1, \ldots, a_n$.

Theorem 4. *If $a_1 \geqq 0, \ldots, a_n \geqq 0$, then*

$$\sqrt[n]{(a_1 a_2 \ldots a_n)} \leqq \frac{a_1 + \ldots + a_n}{n} \leqq \sqrt{\frac{a_1^2 + \ldots + a_n^2}{n}}.$$

1.6. Complex Numbers

Complex numbers are numbers of the form $\alpha = a + ib$, where a, b are real numbers and i is the so-called *imaginary unit* (in electrical engineering j is often used instead of i) which is such that

$$\mathrm{i}^2 = -1, \quad \mathrm{i}^3 = -\mathrm{i}, \quad \mathrm{i}^4 = 1.$$

Definition 1. The *equality* of two complex numbers α_1, α_2 is defined as follows: The number $\alpha_1 = a_1 + ib_1$ is equal to $\alpha_2 = a_2 + ib_2$ if and only if $a_1 = a_2, b_1 = b_2$.

Definition 2. *Addition* and *multiplication* of complex numbers are defined in the following way:

$$(a_1 + ib_1) + (a_2 + ib_2) = (a_1 + a_2) + \mathrm{i}(b_1 + b_2),$$
$$(a_1 + ib_1)(a_2 + ib_2) = (a_1 a_2 - b_1 b_2) + \mathrm{i}(a_1 b_2 + a_2 b_1),$$

respectively.

Theorem 1. *Addition and multiplication of complex numbers satisfy the rules* 1–9 *of Theorem* 1.2.3 (p. 42). *Complex numbers cannot be ordered in such a way that the rules* A–E *of Theorem* 1.2.1 (p. 41) *hold.*

Division of complex numbers is performed by application of the following theorem:

Theorem 2. *If*

$$\alpha = a + ib \neq 0,$$

then

$$\frac{1}{\alpha} = \alpha^{-1} = \frac{a - ib}{a^2 + b^2}\,.$$

Definition 3. If $\alpha = a + ib$, then the real number a is called the *real part* of the number α (denoted by Re α) and the real number b the *imaginary part* of the number α (denoted by Im α). A number $\alpha = a + ib$ is said to be *pure imaginary* if $a = \text{Re}\,\alpha = 0$ and $\text{Im}\,\alpha \neq 0$.

REMARK 1. Some authors use the symbols R[α] and I[α] or script letters $\mathcal{R}(\alpha)$, $\mathcal{I}(\alpha)$, instead of Re α and Im α, respectively.

Definition 4. The number $a - ib$ is called the *complex conjugate* of the number $\alpha = a + ib$ and is denoted by $\bar{\alpha}$.

Theorem 3. *For conjugates of complex numbers the following relations hold*:

$$\overline{\alpha + \beta} = \bar{\alpha} + \bar{\beta}\,, \quad \overline{\alpha\beta} = \bar{\alpha}\bar{\beta}\,, \quad \overline{\left(\frac{\alpha}{\gamma}\right)} = \frac{\bar{\alpha}}{\bar{\gamma}}$$

for $\gamma \neq 0$. *Further*, $\alpha = \bar{\alpha} \Leftrightarrow \alpha$ *is a real number.*

Definition 5. The *absolute value* (*modulus*) of a complex number $\alpha = a + ib$ is defined to be the real number $|\alpha| = \sqrt{(a^2 + b^2)} \geqq 0$.

Theorem 4. *The relations* $|\alpha + \beta| \leqq |\alpha| + |\beta|$, $|\alpha\beta| = |\alpha|\,|\beta|$, $|\alpha| = |\bar{\alpha}|$, $\big||\alpha| - |\beta|\big| \leqq |\alpha - \beta| \leqq |\alpha| + |\beta|$ *hold.*

REMARK 2. *The Geometrical Representation of Complex Numbers* (the *Argand Diagram*) is shown in Fig. 1.3a,b. Fig. 1.3b illustrates the first of the inequalities of Theorem 4 (the so-called *triangle inequality*).

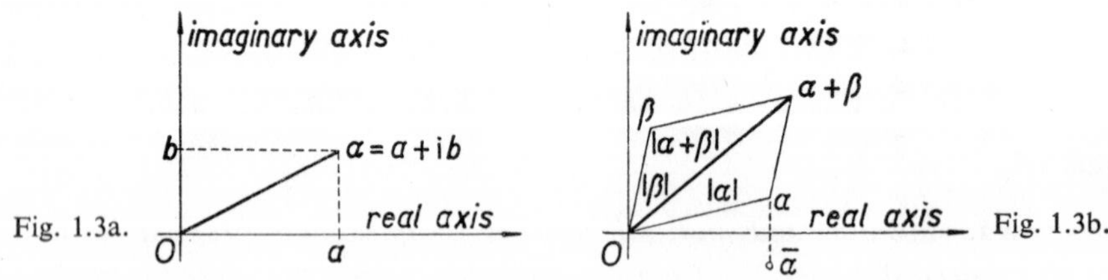

Fig. 1.3a. Fig. 1.3b.

Theorem 5 (*Trigonometric Form of Complex Numbers*). *Every complex number* $\alpha = a + ib \neq 0$ *can be written in the form*

$$\alpha = a + ib = r(\cos\varphi + i\sin\varphi) = re^{i\varphi}\,,$$

where $r = |\alpha|$ *and the angle* φ (*in radian measure*) *is determined apart from an integral multiple of* 2π *by the relations*

$$\cos\varphi = \frac{a}{\sqrt{(a^2+b^2)}}, \quad \sin\varphi = \frac{b}{\sqrt{(a^2+b^2)}};$$

this angle φ *is called the argument (amplitude) of the complex number* α.

The *principal value of the argument* of a complex number α (denoted by arg α) is the (uniquely determined) argument φ for which $-\pi < \varphi \leqq \pi$ (Fig. 1.4a).

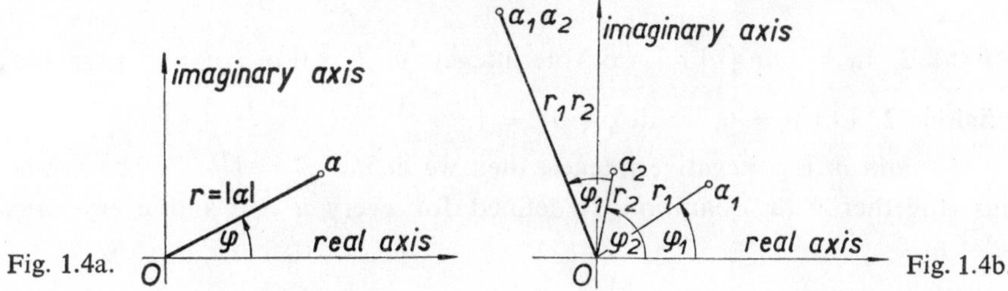

Fig. 1.4a. Fig. 1.4b.

Theorem 6 (*De Moivre's Formula*). *If* $\alpha = r(\cos\varphi + \mathrm{i}\sin\varphi) \neq 0$ *is a complex number, then*

$$\alpha^n = [r(\cos\varphi + \mathrm{i}\sin\varphi)]^n = r^n(\cos n\varphi + \mathrm{i}\sin n\varphi)$$

for every integer n; in particular,

$$(\cos\varphi + \mathrm{i}\sin\varphi)^n = \cos n\varphi + \mathrm{i}\sin n\varphi\,.$$

Theorem 7. *For* $\alpha_1 = r_1(\cos\varphi_1 + \mathrm{i}\sin\varphi_1)$ *and* $\alpha_2 = r_2(\cos\varphi_2 + \mathrm{i}\sin\varphi_2)$, *the following relation holds:*

$$\alpha_1\alpha_2 = r_1r_2[\cos(\varphi_1+\varphi_2) + \mathrm{i}\sin(\varphi_1+\varphi_2)]\,.$$

REMARK 3. Theorems 6 and 7 are used for multiplication, raising to powers and extracting roots of complex numbers. For example, $[\sqrt{(3)} + \mathrm{i}]^3 = [2(\cos 30^\circ + {} + \mathrm{i}\sin 30^\circ)]^3 = 2^3(\cos 90^\circ + \mathrm{i}\sin 90^\circ) = 8\mathrm{i}$. On the use of Theorem 6 for finding roots, see § 1.21 (p. 80). Multiplication of complex numbers in the Argand diagram is performed according to Theorem 7; this can be seen in Fig. 1.4b; the number $|\alpha_1\alpha_2| = r_1r_2$ is usually determined by calculation.

1.7. Powers with Integral Exponents

(a) Powers with a Positive Integral Exponent

REMARK 1. In section (a) m, n denote natural numbers, a, b real or complex numbers.

Definition 1. The n-th *power* of a number a is the number $a^n = aa \dots a$ (n factors a); a is called the *base* and n the *exponent* of the power.

Theorem 1. $a^m a^n = a^{m+n}$, $a^n b^n = (ab)^n$, $(a^m)^n = a^{mn}$.

Theorem 2. $0^n = 0$, $a^n \neq 0$ if $a \neq 0$.

(b) Powers with any Integral Exponent

REMARK 2. In section (b) m, n denote integers, a, b real or complex numbers.

Definition 2. For $a \neq 0$, we define $a^0 = 1$.

If $a \neq 0$ and m is a negative integer, then we define $a^m = 1/a^{-m}$. The symbol a^n is thus (together with Definition 1) defined for every $a \neq 0$ and every *integral* value of n.

Theorem 3. *If* $a \neq 0$, $b \neq 0$ *and* m, n *are integers, then* $a^m a^n = a^{m+n}$, $a^n b^n = (ab)^n$, $(a^m)^n = a^{mn}$, and

$$\frac{a^n}{a^m} = a^{n-m}, \quad \frac{a^n}{b^n} = \left(\frac{a}{b}\right)^n.$$

1.8. Roots of Real Numbers

Definition 1. Let $a > 0$ be a real number, n a natural number. Then there exists exactly one *positive real* number x such that $x^n = a$. The number x is called the *n-th root* of a (denoted by $\sqrt[n]{a}$). Instead of $\sqrt[2]{a}$ we write $\sqrt{a}$.

Example 1. $\sqrt{4} = 2$; the statements $\sqrt{4} = -2$ or $\sqrt{4} = \pm 2$ are not correct.

Definition 2. For $a = 0$, we define $\sqrt[n]{0} = 0$.

Definition 3. For $a < 0$ and an *odd* n we define $\sqrt[n]{a} = -\sqrt[n]{(-a)}$ (since, $-\sqrt[n]{(-a)}$ is the only real number, whose n-th power is a). Thus, e.g. $\sqrt[3]{(-8)} = -\sqrt[3]{8} = -2$.

Theorem 1. *Let* x, y *be positive numbers,* m, n *be natural numbers. Then*

$$\sqrt[n]{(xy)} = \sqrt[n]{x}\sqrt[n]{y}, \quad \sqrt[n]{\frac{x}{y}} = \frac{\sqrt[n]{x}}{\sqrt[n]{y}}, \quad \sqrt[n]{x^k} = (\sqrt[n]{x})^k$$

(k *being an integer*),

$$\sqrt[m]{(\sqrt[n]{x})} = \sqrt[mn]{x}, \quad (\sqrt[n]{x})^n = x.$$

Theorem 2. *For any real number x and any even number n, we have* $\sqrt[n]{x^n} = |x|$. Thus, e.g. $\sqrt{x^2} = |x|$, but in general $\sqrt{x^2} \neq x$.

REMARK 1. On roots of complex numbers see § 1.21, p. 80.

1.9. General Powers of Real Numbers

(a) Power with a Rational Exponent

Definition 1. Let $x > 0$ be a real number, r a rational number. Then we define $x^r = \sqrt[q]{x^p}$, where p and q are integers such that $q > 0$ and $r = p/q$. Thus, if n is a natural number, then

$$x^{1/n} = \sqrt[n]{x}\,, \quad x^{-1/n} = \sqrt[n]{(1/x)} = 1/\sqrt[n]{x}\,.$$

REMARK 1. The rules for operations with powers with a rational exponent are the same as those in Theorem 1 in the next section (b).

(b) General Powers

Definition 2. For a *positive* real number x and for an arbitrary real number a, the *general power* x^a is defined as the limit of a sequence (see Definition 10.1.2, p. 374) $\{x^{a_n}\}$, where $\{a_n\}$ is an (arbitrary) sequence of rational numbers a_n such that its limit is the number a. (If, in particular, a is rational, then this definition evidently coincides with that of Definition 1 and thus x^a is the same real number according to both definitions.)

Theorem 1 (*Properties of General Powers*). *Let x, y be positive real numbers and a, b real numbers. Then the following rules hold:*

1. $1^a = 1$,

2. $x^a y^a = (xy)^a; \quad \dfrac{x^a}{y^a} = \left(\dfrac{x}{y}\right)^a; \quad \dfrac{1}{x^a} = \left(\dfrac{1}{x}\right)^a;$

3. $x^a x^b = x^{a+b}; \quad \dfrac{x^a}{x^b} = x^{a-b}; \quad x^{-a} = \dfrac{1}{x^a};$

4. $(x^a)^b = x^{ab}$.

Theorem 2 (*Inequalities*). *Let x, y be positive real numbers and a, b real numbers. Then*

1. $x^a > 0,\ x^0 = 1$;
2. $x < y,\ a > 0 \Rightarrow x^a < y^a$;
3. $x < y,\ a < 0 \Rightarrow x^a > y^a$;
4. $x > 1,\ a < b \Rightarrow x^a < x^b$;
5. $x < 1,\ a < b \Rightarrow x^a > x^b$.

Definition 3. For $a > 0$, we define $0^a = 0$.

1.10. Logarithms

(a) The Concept and Properties of Logarithms

Definition 1. Let x, a be *positive* numbers, $a \neq 1$. Then there exists a unique real number y such that $a^y = x$; the number y is called the *logarithm of the number x to the base a* (in symbols, $\log_a x$). Thus the logarithm of a number $x > 0$ to a base a is the number $y = \log_a x$ for which $a^{\log_a x} = x$.

Theorem 1 (*Properties of Logarithms*). *Let a, b, c, x, y be real numbers, $0 < a \neq$ $\neq 1$, $0 < b \neq 1$, $x > 0$, $y > 0$. Then*

1. $a^{\log_a x} = x$;
2. $\log_a a = 1,\quad \log_a 1 = 0$;
3. $\log_a xy = \log_a x + \log_a y,\quad \log_a (x/y) = \log_a x - \log_a y$;

   $\log_a \dfrac{1}{x} = -\log_a x;\quad \log_a x^c = c \log_a x$;

4. $\log_b x = \dfrac{\log_a x}{\log_a b}$; *in particular*

   $\log_{10} x \approx 0{\cdot}434\,294 \ln x$,

   $\ln x \approx 2{\cdot}302\,585 \log_{10} x$ (cf. Definition 2).

5. *For $a > 1$ and $x < y$,* $\log_a x < \log_a y$;
   for $a < 1$ and $x < y$, $\log_a x > \log_a y$.
6. *For $a > 1$ and $x > 1$,* $\log_a x > 0$;
   for $a > 1$ and $x < 1$, $\log_a x < 0$.

Definition 2. Logarithms to the base $e = 2{\cdot}718\,28\ldots$ [$e = \lim_{n\to\infty}(1 + 1/n)^n$, see Theorem 10.1.11] are called *natural* or *Napierian logarithms.* Instead of $\log_e x$ we usually write $\ln x$. Then $\log_e 10 = \ln 10 \doteq 2{\cdot}302\,585$. Logarithms to the base 10 are called *common* or *Briggs logarithms.* The value of $\log_{10} e$ is approximately $0{\cdot}434\,294$.

REMARK 1. The use of logarithms for calculating the product and the quotient of two positive numbers, or for calculating the powers of a positive number, is apparent from the property 3. The practical procedure is described in every table of logarithms.

REMARK 2. In the following simple examples, methods of solutions of some exponential and logarithmic equations will be shown.

(b) Exponential Equations

Example 1. Solve the equation $2^{4x} . 2^{x^2} = \frac{1}{16}$.

We arrange the equation in the form $2^{4x+x^2} = 2^{-4}$ and deduce (comparing the exponents) that $4x + x^2 = -4$; the problem is thus reduced to the solution of a quadratic equation (the solution is $x = -2$).

Example 2. Solve the equation $2^x = 3^{x-2} . 5^x$. Taking the logarithm of each term we get $x \log_{10} 2 = (x - 2)\log_{10} 3 + x \log_{10} 5$. Hence

$$x = \frac{-2\log_{10} 3}{\log_{10} 2 - \log_{10} 3 - \log_{10} 5}\,.$$

(c) Logarithmic Equations

Example 3. Solve the equation

$$[\log_{10}(x^2 + 2)]^2 - 5\log_{10}(x^2 + 2) + 6 = 0\,.$$

We put $y = \log_{10}(x^2 + 2)$ and solve the equation $y^2 - 5y + 6 = 0$; this equation has two roots $y_1 = 2$, $y_2 = 3$. Thus the solution consists of those x for which either $\log_{10}(x^2 + 2) = 2$, i.e. $x^2 + 2 = 10^2$, or $\log_{10}(x^2 + 2) = 3$, i.e. $x^2 + 2 = 10^3$, that is $x = \pm\sqrt{98}$ or $x = \pm\sqrt{998}$.

Example 4. Solve the equation

$$2\log_{10}(2x + 3) - \log_{10}(x - 2) - 1 = 0\,.$$

We arrange the equation in the form

$$\log_{10}(2x+3)^2 - \log_{10}(x-2) = 1\,, \quad \log_{10}\frac{(2x+3)^2}{x-2} = 1 = \log_{10} 10\,.$$

Hence $(2x+3)^2/(x-2) = 10$; the problem has thus been reduced to the solution of a quadratic equation.

In more complicated cases, numerical methods are employed (see Chap. 31).

1.11. Arithmetic and Geometric Sequences. Sums of Powers of Natural Numbers; Formulae for $a^n \pm b^n$

Definition 1. An *arithmetic sequence* is a sequence (see Definition 10.1.1, p. 374) of real or complex numbers $a_1, a_2, a_3, \ldots, a_n, \ldots$, such that $a_2 - a_1 = a_3 - a_2 = \ldots = a_{n+1} - a_n = d$ $(n = 1, 2, \ldots)$.

Theorem 1. *The relations*

$$a_n = a_1 + (n-1)\,d\,; \quad s_n = \tfrac{1}{2}n(a_1 + a_n)$$

hold, where $s_n = \sum_{i=1}^{n} a_i$ *is the sum of the first n terms.*

Definition 2. *A geometric sequence* is a sequence of real or complex numbers $b_1, b_2, \ldots, b_n, \ldots$, such that there exists a number q with the property that the relations $b_2 = b_1 q$, $b_3 = b_2 q, \ldots, b_{n+1} = b_n q$ $(n = 1, 2, \ldots)$ hold.

Theorem 2. *For* $q \neq 1$, *the relations*

$$b_n = b_1 q^{n-1}\,, \quad S_n = b_1(q^n - 1)/(q - 1)$$

hold, where $S_n = \sum_{i=1}^{n} b_i$.

Theorem 3. *Sums of powers of natural numbers.*

1. $1 + 2 + \ldots + n = \dfrac{n(n+1)}{2}$;

2. $1^2 + 2^2 + \ldots + n^2 = \dfrac{n(n+1)(2n+1)}{6}$;

3. $1^3 + 2^3 + \ldots + n^3 = \dfrac{n^2(n+1)^2}{4}$;

4. $1^4 + 2^4 + \ldots + n^4 = \dfrac{n(n+1)(2n+1)(3n^2+3n-1)}{30}$;

5. $1^2 + 3^2 + 5^2 + \ldots + (2n-1)^2 = \dfrac{n(4n^2-1)}{3}$;

6. $1^3 + 3^3 + 5^3 + \ldots + (2n-1)^3 = n^2(2n^2-1)$.

Theorem 4. *Formulae for* $a^n \pm b^n$ *(n, k being natural numbers):*

1. $a^2 - b^2 = (a+b)(a-b)$, $a^2 + b^2 = (a+ib)(a-ib)$;
2. $a^3 \pm b^3 = (a \pm b)(a^2 \mp ab + b^2)$;
3. $a^n - b^n = (a-b)(a^{n-1} + a^{n-2}b + a^{n-3}b^2 + \ldots + ab^{n-2} + b^{n-1})$;
4. $a^{2k} - b^{2k} = (a+b)(a^{2k-1} - a^{2k-2}b + a^{2k-3}b^2 - \ldots - b^{2k-1})$;
5. $a^{2k+1} + b^{2k+1} = (a+b)(a^{2k} - a^{2k-1}b + a^{2k-2}b^2 - \ldots + b^{2k})$.

1.12. Permutations and Combinations

Definition 1. Every ordered n-tuple formed from n given mutually different elements is called a *permutation* of these elements.

Example 1. The permutations of three elements a, b, c are the ordered arrangements (a, b, c), (a, c, b), (b, a, c), (b, c, a), (c, a, b), (c, b, a).

Theorem 1. *The number of all (different) permutations of a collection of n elements is*

$$P_n = 1 \,.\, 2 \,.\, 3 \,.\, \ldots \,.\, (n-1)\, n = n!$$

REMARK 1. The symbol $n!$ is read *factorial n*. For $n = 0$, $0!$ is defined as having the value 1.

Definition 2. Let $(i_1, i_2, \ldots, i_n)$ be a permutation of the numbers $1, 2, 3, \ldots, n$. We say that the numbers i_j, i_k, when $j < k$ $(1 \leqq j \leqq n,\ 1 \leqq k \leqq n)$ form an *inversion* in this permutation if $i_j > i_k$. A permutation possessing an odd, or even number of inversions is called *odd*, or *even*, respectively.

Example 2.

(a) In the permutation $(2, 4, 1, 3)$ of the numbers 1, 2, 3, 4 each of the pairs $(2, 1)$, $(4, 1)$, $(4, 3)$ is an inversion. The permutation possesses 3 inversions; therefore it is odd.

(b) The permutation $(4, 2, 1, 3)$ possesses 4 inversions and is thus even; the permutation $(1, 2, 3, 4)$ possesses no inversion and is thus even.

Theorem 2. *If, among n elements a, b, c, …, a occurs α times, b β times, c γ times, …, then the number of all different ordered n-tuples is*

$$\frac{n!}{\alpha!\,\beta!\,\gamma!\,\ldots}.$$

Definition 3. By *combinations of n different elements taken k at a time* we mean all possible selections consisting of k different elements chosen from the n given elements, without regard to the order of selection.

Theorem 3. *The number of all combinations of n different elements taken k at a time is*

$$\binom{n}{k} = \frac{n(n-1)\dots(n-k+1)}{1.2.\dots.k} = \frac{n!}{k!\,(n-k)!}.$$

REMARK 2. Besides $\binom{n}{k}$, the symbols $C(n, k)$, C_n^k, ${}_nC_k$, ${}^nC_k$ and C_k^n are also employed.

For $k = 0$, $\binom{n}{0}$ is defined as having the value 1.

Example 3. Find the number of chess matches required if there are 10 players and every player is to play once with each other.

The number of matches P is equal to the number of pairs formed out of 10 elements, i.e. it is equal to the number of combinations of 10 elements taken 2 at a time. Hence $P = \binom{10}{2} = \frac{10.9}{2.1} = 45.$

Theorem 4. *The symbol* $\binom{n}{k}$ *satisfies the relations*

1. $\binom{n}{k} = \binom{n}{n-k}$;

2. $\binom{n}{1} = \binom{n}{n-1} = n$, $\binom{n}{n} = \binom{n}{0} = 1$;

3. $\binom{n+1}{k} = \binom{n}{k} + \binom{n}{k-1}$;

4. $\binom{n+1}{k+1} = \binom{n}{k} + \binom{n-1}{k} + \dots + \binom{k}{k}$.

REMARK 3. By *combinations with repetitions of n different elements taken k at a time* we understand all possible selections consisting of k elements chosen from n given elements (without regard to the order of selection) such that each element can be repeated any number of times. The number of these combinations with repetitions is $\binom{n+k-1}{k}$. For example, all combinations with repetitions of the elements 1, 2, 3 taken 2 at a time are (1, 1), (2, 2), (3, 3), (1, 2), (1, 3), (2, 3).

Definition 4. By *permutations of n different elements taken k at a time* is meant all possible ordered arrangements consisting of k different elements chosen from the n given elements.

Theorem 5. *The number of all permutations of n different elements taken k at a time is*

$$P_{n/k} = n(n-1)\dots(n-k+1) = \frac{n!}{(n-k)!}.$$

Example 4. All permutations of the three elements 1, 2, 3 taken 2 at a time are (1, 2), (1, 3), (2, 1), (2, 3), (3, 1), (3, 2). Their number is $6 = 3!/(3-2)!$.

REMARK 4. The *permutations with repetitions of n different elements taken k at a time* are all possible ordered arrangements consisting of k elements chosen from the n given elements such that each element can be repeated any number of times. The number of these permutations with repetitions is n^k.

For example, all permutations with repetition of the elements 1, 2 taken 2 at a time are (1, 1), (2, 2), (1, 2), (2, 1).

1.13. Binomial Theorem

Theorem 1. *Let n be a natural number and let a, b be real or complex numbers. Then the following (Newton's) formula holds:*

$$(a \pm b)^n = \sum_{k=0}^{n} (\pm 1)^k \binom{n}{k} a^{n-k} b^k = a^n \pm \binom{n}{1} a^{n-1} b + \binom{n}{2} a^{n-2} b^2 + \dots + (\pm 1)^n b^n.$$

In particular

1. $(a \pm b)^2 = a^2 \pm 2ab + b^2$;
2. $(a \pm b)^3 = a^3 \pm 3a^2 b + 3ab^2 \pm b^3$.

REMARK 1. The *binomial coefficients* $\binom{n}{k}$ can be readily determined by means of *Pascal's triangle*:

n	Binomial coefficients
0	1
1	1 1
2	1 2 1
3	1 3 3 1
4	1 4 6 4 1
5	1 5 10 10 5 1
	. . . .

REMARK 2. The case where n is not a natural number is treated in Theorem 15.5.2, p. 682.

1.14. Polynomials

Definition 1. Let n be a natural number and let $a_0, a_1, \ldots, a_n$ be real or complex numbers. The function $P(x)$ which may be defined for all (real or complex) numbers x by the formula

$$P(x) = a_0x^n + a_1x^{n-1} + \ldots + a_{n-1}x + a_n = \sum_{i=0}^{n} a_ix^{n-i} \tag{1}$$

is called a *polynomial* (in one variable x with real, or complex, coefficients). Besides the term polynomial the expression *rational integral function* is also used. The numbers $a_0, a_1, \ldots, a_n$ are called the *coefficients of the polynomial* $P(x)$.

Definition 2. Two polynomials $P(x)$ and $Q(x)$ are *equal* [in symbols, $P(x) = Q(x)$ or, more precisely, $P(x) \equiv Q(x)$] if, for every number a, the equality $P(a) = Q(a)$ holds.

Definition 3. The highest power of the variable x with a non-zero coefficient in the expression (1) is called the *degree of the polynomial* $P(x)$. If $a_0 \neq 0$ in (1), then $P(x)$ has degree n. (See also Theorem 1.)

Definition 4. The polynomial, all the coefficients of which are equal to zero, is called a *zero polynomial*. A zero polynomial has no degree. If $P(x)$ is a zero polynomial, we write $P(x) = 0$ or, more precisely, $P(x) \equiv 0$. Otherwise, we write $P(x) \neq 0$ or $P(x) \not\equiv 0$.

Theorem 1. *Two polynomials are equal if and only if their difference is a zero polynomial,* i.e. *if the coefficients of the corresponding powers of the variable x are identical.*

Theorem 2. *The sum and the difference of two polynomials of degrees m and n are polynomials of degree less than or equal to the number* max (m, n) (*or zero polynomials*).

Theorem 3. *The product of two polynomials of degrees m and n is a polynomial of degree $m + n$.*

Theorem 4. *The product of non-zero polynomials is a non-zero polynomial.*

Theorem 5. *The quotient of two polynomials need not always be a polynomial.*

How to procced in dividing a polynomial by another polynomial is shown in Example 1.

Example 1.

$$
\begin{array}{l}
(x^3 - 2x^2 + x - 1) : (x^2 - 3x + 2) = x + 1 \text{ (partial quotient)} \\
\underline{x^3 - 3x^2 + 2x} \\
\qquad x^2 - x - 1 \\
\qquad \underline{x^2 - 3x + 2} \\
\qquad\qquad 2x - 3 \quad \text{(remainder)}.
\end{array}
$$

Hence, $x^3 - 2x^2 + x - 1 = (x^2 - 3x + 2)(x + 1) + 2x - 3$.

Definition 5. If the remainder on dividing a polynomial $P(x)$ by a polynomial $Q(x)$ ($Q(x) \not\equiv 0$) equals zero, then the polynomial $P(x)$ is said to be *divisible* by the polynomial $Q(x)$; $Q(x)$ is called a *divisor* of the polynomial $P(x)$.

An important result concerning the process of dividing is contained in

Theorem 6. *For every two polynomials $P(x)$ and $Q(x) \not\equiv 0$, there exist uniquely determined polynomials $S(x)$ and $R(x)$, such that*

1. $P(x) = Q(x)\,S(x) + R(x)$;

2. *$R(x)$ is either a zero polynomial or a polynomial of lower degree than the polynomial $Q(x)$.*

Definition 6. A common divisor, with highest possible degree, of the polynomials $P(x)$ and $Q(x)$ is called the *greatest common divisor* of the polynomials $P(x)$ and $Q(x)$; the polynomials $P(x)$ and $Q(x)$ are said to be *relatively prime* if their greatest common divisor has degree zero.

Theorem 7 (*Euclidean Algorithm*). *The greatest common divisor of two (non-zero) polynomials $P(x)$ and $Q(x)$ can be found in the following way:*

(i) *in accordance with Theorem 6, divide $P(x)$ by the polynomial $Q(x)$,* i.e. $P(x) = Q(x)\,S_1(x) + R_1(x)$ (*where $R_1(x)$ is the remainder*) ;

(ii) *divide $Q(x)$ by the polynomial $R_1(x)$,* i.e. $Q(x) = R_1(x)\,S_2(x) + R_2(x)$, *then $R_1(x)$ by the polynomial $R_2(x)$,* i.e. $R_1(x) = R_2(x)\,S_3(x) + R_3(x)$ *etc., the last remainder $R_k(x) \neq 0$ is the required greatest common divisor.*

Definition 7. The number α (in general complex) is called a *zero* of the polynomial $P(x) = \sum_{i=0}^{n} a_i x^{n-i}$ if $P(\alpha) = \sum_{i=0}^{n} a_i \alpha^{n-i} = 0$.

Theorem 8 (*The Fundamental Theorem of Algebra*). *Every polynomial of degree $n \geqq 1$ has at least one zero.*

Theorem 9. *If a polynomial $P(x)$ has a zero α, then $P(x)$ is divisible by the linear polynomial $x - \alpha$ and vice versa.* [$x - \alpha$ is a so-called *linear factor of the polynomial $P(x)$.*]

Theorem 10 (*The Factorisation of a Polynomial into Linear Factors*). *Every polynomial* $P(x) = \sum_{i=0}^{n} a_i x^{n-i}$, $n \geqq 1$, *can be uniquely written as a product of linear factors*:

$$P(x) = a_0(x - \alpha_1)^{k_1} (x - \alpha_2)^{k_2} \dots (x - \alpha_r)^{k_r}, \quad k_1 + k_2 + \dots + k_r = n .$$

The numbers $\alpha_1, \dots, \alpha_r$ *are all distinct zeros of the polynomial* $P(x)$. [α_1 is called a k_1*-fold zero*, ..., α_r *a* k_r*-fold zero* of the polynomial $P(x)$. *If* $k_j = 1$, α_j is called a *simple zero* of $P(x)$.]

Theorem 11. α *is a k-fold zero of a polynomial* $P(x)$ *if and only if it is also a zero of the first, the second, ..., the* $(k - 1)$*-th derivatives of the polynomial* $P(x)$, *but is not a zero of its k-th derivative*:

$$P(\alpha) = P'(\alpha) = \dots = P^{(k-1)}(\alpha) = 0, \quad P^{(k)}(\alpha) \neq 0 .$$

(For the derivative, see § 11.5.)

Example 2. Let us consider the polynomial $P(x) = x^3 - 3x^2 + 4$. We get $P'(x) = 3x^2 - 6x$, $P''(x) = 6x - 6$, $P'''(x) = 6$. It is easy to check that $P(2) = 0$, $P'(2) = 0$, $P''(2) \neq 0$. Thus $\alpha = 2$ is a double zero of the polynomial $P(x)$. Indeed, $P(x) = (x - 2)^2 (x + 1)$.

Theorem 12 (*Polynomials with Real Coefficients*).

(i) *If the polynomial* $P(x) = \sum_{i=0}^{n} a_i x^{n-i}$ *with real coefficients* a_i *has a k-fold zero* $\alpha = a + ib$, *it has also the k-fold zero* $\bar{\alpha} = a - ib$.

(ii) *A polynomial* $P(x)$ *can be uniquely factorised into linear and quadratic polynomials with real coefficients*:

$$P(x) = a_0(x - \alpha_1)^{r_1} \dots (x - \alpha_i)^{r_i} (x^2 + p_1 x + q_1)^{s_1} \dots (x^2 + p_j x + q_j)^{s_j},$$

where $r_1 + \dots + r_i + 2s_1 + \dots + 2s_j = n$ *and* $p_k^2 - 4q_k < 0$ $(k = 1, 2, \dots, j)$, *so that* $x^2 + p_k x + q_k$ *has no real zeros* (Cf. § 13.3, p. 495.)

Remark 1 (*Horner's Method*). Horner's method is used:

(i) to find the value $P(a)$ of a polynomial $P(x)$ and its derivatives at a given point a;
(ii) to divide a polynomial $P(x)$ by a linear polynomial $x - a$;
(iii) to transform a polynomial $P(x)$ by a substitution $y = x - a$.

Let $P(x) = a_0 x^n + \dots + a_n$; let a be a real or complex number. We then construct the following (Horner's) scheme:

	a_0	a_1	a_2	$\dots$	a_n
		aa_0	ab_1	$\dots$	ab_{n-1}
a	a_0	b_1	b_2	$\dots$	$b_n = P(a)$

Write all coefficients (zero coefficients included) in the first row, leaving for a moment the second row open; in the third row under the number a_0 write a_0 again, then, under a_1, write aa_0 in the second row and $b_1 = a_1 + aa_0$ in the third row. Similarly, under a_2, write ab_1 and $b_2 = a_2 + ab_1$ in the second and third rows respectively, etc. The last number b_n is then the value $P(a)$. Moreover $P(x) = (x - a) \cdot (a_0x^{n-1} + b_1x^{n-2} + \ldots + b_{n-1}) + b_n$ so that the third row determines the quotient and the remainder on dividing the polynomial $P(x)$ by the linear polynomial $x - a$.

Applying Horner's scheme for a number a to the polynomial $a_0x^{n-1} + b_1x^{n-2} + \ldots + b_{n-1}$, we get, as the last number in the scheme, $c_{n-1} = P'(a)/1!$ Continuing in this way, we get $P''(a)/2!, \ldots, P^{(n)}(a)/n!$, successively. The following Example 3 illustrates the procedure.

Example 3. $P(x) = 5x^4 + 10x^3 + x - 1$; $a = -2$.

	5	10	0	1	−1
		−10	0	0	−2
−2	5	0	0	1	$-3 = P(-2)$
		−10	20	−40	
−2	5	−10	20	$-39 = \frac{1}{1!} P'(-2)$	
		−10	40		
−2	5	−20	$60 = \frac{1}{2!} P''(-2)$		
		−10			
−2	5	$-30 = \frac{1}{3!} P'''(-2)$			
−2	$5 = \frac{1}{4!} P^{(4)}(-2)$				

Further, by Taylor's formula for a polynomial of degree n (see § 11.10, p. 434),

$$P(x) = P(a) + \frac{1}{1!} P'(a)(x - a) + \ldots + \frac{1}{n!} P^{(n)}(a)(x - a)^n$$

so that we have in our example

$$P(x) = -3 - 39(x + 2) + 60(x + 2)^2 - 30(x + 2)^3 + 5(x + 2)^4 .$$

By the substitution $y = x + 2$, $P(x)$ is transformed into the polynomial

$$5y^4 - 30y^3 + 60y^2 - 39y - 3 .$$

From the third row of Horner's scheme we get

$$P(x) = (x + 2)(5x^3 + 1) - 3 .$$

1.15. Vectors in Algebra

Definition 1. Let n be a fixed natural number. Then, by an *n-component* (*n-coordinate*) *complex vector* (n-vector for short) $\boldsymbol{a} = (a_1, a_2, \ldots, a_n)$ we understand in algebra an ordered n-tuple of complex numbers $a_1, a_2, \ldots, a_n$. [Besides $\boldsymbol{a} = (a_1, a_2, \ldots \ldots, a_n)$ also the notation $\boldsymbol{a}(a_1, a_2, \ldots, a_n)$ is used.] All these n-component vectors (i.e. the set of all ordered n-tuples of complex numbers) form a so-called *n-dimensional vector space* V_n (over the complex numbers).

The vectors $\boldsymbol{a} = (a_1, \ldots, a_n)$, $\boldsymbol{b} = (b_1, \ldots, b_n)$ are said to be *equal* if and only if $a_1 = b_1, a_2 = b_2, \ldots, a_n = b_n$.

REMARK 1. In the same way as for complex vectors one can define *n-component real vectors* (real n-vectors); their *components*, i.e. the numbers $a_1, \ldots, a_n$ being real numbers. In the following text all concepts and theorems formulated for complex vectors are valid also for real vectors.

Addition and multiplication by a (scalar) number of n-component vectors of V_n is performed in accordance with the following definition:

Definition 2. 1. The *sum of the vectors* $\boldsymbol{a} = (a_1, \ldots, a_n)$ and $\boldsymbol{b} = (b_1, \ldots, b_n)$ is the vector $\boldsymbol{a} + \boldsymbol{b} = (a_1 + b_1, \ldots, a_n + b_n)$.

2. The *product of the vector* $\boldsymbol{a} = (a_1, \ldots, a_n)$ *and the number* c is the vector $c\boldsymbol{a} = (ca_1, \ldots, ca_n)$.

REMARK 2. We write $-\boldsymbol{a}$ instead of $(-1)\,\boldsymbol{a}$; thus $-\boldsymbol{a} = (-a_1, \ldots, -a_n)$.

Example 1. The sum of the vectors $\boldsymbol{a} = (1, 0, -2)$ and $\boldsymbol{b} = (3, 2, 0)$ is $\boldsymbol{a} + \boldsymbol{b} = (4, 2, -2)$; also $3\boldsymbol{a} = (3, 0, -6)$.

Theorem 1. *For the operations on vectors introduced in Definition 2, the following rules hold*:

1. $\boldsymbol{a} + \boldsymbol{b} = \boldsymbol{b} + \boldsymbol{a}\,, \quad \boldsymbol{a} + (\boldsymbol{b} + \boldsymbol{c}) = (\boldsymbol{a} + \boldsymbol{b}) + \boldsymbol{c}\,;$
2. *there exists a vector* [*the so-called zero vector* $\boldsymbol{0} = (0, \ldots, 0)$] *such that* $\boldsymbol{a} + \boldsymbol{0} = \boldsymbol{a}\,;$
3. *for every* $\boldsymbol{a} = (a_1, \ldots, a_n)$ *and* $\boldsymbol{b} = (b_1, \ldots, b_n)$ *there exists a vector* $\boldsymbol{x}$ such *that* $\boldsymbol{a} + \boldsymbol{x} = \boldsymbol{b}$; $\boldsymbol{x} = \boldsymbol{b} - \boldsymbol{a} = (b_1 - a_1, \ldots, b_n - a_n)\,;$
4. $c(\boldsymbol{a} + \boldsymbol{b}) = c\boldsymbol{a} + c\boldsymbol{b}\,;$
5. $(c + d)\,\boldsymbol{a} = c\boldsymbol{a} + d\boldsymbol{a}\,;$
6. $c(d\boldsymbol{a}) = (cd)\,\boldsymbol{a}\,; \quad 0\boldsymbol{a} = \boldsymbol{0}\,, \quad c\boldsymbol{0} = \boldsymbol{0}\,;$
7. *the equality* $c\boldsymbol{a} = \boldsymbol{0}$ *holds if and only if* $c = 0$ *or* $\boldsymbol{a} = \boldsymbol{0}\,;$
8. $-(c\boldsymbol{a}) = (-c)\,\boldsymbol{a} = c(-\boldsymbol{a})\,.$

Definition 3. We say that the vectors $\boldsymbol{a}_1, \ldots, \boldsymbol{a}_k$ of V_n are *linearly dependent* if there exist complex numbers $c_1, \ldots, c_k$, *which are not all zero*, such that $c_1\boldsymbol{a}_1 + c_2\boldsymbol{a}_2 + \ldots + c_k\boldsymbol{a}_k = \boldsymbol{0}$.

If the vectors $\boldsymbol{a}_1, \ldots, \boldsymbol{a}_k$ are not linearly dependent, we say that they are *linearly independent*.

Example 2. The vectors $\boldsymbol{a} = (1, -1, 0)$, $\boldsymbol{b} = (0, -2, 1)$. $\boldsymbol{c} = (2, 4, -3)$ are linearly dependent, for $2\boldsymbol{a} + (-3)\,\boldsymbol{b} + (-1)\,\boldsymbol{c} = \boldsymbol{0}$. The vectors $\mathbf{e}_1 = (1, 0, 0)$, $\mathbf{e}_2 = (0, 1, 0)$, $\mathbf{e}_3 = (0, 0, 1)$ are linearly independent.

Definition 4. A vector $\boldsymbol{a} \in V_n$ is said to be a *linear combination of the vectors* $\boldsymbol{a}_1, \ldots, \boldsymbol{a}_k$ of V_n if complex numbers $d_1, \ldots, d_k$ exist such that

$$\boldsymbol{a} = d_1\boldsymbol{a}_1 + \ldots + d_k\boldsymbol{a}_k .$$

Theorem 2. *Vectors* $\boldsymbol{a}_1, \ldots, \boldsymbol{a}_p$ *of* V_n *are linearly dependent if and only if at least one of them can be expressed as a linear combination of the others.*

Example 3. The vectors $\boldsymbol{a}_1 = (3, 1, 2)$, $\boldsymbol{a}_2 = (-1, 0, 2)$, $\boldsymbol{a}_3 = (7, 2, 2)$ are linearly dependent, for $2\boldsymbol{a}_1 - \boldsymbol{a}_2 - \boldsymbol{a}_3 = 0$. From this equation it follows that

$$\boldsymbol{a}_1 = \tfrac{1}{2}\boldsymbol{a}_2 + \tfrac{1}{2}\boldsymbol{a}_3\,, \quad \boldsymbol{a}_2 = 2\boldsymbol{a}_1 - \boldsymbol{a}_3\,, \quad \boldsymbol{a}_3 = 2\boldsymbol{a}_1 - \boldsymbol{a}_2$$

so that each of them is a linear combination of the other two.

Definition 5. We say that a system $\{\boldsymbol{a}_1, \ldots, \boldsymbol{a}_k\}$ of vectors of V_n has the *rank h* if there are h linearly independent vectors among the vectors $\boldsymbol{a}_1, \ldots, \boldsymbol{a}_k$ but any $h+1$ vectors of $\boldsymbol{a}_1, \ldots, \boldsymbol{a}_k$ are always linearly dependent. (Then h is the maximal number of linearly independent vectors of the given system.)

Example 4. The rank of the system $\{\boldsymbol{a}, \boldsymbol{b}, \boldsymbol{c}\}$ of the vectors of Example 2 is equal to two, for $\boldsymbol{a}, \boldsymbol{b}$ are linearly independent while $\boldsymbol{a}, \boldsymbol{b}, \boldsymbol{c}$ are linearly dependent.

Theorem 3. *Every system of n-component vectors is of rank* $h \leqq n$.

Theorem 4. *The rank of a system of n-component vectors does not change if*

1. *we change the order of the vectors in the system;*
2. *we multiply one of the vectors of the system by a non-zero number;*
3. *we add to one of the vectors a linear combination of the remaining vectors;*
4. *we drop a vector which is a linear combination of the remaining vectors of the system.*

REMARK 3. Theorem 4 is useful in determining the rank of a given system of vectors. In practice, we can find the rank also by determining the rank of the matrix whose rows are the vectors of the given system (see Remark 1.16.2 and Example 1.16.2 on p. 65).

REMARK 4. On vectors in *three-dimensional* space (scalar product, vector product, etc.) see also Chap. 7.

1.16. Matrices

Definition 1. A rectangular array $\mathbf{A}$ of mn real or complex numbers $a_{11}, a_{12}, \ldots, a_{mn}$ arranged in m rows and n columns is called an *m by n matrix*:

$$\mathbf{A} = \begin{Vmatrix} a_{11}, & a_{12}, & a_{13}, & \ldots, & a_{1n} \\ a_{21}, & a_{22}, & a_{23}, & \ldots, & a_{2n} \\ \ldots & \ldots & \ldots & \ldots & \ldots \\ a_{m1}, & a_{m2}, & a_{m3}, & \ldots, & a_{mn} \end{Vmatrix}.$$

If $m = n$, we call $\mathbf{A}$ a *square matrix of order n*, or an *n-rowed square matrix*. The elements $a_{11}, a_{22}, a_{33}, \ldots$ of the matrix $\mathbf{A}$ form its *principal diagonal*, the elements $a_{1n}, a_{2,n-1}, a_{3,n-2}, \ldots$ of $\mathbf{A}$ form its *secondary diagonal*. The matrix, all the elements of which are equal to zero, is called a *zero matrix*.

Definition 2. The *rank of a matrix* is the rank of the system of all vectors formed by the rows of the matrix (see Definition 1.15.5, p. 63). (Cf. Theorem 2.)

Thus, a matrix $\mathbf{A}$ is of rank h if there are h linearly independent rows among its rows, every futher row of the matrix being a linear combination of these h rows.

Example 1. The matrix

$$\begin{Vmatrix} 1, & -1, & 0 \\ 0, & -2, & 1 \\ 2, & 4, & -3 \end{Vmatrix}$$

is of rank 2, for the system of the vectors $\mathbf{a} = (1, -1, 0)$, $\mathbf{b} = (0, -2, 1)$, $\mathbf{c} = (2, 4, -3)$ is of rank 2 (Example 1.15.4, p. 63).

Theorem 1. *For the rank h of an m by n matrix* $\mathbf{A}$, *the inequality*

$$h \leqq \min(m, n)$$

holds.

Definition 3. The matrix

$$\mathbf{A}' = \begin{Vmatrix} a_{11}, & a_{21}, & \ldots, & a_{m1} \\ a_{12}, & a_{22}, & \ldots, & a_{m2} \\ \ldots & \ldots & \ldots & \ldots \\ a_{1n}, & a_{2n}, & \ldots, & a_{mn} \end{Vmatrix},$$

formed from the matrix $\mathbf{A}$ by a transposition of its elements with respect to the principal diagonal (i.e. by an interchange of its rows and columns) is called the *transpose of the matrix* $\mathbf{A}$ and is an n by m matrix.

Theorem 2. *The rank of a matrix* $\mathbf{A}$ *and that of its transpose* $\mathbf{A}'$ *are equal.*

Theorem 3. *The rank of a matrix does not change if*

1. *we change the order of the rows of the matrix;*

2. *we multiply one of the rows by a non-zero number;*

3. *we add to one of the rows a linear combination of the remaining rows;*

4. *we drop a row of the matrix which is a linear combination of the remaining rows of the matrix.*

Thus, if we apply one of these operations, to a matrix, then the resulting matrix has the same rank as the original matrix.

REMARK 1. According to Theorem 2 we can apply the operations of Theorem 3 also to the columns without affecting the rank of the given matrix.

Theorem 4. *The matrix*

$$\mathbf{B} = \begin{Vmatrix} b_{11}, & b_{12}, & b_{13}, & \dots, & & & b_{1n} \\ 0, & b_{22}, & b_{23}, & \dots, & & & b_{2n} \\ 0, & 0, & b_{33}, & \dots, & & & b_{3n} \\ \dots & \dots & \dots & \dots & \dots & \dots & \dots \\ 0, & 0, & \dots, & 0, & b_{kk}, & \dots, & b_{kn} \end{Vmatrix}, \tag{1}$$

where $b_{11}b_{22} \dots b_{kk} \neq 0$ and where all elements below the principal diagonal are equal to zero, is of rank k.

REMARK 2. Theorems 3 and 4 can be used in practice to determine the rank of a given matrix: By means of the operations 1–4 of Theorem 3 and by permutation of the columns we transform the given matrix to a matrix of the same rank and of the form (1) and then apply Theorem 4.

Example 2.

$$\mathbf{A} = \begin{Vmatrix} 1, & 0, & 2, & 3 \\ -2, & 1, & 0, & -1 \\ -1, & 1, & 2, & 2 \\ -1, & 2, & 6, & 7 \end{Vmatrix}.$$

The third row is the sum of the first and second rows; if we drop it, we get the matrix

$$\mathbf{A}_1 = \begin{Vmatrix} 1, & 0, & 2, & 3 \\ -2, & 1, & 0, & -1 \\ -1, & 2, & 6, & 7 \end{Vmatrix}.$$

Applying operations 2 and 3 we can get a matrix in which all elements of the first column of the matrix except the first are zero: First, we add twice the first row to the second row and then we add the first to the third row. We thus obtain the matrix

$$\mathbf{A}_2 = \begin{Vmatrix} 1, & 0, & 2, & 3 \\ 0, & 1, & 4, & 5 \\ 0, & 2, & 8, & 10 \end{Vmatrix}.$$

Now we adjust the second column so as to get zero below the second element: We subtract twice the second row from the third row and thus obtain

$$\mathbf{A}_3 = \begin{Vmatrix} 1, & 0, & 2, & 3 \\ 0, & 1, & 4, & 5 \\ 0, & 0, & 0, & 0 \end{Vmatrix}.$$

In accordance with Theorem 3 we can drop the last row of this matrix. We get a matrix the rank of which is 2, according to Theorem 4. Hence the rank of $\mathbf{A}$ is also 2.

Definition 4. The determinant of order k (see Definition 1.17.1, p. 67) formed by the elements in the intersections of arbitrary k rows and k columns of a matrix

$$\mathbf{A} = \begin{Vmatrix} a_{11}, & a_{12}, & \ldots, & a_{1n} \\ \ldots & \ldots & \ldots & \ldots \\ a_{m1}, & a_{m2}, & \ldots, & a_{mn} \end{Vmatrix},$$

is called a *minor of order k of the matrix* $\mathbf{A}$ $[1 \leqq k \leqq \min(m, n)]$.

Theorem 5. *A matrix* $\mathbf{A}$ *is of rank h if and only if there exists a minor of* $\mathbf{A}$ *of order h different from zero, any minor of* $\mathbf{A}$ *of order higher than h being equal to zero.*

Example 3. Consider the matrix

$$\mathbf{A} = \begin{Vmatrix} 1, & 0, & 2, & 3 \\ -2, & 1, & 0, & -1 \\ -1, & 2, & 6, & 7 \end{Vmatrix}.$$

All its minors of order 3, namely

$$\begin{vmatrix} 1, & 0, & 2 \\ -2, & 1, & 0 \\ -1, & 2, & 6 \end{vmatrix}, \quad \begin{vmatrix} 1, & 0, & 3 \\ -2, & 1, & -1 \\ -1, & 2, & 7 \end{vmatrix}, \quad \begin{vmatrix} 1, & 2, & 3 \\ -2, & 0, & -1 \\ -1, & 6, & 7 \end{vmatrix}, \quad \begin{vmatrix} 0, & 2, & 3 \\ 1, & 0, & -1 \\ 2, & 6, & 7 \end{vmatrix}$$

are equal to zero, while the minor of order 2,

$$\begin{vmatrix} 1, & 0 \\ -2, & 1 \end{vmatrix} = 1 \neq 0.$$

Hence the rank of $\mathbf{A}$ is 2, in accordance with Example 2.

Remark 3. For further results on matrices see § 1.25, p. 87.

1.17. Determinants

Definition 1. The *determinant of order n* of a square matrix

$$\mathbf{A} = \left\| \begin{matrix} a_{11}, & a_{12}, & \ldots, & a_{1n} \\ a_{21}, & a_{22}, & \ldots, & a_{2n} \\ \ldots & \ldots & \ldots & \ldots \\ a_{n1}, & a_{n2}, & \ldots, & a_{nn} \end{matrix} \right\|$$

is defined as the number

$$A = \sum (-1)^r \, a_{1k_1} a_{2k_2} \ldots a_{nk_n} \,,$$

where the symbol $\sum$ indicates the sum of all terms for all possible permutations $(k_1, k_2, \ldots, k_n)$ of the numbers $1, 2, \ldots, n$, the integer r being the number of inversions (Definition 1.12.2, p. 55) in the permutation $(k_1, k_2, \ldots, k_n)$; we write

$$A = \begin{vmatrix} a_{11}, & a_{12}, & \ldots, & a_{1n} \\ a_{21}, & a_{22}, & \ldots, & a_{2n} \\ \ldots & \ldots & \ldots & \ldots \\ a_{n1}, & a_{n2}, & \ldots, & a_{nn} \end{vmatrix} . \tag{1}$$

Example 1.

$$\begin{vmatrix} a_{11}, & a_{12} \\ a_{21}, & a_{22} \end{vmatrix} = (-1)^0 \, a_{11} a_{22} + (-1)^1 \, a_{12} a_{21} = a_{11} a_{22} - a_{12} a_{21} \,,$$

since the permutations (1, 2), and (2, 1) of the numbers 1, 2 have no inversions and 1 inversion, respectively.

Theorem 1. *The value of a determinant remains unaltered if its columns and rows are interchanged*:

$$\begin{vmatrix} a_{11}, & a_{12}, & \ldots, & a_{1n} \\ a_{21}, & a_{22}, & \ldots, & a_{2n} \\ \ldots & \ldots & \ldots & \ldots \\ a_{n1}, & a_{n2}, & \ldots, & a_{nn} \end{vmatrix} = \begin{vmatrix} a_{11}, & a_{21}, & \ldots, & a_{n1} \\ a_{12}, & a_{22}, & \ldots, & a_{n2} \\ \ldots & \ldots & \ldots & \ldots \\ a_{1n}, & a_{2n}, & \ldots, & a_{nn} \end{vmatrix} .$$

Hence, all properties of determinants expressed in the following text for rows hold also for columns and vice versa.

Theorem 2. *The value of a determinant is unaltered if, to one of its rows, a linear combination of the remaining rows is added.*

Theorem 3. *If one of the rows is a linear combination of the remaining rows, then the value of the determinant is zero.*

Theorem 4. *The value of the determinant changes its sign if we interchange two of its rows.*

Definition 2. The determinant

$$A_{ij} = \begin{vmatrix} a_{11}, & a_{12}, & \ldots, & a_{1,j-1}, & a_{1,j+1}, & \ldots, & a_{1n} \\ a_{21}, & a_{22}, & \ldots, & a_{2,j-1}, & a_{2,j+1}, & \ldots, & a_{2n} \\ \ldots & \ldots & \ldots & \ldots & \ldots & \ldots & \ldots \\ a_{i-1,1}, & a_{i-1,2}, & \ldots, & a_{i-1,j-1}, & a_{i-1,j+1}, & \ldots, & a_{i-1,n} \\ a_{i+1,1}, & a_{i+1,2}, & \ldots, & a_{i+1,j-1}, & a_{i+1,j+1}, & \ldots, & a_{i+1,n} \\ \ldots & \ldots & \ldots & \ldots & \ldots & \ldots & \ldots \\ a_{n1}, & a_{n2}, & \ldots, & a_{n,j-1}, & a_{n,j+1}, & \ldots, & a_{nn} \end{vmatrix}, \qquad (2)$$

originating from the determinant A by omitting the i-th row and j-th column is called the *minor of order $n - 1$ of the determinant A belonging to the element a_{ij}.*

The *cofactor* A_{ij} *of the element a_{ij} in the determinant A* is defined as the minor A_{ij} equipped with the sign $(-1)^{i+j}$; thus, $\mathrm{A}_{ij} = (-1)^{i+j} A_{ij}$.

Theorem 5 (*The Expansion of a Determinant According to the i-th Row*). *For the determinant* (1) *the following expansion holds*:

$$A = a_{i1}\mathrm{A}_{i1} + a_{i2}\mathrm{A}_{i2} + \ldots + a_{in}\mathrm{A}_{in} =$$
$$= (-1)^{i+1} a_{i1}A_{i1} + (-1)^{i+2} a_{i2}A_{i2} + \ldots + (-1)^{i+n} a_{in}A_{in} .$$

Theorem 6. *For $i \neq j$,*

$$a_{i1}\mathrm{A}_{j1} + a_{i2}\mathrm{A}_{j2} + \ldots + a_{in}\mathrm{A}_{jn} = 0 .$$

Theorem 7 (*The Addition Rule*). *The relation*

$$\begin{vmatrix} a_1 + b_1, & a_{12}, & \ldots, & a_{1n} \\ a_2 + b_2, & a_{22}, & \ldots, & a_{2n} \\ \ldots & \ldots & \ldots & \ldots \\ a_n + b_n, & a_{n2}, & \ldots, & a_{nn} \end{vmatrix} = \begin{vmatrix} a_1, & a_{12}, & \ldots, & a_{1n} \\ a_2, & a_{22}, & \ldots, & a_{2n} \\ \ldots & \ldots & \ldots & \ldots \\ a_n, & a_{n2}, & \ldots, & a_{nn} \end{vmatrix} + \begin{vmatrix} b_1, & a_{12}, & \ldots, & a_{1n} \\ b_2, & a_{22}, & \ldots, & a_{2n} \\ \ldots & \ldots & \ldots & \ldots \\ b_n, & a_{n2}, & \ldots, & a_{nn} \end{vmatrix}$$

holds, and similarly for other columns.

Theorem 8 (*The Multiplication of a Determinant by a Number*). *The relation*

$$\begin{vmatrix} ca_{11}, & a_{12}, & \ldots, & a_{1n} \\ ca_{21}, & a_{22}, & \ldots, & a_{2n} \\ \ldots & \ldots & \ldots & \ldots \\ ca_{n1}, & a_{n2}, & \ldots, & a_{nn} \end{vmatrix} = c \begin{vmatrix} a_{11}, & a_{12}, & \ldots, & a_{1n} \\ a_{21}, & a_{22}, & \ldots, & a_{2n} \\ \ldots & \ldots & \ldots & \ldots \\ a_{n1}, & a_{n2}, & \ldots, & a_{nn} \end{vmatrix}$$

holds, and similarly for other columns.

In other words: *We multiply a determinant by a number c if we multiply by this number all the elements of a row or of a column.*

Theorem 9 (*The Multiplication of Determinants*). *The relation*

$$\begin{vmatrix} a_{11}, & a_{12}, & \dots, & a_{1n} \\ a_{21}, & a_{22}, & \dots, & a_{2n} \\ \dots & \dots & \dots & \dots \\ a_{n1}, & a_{n2}, & \dots, & a_{nn} \end{vmatrix} \begin{vmatrix} b_{11}, & b_{12}, & \dots, & b_{1n} \\ b_{21}, & b_{22}, & \dots, & b_{2n} \\ \dots & \dots & \dots & \dots \\ b_{n1}, & b_{n2}, & \dots, & b_{nn} \end{vmatrix} = \begin{vmatrix} c_{11}, & c_{12}, & \dots, & c_{1n} \\ c_{21}, & c_{22}, & \dots, & c_{2n} \\ \dots & \dots & \dots & \dots \\ c_{n1}, & c_{n2}, & \dots, & c_{nn} \end{vmatrix},$$

holds, where

$$c_{ik} = a_{i1}b_{1k} + a_{i2}b_{2k} + \dots + a_{in}b_{nk} \quad (i, k = 1, 2, \dots, n).$$

Remark 1 (*Evaluation of a determinant*).

1. 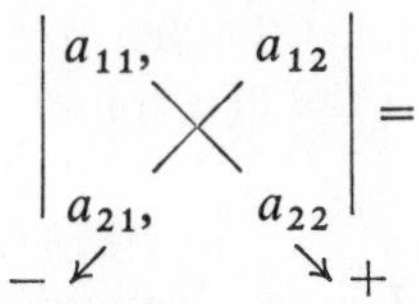
$$\begin{vmatrix} a_{11}, & a_{12} \\ a_{21}, & a_{22} \end{vmatrix} = a_{11}a_{22} - a_{12}a_{21}.$$

2. *Sarrus's rule* for the evaluation of a determinant of the third order:

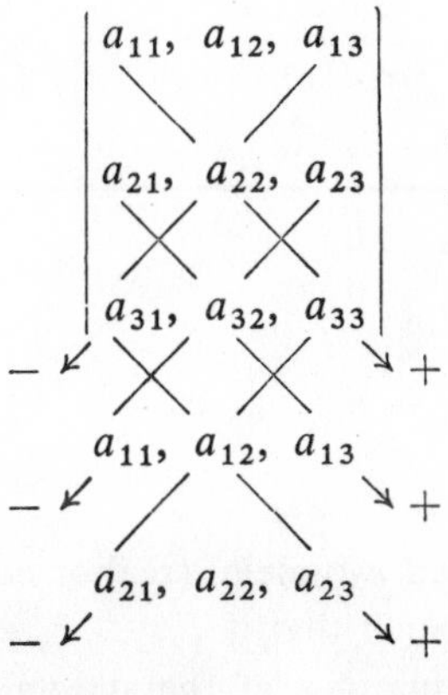

$$\begin{vmatrix} a_{11}, & a_{12}, & a_{13} \\ a_{21}, & a_{22}, & a_{23} \\ a_{31}, & a_{32}, & a_{33} \end{vmatrix} = a_{11}a_{22}a_{33} + a_{21}a_{32}a_{13} + a_{31}a_{12}a_{23} - a_{13}a_{22}a_{31} - a_{23}a_{32}a_{11} - a_{33}a_{12}a_{21}.$$

(The scheme repeats the rows a_{11}, a_{12}, a_{13} and a_{21}, a_{22}, a_{23} below the determinant; products along the diagonals ↘ are taken with $+$, those along ↙ with $-$.)

3. The evaluation of a determinant of order n for $n \geqq 3$ can be reduced, according to Theorem 5, to the evaluation of a determinant of order $n - 1$. First, it is often advantageous to arrange the original determinant by means of Theorem 2 or Theorem 8 in order to get, in a certain row or column, as many zeros as possible. Then, we expand the determinant according to this row or column (Theorem 5).

Example 2.

$$\begin{vmatrix} 1, & -1, & 2, & 4 \\ 0, & 1, & -1, & 2 \\ 3, & -1, & 2, & 0 \\ -1, & 0, & 3, & 2 \end{vmatrix} = 2 \begin{vmatrix} 1, & -1, & 2, & 2 \\ 0, & 1, & -1, & 1 \\ 3, & -1, & 2, & 0 \\ -1, & 0, & 3, & 1 \end{vmatrix} = 2 \begin{vmatrix} 1, & -1, & 2, & 2 \\ 0, & 1, & -1, & 1 \\ 3, & -1, & 2, & 0 \\ 0, & -1, & 5, & 3 \end{vmatrix} =$$

$$= 2 \left(1 \begin{vmatrix} 1, & -1, & 1 \\ -1, & 2, & 0 \\ -1, & 5, & 3 \end{vmatrix} - 0 \begin{vmatrix} -1, & 2, & 2 \\ -1, & 2, & 0 \\ -1, & 5, & 3 \end{vmatrix} + 3 \begin{vmatrix} -1, & 2, & 2 \\ 1, & -1, & 1 \\ -1, & 5, & 3 \end{vmatrix} - 0 \begin{vmatrix} -1, & 2, & 2 \\ 1, & -1, & 1 \\ -1, & 2, & 0 \end{vmatrix} \right) =$$

$$= 2 \begin{vmatrix} 1, & -1, & 1 \\ -1, & 2, & 0 \\ -1, & 5, & 3 \end{vmatrix} + 6 \begin{vmatrix} -1, & 2, & 2 \\ 1, & -1, & 1 \\ -1, & 5, & 3 \end{vmatrix} = 48\,.$$

We proceeded in the above evaluation as follows: First, a common factor 2 was removed from the last column; then we added the first row to the last row, finally, we expanded the determinant according to the first column.

1.18. Systems of Linear Equations

(a) Definition and Properties of Systems of Linear Equations

Definition 1. By a *system of m linear equations in n unknowns* $x_1, x_2, \ldots, x_n$ we understand the system

$$\begin{aligned} a_{11}x_1 + a_{12}x_2 + \ldots + a_{1n}x_n &= b_1\,, \\ a_{21}x_1 + a_{22}x_2 + \ldots + a_{2n}x_n &= b_2\,, \\ \ldots\ldots\ldots\ldots\ldots\ldots\ldots\ldots\ldots \\ a_{m1}x_1 + a_{m2}x_2 + \ldots + a_{mn}x_n &= b_m \end{aligned} \tag{1}$$

($a_{11}, \ldots, a_{mn}, b_1, \ldots, b_m$ being given real or complex numbers).

By a *solution of the system* (1) we mean any ordered n-tuple of (real or complex) numbers $(\xi_1, \xi_2, \ldots, \xi_n)$, i.e. an n-component vector such that if $\xi_1, \ldots, \xi_n$ are substituted for the unknowns $x_1, \ldots, x_n$, then all the equations of the system (1) are satisfied. Two systems of linear equations (in the same number of unknowns $x_1, \ldots, x_n$) are said to be *equivalent systems of linear equations* if every solution of the first system is also a solution of the second system and vice versa. The matrix

$$\mathbf{A} = \left\| \begin{matrix} a_{11}, & a_{12}, & \ldots, & a_{1n} \\ a_{21}, & a_{22}, & \ldots, & a_{2n} \\ \ldots & \ldots & \ldots & \ldots \\ a_{m1}, & a_{m2}, & \ldots, & a_{mn} \end{matrix} \right\|$$

is called the *matrix of the system* (1). The matrix

$$\mathbf{B} = \left\| \begin{matrix} a_{11}, & a_{12}, & \ldots, & a_{1n}, & b_1 \\ a_{21}, & a_{22}, & \ldots, & a_{2n}, & b_2 \\ \ldots & \ldots & \ldots & \ldots & \ldots \\ a_{m1}, & a_{m2}, & \ldots, & a_{mn}, & b_m \end{matrix} \right\|$$

is the so-called *augmented matrix of the system* (1).

Theorem 1 (*Theorem of Frobenius*). *The system* (1) *is solvable if and only if the rank of the matrix of the system is equal to the rank of the augmented matrix of the system.*

Theorem 2. *The system of m homogeneous equations in n unknowns*

$$\begin{aligned} a_{11}x_1 + \ldots + a_{1n}x_n &= 0\,, \\ \ldots\ldots\ldots\ldots\ldots\ldots& \\ a_{m1}x_1 + \ldots + a_{mn}x_n &= 0 \end{aligned} \qquad (2)$$

has always the (trivial) zero solution $\mathbf{0} = (0, 0, \ldots, 0)$.

Theorem 3. *If the system* (2) *has a solution* $\xi = (\xi_1, \ldots, \xi_n)$, *then it has also the solution* $\alpha\xi = (\alpha\xi_1, \ldots, \alpha\xi_n)$, *where* α *is an arbitrary real or complex number.*

If the vectors $\xi^{(1)} = (\xi_1^{(1)}, \ldots, \xi_n^{(1)}), \ldots, \xi^{(k)} = (\xi_1^{(k)}, \ldots, \xi_n^{(k)})$ *are solutions of the system* (2), *then every linear combination of the form* $\alpha_1\xi^{(1)} + \alpha_2\xi^{(2)} + \ldots + \alpha_k\xi^{(k)}$ (see Definition 1.15.4) *is also a solution of the system* (2).

Theorem 4. *If the rank of the matrix of the system* (2) *of homogeneous equations is h, then the system* (2) *has* $n - h$ *linearly independent solutions* [in the sense of linear independence of vectors (see Definition 1.15.3)] *and every solution of the system* (2) *is a linear combination of these* $n - h$ *solutions.*

In particular, if $h = n$, then the system has only the trivial solution $\mathbf{0} = (0, \ldots, 0)$. If $m = n$ in (2), then the system has a nontrivial solution if and only if the determinant of the system is zero.

Theorem 5. *Let the rank of the matrix of the system* (1) *be h, let* $\boldsymbol{\eta} = (\eta_1, \ldots, \eta_n)$ *be a solution of the system* (1) *and let* $\xi^{(1)}, \xi^{(2)}, \ldots, \xi^{(n-h)}$ *be* $n - h$ *linearly independent solutions of the system* (2). *Then every solution of the system* (1) *is the sum of a solution* $\alpha_1\xi^{(1)} + \ldots + \alpha_{n-h}\xi^{(n-h)}$ *of the homogeneous system* (2) *and the solution* $\boldsymbol{\eta}$ *of the system* (1); thus, the form of every solution of the system (1) is $\alpha_1\xi^{(1)} + \ldots + \alpha_{n-h}\xi^{(n-h)} + \boldsymbol{\eta}$, where $\alpha_1, \ldots, \alpha_{n-h}$ are real or complex numbers.

(b) Solution of Systems of Linear Equations without the Use of Determinants

Theorem 6. *The augmented matrix* $\mathbf{B}$ *of the system* (1) *can be transformed by the operations* 1–4 *of Theorem* 1.16.3, p. 65 (see Example 1.16.2, p. 65), *to the* *matrix* $\mathbf{B}'$ *which has only zeros below the principal diagonal. The system of the equations in n unknowns whose augmented matrix is the matrix* $\mathbf{B}'$ *is equivalent to the system* (1). In this way, the solution of the system (1) is transformed to the solution of a system which can be easily solved.

Example 1.

(a)

$$\begin{aligned} x - 2y + 3z &= 2\,, \\ 3x - y + z &= 0\,, \\ 3x + 4y - 7z &= -6\,, \\ 5y - 8z &= -6\,; \end{aligned} \qquad \mathbf{B} = \begin{Vmatrix} 1, & -2, & 3, & 2 \\ 3, & -1, & 1, & 0 \\ 3, & 4, & -7, & -6 \\ 0, & 5, & -8, & -6 \end{Vmatrix}.$$

The matrix $\mathbf{B}$ can be arranged as follows: The fourth row is a linear combination of the second and third rows and therefore can be omitted; also the third row is a linear combination of the first and second rows (namely, it is the difference of twice the second row and three times the first row) and thus can also be omitted. It is sufficient to consider the matrix

$$\begin{Vmatrix} 1, & -2, & 3, & 2 \\ 3, & -1, & 1, & 0 \end{Vmatrix}.$$

Here, we subtract three times the first row from the second one and get the matrix

$$\mathbf{B}' = \begin{Vmatrix} 1, & -2, & 3, & 2 \\ 0, & 5, & -8, & -6 \end{Vmatrix}.$$

Thus, we solve the system

$$\begin{aligned} x - 2y + 3z &= 2\,, \\ 5y - 8z &= -6\,. \end{aligned}$$

We get $y = \frac{1}{5}(8z - 6)$, $x = \frac{1}{5}(z - 2)$. Hence, we can choose an arbitrary (complex) number for z. The system has an infinite number of solutions $x = \frac{1}{5}(\alpha - 2)$, $y = \frac{1}{5}(8\alpha - 6)$, $z = \alpha$ (α being arbitrary).

(b)

$$\begin{aligned} x - 2y + 3z &= 2\,, \\ 3x - y + 19z &= 0\,, \\ 3x + 4y - 7z &= 1\,, \\ 3y - 6z &= -6\,. \end{aligned}$$

From the matrix

$$\mathbf{B} = \begin{Vmatrix} 1, & -2, & 3, & 2 \\ 3, & -1, & 19, & 0 \\ 3, & 4, & -7, & 1 \\ 0, & 3, & -6, & -6 \end{Vmatrix}$$

we get successively

$$\begin{Vmatrix} 1, & -2, & 3, & 2 \\ 0, & 5, & 10, & -6 \\ 0, & 10, & -16, & -5 \\ 0, & 3, & -6, & -6 \end{Vmatrix}, \quad \begin{Vmatrix} 1, & -2, & 3, & 2 \\ 0, & 5, & 10, & -6 \\ 0, & 10, & -16, & -5 \\ 0, & 1, & -2, & -2 \end{Vmatrix}, \quad \begin{Vmatrix} 1, & -2, & 3, & 2 \\ 0, & 5, & 10, & -6 \\ 0, & 0, & -36, & 7 \\ 0, & 0, & 5, & 1 \end{Vmatrix},$$

$$\mathbf{B}' = \begin{Vmatrix} 1, & -2, & 3, & 2 \\ 0, & 5, & 10, & -6 \\ 0, & 0, & -36, & 7 \\ 0, & 0, & 0, & 71 \end{Vmatrix}.$$

To get the solution we solve the system

$$\begin{aligned} x - 2y + 3z &= 2, \\ 5x + 10z &= -6, \\ -36z &= 7, \\ 0 &= 71; \end{aligned}$$

however, this system has no solution, for the last equation cannot be satisfied. Hence, the given system is not solvable.

REMARK 1. When rearranging the matrix $\mathbf{B}$ of Theorem 6 it is sometimes advantageous to interchange two columns. This can be done, provided neither of them is the last column; however, we must then interchange the unknowns in the resulting system corresponding to the interchanged columns. The procedure is obvious from Example 2.

Example 2.

$$\begin{aligned} 3x + y + 3z &= 2, \\ -x \quad\quad + 3z &= 3, \\ 4x \quad\quad - z &= 0. \end{aligned} \qquad \mathbf{B} = \begin{Vmatrix} 3, & 1, & 3, & 2 \\ -1, & 0, & 3, & 3 \\ 4, & 0, & -1, & 0 \end{Vmatrix};$$

we interchange the first and second columns:

$$\begin{Vmatrix} 1, & 3, & 3, & 2 \\ 0, & -1, & 3, & 3 \\ 0, & 4, & -1, & 0 \end{Vmatrix}, \quad \mathbf{B}' = \begin{Vmatrix} 1, & 3, & 3, & 2 \\ 0, & -1, & 3, & 3 \\ 0, & 0, & 11, & 12 \end{Vmatrix}.$$

The solution to be found is then the solution of the system

$$\begin{aligned} y + 3x + 3z &= 2, \\ -x + 3z &= 3, \\ 11z &= 12. \end{aligned}$$

(c) Solution of Systems of Linear Equations by Means of Determinants

Theorem 7 (*Cramer's Rule*). *The system of n equations in n unknowns*

$$\begin{array}{c} a_{11}x_1 + \ldots + a_{1n}x_n = b_1 \,, \\ \ldots\ldots\ldots\ldots\ldots\ldots\ldots\ldots \\ a_{n1}x_1 + \ldots + a_{nn}x_n = b_n \end{array} \tag{3}$$

with a non-zero determinant of the system

$$D = \begin{vmatrix} a_{11}, \ldots, a_{1n} \\ \ldots\ldots\ldots\ldots \\ a_{n1}, \ldots, a_{nn} \end{vmatrix} \neq 0 \,,$$

has a unique solution $(x_1, \ldots, x_n)$, *where* $x_i = D_i/D$; *here,* D_i *is the determinant obtained by replacing the i-th column of D by the column of elements forming the right-hand sides of equations* (3).

Example 3.

$$\begin{aligned} 3x_1 - 2x_2 + x_3 &= 1 \,, \\ x_1 + x_2 - x_3 &= -2 \,, \\ 2x_1 \qquad - 3x_3 &= 0 \,. \end{aligned}$$

$$D = \begin{vmatrix} 3, & -2, & 1 \\ 1, & 1, & -1 \\ 2, & 0, & -3 \end{vmatrix} = -13 \neq 0 \,; \quad x_1 = -\tfrac{1}{13} \begin{vmatrix} 1, & -2, & 1 \\ -2, & 1, & -1 \\ 0, & 0, & -3 \end{vmatrix} = -\tfrac{9}{13} \,,$$

$$x_2 = -\tfrac{1}{13} \begin{vmatrix} 3, & 1, & 1 \\ 1, & -2, & -1 \\ 2, & 0, & -3 \end{vmatrix} = -\tfrac{23}{13} \,, \quad x_3 = -\tfrac{1}{13} \begin{vmatrix} 3, & -2, & 1 \\ 1, & 1, & -2 \\ 2, & 0, & 0 \end{vmatrix} = -\tfrac{6}{13} \,.$$

Theorem 8 (*The System of m Equations in n Unknowns*). *Let the matrix of a system and the augmented matrix of the system have the same rank h. Solution: In the matrix, we find a minor* $D_h \neq 0$ *of order h. In the h equations of the given system* (1) *containing the elements of the determinant* D_h, *we leave on the left-hand side those unknowns whose coefficients belong to* D_h. *We choose arbitrary values for the remaining unknowns, transfer them to the right-hand side and solve this system of h equations in h unknowns by Cramer's Rule. We can always proceed this way, both for homogeneous and non-homogeneous systems.*

Example 4.

(a)

$$\begin{aligned} 3x_1 - 2x_2 + x_3 - x_4 &= 2 \,, \\ -x_1 \qquad + 3x_3 + x_4 &= -1 \,, \\ x_2 + 3x_3 + 2x_4 &= 3 \,. \end{aligned}$$

The matrix of the system and the augmented matrix have rank 3,

$$\begin{vmatrix} 3, & -2, & 1 \\ -1, & 0, & 3 \\ 0, & 1, & 3 \end{vmatrix} = -16 .$$

Transform the system to the form

$$\begin{aligned} 3x_1 - 2x_2 + \ x_3 &= \ \ 2 + \ x_4 , \\ -x_1 \qquad\quad + 3x_3 &= -1 - \ x_4 , \\ x_2 + 3x_3 &= \ \ 3 - 2x_4 . \end{aligned}$$

$$x_1 = -\tfrac{1}{16}\begin{vmatrix} 2 + \ x_4, & -2, & 1 \\ -1 - \ x_4, & 0, & 3 \\ 3 - 2x_4, & 1, & 3 \end{vmatrix} = -\tfrac{1}{16}\left[\begin{vmatrix} 2, & -2, & 1 \\ -1, & 0, & 3 \\ 3, & 1, & 3 \end{vmatrix} + \right.$$

$$\left. + x_4 \begin{vmatrix} 1, & -2, & 1 \\ -1, & 0, & 3 \\ -2, & 1, & 3 \end{vmatrix}\right] = -\tfrac{1}{16}[-31 + 2x_4] ,$$

$$x_2 = -\tfrac{1}{16}\begin{vmatrix} 3, & 2 + \ x_4, & 1 \\ -1, & -1 - \ x_4, & 3 \\ 0, & 3 - 2x_4, & 3 \end{vmatrix} = -\tfrac{1}{16}[-33 + 14x_4] ,$$

$$x_3 = -\tfrac{1}{16}\begin{vmatrix} 3, & -2, & 2 + \ x_4 \\ -1, & 0, & -1 - \ x_4 \\ 0, & 1, & 3 - 2x_4 \end{vmatrix} = -\tfrac{1}{16}[-5 + 6x_4]$$

(x_4 is arbitrary).

(b) *The system of two homogeneous equations in three unknowns*

$$\begin{aligned} a_{11}x_1 + a_{12}x_2 + a_{13}x_3 &= 0 , \\ a_{21}x_1 + a_{22}x_2 + a_{23}x_3 &= 0 \end{aligned}$$

is as follows.

If the rank of the matrix of the system is 2, then the solution is

$$x_1 : x_2 : x_3 = \begin{vmatrix} a_{12}, & a_{13} \\ a_{22}, & a_{23} \end{vmatrix} : \begin{vmatrix} a_{13}, & a_{11} \\ a_{23}, & a_{21} \end{vmatrix} : \begin{vmatrix} a_{11}, & a_{12} \\ a_{21}, & a_{22} \end{vmatrix} .$$

REMARK 2. On the numerical solution of systems of linear equations see Chap. 30.

1.19. Algebraic Equations of Higher Degree. General Properties

Definition 1. An equation

$$a_0x^n + a_1x^{n-1} + \ldots + a_n = 0 , \quad a_0 \neq 0 , \tag{1}$$

where $a_0, a_1, \ldots, a_n$ are real or complex numbers, is called an *algebraic equation of degree n.*

REMARK 1. On the concept of a *root*, its *multiplicity* and theorems on the number of roots see § 1.14, p. 58.

Theorem 1 (*Properties of Roots*). *The roots $x_1, x_2, \ldots, x_n$ of the equation*

$$x^n + a_1 x^{n-1} + \ldots + a_n = 0 \tag{2}$$

satisfy the relations

$$a_1 = -(x_1 + x_2 + \ldots + x_n) = -\sum_{i=1}^{n} x_i ,$$

$$a_2 = x_1 x_2 + x_1 x_3 + \ldots + x_{n-1} x_n = \sum_{\substack{i,j=1\\ i<j}}^{n} x_i x_j ,$$

$$a_3 = -(x_1 x_2 x_3 + x_1 x_2 x_4 + \ldots + x_{n-2} x_{n-1} x_n) = -\sum_{\substack{i,j,k=1\\ i<j<k}}^{n} x_i x_j x_k ,$$

$$\cdots\cdots\cdots\cdots\cdots\cdots\cdots\cdots\cdots\cdots\cdots\cdots$$

$$a_n = (-1)^n x_1 x_2 \ldots x_n .$$

REMARK 2. The expressions

$$y_1 = \sum_{i=1}^{n} x_i , \quad y_2 = \sum_{\substack{i,j=1\\ i<j}}^{n} x_i x_j , \ \ldots , \ y_n = x_1 x_2 \ldots x_n$$

are called *elementary symmetric functions* of the variables $x_1, x_2, \ldots, x_n$.

REMARK 3. On the numerical solution of algebraic equations see Chap. 31.

Definition 2. The *resultant of two algebraic equations*

$$\begin{aligned} a_0 x^m + a_1 x^{m-1} + \ldots + a_m &= 0 , \quad a_0 \neq 0 , \\ b_0 x^n + b_1 x^{n-1} + \ldots + b_n &= 0 , \quad b_0 \neq 0 \end{aligned} \tag{3}$$

is defined as the determinant

$$\left| \begin{array}{cccccccc} a_0, & a_1, & \ldots, & a_{m-1}, & a_m, & 0, & \ldots, & 0 \\ 0, & a_0, & a_1, & \ldots, & a_{m-1}, & a_m, & 0, \ldots, & 0 \\ \cdots & \cdots & \cdots & \cdots & \cdots & \cdots & \cdots & \cdots \\ 0, & \ldots, 0, & a_0, & a_1, & \ldots, & & a_{m-1}, & a_m \\ b_0, & b_1, & \ldots, & b_{n-1}, & b_n, & 0, & \ldots, & 0 \\ 0, & b_0, & b_1, & \ldots, & b_{n-1}, & b_n, & 0, \ldots, & 0 \\ \cdots & \cdots & \cdots & \cdots & \cdots & \cdots & \cdots & \cdots \\ 0, & \ldots, 0, & b_0, & b_1, & \ldots, & & b_{n-1}, & b_n \end{array} \right| \begin{array}{l} \left.\begin{array}{c} \\ \\ \\ \\ \end{array}\right\} n \text{ rows} \\ \left.\begin{array}{c} \\ \\ \\ \\ \end{array}\right\} m \text{ rows} . \end{array}$$

Theorem 2. *The equations* (3) *have a common root if and only if their resultant is equal to zero.*

1.20. Quadratic, Cubic and Biquadratic Equations

(a) A quadratic equation is of the form

(a) $ax^2 + bx + c = 0 \quad (a \neq 0)$ or
(b) $x^2 + px + q = 0$ (*reduced form*).

Definition 1. The *discriminant* of the equation (a) is the number $D = b^2 - 4ac$ and that of the equation (b) is the number $D = p^2 - 4q$.

Theorem 1.

For $D \neq 0$, the equation has two distinct roots;
for $D = 0$, the equation has one double root.

If the coefficients of the equation are real, then
for $D > 0$, it has two distinct real roots;
for $D < 0$, it has two complex conjugate roots;
for $D = 0$, it has only one real (double) root.

The solution can be found:

1. by factorization into linear factors:

$$\begin{aligned} ax^2 + bx + c &= a(x - x_1)(x - x_2) \\ a(x_1 + x_2) &= -b, \\ a(x_1 x_2) &= c \end{aligned} \quad \text{or} \quad \begin{aligned} x^2 + px + q &= (x - x_1)(x - x_2), \\ x_1 + x_2 &= -p, \\ x_1 x_2 &= q \end{aligned}$$

[e.g. $x^2 - 5x + 6 = 0$, $(x - 2)(x - 3) = 0$, $x_1 = 2$, $x_2 = 3$];

2. in the case of the equation $ax^2 + bx + c = 0$ by the formula

$$x_{1,2} = \frac{-b \pm \sqrt{(b^2 - 4ac)}}{2a};$$

3. in the case of the equation $x^2 + px + q = 0$ by the formula

$$x_{1,2} = -\frac{p}{2} \pm \sqrt{\left(\frac{p^2}{4} - q\right)}.$$

(b) A cubic equation is of the form

$$ax^3 + bx^2 + cx + d = 0, \quad a \neq 0. \tag{1}$$

Theorem 2. *By the substitution $x = y - b/3a$ and dividing by a, the equation* (1) *becomes*

$$y^3 + 3py + 2q = 0\,, \tag{2}$$

where

$$3p = \frac{3ac - b^2}{3a^2}\,, \quad 2q = \frac{2b^3}{27a^3} - \frac{bc}{3a^2} + \frac{d}{a}\,.$$

Definition 2. The *discriminant* of the equation (2) is the number $D = -p^3 - q^2$.

Theorem 3.

For $D \neq 0$, the equation (2) *has three distinct roots;*
for $D = 0$, the equation (2) *has either a double root (if $p^3 = -q^2 \neq 0$) or a triple zero root (if $p = q = 0$).*

If the coefficients of the equation (2) *are real, then*
for $D \geqq 0$, it has three real roots which are distinct if $D > 0$;
for $D < 0$, it has one real and two complex conjugate roots.

Solution (see also Chap. 31):

1. By factorization into linear factors:

$$ax^3 + bx^2 + cx + d = 0\,; \quad a(x - x_1)(x - x_2)(x - x_3) = 0$$

(x_1, x_2, x_3 are the roots);

$$x_1 + x_2 + x_3 = -\frac{b}{a}\,, \quad x_1x_2 + x_1x_3 + x_2x_3 = \frac{c}{a}\,, \quad x_1x_2x_3 = -\frac{d}{a}$$

[e.g. $x^3 + 5x^2 + 6x = 0$; $x(x^2 + 5x + 6) = x(x + 2)(x + 3)$; the roots are $x_1 = 0$, $x_2 = -2$, $x_3 = -3$].

2. *The algebraic solution* (*Tartaglia's or Cardan's Formulae*). The roots y_1, y_2, y_3 of equation (2) are

$$y_1 = u + v\,, \quad y_2 = \varepsilon_1 u + \varepsilon_2 v\,, \quad y_3 = \varepsilon_2 u + \varepsilon_1 v\,,$$

where

$$\varepsilon_{1,2} = -\frac{1}{2} \pm i\frac{\sqrt{3}}{2}\,, \quad u = \sqrt[3]{[-q + \sqrt{(q^2 + p^3)}]}\,, \quad v = \sqrt[3]{[-q - \sqrt{(q^2 + p^3)}]}\,;$$

here we choose the cube roots (see § 1.21, p. 80) so that $uv = -p$. This method is not suitable if (2) has real coefficients and $D > 0$, since the real roots y_1, y_2, y_3 are expressed in terms of roots of complex numbers (the irreducible case).

3. *The trigonometric solution.* Let the coefficients p, q of the equation (2) be real and different from zero. Denote the roots by y_1, y_2, y_3. Put $r = \varepsilon\sqrt{|p|}$, where

TABLE 1.1

$p < 0$		$p > 0$	Check
$p^3 + q^2 \leqq 0$	$p^3 + q^2 > 0$		
$\cos \varphi = \frac{q}{r^3}$ *	$\cosh \varphi = \frac{q}{r^3}$	$\sinh \varphi = \frac{b}{r^3}$	
$y_1 = -2r \cos \frac{\varphi}{3}$	$y_1 = -2r \cosh \frac{\varphi}{3}$	$y_1 = -2r \sinh \frac{\varphi}{3}$	
$y_2 = 2r \cos \left(60° - \frac{\varphi}{3}\right)$	$y_2 = r \cosh \frac{\varphi}{3} + i \sqrt{(3)}\, r \sinh \frac{\varphi}{3}$	$y_2 = r \sinh \frac{\varphi}{3} + i \sqrt{(3)}\, r \cosh \frac{\varphi}{3}$	$y_1 + y_2 + y_3 = 0$
$y_3 = 2r \cos \left(60° + \frac{\varphi}{3}\right)$	$y_3 = r \cosh \frac{\varphi}{3} - i \sqrt{(3)}\, r \sinh \frac{\varphi}{3}$	$y_3 = r \sinh \frac{\varphi}{3} - i \sqrt{(3)}\, r \cosh \frac{\varphi}{3}$	

* φ is in the interval (0°, 90°), $r = \varepsilon \sqrt{|p|}$ (see above).

$\varepsilon = 1$ if $q > 0$ and $\varepsilon = -1$ if $q < 0$. Then the roots can be determined by means of the trigonometric or hyperbolic functions according to Table 1.1.

If $q = 0$ in equation (2), then the equation has the common factor y and can be solved easily.

If $p = 0$ in equation (2), then (2) is a binomial equation (see § 1.21, p. 80).

(c) A biquadratic (or quartic) equation has the form:

$$ax^4 + bx^3 + cx^2 + dx + e = 0\,, \quad a \neq 0\,. \tag{3}$$

Theorem 4. *By the substitution $x = y - b/4a$ and dividing by a, the equation* (5) *becomes*

$$x^4 + py^2 + qy + r = 0 \tag{4}$$

where

$$p = -\frac{3b^2}{8a^2} + \frac{c}{a}\,, \quad q = \frac{b^3}{8a^3} - \frac{bc}{2a^2} + \frac{d}{a}\,, \quad r = -\frac{3b^4}{256a^4} + \frac{b^2c}{16a^3} - \frac{bd}{4a^2} + \frac{e}{a}\,.$$

Solution:

1. By factorization into linear factors:

$$ax^4 + bx^3 + cx^2 + dx + e = 0\,; \quad a(x - x_1)(x - x_2)(x - x_3)(x - x_4) = 0$$

(x_1, x_2, x_3, x_4 are the roots);

$$x_1 + x_2 + x_3 + x_4 = -\frac{b}{a}; \quad x_1x_2 + x_1x_3 + x_1x_4 + x_2x_3 + x_2x_4 + x_3x_4 = \frac{c}{a};$$

$$x_1x_2x_3 + x_1x_2x_4 + x_1x_3x_4 + x_2x_3x_4 = -\frac{d}{a}; \quad x_1x_2x_3x_4 = \frac{e}{a}.$$

2. *The algebraic solution.* The roots y_1, y_2, y_3, y_4 of the equation (4) are

$$y_1 = \sqrt{z_1} + \sqrt{z_2} + \sqrt{z_3}, \quad y_2 = \sqrt{z_1} - \sqrt{z_2} - \sqrt{z_3},$$
$$y_3 = -\sqrt{z_1} + \sqrt{z_2} - \sqrt{z_3}, \quad y_4 = -\sqrt{z_1} - \sqrt{z_2} + \sqrt{z_3},$$

where z_1, z_2, z_3 are the roots of the equation (*the reducing cubic*)

$$z^3 + \frac{p}{2}z^2 + \left(\frac{p^2}{16} - \frac{r}{4}\right)z - \frac{q^2}{64} = 0;$$

here, the roots $\sqrt{z_1}, \sqrt{z_2}, \sqrt{z_3}$ should be chosen (see § 1.21) such that

$$\sqrt{(z_1)}\sqrt{(z_2)}\sqrt{(z_3)} = -\frac{q}{8}.$$

REMARK 1. This method is not suitable for numerical solution (see Chap. 31).

1.21. Binomial Equations

Definition 1. An equation of the form

$$x^n - \alpha = 0, \tag{1}$$

where α is a non-zero complex number, is called a *binomial equation.*

Definition 2. The roots of equation (1) are said to be the *n-th roots of the number* α and are denoted in the theory of algebraic equations by the symbol $\sqrt[n]{\alpha}$; thus, in this case (in contrast to § 1.8, p. 50) $\sqrt[n]{\alpha}$ stands for any of the n roots of the equation (1).

Theorem 1. *Equation* (1) *has n simple roots* $x_1, \ldots, x_n$ *given by*

$$x_{k+1} = \sqrt[n]{(r)}\left(\cos\frac{\varphi + 2k\pi}{n} + \mathrm{i}\sin\frac{\varphi + 2k\pi}{n}\right) \quad (k = 0, 1, \ldots, n - 1),$$

where $\alpha = r(\cos\varphi + \mathrm{i}\sin\varphi)$ *is the trigonometric form of the number* α, $\sqrt[n]{(r)} > 0$.

REMARK 1. By means of Theorem 1 we easily find all the n-th roots of any complex number.

Example 1. (a) $x^3 - 2 = 0$. First, $2 = 2(\cos 0 + i \sin 0)$. Hence

$$x_1 = \sqrt[3]{(2)}\,(\cos 0 + i \sin 0) = \sqrt[3]{(2)} \doteq 1{\cdot}260\,,$$

$$x_2 = \sqrt[3]{(2)}\,(\cos \tfrac{2}{3}\pi + i \sin \tfrac{2}{3}\pi) = \sqrt[3]{(2)}\left(-\frac{1}{2} + i\frac{\sqrt{3}}{2}\right),$$

$$x_3 = \sqrt[3]{(2)}\,(\cos \tfrac{4}{3}\pi + i \sin \tfrac{4}{3}\pi) = \sqrt[3]{(2)}\left(-\frac{1}{2} - i\frac{\sqrt{3}}{2}\right).$$

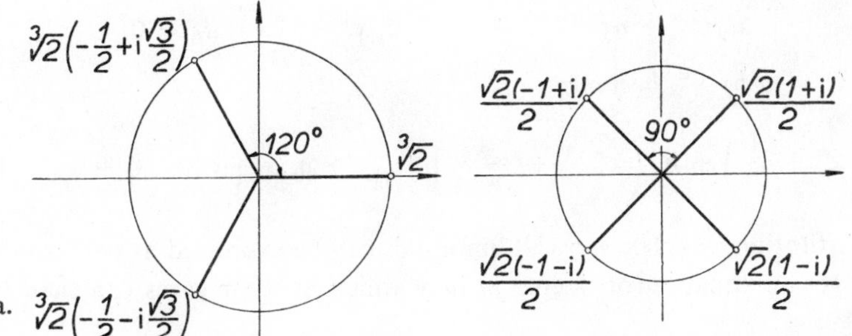

Fig. 1.5a. Fig. 1.5b.

(b) $x^4 + 1 = 0$. We have $-1 = \cos \pi + i \sin \pi$. Hence

$$x_1 = \cos \tfrac{1}{4}\pi + i \sin \tfrac{1}{4}\pi = \frac{\sqrt{2}}{2}(1 + i)\,,$$

$$x_2 = \cos \tfrac{3}{4}\pi + i \sin \tfrac{3}{4}\pi = \frac{\sqrt{2}}{2}(-1 + i)\,,$$

$$x_3 = \cos \tfrac{5}{4}\pi + i \sin \tfrac{5}{4}\pi = \frac{\sqrt{2}}{2}(-1 - i)\,,$$

$$x_4 = \cos \tfrac{7}{4}\pi + i \sin \tfrac{7}{4}\pi = \frac{\sqrt{2}}{2}(1 - i)\,.$$

REMARK 2. The roots of the equation $x^n - \alpha = 0$ $(\alpha \neq 0)$ form, in the Argand diagram, the vertices of an n-sided regular polygon inscribed in the circle with centre at the origin and radius $\sqrt[n]{|\alpha|} > 0$. Fig. 1.5 illustrates the roots of the equations $x^3 - 2 = 0$, $x^4 + 1 = 0$. One sees from the figure that the n-th roots can easily be constructed geometrically.

1.22. Reciprocal Equations

Definition 1. By a *reciprocal equation* we understand an equation of the form

$$a_0 x^n + a_1 x^{n-1} + \ldots + a_n = 0$$

where (a) $a_i = a_{n-i}$ $(i = 0, 1, \ldots, n)$ (*positively reciprocal equation*) or
(b) $a_i = -a_{n-i}$ $(i = 0, 1, \ldots, n)$ (*negatively reciprocal equation*).

Theorem 1. *Every positively reciprocal equation of odd degree and every negatively reciprocal equation of even degree has the root* -1.

Theorem 2. *Every negatively reciprocal equation has the root* $+1$.

Theorem 3. *Reducing a reciprocal equation by linear factors of the form* $x - 1$, $x + 1$, *we get a positively reciprocal equation of even degree*

$$a_0x^{2m} + a_1x^{2m-1} + \ldots + a_mx^m + \ldots + a_0 = 0$$

which, if divided by x^m, *becomes*

$$a_0\left(x^m + \frac{1}{x^m}\right) + a_1\left(x^{m-1} + \frac{1}{x^{m-1}}\right) + \ldots + a_{m-1}\left(x + \frac{1}{x}\right) + a_m = 0\,.$$

By the substitution $x + 1/x = y$, all binomials can be expressed as polynomials in y; hence we get an equation of degree m in y which in some cases can then be easily solved.

Example 1. The equation $x^6 + x^5 - 5x^4 + 5x^2 - x - 1 = 0$ is a negatively reciprocal equation of even degree; thus it has the roots $\xi_1 = -1$, $\xi_2 = +1$. Dividing by $(x + 1)(x - 1)$, we get a positively reciprocal equation of even degree

$$x^4 + x^3 - 4x^2 + x + 1 = 0\,, \quad \text{i.e.} \quad \left(x^2 + \frac{1}{x^2}\right) + \left(x + \frac{1}{x}\right) - 4 = 0\,.$$

By the substitution $x + 1/x = y$, we transform the equation into the form $y^2 + y - 6 = 0$ (since $x^2 + 1/x^2 = (x + 1/x)^2 - 2$) with roots $y_1 = 2$, $y_2 = -3$. Hence, the remaining four roots of the original equation are the roots of the quadratic equations

$$x + \frac{1}{x} = 2\,, \quad x + \frac{1}{x} = -3\,.$$

1.23. The Concept of a Set and the Concept of a Mapping

A *set* is a collection of certain objects, called the *elements* of the set. A set is completely determined by its elements. Thus, if the sets A, B consist of the same elements, we say that they are *equal* and write $A = B$.

Examples of sets:

(a) the set of all even numbers;
(b) the set of all points on the circumference of a given circle;
(c) the set of the numbers 1, 2, 3 [we denote it by either $\{1, 2, 3\}$ or $(1, 2, 3)$].

The *empty* (or *void*) *set* (denoted by $\emptyset$) contains no elements at all. For example, the set of all even numbers greater than 0 and less than 2 is empty.

If x is an element of the set M, we write $x \in M$; if x is not an element of this set, then we write either $x \notin M$ or x non $\in M$.

Definition 1. The set A is called a *subset* of the set B (in symbols, $A \subset B$) if every element x of the set A is also an element of the set B, i.e. if $x \in A \Rightarrow x \in B$.

REMARK 1. For the sets A, B the equality $A = B$ holds if and only if both $A \subset B$ and $B \subset A$ hold.

Definition 2. The *union* (or *sum*) *of the sets* A, B (in symbols, $A \cup B$ or $A + B$) is the set of those elements which belong to at least one of the sets. [Similarly, for a greater (even infinite) number of sets.]

Definition 3. The *intersection* (or *product*) *of the sets* A, B (in symbols, $A \cap B$ or $A \,.\, B$ or AB) is the set of elements belonging simultaneously to both A and B. (Similarly, for a greater (even infinite) number of sets.) If $A \cap B = \emptyset$, we say that A and B are *disjoint sets* (they have no common element).

Definition 4. The *difference* (or *relative complement*) *of the sets* A, B (in symbols, $A \dot{-} B$ or $A - B$ or $A \setminus B$) is the set of those elements of A which do not belong to B.

Example 1. If A is the set of real numbers x satisfying $1 \leqq x \leqq 10$ (i.e. $A = [1, 10]$) and if, similarly $B = [5, 15]$, then $A \cup B = [1, 15]$, $A \cap B = [5, 10]$, $A \dot{-} B = [1, 5)$. If $C = [1, 2]$, then $C \subset A$, $C \cap B = \emptyset$.

Definition 5. The set of all *ordered* pairs (x, y), where $x \in A$, $y \in B$, is called the *cartesian product of the sets* A, B (denoted by $A \times B$).

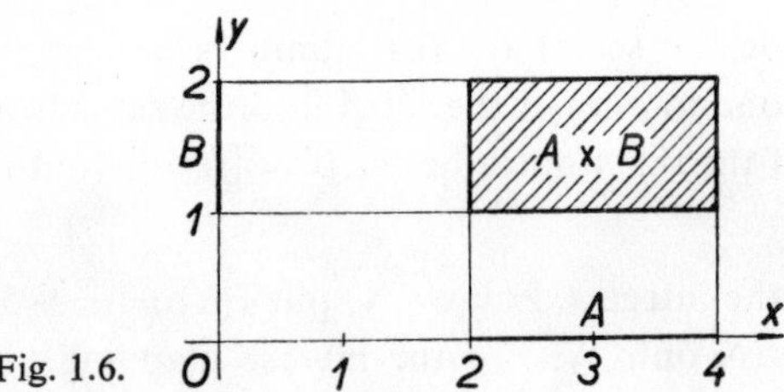

Fig. 1.6.

Example 2. If $A = [2, 4]$, $B = [1, 2]$, then $A \times B$ is the set of the ordered pairs (x, y), where $2 \leqq x \leqq 4$, $1 \leqq y \leqq 2$. If we illustrate (Fig. 1.6) the sets A and B in the plane of the coordinate axes x, y as the intervals $[2, 4]$ of the x axis and as $[1, 2]$ of the y axis, respectively, then $A \times B$ is represented by the rectangle with the vertices $(2, 1)$, $(2, 2)$, $(4, 1)$, $(4, 2)$.

REMARK 2. Let A_α be a system of sets with the index α running through a set M. Then the union and intersection of all the sets A_α are denoted by the symbols $\bigcup_{\alpha\in M} A_\alpha$ and $\bigcap_{\alpha\in M} A_\alpha$, respectively. If M is the set of the natural numbers, then the union and intersection of the sets $A_1, A_2, A_3, \ldots$ are often denoted by $\bigcup_{i=1}^{\infty} A_i$ and $\bigcap_{i=1}^{\infty} A_i$, respectively. Similarly, we write $\bigcup_{k=1}^{n} A_k$ instead of $A_1 \cup A_2 \cup \ldots \cup A_n$ and correspondingly for the intersection.

Theorem 1 (*De Morgan's Formulae*). *The relations*

$$\bigcap_{\alpha\in M} (B \cup A_\alpha) = B \cup \Big(\bigcap_{\alpha\in M} A_\alpha\Big); \quad \bigcup_{\alpha\in M} (B \cap A_\alpha) = B \cap \Big(\bigcup_{\alpha\in M} A_\alpha\Big);$$

$$B \dot{-} \bigcup_{\alpha\in M} A_\alpha = \bigcap_{\alpha\in M} (B \dot{-} A_\alpha); \quad B \dot{-} \bigcap_{\alpha\in M} A_\alpha = \bigcup_{\alpha\in M} (B \dot{-} A_\alpha)$$

hold.

Definition 6. A *mapping f of a set A into a set B* is a rule which assigns to every element $x \in A$ a definite element $y \in B$ (uniquely determined by the element x). The element y is denoted by the symbol $f(x)$ and is called the *image* of the element x. The element x is said to be the *original* or *inverse image* of the element $f(x)$. The set A is called the *domain of the mapping f*.

Definition 7. The mapping f of Definition 6 is a *mapping of the set A onto the set B* if, for every $y \in B$, there exists at least one $x \in A$ such that $y = f(x)$.

Definition 8. The mapping f is said to be *one-to-one*, if $x_1 \neq x_2 \Rightarrow f(x_1) \neq f(x_2)$.

Definition 9. Let f be a one-to-one mapping of a set A onto a set B. The mapping f^{-1} which assigns to every $y \in B$ the element $f^{-1}(y) = x \in A$ such that $f(x) = y$, is called the *inverse mapping to f*.

Example 3. (a) Let A be the set of the real numbers. For $x \in A$, put $f(x) = x^2$. Then f is a mapping (not one-to-one) of the set A into the set A (not a mapping of A onto A); f is a mapping of the set A onto the set $B = [0, \infty)$ (onto the set of all real *non-negative* numbers).

(b) Let N be the set of the integers. For $x \in N$, put $f(x) = x + 5$. Then f is a one-to-one mapping of the set N onto N. For the inverse mapping f^{-1} to f, $f^{-1}(y) = y - 5$ holds.

REMARK 3. Besides the term "mapping" the terms *transformation, correspondence, operation, operator, map, functional, function* are also used, in cases where the sets A, B are in some way specialized.

REMARK 4. On the concept of a function of one real variable x see § 11.1. This function is usually denoted by $f(x)$, in contrast to mere f as in Definition 6. [In theoretical considerations, it is often more advantageous to write only f instead of $f(x)$, for there can be no misunderstanding as to whether $f(x)$ is a function or the value of the function at the point x.]

1.24. Groups, Rings, Division Rings, Fields

Definition 1. A *group* is a non-empty set G in which multiplication is determined in some way, i.e. a rule is given which assigns to each ordered pair a, b of G a unique element $c = ab \in G$, their product. Moreover, the multiplication satisfies the following rules (laws, axioms):

1. $(ab)\,c = a(bc)$ (*associative law*).

2. For each two elements $a, b \in G$ there exist elements $x, y \in G$ such that $ax = b$ and $ya = b$.

REMARK 1. The axioms 1, 2 immediately imply that there is a unique *identity element* e in the group G such that $ea = ae = a$ for every $a \in G$. Furthermore, for each element $a \in G$, there exists a *unique inverse* a^{-1} of a such that $aa^{-1} = a^{-1}a = e$.

Definition 2. A group is called *abelian*, or *commutative*, if, for every two of its elements a, b, the relation $ab = ba$ holds.

REMARK 2. If G is an abelian group, then we frequently use additive notation, i.e. we write $a + b$ instead of ab. The identity element is denoted by 0 (zero element); the inverse of a is denoted by $-a$.

Example 1. (a) The set of all non-zero rational numbers is, with respect to multiplication, an abelian group; the number 1 is its identity element.

(b) The set of all integers is an abelian group with respect to addition; the number 0 is its identity element, the number $-a$ is the inverse of the number a.

(c) The set of all regular matrices of order n is a (non-commutative) group with respect to matrix multiplication (see Definition 1.25.3, p. 87).

Definition 3. By a *ring* (more exactly an *associative ring*) is meant a non-empty set R, in which addition and multiplication are determined in some way, i.e. rules are given which assign to each ordered pair $a, b \in R$ a unique element $a + b \in R$ (their sum) and a unique element $ab \in R$ (their product). Moreover, this addition and multiplication satisfy the following rules (laws, axioms):

1. The set R is, with respect to addition, an abelian group, i.e. for every three elements the relations $(a + b) + c = a + (b + c)$ and $a + b = b + a$ hold and there exists an element x such that $a + x = b$ (the zero element is denoted by 0).

2. Multiplication is associative and is distributive with respect to addition, i.e. for every three elements $a, b, c \in R$

$$(ab)\,c = a(bc)\,,$$
$$(a + b)\,c = ac + bc\,, \quad a(b + c) = ab + ac\,.$$

holds.

REMARK 3. In a ring R, the relation $a0 = 0a = 0$ holds for every element $a \in R$. In R, non-zero elements a, b may exist such that their product is zero: $ab = 0$. Such elements are said to be *zero-divisors*. If there is an element e in R such that $ae = ea = a$ for every $a \in R$, we say that R is a *ring with identity*. If an identity element exists in R, then it is uniquely determined.

Definition 4. A ring R is called a *commutative ring* if, for each $a, b \in R$, the relation $ab = ba$ holds.

Example 2. (a) The set of all integers is a (commutative) ring with identity with respect to addition and multiplication.

(b) The set of all even numbers is a (commutative) ring without identity with respect to addition and multiplication.

(c) The set of all square matrices of order n is a (non-commutative) ring with zero-divisors with respect to matrix addition and multiplication (see Theorem 1.25.3, p. 88).

Definition 5. A *division ring* (*skew field* or *s-field*) D is a ring with identity $e \neq 0$ such that, for every $a \in D$, $a \neq 0$, there exists an inverse $a^{-1} \in D$ such that $aa^{-1} = a^{-1}a = e$. If, moreover, the ring is commutative, then it is called a *field*.

REMARK 4. The identity element e of the division ring D is usually denoted by 1. A division ring has no zero-divisors: If the product of two elements of a division ring is equal to zero, then at least one of the elements is zero. In a division ring, to any non-zero element there corresponds a unique inverse. The set of all non-zero elements of a division ring is a group with respect to multiplication.

Example 3. (a) The set of all rational numbers* is a field (the so-called *field of rational numbers*).

(b) The set of all real numbers* is a field (the so-called *field of real numbers*).

(c) The set of all complex numbers* is a field (the so-called *field of complex numbers*).

---

* With operations of addition and multiplication defined in the usual way.

1.25. Matrices (continued). Operations on Matrices

REMARK 1. The elements of the matrices under consideration – unless otherwise stated – will always be real or complex numbers.

Definition 1. The matrix $\alpha\mathbf{A}$ obtained from a matrix $\mathbf{A}$ by multiplication of all its elements by a number α is called the (*scalar*) *product of the matrix* $\mathbf{A}$ *and the number* α:

$$\alpha\mathbf{A} = \alpha \begin{Vmatrix} a_{11}, \ldots, a_{1n} \\ \ldots\ldots\ldots\ldots \\ a_{m1}, \ldots, a_{mn} \end{Vmatrix} = \begin{Vmatrix} \alpha a_{11}, \ldots, \alpha a_{1n} \\ \ldots\ldots\ldots\ldots \\ \alpha a_{m1}, \ldots, \alpha a_{mn} \end{Vmatrix}.$$

Definition 2. The *sum* $\mathbf{A} + \mathbf{B}$ *of two m by n matrices* $\mathbf{A}$, $\mathbf{B}$ is the m by n matrix whose elements are the sums of the corresponding elements:

$$\begin{Vmatrix} a_{11}, \ldots, a_{1n} \\ \ldots\ldots\ldots\ldots \\ a_{m1}, \ldots, a_{mn} \end{Vmatrix} + \begin{Vmatrix} b_{11}, \ldots, b_{1n} \\ \ldots\ldots\ldots\ldots \\ b_{m1}, \ldots, b_{mn} \end{Vmatrix} = \begin{Vmatrix} a_{11} + b_{11}, \ldots, a_{1n} + b_{1n} \\ \ldots\ldots\ldots\ldots\ldots\ldots \\ a_{m1} + b_{m1}, \ldots, a_{mn} + b_{mn} \end{Vmatrix}.$$

Theorem 1. *The scalar multiplication and addition of m by n matrices satisfy the following rules*:

1. $\mathbf{A} + (\mathbf{B} + \mathbf{C}) = (\mathbf{A} + \mathbf{B}) + \mathbf{C}$.
2. $\mathbf{A} + \mathbf{B} = \mathbf{B} + \mathbf{A}$.
3. $\mathbf{A} + \mathbf{0} = \mathbf{A}$, *where*

$$\mathbf{0} = \begin{Vmatrix} 0, \ldots, 0 \\ \ldots\ldots \\ 0, \ldots, 0 \end{Vmatrix}$$

is the so-called zero-matrix.

4. *For two matrices* $\mathbf{A}$, $\mathbf{B}$ *there exists a matrix* $\mathbf{X}$ *such that* $\mathbf{A} + \mathbf{X} = \mathbf{B}$; *it is the matrix* $\mathbf{X} = \mathbf{B} + (-1)\,\mathbf{A} = \mathbf{B} - \mathbf{A}$.
5. $\alpha(\mathbf{A} + \mathbf{B}) = \alpha\mathbf{A} + \alpha\mathbf{B}$; $(\alpha + \beta)\,\mathbf{A} = \alpha\mathbf{A} + \beta\mathbf{A}$.

Definition 3. The *product* $\mathbf{AB}$ *of an m by n matrix* $\mathbf{A}$ *and an n by p matrix* $\mathbf{B}$ is the m by p matrix $\mathbf{C}$ defined as follows: If

$$\mathbf{A} = \begin{Vmatrix} a_{11}, \ldots, a_{1n} \\ \ldots\ldots\ldots\ldots \\ a_{m1}, \ldots, a_{mn} \end{Vmatrix}, \quad \mathbf{B} = \begin{Vmatrix} b_{11}, \ldots, b_{1p} \\ \ldots\ldots\ldots\ldots \\ b_{n1}, \ldots, b_{np} \end{Vmatrix}, \quad \text{then} \quad \mathbf{C} = \begin{Vmatrix} c_{11}, \ldots, c_{1p} \\ \ldots\ldots\ldots \\ c_{m1}, \ldots, c_{mp} \end{Vmatrix},$$

where $c_{ij} = a_{i1}b_{1j} + a_{i2}b_{2j} + \ldots + a_{in}b_{nj}$ $(i = 1, 2, \ldots, m;\ j = 1, 2, \ldots, p)$. (In words: The *rows* of the matrix $\mathbf{A}$ are multiplied by the *columns* of the matrix $\mathbf{B}$. The number of columns of the first matrix must be equal to the number of rows of the second.)

Theorem 2. *The multiplication of matrices* **A**, **B**, **C** *satisfies the relations*

1. $(\mathbf{AB})\,\mathbf{C} = \mathbf{A}(\mathbf{BC})$,
2. $(\mathbf{A} + \mathbf{B})\mathbf{C} = \mathbf{AC} + \mathbf{BC}$, $\mathbf{A}(\mathbf{B} + \mathbf{C}) = \mathbf{AB} + \mathbf{AC}$,

if the sums and product of the matrices considered are defined (*i.e. if the matrices* **A**, **B**, **C** *are of prescribed type*).

REMARK 2. In the following text, we restrict ourselves to *square matrices of order n*, i.e. to n by n matrices.

Theorem 3. *The set of all square matrices of order n whose elements are real or complex numbers constitutes a ring with respect to matrix addition and multiplication* (see 1.24, p. 85), *the so-called ring of square real or complex matrices. For $n > 1$, this ring is non-commutative* (*i.e., in general,* $\mathbf{AB} \neq \mathbf{BA}$); *it has an identity element – the identity matrix*

$$\mathbf{E} = \begin{Vmatrix} 1, & 0, & \ldots, & 0 \\ 0, & 1, & \ldots, & 0 \\ \hdotsfor{4} \\ 0, & 0, & \ldots, & 1 \end{Vmatrix}.$$

Furthermore, this ring has zero-divisors, i.e. there are pairs of non-zero matrices **A**, **B** *such that their product is the zero matrix.*

Definition 4. A square matrix $\mathbf{A} = \|a_{ij}\|$ of order n is said to be *regular* or *non-singular* if its determinant $|a_{ij}|$ is different from zero (i.e. if **A** is of rank n); a matrix which is not regular is called *singular*.

Theorem 4. *The determinant of the matrix* **AB** *– the product of square matrices* **A**, **B** *of the same order – is equal to the product of the determinants of the matrices* **A**, **B**.

Theorem 5. *The product of regular matrices of the same order is again a regular matrix.*

Definition 5. The *inverse of a square matrix* **A** *of order n is a square matrix* $\mathbf{A}^{-1}$ of order n such that $\mathbf{AA}^{-1} = \mathbf{A}^{-1}\mathbf{A} = \mathbf{E}$, where **E** is the identity matrix.

Theorem 6. *The inverse* $\mathbf{A}^{-1}$ *of a square matrix* **A** *exists if and only if* **A** *is regular. If*

$$\mathbf{A} = \begin{Vmatrix} a_{11}, & a_{12}, & \ldots, & a_{1n} \\ a_{21}, & a_{22}, & \ldots, & a_{2n} \\ \hdotsfor{4} \\ a_{n1}, & a_{n2}, & \ldots, & a_{nn} \end{Vmatrix}, \quad \text{then} \quad \mathbf{A}^{-1} = \begin{Vmatrix} \mathrm{A}_{11}A^{-1}, & \mathrm{A}_{21}A^{-1}, & \ldots, & \mathrm{A}_{n1}A^{-1} \\ \mathrm{A}_{12}A^{-1}, & \mathrm{A}_{22}A^{-1}, & \ldots, & \mathrm{A}_{n2}A^{-1} \\ \hdotsfor{4} \\ \mathrm{A}_{1n}A^{-1}, & \mathrm{A}_{2n}A^{-1}, & \ldots, & \mathrm{A}_{nn}A^{-1} \end{Vmatrix},$$

where A is the determinant of the matrix **A** *and* A_{ij} *is the cofactor belonging to the element* a_{ij} *in the determinant* A (see Definition 1.17.2, p. 68).

REMARK 3 (*A System of Linear Equations in Matrix Form*). Let

$$\begin{aligned} a_{11}x_1 + a_{12}x_2 + \ldots + a_{1n}x_n &= b_1 , \\ \ldots\ldots\ldots\ldots\ldots\ldots\ldots\ldots & \\ a_{n1}x_1 + a_{n2}x_2 + \ldots + a_{nn}x_n &= b_n \end{aligned} \tag{1}$$

be a system of n equations in n unknowns. Put

$$\mathbf{A} = \left\| \begin{matrix} a_{11}, a_{12}, \ldots, a_{1n} \\ \ldots\ldots\ldots\ldots \\ a_{n1}, a_{n2}, \ldots, a_{nn} \end{matrix} \right\| , \quad \mathbf{x} = \left\| \begin{matrix} x_1 \\ x_2 \\ \vdots \\ x_n \end{matrix} \right\| , \quad \mathbf{b} = \left\| \begin{matrix} b_1 \\ b_2 \\ \vdots \\ b_n \end{matrix} \right\| .$$

Then the system of equations (1) can be rewritten in the matrix form

$$\mathbf{Ax} = \mathbf{b} . \tag{2}$$

If the determinant of the system is non-zero, then the matrix $\mathbf{A}$ is regular, and thus, its inverse $\mathbf{A}^{-1}$ exists. Multiplying (2) by this matrix $\mathbf{A}^{-1}$, we get

$$\mathbf{A}^{-1}\mathbf{Ax} = \mathbf{A}^{-1}\mathbf{b} , \quad \text{i.e.} \quad \mathbf{x} = \mathbf{A}^{-1}\mathbf{b} . \tag{3}$$

If the inverse $\mathbf{A}^{-1}$ of the matrix $\mathbf{A}$ is known, then we can, in accordance with (3), immediately write down the solution of the system (1).

Theorem 7. *The inverse of the product* $\mathbf{AB}$ *of regular matrices* $\mathbf{A}$, $\mathbf{B}$ *is equal to the product of the inverses of the matrices* $\mathbf{A}$ *and* $\mathbf{B}$ *taken in reverse order*: $(\mathbf{AB})^{-1} = \mathbf{B}^{-1}\mathbf{A}^{-1}$.

REMARK 4. In what follows, $\mathbf{A}'$ denotes the transpose of the matrix $\mathbf{A}$ (see Definition 1.16.3, p. 64).

Theorem 8. *The transpose of the product* $\mathbf{AB}$ *of two matrices* $\mathbf{A}$, $\mathbf{B}$ *is equal to the product of the transposes of the matrices* $\mathbf{A}$ *and* $\mathbf{B}$ *taken in the reverse order*: $(\mathbf{AB})' = \mathbf{B}'\mathbf{A}'$. *Furthermore* $(\mathbf{A} + \mathbf{B})' = \mathbf{A}' + \mathbf{B}'$.

Definition 6. A matrix is called *symmetric* or *skew-symmetric* respectively, if $\mathbf{A} = \mathbf{A}'$, or $\mathbf{A} = -\mathbf{A}'$, i.e. if $a_{ij} = a_{ji}$, or $a_{ij} = -a_{ji}$, for $i, j = 1, 2, \ldots, n$, respectively.

REMARK 5. The diagonal elements of a skew-symmetric matrix are zero.

Theorem 9 (*A matrix expressed as a sum of a symmetric and a skew-symmetric matrix*). *A matrix* $\mathbf{A}$ *is a sum of the symmetric matrix* $\frac{1}{2}(\mathbf{A} + \mathbf{A}')$ *and the skew-symmetric matrix* $\frac{1}{2}(\mathbf{A} - \mathbf{A}')$; *hence* $\mathbf{A} = \frac{1}{2}(\mathbf{A} + \mathbf{A}') + \frac{1}{2}(\mathbf{A} - \mathbf{A}')$.

Theorem 10. *The product of two symmetric matrices* $\mathbf{A}$, $\mathbf{B}$ *is a symmetric matrix if and only if the matrices commute, i.e. if* $\mathbf{AB} = \mathbf{BA}$.

Theorem 11. *The rank of a skew-symmetric matrix is always an even number.*

Definition 7. A matrix $\mathbf{A}$ is called *orthogonal* if $\mathbf{AA}' = \mathbf{E}$, i.e. if $\mathbf{A}' = \mathbf{A}^{-1}$.

Theorem 12. *Let* $\mathbf{A} = \|a_{ij}\|$ *be an orthogonal matrix of order n. Then the relations*

$$\sum_{j=1}^{n} a_{ij}^2 = 1\,, \quad \sum_{j=1}^{n} a_{ij}a_{kj} = 0 \quad (i \neq k) \tag{4}$$

hold. In words: *In an orthogonal matrix, the sum of the products of the elements of an arbitrary row and of the corresponding elements of another row is zero and the sum of the squares of the elements of an arbitrary row is unity.*

A similar statement holds for the columns:

$$\sum_{j=1}^{n} a_{ji}^2 = 1\,, \quad \sum_{j=1}^{n} a_{ji}a_{jk} = 0 \quad (i \neq k)\,. \tag{5}$$

Conversely, if the elements of a matrix $\mathbf{A} = \|a_{ij}\|$ *satisfy* (4) *or* (5), *then* $\mathbf{A}$ *is orthogonal.*

Example 1. The rotation of the rectangular axes in a plane through an angle α is expressed by the following equations (cf. Theorem 5.13.3, p. 224):

$$x' = x \cos \alpha + y \sin \alpha\,,$$
$$y' = -x \sin \alpha + y \cos \alpha\,.$$

The matrix of this transformation, i.e. the matrix

$$\left\| \begin{matrix} \cos \alpha, & \sin \alpha \\ -\sin \alpha, & \cos \alpha \end{matrix} \right\|$$

is orthogonal, as can easily be checked using Theorem 12.

Theorem 13. *The determinant of an orthogonal matrix is equal to* 1 *or to* -1.

Theorem 14. *The product of orthogonal matrices is an orthogonal matrix.*

Theorem 15. *The inverse of an orthogonal matrix is an orthogonal matrix.*

Definition 8. *The* (*complex*) *conjugate* $\overline{\mathbf{A}}$ *of a matrix* $\mathbf{A}$ (whose elements are complex numbers) is the matrix obtained from $\mathbf{A}$ by replacing every element a_{ij} of $\mathbf{A}$ by its conjugate $\overline{a_{ij}}$.

Theorem 16. *The relations*

$$\overline{\alpha\mathbf{A} + \beta\mathbf{B}} = \bar{\alpha}\overline{\mathbf{A}} + \bar{\beta}\overline{\mathbf{B}}\,; \quad \overline{\mathbf{AB}} = \overline{\mathbf{A}}\overline{\mathbf{B}}\,;$$
$$\overline{\mathbf{A}'} = (\overline{\mathbf{A}})'\,; \quad \overline{\mathbf{A}^{-1}} = (\overline{\mathbf{A}})^{-1}\,.$$

hold.

Definition 9. A matrix $\mathbf{A}$ is called *Hermitian*, or *skew-Hermitian*, if

$$\mathbf{A} = \overline{\mathbf{A}'}\,, \quad \text{or} \quad \mathbf{A} = -\overline{\mathbf{A}'}\,, \quad \text{respectively}\,.$$

Definition 10. A matrix $\mathbf{A}$ such that

$$\mathbf{A}\overline{\mathbf{A}'} = \mathbf{E}\,, \quad \text{i.e.} \quad \overline{\mathbf{A}'} = \mathbf{A}^{-1}\,,$$

is called *unitary*.

Theorem 17. *A matrix* $\mathbf{A} = \|a_{ij}\|$ *of order n is unitary if and only if the relations*

$$\sum_{j=1}^{n} a_{ij}\bar{a}_{ij} = 1\,, \quad \sum_{j=1}^{n} a_{ij}\bar{a}_{kj} = 0 \quad (i \neq k) \tag{6}$$

or

$$\sum_{j=1}^{n} a_{ji}\bar{a}_{ji} = 1\,, \quad \sum_{j=1}^{n} a_{ji}\bar{a}_{jk} = 0 \quad (i \neq k) \tag{7}$$

hold (cf. Theorem 12).

Theorem 18. *The product of unitary matrices is a unitary matrix.*

Theorem 19. *The inverse of a unitary matrix is again unitary.*

Theorem 20. *The absolute value of the determinant of a unitary matrix is* 1.

Definition 11. By the *trace of the square matrix*

$$\begin{Vmatrix} a_{11}, & \ldots, & a_{1n} \\ \ldots & \ldots & \ldots \\ a_{n1}, & \ldots, & a_{nn} \end{Vmatrix}$$

is meant the sum $a_{11} + a_{22} + \ldots + a_{nn}$ of the diagonal elements of the matrix.

1.26. Matrices Partitioned into Blocks and Operations on Them; Triangular and Diagonal Matrices

Definition 1 (*A Matrix Partitioned into Blocks*). Let $\mathbf{A}$ be an m by n matrix. Divide it into parts by drawing lines between certain rows and certain columns. These parts (so-called *blocks*) are again matrices and the matrix $\mathbf{A}$ is formed from these blocks which constitute its elements. We say that the matrix $\mathbf{A}$ is *partitioned into blocks.*

Example 1.

$$\mathbf{A} = \begin{Vmatrix} 1, & 4, & 2, & 3 \\ 0, & 1, & 2, & 3 \\ 1, & -1, & 0, & 1 \end{Vmatrix}.$$

Thus, for example,

$$\mathbf{A} = \begin{Vmatrix} \mathbf{A}_{11}, & \mathbf{A}_{12} \\ \mathbf{A}_{21}, & \mathbf{A}_{22} \end{Vmatrix},$$

where the individual blocks are the matrices

$$\mathbf{A}_{11} = \begin{Vmatrix} 1, & 4 \\ 0, & 1 \end{Vmatrix}, \qquad \mathbf{A}_{12} = \begin{Vmatrix} 2, & 3 \\ 2, & 3 \end{Vmatrix},$$

$$\mathbf{A}_{21} = \begin{Vmatrix} 1, & -1 \end{Vmatrix}, \quad \mathbf{A}_{22} = \begin{Vmatrix} 0, & 1 \end{Vmatrix}.$$

Theorem 1 (*Multiplication by a Scalar*). *Let a matrix* $\mathbf{A}$ *be partitioned into blocks* $\mathbf{A}_{ij}$:

$$\mathbf{A} = \begin{Vmatrix} \mathbf{A}_{11}, & \mathbf{A}_{12}, & \ldots, & \mathbf{A}_{1n} \\ \cdots & \cdots & \cdots & \cdots \\ \mathbf{A}_{m1}, & \mathbf{A}_{m2}, & \ldots, & \mathbf{A}_{mn} \end{Vmatrix}. \tag{1}$$

Let α *be a real or complex number. Then*

$$\alpha\mathbf{A} = \begin{Vmatrix} \alpha\mathbf{A}_{11}, & \alpha\mathbf{A}_{12}, & \ldots, & \alpha\mathbf{A}_{1n} \\ \cdots & \cdots & \cdots & \cdots \\ \alpha\mathbf{A}_{1m}, & \alpha\mathbf{A}_{m2}, & \ldots, & \alpha\mathbf{A}_{mn} \end{Vmatrix}.$$

Theorem 2 (*Addition*). *Let the matrix*

$$\mathbf{B} = \begin{Vmatrix} \mathbf{B}_{11}, & \ldots, & \mathbf{B}_{1n} \\ \cdots & \cdots & \cdots \\ \mathbf{B}_{m1}, & \ldots, & \mathbf{B}_{mn} \end{Vmatrix}$$

be partitioned into blocks of the same type as the matrix (1). *Then*

$$\mathbf{A} + \mathbf{B} = \begin{Vmatrix} \mathbf{A}_{11} + \mathbf{B}_{11}, & \ldots, & \mathbf{A}_{1n} + \mathbf{B}_{1n} \\ \cdots & \cdots & \cdots \\ \mathbf{A}_{m1} + \mathbf{B}_{m1}, & \ldots, & \mathbf{A}_{mn} + \mathbf{B}_{mn} \end{Vmatrix}.$$

Theorem 3 (*Product*). *Let two matrices* $\mathbf{C}$, $\mathbf{D}$ *be partitioned into blocks*

$$\mathbf{C} = \begin{Vmatrix} \mathbf{C}_{11}, & \ldots, & \mathbf{C}_{1n} \\ \cdots & \cdots & \cdots \\ \mathbf{C}_{m1}, & \ldots, & \mathbf{C}_{mn} \end{Vmatrix} \qquad \mathbf{D} = \begin{Vmatrix} \mathbf{D}_{11}, & \ldots, & \mathbf{D}_{1p} \\ \cdots & \cdots & \cdots \\ \mathbf{D}_{n1}, & \ldots, & \mathbf{D}_{np} \end{Vmatrix}$$

in such a way that the number of columns of the matrix $\mathbf{C}_{ij}$ *is equal to the number of rows of the matrix* $\mathbf{D}_{jk}$ $(i = 1, \ldots, m;\ k = 1, \ldots, p)$. *Then*

$$\mathbf{CD} = \begin{Vmatrix} \mathbf{F}_{11}, & \ldots, & \mathbf{F}_{1p} \\ \cdots & \cdots & \cdots \\ \mathbf{F}_{m1}, & \ldots, & \mathbf{F}_{mp} \end{Vmatrix},$$

where $\mathbf{F}_{ik} = \mathbf{C}_{i1}\mathbf{D}_{1k} + \mathbf{C}_{i2}\mathbf{D}_{2k} + \ldots + \mathbf{C}_{in}\mathbf{D}_{nk}$.

Hence: The blocks of the matrix $\mathbf{CD}$ are the sums of the products of the blocks forming the elements of the rows of the matrix $\mathbf{C}$ and the blocks forming the elements of the columns of the matrix $\mathbf{D}$.

REMARK 1. The products $\mathbf{C}_{i1}\mathbf{D}_{1k}, \mathbf{C}_{i2}\mathbf{D}_{2k}, \ldots$ are defined, since, according to our assumption, the number of columns of the matrix $\mathbf{C}_{ij}$ equals the number of rows of the matrix $\mathbf{D}_{jk}$.

Example 2. The relation

$$\begin{Vmatrix} \mathbf{C}_{11}^{(2,3)}, & \mathbf{C}_{12}^{(2,1)} \\ \mathbf{C}_{21}^{(4,3)}, & \mathbf{C}_{22}^{(4,1)} \end{Vmatrix} \begin{Vmatrix} \mathbf{D}_{11}^{(3,4)}, & \mathbf{D}_{12}^{(3,2)} \\ \mathbf{D}_{21}^{(1,4)}, & \mathbf{D}_{22}^{(1,2)} \end{Vmatrix} = \begin{Vmatrix} \mathbf{F}_{11}^{(2,4)}, & \mathbf{F}_{12}^{(2,2)} \\ \mathbf{F}_{21}^{(4,4)}, & \mathbf{F}_{22}^{(4,2)} \end{Vmatrix},$$

holds, where the upper indices indicate the type of the corresponding matrices.

Definition 2. A matrix A partitioned into *square* blocks of the form

$$\mathbf{A} = \begin{Vmatrix} \mathbf{A}_{11}, & \mathbf{0}, & \ldots, & & \mathbf{0} \\ \mathbf{0}, & \mathbf{A}_{22}, & \mathbf{0}, & \ldots, & \mathbf{0} \\ \ldots & \ldots & \ldots & \ldots & \ldots \\ \mathbf{0}, & \ldots, & & \mathbf{0}, & \mathbf{A}_{nn} \end{Vmatrix},$$

where the symbols $\mathbf{0}$ denote zero matrices (and which are, for the sake of brevity, omitted in the formulation of the following Theorem 4), is called the *matrix decomposed into diagonal blocks.*

Theorem 4. *The sum and the product of matrices decomposed into diagonal blocks (where corresponding blocks have the same order) is a matrix decomposed into diagonal blocks; these blocks are sums, or products, of the corresponding blocks of the given matrices, respectively*:

$$\begin{Vmatrix} \mathbf{A}_{11} & & \\ & \ddots & \\ & & \mathbf{A}_{nn} \end{Vmatrix} + \begin{Vmatrix} \mathbf{B}_{11} & & \\ & \ddots & \\ & & \mathbf{B}_{nn} \end{Vmatrix} = \begin{Vmatrix} \mathbf{A}_{11} + \mathbf{B}_{11} & & \\ & \ddots & \\ & & \mathbf{A}_{nn} + \mathbf{B}_{nn} \end{Vmatrix},$$

$$\begin{Vmatrix} \mathbf{A}_{11} & & \\ & \ddots & \\ & & \mathbf{A}_{nn} \end{Vmatrix} \begin{Vmatrix} \mathbf{B}_{11} & & \\ & \ddots & \\ & & \mathbf{B}_{nn} \end{Vmatrix} = \begin{Vmatrix} \mathbf{A}_{11}\mathbf{B}_{11} & & \\ & \ddots & \\ & & \mathbf{A}_{nn}\mathbf{B}_{nn} \end{Vmatrix}.$$

Definition 3. An upper *triangular matrix* is a square matrix of the form

$$\begin{Vmatrix} a_{11}, & a_{12}, & \ldots, & a_{1n} \\ 0, & a_{22}, & \ldots, & a_{2n} \\ \ldots & \ldots & \ldots & \ldots \\ 0, & 0, & \ldots, & a_{nn} \end{Vmatrix},$$

where the elements below the principal diagonal are zero.

Theorem 5. *The sum and the product of upper triangular matrices of the same order is again an upper triangular matrix. The determinant of an upper triangular matrix is equal to the product of the elements in the principal diagonal.*

Definition 4. A *diagonal matrix* is a square matrix of the form

$$\begin{Vmatrix} a_{11}, & 0, & \ldots, & 0 \\ 0, & a_{22}, & \ldots, & 0 \\ \ldots & \ldots & \ldots & \ldots \\ 0, & 0, & \ldots, & a_{nn} \end{Vmatrix}.$$

Theorem 6. *The sum (product) of diagonal matrices of the same order is again a diagonal matrix; the elements of its principal diagonal are the sums (products) of the corresponding diagonal elements of the given matrices.*

1.27. λ-matrices, Equivalence of λ-matrices

Definition 1. A λ-matrix $\mathbf{A}(\lambda)$ is a square matrix whose elements are polynomials in the variable λ with real or complex coefficients.

REMARK 1. Addition and multiplication of λ-matrices and the rank of a λ-matrix are defined in the same way as in § § 1·25 and 1.16, pp. 87 and 64.

Example 1. The matrix

$$\begin{Vmatrix} 3 - \lambda, & 1 + \lambda \\ 1, & 5 - \lambda \end{Vmatrix}$$

is a λ-matrix. Its rank is 2, for its determinant is a non-zero polynomial $(3 - \lambda) . (5 - \lambda) - (\lambda + 1) \neq 0$. However, if we substitute for λ a particular numerical value, then we obtain another matrix (no longer the original λ-matrix) whose rank can be smaller. For example, if $\lambda = 2$, we get a matrix of rank 1.

REMARK 2. λ-matrices include also ordinary matrices as a particular case where the elements are polynomials of zero degree or zero polynomials.

Definition 2. By an *elementary transformation of a given λ-matrix* $\mathbf{A}(\lambda)$ we understand one of the following rearrangements of the matrix:

1. an interchange of two rows or two columns of the matrix;
2. multiplication of a row or a column by a non-zero number;
3. addition of a row, or a column, multiplied by a polynomial $\varphi(\lambda)$ to another row, or column, respectively.

Definition 3 (*The Equivalence of λ-matrices*). A λ-matrix $\mathbf{A}(\lambda)$ is said to be *equivalent* to a λ-matrix $\mathbf{B}(\lambda)$ if the matrix $\mathbf{B}(\lambda)$ can be obtained from $\mathbf{A}(\lambda)$ by a finite number of elementary transformations. In this case, we write $\mathbf{A}(\lambda) \sim \mathbf{B}(\lambda)$.

Theorem 1. *λ-matrices $\mathbf{A}(\lambda)$, $\mathbf{B}(\lambda)$ of order n are equivalent if and only if there exist λ-matrices $\mathbf{C}(\lambda)$, $\mathbf{D}(\lambda)$ of order n such that their determinants are non-zero (real or complex) numbers and*

$$\mathbf{B}(\lambda) = \mathbf{C}(\lambda)\,\mathbf{A}(\lambda)\,\mathbf{D}(\lambda)\,.$$

Theorem 2. *Equivalent λ-matrices have the same rank* (the converse does not hold – see Theorem 6).

Theorem 3. *Two matrices $\mathbf{A}$, $\mathbf{B}$ of the same order whose elements are real or complex numbers are equivalent if and only if they have the same rank.*

Theorem 4. *A λ-matrix $\mathbf{A}(\lambda)$ of order n is equivalent to one and only one of the λ-matrices of the form*

$$\begin{Vmatrix} E_1(\lambda), & 0, & 0, & \ldots, & 0 \\ 0, & E_2(\lambda), & 0, & \ldots, & 0 \\ 0, & 0, & E_3(\lambda), & \ldots, & 0 \\ \multicolumn{5}{c}{\dotfill} \\ 0, & 0, & 0, & \ldots, & E_n(\lambda) \end{Vmatrix}, \tag{1}$$

where the polynomials $E_i(\lambda)$ are either zero polynomials or have the coefficient of the highest power of λ equal to 1, *and the polynomial $E_{j+1}(\lambda)$ is divisible by the polynomial $E_j(\lambda)$ $(j = 1, 2, \ldots, n-1)$. If the rank of $\mathbf{A}(\lambda)$ is h, then $E_1(\lambda)\,E_2(\lambda)\ldots$ $\ldots E_h(\lambda) \neq 0$, $E_{h+1}(\lambda) = E_{h+2}(\lambda) = \ldots = E_n(\lambda) = 0$.*

Definition 4. The polynomials $E_1(\lambda), \ldots, E_n(\lambda)$ are called *invariant factors of the matrix $\mathbf{A}(\lambda)$*; the form (1) is called the *rational canonical form of the matrix $\mathbf{A}(\lambda)$*.

Example 2. Rearrangements transforming the matrix of Example 1 to the rational canonical form reduce the given matrix successively to the following matrices:

$$\begin{Vmatrix} 3-\lambda, & 1+\lambda \\ 1, & 5-\lambda \end{Vmatrix} \sim \begin{Vmatrix} 1, & 5-\lambda \\ 3-\lambda, & 1+\lambda \end{Vmatrix} \sim \begin{Vmatrix} 1, & 5-\lambda \\ 0, & (1+\lambda)+(5-\lambda)(-3+\lambda) \end{Vmatrix} =$$

$$= \begin{Vmatrix} 1, & 5-\lambda \\ 0, & -\lambda^2+9\lambda-14 \end{Vmatrix} \sim \begin{Vmatrix} 1, & 0 \\ 0, & -\lambda^2+9\lambda-14 \end{Vmatrix} \sim \begin{Vmatrix} 1, & 0 \\ 0, & \lambda^2-9\lambda+14 \end{Vmatrix}.$$

First, we interchanged the rows; then we added the first row multiplied by $-(3-\lambda)$ to the second one; then we added to the second column the first one multiplied by $\lambda - 5$ and, finally, we multiplied the second row by the number -1. Thus, the invariant factors of the matrix are the polynomials $E_1(\lambda) = 1$, $E_2(\lambda) = \lambda^2 - 9\lambda + 14$.

Theorem 5. *The greatest common divisor $D_i(\lambda)$ (the so-called i-th determinant divisor) of all the i-rowed minors of a matrix $\mathbf{A}(\lambda)$ satisfies the relation*

$$D_i(\lambda) = cE_1(\lambda)\,E_2(\lambda)\ldots E_i(\lambda)\,,$$

where $E_j(\lambda)$ are the invariant factors of $\mathbf{A}(\lambda)$ *and c is a non-zero real or complex number.*

Theorem 6. *Two λ-matrices* $\mathbf{A}(\lambda)$, $\mathbf{B}(\lambda)$ *of the same order are equivalent if and only if they have the same invariant factors.*

Definition 5 (*Elementary Divisors of a λ-matrix*). Let a matrix $\mathbf{A}(\lambda)$ have the invariant factors $E_1(\lambda), \ldots, E_n(\lambda)$. We can factorize each of these polynomials in the variable λ into the product of powers of distinct linear factors $(\lambda - \alpha)^k$. Then every such power of a linear factor $(\lambda - \alpha)^k$ is called an *elementary divisor of the matrix* $\mathbf{A}(\lambda)$. The elementary divisors of the matrix $\mathbf{A}(\lambda)$ form the so-called *system of elementary divisors of the matrix* $\mathbf{A}(\lambda)$ (see Examples 3–5).

Example 3. Let a matrix $\mathbf{A}(\lambda)$ of order 5 have the invariant factors $E_1(\lambda) = 1$, $E_2(\lambda) = \lambda$, $E_3(\lambda) = \lambda(\lambda + 1)^2$, $E_4(\lambda) = \lambda^2(\lambda + 1)^2$, $E_5(\lambda) = 0$. Then the system of its elementary divisors is $\lambda, \lambda, \lambda^2, (\lambda + 1)^2, (\lambda + 1)^2$.

Example 4. The matrix of Example 2 has the elementary divisors $\lambda - 2$, $\lambda - 7$.

Theorem 7. *The invariant factors of a given matrix are uniquely determined by the order, rank and system of the elementary divisors.*

Example 5. Let us determine invariant factors of a matrix $\mathbf{A}(\lambda)$ of order 5 and rank 4 if the system of its elementary divisors is $\lambda - 1, \lambda - 1, (\lambda - 1)^2, (\lambda + 1)^2$.

In order to find the invariant factors, let us use Theorem 4. Since $h = 4$, we have $E_5(\lambda) = 0$. Now $E_1(\lambda)\,E_2(\lambda)\,E_3(\lambda)\,E_4(\lambda) = (\lambda - 1)(\lambda - 1)(\lambda - 1)^2(\lambda + 1)^2$. Since $E_1(\lambda)$, $E_2(\lambda)$, $E_3(\lambda)$ are divisors of $E_4(\lambda)$, we get immediately (using Definition 5) $E_4(\lambda) = (\lambda - 1)^2(\lambda + 1)^2$. Now $E_1(\lambda)\,E_2(\lambda)\,E_3(\lambda) = (\lambda - 1)(\lambda - 1)$, $E_1(\lambda)$, $E_2(\lambda)$ are divisors of $E_3(\lambda)$. Hence, $E_3(\lambda) = \lambda - 1$. Similarly, we find that $E_2(\lambda) = \lambda - 1$, and, finally, $E_1(\lambda) = 1$.

Theorem 8. *Two λ-matrices* $\mathbf{A}(\lambda)$, $\mathbf{B}(\lambda)$ *of order n are equivalent if and only if they have the same rank and the same system of elementary divisors.*

Theorem 9. *Let a λ-matrix* $\mathbf{A}(\lambda)$ *be partitioned into diagonal blocks:*

$$\mathbf{A}(\lambda) = \begin{Vmatrix} \mathbf{A}_{11}(\lambda), & \mathbf{0}, & \ldots, & \mathbf{0} \\ \mathbf{0}, & \mathbf{A}_{22}(\lambda), & \ldots, & \mathbf{0} \\ \cdots & \cdots & \cdots & \cdots \\ \mathbf{0}, & \mathbf{0}, & \ldots, & \mathbf{A}_{nn}(\lambda) \end{Vmatrix}.$$

Then the system of elementary divisors of the matrix $\mathbf{A}(\lambda)$ *is the collection of the systems of all elementary divisors of the diagonal blocks, i.e. of the λ-matrices* $\mathbf{A}_{11}(\lambda), \ldots, \mathbf{A}_{nn}(\lambda)$.

1.28. Similar Matrices; the Characteristic Matrix and Characteristic Polynomial of a Matrix

Definition 1. We define two square matrices $\mathbf{A}$, $\mathbf{B}$ of order n, the elements of which are real or complex numbers, to be *similar* if a regular matrix $\mathbf{P}$ of order n exists, the elements of which are real or complex numbers respectively, for which the relation

$$\mathbf{B} = \mathbf{P}^{-1}\mathbf{A}\mathbf{P}$$

holds.

Definition 2. By the *characteristic matrix* of a square matrix $\mathbf{A}$, we understand the λ-matrix $\lambda\mathbf{E} - \mathbf{A}$, where $\mathbf{E}$ is the identity matrix (p. 88). Thus, if

$$\mathbf{A} = \begin{Vmatrix} a_{11}, & a_{12}, & \ldots, & a_{1n} \\ a_{21}, & a_{22}, & \ldots, & a_{2n} \\ \ldots & \ldots & \ldots & \ldots \\ a_{n1}, & a_{n2}, & \ldots, & a_{nn} \end{Vmatrix},$$

then

$$\lambda\mathbf{E} - \mathbf{A} = \begin{Vmatrix} \lambda - a_{11}, & -a_{12}, & \ldots, & -a_{1n} \\ -a_{21}, & \lambda - a_{22}, & \ldots, & -a_{2n} \\ \ldots & \ldots & \ldots & \ldots \\ -a_{n1}, & -a_{n2}, & \ldots, & \lambda - a_{nn} \end{Vmatrix}.$$

Definition 3. The determinant of the matrix $\lambda\mathbf{E} - \mathbf{A}$ is said to be the *characteristic polynomial of the matrix* $\mathbf{A}$. Its zeros are called the *eigenvalues* (or *characteristic values* or *characteristic numbers*) *of the matrix* $\mathbf{A}$.

Example 1. The characteristic polynomial of the matrix

$$\begin{Vmatrix} 1, & 2 \\ -1, & 1 \end{Vmatrix}$$

is

$$f(\lambda) = \begin{vmatrix} \lambda - 1, & -2 \\ 1, & \lambda - 1 \end{vmatrix} = (\lambda - 1)(\lambda - 1) + 2 = \lambda^2 - 2\lambda + 3 ;$$

its zeros $1 \pm i\sqrt{2}$ are the eigenvalues of the given matrix. (On numerical methods for evaluation of eigenvalues see Chap. 30.)

Theorem 1. *The eigenvalues of an upper triangular matrix*

$$\mathbf{A} = \begin{Vmatrix} a_{11}, & a_{12}, & \ldots, & a_{1n} \\ 0, & a_{22}, & \ldots, & a_{2n} \\ \ldots & \ldots & \ldots & \ldots \\ 0, & 0, & \ldots, & a_{nn} \end{Vmatrix}$$

are equal to the elements in the principal diagonal, i.e. to the numbers $a_{11}, a_{22}, \ldots, a_{nn}$.

Theorem 2. *If the characteristic polynomial of a matrix* $\mathbf{A}$ *of order* n *has* n *simple zeros* $\alpha_1, \ldots, \alpha_n$, *then the system of elementary divisors of the characteristic matrix* $\lambda\mathbf{E} - \mathbf{A}$ *is* $\lambda - \alpha_1, \ldots, \lambda - \alpha_n$.

Theorem 3. *The product of all elementary divisors of the characteristic matrix* $\lambda\mathbf{E} - \mathbf{A}$ *of a given matrix* $\mathbf{A}$ *is equal to the characteristic polynomial of the matrix* $\mathbf{A}$.

Theorem 4. *Two matrices* $\mathbf{A}$, $\mathbf{B}$ *of the same order are similar if and only if their characteristic matrices* $\lambda\mathbf{E} - \mathbf{A}$, $\lambda\mathbf{E} - \mathbf{B}$ *are equivalent, i.e. if they have the same elementary divisors.*

Theorem 5. *Similar matrices have the same characteristic polynomial, and thus also the same eigenvalues.*

Theorem 6. *Similar matrices have the same traces* (see Definition 1.25.11, p. 91).

Definition 4. By a *Jordan block* (*of order* k) is meant a matrix of order k of the form

$$\begin{Vmatrix} \varrho, & 1, & 0, & \ldots, & 0, & 0 \\ 0, & \varrho, & 1, & \ldots, & 0, & 0 \\ \multicolumn{6}{c}{\dotfill} \\ 0, & 0, & 0, & \ldots, & \varrho, & 1 \\ 0, & 0, & 0, & \ldots, & 0, & \varrho \end{Vmatrix}, \tag{1}$$

where ϱ is a real or complex number.

A matrix decomposed into diagonal blocks (see Definition 1.26.2) which are Jordan blocks is called a *Jordan matrix*.

Theorem 7. *The characteristic matrix of the Jordan block* (1) *has the single elementary divisor* $(\lambda - \varrho)^k$. *Hence, using Theorem* 1.27.9, *one can determine the system of elementary divisors of the characteristic matrix of a given Jordan matrix.*

Example 2. The matrix

$$\mathbf{A} = \begin{Vmatrix} 1, & 0, & 0 \\ 0, & 2, & 1 \\ 0, & 0, & 2 \end{Vmatrix}$$

is a Jordan matrix; its Jordan blocks are the matrices $\|1\|$ and $\begin{Vmatrix} 2, & 1 \\ 0, & 2 \end{Vmatrix}$. Hence, the elementary divisors of the characteristic matrix $\lambda\mathbf{E} - \mathbf{A}$ are $\lambda - 1$, $(\lambda - 2)^2$.

Theorem 8. *Every square matrix* $\mathbf{A}$ *of order n is similar to a Jordan matrix of order n; if* $\mathbf{A}$ *is similar to two Jordan matrices, then these matrices differ only in the order of arrangement of their diagonal blocks.*

REMARK 1. The following two examples indicate the method if determining (at least theoretically) a Jordan matrix which is similar to a given matrix $\mathbf{A}$.

Example 3. Let

$$\mathbf{A} = \left\| \begin{matrix} 1, & 2, & 0 \\ 0, & 2, & 0 \\ -2, & -2, & -1 \end{matrix} \right\| .$$

The characteristic polynomial is

$$\begin{vmatrix} \lambda - 1, & -2, & 0 \\ 0, & \lambda - 2, & 0 \\ 2, & 2, & \lambda + 1 \end{vmatrix} = (\lambda - 1)(\lambda + 1)(\lambda - 2);$$

this polynomial has simple zeros $1, -1, 2$, and therefore, by Theorem 2, the system of elementary divisors of the characteristic matrix is $\lambda - 1$, $\lambda + 1$, $\lambda - 2$. According to Theorem 4, the characteristic matrix of the required Jordan matrix $\mathbf{B}$ has the same system of elementary divisors; hence we can find this Jordan matrix $\mathbf{B}$ by Theorem 7. The matrix $\mathbf{B}$ is decomposed into three Jordan blocks of order 1 corresponding to the elementary divisors $(\lambda - 1)$, $(\lambda + 1)$, $(\lambda - 2)$. Hence

$$\mathbf{B} = \left\| \begin{matrix} 1, & 0, & 0 \\ 0, & -1, & 0 \\ 0, & 0, & 2 \end{matrix} \right\| .$$

Example 4. Let

$$\mathbf{A} = \left\| \begin{matrix} 3, & 1, & -3 \\ -7, & -2, & 9 \\ -2, & -1, & 4 \end{matrix} \right\| .$$

The characteristic polynomial is

$$\begin{vmatrix} \lambda - 3, & -1, & 3 \\ 7, & \lambda + 2, & -9 \\ 2, & 1, & \lambda - 4 \end{vmatrix} = (\lambda - 1)(\lambda - 2)^2 .$$

Since it does not have simple zeros, Theorem 2 cannot be applied. We therefore first determine invariant factors $E_1(\lambda)$, $E_2(\lambda)$, $E_3(\lambda)$ of the matrix $\lambda\mathbf{E} - \mathbf{A}$. Since their product is equal to the product of all the elementary divisors, Theorem 3 shows that it is equal to $(\lambda - 1)(\lambda - 2)^2$. Hence either $\mathbf{E}_3(\lambda) = (\lambda - 1)(\lambda - 2)^2$, $\mathbf{E}_2(\lambda) = \mathbf{E}_1(\lambda) = 1$, or $\mathbf{E}_3(\lambda) = (\lambda - 1)(\lambda - 2)$, $\mathbf{E}_2(\lambda) = \lambda - 2$, $\mathbf{E}_1(\lambda) = 1$ [see Example 1.27.5, p. 96]. Now, $\mathbf{E}_1(\lambda)\,\mathbf{E}_2(\lambda)$ is the greatest common divisor of all minors of order 2 of the matrix

$$\lambda\mathbf{E} - \mathbf{A} = \left\|\begin{matrix} \lambda - 3, & -1, & 3 \\ 7, & \lambda + 2, & -9 \\ 2, & 1, & \lambda - 4 \end{matrix}\right\| .$$

One of these minors is $\left|\begin{matrix} -1, & 3 \\ 1, & \lambda - 4 \end{matrix}\right| = -\lambda + 1$. Thus, $E_2(\lambda)$ cannot be equal to $\lambda - 2$, for $\lambda - 2$ is not a factor of the binomial $\lambda - 1$. So $E_2(\lambda) = 1$ and the system of elementary divisors of the matrix $\lambda\mathbf{E} - \mathbf{A}$ consists of the polynomials $\lambda - 1$, $(\lambda - 2)^2$. The corresponding Jordan matrix is therefore

$$\left\|\begin{matrix} 1, & 0, & 0 \\ 0, & 2, & 1 \\ 0, & 0, & 2 \end{matrix}\right\| .$$

Alternative method: By elementary transformations (Definition 1.27.2, p. 94), we bring the matrix $\lambda\mathbf{E} - \mathbf{A}$ to the rational canonical form and then, by Theorem 1.27.9, we determine its elementary divisors.

Theorem 9. *The eigenvalues of a Hermitian matrix are real numbers.*

Theorem 10. *Let* $\mathbf{A}$ *be a symmetric matrix whose elements are real numbers. Then its eigenvalues are real numbers.*

Theorem 11. *Let* $\mathbf{A}$ *be a Hermitian matrix. Then there exists a unitary matrix* $\mathbf{U}$ *such that the matrix* $\mathbf{U}^{-1}\mathbf{A}\mathbf{U}$ *is diagonal (and real).* $\mathbf{U}^{-1}\mathbf{A}\mathbf{U}$ *is the Jordan form of the matrix* $\mathbf{A}$.

Theorem 12. *Let* $\mathbf{A}$ *be a symmetric matrix the elements of which are real numbers. Then there exists a real orthogonal matrix* $\mathbf{P}$ *such that the matrix* $\mathbf{P}^{-1}\mathbf{A}\mathbf{P}$ *is diagonal.* $\mathbf{P}^{-1}\mathbf{A}\mathbf{P}$ *is the Jordan form of the matrix* $\mathbf{A}$.

REMARK 2. A method for finding the matrix $\mathbf{P}$ of Theorem 12 is given in Example 1.29.3, p. 103.

Theorem 13. *Let* $\mathbf{U}$ *be a unitary matrix. Then there exists a unitary matrix* $\mathbf{V}$ *such that* $\mathbf{V}^{-1}\mathbf{U}\mathbf{V}$ *is diagonal and the absolute value of each of its elements in the principal diagonal is* 1. $\mathbf{V}^{-1}\mathbf{U}\mathbf{V}$ *is the Jordan form of the matrix* $\mathbf{U}$.

1.29. Quadratic and Hermitian Forms

Definition 1. A *quadratic form in n variables* $x_1, x_2, \ldots, x_n$ is a polynomial of the form

$$\begin{aligned} f(x_1, \ldots, x_n) = a_{11}x_1^2 + 2a_{12}x_1x_2 + & \ldots + 2a_{1n}x_1x_n + \\ + \ a_{22}x_2^2 \ + 2a_{23}x_2x_3 + & \ldots + 2a_{2n}x_2x_n + \\ + \ \ldots\ldots\ldots\ldots\ldots\ldots & \ldots\ldots + a_{nn}x_n^2 , \end{aligned}$$

briefly

$$f(x_1, \ldots, x_n) = \sum_{i,j=1}^{n} a_{ij}x_ix_j \quad (a_{ij} = a_{ji}), \tag{1}$$

where a_{ij} are real or complex numbers. In the case, where the a_{ij} are real, we say that the form $f(x_1, \ldots, x_n)$ is *real*.

Definition 2. The symmetric matrix of order n

$$\mathbf{A} = \begin{Vmatrix} a_{11}, & a_{12}, & \ldots, & a_{1n} \\ a_{21}, & a_{22}, & \ldots, & a_{2n} \\ \ldots & \ldots & \ldots & \ldots \\ a_{n1}, & a_{n2}, & \ldots, & a_{nn} \end{Vmatrix}$$

is called the *matrix of the quadratic form* (1), its rank being the *rank of the quadratic form* (1).

Definition 3 (*Linear Mapping*). The mapping

$$\begin{aligned} y_1 &= q_{11}x_1 + q_{12}x_2 + \ldots + q_{1n}x_n, \\ y_2 &= q_{21}x_1 + q_{22}x_2 + \ldots + q_{2n}x_n, \\ &\ldots\ldots\ldots\ldots\ldots\ldots\ldots\ldots \\ y_n &= q_{n1}x_1 + q_{n2}x_2 + \ldots + q_{nn}x_n, \end{aligned}$$

briefly

$$y_i = \sum_{j=1}^{n} q_{ij}x_j \quad (i = 1, 2, \ldots, n) \tag{2}$$

where q_{ij} are fixed real or complex numbers, assigns to every ordered n-tuple (i.e. n-component vector) $\mathbf{x} = (x_1, \ldots, x_n)$ an ordered n-tuple (i.e. n-component vector) $\mathbf{y} = (y_1, \ldots, y_n)$ and is called the *linear mapping* (*of the n-dimensional vector space V_n into itself*).

The matrix

$$\mathbf{Q} = \begin{Vmatrix} q_{11}, & q_{12}, & \ldots, & q_{1n} \\ \ldots & \ldots & \ldots & \ldots \\ q_{n1}, & q_{n2}, & \ldots, & q_{nn} \end{Vmatrix}$$

is called the *matrix of the linear mapping* (2). The mapping (2) is said to be *regular* if the matrix $\mathbf{Q}$ is regular.

Theorem 1. *If the mapping* (2) *is regular, then there exists an inverse linear mapping* $x_i = \sum_{j=1}^{n} p_{ij}y_j$ $(i = 1, \ldots, n)$, *whose matrix* $\mathbf{P}$ *is inverse to the matrix* $\mathbf{Q}$.

Theorem 2. *The composition of two linear mappings* $z_i = \sum_{j=1}^{n} r_{ij}y_j$, $y_i = \sum_{j=1}^{n} s_{ij}x_j$

with matrices **R**, **S** *is a linear mapping* $z_i = \sum_{j=1}^{n} t_{ij} x_j$ *with matrix* $\mathbf{T} = \mathbf{RS}$. *If both mappings are regular, then the composite mapping is also regular.*

Definition 4. If in a quadratic form $f(x_1, \ldots, x_n)$ we substitute for the variables $x_1, \ldots, x_n$ the variables $y_1, \ldots, y_n$ by means of a linear mapping

$$x_i = \sum_{j=1}^{n} p_{ij} y_j \quad (i = 1, 2, \ldots, n), \tag{3}$$

we say that we apply the *linear substitution* (3) to $f(x_1, \ldots, x_n)$. If the mapping (3) is regular, then the corresponding linear substitution is said to be *regular*. If the numbers p_{ij} are real, then the substitution (3) is *real*.

REMARK 1 (*Matrix Notation for Linear Mappings and Quadratic Forms*). If we denote by $\mathbf{x}$ the n by 1 matrix $\begin{Vmatrix} x_1 \\ x_2 \\ \vdots \\ x_n \end{Vmatrix}$ and by $\mathbf{y}$ the matrix $\begin{Vmatrix} y_1 \\ y_2 \\ \vdots \\ y_n \end{Vmatrix}$, then the linear mapping (3) can be written in the matrix form $\mathbf{x} = \mathbf{Py}$, where $\mathbf{P}$ is the matrix (of order n) of the mapping (3). Similarly, the quadratic form (1) can be written in the matrix form, $f(x_1, \ldots, x_n) = \mathbf{x}'\mathbf{Ax}$ where $\mathbf{x}' = \|x_1, x_2, \ldots, x_n\|$.

Example 1. In matrix notation, the form $x_1^2 - 4x_1x_2 + 2x_2^2$ is written

$$\mathbf{x}' \begin{Vmatrix} 1, & -2 \\ -2, & 2 \end{Vmatrix} \mathbf{x}, \quad \text{where} \quad \mathbf{x} = \begin{Vmatrix} x_1 \\ x_2 \end{Vmatrix},$$

since

$$\| x_1, x_2 \| \begin{Vmatrix} 1, & -2 \\ -2, & 2 \end{Vmatrix} \begin{Vmatrix} x_1 \\ x_2 \end{Vmatrix} = \| x_1 - 2x_2, \ -2x_1 + 2x_2 \| \begin{Vmatrix} x_1 \\ x_2 \end{Vmatrix} =$$

$$= x_1^2 - 2x_1x_2 - 2x_1x_2 + 2x_2^2 .$$

Theorem 3. *A quadratic form* $f(x_1, \ldots, x_n) = \sum_{i,j=1}^{n} a_{ij} x_i x_j$ *with matrix* **A** *is transformed by a linear substitution* (3) (*i.e. by a substitution* $\mathbf{x} = \mathbf{Py}$) *into the form*

$$g(y_1, \ldots, y_n) = (\mathbf{Py})' \, \mathbf{A}(\mathbf{Py}) = \mathbf{y}'(\mathbf{P}'\mathbf{AP}) \, \mathbf{y} = \sum_{i,j=1}^{n} b_{ij} y_i y_j$$

with matrix $\mathbf{B} = \mathbf{P}'\mathbf{AP}$, *where* **P** *is the matrix of the linear substitution* (3).

REMARK 2. Square matrices **A**, **B** of order n are said to be *congruent* if there exists a regular matrix **P** such that $\mathbf{B} = \mathbf{P}'\mathbf{AP}$.

Theorem 4. *The quadratic form $g(y_1, \ldots, y_n)$ obtained from a form $f(x_1, \ldots, x_n)$ by a regular linear substitution $x_i = \sum_{j=1}^{n} p_{ij} y_j$ $(i = 1, \ldots, n)$ has the same rank as the form $f(x_1, \ldots, x_n)$.*

Theorem 5. *For every (for every real) quadratic form $f(x_1, \ldots, x_n)$ of rank h there exists a regular linear substitution (3) [a real regular linear substitution (3)] which transforms the form $f(x_1, \ldots, x_n)$ into the form*

$$g(y_1, \ldots, y_n) = c_1 y_1^2 + c_2 y_2^2 + \ldots + c_n y_n^2 ,$$

where $c_1, \ldots, c_n$ are complex (real) numbers, precisely h of which are non-zero.

Example 2. A method of finding such a substitution will be shown in the following example: Let $f(x_1, x_2, x_3) = x_1 x_2 + 4x_2 x_3 - 2x_1 x_3$. Since the form does not contain the square of any variable, we transform it, first, by the regular linear substitution $x_1 = z_1 + z_2$, $x_2 = z_1 - z_2$, $x_3 = z_3$ into the form $h(z_1, z_2, z_3) = z_1^2 - z_2^2 + 2z_1 z_3 - 6z_2 z_3$. This can be rewritten in the form $h(z_1, z_2, z_3) = (z_1 + z_3)^2 - z_3^2 - z_2^2 - 6z_2 z_3$. We then apply the regular linear transformation $t_1 = z_1 + z_3$, $t_2 = z_2$, $t_3 = z_3$ thus obtaining the form $k(t_1, t_2, t_3) = t_1^2 - t_2^2 - t_3^2 - 6t_2 t_3 = t_1^2 - (t_2 + 3t_3)^2 + 9t_3^2 - t_3^2$ which, by means of the regular linear substitution $y_1 = t_1$, $y_2 = t_2 + 3t_3$, $y_3 = t_3$ is transformed into the form $g(y_1, y_2, y_3) = y_1^2 - y_2^2 + 8y_3^2$ as required. By combining the applied substitutions, we find that $f(x_1, x_2, x_3)$ is transformed into this final form by the regular linear substitution $x_1 = y_1 + y_2 - 4y_3$, $x_2 = y_1 - y_2 + 2y_3$, $x_3 = y_3$.

Theorem 6. *For every real quadratic form $f(x_1, \ldots, x_n)$ with a matrix $\mathbf{A}$, there exists a (real) regular linear substitution (3) with an orthogonal matrix $\mathbf{P}$ which transforms the form f into the form*

$$g(y) = \alpha_1 y_1^2 + \alpha_2 y_2^2 + \ldots + \alpha_n y_n^2 , \tag{4}$$

where $\alpha_1, \ldots, \alpha_n$ are the eigenvalues of the matrix $\mathbf{A}$ (and are real – see Theorem 1.28.10, p. 100).

Example 3. The problem is to find a transformation $x_i = \sum_{j=1}^{3} p_{ij} y_j$ $(i = 1, 2, 3)$ with an orthogonal matrix $\mathbf{P} = \|p_{ij}\|$ which brings the given quadratic form $f(x_1, x_2, x_3) = 2x_1^2 + x_2^2 - 4x_1 x_2 - 4x_2 x_3$ into the form $g(y_1, y_2, y_3)$ given by (4). The matrix of the form $f(x_1, x_2, x_3)$ is

$$\mathbf{A} = \|a_{ij}\| = \begin{Vmatrix} 2, & -2, & 0 \\ -2, & 1, & -2 \\ 0, & -2, & 0 \end{Vmatrix} ;$$

its eigenvalues, i.e. the roots of the equation

$$\begin{vmatrix} \lambda - 2, & 2, & 0 \\ 2, & \lambda - 1, & 2 \\ 0, & 2, & \lambda \end{vmatrix} = 0$$

are 1, −2, 4. According to Theorem 6, there exists a transformation $x_i = \sum_{j=1}^{3} p_{ij} y_j$ with an orthogonal matrix $\mathbf{P} = \|p_{ij}\|$ which brings the form $f(x_1, x_2, x_3)$ into the form $g(y_1, y_2, y_3) = y_1^2 - 2y_2^2 + 4y_3^2$. The matrix $\mathbf{P}$ (and, thus, the required transformation) can be found as follows: If we denote by $\mathbf{B} = \|b_{ij}\|$ the matrix of the form $g(y_1, y_2, y_3)$, i.e.

$$\mathbf{B} = \begin{Vmatrix} 1, & 0, & 0 \\ 0, & -2, & 0 \\ 0, & 0, & 4 \end{Vmatrix},$$

then, by Theorem 3, $\mathbf{B} = \mathbf{P}'\mathbf{AP}$. Since $\mathbf{P}$ is orthogonal, i.e. by Definition 1.25.7, $\mathbf{P}' = \mathbf{P}^{-1}$, the equality $\mathbf{B} = \mathbf{P}'\mathbf{AP}$ can be written in the form $\mathbf{PB} = \mathbf{AP}$. This means that $\sum_{l=1}^{3} a_{il} p_{lk} = \sum_{j=1}^{3} p_{ij} b_{jk}$ $(i, k = 1, 2, 3)$. In our problem, we thus get the following equations:

$$\text{(a)}\quad \begin{aligned} 2p_{11} - 2p_{21} \qquad\quad &= p_{11}, \\ -2p_{11} + p_{21} - 2p_{31} &= p_{21}, \\ - 2p_{21} \qquad\quad &= p_{31}; \end{aligned} \qquad \text{(b)}\quad \begin{aligned} 2p_{12} - 2p_{22} \qquad\quad &= -2p_{12}, \\ -2p_{12} + p_{22} - 2p_{23} &= -2p_{22}, \\ - 2p_{22} \qquad\quad &= -2p_{32}; \end{aligned}$$

$$\text{(c)}\quad \begin{aligned} 2p_{13} - 2p_{23} \qquad\quad &= 4p_{13}, \\ -2p_{13} + p_{23} - 2p_{33} &= 4p_{23}, \\ - 2p_{23} \qquad\quad &= 4p_{33}. \end{aligned}$$

These are three systems of homogeneous equations, each of them being of rank 2. Solving, for example, the system a) we can confine ourselves to the first and third equations from which there follows

$$p_{11} : p_{21} : p_{31} = \begin{vmatrix} -2, & 0 \\ -2, & -1 \end{vmatrix} : \begin{vmatrix} 0, & 1 \\ -1, & 0 \end{vmatrix} : \begin{vmatrix} 1, & -2 \\ 0, & -2 \end{vmatrix} = 2 : 1 : -2\,.$$

Since $p_{11}^2 + p_{21}^2 + p_{31}^2 = 1$ (see Theorem 1.25.12, p. 90), we get

$$p_{11} = \frac{\pm 2}{\sqrt{(2^2 + 1^2 + 2^2)}} = \pm\tfrac{2}{3}, \quad p_{21} = \pm\tfrac{1}{3}, \quad p_{31} = \mp\tfrac{2}{3}\,.$$

Similarly, we find that $p_{12} = \pm\frac{1}{3}$, $p_{22} = \pm\frac{2}{3}$, $p_{32} = \pm\frac{2}{3}$, $p_{13} = \pm\frac{2}{3}$, $p_{23} = \mp\frac{2}{3}$, $p_{33} = \pm\frac{1}{3}$. Thus, the matrix $\mathbf{P}$ can be chosen as follows:

$$\mathbf{P} = \begin{Vmatrix} \frac{2}{3}, & \frac{1}{3}, & \frac{2}{3} \\ \frac{1}{3}, & \frac{2}{3}, & -\frac{2}{3} \\ -\frac{2}{3}, & \frac{2}{3}, & \frac{1}{3} \end{Vmatrix}.$$

Theorem 7 (*Sylvester's Law of Inertia*). *Any real quadratic form $f(x_1, \ldots, x_n)$ of rank h can be transformed by a real regular linear substitution into the form*

$$g(y_1, \ldots, y_n) = y_1^2 + y_2^2 + \ldots + y_{s_1}^2 - y_{s_1+1}^2 - \ldots - y_{s_1+s_2}^2 \quad (s_1 + s_2 = h). \quad (5)$$

The substitution transforming $f(x_1, \ldots, x_n)$ into the form (5) is not unique; however, the number s_1 of positive signs as well as the number $s_2 = h - s_1$ of the negative sings in the resulting form is always the same.

Definition 5. The number $s_1 - s_2$ in Theorem 7 is called the *signature of the form* $f(x_1, \ldots, x_n)$.

Theorem 8. *A real quadratic form $f(x_1, \ldots, x_n)$ can be transformed by a real regular linear substitution into a form $g(y_1, \ldots, y_n)$ if and only if both forms have the same rank and signature.*

Definition 6. Let $f(x_1, \ldots, x_n)$ be a real quadratic form.

(a) The form $f(x_1, \ldots, x_n)$ is called *positive* (or *negative*) *definite* if, for every non-zero n-tuple $(\alpha_1, \ldots, \alpha_n)$ of real numbers $\alpha_1, \ldots, \alpha_n$ (briefly: for any real non-zero n-tuple), the number $f(\alpha_1, \ldots, \alpha_n)$ is positive (or negative).

(b) The form $f(x_1, \ldots, x_n)$ is called *positive* (or *negative*) *semidefinite* if, for every non-zero real n-tuple $(\alpha_1, \ldots, \alpha_n)$, the inequality $f(\alpha_1, \ldots, \alpha_n) \geqq 0$ [or $f(\alpha_1, \ldots, \alpha_n) \leqq \leqq 0$] holds and at the same time there exist non-zero real n-tuples $(\beta_1, \ldots, \beta_n)$ such that $f(\beta_1, \ldots, \beta_n) = 0$.

(c) The form $f(x_1, \ldots, x_n)$ is said to be *indefinite* if there are non-zero real n-tuples $(\alpha_1, \ldots, \alpha_n)$ and $(\beta_1, \ldots, \beta_n)$ such that $f(\alpha_1, \ldots, \alpha_n) > 0$ and $f(\beta_1, \ldots, \beta_n) < 0$.

REMARK 3. The matrix of a positive, or negative, definite quadratic form is called *positive* (or *negative*) *definite*. In the following theorems, some conditions for a symmetric matrix to be positive definite are introduced.

Theorem 9. *Let $f(x_1, \ldots, x_n)$ be a real quadratic form of rank h and signature s.*

1. *$f(x_1, \ldots, x_n)$ is positive (or negative) definite if and only if $h = n$ and $s = n$ (or $s = -n$); the form can be transformed by a real regular linear substitution into a sum of positive (or negative) squares of all n variables.*

2. *$f(x_1, \ldots, x_n)$ is positive (or negative) semidefinite if and only if $h < n$ and $s = h$ (or $s = -h$).*

3. *$f(x_1, \ldots, x_n)$ is indefinite if $-h < s < h$.*

REMARK 4. If a form $f(x_1, \ldots, x_n)$ is positive definite, or semidefinite, then the form $-f(x_1, \ldots, x_n)$ is obviously negative definite, or semidefinite, respectively. Therefore, we can restrict our consideration to positive definite or positive semidefinite forms.

Theorem 10. *Let $f(x_1, \ldots, x_n)$ be a real quadratic form and let* $\mathbf{A}$ *be its matrix. The form $f(x_1, \ldots, x_n)$ is positive definite, or semidefinite, if and only if all the eigenvalues of the matrix* $\mathbf{A}$ *are positive, or non-negative, respectively.*

Theorem 11. *A real quadratic form $f(x_1, \ldots, x_n)$ with a matrix*

$$\begin{Vmatrix} a_{11}, & a_{12}, & \ldots, & a_{1n} \\ \ldots & \ldots & \ldots & \ldots \\ a_{n1}, & a_{n2}, & \ldots, & a_{nn} \end{Vmatrix}$$

is positive definite if and only if all the principal minors

$$|a_{11}|, \quad \begin{vmatrix} a_{11}, & a_{12} \\ a_{21}, & a_{22} \end{vmatrix}, \quad \begin{vmatrix} a_{11}, & a_{12}, & a_{13} \\ a_{21}, & a_{22}, & a_{23} \\ a_{31}, & a_{32}, & a_{33} \end{vmatrix}, \quad \ldots, \quad \begin{vmatrix} a_{11}, & \ldots, & a_{1n} \\ \ldots & \ldots & \ldots \\ a_{n1}, & \ldots, & a_{nn} \end{vmatrix}$$

of the matrix $\mathbf{A}$ *are positive.*

Example 4. *The form* $a_{11}x_1^2 + 2a_{12}x_1x_2 + a_{22}x_2^2$ is positive definite if and only if $a_{11} > 0$, $a_{11}a_{22} - a_{12}^2 > 0$; it is semidefinite if $a_{11}a_{22} - a_{12}^2 = 0$; it is indefinite, if $a_{11}a_{22} - a_{12}^2 < 0$.

Definition 7. A *Hermitian (quadratic) form in n variables* $x_1, \ldots, x_n$ is a polynomial of the form

$$f(x_1, x_2, \ldots, x_n) = \sum_{i,j=1}^{n} a_{ij}x_i\bar{x}_j\,, \quad a_{ij} = \bar{a}_{ji} \quad (i, j = 1, \ldots, n)\,, \tag{6}$$

where the bar indicates a conjugate complex number. The matrix

$$\begin{Vmatrix} a_{11}, & \ldots, & a_{1n} \\ \ldots & \ldots & \ldots \\ a_{n1}, & \ldots, & a_{nn} \end{Vmatrix}$$

of a Hermitian form is a Hermitian matrix, i.e. $\mathbf{A} = \bar{\mathbf{A}}'$ holds.

Theorem 12. *A Hermitian form with a matrix* $\mathbf{A}$ *is transformed by a linear substitution* $x_i = \sum p_{ij}y_j$ *into the Hermitian form with the matrix* $\mathbf{B} = \mathbf{P}'\mathbf{A}\bar{\mathbf{P}}$.

REMARK 5. Matrices $\mathbf{A}$, $\mathbf{B}$ are said to be *conjunctive (Hermitian congruent)* if there exists a regular matrix $\mathbf{P}$ such that $\mathbf{B} = \mathbf{P}'\mathbf{A}\bar{\mathbf{P}}$.

Theorem 13. *If $(\alpha_1, \ldots, \alpha_n)$ is an arbitrary n-tuple of real or complex numbers and if $f(x_1, \ldots, x_n)$ is a Hermitian form, then the number $f(\alpha_1, \ldots, \alpha_n)$ is real.*

REMARK 6. In the same way as for real quadratic forms, we define the rank and signature of a Hermitian form, and also positive (negative) definite, semidefinite and indefinite Hermitian forms (see Definitions 2, 5 and 6). Theorems formulated for real quadratic forms hold also for Hermitian forms; in such formulation, instead of real regular linear substitutions we have complex regular linear substitutions and in Theorem 6 we must replace "orthogonal matrix $\mathbf{P}$" by "unitary matrix $\mathbf{P}$".

2. TRIGONOMETRIC AND INVERSE TRIGONOMETRIC FUNCTIONS. HYPERBOLIC AND INVERSE HYPERBOLIC FUNCTIONS

By VÁCLAV VILHELM

References: [2], [48], [93], [144], [220], [265], [303], [440].

2.1. Measurement of Angles (Measurement by Degrees and Circular Measure)

If theoretical problems are under consideration, angles are not measured in degrees, but in radians (circular measure): The magnitude of an angle α is given by the length l of the arc, intercepted by the arms of the angle α on the unit circle with centre at the vertex of the angle (Fig. 2.1). We shall denote the magnitude of the angle α in circular measure again by α; sometimes, instead of α, the notation arc α° is employed, α° denoting the magnitude of the angle α expressed in degrees (in the sexagesimal system).

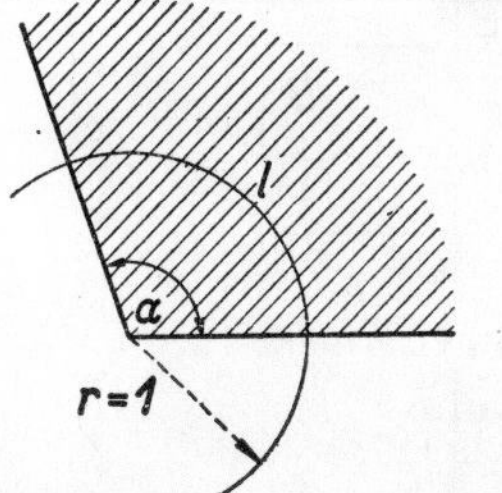

Fig. 2.1.

Theorem 1. *The relationship between circular measure and degrees is*

$$\alpha = \text{arc}\,\alpha^\circ = \frac{\pi}{180^\circ}\,\alpha^\circ\,.$$

Example 1. The angle of 90° is in circular measure

$$\alpha = \frac{\pi}{180^\circ}\,.\,90^\circ = \tfrac{1}{2}\pi\,.$$

Definition 1. The angle ϱ, whose circular measure is 1, is called the *radian*; its magnitude measured in degrees (in the sexagesimal system) is

$$\varrho^\circ = 180^\circ/\pi \doteq 57{\cdot}295\,779\,5^\circ = 57^\circ 17' 44{\cdot}806'' .$$

In centesimal measure,

$$\varrho^{g} = 400^{g}/2\pi \doteq 63{\cdot}661\,977^{g}\,(\text{grades}) .$$

In particular,

$$360^\circ = 2\pi\,,\ 180^\circ = \pi\,,\ 90^\circ = \tfrac{1}{2}\pi\,,\ 60^\circ = \tfrac{1}{3}\pi\,,\ 45^\circ = \tfrac{1}{4}\pi\,,\ 30^\circ = \tfrac{1}{6}\pi\,.$$

2.2. Definition of Trigonometric Functions

Definition 1. The trigonometric functions of an angle α in the interval $[0, 2\pi)$ are defined by means of a unit circle or (for acute angles) by means of a right angled triangle (Fig. 2.2) as follows:

$$\begin{aligned}
&\sin\alpha = \widetilde{AB} && (\text{for } 0 < \alpha < \tfrac{1}{2}\pi,\ \sin\alpha = a/c)\,,\\
&\cos\alpha = \widetilde{OA} && (\text{for } 0 < \alpha < \tfrac{1}{2}\pi,\ \cos\alpha = b/c)\,,\\
&\tan\alpha = \sin\alpha/\cos\alpha,\ \alpha \neq \tfrac{1}{2}\pi,\ \tfrac{3}{2}\pi && (\text{for } 0 < \alpha < \tfrac{1}{2}\pi,\ \tan\alpha = \widetilde{CD} = a/b)\,,\\
&\cot\alpha = \cos\alpha/\sin\alpha,\ \alpha \neq 0,\ \pi && (\text{for } 0 < \alpha < \tfrac{1}{2}\pi,\ \cot\alpha = \widetilde{EF} = b/a)\,,\\
&\sec\alpha = 1/\cos\alpha,\ \alpha \neq \tfrac{1}{2}\pi,\ \tfrac{3}{2}\pi && (\text{for } 0 < \alpha < \tfrac{1}{2}\pi,\ \sec\alpha = c/b)\,,\\
&\operatorname{cosec}\alpha = 1/\sin\alpha,\ \alpha \neq 0,\ \pi && (\text{for } 0 < \alpha < \tfrac{1}{2}\pi,\ \operatorname{cosec}\alpha = c/a)\,.
\end{aligned}$$

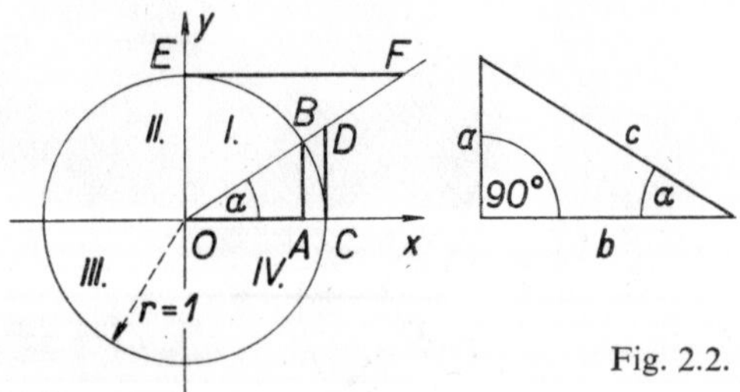

Fig. 2.2.

Here, $\widetilde{AB}$ is the directed length of the segment AB, i.e. $\widetilde{AB} > 0$ if AB is in the same direction and $\widetilde{AB} < 0$ if AB is in the opposite direction to the positive direction of the y-axis. The other lengths are used with a similar meaning (for example $\widetilde{OA} > 0$ if OA is in the same direction as the positive direction of the x-axis).

Further we define:

Definition 2.

$$\sin(2k\pi + \alpha) = \sin\alpha\,,\quad \cos(2k\pi + \alpha) = \cos\alpha\,, \tag{1}$$

$$\tan(k\pi + \alpha) = \tan\alpha\,,\quad \cot(k\pi + \alpha) = \cot\alpha \tag{2}$$

for an arbitrary integer k. In this way, the functions $\sin \alpha$ and $\cos \alpha$ are defined for all real α, the function $\tan \alpha$ for all real α different from $\frac{1}{2}\pi + k\pi$ and the function $\cot \alpha$ for all real α different from $k\pi$.

REMARK 1. The functions $\sin \alpha$ and $\cos \alpha$ are periodic functions with period 2π; the functions $\tan \alpha$ and $\cot \alpha$ are periodic functions with period π.

REMARK 2. In the case where an angle is measured in radians, instead of α we often write the letter x as is usual in the case of functions, where the letter x stands for the independent variable; we thus speak about the functions $\sin x$, $\cos x$, $\tan x$, $\cot x$.

2.3. Behaviour of Trigonometric Functions. Their Fundamental Properties

REMARK 1. In Fig. 2.3, x denotes the angle measured in radians; the figure represents the graph of the functions $\sin x$, $\cos x$, $\tan x$, $\cot x$ for x in the interval $[-\pi, 2\pi]$.

Fundamental properties:

1.

$$\begin{aligned} -1 &\leqq \sin \alpha \leqq 1\,, \\ -1 &\leqq \cos \alpha \leqq 1\,, \\ -\infty &< \tan \alpha < +\infty\,, \\ -\infty &< \cot \alpha < +\infty\,. \end{aligned}$$

2.

$$\begin{aligned} \sin(-\alpha) &= -\sin \alpha\,; \\ \cos(-\alpha) &= \cos \alpha\,; \\ \tan(-\alpha) &= -\tan \alpha\,; \\ \cot(-\alpha) &= -\cot \alpha\,. \end{aligned}$$

2.4. Relations Among Trigonometric Functions of the Same Angle

1.

$$\sin^2 \alpha + \cos^2 \alpha = 1\,; \quad \tan \alpha = \frac{\sin \alpha}{\cos \alpha}\,, \quad \cot \alpha = \frac{\cos \alpha}{\sin \alpha}\,,$$

$$\sec \alpha = \frac{1}{\cos \alpha}\,, \quad \operatorname{cosec} \alpha = \frac{1}{\sin \alpha}\,,$$

$$1 + \tan^2 \alpha = \frac{1}{\cos^2 \alpha}\,, \quad 1 + \cot^2 \alpha = \frac{1}{\sin^2 \alpha}\,.$$

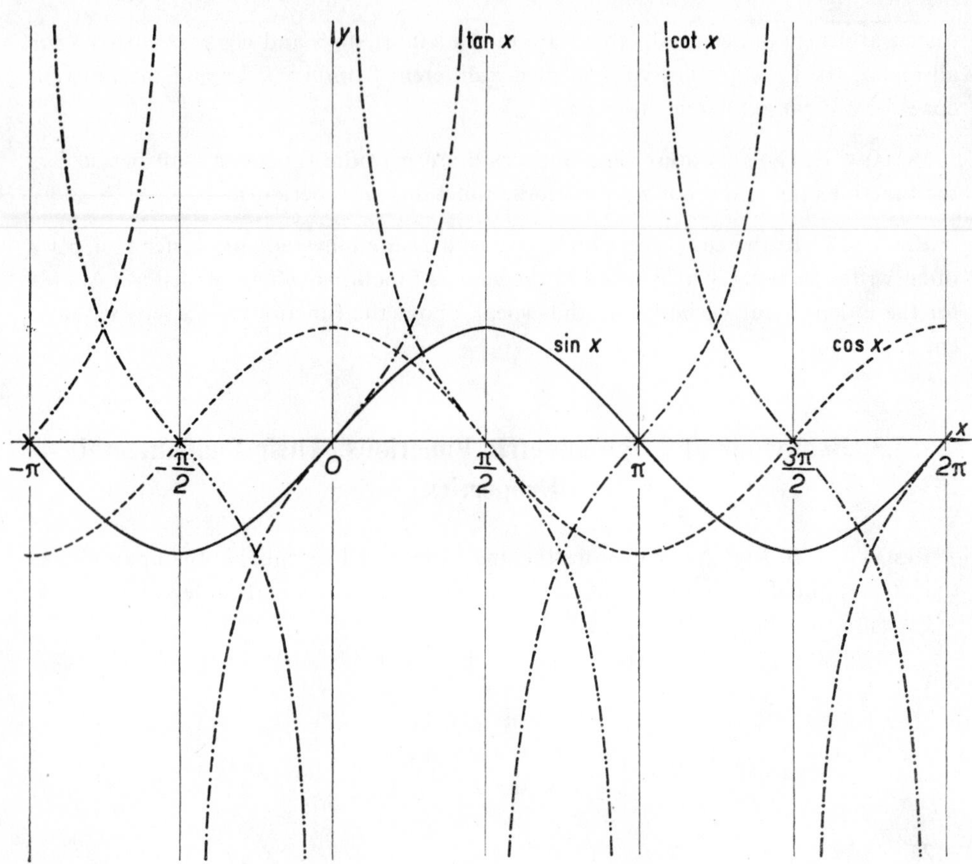

Fig. 2.3.

TABLE 2.1

Signs of trigonometric functions in individual quadrants

Function	Quadrant			
	I	II	III	IV
sin α	+	+	−	−
cos α	+	−	−	+
tan α	+	−	+	−
cot α	+	−	+	−

TABLE 2.2 *Values of trigonometric functions for some special angles*

Degrees Radians	0° 0	30° $\frac{1}{6}\pi$	45° $\frac{1}{4}\pi$	60° $\frac{1}{3}\pi$	90° $\frac{1}{2}\pi$	120° $\frac{2}{3}\pi$	135° $\frac{3}{4}\pi$	150° $\frac{5}{6}\pi$
sin α	0	$\frac{1}{2}$	$\frac{1}{2}\sqrt{2}$	$\frac{1}{2}\sqrt{3}$	1	$\frac{1}{2}\sqrt{3}$	$\frac{1}{2}\sqrt{2}$	$\frac{1}{2}$
cos α	1	$\frac{1}{2}\sqrt{3}$	$\frac{1}{2}\sqrt{2}$	$\frac{1}{2}$	0	$-\frac{1}{2}$	$-\frac{1}{2}\sqrt{2}$	$-\frac{1}{2}\sqrt{3}$
tan α	0	$\frac{1}{3}\sqrt{3}$	1	$\sqrt{3}$		$-\sqrt{3}$	-1	$-\frac{1}{3}\sqrt{3}$
cot α		$\sqrt{3}$	1	$\frac{1}{3}\sqrt{3}$	0	$-\frac{1}{3}\sqrt{3}$	-1	$-\sqrt{3}$

Degrees Radians	180° π	210° $\frac{7}{6}\pi$	225° $\frac{5}{4}\pi$	240° $\frac{4}{3}\pi$	270° $\frac{3}{2}\pi$	300° $\frac{5}{3}\pi$	315° $\frac{7}{4}\pi$	330° $\frac{11}{6}\pi$
sin α	0	$-\frac{1}{2}$	$-\frac{1}{2}\sqrt{2}$	$-\frac{1}{2}\sqrt{3}$	-1	$-\frac{1}{2}\sqrt{3}$	$-\frac{1}{2}\sqrt{2}$	$-\frac{1}{2}$
cos α	-1	$-\frac{1}{2}\sqrt{3}$	$-\frac{1}{2}\sqrt{2}$	$-\frac{1}{2}$	0	$\frac{1}{2}$	$\frac{1}{2}\sqrt{2}$	$\frac{1}{2}\sqrt{3}$
tan α	0	$\frac{1}{3}\sqrt{3}$	1	$\sqrt{3}$		$-\sqrt{3}$	-1	$-\frac{1}{3}\sqrt{3}$
cot α		$\sqrt{3}$	1	$\frac{1}{3}\sqrt{3}$	0	$-\frac{1}{3}\sqrt{3}$	-1	$-\sqrt{3}$

TABLE 2.3 *Reduction of trigonometric functions to the first quadrant*

Function	$\beta = 90° \pm \alpha$	$\beta = 180° \pm \alpha$	$\beta = 270° \pm \alpha$	$\beta = 360° \pm \alpha$
sin β	$+\cos\alpha$	$\mp\sin\alpha$	$-\cos\alpha$	$\pm\sin\alpha$
cos β	$\mp\sin\alpha$	$-\cos\alpha$	$\pm\sin\alpha$	$+\cos\alpha$
tan β	$\mp\cot\alpha$	$\pm\tan\alpha$	$\mp\cot\alpha$	$\pm\tan\alpha$
cot β	$\mp\tan\alpha$	$\pm\cot\alpha$	$\mp\tan\alpha$	$\pm\cot\alpha$

2.

$$|\sin\alpha| = \sqrt{(1-\cos^2\alpha)} = \frac{|\tan\alpha|}{\sqrt{(1+\tan^2\alpha)}} = \frac{1}{\sqrt{(1+\cot^2\alpha)}};$$

$$|\cos\alpha| = \sqrt{(1-\sin^2\alpha)} = \frac{1}{\sqrt{(1+\tan^2\alpha)}} = \frac{|\cot\alpha|}{\sqrt{(1+\cot^2\alpha)}};$$

$$|\tan\alpha| = \frac{|\sin\alpha|}{\sqrt{(1-\sin^2\alpha)}} = \frac{\sqrt{(1-\cos^2\alpha)}}{|\cos\alpha|} = \frac{1}{|\cot\alpha|};$$

$$\tan\alpha = \frac{1}{\cot\alpha}; \quad \cot\alpha = \frac{1}{\tan\alpha}.$$

REMARK 1. The absolute value must be used in relations 2, since, for example, $\sin 30° = \sqrt{(1-\cos^2 30°)}$, but $\sin 270° = -\sqrt{(1-\cos^2 270°)}$. For a definite α we have $\sin\alpha = \sqrt{(1-\cos^2\alpha)}$ or $\sin\alpha = -\sqrt{(1-\cos^2\alpha)}$ according to the sign of $\sin\alpha$ in the corresponding quadrant (Table 2.1).

Similarly for the other formulae in which the absolute values occur.

2.5. The Addition Formulae, the Multiple-angle and Half-angle Formulae

1.

$$\sin(\alpha\pm\beta) = \sin\alpha\cos\beta \pm \cos\alpha\sin\beta;$$

$$\cos(\alpha\pm\beta) = \cos\alpha\cos\beta \mp \sin\alpha\sin\beta;$$

$$\tan(\alpha\pm\beta) = \frac{\tan\alpha\pm\tan\beta}{1\mp\tan\alpha\tan\beta};$$

$$\cot(\alpha\pm\beta) = \frac{\cot\alpha\cot\beta\mp 1}{\cot\beta\pm\cot\alpha}.$$

2. $\sin n\alpha$, $\cos n\alpha$, for n a natural number can be determined by De Moivre's theorem (Theorem 1.6.6, p. 49),

$$\cos n\alpha + \mathrm{i}\sin n\alpha = (\cos\alpha + \mathrm{i}\sin\alpha)^n = \sum_{k=0}^{n}\binom{n}{k}\cos^k\alpha\,(\mathrm{i}\sin\alpha)^{n-k}.$$

3.

$$\sin 2\alpha = 2\sin\alpha\cos\alpha; \quad \sin 3\alpha = 3\sin\alpha - 4\sin^3\alpha;$$

$$\sin n\alpha = n\sin\alpha\cos^{n-1}\alpha - \binom{n}{3}\sin^3\alpha\cos^{n-3}\alpha + \binom{n}{5}\sin^5\alpha\cos^{n-5}\alpha - \ldots.$$

4.

$$\cos 2\alpha = \cos^2 \alpha - \sin^2 \alpha\,; \quad \cos 3\alpha = 4\cos^3 \alpha - 3\cos \alpha\,;$$

$$\cos n\alpha = \cos^n \alpha - \binom{n}{2} \sin^2 \alpha \cos^{n-2} \alpha + \binom{n}{4} \sin^4 \alpha \cos^{n-4} \alpha - \dots .$$

5.

$$\tan 2\alpha = \frac{2\tan \alpha}{1 - \tan^2 \alpha}\,; \quad \tan 3\alpha = \frac{3\tan \alpha - \tan^3 \alpha}{1 - 3\tan^2 \alpha}\,;$$

$$\tan n\alpha = \frac{n \tan \alpha - \binom{n}{3} \tan^3 \alpha + \binom{n}{5} \tan^5 \alpha - \dots}{1 - \binom{n}{2} \tan^2 \alpha + \binom{n}{4} \tan^4 \alpha - \binom{n}{6} \tan^6 \alpha + \dots}.$$

6.

$$\cot 2\alpha = \frac{\cot^2 \alpha - 1}{2\cot \alpha}\,; \quad \cot 3\alpha = \frac{\cot^3 \alpha - 3\cot \alpha}{3\cot^2 \alpha - 1}\,;$$

$$\cot n\alpha = \frac{\cot^n \alpha - \binom{n}{2} \cot^{n-2} \alpha + \binom{n}{4} \cot^{n-4} \alpha - \dots}{n \cot^{n-1} \alpha - \binom{n}{3} \cot^{n-3} \alpha + \binom{n}{5} \cot^{n-5} \alpha - \dots}.$$

7.

$$\left|\sin \frac{\alpha}{2}\right| = \sqrt{[\tfrac{1}{2}(1 - \cos \alpha)]}\,; \quad \left|\tan \frac{\alpha}{2}\right| = \sqrt{\frac{1 - \cos \alpha}{1 + \cos \alpha}}\,;$$

$$\tan \frac{\alpha}{2} = \frac{1 - \cos \alpha}{\sin \alpha} = \frac{\sin \alpha}{1 + \cos \alpha}\,;$$

$$\left|\cos \frac{\alpha}{2}\right| = \sqrt{[\tfrac{1}{2}(1 + \cos \alpha)]}\,; \quad \left|\cot \frac{\alpha}{2}\right| = \sqrt{\frac{1 + \cos \alpha}{1 - \cos \alpha}}\,;$$

$$\cot \frac{\alpha}{2} = \frac{1 + \cos \alpha}{\sin \alpha} = \frac{\sin \alpha}{1 - \cos \alpha}.$$

8.

$$\sin \alpha = \frac{2\tan \frac{1}{2}\alpha}{1 + \tan^2 \frac{1}{2}\alpha}\,; \quad \cos \alpha = \frac{1 - \tan^2 \frac{1}{2}\alpha}{1 + \tan^2 \frac{1}{2}\alpha}.$$

2.6. Sum, Difference, Product of Trigonometric Functions, Powers of Trigonometric Functions

1.

$$\sin\alpha + \sin\beta = 2\sin\frac{\alpha+\beta}{2}\cos\frac{\alpha-\beta}{2};$$

$$\sin\alpha - \sin\beta = 2\cos\frac{\alpha+\beta}{2}\sin\frac{\alpha-\beta}{2};$$

$$\cos\alpha + \cos\beta = 2\cos\frac{\alpha+\beta}{2}\cos\frac{\alpha-\beta}{2};$$

$$\cos\alpha - \cos\beta = -2\sin\frac{\alpha+\beta}{2}\sin\frac{\alpha-\beta}{2};$$

$$\tan\alpha \pm \tan\beta = \frac{\sin(\alpha\pm\beta)}{\cos\alpha\cos\beta};$$

$$\cot\alpha \pm \cot\beta = \frac{\sin(\beta\pm\alpha)}{\sin\alpha\sin\beta};$$

$$\tan\alpha \pm \cot\beta = \pm\frac{\cos(\alpha\mp\beta)}{\cos\alpha\sin\beta}.$$

2.

$$\sin\alpha\sin\beta = \tfrac{1}{2}[\cos(\alpha-\beta) - \cos(\alpha+\beta)];$$

$$\cos\alpha\cos\beta = \tfrac{1}{2}[\cos(\alpha-\beta) + \cos(\alpha+\beta)];$$

$$\sin\alpha\cos\beta = \tfrac{1}{2}[\sin(\alpha-\beta) + \sin(\alpha+\beta)];$$

$$\tan\alpha\tan\beta = \frac{\tan\alpha+\tan\beta}{\cot\alpha+\cot\beta}; \quad \cot\alpha\cot\beta = \frac{\cot\alpha+\cot\beta}{\tan\alpha+\tan\beta};$$

$$\tan\alpha\cot\beta = \frac{\tan\alpha+\cot\beta}{\tan\beta+\cot\alpha}.$$

3.

$$\sin^2\alpha = \tfrac{1}{2}(1-\cos 2\alpha); \quad \sin^3\alpha = \tfrac{1}{4}(3\sin\alpha - \sin 3\alpha);$$

$$\cos^2\alpha = \tfrac{1}{2}(1+\cos 2\alpha); \quad \cos^3\alpha = \tfrac{1}{4}(\cos 3\alpha + 3\cos\alpha);$$

$$\sin^4\alpha = \tfrac{1}{8}(\cos 4\alpha - 4\cos 2\alpha + 3); \quad \cos^4\alpha = \tfrac{1}{8}(\cos 4\alpha + 4\cos 2\alpha + 3).$$

REMARK 1. Higher powers can be found by De Moivre's theorem (see relations 2, 3 and 4 of the previous § 2.5).

2.7. Trigonometric Sums

Theorem 1. For an arbitrary real α, an arbitrary real $x \neq 2k\pi$ (k being an integer) and for n a natural number we have

$$\sin x + \sin 2x + \ldots + \sin nx = \frac{\sin \frac{1}{2}nx}{\sin \frac{1}{2}x} \sin \tfrac{1}{2}(n+1)\,x\,;$$

$$\cos x + \cos 2x + \ldots + \cos nx = \frac{\sin \frac{1}{2}nx}{\sin \frac{1}{2}x} \cos \tfrac{1}{2}(n+1)\,x\,;$$

$$\sum_{j=1}^{n} \sin(\alpha + jx) = \frac{\sin \frac{1}{2}nx}{\sin \frac{1}{2}x} \sin\left[\alpha + \tfrac{1}{2}(n+1)\,x\right];$$

$$\sum_{j=1}^{n} \cos(\alpha + jx) = \frac{\sin \frac{1}{2}nx}{\sin \frac{1}{2}x} \cos\left[\alpha + \tfrac{1}{2}(n+1)\,x\right].$$

2.8. Trigonometric Equations

Trigonometric equations are equations in the unknown x of the form

$$f(\cos x, \sin x, \tan x, \cot x, x) = 0\,. \tag{1}$$

A trigonometric equation can be solved either by employing numerical methods (see Chap. 31), or, in some simple cases, by rearranging the equation using suitable formulae, to contain only one trigonometric function; then we solve the equation for this function.

Example 1. $\sin x - \cos^2 x + \frac{1}{4} = 0$; we rearrange the equation by means of the relation $\cos^2 x = 1 - \sin^2 x$ and put $y = \sin x$. We thus obtain the equation $y^2 + y - \frac{3}{4} = 0$ with the roots $y_1 = \frac{1}{2}$, $y_2 = -\frac{3}{2}$. There is no real solution corresponding to the root y_2 (since $|\sin x| \leqq 1$); the root $y_1 = \frac{1}{2}$ gives the solutions $x = \frac{1}{6}\pi + 2k\pi$, $x = \frac{5}{6}\pi + 2k\pi$ (k being any integer).

Example 2. $a \cos x + b \sin x = c$ $(ab \neq 0)$. We put $a = r \cos \lambda$, $b = r \sin \lambda$, $r > 0$. Then $\tan \lambda = b/a$, $r = a/\cos \lambda = b/\sin \lambda$. The angle λ is determined to within an integral multiple of 2π. The equation is transformed into the form $r \cos x \cos \lambda + r \sin x \sin \lambda = c$, i.e. $\cos(x - \lambda) = c/r$. We get, in general, two values for $x - \lambda$ which are determined to within an integral multiple of 2π (provided, of course, that $|c/r| \leqq 1$).

REMARK 1. If relation (1) is satisfied for all real x for which the expression* $f(\cos x, \sin x, \tan x, \cot x, x)$ has a meaning, then it is called a *trigonometric identity*.

---

* Other angles $y, z, \ldots$ may also be contained in this expression.

Example 3. Let us decide whether the relation

$$\frac{\sin x + \sin y}{\cos x + \cos y} = \tan \frac{x + y}{2} \tag{2}$$

is a trigonometric identity.

We try to arrange the left-hand side in the form $\tan \frac{1}{2}(x + y)$. Applying formulae 1 of § 2.6 (p. 114) we get the left-hand side in the form

$$\frac{2 \sin \frac{1}{2}(x + y) \cos \frac{1}{2}(x - y)}{2 \cos \frac{1}{2}(x + y) \cos \frac{1}{2}(x - y)} = \tan \frac{1}{2}(x + y)\,;$$

this means that the relation (2) is a trigonometric identity.

2.9. Plane Trigonometry

(a) Right-angled Triangle (Fig. 2.4a)

REMARK 1. In this section the following symbols for the elements of a right-angled triangle will be used: a, b enclose the right angle, c is the hypotenuse; A, B, C are the vertices opposite to the sides a, b, c, respectively; α, β, $90°$ are the interior angles corresponding to the vertices A, B, C, respectively; h is the altitude; P-the area.

TABLE 2.4

Formulae for determining the remaining elements of a right-angled triangle if two elements are given

The given elements	The other elements of the triangle				
a, α	$\beta = 90° - \alpha$	$b = a \cot \alpha$	$c = \dfrac{a}{\sin \alpha}$	$h = a \cos \alpha$	$P = \frac{1}{2}a^2 \cot \alpha$
c, α	$\beta = 90° - \alpha$	$a = c \sin \alpha$	$b = c \cos \alpha$	$h = \frac{1}{2}c \sin 2\alpha$	$P = \frac{1}{4}c^2 \sin 2\alpha$
a, b	$\tan \alpha = \dfrac{a}{b}$	$\tan \beta = \dfrac{b}{a}$	$c = \dfrac{a}{\sin \alpha} =$ $= \sqrt{(a^2 + b^2)}$	$h = a \cos \alpha =$ $= b \sin \alpha$	$P = \frac{1}{2}ab$
a, c	$\sin \alpha = \dfrac{a}{c}$	$\cos \beta = \dfrac{a}{c}$	$b = c \cos \alpha =$ $= c \sin \beta =$ $= \sqrt{(c^2 - a^2)}$	$h = a \cos \alpha =$ $= a \sin \beta$	$P = \frac{1}{4}c^2 \sin 2\alpha =$ $= \frac{1}{2}a^2 \tan \beta$

(b) General (Scalene) Triangle (Fig. 2.4b)

REMARK 2. In this section the following symbols for the elements of a triangle will be used: a, b, c are the sides; A, B, C the vertices opposite to the sides a, b, c, respectively; α, β, γ the interior angles corresponding to the vertices A, B, C, respectively; r is the radius of the inscribed circle; R the radius of the circumscribed circle; h_a, h_b, h_c are the altitudes corresponding to the vertices A, B, C or to the sides a, b, c, respectively, P the area and $s = \frac{1}{2}(a + b + c)$.

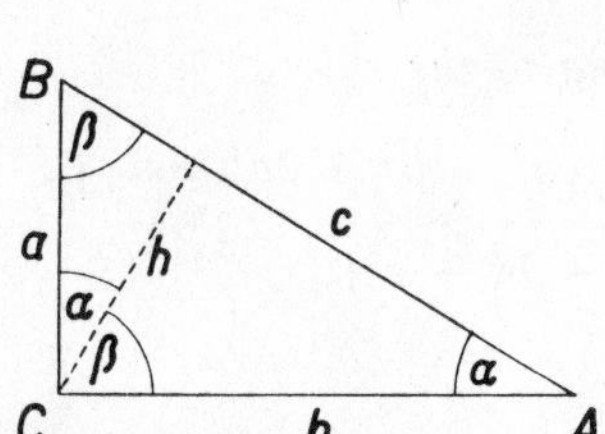

Fig. 2.4a.

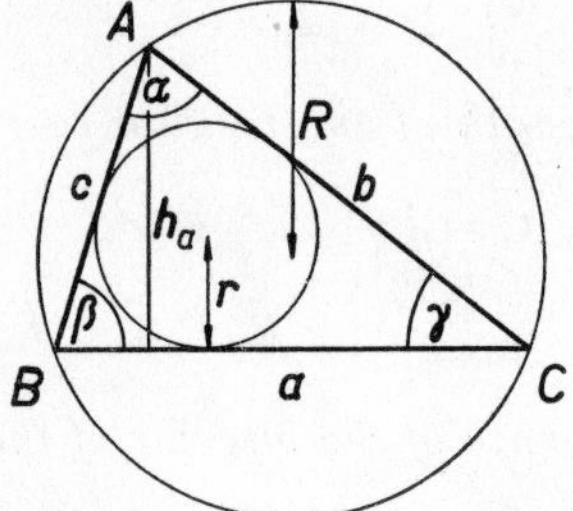

Fig. 2.4b.

Theorem 1. *Fundamental relations:*

1. $$\frac{a}{\sin\alpha} = \frac{b}{\sin\beta} = \frac{c}{\sin\gamma} \; (= 2R) \quad \textit{(the Sine Theorem)}.$$

2. $$a^2 = b^2 + c^2 - 2bc\cos\alpha \quad \textit{(the Cosine Theorem)}.$$

3. $$\frac{a+b}{a-b} = \frac{\tan\frac{1}{2}(\alpha+\beta)}{\tan\frac{1}{2}(\alpha-\beta)} \quad \textit{(the Tangent Theorem)}.$$

Theorem 2. *Further relations:*

4. $$a = b\cos\gamma + c\cos\beta.$$

5. $$\frac{a+b}{c} = \frac{\cos\frac{1}{2}(\alpha-\beta)}{\cos\frac{1}{2}(\alpha+\beta)}; \quad \frac{a-b}{c} = \frac{\sin\frac{1}{2}(\alpha-\beta)}{\sin\frac{1}{2}(\alpha+\beta)}.$$

6. $$\sin\frac{\alpha}{2} = \sqrt{\frac{(s-b)(s-c)}{bc}}; \quad \cos\frac{\alpha}{2} = \sqrt{\frac{s(s-a)}{bc}}.$$

7. $$\tan\frac{\alpha}{2} = \frac{P}{s(s-a)}; \quad \tan\frac{\alpha}{2} = \frac{r}{s-a}.$$

8. $$\tan\alpha = \frac{a\sin\gamma}{b - a\cos\gamma}.$$

9. $r = s \tan \frac{\alpha}{2} \tan \frac{\beta}{2} \tan \frac{\gamma}{2}$.

10. $a = 2R \sin \alpha$; $R = \frac{abc}{4P}$.

11. $h_a = b \sin \gamma = c \sin \beta$; $\frac{1}{h_a} : \frac{1}{h_b} : \frac{1}{h_c} = a : b : c$; $\frac{1}{h_a} + \frac{1}{h_b} + \frac{1}{h_c} = \frac{1}{r}$.

12. $s = 4R \cos \frac{\alpha}{2} \cos \frac{\beta}{2} \cos \frac{\gamma}{2}$.

13. *The length of the median corresponding to the side c*:

$$t_c = \tfrac{1}{2}\sqrt{[2(a^2 + b^2) - c^2]} = \tfrac{1}{2}\sqrt{[a^2 + b^2 + 2ab \cos \gamma]} ;$$

$$t_a^2 + t_b^2 + t_c^2 = \tfrac{3}{4}(a^2 + b^2 + c^2) .$$

14. *The length of the bisector of the angle* γ:

$$u_\gamma = \frac{2\sqrt{[abs(s - c)]}}{a + b} = \frac{\sqrt{[ab[(a + b)^2 - c^2]]}}{a + b} = \frac{2ab \cos \frac{1}{2}\gamma}{a + b} .$$

15. *The radius of the circumscribed circle*:

$$R = \frac{a}{2 \sin \alpha} .$$

16. *The radius of the inscribed circle*:

$$r = 4R \sin \frac{\alpha}{2} \sin \frac{\beta}{2} \sin \frac{\gamma}{2} = \frac{abc}{4Rs} = \sqrt{\frac{(s - a)(s - b)(s - c)}{s}} .$$

17. *The area of the triangle*:

$$P = \tfrac{1}{2}ab \sin \gamma = a^2 \frac{\sin \beta \sin \gamma}{2 \sin \alpha} = r^2 \cot \frac{\alpha}{2} \cot \frac{\beta}{2} \cot \frac{\gamma}{2} = 2R^2 \sin \alpha \sin \beta \sin \gamma ,$$

$$P = \sqrt{[s(s - a)(s - b)(s - c)]} \quad (\textit{Heron's Formula}) .$$

Theorem 3. *Solution of a general triangle*:

1. *Given the elements* a, β, γ $(\beta + \gamma < 180°)$:

$$\alpha = 180° - (\beta + \gamma) ; \quad b = \frac{a \sin \beta}{\sin \alpha} ; \quad c = \frac{a \sin \gamma}{\sin \alpha} .$$

2. *Given the elements* a, b, γ:

$$\tfrac{1}{2}(\alpha + \beta) = 90° - \tfrac{1}{2}\gamma ; \quad \tan \tfrac{1}{2}(\alpha - \beta) = \frac{a - b}{a + b} \cot \tfrac{1}{2}\gamma ;$$

hence we determine the angles α, β:

$$\alpha = \tfrac{1}{2}(\alpha + \beta) + \tfrac{1}{2}(\alpha - \beta)\,; \quad \beta = \tfrac{1}{2}(\alpha + \beta) - \tfrac{1}{2}(\alpha - \beta)\,;$$

$$c = \sqrt{(a^2 + b^2 - 2ab \cos \gamma)}\,.$$

An alternative method:

$$\tan \alpha = \frac{a \sin \gamma}{b - a \cos \gamma}\,; \quad \tan \beta = \frac{b \sin \gamma}{a - b \cos \gamma}\,;$$

$$c = \frac{a \sin \gamma}{\sin \alpha} = \frac{a - b \cos \gamma}{\cos \beta}\,.$$

3. *Given the elements a, b, α:*

$$\sin \beta = \frac{b \sin \alpha}{a}\,; \quad \gamma = 180° - (\alpha + \beta)\,;$$

$$c = \frac{a \sin \gamma}{\sin \alpha}\,; \quad P = \tfrac{1}{2}ab \sin \gamma\,.$$

If $a > b$, then $\beta < 90°$ and there exists a single solution.
If $a = b$, there exists a single solution for $\alpha < 90°$.
If $a < b$ (and, $\alpha < 90°$, of course), then

1° *for $b \sin \alpha < a$ there exist two solutions (there are two angles β, satisfying the relation $\beta_2 = 180° - \beta_1$)*;
2° *for $b \sin \alpha = a$ there exists a single solution* $(\beta = 90°)$;
3° *there is no solution for $b \sin \alpha > a$.*

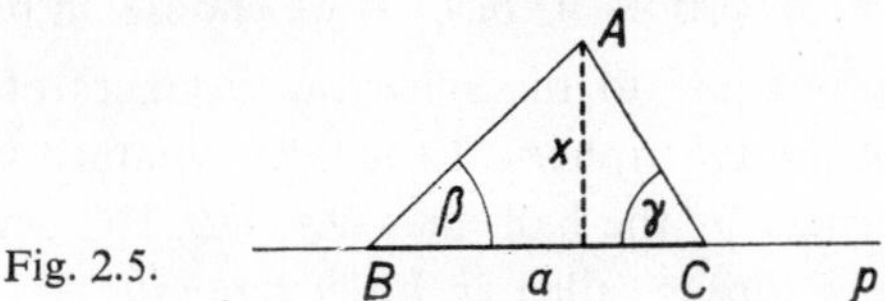

Fig. 2.5.

4. *Given the elements a, b, c:*

If the sum of any two sides is greater than the third side then a single solution exists and is given by

$$\cos \alpha = \frac{b^2 + c^2 - a^2}{2bc}\,, \quad \tan \frac{\alpha}{2} = \frac{P}{s(s - a)}\,, \quad P = \sqrt{[s(s - a)(s - b)(s - c)]}$$

and similarly for the angles β, γ.

Example 1. The problem is to find the distance x of an inaccessible point A from a straight road p (see Fig. 2.5).

On the road, the points B, C have been chosen, the distance a between them determined and the angles $ABC = \beta$, $ACB = \gamma$ measured. By Theorem 2 (formula 17), the area P of the triangle $\triangle ABC$ is

$$P = \frac{a^2 \sin\beta \sin\gamma}{2 \sin\alpha}, \quad \alpha = 180^\circ - (\beta + \gamma).$$

In addition, $P = \frac{1}{2}ax$. Hence $x = a \sin\beta \sin\gamma / \sin\alpha$.

2.10. Spherical Trigonometry

(a) Great Circle on a Sphere; Spherical (Euler's) Triangle

Definition 1. By a *great circle* on a given sphere we mean any circle lying on this sphere, whose centre coincides with the centre of the sphere. Through two points A, B on a sphere, which do not lie on the same diameter, one and only one great circle can be drawn; the smaller of the two arcs cut off by the points A, B on this circle has the shortest length d of all the curves on the given sphere joining the points A, B. This number d is called the *spherical distance* of the points A, B. The spherical distance of opposite points on a sphere equals the semi-circumference of a great circle.

Definition 2 (*Spherical Triangle*). Let A, B, C be three points on a sphere which do not lie on the same great circle. If we draw the three arcs $\overparen{AB}$, $\overparen{AC}$, $\overparen{BC}$ of the great circles which do not intersect except at the points A, B, C, the spherical surface splits into two spherical triangles with vertices A, B, C. If we choose, in particular, the arcs $\overparen{AB}$, $\overparen{AC}$, $\overparen{BC}$ to be of lengths equal to the spherical distances of their end points A, B, C, then the smaller of the two spherical triangles obtained (i.e. the one lying inside the trihedral angle formed by the half-lines OA, OB, OC emanating from the centre O of the sphere, see Fig. 2.6) is called an *Euler triangle.*

REMARK 1. In what follows we deal only with Euler triangles; moreover, we choose (except in Theorem 3) the radius of the sphere $r = 1$.

Definition 3. The lengths a, b, c of the corresponding arcs $\overparen{BC}$, $\overparen{AC}$, $\overparen{AB}$ of the great circles are called the *sides* of the spherical triangle $\triangle ABC$. Thus, they are determined by the angles BOC, AOC, AOB of the half-lines OA, OB, OC and are measured in radians or in degrees (Fig. 2.6).

Definition 4. The interior angles of the faces of the trihedral $OABC$ are called the angles α, β, γ of the spherical triangle $\triangle ABC$. They are measured in radians or in degrees (Fig. 2.6).

REMARK 2. The half-lines joining the centre of the sphere with the vertices of the spherical triangle $\triangle ABC$ form the basic trihedral $OABC$. The so-called *polar trihedral* $OA'B'C'$ has its edges normal to the faces of the basic trihedral and defines on the sphere an Euler spherical triangle $\triangle A'B'C'$, which is polar to $\triangle ABC$. The sides of the polar triangle are $a = 180° - \alpha$, $b = 180° - \beta$, $c = 180° - \gamma$; its angles are $\alpha = 180° - a$, $\beta = 180° - b$, $\gamma = 180° - c$. Thus, substituting the supplements of the angles for the sides and the supplements of the sides for the angles in any formula, we get a new formula.

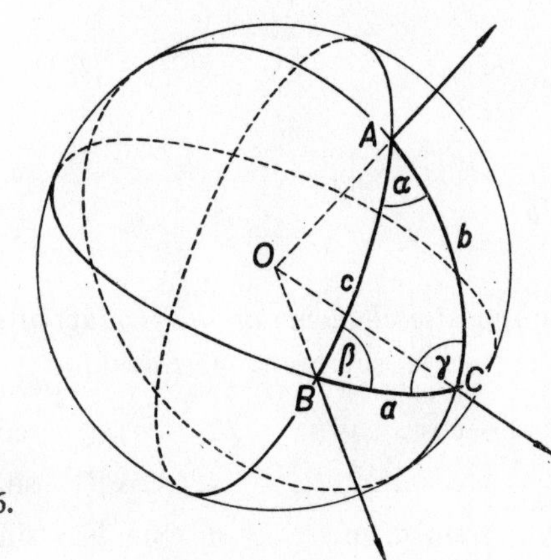

Fig. 2.6.

Fundamental properties of spherical triangles:

Theorem 1. *The sides and the angles of an Euler triangle are less than* 180° (*less than* π).

Theorem 2. *The sum of the angles* α, β, γ *of a spherical triangle is always greater than* 180°.

Definition 5. The number

$$\varepsilon° = \alpha° + \beta° + \gamma° - 180°$$

is called the *spherical excess of a spherical triangle.*

Theorem 3. *The area of a spherical triangle is*

$$P = \frac{\varepsilon°}{180°} \pi r^2 ,$$

where r *is the radius of the sphere and* $\varepsilon°$ *the excess of the triangle expressed in degrees.*

(b) Right-angled Spherical Triangle

REMARK 3. In this section, c denotes the hypotenuse, a, b the sides "enclosing" the right angle and α, β the angles opposite to the sides a, b, respectively (Fig. 2.7).

Theorem 4 (*Napier's Rule*). *We ascribe the hypotenuse c, the angles α, β and the complements of the sides $90° - a$, $90° - b$ to the vertices of a pentagon in the order indicated in Fig. 2.8. Then, the cosine of an arbitrary element equals the product*

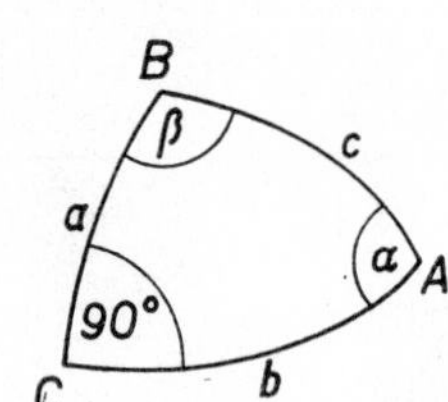

Fig. 2.7.

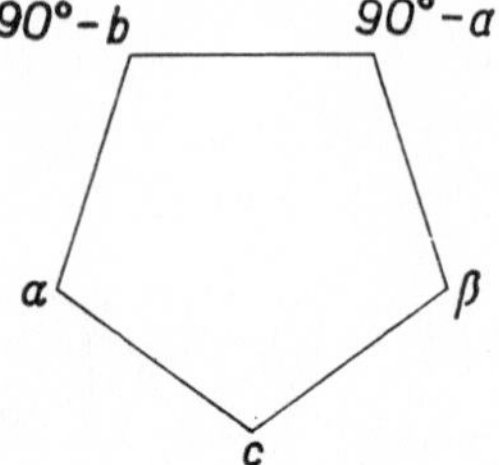

Fig. 2.8.

of the sines of the two opposite elements or the product of the cotangents of the two adjacent elements. In this way, we obtain the formulae

$$
\begin{array}{llll}
1. & \cos c = \cos a \cos b\,, & 2. & \cos c = \cot\alpha \cot\beta\,,\\
3. & \cos\alpha = \cos a \sin\beta\,, & 4. & \cos\beta = \sin\alpha \cos b\,,\\
5. & \sin a = \sin\alpha \sin c\,, & 6. & \sin b = \sin\beta \sin c\,,\\
7. & \cos\alpha = \tan b \cot c\,, & 8. & \cos\beta = \tan a \cot c\,,\\
9. & \sin a = \tan b \cot\beta\,, & 10. & \sin b = \tan a \cot\alpha\,.
\end{array}
$$

Theorem 5. *Spherical excess*:

$$\tan\frac{\varepsilon}{2} = \tan\frac{a}{2}\tan\frac{b}{2}\,.$$

Theorem 6. *The solution of a right-angled spherical triangle* (Table 2.5):

TABLE 2.5

Given elements	Number in parentheses denotes the corresponding formula of Theorem 4
a, b	c (1), α (10), β (9)
a, c	b (1), α (5), β (8)
a, α	b (10), c (5), β (3)
a, β	b (9), c (8), α (3)
c, α	a (5), b (7), β (2)
α, β	a (3), b (4), c (2)

(c) General (Oblique) Spherical Triangle

REMARK 4. Let us denote the sides of the triangle $\triangle ABC$ by a, b, c and the angles corresponding to the vertices A, B, C by α, β, γ, respectively (Fig. 2.9).

Theorem 7. *Fundamental formulae for an Euler triangle*:

1. $$\frac{\sin a}{\sin \alpha} = \frac{\sin b}{\sin \beta} = \frac{\sin c}{\sin \gamma} \quad (\textit{the Sine Theorem}) .$$

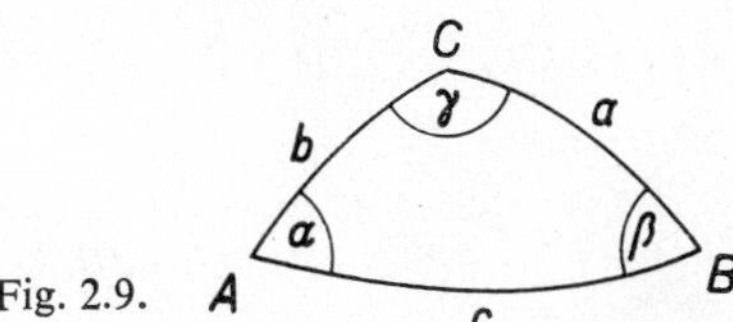

Fig. 2.9.

2. $\cos a = \cos b \cos c + \sin b \sin c \cos \alpha$ (*the Cosine Theorem for the sides*).

3. $\cos \alpha = -\cos \beta \cos \gamma + \sin \beta \sin \gamma \cos a$ (*the Cosine Theorem for the angles*).

4. (a) $\cos a \sin b = \sin a \cos b \cos \gamma + \sin c \cos \alpha$;
   (b) $\cot a \sin b = \sin \gamma \cot \alpha + \cos \gamma \cos b$.

5. (a) $\cos \alpha \sin \beta = \sin \gamma \cos a - \sin \alpha \cos \beta \cos c$;
   (b) $\cot \alpha \sin \beta = \sin c \cot a - \cos c \cos \beta$.

Theorem 8. *Further formulae for an Euler triangle*:

6. $$\tan \tfrac{1}{2}(a + b) = \frac{\cos \frac{1}{2}(\alpha - \beta)}{\cos \frac{1}{2}(\alpha + \beta)} \tan \tfrac{1}{2}c .$$

7. $$\tan \tfrac{1}{2}(a - b) = \frac{\sin \frac{1}{2}(\alpha - \beta)}{\sin \frac{1}{2}(\alpha + \beta)} \tan \tfrac{1}{2}c .$$

8. $$\tan \tfrac{1}{2}(\alpha + \beta) = \frac{\cos \frac{1}{2}(a - b)}{\cos \frac{1}{2}(a + b)} \cot \tfrac{1}{2}\gamma .$$

9. $$\tan \tfrac{1}{2}(\alpha - \beta) = \frac{\sin \frac{1}{2}(a - b)}{\sin \frac{1}{2}(a + b)} \cot \tfrac{1}{2}\gamma .$$

10. $\cos \tfrac{1}{2}(\alpha + \beta) \cos \tfrac{1}{2}c = \cos \tfrac{1}{2}(a + b) \sin \tfrac{1}{2}\gamma$.

11. $\sin \tfrac{1}{2}(\alpha + \beta) \cos \tfrac{1}{2}c = \cos \tfrac{1}{2}(a - b) \cos \tfrac{1}{2}\gamma$.

12. $\cos \tfrac{1}{2}(\alpha - \beta) \sin \tfrac{1}{2}c = \sin \tfrac{1}{2}(a + b) \sin \tfrac{1}{2}\gamma$.

13. $\sin \tfrac{1}{2}(\alpha - \beta) \sin \tfrac{1}{2}c = \sin \tfrac{1}{2}(a - b) \cos \tfrac{1}{2}\gamma$.

In formulae 14 *and* 15 *the notation*

$$s = \tfrac{1}{2}(a + b + c)\,, \quad s_1 = s - a\,, \quad s_2 = s - b\,, \quad s_3 = s - c\,,$$
$$\sigma = \tfrac{1}{2}(\alpha + \beta + \gamma)\,, \quad \sigma_1 = \sigma - \alpha\,, \quad \sigma_2 = \sigma - \beta\,, \quad \sigma_3 = \sigma - \gamma$$

is used.

14. $$\cot \frac{a}{2} = \frac{1}{\cos \sigma_1} \sqrt{\frac{\cos \sigma_1 \cos \sigma_2 \cos \sigma_3}{-\cos \sigma}}\,.$$

15. $$\tan \frac{\alpha}{2} = \frac{1}{\sin s_1} \sqrt{\frac{\sin s_1 \sin s_2 \sin s_3}{\sin s}}\,.$$

Theorem 9. *The solution of a general spherical triangle is shown in Table* 2.6.

TABLE 2.6

Given elements	Number in parentheses denotes the corresponding formula of Theorems 7 and 8
a, b, γ	$\frac{\alpha+\beta}{2}$ (8), $\frac{\alpha-\beta}{2}$ (9), c (e. g. 10 or 12)
α, β, c	$\frac{a+b}{2}$ (6), $\frac{a-b}{2}$ (7), γ (e.g. 11 or 12)
a, b, c	α (15), similarly β and γ
α, β, γ	a (14), similarly b and c
a, b, α*	β (1), γ (9), c (7)
α, β, a**	b (1), c (7), γ (9)

2.11. Inverse Trigonometric Functions

Inverse trigonometric functions are the functions $\arcsin x$ (or $\sin^{-1} x$), $\arccos x$ ($\cos^{-1} x$), $\arctan x$ ($\tan^{-1} x$), $\operatorname{arccot} x$ ($\cot^{-1} x$), which are inverse (see § 11.1, p. 400) to the trigonometric functions.

REMARK 1. In this section, the angles are expressed in circular measure.

* If $\sin b \sin \alpha > \sin a$, then no solution exists. If $\sin b \sin \alpha = \sin a$ there is a single solution (the triangle is right-angled). If $\sin b \sin \alpha < \sin a$, it is necessary to distinguish two cases: 1° if a is nearer to 90° than b, then there exists one solution (β and b are of the same kind, i.e. both either acute or obtuse); 2° if b is nearer to 90° than a, then there are two solutions or no solution according to whether a and α are of the same or of different kinds, acute or obtuse, respectively.

** The discussion of this case can be obtained from the discussion of the case* by substituting throughout the sides a, b, c for the angles α, β, γ and vice versa.

Definition 1. The function $y = \arcsin x$ is inverse to the function $x = \sin y$ $(-\frac{1}{2}\pi \leqq y \leqq \frac{1}{2}\pi)$; it is defined for x in the interval $[-1, 1]$. Thus: If $-1 \leqq x \leqq 1$, then $\arcsin x$ is the unique angle y in the range $[-\frac{1}{2}\pi, \frac{1}{2}\pi]$ such that $\sin y = x$.*

Definition 2. The function $y = \arccos x$ is inverse to the function $x = \cos y$ $(0 \leqq y \leqq \pi)$; it is defined for x in the interval $[-1, 1]$. Thus: If $-1 \leqq x \leqq 1$, then $\arccos x$ is the unique angle y in the range $[0, \pi]$ such that $\cos y = x$.

Definition 3. The function $y = \arctan x$ is inverse to the function $x = \tan y$ $(-\frac{1}{2}\pi < y < \frac{1}{2}\pi)$; it is defined for all real x. Thus, if x is a real number, then $\arctan x$ is the unique angle y in the range $(-\frac{1}{2}\pi, \frac{1}{2}\pi)$ such that $\tan y = x$.

Definition 4. The function $y = \operatorname{arccot} x$ is inverse to the function $x = \cot y$ $(0 < y < \pi)$; it is defined for all real x. Thus, if x is a real number, then $\operatorname{arccot} x$ is the unique angle y in the range $(0, \pi)$ such that $\cot y = x$. (The range $(-\frac{1}{2}\pi, \frac{1}{2}\pi)$ is sometimes used.)

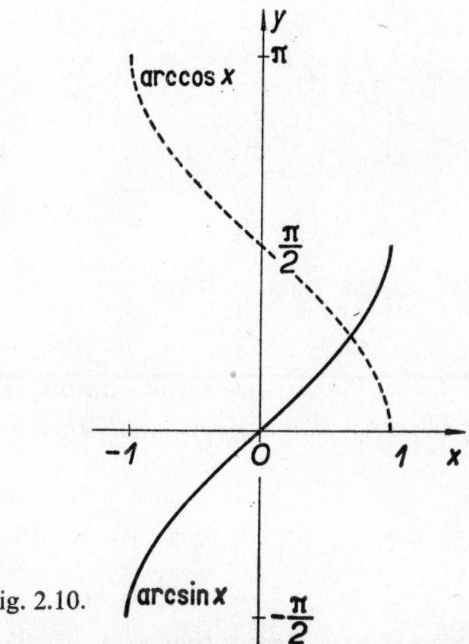

Fig. 2.10.

REMARK 2. The graphs of the functions $\arcsin x$, $\arccos x$, $\arctan x$, $\operatorname{arccot} x$ are illustrated in Fig. 2.10 and 2.11.

Theorem 1. *The values of the inverse trigonometric functions at some special points:*

$$\arcsin 0 = 0\,, \quad \arcsin \tfrac{1}{2} = \tfrac{1}{6}\pi\,, \quad \arcsin 1 = \tfrac{1}{2}\pi\,, \quad \arcsin(-1) = -\tfrac{1}{2}\pi\,;$$

*) In English literature this function is more usually called *the principal value of* $\arcsin x$, the general function $\arcsin x$ being the (multi-valued) function inverse to $x = \sin y$, and similarly for the other inverse functions.

$$\arccos 0 = \tfrac{1}{2}\pi\,,\quad \arccos \tfrac{1}{2} = \tfrac{1}{3}\pi\,,\quad \arccos 1 = 0\,,\quad \arccos(-1) = \pi\,;$$

$$\arctan 0 = 0\,,\quad \arctan 1 = \tfrac{1}{4}\pi\,,\quad \lim_{x\to+\infty} \arctan x = \tfrac{1}{2}\pi\,,\quad \lim_{x\to-\infty} \arctan x = -\tfrac{1}{2}\pi\,;$$

$$\operatorname{arccot} 0 = \tfrac{1}{2}\pi\,,\quad \operatorname{arccot} 1 = \tfrac{1}{4}\pi\,,\quad \lim_{x\to+\infty} \operatorname{arccot} x = 0\,,\quad \lim_{x\to-\infty} \operatorname{arccot} x = \pi\,.$$

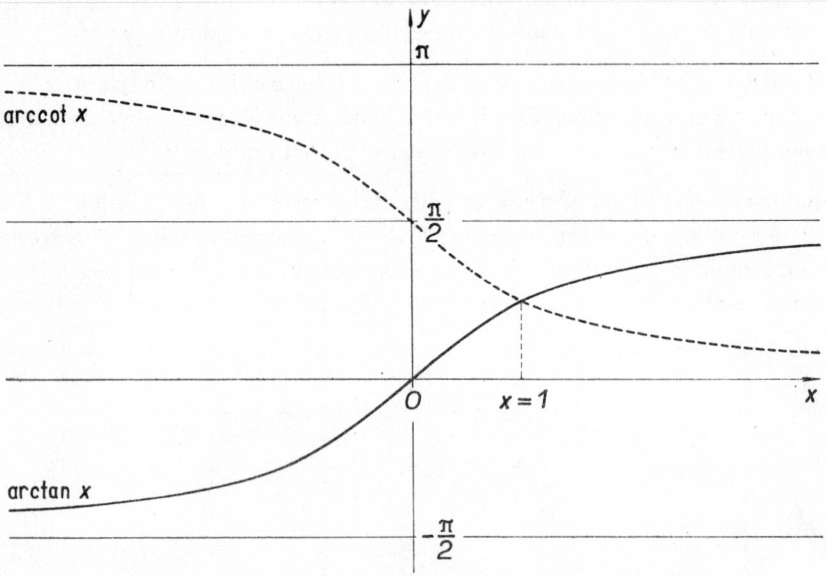

Fig. 2.11.

Theorem 2. *Fundamental formulae and relations among inverse trigonometric functions (if the domain of validity is not mentioned, then the formula holds for all x):*

1. $\arcsin(\sin x) = x \quad (|x| \leqq \tfrac{1}{2}\pi)\,,\quad \arccos(\cos x) = x \quad (0 \leqq x \leqq \pi)\,.$
2. $\sin(\arcsin x) = x\,,\quad \cos(\arccos x) = x \quad (|x| \leqq 1)\,.$
3. $\arctan(\tan x) = x \quad (|x| < \tfrac{1}{2}\pi)\,,\quad \operatorname{arccot}(\cot x) = x \quad (0 < x < \pi)\,.$
4. $\tan(\arctan x) = x\,.$
5. $\cot(\operatorname{arccot} x) = x\,.$
6. $\arcsin x + \arccos x = \tfrac{1}{2}\pi \quad (|x| \leqq 1)\,.$
7. $\arctan x + \operatorname{arccot} x = \tfrac{1}{2}\pi\,.$
8. $\arcsin(-x) = -\arcsin x \quad (|x| \leqq 1)\,.$
9. $\arccos(-x) = \pi - \arccos x \quad (|x| \leqq 1)\,.$
10. $\arctan(-x) = -\arctan x\,.$

11. $\operatorname{arccot}(-x) = \pi - \operatorname{arccot} x\,.$

12. $\arcsin x = \arctan \dfrac{x}{\sqrt{(1 - x^2)}} \quad (|x| < 1)\,.$

13. $\arccos x = \operatorname{arccot} \dfrac{x}{\sqrt{(1 - x^2)}} \quad (|x| < 1)\,.$

14. $\arctan x = \arcsin \dfrac{x}{\sqrt{(1 + x^2)}}\,.$

15. $\operatorname{arccot} x = \arccos \dfrac{x}{\sqrt{(1 + x^2)}}\,.$

16. $\arctan x = \operatorname{arccot} \dfrac{1}{x} \quad (x > 0)\,.$

17. $\arcsin x = \arccos \sqrt{(1 - x^2)}\,, \quad \arccos x = \arcsin \sqrt{(1 - x^2)} \quad (0 \leqq x \leqq 1)\,.$

18. $\arcsin x + \arcsin y =$
$= \arcsin [x \sqrt{(1 - y^2)} + y \sqrt{(1 - x^2)}] \quad (xy \leqq 0 \text{ or } x^2 + y^2 \leqq 1)\,,$
$= \pi - \arcsin [x \sqrt{(1 - y^2)} + y \sqrt{(1 - x^2)}] \quad (x > 0,\, y > 0 \text{ and } x^2 + y^2 > 1)\,,$
$= -\pi - \arcsin [x \sqrt{(1 - y^2)} + y \sqrt{(1 - x^2)}] \quad (x < 0,\, y < 0 \text{ and } x^2 + y^2 > 1)\,.$

19. $\arcsin x - \arcsin y =$
$= \arcsin [x \sqrt{(1 - y^2)} - y \sqrt{(1 - x^2)}] \quad (xy \geqq 0 \text{ or } x^2 + y^2 \leqq 1)\,,$
$= \pi - \arcsin [x \sqrt{(1 - y^2)} - y \sqrt{(1 - x^2)}] \quad (x > 0,\, y < 0 \text{ and } x^2 + y^2 > 1)\,,$
$= -\pi - \arcsin [x \sqrt{(1 - y^2)} - y \sqrt{(1 - x^2)}] \quad (x < 0,\, y > 0 \text{ and } x^2 + y^2 > 1)\,.$

20. $\arccos x + \arccos y = \arccos [xy - \sqrt{(1 - x^2)} \sqrt{(1 - y^2)}] \quad (x + y \geqq 0)\,,$
$= 2\pi - \arccos [xy - \sqrt{(1 - x^2)} \sqrt{(1 - y^2)}] \quad (x + y < 0)\,.$

21. $\arccos x - \arccos y = -\arccos [xy + \sqrt{(1 - x^2)} \sqrt{(1 - y^2)}] \quad (x \geqq y)\,,$
$= \arccos [xy + \sqrt{(1 - x^2)} \sqrt{(1 - y^2)}] \quad (x < y)\,.$

22. $\arctan x + \arctan y = \arctan \dfrac{x + y}{1 - xy} \quad (xy < 1)\,,$

$$= \pi + \arctan \frac{x + y}{1 - xy} \quad (xy > 1,\, x > 0)\,,$$

$$= -\pi + \arctan \frac{x + y}{1 - xy} \quad (xy > 1,\, x < 0)\,.$$

23. $\arctan x - \arctan y = \arctan \dfrac{x - y}{1 + xy} \quad (xy > -1)\,,$

$$= \pi + \arctan \frac{x - y}{1 + xy} \quad (xy < -1, x > 0),$$

$$= -\pi + \arctan \frac{x - y}{1 + xy} \quad (xy < -1, x < 0).$$

2.12. Hyperbolic Functions

Definition 1. The functions $\sinh x$ (*hyperbolic sine*), $\cosh x$ (*hyperbolic cosine*) and $\tanh x$ (*hyperbolic tangent*) are defined for all real x as follows:

$$\sinh x = \tfrac{1}{2}(e^x - e^{-x}), \quad \cosh x = \tfrac{1}{2}(e^x + e^{-x}),$$

$$\tanh x = \frac{e^x - e^{-x}}{e^x + e^{-x}} = \frac{\sinh x}{\cosh x}.$$

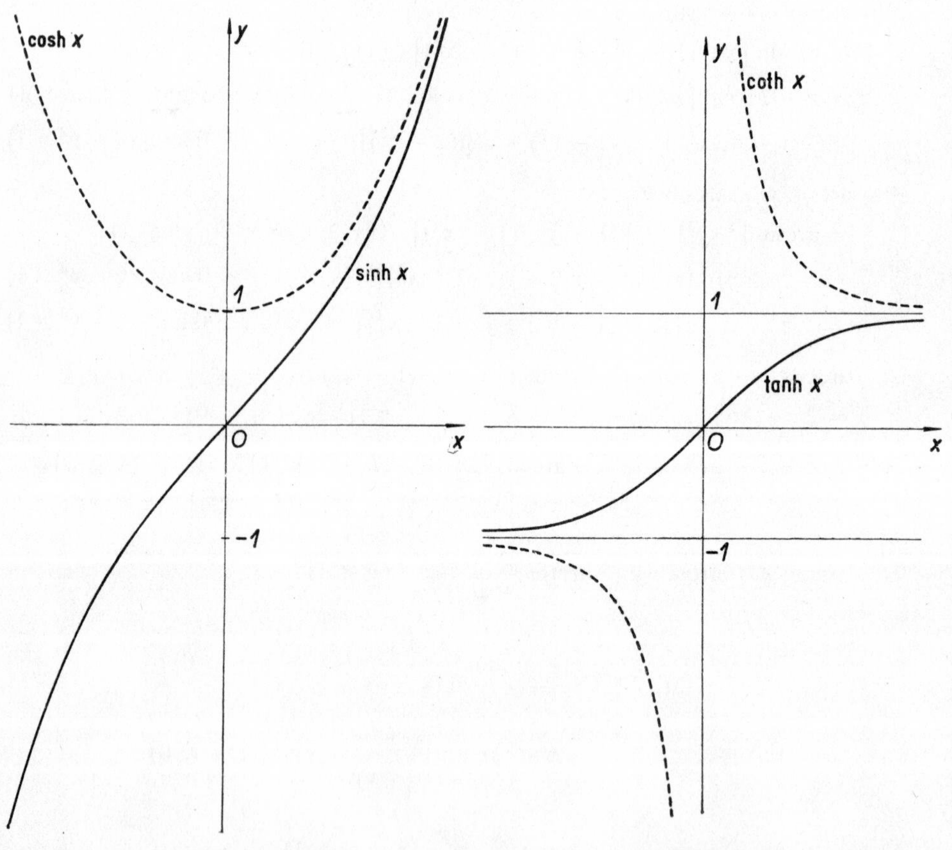

Fig. 2.12a. Fig. 2.12b.

For $x \neq 0$, the function $\coth x$ (*hyperbolic cotangent*) is defined by the relation

$$\coth x = \frac{e^x + e^{-x}}{e^x - e^{-x}} = \frac{1}{\tanh x}.$$

Further, the following functions are defined

$$\operatorname{sech} x = \frac{1}{\cosh x} \quad (\textit{hyperbolic secant})$$

$$\operatorname{cosech} x = \frac{1}{\sinh x} \quad \text{for} \quad x \neq 0 \quad (\textit{hyperbolic cosecant}).$$

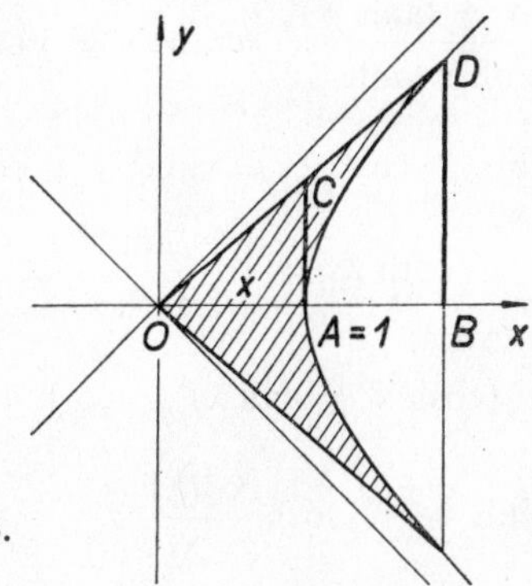

Fig. 2.13.

REMARK 1. The behaviour of the hyperbolic functions can be seen in Fig. 2.12a,b.

REMARK 2. The hyperbolic functions stand in a similar relation to an equiangular hyperbola with semi-axis of length 1 as do the trigonometric functions to a unit circle; the independent variable (argument) $x \geqq 0$ denotes the area of the hyperbolic sector (the shaded area in Fig. 2.13). Here,

$$\sinh x = \widetilde{BD}, \quad \cosh x = \widetilde{OB}, \quad \tanh x = \widetilde{AC}.$$

Theorem 1. *Relations between hyperbolic functions:*

1. $\cosh^2 x - \sinh^2 x = 1$,

2. $\cosh x + \sinh x = e^x$,

   $\cosh x - \sinh x = e^{-x}$.

3. $\sinh(-x) = -\sinh x$,

   $\cosh(-x) = \cosh x$,

   $\tanh(-x) = -\tanh x$,

   $\coth(-x) = -\coth x$.

4. $|\sinh x| = \sqrt{(\cosh^2 x - 1)} = \dfrac{|\tanh x|}{\sqrt{(1 - \tanh^2 x)}} = \dfrac{1}{\sqrt{(\coth^2 x - 1)}}$.

5. $\cosh x = \sqrt{(\sinh^2 x + 1)} = \dfrac{|\coth x|}{\sqrt{(\coth^2 x - 1)}} = \dfrac{1}{\sqrt{(1 - \tanh^2 x)}}$.

6. $\tanh x = \dfrac{\sinh x}{\sqrt{(\sinh^2 x + 1)}}$.

7. $\sinh (x \pm y) = \sinh x \cosh y \pm \cosh x \sinh y$,

$\cosh (x \pm y) = \cosh x \cosh y \pm \sinh x \sinh y$,

$\tanh (x \pm y) = \dfrac{\tanh x \pm \tanh y}{1 \pm \tanh x \tanh y}$, $\coth (x \pm y) = \dfrac{1 \pm \coth x \coth y}{\coth x \pm \coth y}$.

8. $\sinh 2x = 2 \sinh x \cosh x$, $\cosh 2x = \sinh^2 x + \cosh^2 x$,

$\tanh 2x = \dfrac{2 \tanh x}{1 + \tanh^2 x}$, $\coth 2x = \dfrac{1 + \coth^2 x}{2 \coth x}$.

9. *De Moivre's Theorem*: $(\cosh x \pm \sinh x)^n = \cosh nx \pm \sinh nx$.

10. $\sinh x \pm \sinh y = 2 \sinh \dfrac{x \pm y}{2} \cosh \dfrac{x \mp y}{2}$,

$\cosh x + \cosh y = 2 \cosh \dfrac{x + y}{2} \cosh \dfrac{x - y}{2}$,

$\cosh x - \cosh y = 2 \sinh \dfrac{x + y}{2} \sinh \dfrac{x - y}{2}$,

$\tanh x \pm \tanh y = \dfrac{\sinh (x \pm y)}{\cosh x \cosh y}$.

11. *Relations between hyperbolic and trigonometric functions* (see Remark 20.4.4):

$$\sin \mathrm{i}x = \mathrm{i} \sinh x, \quad \cos \mathrm{i}x = \cosh x,$$

$$\tan \mathrm{i}x = \mathrm{i} \tanh x, \quad \cot \mathrm{i}x = -\mathrm{i} \coth x.$$

2.13. Inverse Hyperbolic Functions

Inverse hyperbolic functions are the functions arsinh x ($\sinh^{-1} x$), arcosh x ($\cosh^{-1} x$), artanh x ($\tanh^{-1} x$), arcoth x ($\coth^{-1} x$) which are inverse (see 11.1, p. 400) to the hyperbolic functions.

Definition 1. The function $y = \operatorname{arsinh} x$ is inverse to the function $x = \sinh y$; it is defined for all real x. Thus: If x is a real number, then $\operatorname{arsinh} x$ is the unique number y such that $\sinh y = x$.

Definition 2. The function $y = \operatorname{arcosh} x$ is inverse to the function $x = \cosh y$ considered only in the interval $[0, \infty)$; it is defined for every x in the interval $[1, \infty)$. Thus: If $1 \leqq x < +\infty$, then $\operatorname{arcosh} x$ is the unique number y in the interval $[0, \infty)$ such that $\cosh y = x$.*)

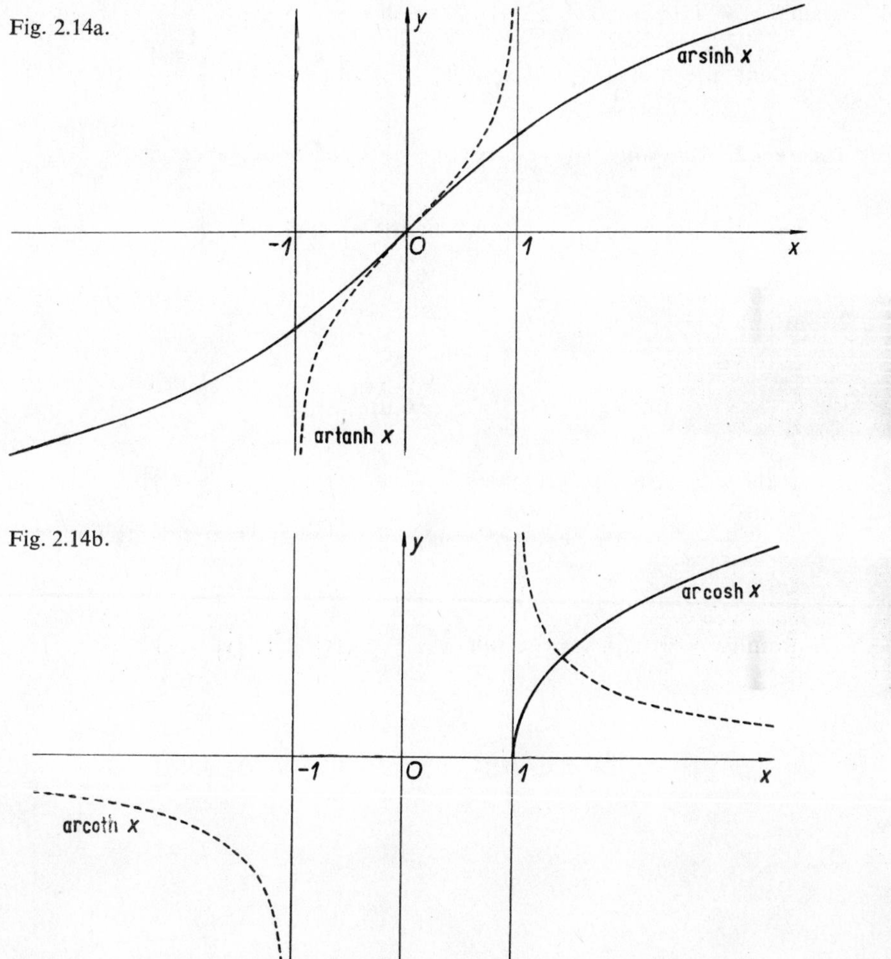

Fig. 2.14a.

Fig. 2.14b.

Definition 3. The function $y = \operatorname{artanh} x$ is inverse to the function $x = \tanh y$; it is defined for all x in the interval $(-1, 1)$. Thus: If $-1 < x < 1$, then $\operatorname{artanh} x$ is the unique number y such that $\tanh y = x$.

*) In English literature the function $\operatorname{arcosh} x$ is more usually defined as the two-valued function inverse to $x = \cosh y$.

Definition 4. The function $y = \operatorname{arcoth} x$ is inverse to the function $x = \coth y$; it is defined for all x satisfying $|x| > 1$. Thus: If $|x| > 1$, then $\operatorname{arcoth} x$ is the unique number y such that $\coth y = x$.

REMARK 1. The graphs of the inverse hyperbolic functions are illustrated in Fig. 2.14a,b.

Theorem 1. *Inverse hyperbolic functions expressed by means of logarithms*:

$$\operatorname{arsinh} x = \ln\left[x + \sqrt{(x^2 + 1)}\right], \quad \operatorname{arcosh} x = \ln\left[x + \sqrt{(x^2 - 1)}\right] \quad (x \geqq 1),$$

$$\operatorname{artanh} x = \tfrac{1}{2} \ln \frac{1 + x}{1 - x} \quad (|x| < 1), \quad \operatorname{arcoth} x = \tfrac{1}{2} \ln \frac{x + 1}{x - 1} \quad (|x| > 1).$$

Theorem 2. *Relations between the inverse hyperbolic functions*:

1. $$\operatorname{arsinh} x = \operatorname{artanh} \frac{x}{\sqrt{(x^2 + 1)}}, \quad |\operatorname{arsinh} x| = \operatorname{arcosh} \sqrt{(x^2 + 1)}.$$

2. $$\operatorname{artanh} x = \operatorname{arsinh} \frac{x}{\sqrt{(1 - x^2)}} \quad (|x| < 1),$$

$$= \operatorname{arcoth} \frac{1}{x} \quad (|x| < 1,\ x \neq 0).$$

3. $$\operatorname{arsinh} x \pm \operatorname{arsinh} y = \operatorname{arsinh}\left[x \sqrt{(1 + y^2)} \pm y \sqrt{(1 + x^2)}\right],$$

$$|\operatorname{arcosh} x \pm \operatorname{arcosh} y| = \operatorname{arcosh} [\![xy \pm \sqrt{[(x^2 - 1)(y^2 - 1)]}]\!] \quad (x \geqq 1,\ y \geqq 1),$$

$$\operatorname{artanh} x \pm \operatorname{artanh} y = \operatorname{artanh} \frac{x \pm y}{1 \pm xy} \quad (|x| < 1,\ |y| < 1).$$

3. SOME FORMULAE (AREAS CIRCUMFERENCES, VOLUMES, SURFACES, CENTROIDS, MOMENTS OF INERTIA)

By VÁCLAV VILHELM

References: [1], [2], [48], [49], [90], [144]

3.1. Area, Circumference, Centroid and Moments of Inertia of Plane Figures

REMARK 1. For the calculation of areas and circumferences of plane figures by means of integrals see § 14.9.

(a) The Triangle (Fig. 3.1). Consider a triangle ABC, denoting its sides by a, b, c, interior angles by α, β, γ, altitudes by h_a, h_b, h_c, radius of the inscribed circle by r, radius of the circumscribed circle by R, area by P, semi-perimeter by $s = \frac{1}{2}(a + b + c)$, medians by t_a, t_b, t_c, centroid by T. The following relations hold:

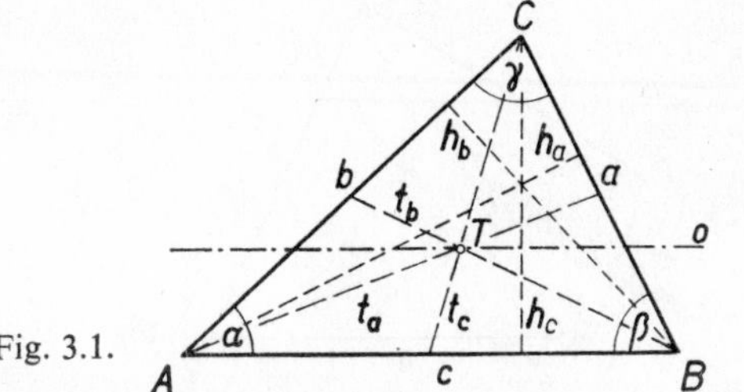

Fig. 3.1.

$$P = \tfrac{1}{2}ah_a = \tfrac{1}{2}bh_b = \tfrac{1}{2}ch_c\,, \tag{1}$$

$$= \sqrt{[s(s-a)(s-b)(s-c)]} \quad (\textit{Heron's formula})\,, \tag{2}$$

$$= \frac{abc}{4R} = 2R^2 \sin\alpha \sin\beta \sin\gamma\,, \tag{3}$$

$$= rs = r^2 \cot\frac{\alpha}{2} \cot\frac{\beta}{2} \cot\frac{\gamma}{2}\,, \tag{4}$$

$$= \tfrac{1}{2}ab \sin\gamma\,. \tag{5}$$

If $x_1, y_1; x_2, y_2; x_3, y_3$ are the coordinates of the vertices A, B, C of a triangle in a cartesian coordinate system, then

$$\pm P = \frac{1}{2}\begin{vmatrix} x_1, & y_1, & 1 \\ x_2, & y_2, & 1 \\ x_3, & y_3, & 1 \end{vmatrix} = \frac{1}{2}\begin{vmatrix} x_2 - x_1, & y_2 - y_1 \\ x_3 - x_1, & y_3 - y_1 \end{vmatrix}; \tag{6}$$

the minus sign relates to the case where the determinant is negative.

The coordinates of the centroid T (the point of intersection of the medians t_a, t_b, t_c) are

$$x_T = \tfrac{1}{3}(x_1 + x_2 + x_3), \quad y_T = \tfrac{1}{3}(y_1 + y_2 + y_3). \tag{7}$$

The moment of inertia about a median axis o, i.e. an axis through the centroid parallel to the side c, or about the side c is

$$I_0 = \tfrac{1}{36}ch_c^3, \quad \text{or} \quad I_c = \tfrac{1}{12}ch_c^3, \quad \text{respectively}. \tag{8}$$

The area of a *right-angled* triangle ABC with hypotenuse c (hence, $\gamma = 90°$) is

$$P = \tfrac{1}{2}ab = \tfrac{1}{2}a^2 \tan \beta = \tfrac{1}{4}c^2 \sin 2\alpha. \tag{9}$$

REMARK 2. For trigonometric formulae concerning a triangle see § 2.9, p. 116.

(b) The Quadrilateral (Fig. 3.2). Consider a quadrilateral with sides a, b, c, d and with vertices A, B, C, D (the sides intersecting only at the vertices). Let u_1, u_2 be its diagonals, φ the angle between them, and h_1, h_2, the altitudes of the triangles

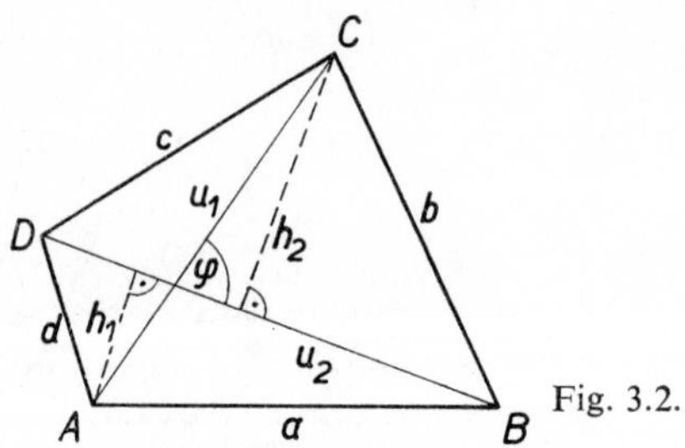

Fig. 3.2.

ABD, BDC, dropped from the points A, C, respectively. Then the area P of the quadrilateral is

$$P = \tfrac{1}{2}u_1u_2 \sin \varphi = \tfrac{1}{2}(h_1 + h_2)\, u_2 \tag{10}$$

(if the quadrilateral is not convex, then u_2 in (10) is the inner diagonal).

If the vertices of a convex quadrilateral lie on a circle then the area of the quadrilateral is

$$P = \sqrt{[(s-a)(s-b)(s-c)(s-d)]} \tag{11}$$

where $s = \tfrac{1}{2}(a + b + c + d)$.

A *trapezium* (Fig. 3.3a) is a (convex) quadrilateral two opposite sides of which are parallel. The area P is given by the formulae (10), (11) and

$$P = \tfrac{1}{2}(a + c)\, h\,. \tag{12}$$

The centroid T lies on the segment MN, where M, N, are the mid-points of the sides a, c, respectively, at distance $h_a = h(a + 2c)/[3(a + c)]$ from the side a.

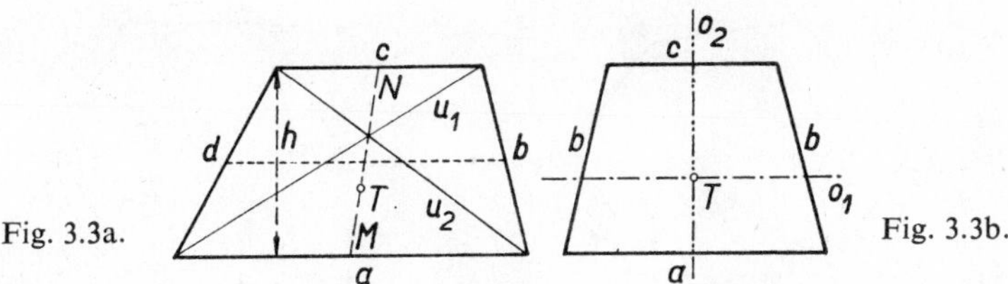

Fig. 3.3a. Fig. 3.3b.

The moments of inertia of an isosceles trapezium of altitude h about the median axes o_1, o_2 (Fig. 3.3b) are

$$I_{o_1} = \frac{h^3(a^2 + 4ac + c^2)}{36(a + c)}\,; \quad I_{o_2} = \frac{h(a^4 - c^4)}{48(a - c)}\,. \tag{13}$$

A *parallelogram* (Fig. 3.4) is a quadrilateral the opposite sides of which are parallel and, consequently, of the same length. If $\gamma = 90°$, we get a rectangle or a square. The area P of a parallelogram is given by formulae (10), (11), (12) (where $a = c$, $b = d$) and

$$P = ab \sin \gamma\,. \tag{14}$$

A *rhombus* is a parallelogram with $a = b$. Then, $\varphi = 90°$ and

$$P = a^2 \sin \gamma = \tfrac{1}{2}u_1 u_2\,. \tag{15}$$

A square is a rhombus with $\gamma = 90°$.

The centroid T of a parallelogram lies at the point of intersection of the diagonals.

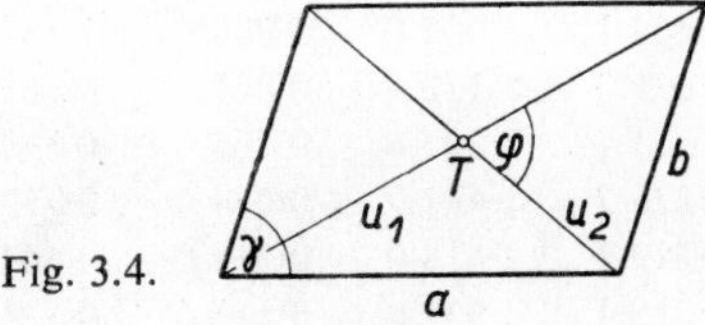

Fig. 3.4.

The moment of inertia of a parallelogram about the diagonal u_1 is

$$I_{u_1} = \tfrac{1}{48}u_1 u_2^3 \sin^3 \varphi = \tfrac{1}{24}P u_2^2 \sin^2 \varphi\,. \tag{16}$$

The moment of inertia of a rectangle with sides a, b about a median axis o parallel to the side a is

$$I_0 = \tfrac{1}{12}ab^3\,. \tag{17}$$

(c) The Polygon. The area can be determined by dividing the polygon into simple figures, for example into triangles (see Fig. 3.5a).

A regular polygon (Fig. 3.5b) has all its sides and all its angles equal. Let n be the number of sides, a their common length, $\alpha = 360°/n$ the central angle, r the radius of the inscribed circle, R the radius of the circumscribed circle, P the area, C the circumference of the regular polygon. Then

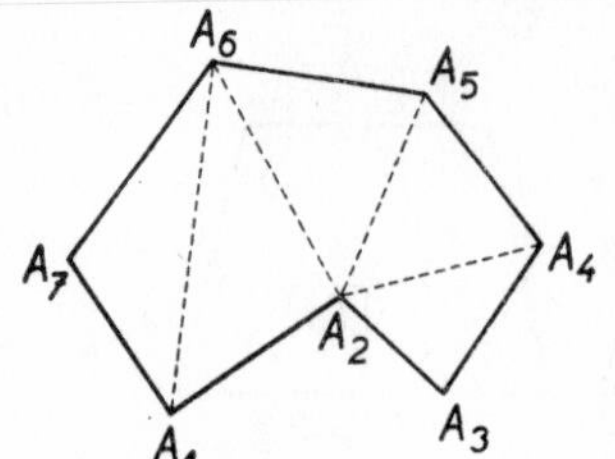

Fig. 3.5a.

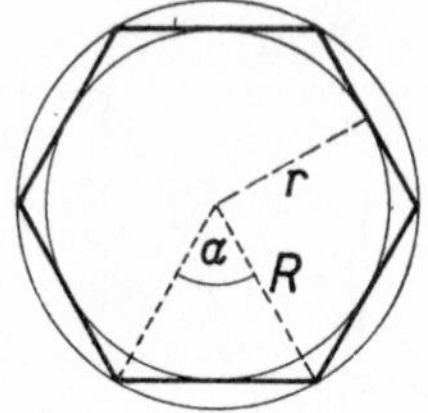

Fig. 3.5b.

$$P = \tfrac{1}{2}nar = \tfrac{1}{4}na^2 \cot\frac{\alpha}{2} = nr^2 \tan\frac{\alpha}{2} = \tfrac{1}{2}nR^2 \sin\alpha\,, \tag{18}$$

$$C = na = 2nR \sin\frac{\alpha}{2} = 2nr \tan\frac{\alpha}{2}\,. \tag{19}$$

TABLE 3.1

Calculation of the elements of regular polygons

n	$\frac{P}{a^2}$	$\frac{P}{R^2}$	$\frac{P}{r^2}$	$\frac{R}{a}$	$\frac{a}{R}$	$\frac{a}{r}$	$\frac{R}{r}$	$\frac{r}{R}$
3	0·433 0	1·299 0	5·196 2	0·577 4	1·732 1	3·464 1	2·000 0	0·500 0
4	1·000 0	2·000 0	4·000 0	0·707 1	1·414 2	2·000 0	1·414 2	0·707 1
5	1·720 5	2·377 6	3·632 7	0·850 7	1·175 6	1·453 1	1·236 1	0·809 0
6	2·598 1	2·598 1	3·464 1	1·000 0	1·000 0	1·154 7	1·154 7	0·866 0
7	3·633 9	2·736 4	3·371 0	1·152 4	0·867 8	0·963 1	1·109 9	0·901 0
8	4·828 4	2·828 4	3·313 7	1·306 6	0·765 4	0·828 4	1·082 4	0·923 9
9	6·181 8	2·892 5	3·275 7	1·461 9	0·684 0	0·727 9	1·064 2	0·939 7
10	7·694 2	2·938 9	3·249 2	1·618 0	0·618 0	0·649 8	1·051 5	0·951 1
12	11·196 2	3·000 0	3·215 4	1·931 9	0·517 6	0·535 9	1·035 3	0·965 9
15	17·642 4	3·050 5	3·188 3	2·404 9	0·415 8	0·425 1	1·022 3	0·978 1
16	20·109 4	3·061 5	3·182 6	2·562 9	0·390 2	0·397 8	1·019 6	0·980 8
20	31·568 8	3·090 2	3·167 7	3·196 2	0·312 9	0·316 8	1·012 5	0·987 7
24	45·574 5	3·105 8	3·159 7	3·830 6	0·261 1	0·263 3	1·008 6	0·991 4
32	81·225 4	3·121 4	3·151 7	5·101 1	0·196 0	0·197 0	1·004 8	0·995 2
48	183·084 6	3·132 6	3·146 1	7·644 9	0·130 8	0·131 1	1·002 1	0·997 9
64	325·687 5	3·136 5	3·144 1	10·190 0	0·098 1	0·098 3	1·001 2	0·998 8

The centroid of a regular polygon of n vertices is at its centre, the moment of inertia about an arbitrary axis o passing through the centre is

$$I_0 = \tfrac{1}{96}nar(12r^2 + a^2) = \tfrac{1}{24}P(6R^2 - a^2). \tag{20}$$

(d) The Circle. Let r denote the radius, d the diameter, P the area and C the circumference of the circle. Then

$$P = \pi r^2 = \tfrac{1}{4}\pi d^2 = \tfrac{1}{4}Cd \approx 0{\cdot}785\,4d^2 ; \tag{21}$$

$$C = 2\pi r = \pi d \approx 3{\cdot}141\,59d . \tag{22}$$

The centroid of a circle is at its centre.

The moment of inertia about an axis o passing through the centre is

$$I_0 = \tfrac{1}{4}\pi r^4 . \tag{23}$$

REMARK 3. For the measurement of angles and for conversion of angles measured in degrees into radians and vice versa, see § 2.1, p. 107.

The length l of a circular arc of radius r, corresponding to the central angle α (Fig. 3.6):

$$l = r \operatorname{arc} \alpha \text{ (arc } \alpha \text{ denotes the magnitude of the angle } \alpha \text{ in radians)}, \tag{24}$$

$$l = \frac{\pi r \alpha}{180} \approx 0{\cdot}017\,453 r\alpha \text{ (the angle in degrees)}, \tag{25}$$

$$l \approx \sqrt{(t^2 + \tfrac{16}{3}h^2)} . \tag{26}$$

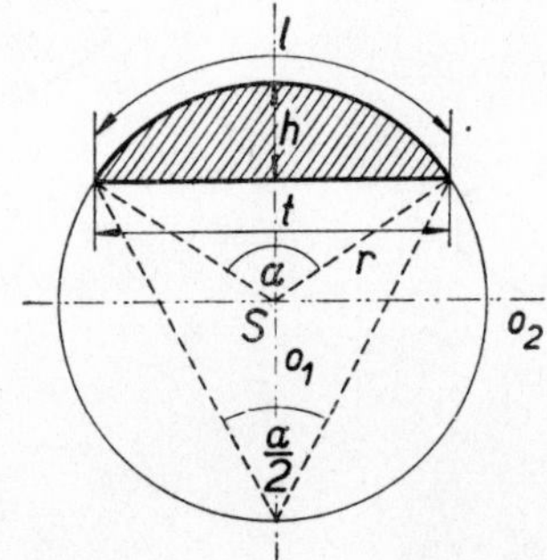

Fig. 3.6.

A *segment of a circle* (Fig. 3.6, the shaded area). Let r be the radius, l the length of the arc, t the length of the chord, α the central angle (in degrees), h the altitude of the segment, P the area of the segment. Then

$$t = 2\sqrt{(2hr - h^2)} = 2r \sin\frac{\alpha}{2}, \quad h = \frac{t}{2}\tan\frac{\alpha}{4}, \tag{27}$$

$$P = \tfrac{1}{2}r^2\left(\frac{\pi\alpha}{180} - \sin\alpha\right) = \tfrac{1}{2}[lr - t(r - h)]\,. \tag{28}$$

The centroid T lies on the bisector o_1 of the central angle (Fig. 3.6); its distance from the centre S is

$$\overline{TS} = \frac{4r\sin^3\frac{1}{2}\alpha}{3(\frac{1}{180}\pi\alpha - \sin\alpha)}\,. \tag{29}$$

The moments of inertia about the axes o_1, o_2 (Fig. 3.6) are

$$I_{o_1} = \tfrac{1}{48}r^4\left(\frac{\pi\alpha}{30} - 8\sin\alpha + \sin 2\alpha\right), \quad I_{o_2} = \tfrac{1}{16}r^4\left(\frac{\pi\alpha}{90} - \sin 2\alpha\right). \tag{30}$$

A *sector of a circle* (Fig. 3.7). The area

$$P = \frac{\pi r^2\alpha}{360} = \tfrac{1}{2}rl\,, \tag{31}$$

where α stands for the magnitude of the central angle in degrees.

The centroid T lies on the bisector o_1 of the central angle; its distance from the centre S (Fig. 3.7) is

$$\overline{TS} = \frac{240r\sin\frac{1}{2}\alpha}{\pi\alpha}\,. \tag{32}$$

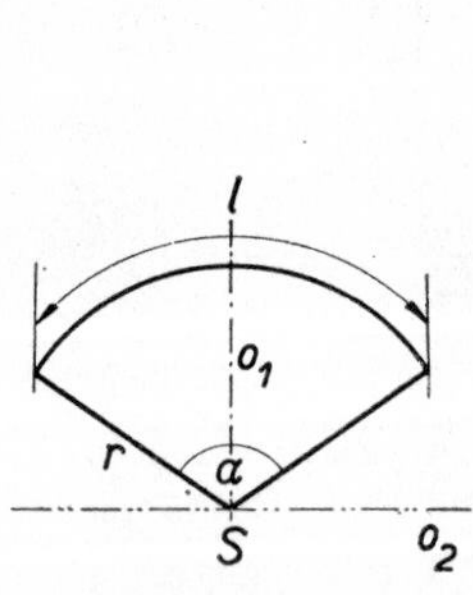

Fig. 3.7.

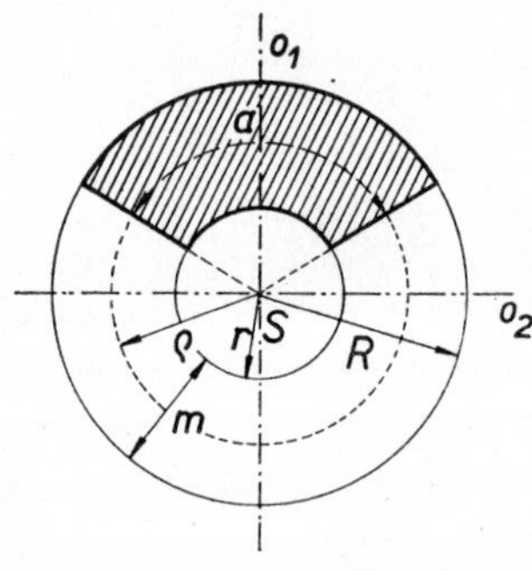

Fig. 3.8.

The moments of inertia about the axes o_1, o_2 (Fig. 3.7) are

$$I_{o_1} = \tfrac{1}{8}r^4\left(\frac{\pi\alpha}{180} - \sin\alpha\right),$$

$$I_{o_2} = \tfrac{1}{8}r^4\left(\frac{\pi\alpha}{180} + \sin\alpha\right). \tag{33}$$

An *annulus* (Fig. 3.8). Let r be the radius of the inner circle, R the radius of the outer circle, $\varrho = \frac{1}{2}(r + R)$, $m = R - r$, and P the area. Then

$$P = \pi(R^2 - r^2) = 2\pi\varrho m\,. \tag{34}$$

The centroid lies at the centre S.

The moment of inertia about the median axis o_1 is

$$I_{o_1} = \tfrac{1}{4}\pi(R^4 - r^4)\,. \tag{35}$$

A *sector of an annulus* (with central angle α in degrees, see Fig. 3.8, the shaded area). The area P is given by

$$P = \frac{\pi\alpha}{360}(R^2 - r^2) = am\,. \tag{36}$$

The centroid T lies on the bisector o_1 of the central angle; its distance from the centre S is

$$\overline{TS} = \frac{4}{3}\frac{R^3 - r^3}{R^2 - r^2}\frac{\sin\frac{1}{2}\alpha}{\frac{1}{180}\pi\alpha}\,. \tag{37}$$

The moments of inertia about the axes o_1, o_2 (Fig. 3.8) are

$$I_{o_1} = \tfrac{1}{8}(R^4 - r^4)\left(\frac{\pi\alpha}{180} - \sin\alpha\right), \quad I_{o_2} = \tfrac{1}{8}(R^4 - r^4)\left(\frac{\pi\alpha}{180} + \sin\alpha\right). \tag{38}$$

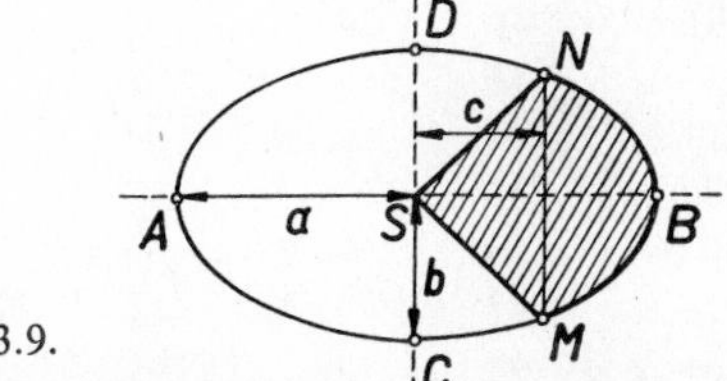

Fig. 3.9.

(e) The Ellipse (Fig. 3.9). Let $\overline{SA} = a$ be the semi-major axis, $\overline{SC} = b$ the semi-minor axis, $e = \sqrt{(a^2 - b^2)}/a$ the eccentricity of the ellipse, C its circumference, P its area. The following relations hold:

$$P = \pi ab\,, \tag{39}$$

$$C = 4aE(e, \tfrac{1}{2}\pi) \tag{40}$$

where $E(e, \frac{1}{2}\pi) = \int_0^{\pi/2} \sqrt{(1 - e^2 \sin^2\varphi)}\,d\varphi$ is the so-called complete elliptic integral of the second kind (see § 13.12). The following approximate formulae hold for the circumference of an ellipse:

$$C \approx \pi[1{\cdot}5(a + b) - \sqrt{(ab)}]\,, \quad C \approx \pi(a + b)\frac{64 - 3l^4}{64 - 16l^2}\,, \quad \text{where } l = \frac{a - b}{a + b}\,. \tag{41}$$

The circumference of an ellipse with semi-axes a, b can be calculated by using Table 3.2. The circumference C is given by the formula $C = ak$.

The centroid of an ellipse lies at the centre S.

The moments of inertia about the axes a, b are as follows:

$$I_a = \tfrac{1}{4}\pi ab^3\,, \quad I_b = \tfrac{1}{4}\pi a^3 b\,. \tag{42}$$

An *elliptic sector* (Fig. 3.9, the shaded area) has the area

$$P = ab \arccos \frac{c}{a}\,. \tag{43}$$

REMARK 4. For further properties of the ellipse see §§ 4.2 and 5.10.

TABLE 3.2

$\frac{b}{a}$	k	$\frac{b}{a}$	k	$\frac{b}{a}$	k	$\frac{b}{a}$	k	$\frac{b}{a}$	k
0·00	4·000 0	0·20	4·202 0	0·40	4·602 6	0·60	5·105 4	0·80	5·672 3
01	001 1	21	218 6	41	625 8	61	132 4	81	702 0
02	003 7	22	235 6	42	649 2	62	159 6	82	731 7
03	007 8	23	253 1	43	672 8	63	187 0	83	761 5
04	013 1	24	271 0	44	696 6	64	214 5	84	791 5
05	019 4	25	289 2	45	720 7	65	242 1	85	821 5
06	026 7	26	307 8	46	745 0	66	269 9	86	851 6
07	034 8	27	326 8	47	769 5	67	297 8	87	781 9
08	043 8	28	346 2	48	794 2	68	325 9	88	912 2
09	053 5	29	365 9	49	819 1	69	354 1	89	942 6
0·10	064 0	0·30	385 9	0·50	844 2	0·70	382 4	0·90	973 2
11	075 2	31	406 2	51	869 5	71	410 8	91	6·003 8
12	087 0	32	426 9	52	895 0	72	439 4	92	034 5
13	099 4	33	447 9	53	920 7	73	468 1	93	065 3
14	112 5	34	469 2	54	946 6	74	496 9	94	096 2
15	126 1	35	490 8	55	972 6	75	525 8	95	127 1
16	140 3	36	512 6	56	998 8	76	554 9	96	158 2
17	155 0	37	534 7	57	5·025 2	77	584 1	97	189 3
18	170 2	38	557 1	58	051 8	78	613 4	98	220 5
19	185 9	39	579 7	59	078 5	79	642 8	99	251 8

(f) The Hyperbola (Fig. 3.10). Let $OA = a$ be the semi-major axis, $OC = b$ be the semi-minor axis, $e = \sqrt{(a^2 + b^2)}/a$ the eccentricity.

A *segment MBN of the hyperbola* (Fig. 3.10, the shaded area) has the area

$$P = xy - ab \ln\left(\frac{x}{a} + \frac{y}{b}\right) = xy - ab \operatorname{arcosh} \frac{x}{a}, \tag{44}$$

where $y = (b/a)\sqrt{(x^2 - a^2)}$.

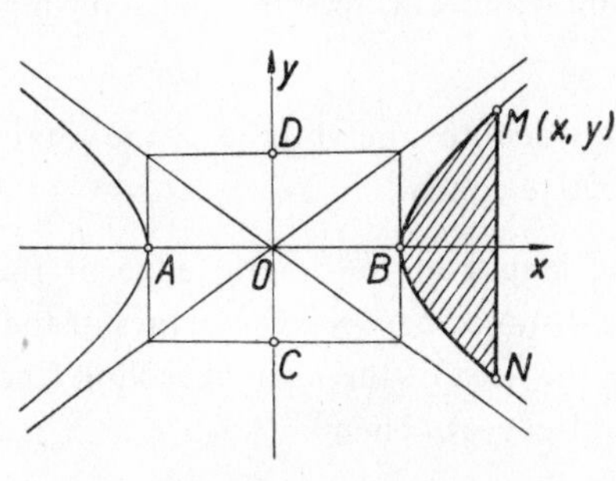

Fig. 3.10.

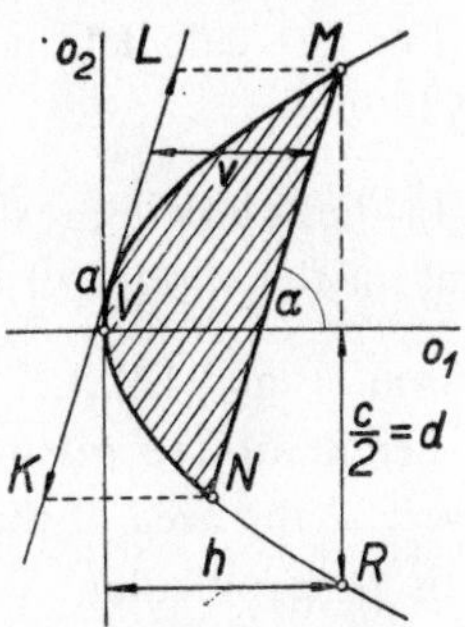

Fig. 3.11.

REMARK 5. For further properties of the hyperbola see § 4.3, 5.11.

(g) The Parabola (Fig. 3.11). The area of a *segment MVN of the parabola* (the shaded area) is

$$P = \tfrac{2}{3}av \sin \alpha\,; \tag{45}$$

it is thus equal to two-thirds of the area of the parallelogram *KLMN*.

The length l of an arc *MVR* of a parabola is

$$l = \tfrac{1}{2}\sqrt{[c^2 + (4h)^2]} + \frac{c^2}{8h} \ln\left[\frac{4h}{c} + \frac{1}{c}\sqrt{[c^2 + (4h)^2]}\right]; \tag{46}$$

the following relation holds approximately (for small h/c):

$$l \approx c\left[1 + \frac{8}{3}\left(\frac{h}{c}\right)^2 - \frac{32}{5}\left(\frac{h}{c}\right)^4\right]. \tag{47}$$

The centroid T of a parabolic segment *MVR* (Fig. 3.11) lies on the axis o_1 of the parabola; its distance from the vertex V is

$$\overline{TV} = \tfrac{3}{5}h\,. \tag{48}$$

The moments of inertia of a parabolic segment MVR about the axes o_1, o_2 (Fig. 3.11) are

$$I_{o_1} = \tfrac{4}{15}hd^3\,, \quad I_{o_2} = \tfrac{4}{7}h^3d\,. \tag{49}$$

REMARK 6. For further properties of the parabola see § 4.4, 5.12.

3.2. Volume, Surface, Centroid and Moments of Inertia of Solids

REMARK 1. For the calculation of volumes and surfaces of solids by means of integrals see § 14.9.

REMARK 2. In the following text, V always denotes the volume, S the total and Q the lateral area* of the surface of the respective solids.

(a) The Prism (Fig. 3.12). Let a be the length of the lateral edge or the slant height, h the height of the prism (i.e. the distance between the planes of the upper and lower bases), P the area of the base, N the area of the normal section (the plane section which is perpendicular to the lateral edges). Then

$$V = Ph = Na\,, \tag{1}$$

$$Q = C_N a\,, \quad S = 2P + C_N a \tag{2}$$

where C_N is the circumference of the normal section.

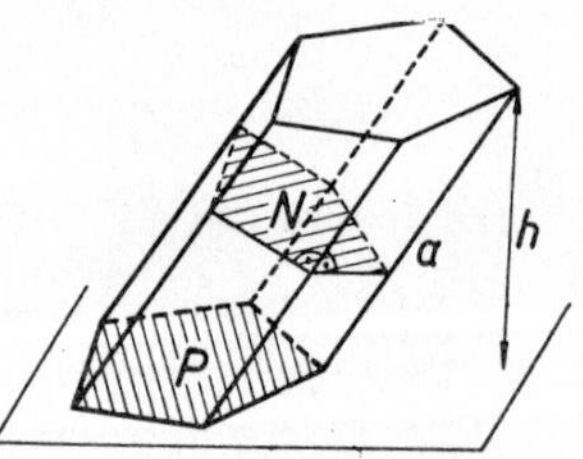

Fig. 3.12.

The centroid lies at the mid-point of the segment connecting the centroids of the two bases of the prism.

A *truncated triangular prism* (i.e. cut off by a plane non-parallel to the plane of the base; Fig. 3.13), whose lateral edges are of lengths a, b, c, has the volume

$$V = \tfrac{1}{3}N(a + b + c)\,. \tag{3}$$

A *parallelepiped* is a prism, the base of which is a parallelogram.

* i.e. area of the slant faces or of the curved surface.

A *right parallelepiped* (i.e. a right prism, the base of which is a rectangle or a square), whose edges are of lengths a, b, c, has the volume

$$V = abc \tag{4}$$

and the surface area

$$S = 2(ab + ac + bc)\,. \tag{5}$$

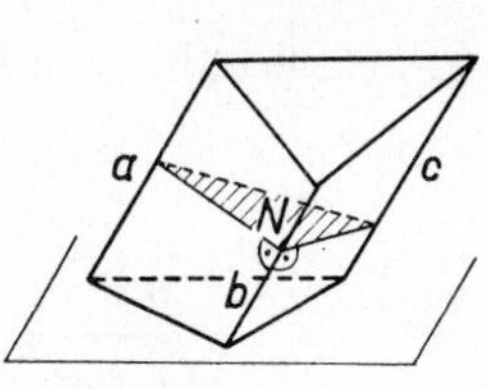

Fig. 3.13.

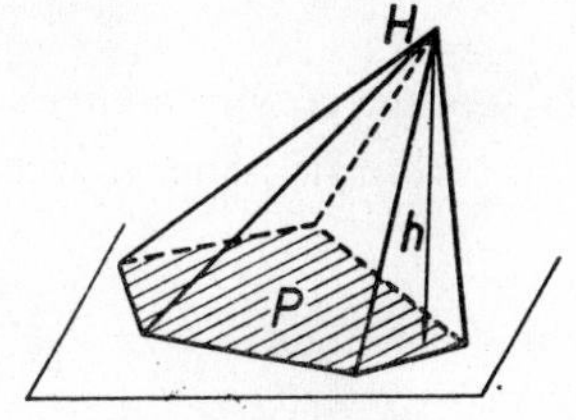

Fig. 3.14.

The moment of inertia about the median axis o which is parallel to the edge c is

$$I_o = \tfrac{1}{12}abc(a^2 + b^2)\,.$$

A *cube* is a right parallelepiped whose edges are of the same length a; the volume and surface are given by

$$V = a^3\,, \quad S = 6a^2\,. \tag{6}$$

(b) The Pyramid (Fig. 3.14). Let h be the height of the pyramid (the distance of the apex H from the plane of the base), P the area of the base. Then

$$V = \tfrac{1}{3}Ph\,. \tag{7}$$

The centroid lies on the segment connecting the apex and the centroid of the base; its distance from the base is $\frac{1}{4}h$.

A *triangular pyramid* with one vertex at the origin of the cartesian coordinate system, the other three vertices being (x_i, y_i, z_i) $(i = 1, 2, 3)$, has the volume equal to one-sixth of the absolute value of the determinant

$$D = \begin{vmatrix} x_1, & y_1, & z_1 \\ x_2, & y_2, & z_2 \\ x_3, & y_3, & z_3 \end{vmatrix}, \quad \text{i.e.} \quad V = \tfrac{1}{6}|D|\,.$$

A *regular pyramid* (i.e. a pyramid whose base is a regular polygon, and whose altitude passes through the centre of the base). The lateral area

$$Q = \tfrac{1}{2}Cl \tag{8}$$

where C is the circumference of the base and l the length of the perpendicular from the apex to (any) of the edges of the base.

A *frustum of a pyramid* (Fig. 3.15) (the bases lie in parallel planes). Let P_1, P_2 be the areas of the bases and h the height (the distance between the two bases). Then

$$V = \tfrac{1}{3}h[P_1 + P_2 + \sqrt{(P_1P_2)}]\,. \tag{9}$$

If a frustum of a pyramid is *regular*, then the lateral area is

$$Q = \tfrac{1}{2}(C_1 + C_2)\,l \tag{10}$$

where C_1, C_2 are the circumferences of the bases and l is the altitude of the trapezoid formed by an (arbitrary) lateral face.

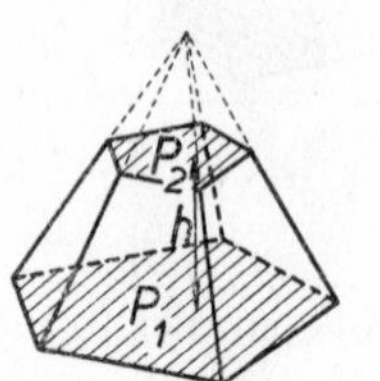

Fig. 3.15.

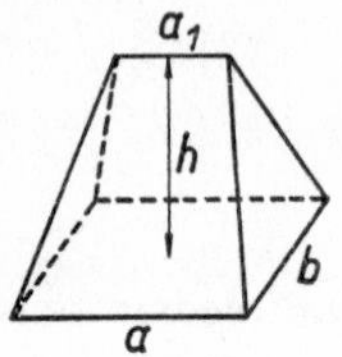

Fig. 3.16a.

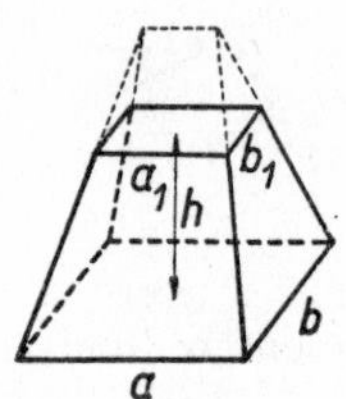

Fig. 3.16b.

REMARK 3. (a) A *dihedral angle* (Fig. 3.16a) (the base is a rectangle with sides a, b, one pair of opposite slant faces is formed by two congruent isosceles triangles, the other pair by two congruent isosceles trapezia). The volume:

$$V = \tfrac{1}{6}(2a + a_1)\,bh\,. \tag{11}$$

The centroid lies on the segment connecting the centre of the upper edge a_1 and the centre of the base; its distance from the base is

$$z = \frac{h(a + a_1)}{2(2a + a_1)}\,. \tag{12}$$

(b) An *obelisk* (Fig. 3.16b) (the bases are rectangles with sides a, b and a_1, b_1, the opposite slant faces make the same angle with the base, but they do not intersect at one point). The volume

$$V = \tfrac{1}{6}h[(2a + a_1)\,b + (2a_1 + a)\,b_1]\,. \tag{13}$$

The centroid lies on the segment connecting the centres of the two bases; its distance from the lower base is

$$z = \frac{h}{2}\,\frac{a(b + b_1) + a_1(b + 3b_1)}{a(2b + b_1) + a_1(b + 2b_1)}\,. \tag{14}$$

(c) The Cylinder (Fig. 3.17). Let h be the height, l the length of the side, P the area of the base, N the area of the normal section (plane section perpendicular to the sides) of the cylinder. Then, the volume V and the lateral area Q of the cylinder are given by:

$$V = Ph = Nl\,, \quad Q = C_P h = C_N l\,, \tag{15}$$

where C_P, and C_N, are the circumferences of the base, and of the normal section, respectively.

The centroid lies at the mid-point of the segment connecting the centroids of upper and lower bases of the cylinder.

A *right circular cylinder*. The base is a circle of radius r, lying in a plane which is perpendicular to the side of the cylinder, h is the height. Then

$$V = \pi r^2 h\,, \quad Q = 2\pi r h\,,$$

$$S = 2\pi r(r + h)\,. \tag{16}$$

The moment of inertia about the axis of revolution o is

$$I_o = \tfrac{1}{2}\pi r^4 h\,. \tag{17}$$

A *truncated right circular cylinder* (Fig. 3.18). Let h_1 be the shortest and h_2 the longest side of the cylinder. Then

$$V = \pi r^2 \frac{h_1 + h_2}{2}\,, \quad Q = \pi r(h_1 + h_2)\,,$$

$$S = \pi r \left[h_1 + h_2 + r + \sqrt{\left[r^2 + \left(\frac{h_2 - h_1}{2}\right)^2\right]}\right]. \tag{18}$$

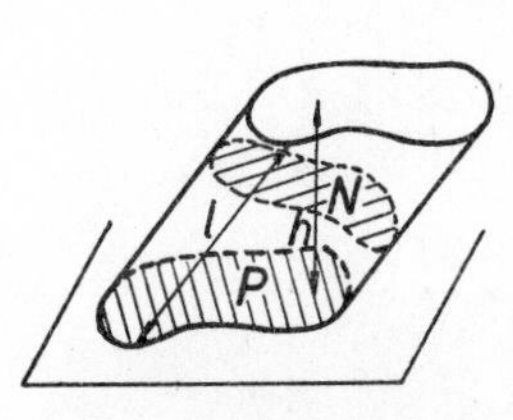

Fig. 3.17.

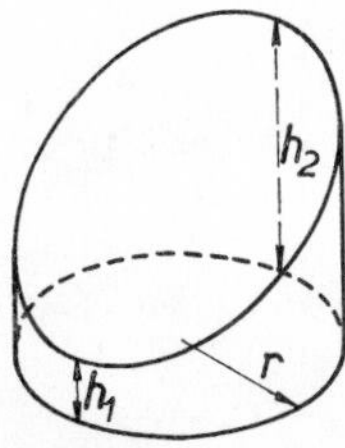

Fig. 3.18.

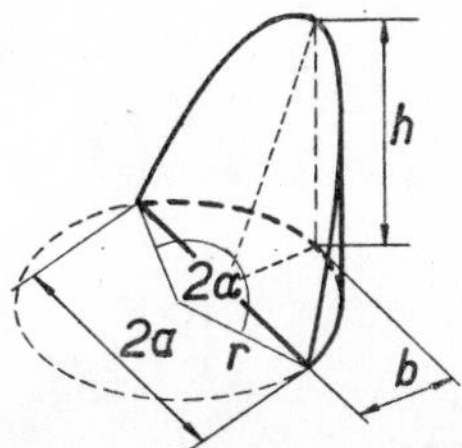

Fig. 3.19.

A *segment of a right circular cylinder* – a *cylindrical angle* (Fig. 3.19). Using the notation of Fig. 3.19, we have

$$V = \frac{h}{3b}\left[a(3r^2 - a^2) + 3r^2(b - r)\,\alpha\right] = \frac{hr^3}{b}(\sin\alpha - \tfrac{1}{3}\sin^3\alpha - \alpha\cos\alpha)\,,$$

$$Q = \frac{2rh}{b}\left[(b - r)\,\alpha + a\right]\,, \tag{19}$$

with the angle α measured in radians $(0 < \alpha \leqq \pi)$. For $\alpha = \frac{1}{2}\pi$, we have $a = b = r$ and

$$V = \tfrac{2}{3}r^2h\,, \quad Q = 2rh\,. \tag{20}$$

A *hollow right circular cylinder* – a *tube* (Fig. 3.20). Let r be the inner radius, R the outer radius, $a = R - r$ the thickness, $\varrho = \frac{1}{2}(r + R)$ the mean radius, h the height. Then

$$V = \pi(R^2 - r^2)\,h = \pi ah(2R - a) = \pi ah(2r + a) = 2\pi\varrho ah\,. \tag{21}$$

The moment of inertia about the axis of revolution o is

$$I_o = \tfrac{1}{2}\pi h(R^4 - r^4)\,. \tag{22}$$

(d) The Cone (Fig. 3.21). Let h be the height, P the area of the base. Then

$$V = \tfrac{1}{3}Ph\,. \tag{23}$$

The centroid lies on the segment connecting the apex and the centroid of the base; its distance from the base is $\frac{1}{4}h$.

A *right circular cone*. Its base is a circle of radius r, and the line passing through the apex and through the centre of the base (the axis of the cone) is perpendicular to the plane of the base; let h be the height. Then

$$V = \tfrac{1}{3}\pi r^2h\,, \quad Q = \pi rl\,, \quad S = \pi r(r + l) \tag{24}$$

where $l = \sqrt{(r^2 + h^2)}$ is the length of the side of the cone.

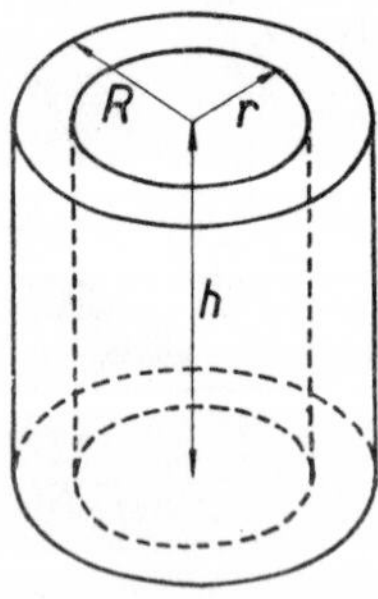

Fig. 3.20.

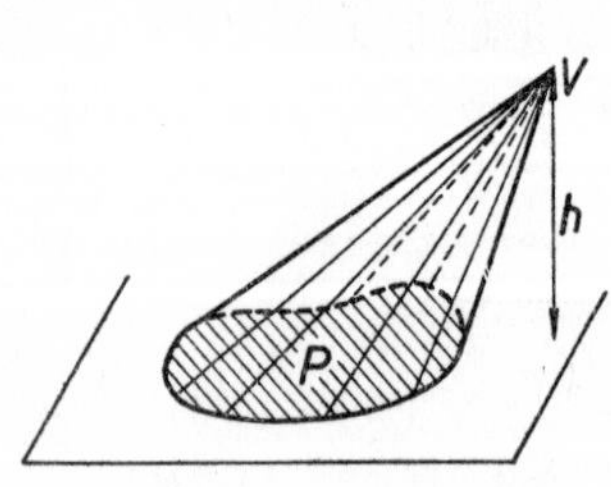

Fig. 3.21.

The moment of inertia about the axis of revolution o is

$$I_o = \tfrac{1}{10}\pi r^4h\,. \tag{25}$$

A *frustum of a right circular cone* (Fig. 3.22). Using the notation of Fig. 3.22,

we have

$$V = \tfrac{1}{3}\pi h(R^2 + Rr + r^2)\,, \quad Q = \pi(R + r)\,a \tag{26}$$

where $a = \sqrt{[h^2 + (R - r)^2]}$ is the length of the side of the frustum.

The centroid lies on the axis of revolution o; its distance from the lower base (of radius R) is

$$z = \frac{h(R^2 + 2Rr + 3r^2)}{4(R^2 + Rr + r^2)}\,. \tag{27}$$

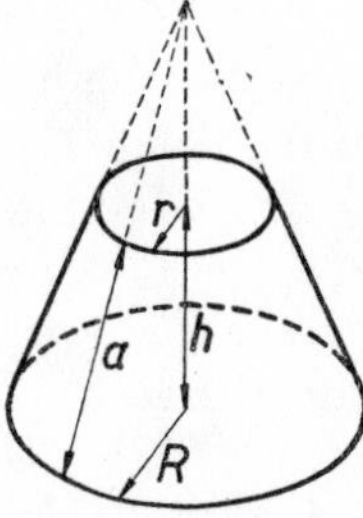

Fig. 3.22.

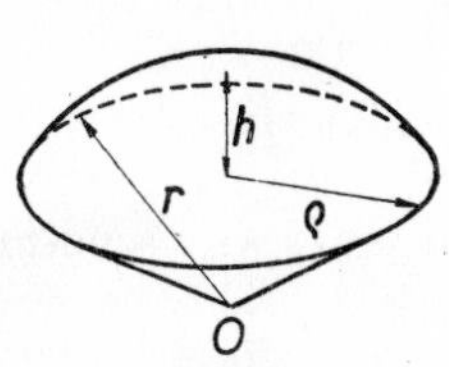

Fig. 3.23.

The moment of inertia about the axis of revolution o is

$$I_o = \frac{\pi h(R^5 - r^5)}{10(R - r)}\,. \tag{28}$$

(e) The Sphere. If r is the radius of the sphere, then

$$V = \tfrac{4}{3}\pi r^3 \approx 4{\cdot}188\,8r^3\,, \quad S = 4\pi r^2 \approx 12{\cdot}566r^2\,. \tag{29}$$

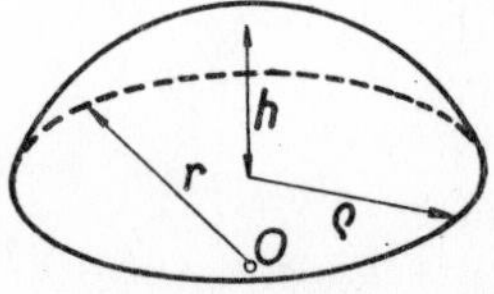

Fig. 3.24.

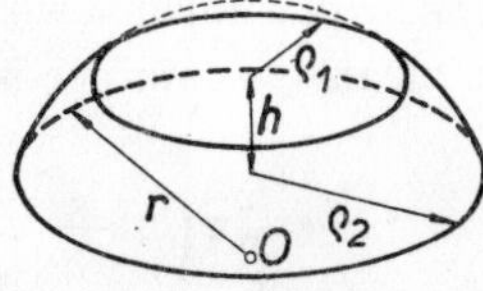

Fig. 3.25.

The moment of inertia about the axis o passing through the centre of the sphere

$$I_o = \tfrac{8}{15}\pi r^5\,. \tag{30}$$

A *sector of a sphere* (Fig. 3.23). Using the notation of Fig. 3.23, we have

$$V = \tfrac{2}{3}\pi r^2 h\,, \quad S = \pi r(2h + \varrho)\,. \tag{31}$$

A *segment of a sphere* (Fig. 3.24). Using the notation of Fig. 3.24, we have

$$V = \tfrac{1}{6}\pi h(3\varrho^2 + h^2) = \tfrac{1}{3}\pi h^2(3r - h)\,,$$
$$S = 2\pi rh + \pi\varrho^2\,, \quad Q = 2\pi rh\,. \tag{32}$$

A *spherical layer* (Fig. 3.25). Using the notation of Fig. 3.25, we have

$$V = \tfrac{1}{6}\pi h(3\varrho_2^2 + 3\varrho_1^2 + h^2)\,,$$
$$S = \pi(2rh + \varrho_1^2 + \varrho_2^2)\,, \quad Q = 2\pi rh\,. \tag{33}$$

A *spherical ring* is the part of a spherical layer, obtained by removing from it an inscribed frustrum of a cone (or of a cylinder). If a is the length of the side of the inscribed frustrum of cone (or of the cylinder), then the volume V of the spherical ring is

$$V = \tfrac{1}{6}\pi h a^2\,. \tag{34}$$

(f) The Ellipsoid with semi-axes a, b, c has the lateral area

$$S = 2\pi c^2 + \frac{2\pi b}{\sqrt{(a^2 - c^2)}}\left[c^2 F(k, \varphi) + (a^2 - c^2)\, E(k, \varphi)\right], \tag{35}$$

where

$$k = \frac{a}{b}\sqrt{\frac{b^2 - c^2}{a^2 - c^2}}\,, \quad \varphi = \arccos\frac{c}{a}$$

and $F(k, \varphi)$, $E(k, \varphi)$ are the elliptic integrals of the first and second kinds (see § 13.12, p. 590).

The volume of an ellipsoid

$$V = \tfrac{4}{3}\pi abc\,. \tag{36}$$

A *prolate spheroid* is formed when an ellipse with semi-axes a, b $(a > b)$ is rotated around its major axis; its surface is

$$S = 2\pi\left(b^2 + ab\,\frac{\arcsin e}{e}\right), \quad e = \frac{\sqrt{(a^2 - b^2)}}{a}\,. \tag{37}$$

An *oblate spheroid* is formed when an ellipse with semi-axes a, b $(a > b)$ is rotated around the minor axis; its surface is

$$S = 2\pi\left(a^2 + \frac{b^2}{2e}\ln\frac{1 + e}{1 - e}\right), \quad e = \frac{\sqrt{(a^2 - b^2)}}{a}\,. \tag{38}$$

The moment of inertia of a spheroid about the semi-axis a is

$$I_a = \tfrac{8}{15}\pi ab^4\,. \tag{39}$$

(g) The Paraboloid of Revolution (Fig. 3.26). The volume bounded by a paraboloid of revolution and by a plane perpendicular to its axis at distance h from the vertex O (the radius of the base being r) is

$$V = \tfrac{1}{2}\pi r^2 h \tag{40}$$

the lateral area is

$$Q = \frac{\pi r}{6h^2}\left[(r^2 + 4h^2)^{3/2} - r^3\right]. \tag{41}$$

The centroid lies on the axis of revolution o; its distance from the vertex O of the paraboloid is $\frac{2}{3}h$.

The moment of inertia about the axis o is

$$I_o = \tfrac{1}{6}\pi r^4 h . \tag{42}$$

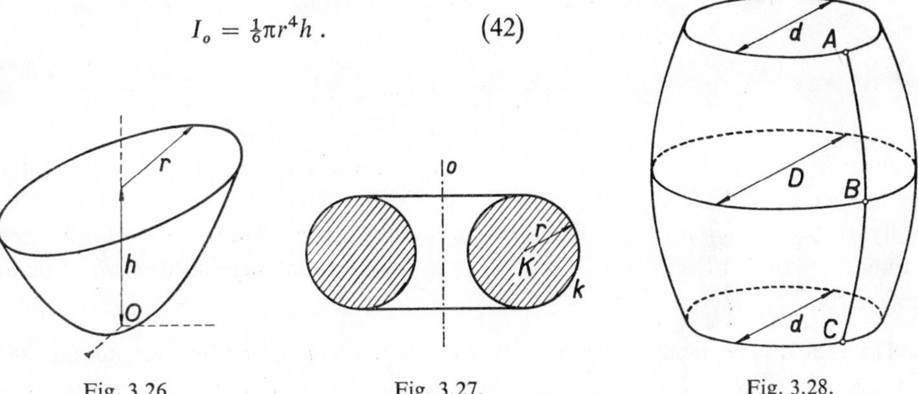

Fig. 3.26. Fig. 3.27. Fig. 3.28.

(h) The Torus (annuloid, ring) (Fig. 3.27) is formed by rotation of a circle k of radius r, with centre K around the axis o, lying in the plane of the circle at distance R $(R > r)$ from the centre K.

$$V = 2\pi^2 Rr^2 \approx 19{\cdot}739Rr^2 , \tag{43}$$

$$S = 4\pi^2 Rr \approx 39{\cdot}478Rr . \tag{44}$$

The moment of inertia of a torus about the axis of revolution o is

$$I_o = \tfrac{1}{2}\pi^2 Rr^2(4R^2 + 3r^2) . \tag{45}$$

(i) The Cask (Fig. 3.28). The diameter of the upper and lower bases is d, the diameter of the central section is D, the height is h.

For a *circular* shape (ABC being an arc of a circle)

$$V \approx 0{\cdot}262h(2D^2 + d^2) . \tag{46}$$

For a *parabolic* shape (ABC being an arc of a parabola)

$$V \approx 0{\cdot}052\,36h(8D^2 + 4Dd + 3d^2) . \tag{47}$$

4. PLANE CURVES AND CONSTRUCTIONS

By Karel Drábek

References: [18], [94], [164], [261], [264], [327], [359], [378], [396], [431], [441].

4.1. The Circle

A circle (for the definition see § 5.9) with centre S and radius r will be denoted by $k(S, r)$.

By the *construction of a circle* we mean the determination of its centre and radius from certain given conditions (with the help of fundamental theorems of plane geometry).

Theorem 1. *The circle is axially symmetrical about any line passing through its centre S (and called a diameter) and, hence, it is radially symmetrical about its centre S* (Fig. 4.1).

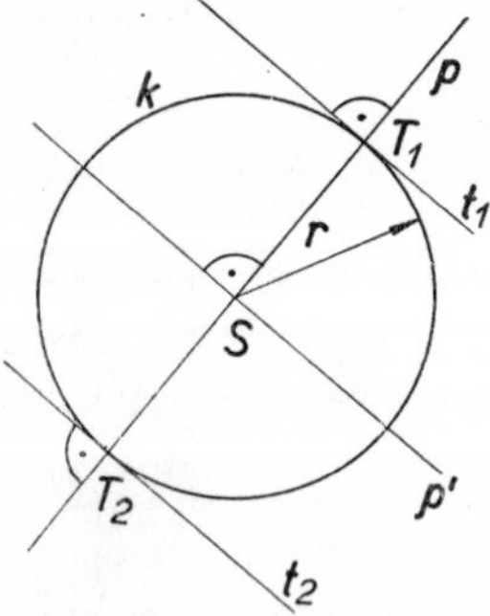

Fig. 4.1.

Theorem 2. *The tangent at a point of a circle is perpendicular to the line connecting this point and the centre of the given circle; consequently, all the normals of a circle pass through the centre of the circle.*

Theorem 3. *The tangents at the points of intersection of a circle and a diameter are parallel.*

In what follows the term diameter will normally be used in the sense of a so-called *bounded diameter*, i.e. the segment determined by the points of intersection of the diameter and the circle (ellipse, hyperbola, etc.), or its length.

Definition 1. The diameter of a circle parallel to the tangents at the end points of a given diameter is called the *conjugate diameter* to the original diameter.

Hence, conjugate diameters of a circle are perpendicular.

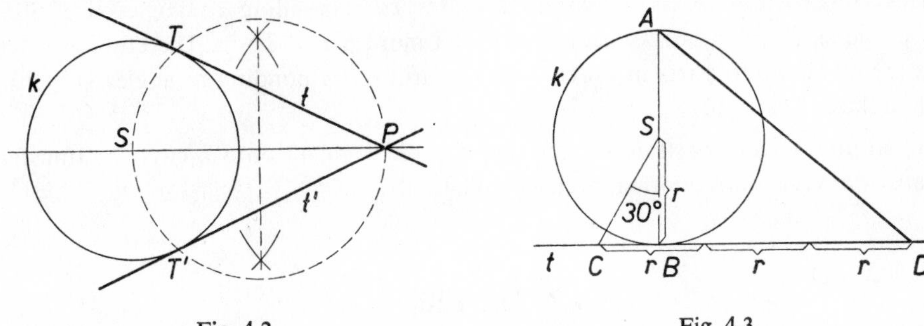

Fig. 4.2. Fig. 4.3.

Construction 1 of the tangents to a circle $k(S, r)$ from an external point P (i.e. from a point whose distance from the centre S of the circle k is $d > r$): the circle constructed on the diameter PS, the so-called *Thalet's circle*, meets the given circle k at two points T, T' (Fig. 4.2) which are the points of contact of the tangents $t \equiv PT$, $t' \equiv PT'$ from the given point P.

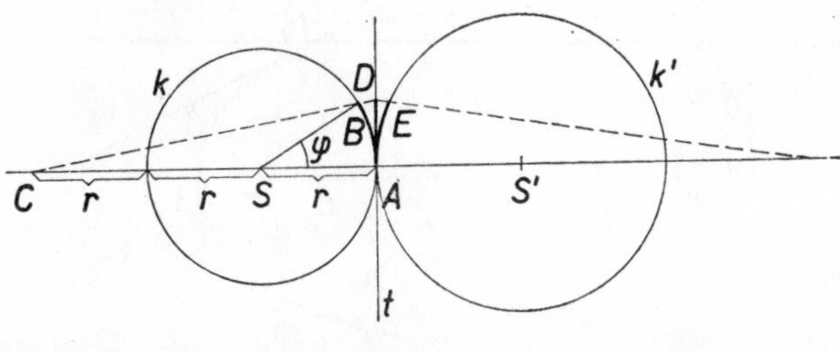

Fig. 4.4.

Definition 2. The construction of a segment equal in length to the circumference of a circle, or of a circular arc, is called the *rectification of the circle*, or of the circular arc, respectively.

In practice, i.e. using a ruler and a pair of compasses, these constructions for a circle are only approximate.

Construction 2 (*Kochański's rectification of a circle* (Fig. 4.3)). At the point B of a diameter AB we construct the tangent t and determine the point C of intersection of t and of the other arm of the angle $BSC = 30°$. On CB produced we find the point D such that $\overline{CD} = 3r$. Then $\overline{AD} \approx \pi r$.

Since the error reaches only 1 mm for $r \approx 17$ m, it need not be taken into account in our constructions.

Construction 3 (*Sobotka's rectification of a circular arc* $\widehat{AB}$ (Fig. 4.4)). We determine the point C on the half-line AS such that $\overline{AC} = 3r$. The line CB meets the tangent t constructed at the point A of the circle k at a point D. Then $\overline{AD} \approx \widehat{AB}$.

This construction is very accurate for arcs corresponding to angles $\varphi \leqq 30°$. For example, for $\varphi = 30°$ we get an error of 1 mm for $r \approx 2{\cdot}5$ m. Therefore, greater arcs are divided into parts in order to rectify arcs corresponding to angles $\varphi \leqq 30°$ with sufficient accuracy.

By an inverse construction we can wind a given segment onto a circle or transfer an arc of a circle onto another circle (Fig. 4.4).

4.2. The Ellipse

For the definition of the ellipse see § 5.10 (p. 221). We denote the foci by F_1, F_2 (Fig. 4.5); the line connecting a point of the ellipse and a focus is called a *focal radius*.

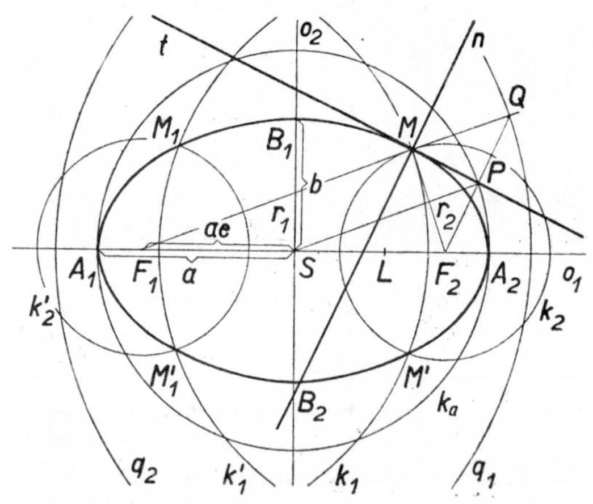

Fig. 4.5.

Theorem 1. *The ellipse is a curve symmetrical about the axis connecting both foci (the major axis) and about the perpendicular bisector of the segment* F_1F_2 *(the minor axis) and hence it is radially symmetrical about the point* S *of intersection of the axes of the ellipse (the centre of the ellipse).*

The points A_1, A_2 of the ellipse on the major axis are called the *major vertices*, the points B_1, B_2 on the minor axis the *minor vertices*. The length $\overline{A_1S} = \overline{A_2S} = a$ is called the *semi-major axis*, and the length $\overline{B_1S} = \overline{B_2S} = b$ the *semi-minor axis*. Let the length $F_1S = F_2S = ae$, so that the ratio $F_1S/A_1S = e$. Then e is called the *eccentricity* of the ellipse.

Theorem 2. *Between the lengths a, b and the eccentricity e, the relation* $a^2e^2 = a^2 - b^2$ *holds.*

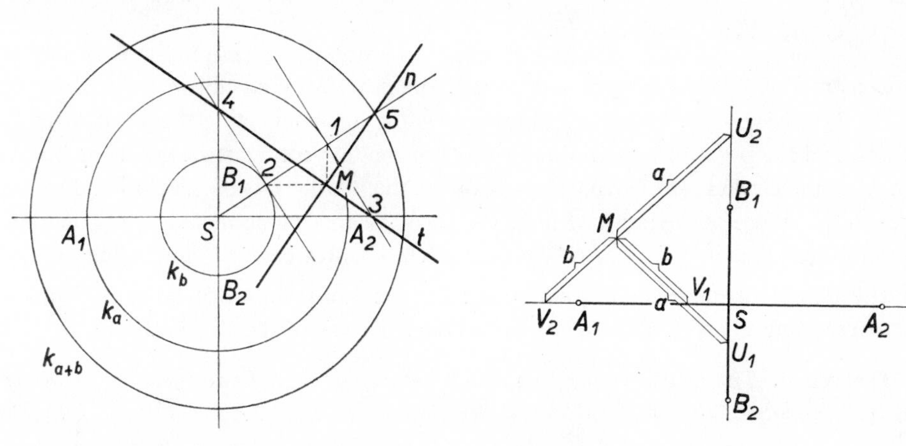

Fig. 4.6. Fig. 4.7.

Construction 1 of points of an ellipse with semi-axes a, b by means of its definition (Fig. 4.5): We determine the major and minor vertices of the ellipse and, by Theorem 2, we determine the foci F_1, F_2 on the major axis. We choose an arbitrary point L between the points F_1, F_2 and describe circles of radii $\overline{A_1L}$, about one focus and $\overline{A_2L}$ about the other. The points of intersection M, M' of these circles k_1, k_2 are points of the ellipse. By an interchange of the foci as centres of the constructed circles, we get two further points M_1, M_1' of the ellipse. This construction is not accurate in the vicinity of the major vertices A_1, A_2.

Construction 2 of points of an ellipse with given semi-axes using affinity with a circle (Fig. 4.6): Let the (*vertex*) circles k_a, k_b with centres at the point S and radii a, b be cut by a radius from the point S at the points *1*, *2*. The line through the point *1* parallel to the minor axis and the line through the point *2* parallel to the major axis intersect at a point M of the ellipse. The construction is always accurate, for the auxiliary lines intersect at right angles.

Construction 3 of an ellipse with given semi-axes a, b (Fig. 4.7).

(a) By means of the difference of the semi-axes: If the segment $\overline{U_1V_1} = a - b$ is moved along two perpendicular lines, then the point M (exterior to the segment $\overline{U_1V_1}$) describes the ellipse with semi-axes $\overline{MU_1} = a$, $\overline{MV_1} = b$.

(b) By means of the sum of the semi-axes: If the segment $\overline{U_2V_2} = a + b$ is moved along two perpendicular lines, then the point M (interior to the segment $\overline{U_2V_2}$) describes the ellipse with semi-axes $\overline{MU_2} = a$, $\overline{MV_2} = b$.

This construction is often used to determine the length of one of the semi-axes, given the other semi-axis, the position of the axes and a point of the ellipse.

Theorem 3. *The tangent, or the normal, at a given point of an ellipse bisects the angle between the focal radii which contains, or does not contain, a major vertex of the ellipse, respectively.*

Construction 4 of the tangent and normal at a point M of an ellipse using the ciclres k_a, k_b, k_{a+b} (Fig. 4.6): The required tangent is the line connecting the point M of the ellipse and the point of intersection *3* (or *4*) of the tangent constructed at the point *1* (or 2) of the circle k_a (or k_b) and the major (or the minor) axis of the ellipse. The normal to the ellipse at the point M joins the point M and the point *5* which is the point of intersection of the half line *S1* and the circle k_{a+b} (of centre S and radius $a + b$).

The following theorems are important for the construction of tangents from an external point of an ellipse and for some constructions of the ellipse (Fig. 4.5).

Theorem 4. *The locus of points Q which are reflections of one focus of an ellipse in its tangents is the circle q having its centre at the other focus and radius $2a$.*

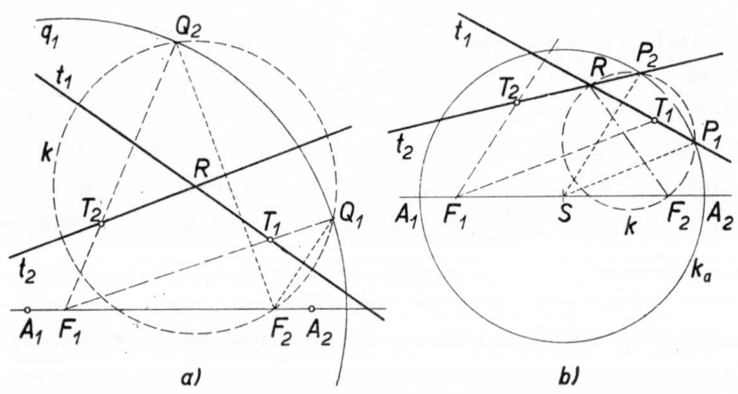

Fig. 4.8.

Theorem 5. *The locus of the feet P of perpendicular lines dropped from the foci of an ellipse to its tangents is the vertex circle $k_a(S, a)$.*

Theorem 6. *The locus of the centres of circles touching the circle $q_2(F_2, 2a)$ and passing through its internal point F_1 is the ellipse with foci F_1, F_2 and with its major axis of length $2a$.*

Theorem 7. *Let the vertex of a right angle move along the circle $k_a(S, a)$ so that one of its arms passes through an internal point F_1 of the circle k_a; then the other arm is a tangent to the ellipse with focus F_1, centre S and semi-major axis of length a.*

Construction 5 of tangents to an ellipse from an external point R:

(a) By means of the circle q_1 (Fig. 4.8a): We determine the points of intersection Q_1, Q_2 of the circles $k(R, \overline{RF_2})$ and $q_1(F_1, 2a)$. The perpendicular bisectors of the segments $\overline{Q_1F_2}$, $\overline{Q_2F_2}$ are the tangents t_1, t_2 from the point R to the ellipse. The

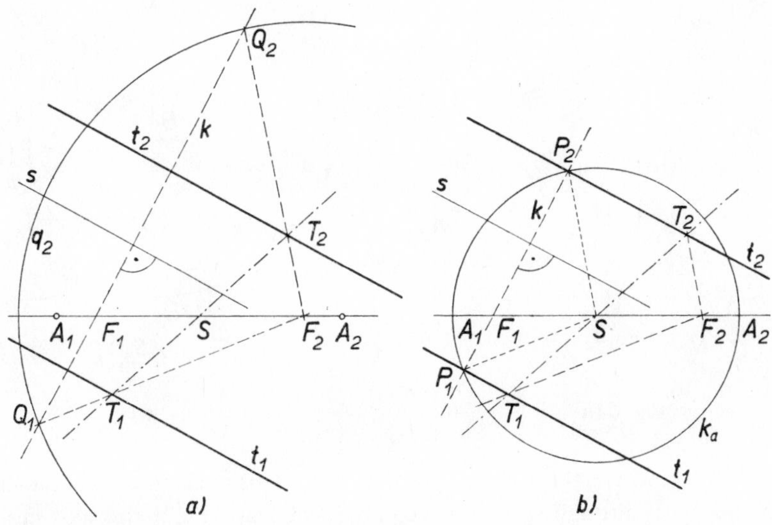

Fig. 4.9.

points of contact T_1, T_2 are the points of intersection of the tangents t_1, t_2 and the lines connecting the points Q_1, Q_2 and the focus F_1 (i.e. the focus about which the circle q_1 is described).

(b) By means of the vertex circle $k_a(S, a)$ (Fig. 4.8b): We determine the points of intersection P_1, P_2 of the circle k_a and the Thalet circle drawn on the diameter RF_2. The lines connecting P_1 and P_2 and the point R are tangents t_1, t_2 of the ellipse. The points of contact T_1, T_2 are the points of intersection of the tangents t_1, t_2 and the lines through the focus F_1 parallel to SP_1, SP_2, respectively.

Construction 6 of tangents to an ellipse, which are parallel to a given direction s:

(a) By means of the circle q_2 (Fig. 4.9a): The line k through the point F_1 perpendicular to the direction s intersects the circle $q_2(F_2, 2a)$ at points Q_1, Q_2; the perpendicular bisectors of the segments Q_1F_1, Q_2F_1 are the required tangents t_1, t_2.

(b) By means of the vertex circle k_a (Fig. 4.9b): The line k through the point F_1 perpendicular to the direction s intersects the circle $k_a(S, a)$ at points P_1, P_2: then the required tangents t_1, t_2 pass through P_1, P_2 and are parallel to the direction s.

The line connecting the points of contact T_1, T_2 of the parallel tangents t_1, t_2 passes through the centre S of the ellipse and is called the *conjugate diameter to the direction s*. Tangents to an ellipse parallel to a given direction a always exist.

Construction 7 (*the Rytz construction*) of the axes of an ellipse given by conjugate diameters M_1M_2, N_1N_2: On the perpendicular erected to one of the diameters, say M_1M_2, at the centre S (Fig. 4.10), we draw the segment $\overline{SR} = \overline{M_1S}$, join the points R and N_1 and describe a circle through the centre of the ellipse about the

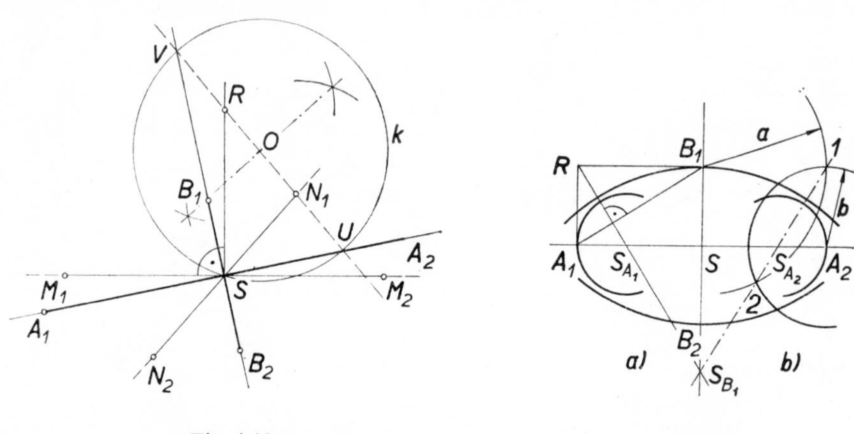

Fig. 4.10.

Fig. 4.11.

point O as centre, where O is the mid-point of RN_1. This circle intersects the line RN_1 in two points U, V through which the required axes pass (the major axis lies always within the acute angle made by the given conjugate diameters). Furthermore, $a = \overline{RU} = \overline{N_1V}$, $b = \overline{RV} = \overline{N_1U}$.

Construction 8 of the centres of curvature at the vertices of an ellipse:

(a) A perpendicular dropped from the vertex R of the rectangle SA_1RB_1 (Fig. 4.11a) to its diagonal A_1B_1 intersects the major, or minor axis at the centre of curvature corresponding to the major, or minor vertex of the ellipse, respectively.

(b) The line connecting the points of intersection *1* and *2* of the circles $k_1(A_2, b)$ $k_2(B_1, a)$ intersects the major, or minor axis at the required centres of curvature (Fig. 4.11b).

The circle with its centre at a centre of curvature constructed as above, which passes through the corresponding vertex (the *osculating circle* of the vertex) approximates to the given ellipse in the neighbourhood of the vertex.

4.3. The Hyperbola

For the definition of the hyperbola see § 5.11 (p. 222). We denote the foci by F_1, F_2 (Fig. 4.12); by a *focal radius*, denoted by r_1, r_2, we shall again mean a line connecting a point of the hyperbola and a focus.

Theorem 1. *The hyperbola is a curve symmetrical about the axis connecting both foci (the major axis) and about their perpendicular bisector (the minor axis) and*

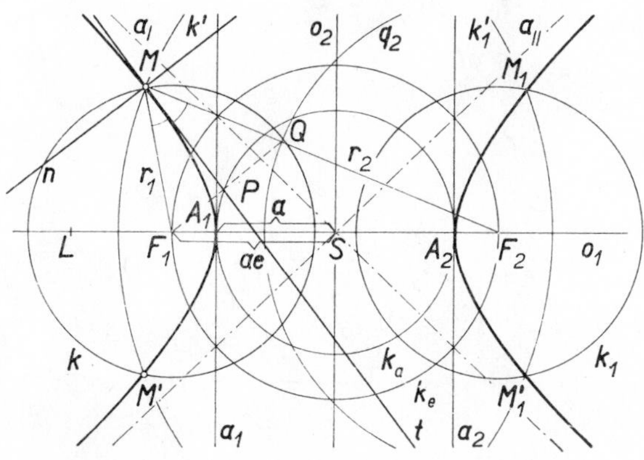

Fig. 4.12.

hence it is radially symmetrical about the point S of intersection of the axes of the hyperbola (the centre of the hyperbola).

The points A_1, A_2 of the hyperbola on the major axis are called the *major vertices*. The length $\overline{A_1S} = \overline{A_2S} = a$ is called the *semi-major axis*. Let the length $F_1S = F_2S = ae$, so that the ratio $F_1S/A_1S = e$. Then e is called the *eccentricity of the hyperbola*.

Construction 1 of points of a hyperbola given by the semi-major axis and focal distance ae (Fig. 4.12): We chose an arbitrary point L outside the segment F_1F_2 and describe circles of radii A_1L about one focus and A_2L about the other. The points of intersection M, M' of these circles are points of the hyperbola. Interchanging the foci as centres of the constructed circles, we get two further points M_1, M'_1 of the hyperbola.

From Construction 1, it is evident that the points of a hyperbola lie on two branches. The points of one branch satisfy the relation $r_1 - r_2 = 2a$ while the points of the other branch satisfy $r_2 - r_1 = 2a$. All points of a hyperbola (excepting the major vertices) lie outside the strip bounded by the lines a_1, a_2 parallel to the minor axis and passing through the points A_1, A_2.

Theorem 2. *The tangent, or the normal, at a point of a hyperbola bisects the angle between the focal radii which contains, or does not contain, the major vertices respectively.*

The following theorems are important for the construction of tangents from an external point of a hyperbola (i.e. from a point for which the absolute value of the difference of the focal radii is less than $2a$) and for some constructions of the hyperbola (Fig. 4.12):

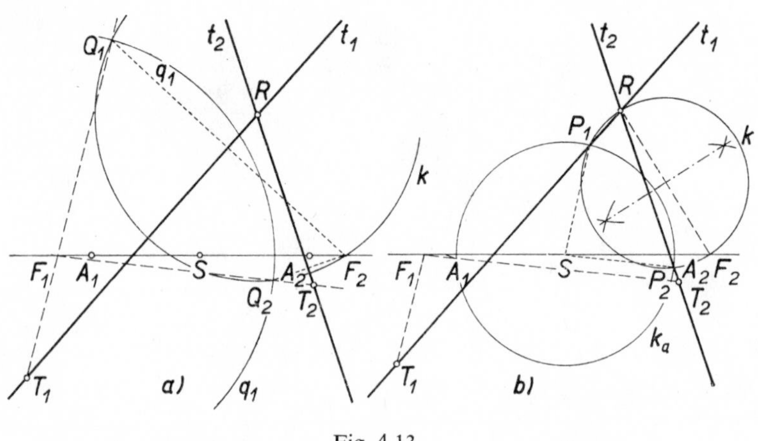

Fig. 4.13.

Theorem 3. *The locus of points Q which are reflections of one focus of a hyperbola in its tangents is the circle q having its centre at the other focus and radius equal to the length of the major axis $2a$.*

Theorem 4. *The locus of the feet P of perpendicular lines dropped from the foci of a hyperbola to its tangents is the (vertex) circle $k_a(S, a)$.*

Theorem 5. *The locus of the centres of circles touching the circle $q_2(F_2, 2a)$ and passing through the external point F_1 of the circle is the hyperbola with foci F_1, F_2 and with its major axis of length $2a$.*

Theorem 6. *Let the vertex of a right angle move along the circle $k_a(S, a)$ so that one of its arms passes through an external point F_1 of the circle k_a; then the other arm is a tangent to the hyperbola with focus F_1 and vertex circle $k_a(S, a)$.*

Construction 2 of tangents to a hyperbola from an external point R (Fig. 4.13a,b):

(a) By means of the circle q: We determine the points of intersection Q_1, Q_2 of the circles $k(R, \overline{RF_2})$ and $q_1(F_1, 2a)$. The perpendicular bisectors of the segments $\overline{Q_1F_2}$, $\overline{Q_2F_2}$ are the tangents t_1, t_2 from the point R to the hyperbola. The points of contact T_1, T_2 are the points of intersection of the tangents t_1, t_2 and the lines connecting the points Q_1, Q_2 and the focus F_1 (about which the circle q_1 is described).

(b) By means of the vertex circle k_a: We determine the points of intersection P_1, P_2 of the Thalet circle k drawn on the diameter RF_2 and the vertex circle k_a. The lines connecting P_1 and P_2 and the point R are tangents t_1, t_2 of the hyperbola. The points of contact T_1, T_2 are the points of intersection of the tangents t_1, t_2 and the lines through the focus F_1 parallel to SP_1, SP_2.

When constructing the tangents from the centre S of a hyperbola, we obtain the points of contact on these tangents a_{I}, a_{II} as points at infinity. We usually extend the locus defining the hyperbola to include these points.

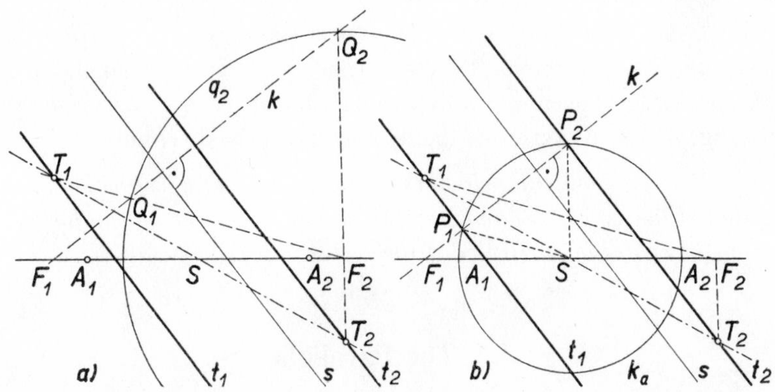

Fig. 4.14.

Definition 1. The tangents a_{I}, a_{II} from the centre S to a hyperbola are called the *asymptotes*; their directions ${}^{s}a_{\text{I}}$, ${}^{s}a_{\text{II}}$ which determine the points of the hyperbola at infinity are called the *directions of the asymptotes.*

Construction 3 of tangents parallel to a given direction s:

(a) By means of the circle q_2 (Fig. 4.14a): The line k through the point F_1 perpendicular to the direction s intersects the circle $q_2(F_2, 2a)$ at points Q_1, Q_2; the perpendicular bisectors of the segments Q_1F_1, Q_2F_1 are the required tangents t_1, t_2 which are parallel to s.

(b) By means of the vertex circle k_a (Fig. 4.14b): The line k through the point F_1 perpendicular to the direction s intersects the circle $k_a(S, a)$ at points P_1, P_2 through which pass the required tangents t_1, t_2 which are parallel to s.

The line connecting the points of contact T_1, T_2 of parallel tangents t_1, t_2 passes through the centre S of a hyperbola and is called the *conjugate diameter to the direction s.*

If φ is the acute angle between an asymptote and the major axis of a hyperbola, and if ψ is the acute angle between the direction s and the major axis, then tangents parallel to the given direction s exist only for $\psi > \varphi$.

The following theorems can be advantageously used when constructing a hyperbola:

Theorem 7. *The segments on an arbitrary secant of a hyperbola (intersecting the asymptotes) between the points of the hyperbola and the asymptotes are equal. In particular: The point of contact of a tangent to a hyperbola bisects its segment between the asymptotes.*

Theorem 8. *The parallelograms formed by the asymptotes of a hyperbola and by the lines constructed through points of the hyperbola parallel to the asymptotes are of a constant area. In particular: The triangles formed by the asymptotes and the tangents of a hyperbola are of a constant area.*

By means of variable parallelograms of constant area we can construct points of a hyperbola with given asymptotes. Further, by means of variable triangles of constant area we can construct tangents to a hyperbola with given asymptotes; in particular the vertex tangent and the vertex of the hyperbola can be determined.

Theorem 9. *The perpendicular drawn to an asymptote of a hyperbola at the point of intersection of the asymptote and a vertex tangent intersects the major axis of the hyperbola at the centre of curvature of the vertex.*

4.4. The Parabola

For the definition of the parabola see § 5.12 (p. 223). The focus will be denoted by F, the directrix by f (Fig. 4.15). The point F does not lie on the line f.

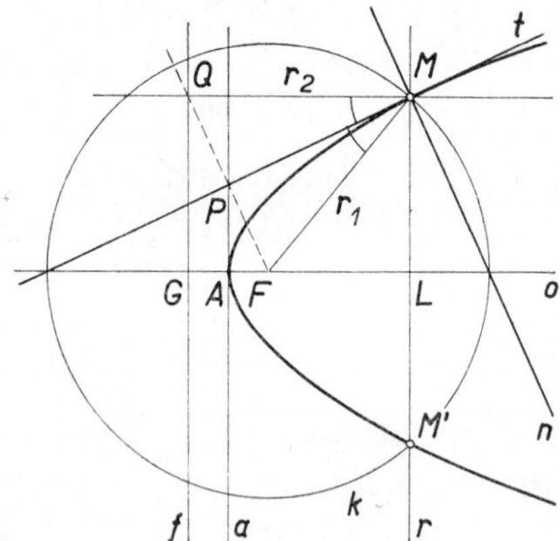

Fig. 4.15.

If M is a point of the parabola, then $MF = r_1$ is one of the focal radii of the point M; as the second (focal) radius the line through the point M perpendicular to the directrix f is to be understood.

The point A of the parabola bisecting the distance of the focus F from the directrix f (this distance is called the *parameter*) is said to be the *vertex of the parabola*; the tangent a at the point A is called the *vertex tangent*.

Theorem 1. *The parabola is a curve symmetrical about the axis, i.e. the line perpendicular to the directrix through the focus F.*

Construction 1 of points of the parabola given by a focus F and a directrix f: Through an arbitrary point L on AF produced we construct the line r parallel to the directrix f. If G is the point of intersection of the axis o and the directrix f, then the points of intersection of the circle $k(F, \overline{GL})$ and the line r are points M, M' of the parabola.

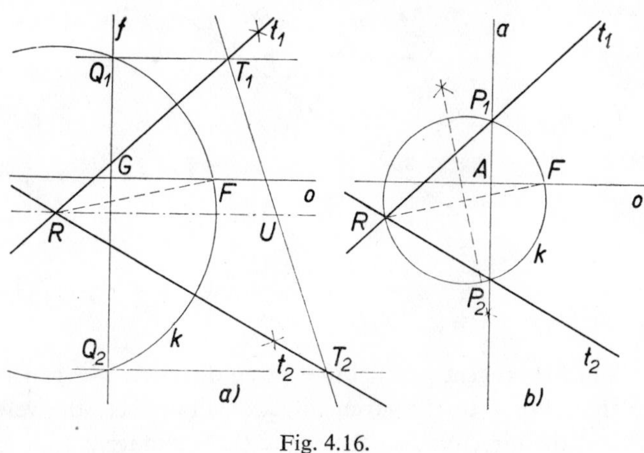

Fig. 4.16.

Theorem 2. *The tangent, or the normal, at a point of the parabola bisects the angle between the focal radii in which the vertex of the parabola lies, or does not lie, respectively.*

Theorem 3. *The locus of points Q which are reflections of the focus F of the parabola in its tangents is the directrix f.*

Theorem 4. *The locus of the feet P of perpendicular lines dropped from the focus F of the parabola to its tangents is the vertex tangent a.*

Theorem 5. *The locus of centres of the circles, touching a line f and passing through a point F (which does not lie on f) is a parabola with focus F and directrix f.*

Theorem 6. *Let the vertex of a right angle move along a straight line a so that one of its arms passes through a point F (which does not lie on a); then the other arm is a tangent to the parabola with focus F and vertex tangent a.*

Construction 2 of tangents to a parabola from an external point R whose distance from the focus F is greater than the distance from the directrix f:

(a) By means of the directrix f (Fig. 4.16a): The circle $k(R, \overline{RF})$ intersects the directrix f at the points Q_1, Q_2; the perpendicular bisectors of the segments Q_1F, Q_2F

are the required tangents t_1, t_2. The points of contact T_1, T_2 are the points of intersection of the tangents t_1, t_2 and the lines through Q_1, Q_2 parallel to the axis of the parabola.

(b) By means of the vertex tangent a (Fig. 4.16b): The circle on the diameter RF intersects the vertex tangent a at the points P_1, P_2 through which pass the required tangents $t_1 \equiv RP_1$, $t_2 \equiv RP_2$. The points of contact should be determined as in (a); consequently, the construction (b) is not convenient in this case.

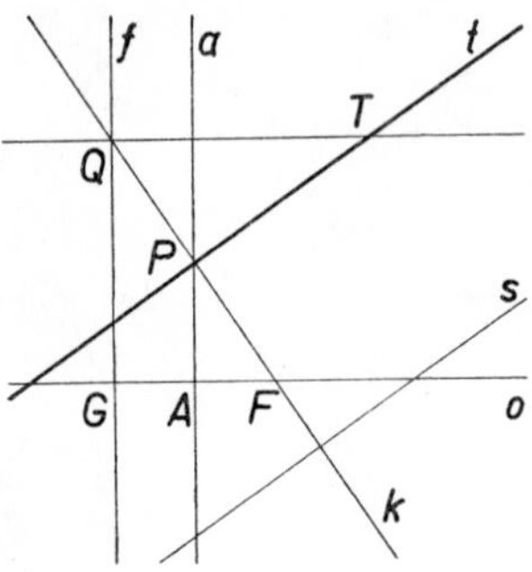

Fig. 4.17.

Construction 3 of the tangent parallel to a given direction s (Fig. 4.17): A perpendicular k from the focus F to the given direction s intersects the vertex tangent a at the point P and the directrix f at the point Q. The tangent $t \parallel s$ passes through the point P (perpendicularly to FQ) and its point of contact T is the point of intersection of the tangent t and the line through Q parallel to the axis of the parabola.

Definition 1. The distance between the point of contact T and the point of intersection of a tangent, or a normal, of the parabola and its axis is called the *length of the tangent*, or the *length of the normal*, briefly the *tangent*, or the *normal*, respectively. The rectangular projection of the tangent, or the normal, onto the axis of the parabola is called the *sub-tangent*, or the *sub-normal*, respectively.

Theorem 7. *A sub-tangent is bisected by the vertex. A sub-normal is of constant length equal to the parameter. The segment which is the sum of the sub-tangent and the sub-normal is bisected by the focus.*

Theorem 8. *The line connecting the point of intersection of two tangents to a parabola and the midpoint of the corresponding chord of contact is parallel to the axis of the parabola (and is called a diameter of the parabola).*

From this theorem, it follows that all diameters of a parabola are parallel.

Theorem 9. *The circle circumscribed about a triangle formed by three tangents to a parabola passes through its focus.*

Theorem 10. *The radius of curvature at the vertex of a parabola is equal to the parameter.*

Construction 4 of a parabola given by two tangents t_1, t_2 with points of contact T_1, T_2: We determine (Fig. 4.18) the diameter p of the parabola by means of the point of intersection R of the tangents t_1, t_2 and the mid-point U of the chord T_1T_2. Denote by *1, 2* the points of intersection of the lines through the points T_1, T_2 parallel to the diameter p and a line r (arbitrarily chosen) through the point R, respectively. The diagonals $1T_1$, $2T_2$ of the constructed trapezium $12T_1T_2$ meet at a point T of the parabola; the tangent t at T is parallel to r.

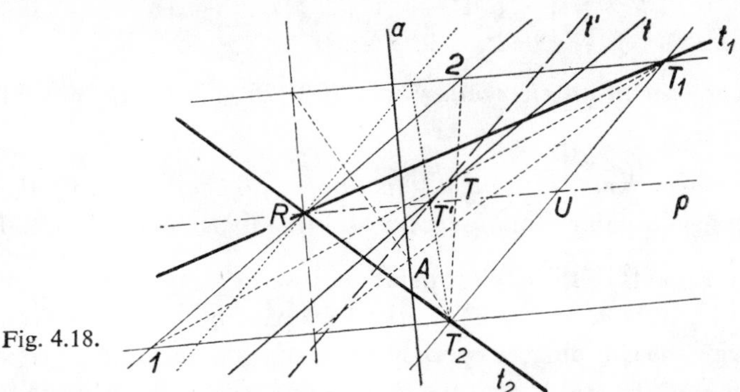

Fig. 4.18.

In particular: If $r \perp p$, we get the vertex A and the vertex tangent a. If $r \parallel T_1T_2$ (then the trapezium becomes a parallelogram) we get a tangent t' parallel to the chord T_1T_2 whose point of contact T' bisects the segment RU.

4.5. Parabolas and Hyperbolas of Higher Degree (Power Curves)

Definition 1. A curve given by the equation

$$y = ax^n \tag{1}$$

is called a *power curve* (a being constant, n rational, x positive, in general). For $n > 1$, we get the so-called *parabolas of higher degree*, for $n < -1$ we get the *hyperbolas of higher degree*.

If $|n| \in (0, 1)$ and $a > 0$, in general, we can write (interchanging the role of the coordinates)

$$x = by^{1/n}, \quad \text{i.e.} \quad x = by^m, \quad \text{where} \quad b = a^{-1/n}, \quad m = 1/n > 1 \quad \text{or} \quad < -1. \tag{2}$$

Theorem 1. *A tangent t (at a given point $P(x_0, y_0)$) cuts off on the y-axis an intercept equal to $(1 - n)\, y_0$. The length of a sub-tangent s_t^x on the x-axis, or s_t^y on the y-axis, is $|x_0/n|$, or $|ny_0|$, respectively.*

Theorem 2a. *The tangent to the parabola* (1), *or* (2), *at the origin* O *is the* x*-axis, or* y*-axis, respectively.*

Theorem 2b. *The asymptotes of the hyperbola* (1) *are the* x*-axis and* y*-axis.*

Theorem 3. *The length of an arc of the parabola* (1) *from the point* O *to the point* $P(x_0, y_0)$ *is given by the integral*

$$s = \int_0^{|x_0|} \sqrt{(1 + a^2 n^2 x^{2n-2})}\, dx ,$$

which can be expressed in an elementary way if $1/(2n - 2)$ *or* $1/(2n - 2) + \frac{1}{2}$ *is an integer.*

Theorem 4. *The area bounded by a parabola of a higher degree, by the* x*-axis and by the ordinate of a point with abscissa* x_0 *is given by* $P = |x_0 y_0|/(n + 1)$.

Construction 1 of points of the cubical parabola $y = ax^3$ (Fig. 4.19a), or of points of the semicubical parabola $y^2 = ax^3$ (Fig. 4.19b) passing through a given point $P(x_0, y_0)$: We divide the coordinates x_0, y_0 of the point $P(x_0, y_0)$ into an equal number of parts of the same length. If M is the foot of a perpendicular dropped from the point P to the x-axis, we describe a semicircle on MP and erect perpendiculars to the x-axis at the points of subdivision of the segment OM.

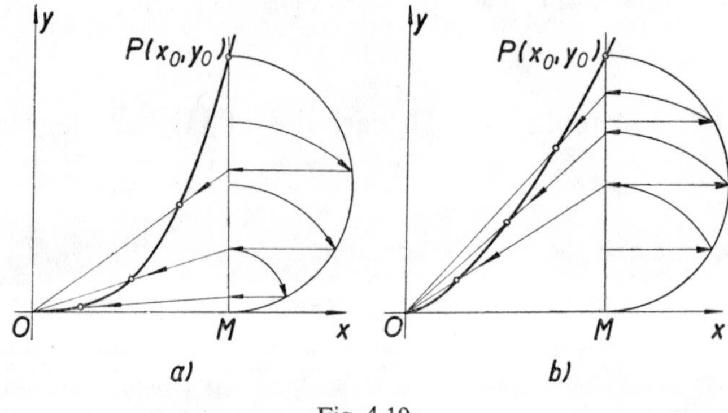

Fig. 4.19.

(a) Points of the *cubical* parabola: The circles with the centre M passing through the points of subdivision of the ordinate MP meet the semicircle on MP at points which we project rectangularly back onto MP. The lines connecting these projections and the origin O intersect the perpendiculars constructed at the points of subdivision of the segment OM at points of a cubical parabola.

(b) Points of the *semicubical* (Neil's) parabola: We project the points of subdivision of the ordinate y_0 parallel to the x-axis onto the semicircle on MP; we turn the points of intersection obtained in this way back onto the segment MP by circles with centre M. The lines connecting the points so obtained and the origin O intersect the perpendiculars at the points of subdivision of the segment OM at points of a semicubical parabola.

4.6. The Cyclic Curves

Definition 1. By rolling a curve h (the *generating curve* or *moving polhode*), without slipping, along a fixed curve p (the *basic curve* or *fixed polhode*) each point of a plane moving with the curve h describes a curve called a *trochoid.*

Theorem 1. *The fixed polhode is the locus of points which are instantaneous centres of rotation for the respective stages of the motion. The moving polhode is the locus of points which become instantaneous centres of rotation during the motion. Both polhodes always touch, at a point which is an instantaneous centre of rotation.*

Theorem 2. *The normal at a point of a trochoid passes through the instantaneous centre of rotation.*

In what follows we consider only the cases in which both polhodes are circles, or one of them is a circle and the other a straight line.

(a) The cycloids

Definition 2. By rolling a circle h along a straight line p without slipping, each point of the circle describes a *simple (general, normal) cycloid.* If the original position of the generating point coinciding with the point of contact of the circle h and the straight line p is the origin O and the straight line p is the x-axis, then

$$x = r(t - \sin t), \quad y = r(1 - \cos t) \tag{1}$$

are parametric equations of this simple cycloid; here, r is the radius of the generating circle h and t is the angle through which the rolling circle has turned at any instant.

Construction 1 of points of a simple cycloid (Fig. 4.20). We divide the circumference of the circle h and its rectified length on the tangent p at the point A into the same number of equal parts (there are 12 in Fig. 4.20). Consequently, $\widehat{A1} = \overline{AI}$, $\widehat{12} = \overline{I\,II}, \ldots$. Perpendiculars constructed through the points on the straight line p determine on the line through the point H parallel to the straight line p (i.e. on the path of the point H) the centres $H_1, H_2, \ldots$ of circles $h_1, h_2, \ldots$. Lines through the points of subdivision $1, 2, 3, \ldots$ of the circle h, parallel to the straight line p, meet the circles $h_1, h_2, \ldots$ at the points $A_1, A_2, \ldots$ of the cycloid.

Theorem 3. *The normal to a simple cycloid at a given point passes through the corresponding point of contact of the generating circle h on the given straight line p (i.e. through the instantaneous centre of rotation). The tangent to a simple cycloid at a given point passes through the point of the circle h which is diametrically opposite to the instantaneous centre of rotation.*

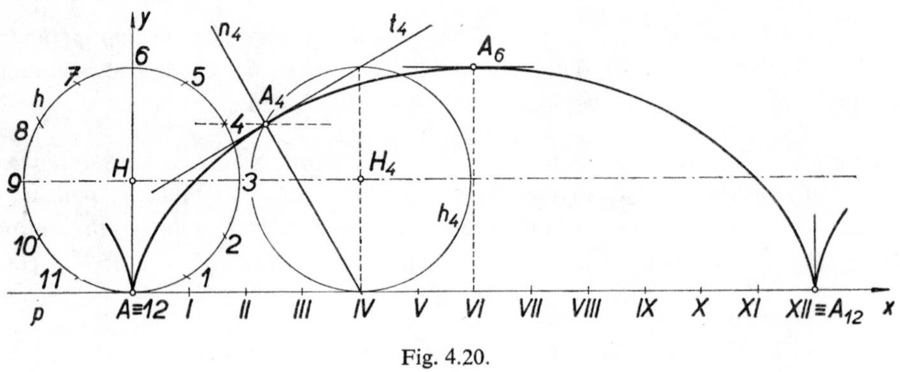

Fig. 4.20.

Theorem 4. *The length of a normal is*

$$n = 2r \sin \frac{t}{2} = \sqrt{(2ry)}\ .$$

Theorem 5. *The radius of curvature at a point other than the cuspidal point of a simple cycloid is*

$$R = 4r \sin \frac{t}{2} = 2\sqrt{(2ry)} = 2n\ ;$$

thus, at the vertex

$$R = 4r\ .$$

Theorem 6. *The length of arc (on a single branch) of a simple cycloid measured from the cuspidal point to the point P(x, y) is*

$$s = 4r\left(1 - \cos \frac{t}{2}\right);$$

thus, the length of the entire branch is

$$s = 8r\ .$$

Theorem 7. *The area bounded by the x-axis and by a branch of a simple cycloid is*

$$P = 3\pi r^2\ .$$

Definition 3. If a circle h rolls along a fixed straight line p, without slipping, an internal or external point moving with the circle h describes a *curtate*, or a *prolate*, *cycloid*, respectively.

Theorem 8. *If t is the angle through which the generating circle h has rolled and if $d \lesseqgtr r$ is the distance between the moving point P and the centre H of the circle h, the parametric equations of the curtate, or prolate, cycloid traced out by P are given by the equations*

$$x = rt - d \sin t\,, \quad y = r - d \cos t\,. \tag{2}$$

Construction 2 of points of a curtate, or a prolate, cycloid (Fig. 4.21): We attach to the generating circle h a concentric circle h' of radius $r' = \overline{HB} < r$ or h'' of radius $r'' = \overline{HC} > r$, respectively. Then, on the appropriate radius of an instantaneous position of the generating circle h_k with centre H_k $(k = 1, 2, \ldots)$ we determine the position of the circle h'_k, or h''_k, and, consequently, get the point B_k, or C_k of a curtate or prolate cycloid respectively.

Theorem 9. *The normal at a point of a curtate (prolate) cycloid passes through the point of contact of an instantaneous position of the generating circle h and the straight line p.*

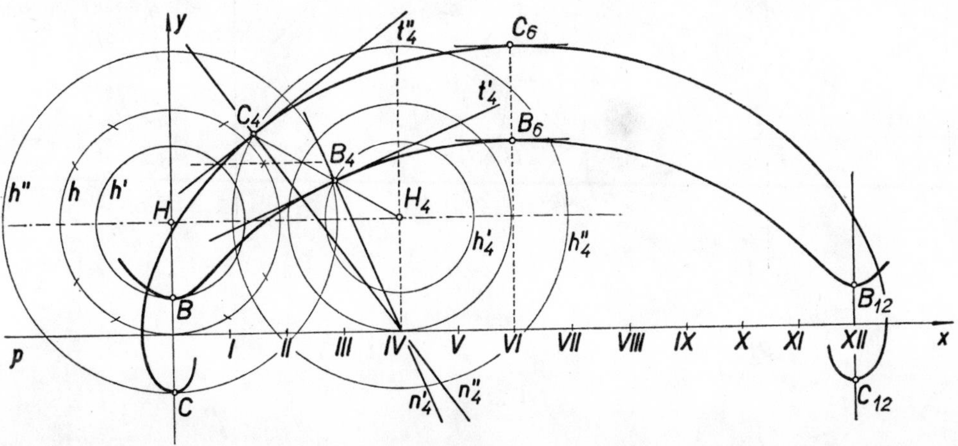

Fig. 4.21.

Theorem 10. *The radius of curvature at the points of a minimum or a maximum of a curtate (prolate) cycloid is*

$$R = \frac{(r-d)^2}{d}\,, \quad \textit{or} \quad R = \frac{(r+d)^2}{d}\,, \quad \textit{respectively}\,.$$

A simple cycloid has an infinite number of *cuspidal points*, a curtate cycloid has an infinite number of *points of inflexion* and a prolate cycloid an infinite number of *double points* (so-called *nodes*).

(b) The epicycloids and hypocycloids

Definition 4. If a generating circle h of radius r rolls along the exterior, or the interior circumference of a fixed circle p of radius $\bar{r}$, then each point of the circle h describes a *simple* (*general, normal*) *epicycloid*, or *hypocycloid*, respectively.

Theorem 11. *The equations*

$$x = (\bar{r} \pm r)\cos t \mp r\cos\frac{\bar{r} \pm r}{r}t\,, \quad y = (\bar{r} \pm r)\sin t - r\sin\frac{\bar{r} \pm r}{r}t \tag{3}$$

are parametric equations of a simple epicycloid (*the upper sign*) *or hypocycloid* (*the lower sign*).

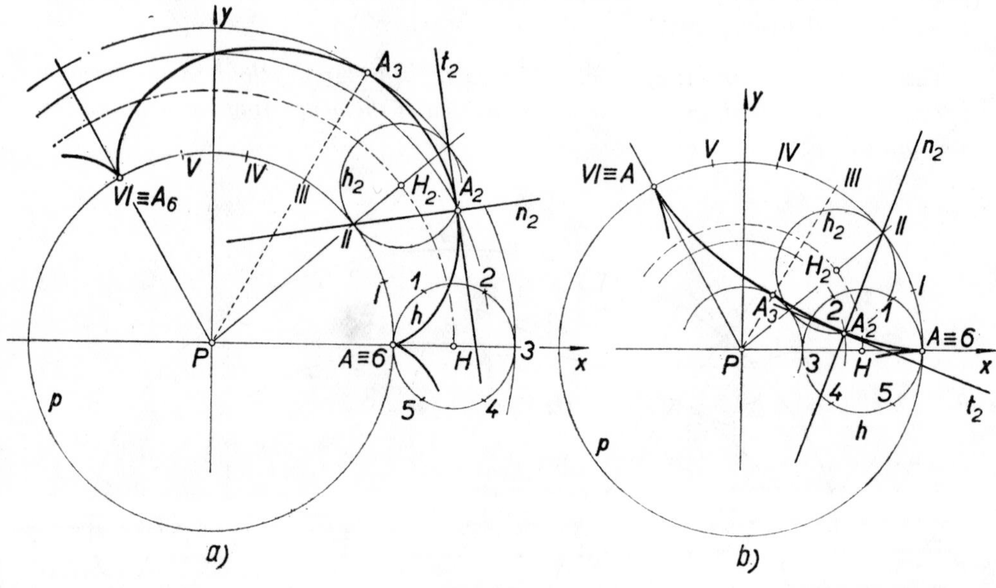

Fig. 4.22.

If $\lambda = \bar{r}/r$ is an integer, then λ denotes the number of branches of the curve formed by a single rotation of the circle h around the circle p. If $\lambda = p/q$ is a rational number, then the curve consists of p branches which are formed by q rotations of the circle h around the circle p. For an irrational λ, the curve contains an infinite number of branches.

Construction 3 of points of a simple epicycloid, or hypocycloid: We divide the circumference of the circle h into a certain number of equal parts (Fig. 4.22a,b), to give the points A, *1*, *2*, ...

On the circle p we determine an arc whose length is equal to the circumference of the circle h, and divide it by points $A, I, II, \ldots$ into an equal number of parts of the same length as those on h. We describe concentric circles through the points $1, 2, \ldots$ about the centre P of the circle p and find the points $H_1, H_2, \ldots$ of intersection of the radii $PI, PII, \ldots$ with the circle described about P through the centre H of the circle h. Then the circle $h_1(H_1, r)$ meets the circle $k_1(P, \overline{PI})$ at a point A_1 of a simple epicycloid, or hypocycloid, respectively, etc.

Theorem 12. *The radius of curvature at a point (other than a cuspidal point) of a simple epicycloid, or hypocycloid, is*

$$R = \left| \frac{4r(\bar{r} \pm r)}{\bar{r} \pm 2r} \sin \frac{\bar{r}t}{2r} \right| ;$$

the length of arc (on the same branch) from the point $t = 0$ to a point t is

$$s = \frac{8r(\bar{r} \pm r)}{\bar{r}} \sin^2 \frac{\bar{r}t}{4r}$$

($r < \bar{r}$ is assumed for the hypocycloid). In particular: The radius of curvature at a vertex is

$$R = \left| \frac{4r(\bar{r} \pm r)}{\bar{r} \pm 2r} \right|$$

and the length of one branch is

$$s = \frac{8r(\bar{r} \pm r)}{\bar{r}} .$$

Here the positive sign holds for an epicycloid, the negative sign for a hypocycloid.

Definition 5. The curve described by an internal, or an external, point rotating with the generating circle h is called a *curtate*, or a *prolate, epicycloid (hypocycloid)*, respectively.

Theorem 13. *If $d \lesseqgtr r$ is the distance between the generating point and the centre H of the circle h, then*

$$x = (\bar{r} \pm r) \cos t \mp d \cos \frac{\bar{r} \pm r}{r} t , \quad y = (\bar{r} \pm r) \sin t - d \sin \frac{\bar{r} \pm r}{r} t$$

are parametric equations of the curves of Definition 5 (the upper signs refer to an epicycloid, the lower ones to a hypocycloid).

For the construction of points of a curtate or a prolate epicycloid (hypocycloid) we employ again the concentric circle h', or h'', attached to the circle h, as in Construction 2 above.

Example 1. If $\bar{r} = r$, $d > r$, we get a prolate epicycloid (the *limaçon of Pascal*) of parametric equations (Fig. 4.23)

$$x = 2r\cos t - d\cos 2t\,,$$

$$y = 2r\sin t - d\sin 2t\,,$$

whose equation in rectangular cartesian coordinates (by elimination of the parameter t and translation of the origin to the point $(d, 0)$) is

$$(x^2 + y^2 + 2dx)^2 = 4r^2(x^2 + y^2)\,.$$

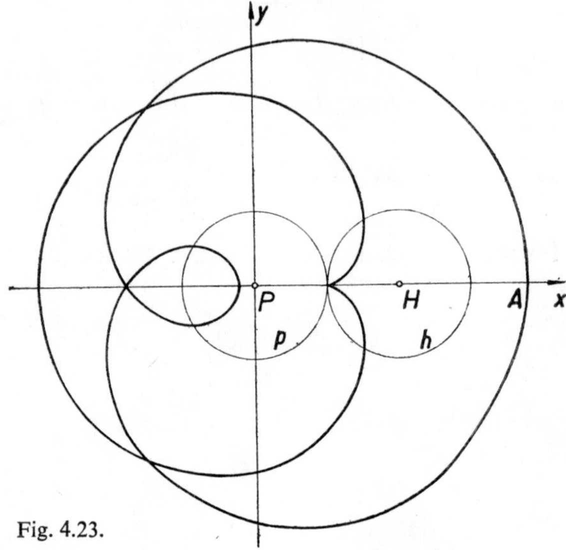

Fig. 4.23.

Example 2. For $r = \frac{1}{2}\bar{r}$, the equations of a simple hypocycloid are

$$x = \bar{r}\cos t\,, \quad y = 0\,;$$

hence, it is a segment of length $2\bar{r}$ on the x-axis.

The equations of a curtate hypocycloid are

$$x = \left(\frac{\bar{r}}{2} + d\right)\cos t\,, \quad y = \left(\frac{\bar{r}}{2} - d\right)\sin t\,;$$

hence it is an ellipse with a semi-major axis of length $r + d$ on the x-axis and a semi-minor axis of length $r - d$ on the y-axis.

Example 3. For $r = \bar{r}$ (Fig. 4.23) the parametric equations of a (simple) epicycloid (the *cardioid*) are

$$x = r(2\cos t - \cos 2t)\,, \quad y = r(2\sin t - \sin 2t)\,.$$

If the origin of cartesian coordinates is at the centre of the fixed curve and the cuspidal point lies on the x-axis, then we get the equation of the curve in the form

$$(x^2 + y^2)^2 - 6r^2(x^2 + y^2) - 8r^3x - 3r^4 = 0\,;$$

if the origin is at the double point (and the x-axis is the axis of symmetry), then we get the equation

$$(x^2 + y^2 - 2rx)^2 - 4r^2(x^2 + y^2) = 0\,.$$

The equation of the cardioid in polar coordinates is

$$\varrho = 2r(1 + \cos\varphi)\,.$$

A cardioid can also be obtained as an orthogonal pedal curve (see Definition 9.10.1, p. 341) of a circle for a pole on the circle.

Example 4. For $r = \frac{1}{2}\bar{r}$ (Fig. 4.24), the parametric equations of a simple epicycloid (the *nephroid*) are

$$x = r(3\cos t - \cos 3t)\,, \quad y = r(3\sin t - \sin 3t)\,;$$

the equation of the curve in cartesian coordinates is

$$(x^2 + y^2 - 4r^2)^3 - 108r^4y^2 = 0\,.$$

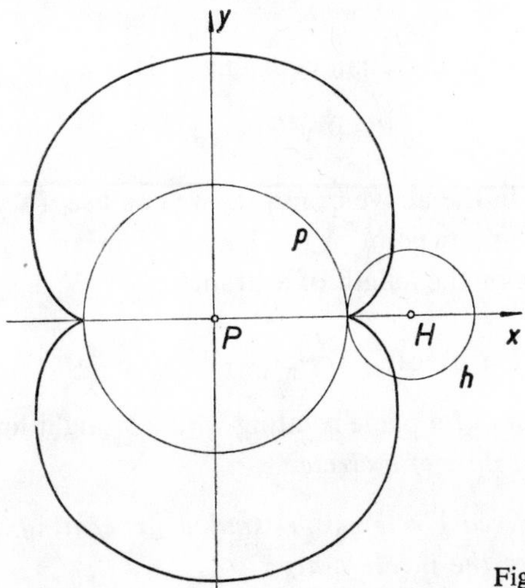

Fig. 4.24.

Example 5. For $r = \frac{1}{3}\bar{r}$ (Fig. 4.25), the parametric equations of a (simple) hypocycloid (*Steiner's hypocycloid*) are

$$x = r(2\cos t + \cos 2t)\,,$$
$$y = r(2\sin t - \sin 2t)\,;$$

the equation of the curve in cartesian coordinates is

$$(x^2 + y^2)^2 + 8rx(3y^2 - x^2) + 18r^2(x^2 + y^2) - 27r^4 = 0 .$$

Example 6. For $r = \frac{1}{4}\bar{r}$, the parametric equations of a (simple) hypocycloid (the *astroid*, Fig. 4.26) are

$$x = r(3 \cos t + \cos 3t) , \quad y = r(3 \sin t - \sin 3t) ;$$

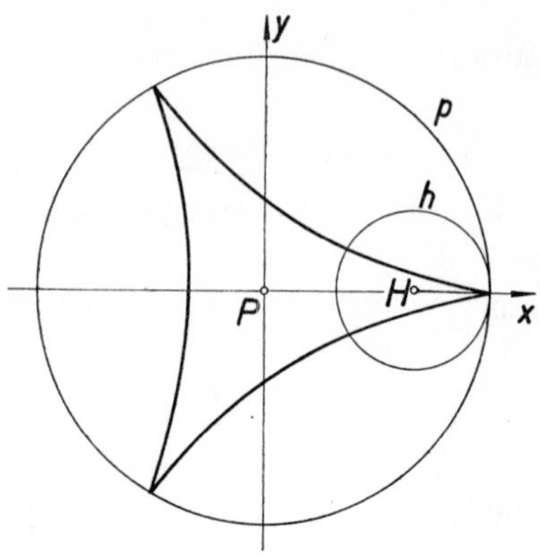

Fig. 4.25.

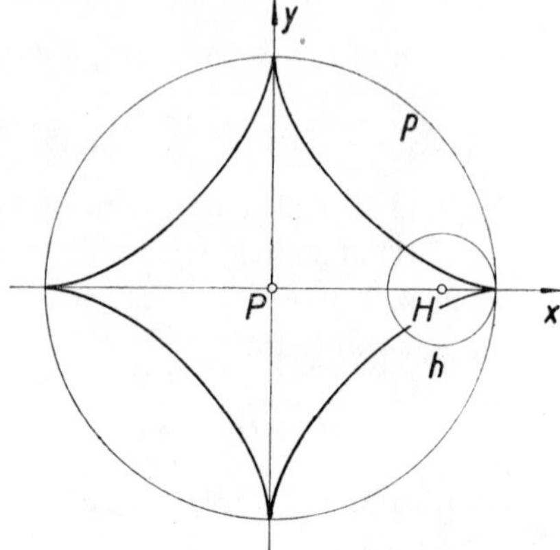

Fig. 4.26.

the equation of the curve in cartesian coordinates is

$$x^{2/3} + y^{2/3} = (4r)^{2/3} .$$

For the curves given in the above examples we can use Theorem 12 to determine the radius of curvature at any point or the length of an arc, in particular the radius of curvature at a vertex or the length of a branch.

(c) The involute of a circle

Definition 6. Any point of a plane rotating with a straight line h, rolling on a fixed circle p, describes an *involute of a circle.*

Theorem 14. *For a fixed circle* $p(0, r)$ *and a generating point* $A(r + d, 0)$ *the parametric equations of the involute are*

$$x = (r + d) \cos t + rt \sin t , \quad y = (r + d) \sin t - rt \cos t ,$$

where t *is the angle between the* x*-axis and the radius of the circle* p *perpendicular to the position of the straight line* h.

Definition 7. For $d = 0$, the (simple, general, normal) *circular involute* is generated, for $d > 0$ (the generating point and the circle p are on opposite sides of the straight line h) a *curtate involute* is generated, for $d < 0$ (the generating point and the circle p lie on the same side of the straight line h) we get a *prolate involute.*

Construction 4 of points of a (simple) circular involute (Fig. 4.27): We divide the circumference of the given circle p into a certain number of equal parts (for example, into 12) by the points *A, 1, 2, …*; we rectify the arc corresponding to one part and then we determine on the tangent to the circle p at every point of subdivision the

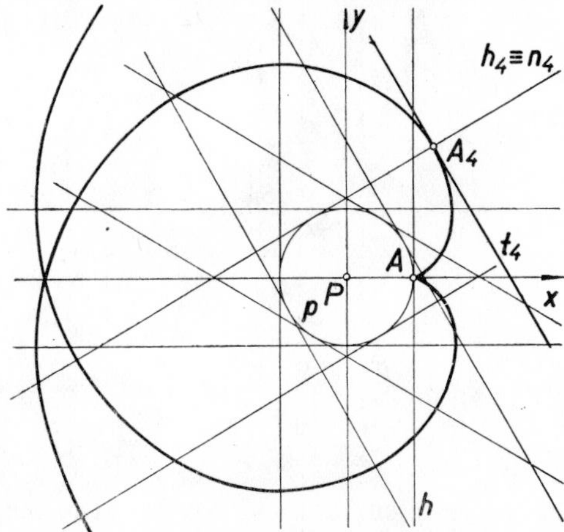

Fig. 4.27.

point at a distance equal to the length of the corresponding number of arcs: at the point *1* at a distance of one arc, at the point *2* of two arcs, etc.

Theorem 15. *The normals to a circular involute are tangents to the fixed circle p which is therefore an involute of the given curve* (see Definition 9.8.3, p. 335) (*and consequently, the locus of centres of curvature*).

Theorem 16. *For the radius of curvature of a circular involute we have*

$$R = rt\,,$$

and for its length of arc

$$s = \tfrac{1}{2}rt^2\,.$$

Example 7. For $d = -r$, a prolate involute, the *spiral of Archimedes* (see § 4.7, Fig. 4.29) is generated; its parametric equations are

$$x = rt\sin t\,, \quad y = -rt\cos t\,;$$

the equation in polar coordinates is

$$\varrho = r\varphi\,.$$

(d) Construction of centres of curvature of cyclic curves

Construction 5 at a point M which is not a vertex of the curve (Fig. 4.28): The centre of curvature S_M lies on the normal $n \equiv SM$. We construct a perpendicular k through the point S to the normal n and find the point of intersection *1* of the lines k, MH. Then S_M is the intersection of n and $1P$.

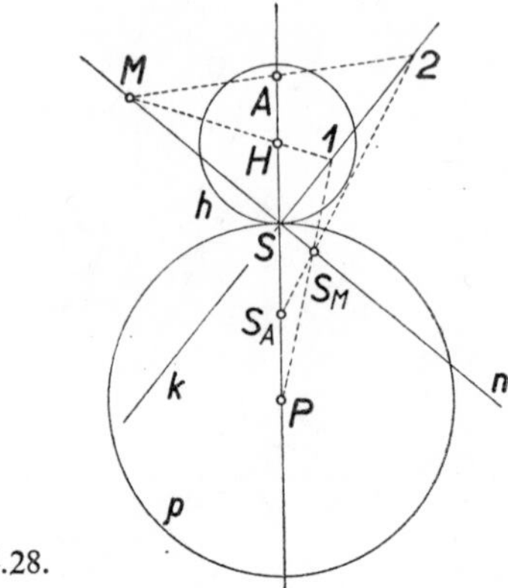

Fig. 4.28.

Construction 6 of the centre of curvature at a vertex A of the curve (i.e. at a point lying on PH) (Fig. 4.28): Using Construction 5, we construct the centre of curvature S_M for an arbitrary point M of the curve. We determine the point of intersection *2* of k and MA, and then S_A is the intersection of $2S_M$ and HP.

4.7. Spirals

Definition 1. A curve generated by a point moving uniformly along a polar radius rotating uniformly around its pole is called a *spiral of Archimedes* (Fig. 4.29).

Construction 1 of points of a spiral of Archimedes (Fig. 4.29): After one revolution, the distance of the moving point M from the origin O is equal to r_0. We divide the angle 2π and the segment r_0 (in the figure, $r_0 = \overline{OM_{12}}$) into n (say, 12) equal parts. Starting at the origin O, we successively mark off segments of lengths r_0/n, $2r_0/n$, ... on the corresponding polar radii. The end points of the segments are points of a spiral of Archimedes.

Theorem 1. *The equation of a spiral of Archimedes in polar coordinates is*

$$\varrho = \frac{r_0}{2\pi}\,\varphi = a\varphi$$

(*where* r_0, *and hence* $a = r_0/2\pi$, *is a given constant*).

The acute angle between a tangent to the curve and the polar radius at the point of contact increases with the polar radius and converges to the value $\frac{1}{2}\pi$.

Theorem 2. *The length of a polar sub-normal* s_n *is constant and equal to* a. (*The polar sub-normal is the segment between the pole and the point N of intersection of the normal n at the point M under consideration with the perpendicular constructed at the pole O to the polar radius.*)

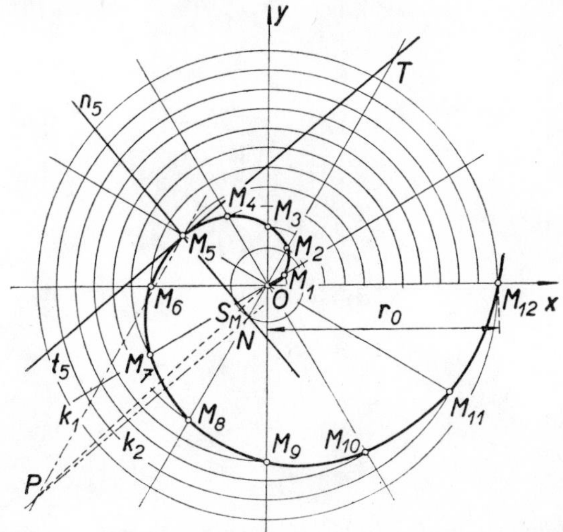

Fig. 4.29.

At a given point of a spiral of Archimedes we construct its normal and tangent by using Theorem 2.

The segment $\overline{OT} = s_t$, where T is the point of intersection of the tangent t at a point M of the curve and a perpendicular constructed at the pole to the polar radius of the point, is called a *polar sub-tangent*. Thus, for a spiral of Archimedes the equation

$$s_t = \frac{\varrho^2}{s_n} = \frac{\varrho^2}{a} = a\varphi^2$$

holds.

Construction 2 of the centre of curvature at a point of a spiral of Archimedes (Fig. 4.29): The perpendicular k_1 constructed at the point M to its polar radius, meets the perpendicular k_2 constructed at the point N to the normal n, at a point P. The centre of curvature S_M is the intersection of n and PO.

Definition 2. The arc of a spiral of Archimedes for which $2(n-1)\pi \leqq \varphi < 2n\pi$, is called the *n-th coil of the curve.*

Theorem 3. *The individual coils of a spiral of Archimedes are equidistant curves.*

Since the motion of a generating point on the polar radius can be accomplished in two directions, a spiral of Archimedes has two branches symmetrical about the x-axis.

Definition 3. A curve for which (in polar coordinates) the product of the length of the polar radius and the argument is constant, is called a *hyperbolic spiral or reciprocal spiral* (Fig. 4.30).

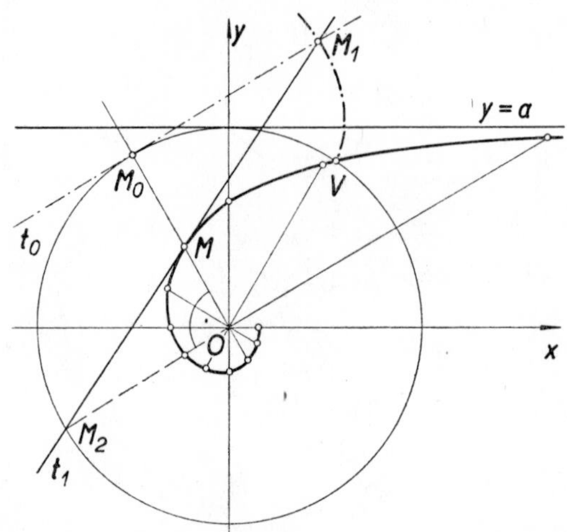

Fig. 4.30.

Theorem 4. *The equation of a hyperbolic spiral in polar coordinates is*

$$\varrho = \frac{a}{\varphi} \quad (\textit{where } a \textit{ is constant}) .$$

If $a < 0$, then also $\varphi < 0$ and we get the second branch (not illustrated in the figure) which is symmetrical to the first one about the x-axis.

Theorem 5. *The straight line $y = a$ is an asymptote of the hyperbolic spiral; the pole O is an asymptotic point.*

Theorem 6. *For a hyperbolic spiral, the length of the polar sub-tangent s_t is constant and equal to a.*

Construction 3 of points and of tangents to a hyperbolic spiral (Fig. 4.30): By Theorem 6, the end points of polar sub-tangents lie on the circle $k(O, a)$. The point V of the circle k, where the polar radius makes with the axis an angle $\varphi = 1$ (in circular measure) is also a point of the hyperbolic spiral. On an (arbitrary) radius OM_0 we find a point M of the hyperbolic spiral and the tangent in the following way: We connect the point of intersection M_1 of the tangent t_0 to the circle k at the point M_0 and the evolute e of the point V of the circle k, with the point M_2 of the polar sub-tangent and thus get a tangent t to the hyperbolic spiral, which meets the polar radius OM_0 at the point of contact M.

Definition 4. A curve making a constant angle with the polar radii at its points is called a *logarithmic* (*equiangular*, *logistic*) *spiral* (Fig. 4.31). It can also be characterized as the curve whose length of arc between a fixed and a variable point is proportional to the polar radius of the latter point.

Theorem 7. *The equation of a logarithmic spiral in polar coordinates is*

$$\varrho = a\mathrm{e}^{b\varphi} ,$$

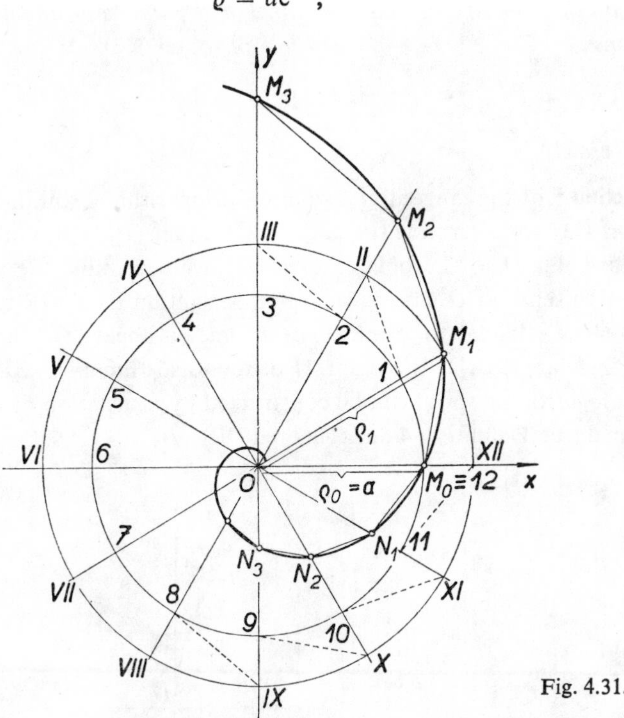

Fig. 4.31.

where $a, b > 0$ *are constant,* φ *is the angle (in radians) between the polar radius and the polar axis and* e *is the base of natural logarithms.*

Theorem 8. *If the angles* φ *form an arithmetic progression, then the corresponding polar radii* ϱ *form a geometric progression.*

Theorem 9. *The pole O is an asymptotic point of a logarithmic spiral. For* $\varphi = 0$, $\varrho_0 = a$.

Construction 4 of points of a logarithmic spiral (Fig. 4.31): We divide the angle 2π into n (say, 12) equal angles (Fig. 4.31) and calculate two adjacent polar radii $\varrho_0 = a$, $\varrho_1 = a\mathrm{e}^{b(\pi/6)}$. The triangles OM_0M_1, OM_1M_2, ... are similar. We describe circles k_0, k_1 with radii ϱ_0, ϱ_1 about the pole O and mark on them the points M_0, *1*, *2*, ..., and *XII*, M_1, *II*, ..., determined by the polar radii. Then, the line through the point M_1 parallel to the line joining 1, *II* meets the polar radius ϱ_2 at the point M_2 of the logarithmic spiral, the line through the point M_2 parallel to the line joining *2*, *III* meets the polar radius ϱ_3 at the point M_3, etc. Similarly, the point N_1 of the polar radius corresponding to the angle $\varphi = -\frac{1}{6}\pi$ can be obtained as the point of intersection of the polar radius and the straight line through the point M_0 drawn parallel to the line joining *XII*, *11*, etc.

Theorem 10. *The tangent at a point of a logarithmic spiral makes with its polar radius an angle* ϑ *satisfying* $\tan \vartheta = 1/b$. *For a polar sub-tangent, or sub-normal, the relations*

$$s_t = \frac{\varrho}{b}, \quad s_n = b\varrho$$

hold, respectively.

Construction 5 of the tangent at a point of a logarithmic spiral (Fig. 4.32): On the polar radius OM we determine the point Q such that $\overline{OQ} = 1$. On the perpendicular through the point O to the polar radius OM we determine the point R such that $\overline{OR} = 1/b$ (the sense of OR being such that a rotation from OR to OM is positive). The angle OQR is equal to ϑ and hence the line parallel to QR through the point M is the required tangent t at the point M of the logarithmic spiral. Other tangents at points of a logarithmic spiral can be constructed by translation of the constant angle ϑ so obtained (see Definition 4 and Theorem 10).

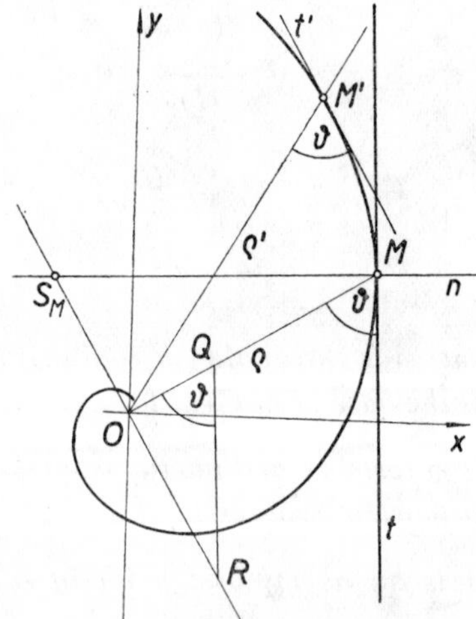

Fig. 4.32.

Theorem 11. *The radius of curvature at a point of a logarithmic spiral is*

$$R = \varrho \sqrt{(1 + b^2)}$$

and it is equal to the length of the polar normal. The centre of curvature lies at the point of intersection of the normal and the perpendicular through the point O to the polar radius of the point.

Definition 5. Curves satisfying the polar equation

$$\varrho^m = a^m \sin m\varphi$$

are called *sinusoidal spirals.*

Theorem 12. *By a rotation of the coordinate system through an angle* $\psi = (\frac{1}{2}\pi/m) - \varphi$ *the equation of a sinusoidal spiral becomes*

$$\varrho^m = a^m \cos m\psi \, .$$

Theorem 13. *For rational m, sinusoidal spirals are algebraic curves; for irrational m, they are transcendental curves.*

Example 1. For special values of m, we get the following sinusoidal spirals:

(a) $m = 1$; $\varrho = a \cos \varphi$, the sinusoidal spiral is a circle given by the equation $x^2 + y^2 = ax$;

(b) $m = 2$; $\varrho^2 = a^2 \cos 2\varphi$, the sinusoidal spiral is a *lemniscate of Bernoulli* satisfying the equation $(x^2 + y^2)^2 = a^2(x^2 - y^2)$ (see § 4.11);

(c) $m = -1$; $\varrho = a/\cos \varphi$, the sinusoidal spiral is a straight line given by the equation $x = a$;

(d) $m = -2$; $\varrho^2 = a^2/\cos 2\varphi$, the sinusoidal spiral is a rectangular hyperbola satisfying the equation $x^2 - y^2 = a^2$;

(e) $m = \frac{1}{2}$; $\varrho = a \cos^2 \frac{1}{2}\varphi$, the sinusoidal spiral is a cardioid given by the equation $\varrho = 2r(1 + \cos \varphi)$, which can be obtained by use of the relation $2 \cos^2 \frac{1}{2}\varphi = 1 + \cos \varphi$ on putting $a = 4r$.

(f) $m = -\frac{1}{2}$; $\varrho = a/\cos^2 \frac{1}{2}\varphi$, the sinusoidal spiral is a parabola satisfying the equation $y^2 = 4a(a - x)$.

4.8. The Clothoid (Cornu Spiral)

Definition 1. A curve whose radius of curvature R at a point M is inversely proportional to the length s of the arc between this point and a fixed point O is called the *clothoid or Cornu Spiral* (Fig. 4.33).

Theorem 1. *The intrinsic equation of a clothoid* (see Definition 9.4.3 and Remark 9.4.10, p. 319) *is*

$$R = \frac{a^2}{s} \, .$$

Theorem 2. *Parametric equations of a clothoid with the arc s as a parameter are given by the Fresnel integrals* (§ 13.12)

$$x = \int_0^s \cos \frac{s^2}{2a^2} \, ds \, , \quad y = \int_0^s \sin \frac{s^2}{2a^2} \, ds \, .$$

If the angle $\varphi = \frac{1}{2}s^2/a^2$ of the tangent at the point under consideration is taken as a parameter, then the equations are of the form

$$x = \frac{a}{\sqrt{2}} \int_0^{\varphi} \frac{\cos \varphi}{\sqrt{\varphi}} \, d\varphi \,, \quad y = \frac{a}{\sqrt{2}} \int_0^{\varphi} \frac{\sin \varphi}{\sqrt{\varphi}} \, d\varphi \,.$$

If $\varphi = \frac{1}{2}\pi t^2$, then the parametric equations have the form

$$x = a \sqrt{\pi} \int_0^t \cos \frac{\pi t^2}{2} \, dt \,, \quad y = a \sqrt{\pi} \int_0^t \sin \frac{\pi t^2}{2} \, dt \,.$$

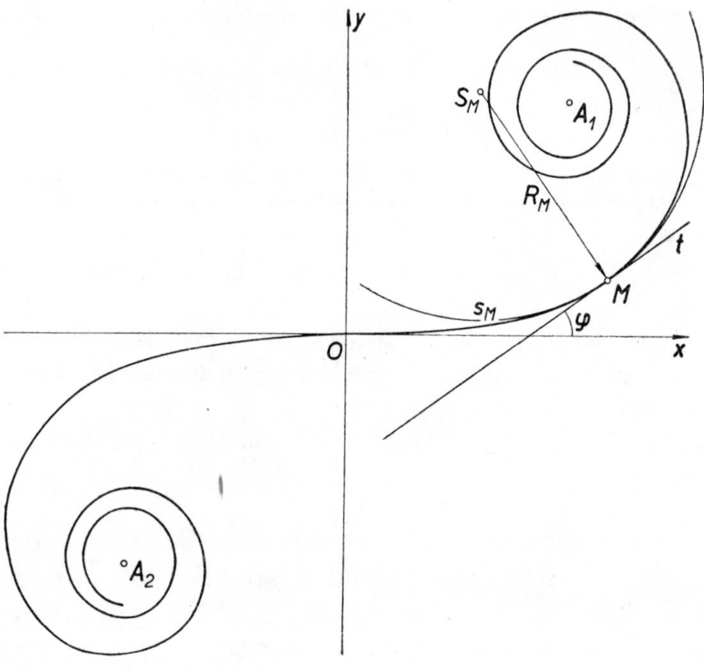

Fig. 4.33.

Theorem 3. *A clothoid is symmetrical about the point O which is a point of inflexion and it touches the x-axis at this point.*

Theorem 4. *The points $(\frac{1}{2}a\sqrt{\pi}, \frac{1}{2}a\sqrt{\pi})$, $(-\frac{1}{2}a\sqrt{\pi}, -\frac{1}{2}a\sqrt{\pi})$ are asymptotic points for a clothoid.*

Theorem 5. *The tangents of a clothoid at points for which*

$$\frac{s^2}{2a^2} = k\pi \quad (k = 0, 1, 2, \ldots)$$

are parallel to the x-axis, and tangents at the points for which

$$\frac{s^2}{2a^2} = \frac{2k+1}{2}\pi \quad (k = 0, 1, 2, \ldots)$$

are parallel to the y-axis.

Theorem 6. *For the angle φ made by a tangent to the curve and the x-axis, Theorem 2 and Definition 1 yield*

$$\varphi = \frac{s}{2R}.$$

Theorem 7. *The following relations hold between the quantities a, s, R, φ (in circular measure):*

$$\text{(a)} \quad a = \sqrt{(sR)} = \frac{s}{\sqrt{(2\varphi)}} = R\sqrt{(2\varphi)};$$

$$\text{(b)} \quad s = \frac{a^2}{R} = 2\varphi R = a\sqrt{(2\varphi)};$$

$$\text{(c)} \quad R = \frac{a^2}{s} = \frac{s}{2\varphi} = \frac{a}{\sqrt{(2\varphi)}};$$

$$\text{(d)} \quad \varphi = \frac{s}{2R} = \frac{s^2}{2a^2} = \frac{a^2}{2R^2}.$$

For practical use, the Fresnel integrals are tabulated. The constant a is the *parameter* determining the relative magnitude of the curve. If, for example, $a = 200$, then all the longitudinal values of the corresponding clothoid are double the values for the parameter $a = 100$.

4.9. The Exponential Curve

Definition 1. The curve whose equation in cartesian coordinates is

$$y = ab^{cx} \left(\text{or} \quad x = \frac{1}{c}\log_b \frac{y}{a}\right),$$

where $a > 0$, $b > 0$, c are constant, is called the *exponential curve.*

For $b = e$, we get

$$y = ae^{cx} \left(\text{or} \quad x = \frac{1}{c}\ln \frac{y}{a}\right).$$

Theorem 1. *The curves $y = ab^{cx}$, $y = ab^{-cx}$ are symmetrical about the y-axis. Both curves have only positive ordinates y and pass through the point $A(0, a)$. The x-axis is their common asymptote.*

Theorem 2. *Three points $P(x, y)$, $P_1(x_1, y_1)$, $P_2(x_2, y_2)$ of the curve satisfy the following relation:*

$$\left(\frac{y}{y_1}\right)^{x_1 - x_2} = \left(\frac{y_1}{y_2}\right)^{x - x_1}.$$

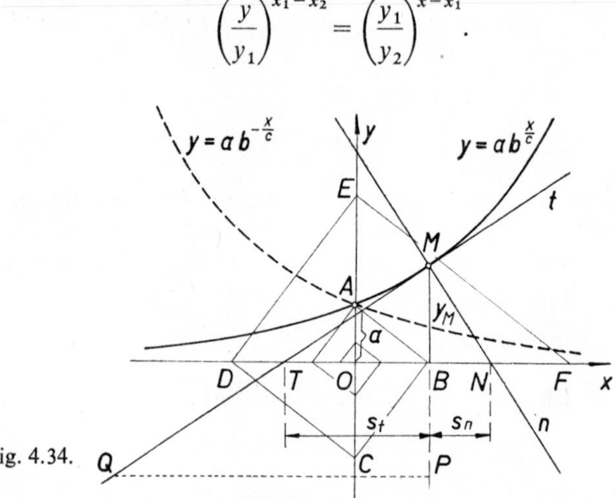

Fig. 4.34.

Construction 1 of points of the curve (Fig. 4.34): The ordinates y form a geometric progression if the abscissae x form an arithmetic progression. Hence we construct the points $A(0, a)$ and $B(ab^c, 0)$, and then draw successive perpendiculars $BC \perp AB$, $CD \perp BC, \ldots$; the segments $\overline{OA}$, $\overline{OB}$, $\overline{OC}, \ldots$ are now the ordinates of the points whose abscissae are $x = 0, 1, 2, \ldots$. By a reverse procedure we obtain the points of the curve for $x = -1, -2, \ldots$.

Theorem 3. *The sub-tangent of an exponential curve $y = ae^{cx}$ (with respect to the x-axis) has the constant value*

$$s_t = -\frac{1}{c}.$$

For the sub-normal we have $s_n = cy^2$. The length of the tangent is $t = \sqrt{(y^2 + 1/c^2)}$, and of the normal is $n = y\sqrt{(c^2y^2 + 1)}$.

For a curve $y = ab^{cx}$, the sub-tangent is $s_t = -1/(c \ln b)$.

Theorem 4. *The radius of curvature R of an exponential curve is given by the expression*

$$R = \frac{\sqrt{(y^2 + (1/c^2))^3}}{y}\, c = \frac{n^3}{c^2y^4} = \frac{n^3}{s_n^2}.$$

For the point for which $y = \sqrt{(2)}/2c$, the radius R is minimal and equal to $3\sqrt{(3)}/2c$.

Construction 2 of the radius of curvature of the curve $y = ab^{cx}$ at a point M (Fig. 4.34): Since $y/t = (s_n + s_t)/R$, we construct on the perpendicular through the point M to the x-axis, the point P such that $\overline{MP} = (s_n + s_t)$. The tangent t and the perpendicular to the y-axis at the point P meet at the point Q; then $R = \overline{MQ} = \overline{MS_M}$.

4.10. The Catenaries (Chainettes)

(a) The general catenary

Definition 1. A curve satisfying the equation

$$y = \tfrac{1}{2}a(e^{x/a} + e^{-x/a}), \quad \text{i.e.} \quad y = a \cosh \frac{x}{a}$$

is called a *general catenary* (Fig. 4.35).

A heavy homogeneous perfectly flexible cable suspended by two points assumes the form of a general catenary.

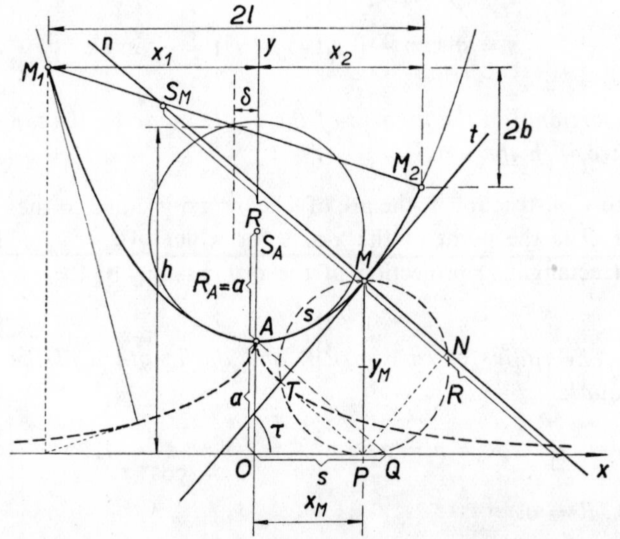

Fig. 4.35.

Theorem 1. *A general catenary is symmetrical about the y-axis, on which it has its vertex A at a distance a from the origin O; the value a is called the parameter of the catenary.*

Theorem 2. *A general catenary and the parabola $y = a + x^2/2a$ have three-point contact at the vertex $A(0, a)$. Also an ellipse with centre $S(0, 4a)$, major axis of length $6a$ on the y-axis and semi-minor axis $a\sqrt{3}$ has three-point contact with the general catenary at the common vertex A.*

Theorem 3. *By a translation of the origin of the coordinate system to the point of suspension M_1 (with the abscissa $-m$ in the original system) the equation of a general catenary becomes*

$$y = a\left(\cosh\frac{x-m}{a} - \cosh\frac{m}{a}\right).$$

Theorem 4. *The angle τ between a tangent and the x-axis satisfies (in the original system)*

$$\tan\tau = \sinh\frac{x}{a} = \frac{1}{a}\sqrt{(y^2-a^2)}\,,\quad \cos\tau = \frac{a}{y}\,.$$

Construction 1 of the tangent and the normal at a point M of a general catenary (Fig. 4.35): The circle on the ordinate MP of the point M and the circular arc about the point M, or P, of radius $r = a$ intersect at the points N, or T, respectively. The point T is a point of the tangent, the point N is a point of the normal at the given point M.

Theorem 5. *The arc s of a general catenary (measured from its vertex A) is*

$$s = a\sinh\frac{x}{a} = \sqrt{(y^2-a^2)} = a\tan\tau\,;$$

thus it is proportional to the tangent of the angle made by the tangent at the end point of the arc with the x-axis.

According to Construction 1, the arc of Theorem 5 is equal to the segment $\overline{MT} = \overline{OQ}$ (where Q is the point of the x-axis for which $AQ = y_M$); thus the arc is equal to the (rectangular) projection of the ordinate y_M of the point M onto the tangent.

Theorem 6. *The radius of curvature R and the length n of the corresponding normal are equal:*

$$R = n = a\cosh^2\frac{x}{a} = \frac{y^2}{a} = \frac{a}{\cos^2\tau}\,.$$

For the vertex, $R = a$.

Theorem 7. *The area enclosed by the x-axis, the y-axis, the arc of a general catenary and the ordinate of a given point is given by*

$$P = a^2\sinh\frac{x}{a} = as\,.$$

Example 1. The determination of the parameter a and the position of the axes, given the length of the cable $2s$, the horizontal distance between the points of suspension $2l$ and the difference of the heights $2b$ (Fig. 4.35).

To solve the problem we use the relations

$$\frac{\sqrt{(s^2 - b^2)}}{l} = \frac{\sinh u}{u}\,; \quad a = \frac{l}{u}\,.$$

Putting $c = (1/l)\sqrt{(s^2 - b^2)}$, we determine u from the equation $\sinh u = cu$ (for example, by using the tables of the function $\Theta(u) = \sinh u/u$) and hence the parameter a.

Then, the distance of the x-axis from the centre of the segment joining the points of suspension is

$$h = s \coth u\,.$$

The displacement of the y-axis in the direction of the lower point of suspension is $\delta = av$, where $\tanh v = b/s$.

The angles between the tangents at the points of suspension M_1, M_2 and the x-axis are $\tan \alpha_i = \sinh x_i/a$ $(i = 1, 2)$, where x_1, x_2 are the abscissae of the given points of suspension.

Theorem 8. *The involute of a catenary, called the tractrix, has the equation*

$$x = a \ln \frac{a - \sqrt{(a^2 - y^2)}}{y} + \sqrt{(a^2 - y^2)}\,.$$

Its tangent is of a constant length a.

The points of a general catenary can be constructed using tables of the hyperbolic cosine.

(b) The catenary of constant strength

Definition 2. The curve satisfying the equation

$$e^{y/a} \cos \frac{x}{a} = 1\,, \quad \text{i.e.} \quad y = -a \ln \cos \frac{x}{a}\,,$$

where $a > 0$ and where x satisfies the inequalities

$$a(4k - 1)\frac{\pi}{2} < x < a(4k + 1)\frac{\pi}{2} \quad (k \text{ an integer})\,,$$

is called a *catenary of constant strength.*

A heavy perfectly flexible and inelastic cable whose cross-section varies in such a manner that its resistance to breakage is constant, assumes, after being suspended, the form of a catenary of constant strength (Fig. 4.36).

Theorem 9. *A catenary of constant strength consists of an infinite number of congruent branches, touching the x-axis at the points $x = 2k\pi a$ and having the straight lines $x = a(4k \pm 1)\,\pi/2$ (k an integer) as asymptotes.*

Theorem 10. *The angle τ between a tangent at a point of a catenary of constant strength and the x-axis is proportional to the abscissa of the point of contact:*

$$\tau = \frac{x}{a}\,.$$

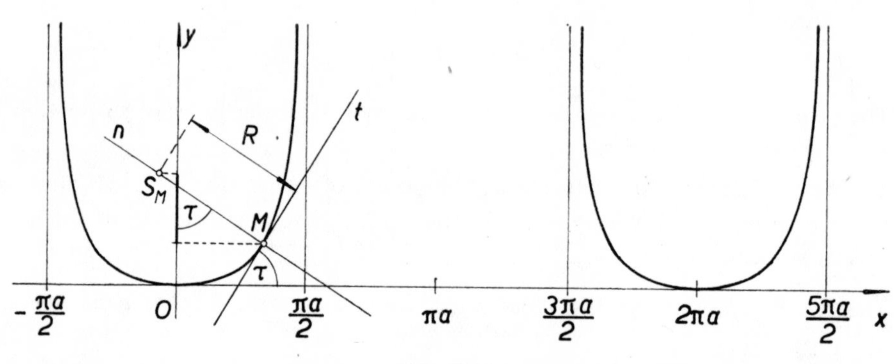

Fig. 4.36.

Theorem 11. *The radius of curvature at a point of the curve under consideration is*

$$R = \frac{a}{\cos \dfrac{x}{a}} = \frac{a}{\cos \tau}\,;$$

thus, since $R \cos \tau = a$, the rectangular projection of the radius of curvature on the y-axis is constant.

Theorem 12. *For an arc s of a catenary of constant strength the relation*

$$s = a \ln \tan\left(\frac{x}{2a} + \frac{\pi}{4}\right).$$

holds.

Theorem 13. *The area enclosed by a branch of a catenary of constant strength, by both asymptotes and by the x-axis is*

$$P = \pi a^2 \ln 2\,.$$

4.11. Examples of Some Algebraic Curves

Example 1 (The *cissoid of Diocles*; Fig. 4.37). We construct the tangent $t \parallel y$ to the circle k of diameter a, with the centre on the x-axis and passing through the origin O. We draw lines through O intersecting k at the points *1*, *2*, ..., and t at the points *1'*, *2'*, ... On every such line, we mark the point whose distance from the origin is

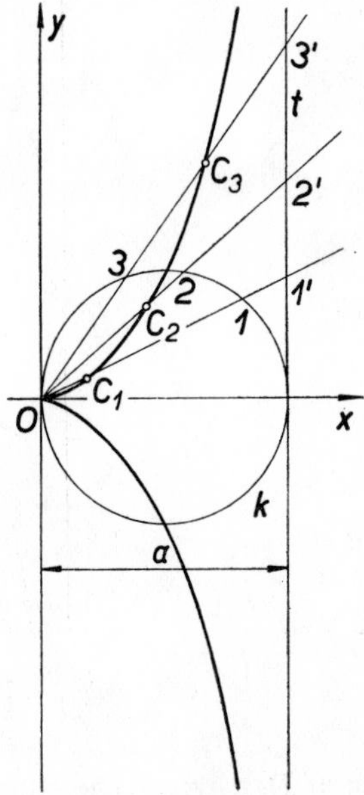

Fig. 4.37.

equal to the length of the segment determined by the points of intersection of the line with the circle and the tangent, i.e. $OC_1 = 11'$, $OC_2 = 22'$, ... Then $C_1, C_2, \ldots$ are the points of a cissoid of Diocles.

The equation of the curve in polar coordinates is

$$\varrho = a \sin \varphi \tan \varphi = a \frac{\sin^2 \varphi}{\cos \varphi},$$

and in cartesian coordinates

$$x(x^2 + y^2) - ay^2 = 0 \quad \left(\text{or} \quad y^2 = \frac{x^3}{a - x}\right).$$

The parametric equations are

$$x = \frac{at^2}{1 + t^2}, \quad y = \frac{at^3}{1 + t^2}.$$

A cissoid of Diocles is an algebraic curve of the third degree. The tangent t (of equation $x = a$) to the circle k is an asymptote of the curve.

A cissoid of Diocles is the orthogonal pedal curve of a parabola for a pole at the vertex of the parabola.

Example 2 (The *folium of Descartes*, Fig. 4.38). A cissoid of the ellipse $x^2 - xy + y^2 + a(x + y) = 0$, $a > 0$, with regard to the straight line $x + y + a = 0$ for

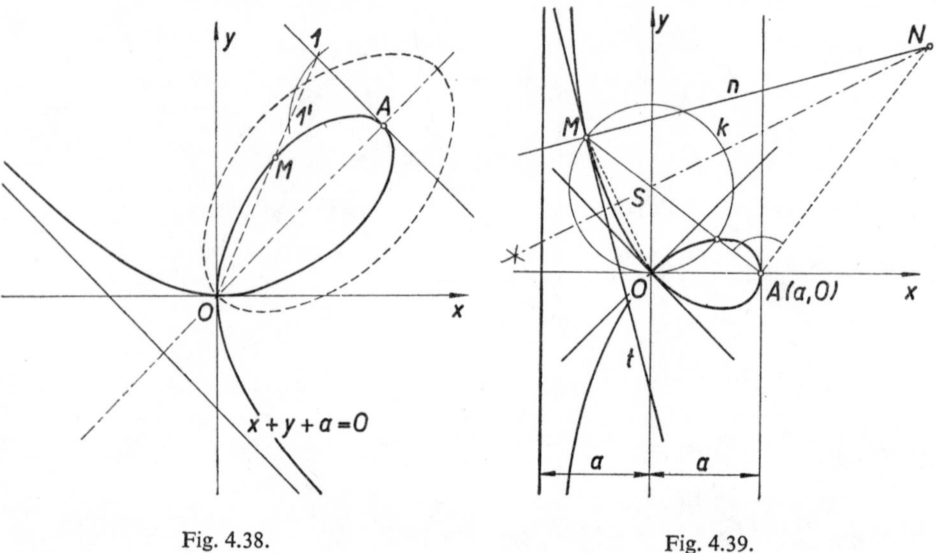

Fig. 4.38. Fig. 4.39.

the pole O is called the *folium of Descartes*. The equation of this curve in cartesian coordinates is

$$x^3 + y^3 - 3axy = 0\,,$$

and in polar coordinates

$$\varrho = \frac{3a \sin \varphi \cos \varphi}{\sin^3 \varphi + \cos^3 \varphi}\,;$$

the parametric equations are

$$x = \frac{3at}{1 + t^3}\,, \quad y = \frac{3at^2}{1 + t^3}\,.$$

A folium of Descartes is a curve of the third degree symmetrical about the straight line $y = x$; at the point O, it has a node with the x-axis and y-axis as tangents; at the point $A(\frac{3}{2}a, \frac{3}{2}a)$, it has a vertex; the straight line $x + y + a = 0$ is its asymptote.

Construction 1 of points of a folium of Descartes: We draw a line through the pole O to meet the tangent constructed at the vertex A at the point 1. On this line we determine the point $1'$ such that $A1 = A1'$; we then construct the harmonic point M to the point 1 with regard to O and $1'$ (i.e. the cross-ratio $(O, 1', 1, M) = -1$). The point M is a point of the folium of Descartes.

Example 3 (The *strophoid*; Fig. 4.39). We intersect the pencil of circles having the x-axis as a common tangent, with the common point of contact at the origin, by the diameters drawn through the point $A(a, 0)$. The end points of the diameters lie on a (straight) strophoid, whose equation in polar coordinates is

$$\varrho = a \frac{\cos 2\varphi}{\cos \varphi},$$

and in cartesian coordinates is

$$x(x^2 + y^2) - a(x^2 - y^2) = 0 \quad \left(\text{or} \quad y^2 = x^2 \frac{a - x}{a + x}\right);$$

the parametric equations are

$$x = \frac{a(1 - t^2)}{1 + t^2}, \quad y = \frac{at(1 - t^2)}{1 + t^2}.$$

The curve is symmetrical about the x-axis, it has a node with the tangents $y = \pm x$ at the origin O and the straight line $x + a = 0$ is its asymptote.

Construction 2 of the normal and tangent at a point M of a strophoid: The perpendicular bisector of the segment $\overline{OM}$ intersects the line through the point A perpendicular to AM at the point N of the normal n; having obtained the normal, the tangent at M can be determined.

Example 4 (The *lemniscate of Bernoulli*; Fig. 4.40). This curve is a rectangular pedal curve of the rectangular hyperbola $x^2 - y^2 = a^2$. Its equation is

$$(x^2 + y^2)^2 = a^2(x^2 - y^2),$$

or in the polar form

$$\varrho^2 = a^2 \cos 2\varphi.$$

Its parametric equations are

$$x = \frac{at(1 + t^2)}{1 + t^4}, \quad y = \frac{at(1 - t^2)}{1 + t^4}.$$

The vertices A_1, A_2 of the rectangular hyperbola are the vertices of the lemniscate, at the point O there is a double point (of inflexion) with the tangents $y = \pm x$ (which are the asymptotes of the hyperbola).

The lemniscate of Bernoulli is one of the *Cassinian ovals*, i.e. its points have a constant product (equal to $\frac{1}{2}a^2$) of their distances from two fixed points $(\pm\frac{1}{2}a\sqrt{2}, 0)$.

Construction 3 of points of a lemniscate of Bernoulli: We intersect the circle with centre O and radius $\frac{1}{2}a\sqrt{2}$ by a line, for example, from the vertex A_1, at the points C_1, C_2. Then $r_1 = \overline{A_1C_1}$, $r_2 = \overline{A_1C_2}$ are the focal radii of a point of the lemniscate of Bernoulli.

The polar form shows that φ is restricted to the intervals $(-\frac{1}{4}\pi, \frac{1}{4}\pi)$, $(\frac{3}{4}\pi, \frac{5}{4}\pi)$, and thus the curve lies within the right angles made by the tangents at the point O, containing the x-axis.

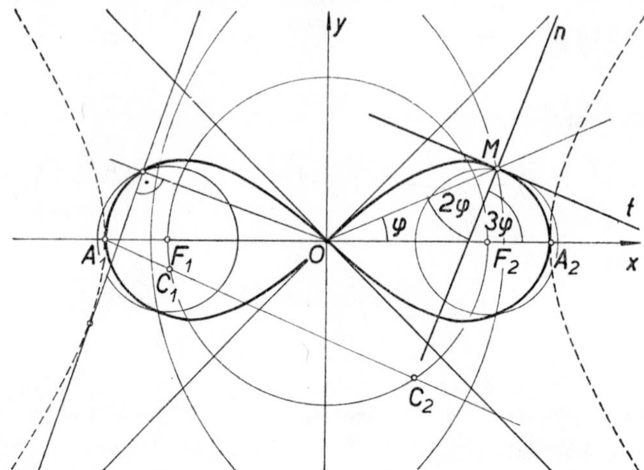

Fig. 4.40.

A lemniscate of Bernoulli has two axes of symmetry and, hence, it is radially symmetrical about their point of intersection. At the points whose coordinates are $(\pm\frac{1}{4}a\sqrt{6}, \frac{1}{4}a\sqrt{2})$, i.e. for which $\varphi = \pm\frac{1}{6}\pi, \frac{5}{6}\pi, \frac{7}{6}\pi$ and $\varrho = \frac{1}{2}a\sqrt{2}$, the tangents are parallel to the x-axis.

The angle between a tangent and the polar axis is equal to $\pm\frac{1}{2}\pi + 3\varphi$, the angle between a normal and the polar axis is equal to 3φ, and the angle between a normal and the polar radius is equal to 2φ.

Example 5 (The *conchoid of Nicomedes*; Fig. 4.41). We intersect a fixed straight line $x = a$ by a pencil of straight lines with vertex (pole) at O. On each line of the pencil we mark off segments of a constant length b on both sides of the point of intersection with the fixed straight line. The end point of the segments lie on a conchoid of Nicomedes whose equation in polar coordinates is

$$\varrho = \frac{a}{\cos\varphi} \pm b\,,$$

and in rectangular coordinates is

$$(x^2 + y^2)(x - a)^2 - b^2x^2 = 0\,.$$

A conchoid of Nicomedes consists of two branches; it is symmetrical about the x-axis, and the straight line $x = a$ is its asymptote. If $b > a$, then a branch has a node at O; if $b = a$, then it has a cusp at O; if $b < a$, then O is an isolated point.

The normals of all conchoids corresponding to the points of a given polar radius pass through a point N which is the point of intersection of the perpendicular to the polar radius at the pole O with the line parallel to the x-axis through the point of intersection P of the polar radius and the straight line $x = a$.

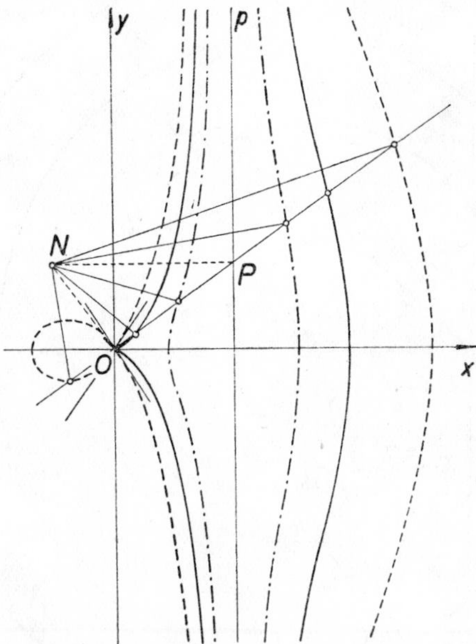

Fig. 4.41.

Example 6 (The *conchoid of a circle*; Fig. 4.42). We intersect the circle $\varrho = a \cos \varphi$ by a pencil of straight lines with the vertex (pole) at O and mark off segments of a constant length b on both sides from the point of intersection of a straight line of the pencil and the circle. The end points of the segments lie on a conchoid of the circle whose equation in polar coordinates is

$$\varrho = a \cos \varphi \pm b ,$$

and in rectangular coordinates is

$$(x^2 + y^2 - ax)^2 - b^2(x^2 + y^2) = 0 .$$

The conchoid of a circle (the *limaçon of Pascal*) is symmetrical about the x-axis, and has a double point at the pole O. (For $b < a$ the double point is a node, for $b = a$ it is a cuspidal point (the curve is a cardioid) and for $b > a$ it is an isolated point.) The equations of the tangents at the pole are

$$x\sqrt{(a^2 - b^2)} \pm by = 0\,;$$

the equation of the double tangent is

$$x + \frac{b^2}{4a} = 0\,;$$

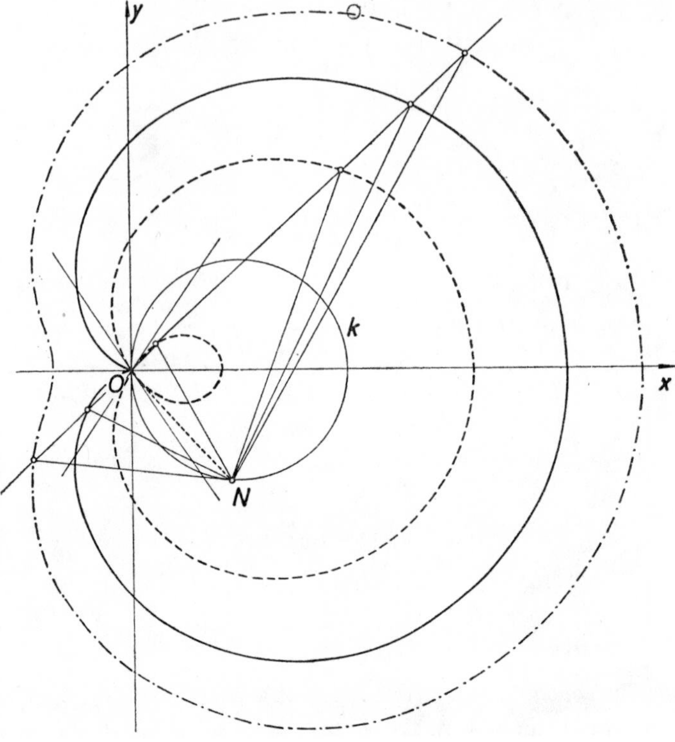

Fig. 4.42.

the points of contact have the ordinates

$$y = \frac{\pm b\sqrt{(4a^2 - b^2)}}{4a}.$$

The normals of all conchoids corresponding to the points of a given polar radius pass through a point N which is the point of intersection of the perpendicular erected to the polar radius at the pole O and the fixed circle.

4.12. The Sine Curves

To express periodical phenomena that repeat, without change, after a certain time, we use the *sine functions*. The least value of the constant, which, being added to the argument, does not change the value of the function, is called a *primitive period.*

Example 1. $y = \sin x$ (Fig. 4.43). The primitive period is 2π, the zero points $0, \pm\pi, \pm 2\pi, \ldots$ are points of inflexion of the curve and the tangents at these points make an angle of $\pm\frac{1}{4}\pi$ with the x-axis.

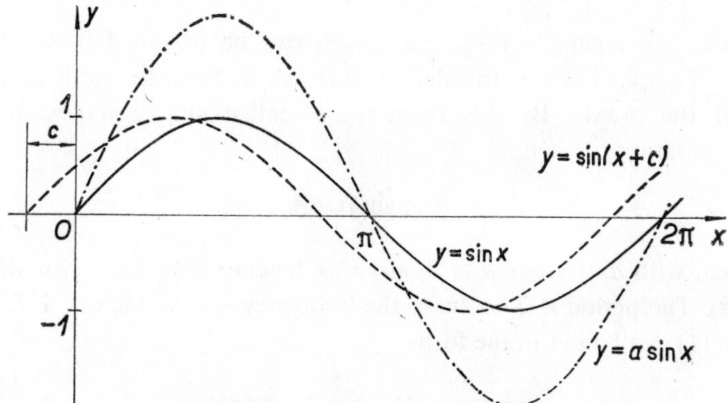

Fig. 4.43.

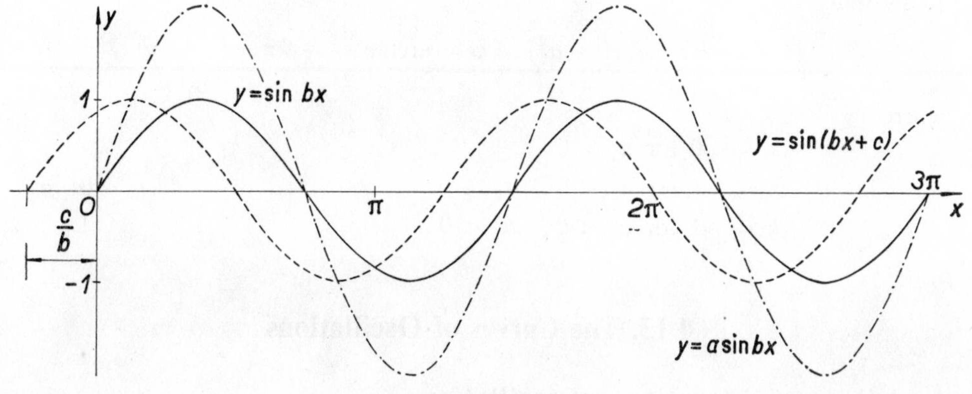

Fig. 4.44.

Example 2. $y = \sin(x + c)$. The graph can be obtained from the graph of Example 1 (Fig. 4.43) by a translation through a distance $-c$ in the direction of the x-axis.

Example 3. $y = a \sin x$ (Fig. 4.43). The graph can be obtained from the graph of Example 1 by multiplying the ordinates y by a.

Example 4. $y = \sin bx$ (Fig. 4.44). The primitive period is $2\pi/b$ and the zero points (of inflexion) are $x = \pm k\pi/b$ $(k = 0, 1, 2, \ldots)$. The tangents at these points have the direction of the hypotenuse of the right-angled triangle one side of which is on the x-axis and is of unit length, the other side being of length b. The graph can be obtained from the graph of Example 1 by a change of the x-coordinates in the ratio $1/b$. The coefficient b indicates the number of waves coming into the length 2π and is called the *circular (angular) frequency*.

Example 5. $y = a \sin bx$ (Fig. 4.44). The graph can be obtained from the graph of Example 4 by multiplying the ordinates y by a.

Example 6. $y = a \sin (bx + c)$. The graph can be obtained from the graph of Example 5 (Fig. 4.44) by a translation through a distance $-c/b$ in the positive direction of the x-axis. By this function, so-called simple *harmonic motion* is given. The notation

$$y = a \sin (\omega t + \varphi) \tag{1}$$

is often used, with *amplitude* $a > 0$, *circular frequency* ω and *phase displacement* φ $(|\varphi| < \pi)$. The period is $T = 2\pi/\omega$; the frequency is $n = \frac{1}{2}\omega/\pi = 1/T$.

Equation (1) can be put in the form

$$y = a_1 \sin \omega t + a_2 \cos \omega t\,, \tag{2}$$

where $a_1 = a \cos \varphi$, $a_2 = a \sin \varphi$. Conversely, if the form (2) is given, we get (1) by putting

$$a = \sqrt{(a_1^2 + a_2^2)}\,, \quad \varphi = \arctan \frac{a_2}{a_1} + k\pi\,,$$

where

$$\begin{aligned} k &= 0 \text{ for } a_1 > 0\,, \\ k &= 1 \text{ for } a_1 < 0\,, \quad a_2 > 0\,, \\ k &= -1 \text{ for } a_1 < 0\,, \quad a_2 < 0\,. \end{aligned}$$

4.13. The Curves of Oscillations

(a) Undamped (continuous) oscillations

(α) *Free undamped oscillations* are accomplished by a particle of mass m, on which a force Cy proportional to the displacement y from an equilibrium position is exerted; in dynamics, C is called the *spring constant*. The motion is given by the following differential equation:

$$\ddot{y} + \frac{C}{m} y = 0\,.$$

The solution is (see § 17·13 and equation (4.12.2)):

$$y = a \sin(\omega_0 t + \varphi), \tag{1}$$

where $\omega_0 = \sqrt{(C/m)}$ and a, φ are constants given by the initial conditions of the motion.

The composition of two harmonic motions

I. IDENTICAL CIRCULAR FREQUENCIES ω

$$a \sin(\omega t + \varphi_1) + b \sin(\omega t + \varphi_2) = A_1 \sin \omega t + A_2 \cos \omega t = A \sin(\omega t + \varphi), \tag{2}$$

where

$$A_1 = a \cos \varphi_1 + b \cos \varphi_2, \quad A_2 = a \sin \varphi_1 + b \sin \varphi_2, \quad A = \sqrt{(A_1^2 + A_2^2)},$$

$$\tan \varphi = \frac{A_2}{A_1}.$$

In particular, for equal amplitudes $b = a$ we get

$$a \sin(\omega t + \varphi_1) + a \sin(\omega t + \varphi_2) = 2a \cos \frac{\varphi_1 - \varphi_2}{2} \sin(\omega t + \varphi),$$

where

$$\varphi = \frac{\varphi_1 + \varphi_2}{2}. \tag{3}$$

Thus, in the case of equal frequencies the sum is a harmonic motion of the same frequency.

II. IDENTICAL AMPLITUDES, DIFFERENT FREQUENCIES

$$a \sin \omega_1 t + a \sin \omega_2 t = 2a \cos \frac{\omega_1 - \omega_2}{2} t \sin \frac{\omega_1 + \omega_2}{2} t. \tag{4}$$

(β) *Forced undamped (continuous) oscillation* is the motion of a particle of mass m under a periodically varying force $P \sin \omega t$ in addition to the force Cy. This motion satisfies the differential equation

$$\ddot{y} + \frac{C}{m} y = \frac{P}{m} \sin \omega t.$$

The solution is (see § 17.14):

$$y = Y \sin \omega t + a \sin (\omega_0 t + \varphi) \quad (\omega \neq \omega_0) \tag{5}$$

where $\omega_0 = \sqrt{(C/m)}$, a, φ are constants given by the initial conditions and

$$Y = \frac{P}{m(\omega_0^2 - \omega^2)}. \tag{6}$$

For $\omega = \omega_0$ (the case of *resonance*),

$$y = -\frac{P}{2m\omega_0} t \cos \omega_0 t + a \sin (\omega_0 t + \varphi). \tag{7}$$

As a rule, m, ω_0, P are fixed constants. The dependence of Y on ω expressed in (6) is illustrated in Fig. 4.45 (the *resonance curve*).

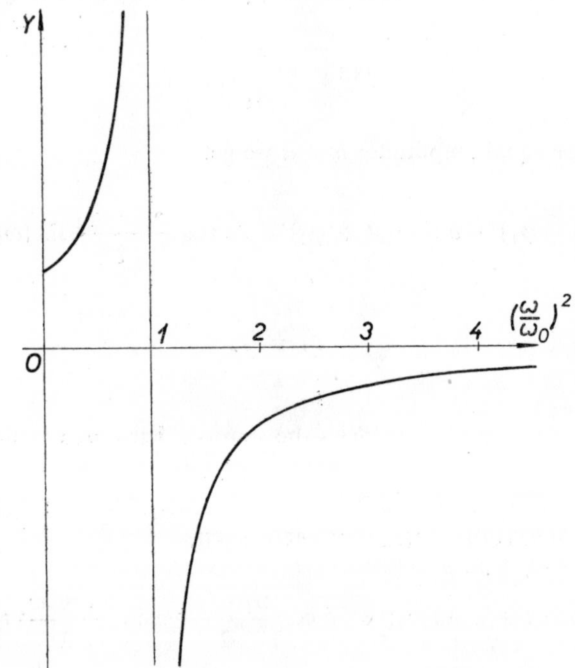

Fig. 4.45

(b) Damped oscillations. The motion is retarded by a force F proportional to the velocity ($F = -k\dot{y}$).

(α) *Free damped oscillations.* The differential equation of the motion is

$$\ddot{y} + 2b\dot{y} + \omega_0^2 y = 0 \tag{8}$$

where $\omega_0 = \sqrt{(C/m)}$, $b = k/(2m)$. The solution depends on the roots of the auxiliary equation (cf. § 17.13)

$$\alpha^2 + 2b\alpha + \omega_0^2 = 0. \tag{9}$$

1. $\alpha_1 = \alpha_2 = -\omega_0 = -b$. The general solution is

$$y = e^{-\omega_0 t}(C_1 + C_2 t).$$

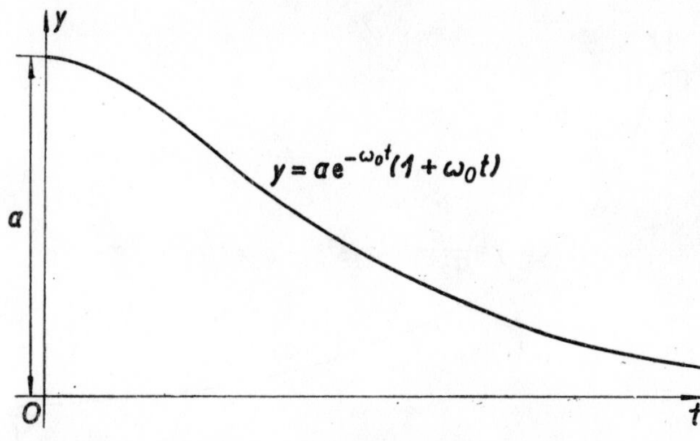

Fig. 4.46.

If, for $t = 0$, the initial conditions are $y = a$, $\dot{y} = 0$, then the solution is

$$y = ae^{-\omega_0 t}(1 + \omega_0 t).$$

This is the case of *critical damping*. For $t \to +\infty$ we have $y \to 0$ (Fig. 4.46).

2. If $b > \omega_0$, then the roots

$$\alpha_1 = -b + \sqrt{(b^2 - \omega_0^2)}, \quad \alpha_2 = -b - \sqrt{(b^2 - \omega_0^2)}$$

are real and distinct. The general solution is

$$y = C_1 e^{\alpha_1 t} + C_2 e^{\alpha_2 t};$$

if for $t = 0$,

$$y = a, \quad \dot{y} = 0,$$

then

$$y = \frac{a}{\alpha_1 - \alpha_2}(\alpha_1 e^{\alpha_2 t} - \alpha_2 e^{\alpha_1 t}).$$

This case is referred to as *supercritical damping* (Fig. 4.47). The motions 1, 2 are called *aperiodic*.

3. If $b < \omega_0$, then, writing

$$\omega_1 = \sqrt{(\omega_0^2 - b^2)},$$

the general solution is

$$y = e^{-bt}(C_1 \cos \omega_1 t + C_2 \sin \omega_1 t)\,.$$

If, for $t = 0$, the initial conditions are $y = a$, $\dot{y} = 0$, then

$$y = e^{-bt}\left(a \cos \omega_1 t + \frac{ab}{\omega_1} \sin \omega_1 t\right) = Ae^{-bt} \sin(\omega_1 t + \varphi) \qquad (10)$$

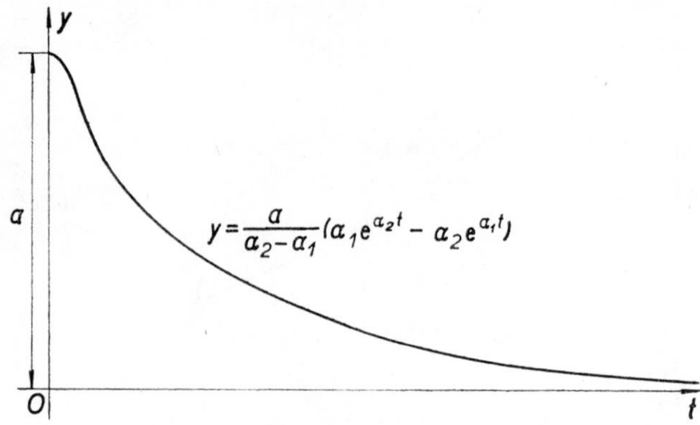

Fig. 4.47.

(cf. equation (4.12.2)). Here, the period $T = 2\pi/\omega_1$ is longer than in the case of an undamped oscillation. The ratio of the displacements y_1, y_2 at instants t_1, $t_1 + T$ is

$$\frac{y_1}{y_2} = e^{2\pi b/\omega_1}\,;$$

its natural logarithm $\vartheta = 2\pi b/\omega_1 = bT$ is called the *logarithmic decrement* of the motion.

The zero points of the curve (10) are obtained for $t_n = (n\pi - \varphi)/\omega_1$, the vertices for

$$t_n = \frac{\arctan(\omega_1/b) - \varphi + n\pi}{\omega_1}\,, \quad y_n = \pm \frac{A\omega_1}{\sqrt{(\omega_1^2 + b^2)}} e^{-bt_n}\,.$$

The curve (10) can be constructed (Fig. 4.48) by means of the enveloping curves

$$y_1 = Ae^{-bt} \quad \text{and} \quad \bar{y}_1 = -Ae^{-bt}$$

(see Construction 4.9.1) and by the curve

$$y_2 = \sin(\omega_1 t + \varphi)\,,$$

using the proportion

$$1 : y_1 = y_2 : y\,.$$

(β) *Forced damped oscillations.* The differential equation is

$$\ddot{y} + 2b\dot{y} + \omega_0^2 y = \frac{P}{m} \sin \omega t \tag{11}$$

(the notation as in equation (8)). The general solution (provided equation (9) has complex roots) is

$$\begin{aligned} y &= A_1 \sin \omega t + A_2 \cos \omega t + e^{-bt}(C_1 \cos \omega_1 t + C_2 \sin \omega_1 t) = \\ &= A \sin(\omega t + \varphi) + e^{-bt}(C_1 \cos \omega_1 t + C_2 \sin \omega_1 t) \end{aligned} \tag{12}$$

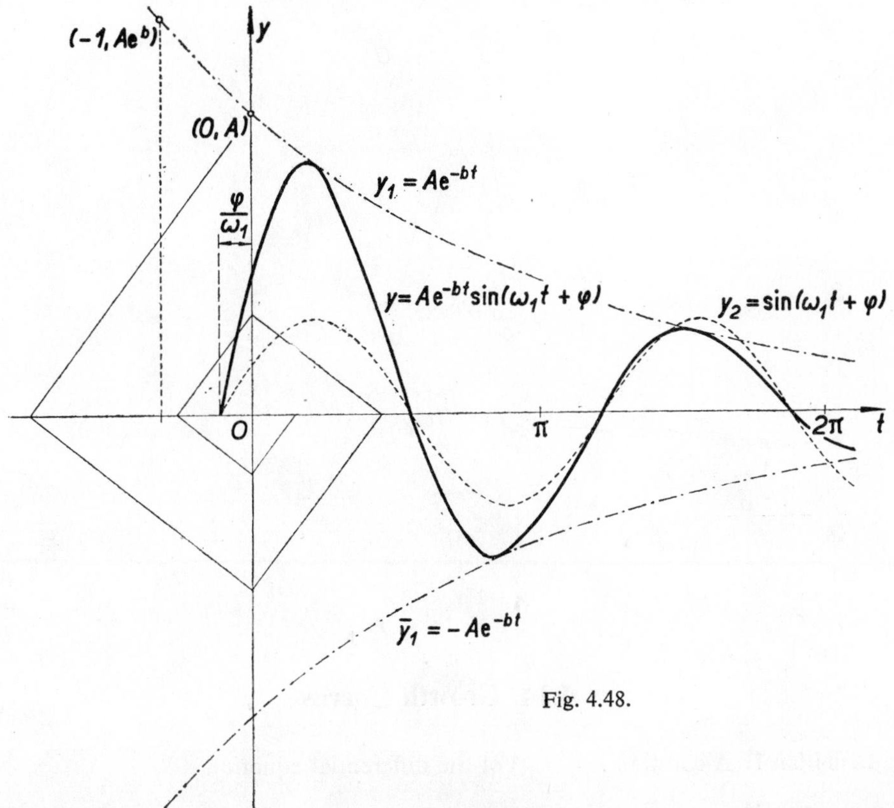

Fig. 4.48.

(cf. (4.12.2)). Here

$$\omega_1 = \sqrt{(\omega_0^2 - b^2)}, \quad A_1 = \frac{P(\omega_0^2 - \omega^2)}{m[(\omega_0^2 - \omega^2)^2 + 4b^2\omega^2]},$$

$$A_2 = \frac{-2Pb\omega}{m[(\omega_0^2 - \omega^2)^2 + 4b^2\omega^2]}, \tag{13}$$

$$A = \frac{P}{m\sqrt{[(\omega_0^2 - \omega^2)^2 + 4b^2\omega^2]}}.$$

Since $b > 0$, the second term of (12) becomes negligible, after a certain time, with regard to the first one, so that the motion is characterized only by the first term with the amplitude A. The magnitude of this amplitude depends on ω (P, m, ω_0, b are constants) and is illustrated in Fig. 4.49 (the *resonance curve*). The case $\omega = \omega_0$ is said to give *resonance*. For small b ($b \ll \omega_0$) A attains its maximum for $\omega \approx \omega_0$ (more precisely, if $\omega_0^2 > 2b^2$, for $\omega = \sqrt{(\omega_0^2 - 2b^2)}$). The first and second terms of the oscillation (12) are often called the *steady-state* and *transient oscillations*, respectively.

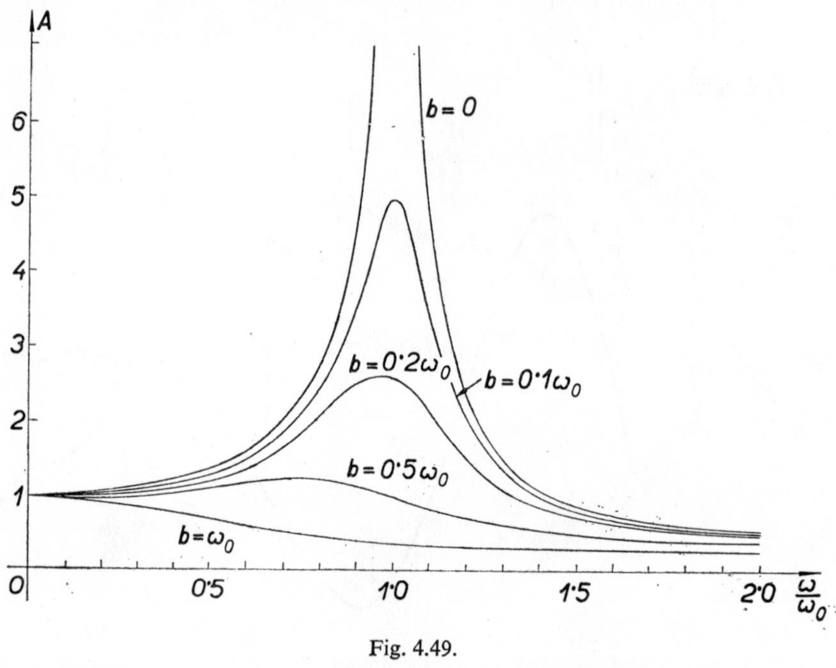

Fig. 4.49.

4.14. Growth Curves

Definition 1. A solution $x = F(t)$ of the differential equation

$$\frac{dx}{dt} = f(x) \tag{1}$$

is called the *law of growth* which is assigned to any phenomenon observed to satisfy the equation.

We make the following assumptions:

(a) the necessary parameters involved in $f(x)$ have been established for the phenomenon under consideration on the basis of statistical data;

(b) the growth of the quantity x in time t takes place without any external intervention;

(c) the initial condition $t = 0$, $x = x_0$ holds.

Example 1. If $f(x) = m = \text{const.}$, the solution of the differential equation (1) is

$$x = mt + x_0 \tag{2}$$

and the growth curve is the straight line of gradient $\tan\alpha = m$; this line passes through the point $(0, x_0)$.

Example 2. If $f(x) = ax + b$, $a \neq 0$, the solution is

$$x = -\frac{b}{a} + \left(x_0 + \frac{b}{a}\right) e^{at} \tag{3}$$

and the law of growth is given by the exponential curve passing through the point $(0, x_0)$.

If $b = 0$, $x_0 \neq 0$, the law (3) assumes the form (Fig. 4.50)

$$x = x_0 e^{at}, \quad \text{or (writing } c = -a \text{ when } a < 0) \quad x = x_0 e^{-ct}. \tag{4}$$

If $b \neq 0$, $x_0 = 0$, the law (3) takes the form (Fig. 4.51)

$$x = A(e^{at} - 1) \quad \text{or} \quad x = A(1 - e^{-ct}), \tag{5}$$

where $c = -a$ (if $a < 0$) and $A = b/a$ or $A = b/c$, respectively.

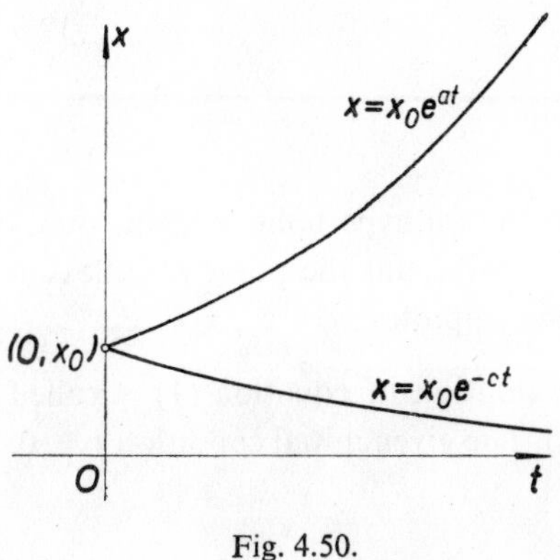

Fig. 4.50.

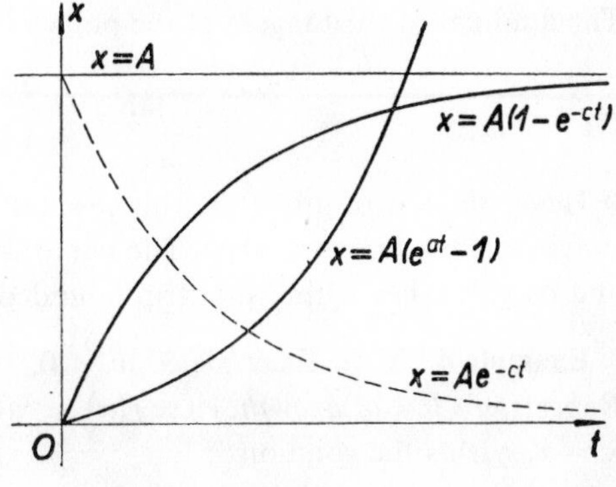

Fig. 4.51.

In Fig. 4.51 the auxiliary curve $x = Ae^{-ct}$ is also shown. The line $x = A = b/a$ is the asymptote of the second of curves (5) and it determines the limit of the evolution.

Example 3. If $f(x) = m + ax - bx^2 = b(x - x_2)(x_1 - x)$, $a > 0$, $b > 0$, $m > 0$ so that x_1, x_2 are the roots of the quadratic equation $f(x) = 0$, then the general solution of equation (1) is

$$\frac{x - x_2}{x_1 - x} = e^{(x_1 - x_2)(bt + c)}, \quad c = \text{const}. \tag{6}$$

The solution can be written in the form

$$x = x_2 + \frac{x_1 - x_2}{1 + Ce^{-(x_1 - x_2)bt}} \tag{7}$$

or

$$x = \frac{x_1 + x_2}{2} + \frac{x_1 - x_2}{2} \tanh \frac{x_1 - x_2}{2} (bt + c). \tag{8}$$

The straight lines $x = x_1$, $x = x_2$ are the asymptotes of integral curves (6). The ordinate of the point of inflexion of the curve is $\xi = \frac{1}{2}(x_1 + x_2)$; the abscissa τ of this point satisfies $b\tau + c = 0$. Using the given initial condition $x = x_0$, we get

$$\tau = \frac{1}{b(x_1 - x_2)} \ln \frac{x_1 - x_0}{x_0 - x_2}.$$

By translation of the origin to the point of inflexion (by means of the equations $X = x - \xi$, $T = t - \tau$) equation (8) assumes the form

$$X = \frac{x_1 - x_2}{2} \tanh \frac{x_1 - x_2}{2} bT. \tag{9}$$

The gradient of the tangent at the point of inflexion is

$$\tan \alpha = \frac{b}{4}(x_1 - x_2)^2.$$

Hence the law of growth in this case takes the form of a hyperbolic tangent, sometimes called a *logistic curve*. The curve is symmetrical about the point of inflexion and its graph lies within the strip bounded by the asymptotes $x = x_1$, $x = x_2$.

Example 4. If, in Example 3, $m = 0$, then the solution of equation (1) is called *Robertson's law of growth*. Here $f(x) = x(a - bx)$. The given initial condition $t = 0$, $x = x_0$ yields the solution

$$x = \frac{a}{b(1 + Ce^{-at})}, \quad \text{where} \quad C = \frac{a - bx_0}{bx_0}. \tag{10}$$

Using the coordinates $\xi = a/2b$, $\tau = (1/a) \ln [(a - bx_0)/bx_0]$ of the point of inflexion we get the form

$$x = \frac{a}{2b} + \frac{a}{2b} \tanh \frac{a}{2}(t - \tau); \tag{11}$$

on translating the origin we get

$$X = \frac{a}{2b} \tanh \frac{a}{2} T. \tag{12}$$

The asymptotes are $x = a/b$, $x = 0$; the gradient of the tangent at the point of inflexion is

$$\tan \alpha = \frac{a^2}{4b}.$$

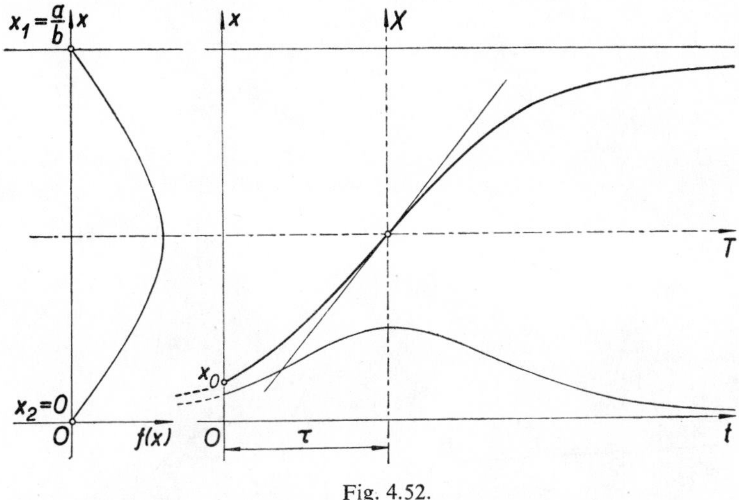

Fig. 4.52.

In Fig. 4.52 Robertson's law of growth together with the parabola $f(x) = ax - bx^2$ and the determination of the coordinates of the point of inflexion is illustrated; the curve is constructed in the coordinates T, X. The curve

$$\frac{dX}{dT} = \frac{a^2}{4b \cosh^2 \frac{1}{2}aT},$$

(also illustrated), shows the speed of growth.

4.15. Some Approximate Constructions

Construction 1 of the tangent t at a given point T of a curve k (Fig. 4.53): We draw secants through the point T meeting the curve in the points $\ldots, -2, -1, 1, 2, \ldots$ near T, and an arbitrary straight line q (not passing through T) in the points $\ldots, -2'$, $-1', 1', 2', \ldots$. Now, on the secants corresponding to the points $1, 2, \ldots$ we determine the points at the distances $T1$, $T2$, $\ldots$ from $1'$, $2'$, $\ldots$ on one side of the straight line q;

in a similar manner, we determine the points on the secants corresponding to -1, $-2, \ldots$ on the other side of q. These points determine the curve s intersecting the straight line q at the point O'. The tangent t is then the line $O'T$.

Construction 2 of the point of contact T of a tangent t constructed (by means of a ruler) from a point R to a curve k (Fig. 4.54): From the point R we construct a pencil of secants, which intersect the curve k in the neighbourhood of the required point T

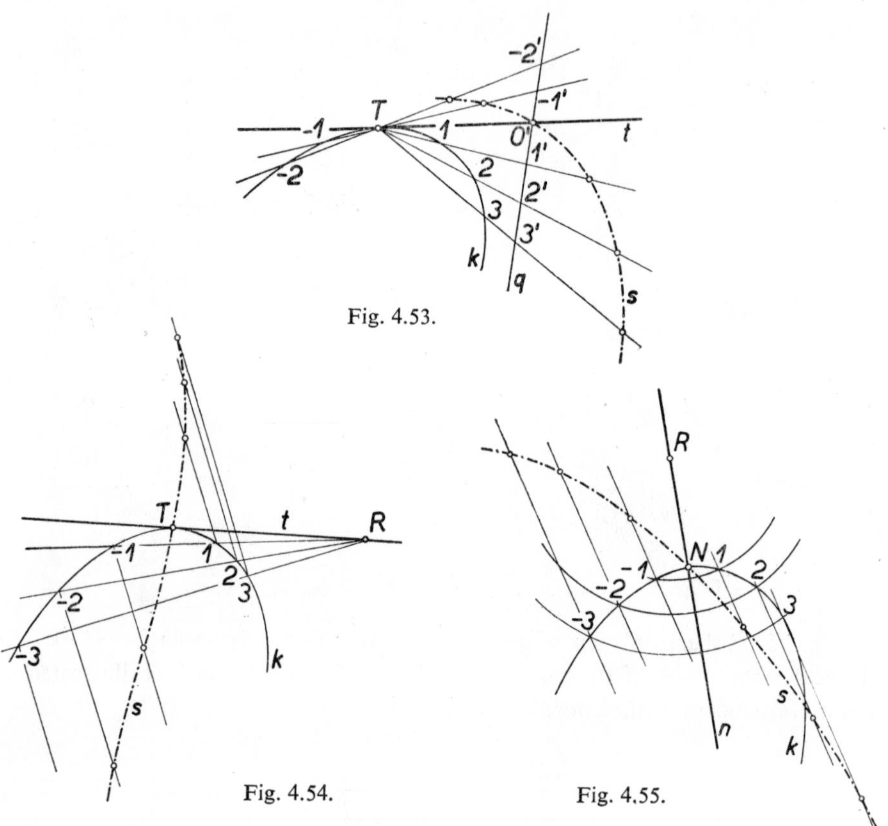

Fig. 4.53.

Fig. 4.54.

Fig. 4.55.

in pairs of points $-1, +1; -2, +2; \ldots$. On parallel lines drawn through these points we mark off points at distances equal to the length of the corresponding chords; on one side of the curve for the points marked by $+$ and on the other side for the points marked by $-$. If we join those points, the resulting curve s intersects the tangent t (and thus, also the curve k) at the point of contact T of the tangent t.

Construction 3 of a normal n from a point R to a curve k (Fig. 4.55): From the point R we describe concentric circles, each intersecting the curve at two points $-1, +1; -2, +2; \ldots$ in the neighbourhood of the foot N of the required normal. The curve s constructed in the same way as in Construction 2 intersects the curve k at the foot N of the required normal n from the point R.

5. PLANE ANALYTIC GEOMETRY

By Miloslav Zelenka

References: [19], [190], [263], [347], [360], [368], [378], [384], [394]

5.1. Coordinates of a Point on a Straight Line and in a Plane. Distance between Two Points

Definition 1. Let us divide a straight line p by a point O into two half-lines $+p$ and $-p$ (Fig. 5.1). Let us choose on p a unit of length. The *coordinate* x of a point M is defined to be the distance of the point M from the point O prefixed by a sign, plus or minus (the so-called *directed distance*) according as M belongs to $+p$ or $-p$, respectively. We write $M(x)$.

Remark 1. The position of a point M on a line p is uniquely determined by the coordinate x (and vice versa). We say that a *coordinate system* has been introduced on the line p. The point O is called the *origin of the coordinate system.*

N(-2) O M(3;6)
-4 -3 -2 -1 0 1 2 3 4 p

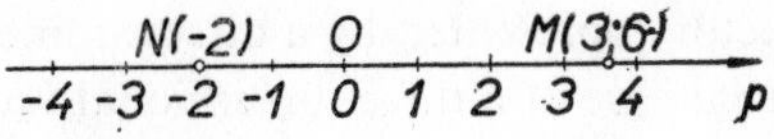

Fig. 5.1.

Theorem 1. *The distance d between two points $A(x_1)$ and $B(x_2)$ on a line is equal to*

$$d = |x_2 - x_1| . \tag{1}$$

Remark 2. In a similar way, a coordinate system can be introduced in a plane (Fig. 5.2): We select units of length on two intersecting lines, called *axes of coordinates*; the intersection of the lines is taken as the origin on each of them; we denote it by O and call it the *origin of the coordinate system in the plane.* A point M in the plane is then uniquely determined by its coordinates x, y (see Fig. 5.2), and vice versa.

If the axes x, y are mutually perpendicular, the coordinate system is called *rectangular.* As in the case of coordinates on a line, the *coordinate* x or y gives the *directed distance* of the point $M(x, y)$ from the coordinate axis y or x, respectively.

A rectangular system in which both axes have the same unit of length is called *cartesian*. Throughout this chapter – unless otherwise stated – we use the cartesian coordinate system.

The plane is divided by the coordinate axes into four parts called the *quadrants* (Fig. 5.2).

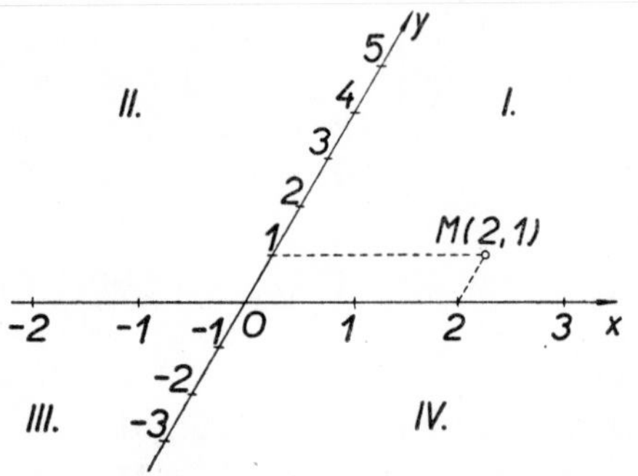

Fig. 5.2.

Theorem 2. *The distance between two points $A(x_1, y_1)$ and $B(x_2, y_2)$ in a cartesian coordinate system is equal to*

$$d = \sqrt{[(x_2 - x_1)^2 + (y_2 - y_1)^2]} . \tag{2}$$

REMARK 3. In § 7.1, the concept of a three-dimensional (real) vector is introduced. Similarly, a two-dimensional vector $\mathbf{a}$ with two components a_1, a_2 can be defined; the notation $\mathbf{a}(a_1, a_2)$ or $\mathbf{a} = (a_1, a_2)$ is used. As in the case of three-dimensional vectors, two-dimensional vectors can also be represented by directed line segments. If two-dimensional vectors are used in problems of analytical geometry in a plane, they are, of course, represented by directed segments lying in the plane.

A two-dimensional vector represented by a directed line segment in a plane xy is often considered as a special case of a three-dimensional vector, the third component of which is zero (although this is not written explicitly). Then we can, without any alterations, apply the definitions and theorems of Chap. 7 concerning operations on vectors in three-dimensional space. For example, the formula

$$\mathbf{a} . \mathbf{b} = a_1 b_1 + a_2 b_2$$

for the scalar product of two vectors $\mathbf{a}(a_1, a_2)$ and $\mathbf{b}(b_1, b_2)$ holds.

The number $\sqrt{(a_1^2 + a_2^2)}$ is called the *length* (or *magnitude*) *of the vector* $\mathbf{a}$ and is denoted by either $|\mathbf{a}|$ or a.

5.2. Division of a Line Segment in a Given Ratio. Area of a Triangle and Polygon

Theorem 1. *An arbitrary point $M(x, y)$ on the line segment with the end points $M_1(x_1, y_1)$ and $M_2(x_2, y_2)$ can be represented in vector form by*

$$\mathbf{m} = \mathbf{m}_1 + t(\mathbf{m}_2 - \mathbf{m}_1) \quad (0 \leqq t \leqq 1), \tag{1}$$

where $\mathbf{m}$, $\mathbf{m}_1$ and $\mathbf{m}_2$ are the radius vectors of the points M, M_1 and M_2, respectively, or in coordinate form by

$$x = x_1 + t(x_2 - x_1), \quad y = y_1 + t(y_2 - y_1) \tag{2}$$

$(0 \leqq t \leqq 1)$.

If the point $M(x, y)$ divides the line segment M_1M_2 in the ratio $M_1M : M_2M = \lambda < 0$, then, putting $t = -\lambda/(1 - \lambda)$, we obtain

$$x = \frac{x_1 - \lambda x_2}{1 - \lambda}, \quad y = \frac{y_1 - \lambda y_2}{1 - \lambda}. \tag{3}$$

If $\lambda = -1$, M is the midpoint of the line segment M_1M_2, and formulae (3) become

$$x = \frac{x_1 + x_2}{2}, \quad y = \frac{y_1 + y_2}{2}. \tag{4}$$

Theorem 2. *The area of a polygon with vertices $A_1(x_1, y_1)$, $A_2(x_2, y_2)$, ..., ..., $A_n(x_n, y_n)$ occuring in that order, is*

$$P = \tfrac{1}{2}\left|\begin{vmatrix} x_1, & x_2 \\ y_1, & y_2 \end{vmatrix} + \begin{vmatrix} x_2, & x_3 \\ y_2, & y_3 \end{vmatrix} + \ldots + \begin{vmatrix} x_n, & x_1 \\ y_n, & y_1 \end{vmatrix}\right|. \tag{5}$$

In particular, the area of the triangle with vertices $A_1(x_1, y_1)$, $A_2(x_2, y_2)$ and $A_3(x_3, y_3)$ is

$$P = \tfrac{1}{2}\left|\begin{vmatrix} x_1, & y_1, & 1 \\ x_2, & y_2, & 1 \\ x_3, & y_3, & 1 \end{vmatrix}\right|. \tag{6}$$

5.3. The Equation of a Curve as the Locus of a Point

Definition 1. The *equation of a curve* is the name given to the relation (equation) which is satisfied by the coordinates x, y of all the points lying on the given curve (and only those points).

In order to obtain the equation of a curve as a locus of a point having a given property, we proceed as follows:

1. we choose an arbitrary point M of the curve and denote its coordinates by (x, y);
2. we express the required property of points on the locus by an equation between x and y;
3. we arrange the equation in a simpler form, if possible, at the same time expressing all the quantities involved in terms of x, y and the given elements (constants).

Example 1. Let us obtain the equation of the locus of the point in a plane, which is always at a distance $d = 3$ from the point $S(-2, 1)$.

Thus,

1. $$M(x, y)\,; \tag{1}$$

2. $$\sqrt{[(x + 2)^2 + (y - 1)^2]} = 3\,; \tag{2}$$

3. we square equation (2) and obtain the equation $(x + 2)^2 + (y - 1)^2 = 9$, which needs no further rearrangement; it is the equation of a circle.

5.4. The Gradient, Intercept, General and Vector Forms of the Equation of a Straight Line. Parametric Equations of a Straight Line. Equation of the Straight Line through Two Given Points. The Point of Intersection of Two Straight Lines. Equation of a Pencil of Lines

$$y = kx + q \quad (\textit{the gradient form of the equation of a straight line})\,; \tag{1}$$

$$\frac{x}{p} + \frac{y}{q} = 1 \quad (p \neq 0,\ q \neq 0) \quad (\textit{the intercept form of the equation of a straight line})\,; \tag{2}$$

$$ax + by + c = 0 \quad (a^2 + b^2 > 0) \quad (\textit{the general equation of a straight line})\,; \tag{3}$$

$$\mathbf{m} = \mathbf{m}_1 + t\mathbf{a} \quad (\mathbf{a} \neq \mathbf{0}) \quad (\textit{the vector equation of a straight line})\,; \tag{4}$$

$$\left.\begin{aligned} x &= x_1 + a_1 t \\ y &= y_1 + a_2 t \end{aligned}\right\} \quad (\textit{the parametric equations of a straight line})\,. \tag{5}$$

The geometrical meaning of the constants involved in these equations can be seen from Fig. 5.3, 5.4, 5.5, 5.6; $k = \tan\varphi$ is the so-called *slope* (or *gradient*) of the line. For $k = 0$ the line is parallel to the x-axis. For $q = 0$ the line passes through the origin. The equation of the y-axis or a line parallel to the y-axis (i.e. if $\varphi = \frac{1}{2}\pi$) cannot be written in the form (1).

The numbers, p, q, in equation (2) (which may be positive or negative) are the so-called *intercepts on the axes*. A straight line which passes through the origin or is parallel to a coordinate axis cannot be written in the form (2). The constants a, b in equation (3) determine the vector $\mathbf{n}(a, b)$ perpendicular to the line (3). If $a = 0$, the line is parallel to the x-axis; if $b = 0$, it is parallel to the y-axis. The third parameter c is related to the distance of the line from the origin (see § 5.6); if $c = 0$, the line passes through the origin.

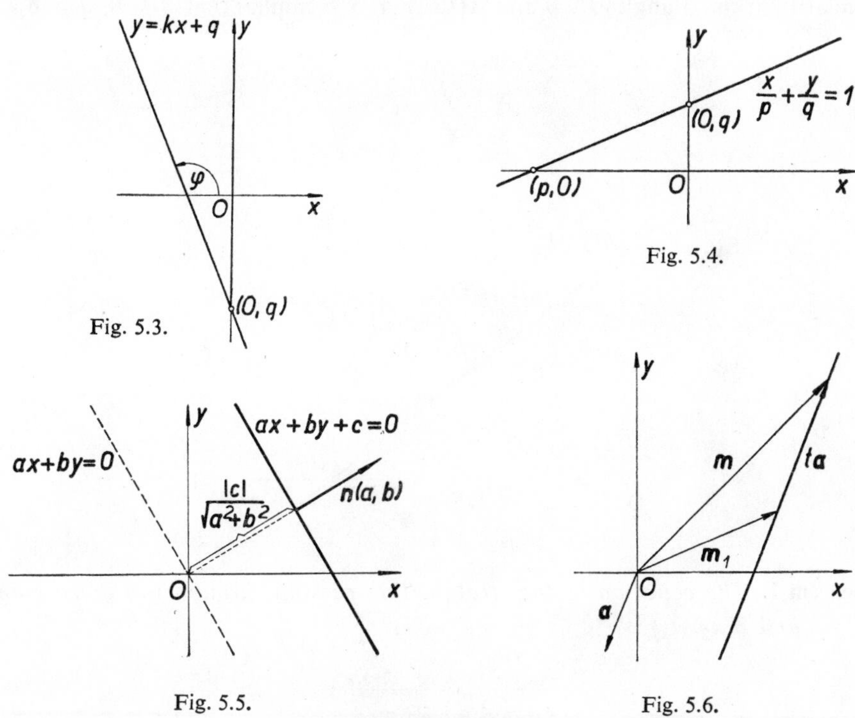

Fig. 5.3.

Fig. 5.4.

Fig. 5.5.

Fig. 5.6.

The line (4) passes through the point $M_1(x_1, y_1)$, the radius vector of which is denoted by $\mathbf{m}_1$; its direction is determined by the vector $\mathbf{a}(a_1, a_2)$ and t is a variable parameter $(-\infty < t < +\infty)$. To each particular value of t there corresponds a particular point $M(x, y)$ whose radius vector is $\mathbf{m}$ (Fig. 5.6). The vector equation is, in fact, a more concise version of the parametric equations (5).

Example 1. The straight line given by the parametric equations

$$x = 3 + 2t\,,$$

$$y = 1 - 3t$$

is to be expressed in the form (3).

Eliminating t from the parametric equations we obtain the required relation between x and y: adding three times the first equation to twice the second we obtain

$$3x + 2y = 11\,,$$

i.e.

$$3x + 2y - 11 = 0 .$$

Example 2. Find the equation of the straight line whose segment AB, intercepted by the positive semi-axes x and y, is bisected by the point $P(4, 3)$.

Similarity of the triangles PCB and AOB (Fig. 5.7) implies that $p = 8$, $q = 6$ and thus, by (2) the required equation is

$$\frac{x}{8} + \frac{y}{6} = 1 .$$

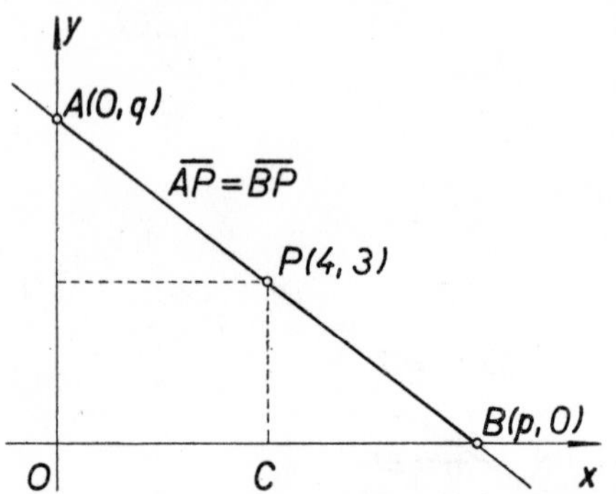

Fig. 5.7.

Theorem 1. *The equation of the straight line passing through two given points* $A(x_1, y_1)$ *and* $B(x_2, y_2)$ *is*

$$\frac{x - x_1}{x_2 - x_1} = \frac{y - y_1}{y_2 - y_1} ,$$

i.e.

$$y - y_1 = \frac{y_2 - y_1}{x_2 - x_1}(x - x_1) .$$

If $x_2 = x_1$, or $y_2 = y_1$, then the equation of the straight line is $x = x_1$, or $y = y_1$, respectively.

Example 3. The line passing through the points $A(-1, 5)$, $B(3, 7)$ has the equation

$$y - 5 = \frac{7 - 5}{3 + 1}(x + 1) , \quad \text{i.e.} \quad x - 2y + 11 = 0 .$$

The line passing through the points $A(3, 3)$, $B(3, 8)$ has, of course, the equation $x = 3$ (and is a line parallel to the y-axis).

Theorem 2. *The point of intersection* $P(x_0, y_0)$ *of two intersecting straight lines given by equations* $a_1x + b_1y + c_1 = 0$ *and* $a_2x + b_2y + c_2 = 0$ *can be found* *by solving these equations simultaneously; hence,*

$$x_0 = \frac{\begin{vmatrix} -c_1, & b_1 \\ -c_2, & b_2 \end{vmatrix}}{\begin{vmatrix} a_1, & b_1 \\ a_2, & b_2 \end{vmatrix}}, \quad y_0 = \frac{\begin{vmatrix} a_1, & -c_1 \\ a_2, & -c_2 \end{vmatrix}}{\begin{vmatrix} a_1, & b_1 \\ a_2, & b_2 \end{vmatrix}}, \quad \text{provided that} \quad \begin{vmatrix} a_1, & b_1 \\ a_2, & b_2 \end{vmatrix} \neq 0. \tag{6}$$

If $\begin{vmatrix} a_1, & b_1 \\ a_2, & b_2 \end{vmatrix} = 0$, *the lines are parallel or they coincide.*

Definition 1. The set of straight lines in a plane, all of which pass through one point S, is called a *pencil of lines.* The point S is called the *centre* (or *vertex*) *of the pencil of lines.*

Theorem 3. *The straight lines belonging to the pencil with centre* $S(x_0, y_0)$ *have as their equation either*

$$y - y_0 = k_1(x - x_0) \tag{7}$$

or

$$x - x_0 = k_2(y - y_0), \tag{7'}$$

where k_i $(i = 1, 2)$ *is a variable parameter; to each line of the pencil, there corresponds a unique value of* k_1 *or* k_2, *and conversely.*

Theorem 4. *The straight lines belonging to the pencil determined by the intersecting lines*

$$a_1x + b_1y + c_1 = 0 \quad \textit{and} \quad a_2x + b_2y + c_2 = 0$$

have as their equations:

$$\lambda_1(a_1x + b_1y + c_1) + \lambda_2(a_2x + b_2y + c_2) = 0, \tag{8}$$

where λ_i $(i = 1, 2)$ *are variable parameters not simultaneously equal to zero; to each line of the pencil there corresponds a unique ratio* λ_1/λ_2 or λ_2/λ_1, *and conversely.*

Definition 2. Equations (7), (7′) or (8) are called the *equations of a pencil of lines.*

Example 4. Let us find the equation of the straight line which passes through the point of intersection of the lines

$$2x - y + 3 = 0, \quad x + 3y - 1 = 0$$

and through the point $P(2, 1)$.

The equation of the line will be, by (8), of the form

$$\lambda_1(2x - y + 3) + \lambda_2(x + 3y - 1) = 0. \tag{9}$$

The point $P(2, 1)$ must lie on the line (9) and hence

$$\lambda_1(2\,.\,2 - 1 + 3) + \lambda_2(2 + 3\,.\,1 - 1) = 0\,,$$
$$6\lambda_1 + 4\lambda_2 = 0\,; \tag{10}$$

in order to satisfy equation (10), it suffices to put $\lambda_1 = 2$, $\lambda_2 = -3$. Then, by (9) the required equation is

$$2(2x - y + 3) - 3(x + 3y - 1) = 0\,, \quad \text{i.e.} \quad x - 11y + 9 = 0\,.$$

Check: We can calculate the coordinates of Q, the point of intersection of the two lines (by Theorem 2) and then verify that the points P and Q satisfy the equation $x - 11y + 9 = 0$.

5.5. Directed (Oriented) Straight Line. Direction Cosines. The Angle between Two Straight Lines

Definition 1. A straight line p is said to be *directed* (or *oriented*), if, for every pair of points A, B $(A \neq B)$ on this line, one can decide by means of a given rule, whether A lies before B (notation $A \prec B$) or B lies before A; (the two relations $A \prec B$ and $B \prec C$ together imply $A \prec C$). We say that on p the so-called *positive sense* and *negative sense* of orientation are given. It is customary to mark the direction (orientation) of a line in diagrams by an arrow showing its positive sense.

In a similar way, a *directed half-line* and *directed line segment* are defined. In the case of a directed half-line we speak of its initial point; in the case of a directed line segment we speak of its initial and end points. If we choose a point O on a directed line, we divide it into the so-called *positive part* (*positive half-line*) $+p$ and the *negative part* (*negative half-line*) $-p$ by this point.

Definition 2. If $A(a_1, a_2)$ and $B(b_1, b_2)$ are two points on a directed line p such that A lies before B, then the expressions

$$\frac{b_1 - a_1}{\sqrt{[(b_1 - a_1)^2 + (b_2 - a_2)^2]}}$$

and

$$\frac{b_2 - a_2}{\sqrt{[(b_1 - a_1)^2 + (b_2 - a_2)^2]}}$$

are called the *direction cosines* of the directed line p; we denote them by $\cos \alpha_1$, $\cos \alpha_2$. (The unit-vector with components $\cos \alpha_1$, $\cos \alpha_2$ lies on the line p.)

Theorem 1. *The expressions introduced in Definition 2 and denoted by* $\cos \alpha_1$, $\cos \alpha_2$ *are cosines of the undirected angles* α_1, α_2 $(0 \leqq \alpha_1, \alpha_2 \leqq 180°)$ *between the* *positive part of the line p and the positive parts of the coordinate axes x and y, respectively.*

Example 1. Let us choose the points $A(1, 2)$ and $B(0, -1)$ on the straight line $y = 3x - 1$ (Fig. 5.8); if this line is directed from A to B then its direction cosines are

$$\cos \alpha_1 = \frac{0 - 1}{\sqrt{[(0 - 1)^2 + (-1 - 2)^2]}} = \frac{-1}{\sqrt{10}},$$

$$\cos \alpha_2 = \frac{-1 - 2}{\sqrt{10}} = \frac{-3}{\sqrt{10}}.$$

The corresponding direction angles are

$$\alpha_1 \doteq 108°26',$$

$$\alpha_2 \doteq 161°34'.$$

Theorem 2. *If φ is the acute angle between the lines $y = k_1 x + q_1$ and $y = k_2 x + q_2$ then*

$$\tan \varphi = \left|\frac{k_1 - k_2}{1 + k_1 k_2}\right| \quad \left(k_1 \neq -\frac{1}{k_2}\text{; see Theorem 4}\right). \tag{1}$$

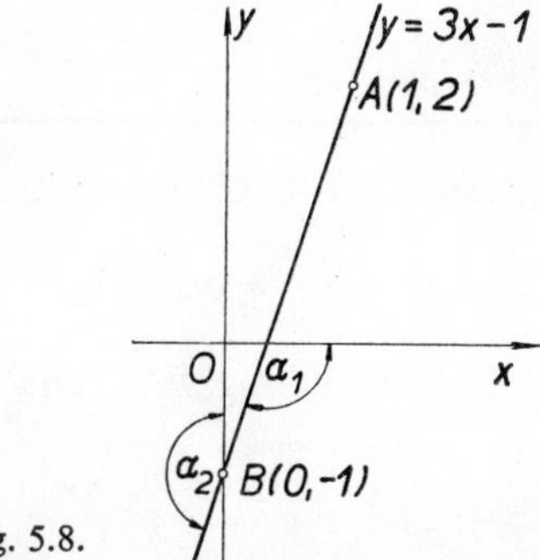

Fig. 5.8.

If the lines are given in the form $a_1 x + b_1 y + c_1 = 0$ and $a_2 x + b_2 y + c_2 = 0$, then

$$\tan \varphi = \frac{|a_1 b_2 - a_2 b_1|}{|a_1 a_2 + b_1 b_2|} \quad (a_1 a_2 + b_1 b_2 \neq 0\text{; see Theorem 4}). \tag{1'}$$

Theorem 3. *The condition for the straight lines of Theorem 2 to be parallel is*

$$k_1 = k_2, \quad or \quad a_1 b_2 - a_2 b_1 = 0.$$

Theorem 4. *The condition for the straight lines of Theorem 2 to be perpendicular is*

$$k_2 = -\frac{1}{k_1}, \quad or \quad a_1a_2 + b_1b_2 = 0\,.$$

Example 2. Let us determine the angles between the following straight lines: a) $y = 3x - 1$, $y = -x + 7$; b) $2x + 3y - 5 = 0$, $3x - 2y + 1 = 0$; c) $y = 2x + 3$, $4x - 2y + 1 = 0$.

Solution: a) $k_1 = 3$, $k_2 = -1$. Thus, $\tan \varphi = -4/(1 - 3) = 2$, $\varphi \doteq 63°26'$. b) $a_1a_2 + b_1b_2 = 2\,.\,3 - 3\,.\,2 = 0$; the lines are perpendicular. c) First, we put the equations of both lines in the same form: $2x - y + 3 = 0$, $4x - 2y + 1 = 0$. Clearly, $a_1b_2 - a_2b_1 = 2\,.\,(-2) - (-1)\,.\,4 = 0$; the lines are parallel.

REMARK 1. The equation of any line perpendicular to the straight line

$$a_1x + b_1y + c_1 = 0 \tag{2}$$

can be written in the form

$$b_1x - a_1y + c_2 = 0 \tag{3}$$

(since, the coefficients of the variables x and y in equations (2) and (3) satisfy the conditions of Theorem 4, namely $a_1b_1 + b_1(-a_1) = 0$).

Example 3. Find the equation of the line p passing through the point $P(1, 4)$ and perpendicular to the line

$$2x + 3y + 5 = 0\,. \tag{4}$$

By (3), the equation of the line p can be written in the form

$$3x - 2y + c_2 = 0\,. \tag{5}$$

Substituting the coordinates 1 and 4 of the point P for x and y respectively into (5), we obtain

$$3\,.\,1 - 2\,.\,4 + c_2 = 0\,,$$

i.e. $c_2 = 5$ and thus the equation of the line p is

$$3x - 2y + 5 = 0\,.$$

Example 4. Find the equation of the straight line which passes through the point of intersection of the lines

$$x - 2y + 3 = 0\,,$$

$$3x + 5y - 2 = 0$$

and which is perpendicular to the line

$$4x + y - 7 = 0 . \tag{6}$$

By (5.4.8), p. 211, the equation of the required line is of the form

$$\lambda_1(x - 2y + 3) + \lambda_2(3x + 5y - 2) = 0 ,$$

i.e.

$$(\lambda_1 + 3\lambda_2)\,x + (-2\lambda_1 + 5\lambda_2)\,y + (3\lambda_1 - 2\lambda_2) = 0 . \tag{7}$$

The condition for the lines (6) and (7) to be perpendicular is, by Theorem 4,

$$4(\lambda_1 + 3\lambda_2) + (-2\lambda_1 + 5\lambda_2) = 0 ,$$

i.e.

$$2\lambda_1 + 17\lambda_2 = 0 .$$

Hence, it suffices to choose $\lambda_2 = 2$, $\lambda_1 = -17$. Substituting these values into (7), we obtain the required equation in the form

$$-11x + 44y - 55 = 0 ,$$

i.e.

$$x - 4y + 5 = 0 .$$

5.6. The Normal Equation of a Straight Line. Distance of a Point from a Straight Line. The Equations of the Bisectors of the Angles between Two Straight Lines

Definition 1. The equation

$$\frac{a}{\pm\sqrt{(a^2 + b^2)}}\,x + \frac{b}{\pm\sqrt{(a^2 + b^2)}}\,y + \frac{c}{\pm\sqrt{(a^2 + b^2)}} = 0 , \tag{1}$$

where a, b, c are three arbitrary numbers $(a^2 + b^2 > 0)$ and the sign of the denominators is the opposite of that of the number c, is called the *normal equation of a straight line.*

The geometrical meanings of the coefficients are:

1. $\left(\dfrac{a}{\pm\sqrt{(a^2 + b^2)}}, \dfrac{b}{\pm\sqrt{(a^2 + b^2)}}\right)$ is a unit-vector perpendicular to the straight line (directed from the origin of coordinates to the line);

2. $\left|\dfrac{c}{\pm\sqrt{(a^2 + b^2)}}\right|$ is the length d of the perpendicular from the origin to the straight line.

Denoting the direction cosines of the above-mentioned vector by $\cos\alpha$ and $\cos\beta$, (α, β are the magnitudes of the angles which it makes with the positive parts of the axes x, y; see Theorem 5.5.1, p. 212), then

$$\cos\alpha = \frac{a}{\pm\sqrt{(a^2+b^2)}}, \quad \cos\beta = \frac{b}{\pm\sqrt{(a^2+b^2)}},$$

and the equation (1) can be rewritten in the form

$$x\cos\alpha + y\cos\beta - d = 0 .$$

Theorem 1. *The distance d of a point $A(x_0, y_0)$ from a straight line $ax + by + c = 0$ is given by*

$$d = \left|\frac{ax_0 + by_0 + c}{\sqrt{(a^2+b^2)}}\right| . \tag{2}$$

Theorem 2. *The equations of the bisectors of the angles between the two straight lines $a_1x + b_1y + c_1 = 0$ and $a_2x + b_2y + c_2 = 0$ are*

$$\frac{a_1x + b_1y + c_1}{\sqrt{(a_1^2+b_1^2)}} + \frac{a_2x + b_2y + c_2}{\sqrt{(a_2^2+b_2^2)}} = 0 \tag{3}$$

and

$$\frac{a_1x + b_1y + c_1}{\sqrt{(a_1^2+b_1^2)}} - \frac{a_2x + b_2y + c_2}{\sqrt{(a_2^2+b_2^2)}} = 0 . \tag{3'}$$

REMARK 1. In order to decide which of the two bisectors (3), (3′) passing through a given vertex of a triangle is the internal bisector of the angle, it is sufficient to find which one meets the opposite side of the triangle.

5.7. Polar Coordinates

The position of a point M may be determined by polar coordinates ϱ, φ: the coordinate ϱ is the distance of the point M from the origin (or *pole*) O, the coordinate φ is the directed angle between the segment OM and a fixed half-line p (with initial

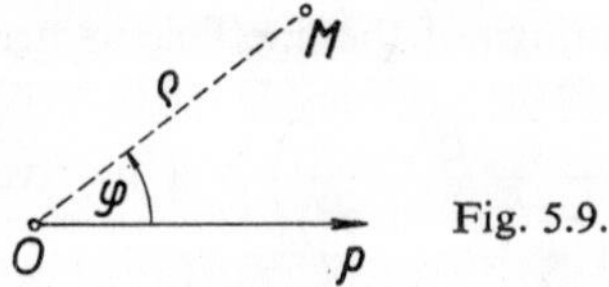

Fig. 5.9.

point O) called the *polar semi-axis* (or *initial-line* (Fig. 5.9.)). Here $\varrho \geqq 0$, $0 \leqq \varphi < 2\pi$. It is necessary to restrict the coordinates ϱ and φ in some way in order to establish a one-to-one correspondence between the points of a plane and the pairs of numbers (ϱ, φ) (with the exception of the pole), and we choose this particular way. However, sometimes φ is not restricted to this, or even any interval (and occasionally even $\varrho < 0$ is used, especially in the equations of spirals etc.).

The relations between the cartesian and polar coordinates in the case where the pole is at the origin of the cartesian system and the polar semi-axis coincides with the positive part of the x-axis are:

$$x = \varrho \cos \varphi\,, \quad y = \varrho \sin \varphi\,; \tag{1}$$

conversely

$$\begin{aligned}
&\varrho = \sqrt{(x^2 + y^2)}\,, && \\
&\varphi = \arctan \frac{y}{x} && \text{for} \quad x > 0,\, y > 0\,; \\
&\varphi = \tfrac{1}{2}\pi && \text{for} \quad x = 0,\, y > 0\,; \\
&\varphi = \pi + \arctan \frac{y}{x} && \text{for} \quad x < 0,\, y \text{ arbitrary}\,; \\
&\varphi = \tfrac{3}{2}\pi && \text{for} \quad x = 0,\, y < 0\,; \\
&\varphi = 2\pi + \arctan \frac{y}{x} && \text{for} \quad x > 0,\, y < 0\,.
\end{aligned} \tag{1'}$$

Example 1. The point M, the cartesian coordinates of which are $(-2, 1)$, has polar coordinates $(\sqrt{5}, 2{\cdot}68)$ (the angle being measured in radians). The point N, the polar coordinates of which are $(4, \frac{5}{3}\pi)$, has cartesian coordinates $(2, -2\sqrt{3})$ (Fig. 5.10).

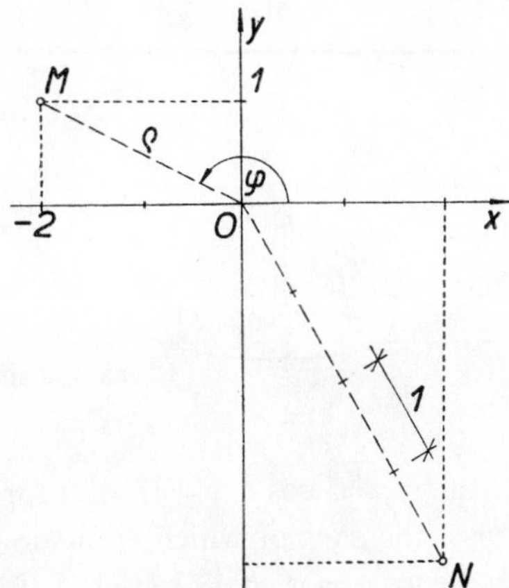

Fig. 5.10.

The equation of a curve in polar coordinates is – as in the case of cartesian co-ordinates – a relation which is satisfied by the coordinates of all the points of the curve (and only those points). The equations of some curves have a particularly simple form in polar coordinates.

Example 2. The cartesian equation of the circle whose centre is at the origin and whose radius is 7 is $x^2 + y^2 = 49$. If the coordinates are changed to polars using (1), the equation, after a small simplification, takes the form $\varrho = 7$, which is the equation of the same circle. (This is obvious geometrically.)

Example 3. If we choose as pole the focus of an ellipse, hyperbola, or parabola, and if that part of the focal axis of symmetry which does not contain the nearer vertex, be chosen as polar semi-axis, then all these curves have an equation of the form:

$$\varrho = \frac{p}{1 - e \cos \varphi}, \tag{2}$$

where e is the eccentricity (see §§ 5.10, 5.11, 5.12) and $2p$ is the *latus rectum* (i.e. the focal chord perpendicular to the focal axis of symmetry). In the case of the ellipse and hyperbola, $p = b^2/a$.

5.8. Parametric Equations of a Curve in a Plane

The equations $x = x(t)$, $y = y(t)$, where t is a variable *parameter*, are called the *parametric equations of a curve.*

Here $x(t)$, $y(t)$ are, as a rule, differentiable functions of t within an interval I. If t ranges over this interval, the point $M(x, y)$ moves along the curve.

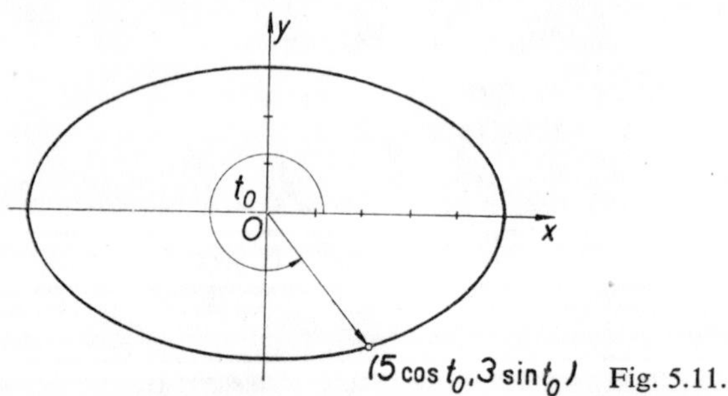

Fig. 5.11.

Example 1. a) The equations $x = 5 \cos t$, $y = 3 \sin t$ for $0 \leqq t < 2\pi$ are the parametric equations of an ellipse, the axes of which coincide with the coordinate axes, the lengths of the semi-axes being $a = 5$, $b = 3$ (Fig. 5.11).

b) If we eliminate t from these equations, we obtain the equation of the ellipse in cartesian coordinates (see § 10):

$$\frac{x}{5} = \cos t\,, \quad \frac{y}{3} = \sin t\,;$$

hence

$$\left(\frac{x}{5}\right)^2 + \left(\frac{y}{3}\right)^2 = \cos^2 t + \sin^2 t = 1\,,$$

i.e.

$$\frac{x^2}{25} + \frac{y^2}{9} = 1\,.$$

REMARK 1. Some curves can only be expressed in a simple form parametrically, (either we cannot eliminate t from the parametric equations or it is inconvenient to do so; for example, in the case of a cycloid $x = t - \sin t$, $y = 1 - \cos t$); some curves can be expressed simply in both ways. Obviously, we use whichever form is more convenient.

5.9. The Circle (see also § 4.1, p. 150)

Definition 1. A *circle* is the locus of a point $X(x, y)$ in a plane which moves so that its distance from a fixed point – the *centre S* – is constant.

Theorem 1. *The equation of the circle, whose centre in cartesian coordinates is* $S(x_0, y_0)$ *and whose radius is* r, *is*

$$(x - x_0)^2 + (y - y_0)^2 = r^2\,. \tag{1}$$

Example 1. The circle with centre $S(-2, 1)$ and radius $r = 3$ has the equation

$$(x + 2)^2 + (y - 1)^2 = 9\,.$$

If we remove the brackets in equation (1), we obtain an equation of the form

$$x^2 + y^2 + mx + ny + p = 0. \tag{2}$$

If we want to obtain an equation of the form (1) from equation (2), we "complete the squares" on the left-hand side of equation (2) and obtain

$$\left(x + \frac{m}{2}\right)^2 + \left(y + \frac{n}{2}\right)^2 = \frac{m^2}{4} + \frac{n^2}{4} - p\,. \tag{3}$$

Comparing this with equation (1) we can see that the expression on the right-hand side of equation (3) must be positive in order to get a real circle; also $(-\frac{1}{2}m, -\frac{1}{2}n)$ are the coordinates of the centre of this circle.

Theorem 2. *The parametric equations of a circle are*

$$x = x_0 + r\cos t\,,$$

$$y = y_0 + r\sin t\,,$$

where the point $S(x_0, y_0)$ *is the centre,* r *is the radius,* (x, y) *are the coordinates of a general point* X *on the circle and* t $(0 \leqq t < 2\pi)$ *is a variable parameter, the geometrical significance of which is that it is the angle formed by the half-line* SX *and the positive semi-axis* $+x$.

Theorem 3. *The equation of a circle of radius* a *in polar coordinates is* $\varrho = a$, *if* S *coincides with* O (see Example 5.7.2, p. 218), *and* $\varrho = 2a \cos \varphi$ $(-\frac{1}{2}\pi < \varphi \leqq \frac{1}{2}\pi)$, *if* S *lies on the polar semi-axis and the circle passes through the pole.*

Example 2. Find the coordinates of the points of intersection P_1, P_2 of the line

$$4x - 3y + 4 = 0 \tag{4}$$

with the circle whose centre is at the point $(2, 4)$ and whose radius r is 5.

The equation of the circle is, by (1),

$$(x - 2)^2 + (y - 4)^2 = 25\,. \tag{5}$$

The coordinates of the common points of the line and the circle satisfy simultaneously equations (4) and (5); hence, they are given by solving the equations (4) and (5). From (4), it follows that

$$y = \tfrac{4}{3}(x + 1)\,. \tag{6}$$

Substituting (6) in (5), we obtain the quadratic equation

$$25(x^2 - 4x + 4) = 225\,,$$

i.e.

$$x^2 - 4x - 5 = 0\,,$$

for the x-coordinates of the points of intersection, the roots of this equation being

$$x_1 = 5\,, \quad x_2 = -1\,. \tag{7}$$

The corresponding values for y_1, y_2 are found by substituting (7) into (6) (not into (5)!):

$$y_1 = 8\,, \quad y_2 = 0\,.$$

The required points of intersection are $P_1(5, 8)$, $P_2(-1, 0)$.

REMARK 1. The problem of finding the points of intersection of a straight line and a circle reduces therefore to the solution of a quadratic equation. If this equation possesses two real roots, or a double root, or two conjugate complex roots, then the straight line is a secant (chord) of the circle, or a tangent to the circle, or it does not intersect the circle at all, respectively.

We proceed in the same way (and the same conclusion holds) when finding the points of intersection of a straight line and other conics. The only exceptions are the lines parallel to the axis of a parabola and to the asymptotes of a hyperbola.

5.10. The Ellipse (see also § 4.2, p. 152)

Definition 1. An *ellipse* is the locus of a point $X(x, y)$ which moves in a plane such that the sum of its distances from two fixed points $F_1(x_1, y_1)$ and $F_2(x_2, y_2)$ – the *foci* – is equal to a constant which is usually denoted by $2a$ if the foci both lie on the x-axis.

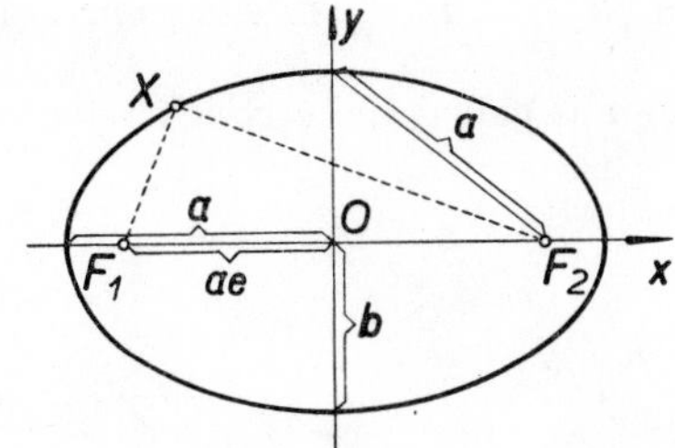

Fig. 5.12.

Clearly $\overline{F_1F_2} < 2a$ (i.e. $\overline{F_1F_2} = 2ae$ where e is some positive number less than unity).

Definition 2. The number e is called the *eccentricity of the ellipse.*

Theorem 1. *The standard equation of an ellipse (for the case where the axes of the ellipse coincide with the coordinate axes, the foci lying on the x-axis (Fig. 5.12)) is*

$$\frac{x^2}{a^2} + \frac{y^2}{b^2} = 1\,, \tag{1}$$

where $b^2 = a^2(1 - e^2)$.

In fact a is the length of the semi-major axis and b the length of the semi-minor axis of the ellipse.

REMARK 1. If the axes of an ellipse are parallel to the cooordinate axes and if the centre is at the point $S(x_0, y_0)$, then (1) becomes

$$\frac{(x - x_0)^2}{a^2} + \frac{(y - y_0)^2}{b^2} = 1\,. \tag{1'}$$

REMARK 2. If the foci lie on the y-axis the sum of the focal distances is denoted by $2b$, $\overline{F_1F_2} = 2be$ and a is now defined by $a^2 = b^2(1 - e^2)$. The equation is the same as (1).

For the equation of an ellipse in polar coordinates see Example 5.7.3, p. 218; for parametric equations of an ellipse see Example 5.8.1, p. 218.

5.11. The Hyperbola (see also § 4.3, p. 157)

Definition 1. A hyperbola is the locus of a point $X(x, y)$ which moves in a plane such that the difference of its distances from two fixed points $F_1(x_1, y_1)$ and $F_2(x_2, y_2)$ – the foci – is in absolute value equal to a constant which is usually denoted by $2a$ if the foci both lie on the x-axis.

Clearly $\overline{F_1F_2} > 2a$ (i.e. $\overline{F_1F_2} = 2ae$ where e is some number greater than unity).

Definition 2. The number e is called the *eccentricity of the hyperbola.*

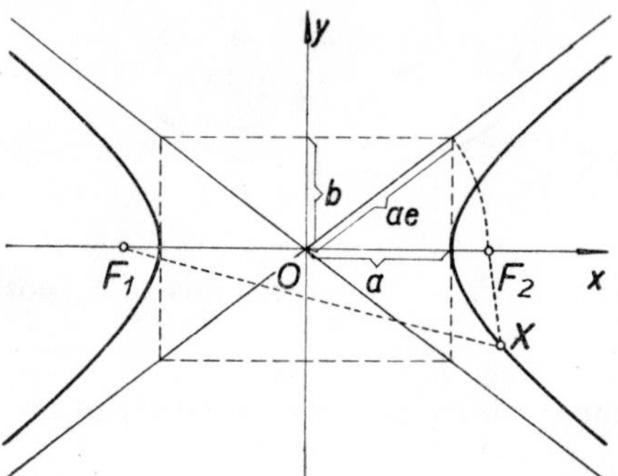

Fig. 5.13.

Theorem 1. *The standard equation of a hyperbola (for the case where the axes of the hyperbola coincide with the coordinate axes, the foci lying on the x-axis* (Fig. 5.13)) *is*

$$\frac{x^2}{a^2} - \frac{y^2}{b^2} = 1, \tag{1}$$

where $b^2 = a^2(e^2 - 1)$.

In fact a is the length of the (real) semi-major axis and b the length of the (imaginary) semi-minor axis of the hyperbola.

Remark 1. If the foci lie on the y-axis the absolute value of the difference of the focal distances is denoted by $2b$, $F_1F_2 = 2be$ and a is now defined by $a^2 = b^2(e^2 - 1)$. Equation (1) becomes

$$\frac{x^2}{a^2} - \frac{y^2}{b^2} = -1. \tag{1'}$$

Remark 2. If the axes of the hyperbola are parallel to the coordinate axes and the centre is at the point $S(x_0, y_0)$, then (1) and (1′) become

$$\frac{(x - x_0)^2}{a^2} - \frac{(y - y_0)^2}{b^2} = \pm 1. \tag{1''}$$

Theorem 2. *The lines* $y = \pm bx/a$ *are the asymptotes of the hyperbolas* (1) *and* (1'). *Their equations can be combined thus:*

$$\frac{x^2}{a^2} - \frac{y^2}{b^2} = 0\,.$$

REMARK 3. If $a = b$ the hyperbola is called *rectangular*; its equation is $x^2 - y^2 = a^2$ (or $y^2 - x^2 = a^2$), the equations of its asymptotes being $y = \pm x$.

REMARK 4. The hyperbolas $x^2/a^2 - y^2/b^2 = 1$ and $y^2/b^2 - x^2/a^2 = 1$ are called *conjugate*. They have the same asymptotes; the first of these hyperbolas has real points of intersection with the x-axis, the second one with the y-axis.

5.12. The Parabola (see also § 4.4, p. 160)

Definition 1. A *parabola* is the locus of a point $X(x, y)$ in a plane, equidistant from a fixed point $F(x_1, y_1)$ – the *focus* – and from a fixed line d – the *directrix*.

Theorem 1. *The parabola whose vertex is at the origin of the coordinate system* (Fig. 5.14) *and whose axis coincides with the x-axis, or y-axis, has the cartesian equation*

$$y^2 = 2px\,, \quad \text{or} \quad x^2 = 2py\,, \quad \textit{respectively}\,. \tag{1}$$

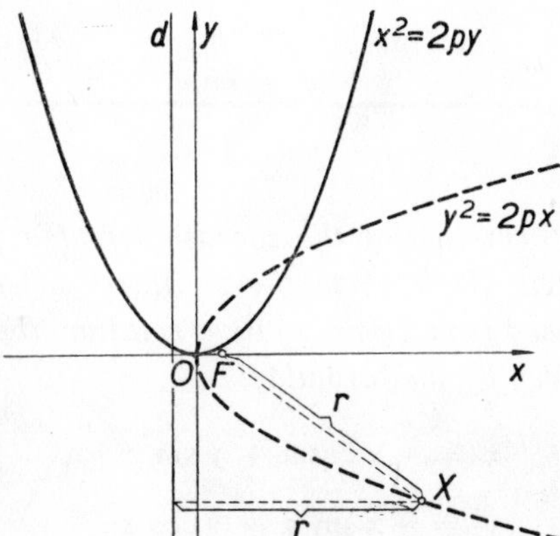

Fig. 5.14.

Theorem 2. *The parabola* $y^2 = 2px$ *has the focus* $F(\frac{1}{2}p, 0)$, *and the directrix* $x = -\frac{1}{2}p$. *The parabola* $x^2 = 2py$ *has the focus* $F_1(0, \frac{1}{2}p)$ *and the directrix* $y = -\frac{1}{2}p$.

(In the English literature the standard form of the equation of a parabola is generally taken as $y^2 = 4ax$, so that the focus is $(a, 0)$ and the directrix is $x = -a$.)

REMARK 1. If the axis of the parabola is parallel to the x-axis, or to the y-axis and its vertex is at the point $V(x_0, y_0)$, equations (1) become

$$(y - y_0)^2 = 2p(x - x_0)\,, \quad \text{or} \quad (x - x_0)^2 = 2p(y - y_0)\,, \quad \text{respectively}\,. \tag{2}$$

5.13. Congruent Transformations of Cartesian Coordinates in a Plane

We assume that all the different cartesian coordinate systems considered in this paragraph have the same unit of length.

Theorem 1. *Any change from one cartesian coordinate system to another can be performed by one translation and one rotation provided that both systems have the same orientation (i.e. both are right-handed or both left-handed).*

Theorem 2. *The transformation of the coordinates of a point X when passing from the cartesian system $(O;\, x, y)$ to the similarly oriented cartesian system $(O';\, x', y')$, in which the axes x', y', are parallel to the axes x, y, respectively, is given by the formulae:*

$$\begin{aligned} x' &= x - m\,, \\ y' &= y - n\,, \end{aligned} \tag{1}$$

where (m, n) are the coordinates of the new origin O' in the original system $(O;\, x, y)$.

Conversely:

$$\begin{aligned} x &= x' + m\,, \\ y &= y' + n\,. \end{aligned} \tag{1'}$$

Theorem 3. *The transformation of the coordinates of a point X when passing from the cartesian system $(O;\, x, y)$ to the similarly oriented cartesian system $(O;\, x', y')$ which is obtained from $(O;\, x, y)$ by a rotation about the common origin through an angle α, is given by the formulae:*

$$\begin{aligned} x' &= x \cos \alpha + y \sin \alpha\,, \\ y' &= -x \sin \alpha + y \cos \alpha\,. \end{aligned} \tag{2}$$

Conversely:

$$\begin{aligned} x &= x' \cos \alpha - y' \sin \alpha\,, \\ y &= x' \sin \alpha + y' \cos \alpha\,. \end{aligned} \tag{2'}$$

Theorem 4. *In any change from a cartesian coordinate system $(O;\, x, y)$ to a similarly oriented cartesian coordinate system $(O';\, x', y')$ the coordinates are trans-*

formed according to the formulae:

$$\begin{aligned} x' &= \quad x \cos \alpha + y \sin \alpha - m' , \\ y' &= -x \sin \alpha + y \cos \alpha - n' . \end{aligned} \tag{3}$$

Conversely:

$$\begin{aligned} x &= x' \cos \alpha - y' \sin \alpha + m , \\ y &= x' \sin \alpha + y' \cos \alpha + n . \end{aligned} \tag{3'}$$

The numbers m, n are the coordinates of the origin O' in the system $(O; x, y)$, $m' = m \cos \alpha + n \sin \alpha$, $n' = -m \sin \alpha + n \cos \alpha$, and α is the angle of rotation of the coordinate axes.

REMARK 1. The determinant consisting of the coefficients of x, y in equations (2) or (3) equals unity. This fact is characteristic of congruent transformations of cartesian coordinates in a plane; it is also a sufficient condition for the equations (2) or (3) to be solvable in x, y (to give the inverse transformations (2') and (3'); the derivation of (1') from (1) is obvious).

REMARK 2. Equations (1), (2), (3) represent the relationship between the coordinates of a fixed point in a plane with respect to two different coordinate systems; but they may also be interpreted as showing the relationship between the coordinates of two different points in a plane with respect to the same coordinate system.

5.14. Homogeneous Coordinates

Definition 1. The three ordered numbers (ξ_0, ξ_1, ξ_2) $(\xi_0 \neq 0)$ are called the *rectangular homogeneous coordinates* of a point M in a plane, if $\xi_1/\xi_0 = x$, $\xi_2/\xi_0 = y$, where (x, y) *are the cartesian coordinates of the point M. We write $M(\xi_0, \xi_1, \xi_2)$.* [The coordinate ξ_0 is frequently placed last in the group of these numbers viz. (ξ_1, ξ_2, ξ_0)].

Theorem 1. *If we use homogeneous coordinates then the equations of algebraic curves in a plane are homogeneous.*

Example 1. If we transform the cartesian equation of a line $ax + by + c = 0$ by means of the formulae in Definition 1, we obtain the linear homogeneous equation:

$$a\xi_1 + b\xi_2 + c\xi_0 = 0 . \tag{1}$$

REMARK 1. In contrast to the case of the general equation of a line in cartesian coordinates, the numbers a and b in equation (1) may both be equal to zero, so that the equation can be of the form

$$\xi_0 = 0 ;$$

this is the equation of the so-called *line at infinity* (*improper line*), by which the plane has been extended due to the introduction of homogeneous coordinates.

Example 2. The equation of an ellipse

$$\frac{(x - \bar{x})^2}{a^2} + \frac{(y - \bar{y})^2}{b^2} = 1$$

[whose centre is at the point $(\bar{x}, \bar{y})$] becomes, on substitution of homogeneous coordinates and after some simplification:

$$(a^2\bar{\xi}_2^2 + b^2\bar{\xi}_1^2)\,\xi_0^2 + b^2\bar{\xi}_0^2\xi_1^2 + a^2\bar{\xi}_0^2\xi_2^2 - 2b^2\bar{\xi}_0\bar{\xi}_1\xi_0\xi_1 - 2a^2\bar{\xi}_0\bar{\xi}_2\xi_0\xi_2 =$$
$$= a^2b^2\bar{\xi}_0^2\xi_0^2\,;$$

this is a homogeneous equation of the second degree in the variables ξ_0, ξ_1, ξ_2.

5.15. General Equation of a Conic

Theorem 1. *The general equation of a curve of the second degree in cartesian coordinates is*

$$a_{11}x^2 + 2a_{12}xy + a_{22}y^2 + 2a_{13}x + 2a_{23}y + a_{33} = 0\,; \tag{1}$$

this equation may represent an ellipse (a circle as a special case), a hyperbola, a parabola, a pair of straight lines (which may coincide), a point, or it may not be satisfied by any (real) point at all.

Dafinition 1. Let us form two determinants from the coefficients of equation (1):

$$\Delta = \begin{vmatrix} a_{11}, & a_{12}, & a_{13} \\ a_{12}, & a_{22}, & a_{23} \\ a_{13}, & a_{23}, & a_{33} \end{vmatrix}, \quad \delta = \begin{vmatrix} a_{11}, & a_{12} \\ a_{12}, & a_{22} \end{vmatrix};$$

the number Δ is called the *discriminant* of the conic section (1), δ – the *discriminant of the quadratic members.*

Theorem 2. *The curves of the second degree can be classified in terms of Δ and δ, as shown in Tab.* 5.1, p. 227.

Theorem 3. *By means of a rotation of the coordinate system through an angle φ it is possible to make the axes (or the axis) of a regular conic parallel to the coordinate axes; the angle φ can be found from the relation*

$$\tan 2\varphi = \frac{2a_{12}}{a_{11} - a_{22}}\,. \tag{2}$$

If $a_{11} = a_{22}$ we can choose $\varphi = \frac{1}{4}\pi$ (see Example 5.17.1, p. 231).

TABLE 5.1

	Regular (non-singular) conic sections ($\Delta \neq 0$)	Singular conic sections ($\Delta = 0$)
$\delta > 0$	an ellipse (real or imaginary)	two imaginary lines with a real point of intersection
$\delta < 0$	a hyperbola	two intersecting lines
$\delta = 0$	a parabola	two parallel lines (real or imaginary, different or coincident)

Theorem 4. *If the position of a regular conic is such that its axes are parallel to the coordinate axes (or its axis is parallel to one of the coordinate axes), then its equation does not contain the term involving xy, i.e. $a_{12} = 0$ (and conversely). The nature of the conic can then be determined as follows:*

a) $a_{11}a_{22} > 0$ – *an ellipse (a circle if $a_{11} = a_{22}$),*
b) $a_{11}a_{22} < 0$ – *a hyperbola (a rectangular hyperbola if $a_{11} = -a_{22}$),*
c) $a_{11}a_{22} = 0$ – *a parabola.*

REMARK 1. In the case mentioned in Theorem 4, i.e. when the equation does not contain any xy term, we can easily find the type of the conic section and at the same time find its centre (or vertex) and semi-axes by the method of "completing the squares", as in the case of the circle (Equation (5.9.3), p. 219).

5.16. Affine and Projective Transformations

Definition 1. *The affine (position) ratio* of a point M on a straight line with respect to two base points P, Q of the line is the ratio of the distances of the point M from the two points P, Q; if M is an inner point of the line-segment PQ, the ratio is negative, if it is an external point, the ratio is positive. We use the notation

$$(PQM) = \frac{PM}{QM} \quad (M \neq Q).$$

Definition 2. *The cross ratio of four points P, Q, M, N* on a line (the order in which they are written is important) is the quotient of the affine ratios of the points M and N

with respect to P and Q. We write

$$(PQMN) = \frac{(PQM)}{(PQN)} = \frac{PM}{QM} \cdot \frac{QN}{PN} \quad (M \neq Q, N \neq P).$$

Definition 3. By an *affine transformation* of a plane we mean a transformation which carries the point $M(x, y)$ into the point $M'(x', y')$ according to the equations

$$\begin{aligned} x' &= a_1 x + b_1 y + c_1 , \\ y' &= a_2 x + b_2 y + c_2 \end{aligned} \tag{1}$$

where

$$\begin{vmatrix} a_1, & b_1 \\ a_2, & b_2 \end{vmatrix} \neq 0 .$$

Theorem 1. *An affine transformation preserves the affine ratio of a point on a line with respect to any two points on the line; the line at infinity is transformed into itself (i.e. parallelism is preserved).*

Theorem 2. *An affine transformation of a plane into which a homogeneous coordinate system is introduced, is given by the equations:*

$$\begin{aligned} \xi_0' &= a_0 \xi_0 , \\ \xi_1' &= a_1 \xi_0 + b_1 \xi_1 + c_1 \xi_2 , \\ \xi_2' &= a_2 \xi_0 + b_2 \xi_1 + c_2 \xi_2 , \end{aligned} \tag{2}$$

where

$$\begin{vmatrix} a_0, & 0, & 0 \\ a_1, & b_1, & c_1 \\ a_2, & b_2, & c_2 \end{vmatrix} \neq 0 .$$

Theorem 3. *Every congruent transformation is a particular case of an affine transformation.*

Theorem 4. *By an affine transformation, a conic section is transformed into a conic section of the same type, i.e. an ellipse into an ellipse, a hyperbola into a hyperbola and a parabola into a parabola.*

Definition 4. By a *projective transformation* of a plane we mean a transformation which carries the point $M(x, y)$ into the point $M'(x', y')$ according to the equations:

$$\begin{aligned} x' &= \frac{a_{11}x + a_{12}y + a_{13}}{a_{31}x + a_{32}y + a_{33}} , \\ y' &= \frac{a_{21}x + a_{22}y + a_{23}}{a_{31}x + a_{32}y + a_{33}} , \end{aligned} \tag{3}$$

where

$$\begin{vmatrix} a_{11}, & a_{12}, & a_{13} \\ a_{21}, & a_{22}, & a_{23} \\ a_{31}, & a_{32}, & a_{33} \end{vmatrix} \neq 0 .$$

Theorem 5. *A projective transformation preserves the cross ratio of any four points on a line.*

Theorem 6. *A projective transformation of a plane into which a homogeneous coordinate system is introduced is given by the equations:*

$$\begin{aligned} \xi_0' &= a_{11}\xi_0 + a_{12}\xi_1 + a_{13}\xi_2 , \\ \xi_1' &= a_{21}\xi_0 + a_{22}\xi_1 + a_{23}\xi_2 , \\ \xi_2' &= a_{31}\xi_0 + a_{32}\xi_1 + a_{33}\xi_2 , \end{aligned} \tag{4}$$

where

$$\begin{vmatrix} a_{11}, & a_{12}, & a_{13} \\ a_{21}, & a_{22}, & a_{23} \\ a_{31}, & a_{32}, & a_{33} \end{vmatrix} \neq 0 .$$

Theorem 7. *Every affine transformation is a particular case of a projective transformation.*

Theorem 8. *By a projective transformation a regular conic section is transformed into a regular conic section (not necessarily of the same type), a singular conic section is transformed into a singular conic section (of the same type in the projective sense; i.e. the properties of being real, imaginary, distinct or coincident are preserved).*

REMARK 1. Since the determinants of the systems (1)–(4) are different from zero, the undashed coordinates can be expressed by means of the dashed coordinates in each of the systems, i.e. there exists an inverse transformation for each of the transformations under consideration.

5.17. Pole, Polar, Centre, Conjugate Diameters and Tangents of a Conic Section

Definition 1. If the cross ratio of four points A, B, C, D is equal to -1, i.e. $(ABCD) = -1$, we say that these points form a *harmonic set* (*range*).

Theorem 1. *Let us consider a pencil of lines passing through a point P chosen in the plane of a regular conic, the individual lines intersecting the conic in pairs of points M_1, N_1; M_2, N_2 etc.* (Fig. 5.15). *Then, the locus of a point Q_i, which forms a harmonic set with the point P and the points M_i, N_i on every line of the pencil* (i.e. $(M_iN_iPQ_i) = -1$), *is a straight line p called the polar of the point P with respect to the conic. The point P is called the pole of the line p with respect to the conic.*

Theorem 2. *The equation of the polar p of a point $P(x_0, y_0)$ with respect to the regular conic*

$$a_{11}x^2 + 2a_{12}xy + a_{22}y^2 + 2a_{13}x + 2a_{23}y + a_{33} = 0 \tag{1}$$

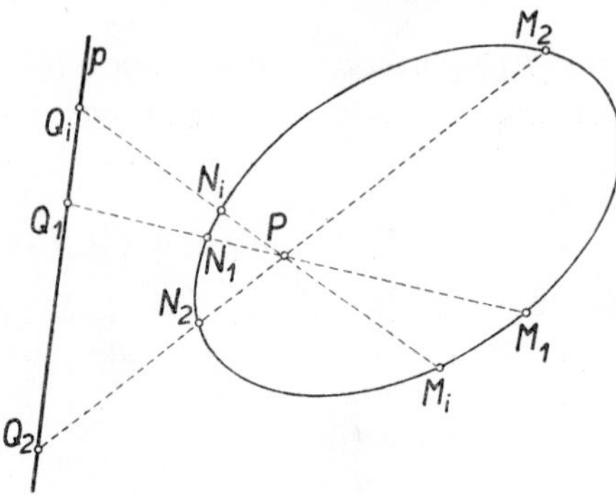

Fig. 5.15.

is

$$(a_{11}x_0 + a_{12}y_0 + a_{13})\,x + (a_{12}x_0 + a_{22}y_0 + a_{23})\,y + (a_{13}x_0 + a_{23}y_0 + a_{33}) = 0\,. \tag{2}$$

Theorem 3. *The polar of a point $T(x_0, y_0)$ of a regular conic with respect to this conic passes through the point T and is the tangent to the conic at this point. Its equation is*

$$\frac{xx_0}{a^2} + \frac{yy_0}{b^2} = 1 \quad \textit{for an ellipse whose equation is in standard form}\,,$$

$$\frac{xx_0}{a^2} - \frac{yy_0}{b^2} = 1 \quad \textit{for a hyperbola whose equation is in standard form}\,,$$

$$yy_0 = p(x + x_0) \quad \textit{for the parabola } y^2 = 2px\,.$$

Theorem 4. *The tangents from a point P to a regular conic (if they exist) pass through the points of intersection of the polar of P and the conic.*

Theorem 5. *The mid-points of all parallel chords of a regular conic lie on a line.*

Definition 2. The line on which all the mid-points of parallel chords of a regular conic lie is called a *diameter of the conic.*

Definition 3. The common *direction* of parallel chords of a regular conic is said to be *conjugate to the direction* of the diameter which passes through the mid-points of those chords.

Theorem 6. a) *All diameters of a parabola are parallel.*

b) *All diameters of an ellipse (or hyperbola) pass through a common point called the centre of the ellipse (or hyperbola).*

Theorem 7. *The coordinates of the centre of the conic* (1) *are given by the solution of the equations:*

$$\begin{aligned} a_{11}x + a_{12}y + a_{13} &= 0\,, \\ a_{12}x + a_{22}y + a_{23} &= 0\,; \end{aligned} \tag{3}$$

(*the left-hand sides of equations* (3) *are in fact half the partial derivatives of the left-hand side of equation* (1) *with respect to x and y, respectively.*)

Theorem 8. *If the direction* (s_1) *is conjugate to the direction* (s_2) *with respect to a regular central conic, then the direction* (s_2) *is conjugate to the direction* (s_1) *with respect to the same conic.* Thus, such directions are called *conjugate directions with respect to the conic.*

Definition 4. Two diameters of a regular conic whose directions are conjugate with respect to this conic are said to be *conjugate diameters* of this conic.

Definition 5. The two conjugate diameters of a central conic which are perpendicular are called the *axes of the conic.*

Theorem 9. *The slopes* k_1, k_2 *of conjugate directions satisfy the relation*

$$a_{11} + a_{12}(k_1 + k_2) + a_{22}k_1k_2 = 0\,, \tag{4}$$

i.e.

$$k_1k_2 = -\frac{b^2}{a^2} \quad \textit{for an ellipse whose equation is in standard form}\,,$$

$$k_1k_2 = \frac{b^2}{a^2} \quad \textit{for a hyperbola whose equation is in standard form}\,.$$

Theorem 10. *The equation of the diameter conjugate to the direction whose slope is k, is*

$$a_{11}x + a_{12}y + a_{13} + k(a_{12}x + a_{22}y + a_{23}) = 0\,, \tag{5}$$

i.e.

$$y = -\frac{b^2}{a^2k}x \quad \textit{for an ellipse whose equation is in standard form}\,,$$

$$y = \frac{b^2}{a^2k}x \quad \textit{for a hyperbola whose equation is in standard form}\,,$$

$$y = \frac{p}{k} \quad \textit{for the parabola } y^2 = 2px\,.$$

Example 1. Let us investigate the curve of the second degree which is given by the equation

$$3x^2 - 2xy + 3y^2 + 4x + 4y - 4 = 0 \tag{6}$$

and draw the tangents from the point $P(3, 1)$ to it.

Solution: Since

$$\Delta = \begin{vmatrix} 3, & -1, & 2 \\ -1, & 3, & 2 \\ 2, & 2, & -4 \end{vmatrix} = -64 \neq 0\,, \quad \delta = \begin{vmatrix} 3, & -1 \\ -1, & 3 \end{vmatrix} = 8 > 0\,,$$

either the curve (6) is an ellipse or it contains no real points (see Theorem 5.15.2). Further, from the formula (5.15.2) we can see that $\varphi = \frac{1}{4}\pi$; thus, $\sin \varphi = \cos \varphi = \frac{1}{2}\sqrt{2}$ so that by successive substitution into equations (5.13.2′) and from them into (6), we obtain the equation

$$x'^2 + 2y'^2 + 2\sqrt{(2)}\,x' - 2 = 0\,,$$

i.e.

$$\frac{(x' + \sqrt{2})^2}{4} + \frac{y'^2}{2} = 1\,.$$

Hence, (6) is the ellipse whose centre is at the point $(-\sqrt{2}, 0)$ and whose semi-axes a (of length 2) and b (of length $\sqrt{2}$) make an angle φ $(= \frac{1}{4}\pi)$ with the coordinate axes. The coordinates of the centre S are expressed in the transformed coordinates; its coordinates in the original system can be found, for instance, by equations (3):

$$3x - y + 2 = 0\,,$$
$$-x + 3y + 2 = 0\,;$$

hence $x_0 = -1$, $y_0 = -1$, and so S is the point $(-1, -1)$.

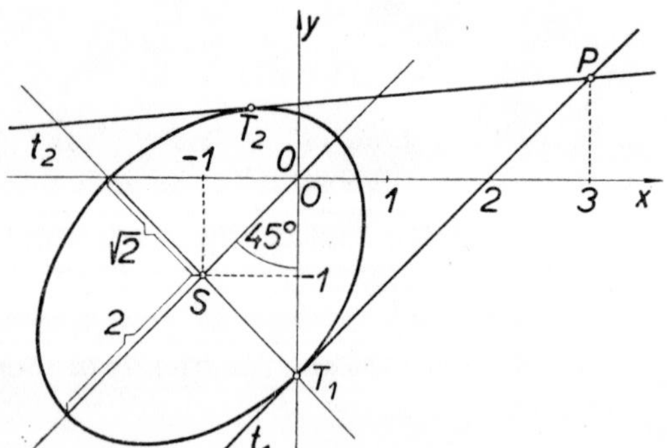

Fig. 5.16.

In order to find the equation of a tangent from the point P to the ellipse, we first find its polar; by (2) the equation of the polar is $5x + y + 2 = 0$. The coordinates of the points of contact T_1, T_2 are given by the simultaneous solution of this equation and equation (6): $T_1(0, -2)$, $T_2(-\frac{6}{11}, \frac{8}{11})$. Then, by means of the coordinates of P, T_1 and T_2, we can easily find the equations of the tangent lines: $x - y - 2 = 0$, $x - 13y + 10 = 0$.

6. SOLID ANALYTIC GEOMETRY

By František Kejla

References: [30], [159], [276], [369], [379].

6.1. Coordinate Systems

The position of an arbitrary point in three dimensional space is usually determined so that, to each point of the space an ordered triplet of real numbers (called *coordinates*) is assigned and conversely, to each ordered triplet of real numbers there corresponds a certain unique point of the space.

Definition 1. Surfaces consisting of those points which have one particular coordinate constant are called *coordinate surfaces.*

Various coordinate systems can be established, the most important being:

a) Rectangular coordinate system. This system is introduced in a manner similar to that for a rectangular coordinate system in a plane (see Remark 5.1.2, p. 205): We choose three mutually perpendicular directed lines x, y, z in space passing through a common point O and three units of length, one on each line. The directed distances x, y, z (cf. Remark 5.1.2, p. 205) of an arbitrary point M from the planes yz, xz, xy respectively, are called the *rectangular coordinates* of the point M. We write $M(x, y, z)$ to denote this. If the units of length are identical, the coordinates are called *cartesian.* In what follows we shall always be referring to these coordinates (unless otherwise stated).

The point $M_1(x, y, 0)$ is the orthogonal projection (top view) of the point M onto the plane xy, so that x, y are the cartesian coordinates of the projection of the point M in the system $(O; x, y)$ in the sense of plane analytical geometry (see Remark 5.1.2, p. 205). The situation is similar for the point $M_2(0, y, z)$ – front view of the point M – and for the point $M_3(x, 0, z)$ – side view of the point M.

The lines x, y, z are called the *coordinate axes,* the point O – the *origin of the coordinate system,* and the planes yz, xz, xy – the *coordinate planes.*

Definition 2. If when viewed from an arbitrary point on the positive semi-axis $+z$, the positive semi-axis $+x$ is carried by counter-clockwise rotation through a right angle into the positive semi-axis $+y$, the coordinate system $(O; x, y, z)$ is said to be *positively oriented* (*right-handed*). Otherwise the system is said to be negatively *oriented* (*left-handed*) – see Fig. 6.1a,b.

Theorem 1. *The coordinate surfaces in a cartesian coordinate system are planes parallel to the coordinate planes (perpendicular to the corresponding coordinate axes).*

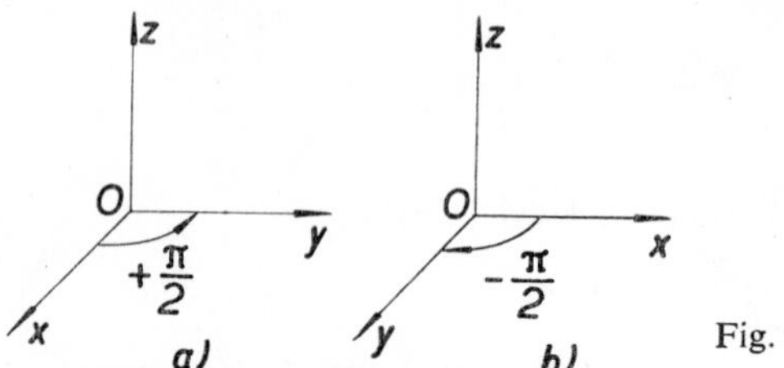

Fig. 6.1.

b) Cylindrical (semi-polar) coordinate system. This system is determined by a coordinate plane xy into which polar coordinates ϱ, φ are introduced (see p. 216) and by a directed z-axis, passing through the pole of the system, perpendicular to the plane. An arbitrary point M is determined by an ordered triplet of numbers (ϱ, φ, z), where ϱ, φ are the polar coordinates of the orthogonal projection M_1 of the point M onto the plane xy and z is the directed distance of the point M from the plane xy (see Fig. 6.2).

Theorem 2. *The coordinate surfaces in a cylindrical system are*

a) *half-planes passing through the z-axis* (φ = const.);

b) *cylinders of revolution with axis coinciding with the z-axis* (ϱ = const.; $\varrho = 0$ corresponds to the z-axis);

c) *planes perpendicular to the z-axis* (z = const.).

c) Spherical (polar) coordinate system. This system is determined by a coordinate plane xy into which polar coordinates ϱ, φ are introduced (see p. 216) and by a directed half-line, passing through the pole of the system perpendicular to the plane. The following coordinates determine the position of an arbitrary point M in this coordinate system:

a) the distance r of the point M from the origin (pole) O of the system;

b) the magnitude φ of the angle between the half-line OM_1 and the positive semi-axis $+x$, where M_1 is the orthogonal projection of the point M onto the plane xy;

c) the magnitude ϑ of the angle between the half-line OM and the positive semi-axis $+z$ (the axes x, y, z are at the same time the axes of a cartesian coordinate system – the so-called *adjoined system*; see Fig. 6.3).

The coordinate r is never negative ($r \geqq 0$), the coordinate φ ranges over the interval $[0, 2\pi)$, the coordinate ϑ ranges over the interval $[0, \pi]$. (Sometimes the interval $(-\pi, \pi]$ is used for φ.)

Theorem 3. *The coordinate surfaces in a spherical system are*

a) *spheres whose centres are at the pole of the system* ($r = \text{const.}$);
b) *half-planes passing through the z-axis* ($\varphi = \text{const.}$);
c) *cones* (or *half-cones*, more precisely) *of revolution, the vertices of which are at the pole and the axes of which coincide with the z-axis* ($\vartheta = \text{const.}$).

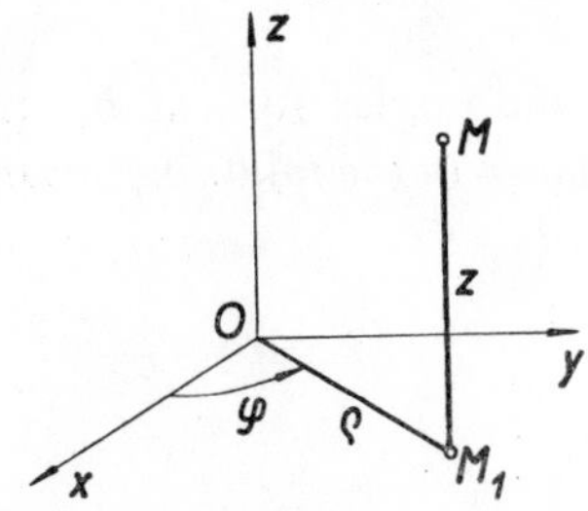

Fig. 6.2.

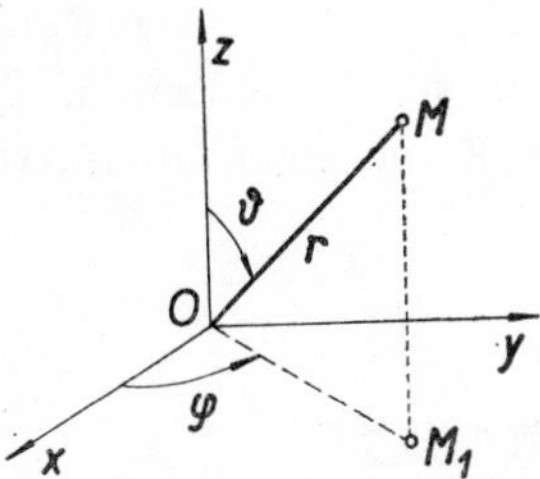

Fig. 6.3.

In particular, $r = 0$ gives one point (the pole), $\vartheta = \frac{1}{2}\pi$ – the plane xy, $\vartheta = 0$, π – the half-lines $+z$, $-z$, respectively.

Theorem 4. *The cartesian coordinates x, y, z, the cylindrical coordinates ϱ, φ, z and the spherical coordinates r, ϑ, φ of the same point satisfy the following relations:*

a) $x = \varrho \cos \varphi\,, \quad y = \varrho \sin \varphi\,, \quad z = z\,,$

$$\varrho = \sqrt{(x^2 + y^2)}\,, \quad \sin \varphi = \frac{y}{\sqrt{(x^2 + y^2)}}\,, \quad \cos \varphi = \frac{x}{\sqrt{(x^2 + y^2)}}\,, \quad \tan \varphi = \frac{y}{x}\,;$$

b) $x = r \sin \vartheta \cos \varphi\,, \quad y = r \sin \vartheta \sin \varphi\,, \quad z = r \cos \vartheta\,,$

$$r = \sqrt{(x^2 + y^2 + z^2)}\,, \quad \sin \vartheta = \sqrt{\frac{x^2 + y^2}{x^2 + y^2 + z^2}}\,,$$

$$\cos \vartheta = \frac{z}{\sqrt{(x^2 + y^2 + z^2)}}\,, \quad \tan \vartheta = \frac{\sqrt{(x^2 + y^2)}}{z}\,,$$

$$\sin \varphi = \frac{y}{\sqrt{(x^2 + y^2)}}\,, \quad \cos \varphi = \frac{x}{\sqrt{(x^2 + y^2)}}\,, \quad \tan \varphi = \frac{y}{x}\,.$$

REMARK 1. The correspondence between the sets of coordinates and the points themselves is one-to-one without exception, only in the case of a rectangular system.

It is not so in the other two systems. All points on the z-axis are so-called *singular points* of those systems – the coordinate φ may be chosen quite arbitrarily; however these points are uniquely determined by their remaining two coordinates. In calculations though, care is sometimes required.

Theorem 5 (*Transformation of a Cartesian Coordinate System*).

a) Translation. *If x, y, z denote the coordinates in the original system, X, Y, Z – the coordinates of the same point in the new system, x_0, y_0, z_0 – the coordinates of the new origin in the original system, then*

$$X = x - x_0\,, \quad Y = y - y_0\,, \quad Z = z - z_0\,.$$

b) Rotation and reflection. *If the cosines of the angles formed by the new axes X, Y, Z and the original axes x, y, z are as shown in the following scheme*

	X	Y	Z
x	a_1	a_2	a_3
y	b_1	b_2	b_3
z	c_1	c_2	c_3

then the following relations hold between the original and new coordinates of the same point:

$$x = a_1X + a_2Y + a_3Z\,, \quad X = a_1x + b_1y + c_1z\,,$$

$$y = b_1X + b_2Y + b_3Z\,, \quad Y = a_2x + b_2y + c_2z\,,$$

$$z = c_1X + c_2Y + c_3Z\,; \quad Z = a_3x + b_3y + c_3z\,.$$

Theorem 6. *The cosines listed in the table above satisfy the relations:*

$$a_1^2 + a_2^2 + a_3^2 = 1\,, \quad a_1^2 + b_1^2 + c_1^2 = 1\,,$$

$$b_1^2 + b_2^2 + b_3^2 = 1\,, \quad a_2^2 + b_2^2 + c_2^2 = 1\,,$$

$$c_1^2 + c_2^2 + c_3^2 = 1\,; \quad a_3^2 + b_3^2 + c_3^2 = 1\,;$$

$$a_1b_1 + a_2b_2 + a_3b_3 = 0\,, \quad a_1a_2 + b_1b_2 + c_1c_2 = 0\,,$$

$$a_1c_1 + a_2c_2 + a_3c_3 = 0\,, \quad a_1a_3 + b_1b_3 + c_1c_3 = 0\,,$$

$$b_1c_1 + b_2c_2 + b_3c_3 = 0\,; \quad a_2a_3 + b_2b_3 + c_2c_3 = 0\,.$$

Theorem 7.

$$\Delta = \begin{vmatrix} a_1, & a_2, & a_3 \\ b_1, & b_2, & b_3 \\ c_1, & c_2, & c_3 \end{vmatrix} = \pm 1 \quad \text{(the so-called } \textit{determinant of the transformation}\text{)},$$

where the $\left\{\begin{matrix}\textit{upper}\\ \textit{lower}\end{matrix}\right\}$ *sign is valid according as the two systems have* $\left\{\begin{matrix}\textit{the same}\\ \textit{different}\end{matrix}\right\}$ *orientations* (see Definition 2).

If the orientation of the two systems is the same, then each element of the determinant above is equal to its complement (cofactor). In the case of different orientations of the systems each element of the determinant of transformation is equal to minus its complement.

6.2. Linear Concepts

In § 6.2 we confine ourselves to relations expressed in cartesian coordinates. (For the meaning at any vector terminology mentioned see Chap. 7.)

Theorem 1. *The distance d between two points $M_1(x_1, y_1, z_1)$, $M_2(x_2, y_2, z_2)$ is equal to the length of the vector $\overrightarrow{M_1M_2}$*; i.e.

$$d = |\overrightarrow{M_1M_2}| = |\mathbf{r}_2 - \mathbf{r}_1| = \sqrt{[(x_2 - x_1)^2 + (y_2 - y_1)^2 + (z_2 - z_1)^2]}\,,$$

where $\mathbf{r}_1$, $\mathbf{r}_2$ are radius vectors of the points M_1, M_2.

Theorem 2. *A point M lying on the line which joins the points M_1, M_2 is determined by the ratio $\lambda = \overline{M_1M}/\overline{M_2M}$* (cf. § 5.2). *If $\mathbf{r}_1$, $\mathbf{r}_2$ are the radius vectors of the points M_1, M_2, then the radius vector $\mathbf{r}$ of the point M is given by*

$$\mathbf{r} = \frac{\mathbf{r}_1 - \lambda \mathbf{r}_2}{1 - \lambda},$$

i.e. the coordinates x, y, z of the point M are given by

$$x = \frac{x_1 - \lambda x_2}{1 - \lambda}, \quad y = \frac{y_1 - \lambda y_2}{1 - \lambda}, \quad z = \frac{z_1 - \lambda z_2}{1 - \lambda}.$$

In particular, if M is the mid-point of the line segment, we have

$$\mathbf{r} = \frac{\mathbf{r}_1 + \mathbf{r}_2}{2}, \quad \text{i.e.} \quad x = \frac{x_1 + x_2}{2}, \quad y = \frac{y_1 + y_2}{2}, \quad z = \frac{z_1 + z_2}{2}.$$

Theorem 3. *The centroid (centre of mass – if the mass is uniformly distributed) of a triangle is given by*

$$\mathbf{r} = \frac{\mathbf{r}_1 + \mathbf{r}_2 + \mathbf{r}_3}{3}, \quad \text{i.e.} \quad x = \frac{x_1 + x_2 + x_3}{3}, \quad y = \frac{y_1 + y_2 + y_3}{3},$$

$$z = \frac{z_1 + z_2 + z_3}{3},$$

where the $\mathbf{r}_k$ $(k = 1, 2, 3)$ *denote the radius vectors, and the* x_k, y_k, z_k $(k = 1, 2, 3)$ *the coordinates, of the vertices of the triangle.*

REMARK 1. The radius vector of the centre of mass of a system of particles $M_k(\mathbf{r}_k)$ with masses m_k $(k = 1, 2, \ldots, n)$ is given by

$$\mathbf{r} = \frac{\sum_{k=1}^{n} m_k \mathbf{r}_k}{\sum_{k=1}^{n} m_k}.$$

Theorem 4. *The volume* V_4, *of a tetrahedron with vertices* $M_k(x_k, y_k, z_k)$ $(k = = 1, 2, 3, 4)$ *is given by*

$$\pm V_4 = \frac{1}{6} \begin{vmatrix} x_2 - x_1, & y_2 - y_1, & z_2 - z_1 \\ x_3 - x_1, & y_3 - y_1, & z_3 - z_1 \\ x_4 - x_1, & y_4 - y_1, & z_4 - z_1 \end{vmatrix} = \frac{1}{6} \begin{vmatrix} x_1, & y_1, & z_1, & 1 \\ x_2, & y_2, & z_2, & 1 \\ x_3, & y_3, & z_3, & 1 \\ x_4, & y_4, & z_4, & 1 \end{vmatrix}.$$

(Obviously the positive value of V_4 is taken).

REMARK 2. All the four points M_k lie in the same plane if and only if the above determinant is equal to zero.

Theorem 5. *The equation of a plane can be written:*

a) *in general form:* $Ax + By + Cz + D = 0$ (*at least one of the numbers* A, B, C *being non-zero*);

b) *in vector form:* $\mathbf{r} \cdot \mathbf{n} + D = 0$
[*the vector* $\mathbf{n} = (A, B, C)$ *is perpendicular to the plane,* and is a so-called *normal vector to the plane*];

c) *in normal form:* $x \cos \alpha + y \cos \beta + z \cos \gamma - d = 0$, i.e. $\mathbf{r} \cdot \mathbf{n}^0 - d = 0$ ($d \geq 0$ *is the distance of the plane from the origin;* α, β, γ *are the magnitudes of the angles formed by that normal to the plane which is directed from the origin, and the coordinate axes;* $\mathbf{n}^0$ *is the unit vector in the direction of the normal; if* $d = 0$ *then the orientation of the normal is not uniquely determined*);

d) *in intercept form*:

$$\frac{x}{p} + \frac{y}{q} + \frac{z}{r} = 1$$

(p, q, r *are the intercepts cut on the coordinate axes by the plane, due regard being paid to their orientation*; for instance $p = -3$ means that the plane cuts the negative semi-axis $-x$ at a distance 3 from the origin, i.e. at the point $(-3, 0, 0)$).

REMARK 3. The equation of any plane can be written in the forms a), b), c) but it is impossible to write the equation of planes which are parallel to one of the coordinate axes or which pass through the origin in the intercept form d).

Theorem 6. *The general equation of a plane can be turned into the normal form by dividing throughout by the number* $\pm\sqrt{(A^2 + B^2 + C^2)}$. *The sign in front of the root is the opposite of that of the constant term in the general equation. (The normal of the plane is thereby directed from the origin to the plane.)*

Example 1. $2x - 3y - 6z + 21 = 0| : [-\sqrt{(2^2 + 3^2 + 6^2)}] = -7$, and so we obtain

$$\frac{2}{-7}x + \frac{-3}{-7}y + \frac{-6}{-7}z - 3 = 0\,, \quad \text{i.e.} \quad \frac{-2}{7}x + \frac{3}{7}y + \frac{6}{7}z - 3 = 0\,.$$

To construct the given plane, first construct the vector

$$\mathbf{n} = 3\mathbf{n}^0 = 3(-\tfrac{2}{7}, \tfrac{3}{7}, \tfrac{6}{7}) = (-\tfrac{6}{7}, \tfrac{9}{7}, \tfrac{18}{7})$$

whose initial point is at the origin, and then the plane which is perpendicular to, and which contains the terminal point of, $\mathbf{n}$.

Theorem 7. *The general equation of a plane can be turned into the intercept form by dividing throughout by minus the constant term* (i.e. by the number $-D$).

Example 2. $2x - 3y - 6z + 21 = 0\,|:(-21)$,

$$\frac{x}{-\frac{21}{2}} + \frac{y}{7} + \frac{z}{\frac{7}{2}} = 1\,.$$

Theorem 8. *The equation of the plane which is perpendicular to the vector* $\mathbf{a}(a_1, a_2, a_3)$ *and which passes through the point* $M(x_1, y_1, z_1)$, *is*

$$a_1(x - x_1) + a_2(y - y_1) + a_3(z - z_1) = 0\,.$$

Theorem 9. *The equation of the plane containing the three non-collinear points* $M_k(x_k, y_k, z_k)$ $(k = 1, 2, 3)$, *is*

$$\begin{vmatrix} x, & y, & z, & 1 \\ x_1, & y_1, & z_1, & 1 \\ x_2, & y_2, & z_2, & 1 \\ x_3, & y_3, & z_3, & 1 \end{vmatrix} = 0\,;$$

or, in vector form:

$$[(\mathbf{r} - \mathbf{r}_1)(\mathbf{r}_2 - \mathbf{r}_1)(\mathbf{r}_3 - \mathbf{r}_1)] = 0 .$$

($[\mathbf{a}\,\mathbf{b}\,\mathbf{c}]$ is the mixed or triple scalar product of the vectors $\mathbf{a}$, $\mathbf{b}$, $\mathbf{c}$; see Definition 7.1.13, p. 268.)

Theorem 10. *The equation of the plane which contains the two points* $M_k(x_k, y_k, z_k)$ $(k = 1, 2)$, *and which is parallel to the vector* $\mathbf{a}(a_1, a_2, a_3)$, *is*

$$\begin{vmatrix} x - x_1, & y - y_1, & z - z_1 \\ x_2 - x_1, & y_2 - y_1, & z_2 - z_1 \\ a_1, & a_2, & a_3 \end{vmatrix} = 0 ;$$

or, in vector form:

$$[\mathbf{a}(\mathbf{r} - \mathbf{r}_1)(\mathbf{r}_2 - \mathbf{r}_1)] = 0 .$$

Theorem 11. *The equation of the plane which contains the point* $M(x_1, y_1, z_1)$ *and which is parallel to the two vectors* $\mathbf{a}(a_1, a_2, a_3)$, $\mathbf{b}(b_1, b_2, b_3)$ *is*

$$\begin{vmatrix} x - x_1, & y - y_1, & z - z_1 \\ a_1, & a_2, & a_3 \\ b_1, & b_2, & b_3 \end{vmatrix} = 0 ;$$

or, in vector form:

$$[\mathbf{a}\,\mathbf{b}(\mathbf{r} - \mathbf{r}_1)] = 0 .$$

Theorem 12. *The distance of the point* $M(x_1, y_1, z_1)$ *from the plane* $Ax + By + Cz + D = 0$ *is equal to the absolute value of the result of substituting the coordinates of the point in the left-hand side of the normal equation of the plane,* i.e.

$$d = \left| \frac{Ax_1 + By_1 + Cz_1 + D}{\sqrt{(A^2 + B^2 + C^2)}} \right| .$$

Theorem 13. *The angle between the two planes* $A_1x + B_1y + C_1z + D_1 = 0$, $A_2x + B_2y + C_2z + D_2 = 0$ *is equal to the angle between their normals (which are represented by the vectors* $\mathbf{n}_1(A_1, B_1, C_1)$, $\mathbf{n}_2(A_2, B_2, C_2)$*). Thus considering* $0 \leqq \varphi \leqq \frac{1}{2}\pi$,

$$\cos\varphi = \frac{|\mathbf{n}_1 \cdot \mathbf{n}_2|}{|\mathbf{n}_1|\,|\mathbf{n}_2|} = \frac{|A_1A_2 + B_1B_2 + C_1C_2|}{\sqrt{[(A_1^2 + B_1^2 + C_1^2)(A_2^2 + B_2^2 + C_2^2)]}} .$$

In particular, a necessary and sufficient condition for the two planes to be perpendicular is that

$$\mathbf{n}_1 \cdot \mathbf{n}_2 = A_1A_2 + B_1B_2 + C_1C_2 = 0 .$$

Further, since the normals to two parallel planes are parallel, we have the following theorem:

Theorem 14. *A necessary and sufficient condition for the two planes*

$$A_1x + B_1y + C_1z + D_1 = 0\,, \quad A_2x + B_2y + C_2z + D_2 = 0$$

to be parallel is that

$$A_1 : A_2 = B_1 : B_2 = C_1 : C_2\,, \quad \text{i.e.} \quad \mathbf{n}_1 = \lambda \mathbf{n}_2\,.$$

REMARK 4. The equations of two parallel planes can, therefore, always be modified so that the coefficients of the variables are the same for both planes and the equations may differ only in the constant term. This is especially useful when calculating the distance between two parallel planes:

Theorem 15. *The distance d between the two parallel planes $Ax + By + Cz + D_1 = 0$, $Ax + By + Cz + D_2 = 0$ is given by*

$$d = \left| \frac{D_2 - D_1}{\sqrt{(A^2 + B^2 + C^2)}} \right|.$$

Example 3. The distance between the planes $4x - 2y - 4z + 11 = 0$, $-2x + y + 2z + 5 = 0$ can be calculated by the foregoing formula thus: we first multiply the second equation by the number -2. Then, $A = 4$, $B = -2$, $C = -4$, $D_1 = 11$, $D_2 = -10$ and so

$$d = \left| \frac{-10 - 11}{\sqrt{(16 + 4 + 16)}} \right| = \left| \frac{-21}{6} \right| = 3{\cdot}5\,.$$

Definition 1. The set of all planes which pass through a fixed line or the set of all planes parallel to a particular plane is called an (*axial*) *pencil of planes* (*sheaf of planes*).

REMARK 5. We often speak in geometry about *points, lines and planes at infinity.* Accordingly, in the preceding definition it is sufficient to refer to the set of all planes which have a line in common (at infinity in the case of parallel planes).

Theorem 16. *The planes belonging to a pencil, two of whose members are the planes $A_1x + B_1y + C_1z + D_1 = 0$ and $A_2x + B_2y + C_2z + D_2 = 0$, have as their equations:*

$$\lambda_1(A_1x + B_1y + C_1z + D_1) + \lambda_2(A_2x + B_2y + C_2z + D_2) = 0\,, \tag{1}$$

where λ_1, λ_2 are variable parameters, at least one of them being non-zero; clearly, only their ratio is significant.

REMARK 6. The above equation is especially useful when solving problems in which the equation of a plane passing through the line of intersection of two given planes and satisfying some additional condition is to be found.

Example 4. A plane is determined by the point $M(2, -1, 3)$ and the line of intersection of the planes whose equations are $6x + 2y - z - 3 = 0$, $3x + 4y - 2z - 2 = 0$. Find its equation.

If we substitute the coordinates of the given point into equation (1) which represents this particular pencil, we obtain the condition $4\lambda_1 - 6\lambda_2 = 0$ for λ_1, λ_2. This condition will be satisfied, if we choose, for instance $\lambda_1 = 3$, $\lambda_2 = 2$. The equation of the required plane is then

$$3(6x + 2y - z - 3) + 2(3x + 4y - 2z - 2) = 0\,,$$
$$\text{i.e.}\quad 24x + 14y - 7z - 13 = 0\,.$$

Theorem 17. *The equations of the planes which bisect the angles between two intersecting planes $\varrho_1 = 0$, $\varrho_2 = 0$ can be obtained if we add and subtract the normal equations of the two given planes:*

$$\frac{A_1x + B_1y + C_1z + D_1}{\sqrt{(A_1^2 + B_1^2 + C_1^2)}} \pm \frac{A_2x + B_2y + C_2z + D_2}{\sqrt{(A_2^2 + B_2^2 + C_2^2)}} = 0\,.$$

Example 5. $\varrho_1 \equiv 2x - y - 2z + 3 = 0$, $\varrho_2 \equiv 3x + 2y + 6z - 1 = 0$. The normal equations of these planes are

$$\frac{2x - y - 2z + 3}{-3} = 0 \quad \text{and} \quad \frac{3x + 2y + 6z - 1}{7} = 0\,,$$

respectively. Adding both equations and simplifying we obtain

$$5x - 13y - 32z + 24 = 0\,.$$

Subtracting we obtain, similarly,

$$23x - y + 4z + 18 = 0\,.$$

Definition 2. The set of all planes which pass through a fixed point (see Remark 5) is called a *bundle of planes* (*star of planes*).

Theorem 18. *The planes belonging to a bundle, three of whose members are the planes $A_1x + B_1y + C_1z + D_1 = 0$, $A_2x + B_2y + C_2z + D_2 = 0$, $A_3x + B_3y + C_3z + D_3 = 0$, have as their equations:*

$$\lambda_1(A_1x + B_1y + C_1z + D_1) + \lambda_2(A_2x + B_2y + C_2z + D_2) + \\ + \lambda_3(A_3x + B_3y + C_3z + D_3) = 0\,,$$

where λ_1, λ_2, λ_3 are variable parameters, at least one of them being non-zero; clearly, only their ratios are significant.

REMARK 7. The relative positions of several planes can be decided by a detailed analysis of the solution of the system of linear equations which represent these planes (see § 1.18).

The position of two points relative to each other with respect to a given plane can easily be decided on the basis of the result of substituting the coordinates of these points into the equation of the plane:

Theorem 19. *If the results of substituting the coordinates of two points into the left-hand side of the general equation of a plane are of the same sign, then both points lie in the same half-space determined by the plane (i.e. on the same side of the plane); if they are of different sign, then the two points lie in different half-spaces (i.e. on different sides of the plane). (If the result equals zero, the point lies in the plane, of course.)*

Theorem 20. *The equations of a straight line:*

a) *the general equations are*

$$A_1x + B_1y + C_1z + D_1 = 0\,,$$
$$A_2x + B_2y + C_2z + D_2 = 0$$

provided that the two planes represented by these equations intersect, i.e.

$$A_1 : B_1 : C_1 \neq A_2 : B_2 : C_2\,;$$

b) *the vector equation is*

$$\mathbf{r} = \mathbf{r}_0 + t\mathbf{a}$$

where $\mathbf{r}_0$ is the radius vector of a fixed point on the line, $\mathbf{a}$ – the direction of the line, i.e. a vector parallel to the given line, and t is a variable parameter;

c) *the parametric equations (merely a paraphrase of the vector equation):*

$$x = x_0 + a_1t\,, \quad y = y_0 + a_2t\,, \quad z = z_0 + a_3t\,;$$

where (x_0, y_0, z_0) is a fixed point on the line, $\mathbf{a} = (a_1, a_2, a_3)$ – the direction vector of the line (a_1, a_2, a_3 are so-called direction parameters of the line);

d) *the reduced equations:*

$$x = mz + p\,, \quad y = nz + q\,.$$

The equations of lines parallel to the plane xy cannot be written in this form. These equations are a particular case of the general equations, the reference planes being those planes which contain the line and its projection on the xz and yz planes respectively. They are, however, also a particular case of the parametric equations, in which the coordinate z is chosen as the parameter, i.e. $z = t$. The point (p, q) is then the point of intersection of the given line and the plane xy. Choosing the coordinate x, or y, as the parameter we obtain the other pairs of reduced equations of the line – provided that the line is not parallel to the plane yz, or xz, respectively.

Theorem 21. *The equation of a line determined by*

a) *a point* (x_0, y_0, z_0) *and a direction vector* $\boldsymbol{a} = (a_1, a_2, a_3)$: *in canonical form*

$$\frac{x - x_0}{a_1} = \frac{y - y_0}{a_2} = \frac{z - z_0}{a_3}; \tag{2}$$

in parametric form

$$x = x_0 + a_1 t, \quad y = y_0 + a_2 t, \quad z = z_0 + a_3 t; \tag{3}$$

b) *two points* (x_1, y_1, z_1), (x_2, y_2, z_2):

$$\frac{x - x_1}{x_2 - x_1} = \frac{y - y_1}{y_2 - y_1} = \frac{z - z_1}{z_2 - z_1}$$

or (in parametric form)

$$x = x_1 + (x_2 - x_1)\,t, \quad y = y_1 + (y_2 - y_1)\,t, \quad z = z_1 + (z_2 - z_1)\,t.$$

REMARK 8. Equations (2) can be obtained from the parametric equations (3) by eliminating the parameter. If zero occurs in the denominator of some of the above fractions we consider those equations which are involved as a mere formal notation and, as a rule, use another form of the equations.

Theorem 22. *The direction vector of a line is parallel to the vector product of the normal vectors of any two planes which contain the line,* i.e.

$$a_1 : a_2 : a_3 = \begin{vmatrix} B_1, & C_1 \\ B_2, & C_2 \end{vmatrix} : \begin{vmatrix} C_1, & A_1 \\ C_2, & A_2 \end{vmatrix} : \begin{vmatrix} A_1, & B_1 \\ A_2, & B_2 \end{vmatrix}.$$

Example 6. Obtain the parametric equations of the line given by the general equations

$$3x + 4y + 5z - 3 = 0,$$
$$x - 2y - 3z + 4 = 0.$$

The direction vector of the line can be found by Theorem 22:

$$a_1 : a_2 : a_3 = \begin{vmatrix} 4, & 5 \\ -2, & -3 \end{vmatrix} : \begin{vmatrix} 5, & 3 \\ -3, & 1 \end{vmatrix} : \begin{vmatrix} 3, & 4 \\ 1, & -2 \end{vmatrix} =$$
$$= (-2) : 14 : (-10) = 1 : (-7) : 5.$$

In order to find the parametric equations of this line we determine one of its points. We choose $z_0 = 0$ say and find x_0, y_0, from the equations

$$3x + 4y - 3 = 0\,,$$
$$x - 2y + 4 = 0\,.$$

We obtain $x_0 = -1$, $y_0 = 1{\cdot}5$. Thus, the required equations are

$$x = -1 + t\,, \quad y = 1{\cdot}5 - 7t\,, \quad z = 5t\,.$$

Theorem 23. *The distance d of a point $P(x_0, y_0, z_0)$ from the line*

$$\frac{x - x_1}{a_1} = \frac{y - y_1}{a_2} = \frac{z - z_1}{a_3}$$

is given by

$$d = \frac{|\boldsymbol{u} \times \boldsymbol{a}|}{|\boldsymbol{a}|}\,, \quad \textit{where} \quad \boldsymbol{u} = (x_0 - x_1, y_0 - y_1, z_0 - z_1)\,, \quad \boldsymbol{a} = (a_1, a_2, a_3)\,.$$

Example 7. Find the distance of the point $M(3, -1, 2)$ from the line

$$\frac{x - 2}{2} = \frac{y}{1} = \frac{z + 1}{-2}\,.$$

Here,

$$\boldsymbol{u} = (1, -1, 3)\,, \quad \boldsymbol{a} = (2, 1, -2)\,, \quad \boldsymbol{u} \times \boldsymbol{a} = \left(\begin{vmatrix} -1, & 3 \\ 1, & -2 \end{vmatrix}, \begin{vmatrix} 3, & 1 \\ -2, & 2 \end{vmatrix}, \begin{vmatrix} 1, & -1 \\ 2, & 1 \end{vmatrix} \right) =$$
$$= (-1, 8, 3)\,,$$
$$|\boldsymbol{u} \times \boldsymbol{a}| = \sqrt{[(-1)^2 + 8^2 + 3^2]} = \sqrt{74}\,, \quad |\boldsymbol{a}| = \sqrt{[2^2 + 1^2 + (-2)^2]} = 3\,;$$
$$d = \frac{\sqrt{74}}{3} = 2{\cdot}867\ldots\,.$$

REMARK 9. In the same way, the distance between two parallel lines can be calculated: we choose a particular point on one of them and then find the distance of this point from the other line.

Theorem 24. *The distance d between two skew lines ${}^1p \equiv \boldsymbol{r} = \boldsymbol{r}_1 + \boldsymbol{a}t$, ${}^2p \equiv \boldsymbol{r} = \boldsymbol{r}_2 + \boldsymbol{b}t'$ is given by*

$$d = \frac{|[(\boldsymbol{r}_2 - \boldsymbol{r}_1)\, \boldsymbol{a}\boldsymbol{b}]|}{|\boldsymbol{a} \times \boldsymbol{b}|}\,.$$

Example 8. Find the distance between the skew lines

$${}^1p \equiv \frac{x - 1}{2} = \frac{y + 2}{2} = \frac{z + 3}{-1}\,, \quad {}^2p \equiv \frac{x - 2}{1} = \frac{y + 1}{-2} = \frac{z - 1}{-2}\,.$$

Here $\mathbf{r}_1 = (1, -2, -3)$, $\mathbf{r}_2 = (2, -1, 1)$, $\mathbf{a} = (2, 2, -1)$, $\mathbf{b} = (1, -2, -2)$, and thus $\mathbf{r}_2 - \mathbf{r}_1 = (1, 1, 4)$,

$$\mathbf{a} \times \mathbf{b} = \left(\begin{vmatrix} 2, & -1 \\ -2, & -2 \end{vmatrix}, \begin{vmatrix} -1, & 2 \\ -2, & 1 \end{vmatrix}, \begin{vmatrix} 2, & 2 \\ 1, & -2 \end{vmatrix} \right) = (-6, 3, -6),$$

$$|\mathbf{a} \times \mathbf{b}| = 9, \quad [(\mathbf{r}_2 - \mathbf{r}_1)\mathbf{ab}] = (\mathbf{r}_2 - \mathbf{r}_1) . (\mathbf{a} \times \mathbf{b}) = -6 + 3 - 24 = -27 .$$

Hence

$$d = \frac{|-27|}{9} = 3 .$$

REMARK 10. Two lines ${}^1p \equiv \mathbf{r} = \mathbf{r}_1 + \mathbf{a}t$, ${}^2p \equiv \mathbf{r} = \mathbf{r}_2 + \mathbf{b}t'$ lie in the same plane if and only if the mixed product $[(\mathbf{r}_2 - \mathbf{r}_1)\mathbf{ab}]$ equals zero. If, in addition, $\mathbf{a}$ is not parallel to $\mathbf{b}$, then they intersect.

Theorem 25. *The angle between two lines is equal to the angle between their direction vectors* $\mathbf{a}$, $\mathbf{b}$, i.e.

$$\cos \varphi = \frac{\mathbf{a} . \mathbf{b}}{|\mathbf{a}| . |\mathbf{b}|} = \frac{a_1 b_1 + a_2 b_2 + a_3 b_3}{\sqrt{(a_1^2 + a_2^2 + a_3^2)} \sqrt{(b_1^2 + b_2^2 + b_3^2)}} .$$

Theorem 26. a) *A necessary and sufficient condition for two straight lines whose direction vectors are* $\mathbf{a}$, $\mathbf{b}$ *to be perpendicular is*

$$\mathbf{a} . \mathbf{b} = 0 ; \quad \text{i.e.} \quad a_1 b_1 + a_2 b_2 + a_3 b_3 = 0 .$$

b) *A necessary and sufficient condition for two straight lines whose direction vectors are* $\mathbf{a}$, $\mathbf{b}$ *to be parallel is*

$$\mathbf{a} \parallel \mathbf{b} , \quad \text{i.e.} \quad a_1 : a_2 : a_3 = b_1 : b_2 : b_3 .$$

Theorem 27. *The angle* φ *between a line and a plane is equal to the complement of the angle between the direction vector of the line and a normal vector to the plane; thus*

$$\sin \varphi = \frac{|\mathbf{a} . \mathbf{n}|}{|\mathbf{a}| . |\mathbf{n}|} ,$$

i.e. *if* $\mathbf{r} = \mathbf{r}_0 + \mathbf{a}t$ *and* $Ax + By + Cz + D = 0$ *are the equations of the line and the plane, respectively, then*

$$\sin \varphi = \frac{|a_1 A + a_2 B + a_3 C|}{\sqrt{(a_1^2 + a_2^2 + a_3^2)} \sqrt{(A^2 + B^2 + C^2)}} .$$

A necessary and sufficient condition for a line and a plane to be perpendicular is

$$\mathbf{a} \parallel \mathbf{n} ; \quad \text{i.e.} \quad a_1 : a_2 : a_3 = A : B : C .$$

A necessary and sufficient condition for a line and a plane to be parallel is

$$\mathbf{a} \cdot \mathbf{n} = 0 ; \quad \text{i.e.} \quad a_1 A + a_2 B + a_3 C = 0 .$$

6.3. Quadrics (Surfaces of the Second Order)

REMARK 1. In this section, a *surface* is defined as the locus of a point whose rectangular coordinates satisfy the equation $F(x, y, z) = 0$, where F is a function having continuous partial derivatives of at least the first order at every point. The points of a surface at which at least one of these partial derivatives differs from zero are called *regular points* of the surface, whereas the points at which all the first partial derivatives vanish are called *singular points* of the surface (for example the vertex of a cone).

(For a more detailed treatment see Chap. 9.)

Theorem 1. *The equation of the sphere with centre $S(x_0, y_0, z_0)$ and radius r is*

$$(x - x_0)^2 + (y - y_0)^2 + (z - z_0)^2 = r^2 .$$

If we perform the operations indicated in this equation, we obtain the *general equation of a sphere* in the form

$$x^2 + y^2 + z^2 + mx + ny + pz + q = 0 .$$

It should be noticed that the products xy, xz, yz do not occur and that the coefficients of the squared variables are all equal.

The coordinates of the centre, and the radius, of a sphere given by the general equation can be found by completing the squares:

$$\left(x + \frac{m}{2}\right)^2 + \left(y + \frac{n}{2}\right)^2 + \left(z + \frac{p}{2}\right)^2 = \frac{m^2 + n^2 + p^2}{4} - q .$$

If the right-hand side of this modified equation is a positive number, then the general equation represents the so-called *real sphere* with centre $S(-\frac{1}{2}m, -\frac{1}{2}n, -\frac{1}{2}p)$ and radius $r = \sqrt{[\frac{1}{4}(m^2 + n^2 + p^2) - q]}$; if the right-hand side equals zero, then only one real point (the centre of the sphere of zero radius) satisfies the general equation; if the right-hand side is a negative number, then no real point in space satisfies the general equation. (In this case we speak about a *virtual sphere*.)

Theorem 2. *The equation of the general ellipsoid with centre at the origin and the semi-axes a, b, c, coincident with the x, y and z axes, respectively, is*

$$\frac{x^2}{a^2} + \frac{y^2}{b^2} + \frac{z^2}{c^2} = 1 .$$

Particular cases are:

a) $a = b > c$ (*an oblate spheroid*; Fig. 6.4) ;
b) $a = b < c$ (*a prolate spheroid*; Fig. 6.5) ;
c) $a = b = c$ (*a sphere of radius a*) .

In cases a,) b) *the z-axis is the axis of revolution.*

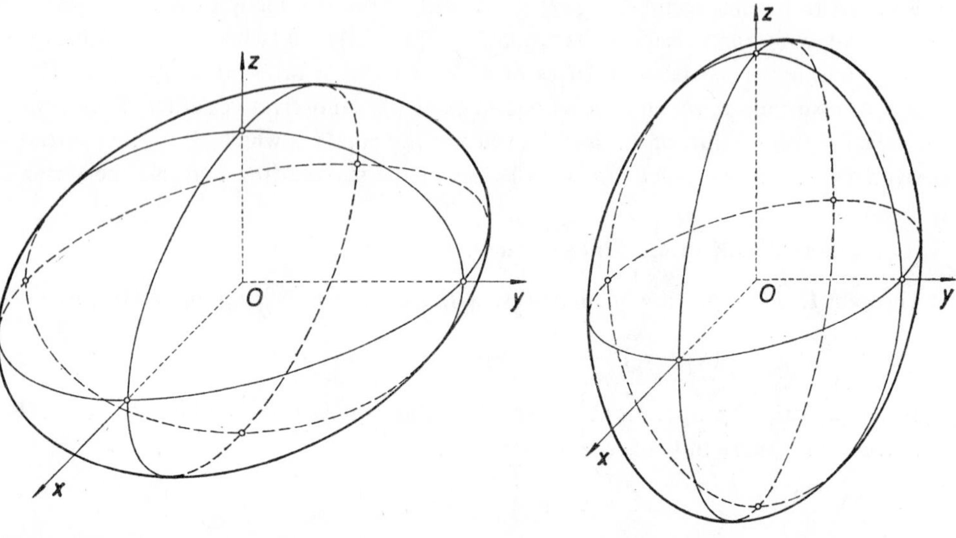

Fig. 6.4. Fig. 6.5.

Theorem 3. *Hyperboloids with centre at the origin and semi-axes a, b, c coincident with the axes x, y, z, respectively are of two types:*

a) *a hyperboloid of one sheet* (Fig. 6.6), *having the equation*

$$\frac{x^2}{a^2} + \frac{y^2}{b^2} - \frac{z^2}{c^2} = 1$$

(*a, b being its real semi-axes, and c its imaginary semi-axis*),

b) *a hyperboloid of two sheets* (Fig. 6.7), *having the equation*

$$\frac{x^2}{a^2} + \frac{y^2}{b^2} - \frac{z^2}{c^2} = -1$$

(*a, b being its imaginary semi-axes, and c the real semi-axis*).

If, in either case, $a = b$, then the hyperboloid is a hyperboloid of revolution and the z-axis is the axis of revolution.

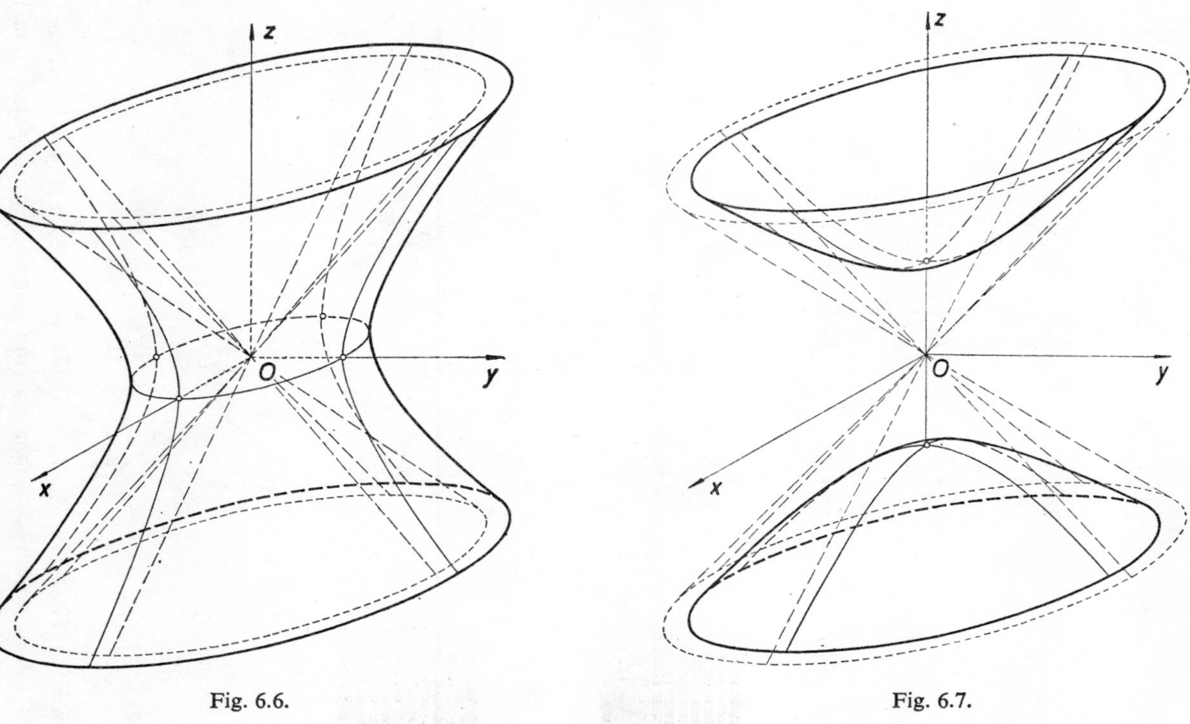

Fig. 6.6.

Fig. 6.7.

Theorem 4. *On a hyperboloid of one sheet there exist two sets of straight lines. Every line of the first set intersects every line of the other set, while no two lines of the same set intersect. The equations of the two sets of lines on the hyperboloid*

$$\frac{x^2}{a^2} + \frac{y^2}{b^2} - \frac{z^2}{c^2} = 1$$

are

a) $$k_1 \left(\frac{x}{a} + \frac{z}{c} \right) = k_2 \left(1 - \frac{y}{b} \right),$$

$$k_2 \left(\frac{x}{a} - \frac{z}{c} \right) = k_1 \left(1 + \frac{y}{b} \right);$$

b) $$k_1 \left(\frac{x}{a} + \frac{z}{c} \right) = k_2 \left(1 + \frac{y}{b} \right),$$

$$k_2 \left(\frac{x}{a} - \frac{z}{c} \right) = k_1 \left(1 - \frac{y}{b} \right),$$

where k_1, k_2 are arbitrary real numbers (not both equal to zero); clearly, only their ratio $k_1 : k_2$ is significant.

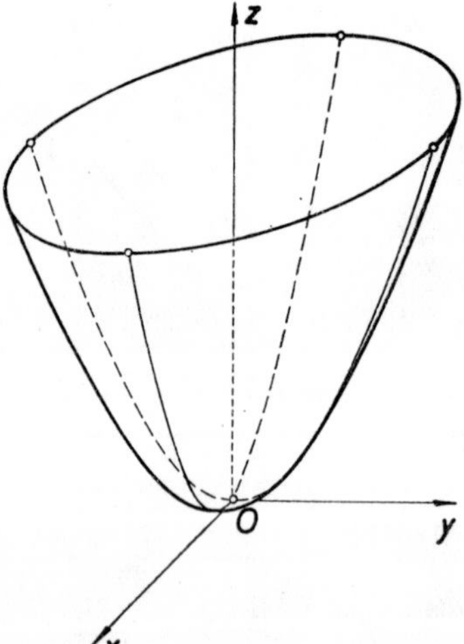

Fig. 6.8.

Theorem 5. a) *The equation of an elliptic paraboloid* (Fig. 6.8) *with vertex at the origin and whose planes of symmetry (the so-called principal sections) coincide with the planes $x = 0$, $y = 0$, is*

$$z = \frac{x^2}{2p} + \frac{y^2}{2q}, \quad pq > 0 .$$

b) *The equation of a hyperbolic paraboloid* (Fig. 6.9) *with vertex at the origin and whose planes of symmetry (principal sections) coincide with the planes* $x = 0$, $y = 0$, *is*

$$z = \frac{x^2}{2p} - \frac{y^2}{2q}, \quad pq > 0 .$$

If, in the case of an elliptic paraboloid, $p = q$, then the paraboloid is a surface of revolution and the z-axis is the axis of revolution.

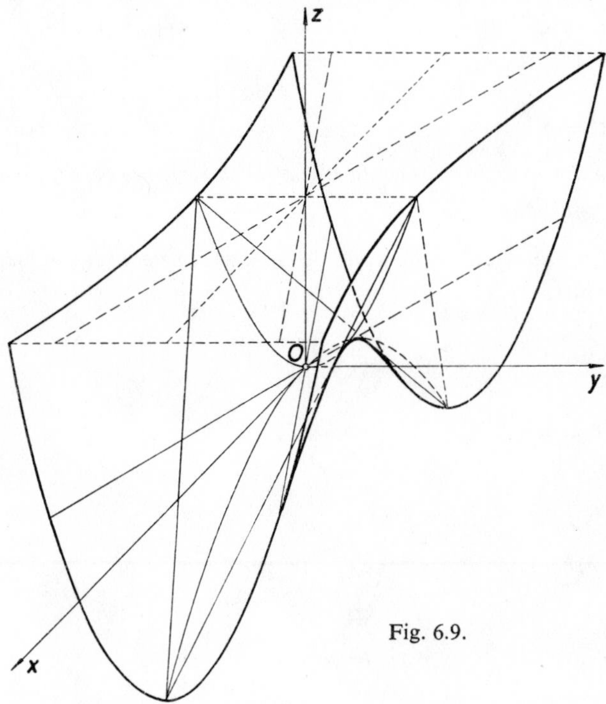

Fig. 6.9.

REMARK 2. If, in the equations of surfaces of Theorems 2 and 3, we replace x, y, z by $x - x_0$, $y - y_0$, $z - z_0$ respectively, we obtain the equations of the same surfaces with their centres translated to the point (x_0, y_0, z_0) and their axes parallel to the coordinate axes.

The same change in the equations of Theorem 5 gives the equations of paraboloids whose vertices are at the point (x_0, y_0, z_0) and whose planes of symmetry are parallel to the planes $x = 0$, $y = 0$.

In all these equations (after removing brackets) the products xy, xz, yz of the variables are missing. By similar re-arrangements as in the case of a sphere (see p. 247) the position of the centre, or the vertex, as well as the quantities a, b, c or p, q, can be found.

Theorem 6. *On a hyperbolic paraboloid there exist two sets of straight lines. Every line of the first set intersects every line of the other set, while no two lines of the same set intersect. The equations of the two sets of lines on the hyperbolic paraboloid*

$$z = \frac{x^2}{2p} - \frac{y^2}{2q} \quad (pq > 0)$$

are

$$\text{a)}\ k_1\left(\frac{x}{\sqrt{(2|p|)}} + \frac{y}{\sqrt{(2|q|)}}\right) = \frac{p}{|p|}k_2 z, \quad \text{b)}\ k_1\left(\frac{x}{\sqrt{(2|p|)}} - \frac{y}{\sqrt{(2|q|)}}\right) = \frac{p}{|p|}k_2 z,$$

$$k_2\left(\frac{x}{\sqrt{(2|p|)}} - \frac{y}{\sqrt{(2|q|)}}\right) = k_1; \qquad k_2\left(\frac{x}{\sqrt{(2|p|)}} + \frac{y}{\sqrt{(2|q|)}}\right) = k_1,$$

where k_1, k_2 are arbitrary real numbers (not both equal to zero); clearly only their ratio $k_1 : k_2$ is significant.

Theorem 7. *The equation of the quadric cone with vertex at the origin and whose directrix* is the ellipse*

$$\frac{x^2}{a^2} + \frac{y^2}{b^2} = 1$$

in the plane $z = c$, is

$$\frac{x^2}{a^2} + \frac{y^2}{b^2} = \frac{z^2}{c^2}.$$

This cone is also the asymptotic cone of the two hyperboloids

$$\frac{x^2}{a^2} + \frac{y^2}{b^2} - \frac{z^2}{c^2} = \pm 1.$$

If $a = b$, the cone is a cone of revolution and the z-axis is the axis of revolution.

Theorem 8. *Quadric cylinders are of three types:*

a) *elliptic:*

$$\frac{x^2}{a^2} + \frac{y^2}{b^2} = 1$$

(its directrix is an ellipse of the same equation in the plane $z = 0$ and the generating lines are parallel to the z-axis; if $a = b$, the cylinder is a cylinder of revolution and the z-axis is the axis of revolution);

* For the meaning of the term *directrix*, as used here, see Definition 6.4.4, p. 259.

b) *hyperbolic:*

$$\frac{x^2}{a^2} - \frac{y^2}{b^2} = 1$$

(*its directrix is a hyperbola of the same equation in the plane* $z = 0$ *and the generating lines are parallel to the z-axis*);

c) *parabolic:*

$$y^2 = 2px$$

(*its directrix is a parabola of the same equation in the plane* $z = 0$ *and the generating lines are parallel to the z-axis*).

Theorem 9. *The general equation of a quadric is*

$$a_{11}x^2 + a_{22}y^2 + a_{33}z^2 + 2a_{12}xy + 2a_{13}xz + 2a_{23}yz + 2a_{14}x + 2a_{24}y + \\ + 2a_{34}z + a_{44} = 0\,;$$

it has four so-called orthogonal invariants (i.e. functions of its coefficients whose values do not alter under translation or rotation of the coordinate system):

a) *the discriminant of the quadric*

$$A = \begin{vmatrix} a_{11}, & a_{12}, & a_{13}, & a_{14} \\ a_{12}, & a_{22}, & a_{23}, & a_{24} \\ a_{13}, & a_{23}, & a_{33}, & a_{34} \\ a_{14}, & a_{24}, & a_{34}, & a_{44} \end{vmatrix};$$

b) *the minor* A_{44} *of the discriminant*

$$A_{44} = \begin{vmatrix} a_{11}, & a_{12}, & a_{13} \\ a_{12}, & a_{22}, & a_{23} \\ a_{13}, & a_{23}, & a_{33} \end{vmatrix};$$

c) *the quadratic invariant*

$$I_2 = \begin{vmatrix} a_{11}, & a_{12} \\ a_{12}, & a_{22} \end{vmatrix} + \begin{vmatrix} a_{11}, & a_{13} \\ a_{13}, & a_{33} \end{vmatrix} + \begin{vmatrix} a_{22}, & a_{23} \\ a_{23}, & a_{33} \end{vmatrix};$$

d) *the linear invariant*

$$I_1 = a_{11} + a_{22} + a_{33}\,.$$

In addition, the above equation has two so-called semi-invariants whose values do not alter under rotation of the coordinate system:

e)

$$S_2 = \begin{vmatrix} a_{11}, & a_{14} \\ a_{14}, & a_{44} \end{vmatrix} + \begin{vmatrix} a_{22}, & a_{24} \\ a_{24}, & a_{44} \end{vmatrix} + \begin{vmatrix} a_{33}, & a_{34} \\ a_{34}, & a_{44} \end{vmatrix};$$

TABLE 6.1

*Determination of the type of a quadric given by the general equation in rectangular coordinates**

		Type of surface	Transformed equation	Canonical equation
	$A < 0$ $I_2 > 0$ $I_1 A_{44} > 0$	real ellipsoid		$\frac{x^2}{a^2} + \frac{y^2}{b^2} + \frac{z^2}{c^2} = 1$
	$A > 0$ $I_2 > 0$ $I_1 A_{44} > 0$	virtual ellipsoid		$\frac{x^2}{a^2} + \frac{y^2}{b^2} + \frac{z^2}{c^2} = -1$
	$A = 0$ $I_2 > 0$ $I_1 A_{44} > 0$	virtual cone (real point)		$\frac{x^2}{a^2} + \frac{y^2}{b^2} + \frac{z^2}{c^2} = 0$
$A_{44} \neq 0$	$A > 0$; at least one of the numbers I_2, $I_1 A_{44}$ negative	hyperboloid of one sheet	$k_1 x^2 + k_2 x^2 + k_3 z^2 + \frac{A}{A_{44}} = 0$	$\frac{x^2}{a^2} + \frac{y^2}{b^2} - \frac{z^2}{c^2} = 1$
	$A < 0$, at least one of the numbers I_2, $I_1 A_{44}$ negative	hyperboloid of two sheets		$\frac{x^2}{a^2} + \frac{y^2}{b^2} - \frac{z^2}{c^2} = -1$
	$A = 0$, at least one of the numbers I_2, $I_1 A_{44}$ negative	real cone		$\frac{x^2}{a^2} + \frac{y^2}{b^2} - \frac{z^2}{c^2} = 0$

* The entries of the first and second columns of Table 6.1 are necessary and sufficient conditions for the type of the surface stated in the third column. This table thus replaces a number of theorems, e.g.: a necessary and sufficient condition for a quadric to be a paraboloid is $A \neq 0$, $A_{44} = 0$; or, a necessary and sufficient condition for a quadric to be degenerate is $A = 0$, $A_{44} = 0$, $S_3 = 0$.

Table 6.1

		Type of surface	Transformed equation	Canonical equation
$A_{44} = 0$	$A < 0$	elliptic paraboloid	$k_1 x^2 + k_2 y^2 \pm 2\sqrt{\left(-\frac{A}{I_2}\right)} z = 0$	$\frac{x^2}{p} + \frac{y^2}{q} - 2z = 0$
$A \neq 0$	$A > 0$	hyperbolic paraboloid		$\frac{x^2}{p} - \frac{y^2}{q} - 2z = 0$
	$I_2 > 0$, $I_1 S_3 < 0$	real elliptic cylinder		$\frac{x^2}{a^2} + \frac{y^2}{b^2} = 1$
	$I_2 > 0$, $I_1 S_3 > 0$	virtual elliptic cylinder		$\frac{x^2}{a^2} + \frac{y^2}{b^2} = -1$
$A_{44} = 0$, $A = 0$, $I_2 \neq 0$	$I_2 > 0$, $S_3 = 0$	two intersecting virtual planes (a real line)	$k_1 x^2 + k_2 y^2 + \frac{S_3}{I_2} = 0$	$\frac{x^2}{a^2} + \frac{y^2}{b^2} = 0$
	$I_2 < 0$, $S_3 \neq 0$	hyperbolic cylinder		$\frac{x^2}{a^2} - \frac{y^2}{b^2} = 1$
	$I_2 < 0$, $S_3 = 0$	two intersecting real planes		$\frac{x^2}{a^2} - \frac{y^2}{b^2} = 0$
$A_{44} = 0$, $A = 0$, $I_2 = 0$	$S_3 \neq 0$	parabolic cylinder	$k_1 x^2 \pm 2\sqrt{\left(-\frac{S_3}{I_1}\right)} y = 0$	$x^2 - 2py = 0$
	$S_2 < 0$	two parallel real planes		$x^2 - a^2 = 0$
$A_{44} = 0$, $A = 0$, $I_2 = 0$, $S_3 = 0$	$S_2 > 0$	two parallel virtual planes	$k_1 x^2 + \frac{S_2}{I_1} = 0$	$x^2 + a^2 = 0$
	$S_2 = 0$	single plane (two coinciding planes)		$x^2 = 0$

f)

$$S_3 = \begin{vmatrix} a_{11}, a_{12}, a_{14} \\ a_{12}, a_{22}, a_{24} \\ a_{14}, a_{24}, a_{44} \end{vmatrix} + \begin{vmatrix} a_{11}, a_{13}, a_{14} \\ a_{13}, a_{33}, a_{34} \\ a_{14}, a_{34}, a_{44} \end{vmatrix} + \begin{vmatrix} a_{22}, a_{23}, a_{24} \\ a_{23}, a_{33}, a_{34} \\ a_{24}, a_{34}, a_{44} \end{vmatrix}.$$

Using these invariants and semi-invariants, the type of a quadric given by the general equation introduced above can be determined (Table 6.1). It should be noticed that a general quadratic equation in three variables with real coefficients need not always represent an equation of a quadric in the proper sense, i.e. ellipsoids, hyperboloids, paraboloids, cones or cylinders. For example, the equation $x^2 - y^2 = 0$ is satisfied by the points of two planes $x - y = 0$ and $x + y = 0$ (in this case, we speak about a *degenerate quadric*); similarly there is no point in space satisfying the equation $x^2 + y^2 + z^2 = -1$ and in this case we speak about a *virtual quadric.*

Theorem 10. *The coordinates of the centre of a quadric given by the general equation satisfy the system of linear equations*

$$\begin{aligned} a_{11}x + a_{12}y + a_{13}z + a_{14} &= 0\,, \\ a_{12}x + a_{22}y + a_{23}z + a_{24} &= 0\,, \\ a_{13}x + a_{23}y + a_{33}z + a_{34} &= 0\,. \end{aligned}$$

Depending upon the number of solutions of the above system, the quadric has no centre (quadric without centre), a single centre, an entire line of centres, or, finally, an entire plane of centres. (For a detailed analysis of the solution of a system of linear equations see § 1.18.)

To convert the general equation of a quadric to the standard form (so-called *canonical equation*) used in Theorems 1, 2, 3, 5, 7 and 8, the so-called *discriminating cubic* is important.

Definition 1. The equation $k^3 - I_1k^2 + I_2k - A_{44} = 0$ is called the *discriminating cubic of the quadric.* Its roots k_1, k_2, k_3 are all real.

Example 1. Determine the type and the canonical equation of the surface $2xy - 2xz + 2yz - 4x + 1 = 0$.

$$A = \begin{vmatrix} 0, & 1, & -1, & -2 \\ 1, & 0, & 1, & 0 \\ -1, & 1, & 0, & 0 \\ -2, & 0, & 0, & 1 \end{vmatrix} = 2 > 0\,; \quad A_{44} = \begin{vmatrix} 0, & 1, & -1 \\ 1, & 0, & 1 \\ -1, & 1, & 0 \end{vmatrix} = -2 \neq 0\,.$$

It follows immediately (according to Table 6.1) that the surface is either a virtual ellipsoid or a hyperboloid of one sheet. Also,

$$I_1 = 0\,; \quad I_2 = \begin{vmatrix} 0, & 1 \\ 1, & 0 \end{vmatrix} + \begin{vmatrix} 0, & -1 \\ -1, & 0 \end{vmatrix} + \begin{vmatrix} 0, & 1 \\ 1, & 0 \end{vmatrix} = -3 < 0\,.$$

Thus, the given surface is a hyperboloid of one sheet. Solving the discriminating cubic

$$k^3 - 0 \,.\, k^2 + (-3) \,.\, k - (-2) = 0\,,$$

i.e.

$$k^3 - 3k + 2 = 0\,,$$

we find that the roots are $k_1 = k_2 = 1$, $k_3 = -2$. Further, since $A/A_{44} = -1$, the equation of the surface after translation and rotation of the coordinate system is

$$x^2 + y^2 - 2z^2 - 1 = 0\,,$$

i.e.

$$x^2 + y^2 - 2z^2 = 1\,.$$

Hence, it is a hyperboloid of revolution. Finally, let us find the position of the centre of the surface in the original coordinate system:

$$\begin{aligned} y - z - 2 &= 0\,, \\ x \quad + z \quad &= 0\,, \\ -x + y \quad &= 0\,. \end{aligned}$$

Solving these equations we obtain $x = 1$, $y = 1$, $z = -1$; these are the coordinates of the centre of the given surface.

6.4. Surfaces of Revolution and Ruled Surfaces

Definition 1. In this section a *curve* is defined as a (one-parameter) set of points such that their rectangular coordinates satisfy equations of the form $x = x(t)$, $y = y(t)$, $z = z(t)$, where $x(t)$, $y(t)$, $z(t)$ are functions which are defined in a given domain (for example, in an interval I). These functions are assumed to possess, at every point of the domain considered, first derivatives which are not all zero. The above equations can be replaced by a single vector equation $\boldsymbol{r} = \boldsymbol{r}(t)$.

(For a more detailed treatment see Chap. 9.)

REMARK 1. A curve is often given in space as the intersection of two surfaces (not having a common two-parameter part), for example $z = f(x, y)$ and $z = g(x, y)$, or $F(x, y, z) = 0$ and $G(x, y, z) = 0$. It is usually possible, by a suitable choice of the parameter, to obtain the parametric equations of Definition 1.

Definition 2. A surface generated by a curve rotating about a fixed straight line (axis of revolution) is called a *surface of revolution.*

Theorem 1. *The equation of the surface generated by the rotation about the z-axis of the curve* $y = f(z)$ in the plane $x = 0$ *is*

$$x^2 + y^2 = [f(z)]^2 .$$

Theorem 2. *The equation of the surface generated by the rotation about the z-axis of the curve* $f(y, z) = 0$ $(y \geqq 0)$ *in the plane* $x = 0$ *is*

$$f(\sqrt{(x^2 + y^2)}, z) = 0 .$$

Example 1. The equation of the torus (anchor-ring) generated by the circle $x = 0$, $(y - a)^2 + z^2 = r^2$ $(0 < r \leqq a)$ rotating about the z-axis is

$$(\sqrt{(x^2 + y^2)} - a)^2 + z^2 = r^2 ,$$

which can be put (on removing the radicals) into the form

$$(x^2 + y^2 + z^2 + a^2 - r^2)^2 = 4a^2(x^2 + y^2) .$$

REMARK 2. The condition $y \geqq 0$ stated in Theorem 2 is usually not essential, since the equation $f(y, z) = 0$ can often be put into the form $g(y^2, z) = 0$. Then, the equation of the corresponding surface of revolution is $g(x^2 + y^2, z) = 0$. For example, the equation of the spheroid generated by the ellipse

$$x = 0 , \quad \frac{y^2}{b^2} + \frac{z^2}{c^2} - 1 = 0$$

rotating about the z-axis is

$$\frac{x^2 + y^2}{b^2} + \frac{z^2}{c^2} - 1 = 0 .$$

REMARK 3. In the general case, when the surface of revolution is generated by the curve $x = x(t)$, $y = y(t)$, $z = z(t)$ rotating about the straight line

$$\frac{x - x_0}{a} = \frac{y - y_0}{b} = \frac{z - z_0}{c} ,$$

we derive the equation of the surface in the following way:

We write down the equations of the circle traced by a general point of the rotating curve and then eliminate the parameter of this general point from these equations. This circle is the intersection of the plane in which the general point moves (this plane being perpendicular to the axis of rotation) and a sphere whose centre is at a point on the axis of rotation – for example the point (x_0, y_0, z_0) – and passing through the general point of the rotating curve, as shown in Example 2.

Example 2. The line

$$x = t\sqrt{2}, \quad y = \tfrac{1}{2} + t, \quad z = -1 + t(2 + \sqrt{2})$$

rotates about the axis

$$\frac{x-1}{1} = \frac{y-1}{-1} = \frac{z+1}{1}.$$

Derive the equation of the surface of revolution.

1. The plane of rotation passes through the point $M(t\sqrt{2}, \frac{1}{2} + t, -1 + t(2 + \sqrt{2}))$ and is perpendicular to the vector $(1, -1, 1)$. Its equation is

$$x - y + z - t(2\sqrt{(2)} + 1) + \tfrac{3}{2} = 0.$$

2. The equation of the sphere with centre at the point $S(1, 1, -1)$ passing through the point M is

$$x^2 + y^2 + z^2 - 2x - 2y + 2z + \tfrac{7}{4} = (2\sqrt{(2)} + 1)^2 t^2 - t(2\sqrt{(2)} + 1).$$

In order to eliminate the parameter t we obtain an explicit expression for t from the first equation

$$t(2\sqrt{(2)} + 1) = x - y + z + \tfrac{3}{2}$$

and substitute it into the second. On simplification we obtain

$$2xy - 2xz + 2yz - 4x + 1 = 0;$$

this is the equation of a hyperboloid of revolution of one sheet (see Example 6.3.1).

Definition 3. A surface through every point of which it is possible to draw a straight line lying entirely on the surface is called a *ruled surface.*

Particular examples of ruled surfaces are the (general) conical and (general) cylindrical surfaces.

Definition 4. The set of all straight lines passing through a fixed point V and intersecting a fixed curve c is called a *(general) conical surface* (with the exception of the case when c is a straight line passing through the point V). The point V is the *vertex,* the curve c is *the directrix*, the lines of the surface are the *ruling (generating) lines*, or *generators.*

Definition 5. The aggregate of all straight lines parallel to a given direction (i.e. to a vector or a line) and intersecting a fixed curve c is called a *(general) cylindrical surface.* The curve c is the *directrix*, the lines of the surface are the *ruling (generating) lines* or *generators.*

The equations of conical and cylindrical surfaces can be derived, for example, in the following way: express the directrix parametrically and obtain the equation of the line joining the vertex (at infinity in the case of a cylinder) to a general point of the directrix. Then, eliminate from these equations the parameter of the general point.

Example 3. A conical surface has directrix

$$c \equiv (x^2 + y^2 + z^2 = r^2 \,,\ x^2 + y^2 = rx)$$

and its vertex is at the origin of the coordinate system.

The parametric equations of the directrix can be written in the form $x = r\cos^2 t$, $y = r\sin t\cos t$, $z = r\sin t$. A general line of the surface is therefore

$$\frac{x}{r\cos^2 t} = \frac{y}{r\sin t\cos t} = \frac{z}{r\sin t}$$

and can be re-written in the form

$$x\sin t = y\cos t \,,$$
$$y = z\cos t \,.$$

The easiest way to eliminate the parameter is to express $\cos t$ and $\sin t$ in terms of x, y, z from the latter equations, and then to square and add, giving

$$\cos t = \frac{y}{z}\,,$$

$$\sin t = \frac{y^2}{xz}$$

$$1 = \frac{y^2}{z^2} + \frac{y^4}{x^2z^2}\,.$$

Hence we obtain, on simplification, the required equation of the surface:

$$x^2z^2 = y^2(x^2 + y^2)\,.$$

Theorem 3. a) *The equation of the cylindrical surface whose directrix in the plane* $z = 0$ *is* $f(x, y) = 0$ *and whose ruling lines are parallel to the z-axis is*

$$f(x, y) = 0\,.$$

b) *If, in the equation of a surface, one of the variables is missing, then this equation represents a cylindrical surface such that the ruling lines are parallel to the axis which is denoted by the missing variable, and the equation of the directrix is identical with that of the given surface and lies in the coordinate plane perpendicular to the ruling lines.*

Example 4. The equation $(x^2 + y^2)^2 = a^2(x^2 - y^2)$ is the equation of the cylindrical surface whose lines are parallel to the z-axis and whose directrix is the lemniscate of Bernoulli (see § 4.11) in the plane xy.

Theorem 4. *The equation of a cylindrical surface whose ruling lines are parallel to the vector (a_1, a_2, a_3) can always be put into the form*

$$F(a_3x - a_1z, a_3y - a_2z) = 0 .$$

Theorem 5. *The equation of a conical surface with vertex $V(x_0, y_0, z_0)$ can always be put into the form*

$$F\left(\frac{x - x_0}{z - z_0}, \frac{y - y_0}{z - z_0}\right) = 0 .$$

The equation of the conical surface derived in Example 3 can, for example, be written in the form

$$\frac{y^2}{z^2} + \frac{y^4}{z^4}\frac{z^2}{x^2} - 1 = 0 .$$

More general ruled surfaces are usually determined by three directrices and by the condition that the surface is formed by lines intersecting all the directrices. The equations of such surfaces can be derived in a way similar to that used in the case of the equations of cylindrical and conical surfaces: express one of the directrices parametrically, project the other two from a general point of this directrix and obtain, in this way, the equations of two conical surfaces with a common vertex. From these equations, eliminate the parameter of the variable point of the first directrix and so obtain the required equation of the surface.

In some ruled surfaces, one of the directrices may be a *plane*; the ruling lines are parallel to it.

Definition 6. A ruled surface determined by three directrices which consist of a curve, a straight line and a plane, is called a *conoid*.

Example 5. The conoid determined by the directrices: the curve $c \equiv (x = a \cos t, y = a \sin t, z = bt)$, the line $p \equiv (x = 0, y = 0)$ and the plane $z = 0$ is a so-called *helicoid*.)

Through a general point of the directrix-curve we draw on the one hand a plane parallel to the directrix-plane and on the other a plane containing the directrix-line. Eliminating the parameter, we obtain the required equation of the surface in the form

$$y = x \tan \frac{z}{b} .$$

Theorem 6. *The equation of the circular conoid determined by the directrix-curve* $x = 0, y^2 + z^2 = r^2$, *the directrix-line* $z = 0, x = a$ *and the directrix-plane* $y = 0$ *is*

$$a^2z^2 = (r^2 - y^2)(a - x)^2 .$$

Theorem 7. *The equation of Plücker's conoid determined by the directrix curve* $x^2 + y^2 = rx$, $z = x \tan \alpha$, *the directrix-line* $x = y = 0$ *and the directrix-plane* $z = 0$ *is*

$$z(x^2 + y^2) = rx^2 \tan \alpha .$$

Theorem 8. *The equation of Küpper's conoid determined by the directrix-curve* $x^2 + y^2 = rx$, $z = 0$, *the directrix-line* $x = y = 0$ *and the directrix-plane* $z = x$ *is*

$$rx^2 = (x^2 + y^2)(x - z) .$$

Theorem 9. *The equation of the Montpellier conoid determined by the directrix-circle* $y^2 + z^2 = r^2$, $x = 0$ *and two directrix-lines* ${}^1p \equiv (y = z = 0)$, ${}^2p \equiv (x = a, z = b)$ *is*

$$r^2z^2(a - x)^2 = (az - bx)^2 (y^2 + z^2) .$$

7. VECTOR CALCULUS

References: [16], [34], [44], [63], [76], [208], [215], [302], [339], [344], [386], [421], [423].

A. VECTOR ALGEBRA

By FRANTIŠEK KEJLA

7.1. Vector Algebra; Scalar (Inner), Vector (Cross), Mixed and Treble Products

There is a great advantage in using vector calculus when solving various problems in applied mathematics. This advantage consists on the one hand in a special notation facilitating a very simple description of many relations which would otherwise be expressed by awkward and incomprehensible formulae, on the other in the possibility of expressing many laws and formulae in a form independent of the coordinate system.

Convention 1. Throughout this chapter, by the term "vector" a three-component vector, i.e. a vector in three-dimensional space will be understood (for general definition of a vector see § 1.15).

Definition 1. Ordered triplets of real numbers for which

a) equality: $(a_1, a_2, a_3) = (b_1, b_2, b_3)$ if and only if $a_1 = b_1$, $a_2 = b_2$, $a_3 = b_3$;

b) sum: $(a_1, a_2, a_3) + (b_1, b_2, b_3) = (a_1 + b_1, a_2 + b_2, a_3 + b_3)$;

c) product of a triple and a number: $k(a_1, a_2, a_3) = (ka_1, ka_2, ka_3)$ are defined, are called *vectors*. We denote them usually by bold letters, i.e. $\boldsymbol{a} = (a_1, a_2, a_3)$ or $\boldsymbol{a}(a_1, a_2, a_3)$. The numbers a_1, a_2, a_3 are said to be the *components of the vector* $\boldsymbol{a}$.

Definition 2. The vector $(0, 0, 0)$ is called the *zero vector* or *null vector;* we denote it by $\mathbf{0}$.

Definition 3. By the *vector opposite* to a vector $\boldsymbol{a}(a_1, a_2, a_3)$ we mean the vector $(-a_1, -a_2, -a_3)$ and denote it by $-\boldsymbol{a}$.

Theorem 1. *Vectors satisfy*

(i) *the commutative law:* $\boldsymbol{a} + \boldsymbol{b} = \boldsymbol{b} + \boldsymbol{a}$;
(ii) *the associative law:* $(\boldsymbol{a} + \boldsymbol{b}) + \boldsymbol{c} = \boldsymbol{a} + (\boldsymbol{b} + \boldsymbol{c})$;
(iii) *distributive laws:* $k(\boldsymbol{a} + \boldsymbol{b}) = k\boldsymbol{a} + k\boldsymbol{b}$, $(k_1 + k_2)\,\boldsymbol{a} = k_1\boldsymbol{a} + k_2\boldsymbol{a}$, where k, k_1, k_2 are numbers.

For further properties of vectors see § 1.15.

Some quantities in physics (force, velocity etc.) are known to be of vector character. They are customarily represented by *directed segments,* i.e. by a segment having a certain length and direction. In what follows, we assume that a fixed cartesian coordinate system in space has been chosen. In this system, we establish the above-mentioned representation of vectors by the following definition:

Definition 4. Every ordered pair of points $A(a_1, a_2, a_3)$, $B(b_1, b_2, b_3)$ determines a vector $(b_1 - a_1, b_2 - a_2, b_3 - a_3)$. This vector is denoted by $\overrightarrow{AB}$. By the arrow the direction of the vector is marked; A is called the *initial* (*starting*) *point, B the end* (*terminal*) *point of the vector* $\overrightarrow{AB}$.

In figures a vector $\overrightarrow{AB}$ is illustrated by a segment AB with an arrow at the end point B (see Fig. 7.1).

Convention 2. In the following text we shall use the term "vector" also for the graphical illustration of a vector.

Theorem 2. *If a point* $A(a_1, a_2, a_3)$ *and a vector* $\mathbf{u}(u_1, u_2, u_3)$ *are given, then there exists a unique point B such that* $\overrightarrow{AB} = \mathbf{u}$. *The coordinates of the point B are* $a_1 + u_1$, $a_2 + u_2$, $a_3 + u_3$.

Theorem 3. *Two vectors* $\overrightarrow{AB}$, $\overrightarrow{CD}$ *are equal if and only if the equations* $b_1 - a_1 = d_1 - c_1$, $b_2 - a_2 = d_2 - c_2$, $b_3 - a_3 = d_3 - c_3$ *hold.*

Fig. 7.1. Fig. 7.2.

REMARK 1. Theorem 3 states that a vector can be arbitrarily placed in space by a choice of its initial point. Its end point is then determined uniquely. We speak of so-called *free vectors.*

A vector with its initial point at the origin and end point at the given point P is called the *radius* (*position*) *vector of the point P.*

Theorem 4. *Let* $\mathbf{a}$, $\mathbf{b}$ *be two vectors. If we place the vector* $\mathbf{b}$ *so that its initial point coincides with the end point of the vector* $\mathbf{a}$, *then the vector* $\mathbf{c}$ *determined by the initial point of the vector* $\mathbf{a}$ *and by the end point of the vector* $\mathbf{b}$ *(and directed in this sense) equals the sum of the vectors* $\mathbf{a}$, $\mathbf{b}$, *i.e.* $\mathbf{c} = \mathbf{a} + \mathbf{b}$ (see Fig. 7.2).

Definition 5. By the *length of a vector* $\mathbf{a}(a_1, a_2, a_3)$ we understand the non-negative number $\sqrt{(a_1^2 + a_2^2 + a_3^2)}$; we denote it by $|\mathbf{a}|$ or a. A vector whose length equals unity is called a *unit vector.*

REMARK 2. Instead of "the length of a vector" the terms *modulus, magnitude, norm* or *absolute value of a vector* are also used. *The length of a vector is equal to the length of a segment representing the given vector.*

Theorem 5. *For every non-zero vector* $\mathbf{a}$ *there exists a unit vector* $\mathbf{a}^0$ *conformably parallel to the vector* $\mathbf{a}$ (Theorem 7). *It is (uniquely) determined by the relation*

$$\mathbf{a}^0 = \frac{\mathbf{a}}{|\mathbf{a}|}.$$

(The notation $\hat{\mathbf{a}}$ is often used for a unit vector.)

Definition 6. The linearly independent vectors $\mathbf{i}(1, 0, 0)$, $\mathbf{j}(0, 1, 0)$, $\mathbf{k}(0, 0, 1)$ (Fig. 7.3) are called *principal* or *coordinate vectors.* (For the concept of linear dependence and independence see Definition 1.15.3, p. 62.)

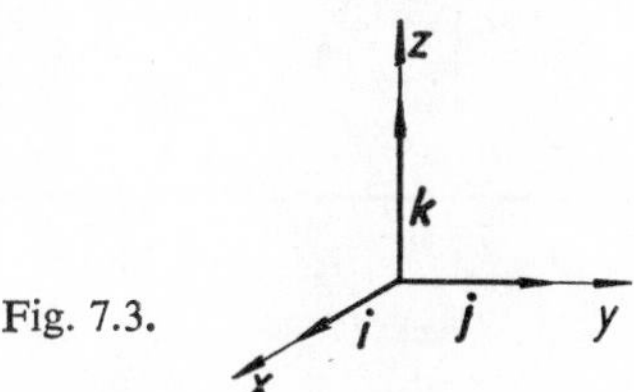

Fig. 7.3.

Theorem 6. *Any four vectors are linearly dependent. Thus every vector* $\mathbf{a}(a_1, a_2, a_3)$ *can be expressed as a linear combination of three linearly independent vectors, in particular as a linear combination of the principal vectors* $\mathbf{i}$, $\mathbf{j}$, $\mathbf{k}$:

$$\mathbf{a} = a_1\mathbf{i} + a_2\mathbf{j} + a_3\mathbf{k}.$$

Definition 7. Two linearly dependent vectors are called *collinear (parallel)*; three linearly dependent vectors are called *coplanar.*

Theorem 7. *Two vectors* $\mathbf{a}, \mathbf{b}$ *are linearly dependent (parallel) if and only if one of them is a multiple of the other, i.e. if there is a number k such that either* $\mathbf{a} = k\mathbf{b}$ or $\mathbf{b} = k\mathbf{a}$. *In a graphical illustration, they are represented by two parallel segments (unless one of them is the zero vector) whose lengths satisfy* $|\mathbf{a}| = |k|\,|\mathbf{b}|$, or $|\mathbf{b}| = |k|\,|\mathbf{a}|$. If $k > 0$, the vectors $\mathbf{a}$, $\mathbf{b}$ are said to be *conformably collinear (con-*

formably parallel); if $k < 0$, they are said to be *unconformably collinear* (*unconformably parallel*).

Theorem 8. *Coplanar vectors are parallel to a common plane (they can be placed in the same plane).*

Definition 8. The *angle between two non-zero vectors* $\boldsymbol{a}$, $\boldsymbol{b}$ is the angle φ ($0 \leqq \varphi \leqq \pi$) between the directed segments representing both vectors.

Definition 9. The *scalar* (*inner, dot*) *product* ($\boldsymbol{a} . \boldsymbol{b}$ or $\boldsymbol{ab}$, in symbols) of vectors $\boldsymbol{a}(a_1, a_2, a_3)$, $\boldsymbol{b}(b_1, b_2, b_3)$ is the (scalar) number $a_1 b_1 + a_2 b_2 + a_3 b_3$.

Theorem 9. *If* $\boldsymbol{a}$, $\boldsymbol{b}$ *are non-zero vectors and* φ *the angle between them, then the following relation holds for their scalar product:*

$$\boldsymbol{a} . \boldsymbol{b} = |\boldsymbol{a}| \, |\boldsymbol{b}| \cos \varphi .$$

Example 1.

$$\boldsymbol{i} . \boldsymbol{i} = 1 , \quad \boldsymbol{j} . \boldsymbol{j} = 1 , \quad \boldsymbol{k} . \boldsymbol{k} = 1 ,$$
$$\boldsymbol{i} . \boldsymbol{j} = 0 , \quad \boldsymbol{j} . \boldsymbol{k} = 0 , \quad \boldsymbol{k} . \boldsymbol{i} = 0 ;$$

these results can easily be verified by Definition 9 or Theorem 9 (see Fig. 7.3).

REMARK 3. Theorem 9 is very often used to compute the angle between two vectors which are given by their components.

Example 2. For $\boldsymbol{a}(2, 1, 2)$, $\boldsymbol{b}(1, -1, 4)$

$$\cos \varphi = \frac{\boldsymbol{a} . \boldsymbol{b}}{|\boldsymbol{a}| \, |\boldsymbol{b}|} = \frac{2 . 1 + 1 . (-1) + 2 . 4}{\sqrt{(2^2 + 1^2 + 2^2)} \sqrt{[1^2 + (-1)^2 + 4^2]}} = \frac{9}{3 . \sqrt{18}} = \frac{1}{\sqrt{2}} ,$$

and thus $\varphi = \frac{1}{4}\pi$.

Theorem 10. *Two non-zero vectors* $\boldsymbol{a}$, $\boldsymbol{b}$ *are perpendicular if and only if*

$$\boldsymbol{a} . \boldsymbol{b} = 0 .$$

Theorem 11. *The scalar product of vectors satisfies the relations:*

a) $\boldsymbol{a} . \boldsymbol{b} = \boldsymbol{b} . \boldsymbol{a}$;
b) $(\boldsymbol{a} + \boldsymbol{b}) . \boldsymbol{c} = \boldsymbol{a} . \boldsymbol{c} + \boldsymbol{b} . \boldsymbol{c}$;
c) $\boldsymbol{a} . \boldsymbol{a} = |\boldsymbol{a}|^2$.

REMARK 4. Instead of $\boldsymbol{a} . \boldsymbol{a}$ we often write $\boldsymbol{a}^2$.

Definition 10. The angles which a non-zero vector makes with the principal vectors (and thus with the coordinate axes) are called the *direction angles* and their cosines the *direction cosines* of the given vector.

Theorem 12. *Denoting by* α, β, γ *the direction angles of a non-zero vector* $\boldsymbol{a}(a_1, a_2, a_3)$, *the following relations hold for the direction cosines of the vector* $\boldsymbol{a}$:

a) $\cos \alpha = \dfrac{a_1}{|\boldsymbol{a}|}$; $\cos \beta = \dfrac{a_2}{|\boldsymbol{a}|}$; $\cos \gamma = \dfrac{a_3}{|\boldsymbol{a}|}$;

b) $\cos^2 \alpha + \cos^2 \beta + \cos^2 \gamma = 1$.

Definition 11. Three linearly independent vectors $\boldsymbol{a}$, $\boldsymbol{b}$, $\boldsymbol{c}$ with a common initial point determine a trihedral angle $(\boldsymbol{a}, \boldsymbol{b}, \boldsymbol{c})$. It is said to be *positively oriented* if the determinant of the coordinates of the vectors (in the given order), i.e. the determinant

$$\begin{vmatrix} a_1, & a_2, & a_3 \\ b_1, & b_2, & b_3 \\ c_1, & c_2, & c_3 \end{vmatrix}$$

is positive. If it is negative, then we say that the trihedral angle is *negatively oriented.*

Example 3. The trihedral angle defined by the vectors $\boldsymbol{i}$, $\boldsymbol{j}$, $\boldsymbol{k}$ (in the given order) is positively oriented, since

$$\begin{vmatrix} 1, & 0, & 0 \\ 0, & 1, & 0 \\ 0, & 0, & 1 \end{vmatrix} = 1 > 0 .$$

Definition 12. The *vector (cross, outer) product* ($\boldsymbol{a} \times \boldsymbol{b}$, or $\boldsymbol{a} \wedge \boldsymbol{b}$, in symbols) of vectors $\boldsymbol{a}(a_1, a_2, a_3)$, $\boldsymbol{b}(b_1, b_2, b_3)$ is the vector

$$\boldsymbol{w}\left(\begin{vmatrix} a_2, & a_3 \\ b_2, & b_3 \end{vmatrix}, \begin{vmatrix} a_3, & a_1 \\ b_3, & b_1 \end{vmatrix}, \begin{vmatrix} a_1, & a_2 \\ b_1, & b_2 \end{vmatrix}\right).$$

Theorem 13. *The vector product satisfies the relations*:

a) $\boldsymbol{a} \times \boldsymbol{b} = -(\boldsymbol{b} \times \boldsymbol{a})$;
b) $k\boldsymbol{a} \times \boldsymbol{b} = k(\boldsymbol{a} \times \boldsymbol{b})$;
c) $\boldsymbol{a} \times (\boldsymbol{b} + \boldsymbol{c}) = \boldsymbol{a} \times \boldsymbol{b} + \boldsymbol{a} \times \boldsymbol{c}$.

REMARK 5. Hence, the commutative law does not hold for the vector product.

Theorem 14. *The vector product of two linearly dependent vectors is the zero vector.*

Theorem 15. *The vector product* $\boldsymbol{w} = \boldsymbol{a} \times \boldsymbol{b}$ *of two linearly independent vectors* $\boldsymbol{a}$, $\boldsymbol{b}$ *possesses the following properties*:

a) *It is perpendicular to both given vectors* $\boldsymbol{a}$, $\boldsymbol{b}$, i.e. $\boldsymbol{w} . \boldsymbol{a} = 0$, $\boldsymbol{w} . \boldsymbol{b} = 0$;

b) *its length is numerically equal to the area of the parallelogram of which the vectors* $\boldsymbol{a}$, $\boldsymbol{b}$ *are concurrent sides*: i.e. $|\boldsymbol{w}| = |\boldsymbol{a}| . |\boldsymbol{b}| . \sin \varphi$, *where* φ *is the angle between the vectors* $\boldsymbol{a}$, $\boldsymbol{b}$;

c) *the trihedral angle* $(\boldsymbol{a}, \boldsymbol{b}, \boldsymbol{w})$ *is positively oriented.*

Example 4. In view of Definition 12, or Theorem 15, the following relations can readily be established:

$$\boldsymbol{i} \times \boldsymbol{j} = \boldsymbol{k}, \quad \boldsymbol{j} \times \boldsymbol{k} = \boldsymbol{i}, \quad \boldsymbol{k} \times \boldsymbol{i} = \boldsymbol{j},$$

$$\boldsymbol{i} \times \boldsymbol{i} = \boldsymbol{0}, \quad \boldsymbol{j} \times \boldsymbol{j} = \boldsymbol{0}, \quad \boldsymbol{k} \times \boldsymbol{k} = \boldsymbol{0}$$

(see Fig. 7.3).

REMARK 6. The properties listed in Theorem 15 are sometimes used (in physics, for example) to define the vector product.

Theorem 16. *The vector product of vectors* $\boldsymbol{a}$, $\boldsymbol{b}$ *can be written by means of the principal vectors in the form*

$$\boldsymbol{a} \times \boldsymbol{b} = \begin{vmatrix} \boldsymbol{i}, & \boldsymbol{j}, & \boldsymbol{k} \\ a_1, & a_2, & a_3 \\ b_1, & b_2, & b_3 \end{vmatrix}.$$

Definition 13. *The mixed product or triple scalar product of three vectors* $\boldsymbol{a}$, $\boldsymbol{b}$, $\boldsymbol{c}$ is the number $\boldsymbol{a} \cdot (\boldsymbol{b} \times \boldsymbol{c})$, denoted by $[\boldsymbol{abc}]$ or $\boldsymbol{abc}$.

REMARK 7. The mixed product of three vectors is sometimes called also a *trivector.*

Theorem 17. *The following relations hold for the mixed product of vectors* $\boldsymbol{a}$, $\boldsymbol{b}$, $\boldsymbol{c}$:

$$[\boldsymbol{abc}] = [\boldsymbol{bca}] = [\boldsymbol{cab}] = -[\boldsymbol{acb}] = -[\boldsymbol{cba}] = -[\boldsymbol{bac}] = \begin{vmatrix} a_1, & a_2, & a_3 \\ b_1, & b_2, & b_3 \\ c_1, & c_2, & c_3 \end{vmatrix}.$$

Theorem 18. *The absolute value of the mixed product is equal to the volume of the parallelepiped of which* $\boldsymbol{a}$, $\boldsymbol{b}$, $\boldsymbol{c}$ *are concurrent edges.*

Theorem 19. *Three vectors* $\boldsymbol{a}$, $\boldsymbol{b}$, $\boldsymbol{c}$ *are coplanar if and only if*

$$[\boldsymbol{abc}] = 0.$$

Definition 14. The vector $\boldsymbol{a} \times (\boldsymbol{b} \times \boldsymbol{c})$ is said to be the *triple vector product of vectors* $\boldsymbol{a}$, $\boldsymbol{b}$, $\boldsymbol{c}$ (in the given order).

Theorem 20. *The triple vector product of vectors* $\boldsymbol{a}$, $\boldsymbol{b}$, $\boldsymbol{c}$ *can be expressed without using vector multiplication*: $\boldsymbol{a} \times (\boldsymbol{b} \times \boldsymbol{c}) = (\boldsymbol{a} \cdot \boldsymbol{c})\, \boldsymbol{b} - (\boldsymbol{a} \cdot \boldsymbol{b})\, \boldsymbol{c}$.

REMARK 8. In general, $\boldsymbol{a} \times (\boldsymbol{b} \times \boldsymbol{c}) \neq (\boldsymbol{a} \times \boldsymbol{b}) \times \boldsymbol{c}$.

Theorem 21. $(\boldsymbol{a} \times \boldsymbol{b}) \cdot (\boldsymbol{c} \times \boldsymbol{d}) = (\boldsymbol{a} \cdot \boldsymbol{c})(\boldsymbol{b} \cdot \boldsymbol{d}) - (\boldsymbol{b} \cdot \boldsymbol{c})(\boldsymbol{a} \cdot \boldsymbol{d})$ (the so-called *Lagrange identity*).

REMARK 9. In particular, Theorem 21 yields: $(\boldsymbol{a} \times \boldsymbol{b})^2 = \boldsymbol{a}^2\boldsymbol{b}^2 - (\boldsymbol{a} \cdot \boldsymbol{b})^2$.

Theorem 22.

$$[\boldsymbol{abc}] \cdot [\boldsymbol{def}] = \begin{vmatrix} \boldsymbol{a} \cdot \boldsymbol{d}, & \boldsymbol{a} \cdot \boldsymbol{e}, & \boldsymbol{a} \cdot \boldsymbol{f} \\ \boldsymbol{b} \cdot \boldsymbol{d}, & \boldsymbol{b} \cdot \boldsymbol{e}, & \boldsymbol{b} \cdot \boldsymbol{f} \\ \boldsymbol{c} \cdot \boldsymbol{d}, & \boldsymbol{c} \cdot \boldsymbol{e}, & \boldsymbol{c} \cdot \boldsymbol{f} \end{vmatrix}.$$

B. VECTOR ANALYSIS

By KAREL REKTORYS

7.2. Derivative of a Vector. Scalar and Vector Fields. Gradient, Divergence, Curl (Rotation). Operator ∇, Laplace Operator. Transformation to Cylindrical and Spherical Coordinates

In physical and geometrical considerations we often have to deal with the case where the components of a vector are functions of a scalar variable t,

$$\boldsymbol{a}(t) = a_1(t)\,\boldsymbol{i} + a_2(t)\,\boldsymbol{j} + a_3(t)\,\boldsymbol{k}\,. \tag{1}$$

Thus, for every t of the domain under consideration we get, in general, a different vector; we speak about the *vector field*, or briefly about the *vector* $\boldsymbol{a}(t)$. The components of a vector can also be functions of several variables.

We define the derivative of a vector $\boldsymbol{a}(t)$. (The corresponding partial derivatives, when the components of $\boldsymbol{a}$ are functions of several variables, follow in a similar way.)

Derivative of a vector $\boldsymbol{a}(t)$:

$$\boldsymbol{a}'(t) = \frac{\mathrm{d}\boldsymbol{a}(t)}{\mathrm{d}t} = \lim_{\Delta\to 0} \frac{\boldsymbol{a}(t+\Delta t) - \boldsymbol{a}(t)}{\Delta t} = a_1'(t)\,\boldsymbol{i} + a_2'(t)\,\boldsymbol{j} + a_3'(t)\,\boldsymbol{k}\,. \tag{2}$$

Similarly

$$\boldsymbol{a}''(t) = \lim_{\Delta\to 0} \frac{\boldsymbol{a}'(t+\Delta t) - \boldsymbol{a}'(t)}{\Delta t} = a_1''(t)\,\boldsymbol{i} + a_2''(t)\,\boldsymbol{j} + a_3''(t)\,\boldsymbol{k} \tag{3}$$

etc.

Theorem 1.

$$(\boldsymbol{a} \cdot \boldsymbol{b})' = \boldsymbol{a}' \cdot \boldsymbol{b} + \boldsymbol{a} \cdot \boldsymbol{b}'\,, \quad (\boldsymbol{a} \cdot \boldsymbol{b})'' = \boldsymbol{a}'' \cdot \boldsymbol{b} + 2\boldsymbol{a}' \cdot \boldsymbol{b}' + \boldsymbol{a} \cdot \boldsymbol{b}''\,, \tag{4}$$

$$(\boldsymbol{a} \times \boldsymbol{b})' = \boldsymbol{a}' \times \boldsymbol{b} + \boldsymbol{a} \times \boldsymbol{b}'\,, \quad (\boldsymbol{a} \times \boldsymbol{b})'' = \boldsymbol{a}'' \times \boldsymbol{b} + 2\boldsymbol{a}' \times \boldsymbol{b}' + \boldsymbol{a} \times \boldsymbol{b}''\,. \tag{5}$$

Example 1. If the length of a vector $\boldsymbol{a}(t)$ is constant and equal to k then from the equation

$$\boldsymbol{a}^2(t) = \boldsymbol{a}(t) \,.\, \boldsymbol{a}(t) = k^2$$

the relation

$$\boldsymbol{a}'(t)\, \boldsymbol{a}(t) + \boldsymbol{a}(t)\, \boldsymbol{a}'(t) = 2\boldsymbol{a}'(t)\, \boldsymbol{a}(t) = 0$$

follows and thus (for every t at which the derivative $\boldsymbol{a}'(t)$ exists and $\boldsymbol{a}(t) \neq \mathbf{0}$, $\boldsymbol{a}'(t) \neq \mathbf{0}$)

$$\boldsymbol{a}'(t) \perp \boldsymbol{a}(t)\,.$$

REMARK 1. A case of particular importance is that of the space curve described by the end-point of the radius vector

$$\boldsymbol{r}(s) = x(s)\, \boldsymbol{i} + y(s)\, \boldsymbol{j} + z(s)\, \mathbf{k}\,;$$

here s denotes the length of the curve, measured from a fixed point on the curve (see § 9.2, p. 303). Then, the vector

$$\frac{\mathrm{d}\boldsymbol{r}}{\mathrm{d}s} = \mathbf{t} \tag{6}$$

is a unit vector and is called the *tangent (unit) vector* of the curve under consideration. The unit vector

$$\frac{\dfrac{\mathrm{d}\mathbf{t}}{\mathrm{d}s}}{\left|\dfrac{\mathrm{d}\mathbf{t}}{\mathrm{d}s}\right|} = \boldsymbol{n} \tag{7}$$

is called the *principal normal (unit) vector* of the curve. The vector

$$\mathbf{b} = \mathbf{t} \times \boldsymbol{n} \tag{8}$$

is called the *binormal vector* of this curve. The vectors $\mathbf{t}$, $\boldsymbol{n}$, $\mathbf{b}$ are mutually orthogonal and form the so-called *moving trihedral* of the curve (for more details see § 9.3).

GRADIENT. By means of a function

$$u = f(x, y, z)\,,$$

a *scalar field* is given in the region O in which the function is defined. The surfaces $u = \text{const.}$ are the *level (equipotential) surfaces* of this scalar field.

Definition 1. The vector

$$\operatorname{grad} u = \frac{\partial u}{\partial x}\, \boldsymbol{i} + \frac{\partial u}{\partial y}\, \boldsymbol{j} + \frac{\partial u}{\partial z}\, \mathbf{k} \tag{9}$$

is said to be the *gradient* of the given scalar field.

REMARK 2. Thus, the gradient of a scalar field is a vector. *At a fixed point* $(x_0, y_0, z_0) \in O$, *this vector is perpendicular to the level surface passing through this point* .

An example of a scalar field is an electrostatic potential field. Its level surfaces are called *equipotential surfaces.* The gradient defines the *vector field* characterizing (at every point) the intensity of the given electrostatic field. The curves touching this field at every point (i.e. curves such that the gradient at every point of a curve is a tangent vector of this curve) are called the *lines of force.*

If a vector (vector field) $\boldsymbol{a}(x, y, z)$ is given in a region O and if there is a (univalent) function $u = f(x, y, z)$ in O such that this vector is the gradient of the function u in O, i.e.

$$\boldsymbol{a}(x, y, z) = \operatorname{grad} u(x, y, z)\,, \tag{10}$$

then this vector field is called *irrotational* (*conservative*) and u is the *scalar potential.* In an irrotational field the work done by the force $\boldsymbol{a} = \operatorname{grad} u$ along a curve c lying in O and connecting two points A, B of this field does not depend on the form of this curve. In particular, the work along a closed curve is zero:

$$\oint_c \operatorname{grad} u \,.\, \mathrm{d}\boldsymbol{s} = 0 \tag{11}$$

(see equation (7.3.4)); $\mathrm{d}\boldsymbol{s} = \boldsymbol{i}\,\mathrm{d}x + \boldsymbol{j}\,\mathrm{d}y + \boldsymbol{k}\,\mathrm{d}z$.

Theorem 2. *The relation*

$$\mathrm{d}u = \operatorname{grad} u \,.\, \mathrm{d}\boldsymbol{s} \tag{12}$$

holds.

Roughly speaking (replacing the increment by the differential): *The increment of the potential along the path characterized by a small vector* d$\boldsymbol{s}$ *is given by the scalar product* (12).

REMARK 3. At a fixed point and for a fixed length ds of the vector d$\boldsymbol{s}$ the increment (differential, more precisely) of the potential u is (in accordance to Theorem 7.1.9) the greatest in the direction of the gradient. Thus, *the gradient determines at every point the greatest descent in the field.*

Theorem 3.

$$\operatorname{grad}(u_1 + u_2 + \ldots + u_n) = \operatorname{grad} u_1 + \operatorname{grad} u_2 + \ldots + \operatorname{grad} u_n\,, \tag{13}$$

$$\operatorname{grad}(uv) = u \operatorname{grad} v + v \operatorname{grad} u\,, \tag{14}$$

$\operatorname{grad} r = \dfrac{\boldsymbol{r}}{r} = \hat{\boldsymbol{r}}$ (*$\boldsymbol{r}$ is the radius vector of the point (x, y, z), $\hat{\boldsymbol{r}}$ is the unit vector in the direction of* $\boldsymbol{r}$), (15)

$$\operatorname{grad} f(u) = f'(u) \operatorname{grad} u\,; \tag{16}$$

in particular

$$\operatorname{grad}\frac{1}{r} = -\frac{1}{r^2}\operatorname{grad} r = -\frac{\boldsymbol{r}}{r^3} \tag{17}$$

(*field of force of a unit charge lying at the origin of the coordinate system*);

$$|\operatorname{grad} u| = \sqrt{\left[\left(\frac{\partial u}{\partial x}\right)^2 + \left(\frac{\partial u}{\partial y}\right)^2 + \left(\frac{\partial u}{\partial z}\right)^2\right]}. \tag{18}$$

REMARK 4. For the gradient the following notation is used:

$$\operatorname{grad} u = \nabla u = \left(\boldsymbol{i}\frac{\partial}{\partial x} + \boldsymbol{j}\frac{\partial}{\partial y} + \boldsymbol{k}\frac{\partial}{\partial z}\right) u\,, \tag{19}$$

where

$$\nabla = \boldsymbol{i}\frac{\partial}{\partial x} + \boldsymbol{j}\frac{\partial}{\partial y} + \boldsymbol{k}\frac{\partial}{\partial z} \tag{20}$$

is the so-called *Hamilton nabla operator*, often called "del". Formula (12) is then written in the form

$$du = \nabla u \,.\, \mathrm{d}\boldsymbol{s}\,. \tag{21}$$

DIVERGENCE AND CURL OF A VECTOR FIELD. Consider a vector field given by the vector

$$\boldsymbol{a}(x, y, z) = a_1(x, y, z)\,\boldsymbol{i} + a_2(x, y, z)\,\boldsymbol{j} + a_3(x, y, z)\,\boldsymbol{k} \tag{22}$$

which is thus a vector function of a point (x, y, z).

Definition 2. The *divergence of the vector* $\boldsymbol{a}$ is the scalar

$$\operatorname{div}\boldsymbol{a} = \nabla\boldsymbol{a} = \frac{\partial a_1}{\partial x} + \frac{\partial a_2}{\partial y} + \frac{\partial a_3}{\partial z}\,. \tag{23}$$

In English mathematical literature it is more common to write $\nabla \,.\, \boldsymbol{a}$, rather than $\nabla\boldsymbol{a}$, for div $\boldsymbol{a}$.

REMARK 5. Let us consider a steady flow of fluid characterized (in a region O) by a velocity vector $\boldsymbol{a}(x, y, z)$. The divergence of the vector $\boldsymbol{a}$ measures the (volume) quantity of fluid produced in a unit volume in unit time.

For an incompressible fluid div $\boldsymbol{a} = 0$. Such a vector field (i.e. a field for which div $\boldsymbol{a} = 0$) is called *solenoidal* (*sourceless*). The flux of such a field through a closed surface equals zero; the quantity leaving the surface is the same as that entering it (see equation (7.3.7)).

Theorem 4. *The following relations hold*:

$$\operatorname{div}(\boldsymbol{a} + \boldsymbol{b}) = \operatorname{div} \boldsymbol{a} + \operatorname{div} \boldsymbol{b}\,, \quad \operatorname{div}(u\boldsymbol{a}) = u \operatorname{div} \boldsymbol{a} + \boldsymbol{a} \operatorname{grad} u\,, \tag{24}$$

$$\operatorname{div} \boldsymbol{r} = 3\,, \quad \operatorname{div} \hat{\boldsymbol{r}} = \frac{2}{r}\,, \quad \operatorname{div} \frac{\hat{\boldsymbol{r}}}{r^2} = 0\,, \tag{25}$$

where $\hat{\boldsymbol{r}} = \dfrac{\boldsymbol{r}}{r}$ *(*$\boldsymbol{r}$ *is the radius vector of the point* (x, y, z)*)*.

Definition 3. The *curl of a vector* $\boldsymbol{a}$ is the vector

$$\operatorname{curl} \boldsymbol{a} = \nabla \times \boldsymbol{a} = \left(\frac{\partial a_3}{\partial y} - \frac{\partial a_2}{\partial z}\right)\boldsymbol{i} + \left(\frac{\partial a_1}{\partial z} - \frac{\partial a_3}{\partial x}\right)\boldsymbol{j} + \left(\frac{\partial a_2}{\partial x} - \frac{\partial a_1}{\partial y}\right)\boldsymbol{k}$$

$$= \begin{vmatrix} \boldsymbol{i}, & \boldsymbol{j}, & \boldsymbol{k} \\ \dfrac{\partial}{\partial x}, & \dfrac{\partial}{\partial y}, & \dfrac{\partial}{\partial z} \\ a_1, & a_2, & a_3 \end{vmatrix}. \tag{26}$$

The symbol rot $\boldsymbol{a}$ is often used instead of curl $\boldsymbol{a}$.

REMARK 6. If $\boldsymbol{a}$ is the velocity of a fluid, then the direction of curl $\boldsymbol{a}$ indicates the direction of the axis about which the fluid rotates in a "small" neighbourhood of the point under consideration. The length of the vector $\frac{1}{2}$ curl $\boldsymbol{a}$ determines the speed of rotation (in circular measure).

Theorem 5. *The following relations hold:*

$$\operatorname{curl}(\boldsymbol{a} + \boldsymbol{b}) = \operatorname{curl} \boldsymbol{a} + \operatorname{curl} \boldsymbol{b}\,, \quad \operatorname{curl}(u\boldsymbol{a}) = u \operatorname{curl} \boldsymbol{a} - \boldsymbol{a} \times \operatorname{grad} u\,, \tag{27}$$

$$\operatorname{curl} \boldsymbol{r} = \boldsymbol{0}\,, \quad \operatorname{curl} f(r)\,\boldsymbol{r} = 0 \quad (\boldsymbol{r} \textit{ is the radius vector of the point } (x, y, z))\,. \tag{28}$$

REMARK 7. The field in which curl $\boldsymbol{a} = \boldsymbol{0}$ holds is called *irrotational.* A vector field constructed as the gradient of a scalar field $u(x, y, z)$, is irrotational. Conversely, every irrotational field in a simply connected region can be represented as the gradient of a scalar field.

REMARK 8. The scalar product of the operator ∇ with itself gives the so-called *Laplacian operator* Δ *(delta)*:

$$\Delta = \nabla\nabla = \nabla^2 = \frac{\partial^2}{\partial x^2} + \frac{\partial^2}{\partial y^2} + \frac{\partial^2}{\partial z^2}\,. \tag{29}$$

Theorem 6. *The above-defined operations expressed in terms of*

a) *cylindrical* (*polar*) *coordinates* $x = \varrho \cos \varphi$, $y = \varrho \sin \varphi$, $z = z$ give

$$\operatorname{div} \boldsymbol{a} = \frac{1}{\varrho} \frac{\partial}{\partial \varrho}(\varrho a_\varrho) + \frac{1}{\varrho} \frac{\partial a_\varphi}{\partial \varphi} + \frac{\partial a_z}{\partial z}, \tag{30}$$

$$\Delta u = \frac{\partial^2 u}{\partial \varrho^2} + \frac{1}{\varrho} \frac{\partial u}{\partial \varrho} + \frac{1}{\varrho^2} \frac{\partial^2 u}{\partial \varphi^2} + \frac{\partial^2 u}{\partial z^2}; \tag{31}$$

the components of the vector grad u *in the directions of* ϱ, φ, z:

$$\frac{\partial u}{\partial \varrho}, \quad \frac{1}{\varrho} \frac{\partial u}{\partial \varphi}, \quad \frac{\partial u}{\partial z}: \tag{32}$$

the components of the vector curl $\boldsymbol{a}$ *in the directions of* ϱ, φ, z:

$$\frac{1}{\varrho} \frac{\partial a_z}{\partial \varphi} - \frac{\partial a_\varphi}{\partial z}, \quad \frac{\partial a_\varrho}{\partial z} - \frac{\partial a_z}{\partial \varrho}, \quad \frac{1}{\varrho} \frac{\partial}{\partial \varrho}(\varrho a_\varphi) - \frac{1}{\varrho} \frac{\partial a_\varrho}{\partial \varphi}; \tag{33}$$

b) *spherical coordinates* $x = r \sin \vartheta \cos \varphi$, $y = r \sin \vartheta \sin \varphi$, $z = r \cos \vartheta$ give

$$\operatorname{div} \boldsymbol{a} = \frac{1}{r^2} \frac{\partial}{\partial r}(r^2 a_r) + \frac{1}{r \sin \vartheta} \frac{\partial}{\partial \vartheta}(a_\vartheta \sin \vartheta) + \frac{1}{r \sin \vartheta} \frac{\partial a_\varphi}{\partial \varphi}, \tag{34}$$

$$\Delta u = \frac{\partial^2 u}{\partial r^2} + \frac{2}{r} \frac{\partial u}{\partial r} + \frac{1}{r^2} \frac{\partial^2 u}{\partial \vartheta^2} + \frac{1}{r^2} \cot \vartheta \frac{\partial u}{\partial \vartheta} + \frac{1}{r^2 \sin^2 \vartheta} \frac{\partial^2 u}{\partial \varphi^2}; \tag{35}$$

the components of the vector grad u *in the directions of* r, ϑ, φ:

$$\frac{\partial u}{\partial r}, \quad \frac{1}{r} \frac{\partial u}{\partial \vartheta}, \quad \frac{1}{r \sin \vartheta} \frac{\partial u}{\partial \varphi};$$

the components of the vector curl $\boldsymbol{a}$ *in the directions of* r, ϑ, φ:

$$-\frac{1}{r \sin \vartheta}\left[\frac{\partial a_\vartheta}{\partial \varphi} - \frac{\partial}{\partial \vartheta}(a_\varphi \sin \vartheta)\right], \quad -\frac{1}{r}\left[\frac{\partial}{\partial r}(r a_\varphi) - \frac{1}{\sin \vartheta} \frac{\partial a_r}{\partial \varphi}\right],$$
$$-\left[\frac{1}{r} \frac{\partial a_r}{\partial \vartheta} - \frac{1}{r} \frac{\partial}{\partial r}(r a_\vartheta)\right]. \tag{36}$$

Theorem 7 (*Some Formulae for Calculation with the Operators* ∇ *and* Δ).

1. $\nabla(uv) = \operatorname{grad}(uv) = u \operatorname{grad} v + v \operatorname{grad} u$,
2. $\nabla(u\boldsymbol{a}) = \operatorname{div}(u\boldsymbol{a}) = u \operatorname{div} \boldsymbol{a} + \boldsymbol{a} \operatorname{grad} u$,
3. $\nabla \times (u\boldsymbol{a}) = \operatorname{curl}(u\boldsymbol{a}) = u \operatorname{curl} \boldsymbol{a} - \boldsymbol{a} \times \operatorname{grad} u$,

4. $\nabla(\boldsymbol{a} \,.\, \boldsymbol{b}) = \text{grad}\,(\boldsymbol{a} \,.\, \boldsymbol{b}) = \boldsymbol{a} \times \text{curl}\,\boldsymbol{b} + \boldsymbol{b} \times \text{curl}\,\boldsymbol{a} +$
$$+\left(a_1 \frac{\partial \boldsymbol{b}}{\partial x} + a_2 \frac{\partial \boldsymbol{b}}{\partial y} + a_3 \frac{\partial \boldsymbol{b}}{\partial z}\right) + \left(b_1 \frac{\partial \boldsymbol{a}}{\partial x} + b_2 \frac{\partial \boldsymbol{a}}{\partial y} + b_3 \frac{\partial \boldsymbol{a}}{\partial z}\right),$$
5. $\nabla(\boldsymbol{a} \times \boldsymbol{b}) = \text{div}\,(\boldsymbol{a} \times \boldsymbol{b}) = \boldsymbol{b} \,.\, \text{curl}\,\boldsymbol{a} - \boldsymbol{a} \,.\, \text{curl}\,\boldsymbol{b}\,,$
6. $\nabla \times (\boldsymbol{a} \times \boldsymbol{b}) = \text{curl}\,(\boldsymbol{a} \times \boldsymbol{b}) = \boldsymbol{a}\,\text{div}\,\boldsymbol{b} - \boldsymbol{b}\,\text{div}\,\boldsymbol{a} +$
$$+\left(b_1 \frac{\partial \boldsymbol{a}}{\partial x} + b_2 \frac{\partial \boldsymbol{a}}{\partial y} + b_3 \frac{\partial \boldsymbol{a}}{\partial z}\right) - \left(a_1 \frac{\partial \boldsymbol{b}}{\partial x} + a_2 \frac{\partial \boldsymbol{b}}{\partial y} + a_3 \frac{\partial \boldsymbol{b}}{\partial z}\right),$$
7. $\nabla^2 u = \nabla\nabla u = \text{div grad}\,u = \Delta u\,,$
8. $\nabla \times (\nabla u) = \text{curl grad}\,u = \boldsymbol{0}\,,$
9. $\nabla(\nabla \boldsymbol{a}) = \text{grad div}\,\boldsymbol{a} = \text{curl curl}\,\boldsymbol{a} + \Delta \boldsymbol{a}\,,$
10. $\nabla(\nabla \times \boldsymbol{a}) = \text{div curl}\,\boldsymbol{a} = 0\,.$

Theorem 8 (*Some Properties of the Laplacian Operator*).

1. $\Delta(u + v) = \Delta u + \Delta v\,, \quad \Delta(uv) = u\,\Delta v + v\,\Delta u + 2\,\text{grad}\,u \,.\, \text{grad}\,v\,,$
2. $\Delta(\boldsymbol{a} + \boldsymbol{b}) = \Delta \boldsymbol{a} + \Delta \boldsymbol{b}\,, \quad \Delta\,\text{grad}\,u = \text{grad}\,(\Delta u)\,, \quad \Delta\,\text{curl}\,\boldsymbol{a} = \text{curl}\,\Delta \boldsymbol{a}\,,$
3. $\Delta \dfrac{1}{r} = 0\,.$

REMARK 9. In accordance with the definition of the gradient of a *scalar*, the divergence of a *vector* and the curl of a *vector*, we read, of course,

$$\nabla u = \text{grad}\,u\,, \quad \nabla \boldsymbol{a} = \text{div}\,\boldsymbol{a}\,, \quad \nabla \times \boldsymbol{a} = \text{curl}\,\boldsymbol{a}$$

and not e.g. $\nabla u = \text{div}\,u$, for the operator "divergence" can be applied to a *vector* only, not to a *scalar*, etc.

The formulae stated in Theorem 7 can often easily be formally deduced if we note that the operator ∇ is given in vector form by (20); for example,

$$\nabla^2 = \nabla \,.\, \nabla = \left(\boldsymbol{i}\frac{\partial}{\partial x} + \boldsymbol{j}\frac{\partial}{\partial y} + \boldsymbol{k}\frac{\partial}{\partial z}\right) . \left(\boldsymbol{i}\frac{\partial}{\partial x} + \boldsymbol{j}\frac{\partial}{\partial y} + \boldsymbol{k}\frac{\partial}{\partial z}\right) =$$
$$= \frac{\partial^2}{\partial x^2} + \frac{\partial^2}{\partial y^2} + \frac{\partial^2}{\partial z^2} = \Delta\,.$$

Similarly, also

$$\nabla(u\boldsymbol{a}) = (\nabla u) \,.\, \boldsymbol{a} + u\,\nabla \boldsymbol{a} = \text{grad}\,u \,.\, \boldsymbol{a} + u\,\text{div}\,\boldsymbol{a}\,.$$

The last two expressions in formula 4 of Theorem 7 can be symbolically written as

$$\left(a_1 \frac{\partial \boldsymbol{b}}{\partial x} + a_2 \frac{\partial \boldsymbol{b}}{\partial y} + a_3 \frac{\partial \boldsymbol{b}}{\partial z}\right) + \left(b_1 \frac{\partial \boldsymbol{a}}{\partial x} + b_2 \frac{\partial \boldsymbol{a}}{\partial y} + b_3 \frac{\partial \boldsymbol{a}}{\partial z}\right) = (\boldsymbol{a}\nabla)\,\boldsymbol{b} + (\boldsymbol{b}\nabla)\,\boldsymbol{a}\,,$$

where

$$\boldsymbol{a}\nabla = \boldsymbol{a}\,\text{grad} = (a_1\boldsymbol{i} + a_2\boldsymbol{j} + a_3\mathbf{k}) \,.\, \left(\boldsymbol{i}\frac{\partial}{\partial x} + \boldsymbol{j}\frac{\partial}{\partial y} + \mathbf{k}\frac{\partial}{\partial z}\right) =$$

$$= a_1\frac{\partial}{\partial x} + a_2\frac{\partial}{\partial y} + a_3\frac{\partial}{\partial z}\,, \text{ etc.}$$

The corresponding expressions in formula 6 can be written in a similar way.

7.3. Curvilinear and Surface Integrals of a Vector. Vector Notation for the Theorems of Stokes, Gauss and Green

Let c be a sectionally smooth oriented curve in space (see p. 603). Denote $\mathbf{t}\,\mathrm{d}s$ by $\mathrm{d}\mathbf{s}$, where s is arc-length and $\mathbf{t}$ the tangent unit vector at the point of the curve under consideration. We define

$$\int_c \boldsymbol{a}\,.\,\mathrm{d}\mathbf{s} = \int_c \boldsymbol{a}\,.\,\mathbf{t}\,\mathrm{d}s = \int_c (a_1\,\mathrm{d}x + a_2\,\mathrm{d}y + a_3\,\mathrm{d}z)\,. \tag{1}$$

If c is a closed curve, we usually write

$$\oint_c \boldsymbol{a}\,.\,\mathrm{d}\mathbf{s} \tag{2}$$

(this is the *circulation of the vector* $\boldsymbol{a}$ *along the closed curve* c).

If the vector $\boldsymbol{a}$ denotes a force, then integral (1) represents the work done by this force along the curve c.

If $\boldsymbol{a} = \text{grad}\, u$, then

$$\oint_A^B \boldsymbol{a}\,.\,\mathrm{d}\mathbf{s} = u(B) - u(A)\,, \tag{3}$$

where A is the initial and B the end point of the curve. Thus, in an irrotational field the integral (1) depends on the initial and end points of the curve and not on the shape of the curve. In particular, the integral along a closed curve in an irrotational field $\boldsymbol{a}$ is equal to zero:

$$\oint_c \boldsymbol{a}\,.\,\mathrm{d}\mathbf{s} = \oint_c \text{grad}\, u\,.\,\mathrm{d}\mathbf{s} = 0\,. \tag{4}$$

If a vector (vector field) $\boldsymbol{a}$ is irrotational in a simply connected region O, i.e. if $\text{curl}\,\boldsymbol{a} = \mathbf{0}$ in O, then the vector $\boldsymbol{a}$ can be expressed as the gradient of a scalar u (cf. Remark 7.2.7); the integral (1) does not depend on the shape of the curve but merely on its initial and end points; the integral along a closed curve is equal to zero.

In a similar way the surface integral of a vector $\boldsymbol{a}$ (over a sectionally smooth oriented surface, with unit normal vector $\boldsymbol{n}$ can be defined:

$$\iint_S \boldsymbol{a} \,.\, \mathrm{d}\boldsymbol{S} = \int_S \boldsymbol{a} \,.\, \boldsymbol{n} \,\mathrm{d}S \,. \tag{5}$$

In vector notation some theorems of integral calculus (see § 14·8) can be written in a simple form:

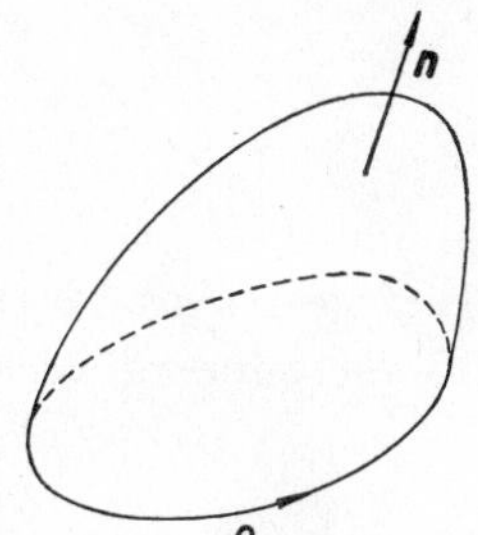

Fig. 7.4.

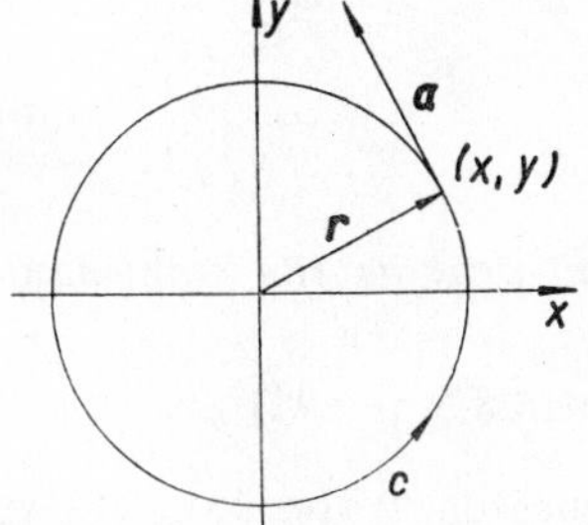

Fig. 7.5.

1. *Stokes's Theorem*:

$$\iint_S \operatorname{curl} \boldsymbol{a} \,.\, \mathrm{d}\boldsymbol{S} = \oint_c \boldsymbol{a} \,.\, \mathrm{d}\boldsymbol{s} \tag{6}$$

where S is a surface bounded by a curve c (Fig. 7.4); the orientation of the surface and the curve is to be seen in Fig. 14.20 (Theorem 14.8.6, p. 643).

The physical interpretation of equation (6): The flux of the vector curl $\boldsymbol{a}$ through a surface S equals the circulation of the vector $\boldsymbol{a}$ along the bounding curve c of S.

Example 1. Consider a two dimensional vector field (in the xy plane) given by the vector

$$\boldsymbol{a} = -y\boldsymbol{i} + x\boldsymbol{j}$$

(at every point (x, y) the vector $\boldsymbol{a}$ is perpendicular to the corresponding radius vector $\boldsymbol{r}$ of this point and its length is r (see Fig. 7.5)). Let c be the circle of radius r with centre at the origin, positively oriented with respect to its interior S. Thus, the normal $\boldsymbol{n}$ of the surface S is directed upwards (in the positive sense of the z-axis). Evidently

$$\int_c \boldsymbol{a} \,.\, \mathrm{d}\boldsymbol{s} = r \,.\, 2\pi r = 2\pi r^2 \,.$$

Further

$$\operatorname{curl} \boldsymbol{a} = \begin{vmatrix} \boldsymbol{i}, & \boldsymbol{j}, & \boldsymbol{k} \\ \dfrac{\partial}{\partial x}, & \dfrac{\partial}{\partial y}, & \dfrac{\partial}{\partial z} \\ -y, & x, & 0 \end{vmatrix} = 2\boldsymbol{k} \,,$$

and thus

$$\iint_S \operatorname{curl} \boldsymbol{a} \,.\, \mathrm{d}\boldsymbol{S} = 2 \,.\, \pi r^2 = 2\pi r^2$$

in accordance with (6).

If $\operatorname{curl} \boldsymbol{a} = \boldsymbol{0}$, then equation (6) yields $\oint_c \boldsymbol{a} \,.\, \mathrm{d}\boldsymbol{s} = 0$ in accordance with (4).

2. *Gauss's Theorem*:

$$\iiint_V \operatorname{div} \boldsymbol{a} \,\mathrm{d}V = \iint_S \boldsymbol{a} \,.\, \mathrm{d}\boldsymbol{S}\,, \tag{7}$$

where the integral on the right-hand side is the surface integral over a closed surface whose interior is V; $\mathrm{d}\boldsymbol{S} = \boldsymbol{n}\,\mathrm{d}S$, where $\boldsymbol{n}$ is the outward normal vector (see Theorem14.8.5, p. 642).

Physical meaning: The flux of a vector $\boldsymbol{a}$ through a closed surface is equal to the volume integral of the divergence of the vector $\boldsymbol{a}$.

Similarly, the relations

$$\iiint_V \operatorname{grad} u \,\mathrm{d}V = \iint_S u \,\mathrm{d}\boldsymbol{S}\,, \quad \iiint_V \operatorname{curl} \boldsymbol{a} \,\mathrm{d}V = -\iint_S \boldsymbol{a} \times \mathrm{d}\boldsymbol{S} \tag{8}$$

hold.

REMARK 1. On the basis of the relations (7), (8), the operators grad, div, curl can be defined, without use of a special coordinate system:

$$\operatorname{grad} u = \lim_{V \to 0} \frac{1}{V} \iint_S u \,\mathrm{d}\boldsymbol{S}\,, \quad \operatorname{div} \boldsymbol{a} = \lim_{V \to 0} \frac{1}{V} \iint_S \boldsymbol{a} \,.\, \mathrm{d}\boldsymbol{S}\,, \tag{9}$$

$$\operatorname{curl} \boldsymbol{a} = -\lim_{V \to 0} \frac{1}{V} \iint_S \boldsymbol{a} \times \mathrm{d}\boldsymbol{S}\,. \tag{10}$$

3. *Green's Theorems*:

$$\iiint_V (\operatorname{grad} u \,.\, \operatorname{grad} v) \,\mathrm{d}V + \iiint_V u \,\Delta v \,\mathrm{d}V = \iint_S u \frac{\partial v}{\partial n} \,\mathrm{d}S\,, \tag{11}$$

$$\iiint_V (u \,\Delta v - v \,\Delta u) \,\mathrm{d}V = \iint_S \left(u \frac{\partial v}{\partial n} - v \frac{\partial u}{\partial n} \right) \mathrm{d}S\,, \tag{12}$$

where n is the outward unit normal (see p. 644). If we put $v = u$ in (11) and if, moreover, u is harmonic ($\Delta u = 0$), then

$$\iiint_V (\nabla u)^2 \, dV = \iiint_V \left[\left(\frac{\partial u}{\partial x}\right)^2 + \left(\frac{\partial u}{\partial y}\right)^2 + \left(\frac{\partial u}{\partial z}\right)^2\right] dV = \iint_V u \frac{\partial u}{\partial n} \, dS . \tag{13}$$

4. *Let the point $Q(x_0, y_0, z_0)$ be inside S, u be harmonic inside S, $\partial u/\partial n$ continuously extensible on S and $r = \sqrt{[(x - x_0)^2 + (y - y_0)^2 + (z - z_0)^2]}$. Then*

$$\iint_S \left[u \frac{\partial}{\partial n}\left(\frac{1}{r}\right) - \frac{1}{r}\frac{\partial u}{\partial n}\right] dS = -4\pi u_0 , \tag{14}$$

where u_0 is the value of the function u at the point Q.

8. TENSOR CALCULUS

By Václav Vilhelm

References: [34], [44], [63], [76], [185], [202], [258], [279], [355], [377], [385], [390], [393], [424].

8.1. Contravariant and Covariant Coordinates of a Vector and their Transformation by a Change of the Coordinate System

If $\mathbf{e}_1$, $\mathbf{e}_2$, $\mathbf{e}_3$ are three arbitrary non-coplanar vectors then they define a coordinate system $(\mathbf{e}_1, \mathbf{e}_2, \mathbf{e}_3)$ in space in the sense that every vector $\boldsymbol{a}$ can be written uniquely in the form

$$\boldsymbol{a} = a^1\mathbf{e}_1 + a^2\mathbf{e}_2 + a^3\mathbf{e}_3\,, \tag{1}$$

where a^1, a^2, a^3 are real numbers. (We mention explicitly that a^i does *not* denote the i-th power of a but a number a^i with the upper index i.)

Definition 1. The numbers a^1, a^2, a^3 are called the *contravariant coordinates of the vector* $\boldsymbol{a}$ in the coordinate system $(\mathbf{e}_1, \mathbf{e}_2, \mathbf{e}_3)$.

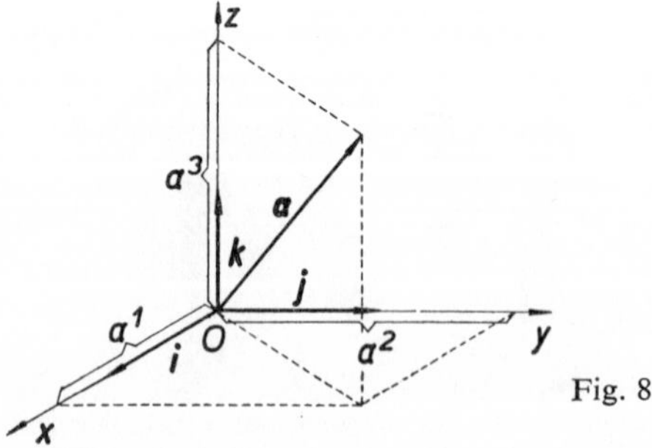

Fig. 8.1.

Example 1. Let us choose three mutually perpendicular unit vectors in space and denote them by $\mathbf{e}_1 = \boldsymbol{i}$, $\mathbf{e}_2 = \boldsymbol{j}$, $\mathbf{e}_3 = \boldsymbol{k}$ (Fig. 8.1). If $\boldsymbol{a}$ is an arbitrary vector, then $\boldsymbol{a} = a^1\boldsymbol{i} + a^2\boldsymbol{j} + a^3\boldsymbol{k} = a^i\mathbf{e}_i$ (see Remark 2); the numbers a^1, a^2, a^3 are its contravariant coordinates in the coordinate system $(\boldsymbol{i}, \boldsymbol{j}, \boldsymbol{k})$.

REMARK 1. From (1), a vector is obviously uniquely determined by its contravariant coordinates in the given coordinate system.

REMARK 2. Equation (1) can be written in the form $\boldsymbol{a} = \sum_{i=1}^{3} a^i \mathbf{e}_i$. It is customary in tensor algebra to adopt the "summation convention" by which we omit the sum symbol $\sum$ and write simply $\boldsymbol{a} = a^i \mathbf{e}_i$. In this convention it is understood that whenever an index is *repeated* (as in $a^i \mathbf{e}_i$) we sum over the values $i = 1, 2, 3$. Thus, $a^i \mathbf{e}_i$ stands for the sum $a^1 \mathbf{e}_1 + a^2 \mathbf{e}_2 + a^3 \mathbf{e}_3$. In what follows we shall normally use this brief notation.

If $\mathbf{e}'_1, \mathbf{e}'_2, \mathbf{e}'_3$ denote three other *non-coplanar* vectors in space, then

$$\begin{aligned} \mathbf{e}'_1 &= e_1^1 \mathbf{e}_1 + e_1^2 \mathbf{e}_2 + e_1^3 \mathbf{e}_3 \,, \\ \mathbf{e}'_2 &= e_2^1 \mathbf{e}_1 + e_2^2 \mathbf{e}_2 + e_2^3 \mathbf{e}_3 \,, \\ \mathbf{e}'_3 &= e_3^1 \mathbf{e}_1 + e_3^2 \mathbf{e}_2 + e_3^3 \mathbf{e}_3 \,, \end{aligned} \tag{2}$$

briefly this may be written as

$$\mathbf{e}'_i = e_i^j \mathbf{e}_j \quad (i = 1, 2, 3) \,,$$

since j is a repeatable index.

Definition 2. The matrix $\mathbf{A} = \|e_j^i\|$ (the upper index refers to the columns, the lower to the rows) is called the *transformation matrix of the coordinate system* $(\mathbf{e}_1, \mathbf{e}_2, \mathbf{e}_3)$ *to the coordinate system* $(\mathbf{e}'_1, \mathbf{e}'_2, \mathbf{e}'_3)$.

Theorem 1. *The determinant of the transformation matrix is different from zero; hence we may write*

$$\mathbf{e}_i = f_i^j \mathbf{e}'_j \quad (i = 1, 2, 3) \,, \tag{3}$$

where the matrix $\|f_j^i\|$ *is the inverse of* $\|e_j^i\|$ (see Example 2).

Theorem 2 (*Transformation of the Contravariant Coordinates of a Vector*). *If the contravariant coordinates of a vector* $\boldsymbol{a}$ *in the coordinate system* $(\mathbf{e}_1, \mathbf{e}_2, \mathbf{e}_3)$ *are* a^1, a^2, a^3 *and those in the coordinate system* $(\mathbf{e}'_1, \mathbf{e}'_2, \mathbf{e}'_3)$ *are* a'^1, a'^2, a'^3, *then the following relation holds between these coordinates:*

$$a'^i = f_j^i a^j \,, \quad a^i = e_j^i a'^j \quad (i = 1, 2, 3) \,. \tag{4}$$

Here the matrix

$$\left\| \begin{matrix} e_1^1, & e_2^1, & e_3^1 \\ e_1^2, & e_2^2, & e_3^2 \\ e_1^3, & e_2^3, & e_3^3 \end{matrix} \right\|$$

is the transpose $\mathbf{A}'$ *of the transformation matrix* $\mathbf{A}$ *of the system* $(\mathbf{e}_1, \mathbf{e}_2, \mathbf{e}_3)$ *to the system* $(\mathbf{e}'_1, \mathbf{e}'_2, \mathbf{e}'_3)$ *and the matrix*

$$\left\| \begin{matrix} f_1^1, & f_2^1, & f_3^1 \\ f_1^2, & f_2^2, & f_3^2 \\ f_1^3, & f_2^3, & f_3^3 \end{matrix} \right\|$$

is the inverse of the matrix $\mathbf{A}'$, *i.e.*

$$\sum_{i=1}^{3} e_j^i f_i^k = \delta_j^k \quad (\textit{briefly} \quad e_j^i f_i^k = \delta_j^k)\,, \tag{5}$$

where $\delta_j^k = \begin{cases} 1 \textit{ for } k = j \\ 0 \textit{ for } k \neq j \end{cases}$ (see Example 2).

REMARK 3. δ_j^k in Theorem 2 is known as the *Kronecker delta*.

Theorem 3. *Let the contravariant coordinates of vectors* $\mathbf{a}$ *and* $\mathbf{b}$ *in the coordinate system* $(\mathbf{e}_1, \mathbf{e}_2, \mathbf{e}_3)$ *be* a^i *and* b^i, *respectively. Then, for the scalar product* $\mathbf{a} \,.\, \mathbf{b}$ *we have the relation*

$$\mathbf{a} \,.\, \mathbf{b} = \sum_{i=1}^{3} \sum_{j=1}^{3} g_{ij} a^i b^j$$

or briefly

$$\mathbf{a} \,.\, \mathbf{b} = g_{ij} a^i b^j\,, \tag{6}$$

where

$$g_{ij} = \mathbf{e}_i \,.\, \mathbf{e}_j \quad (i, j = 1, 2, 3)$$

(see Example 2 below).

Definition 3. The numbers

$$\sum_{j=1}^{3} g_{1j} a^j\,, \quad \sum_{j=1}^{3} g_{2j} a^j\,, \quad \sum_{j=1}^{3} g_{3j} a^j$$

(see Theorem 4) are called the *covariant coordinates* of the vector $\mathbf{a}$ in the coordinate system $(\mathbf{e}_1, \mathbf{e}_2, \mathbf{e}_3)$ and are denoted by a_1, a_2, a_3. Thus, $a_i = g_{ij} a^j$.

REMARK 4. Since the numbers a^1, a^2, a^3 can be determined from the equations $a_i = g_{ij} a^j$ uniquely (because the determinant of the matrix $\|g_{ij}\|$ is different from zero), a vector is, according to Remark 1, uniquely determined by its covariant coordinates in the given coordinate system. In the system $(\mathbf{i}, \mathbf{j}, \mathbf{k})$ of Example 1, $a^i = a_i$.

Theorem 4 (*Transformation of the Covariant Coordinates of a Vector*). *If the covariant coordinates of a vector* $\mathbf{a}$ *in the coordinate system* $(\mathbf{e}_1, \mathbf{e}_2, \mathbf{e}_3)$ *are* a_1, a_2, a_3 *and those in the coordinate system* $(\mathbf{e}'_1, \mathbf{e}'_2, \mathbf{e}'_3)$ *are* a'_1, a'_2, a'_3, *then the following relation holds between these coordinates:*

$$a'_i = e_i^j a_j\,, \quad a_i = f_i^j a'_j\,; \tag{7}$$

the numbers e_i^j, f_i^j *having the same meaning as in Theorem 2.*

Theorem 5. *Let a^i, b^i be the contravariant and a_i, b_i the covariant coordinates of vectors* $\mathbf{a}$, $\mathbf{b}$ *in the given coordinate system, respectively. Then the scalar product* $\mathbf{a} \, . \, \mathbf{b}$ *satisfies*

$$\mathbf{a} \, . \, \mathbf{b} = a^i b_i = a_i b^i$$

(*where again*, $a^i b_i = a^1 b_1 + a^2 b_2 + a^3 b_3$, $a_i b^i = a_1 b^1 + a_2 b^2 + a_3 b^3$; see Example 2).

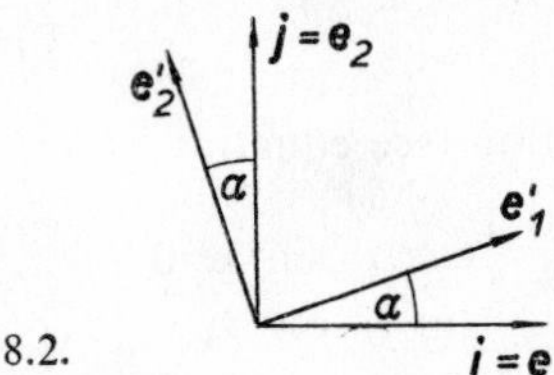

Fig. 8.2.

Example 2. Let $(\mathbf{e}_1, \mathbf{e}_2, \mathbf{e}_3) = (\mathbf{i}, \mathbf{j}, \mathbf{k})$ be the coordinate system of Example 1 and let us choose another three linearly independent vectors $\mathbf{e}'_1$ $\mathbf{e}'_2$, $\mathbf{e}'_3$ such that the vector $\mathbf{e}'_i$ is obtained from the vector $\mathbf{e}_i$ $(i = 1, 2)$ by rotation through an angle α in the plane of the vectors $\mathbf{e}_1$, $\mathbf{e}_2$ (see Fig. 8.2), and $\mathbf{e}'_3 = 2\mathbf{e}_3$. Then equations (2) take the form

$$\begin{aligned} \mathbf{e}'_1 &= \mathbf{e}_1 \cos \alpha + \mathbf{e}_2 \sin \alpha \, , \\ \mathbf{e}'_2 &= -\mathbf{e}_1 \sin \alpha + \mathbf{e}_2 \cos \alpha \, , \\ \mathbf{e}'_3 &= 2\mathbf{e}_3 \, . \end{aligned}$$

The transformation matrix of the coordinate system $(\mathbf{e}_1, \mathbf{e}_2, \mathbf{e}_3)$ to the coordinate system $(\mathbf{e}'_1, \mathbf{e}'_2, \mathbf{e}'_3)$ is

$$\mathbf{A} = \begin{Vmatrix} \cos \alpha, & \sin \alpha, & 0 \\ -\sin \alpha, & \cos \alpha, & 0 \\ 0, & 0, & 2 \end{Vmatrix} .$$

We can easily show that equations (3) have the form

$$\begin{aligned} \mathbf{e}_1 &= \mathbf{e}'_1 \cos \alpha - \mathbf{e}'_2 \sin \alpha \, , \\ \mathbf{e}_2 &= \mathbf{e}'_1 \sin \alpha + \mathbf{e}'_2 \cos \alpha \, , \\ \mathbf{e}_3 &= \tfrac{1}{2}\mathbf{e}'_3 \end{aligned}$$

and, thus, the transformation matrix of the coordinate system $(\mathbf{e}'_1, \mathbf{e}'_2, \mathbf{e}'_3)$ to the system $(\mathbf{e}_1, \mathbf{e}_2, \mathbf{e}_3)$ is

$$\mathbf{B} = \begin{Vmatrix} \cos \alpha, & -\sin \alpha, & 0 \\ \sin \alpha, & \cos \alpha, & 0 \\ 0, & 0, & \tfrac{1}{2} \end{Vmatrix} ,$$

i.e. the inverse of $\mathbf{A}$.

Consider the vector $\boldsymbol{a}$, the contravariant coordinates of which in the coordinate system $(\mathbf{e}_1, \mathbf{e}_2, \mathbf{e}_2)$ are a^1, a^2, a^3, i.e. $\boldsymbol{a} = a^i\mathbf{e}_i$. Hence, $\boldsymbol{a} = a^1(\mathbf{e}'_1 \cos\alpha - \mathbf{e}'_2 \sin\alpha) + a^2(\mathbf{e}'_1 \sin\alpha + \mathbf{e}'_2 \cos\alpha) + a^3 \frac{1}{2}\mathbf{e}'_3 = (a^1 \cos\alpha + a^2 \sin\alpha)\,\mathbf{e}'_1 + (-a^1 \sin\alpha + a^2 \cos\alpha)\,\mathbf{e}'_2 + \frac{1}{2}a^3\mathbf{e}'_3$. Thus, the contravariant coordinates of the vector $\boldsymbol{a}$ in the coordinate system $(\mathbf{e}'_1, \mathbf{e}'_2, \mathbf{e}'_3)$ are

$$\begin{aligned} a'^1 &= a^1 \cos\alpha + a^2 \sin\alpha\,, \\ a'^2 &= -a^1 \sin\alpha + a^2 \cos\alpha\,, \\ a'^3 &= \tfrac{1}{2}a^3\,. \end{aligned}$$

The matrix of this transformation (see equation (4)) is

$$\begin{Vmatrix} \cos\alpha, & \sin\alpha, & 0 \\ -\sin\alpha, & \cos\alpha, & 0 \\ 0, & 0, & \frac{1}{2} \end{Vmatrix},$$

which is the inverse of the matrix $\boldsymbol{A}'$ (see Theorem 2).

The covariant coordinates of the vector $\boldsymbol{a}$ in the coordinate system $(\mathbf{e}_1, \mathbf{e}_2, \mathbf{e}_3)$ are (Theorem 3) $a_i = (\mathbf{e}_i \,.\, \mathbf{e}_j)\, a^j = \delta^i_j a^j = a^i$ (i.e. the same as the contravariant coordinates), while the covariant coordinates of the vector $\boldsymbol{a}$ in the coordinate system $(\mathbf{e}'_1, \mathbf{e}'_2, \mathbf{e}'_3)$ are $a'_1 = (\mathbf{e}'_1 .\ \mathbf{e}'_j)\, a'^j = a'^1$, $a'_2 = (\mathbf{e}'_2 \,.\, \mathbf{e}'_j)\, a'^j = a'^2$, $a'_3 = (\mathbf{e}'_3 \,.\, \mathbf{e}'_j)\, a'^j = 4a'^3$ (since $\mathbf{e}'_i \,.\, \mathbf{e}'_j = 0$ for $i \neq j$, $\mathbf{e}'_1 \,.\, \mathbf{e}'_1 = \mathbf{e}'_2 \,.\, \mathbf{e}'_2 = 1$, $\mathbf{e}'_3 \,.\, \mathbf{e}'_3 = 4$).

The scalar product $\boldsymbol{a} \,.\, \boldsymbol{a}$ is (Theorem 5)

$$\boldsymbol{a} \,.\, \boldsymbol{a} = a^i a_i = (a_1)^2 + (a_2)^2 + (a_3)^2 = a'^i a'_i = (a^1 \cos\alpha + a^2 \sin\alpha)^2 + (-a^1 \sin\alpha + a^2 \cos\alpha)^2 + \tfrac{1}{2}a^3 \,.\, 2a^3\,.$$

8.2. The Concept of a Tensor in Space

We have shown in § 1 that in every coordinate system, a vector is determined by an ordered triplet of numbers – by its contravariant or covariant coordinates. In changing from one coordinate system to another, this system of numbers defining the vector transforms in a certain way. The transformation formulae for contravariant and covariant coordinates are different (see Theorem 8.1.2 and Theorem 8.1.4). On the other hand, if to every coordinate system we assign three numbers a^1, a^2, a^3 or b_1, b_2, b_3 in such a way that when changing from one coordinate system to another, these numbers are transformed according to the formulae $a'^i = f^i_j a^j$ (or $b'_i = e^j_i b_j$), where $\|e^i_j\|$ is the corresponding transformation matrix and $\|f^i_j\|$ is the transpose of the inverse of the matrix $\|e^j_i\|$, then these numbers can be understood to be the contravariant, or covariant coordinates of the vector $\boldsymbol{a}$, or $\boldsymbol{b}$, respectively. This follows from Theorems 8.1.2 and 8.1.4 and, thus, these numbers define the vectors $\boldsymbol{a}$ and $\boldsymbol{b}$. This idea is exploited in the following definition of a tensor.

Definition 1. We say that a *tensor* is defined in space if, to every coordinate system, there correspond 3^{p+q} numbers $a^{ijk\dots}_{rst\dots}$ (the number of upper indices is p, the number of lower indices q) such that they are transformed according to the formulae

$$a'^{ijk\dots}_{rst\dots} = a^{lmn\dots}_{uvw\dots} e^u_r e^v_s e^w_t \dots f^i_l f^j_m f^k_n \dots \tag{1}$$

by any change from one coordinate system to another (in the right-hand side of formulae (1) we sum (from one to three) over all indices which appear twice there). Here, $\|e^i_j\|$ is the transformation matrix and $\|f^i_j\|$ the transpose of the inverse of the matrix $\|e^i_j\|$. The tensor so defined is said to be *p-times contravariant* and *q-times covariant*. The number $p + q$ is called the *rank of the tensor*, the numbers $a^{ijk\dots}_{rst\dots}$ are called the *coordinates of the tensor*.

REMARK 1. Instead of "tensor of rank two" the term "*quadratic tensor*" is used. A quadratic tensor once covariant and once contravariant is called a *mixed quadratic tensor*. A tensor satisfying $q = 0$, or $p = 0$, is called a *contravariant*, or *covariant*, *tensor*, respectively.

Example 1 (*a scalar*). If to every coordinate system $(\mathbf{e}_1, \mathbf{e}_2, \mathbf{e}_3)$ there corresponds the same number a, a tensor of rank zero ($p = q = 0$), called a *scalar*, is defined.

Example 2 (*a contravariant vector*). If a^i are contravariant coordinates of a vector, then, by a change of the coordinate system, they are transformed according to the formulae $a'^i = f^i_j a^j$; this is a particular case of the formulae (1) for $p = 1$, $q = 0$. Thus a^i are the coordinates of a contravariant tensor of rank 1, called a *contravariant vector*.

Example 3 (*a covariant vector*). If a_i are covariant coordinates of a vector, then by a change of the coordinate system, they are transformed (see Theorem 8.1.4) according to formulae (1), where $p = 0$, $q = 1$. Thus, a_i are the coordinates of a covariant tensor of rank 1, called a *covariant vector*.

Example 4. The coordinates of a contravariant tensor a^{ij} are transformed, using (1), as follows:

$$a'^{ij} = a^{lm} f^i_l f^j_m .$$

Hence, in the transformation formulae for a contravariant (or covariant) tensor only the elements of the matrix $\|f^i_j\|$ (or $\|e^i_j\|$) appear.

Example 5 (*a metric tensor of the space*). If to every coordinate system $(\mathbf{e}_1, \mathbf{e}_2, \mathbf{e}_3)$ we assign the numbers $g_{ij} = \mathbf{e}_i \cdot \mathbf{e}_j$ (see Theorem 8.1.3), we can easily check that these numbers are transformed, by a change of the coordinate system, according to the formulae $g'_{ij} = g_{lm} e^l_i e^m_j$. Thus, g_{ij} are the coordinates of a double covariant tensor of rank 2 (i.e. a quadratic double covariant tensor), called the (*covariant*) *metric* *tensor*. The coordinates g_{ij} can be written in the form of a matrix

$$\begin{Vmatrix} g_{11}, g_{12}, g_{13} \\ g_{21}, g_{22}, g_{23} \\ g_{31}, g_{32}, g_{33} \end{Vmatrix}.$$

If

$$\begin{Vmatrix} g^{11}, g^{12}, g^{13} \\ g^{21}, g^{22}, g^{23} \\ g^{31}, g^{32}, g^{33} \end{Vmatrix}$$

is its inverse (i.e. $g_{ij}g^{jk} = \delta_j^k$ – see Remark 8.1.3), then the numbers g^{ij} are the coordinates of a double contravariant tensor of rank 2, called the (*contravariant*) *metric tensor*. If the contravariant (or covariant) coordinates of vectors $\boldsymbol{a}$, $\boldsymbol{b}$ in the given coordinate system are a^i, b^i (or a_i, b_i) and the covariant (or contravariant) coordinates of a metric tensor in this system are g_{ij} (or g^{ij}), then $\boldsymbol{a} \cdot \boldsymbol{b} = g_{ij}a^ib^j = g^{ij}a_ib_j$. This justifies the term "metric tensor".

Example 6. If to every coordinate system we assign the numbers

$$\delta_j^i = \begin{cases} 1 \text{ for } i = j , \\ 0 \text{ for } i \neq j , \end{cases}$$

then $\delta_j'^i = \delta_s^r e_j^s f_r^i = e_j^r f_r^i = \delta_j^i$ (see Theorem 8.1.2). Thus, δ_j^i are the coordinates of a once covariant and once contravariant tensor of rank 2 (i.e. a mixed quadratic tensor). These coordinates are the same in all coordinate systems.

Example 7. Let us choose a coordinate system in space and assign to every vector $\boldsymbol{a}$, the contravariant coordinates of which are a^i, the vector $\boldsymbol{b}$, the contravariant coordinates b^i of which are defined by the equations

$$b^i = c_j^i a^j , \tag{2}$$

i.e.

$$\begin{aligned} b^1 &= c_1^1 a^1 + c_2^1 a^2 + c_3^1 a^3 , \\ b^2 &= c_1^2 a^1 + c_2^2 a^2 + c_3^2 a^3 , \\ b^3 &= c_1^3 a^1 + c_2^3 a^2 + c_3^3 a^3 . \end{aligned}$$

If we change the given coordinate system to a new one in which the coordinates of the vector $\boldsymbol{a}$, or the vector $\boldsymbol{b}$, are a'^i, or b'^i, respectively, then the following relation between these coordinates holds:

$$b'^i = c_j'^i a'^j ,$$

where $c_j'^i = c_r^s e_j^r f_s^i$ ($\|e_j^i\|$ is the transformation matrix of the original coordinate system to the new one). Thus, c_j^i are the coordinates of a mixed quadratic tensor.

In particular, considering the so-called *small deformations of a solid* whereby the vector a^i is transformed into the vector $\bar{a}^i$, then relations (2) hold between the vector a^i and the vector $b^i = \bar{a}^i - a^i$. The coefficients c^i_j are the coordinates of the so-called *deformation tensor* (see Example 8.4.4, p. 294).

8.3. A Tensor on a Surface

Definition 1. Let π be a smooth surface in space defined by the radius vector $\mathbf{r}(u_1, u_2)$ (see equations (9.11.1), (9.11.6) where u, v are written instead of u_1, u_2).

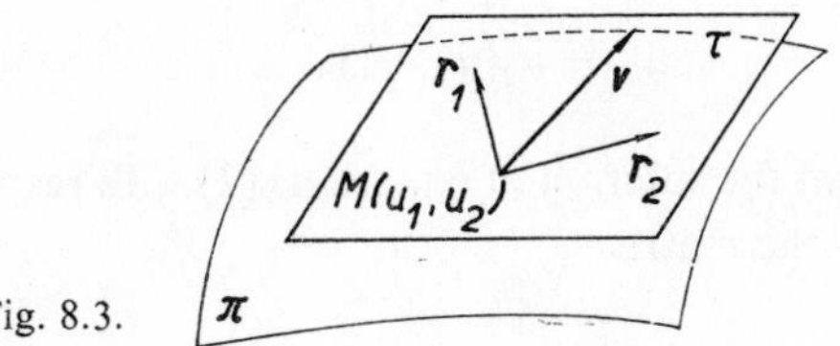

Fig. 8.3.

If, to every point M of the surface π, the coordinates of which are u_1, u_2, there corresponds a vector $\mathbf{v}(u_1, u_2)$ having initial point $M(u_1, u_2)$ and lying in the tangent plane of the surface at this point, we say that a (*tangent*) *vector field*, or briefly a (*tangent*) *vector* $\mathbf{v}(u_1, u_2)$ is given on the surface π (see Fig. 8.3).

REMARK 1. It is known (see § 9·12) that the vectors

$$\mathbf{r}_1(u_1, u_2) = \frac{\partial}{\partial u_1} \mathbf{r}(u_1, u_2),$$

$$\mathbf{r}_2(u_1, u_2) = \frac{\partial}{\partial u_2} \mathbf{r}(u_1, u_2)$$

lie in the tangent plane of the surface $\mathbf{r}(u_1, u_2)$ at the point $M(u_1, u_2)$ and are non-collinear. Therefore, they can be taken as the coordinate vectors in the tangent plane at the point $M(u_1, u_2)$. Every vector $\mathbf{v}(u_1, u_2)$ on the surface π can then be uniquely written in the form

$$\mathbf{v}(u_1, u_2) = v^1(u_1, u_2)\, \mathbf{r}_1(u_1, u_2) + v^2(u_1, u_2)\, \mathbf{r}_2(u_1, u_2), \tag{1}$$

briefly

$$\mathbf{v} = v^i \mathbf{r}_i .$$

Definition 2. $v^1(u_1, u_2)$, $v^2(u_1, u_2)$ (briefly v^i) are the so-called *contravariant coordinates of the vector* $\mathbf{v}(u_1, u_2)$ *on the surface* $\mathbf{r}(u_1, u_2)$ (in the local coordinate system $(\mathbf{r}_1(u_1, u_2), \mathbf{r}_2(u_1, u_2))$ of the point $M(u_1, u_2)$ of the surface).

REMARK 2. If $\mathbf{r}'(u'_1, u'_2)$ is another parametric expression of the surface π of Definition 1 in which the point M with original coordinates u_1, u_2 has coordinates u'_1, u'_2, then we shall always assume that u'_1, u'_2 are continuously differentiable functions of the variables u_1, u_2:

$$u'_1 = u'_1(u_1, u_2), \qquad (2)$$
$$u'_2 = u'_2(u_1, u_2).$$

Similarly, we shall assume that u_1, u_2 are continuously differentiable functions of the variables u'_1, u'_2;

$$u_1 = u_1(u'_1, u'_2), \qquad (3)$$
$$u_2 = u_2(u'_1, u'_2).$$

Here, equations (3) represent the solution of equations (2) with respect to the variables u_1, u_2. The determinant of the matrix

$$\left\| \begin{matrix} \dfrac{\partial u'_1}{\partial u_1}, & \dfrac{\partial u'_1}{\partial u_2} \\ \dfrac{\partial u'_2}{\partial u_1}, & \dfrac{\partial u'_2}{\partial u_2} \end{matrix} \right\| \quad \left(\text{briefly } \left\| \frac{\partial u'_i}{\partial u_j} \right\| \right)$$

is different from zero and the matrix $\left\| \dfrac{\partial u_i}{\partial u'_j} \right\|$ (formed by the partial derivatives of the functions (3)) is the inverse of the matrix $\left\| \dfrac{\partial u'_i}{\partial u_j} \right\|$, hence

$$\frac{\partial u'_i}{\partial u_j} \frac{\partial u_j}{\partial u'_k} = \delta^i_k, \quad \frac{\partial u_i}{\partial u'_j} \frac{\partial u'_j}{\partial u_k} = \delta^i_k,$$

where δ^i_k is the Kronecker delta (see Remark 8.1.3).

Theorem 1. *If we transform the parametric expression $\mathbf{r}(u_1, u_2)$ of a surface to a new parametric expression $\mathbf{r}'(u'_1, u'_2)$ by means of equations (3), then the following relations hold between the local coordinate vectors $\mathbf{r}_1(u_1, u_2)$, $\mathbf{r}_2(u_1, u_2)$ in the original expression and the local coordinate vectors $\mathbf{r}'_1(u'_1, u'_2)$, $\mathbf{r}'_2(u'_1, u'_2)$:*

$$\mathbf{r}'_i(u'_1, u'_2) = \sum_{j=1}^{2} \frac{\partial u_j}{\partial u'_i}(u'_1, u'_2)\, \mathbf{r}_j(u_1, u_2) \quad \left(\textit{briefly } \mathbf{r}'_i = \frac{\partial u_j}{\partial u'_i} \mathbf{r}_j\right), \qquad (4)$$

$$\mathbf{r}_i = \frac{\partial u'_j}{\partial u_i} \mathbf{r}'_j. \qquad (5)$$

REMARK 3. From (4), we see that the transformation matrix of the coordinate system $(\mathbf{r}_1, \mathbf{r}_2)$ to the coordinate system $(\mathbf{r}'_1, \mathbf{r}'_2)$ is $\left\|\frac{\partial u_j}{\partial u'_i}\right\|$ (cf. Definition 8.1.2).

Theorem 2 (*Transformation of the Contravariant Coordinates of a Vector on a Surface*). *If the contravariant coordinates of a tangent vector* $\mathbf{v}$ *at a point* M *on a surface* π *in the local coordinate system* $(\mathbf{r}_1, \mathbf{r}_2)$ *are* v^1, v^2 *and those in the local coordinate system* $(\mathbf{r}'_1, \mathbf{r}'_2)$ *(which has resulted from the original system by the change of the parametric expression of the surface defined by equations* (2) *and* (3)) *are* v'^1, v'^2 *then*

$$v'^i = \frac{\partial u'_i}{\partial u_j} v^j, \quad v^i = \frac{\partial u_i}{\partial u'_j} v'^j .$$

REMARK 4. The coordinates v^1, v^2 in Theorem 2 are naturally functions of the variables u_1, u_2; similarly the coordinates v'^1, v'^2 are functions of the variables u'_1, u'_2.

Theorem 3. *Let the contravariant coordinates of vectors* $\mathbf{a}$ *and* $\mathbf{b}$ *on a surface in the local coordinate system* $(\mathbf{r}_1, \mathbf{r}_2)$ *be* a^i *and* b^i, *respectively. Then*

$$\mathbf{a} \cdot \mathbf{b} = \sum_{i=1}^{2} \sum_{j=1}^{2} g_{ij} a^i b^j = g_{ij} a^i b^j , \tag{6}$$

where $g_{ij} = g_{ij}(u_1, u_2) = \mathbf{r}_i(u_1, u_2) \cdot \mathbf{r}_j(u_1, u_2)$.

Definition 3. The numbers $a_i = g_{ij} a^j$ are said to be the *covariant coordinates of the vector* $\mathbf{a}$ in the coordinate system $(\mathbf{r}_1, \mathbf{r}_2)$.

Theorem 4 (*Transformation of the Covariant Coordinates of a Vector on a Surface*). *If the covariant coordinates of a vector* $\mathbf{v}$ *on a surface* $\mathbf{r}(u_1, u_2)$ *in the local coordinate system* $(\mathbf{r}_1, \mathbf{r}_2)$ *are* v_i *and those in the local coordinate system* $(\mathbf{r}'_1, \mathbf{r}'_2)$ *(which has resulted from the original system by the change of the parametric expression of the surface according to equations* (2) *and* (3)) *are* v'_i, *then*

$$v'_i = \frac{\partial u_j}{\partial u'_i} v_j , \quad v_i = \frac{\partial u'_j}{\partial u_i} v'_j .$$

Theorem 5. *Let* a^i, b^i *be the contravariant and* a_i, b_i *the covariant coordinates of vectors* $\mathbf{a}$, $\mathbf{b}$ *on a surface* $\mathbf{r}(u_1, u_2)$, *respectively. Then* $\mathbf{a} \cdot \mathbf{b} = a^i b_i = a_i b^i$ *(here,* i *runs from* 1 *to* 2).

Definition 4 (*Definition of a Tensor on a Surface*; cf. Definition 8.2.1). We say a *tensor field* (briefly a *tensor*) is defined on a surface π if, to every local coordinate system $(\mathbf{r}_1(u_1, u_2), \mathbf{r}_2(u_1, u_2))$ defined by the corresponding parametric expression $\mathbf{r}(u_1, u_2)$ of the surface π, there correspond 2^{p+q} numbers (depending on the point of the surface) $a^{ij\dots}_{kl\dots}$ (the number of upper indices is p, the number of lower indices q; $i, j, k, l, \dots = 1, 2$) such that they are transformed according to the formulae

$$a'^{ij\dots}_{rs\dots} = a^{lm\dots}_{tv\dots} \frac{\partial u'_i}{\partial u_l} \frac{\partial u'_j}{\partial u_m} \dots \frac{\partial u_t}{\partial u'_r} \frac{\partial u_v}{\partial u'_s} \dots \tag{7}$$

by any change from the coordinate system $(\mathbf{r}_1(u_1, u_2), \mathbf{r}_2(u_1, u_2))$ to the coordinate system $(\mathbf{r}'_1(u'_1, u'_2), \mathbf{r}'_2(u'_1, u'_2))$ which has resulted from the original system by the change of the parametric expression of the surface according to equations (2) and (3). This tensor is said to be p-times contravariant and q-times covariant. The number $p + q$ is called the *rank of the tensor*, the numbers $a^{ij\dots}_{rs\dots}$ are called the *coordinates of the tensor.*

REMARK 5. The coordinates $a^{ij\dots}_{rs\dots}$ of a tensor on a surface π in the local coordinate system $(\mathbf{r}_1, \mathbf{r}_2)$ defined by the parametric expression $\mathbf{r}(u_1, u_2)$ of the surface π are evidently functions of the variables u_1, u_2 (see Remark 4).

Example 1 (*A Scalar on a Surface*). If to every point of a surface π there corresponds a certain fixed number a, then a tensor field of rank zero ($p = q = 0$), called a *scalar field*, briefly a *scalar*, is determined. In the coordinate system defined by the parametric expression $\mathbf{r}(u_1, u_2)$ of the surface π, a is a function of the variables u_1, u_2: $a = a(u_1, u_2)$. For a different expression $\mathbf{r}'(u'_1, u'_2)$ of the surface π in which the point with original curvilinear coordinates u_1, u_2 has curvilinear coordinates u'_1, u'_2, we naturally have $a = a(u'_1, u'_2) = a(u_1, u_2)$.

Example 2 (*A Contravariant and Covariant Vector on a Surface*). Let v^i and v_i be respectively the contravariant and covariant coordinates of a vector $\mathbf{v}$ on a surface Then, comparing the transformation formulae of Theorems 2 and 4 and the formulae of Definition 4, v^i and v_i are easily seen to be the coordinates of a once contravariant, or a once covariant tensor of rank 1, the so-called *contravariant*, or *covariant, vector on a surface*, respectively.

Example 3 (*The Metric Tensor of a Surface (the first fundamental tensor of a surface)*). If we assign to every local coordinate system $(\mathbf{r}_1, \mathbf{r}_2)$ defined by the expression $\mathbf{r}(u_1, u_2)$ the numbers $g_{ij}(u_1, u_2) = \mathbf{r}_1(u_1, u_2) \cdot \mathbf{r}_2(u_1, u_2)$, we easily verify that these numbers under any change of the coordinate system satisfy the transformation formulae

$$g'_{ij}(u'_1, u'_2) = \frac{\partial u_a}{\partial u'_i} \frac{\partial u_b}{\partial u'_j} g_{ab}(u_1, u_2) \,.$$

Hence, g_{ij} are the coordinates of a twice covariant quadratic tensor, the so-called (*covariant*) *metric* (or the *first fundamental*) *tensor of the surface*. The determinant of the matrix $\|g_{ij}\|$ is different from zero and thus there exists the inverse matrix

$\|g^{ij}\|$ (i.e. $g_{ij}g^{jk} = \delta_i^k$; see Remark 8.1.3). The numbers $g^{ij}(u_1, u_2)$ are the coordinates of a twice contravariant quadratic tensor, the so-called (*contravariant*) *metric tensor of the surface*. If a^i, b^i, or a_i, b_i, are the contravariant, or covariant, coordinates, respectively, of the vectors $\boldsymbol{a}$, $\boldsymbol{b}$ on a surface in the coordinate system in which the coordinates of the metric tensor are g_{ij}, then $\boldsymbol{a} \,.\, \boldsymbol{b} = g_{ij}a^ib^j = g^{ij}a_ib_j$, $\boldsymbol{a} \,.\, \boldsymbol{a} = g_{ij}a^ia^j = g^{ij}a_ia_j$. This justifies the term "metric tensor": by means of it we measure the lengths of vectors on a surface, and angles between them. The coordinates $g_{ij}(u_1, u_2)$ are called the *coefficients of the first fundamental form of the surface* (§ 9.14).

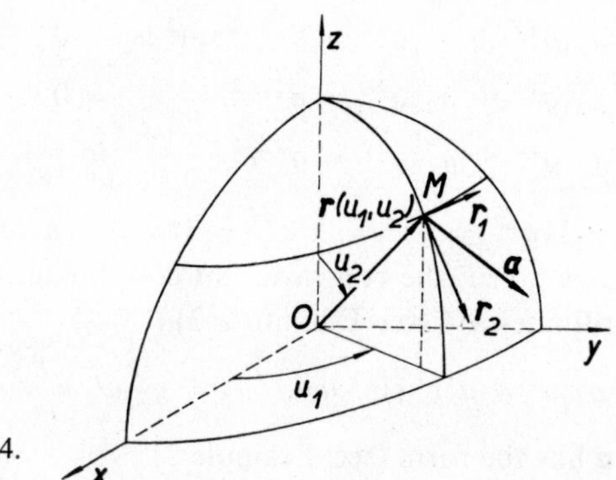

Fig. 8.4.

Example 4 (*The Second Fundamental Tensor of a Surface*). Let π be a surface given by the parametric expression $\boldsymbol{r}(u_1, u_2)$ and $h_{11}(u_1, u_2)\, du_1^2 + 2h_{12}(u_1, u_2) \,.\, du_1\, du_2 + h_{22}(u_1, u_2)\, du_2^2$, briefly $h_{ij}\, du_i\, du_j$, be its second fundamental form (see § 9.15). Then h_{ij} are the coordinates of a quadratic twice covariant tensor, the so-called *second fundamental tensor of the surface*.

Example 5. In a rectangular coordinate system, let a spherical surface with centre at the origin of the coordinate system and radius r be given (see Fig. 8.4). Let us choose the parametric expression $\boldsymbol{r}(u_1, u_2)$ in such a way that the coordinates of $\boldsymbol{r}(u_1, u_2)$ are

$$x = r \cos u_1 \sin u_2\,,$$
$$y = r \sin u_1 \sin u_2\,, \quad 0 \leqq u_1 < 2\pi\,, \quad 0 < u_2 < \pi\,.$$
$$z = r \cos u_2\,,$$

The coordinates of the local coordinate vectors $\boldsymbol{r}_1(u_1, u_2)$, $\boldsymbol{r}_2(u_1, u_2)$ (see Remark 1) are

$$\boldsymbol{r}_1(u_1, u_2) = (-r \sin u_1 \sin u_2, r \cos u_1 \sin u_2, 0)\,,$$
$$\boldsymbol{r}_2(u_1, u_2) = (\ \ r \cos u_1 \cos u_2, r \sin u_1 \cos u_2, -r \sin u_2)\,.$$

The covariant coordinates $g_{ij}(u_1, u_2)$ of the metric tensor (see Example 3) are

$$g_{11} = \mathbf{r}_1 . \mathbf{r}_1 = r^2 \sin^2 u_1 \sin^2 u_2 + r^2 \cos^2 u_1 \sin^2 u_2 = r^2 \sin^2 u_2 ,$$

$$g_{12} = g_{21} = \mathbf{r}_1 . \mathbf{r}_2 = -r^2 \sin u_1 \sin u_2 \cos u_1 \cos u_2 + \\ + r^2 \cos u_1 \sin u_2 \sin u_1 \cos u_2 = 0 ,$$

$$g_{22} = \mathbf{r}_2 . \mathbf{r}_2 = r^2 \cos^2 u_1 \cos^2 u_2 + r^2 \sin^2 u_1 \cos^2 u_2 + r^2 \sin^2 u_2 = r^2 .$$

The contravariant coordinates $g^{ij}(u_1, u_2)$ of the metric tensor satisfy $g_{ij}g^{jk} = \delta^k_i$, i.e.

$$g_{11}g^{11} + g_{12}g^{21} = g^{11}r^2 \sin^2 u_2 = 1 ,$$

$$g_{11}g^{12} + g_{12}g^{22} = g^{12}r^2 \sin^2 u_2 = 0 ,$$

$$g_{21}g^{11} + g_{22}g^{21} = g^{21}r^2 = 0 ,$$

$$g_{21}g^{12} + g_{22}g^{22} = g^{22}r^2 = 1 .$$

Thus, $g^{11} = 1/(r^2 \sin^2 u_2)$, $g^{12} = g^{21} = 0$, $g^{22} = 1/r^2$. Let a vector $\mathbf{a}$ at a point M of the spherical surface be given, the contravariant coordinates of which are a^1, a^2. Then its covariant coordinates are (see Definition 3)

$$a_1 = g_{1j}a^j = a^1r^2 \sin^2 u_2 , \quad a_2 = g_{2j}a^j = a^2r^2 .$$

The scalar product $\mathbf{a} . \mathbf{a}$ has the form (see Example 3)

$$\mathbf{a} . \mathbf{a} = g_{ij}a^ia^j = (a^1)^2 \, r^2 \sin^2 u_2 + (a^2)^2 \, r^2 .$$

8.4. Basic Algebraic Operations on Tensors

REMARK 1. By the term "tensor" we understand here both tensor in space and tensor on a surface. It is necessary to bear in mind that the indices of the coordinates of a tensor in space assume the values 1, 2, 3 while those of the coordinates of a tensor on a surface only the values 1, 2.

Definition 1 (*Equality of Tensors*). We say that *two tensors are equal* if they are both p-times contravariant and q-times covariant and their coordinates are equal in at least one coordinate system. (Then, the coordinates are equal in every coordinate system.)

Definition 2 (*Addition of Tensors*). If $a^{ij\ldots}_{rs\ldots}$, $b^{ij\ldots}_{rs\ldots}$ are the coordinates of two tensors of the same type (i.e. if both are p-times contravariant and q-times covariant), then the numbers

$$c^{ij\ldots}_{rs\ldots} = a^{ij\ldots}_{rs\ldots} + b^{ij\ldots}_{rs\ldots}$$

are the coordinates of a tensor which is said to be the *sum* of these tensors (and is of the same type).

Definition 3 (*Multiplication of Tensors*). If $a^{ij\ldots}_{pq\ldots}$ are the coordinates of a p_1-times contravariant and q_1-times covariant tensor and $b^{kl\ldots}_{rs\ldots}$ the coordinates of a p_2-times contravariant and q_2-times covariant tensor, then the numbers

$$c^{ij\ldots kl\ldots}_{pq\ldots rs\ldots} = a^{ij\ldots}_{pq\ldots} b^{kl\ldots}_{rs\ldots}$$

are the coordinates of a $(p_1 + p_2)$-times contravariant and $(q_1 + q_2)$-times covariant tensor which is said to be the *product* of these tensors.

Definition 4 (*Contraction of Tensors*). Let $a^{ij\ldots}_{rs\ldots}$ be the coordinates of a p-times contravariant and q-times covariant tensor. Consider the sums

$$c^{j\ldots}_{r\ldots} = \sum_i a^{ij\ldots}_{ir\ldots} = a^{ij\ldots}_{ir\ldots}\,.$$

Then $c^{j\ldots}_{r\ldots}$ are the coordinates of a $(p - 1)$-times contravariant and $(q - 1)$-times covariant tensor. The tensor $c^{j\ldots}_{r\ldots}$ is called a *contraction of the tensor* $a^{ij\ldots}_{rs\ldots}$. Contraction can be performed not only on the first upper and the first lower indices but also on arbitrary k upper and k lower indices. For example, the contraction of a tensor a^{ij}_{rs} performed on both upper and both lower indices is the scalar $a = a^{ij}_{ij} = \sum_i \sum_j a^{ij}_{ij}$.

Example 1. Let v^i be a contravariant vector, g_{ij} a (covariant) metric tensor. By multiplication, we get the tensor $g_{ij}v^k$ of rank three; the covariant vector $v_i = g_{ij}v^j$ is its contraction.

Definition 5 (*Lowering and Raising of Indices*). For every p-times contravariant and q-times covariant tensor $a^{ij\ldots}_{rs\ldots}$ a new $(p + q)$-times covariant tensor $a_{ij\ldots rs\ldots} = g_{ki}g_{lj}\ldots a^{kl\ldots}_{rs\ldots}$ can be constructed, where g_{ij} are the coordinates of the metric tensor. We say that the tensor $a_{ij\ldots rs\ldots}$ was obtained from the tensor $a^{ij\ldots}_{rs\ldots}$ by *lowering of indices*.

Similarly a new $(p + q)$-times contravariant tensor $a^{rs\ldots ij\ldots} = g^{rk}g^{sl}\ldots a^{ij\ldots}_{kl\ldots}$ can be constructed from a tensor $a^{ij\ldots}_{rs\ldots}$. The tensor $a^{rs\ldots ij\ldots}$ was obtained from $a^{ij\ldots}_{rs\ldots}$ by *raising of indices*.

REMARK 2. By raising some of the indices of a covariant tensor we again get a tensor; however, it is necessary to indicate those indices which have been raised. This may be done by means of dots which indicate the place of the raised indices, as illustrated in the following examples:

$$a^{i}_{.jk} = g^{il}a_{ljk}\,, \quad a_{i.k}^{\ j} = g^{lj}a_{ilk}\,, \quad a^{ij}_{..k} = g^{il}g^{jp}a_{lpk}\,.$$

A similar notation is used when lowering indices, e.g. $a^{i.k}_{\ j} = g_{lj}a^{ilk}$.

Example 2. By lowering the contravariant coordinates v^i of a vector we get its covariant coordinates $v_i = g_{ij}v^j$.

Definition 6. A tensor is said to be *symmetric with respect to given upper* (*or lower*) *indices* if its coordinates do not alter by an arbitrary permutation of these indices. For example, a tensor a^l_{ijk} is symmetric with respect to the first two lower indices if $a^l_{ijk} = a^l_{jik}$.

Example 3. The metric tensor g_{ij} is symmetric for $g_{ij} = \mathbf{r}_i \cdot \mathbf{r}_j = \mathbf{r}_j \cdot \mathbf{r}_i = g_{ji}$. Also the tensor g^{ij} and the second fundamental tensor h_{ij} of a surface (see Example 8.3.4) are symmetric.

Definition 7. A tensor is said to be *skew-symmetric* (*alternating*) *with respect to a given group of upper* (*lower*) *indices* if the sign changes with every interchange of two arbitrary indices of the group. For example, a tensor a_{ij} is skew-symmetric if $a_{ij} = -a_{ji}$.

Definition 8 (*Operation of Symmetrization*). For every tensor, a tensor symmetric with respect to a given group of indices can be constructed. For example, by symmetrization of a tensor $a_{ijkl\ldots}$ with respect to the first three indices we get the tensor

$$a_{(ijk)l\ldots} = \frac{1}{3!}(a_{ijkl\ldots} + a_{ikjl\ldots} + a_{jikl\ldots} + a_{jkil\ldots} + a_{kijl\ldots} + a_{kjil\ldots}) \,. \qquad (1)$$

The tensor $a_{(ijk)}$ is the so-called *symmetric part of the tensor* a_{ijk}.

Definition 9 (*Operation of Skew-symmetrization*). For every tensor, a tensor skew-symmetric with respect to a given group of indices can be constructed. For example, by skew-symmetrization of a tensor $a_{ijkl\ldots}$ with respect to the first three indices we get the tensor

$$a_{[ijk]l\ldots} = \frac{1}{3!}(a_{ijkl\ldots} - a_{ikjl\ldots} - a_{jikl\ldots} + a_{jkil\ldots} + a_{kijl\ldots} - a_{kjil\ldots}) \,. \qquad (2)$$

(Here, we choose the plus sign with an even and minus with an odd permutation of the indices i, j, k.)

The tensor $a_{[ijk]}$ is the so-called *skew-symmetric part of the tensor* a_{ijk}.

REMARK 3. A quadratic tensor is the sum of its symmetric and skew-symmetric parts: $a_{ij} = a_{(ij)} + a_{[ij]}$.

Example 4. If $a_{ij} = c^k_j g_{ik}$ are the covariant coordinates of the deformation tensor of Example 8.2.7, then its symmetric part $a_{(ij)} = \frac{1}{2}(a_{ij} + a_{ji})$ is the so-called *tensor of a pure deformation*, its skew-symmetric part $a_{[ij]} = \frac{1}{2}(a_{ij} - a_{ji})$ is the so-called *tensor of rotation* (it represents, roughly speaking, the rotation of the body).

8.5. Symmetric Quadratic Tensors

Definition 1. On a surface defined by a parametric expression $\mathbf{r}(u_1, u_2)$, let a quadratic symmetric (non-zero) tensor be given. According to Definition 8.4.5 we can assume that it is covariant and its coordinates are $a_{ij}(u_1, u_2)$ $(a_{ij} = a_{ji})$. Let us choose a point O on the surface whose curvilinear coordinates are u_1, u_2 and construct in the tangent plane at this point the locus of terminal points of vectors t^i on the surface with the initial point at O which satisfy whichever of the equations

$$a_{ij}t^i t^j = a_{11}(t^1)^2 + 2a_{12}t^1t^2 + a_{22}(t^2)^2 = \pm 1 \,. \tag{1}$$

This locus is called the *indicatrix of the tensor a_{ij} at the point (u_1, u_2)*.

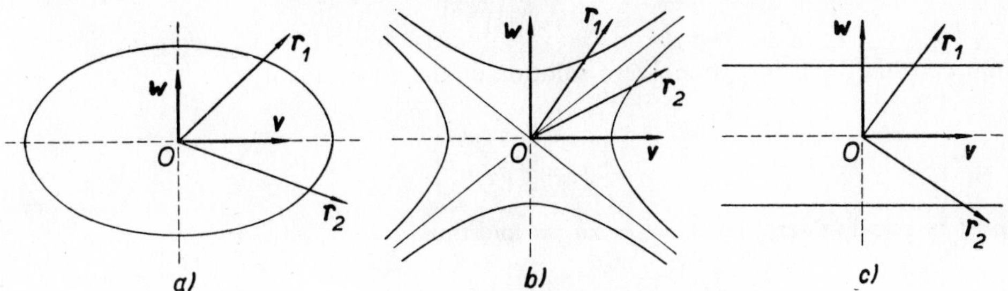

Fig. 8.5.

Theorem 1. *The indicatrix of the tensor a_{ij} at a point O is*
1. *an ellipse (or a circle), if the determinant*

$$\begin{vmatrix} a_{11}, & a_{12} \\ a_{21}, & a_{22} \end{vmatrix} \tag{2}$$

is positive (i.e. the form $a_{ij}t^it^j$ is definite) (see Fig. 8.5a);

2. *a pair of hyperbolas with common asymptotes and centre at the point O* (see Fig. 8.5b), *if the determinant* (2) *is negative (i.e. the form $a_{ij}t^it^j$ is indefinite)*;

3. *a pair of parallel lines, if the determinant* (2) *equals zero* (see Fig. 8.5c)

REMARK 1. In what follows we shall restrict our considerations to the case when the determinant (2) is non-zero.

Definition 2. The directions of conjugate diameters of the indicatrix of a tensor are called *conjugate directions* of the tensor; the directions of the axes are the *principal directions* of the tensor.

Theorem 2. *Vectors v^i, w^j lie in conjugate directions of a tensor a_{ij} if and only if $a_{ij}v^iw^j = 0$.*

REMARK 2. *Determination of the principal directions of a quadratic tensor:*

1. If a_{ij} is a multiple of the metric tensor (i.e. $a_{ij} = \lambda g_{ij}$), then the indicatrix is a circle and any direction is principal;

2. Let a_{ij} not be a multiple of g_{ij}; let $\mathbf{v}$, $\mathbf{w}$ be vectors lying in the principal directions of the tensor a_{ij} (see Fig. 8.5a,b), v^i, w^i their contravariant coordinates. Then

$$\begin{aligned} a_{ij}v^i w^j &= 0\,, \\ \mathbf{v}\,.\,\mathbf{w} = g_{ij}v^i w^j &= 0\,. \end{aligned} \tag{3}$$

Re-write the equations (3) in the form

$$\begin{aligned} v^i(a_{ij}w^j) &= 0\,, \\ v^i(g_{ij}w^j) &= 0\,. \end{aligned}$$

In order that there exist a non-zero solution v^i, the determinant

$$\begin{vmatrix} a_{1j}w^j, & a_{2j}w^j \\ g_{1j}w^j, & g_{2j}w^j \end{vmatrix}$$

must be equal to zero, i.e. $a_{ij}w^j = \lambda g_{ij}w^j$ and thus

$$(a_{ij} - \lambda g_{ij})\, w^j = 0\,. \tag{4}$$

In order that equations (4) have a non-zero solution for w^j, it is necessary that

$$\begin{vmatrix} a_{11} - \lambda g_{11}, & a_{12} - \lambda g_{12} \\ a_{21} - \lambda g_{21}, & a_{22} - \lambda g_{22} \end{vmatrix} = 0\,. \tag{5}$$

The roots λ_1, λ_2 of the quadratic equation (5) are the so-called *characteristic numbers of the tensor*. Substituting them successively into equations (4) we can determine the required vectors w^i, v^i (see Example 1).

REMARK 3. For a quadratic symmetric tensor in space we can obtain results similar to those just introduced. However, the indicatrix is then a quadratic surface (or a pair of such surfaces) and there are, in general, three principal directions.

Example 1. Let the tensor of the membrane stresses in the middle surface of a spherical shell of radius r be given by the covariant coordinates $\sigma_{ij}(u_1, u_2)$, with respect to the coordinate system of Example 8.3.5. Let us find the directions of the principal stresses in the middle surface.

The tensor of the membrane stresses is symmetric and its principal directions coincide with the directions of the principal stresses. In order to determine the principal directions of the tensor σ_{ij} we substitute in equation (5) above (where $g_{11} = r^2 \sin^2 u_2$, $g_{12} = g_{21} = 0$, $g_{22} = r^2$), giving

$$\begin{vmatrix} \sigma_{11} - \lambda r^2 \sin^2 u_2, & \sigma_{12} \\ \sigma_{12}, & \sigma_{22} - \lambda r^2 \end{vmatrix} = 0\,,$$

i.e.

$$\lambda^2 r^4 \sin^2 u_2 - \lambda r^2(\sigma_{11} + \sigma_{22} \sin^2 u_2) + \sigma_{11}\sigma_{22} - \sigma_{12}^2 = 0\,.$$

Denoting the roots of this equation by λ_1, λ_2, we find vectors v^i, w^i lying in the principal directions, from equations $(a_{ij} - \lambda_1 g_{ij})v^j = 0$, $(a_{ij} - \lambda_2 g_{ij})w^j = 0$, i.e.

$$(\sigma_{11} - \lambda_1 r^2 \sin^2 u_2)\, v^1 + \sigma_{12} v^2 = 0\,, \quad (\sigma_{11} - \lambda_2 r^2 \sin^2 u_2)\, w^1 + \sigma_{12} w^2 = 0\,.$$

Thus, $v^1/v^2 = \sigma_{12}/(\lambda_1 r^2 \sin^2 u_2 - \sigma_{11})$ and similarly $w^1/w^2 = \sigma_{12}/(\lambda_2 r^2 \sin^2 u_2 - \sigma_{11})$. Here, the ratio v^1/v^2 (or w^1/w^2) represents the tangent of the angle which the principal direction (axis of the indicatrix) makes with the corresponding parallel line on the middle surface.

REMARK 4. In tensor calculus, tensors may also be introduced by means of the concept of a dyad (see e.g. [208]). For *tensor analysis* (*covariant derivative* etc.) see e.g. [421]).

9. DIFFERENTIAL GEOMETRY

By Bořivoj Kepr

References: [27], [38], [95], [127], [157], [185], [230], [239], [248], [341], [353], [422].

9.1. Introduction

Differential geometry is the study of curves (both plane and space curves) and surfaces by means of the calculus. When investigating geometric configurations (on the basis of their equations) in differential geometry, we aim mostly at the study of *invariant properties*, i.e., properties independent of the choice of the coordinate system and so belonging directly to the curve or surface (e.g. the points of inflexion, the curvature and so on). But we also study those properties of geometric configurations that depend on the choice of the coordinate system (e.g. the sections of a surface by the coordinate planes, the slope of the tangent and so on). Differential geometry studies mostly the *local properties* of curves and surfaces, i.e. those which pertain to sufficiently small portions of the curve or the surface; so it is essentially "geometry in the small". But differential geometry also investigates those properties of curves and surfaces which pertain to the configuration as a whole (e.g. the length of a curve, the number of vertices and so on).

A. CURVES

9.2. Definition and Equations of a Curve, Length of Arc and Tangent Line

Definition 1. A *piecewise smooth space curve, defined parametrically*, is a set of points (x, y, z) given by the equations

$$x = x(t)\,, \quad y = y(t)\,, \quad z = z(t)\,, \tag{1}$$

where the functions $x(t)$, $y(t)$, $z(t)$, defined in some interval I (most often in a closed interval $[a, b]$ or in the interval $(-\infty, +\infty)$),

1. are continuous in the interval I,

2. have, in I, piecewise continuous derivatives $\dot{x}(t)$, $\dot{y}(t)$, $\dot{z}(t)$ (we write $dx/dt = \dot{x}$, etc.), while with the exception of at most a finite number of points t_k from the interval I the relation $\dot{x}^2 + \dot{y}^2 + \dot{z}^2 > 0$ holds (i.e. at least one of the derivatives $\dot{x}$, $\dot{y}$, $\dot{z}$ is non-zero).

REMARK 1. If $I = [a, b]$ and if

$$x(a) = x(b), \quad y(a) = y(b), \quad z(a) = z(b)$$

the curve is said to be *closed*. If the functions $\dot{x}$, $\dot{y}$, $\dot{z}$ are continuous in I (cf. Definition 1) (and in the case of a closed curve the values of the right-hand derivatives of functions $\dot{x}$, $\dot{y}$, $\dot{z}$ at the point a and the left-hand derivatives of these functions at the point b are equal) and if everywhere in I (in the case of a closed curve also at the points a, b) $\dot{x}^2 + \dot{y}^2 + \dot{z}^2 > 0$, the curve is said to be *smooth*. *In the following text the word curve will stand for a smooth or piecewise smooth curve. As is customary in differential geometry, we shall suppose that the functions* $x(t)$, $y(t)$, $z(t)$ *also possess (continuous) derivatives of an order*, r, *higher than one (according as the problem under investigation may require) without stating explicitly this condition.*
(A similar remark holds for Definition 2.)

The argument t in equations (1) is called the *parameter of the curve* and equations (1) are called the *parametric equations of the curve*. Every number t from the interval I is called a *point of the curve* (to every value $t \in I$ there corresponds a point (x, y, z) on the curve), namely a *regular point* when $\dot{x}^2 + \dot{y}^2 + \dot{z}^2 > 0$ and when no other value of $t \in I$ corresponds to the considered point (x, y, z). (An exception may occur at the points a, b in the case of a closed curve.) Every other point is called a *singular point* of the curve.

REMARK 2. Whether a point t is a regular or singular point of a curve may, in the general case, depend on the chosen parametric representation of the curve. For the parabola, represented parametrically by the equations $x = t$, $y = t^2$, $z \equiv 0$, $t \in (-\infty, +\infty)$, the point $t = 0$ is not a singular point; but if it is represented parametrically by equations $x = t^3$, $y = t^6$, $z \equiv 0$, $t \in (-\infty, +\infty)$ (both pairs of equations define the same set of points in the cartesian coordinate system $(O; x, y, z)$), then the point $t = 0$ is a singular point with this parametric representation because $\dot{x}^2 + \dot{y}^2 + \dot{z}^2 = 0$. In these cases we speak of an *removable singular point* of the given curve.

REMARK 3. Of course, we obtain the curve as the set of all points (1) for every t from the interval I. If to several different values of t from I there corresponds a single point $P(x_0, y_0, z_0)$, then such a singular point is called a *multiple point* of the curve (Fig. 9.1).

In the following exposition the word "point" will mean a regular point of the curve. At such a point there exists a unique tangent.

REMARK 4. If $\boldsymbol{r}$ is the *radius vector* of a point on the curve (1) (where the coordinates are $x(t)$, $y(t)$, $z(t)$ and the starting point is at the origin), then we can use a single symbolic equation of the curve (the *vector equation*)

$$\boldsymbol{r} = \boldsymbol{r}(t) \equiv \boldsymbol{i}\, x(t) + \boldsymbol{j}\, y(t) + \boldsymbol{k}\, z(t) \tag{2}$$

where $\boldsymbol{i}$, $\boldsymbol{j}$ and $\boldsymbol{k}$ are the unit vectors along the positive axes; $\boldsymbol{r}$ is the so-called *radius vector* (the *position vector*) of the point (x, y, z) on the curve. As t runs through the interval I, the end point of the radius vector describes the given curve (Fig. 9.2).

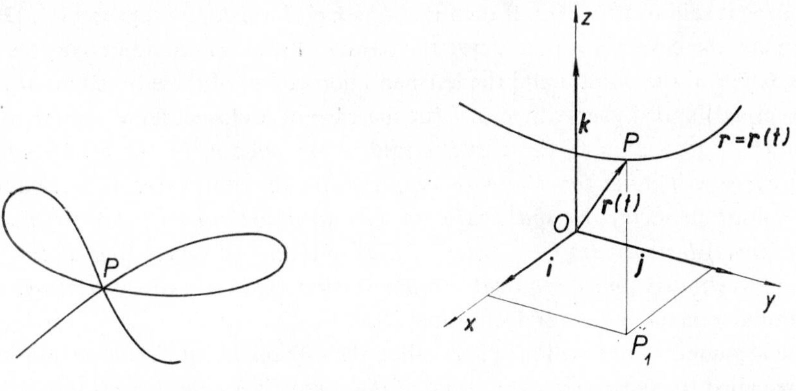

Fig. 9.1. Fig. 9.2.

Example 1. For the *circular helix* (see Example 9.3.1)

$$x = a \cos t\,, \quad y = a \sin t\,, \quad z = bt\,, \quad t \in (-\infty, \infty)\,,$$

the vector equation is of the form

$$\boldsymbol{r} = \boldsymbol{i}a \cos t + \boldsymbol{j}a \sin t + \boldsymbol{k}bt\,, \quad t \in (-\infty, \infty)\,.$$

Definition 2. Let the functions $\varphi(x, y, z)$, $\psi(x, y, z)$, defined in the (three-dimensional) domain O, have continuous partial derivatives of the first order. By a *curve, defined in space implicitly by the equations*

$$\varphi(x, y, z) = 0\,, \quad \psi(x, y, z) = 0\,, \tag{3}$$

is meant the set of points whose cartesian coordinates (x, y, z) satisfy simultaneously both equations (3); we suppose that the matrix

$$\left\| \begin{matrix} \dfrac{\partial \varphi}{\partial x}, & \dfrac{\partial \varphi}{\partial y}, & \dfrac{\partial \varphi}{\partial z} \\[2ex] \dfrac{\partial \psi}{\partial x}, & \dfrac{\partial \psi}{\partial y}, & \dfrac{\partial \psi}{\partial z} \end{matrix} \right\| \tag{4}$$

is of rank 2 at every such point (x, y, z), with the exception of at most a finite number of points.

The curve so defined is the *intersection* of the surfaces (3). Equations (3) are called the *implicit equations of this curve.*

Example 2. The circle with centre at the origin and radius a which lies in the plane $z = x$ may be expressed as the intersection of the considered plane and a sphere with the centre at the origin in the form (3) as follows:

$$z - x = 0\,, \quad x^2 + y^2 + z^2 - a^2 = 0\,.$$

REMARK 5. If the equations (3) are algebraic (or if they can be transformed so as to become algebraic; for example, φ and ψ are polynomials), the curve is called *algebraic*; if the equations (3) are not algebraic (and if they cannot be transformed so that they become algebraic), the curve is said to be *transcendent.* If the functions (1) are rational functions of the variable t (or if they can be so reshaped), then the curve (1) is said to be *rational.* The parameter t is the *coordinate on the curve.* The form (1) is often used in mechanics in the study of the movement of a particle, and in this case t represents the time.

Definition 3. A curve (1) is said to be a *plane curve* if constants A, B, C, D (at least one of which is non-zero) can be found such that

$$A\,x(t) + B\,y(t) + C\,z(t) + D = 0 \tag{5}$$

holds for every t from I. (The whole curve lies in the plane $Ax + By + Cz + D = 0$.) In the contrary case the curve (1) is a *space curve* (a *skew* or *twisted curve*).

REMARK 6. The curve (3) is said to be a *plane curve* if at least one of equations (3) is the equation of a plane or may be replaced by the equation of a plane (cf. Example 2). Otherwise the curve (3) is a *space curve.*

Example 3. The curve

$$x = t\,, \quad y = t^2\,, \quad z = t^3 \tag{6}$$

is a space curve because, as is well known from algebra, the equation

$$At + Bt^2 + Ct^3 + D = 0$$

can be satisfied identically only in the case where $A = B = C = D = 0$.

If we eliminate t from equations (6), we obtain an implicit representation of the curve

$$y = x^2\,, \quad z = x^3\,.$$

Example 4. If the equations

$$x = t\,, \quad y = y(t)\,, \quad z = 0 \tag{7}$$

represent a curve, then it is a plane curve since the identity

$$At + B\,y(t) + C\,.\,0 + D = 0$$

is satisfied by $A = B = D = 0$, C being an arbitrary number. The plane curve (7) is often expressed in the short form

$$y = y(x) \quad \text{or} \quad y = f(x)\,. \tag{8}$$

(The equation $z = 0$ is assumed.) Equation (8) is the so-called *explicit equation of a (plane) curve.*

Example 5. If the equations

$$\varphi(x, y) = 0\,, \quad z = 0 \tag{9}$$

represent a curve, then it is a plane curve because it lies in the coordinate plane xy. We write the plane curve (9) in the form

$$F(x, y) = 0 \text{ or } f(x, y) = 0 \text{ or } \varphi(x, y) = 0 \text{ or in some similar way}\,. \tag{10}$$

(The equation $z = 0$ is assumed.) Equations (10) are called the *implicit equations of a (plane) curve.*

Definition 4. A point (x_0, y_0) on a plane curve $F(x, y) = 0$ is called a *regular (ordinary) point of the curve* if at least one of

$$\frac{\partial F(x_0, y_0)}{\partial x}\,, \quad \frac{\partial F(x_0, y_0)}{\partial y}$$

is non-zero. Every other point on this curve is said to be *singular.* (Cf., however, Remark 2.)

Definition 5. An equation

$$t = t(\bar{t}) \tag{11}$$

expresses a so-called *admissible transformation of the parameter* in an interval $\bar{I}$ if the function (11), defined in $\bar{I}$, possesses the following properties:

1. it is a continuous function and has a continuous derivative (or continuous derivatives up to the order r- cf. Remark 1),
2. $\mathrm{d}t/\mathrm{d}\bar{t} \neq 0$.

The parameter $\bar{t}$ introduced in place of the parameter t by transformation (11) is called an *admissible parameter.*

Remark 7. All admissible parameters (and these only) are mutually equivalent. A very convenient transformation of the parameter is that one which introduces the arc s of the curve instead of a general parameter t (in terms of the arc s many results often become of a much simpler form):

Definition 6. The expression

$$s = s(t) = \int_{t_0}^{t} \mathrm{d}s = \int_{t_0}^{t} \sqrt{(\dot{x}^2 + \dot{y}^2 + \dot{z}^2)}\, \mathrm{d}t =$$

$$= \int_{t_0}^{t} \sqrt{\left(\frac{\mathrm{d}\boldsymbol{r}}{\mathrm{d}t} \cdot \frac{\mathrm{d}\boldsymbol{r}}{\mathrm{d}t}\right)}\, \mathrm{d}t = \int_{t_0}^{t} \sqrt{(\dot{\boldsymbol{r}} \cdot \dot{\boldsymbol{r}})}\, \mathrm{d}t = \int_{t_0}^{t} \sqrt{(\mathrm{d}\boldsymbol{r} \cdot \mathrm{d}\boldsymbol{r})} \tag{12}$$

is called the *arc of the curve* from the point $t_0 \in I$ to the point $t \in I$.

REMARK 8. It is known, from Integral Calculus, that the expression (12) is the *length of the arc of the curve* between the points t_0 and t. The differential

$$\mathrm{d}s = \sqrt{\left(\frac{\mathrm{d}\boldsymbol{r}}{\mathrm{d}t} \cdot \frac{\mathrm{d}\boldsymbol{r}}{\mathrm{d}t}\right)}\, \mathrm{d}t = \sqrt{(\dot{\boldsymbol{r}} \cdot \dot{\boldsymbol{r}})}\, \mathrm{d}t = \sqrt{(\mathrm{d}\boldsymbol{r} \cdot \mathrm{d}\boldsymbol{r})} =$$

$$= \sqrt{(\dot{x}^2 + \dot{y}^2 + \dot{z}^2)}\, \mathrm{d}t = \sqrt{(\mathrm{d}x^2 + \mathrm{d}y^2 + \mathrm{d}z^2)}$$

is called the *element of length*, or *linear element*, *of the curve*. Instead of "the length of the arc s" we say only "the arc s". If $\dot{x}^2 + \dot{y}^2 + \dot{z}^2 > 0$ everywhere in I, the arc s is an admissible parameter of the curve. In this case we can write the equaion of the curve in the form

$$\boldsymbol{r} = \boldsymbol{r}(s) . \tag{13}$$

The *length of the curve* in the interval $[a, b]$ is

$$l = \int_{a}^{b} \sqrt{(\dot{x}^2 + \dot{y}^2 + \dot{z}^2)}\, \mathrm{d}t . \tag{14}$$

Theorem 1. *The parameter t represents the arc length s of the curve if the value $t = 0$ corresponds to the starting point of the curve and the relation*

$$\frac{\mathrm{d}\boldsymbol{r}}{\mathrm{d}t} \cdot \frac{\mathrm{d}\boldsymbol{r}}{\mathrm{d}t} = \dot{\boldsymbol{r}} \cdot \dot{\boldsymbol{r}} = \dot{x}^2 + \dot{y}^2 + \dot{z}^2 = 1 \tag{15}$$

holds good for every $t \in I$.

REMARK 9. We shall denote derivatives with respect to the arc s by primes,

$$x' = \frac{\mathrm{d}x}{\mathrm{d}s}, \quad x'' = \frac{\mathrm{d}^2x}{\mathrm{d}s^2} \quad \text{etc.}$$

The radius vector $\boldsymbol{r}$ of a point on a curve corresponding to the value of the parameter $t = t_0$ is denoted by $\boldsymbol{r}_0$ (we say shortly the point $\boldsymbol{r}_0$) and so on.

Definition 7. The *tangent vector* of the curve $\mathbf{r} = \mathbf{r}(t)$ at its point $\mathbf{r}_0$ (that is, at the point $t = t_0$ or (x_0, y_0, z_0)) is the vector

$$\dot{\mathbf{r}}_0 = \left(\frac{\mathrm{d}\mathbf{r}}{\mathrm{d}t}\right)_0 \tag{16}$$

the coordinates of which are $\dot{x}_0$, $\dot{y}_0$, $\dot{z}_0$ and its starting point at the point $\mathbf{r}_0$ (i.e. the point (x_0, y_0, z_0)). The straight line that contains this vector is called the *tangent to the curve* at the point $\mathbf{r}_0$, which is called the *point of contact* (*contact point*)

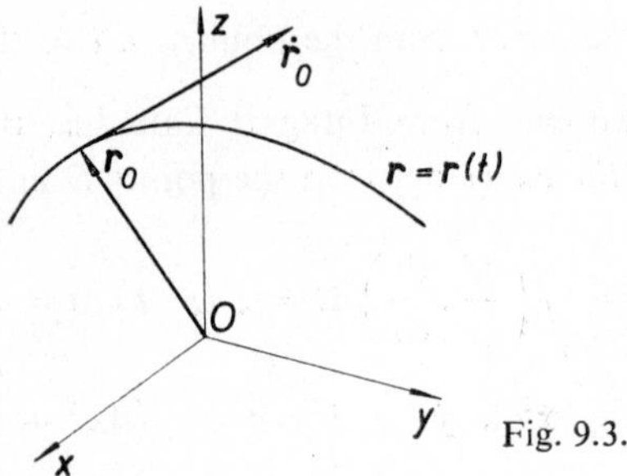

Fig. 9.3.

of the tangent (Fig. 9.3). (The tangent defined in this way is the limiting position of the secant line when its two points of intersection with the curve coincide in a single point of contact.)

REMARK 10. To *norm a tangent vector* $\dot{\mathbf{r}}_0$ means to represent it in terms of a unit vector of the same direction and sense (orientation) and with the starting point $\mathbf{r}_0$.

Theorem 2. *The vector*

$$\mathbf{t} = c\dot{\mathbf{r}}, \quad \textit{where} \quad c = \frac{1}{\sqrt{(\dot{\mathbf{r}} \cdot \dot{\mathbf{r}})}}, \tag{17}$$

with the coordinates (the direction cosines of the tangent)

$$t_x = c\dot{x}, \quad t_y = c\dot{y}, \quad t_z = c\dot{z} \tag{18}$$

is the so-called unit tangent vector (its length is equal to 1).

REMARK 11. If the parameter is the arc s, then the modulus of the tangent vector $\mathbf{r}'$ is 1. We write

$$\mathbf{r}' = \mathbf{t}, \tag{19a}$$

where $\mathbf{t}$ is the unit tangent vector. Its coordinates

$$t_x = x', \quad t_y = y', \quad t_z = z' \tag{19b}$$

are the direction cosines of the tangent vector.

For a general parameter t we obtain the direction cosines of the tangent to the curve in the form

$$t_x = \frac{\dot{x}}{\dot{s}}, \quad t_y = \frac{\dot{y}}{\dot{s}}, \quad t_z = \frac{\dot{z}}{\dot{s}} \quad \text{or } \mathbf{t} = \frac{\dot{\mathbf{r}}}{\dot{s}} \quad \left(\dot{s} = \frac{\mathrm{d}s}{\mathrm{d}t} = \sqrt{(\dot{x}^2 + \dot{y}^2 + \dot{z}^2)}\right) \tag{20}$$

and the following relations hold

$$\mathbf{r}' = \frac{\mathrm{d}\mathbf{r}}{\mathrm{d}s} = \frac{\mathrm{d}\mathbf{r}}{\mathrm{d}t} \cdot \frac{\mathrm{d}t}{\mathrm{d}s} = \dot{\mathbf{r}} \frac{\mathrm{d}t}{\mathrm{d}s} = \dot{\mathbf{r}} \frac{1}{\mathrm{d}s/\mathrm{d}t} = \frac{\dot{\mathbf{r}}}{\dot{s}} = \frac{\dot{\mathbf{r}}}{\sqrt{(\dot{x}^2 + \dot{y}^2 + \dot{z}^2)}}\,. \tag{21}$$

Let us denote by $\mathbf{R}$ the radius-vector of a current point (X, Y, Z) in space.

Theorem 3. *The equations of the tangent of the curve* $\mathbf{r} = \mathbf{r}(t)$ *at its point* $\mathbf{r}_0$ *are:*

$$\mathbf{R} = \mathbf{r}_0 + u\dot{\mathbf{r}}_0\,, \text{ i.e. } X = x_0 + u\dot{x}_0\,, \; Y = y_0 + u\dot{y}_0\,, \; Z = z_0 + u\dot{z}_0 \tag{22}$$

or

$$\frac{X - x_0}{\dot{x}_0} = \frac{Y - y_0}{\dot{y}_0} = \frac{Z - z_0}{\dot{z}_0} \quad \text{or} \quad \frac{X - x_0}{(\mathrm{d}x)_0} = \frac{Y - y_0}{(\mathrm{d}y)_0} = \frac{Z - z_0}{(\mathrm{d}z)_0} \tag{23}$$

(where (x_0, y_0, z_0) *is the point of contact,* $(\dot{x}_0, \dot{y}_0, \dot{z}_0)$ *are the direction ratios (direction parameters) of the tangent,* $\dot{x}_0 = \dot{x}(t_0)$ *etc. and* u *is the variable parameter on the tangent).*

REMARK 12. If a curve is given implicitly by the equations $F(x, y, z) = 0$, $G(x, y, z) = 0$, then the *direction ratios* $(\mathrm{d}x, \mathrm{d}y, \mathrm{d}z)$ of its *tangent* satisfy the relation

$$\mathrm{d}x : \mathrm{d}y : \mathrm{d}z = \begin{vmatrix} \frac{\partial F}{\partial y}, & \frac{\partial F}{\partial z} \\ \frac{\partial G}{\partial y}, & \frac{\partial G}{\partial z} \end{vmatrix} : \begin{vmatrix} \frac{\partial F}{\partial z}, & \frac{\partial F}{\partial x} \\ \frac{\partial G}{\partial z}, & \frac{\partial G}{\partial x} \end{vmatrix} : \begin{vmatrix} \frac{\partial F}{\partial x}, & \frac{\partial F}{\partial y} \\ \frac{\partial G}{\partial x}, & \frac{\partial G}{\partial y} \end{vmatrix}\,. \tag{24}$$

REMARK 13. The *equation of the tangent* to a plane curve $x = \varphi(t)$, $y = \psi(t)$ at its point (x_0, y_0) can be written in the form

$$(X - \varphi_0)\,\dot{\psi}_0 = (Y - \psi_0)\,\dot{\varphi}_0$$

or

$$\frac{X - \varphi_0}{\dot{\varphi}_0} = \frac{Y - \psi_0}{\dot{\psi}_0} \quad (\dot{\varphi}_0 \neq 0, \dot{\psi}_0 \neq 0)\,. \tag{25}$$

The *equation of the tangent* to a plane curve $y = f(x)$ at its point (x_0, y_0) is of the form

$$Y - y_0 = \dot{y}_0(X - x_0) \quad \left(\dot{y}_0 = \left(\frac{\mathrm{d}f}{\mathrm{d}x}\right)_0\right). \tag{26}$$

The *equation of the tangent* to a plane curve $F(x, y) = 0$ at its point (x_0, y_0) is of the form

$$(X - x_0) F_x^0 + (Y - y_0) F_y^0 = 0 \quad \left(F_x^0 = \frac{\partial F}{\partial x}(x_0, y_0), \quad F_y^0 = \frac{\partial F}{\partial y}(x_0, y_0)\right). \quad (27)$$

(The derivatives

$$\dot{y} = \frac{\mathrm{d}y}{\mathrm{d}x} \quad (\text{see } (26)), \quad \frac{\mathrm{d}y}{\mathrm{d}x} = \frac{\dot{\psi}(t)}{\dot{\varphi}(t)} \quad (\text{see } (25)), \quad \frac{\mathrm{d}y}{\mathrm{d}x} = -\frac{\partial F/\partial x}{\partial F/\partial y} \quad (\text{see } (27))$$

denote the tangent of the angle between the tangent at the given point of the curve and the positive x-axis.)

9.3. The Moving Trihedron and the Frenet Formulae

Definition 1. The unit vector $\boldsymbol{n}$ with the same direction and sense as the vector $\mathbf{t}' = \mathrm{d}\mathbf{t}/\mathrm{d}s$ and with starting point at the point $\boldsymbol{r}$ on the curve is called the *principal normal unit vector* or briefly the *principal normal* at the considered point.

REMARK 1. The straight line containing the vector $\boldsymbol{n}$ is also called the *principal normal* of the curve at its point. A principal normal is defined at a point of the curve if $\boldsymbol{r}'' \neq 0$ (i.e. if the coordinates (x'', y'', z'') of this vector are not all simultaneously equal to zero).

Theorem 1. *We have that*

$$\boldsymbol{n} = \frac{\mathbf{t}'}{k_1} = \frac{\boldsymbol{r}''}{k_1}, \quad \textit{its coordinates being} \quad n_x = \frac{x''}{k_1}, \quad n_y = \frac{y''}{k_1}, \quad n_z = \frac{z''}{k_1} \quad (1)$$

where

$$k_1 = \sqrt{(\mathbf{t}' \cdot \mathbf{t}')} = \sqrt{(\boldsymbol{r}'' \cdot \boldsymbol{r}'')} = \sqrt{(x''^2 + y''^2 + z''^2)}, \quad x'' = \frac{\mathrm{d}^2 x}{\mathrm{d}s^2} \quad \textit{etc.}$$

The vectors $\mathbf{t}$ *and* $\boldsymbol{n}$ *are perpendicular* $(\mathbf{t} \cdot \mathbf{n} = 0)$.

REMARK 2. If the curve is given by its parametric representation with a general parameter t, then

$$\boldsymbol{n} = \frac{\ddot{\boldsymbol{r}}c + \dot{\boldsymbol{r}}\dot{c}}{\sqrt{[(\ddot{\boldsymbol{r}}c + \dot{\boldsymbol{r}}\dot{c}) \cdot (\ddot{\boldsymbol{r}}c + \dot{\boldsymbol{r}}\dot{c})]}} \quad \left(c = \frac{1}{\sqrt{(\dot{\boldsymbol{r}} \cdot \dot{\boldsymbol{r}})}} = \frac{1}{\dot{s}}, \quad \ddot{\boldsymbol{r}} \neq \mathbf{0}, \quad \ddot{\boldsymbol{r}}c + \dot{\boldsymbol{r}}\dot{c} \neq \mathbf{0}\right).$$

The *direction cosines of the principal normal* are

$$n_x = \frac{\dot{s}(\dot{x}/\dot{s})^{\cdot}}{\sqrt{(\ddot{x}^2 + \ddot{y}^2 + \ddot{z}^2 - \ddot{s}^2)}}, \quad n_y = \frac{\dot{s}(\dot{y}/\dot{s})^{\cdot}}{\sqrt{(\ddot{x}^2 + \ddot{y}^2 + \ddot{z}^2 - \ddot{s}^2)}},$$

$$n_z = \frac{\dot{s}(\dot{z}/\dot{s})^{\cdot}}{\sqrt{(\ddot{x}^2 + \ddot{y}^2 + \ddot{z}^2 - \ddot{s}^2)}} \quad \left(\dot{s} = \frac{\mathrm{d}s}{\mathrm{d}t}, \quad \ddot{s} = \frac{\mathrm{d}^2 s}{\mathrm{d}t^2}\right) \tag{2a}$$

or

$$n_x = \frac{(\dot{x}/\dot{s})^{\cdot}}{k_1 \dot{s}}, \quad n_y = \frac{(\dot{y}/\dot{s})^{\cdot}}{k_1 \dot{s}}, \quad n_z = \frac{(\dot{z}/\dot{s})^{\cdot}}{k_1 \dot{s}} \quad \left(k_1 = \frac{1}{\dot{s}^2}\sqrt{(\ddot{x}^2 + \ddot{y}^2 + \ddot{z}^2 - \ddot{s}^2)}\right). \tag{2b}$$

REMARK 3. The direction ratios of the principal normal (when the parameter is the arc s of the curve) are given by

$$n_x : n_y : n_z = x'' : y'' : z''$$

or (when the parameter of the curve is a general parameter t) by

$$n_x : n_y : n_z = \left(\frac{\dot{x}}{\dot{s}}\right)^{\cdot} : \left(\frac{\dot{y}}{\dot{s}}\right)^{\cdot} : \left(\frac{\dot{z}}{\dot{s}}\right)^{\cdot}.$$

REMARK 4. The equation of the principal normal of a plane curve $F(x, y) = 0$, at its point (x_0, y_0) is

$$(X - x_0)\, F_y^0 - (Y - y_0)\, F_x^0 = 0 \quad \text{or} \quad \frac{X - x_0}{F_x^0} = \frac{Y - y_0}{F_y^0} \tag{3}$$

$$(\text{provided } F_x^0 \neq 0\,,\ F_y^0 \neq 0)\,,$$

for the curve $x = \varphi(t)$, $y = \psi(t)$ at its point t_0

$$(X - \varphi_0)\, \dot{\varphi}_0 + (Y - \psi_0)\, \dot{\psi}_0 = 0\,, \tag{4}$$

and for the curve $y = f(x)$

$$Y - y_0 = -\left(\frac{\mathrm{d}x}{\mathrm{d}y}\right)_0 (X - x_0)\,. \tag{5}$$

Definition 2. A unit vector $\mathbf{b}$ with its starting point at the point $\mathbf{r}$ on the curve and oriented so that it forms, with the vectors $\mathbf{t}$ and $\mathbf{n}$, a positively oriented normed rectangular trihedron (hence $\mathbf{b} = \mathbf{t} \times \mathbf{n}$, see Theorem 7.1.15), is called the *binormal unit vector* (or for short the *binormal*) to the curve at its point.

REMARK 5. The straight line containing the vector $\mathbf{b}$ is also called the *binormal* to the curve at its point. All perpendiculars to the tangent line at its point of contact are called *normals* to the curve. Among them the principal normal and the binormal are of fundamental importance. In the case of a plane curve the principal normal is that normal which lies in the plane of the curve. This principal normal is briefly called the *normal*.

Theorem 2. *The relations*

$$\mathbf{b} = \mathbf{t} \times \mathbf{n}\,, \quad \mathbf{n} = \mathbf{b} \times \mathbf{t}\,, \quad \mathbf{t} = \mathbf{n} \times \mathbf{b} \tag{6}$$

hold (*where* $\mathbf{t} \times \mathbf{n}$ *etc. are vector products*, see § 7.1, p. 267).

The following relations hold for the direction cosines (b_x, b_y, b_z) *of the binormal* $\mathbf{b}$ *at a point of the curve* $\mathbf{r} = \mathbf{r}(s)$ (*if the parameter is the arc* s):

$$b_x = \frac{\begin{vmatrix} y', & z' \\ y'', & z'' \end{vmatrix}}{k_1}\,, \quad b_y = \frac{\begin{vmatrix} z', & x' \\ z'', & x'' \end{vmatrix}}{k_1}\,, \quad b_z = \frac{\begin{vmatrix} x', & y' \\ x'', & y'' \end{vmatrix}}{k_1} \quad (\text{for } k_1 \text{ see Theorem 1})\,, \tag{7a}$$

in the case of a general parameter t $(\mathbf{r} = \mathbf{r}(t))$:

$$b_x = \frac{\begin{vmatrix} \dot{y}, & \dot{z} \\ \ddot{y}, & \ddot{z} \end{vmatrix}}{\dot{s}^3 k_1}\,, \quad b_y = \frac{\begin{vmatrix} \dot{z}, & \dot{x} \\ \ddot{z}, & \ddot{x} \end{vmatrix}}{\dot{s}^3 k_1}\,, \quad b_z = \frac{\begin{vmatrix} \dot{x}, & \dot{y} \\ \ddot{x}, & \ddot{y} \end{vmatrix}}{\dot{s}^3 k_1} \quad (\text{for } k_1 \text{ see (2b)})\,. \tag{7b}$$

REMARK 6. The *equations of the binormal* to the curve $\mathbf{r} = \mathbf{r}(t)$ at its point (x, y, z) are

$$\frac{X - x}{[\dot{y}, \ddot{z}]} = \frac{Y - y}{[\dot{z}, \ddot{x}]} = \frac{Z - z}{[\dot{x}, \ddot{y}]} \tag{8}$$

(X, Y, Z are the orthogonal coordinates of the running point on the binormal, $[\dot{y}, \ddot{z}] = \begin{vmatrix} \dot{y}, & \dot{z} \\ \ddot{y}, & \ddot{z} \end{vmatrix}$ etc.)

Definition 3. The normed orthogonal and right-handed (positively oriented) trihedron formed by the vectors $\mathbf{t}$, $\mathbf{n}$, $\mathbf{b}$ at a point of a curve is called the *moving trihedron* (*moving trihedral*) of the curve.

Theorem 3 (*The Frenet or Serret-Frenet Formulae*).

a) *For the curve* $\mathbf{r} = \mathbf{r}(s)$ (*the parameter is the arc* s)

$$\begin{aligned} \mathbf{t}' &= \quad\;\; +k_1\mathbf{n} \\ \mathbf{n}' &= -k_1\mathbf{t} \qquad\quad + k_2\mathbf{b} \\ \mathbf{b}' &= \quad\;\; -k_2\mathbf{n} \end{aligned}$$

(for k_1 see Theorem 1,

$$k_2 = \frac{[x', y'', z''']}{x''^2 + y''^2 + z''^2}\,,$$

b) *For the curve* $\mathbf{r} = \mathbf{r}(t)$ (*with a general parameter* t)

$$\begin{aligned} \dot{\mathbf{t}} &= \quad\;\; +k_1\dot{s}\mathbf{n} \\ \dot{\mathbf{n}} &= -k_1\dot{s}\mathbf{t} \qquad\quad + k_2\dot{s}\mathbf{b} \\ \dot{\mathbf{b}} &= \quad\;\; -k_2\dot{s}\mathbf{n} \end{aligned} \tag{9}$$

(for k_1 see (2b),

$$k_2 = \frac{[\dot{x}, \ddot{y}, \dddot{z}]}{\dot{s}^2(\ddot{x}^2 + \ddot{y}^2 + \ddot{z}^2 - \ddot{s}^2)}\,,$$

where

$$[x', y'', z'''] = \begin{vmatrix} x', & y', & z' \\ x'', & y'', & z'' \\ x''', & y''', & z''' \end{vmatrix}\Big) .$$

where

$$[\dot{x}, \ddot{y}, \dddot{z}] = \begin{vmatrix} \dot{x}, & \dot{y}, & \dot{z} \\ \ddot{x}, & \ddot{y}, & \ddot{z} \\ \dddot{x}, & \dddot{y}, & \dddot{z} \end{vmatrix}\Big) .$$

REMARK 7. The Frenet formulae define the relations between the direction cosines (or the direction ratios) of the tangent, the principal normal and the binormal at a general point of the curve, and their derivatives. The numbers k_1 and k_2 in Theorem 1, Remark 2 and Theorem 3 are called the *first curvature* (briefly the *curvature*), and the *second curvature* (briefly the *torsion*), respectively. For a detailed treatment see § 9.4.

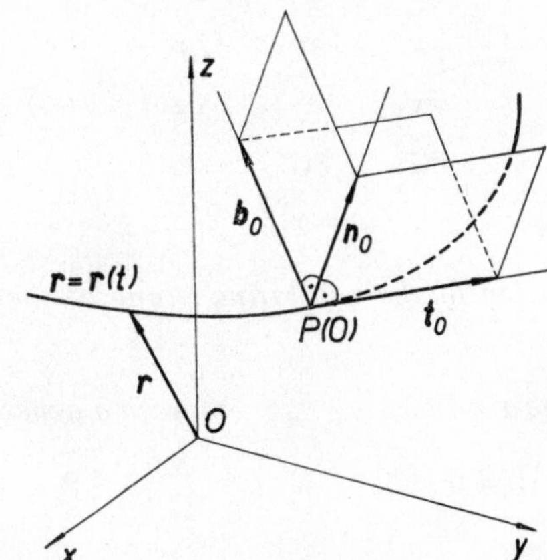

Fig. 9.4.

Definition 4. The plane determined by the principal normal and the binormal (at the point considered) is called the *normal plane*, the plane determined by the binormal and the tangent is called the *rectifying plane* and the plane determined by the tangent and the principal normal is the *osculating plane* (Fig. 9.4).

Theorem 4. *The equations of the normal plane, the rectifying plane and the osculating plane, respectively, at a point* $\mathbf{r}_0$ *of a curve* $\mathbf{r} = \mathbf{r}(t)$ *(or* $\mathbf{r} = \mathbf{r}(s)$*):*

The normal plane (perpendicular to the tangent line):

$$(\mathbf{R} - \mathbf{r}_0) \, . \, \mathbf{t}_0 = 0 \text{ or } (\mathbf{R} - \mathbf{r}_0) \, . \, \dot{\mathbf{r}}_0 = 0 \text{ or } (\mathbf{R} - \mathbf{r}_0) \, . \, \mathbf{r}'_0 = 0 \, , \tag{10}$$

for example,

$$(X - x_0)(\mathrm{d}x)_0 + (Y - y_0)(\mathrm{d}y)_0 + (Z - z_0)(\mathrm{d}z)_0 = 0 \tag{11}$$

((x_0, y_0, z_0) being a point on the curve).

The rectifying plane (perpendicular to the principal normal):

$$(\mathbf{R} - \mathbf{r}_0) \, . \, \mathbf{n}_0 = 0 \quad \text{or} \quad (\mathbf{R} - \mathbf{r}_0) \, . \, \mathbf{r}''_0 = 0 \, , \tag{12}$$

for example,

$$(X - x_0)(n_x)_0 + (Y - y_0)(n_y)_0 + (Z - z_0)(n_z)_0 = 0 .$$

The osculating plane (perpendicular to the binormal):

$$(\mathbf{R} - \mathbf{r}_0) . \mathbf{b}_0 = 0 \tag{13}$$

(**R** *being the radius vector of the running point* (X, Y, Z) *of the plane*).

REMARK 8. If the curve is given by the equations $F(x, y, z) = 0$, $G(x, y, z) = 0$, the *equation of the normal plane* at the point (x, y, z) of the curve is

$$\begin{vmatrix} X - x, & Y - y, & Z - z \\ \dfrac{\partial F}{\partial x}, & \dfrac{\partial F}{\partial y}, & \dfrac{\partial F}{\partial z} \\ \dfrac{\partial G}{\partial x}, & \dfrac{\partial G}{\partial y}, & \dfrac{\partial G}{\partial z} \end{vmatrix} = 0 . \tag{14}$$

Theorem 5. *The equation of the osculating plane at the point* (x, y, z) *with the radius vector* **r**

a) *if the parameter is the arc* s:

$$[\mathbf{R} - \mathbf{r}, \mathbf{r}', \mathbf{r}''] = 0 ,$$

i.e.

$$\begin{vmatrix} X - x, & Y - y, & Z - z \\ x', & y', & z' \\ x'', & y'', & z'' \end{vmatrix} = 0$$

b) *with a general parameter* t:

$$[\mathbf{R} - \mathbf{r}, \dot{\mathbf{r}}, \ddot{\mathbf{r}}] = 0 ,$$

i.e.

$$\begin{vmatrix} X - x, & Y - y, & Z - z \\ \dot{x}, & \dot{y}, & \dot{z} \\ \ddot{x}, & \ddot{y}, & \ddot{z} \end{vmatrix} = 0 \tag{15a}$$

or

$$\begin{vmatrix} X - x, & Y - y, & Z - z \\ \mathrm{d}x, & \mathrm{d}y, & \mathrm{d}z \\ \mathrm{d}^2x, & \mathrm{d}^2y, & \mathrm{d}^2z \end{vmatrix} = 0 . \tag{15b}$$

REMARK 9. Every plane passing through the tangent of a (space) curve is called a *tangent plane* to the curve at the corresponding point of contact. At this point the contact of the tangent plane with the curve is at least a two-point contact. The osculating plane of the curve at its point is an important tangent plane because its contact with the curve at this point is at least a three-point contact. At a point of the curve where the equation $\mathbf{r}'' = \mathbf{0}$ (with the parameter s) or $\ddot{\mathbf{r}} = \lambda\dot{\mathbf{r}}$ (with general parameter t; λ being a real number) is satisfied, the osculating plane is indefinite. At such points on the curve, e.g. at so-called *points of inflexion*, the equation of the osculating plane is satisfied identically. In the case of a plane curve, the osculating plane at each of its points is the plane of the curve.

Example 1. The curve given by the equations

$$x = a \cos t\,, \ y = a \sin t\,, \ z = bt \ (a > 0, b \neq 0 \text{ being real constants}) \tag{16}$$

is called a *circular helix*. It is one of the more important curves in applications. It lies on the circular cylinder (Fig. 9.5)

$$x^2 + y^2 = a^2 \tag{17}$$

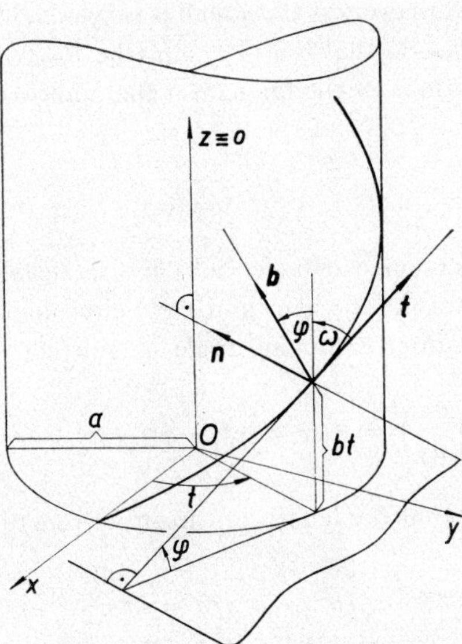

Fig. 9.5.

(as we can find by elimination of t from the first two equations (16)), and the axis of the cylinder is the coordinate z-axis. The axis of the cylinder on which the circular helix lies is called the *axis of the helix*. From the first and second of equations (16) we obtain in the first quadrant (for $0 \leqq t < \frac{1}{2}\pi$) $t = \tan^{-1}(y/x)$ and by the third of equations (16)

$$z = b \tan^{-1} \frac{y}{x}\,, \tag{18}$$

i.e. the equation of the *right helicoid* (see Example 9.12.2, p. 354). For $\frac{1}{2}\pi < t < \frac{3}{2}\pi$ we have $z = b[\tan^{-1}(y/x) + \pi]$ etc. The given helix is the intersection of the conoid (18) and the circular cylinder (17). From (16) it follows that

$$\dot{x} = -a \sin t\,, \quad \dot{y} = a \cos t\,, \quad \dot{z} = b\,, \tag{19}$$

hence

$$\frac{1}{\sqrt{(\dot{\mathbf{r}} \cdot \dot{\mathbf{r}})}} = \frac{1}{\sqrt{(a^2 + b^2)}}$$

and for the direction cosines (t_x, t_y, t_z) of the tangent line we obtain (see Theorem 9.2.2)

$$t_x = -\frac{a \sin t}{\sqrt{(a^2 + b^2)}}, \quad t_y = \frac{a \cos t}{\sqrt{(a^2 + b^2)}}, \quad t_z = \frac{b}{\sqrt{(a^2 + b^2)}}. \tag{20}$$

Thus, the tangents to a circular helix make a constant angle ω with its axis, where $\cos \omega = b/\sqrt{(a^2 + b^2)}$, and therefore the tangent makes a constant angle φ (the *gradient of the helix*), with every plane which is perpendicular to the axis of the helix. Moreover, $\sin \varphi = \cos \omega$, so that $\tan \varphi = b/a$ (the *slope of the helix*). By (20) and Theorem 9.2.3 the equations of the tangent at the point t of a helix are of the form

$$-\frac{X - a \cos t}{a \sin t} = \frac{Y - a \sin t}{a \cos t} = \frac{Z - bt}{b}. \tag{21}$$

If the circular cylinder, on which the helix lies, is developed upon a plane, then at the same time, every turn of the helix and every circle on the cylinder are developed into two segments that intersect at an angle φ. Further, from (19), we have that

$$\dot{s}^2 = \left(\frac{ds}{dt}\right)^2 = \dot{\boldsymbol{r}} \cdot \dot{\boldsymbol{r}} = \dot{x}^2 + \dot{y}^2 + \dot{z}^2 = a^2 + b^2$$

(see Definition 9.2.6) so that the length of the arc of one turn of the helix is

$$s = \int_0^{2\pi} \sqrt{(a^2 + b^2)}\, dt = 2\pi \sqrt{(a^2 + b^2)}$$

(as can be seen immediately from the development in Fig. 9.6). The length of the arc of the helix from the point $t = 0$ to the point t, is $s(t) = t\sqrt{(a^2 + b^2)} = ct$. If we substitute in equation (16) the length of arc s as parameter instead of the general parameter t, we obtain the equations of the helix in the form

$$x = a \cos \frac{s}{c}, \quad y = a \sin \frac{s}{c}, \quad z = b\frac{s}{c} \quad (c = \sqrt{(a^2 + b^2)}).$$

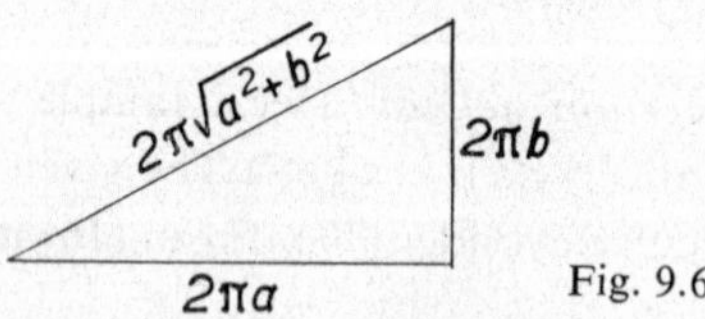

Fig. 9.6.

From (16) we obtain

$$\ddot{x} = -a \cos t, \quad \ddot{y} = -a \sin t, \quad \ddot{z} = 0, \quad \ddot{x}^2 + \ddot{y}^2 + \ddot{z}^2 = a^2 \tag{22}$$

and further $\ddot{s} = 0$. Substituting in (2) we obtain for the direction cosines of the principal normal the expressions

$$n_x = -\cos t\,, \quad n_y = -\sin t\,, \quad n_z = 0\,. \tag{23}$$

Thus, the principal normals of the helix are perpendicular to the axis of the helix (its direction cosines are (0, 0, 1)). From (7) we obtain for the direction cosines of the binormal the expressions

$$b_x = \frac{b \sin t}{\sqrt{(a^2 + b^2)}}\,, \quad b_y = -\frac{b \cos t}{\sqrt{(a^2 + b^2)}}\,, \quad b_z = \frac{a}{\sqrt{(a^2 + b^2)}}\,. \tag{24}$$

Thus, the binormals of the helix are inclined at a constant angle φ to the axis of the helix, such that $\cos \varphi = a/\sqrt{(a^2 + b^2)}$.

By (15a), (19) and (22) we obtain the equation of the osculating plane at the point t of the helix

$$\begin{vmatrix} X - a\cos t, & Y - a \sin t, & Z - bt \\ -a \sin t, & a \cos t, & b \\ -a \cos t, & -a \sin t, & 0 \end{vmatrix} = Xb \sin t - Yb \cos t + Za - abt = 0\,.$$

Further, it follows from (24) that

$$\dot{b}_x = \frac{b \cos t}{\sqrt{(a^2 + b^2)}}\,, \quad \dot{b}_y = \frac{b \sin t}{\sqrt{(a^2 + b^2)}}\,, \quad \dot{b}_z = 0\,.$$

Putting $\dot{s} = \sqrt{(a^2 + b^2)}$ in the third of the Frenet formulae (9) and using (23), we obtain for the torsion k_2 at the point t of the helix, the expression

$$k_2 = \frac{b}{a^2 + b^2}\,. \tag{25}$$

Similarly, from the first of the Frenet formulae (9) it follows that at the point t of the helix the curvature k_1 is given by the expression

$$k_1 = \frac{a}{a^2 + b^2}\,. \tag{26}$$

Thus, the circular helix (and the circular helix alone from among all space curves) has both curvatures constant (and not vanishing at all its points). The curvature k_1 and the torsion k_2 can be computed, of course, by formulae (9.4.1) and (9.4.4). The sign of k_2 agrees with the sign of the constant b. The helix is *right-handed* or *left-handed* according as $b > 0$ (and then $k_2 > 0$) or $b < 0$ respectively.

If k_1 and k_2 in (25) and (26) stand for the curvature and the torsion, respectively, at any point of a space curve and if (25) and (26) are solved for a and b, then these numbers a and b define a circular helix (generally turned round the axis with regard to the helix (16)); this helix has the same curvature and the same torsion as the given curve at the point considered. Such a helix plays a similar role for the given curve as the osculating circle does for a plane curve, and the contact of this helix with the given space curve at the point considered is of the fourth order at least (cf. § 9.5).

Example 2 (*Components of the Vector of Acceleration*). Let a particle move along a space curve

$$\mathbf{r} = \mathbf{i}\,x(t) + \mathbf{j}\,y(t) + \mathbf{k}\,z(t)$$

(t denoting time).

The velocity vector is

$$\mathbf{v} = \frac{d\mathbf{r}}{dt} = \frac{d\mathbf{r}}{ds} \cdot \frac{ds}{dt} = \frac{ds}{dt}\mathbf{r}' = \frac{ds}{dt}\mathbf{t}\,.$$

The acceleration vector is

$$\mathbf{a} = \frac{d\mathbf{v}}{dt} = \frac{d^2 s}{dt^2}\mathbf{t} + \frac{ds}{dt}\frac{d\mathbf{t}}{dt} = \frac{d^2 s}{dt^2}\mathbf{t} + \frac{ds}{dt}\frac{d\mathbf{t}}{ds}\frac{ds}{dt} = \frac{d^2 s}{dt^2}\mathbf{t} + \left(\frac{ds}{dt}\right)^2 \mathbf{t}'\,.$$

Using the first Frenet formula $\mathbf{t}' = k_1 \mathbf{n}$ and writing

$$k_1 = \frac{1}{r_1}$$

(r_1 is the so-called *radius of curvature*), we obtain

$$\mathbf{a} = \frac{d^2 s}{dt^2}\mathbf{t} + \frac{\left(\dfrac{ds}{dt}\right)^2}{r_1}\mathbf{n}\,.$$

We can see from the result that the acceleration vector resolves into two components, one in the direction of the tangent vector and the other in the direction of the principal normal vector. They therefore lie in the osculating plane. The normal component,

$$\frac{\left(\dfrac{ds}{dt}\right)^2}{r_1}\mathbf{n}\,,$$

is called the *normal acceleration*. For the case of circular motion, this represents a well-known formula of physics.

REMARK 10. When considering plane curves, we often speak of a *subtangent* s_t or a *subnormal* s_n; they are the (oriented) orthogonal projections on the x-axis of the segments of the tangent t, or the normal n, from the contact point P up to the point of intersection of the tangent, or the normal, with the x-axis, respectively. If we suppose that the derivative $\dot{y} = dy/dx \neq 0$ at the point P is finite, then it follows from Fig. 9.7 that

$$s_t = -\frac{y}{\dot{y}}, \quad s_n = y\dot{y}.$$

(Often only the absolute values of s_t and s_n are considered.)

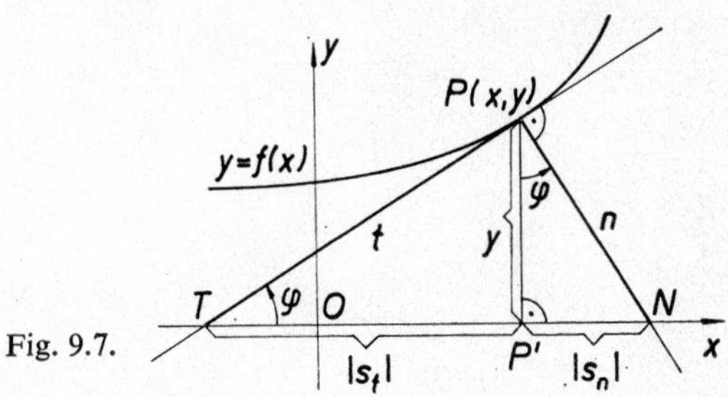

Fig. 9.7.

For the *length of the tangent* t *and the normal* n we obtain

$$t = \left|\frac{y\dot{s}}{\dot{y}}\right|, \quad n = |y\dot{s}| \quad \left(\dot{s} = \frac{ds}{dx} = \sqrt{(1 + \dot{y}^2)}\right).$$

9.4. First and Second Curvatures, Natural Equations of a Curve

Definition 1. The expression k_1 given by the equation

a) for the parameter s:

$$k_1 = \sqrt{(\mathbf{t}' . \mathbf{t}')} = \sqrt{(\mathbf{r}'' . \mathbf{r}'')} = \sqrt{(x''^2 + y''^2 + z''^2)}$$

(see (9.3.1) and (9.3.9))

b) for a general parameter t:

$$k_1 = \frac{1}{\dot{s}^2} \sqrt{(\ddot{x}^2 + \ddot{y}^2 + \ddot{z}^2 - \ddot{s}^2)} \quad (1)$$

(see (9.3.2b) and (9.3.9))

is called the *first curvature* (briefly: the *curvature*) of the curve at the point considered; its reciprocal value $r_1 = 1/k_1$ is the so-called *radius of curvature* (cf. Examples 9.3.1 and 9.3.2).

REMARK 1. A necessary and sufficient condition that a curve be a straight line is that the equation $k_1 = 0$ (i.e. $\mathbf{r}'' = \mathbf{0}$) holds at every point of the curve. The case where the curvature vanishes only at individual points of the curve (the case of points of inflexion) is considered in § 9.5.

Theorem 1 (*The Geometric Interpretation of the Curvature k_1 of a Curve*). *Let $\varphi = \varphi(s, 0)$ be the angle between the tangent lines $\mathbf{t}(s)$ and $\mathbf{t}_0$ $(s > 0)$ at the* *points* $Q(s)$ and $P(0)$, *respectively on the curve* $\mathbf{r} = \mathbf{r}(s)$. *Then* (Fig. 9.8)

$$\lim_{s \to 0} \frac{\varphi}{s} = \left(\frac{d\varphi}{ds}\right)_0 = (k_1)_0 \,. \tag{2}$$

REMARK 2. If $\Delta s = \widehat{PQ}$ is the arc between two "neighbouring" points P and Q on curve k and if $\Delta\varphi$ is the acute angle between the tangents to the curve constructed at these points, then the ratio $\Delta\varphi/\Delta s$ is called the *mean curvature* of the curve on the

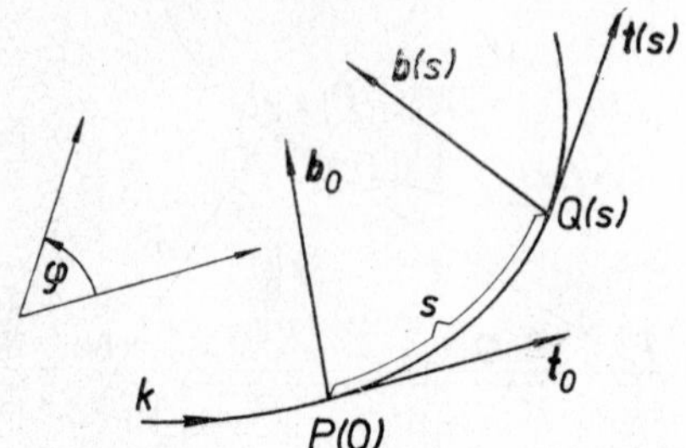

Fig. 9.8.

arc Δs. The acute angle between the tangents $\mathbf{t}_0$ and $\mathbf{t}(\Delta s)$ at the points $P(0)$ and $Q(\Delta s)$ is called the *angle of contingence.*

REMARK 3. In the case of a plane curve $y = f(x)$ the curvature at the point (x, y) is

$$k_1 = \left|\frac{\ddot{y}}{(1 + \dot{y}^2)^{3/2}}\right| \quad \text{or} \quad r_1 = \left|\frac{(1 + \dot{y}^2)^{3/2}}{\ddot{y}}\right| \quad (\ddot{y} \neq 0) \,. \tag{3}$$

For example, for the circle

$$y = \sqrt{(r^2 - x^2)}$$

we obtain $k_1 = 1/r$ and so the radius of curvature of the circle is, at all its points, constant and has the same value as its radius (this property holds only for the circle).

Definition 2. The expression k_2, given by the equation

a) for the parameter s:

$$k_2 = \frac{\begin{vmatrix} x', & y', & z' \\ x'', & y'', & z'' \\ x''', & y''', & z''' \end{vmatrix}}{x''^2 + y''^2 + z''^2}$$

b) for a general parameter t:

$$k_2 = \frac{\begin{vmatrix} \dot{x}, & \dot{y}, & \dot{z} \\ \ddot{x}, & \ddot{y}, & \ddot{z} \\ \dddot{x}, & \dddot{y}, & \dddot{z} \end{vmatrix}}{\dot{s}^2(\ddot{x}^2 + \ddot{y}^2 + \ddot{z}^2 - \ddot{s}^2)} \tag{4}$$

(see (9.3.9)), is called the *second curvature* (the *torsion*) of the curve at the point considered, and its reciprocal value $r_2 = 1/k_2$ is called the *radius of torsion* (cf. Example 9.3.1).

Theorem 2. *A necessary and sufficient condition that a curve be a plane curve is that*

$$k_2 = 0\,, \quad \text{i.e.} \quad [x', y'', z'''] \text{ or } [\dot{x}, \ddot{y}, \dddot{z}] = 0 \quad (\text{see } (4))$$

at all points of the curve.

REMARK 4. Space curves ($k_1 \neq 0, k_2 \neq 0$) are sometimes called curves of two curvatures or curves with torsion. In the case of plane curves the Frenet formulae reduce (see (9.3.9)) to

$$\mathbf{t}' = k_1 \mathbf{n}\,, \qquad \mathbf{n}' = -k_1 \mathbf{t} \quad (\text{for the parameter } s) \qquad \Bigg| \qquad \dot{\mathbf{t}} = k_1 \dot{s} \mathbf{n}\,, \qquad \dot{\mathbf{n}} = -k_1 \dot{s} \mathbf{t} \quad (\text{for the general parameter } t).$$

Theorem 3. (*The Geometric Interpretation of the Torsion k_2 of a Curve*). *Let $\psi = \psi(s, 0)$ be the angle between the binormals $\mathbf{b}(s)$ and $\mathbf{b}_0$ $(s > 0)$ at the points $Q(s)$ and $P(0)$, respectively, on the curve $\mathbf{r} = \mathbf{r}(s)$. Then* (Fig. 9.8)

$$\lim_{s \to 0} \frac{\psi}{s} = \left(\frac{\mathrm{d}\psi}{\mathrm{d}s}\right)_0 = |(k_2)_0|\,. \tag{5}$$

REMARK 5. If $\Delta s = \widehat{PQ}$ is the arc between two neighbouring points $P(0)$ and $Q(\Delta s)$ on the curve and $\Delta\psi$ is the acute angle between the binormals at these points, then the ratio $\Delta\psi/\Delta s$ is called the *mean torsion* of the curve on the arc Δs. For plane curves, $\mathbf{b}'(s) = \dot{\mathbf{b}}(t) = \mathbf{0}$ identically.

REMARK 6. If we introduce a system of coordinates such that at the point $s = 0$ the vectors $\mathbf{t}, \mathbf{n}, \mathbf{b}$ of the moving trihedron correspond to the half-axes $+x, +y, +z$ respectively, then we can write, for the cartesian coordinates $x(s), y(s), z(s)$ of a point on the curve in a neighbourhood of the point $s = 0$, the series

$$\begin{aligned} x(s) &= s - \frac{s^3}{3!}(k_1^2)_0 - \ldots\,, \\ y(s) &= \frac{s^2}{2!}(k_1)_0 + \frac{s^3}{3!}(k_1')_0 + \ldots\,, \\ z(s) &= \frac{s^3}{3!}(k_1 k_2)_0 + \ldots\,. \end{aligned} \tag{6}$$

Equations (6) are called the *canonical equations* (or the *canonical representation*) *of the curve*. We obtain from (6) (when we use only the first term of each series) the equations of the simplest algebraic space curve (the so-called *cubical parabola*),

$$x(s) = s\,, \quad y(s) = \frac{(k_1)_0}{2} s^2\,, \quad z(s) = \frac{(k_1 k_2)_0}{6} s^3\,.$$

This parabola approximates the given curve at the point $s = 0$.

REMARK 7. A curve is said to be *right-handed* or *left-handed* if at any point of the curve $k_2 > 0$ or $k_2 < 0$, respectively (cf. Example 9.3.1).

REMARK 8. If two continuous functions $k_1 = k_1(s) > 0$, and $k_2 = k_2(s)$ are given, then it is possible to construct a curve for which k_1 is its curvature, k_2 its torsion and the parameter s its arc. This curve is thus determined uniquely except for its position in space (if no definite special conditions are given).

Theorem 4. *Let two continuous functions*

$$k_1 = k_1(s) > 0\,, \quad k_2 = k_2(s) \tag{7}$$

be given (k_1 having a continuous second derivative and k_2 having a continuous first derivative). Then there exists a unique curve having the following properties:

1. *its arc is s, its curvature is k_1 and its torsion is k_2;*
2. *it passes through an arbitrary given point s_0;*
3. *three arbitrary mutually perpendicular unit vectors $\mathbf{t}_0$, $\mathbf{n}_0$, $\mathbf{b}_0$ are the tangent unit vector, the principal normal unit vector, the binormal unit vector, respectively, at the point s_0.*

Theorem 5. *It is always possible to write the equations of a plane curve with given curvature $k_1(s)$ (the torsion k_2 being equal to zero identically) in the form*

$$\left.\begin{aligned} x &= \int \cos\left(\int k_1\,\mathrm{d}s + c\right)\mathrm{d}s + a\,, \\ y &= \pm \int \sin\left(\int k_1\,\mathrm{d}s + c\right)\mathrm{d}s + b\,, \\ z &= 0 \end{aligned}\right\} \tag{8}$$

(a, b, c are arbitrary real constants).

REMARK 9. All plane curves (in the plane $z = 0$) with the same curvature $k_1(s)$ at the point s may be obtained from the rotation of the curve

$$x = \int \cos\left(\int k_1\,\mathrm{d}s\right)\mathrm{d}s\,, \quad y = \pm \int \sin\left(\int k_1\,\mathrm{d}s\right)\mathrm{d}s\,, \quad z = 0$$

through an angle c and by the translation of the curve along the oriented segment given by the vector with the coordinates (a, b) (for the constants a, b, c see Theorem 5).

Definition 3. The quantities s, k_1 and k_2 are called the *natural coordinates* and the relations

$$k_1 = k_1(s)\,, \quad k_2 = k_2(s) \tag{9}$$

between k_1, k_2 and s are called the *natural equations* (*intrinsic equations*) *of the curve.*

REMARK 10. The natural coordinates of a curve are independent of any coordinate system. The natural equations of a curve, which express the curve independently of the choice of the coordinate system, are suitable for the investigation of those properties of curves that are not dependent on the coordinates. For example, the natural equation

$$k_1 = \frac{1}{a} \quad (k_2 = 0,\ a \text{ is a positive constant})$$

is the equation of *all circles* of radius $r = a$ in the plane $z = 0$, the implicit equation of which is

$$(x - m)^2 + (y - n)^2 = a^2 \quad (m, n \text{ being arbitrary real constants}) .$$

The plane curve having the property that its curvature k_1 is directly proportional to the length of the arc s is called a *clothoid* (cf. § 4.8). This property is expressed by its natural equation

$$k_1 = \frac{1}{a^2} s \quad (a \text{ is a real constant}) .$$

By the proper choice of the coordinate system, the cartesian coordinates of the points of a clothoid may be expressed with the help of the Fresnel integrals,

$$x = \frac{a}{\sqrt{2}} \int_0^{\varphi} \frac{\cos \varphi}{\sqrt{\varphi}} \, d\varphi , \quad y = \frac{a}{\sqrt{2}} \int_0^{\varphi} \frac{\sin \varphi}{\sqrt{\varphi}} \, d\varphi ,$$

where $\varphi = s^2/2a^2$ is the angle between the tangent line at its point (x, y) and the half-axis $+x$.

9.5. Contact of Curves, Osculating Circle

Let

$$^1\mathbf{r} = {}^1\mathbf{r}(s) , \quad ^2\mathbf{r} = {}^2\mathbf{r}(s) \tag{1}$$

be the equations of two curves ^1k and ^2k represented in terms of the same parameter s, which is the length of the arc on both curves. Let the curves have a common (regular) point $s = 0$ (i.e. $^1\mathbf{r}_0 = {}^2\mathbf{r}_0$) from which we shall measure the common parameter along both curves. Let us consider a point on each curve corresponding to the same value of the parameter s (Fig. 9.9) and let us investigate the mutual position of both curves (1) in a sufficiently small neighbourhood of their common point $s = 0$.

Definition 1. Two curves $^1\mathbf{r} = {}^1\mathbf{r}(s)$, $^2\mathbf{r} = {}^2\mathbf{r}(s)$ are said to have, at their common point $^1\mathbf{r}_0 = {}^2\mathbf{r}_0$, *contact of order* q *at least* i.e. *at least* $(q + 1)$*-point contact*, provided that the equations

$$\lim_{s\to 0} \frac{\mathbf{d}(s)}{s^p} = \mathbf{0} \quad (p = 0, 1, 2, \ldots, q),$$

are satisfied, where

$$\mathbf{d}(s) = {}^1\mathbf{r}(s) - {}^2\mathbf{r}(s). \tag{2}$$

Theorem 1. *The necessary and sufficient condition that the curves* (1) *have contact of order q at least, i.e. at least* $(q + 1)$*-point contact, at their common point is that the equations*

$${}^1\mathbf{r}_0 = {}^2\mathbf{r}_0, \quad {}^1\mathbf{r}'_0 = {}^2\mathbf{r}'_0, \quad \ldots, \quad {}^1\mathbf{r}_0^{(q)} = {}^2\mathbf{r}_0^{(q)}. \tag{3}$$

are satisfied. (*The existence of a sufficient number of derivatives is assumed.*)

REMARK 1. The phrase "at least $(q + 1)$-point contact" corresponds to the conception that at the common point of contact of two curves, both curves have at least $(q + 1)$ coincident points of intersection. For example if ${}^1\mathbf{r}_0 = {}^2\mathbf{r}_0$, and ${}^1\mathbf{r}'_0 \neq {}^2\mathbf{r}'_0$, then at the point $s = 0$ the curves have contact of order 0 exactly, i.e. exactly one-point contact (they intersect at the point $s = 0$).

REMARK 2. For the plane curves ${}^1y = {}^1y(x)$ and ${}^2y = {}^2y(x)$ we may take, instead of equations (2), the equations

$$\lim_{x\to 0} \frac{d(x)}{x^p} = 0 \quad (p = 0, 1, 2, \ldots, q), \quad d(x) = {}^1y(x) - {}^2y(x)$$

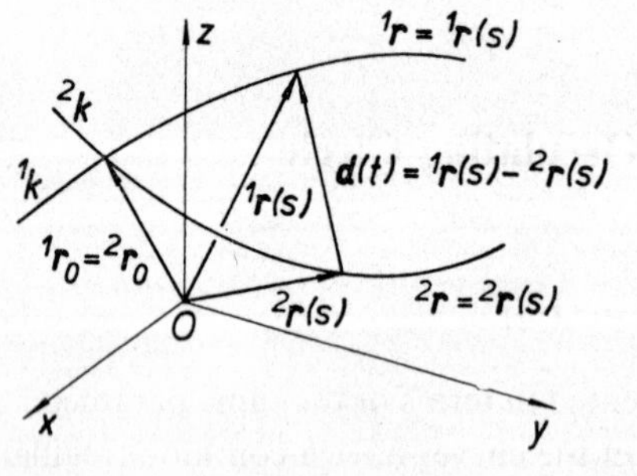

Fig. 9.9.

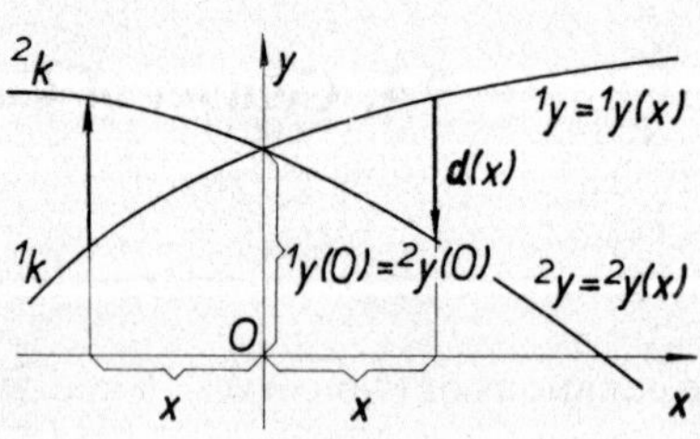

Fig. 9.10.

assuming that no curve possesses the tangent $x = 0$ at their common point $x = 0$ (see Fig. 9.10, where ${}^1y(0) = {}^2y(0)$, but ${}^1\dot{y}(0) \neq {}^2\dot{y}(0)$). Instead of equations (3) we have, in the case of the curves ${}^1y = {}^1y(x)$ and ${}^2y = {}^2y(x)$,

$${}^1y_0 = {}^2y_0, \quad {}^1\dot{y}_0 = {}^2\dot{y}_0, \quad \ldots, \quad {}^1y_0^{(q)} = {}^2y_0^{(q)}. \tag{4}$$

(We use the notation

$$ {}^i\dot{y}(x) = \frac{\mathrm{d}^i y}{\mathrm{d}x} $$

etc.)

REMARK 3. If two curves have, at their common point, the same tangent (in the case of space curves also the same tangent plane) then at that point they have contact of order 1 at least, i.e. at least two-point contact, and conversely. If at their common point two curves have the same tangent, principal normal and curvature, they have contact of order 2 at least, i.e. at least three-point contact, and conversely. Two space curves that have, at their common point, contact at least of order 2 (at least three-point contact) have at this point a common osculating plane (provided that the osculating planes exist at this point). The plane that has contact of the first order (two-point contact) or contact of the second order (three-point contact) with a given curve is the tangent plane or the osculating plane, respectively, at the point considered.

REMARK 4. If ${}^1\mathbf{r}_0 = {}^2\mathbf{r}_0$, ${}^1\mathbf{r}'_0 = {}^2\mathbf{r}'_0$, but ${}^1\mathbf{r}''_0 \neq {}^2\mathbf{r}''_0$ (or ${}^1y_0 = {}^2y_0$, ${}^1\dot{y}_0 = {}^2\dot{y}_0$, but ${}^1\ddot{y}_0 \neq {}^2\ddot{y}_0$), then we say that both curves have *contact of exactly the first order* or *exactly two-point contact.* Similarly for the contact of any order. In the example considered, both curves touch each other at their common point and their contact is a so-called ordinary one. The curve ${}^2y = {}^2y(x)$ lies on the same side of the curve ${}^1y = {}^1y(x)$ in a neighbourhood of their common point (Fig. 9.11). If we replace the curve ${}^2y$ by a straight line, then the straight line which has contact of the first order at least (at least two-point contact) with the curve is the tangent at this point.

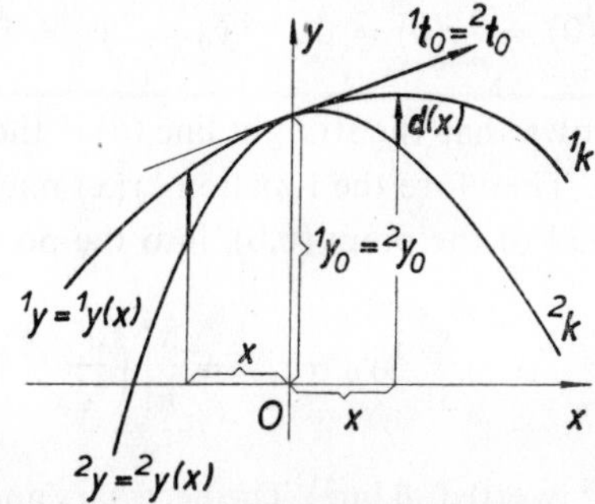

Fig. 9.11.

Definition 2. A curve ${}^2k$ of a given type (e.g. a circle) that has contact of highest possible order with the curve ${}^1k$ at their common point is said to be an *osculating curve* of the curve ${}^1k$ at the point considered. We say that the curve ${}^2k$ *osculates* the curve ${}^1k$ at this point.

REMARK 5. An osculating curve of the given curve is generally completely determined by the condition of osculation. In special cases, i.e. at some special points on the given curve (e.g. at the points of inflexion, at the vertices, etc.) the osculating curve may have contact of an order that is higher than is the highest possible at an *ordinary* point.

Definition 3. The curve ${}^2k$ that has contact with the curve ${}^1k$ at their common point of an order higher than is generally the highest possible order, is said to be the *superosculating curve* of the curve ${}^1k$ at such point.

Example 1. In the general case a straight line and a curve may have, at a regular point of the given curve, contact of the first order at most (i.e. two-point contact). Of course, this osculating straight line is the tangent to the given curve. If a further condition regarding contact of a higher order is imposed, it relates only to the properties of the given curve and may be fulfilled, for example, at its points of inflexion. The *tangent at a point of inflexion* is also a superosculating straight line and its contact with the given curve is at least a three-point contact. We may proceed to add further conditions as far as the given curve possesses derivatives of higher order.

Example 2. Let us investigate the contact of the (plane) curve ${}^1k$,

$$ {}^1x = t\,,\ {}^1y = {}^1y(t)\,,\ {}^1z = 0 \ \text{(i.e. the curve } {}^1y = {}^1y(x)) \tag{5}$$

with the straight line

$$ {}^2x = t\,,\quad {}^2y = 0\,,\quad {}^2z = 0 \tag{6}$$

(i.e. with the x-axis). The necessary and sufficient condition that the curve (5) and the straight line (6) have contact of the first order at least ($q = 1$), at the point $t = 0$ (which is assumed to be their common point), is

$$ {}^1y(0) = {}^2y(0) = 0\,,\quad {}^1\dot{y}_0 = {}^2\dot{y}_0 = 0\,. \tag{7}$$

From this condition it follows that the straight line (6) is the *tangent* to the curve (5) at the contact point (0, 0). Therefore the function ${}^1y(x)$ may be expanded, in a sufficiently small neighbourhood of the point (0,0), into the power series

$$ {}^1y = \frac{x^2}{2!}\,{}^1\ddot{y}_0 + \frac{x^3}{3!}\,{}^1\dddot{y}_0 + \ldots + \frac{x^n}{n!}\left(\frac{\mathrm{d}^n\,{}^1y}{\mathrm{d}x^n}\right)_0 + \ldots \tag{8}$$

(see Fig. 9.12b for the case ${}^1\ddot{y}_0 \neq 0$, full line). The necessary and sufficient condition that the curve (5) and the tangent (6) have, at the point (0, 0), contact of the second order at least (see (4)), is that in addition to (7), at least ${}^1\ddot{y}_0 = 0$. From (8) it then follows that if ${}^1\dddot{y}_0 \neq 0$, the curve crosses the tangent ${}^2y = 0$ at the point (0,0) (Fig. 9.12a, full line). The point (0, 0) is called the *point of inflexion* of the curve (it is an *ordinary inflexion*, an *inflexion of the first order*). Similarly it can be shown that the vanishing of all derivatives of the function ${}^1y$ with respect to x up to the q-th derivative inclusive is a necessary and sufficient condition for contact at least of order q between the given curve and the tangent at the point (0, 0). If q is even, $q > 2$ and

$(\mathrm{d}^{q+1}\,{}^1y/\mathrm{d}x^{q+1})_0 \neq 0$, the curve crosses the tangent line at the point $(0, 0)$ (Fig. 9.12a, the dashed line) and we speak of a *higher inflexion* (an *inflexion of higher order*, an inflexion of the order q). If q is odd, $q > 1$, the curve remains in a sufficiently small neighbourhood of the point $(0, 0)$, on the same side of the tangent (Fig. 9.12b, the dashed line). In this case the contact point $(0, 0)$ is called a *flat point*.

Example 3. Let us consider two (plane) curves ${}^1k$ and ${}^2k$,

$$ {}^1y = {}^1y(x)\,, \quad {}^2y = {}^2y(x)\,, \tag{9} $$

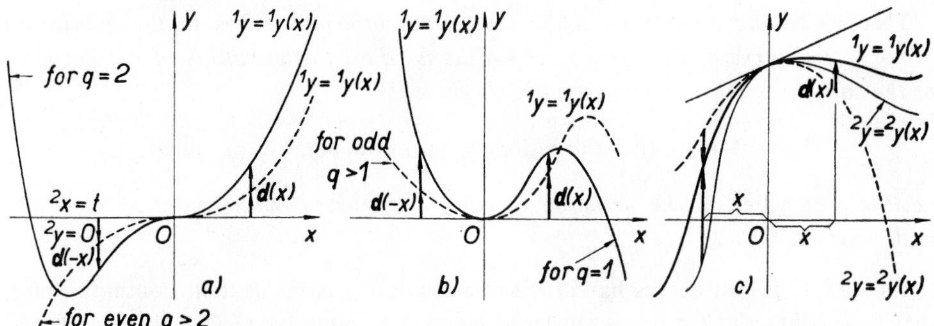

Fig. 9.12.

for which the relations

$$ {}^1y_0 = {}^2y_0\,, \quad {}^1y_0^{(p)} = {}^2y_0^{(p)} \quad (p = 1, 2, \ldots, q) $$

hold and where

$$ {}^1y_0^{(q+1)} \neq {}^2y_0^{(q+1)}\,. $$

In a sufficiently small neighbourhood of the point $x = 0$ the difference $d(x) = {}^1y(x) - {}^2y(x)$ may be represented by the expansion

$$ d(x) = \frac{x^{q+1}}{(q+1)!}\left({}^1y_0^{(q+1)} - {}^2y_0^{(q+1)}\right) + \frac{x^{q+2}}{(q+2)!}\left({}^1y_0^{(q+2)} - {}^2y_0^{(q+2)}\right) + \ldots. \tag{10} $$

If q is *even*, then $d(x)$ changes its sign at the point $x = 0$ (for sufficiently small x) and the curve ${}^2k$ *crosses* the curve ${}^1k$ at the point $x = 0$ (Fig. 9.12c, the curve ${}^2k$ is the dashed line). If q is odd, then $d(x)$ keeps its sign (for sufficiently small x) and the curve ${}^2k$ lies, in a neighbourhood of the point $x = 0$, *on the same side* of the curve ${}^1k$ (Fig. 9.12c, the curve ${}^2k$ is drawn as a narrow line).

Definition 4. A circle which passes through a point of a (space) curve (at which $k_1 \neq 0$) and has there contact of the second order at least (three-point contact) is called the *osculating circle* (the *circle of curvature*) of the curve at that point. The centre of this circle is called the *centre of curvature* and its radius is the *radius of curvature* at that point.

REMARK 6. In general, contact of the second order exactly (a three-point contact), of a curve and a circle (or a plane) is contact of highest possible order. This means that, at a general point of a curve, its osculating circle (or its osculating plane) is uniquely determined and we cannot construct a circle (or a plane) with contact of the third or higher order at that point. But there may be such points on the curve (at which $k_2 = 0$) that admit the existence of a circle (or a plane) with contact of a higher order.

Theorem 2. *The osculating circle at a point of a curve lies in the osculating plane of the curve at that point, its radius is equal to the radius of curvature r_1 of the curve and the radius vector of its centre is*

$$\bar{\mathbf{r}} = \mathbf{r} + r_1\mathbf{n} \quad (\mathbf{r} \textit{ is the radius vector of the point of the curve}),$$

i.e. *the centre lies on the principal normal constructed at the point of the curve under consideration.*

REMARK 7. If two curves have the same osculating circle at their common point, they have contact of the second order at least (three-point contact) at that point, and conversely.

Theorem 3. *The orthogonal projection of a curve into the osculating plane (at a given point of the curve) is a plane curve which has a common osculating circle with the curve at the point mentioned.*

Theorem 4. *A curve $y = y(x)$ has at every point (x, y) for which $\ddot{y} \neq 0$* (i.e., for example, the points of inflexion are excluded) *a unique osculating circle; its radius (the radius of curvature) r_1 and the coordinates (m, n) of its centre S are given by the expressions*

$$r_1 = \frac{(1 + \dot{y}^2)^{3/2}}{|\ddot{y}|}, \quad m = x - \dot{y}\,\frac{1 + \dot{y}^2}{\ddot{y}}, \quad n = y + \frac{1 + \dot{y}^2}{\ddot{y}}. \tag{11}$$

REMARK 8. If the osculating circle of a curve at its point has *exactly* three-point contact with this curve, then it crosses the curve at the contact point.

Theorem 5. *In the case of the curve $F(x, y) = 0$, the radius r_1 (the radius of curvature) and the coordinates (m, n) of the centre S of the osculating circle are given by the expressions*

$$r_1 = \frac{(F_x^2 + F_y^2)^{3/2}}{|J|}, \quad m = x - F_x\,\frac{F_x^2 + F_y^2}{J}, \quad n = y - F_y\,\frac{F_x^2 + F_y^2}{J}, \tag{12}$$

where

$$J = F_{xx}F_y^2 - 2F_{xy}F_xF_y + F_{yy}F_x^2 \neq 0. \tag{13}$$

(It is assumed that the tangent at the point (x, y) *of the curve is not parallel to the* y*-axis.)*

Theorem 6. *A curve* $x = \varphi(t)$, $y = \psi(t)$ *has at every point* t, *for which* $\dot{\varphi}\ddot{\psi} - \ddot{\varphi}\dot{\psi} \neq 0$, *a unique osculating circle; its radius (the radius of curvature)* r_1 *and the coordinates* m, n *of its centre* S *are given by the expressions*

$$r_1 = \frac{(\dot{\varphi}^2 + \dot{\psi}^2)^{3/2}}{|\dot{\varphi}\ddot{\psi} - \ddot{\varphi}\dot{\psi}|}, \quad m = \varphi - \dot{\psi}\,\frac{\dot{\varphi}^2 + \dot{\psi}^2}{\dot{\varphi}\ddot{\psi} - \ddot{\varphi}\dot{\psi}}, \quad n = \psi + \dot{\varphi}\,\frac{\dot{\varphi}^2 + \dot{\psi}^2}{\dot{\varphi}\ddot{\psi} - \ddot{\varphi}\dot{\psi}}. \tag{14}$$

REMARK 9. To find the *points of inflexion* of a curve $y = f(x)$ we solve the equation of the curve and the equation $\ddot{y} = 0$. Such points are really points of inflexion if the order of the highest non-vanishing derivative is odd ($y^{(p)} \neq 0$, $p > 2$, p odd). The inflexion points at which the tangent is perpendicular to the x-axis must be found separately. The coordinates of a point of inflexion of a curve $F(x, y) = 0$ may be found from the common solution of the equation of the curve and the equation

$$J \equiv F_{xx}F_y^2 - 2F_{xy}F_xF_y + F_{yy}F_x^2 = 0\,,$$

which is satisfied by the coordinates of the point of inflexion (this condition is only a necessary one). But the equation $J = 0$ is also satisfied by the *coordinates of all singular points* of the curve ($F_x = F_y = 0$). Thus, among the points obtained, there will be the points of inflexion and also the singular points.

The coordinates of a point of inflexion of a curve $x = \varphi(t)$, $y = \psi(t)$ may be determined from the equation $\dot{\varphi}\ddot{\psi} - \ddot{\varphi}\dot{\psi} = 0$, which is satisfied by the parameter of a point of inflexion (again, this is only a necessary condition because the last equation is satisfied by *all singular points* of the given curve).

Definition 5. If a point of a curve admits the existence of a circle which at this point has contact of the third order at least (four-point contact) with the curve, then such a circle is called the *superosculating circle* of the curve at that point.

Theorem 7. *At a point* (x, y) *of a curve* $y = y(x)$ *for which*

$$3\dot{y}\ddot{y}^2 - (1 + \dot{y}^2)\,\dddot{y} = 0 \tag{15}$$

the osculating circle has with the curve contact of the third order at least (four-point contact), i.e. *the circle is the superosculating circle.*

REMARK 10. The superosculating circle that has exactly four-point contact with a given plane curve at its point (x, y) lies, in a certain neighbourhood of the point of contact, on the same side of the curve.

Theorem 8. *The radius of a superosculating circle is a stationary value of the radius of curvature along a given curve (we assume that* $r_1(t)$ *has a sufficient number of derivatives).*

Definition 6. A point on a plane curve at which the radius of curvature attains its (relative) extreme value is called an *apex* of the curve; it is a point at which the plane curve has contact of an odd order with the superosculating circle.

Definition 7. The straight line through the centre of the osculating circle at a point of a (space) curve and perpendicular to the corresponding osculating plane is called the *polar line* associated with that point.

REMARK 11. The polar line associated with a point of a curve is parallel to the binormal at that point.

REMARK 12. A curve, for which the equation

$$\frac{r_1}{r_2} + (r_1' r_2)' = 0 \quad \left(r_1' = \frac{dr_1}{ds}, \quad r_2 \neq 0\right)$$

is satisfied at every of its points, lies entirely on a sphere and is called a *spherical curve*. The osculating circle at any point of such curve is the intersection of the sphere, on which the curve lies, with the osculating plane at that point.

9.6. Asymptotes. Singular Points of Plane Curves

Definition 1. Let a point T be given on a plane curve and let v be the distance of the point T from a straight line p. If $\lim v = 0$, when at least one of the coordinates of the point T tends to $\pm\infty$, then the straight line p is called an *asymptote* of the given curve.

There may be two kinds of asymptotes of the curve $y = f(x)$: If $\lim f(x) = \pm\infty$, when $x \to c$ (also only when $x \to c+$, or $x \to c-$, respectively), then the straight line $x = c$ (parallel to the y-axis) is an asymptote of the curve. For asymptotes that are not parallel to the y-axis, the following theorem holds:

Theorem 1. *If a curve $y = f(x)$ is defined in the interval $[a, +\infty)$ and the limits*

$$k = \lim_{x \to +\infty} \frac{f(x)}{x}, \tag{1}$$

$$q = \lim_{x \to +\infty} (f(x) - kx) \tag{2}$$

exist, then the curve has an asymptote; its equation is

$$y = kx + q\,. \tag{3}$$

REMARK 1. An entirely similar theorem may be formulated for the interval $(-\infty, a]$.

The tangent to a curve at infinity is an asymptote of this curve. But there are asymptotes of another kind (see Theorem 2):

Theorem 2. *If the equation of the given curve is of the form*

$$y = kx + q + \mu(x) \tag{4}$$

where $\lim_{x \to +\infty} \mu(x) = 0$, *then the curve has the asymptote* $y = kx + q$.

Theorem 3. *If, for a plane curve* $y = f(x)$,

$$\lim_{|x| \to +\infty} y = q \quad \text{or} \quad \lim_{|y| \to +\infty} x = c\,,$$

then the curve has the asymptote

$$y = q \quad \text{or} \quad x = c\,, \tag{5}$$

respectively.

Theorem 4. *If the equation of an algebraic curve can be put in the form*

$$x^n\,\varphi(y) + x^{n-1}\,\varphi_1(y) + x^{n-2}\,\varphi_2(y) + \ldots = 0\,,$$

or

$$y^m\,\psi(x) + y^{m-1}\,\psi_1(x) + y^{m-2}\,\psi_2(x) + \ldots = 0$$

(*we suppose that the polynomials* $\varphi(y)$, $\varphi_1(y)$, …, *or* $\psi(x)$, $\psi_1(x)$, …, *respectively, have no common factor*) *then the curve has asymptotes parallel to the x-axis or to the y-axis, given by*

$$\varphi(y) = 0 \quad or \quad \psi(x) = 0\,, \quad respectively\,. \tag{6}$$

Theorem 5. *If an algebraic curve of degree* $n \geqq 2$,

$$F(x, y) \equiv \varphi_n(x, y) + \varphi_{n-1}(x, y) + \ldots = 0\,,$$

where $\varphi_h(x, y)$ *denotes the sum of terms of degree h, has the asymptote* $y = kx + q$, *then k satisfies the equation*

$$\varphi_n(1, k) = 0 \tag{7}$$

and q the equation

$$q = -\frac{\varphi_{n-1}(1, k)}{\varphi_n'(1, k)} \quad \left(\varphi_n'(1, k) = \frac{\partial \varphi_n}{\partial y}(1, k) \neq 0\right), \tag{8}$$

respectively.

REMARK 2. Theorem 5 thus states the following: Substitute $y = kx + q$ into the equation of the given algebraic curve and equate the coefficient of the two highest powers of the variable x to zero. From the first condition we compute the values k_i

$(i = 1, 2, \ldots, n)$ and substitute these values into the second condition to determine the q_i corresponding to the particular k_i. In this way we find also the asymptotes parallel to the x-axis (but not those parallel to the y-axis). If $\varphi_n'(1, k) = \varphi_{n-1}(1, k) = 0$, then we find q from the equation

$$\tfrac{1}{2}q^2 \, \varphi_n''(1, k) + q \, \varphi_{n-1}'(1, k) + \varphi_{n-2}(1, k) = 0 \, .$$

Example 1. The curve

$$x^3 + y^3 - 3axy = 0$$

(folium of Descartes) is an algebraic curve. We substitute $y = kx + q$ and equate the coefficients of x^3 and x^2 to zero, so that

$$x^3 + k^3x^3 + 3k^2qx^2 + 3kq^2x + q^3 - 3akx^2 - 3aqx = 0$$

giving

$$1 + k^3 = 0 \quad \text{and} \quad 3k^2q - 3ak = 0 \, .$$

The only real solution to the first condition is $k = -1$, which when substituted into the second condition gives

$$3q + 3a = 0 \, .$$

Hence $q = -a$ and the asymptote is

$$y = -x - a \, .$$

Theorem 6. *There exists only one straight line passing through a regular point of a curve (it is the tangent) such that at this point, the line and the curve have two-point contact at least. For all other straight lines there is only one-point contact.*

Definition 2. A singular point (x_0, y_0) of the curve $F(x, y) = 0$ is called a *double point* of the curve if $F_x^0 = F_y^0 = 0$ and if, at the same time, not all partial derivatives of the second order of $F(x, y)$ vanish at this point.

Theorem 7. *There are at most two (real) straight lines passing through a double point (x_0, y_0) of a curve $F(x, y) = 0$ such that the point (x_0, y_0) is their point of intersection with the curve and has a multiplicity at least three.*

The slopes k of these straight lines satisfy the equation

$$F_{xx}^0 + 2F_{xy}^0 k + F_{yy}^0 k^2 = 0 \, . \tag{9}$$

There are two or one or no such straight lines according as to whether

$$F_{xx}^0 F_{yy}^0 - (F_{xy}^0)^2 < 0 \, , \quad or \quad = 0 \, , \quad or \quad > 0 \, , \tag{10}$$

respectively.

Definition 3. The straight lines that pass through a double point of a curve $F(x, y) = 0$ and the slopes of which satisfy equation (9) are called the *tangents to the curve, at its double point.*

Definition 4. A double point of a curve $F(x, y) = 0$ for which

$$\text{a)}\quad F_{xx}F_{yy} - F_{xy}^2 < 0\,,$$

is called a *double point with (two) distinct tangents* or a *node* (the curve has two branches through the singular point, each of them touching the corresponding tangent (Fig. 9.13a));

$$\text{b)}\quad F_{xx}F_{yy} - F_{xy}^2 > 0\,,$$

is called an *isolated point* (the curve has no (real) tangent at this point (Fig. 9.13c, point P));

$$\text{c)}\quad F_{xx}F_{yy} - F_{xy}^2 = 0\,,$$

is called a *cusp* (the curve has two branches which tend to the singular point from one side only and lie on opposite sides of the tangent (Fig. 9.13b)).

REMARK 3. In case c) the curve may have a *point of self-tangency* (a double cusp) (Fig. 9.13d).

REMARK 4. If at a point (x_0, y_0) of a curve $F(x, y) = 0$, in addition to $F_x^0 = F_y^0 = 0$, $F_{xx}^0 = F_{yy}^0 = F_{xy}^0 = 0$ and at least one of the derivatives of the third order of $F(x, y)$ at that point is non-zero, then the curve has a *triple point* (a singular point of multiplicity three) at that point etc. Besides the multiple points, singular points of other kind may also exist, e.g. the *end point* (on the curve $y = e^{1/x}$, see Fig. 9·14a) or the *angular point* (Fig. 9.14b).

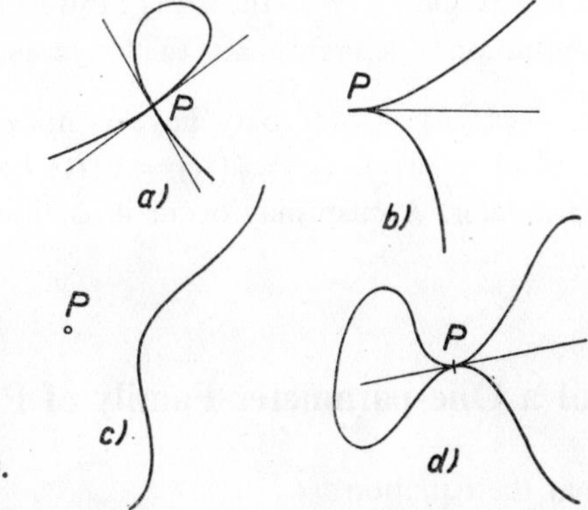

Fig. 9.13.

Theorem 8. *The coordinates (X, Y) of a current point on the tangents at the double point $(0, 0)$ or (x_0, y_0) of a given curve $F(x, y) = 0$ satisfy the equations*

$$F_{xx}^0 X^2 + 2F_{xy}^0 XY + F_{yy}^0 Y^2 = 0$$

(the derivatives being computed at the point $(0, 0)$) (11)

or

$$F^0_{xx}(X - x_0)^2 + 2F^0_{xy}(X - x_0)(Y - y_0) + F^0_{yy}(Y - y_0)^2 = 0$$
$$\text{(the derivatives being computed at the point } (x_0, y_0)) . \tag{12}$$

REMARK 5. A curve $F(x, y) = 0$ has a singular (multiple) point at points whose coordinates satisfy the equations

$$F(x, y) = 0 , \quad F_x = 0 , \quad F_y = 0 .$$

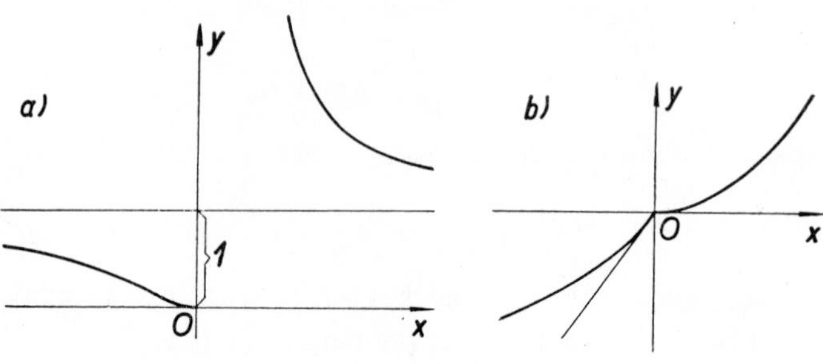

Fig. 9.14.

A deeper investigation of these points can be made by means of the analysis of higher derivatives. In the case of algebraic curves we may proceed as follows: If an algebraic curve passes through the origin and if the origin is a regular point, then we obtain the equation of the tangent at this point by equating to zero the sum of terms of the first order of the equation of the curve. If the lowest terms of the equation of an algebraic curve are of degree at least two then the origin is its singular point. The lowest degree of terms of the equation of the curve gives the multiplicity of the singular point (at the origin). If we equate to zero the sum of lowest terms of the equation of the curve we obtain an equation which represents the tangents at the singular point.

REMARK 6. A curve $x = \varphi(t)$, $y = \psi(t)$ may have a node at a point (x_0, y_0) when the relations $x_0 = \varphi(t_1) = \varphi(t_2)$, $y_0 = \psi(t_1) = \psi(t_2)$ hold for two different values t_1 and t_2 of the parameter. A cusp may occur if $\dot{\varphi}(t) = \dot{\psi}(t) = 0$ holds for a value of the parameter.

9.7. Envelopes of a One-parameter Family of Plane Curves

Definition 1. We say that the equation

$$F(x, y, c) = 0 \tag{1}$$

defines a *one-parameter family of curves* in a domain O of the xy-plane if

1. the function $F(x, y, c)$ is a continuous function of the variables x, y, c for $(x, y) \in O$ and c of an interval I,

2. for every $c \in I$, equation (1) defines a certain curve in the domain O (cf. § 9.2) in such a manner that to any two different values of $c \in I$ there correspond two different curves.

Example 1. The equation

$$y - cx^2 = 0 \tag{2}$$

defines a one-parameter family of parabolas passing through the origin (for $c = 0$ the parabola reduces to the straight line $y = 0$).

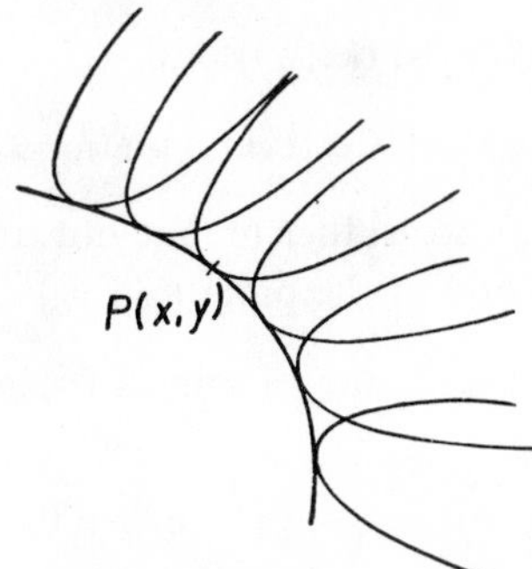

Fig. 9.15.

Definition 2. A curve is called (if such a curve acutally exists) the *envelope* of the family (1), if it touches every curve of the given family and, conversely, if every point $P(x, y)$ of the curve is the point of contact with a curve of the family (Fig. 9.15).

Theorem 1. *Let the function $F(x, y, c)$ have continuous partial derivatives*

$$\frac{\partial F}{\partial x}, \quad \frac{\partial F}{\partial y}, \quad \frac{\partial F}{\partial c}, \quad \frac{\partial^2 F}{\partial c \, \partial x}, \quad \frac{\partial^2 F}{\partial c \, \partial y}, \quad \frac{\partial^2 F}{\partial c^2} \tag{3}$$

in a neighbourhood of the point (x_0, y_0, c_0). At the point (x_0, y_0, c_0) let the following relations be fulfilled:

$$F(x_0, y_0, c_0) = 0, \quad \frac{\partial F}{\partial c}(x_0, y_0, c_0) = 0, \tag{4}$$

$$\frac{\partial^2 F}{\partial c^2} \neq 0, \quad \begin{vmatrix} \dfrac{\partial F}{\partial x}, & \dfrac{\partial F}{\partial y} \\ \dfrac{\partial^2 F}{\partial c \, \partial x}, & \dfrac{\partial^2 F}{\partial c \, \partial y} \end{vmatrix} \neq 0. \tag{5}$$

Then in a certain neighbourhood U of the point (x_0, y_0) and for c from a definite neighbourhood V of the point c_0, there exists an envelope of the family (1).

REMARK 1. We may obtain the equation of the envelope from the equations

$$F(x, y, c) = 0\,, \quad \frac{\partial F}{\partial c}(x, y, c) = 0 \tag{6}$$

by expressing x and y as functions of the variable c (this can be done because of the second condition (5)) or by expressing c as a function of the variables x, y (this can be done because of the first condition (5)) and substituting for c into the first equation (6),

$$F(x, y, c(x, y)) = 0 \tag{7}$$

(i.e. the "elimination" of the parameter c from equations (6)).

REMARK 2. The conditions (5) are sufficient but not necessary for the existence of the envelope in a neighbourhood of the point (x_0, y_0):

Example 2. The envelope of the family of curves (occasionally called parabolas of the fourth order)

$$F(x, y, c) \equiv y - (x - c)^4 = 0\,, \tag{8}$$

is evidently the x-axis; we easily obtain its equation $y = 0$ from equations (6),

$$y - (x - c)^4 = 0\,, \quad 4(x - c)^3 = 0\,. \tag{9}$$

However, at every point of this envelope, both expressions (5) are zero:

$$\frac{\partial^2 F}{\partial c^2} = -12(x - c)^2\,, \quad \begin{vmatrix} \dfrac{\partial F}{\partial x}, & \dfrac{\partial F}{\partial y} \\ \dfrac{\partial^2 F}{\partial c\, \partial x}, & \dfrac{\partial^2 F}{\partial c\, \partial y} \end{vmatrix} = \begin{vmatrix} -4(x - c)^3, & 1 \\ 12(x - c)^2, & 0 \end{vmatrix} = -12(x - c)^2$$

(in consequence of (8) and of the equation $y = 0$).

If conditions (5) are not fulfilled, equations (6) need not define the envelope:

Example 3. Let us consider the family of curves

$$F(x, y, c) \equiv (y - c)^2 - (x - c)^3 = 0\,. \tag{10}$$

(Fig. 9.16.) From (10) it follows that

$$\frac{\partial F}{\partial c} = -2(y - c) + 3(x - c)^2 = 0\,,$$

i.e.

$$y - c = \tfrac{3}{2}(x - c)^2\,. \tag{11}$$

Substituting (11) into (10) we obtain

$$\tfrac{9}{4}(x-c)^3\,(x-c-\tfrac{4}{9})=0\,.$$

Hence either $x-c=0$ or $x-c-\frac{4}{9}=0$.

1. $x-c=0$. Then from (11) it follows that $y-c=0$. Equations $x=c$, $y=c$ are parametric equations of the straight line $y=x$.

2. $x-c=\frac{4}{9}$. Then, by (11), $y-c=\frac{8}{27}$.

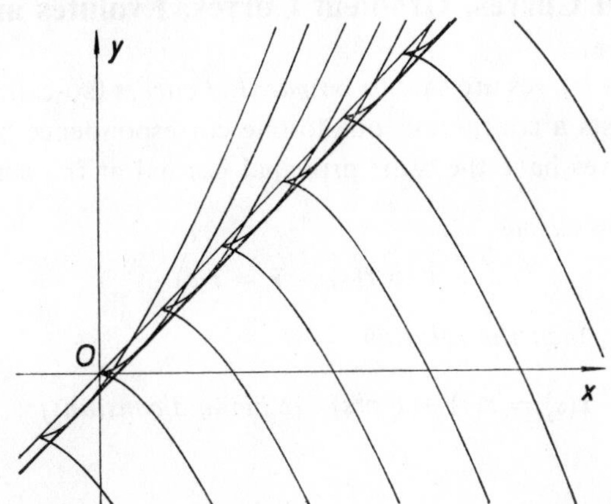

Fig. 9.16.

Equations

$$x=c+\tfrac{4}{9}\,,\quad y=c+\tfrac{8}{27} \tag{12}$$

are parametric equations of the straight line

$$y=x-\tfrac{4}{27}\,. \tag{13}$$

Equations (10) and (11) provide two straight lines, the straight line (13) and the straight line $y=x$. At every point of the former straight line the relation

$$\frac{\partial^2 F}{\partial c^2}\equiv 2-6(x-c)=2-\tfrac{8}{3}=-\tfrac{2}{3}$$

holds (using the first equation (12)) and

$$\begin{vmatrix} \dfrac{\partial F}{\partial x}, & \dfrac{\partial F}{\partial y} \\ \dfrac{\partial^2 F}{\partial c\,\partial x}, & \dfrac{\partial^2 F}{\partial c\,\partial y} \end{vmatrix} \equiv \begin{vmatrix} -3(x-c)^2, & 2(y-c) \\ 6(x-c), & -2 \end{vmatrix} = 6(x-c)[(x-c)-2(y-c)] =$$

$$=\tfrac{8}{3}(\tfrac{4}{9}-\tfrac{16}{27})=-\tfrac{32}{81}$$

(also using equations (12)). Thus, both expressions (5) are non-zero at all points of the straight line (13), which is the envelope of the family (10), since the other conditions of Theorem 1 are evidently fulfilled. The straight line $y = x$ (see Fig. 9.16) is the locus of singular points (the cusps) and evidently it is not an envelope of the curves (10). The reader may easily verify that the determinant (5) vanishes at every point of this straight line.

9.8. Parallel Curves, Gradient Curves, Evolutes and Involutes

Definition 1. Two curves are said to be *parallel curves* (so-called *pair of Bertrand curves*) if there exists a continuous one-to-one correspondence between their points such that both curves have the same principal normal at the corresponding points.

Theorem 1. *If two curves*

$$\mathbf{r} = \mathbf{r}(s)\,, \quad \bar{\mathbf{r}} = \bar{\mathbf{r}}(s)$$

are parallel curves, then the relation

$$\bar{\mathbf{r}}(s) = \mathbf{r}(s) + c\,\mathbf{n}(s) \quad (c \textit{ being a constant}) \tag{1}$$

holds.

REMARK 1. There *need not* exist a parallel curve to each *space* curve. The constant c in (1) determines the distance between corresponding points on the common principal normals. Every *plane* curve possesses an infinite number of parallel curves (for different values of the real constant c in (1)).

A (plane) curve k,

$$x = \varphi(t)\,, \quad y = \psi(t)$$

has a parallel curve $\bar{k}$

$$\bar{x} = \varphi(t) + c\,\frac{\dot{\psi}}{\sqrt{(\dot{\varphi}^2 + \dot{\psi}^2)}}\,, \quad \bar{y} = \psi(t) - c\,\frac{\dot{\varphi}}{\sqrt{(\dot{\varphi}^2 + \dot{\psi}^2)}} \tag{2}$$

(c being an arbitrary real constant). To a given plane curve k we may construct a parallel curve $\bar{k}$ in this manner: We mark off a segment of constant length c on the normal at every point of the given curve, always on the same side of the curve, the starting point of the segment being the point on the given curve (Fig. 9.17). The end points of these segments form a plane curve $\bar{k}$ parallel to the given curve. If the parameter t can be eliminated from equations (2), we obtain the equation of a one-parameter family of curves (with the parameter c) parallel to the given curve. Any two curves of this family are mutually parallel. A plane curve $\bar{k}$ parallel to a given plane curve k is said to be an *equidistant curve* of the curve k (the curves k and $\bar{k}$ are said to be the *curves of equidistance* or *equidistant curves*). The equation of the equidistant curve to the curve $F(x, y) = 0$ (in the variables (X, Y)) may be obtained by the elimination of (x, y) from the equations

$$F(x, y) = 0\,, \quad Y - y = -\frac{\mathrm{d}x}{\mathrm{d}y}(X - x)\,, \quad (X - x)^2 + (Y - y)^2 = c^2\,.$$

The radii of curvature at points on the common normal of two parallel curves differ by the constant length c (the centre of curvature is common). A circular helix is the only

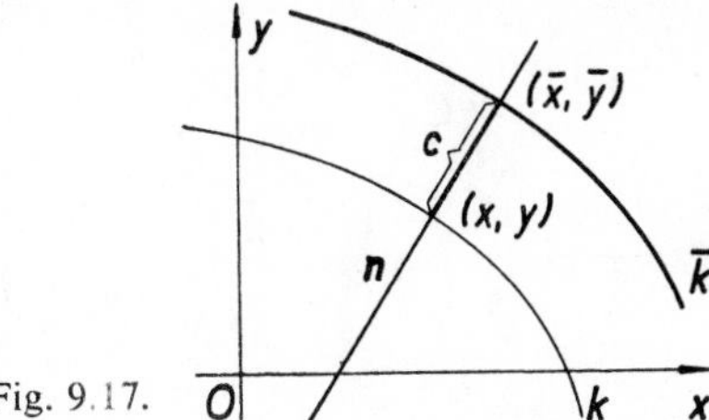

Fig. 9.17.

space curve that also has an infinite number of parallel curves, each of them being a circular helix (lying on a circular cylinder with the same axis as that of the given helix).

Definition 2. A *gradient curve with respect to a fixed direction* is a curve, the tangents to which make a constant angle ω with that fixed direction.

Theorem 2. *A necessary and sufficient condition that a curve be a gradient curve with respect to a given direction is that the relation*

$$k_2 \sin \omega - k_1 \cos \omega = 0 \tag{3a}$$

holds at every point of the curve (k_1 and k_2 being the curvature and torsion respectively).

REMARK 2. Gradient curves are sometimes called *cylindrical helices.* Their characteristic feature is that the equation

$$\frac{k_1}{k_2} = \tan \omega = k \quad (k = \text{const}). \tag{3b}$$

holds along the entire curve. The circular helix is a gradient curve with respect to its axis.

Definition 3. An orthogonal trajectory $\bar{k}$ of the tangents of a curve k is called an *involute* of the curve k. The curve k is called an *evolute* of $\bar{k}$.

REMARK 3. An *orthogonal trajectory* of the tangents of a curve is a curve that intersects the tangents to the given curve at right angles.

Theorem 3. *The equation of an involute $\bar{k}$ of the given evolute $k \equiv \mathbf{r} = \mathbf{r}(s)$ is*

$$\bar{\mathbf{r}} = \mathbf{r} - (s + c)\,\mathbf{t} \quad (c \text{ is a constant; Fig. 9.18}) . \tag{4}$$

REMARK 4. If a curve k (the evolute) is a plane curve, then its involutes (for different values of the constant c) are also plane curves. Every two of these involutes have common normals in the tangents to the curve k (the evolute) and are parallel curves (Fig. 9.18).

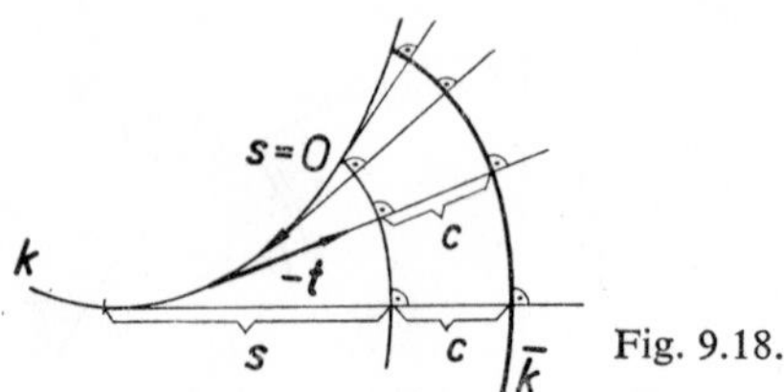

Fig. 9.18.

REMARK 5. To every curve there exists an infinite number of evolutes (forming a one-parameter family of curves) and of involutes (c in (4) is the parameter of the family of involutes).

Theorem 4. *The evolutes of a plane curve $\bar{k}$ are gradient curves (with respect to the perpendicular to the plane of the given curve). Among them, there is one plane evolute k formed by the centres of curvature of the given curve. The equation of this plane evolute is*

$$\mathbf{r} = \bar{\mathbf{r}} + \bar{r}_1 \bar{\mathbf{n}} \tag{5}$$

($\bar{r}_1$ is the radius of curvature at the variable point of the given plane curve $\bar{k}$).

REMARK 6. The *plane evolute* of a curve $x = \varphi(t)$, $y = \psi(t)$ is expressed parametrically by equations (9.5.14) in Theorem 9.5.6, p. 325, defining the coordinates (m, n) of the centre of curvature. By the elimination of t from these equations we obtain the equation of the evolute in the form $F(m, n) = 0$. If a curve is given by an equation $y = f(x)$ or $F(x, y) = 0$, we find the equation of its evolute by the elimination of the variables (x, y) from equations (9.5.11) (Theorem 9.5.4) and $y = f(x)$, or equations (9.5.12) and $F(x, y) = 0$, respectively. If a curve is given by an equation $y = f(x)$, we may retain the parametric expression for an evolute if we choose, for example, x as the parameter.

Example 1. For the parabola $y = x^2$, we have

$$\dot{y} = \frac{\mathrm{d}y}{\mathrm{d}x} = 2x\,, \quad \ddot{y} = \frac{\mathrm{d}^2y}{\mathrm{d}x^2} = 2$$

and by (9.5.11)

$$m = x - 2x\,\frac{1 + 4x^2}{2} = -4x^3\,, \quad n = x^2 + \frac{1 + 4x^2}{2} = \tfrac{1}{2} + 3x^2\,.$$

From the first equation it follows that

$$x = -\sqrt[3]{\frac{m}{4}};$$

on substituting for x into the second equation, we obtain the equation of the evolute of the given parabola,

$$n = \tfrac{1}{2} + 3\sqrt[3]{\frac{m^2}{16}} \quad \text{or} \quad (n - \tfrac{1}{2})^3 = \tfrac{27}{16}m^2 .$$

Conversely the parabola $y = x^2$ is the involute of this curve.

REMARK 7. Eliminating m and n from the equations

$$m = x - \dot{y}\,\frac{1 + \dot{y}^2}{\ddot{y}}, \quad n = y + \frac{1 + \dot{y}^2}{\ddot{y}}, \quad F(m, n) = 0 ,$$

(see (9.5.11)) we obtain the *differential equation of the involutes* of the given curve $F(m, n) = 0$.

REMARK 8. The plane evolute of a given plane curve is the envelope of normals of this curve. The centre of curvature at a fixed point of a given plane curve is the limiting point of the point of intersection of the normal at that point and the normal at a point which tends to the given point along the curve. A normal of a given plane curve is a tangent to its plane evolute with the point of contact at the corresponding centre of curvature of the given plane curve.

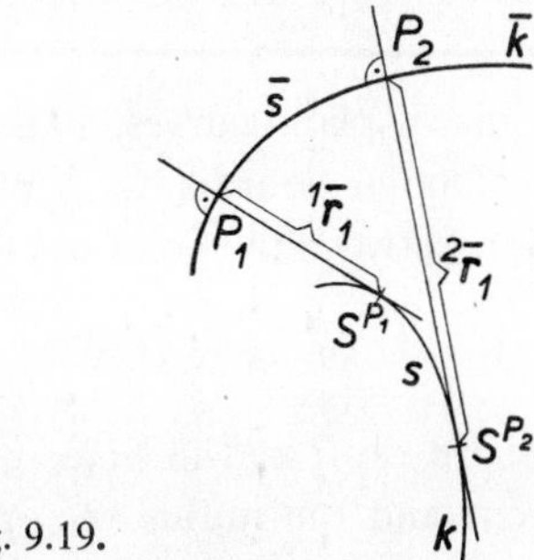

Fig. 9.19.

Theorem 5. *The length of the arc s on the plane evolute k of a given plane curve $\bar{k}$ is equal (in absolute value) to the difference in length of the radii of curvature of the curve $\bar{k}$ that lie on the normals touching the evolute at the end points of the arc s* (Fig. 9.19), i.e.

$$|s| = |{}^2\bar{r}_1 - {}^1\bar{r}_1| \tag{6}$$

($s = \widehat{S^{P_1}S^{P_2}}$ *is the arc of the evolute k, S^{P_1} and S^{P_2} are the centres of curvature* *corresponding to the points P_1 and P_2 on the given curve $\bar{k}$ and ${}^1\bar{r}_1$ and ${}^1\bar{r}_2$ are the corresponding radii of curvature*).

REMARK 9. It is assumed in Theorem 5 that there are no points of inflexion or singular points on the arc $\bar{s} = \widehat{P_1P_2}$ of the curve $\bar{k}$ and that no inner point of this arc is a summit of the given curve $\bar{k}$.

REMARK 10. To every curve k it is possible to construct an infinite number of involutes as follows: we choose a fixed point on the curve k and then mark off on every tangent of the curve k a segment whose length is equal to the length of the arc of the curve k measured from the fixed point to the point of contact of the tangent; the starting point of the segment is the point of contact of the tangent, the end point of the segment is a point of the involute so constructed. The plane involutes of the same family have a single plane evolute in common. To a regular point (which is not a point of inflexion) of a plane curve with maximal or minimal curvature (i.e., to a summit of the curve) there corresponds a cusp of its plane evolute. The normal at a point of inflexion of a plane curve is an asymptote of its plane evolute. The plane evolute passes through the cusps of its plane involute having there its tangent perpendicular to the double tangent at the cusp of the involute (cf. Example 1). Using equation (6), we may determine the length of the arc of the evolute (or of any curve that may be considered as an evolute) if the radii of curvature of its involute are known.

9.9. Direction of the Tangent, Curvature and Asymptotes of Plane Curves in Polar Coordinates

For analytic representation of many plane curves, it is often convenient to use polar coordinates. In the system of polar coordinates a plane curve is represented by an equation expressing a relation between the polar coordinates (ϱ, φ):

$$F(\varrho, \varphi) = 0 \quad \text{or} \quad \varrho = f(\varphi).$$

The direction of a tangent to a curve expressed in polar coordinates is determined by the angle ϑ between the tangent and the radius vector of its point of contact; this angle ϑ is called the *direction angle of the tangent.*

Theorem 1. *The direction angle* ϑ (Fig. 9.20) *is determined by the relation*

$$\tan \vartheta = \varrho/\varrho' \quad (0 < \vartheta < \pi,\ \vartheta \neq \tfrac{1}{2}\pi,\ \varrho' = \mathrm{d}\varrho/\mathrm{d}\varphi \neq 0). \tag{1}$$

REMARK 1. The tangent to the curve at the point $P(\varrho, \varphi)$ makes an angle $\varphi + \vartheta$ with the polar axis o, the normal at the point P on the curve and the radius vector of the point P are inclined at an angle $\vartheta + \frac{1}{2}\pi$, while the same normal and the polar axis o are inclined at an angle $\varphi + \vartheta + \frac{1}{2}\pi$. The distance v of the tangent t from the origin O is given by the relation

$$v = \frac{\varrho^2}{\sqrt{(\varrho^2 + \varrho'^2)}} \quad (\varrho \text{ being the radius vector of the point of contact}) .$$

REMARK 2. In order to construct the tangent or the normal at the point P or to determine their mutual position, it may sometimes be convenient to find the (oriented) length of a segment on the tangent (the *length of the polar tangent* t) and on the

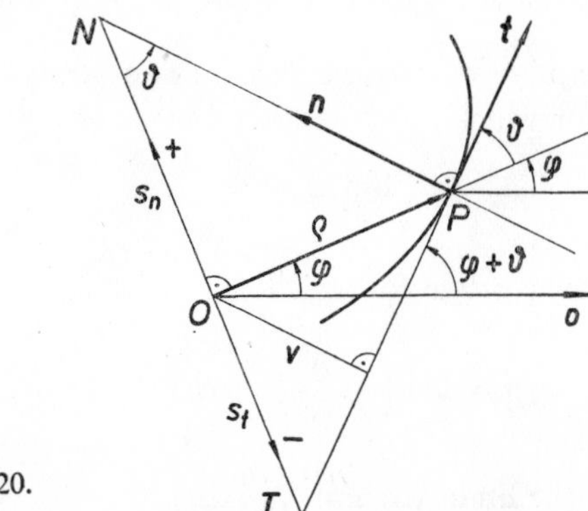

Fig. 9.20.

normal (*the length of the polar normal* n). These lengths are measured from the point of contact P to their respective points of intersection with the perpendicular to the radius vector of the point P through the origin O. We may also need the orthogonal projections of these two segments on the above-mentioned perpendicular, the so-called *polar subtangent* s_t and *subnormal* s_n. The relations

$$s_t = \varrho \tan \vartheta = \frac{\varrho^2}{\varrho'}, \quad s_n = \frac{\varrho}{\tan \vartheta} = \varrho' \tag{2}$$

hold for s_t and s_n (Fig. 9.20). (The segments s_t and s_n are taken either oriented as shown in Fig. 9.20, or in absolute values.) For the length t of the polar tangent and the length n of the polar normal the following relations hold:

$$t = \left|\frac{\varrho}{\varrho'}\right| s', \quad n = s' \quad \left(s' = \frac{\mathrm{d}s}{\mathrm{d}\varphi} = \sqrt{(\varrho^2 + \varrho'^2)}, \text{ d}s \text{ being the differential of the arc}\right). \tag{3}$$

Theorem 2. *The radius of the circle of curvature of the curve* $F(\varphi, \varrho) = 0$ *is given by the expression*

$$r_1 = \frac{(\varrho^2 + \varrho'^2)^{3/2}}{\varrho^2 + 2\varrho'^2 - \varrho\varrho''} . \tag{4}$$

The coordinates ϱ_0, φ_0 *of the centre of the circle of curvature satisfy the equations*

$$\varrho_0 \cos(\varphi_0 - \varphi) = \frac{\varrho(\varrho'^2 - \varrho\varrho'')}{\varrho^2 + 2\varrho'^2 - \varrho\varrho''},$$

and (5)

$$\varrho_0 \sin(\varphi_0 - \varphi) = \frac{\varrho'(\varrho^2 + \varrho'^2)}{\varrho^2 + 2\varrho'^2 - \varrho\varrho''}.$$

REMARK 3. If we eliminate the variables φ and ϱ from equations (5) and the equation $F(\varrho, \varphi) = 0$, we obtain the equation of the plane evolute of the given curve.

Definition 1. An *asymptote* of a curve $F(\varrho, \varphi)$ is a straight line intersecting the polar axis at an angle, for which

$$\lim_{\varrho \to +\infty} \varphi = \alpha \tag{6}$$

and at a distance v from the origin for which

$$v = \lim_{\varphi \to \alpha} \left(- \frac{\varrho^2}{\varrho'} \right) \quad \text{(see (2))} \tag{7}$$

(on the assumption that the limits (6) and (7) exist).

REMARK 4. According to Definition 1, we find an asymptote (if it exists) of a curve as follows: First we determine its direction defined by the angle α, i.e. the angle of inclination to the polar axis and then its distance v from the origin. This distance is the limiting value of the polar subtangent (see (2)) for $\varphi \to \alpha$.

Example 1. For the hyperbolic spiral (reciprocal spiral)

$$\varrho = \frac{a}{\varphi} \quad (a > 0, \varphi > 0),$$

the relation $\varrho \to +\infty$ yields

$$\varphi \to 0, \quad \text{hence} \quad \alpha = 0.$$

Further,

$$v = \lim_{\varphi \to 0} \left(- \frac{\varrho^2}{\varrho'} \right) = \lim_{\varphi \to 0} \frac{a^2/\varphi^2}{a/\varphi^2} = a.$$

Thus the asymptote is a line parallel to, and at a distance a from the polar axis.

REMARK 5. Besides the asymptotes there may exist, with some plane curves, so-called *asymptotic points*. An asymptotic point of a plane curve is a point which does not lie on the curve but to which a point moving along the curve approaches to within an arbitrarily small distance. (This occurs, for example, in the case of the clothoid; cf. § 4.8.) Of a similar meaning is the concept of an *asymptotic curve* (so-called *curvilinear asymptote*). For example, for the curve

$$\varrho = \frac{a\varphi}{\varphi - 1} \quad (a \neq 0 \text{ is a real constant})$$

we have $\lim \varrho = a$ when $|\varphi| \to +\infty$, thus the curve has an *asymptotic circle* of radius a and the curve considered approaches this circle asymptotically from outside and inside (for $\varphi \to +\infty$ and $\varphi \to -\infty$, respectively).

9.10. Supplementary Notes to Part A

a) To determine the *equations of the tangents drawn to the curve* $F(x, y) = 0$ *from an arbitrary point* (x_0, y_0), we find first the coordinates of their points of contact by solving the equations

$$\left.\begin{aligned} &F(x, y) = 0 \\ \text{and} \quad &(x_0 - x)\frac{\partial F}{\partial x} + (y_0 - y)\frac{\partial F}{\partial y} = 0 \end{aligned}\right\} \tag{1}$$

and then substitute these coordinates into the equation of the tangent to the given curve.

b) To determine the *equations of the normals drawn to the curve* $F(x, y) = 0$ *from an arbitrary point* (x_0, y_0), we find those points at which the normals cut the given curve orthogonally (so-called *feet of the normals*), by solving the equations

$$F(x, y) = 0 \quad \text{and} \quad (x_0 - x)\frac{\partial F}{\partial y} - (y_0 - y)\frac{\partial F}{\partial x} = 0\,. \tag{2}$$

If a given algebraic curve is of degree n, then at most n^2 normals can be drawn from a given point to the curve.

Definition 1. The locus of the foot of the perpendicular from a given point (in the plane of the curve) on the tangent to a given curve is called the *pedal curve* of the given curve with respect to the given point (the *pole*).

c) We can find the *equation of the pedal curve* of the curve $F(x, y) = 0$ with respect to the pole (x_0, y_0) by elimination of the variables (x, y) from the equations

$$(X - x)\frac{\partial F}{\partial x} + (Y - y)\frac{\partial F}{\partial y} = 0\,, \quad (X - x_0)\frac{\partial F}{\partial y} - (Y - y_0)\frac{\partial F}{\partial x} = 0\,, \quad F(x, y) = 0\,.$$

The equation of the pedal curve will thus be expressed in the coordinates (X, Y).

d) *The angle between two curves* $y = f(x)$, $y = g(x)$ or $u(x, y) = 0$, $v(x, y) = 0$ at their common point is given by the equation

$$\tan \omega = \frac{\dfrac{dg}{dx} - \dfrac{df}{dx}}{1 + \dfrac{df}{dx}\dfrac{dg}{dx}}, \quad \text{or } \tan \omega = \frac{\dfrac{\partial u}{\partial x}\dfrac{\partial v}{\partial y} - \dfrac{\partial u}{\partial y}\dfrac{\partial v}{\partial x}}{\dfrac{\partial u}{\partial x}\dfrac{\partial v}{\partial x} + \dfrac{\partial u}{\partial y}\dfrac{\partial v}{\partial y}},$$

respectively. (The derivatives are to be evaluated for the coordinates of the point of intersection of the curves.)

The curves $u(x, y) = 0$, $v(x, y) = 0$ *intersect at right angles* if

$$\frac{\partial u}{\partial x}\frac{\partial v}{\partial x} + \frac{\partial u}{\partial y}\frac{\partial v}{\partial y} = 0 .$$

e) The curves which intersect a one-parameter family $F(x, y, c) = 0$ of curves at right angles are called the *orthogonal trajectories* of this family and they form a one-parameter family of curves. Both families form an *orthogonal* net (provided exactly one curve of each family passes through each point). By eliminating the parameter c from the equations

$$\frac{\partial F}{\partial x} + \frac{\partial F}{\partial y}\frac{dy}{dx} = 0, \quad F(x, y, c) = 0,$$

we obtain the differential equation $f(x, y, \dot{y}) = 0$ of the given family. By putting $-1/\dot{y}$ for $\dot{y}$ we obtain the differential equation

$$f\left(x, y, -\frac{1}{\dot{y}}\right) = 0$$

of the family of orthogonal trajectories (cf. § 17·6).

In polar coordinates, by eliminating the parameter c from the equations

$$\frac{\partial F}{\partial \varphi} + \frac{\partial F}{\partial \varrho}\frac{d\varrho}{d\varphi} = 0, \quad F(\varrho, \varphi, c) = 0,$$

we obtain the differential equation $f(\varrho, \varphi, \varrho') = 0$ of the family of curves $F(\varrho, \varphi, c) =$ $= 0$. Writting $-\varrho^2/\varrho'$ for ϱ', we obtain the differential equation

$$f\left(\varrho, \varphi, -\frac{\varrho^2}{\varrho'}\right) = 0$$

of the orthogonal trajectories.

The curves which intersect a one-parameter family of curves at a constant angle $\omega \neq \frac{1}{2}\pi$ are called the *isogonal trajectories* of this family. If $f(x, y, \dot{y}) = 0$ is the differential equation of the given family, then

$$f\left(x, y, \frac{\dot{y} - k}{1 + k\dot{y}}\right) = 0 \quad (k = \tan \omega)$$

is the differential equation of the isogonal trajectories. If, in polar coordinates, $f(\varrho, \varphi, \varrho') = 0$ is the differential equation of a family of curves, then

$$f\left(\varrho, \varphi, \frac{k\varrho^2 + \varrho\varrho'}{\varrho - k\varrho'}\right) = 0 \quad (k = \tan \omega)$$

is the differential equation of the isogonal trajectories.

Fig. 9.21.

Definition 2. The locus of the end point of a segment of constant length c marked off on the tangent of a given curve, with the starting point at the point of contact, is called the *equitangential curve* of the given curve (Fig. 9.21).

f) We find the *equation of an equitangential curve* of the given curve $F(x, y) = 0$ by eliminating the variables (x, y) from the equations

$$F(x, y) = 0\,, \quad Y - y = \frac{dy}{dx}(X - x)\,, \quad (X - x)^2 + (Y - y)^2 = c^2\,.$$

The equation of the equitangential curve is then given in the coordinates (X, Y). The *equations of an equitangential curve* to the curve $x = \varphi(t)$, $y = \psi(t)$ are

$$X = \varphi + \frac{c\dot{\varphi}}{\sqrt{(\dot{\varphi}^2 + \dot{\psi}^2)}}\,, \quad Y = \psi + \frac{c\dot{\psi}}{\sqrt{(\dot{\varphi}^2 + \dot{\psi}^2)}}\,.$$

B. SURFACES

9.11. Definition and Equations of a Surface; Coordinates on a Surface

Definition 1. A *finite piecewise smooth surface*, defined *parametrically*, is a set of points (x, y, z) given by the equations

$$x = x(u, v)\,, \quad y = y(u, v)\,, \quad z = z(u, v)\,; \tag{1}$$

we suppose that the functions $x(u, v)$, $y(u, v)$, $z(u, v)$ are defined in a domain I which is a region of the type A (p. 603) containing, possibly, its boundary or a part of its boundary, and possess the following properties:

1. They are continuous and have piecewise continuous derivatives (Remark 12.1.8, p. 443) of the first order in I. (On the curves of discontinuity or at the boundary points, a derivative is taken to be the value of the corresponding continuous extension.)

2. The matrix

$$\mu = \left\| \begin{matrix} \dfrac{\partial x}{\partial u}, & \dfrac{\partial y}{\partial u}, & \dfrac{\partial z}{\partial u} \\[2ex] \dfrac{\partial x}{\partial v}, & \dfrac{\partial y}{\partial v}, & \dfrac{\partial z}{\partial v} \end{matrix} \right\| \tag{2}$$

is everywhere – with the exception of at most a finite number of points – of rank $h = 2$ (i.e. at least one of its determinants of order two is non-vanishing).

REMARK 1. In Remark 9.2.1, we defined a closed curve and a smooth curve; in a somewhat similar fashion we can define a closed surface and a smooth surface. Briefly: a smooth surface has at every point a definite normal (§ 9.12) which varies continuously with the point on the surface. When a surface is considered in this chapter, a smooth surface or a piecewise smooth surface will be assumed. Furthermore, in the case of a surface, we often suppose (as the nature of the problem may require) that the functions $x(u, v)$, $y(u, v)$, $z(u, v)$ have derivatives of an order $r > 1$, without explicitly pointing out this condition. (The same remark applies to the surface (4).)

The arguments (u, v) in equations (1) are called the *parameters of the surface* and equations (1) are called the *parametric equations of the surface.* Every pair of numbers (u, v) from I is called a *point of the surface* (because to every pair of numbers (u, v) there corresponds a definite point (x, y, z) on the surface); such a point is said to be a *regular (general, ordinary) point of the surface* if, at this point, the functions (1) possess continuous partial derivatives and if the matrix (2) is of rank $h = 2$. Otherwise it is called a *singular point of the surface.* (However, see also Definition 9.12.2 and Remark 9.12.1.)

REMARK 2. If equations (1) are of the form

$$x = u\,, \quad y = v\,, \quad z = z(u, v)\,,$$

the equation

$$z = z(x, y) \quad \text{or} \quad z = f(x, y) \tag{3}$$

is called the *explicit equation of the surface.*

REMARK 3. A surface may be defined *implicitly* as a set of points satisfying an equation

$$F(x, y, z) = 0\,. \tag{4}$$

When we speak, in the following text, of a *surface given implicitly*, then we shall suppose that the function $F(x, y, z)$ is continuous and has continuous or piecewise continuous partial derivatives of the first order in the domain considered and that, with the exception of a finite number of points on the surface, at least one of them

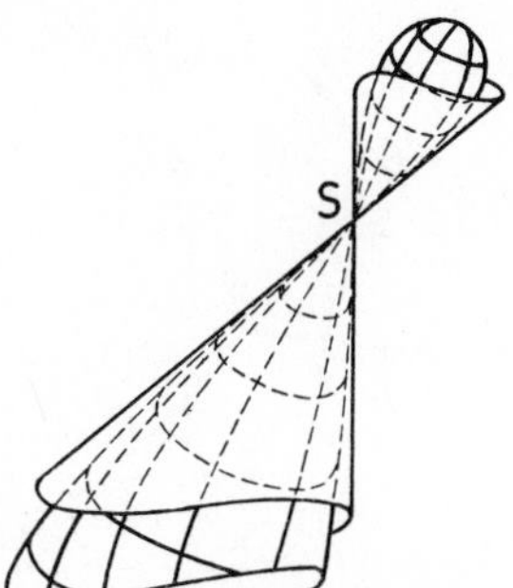

Fig. 9.22.

is non-zero. (In the case of piecewise continuous derivatives, the derivatives are to be understood in the sense of a continuous extension – cf. Definition 1.) If the relations

$$\frac{\partial F}{\partial x} = 0\,, \quad \frac{\partial F}{\partial y} = 0\,, \quad \frac{\partial F}{\partial z} = 0 \tag{5}$$

hold at a point (x, y, z) on the surface (4), then such a point is said to be a *singular point on the surface*. A point on the surface (4) at which at least one of the partial derivatives (5) is non-zero is called a *regular* (*general, ordinary*) *point of the surface* (4). (Cf., however, Remark 9.12.1 which may be applied in a similar form to equation (4).)

REMARK 4. A singular point on the surface (4) at which at least one partial derivative of the second order of the function F is non-zero is called a *conical point of the surface* (the point S in Fig. 9.22).

We shall suppose in the following text that the surface does not intersect itself at its regular points.

If we denote by $\boldsymbol{r}$ the radius vector of a point on the surface, the coordinates of which are, in the parametric representation (1), $x(u, v)$, $y(u, v)$, $z(u, v)$, we can use, instead of equations (1), the single symbolic equation

$$\boldsymbol{r} = \boldsymbol{r}(u, v) = \boldsymbol{i}x(u, v) + \boldsymbol{j}y(u, v) + \boldsymbol{k}z(u, v) \tag{6}$$

for the equation of the surface (the so-called *vector equation*).

Definition 2. The set of points (u_0, v) from I (where u_0 is fixed) for which the derivatives

$$\frac{\partial x(u_0, v)}{\partial v}, \quad \frac{\partial y(u_0, v)}{\partial v}, \quad \frac{\partial z(u_0, v)}{\partial v}$$

do not vanish simultaneously, is called the *parametric u-curve* (briefly *u-curve*) on the surface (1). The *parametric v-curve* (briefly the *v-curve*) is defined in a similar way.

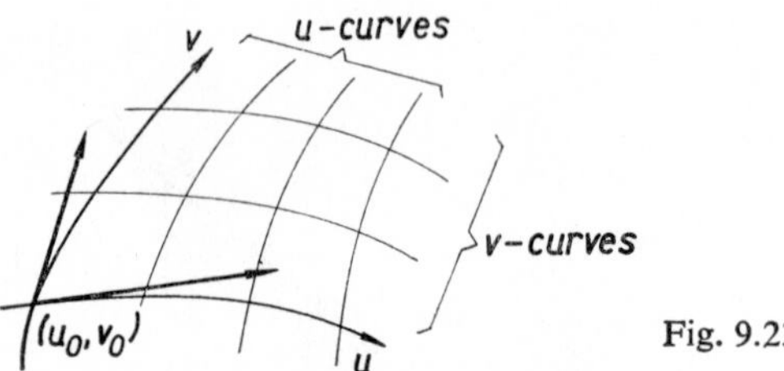

Fig. 9.23.

REMARK 5. If one of the parameters in equations (1) is kept constant while the other varies continuously, we obtain two one-parameter families of curves, one being formed by the u-curves and the other by the v-curves. These curves are called the *coordinate curves* (Fig. 9.23). Their equations on the surface are:

$$u = u_0 = \text{const.}, \quad v = t, \tag{7a}$$

(the u-curve; only the parameter v varies along the curve);

$$u = \bar{t}, \quad v = v_0 = \text{const.}, \tag{7b}$$

(the v-curve; only the parameter u varies along the curve).

The parameters u and v constitute the so-called *curvilinear coordinates* (u, v) of a point on the surface (1) in such a domain where

1. no two curves from the same family intersect,

2. one and only one curve from each family passes through every point of the surface.

The parametric equations of a u-curve (with variable parameter v) or the parametric equations of a v-curve (with variable parameter u) – in space – are

$$x = x(u_0, v), \quad y = y(u_0, v), \quad z = z(u_0, v), \tag{8a}$$

or

$$x = x(u, v_0), \quad y = y(u, v_0), \quad z = z(u, v_0), \tag{8b}$$

respectively.

The *direction cosines of the tangent* of these coordinate curves are proportional to the numbers

$$\frac{\partial x(u_0, v)}{\partial v}, \quad \frac{\partial y(u_0, v)}{\partial v}, \quad \frac{\partial z(u_0, v)}{\partial v} \tag{9a}$$

or

$$\frac{\partial x(u, v_0)}{\partial u}, \quad \frac{\partial y(u, v_0)}{\partial u}, \quad \frac{\partial z(u, v_0)}{\partial u}, \tag{9b}$$

respectively.

Theorem 1. *The tangents of a u-curve and a v-curve constructed at their common regular point are different.* (See Fig. 9.23, the point (u_0, v_0).)

9.12. Curves on Surfaces, Tangent Planes and Normal Lines

Definition 1. The equations

$$u^* = u^*(u, v), \quad v^* = v^*(u, v) \tag{1}$$

express a (*regular*) *transformation of parameters* in a two-dimensional domain I^* from I, if the functions (1), defined in I^*, possess the following properties:

1. They are continuous and have continuous derivatives of order at least $m \geqq r$ (cf. Remark 9.11.1);

2. the determinant

$$\Delta = \begin{vmatrix} \dfrac{\partial u^*}{\partial u}, & \dfrac{\partial u^*}{\partial v} \\ \dfrac{\partial v^*}{\partial u}, & \dfrac{\partial v^*}{\partial v} \end{vmatrix} \tag{2}$$

is non-vanishing in I^*.

Definition 2. Let (u_0, v_0) be a singular point on a surface. Suppose there exist two functions $u^* = u^*(u, v)$, $v^* = v^*(u, v)$, continuous in a neighbourhood of the point (x_0, y_0) and having there continuous partial derivatives of the order $m \geqq r$ (see Definition 1) with $\Delta \neq 0$. Moreover, let the point (u_0, v_0) be not a singular point with respect to the new parameters. Then the point (u_0, v_0) is called an *unsubstantially singular point* (a *pole*) with respect to the parameters (u, v). Every singular point which is not a pole is called *substantially singular*.

REMARK 1. Thus, an unsubstantially singular point is a singular point of a surface only with respect to a certain system of coordinates on the surface. For example on a sphere, if we choose the parallels of latitude for the u-curves, the meridians (parallels of longitude) for the v-curves, then, according to Remark 9.11.1, p. 344, the "north" and "south" poles are singular points of the sphere. But according to Definition 2 these points are unsubstantially singular (because they may be regular in some other system of coordinates.)

The notion of a general curve on a surface is introduced by the following definition:

Definition 3. The equations

$$u = u(t), \quad v = v(t), \quad t \in i, \tag{3}$$

express parametrically a *curve on a surface* (9.11.1) provided that functions (3), defined in an interval i, have the following properties in the interval i:

1. They are continuous and possess continuous first derivatives which do not vanish simultaneously;
2. the points (3) lie for all $t \in i$ in the domain I (see Definition 11.1);
3. the elements of the matrix (9.11.2) vanish simultaneously at a finite number of points at most.

Equations (3) are called the *parametric equations of a curve on the surface* (9.11.1).

REMARK 2. If equations (3) are of the form

$$u = t, \quad v = v(t),$$

then the equation

$$v = v(u) \tag{4}$$

is called the *explicit equation of a curve on the surface* (9.11.1). Equations (3) and (4) are the equations of a curve expressed in the coordinates on the surface. The equations

$$x = x(u(t), v(t)), \quad y = y(u(t), v(t)), \quad z = z(u(t), v(t)) \tag{5a}$$

or briefly

$$\boldsymbol{r} = \boldsymbol{r}(u(t), v(t)) \quad (\boldsymbol{r} \text{ being the radius vector of a point on the curve}) \tag{5b}$$

give the parametric representation of the same curve, which lies on the surface considered, in orthogonal cartesian coordinates (i.e. parametric representation of a space curve).

Definition 4. We also say that the equation

$$f(u, v) = 0 \tag{6}$$

expresses, in a definite neighbourhood of a certain (regular) point of a surface, a *curve on the surface defined implicitly* in I, if $f(u, v)$ is defined in I and has continuous first partial derivatives in I such that at least one of them is non-zero at every point. The equation (6) is called the *implicit equation of a curve on the surface.*

REMARK 3. The functional relation between the curvilinear coordinates defines (under the above-mentioned conditions) a curve on the surface (and conversely).

Theorem 1. *Let us consider all the curves that lie on a surface* $\mathbf{r} = \mathbf{r}(u, v)$ *and pass through a regular point* (u_0, v_0) *of this surface, and themselves have a regular point at this point. Then the tangent lines at* (u_0, v_0) *of all these curves lie in a plane* (Fig. 9.24).

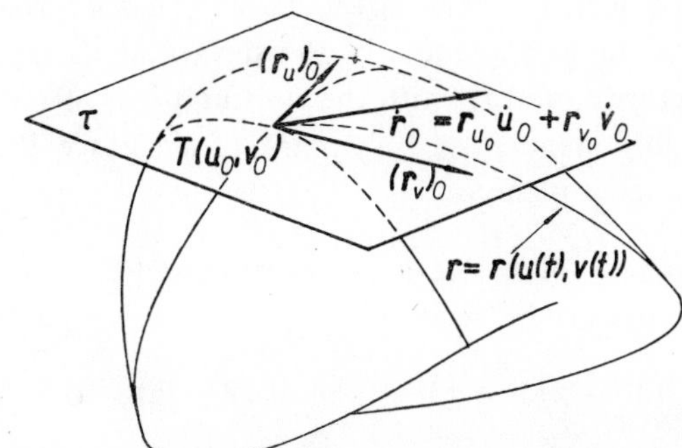

Fig. 9.24.

Definition 5. The plane mentioned in Theorem 1 is called the *tangent plane of the surface* $\mathbf{r} = \mathbf{r}(u, v)$, at its regular point (u_0, v_0). The point (u_0, v_0) is called the *point of contact* of this plane (the plane τ in Fig. 9.24).

REMARK 4. The tangential vector of a curve on a surface is the so-called *tangential vector of the surface* (shortly the *vector of the surface*) at the corresponding point of contact.

Theorem 2. *At a regular point* (u_0, v_0) *of a surface* $\mathbf{r} = \mathbf{r}(u, v)$ *there exists just one tangent plane and its equation in orthogonal cartesian coordinates* (X, Y, Z) *is*

$$[\mathbf{R} - \mathbf{r}_0, (\mathbf{r}_u)_0, (\mathbf{r}_v)_0] \equiv$$

$$\equiv \begin{vmatrix} X - x_0, & Y - y_0, & Z - z_0 \\ \left(\dfrac{\partial x}{\partial u}\right)_0, & \left(\dfrac{\partial y}{\partial u}\right)_0, & \left(\dfrac{\partial z}{\partial u}\right)_0 \\ \left(\dfrac{\partial x}{\partial v}\right)_0, & \left(\dfrac{\partial y}{\partial v}\right)_0, & \left(\dfrac{\partial z}{\partial v}\right)_0 \end{vmatrix} \equiv \begin{vmatrix} X - x_0, & Y - y_0, & Z - z_0 \\ (x_u)_0, & (y_u)_0, & (z_u)_0 \\ (x_v)_0, & (y_v)_0, & (z_v)_0 \end{vmatrix} = 0 \qquad (7)$$

(*where* $\mathbf{R}$ *is the vector of the variable point on the plane,* $(\mathbf{r}_u)_0 = (\partial\mathbf{r}/\partial u)_0$, $(\mathbf{r}_v)_0 = (\partial\mathbf{r}/\partial v)_0$).

REMARK 5. The linearly independent (i.e. non-collinear) vectors $\boldsymbol{r}_u = \partial\boldsymbol{r}/\partial u$ and $\boldsymbol{r}_v = \partial\boldsymbol{r}/\partial v$, with the starting point at a regular point (u, v) of a surface, form the tangential vectors of the corresponding v-curve and u-curve at this point, respectively. The tangential vector $\dot{\boldsymbol{r}}$ of a curve $\boldsymbol{r} = \boldsymbol{r}(u(t), v(t))$ on the surface passing through its regular point (u, v) is given by the linear combination of the vectors $\boldsymbol{r}_u$ and $\boldsymbol{r}_v$,

$$\dot{\boldsymbol{r}} = \boldsymbol{r}_u\dot{u} + \boldsymbol{r}_v\dot{v}\,, \quad \dot{u} = \frac{du}{dt}\,, \quad \dot{v} = \frac{dv}{dt}\,.$$

The ratio $\dot{u}/\dot{v}$ defines, at the point (u, v) of the surface, the direction on the surface in which the curve $u = u(t)$, $v = v(t)$ on the surface passes through the point considered (Fig. 9.24). The tangent plane τ of the surface at its regular point (u, v) is determined by the vectors $\boldsymbol{r}_u$ and $\boldsymbol{r}_v$, with the starting point of both vectors at (u, v). A regular (or removably singular, p. 347) point of a surface is a point at which there exists just one tangent plane.

Theorem 3. *The tangent plane at a regular point (x, y, z) of a surface $z = z(x, y)$ has the equation*

$$(X - x)\,p + (Y - y)\,q - (Z - z) = 0 \tag{8}$$

(where $p = \partial z/\partial x$, $q = \partial z/\partial y$, and (X, Y, Z) are the coordinates of the variable point on the plane).

Theorem 4. *The tangent plane at a regular point (x, y, z) of a surface $F(x, y, z) = 0$ has the equation*

$$\frac{\partial F}{\partial x}(X - x) + \frac{\partial F}{\partial y}(Y - y) + \frac{\partial F}{\partial z}(Z - z) = 0 \tag{9}$$

((X, Y, Z) being the coordinates of the variable point on the plane).

Definition 6. The *normal* $\boldsymbol{n} = \boldsymbol{n}(u, v)$ *to a surface* $\boldsymbol{r} = \boldsymbol{r}(u, v)$, at a regular point (u, v), is the unit vector with its starting point at (u, v), perpendicular to the tangent plane at (u, v) and oriented so that

$$[\boldsymbol{n}\boldsymbol{r}_u\boldsymbol{r}_v] > 0\,,$$

where $[\boldsymbol{n}\boldsymbol{r}_u\boldsymbol{r}_v]$ is the mixed product (scalar triple product) of the vectors $\boldsymbol{n}, \boldsymbol{r}_u, \boldsymbol{r}_v$ (Fig. 9.25).

Theorem 5. *The unit vector $\boldsymbol{n}$ of the normal to the surface $x = x(u, v)$, $y = y(u, v)$, $z = z(u, v)$ is*

$$\boldsymbol{n} = \frac{\boldsymbol{r}_u \times \boldsymbol{r}_v}{\sqrt{(EG - F^2)}}\,, \tag{10}$$

where

$$E = \boldsymbol{r}_u \,.\, \boldsymbol{r}_u = \left(\frac{\partial x}{\partial u}\right)^2 + \left(\frac{\partial y}{\partial u}\right)^2 + \left(\frac{\partial z}{\partial u}\right)^2,$$

$$F = \boldsymbol{r}_u \,.\, \boldsymbol{r}_v = \frac{\partial x}{\partial u}\frac{\partial x}{\partial v} + \frac{\partial y}{\partial u}\frac{\partial y}{\partial v} + \frac{\partial z}{\partial u}\frac{\partial z}{\partial v}, \tag{11}$$

$$G = \boldsymbol{r}_v \,.\, \boldsymbol{r}_v = \left(\frac{\partial x}{\partial v}\right)^2 + \left(\frac{\partial y}{\partial v}\right)^2 + \left(\frac{\partial z}{\partial v}\right)^2.$$

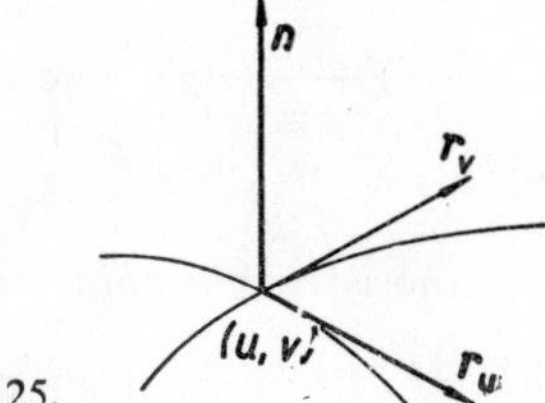

Fig. 9.25.

REMARK 6. At a regular point of a surface the relation $D^2 = EG - F^2 > 0$ always holds. The expression

$$D = \sqrt{(EG - F^2)} = |\boldsymbol{r}_u \times \boldsymbol{r}_v| = [\boldsymbol{n}\boldsymbol{r}_u\boldsymbol{r}_v] > 0 \tag{12}$$

is called the *discriminant of the surface.* We have

$$D^2 = \begin{vmatrix} E, & F \\ F, & G \end{vmatrix}. \tag{13}$$

The lengths of the tangential vectors $\boldsymbol{r}_u$ and $\boldsymbol{r}_v$ to a surface are

$$|\boldsymbol{r}_u| = \sqrt{E}, \quad |\boldsymbol{r}_v| = \sqrt{G}.$$

Theorem 6. *The direction cosines (n_x, n_y, n_z) of the normal $\boldsymbol{n}$ to a surface $\boldsymbol{r} = \boldsymbol{r}(u, v)$, are given by the expressions*

$$n_x = \frac{\begin{vmatrix} \dfrac{\partial y}{\partial u}, & \dfrac{\partial z}{\partial u} \\ \dfrac{\partial y}{\partial v}, & \dfrac{\partial z}{\partial v} \end{vmatrix}}{D}, \quad n_y = \frac{\begin{vmatrix} \dfrac{\partial z}{\partial u}, & \dfrac{\partial x}{\partial u} \\ \dfrac{\partial z}{\partial v}, & \dfrac{\partial x}{\partial v} \end{vmatrix}}{D}, \quad n_z = \frac{\begin{vmatrix} \dfrac{\partial x}{\partial u}, & \dfrac{\partial y}{\partial u} \\ \dfrac{\partial x}{\partial v}, & \dfrac{\partial y}{\partial v} \end{vmatrix}}{D}. \tag{14}$$

Theorem 7. *The direction cosines of the normal to a surface $z = f(x, y)$ are given by the expressions*

$$n_x = \frac{-p}{\sqrt{(p^2 + q^2 + 1)}}, \quad n_y = \frac{-q}{\sqrt{(p^2 + q^2 + 1)}}, \quad n_z = \frac{1}{\sqrt{(p^2 + q^2 + 1)}} \tag{15}$$

(where $p = \partial z/\partial x$, $q = \partial z/\partial y$).

Theorem 8. *The direction cosines of the normal to a surface* $F(x, y, z) = 0$ *are given by the expressions*

$$n_x = \frac{F_x}{J}, \quad n_y = \frac{F_y}{J}, \quad n_z = \frac{F_z}{J} \tag{16}$$

(*where* $F_x = \partial F/\partial x$, $F_y = \partial F/\partial y$, $F_z = \partial F/\partial z$ *and* $J = \sqrt{(F_x^2 + F_y^2 + F_z^2)}$).

The equations of a normal to the surface $F(x, y, z) = 0$ *at its regular point* (x, y, z) *are*

$$\frac{X - x}{F_x} = \frac{Y - y}{F_y} = \frac{Z - z}{F_z} \tag{17}$$

((X, Y, Z) *being the coordinates of the variable point on the normal*).

REMARK 7. The direction ratios of the normal to a surface $F(x, y, z) = 0$, or $z = f(x, y)$ are (F_x, F_y, F_z), or $(\partial f/\partial x, \partial f/\partial y, -1)$, respectively.

Example 1. If we choose the spherical coordinates u and v (polar coordinates in space) (Fig. 9.26) for the representation of a spherical surface of radius r with its centre at the origin of the cartesian coordinate system, then the equations

$$x = r \sin u \cos v\,, \quad y = r \sin u \sin v\,, \quad z = r \cos u \tag{18}$$
$$(0 \leqq u \leqq \pi, \quad 0 \leqq v < 2\pi)$$

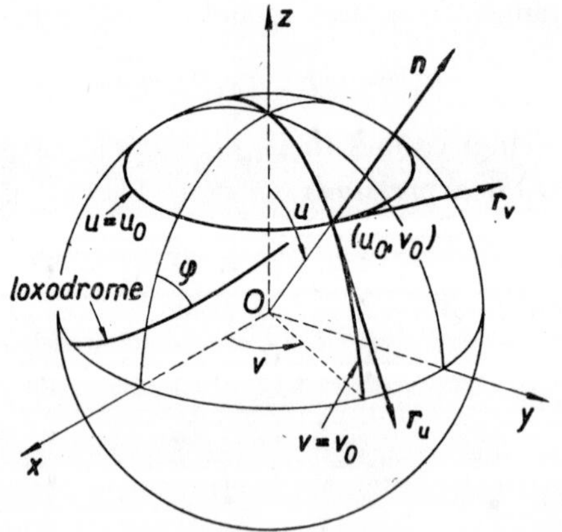

Fig. 9.26.

are the parametric equations of this spherical surface. The geometrical interpretation of the parameters u and v may be seen in Fig. 9.26. The net of the coordinate curves is formed by the parallel circles (u-curves) and the half-meridians (v-curves) of the spherical surface. For the vector $\mathbf{r}_u$ (the direction of the tangent to a meridian) and the vector $\mathbf{r}_v$ (the direction of the tangent to a parallel circle), at the point (u, v), we obtain from (18) (for the sake of clarity we use here the notation $\boldsymbol{a} = \begin{cases} a_1 \\ a_2 \\ a_3 \end{cases}$ instead of $\boldsymbol{a} = (a_1, a_2, a_3)$)

$$\boldsymbol{r}_u(u, v) = \begin{cases} r\cos u \cos v\,, \\ r\cos u \sin v\,, \\ -r\sin u\,, \end{cases} \quad \boldsymbol{r}_v(u, v) = \begin{cases} -r\sin u \sin v\,, \\ r\sin u \cos v\,, \\ 0\,. \end{cases}$$

For the direction ratios of the vector of the normal we obtain

$$\boldsymbol{r}_u \times \boldsymbol{r}_v = \begin{cases} r^2 \sin^2 u \cos v\,, \\ r^2 \sin^2 u \sin v\,, \\ r^2 \sin u \cos u\,, \end{cases} \quad \text{i.e.} \quad \boldsymbol{n} = \begin{cases} \sin u \cos v\,, \\ \sin u \sin v\,, \\ \cos u \end{cases}$$

and, using (18), we obtain the equation of the tangent plane at the point (u, v) in the form

$$X \sin u \cos v + Y \sin u \sin v + Z \cos u - r = 0\,.$$

Example 2. Let the equations

$$x = u\cos v\,, \quad y = u\sin v\,, \quad z = f(v) \quad (u \in (-\infty, +\infty), \quad v \in (-\tfrac{1}{2}\pi, \tfrac{1}{2}\pi)) \tag{19}$$

represent a surface. By the elimination of u and v from (19) we obtain the explicit equation of the surface in the form

$$z = f\left(\tan^{-1}\frac{y}{x}\right). \tag{20}$$

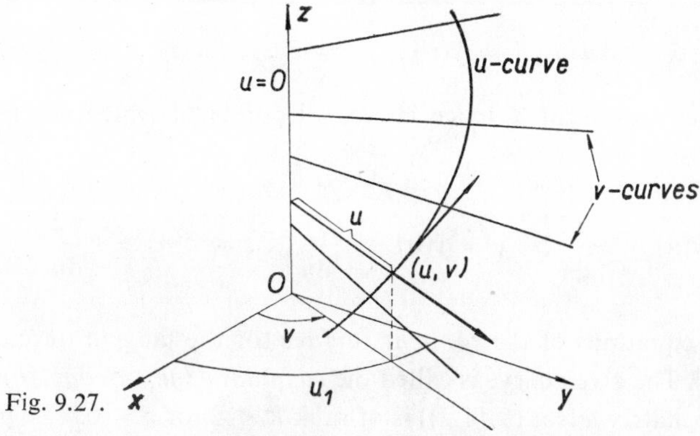

Fig. 9.27.

For $u = u_0 \neq 0$ the parametric u-curve is the intersection of the surface (20) and the circular cylinder $x^2 + y^2 = u_0^2$. The equation $u = 0$ denotes the z-axis. The v-curves are the sections of the surface (20) cut by the planes $y = x \tan v_0$ and therefore constitute the straight lines which intersect the z-axis at right angles (Fig. 9.27). This surface (which contains a one-parameter family of straight lines) is a *ruled surface* (in this case a *skew surface* or a *scroll*, i.e. a non-developable surface), in fact a *right conoid*. The equation of the tangent plane at the point (u, v) is, by (7),

$$\begin{vmatrix} X - u\cos v, & Y - u\sin v, & Z - f(v) \\ \cos v, & \sin v, & 0 \\ -u\sin v, & u\cos v, & \mathrm{d}f/\mathrm{d}v \end{vmatrix} =$$

$$= (X\sin v - Y\cos v)\frac{\mathrm{d}f}{\mathrm{d}v} + [Z - f(v)]\,u = 0\,.$$

In particular, if we choose for the function $z = f(v)$, the function $z = cv$ ($c \neq 0$ being a real constant), we obtain a right conoid and screw surface, known as a *helicoid*; its equation is

$$z = c\tan^{-1}\frac{y}{x}$$

(the u-curves are formed by coaxial helices). At the point (x, y, z) we obtain, from (8), the equation of the tangent plane of this screw surface

$$cyX - cxY + (x^2 + y^2)\,Z = (x^2 + y^2)\,z$$

and, from (15), the equations of the normal line

$$\frac{X - x}{cy} = \frac{Y - y}{-cx} = \frac{Z - z}{x^2 + y^2}\,.$$

Example 3. Let

$$\bar{x} = \bar{x}(u)\,, \quad \bar{y} = \bar{y}(u)\,, \quad \bar{z} = \bar{z}(u)\,, \quad u \in I\,, \tag{21}$$

be the parametric equations of a space curve, all points of which are regular for $u \in I$. The equations

$$x = \bar{x}(u) + v\frac{\mathrm{d}\bar{x}}{\mathrm{d}u}\,, \quad y = \bar{y}(u) + v\frac{\mathrm{d}\bar{y}}{\mathrm{d}u}\,, \quad z = \bar{z}(u) + v\frac{\mathrm{d}\bar{z}}{\mathrm{d}u} \tag{22}$$

are the parametric equations of the *tangent surface* (or the tangent developable) of the space curve (21). The given curve is called the *cuspidal edge*, or *edge of regression* of this surface. The matrix μ (see (9.11.2)) is of rank $h < 2$ for $v = 0$. The parametric curves are on the one hand the curves $u =$ const., i.e. the tangents to the curve (21), and on the other $v =$ const., i.e. the curves which have the same distance v, measured along the tangent to the given curve, from the given curve (i.e. from the curve $v = 0$). Every point of the given curve is a singular point of the surface (22) (see Fig. 9.28). The surface (22) which contains a one-parameter family of straight lines (the u-curves) is a ruled surface, but is *developable*. The equation of the tangent plane at a regular point (u, v) (for which $v \neq 0$) is

$$\left[\boldsymbol{R} - \bar{\boldsymbol{r}}, \frac{d^2\bar{\boldsymbol{r}}}{du^2}, \frac{d\bar{\boldsymbol{r}}}{du}\right] = 0 .$$

This equation (which is independent of v) is, however, an equation of the osculating plane of the curve $\bar{\boldsymbol{r}} = \bar{\boldsymbol{r}}(u)$ at its point (u). Any tangent plane of the given surface is a tangent plane at every point $v \neq 0$ to a fixed u-curve, i.e. to a definite straight

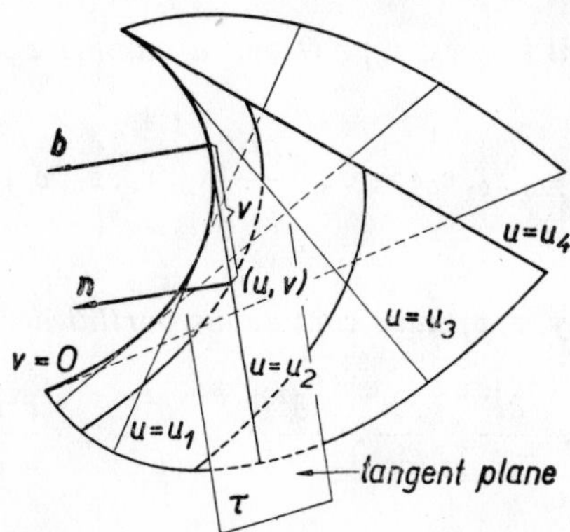

Fig. 9.28.

line on the surface (the so-called *generator*, or *ruling of the surface*). The normal $\boldsymbol{n}$ to the surface does not change along this straight line and is parallel to the binormal $\boldsymbol{b}$ to the given curve at the corresponding point of contact $(u, 0)$.

Example 4. According to (9), the surface

$$xyz = a^3 \quad (a \neq 0 \text{ being a real constant})$$

has at the point (x, y, z) the tangent plane with the equation

$$\frac{X}{x} + \frac{Y}{y} + \frac{Z}{z} = 3 .$$

9.13. Envelope of a One-parameter Family of Surfaces, Ruled Surfaces (Torses and Scrolls)

Definition 1. We say that an equation

$$F(x, y, z, c) = 0 \tag{1}$$

defines a *one-parameter family of surfaces* in a domain O of the three-dimensional space xyz if

1. the function $F(x, y, z, c)$ is a continuous function of the variables (x, y, z, c) for $(x, y, z) \in O$ and for c from an interval I,

2. for every $c \in I$, equation (1) defines in the domain O a certain surface (cf. § 9.11) and to every two different values of $c \in I$ there correspond two different surfaces in the domain O.

Definition 2. A surface is said to be the *envelope* of the family (1) (provided such a surface exists) if it touches every surface of the given family, and conversely, if its every point (x, y, z) is a point of contact with a surface of the family (1).

Theorem 1. *Let a point $(x_0, y_0, z_0) \in O$ and a number $c_0 \in I$ exist such that the equations*

$$F(x_0, y_0, z_0, c_0) = 0\,, \quad \frac{\partial F}{\partial c}(x_0, y_0, z_0, c_0) = 0 \tag{2}$$

hold.

Let the function $F(x, y, z, c)$ have continuous partial derivatives

$$\frac{\partial F}{\partial x}, \frac{\partial F}{\partial y}, \frac{\partial F}{\partial z}, \frac{\partial F}{\partial c}, \frac{\partial^2 F}{\partial c\,\partial x}, \frac{\partial^2 F}{\partial c\,\partial y}, \frac{\partial^2 F}{\partial c\,\partial z}, \frac{\partial^2 F}{\partial c^2} \tag{3}$$

in a certain neighbourhood of the point (x_0, y_0, z_0, c_0). Further at the point (x_0, y_0, z_0, c_0), let

$$\frac{\partial^2 F}{\partial c^2} \neq 0 \tag{4}$$

and let the matrix

$$\left\| \begin{matrix} \dfrac{\partial F}{\partial x}, & \dfrac{\partial F}{\partial y}, & \dfrac{\partial F}{\partial z} \\ \dfrac{\partial^2 F}{\partial c\,\partial x}, & \dfrac{\partial^2 F}{\partial c\,\partial y}, & \dfrac{\partial^2 F}{\partial c\,\partial z} \end{matrix} \right\| \tag{5}$$

be of rank

$$h = 2\,. \tag{6}$$

Then, in a certain neighbourhood U of the point (x_0, y_0, z_0) and for c from a certain neighbourhood V of the point c_0, there exists an envelope of the family (1).

REMARK 1. The equation of the envelope may be obtained, for example, from the equations

$$F(x, y, z, c) = 0\,, \quad \frac{\partial F}{\partial c}(x, y, z, c) = 0 \tag{7}$$

by expressing c from the second equation as a function of (x, y, z) and substituting into the first equation; by eliminating c we obtain

$$F(x, y, z, c(x, y, z)) = 0. \tag{8}$$

For a fixed c, equations (7) give the so-called *characteristic curve*, or simply the *characteristic*, of the family (1), along which the envelope touches the corresponding surfaces of the given family.

REMARK 2. By considering (4) and (6), we can make a remark, similar to Remark 9.7.1, on the envelope of a given family of curves. In the next example, the reader may verify the fulfilment of the above-mentioned assumptions:

Example 1. The equation

$$x^2 + y^2 + (z - c)^2 = r^2(c) \tag{9}$$

is the equation of a one-parameter family of spheres, whose centres are on the z-axis and whose radii depend on the distance c between the centre of the sphere and the origin. Equation (9), together with the second equation (7),

$$-(z - c) = r(c)\frac{dr(c)}{dc} \tag{10}$$

express (c being fixed) the characteristic of a sphere from the family investigated. By using (9) and (10), the equation of this characteristic may be expressed in the form

$$x^2 + y^2 = r^2(c)\left[1 - \left(\frac{dr(c)}{dc}\right)^2\right]; \quad z = c - r(c)\frac{dr(c)}{dc}. \tag{11}$$

The first of equations (11) represents a circular cylinder with its axis along the z-axis; the second of equations (11) represents a plane which is perpendicular to the z-axis, so that the characteristic is a circle lying in this plane and having its centre on the z-axis. We obtain the equation of the envelope of the family (9) by eliminating the parameter c from equations (7) in the form

$$x^2 + y^2 = f(z), \tag{12}$$

which is the equation of a surface of revolution with its axis of revolution in the z-axis. If, for example, $r^2(c) = 2c - 1$, equations (11) reduce, after rearrangement, to

$$x^2 + y^2 = 2(c - 1), \quad z = c - 1.$$

The equation of the envelope of the family $x^2 + y^2 + (z - c)^2 = 2c - 1$ will be obtained by the elimination of the parameter c from the two preceding equations, in the form

$$x^2 + y^2 = 2z;$$

this is the equation of a paraboloid of revolution generated by the revolution of the parabola $2z = x^2$ about the z-axis.

REMARK 3. Surfaces of revolution are examples of envelopes of a one-parameter family of spheres. The characteristics are circles.

Definition 3. If the equations

$$F(x, y, z, c) = 0\,, \quad \frac{\partial F}{\partial c} = 0\,, \quad \frac{\partial^2 F}{\partial c^2} = 0 \quad \left(\frac{\partial^3 F}{\partial c^3} \neq 0 \text{ for all } c \text{ from } I\right) \tag{13}$$

determine a curve, then this curve is called the *edge of regression* of the given one-parameter family of surfaces.

REMARK 4. We also speak of an edge of regression in the case where equations (13) define a finite number of points.

Theorem 2. *If the edge of regression is a curve, then at each of its points* (x_0, y_0, z_0) *the characteristic corresponding to a parameter* c_0 *and the edge of regression are tangential to each other*, i.e. *every characteristic of the surface of the family is a tangent to the edge of regression* (*at the so-called focal point*).

Theorem 3. *If the edge of regression of a one-parameter family of planes is a space curve, then this family of planes is formed by the osculating planes of the edge of regression and the characteristics are the tangents to the edge of regression.*

Theorem 4. *The envelope of a one-parameter family of planes, whose edge of regression is a space curve, is the tangent surface of this space curve.*

REMARK 5. The envelopes of one-parameter families of planes are developable ruled surfaces. The straight lines of the surface (so-called *rulings* or *generators*) are the characteristics of the family (of planes). The tangent surfaces of space curves, all cones and cylinders belong to this group. The tangent surfaces of gradient curves (Definition 9.8.2) are called the *gradient surfaces*. All tangent planes of such a surface make a constant angle with a fixed direction.

REMARK 6. In Examples 9.12.2 and 9.12.3 two substantially different cases of ruled surfaces were shown. The ruled surfaces are of great importance in technical applications.

Definition 4. A *ruled surface* is a surface such that, through every point of it, there passes at least one straight line lying entirely on it.

REMARK 7. We may also represent a ruled surface as a locus of a straight line moving continuously in space, i.e. a ruled surface is a one-parameter family of straight lines where the corresponding parameter changes continuously through an interval.

Theorem 5. *The parametric equations of a ruled surface are*

$$\begin{aligned} x(u, v) &= \bar{x}(u) + v \,.\, a(u)\,, \\ y(u, v) &= \bar{y}(u) + v \,.\, b(u)\,, \quad \text{i.e.} \quad \mathbf{r}(u, v) = \bar{\mathbf{r}}(u) + v \,.\, \mathbf{p}(u)\,, \\ z(u, v) &= \bar{z}(u) + v \,.\, c(u)\,, \end{aligned} \tag{14}$$

where $\bar{x} = \bar{x}(u)$, $\bar{y} = \bar{y}(u)$, $\bar{z} = \bar{z}(u)$ *are the parametric equations of the so-called director curve, and* $a(u)$, $b(u)$, $c(u)$ *are continuous functions of the argument* u *and determine the direction ratios of the straight lines (so-called rulings or generators) on the surface.*

Theorem 6. *If on a ruled surface* (14), *it is not possible to determine a value of* u *for which the ruling is such that the tangent planes at the points along that ruling form a pencil of planes, the ruled surface is a cylinder, cone, or tangent surface of a (space) curve. The characteristic feature of these surfaces is that every tangent plane touches the surface along an entire ruling.*

Definition 5. A ruled surface of the type mentioned in Theorem 6 is called a *developable surface* or a *torse.* A ruled surface which possesses a property such that the tangent planes at the points along its rulings (with the possible exception of a finite number of rulings called *torsal lines*) form a pencil of planes, is called an *undevelopable ruled surface*, a *skew surface* or a *scroll.* The ruling of an undevelopable ruled surface is called the *torsal line* if it has the same tangent plane at all its regular points.

REMARK 8. Every ruling on a cone, cylinder or tangent surface of a space curve (i.e. on a developable ruled surface) is a torsal line. The skew conicoids (the hyperbolic paraboloid and the hyperboloid of one sheet) and the helicoid from Example 9.12.2 serve as examples of skew surfaces that have no torsal lines.

Theorem 7 (*Chasles' Theorem*). *The tangent planes of a skew surface along its rulings form pencils of planes (their axes are the rulings). There is a projectivity between the pencil of tangent planes and the range of their points of contact on a ruling,* i.e. *the cross-ratio of four tangent planes to a skew surface at points of a ruling is equal to the cross-ratio of the points of contact.*

Theorem 8. *A necessary and sufficient condition that a ruled surface* (14) *be developable is*

$$[\dot{\bar{\mathbf{r}}}, \mathbf{p}, \dot{\mathbf{p}}] = 0\,, \tag{15}$$

where $\dot{\bar{\mathbf{r}}} = \dfrac{\mathrm{d}\bar{\mathbf{r}}}{\mathrm{d}u}$, $\dot{\mathbf{p}} = \dfrac{\mathrm{d}\mathbf{p}}{\mathrm{d}u}$.

Theorem 9. *If a ruled surface is given by the equations*

$$x = a(t)\, z + m(t)\,, \quad y = b(t)\, z + n(t)$$

(a, b, m *and* n *being functions of the same parameter* t*), then a necessary and*

sufficient condition that this surface be developable is

$$\dot{a}\dot{n} = \dot{b}\dot{m} \quad \left(\dot{a} = \frac{\mathrm{d}a}{\mathrm{d}t}, \quad \dot{n} = \frac{\mathrm{d}n}{\mathrm{d}t}, \ etc.\right).$$

Theorem 10. *The equation*

$$f_{xx}f_{yy} - f_{xy}^2 = 0 \quad \left(f_{xx} = \frac{\partial^2 f}{\partial x^2}, \quad f_{yy} = \frac{\partial^2 f}{\partial y^2}, \quad f_{xy} = \frac{\partial^2 f}{\partial x\,\partial y}\right) \tag{16}$$

is the differential equation of developable surfaces (for surfaces expressed by the equation $z = f(x, y)$).

9.14. First Fundamental Form of the Surface

Theorem 1. *The square of the differential of the arc of a curve $u = u(t)$, $v = v(t)$ on a surface $\mathbf{r} = \mathbf{r}(u, v)$ is given by the formula*

$$\mathrm{d}s^2 = E\,\mathrm{d}u^2 + 2F\,\mathrm{d}u\,\mathrm{d}v + G\,\mathrm{d}v^2 \equiv I. \tag{1}$$

Definition 1. The quadratic differential form (1) is called the *first fundamental* (or *metric*) *form of a surface*, ds is called the *linear element* (the *element of the arc*) on the surface and E, F, G are called the *first fundamental coefficients* (see Theorem 9.12.5).

Theorem 2. *If φ is the angle between the curves $u = u(t)$, $v = v(t)$ and $\bar{u} = \bar{u}(t)$, $\bar{v} = \bar{v}(t)$ which lie on a surface $\mathbf{r} = \mathbf{r}(u, v)$* and *pass through any of its regular points, then*

$$\cos\varphi = \frac{E\dot{u}\dot{\bar{u}} + F(\dot{u}\dot{\bar{v}} + \dot{\bar{u}}\dot{v}) + G\dot{v}\dot{\bar{v}}}{\sqrt{(E\dot{u}^2 + 2F\dot{u}\dot{v} + G\dot{v}^2)}\,.\,\sqrt{(E\dot{\bar{u}}^2 + 2F\dot{\bar{u}}\dot{\bar{v}} + G\dot{\bar{v}}^2)}}. \tag{2}$$

REMARK 1. For the parametric curves of a surface we have that $\mathrm{d}s^2 = G\,\mathrm{d}v^2$ (for $u = \text{const.}$) and $\mathrm{d}s^2 = E\,\mathrm{d}u^2$ (for $v = \text{const.}$).

If the second curve of Theorem 2 is a parametric curve, then

$$\cos\varphi = \frac{E\dot{u} + F\dot{v}}{\sqrt{(E)}\sqrt{(E\dot{u}^2 + 2F\dot{u}\dot{v} + G\dot{v}^2)}} \quad \text{or} \quad \cos\varphi = \frac{F\dot{u} + G\dot{v}}{\sqrt{(G)}\sqrt{(E\dot{u}^2 + 2F\dot{u}\dot{v} + G\dot{v}^2)}} \tag{3}$$

for $\bar{v}(t) = \bar{v}_0$ or $\bar{u}(t) = \bar{u}_0$, respectively (the curve is oriented in the sense of the increasing parameter). If the curves of Theorem 2 intersect at right angles, then

$$(E\,\mathrm{d}u + F\,\mathrm{d}v)\,\mathrm{d}\bar{u} + (F\,\mathrm{d}u + G\,\mathrm{d}v)\,\mathrm{d}\bar{v} = 0.$$

Theorem 3. *For the angle made by a parametric v-curve with a parametric u-curve (in this order) the following relations hold:*

$$\cos \varphi = \frac{F}{\sqrt{(EG)}}, \quad \sin \varphi = \frac{D}{\sqrt{(EG)}} \quad (D = \sqrt{(EG - F^2)}) . \tag{4}$$

Theorem 4. *A necessary and sufficient condition that the parametric curves intersect at right angles at every point of a surface, i.e., that the parametric net on a surface be orthogonal, is*

$$F(u, v) \equiv 0 \quad (\textit{identically}) . \tag{5}$$

Example 1. For a *sphere* (Example 9.12.1, p. 352)

$$x = r \sin u \cos v , \quad y = r \sin u \sin v , \quad z = r \cos u$$

we obtain (using (9.12.11))

$$E = r^2 , \quad F = 0 , \quad G = r^2 \sin^2 u .$$

Hence the metric form is

$$ds^2 = r^2(du^2 + \sin^2 u \, dv^2) ,$$

and this relation holds for the element of arc of every curve on the sphere. A curve which makes a constant angle φ $(0 < \varphi < \frac{1}{2}\pi)$ with the meridians of a sphere is called a *loxodrome* on the sphere (Fig. 9.26). Its equation in the coordinates u and v is

$$v + \tan \varphi \ln \tan \frac{u}{2} = c \quad (c \text{ being a real constant}) .$$

The length of its arc s is given by

$$s = r \int \sqrt{\left[1 + \sin^2 u \left(\frac{dv}{du}\right)^2\right]} du .$$

By differentiation of the equation of the loxodrome we obtain its differential equation

$$\frac{dv}{du} = \frac{-\tan \varphi}{\sin u}$$

and for the length of arc s, e.g. between $u = \frac{1}{4}\pi$ and $u = \frac{1}{2}\pi$

$$s = r \int_{\pi/4}^{\pi/2} \sqrt{(1 + \tan^2 \varphi)} \, du = r \sqrt{(1 + \tan^2 \varphi)} \, . \, [u]_{\pi/4}^{\pi/2} = \frac{\pi r}{4} \sqrt{(1 + \tan^2 \varphi)} .$$

Theorem 5. *The area of the parallelogram, two adjacent sides of which are tangent vectors of the surface, is* (Fig. 9.29)

$$\mathrm{d}P = D \begin{vmatrix} \mathrm{d}u, & \mathrm{d}v \\ \mathrm{d}\bar{u}, & \mathrm{d}\bar{v} \end{vmatrix} . \tag{6a}$$

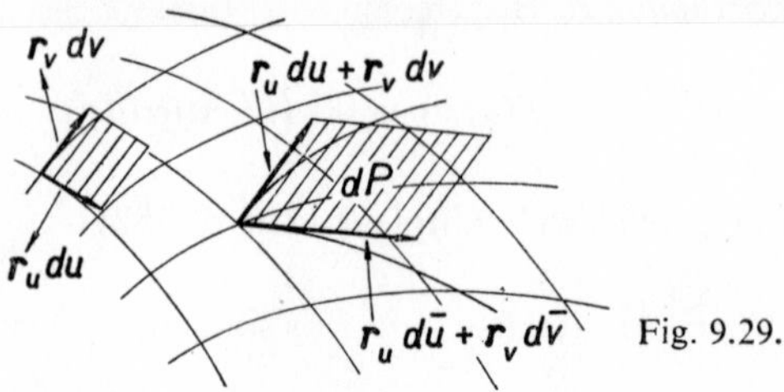

Fig. 9.29.

Definition 2. The expression (6a) for dP is called the *element of area* of the surface $\boldsymbol{r} = \boldsymbol{r}(u, v)$.

REMARK 2. If for the tangent vectors of Theorem 5 we choose the tangent vector of parametric curves (Fig. 9.29), then equation (6a) reduces to

$$\mathrm{d}P = D\,\mathrm{d}u\,\mathrm{d}v\,. \tag{6b}$$

The *area of a region of a surface* $x = x(u, v)$, $y = y(u, v)$, $z = z(u, v)$, over a domain O, may be found by evaluating the double integral

$$P = \iint_O \sqrt{(EG - F^2)}\,\mathrm{d}u\,\mathrm{d}v\,. \tag{7}$$

REMARK 3. For a surface $z = z(x, y)$ we obtain the first fundamental coefficients in the form

$$E = 1 + \left(\frac{\partial z}{\partial x}\right)^2, \quad F = \frac{\partial z}{\partial x}\frac{\partial z}{\partial y}, \quad G = 1 + \left(\frac{\partial z}{\partial y}\right)^2,$$

the *discriminant* $D = \sqrt{[1 + (\partial z/\partial x)^2 + (\partial z/\partial y)^2]}$ and the *element of area*

$$\mathrm{d}P = \sqrt{\left[1 + \left(\frac{\partial z}{\partial x}\right)^2 + \left(\frac{\partial z}{\partial y}\right)^2\right]}\,\mathrm{d}x\,\mathrm{d}y = \sqrt{(1 + p^2 + q^2)}\,\mathrm{d}x\,\mathrm{d}y\,.$$

For the *differential* ds *of the arc* of a curve on the surface $z = f(x, y)$, we obtain

$$\mathrm{d}s = \sqrt{[(1 + p^2)\,\mathrm{d}x^2 + 2pq\,\mathrm{d}x\,\mathrm{d}y + (1 + q^2)\,\mathrm{d}y^2]}\,.$$

9.15. Second Fundamental Form of the Surface, Shape of the Surface with Respect to its Tangent Plane

Theorem 1. *Along a curve* $u = u(s)$, $v = v(s)$ *(s being the arc of the curve) on the surface* $\mathbf{r} = \mathbf{r}(u, v)$ *the following equation holds:*

$$-\mathrm{d}\mathbf{r} \cdot \mathrm{d}\mathbf{n} = L\,\mathrm{d}u^2 + 2M\,\mathrm{d}u\,\mathrm{d}v + N\,\mathrm{d}v^2 \equiv II\,, \tag{1}$$

where

$$L = -\mathbf{r}_u \cdot \mathbf{n}_u\,, \quad 2M = -(\mathbf{r}_u \cdot \mathbf{n}_v + \mathbf{r}_v \cdot \mathbf{n}_u)\,, \quad N = -\mathbf{r}_v \cdot \mathbf{n}_v \tag{2}$$

($\mathbf{n}_u = \partial\mathbf{n}/\partial u$, $\mathbf{n}_v = \partial\mathbf{n}/\partial v$, and $\mathbf{n}$ *is the unit normal vector of the surface*).

Definition 1. The form (1) is called the *second fundamental form of the surface*, while the coefficients L, M, N are called the *second fundamental coefficients of the surface.*

Theorem 2. *The following relations hold:*

$$L = \frac{[\mathbf{r}_{uu}\mathbf{r}_u\mathbf{r}_v]}{\sqrt{(EG - F^2)}}\,, \quad M = \frac{[\mathbf{r}_{uv}\mathbf{r}_u\mathbf{r}_v]}{\sqrt{(EG - F^2)}}\,, \quad N = \frac{[\mathbf{r}_{vv}\mathbf{r}_u\mathbf{r}_v]}{\sqrt{(EG - F^2)}}\,, \tag{3}$$

where

$$\mathbf{r}_{uu} = \frac{\partial^2\mathbf{r}}{\partial u^2}\,, \quad \mathbf{r}_{uv} = \frac{\partial^2\mathbf{r}}{\partial u\,\partial v}\,, \quad \mathbf{r}_{vv} = \frac{\partial^2\mathbf{r}}{\partial v^2}\,.$$

Theorem 3. *For a surface* $z = f(x, y)$ *we have*

$$L = \frac{r}{\sqrt{(1 + p^2 + q^2)}}\,, \quad M = \frac{s}{\sqrt{(1 + p^2 + q^2)}}\,, \quad N = \frac{t}{\sqrt{(1 + p^2 + q^2)}} \tag{4}$$

$$\left(r = \frac{\partial^2 z}{\partial x^2}\,, \quad s = \frac{\partial^2 z}{\partial x\,\partial y}\,, \quad t = \frac{\partial^2 z}{\partial y^2}\,, \quad p = \frac{\partial z}{\partial x}\,, \quad q = \frac{\partial z}{\partial y}\right).$$

Definition 2. A regular point of a surface at which

$$LN - M^2 > 0 \quad \text{or} \quad LN - M^2 = 0 \quad \text{or} \quad LN - M^2 < 0 \tag{5}$$

is called an *elliptic, parabolic* or *hyperbolic point of the surface*, respectively.

Theorem 4. *In a sufficiently small neighbourhood of a regular point P which is an elliptic, or hyperbolic point, the surface lies on one side, or on both sides of the tangent plane* τ *at P, respectively. The tangent plane* τ *at an elliptic, parabolic or hyperbolic point P of a surface cuts the surface in a curve for which the point P is a double point with imaginary conjugate, real coincident, or real distinct tangents, respectively* (Fig. 9.30).

REMARK 1. If a surface is given by the equation $z = f(x, y)$, then at an *elliptic point* $f_{xx}f_{yy} - f_{xy}^2 > 0$, at a *hyperbolic point* $f_{xx}f_{yy} - f_{xy}^2 < 0$, and at a *parabolic point* $f_{xx}f_{yy} - f_{xy}^2 = 0$. Developable surfaces have only parabolic points (with the exception of singular points).

Example 1. Every point of an elliptic paraboloid is elliptic; a hyperbolic paraboloid and a skew helicoid consist exclusively of hyperbolic points.

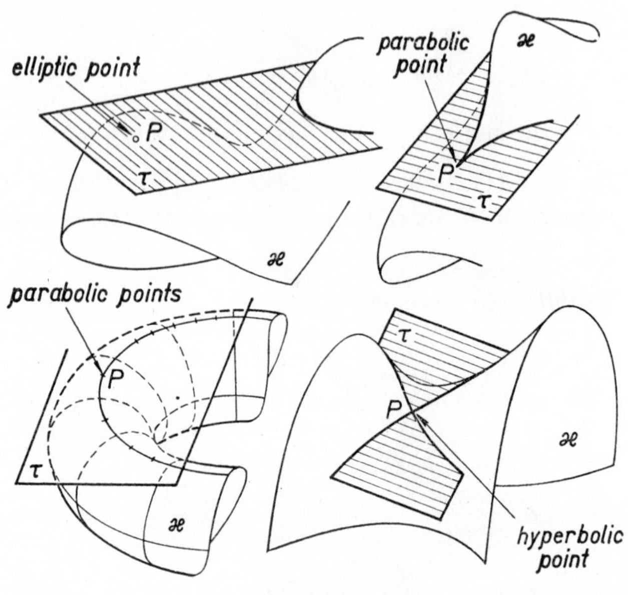

Fig. 9.30.

REMARK 2. The directions dv/du of the tangent vectors at a point of a surface which are tangent vectors of the section of the surface by the tangent plane at this point, are called the *asymptotic directions* and they satisfy the equation

$$L\, du^2 + 2M\, du\, dv + N\, dv^2 = 0\,. \tag{6}$$

If the equation of a surface is of the form $z = f(x, y)$, then equation (6) for the asymptotic directions reduces to

$$f_{xx} + 2f_{xy}k + f_{yy}k^2 = 0 \quad \left(k = \frac{dy}{dx}\right). \tag{7}$$

9.16. Curvature of a Surface

Theorem 1. *All curves on a surface which pass through a regular point P of the surface and have the same osculating plane at P have also the same curvature at P.*

REMARK 1. The radius of curvature of the curve of the section cut by a plane passing through a point P of the surface $\mathbf{r} = \mathbf{r}(u, v)$ is

$$r = \frac{E\,du^2 + 2F\,du\,dv + G\,dv^2}{|L\,du^2 + 2M\,du\,dv + N\,dv^2|}\cos\vartheta \quad (II \neq 0;\ \text{cf. Theorem 9.15.1}) \qquad (1)$$

(ϑ being the angle between the plane of section and the normal to the surface at P; Fig. 9.31). Formula (1) also holds for space curves on the surface (in this case we

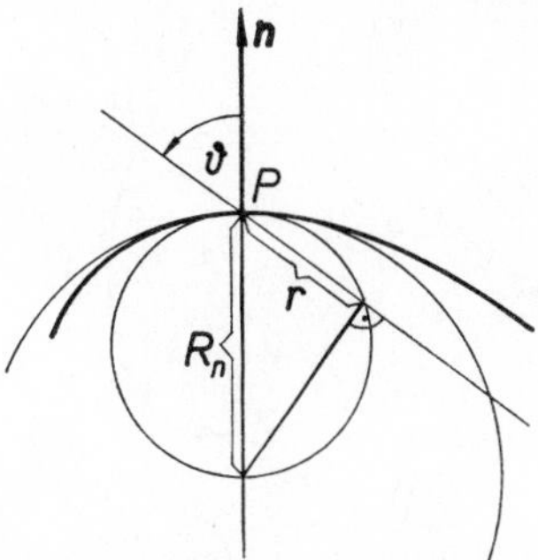

Fig. 9.31.

consider the corresponding osculating plane in place of the plane of section). A curve cut on a surface by a plane that contains a normal to the surface is called a *curve of normal section* (Fig. 9.31).

Theorem 2 (*Theorem of Meusnier*). *A curve of section passing through a regular point P on a surface, has at P a radius of curvature which is the orthogonal projection of the radius of curvature R_n of the curve of normal section (into the osculating plane of the first curve at P), while both curves of section have a common tangent line at P* (Fig. 9.31, schematic), i.e.

$$r = R_n \cos\vartheta\,. \qquad (2)$$

REMARK 2. The circles of curvature of all the curves of section on a surface, through a regular point P and with the same tangent line t at P, lie on a sphere of radius R_n (the radius of curvature of the curve of normal section through the common tangent t) with its centre on the normal to the surface at P. Theorem 2 holds also for space curves on a surface.

REMARK 3. If, for example, we consider a regular point P of a surface of revolution (Fig. 9.32, schematic), then the centre of curvature S of the curve of normal section lying in the plane through the tangent line to the corresponding parallel circle at P is on the axis of the surface.

REMARK 4. The *normal curvature* $1/R$ in a given direction at a point P of a surface $\boldsymbol{r} = \boldsymbol{r}(u, v)$ is

$$\frac{1}{R} = \frac{L\,du^2 + 2M\,du\,dv + N\,dv^2}{E\,du^2 + 2F\,du\,dv + G\,dv^2} = \frac{\varepsilon}{R_n} \quad (\varepsilon = \pm 1)\,; \tag{3a}$$

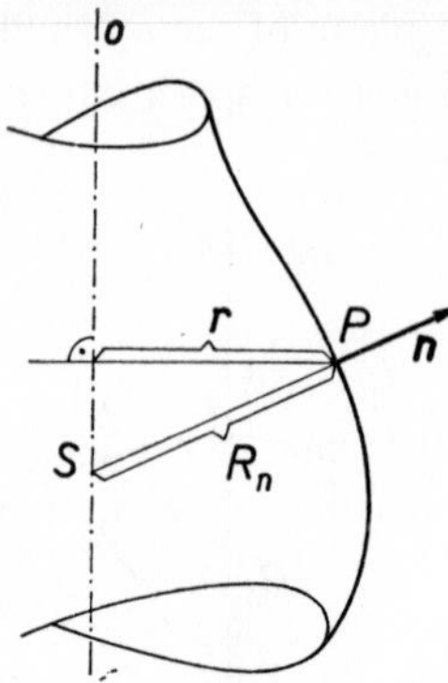

Fig. 9.32.

if a surface is given by an equation $z = f(x, y)$, then the following relation holds for the *radius of curvature of a curve of normal section* at a point of the surface:

$$\varepsilon R_n = \frac{\sqrt{(p^2 + q^2 + 1)}}{f_{xx}\cos^2\alpha + 2f_{xy}\cos\alpha\cos\beta + f_{yy}\cos^2\beta} \tag{3b}$$

(where α, β are the angles which the tangent of the curve of normal section at the point considered makes with the axes x and y).

If a regular point P of a surface $z = f(x, y)$ is the origin of the coordinate system and the normal to the surface at the point P is the z-axis (i.e. the tangent plane of the surface at P is the coordinate plane xy), then the expression for the curvature of a curve of normal section at P is of the form

$$\frac{\varepsilon}{R_n} = (f_{xx})_0\cos^2\varphi + 2(f_{xy})_0\sin\varphi\cos\varphi + (f_{yy})_0\sin^2\varphi \tag{4a}$$

or

$$\frac{\varepsilon}{R_n} = \tfrac{1}{2}(f_{xx} + f_{yy})_0 + \tfrac{1}{2}(f_{xx} - f_{yy})_0\cos 2\varphi + (f_{xy})_0\sin 2\varphi \tag{4b}$$

(φ being the angle between the tangent vector of the curve of normal section at P and the x-axis).

Theorem 3. *Among the curves of normal section at a point P of a surface there exist at least two curves in mutually perpendicular planes such that the normal* *curvature of one curve has a maximum value and that of the other curve a minimum value at P.*

Definition 1. The curves of normal section of a surface for which the corresponding normal curvatures have extreme values are called the *principal curves of normal section*, and their radii of curvature R_1 and R_2 the *principal radii of curvature*, at the point considered on the surface.

Theorem 4 (*Euler's Theorem*). *The curvature* $1/R_n$ *of a curve of normal section at a regular point of a surface is given by the formula*

$$\frac{1}{R_n} = \frac{\cos^2 \delta}{\varepsilon R_1} + \frac{\sin^2 \delta}{\varepsilon R_2} \quad (\varepsilon = \pm 1) \tag{5}$$

(δ *being the angle between the plane of the curve of normal section and the plane of the first principal curve of normal section*).

REMARK 5. Let us introduce in the tangent plane at a point P of a surface, cartesian coordinates such that the x- or y-axis is in the tangent line of the first, or second, principal curve of normal section at P, respectively. If the point P is elliptic, hyperbolic, or parabolic (in this case, let the curvature of the second curve of normal section be equal to zero), let us construct, in the tangent plane at P, the ellipse, two hyperbolas, or two parallel straight lines given by the equations

$$\frac{x^2}{R_1} + \frac{y^2}{R_2} = 1\,, \quad \text{or} \quad \pm\frac{x^2}{R_1} \mp \frac{y^2}{R_2} = 1\,, \quad \text{or} \quad \frac{x^2}{R_1} = 1\,, \quad \text{respectively}, \tag{6}$$

(Fig. 9.33).

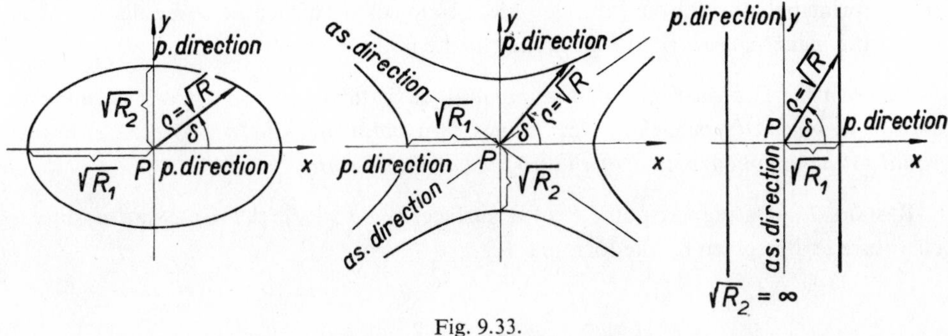

Fig. 9.33.

Then the length of radius vector ϱ of each point of the ellipse, hyperbolas, or the straight lines is the square root of the radius of curvature of the curve of normal section whose plane passes through the radius vector ϱ at the point P. The angle δ between ϱ and the first principal direction is the angle δ from equation (5).

Definition 2. The ellipse, two hyperbolas, or two parallel straight lines (6) in the tangent plane at a point of a surface is called the *indicatrix of Dupin.*

Definition 3. An elliptic point of a surface for which $R_1 = R_2$ is called an *umbilical point,* an *umbilic,* or a *circular point* of the surface.

REMARK 6. The indicatrix of Dupin at an umbilical point of a surface is a circle. Spheres are the only surfaces which have every point an umbilic.

Theorem 5. *The following relations hold:*

$$\pm \frac{1}{R_1} \cdot \frac{1}{R_2} = \frac{LN - M^2}{EG - F^2}, \quad \frac{1}{\varepsilon R_1} + \frac{1}{\varepsilon R_2} = \frac{EN - 2FM + GL}{EG - F^2} \quad (\varepsilon = \pm 1). \tag{7}$$

The principal radii R_1 and R_2 are the roots of the equation

$$(EG - F^2)\frac{1}{R^2} - (EN - 2FM + GL)\frac{1}{R} + (LN - M^2) = 0. \tag{8}$$

Definition 4. The product

$$K = \pm \frac{1}{R_1} \cdot \frac{1}{R_2} = \frac{LN - M^2}{EG - F^2} = \frac{W^2}{D^2} \qquad (LN - M^2 = W^2) \tag{9}$$

of the principal normal curvatures $1/\varepsilon R_1$, $1/\varepsilon R_2$ at a regular point of a surface is called the *total curvature* or the *Gaussian curvature* at the point of the surface.

The average

$$H = \frac{1}{2}\left(\frac{1}{\varepsilon R_1} + \frac{1}{\varepsilon R_2}\right) = \frac{EN - 2FM + GL}{2D^2} \qquad (\varepsilon = \pm 1) \tag{10}$$

of the principal normal curvatures $1/\varepsilon R_1$, $1/\varepsilon R_2$ of a surface at a regular point is called the *mean curvature of the surface* at the point considered.

Theorem 6. *The average of the normal curvatures of two curves of normal section in mutually perpendicular planes at a point of a surface is constant, and equal to the mean curvature of the surface at that point.*

REMARK 7. At a regular point P of a surface $z = f(x, y)$, the Gaussian or mean curvature at P is given by the formula

$$K = \frac{f_{xx}f_{yy} - f_{xy}^2}{(f_x^2 + f_y^2 + 1)^2} \tag{11}$$

or

$$H = \frac{1}{2}\frac{(1 + f_y^2)f_{xx} - 2f_xf_yf_{xy} + (1 + f_x^2)f_{yy}}{(f_x^2 + f_y^2 + 1)^{3/2}}, \tag{12}$$

respectively.

REMARK 8. If P is an elliptic, parabolic or hyperbolic point of the surface, then $K > 0$ (i.e. $LN - M^2 > 0$), or $K = 0$ (i.e. $LN - M^2 = 0$ or at least one of the principal curvatures is zero), or $K < 0$ (i.e. $LN - M^2 < 0$), respectively. For developable surfaces (and for planes) $K = 0$ identically, and conversely. For so-called *minimal surfaces* the equation $H = 0$ holds identically.

9.17. Lines of Curvature

Definition 1. A curve on a surface whose tangent line at every point lies in the principal direction on the surface (Definition 9.16.1) is called a *line of curvature.*

Theorem 1. *The differential equation of the lines of curvature is of the form*

$$(LF - ME)\,du^2 + (LG - NE)\,du\,dv + (MG - NF)\,dv^2 = 0\,. \tag{1}$$

REMARK 1. Through a regular (not umbilical) point of a surface there pass two real lines of curvature intersecting at right angles. The lines of curvature constitute an *orthogonal conjugate net* on a surface.

Theorem 2. *A necessary and sufficient condition for a point of a surface* $\mathbf{r} = \mathbf{r}(u, v)$ *to be an umbilical point is*

$$L = \lambda E\,, \quad M = \lambda F\,, \quad N = \lambda G \quad (\lambda \neq 0)\,. \tag{2a}$$

If a surface is given by an equation $z = f(x, y)$, *then the equations*

$$\frac{1 + f_x^2}{f_{xx}} = \frac{f_x f_y}{f_{xy}} = \frac{1 + f_y^2}{f_{yy}} \tag{2b}$$

express the conditions for an umbilical point of the surface.

REMARK 2. For the radii of curvature of curves of normal section at an umbilical point of a surface $z = f(x, y)$ the following relation holds:

$$R_n = \frac{1 + f_x^2}{f_{xx}} \sqrt{(1 + f_x^2 + f_y^2)}\,. \tag{3}$$

Theorem 3. *A curve on a surface is a line of curvature of the surface if and only if the ruled surface of normals to the surface at points of the curve is a developable surface.*

REMARK 3. Every curve on spheres and planes is a line of curvature.

9.18. Asymptotic Curves

Definition 1. An *asymptotic curve* (or *asymptotic line*) is a curve on the surface, to which the tangent line, at every point of it, has asymptotic direction (Remark 9.15.2).

Theorem 1. *The differential equation of asymptotic curves is of the form*

$$L\,\mathrm{d}u^2 + 2M\,\mathrm{d}u\,\mathrm{d}v + N\,\mathrm{d}v^2 = 0\,. \tag{1}$$

REMARK 1. Only the plane has the property that every curve in it is an asymptotic curve.

Theorem 2. *Real asymptotic curves exist only on that part of a surface where all points are hyperbolic* ($K < 0$) *or parabolic* ($K = 0$). *Through every hyperbolic point of a surface there pass two real distinct asymptotic curves. Through every parabolic point of a surface there passes exactly one real asymptotic curve.*

Theorem 3. *If a curve of section of a surface by its tangent plane has a double point at the corresponding point of contact, then the tangent lines of the curve at the double point are in asymptotic directions.*

Theorem 4. *An asymptotic curve on a surface is a curve such that the osculating plane at every point of the curve coincides with the tangent plane of the surface at the same point.*

REMARK 2. If a surface is given by the equation $z = f(x, y)$, then the equation

$$f_{xx}\,\mathrm{d}x^2 + 2f_{xy}\,\mathrm{d}x\,\mathrm{d}y + f_{yy}\,\mathrm{d}y^2 = 0$$

is the differential equation of the orthogonal projection of asymptotic curves of the surface onto the coordinate plane xy.

The curves of normal section of a surface which touch an asymptotic curve have zero curvature at their common point. An asymptotic curve follows a direction of zero normal curvature on the surface. The tangent lines of asymptotic curves at every point of the surface are identical with the asymptotes of the corresponding indicatrix of Dupin. On a ruled surface one family of asymptotic curves is constituted by the generators of the surface.

9.19. Fundamental Equations of Weingarten, Gauss and Codazzi

The formulae of the following theorems give the relations between the vectors $\boldsymbol{n}_u$, $\boldsymbol{n}_v$ and the vectors $\boldsymbol{r}_u$, $\boldsymbol{r}_v$ ($\partial\boldsymbol{n}/\partial u = \boldsymbol{n}_u$, $\partial\boldsymbol{r}/\partial u = \boldsymbol{r}_u$, etc.) corresponding to the Frenet formulae of the theory of curves.

Theorem 1 (*The Weingarten Equations*). *The following relations exist between the vectors* $\mathbf{n}_u$, $\mathbf{n}_v$ *and* $\mathbf{r}_u$, $\mathbf{r}_v$:

$$\left.\begin{aligned} \mathbf{n}_u &= \frac{FM - GL}{D^2}\mathbf{r}_u + \frac{FL - EM}{D^2}\mathbf{r}_v\,; \quad \mathbf{r}_u = \frac{MF - NE}{W^2}\mathbf{n}_u + \frac{ME - LF}{W^2}\mathbf{n}_v\,; \\ \mathbf{n}_v &= \frac{FN - GM}{D^2}\mathbf{r}_u + \frac{FM - EN}{D^2}\mathbf{r}_v\,; \quad \mathbf{r}_v = \frac{MG - NF}{W^2}\mathbf{n}_u + \frac{MF - LG}{W^2}\mathbf{n}_v \end{aligned}\right\} \quad (1)$$

(*where* $W^2 = LN - M^2$).

Theorem 2 (*The Gauss Equations*). *The following relations exist between the vectors* $\mathbf{r}_{uu}$, $\mathbf{r}_{uv}$, $\mathbf{r}_{vv}$ *and* $\mathbf{r}_u$, $\mathbf{r}_v$, $\mathbf{n}$ (*where* $\partial^2\mathbf{r}/\partial u\,\partial v = \mathbf{r}_{uv}$, *etc.*):

$$\left.\begin{aligned} \mathbf{r}_{uu} &= \frac{GE_u - 2FF_u + FE_v}{2D^2}\mathbf{r}_u + \frac{-FE_u + 2EF_u - EE_v}{2D^2}\mathbf{r}_v + L\mathbf{n}\,, \\ \mathbf{r}_{uv} &= \frac{GE_v - FG_u}{2D^2}\mathbf{r}_u + \frac{EG_u - FE_v}{2D^2}\mathbf{r}_v + M\mathbf{n}\,, \\ \mathbf{r}_{vv} &= \frac{-FG_v + 2GF_v - GG_u}{2D^2}\mathbf{r}_u + \frac{EG_v - 2FF_v + FG_u}{2D^2}\mathbf{r}_v + N\mathbf{n} \end{aligned}\right\} \quad (2)$$

(*where* $E_u = \partial E/\partial u$, $F_v = \partial F/\partial v$, *etc.*).

The following relations express the mutual dependence of the six functions E, F, G; L, M, N which correspond to the same surface:

Theorem 3 (*The Codazzi or Mainardi Equations*):

$$\left.\begin{aligned} &(EG - 2FF + GE)(L_v - M_u) - (EN - 2FM + GL)(E_v - F_u) + \\ &\qquad + \begin{vmatrix} E, & E_u, & L \\ F, & F_u, & M \\ G, & G_u, & N \end{vmatrix} = 0\,, \\ &(EG - 2FF + GE)(M_v - N_u) - (EN - 2FM + GL)(F_v - G_u) + \\ &\qquad + \begin{vmatrix} E, & E_v, & L \\ F, & F_v, & M \\ G, & G_v, & N \end{vmatrix} = 0 \end{aligned}\right\} \quad (3)$$

(*where* $L_v = \partial L/\partial v$, $M_u = \partial M/\partial u$, *etc.*).

Theorem 4 (*The Gauss Theorem Egregium*).

$$K = -\frac{1}{4D^4}\begin{vmatrix} E, & E_u, & E_v \\ F, & F_u, & F_v \\ G, & G_u, & G_v \end{vmatrix} - \frac{1}{2D}\left(\frac{\partial}{\partial v}\frac{E_v - F_u}{D} - \frac{\partial}{\partial u}\frac{F_v - G_u}{D}\right). \quad (4)$$

9.20. Geodesic Curvature, Geodesic Curves and Gradient Curves on a Surface

Definition 1. The *geodesic curvature* k_g of a curve on a surface at a regular point P is defined by the relation

$$k_g = k_1 \sin \vartheta \tag{1}$$

(ϑ being the angle between the principal normal to the curve and the normal to the surface at P and k_1 the curvature of this curve at P).

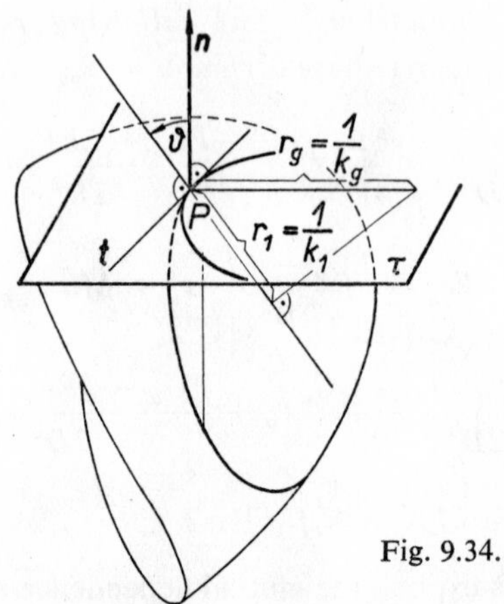

Fig. 9.34.

Theorem 1. *The geodesic curvature of a curve on a surface at a point P is equal to the curvature of the orthogonal projection of the curve onto the tangent plane of the surface at P* (Fig. 9.34).

If a curve $\boldsymbol{r}(s) = \boldsymbol{r}(u(s), v(s))$ on a surface $\boldsymbol{r} = \boldsymbol{r}(u, v)$ is given, then for its geodesic curvature k_g the following relation holds:

Theorem 2.

$$k_g = \frac{1}{r_g} = [\boldsymbol{r}_s \boldsymbol{r}_{ss} \boldsymbol{n}]\,. \tag{2}$$

REMARK 1. If a curve depends on a parameter t different from the arc s, then

$$k_g = \frac{1}{\mathrm{d}s^3} [\mathrm{d}\boldsymbol{r}, \mathrm{d}^2\boldsymbol{r}, \boldsymbol{n}] \tag{3}$$

holds instead of (2).

Definition 2. A curve on a surface is called a *geodesic curve*, or simply a *geodesic* if at every point of the curve its osculating plane contains the corresponding normal to the surface.

Theorem 3. *A necessary and sufficient condition that a curve on a surface is a geodesic is that the geodesic curvature of this curve at every point is zero,* i.e. $k_g = 0$ *at every point of the curve.*

REMARK 2. The geodesics on a surface also satisfy the condition (the differential equation)

$$[\boldsymbol{r}_s \boldsymbol{r}_{ss} \boldsymbol{n}] = 0 \quad \text{or} \quad [\mathrm{d}\boldsymbol{r}, \mathrm{d}^2\boldsymbol{r}, \boldsymbol{n}] = 0 . \tag{4}$$

A straight line on a surface is a geodesic on this surface.

Theorem 4. *On the assumption that the curvilinear coordinates on a surface constitute an orthogonal net, the differential equation of geodesics on the surface can be put in the form*

$$\sqrt{(EG)}\frac{\mathrm{d}\varphi}{\mathrm{d}s} + \tfrac{1}{2}G_u\frac{\mathrm{d}v}{\mathrm{d}u} - \tfrac{1}{2}E_v = 0 \tag{5}$$

($\mathrm{d}s^2 = E\,\mathrm{d}u^2 + G\,\mathrm{d}v^2$,
$\mathrm{d}\varphi = [(EG\,\mathrm{d}u\,\mathrm{d}v^2 - \tfrac{1}{2}G\,\mathrm{d}E\,\mathrm{d}u\,\mathrm{d}v + \tfrac{1}{2}E\,\mathrm{d}G\,\mathrm{d}u\,\mathrm{d}v)/\sqrt{(EG)}]\,\mathrm{d}s^2$, φ *being the angle which the geodesic makes with the parametric v-curve at a point of the surface*).

Theorem 5. *From among all arcs on a given (properly chosen) part of a surface joining two definite points of the surface the shortest arc is on a geodesic.*

REMARK 3. All straight lines on a plane and on a ruled surface are geodesics. The helix is a geodesic on a cylinder, etc. A geodesic on a surface is determined by a point of the surface and its tangent line at that point. Through every regular point of a surface there passes a one-parameter family of geodesics. When a developable surface is developed upon a plane, the arc of every geodesic on the surface becomes a line segment in the plane.

Definition 3. The *gradient curves* (the *curves of greatest slope*) on a surface are the orthogonal trajectories of the level lines of the surface with respect to a given plane.

Theorem 6. *The equation*

$$\frac{\mathrm{d}y}{\mathrm{d}x} = \frac{f_y}{f_x} \tag{6}$$

is the differential equation of the orthogonal projection of the gradient curves of a surface $z = f(x, y)$ *with respect to the coordinate plane xy, onto this plane.*

10. SEQUENCES AND SERIES OF CONSTANT TERMS. INFINITE PRODUCTS

By KAREL REKTORYS

References: [47], [48], [49], [50], [73], [109], [112], [138], [172] [173], [198], [226], [229], [233], [305], [308], [320], [341], [362], [367], [376], [387]

For sequences and series of variable terms see Chapters 15 and 16.

10.1. Sequences of Constant Terms

Definition 1. If to every natural number n we assign a number a_n (which may be real or complex) and order the numbers a_n according to their increasing suffixes, then we say that we have formed a *sequence*. We denote it by $a_1, a_2, a_3, \ldots, a_n, \ldots$ or briefly by $\{a_n\}$.

Example 1. The relation $a_n = (n - 1)/n$ defines the sequence

$$0, \tfrac{1}{2}, \tfrac{2}{3}, \tfrac{3}{4}, \ldots .$$

Definition 2. We say that a sequence $\{a_n\}$ possesses a (*finite*) *limit* a (in other words: *tends* to a number a, *converges* with limit a, is said to converge to a limit a), if, to any arbitrary number $\varepsilon > 0$, there corresponds a number n_0 (depending in

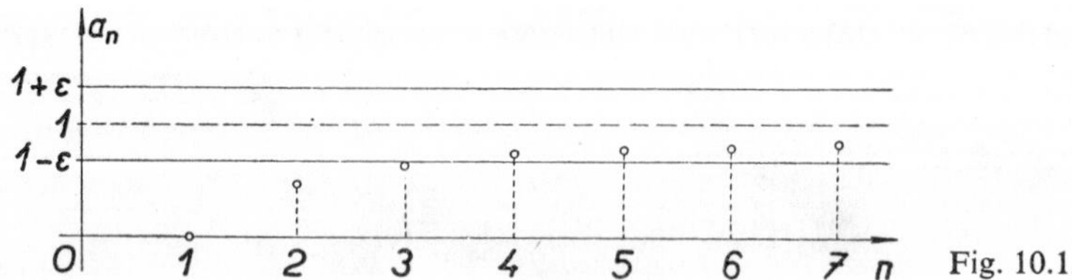

Fig. 10.1.

general on the choice of the number ε), such that $|a_n - a| < \varepsilon$ holds for every $n > n_0$ (Fig 10.1). We then write

$$\lim_{n \to \infty} a_n = a .$$

(When dealing with sequences, it is customary to write simply $n \to \infty$ rather than $n \to +\infty$.)

REMARK 1. Roughly speaking: A sequence $\{a_n\}$ possesses a limit a if the number a_n "approaches closer and closer" to the number a as the suffix n increases.

Definition 3. When a sequence possesses a (finite) limit, it is said to be *convergent*. Otherwise it is said to be *divergent*.

Theorem 1. *A sequence $\{a_n\}$ can have at most one limit.*

Definition 4. We say that $\{a_n\}$ *diverges* to $+\infty$ (has an *infinite limit* $+\infty$, is definitely *divergent* with the limit $+\infty$), if, for any arbitrary number K, there exists a number n_0 (depending on the choice of the number K), such that $a_n > K$ for every $n > n_0$. We then write

$$\lim_{n\to\infty} a_n = +\infty .$$

A corresponding meaning holds if we write

$$\lim_{n\to\infty} a_n = -\infty .$$

Example 2. The sequence given in Example 1 is convergent and the number 1 is its limit. For any given $\varepsilon > 0$, it is sufficient to choose as n_0 any number greater than $1/\varepsilon$ (Fig. 10.1). For, if $n_0 > 1/\varepsilon$ and $n > n_0$, then

$$|a_n - 1| = \left|\frac{n-1}{n} - 1\right| = \frac{1}{n} < \frac{1}{n_0} < \varepsilon .$$

Example 3. The sequence defined by the relation $a_n = 2^n$ (i.e. the sequence 2, 4, 8, ...) diverges to $+\infty$, i.e.

$$\lim_{n\to\infty} 2^n = +\infty .$$

Example 4. The sequence 0, 1, 0, 1, 0, 1, ... is divergent (its terms do not tend to any (single) number). As a rule, sequences of this type are said to *oscillate*.

Theorem 2. *A sequence $\{a_n\}$ is convergent if and only if it fulfils the following (Bolzano-Cauchy) condition: Given any (arbitrarily small) number $\varepsilon > 0$, there exists a number n_0 (depending on the choice of the number ε) such that the relation $|a_m - a_n| < \varepsilon$ is valid for every pair of numbers m, n, such that $m > n_0$, $n > n_0$.*

Theorem 3. *If $\{a_n\}$, $\{b_n\}$ are convergent sequences such that*

$$\lim_{n\to\infty} a_n = a , \quad \lim_{n\to\infty} b_n = b ,$$

then

$$\lim_{n\to\infty} (a_n \pm b_n) = a \pm b , \quad \lim_{n\to\infty} ka_n = ka \quad (k \text{ a constant}) , \quad \lim_{n\to\infty} |a_n| = |a| ,$$

$$\lim_{n\to\infty} a_n b_n = ab , \quad \lim_{n\to\infty} \frac{a_n}{b_n} = \frac{a}{b} \quad \textit{for} \quad b \neq 0 .$$

(*Thus each of the sequences mentioned above is convergent. In the last case, we omit from the sequence* $\{a_n/b_n\}$ *the terms with suffixes for which* $b_n = 0$ *and of which there are a finite number, because* $b \neq 0$.)

Example 5. Find

$$\lim_{n\to\infty} \frac{2n^2 - n + 1}{7n^2 - 5n + 2}.$$

Because the sequences $\{a_n\}$ (in the numerator) and $\{b_n\}$ (in the denominator) are divergent, we cannot apply Theorem 3 directly. However, for every n

$$\frac{2n^2 - n + 1}{7n^2 - 5n + 2} = \frac{2 - \frac{1}{n} + \frac{1}{n^2}}{7 - \frac{5}{n} + \frac{2}{n^2}},$$

so that

$$\lim_{n\to\infty} \frac{2n^2 - n + 1}{7n^2 - 5n + 2} = \lim_{n\to\infty} \frac{2 - \frac{1}{n} + \frac{1}{n^2}}{7 - \frac{5}{n} + \frac{2}{n^2}} = \frac{2}{7},$$

because

$$\lim_{n\to\infty} \frac{1}{n} = 0 \quad \text{and} \quad \lim_{n\to\infty} \frac{1}{n^2} = 0.$$

REMARK 2. Theorem 3 can be generalized to cases where the sequences $\{a_n\}$, $\{b_n\}$ diverge to $+\infty$ or $-\infty$. If, for example,

$$\lim_{n\to\infty} a_n = +\infty \quad \text{and} \quad \lim_{n\to\infty} b_n = -\infty$$

then

$$\lim_{n\to\infty} (a_n - b_n) = +\infty \quad \text{and} \quad \lim_{n\to\infty} a_n b_n = -\infty.$$

But we cannot apply the theorem directly when the computation leads to so-called indeterminate expressions of the type $\infty - \infty$, $0 . \infty$ or ∞/∞.

Theorem 4. *Let*

$$\lim_{n\to\infty} a_n = a, \quad \lim_{n\to\infty} b_n = a$$

and let $\{c_n\}$ *be a sequence such that* $a_n \leqq c_n \leqq b_n$ *holds for every* n. *Then*

$$\lim_{n\to\infty} c_n = a.$$

REMARK 3. In other words: If a sequence $\{c_n\}$ lies between two convergent sequences $\{a_n\}$ and $\{b_n\}$ which have the same limit, then $\{c_n\}$ is also convergent and has this same limit.

Theorem 5. *Let $\{a_n\}$ be a sequence of complex numbers and for every n let $a_n = \alpha_n + i\beta_n$, where α_n, β_n are real. Then $\{a_n\}$ is convergent if and only if both sequences $\{\alpha_n\}$ and $\{\beta_n\}$ are convergent. Moreover, the relation*

$$\lim_{n\to\infty} a_n = \lim_{n\to\infty} \alpha_n + i \lim_{n\to\infty} \beta_n$$

holds.

Theorem 6. *Let $\lim_{n\to\infty} a_n = a$, $\lim_{n\to\infty} b_n = b$. If, for every n, $a_n \leqq b_n$, then $a \leqq b$.*

REMARK 4. If $a_n < b_n$ for every n, then we cannot conclude that $a < b$, but only that $a \leqq b$. For example:

$$\{a_n\} = \left\{\frac{1}{2n}\right\}, \quad \{b_n\} = \left\{\frac{1}{n}\right\}.$$

Then $a_n < b_n$ for every n, but $a = b = 0$.

Definition 5. Let us consider the sequence $\{a_n\}$. A sequence $\{a_{k_n}\}$, where k_n are positive integers, such that $k_1 < k_2 < k_3 < \ldots$, is called a *sub-sequence of the sequence* $\{a_n\}$.

Example 6. If, for the sequence 0, 1, 0, 1, 0, 1, …, we choose $k_1 = 2$, $k_2 = 4$, $k_3 = 6$, …, we obtain the sub-sequence 1, 1, 1, ….

Example 7. Let us form a sub-sequence of the sequence $\{a_n\} = \{(n-1)/n\}$ in such a way that we take only every third term, i.e. $k_1 = 3$, $k_2 = 6$, $k_3 = 9$, … Then the resulting sequence is

$$\tfrac{2}{3}, \tfrac{5}{6}, \tfrac{8}{9}, \ldots .$$

Theorem 7. *If* $\lim_{n\to\infty} a_n = a$, *then* $\lim_{n\to\infty} a_{k_n} = a$ *also.*

REMARK 5. The converse of the Theorem is not true. If a sub-sequence is convergent then the original sequence need not be convergent. For example, the sequence 0, 1, 0, 1, 0, 1, … from Example 6 is divergent but its sub-sequence 1, 1, 1, … is convergent.

Definition 6. A sequence $\{a_n\}$ is said to be *bounded above*, or *bounded below*, or *bounded* if there exists a finite number K_1, or K_2, or M, respectively, such that

$$a_n < K_1\,, \quad \text{or} \quad a_n > K_2\,, \quad \text{or} \quad |a_n| < M\,, \quad \text{respectively}$$

for every n.

Definition 7. A number d is called a *point of accumulation* or *limiting point* of a sequence $\{a_n\}$ if an infinite number of terms of the given sequence lie in every arbitrarily small ε-neighbourhood of the point d (i.e. in the interval $(d - \varepsilon, d + \varepsilon)$). (The term cluster point is also used.)

Theorem 8 (*The Bolzano-Weierstrass Theorem*). *Every bounded sequence $\{a_n\}$ possesses at least one limiting point. There always exist (even when there are infinitely many limiting points) a greatest limiting point and a least limiting point; we denote them by:*

$$\limsup_{n\to\infty} a_n \quad or \quad \overline{\lim_{n\to\infty}}\, a_n\,, \quad \liminf_{n\to\infty} a_n \quad or \quad \underline{\lim_{n\to\infty}}\, a_n\,,$$

respectively.

REMARK 6 The expression

$$\overline{\lim_{n\to\infty}}\, a_n = +\infty\,,$$

which is occasionally to be found in the literature, stands for the assertion "However large a number K be chosen, the sequence $\{a_n\}$ always possesses infinitely many terms such that $a_n > K$." The expression

$$\underline{\lim_{n\to\infty}}\, a_n = -\infty$$

has a corresponding meaning. For example, for the sequence $a_n = (-n)^n$ the relations

$$\overline{\lim_{n\to\infty}}\, a_n = +\infty\,, \quad \underline{\lim_{n\to\infty}}\, a_n = -\infty$$

hold.

Theorem 9. *The sequence $\{a_n\}$ is convergent if, and only if, the numbers $\overline{\lim}_{n\to\infty} a_n$ and $\underline{\lim}_{n\to\infty} a_n$ are finite and*

$$\overline{\lim_{n\to\infty}}\, a_n = \underline{\lim_{n\to\infty}}\, a_n\,,$$

i.e. if the sequence $\{a_n\}$ possesses one and only one limiting point (which is not infinite).

Example 8. The sequence 0, 1, 0, 1, 0, 1, ... in Example 4 possesses two limiting points: the point 0 and the point 1. Thus

$$\overline{\lim_{n\to\infty}}\, a_n = 1\,, \quad \underline{\lim_{n\to\infty}}\, a_n = 0\,.$$

Then the fact that this sequence is not convergent follows from Theorem 9.

Definition 8. A sequence is said to be

$$\left.\begin{array}{l} \textit{strictly increasing} \\ \textit{strictly decreasing} \\ \textit{decreasing} \\ \textit{increasing} \end{array}\right\} \text{ if, for every } n, \quad \begin{array}{l} a_{n+1} > a_n \\ a_{n+1} < a_n \\ a_{n+1} \leqq a_n \\ a_{n+1} \geqq a_n . \end{array}$$

All such sequences will be referred to as *monotonic sequences*, the first two as *strictly monotonic sequences*.

Theorem 10. *Every increasing (and thus every strictly increasing) sequence which is bounded above possesses a limit equal to the l. u. b. of all the values a_n). Similarly every decreasing (and thus every strictly decreasing) sequence which is bounded below is convergent (and its limit is equal to the g. l. b. of all the values a_n).*

Theorem 11. *The sequences*

$$a_n = \left(1 + \frac{1}{n}\right)^n, \quad b_n = \left(1 + \frac{1}{n}\right)^{n+1}$$

(the former being strictly increasing, and the latter strictly decreasing) possess the same limit, the number e *(so-called) which is the base of natural logarithms. Thus*

$$\mathrm{e} = \lim_{n\to\infty}\left(1 + \frac{1}{n}\right)^n = \lim_{n\to\infty}\left(1 + \frac{1}{n}\right)^{n+1} = 2{\cdot}718,281,828,459,0\ldots.$$

Theorem 12. *If $a > 0$, $\lim_{n\to\infty} b_n = b$ (b_n need not be rational*; see § 1.9), *then*

$$\lim_{n\to\infty} a^{b_n} = a^{\lim_{n\to\infty} b_n} = a^b .$$

Example 9.

$$\lim_{n\to\infty} \sqrt[n]{a} = 1 \ (a > 0), \text{ because } \sqrt[n]{a} = a^{1/n} \text{ and } \lim_{n\to\infty} \frac{1}{n} = 0 .$$

Theorem 13. *The sequence*

$$a_1, \ a_1 + d, \ \ldots, \ a_1 + nd, \ \ldots,$$

called an arithmetic progression, is divergent for every $d \neq 0$.

The sequence

$$a_1, a_1q, a_1q^2, \ldots, a_1q^n, \ldots,$$

called a geometric progression, has limit 0 for $|q| < 1$ and limit a_1 for $q = +1$; if $a_1 \neq 0$, then for any other value of q this sequence is divergent.

REMARK 7. It is often convenient to replace the calculation of the limit of a sequence by the calculation of the limit of a suitable function as $x \to +\infty$. If $f(x)$ is a function such that $f(n) = a_n$ for every positive integer n, then from the existence of the limit $\lim_{x \to +\infty} f(x) = A$, it follows that $\lim_{n \to \infty} a_n = A$.

Example 10. To calculate

$$\lim_{n \to \infty} \left(1 + \frac{a}{n}\right)^n$$

we make use of the function

$$f(x) = \left(1 + \frac{a}{x}\right)^x.$$

We have

$$\left(1 + \frac{a}{x}\right)^x = \left[e^{\ln(1+a/x)}\right]^x = e^{x \ln(1+a/x)},$$

while

$$\lim_{x \to +\infty} \left\{x \ln\left(1 + \frac{a}{x}\right)\right\} = \lim_{x \to +\infty} \frac{\ln(1 + a/x)}{1/x} =$$

$$= \lim_{z \to 0+} \frac{\ln(1 + az)}{z} = \lim_{z \to 0+} \frac{a}{1 + az} = a$$

(according to l'Hôpital's Rule, Theorem 11.8.1). Because the exponential function is continuous everywhere we have

$$\lim_{x \to \infty} e^{x \ln(1+a/x)} = e^{\lim\limits_{x \to +\infty} x \ln(1+a/x)} = e^a .$$

Hence

$$\lim_{n \to \infty} \left(1 + \frac{a}{n}\right)^n = e^a .$$

Theorem 14. *Survey of Important Formulae and Limits*

1. $\lim\limits_{n \to \infty} (a_n \pm b_n) = \lim\limits_{n \to \infty} a_n \pm \lim\limits_{n \to \infty} b_n$; $\lim\limits_{n \to \infty} ka_n = k \lim\limits_{n \to \infty} a_n$ (k a constant);

   $\lim\limits_{n \to \infty} |a_n| = \left|\lim\limits_{n \to \infty} a_n\right|$; $\lim\limits_{n \to \infty} a_n b_n = \lim\limits_{n \to \infty} a_n \,.\, \lim\limits_{n \to \infty} b_n$;

   $$\lim_{n \to \infty} \frac{a_n}{b_n} = \frac{\lim\limits_{n \to \infty} a_n}{\lim\limits_{n \to \infty} b_n} \quad (\lim_{n \to \infty} b_n \neq 0; \text{ cf. Theorem 3}).$$

2. $a_n \leqq b_n \Rightarrow \lim\limits_{n \to \infty} a_n \leqq \lim\limits_{n \to \infty} b_n$ (Theorem 6).

3. $\lim_{n\to\infty} a_{k_n} = \lim_{n\to\infty} a_n$ (Theorem 7).

4. $\lim_{n\to\infty}\left(1+\frac{a}{n}\right)^n = \mathrm{e}^a$ *for every a; in particular* $\lim_{n\to\infty}\left(1+\frac{1}{n}\right)^n = \mathrm{e}$.

5. $\lim_{n\to\infty} a^n = 0$ *for* $|a| < 1$.

6. $\lim_{n\to\infty}\left(1+\frac{1}{2}+\frac{1}{3}+\ldots+\frac{1}{n}-\ln n\right) = C = 0{\cdot}577,215,664,9\ldots$ (*Euler's Constant*).

7. $\lim_{n\to\infty}\sqrt[n]{n!} = +\infty$; $\lim_{n\to\infty}\sqrt[n]{\frac{1}{n!}} = 0$; $\lim_{n\to\infty}\frac{\sqrt[n]{n!}}{n} = \frac{1}{\mathrm{e}}$; $\lim_{n\to\infty}\frac{n!}{n^n} = 0$.

8. $\lim_{n\to\infty}\frac{n!}{n^n\mathrm{e}^{-n}\sqrt{n}} = \sqrt{(2\pi)}$ (*Stirling's Formula*).

9. $\lim_{n\to\infty}\left[\frac{2.4.6.\ldots.2n}{1.3.5.\ldots.(2n-1)}\right]^2.\frac{1}{2n} = \frac{\pi}{2}$ (*Wallis's Product*).

10. $\lim_{n\to\infty}\frac{1^k+2^k+\ldots+n^k}{n^{k+1}} = \frac{1}{k+1}$ (*k a positive integer*).

11. $\lim_{n\to\infty}\frac{1^2+3^2+5^2+\ldots+(2n-1)^2}{n^3} = \frac{4}{3}$.

12. *If* $a_n > 0$ *and* $\lim_{n\to\infty}\frac{a_{n+1}}{a_n} = a$ *then* $\lim_{n\to\infty}\sqrt[n]{a_n} = a$ *also*.

13. *If* $\lim_{n\to\infty} a_n = a$ *then* $\lim_{n\to\infty}\frac{a_1+a_2+\ldots+a_n}{n} = a$ *also*.

10.2. Infinite Series (of Constant Terms)

Definition 1. Let the sequence $\{a_n\}$ be given. The symbolical expression

$$a_1 + a_2 + a_3 + \ldots = \sum_{n=1}^{\infty} a_n \tag{1}$$

is called the (*infinite*) *series* corresponding to the given sequence.

The sum

$$s_n = a_1 + a_2 + \ldots + a_n$$

is called a *partial sum of the series* (1).

Definition 2. If the sequence of partial sums $s_1, s_2, s_3, \ldots$ is convergent (see definition 10.1.3) and possesses a (finite) limit s, we say that the series is *convergent* and *has the sum s*. If the sequence $\{s_n\}$ is divergent then we say that the series (1) is *divergent*.

Example 1. For the so-called *geometric* series $1 + q + q^2 + \ldots$ we have

$$s_n = \frac{1 - q^n}{1 - q} \quad (q \neq 1) .$$

If $|q| < 1$ then

$$s = \lim_{n\to\infty} s_n = \frac{1}{1 - q},$$

so the series is convergent. If $q = 1$ then $s_n = n$, so the sequence of partial sums diverges to $+\infty$ (see Definition 10.1.4). (In this case we say that the *series has the sum* $+\infty$.) For $q = -1$, $s_1 = 1$, $s_2 = 0$, $s_3 = 1$, $s_4 = 0, \ldots$ and, the sequence s_n having no limit, the series is divergent (we speak of an *oscillating series*).

Example 2. The *arithmetic* series

$$a_1 + (a_1 + d) + (a_1 + 2d) + \ldots$$

is divergent when at least one of the numbers a_1, d is not zero. For $d > 0$ its sum is $+\infty$, for $d < 0$ its sum is $-\infty$.

Example 3. The series

$$1 + \tfrac{1}{2} + \tfrac{1}{3} + \tfrac{1}{4} + \ldots$$

(the so-called *harmonic series*) is divergent (its sum is $+\infty$, see Examples 6 and 7.)

REMARK 1. The problem of deciding whether a given series is convergent or not is a very important one. When we know that a series is convergent we may (according to the definition) determine an approximation to its sum to any desired degree of accuracy by considering the finite sum of a sufficiently large number of its terms. It is for this reason that in the following text we are so concerned with tests for the convergence of infinite series. To determine exactly the sum of a series is generally a difficult problem, and theorems on the differentiation and integration of power series (§ 15.4) and on double series (see Remark 14) etc. provide an effective help.

Theorem 1. *For the series* $a_1 + a_2 + a_3 + \ldots$ *to be convergent it is necessary that*

$$\lim_{n\to\infty} a_n = 0 .$$

REMARK 2. This condition is not, however, sufficient for the convergence of a given series as we can see in the case of the harmonic series in Example 3.

Theorem 2 (*The Bolzano-Cauchy Condition*). *A necessary and sufficient condition for the convergence of the series $a_1 + a_2 + a_3 + \ldots$ is that, for any given $\varepsilon > 0$, there exists a number n_0 such that for every $n > n_0$ and every positive integer p the relation*

$$|a_n + a_{n+1} + a_{n+2} + \ldots + a_{n+p}| < \varepsilon$$

holds.

Theorem 3. *If the series $a_1 + a_2 + a_3 + \ldots$ and $b_1 + b_2 + b_3 + \ldots$ are convergent with respective sums s and t, then the series whose n-th terms are respectively $a_n \pm b_n$ and ka_n (k constant), are also convergent and*

$$\sum_{n=1}^{\infty} (a_n \pm b_n) = s \pm t\,, \quad \sum_{n=1}^{\infty} ka_n = ks\,.$$

REMARK 3. For the multiplication of infinite series see Theorems 21, 22 and 24.

REMARK 4. If the series $\sum_{n=1}^{\infty} (a_n + b_n)$ is convergent then neither of the series $\sum_{n=1}^{\infty} a_n$, $\sum_{n=1}^{\infty} b_n$ need be convergent. For example, the series $(1 - 1) + (1 - 1) + (1 - 1) + \ldots$ is convergent as each of its terms is zero but neither the series $1 + 1 + 1 + \ldots$ nor the series $-1 - 1 - 1 - \ldots$ is convergent. The series without parentheses is also divergent.

Theorem 4. *If the series*

$$|a_1| + |a_2| + |a_3| + \ldots \tag{2}$$

converges, then the series

$$a_1 + a_2 + a_3 + \ldots \tag{3}$$

also converges.

Definition 3. If the series (2) converges then the series (3) is said to be *absolutely convergent*. If the series (3) converges but the series (2) does not, then the series (3) is said to be *conditionally convergent*.

Example 4. The series whose n-th term is

$$a_n = \frac{(-1)^{n+1}}{n} \tag{4}$$

(that is the series

$$1 - \tfrac{1}{2} + \tfrac{1}{3} - \tfrac{1}{4} + \ldots),$$

converges (see Example 9), but the corresponding series of absolute values (2), that is

$$1 + \tfrac{1}{2} + \tfrac{1}{3} + \ldots$$

is the harmonic series (see Example 3) which is divergent. Hence the series

$$1 - \tfrac{1}{2} + \tfrac{1}{3} - \tfrac{1}{4} + \dots \tag{5}$$

is conditionally convergent.

Theorem 5. *If* (1) *is absolutely convergent then every series that arises from the series* (1) *by a rearrangement of its terms* (*i.e. by a change of the order of its terms*) *is also absolutely convergent and possesses the same sum.*

Theorem 6. *If* (1) *is conditionally convergent then we can obtain from* (1), *by a suitable rearrangement of its terms, a series converging to any given value, or even one which diverges.*

Theorem 7. *If we remove from an infinite series a finite number of its terms we change in general the sum of the series, but alter nothing regarding its convergence or divergence.*

REMARK 5. If we omit in a given series the terms with value zero (finite or infinite in number) then we change neither the convergence nor the sum of the series. *We can thus suppose in the following exposition that the series considered have no zero terms*, so that instead of series with non-negative terms we need consider only *series with positive terms*. This is important in particular in the case of d'Alembert's Ratio Test (Theorem 13) where the term a_n occurs as a divisor.

Theorem 8. *A series of positive terms converges if and only if the sequence* s_n *of its partial sums* (see Definition 1) *is bounded above.*

REMARK 6. When we say that the terms a_n of a given series have some property for *almost every n*, then we mean that they have that property for all n with the exception of a finite number of terms at most. For example, if the relation $a_n < a_{n+1}$ holds for almost every n, it means that this inequality fails for a finite number of indices n at most.

Theorem 9 (*Comparison Test*). *Let*

$$a_1 + a_2 + a_3 + \dots, \tag{6}$$

$$b_1 + b_2 + b_3 + \dots \tag{7}$$

be two series of positive (Remark 5) *terms and suppose* $a_n \leqq b_n$ *for almost every* n (Remark 6). *Then, from the convergence of the series* (7) *follows the convergence of the series* (6), *while from the divergence of the series* (6) *follows the divergence of the series* (7).

(A series such as (7) is called a *majorant* of the series (6).)

Theorem 10. *Let* (6), (7) *be two series of positive terms and* $K \neq 0$ *be a finite number. If the limit*

$$\lim_{n\to\infty} \frac{a_n}{b_n} = K$$

exists then both the series (6), (7) *are simultaneously either convergent or divergent.*

Theorem 11. *If the series of positive terms* $a_1 + a_2 + a_3 + \ldots$ *is convergent and* $c_1, c_2, c_3, \ldots$ *are positive numbers, for which* $c_n < A$ *for almost every* n (*A being a constant*) *then the series*

$$c_1a_1 + c_2a_2 + c_3a_3 + \ldots$$

is also convergent.

Theorem 12. *If* $a_1 + a_2 + a_3 + \ldots$, $b_1 + b_2 + b_3 + \ldots$ *are two series of positive terms, the latter being convergent, and if*

$$\frac{a_{n+1}}{a_n} \leq \frac{b_{n+1}}{b_n}$$

for almost every n (Remark 6), *then the former of these series is also convergent.*

Theorem 13. *Let* $a_1 + a_2 + a_3 + \ldots$ *be a series of positive terms. If*

$$\left.\begin{array}{ll}
\lim\limits_{n\to\infty} \sqrt[n]{a_n} = l & \text{and } l < 1 \\
\lim\limits_{n\to\infty} \sqrt[n]{a_n} = l & \text{and } l > 1 \\
\lim\limits_{n\to\infty} \dfrac{a_{n+1}}{a_n} = k & \text{and } k < 1 \\
\lim\limits_{n\to\infty} \dfrac{a_{n+1}}{a_n} = k & \text{and } k > 1 \\
\lim\limits_{n\to\infty} n\left(\dfrac{a_n}{a_{n+1}} - 1\right) = m & \text{and } m > 1 \\
\lim\limits_{n\to\infty} n\left(\dfrac{a_n}{a_{n+1}} - 1\right) = m & \text{and } m < 1
\end{array}\right\}
\begin{array}{c}\text{then} \\ \text{the series} \\ \text{is}\end{array}
\left\{\begin{array}{ll}
\left.\begin{array}{l}\text{convergent} \\ \text{divergent}\end{array}\right\} & (\textit{Cauchy's Root Test}). \\
\left.\begin{array}{l}\text{convergent} \\ \text{divergent}\end{array}\right\} & (\textit{d'Alembert's Ratio Test}). \\
\left.\begin{array}{l}\text{convergent} \\ \text{divergent}\end{array}\right\} & (\textit{Raabe's Test}).
\end{array}\right.$$

REMARK 7. Some statements of Theorem 13 are valid under more general assumptions. In particular, for a given series to be convergent, it is sufficient that one of the relations

$$\overline{\lim_{n\to\infty}} \sqrt[n]{a_n} < 1\,, \quad \overline{\lim_{n\to\infty}} \frac{a_{n+1}}{a_n} < 1$$

(Theorem 10.1.8) holds. On the other hand, if $\sqrt[n]{a_n} \geqq 1$ or $a_{n+1}/a_n \geqq 1$ for almost every n, then the series is divergent.

REMARK 8. The tests of Cauchy and d'Alembert are inconclusive when $l = 1$ and $k = 1$, respectively. For Raabe's test we have the result: "If the relation

$$n\left(\frac{a_n}{a_{n+1}} - 1\right) \leqq 1$$

holds for almost every n, then the series diverges". Using Cauchy's test we can often conclude that a series converges, even when d'Alembert's test fails (but not vice versa). If Cauchy's test fails, it is often possible to reach a conclusion with the help of Raabe's test:

Example 5. For the series $\sum_{n=1}^{\infty} \frac{1}{n^2}$ neither d'Alembert's nor Cauchy's Test gives a conclusion. Using Raabe's test we have

$$\lim_{n\to\infty} n\left(\frac{a_n}{a_{n+1}} - 1\right) = \lim_{n\to\infty} n\left[\frac{(n+1)^2}{n^2} - 1\right] = \lim_{n\to\infty}\left(2 + \frac{1}{n}\right) = 2,$$

so the series is convergent.

Example 6. According to Remark 8 we can conclude that the harmonic series $\sum_{n=1}^{\infty} \frac{1}{n}$ is divergent, because $n\left(\frac{a_n}{a_{n+1}} - 1\right) = n\left(\frac{n+1}{n} - 1\right) = 1$ for every n.

Theorem 14 (*Integral Test*). *Let $f(x)$ be a non-negative decreasing function defined in an interval $[a, \infty)$ $(a > 0)$ such that $f(n) = a_n$ for every positive integer n. Then the series $a_1 + a_2 + a_3 + \ldots$ and the integral $\int_a^\infty f(x)\,\mathrm{d}x$ both converge or both diverge.*

Example 7. Let us find for which $\alpha > 0$ the series

$$1 + \frac{1}{2^\alpha} + \frac{1}{3^\alpha} + \frac{1}{4^\alpha} + \ldots \tag{8}$$

converges (for $\alpha \leqq 0$ it is evidently divergent). In Theorem 14 let us choose $f(x) = 1/x^\alpha$. The integral

$$\int_1^\infty \frac{\mathrm{d}x}{x^\alpha}$$

is convergent if and only if $\alpha > 1$ (see Theorem 13.8.9, p. 567). Thus, by Theorem 14 the series (8) is convergent for $\alpha > 1$ and divergent for $\alpha \leqq 1$. (In particular, from this result it follows that the harmonic series diverges, since $\alpha = 1$ in this case.)

Theorem 15 (*Cauchy's Theorem*). *Let $a_1 + a_2 + a_3 + \ldots$ be a series of positive decreasing terms $(a_n \geqq a_{n+1} \geqq \ldots > 0)$ and c be a positive integer such that $c > 1$. Then the given series and the series*

$$ca_c + c^2 a_{c^2} + c^3 a_{c^3} + \ldots$$

both converge or both diverge.

Example 8. Let us find for which $\alpha \geqq 0$ the series

$$\frac{1}{2 \ln^\alpha 2} + \frac{1}{3 \ln^\alpha 3} + \frac{1}{4 \ln^\alpha 4} + \ldots = \sum_{n=2}^{\infty} \frac{1}{n \ln^\alpha n} \tag{9}$$

converges. (For $\alpha < 0$ the series is evidently divergent; it is sufficient to compare it with the harmonic series.) The conditions of Theorem 15 are fulfilled, so let us choose $c = 2$. By Theorem 15 our series converges if and only if the series

$$2a_2 + 4a_4 + 8a_8 + \ldots$$

with the general term

$$2^k a_{2^k} = \frac{2^k}{2^k \ln^\alpha 2^k} = \frac{1}{(k \ln 2)^\alpha} = \frac{1}{\ln^\alpha 2} \cdot \frac{1}{k^\alpha}$$

converges. According to Example 7 the series with the general term $1/k^\alpha$ converges if and only if $\alpha > 1$. Thus the series (9) converges for $\alpha > 1$ and diverges for $\alpha \leqq 1$.

REMARK 9. The criteria (tests) expressed in Theorems 13–15 are valid for series of positive (Remark 5) terms. For other series they are useful because we can prove the convergence of the series (3) on the basis of Theorem 4 by proving the convergence of the series (2) for which our tests are applicable. From the divergence of the series (2) the divergence of the series (3) does not follow. Let us remark, however, that from the relations

$$\lim \frac{|a_{n+1}|}{|a_n|} > 1 \quad \text{or} \quad \lim \sqrt[n]{|a_n|} > 1$$

follows not only the divergence of the series (2) but also the divergence of the series (3).

Theorems on differentiation and integration of power series provide further effective means for investigating the convergence (and also for finding the sum) of many series (§ 15.4).

REMARK 10. The series $\sum_{n=1}^{\infty} a_n$ of complex terms $a_n = \alpha_n + i\beta_n$ (α_n, β_n real) is convergent if and only if both series $\sum_{n=1}^{\infty} \alpha_n$, $\sum_{n=1}^{\infty} \beta_n$ converge (and then

$$\sum_{n=1}^{\infty} a_n = \sum_{n=1}^{\infty} \alpha_n + i \sum_{n=1}^{\infty} \beta_n) .$$

To prove the convergence of a series of complex terms we can often with advantage use Theorem 4 on absolute convergence and so transform the problem to the investigation of the convergence of a series with positive terms.

Definition 4. A series

$$a_1 - a_2 + a_3 - a_4 \ldots, \quad a_n \geqq 0 \tag{10}$$

(with alternating positive and negative terms) is called an *alternating series.*

Theorem 16 (*Leibniz's Rule*). *If, for the series* (10), *the relations*

$$a_1 \geqq a_2 \geqq a_3 \geqq \ldots \quad \text{and} \quad \lim a_n = 0$$

hold, then the series (10) *is convergent. The absolute value of the difference* $s - s_n$ *between the sum* s *of this series and the partial sum* s_n *is less than or equal to the number* a_{n+1} *and this difference has the same sign as the* $(n + 1)$*-th term in the series* (10).

Example 9. By Theorem 16 the alternating series

$$1 - \tfrac{1}{2} + \tfrac{1}{3} - \tfrac{1}{4} + \ldots$$

is convergent. If we consider, for instance, the first eight terms, then $0 < s - s_8 < \frac{1}{9}$.

Theorem 17 (*Dirichlet's Test*). *Let two sequences* $\{a_n\}$, $\{b_n\}$ *be given, where* $\{b_n\}$ *is a monotonic sequence* (Definition 10.1.8) *such that* $\lim_{n\to\infty} b_n = 0$. *If* $\left|\sum_{k=1}^{n} a_k\right| \leqq K$ *for every* n (K *constant*), *then the series* $\sum_{k=1}^{\infty} a_k b_k$ *is convergent.*

Theorem 18 (*Abel's Test*). *Let two sequences* $\{a_n\}$, $\{b_n\}$ *be given, where* $\{b_n\}$ *is bounded and monotonic. If the series* $\sum_{k=1}^{\infty} a_k$ *converges, then the series* $\sum_{k=1}^{\infty} a_k b_k$ *is also convergent.*

Example 10. Let $b_1 \geqq b_2 \geqq b_3 \geqq \ldots \geqq 0$ and $\lim_{n\to\infty} b_n = 0$. Then the series

$$b_1 \sin x + b_2 \sin 2x + b_3 \sin 3x + \ldots \tag{11}$$

is convergent for every x: If $x = 2k\pi$ (k integral), the sum of the series is zero. For $x \neq 2k\pi$

$$a_1 + a_2 + \ldots + a_n = \sin x + \sin 2x + \ldots + \sin nx = \frac{\sin \frac{1}{2}nx}{\sin \frac{1}{2}x} \sin \frac{n+1}{2} x$$

(see Theorem 25), so

$$\left|\sum_{k=1}^{n} a_k\right| \leqq \left|\frac{1}{\sin \frac{1}{2}x}\right|$$

for every n, and the convergence of the series (11) for every fixed $x \neq 2k\pi$ follows from Theorem 17.

Definition 5. Let us consider the "two-dimensional" array of (real or complex) numbers

$$\begin{array}{l} a_{11}, a_{12}, a_{13}, a_{14}, \ldots, \\ a_{21}, a_{22}, a_{23}, a_{24}, \ldots, \\ a_{31}, a_{32}, a_{33}, a_{34}, \ldots, \\ \ldots\ldots\ldots\ldots\ldots\ldots \end{array} \tag{12}$$

The expression

$$\lim_{\substack{m \to \infty \\ n \to \infty}} a_{mn} = a$$

means: "For an arbitrarily chosen $\varepsilon > 0$ there exists a number K such that the relation

$$|a_{mn} - a| < \varepsilon$$

holds whenever both $m > K$ and $n > K$."

Definition 6. Let us write

$$s_{mn} = \sum_{i=1}^{m} \sum_{k=1}^{n} a_{ik}\,.$$

If the limit

$$\lim_{\substack{m \to \infty \\ n \to \infty}} s_{mn} = s$$

exists, then we write

$$\sum_{i,k=1}^{\infty} a_{ik} = s \tag{13}$$

and say that the so-called *double series* $\sum_{i,k=1}^{\infty} a_{ik}$ formed from the array (12), is *convergent* and *has the sum s*.

REMARK 11. Let us form from the array (12) an arbitrary series such that it includes every term from the array (12) exactly once. In particular, let us form the series

$$a_{11} + a_{12} + a_{21} + a_{13} + a_{22} + a_{31} + \ldots \tag{14}$$

(we note down successively from the array (12) the terms that have the same sum of indices). If any one of these series converges *absolutely*, in particular the series (14), then all such series converge absolutely and possess (Theorem 5) the same sum s. It is then possible to sum the terms of the array (12) in any order. In particular it follows that $\sum_{i,k=1}^{\infty} a_{ik} = s$ [see (13)]. We speak of the *absolute convergence of the double series* (13). Theorem 19 provides a simple criterion for absolute convergence.

REMARK 12. We meet double series in particular when dealing with Fourier series in two-dimensional domains. For example, the equation

$$\sum_{m,n=1}^{\infty} A_{mn} \sin mx \sin ny = f(x, y) \tag{15}$$

(A_{mn} are the Fourier coefficients of the function $f(x, y)$) means, according to Definition 5: For every arbitrarily small $\varepsilon > 0$ there exists a number K such, that for every pair of numbers M, N, for which both the relations $M > K$, $N > K$ hold, the relation

$$\left|\sum_{m=1}^{M} \sum_{n=1}^{N} A_{mn} \sin mx \sin ny - f(x, y)\right| < \varepsilon \tag{16}$$

holds. If we know (from the theorems on Fourier series and from the properties of the function $f(x, y)$) that the series (15) converges absolutely to $f(x, y)$ at the point (x, y), we can sum the terms of that series in an arbitrary order.

Theorem 19. *Let the series* $\sum_{k=1}^{\infty} |a_{ik}|$ *be convergent for every* i. *Let us write* $\sum_{k=1}^{\infty} |a_{ik}| = \sigma_i$. *If the series* $\sum_{i=1}^{\infty} \sigma_i$ *converges, then the double series* (13) *is absolutely convergent.*

Theorem 20. *Let the double series* (13) *be absolutely convergent. Then the series* $\sum_{k=1}^{\infty} a_{ik} = s_i$ *are absolutely convergent for every* i. *The series* $\sum_{i=1}^{\infty} s_i$ *is also absolutely convergent and its sum* s *is equal to the sum of the double series* (13).

REMARK 13. The summing in theorems 19 and 20 is done first by rows, then by columns. The theorems, however, retain their validity if the summing is done first by columns and then by rows.

REMARK 14. The fact that an absolutely convergent double series can be summed in an arbitrary order, may be used to *improve the rapidity of convergence* of simple series. In particular, the series

$$s = \frac{x}{1-x} + \frac{x^2}{1-x^2} + \frac{x^3}{1-x^3} + \ldots \tag{17}$$

is absolutely convergent for every x for which $|x| < 1$, as follows, for example, from d'Alembert's Test (Theorem 13). But this series converges slowly for $|x|$ in the vicinity of 1. By means of the so-called *Clausen's transformation* this series changes into the series

$$s = x\frac{1+x}{1-x} + x^4\frac{1+x^2}{1-x^2} + x^9\frac{1+x^3}{1-x^3} + x^{16}\frac{1+x^4}{1-x^4} + \dots, \tag{18}$$

which converges much faster. Namely, the i-th term of the series (17) is the sum of the series formed by members of the i-th row of the array

$$\begin{array}{l} x,\ x^2, x^3, x^4,\ \dots, \\ x^2, x^4, x^6, x^8,\ \dots, \\ x^3, x^6, x^9, x^{12}, \dots, \\ \dots\dots\dots\dots\dots\dots \end{array} \tag{19}$$

Each of these series converges absolutely for $|x| < 1$. The series corresponding to the series $\sum_{i=1}^{\infty} \sigma_i$ from Theorem 19 is here the series

$$\frac{|x|}{1-|x|} + \frac{|x|^2}{1-|x|^2} + \frac{|x|^3}{1-|x|^3} + \dots$$

which is convergent (e.g. according to d'Alembert's Test) for $|x| < 1$. Thus, the double series corresponding to the array (19) is absolutely convergent by Theorem 19. We form the sum in the following way:

$$s = U_1 + U_2 + U_3 + \dots$$

where U_1 is the sum of the elements from the first row and the first column in (19), U_2 is the sum of the remaining elements of the second row and the second column and so on. It follows that

$$U_1 = x + 2x^2 + 2x^3 + 2x^4 + \dots = x + 2x^2\frac{1}{1-x} = x\frac{1+x}{1-x},$$

$$U_2 = x^4 + 2x^6 + 2x^8 + 2x^{10} + \dots = x^4 + 2x^6\frac{1}{1-x^2} = x^4\frac{1+x^2}{1-x^2}.$$

In this way, we proceed to the series (18).

REMARK 15. By analogy with the definition of double series we can define *triple series*, etc.

Theorem 21 (*Multiplication or Product of Series*). *Let the series* $\sum_{n=1}^{\infty} a_n$ *and* $\sum_{n=1}^{\infty} b_n$ *be absolutely convergent and have the sums* s *and* t, *respectively. Then the double series*

$$\sum_{i,k=1}^{\infty} a_i b_k$$

is absolutely convergent and has the sum $s \cdot t$ (*i.e. we can multiply the series as we would multiply polynomials and then sum in an arbitrary order*).

Theorem 22 (*Cauchy Product of Series*). *Let* $\sum_{n=1}^{\infty} a_n$, $\sum_{n=1}^{\infty} b_n$ *be convergent series* (*with sums* s, t, *respectively*), *one of which at least is absolutely convergent. Then the series*

$$\sum_{n=1}^{\infty} U_n, \text{ where } U_n = a_1 b_n + a_2 b_{n-1} + \ldots + a_{n-1} b_2 + a_n b_1$$

is convergent and has the sum $s \cdot t$.

Theorem 23. *Let us write in the customary way* $s_n = a_1 + a_2 + \ldots + a_n$. *Let* $\sum_{n=1}^{\infty} a_n$ *be convergent, that is* $\lim_{n \to \infty} s_n = s$. *Then also*

$$\lim_{n \to \infty} S_n = \lim_{n \to \infty} \frac{s_1 + s_2 + \ldots + s_n}{n} = s .$$

REMARK 16. The converse of Theorem 23 does not hold. We say that a series, for which the sequence S_n converges, is *summable by arithmetic means of the first order* [or *Cesàro summable*; we write: *summable* (C, 1)]. This method assigns a "sum" to certain divergent series. For some definitions of the summability and particularly for the applicability of divergent series to asymptotic expansions (representations) see § 15.7. We shall mention one application here:

Theorem 24. *If* $a_1 + a_2 + a_3 + \ldots$ *and* $b_1 + b_2 + b_3 + \ldots$ *are two convergent series with respective sums* s *and* t *then the series*

$$a_1 b_1 + (a_1 b_2 + a_2 b_1) + (a_1 b_3 + a_2 b_2 + a_3 b_1) + \ldots$$

is summable by arithmetic means (Remark 16) *to the sum* $s \cdot t$ (*i.e.* $\lim_{n \to \infty} S_n = s \cdot t$).

Theorem 25 (*Survey of Important Formulae*).

1. $\sum_{n=1}^{\infty} (a_n \pm b_n) = \sum_{n=1}^{\infty} a_n \pm \sum_{n=1}^{\infty} b_n$, $\sum_{n=1}^{\infty} k a_n = k \sum_{n=1}^{\infty} a_n$ (k a constant) (Theorem 3).

2. $$a_n > 0, \quad \lim_{n \to \infty} \frac{a_{n+1}}{a_n} = \begin{Bmatrix} k < 1 \\ \\ k > 1 \end{Bmatrix} \Rightarrow \begin{Bmatrix} \sum_{n=1}^{\infty} a_n \text{ is convergent} \\ \\ \sum_{n=1}^{\infty} a_n \text{ is divergent} \end{Bmatrix} \text{ (Theorem 13);}$$

$$a_n > 0, \quad \lim_{n \to \infty} \sqrt{a_n} = \begin{Bmatrix} l < 1 \\ \\ l > 1 \end{Bmatrix} \Rightarrow \begin{Bmatrix} \sum_{n=1}^{\infty} a_n \text{ is convergent} \\ \\ \sum_{n=1}^{\infty} a_n \text{ is divergent} \end{Bmatrix} \text{ (Theorem 13);}$$

$$a_n > 0\,, \quad \lim_{n\to\infty} n\left(\frac{a_n}{a_{n+1}} - 1\right) = \left\{\begin{matrix} m > 1 \\ \\ m < 1 \end{matrix}\right\} \Rightarrow \left\{\begin{matrix} \sum_{n=1}^{\infty} a_n \textit{ is convergent} \\ \\ \sum_{n=1}^{\infty} a_n \textit{ is divergent} \end{matrix}\right\} \text{(Theorem 13).}$$

3. *The arithmetic series* $a_1 + a_2 + a_3 + \ldots$, *where* $a_n = a_1 + (n-1)\,d$, *is convergent if and only if* $a_1 = 0$ *and* $d = 0$. *The partial sum of the first n terms is*

$$a_1 + a_2 + \ldots + a_n = \tfrac{1}{2}n(a_1 + a_n) = \tfrac{1}{2}n[2a_1 + (n-1)\,d]\,.$$

4. *The geometric series*

$$a + aq + aq^2 + \ldots \quad (a \neq 0)$$

converges if and only if $|q| < 1$. *The sum of the series is*

$$s = \frac{a}{1-q}\,.$$

The partial sum is:

$$s_n = a\,\frac{q^n - 1}{q - 1} \quad (q \neq 1)\,.$$

5. $1 + 2 + 3 + \ldots + n = \tfrac{1}{2}n(n+1)\,;$

$$1^2 + 2^2 + 3^2 + \ldots + n^2 = \tfrac{1}{6}n(n+1)(2n+1)\,;$$

$$1^3 + 2^3 + 3^3 + \ldots + n^3 = \tfrac{1}{4}n^2(n+1)^2\,;$$

$$1^4 + 2^4 + 3^4 + \ldots + n^4 = \tfrac{1}{30}n(n+1)(2n+1)(3n^2 + 3n - 1)\,.$$

6. $\dfrac{1}{1^2} + \dfrac{1}{2^2} + \dfrac{1}{3^2} + \ldots = \dfrac{\pi^2}{6}\,;$

$$\frac{1}{1^4} + \frac{1}{2^4} + \frac{1}{3^4} + \ldots = \frac{\pi^4}{90}\,;$$

$$\frac{1}{1^6} + \frac{1}{2^6} + \frac{1}{3^6} + \ldots = \frac{\pi^6}{945}\,;$$

$$\frac{1}{1^2} + \frac{1}{3^2} + \frac{1}{5^2} + \ldots = \frac{\pi^2}{8}\,.$$

7. $$\left.\begin{aligned} \sin x + \sin 2x + \ldots + \sin nx &= \frac{\sin \frac{1}{2}nx}{\sin \frac{1}{2}x} \sin \frac{1}{2}(n+1)\,x\,, \\ \cos x + \cos 2x + \ldots + \cos nx &= \frac{\sin \frac{1}{2}nx}{\sin \frac{1}{2}x} \cos \frac{1}{2}(n+1)\,x\,, \end{aligned}\right\} \begin{matrix} x \neq 2k\pi\,, \\ k \textit{ an integer.} \end{matrix}$$

8. *Let* $\frac{1}{p} + \frac{1}{q} = 1$. *The following results then hold:*

(a) *If* $a_n \geqq 0$, $b_n \geqq 0$, $p > 1$, *then*

$$\sum_{n=1}^{\infty} a_n b_n \leqq \left(\sum_{n=1}^{\infty} a_n^p\right)^{1/p} \cdot \left(\sum_{n=1}^{\infty} b_n^q\right)^{1/q} \quad \text{(Hölder's Inequality)}$$

(b) *If* $a_n \geqq 0$, $b_n \geqq 0$, $p \geqq 1$, *then*

$$\left[\sum_{n=1}^{\infty} (a_n + b_n)^p\right]^{1/p} \leqq \left(\sum_{n=1}^{\infty} a_n^p\right)^{1/p} + \left(\sum_{n=1}^{\infty} b_n^p\right)^{1/p} \quad \text{(Minkowski's Inequality)}$$

(c) *From* (a) *for* $p = 2$ $(a_n \geqq 0, b_n \geqq 0)$ *it follows that*

$$\left(\sum_{n=1}^{\infty} a_n b_n\right)^2 \leqq \sum_{n=1}^{\infty} a_n^2 \cdot \sum_{n=1}^{\infty} b_n^2 \quad \text{(Schwarz or Schwarz-Cauchy Inequality)}$$

(*so, if the series* $\sum_{n=1}^{\infty} a_n^2$, $\sum_{n=1}^{\infty} b_n^2$ *converge, then the series* $\sum_{n=1}^{\infty} a_n b_n$ *also converges*).

Theorem 26 (*The Sums of Some Series*).

1. $1 + \frac{1}{1!} + \frac{1}{2!} + \frac{1}{3!} + \ldots + \frac{1}{n!} + \ldots = \mathrm{e}\,.$

2. $1 - \frac{1}{1!} + \frac{1}{2!} - \frac{1}{3!} + \ldots + (-1)^n \frac{1}{n!} + \ldots = \frac{1}{\mathrm{e}}\,.$

3. $1 - \frac{1}{2} + \frac{1}{3} - \frac{1}{4} + \ldots + (-1)^{n+1} \frac{1}{n} + \ldots = \ln 2\,.$

4. $1 + \frac{1}{2} + \frac{1}{4} + \frac{1}{8} + \ldots + \frac{1}{2^n} + \ldots = 2\,.$

5. $1 - \frac{1}{2} + \frac{1}{4} - \frac{1}{8} + \ldots + (-1)^n \frac{1}{2^n} + \ldots = \frac{2}{3}\,.$

6. $1 - \frac{1}{3} + \frac{1}{5} - \frac{1}{7} + \ldots + (-1)^{n+1} \frac{1}{2n-1} + \ldots = \frac{\pi}{4}\,.$

7. $\frac{1}{1\,.\,2} + \frac{1}{2\,.\,3} + \frac{1}{3\,.\,4} + \ldots + \frac{1}{n(n+1)} + \ldots = 1\,.$

8. $\frac{1}{1\,.\,3} + \frac{1}{3\,.\,5} + \frac{1}{5\,.\,7} + \ldots + \frac{1}{(2n-1)(2n+1)} + \ldots = \frac{1}{2}\,.$

9. $\dfrac{1}{1.3} + \dfrac{1}{2.4} + \dfrac{1}{3.5} + \ldots + \dfrac{1}{n(n+2)} + \ldots = \dfrac{3}{4}$.

10. $\dfrac{1}{3.5} + \dfrac{1}{7.9} + \dfrac{1}{11.13} + \ldots + \dfrac{1}{(4n-1)(4n+1)} + \ldots = \dfrac{1}{2} - \dfrac{\pi}{8}$.

11. $\dfrac{1}{1.2.3} + \dfrac{1}{2.3.4} + \ldots + \dfrac{1}{n(n+1)(n+2)} + \ldots = \dfrac{1}{4}$.

12. $\dfrac{1}{1.2.\ldots.l} + \dfrac{1}{2.3.\ldots.(l+1)} + \ldots + \dfrac{1}{n.\ldots.(n+l-1)} + \ldots =$

$$= \frac{1}{(l-1)(l-1)!}.$$

REMARK 17. For series with variable terms see Chap. 15 where the application of power series to the numerical summation of series is dealt with. For the summation of infinite series by means of integral transformations see the article "The Summation of Infinite Series by means of Integral Transformations" by D. MAYER and J. NEČAS, *Aplikace matematiky* 1 (1956), No. 3, pp. 165–185. (In Czech, English summary.)

10.3. Infinite Products

Definition 1. Suppose we are given the sequence of (real or complex) numbers $p_1, p_2, p_3, \ldots$. Let us define

$$P_n = p_1 \cdot p_2 \cdot p_3 \cdot \ldots \cdot p_n . \tag{1}$$

The symbol

$$\prod_{n=1}^{\infty} p_n = p_1 \cdot p_2 \cdot p_3 \cdot \ldots \tag{2}$$

is called an *infinite product.* If $\lim_{n\to\infty} P_n$ exists, then this limit is called the *value of the infinite product* (2).

Definition 2. We say that the product (2) is *convergent* if, either (i) the limit $\lim_{n\to\infty} P_n$, finite and *different from zero*, exists, or (ii) in the product (2) there is only a finite number of factors equal to zero and their omission leads again to a finite limit *different from zero.* (In the latter case the infinite product has the value 0 according to Definition 1.)

In every other case we shall say that (2) is *divergent*.

Example 1. The product

$$\frac{1}{1} \cdot \frac{1}{2} \cdot \frac{1}{3} \cdot \ldots$$

is divergent because

$$P_n = \frac{1}{n!} \quad \text{and} \quad \lim_{n \to \infty} P_n = 0 .$$

REMARK 1. The reader's attention is drawn to the fact that there is no uniformity in mathematical literature regarding the definition of the convergence of an infinite product.

REMARK 2. In applications, the investigation of infinite products frequently relates to cases where the factors p_n are of the form $1 + a_n$.

Theorem 1. *Let the product*

$$(1 + |a_1|)(1 + |a_2|)(1 + |a_3|) \ldots \tag{3}$$

be convergent. Then the product

$$(1 + a_1)(1 + a_2)(1 + a_3) \ldots \tag{4}$$

is also convergent.

Definition 3. If the product (3) converges then the product (4) is said to be *absolutely convergent.*

Theorem 2. *If the series $a_1 + a_2 + a_3 + \ldots$ is absolutely convergent* (see Definition 10.2.3) *then the product* (4) *is absolutely convergent* (*and conversely*) *and its value does not depend on the ordering of its factors.* (*We say in this case that its factors may be rearranged.*)

REMARK 3. If the series $a_1 + a_2 + a_3 + \ldots$ is only conditionally convergent (Definition 10.2.3) then the product (4) need not be convergent.

Theorem 3. *For every x* (*real or complex*) *the relations*

$$\sin x = x\left(1 - \frac{x^2}{\pi^2 \cdot 1^2}\right)\left(1 - \frac{x^2}{\pi^2 \cdot 2^2}\right)\left(1 - \frac{x^2}{\pi^2 \cdot 3^2}\right) \ldots = x \prod_{n=1}^{\infty}\left(1 - \frac{x^2}{n^2\pi^2}\right),$$

$$\cos x = \left(1 - \frac{2^2x^2}{\pi^2 \cdot 1^2}\right)\left(1 - \frac{2^2x^2}{\pi^2 \cdot 3^2}\right)\left(1 - \frac{2^2x^2}{\pi^2 \cdot 5^2}\right) \ldots = \prod_{n=1}^{\infty}\left(1 - \frac{2^2x^2}{\pi^2(2n-1)^2}\right)$$

hold. Also

$$\frac{\pi}{2} = \frac{2}{1} \cdot \frac{2}{3} \cdot \frac{4}{3} \cdot \frac{4}{5} \cdot \frac{6}{5} \cdot \frac{6}{7} \ldots \quad (\textit{Wallis's Product}).$$

REMARK 4. For the expression of the Γ function as an infinite product see § 13.11.

11. DIFFERENTIAL CALCULUS OF FUNCTIONS OF A REAL VARIABLE

By Karel Rektorys

References: [16], [27], [32], [43], [48], [50], [57], [73], [110], [112], [138], [155], [172], [226], [229], [242], [246], [269], [270], [307], [308], [341], [362], [367], [376], [394]

11.1. The Concept of a Function. Composite Functions. Inverse Functions

Notation: x is a real number. Instead of "the number x" we often say "the point x".

The *closed interval* $[a, b]$ is the set of all x, for which $a \leqq x \leqq b$;
the *open interval* (a, b) is the set of all x, for which $a < x < b$;
the *semi-closed* (*semi-open*) *interval* $[a, b)$ is the set of all x, for which $a \leqq x < b$;
the *semi-closed* (*semi-open*) *interval* $(a, b]$ is the set of all x, for which $a < x \leqq b$.

The interval $(a, +\infty)$	(briefly (a, ∞))	is the set of all x, for which $x > a$;
the interval $[a, +\infty)$	(briefly $[a, \infty)$)	is the set of all x, for which $x \geqq a$;
the interval $(-\infty, a)$		is the set of all x, for which $x < a$;
the interval $(-\infty, a]$		is the set of all x, for which $x \leqq a$;
the interval $(-\infty, +\infty)$	(briefly $(-\infty, \infty)$)	is the set of all real numbers x.

We shall write the interval I in speaking of an interval without special reference to its end-points. The notation $x \in M$ means: x is an element belonging to the set M. For example, $x \in [a, b]$ means that x is in the interval $[a, b]$.

Definition 1. We say that a *real function* is defined on a set M of real numbers, if a rule (relation) is given by virtue of which to each number $x \in M$ there corresponds *exactly one* real number y. The number x is called the *independent variable* (*argument*), y is called the *dependent variable*. The set M is called the *domain of definition of the function.*

A function is generally denoted by the letters $f, g, \ldots$. The value y of the function corresponding to an arbitrary point $x \in M$ is denoted by $f(x)$, $g(x)$, etc. Instead of "the function f" we often say "the function $f(x)$" or "the function $y = f(x)$".

Example 1. The area y of a square is a function of the length x of its side, $y = x^2$. The domain of definition is the interval $(0, +\infty)$, since the length of the side of the square is always expressed by a positive number.

REMARK 1. The domain of definition of a function is most often an interval, e.g. the interval $[-1, 1]$ for the function $y = \arcsin x$, etc. But every "reasonable" function need not have an interval as its domain of definition. For example, the function $y = \tan x$ is defined in the interval $(-\infty, +\infty)$ from which the points $\pm\frac{1}{2}\pi$, $\pm\frac{3}{2}\pi$, $\pm\frac{5}{2}\pi$, ... are excluded.

REMARK 2. The relationship defining the function need not be given by an equation (i.e. by an analytic formula, from which the value of the dependent variable y can be calculated for a given value of the independent variable x) as was the case in Example 1. Frequently, in applications, the correspondence between the independent variable x and the dependent variable y is established by a graph, expecially in cases where the values of the independent and dependent variables can be read off the graph with adequate accuracy. When we perform different types of measurements we compile a table of measured values. From this table we often try to obtain values of a function for the whole domain of definition (e.g., by interpolation). The function may be given also as the limit of a sequence of functions, as the sum of a series of functions, etc. Often a relationship between x and y is given (most frequently by an equation) from which it is necessary to determine the single-valued correspondence between the dependent variable y and the independent variable x (Fig. 11.1). (E.g. $x^2 + y^2 = 25$; in the neighbourhood of the point $(3, 4)$ the function $y = f(x)$ will be given by the relation $y = \sqrt{(25 - x^2)}$. There we say that the *explicit* function $y = \sqrt{(25 - x^2)}$ is given in a neighbourhood of the point $(3, 4)$ by the *implicit*

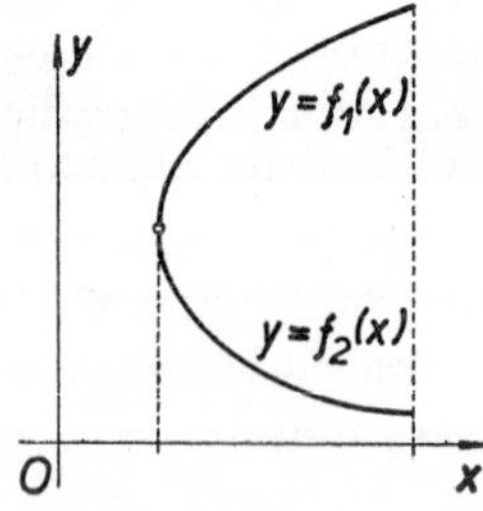

Fig. 11.1. The curve in this figure is the graph of two functions, $y = f_1(x)$ and $y = f_2(x)$; it is not the graphical representation of a single-valued function.

equation $x^2 + y^2 = 25$.) See § 12·9 for more detailed treatment. *Until further notice the concept of a function means the (single-valued) function as defined in Definition 2.*

REMARK 3. If to each real number x from a certain domain M there corresponds a complex number $y = y_1 + iy_2$ we say that $y = f(x) = f_1(x) + if_2(x)$ is a *complex function of the real variable* x. The study of these functions is reduced to that of the real functions $f_1(x)$ and $f_2(x)$ (e.g. the derivative is defined by the relation $f'(x) = f_1'(x) + if_2'(x)$, etc.) so that in the following we shall deal only with *real* functions of a real variable. (For functions where the independent variable is complex see Chap. 20.)

REMARK 4. If the functional relationship is given by an analytical formula then we are interested in those x for which the formula has a sense (in the domain of real numbers). The set of those x is then accepted as the domain of definition of the given function. For example, the domain of definition of the function given by the formula $y = \sqrt{(4 - x^2)}$ is taken to be the interval $[-2, 2]$ (for $|x| > 2$, $\sqrt{(4 - x^2)}$ is no longer a real number).

Definition 2. By the *graph of the function* $y = f(x)$ is understood the set of all points (x, y) in the plane xy (with a cartesian coordinate system $(O; x, y)$) such that $x \in M$, $y = f(x)$. Instead of "the graph of the function $y = f(x)$" we often say "the curve $y = f(x)$". The coordinate x is called the *first coordinate* (the *abscissa*, *x-coordinate*), the coordinate y is called the *second coordinate* (the *ordinate*, *y-coordinate*).

Example 2. Graphs of the trigonometric functions are given on p. 110.

Definition 3. Let the function $y = f(x)$ be defined in the interval M_1. We say that the function $y = f(x)$ maps the interval M_1 *into* (*in*) the interval M_2 if for every $x \in M_1$ it follows that $y \in M_2$.

If in addition it is possible to find, for every $y \in M_2$, at least one $x \in M_1$ such that $y = f(x)$ then we say that the function $y = f(x)$ maps the interval M_1 *onto* (*on*) the interval M_2.

Example 3. The interval $[0, 7\pi]$ is mapped by the function $y = \sin x$ onto the interval $[-1, 1]$, because for every $x \in [0, 7\pi]$, $y = f(x) \in [-1, 1]$ and to any number $y \in [-1, 1]$ we can find even several $x \in [0, 7\pi]$ such that $y = \sin x$. We can say as well that the interval $[0, 7\pi]$ is mapped *into* any interval that includes the interval $[-1, 1]$, e.g. into the interval $[-5, 10]$.

Definition 4 (*of a Composite Function*). Let the interval M_1 be mapped by the function $z = f(x)$ into the interval M_2. Let $y = g(z)$ be a function defined in the domain M_2. The function $y = g(f(x))$ is said to be a *composite function* of the functions $z = f(x)$ and $y = g(z)$.

REMARK 5. Thus, $y = g(f(x))$ has the following meaning (Fig. 11.2): When we choose any number $x \in M_1$ then by means of the relation $z = f(x)$ we can evaluate z. To this number z we can then by means of the relation $y = g(z)$ evaluate y. Thus y is determined uniquely by the choice of the number $x \in M_1$, so that finally $y = g(f(x))$ is a function of the variable x only. (On how to use composite functions for finding derivatives see Theorem 11.5.5, p. 420.)

Example 4. The function $y = \sqrt{(1 - \frac{1}{2}\sin^2 x)}$ may be "decomposed" into the functions $y = \sqrt{z}$, $z = 1 - \frac{1}{2}\sin^2 x$. As the interval M_1 we can take the interval $(-\infty, +\infty)$ because the function $z = 1 - \frac{1}{2}\sin^2 x$ maps the interval $(-\infty, +\infty)$ onto the interval $[\frac{1}{2}, 1]$ and consequently also into the interval $[0, +\infty)$, which is the domain of definition of the function $y = \sqrt{z}$.

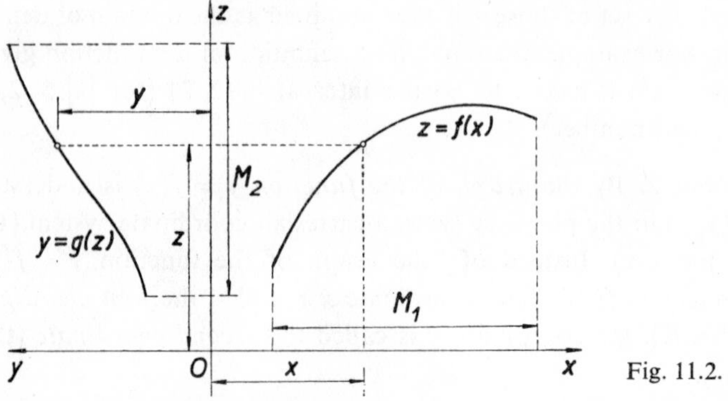

Fig. 11.2.

REMARK 6. In general, we define composite functions (and also mappings) for sets other than intervals. It is then necessary to replace in definitions 3 and 4 the word "interval" by the word "set".

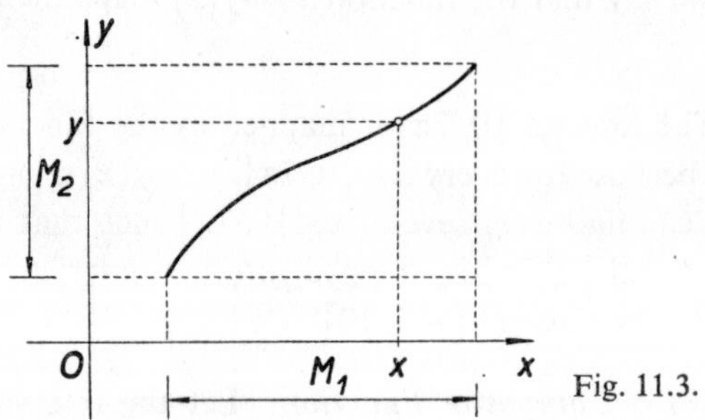

Fig. 11.3.

Definition 5. Let the interval M_1 be mapped by the function $y = f(x)$ onto the interval M_2 (see definition 3) with a *one-to-one correspondence* (Fig. 11.3), which means that not only to every $x \in M_1$ there corresponds exactly one $y \in M_2$, but also to every $y \in M_2$ there corresponds exactly one $x \in M_1$ such that $y = f(x)$. Because to every $y \in M_2$ there corresponds just one $x \in M_1$, a function is defined on the interval M_2 which we denote by $x = \varphi(y)$. This function is called the *inverse function* of the function $y = f(x)$. Conversely, the function $y = f(x)$ is the inverse function of the function $x = \varphi(y)$.

REMARK 7. The one-to-one correspondence is ensured, for example, if $y = f(x)$ is strictly increasing or decreasing in M_1 (or if the function $x = \varphi(y)$ is strictly increasing or decreasing in M_2) (see Fig. 11.3). This case is very common.

REMARK 8. It is possible to add a remark to the definition of inverse function similar to Remark 6.

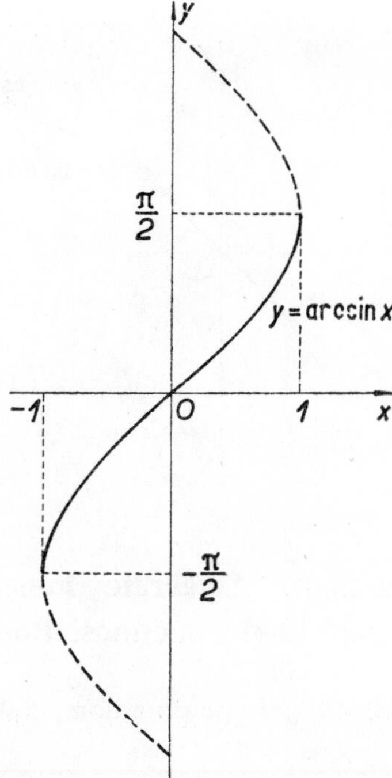

Fig. 11.4.

Example 5. The function $x = \sin y$ is a strictly increasing function of the variable y in the interval $[-\frac{1}{2}\pi, \frac{1}{2}\pi]$. The corresponding interval of the variable x is $[-1, 1]$ (see Fig. 11.4). The inverse function of the function $x = \sin y$ is called *arcussinus x* and is denoted by $y = \arcsin x$ (or $y = \sin^{-1} x$). Thus this function is defined in the interval $[-1, 1]$ (of the independent variable x). (For details see § 2.11.)

Example 6. The interval $(-\infty, +\infty)$ of the independent variable x is mapped by the function $y = x^2$ onto the interval $[0, +\infty)$ of the dependent variable y, but the correspondence between x and y is not one-to-one, because, for example, $f(2) = f(-2) = 4$. But the interval $[0, +\infty)$ of the variable x is mapped by the given function onto the interval $[0, +\infty)$ in a one-to-one correspondence. The corresponding inverse function is then $x = +\sqrt{y}$ as we easily derive from the equation $y = x^2$, and is defined in the interval $0 \leqq y < +\infty$.

REMARK 9. We draw attention to the fact that, according to Definition 5, the function $y = \arcsin x$ $(-1 \leqq x \leqq 1)$ is the inverse of the function $x = \sin y$

$(-\frac{1}{2}\pi \leq y \leq \frac{1}{2}\pi)$ (Example 5). Both equations have the same meaning and both functions have the same *graphical representation* (in the chosen coordinate system xy). When we interchange the variables x and y in one of them (e.g. in Example 5 we write $y = \sin x$ instead of $x = \sin y$) then the graphs of the functions will be symmetrical with respect to the straight line $y = x$ (i.e. the straight line bisecting the angle between the positive x-axis and the positive y-axis). (See Fig. 11.5.)

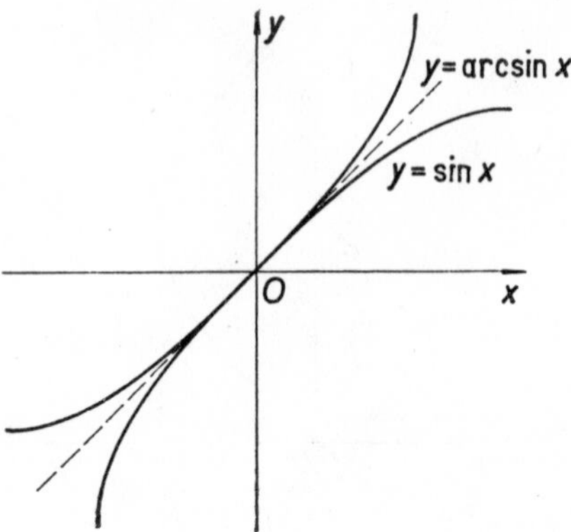

Fig. 11.5.

11.2. Elementary Functions. Algebraic Functions, Transcendental Functions. Even and Odd Functions. Bounded Functions

A function $y = f(x)$ is called *algebraic* (in a domain M) if it satisfies identically an equation

$$F(x, y) = 0 \tag{1}$$

where $F(x, y)$ is a polynomial in the variables x, y. For example, the function $y = \sqrt{(1 - x^2)}$, $x \in [-1, 1]$, is algebraic since it satisfies in the interval mentioned the equation $1 - x^2 - y^2 \equiv 0$ and $1 - x^2 - y^2$ is a polynomial in the variables x, y. Functions which are not algebraic are called *transcendental functions.*

REMARK 1. Algebraic functions include, first, *polynomials* (or *rational integral functions*) and *fractional rational functions* (or briefly *rational functions*), i.e. functions of the form

$$y = \frac{a_n x^n + a_{n-1}x^{n-1} + \ldots + a_0}{b_m x^m + b_{m-1}x^{m-1} + \ldots + b_0},$$

where m, n are non-negative integers. Additional examples are the functions

$$y = \sqrt{x}\ (x \geqq 0), \quad y = \sqrt[3]{(1-x^2)}\ \ (-1 \leqq x \leqq 1), \quad y = \sqrt{(1+x^2)}\ \ (-\infty < x < +\infty),$$

etc. (For more details on polynomials see § 1.14.)

Transcendental functions include the general power $y = x^n$ ($x > 0$, n irrational), trigonometric, hyperbolic and exponential functions and their inverse functions. All these functions are called *elementary transcendental functions*. Further transcendental functions are defined by means of differential equations and integrals (so-called *higher transcendental functions*; e.g. $g(x) = \int_0^x e^{-t^2}\,dt$).

REMARK 2. For the trigonometric and hyperbolic functions and their inverse functions in more detail, see Chap. 2.

In Definition 1.9.2, p. 51, the meaning of the symbol a^b ($a > 0$) for b irrational is explained. If b is a constant and a a variable, we get the *general power* with standard notation: $y = x^n$ ($x > 0$, n any real number). The function $y = x^n$ is continuous and strictly increasing or strictly decreasing or constant in the interval $(0, \infty)$ according as $n > 0$, or $n < 0$, or $n = 0$, respectively. For certain n it is possible to extend the domain of definition of the function $y = x^n$ to values of x other than $x > 0$. For example, the relation $y = x^2$ has sense for all x (the domain of definition is $(-\infty, +\infty)$).

If the exponent b in the expression a^b changes its value and the base a remains constant we get the *exponential function* $y = a^x$ ($a > 0$); its domain of definition is $(-\infty, +\infty)$. For the graph of the function $y = a^x$ see Fig. 11.6. The function is

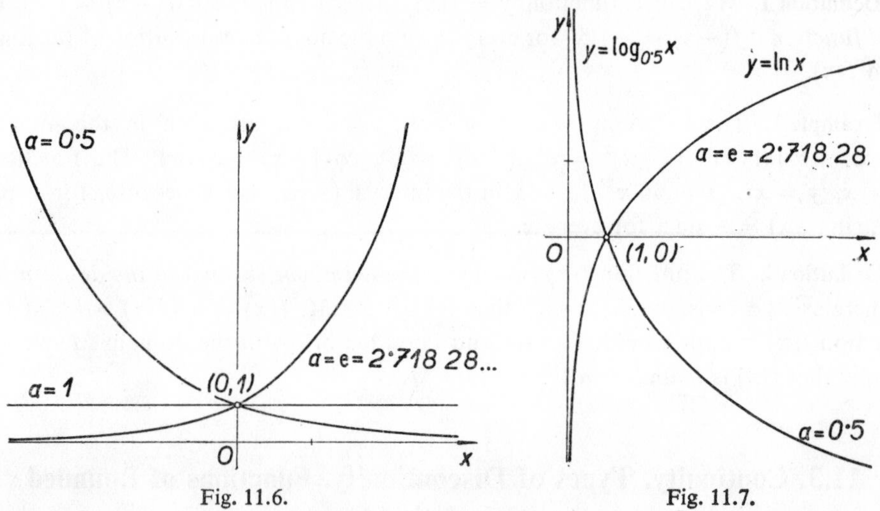

Fig. 11.6. Fig. 11.7.

positive for all x, it is increasing for $a > 1$, constant for $a = 1$ ($y \equiv 1$) and decreasing for $0 < a < 1$. For $a = e = 2{\cdot}718, 281, 828, 459, 0\ldots$ (Theorem 10.1.11) we obtain the important function $y = e^x$ (frequently used notation: $y = \exp(x)$).

If $a > 0$, $a \neq 1$, the function $x = a^y$ maps the interval $-\infty < y < +\infty$ onto the interval $0 < x < +\infty$ in a one-to-one correspondence and so we can define (see Definition 11.1.5) the inverse function of the exponential function, called *logarithm of x to the base a*. Notation: $y = \log_a x$. For the graphical representation see Fig. 11.7.

The function is defined for $x > 0$ and is strictly increasing for $a > 1$ and strictly decreasing for $0 < a < 1$. For $a = \mathrm{e}$ we write $y = \ln x$ (in the literature the notation $\lg x$, $\log x$ is also used) and this function is called the *natural logarithm of x*. This function is the inverse function of the function $x = \mathrm{e}^y$.

REMARK 3. On the differentiation of the elementary functions see § 11.5. All current rules known from elementary mathematics, e.g. $a^{x_1} . a^{x_2} = a^{x_1+x_2}$, $\log_a (x_1 . x_2) = \log_a x_1 + \log_a x_2$ $(x_1 > 0, x_2 > 0)$, etc., are valid for general powers and for the exponential and logarithmic functions (for details see §§ 1.9 and 1.10). Further we have

$$\log_a x = \log_b x . \log_a b ,$$

in particular

$$\log_{10} x = M \log_e x = M \ln x \approx 0{\cdot}434\,294 \ln x ,$$

$$\log_e x = \ln x = \frac{1}{M} \log_{10} x \approx 2{\cdot}302\,585 \log_{10} x .$$

The number M is called the *conversion modulus from natural to common logarithms*, the number $1/M$ is the *conversion modulus from common to natural logarithms.*

Definition 1. We call a function $y = f(x)$ an *even function* if $f(-x) = f(x)$, an *odd function* if $f(-x) = -f(x)$ for every x from the domain of definition of the function $f(x)$.

Example 1. The functions $y = x^2$, $y = x^4$, $y = \cos x$ are even in the interval $(-\infty, +\infty)$ (because $(-x)^2 = x^2$, $(-x)^4 = x^4$, $\cos(-x) = \cos x$). The functions $y = x$, $y = x^3$, $y = \sin x$ are odd in the interval $(-\infty, +\infty)$, because, for example, $\sin(-x) = -\sin x$ for every x.

Definition 2. The function $f(x)$ is called *bounded above (below) in the domain M*, if there exists a constant K (k) such that for all $x \in M$, $f(x) < K$ $(f(x) > k)$. If the function $f(x)$ is both bounded above and bounded below in the domain M, we say simply that $f(x)$ is *bounded in the domain M.*

11.3. Continuity. Types of Discontinuity. Functions of Bounded Variation

Definition 1. By a *δ-neighbourhood of a point a* we mean a set of all points x such that their distance from the point a is smaller than δ (or: such that they lie in the interval $(a - \delta, a + \delta)$ or for which $|x - a| < \delta$; we often denote such a neighbourhood by $U_\delta(a)$).

Definition 2 (*Cauchy's Definition of Continuity*). We say that $f(x)$ is *continuous at the point a* if, an *arbitrary* number $\varepsilon > 0$ being chosen, another number $\delta > 0$

exists (depending in general on the choice of the number ε), such that for *every* x in the δ-neighbourhood of the point a the relation

$$|f(x) - f(a)| < \varepsilon \tag{1}$$

holds.

REMARK 1. Roughly speaking: $f(x)$ is continuous at the point a, if $f(x)$ differs from $f(a)$ by a small enough quantity when x is sufficiently near to the point a (Fig. 11.8). Or also (writing $\Delta y = f(x + \Delta x) - f(x)$) if $\Delta y \to 0$ when $\Delta x \to 0$.

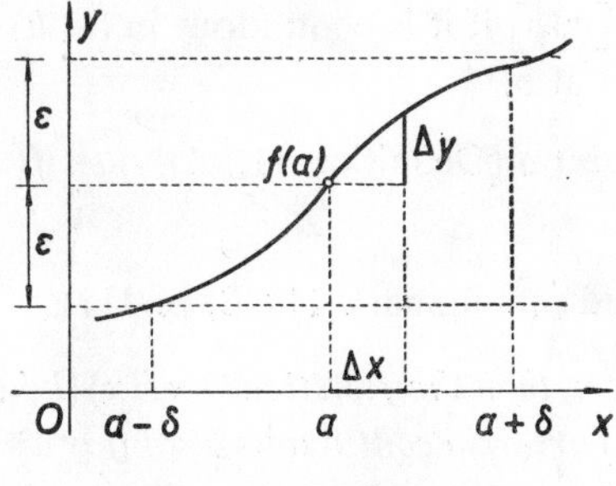

Fig. 11.8.

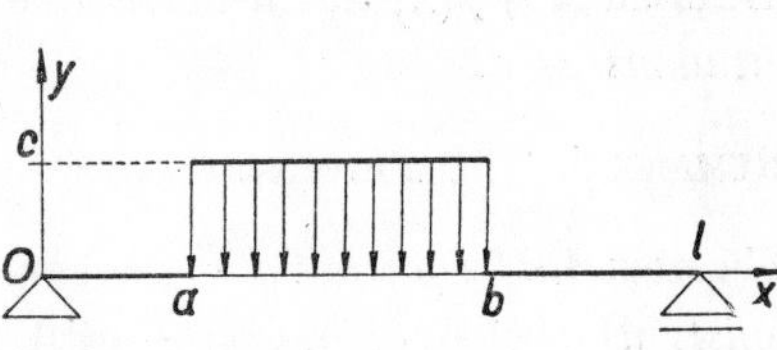

Fig. 11.9.

REMARK 2. It follows from the definition that: Should $f(x)$ be continuous at a point a, it must be defined in a certain neighbourhood of the point a (and so also at the point a itself).

REMARK 3. It is possible to define the continuity of a function $f(x)$ at the point a by means of sequences of x_n tending to the point a the (so-called *Heine definition*).

Definition 3. We say that $f(x)$ is *continuous from (on) the right at the point a* if, an arbitrary number $\varepsilon > 0$ being chosen, another number $\delta > 0$ exists, such that for every $x \geqq a$ in the δ-neighbourhood of the point a the relation

$$|f(x) - f(a)| < \varepsilon$$

holds. Analogous is the definition of *left-hand continuity.* (In these definitions we consider "the right-hand" or "the left-hand" δ-neighbourhood of the point a.)

Example 1. The function $y = \sqrt{(4 - x^2)}$ is continuous from the right (according to Theorems 4 and 5 and Remark 6) at the point $x = -2$ (to the left of the point -2 it is not defined at all). Similarly it is continuous from the left at the point $x = 2$.

REMARK 4. The function in Fig. 11. 9 illustrating the loading of a bar is continuous from the right at the point a, and continuous from the left at the point b, if we define the function at the points a, b by the value c. If we assign to our function the value 0 at these two points it will be continuous from the left at the point a (and discontinuous from the right) and continuous from the right at the point b. (Of course we cannot assign to our function both values 0 and c at the point a; this would be in contradiction to the definition of a *single-valued function*).

Theorem 1. *The function $f(x)$ is continuous at the point a if and only if it is continuous from the right and continuous from the left at this point.*

Theorem 2. *The function $f(x)$ is continuous at the point a if and only if it is defined at the point a and has a limit at the point a* (see Definition 11.4.1) *equal to the number $f(a)$.*

Definition 4. We say that $f(x)$ is *continuous in a domain M*, if it is continuous at every point of this domain. It is continuous in $[a, b]$ if it is continuous in (a, b) and continuous from the right at a and from the left at b.

Theorem 3. *If $f(x)$ has a derivative at the point a* (Definition 11.5.1), *then $f(x)$ is continuous at a.*

REMARK 5. The converse theorem is not valid (cf. Remark 11.5.3, p. 417).

Theorem 4. *If $f(x)$ and $g(x)$ are continuous at a, then the functions $k \,.\, f(x)$ (k being a constant), $f(x) + g(x)$, $f(x) - g(x)$, $f(x) \,.\, g(x)$ are also continuous at a; if $g(a) \neq 0$ then also $f(x)/g(x)$ is continuous at a.* A similar theorem is true on the continuity from the right and from the left.

Theorem 5. *A continuous function composed of continuous functions is also continuous.* Precisely: *If $f(x)$ is continuous at a and $g(z)$ continuous at the corresponding point $z_0 = f(a)$, then the function $y = g(f(x))$* (see Definition 11.1.4) *is continuous at the point a.*

REMARK 6. On the basis of Theorems 4 and 5 it is easy to show that the great majority of functions we meet with in applications are continuous functions. Especially all polynomials and rational functions with non-vanishing denominators are continuous functions. The function $y = \sqrt{x}$ is continuous for $x \geqq 0$. Further, trigonometric functions are continuous (with the exception of points in the neighbourhood of which they are not bounded, e.g. the function $\tan x$ is not continuous at the point $\frac{1}{2}\pi$), inverse trigonometric functions, exponential functions, logarithmic functions and functions generated from them by addition, subtraction, multiplication and division (with non-vanishing divisors) as well as their composite functions are continuous.

Definition 5 (*Points of Discontinuity*). We say that the point a is a *point of discontinuity of the first kind* for the function $f(x)$ (the point a_1 in Fig. 11.10) if there exist a finite right-hand limit and a finite left-hand limit of the function $f(x)$ at the point a (denoted by the symbols $f(a + 0)$ or $f(a - 0)$, respectively, see Remark 11.4.2, p. 410) and if $f(a + 0) \neq f(a - 0)$. We call the number $f(a + 0) - f(a - 0)$ *the jump of the function $f(x)$ at the point a.*

If at least one of the one-sided limits does not exist then we call the point a a *point of discontinuity of the second kind* of the function $f(x)$ (the point a_3 in Fig. 11.10).

If the finite limit $\lim_{x \to a} f(x) = A$ exists but either the function $f(x)$ is not defined at the point a or $f(a) \neq A$, then we say that $f(x)$ has a *removable discontinuity at the point a* (the point a_2 in Fig. 11.10).

Example 2. The function $f(x) = \sin(1/x)$ has a discontinuity of the second kind at the point $x = 0$. The function represented in Fig. 11.9 has a discontinuity of the first kind at the points a and b. The function

$$g(x) = \frac{\sin x}{x}$$

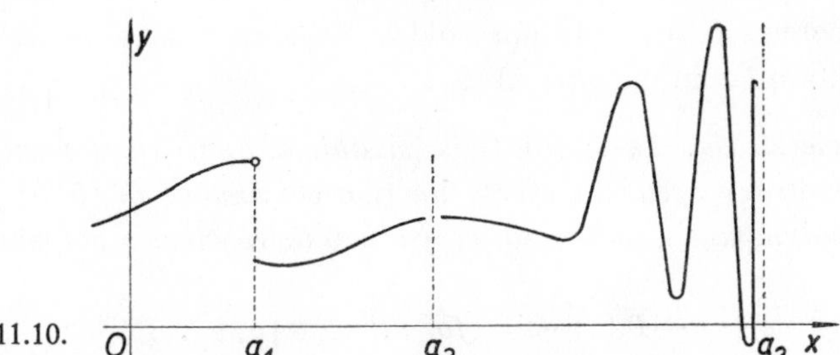

Fig. 11.10.

has a removable discontinuity at the point $x = 0$ because it is not defined at the point $x = 0$ but

$$\lim_{x \to 0} \frac{\sin x}{x} = 1$$

(cf. Theorem 11.4.9, p. 415, formula 2).

Definition 6. A function $f(x)$ defined in the interval $[a, b]$ is called *sectionally* or *piecewise continuous in the interval* $[a, b]$ if it is continuous in $[a, b]$ except at a finite number of points of discontinuity of the first kind.

Example 3. The function illustrated in Fig. 11.9 is sectionally continuous in the interval $[0, l]$.

Theorem 6. *A function $f(x)$ continuous in $[a, b]$ takes on a greatest and a least value in $[a, b]$.* Precisely: *There exist at least one point $x_1 \in [a, b]$ such that $f(x_1) \geqq f(x)$ for all $x \in [a, b]$ and at least one point $x_2 \in [a, b]$ such that $f(x_2) \leqq f(x)$ for all $x \in [a, b]$.*

Remark 7. There may be several such points. For example, the function $y = \sin x$ attains in the interval $[-2\pi, 2\pi]$ its maximum value at the points $-\frac{3}{2}\pi$ and $\frac{1}{2}\pi$, its minimum value at the points $-\frac{1}{2}\pi$ and $\frac{3}{2}\pi$.

Theorem 7. *Let $f(x)$ be continuous in $[a, b]$, $f(a) \neq f(b)$, let c be any number between $f(a)$ and $f(b)$ (i.e. either $f(a) < c < f(b)$ or $f(a) > c > f(b)$). Then there exists at least one point $x_0 \in (a, b)$ such that $f(x_0) = c$.*

REMARK 8. Thus a function continuous in $[a, b]$ assumes every value between $f(a)$ and $f(b)$. Especially, if $f(x)$ is continuous in $[a, b]$ and $f(a) . f(b) < 0$, then $f(x)$ has at least one zero in (a, b).

Theorem 8. *A function $f(x)$, continuous in the interval $[a, b]$, is uniformly continuous in this interval, that is, to any arbitrary $\varepsilon > 0$ there exists $\delta > 0$ (depending only on the choice of the number ε) such that for every two points x_1, x_2 from the interval $[a, b]$, the distance of which is smaller than δ, the relation*

$$|f(x_1) - f(x_2)| < \varepsilon$$

holds.

REMARK 9. Theorems 6 and 8 do not hold in an *open* interval as we can easily verify for the function $1/x$ in the interval $(0, 1)$.

Theorem 9. (*Weierstrass's Theorem*). *It is possible to approximate uniformly in $[a, b]$ with an arbitrary accuracy every function continuous in $[a, b]$ by means of a sequence of polynomials, that is, to every $\varepsilon > 0$ there exists a polynomial $P_n(x)$ such that*

$$|f(x) - P_n(x)| < \varepsilon \quad \textit{for all} \quad x \in [a, b].$$

Definition 7. Let $f(x)$ be defined in $[a, b]$. Let us divide this interval into subintervals by means of the points $a = x_0 < x_1 < x_2 < \ldots < x_{n-1} < x_n = b$ and let us form the sum

$$V = \sum_{k=1}^{n} |f(x_k) - f(x_{k-1})| .$$

V is a non-negative number depending on the choice of the points of division $x_1, x_2, \ldots$ $\ldots, x_{n-1}$. If we choose all possible n and all possible divisions of the interval $[a, b]$, then the numbers V form a set of non-negative numbers. Its lowest upper bound (Definition 1.3.3) is called the *variation* (or more precisely the *total variation*) *of the function $f(x)$ in the interval $[a, b]$*. We denote this by

$$\overset{b}{\underset{a}{V}}(f) .$$

If $\overset{b}{\underset{a}{V}}(f)$ is a finite number, then $f(x)$ is said to be *of bounded variation in* $[a, b]$.

Example 4.

$$\overset{2\pi}{\underset{0}{V}}(\cos x) = 4$$

(it is sufficient to divide the interval $[0, 2\pi]$ into the intervals $[0, \pi]$, $[\pi, 2\pi]$ in which $\cos x$ is decreasing or increasing, respectively.

Theorem 10. *If $f(x)$ has a bounded derivative in $[a, b]$ or if $f(x)$ is monotonic in $[a, b]$ or if $f(x)$ is continuous and attains a finite number of maximum and minimum values in $[a, b]$, then $f(x)$ is of bounded variation in $[a, b]$.*

Theorem 11. *$f(x)$ is a function of bounded variation if and only if it can be expressed as the difference of two non-decreasing functions.*

11.4. Limit. Infinite Limits. Evaluation of Limits. Some Important Limits. Symbols $O(g(x))$, $o(g(x))$

Definition 1. We say that $f(x)$ has the *limit A at the point a* (in more detail *the finite limit A*) if to any arbitrary $\varepsilon > 0$ there exists a $\delta > 0$ (depending in general

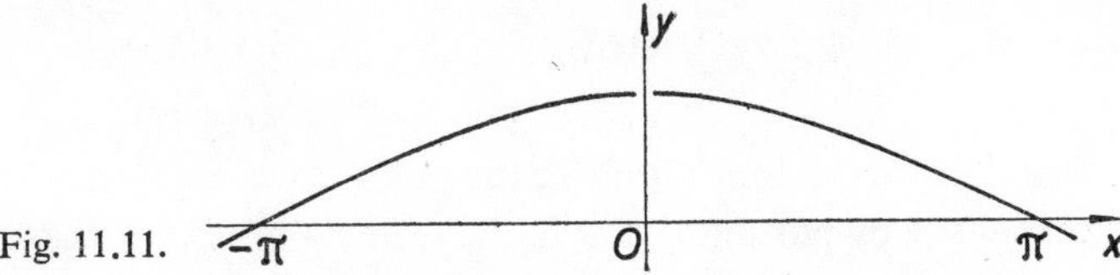

Fig. 11.11.

on the choice of the number ε) such that for all x from the δ-neighbourhood of the point a, *from which we exclude the point a* (i.e. for all x, for which $0 < |x - a| < \delta$) the relation

$$|f(x) - A| < \varepsilon$$

holds.

We write

$$\lim_{x \to a} f(x) = A .$$

REMARK 1. Roughly speaking: $f(x)$ has the limit A at the point a, if $f(x)$ differs from the number A by as little as we please when x is sufficiently near to the point a.

Example 1. The function

$$y = \frac{\sin x}{x}$$

(Fig. 11.11) is not defined at the point $a = 0$ (therefore it cannot be continuous at that point, cf. Remark 11.3.2) but it has a limit at that point equal to 1 (see Theorem 9). This example shows why we exclude the point a (here the point zero) from our consideration, because our function need not be defined at this point at all.

REMARK 2. The *limit from the right* or the *limit from the left* is defined quite analogously as in Definition 1 (we take into consideration only those x that lie in the "right" or "left" neighbourhood of the point a while the point a is excluded). We write

$$\lim_{x \to a+} f(x) = A \quad \text{or} \quad \lim_{x \to a-} f(x) = B$$

or often

$$f(a + 0) = A \quad \text{or} \quad f(a - 0) = B .$$

Example 2. For the function represented in Fig. 11.9 we see that $f(a + 0) = c$, $f(a - 0) = 0$.

Theorem 1. *The function $f(x)$ has a limit at the point a if and only if it has both limits from the right and from the left at this point and if these two limits are equal.*

Theorem 2. *The function $f(x)$ is continuous at the point a if and only if it is defined at this point and if*

$$\lim_{x \to a} f(x) = f(a) .$$

REMARK 3. If the function $f(x)$ is continuous at the point a, we compute the limit of $f(x)$ at that point very easily by putting $x = a$ in the formula of $f(x)$. For example, the function $y = \sin x$ is continuous at the point $a = \frac{1}{3}\pi$, hence

$$\lim_{x \to \pi/3} \sin x = \sin \frac{\pi}{3} = \frac{\sqrt{3}}{2} .$$

Theorem 3. *The function $f(x)$ has a (finite) limit at the point a if and only if the Bolzano–Cauchy condition is satisfied: To any arbitrary $\varepsilon > 0$ there exists a $\delta > 0$ such that for every pair of numbers x_1, x_2, $0 < |x_1 - a| < \delta$, $0 < |x_2 - a| < \delta$ the relation $|f(x_1) - f(x_2)| < \varepsilon$ holds.*

Theorem 4. *If $f(x)$ has the limit A and $g(x)$ the limit B both at the point a, then the functions $k\,.\,f(x)$ (k = const), $f(x) \pm g(x)$, $f(x)\,.\,g(x)$, $f(x)/g(x)$* (if $B \neq 0$) *have limits at the point a and the relations*

$$\lim_{x \to a} k\,.\,f(x) = kA\,, \quad \lim_{x \to a} [f(x) \pm g(x)] = A \pm B$$

$$\lim_{x \to a} f(x)\,g(x) = AB\,, \quad \lim_{x \to a} \frac{f(x)}{g(x)} = \frac{A}{B}$$

hold. A similar theorem holds for the limit from the right or from the left.

REMARK 4. This theorem facilitates the practical computation of limits of many functions; cf. the similar Remark 11.3.6.

Theorem 5 (*Limit of Composite Functions*). *Let*

$$\lim_{x \to a} f(x) = A\,, \quad \lim_{z \to A} g(z) = B$$

and let $\delta > 0$ exist such that for all x, for which

$$0 < |x - a| < \delta, \text{ the relation} f(x) \neq A \text{ holds}. \tag{1}$$

Then

$$\lim_{x \to a} g(f(x)) = B.$$

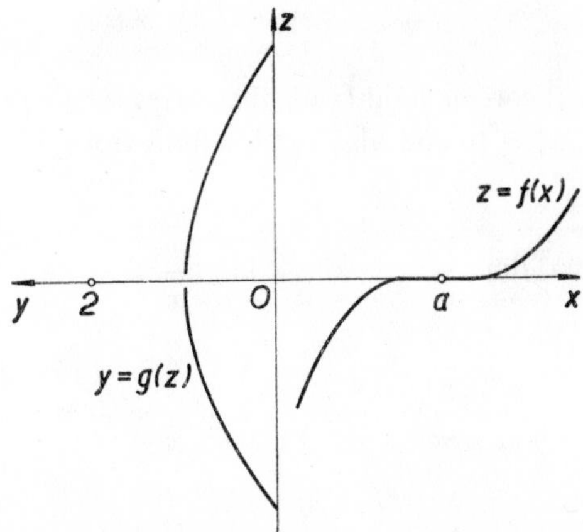

Fig. 11.12.

REMARK 5. If condition (1) is not satisfied, an incorrect result may be obtained as we can see from Fig. 11.12 where $f(x) = 0$ in some neighbourhood of the point a. Because here $A = f(a) = 0$, $g(0) = 2$, we have

$$\lim_{x \to a} g(f(x)) = 2$$

whereas

$$B = \lim_{z \to 0} g(z) = 1.$$

Definition 2 (*Infinite Limit*). We say that $f(x)$ has the *infinite limit* $+\infty$ *at the point* a (we write $\lim_{x \to a} f(x) = +\infty$) if to any (arbitrarily great) number $K > 0$ there exists a $\delta > 0$ such that for all x from the δ-neighbourhood of the point a (except the point a itself) the relation $f(x) > K$ holds.

REMARK 6. We can give similar definitions of the *infinite limit* $-\infty$ and of the *infinite limit from the right* or *from the left*. For example, the infinite limit $-\infty$ from the right is defined in this way:

Definition 3. We say that the function $f(x)$ has the *infinite limit* $-\infty$ *from the right at the point* a if to any (arbitrarily great) number $L > 0$ there exists a $\delta > 0$ such that for all x from the interval $(a, a + \delta)$ the relation

$$f(x) < -L$$

holds.

Theorem 6. *If $f(x)$ is continuous from the right at the point a and $f(a) > 0$, and if for a certain $\delta > 0$ $g(x) > 0$ in the interval $(a, a + \delta)$ and*

$$\lim_{x \to a+} g(x) = 0 ,$$

then

$$\lim_{x \to a+} \frac{f(x)}{g(x)} = +\infty .$$

REMARK 7. Similar theorems hold (with the corresponding sign) for various combinations $f(a) \gtrless 0$, $g(x) \gtrless 0$, and also for the limit from the left and for the limit.

Example 3.

$$\lim_{x \to 0+} \frac{1}{x} = +\infty , \quad \lim_{x \to 0-} \frac{1}{x} = -\infty .$$

For $b > 0$,

$$\lim_{x \to 0+} \frac{b}{x} = +\infty , \quad \lim_{x \to 0-} \frac{b}{x} = -\infty ;$$

for $b < 0$,

$$\lim_{x \to 0+} \frac{b}{x} = -\infty , \quad \lim_{x \to 0-} \frac{b}{x} = +\infty .$$

Example 4.

$$\lim_{x \to 2+} \frac{2x^2 - 5x + 1}{x^2 - x - 2} = \lim_{x \to 2+} \frac{\dfrac{2x^2 - 5x + 1}{x + 1}}{x - 2} = -\infty ,$$

because $x^2 - x - 2 = (x - 2)(x + 1)$, and if we write

$$f(x) = \frac{2x^2 - 5x + 1}{x + 1} , \quad g(x) = x - 2,$$

then

$$f(2) = -\tfrac{1}{3} < 0 , \quad g(x) > 0 \quad \text{for} \quad x > 2 \quad \text{and} \quad \lim_{x \to 2+} g(x) = 0 .$$

The limit from the left at the point $x = 2$ is $+\infty$ because $g(x) < 0$ for $x < 2$.

REMARK 8. If the numerator of the quotient $f(x)/g(x)$ also vanishes or if both $f(x)$ and $g(x)$ become infinite as $x \to a$ (an expression of the type $0/0$ or ∞/∞), then, in many cases, we can conclude whether the quotient has a limit, by means of l'Hospital's Rule (Theorems 11.8.1, 11.8.2).

If, however, $f(x)$ is bounded in the neighbourhood of the point a (i.e. $|f(x)| < M$) and

$$\lim_{x \to a} g(x) = +\infty \quad \text{or} \quad \lim_{x \to a} g(x) = -\infty ,$$

then

$$\lim_{x \to a} \frac{f(x)}{g(x)} = 0$$

(and similarly for the limit from the right or from the left).

Example 5.

$$\lim_{x \to 0+} \frac{\sin \dfrac{1 + x^2}{x}}{e^{1/x}} = 0$$

because

$$\left| \sin \frac{1 + x^2}{x} \right| \leqq 1$$

for all $x \neq 0$ and

$$\lim_{x \to 0+} e^{1/x} = +\infty .$$

Definition 4 (*Limits at the Points at Infinity*). We say that $f(x)$ has the *limit A at the point at infinity* $+\infty$ (or briefly, *at the point* $+\infty$) if to an arbitrary $\varepsilon > 0$ there exists an x_0 such that for all $x > x_0$

$$|f(x) - A| < \varepsilon .$$

We write

$$\lim_{x \to +\infty} f(x) = A .$$

REMARK 9. The definition of the limit at the point at infinity $-\infty$ is analogous. The definitions of the *infinite* limit at the points at infinity $+\infty$ or $-\infty$ are analogous. For example:

Definition 5. We say that the function $f(x)$ has the *infinite limit* $+\infty$ *at the point at infinity* $-\infty$ if to an arbitrary $K > 0$ there exists such a number x_0 that for all $x < x_0$

$$f(x) > K .$$

We write

$$\lim_{x \to -\infty} f(x) = +\infty .$$

Example 6.

$$\lim_{x \to +\infty} \frac{1}{x} = 0 , \quad \lim_{x \to +\infty} x = +\infty , \quad \lim_{x \to -\infty} x^3 = -\infty ,$$

$$\lim_{x \to +\infty} \frac{x^2 - 2x + 3}{2x - 1} = \lim_{x \to +\infty} x \frac{1 - 2/x + 3/x^2}{2 - 1/x} = +\infty$$

(cf. Remark 12).

REMARK 10. We often write ∞ only instead of $+\infty$.

REMARK 11. L'Hospital's rule may also be applied to find limits of the form $0/0$, ∞/∞ at the points at infinity (§ 11.8).

REMARK 12. For computation with finite limits at points at infinity we apply Theorem 4. For computation with infinite limits as $x \to a$ or $x \to +\infty$ or $x \to -\infty$ we may apply Theorem 4 unless the result is of an "indeterminate form" ($0 . \infty$, $\infty - \infty$, etc.). For example,

$$\lim_{x \to 0+} \frac{\sin x}{x^2} = \lim_{x \to 0+} \frac{\sin x}{x} . \frac{1}{x} = 1 . (+\infty) = +\infty .$$

We naturally obtain the same result if we apply Theorem 11.8.1. (Cf. also Example 6.)

Theorem 7. *If $f(x) \leqq g(x)$, for all x in some neighbourhood of the point a (with the possible exception of the point a itself), then $\lim_{x \to a} f(x) \leqq \lim_{x \to a} g(x)$ if both these limits exist. A similar theorem holds for the limit from the right or from the left. (If $a = +\infty$ or $a = -\infty$, then we take into consideration all x greater (or smaller) then a certain number x_0 (instead of a "neighbourhood of the point a").)*

REMARK 13. If $f(x) < g(x)$ for all x in a neighbourhood of the point a, then the relation $\lim_{x \to a} f(x) < \lim_{x \to a} g(x)$ need not hold, the sign of equality may also be valid. For example, in a sufficiently small neighbourhood of the origin $|x| > x^2$ for all x ($x \neq 0$), but $\lim_{x \to 0} |x| = \lim_{x \to 0} x^2 = 0$.

Theorem 8. *If $f(x) \leqq g(x) \leqq h(x)$ for all x in a neighbourhood of the point a (with the possible exception of the point a itself) and if the limits $\lim_{x \to a} f(x) = A$, $\lim_{x \to a} h(x) = A$ exist, then $\lim_{x \to a} g(x)$ exists also and is equal to A.*

REMARK 14. Theorem 8 has a simple geometrical interpretation: If the graph of $g(x)$ lies between the graphs of $f(x)$ and $h(x)$ and if both $f(x)$ and $h(x)$ tend to the same value as $x \to a$, then $g(x)$ also tends to the same value.

Definition 6. We say that $f(x)$ is *of the order $O(g(x))$ in the neighbourhood of the point a if the expression* $|f(x)/g(x)|$ is bounded for all $x \neq a$ from a certain neighbourhood of the point a. If

$$\lim_{x \to a} \left| \frac{f(x)}{g(x)} \right| = 0 ,$$

we say that $f(x)$ is *of the order $o(g(x))$* or *of a smaller order than $g(x)$, in the neighbourhood of the point a.*

REMARK 15. We naturally demand that the expression $|f(x)/g(x)|$ has sense in the neighbourhood of the point a ($x \neq a$), i.e. that $g(x) \neq 0$.

REMARK 16. One can see from the definition that if $f(x) = o(g(x))$, then also $f(x) = O(g(x))$; however, in general, the converse is not true.

REMARK 17. By the point a in Definition 6 we may understand also the point at infinity.

Example 7. $\sin x = O(x)$ in the neighbourhood of the point $x = 0$ because

$$\lim_{x\to 0}\left|\frac{\sin x}{x}\right| = 1\,;$$

$x^3 = o(e^x)$ for $x \to +\infty$ because $\lim\limits_{x\to+\infty} \dfrac{x^3}{e^x} = 0$.

Theorem 9 (*Some Important Limits*).

1. $\lim\limits_{x\to+\infty}\left(1+\dfrac{a}{x}\right)^x = e^a\,,\quad \lim\limits_{x\to-\infty}\left(1+\dfrac{a}{x}\right)^x = e^a\,,$

2. $\lim\limits_{x\to 0}\dfrac{\sin x}{x} = 1\,,\quad \lim\limits_{x\to 0}\dfrac{\tan x}{x} = 1\,,$

3. $\lim\limits_{x\to+\infty} a^x = +\infty\quad (a > 1)\,,\quad \lim\limits_{x\to-\infty} a^x = 0\quad (a > 1)\,,$

4. $\lim\limits_{x\to+\infty} a^x = 0\ (0 < a < 1)\,,\quad \lim\limits_{x\to-\infty} a^x = +\infty\ (0 < a < 1)\,,$

5. $\lim\limits_{x\to 0}\dfrac{e^x - 1}{x} = 1\,,\quad \lim\limits_{x\to 0}\dfrac{a^x - 1}{x} = \ln a\ (a > 0)\,,$

6. $\lim\limits_{x\to+\infty}\dfrac{x^n}{e^{kx}} = 0\quad (k > 0,\ n\ \textit{arbitrary})\,,$

7. $\lim\limits_{x\to+\infty}\dfrac{(\ln x)^n}{x^\alpha} = 0\quad (\alpha > 0,\ n\ \textit{arbitrary})\,,$

   $\lim\limits_{x\to 0+} x^\alpha(-\ln x)^n = 0\quad (\alpha > 0,\ n\ \textit{arbitrary})\,.$

Especially $\lim\limits_{x\to 0+} x \ln x = 0$.

11.5. Derivative. Formulae for Computing Derivatives. Derivatives of Composite and Inverse Functions

Definition 1. If there exists the (finite) limit

$$\lim_{h\to 0}\frac{f(a+h) - f(a)}{h}\,, \tag{1}$$

we say that $f(x)$ has a *derivative at the point a*. The corresponding limit is denoted by $f'(a)$.

REMARK 1. The geometrical representation of the number $f'(a)$ is, as we can see from Fig. 11.13, the slope of the tangent to the curve, given by the equation $y = f(x)$, at the point a (because the tangent at the point a is the limiting position of the chord for $h \to 0$). In dynamics, if x stands for time, y for the length of path traversed by a particle up to the instant x, $y = f(x)$ for the equation of motion, then the meaning of the derivative is the limit of the average velocity, i.e. the instantaneous velocity v at the instant a.

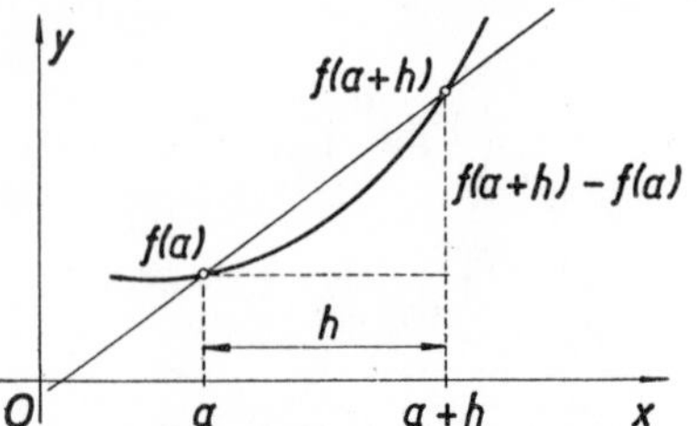

Fig. 11.13.

Definition 2. If there exists in (1) only the limit from the right or from the left, then we say that $f(x)$ has a *right-hand* or *left-hand derivative at the point a*. We write $f'_+(a)$ or $f'_-(a)$.

Theorem 1. *A necessary and sufficient condition that $f(x)$ has a derivative at the point a is that it has a right-hand and left-hand derivative at the point a and* $f'_+(a) = f'_-(a)$.

Definition 3. If (1) is an infinite limit (Definition 11.4.2) or an infinite limit from the right (or from the left), we say that $f(x)$ has an *infinite derivative at the point a* or an *infinite right-hand* (or *left-hand*) *derivative.* (One of the possible geometrical interpretations: The tangent to the curve $y = f(x)$ is vertical.)

REMARK 2. *When we say that $f(x)$ has a derivative at the point a, we shall always mean a finite derivative.*

Definition 4. If $f(x)$ has a derivative at every point $x \in (a, b)$, we say that $f(x)$ is *differentiable in the interval* (a, b) or that it has a *derivative in the interval* (a, b). Current notations for derivatives:

$$f'(x), \quad y'(x), \quad \frac{\mathrm{d}f}{\mathrm{d}x}, \quad \frac{\mathrm{d}y}{\mathrm{d}x}, \quad \frac{\mathrm{d}}{\mathrm{d}x}f(x), \quad \frac{\mathrm{d}}{\mathrm{d}x}y(x), \quad y', \quad f', \quad [f(x)]'.$$

If $f(x)$ is differentiable in (a, b) and if it also has a right-hand derivative at a and a left-hand derivative at b, we say that $f(x)$ is *differentiable in* $[a, b]$ or that $f(x)$ has a *derivative in* $[a, b]$.

Definition 5. A function $y = f(x)$ that has a continuous derivative in $[a, b]$ is called a *smooth function in* $[a, b]$. The curve $y = f(x)$, its graphical representation, is also said to be *smooth.*

Definition 6. If there exists the (finite) limit

$$\lim_{h \to 0} \frac{f'(a + h) - f'(a)}{h},$$

we say that $f(x)$ has a *second derivative at a.* We write it as $f''(a)$. Similarly we define *higher derivatives.* We write them as $f'''(a)$, $f^{(4)}(a)$, $f^{(5)}(a)$, etc.

Theorem 2. *If $f(x)$ has a derivative at the point a, then it is continuous at this point.*

REMARK 3. The converse is not true as we can see when $f(x) = |x|$ (Fig. 11.14), which is continuous at the origin but has no derivative there (it has there the right-hand derivative equal to $+1$ and the left-hand derivative equal to -1).

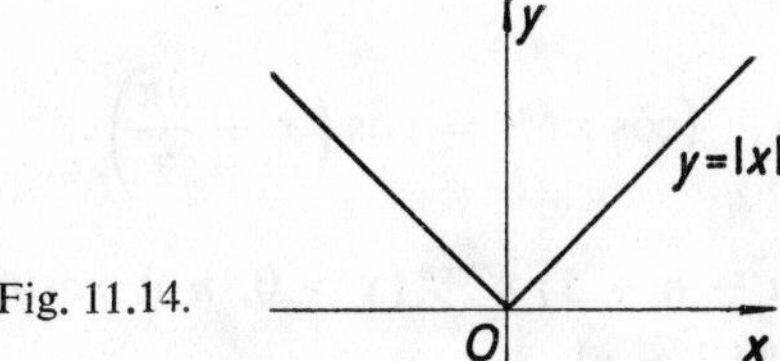

Fig. 11.14.

Theorem 3 (*Fundamental Formulae*). *We denote a derivative by a dash. Unless the contrary is stated, all the formulae are valid for all x and every value of the constants referred to.*

1. $(x^n)' = nx^{n-1}$ ($x > 0$, *n arbitrary*). It is possible to enlarge the domain of validity for the variable x for some n. For example, $(x^3)' = 3x^2$ holds for all x; $(x^{1/3})' = \frac{1}{3}x^{-2/3}$ holds for all $x \neq 0$.

2. If $f(x) = \text{const.}$, *then* $f'(x) = 0$.

3. $(a^x)' = a^x \ln a \quad (a > 0)\,, \quad (e^x)' = e^x\,.$

4. $(\log_a x)' = \dfrac{1}{x \ln a} \quad (a > 0, a \neq 1, x > 0)\,, \quad (\ln x)' = \dfrac{1}{x} \quad (x > 0)\,,$

$[\ln (kx)]' = \dfrac{1}{x}$ (*k arbitrary but such that* $kx > 0$).

5. $(\sin x)' = \cos x\,, \quad (\cos x)' = -\sin x\,.$

6. $(\tan x)' = \dfrac{1}{\cos^2 x} \quad (x \neq \pm\frac{1}{2}\pi,\ \pm\frac{3}{2}\pi, \ldots)\,,$

$$(\cot x)' = -\frac{1}{\sin^2 x} \quad (x \neq 0, \pm\pi, \pm 2\pi, \ldots).$$

7. $$(\arcsin x)' = \frac{1}{\sqrt{(1 - x^2)}} \quad (|x| < 1),$$

$$(\arccos x)' = -\frac{1}{\sqrt{(1 - x^2)}} \quad (|x| < 1).$$

8. $$(\arctan x)' = \frac{1}{1 + x^2}, \quad (\operatorname{arccot} x)' = -\frac{1}{1 + x^2}.$$

9. $$(\sinh x)' = \cosh x, \quad (\cosh x)' = \sinh x.$$

10. $$(\tanh x)' = \frac{1}{\cosh^2 x}, \quad (\coth x)' = -\frac{1}{\sinh^2 x} \quad (x \neq 0).$$

Some other common formulae:

11. $$(\sin x)^{(n)} = \sin\left(x + \frac{n\pi}{2}\right), \quad (\cos x)^{(n)} = \cos\left(x + \frac{n\pi}{2}\right).$$

12. $(x^m)^{(n)} = m(m - 1) \ldots (m - n + 1)\, x^{m-n}$ ($x > 0$, *n a positive integer, m arbitrary*), *in particular* $(x^n)^{(n)} = n!$.

13. $$(a^x)^{(n)} = a^x(\ln a)^n \ (a > 0), \quad (e^x)^{(n)} = e^x,$$

$$(\ln x)^{(n)} = (-1)^{n-1} \frac{(n - 1)!}{x^n} \quad (x > 0).$$

14. $$(\sin ax)' = a \cos ax, \quad (\cos ax)' = -a \sin ax.$$

15. $$\left(\arcsin \frac{x}{a}\right)' = \frac{1}{\sqrt{(a^2 - x^2)}} \quad (a > 0, |x| < a),$$

$$= -\frac{1}{\sqrt{(a^2 - x^2)}} \quad (a < 0, |x| < |a|),$$

$$\left(\arccos \frac{x}{a}\right)' = -\frac{1}{\sqrt{(a^2 - x^2)}} \quad (a > 0, |x| < a),$$

$$= \frac{1}{\sqrt{(a^2 - x^2)}} \quad (a < 0, |x| < |a|).$$

16. $$\left(\arctan \frac{x}{a}\right)' = \frac{a}{a^2 + x^2}, \quad \left(\operatorname{arccot} \frac{x}{a}\right)' = -\frac{a}{a^2 + x^2} \quad (a \neq 0).$$

17. $(\ln |x + \sqrt{(x^2 + a)}|)' = \dfrac{1}{\sqrt{(x^2 + a)}} \quad (a \neq 0, x^2 + a > 0)$.

18. $[\sin (ax + b)]' = a \cos (ax + b), \quad [\cos (ax + b)]' = -a \sin (ax + b)$.

19. $(e^{ax+b})' = ae^{ax+b}$.

20. $[\ln (ax + b)]' = \dfrac{a}{ax + b} \quad (ax + b > 0)$.

21. $(x^x)' = x^x(1 + \ln x) \quad (x > 0)$.

22. $[f(x)^{g(x)}]' = [e^{g(x)\ln f(x)}]' = f(x)^{g(x)} \left[g'(x) \ln f(x) + \dfrac{g(x) f'(x)}{f(x)} \right] \quad (f(x) > 0)$.

23. $(\operatorname{arsinh} x)' = \dfrac{1}{\sqrt{(1 + x^2)}}, \quad (\operatorname{arcosh} x)' = \dfrac{1}{\sqrt{(x^2 - 1)}} \quad (x > 1)$,

$(\operatorname{artanh} x)' = \dfrac{1}{1 - x^2} \quad (-1 < x < 1), \quad (\operatorname{arcoth} x)' = -\dfrac{1}{x^2 - 1} \quad (|x| > 1)$.

24. $\left(\operatorname{arsinh} \dfrac{x}{a}\right)' = \dfrac{1}{\sqrt{(a^2 + x^2)}} \quad (a > 0)$,

$= -\dfrac{1}{\sqrt{(a^2 + x^2)}} \quad (a < 0)$,

$\left(\operatorname{arcosh} \dfrac{x}{a}\right)' = \dfrac{1}{\sqrt{(x^2 - a^2)}} \quad (x > a > 0)$,

$= -\dfrac{1}{\sqrt{(x^2 - a^2)}} \quad (x < a < 0)$.

Theorem 4. *If the functions $f(x)$, $g(x)$ have a derivative at the point a, then also the functions $k \,.\, f(x)$ ($k =$ const.), $f(x) \pm g(x)$, $f(x)\, g(x)$ and, if $g(a) \neq 0$, $f(x)/g(x)$ each have a derivative at the point a. Moreover* (briefly written),

$$(kf)' = kf', \quad (f \pm g)' = f' \pm g', \quad (fg)' = f'g + fg', \quad \left(\frac{f}{g}\right)' = \frac{f'g - fg'}{g^2}.$$

Example 1. Applying the rule for differentiation of a quotient we find that for all $x \neq 0$ the function

$$y = \frac{\sin x}{x}$$

has the derivative

$$y' = \frac{x \cos x - \sin x}{x^2}.$$

Theorem 5 (*Differentiation of a Composite Function or the Chain Rule*). *Let* $y = g(f(x))$ *be a composite function where* $y = g(z)$, $z = f(x)$ (Definition 11.1.4). *If the function* $z = f(x)$ *has a derivative with respect to* x *at the point* a *and if the function* $y = g(z)$ *has a derivative with respect to* z *at the point* $z_0 = f(a)$, *then* $y = g(f(x))$ *has a derivative with respect to* x, *at the point* a, *equal to* $g'(z_0) \, . \, f'(a)$; *in short notation*

$$\frac{dy}{dx} = \frac{dy}{dz} \cdot \frac{dz}{dx} \, . \tag{2}$$

REMARK 4. We therefore compute the derivative of the function $g(f(x))$ with respect to x at the point a by multiplying the derivative of the function $y = g(z)$ (with respect to z) at the point $z_0 = f(a)$ by the derivative of the function $z = f(x)$ (with respect to x) at the point a. A more exact form of equation (2) is

$$\frac{dy}{dx} = \frac{dg}{dz} \cdot \frac{df}{dx} \, . \tag{3}$$

Example 2. We express the function $y = \sin^5 x$ as a composite function as follows: $y = z^5$, $z = \sin x$. From (2), or (3) we have

$$y' = \frac{dy}{dx} = \frac{dy}{dz}\frac{dz}{dx} = 5z^4 \cos x = 5 \sin^4 x \cos x \, .$$

Example 3. If $y = \sin(x^5)$, we choose $y = \sin z$, $z = x^5$; hence

$$y' = \frac{dy}{dx} = \frac{dy}{dz}\frac{dz}{dx} = \cos z \, . \, 5x^4 = 5x^4 \cos(x^5) \, .$$

Theorem 6 (*Derivative of an Inverse Function*). *If* $y = f(x)$ *is the inverse function of the function* $x = g(y)$ (Definition 11.1.5) *and if* $g(y)$ *has a non-zero derivative (with respect to* y*) at the point* y_0, *then the function* $y = f(x)$ *has a derivative (with respect to* x*) at the corresponding point* $x_0 = g(y_0)$ *and the relation*

$$\frac{dy}{dx} = \frac{1}{\dfrac{dx}{dy}} \quad \left(\text{or, more exactly: } \frac{df}{dx} = \frac{1}{\dfrac{dg}{dy}}\right)$$

holds.

Example 4.

$$y = \arcsin x \quad (\text{i.e. } y = \sin^{-1} x) \, , \quad -1 \leqq x \leqq 1 \, ,$$

is the inverse function of $x = \sin y$ $(-\frac{1}{2}\pi \leqq y \leqq \frac{1}{2}\pi)$ (Example 11.1.5, p. 401). Then

$$\frac{dx}{dy} = \cos y \neq 0 \quad \text{for} \quad y \neq \pm\tfrac{1}{2}\pi \, ,$$

hence

$$\frac{dy}{dx} = \frac{1}{\cos y} = \frac{1}{+\sqrt{(1-\sin^2 y)}} = \frac{1}{\sqrt{(1-x^2)}} \quad (-1 < x < 1).$$

(We take the positive root because $\cos y > 0$ when $-\frac{1}{2}\pi < y < \frac{1}{2}\pi$.)

Theorem 7. *The derivatives of a function given parametrically by equations* $x = \varphi(t)$, $y = \psi(t)$:

$$y' = \frac{dy}{dx} = \frac{\psi'(t)}{\varphi'(t)}, \quad y'' = \frac{d}{dx}(y') = \frac{d}{dt}(y')\frac{dt}{dx} = \frac{\psi''(t)\,\varphi'(t) - \psi'(t)\,\varphi''(t)}{[\varphi'(t)]^3},$$

$$y''' = \frac{d^3y}{dx^3} = \frac{\varphi'^2\psi''' - \varphi'\psi'\varphi''' - 3\varphi'\varphi''\psi'' + 3\varphi''^2\psi'}{\varphi'^5} \quad (\varphi'(t) \neq 0).$$

Example 5. For the ellipse $x = a\cos t$, $y = b\sin t$ we have

$$y' = \frac{b\cos t}{-a\sin t} = -\frac{b}{a}\cot t = -\frac{b^2}{a^2}\frac{x}{y}, \quad t \neq k\pi \quad (\text{where } k \text{ is an integer}).$$

REMARK 5. Similarly, for polar coordinates

$$x = r\cos\varphi, \quad y = r\sin\varphi,$$

the relations

$$dx = dr\cos\varphi - r\sin\varphi\,d\varphi, \quad dy = dr\sin\varphi + r\cos\varphi\,d\varphi$$

hold; hence

$$\frac{dy}{dx} = \frac{\sin\varphi\dfrac{dr}{d\varphi} + r\cos\varphi}{\cos\varphi\dfrac{dr}{d\varphi} - r\sin\varphi},$$

etc.

Some formulae that find frequent application:

Theorem 8. *For the derivative of a product of functions* $u_1(x), u_2(x), \ldots, u_n(x)$ *the following rule holds:*

$$(u_1u_2u_3\ldots u_n)' = u_1'u_2u_3\ldots u_n + u_1u_2'u_3\ldots u_n + \ldots + u_1u_2u_3\ldots u_n' =$$

$$= u_1u_2u_3\ldots u_n\left(\frac{u_1'}{u_1} + \frac{u_2'}{u_2} + \frac{u_3'}{u_3} + \ldots + \frac{u_n'}{u_n}\right)$$

(*we apply the last form if* $u_1 \neq 0$, $u_2 \neq 0, \ldots$). *In another form:*

$$\frac{(u_1 u_2 u_3 \ldots u_n)'}{u_1 u_2 u_3 \ldots u_n} = \frac{u_1'}{u_1} + \frac{u_2'}{u_2} + \frac{u_3'}{u_3} + \ldots + \frac{u_n'}{u_n}.$$

Theorem 9 (*Leibniz's Rule*).

$$(uv)^{(n)} = u^{(n)}v^{(0)} + \binom{n}{1} u^{(n-1)}v^{(1)} + \binom{n}{2} u^{(n-2)}v^{(2)} + \ldots + u^{(0)}v^{(n)}.$$

The upper indices in brackets stand for the order of the derivative; $u^{(0)} = u$, $v^{(0)} = v$; $\binom{n}{k}$ *are the binomial coefficients* (§ 1.12, p. 56).

Example 6. For the second derivative of a product of functions $u(x) . v(x)$ the relation

$$(uv)'' = u''v + 2u'v' + uv''$$

holds. For example

$$(x^3 \sin x)'' = 6x \sin x + 6x^2 \cos x - x^3 \sin x.$$

REMARK 6. If the given function $y = h(x)$ is positive and if we can easily differentiate the function $\ln h(x)$, then we frequently use so-called *logarithmic differentiation*. Then $y = h(x) = e^{\ln h(x)}$, $y' = e^{\ln h(x)}[\ln h(x)]' = h(x) . [\ln h(x)]'$.

Example 7. $y = x^x$ $(x > 0)$, $\ln x^x = x \ln x$, $[x \ln x]' = \ln x + x/x = \ln x + 1$, so $y' = x^x (\ln x + 1)$.

More generally:

$$y = f(x)^{g(x)} \quad (f(x) > 0), \quad \ln [f(x)^{g(x)}] = g(x) \ln f(x),$$

$$[g(x) \ln f(x)]' = g'(x) \ln f(x) + g(x) \frac{f'(x)}{f(x)},$$

so

$$y' = f(x)^{g(x)} \left[g'(x) \ln f(x) + g(x) \frac{f'(x)}{f(x)} \right].$$

11.6. Differential. Differences

Definition 1. We say that $f(x)$ is *differentiable* or *has a differential at the point a* if we can express its increment $\Delta f = f(a + h) - f(a)$ in the form

$$\Delta f = f(a + h) - f(a) = Ah + h\,\tau(h) \tag{1}$$

where A is a constant and

$$\lim_{h \to 0} \tau(h) = 0 . \tag{2}$$

Theorem 1. *The function $f(x)$ has a differential at the point a if and only if $f(x)$ has a derivative at the point a. The constant A in (1) is equal to $f'(a)$*, i.e.

$$f(a + h) - f(a) = f'(a)\, h + h\, \tau(h) . \tag{3}$$

Definition 2. The expression $f'(a)\, h$ is called the *differential of the function $f(x)$ at the point a.* We denote it by $\mathrm{d}f(a)$. At a general point we write $\mathrm{d}f(x)$ or $\mathrm{d}y$.

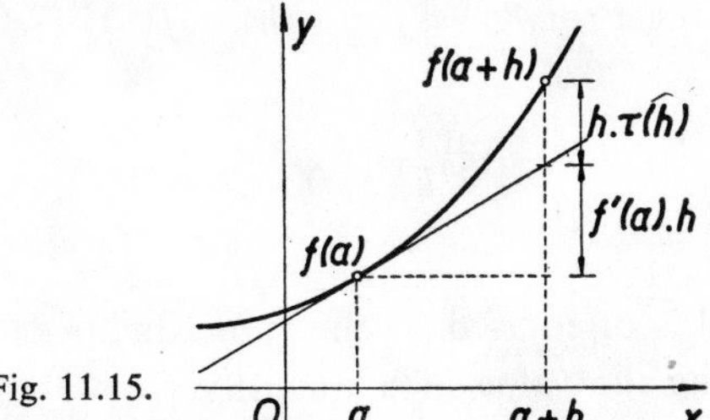

Fig. 11.15.

The geometrical interpretation (Fig. 11.15): If we replace the increment $\Delta f = f(a + h) - f(a)$ by the differential $f'(a)\, h$, then it means that we take only the increment on the tangent $y = f(a) + f'(a)\,(x - a)$ instead of the increment of the function $y = f(x)$. According to (3), the "error" $\Delta f - \mathrm{d}f$ is equal to the function $h\, \tau(h)$ of the variable h; if $h \to 0$, then $\mathrm{d}f \to 0$ and $h\, \tau(h) \to 0$. The condition (2) says that, if $f'(a) \neq 0$, then $h\, \tau(h)$ tends to zero at a "higher order" than $f'(a)\, h$. That is, the smaller is h, the smaller *relative* error we commit in replacing Δf by the differential $\mathrm{d}f$.

At a general point x, $\mathrm{d}f(x) = f'(x)\, h$. For the function $y = x$ we have $\mathrm{d}y = \mathrm{d}x = h$. This justifies the notation of differentials

$$\mathrm{d}f(x) = f'(x)\, \mathrm{d}x \quad \text{or} \quad \mathrm{d}y = f'(x)\, \mathrm{d}x$$

used almost exclusively nowadays.

We call the reader's attention to the fact that here $\mathrm{d}x$ is by no means an "infinitesimally small quantity", but it can assume any value. Sufficient accuracy when replacing the increment of a function by its differential is, of course, secured by (2) only for sufficiently small $\mathrm{d}x$.

Example 1. For the function $y = x^3$, $\mathrm{d}y = 3x^2\, \mathrm{d}x$; $\Delta y = (x + \mathrm{d}x)^3 - x^3 = 3x^2\, \mathrm{d}x + 3x\, \mathrm{d}x^2 + \mathrm{d}x^3$, hence $\tau(\mathrm{d}x) = 3x\, \mathrm{d}x + \mathrm{d}x^2$. (Cf. the notation in Remark 4.) Obviously for every x,

$$\lim_{\mathrm{d}x \to 0} \tau(\mathrm{d}x) = 0 .$$

Further,

$$dy = 2{\cdot}7\,,\quad \Delta y = 2{\cdot}791 \quad \text{when} \quad x = 3\,,\quad dx = 0{\cdot}1\,.$$
$$dy = 0{\cdot}027\,,\quad \Delta y = 0{\cdot}027{,}009{,}001 \quad \text{when} \quad x = 3\,,\quad dx = 0{\cdot}001\,.$$

REMARK 1. In technical subjects we often speak about an *increment of the function* instead of a *differential of the function.*

Taylor's formula gives a better approximation to the increment of a function than the differential does (§ 11.10, p. 434).

REMARK 2. If we replace the increment of a function by its differential, then it follows for the estimation of the error $R = [f(a + h) - f(a)] - f'(a)\,h$ by Taylor's formula that

$$|R| = \frac{|f''(a + \vartheta h)|}{2!}\,h^2\,,\quad 0 < \vartheta < 1\,.$$

REMARK 3. The differential is often used for the approximate determination of the error committed in computing the value of a quantity from the value of another quantity measured with some error. For example, if the radius of a sphere is measured as $a = 4$ cm and if we know that the error of that measurement is 0·1 mm at most, then the maximum error in the determination of the volume $V = \frac{4}{3}\pi x^3 = \frac{4}{3}\pi\,.\,4^3\ \text{cm}^3$ is given approximately by the differential $V'\,.\,h = 4\pi x^2 h = 4\pi\,.\,4^2\,.\,0{\cdot}01\ \text{cm}^3$ (about 0·75 per cent).

Definition 3. Let $f(x)$ have a second derivative which is continuous at the point x. The *second differential of the function* $f(x)$ *at the point* x is the expression

$$d^2f(x) = f''(x)\,dx^2\,.$$

Analogously we define *differentials of higher orders*, $d^{(n)}f(x) = f^{(n)}(x)\,dx^n$; it is supposed that $f(x)$ has a continuous n-th derivative at the point x.

The second (n-th) differential is obtained formally as the differential of the first (or $(n - 1)$-th) differential for the same constant h:

$$d^2f(x) = d[h\,f'(x)] = hf''(x)\,.\,h = h^2\,f''(x) = f''(x)\,dx^2\,.$$

REMARK 4. The notation dx^n is usual for $(dx)^n$, thus it does not stand for $d(x^n)$. Similarly Δx^n is used instead of $(\Delta x)^n$. Thus Δx^n is not $\Delta(x^n)$.

Definition 4. The *first difference* $\Delta f(a)$ *of the function* $f(x)$ *at the point* a is defined as

$$\Delta f(a) = f(a + \Delta x) - f(a)\,.$$

The *second difference* is the difference of the first difference, $\Delta^2 f(a) = \Delta[\Delta f(a)] = [f(a + \Delta x + \Delta x) - f(a + \Delta x)] - [f(a + \Delta x) - f(a)] = f(a + 2\Delta x) - 2f(a + \Delta x) +$

$+ f(a)$. Generally, the *n-th difference* is the difference of the $(n-1)$-th difference,

$$\Delta^n f(a) = f(a + n\Delta x) - \binom{n}{1} f[a + (n-1)\Delta x] + \binom{n}{2} f[a + (n-2)\Delta x] - \\ - \ldots + (-1)^n f(a).$$

Theorem 2. *If $f(x)$ has the n-th derivative which is continuous at the point x, then*

$$f^{(n)}(x) = \lim_{\Delta x \to 0} \frac{\Delta^n f(x)}{\Delta x^n}.$$

For more detailed treatment of differences see Chap. 32.

11.7. General Theorems on Derivatives. Rolle's Theorem. Mean-Value Theorem

Theorem 1 (*Rolle's Theorem*). *If a function $f(x)$ is continuous in $[a, b]$, has a derivative (finite or infinite) in (a, b), and $f(a) = f(b)$, then there exists at least one point $c \in (a, b)$ such that $f'(c) = 0$* (so that tangent to the graph at c is horizontal).

Theorem 2 (*Mean-Value Theorem or Lagrange's Theorem*). *If a function $f(x)$ is continuous in $[a, b]$ and has a derivative (finite or infinite) in (a, b), then there exists at least one point $c \in (a, b)$ such that*

$$f'(c) = \frac{f(b) - f(a)}{b - a} \quad \text{or} \quad f(b) - f(a) = (b - a) f'(c).$$

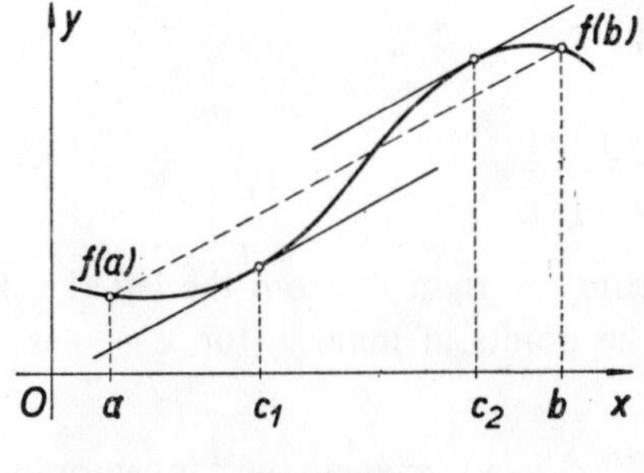

Fig. 11.16.

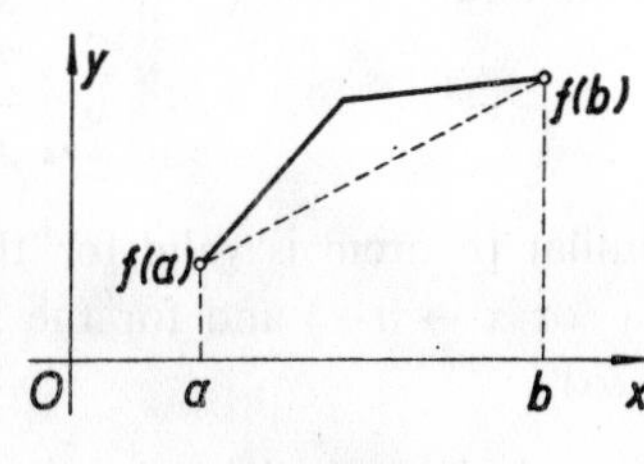

Fig. 11.17.

REMARK 1. Geometrical interpretation: There exists at least one point c (in Fig. 11.16 even two such points, c_1 and c_2) such that the tangent at c is parallel to the straight line joining the points $(a, f(a))$, $(b, f(b))$. For $f(a) = f(b)$ we have Rolle's Theorem. The theorem does not hold if $f(x)$ has a right-hand and left-hand derivative at an interior point, but these are not equal, as we can see in Fig. 11.17.

Theorem 3 (*The Generalized Mean-Value Theorem*). *If the functions $f(x)$ and $g(x)$ are continuous in $[a, b]$ and have derivatives in (a, b) (an infinite derivative of $f(x)$ is admitted), $g'(x) \neq 0$ in (a, b), then there exists at least one point $c \in (a, b)$ such that*

$$\frac{f(b) - f(a)}{g(b) - g(a)} = \frac{f'(c)}{g'(c)}.$$

REMARK 2. For $g(x) = x$ we have the preceding Mean-Value theorem.

Theorem 4. *If $f(x)$ is continuous from the right at the point a, has a finite or infinite derivative when $a < x < a + \delta$ $(\delta > 0)$ and if the finite or infinite limit $\lim_{x \to a+} f'(x)$ exists, then there exists a derivative from the right of $f(x)$ at the point a which is equal to this limit.*

An analogous theorem holds for a left-hand neighbourhood of the point a.

11.8. The Computation of Certain Limits by means of l'Hospital's Rule

Theorem 1 (*Computation of Limits of the Form* 0/0). *If*

$$\lim_{x \to a} f(x) = 0, \quad \lim_{x \to a} g(x) = 0$$

and

$$\lim_{x \to a} \frac{f'(x)}{g'(x)}$$

exists (finite or infinite), then

$$\lim_{x \to a} \frac{f(x)}{g(x)}$$

also exists and

$$\lim_{x \to a} \frac{f(x)}{g(x)} = \lim_{x \to a} \frac{f'(x)}{g'(x)}.$$

A similar theorem is valid for the limit from the right or from the left (i.e. for $x \to a+$ or $x \to a-$) and for the limits at the points at infinity (for $x \to +\infty$ or $x \to -\infty$).

REMARK 1. Frequently it is necessary, if $f'(x)/g'(x)$ is again of "indeterminate form" (i.e. if again $\lim_{x \to a} f'(x) = 0$ and $\lim_{x \to a} g'(x) = 0$), to repeat the application of l'Hospital's rule.

Example 1.

$$\lim_{x \to 0} \frac{1 - \cos x}{x^2} = \lim_{x \to 0} \frac{\sin x}{2x} = \lim_{x \to 0} \frac{\cos x}{2} = \frac{1}{2}.$$

REMARK 2. L'Hospital's Rule cannot be used if one of the functions tends to zero when $x \to a$ while the other does not.

REMARK 3. Note that $f(x)/g(x)$ is not differentiated as a quotient; the functions in the numerator and the denominator are differentiated separately.

Theorem 2 (*Computation of Limits of the Form* ∞/∞). *Let* $\lim_{x\to a} |g(x)| = +\infty$ (*we do not suppose anything about* $\lim_{x\to a} f(x)$, *not even the existence of that limit*). *Then, if*

$$\lim_{x\to a} \frac{f'(x)}{g'(x)}$$

exists (*finite or infinite*) *so does*

$$\lim_{x\to a} \frac{f(x)}{g(x)}$$

and

$$\lim_{x\to a} \frac{f(x)}{g(x)} = \lim_{x\to a} \frac{f'(x)}{g'(x)} .$$

Theorem 2 holds analogously for $x \to a+$, $x \to a-$, $x \to +\infty$, $x \to -\infty$, respectively.

Example 2.

$$\lim_{x\to+\infty} \frac{x^3 + 5x - 2}{x^2 - 1} = \lim_{x\to+\infty} \frac{3x^2 + 5}{2x} = \lim_{x\to+\infty} \frac{6x}{2} = +\infty .$$

Example 3. For $a > 1$, n a positive integer,

$$\lim_{x\to+\infty} \frac{a^x}{x^n} = \lim_{x\to+\infty} \frac{a^x \ln a}{nx^{n-1}} = \ldots = \lim_{x\to+\infty} \frac{a^x (\ln a)^n}{n!} = +\infty .$$

In particular,

$$\lim_{x\to+\infty} \frac{e^x}{x^n} = +\infty .$$

This fact is expressed by the statement that the exponential function increases faster than any power of x, when $x \to +\infty$.

REMARK 4. The computation of limits of the form $0 . \infty$, $\infty - \infty$ is frequently reduced to the preceding forms:

Example 4.

$$\lim_{x\to 0+} x \ln x = \lim_{x\to 0+} \frac{\ln x}{\frac{1}{x}} = \lim_{x\to 0+} \frac{\frac{1}{x}}{-\frac{1}{x^2}} = \lim_{x\to 0+} (-x) = 0 .$$

Example 5.

$$\lim_{x\to 0+}\left(\frac{1}{\sin x}-\frac{1}{x}\right)=\lim_{x\to 0+}\frac{x-\sin x}{x\sin x}=\lim_{x\to 0+}\frac{1-\cos x}{\sin x+x\cos x}=$$

$$=\lim_{x\to 0+}\frac{\sin x}{2\cos x-x\sin x}=0\,.$$

We try to reduce "indeterminate expressions" of other forms to the preceding forms also:

Example 6.

$$\lim_{x\to 0+} x^x=\lim_{x\to 0+} e^{x\ln x}=\exp\left(\lim_{x\to 0+} x\ln x\right)=e^0=1\,.$$

Here we have applied the result of example 4, the relation $x = e^{\ln x}$ (following from the definition of $\ln x$) and the continuity of the exponential function,

$$\lim_{z\to z_0} e^z=\exp\left(\lim_{z\to z_0} z\right).$$

11.9. Investigation of a Function. Graphical Representation. Monotonic Functions. Concavity. Convexity. Points of Inflection. Maxima and Minima

In this paragraph the briefer term "the function" instead of "the graph of the function" is frequently used.

Definition 1. We say that $f(x)$ is *strictly increasing at the point a* if, in a certain neighbourhood of the point a,

$$f(x)>f(a)\quad\text{when}\quad x>a\,, \tag{1}$$
$$f(x)<f(a)\quad\text{when}\quad x<a$$

(the points x_1, x_4 in Fig. 11.18). Analogously we define a function *strictly decreasing at the point a*.

If in (1) $f(x) \geqq f(a)$ when $x > a$ and $f(x) \leqq f(a)$ when $x < a$, we say that $f(x)$ is *increasing at the point a*; analogously we speak of a function being *decreasing at a*. For example the function $f(x) = \text{const.}$ is both increasing and decreasing at a.

Theorem 1. *If $f'(a) > 0$, then $f(x)$ is strictly increasing at the point a; if $f'(a) < 0$, then $f(x)$ is strictly decreasing at the point a.*

Remark 1. If $f(x)$ is strictly increasing at every point of an interval I, we say that it is *strictly increasing in I*. The following definition is equivalent: $f(x)$ is called *strictly increasing* in I if for every pair of points x_1, x_2 of this interval, satisfying $x_1 < x_2$, the relation

$$f(x_1) < f(x_2) \tag{2}$$

holds. *Increasing* (with the sign $\leqq$ in (2)), *strictly decreasing*, and *decreasing functions in I* are defined analogously. All such functions are called *monotonic in I*. Strictly increasing or strictly decreasing functions are called *strictly monotonic in I*. *If $f'(x) > 0$ ($f'(x) < 0$) in I, then $f(x)$ is strictly increasing (decreasing) in I.*

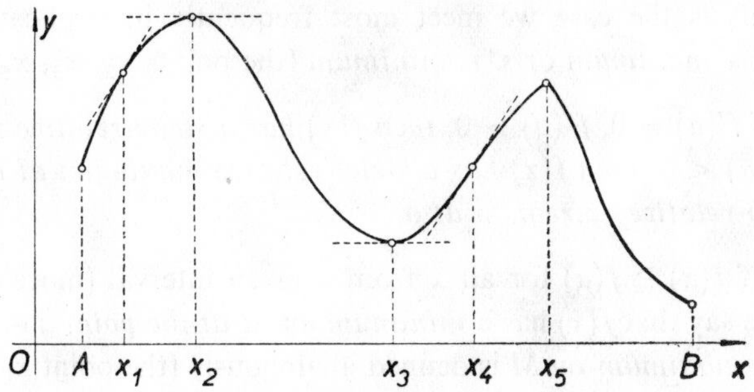

Fig. 11.18.

Definition 2. If, in a certain neighbourhood U of the point a, the graph of the function $f(x)$ lies below the tangent (or on the tangent) drawn at the point $(a, f(a))$ (i.e. if the relation $f(x) \leqq f(a) + (x - a) f'(a)$ holds in U; the points x_1, x_2 in Fig. 11.18), we say that $f(x)$ is *concave at the point a*. If the graph lies above the tangent (or on the tangent), i.e. the relation $f(x) \geqq f(a) + (x - a) f'(a)$ holds in U, we say that $f(x)$ is *convex at the point a* (the point x_3 in Fig. 11.18).

REMARK 2. If in these cases the graph of the function coincides in a certain neighbourhood U of the point a with the tangent *only* at the point of contact (this is the case we meet most frequently in applications), the function is called *strictly concave* or *strictly convex at a*.

Theorem 2. *If $f''(a) > 0$, then $f(x)$ is strictly convex at a; if $f''(a) < 0$, then $f(x)$ is strictly concave at a.*

REMARK 3. If $f(x)$ is convex at every point of an interval I, we say that it is *convex in the interval I*. Analogously a function is defined to be *strictly convex* or *concave* or *strictly concave in an interval I*.

If $f''(x) > 0$ (or $f''(x) < 0$) everywhere in I, then $f(x)$ is strictly convex (or strictly concave) in I.

Definition 3. If the graph of a function crosses, at the point $x = a$, its tangent at this point, we say that $f(x)$ has a *point of inflection at the point a* (the point x_4 in Fig. 11.18).

Theorem 3. *If $f''(a) = 0$, $f'''(a) \neq 0$, then $f(x)$ has a point of inflection at a. Further if $f'''(a) < 0$, the graph of the function crosses its tangent from above* (the point x_4 in Fig. 11.18); *if $f'''(a) > 0$, the graph crosses its tangent from below. If $f''(a) \neq 0$, then $f(x)$ has no point of inflection at a.*

Definition 4. If

$$f(x) \leqq f(a) \quad (\text{or} \quad f(x) \geqq f(a)) \tag{3}$$

in some neighbourhood U of the point a, we say that $f(x)$ has a *relative maximum* (or a *relative minimum*) *at the point a*. If the sign of equality in (3) holds in U at the point a only (this is the case we meet most frequently in applications) we speak of a *strict relative maximum* or *s. r. minimum* (the points x_2, x_3, x_5 in Fig. 11.18).

Theorem 4. *If $f'(a) = 0$, $f''(a) > 0$, then $f(x)$ has a strict relative minimum at a; if $f'(a) = 0$, $f''(a) < 0$, then $f(x)$ has a strict relative maximum at a. If $f'(a) \neq 0$, then $f(x)$ has no relative extremum at a.*

Definition 5. If $f(x) \geqq f(a)$ for all x from a given interval (more generally from a domain) M, we say that $f(x)$ has a *minimum on M at the point $a \in M$* (the point B in Fig. 11.18). A *maximum* on M is defined analogously (the point x_2 in Fig. 11.18). (Both these extremes are often called *total* or *absolute.*)

REMARK 4. As we can observe in Fig. 11.18, absolute extremes on M need not always be at the points of the relative extremes (the point B in Fig. 11.18); it is necessary to investigate also the values of the function at the boundary points of the domain M. In the same figure, we can see that the relative extremes occur not only at the points where $f'(a) = 0$ (Theorem 4) but they may occur also at points at which $f(x)$ has no derivative at all (the point x_5 in Fig. 11.18).

REMARK 5. Even if a function $f(x)$ has a sufficient number of higher derivatives at the point a, Theorems 1–4 may prove to be ineffective when several of the first derivatives vanish. Then the following theorems may be useful:

Theorem 5. *Let $f''(a) = f'''(a) = \ldots = f^{(n-1)}(a) = 0$, $f^{(n)}(a) \neq 0$ $(n \geqq 2)$. Then*

$f(x)$ is strictly convex at a when $f^{(n)}(a) > 0$, n even ,
$f(x)$ is strictly concave at a when $f^{(n)}(a) < 0$, n even ,
$f(x)$ has a point of inflection at a (and crosses the tangent from below) when $f^{(n)}(a) > 0$, n odd ,
$f(x)$ has a point of inflection at a (and crosses the tangent from above) when $f^{(n)}(a) < 0$, n odd .

Theorem 6. *Let $f'(a) = f''(a) = \ldots = f^{(n-1)}(a) = 0$, $f^{(n)}(a) \neq 0$ $(n \geqq 1)$. Then*
$f(x)$ has a strict relative minimum at a when $f^{(n)}(a) > 0$, n even ,
$f(x)$ has a strict relative maximum at a when $f^{(n)}(a) < 0$, n even ,
$f(x)$ is strictly increasing at a when $f^{(n)}(a) > 0$, n odd ,
$f(x)$ is strictly decreasing at a when $f^{(n)}(a) < 0$, n odd .

REMARK 6. Putting $n = 2, 3$ in Theorem 5, we obtain Theorems 2, 3 respectively. Putting $n = 1, 2$ in Theorem 6, we obtain Theorems 1, 4 respectively.

REMARK 7. If $f'(a) = 0$ (the zeros of the derivative are generally known as "*stationary*" points of the function $f(x)$) and if the computation of the second derivative or of higher derivatives is not complicated, then we decide easily whether there is an extremum at the point a (and its type) by Theorem 4 or 6. If the computation of derivatives is rather lengthy, we may use the following theorem:

Theorem 7. *Let $f'(a) = 0$ and $f'(x) > 0$ when $x < a$, $f'(x) < 0$ when $x > a$ in a certain neighbourhood U of the point a.* (We say briefly that the derivative is *changing its sign from positive to negative.*) *Then $f(x)$ has a strict relative maximum at the point a.*

If $f'(a) = 0$ and $f'(x) < 0$ when $x < a$, $f'(x) > 0$ when $x > 0$ in U, then $f(x)$ has a strict relative minimum at the point a.

If $f'(a) = 0$ and $f'(x) > 0$ (or $f'(x) < 0$) when $x < a$ as well as when $x > a$ in U, then $f(x)$ is strictly increasing (strictly decreasing) at the point a.

Similarly we have:

Theorem 8. *If $f''(a) = 0$ and if, in a certain neighbourhood U of the point a, $f''(x) > 0$ when $x < a$ and $f''(x) < 0$ when $x > a$, then $f(x)$ has a point of inflection at the point a and the graph of $f(x)$ crosses the tangent from above. If $f''(a) = 0$ and $f''(x) < 0$ when $x < a$, $f''(x) > 0$ when $x > a$ in U, then $f(x)$ has a point of inflection at the point a and the graph crosses the tangent from below.*

Example 1 (The *Investigation of a Function*). Using the theorems of this paragraph, let us investigate the characteristic features of the function

$$f(x) = x + \frac{4}{x} \tag{4}$$

and plot its graph approximately.

The functional relation (4) is defined for every $x \neq 0$. Also the domain of definition is (Remark 11.1.4, p. 399) the interval $(-\infty, +\infty)$ from which the point $x = 0$ is excluded. The function (4) has derivatives of all orders. Especially

$$f'(x) = 1 - \frac{4}{x^2}, \tag{5}$$

$$f''(x) = \frac{8}{x^3}. \tag{6}$$

From (5) it follows that $f'(x) = 0$ for $x_1 = 2$, $x_2 = -2$. By (6) $f''(x_1) > 0$, $f''(x_2) < 0$, hence by Theorem 4, $f(x)$ has a strict relative minimum at the point x_1 and a strict relative maximum at the point x_2. We easily compute $f(2) = 4$, $f(-2) = -4$ (the points A, B in Fig. 11.19). For $x \neq \pm 2$, $f'(x) \neq 0$ and thus the given function has no other relative extremes. For $|x| > 2$ we have by (5) $f'(x) > 0$ and so (Remark 1) the function $f(x)$ is strictly increasing in the intervals $(2, +\infty)$ and $(-\infty, -2)$. In the intervals $(0, 2)$, $(-2, 0)$ we have $f'(x) < 0$ by (5) and $f(x)$ is strictly decreasing there (Fig. 11.19).

By (6), $f''(x) > 0$ for $x > 0$, hence $f(x)$ is strictly convex in the interval $(0, +\infty)$ (Remark 3). In the interval $(-\infty, 0)$, $f''(x) < 0$ and thus $f(x)$ is strictly concave (Fig. 11.19).

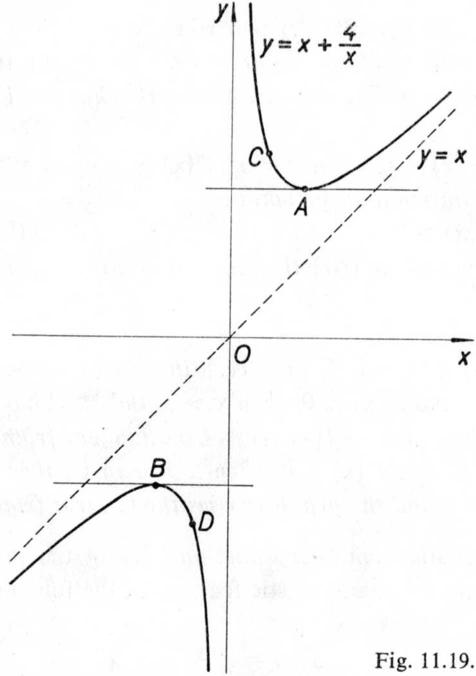

Fig. 11.19.

Because $f''(x) \neq 0$ everywhere in the domain of definition M, $f(x)$ has nowhere a point of inflection (Theorem 3).

For an approximate plotting of the graph it is useful to determine the asymptotes of the graph. Evidently,

$$\lim_{x\to 0+} f(x) = +\infty\,, \quad \lim_{x\to 0-} f(x) = -\infty\,,$$

so that the straight line $x = 0$ is a vertical asymptote of the graph (§ 9.6, p. 326). Further, there exist finite limits for $x \to +\infty$ and $x \to -\infty$,

$$k = \lim \frac{f(x)}{x} = \lim \frac{x + \dfrac{4}{x}}{x} = \lim \left(1 + \frac{4}{x^2}\right) = 1\,,$$

$$q = \lim [f(x) - kx] = \lim \left(x + \frac{4}{x} - x\right) = \lim \frac{4}{x} = 0,$$

so that the straight line

$$y = x$$

is a (non-vertical) asymptote of the graph (Theorem 9.6.1, p. 326).

To facilitate the drawing of the graph, we observe that the investigated function is odd (Definition 11.2.1, p. 404),

$$f(-x) = -x + \frac{4}{-x} = -\left(x + \frac{4}{x}\right) = -f(x),$$

thus its graph is symmetric with respect to the origin. (Moreover, we could have used this property earlier for the investigation of the given function for $x > 0$ only: If $f(x)$ has a strict relative minimum at $x = 2$, then it has a strict relative maximum at $x = -2$, etc.) For the construction of the graph it is advantageous, further, to work out the values of the function itself at some points where their computation is easy, e.g. at the points $x = 1$, $x = -1$ (the points C, D in Fig. 11.19).

Thus the investigation of the properties of the function $f(x)$ and the preparation for the approximate drawing of its graph are finished. Briefly, we say that we have performed the investigation of the given function.

Example 2. Of all rectangles of perimeter 20 cm, to find that with the greatest area.

Denoting the lengths of the sides by x, y, the area is $P = xy$ with the condition $2x + 2y = 20$, hence $y = 10 - x$, and $P = x(10 - x) = 10x - x^2$. We look for that value of $x \in [0, 10]$ for which P assumes its maximum value. Evidently we have to find the *relative* maximum, since $P(0) = P(10) = 0$. Because the function $P(x)$ has a derivative everywhere it may have a maximum only at the point where $P' = 0$ (Theorem 4). From the equation $P'(x) = 0$ or $10 - 2x = 0$ we obtain $x = 5$. Indeed, there is a maximum at the point $x = 5$ (by Theorem 4), because $P''(5) = -2 < 0$. (Thus the square, the length of side of which is 5 cm, has a maximum area of all rectangles of the given perimeter 20 cm.)

Example 3. Of all right circular cylinders of given volume V, to find that with the least surface S.

We have

$$V = \pi r^2 h, \tag{7}$$

$$S = 2\pi r h + 2\pi r^2. \tag{8}$$

If we choose for instance r as independent variable, then we can express h using (7) by

$$h = \frac{V}{\pi r^2}, \tag{9}$$

for V is fixed. Putting (9) in (8),

$$S = \frac{2V}{r} + 2\pi r^2 , \tag{10}$$

which gives S as a function of the single variable r. We will find the minimum of this function for $r \in (0, +\infty)$. If we put $S' = 0$, i.e.

$$-\frac{2V}{r^2} + 4\pi r = 0 , \tag{11}$$

we obtain

$$r_{\min} = \sqrt[3]{\frac{V}{2\pi}} .$$

For this value of r the surface S really attains its minimum value in the interval $(0, +\infty)$ because for $r > r_{\min}$, $S' > 0$ and for $0 < r < r_{\min}$, $S' < 0$ (as follows from the left-hand side of equation (11)). For $r_{\min}$ we obtain from (9)

$$h_{\min} = \sqrt[3]{\frac{4V}{\pi}} = 2r_{\min} .$$

Thus, the resulting cylinder is such that its height is equal to the diameter of its base.

Example 4. Discuss the behaviour of the function

$$f(x) = x^3 e^{-x}$$

at the point $x = 0$. By easy computation we obtain $f(0) = 0$, $f'(0) = 0$, $f''(0) = 0$, $f'''(0) > 0$. By Theorem 6, $f(x)$ is increasing at the point $x = 0$; by Theorem 5, it has a point of inflection there and crosses its tangent from below (its behaviour in the neighbourhood of that point is represented in Fig. 11.20).

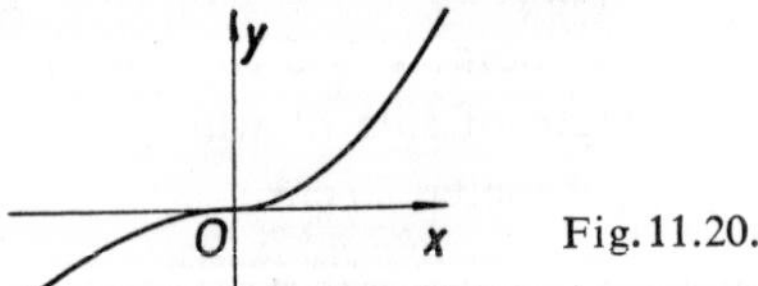

Fig. 11.20.

11.10. Taylor's Theorem

Theorem 1 (*Taylor's Theorem*). *Let $f(x)$ have continuous derivatives up to the n-th order inclusive in $[a, a + h]$ (or in $[a + h, a]$ if h is negative) and a continuous derivative of the $(n + 1)$-th order in $(a, a + h)$ (or in $(a + h, a)$). Then*

$$f(a + h) = f(a) + \frac{f'(a)}{1!} h + \frac{f''(a)}{2!} h^2 + \ldots + \frac{f^{(n)}(a)}{n!} h^n + R_{n+1} \tag{1}$$

where the expression for the remainder R_{n+1} may be put in one of these forms:

$$R_{n+1} = \frac{f^{(n+1)}(a + \vartheta h)}{(n+1)!} h^{n+1} \quad (0 < \vartheta < 1) \qquad \textit{(Lagrange form)}, \quad (2)$$

$$R_{n+1} = \frac{f^{(n+1)}(a + \eta h)}{n!} (1 - \eta)^n h^{n+1} \quad (0 < \eta < 1) \quad \textit{(Cauchy form)}, \quad (3)$$

$$R_{n+1} = \int_a^{a+h} f^{(n+1)}(t) \frac{(a + h - t)^n}{n!} \, dt \qquad \textit{(integral form)}. \quad (4)$$

REMARK 1. If we write $h = x - a$, we obtain the frequently used form

$$f(x) = f(a) + \frac{f'(a)}{1!}(x - a) + \frac{f''(a)}{2!}(x - a)^2 + \ldots + \frac{f^{(n)}(a)}{n!}(x - a)^n + R_{n+1} \quad (5)$$

where

$$R_{n+1} = \frac{f^{(n+1)}[a + \vartheta(x - a)]}{(n+1)!}(x - a)^{n+1} \quad (0 < \vartheta < 1) \quad (6)$$

or

$$R_{n+1} = \frac{f^{(n+1)}[a + \eta(x - a)]}{n!}(1 - \eta)^n (x - a)^{n+1} \quad (0 < \eta < 1), \quad (7)$$

or

$$R_{n+1} = \int_a^x f^{(n+1)}(t) \frac{(x - t)^n}{n!} \, dt. \quad (8)$$

REMARK 2. In particular, if $a = 0$, then

$$f(x) = f(0) + \frac{f'(0)}{1!} x + \frac{f''(0)}{2!} x^2 + \ldots + \frac{f^{(n)}(0)}{n!} x^n + R_{n+1}, \quad (9)$$

where it is sufficient to put $a = 0$ in (6), (7), (8), respectively, to obtain the expression for the remainder R_{n+1}. Formula (9) is called *Maclaurin's formula*.

REMARK 3. We use Taylor's or Maclaurin's Theorem in order to compute the values of a function or to express approximately a given function by means of a polynomial in the neighbourhood of the point a (or zero). Because we know the values of several functions at the origin and can then easily compute the values of the derivatives at this point, formula (9) finds the most frequent application; note, that x (or h in (1)) may be negative.

REMARK 4. If we consider only the first two terms on the right of (1), then we obtain (except for the remainder R_2) the replacing of the difference $f(a + h) - f(a)$ by the differential $hf'(a)$. For $n = 0$ we obtain from (1) the Mean-Value Theorem.

Example 1. Let us try to approximate the function $y = \sin x$ in the neighbourhood of the origin by means of a polynomial of degree 4 and let us estimate the error.

We apply (9); $f(x) = \sin x$, $f(0) = 0$, $f'(0) = 1$, $f''(0) = 0$, $f'''(0) = -1$, $f^{(4)}(0) = 0$, hence

$$\sin x = x - \frac{x^3}{3!} + R_5 . \tag{10}$$

According to formula 11, p. 418, $(\sin x)^{(5)} = \sin(x + \frac{5}{2}\pi) = \cos x$. According to (6), for $a = 0$

$$R_5 = \frac{\cos \vartheta x}{5!} x^5 \quad (0 < \vartheta < 1) . \tag{11}$$

Let us consider the interval $[-\frac{1}{10}, \frac{1}{10}]$. Because $|\cos \vartheta x| \leqq 1$, we shall have for all x in this interval $|R_5| \leqq (\frac{1}{10})^5/5! = 1/12{,}000{,}000$.

Example 2. Let us compute the approximate value of $\sin 3°$.

In radians

$$x = \frac{2\pi}{360} . 3 = \frac{\pi}{60} \doteq 0{\cdot}052{,}359{,}878 .$$

Further

$$\frac{1}{3!}\left(\frac{\pi}{60}\right)^3 \doteq 0{\cdot}000{,}023{,}925 ,$$

thus by (10)

$$\sin 3° \approx \frac{\pi}{60} - \frac{1}{3!}\left(\frac{\pi}{60}\right)^3 \doteq 0{\cdot}052{,}335{,}953$$

with an error (by (11)) of less than

$$\frac{1}{5!}\left(\frac{\pi}{60}\right)^5 \doteq 3{\cdot}3 \times 10^{-9} .$$

Theorem 2. *Let $f(x)$ have derivatives of all orders in $[a, x]$ (or in $[x, a]$ if $x < a$). Then a necessary and sufficient condition that the series*

$$f(a) + \frac{f'(a)}{1!}(x - a) + \frac{f''(a)}{2!}(x - a)^2 + \dots$$

converges and has the sum $f(x)$ is that $\lim_{n\to\infty} R_{n+1} = 0$ *(for the specified x).*

11.11. Approximate Expressions. Computation with Small Numbers

REMARK 1. If we take into consideration only the first few terms of the right-hand side of equation (11.10.1), p. 434, we obtain an approximating formula for the evaluation of the number $f(a + h)$. If need be, we can use the formulae (11.10.2)–(11.10.4) or (11.10.6)–(11.10.8) for the eventual estimation of the error. Some frequently used approximations (we denote the approximations by $\approx$, ε is a relatively small number (in absolute value) not necessarily positive) are:

Theorem 1.

1. $(1 \pm \varepsilon)^n \approx 1 \pm n\varepsilon\,, \quad (1 \pm \varepsilon)^2 \approx 1 \pm 2\varepsilon\,, \quad \sqrt{(1 \pm \varepsilon)} \approx 1 \pm \tfrac{1}{2}\varepsilon\,,$

$$\frac{1}{1 \pm \varepsilon} \approx 1 \mp \varepsilon\,, \quad \frac{1}{(1 \pm \varepsilon)^2} \approx 1 \mp 2\varepsilon\,, \quad \frac{1}{\sqrt{(1 \pm \varepsilon)}} \approx 1 \mp \tfrac{1}{2}\varepsilon\,.$$

2. $a^\varepsilon \approx 1 + \varepsilon \ln a\,, \quad e^\varepsilon \approx 1 + \varepsilon\,.$

3. $(1 \pm \varepsilon)(1 \pm \delta)(1 \pm \eta) \approx 1 \pm \varepsilon \pm \delta \pm \eta\,,$

$$\frac{(1 \pm \varepsilon)(1 \pm \delta)}{(1 \pm \eta)(1 \pm \varkappa)} \approx 1 \pm \varepsilon \pm \delta \mp \eta \mp \varkappa\,.$$

4. *For positive numbers $p \approx q$ the relation $\sqrt{(pq)} \approx \frac{1}{2}(p + q)$ holds.*

5. $\sin \varepsilon \approx \varepsilon\,, \quad \cos \varepsilon \approx 1\,, \quad \tan \varepsilon \approx \varepsilon\,,$

$$\sin (x + \varepsilon) \approx \sin x + \varepsilon \cos x\,, \quad \cos (x + \varepsilon) \approx \cos x - \varepsilon \sin x\,,$$

$$\tan (x + \varepsilon) = \tan x + \frac{\varepsilon}{\cos^2 x}\,,$$

$$e^\varepsilon \approx 1 + \varepsilon\,, \quad \ln (1 + \varepsilon) \approx \varepsilon\,, \quad \ln (x + \varepsilon) \approx \ln x + \frac{\varepsilon}{x}\,,$$

$$\ln \frac{x + \varepsilon}{x - \varepsilon} \approx \frac{2\varepsilon}{x}\,.$$

first approximation,

6. $\sin \varepsilon \approx \varepsilon - \dfrac{\varepsilon^3}{3!}\,, \quad \cos \varepsilon \approx 1 - \dfrac{\varepsilon^2}{2!}\,, \quad \tan \varepsilon \approx \varepsilon + \dfrac{\varepsilon^3}{3}\,,$

$$e^\varepsilon \approx 1 + \varepsilon + \frac{\varepsilon^2}{2!} + \frac{\varepsilon^3}{3!}\,,$$

$$\ln (1 + \varepsilon) \approx \varepsilon - \tfrac{1}{2}\varepsilon^2 + \tfrac{1}{3}\varepsilon^3\,,$$

$$\ln (x + \varepsilon) \approx \ln x + \frac{\varepsilon}{x} - \frac{1}{2}\frac{\varepsilon^2}{x^2} + \frac{1}{3}\frac{\varepsilon^3}{x^3}\,,$$

$$\ln \frac{x + \varepsilon}{x - \varepsilon} \approx 2\left(\frac{\varepsilon}{x} + \frac{\varepsilon^3}{3x^3}\right).$$

higher approximations.

(For higher approximations we have considered the first four terms of the right-hand side of equation (11.10.1)).

Example 1. Let us compute $\sqrt{9{,}986}$.

Using the relation $\sqrt{(1 - \varepsilon)} \approx 1 - \frac{1}{2}\varepsilon$, we have:

$$\sqrt{9{,}986} = \sqrt{(10{,}000 - 14)} = 100 \sqrt{\left(1 - \frac{14}{10{,}000}\right)} \approx 100\left(1 - \frac{7}{10{,}000}\right) = 99{\cdot}93\,.$$

Estimation of error: If we take $f(x) = \sqrt{(1 - x)}$, then by (11.10.2) ($a = 0$, $h = 0{\cdot}001{,}4$, $0 < \vartheta < 1$)

$$|R_2| = \left|\frac{[\sqrt{(1-x)}]''_{x=0.001,4\vartheta}}{2!}\, 0{\cdot}001{,}4^2\right| = \frac{0{\cdot}001{,}4^2}{4\,.\,2!\,[\sqrt{(1-x)}]^3_{x=0.001,4\vartheta}} < \frac{1}{4\,.\,10^6}\,.$$

Because the whole expression $\sqrt{(1 - 0{\cdot}001{,}4)}$ is multiplied by the number 100, we obtain the result $\sqrt{9{,}986} \doteq 99{\cdot}93$ with an error less than $10^{-4}/4$.

11.12. Survey of Some Important Formulae from Chapter 11

(Cf. also Theorems 11.4.9, 11.5.3, 11.5.8, 11.5.9.)

1. $\lim\limits_{x\to a} [f(x) \pm g(x)] = \lim\limits_{x\to a} f(x) \pm \lim\limits_{x\to a} g(x)\,, \quad \lim\limits_{x\to a} k\,f(x) = k \lim\limits_{x\to a} f(x)\,,$

$$\lim_{x\to a} f(x)\,g(x) = \lim_{x\to a} f(x) \lim_{x\to a} g(x)\,,$$

$$\lim_{x\to a} \frac{f(x)}{g(x)} = \frac{\lim\limits_{x\to a} f(x)}{\lim\limits_{x\to a} g(x)} \quad (\lim_{x\to a} g(x) \neq 0) \quad \text{(Theorem 11.4.4)}\,.$$

2. $\lim\limits_{x\to 0+} \dfrac{1}{x} = +\infty\,, \quad \lim\limits_{x\to 0-} \dfrac{1}{x} = -\infty\,,$

$$\lim_{x\to 0+} \frac{a}{x} = \begin{cases} +\infty & \text{when } a > 0 \\ -\infty & \text{when } a < 0\,, \end{cases} \qquad \lim_{x\to 0-} \frac{a}{x} = \begin{cases} -\infty & \text{when } a > 0\,, \\ +\infty & \text{when } a < 0\,. \end{cases}$$

3. $\lim\limits_{x\to a} \dfrac{f(x)}{g(x)} = \lim\limits_{x\to a} \dfrac{f'(x)}{g'(x)}\,,$ *if at the same time* $\lim\limits_{x\to a} f(x) = \lim\limits_{x\to a} g(x) = 0\,,$

or if $\lim\limits_{x\to a} |g(x)| = +\infty$ (Theorems 11.8.1, 11.8.2).

4. $[f(x) \pm g(x)]' = f'(x) \pm g'(x)\,, \quad [f(x)\,g(x)]' = f'(x)\,g(x) + f(x)\,g'(x)\,,$

$$\left[\frac{f(x)}{g(x)}\right]' = \frac{f'(x)\,g(x) - f(x)\,g'(x)}{g^2(x)} \quad \text{(Theorem 11.5.4)}\,,$$

$$\frac{dy}{dx} = \frac{dy}{dz}\frac{dz}{dx}$$ (*differentiation of composite functions*, Theorem 11.5.5).

5. $$\frac{dy}{dx} = \frac{1}{dx/dy}$$ (*differentiation of inverse functions*, Theorem 11.5.6).

6. $$(uv)^{(n)} = u^{(n)}v^{(0)} + \binom{n}{1} u^{(n-1)}v^{(1)} + \binom{n}{2} u^{(n-2)}v^{(2)} + \ldots + u^{(0)}v^{(n)}.$$

7. $$\frac{(u_1 u_2 \ldots u_n)'}{u_1 u_2 \ldots u_n} = \frac{u_1'}{u_1} + \frac{u_2'}{u_2} + \ldots + \frac{u_n'}{u_n}, \quad u_1 u_2 \ldots u_n \neq 0.$$

8. $$[f(x)^{g(x)}]' = f(x)^{g(x)} \left[g'(x) \ln f(x) + g(x) \frac{f'(x)}{f(x)} \right]$$ ($f(x) > 0$; Example 11.5.7).

9. $\mathrm{d}f(x) = f'(x)\,\mathrm{d}x$ (§ 11.6).

10. $f'(a) = 0$, $f''(a) > 0 \Rightarrow f(x)$ *has a strict relative minimum at the point* a,

$f'(a) = 0$, $f''(a) < 0 \Rightarrow f(x)$ *has a strict relative maximum at the point* a (Theorem 11.9.4).

11. $$f(x) = f(a) + \frac{f'(a)}{1!}(x - a) + \frac{f''(a)}{2!}(x - a)^2 + \ldots + \frac{f^{(n)}(a)}{n!}(x - a)^n + R_{n+1},$$

where e.g.

$$R_{n+1} = \frac{f^{(n+1)}[a + \vartheta(x - a)]}{(n + 1)!}(x - a)^{n+1} \quad (0 < \vartheta < 1)$$

(*Taylor's formula*, Theorem 11.10.1, Remark 11.10.1).

12. $$e^x = 1 + \frac{x}{1!} + \frac{x^2}{2!} + \frac{x^3}{3!} + \ldots + \frac{x^n}{n!} + \frac{x^{n+1}}{(n + 1)!} e^{\vartheta x}, \quad 0 < \vartheta < 1;$$

$$\sin x = x - \frac{x^3}{3!} + \frac{x^5}{5!} - \ldots + (-1)^{n-1} \frac{x^{2n-1}}{(2n - 1)!} + (-1)^n \frac{x^{2n+1}}{(2n + 1)!} \cos \vartheta x, \quad 0 < \vartheta < 1;$$

$$\cos x = 1 - \frac{x^2}{2!} + \frac{x^4}{4!} - \ldots + (-1)^n \frac{x^{2n}}{(2n)!} + (-1)^{n+1} \frac{x^{2n+2}}{(2n + 2)!} \cos \vartheta x, \quad 0 < \vartheta < 1.$$

All three formulae hold for every x.

13. $1 + x > 0$, $x \neq 0 \Rightarrow (1 + x)^n > 1 + nx$, n *any positive integer greater than* 1.

12. FUNCTIONS OF TWO OR MORE VARIABLES

By KAREL REKTORYS

References: [16], [27], [48], [57], [73], [110], [112], [118], [155], [215], [226], [229], [302], [308], [320], [341], [367], [376], [389].

12.1. Functions of Several Variables. Composite Functions. Limit, Continuity

Definition 1. Let us consider a set M of points (x, y) in the xy-plane (this set is most often a *region*; see Remark 1). We say that a *real function of real variables* x,y *is defined in* (*on* or *over*) *the set* M if a rule is given according to which *exactly one* real number z is assigned to each point (x, y) of M. The set M is called the *domain of definition of the function*. Similarly a function of n variables

$$z = f(x_1, x_2, \ldots, x_n)$$

may be defined.

Usually we denote functions by letters f, g, etc. At an arbitrary point $(x, y) \in M$ or $(x_1, x_2, \ldots, x_n) \in M$ we then write

$$z = f(x, y) \quad \text{or} \quad z = g(x_1, x_2, \ldots, x_n),$$

etc. Cf. also Remark to Definition 11.1.1, p. 397.

REMARK 1. The domain of definition is most often a *region* or a *closed region* (that is, a region with its boundary included). Both these concepts are defined in § 22.1. (Thus, for example, the interior of a circle, the interior of an ellipse, the whole xy-plane, etc., are regions. An example of a closed region is a circle with its circumference included, the so-called *closed circle*.)

Example 1. The function $z = x^2y$ is defined in the whole plane. The function

$$z = \sqrt{(1 - x^2 - y^2)}$$

is defined at all points for which

$$1 - x^2 - y^2 \geqq 0 \quad \text{or} \quad x^2 + y^2 \leqq 1 .$$

For its domain of definition the closed region $x^2 + y^2 \leqq 1$ may be taken, i.e. the closed circle with its centre at the origin and radius equal to 1.

Similarly the function

$$z = \sqrt{(1 - x_1^2 - x_2^2 - \ldots - x_n^2)}$$

is defined in a "closed n-dimensional sphere"

$$x_1^2 + x_2^2 + \ldots + x_n^2 \leqq 1 .$$

REMARK 2. The geometrical interpretation of a function $z = f(x, y)$ (in so far as the function is a "reasonable" one) is a surface in three-dimensional space. Functions of more than three variables can no longer be represented in such a simple way.

REMARK 3. The function

$$z = h(f(x, y), g(x, y)) \tag{1}$$

is called a *composite function*, composed of the functions

$$u = f(x, y), \quad v = g(x, y), \tag{2}$$

$$z = h(u, v). \tag{3}$$

The functions (2) are defined in the set M, the function (3) in the set N and it is required that for every point $(x, y) \in M$ the relation $(u, v) \in N$ be satisfied. For a given point $(x, y) \in M$ we can then compute by (2) the values of u and v and by (3) the corresponding value of z.

Example 2. The function $z = (1 + x^2 + y^2)^{\sin xy}$ may be considered as a composite function by means of the relations $z = u^v$, $u = 1 + x^2 + y^2$, $v = \sin xy$. For the set M the whole plane xy may be taken because the function $z = u^v$ is defined at all points (u, v), where $u > 0$, and the function $u = 1 + x^2 + y^2$ is positive for all x, y.

Example 3. The function $z = \sqrt{(1 - x^2 \sin x)}$ may be considered as a function composed of the functions $z = \sqrt{(1 - uv)}$, $u = x^2$, $v = \sin x$. (This function is a function of only one variable x.)

Definition 2. The *distance between two points* (x_1, y_1), (x_2, y_2) is, by definition, the number

$$d = \sqrt{[(x_2 - x_1)^2 + (y_2 - y_1)^2]}$$

(we use cartesian coordinates unless otherwise stated). Similarly, the distance between the points $(x_1, x_2, \ldots, x_n)$, $(y_1, y_2, \ldots, y_n)$ is defined as the number

$$d = \sqrt{[(y_1 - x_1)^2 + (y_2 - x_2)^2 + \ldots + (y_n - x_n)^2]}\,.$$

Definition 3. A set of all points, the distance of which from a point P is smaller than δ, is called a *δ-neighbourhood of the point P*. (Note that the point P itself belongs to the δ-neighbourhood of the point P.) It is often denoted by $U_\delta(P)$.

REMARK 4. In the xy-plane a δ-neighbourhood of a point P is formed by all points that lie inside the circle with centre at the point P and radius δ. In three-dimensional space a δ-neighbourhood is the interior of a sphere, etc.

Definition 4. We say that a function $z = f(x, y)$ has a *limit A at a point $P(x_0, y_0)$* if to every (arbitrarily small) number $\varepsilon > 0$, there exists a number $\delta > 0$ (depending, in general, on the choice of the number ε) such that for all points $(x, y) \neq P$ in the δ-neighbourhood of the point P the relation

$$|f(x, y) - A| < \varepsilon$$

holds.

The notations

$$\lim_{(x,y)\to P} f(x, y) = A\,, \qquad \lim_{(x,y)\to(x_0,y_0)} f(x, y) = A\,, \qquad \lim_{\substack{x\to x_0\\ y\to y_0}} f(x, y) = A$$

are used.

The limit of a function of several variables is defined similarly.

REMARK 5. The intuitive meaning of Definition 4: $f(x, y)$ has a limit A at a point P, if $f(x, y)$ is sufficiently close to the value A for all points (x, y) that are sufficiently close to the point P, except possibly P itself.

Definition 5. We say that $f(x, y)$ is *continuous at a point* (x_0, y_0) if it is defined at this point and if, corresponding to an arbitrary $\varepsilon > 0$, there exists a $\delta > 0$, such that for all points (x, y) in the δ-neighbourhood of the point (x_0, y_0) the relation $|f(x, y) - f(x_0, y_0)| < \varepsilon$ holds. The continuity of a function of several variables is defined similarly.

REMARK 6. It follows from the definition of continuity that a function $f(x, y)$ continuous at the point (x_0, y_0) is defined in a definite neighbourhood of the point (x_0, y_0) (this point included).

Theorem 1. *Let a function $f(x, y)$ be defined at a point (x_0, y_0). Then it is continuous at that point if, and only if,* $\lim_{(x,y)\to(x_0,y_0)} f(x, y) = f(x_0, y_0)$. *Similarly for functions of several variables.*

Definition 6. If a function is continuous at every point of a region (more generally: of a set) M we say that it is *continuous in* (or *on*) M.

REMARK 7. When we say that $f(x, y)$ is continuous in a closed region $\bar{O}$, we mean that it is continuous in O and that on the boundary it is *continuous with regard to the points from* $\bar{O}$ (i.e. if the point (x_0, y_0) lies on the boundary, then we consider in Definition 5 only those points from the δ-neighbourhood of the point (x_0, y_0) that belong to $\bar{O}$). The same remark holds for functions of several variables.

Example 4. The function $z = \sqrt{(1 - x^2 - y^2)}$ is continuous in the closed region given by $x^2 + y^2 \leqq 1$, i.e. in a closed circle.

REMARK 8. We often have to deal with the case where $f(x, y)$ is defined and continuous in the region O and at the same time can be defined on the boundary h of this region in such a manner that the extended function is continuous in $\bar{O}$. Then we say that $f(x, y)$ is *continuously extensible on the boundary* h.

In a similar way we define continuous extensibility on the boundary for functions of several variables.

We also often come across the case where the given function is continuous in a closed region $\bar{O}$ and has in $\bar{O}$ continuous partial derivatives of the first order. It is to be understood that these derivatives are continuous in O and continuously extensible on the boundary. Similarly, we may speak of the continuity of higher derivatives in $\bar{O}$.

When we say that a function $f(x, y)$ is piecewise continuous in a region O of the type A (Definition 14.1.2, p. 603) we shall mean that it is possible to divide the region O by means of a finite number of simple finite piecewise smooth curves (Definition 14.1.1) into a (finite) number of regions O_n of the type A so that $f(x, y)$ is continuous in every region O_n and continuously extensible on its boundary. (For example, the function considered in Example 14.1.2, p. 604, is piecewise continuous in O.)

Similarly we define a piecewise continuous function in a three-dimensional region of the type A.

A function is said to be *piecewise smooth in* O when the function and its partial derivatives of the first order are piecewise continuous in O.

Further definitions and theorems (similar to those given in § 11.3, 11.4 for functions of one variable) can be formulated for functions of two or more variables.

Of most frequent application are the following theorems:

Theorem 2. *If* $f(x, y)$ *and* $g(x, y)$ *possess the limits* A *and* B, *respectively, at the point* (x_0, y_0), *then the functions* $k \,.\, f(x, y)$ $(k = \text{const.})$, $f(x, y) \pm g(x, y)$, $f(x, y) \,.\, g(x, y)$ *and (if* $B \neq 0$*)* $f(x, y)/g(x, y)$ *also possess a limit at the point* (x_0, y_0) *and the relations*

$$\lim_{(x,y)\to(x_0,y_0)} k f(x, y) = kA \; (k \text{ a constant}), \qquad \lim_{(x,y)\to(x_0,y_0)} [f(x, y) \pm g(x, y)] = A \pm B,$$

$$\lim_{(x,y)\to(x_0,y_0)} f(x, y)\, g(x, y) = AB, \qquad \lim_{(x,y)\to(x_0,y_0)} \frac{f(x, y)}{g(x, y)} = \frac{A}{B}.$$

hold.

A similar theorem holds in the n-dimensional case. A similar statement holds also for continuity.

Theorem 3. *A continuous function of continuous functions is itself continuous. In more detail (for functions of two variables): If* $u = f(x, y)$, $v = g(x, y)$ *are continuous at the point* (x_0, y_0) *and if* $z = h(u, v)$ *is continuous at the corresponding point* (u_0, v_0), *then* $z = h(f(x, y), g(x, y))$ *is continuous (considered as a function of the variables* x, y*) at the point* (x_0, y_0).

REMARK 9. On the basis of the last two theorems we may decide on the continuity of many functions which we come across in applications. In particular all polynomials in x and y, all rational functions (provided the denominator is non-zero), all functions composed of continuous functions (e.g., the function $z = xy \sin^2 x$), etc., are continuous.

Theorem 4. *If* $f(x, y)$ *is continuous in a region* O *and if* (x_1, y_1), (x_2, y_2) *are any two points in this region, then* $f(x, y)$ *takes on in* O *every value between* $f(x_1, y_1)$ *and* $f(x_2, y_2)$.

Theorem 5. *A function that is continuous in a bounded closed region* $\bar{O}$ *takes on a greatest value at least at one point* $(x_0, y_0) \in \bar{O}$ *(that is:* $f(x_0, y_0) \geqq f(x, y)$ *for all points* $(x, y) \in \bar{O}$*) and a least value at least at one point* $(x_1, y_1) \in \bar{O}$.

Theorem 6. *A function* $f(x, y)$ *that is continuous in a bounded closed domain* $\bar{O}$ *is uniformly continuous there. This means: To an arbitrary* $\varepsilon > 0$ *there exists* $\delta > 0$ *depending only on the choice of the number* ε *(thus the same for the whole closed region* $\bar{O}$*) such that*

$$|f(x_2, y_2) - f(x_1, y_1)| < \varepsilon$$

for every pair of points $(x_1, y_1) \in \bar{O}$, $(x_2, y_2) \in \bar{O}$, *the distance between which is smaller than* δ.

REMARK 10. Theorems similar to Theorems 4, 5, 6 hold for functions of several variables.

12.2. Partial Derivatives. Change of Order of Differentiation

Definition 1. We say that a function $z = f(x, y)$ has a *partial derivative with respect to* x at the point (x_0, y_0) if the (finite) limit

$$\lim_{h \to 0} \frac{f(x_0 + h, y_0) - f(x_0, y_0)}{h}$$

exists.

The following notations are used:

$$\frac{\partial f}{\partial x}(x_0, y_0), \quad \frac{\partial z}{\partial x}(x_0, y_0), \quad f_x(x_0, y_0), \quad f'_x(x_0, y_0).$$

Similarly

$$\frac{\partial f}{\partial y}(x_0, y_0) = \frac{\partial z}{\partial y}(x_0, y_0) = f_y(x_0, y_0) = f'_y(x_0, y_0) = \lim_{k\to 0}\frac{f(x_0, y_0 + k) - f(x_0, y_0)}{k}.$$

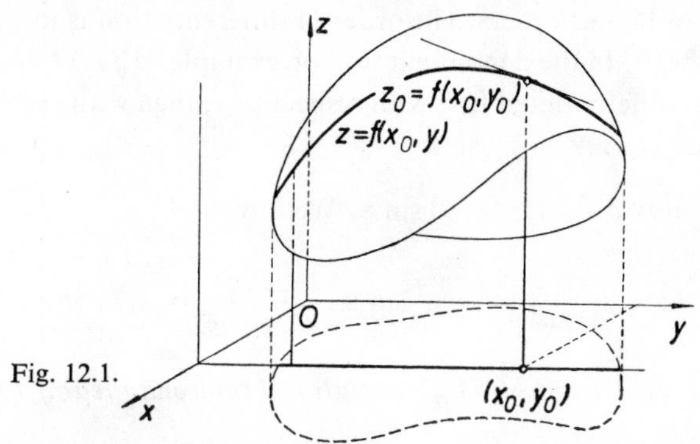

Fig. 12.1.

REMARK 1. The geometrical interpretation of the partial derivative is given in Fig. 12.1. The section of the surface $z = f(x, y)$ by the plane $x = x_0$ (a plane parallel to the coordinate plane $x = 0$) is the curve $z = f(x_0, y)$ (z depends only on y if x_0 is constant); $\partial f/\partial y(x_0, y_0)$ is the slope of the tangent to this curve at the point (x_0, y_0, z_0) indicated in Fig. 12.1.

REMARK 2. *Partial derivatives of functions of several variables* are defined in a similar manner. For example,

$$\frac{\partial f}{\partial x_2}(x_1, x_2, \ldots, x_n) = \lim_{h_2\to 0}\frac{f(x_1, x_2 + h_2, x_3, \ldots, x_n) - f(x_1, x_2, x_3, \ldots, x_n)}{h_2}.$$

REMARK 3. To compute a partial derivative, we differentiate the given function regarding it as a function of the *single* variable, with respect to which the derivative is required. The other variables are treated as though they were constants.

Example 1.

$$z = x^3 y, \quad \frac{\partial z}{\partial x} = 3x^2 y$$

(during this differentiation y is kept constant);

$$z = \sin(xy), \quad \frac{\partial z}{\partial y} = x\cos(xy)$$

(during this differentiation x is constant).

REMARK 4. $\partial f/\partial x$, $\partial f/\partial y$ are again functions of x, y. *Derivatives of the second order* are defined by the relations:

$$\frac{\partial^2 f}{\partial x^2} = \frac{\partial}{\partial x}\left(\frac{\partial f}{\partial x}\right), \quad \frac{\partial^2 f}{\partial y^2} = \frac{\partial}{\partial y}\left(\frac{\partial f}{\partial y}\right), \quad \frac{\partial^2 f}{\partial x\, \partial y} = \frac{\partial}{\partial x}\left(\frac{\partial f}{\partial y}\right), \quad \frac{\partial^2 f}{\partial y\, \partial x} = \frac{\partial}{\partial y}\left(\frac{\partial f}{\partial x}\right).$$

The last two derivatives are called *mixed.*

The derivatives of second order of functions of several variables are defined similarly and so are those of higher orders. The order of differentiation is indicated by the order of the symbols ∂x, ∂y in the denominator. For example, $\partial^3 f/\partial x^2\, \partial y$ means that the function z has been differentiated first with respect to y, then with respect to x and then again with respect to x.

Example 2. Find $\partial^3 z/\partial y\, \partial x^2$, if $z = y^2 \sin x$. We have

$$\frac{\partial z}{\partial x} = y^2 \cos x\,, \quad \frac{\partial^2 z}{\partial x^2} = -y^2 \sin x\,, \quad \frac{\partial^3 z}{\partial y\, \partial x^2} = -2y \sin x\,.$$

Theorem 1. *Change of Order of Differentiation (Interchangeability of Mixed Derivatives). If*

$$\frac{\partial^2 f}{\partial x\, \partial y}, \quad \frac{\partial^2 f}{\partial y\, \partial x}$$

are both continuous at the point (x_0, y_0), then they are equal at this point, i.e. the relation

$$\frac{\partial^2 f}{\partial x\, \partial y} = \frac{\partial^2 f}{\partial y\, \partial x}$$

holds.

REMARK 5. Under similar assumptions theorems on the interchangeability of mixed derivatives of higher orders or on the interchangeability of mixed partial derivatives of functions of several variables also hold. If, for instance, $\partial^4 f/\partial y^2\, \partial z\, \partial x$ and $\partial^4 f/\partial x\, \partial y^2\, \partial z$ are both continuous at the point (x_0, y_0, z_0), then they are equal at that point. If the equality holds at every point of a given domain then, of course, the corresponding derivatives are equal in the whole domain.

Example 3. Let us consider the function $z = y^2 \sin x$ from Example 2. Then

$$\frac{\partial z}{\partial y} = 2y \sin x\,, \quad \frac{\partial^2 z}{\partial x\, \partial y} = 2y \cos x\,, \quad \frac{\partial^3 z}{\partial x^2\, \partial y} = -2y \sin x\,,$$

hence

$$\frac{\partial^3 z}{\partial x^2\, \partial y} = \frac{\partial^3 z}{\partial y\, \partial x^2}\,.$$

12.3. Total Differential

Definition 1. The function $z = f(x, y)$ is said to be *differentiable at the point* (x_0, y_0) when its increment $\Delta z = f(x_0 + h, y_0 + k) - f(x_0, y_0)$ can be expressed, in a certain neighbourhood of the point (x_0, y_0), in the form

$$\Delta z \equiv f(x_0 + h, y_0 + k) - f(x_0, y_0) = Ah + Bk + \varrho\tau(h, k)\,, \tag{1}$$

where A, B are constants, $\varrho = \sqrt{(h^2 + k^2)}$ and

$$\lim_{\substack{h\to 0\\ k\to 0}} \tau(h, k) = 0\,. \tag{2}$$

Note that in general τ contains h, k, x_0, y_0; x_0, y_0 are treated as constants. (See Definition 11.6.1 and Remark to Definition 11.6.2, p. 423.)

Theorem 1. *If $f(x, y)$ is differentiable at the point (x_0, y_0), then it possesses partial derivatives at (x_0, y_0) and the relations*

$$A = \frac{\partial f}{\partial x}(x_0, y_0)\,, \quad B = \frac{\partial f}{\partial y}(x_0, y_0)$$

hold.

REMARK 1. If we pass to the customary notation $h = \mathrm{d}x$, $k = \mathrm{d}y$, we have

$$\Delta z = \frac{\partial f}{\partial x}\,\mathrm{d}x + \frac{\partial f}{\partial y}\,\mathrm{d}y + \varrho\tau(\mathrm{d}x, \mathrm{d}y)\,. \tag{3}$$

Definition 2. If $f(x, y)$ is differentiable at (x_0, y_0), then the expression

$$\mathrm{d}z = \frac{\partial f}{\partial x}\,\mathrm{d}x + \frac{\partial f}{\partial y}\,\mathrm{d}y$$

is called the *total differential* of the function $z = f(x, y)$.

Theorem 2. *If $f(x, y)$ is differentiable at the point (x_0, y_0), then it is continuous at that point.*

REMARK 2. If $f(x, y)$ has partial derivatives of the first order at the point (x_0, y_0) then it need not be continuous at that point as a function of both variables x, y (it is continuous only as a function of the variable x on the straight line parallel to the x-axis drawn through the point (x_0, y_0) and as a function of the variable y on the straight line parallel to y-axis). This can be shown, for example, by the function

$$z(x, y) = \frac{xy}{x^2 + y^2} \quad \text{for} \quad (x, y) \neq (0, 0)\,, \quad z(0, 0) = 0$$

which possesses derivatives of the first order at the origin (equal to zero) but is not continuous at this point. This example shows (see Theorem 2) that a function which possesses partial derivatives of the first order at (x_0, y_0) need not be differentiable at that point. The following theorem, however, holds:

Theorem 3. *If*

$$\frac{\partial f}{\partial x}, \quad \frac{\partial f}{\partial y}$$

are continuous at (x_0, y_0), then $f(x, y)$ is differentiable at (x_0, y_0) (and therefore also continuous).

Theorem 4. *If $f(x, y)$ is differentiable at (x_0, y_0), then the surface $z = f(x, y)$ possesses a tangent plane at the point (x_0, y_0, z_0) (where $z_0 = f(x_0, y_0)$). Its equation is*

$$z - z_0 = \left(\frac{\partial f}{\partial x}\right)_0 (x - x_0) + \left(\frac{\partial f}{\partial y}\right)_0 (y - y_0),$$

where

$$\left(\frac{\partial f}{\partial x}\right)_0 = \frac{\partial f}{\partial x}(x_0, y_0), \quad \left(\frac{\partial f}{\partial y}\right)_0 = \frac{\partial f}{\partial y}(x_0, y_0).$$

REMARK 3. In the same way as we approximate, in the case of a function of *one* variable, the increment of the function by its differential (geometrically: we substitute the increment on the tangent for the increment of the function) so here we approximate the increment of the function by its total differential (geometrically: we substitute the increment on the tangent plane for the increment of the function).

Example 1. $f(x, y) = x^3 + 4y^3$. Let us find the approximate value of $f(1{\cdot}11; 0{\cdot}58)$.

First, we have $f(1; 0{\cdot}5) = 1{\cdot}5$. If we compute the total differential for $dx = 0{\cdot}11$, $dy = 0{\cdot}08$ at this point (i.e. at the point $x_0 = 1$, $y_0 = 0{\cdot}5$), we obtain: $dz = 3x^2\,dx + 12y^2\,dy = 3 \cdot 0{\cdot}11 + 12 \cdot 0{\cdot}5^2 \cdot 0{\cdot}08 = 0{\cdot}57$. Hence $f(1{\cdot}11; 0{\cdot}58) \approx f(1; 0{\cdot}5) + dz = 1{\cdot}5 + 0{\cdot}57 = 2{\cdot}07$. (The exact result is $f(1{\cdot}11; 0{\cdot}58) = 2{\cdot}148\,079$.)

Theorem 5. *If the functions*

$$P(x, y), \quad Q(x, y), \quad \frac{\partial P}{\partial y}(x, y), \quad \frac{\partial Q}{\partial x}(x, y)$$

are continuous in a simply-connected region O, then a necessary and sufficient condition that the expression

$$P(x, y)\,dx + Q(x, y)\,dy$$

be the total differential of a function $f(x, y)$ in O is that the relation

$$\frac{\partial P}{\partial y} = \frac{\partial Q}{\partial x}$$

holds in O.

The conditions for a similar expression

$$P(x, y, z)\,\mathrm{d}x + Q(x, y, z)\,\mathrm{d}y + R(x, y, z)\,\mathrm{d}z$$

to be the total differential of a function $f(x, y, z)$ are

$$\frac{\partial P}{\partial y} = \frac{\partial Q}{\partial x}, \quad \frac{\partial P}{\partial z} = \frac{\partial R}{\partial x}, \quad \frac{\partial Q}{\partial z} = \frac{\partial R}{\partial y} \quad \textit{(simultaneously) in } O\,.$$

(For a more detailed treatment see § 14.7.)

Definition 3. Let $z = f(x, y)$ have a total differential in a neighbourhood of a point (x_0, y_0) and let the partial derivatives

$$\frac{\partial f}{\partial x}(x, y), \quad \frac{\partial f}{\partial y}(x, y)$$

have a total differential at the point (x_0, y_0). Then we say that $f(x, y)$ has a *total differential of the second order* (briefly a *second differential*). By this differential we understand the expression

$$\mathrm{d}^2 z = h^2 \frac{\partial^2 f}{\partial x^2}(x_0, y_0) + 2hk \frac{\partial^2 f}{\partial x\,\partial y}(x_0, y_0) + k^2 \frac{\partial^2 f}{\partial y^2}(x_0, y_0)\,.$$

Instead of h, k we often write $\mathrm{d}x$, $\mathrm{d}y$.

REMARK 4. Formally, we obtain the second differential as the differential of the first differential at the point (x_0, y_0) regarding h and k as constants (cf. Definition 11.6.3):

$$\mathrm{d}^2 z = \mathrm{d}\left(\frac{\partial f}{\partial x} h + \frac{\partial f}{\partial y} k\right) = h\,\mathrm{d}\left(\frac{\partial f}{\partial x}\right) + k\,\mathrm{d}\left(\frac{\partial f}{\partial y}\right) =$$

$$= h\left(h \frac{\partial^2 f}{\partial x^2} + k \frac{\partial^2 f}{\partial y\,\partial x}\right) + k\left(h \frac{\partial^2 f}{\partial x\,\partial y} + k \frac{\partial^2 f}{\partial y^2}\right) =$$

$$= h^2 \frac{\partial^2 f}{\partial x^2} + 2hk \frac{\partial^2 f}{\partial x\,\partial y} + k^2 \frac{\partial^2 f}{\partial y^2} = \left(h \frac{\partial}{\partial x} + k \frac{\partial}{\partial y}\right)^2 f\,.$$

We have used here the symbolic operator notation which is convenient especially in the case of higher differentials.

Similarly we can also define higher differentials:

Definition 4. Let $f(x, y)$ and all its partial derivatives up to the $(n-2)$-th order have a total differential in a neighbourhood of the point (x_0, y_0). Let the partial derivatives of the $(n-1)$-th order have a total differential at the point (x_0, y_0). Then we say that the function $f(x, y)$ has *at the point* (x_0, y_0) *a total differential of the n-th order* (briefly an *n-th differential*) and by this differential we mean the expression

$$\mathrm{d}^n z = \left(h \frac{\partial}{\partial x} + k \frac{\partial}{\partial y}\right)^n f =$$

$$= h^n \frac{\partial^n f}{\partial x^n} + \binom{n}{1} h^{n-1} k \frac{\partial^n f}{\partial x^{n-1} \partial y} + \ldots + \binom{n}{n-1} h k^{n-1} \frac{\partial^n f}{\partial x \partial y^{n-1}} + k^n \frac{\partial^n f}{\partial y^n}.$$

REMARK 5. In a similar way we define a *differentiable function of n variables.* Its m-th total differential is given by the formula

$$\mathrm{d}^m z = \left(h_1 \frac{\partial}{\partial x_1} + h_2 \frac{\partial}{\partial x_2} + \ldots + h_n \frac{\partial}{\partial x_n}\right)^m f.$$

REMARK 6. In contradistinction to the total differential we often speak of the *partial differential of the function* $z = f(x, y)$ *with respect to* x or y:

$$\mathrm{d}_x z = \frac{\partial f}{\partial x} \mathrm{d}x \quad \text{or} \quad \mathrm{d}_y z = \frac{\partial f}{\partial y} \mathrm{d}y.$$

12.4. Differentiation of Composite Functions

Theorem 1. *Let the functions* $u = f(x, y)$, $v = g(x, y)$ *be differentiable at the point* (x_0, y_0). *(In order that the functions be differentiable it is sufficient by Theorem* 12.3.3 *that they have continuous partial derivatives at that point.) Let the function* $z = h(u, v)$ *be differentiable at the corresponding point* (u_0, v_0) (where $u_0 = f(x_0, y_0)$, $v_0 = g(x_0, y_0)$). *Then the composite function* (see Remark 12.1.3)

$$z = h(f(x, y), g(x, y))$$

is a differentiable function (as a function of the variables x, y*) at the point* (x_0, y_0) *and*

$$\frac{\partial z}{\partial x} = \frac{\partial z}{\partial u}\frac{\partial u}{\partial x} + \frac{\partial z}{\partial v}\frac{\partial v}{\partial x}, \quad \frac{\partial z}{\partial y} = \frac{\partial z}{\partial u}\frac{\partial u}{\partial y} + \frac{\partial z}{\partial v}\frac{\partial v}{\partial y}. \tag{1}$$

In somewhat more precise notation:

$$\frac{\partial z}{\partial x} = \frac{\partial h}{\partial u}\frac{\partial f}{\partial x} + \frac{\partial h}{\partial v}\frac{\partial g}{\partial x}, \quad \frac{\partial z}{\partial y} = \frac{\partial h}{\partial u}\frac{\partial f}{\partial y} + \frac{\partial h}{\partial v}\frac{\partial g}{\partial y}, \tag{2}$$

where the derivatives of the functions f, g are computed at the point (x_0, y_0) and those of the function h at the point (u_0, v_0).

Under similar assumptions regarding the functions

$$z = h(u_1, u_2, \ldots, u_n), \quad u_1 = f_1(x_1, x_2, \ldots, x_m), \quad \ldots, \quad u_n = f_n(x_1, x_2, \ldots, x_m)$$

the relation

$$\frac{\partial z}{\partial x_k} = \frac{\partial z}{\partial u_1}\frac{\partial u_1}{\partial x_k} + \frac{\partial z}{\partial u_2}\frac{\partial u_2}{\partial x_k} + \ldots + \frac{\partial z}{\partial u_n}\frac{\partial u_n}{\partial x_k}, \quad (k = 1, 2, \ldots, m)$$

holds.

Example 1. $z = (y \sin x)^{e^{x^2y}}$ $(y \sin x > 0)$. Let us put

$$z = u^v, \quad u = y \sin x, \quad v = e^{x^2y}.$$

By (1) or (2)

$$\frac{\partial z}{\partial x} = vu^{v-1}\, y \cos x + u^v \ln u\; 2xy\, e^{x^2y}$$

$$= u^v \left(\frac{e^{x^2y}}{u}\, y \cos x + 2xy\, e^{x^2y} \ln u\right)$$

$$= (y \sin x)^{e^{x^2y}} . e^{x^2y} . [\cot x + 2xy \ln (y \sin x)] .$$

REMARK 1. When computing the *second* derivatives it is necessary to bear in mind that the functions $\partial h/\partial u$ and $\partial h/\partial v$ are functions of u, v and thus *they are to be differentiated with respect to x or y as composite functions* (according to (1), (2)). For example, using (2):

$$\frac{\partial^2 z}{\partial x^2} = \frac{\partial}{\partial x}\left(\frac{\partial z}{\partial x}\right) = \frac{\partial}{\partial x}\left[\frac{\partial h}{\partial u}\frac{\partial f}{\partial x} + \frac{\partial h}{\partial v}\frac{\partial g}{\partial x}\right] =$$

$$= \frac{\partial}{\partial x}\left(\frac{\partial h}{\partial u}\right)\frac{\partial f}{\partial x} + \frac{\partial h}{\partial u}\frac{\partial}{\partial x}\left(\frac{\partial f}{\partial x}\right) + \frac{\partial}{\partial x}\left(\frac{\partial h}{\partial v}\right)\frac{\partial g}{\partial x} + \frac{\partial h}{\partial v}\frac{\partial}{\partial x}\left(\frac{\partial g}{\partial x}\right) =$$

$$= \left[\frac{\partial}{\partial u}\left(\frac{\partial h}{\partial u}\right)\frac{\partial f}{\partial x} + \frac{\partial}{\partial v}\left(\frac{\partial h}{\partial u}\right)\frac{\partial g}{\partial x}\right]\frac{\partial f}{\partial x} + \frac{\partial h}{\partial u}\frac{\partial^2 f}{\partial x^2} +$$

$$+ \left[\frac{\partial}{\partial u}\left(\frac{\partial h}{\partial v}\right)\frac{\partial f}{\partial x} + \frac{\partial}{\partial v}\left(\frac{\partial h}{\partial v}\right)\frac{\partial g}{\partial x}\right]\frac{\partial g}{\partial x} + \frac{\partial h}{\partial v}\frac{\partial^2 g}{\partial x^2} =$$

$$= \frac{\partial^2 h}{\partial u^2}\left(\frac{\partial f}{\partial x}\right)^2 + 2\frac{\partial^2 h}{\partial u\,\partial v}\frac{\partial f}{\partial x}\frac{\partial g}{\partial x} + \frac{\partial^2 h}{\partial v^2}\left(\frac{\partial g}{\partial x}\right)^2 + \frac{\partial h}{\partial u}\frac{\partial^2 f}{\partial x^2} + \frac{\partial h}{\partial v}\frac{\partial^2 g}{\partial x^2}.$$

(We have already applied the interchangeability of the order of differentiation, $\partial^2 h/\partial u\,\partial v = \partial^2 h/\partial v\,\partial u$.) Similarly,

$$\frac{\partial^2 z}{\partial y^2} = \frac{\partial^2 h}{\partial u^2}\left(\frac{\partial f}{\partial y}\right)^2 + 2\frac{\partial^2 h}{\partial u\,\partial v}\frac{\partial f}{\partial y}\frac{\partial g}{\partial y} + \frac{\partial^2 h}{\partial v^2}\left(\frac{\partial g}{\partial y}\right)^2 + \frac{\partial h}{\partial u}\frac{\partial^2 f}{\partial y^2} + \frac{\partial h}{\partial v}\frac{\partial^2 g}{\partial y^2},$$

$$\frac{\partial^2 z}{\partial x\,\partial y} = \frac{\partial^2 h}{\partial u^2}\frac{\partial f}{\partial x}\frac{\partial f}{\partial y} + \frac{\partial^2 h}{\partial u\,\partial v}\left(\frac{\partial f}{\partial x}\frac{\partial g}{\partial y} + \frac{\partial f}{\partial y}\frac{\partial g}{\partial x}\right) +$$

$$+ \frac{\partial^2 h}{\partial v^2}\frac{\partial g}{\partial x}\frac{\partial g}{\partial y} + \frac{\partial h}{\partial u}\frac{\partial^2 f}{\partial x\,\partial y} + \frac{\partial h}{\partial v}\frac{\partial^2 g}{\partial x\,\partial y}.$$

REMARK 2. We often come across the case where the composite function is given as follows:

$$z = h(x, y, u, v), \quad u = f(x, y), \quad v = g(x, y). \tag{3}$$

Here the function h contains the variable x partly directly, partly through the functions u, v so that

$$\frac{\partial z}{\partial x} = \frac{\partial h}{\partial x} + \frac{\partial h}{\partial u}\frac{\partial u}{\partial x} + \frac{\partial h}{\partial v}\frac{\partial v}{\partial x}, \quad \frac{\partial z}{\partial y} = \frac{\partial h}{\partial y} + \frac{\partial h}{\partial u}\frac{\partial u}{\partial y} + \frac{\partial h}{\partial v}\frac{\partial v}{\partial y}. \tag{4}$$

(In this case we cannot apply notation (1) and write $\partial z/\partial x$ instead of $\partial h/\partial x$.)

Example 2. $z = xu^2 + 2y^2v$, $u = y \sin x$, $v = x \ln y$. By (4)

$$\frac{\partial z}{\partial x} = u^2 + 2xuy \cos x + 2y^2 \ln y = y^2 \sin^2 x + 2xy^2 \sin x \cos x + 2y^2 \ln y.$$

REMARK 3. The preceding differentiation could also be carried out directly after putting the expressions $y \sin x$ and $x \ln y$ for u and v into the formula $z = xu^2 + 2y^2v$. But, if one (or more) of the functions (3) is given implicitly, then the substitution cannot (at least in the general case) be carried out. Then the application of formulae for differentiation of composite functions is necessary (cf. § 12.9).

12.5. Taylor's Theorem, the Mean-Value Theorem. Differentiation in a Given Direction

Theorem 1 (*Taylor's Theorem*). *Let the function $z = f(x, y)$ possess total differentials (§ 12.3) up to order $n + 1$ at every point of the closed segment u joining the points (x_0, y_0), $(x_0 + h, y_0 + k)$. Then*

$$f(x_0 + h, y_0 + k) = f(x_0, y_0) + \frac{\left(h\dfrac{\partial}{\partial x} + k\dfrac{\partial}{\partial y}\right) f_{x_0, y_0}}{1!} + \ldots +$$

$$+\frac{\left(h\frac{\partial}{\partial x}+k\frac{\partial}{\partial y}\right)^n f_{x_0,y_0}}{n!}+R_{n+1}\,, \tag{1}$$

where the most often applied form of the remainder is Lagrange's form, namely

$$R_{n+1}=\frac{\left(h\frac{\partial}{\partial x}+k\frac{\partial}{\partial y}\right)^{n+1} f_{c,d}}{(n+1)!}\,.$$

Here, the point (c, d) is an interior point of the above-mentioned segment u. The suffixes x_0, y_0 or c, d denote the point at which the derivatives are to be computed.

REMARK 1. For $n = 0$, the relation (1) reduces to the *mean-value theorem:*

$$f(x_0+h, y_0+k)-f(x_0, y_0)=h\frac{\partial f}{\partial x}(c, d)+k\frac{\partial f}{\partial y}(c, d)\,.$$

REMARK 2. The generalization of *Taylor's Theorem* (or of the mean-value theorem) *to functions of several variables* is immediate:

$$f(x_1+h_1, x_2+h_2, \ldots, x_n+h_n)=f(x_1, x_2, \ldots, x_n)+$$

$$+\frac{\left(h_1\frac{\partial}{\partial x_1}+\ldots+h_n\frac{\partial}{\partial x_n}\right) f_{x_1,x_2,\ldots,x_n}}{1!}+\ldots+$$

$$+\frac{\left(h_1\frac{\partial}{\partial x_1}+\ldots+h_n\frac{\partial}{\partial x_n}\right)^m f_{x_1,x_2,\ldots,x_n}}{m!}+R_{m+1}\,,$$

where

$$R_{m+1}=\frac{\left(h_1\frac{\partial}{\partial x_1}+\ldots+h_n\frac{\partial}{\partial x_n}\right)^{m+1} f_{c_1,c_2,\ldots,c_n}}{(m+1)!}\,,$$

$c_k = x_k + \vartheta h_k$ and $0 < \vartheta < 1$. (For $n = 1$ we obtain Taylor's Theorem for functions of one variable, § 11.10.)

Definition 1. Let $\cos\alpha$, $\cos\beta$ be the direction-cosines of the oriented segment u joining the points (x_0, y_0), $(x_0 + h, y_0 + k)$. If $s = \sqrt{(h^2 + k^2)}$ denotes the length of this segment, then $h = s\cos\alpha$, $k = s\cos\beta$ (Fig. 12.2). The limit

$$\lim_{s\to 0}\frac{f(x_0+s\cos\alpha, y_0+s\cos\beta)-f(x_0, y_0)}{s} \tag{2}$$

(if it exists) is called the *derivative of the function* $f(x, y)$ *in the direction* $(\cos \alpha, \cos \beta)$ *at the point* (x_0, y_0) and is denoted by df/ds.

Theorem 2. *If* $f(x, y)$ *is differentiable at* (x_0, y_0), *then*

$$\frac{\partial f}{\partial s} = \frac{\partial f}{\partial x}(x_0, y_0) \cos \alpha + \frac{\partial f}{\partial y}(x_0, y_0) \sin \alpha .$$

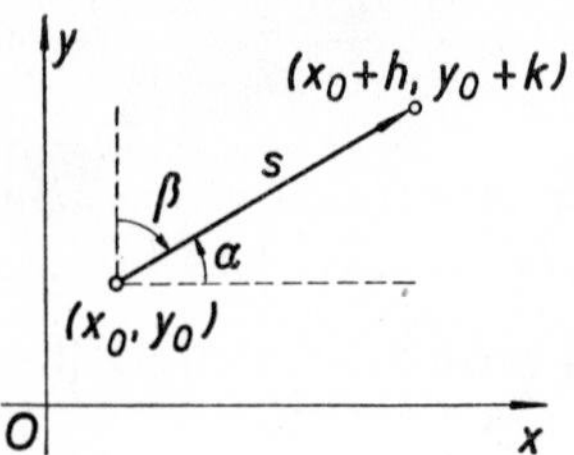

Fig. 12.2.

12.6. Euler's Theorem on Homogeneous Functions

Definition 1. The function $z = f(x, y)$ is said to be *homogeneous of degree n in a region O* if the relation

$$f(tx, ty) = t^n f(x, y)$$

holds identically for every point $(x, y) \in O$ and for every t from a certain neighbourhood of the point $t = 1$ (depending, in general, on the point (x, y)).

Example 1. The function

$$z = x^2 + y^2$$

is a homogeneous function of the second degree in the whole plane because (for every t)

$$(tx)^2 + (ty)^2 = t^2(x^2 + y^2) .$$

The function

$$z = \frac{1}{\sqrt{(x - y)}} \quad (x > y)$$

is, as we see in a similar way, a homogeneous function of degree $n = -\frac{1}{2}$ in the half-plane $x > y$.

Theorem 1 (*Euler's Theorem on Homogeneous Functions*). *If a function* $z = f(x, y)$, *homogeneous of degree n in a region O, has a total differential in O, then* *the relation*

$$x \frac{\partial f}{\partial x} + y \frac{\partial f}{\partial y} = nf(x, y) \tag{1}$$

holds in O.

Example 2. The function $z = x^2 + y^2$ satisfies the relation

$$x \frac{\partial f}{\partial x} + y \frac{\partial f}{\partial y} = x \,.\, 2x + y \,.\, 2y = 2(x^2 + y^2)\,.$$

REMARK 1. Under similar assumptions the relation

$$x_1 \frac{\partial f}{\partial x_1} + \ldots + x_n \frac{\partial f}{\partial x_n} = m\, f(x_1, x_2, \ldots, x_n)$$

holds for a homogeneous function of degree m of n variables.

REMARK 2. The converse of Theorem 1 is also true: If a function $f(x, y)$ possesses a total differential in the region O and if equation (1) is satisfied everywhere in O, then $f(x, y)$ is a homogeneous function of degree n in O. A similar assertion holds for functions of several variables.

12.7. Regular Mappings. Functional Determinants

Definition 1. Let us consider m functions

$$\begin{aligned} y_1 &= f_1(x_1, x_2, \ldots, x_n)\,, \\ y_2 &= f_2(x_1, x_2, \ldots, x_n)\,, \\ &\ldots\ldots\ldots\ldots\ldots\ldots\ldots \\ y_m &= f_m(x_1, x_2, \ldots, x_n) \end{aligned} \tag{1}$$

defined in an n-dimensional domain M. By system (1), to every point $X\,(x_1, x_2, \ldots, x_n$ of M, there corresponds a certain point $Y(y_1, y_2, \ldots, y_m)$ of the m-dimensional space E_m. This correspondence is called a *mapping*, or *transformation* (*of M into* E_m). The point X is called the *original* (or the *model* or the *inverse image*), the point Y is the *image* (or the *transform*). (On the simplest case $m = n = 1$ see Definition 11.1.3.)

Example 1. The parametric representation of a surface

$$x = f_1(u, v)\,, \quad y = f_2(u, v)\,, \quad z = f_3(u, v)$$

is a mapping, where the points $X(u, v)$ of the plane uv are the originals and the points $Y(x, y, z)$ of the space E_3 are the images.

Definition 2. The mapping

$$\begin{aligned} y_1 &= f_1(x_1, x_2, \ldots, x_n), \\ y_2 &= f_2(x_1, x_2, \ldots, x_n), \\ &\ldots\ldots\ldots\ldots\ldots\ldots \\ y_n &= f_n(x_1, x_2, \ldots, x_n) \end{aligned} \tag{2}$$

(where the number of functions is the same as the number of variables) is said to be *regular in a region M* if each of the functions $y_1, y_2, \ldots, y_n$ possesses continuous partial derivatives of the first order in M and if the determinant

$$\begin{vmatrix} \dfrac{\partial f_1}{\partial x_1}, & \dfrac{\partial f_1}{\partial x_2}, & \ldots, & \dfrac{\partial f_1}{\partial x_n} \\ \dfrac{\partial f_2}{\partial x_1}, & \dfrac{\partial f_2}{\partial x_2}, & \ldots, & \dfrac{\partial f_2}{\partial x_n} \\ \ldots & \ldots & \ldots & \ldots \\ \dfrac{\partial f_n}{\partial x_1}, & \dfrac{\partial f_n}{\partial x_2}, & \ldots, & \dfrac{\partial f_n}{\partial x_n} \end{vmatrix} \tag{3}$$

is different from zero in M.

REMARK 1. The determinant (3) is called the *functional determinant* of the given mapping or the *Jacobian*. We denote it by

$$\frac{\partial(y_1, y_2, \ldots, y_n)}{\partial(x_1, x_2, \ldots, x_n)}.$$

The Jacobian is continuous and non-zero in M; hence its sign does not change in M.

Definition 3. The mapping (1) is called *continuous at the point* $A(x_1, x_2, \ldots, x_n)$ if all the functions $f_1, f_2, \ldots, f_n$ are continuous at the point A.

REMARK 2. Evidently, every regular mapping is continuous but the converse is not true in general.

Definition 4. Let M and E_m have the same meaning as in Definition 1 and let N be a subregion of M; we do not exclude the possibility that $N = M$. Denote by Q the set of all points Y in E_m which correspond to all points $X \in N$ according to equations (1). We then say that the set N is transformed (or mapped) by equations (1) *onto* the set Q. If this correspondence has the further property that to each $Y \in Q$ there corresponds a unique original $X \in N$, then we say the *correspondence* is *one-to-one*. The so obtained mapping from Q to N (denoted by $Q \to N$), where thus Y is the original and X is the image, is called the *inverse mapping* to the original mapping $N \to Q$.

Theorem 1. *If a mapping is regular in M, then it is a one-to-one mapping in a sufficiently small neighbourhood of every interior point $X_0 \in M$.*

REMARK 3. Thus, this means that in a neighbourhood of every interior point $X_0 \in M$ there exists an inverse mapping, i.e. we can compute $x_1, x_2, \ldots, x_n$ from the system (2) as functions of the variables $y_1, y_2, \ldots, y_n$. The theorem has, however, a *local* character, i.e. a mapping that is regular in a region need not be a one-to-one mapping in the whole region.

Example 2. The Jacobian of the mapping $x = \varrho \cos \varphi$, $y = \varrho \sin \varphi$ (where ϱ and φ are the polar coordinates of the point (x, y)), namely

$$\frac{\partial(x, y)}{\partial(\varrho, \varphi)} = \begin{vmatrix} \dfrac{\partial x}{\partial \varrho}, & \dfrac{\partial x}{\partial \varphi} \\ \dfrac{\partial y}{\partial \varrho}, & \dfrac{\partial y}{\partial \varphi} \end{vmatrix} = \begin{vmatrix} \cos \varphi, & -\varrho \sin \varphi \\ \sin \varphi, & \varrho \cos \varphi \end{vmatrix} = \varrho\,,$$

is non-vanishing when $\varrho > 0$ (ϱ is always non-negative) but, evidently, x and y are the same, for instance, when $\varrho = 1$, $\varphi = \frac{1}{2}\pi$ and $\varrho = 1$, $\varphi = \frac{5}{2}\pi$. If we restrict the values of φ, say, to the interval $0 \leqq \varphi < 2\pi$, then the mapping will be one-to-one for $\varrho > 0$. (But not for $\varrho = 0$ because, for example, for $\varrho = 0$, $\varphi = \frac{1}{2}\pi$ and $\varrho = 0$, $\varphi = \pi$ we obtain the same values $x = 0$, $y = 0$.)

REMARK 4. If $n = 1$ in (2), then we obtain the mapping $y = f(x)$. Its Jacobian is $f'(x)$. If, in a certain interval, $f'(x) \neq 0$, then the function $y = f(x)$ is strictly increasing or strictly decreasing in the whole interval and the inverse mapping exists in the whole interval.

Theorem 2 (*Theorem on the Preservation of the Region*). *The image of a region by a one-to-one regular mapping is again a region.*

Theorem 3. *If the mapping* (2) *is regular in a neighbourhood of the point $X_0 \in M$, then the inverse mapping* (see Remark 3) *is also regular in a neighbourhood of the corresponding point Y_0 and the values of the corresponding Jacobians are reciprocal,*

$$\frac{\partial(y_1, y_2, \ldots, y_n)}{\partial(x_1, x_2, \ldots, x_n)} = \frac{1}{\dfrac{\partial(x_1, x_2, \ldots, x_n)}{\partial(y_1, y_2, \ldots, y_n)}}\,.$$

Example 3. Let us consider the mapping

$$x = \varrho \cos \varphi\,, \quad y = \varrho \sin \varphi \tag{4}$$

(Example 2). Let us choose a point (x_0, y_0), for example so that $x_0 > 0$, $y_0 > 0$. Squaring equations (4) and adding them we obtain $x^2 + y^2 = \varrho^2$. Dividing them we have $y/x = \tan\varphi$. So we obtain the inverse mapping (in a neighbourhood of the chosen point)

$$\varrho = \sqrt{(x^2 + y^2)}, \quad \varphi = \arctan\frac{y}{x}. \tag{5}$$

Its Jacobian is

$$\frac{\partial(\varrho, \varphi)}{\partial(x, y)} = \begin{vmatrix} \dfrac{x}{\sqrt{(x^2 + y^2)}}, & \dfrac{y}{\sqrt{(x^2 + y^2)}} \\ -\dfrac{y}{x^2}\dfrac{1}{1 + y^2/x^2}, & \dfrac{1}{x}\dfrac{1}{1 + y^2/x^2} \end{vmatrix} =$$

$$= \frac{1}{\sqrt{(x^2 + y^2)}\,(1 + y^2/x^2)} \begin{vmatrix} x, & y \\ -y/x^2, & 1/x \end{vmatrix} = \frac{1}{\sqrt{(x^2 + y^2)}} = \frac{1}{\varrho}$$

in accordance with Theorem 2 (the value of the Jacobian of the mapping (4) is ϱ (Example 2)).

Theorem 4. *The mapping which is the resultant mapping obtained by combining two regular mappings is again a regular mapping. Its Jacobian is equal to the product of the Jacobians of the individual mappings:*

$$\frac{\partial(y_1, y_2, \ldots, y_n)}{\partial(x_1, x_2, \ldots, x_n)} = \frac{\partial(y_1, y_2, \ldots, y_n)}{\partial(u_1, u_2, \ldots, u_n)} \cdot \frac{\partial(u_1, u_2, \ldots, u_n)}{\partial(x_1, x_2, \ldots, x_n)}.$$

12.8. Dependence of Functions

Let us consider m functions defined in an n-dimensional region O,

$$\begin{aligned} y_1 &= f_1(x_1, x_2, \ldots, x_n), \\ y_2 &= f_2(x_1, x_2, \ldots, x_n), \\ &\ldots\ldots\ldots\ldots\ldots\ldots \\ y_m &= f_m(x_1, x_2, \ldots, x_n), \end{aligned} \tag{1}$$

having continuous partial derivatives of the first order in O.

Theorem 1. *Let the so-called Jacobian matrix of the functions* (1),

$$\left\| \begin{matrix} \frac{\partial f_1}{\partial x_1}, & \frac{\partial f_1}{\partial x_2}, & \ldots, & \frac{\partial f_1}{\partial x_n} \\ \frac{\partial f_2}{\partial x_1}, & \frac{\partial f_2}{\partial x_2}, & \ldots, & \frac{\partial f_2}{\partial x_n} \\ \ldots & \ldots & \ldots & \ldots \\ \frac{\partial f_m}{\partial x_1}, & \frac{\partial f_m}{\partial x_2}, & \ldots, & \frac{\partial f_m}{\partial x_n} \end{matrix} \right\| , \tag{2}$$

be of rank h $(0 < h < m)$ *in a neighbourhood* U *of the point* $A(a_1, a_2, \ldots, a_n) \in O$. *Thus, at the point* A *(and, hence, owing to the continuity of the functions* $\partial f_i/\partial x_k$ *also in a neighbourhood of the point* A*) at least one minor of order* h *is non-zero; let it be, for example, the minor*

$$\frac{\partial(y_1, y_2, \ldots, y_h)}{\partial(x_1, x_2, \ldots, x_h)} ,$$

Let us denote by $B(b_1, b_2, \ldots, b_h)$ *the point with the coordinates*

$$b_k = f_k(a_1, a_2, \ldots, a_n) , \quad k = 1, 2, \ldots, h .$$

Then there exist functions

$$F_1(y_1, y_2, \ldots, y_h) , \; F_2(y_1, y_2, \ldots, y_h) , \; \ldots , \; F_{m-h}(y_1, y_2, \ldots, y_h) , \tag{3}$$

with continuous partial derivatives of the first order in a sufficiently small neighbourhood of the point B, *such that at every point* $X(x_1, x_2, \ldots, x_n)$ *from a sufficiently small neighbourhood* Ω *of the point* A *the equations*

$$\begin{aligned} y_{h+1} &= F_1(y_1, y_2, \ldots, y_h) , \\ y_{h+2} &= F_2(y_1, y_2, \ldots, y_h) , \\ &\ldots\ldots\ldots\ldots\ldots \\ y_m &= F_{m-h}(y_1, y_2, \ldots, y_h) \end{aligned} \tag{4}$$

hold.

REMARK 1. The functions (1) are said in this case (i.e. when some of them can be expressed as functions of the others) to be *dependent in* Ω. In the opposite case we say that the functions (1) are *independent* in Ω. This case occurs when $h = m$. In particular the functions defining a regular mapping (Definition 12.7.2) are independent.

REMARK 2. If the matrix (2) is of rank h in the whole domain O, we cannot, in general, affirm that the conclusion of Theorem 1 holds in the whole domain O;

Theorem 1 has a *local* character (therefore we often speak of *local dependency* of functions). It may, of course, happen that equations (4) hold in the whole domain O.

Example 1. The rank of the Jacobian of the system of functions

$$u_1 = x^2 + y^2 + z^2\,, \quad u_2 = x + y + z\,, \quad u_3 = xy + xz + yz \tag{5}$$

is less than 3 because the corresponding determinant

$$\begin{vmatrix} 2x, & 2y, & 2z \\ 1, & 1, & 1 \\ y + z, & x + z, & x + y \end{vmatrix}$$

is zero in the whole three-dimensional space xyz (as may be easily verified). Hence the functions (5) are locally dependent. It is easy to find that $u_3 = \frac{1}{2}(u_2^2 - u_1)$. Therefore, the functions (5) are dependent in the whole three-dimensional space.

Example 2. The functions $y_1 = \sin x$, $y_2 = \cos x$ are dependent, for instance, in the interval $[0, \pi]$ because the relation $y_1 = \sqrt{(1 - y_2^2)}$ holds for every x from this interval (see, however, Example 4 and Remark 4).

Definition 1. We say that the function $y_r(x_1, x_2, \ldots, x_n)$ is a *linear combination of the functions* $y_1(x_1, x_2, \ldots, x_n), \ldots, y_{r-1}(x_1, x_2, \ldots, x_n)$ *in the region* O if it is possible to find constants $c_1, c_2, \ldots, c_{r-1}$ such that

$$y_r = c_1 y_1 + c_2 y_2 + \ldots + c_{r-1} y_{r-1} \text{ identically in } O\,.$$

Definition 2. The functions (1) are said to be *linearly dependent in* O if at least one of them can be expressed as a linear combination of the others. In the opposite case the functions (1) are said to be *linearly independent in* O.

REMARK 3. The following definition is equivalent to Definition 2:

Definition 3. The functions (1) are said to be *linearly dependent in* O if m constants $c_1, c_2, \ldots, c_m$, *at least one of which is different from zero,* exist such that

$$c_1 y_1 + c_2 y_2 + \ldots + c_m y_m = 0 \text{ identically in } O\,. \tag{6}$$

If the identity (6) holds only when all c_k in (6) are equal to zero, the functions (1) are said to be *linearly independent in* O.

Example 3. The functions $1, x, x^2$ are linearly independent in every interval I because (as is well known from algebra) the equation

$$c_1 + c_2 x + c_3 x^2 \equiv 0$$

holds in I only when $c_1 = c_2 = c_3 = 0$. The functions

$$f_1(x) = \sin^2 x\,, \quad f_2(x) = 4\,, \quad f_3(x) = \cos^2 x$$

are linearly dependent in every interval because

$$4 \sin^2 x - 4 + 4 \cos^2 x \equiv 0 .$$

Theorem 2. *Let the functions* (1) *be square-integrable in* O (§ 16.1.1) *and let their number be* n. *Let us denote* (cf. Definition 16.2.1 and Remark 16.2.17) *by* (f_i, f_k) *the scalar product of functions* $f_i(x_1, x_2, \ldots, x_n), f_k(x_1, x_2, \ldots, x_n)$, i.e.

$$(f_i, f_k) = \int_O f_i(x_1, x_2, \ldots, x_n) f_k(x_1, x_2, \ldots, x_n) \, dx_1 \, dx_2 \ldots dx_n . \tag{7}$$

Then the necessary and sufficient condition for the functions (1) *to be linearly dependent in* O *is the vanishing of the so-called Gram determinant,*

$$G = \begin{vmatrix} (f_1, f_1), & (f_1, f_2), & \ldots, & (f_1, f_n) \\ (f_2, f_1), & (f_2, f_2), & \ldots, & (f_2, f_n) \\ \ldots & \ldots & \ldots & \ldots \\ (f_n, f_1), & (f_n, f_2), & \ldots, & (f_n, f_n) \end{vmatrix} = 0 . \tag{8}$$

Example 4. Let us apply (8) to the investigation of the linear dependence of the functions $f_1(x) = \sin x$, $f_2(x) = \cos x$ in the interval $[0, \pi]$. We have

$$(f_1, f_1) = \int_0^\pi \sin^2 x \, dx = \tfrac{1}{2}\pi , \quad (f_2, f_2) = \int_0^\pi \cos^2 x \, dx = \tfrac{1}{2}\pi ,$$

$$(f_1, f_2) = (f_2, f_1) = \int_0^\pi \sin x \cos x = 0 .$$

So

$$G = \begin{vmatrix} \tfrac{1}{2}\pi, & 0 \\ 0, & \tfrac{1}{2}\pi \end{vmatrix} = \tfrac{1}{4}\pi^2 \neq 0$$

and the functions under consideration are linearly independent in O. (Note that, by Example 2, the functions are dependent in the sense defined in Remark 1.)

REMARK 4. If the functions (1) are *linearly* dependent, then they are naturally dependent in the sense of Remark 1. If they are independent in accordance with Remark 1, then they are also *linearly* independent.

For a simple criterion (the so-called *Wronski determinant* or *Wronskian*) in the investigation of the linear independence of functions that are integrals of a linear differential equation see § 17.11, pp. 776, 777.

12.9. Theorem on Implicit Functions. Equations $f(x, y) = 0$, $f(x, y, z) = 0$

Definition 1. Let an equation $f(x, y) = 0$ be given. We say that the function $y = \varphi(x)$ is a *solution of this equation in the domain* M if the relation $f(x, \varphi(x)) = 0$

holds identically in M. The functions defined in this sense by the equation $f(x, y) = 0$ are said to be *given implicitly*, or are briefly called *implicit functions*.

Example 1. Consider $2 \ln x - x^2 + e^y - y = 0$. In the interval $(0, +\infty)$ the function $y = 2 \ln x$ is a solution of this equation.

Theorem 1. *Let us consider an equation*

$$f(x, y) = 0 \tag{1}$$

and a point (x_0, y_0) *such that* $f(x_0, y_0) = 0$. *Let* $f(x, y)$ *have continuous partial derivatives of the first order in a neighbourhood of this point and let*

$$\frac{\partial f}{\partial y}(x_0, y_0) \neq 0 .$$

Then, in a certain neighbourhood of the point x_0, *there exists a unique continuous solution* $y = \varphi(x)$ *of equation* (1) *that satisfies the condition* $\varphi(x_0) = y_0$. *The function* $y = \varphi(x)$ *has a continuous derivative* $y' = \varphi'(x)$ *in a neighbourhood of the point* x_0. *This derivative may be computed from the equation*

$$\frac{\partial f}{\partial x} + \frac{\partial f}{\partial y} y' = 0 \tag{2}$$

or

$$y' = -\frac{\partial f / \partial x}{\partial f / \partial y} . \tag{3}$$

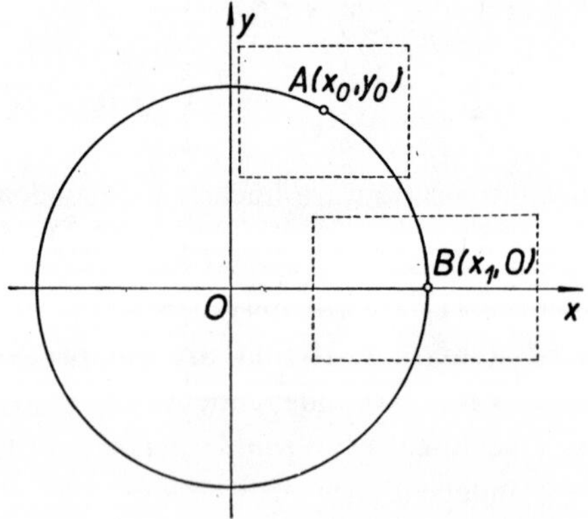

Fig. 12.3.

REMARK 1. *Geometrical interpretation*: If $f(x, y)$ is a "reasonable" function, then by equation (1) a curve in the plane xy is given; e.g. the equation

$$x^2 + y^2 - 25 = 0$$

represents a circle (Fig. 12.3). If we choose a point (x_0, y_0) on the curve, then the question arises if it is possible to express *all* points of this curve in a neighbourhood of the point (x_0, y_0) by an explicit single valued function $y = \varphi(x)$. It can be seen from Fig. 12.3 that, for example, in the neighbourhood of the point A it certainly is possible (indeed, here this function can be found directly because y may be computed from the given equation, $y = +\sqrt{(25 - x^2)}$), but this is not so at the point B: However small we choose the neighbourhood of the point B, two values of y will always correspond to a single value of $x \in (x_1 - \delta, x_1)$ (if *all* points of the circle in

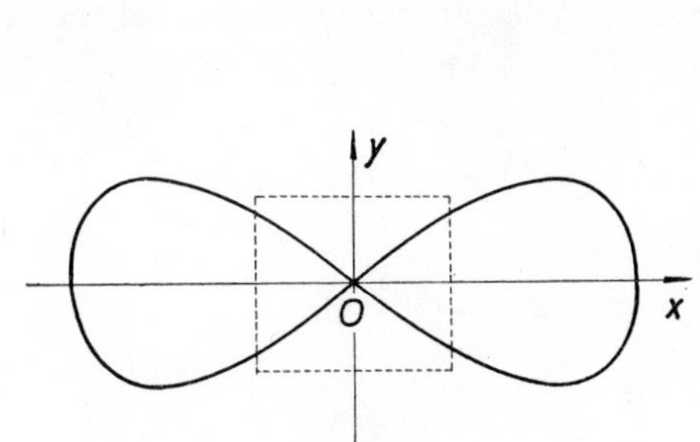

Fig. 12.4.

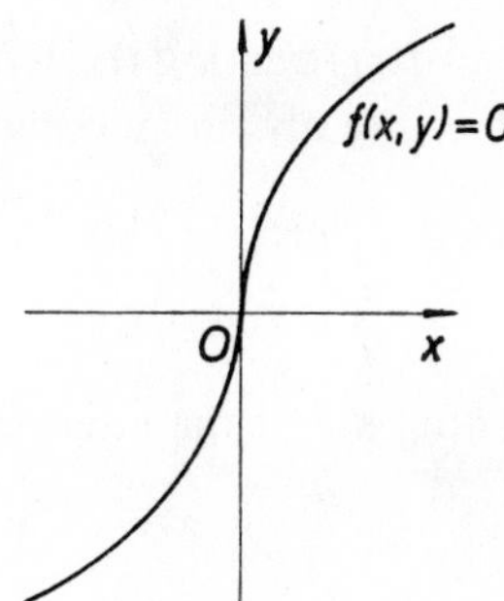

Fig. 12.5. $x - y^3 = 0$; at the point (0, 0) we have $\partial f/\partial y = 0$.

the neighbourhood of the point B are to be considered), and not a single value as is required by the definition of a function (Definition 11.1.1). We note that

$$\frac{\partial f}{\partial y} = 2y = 0$$

at the point $B(x_1, 0)$, so that even in this simple case we can see the importance of the condition $\dfrac{\partial f}{\partial y}(x_0, y_0) \neq 0$.

Likewise it can be seen that, in the case of the lemniscate $(x^2 + y^2)^2 + 2c^2(y^2 - x^2) = 0$ (Fig. 12.4), it is not possible to express *all* its points by a unique function $y = \varphi(x)$ in any (non-zero) neighbourhood of the origin – however small this neighbourhood is chosen. Here also we note that $\dfrac{\partial f}{\partial y}(0, 0) = 0$.

The condition $\dfrac{\partial f}{\partial y}(x_0, y_0) \neq 0$ is not, however, a necessary condition – as can be seen from Fig. 12.5.

REMARK 2. Since $\dfrac{\partial f}{\partial y}$ is continuous in a neighbourhood of (x_0, y_0), and since $\dfrac{\partial f}{\partial y} \neq 0$ at (x_0, y_0), it is possible to compute y' from (2) in a certain neighbourhood of the point (x_0, y_0). If $\dfrac{\partial f}{\partial y}(x_0, y_0) = 0$ and simultaneously $\dfrac{\partial f}{\partial x}(x_0, y_0) \neq 0$, then this means geometrically that the tangent to the curve $f(x, y) = 0$ at the point (x_0, y_0) is parallel to the y-axis.

REMARK 3. Equation (2) results from (1) formally by differentiating equation (1) with respect to x as a composite function of a single variable x (i.e. when considering y as a function of x, $y = \varphi(x)$). Similarly, by differentiating equation (2) with respect to x (e.g. under the condition that $f(x, y)$ has continuous partial derivatives of the second order), treating the left-hand side of (2) as a composite function of the single variable x, we obtain an equation from which we can compute y'', namely

$$\frac{\partial^2 f}{\partial x^2} + \frac{\partial^2 f}{\partial y\, \partial x} y' + \left(\frac{\partial^2 f}{\partial x\, \partial y} + \frac{\partial^2 f}{\partial y^2} y'\right) y' + \frac{\partial f}{\partial y} y'' = 0 .$$

Or (making use of the interchangeability of mixed derivatives),

$$\frac{\partial^2 f}{\partial x^2} + 2 \frac{\partial^2 f}{\partial x\, \partial y} y' + \frac{\partial^2 f}{\partial y^2} y'^2 + \frac{\partial f}{\partial y} y'' = 0 . \tag{4}$$

By further differentiation of this equation we obtain an equation for y''', etc.

Example 2. Let us compute y' and y'' for the function given implicitly by the equation

$$x^2 + y^2 - 25 = 0 \tag{5}$$

(a circle) at the point $(3, 4)$.

The chosen point does satisfy equation (5) (i.e. lies on the circle (5)) and

$$\frac{\partial f}{\partial y} = 2y = 8 \neq 0 .$$

Hence the condition $\partial f/\partial y \neq 0$ is fulfilled. From (5) we have (because $\partial f/\partial x = 2x$)

$$2x + 2yy' = 0 \tag{6}$$

or (when we put $x = 3$, $y = 4$) $y' = -\frac{3}{4}$. Further differentiation of equation (6) (divided by 2) gives

$$1 + y'y' + yy'' = 0 .$$

We substitute $y = 4$ and for y' the computed value $y' = -\frac{3}{4}$:

$$1 + (\tfrac{3}{4})^2 + 4y'' = 0 ,$$

$$y'' = -\tfrac{25}{64} .$$

Hence, the function $y = \varphi(x)$ given by equation (5) and such that $\varphi(3) = 4$, has the following values of derivatives for $x = 3$: $y' = -\frac{3}{4}$, $y'' = -\frac{25}{64}$.

REMARK 4. The foregoing example is a very simple one and was used here only as an illustration; y' and y'' could have been computed directly by differentiating the equation $y = \sqrt{(25 - x^2)}$. But for implicit equations, where $y = \varphi(x)$ cannot be computed on the basis of current elementary functions (for example, in the case of the equation $e^y \sin x + x^2y^2 - x \ln y + 1 = 0$), the application of the theorem considered is essential.

Theorem 2. *Let us consider the equation*

$$f(x, y, z) = 0\,. \tag{7}$$

Let

$$f(x_0, y_0, z_0) = 0 \tag{8}$$

and let $f(x, y, z)$ have continuous partial derivatives of the first order in a neighbourhood of the point (x_0, y_0, z_0). Further, let

$$\frac{\partial f}{\partial z}(x_0, y_0, z_0) \neq 0\,. \tag{9}$$

Then, in a certain neighbourhood of the point (x_0, y_0), there exists a unique continuous solution

$$z = \varphi(x, y) \tag{10}$$

of equation (7) that satisfies the condition $\varphi(x_0, y_0) = z_0$. The function $z = \varphi(x, y)$ has continuous partial derivatives of the first order in a neighbourhood of the point (x_0, y_0) and these derivatives can be determined from the relations

$$\frac{\partial f}{\partial x} + \frac{\partial f}{\partial z}\frac{\partial z}{\partial x} = 0\,, \quad \frac{\partial f}{\partial y} + \frac{\partial f}{\partial z}\frac{\partial z}{\partial y} = 0\,. \tag{11}$$

REMARK 5. Equations (11) result formally from equation (7) if we differentiate this equation partially with respect to x (or with respect to y) considering z as a function of x and y. In the same way as was shown in Remark 3 we can obtain higher derivatives by differentiation of (11). For example, if we differentiate the first equation (11) with respect to x, we obtain an equation for $\partial^2 z/\partial x^2$:

$$\frac{\partial^2 f}{\partial x^2} + \frac{\partial^2 f}{\partial z\, \partial x}\frac{\partial z}{\partial x} + \left(\frac{\partial^2 f}{\partial x\, \partial z} + \frac{\partial^2 f}{\partial z^2}\frac{\partial z}{\partial x}\right)\frac{\partial z}{\partial x} + \frac{\partial f}{\partial z}\frac{\partial^2 z}{\partial x^2} = 0\,,$$

or, after rearrangement,

$$\frac{\partial^2 f}{\partial x^2} + 2\frac{\partial^2 f}{\partial x\, \partial z}\frac{\partial z}{\partial x} + \frac{\partial^2 f}{\partial z^2}\left(\frac{\partial z}{\partial x}\right)^2 + \frac{\partial f}{\partial z}\frac{\partial^2 z}{\partial x^2} = 0\,.$$

Example 3. Let us find the equation of the tangent plane of the ellipsoid

$$\frac{x^2}{12} + \frac{y^2}{27} + \frac{z^2}{3} - 1 = 0 \tag{12}$$

at the point (2, 3, 1).

This point does satisfy equation (12) and at this point

$$\frac{\partial f}{\partial z} = \frac{2z}{3} = \frac{2}{3} \neq 0 .$$

Condition (9) is thus satisfied. The equation of the tangent plane is (Theorem 12.3.4, p. 448)

$$z - z_0 = \left(\frac{\partial f}{\partial x}\right)_0 (x - x_0) + \left(\frac{\partial f}{\partial y}\right)_0 (y - y_0) .$$

By (11)

$$\frac{2x}{12} + \frac{2z}{3}\frac{\partial z}{\partial x} = 0 , \quad \frac{2y}{27} + \frac{2z}{3}\frac{\partial z}{\partial y} = 0 .$$

If we insert the values $x_0 = 2$, $y_0 = 3$, $z_0 = 1$, we get

$$\left(\frac{\partial f}{\partial x}\right)_0 = \frac{\partial z}{\partial x}(2, 3) = -\frac{1}{2} , \quad \left(\frac{\partial f}{\partial y}\right)_0 = \frac{\partial z}{\partial y}(2, 3) = -\frac{1}{3} .$$

Thus, the equation of the tangent plane is

$$z - 1 = -\tfrac{1}{2}(x - 2) - \tfrac{1}{3}(y - 3)$$

or

$$3x + 2y + 6z - 18 = 0 .$$

Example 4. By the equations

$$uz - 2e^{vz} = 0 , \quad u = x^2 + y^2 , \quad v^2 - xy \ln v - 1 = 0 \tag{13}$$

z is given as a composite function of the variables x, y (by means of the functions u, v). We wish to find $\partial z/\partial x$ at the point $x = 0$, $y = e$ (the corresponding values of u, v, z are $u = e^2$, $v = 1$, $z = 2$). According to the rule for differentiation of composite functions (Theorem 12.4.1) we obtain

$$\frac{\partial z}{\partial x} = \frac{\partial z}{\partial u}\frac{\partial u}{\partial x} + \frac{\partial z}{\partial v}\frac{\partial v}{\partial x} . \tag{14}$$

Let us denote $uz - 2e^{vz}$ by $f(u, v, z)$. Then

$$\frac{\partial f}{\partial z} = u - 2v\, e^{vz} , \quad \frac{\partial f}{\partial z}(e^2, 1, 2) = -e^2 \neq 0 .$$

Thus, according to (11),

$$\frac{\partial f}{\partial u} + \frac{\partial f}{\partial z}\frac{\partial z}{\partial u} = 0\,, \quad z + (u - 2v\,e^{vz})\frac{\partial z}{\partial u} = 0\,, \quad 2 - e^2\frac{\partial z}{\partial u} = 0$$

and hence

$$\frac{\partial z}{\partial u} = \frac{2}{e^2}\,.$$

Similarly

$$\frac{\partial f}{\partial v} + \frac{\partial f}{\partial z}\frac{\partial z}{\partial v} = 0\,, \quad -2z\,e^{vz} + (u - 2v\,e^{vz})\frac{\partial z}{\partial v} = 0\,,$$

$$-4e^2 - e^2\frac{\partial z}{\partial v} = 0\,, \quad \frac{\partial z}{\partial v} = -4\,.$$

Further (after substitution of the values $x = 0$, $y = e$, $v = 1$)

$$\frac{\partial u}{\partial x} = 2x = 0\,.$$

If

$$g(x, y, v) = v^2 - xy\ln v - 1\,, \quad \text{then} \quad \frac{\partial g}{\partial v} = 2v - \frac{xy}{v} = 2 \neq 0\,.$$

Hence

$$\frac{\partial g}{\partial x} + \frac{\partial g}{\partial v}\frac{\partial v}{\partial x} = 0\,, \quad -y\ln v + \left(2v - \frac{xy}{v}\right)\frac{\partial v}{\partial x} = 0\,, \quad 0 + 2\frac{\partial v}{\partial x} = 0\,, \quad \frac{\partial v}{\partial x} = 0\,.$$

If we insert all partial results into (14) we obtain:

$$\frac{\partial z}{\partial x}(0, e) = \frac{2}{e^2}\,.\,0 - 4\,.\,0 = 0\,.$$

REMARK 6. Instead of using equation (14) we may compute $\partial z/\partial x$ by differentiation of the first equation of (13) with respect to x taking into consideration the fact that u, v and z contain x:

$$\frac{\partial u}{\partial x}z + u\frac{\partial z}{\partial x} - 2e^{vz}\left(\frac{\partial v}{\partial x}z + v\frac{\partial z}{\partial x}\right) = 0$$

($\partial u/\partial x$ and $\partial v/\partial x$ have naturally to be found from the second and third equations of (13)). We arrive at the same result. The advantage of this method is that it can be applied without difficulty even in the case where the function f contains explicitly not only u and v but also x and y, when equation (14) cannot be applied (cf. Remark 12.4.2).

REMARK 7. The problem becomes somewhat complicated when the second and third equations of (13) do not contain u and v separately but when u and v are given (as functions of x and y) by the equations

$$g(x, y, u, v) = 0\,, \quad h(x, y, u, v) = 0\,. \tag{15}$$

The required derivatives $\partial u/\partial x$, $\partial v/\partial x$ are found by solving the equations derived from (15), by differentiating with respect to x:

$$\frac{\partial g}{\partial x} + \frac{\partial g}{\partial u}\frac{\partial u}{\partial x} + \frac{\partial g}{\partial v}\frac{\partial v}{\partial x} = 0\,, \quad \frac{\partial h}{\partial x} + \frac{\partial h}{\partial u}\frac{\partial u}{\partial x} + \frac{\partial h}{\partial v}\frac{\partial v}{\partial x} = 0$$

(see § 12.10).

12.10. Theorem on Implicit Functions. General Case

Definition 1. Let us have a system of equations

$$\begin{aligned} F_1(x_1, x_2, \ldots, x_n, y_1, y_2, \ldots, y_m) &= 0\,, \\ F_2(x_1, x_2, \ldots, x_n, y_1, y_2, \ldots, y_m) &= 0\,, \\ \ldots\ldots\ldots\ldots\ldots\ldots\ldots\ldots\ldots\ldots& \\ F_m(x_1, x_2, \ldots, x_n, y_1, y_2, \ldots, y_m) &= 0\,. \end{aligned} \tag{1}$$

By a solution of system (1) (in a region O) we understand the functions

$$\begin{aligned} y_1 &= f_1(x_1, x_2, \ldots, x_n)\,, \\ y_2 &= f_2(x_1, x_2, \ldots, x_n)\,, \\ &\ldots\ldots\ldots\ldots\ldots\ldots \\ y_m &= f_m(x_1, x_2, \ldots, x_n) \end{aligned} \tag{2}$$

such that if we substitute f_1 for y_1, etc., into (1), all the equations of the system become identities in $x_1, x_2, \ldots, x_n$ in the region O. The functions (2) are said to be *given implicitly by equations* (1). Briefly they are *implicit functions.*

Theorem 1. *Let*

1. *the point* $(x_1^0, x_2^0, \ldots, x_n^0, y_1^0, y_2^0, \ldots, y_m^0)$ *satisfy all the equations of system* (1),

2. *the functions* (1) *possess continuous partial derivatives of the first order with respect to all variables in a neighbourhood of this point,*

3. *the determinant*

$$\frac{\partial(F_1, F_2, \ldots, F_m)}{\partial(y_1, y_2, \ldots, y_m)} = \begin{vmatrix} \dfrac{\partial F_1}{\partial y_1}, & \dfrac{\partial F_1}{\partial y_2}, & \ldots, & \dfrac{\partial F_1}{\partial y_m} \\ \dfrac{\partial F_2}{\partial y_1}, & \dfrac{\partial F_2}{\partial y_2}, & \ldots, & \dfrac{\partial F_2}{\partial y_m} \\ \cdots & \cdots & \cdots & \cdots \\ \dfrac{\partial F_m}{\partial y_1}, & \dfrac{\partial F_m}{\partial y_2}, & \ldots, & \dfrac{\partial F_m}{\partial y_m} \end{vmatrix} \tag{3}$$

be non-zero at the point $(x_1^0, x_2^0, \ldots, x_n^0, y_1^0, y_2^0, \ldots, y_m^0)$.

Then, in a certain neighbourhood of the point $(x_1^0, x_2^0, \ldots, x_n^0)$, *there exists a unique system of continuous functions* (2) *which is the solution of equations* (1) *in this neighbourhood and for which the equations*

$$\begin{aligned} f_1(x_1^0, x_2^0, \ldots, x_n^0) &= y_1^0, \\ f_2(x_1^0, x_2^0, \ldots, x_n^0) &= y_2^0, \\ &\cdots\cdots \\ f_m(x_1^0, x_2^0, \ldots, x_n^0) &= y_m^0 \end{aligned}$$

hold simultaneously. Each of the functions (2) *possesses continuous partial derivatives of the first order with respect to all variables* $x_1, x_2, \ldots, x_n$ *in the neighbourhood of the point considered.*

REMARK 1. The formulae for the computation of derivatives are rather cumbersome, expecially those for the computation of derivatives of higher orders (existence of these derivatives is established if, for example, all the functions (1) possess continuous derivatives of the order considered in a neighbourhood of the point $(x_1^0, x_2^0, \ldots$ $\ldots, x_n^0, y_1^0, y_2^0, \ldots, y_m^0)$). The computation of derivatives is carried out practically by differentiating the equations (1) with respect to the corresponding variable x_k and considering simultaneously $y_1, y_2, \ldots, y_m$ as functions of the variables $x_1, x_2, \ldots$ $\ldots, x_n$. For example, to compute the derivatives with respect to x_2 we have:

$$\begin{gathered} \frac{\partial F_1}{\partial x_2} + \frac{\partial F_1}{\partial y_1}\frac{\partial y_1}{\partial x_2} + \frac{\partial F_1}{\partial y_2}\frac{\partial y_2}{\partial x_2} + \ldots + \frac{\partial F_1}{\partial y_m}\frac{\partial y_m}{\partial x_2} = 0, \\ \cdots\cdots\cdots\cdots \\ \frac{\partial F_m}{\partial x_2} + \frac{\partial F_m}{\partial y_1}\frac{\partial y_1}{\partial x_2} + \frac{\partial F_m}{\partial y_2}\frac{\partial y_2}{\partial x_2} + \ldots + \frac{\partial F_m}{\partial y_m}\frac{\partial y_m}{\partial x_2} = 0. \end{gathered} \tag{4}$$

This system of equations for $\partial y_1/\partial x_2, \partial y_2/\partial x_2, \ldots, \partial y_m/\partial x_2$ is uniquely solvable because the determinant (3) is non-zero.

Example 1. Let y_1 and y_2 be given implicitly by the equations

$$x_1 e^{y_2} + y_1 \ln x_2 - e = 0\,, \quad x_1 y_1 + x_2 e^{y_2} - (2 + e) = 0\,. \tag{5}$$

Let us compute $\partial y_1/\partial x_1$, $\partial y_2/\partial x_1$ at the point $x_1 = 1$, $x_2 = 1$.

From equations (5) it follows (at the point $x_1 = 1$, $x_2 = 1$) that $y_2 = 1$, $y_1 = 2$. (In the general case, when there are more possibilities, it is necessary to state in advance for what values y_1, y_2 the problem is to be solved.) We differentiate equations (5) with respect to x_1 considering (see Remark 1) y_1 and y_2 dependend on x_1:

$$e^{y_2} + x_1 e^{y_2} \frac{\partial y_2}{\partial x_1} + \frac{\partial y_1}{\partial x_1} \ln x_2 = 0\,, \quad y_1 + x_1 \frac{\partial y_1}{\partial x_1} + x_2 e^{y_2} \frac{\partial y_2}{\partial x_1} = 0\,. \tag{6}$$

When the numerical values of x_1, x_2, y_1, y_2 are substituted into (6), we obtain

$$0 \,.\, \frac{\partial y_1}{\partial x_1} + e \frac{\partial y_2}{\partial x_1} + e = 0\,,$$

$$\frac{\partial y_1}{\partial x_1} + e \frac{\partial y_2}{\partial x_1} + 2 = 0\,,$$

hence

$$\frac{\partial y_2}{\partial x_1}(1, 1) = -1\,, \quad \frac{\partial y_1}{\partial x_1}(1, 1) = e - 2\,.$$

12.11. Introduction of New Variables. Transformations of Differential Equations and Differential Expressions (Especially into Polar, Spherical and Cylindrical Coordinates)

In the course of the solution of differential equations and in many other problems it may be convenient to introduce new variables. In this section we shall consider the most frequently occurring cases of introducing new variables, independent as well as dependent, and shall show how to express the given differential expressions with the help of the new variables.

REMARK 1. Throughout this section we suppose that the functions considered possess all the necessary (continuous) derivatives and that the correspondence between the original and the new variables is one-to-one.

(a) CASE OF ONE VARIABLE

(α) *Introduction of a new Independent Variable.* By introducing a new variable $x = \varphi(t)$, the function $y = f(x)$ becomes a composite function of the variable t.

We have to find the relation between $y' = \mathrm{d}y/\mathrm{d}x$ and $\dot{y} = \mathrm{d}y/\mathrm{d}t$ and, similarly, between higher derivatives.

By Theorem 11.5.5 (assuming $\dot{x} = \mathrm{d}x/\mathrm{d}t \neq 0$ and applying Theorem 11.5.6) we get

$$y' = \frac{\mathrm{d}y}{\mathrm{d}x} = \frac{\mathrm{d}y}{\mathrm{d}t} \cdot \frac{\mathrm{d}t}{\mathrm{d}x} = \dot{y} \cdot \frac{1}{\mathrm{d}x/\mathrm{d}t} = \dot{y} \cdot \frac{1}{\dot{x}}, \tag{1}$$

$$y'' = \frac{\mathrm{d}y'}{\mathrm{d}x} = \frac{\mathrm{d}}{\mathrm{d}x}\left(\dot{y} \cdot \frac{1}{\dot{x}}\right) = \frac{\mathrm{d}}{\mathrm{d}t}\left(\dot{y} \cdot \frac{1}{\dot{x}}\right)\frac{\mathrm{d}t}{\mathrm{d}x} = \left(\ddot{y}\,\frac{1}{\dot{x}} - \dot{y}\,\frac{\ddot{x}}{\dot{x}^2}\right)\frac{1}{\dot{x}}, \tag{2}$$

$$y''' = \frac{\mathrm{d}y''}{\mathrm{d}x} = \frac{\mathrm{d}}{\mathrm{d}x}\left[\ddot{y}\,\frac{1}{\dot{x}^2} - \dot{y}\,\frac{\ddot{x}}{\dot{x}^3}\right] = \frac{\mathrm{d}}{\mathrm{d}t}\left[\ddot{y}\,\frac{1}{\dot{x}^2} - \dot{y}\,\frac{\ddot{x}}{\dot{x}^3}\right]\frac{\mathrm{d}t}{\mathrm{d}x} =$$

$$= \left\{\dddot{y}\,\frac{1}{\dot{x}^2} - 3\ddot{y}\,\frac{\ddot{x}}{\dot{x}^3} - \dot{y}\,\frac{\dddot{x}\dot{x}^3 - 3\dot{x}\dot{x}^2\ddot{x}}{\dot{x}^6}\right\}\frac{1}{\dot{x}} = \dddot{y}\,\frac{1}{\dot{x}^3} - 3\ddot{y}\,\frac{\ddot{x}}{\dot{x}^4} - \dot{y}\,\frac{\dddot{x}\dot{x} - 3\ddot{x}^2}{\dot{x}^5}. \tag{3}$$

To compute higher derivatives, we proceed in a similar way. Derivatives with respect to t expressed in terms of derivatives with respect to x may be obtained from equations (1), (2), (3), or directly:

$$\dot{y} = \frac{\mathrm{d}y}{\mathrm{d}t} = \frac{\mathrm{d}y}{\mathrm{d}x} \cdot \frac{\mathrm{d}x}{\mathrm{d}t} = y' \cdot \dot{x}, \tag{4}$$

$$\ddot{y} = \frac{\mathrm{d}^2y}{\mathrm{d}t^2} = \frac{\mathrm{d}}{\mathrm{d}t}(y' \cdot \dot{x}) = \dot{x}\,\frac{\mathrm{d}}{\mathrm{d}t}(y') + y' \cdot \frac{\mathrm{d}}{\mathrm{d}t}(\dot{x}) =$$

$$= \dot{x}\,\frac{\mathrm{d}}{\mathrm{d}x}(y') \cdot \frac{\mathrm{d}x}{\mathrm{d}t} + y' \cdot \ddot{x} = \dot{x}^2 y'' + \ddot{x}y', \quad \text{etc.} \tag{5}$$

Example 1. Let us transform the left-hand side of the differential equation

$$x^2y'' + 4xy' - 2y = 0 \tag{6}$$

by introducing a new independent variable t by the relation $x = \mathrm{e}^t$ $(x > 0)$.

Evidently, the correspondence is one-to-one for all t and the function e^t possesses derivatives of all orders. We may proceed according to (1), (2) or differentiate directly:

$$y' = \dot{y} \cdot \frac{1}{\dot{x}} = \dot{y} \cdot \frac{1}{\mathrm{e}^t} = \dot{y}\mathrm{e}^{-t}, \quad y'' = \frac{\mathrm{d}}{\mathrm{d}t}(\dot{y}\mathrm{e}^{-t})\frac{\mathrm{d}t}{\mathrm{d}x} =$$

$$= (\ddot{y}\mathrm{e}^{-t} - \dot{y}\mathrm{e}^{-t})\,\mathrm{e}^{-t} = \ddot{y}\mathrm{e}^{-2t} - \dot{y}\mathrm{e}^{-2t}.$$

After substituting into (6) we obtain

$$\mathrm{e}^{2t}(\ddot{y}\mathrm{e}^{-2t} - \dot{y}\mathrm{e}^{-2t}) + 4\mathrm{e}^t \cdot \dot{y}\mathrm{e}^{-t} - 2y = 0$$

or

$$\ddot{y} + 3\dot{y} - 2y = 0$$

(see § 17.13, Euler's differential equation).

(β) *Introduction of a New Dependent Variable.* Instead of the dependent variable y we shall introduce a new dependent variable z by the relation $y = \varphi(z)$ or $z = \psi(y)$. In the same way as in (4), (5) we get

$$y' = \frac{dy}{dx} = \frac{dy}{dz} \cdot \frac{dz}{dx} = \frac{dy}{dz} \cdot z' , \tag{7}$$

$$y'' = \frac{d^2y}{dx^2} = \frac{d}{dx}\left(\frac{dy}{dz} z'\right) = z' \frac{d}{dz}\left(\frac{dy}{dz}\right)\frac{dz}{dx} + \frac{dy}{dz}\frac{d}{dx}(z') =$$

$$= \frac{d^2y}{dz^2} z'^2 + \frac{dy}{dz} z'' , \quad \text{etc} . \tag{8}$$

Similarly for the relation $z = \psi(y)$:

$$z' = \frac{dz}{dx} = \frac{dz}{dy} y' , \quad z'' = \frac{d^2z}{dx^2} = \frac{d^2z}{dy^2} y'^2 + \frac{dz}{dy} y'' , \quad \text{etc} . \tag{9}$$

Example 2. If we introduce into the differential equation

$$y' e^x \cos y - x \sin y = \ln x \tag{10}$$

a new dependent variable by the relation $\sin y = z$, then $\cos y \,.\, y' = z'$. Substituting into (10) we obtain the linear differential equation

$$e^x z' - xz = \ln x \tag{11}$$

for the new unknown function $z(x)$.

Remark 2. In the majority of cases we need not apply the general transformation formulae because simpler methods are available as we have seen in Examples 1 and 2. A very simple treatment is also possible in the case of so-called *homogeneous differential equations of the first order*, where (see § 17.3) we introduce instead of $y(x)$ a new dependent variable $z(x)$ by means of the relation $y = xz$. Then $y' = z + xz'$.

Remark 3. It is relatively seldom that we introduce simultaneously both a new independent variable and a dependent variable. As a rule, the transformation may be carried out successively according to α) and β).

(b) Case of Two or More Variables

If we introduce new independent variables

$$x = x(u, v) , \quad y = y(u, v) , \tag{12}$$

then the function $z = f(x, y)$ becomes a composite function of the variables u, v, $z = f(x(u, v), y(u, v))$. By Theorem 12.4.1 we have

$$\frac{\partial z}{\partial u} = \frac{\partial z}{\partial x}\frac{\partial x}{\partial u} + \frac{\partial z}{\partial y}\frac{\partial y}{\partial u}, \quad \frac{\partial z}{\partial v} = \frac{\partial z}{\partial x}\frac{\partial x}{\partial v} + \frac{\partial z}{\partial y}\frac{\partial y}{\partial v}. \tag{13}$$

Higher derivatives are computed similarly (for a more detailed treatment see Remark 12.4.1 where second derivatives are computed directly; it is necessary, however, in the formulae of that Remark to interchange (x, y) and (u, v) because in the present case $z = z(x, y)$, $x = x(u, v)$, $y = y(u, v)$, which is different from the notation of Theorem 12.4.1).

Solving (13) for $\partial z/\partial x$, $\partial z/\partial y$ we obtain (if $\partial(x, y)/\partial(u, v) \neq 0$; Remark 12.7.1) the derivatives $\partial z/\partial x$, $\partial z/\partial y$ expressed in terms of the derivatives $\partial z/\partial u$, $\partial z/\partial v$.

Example 3. Let us transform the differential equation

$$\frac{\partial^2 z}{\partial x^2} + \frac{\partial^2 z}{\partial y^2} = 0 \tag{14}$$

by introducing the polar coordinates

$$x = r\cos\varphi, \quad y = r\sin\varphi \quad (r > 0,\ 0 \leqq \varphi < 2\pi). \tag{15}$$

By (13) (writing $u = r$, $v = \varphi$) we have

$$\frac{\partial z}{\partial r} = \frac{\partial z}{\partial x}\cos\varphi + \frac{\partial z}{\partial y}\sin\varphi, \quad \frac{\partial z}{\partial \varphi} = \frac{\partial z}{\partial x}(-r\sin\varphi) + \frac{\partial z}{\partial y} r\cos\varphi. \tag{16}$$

Solving (16) for $\partial z/\partial x$, $\partial z/\partial y$ we obtain

$$\frac{\partial z}{\partial x} = \frac{\partial z}{\partial r}\cos\varphi - \frac{\partial z}{\partial \varphi}\frac{1}{r}\sin\varphi, \quad \frac{\partial z}{\partial y} = \frac{\partial z}{\partial r}\sin\varphi + \frac{\partial z}{\partial \varphi}\frac{1}{r}\cos\varphi. \tag{17}$$

If we carry out these operations with the function $\partial z/\partial x$ instead of the function z, we obtain by (17)

$$\frac{\partial}{\partial x}\left(\frac{\partial z}{\partial x}\right) = \frac{\partial\left(\dfrac{\partial z}{\partial x}\right)}{\partial r}\cos\varphi - \frac{\partial\left(\dfrac{\partial z}{\partial x}\right)}{\partial \varphi}\frac{1}{r}\sin\varphi. \tag{18}$$

Substituting the right-hand side of the first equation (17) into the right-hand side of (18) we obtain

$$\frac{\partial^2 z}{\partial x^2} = \cos\varphi\frac{\partial}{\partial r}\left[\frac{\partial z}{\partial r}\cos\varphi - \frac{\partial z}{\partial \varphi}\frac{1}{r}\sin\varphi\right] - \frac{1}{r}\sin\varphi\frac{\partial}{\partial \varphi}\left[\frac{\partial z}{\partial r}\cos\varphi - \frac{\partial z}{\partial \varphi}\frac{1}{r}\sin\varphi\right] =$$

$$= \frac{\partial^2 z}{\partial r^2}\cos^2\varphi - \frac{\partial^2 z}{\partial r\,\partial\varphi}\frac{2}{r}\cos\varphi\sin\varphi + \frac{\partial^2 z}{\partial\varphi^2}\frac{1}{r^2}\sin^2\varphi +$$

$$+ \frac{1}{r}\frac{\partial z}{\partial r}\sin^2\varphi + \frac{\partial z}{\partial\varphi}\frac{2}{r^2}\sin\varphi\cos\varphi. \tag{19}$$

Similarly, by application of the second equation (17) we get

$$\frac{\partial^2 z}{\partial y^2} = \frac{\partial^2 z}{\partial r^2} \sin^2 \varphi + \frac{\partial^2 z}{\partial r\, \partial \varphi} \frac{2}{r} \sin \varphi \cos \varphi + \frac{\partial^2 z}{\partial \varphi^2} \frac{1}{r^2} \cos^2 \varphi +$$

$$+ \frac{1}{r} \frac{\partial z}{\partial r} \cos^2 \varphi - \frac{\partial z}{\partial \varphi} \frac{2}{r^2} \sin \varphi \cos \varphi \,. \tag{20}$$

Substituting (19), (20) into (14) we obtain the transformed equation

$$\frac{\partial^2 z}{\partial r^2} + \frac{1}{r^2} \frac{\partial^2 z}{\partial \varphi^2} + \frac{1}{r} \frac{\partial z}{\partial r} = 0 \,. \tag{21}$$

REMARK 4. Without making use of the above formulae we can – by a rather more laborious process – reach the same result directly, starting out from the transformation inverse to (15),

$$r = \sqrt{(x^2 + y^2)}, \quad \varphi = \arctan \frac{y}{x} \quad (+\text{ const.}) \tag{22}$$

and applying the equations

$$\frac{\partial z}{\partial x} = \frac{\partial z}{\partial r} \frac{\partial r}{\partial x} + \frac{\partial z}{\partial \varphi} \frac{\partial \varphi}{\partial x}, \quad \frac{\partial z}{\partial y} = \frac{\partial z}{\partial r} \frac{\partial r}{\partial y} + \frac{\partial z}{\partial \varphi} \frac{\partial \varphi}{\partial y} \,.$$

The constant in the second equation (22) depends on the quadrant in which the point (x, y) lies. It is irrelevant when differentiating.

Another method, making use of the simplicity of equation (14) and the substitution (15), is the following: Differentiating the first equation (16) with respect to r, we obtain

$$\frac{\partial^2 z}{\partial r^2} = \left(\frac{\partial^2 z}{\partial x^2} \cos \varphi + \frac{\partial^2 z}{\partial x\, \partial y} \sin \varphi \right) \cos \varphi + \left(\frac{\partial^2 z}{\partial y\, \partial x} \cos \varphi + \frac{\partial^2 z}{\partial y^2} \sin \varphi \right) \sin \varphi \,,$$

and differentiating the second equation with respect to φ we have

$$\frac{\partial^2 z}{\partial \varphi^2} = \left(-\frac{\partial^2 z}{\partial x^2} r \sin \varphi + \frac{\partial^2 z}{\partial x\, \partial y} r \cos \varphi \right) (-r \sin \varphi) - \frac{\partial z}{\partial x} r \cos \varphi +$$

$$+ \left(-\frac{\partial^2 z}{\partial y\, \partial x} r \sin \varphi + \frac{\partial^2 z}{\partial y^2} r \cos \varphi \right) r \cos \varphi - \frac{\partial z}{\partial y} r \sin \varphi \,.$$

Then

$$\frac{\partial^2 z}{\partial r^2} + \frac{1}{r^2} \frac{\partial^2 z}{\partial \varphi^2} + \frac{1}{r} \frac{\partial z}{\partial r} = \frac{\partial^2 z}{\partial x^2} + \frac{\partial^2 z}{\partial y^2} \,.$$

REMARK 5. The method of transformation in the case of several variables is quite similar. For example, the equation

$$\frac{\partial^2 u}{\partial x^2} + \frac{\partial^2 u}{\partial y^2} + \frac{\partial^2 u}{\partial z^2} = 0 \tag{23}$$

s converted (i) by introducing cylindrical coordinates

$$x = r\cos\varphi\,, \quad y = r\sin\varphi\,, \quad z = z \tag{24}$$

into the equation

$$\frac{\partial^2 u}{\partial r^2} + \frac{1}{r^2}\frac{\partial^2 u}{\partial \varphi^2} + \frac{1}{r}\frac{\partial u}{\partial r} + \frac{\partial^2 u}{\partial z^2} = 0\,, \tag{25}$$

and (ii) by introducing spherical coordinates

$$x = r\sin\vartheta\cos\varphi\,, \quad y = r\sin\vartheta\sin\varphi\,, \quad z = r\cos\vartheta \tag{26}$$

into the equation

$$\frac{\partial^2 u}{\partial r^2} + \frac{1}{r^2}\frac{\partial^2 u}{\partial \vartheta^2} + \frac{1}{r^2\sin^2\vartheta}\frac{\partial^2 u}{\partial \varphi^2} + \frac{2}{r}\frac{\partial u}{\partial r} + \frac{1}{r^2}\cot\vartheta\,\frac{\partial u}{\partial \vartheta} = 0\,. \tag{27}$$

The expression

$$\left(\frac{\partial z}{\partial x}\right)^2 + \left(\frac{\partial z}{\partial y}\right)^2$$

is converted by the transformation (15) into the expression

$$\left(\frac{\partial z}{\partial r}\right)^2 + \frac{1}{r^2}\left(\frac{\partial z}{\partial \varphi}\right)^2.$$

The expression

$$\left(\frac{\partial u}{\partial x}\right)^2 + \left(\frac{\partial u}{\partial y}\right)^2 + \left(\frac{\partial u}{\partial z}\right)^2$$

is converted by the transformation (24) into the expression

$$\left(\frac{\partial u}{\partial r}\right)^2 + \frac{1}{r^2}\left(\frac{\partial u}{\partial \varphi}\right)^2 + \left(\frac{\partial u}{\partial z}\right)^2,$$

and by the transformation (26) into the expression

$$\left(\frac{\partial u}{\partial r}\right)^2 + \frac{1}{r^2\sin^2\vartheta}\left(\frac{\partial u}{\partial \varphi}\right)^2 + \frac{1}{r^2}\left(\frac{\partial u}{\partial \vartheta}\right)^2.$$

REMARK 6. The transformation of a dependent variable may be carried out in the same way as in (α), (β). If, for example, $z = \varphi(t)$, then

$$\frac{\partial z}{\partial x} = \frac{\mathrm{d}z}{\mathrm{d}t}\frac{\partial t}{\partial x}, \quad \frac{\partial z}{\partial y} = \frac{\mathrm{d}z}{\mathrm{d}t}\frac{\partial t}{\partial y},$$

whence we obtain $\partial t/\partial x$ and $\partial t/\partial y$. Further,

$$\frac{\partial^2 z}{\partial x^2} = \frac{\partial}{\partial x}\left(\frac{\mathrm{d}z}{\mathrm{d}t}\frac{\partial t}{\partial x}\right) = \left[\frac{\mathrm{d}}{\mathrm{d}t}\left(\frac{\mathrm{d}z}{\mathrm{d}t}\right)\frac{\partial t}{\partial x}\right]\frac{\partial t}{\partial x} + \frac{\mathrm{d}z}{\mathrm{d}t}\frac{\partial^2 t}{\partial x^2} = \frac{\mathrm{d}^2 z}{\mathrm{d}t^2}\left(\frac{\partial t}{\partial x}\right)^2 + \frac{\mathrm{d}z}{\mathrm{d}t}\frac{\partial^2 t}{\partial x^2}, \quad \text{etc.}$$

12.12. Extremes of Functions of Several Variables. Constrained Extremes. Lagrange's Method of Undetermined Coefficients. Extremes of Implicit Functions

Definition 1. The function $z = f(x, y)$ is said to have a *relative (local) maximum at the point* (x_0, y_0) if there exists a neighbourhood O of the point (x_0, y_0) such that the relation

$$f(x, y) \leqq f(x_0, y_0) \tag{1}$$

holds at every point of this neighbourhood. A *relative minimum* is defined similarly and so are *relative extremes* for functions of several variables.

If, in the neighbourhood O of the point (x_0, y_0), the equality in (1) occurs *only* at the point (x_0, y_0) we speak of a *strict relative maximum.* (Similarly for a minimum and for functions of several variables.)

Definition 2. The function $z = f(x, y)$ is said to have at the point (x_0, y_0) a *maximum in the region (set) M* (a so-called *absolute maximum*) if the relation $f(x, y) \leqq$ $\leqq f(x_0, y_0)$ holds at every point (x, y) of M.

REMARK 1. An *(absolute) minimum on M* is defined similarly. The definition of absolute extremes for functions of several variables is also similar.

Theorem 1. *If the function $z = f(x, y)$ possesses partial derivatives of the first order in the region O, then it may attain a relative extreme only at a point (x_0, y_0) for which*

$$\frac{\partial f}{\partial x}(x_0, y_0) = 0, \quad \frac{\partial f}{\partial y}(x_0, y_0) = 0. \tag{2}$$

If, moreover, $z = f(x, y)$ possesses a second total differential at this point (a sufficient condition for this is the continuity of the second partial derivatives) and if

$$AC - B^2 > 0, \tag{3}$$

where

$$A = \frac{\partial^2 f}{\partial x^2}(x_0, y_0), \quad B = \frac{\partial^2 f}{\partial x\, \partial y}(x_0, y_0), \quad C = \frac{\partial^2 f}{\partial y^2}(x_0, y_0),$$

then there is in fact a relative extreme at the point (x_0, y_0), *namely a strict relative maximum when* $A < 0$, *a strict relative minimum when* $A > 0$. *If*

$$AC - B^2 < 0, \tag{4}$$

then there is no extreme at the point (x_0, y_0). *If this expression vanishes at the point* (x_0, y_0), *then there may be, but need not be, a relative extreme at the point* (x_0, y_0) *and further investigation is necessary.*

REMARK 2. As a rule, this further investigation is done by considering straight lines $y - y_0 = k(x - x_0)$ drawn through the point (x_0, y_0) and investigating the given function only on these lines as a function of the single variable x. (But if the given function has, for example, a strict relative maximum on every straight line going through the point (x_0, y_0), then generally it does not follow that it has a strict relative maximum at this point by Definition 1 when considered as a function of the variables x, y.) In addition: If $AC - B^2 = 0$ at the point (x_0, y_0) and $AC - B^2 > 0$, or $AC - B^2 < 0$ holds in a certain neighbourhood of the point (x_0, y_0) from which the point (x_0, y_0) is excluded, then $f(x, y)$ has, or has not a strict relative extreme at that point, respectively. (Here $A = \dfrac{\partial^2 f}{\partial x^2}(x, y)$, etc.)

Theorem 2. *If* $z = f(x_1, x_2, \ldots, x_n)$ *possesses partial derivatives of the first order in the region* O, *then necessary conditions for the existence of a relative extreme at a point* $P(x_1^0, x_2^0, \ldots, x_n^0) \in O$ *are*

$$\frac{\partial f}{\partial x_1} = 0, \quad \frac{\partial f}{\partial x_2} = 0, \quad \ldots, \quad \frac{\partial f}{\partial x_n} = 0, \tag{5}$$

at that point.

Sufficient conditions: Let, moreover, $f(x_1, x_2, \ldots, x_n)$ have continuous derivatives of the second order in a neighbourhood of the point P. Let

$$\frac{\partial^2 f}{\partial x_i\, \partial x_k}(x_1^0, x_2^0, \ldots, x_n^0) = A_{ik}, * \quad i = 1, 2, \ldots, n, \quad k = 1, 2, \ldots, n$$

form the matrix

$$\mathbf{A} = \left\| \begin{matrix} A_{11}, & A_{12}, & \ldots, & A_{1n} \\ A_{21}, & A_{22}, & \ldots, & A_{2n} \\ \ldots & \ldots & \ldots & \ldots \\ A_{n1}, & A_{n2}, & \ldots, & A_{nn} \end{matrix} \right\| .$$

* $A_{ik} = A_{ki}$ under the assumptions considered.

If this matrix is positive definite, i.e. if all its principal minors

$$A_{11}, \quad \begin{vmatrix} A_{11}, & A_{12} \\ A_{21}, & A_{22} \end{vmatrix}, \ldots,$$

(p. 106) are positive, then $f(x_1, x_2, \ldots, x_n)$ has a strict relative minimum at the point $(x_1^0, x_2^0, \ldots, x_n^0)$.

If $\boldsymbol{A}$ is negative definite, i.e. if its principal minors of odd and even order are negative and positive, respectively, then $f(x_1, x_2, \ldots, x_n)$ has a strict relative maximum at the point $(x_1^0, x_2^0, \ldots, x_n^0)$.

REMARK 3. The solution of equations (2), and even more so that of equations (5), is difficult in the general case. (In the case of equations (2) we have to find – geometrically speaking – the points of intersection of the curves (2).) In applications, however, there are usually no difficulties in solving those equations. See also Example 31.5.1.

Example 1. Let us find the dimensions of a rectangular water-tank of volume 32 m^3 with minimal area of its base and vertical walls.

If we denote the dimensions of the base by x and y and the height of the tank by z, then we have for the volume

$$V = xyz \tag{6}$$

and for the area of the walls and base

$$P = xy + 2xz + 2yz \,. \tag{7}$$

P is a function of x, y and z but, in fact, it depends only upon two variables, because x, y and z are subject to the relation (6) where V is a given constant, so that (after substituting $z = V/xy$) we have

$$P = xy + \frac{2V}{y} + \frac{2V}{x} \,. \tag{8}$$

Equations (2) now give

$$\frac{\partial P}{\partial x} = 0 \,, \quad \frac{\partial P}{\partial y} = 0 \text{ or } y - \frac{2V}{x^2} = 0 \,, \quad x - \frac{2V}{y^2} = 0 \,. \tag{9}$$

From (9) it follows that

$$x^2 y = 2V, \quad xy^2 = 2V; \tag{10}$$

by dividing the first of these equations by the second we get $x/y = 1$ or $x = y$. If we put $y = x$ in the first equation (10) we obtain $x = \sqrt[3]{(2V)}$ and then $y = \sqrt[3]{(2V)}$. From (6) it then follows that $z = \sqrt[3]{(V/4)}$. Substituting the numerical values we obtain $x = y = 4$ m, $z = 2$ m.

We have really found a (strict) relative extreme because

$$\frac{\partial^2 P}{\partial x^2} = \frac{4V}{x^3} = \frac{4V}{2V} = 2\,, \quad \frac{\partial^2 P}{\partial y^2} = \frac{4V}{y^3} = 2\,, \quad \frac{\partial^2 P}{\partial x\,\partial y} = 1$$

so that $AC - B^2 = 4 - 1 = 3 > 0$ in accordance with (3); also, $\partial^2 P/\partial x^2 = 2$ so that a minimum is assured. It is easy to verify that this relative minimum is a minimum in the whole region $x > 0$, $y > 0$.

REMARK 4. The *extremes* of a given function *in a domain M* can be expected at the points, where (2) or (5) are fulfilled, or where these derivatives do not exist, or, finally, on the boundary of the domain M. The last problem leads to the so-called *constrained extremes* (or *extremes with subsidiary conditions*).

Definition 3. We say that the function $z = f(x, y)$ has a *relative maximum* (*or minimum*) *on the curve* $g(x, y) = 0$ *at the point* (x_0, y_0) for which $g(x_0, y_0) = 0$ (so-called *relative constrained extreme*) if the relation $f(x_0, y_0) \geqq f(x, y)$ (or $f(x_0, y_0) \leqq f(x, y)$) holds at every point (x, y) on the curve $g(x, y) = 0$ in a certain neighbourhood of the point (x_0, y_0). If in this neighbourhood the equality occurs *only* at the point (x_0, y_0), then we speak of a *strict relative constrained extreme.*

If $f(x_0, y_0) \geqq f(x, y)$ or $f(x_0, y_0) \leqq f(x, y)$ holds at *every* point (x, y) of the curve $g(x, y) = 0$, then we speak of an (*absolute*) *extreme of the function* $f(x, y)$ *on the curve* $g(x, y) = 0$ (or of an *absolute constrained extreme*) at the point (x_0, y_0). The curve $g(x, y)$ is often called a *constraint.*

REMARK 5. In applications, we find the constrained extremes (at least on a certain part of the given curve) either by computing, for example, y as a function of x, substituting into $z = f(x, y)$ and thus reducing the problem to one of finding extremes of a function of a single variable (Example 2) or by the application of Lagrange's method (Example 3).

Example 2. Let us find the (absolute) maximum and minimum of the function

$$z = x^2 + y^2 - 2x - 4y + 1 \tag{11}$$

in a closed triangle (i.e. its boundary included) with vertices (0, 0), (3, 0), (0, 5).

First, we find the relative extremes by (2):

$$2x - 2 = 0\,, \quad 2y - 4 = 0\,, \quad \text{hence} \quad x = 1\,, \quad y = 2\,. \tag{12}$$

By Theorem 1 we easily verify that at the point (1, 2) there is a (strict) relative minimum, $z = -4$; this point is an interior point of the given triangle.

Constrained extremes: The equations of the sides of the triangle are (§ 5.4, p. 208):

$$y = 0\,, \quad x = 0\,, \quad 5x + 3y - 15 = 0\,. \tag{13}$$

On the segment $y = 0, 0 \leqq x \leqq 3$, the function (11) is of the form $z = x^2 - 2x + 1$. From $dz/dx = 2x - 2 = 0$ it follows that $x = 1$. Because $d^2z/dx^2(1, 0) = 2$, there is a (strict) relative minimum, $z = 0$, at the point $x = 1$. There are no other relative extremes of the function $z = x^2 - 2x + 1$. At the first end point $x = 0$, we have $z = 1$, at the second end point $x = 3$, $z = 4$. Thus the function $z = x^2 - 2x + 1$ attains its maximum value on the segment concerned at the point $x = 3$. The absolute extremes of the function (11) on the segment $y = 0$, $0 \leqq x \leqq 3$ are therefore at the points $(1, 0)$, $(3, 0)$.

Similarly, we find that the absolute extremes of the function (11) on the segment $x = 0$, $0 \leqq y \leqq 5$ are at the points $(0, 2)$ (minimum, $z = -3$), $(0, 5)$ (maximum, $z = 6$).

The third side of the triangle is the segment $5x + 3y - 15 = 0$, $0 \leqq x \leqq 3$. Hence, $y = (15 - 5x)/3$. If we put y into (11) we obtain after rearrangement $9z = 34x^2 - 108x + 54$. If we put the first derivative equal to zero, i.e. $68x - 108 = 0$, we get $x = \frac{27}{17}$, $y = \frac{40}{17}$. The second derivative with respect to x is positive; so we have a (strict) relative minimum, $z \doteq -3{\cdot}6$, at this point. At the end points of the segment, $(3, 0)$, $(0, 5)$ we have $z = 4$, $z = 6$, respectively. The absolute extremes on the investigated segment are therefore at the points $(\frac{27}{17}, \frac{40}{17})$ and $(0, 5)$.

The result. The function (11) attains its (absolute) minimum, $z = -4$ at the point $(1, 2)$ and its (absolute) maximum $z = 6$, at the point $(0, 5)$ of the triangle concerned.

Theorem 3 (*Lagrange's Method of Undetermined Coefficients (Multipliers)*). *Let $f(x, y)$ and $g(x, y)$ have total differentials in a neighbourhood of the points of the curve $g(x, y) = 0$. Let at least one of the derivatives $\partial g/\partial x$, $\partial g/\partial y$ be non-zero at every point of the curve $g(x, y) = 0$. If the function $z = f(x, y)$ has a relative extreme on the curve $g(x, y) = 0$ at a point (x_0, y_0) of this curve, then there exists a constant λ such that for the function*

$$F(x, y) = f(x, y) + \lambda g(x, y) \tag{14}$$

the equations

$$\frac{\partial F}{\partial x}(x_0, y_0) = 0\,, \quad \frac{\partial F}{\partial y}(x_0, y_0) = 0 \quad (\textit{and also } g(x_0, y_0) = 0) \tag{15}$$

are satisfied at the point (x_0, y_0).

REMARK 6. Constrained extremes can thus be found by constructing the function (14) and then solving equations (15) for the unknowns x_0, y_0, λ. We observe that equations (15) are *necessary* conditions for the existence of a constrained extreme.

REMARK 7. *Sufficient conditions*: Let us construct the second differential of the function (14) at the above-mentioned point (x_0, y_0),

$$d^2F(x_0, y_0) = \frac{\partial^2 F}{\partial x^2}(x_0, y_0)\,dx^2 + 2\frac{\partial^2 F}{\partial x\,\partial y}(x_0, y_0)\,dx\,dy + \frac{\partial^2 F}{\partial y^2}(x_0, y_0)\,dy^2\,. \tag{16}$$

If for all points $(x_0 + \mathrm{d}x, y_0 + \mathrm{d}y)$ from a certain neighbourhood of the point (x_0, y_0), and such that $g(x_0 + \mathrm{d}x, y_0 + \mathrm{d}y) = 0$ and $\mathrm{d}x$ and $\mathrm{d}y$ are not simultaneously zero, the differential (16) is positive (or negative), then there is a constrained relative minimum (or maximum) at the point (x_0, y_0).

REMARK 8. For this condition to be fulfilled it is sufficient that the form (16) be positive (or negative) definite at the point (x_0, y_0) (§ 1.29, p. 105). For this it is again sufficient that

$$AC - B^2 > 0\,, \tag{17}$$

where

$$A = \frac{\partial^2 F}{\partial x^2}(x_0, y_0)\,, \quad B = \frac{\partial^2 F}{\partial x\,\partial y}(x_0, y_0)\,, \quad C = \frac{\partial^2 F}{\partial y^2}(x_0, y_0)\,,$$

and that

$$\frac{\partial^2 F}{\partial x^2}(x_0, y_0) > 0 \quad \left(\text{or } \frac{\partial^2 F}{\partial x^2}(x_0, y_0) < 0\right). \tag{18}$$

Example 3. Let us find the constrained extremes of the function

$$z = x^2 + y^2 \tag{19}$$

on the curve

$$x^2 + 4y^2 - 1 = 0\,.$$

At least one of the derivatives with respect to x or y is non-zero at every point of the given ellipse. We construct the function (14),

$$F(x, y) = x^2 + y^2 + \lambda(x^2 + 4y^2 - 1) = x^2(1 + \lambda) + y^2(1 + 4\lambda) - \lambda\,. \tag{20}$$

Equations (15) are now

$$2x_0(1 + \lambda) = 0\,, \tag{21}$$
$$2y_0(1 + 4\lambda) = 0\,, \tag{22}$$
$$x_0^2 + 4y_0^2 - 1 = 0\,. \tag{23}$$

If $\lambda \neq -1$ and also $\lambda \neq -\frac{1}{4}$, then by (21) and (22) $x_0 = y_0 = 0$ and (23) is not fulfilled. It follows that either $\lambda = -1$ or $\lambda = -\frac{1}{4}$. For $\lambda = -1$ we obtain from (22) $y_0 = 0$ and from (23) $x_0 = \pm 1$. For $\lambda = -\frac{1}{4}$ we obtain similarly $x_0 = 0$, $y_0 = \pm 0{\cdot}5$. So we have four points:

$$\text{For } \lambda = -1\text{: } (1, 0), (-1, 0)\,; \quad \text{for } \lambda = -\tfrac{1}{4}\text{: } \quad (0, 0{\cdot}5), (0, -0{\cdot}5)\,. \tag{24}$$

For the case $\lambda = -1$ the second differential (16) is (see (20))

$$\mathrm{d}^2 F(\pm 1; 0) = -6\,\mathrm{d}y^2\,, \tag{25}$$

for $\lambda = -\frac{1}{4}$

$$\mathrm{d}^2 F(0; \pm 0{\cdot}5) = +\tfrac{3}{2}\,\mathrm{d}x^2\,. \tag{26}$$

From the form of the right-hand member of (25) it can be seen that the condition of Remark 8 or 9 is fulfilled at the points $(\pm 1, 0)$ and a constrained relative maximum, $z = 1$, is attained there. Similarly, at the points $(0, \pm 0{\cdot}5)$ a constrained relative minimum, $z = 0{\cdot}25$ is attained. (Evidently, these extremes are also absolute constrained extremes.)

REMARK 9. We could not apply (17) here, because $AC - B^2 = 4(1 + \lambda)(1 + 4\lambda) = 0$ at the points considered.

REMARK 10. The problem of finding the constrained extremes of a function of several variables is solved similarly. For example, a necessary condition for the existence of a relative extreme of the function

$$w = f(x, y, z, u, v)$$

with subsidiary conditions

$$F_1(x, y, z, u, v) = 0\,, \quad F_2(x, y, z, u, v) = 0\,,$$

is that the equations

$$\frac{\partial G}{\partial x} = 0\,, \quad \frac{\partial G}{\partial y} = 0\,, \quad \frac{\partial G}{\partial z} = 0\,, \quad \frac{\partial G}{\partial u} = 0\,, \quad \frac{\partial G}{\partial v} = 0\,, \quad F_1 = 0\,, \quad F_2 = 0\,,$$

where

$$G = f + \lambda_1 F_1 + \lambda_2 F_2\,,$$

be satisfied. (We suppose that at least one of the functional determinants

$$\frac{\partial(F_1, F_2)}{\partial(x, y)}\,, \quad \frac{\partial(F_1, F_2)}{\partial(x, z)}\,, \quad \ldots\,, \quad \frac{\partial(F_1, F_2)}{\partial(u, v)}$$

is non-zero at the points of the hypersurfaces $F_1 = 0$, $F_2 = 0$.) This gives a system of seven equations for seven unknowns $x_0, y_0, z_0, u_0, v_0, \lambda_1, \lambda_2$.

The sufficient conditions are similar to those in Remarks 7 and 8. In particular, we obtain a constrained relative minimum if the second differential of the function G at the point $(x_0, y_0, z_0, u_0, v_0)$ is a positive definite form and a constrained relative maximum if it is a negative definite form.

REMARK 11. *Relative extremes of implicit functions* are found from equations (12.9.2) or (12.9.11), pp. 462 and 465. Since for a function of one variable we require that $y' = 0$, equation (12.9.2) gives the condition

$$\frac{\partial f}{\partial x} = 0\,, \tag{27}$$

while at the same time $f(x, y) = 0$ has to be fulfilled, as the required point (x_0, y_0) has to lie on the curve $f(x, y) = 0$, and

$$\frac{\partial f}{\partial y} \neq 0 .$$

It then follows from equation (12.9.4) (since $y' = 0$) that $y'' \neq 0$ if

$$\frac{\partial^2 f}{\partial x^2} \neq 0 .$$

Generally: The relative extremes of the function $y = \varphi(x)$ given implicitly by the equation $f(x, y) = 0$ (more exactly: of all functions defined implicitly by the equation $f(x, y) = 0$) are to be found at the points at which $\partial f/\partial x = 0$ and simultaneously $f(x, y) = 0$. If at such a point $\partial f/\partial y \neq 0$, then the existence of a relative extreme is assured if the first non-zero derivative in the sequence

$$\frac{\partial f}{\partial x}, \quad \frac{\partial^2 f}{\partial x^2}, \quad \frac{\partial^3 f}{\partial x^3}, \quad \ldots$$

is of an even order. If its sign is the same as that of $\partial f/\partial y$, we obtain a strict relative maximum, if its sign is opposite, we obtain a strict relative minimum.

The case of functions of several variables can be dealt with in a similar manner. For example, the extremes of a function $z = \varphi(x, y)$ given by $f(x, y, z) = 0$ are to be found at the points where, simultaneously,

$$f(x, y, z) = 0 , \quad \frac{\partial f}{\partial x} = 0 , \quad \frac{\partial f}{\partial y} = 0$$

(see equation (12.9.11)) under the assumption that

$$\frac{\partial f}{\partial z} \neq 0 .$$

Example 4. Let us find the extremes of the function (more exactly: of the functions defined implicitly by the equation)

$$x^2 + y^2 - 25 = 0 .$$

According to (27) we solve the system

$$2x = 0 , \quad x^2 + y^2 - 25 = 0 .$$

Solving it, we get two points: (0, 5), (0, −5). At both points $\partial f/\partial y \neq 0$, since

$$\frac{\partial f}{\partial y}(0, 5) = 10 , \quad \frac{\partial f}{\partial y}(0, -5) = -10 .$$

Further

$$\frac{\partial^2 f}{\partial x^2} = 2 .$$

At the point $(0, 5)$ there is a strict relative maximum because

$$\frac{\partial^2 f}{\partial x^2} \quad \text{and} \quad \frac{\partial f}{\partial y}$$

are of the same sign at that point. At the point $(0, -5)$ there is a strict relative minimum because the derivatives are of opposite signs. (This example is only an illustrative one – geometrically the problem is self-evident.)

REMARK 12. From equations (12.9.2) and (12.9.4) and from the corresponding equations for higher derivatives conclusions may also be drawn (on the basis of theorems from § 11.9) on other properties of implicit functions (on convexity, on points of inflexion, etc.), not only on extremes. Implicit functions of several variables may also be studied by means of equations of the type (12.9.11) and of similar equations for higher derivatives.

12.13. Survey of Some Important Formulae from Chapter 12

1. $$\lim_{\substack{x \to x_0 \\ y \to y_0}} [f(x, y) \pm g(x, y)] = \lim_{\substack{x \to x_0 \\ y \to y_0}} f(x, y) \pm \lim_{\substack{x \to x_0 \\ y \to y_0}} g(x, y) ,$$

$$\lim_{\substack{x \to x_0 \\ y \to y_0}} kf(x, y) = k \lim_{\substack{x \to x_0 \\ y \to y_0}} f(x, y) \quad (k \text{ a constant}) ,$$

$$\lim_{\substack{x \to x_0 \\ y \to y_0}} [f(x, y)\, g(x, y)] = \lim_{\substack{x \to x_0 \\ y \to y_0}} f(x, y) \lim_{\substack{x \to x_0 \\ y \to y_0}} g(x, y) ,$$

$$\lim_{\substack{x \to x_0 \\ y \to y_0}} \frac{f(x, y)}{g(x, y)} = \frac{\lim\limits_{\substack{x \to x_0 \\ y \to y_0}} f(x, y)}{\lim\limits_{\substack{x \to x_0 \\ y \to y_0}} g(x, y)} \quad \text{if} \quad \lim_{\substack{x \to x_0 \\ y \to y_0}} g(x, y) \neq 0 \quad \text{(Theorem 12.1.2)}.$$

2. $$\frac{\partial f}{\partial x}(x_0, y_0) = \lim_{h \to 0} \frac{f(x_0 + h, y_0) - f(x_0, y_0)}{h} ,$$

$$\frac{\partial f}{\partial y}(x_0, y_0) = \lim_{k = 0} \frac{f(x_0, y_0 + k) - f(x_0, y_0)}{k} \quad \text{(Definition 12.2.1)} .$$

3. $$\frac{\partial^2 f}{\partial x\, \partial y} = \frac{\partial^2 f}{\partial y\, \partial x} \quad \text{(Theorem 12.2.1)} .$$

4. $dz = \frac{\partial z}{\partial x} dx + \frac{\partial z}{\partial y} dy\,, \quad d^{(m)} z = \left(h \frac{\partial}{\partial x} + k \frac{\partial}{\partial y}\right)^m z$

(Definition 12.3.2, Remark 12.3.4),

$$\frac{\partial z}{\partial x} = \frac{\partial z}{\partial u} \frac{\partial u}{\partial x} + \frac{\partial z}{\partial v} \frac{\partial v}{\partial x}\,, \quad \frac{\partial z}{\partial y} = \frac{\partial z}{\partial u} \frac{\partial u}{\partial y} + \frac{\partial z}{\partial v} \frac{\partial v}{\partial y}$$

(differentiation of composite functions, Theorem 12.4.1; see also Remarks 12.4.1 and 12.4.2).

5. $z - z_0 = \left(\frac{\partial f}{\partial x}\right)_0 (x - x_0) + \left(\frac{\partial f}{\partial y}\right)_0 (y - y_0)$

(equation of a tangent plane, Theorem 12.3.4).

6. $\frac{\partial f}{\partial s} = \left(\frac{\partial f}{\partial x}\right)_0 \cos \alpha + \left(\frac{\partial f}{\partial y}\right)_0 \sin \alpha$

(differentiation in a given direction, Theorem 12.5.2).

7. $x \frac{\partial f}{\partial x} + y \frac{\partial f}{\partial y} = nf(x, y)$ (Theorem 12.6.1 on homogeneous functions).

8. $y' = -\frac{\partial f / \partial x}{\partial f / \partial y}$

(differentiation of a function y given implicitly by an equation $f(x, y) = 0$, Theorem 12.9.1).

9. Extremes of a function $z = f(x, y)$: If

$$\frac{\partial f}{\partial x}(x_0, y_0) = 0\,, \quad \frac{\partial f}{\partial y}(x_0, y_0) = 0\,,$$

$$\frac{\partial^2 f}{\partial x^2}(x_0, y_0) \frac{\partial^2 f}{\partial y^2}(x_0, y_0) - \left[\frac{\partial^2 f}{\partial x\, \partial y}(x_0, y_0)\right]^2 > 0\,,$$

then for $\frac{\partial^2 f}{\partial x^2}(x_0, y_0) > 0$ the function has a strict relative minimum at the point (x_0, y_0), for $\frac{\partial^2 f}{\partial x^2}(x_0, y_0) < 0$ a strict relative maximum (Theorem 12.12.1).

13. INTEGRAL CALCULUS OF FUNCTIONS OF ONE VARIABLE

By KAREL REKTORYS

References: [16], [27], [32], [48], [73], [112], [155], [156], [169], [175], [182], [184], [188], [210], [224], [226], [246], [280], [294], [299], [308], [338], [341], [346], [362], [367], [376], [438], [445].

13.1. Primitive Function (Indefinite Integral). Basic Integrals

Definition 1. The function $F(x)$ is said to be a *primitive* (*primitive function, indefinite integral*) of the function $f(x)$ in the interval (a, b), if the relation $F'(x) = f(x)$ holds for all $x \in (a, b)$.

Theorem 1. *For each function $f(x)$ which is continuous in (a, b) there exists a primitive. In fact, there exists an infinite number of them. If $F(x)$ is a primitive, then all others are of the form*

$$F(x) + C\,, \tag{1}$$

where C is an arbitrary constant.

We write

$$\int f(x)\,\mathrm{d}x = F(x) + C\,. \tag{2}$$

The function $f(x)$ itself is called the *integrand* of the integral (2).

REMARK 1. According to the definition we have, keeping C fixed in (2),

$$\frac{\mathrm{d}}{\mathrm{d}x}\int f(x)\,\mathrm{d}x = f(x)\,.$$

(In this sense differentiation and integration may be regarded as inverse operations.)

Example 1. We have

$$\int x^2\,\mathrm{d}x = \frac{x^3}{3} + C \tag{3}$$

(where C is an arbitrary constant, e.g. 7). For, if we differentiate the function on the right-hand side of equation (3) (keeping C fixed), we obtain x^2. According to Theorem 1, functions of the form $x^3/3 + C$ are the only ones which possess this property.

Theorem 2 (*Standard Integrals*). (*Unless the contrary is stated, the formulae are valid for all real x and for all values of constants involved.*)

1. $$\int x^n \, dx = \frac{x^{n+1}}{n+1} + C \quad (x > 0, n \text{ real}, n \neq -1).$$

(For some n, the range of validity may be extended. For instance, if k is a positive integer, the relation

$$\int x^k \, dx = \frac{x^{k+1}}{k+1} + C$$

holds for all x.)

2. $$\int \frac{dx}{x} = \ln |x| + C \quad (x \neq 0), \quad \text{or} \quad \int \frac{dx}{x} = \ln kx \quad (kx > 0).$$

3. $$\int e^x \, dx = e^x + C, \qquad \int a^x \, dx = \frac{a^x}{\ln a} + C \quad (a > 0).$$

4. $$\int \sin x \, dx = -\cos x + C, \qquad \int \cos x \, dx = \sin x + C.$$

5. $$\int \frac{dx}{\sin^2 x} = -\cot x + C \quad (x \neq n\pi, \ n \text{ an integer}),$$

$$\int \frac{dx}{\cos^2 x} = \tan x + C \quad (x \neq \tfrac{1}{2}\pi + n\pi, \quad n \text{ an integer}).$$

6. $$\int \frac{dx}{1 + x^2} = \arctan x + C = -\operatorname{arccot} x + k.$$

7. $$\int \frac{dx}{\sqrt{(1 - x^2)}} = \arcsin x + C = -\arccos x + k \quad (-1 < x < 1).$$

8. $$\int \frac{dx}{1 - x^2} = \tfrac{1}{2} \ln \left| \frac{1 + x}{1 - x} \right| + C \quad \text{for} \quad |x| \neq 1,$$

$$= \operatorname{artanh} x + C \quad \text{for} \quad |x| < 1,$$

$$= \operatorname{arcoth} x + C \quad \text{for} \quad |x| > 1.$$

9. $$\int \sinh x \, dx = \cosh x + C, \qquad \int \cosh x \, dx = \sinh x + C.$$

10. $\displaystyle\int \frac{dx}{\sinh^2 x} = -\coth x + C \quad (x \neq 0), \quad \int \frac{dx}{\cosh^2 x} = \tanh x + C.$

11. $\displaystyle\int \frac{dx}{\sqrt{(x^2 + 1)}} = \ln [x + \sqrt{(x^2 + 1)}] + C = \operatorname{arsinh} x + C.$

12. $\displaystyle\int \frac{dx}{\sqrt{(x^2 - 1)}} = \ln |x + \sqrt{(x^2 - 1)}| + C \quad (|x| > 1),$

$$= \operatorname{arcosh} x + C \quad (x > 1).$$

REMARK 2. $\operatorname{arsinh} x$, $\operatorname{arcosh} x$, $\operatorname{artanh} x$, $\operatorname{arcoth} x$ are the functions inverse to the hyperbolic functions ($x = \sinh y$ etc., see § 2.13).

REMARK 3. We have given two expressions for the integral $\int dx/x$. In the domain where x is negative, $\int dx/x = \ln(-x) + C$ (for there we have $|x| = -x$), or $\int dx/x = \ln kx$, where $k < 0$. The fact that $\ln kx$ is the primitive of the function $1/x$, follows from the chain rule (Theorem 11.5.5), since $(\ln kx)' = (1/kx) \cdot k = 1/x$.

Two forms of the indefinite integral do not contradict Theorem 1, for one can be reduced to the other. For instance, $\ln kx = \ln x + \ln k$ holds for $x > 0$ and $k > 0$, and the right-hand side is already of the form $\ln |x| + C$. This situation occurs very frequently. If two forms of an indefinite integral are found, it is always possible to reduce one to the other.

REMARK 4. Even though for each continuous function there exists a primitive, in many cases it is not possible to express it in terms of elementary functions (i.e. algebraic functions and elementary transcendental functions, see § 11.2). An example of such a primitive is the indefinite integral of the function e^{x^2}. Hence a new transcendental function is defined by the integral

$$\int e^{x^2} dx.$$

Similarly, new transcendental functions are defined by the integrals

$$\operatorname{Si} x = \int_0^x \frac{\sin t}{t} dt \quad (\textit{sine integral}), \quad \operatorname{Ci} x = -\int_x^\infty \frac{\cos t}{t} dt \quad (\textit{cosine integral}),$$

$$\operatorname{li} x = \int_0^x \frac{dt}{\ln t} \quad (\textit{logarithm integral}), \quad \operatorname{Ei}(x) = \int_{-\infty}^x \frac{e^t}{t} dt.$$

(At singular points of the integrands, the integrals are understood in the sense of the Cauchy principal value, see Remark 13.8.3.) See also Theorem 13.12.1, p. 588. On elliptic integrals see § 13.12.

13.2. Methods of Integration. Integration by Parts. Method of Substitution. Method of Differentiation with Respect to a Parameter

Theorem 1. *If there exist primitives of the functions* $f_1(x)$, $f_2(x)$, $f(x)$, *then*

$$\int [f_1(x) \pm f_2(x)]\,\mathrm{d}x = \int f_1(x)\,\mathrm{d}x \pm \int f_2(x)\,\mathrm{d}x\,,$$

$$\int k\,f(x)\,\mathrm{d}x = k \int f(x)\,\mathrm{d}x \quad (k = \text{const.})\,.$$

(*The relation* $\int f(x)\,g(x)\,\mathrm{d}x = \int f(x)\,\mathrm{d}x\,.\int g(x)\,\mathrm{d}x$ *does not hold in general.*)

Example 1.

$$\int \frac{4x - 1}{\sqrt{x}}\,\mathrm{d}x = \int \left(4\sqrt{(x)} - \frac{1}{\sqrt{(x)}}\right)\mathrm{d}x =$$

$$= 4\int x^{1/2}\,\mathrm{d}x - \int x^{-1/2}\,\mathrm{d}x = \tfrac{8}{3}x^{3/2} - 2x^{1/2} + C = \tfrac{8}{3}\sqrt{(x^3)} - 2\sqrt{(x)} + C \quad (x > 0)\,.$$

Theorem 2 (*Integration by Parts*). *Let us assume that the functions* $u(x)$ *and* $v(x)$ *possess continuous derivatives in the interval* (a, b) (*hence these functions are also continuous in the interval* (a, b) by Theorem 11.5.2). *Then the relation*

$$\int u'v\,\mathrm{d}x = uv - \int uv'\,\mathrm{d}x \tag{1}$$

holds in the interval (a, b).

REMARK 1. Equation (1) is often written in the form

$$\int v\,\mathrm{d}u = uv - \int u\,\mathrm{d}v\,.$$

REMARK 2. Integration by parts is convenient when integrating functions of the type

$$x^n \sin x\,, \quad x^n \cos x\,, \quad x^n \mathrm{e}^x\,, \quad \ln x\,, \quad \arctan x\,,$$

etc. (In the two last cases we put $u' = 1$.) See Examples 2, 3, 4.

Example 2. To evaluate the integral

$$I = \int x \sin x\,\mathrm{d}x$$

we put

$$u' = \sin x\,, \quad v = x\,, \quad \text{so that} \quad u = -\cos x\,, \quad v' = 1\,;$$

hence by (1)

$$\int x \sin x \,\mathrm{d}x = -x \cos x + \int \cos x \,\mathrm{d}x = -x \cos x + \sin x + C.$$

Example 3.

$$\int \ln x \,\mathrm{d}x = x \ln x - \int x \frac{1}{x} \,\mathrm{d}x = x \ln x - x + C = x(\ln x - 1) + C \quad (x > 0);$$

$$u' = 1, \quad v = \ln x,$$

$$u = x, \quad v' = \frac{1}{x}.$$

Example 4.

$$I = \int \mathrm{e}^{ax} \sin bx \,\mathrm{d}x.$$

First we put

$$u' = \mathrm{e}^{ax}, \quad v = \sin bx \quad \text{hence} \quad u = \frac{1}{a} \mathrm{e}^{ax}, \quad v' = b \cos bx,$$

thus reducing the given integral to the integral of the function $\mathrm{e}^{ax} \cos bx$. Again integrating by parts, with

$$u' = \mathrm{e}^{ax}, \quad v = \cos bx, \quad u = \frac{1}{a} \mathrm{e}^{ax}, \quad v' = -b \sin bx,$$

we obtain the original integral with the opposite sign (and a different constant). From the equation so obtained the required integral may then be easily evaluated:

$$I = \int \mathrm{e}^{ax} \sin bx \,\mathrm{d}x = \frac{1}{a} \mathrm{e}^{ax} \sin bx - \frac{b}{a} \int \mathrm{e}^{ax} \cos bx \,\mathrm{d}x =$$

$$= \frac{1}{a} \mathrm{e}^{ax} \sin bx - \frac{b}{a} \left(\frac{1}{a} \mathrm{e}^{ax} \cos bx + \frac{b}{a} \int \mathrm{e}^{ax} \sin bx \,\mathrm{d}x \right) =$$

$$= \frac{1}{a} \mathrm{e}^{ax} \sin bx - \frac{b}{a^2} \mathrm{e}^{ax} \cos bx - \frac{b^2}{a^2} I.$$

From this equation (to the right-hand side, we can assign an arbitrary constant) we obtain

$$I = \frac{1}{a^2 + b^2} (a\mathrm{e}^{ax} \sin bx - b\mathrm{e}^{ax} \cos bx) + C.$$

This method of procedure is often used.

Method of Substitution

A. Substitution $h(x) = z$:

Theorem 3. *Let $f(x)$ be of the form $f(x) = g(h(x))\,h'(x)$ in the interval (a, b), where $h'(x)$ is a continuous function in (a, b) and $g(z)$ is continuous for all $z = h(x)$ when x runs through the interval (a, b). Then*

$$\int f(x)\,\mathrm{d}x = \int g(h(x))\,h'(x)\,\mathrm{d}x = \int g(z)\,\mathrm{d}z = G(z) + C\,, \tag{2}$$

where $h(x)$ is to be substituted for z in the result.

B. Substitution $x = \varphi(z)$:

Theorem 4. *Let $f(x)$ be continuous in the interval (a, b). Let $x = \varphi(z)$ be a function (of the variable z), which is strictly increasing or strictly decreasing in the interval (α, β) and possesses a continuous derivative $\varphi'(z)$. Let us denote by $z = \psi(x)$ the function inverse to the function $x = \varphi(z)$. If, further, $a < \varphi(z) < b$ holds for $z \in (\alpha, \beta)$, then*

$$\int f(x)\,\mathrm{d}x = \int f(\varphi(z))\,\varphi'(z)\,\mathrm{d}z = H(z) + C\,, \tag{3}$$

where $\psi(x)$ is to be substituted for z in the result.

REMARK 3. Application of the method of substitution requires a certain amount of experience, in order to foresee the form of the resulting integral or in order to see in the function $f(x)$ the form $g(h(x))\,h'(x)$. Note that the existence of the inverse function to $h(x) = z$ is not assumed in Theorem 3.

REMARK 4. Note that the right-hand side of equation (3) is formally obtained by substituting $\varphi(z)$ for x and $\varphi'(z)\,\mathrm{d}z$ (which arises as a result of differentiation of the right-hand side of the equation $x = \varphi(z)$) for $\mathrm{d}x$. Similarly in (2).

In the following examples, the range of validity of the result is mentioned only in those cases where it is not evident.

Example 5.

$$\int \sin^2 x \cos x\,\mathrm{d}x = \int z^2\,\mathrm{d}z = \frac{z^3}{3} + C = \frac{\sin^3 x}{3} + C\,.$$

We have made use of the substitution $\sin x = z$ (from which it follows that $\cos x\,\mathrm{d}x = \mathrm{d}z$) and substituted $z = \sin x$ in the result (Theorem 3).

Example 6. Using the substitution

$$\tan x = z\,, \quad \frac{\mathrm{d}x}{\cos^2 x} = \mathrm{d}z \quad \text{or} \quad (1 + \tan^2 x)\,\mathrm{d}x = \mathrm{d}z\,,$$

we obtain (in every interval $(\frac{1}{2}k\pi, \frac{1}{2}(k+1)\pi)$, where k is an integer)

$$\int \tan^4 x \,\mathrm{d}x = \int \frac{z^4}{1+z^2}\,\mathrm{d}z = \int \frac{(z^4+z^2)-(z^2+1)+1}{z^2+1}\,\mathrm{d}z =$$

$$= \int\left(z^2 - 1 + \frac{1}{z^2+1}\right)\mathrm{d}z = \frac{z^3}{3} - z + \arctan z + C = \frac{\tan^3 x}{3} - \tan x + x + k$$

(since $\arctan \tan x = x + \text{const.}$).

Example 7 (using Theorem 3).

$$\int \frac{x^2}{\sqrt{(1-x^6)}}\,\mathrm{d}x = \frac{1}{3}\int \frac{\mathrm{d}z}{\sqrt{(1-z^2)}} = \tfrac{1}{3}\arcsin z + C = \tfrac{1}{3}\arcsin x^3 + C \quad (|x| < 1),$$

where

$$x^3 = z, \quad 3x^2\,\mathrm{d}x = \mathrm{d}z.$$

Example 8. Using the substitution

$$f(x) = z, \quad f'(x)\,\mathrm{d}x = \mathrm{d}z$$

we obtain

$$\int \frac{f'(x)}{f(x)}\,\mathrm{d}x = \ln|f(x)| + C \quad (f(x) \neq 0).$$

For example

$$\int \frac{5x-1}{x^2+1}\,\mathrm{d}x = \frac{5}{2}\int \frac{2x}{x^2+1}\,\mathrm{d}x - \int \frac{\mathrm{d}x}{x^2+1} = \frac{5}{2}\ln(x^2+1) - \arctan x + C.$$

Example 9. Using the substitution

$$ax + b = z, \quad a\,\mathrm{d}x = \mathrm{d}z$$

we have (if $F(z)$ denotes a primitive of $f(z)$)

$$\int f(ax+b)\,\mathrm{d}x = \frac{1}{a}\int f(z)\,\mathrm{d}z = \frac{1}{a}F(z) + C = \frac{1}{a}F(ax+b) + C.$$

For example

$$\int \cos(3x+5)\,\mathrm{d}x = \tfrac{1}{3}\int \cos z\,\mathrm{d}z = \tfrac{1}{3}\sin z + C = \tfrac{1}{3}\sin(3x+5) + C.$$

Example 10. Using the substitution

$$x = \sin z, \quad \mathrm{d}x = \cos z\,\mathrm{d}z$$

we obtain

$$\int \sqrt{(1 - x^2)}\, dx = \int \sqrt{(1 - \sin^2 z)} \cos z\, dz = \int \cos^2 z\, dz = \int \tfrac{1}{2}(1 + \cos 2z)\, dz =$$

$$= \tfrac{1}{2}z + \tfrac{1}{4} \sin 2z + C = \tfrac{1}{2}z + \tfrac{1}{2} \sin z \cos z + C = \tfrac{1}{2} \arcsin x +$$

$$+ x \sqrt{(1 - x^2)} + C$$

(Theorem 4). The interval $(-1, 1)$ for x corresponds to the interval $(-\frac{1}{2}\pi, \frac{1}{2}\pi)$ for z, which was denoted by (α, β) in Theorem 4. (The same considerations hold also for closed intervals.) The inverse function to $x = \sin z$ is $z = \arcsin x$; $\sqrt{(1 - \sin^2 z)} = + \cos z$, for $\cos z > 0$ if z belongs to the interval $(-\frac{1}{2}\pi, \frac{1}{2}\pi)$. When integrating $\cos 2z$ we employ the substitution $2z = t$; cf. Example 9.

The integral

$$\int \sqrt{(a^2 - x^2)}\, dx$$

can be evaluated either by the substitution $x = a \sin z$ or may be reduced by the substitution $x = at$ to the previous case.

For some further typical examples on the method of substitution see § 13.4.

REMARK 5. Theorems on integration by substitution and by parts may be formulated under rather weaker restrictions than those introduced in Theorems 2 and 3. In practice the two methods are often combined:

Example 11. Integrating by parts,

$$u' = 1\,, \quad v = \arctan x\,, \quad u = x\,, \quad v' = \frac{1}{1 + x^2}$$

and using the result of Example 8, we obtain

$$\int \arctan x\, dx = x \arctan x - \int \frac{x}{x^2 + 1}\, dx = x \arctan x - \tfrac{1}{2} \ln (x^2 + 1) + C\,.$$

REMARK 6 (*Method of Differentiation of Integrals with Respect to a Parameter*). We have (Example 13.4.6)

$$\int \frac{dx}{\sqrt{(x^2 + a)}} = \ln [x + \sqrt{(x^2 + a)}] + C \quad (a > 0)\,.$$

This equation, *formally* differentiated with respect to a (not with respect to x!), gives

$$-\int \frac{dx}{2\sqrt{(x^2 + a)^3}} = \frac{1}{x + \sqrt{(x^2 + a)}}\, \frac{1}{2\sqrt{(x^2 + a)}}\,.$$

Hence, if differentiation with respect to a under the integral sign is "permissible", then this procedure gives the primitive of the function $1/\sqrt{(x^2 + a)^3}$. For the conditions under which this method leads to correct results we refer the reader to § 13.9. Here, we shall treat only the simplest case.

Theorem 5. *Let us denote*

$$\int f(x, a)\,\mathrm{d}x = F(x, a) + C\,. \tag{4}$$

If the functions $f(x, a)$, $\dfrac{\partial f}{\partial a}(x, a)$, are continuous as functions of two variables (in the region considered), then

$$\int \frac{\partial f}{\partial a}(x, a)\,\mathrm{d}x = \frac{\partial F}{\partial a}(x, a) + k\,. \tag{5}$$

Example 12. Let us determine

$$\int \frac{\mathrm{d}x}{(a + bx^2)^2} \quad (a > 0, b > 0)\,.$$

We use the relation (see e.g. § 13.5, formula 33)

$$\int \frac{\mathrm{d}x}{a + bx^2} = \frac{1}{\sqrt{(ab)}} \arctan \sqrt{\left(\frac{b}{a}\right)}\, x + C \quad (a > 0, b > 0)\,. \tag{6}$$

Obviously, the function $1/(a + bx^2)$ as well as the function $-1/(a + bx^2)^2$ are continuous functions of x, a, b for all x (since $a > 0$, $b > 0$), hence (4) and (5) are applicable. Integrating (6) with respect to a, we have

$$-\int \frac{\mathrm{d}x}{(a + bx^2)^2} = -\frac{1}{2\sqrt{(a^3 b)}} \arctan \sqrt{\left(\frac{b}{a}\right)}\, x + \frac{1}{\sqrt{(ab)}}\, \frac{1}{1 + \dfrac{b}{a}x^2} \left(-\frac{\sqrt{b}}{2\sqrt{a^3}}\, x\right) + k\,;$$

hence on rearranging

$$\int \frac{\mathrm{d}x}{(a + bx^2)^2} = \frac{1}{2a\sqrt{(ab)}} \arctan \sqrt{\left(\frac{b}{a}\right)}\, x + \frac{1}{2a}\, \frac{x}{a + bx^2} + k \quad (a > 0, b > 0)\,.$$

REMARK 7. *Graphical integration* is based on the following idea: If $F(x)$ denotes the primitive of $f(x)$, then $F'(x) = f(x)$. Hence (see § 11.6)

$$F(a + h) - F(a) = h\,f(a) + h\,\tau(h)\,, \quad \text{where} \quad \tau(h) \to 0 \quad \text{for} \quad h \to 0\,.$$

Neglecting the second term, one gets

$$F(a + h) \approx F(a) + h\,f(a)\,.$$

Similarly

$$F(a + 2h) - F(a + h) \approx h\,f(a + h)\,,$$

whence $F(a + 2h)$ is determined, etc. For details and for some modifications for practical purposes see § 32.6.

13.3. Integration of Rational Functions

In this paragraph, we shall deal with integrals of the form

$$\int \frac{P(x)}{Q(x)}\, dx\,, \tag{1}$$

where $P(x)$ and $Q(x)$ are polynomials (we shall assume throughout that $P(x)$ and $Q(x)$ have *real coefficients*). Basically the method is to split the integrand into the sum of simple functions, which can be integrated directly.

If the degree m of the polynomial $P(x)$ is greater than (or equal to) the degree n of the polynomial $Q(x)$, then the fraction $P(x)/Q(x)$ can be reduced to a sum of a polynomial (of order $m - n$) and of a proper fraction

$$\frac{R(x)}{Q(x)}\,, \tag{2}$$

where the degree r of the polynomial $R(x)$ is less than the degree n of the polynomial $Q(x)$ (or $R(x) \equiv 0$). The mechanism of division is to be seen from the following example:

Example 1.

$$(x^3 + 4x^2 - x + 2)/(x^2 + x - 3) = x + 3 + \frac{-x + 11}{x^2 + x - 3}$$

$$\begin{array}{r} -x^3 \pm x^2 \mp 3x \\ \hline 3x^2 + 2x + 2 \\ -3x^2 \pm 3x \mp 9 \\ \hline -x + 11 \end{array}$$

Theorem 1. *Every polynomial*

$$Q(x) = a_n x^n + a_{n-1} x^{n-1} + \ldots + a_1 x + a_0$$

of the n-th order, with real coefficients can be reduced in a unique way, except for the ordering of the factors, into a product of the form (see Example 2)

$$Q(x) = a_n(x - \alpha_1)^{k_1} (x - \alpha_2)^{k_2} \ldots (x - \alpha_i)^{k_i} (x^2 + p_1 x + q_1)^{l_1} (x^2 + p_2 x + q_2)^{l_2} \ldots \ldots (x^2 + p_j x + q_j)^{l_j}\,, \tag{3}$$

where the numbers $\alpha_1, \alpha_2, \ldots, \alpha_i, p_1, q_1, p_2, q_2, \ldots$ *are real and the quadratic expressions in* (3) *are irreducible to real linear factors* (*i.e. they have negative discriminants,*

$$\frac{p^2}{4} - q < 0) . \tag{4}$$

REMARK 1. The reduction (3) is obtained from the well-known method of factorisation in linear factors in the following way: If α is a real zero of $Q(x)$, then the reduction contains the factor $x - \alpha$. If β ($\beta = \beta_1 + i\beta_2$) is a complex zero of $Q(x)$, then the complex conjugate $\bar{\beta} = \beta_1 - i\beta_2$ is also a zero of $Q(x)$. The product of corresponding factors

$$[x - (\beta_1 + i\beta_2)]\,[x - (\beta_1 - i\beta_2)] = (x - \beta_1)^2 + \beta_2^2 =$$
$$= x^2 - 2\beta_1 x + \beta_1^2 + \beta_2^2$$

gives a quadratic expression with *real* coefficients. If the multiplicity of β is l, then $\bar{\beta}$ is of the same multiplicity and we obtain in (3) the factor

$$(x^2 + px + q)^l .$$

REMARK 2. On the practical determination of zeros of a polynomial of the n-th degree see Chaps. 1 and 31. It is often possible to find a zero of the given polynomial by inspection, especially if the polynomial has integral coefficients. Dividing by the corresponding linear factor $x - \alpha$ we reduce the order of the polynomial, as in the following example.

Example 2. The polynomial $Q(x) = x^4 - x^3 - x^2 - x - 2$ has obviously the zero $\alpha_1 = -1$. We have

$$\begin{array}{l}
(x^4 - x^3 - x^2 - x - 2)/(x + 1) = x^3 - 2x^2 + x - 2 . \\
\underline{-x^4 \pm x^3} \\
\quad - 2x^3 - x^2 \\
\quad \underline{\mp 2x^3 \mp 2x^2} \\
\qquad\quad x^2 - x \\
\qquad \underline{- x^2 \pm x} \\
\qquad\quad - 2x - 2
\end{array}$$

The resulting polynomial has the zero $\alpha_2 = 2$. Dividing by the linear factor $(x - 2)$ we obtain the polynomial $x^2 + 1$, which is irreducible to real linear factors (see (4); $p = 0$, $q = 1$). Hence

$$Q(x) \equiv x^4 - x^3 - x^2 - x - 2 = (x + 1)(x - 2)(x^2 + 1)$$

and this factorisation is of the form (3).

Theorem 2. *Let*

$$\frac{P(x)}{Q(x)} \tag{5}$$

be a proper rational fraction with real coefficients and let $Q(x)$ be of the form (3), i.e.

$$Q(x) = a_n(x - \alpha_1)^{k_1} (x - \alpha_2)^{k_2} \dots (x^2 + p_1x + q_1)^{l_1} (x^2 + p_2x + q_2)^{l_2} \dots .$$

Then there exist real numbers $A_1, A_2, \dots, A_{k_1}, \dots; C_1, C_2, \dots, C_{l_1}; D_1, D_2, \dots, D_{l_1} \dots$ (uniquely determined by the function (5)), *such that for all x different from the zeros of $Q(x)$ the relation*

$$\frac{P(x)}{Q(x)} = \frac{A_1}{x - \alpha_1} + \frac{A_2}{(x - \alpha_1)^2} + \dots + \frac{A_{k_1}}{(x - \alpha_1)^{k_1}} +$$

$$+ \frac{B_1}{x - \alpha_2} + \frac{B_2}{(x - \alpha_2)^2} + \dots + \frac{B_{k_2}}{(x - \alpha_2)^{k_2}} + \dots +$$

$$+ \frac{C_1x + D_1}{x^2 + p_1x + q_1} + \frac{C_2x + D_2}{(x^2 + p_1x + q_1)^2} + \dots + \frac{C_{l_1}x + D_{l_1}}{(x^2 + p_1x + q_1)^{l_1}} +$$

$$+ \frac{E_1x + F_1}{x^2 + p_2x + q_2} + \frac{E_2x + F_2}{(x^2 + p_2x + q_2)^2} + \dots + \frac{E_{l_2}x + F_{l_2}}{(x^2 + p_2x + q_2)^{l_2}} + \dots \tag{6}$$

holds.

REMARK 3. This means that if α_1 is a real k_1-fold root of $Q(x) = 0$, then all fractions with denominators $x - \alpha_1, (x - \alpha_1)^2, \dots, (x - \alpha_1)^{k_1}$ occur in the reduction. If α_1 is a simple root, then only one fraction in the reduction (6) corresponds to it; similarly for all other real roots. The situation for expressions of the form $x^2 + px + q$ is similar; the linear binomials of the form $Cx + D$, however, should be written in the numerators instead of constants.

REMARK 4. The unknown constants $A_1, A_2, \dots$ can be determined by several methods, only two of which will be mentioned here: the *method of undetermined coefficients* and the *substitution method:*

Example 3. Let us reduce the function

$$\frac{x^2 - 2}{x^4 - 2x^3 + 2x^2} \tag{7}$$

according to Theorem 2.

The function (7) is already a proper fraction, hence it is not necessary to perform the preliminary division as in Example 1. The denominator can be written in the form

$$Q(x) \equiv x^4 - 2x^3 + 2x^2 = x^2(x^2 - 2x + 2),$$

where the quadratic expression $x^2 - 2x + 2$ has a negative discriminant (4). Hence, the reduction (6) is of the form ($\alpha = 0$)

$$\frac{x^2 - 2}{x^4 - 2x^3 + 2x^2} = \frac{A_1}{x} + \frac{A_2}{x^2} + \frac{Bx + C}{x^2 - 2x + 2}. \tag{8}$$

1. Determination of the constants A_1, A_2, B, C by the *method of undetermined coefficients*. We multiply equation (8) by $x^2(x^2 - 2x + 2)$, giving

$$x^2 - 2 = A_1 x(x^2 - 2x + 2) + A_2(x^2 - 2x + 2) + (Bx + C)\,x^2 \tag{9}$$

or

$$x^2 - 2 = (A_1 + B)\,x^3 + (-2A_1 + A_2 + C)\,x^2 + (2A_1 - 2A_2)\,x + 2A_2\,. \tag{10}$$

Since equation (8) should be valid for infinitely many values of x, the same is true for equations (9) and (10). Hence, it follows (see Theorem 1.14.1) that coefficients of the same powers of x are equal. Comparing coefficients, we obtain the equations $A_1 + B = 0$, $-2A_1 + A_2 + C = 1$, $2A_1 - 2A_2 = 0$, $2A_2 = -2$, from which

$$A_1 = -1\,, \quad A_2 = -1\,, \quad B = 1\,, \quad C = 0\,.$$

2. *Method of substitution.* Equation (9) is valid for an infinite number of values of x; hence it is valid for all x, in particular for zeros of the polynomial $Q(x)$. If $x = 0$ is substituted into (9), then equation (9) yields

$$-2 = 2A_2 \Rightarrow A_2 = -1\,.$$

To evaluate A_1 we differentiate (9) with respect to x and get

$$2x = A_1(x^2 - 2x + 2) + A_1 x(2x - 2) + A_2(2x - 2) + 2x(Bx + C) + Bx^2\,.$$

If we put $x = 0$ and $A_2 = -1$ (which has been already evaluated), we obtain $0 = 2A_1 + 2 \Rightarrow A_1 = -1$.

If $x = 0$ were a three-fold root of the equation $Q(x) = 0$, then we would determine the corresponding third constant by repeated differentiation of equation (9) (and then substituting $x = 0$). This procedure always leads to the required result.

Next, we substitute $x = 1 + \mathrm{i}$ (which is one of the zeros of the quadratic polynomial $x^2 - 2x + 2$), and obtain (note that $(1 + \mathrm{i})^2 = 2\mathrm{i}$)

$$2\mathrm{i} - 2 = [B(1 + \mathrm{i}) + C]\,2\mathrm{i}$$

or

$$-2 + 2\mathrm{i} = -2B + (2B + 2C)\,\mathrm{i}\,.$$

Comparing the real and the imaginary parts we get $B = 1$, $C = 0$.

(If the expression $x^2 - 2x + 2$ had appeared squared in the reduction, we would have evaluated further coefficients in this case also by differentiation of equation (9) and substitution of $x = 1 + \mathrm{i}$.)

Hence

$$\frac{x^2 - 2}{x^4 - 2x^3 + 2x^2} = \frac{-1}{x} + \frac{-1}{x^2} + \frac{x}{x^2 - 2x + 2} \tag{11}$$

for all x other than 0 and $1 \pm i$.

REMARK 5. The numbers A_1, A_2, B, C may alternatively be determined by substituting any four values for x and solving the resulting four equations for the four unknown constants A_1, A_2, B, C.

REMARK 6. We draw the reader's attention to the fact that a rational function can be reduced to a sum of fractions (6) only if the degree of the polynomial in the numerator is less than that in the denominator. Otherwise division as in Example 1 is first to be performed.

REMARK 7 (concerning practical evaluation of constants). If α is a simple zero of the polynomial $Q(x)$ then the corresponding constant A can be evaluated by the formula

$$A = \frac{P(\alpha)}{Q'(\alpha)}.$$

Example 4. Using (6), we may write

$$\frac{1}{x^3 - 3x^2 + 4} = \frac{1}{(x-2)^2 (x+1)} = \frac{A_1}{x-2} + \frac{A_2}{(x-2)^2} + \frac{B}{x+1}.$$

Then

$$B = \frac{P(-1)}{Q'(-1)} = \frac{1}{(3x^2 - 6x)_{x=-1}} = \frac{1}{9},$$

since $\alpha = -1$ is a simple zero of the polynomial $Q(x)$.

REMARK 8 (*Integration after Reduction*). Every proper rational function with real coefficients can be reduced by Theorem 2 in a unique way into the sum (6) (of so-called *partial fractions*). Each term on the right-hand side of equation (6) can be integrated by elementary methods:

$$\int \frac{A_1}{x - \alpha_1}\,\mathrm{d}x = A_1 \ln |x - \alpha_1| + C \quad (\text{by substitution } x - \alpha_1 = z),$$

$$\int \frac{A_k}{(x - \alpha_1)^k}\,\mathrm{d}x = \frac{A_k}{-k+1} \frac{1}{(x - \alpha_1)^{k-1}} + C \quad (k \neq 1, \text{ substitution } x - \alpha_1 = z).$$

Terms of the type

$$\frac{Bx + C}{(x^2 + px + q)^k}$$

are integrated as follows (we assume the quadratic polynomial $x^2 + px + q$ has a negative discriminant $p^2/4 - q < 0$):

$$\frac{Bx + C}{(x^2 + px + q)^k} = \frac{B}{2} \frac{2x + p}{(x^2 + px + q)^k} + \left(C - \frac{Bp}{2}\right) \frac{1}{[(x + \frac{1}{2}p)^2 + q - (\frac{1}{2}p)^2]^k}.$$

Putting $x^2 + px + q = z$, $(2x + p)\,\mathrm{d}x = \mathrm{d}z$ we have:

$$\int \frac{2x + p}{(x^2 + px + q)^k}\,\mathrm{d}x = \frac{1}{-k + 1} \frac{1}{(x^2 + px + q)^{k-1}} + C, \quad \text{if} \quad k \neq 1;$$

$$\int \frac{2x + p}{x^2 + px + q}\,\mathrm{d}x = \ln (x^2 + px + q) + C, \quad \text{if} \quad k = 1.$$

Next on substituting $x + \frac{1}{2}p = z\sqrt{[q - (\frac{1}{2}p)^2]}$, $\mathrm{d}x = \sqrt{[q - (\frac{1}{2}p)^2]}\,\mathrm{d}z$,

$$\int \frac{\mathrm{d}x}{[(x + \frac{1}{2}p)^2 + q - (\frac{1}{2}p)^2]^k} = \frac{1}{[q - (\frac{1}{2}p)^2]^{k-\frac{1}{2}}} \int \frac{\mathrm{d}z}{(z^2 + 1)^k}.$$

For $k = 1$, we have

$$\int \frac{\mathrm{d}z}{z^2 + 1} = \arctan z + C = \arctan \frac{x + \frac{1}{2}p}{\sqrt{[q - (\frac{1}{2}p)^2]}} + C.$$

For $k > 1$ the reduction formula

$$I_{k+1} = \frac{1}{2k} \frac{z}{(1 + z^2)^k} + \frac{2k - 1}{2k} I_k, \tag{12}$$

where

$$I_k = \int \frac{\mathrm{d}z}{(1 + z^2)^k}, \quad I_{k+1} = \int \frac{\mathrm{d}z}{(1 + z^2)^{k+1}},$$

is valid.

Example 5. Using (12) we have

$$\int \frac{\mathrm{d}z}{(1 + z^2)^2} = \frac{1}{2} \frac{z}{1 + z^2} + \frac{2 - 1}{2} \int \frac{\mathrm{d}z}{1 + z^2} = \frac{1}{2} \frac{z}{1 + z^2} + \frac{1}{2} \arctan z + C.$$

Example 6. Using (11) and Remark 8 we get

$$\int \frac{x^2 - 2}{x^4 - 2x^3 + 2x^2}\,\mathrm{d}x = \int \left(\frac{-1}{x} + \frac{-1}{x^2} + \frac{x}{x^2 - 2x + 2}\right) \mathrm{d}x =$$

$$= -\ln |x| + \frac{1}{x} + \frac{1}{2} \ln (x^2 - 2x + 2) + \arctan (x - 1) + C.$$

13.4. Integrals which can be rationalized

Some types of integral can be reduced to integrals of rational functions, which are then integrated according to § 13.3. In particular, the following types are considered.

I. $$\int R\left(x, \sqrt[n]{\frac{ax+b}{cx+d}}\right) dx,$$

where $R(x, t)$ is a rational function of the variables x, t.

Integrals like this can be transformed into integrals of rational functions by the substitution

$$\frac{ax+b}{cx+d} = z^n \quad \text{or} \quad x = \frac{dz^n - b}{a - cz^n}.$$

Example 1. The function

$$\int \frac{3x + \sqrt{(2x-1)}}{x - \sqrt{(2x-1)^3}} dx \quad (2x - 1 \geqq 0, \quad x - \sqrt{(2x-1)^3} \neq 0) \tag{1}$$

is of the type mentioned, where

$$R(x, t) = \frac{3x + t}{x - t^3}, \quad t = \sqrt{(2x-1)}.$$

We make the substitution

$$2x - 1 = z^2 \quad (z > 0),$$

from which it follows that $dx = z\,dz$ and

$$\int \frac{3x + \sqrt{(2x-1)}}{x - \sqrt{(2x-1)^3}} dx = \int \frac{3\,\dfrac{z^2+1}{2} + z}{\dfrac{z^2+1}{2} - z^3} z\,dz = -\int \frac{3z^3 + 2z^2 + 3z}{2z^3 - z^2 - 1} dz,$$

and this is now an integral of a rational function. After integration we substitute again $z = \sqrt{(2x-1)}$.

Remark 1. The fraction $(ax+b)/(cx+d)$ may occur with different (rational) exponents. For instance, the integrand may be a rational function $R(x, t, u)$, where

$$t = \left(\frac{ax+b}{cx+d}\right)^{1/r}, \quad u = \left(\frac{ax+b}{cx+d}\right)^{1/s},$$

is to be substituted. Such an integral can be rationalized by the substitution $(ax+b) : : (cx+d) = t^p$, where p stands for the least common multiple of the numbers r and s. The procedure is similar when several roots are involved.

Example 2. The integral

$$\int \frac{\mathrm{d}x}{\sqrt{(x+1)} - \sqrt[3]{(x+1)}}$$

is rationalized by the substitution $x + 1 = t^6$ $(t > 0)$; we get

$$\int \frac{\mathrm{d}x}{\sqrt{(x+1)} - \sqrt[3]{(x+1)}} = \int \frac{6t^5\,\mathrm{d}t}{t^3 - t^2} = 6\int \frac{t^3\,\mathrm{d}t}{t-1} =$$

$$= 6\int \frac{(t^3 - t^2) + (t^2 - t) + (t - 1) + 1}{t - 1}\,\mathrm{d}t = 6\left(\frac{t^3}{3} + \frac{t^2}{2} + t + \ln|t - 1|\right) + C =$$

$$= 2\sqrt{(x+1)} + 3\sqrt[3]{(x+1)} + 6\sqrt[6]{(x+1)} + 6\ln|\sqrt[6]{(x+1)} - 1| + C\,.$$

II. *Binomial integrals* are those of the form

$$\int x^m(a + bx^n)^p\,\mathrm{d}x \quad (a \neq 0,\ b \neq 0,\ n \neq 0,\ p \neq 0)\,, \tag{2}$$

where m, n, p are rational numbers.

Theorem 1. *Integrals* (2) *can be expressed in terms of elementary functions (algebraic functions and elementary transcendental functions,* see § 11.2) *if and only if one of the numbers*

$$p, \frac{m+1}{n}, \frac{m+1}{n} + p$$

is an integer.

REMARK 2. (α) If p is a positive integer we evaluate the expression $(a + bx^n)^p$ by the Binomial Theorem and we obtain an integral of a sum of powers of x.

If p is a negative integer and

$$m = \frac{r}{s}, \quad n = \frac{u}{v}$$

where r, s $(s > 0)$ are integers without a common factor, and similarly for u and v, then the substitution

$$x = z^t\,,$$

where t is the least common multiple of the numbers s, v, will rationalize the integral (2).

(β) If

$$\frac{m+1}{n}$$

is an integer, then the substitution

$$a + bx^n = z$$

will reduce the integral (2) to the previous case.

(γ) If

$$\frac{m+1}{n} + p$$

is an integer then we reduce the integration of (2) to the case (α) by the substitution

$$ax^{-n} + b = z. \tag{3}$$

Example 3.

$$\int \frac{(1+\sqrt[6]{x})^3}{\sqrt[3]{x^2}}\,dx = \int x^{-2/3}(1+x^{1/6})^3\,dx.$$

Case (α), p is a positive integer. Making use of the Binomial Theorem, we have

$$\int x^{-2/3}(1+x^{1/6})^3\,dx = \int x^{-2/3}(1+3x^{1/6}+3x^{2/6}+x^{3/6})\,dx =$$

$$= \int (x^{-2/3}+3x^{-1/2}+3x^{-1/3}+x^{-1/6})\,dx = 3x^{1/3}+6x^{1/2}+\tfrac{9}{2}x^{2/3}+\tfrac{6}{5}x^{5/6}+C.$$

Example 4.

$$\int \frac{dx}{\sqrt{(a+bx^2)^3}} = \int (a+bx^2)^{-3/2}\,dx.$$

Here

$$\frac{m+1}{n} + p = \tfrac{1}{2} - \tfrac{3}{2} = -1$$

(case γ). Putting

$$\frac{a}{x^2} + b = z, \quad -\frac{2a}{x^3}\,dx = dz$$

we obtain (for $x > 0$)

$$\int \frac{dx}{\sqrt{(a+bx^2)^3}} = \int \frac{1}{x^3}\frac{dx}{\sqrt{\left(\frac{a}{x^2}+b\right)^3}} = -\frac{1}{2a}\int \frac{dz}{z^{3/2}} =$$

$$= \frac{1}{a}z^{-1/2} + C = \frac{1}{a}\,\frac{1}{\sqrt{\left(\frac{a}{x^2}+b\right)}} + C.$$

Example 5. The integral

$$\int x^{-3/4}(1 - x^{1/6})^{-2}\,dx$$

(case α, p is a negative integer) is rationalized by the substitution $x = z^{12}$.

III. Integrals of the form

$$\int R(x, \sqrt{(ax^2 + bx + c)})\,dx \quad (ax^2 + bx + c > 0)\,, \tag{4}$$

where $R(x, t)$ is a rational function of the variables x, t, are rationalized as follows:

(α) If $a > 0$ we make use of the substitution

$$\sqrt{(ax^2 + bx + c)} + \sqrt{(a)}\,x = z\,, \tag{5}$$

from which it follows that

$$ax^2 + bx + c = ax^2 - 2\sqrt{(a)}\,xz + z^2 \quad \text{or} \quad x = \frac{z^2 - c}{b + 2\sqrt{(a)}\,z}\,,$$

$$dx = 2\frac{c\sqrt{(a)} + bz + \sqrt{(a)}\,z^2}{[b + 2\sqrt{(a)}\,z]^2}\,dz\,, \quad \sqrt{(ax^2 + bx + c)} = \frac{c\sqrt{(a)} + bz + \sqrt{(a)}\,z^2}{b + 2\sqrt{(a)}\,z}\,,$$

$$\frac{dx}{\sqrt{(ax^2 + bx + c)}} = \frac{2dz}{b + 2\sqrt{(a)}\,z}\,. \tag{6}$$

(β) If $a < 0$, then there are two distinct zeros x_1 and x_2 of the polynomial $ax^2 + bx + c$ (since we are given that $ax^2 + bx + c > 0$ for some x and the latter polynomial tends to $-\infty$ as $x \to \pm\infty$), hence $ax^2 + bx + c = a(x - x_1)(x - x_2)$. If $x_1 < x_2$, hen

$$\sqrt{(ax^2 + bx + c)} = \sqrt{(-a)}\,\sqrt{[(x - x_1)(x_2 - x)]} = \sqrt{(-a)}\,(x - x_1)\sqrt{\frac{x_2 - x}{x - x_1}}\,.$$

We now proceed as in case I, i.e. we employ the substitution

$$\frac{x_2 - x}{x - x_1} = z^2\,.$$

The given integral can often be transformed to the integral

$$\int \frac{dt}{\sqrt{(1 - t^2)}} \quad \text{or} \quad \int \sqrt{(1 - t^2)}\,dt\,.$$

The first integral is a standard integral and the second one is easily evaluated by the substitution $t = \sin z$ (Example 13.2.10, p. 492).

On some other methods for more complicated cases see e.g. [112].

Example 6. To evaluate

$$\int \frac{\mathrm{d}x}{\sqrt{(x^2 + k)}} \quad (k \neq 0\,,\ x^2 + k > 0)\,.$$

Making use of substitution (5) and equation (6), where $a = 1$, $b = 0$, $c = k$, we get

$$\int \frac{\mathrm{d}x}{\sqrt{(x^2 + k)}} = \int \frac{\mathrm{d}z}{z} = \ln |z| + C = \ln |\sqrt{(x^2 + k)} + x| + C\,.$$

See also § 13.1, formula 11 (p. 488).

IV. Integrals of the form

$$\int R(\mathrm{e}^{ax})\,\mathrm{d}x\,,$$

where $R(t)$ is a rational function of the variable t and a is a real constant, are rationalized by the substitution

$$\mathrm{e}^{ax} = z\,.$$

Integrals

$$\int R(\ln x)\frac{\mathrm{d}x}{x}\,, \quad x > 0\,,$$

where $R(t)$ is again a rational function, are rationalized by putting

$$\ln x = z\,.$$

V. Integrals of the type

$$\int R(\cos x, \sin x)\,\mathrm{d}x\,,$$

where $R(u, t)$ is a rational function of the variables u and t, can always be rationalized by the substitution

$$\tan \tfrac{1}{2}x = z\,.$$

From this equation it follows (making use of the relation $\cos^2 \frac{1}{2}x = 1/(1 + \tan^2 \frac{1}{2}x)$) that

$$\cos x = \frac{1 - z^2}{1 + z^2}\,, \quad \sin x = \frac{2z}{1 + z^2}\,, \quad \mathrm{d}x = \frac{2\,\mathrm{d}z}{1 + z^2}\,.$$

Example 7. The integral

$$\int \frac{1 - \sin x + \cos x}{5 \sin x \cos x} \, dx$$

is transformed by the above substitution into the integral

$$\int \frac{1 - \dfrac{2z}{1 + z^2} + \dfrac{1 - z^2}{1 + z^2}}{5 \dfrac{2z}{1 + z^2} \cdot \dfrac{1 - z^2}{1 + z^2}} \cdot \frac{2dz}{1 + z^2} = \frac{2}{5} \int \frac{1 - z}{z(1 - z^2)} \, dz = \frac{2}{5} \int \frac{dz}{z(1 + z)} .$$

Hence

$$\int \frac{1 - \sin x + \cos x}{5 \sin x \cos x} \, dx = \frac{2}{5} \int \frac{dz}{z(1 + z)} = \frac{2}{5} \int \left(\frac{1}{z} - \frac{1}{1 + z} \right) dz =$$

$$= \frac{2}{5} \ln \left| \frac{z}{1 + z} \right| + C = \frac{2}{5} \ln \left| \frac{\tan \frac{1}{2}x}{1 + \tan \frac{1}{2}x} \right| + C .$$

REMARK 3. One can often choose substitutions which are simpler than the substitution $\tan \frac{1}{2}x = z$. For example, the integral

$$\int \frac{1 + \cos^2 x}{\cos^4 x} \, dx$$

is transformed by the substitution

$$\tan x = z , \quad \text{from which we have} \quad dx/(\cos^2 x) = dz$$

into the integral (making use of the relation $\cos^2 x = 1/(1 + \tan^2 x)$)

$$\int (1 + z^2 + 1) \, dz = \int (2 + z^2) \, dz .$$

Hence

$$\int \frac{1 + \cos^2 x}{\cos^4 x} \, dx = \int (2 + z^2) \, dz = 2z + \frac{z^3}{3} + C = 2 \tan x + \frac{\tan^3 x}{3} + C .$$

In general if the function $R(\cos x, \sin x)$ is odd with respect to the function $\sin x$, i.e. if $R(\cos x, \sin x) = -R(\cos x, -\sin x)$, then it is possible to use the substitution $\cos x = z$ to rationalize the integral $\int R(\cos x, \sin x) \, dx$.

If $R(\cos x, \sin x)$ is odd with respect to $\cos x$, i.e. if $R(\cos x, \sin x) = -R(-\cos x, \sin x)$, the substitution $\sin x = z$ can be used.

If the function $R(\cos x, \sin x)$ is even with respect to both functions, i.e. if $R(\cos x, \sin x) = R(-\cos x, -\sin x)$, the substitution $\tan x = z$ may be employed (see the previous example).

Integrals of the type

$$\int \sin^n x \cos^m x \, dx \tag{7}$$

(where m, n are integers) are evaluated, if n or m is an odd number, by the substitution

$$\cos x = z \quad \text{or} \quad \sin x = z .$$

Example 8 (substitution $\sin x = z$).

$$\int \frac{dx}{\cos x} = \int \frac{\cos x \, dx}{\cos^2 x} = \int \frac{\cos x \, dx}{1 - \sin^2 x} = \int \frac{dz}{1 - z^2} = \frac{1}{2} \int \left(\frac{1}{1 - z} + \frac{1}{1 + z} \right) dz =$$

$$= \tfrac{1}{2} \ln \left| \frac{1 + z}{1 - z} \right| + C = \tfrac{1}{2} \ln \left| \frac{1 + \sin x}{1 - \sin x} \right| + C .$$

(By first transforming the integrand as indicated below this integral can be solved by the substitution $\tan(\frac{1}{4}\pi + \frac{1}{2}x) = z$. Thus:

$$\frac{1}{\cos x} = \frac{1}{\sin(\frac{1}{2}\pi + x)} = \frac{1}{2 \sin(\frac{1}{4}\pi + \frac{1}{2}x) \cos(\frac{1}{4}\pi + \frac{1}{2}x)} =$$

$$= \frac{1}{2 \tan(\frac{1}{4}\pi + \frac{1}{2}x) \cos^2(\frac{1}{4}\pi + \frac{1}{2}x)} ,$$

hence

$$\int \frac{dx}{\cos x} = \ln \left| \tan(\tfrac{1}{4}\pi + \tfrac{1}{2}x) \right| + C .$$

We can show that these two results are equivalent as follows:

$$\tfrac{1}{2} \ln \left| \frac{1 + \sin x}{1 - \sin x} \right| = \tfrac{1}{2} \ln \left| \frac{1 + 2 \sin \frac{1}{2}x \cos \frac{1}{2}x}{1 - 2 \sin \frac{1}{2}x \cos \frac{1}{2}x} \right| =$$

$$= \tfrac{1}{2} \ln \left| \frac{\dfrac{1}{\cos^2 \frac{1}{2}x} + 2 \tan \frac{1}{2}x}{\dfrac{1}{\cos^2 \frac{1}{2}x} - 2 \tan \frac{1}{2}x} \right| = \tfrac{1}{2} \ln \left| \frac{1 + \tan^2 \frac{1}{2}x + 2 \tan \frac{1}{2}x}{1 + \tan^2 \frac{1}{2}x - 2 \tan \frac{1}{2}x} \right| =$$

$$= \tfrac{1}{2} \ln \left(\frac{1 + \tan \frac{1}{2}x}{1 - \tan \frac{1}{2}x} \right)^2 = \ln \left| \frac{1 + \tan \frac{1}{2}x}{1 - \tan \frac{1}{2}x} \right| = \ln \left| \tan(\tfrac{1}{4}\pi + \tfrac{1}{2}x) \right| .)$$

Example 9 (substitution $\sin x = z$):

$$\int \sin^3 x \cos^5 x \, dx = \int z^3(1 - z^2)^2 \, dz = \frac{z^4}{4} - \frac{z^6}{3} + \frac{z^8}{8} + C =$$

$$= \frac{\sin^4 x}{4} - \frac{\sin^6 x}{3} + \frac{\sin^8 x}{8} + C .$$

If both m and n in (7) are even and non-negative, then making use of the formulae $\cos^2 x = \frac{1}{2}(1 + \cos 2x)$, $\sin^2 x = \frac{1}{2}(1 - \cos 2x)$ and if necessary by their further application ($\cos^2 2x = \frac{1}{2}(1 + \cos 4x)$, etc.) we can reduce the degree and hence reduce the integration to the previous case.

Example 10.

$$\int \sin^2 x \, dx = \frac{1}{2} \int (1 - \cos 2x) \, dx = \frac{1}{2}(x - \tfrac{1}{2}\sin 2x) + C =$$

$$= \tfrac{1}{2}(x - \sin x \cos x) + C .$$

Example 11.

$$\int \cos^4 x \, dx = \frac{1}{4} \int (1 + \cos 2x)^2 \, dx = \tfrac{1}{4}x + \tfrac{1}{4} \sin 2x + \tfrac{1}{8} \int (1 + \cos 4x) \, dx =$$

$$= \tfrac{3}{8}x + \tfrac{1}{4} \sin 2x + \tfrac{1}{32} \sin 4x + C .$$

13.5. Table of Indefinite Integrals

See, in particular, [48]. (For standard integrals see § 13.1, p. 487.) Constants of integration are omitted in the Table; m, n are integers, r stands for an arbitrary real number.

The range of validity is indicated only in non-trivial cases. (For example, if the expression $xX = x(ax + b)$ appears in the denominator, we do not draw the reader's attention to the fact that the range of validity is determined by the conditions $x \neq 0$, $ax + b \neq 0$; if the root of the expression $X = ax + b$ is considered, then, of course, the range of validity follows from the condition $ax + b \geqq 0$, i.e. $x \geqq -b/a$ for $a > 0$ and $x \leqq -b/a$ for $a < 0$, etc.)

If $a^2 + x^2$ appears in the integral, then $a > 0$ is assumed, since this case is most often met in applications. The case $a < 0$ is easily reduced to the previous one by putting $a = -b$, $b > 0$.

(a) ***Rational Functions*** (see also Remark 1, p. 548).

Notation: $X = ax + b \ (a \neq 0, b \neq 0)$.

1. $\int X^n \, dx = \frac{1}{a(n+1)} X^{n+1}$.

2. $\int \frac{dx}{X^n} = \frac{1}{a(1-n)} \cdot \frac{1}{X^{n-1}} \quad (n \neq 1)$.

3. $\int \frac{dx}{X} = \frac{1}{a} \ln |X|$.

4. $\int x X^n \, dx = \frac{1}{a^2(n+2)} X^{n+2} - \frac{b}{a^2(n+1)} X^{n+1}$.

5. $\int x^m X^n \, dx$ (see Remark 13.4.2, case (α), p. 502).

6. $\int \frac{x \, dx}{X} = \frac{x}{a} - \frac{b}{a^2} \ln |X|$.

7. $\int \frac{x \, dx}{X^2} = \frac{b}{a^2 X} + \frac{1}{a^2} \ln |X|$.

8. $\int \frac{x \, dx}{X^3} = \frac{1}{a^2} \left(-\frac{1}{X} + \frac{b}{2X^2} \right)$.

9. $\int \frac{x \, dx}{X^n} = \frac{1}{a^2} \left(\frac{-1}{(n-2) X^{n-2}} + \frac{b}{(n-1) X^{n-1}} \right) \quad (n \neq 1, n \neq 2)$.

10. $\int \frac{x^2 \, dx}{X} = \frac{1}{a^3} \left(\tfrac{1}{2} X^2 - 2bX + b^2 \ln |X| \right)$.

11. $\int \frac{x^2 \, dx}{X^2} = \frac{1}{a^3} \left(X - 2b \ln |X| - \frac{b^2}{X} \right)$.

12. $\int \frac{x^2 \, dx}{X^3} = \frac{1}{a^3} \left(\ln |X| + \frac{2b}{X} - \frac{b^2}{2X^2} \right)$.

13. $\int \frac{x^2 \, dx}{X^n} = \frac{1}{a^3} \left[\frac{-1}{(n-3) X^{n-3}} + \frac{2b}{(n-2) X^{n-2}} - \frac{b^2}{(n-1) X^{n-1}} \right]$
$(n \neq 1, \; n \neq 2, \; n \neq 3)$.

14. $\int \frac{dx}{xX} = -\frac{1}{b} \ln \left| \frac{X}{x} \right|$.

15. $\int \frac{dx}{xX^2} = -\frac{1}{b^2} \left(\ln \left| \frac{X}{x} \right| + \frac{ax}{X} \right)$.

16. $\displaystyle\int \frac{dx}{xX^3} = -\frac{1}{b^3}\left(\ln\left|\frac{X}{x}\right| + \frac{2ax}{X} - \frac{a^2x^2}{2X^2}\right).$

17. $\displaystyle\int \frac{dx}{xX^n} = -\frac{1}{b^n}\left[\ln\left|\frac{X}{x}\right| - \sum_{k=1}^{n-1}\binom{n-1}{k}\frac{(-a)^k x^k}{kX^k}\right] \quad (n > 1).$

18. $\displaystyle\int \frac{dx}{x^2X} = -\frac{1}{bx} + \frac{a}{b^2}\ln\left|\frac{X}{x}\right|.$

19. $\displaystyle\int \frac{dx}{x^2X^2} = -a\left[\frac{1}{b^2X} + \frac{1}{ab^2x} - \frac{2}{b^3}\ln\left|\frac{X}{x}\right|\right].$

20. $\displaystyle\int \frac{dx}{x^2X^3} = -a\left[\frac{1}{2b^2X^2} + \frac{2}{b^3X} + \frac{1}{ab^3x} - \frac{3}{b^4}\ln\left|\frac{X}{x}\right|\right].$

21. $\displaystyle\int \frac{dx}{x^2X^n} = \frac{1}{b^{n+1}}\left[an\ln\left|\frac{X}{x}\right| - \frac{X}{x} + \sum_{k=2}^{n}\binom{n}{k}\frac{(-a)^k x^{k-1}}{(k-1)X^{k-1}}\right] \quad (n \geqq 2).$

22. $\displaystyle\int \frac{dx}{x^mX^n} = -\frac{1}{b^{m+n-1}}\sum_{k=0}^{m+n-2}\binom{m+n-2}{k}\frac{(-a)^k X^{m-k-1}}{(m-k-1)x^{m-k-1}};$

the term for which $m - k - 1 = 0$ is to be replaced in the sum by the term

$$\binom{m+n-2}{n-1}(-a)^{m-1}\ln\left|\frac{X}{x}\right|.$$

Notation: $\Delta = bf - ag$.

23. $\displaystyle\int \frac{ax+b}{fx+g}\,dx = \frac{ax}{f} + \frac{\Delta}{f^2}\ln|fx+g| \quad (f \neq 0).$

24. $\displaystyle\int \frac{dx}{(ax+b)(fx+g)} = \frac{1}{\Delta}\ln\left|\frac{fx+g}{ax+b}\right| \quad (\Delta \neq 0).$

25. $\displaystyle\int \frac{x\,dx}{(ax+b)(fx+g)} = \frac{1}{\Delta}\left(\frac{b}{a}\ln|ax+b| - \frac{g}{f}\ln|fx+g|\right)$

$(a \neq 0,\ f \neq 0,\ \Delta \neq 0).$

26. $\displaystyle\int \frac{dx}{(ax+b)^2(fx+g)} = \frac{1}{\Delta}\left(\frac{1}{ax+b} + \frac{f}{\Delta}\ln\left|\frac{fx+g}{ax+b}\right|\right) \quad (\Delta \neq 0).$

27. $\displaystyle\int \frac{x\,dx}{(a+x)(b+x)^2} = \frac{b}{(a-b)(b+x)} - \frac{a}{(a-b)^2}\ln\left|\frac{a+x}{b+x}\right| \quad (a \neq b).$

28. $$\int \frac{x^2\,dx}{(a+x)(b+x)^2} =$$
$$= \frac{b^2}{(b-a)(b+x)} + \frac{a^2}{(b-a)^2}\ln|a+x| + \frac{b^2-2ab}{(b-a)^2}\ln|b+x| \quad (a \neq b).$$

29. $$\int \frac{dx}{(a+x)^2(b+x)^2} = \frac{-1}{(a-b)^2}\left(\frac{1}{a+x}+\frac{1}{b+x}\right) + \frac{2}{(a-b)^3}\ln\left|\frac{a+x}{b+x}\right| \quad (a \neq b).$$

30. $$\int \frac{dx}{(a+x)^m(b+x)^n} = \int \frac{dz}{z^m(z+c)^n} \quad (\text{see 22; } z = x+a,\ c = b-a,\ a \neq b).$$

31. $$\int \frac{x\,dx}{(a+x)^2(b+x)^2} =$$
$$= \frac{1}{(a-b)^2}\left(\frac{a}{a+x}+\frac{b}{b+x}\right) + \frac{a+b}{(a-b)^3}\ln\left|\frac{a+x}{b+x}\right| \quad (a \neq b).$$

32. $$\int \frac{x^2\,dx}{(a+x)^2(b+x)^2} =$$
$$= \frac{-1}{(a-b)^2}\left(\frac{a^2}{a+x}+\frac{b^2}{b+x}\right) + \frac{2ab}{(a-b)^3}\ln\left|\frac{a+x}{b+x}\right| \quad (a \neq b).$$

Notation: $X = ax^2 + bx + c$, $\Delta = 4ac - b^2$ $(a \neq 0, \Delta \neq 0)$.

33. $$\int \frac{dx}{X} = \frac{2}{\sqrt{\Delta}}\arctan\frac{2ax+b}{\sqrt{\Delta}} \quad (\Delta > 0),$$
$$= \frac{1}{\sqrt{(-\Delta)}}\ln\left|\frac{2ax+b-\sqrt{(-\Delta)}}{2ax+b+\sqrt{(-\Delta)}}\right| \quad (\Delta < 0).$$

34. $$\int \frac{dx}{X^2} = \frac{2ax+b}{\Delta X} + \frac{2a}{\Delta}\int\frac{dx}{X} \quad (\text{see 33}).$$

35. $$\int \frac{dx}{X^3} = \frac{2ax+b}{\Delta}\left(\frac{1}{2X^2}+\frac{3a}{\Delta X}\right) + \frac{6a^2}{\Delta^2}\frac{dx}{X} \quad (\text{see 33}).$$

36. $$\int \frac{dx}{X^n} = \frac{2ax+b}{(n-1)\Delta X^{n-1}} + \frac{(2n-3)\,2a}{(n-1)\Delta}\int\frac{dx}{X^{n-1}} \quad (n > 1).$$

37. $\displaystyle\int \frac{x\,dx}{X} = \frac{1}{2a}\ln|X| - \frac{b}{2a}\int\frac{dx}{X}$ (see 33).

38. $\displaystyle\int \frac{x\,dx}{X^2} = -\frac{bx+2c}{\Delta X} - \frac{b}{\Delta}\int\frac{dx}{X}$ (see 33).

39. $\displaystyle\int \frac{x\,dx}{X^n} = -\frac{bx+2c}{(n-1)\,\Delta X^{n-1}} - \frac{b(2n-3)}{(n-1)\,\Delta}\int\frac{dx}{X^{n-1}}$ $(n > 1)$.

40. $\displaystyle\int \frac{x^2\,dx}{X} = \frac{x}{a} - \frac{b}{2a^2}\ln|X| + \frac{b^2-2ac}{2a^2}\int\frac{dx}{X}$ (see 33).

41. $\displaystyle\int \frac{x^2\,dx}{X^2} = \frac{(b^2-2ac)\,x + bc}{a\,\Delta X} + \frac{2c}{\Delta}\int\frac{dx}{X}$ (see 33).

42. $\displaystyle\int \frac{x^2\,dx}{X^n} = \frac{-x}{(2n-3)\,aX^{n-1}} + \frac{c}{(2n-3)\,a}\int\frac{dx}{X^n} - \frac{(n-2)\,b}{(2n-3)\,a}\int\frac{x\,dx}{X^n}$

(see 36 and 39).

43. $\displaystyle\int \frac{x^m\,dx}{X^n} = -\frac{x^{m-1}}{(2n-m-1)\,aX^{n-1}} + \frac{(m-1)\,c}{(2n-m-1)\,a}\int\frac{x^{m-2}\,dx}{X^n} -$

$\displaystyle -\frac{(n-m)\,b}{(2n-m-1)\,a}\int\frac{x^{m-1}\,dx}{X^n}$ $(m \neq 2n-1;$ for $m = 2n-1$ see 44).

44. $\displaystyle\int \frac{x^{2n-1}\,dx}{X^n} = \frac{1}{a}\int\frac{x^{2n-3}\,dx}{X^{n-1}} - \frac{c}{a}\int\frac{x^{2n-3}\,dx}{X^n} - \frac{b}{a}\int\frac{x^{2n-2}\,dx}{X^n}$

$(n > 1;$ for $n = 1$ see 37).

45. $\displaystyle\int \frac{dx}{xX} = \frac{1}{2c}\ln\frac{x^2}{|X|} - \frac{b}{2c}\int\frac{dx}{X}$ (see 33; $c \neq 0$).

46. $\displaystyle\int \frac{dx}{xX^n} = \frac{1}{2c(n-1)\,X^{n-1}} - \frac{b}{2c}\int\frac{dx}{X^n} + \frac{1}{c}\int\frac{dx}{xX^{n-1}}$ $(c \neq 0, n > 1)$.

47. $\displaystyle\int \frac{dx}{x^2X} = \frac{b}{2c^2}\ln\frac{|X|}{x^2} - \frac{1}{cx} + \left(\frac{b^2}{2c^2} - \frac{a}{c}\right)\int\frac{dx}{X}$ (see 33; $c \neq 0$).

48. $\displaystyle\int \frac{dx}{x^mX^n} = -\frac{1}{(m-1)\,cx^{m-1}X^{n-1}} - \frac{(2n+m-3)\,a}{(m-1)\,c}\int\frac{dx}{x^{m-2}X^n} -$

$\displaystyle -\frac{(n+m-2)\,b}{(m-1)\,c}\int\frac{dx}{x^{m-1}X^n}$ $(c \neq 0, m > 1)$.

49. $$\int \frac{dx}{(fx+g)X} = \frac{f}{2(cf^2 - gbf + g^2a)} \ln \frac{(fx+g)^2}{|X|} + \frac{2ga - bf}{2(cf^2 - gbf + g^2a)} \int \frac{dx}{X} \quad (\text{see 33; } cf^2 - gbf + g^2a \neq 0).$$

Notation: $X = a^2 \pm x^2 \quad (a > 0)$,

$Y = \arctan \frac{x}{a}$ for the positive sign,

$$Y = \tfrac{1}{2} \ln \left|\frac{x+a}{x-a}\right| = \begin{cases} \operatorname{artanh}(x/a) & \text{for the negative sign and for } |x| < a, \\ \operatorname{arcoth}(x/a) & \text{for the negative sign and for } |x| > a. \end{cases}$$

If both signs occur in a formula, then the upper or lower sign corresponds to the case $X = a^2 + x^2$, or $X = a^2 - x^2$, respectively.

50. $$\int \frac{dx}{X} = \frac{1}{a} Y.$$

51. $$\int \frac{dx}{X^2} = \frac{x}{2a^2X} + \frac{1}{2a^3} Y.$$

52. $$\int \frac{dx}{X^3} = \frac{x}{4a^2X^2} + \frac{3x}{8a^4X} + \frac{3}{8a^5} Y.$$

53. $$\int \frac{dx}{X^{n+1}} = \frac{x}{2na^2X^n} + \frac{2n-1}{2na^2} \int \frac{dx}{X^n} \quad (n \neq 0).$$

54. $$\int \frac{x\,dx}{X} = \pm\tfrac{1}{2} \ln |X|.$$

55. $$\int \frac{x\,dx}{X^2} = \mp \frac{1}{2X}.$$

56. $$\int \frac{x\,dx}{X^3} = \mp \frac{1}{4X^2}.$$

57. $$\int \frac{x\,dx}{X^{n+1}} = \mp \frac{1}{2nX^n} \quad (n \neq 0).$$

58. $$\int \frac{x^2\,dx}{X} = \pm x \mp aY.$$

59. $$\int \frac{x^2\,dx}{X^2} = \mp \frac{x}{2X} \pm \frac{1}{2a} Y.$$

60. $$\int \frac{x^2\,dx}{X^3} = \mp \frac{x}{4X^2} \pm \frac{x}{8a^2X} \pm \frac{1}{8a^3} Y.$$

61. $$\int \frac{x^2\,dx}{X^{n+1}} = \mp \frac{x}{2nX^n} \pm \frac{1}{2n}\int \frac{dx}{X^n} \quad (n \neq 0).$$

62. $$\int \frac{dx}{xX} = \frac{1}{2a^2} \ln \frac{x^2}{|X|}.$$

63. $$\int \frac{dx}{xX^2} = \frac{1}{2a^2X} + \frac{1}{2a^4} \ln \frac{x^2}{|X|}.$$

64. $$\int \frac{dx}{xX^3} = \frac{1}{4a^2X^2} + \frac{1}{2a^4X} + \frac{1}{2a^6} \ln \frac{x^2}{|X|}.$$

65. $$\int \frac{dx}{x^2X} = -\frac{1}{a^2x} \mp \frac{1}{a^3} Y.$$

66. $$\int \frac{dx}{x^2X^2} = -\frac{1}{a^4x} \mp \frac{x}{2a^4X} \mp \frac{3}{2a^5} Y.$$

67. $$\int \frac{dx}{x^2X^3} = -\frac{1}{a^6x} \mp \frac{x}{4a^4X^2} \mp \frac{7x}{8a^6X} \mp \frac{15}{8a^7} Y.$$

68. $$\int \frac{dx}{(b+cx)X} = \frac{1}{a^2c^2 \pm b^2}\left[c \ln|b+cx| - \frac{c}{2}\ln|X| \pm \frac{b}{a} Y\right] \quad (a^2c^2 \pm b^2 \neq 0).$$

Notation: $X = a^3 \pm x^3$ $(a \neq 0)$.
If both signs occur in a formula, then the upper or lower sign corresponds to the case $X = a^3 + x^3$, or $X = a^3 - x^3$, respectively.

69. $$\int \frac{dx}{X} = \pm \frac{1}{6a^2} \ln \frac{(a \pm x)^2}{a^2 \mp ax + x^2} + \frac{1}{a^2\sqrt{3}} \arctan \frac{2x \mp a}{a\sqrt{3}}.$$

70. $$\int \frac{dx}{X^2} = \frac{x}{3a^3X} + \frac{2}{3a^3}\int \frac{dx}{X} \quad \text{(see 69)}.$$

71. $$\int \frac{x\,dx}{X} = \frac{1}{6a} \ln \frac{a^2 \mp ax + x^2}{(a \pm x)^2} \pm \frac{1}{a\sqrt{3}} \arctan \frac{2x \mp a}{a\sqrt{3}}.$$

72. $\displaystyle\int \frac{x\,dx}{X^2} = \frac{x^2}{3a^3X} + \frac{1}{3a^3}\int \frac{x\,dx}{X}$ (see 71).

73. $\displaystyle\int \frac{x^2\,dx}{X} = \pm \tfrac{1}{3} \ln |X|$.

74. $\displaystyle\int \frac{x^2\,dx}{X^2} = \mp \frac{1}{3X}$.

75. $\displaystyle\int \frac{dx}{xX} = \frac{1}{3a^3} \ln \left|\frac{x^3}{X}\right|$.

76. $\displaystyle\int \frac{dx}{xX^2} = \frac{1}{3a^3X} + \frac{1}{3a^6} \ln \left|\frac{x^3}{X}\right|$.

77. $\displaystyle\int \frac{dx}{x^2X} = -\frac{1}{a^3x} \mp \frac{1}{a^3}\int \frac{x\,dx}{X}$ (see 71).

78. $\displaystyle\int \frac{dx}{x^2X^2} = -\frac{1}{a^6x} \mp \frac{x^2}{3a^6X} \mp \frac{4}{3a^6}\int \frac{x\,dx}{X}$ (see 71).

79. $\displaystyle\int \frac{dx}{a^4 + x^4} = \frac{1}{4a^3\sqrt{2}} \ln \frac{x^2 + ax\sqrt{(2)} + a^2}{x^2 - ax\sqrt{(2)} + a^2} + \frac{1}{2a^3\sqrt{2}} \arctan \frac{ax\sqrt{2}}{a^2 - x^2}$ $(a \neq 0)$.

80. $\displaystyle\int \frac{x\,dx}{a^4 + x^4} = \frac{1}{2a^2} \arctan \frac{x^2}{a^2}$ $(a \neq 0)$.

81. $\displaystyle\int \frac{x^2\,dx}{a^4 + x^4} = -\frac{1}{4a\sqrt{2}} \ln \frac{x^2 + ax\sqrt{(2)} + a^2}{x^2 - ax\sqrt{(2)} + a^2} + \frac{1}{2a\sqrt{2}} \arctan \frac{ax\sqrt{2}}{a^2 - x^2}$ $(a \neq 0)$.

82. $\displaystyle\int \frac{x^3\,dx}{a^4 + x^4} = \tfrac{1}{4} \ln |a^4 + x^4|$.

83. $\displaystyle\int \frac{dx}{a^4 - x^4} = \frac{1}{4a^3} \ln \left|\frac{a + x}{a - x}\right| + \frac{1}{2a^3} \arctan \frac{x}{a}$ $(a \neq 0)$.

84. $\displaystyle\int \frac{x\,dx}{a^4 - x^4} = \frac{1}{4a^2} \ln \left|\frac{a^2 + x^2}{a^2 - x^2}\right|$ $(a \neq 0)$.

85. $\displaystyle\int \frac{x^2\,dx}{a^4 - x^4} = \frac{1}{4a} \ln \left|\frac{a + x}{a - x}\right| - \frac{1}{2a} \arctan \frac{x}{a}$ $(a \neq 0)$.

86. $\displaystyle\int \frac{x^3\,dx}{a^4 - x^4} = -\tfrac{1}{4}\ln|a^4 - x^4|\,.$

(b) ***Irrational Functions.***

Notation: $X = a^2 \pm b^2x \; (a > 0, b > 0)$,

$Y = \arctan \dfrac{b\sqrt{x}}{a}$ for the positive sign,

$Y = \frac{1}{2}\ln\left|\dfrac{a + b\sqrt{x}}{a - b\sqrt{x}}\right|$ for the negative sign.

The upper or lower sign in the formulae corresponds to the case $X = a^2 + b^2x$, or $X = a^2 - b^2x$, respectively.

87. $\displaystyle\int \frac{\sqrt{(x)}\,dx}{X} = \pm\frac{2\sqrt{x}}{b^2} \mp \frac{2a}{b^3}\,Y\,.$

88. $\displaystyle\int \frac{\sqrt{(x)}\,dx}{X^2} = \mp\frac{\sqrt{x}}{b^2X} \pm \frac{1}{ab^3}\,Y\,.$

89. $\displaystyle\int \frac{dx}{X\sqrt{x}} = \frac{2}{ab}\,Y\,.$

90. $\displaystyle\int \frac{dx}{X^2\sqrt{x}} = \frac{\sqrt{x}}{a^2X} + \frac{1}{a^3b}\,Y\,.$

91. $\displaystyle\int \frac{\sqrt{(x)}\,dx}{a^4 + x^2} = -\frac{1}{2a\sqrt{2}}\ln\frac{x + a\sqrt{(2x)} + a^2}{x - a\sqrt{(2x)} + a^2} + \frac{1}{a\sqrt{2}}\arctan\frac{a\sqrt{(2x)}}{a^2 - x}$

$(a \neq 0)\,.$

92. $\displaystyle\int \frac{dx}{(a^4 + x^2)\sqrt{x}} = \frac{1}{2a^3\sqrt{2}}\ln\frac{x + a\sqrt{(2x)} + a^2}{x - a\sqrt{(2x)} + a^2} + \frac{1}{a^3\sqrt{2}}\arctan\frac{a\sqrt{(2x)}}{a^2 - x}$

$(a \neq 0)\,.$

93. $\displaystyle\int \frac{\sqrt{(x)}\,dx}{a^4 - x^2} = \frac{1}{2a}\ln\left|\frac{a + \sqrt{x}}{a - \sqrt{x}}\right| - \frac{1}{a}\arctan\frac{\sqrt{x}}{a} \quad (a \neq 0)\,.$

94. $\displaystyle\int \frac{dx}{(a^4 - x^2)\sqrt{x}} = \frac{1}{2a^3}\ln\left|\frac{a + \sqrt{x}}{a - \sqrt{x}}\right| + \frac{1}{a^3}\arctan\frac{\sqrt{x}}{a} \quad (a \neq 0)\,.$

Notation: $X = ax + b$ $(a \neq 0,\ b \neq 0)$.

95. $$\int X^r \,\mathrm{d}x = \frac{1}{(r+1)\,a} X^{r+1} \quad (r \neq -1;\ \text{for } r = -1 \text{ see } 3).$$

96. $$\int \sqrt{(X)} \,\mathrm{d}x = \frac{2}{3a} \sqrt{X^3}.$$

97. $$\int x \sqrt{(X)} \,\mathrm{d}x = \frac{2(3ax - 2b)\sqrt{X^3}}{15a^2}.$$

98. $$\int x^2 \sqrt{(X)} \,\mathrm{d}x = \frac{2(15a^2x^2 - 12abx + 8b^2)\sqrt{X^3}}{105a^3}.$$

99. $$\int \frac{\mathrm{d}x}{\sqrt{X}} = \frac{2\sqrt{X}}{a}.$$

100. $$\int \frac{x \,\mathrm{d}x}{\sqrt{X}} = \frac{2(ax - 2b)}{3a^2} \sqrt{X}.$$

101. $$\int \frac{x^2 \,\mathrm{d}x}{\sqrt{X}} = \frac{2(3a^2x^2 - 4abx + 8b^2)\sqrt{X}}{15a^3}.$$

102. $$\int \frac{\mathrm{d}x}{x\sqrt{X}} = \begin{cases} \dfrac{1}{\sqrt{b}} \ln \left| \dfrac{\sqrt{X} - \sqrt{b}}{\sqrt{X} + \sqrt{b}} \right| & \text{for } b > 0, \\[2ex] \dfrac{2}{\sqrt{(-b)}} \arctan \sqrt{\dfrac{X}{-b}} & \text{for } b < 0. \end{cases}$$

103. $$\int \frac{\sqrt{X}}{x} \,\mathrm{d}x = 2\sqrt{X} + b \int \frac{\mathrm{d}x}{x\sqrt{X}} \quad (\text{see } 102).$$

104. $$\int \frac{\mathrm{d}x}{x^2\sqrt{X}} = -\frac{\sqrt{X}}{bx} - \frac{a}{2b} \int \frac{\mathrm{d}x}{x\sqrt{X}} \quad (\text{see } 102).$$

105. $$\int \frac{\sqrt{X}}{x^2} \,\mathrm{d}x = -\frac{\sqrt{X}}{x} + \frac{a}{2} \int \frac{\mathrm{d}x}{x\sqrt{X}} \quad (\text{see } 102).$$

106. $$\int \frac{\mathrm{d}x}{x^n\sqrt{X}} = -\frac{\sqrt{X}}{(n-1)\,bx^{n-1}} - \frac{(2n-3)\,a}{(2n-2)\,b} \int \frac{\mathrm{d}x}{x^{n-1}\sqrt{X}} \quad (n > 1).$$

107. $$\int \sqrt{(X^3)}\,\mathrm{d}x = \frac{2\sqrt{X^5}}{5a}.$$

108. $$\int x\sqrt{(X^3)}\,\mathrm{d}x = \frac{2}{35a^2}\left[5\sqrt{(X^7)} - 7b\sqrt{(X^5)}\right].$$

109. $$\int x^2\sqrt{(X^3)}\,\mathrm{d}x = \frac{2}{a^3}\left(\frac{\sqrt{X^9}}{9} - \frac{2b\sqrt{X^7}}{7} + \frac{b^2\sqrt{X^5}}{5}\right).$$

110. $$\int \frac{\sqrt{X^3}}{x}\,\mathrm{d}x = \frac{2\sqrt{X^3}}{3} + 2b\sqrt{(X)} + b^2\int\frac{\mathrm{d}x}{x\sqrt{X}} \quad \text{(see 102)}.$$

111. $$\int \frac{x\,\mathrm{d}x}{\sqrt{X^3}} = \frac{2}{a^2}\left(\sqrt{(X)} + \frac{b}{\sqrt{X}}\right).$$

112. $$\int \frac{x^2\,\mathrm{d}x}{\sqrt{X^3}} = \frac{2}{a^3}\left(\frac{\sqrt{X^3}}{3} - 2b\sqrt{X} - \frac{b^2}{\sqrt{X}}\right).$$

113. $$\int \frac{\mathrm{d}x}{x\sqrt{X^3}} = \frac{2}{b\sqrt{X}} + \frac{1}{b}\int\frac{\mathrm{d}x}{x\sqrt{X}} \quad \text{(see 102)}.$$

114. $$\int \frac{\mathrm{d}x}{x^2\sqrt{X^3}} = -\frac{1}{bx\sqrt{X}} - \frac{3a}{b^2\sqrt{X}} - \frac{3a}{2b^2}\int\frac{\mathrm{d}x}{x\sqrt{X}} \quad \text{(see 102)}.$$

115. $$\int X^{\pm n/2}\,\mathrm{d}x = \frac{2X^{\frac{1}{2}(2\pm n)}}{a(2\pm n)} \quad (n \pm 2 \neq 0).$$

116. $$\int xX^{\pm n/2}\,\mathrm{d}x = \frac{2}{a^2}\left(\frac{X^{\frac{1}{2}(4\pm n)}}{4\pm n} - \frac{bX^{\frac{1}{2}(2\pm n)}}{2\pm n}\right) \quad (n \pm 2 \neq 0,\ n \pm 4 \neq 0).$$

117. $$\int x^2X^{\pm n/2}\,\mathrm{d}x = \frac{2}{a^3}\left(\frac{X^{\frac{1}{2}(6\pm n)}}{6\pm n} - \frac{2bX^{\frac{1}{2}(4\pm n)}}{4\pm n} + \frac{b^2X^{\frac{1}{2}(2\pm n)}}{2\pm n}\right)$$
$$(n \pm 2 \neq 0,\ n \pm 4 \neq 0,\ n \pm 6 \neq 0).$$

118. $$\int \frac{X^{n/2}\,\mathrm{d}x}{x} = \frac{2X^{n/2}}{n} + b\int\frac{X^{\frac{1}{2}(n-2)}}{x}\,\mathrm{d}x \quad (n \neq 0).$$

119. $$\int \frac{\mathrm{d}x}{xX^{n/2}} = \frac{2}{(n-2)\,bX^{\frac{1}{2}(n-2)}} + \frac{1}{b}\int\frac{\mathrm{d}x}{xX^{\frac{1}{2}(n-2)}} \quad (n \neq 2).$$

120. $$\int \frac{\mathrm{d}x}{x^2X^{n/2}} = -\frac{1}{bxX^{\frac{1}{2}(n-2)}} - \frac{na}{2b}\int\frac{\mathrm{d}x}{xX^{n/2}}.$$

Notation: $X = ax + b,\ Y = fx + g,\ \Delta = bf - ag$ $(a \neq 0,\ f \neq 0,\ \Delta \neq 0)$.

121. $$\int \frac{dx}{\sqrt{(XY)}} \begin{cases} = \dfrac{2}{\sqrt{(-af)}} \arctan \sqrt{-\dfrac{fX}{aY}} & \text{for } af < 0 \quad (aY > 0), \\ = \dfrac{2}{\sqrt{(af)}} \operatorname{artanh} \sqrt{\dfrac{fX}{aY}} = \dfrac{2}{\sqrt{(af)}} \ln |\sqrt{(aY)} + \sqrt{(fX)}| \\ \text{for } af > 0 \quad (aY > 0). \end{cases}$$

122. $$\int \frac{x\,dx}{\sqrt{(XY)}} = \frac{\sqrt{(XY)}}{af} - \frac{ag + bf}{2af} \int \frac{dx}{\sqrt{(XY)}} \quad \text{(see 121)}.$$

123. $$\int \frac{dx}{\sqrt{(X)}\sqrt{Y^3}} = -\frac{2\sqrt{X}}{\Delta\sqrt{Y}}.$$

124. $$\frac{dx}{Y\sqrt{X}} = \begin{cases} \dfrac{2}{\sqrt{(-\Delta f)}} \arctan \dfrac{f\sqrt{X}}{\sqrt{(-\Delta f)}} & \text{for } \Delta f < 0, \\ \dfrac{1}{\sqrt{(\Delta f)}} \ln \left| \dfrac{f\sqrt{(X)} - \sqrt{(\Delta f)}}{f\sqrt{(X)} + \sqrt{(\Delta f)}} \right| & \text{for } \Delta f > 0. \end{cases}$$

125. $$\int \sqrt{(XY)}\,dx = \frac{\Delta + 2aY}{4af} \sqrt{(XY)} - \frac{\Delta^2}{8af} \int \frac{dx}{\sqrt{(XY)}} \quad \text{(see 121)}.$$

126. $$\int \sqrt{\left(\frac{Y}{X}\right)}\,dx = \frac{1}{a} \sqrt{(XY)} - \frac{\Delta}{2a} \int \frac{dx}{\sqrt{(XY)}} \quad \text{(see 121; } Y > 0\text{)}.$$

127. $$\int \frac{\sqrt{(X)}\,dx}{Y} = \frac{2\sqrt{X}}{f} + \frac{\Delta}{f} \int \frac{dx}{Y\sqrt{X}} \quad \text{(see 124)}.$$

128. $$\int \frac{Y^n\,dx}{\sqrt{X}} = \frac{2}{(2n+1)a} \left(\sqrt{(X)}\,Y^n - n\Delta \int \frac{Y^{n-1}\,dx}{\sqrt{X}} \right).$$

129. $$\int \frac{dx}{\sqrt{(X)}\,Y^n} = -\frac{1}{(n-1)\Delta} \left\{ \frac{\sqrt{X}}{Y^{n-1}} + \left(n - \frac{3}{2}\right) a \int \frac{dx}{\sqrt{(X)}\,Y^{n-1}} \right\} \quad (n > 1).$$

130. $$\int \sqrt{(X)}\,Y^n\,dx = \frac{1}{(2n+3)f} \left(2\sqrt{(X)}\,Y^{n+1} + \Delta \int \frac{Y^n\,dx}{\sqrt{X}} \right) \quad \text{(see 128)}.$$

131. $$\int \frac{\sqrt{(X)}\,dx}{Y^n} = \frac{1}{(n-1)f} \left(-\frac{\sqrt{X}}{Y^{n-1}} + \frac{a}{2} \int \frac{dx}{\sqrt{(X)}\,Y^{n-1}} \right) \quad (n > 1).$$

Notation: $X = a^2 - x^2 \ (a > 0)$.

132. $\displaystyle\int \sqrt{(X)}\,dx = \frac{1}{2}\left(x\sqrt{(X)} + a^2 \arcsin\frac{x}{a}\right).$

133. $\displaystyle\int x\sqrt{(X)}\,dx = -\tfrac{1}{3}\sqrt{X^3}.$

134. $\displaystyle\int x^2\sqrt{(X)}\,dx = -\frac{x}{4}\sqrt{(X^3)} + \frac{a^2}{8}\left(x\sqrt{(X)} + a^2 \arcsin\frac{x}{a}\right).$

135. $\displaystyle\int \frac{\sqrt{X}}{x}\,dx = \sqrt{(X)} - a\ln\left|\frac{a+\sqrt{X}}{x}\right|.$

136. $\displaystyle\int \frac{\sqrt{X}}{x^2}\,dx = -\frac{\sqrt{X}}{x} - \arcsin\frac{x}{a}.$

137. $\displaystyle\int \frac{dx}{\sqrt{X}} = \arcsin\frac{x}{a}.$

138. $\displaystyle\int \frac{x\,dx}{\sqrt{X}} = -\sqrt{X}.$

139. $\displaystyle\int \frac{x^2\,dx}{\sqrt{X}} = -\frac{x}{2}\sqrt{(X)} + \frac{a^2}{2}\arcsin\frac{x}{a}.$

140. $\displaystyle\int \frac{dx}{x\sqrt{X}} = -\frac{1}{a}\ln\left|\frac{a+\sqrt{X}}{x}\right|.$

141. $\displaystyle\int \frac{dx}{x^2\sqrt{X}} = -\frac{\sqrt{X}}{a^2 x}.$

142. $\displaystyle\int \sqrt{(X^3)}\,dx = \frac{1}{4}\left(x\sqrt{(X^3)} + \frac{3a^2 x}{2}\sqrt{(X)} + \frac{3a^4}{2}\arcsin\frac{x}{a}\right).$

143. $\displaystyle\int x\sqrt{(X^3)}\,dx = -\tfrac{1}{5}\sqrt{X^5}.$

144. $\displaystyle\int x^2\sqrt{(X^3)}\,dx = -\frac{x\sqrt{X^5}}{6} + \frac{a^2 x\sqrt{X^3}}{24} + \frac{a^4 x\sqrt{X}}{16} + \frac{a^6}{16}\arcsin\frac{x}{a}.$

145. $\displaystyle\int \frac{\sqrt{X^3}}{x}\,dx = \frac{\sqrt{X^3}}{3} + a^2\sqrt{(X)} - a^3\ln\left|\frac{a+\sqrt{X}}{x}\right|.$

146. $\int \frac{\sqrt{X^3}}{x^2} \, dx = -\frac{\sqrt{X^3}}{x} - \frac{3}{2} x \sqrt{X} - \frac{3}{2} a^2 \arcsin \frac{x}{a}$

147. $\int \frac{dx}{\sqrt{X^3}} = \frac{x}{a^2 \sqrt{X}}.$

148. $\int \frac{x \, dx}{\sqrt{X^3}} = \frac{1}{\sqrt{X}}.$

149. $\int \frac{x^2 \, dx}{\sqrt{X^3}} = \frac{x}{\sqrt{X}} - \arcsin \frac{x}{a}.$

150. $\int \frac{dx}{x \sqrt{X^3}} = \frac{1}{a^2 \sqrt{X}} - \frac{1}{a^3} \ln \left| \frac{a + \sqrt{X}}{x} \right|.$

151. $\int \frac{dx}{x^2 \sqrt{X^3}} = \frac{1}{a^4} \left(-\frac{\sqrt{X}}{x} + \frac{x}{\sqrt{X}} \right).$

Notation: $X = a^2 + x^2, \; a > 0.$

152. $\int \sqrt{(X)} \, dx = \frac{1}{2}\left(x \sqrt{(X)} + a^2 \operatorname{arsinh} \frac{x}{a} \right) + C =$

$= \frac{1}{2}(x \sqrt{(X)} + a^2 \ln |x + \sqrt{X}|) + C_1.$

153. $\int x \sqrt{(X)} \, dx = \frac{1}{3} \sqrt{X^3}.$

154. $\int x^2 \sqrt{(X)} \, dx = \frac{x}{4} \sqrt{(X^3)} - \frac{a^2}{8} \left(x \sqrt{(X)} + a^2 \operatorname{arsinh} \frac{x}{a} \right) + C =$

$= \frac{x}{4} \sqrt{(X^3)} - \frac{a^2}{8} (x \sqrt{(X)} + a^2 \ln |x + \sqrt{X}|) + C_1.$

155. $\int x^3 \sqrt{(X)} \, dx = \frac{\sqrt{X^5}}{5} - \frac{a^2 \sqrt{X^3}}{3}.$

156. $\int \frac{\sqrt{X}}{x} \, dx = \sqrt{(X)} - a \ln \left| \frac{a + \sqrt{X}}{x} \right|.$

157. $\int \frac{\sqrt{X}}{x^2} \, dx = -\frac{\sqrt{X}}{x} + \operatorname{arsinh} \frac{x}{a} + C = -\frac{\sqrt{X}}{x} + \ln |x + \sqrt{X}| + C_1.$

158. $\int \frac{\sqrt{X}}{x^3} \, dx = -\frac{\sqrt{X}}{2x^2} - \frac{1}{2a} \ln \left| \frac{a + \sqrt{X}}{x} \right|.$

159. $\displaystyle\int \frac{dx}{\sqrt{X}} = \operatorname{arsinh} \frac{x}{a} + C = \ln |x + \sqrt{X}| + C_1 .$

160. $\displaystyle\int \frac{x\, dx}{\sqrt{X}} = \sqrt{X} .$

161. $\displaystyle\int \frac{x^2\, dx}{\sqrt{X}} = \frac{x}{2} \sqrt{(X)} - \frac{a^2}{2} \operatorname{arsinh} \frac{x}{a} + C = \frac{x}{2} \sqrt{(X)} - \frac{a^2}{2} \ln |x + \sqrt{X}| + C_1 .$

162. $\displaystyle\int \frac{x^3\, dx}{\sqrt{X}} = \frac{\sqrt{(X^3)}}{3} - a^2 \sqrt{X} .$

163. $\displaystyle\int \frac{dx}{x \sqrt{X}} = - \frac{1}{a} \ln \left| \frac{a + \sqrt{X}}{x} \right| .$

164. $\displaystyle\int \frac{dx}{x^2 \sqrt{X}} = - \frac{\sqrt{X}}{a^2 x} .$

165. $\displaystyle\int \frac{dx}{x^3 \sqrt{X}} = - \frac{\sqrt{X}}{2a^2 x^2} + \frac{1}{2a^3} \ln \left| \frac{a + \sqrt{X}}{x} \right| .$

166. $$\int \sqrt{(X^3)}\, dx = \frac{1}{4} \left(x \sqrt{(X^3)} + \frac{3a^2 x}{2} \sqrt{(X)} + \frac{3a^4}{2} \operatorname{arsinh} \frac{x}{a} \right) + C =$$
$$= \frac{1}{4} \left(x \sqrt{(X^3)} + \frac{3a^2 x}{2} \sqrt{(X)} + \frac{3a^4}{2} \ln |x + \sqrt{X}| \right) + C_1 .$$

167. $\displaystyle\int x \sqrt{(X^3)}\, dx = \tfrac{1}{5} \sqrt{X^5} .$

168. $$\int x^2 \sqrt{(X^3)}\, dx = \frac{x \sqrt{X^5}}{6} - \frac{a^2 x \sqrt{X^3}}{24} - \frac{a^4 x \sqrt{X}}{16} - \frac{a^6}{16} \operatorname{arsinh} \frac{x}{a} + C =$$
$$= \frac{x \sqrt{X^5}}{6} - \frac{a^2 x \sqrt{X^3}}{24} - \frac{a^4 x \sqrt{X}}{16} - \frac{a^6}{16} \ln |x + \sqrt{X}| + C_1 .$$

169. $\displaystyle\int x^3 \sqrt{(X^3)}\, dx = \frac{\sqrt{X^7}}{7} - \frac{a^2 \sqrt{X^5}}{5} .$

170. $\displaystyle\int \frac{\sqrt{X^3}}{x}\, dx = \frac{\sqrt{X^3}}{3} + a^2 \sqrt{(X)} - a^3 \ln \left| \frac{a + \sqrt{X}}{x} \right| .$

171. $$\int \frac{\sqrt{X^3}}{x^2}\, dx = - \frac{\sqrt{X^3}}{x} + \tfrac{3}{2} x \sqrt{(X)} + \tfrac{3}{2} a^2 \operatorname{arsinh} \frac{x}{a} + C =$$
$$= - \frac{\sqrt{X^3}}{x} + \tfrac{3}{2} x \sqrt{(X)} + \tfrac{3}{2} a^2 \ln |x + \sqrt{X}| + C_1 .$$

172. $\int \frac{\sqrt{X^3}}{x^3}\,dx = -\frac{\sqrt{X^3}}{2x^2} + \frac{3}{2}\sqrt{(X)} - \frac{3}{2}a \ln\left|\frac{a+\sqrt{X}}{x}\right|.$

173. $\int \frac{dx}{\sqrt{X^3}} = \frac{x}{a^2\sqrt{X}}.$

174. $\int \frac{x\,dx}{\sqrt{X^3}} = -\frac{1}{\sqrt{X}}.$

175. $\int \frac{x^2\,dx}{\sqrt{X^3}} = -\frac{x}{\sqrt{X}} + \operatorname{arsinh}\frac{x}{a} + C = -\frac{x}{\sqrt{X}} + \ln|x + \sqrt{X}| + C_1.$

176. $\int \frac{x^3\,dx}{\sqrt{X^3}} = \sqrt{(X)} + \frac{a^2}{\sqrt{X}}.$

177. $\int \frac{dx}{x\sqrt{X^3}} = \frac{1}{a^2\sqrt{X}} - \frac{1}{a^3}\ln\left|\frac{a+\sqrt{X}}{x}\right|.$

178. $\int \frac{dx}{x^2\sqrt{X^3}} = -\frac{1}{a^4}\left(\frac{\sqrt{X}}{x} + \frac{x}{\sqrt{X}}\right).$

179. $\int \frac{dx}{x^3\sqrt{X^3}} = -\frac{1}{2a^2x^2\sqrt{X}} - \frac{3}{2a^4\sqrt{X}} + \frac{3}{2a^5}\ln\left|\frac{a+\sqrt{X}}{x}\right|.$

Notation: $X = x^2 - a^2$, $a > 0$.
If arcosh (x/a) occurs in a formula, x belonging to the interval $[a, \infty)$ is assumed. For $x \in (-\infty, -a]$ the function arcosh (x/a) is to be replaced by $-\operatorname{arcosh}(-x/a)$.

180. $\int \sqrt{(X)}\,dx = \frac{1}{2}\left(x\sqrt{(X)} - a^2 \operatorname{arcosh}\frac{x}{a}\right) + C =$

$= \frac{1}{2}(x\sqrt{(X)} - a^2 \ln|x + \sqrt{X}|) + C_1.$

181. $\int x\sqrt{(X)}\,dx = \frac{1}{3}\sqrt{X^3}.$

182. $\int x^2\sqrt{(X)}\,dx = \frac{x}{4}\sqrt{(X^3)} + \frac{a^2}{8}\left(x\sqrt{(X)} - a^2 \operatorname{arcosh}\frac{x}{a}\right) + C =$

$= \frac{x}{4}\sqrt{(X^3)} + \frac{a^2}{8}[x\sqrt{(X)} - a^2 \ln|x + \sqrt{X}|] + C_1.$

183. $\int x^3 \sqrt{(X)}\,dx = \frac{\sqrt{X^5}}{5} + \frac{a^2\sqrt{X^3}}{3}.$

184. $\int \frac{\sqrt{X}}{x}\,dx = \sqrt{(X)} - a \arccos \frac{a}{|x|}.$

185. $\int \frac{\sqrt{X}}{x^2}\,dx = -\frac{\sqrt{X}}{x} + \operatorname{arcosh}\frac{x}{a} + C = -\frac{\sqrt{X}}{x} + \ln|x + \sqrt{X}| + C_1.$

186. $\int \frac{\sqrt{X}}{x^3}\,dx = -\frac{\sqrt{X}}{2x^2} + \frac{1}{2a}\arccos\frac{a}{|x|}.$

187. $\int \frac{dx}{\sqrt{X}} = \operatorname{arcosh}\frac{x}{a} + C = \ln|x + \sqrt{X}| + C_1.$

188. $\int \frac{x\,dx}{\sqrt{X}} = \sqrt{X}.$

189. $\int \frac{x^2\,dx}{\sqrt{X}} = \frac{x}{2}\sqrt{(X)} + \frac{a^2}{2}\operatorname{arcosh}\frac{x}{a} + C = \frac{x}{2}\sqrt{(X)} + \frac{a^2}{2}\ln|x + \sqrt{X}| + C_1.$

190. $\int \frac{x^3\,dx}{\sqrt{X}} = \frac{\sqrt{X^3}}{3} + a^2\sqrt{X}.$

191. $\int \frac{dx}{x\sqrt{X}} = \frac{1}{a}\arccos\frac{a}{|x|}.$

192. $\int \frac{dx}{x^2\sqrt{X}} = \frac{\sqrt{X}}{a^2x}.$

193. $\int \frac{dx}{x^3\sqrt{X}} = \frac{\sqrt{X}}{2a^2x^2} + \frac{1}{2a^3}\arccos\frac{a}{|x|}.$

194. $\int \sqrt{(X^3)}\,dx = \frac{1}{4}\left(x\sqrt{(X^3)} - \frac{3a^2x}{2}\sqrt{(X)} + \frac{3a^4}{2}\operatorname{arcosh}\frac{x}{a}\right) + C =$

$$= \frac{1}{4}\left(x\sqrt{(X^3)} - \frac{3a^2x}{2}\sqrt{(X)} + \frac{3a^4}{2}\ln|x + \sqrt{X}|\right) + C_1.$$

195. $\int x\sqrt{(X^3)}\,dx = \frac{1}{5}\sqrt{X^5}.$

196. $\int x^2\sqrt{(X^3)}\,dx = \frac{x\sqrt{X^5}}{6} + \frac{a^2x\sqrt{X^3}}{24} - \frac{a^4x\sqrt{X}}{16} + \frac{a^6}{16}\operatorname{arcosh}\frac{x}{a} + C =$

$$= \frac{x\sqrt{X^5}}{6} + \frac{a^2x\sqrt{X^3}}{24} - \frac{a^4x\sqrt{X}}{16} + \frac{a^6}{16}\ln|x + \sqrt{X}| + C_1.$$

197. $$\int x^3 \sqrt{(X^3)}\,dx = \frac{\sqrt{X^7}}{7} + \frac{a^2\sqrt{X^5}}{5}.$$

198. $$\int \frac{\sqrt{X^3}}{x}\,dx = \frac{\sqrt{X^3}}{3} - a^2\sqrt{(X)} + a^3 \arccos\frac{a}{|x|}.$$

199. $$\int \frac{\sqrt{X^3}}{x^2}\,dx = -\frac{\sqrt{X^3}}{x} + \tfrac{3}{2}x\sqrt{(X)} - \tfrac{3}{2}a^2 \operatorname{arcosh}\frac{x}{a} + C =$$
$$= -\frac{\sqrt{X^3}}{x} + \tfrac{3}{2}x\sqrt{(X)} - \tfrac{3}{2}a^2 \ln|x + \sqrt{X}| + C_1.$$

200. $$\int \frac{\sqrt{X^3}}{x^3}\,dx = -\frac{\sqrt{X^3}}{2x^2} + \frac{3\sqrt{X}}{2} - \tfrac{3}{2}\,a \arccos\frac{a}{|x|}.$$

201. $$\int \frac{dx}{\sqrt{X^3}} = -\frac{x}{a^2\sqrt{X}}.$$

202. $$\int \frac{x\,dx}{\sqrt{X^3}} = -\frac{1}{\sqrt{X}}.$$

203. $$\int \frac{x^2\,dx}{\sqrt{X^3}} = -\frac{x}{\sqrt{X}} + \operatorname{arcosh}\frac{x}{a} + C = -\frac{x}{\sqrt{X}} + \ln|x + \sqrt{X}| + C_1.$$

204. $$\int \frac{x^3\,dx}{\sqrt{X^3}} = \sqrt{(X)} - \frac{a^2}{\sqrt{X}}.$$

205. $$\int \frac{dx}{x\sqrt{X^3}} = -\frac{1}{a^2\sqrt{X}} - \frac{1}{a^3}\arccos\frac{a}{|x|}.$$

206. $$\int \frac{dx}{x^2\sqrt{X^3}} = -\frac{1}{a^4}\left(\frac{\sqrt{X}}{x} + \frac{x}{\sqrt{X}}\right).$$

207. $$\int \frac{dx}{x^3\sqrt{X^3}} = \frac{1}{2a^2x^2\sqrt{X}} - \frac{3}{2a^4\sqrt{X}} - \frac{3}{2a^5}\arccos\frac{a}{|x|}.$$

Notation: $X = ax^2 + bx + c$, $\Delta = 4ac - b^2$, $k = 4a/\Delta$, $a \neq 0$, $\Delta \neq 0$.

If $\Delta > 0$, then $ax^2 + bx + c$ has the same sign for all x and $\sqrt{X}$ is real either for all x, if $a > 0$, or for no x, if $a < 0$.

If $\Delta < 0$, then the equation $ax^2 + bx + c = 0$ has two distinct real roots $\alpha_1 < \alpha_2$ and $\sqrt{X}$ is real either for $x \in [\alpha_1, \alpha_2]$ if $a < 0$, or for $x \in (-\infty, \alpha_1]$ and $x \in [\alpha_2, \infty)$ if $a > 0$.

208. $$\int \frac{dx}{\sqrt{X}} = \begin{cases} \frac{1}{\sqrt{a}} \ln |2\sqrt{(aX)} + 2ax + b| + C & \text{for} \quad a > 0\,, \\ \frac{1}{\sqrt{a}} \operatorname{arsinh} \frac{2ax + b}{\sqrt{\Delta}} + C_1 & \text{for} \quad a > 0,\ \Delta > 0\,, \\ -\frac{1}{\sqrt{(-a)}} \arcsin \frac{2ax + b}{\sqrt{(-\Delta)}} & \text{for} \quad a < 0,\ \Delta < 0\,. \end{cases}$$

209. $$\int \frac{dx}{X\sqrt{X}} = \frac{2(2ax + b)}{\Delta\sqrt{X}}\,.$$

210. $$\int \frac{dx}{X^2\sqrt{X}} = \frac{2(2ax + b)}{3\Delta\sqrt{X}}\left(\frac{1}{X} + 2k\right).$$

211. $$\int \frac{dx}{X^{\frac{1}{2}(2n+1)}} = \frac{2(2ab + b)}{(2n - 1)\,\Delta X^{\frac{1}{2}(2n-1)}} + \frac{2k(n - 1)}{2n - 1}\int \frac{dx}{X^{\frac{1}{2}(2n-1)}}\,.$$

212. $$\int \sqrt{(X)}\,dx = \frac{(2ax + b)\sqrt{X}}{4a} + \frac{1}{2k}\int \frac{dx}{\sqrt{X}} \quad \text{(see 208)}\,.$$

213. $$\int X\sqrt{(X)}\,dx = \frac{(2ax + b)\sqrt{X}}{8a}\left(X + \frac{3}{2k}\right) + \frac{3}{8k^2}\int \frac{dx}{\sqrt{X}} \quad \text{(see 208)}\,.$$

214. $$\int X^2\sqrt{(X)}\,dx = \frac{(2ax + b)\sqrt{X}}{12a}\left(X^2 + \frac{5X}{4k} + \frac{15}{8k^2}\right) + \frac{5}{16k^3}\int \frac{dx}{\sqrt{X}} \quad \text{(see 208)}\,.$$

215. $$\int X^{\frac{1}{2}(2n+1)}\,dx = \frac{(2ax + b)\,X^{\frac{1}{2}(2n+1)}}{4a(n + 1)} + \frac{2n + 1}{2k(n + 1)}\int X^{\frac{1}{2}(2n-1)}\,dx$$
(see 208 and 212).

216. $$\int \frac{x\,dx}{\sqrt{X}} = \frac{\sqrt{X}}{a} - \frac{b}{2a}\int \frac{dx}{\sqrt{X}} \quad \text{(see 208)}\,.$$

217. $$\int \frac{x\,dx}{X\sqrt{X}} = -\frac{2(bx + 2c)}{\Delta\sqrt{X}}\,.$$

218. $$\int \frac{x\,dx}{X^{\frac{1}{2}(2n+1)}} = -\frac{1}{(2n - 1)\,aX^{\frac{1}{2}(2n-1)}} - \frac{b}{2a}\int \frac{dx}{X^{\frac{1}{2}(2n+1)}} \quad \text{(see 211)}\,.$$

219. $$\int \frac{x^2\,dx}{\sqrt{X}} = \left(\frac{x}{2a} - \frac{3b}{4a^2}\right)\sqrt{(X)} + \frac{3b^2 - 4ac}{8a^2}\int \frac{dx}{\sqrt{X}} \quad \text{(see 208)}\,.$$

220. $$\int \frac{x^2\,dx}{X\sqrt{X}} = \frac{(2b^2 - 4ac)\,x + 2bc}{a\Delta\sqrt{X}} + \frac{1}{a}\int \frac{dx}{\sqrt{X}} \quad \text{(see 208)}\,.$$

221. $\displaystyle\int x\sqrt{(X)}\,dx = \frac{X\sqrt{X}}{3a} - \frac{b(2ax+b)}{8a^2}\sqrt{(X)} - \frac{b}{4ak}\int\frac{dx}{\sqrt{X}}$ (see 208).

222. $\displaystyle\int xX\sqrt{(X)}\,dx = \frac{X^2\sqrt{X}}{5a} - \frac{b}{2a}\int X\sqrt{(X)}\,dx$ (see 213).

223. $\displaystyle\int xX^{\frac{1}{2}(2n+1)}\,dx = \frac{X^{\frac{1}{2}(2n+3)}}{(2n+3)\,a} - \frac{b}{2a}\int X^{\frac{1}{2}(2n+1)}\,dx$ (see 215).

224. $\displaystyle\int x^2\sqrt{(X)}\,dx = \left(x - \frac{5b}{6a}\right)\frac{X\sqrt{X}}{4a} + \frac{5b^2-4ac}{16a^2}\int\sqrt{(X)}\,dx$ (see 212).

225. $$\int\frac{dx}{x\sqrt{X}} = \begin{cases} -\dfrac{1}{\sqrt{c}}\ln\left|\dfrac{2\sqrt{(cX)}}{x} + \dfrac{2c}{x} + b\right| + C & \text{for } c > 0\,, \\ -\dfrac{1}{\sqrt{c}}\operatorname{arsinh}\dfrac{bx+2c}{x\sqrt{\Delta}} + C_1 & \text{for } c > 0,\ \Delta > 0\,, \\ \dfrac{1}{\sqrt{(-c)}}\arcsin\dfrac{bx+2c}{x\sqrt{(-\Delta)}} & \text{for } c < 0,\ \Delta < 0 \end{cases}$$
(for $c = 0$ see 231).

226. $\displaystyle\int\frac{dx}{x^2\sqrt{X}} = -\frac{\sqrt{X}}{cx} - \frac{b}{2c}\int\frac{dx}{x\sqrt{X}}$ (see 225).

227. $\displaystyle\int\frac{\sqrt{(X)}\,dx}{x} = \sqrt{(X)} + \frac{b}{2}\int\frac{dx}{\sqrt{X}} + c\int\frac{dx}{x\sqrt{X}}$ (see 208 and 225).

228. $\displaystyle\int\frac{\sqrt{(X)}\,dx}{x^2} = -\frac{\sqrt{(X)}}{x} + a\int\frac{dx}{\sqrt{X}} + \frac{b}{2}\int\frac{dx}{x\sqrt{X}}$ (see 208 and 225).

229. $\displaystyle\int\frac{X^{\frac{1}{2}(2n+1)}}{x}\,dx = \frac{X^{\frac{1}{2}(2n+1)}}{2n+1} + \frac{b}{2}\int X^{\frac{1}{2}(2n-1)}\,dx + c\int\frac{X^{\frac{1}{2}(2n-1)}}{x}\,dx$

(see 212 and 227).

230. $$\int\frac{a_nx^n + a_{n-1}x^{n-1} + \ldots + a_1x + a_0}{\sqrt{X}}\,dx =$$
$$= (A_0 + A_1x + \ldots + A_{n-1}x^{n-1})\sqrt{(X)} + A_n\int\frac{dx}{\sqrt{X}}\,;$$

the constants $A_0, A_1, \ldots, A_n$ can be determined by differentiation and then by the method of undetermined coefficients (by equating coefficients).

231. $\displaystyle\int\frac{dx}{x\sqrt{(ax^2+bx)}} = -\frac{2}{bx}\sqrt{(ax^2+bx)}$ $(b \neq 0)$.

232. $$\int \frac{dx}{\sqrt{(2ax - x^2)}} = \arcsin \frac{x - a}{|a|} \quad (a \neq 0).$$

233. $$\int \frac{x\,dx}{\sqrt{(2ax - x^2)}} = -\sqrt{(2ax - x^2)} + a \arcsin \frac{x - a}{|a|} \quad (a \neq 0).$$

234. $$\int \sqrt{(2ax - x^2)}\,dx = \frac{x - a}{2}\sqrt{(2ax - x^2)} + \frac{a^2}{2} \arcsin \frac{x - a}{|a|} \quad (a \neq 0).$$

235. $$\int \frac{dx}{(ax^2 + b)\sqrt{(fx^2 + g)}} = \frac{1}{\sqrt{(b)}\sqrt{(ag - bf)}} \arctan \frac{x\sqrt{(ag - bf)}}{\sqrt{(b)}\sqrt{(fx^2 + g)}} \quad (ag - bf > 0),$$

$$= \frac{1}{2\sqrt{(b)}\sqrt{(bf - ag)}} \ln \left| \frac{\sqrt{(b)}\sqrt{(fx^2 + g)} + x\sqrt{(bf - ag)}}{\sqrt{(b)}\sqrt{(fx^2 + g)} - x\sqrt{(bf - ag)}} \right| \quad (ag - bf < 0).$$

236. $$\int \sqrt[n]{(ax + b)}\,dx = \frac{n(ax + b)}{(n + 1)a} \sqrt[n]{(ax + b)} \quad (a \neq 0).$$

237. $$\int \frac{dx}{\sqrt[n]{(ax + b)}}\,dx = \frac{n(ax + b)}{(n - 1)a} \frac{1}{\sqrt[n]{(ax + b)}} \quad (n \neq 1;\ a \neq 0).$$

238. $$\int \frac{dx}{x\sqrt{(x^n + a^2)}} = -\frac{2}{na} \ln \left| \frac{a + \sqrt{(x^n + a^2)}}{\sqrt{x^n}} \right| \quad (x > 0).$$

239. $$\int \frac{dx}{x\sqrt{(x^n - a^2)}} = \frac{2}{na} \arccos \frac{a}{\sqrt{x^n}} \quad (x > \sqrt[n]{a^2}).$$

240. $$\int \frac{\sqrt{(x)}\,dx}{\sqrt{(a^3 - x^3)}} = \tfrac{2}{3} \arcsin \sqrt{\left(\frac{x}{a}\right)^3} \quad (0 \leqq x < a).$$

Reduction Formulae for Binomial Integrals

241. $$\int x^m(ax^n + b)^p\,dx =$$

$$= \frac{1}{m + np + 1}\left[x^{m+1}(ax^n + b)^p + npb \int x^m(ax^n + b)^{p-1}\,dx\right] \quad (m + np + 1 \neq 0),$$

$$= \frac{1}{bn(p + 1)}\left[-x^{m+1}(ax^n + b)^{p+1} + (m + n + np + 1)\int x^m(ax^n + b)^{p+1}\,dx\right] \quad (bn(p + 1) \neq 0),$$

$$= \frac{1}{(m+1)\,b}\left[x^{m+1}(ax^n+b)^{p+1} - a(m+n+np+1)\int x^{m+n}(ax^n+b)^p\,dx\right]$$

$$((m+1)\,b \neq 0)\,,$$

$$= \frac{1}{a(m+np+1)}\left[x^{m-n+1}(ax^n+b)^{p+1} - (m-n+1)\,b\int x^{m-n}(ax^n+b)^p\,dx\right]$$

$$(a(m+np+1) \neq 0)\,.$$

(In 241 m, n, p are rational numbers, $x > 0$.)

(c) *Trigonometric Functions.*

$a \neq 0$ is assumed in all cases .

(See also 402–405, 417–422, 440–444.)

(α) *Integrals Containing the Sine Only.*

242. $\int \sin ax\,dx = -\frac{1}{a}\cos ax\,.$

243. $\int \sin^2 ax\,dx = \frac{1}{2}x - \frac{1}{4a}\sin 2ax\,.$

244. $\int \sin^3 ax\,dx = -\frac{1}{a}\cos ax + \frac{1}{3a}\cos^3 ax\,.$

245. $\int \sin^4 ax\,dx = \frac{3}{8}x - \frac{1}{4a}\sin 2ax + \frac{1}{32a}\sin 4ax\,.$

246. $\int \sin^n ax\,dx = -\frac{\sin^{n-1} ax\cos ax}{na} + \frac{n-1}{n}\int \sin^{n-2} ax\,dx\,.$

247. $\int x\sin ax\,dx = \frac{\sin ax}{a^2} - \frac{x\cos ax}{a}\,.$

248. $\int x^2\sin ax\,dx = \frac{2x}{a^2}\sin ax - \left(\frac{x^2}{a} - \frac{2}{a^3}\right)\cos ax\,.$

249. $\int x^3\sin ax\,dx = \left(\frac{3x^2}{a^2} - \frac{6}{a^4}\right)\sin ax - \left(\frac{x^3}{a} - \frac{6x}{a^3}\right)\cos ax\,.$

250. $\int x^n \sin ax \, dx = -\frac{x^n}{a}\cos ax + \frac{n}{a}\int x^{n-1}\cos ax \, dx$ (see 289).

251. $\int \frac{\sin ax}{x}\, dx = ax - \frac{(ax)^3}{3 \cdot 3!} + \frac{(ax)^5}{5 \cdot 5!} - \frac{(ax)^7}{7 \cdot 7!} + \ldots$

(the series is convergent for all x; see also §§ 13.12 and 15.7).

252. $\int \frac{\sin ax}{x^2}\, dx = -\frac{\sin ax}{x} + a\int \frac{\cos ax \, dx}{x}$ (see 290).

253. $\int \frac{\sin ax}{x^n}\, dx = -\frac{1}{n-1}\frac{\sin ax}{x^{n-1}} + \frac{a}{n-1}\int \frac{\cos ax}{x^{n-1}}\, dx$ (see 292; $n > 1$).

254. $\int \frac{dx}{\sin ax} = \frac{1}{a}\ln\left|\tan\frac{ax}{2}\right|$.

255. $\int \frac{dx}{\sin^2 ax} = -\frac{1}{a}\cot ax$.

256. $\int \frac{dx}{\sin^3 ax} = -\frac{\cos ax}{2a\sin^2 ax} + \frac{1}{2a}\ln\left|\tan\frac{ax}{2}\right|$.

257. $\int \frac{dx}{\sin^n ax} = -\frac{1}{a(n-1)}\frac{\cos ax}{\sin^{n-1} ax} + \frac{n-2}{n-1}\int \frac{dx}{\sin^{n-2} ax}$ $(n > 1)$.

258. $\int \frac{x\, dx}{\sin ax} = \frac{1}{a^2}\left(ax + \frac{(ax)^3}{3 \cdot 3!} + \frac{7(ax)^5}{3 \cdot 5 \cdot 5!} + \frac{31(ax)^7}{3 \cdot 7 \cdot 7!} + \frac{127(ax)^9}{3 \cdot 5 \cdot 9!} + \ldots + \right.$

$\left. + \frac{2(2^{2n-1}-1)}{(2n+1)!} B_n \cdot (ax)^{2n+1} + \ldots\right)$ ($|x| < \pi/a$; see Remark 3, p. 549).

259. $\int \frac{x\, dx}{\sin^2 ax} = -\frac{x}{a}\cot ax + \frac{1}{a^2}\ln|\sin ax|$.

260. $\int \frac{x\, dx}{\sin^n ax} = -\frac{x\cos ax}{(n-1)\, a\sin^{n-1} ax} - \frac{1}{(n-1)(n-2)\, a^2\sin^{n-2} ax} +$

$+ \frac{n-2}{n-1}\int \frac{x\, dx}{\sin^{n-2} ax}$ $(n > 2)$.

261. $\int \frac{dx}{1 + \sin ax} = -\frac{1}{a}\tan\left(\frac{\pi}{4} - \frac{ax}{2}\right)$.

262. $\int \frac{dx}{1 - \sin ax} = \frac{1}{a}\tan\left(\frac{\pi}{4} + \frac{ax}{2}\right)$.

263. $\displaystyle\int \frac{x\,dx}{1+\sin ax} = -\frac{x}{a}\tan\left(\frac{\pi}{4}-\frac{ax}{2}\right)+\frac{2}{a^2}\ln\left|\cos\left(\frac{\pi}{4}-\frac{ax}{2}\right)\right|.$

264. $\displaystyle\int \frac{x\,dx}{1-\sin ax} = \frac{x}{a}\cot\left(\frac{\pi}{4}-\frac{ax}{2}\right)+\frac{2}{a^2}\ln\left|\sin\left(\frac{\pi}{4}-\frac{ax}{2}\right)\right|.$

265. $\displaystyle\int \frac{\sin ax\,dx}{1\pm\sin ax} = \pm x+\frac{1}{a}\tan\left(\frac{\pi}{4}\mp\frac{ax}{2}\right).$

266. $\displaystyle\int \frac{dx}{\sin ax(1\pm\sin ax)} = \frac{1}{a}\tan\left(\frac{\pi}{4}\mp\frac{ax}{2}\right)+\frac{1}{a}\ln\left|\tan\frac{ax}{2}\right|.$

267. $\displaystyle\int \frac{dx}{(1+\sin ax)^2} = -\frac{1}{2a}\tan\left(\frac{\pi}{4}-\frac{ax}{2}\right)-\frac{1}{6a}\tan^3\left(\frac{\pi}{4}-\frac{ax}{2}\right).$

268. $\displaystyle\int \frac{dx}{(1-\sin ax)^2} = \frac{1}{2a}\cot\left(\frac{\pi}{4}-\frac{ax}{2}\right)+\frac{1}{6a}\cot^3\left(\frac{\pi}{4}-\frac{ax}{2}\right).$

269. $\displaystyle\int \frac{\sin ax\,dx}{(1+\sin ax)^2} = -\frac{1}{2a}\tan\left(\frac{\pi}{4}-\frac{ax}{2}\right)+\frac{1}{6a}\tan^3\left(\frac{\pi}{4}-\frac{ax}{2}\right).$

270. $\displaystyle\int \frac{\sin ax\,dx}{(1-\sin ax)^2} = -\frac{1}{2a}\cot\left(\frac{\pi}{4}-\frac{ax}{2}\right)+\frac{1}{6a}\cot^3\left(\frac{\pi}{4}-\frac{ax}{2}\right).$

271. $\displaystyle\int \frac{dx}{1+\sin^2 ax} = \frac{1}{2\sqrt{(2)}\,a}\arcsin\left(\frac{3\sin^2(ax)-1}{\sin^2(ax)+1}\right) \quad (\sin 2ax > 0)\,.$

272. $\displaystyle\int \frac{dx}{1-\sin^2 ax} = \int\frac{dx}{\cos^2 ax} = \frac{1}{a}\tan ax\,.$

273. $\displaystyle\int \sin ax\sin bx\,dx = \frac{\sin(a-b)x}{2(a-b)}-\frac{\sin(a+b)x}{2(a+b)}$

$(|a| \neq |b|;$ for $|a| = |b|$ see 243$)$.

274. $\displaystyle\int \frac{dx}{b+c\sin ax} = \frac{2}{a\sqrt{(b^2-c^2)}}\arctan\frac{b\tan(\frac{1}{2}ax)+c}{\sqrt{(b^2-c^2)}} \quad (\text{for } b^2 > c^2)\,,$

$$= \frac{1}{a\sqrt{(c^2-b^2)}}\ln\left|\frac{b\tan(\frac{1}{2}ax)+c-\sqrt{(c^2-b^2)}}{b\tan(\frac{1}{2}ax)+c+\sqrt{(c^2-b^2)}}\right| \quad (\text{for } b^2 < c^2)\,.$$

275. $\displaystyle\int \frac{\sin ax\,dx}{b+c\sin ax} = \frac{x}{c}-\frac{b}{c}\int\frac{dx}{b+c\sin ax} \quad (\text{see } 274)\,.$

276. $\displaystyle\int \frac{dx}{\sin ax(b+c\sin ax)} = \frac{1}{ab}\ln\left|\tan\frac{ax}{2}\right|-\frac{c}{b}\int\frac{dx}{b+c\sin ax} \quad (\text{see } 274)\,.$

277. $$\int \frac{dx}{(b + c\sin ax)^2} = \frac{c\cos ax}{a(b^2 - c^2)(b + c\sin ax)} + \frac{b}{b^2 - c^2}\int \frac{dx}{b + c\sin ax} \quad \text{(see 274)}.$$

278. $$\int \frac{\sin ax\, dx}{(b + c\sin ax)^2} = \frac{b\cos ax}{a(c^2 - b^2)(b + c\sin ax)} + \frac{c}{c^2 - b^2}\int \frac{dx}{b + c\sin ax} \quad \text{(see 274)}.$$

279. $$\int \frac{dx}{b^2 + c^2\sin^2 ax} = \frac{1}{ab\sqrt{(b^2 + c^2)}}\arctan \frac{\sqrt{(b^2 + c^2)}\tan ax}{b} \quad (b \neq 0).$$

280. $$\int \frac{dx}{b^2 - c^2\sin^2 ax} = \frac{1}{ab\sqrt{(b^2 - c^2)}}\arctan \frac{\sqrt{(b^2 - c^2)}\tan ax}{b} \quad (b^2 > c^2,\ b \neq 0),$$

$$= \frac{1}{2ab\sqrt{(c^2 - b^2)}}\ln\left|\frac{\sqrt{(c^2 - b^2)}\tan(ax) + b}{\sqrt{(c^2 - b^2)}\tan(ax) - b}\right| \quad (c^2 > b^2,\ b \neq 0).$$

(β) *Integrals Containing the Cosine Only.*

281. $$\int \cos ax\, dx = \frac{1}{a}\sin ax.$$

282. $$\int \cos^2 ax\, dx = \frac{1}{2}x + \frac{1}{4a}\sin 2ax.$$

283. $$\int \cos^3 ax\, dx = \frac{1}{a}\sin ax - \frac{1}{3a}\sin^3 ax.$$

284. $$\int \cos^4 ax\, dx = \frac{3}{8}x + \frac{1}{4a}\sin 2ax + \frac{1}{32a}\sin 4ax.$$

285. $$\int \cos^n ax\, dx = \frac{\cos^{n-1} ax\sin ax}{na} + \frac{n-1}{n}\int \cos^{n-2} ax\, dx.$$

286. $$\int x\cos ax\, dx = \frac{\cos ax}{a^2} + \frac{x\sin ax}{a}.$$

287. $$\int x^2\cos ax\, dx = \frac{2x}{a^2}\cos ax + \left(\frac{x^2}{a} - \frac{2}{a^3}\right)\sin ax.$$

288. $\int x^3 \cos ax \, dx = \left(\frac{3x^2}{a^2} - \frac{6}{a^4} \right) \cos ax + \left(\frac{x^3}{a} - \frac{6x}{a^3} \right) \sin ax \, .$

289. $\int x^n \cos ax \, dx = \frac{x^n \sin ax}{a} - \frac{n}{a} \int x^{n-1} \sin ax \, dx$ (see 250) .

290. $\int \frac{\cos ax}{x} \, dx = \ln |ax| - \frac{(ax)^2}{2 \, . \, 2!} + \frac{(ax)^4}{4 \, . \, 4!} - \frac{(ax)^6}{6 \, . \, 6!} + \ldots$

(the series is convergent for all x; see also §§ 13.12 and 15.7).

291. $\int \frac{\cos ax}{x^2} \, dx = - \frac{\cos ax}{x} - a \int \frac{\sin ax \, dx}{x}$ (see 251) .

292. $\int \frac{\cos ax}{x^n} \, dx = - \frac{\cos ax}{(n-1) \, x^{n-1}} - \frac{a}{n-1} \int \frac{\sin ax \, dx}{x^{n-1}}$ ($n > 1$; see 253) .

293. $\int \frac{dx}{\cos ax} = \frac{1}{a} \ln \left| \tan \left(\frac{ax}{2} + \frac{\pi}{4} \right) \right| .$

294. $\int \frac{dx}{\cos^2 ax} = \frac{1}{a} \tan ax \, .$

295. $\int \frac{dx}{\cos^3 ax} = \frac{\sin ax}{2a \cos^2 ax} + \frac{1}{2a} \ln \left| \tan \left(\frac{\pi}{4} + \frac{ax}{2} \right) \right| .$

296. $\int \frac{dx}{\cos^n ax} = \frac{1}{a(n-1)} \frac{\sin ax}{\cos^{n-1} ax} + \frac{n-2}{n-1} \int \frac{dx}{\cos^{n-2} ax}$ ($n > 1$) .

297. $\int \frac{x \, dx}{\cos ax} = \frac{1}{a^2} \left(\frac{(ax)^2}{2} + \frac{(ax)^4}{4 \, . \, 2!} + \frac{5(ax)^6}{6 \, . \, 4!} + \frac{61(ax)^8}{8. \, 6!} + \frac{1{,}385(ax)^{10}}{10 \, . \, 8!} + \ldots \right.$

$$\left. \ldots + \frac{E_n \, . \, (ax)^{2n+2}}{(2n+2)(2n!)} + \ldots \right) \quad \left(|x| < \frac{\pi}{2|a|} ; \text{ see Remark 4, p. 549} \right).$$

298. $\int \frac{x \, dx}{\cos^2 ax} = \frac{x}{a} \tan ax + \frac{1}{a^2} \ln |\cos ax| \, .$

299. $\int \frac{x \, dx}{\cos^n ax} = \frac{x \sin ax}{(n-1) \, a \cos^{n-1} ax} - \frac{1}{(n-1)(n-2) \, a^2 \cos^{n-2} ax} +$

$$+ \frac{n-2}{n-1} \int \frac{x \, dx}{\cos^{n-2} ax} \quad (n > 2) \, .$$

300. $\int \frac{dx}{1 + \cos ax} = \frac{1}{a} \tan \frac{ax}{2} \, .$

301. $\displaystyle\int \frac{dx}{1 - \cos ax} = -\frac{1}{a}\cot\frac{ax}{2}.$

302. $\displaystyle\int \frac{x\,dx}{1 + \cos ax} = \frac{x}{a}\tan\frac{ax}{2} + \frac{2}{a^2}\ln\left|\cos\frac{ax}{2}\right|.$

303. $\displaystyle\int \frac{x\,dx}{1 - \cos ax} = -\frac{x}{a}\cot\frac{ax}{2} + \frac{2}{a^2}\ln\left|\sin\frac{ax}{2}\right|.$

304. $\displaystyle\int \frac{\cos ax\,dx}{1 + \cos ax} = x - \frac{1}{a}\tan\frac{ax}{2}.$

305. $\displaystyle\int \frac{\cos ax\,dx}{1 - \cos ax} = -x - \frac{1}{a}\cot\frac{ax}{2}.$

306. $\displaystyle\int \frac{dx}{\cos ax(1 + \cos ax)} = \frac{1}{a}\ln\left|\tan\left(\frac{\pi}{4} + \frac{ax}{2}\right)\right| - \frac{1}{a}\tan\frac{ax}{2}.$

307. $\displaystyle\int \frac{dx}{\cos ax(1 - \cos ax)} = \frac{1}{a}\ln\left|\tan\left(\frac{\pi}{4} + \frac{ax}{2}\right)\right| - \frac{1}{a}\cot\frac{ax}{2}.$

308. $\displaystyle\int \frac{dx}{(1 + \cos ax)^2} = \frac{1}{2a}\tan\frac{ax}{2} + \frac{1}{6a}\tan^3\frac{ax}{2}.$

309. $\displaystyle\int \frac{dx}{(1 - \cos ax)^2} = -\frac{1}{2a}\cot\frac{ax}{2} - \frac{1}{6a}\cot^3\frac{ax}{2}.$

310. $\displaystyle\int \frac{\cos ax\,dx}{(1 + \cos ax)^2} = \frac{1}{2a}\tan\frac{ax}{2} - \frac{1}{6a}\tan^3\frac{ax}{2}.$

311. $\displaystyle\int \frac{\cos ax\,dx}{(1 - \cos ax)^2} = \frac{1}{2a}\cot\frac{ax}{2} - \frac{1}{6a}\cot^3\frac{ax}{2}.$

312. $\displaystyle\int \frac{dx}{1 + \cos^2 ax} = \frac{1}{2\sqrt{(2)}\,a}\arcsin\left(\frac{1 - 3\cos^2 ax}{1 + \cos^2 ax}\right) \quad (\sin ax \cos ax > 0).$

313. $\displaystyle\int \frac{dx}{1 - \cos^2 ax} = \int \frac{dx}{\sin^2 ax} = -\frac{1}{a}\cot ax.$

314. $\displaystyle\int \cos ax \cos bx\,dx = \frac{\sin(a - b)x}{2(a - b)} + \frac{\sin(a + b)x}{2(a + b)}$

$(|a| \neq |b|;$ for $|a| = |b|$ see 282$)$.

315. $$\int \frac{dx}{b + c\cos ax} = \frac{2}{a\sqrt{(b^2 - c^2)}} \arctan \frac{(b - c)\tan \frac{1}{2}ax}{\sqrt{(b^2 - c^2)}} \quad (\text{for } b^2 > c^2),$$
$$= \frac{1}{a\sqrt{(c^2 - b^2)}} \ln \left| \frac{(c - b)\tan \frac{1}{2}ax + \sqrt{(c^2 - b^2)}}{(c - b)\tan \frac{1}{2}ax - \sqrt{(c^2 - b^2)}} \right| \quad (\text{for } b^2 < c^2).$$

316. $$\int \frac{\cos ax\, dx}{b + c\cos ax} = \frac{x}{c} - \frac{b}{c}\int \frac{dx}{b + c\cos ax} \quad (\text{see } 315).$$

317. $$\int \frac{dx}{\cos ax(b + c\cos ax)} = \frac{1}{ab}\ln\left|\tan\left(\frac{ax}{2} + \frac{\pi}{4}\right)\right| - \frac{b}{c}\int \frac{dx}{b + c\cos ax} \quad (\text{see } 315).$$

318. $$\int \frac{dx}{(b + c\cos ax)^2} = \frac{c\sin ax}{a(c^2 - b^2)(b + c\cos ax)} - \frac{b}{c^2 - b^2}\int \frac{dx}{b + c\cos ax}$$
$$(\text{see } 315) \quad (b^2 \neq c^2).$$

319. $$\int \frac{\cos ax\, dx}{(b + c\cos ax)^2} = \frac{b\sin ax}{a(b^2 - c^2)(b + c\cos ax)} - \frac{c}{b^2 - c^2}\int \frac{dx}{b + c\cos ax}$$
$$(\text{see } 315) \quad (b^2 \neq c^2).$$

320. $$\int \frac{dx}{b^2 + c^2\cos^2 ax} = \frac{1}{ab\sqrt{(b^2 + c^2)}} \arctan \frac{b\tan ax}{\sqrt{(b^2 + c^2)}} \quad (b > 0).$$

321. $$\int \frac{dx}{b^2 - c^2\cos^2 ax} = \frac{1}{ab\sqrt{(b^2 - c^2)}} \arctan \frac{b\tan ax}{\sqrt{(b^2 - c^2)}} \quad (b^2 > c^2 > 0),$$
$$= \frac{1}{2ab\sqrt{(c^2 - b^2)}} \ln\left|\frac{b\tan ax - \sqrt{(c^2 - b^2)}}{b\tan ax + \sqrt{(c^2 - b^2)}}\right| \quad (c^2 > b^2 > 0).$$

(γ) *Integrals Containing both Sine and Cosine.*

322. $$\int \sin ax\cos ax\, dx = \frac{1}{2a}\sin^2 ax.$$

323. $$\int \sin^2 ax\cos^2 ax\, dx = \frac{x}{8} - \frac{\sin 4ax}{32a}.$$

324. $$\int \sin^r ax\cos ax\, dx = \frac{1}{a(r + 1)}\sin^{r+1} ax \quad (r \neq -1).$$

325. $$\int \sin ax\cos^r ax\, dx = -\frac{1}{a(r + 1)}\cos^{r+1} ax \quad (r \neq -1).$$

326. $$\int \sin^n ax \cos^m ax \, dx = -\frac{\sin^{n-1} ax \cos^{m+1} ax}{a(n+m)} + \frac{n-1}{n+m} \int \sin^{n-2} ax \cos^m ax \, dx =$$
$$= \frac{\sin^{n+1} ax \cos^{m-1} ax}{a(n+m)} + \frac{m-1}{n+m} \int \sin^n ax \cos^{m-2} ax \, dx \, .$$

327. $$\int \frac{dx}{\sin ax \cos ax} = \frac{1}{a} \ln |\tan ax| \, .$$

328. $$\int \frac{dx}{\sin^2 ax \cos ax} = \frac{1}{a} \left[\ln \left| \tan \left(\frac{\pi}{4} + \frac{ax}{2} \right) \right| - \frac{1}{\sin ax} \right].$$

329. $$\int \frac{dx}{\sin ax \cos^2 ax} = \frac{1}{a} \left(\ln \left| \tan \frac{ax}{2} \right| + \frac{1}{\cos ax} \right).$$

330. $$\int \frac{dx}{\sin^3 ax \cos ax} = \frac{1}{a} \left(\ln |\tan ax| - \frac{1}{2\sin^2 ax} \right).$$

331. $$\int \frac{dx}{\sin ax \cos^3 ax} = \frac{1}{a} \left(\ln |\tan ax| + \frac{1}{2\cos^2 ax} \right).$$

332. $$\int \frac{dx}{\sin^2 ax \cos^2 ax} = -\frac{2}{a} \cot 2ax \, .$$

333. $$\int \frac{dx}{\sin ax \cos^n ax} = \frac{1}{a(n-1) \cos^{n-1} ax} + \int \frac{dx}{\sin ax \cos^{n-2} ax}$$

(see 327, 329, 331; $n > 1$).

334. $$\int \frac{dx}{\sin^n ax \cos ax} = -\frac{1}{a(n-1) \sin^{n-1} ax} + \int \frac{dx}{\sin^{n-2} ax \cos ax}$$

(see 327, 328, 330; $n > 1$).

335. $$\int \frac{dx}{\sin^n ax \cos^m ax} =$$

$$= -\frac{1}{a(n-1)} \cdot \frac{1}{\sin^{n-1} ax \cos^{m-1} ax} + \frac{n+m-2}{n-1} \int \frac{dx}{\sin^{n-2} ax \cos^m ax}$$

($n > 1$),

$$= \frac{1}{a(m-1)} \cdot \frac{1}{\sin^{n-1} ax \cos^{m-1} ax} + \frac{n+m-2}{m-1} \int \frac{dx}{\sin^n ax \cos^{m-2} ax}$$

($m > 1$).

336. $\displaystyle\int \frac{\sin ax\,dx}{\cos^2 ax} = \frac{1}{a\cos ax}$.

337. $\displaystyle\int \frac{\sin ax\,dx}{\cos^3 ax} = \frac{1}{2a\cos^2 ax} + C = \frac{1}{2a}\tan^2 ax + C_1$.

338. $\displaystyle\int \frac{\sin ax\,dx}{\cos^n ax} = \frac{1}{a(n-1)\cos^{n-1} ax}$ $(n > 1;$ for $n = 1$ see 364).

339. $\displaystyle\int \frac{\sin^2 ax\,dx}{\cos ax} = -\frac{1}{a}\sin ax + \frac{1}{a}\ln\left|\tan\left(\frac{\pi}{4} + \frac{ax}{2}\right)\right|$.

340. $\displaystyle\int \frac{\sin^2 ax\,dx}{\cos^3 ax} = \frac{1}{a}\left[\frac{\sin ax}{2\cos^2 ax} - \frac{1}{2}\ln\left|\tan\left(\frac{\pi}{4} + \frac{ax}{2}\right)\right|\right]$.

341. $\displaystyle\int \frac{\sin^2 ax\,dx}{\cos^n ax} = \frac{\sin ax}{a(n-1)\cos^{n-1} ax} - \frac{1}{n-1}\int \frac{dx}{\cos^{n-2} ax}$

(see 293–296; $n > 1$).

342. $\displaystyle\int \frac{\sin^n ax}{\cos ax} = -\frac{\sin^{n-1} ax}{a(n-1)} + \int \frac{\sin^{n-2} ax\,dx}{\cos ax}$ $(n > 1;$ for $n = 1$ see 364).

343. $\displaystyle\int \frac{\sin^n ax}{\cos^m ax}\,dx = \frac{\sin^{n+1} ax}{a(m-1)\cos^{m-1} ax} - \frac{n-m+2}{m-1}\int \frac{\sin^n ax}{\cos^{m-2} ax}\,dx$ $(m > 1)$,

$$= -\frac{\sin^{n-1} ax}{a(n-m)\cos^{m-1} ax} + \frac{n-1}{n-m}\int \frac{\sin^{n-2} ax\,dx}{\cos^m ax}$$

($m \neq n$; for $m = n$ see 367),

$$= \frac{\sin^{n-1} ax}{a(m-1)\cos^{m-1} ax} - \frac{n-1}{m-1}\int \frac{\sin^{n-2} ax\,dx}{\cos^{m-2} ax} \quad (m > 1).$$

344. $\displaystyle\int \frac{\cos ax\,dx}{\sin^2 ax} = -\frac{1}{a\sin ax}$.

345. $\displaystyle\int \frac{\cos ax\,dx}{\sin^3 ax} = -\frac{1}{2a\sin^2 ax} + C = -\frac{\cot^2 ax}{2a} + C_1$.

346. $\displaystyle\int \frac{\cos ax\,dx}{\sin^n ax} = -\frac{1}{a(n-1)\sin^{n-1} ax}$ $(n > 1;$ for $n = 1$ see 373).

347. $\displaystyle\int \frac{\cos^2 ax\,dx}{\sin ax} = \frac{1}{a}\left(\cos ax + \ln\left|\tan\frac{ax}{2}\right|\right)$.

348. $$\int \frac{\cos^2 ax \, dx}{\sin^3 ax} = -\frac{1}{2a}\left(\frac{\cos ax}{\sin^2 ax} - \ln\left|\tan\frac{ax}{2}\right|\right).$$

349. $$\int \frac{\cos^2 ax \, dx}{\sin^n ax} = -\frac{1}{(n-1)}\left(\frac{\cos ax}{a \sin^{n-1} ax} + \int \frac{dx}{\sin^{n-2} ax}\right)$$

(see 254–257; $n > 1$).

350. $$\int \frac{\cos^n ax}{\sin ax} dx = \frac{\cos^{n-1} ax}{a(n-1)} + \int \frac{\cos^{n-2} ax \, dx}{\sin ax} \quad (n > 1;\ \text{for } n = 1 \text{ see } 373).$$

351. $$\int \frac{\cos^n ax \, dx}{\sin^m ax} =$$

$$= -\frac{\cos^{n+1} ax}{a(m-1)\sin^{m-1} ax} - \frac{n-m+2}{m-1}\int \frac{\cos^n ax \, dx}{\sin^{m-2} ax} \quad (m > 1),$$

$$= \frac{\cos^{n-1} ax}{a(n-m)\sin^{m-1} ax} + \frac{n-1}{n-m}\int \frac{\cos^{n-2} ax \, dx}{\sin^m ax}$$

($m \neq n$; for $m = n$ see 376),

$$= -\frac{\cos^{n-1} ax}{a(m-1)\sin^{m-1} ax} - \frac{n-1}{m-1}\int \frac{\cos^{n-2} ax \, dx}{\sin^{m-2} ax} \quad (m > 1).$$

352. $$\int \frac{dx}{\sin ax(1 \pm \cos ax)} = \pm\frac{1}{2a(1 \pm \cos ax)} + \frac{1}{2a}\ln\left|\tan\frac{ax}{2}\right|.$$

353. $$\int \frac{dx}{\cos ax(1 \pm \sin ax)} = \mp\frac{1}{2a(1 \pm \sin ax)} + \frac{1}{2a}\ln\left|\tan\left(\frac{\pi}{4} + \frac{ax}{2}\right)\right|.$$

354. $$\int \frac{\sin ax \, dx}{\cos ax(1 \pm \cos ax)} = \frac{1}{a}\ln\left|\frac{1 \pm \cos ax}{\cos ax}\right|.$$

355. $$\int \frac{\cos ax \, dx}{\sin ax(1 \pm \sin ax)} = -\frac{1}{a}\ln\left|\frac{1 \pm \cos ax}{\sin ax}\right|.$$

356. $$\int \frac{\sin ax \, dx}{\cos ax(1 \pm \sin ax)} = \frac{1}{2a(1 \pm \sin ax)} \pm \frac{1}{2a}\ln\left|\tan\left(\frac{\pi}{4} + \frac{ax}{2}\right)\right|.$$

357. $$\int \frac{dx}{b \sin ax + c \cos ax} = \frac{1}{a\sqrt{(b^2 + c^2)}}\ln\left|\tan\frac{ax + \vartheta}{2}\right|,$$

where $\sin\vartheta = \dfrac{c}{\sqrt{(b^2 + c^2)}}$, $\cos\vartheta = \dfrac{b}{\sqrt{(b^2 + c^2)}}$.

358. $\displaystyle\int \frac{\sin ax \, dx}{b + c \cos ax} = -\frac{1}{ac} \ln |b + c \cos ax| .$

359. $\displaystyle\int \frac{\cos ax \, dx}{b + c \sin ax} = \frac{1}{ac} \ln |b + c \sin ax| .$

360. $\displaystyle\int \frac{dx}{b + c \cos ax + f \sin ax} = \int \frac{d(x + \vartheta/a)}{b + \sqrt{(c^2 + f^2)} \sin (ax + \vartheta)},$

where $\displaystyle\sin \vartheta = \frac{c}{\sqrt{(c^2 + f^2)}}, \quad \cos \vartheta = \frac{f}{\sqrt{(c^2 + f^2)}}$ (see 274).

361. $\displaystyle\int \frac{dx}{b^2 \cos^2 ax + c^2 \sin^2 ax} = \frac{1}{abc} \arctan \left(\frac{c}{b} \tan ax \right).$

362. $\displaystyle\int \frac{dx}{b^2 \cos^2 ax - c^2 \sin^2 ax} = \frac{1}{2abc} \ln \left| \frac{c \tan (ax) + b}{c \tan (ax) - b} \right| .$

363. $\displaystyle\int \sin ax \cos bx \, dx = -\frac{\cos (a + b) x}{2(a + b)} - \frac{\cos (a - b) x}{2(a - b)}$

($a^2 \neq b^2$, for $a = b$ see 322).

(δ) *Integrals Containing the Tangent and Cotangent.*

364. $\displaystyle\int \tan ax \, dx = -\frac{1}{a} \ln |\cos ax| .$

365. $\displaystyle\int \tan^2 ax \, dx = \frac{\tan ax}{a} - x .$

366. $\displaystyle\int \tan^3 ax \, dx = \frac{1}{2a} \tan^2 ax + \frac{1}{a} \ln |\cos ax| .$

367. $\displaystyle\int \tan^n ax \, dx = \frac{1}{a(n-1)} \tan^{n-1} ax - \int \tan^{n-2} ax \, dx .$

368. $\displaystyle\int x \tan ax \, dx =$

$$= \frac{ax^3}{3} + \frac{a^3x^5}{15} + \frac{2a^5x^7}{105} + \frac{17a^7x^9}{2,835} + \ldots + \frac{2^{2n}(2^{2n} - 1) B_n a^{2n-1} x^{2n+1}}{(2n + 1)!} + \ldots$$

$\left(|x| < \frac{\pi}{2|a|} ;\right.$ see Remark 3, p. 549$\left.\right)$.

369. $$\int \frac{\tan ax\,dx}{x} =$$

$$= ax + \frac{(ax)^3}{9} + \frac{2(ax)^5}{75} + \frac{17(ax)^7}{2,205} + \dots + \frac{2^{2n}(2^{2n}-1)\,B_n\,.\,(ax)^{2n-1}}{(2n-1)\,(2n)!} + \dots$$

$$\left(|x| < \frac{\pi}{2|a|}\ ;\ \text{see Remark 3, p. 549}\right).$$

370. $$\int \frac{\tan^n ax}{\cos^2 ax}\,dx = \frac{1}{a(n+1)}\tan^{n+1} ax\,.$$

371. $$\int \frac{dx}{\tan ax \pm 1} = \pm\frac{x}{2} + \frac{1}{2a}\ln|\sin ax \pm \cos ax|\,.$$

372. $$\int \frac{\tan ax\,dx}{\tan(ax) \pm 1} = \frac{x}{2} \mp \frac{1}{2a}\ln|\sin ax \pm \cos ax|\,.$$

373. $$\int \cot ax\,dx = \frac{1}{a}\ln|\sin ax|\,.$$

374. $$\int \cot^2 ax\,dx = -\frac{\cot ax}{a} - x\,.$$

375. $$\int \cot^3 ax\,dx = -\frac{1}{2a}\cot^2 ax - \frac{1}{a}\ln|\sin ax|\,.$$

376. $$\int \cot^n ax\,dx = -\frac{1}{a(n-1)}\cot^{n-1} ax - \int \cot^{n-2} ax\,dx \quad (n \neq 1)\,.$$

377. $$\int x\cot ax\,dx = \frac{x}{a} - \frac{ax^3}{9} - \frac{a^3x^5}{225} - \dots - \frac{2^{2n}B_n a^{2n-1}x^{2n+1}}{(2n+1)!} - \dots$$

$$(|x| < \pi/|a|;\ \text{see Remark 3, p. 549})\,.$$

378. $$\int \frac{\cot ax\,dx}{x} = -\frac{1}{ax} - \frac{ax}{3} - \frac{(ax)^3}{135} - \frac{2(ax)^5}{4,725} - \dots - \frac{2^{2n}B_n\,.\,(ax)^{2n-1}}{(2n-1)\,(2n)!} - \dots$$

$$(|x| < \pi/|a|;\ x \neq 0;\ \text{see Remark 3, p. 549})\,.$$

379. $$\int \frac{\cot^n ax}{\sin^2 ax}\,dx = -\frac{1}{a(n+1)}\cot^{n+1} ax\,.$$

380. $$\int \frac{dx}{1 \pm \cot ax} = \int \frac{\tan ax\,dx}{\tan(ax) \pm 1} \quad (\text{see 372})\,.$$

381. $\displaystyle\int \frac{\tan^r ax}{\cos^2 ax} = \frac{1}{a(r+1)} \tan^{r+1} ax \quad (r \neq -1)\,.$

382. $\displaystyle\int \frac{\cot^r ax}{\sin^2 ax} = -\frac{1}{a(r+1)} \cot^{r+1} ax \quad (r \neq -1)\,.$

(d) ***Other Transcendental Functions.***

$a \neq 0$ is assumed

(α) *Hyperbolic Functions.*

383. $\displaystyle\int \sinh ax \, dx = \frac{1}{a} \cosh ax\,.$

384. $\displaystyle\int \cosh ax \, dx = \frac{1}{a} \sinh ax\,.$

385. $\displaystyle\int \sinh^2 ax \, dx = \frac{1}{2a} \sinh ax \cosh ax - \tfrac{1}{2}x\,.$

386. $\displaystyle\int \cosh^2 ax \, dx = \frac{1}{2a} \sinh ax \cosh ax + \tfrac{1}{2}x\,.$

387. $\displaystyle\int \sinh^n ax \, dx = \frac{1}{an} \sinh^{n-1} ax \cosh ax - \frac{n-1}{n} \int \sinh^{n-2} ax \, dx\,.$

388. $\displaystyle\int \frac{dx}{\sinh^n ax} = \frac{\cosh ax}{a(1-n)\sinh^{n-1} ax} - \frac{2-n}{1-n} \int \frac{dx}{\sinh^{n-2} ax} \quad (n \neq 1)\,.$

389. $\displaystyle\int \cosh^n ax \, dx = \frac{1}{an} \sinh ax \cosh^{n-1} ax + \frac{n-1}{n} \int \cosh^{n-2} ax \, dx\,.$

390. $\displaystyle\int \frac{dx}{\cosh^n ax} = -\frac{\sinh ax}{a(1-n)\cosh^{n-1} ax} + \frac{2-n}{1-n} \int \frac{dx}{\cosh^{n-2} ax} \quad (n \neq 1)\,.$

391. $\displaystyle\int \frac{dx}{\sinh ax} = \frac{1}{a} \ln \left| \tanh \frac{ax}{2} \right|.$

392. $\displaystyle\int \frac{dx}{\cosh ax} = \frac{2}{a} \arctan e^{ax}\,.$

393. $\int x \sinh ax \, dx = \frac{1}{a} x \cosh ax - \frac{1}{a^2} \sinh ax .$

394. $\int x \cosh ax \, dx = \frac{1}{a} x \sinh ax - \frac{1}{a^2} \cosh ax .$

395. $\int \tanh ax \, dx = \frac{1}{a} \ln \cosh ax .$

396. $\int \coth ax \, dx = \frac{1}{a} \ln |\sinh ax| .$

397. $\int \tanh^2 ax \, dx = x - \frac{\tanh ax}{a} .$

398. $\int \coth^2 ax \, dx = x - \frac{\coth ax}{a} .$

399. $\int \sinh ax \sinh bx \, dx = \frac{1}{a^2 - b^2} (a \cosh ax \sinh bx - b \sinh ax \cosh bx)$

400. $\int \cosh ax \cosh bx \, dx = \frac{1}{a^2 - b^2} (a \sinh ax \cosh bx - b \cosh ax \sinh bx)$ $(a^2 \neq b^2)$.

401. $\int \cosh ax \sinh bx \, dx = \frac{1}{a^2 - b^2} (a \sinh ax \sinh bx - b \cosh ax \cosh bx)$

$$\int \cosh ax \sinh ax \, dx = \frac{\cosh^2 ax}{2a} + C = \frac{\sinh^2 ax}{2a} + C_1 .$$

402. $\int \sinh ax \sin ax \, dx = \frac{1}{2a} (\cosh ax \sin ax - \sinh ax \cos ax) .$

403. $\int \cosh ax \cos ax \, dx = \frac{1}{2a} (\sinh ax \cos ax + \cosh ax \sin ax) .$

404. $\int \sinh ax \cos ax \, dx = \frac{1}{2a} (\cosh ax \cos ax + \sinh ax \sin ax) .$

405. $\int \cosh ax \sin ax \, dx = \frac{1}{2a} (\sinh ax \sin ax - \cosh ax \cos ax) .$

(β) *Exponential Functions.*

406. $\int e^{ax}\,dx = \frac{1}{a}\,e^{ax}.$

407. $\int xe^{ax}\,dx = \frac{e^{ax}}{a^2}(ax - 1).$

408. $\int x^2 e^{ax}\,dx = e^{ax}\left(\frac{x^2}{a} - \frac{2x}{a^2} + \frac{2}{a^3}\right).$

409. $\int x^n e^{ax}\,dx = \frac{1}{a}\,x^n e^{ax} - \frac{n}{a}\int x^{n-1} e^{ax}\,dx.$

410. $\int \frac{e^{ax}}{x}\,dx = \ln|x| + \frac{ax}{1\,.\,1!} + \frac{(ax)^2}{2\,.\,2!} + \frac{(ax)^3}{3\,.\,3!} + \ldots \qquad (x \neq 0;\ \text{see also § 15.7}).$

411. $\int \frac{e^{ax}}{x^n}\,dx = \frac{1}{n-1}\left(-\frac{e^{ax}}{x^{n-1}} + a\int \frac{e^{ax}}{x^{n-1}}\,dx\right) \quad (n > 1).$

412. $\int \frac{dx}{b + ce^{ax}} = \frac{x}{b} - \frac{1}{ab}\ln|b + ce^{ax}| \quad (b \neq 0).$

413. $\int \frac{e^{ax}\,dx}{b + ce^{ax}} = \frac{1}{ac}\ln|b + ce^{ax}| \quad (c \neq 0).$

414. $\int \frac{dx}{be^{ax} + ce^{-ax}} = \frac{1}{a\sqrt{(bc)}}\arctan\left(e^{ax}\sqrt{\frac{b}{c}}\right) \quad (b > 0,\ c > 0),$

$$= \frac{1}{2a\sqrt{(-bc)}}\ln\left|\frac{c + e^{ax}\sqrt{(-bc)}}{c - e^{ax}\sqrt{(-bc)}}\right| \quad (bc < 0).$$

415. $\int \frac{xe^{ax}\,dx}{(1 + ax)^2} = \frac{e^{ax}}{a^2(1 + ax)}.$

416. $\int e^{ax}\ln x\,dx = \frac{e^{ax}\ln x}{a} - \frac{1}{a}\int \frac{e^{ax}}{x}\,dx \quad (x > 0;\ \text{see 410}).$

417. $\int e^{ax}\sin bx\,dx = \frac{e^{ax}}{a^2 + b^2}(a\sin bx - b\cos bx).$

418. $\int e^{ax}\cos bx\,dx = \frac{e^{ax}}{a^2 + b^2}(a\cos bx + b\sin bx).$

419. $$\int e^{ax} \sin^n x \, dx = \frac{e^{ax} \sin^{n-1} x}{a^2 + n^2}(a \sin x - n \cos x) + \\ + \frac{n(n-1)}{a^2 + n^2} \int e^{ax} \sin^{n-2} x \, dx \quad \text{(see 406 and 417)}.$$

420. $$\int e^{ax} \cos^n x \, dx = \frac{e^{ax} \cos^{n-1} x}{a^2 + n^2}(a \cos x + n \sin x) + \\ + \frac{n(n-1)}{a^2 + n^2} \int e^{ax} \cos^{n-2} x \, dx \quad \text{(see 406 and 418)}.$$

421. $$\int x e^{ax} \sin bx \, dx = \frac{x e^{ax}}{a^2 + b^2}(a \sin bx - b \cos bx) - \\ - \frac{e^{ax}}{(a^2 + b^2)^2}\left[(a^2 - b^2) \sin bx - 2ab \cos bx\right].$$

422. $$\int x e^{ax} \cos bx \, dx = \frac{x e^{ax}}{a^2 + b^2}(a \cos bx + b \sin bx) - \\ - \frac{e^{ax}}{(a^2 + b^2)^2}\left[(a^2 - b^2) \cos bx + 2ab \sin bx\right].$$

423. $$\int b^{ax} \, dx = \frac{b^{ax}}{a \ln b} \quad (b > 0,\ b \neq 1).$$

424. $$\int x b^{ax} \, dx = \frac{x b^{ax}}{a \ln b} - \frac{b^{ax}}{a^2 (\ln b)^2} \quad (b > 0,\ b \neq 1).$$

(γ) *Logarithmic Functions.*

$x > 0$ is assumed

425. $$\int \ln x \, dx = x \ln x - x.$$

426. $$\int (\ln x)^2 \, dx = x(\ln x)^2 - 2x \ln x + 2x.$$

427. $$\int (\ln x)^3 \, dx = x(\ln x)^3 - 3x(\ln x)^2 + 6x \ln x - 6x.$$

428. $\int (\ln x)^n \, dx = x(\ln x)^n - n \int (\ln x)^{n-1} \, dx$.

429. $\int \frac{dx}{\ln x} = \ln |\ln x| + \ln x + \frac{(\ln x)^2}{2 \cdot 2!} + \frac{(\ln x)^3}{3 \cdot 3!} + \ldots$

($x > 0$, $x \neq 1$; see also § 13.12 and 15.7) .

430. $\int \frac{dx}{(\ln x)^n} = -\frac{x}{(n-1)(\ln x)^{n-1}} + \frac{1}{n-1} \int \frac{dx}{(\ln x)^{n-1}}$ ($n > 1$; see 429) .

431. $\int x^r \ln x \, dx = x^{r+1} \left[\frac{\ln x}{r+1} - \frac{1}{(r+1)^2} \right]$ ($r \neq -1$) .

432. $\int x^r (\ln x)^n \, dx = \frac{x^{r+1}(\ln x)^n}{r+1} - \frac{n}{r+1} \int x^r (\ln x)^{n-1} \, dx$ ($r \neq -1$; see 431) .

433. $\int \frac{(\ln x)^n}{x} \, dx = \frac{(\ln x)^{n+1}}{n+1}$.

434. $\int \frac{x^r \, dx}{\ln x} = \int \frac{e^{-y}}{y} \, dy$, where $y = -(r+1) \ln x$ ($r \neq -1$; see 410, 436) .

435. $\int \frac{x^r \, dx}{(\ln x)^n} = -\frac{x^{r+1}}{(n-1)(\ln x)^{n-1}} + \frac{r+1}{n-1} \int \frac{x^r \, dx}{(\ln x)^{n-1}}$ ($n > 1$) .

436. $\int \frac{dx}{x \ln x} = \ln |\ln x|$.

437. $\int \frac{dx}{x^n \ln x} = \ln |\ln x| - (n-1) \ln x + \frac{(n-1)^2 (\ln x)^2}{2 \cdot 2!} -$

$- \frac{(n-1)^3 (\ln x)^3}{3 \cdot 3!} + \ldots$ ($x > 0$, $x \neq 1$) .

438. $\int \frac{dx}{x(\ln x)^n} = \frac{-1}{(n-1)(\ln x)^{n-1}}$ ($n > 1$) .

439. $\int \frac{dx}{x^r (\ln x)^n} = \frac{-1}{x^{r-1}(n-1)(\ln x)^{n-1}} - \frac{r-1}{n-1} \int \frac{dx}{x^r (\ln x)^{n-1}}$ ($n > 1$) .

440. $\int \ln \sin x \, dx = x \ln x - x - \frac{x^3}{18} - \frac{x^5}{900} - \ldots - \frac{2^{2n-1} B_n x^{2n+1}}{n(2n+1)!} - \ldots$

($0 < x < \pi$; see Remark 3, p. 549)

441. $$\int \ln \cos x \, dx = -\frac{x^3}{6} - \frac{x^5}{60} - \frac{x^7}{315} - \dots - \frac{2^{2n-1}(2^{2n}-1) B_n x^{2n+1}}{n(2n+1)!} - \dots$$
$(-\tfrac{1}{2}\pi < x < \tfrac{1}{2}\pi$; see Remark 3, p. 549).

442. $$\int \ln \tan x \, dx = x \ln x - x + \frac{x^3}{9} + \frac{7x^5}{450} + \dots + \frac{2^{2n}(2^{2n-1}-1) B_n x^{2n+1}}{n(2n+1)!} + \dots$$
$(0 < x < \tfrac{1}{2}\pi$; see Remark 3, p. 549).

443. $$\int \sin \ln x \, dx = \frac{x}{2}(\sin \ln x - \cos \ln x).$$

444. $$\int \cos \ln x \, dx = \frac{x}{2}(\sin \ln x + \cos \ln x).$$

445. $$\int e^{ax} \ln x \, dx = \frac{1}{a} e^{ax} \ln x - \frac{1}{a}\int \frac{e^{ax}}{x} dx \quad \text{(see 410)}.$$

(δ) *Inverse Trigonometric Functions.*

446. $$\int \arcsin \frac{x}{a} dx = x \arcsin \frac{x}{a} + \sqrt{(a^2 - x^2)} \quad (a > 0).$$

447. $$\int x \arcsin \frac{x}{a} dx = \left(\frac{x^2}{2} - \frac{a^2}{4}\right) \arcsin \frac{x}{a} + \frac{x}{4}\sqrt{(a^2 - x^2)} \quad (a > 0).$$

448. $$\int x^2 \arcsin \frac{x}{a} dx = \frac{x^3}{3} \arcsin \frac{x}{a} + \frac{1}{9}(x^2 + 2a^2)\sqrt{(a^2 - x^2)} \quad (a > 0).$$

449. $$\int \frac{\arcsin \frac{x}{a} dx}{x} = \frac{x}{a} + \frac{1}{2.3.3}\frac{x^3}{a^3} + \frac{1.3}{2.4.5.5}\frac{x^5}{a^5} + \frac{1.3.5}{2.4.6.7.7}\frac{x^7}{a^7} + \dots \quad (|x| \leqq |a|).$$

450. $$\int \frac{\arcsin \frac{x}{a} dx}{x^2} = -\frac{1}{x} \arcsin \frac{x}{a} - \frac{1}{a} \ln \left| \frac{a + \sqrt{(a^2 - x^2)}}{x} \right| \quad (a > 0).$$

451. $$\int \arccos \frac{x}{a} dx = x \arccos \frac{x}{a} - \sqrt{(a^2 - x^2)} \quad (a > 0).$$

452. $\int x \arccos \frac{x}{a} \, dx = \left(\frac{x^2}{2} - \frac{a^2}{4} \right) \arccos \frac{x}{a} - \frac{x}{4} \sqrt{(a^2 - x^2)} \quad (a > 0) .$

453. $\int x^2 \arccos \frac{x}{a} \, dx = \frac{x^3}{3} \arccos \frac{x}{a} - \frac{1}{9} (x^2 + 2a^2) \sqrt{(a^2 - x^2)} \quad (a > 0) .$

454. $$\int \frac{\arccos \frac{x}{a} \, dx}{x} = \frac{\pi}{2} \ln |x| - \frac{x}{a} - \frac{1}{2 . 3 . 3} \frac{x^3}{a^3} - \frac{1 . 3}{2 . 4 . 5 . 5} \frac{x^5}{a^5} - $$
$$- \frac{1 . 3 . 5}{2 . 4 . 6 . 7 . 7} \frac{x^7}{a^7} - \dots \quad (|x| \leqq |a|, \; x \neq 0) .$$

455. $\int \frac{\arccos \frac{x}{a} \, dx}{x^2} = - \frac{1}{x} \arccos \frac{x}{a} + \frac{1}{a} \ln \left| \frac{a + \sqrt{(a^2 - x^2)}}{x} \right| \quad (a > 0) .$

456. $\int \arctan \frac{x}{a} \, dx = x \arctan \frac{x}{a} - \frac{a}{2} \ln (a^2 + x^2) .$

457. $\int x \arctan \frac{x}{a} \, dx = \tfrac{1}{2}(x^2 + a^2) \arctan \frac{x}{a} - \frac{ax}{2} .$

458. $\int x^2 \arctan \frac{x}{a} \, dx = \frac{x^3}{3} \arctan \frac{x}{a} - \frac{ax^2}{6} + \frac{a^3}{6} \ln (a^2 + x^2) .$

459. $\int x^n \arctan \frac{x}{a} \, dx = \frac{x^{n+1}}{n + 1} \arctan \frac{x}{a} - \frac{a}{n + 1} \int \frac{x^{n+1} \, dx}{a^2 + x^2} .$

460. $\int \frac{\arctan \frac{x}{a} \, dx}{x} = \frac{x}{a} - \frac{x^3}{3^2 a^3} + \frac{x^5}{5^2 a^5} - \frac{x^7}{7^2 a^7} + \dots \quad (|x| \leqq |a|) .$

461. $\int \frac{\arctan \frac{x}{a} \, dx}{x^2} = - \frac{1}{x} \arctan \frac{x}{a} - \frac{1}{2a} \ln \frac{a^2 + x^2}{x^2} .$

462. $\int \frac{\arctan \frac{x}{a} \, dx}{x^n} = - \frac{1}{(n - 1) \, x^{n-1}} \arctan \frac{x}{a} + \frac{a}{n - 1} \int \frac{dx}{x^{n-1}(a^2 + x^2)} \quad (n > 1) .$

463. $\int \operatorname{arccot} \frac{x}{a} \, dx = x \operatorname{arccot} \frac{x}{a} + \frac{a}{2} \ln (a^2 + x^2) .$

464. $\int x \operatorname{arccot} \frac{x}{a} \, dx = \frac{1}{2}(x^2 + a^2) \operatorname{arccot} \frac{x}{a} + \frac{ax}{2}$.

465. $\int x^2 \operatorname{arccot} \frac{x}{a} \, dx = \frac{x^3}{3} \operatorname{arccot} \frac{x}{a} + \frac{ax^2}{6} - \frac{a^3}{6} \ln (a^2 + x^2)$.

466. $\int x^n \operatorname{arccot} \frac{x}{a} \, dx = \frac{x^{n+1}}{n+1} \operatorname{arccot} \frac{x}{a} + \frac{a}{n+1} \int \frac{x^{n+1} \, dx}{a^2 + x^2}$.

467. $\int \frac{\operatorname{arccot} \frac{x}{a} \, dx}{x} = \frac{\pi}{2} \ln |x| - \frac{x}{a} + \frac{x^3}{3^2 a^3} - \frac{x^5}{5^2 a^5} + \frac{x^7}{7^2 a^7} - \dots \quad (|x| \leqq |a|, \; x \neq 0)$.

468. $\int \frac{\operatorname{arccot} \frac{x}{a} \, dx}{x^2} = -\frac{1}{x} \arctan \frac{x}{a} + \frac{1}{2a} \ln \frac{a^2 + x^2}{x^2}$.

469. $\int \frac{\operatorname{arccot} \frac{x}{a} \, dx}{x^n} = -\frac{1}{(n-1)\, x^{n-1}} \operatorname{arccot} \frac{x}{a} - \frac{a}{n-1} \int \frac{dx}{x^{n-1}(a^2 + x^2)} \quad (n > 1)$.

(ε) *Inverse Hyperbolic Functions.*

470. $\int \operatorname{arsinh} \frac{x}{a} \, dx = x \operatorname{arsinh} \frac{x}{a} - \sqrt{(x^2 + a^2)} \quad (a > 0)$.

471. $\int \operatorname{arcosh} \frac{x}{a} \, dx = x \operatorname{arcosh} \frac{x}{a} - \sqrt{(x^2 - a^2)} \quad (x > a > 0)$.

472. $\int \operatorname{artanh} \frac{x}{a} \, dx = x \operatorname{artanh} \frac{x}{a} + \frac{a}{2} \ln (a^2 - x^2) \quad (|x| < |a|)$.

473. $\int \operatorname{arcoth} \frac{x}{a} \, dx = x \operatorname{arcoth} \frac{x}{a} + \frac{a}{2} \ln (x^2 - a^2) \quad (|x| > |a|)$.

REMARK 1. Some simple reductions of rational functions to partial fractions:

$$\frac{1}{(a + bx)(f + gx)} = \frac{1}{bf - ag} \left(\frac{b}{a + bx} - \frac{g}{f + gx} \right);$$

$$\frac{1}{(x + a)(x + b)(x + c)} = \frac{A}{x + a} + \frac{B}{x + b} + \frac{C}{x + c},$$

where

$$A = \frac{1}{(b-a)(c-a)}, \quad B = \frac{1}{(a-b)(c-b)}, \quad C = \frac{1}{(a-c)(b-c)};$$

$$\frac{1}{(x+a)(x+b)(x+c)(x+d)} = \frac{A}{x+a} + \frac{B}{x+b} + \frac{C}{x+c} + \frac{D}{x+d},$$

where

$$A = \frac{1}{(b-a)(c-a)(d-a)}, \quad B = \frac{1}{(a-b)(c-b)(d-b)},$$

$$C = \frac{1}{(a-c)(b-c)(d-c)}, \quad D = \frac{1}{(a-d)(b-d)(c-d)};$$

$$\frac{1}{(a+bx^2)(f+gx^2)} = \frac{1}{bf-ag}\left(\frac{b}{a+bx^2} - \frac{g}{f+gx^2}\right).$$

REMARK 2. On integrals of the type

$$\int x^m(a+bx^n)^p \, dx$$

(binomial integrals) see § 13.4.

REMARK 3. The Bernoulli coefficients B_n:

B_1	$\frac{1}{6}$	B_4	$\frac{1}{30}$	B_7	$\frac{7}{6}$	B_{10}	$\frac{174,611}{330}$
B_2	$\frac{1}{30}$	B_5	$\frac{5}{66}$	B_8	$\frac{3,617}{510}$	B_{11}	$\frac{854,513}{138}$
B_3	$\frac{1}{42}$	B_6	$\frac{691}{2,730}$	B_9	$\frac{43,867}{798}$		

REMARK 4. The Euler coefficients E_n:

E_1	1	E_3	61	E_5	50,521	E_7	199,360,981
E_2	5	E_4	1,385	E_6	2,702,765		

13.6. Definite Integrals. Cauchy–Riemann Definition. Basic Properties. Mean Value Theorems. Evaluation of a Definite Integral

Suppose we are given a function $y = f(x)$ which is continuous in the interval $[a, b]$. Let us divide the interval $[a, b]$ at the points $x_1, x_2, \ldots, x_{n-1}$ into n (closed) subintervals $\Delta x_1, \Delta x_2, \ldots, \Delta x_n$ (Fig. 13.1) which need not be equally long. Since $f(x)$ is

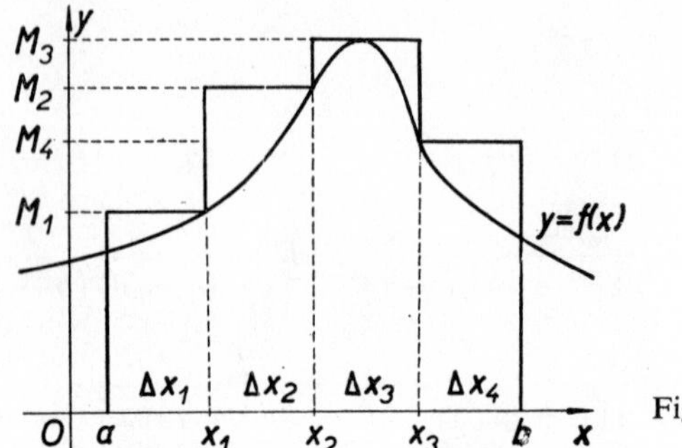

Fig. 13.1.

continuous in $[a, b]$, it assumes its maximum value M and minimum value m in $[a, b]$, and it takes on also its maximum value M_i and minimum value m_i ($M_i \leqq M$, $m_i \geqq m$) in each of the subintervals Δx_i. Let us denote by d the chosen partition of the interval $[a, b]$ and write

$$S(d) = \sum_{i=1}^{n} M_i \, \Delta x_i \,, \quad s(d) = \sum_{i=1}^{n} m_i \, \Delta x_i \,. \tag{1}$$

The numbers $S(d)$ and $s(d)$ are called the *upper Darboux sum* and the *lower Darboux sum* (corresponding to the function $f(x)$ and to the chosen partition), respectively. The geometric meaning of the upper Darboux sum is apparent from Fig. 13.1. The geometric meaning of the lower Darboux sum is similar. If another partition d of $[a, b]$ is chosen, then, generally speaking, other Darboux sums $S(d)$ and $s(d)$ correspond to it. The greatest lower bound (Definition 1.3.3, p. 43) of the values of all upper Darboux sums is called the *upper integral* of the function $f(x)$ in the interval $[a, b]$,

$$\inf S(d) = \overline{\int_a^b} f(x) \, dx \,; \tag{2}$$

similarly the least upper bound of the values of all lower Darboux sums is called the *lower integral* of $f(x)$ in $[a, b]$,

$$\sup s(d) = \underline{\int_a^b} f(x) \, dx \,. \tag{3}$$

REMARK 1. We need not assume the continuity of $f(x)$ in order to be able to define the upper and lower integral; it suffices to assume that $f(x)$ is *bounded* in

$[a, b]$. Instead of maxima and minima of $f(x)$, greatest lower bounds and least upper bounds in the corresponding intervals are then to be considered.

Definition 1. If

$$\overline{\int_a^b} f(x) \, dx = \underline{\int_a^b} f(x) \, dx ,$$

then the common value of these integrals is called the *(definite) integral of the function $f(x)$ over the interval $[a, b]$* and the function $f(x)$ is said to be *integrable in $[a, b]$ according to the Cauchy–Riemann definition.* We write

$$\int_a^b f(x) \, dx .$$

Theorem 1. *Any function piecewise continuous in $[a, b]$* (Definition 11.3.6) *is integrable in $[a, b]$. In particular, any function continuous in $[a, b]$ is integrable in $[a, b]$.* (In § 13.14 an example of a function is given which is not integrable according to the Cauchy–Riemann definition.)

REMARK 2. The Lebesgue and Stieltjes definitions of an integral are briefly mentioned in § 13.14, p. 595.

REMARK 3. The following sums are often considered instead of Darboux sums (cf. Theorem 2): Let us choose, for a fixed partition d in the interval $[a, b]$, an arbitrary point c_k in each interval Δx_k and let us write

$$\sigma(d) = \sum_{k=1}^{n} f(c_k) \, \Delta x_k \tag{4}$$

(this sum depends on how we have chosen d and the points c_k in Δx_k). Obviously

$$s(d) \leqq \sigma(d) \leqq S(d) .$$

The greatest of the lengths of the intervals Δx_i is called the *norm $n(d)$ of the partition d.*

Theorem 2. *Let us consider a sequence of partitions $d_1, d_2, d_3, \ldots$ such that* $\lim_{k\to\infty} n(d_k) = 0$. *If $f(x)$ is integrable in $[a, b]$, then*

$$\int_a^b f(x) \, dx = \lim_{k\to\infty} S(d_k) = \lim_{k\to\infty} s(d_k) = \lim_{k\to\infty} \sigma(d_k) .$$

REMARK 4. Briefly speaking: The integral of the function $f(x)$ in $[a, b]$ is the limit of upper Darboux sums provided that the norms of partitions tend to zero. Similar statements are valid concerning lower Darboux sums and sums (4).

Theorem 3. *If $f_1(x)$ and $f_2(x)$ are integrable in $[a, b]$, then the same is true for the functions $k_1 f_1(x) + k_2 f_2(x)$, $f_1(x) f_2(x)$, $|f_1(x)|$ (and of course for $|f_2(x)|$), and* *the relations*

$$\int_a^b [k_1 f_1(x) + k_2 f_2(x)]\, dx = k_1 \int_a^b f_1(x)\, dx + k_2 \int_a^b f_2(x)\, dx\,,$$

$$\left| \int_a^b f_1(x)\, dx \right| \leqq \int_a^b |f_1(x)|\, dx$$

hold. (The equation

$$\int_a^b f(x)\, g(x)\, dx = \int_a^b f(x)\, dx \int_a^b g(x)\, dx$$

is not, in general, valid! For the function

$$\frac{f_1(x)}{f_2(x)}$$

to be integrable it is sufficient that f_1 and f_2 be integrable and

$$0 < m \leqq f_2(x) \quad \text{or} \quad f_2(x) \leqq M < 0$$

in $[a, b]$, i.e. $f_2(x)$ is bounded below by a positive constant or bounded above by a negative constant.)

Theorem 4. *If $f(x)$ is integrable in $[a, b]$ and if $a < c < b$, then $f(x)$ is integrable in both $[a, c]$ and $[c, b]$ and*

$$\int_a^b f(x)\, dx = \int_a^c f(x)\, dx + \int_c^b f(x)\, dx$$

(and conversely).

Definition 2. For $b < a$ the integral

$$\int_a^b f(x)\, dx$$

is defined by the equation

$$\int_a^b f(x)\, dx = - \int_b^a f(x)\, dx\,.$$

REMARK 5. *In the sequel* (Theorems 5, 7, 8, 10, 11, 12) *the functions considered are assumed to be integrable in $[a, b]$.*

Theorem 5. *If*

$$f(x) \geqq 0 \quad in \quad [a, b]\,,$$

then

$$\int_a^b f(x)\,\mathrm{d}x \geqq 0\,.$$

If, moreover, $f(x)$ is continuous at least at one point c of that interval and if $f(c) > 0$, then

$$\int_a^b f(x)\,\mathrm{d}x > 0\,.$$

Theorem 6. *If $f(x)$ is continuous in $[a, b]$ and $\int_a^b f^2(x)\,\mathrm{d}x = 0$, then $f(x) \equiv 0$ in $[a, b]$.*

Theorem 7. *If $f(x) \geqq g(x)$ in $[a, b]$, then $\int_a^b f(x)\,\mathrm{d}x \geqq \int_a^b g(x)\,\mathrm{d}x$.*

Theorem 8. *Let*

$$m \leqq f(x) \leqq M \quad \textit{and} \quad g(x) \geqq 0 \quad \textit{in} \quad [a, b]\,.$$

Then

$$m \int_a^b g(x)\,\mathrm{d}x \leqq \int_a^b f(x)\,g(x)\,\mathrm{d}x \leqq M \int_a^b g(x)\,\mathrm{d}x\,. \tag{5}$$

REMARK 6. In particular, if $g(x) \equiv 1$, then

$$m(b - a) \leqq \int_a^b f(x)\,\mathrm{d}x \leqq M(b - a)\,. \tag{6}$$

If

$$|f(x)| \leqq K \quad \text{in} \quad [a, b]\,,$$

then

$$\left|\int_a^b f(x)\,\mathrm{d}x\right| \leqq K(b - a)\,. \tag{7}$$

REMARK 7. The inequality (5) is convenient for estimating the integral

$$\int_a^b f(x)\,g(x)\,\mathrm{d}x$$

(for example in the case where the integration of the product $f(x)\,g(x)$ is rather complicated).

The inequality (6) can also be used to estimate an integral.

Example 1. Let us estimate

$$\int_0^1 \frac{\mathrm{d}x}{10 + \sqrt{(x^2 + 3)} - 0{\cdot}1 \cos^7 x - x^4}\,.$$

In the interval considered the inequalities

$$\frac{1}{10+\sqrt{(1+3)}} < f(x) < \frac{1}{10+\sqrt{(3)}-0\cdot1-1} < \frac{1}{10}$$

obviously hold. Hence

$$\frac{1}{12} < \int_0^1 \frac{\mathrm{d}x}{10+\sqrt{(x^2+3)}-0\cdot1\cos^7 x - x^4} < \frac{1}{10}.$$

Theorem 9 (*The First Mean Value Theorem*). *If $f(x)$ is continuous in $[a, b]$, then there is at least one point $c \in (a, b)$ such that*

$$\int_a^b f(x)\,\mathrm{d}x = (b-a)\,.\,f(c)\,. \tag{8}$$

(The value $f(c)$ defined by equation (8) is called the *mean value of the function $f(x)$ in the interval $[a, b]$*).

More generally: If $f(x)$ is continuous in $[a, b]$, $g(x)$ integrable in $[a, b]$ and $g(x) \geqq 0$ or $g(x) \leqq 0$, then there exists at least one point $c \in (a, b)$ such that

$$\int_a^b f(x)\,g(x)\,\mathrm{d}x = f(c)\int_a^b g(x)\,\mathrm{d}x\,.$$

Theorem 10 (*The Second Mean Value Theorem*). *Let $g(x)$ be monotonic (i.e. either increasing or decreasing) in $[a, b]$. Then there is at least one point $c \in (a, b)$ such that*

$$\int_a^b f(x)\,g(x)\,\mathrm{d}x = g(a)\int_a^c f(x)\,\mathrm{d}x + g(b)\int_c^b f(x)\,\mathrm{d}x\,.$$

Theorem 11 (*The Schwarz or Schwarz–Cauchy inequality*).

$$\left[\int_a^b f(x)\,g(x)\,\mathrm{d}x\right]^2 \leqq \int_a^b f^2(x)\,\mathrm{d}x\int_a^b g^2(x)\,\mathrm{d}x\,.$$

Theorem 12. *The function*

$$G(x) = \int_a^x f(t)\,\mathrm{d}t$$

is a continuous function of the variable x in the interval $[a, b]$. $G(x)$ possesses a derivative at every point for which $f(x)$ is continuous, and the derivative equals the value of the function $f(x)$ at that point, i.e.

$$\frac{\mathrm{d}G}{\mathrm{d}x} = \frac{\mathrm{d}}{\mathrm{d}x}\int_a^x f(t)\,\mathrm{d}t = f(x)\,.$$

REMARK 8. If $f(x)$ is continuous in (a, b), it follows that $G(x)$ is a primitive of this function in (a, b); under the same assumptions,

$$\frac{d}{dx}\int_x^b f(t)\,dt = -f(x)\,.$$

Theorem 13. *If $f(x)$ is continuous in $[a, b]$ and if $F(x)$ is a primitive of $f(x)$ in (a, b) (continuous in $[a, b]$), then*

$$\int_a^b f(x)\,dx = F(b) - F(a)\,. \tag{9}$$

REMARK 9. This equation is of fundamental importance for the evaluation of the definite integral. In applications, the right-hand side of equation (9) is usually denoted by

$$[F(x)]_a^b \quad \text{or} \quad F(x)\Big|_a^b\,, \quad \text{hence} \quad \int_a^b f(x)\,dx = [F(x)]_a^b = F(x)\Big|_a^b\,. \tag{10}$$

Example 2.

$$\int_{-2}^5 x^2\,dx = \left[\frac{x^3}{3}\right]_{-2}^5 = \frac{5^3}{3} - \frac{(-2)^3}{3} = \frac{133}{3} = 44\tfrac{1}{3}\,.$$

Theorem 14. *If $f(x)$ is continuous in $[a, b]$ and $f(x) \geqq 0$ in $[a, b]$, then the integral*

$$\int_a^b f(x)\,dx$$

is equal to the area of the region bounded by the x-axis, by the graph of the function $y = f(x)$ and by the lines parallel to the y-axis through the points $x = a$, $x = b$.

REMARK 10. The area is always positive (or zero). If $f(x) < 0$ holds in the interval $[c, d]$ which is a subinterval of $[a, b]$, then the integral over $[c, d]$ is negative. If we want to determine the area of the region in question, we have to change the sign of the integral in this subinterval. Hence, if the graph of the function $y = f(x)$ crosses the x-axis in the interval $[a, b]$, we determine the coordinates of the intersections and divide the interval $[a, b]$ into intervals in which $f(x)$ is either negative, or positive.

Example 3. To find the area of the region shaded in Fig. 13.2, we write $P = P_1 + P_2$, where

$$P_1 = \int_0^\pi \sin x\,dx = [-\cos x]_0^\pi = -(-1 - 1) = 2\,,$$

$$P_2 = -\int_\pi^{2\pi} \sin x\,dx = [\cos x]_\pi^{2\pi} = 1 - (-1) = 2\,,$$

hence $P = 4$. The second integral had to be taken with negative sign, since $\sin x \leqq 0$ in $[\pi, 2\pi]$. Direct calculation over the whole interval 0 to 2π gives zero:

$$\int_0^{2\pi} \sin x \, dx = [-\cos x]_0^{2\pi} = -(1-1) = 0 .$$

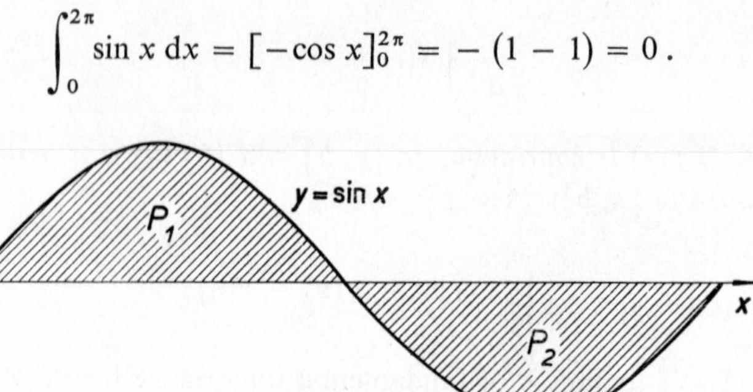

Fig. 13.2.

REMARK 11. If $f(x)$ is an (integrable) *even* function, i.e. if $f(-x) = f(x)$, then $\int_{-a}^{a} f(x)\,dx = 2\int_0^a f(x)\,dx$; if $f(x)$ is *odd*, i.e. if $f(-x) = -f(x)$, then $\int_{-a}^{a} f(x)\,dx = 0$.

Example 4.

$$\int_{-3}^{3} x^2 \, dx = 2\int_0^3 x^2 \, dx = 2\left[\frac{x^3}{3}\right]_0^3 = 18 ;$$

$$\int_{-\pi/4}^{\pi/4} \tan x \, dx = 0 .$$

REMARK 12. The so-called *Newton's definite integral* is defined by equation (9). Hence, Newton's definition assumes (only) the existence of a primitive. The equivalence of Riemann's and Newton's definition for the case where f is continuous follows from Theorem 13.

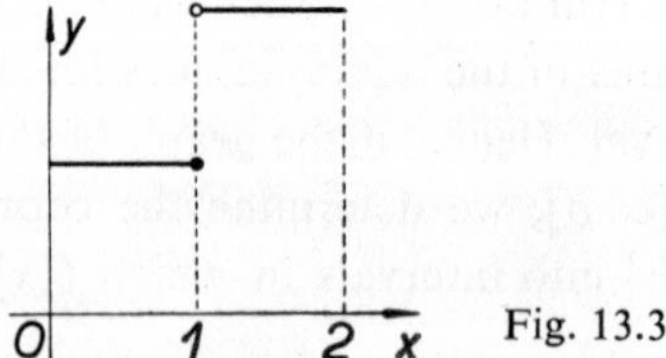

Fig. 13.3.

REMARK 13. The function remains integrable and the value of the integral does not change if the values of the function are changed at a finite number of points of $[a, b]$. For example, the integral of the function

$$f(x) = \begin{cases} 1 & \text{for} \quad 0 \leqq x \leqq 1 , \\ 2 & \text{for} \quad 1 < x \leqq 2 \end{cases}$$

(Fig. 13.3) is calculated as follows:

$$\int_0^2 f(x)\,\mathrm{d}x = \int_0^1 1\,.\,\mathrm{d}x + \int_1^2 2\,.\,\mathrm{d}x = 1 + 2 = 3\,.$$

The second integral is evaluated in the same way as if $f(1)$ were equal to 2.

REMARK 14. On geometrical and physical applications of the definite integral see § 14.9.

13.7. Substitution and Integration by Parts for Definite Integrals

Theorem 1 (*Integration by Parts*). *If $u'(x)$ and $v'(x)$ are continuous in $[a, b]$ (then also $u(x)$ and $v(x)$ are continuous in $[a, b]$,* Theorem 11.5.2), *then*

$$\int_a^b u'v\,\mathrm{d}x = [uv]_a^b - \int_a^b uv'\,\mathrm{d}x\,.$$

In another form

$$\int_a^b v\,\mathrm{d}u = [uv]_a^b - \int_a^b u\,\mathrm{d}v\,.$$

(The notation $[uv]_a^b$ means $u(b)\,v(b) - u(a)\,v(a)$.)

Example 1.

$$\int_0^\pi x \sin x\,\mathrm{d}x = [-x\cos x]_0^\pi + \int_0^\pi \cos x\,\mathrm{d}x = \pi + [\sin x]_0^\pi = \pi\,.$$

REMARK 1. If $u(x)$, $v(x)$, $u'(x)$, $v'(x)$ are only *piecewise* continuous in $[a, b]$ (Definition 11.3.6) and if c_k $(k = 1, 2, \ldots, n)$ denote the points of discontinuity of the functions $u(x)$, $v(x)$ in (a, b), then

$$\int_a^b u'v\,\mathrm{d}x = [uv]_a^b + \sum_{k=1}^n [uv]_{c_k+0}^{c_k-0} - \int_a^b uv'\,\mathrm{d}x = \qquad (1)$$

$$= [uv]_a^{c_1-0} + \sum_{k=1}^{n-1} [uv]_{c_k+0}^{c_{k+1}-0} + [uv]_{c_n+0}^b - \int_a^b uv'\,\mathrm{d}x\,, \qquad (2)$$

where

$$u(c_k - 0) = \lim_{x\to c_k-} u(x)\,, \quad u(c_k + 0) = \lim_{x\to c_k+} u(x)$$

(Remark 11.4.2) and similarly for the function v. We can alternatively first break up the integral into a sum of integrals

$$\int_a^b = \int_a^{c_1} + \int_{c_1}^{c_2} + \ldots + \int_{c_n}^b\,,$$

and then apply the integration by parts to individual integrals. One has to apply the correct limits to the corresponding functions. For instance

$$\int_{c_1}^{c_2} u'v \, dx = [uv]_{c_1+0}^{c_2-0} - \int_{c_1}^{c_2} uv' \, dx \quad \text{etc.} \tag{3}$$

Example 2. Let

$$f(x) = \begin{cases} x \sin x & \text{for} \quad 0 \leqq x \leqq \frac{1}{2}\pi \,, \\ 2x \sin x & \text{for} \quad \frac{1}{2}\pi < x \leqq \pi \,; \end{cases}$$

Then

$$\int_0^\pi f(x) \, dx = \int_0^{\pi/2} x \sin x \, dx + \int_{\pi/2}^{\pi} 2x \sin x \, dx =$$

$$= [-x \cos x]_0^{\pi/2-0} + \int_0^{\pi/2} \cos x \, dx + [-2x \cos x]_{\pi/2+0}^{\pi} + 2\int_{\pi/2}^{\pi} \cos x \, dx =$$

$$= 0 + 1 + 2\pi - 2 = 2\pi - 1 \,.$$

Theorem 2 (*Method of Substitution, Case* (a), *Substitution* $h(x) = z$). *Let* $f(x)$ *be of the form* $f(x) = g(h(x)) \, h'(x)$, *where* $h'(x)$ *is continuous in* $[a, b]$ *and* $g(z)$ *is continuous for all* $z = h(x)$ *if* $x \in [a, b]$. *Then*

$$\int_a^b f(x) \, dx = \int_a^b g(h(x)) \, h'(x) \, dx = \int_{h(a)}^{h(b)} g(z) \, dz \,. \tag{4}$$

Theorem 3 (*Method of Substitution, Case* (b), *Substitution* $x = \varphi(z)$). *Let* $A \leqq a < b \leqq B$, *let* $f(x)$ *be continuous in* $[A, B]$, $\varphi'(z)$ *continuous in* $[\alpha, \beta]$ *and for* $z \in [\alpha, \beta]$ *let* $x = \varphi(z)$ *belong to the interval* $[A, B]$, $\varphi(\alpha) = a$, $\varphi(\beta) = b$. *Then*

$$\int_a^b f(x) \, dx = \int_\alpha^\beta f(\varphi(z)) \, \varphi'(z) \, dz \,. \tag{5}$$

REMARK 2. There are often several possible ways of satisfying the equations $\varphi(\alpha) = a$, $\varphi(\beta) = b$ when applying Theorem 3. It is immaterial which one is chosen as long as the conditions of the theorem are satisfied. This point is illustrated in Example 3.

Example 3. Using the substitution $x = z^2$, we obtain

$$\int_1^4 x \, dx = 2\int_1^2 z^3 \, dz = 2\int_{-1}^2 z^3 \, dz = 2\int_1^{-2} z^3 \, dz = 2\int_{-1}^{-2} z^3 \, dz \,,$$

as can be immediately verified by calculation. Either of the roots of the equation $z^2 = 1$ or $z^2 = 4$, respectively, may be chosen as a new limit of integration.

Example 4. Let us evaluate $\int_{-1}^{1} \sqrt{(1 - x^2)} \, dx$.

We make use of the substitution $x = \sin z$ (Theorem 3). Let us choose $\alpha = -\frac{1}{2}\pi$, $\beta = \frac{1}{2}\pi$. Obviously $\varphi(\alpha) = a$, $\varphi(\beta) = b$, since $\sin(-\frac{1}{2}\pi) = -1$, $\sin\frac{1}{2}\pi = 1$. The conditions of Theorem 3 are satisfied and we have

$$\int_{-1}^{1} \sqrt{(1 - x^2)}\, dx = \int_{-\pi/2}^{\pi/2} \sqrt{(1 - \sin^2 z)} \cos z\, dz = \int_{-\pi/2}^{\pi/2} \cos^2 z\, dz =$$

$$= \frac{1}{2}\int_{-\pi/2}^{\pi/2} (1 + \cos 2z)\, dz = \frac{1}{2}\left[z + \frac{\sin 2z}{2}\right]_{-\pi/2}^{\pi/2} = \tfrac{1}{2}\pi\,.$$

($\sqrt{(1 - \sin^2 z)} = +\cos z$, because $\sqrt{(1 - \sin^2 z)}$ is non-negative and $\cos z \geqq 0$ for $z \in [-\frac{1}{2}\pi, \frac{1}{2}\pi]$.)

Example 5. Evaluate $\int_0^{\pi/2} \sin^3 x \cos x\, dx$.

We use Theorem 2, for the function $\sin^3 x \cos x$ is of the form $g(h(x))\, h'(x)$, where $h(x) = \sin x$, $g(z) = z^3$. Hence, by the substitution $\sin x = z$ we get

$$\int_0^{\pi/2} \sin^3 x \cos x\, dx = \int_0^1 z^3\, dz = \left[\frac{z^4}{4}\right]_0^1 = \frac{1}{4}\,.$$

REMARK 3. The following examples indicate common errors in integration by substitution.

Example 6. The integral $\int_1^4 \sqrt{(x^2 - 1)}\, dx$ may not be integrated by the substitution $x = \sin z$ since $x = \sin z$ could not run through the interval $[1, 4]$, when z runs through any interval $[\alpha, \beta]$. However, the given integral may be evaluated by substitution $\sqrt{(x^2 - 1)} = z - x$ (§ 13.4 substitution (5)). In this case we have $1 \leqq x \leqq 4$, $1 \leqq z \leqq 4 + \sqrt{15}$.

Example 7. Evaluate the integral

$$\int_0^{\pi} \frac{1 + \tan^2 x}{1 + k^2 \tan^2 x}\, dx\,, \quad k > 0\,,\ k \neq 1\,.$$

Substitution:

$$\tan x = z\,, \quad \frac{1}{\cos^2 x}\, dx = dz\,, \quad (1 + \tan^2 x)\, dx = dz\,; \quad \tan(0) = 0\,, \quad \tan(\pi) = 0.$$

Hence

$$\int_0^{\pi} \frac{1 + \tan^2 x}{1 + k^2 \tan^2 x}\, dx = \int_0^0 \frac{dz}{1 + k^2 z^2} = 0\,.$$

Obviously, the result is *wrong*, for an integral of a positive function is positive (Theorem 13.6.5) and we cannot obtain zero as a result. The mistake lies in the substitution $\tan x = z$ and in the fact that $\tan x$ is discontinuous at the point $x = \frac{1}{2}\pi$ in the interval $[0, \pi]$. For the correct solution see Example 13.8.14, p. 571.

13.8. Improper Integrals

Improper integrals are a generalization of the Cauchy–Riemann integral (§ 13.6) which has been defined for a *bounded* function and a *finite* interval.

Definition 1. Let $f(x)$ be integrable according to the Cauchy–Riemann definition in any interval $[a, c]$, $a < c < b$. (We do not require that $f(x)$ be bounded in the whole interval $[a, b]$; it may be unbounded in the (left) neighbourhood of the point b.) If there exists the *finite* limit

$$\lim_{c \to b-} \int_a^c f(x)\,\mathrm{d}x = A\,, \tag{1}$$

we say that the *integral*

$$\int_a^b f(x)\,\mathrm{d}x \tag{2}$$

is convergent (*converges, exists*), and we write

$$\int_a^b f(x)\,\mathrm{d}x = A\,.$$

If the limit (1) does not exist or if it is infinite, integral (2) is said to be *divergent* or to *diverge* (or we say that it *does not exist*).

Example 1.

$$\int_0^1 \frac{\mathrm{d}x}{\sqrt{(1-x)}} = \lim_{c \to 1-} \int_0^c \frac{\mathrm{d}x}{\sqrt{(1-x)}} = \lim_{c \to 1-} \left[-2\sqrt{(1-x)}\right]_0^c =$$

$$= \lim_{c \to 1-} \left[-2\sqrt{(1-c)} + 2\right] = 2\,.$$

Hence this integral is convergent.

Example 2.

$$\int_0^1 \frac{\mathrm{d}x}{1-x} = \lim_{c \to 1-} \int_0^c \frac{\mathrm{d}x}{1-x} = \lim_{c \to 1-} \left[-\ln(1-x)\right]_0^c = \lim_{c \to 1-} \left[-\ln(1-c)\right] = +\infty\,.$$

The integral is divergent. (We say in this case that its value is $+\infty$.)

REMARK 1. Similarly, if $f(x)$ is integrable in any interval $[c, b]$, $a < c < b$ (in (a, c) it need not be bounded), and if there exists the *finite* limit

$$\lim_{c \to a+} \int_c^b f(x)\,\mathrm{d}x = B\,, \tag{3}$$

then the integral

$$\int_a^b f(x)\,\mathrm{d}x$$

is said to be *convergent* (to *converge*, to *exist*); we write

$$\int_a^b f(x)\,\mathrm{d}x = B\,.$$

Otherwise the integral is said to be *divergent*, or to *diverge* (or we say that it *does not exist*).

Definition 2. If d is a fixed point of the interval (a, b) and if $f(x)$ is integrable in arbitrary intervals $[a, a']$, $[b', b]$ with $a < a' < d < b' < b$ (in the neighbourhood of the point d $f(x)$ need not be bounded), then the integral

$$\int_a^b f(x)\,\mathrm{d}x \tag{4}$$

is said to be *convergent* (to *converge*, to *exist*) if the integrals

$$\int_a^d f(x)\,\mathrm{d}x\,, \quad \int_d^b f(x)\,\mathrm{d}x \tag{5}$$

both converge. Their sum is then called the *value of the given integral.* If at least one of the integrals (5) is divergent, the integral (4) is said to be *divergent.*

REMARK 2. If $f(x)$ is not bounded in the neighbourhood of the point a as well as in the neighbourhood of the point b, we consider the integrals

$$\int_a^c f(x)\,\mathrm{d}x\,, \quad \int_c^b f(x)\,\mathrm{d}x\,,$$

where c is an arbitrary point of the interval (a, b). We proceed in a similar way, if $f(x)$ is not bounded in the vicinity of a finite number of points of the interval $[a, b]$.

Example 3.

$$\int_{-1}^2 \frac{\mathrm{d}x}{x} = \int_{-1}^0 \frac{\mathrm{d}x}{x} + \int_0^2 \frac{\mathrm{d}x}{x}$$

provided that the integrals on the right-hand side are both convergent. However, they are both divergent:

$$\int_{-1}^0 \frac{\mathrm{d}x}{x} = \lim_{c\to 0-} \int_{-1}^c \frac{\mathrm{d}x}{x} = \lim_{c\to 0-} [\ln |x|]_{-1}^c = \lim_{c\to 0-} \ln |c| = -\infty$$

and similarly

$$\int_0^2 \frac{dx}{x} = +\infty .$$

Hence the given integral is divergent.

REMARK 3. It is sometimes convenient to deal with the so-called *Cauchy principal value* of the integral. If $f(x)$ is not bounded in the neighbourhood of the point d $(a < d < b)$, then the Cauchy principal value is defined as follows:

$$\int_a^b f(x)\,dx = \lim_{\delta \to 0} \left(\int_a^{d-\delta} f(x)\,dx + \int_{d+\delta}^b f(x)\,dx \right).$$

Thus, symmetric δ-neighbourhoods of the point d are considered and the limit process for $\delta \to 0$ is carried out. If the improper integral exists, then it also exists considered as the Cauchy principal value, but not conversely, in general.

Example 4.

$$\int_{-1}^2 \frac{dx}{x} = \lim_{\delta \to 0+} \left(\int_{-1}^{-\delta} \frac{dx}{x} + \int_{\delta}^2 \frac{dx}{x} \right) = \lim_{\delta \to 0+} (\ln|-\delta| - \ln|-1| + \ln 2 - \ln \delta) =$$

$$= \lim_{\delta \to 0+} (\ln \delta - 0 + \ln 2 - \ln \delta) = \ln 2 .$$

The given integral is convergent considered as the Cauchy principal value, but it is not convergent if taken in the usual sense.

REMARK 4. If we can determine a primitive of a given function, then we can – as a rule – easily determine the limit (1) or (3) and thus evaluate immediately the integral (Examples 1 and 2). In some cases it may be difficult to find the primitive. We then try to evaluate the integral approximately. To do this, we must first know whether the given integral is convergent or not. The following tests enable us to decide this question. (The corresponding Theorems 1–7 are stated *for the case where $f(x)$ is unbounded only in the neighbourhood of the point b and is integrable in any interval $[a, c]$, $a < c < b$*; other cases are treated similarly.)

Theorem 1 (*The Bolzano–Cauchy Condition*). *The integral*

$$\int_a^b f(x)\,dx \tag{6}$$

is convergent if and only if, for arbitrary $\varepsilon > 0$, a $\delta > 0$ can be found such that for every pair of positive numbers δ_1, δ_2, satisfying $\delta_1 < \delta$, $\delta_2 < \delta$, the inequality

$$\left| \int_{b-\delta_1}^{b-\delta_2} f(x)\,dx \right| < \varepsilon$$

holds.

Theorem 2. *If*

$$\int_a^b |f(x)| \,dx$$

is convergent, then so is integral (6). In this case integral (6) is said to be *absolutely convergent.*

Theorem 3. *Let us assume* $0 \leqq \psi(x) \leqq \varphi(x)$ *in* $[a, b)$. *If the integral*

$$\int_a^b \varphi(x) \,dx \tag{7}$$

is convergent, then the same is true for the integral

$$\int_a^b \psi(x) \,dx \,. \tag{8}$$

If integral (8) *is divergent, then integral* (7) *is also divergent.*

Theorem 4. *If the inequality* $|f(x)| \leqq \varphi(x)$ *holds in* $[a, b)$ *and if integral* (7) *is convergent, then integral* (6) *is also convergent (and its convergence is absolute).* (The function $\varphi(x)$ is called a *majorant of* $f(x)$ *in* $[a, b)$.)

Example 5. The integral

$$\int_0^1 \frac{\sin x}{\sqrt{(1 - x)}} \,dx$$

is convergent, for

$$\left|\frac{\sin x}{\sqrt{(1 - x)}}\right| \leqq \frac{1}{\sqrt{(1 - x)}}$$

and the integral of the right-hand side is convergent by Example 1.

Theorem 5. *Let the finite limit*

$$\lim_{x \to b-} \frac{f(x)}{\varphi(x)} = l$$

exist. If

$$\int_a^b |\varphi(x)| \,dx \tag{9}$$

is convergent, then also

$$\int_a^b |f(x)| \,dx \tag{10}$$

is convergent (and hence the same is true for integral (6)). If $l \neq 0$, then if (9) is divergent so also is (10). (At the same time integral (6) may be convergent.)

Theorem 6. *If integral (6) is convergent and $g(x)$ is a bounded monotonic function in $[a, b]$, then*

$$\int_a^b f(x)\, g(x)\, dx \tag{11}$$

is also convergent. Further: If for every $c \in (a, b)$ the inequality

$$\left| \int_a^c f(x)\, dx \right| < K$$

holds and if $g(x)$ is monotonic in $[a, b]$ and

$$\lim_{x \to b-} g(x) = 0\,,$$

then integral (11) is convergent.

Theorem 7. *Let the inequality*

$$|f(x)| \leqq \frac{M}{(b - x)^\alpha} \tag{12}$$

hold in a neighbourhood of the point b (for $x < b$), where M is a constant and $\alpha < 1$. Then the integral (6) is (absolutely) convergent. If

$$|f(x)| \geqq \frac{M}{(b - x)^\alpha} \quad (M > 0, \alpha \geqq 1)\,, \tag{13}$$

then the integral

$$\int_a^b |f(x)|\, dx$$

is divergent (while, however, integral (6) may be convergent).

REMARK 5. In the case where $f(x)$ is unbounded in the neighbourhood of the point a (instead of the point b) (cf. the end of Remark 4), then, of course, we write $x - a$ in place of $b - x$ in (12) and (13).

Example 6. Let us consider the convergence of the integral

$$\int_0^1 \ln^2 x\, dx\,.$$

The function $f(x) = \ln^2 x$ is not bounded in the neighbourhood of the point $x = 0$.

However (by l'Hospital's rule, see Example 11.8.4), if $0 < \alpha < 1$, we have

$$\lim_{x\to 0+} x^\alpha \ln^2 x = \lim_{x\to 0+} \frac{\ln^2 x}{\frac{1}{x^\alpha}} = \lim_{x\to 0+} \frac{\frac{2\ln x}{x}}{-\frac{\alpha}{x^{1+\alpha}}} = -\frac{2}{\alpha}\lim_{x\to 0+} x^\alpha \ln x =$$

$$= -\frac{2}{\alpha}\lim_{x\to 0+} \frac{\frac{1}{x}}{-\frac{\alpha}{x^{1+\alpha}}} = \frac{2}{\alpha^2}\lim_{x\to 0+} x^\alpha = 0\,.$$

Hence the function $x^\alpha \ln^2 x$ is bounded in a (right) neighbourhood of the point $x = 0$, i.e.

$$x^\alpha \ln^2 x < M\,, \quad \ln^2 x < \frac{M}{x^\alpha} \quad (\alpha < 1)\,,$$

and therefore the integral considered is convergent by Theorem 7.

REMARK 6. For details and for many examples see e.g. [112], [367].

REMARK 7. We also speak about improper integrals in the case where the integration is carried out in an *infinite interval*.

Definition 3. Let $f(x)$ be integrable in every interval $[a, b]$, $b > a$. If there exists the finite limit

$$\lim_{b\to+\infty} \int_a^b f(x)\,\mathrm{d}x = A\,, \tag{14}$$

we say that the *integral*

$$\int_a^\infty f(x)\,\mathrm{d}x \tag{15}$$

is convergent (*converges, exists*) and write

$$\int_a^\infty f(x)\,\mathrm{d}x = A\,.$$

If the limit (14) does not exist or is infinite, we say that integral (15) is *divergent* (*diverges, does not exist*).

Similarly, the integral

$$\int_{-\infty}^a f(x)\,\mathrm{d}x$$

is defined.

Example 7.

$$\int_2^{\infty} \frac{dx}{x^3} = \lim_{b\to+\infty} \int_2^b \frac{dx}{x^3} = \lim_{b\to+\infty} \left[-\frac{1}{2x^2} \right]_2^b = \lim_{b\to+\infty} \left(-\frac{1}{2b^2} + \frac{1}{2 \,.\, 2^2} \right) = \frac{1}{8} \,.$$

The integral is convergent.

Definition 4. Let us assume $-\infty < a < +\infty$. If both integrals

$$\int_a^{\infty} f(x)\, dx \,, \quad \int_{-\infty}^{a} f(x)\, dx \tag{16}$$

are convergent, then the integral

$$\int_{-\infty}^{\infty} f(x)\, dx \tag{17}$$

is said to be *convergent* and the sum of integrals (16) to be its *sum*. If at least one of integrals (16) is divergent, integral (17) is said to be *divergent*.

REMARK 8. If the integral (17) is divergent in the sense of Definition 4, it may be convergent as the Cauchy principal value,

$$\int_{-\infty}^{\infty} f(x)\, dx = \lim_{a\to+\infty} \int_{-a}^{a} f(x)\, dx \quad (a > 0) \tag{18}$$

(assuming the limit (18) exists and is finite).

Example 8. The integral

$$\int_{-\infty}^{\infty} x\, dx$$

is divergent in the common sense, for e.g.

$$\int_0^{\infty} x\, dx = \lim_{b\to+\infty} \int_0^b x\, dx = \lim_{b\to+\infty} \frac{b^2}{2} = +\infty \,.$$

However, if we take

$$\int_{-\infty}^{\infty} x\, dx = \lim_{a\to+\infty} \int_{-a}^{a} x\, dx = \lim_{a\to+\infty} \left(\frac{a^2}{2} - \frac{a^2}{2} \right) = 0 \,,$$

it is convergent as the Cauchy principal value.

REMARK 9. Concerning improper integrals in infinite intervals, a remark similar to Remark 4 may be added. Theorems similar to Theorems 1–7 may be used to decide on the convergence or divergence. *In what follows the integrability in every finite interval* $[a, b]$ *is assumed*. Theorems 2–6 keep exactly the same form, the only difference being in writing ∞ in place of b (the analogy of the second assertion of Theorem 6 will be stated additionally in Theorem 10). Theorems 1 and 7 need a slight modification:

Theorem 8 (*The Bolzano–Cauchy Condition*). *The integral*

$$\int_a^\infty f(x)\,\mathrm{d}x$$

is convergent if and only if, for arbitrary $\varepsilon > 0$, *there is a number* B *such that, if* $b_1 > B$, $b_2 > B$, *then the inequality*

$$\left|\int_{b_1}^{b_2} f(x)\,\mathrm{d}x\right| < \varepsilon$$

holds.

Theorem 9. *Let*

$$|f(x)| \leqq \frac{M}{x^\alpha}, \quad M = \text{const.}, \quad \alpha > 1,$$

hold for every $x \geqq a$. *Then the integral*

$$\int_a^\infty f(x)\,\mathrm{d}x \tag{19}$$

is convergent. If

$$|f(x)| \geqq \frac{M}{x^\alpha}, \quad M = \text{const.} > 0, \quad \alpha \leqq 1,$$

then

$$\int_a^\infty |f(x)|\,\mathrm{d}x$$

is divergent (*while integral* (19) *may be convergent*).

REMARK 10. If $f(x)$ is not integrable in every interval $[a, b]$ $(b > a)$ (for example, if it is not bounded in a (right) neighbourhood of the point a), we define

$$\int_a^\infty f(x)\,\mathrm{d}x = \int_a^c f(x)\,\mathrm{d}x + \int_c^\infty f(x)\,\mathrm{d}x$$

(where c is an arbitrary point, $c > a$) provided that the integrals on the right-hand side are both convergent. If at least one of these integrals is divergent, the given integral is said to be divergent.

If the limits of integration are improper and if, moreover, $f(x)$ is not bounded in the vicinity of points $a_1 < a_2 < \dots < a_n$, we define

$$\int_{-\infty}^{\infty} f(x)\,dx = \int_{-\infty}^{a_1} f(x)\,dx + \int_{a_1}^{a_2} f(x)\,dx + \dots + \int_{a_n}^{\infty} f(x)\,dx \tag{20}$$

provided all integrals on the right-hand side of equation (20) are convergent. If at least one is divergent, the integral $\int_{-\infty}^{\infty} f(x)\,dx$ is said to be *divergent.*

Example 9. The integral

$$\int_0^{\infty} \frac{dx}{x^\alpha} \tag{21}$$

is divergent for every α. To show this, let us choose $a > 0$. Integral (21) (see Remark 10) is convergent if and only if integrals

$$\int_0^{a} \frac{dx}{x^\alpha}, \tag{22}$$

$$\int_a^{\infty} \frac{dx}{x^\alpha} \tag{23}$$

are both convergent. By Theorem 9, integral (23) is convergent if $\alpha > 1$ and divergent if $\alpha \leqq 1$. However, if $\alpha > 1$, then by (13) and by Remark 5 integral (22) is divergent.

Theorem 10. *Let us assume that $f(x)$ has a bounded primitive $F(x)$ for $x > a$. (Hence $|F(x)| < K$ holds for all $x > a$). Let $g(x)$ be a monotonic function for $x > a$ such that $\lim_{x\to\infty} g(x) = 0$. Then the integral*

$$\int_a^{\infty} f(x)\,g(x)\,dx$$

is convergent.

Example 10. The integral

$$\int_0^{\infty} \frac{\sin x}{x}\,dx \tag{24}$$

is convergent. First let the function $h(x) = \sin x/x$ be defined at the point $x = 0$ by the equation $h(0) = 1$, then it is continuous at $x = 0$ (see Theorem 11.4.9); hence the point $x = 0$ does not cause difficulty. For all x there is a bounded primitive $F(x) = -\cos x$ of $f(x) = \sin x$; the function $g(x) = 1/x$ is monotonic for $x > 0$ and has zero as its limit as $x \to +\infty$. Hence, by Theorem 10, the integral (24) is convergent. (Its value is $\pi/2$, see Example 13.9.5.)

REMARK 11. Making use of the inequality $|\sin x| \leq |x|$ in the neighbourhood of the point $x = 0$, and of Theorems 7 and 10, it can be shown that the integral

$$\int_0^\infty \frac{\sin x}{x^\alpha}\,\mathrm{d}x$$

is convergent for all α such that $0 < \alpha < 2$.

REMARK 12. It follows directly from the definitions of improper integrals, that if the integrals of the functions $f_1(x)$ and $f_2(x)$ are convergent, then the same is true for the integral of their sum and of their difference, and also of the functions $k f_1(x)$, $k f_2(x)$, where k is a constant.

REMARK 13. The rules of substitution and integration by parts may often be employed with success for the evaluation of improper integrals.

We state corresponding theorems for the case where the functions considered are unbounded only in the neighbourhood of the point b ($b = +\infty$ is also admitted). The other cases are similar.

Theorem 11. *The equation*

$$\int_a^b f'(x)\, g(x)\,\mathrm{d}x = [f(x)\, g(x)]_a^b - \int_a^b f(x)\, g'(x)\,\mathrm{d}x \tag{25}$$

is valid provided that the existence of at least two members of this equation is ensured. The continuity of $f'(x)$ and $g'(x)$ in $[a, b)$ is assumed.

REMARK 14. The expression $[f(x)\, g(x)]_a^b$ is to be understood as the limit

$$\lim_{c \to b-} f(c)\, g(c) - f(a)\, g(a)\,. \tag{26}$$

The existence of the central member of equation (25) is understood to mean the existence of the finite limit (26).

Example 11.

$$\int_0^\infty x\mathrm{e}^{-x}\,\mathrm{d}x = -[x\mathrm{e}^{-x}]_0^\infty + \int_0^\infty \mathrm{e}^{-x}\,\mathrm{d}x = 0 - [\mathrm{e}^{-x}]_0^\infty = 0 - 0 + 1 = 1\,,$$

for $\lim\limits_{x \to +\infty} x\mathrm{e}^{-x} = 0$ (see Theorem 11.4.9).

Example 12.

$$\int_0^\infty \frac{\sin x}{x}\,\mathrm{d}x = -\left[\frac{\cos x}{x}\right]_0^\infty - \int_0^\infty \frac{\cos x}{x^2}\,\mathrm{d}x\,. \tag{27}$$

This equation is nonsense, for

$$\lim_{x\to 0+} \frac{\cos x}{x} = +\infty \quad \text{and the relation} \quad \frac{\cos x}{x^2} > \frac{\frac{1}{2}}{x^2}$$

holds in a sufficiently small neighbourhood of the origin, hence the second integral is divergent by Theorem 7. In spite of this, the integral on the left-hand side of equation (27) is convergent (see Example 10). This example shows how formal use of the method of integration by parts may fail for the evaluation of improper integrals.

Theorem 12. *Let $f(x)$ be continuous in $[a, b)$. Let $x = \varphi(z)$ be an increasing function in the interval (α, β), having a continuous derivative $\varphi'(z)$ in (α, β); further, let $\lim_{z\to\alpha+} \varphi(z) = a$, $\lim_{z\to\beta-} \varphi(z) = b$ (or $\lim_{z\to\beta-} \varphi(z) = +\infty$, if $b = +\infty$). Then the equation*

$$\int_a^b f(x)\,dx = \int_\alpha^\beta f(\varphi(z))\,\varphi'(z)\,dz \tag{28}$$

holds, provided at least one of integrals (28) is convergent. If one of them is divergent, so is the second.

Theorem 12 may be stated for other similar cases, for instance for the case when $\varphi(z)$ is decreasing in (α, β), $\lim_{x\to\alpha+} \varphi(z) = b$, $\lim_{z\to\beta-} \varphi(z) = a$ while e.g. $b = +\infty$ may be admitted. See the following example.

Example 13. The integrals

$$\int_0^\infty \frac{dx}{1 + x^4}, \quad \int_0^\infty \frac{x^2\,dx}{1 + x^4} \tag{29}$$

are convergent by Theorem 9, since for x sufficiently large the integrands are both smaller than $1/x^2$. By the substitution

$$x = \frac{1}{z}, \quad dx = -\frac{dz}{z^2}$$

we obtain

$$\int_0^\infty \frac{x^2}{1 + x^4}\,dx = -\int_\infty^0 \frac{1/z^2}{1 + 1/z^4} \cdot \frac{dz}{z^2} = \int_0^\infty \frac{dz}{1 + z^4}. \tag{30}$$

(Determination of limits: If $x \to 0+$, then $z \to +\infty$; if $x \to +\infty$, then $z \to 0+$.) Hence, both integrals (29) have the same value. This fact may be used for their evaluation. Forming the sum of both integrals, we get

$$\int_0^\infty \frac{dx}{1 + x^4} = \frac{1}{2}\int_0^\infty \frac{1 + x^2}{1 + x^4}\,dx = \frac{1}{2}\int_0^\infty \frac{1 + 1/x^2}{x^2 + 1/x^2}\,dx.$$

By the substitution $x - 1/x = t$, $(1 + 1/x^2)\,dx = dt$, $x^2 + 1/x^2 = t^2 + 2$ we obtain

$$\int_0^\infty \frac{1 + 1/x^2}{x^2 + 1/x^2}\,dx = \int_{-\infty}^\infty \frac{dt}{t^2 + 2} = \left[\frac{1}{\sqrt{2}} \arctan \frac{t}{\sqrt{2}}\right]_{-\infty}^\infty = \frac{1}{\sqrt{2}}\left[\frac{\pi}{2} - \left(-\frac{\pi}{2}\right)\right] = \frac{\pi}{\sqrt{2}}.$$

(Determination of limits: If $x \to 0+$; then $x - 1/x \to -\infty$; if $x \to +\infty$, then $x - 1/x \to +\infty$.) Hence

$$\int_0^\infty \frac{dx}{1 + x^4} = \int_0^\infty \frac{x^2}{1 + x^4}\,dx = \frac{\pi}{2\sqrt{2}}.$$

Example 14. Let us evaluate the integral

$$\int_0^\pi \frac{1 + \tan^2 x}{1 + k^2 \tan^2 x}\,dx, \quad k > 0, \quad k \neq 1.$$

We have

$$I = \int_0^\pi = \int_0^{\pi/2} + \int_{\pi/2}^\pi = I_1 + I_2.$$

By the substitution

$$\tan x = z, \quad \frac{1}{\cos^2 x}\,dx = dz, \quad (1 + \tan^2 x)\,dx = dz$$

and by the further substitution $kz = t$ we obtain

$$I_1 = \int_0^\infty \frac{dz}{1 + k^2 z^2} = \frac{1}{k}\int_0^\infty \frac{dt}{1 + t^2} = \frac{1}{k}\left[\lim_{t \to +\infty} \arctan t - \arctan 0\right] = \frac{1}{k} \cdot \frac{\pi}{2}.$$

In a similar way we obtain (since $\lim_{x \to \pi/2-} \tan x = -\infty$)

$$I_2 = \int_{-\infty}^0 \frac{dz}{1 + k^2 z^2} = \frac{1}{k}\int_{-\infty}^0 \frac{dt}{1 + t^2} = \frac{1}{k}\left[\arctan 0 - \lim_{t \to -\infty} \arctan t\right] =$$

$$= \frac{1}{k}\left[0 - \left(-\frac{\pi}{2}\right)\right] = \frac{1}{k} \cdot \frac{\pi}{2}.$$

Hence

$$I = \frac{\pi}{k}.$$

REMARK 15. The *Schwarz* (or *Cauchy–Schwarz*) *inequality*

$$\left(\int_a^b f(x)\,g(x)\,dx\right)^2 \leqq \int_a^b f^2(x)\,dx \,.\, \int_a^b g^2(x)\,dx$$

is often useful when deciding on the convergence of improper integrals. The limits a and b need not be finite. If the integrals on the right-hand side are both convergent, then the integral on the left-hand side is also convergent.

13.9. Integrals Involving a Parameter

It is often convenient to consider integrals depending on a parameter (cf. Remark 13.2.6 where a primitive was found by differentiating an integral with respect to a parameter).

Formal differentiation with respect to a parameter does not always lead to correct results. It can be shown (see Example 5 below) that

$$\int_0^\infty \frac{\sin \alpha x}{x}\,dx = \frac{\pi}{2} \quad (\alpha \neq 0)\,. \tag{1}$$

If we differentiate equation (1) *with respect to* α, we get

$$\int_0^\infty \cos \alpha x\,dx = 0 \tag{2}$$

and this is obviously wrong because the integral on the left-hand side of equation (2) is not convergent. (The limit

$$\lim_{b\to+\infty} \int_0^b \cos \alpha x\,dx = \lim_{b\to+\infty} \left[\frac{1}{\alpha}\sin \alpha x\right]_0^b = \lim_{b\to+\infty} \frac{1}{\alpha}\sin \alpha b$$

does not exist.) However, the following theorems are valid:

Theorem 1. *Let $f(x, \alpha)$ be continuous (as a function of two variables) in the rectangle $O(a \leqq x \leqq b,\ \alpha_1 \leqq \alpha \leqq \alpha_2;\ a,\ b,\ \alpha_1,\ \alpha_2$ are finite numbers). Then the function*

$$g(\alpha) = \int_a^b f(x, \alpha)\,dx \tag{3}$$

is a continuous function of the variable α in the interval $[\alpha_1, \alpha_2]$ (at α_1 it is continuous from the right, at α_2 from the left), i.e. the relation

$$\lim_{\alpha\to\alpha_0} \int_a^b f(x, \alpha)\,dx = \int_a^b \lim_{\alpha\to\alpha_0} f(x, \alpha)\,dx = \int_a^b f(x, \alpha_0)\,dx \tag{4}$$

holds for every $\alpha_0 \in [\alpha_1, \alpha_2]$.

Theorem 2. *If, in addition, $\partial f/\partial\alpha$ is continuous in O, then the function $g(\alpha)$ possesses a derivative in $[\alpha_1, \alpha_2]$ (at α_1 the right-hand derivative and at α_2 the left-hand derivative) and*

$$\frac{dg}{d\alpha} = \int_a^b \frac{\partial f}{\partial \alpha}(x, \alpha)\,dx\,, \tag{5}$$

i.e.

$$\frac{d}{d\alpha}\int_a^b f(x, \alpha)\,dx = \int_a^b \frac{\partial f}{\partial \alpha}(x, \alpha)\,dx\,. \tag{6}$$

Theorem 3. *If $f(x, \alpha)$ is continuous in O, then the relation*

$$\int_{\alpha_1}^{\alpha_0} g(\alpha)\, d\alpha = \int_a^b \left(\int_{\alpha_1}^{\alpha_0} f(x, \alpha)\, d\alpha \right) dx\,, \tag{7}$$

i.e.

$$\int_{\alpha_1}^{\alpha_0} \left(\int_a^b f(x, \alpha)\, dx \right) d\alpha = \int_a^b \left(\int_{\alpha_1}^{\alpha_0} f(x, \alpha)\, d\alpha \right) dx \tag{8}$$

holds for every $\alpha_0 \in [\alpha_1, \alpha_2]$.

REMARK 1. The assertion of Theorem 3 remains valid under far more general assumptions (see Theorem 14.3.1).

Theorem 4 (*The Limits of Integration Depending on a Parameter*). *Let* $x = \varphi_1(\alpha)$, $x = \varphi_2(\alpha)$ *be functions having continuous derivatives in* $[\alpha_1, \alpha_2]$. *Let us denote by* P *the region* $\alpha_1 \leqq \alpha \leqq \alpha_2$, $\varphi_1(\alpha) \leqq x \leqq \varphi_2(\alpha)$, $\varphi_1(\alpha) < \varphi_2(\alpha)$ (see Fig. 13.4). *If* $f(x, \alpha)$ *and* $\partial f / \partial \alpha(x, \alpha)$ *are continuous in* P, *then the function*

$$g(\alpha) = \int_{\varphi_1(\alpha)}^{\varphi_2(\alpha)} f(x, \alpha)\, dx \tag{9}$$

has a derivative in $[\alpha_1, \alpha_2]$ (*at* α_1 *the right-hand derivative, at* α_2 *the left-hand derivative*) *and the relation*

$$\frac{\partial g}{\partial \alpha} = \int_{\varphi_1(\alpha)}^{\varphi_2(\alpha)} \frac{\partial f}{\partial \alpha}(x, \alpha)\, dx + \varphi_2'(\alpha) \,.\, f(\varphi_2(\alpha), \alpha) - \varphi_1'(\alpha) \,.\, f(\varphi_1(\alpha), \alpha) \tag{10}$$

holds (see Example 2).

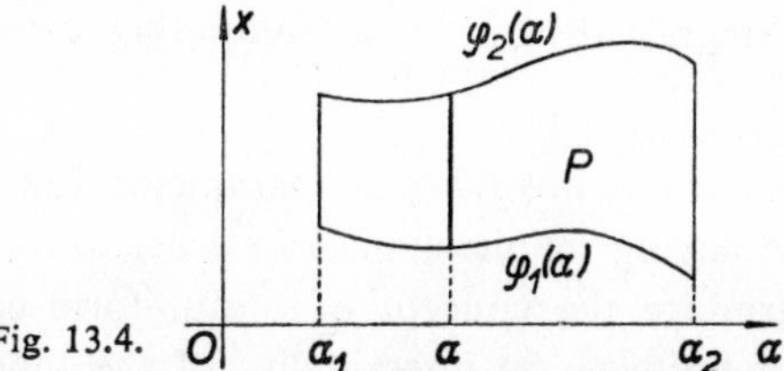

Fig. 13.4.

REMARK 2. The theorems mentioned above, particularly Theorem 2, may advantageously be used for the evaluation of definite integrals. (For the determination of primitives, using the method of a parameter, see Theorem 13.2.5.)

Example 1. We have (§ 13.10, formula 14)

$$\int_0^{\pi/2} \frac{dx}{a^2 \cos^2 x + b^2 \sin^2 x} = \frac{\pi}{2ab} \quad (a > 0, b > 0)\,.$$

Differentiation with respect to a or b yields formulae for more complicated integrals (the assumptions of Theorem 2 are obviously satisfied for $a > 0$, $b > 0$):

$$-\int_0^{\pi/2} \frac{2a \cos^2 x}{(a^2 \cos^2 x + b^2 \sin^2 x)^2} \, dx = -\frac{\pi}{2a^2 b},$$

$$-\int_0^{\pi/2} \frac{2b \sin^2 x}{(a^2 \cos^2 x + b^2 \sin^2 x)^2} \, dx = -\frac{\pi}{2ab^2}. \tag{11}$$

Dividing the first equation by $-2a$ and the second by $-2b$ and summing, we get (since $\cos^2 x + \sin^2 x = 1$)

$$\int_0^{\pi/2} \frac{dx}{(a^2 \cos^2 x + b^2 \sin^2 x)^2} = \frac{\pi}{4ab}\left(\frac{1}{a^2} + \frac{1}{b^2}\right). \tag{12}$$

Example 2.

$$g(\alpha) = \int_\alpha^{\alpha^2+1} (x^3 + \alpha x) \, dx .$$

The assumptions of Theorem 4 are obviously satisfied, hence we obtain by (10)

$$g'(\alpha) = \int_\alpha^{\alpha^2+1} x \, dx + 2\alpha[(\alpha^2 + 1)^3 + \alpha(\alpha^2 + 1)] - 1 . [\alpha^3 + \alpha^2] =$$

$$= \frac{(\alpha^2 + 1)^2}{2} - \frac{\alpha^2}{2} + 2\alpha[(\alpha^2 + 1)^3 + \alpha(\alpha^2 + 1)] - [\alpha^3 + \alpha^2] .$$

This example is only an illustrative one. The same result may be established by direct evaluation of the integral $g(\alpha)$ and by differentiating the result with respect to α.

REMARK 3. *Improper integrals* involving a parameter are of considerable importance, especially those having infinite limits. In order to formulate the corresponding theorems, we introduce the concept of the uniform convergence of an (improper) integral (in what follows, the integrability of the functions considered in every finite interval is assumed):

Definition 1. The integral

$$\int_a^\infty f(x, \alpha) \, dx \tag{13}$$

is said to be *uniformly convergent in the interval* $\alpha_1 \leqq \alpha \leqq \alpha_2$ (in the case where, for example, $\alpha_2 = +\infty$, we shall consider the semi-open interval $\alpha_1 \leqq \alpha < \alpha_2$), if to every $\varepsilon > 0$ there exists a number B_0 (depending, in general, on the choice of ε but independent of α) such that the inequality

$$\left| \int_B^{\infty} f(x, \alpha) \, \mathrm{d}x \right| < \varepsilon \tag{14}$$

holds for any $B > B_0$.

Theorem 5. *If for all $\alpha \in [\alpha_1, \alpha_2]$ the inequality $|f(x, \alpha)| \leqq \varphi(x)$ holds and if $\int_a^\infty \varphi(x) \, \mathrm{d}x$ converges, then the integral (13) is uniformly convergent in the interval $\alpha_1 \leqq \alpha \leqq \alpha_2$.*

Example 3. According to Theorem 5 the integral

$$\int_0^{\infty} \mathrm{e}^{-\alpha x} \sin x \, \mathrm{d}x \tag{15}$$

is uniformly convergent in every interval $[\delta, \infty)$, $\delta > 0$. For, if $\alpha \geqq \delta$, then $|\mathrm{e}^{-\alpha x} \sin x| \leqq \mathrm{e}^{-\delta x}$ and the integral $\int_0^\infty \mathrm{e}^{-\delta x} \, \mathrm{d}x$ converges. (However, $\delta = 0$ may not be admitted, because $\int_0^\infty \sin x \, \mathrm{d}x$ does not converge at all!)

Theorem 6. *Let $g(x, \alpha)$ be continuous for $x \geqq a$, $\alpha_1 \leqq \alpha \leqq \alpha_2$ and let $h(x)$ be continuous and monotonic for $x \geqq a$,* and $\lim_{x \to \infty} h(x) = 0$. *Let $G(x, \alpha)$ be a primitive (with respect to the variable x) of the function $g(x, \alpha)$. If $G(x, \alpha)$ is bounded (i.e. $|G(x, \alpha)| \leqq K$ for all $\alpha_1 \leqq \alpha \leqq \alpha_2$, $x \geqq a$, K being a constant), then the integral*

$$\int_a^{\infty} g(x, \alpha) \, h(x) \, \mathrm{d}x$$

is uniformly convergent in the interval $[\alpha_1, \alpha_2]$.

Example 4. We shall prove that the integral

$$\int_a^{\infty} \mathrm{e}^{-\alpha x} \frac{\sin x}{x} \, \mathrm{d}x \tag{16}$$

is uniformly convergent in the interval $0 \leqq \alpha < +\infty$. (For $\alpha < 0$ it is obviously divergent. For $\alpha = 0$ it is convergent by Example 13.8.10.)

It suffices to examine the convergence of the integral

$$\int_a^{\infty} \mathrm{e}^{-\alpha x} \frac{\sin x}{x} \, \mathrm{d}x \,, \quad a > 0 \,, \tag{17}$$

for if we define the function $(\sin x)/x$ at the point $x = 0$ to be equal to 1, then the function $\mathrm{e}^{-\alpha x}(\sin x)/x$ is everywhere continuous.

Let us write $g(x, \alpha) = e^{-\alpha x} \sin x$, $h(x) = 1/x$. Then

$$G(x, \alpha) = -\frac{e^{-\alpha x}(\alpha \sin x + \cos x)}{1 + \alpha^2}$$

(Example 13.2.4). The functions $g(x, \alpha)$ and $h(x)$ are continuous if $\alpha \geqq 0$, $x \geqq a > 0$, $h(x)$ is decreasing and $\lim_{x\to\infty} h(x) = 0$. Further, obviously, if $\alpha \geqq 0$ and $x \geqq a$, then $|G(x, \alpha)| < 2$. Hence, by Theorem 6, integral (17), and hence also integral (16) is uniformly convergent for all $\alpha \geqq 0$.

Theorem 7. *Let $f(x, \alpha)$ be continuous in the semi-infinite strip $\alpha_1 \leqq \alpha \leqq \alpha_2$, $x \geqq a$. Assume further that the integral*

$$g(\alpha) = \int_a^\infty f(x, \alpha)\, dx$$

is uniformly convergent for all $\alpha \in [\alpha_1, \alpha_2]$. Then

(a) *$g(\alpha)$ is continuous in $[\alpha_1, \alpha_2]$ (at the point α_1 from the right, at the point α_2 from the left).*

(b) *The relation*

$$\int_{\alpha_1}^{\alpha_0} g(\alpha)\, d\alpha = \int_a^\infty \left(\int_{\alpha_1}^{\alpha_0} f(x, \alpha)\, d\alpha \right) dx\,, \tag{18}$$

i.e.

$$\int_{\alpha_1}^{\alpha_0} \left(\int_a^\infty f(x, \alpha)\, dx \right) d\alpha = \int_a^\infty \left(\int_{\alpha_1}^{\alpha_0} f(x, \alpha)\, d\alpha \right) dx \tag{19}$$

holds for all $\alpha_0 \in [\alpha_1, \alpha_2]$. (In the case (a) the interval for α need not be finite.)

Theorem 8. *Let $f(x, \alpha)$ be continuous in the region $x \geqq a$, $\alpha \geqq \alpha_1$. Let the integrals*

$$g(\alpha) = \int_a^\infty f(x, \alpha)\, dx\,, \quad h(x) = \int_{\alpha_1}^\infty f(x, \alpha)\, d\alpha$$

be both uniformly convergent, the first with respect to α, the second with respect to x, from arbitrary finite intervals $[\alpha_1, \alpha_2]$ or $[a, b]$ respectively.

Let at least one of the integrals

$$\int_a^\infty \left(\int_{\alpha_1}^\infty |f(x, \alpha)|\, d\alpha \right) dx\,, \quad \int_{\alpha_1}^\infty \left(\int_a^\infty |f(x, \alpha)|\, dx \right) d\alpha$$

be convergent. Then both the integrals

$$\int_a^\infty h(x)\, dx\,, \quad \int_{\alpha_1}^\infty g(\alpha)\, d\alpha$$

are convergent and the equality

$$\int_a^\infty \left(\int_{\alpha_1}^\infty f(x, \alpha) \, d\alpha \right) dx = \int_{\alpha_1}^\infty \left(\int_a^\infty f(x, \alpha) \, dx \right) d\alpha \tag{20}$$

holds.

Theorem 9. *Let the functions $f(x, \alpha)$ and $\partial f/\partial\alpha(x, \alpha)$ be continuous in the semi-infinite rectangle $\alpha_1 \leqq \alpha \leqq \alpha_2$, $x \geqq a$. If the integral*

$$g(\alpha) = \int_a^\infty f(x, \alpha) \, dx$$

exists for all $\alpha_1 \leqq \alpha \leqq \alpha_2$ and if the integral

$$\int_a^\infty \frac{\partial f}{\partial \alpha}(x, \alpha) \, dx$$

is uniformly convergent for $\alpha \in [\alpha_1, \alpha_2]$, then the function $g(\alpha)$ has a derivative in $[\alpha_1, \alpha_2]$ (at α_1 from the right, at α_2 from the left) and the relation

$$\frac{dg}{d\alpha} = \int_a^\infty \frac{\partial f}{\partial \alpha}(x, \alpha) \, dx , \tag{21}$$

i.e.

$$\frac{d}{d\alpha} \int_a^\infty f(x, \alpha) \, dx = \int_a^\infty \frac{\partial f}{\partial \alpha}(x, \alpha) \, dx , \tag{22}$$

holds.

Example 5. We have to evaluate the integral

$$\int_0^\infty \frac{\sin x}{x} \, dx .$$

We shall use the result concerning integral (16). If the function $\varphi(x) = \sin x/x$ is defined at the origin by the equation $\varphi(0) = 1$, then the function $e^{-\alpha x} \sin x/x$ is continuous for all x and α. The function

$$\frac{\partial}{\partial \alpha} \left(e^{-\alpha x} \frac{\sin x}{x} \right) = - e^{-\alpha x} \sin x$$

is also continuous for all x and α. According to Example 3, the integral

$$\int_0^\infty e^{-\alpha x} \sin x \, dx$$

is uniformly convergent for all $\alpha \geqq \delta > 0$ and is equal to $1/(1 + \alpha^2)$, as can be easily verified using integration by parts. Hence, by Theorem 9, we have

$$\frac{d}{d\alpha} \int_0^\infty e^{-\alpha x} \frac{\sin x}{x} \, dx = - \int_0^\infty e^{-\alpha x} \sin x \, dx = - \frac{1}{1 + \alpha^2} \tag{23}$$

for all $\alpha \geqq \delta$. Consequently,

$$\int_0^\infty e^{-\alpha x} \frac{\sin x}{x} \, dx = - \arctan \alpha + \frac{\pi}{2} \quad (\alpha \geqq \delta) . \tag{24}$$

The value $\frac{1}{2}\pi$ as the constant of integration follows from the relation

$$\int_0^\infty e^{-\alpha x} \frac{\sin x}{x} \, dx \to 0 \quad \text{as} \quad \alpha \to +\infty$$

(which is true since

$$\left| \frac{\sin x}{x} \right| \leqq 1$$

and thus

$$\left| \int_0^\infty e^{-\alpha x} \frac{\sin x}{x} \, dx \right| \leqq \int_0^\infty \left| e^{-\alpha x} \frac{\sin x}{x} \right| dx \leqq \int_0^\infty e^{-\alpha x} \, dx = \frac{1}{\alpha} \bigg) .$$

By Example 4 the integral

$$g(\alpha) = \int_0^\infty e^{-\alpha x} \frac{\sin x}{x} \, dx \tag{25}$$

is uniformly convergent for all $\alpha \geqq 0$. Hence, by Theorem 7, the function $g(\alpha)$ is continuous from the right at the point $\alpha = 0$, i.e.

$$\lim_{\alpha \to 0+} \int_0^\infty e^{-\alpha x} \frac{\sin x}{x} \, dx = g(0) = \int_0^\infty \frac{\sin x}{x} \, dx . \tag{26}$$

It follows from (26) and (24) that

$$\int_0^\infty \frac{\sin x}{x} \, dx = \frac{\pi}{2} . \tag{27}$$

REMARK 4. We had to divide the procedure just carried out into two steps: By differentiation with respect to the parameter α we were led to equation (24) (Theorem 9) and then we made use of the continuity (from the right) of the function $g(\alpha)$ at the point $\alpha = 0$ (equation (26)). Theorem 9 was not directly applicable in view of the divergence of the integral $\int_0^\infty e^{-\alpha x} \sin x \, dx$ for $\alpha = 0$.

REMARK 5. Sometimes we meet the case where the limits of integration are finite but the function $f(x, \alpha)$ or $\dfrac{\partial f}{\partial \alpha}(x, \alpha)$ is *not bounded* in the neighbourhood of the segment $x = b$, $\alpha_1 \leqq \alpha \leqq \alpha_2$. In the same way as in Definition 1, the integral

$$g(\alpha) = \int_a^b f(x, \alpha)\,dx \tag{28}$$

is said to be *uniformly convergent for all* $\alpha \in [\alpha_1, \alpha_2]$ if for every $\varepsilon > 0$ there exists $\delta_0 > 0$ (the same for all $\alpha \in [\alpha_1, \alpha_2]$) such that the inequality

$$\left| \int_{b-\delta}^b f(x, \alpha)\,dx \right| < \varepsilon$$

holds for every δ, $0 < \delta < \delta_0$.

In this case, Theorems 5, 7 and 9 are quite similar; one has only to replace ∞ by b and to examine the continuity of the functions $f(x, \alpha)$ and $\dfrac{\partial f}{\partial \alpha}(x, \alpha)$ in the domain $\alpha_1 \leqq \alpha \leqq \alpha_2$, $a \leqq x < b$.

These theorems may be generalized for the case where $f(x, \alpha)$ or $\dfrac{\partial f}{\partial \alpha}(x, \alpha)$ is not bounded in the neighbourhood of the curve $x = \varphi(\alpha)$ or in the neighbourhood of several such curves. If the limits of integration are infinite and at the same time the functions $f(x, \alpha)$ or $\dfrac{\partial f}{\partial \alpha}(x, \alpha)$ are not bounded on some curves, then we divide the given integral into two or more integrals and examine each of them separately, by a method similar to that of Remark 13.8.10.

13.10. Table of Definite Integrals

Throughout this paragraph m, n are integers, r a real number, $C = 0{\cdot}577, 215, 664, 9\ldots$ is the so-called *Euler's constant*, $\Gamma(x)$ is the *gamma function*,

$$\Gamma(x) = \int_0^\infty e^{-t} t^{x-1}\,dt \quad (x > 0),$$

$B(p, q)$ is the *beta function*,

$$B(p, q) = \frac{\Gamma(p)\,\Gamma(q)}{\Gamma(p + q)} \quad (p > 0,\ q > 0)$$

(see § 13.11).

In particular, for $x = n$

$$\Gamma(n) = (n - 1)!$$

holds and for $0 < x < 1$ we have

$$\Gamma(x)\,\Gamma(1 - x) = \frac{\pi}{\sin \pi x}\,.$$

1. $\displaystyle\int_0^\infty x^r e^{-ax}\,dx = \frac{\Gamma(r + 1)}{a^{r+1}}$ for $a > 0,\ r > -1$.

In particular for $r = n$ (n is a positive integer) this integral is equal to $n!/a^{n+1}$.

2. $\displaystyle\int_0^\infty x^r e^{-ax^2}\,dx = \frac{\Gamma(\frac{1}{2}r + \frac{1}{2})}{2a^{\frac{1}{2}(r+1)}}$ for $a > 0,\ r > -1$.

In particular, for r even ($r = 2k$) and positive this integral is equal to

$$\frac{1\,.\,3\,.\,\ldots\,.\,(2k - 1)\sqrt{\pi}}{2^{k+1}a^{(2k+1)/2}}\,,$$

for r odd ($r = 2k + 1$) it is equal to $k!/2a^{k+1}$.

3. $\displaystyle\int_0^\infty e^{-a^2x^2}\,dx = \frac{\sqrt{\pi}}{2a}$ $(a > 0)$ (*Laplace–Gauss integral*).

4. $\displaystyle\int_{-\infty}^\infty e^{-a^2x^2+bx}\,dx = \frac{\sqrt{\pi}}{a}\,e^{b^2/4a^2}$ $(a > 0)$.

5. $\displaystyle\int_0^\infty x^2 e^{-a^2x^2}\,dx = \frac{\sqrt{\pi}}{4a^3}$ $(a > 0)$.

6. $\displaystyle\int_0^\infty e^{-a^2x^2}\cos bx\,dx = \frac{\sqrt{\pi}}{2a}\,e^{-b^2/(4a^2)}$ $(a > 0)$.

7. $\displaystyle\int_0^\infty \frac{x\,dx}{e^x - 1} = \frac{\pi^2}{6}\,.$

8. $\displaystyle\int_0^\infty \frac{x\,dx}{e^x + 1} = \frac{\pi^2}{12}\,.$

9. $\displaystyle\int_0^\infty e^{-ax}\cos bx = \frac{a}{a^2 + b^2}$ $(a > 0)$.

10. $\displaystyle\int_0^\infty e^{-ax}\sin bx = \frac{b}{a^2 + b^2}$ $(a > 0)$.

11. $$\int_0^\infty \frac{e^{-ax} \sin bx}{x}\, dx = \arctan \frac{b}{a} \quad (a > 0)\,.$$

12. $$\int_0^\infty e^{-x} \ln x\, dx = -C = -0{\cdot}577, 215, 664, 9\ldots\,.$$

13. $$\int_0^{\pi/2} \sin^{2\alpha+1} x \cos^{2\beta+1} x\, dx = \frac{\Gamma(\alpha+1)\,\Gamma(\beta+1)}{2\Gamma(\alpha+\beta+2)} = \tfrac{1}{2}B(\alpha+1,\ \beta+1)\,.$$

This formula can be used, for example, to evaluate the integrals

$$\int_0^{\pi/2} \sqrt{(\sin x)}\, dx\,, \quad \int_0^{\pi/2} \sqrt[3]{(\sin x)}\, dx\,, \quad \int_0^{\pi/2} \frac{dx}{\sqrt[3]{(\cos x)}}, \quad \text{etc}\,.$$

If α and β are positive integers, this integral is equal to

$$\frac{\alpha!\,\beta!}{2(\alpha+\beta+1)!} \quad (\text{see } 24)\,.$$

14. $$\int_0^{\pi/2} \frac{dx}{a^2 \cos^2 x + b^2 \sin^2 x} = \frac{\pi}{2ab}\,.$$

15. $$\int_0^{\pi} \sin mx \sin nx\, dx = \int_0^{\pi} \cos mx \cos nx\, dx = \begin{cases} 0 & \text{for} \quad m \neq n\,, \\ \tfrac{1}{2}\pi & \text{for} \quad m = n\,. \end{cases}$$

16. $$\int_0^{\pi} \sin mx \cos nx\, dx = \begin{cases} 0 & \text{for} \quad m - n \quad \text{even}\,, \\ 2m/(m^2 - n^2) & \text{for} \quad m - n \quad \text{odd}\,. \end{cases}$$

17. $$\int_{-\pi}^{\pi} \sin mx \sin nx\, dx = \int_0^{2\pi} = \begin{cases} 0 & \text{for} \quad m \neq n\,, \\ \pi & \text{for} \quad m = n\,. \end{cases}$$

18. $$\int_{-\pi}^{\pi} \cos mx \cos nx\, dx = \int_0^{2\pi} = \begin{cases} 0 & \text{for} \quad m \neq n\,, \\ \pi & \text{for} \quad m = n\,. \end{cases}$$

19. $$\int_{-\pi}^{\pi} \sin mx \cos nx\, dx = \int_0^{2\pi} = 0\,.$$

20. $$\int_0^{\pi/2} \sin^2 x\, dx = \int_0^{\pi/2} \cos^2 x dx = \frac{\pi}{4}\,.$$

21. $$\int_0^{\pi/2} \sin^{2n} x\, dx = \int_0^{\pi/2} \cos^{2n} x\, dx = \frac{1\,.\,3\,.\,5\,.\ldots.\,(2n-1)}{2\,.\,4\,.\,6\,.\ldots.\,2n}\,\frac{\pi}{2}\,.$$

22. $$\int_0^{\pi/2} \sin^{2n+1} x\, dx = \int_0^{\pi/2} \cos^{2n+1} x\, dx = \frac{2\,.\,4\,.\,6\,.\ldots.\,2n}{1\,.\,3\,.\,5\,.\ldots.\,(2n+1)}\,.$$

23. $$\int_0^{\pi/2} \sin^{2m} x \cos^{2n} x \, dx = \frac{1.3.\ldots.(2m-1).1.3.\ldots.(2n-1)}{2.4.\ldots.(2m+2n)} \frac{\pi}{2}.$$

24. $$\int_0^{\pi/2} \sin^{2m+1} x \cos^{2n+1} x \, dx = \frac{1}{2} \frac{m!\, n!}{(m+n+1)!}.$$

25. $$\int_0^{\infty} \frac{\sin ax}{x} dx = \begin{cases} \frac{1}{2}\pi & (a > 0), \\ -\frac{1}{2}\pi & (a < 0). \end{cases}$$

26. $$\int_0^{\alpha} \frac{\cos ax}{x} dx = +\infty \quad (\alpha > 0 \text{ arbitrary; the integral is divergent}).$$

27. $$\int_0^{\infty} \frac{\tan ax \, dx}{x} = \begin{cases} \frac{1}{2}\pi & (a > 0) \\ -\frac{1}{2}\pi & (a < 0) \end{cases} \quad \text{(taken as the Cauchy principal value)}.$$

28. $$\int_0^{\infty} \frac{\cos ax - \cos bx}{x} dx = \ln \frac{b}{a} \ (a > 0,\ b > 0).$$

29. $$\int_0^{\infty} \frac{\sin x \cos ax}{x} dx = \begin{cases} \frac{1}{2}\pi & \text{for } |a| < 1, \\ \frac{1}{4}\pi & \text{for } |a| = 1, \\ 0 & \text{for } |a| > 1. \end{cases}$$

30. $$\int_0^{\infty} \frac{\sin x}{\sqrt{x}} dx = \int_0^{\infty} \frac{\cos x}{\sqrt{x}} dx = \sqrt{\frac{\pi}{2}} \quad \textit{(Fresnel's integrals)}.$$

31. $$\int_0^{\infty} \frac{x \sin bx}{a^2 + x^2} dx = \pm \frac{\pi}{2} e^{-|ab|} \quad \text{(the sign is to be taken to agree with that of } b\text{)}.$$

32. $$\int_0^{\infty} \frac{\cos ax}{1 + x^2} dx = \frac{\pi}{2} e^{-|a|}.$$

33. $$\int_0^{\infty} \frac{\sin^2 ax}{x^2} dx = \frac{\pi}{2} |a|.$$

34. $$\int_{-\infty}^{+\infty} \sin(x^2) \, dx = \int_{-\infty}^{+\infty} \cos(x^2) \, dx = \sqrt{\frac{\pi}{2}} \quad \textit{(Fresnel's integrals)}.$$

35. $$\int_0^{\pi/2} \frac{\sin x \, dx}{\sqrt{(1 - k^2 \sin^2 x)}} = \frac{1}{2k} \ln \frac{1+k}{1-k} \quad (|k| < 1).$$

36. $$\int_0^{\pi/2} \frac{\cos x \, dx}{\sqrt{(1 - k^2 \sin^2 x)}} = \frac{1}{k} \arcsin k \quad (|k| < 1).$$

37. $$\int_0^{\pi/2} \frac{\sin^2 x \, dx}{\sqrt{(1 - k^2 \sin^2 x)}} = \frac{1}{k^2} (\mathrm{K} - \mathrm{E}) \quad (|k| < 1;\ \mathrm{E}, \mathrm{K} \text{ see § 13.12, p. 590}).$$

38. $\displaystyle\int_0^{\pi/2} \frac{\cos^2 x\,dx}{\sqrt{(1 - k^2 \sin^2 x)}} = \frac{1}{k^2}\left[E - (1 - k^2) K\right]$ ($|k| < 1$; E, K, see § 13.12, p. 590).

39. $\displaystyle\int_0^{\pi} \frac{\cos ax\,dx}{1 - 2b\cos x + b^2} = \frac{\pi b^a}{1 - b^2}$ (a is a non-negative integer, $|b| < 1$).

40. $\displaystyle\int_0^1 \ln|\ln x|\,dx = -C = -0{\cdot}577,215,664,9\ldots$.

41. $\displaystyle\int_0^1 \frac{\ln x}{x - 1}\,dx = \frac{\pi^2}{6}$.

42. $\displaystyle\int_0^1 \frac{\ln x}{x + 1}\,dx = -\frac{\pi^2}{12}$.

43. $\displaystyle\int_0^1 \frac{\ln x}{x^2 - 1}\,dx = \frac{\pi^2}{8}$.

44. $\displaystyle\int_0^1 \frac{\ln(1 + x)}{x^2 + 1}\,dx = \frac{\pi}{8}\ln 2$.

45. $\displaystyle\int_0^1 \left(\ln\frac{1}{x}\right)^a dx = \Gamma(a + 1)$ $(-1 < a < \infty)$.

46. $\displaystyle\int_0^{\pi/2} \ln\sin x\,dx = \int_0^{\pi/2} \ln\cos x\,dx = -\frac{\pi}{2}\ln 2$.

47. $\displaystyle\int_0^{\pi} x\ln\sin x\,dx = -\frac{\pi^2\ln 2}{2}$.

48. $\displaystyle\int_0^{\pi/2} \sin x\ln\sin x\,dx = \ln 2 - 1$.

49. $\displaystyle\int_0^{\pi} \ln(a \pm b\cos x)\,dx = \pi\ln\frac{a + \sqrt{(a^2 - b^2)}}{2}$ for $a \geqq b > 0$.

50. $\displaystyle\int_0^{\pi} \ln(a^2 - 2ab\cos x + b^2)\,dx = \begin{cases} 2\pi\ln a & (a \geqq b > 0), \\ 2\pi\ln b & (b \geqq a > 0). \end{cases}$

51. $\displaystyle\int_0^{\pi/2} \ln\tan x\,dx = 0$.

52. $\displaystyle\int_0^{\pi/4} \ln(1 + \tan x)\,dx = \frac{\pi}{8}\ln 2$.

53. $$\int_0^1 x^\alpha(1-x)^\beta\,dx = 2\int_0^1 x^{2\alpha+1}(1-x^2)^\beta\,dx =$$

$$= \frac{\Gamma(\alpha+1)\,\Gamma(\beta+1)}{\Gamma(\alpha+\beta+2)} = B(\alpha+1,\ \beta+1) \quad \text{(see 13)}.$$

54. $$\int_0^\infty \frac{dx}{(1+x)\,x^a} = \frac{\pi}{\sin a\pi} \quad (0 < a < 1).$$

55. $$\int_0^\infty \frac{x^{a-1}}{1+xb}\,dx = \frac{\pi}{b\sin\dfrac{a\pi}{b}} \quad (0 < a < b).$$

56. $$\int_0^1 \frac{dx}{\sqrt{(1-x^a)}} = \frac{\sqrt{(\pi)}\,\Gamma\left(\dfrac{1}{a}\right)}{a\Gamma\left(\dfrac{2+a}{2a}\right)} \quad (a \neq 0).$$

57. $$\int_0^1 \frac{dx}{1+2x\cos a + x^2} = \frac{a}{2\sin a} \quad (0 < a < \tfrac{1}{2}\pi).$$

58. $$\int_0^\infty \frac{dx}{1+2x\cos a + x^2} = \frac{a}{\sin a} \quad (0 < a < \tfrac{1}{2}\pi).$$

13.11. Euler's Integrals, the Gamma Function, the Beta Function. The Gauss Function. Stirling's Formula

Definition 1. The function

$$\Gamma(x) = \int_0^\infty e^{-t} t^{x-1}\,dt \tag{1}$$

is called the *gamma function* or *Euler's integral (function) of the second kind.*

Theorem 1. *The gamma function is defined by integral* (1) *for* $x > 0$. *If* $x \leqq 0$ *the integral* (1) *is divergent. The function* (1) *is continuous for* $x > 0$. *It has derivatives of all orders; these derivatives are obtained by formal differentiation with respect to* x *under the integral sign:*

$$\Gamma^{(n)}(x) = \int_0^\infty e^{-t} t^{x-1} \ln^n t\,dt. \tag{2}$$

Further, the relations

$$\lim_{x\to+\infty} \Gamma(x) = +\infty\,, \qquad \lim_{x\to 0+} \Gamma(x) = +\infty \tag{3}$$

hold. The gamma function has its local minimum between the points $x = 1$ *and* $x = 2$ $(x \doteq 1{\cdot}46)$.

Theorem 2. *Basic relations:*

$$\Gamma(x + 1) = x\,\Gamma(x)\,; \tag{4}$$

$$\Gamma(x)\,\Gamma(1 - x) = \frac{\pi}{\sin \pi x} \quad (0 < x < 1)\,;$$

$$\Gamma(x)\,\Gamma(x + \tfrac{1}{2}) = \frac{\sqrt{\pi}}{2^{2x-1}}\,\Gamma(2x)\,; \tag{5}$$

$\Gamma(1) = 1$, $\Gamma(2) = 1$, *in general*

$$\Gamma(n) = (n - 1)! \quad \textit{for every natural } n\,. \tag{6}$$

(With the aid of the gamma function we often extend the definition of the factorial function by the equation $x! = \Gamma(x + 1)$ for positive real numbers other than natural.)

$$\Gamma(\tfrac{1}{2}) = \sqrt{\pi}\,, \quad \Gamma(\tfrac{3}{2}) = \tfrac{1}{2}\sqrt{\pi}\,, \quad \Gamma(\tfrac{5}{2}) = \tfrac{3}{2}\,.\,\tfrac{1}{2}\sqrt{\pi}\,, \quad \ldots\,; \tag{7}$$

$$\ln \Gamma(x) = (x - \tfrac{1}{2}) \ln x - x + \ln \sqrt{(2\pi)} + \frac{\vartheta}{4x} \quad (x > 0,\ 0 < \vartheta < 1)\,; \tag{8}$$

$$\Gamma(x) = \lim_{n\to\infty} \frac{(n - 1)!}{x(x + 1)(x + 2)\ldots(x + n - 1)}\, n^x\,; \tag{9}$$

$$\Gamma(x) = \frac{1}{x\,\mathrm{e}^{Cx}} \cdot \frac{1}{\prod_{n=1}^{\infty} \mathrm{e}^{-x/n}\left(1 + \dfrac{x}{n}\right)}\,, \tag{10}$$

where C is the so-called Euler's constant; $C = 0{\cdot}577{,}215{,}664{,}9\ldots.$

REMARK 1. The limit on the right-hand side of equation (9) exists (for the infinite product on the right-hand side of equation (10) is convergent and its value is different from zero) not only for positive x, but for all x other than $0, -1, -2, \ldots$. The function $\Gamma(x)$ is thus defined for all x other than $0, -1, -2, \ldots$. (Sometimes, this extension of the gamma function is denoted by the symbol $\Pi(x - 1)$.)

For the gamma function extended in this way the basic relations given above remain valid for all x, with the exception of those values for which the corresponding expressions have no meaning.

The graph of the gamma function is plotted in Fig. 13.5. Some values of the gamma function for $0 < x \leqq 2$ can be found in Table 13.1. By equation (9) or (10), the function $\Gamma(x)$ is defined also for complex values of x other than $0, -1, -2, \ldots$ The gamma function, thus considered as a function of a complex variable, is holomorphic

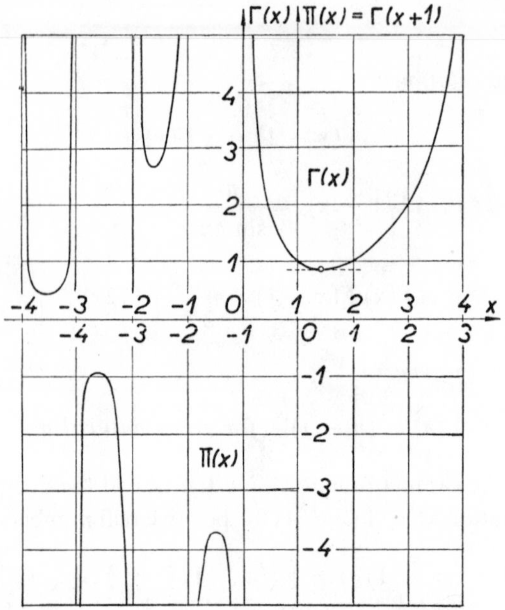

Fig. 13.5.

TABLE 13.1

Table of the gamma function for $0 < x \leqq 2$

x	$\Gamma(x)$	x	$\Gamma(x)$	x	$\Gamma(x)$	x	$\Gamma(x)$
0·00	undefined	0·50	1·772 45	1·00	1·000 00	1·50	0·886 23
05	19·470 09	55	1·616 12	05	0·973 50	55	888 87
10	9·513 51	60	1·489 19	10	951 35	60	893 52
15	6·220 27	65	1·384 80	15	933 04	65	900 12
20	4·590 84	70	1·298 06	20	918 17	70	908 64
0·25	3·625 61	0·75	1·225 42	1·25	0·906 40	1·75	0·919 06
30	2·991 57	80	1·164 23	30	897 47	80	931 38
35	2·546 15	85	1·112 48	35	891 15	85	945 61
40	2·218 16	90	1·068 63	40	887 26	90	961 77
45	1·968 14	95	1·031 45	45	885 66	95	979 88
0·50	1·772 45	1·00	1·000 00	1·46	0·885 60 (minimum)	2·00	1·000 00

(regular) in the complex domain with the exception of the points $0, -1, -2, \ldots$, where it has simple poles.

REMARK 2. The evaluation of many integrals leads to the gamma function (see e.g. [112], [367]). For example

$$\int_0^\infty e^{-x^2}\,dx = \tfrac{1}{2}\Gamma(\tfrac{1}{2}) = \tfrac{1}{2}\sqrt{\pi} \quad (\text{by the substitution } x^2 = z);$$

$$\int_0^\infty e^{-x^2} x^{2k}\,dx = \tfrac{1}{2}\Gamma\left(\frac{2k+1}{2}\right) = \frac{1\,.\,3\,.\,5\,.\ldots.\,(2k-1)}{2^{k+1}}\sqrt{\pi} \quad (k = 1, 2, 3, \ldots);$$

$$\int_0^\infty e^{-nt} t^{x-1}\,dt = \frac{\Gamma(x)}{n^x} \quad (x > 0) \quad (\text{by the substitution } nt = z);$$

$$\int_0^{\pi/2} \sin^{r-1} x \cos^{s-1} x\,dx = \frac{1}{2}\frac{\Gamma(\frac{1}{2}r)\,\Gamma(\frac{1}{2}s)}{\Gamma(\frac{1}{2}r + \frac{1}{2}s)} \quad (r > 0, s > 0);$$

$$\int_0^{\pi/2} \sin^{r-1} x\,dx = \int_0^{\pi/2} \cos^{r-1} x\,dx = \frac{\sqrt{(\pi)}\,\Gamma(\frac{1}{2}r)}{2\Gamma(\frac{1}{2}r + \frac{1}{2})}, \quad r > 0.$$

Definition 2. The *beta function* is defined by the integral

$$B(p, q) = \int_0^1 x^{p-1}(1 - x)^{q-1}\,dx = \int_0^\infty \frac{x^{p-1}\,dx}{(1 + x)^{p+q}} \quad (p > 0, q > 0). \tag{11}$$

The beta function is often called *Euler's integral (function) of the first kind.*

Theorem 3. *The relation*

$$B(p, q) = B(q, p), \tag{12}$$

holds.

REMARK 3. This function is also tabulated and the evaluation of some integrals may be reduced to it. The beta function is related to the gamma function by the equation

$$B(p, q) = \frac{\Gamma(p)\,\Gamma(q)}{\Gamma(p + q)}. \tag{13}$$

Example 1. By the substitution $x^m = z$ we have

$$\int_0^1 \frac{1}{\sqrt[n]{(1 - x^m)}}\,dx = \frac{1}{m}\int_0^1 (1 - z)^{-1/n} z^{(1/m)-1}\,dz = \frac{1}{m}\,B\left(\frac{1}{m}, 1 - \frac{1}{n}\right).$$

In particular for $n = 2$ and $m = 4$ we obtain (using (13))

$$\int_0^1 \frac{1}{\sqrt{(1 - x^4)}}\,dx = \tfrac{1}{4}\,B\,(\tfrac{1}{4}, \tfrac{1}{2}) = \frac{1}{4}\frac{\Gamma(\frac{1}{4})\,\Gamma(\frac{1}{2})}{\Gamma(\frac{3}{4})}.$$

Definition 3. The *Gauss function* Ψ is defined by the relation (we write $\Pi(x) = \Gamma(x+1)$)

$$\Psi(x) = \frac{\mathrm{d} \ln \Pi(x)}{\mathrm{d}x} = \frac{\Pi'(x)}{\Pi(x)} \left(= \frac{\Gamma'(x+1)}{\Gamma(x+1)} \right), \quad x > -1 .$$

Theorem 4. *The relation*

$$\Psi(x) = \lim_{n \to \infty} \left(\ln n - \frac{1}{1+x} - \frac{1}{2+x} - \dots - \frac{1}{n+x} \right)$$

holds. In particular

$$\Psi(0) = \lim_{n \to \infty} \left(\ln n - \frac{1}{1} - \frac{1}{2} - \dots - \frac{1}{n} \right) = -C ,$$

where $C = 0{\cdot}577\,215\,664\,9 \dots$ *is Euler's constant.*

$$\Psi(k) = -C + \frac{1}{1} + \frac{1}{2} + \dots + \frac{1}{k} \quad (k \text{ a positive integer}) .$$

Theorem 5. *Stirling's formula:*

$$n! = n^n \sqrt{(2\pi n)}\, \mathrm{e}^{-n+\vartheta/4n} \quad (0 < \vartheta < 1) ;$$

for large n the approximate formula

$$n! \approx n^n \mathrm{e}^{-n} \sqrt{(2\pi n)} = \left(\frac{n}{\mathrm{e}} \right)^n \sqrt{(2\pi n)}$$

holds.

13.12. Series Expansions of Some Important Integrals. Elliptic Integrals, Elliptic Functions

For general theorems on term-by-term integration of infinite series (including applications to the evaluation of integrals) the reader is referred to §§ 15.2, 15.4, 15.7. On applications of asymptotic expansions to the evaluation of integrals see § 15.7.

Theorem 1. *Sine integral*

$$\mathrm{Si}\, x = \int_0^x \frac{\sin t}{t} \mathrm{d}t = x - \frac{1}{3} \frac{x^3}{3!} + \frac{1}{5} \frac{x^5}{5!} - \dots \quad (x \text{ arbitrary}).$$

Cosine integral:

$$\mathrm{Ci}\, x = - \int_x^\infty \frac{\cos t}{t} \mathrm{d}t = C + \ln x - \frac{1}{2} \frac{x^2}{2!} + \frac{1}{4} \frac{x^4}{4!} - \dots \quad (x > 0) .$$

Logarithm integral:

$$\operatorname{li} x = \int_0^x \frac{dt}{\ln t} = C + \ln|\ln x| + \ln x + \frac{1}{2}\frac{(\ln x)^2}{2!} + \frac{1}{3}\frac{(\ln x)^3}{3!} + \ldots \quad (0 < x < 1),$$

C denotes Euler's constant

$$C = \lim_{n\to\infty}\left(1 + \frac{1}{2} + \frac{1}{3} + \ldots + \frac{1}{n} - \ln n\right) = 0{\cdot}577, 215, 664, 9\ldots.$$

Fresnel's integrals:

$$\int_0^x \frac{\cos t}{\sqrt{t}}\,dt = 2\sqrt{(x)}\left(1 - \frac{x^2}{5\,.\,2!} + \frac{x^4}{9\,.\,4!} + \frac{x^6}{13\,.\,6!} + \ldots\right) \quad (x > 0);$$

$$\int_0^x \frac{\sin t}{\sqrt{t}}\,dt = 2\sqrt{(x)}\left(\frac{x}{3} - \frac{x^3}{7\,.\,3!} + \frac{x^5}{11\,.\,5!} - \frac{x^7}{15\,.\,7!} + \ldots\right) \quad (x > 0);$$

$$\int_0^x e^{t^2}\,dt = x + \frac{x^3}{3\,.\,1!} + \frac{x^5}{5\,.\,2!} + \frac{x^7}{7\,.\,3!} + \ldots \quad (x \text{ arbitrary});$$

$$\int_0^x e^{-t^2}\,dt = x - \frac{x^3}{3\,.\,1!} + \frac{x^5}{5\,.\,2!} - \frac{x^7}{7\,.\,3!} + \ldots \quad (x \text{ arbitrary})$$

(*the Laplace–Gauss integral*). *Asymptotic expansion for large x* (for details see § 15.7):

$$\int_0^x e^{-t^2}\,dt = \frac{\sqrt{\pi}}{2} - \frac{e^{-x^2}}{2x}\left(1 - \frac{1}{2x^2} + \frac{1\,.\,3}{(2x^2)^2} - \frac{1\,.\,3\,.\,5}{(2x^2)^3} + \ldots\right).$$

(The notation

$$\operatorname{erf}(x) = \frac{2}{\sqrt{\pi}}\int_0^x e^{-t^2}\,dt, \quad \operatorname{erfc}(x) = 1 - \operatorname{erf}(x) = \frac{2}{\sqrt{\pi}}\int_x^\infty e^{-t^2}\,dt,$$

is often used in the literature.)

Integrals of the type

$$\int R(x, \sqrt{X(x)})\,dx,$$

where $R(x, y)$ is a rational function of the variables x, y and $X(x)$ is a polynomial of the third or fourth order, are called *elliptic integrals*. (If the order of the polynomial is higher than 4, integrals of this type are called *hyperelliptic*.) The evaluation of elliptic integrals may be reduced by suitable transformations to the evaluation of

Legendre integrals in the normal form:

$$\text{(a)}\quad u = \int_0^x \frac{\mathrm{d}x}{\sqrt{[(1-x^2)(1-k^2x^2)]}} = \int_0^\varphi \frac{\mathrm{d}\varphi}{\sqrt{(1-k^2\sin^2\varphi)}} = F(k,\varphi)$$

$$(0 < k < 1)\,. \quad (1)$$

$$\text{(b)}\quad v = \int_0^x \sqrt{\left(\frac{1-k^2x^2}{1-x^2}\right)}\,\mathrm{d}x = \int_0^\varphi \sqrt{(1-k^2\sin^2\varphi)}\,\mathrm{d}\varphi = E(k,\varphi) \quad (0<k<1)\,. \quad (2)$$

Theorem 2. *The relations*

$$F(k,\varphi) = J_0 + \tfrac{1}{2}k^2J_2 + \frac{1\,.\,3}{2\,.\,4}k^4J_4 + \frac{1\,.\,3\,.\,5}{2\,.\,4\,.\,6}k^6J_6 + \ldots,$$

$$E(k,\varphi) = J_0 - \tfrac{1}{2}k^2J_2 - \frac{1\,.\,1}{2\,.\,4}k^4J_4 - \frac{1\,.\,1\,.\,3}{2\,.\,4\,.\,6}k^6J_6 - \ldots$$

hold, where $0 < k < 1$ *and* φ *is arbitrary (it is sufficient, of course, to consider* $0 < \varphi < \frac{1}{2}\pi$) *and*

$$J_{2n} = \int_0^\varphi \sin^{2n}\varphi\,\mathrm{d}\varphi\,, \quad n = 0, 1, 2, \ldots.$$

COMPLETE ELLIPTIC INTEGRALS OF THE FIRST AND SECOND KIND $(0 < k < 1)$:

Theorem 3. *The relations*

$$\mathrm{K} = F(k,\tfrac{1}{2}\pi) = \int_0^{\pi/2} \frac{\mathrm{d}\varphi}{\sqrt{(1-k^2\sin^2\varphi)}} = \int_0^1 \frac{\mathrm{d}x}{\sqrt{[(1-x^2)(1-k^2x^2)]}} =$$

$$= \frac{\pi}{2}\left[1 + \left(\frac{1}{2}\right)^2 k^2 + \left(\frac{1\,.\,3}{2\,.\,4}\right)^2 k^4 + \ldots\right], \quad (3)$$

$$\mathrm{E} = E(k,\tfrac{1}{2}\pi) = \int_0^{\pi/2} \sqrt{(1-k^2\sin^2\varphi)}\,\mathrm{d}\varphi = \int_0^1 \sqrt{\left(\frac{1-k^2x^2}{1-x^2}\right)}\,\mathrm{d}x =$$

$$= \frac{\pi}{2}\left[1 - \left(\frac{1}{2}\right)^2 \frac{k^2}{1} - \left(\frac{1\,.\,3}{2\,.\,4}\right)^2 \frac{k^4}{3} - \ldots\right] \quad (4)$$

hold.

Theorem 4. *For complementary elliptic integrals*

$$\mathrm{K}' = F(k',\tfrac{1}{2}\pi)\,, \quad \mathrm{E}' = E(k',\tfrac{1}{2}\pi)$$

having modulus $k' = \sqrt{(1 - k^2)}$, *the relation*

$$\text{KE}' + \text{K}'\text{E} - \text{KK}' = \tfrac{1}{2}\pi$$

holds.

THE LEGENDRE (JACOBI) ELLIPTIC FUNCTIONS

$$\operatorname{sn} u = u - (1 + k^2)\frac{u^3}{3!} + (1 + 14k^2 + k^4)\frac{u^5}{5!} - \dots, \quad |u| < \text{K}';$$

$$\operatorname{cn} u = 1 - \frac{u^2}{2!} + (1 + 4k^2)\frac{u^4}{4!} - (1 + 44k^2 + 16k^4)\frac{u^6}{6!} + \dots, \quad |u| < \text{K}';$$

$$\operatorname{dn} u = 1 - k^2\frac{u^2}{2!} + k^2(4 + k^2)\frac{u^4}{4!} - k^2(16 + 44k^2 + k^4)\frac{u^6}{6!} + \dots, \quad |u| < \text{K}'.$$

REMARK 1. The function $x = \operatorname{sn} u$, $0 \leqq u \leqq \text{K}$, is the inverse of the function (1), i.e. of the function

$$u = \int_0^x \frac{1}{\sqrt{[(1 - x^2)(1 - k^2x^2)]}}\,\mathrm{d}x,$$

where x runs through the interval $[0, 1]$. The functions $\operatorname{cn} u$, $\operatorname{dn} u$ may then be defined as continuous functions satisfying the relations

$$\operatorname{cn}^2 u = 1 - \operatorname{sn}^2 u, \quad \operatorname{dn}^2 u = 1 - k^2 \operatorname{sn}^2 u$$

and $\operatorname{cn} 0 = 1$, $\operatorname{dn} 0 = 1$. For a detailed treatment (in the complex plane) see e.g. [429].

For $k = 0$ the functions $\operatorname{sn} u$ and $\operatorname{cn} u$ become the common trigonometric functions $\sin u$, $\cos u$.

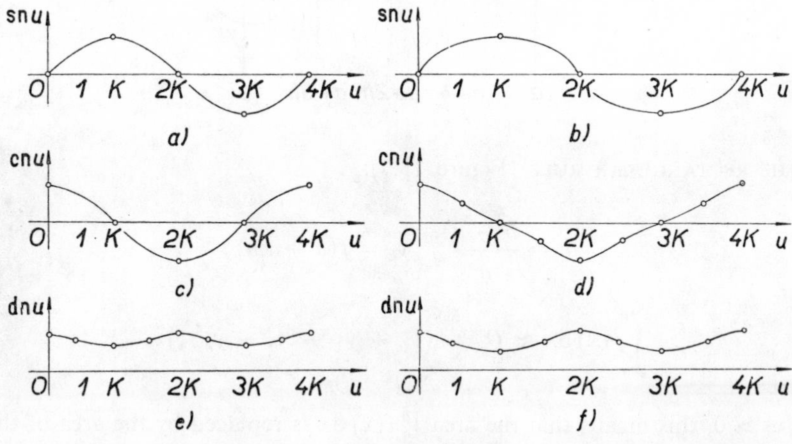

Fig. 13.6.

Basic relations:

$$\operatorname{sn}(-u) = -\operatorname{sn} u\,, \quad \operatorname{cn}(-u) = \operatorname{cn} u\,, \quad \operatorname{dn}(-u) = \operatorname{dn} u\,,$$

$$\operatorname{sn}^2 u + \operatorname{cn}^2 u = 1\,, \quad \operatorname{dn}^2 u - k^2 \operatorname{cn}^2 u = 1 - k^2\,, \quad \operatorname{dn}^2 u + k^2 \operatorname{sn}^2 u = 1\,,$$

$$\frac{\mathrm{d}}{\mathrm{d}u} \operatorname{sn} u = \operatorname{cn} u \operatorname{dn} u\,, \quad \frac{\mathrm{d}}{\mathrm{d}u} \operatorname{cn} u = -\operatorname{sn} u \operatorname{dn} u\,, \quad \frac{\mathrm{d}}{\mathrm{d}u} \operatorname{dn} u = -k^2 \operatorname{sn} u \operatorname{cn} u\,.$$

The graphs of the elliptic functions are shown in Figure 13.6 (for $k^2 = \frac{1}{2}$ in Figures 13.6a, c, e; for $k^2 = \frac{3}{4}$ in Figures 13.6b, d, f).

[The functions sn u, cn u are periodic with period $4K$; the function dn u has period $2K$.]

13.13. Approximate Evaluation of Definite Integrals. Numerical Integration

If a primitive cannot be expressed with the aid of elementary or of tabulated functions or if it is difficult or very laborious to find the primitive, then besides integration by infinite series numerical integration may be used. For this purpose, the integrand is usually replaced by a simpler function (by a polynomial, broken line, etc.).

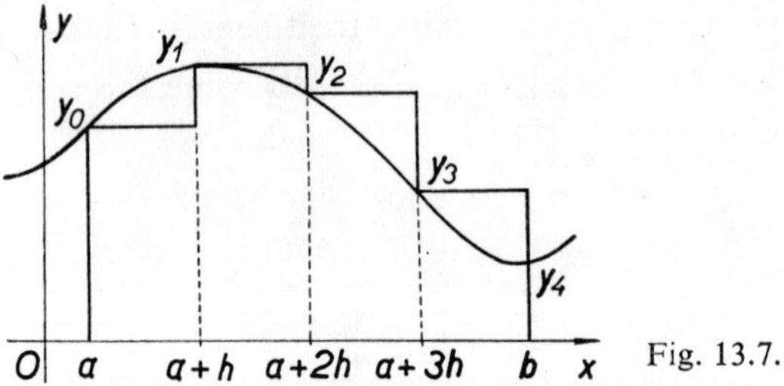

Fig. 13.7.

(a) THE RECTANGULAR RULE (Figure 13.7):

$$h = \frac{b-a}{n}\,, \quad y_k = f(a + kh)\,,$$

$$\int_a^b f(x)\,\mathrm{d}x \approx Q = h(y_0 + y_1 + \dots + y_{n-1})\,.$$

(For $f(x) > 0$, this means that the area $\int_a^b f(x)\,\mathrm{d}x$ is replaced by the area of the rectangular region shown in Fig. 13.7.)

Theorem 1. *If $f(x)$ has the bounded derivative $f'(x)$ in $[a, b]$ and if $m_1 \leqq f'(x) \leqq \leqq M_1$ holds in $[a, b]$, then for an estimate of the error we have*

$$\tfrac{1}{2} m_1 \frac{(b-a)^2}{n} \leqq \int_a^b f(x)\,dx - Q \leqq \tfrac{1}{2} M_1 \frac{(b-a)^2}{n}.$$

(b) THE TRAPEZOIDAL RULE (Figure 13.8):

$$\int_a^b f(x)\,dx \approx R = h(\tfrac{1}{2}y_0 + y_1 + y_2 + \ldots + \tfrac{1}{2}y_n),$$

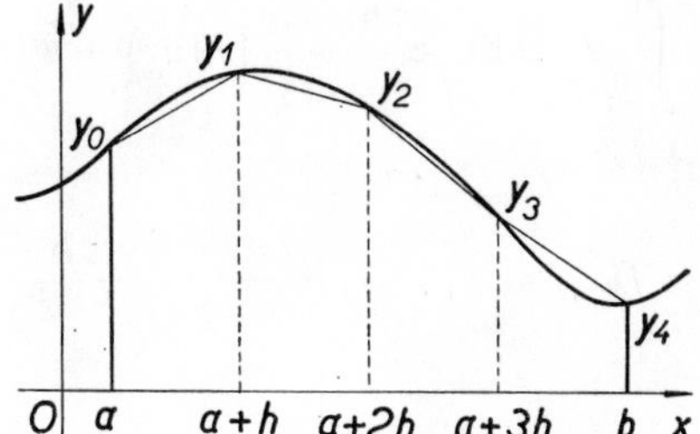

Fig. 13.8.

where again $h = (b - a)/n$, $y_k = f(a + kh)$, $y_0 = f(a)$, $y_n = f(b)$.

Theorem 2. *If $m_2 \leqq f''(x) \leqq M_2$ is satisfied in $[a, b]$, then the estimate*

$$\frac{m_2}{12}\frac{(b-a)^3}{n^2} \leqq R - \int_a^b f(x)\,dx \leqq \frac{M_2}{12}\frac{(b-a)^3}{n^2}$$

holds.

(c) SIMPSON'S RULE

$$h = \frac{b-a}{n}, \quad n \text{ even}, \quad y_k = f(a + kh), \quad y_0 = f(a), \quad y_n = f(b).$$

$$\int_a^b f(x)\,dx \approx S = \tfrac{1}{3}h(y_0 + 4y_1 + 2y_2 + 4y_3 + 2y_4 + \ldots + 4y_{n-1} + y_n).$$

(The function $f(x)$ is approximated by parabolas passing in the interval $[a, a + 2h]$ through the points $(a, f(a))$, $(a + h, f(a + h))$, $(a + 2h, f(a + 2h))$, in the interval $[a + 2h, a + 4h]$ through the points $(a + 2h, f(a + 2h))$, $(a + 3h, f(a + 3h))$, $(a + 4h, f(a + 4h))$, ..., etc.)

Theorem 3. *If $m_4 \leqq f^{(4)}(x) \leqq M_4$ in $[a, b]$, then*

$$m_4 \frac{(b-a)^5}{180n^4} \leqq S - \int_a^b f(x)\,dx \leqq M_4 \frac{(b-a)^5}{180n^4}.$$

REMARK 1. Thus the formula is *exact* if $f(x)$ is a polynomial of at most third degree.

REMARK 2. It is often convenient to have formulae that do not involve the value y_0, y_n at the end points of the interval. We mention here Chebyshev's formula. For other methods and error estimates the reader is referred e.g. to [280].

(d) CHEBYSHEV'S FORMULA:

$$x = \frac{a + b}{2} + \frac{b - a}{2} u ,$$

$$\int_a^b f(x) \, dx = \frac{b - a}{2} \int_{-1}^{1} F(u) \, du \approx \frac{b - a}{n} [F(u_1) + F(u_2) + \ldots + F(u_n)] ,$$

where

$$F(u_i) = f(x_i) = y_i , \quad x_i = \frac{a + b}{2} + \frac{b - a}{2} u_i .$$

We have if

$$n = 2: \quad u_1 = -0{\cdot}577, 350, 3 , \quad u_2 = 0{\cdot}577, 350, 3 ;$$

$$n = 3: \quad u_1 = -0{\cdot}707, 106, 8 , \quad u_2 = 0 , \quad u_3 = +0{\cdot}707, 106, 8$$

(in these two cases the above-mentioned formula is exact for polynomials of the third degree);

$$n = 4: \quad u_1 = -0{\cdot}794, 654, 5 , \quad u_2 = -0{\cdot}187, 592, 5 , \quad u_3 = 0{\cdot}187, 592, 5 , \quad u_4 = 0{\cdot}794, 654, 5 ;$$

$$n = 5: \quad u_1 = -0{\cdot}832, 497, 5 , \quad u_2 = -0{\cdot}374, 541, 4 , \quad u_3 = 0 , \quad u_4 = 0{\cdot}374, 541, 4 , \quad u_5 = 0{\cdot}832, 497, 5$$

(the formula is exact for polynomials of the fifth degree);

$$n = 6: \quad u_1 = -0{\cdot}866, 246, 8 , \quad u_2 = -0{\cdot}422, 518, 7 , \quad u_3 = -0{\cdot}266, 635, 4 , \quad u_4 = 0{\cdot}266, 635, 4 , \quad u_5 = 0{\cdot}422, 518, 7 , \quad u_6 = 0{\cdot}866, 246, 8$$

(the formula is exact for polynomials of the seventh degree).

Example 1. Let us evaluate approximately

$$\int_1^3 f(x) \, dx .$$

(a) Simpson's rule; n must be even. Let us choose $n = 4$. We have $h = (b - a)/n = \frac{1}{2}$ and hence

$$\int_1^3 f(x)\,dx \approx S = \frac{1}{2\,.\,3}\,[f(1) + 4f(1{\cdot}5) + 2f(2) + 4f(2{\cdot}5) + f(3)]\,.$$

(b) Chebyshev's formula; $x = \frac{1}{2}(a + b) + \frac{1}{2}(b - a)\,u$ or $x = 2 + u$. Let us choose $n = 4$. We have $(b - a)/n = \frac{1}{2}$, $x_1 = 2 + u_1 = 2 - 0{\cdot}794{,}654{,}5 = 1{\cdot}205{,}345{,}5$ etc., hence

$$\int_1^3 f(x)\,dx \approx \tfrac{1}{2}[f(1{\cdot}205{,}345{,}5) + f(1{\cdot}812{,}407{,}5) + f(2{\cdot}187{,}592{,}5) + f(2{\cdot}794{,}654{,}5)]\,.$$

13.14. Remark on Lebesgue and Stieltjes Integration

Only relatively simple functions are integrable by the Cauchy–Riemann definition (§ 13·6). For instance, the function f defined in $[0, 1]$ in the following way:

$$f(x) = 1 \quad \text{if } x \text{ is rational}\,,$$

$$f(x) = 0 \quad \text{if } x \text{ is irrational}\,,$$

is not Cauchy–Riemann integrable, since the upper sum is $S(d) = 1$ and the lower sum is $s(d) = 0$ for *every* partition of the interval $[0, 1]$, hence the upper integral equals one, the lower integral equals zero and thus the function is not integrable. Functions may also be met in practice which are not Cauchy–Riemann (or, briefly, Riemann) integrable.

This circumstance and the fact that, when using Riemann integration, some difficulties often arise (for example, when passing to the limit under the integral sign) led to the search for more suitable definitions. One of them is the Lebesgue definition which we shall now briefly mention. For details, we refer e.g. to [175], [210] [299].

First, we define measurable sets. A bounded* set M is said to be *measurable* if, for every $\varepsilon > 0$, an open set N can be found which contains M, while the difference $N - M$ can be covered by a finite or infinite system of open intervals the sum of the lengths of which is less than ε.

To every measurable set M, a (unique) number, called the measure of the set M, is assigned. For instance, the open interval (a, b) or the closed interval $[a, b]$ has the measure $b - a$. Every countable set (and hence also every finite set) is measurable and its measure is zero. (More generally: a set has the measure zero if it can be covered by a finite or infinite system of open intervals the sum of the lengths of which can be made as small as we please.)

* For the purpose of this book, it is sufficient to consider bounded sets only.

Next, the concept of a measurable function is defined: $f(x)$ is said to be *measurable* (in the set M), if for an arbitrary constant K the set of those $x \in M$ for which $f(x) < K$ is measurable.

Let $f(x)$ be a function bounded ($A < f(x) < B$) and measurable in $[a, b]$. Let $A = y_0 < y_1 < y_2 < \ldots < y_{n-1} < y_n = B$. Let us denote by $m(E_k)$ the measure of the set E_k of all x for which $y_{k-1} \leqq f(x) < y_k$. (The set E_3 which consists of two intervals is shown in Figure 13.9.)

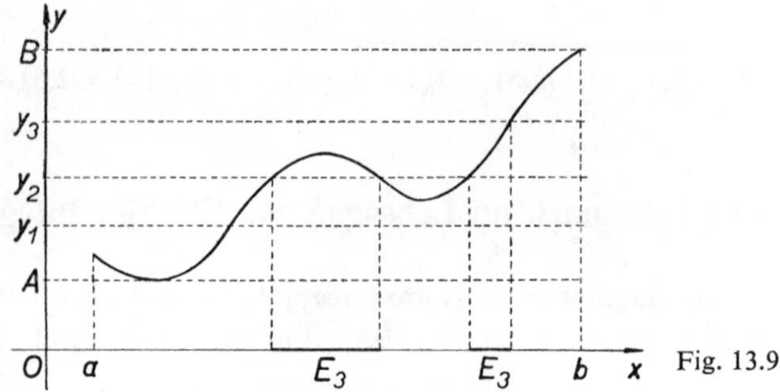

Fig. 13.9.

Let us write

$$S(d) = \sum_{k=1}^{n} y_k\, m(E_k)\,, \quad s(d) = \sum_{k=1}^{n} y_{k-1}\, m(E_k)\,.$$

Now, it can be shown that, if $f(x)$ is measurable, then $\inf S(d) = \sup s(d)$ (where d runs through all possible partitions of the interval $[A, B]$). This common value is called the *Lebesgue integral* of the (bounded) function $f(x)$. Hence:

Theorem 1. *Every function bounded and measurable in the interval* $[a, b]$ *is integrable in Lebesgue's sense in the interval* $[a, b]$.

We shall say L-*integrable* to distinguish Lebesgue integrability from R-*integrability* (i.e. from Riemann integrability).

The definition of the Lebesgue integral of the function $f(x)$ in an arbitrary measurable set M (which need not be an interval) is quite similar. We write

$$\int_M f(x)\,\mathrm{d}x\,. \tag{1}$$

All functions we meet in applications are measurable.

Let the function $f(x)$ be *non-negative* and measurable in $[a, b]$ and generally *not bounded* in $[a, b]$. We choose $K > 0$ and define the function $g(x)$ as follows:

$$g(x) = f(x) \quad \text{if} \quad f(x) \leqq K\,, \quad g(x) = K \quad \text{if} \quad f(x) > K$$

(Fig. 13.10). The function $g(x)$ is measurable and bounded in $[a, b]$, thus the integral

$$\int_a^b g(x)\,dx \tag{2}$$

exists (as the Lebesgue integral). If there exists a finite limit of the integral (2) as $K \to +\infty$, then the function $f(x)$ is said to be L-*integrable* in $[a, b]$ (or the integral $\int_a^b f(x)\,dx$ is said to *exist in the Lebesgue sense*).

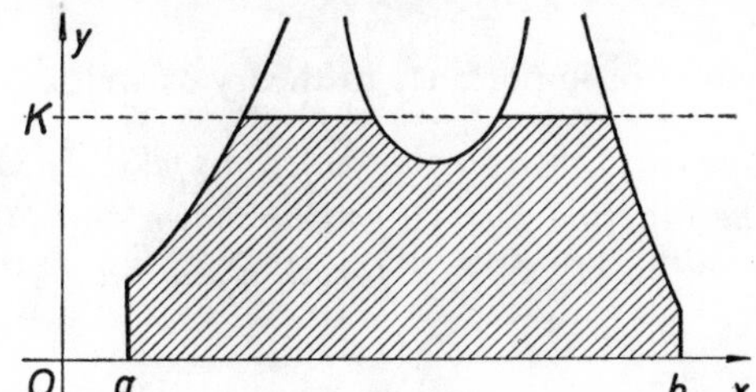

Fig. 13.10.

If we are given a measurable function $f(x)$ (not necessarily bounded and non-negative), we can write

$$f(x) = f_+(x) - f_-(x),$$

where

$$f_+(x) = f(x) \quad \text{if} \quad f(x) \geqq 0\,; \quad f_+(x) = 0 \quad \text{if} \quad f(x) < 0\,;$$
$$f_-(x) = 0 \quad \text{if} \quad f(x) \geqq 0\,; \quad f_-(x) = -f(x) \quad \text{if} \quad f(x) < 0\,.$$

(Intuitively: $f_+(x)$ is that "part" of the function $f(x)$ the graph of which lies above the x-axis etc.)

If the integrals

$$\int_a^b f_+(x)\,dx\,, \quad \int_a^b f_-(x)\,dx$$

both exist, then the function $f(x)$ is said to be *integrable in* $[a, b]$ (*in the Lebesgue sense*) and

$$\int_a^b f(x)\,dx = \int_a^b f_+(x)\,dx - \int_a^b f_-(x)\,dx\,.$$

The definition of the integral of functions not bounded in an arbitrary measurable set M is similar.

An important class of functions consists of functions for which the integrals

$$\int_a^b f(x)\,dx \quad \text{and} \quad \int_a^b f^2(x)\,dx$$

both exist.

These functions constitute the so-called *set of square integrable functions* in $[a, b]$.

The fact that $f(x)$ is square integrable (in Lebesgue's sense) in $[a, b]$ will be briefly denoted as follows: $f \in L_2(a, b)$.

Theorem 2. *Every function bounded in the interval $[a, b]$ and* R-*integrable (i.e. integrable in the Riemann sense) is also* L-*integrable in $[a, b]$ (and the values of the integrals are equal). The same is true for square integrable functions (not necessarily bounded).*

The following theorem is of importance in theory as well as in applications:

Theorem 3. *If the sequence of measurable functions $f_1(x), f_2(x), \ldots$ converges in the measurable set M to the function $F(x)$ (it can be shown that $F(x)$ is then also measurable) and at the same time $|f_n(x)| < K$ for all n and all $x \in M$ (where K is a constant), then the relation*

$$\lim_{n \to \infty} \int_M f_n(x)\, dx = \int_M F(x)\, dx$$

holds (even if the convergence is not uniform).

From Theorem 3 there follows:

Theorem 4. *If the sequence of* R-*integrable functions $f_n(x)$ converges in the interval $[a, b]$ to the function $F(x)$ which is also* R-*integrable and if $|f_n(x)| < K$ for all n and all $x \in [a, b]$, then*

$$\lim_{n \to \infty} \int_a^b f_n(x)\, dx = \int_a^b F(x)\, dx\,.$$

The Lebesgue theory of integration is also of great importance in the theory of orthogonal systems, in functional analysis etc. Measure theory and Lebesgue's integration is dealt with in detail e.g. in [175], [210], [299] etc.

STIELTJES'S INTEGRAL. Let two functions $f(x)$ and $g(x)$ be given in a (finite) interval $[a, b]$. We divide the interval $[a, b]$ into subintervals by inserting the points

$$a = x_0 < x_1 < x_2 < \ldots < x_n = b\,,$$

choose in each interval $[x_k, x_{k+1}]$ an arbitrary point ξ_k and form the sum

$$\sigma = \sum_{k=0}^{n-1} f(\xi_k)\,[g(x_{k+1}) - g(x_k)]\,. \tag{3}$$

The maximal length of all the subintervals is called the *norm* $\nu(d)$ of the given partition d, i.e. $\nu(d) = \max\,(x_{k+1} - x_k)$. If σ tends to a finite number I for $\nu \to 0$, no matter how the interval has been sub-divided and the points ξ_k chosen, then the number I is called the *Stieltjes integral of the function $f(x)$ with respect to the function $g(x)$*. We write

$$I = \int_a^b f(x)\,dg(x)\,. \tag{4}$$

For $g(x) = x$ we have the Cauchy – Riemann integral.

Theorem 5. *If $f(x)$ is continuous in $[a, b]$ and if $g(x)$ is of bounded variation in $[a, b]$* (Definition 11.3.7, p. 408), *then the integral* (4) *exists.*

Theorem 6. *If $f(x)$ is continuous in $[a, b]$ and if $g(x)$ has a continuous derivative in $[a, b]$, then the integral* (4) *exists and the relation*

$$\int_a^b f(x)\,dg(x) = \int_a^b f(x)\,g'(x)\,dx \tag{5}$$

holds.

Theorem 7. *If $f(x)$ is continuous in $[a, b]$ and if $g(x)$ and $g'(x)$ are piecewise continuous, then the integral* (4) *exists. Let us denote by x_k the points of discontinuity of $g(x)$, $a = x_0 < x_1 < x_2 < \ldots < x_p = b$ and s_k the jump of the function $g(x)$ at x_k,* i.e.

$$s_k = g(x_k + 0) - g(x_k - 0)\,,$$

$$\textit{where}\quad g(x_k + 0) = \lim_{x \to x_k+} g(x)\,,\quad g(x_k - 0) = \lim_{x \to x_k-} g(x)\,,$$

$$s_0 = g(a + 0) - g(a)\,,\quad s_p = g(b) - g(b - 0)\,.$$

Then

$$\int_a^b f(x)\,dg(x) = \int_a^b f(x)\,g'(x)\,dx + \sum_{k=0}^{p} f(x_k)\,.\,s_k\,,$$

i.e. *we can easily evaluate the Stieltjes integral as the sum of the above-mentioned Riemann integral and of the jumps of the function $g(x)$ multiplied by the values of the function $f(x)$ at the points x_k.*

Example 1. Let us consider a bar of length l carrying a load $g(x)$ (in Fig. 13.11 a uniform load is illustrated). Denote the total load in the interval $[0, x]$ by $Q(x)$, i.e.

$$Q(x) = \int_0^x q(t)\,dt\,. \tag{6}$$

The total moment about the point O due to this load (the reactions are not considered) can then be written in the form

$$M = \int_0^l x\,dQ(x) \tag{7}$$

for, by Theorem 6,

$$\int_0^l x \, dQ(x) = \int_0^l x \, q(x) \, dx \, .$$

Now, let us consider, in addition, a concentrated load P at the point a (Figure 13.12). The function (6), which describes the total load in the interval $[0, x]$, becomes

$$Q(x) = \int_0^x q(t) \, dt \qquad \text{for} \quad x < a \, ,$$

$$Q(x) = \int_0^x q(t) \, dt + P \quad \text{for} \quad x \geqq a \, .$$

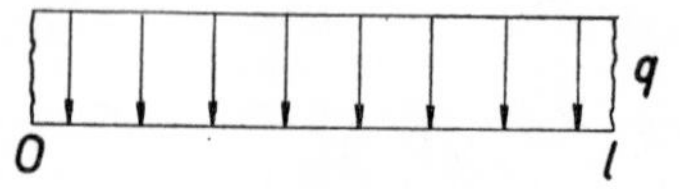

Fig. 13.11.

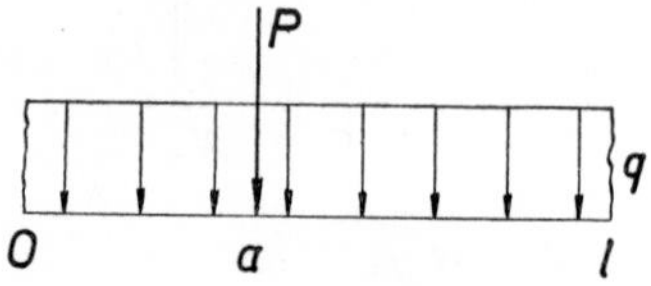

Fig. 13.12.

The Stieltjes integral makes it possible here also to write the expression for M in the closed form (7) (which is often convenient):

$$M = \int_0^l x \, dQ(x) \, ;$$

By Theorems 7 and 6, we have

$$\int_0^l x \, dQ(x) = \int_0^l x \, q(x) \, dx + aP$$

and the right-hand side is the known expression for M.

13.15. Survey of Some Important Formulae from Chapter 13

(See also Theorem 13.1.2, §§ 13.5 and 13.10 and some formulae in §§ 13.11 and 13.12. See also applications to geometry and physics in § 14.9.)

1. $\int [c_1 f_1(x) + c_2 f_2(x)] \, dx = c_1 \int f_1(x) \, dx + c_2 \int f_2(x) \, dx$ (Theorem 13.2.1) .

2. $\int u'v \, dx = uv - \int uv' \, dx$ or $\int v \, du = uv - \int u \, dv$

(integration by parts, Theorem 13.2.2) .

3. $\int g(h(x))\,h'(x)\,\mathrm{d}x = \int g(z)\,\mathrm{d}z\,,\quad \int f(x)\,\mathrm{d}x = \int f(\varphi(z))\,\varphi'(z)\,\mathrm{d}z$

(integration by substitution, Theorems 13.2.3, 13.2.4).

4. $\frac{\mathrm{d}}{\mathrm{d}x}\int_a^x f(t)\,\mathrm{d}t = f(x)\,,\quad \frac{\mathrm{d}}{\mathrm{d}x}\int_x^b f(t)\,\mathrm{d}t = -f(x)$ (Theorem 13.6.12, Remark 13.6.8).

5. $\int_a^b f(x)\,\mathrm{d}x = F(b) - F(a)$ (Theorem 13.6.13).

6. $\int_a^b [c_1 f_1(x) + c_2 f_2(x)] = c_1 \int_a^b f_1(x)\,\mathrm{d}x + c_2 \int_a^b f_2(x)\,\mathrm{d}x$ (Theorem 13.6.3).

7. $\int_a^b u'v\,\mathrm{d}x = [uv]_a^b - \int_a^b uv'\,\mathrm{d}x$ or $\int_a^b v\,\mathrm{d}u = [uv]_a^b - \int_a^b u\,\mathrm{d}v$

(integration by parts, Theorem 13.7.1).

8. $\int_a^b g(h(x))\,h'(x)\,\mathrm{d}x = \int_{h(a)}^{h(b)} g(z)\,\mathrm{d}z\,,$

$\int_a^b f(x)\,\mathrm{d}x = \int_\alpha^\beta f(\varphi(z))\,\varphi'(z)\,\mathrm{d}z\,,$ where $\varphi(\alpha) = a\,,\quad \varphi(\beta) = b$

(integration by substitution, Theorems 13.7.2, 13.7.3).

9. $\int_b^a f(x)\,\mathrm{d}x = -\int_a^b f(x)\,\mathrm{d}x\,.$

10. $\int_a^a f(x)\,\mathrm{d}x = 0\,.$

11. $\int_{-a}^a f(x)\,\mathrm{d}x = 2\int_0^a f(x)\,\mathrm{d}x$ if f is even,

$= 0$ if f is odd (Remark 13.6.11).

12. $\frac{\mathrm{d}}{\mathrm{d}\alpha}\int_a^b f(x,\alpha)\,\mathrm{d}x = \int_a^b \frac{\partial f}{\partial \alpha}(x,\alpha)\,\mathrm{d}x$ (Theorem 13.9.2).

13. $\frac{\mathrm{d}}{\mathrm{d}\alpha}\int_{\varphi_1(\alpha)}^{\varphi_2(\alpha)} f(x,\alpha)\,\mathrm{d}x = \int_{\varphi_1(\alpha)}^{\varphi_2(\alpha)} \frac{\partial f}{\partial \alpha}(x,\alpha)\,\mathrm{d}x + \varphi_2'(\alpha) f(\varphi_2(\alpha),\alpha) -$

$- \varphi_1'(\alpha) f(\varphi_1(\alpha),\alpha)$ (Theorem 13.9.4).

14. $\int_a^b f(x)\,\mathrm{d}g(x) = \int_a^b f(x)\,g'(x)\,\mathrm{d}x + \sum_{k=0}^{p} f(x_k)\,s_k$ (Theorem 13.14.7).

14. INTEGRAL CALCULUS OF FUNCTIONS OF TWO AND MORE VARIABLES

By KAREL REKTORYS

References: [16], [27], [48], [73], [112], [155], [226], [275], [320], [338], [341], [367]

When using the concepts of a *curve*, a *surface*, a *region* and a *function* in this chapter, we suppose that they are of the type defined in § 14.1. Closed domains (i.e. which contain the boundary h) are denoted by a bar: $\overline{M} = M + h$.

14.1. Basic Definitions and Notation

Definition 1. By a *simple finite piecewise smooth curve* in the xy-plane we mean a set of points (x, y) given in parametric form by the equations

$$x = \varphi(t)\,, \quad y = \psi(t) \quad (\alpha \leqq t \leqq \beta)\,, \tag{1}$$

where

1. $\varphi(t)$ and $\psi(t)$ are continuous in $[\alpha, \beta]$ and have a piecewise continuous derivative in $[\alpha, \beta]$;

2. $x'(t)$ and $y'(t)$ do not vanish simultaneously for any $t \in [\alpha, \beta]$ (at points of discontinuity and at the end points of the interval $[\alpha, \beta]$, we understand here by $x'(t)$ and $y'(t)$ the values of their continuous extensions);

3. for any pair $t_1 \neq t_2$ from $[\alpha, \beta]$ (with the possible exception of the pair $t_1 = \alpha$, $t_2 = \beta$), the equations

$$\varphi(t_1) = \varphi(t_2)\,, \quad \psi(t_1) = \psi(t_2)$$

do not simultaneously hold.

REMARK 1. Condition 3 expresses the *simplicity* of the curve, which means that the curve does not intersect itself. If $\varphi(\alpha) = \varphi(\beta)$ and if at the same time $\psi(\alpha) = \psi(\beta)$, we say that the curve is *closed*.

The terms *curve* and *arc* are not uniformly used in the literature. Often the term *curve* is used for a closed curve and the term *arc* is used for an open curve.

The continuity of the functions which is required in condition 1 expresses the fact that the curve is *connected*; the piecewise continuous derivatives express the condition that the curve consists of a finite number of arcs with continuously changing tangent. If the derivatives are *everywhere* continuous, then the curve has everywhere a continuously changing tangent and is said to be *smooth*. Condition 2 excludes various singularities, e.g. the "curve" $x = a$, $y = b$ (degenerating into a point) etc.

The above-mentioned curves are *rectifiable*, i.e. they have a finite length l for which the relation

$$l = \int_\alpha^\beta \sqrt{[\varphi'^2(t) + \psi'^2(t)]}\,dt$$

holds.

Thus, *geometrically*, a simple finite piecewise smooth curve is a curve of finite length not intersecting itself and consisting of a finite number of arcs with continuously changing tangent. In practice we deal almost exclusively with such types of curves (as far as curves of finite length are considered).

Example 1. The circumference of a square is a simple piecewise smooth closed curve. The circle and the ellipse are simple finite smooth closed curves.

In a similar way a *simple finite piecewise smooth curve in three-dimensional space* can be defined. It is given parametrically by the equations

$$x = \varphi(t), \quad y = \psi(t), \quad z = \chi(t) \quad (\alpha \leqq t \leqq \beta).$$

REMARK 2. For $x = t$ we get from (1) $y = \psi(x)$ and the curve is the graph of the function $y = \psi(x)$.

REMARK 3. A rather more general concept is the so-called *Jordan curve* in the plane which is a simple rectifiable closed curve (not necessarily smooth or piecewise smooth). Every Jordan curve divides the plane into two parts; the bounded part, which is said to be the *interior of the curve* (the so-called *Jordan region*), and the unbounded part, called the *exterior of the curve*.

Definition 2. Let O be a bounded region in the plane (see Definitions 22.1.6 and 22.1.10, pp. 995, 996) (O need not be simply connected). If its boundary consists of a finite number of simple piecewise smooth closed curves, then the region O is said to be *of type* A.

We define the *closed region of type* A similarly.

REMARK 4. In practice we almost invariably meet regions of type A only. Examples: the square, the polygon, the interior of an ellipse, the annulus etc.

Definition 3. Let the function $f(x, y)$ be defined in a region O of type A. The function $f(x, y)$ is said to be *of type* B *in* O, if it is bounded in O and continuous with the possible exception of a finite number of points or of points constituting a finite number of simple finite piecewise smooth curves. Similarly we define the *function* $f(x, y)$ *of type* B *in the closed region* O *of type* A.

REMARK 5. If $\bar{O}$ is a closed region of type A, then obviously every function $f(x, y)$ which is continuous in $\bar{O}$ is of type B in $\bar{O}$.

REMARK 6. Clearly, if $f(x, y)$ is of type B in $\bar{O}$, it is of type B also in O. Conversely: If $f(x, y)$ is of type B in O and if we define it on the boundary of the region O in such a way that it remains bounded (otherwise it can be defined arbitrarily), then $f(x, y)$ is of type B in $\bar{O}$.

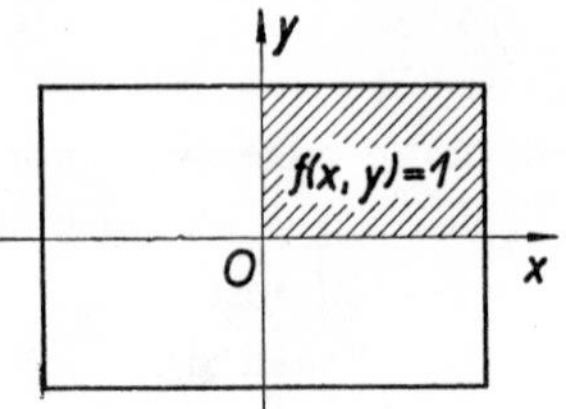

Fig. 14.1.

Example 2. The rectangle $\bar{O}(-a \leqq x \leqq a, \ -b \leqq y \leqq b)$ is a closed region of type A. We define the function $f(x, y)$ as follows (Figure 14.1):

$$f(x, y) = 1 \quad \text{for} \quad 0 \leqq x \leqq a\,, \quad 0 \leqq y \leqq b\,,$$

$$f(x, y) = 0 \quad \text{for the other points of the rectangle } \bar{O}\,.$$

The function $f(x, y)$ is of type B in $\bar{O}$, for it is bounded and continuous everywhere in $\bar{O}$ with exception of the line segments

$$x = 0\,, \quad 0 \leqq y \leqq b \quad \text{and} \quad y = 0\,, \quad 0 \leqq x \leqq a\,.$$

REMARK 7. In a manner similar to that of Definition 1, we define a *simple finite piecewise smooth surface*

$$x = \varphi(u, v)\,, \quad y = \psi(u, v)\,, \quad z = \chi(u, v)\,. \tag{2}$$

The interval $[\alpha, \beta]$ in Definition (1) is replaced here by a closed region $\bar{O}$ of type A (Definition 2) in which the functions (2) are continuous and have piecewise continuous (Remark 12.1.8) partial derivatives of the first order. The condition 2 in Definition 1 is replaced by the condition that at no point (u, v) in the region $\bar{O}$ the determinants

$$\begin{vmatrix} \dfrac{\partial x}{\partial u}, & \dfrac{\partial x}{\partial v} \\ \dfrac{\partial y}{\partial u}, & \dfrac{\partial y}{\partial v} \end{vmatrix}, \quad \begin{vmatrix} \dfrac{\partial x}{\partial u}, & \dfrac{\partial x}{\partial v} \\ \dfrac{\partial z}{\partial u}, & \dfrac{\partial z}{\partial v} \end{vmatrix}, \quad \begin{vmatrix} \dfrac{\partial y}{\partial u}, & \dfrac{\partial y}{\partial v} \\ \dfrac{\partial z}{\partial u}, & \dfrac{\partial z}{\partial v} \end{vmatrix} \tag{3}$$

vanish *simultaneously*.

REMARK 8. *Intuitively:* A piecewise smooth surface may be divided into a finite number of parts which have a continuously changing normal.

REMARK 9. If in (2) $x = u$, $y = v$, we obtain the equation of the surface in the explicit form $z = \chi(x, y)$. This surface will naturally be simple, finite and piecewise smooth, if the function $\chi(x, y)$ is continuous in $\bar{O}$ and has piecewise continuous partial derivatives of the first order in $\bar{O}$.

Definition 4. The bounded *three-dimensional* region O (not necessarily simply connected) is said to be *of type* A if its boundary is formed by a finite number of simple finite piecewise smooth surfaces. Similarly, we define a *closed three-dimensional region of type* A. Instead of the "three-dimensional region of type A" we shall often say the "*solid of type* A".

Example 3. Examples of solids of type A are the cube, the sphere, the ellipsoid etc.

Definition 5. Let the function $u = f(x, y, z)$ be defined in the solid Q of type A. If $f(x, y, z)$ is bounded in Q and at the same time continuous, with the possible exception of a finite number of points or of points constituting a finite number of simple finite piecewise smooth curves or surfaces, we say that $f(x, y, z)$ is *of type* B *in* Q. Similarly, we define the *function* $f(x, y, z)$ *of type* B *in the closed solid* $\bar{Q}$.

REMARK 10. Remarks similar to Remarks 5 and 6 hold in this case also.

REMARK 11. In the following text we use the concepts of a ***curve,*** a ***surface,*** a ***region*** and a ***function*** in the above-mentioned sense (we consider regions of type A, functions of type B etc.).

REMARK 12. If we say that $f(x, y)$ is *continuous on a curve k*, we mean that the function $f(x, y)$ is continuous at each point (x_0, y_0) of the curve k in the usual sense (with ε and δ) with the distinction that in a δ-neighbourhood of the point (x_0, y_0) in question we consider *only* the points of the curve k. Similarly we consider the continuity of the function $f(x, y, z)$ on a surface.

14.2. The Double Integral

Let a continuous function $z = f(x, y)$ (with $m \leqq f(x, y) \leqq M$) be defined in the closed rectangle $\bar{O}(a \leqq x \leqq b,\ c \leqq y \leqq d)$. Let us divide the interval $[a, b]$ into subintervals $\Delta x_1, \ldots, \Delta x_m$, the interval $[c, d]$ into subintervals $\Delta y_1, \ldots, \Delta y_n$ (as in Fig. 14.2, where we have $m = 4$, $n = 3$). We shall denote the rectangle with the base Δx_i and the height Δy_k (including the boundary) by $\bar{O}_{ik}$. Let us construct the so-called *upper and lower sums* (belonging to the given partition p of the rectangle $\bar{O}$),

$$S(p) = \sum_{i,k} M_{ik}\, \Delta x_i\, \Delta y_k\,, \quad s(p) = \sum_{i,k} m_{ik}\, \Delta x_i\, \Delta y_k\,, \tag{1}$$

where M_{ik} and m_{ik} are the maximum and the minimum values of the function $f(x, y)$ in $\bar{O}_{ik}$, respectively. The set of upper sums for all possible partitions p is bounded below (for the inequality $S(p) \geqq m(b - a)(d - c)$ always holds). Its greatest lower bound (Definition 1.3.3) is called the *upper double integral* of the function $f(x, y)$ in $\bar{O}$,

$$\inf S(p) = \overline{\iint_O} f(x, y)\,\mathrm{d}x\,\mathrm{d}y\,;$$

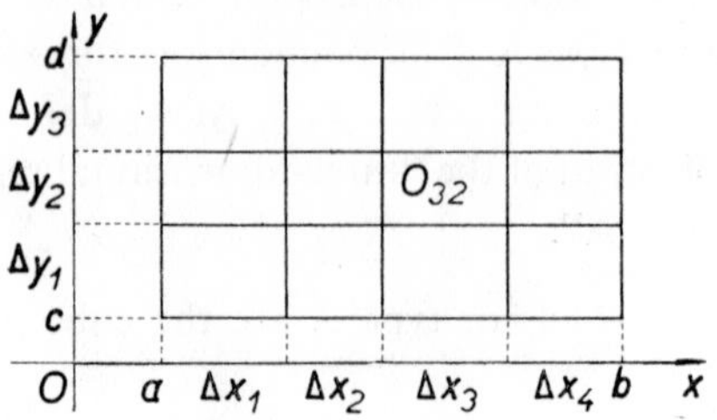

Fig. 14.2.

the least upper bound of the set of the lower sums,

$$\sup s(p) = \underline{\iint_O} f(x, y)\,\mathrm{d}x\,\mathrm{d}y\,,$$

is called the *lower double integral.*

Similarly the upper and the lower integrals are defined in the case where we assume mere *boundedness* instead of continuity of $f(x, y)$ in $\bar{O}$. Then in the sums (1) M_{ik} and m_{ik} need not be the maximum and minimum, respectively, but, in general, the least upper bound, or the greatest lower bound of the values of the function in $\bar{O}_{ik}$, respectively.

Definition 1. If the upper and lower integrals are equal, then their common value is called the *double integral of the function* $f(x, y)$ *in* (or *over*) O,

$$\iint_O f(x, y)\,\mathrm{d}x\,\mathrm{d}y\,, \tag{2}$$

and the function $f(x, y)$ is said to be *integrable in* (or *over*) O *in the Cauchy-Riemann sense.*

REMARK 1. The *geometric meaning* of the integral (2) for the case where $f(x, y)$ is continuous and positive in $\bar{O}$: (2) is the volume of the solid whose lower base is formed by the rectangle $\bar{O}$, upper base by the surface $z = f(x, y)$ above the region $\bar{O}$ and the lateral surface by lines parallel to the z-axis.

Theorem 1. *Every function* $z = f(x, y)$ *which is continuous in* $\bar{O}(a \leqq x \leqq b$, $c \leqq y \leqq d)$ *is integrable in* O.

Theorem 2. *Every function* $f(x, y)$ *which is of type* B *in* $\bar{O}$ (Definition 14.1.3) *is integrable in* O.

REMARK 2. Theorem 2 offers the possibility of defining the integral for other regions than a rectangular one. Let $\overline{M}$ be a closed region of type A (Definition 14.1.2) and let a function $f(x, y)$ of type B be defined in $\overline{M}$. As M is a bounded region, we can construct a closed rectangle $\bar{O}(a \leqq x \leqq b,\ c \leqq y \leqq d)$ in which this region is contained (Fig. 14.3). Let us define the function $g(x, y)$ in $\bar{O}$ as follows:

$$g(x, y) = f(x, y) \quad \text{for} \quad (x, y) \in \overline{M}\,,$$

$$g(x, y) = 0 \text{ for the other points of } \bar{O} \quad \text{(shaded region in Fig. 14.3)}\,.$$

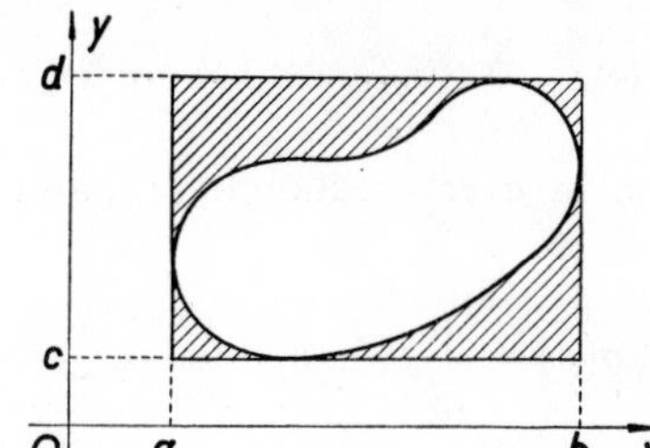

Fig. 14.3.

(Thus, the function $g(x, y)$ coincides with $f(x, y)$ in $\overline{M}$ and at other points of $\bar{O}$ is equal to zero.) We define

$$\iint_M f(x, y)\,\mathrm{d}x\,\mathrm{d}y = \iint_O g(x, y)\,\mathrm{d}x\,\mathrm{d}y\,. \tag{3}$$

REMARK 3. The *geometric meaning* of the integral $\iint_M f(x, y)\,\mathrm{d}x\,\mathrm{d}y$ if $f(x, y)$ is continuous and positive in M is that it is the volume of the body the "lower base" of which is formed by the region $\overline{M}$, the "upper base" by the surface $z = f(x, y)$ and the lateral surface by lines parallel to the z-axis.

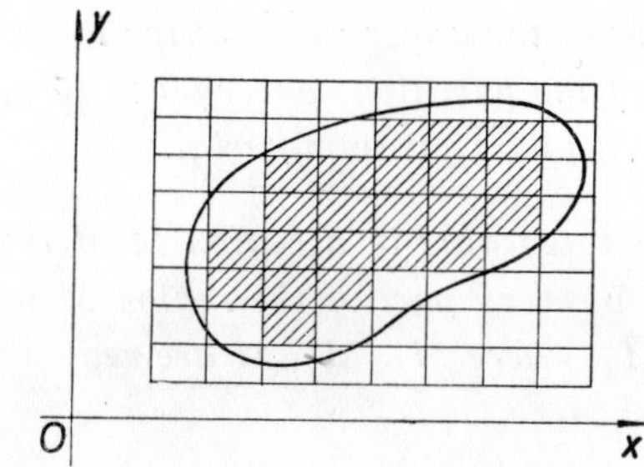

Fig. 14.4.

REMARK 4. The integral

$$\iint_M f(x, y)\,\mathrm{d}x\,\mathrm{d}y$$

may be defined in many other ways different from that used above. Another way of defining the integral is clear from Fig. 14.4. In the sums (1) only such rectangles O_{ik} which lie entirely in the region $\overline{M}$ (shaded region in Fig. 14.4) occur.

Other definitions do not use a rectangular network, but divide the region $\overline{M}$ into (small) regions which are in a certain sense arbitrary.

The definition of the *Lebesgue* integral is based, as in the case of the one-dimensional integral, on the definition of the measure of a two-dimensional region.

We give now the basic theorems for the Cauchy–Riemann integral. First we define:

Definition 2. By the *norm* $v(p)$ *of the partition* p we mean the greatest of the lengths of the intervals Δx_i, Δy_k,

$$v(p) = \max\left(\Delta x_1, \Delta x_2, \ldots, \Delta x_m, \Delta y_1, \Delta y_2, \ldots, \Delta y_n\right).$$

REMARK 5. Let us choose, in every rectangle $\overline{O}_{ik}$, an arbitrary point $P_{ik}(\xi_i, \eta_k)$. Form the sum

$$\sigma(p) = \sum_{i,k} f(\xi_i, \eta_k)\, \Delta x_i\, \Delta y_k$$

depending on the chosen partition p and on the choice of the points (ξ_i, η_i) in $\overline{O}_{ik}$.

Theorem 3. *Let* $p_1, p_2, \ldots$ *be a sequence of partitions such that*

$$\lim_{n\to\infty} v(p_n) = 0\,.$$

Then, if $f(x, y)$ *is integrable in* O, *we have*

$$\iint_O f(x, y)\, \mathrm{d}x\, \mathrm{d}y = \lim_{n\to\infty} S(p_n) = \lim_{n\to\infty} s(p_n) = \lim_{n\to\infty} \sigma(p_n)\,,$$

i.e. *the integral is the limit of the upper sums, of the lower sums, or of the sums* σ, *respectively, if the norm of the partition converges to zero. Regarding the sums* σ, *the choice of the point* (ξ_i, η_k) *in* $\overline{O}_{ik}$ *is immaterial.*

Theorem 4. *If* $f(x, y)$ *is integrable in a region* M *of type* A (Definition 14.1.2), *then it is integrable also in every part of the region* M *which is of type* A. *If, in particular,* $\overline{M} = \overline{M}_1 + \overline{M}_2$, *where* M_1, M_2, M *are regions of type* A *and* M_1, M_2 *have no common points, then*

$$\iint_M f(x, y)\, \mathrm{d}x\, \mathrm{d}y = \iint_{M_1} f(x, y)\, \mathrm{d}x\, \mathrm{d}y + \iint_{M_2} f(x, y)\, \mathrm{d}x\, \mathrm{d}y\,.$$

Theorem 5. *If the functions* $f_1(x, y)$, $f_2(x, y)$ *are integrable in* M, *then the functions*

$$c_1 f_1 + c_2 f_2\,,\quad f_1 f_2\,,\quad |f_1|\,,\quad |f_2|$$

are also integrable in M and the relations

$$\iint_M [c_1 f_1(x, y) + c_2 f_2(x, y)] \, dx \, dy = c_1 \iint_M f_1(x, y) \, dx \, dy + c_2 \iint_M f_2(x, y) \, dx \, dy \,,$$

$$\left| \iint_M f_1(x, y) \, dx \, dy \right| \leqq \iint_M |f_1(x, y)| \, dx \, dy$$

hold.

The equation

$$\iint_M f_1(x, y) f_2(x, y) \, dx \, dy = \iint_M f_1(x, y) \, dx \, dy \,.\, \iint_M f_2(x, y) \, dx \, dy$$

does not hold, in general! A sufficient condition for the integrability of the function

$$\frac{f_1(x, y)}{f_2(x, y)}$$

is that the functions f_1 and f_2 are integrable and either

$$0 < k \leqq f_2(x, y) \,, \quad \text{or} \quad f_2(x, y) \leqq K < 0 \,,$$

i.e. if f_2 is positive and bounded below in M by a positive constant or negative and bounded above by a negative constant, respectively.

Theorem 6. *Let $f(x, y)$, $g(x, y)$ be integrable in M,*

$$k \leqq f(x, y) \leqq K \,, \quad g(x, y) \geqq 0 \quad \text{in } M \,.$$

Then

$$k \iint_M g(x, y) \, dx \, dy \leqq \iint_M f(x, y) \, g(x, y) \, dx \, dy \leqq K \iint_M g(x, y) \, dx \, dy \,.$$

Theorem 7 (*Mean-Value Theorem*). *If the function $f(x, y)$ is continuous in a closed region $\overline{M}$ of type* A, *then there exists at least one point $(x_0, y_0) \in \overline{M}$ such that*

$$\iint_M f(x, y) \, dx \, dy = P f(x_0, y_0) \,,$$

where P is the area of the region $\overline{M}$.

Theorem 8. *Neither the integrability of the function nor the value of the integral is changed if the value of the function $f(x, y)$ is changed at a finite number of points or on a finite number of piecewise smooth finite curves.*

14.3. Evaluation of a Double Integral by Repeated Integration

Theorem 1. *Let $\bar{O}$ be a closed rectangle $a \leqq x \leqq b$, $c \leqq y \leqq d$. If $f(x, y)$ is of type* B *in $\bar{O}$* (see Definition 14.1.3), *then*

$$\iint_O f(x, y)\,\mathrm{d}x\,\mathrm{d}y = \int_a^b \left[\int_c^d f(x, y)\,\mathrm{d}y\right]\mathrm{d}x = \int_c^d \left[\int_a^b f(x, y)\,\mathrm{d}x\right]\mathrm{d}y\,. \tag{1}$$

REMARK 1. Thus, in particular, Theorem 1 holds for functions continuous in $\bar{O}$. It should be noted that in the general case the function

$$F(x) = \int_c^d f(x, y)\,\mathrm{d}y$$

need not be defined for *all* $x \in [a, b]$ (the function $f(x, y)$ need not be integrable as a function of the variable y for all these x). The same holds for the third expression in (1).

REMARK 2. According to (1) the integral

$$\iint_O f(x, y)\,\mathrm{d}x\,\mathrm{d}y \quad \text{is often denoted by} \quad \int_a^b \int_c^d f(x, y)\,\mathrm{d}x\,\mathrm{d}y \quad \text{or} \quad \int_c^d \int_a^b f(x, y)\,\mathrm{d}y\,\mathrm{d}x\,.$$

Instead of

$$\int_a^b \left[\int_c^d f(x, y)\,\mathrm{d}y\right]\mathrm{d}x \quad \text{or} \quad \int_c^d \left[\int_a^b f(x, y)\,\mathrm{d}x\right]\mathrm{d}y$$

we employ the notation

$$\int_a^b \mathrm{d}x \int_c^d f(x, y)\,\mathrm{d}y \quad \text{or} \quad \int_c^d \mathrm{d}y \int_a^b f(x, y)\,\mathrm{d}x\,,$$

respectively.

REMARK 3. *Geometric interpretation* of Theorem 1 (Fig. 14.5; for simplicity let us suppose that $f(x, y)$ is continuous and positive in $\bar{O}$; see Remark 14.2.1). For a constant x, the integral

$$\int_c^d f(x, y)\,\mathrm{d}y = F(x)$$

(where we have integrated only with respect to y) gives the area of the cross-section of the solid shown in Fig. 14.5. For small Δx

$$F(x)\,\Delta x \quad \text{or} \quad \Delta x \int_c^d f(x, y)\,\mathrm{d}y$$

denotes the volume of a small layer of the solid under consideration. Summing up these layers ($\sum F(x_i)\,\Delta x_i$) we get the approximate volume of the solid. By going to the limit, this sum changes into the integral.

We arrive at the same result (see the second of equations (1)) if we "cut" the solid into layers by the planes $y = \text{const.}$

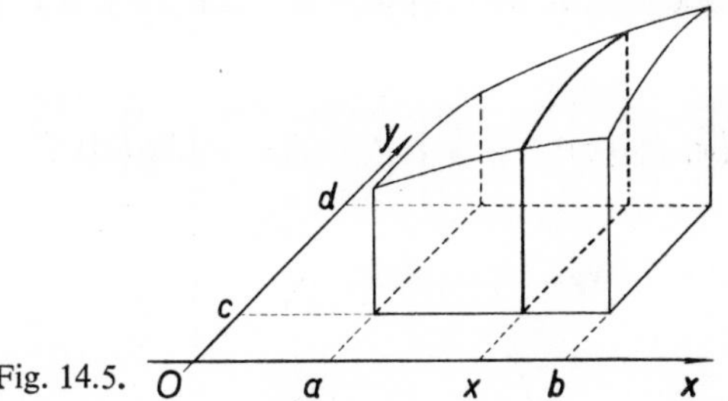

Fig. 14.5.

REMARK 4. The integrals on the right-hand side of equation (1) are called *repeated integrals*. Theorem 1 is a generalization of Theorems 13.9.3.

REMARK 5. According to Remark 14.2.2, Theorem 1 holds for arbitrary closed regions $\overline{M}$ of type A and for functions $f(x, y)$ of type B in $\overline{M}$. Usually we deal with the following simple case: Let $y = h_1(x)$, $y = h_2(x)$ be continuous functions with continuous or piecewise continuous derivatives in $[a, b]$ such that $h_2(x) > h_1(x)$ in (a, b). Let $k = \min h_1(x)$ in $[a, b]$, $K = \max h_2(x)$ in $[a, b]$. Let $\overline{M}$ be the region (obviously of type A) the boundary of which is formed by the functions $y = h_1(x)$, $y = h_2(x)$ and the lines $x = a$, $x = b$ parallel to the y-axis. Denote the rectangle

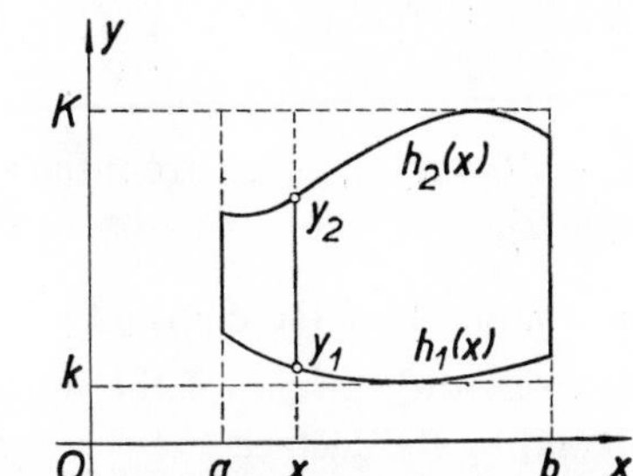

Fig. 14.6.

$a \leqq x \leqq b$, $k \leqq y \leqq K$ by $\overline{O}$ (Fig. 14.6). By equation (3) of Remark 14.2.2 and Theorem 1, we have

$$\iint_M f(x, y)\,\mathrm{d}x\,\mathrm{d}y = \iint_O g(x, y)\,\mathrm{d}x\,\mathrm{d}y = \int_a^b \left[\int_k^K g(x, y)\,\mathrm{d}y\right] \mathrm{d}x\,.$$

The definition of the function $g(x, y)$ (see Remark 14.2.2), however, implies

$$\int_k^K g(x, y)\,\mathrm{d}y = \int_{y_1}^{y_2} f(x, y)\,\mathrm{d}y\,, \quad \text{where} \quad y_1 = h_1(x)\,, \quad y_2 = h_2(x)$$

(Fig. 14.6). Thus

$$\iint_M f(x, y)\,\mathrm{d}x\,\mathrm{d}y = \int_a^b \left[\int_{h_1(x)}^{h_2(x)} f(x, y)\,\mathrm{d}y\right] \mathrm{d}x\,. \tag{2}$$

Similarly: If the boundary of the region M consists of the curves $x = \varphi_1(y)$, $x = \varphi_2(y)$ (with similar properties as before) and of the lines $y = c$, $y = d$ parallel to the x-axis then

$$\iint_M f(x, y)\,\mathrm{d}x\,\mathrm{d}y = \int_c^d \left[\int_{\varphi_1(y)}^{\varphi_2(y)} f(x, y)\,\mathrm{d}x\right] \mathrm{d}y\,. \tag{3}$$

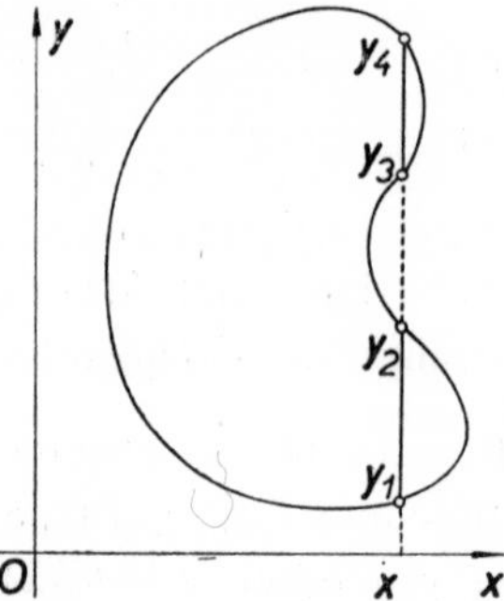

Fig. 14.7.

Similarly if the boundary of the region M is more complicated. For example, for x chosen as in Fig. 14.7, we have

$$\int_c^d g(x, y)\,\mathrm{d}y = \int_{y_1}^{y_2} f(x, y)\,\mathrm{d}y + \int_{y_3}^{y_4} f(x, y)\,\mathrm{d}y\,.$$

In cases like this we evaluate the limits $y_1, y_2, \ldots$ from the equation of the boundary of the region M (see Example 2).

Example 1. Determine the volume V of the ellipsoid with semi-axes a, b, c.

The volume V is equal to twice the volume of the upper half-ellipsoid which is bounded by the plane $z = 0$ and by the surface

$$z = c\sqrt{\left(1 - \frac{x^2}{a^2} - \frac{y^2}{b^2}\right)}, \tag{4}$$

the equation of which is obtained from the equation of the ellipsoid

$$\frac{x^2}{a^2} + \frac{y^2}{b^2} + \frac{z^2}{c^2} = 1$$

for $z \geqq 0$.

The base M of the half-ellipsoid is bounded by the ellipse

$$1 - \frac{x^2}{a^2} - \frac{y^2}{b^2} = 0 \tag{5}$$

the equation of which is obtained from (4) for $z = 0$ (because the base lies in this plane). Thus

$$V = 2 \iint_M c \sqrt{\left(1 - \frac{x^2}{a^2} - \frac{y^2}{b^2}\right)} \, dx \, dy =$$

$$= 2c \int_{-a}^{a} \left[\int_{-b\sqrt{(1-x^2/a^2)}}^{b\sqrt{(1-x^2/a^2)}} \sqrt{\left(1 - \frac{x^2}{a^2} - \frac{y^2}{b^2}\right)} \, dy \right] dx . \tag{6}$$

Using (1) we transformed the double integral into a repeated integral. The limits y_1, y_2 have been determined (for a given x) from (5) (see Fig. 14.8).

In (6), we integrate with respect to y keeping x *fixed*. Writing $\sqrt{(1 - x^2/a^2)} = m \geqq 0$, we obtain (using the substitution $y = bm \sin t$, $dy = bm \cos t \, dt$; see Example 13.2.10)

$$\int_{-b\sqrt{(1-x^2/a^2)}}^{b\sqrt{(1-x^2/a^2)}} \sqrt{\left(1 - \frac{x^2}{a^2} - \frac{y^2}{b^2}\right)} dy = \int_{-bm}^{bm} \sqrt{\left(m^2 - \frac{y^2}{b^2}\right)} dy = bm^2 \int_{-\pi/2}^{\pi/2} \cos^2 t \, dt =$$

$$= \tfrac{1}{2}\pi b m^2 = \tfrac{1}{2}\pi b \left(1 - \frac{x^2}{a^2}\right).$$

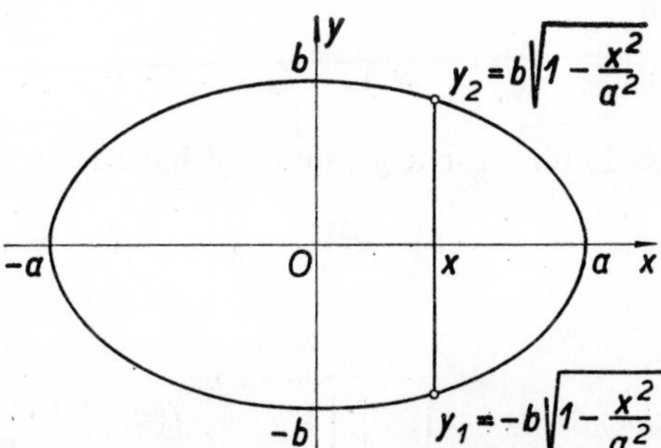

Fig. 14.8.

Substituting into (6), we have

$$V = 2c \int_{-a}^{a} \tfrac{1}{2}\pi b \left(1 - \frac{x^2}{a^2}\right) dx = \pi bc \left[x - \frac{x^3}{3a^2}\right]_{-a}^{a} = \tfrac{4}{3}\pi abc .$$

REMARK 6. It would have been *wrong* to proceed in the following way:

$$\iint_M \sqrt{\left(1 - \frac{x^2}{a^2} - \frac{y^2}{b^2}\right)} dx \, dy = \int_{-a}^{a} \left[\int_{-b}^{b} \sqrt{\left(1 - \frac{x^2}{a^2} - \frac{y^2}{b^2}\right)} dy \right] dx ,$$

for according to Remark 14.2.2 $g(x, y) = 0$ if the point (x, y) does not lie in M. Here, of course, this circumstance is obvious from the fact that at these points the function $f(x, y)$ ceases to have meaning as a real function. For example, at the point with coordinates $x = a$, $y = b$ we would have got

$$\sqrt{\left(1 - \frac{x^2}{a^2} - \frac{y^2}{b^2}\right)} = \sqrt{(1 - 1 - 1)} = \sqrt{(-1)}.$$

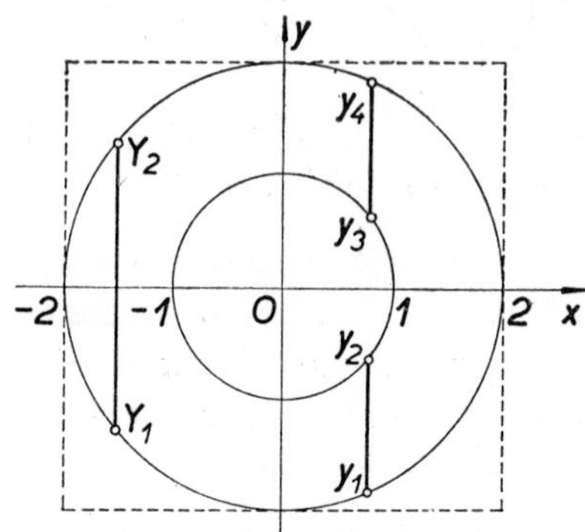

Fig. 14.9.

Example 2. The determination of the limits of integration in the case of the integration of a function $f(x, y)$ of type B in the annulus M shown in Fig. 14.9 is somewhat more difficult. For the inner radius r and outer radius R we have $r = 1$, $R = 2$. If $-2 < x < -1$ or $1 < x < 2$, we obtain, for a chosen x, two values Y_1, Y_2:

$$Y_1 = -\sqrt{(4 - x^2)}, \quad Y_2 = \sqrt{(4 - x^2)}.$$

If, however, $-1 < x < 1$, then for a given x we have four values y_1, y_2, y_3, y_4:

$$y_1 = -\sqrt{(4 - x^2)}, \quad y_2 = -\sqrt{(1 - x^2)}, \quad y_3 = \sqrt{(1 - x^2)}, \quad y_4 = \sqrt{(4 - x^2)}.$$

Thus

$$\iint_M f(x, y)\,\mathrm{d}x\,\mathrm{d}y = \int_{-2}^{-1}\left[\int_{-\sqrt{(4-x^2)}}^{\sqrt{(4-x^2)}} f(x, y)\,\mathrm{d}y\right]\mathrm{d}x +$$

$$+ \int_{-1}^{1}\left[\int_{-\sqrt{(4-x^2)}}^{-\sqrt{(1-x^2)}} f(x, y)\,\mathrm{d}y + \int_{\sqrt{(1-x^2)}}^{\sqrt{(4-x^2)}} f(x, y)\,\mathrm{d}y\right]\mathrm{d}x +$$

$$+ \int_{1}^{2}\left[\int_{-\sqrt{(4-x^2)}}^{\sqrt{(4-x^2)}} f(x, y)\,\mathrm{d}y\right]\mathrm{d}x.$$

14.4. Method of Substitution for Double Integrals

Theorem 1. *Let the closed region* $\bar{N}$ *(of variables* u, v*) be mapped in a one-to-one correspondence by the equations*

$$x = x(u, v), \quad y = y(u, v) \tag{1}$$

on the closed region $\bar{M}$ *(of variables* x, y*). Let* $\bar{M}$ *and* $\bar{N}$ *be of type* A (see Definition 14.1.2) *and* $f(x, y)$ *of type* B *in* $\bar{M}$ (see Definition 14.1.3). *If the functions* $x(u, v)$, $y(u, v)$ *have continuous first partial derivatives in* $\bar{N}$ *and the Jacobian*

$$D(u, v) = \begin{vmatrix} \dfrac{\partial x}{\partial u}, & \dfrac{\partial x}{\partial v} \\ \dfrac{\partial y}{\partial u}, & \dfrac{\partial y}{\partial v} \end{vmatrix} \tag{2}$$

in $\bar{N}$ *is different from zero, then*

$$\iint_M f(x, y)\, dx\, dy = \iint_N f(x(u, v), y(u, v))\, |D(u, v)|\, du\, dv\,. \tag{3}$$

REMARK 1. Note that the *absolute value* of $D(u, v)$ appears in (3). For the most frequently used substitution

$$x = r \cos \varphi\,, \quad y = r \sin \varphi \tag{4}$$

(polar coordinates), the Jacobian is

$$D(r, \varphi) = \begin{vmatrix} \dfrac{\partial x}{\partial r}, & \dfrac{\partial x}{\partial \varphi} \\ \dfrac{\partial y}{\partial r}, & \dfrac{\partial y}{\partial \varphi} \end{vmatrix} = \begin{vmatrix} \cos \varphi, & -r \sin \varphi \\ \sin \varphi, & r \cos \varphi \end{vmatrix} = r \geqq 0 \tag{5}$$

so that, in this case, it is not necessary to pay special attention to the absolute value.

Example 1. Let $f(x, y)$ be of type B in the closed circle M, $x^2 + y^2 \leqq 25$. We make use of the substitution (4); r will run between the limits 0, 5, φ between the limits 0, 2π (Fig. 14.10). Let us investigate if equation (3), i. e. the equation

$$\iint_M f(x, y)\, dx\, dy = \int_0^5 \int_0^{2\pi} f(r \cos \varphi, r \sin \varphi)\, r\, dr\, d\varphi \tag{6}$$

holds. The assumptions of Theorem 1 are not satisfied. For, firstly, if the mapping has to be one-to-one for $r > 0$, then φ has to run through the interval $[0, 2\pi)$ and not through the interval $[0, 2\pi]$ so that $\overline{N}$ is not a closed region. Secondly, to the point $(0, 0) \in \overline{M}$ there corresponds in $\overline{N}$ the entire line segment $r = 0, 0 \leqq \varphi < 2\pi$. In spite of this we can show that (6) holds. To the shaded sector S of the annulus in Fig. 14.11 there corresponds the closed rectangle $\varrho \leqq r \leqq 5, 0 \leqq \varphi \leqq 2\pi - \delta$ ($\varrho > 0$, $\delta > 0$). Now, all the assumptions of Theorem 1 are satisfied, so that the relation

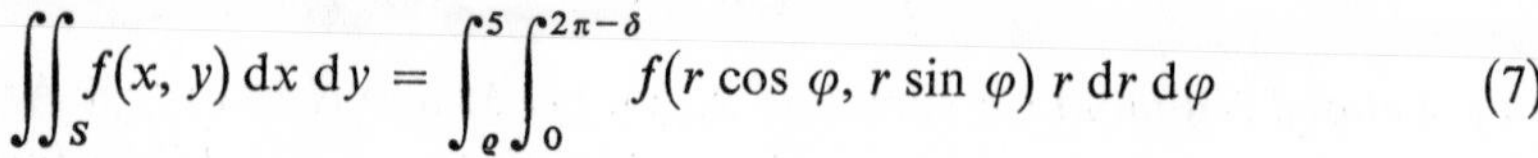

$$\iint_S f(x, y) \, dx \, dy = \int_\varrho^5 \int_0^{2\pi-\delta} f(r \cos \varphi, r \sin \varphi) \, r \, dr \, d\varphi \tag{7}$$

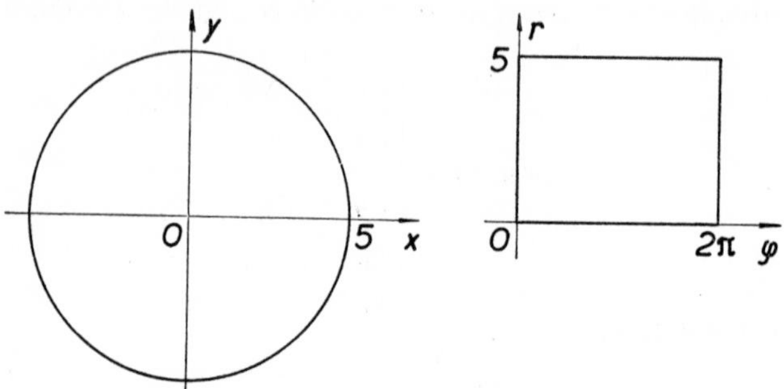

Fig. 14.10.

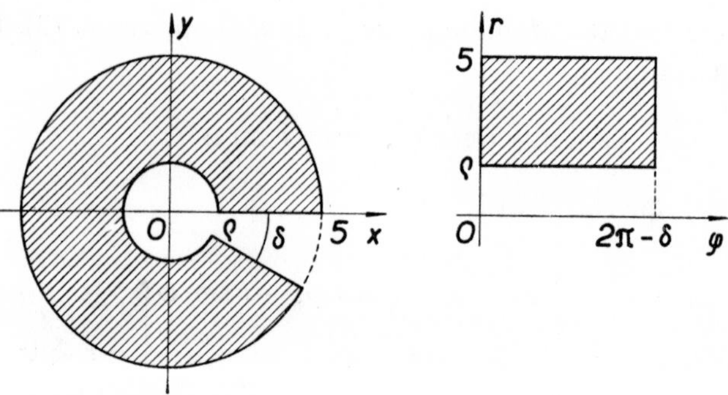

Fig. 14.11.

holds. Since $f(x, y)$ is of type B it is bounded in $\overline{M}$, hence the integral $\iint_S$ differs from $\iint_M$ by as little as we please if δ and ϱ are sufficiently small. Letting $\delta \to 0$ and then $\varrho \to 0$, we obtain (6). Thus, if $f(x, y)$ is of type B, then equation (6) holds.

Example 2. Let us determine the moment of inertia I of the homogeneous sphere of constant density ϱ_0 and of radius R with respect to an axis passing through its centre.

We choose the centre of the sphere as the origin of the coordinate system and the axis mentioned as the z-axis. The equation of the spherical surface is

$$x^2 + y^2 + z^2 = R^2$$

whence, for the upper hemispherical surface, we have

$$z = \sqrt{(R^2 - x^2 - y^2)}\,.$$

The moment of inertia I with respect to the z-axis will be equal to twice the moment of the upper hemisphere. Thus (by formula (14.9.70), p. 657), we have

$$I = 2\varrho_0 \iint_M (x^2 + y^2) \sqrt{(R^2 - x^2 - y^2)}\, \mathrm{d}x\, \mathrm{d}y\,.$$

We integrate over the circle $\overline{M}$ with centre at the origin and with radius R.

Using polar coordinates

$$x = r\cos\varphi\,, \quad y = r\sin\varphi \quad (0 \leqq r \leqq R,\ 0 \leqq \varphi < 2\pi)\,, \tag{8}$$

we have by (6)

$$I = 2\varrho_0 \int_0^R \int_0^{2\pi} r^2 \sqrt{(R^2 - r^2)}\, r\, \mathrm{d}r\, \mathrm{d}\varphi = 2\varrho_0 \int_0^R \left(\int_0^{2\pi} r^2 \sqrt{(R^2 - r^2)}\, r\, \mathrm{d}\varphi \right) \mathrm{d}r =$$

$$= 4\pi\varrho_0 \int_0^R r^2 \sqrt{(R^2 - r^2)}\, r\, \mathrm{d}r = 4\pi\varrho_0 \int_0^R (R^2 - t^2)\, t^2\, \mathrm{d}t =$$

$$= 4\pi\varrho_0 \left[R^2 \frac{t^3}{3} - \frac{t^5}{5} \right]_0^R = \frac{8\pi\varrho_0}{15} R^5\,, \tag{9}$$

where we have used the substitution $\sqrt{(R^2 - r^2)} = t$.

REMARK 2. In Example 2, we integrated over the *circle* and used substitution (8). Often we integrate over the interior of the ellipse

$$\frac{x^2}{a^2} + \frac{y^2}{b^2} = 1\,, \quad a > 0\,, \quad b > 0\,.$$

Then, substitution (8) is not convenient as the upper limit for r turns out to be dependent on φ. In this case we use the substitution

$$x = ar\cos\varphi\,, \quad y = br\sin\varphi \quad (0 \leqq r \leqq 1,\ 0 \leqq \varphi < 2\pi) \tag{10}$$

(r having here a constant upper limit, equal to one, since for $r = 1$ equations (10) give just the parametric equations of the ellipse (9); if $r < 1$, then $x^2/a^2 + y^2/b^2 < 1$ and the point (x, y) lies inside the ellipse, for $r > 1$ outside the ellipse.) For the Jacobian (2) we obtain

$$D = abr\,. \tag{11}$$

The remarks made about the justification of equation (6) apply here also.

Example 3. For the volume V_0 of the upper semiellipsoid with the semi-axes a, b, c we obtain by (10) and (11)

$$V_0 = \iint_M c \sqrt{\left(1 - \frac{x^2}{a^2} - \frac{y^2}{b^2}\right)} \,\mathrm{d}x\,\mathrm{d}y = c \int_0^1 \int_0^{2\pi} \sqrt{(1 - r^2)}\, abr \,\mathrm{d}r\,\mathrm{d}\varphi =$$

$$= abc \,.\, 2\pi \int_0^1 t^2 \,\mathrm{d}t = \tfrac{2}{3}\pi abc \,.$$

(We made use of the substitution $1 - r^2 = t^2$, $t > 0$.) Thus, the total volume V is

$$V = \tfrac{4}{3}\pi abc \,.$$

The evaluation is obviously easier than in Example 14.3.1.

14.5. Triple Integrals

Triple integrals are defined in a way similar to double integrals: Let a bounded function $u = f(x, y, z)$ be given in a rectangular parallelepiped $\overline{H}(a \leqq x \leqq b,\ c \leqq \leqq y \leqq d,\ e \leqq z \leqq f)$. We divide this parallelepiped by planes parallel to the co-ordinate planes into small parallelepipeds and denote the least upper bound and greatest lower bound of the function $f(x, y, z)$ on the parallelepipeds $\Delta x_i\ \Delta y_k\ \Delta z_l$ by M_{ikl} and m_{ikl}, respectively. (If $f(x, y, z)$ is continuous in $\overline{H}$, then the least upper bound is also the maximum and the greatest lower bound is also the minimum of the values of the function.) We construct the upper and lower sums

$$S(p) = \sum M_{ikl}\, \Delta x_i\, \Delta y_k\, \Delta z_l\,, \quad s(p) = \sum m_{ikl}\, \Delta x_i\, \Delta y_k\, \Delta z_l\,,$$

respectively, and sum over all the parallelepipeds into which the parallelepiped H is divided by the chosen partition p. The greatest lower bound of the set of all upper sums (for all possible partitions p) is called the *upper integral*, the least upper bound of the set of all lower sums – the *lower integral* of the function u in (or over) H.

Definition 1. If the upper and lower integrals are equal, then the function $u = f(x, y, z)$ is said to be *integrable in* (or *over*) H *in the Cauchy-Riemann sense* and their common value is called the *triple integral* of the function f in (or over) the parallelepiped H. We write

$$\iiint_H f(x, y, z)\,\mathrm{d}x\,\mathrm{d}y\,\mathrm{d}z\,, \quad \text{often also} \quad \iiint_H f(x, y, z)\,\mathrm{d}V.$$

The integral of a function of more than three variables may be defined like-wise.

REMARK 1. The interpretation of the triple integral if $f(x, y, z) > 0$: the mass of the parallelepiped H the density of which is given by the function $f(x, y, z)$.

REMARK 2. The basic properties of triple integrals (or of higher-dimensional integrals) are similar to those of double integrals (see Theorems 14.2.3 – 14.2.8).

Theorem 1. *Every function of type* B *in* H (Definition 14.1.5) *is integrable in* H.

REMARK 3. In particular, any function continuous in $\bar{H}$ is integrable in H.

REMARK 4. If $f(x, y, z)$ is defined in a region $\bar{O}$ which is a closed region of type A (Definition 14.1.4) but not a parallelepiped, then similarly as in Remark 14.2.2 we define

$$\iiint_O f(x, y, z)\,dx\,dy\,dz = \iiint_H g(x, y, z)\,dx\,dy\,dz\,, \tag{1}$$

where $\bar{H}$ is a parallelepiped containing the region $\bar{O}$, $g(x, y, z) = f(x, y, z)$ for all points $(x, y, z) \in \bar{O}$ and $g(x, y, z) = 0$ for the other points of H.

REMARK 5. The triple integral can be defined also in many other ways. In particular it is possible, using the concept of a *measurable set*, to define the triple (or multi-dimensional) integral in the Lebesque sense.

Theorem 2. *If* $f(x, y, z)$ *is of type* B (Definition 14.1.5) *in the parallelepiped* $\bar{H}(a \leqq x \leqq b,\ c \leqq y \leqq d,\ e \leqq z \leqq f)$, *then*

$$\iiint_H f(x, y, z)\,dx\,dy\,dz = \int_e^f \left[\int_a^b \int_c^d f(x, y, z)\,dx\,dy\right] dz =$$

$$= \int_a^b \int_c^d \left[\int_e^f f(x, y, z)\,dz\right] dx\,dy = \int_a^b \left[\int_c^d \int_e^f f(x, y, z)\,dy\,dz\right] dx =$$

$$= \int_c^d \int_e^f \left[\int_a^b f(x, y, z)\,dx\right] dy\,dz = \int_c^d \left[\int_a^b \int_e^f f(x, y, z)\,dx\,dz\right] dy =$$

$$= \int_a^b \int_e^f \left[\int_c^d f(x, y, z)\,dy\right] dx\,dz = \int_a^b \left\{\int_c^d \left[\int_e^f f(x, y, z)\,dz\right] dy\right\} dx\,, \tag{2}$$

where in the last integral it is again possible to change the order of integration (see, however, also Remark 14.3.1).

REMARK 6. If $f(x, y, z)$ is defined in the domain $\bar{O}$ which need not be a prism, then the evaluation of this integral according to (1) and (2) requires careful determination of the limits of integration in the same way as in Remark 14.3.5. Here also we often meet the following case: In a closed region $\bar{M}$ of type A (Definition 14.1.2) there are given smooth or continuous and piecewise smooth (Remark 12.1.8, p. 443) functions $z = z_1(x, y)$, $z = z_2(x, y)$ such that throughout M we have $z_2(x, y) > z_1(x, y)$. Let $\bar{O}$ be the closed solid with "upper base" $z = z_2(x, y)$ and with "lower base" $z = z_1(x, y)$. The lateral surface is formed by lines parallel to the z-axis passing through the boundary of the domain $\bar{M}$ (Fig. 14.12). Thus, the domain $\bar{O}$ is a closed solid of type A (Definition 14.1.4). Let a function of type B (Definition 14.1.5) be given in $\bar{O}$. Then

$$\iiint_O f(x, y, z)\,dx\,dy\,dz = \iint_M \left[\int_{z_1(x,y)}^{z_2(x,y)} f(x, y, z)\,dz \right] dx\,dy \tag{3}$$

as can easily be derived from (1) and from the second of the relations (2) (see also Fig. 14.12).

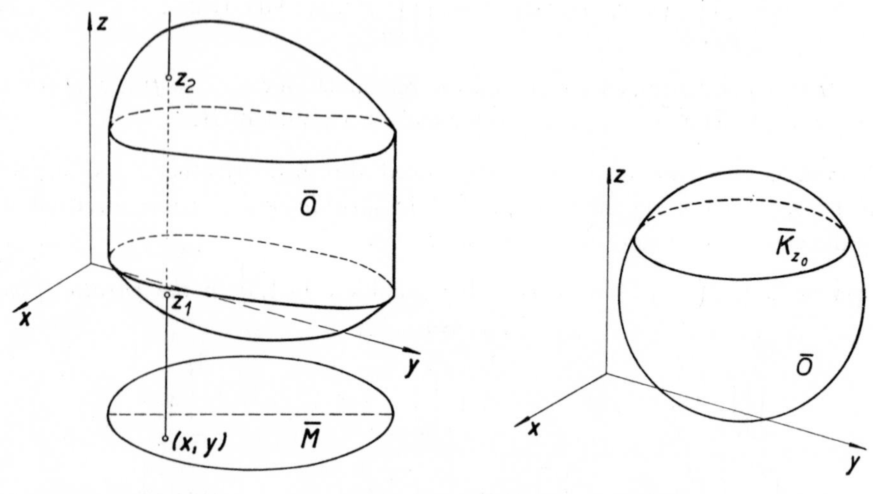

Fig. 14.12. Fig. 14.13.

Another procedure, based on the first of the relations (2), is as follows: $\bar{O}$ lies between the planes $z = e$, $z = f$. Let us denote by $\bar{K}_{z0}$ the cross-section of the solid $\bar{O}$ by the plane $z = z_0$ $(e \leqq z_0 \leqq f)$. Then

$$\iiint_O f(x, y, z)\,dx\,dy\,dz = \int_e^f \left[\iint_{K_z} f(x, y, z)\,dx\,dy \right] dz \tag{4}$$

(Fig. 14.13, where $\bar{O}$ is a sphere).

Example 1. Let us determine the mass m of the sphere $\bar{O}$ whose surface has the equation

$$x^2 + y^2 + z^2 = R^2\,, \tag{5}$$

and whose density increases with the square of the distance from the z-axis, i.e. $\varrho = (x^2 + y^2)\,\varrho_0$, where ϱ_0 is a constant. By Remark 1

$$m = \iiint_O \varrho_0(x^2 + y^2)\,dx\,dy\,dz\,. \tag{6}$$

By (3) (Remark 6) we have

$$\iiint_O \varrho_0(x^2 + y^2)\,dx\,dy\,dz = \iint_M \left(\int_{z_1}^{z_2} \varrho_0(x^2 + y^2)\,dz\right) dx\,dy\,,$$

where $\overline{M}$ is a circle with centre at the origin and radius R, $z_2 = \sqrt{(R^2 - x^2 - y^2)}$, $z_1 = -\sqrt{(R^2 - x^2 - y^2)}$. After integration with respect to z and substitution of the limits we obtain

$$m = \varrho_0 \iint_M (x^2 + y^2) \,.\, 2\sqrt{(R^2 - x^2 - y^2)}\,dx\,dy\,.$$

This integral can be evaluated either by repeated integration in the same way as in Example 14.3.1, or by transformation to polar coordinates as in Example 14.4.2. Result: $m = \frac{8}{15}\pi\varrho_0 R^5$ (see Example 2).

REMARK 7. The given problem is evidently equivalent to that of the determination of the moment of inertia of the homogeneous sphere of density ϱ_0 with respect to the z-axis. Note that in Example 1 ϱ_0 has the dimension g cm^{-5} as follows from the equation $\varrho = (x^2 + y^2)\,\varrho_0$.

Theorem 3 (*The Method of Substitution for Triple Integrals*). *Let $\overline{O}_1$ and $\overline{O}_2$ be closed regions of type* A (Definition 14.1.4) *in variables x, y, z, and u, v, w, respectively. Let there exist a one-to-one mapping between $\overline{O}_1$ and $\overline{O}_2$ expressed by the equations*

$$x = x(u, v, w)\,, \quad y = y(u, v, w)\,, \quad z = z(u, v, w) \qquad (7)$$

and assume that the functions (7) *have continuous partial derivatives of first order in $\overline{O}_2$ and that the Jacobian*

$$D(u, v, w) = \begin{vmatrix} \dfrac{\partial x}{\partial u}, & \dfrac{\partial x}{\partial v}, & \dfrac{\partial x}{\partial w} \\[2ex] \dfrac{\partial y}{\partial u}, & \dfrac{\partial y}{\partial v}, & \dfrac{\partial y}{\partial w} \\[2ex] \dfrac{\partial z}{\partial u}, & \dfrac{\partial z}{\partial v}, & \dfrac{\partial z}{\partial w} \end{vmatrix}$$

is different from zero in $\overline{O}_2$. Further let $f(x, y, z)$ be of type B (Definition 14.1.5) *in $\overline{O}_1$. Then*

$$\iiint_{O_1} f(x, y, z)\,dx\,dy\,dz =$$

$$= \iiint_{O_2} f(x(u, v, w), y(u, v, w), z(u, v, w))\,|D(u, v, w)|\,du\,dv\,dw\,. \qquad (8)$$

REMARK 8. The transformation theorem can be extended under similar assumptions to n-dimensional integrals. The Jacobian will then be n-dimensional.

REMARK 9. The transformation to spherical coordinates given by the equation

$$x = r \sin \vartheta \cos \varphi\,, \quad y = r \sin \vartheta \sin \varphi\,, \quad z = r \cos \vartheta \tag{9}$$

(§ 6.1) is used very frequently. These coordinates are employed when integrating over a sphere with radius R and having centre at the origin (then $0 \leqq r \leqq R, 0 \leqq \varphi < < 2\pi, 0 \leqq \vartheta \leqq \pi$), over a hemisphere (if we deal with the "upper" hemisphere, then $0 \leqq \vartheta \leqq \frac{1}{2}\pi$) etc. The Jacobian is

$$D(r, \varphi, \vartheta) = r^2 \sin \vartheta\,.$$

It is possible to establish by a limiting process similar to that used in Example 14.4.1 the validity of the equation

$$\iiint_K f(x, y, z)\,\mathrm{d}x\,\mathrm{d}y\,\mathrm{d}z =$$

$$= \int_0^R \int_0^{2\pi} \int_0^{\pi} f(r \sin \vartheta \cos \varphi, r \sin \vartheta \sin \varphi, r \cos \vartheta)\, r^2 \sin \vartheta\,\mathrm{d}r\,\mathrm{d}\varphi\,\mathrm{d}\vartheta\,, \tag{10}$$

where K is a sphere with centre at the origin and radius R. We assume that $f(x, y, z)$ is of type B in K.

REMARK 10. If we integrate over the ellipsoid with semi-axes a, b, c, we use instead of (9) the substitution

$$x = ar \sin \vartheta \cos \varphi\,, \quad y = br \sin \vartheta \sin \varphi\,, \quad z = cr \cos \vartheta \tag{11}$$

with the Jacobian $D = abc\, r^2 \sin \vartheta$. In this case $0 \leqq r \leqq 1, 0 \leqq \varphi < 2\pi, 0 \leqq \vartheta \leqq \pi$.

REMARK 11. For substitutions (9) and (11) $D \geqq 0$. In the general case, the absolute value of D appears in (8).

Example 2. By transforming to spherical coordinates we can easily evaluate integral (6) from Example 1. We have

$$x^2 + y^2 = r^2 \sin^2 \vartheta \cos^2 \varphi + r^2 \sin^2 \vartheta \sin^2 \varphi = r^2 \sin^2 \vartheta\,,$$

thus by (10),

$$\iiint_O \varrho_0(x^2 + y^2)\,\mathrm{d}x\,\mathrm{d}y\,\mathrm{d}z = \int_0^R \int_0^{2\pi} \int_0^{\pi} \varrho_0 r^2 \sin^2 \vartheta \,.\, r^2 \sin \vartheta\,\mathrm{d}r\,\mathrm{d}\varphi\,\mathrm{d}\vartheta =$$

$$= 2\pi\varrho_0 \int_0^R r^4 \left(\int_0^{\pi} \sin^3 \vartheta\,\mathrm{d}\vartheta \right) \mathrm{d}r = 2\pi\varrho_0 \,.\, \tfrac{4}{3} \,.\, \tfrac{1}{5}R^5 = \tfrac{8}{15}\pi\varrho_0 R^5\,.$$

14.6. Improper Double and Triple Integrals

By the term *improper* integrals we mean integrals where either the integrand, or the domain of integration is unbounded. The simplest cases are those of improper double integrals, where the integrand is unbounded in the neighbourhood of only one point or where the domain of integration is the entire xy-plane.

Definition 1. Let $f(x, y)$ be defined in the closed region $\bar{O}$ of type A (Definition 14.1.2) and let it become unbounded in the neighbourhood of the point $(x_0, y_0) \in O$. At the point (x_0, y_0) itself f need not be defined. Assume, further, that $f(x, y)$ has the following property: If we remove from $\bar{O}$ an arbitrary (open) region ω of type A ($\bar{\omega} \in O$) containing the point (x_0, y_0), then $f(x, y)$ is of type B (Definition 14.1.3) in $\bar{O} - \omega$ (and thus, the integral

$$I_\omega = \iint_{O-\omega} f(x, y)\,\mathrm{d}x\,\mathrm{d}y \tag{1}$$

exists). If there exists a number I such that for every $\varepsilon > 0$ it is possible to find a rectangle R so small that for all regions ω with the above mentioned property, lying inside R (and containing the point (x_0, y_0)), the inequality

$$|I_\omega - I| < \varepsilon$$

holds, then the integral

$$\iint_O f(x, y)\,\mathrm{d}x\,\mathrm{d}y \tag{2}$$

is said to be *convergent* (or to *exist*) and to have the value I.

REMARK 1. Similarly we define the three-dimensional improper integral (or the m-dimensional improper integral) for the case of one singular point (x_0, y_0, z_0).

Theorem 1. *The integral* (2) *is convergent if and only if* $\iint |f(x, y)|\,\mathrm{d}x\,\mathrm{d}y$ *is convergent.* (*The convergence of the integral is absolute unlike the one-dimensional case of improper integrals.*)

Theorem 2. *If in a given neighbourhood of the point* (x_0, y_0) *we have*

$$|f(x, y)| \leqq \frac{M}{r^\alpha} \tag{3}$$

where M is a positive constant, $r = \sqrt{[(x - x_0)^2 + (y - y_0)^2]}$ (i.e. *the distance of the point (x, y) from the point (x_0, y_0)*) *and $\alpha < 2$, then the integral*

$$I = \iint_O f(x, y)\,\mathrm{d}x\,\mathrm{d}y$$

is convergent and

$$I = \lim_{n \to \infty} \iint_{O - \omega_n} f(x, y) \, dx \, dy \,, \tag{4}$$

where ω_n is an arbitrary sequence of circles contained in O having centres at the point (x_0, y_0) and radius r_n, with $\lim_{n \to \infty} r_n = 0$.

Moreover, under the assumption (3) *the theorem for the replacement of the double integral by the repeated integral* (Theorem 14.3.1) *and the theorem on substitution* (Theorem 14.4.1) *are valid.*

REMARK 2. A similar assertion holds for triple (or m-dimensional) integrals; it is sufficient to require $\alpha < 3$ (or $\alpha < m$, respectively); in the three-dimensional case r is given by the expression

$$r = \sqrt{[(x - x_0)^2 + (y - y_0)^2 + (z - z_0)^2]}$$

(and similarly in the m-dimensional case).

REMARK 3. The advantage of using equation (4) lies in the fact that the integrals (4) can be easily evaluated as the domains ω_n are circles.

REMARK 4. In Theorems 1 and 2 is understood, of course, that the basic assumptions on the existence of the integral, mentioned in Definition 1 (concerning the type B of the function $f(x, y)$ in each region $O - \omega$) are satisfied.

REMARK 5. By the limit (4) the integral in the sense of the *principal value* (see Remark 13.8.3) is defined. The limit (4) may exist even though integral (2) does not.

REMARK 6. The conclusions of Theorem 2 remain valid if condition (3) is replaced by the condition that at least one of the repeated integrals (over the domain O)

$$\int \left[\int |f(x, y)| \, dy \right] dx \,, \quad \int \left[\int |f(x, y)| \, dx \right] dy$$

exists.

Example 1. Let us examine the integral

$$I = \iint_K \frac{dx \, dy}{\sqrt{(x^2 + y^2)}} \,,$$

where K is the circle with centre at the origin and radius $R = 1$.

The singularity is at the origin. Our function is of the form $f(x, y) = 1/r$, thus $M = 1$, $\alpha = 1$ in (3). By Theorem 2, the integral is convergent. Substituting polar coordinates, $x = r \cos \varphi$, $y = r \sin \varphi$, we obtain (since $D(r, \varphi) = r$, see equation

(14.4.5))

$$\iint_K \frac{dx\,dy}{\sqrt{(x^2 + y^2)}} = \int_0^1 \int_0^{2\pi} \frac{1}{r} \cdot r\,dr\,d\varphi = 2\pi\,.$$

REMARK 7. The second important case of improper integrals concerns double integrals where the domain of integration is the entire xy-plane. We write

$$\int_{-\infty}^{\infty} \int_{-\infty}^{\infty} f(x, y)\,dx\,dy\,. \tag{5}$$

The exact meaning is as follows:

Definition 2. Let $f(x, y)$ be of type B in every bounded closed region O of type A (Definition 14.1.2). If there exists a number I such that for every $\varepsilon > 0$ it is possible to find a circle K with centre at the origin and having radius R so large that for every closed region $\bar{O}$ of type A, containing this circle, we have

$$\left| I - \iint_O f(x, y) \right| < \varepsilon\,,$$

then the integral (5) is said to be *convergent* (or to *exist*) and to have the value I.

REMARK 8. We can define improper triple or m-dimensional integrals similarly.

Theorem 3. *Integral* (5) *is convergent if and only if* $\int_{-\infty}^{\infty} \int_{-\infty}^{\infty} |f(x, y)|\,dx\,dy$ *is convergent.* (*The convergence of the integral is absolute.*)

Theorem 4. *If everywhere outside a certain circle with the centre at the origin we have*

$$|f(x, y)| < \frac{M}{r^\alpha}\,, \quad M = \text{const}\,, \quad r = \sqrt{(x^2 + y^2)}\,, \quad \alpha > 2\,, \tag{6}$$

then integral (5) *is convergent and it is possible to use the theorem on repeated integration* (Theorem 14.3.1), *and the theorem on substitution* (Theorem 14.4.1).

In addition,

$$I = \lim_{R \to +\infty} \iint_{K_R} f(x, y)\,dx\,dy\,, \tag{7}$$

where K_R is the circle with centre at the origin and of radius R.

REMARK 9. Remarks similar to Remarks 2–6 are also valid here. In particular, for the m-dimensional integral it is sufficient to require $\alpha > m$ in (6). If using a substitution, we transform the integral (7) and take the limit for $R \to +\infty$.

Example 2. Evaluate the integral

$$A = \int_{-\infty}^{\infty} e^{-x^2} \, dx \, . \tag{8}$$

To this purpose we examine the integral

$$I = \int_{-\infty}^{\infty} \int_{-\infty}^{\infty} e^{-x^2} e^{-y^2} dx \, dy = \int_{-\infty}^{\infty} \int_{-\infty}^{\infty} e^{-(x^2+y^2)} \, dx \, dy \, . \tag{9}$$

The integral (9) is convergent by Theorem 4, since, for sufficiently large r, we have $e^{-r^2} < M/r^k$ ($M > 0$; $k > 0$ may be arbitrary). Thus

$$\int_{-\infty}^{\infty} \int_{-\infty}^{\infty} e^{-x^2} e^{-y^2} \, dx \, dy =$$

$$= \int_{-\infty}^{\infty} \left[\int_{-\infty}^{\infty} e^{-x^2} e^{-y^2} \, dy \right] dx = \int_{-\infty}^{\infty} e^{-y^2} \, dy \int_{-\infty}^{\infty} e^{-x^2} \, dx = A \, . \, A = A^2 \, . \tag{10}$$

Further, by Theorem 4, transformation to polar coordinates $x = r \cos \varphi$, $y = r \sin \varphi$ may be carried out. We obtain

$$\int_{-\infty}^{\infty} \int_{-\infty}^{\infty} e^{-(x^2+y^2)} \, dx \, dy = \lim_{R \to +\infty} \iint_{K_R} e^{-(x^2+y^2)} \, dx \, dy =$$

$$= \lim_{R \to +\infty} \int_0^R \int_0^{2\pi} e^{-r^2} r \, dr \, d\varphi = \lim_{R \to +\infty} 2\pi \, . \, \tfrac{1}{2} \int_0^{R^2} e^{-z} \, dz = \pi \, .$$

Thus, by (10)

$$A = \int_{-\infty}^{\infty} e^{-x^2} \, dx = \sqrt{\pi} \, .$$

Symmetry implies that

$$\int_0^{\infty} e^{-x^2} \, dx = \frac{\sqrt{\pi}}{2} \, .$$

REMARK 10. The definition of the improper integral with infinite domain of integration can be extended to the case where the domain of integration is not the entire plane but an arbitrary (in a certain sense) unbounded region M. In definition 2 one has to take instead of the whole region $\bar{O}$ containing the circle K only that part which the region $\bar{M}$ and the region $\bar{O}$ have in common (i.e. the intersection $\bar{M} \cap \bar{O}$, Fig. 14.14).

REMARK 11. Definition 1 may be extended as well. The following case is of importance: The function $f(x, y)$ is unbounded in $\bar{O}$ in the neighbourhood of a simple finite piecewise smooth curve k. Then the domains ω in Definition 1 are required to contain the curve k and the function $f(x, y)$ is assumed to be of type B in $\bar{O} - \omega$. For this case the following theorems are valid:

Theorem 5. *If at least one of the repeated integrals (over the region O)*

$$\int\left(\int|f(x, y)|\,\mathrm{d}y\right)\mathrm{d}x\,, \quad \int\left(\int|f(x, y)|\,\mathrm{d}x\right)\mathrm{d}y \tag{11}$$

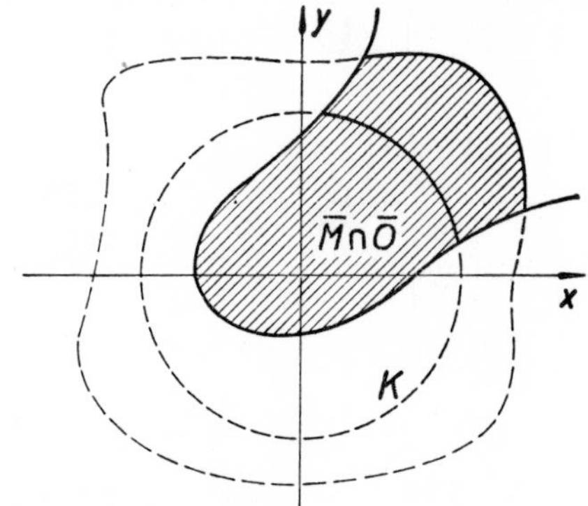

Fig. 14.14.

is convergent, then the integral

$$\iint_O f(x, y)\,\mathrm{d}x\,\mathrm{d}y \tag{12}$$

is also convergent.

Theorem 6. *If integral (12) is convergent and if the integral*

$$\int\left[\int f(x, y)\,\mathrm{d}y\right]\mathrm{d}x \quad \text{or} \quad \int\left[\int f(x, y)\,\mathrm{d}x\right]\mathrm{d}y \tag{13}$$

exists, then this integral is equal to integral (12).

Example 3. It is required to decide whether the integral

$$\iint_M \frac{x}{\sqrt{(x^2 - y^2)}}\,\mathrm{d}x\,\mathrm{d}y\,, \tag{14}$$

where M is the domain shaded in Fig. 14.15, is convergent.

This integral is improper since the integrand is unbounded in the (right) neighbourhood of the line segment $y = x, 0 \leqq x \leqq 2p$. According to Theorem 5 it is sufficient to prove the existence of one of integrals (11), where the sign of absolute value may be omitted for in the domain considered we have $x \geqq 0$ and $x^2 - y^2 \geqq 0$. We get

$$\int_0^{2p}\left(\int_y^{\sqrt{(2py)}} \frac{x}{\sqrt{(x^2-y^2)}}\,dx\right)dy = \int_0^{2p}\left([\sqrt{(x^2-y^2)}]_y^{\sqrt{(2py)}}\right)dy = \int_0^{2p}\sqrt{(2py-y^2)}\,dy\,. \tag{15}$$

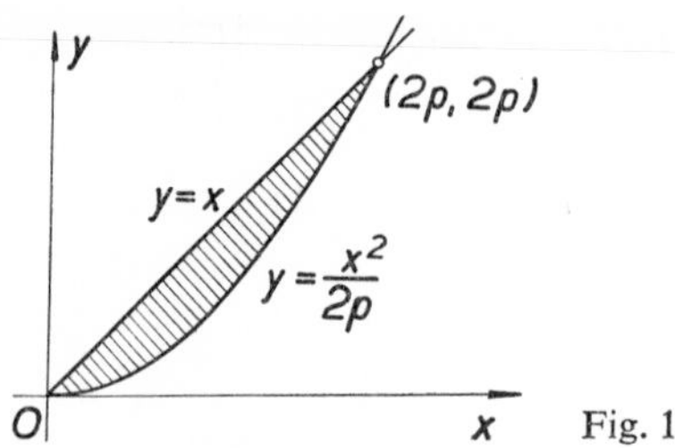

Fig. 14.15.

The last integral is no longer an improper one so that it has a finite value (and thus (14) exists by Theorem 5). Let us evaluate it. By the substitution $y - p = u$ and then $u = pz$ we obtain

$$\int_0^{2p}\sqrt{(2py-y^2)}\,dy = \int_0^{2p}\sqrt{[p^2-(y-p)^2]}\,dy = \int_{-p}^{p}\sqrt{(p^2-u^2)}\,du =$$

$$= p^2\int_{-1}^{1}\sqrt{(1-z^2)}\,dz = \tfrac{1}{2}\pi p^2\,.$$

(Example 13.7.4, p. 558). By Theorem 6 this is also the value of integral (14). If we had evaluated the integral

$$\int_0^{2p}\left(\int_{x^2/2p}^{x}\frac{x}{\sqrt{(x^2-y^2)}}\,dy\right)dx$$

we would have arrived at the same value.

14.7. Curvilinear Integrals. Green's Theorem

REMARK 1. Unless the contrary is stated in this paragraph, the term curve means a *simple finite piecewise smooth curve* according to Definition 14.1.1. If the curve k is the boundary of a region O, then this curve is said to be *positively oriented with respect to O* if O remains on the left-hand side of the point which traces out the curve k in the positive direction. We shall always choose the parametric representation in such a way that if $t_1 < t_2$ then the point A with the parameter t_1 lies on k before the point B with parameter t_2. (Briefly, we say that the curve k is oriented in the sense of increasing parameter).

REMARK 2. We say that in the equations of the curve k

$$x = \varphi(s), \quad y = \psi(s)$$

s denotes the *length of an arc* (or that s is the *parameter of the length of an arc*) if for each s (with the possible exception of a finite number of points)

$$\varphi'^2(s) + \psi'^2(s) = 1 .$$

Geometrically: If points A, B are given by $s = 0$ and $s = s_0$ respectively, then the arc AB has length s_0. For an *arbitrary* parametric representation of the curve k,

$$x = f(t), \quad y = g(t) \quad (\alpha \leqq t \leqq \beta),$$

the length of an arc is given by

$$s_0 = \int_\alpha^{t_0} \sqrt{[f'^2(t) + g'^2(t)]}\, dt .$$

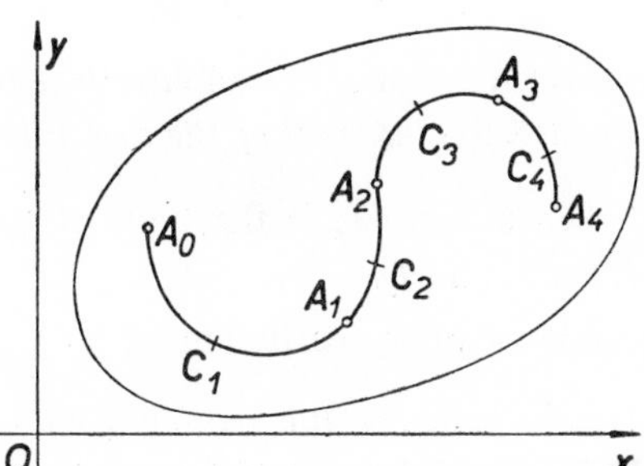

Fig. 14.16.

Definition 1. Let the oriented curve k be given by the equations

$$x = \varphi(t), \quad y = \psi(t), \quad \alpha \leqq t \leqq \beta .$$

Let a function $z = f(x, y)$ be given on the curve k. Let us subdivide k into n arcs $o_1, o_2, \ldots, o_n$ (Fig. 14.16) by points $A_1, A_2, \ldots, A_{n-1}$ with parameters $t_1 < t_2 < \ldots < t_{n-1}$, let us choose arbitrarily on each arc o_k a point C_k (ξ_k, η_k) and write the sums

$$S_x = \sum_{k=1}^{n} f(\xi_k, \eta_k)(x_k - x_{k-1}) = \sum_{k=1}^{n} f(\xi_k, \eta_k)[\varphi(t_k) - \varphi(t_{k-1})], \tag{1}$$

$$S_y = \sum_{k=1}^{n} f(\xi_k, \eta_k)(y_k - y_{k-1}) = \sum_{k=1}^{n} f(\xi_k, \eta_k)[\psi(t_k) - \psi(t_{k-1})], \tag{2}$$

$$S_s = \sum_{k=1}^{n} f(\xi_k, \eta_k)(s_k - s_{k-1}), \tag{3}$$

where

$$s_k = \int_\alpha^{t_k} \sqrt{[\varphi'^2(t) + \psi'^2(t)]}\, dt\,.$$

Let us denote by l_k the length of the arc o_k and by $v(p)$ the greatest of these lengths for the chosen partition p, i.e. $v(p) = \max l_k$ ($v(p)$ is called the *norm of the partition* p).

If there exists a number I_x, or I_y, or I_s such that for an arbitrary $\varepsilon > 0$ it is possible to find a $\delta > 0$ such that

$$|I_x - S_x| < \varepsilon\,, \quad \text{or} \quad |I_y - S_y| < \varepsilon\,, \quad \text{or} \quad |I_s - S_s| < \varepsilon \tag{4}$$

for each partition p for which $v(p) < \delta$ – independently of the choice of the points C_k on o_k – we say that there *exists a curvilinear (line) integral* of the function $f(x, y)$ along the oriented curve k with respect to x, or y, or s, respectively and write

$$I_x = \int_k f(x, y)\, dx\,, \quad \text{or} \quad I_y = \int_k f(x, y)\, dy\,, \quad \text{or} \quad I_s = \int_k f(x, y)\, ds\,. \tag{5}$$

(The first two of the integrals (5) are often called *curvilinear integrals of the second kind*, the third of them is called the *integral of the first kind*.)

REMARK 3. Intuitively: The integral I_x is the limit of the sums (1) as $v(p) \to 0$. A similar assertion is true for I_y and I_s.

For geometric and physical meaning see Remark 7.

Theorem 1. *If $f(x, y)$ is continuous in the region O containing the curve k, then the integrals* (5) *exist. It is even sufficient if $f(x, y)$ is continuous on k* (Remark 14.1.12).

Theorem 2. *If $f(x, y)$ is continuous in O (or on k), then*

$$\int_k f(x, y)\, dx = \int_\alpha^\beta f(\varphi(t), \psi(t))\, \varphi'(t)\, dt\,,$$

$$\int_k f(x, y)\, dy = \int_\alpha^\beta f(\varphi(t), \psi(t))\, \psi'(t)\, dt\,, \tag{6}$$

$$\int_k f(x, y)\, ds = \int_\alpha^\beta f(\varphi(t), \psi(t)) \sqrt{[\varphi'^2(t) + \psi'^2(t)]}\, dt\,.$$

Theorem 3. *If k is given by the equation $y = g(x)$ in $[a, b]$, then*

$$\int_k f(x, y)\, dx = \int_a^b f(x, g(x))\, dx \tag{7}$$

if k is oriented in such a way that $(a, g(a))$ *is its initial point, and*

$$\int_k f(x, y)\,dx = -\int_a^b f(x, g(x))\,dx \tag{8}$$

if $(b, g(b))$ *is its initial point. Furthermore*

$$\int_k f(x, y)\,ds = \int_a^b f(x, g(x))\sqrt{[1 + g'^2(x)]}\,dx \quad (a < b) \tag{9}$$

(*no matter how the curve is oriented*).

Similarly: If k is given by the equation $x = h(y)$ in $[c, d]$, then

$$\int_k f(x, y)\,dy = \pm \int_c^d f(h(y), y)\,dy\,,$$

where the plus or minus sign is taken depending on whether the initial point of the curve k is the point $(h(c), c)$, or the point $(h(d), d)$, respectively. Furthermore

$$\int_k f(x, y)\,ds = \int_c^d f(h(y), y)\sqrt{[1 + h'^2(y)]}dy.$$

REMARK 4. If the curve k is a segment parallel to the x-axis, then $\int_k f(x, y)\,dy = 0$; if it is parallel to the y-axis, then $\int_k f(x, y)\,dx = 0$.

Theorem 4. *If k is composed of two arcs* o_1, o_2, *then*

$$\int_k f(x, y)\,dx = \int_{o_1} f(x, y)\,dx + \int_{o_2} f(x, y)\,dx\,;$$

similarly for

$$\int_k f(x, y)\,dy\,, \quad \int_k f(x, y)\,ds\,.$$

Theorem 5.

$$\int_k [c_1 f_1(x, y) + c_2 f_2(x, y)]\,dx = c_1 \int_k f_1(x, y)\,dx + c_2 \int_k f_2(x, y)\,dx\,,$$

if the integrals on the right-hand side exist; similarly for the line integrals with respect to y and with respect to s.

Theorem 6. *Let us denote by k′ the curve which is inversely oriented to k. Then*

$$\int_{k'} f(x, y)\,dx = -\int_k f(x, y)\,dx\,, \quad \int_{k'} f(x, y)\,dy = -\int_k f(x, y)\,dy\,.$$

REMARK 5. For line integrals with respect to s the sign remains the same.

REMARK 6. By the sum

$$\int_k [f(x, y)\,dx + g(x, y)\,dy]$$

we mean the sum of the integrals

$$\int_k f(x, y)\,dx + \int_k g(x, y)\,dy\,.$$

Example 1. Let k be the circle $x = \cos t$, $y = \sin t$ $(0 \leqq t < 2\pi)$, $f(x, y) = 2 + y$. Then (by (6))

$$\int_k f(x, y)\,dx = \int_0^{2\pi} (2 + \sin t)(-\sin t)\,dt = -\pi\,.$$

This example can also be solved as follows: Let $k = k_1 + k_2$, where k_1 is the upper and k_2 the lower semicircle (Fig. 14.17). On k_1 we have $y = \sqrt{(1 - x^2)}$, on k_2, $y = -\sqrt{(1 - x^2)}$. By Theorem 4 and by (7), (8) we have

$$\int_k (2 + y)\,dx = \int_{k_1} (2 + y)\,dx + \int_{k_2} (2 + y)\,dx =$$

$$= -\int_{-1}^{1} [2 + \sqrt{(1 - x^2)}]\,dx + \int_{-1}^{1} [2 - \sqrt{(1 - x^2)}]\,dx =$$

$$= -2\int_{-1}^{1} \sqrt{(1 - x^2)}\,dx = -\pi\,.$$

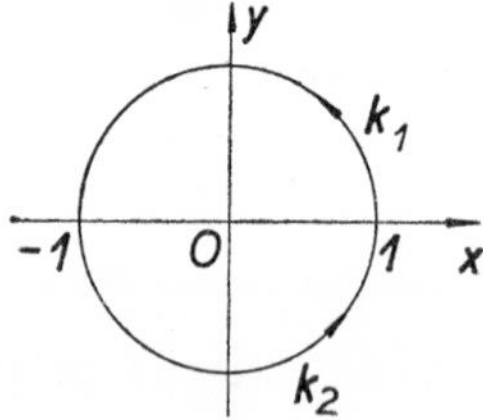

Fig. 14.17.

REMARK 7. The *geometric meaning of a curvilinear integral:* I_s is the area of the surface shown in Fig. 14.18. I_x or I_y is equal to the area of the (oriented) projection of this surface onto the xz-plane, or onto the yz-plane (Fig. 14.18), respectively.

The *physical meaning:* If a field of force $\mathbf{F} = \mathbf{i}\,P(x, y) + \mathbf{j}\,Q(x, y)$ ($\mathbf{i}$, $\mathbf{j}$ are the unit coordinate vectors) is given and if we have to calculate the amount of work done by this field in moving a particle along the oriented curve k, then this work – as we know from physics – is given by the integral (i.e. by the limit of sums, in a certain sense) of the scalar product

$$L = \int_k \mathbf{F}\,\mathrm{d}\mathbf{l} = \int_k [\mathbf{i}\,P(x, y) + \mathbf{j}\,Q(x, y)]\,(\mathbf{i}\,\mathrm{d}x + \mathbf{j}\,\mathrm{d}y) =$$

$$= \int_k P(x, y)\,\mathrm{d}x + \int_k Q(x, y)\,\mathrm{d}y\,. \tag{10}$$

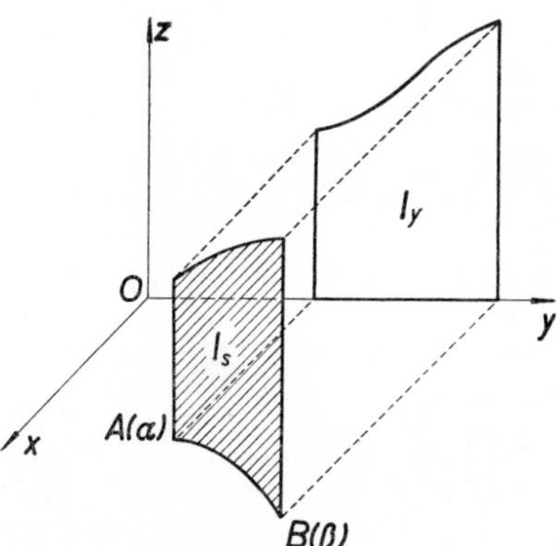

Fig. 14.18.

Thus, the evalution of the amount of work done reduces to the evaluation of two curvilinear integrals.

REMARK 8. The definition of a *curvilinear integral along a curve in space* is similar to that of line integrals in the *plane.* In particular curvilinear integrals of a function continuous in the region O in which the given curve lies exist, and

$$\int_k f(x, y, z)\,\mathrm{d}x = \int_\alpha^\beta f(\varphi(t), \psi(t), \chi(t))\,\varphi'(t)\,\mathrm{d}t\,,$$

$$\int_k f(x, y, z)\,\mathrm{d}y = \int_\alpha^\beta f(\varphi(t), \psi(t), \chi(t))\,\psi'(t)\,\mathrm{d}t\,,$$

$$\int_k f(x, y, z)\,\mathrm{d}z = \int_\alpha^\beta f(\varphi(t), \psi(t), \chi(t))\,\chi'(t)\,\mathrm{d}t\,,$$

$$\int_k f(x, y, z)\,\mathrm{d}s = \int_\alpha^\beta f(\varphi(t), \psi(t), \chi(t))\,\sqrt{[\varphi'^2(t) + \psi'^2(t) + \chi'^2(t)]}\,\mathrm{d}t\,. \tag{11}$$

The physical meaning of the first three integrals is similar to that of (10).

REMARK 9. It can be shown that the value of integrals (11) (as well as that of integrals (6), previously introduced) does not depend on the choice of the parametric representation of the curve k (naturally respecting the convention of Remark 1).

Theorem 7 (*Green's Theorem*). *Let $\bar{O}$ be a closed region of type* A (Definition 14.1.2) *with the boundary k positively oriented with respect to O* (see Remark 1). *Let the functions*

$$P(x, y), \quad Q(x, y), \quad \frac{\partial P}{\partial y}(x, y), \quad \frac{\partial Q}{\partial x}(x, y)$$

be continuous in $\bar{O}$ (i.e. in $O + k$). Then (see Remark 6)

$$\int_k (P\,\mathrm{d}x + Q\,\mathrm{d}y) = \iint_O \left(\frac{\partial Q}{\partial x} - \frac{\partial P}{\partial y}\right) \mathrm{d}x\,\mathrm{d}y\,. \tag{12}$$

If O is a multiply connected region, then $\int_k$ means the sum of the curvilinear integrals along individual parts of the boundary.

REMARK 10. If

$$Q(x, y) \equiv 0 \quad \text{or} \quad P(x, y) \equiv 0\,,$$

we obtain particular cases of Green's theorem

$$\int_k P\,\mathrm{d}x = -\iint_O \frac{\partial P}{\partial y}\,\mathrm{d}x\,\mathrm{d}y\,, \quad \int_k Q\,\mathrm{d}y = \iint_O \frac{\partial Q}{\partial x}\,\mathrm{d}x\,\mathrm{d}y\,. \tag{13}$$

Green's theorem is often written in the following form:

$$\int_k [-P(x, y)\cos\beta + Q(x, y)\cos\alpha]\,\mathrm{d}s = \iint_O \left(\frac{\partial Q}{\partial x} - \frac{\partial P}{\partial y}\right) \mathrm{d}x\,\mathrm{d}y\,,$$

or, writing $-P(x, y) = T(x, y)$, $Q(x, y) = S(x, y)$, in the form

$$\int_k [S(x, y)\cos\alpha + T(x, y)\cos\beta]\,\mathrm{d}s = \iint_O \left(\frac{\partial S}{\partial x} + \frac{\partial T}{\partial y}\right) \mathrm{d}x\,\mathrm{d}y\,,$$

where $\cos\alpha$, $\cos\beta$ are the direction cosines of the outward normal.

Definition 2. Let k be an oriented curve with initial point A and end point B. If the value of the integral

$$\int_k [f(x, y)\,\mathrm{d}x + g(x, y)\,\mathrm{d}y] \tag{14}$$

depends only on the choice of the points A, B, and not on the choice of the curve connecting the points A, B (and naturally, lying in the considered region O), then the integral (14) is said to be *independent of the path of integration in O.*

Theorem 8. *Let*

$$P(x, y), \quad Q(x, y), \quad \frac{\partial P(x, y)}{\partial y}, \quad \frac{\partial Q(x, y)}{\partial x},$$

be continuous functions in a simply connected region O. Then a necessary and sufficient condition that the integral

$$\int_k [P(x, y)\,\mathrm{d}x + Q(x, y)\,\mathrm{d}y] \tag{15}$$

is independent of the path of integration in O is that the equation

$$\frac{\partial P}{\partial y} = \frac{\partial Q}{\partial x} \tag{16}$$

is satisfied in O.

REMARK 11. In particular: If condition (16) is satisfied, then the integral (15) along each *closed* curve k is equal to zero.

Theorem 9. *If P, Q, $\partial P/\partial y$, $\partial Q/\partial x$ are continuous functions in a simply connected region O, then a necessary and sufficient condition that the expression*

$$P\,\mathrm{d}x + Q\,\mathrm{d}y \tag{17}$$

is the total differential of some function $F(x, y)$ in O is that the equation

$$\frac{\partial P}{\partial y} = \frac{\partial Q}{\partial x} \tag{18}$$

holds in O. If this condition is satisfied, then the value of the integral (15) is given by the difference $F(x_2, y_2) - F(x_1, y_1)$, where (x_1, y_1) is the initial and (x_2, y_2) the end point of the curve k.

Theorem 10. *Let $P(x, y)$, $Q(x, y)$ be continuous in O. Then a necessary and sufficient condition that the integral*

$$\int_k (P\,\mathrm{d}x + Q\,\mathrm{d}y)$$

is independent of the path of integration in O is that there exists a function $F(x, y)$ such that the expression

$$P\,\mathrm{d}x + Q\,\mathrm{d}y$$

is its total differential.

Example 2. According to Theorem 9 the expression

$$(x^2 - y^2)\,\mathrm{d}x + (5 - 2xy)\,\mathrm{d}y \tag{19}$$

is a total differential (even in the entire xy-plane). The corresponding function $F(x, y)$ may be found from the condition

$$\mathrm{d}F = \frac{\partial F}{\partial x}\,\mathrm{d}x + \frac{\partial F}{\partial y}\,\mathrm{d}y = P\,\mathrm{d}x + Q\,\mathrm{d}y$$

(which has to be satisfied for each dx, dy):

$$\frac{\partial F}{\partial x} = P = x^2 - y^2 , \quad \frac{\partial F}{\partial y} = Q = 5 - 2xy . \tag{20}$$

From the first equation it follows that

$$F(x, y) = \frac{x^3}{3} - xy^2 + f(y) . \tag{21}$$

By differentiation of (21) with respect to y and by comparison with the second of equations (20), we have

$$-2xy + f'(y) = -2xy + 5 , \quad \text{hence} \quad f(y) = 5y + C .$$

Thus

$$F(x, y) = \frac{x^3}{3} - xy^2 + 5y + C , \tag{22}$$

where C is an arbitrary constant.

The function $F(x, y)$ may also be obtained by the formula

$$F(x, y) = \int_{x_0}^{x} P(x, y_0)\, dx + \int_{y_0}^{y} Q(x, y)\, dy . \tag{23}$$

In our case

$$F(x, y) = \int_{x_0}^{x} (x^2 - y_0^2)\, dx + \int_{y_0}^{y} (5 - 2xy)\, dy =$$

$$= \left(\frac{x^3}{3} - y_0^2 x\right) - \left(\frac{x_0^3}{3} - y_0^2 x_0\right) + (5y - xy^2) - (5y_0 - xy_0^2) =$$

$$= \frac{x^3}{3} + 5y - xy^2 - \frac{x_0^3}{3} + y_0^2 x_0 - 5y_0 \tag{24}$$

and this result is of the form (22).

Remark 12. Theorems 8, 9 and 10 are true also for line integrals along curves in space, with the only difference that instead of integral (15) we have to write the integral

$$\int_k [P(x, y, z)\, dx + Q(x, y, z)\, dy + R(x, y, z)\, dz] \tag{25}$$

and to replace equation (16), or (18) by the equations (which have to be simultaneously fulfilled)

$$\frac{\partial P}{\partial y} = \frac{\partial Q}{\partial x} , \quad \frac{\partial P}{\partial z} = \frac{\partial R}{\partial x} , \quad \frac{\partial Q}{\partial z} = \frac{\partial R}{\partial y} \tag{26}$$

(see Theorem 14.8.6, p. 643). The function $F(x, y, z)$ can be calculated by both the methods of Example 2, in particular

$$F(x, y, z) = \int_{x_0}^{x} P(x, y_0, z_0)\,\mathrm{d}x + \int_{y_0}^{y} Q(x, y, z_0)\,\mathrm{d}y + \int_{z_0}^{z} R(x, y, z)\,\mathrm{d}z\,.$$

REMARK 13 (concerning practical evaluation). From (13) it follows that if P is a function of x only in a simply connected region O, then the integral $\int_k P\,\mathrm{d}x$, along each closed curve k in O, is equal to zero. Similarly, if Q is a function of the variable y only.

REMARK 14. Using (13) it is possible to show that the area p of a finite simply connected region O with the boundary k (positively oriented with respect to O) may be expressed as follows:

$$p = \int_k x\,\mathrm{d}y\,, \quad p = -\int_k y\,\mathrm{d}x$$

or,

$$p = \tfrac{1}{2}\int_k [x\,\mathrm{d}y - y\,\mathrm{d}x]\,. \tag{27}$$

The last of the integrals (27) is often used for the evaluation of the area of a sector (Fig. 14.19). On OA and BO the integral is equal to zero (for on these line segments $x\,\mathrm{d}y = y\,\mathrm{d}x$) so that

$$p = \tfrac{1}{2}\int_{k_0} [x\,\mathrm{d}y - y\,\mathrm{d}x]\,,$$

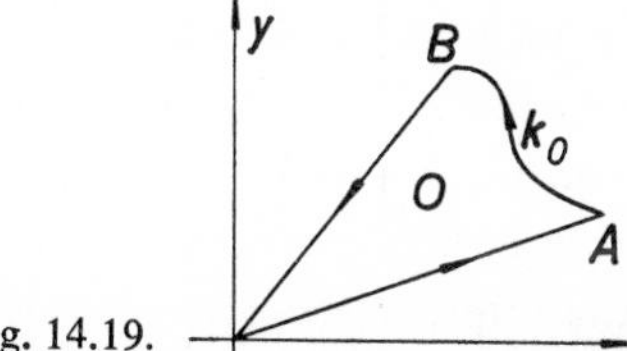

Fig. 14.19.

where $k_0 = \widehat{AB}$. If the equations of the curve k_0 are given in the form

$$x = \varphi(t)\,, \quad y = \psi(t) \quad (\alpha \leqq t \leqq \beta) \tag{28}$$

(with positive orientation in the sense of increasing parameter), then

$$p = \tfrac{1}{2}\int_{\alpha}^{\beta} [\varphi(t)\,\psi'(t) - \psi(t)\,\varphi'(t)]\,\mathrm{d}t\,. \tag{29}$$

14.8. Surface Integrals. The Gauss–Ostrogradski Theorem, Stokes's Theorem, Green's Identities

REMARK 1. In this paragraph we shall deal only with surfaces having some simple properties. If we use the term *surface*, we shall always mean a *simple finite piecewise smooth surface* (Remark 14.1.7).

REMARK 2. A surface is said to be *oriented* if its two sides can be distinguished as the *exterior* and *interior side*, respectively. This concept is intuitively very clear for the case of closed surfaces that constitute boundaries of solids of type A (Definition 14.1.4). In the same sense we speak about the *oriented outward* and *inward normal of a surface.* Of course, for surfaces that constitute boundaries of the above-mentioned solids, the inward normal is oriented inside the solid, while the outward normal has the reverse orientation. (Not all surfaces can be oriented (e.g. the so-called Mobius leaf).

REMARK 3. Let us consider an oriented surface $z = f(x, y)$. The area of the projection of the surface onto the xy-plane, assigned with the plus or minus sign according as the exterior side or the interior side is seen when observing the surface in the negative direction of the z-axis,* is called the *oriented projection of the surface onto the xy-plane.* The value of the oriented projection of a surface onto the xy-plane is (by definition) equal to zero if the surface is formed by parallels to the z-axis.

Similarly we define the oriented projection of the oriented surfaces $x = g(y, z)$ and $y = h(x, z)$ on to the yz-plane and xz-plane, respectively. The plus or minus sign is again chosen according as the exterior or interior side is seen when observing the surface in the negative direction of the x- or y-axis, respectively.

REMARK 4. The concepts introduced in Remarks 2 and 3 are based to a large extent on intuition, and were therefore not given as definitions.

Definition 1. Let two points A, B be given on a surface. Let us connect them with a rectifiable curve k lying on the surface. The greatest lower bound of the lengths of all such curves is called the *distance between the points A, B measured on the given surface.* The least upper bound of the distances between all possible pairs of points on the surface is called the *interior diameter of the surface.*

REMARK 5. Roughly speaking: the distance between the points A, B measured on the surface is given by the length of the shortest curve lying on the surface and connecting these points. The interior diameter of a finite surface is the greatest possible distance between two points measured on this surface.

For example, the spherical surface with radius r has diameter $2r$; however, its *interior* diameter is equal to πr (the length of the meridian between "the north and south poles").

---

* i.e. according as to whether the outward normal makes an angle with the positive z-axis less or greater than $\pi/2$.

Definition 2. Let a function $u = f(x, y, z)$ be given on the surface S (see Remark 1). Let us divide S into n parts S_i and denote the greatest of the interior diameters of the parts S_i (Definition 1) by $\nu(p)$. Choose a point (x_i, y_i, z_i) in each part S_i and write down the sum

$$\sigma = \sum_{i=1}^{n} f(x_i, y_i, z_i)\, s_i\,, \tag{1}$$

where s_i is the area of the part S_i. If there exists a number I_S such that for every $\varepsilon > 0$ there is another number $\delta > 0$ such that for each partition p for which $\nu(p) < \delta$ the inequality

$$|I_S - \sigma| < \varepsilon$$

holds, independently of the choice of the points (x_i, y_i, z_i) in S_i, then we say that the *surface integral*

$$\iint_S f(x, y, z)\, \mathrm{d}S \tag{2}$$

over the surface S exists and that its value is I_S.

REMARK 6. Roughly speaking: (2) is the limit of the sums (1) as $\nu(p) \to 0$.

REMARK 7. The integral (2) is often called the *surface integral of the first kind.*

Definition 3. Let a function $u = f(x, y, z)$ be given on an *oriented* surface S. Let S be divisible into a finite number n of parts S_i which either may be represented in the form $z = f(x, y)$, or are formed by parallels to the z-axis. Let p be such a partition. By the norm $\nu(p)$ of that partition we mean the greatest of the interior diameters of the parts S_i (Definition 1). Let us choose, in each part S_i, a point (x_i, y_i, z_i) and write the sum

$$\sigma_{xy} = \sum_{i=1}^{n} f(x_i, y_i, z_i)\, p_i\,, \tag{3}$$

where p_i is the oriented projection of the part S_i on the xy-plane (Remark 3). If there exists a number I_{xy} such that for an arbitrary $\varepsilon > 0$ there is a $\delta > 0$ such that for each partition p (with the above-mentioned properties) for which $\nu(p) < \delta$ the inequality

$$|I_{xy} - \sigma_{xy}| < \varepsilon$$

holds, independently of the choice of the points (x_i, y_i, z_i) in S_i, then we say that the *surface integral over the oriented surface S with respect to the coordinates x, y,*

$$\iint_S f(x, y, z)\, \mathrm{d}x\, \mathrm{d}y\,, \tag{4}$$

exists and that its value is I_{xy}.

REMARK 8. Roughly speaking: (4) is the limit of the sums (3) as $\nu(p) \to 0$.

REMARK 9. The integrals

$$\iint_S g(x, y, z)\,dy\,dz\,, \quad \iint_S h(x, y, z)\,dz\,dx \tag{5}$$

are defined in a similar way. Integrals (4), (5) are often called *surface integrals of the second kind*. Surface integrals are a generalization of curvilinear integrals in a certain sense. In the same way as in the case of curvilinear integrals of the second kind, the orientation of the surface must be known when considering surface integrals of the second kind.

By the integral

$$\iint_S [f(x, y, z)\,dx\,dy + g(x, y, z)\,dy\,dz + h(x, y, z)\,dz\,dx]$$

we mean the sum of the integrals (4) and (5).

REMARK 10. To ensure the existence of the integrals (2), (4), (5) it is sufficient, under the above-mentioned assumptions regarding the surface S (Remark 1), to assume the continuity of the functions under consideration in the region O in which the surface S lies (in fact continuity on the surface S itself is sufficient).

REMARK 11. The basic properties of surface integrals are similar to those of curvilinear integrals (Theorems 14.7.4, 14.7.5). If the orientation of the surface is changed (i.e. the exterior and the interior sides of the surface are interchanged), then the integrals (4) and (5) change their signs.

Theorem 1. *If the surface S is given in the explicit form $z = \varphi(x, y)$, where $\varphi(x, y)$ is a continuous piecewise smooth function in M, then the integral* (2) *is equal to the integral*

$$\iint_M f(x, y, \varphi(x, y)) \sqrt{\left[1 + \left(\frac{\partial \varphi}{\partial x}\right)^2 + \left(\frac{\partial \varphi}{\partial y}\right)^2\right]}\,dx\,dy\,. \tag{6}$$

Theorem 2. *If the surface S is given parametrically by the equations*

$$x = x(u, v)\,, \quad y = y(u, v)\,, \quad z = z(u, v)\,, \quad (u, v) \in Q$$

(where x, y, z are assumed to be continuous piecewise smooth functions in Q), then the integral (2) *is equal to the integral*

$$\iint_Q f(x(u, v), y(u, v), z(u, v)) \sqrt{(EG - F^2)}\,du\,dv\,. \tag{7}$$

(The functions E, F, G are defined in equations (14.9.75), (14.9.76), p. 658.)

Theorem 3. *If S is given in the form $z = \varphi(x, y)$ for $(x, y) \in M$, then*

$$\iint_S f(x, y, z)\, dx\, dy = \pm \iint_M f(x, y, \varphi(x, y))\, dx\, dy \tag{8}$$

where the plus or minus sign is to be chosen according to Remark 3 (see Example 1).

REMARK 12. Theorem 3 is useful for the practical evaluation of the integral (4). We divide the surface S into surfaces which may be represented in the form $z = \varphi(x, y)$ and surfaces formed by lines parallel to the z-axis. In the second case the corresponding integrals are equal to zero (Remark 3), in the first case they are evaluated using (8).

Theorem 4. *Let the surface S be given parametrically by $x = x(u, v)$, $y = y(u, v)$, $z = z(u, v)$ for $u, v \in Q$. Then*

$$\iint_S f(x, y, z)\, dx\, dy = \pm \iint_Q f(x(u, v), y(u, v), z(u, v)) \frac{D(x, y)}{D(u, v)}\, du\, dv\,, \tag{9}$$

where

$$\frac{D(x, y)}{D(u, v)} = \begin{vmatrix} \dfrac{\partial x}{\partial u}, & \dfrac{\partial x}{\partial v} \\ \dfrac{\partial y}{\partial u}, & \dfrac{\partial y}{\partial v} \end{vmatrix}. \tag{10}$$

The sign in (9) can be determined as follows: We choose a point $(u_0, v_0) \in Q$ such that in its neighbourhood Q', $D(x, y)/D(u, v) \neq 0$. Then the corresponding part S' of the surface S may be represented in the form $z = \varphi(x, y)$. Since S is an oriented surface, we are able, according to Theorem 3 or Remark 3, to determine the correct sign in the equation

$$\iint_{S'} f(x, y, z)\, dx\, dy = \pm \iint_{M'} f(x, y, \varphi(x, y))\, dx\, dy\,. \tag{11}$$

If $D(x, y)/D(u, v) > 0$ in Q', we choose the same sign in (9) as in (11); if $D(x, y)/D(u, v) < 0$ in Q', we choose the opposite sign. (See Example 1.)

REMARK 13. Theorems 3, 4 and Remark 12 are valid in a similar form for the integrals (5). Expression (10) has to be successively replaced by the expressions

$$\frac{D(y, z)}{D(u, v)} = \begin{vmatrix} \dfrac{\partial y}{\partial u}, & \dfrac{\partial y}{\partial v} \\ \dfrac{\partial z}{\partial u}, & \dfrac{\partial z}{\partial v} \end{vmatrix} \quad \text{or} \quad \frac{D(z, x)}{D(u, v)} = \begin{vmatrix} \dfrac{\partial z}{\partial u}, & \dfrac{\partial z}{\partial v} \\ \dfrac{\partial x}{\partial u}, & \dfrac{\partial x}{\partial v} \end{vmatrix}. \tag{12}$$

Example 1. Let us evaluate

$$\iint_S x^2 z \, dx \, dy \,, \tag{13}$$

where S is the sphere with centre at the origin and radius 1. S is a closed surface by which its orientation is thus given. S obviously consists of two surfaces S_1 and S_2 with equations $z = \sqrt{(1 - x^2 - y^2)}$ and $z = -\sqrt{(1 - x^2 - y^2)}$, respectively. If we make use of (8), we choose, according to Theorem 3 and Remark 3, the plus sign for S_1 and the minus sign for S_2. Hence

$$\iint_S x^2 z \, dx \, dy = + \iint_K x^2 \sqrt{(1 - x^2 - y^2)} \, dx \, dy - \iint_K x^2 [-\sqrt{(1 - x^2 - y^2)}] \, dx \, dy,$$

where K is the circle $x^2 + y^2 \leqq 1$. The right-hand side of the equation is easily evaluated by transforming to polar coordinates

$$x = r \cos \varphi \,, \quad y = r \sin \varphi \quad \text{(see Examples 14.4.1, 14.4.2)}\,.$$

The result:

$$\iint_S x^2 z \, dx \, dy = 2 \iint_K x^2 \sqrt{(1 - x^2 - y^2)} \, dx \, dy = \tfrac{4}{15}\pi \,.$$

If S is given in parametric form,

$$x = \sin u \cos v \,, \quad y = \sin u \sin v \,, \quad z = \cos u \quad (0 \leqq u \leqq \pi, \ 0 \leqq v < 2\pi)\,,$$

we use (9). According to Theorem 4, let us choose, for example, $(u_0, v_0) = (\frac{1}{6}\pi, \frac{1}{3}\pi)$. There is obviously a point on S_1 which corresponds to this point so that in (11) we have the plus sign. Furthermore, for $u = \frac{1}{6}\pi$, $v = \frac{1}{3}\pi$ (and – as follows from continuity – also in a certain neighbourhood of this point)

$$\frac{D(x, y)}{D(u, v)} = \begin{vmatrix} \cos u \cos v, & -\sin u \sin v \\ \cos u \sin v, & \sin u \cos v \end{vmatrix} = \sin u \cos u > 0$$

so that according to Theorem 4 we choose in (9) the *same* sign as in (11), i.e. the plus sign. Thus we have

$$\iint_S x^2 z \, dx \, dy = \int_0^\pi \int_0^{2\pi} \sin^2 u \cos^2 v \cos u \sin u \cos u \, du \, dv = \tfrac{4}{15}\pi$$

in accordance with the previous result.

Theorem 5 (*The Gauss–Ostrogradski Theorem*). *Let the functions*

$$P(x, y, z), \quad Q(x, y, z)\,, \quad R(x, y, z) \quad \textit{and} \quad \frac{\partial P}{\partial x}\,, \quad \frac{\partial Q}{\partial y}\,, \quad \frac{\partial R}{\partial z}$$

be continuous in a closed solid $\bar{V}$ *of type* A (Definition 14.1.4) (*which need not necessarily be simply connected*) *with the oriented boundary S so that the outward normal n points out of the solid. Then*

$$\iiint_V \left(\frac{\partial P}{\partial x} + \frac{\partial Q}{\partial y} + \frac{\partial R}{\partial z}\right) \mathrm{d}x\,\mathrm{d}y\,\mathrm{d}z = \iint_S (P\,\mathrm{d}y\,\mathrm{d}z + Q\,\mathrm{d}z\,\mathrm{d}x + R\,\mathrm{d}x\,\mathrm{d}y)\,, \quad (14)$$

or, if we denote the direction cosines of the outward normal by $\cos\alpha$, $\cos\beta$, $\cos\gamma$,

$$\iiint_V \left(\frac{\partial P}{\partial x} + \frac{\partial Q}{\partial y} + \frac{\partial R}{\partial z}\right) \mathrm{d}x\,\mathrm{d}y\,\mathrm{d}z = \iint_S (P\cos\alpha + Q\cos\beta + R\cos\gamma)\,\mathrm{d}S \quad (15)$$

or, if we take into consideration that $\partial x/\partial n = \cos\alpha$ *etc.*

$$\iiint_V \left(\frac{\partial P}{\partial x} + \frac{\partial Q}{\partial y} + \frac{\partial R}{\partial z}\right) \mathrm{d}x\,\mathrm{d}y\,\mathrm{d}z = \iint_S \left(P\frac{\partial x}{\partial n} + Q\frac{\partial y}{\partial n} + R\frac{\partial z}{\partial n}\right) \mathrm{d}S\,. \quad (16)$$

If V is multiply connected, then $\iint_S$ *means the sum of the surface integrals over the individual parts of the boundary.*

Definition 4. Let S be an oriented surface and let k be a *closed* simple finite piecewise smooth (generally spatial) curve on S. The curve k is said to be *positively oriented with respect to S*, if – observed from the side of the outward normal (erected on the exterior part of the surface) – when traversing the curve k in its positive direction, the part of the surface S that is enclosed by the curve k remains on the left-hand side (Fig. 14.20).

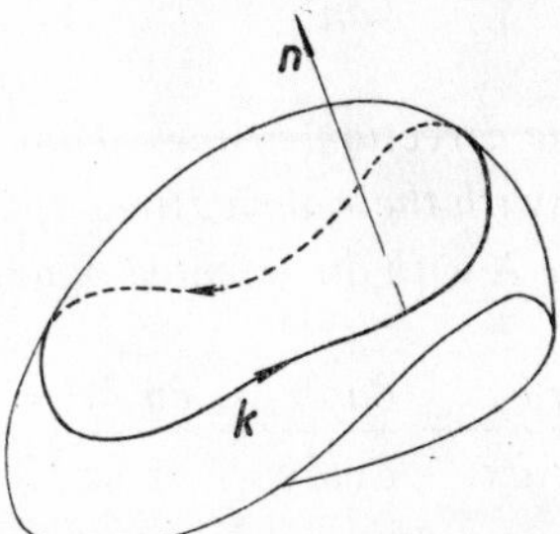

Fig. 14.20.

Theorem 6 (*The Stokes Theorem*). *Let S be an oriented open surface the boundary of which is formed by a closed simple finite piecewise smooth curve k which is positively oriented with respect to S. Then*

$$\int_k (P\,\mathrm{d}x + Q\,\mathrm{d}y + R\,\mathrm{d}z) =$$

$$= -\iint_S \left[\left(\frac{\partial P}{\partial y} - \frac{\partial Q}{\partial x}\right)\mathrm{d}x\,\mathrm{d}y + \left(\frac{\partial Q}{\partial z} - \frac{\partial R}{\partial y}\right)\mathrm{d}y\,\mathrm{d}z + \left(\frac{\partial R}{\partial x} - \frac{\partial P}{\partial z}\right)\mathrm{d}z\,\mathrm{d}x\right]. \quad (17)$$

REMARK 14. Stokes's theorem is a generalization of Green's theorem (Theorem 14.7.7) for simply connected regions in the xy-plane.

From the Gauss–Ostrogradski theorem there easily follows:

Theorem 7. *If the functions P, Q, R, $\partial P/\partial x$, $\partial Q/\partial y$, $\partial R/\partial z$ are continuous in O, then a necessary and sufficient condition that the surface integral* (14) (*or* (15), *or* (16)) *is equal to zero for each surface S, which is the boundary of a closed simply connected solid of type* A *lying in the region O, is given by the equation*

$$\frac{\partial P}{\partial x} + \frac{\partial Q}{\partial y} + \frac{\partial R}{\partial z} \equiv 0 \quad \text{in} \quad O. \tag{18}$$

Theorem 8. *If* (18) *is satisfied, then the surface integral* (14) (*or* (15), *or* (16)) *over the surface S', the boundary of which is a curve k which is positively oriented with respect to S', depends only on k and is independent of the form of the surface S'.*

Theorem 9 (*Green's Identities*). *Let us write, as usual, $\Delta u = \partial^2 u/\partial x^2 + \partial^2 u/\partial y^2 + \partial^2 u/\partial z^2$ and let $\partial u/\partial n$ be the derivative of u in the direction of the outward normal,* i.e.

$$\frac{\partial u}{\partial n} = \frac{\partial u}{\partial x}\frac{\partial x}{\partial n} + \frac{\partial u}{\partial y}\frac{\partial y}{\partial n} + \frac{\partial u}{\partial z}\frac{\partial z}{\partial n},$$

where

$$\frac{\partial x}{\partial n} = \cos\alpha, \quad \frac{\partial y}{\partial n} = \cos\beta, \quad \frac{\partial z}{\partial n} = \cos\gamma$$

and $\cos\alpha$, $\cos\beta$, $\cos\gamma$ *are the direction cosines of the oriented outward normal. Let $u(x, y, z)$, $v(x, y, z)$, together with their derivatives up to the second order, be continuous in a closed solid $\overline{V}$ of type* A *with an oriented boundary S. Then*

$$\iiint_V \left(\frac{\partial u}{\partial x}\frac{\partial v}{\partial x} + \frac{\partial u}{\partial y}\frac{\partial v}{\partial y} + \frac{\partial u}{\partial z}\frac{\partial v}{\partial z}\right) \mathrm{d}x\,\mathrm{d}y\,\mathrm{d}z =$$

$$= -\iiint_V u\,\Delta v\,\mathrm{d}x\,\mathrm{d}y\,\mathrm{d}z + \iint_S u\frac{\partial v}{\partial n}\,\mathrm{d}S = \tag{19}$$

$$= -\iiint_V v\,\Delta u\,\mathrm{d}x\,\mathrm{d}y\,\mathrm{d}z + \iint_S v\frac{\partial u}{\partial n}\,\mathrm{d}S. \tag{20}$$

From (19) *and* (20) *it follows that*

$$\iiint_V (u\,\Delta v - v\,\Delta u)\,\mathrm{d}x\,\mathrm{d}y\,\mathrm{d}z = \iint_S \left(u\frac{\partial v}{\partial n} - v\frac{\partial u}{\partial n}\right)\mathrm{d}S. \tag{21}$$

For $v \equiv 1$ we obtain

$$\iiint_V \Delta u \, dx \, dy \, dz = \iint_S \frac{\partial u}{\partial n} \, dS . \tag{22}$$

REMARK 15. Using the symbolism of vector analysis and the notation $dV = dx\,dy\,dz$, the previous theorems may be rewritten in a rather more concise form:

$$\boldsymbol{A} = \boldsymbol{i}P + \boldsymbol{j}Q + \boldsymbol{k}R , \quad d\boldsymbol{S} = \boldsymbol{i}\,dS \cos\alpha + \boldsymbol{j}\,dS \cos\beta + \boldsymbol{k}\,dS \cos\gamma ,$$

$$d\boldsymbol{s} = \boldsymbol{i}\,dx + \boldsymbol{j}\,dy + \boldsymbol{k}\,dz .$$

Equation (15) (or (14), or (16)):

$$\iiint_V \operatorname{div} \boldsymbol{A} \, dV = \iint_S \boldsymbol{A} \, d\boldsymbol{S} .$$

Equation (17):

$$\int_C \boldsymbol{A} \, d\boldsymbol{s} = \iint_S \operatorname{curl} \boldsymbol{A} \, d\boldsymbol{S} .$$

Equations (19), (20):

$$\iiint_V \operatorname{grad} u \operatorname{grad} v \, dV = -\iiint_V u \, \Delta v \, dV + \iint_S u \operatorname{grad} v \, d\boldsymbol{S} =$$

$$= -\iiint_V v \, \Delta u \, dV + \iint_S v \operatorname{grad} u \, d\boldsymbol{S} .$$

Equation (21):

$$\iiint_V (u \, \Delta v - v \, \Delta u) \, dV = \iint_S (u \operatorname{grad} v - v \operatorname{grad} u) \, d\boldsymbol{S} .$$

Equation (22):

$$\iiint_V \Delta u \, dV = \iint_S \operatorname{grad} u \, d\boldsymbol{S} \quad \text{(see Chap. 7)} .$$

14.9. Applications of the Integral Calculus in Geometry and Physics

(Curves, plane figures, solids, surfaces – lengths, areas, volumes, masses, statical moments, centres of gravity, moments of inertia; the work of a force along a given curve; some special formulae; Guldin's rules; Steiner's theorem; examples.)

REMARK 1. Exact definitions of the length of a curve, of the area of a surface, etc., may be found in many textbooks of integral calculus.

REMARK 2. All formulae given below may easily be obtained from a geometrical or physical conception of the problem. For example, the formula for calculation of the volume of a solid of revolution, the lateral surface of which is obtained by rotating the curve $y = f(x)$ ($f(x) \geqq 0$, $a \leqq x \leqq b$) around the x-axis (Fig. 14.21) can be derived in an intuitive way as follows: We divide the interval $[a, b]$ into n subintervals $\Delta x_1, \Delta x_2, \ldots, \Delta x_n$. Choosing a point ξ_k in each interval Δx_k, we consider the function $y = f(x)$ to be constant in Δx_k, i.e. $y = f(\xi_k)$. By rotating the segment $y = f(\xi_k)$ over Δx_k around the x-axis, we get a circular cylinder with the volume

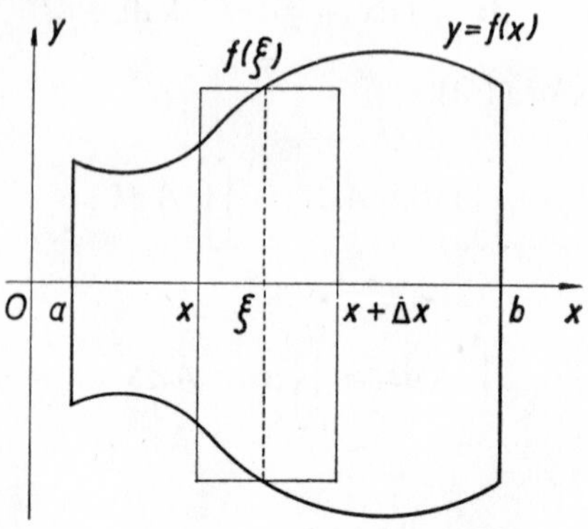

Fig. 14.21.

$\pi f^2(\xi_k)\,\Delta x_k$. For the sum of the volumes of all these cylinders in the interval $[a, b]$ we have

$$\sum_{k=1}^{n} \pi f^2(\xi_k)\,\Delta x_k\,. \tag{1}$$

Now, it seems to be evident that as $n \to \infty$ and $\Delta x_k \to 0$, we get the exact volume of the solid of revolution, while the sum (1) turns into the integral

$$\pi \int_a^b f^2(x)\,dx\,. \tag{2}$$

If the function $f(x)$ is continuous in $[a, b]$, the heuristic consideration just indicated yields – by Theorem 13.6.2 – the correct result. The same is true if upper or lower sums are used instead of the sums (1). The idea just explained may be employed for the derivation of all formulae given below.

In all these formulae we assume the continuity of the function considered. If some of the functions are, for example, piecewise continuous, we proceed according to Remark 13.6.13, p. 556.

NOTE. The majority of formulae are contained in the first four sections, (**a**) curves, (**b**) plane figures, (**c**) solids, (**d**) surfaces. At the beginning of each of these sections, the types of objects are listed for which formulae are presented. For example, in section (**c**) formulae for five types of solids (a) to (e) are given; these types are described at the beginning of section (**c**). Thus, for the volume of solids, five formulae corresponding to types (a) to (e) are given; the formulae for the mass, statical moment, etc., are treated similarly.

(a) Curves.

(α) *Plane curves*. The curve k is given:
(a) by the graph of a function $y = f(x)$, $a \leqq x \leqq b$,
(b) parametrically, $x = \varphi(t)$, $y = \psi(t)$, $t_1 \leqq t \leqq t_2$
(notation: $\mathrm{d}\varphi/\mathrm{d}t = \dot{\varphi}$, $\mathrm{d}\psi/\mathrm{d}t = \dot{\psi}$),
(c) in polar coordinates, $r = r(\varphi)$, $\varphi_1 \leqq \varphi \leqq \varphi_2$, $r \geqq 0$.

The specific density (mass per unit of length, g cm^{-1}) ϱ is given as a function of x or t or φ. (If ϱ is constant, then, of course, ϱ may be put in front of the integral sign.)

LENGTH:

(a) $$l = \int_a^b \sqrt{[1 + f'^2(x)]}\,\mathrm{d}x = \int_a^b \sqrt{(1 + y'^2)}\,\mathrm{d}x\,, \tag{3}$$

(b) $$l = \int_{t_1}^{t_2} \sqrt{[\dot{\varphi}^2(t) + \dot{\psi}^2(t)]}\,\mathrm{d}t = \int_{t_1}^{t_2} \sqrt{(\dot{x}^2 + \dot{y}^2)}\,\mathrm{d}t\,, \tag{4}$$

(c) $$l = \int_{\varphi_1}^{\varphi_2} \sqrt{[r^2(\varphi) + r'^2(\varphi)]}\,\mathrm{d}\varphi\,. \tag{5}$$

MASS:

(a) $$M = \int_a^b \varrho(x) \sqrt{[1 + f'^2(x)]}\,\mathrm{d}x = \int_a^b \varrho \sqrt{(1 + y'^2)}\,\mathrm{d}x\,, \tag{6}$$

(b) $$M = \int_{t_1}^{t_2} \varrho(t) \sqrt{[\dot{\varphi}^2(t) + \dot{\psi}^2(t)]}\,\mathrm{d}t = \int_{t_1}^{t_2} \varrho \sqrt{(\dot{x}^2 + \dot{y}^2)}\,\mathrm{d}t\,, \tag{7}$$

(c) $$M = \int_{\varphi_1}^{\varphi_2} \varrho(\varphi) \sqrt{[r^2(\varphi) + r'^2(\varphi)]}\,\mathrm{d}\varphi\,. \tag{8}$$

STATICAL MOMENT WITH RESPECT TO THE x- OR y-AXIS:

(a) $$S_x = \int_a^b f(x)\,\varrho(x) \sqrt{[1 + f'^2(x)]}\,\mathrm{d}x = \int_a^b y\varrho \sqrt{(1 + y'^2)}\,\mathrm{d}x\,, \tag{9}$$

$$S_y = \int_a^b x\varrho(x) \sqrt{[1 + f'^2(x)]}\,\mathrm{d}x = \int_a^b x\varrho \sqrt{(1 + y'^2)}\,\mathrm{d}x\,, \tag{9'}$$

(b) $$S_x = \int_{t_1}^{t_2} \psi(t)\,\varrho(t)\,\sqrt{[\dot{\varphi}^2(t) + \dot{\psi}^2(t)]}\,\mathrm{d}t = \int_{t_1}^{t_2} y\varrho\,\sqrt{(\dot{x}^2 + \dot{y}^2)}\,\mathrm{d}t\,, \tag{10}$$

$$S_y = \int_{t_1}^{t_2} \varphi(t)\,\varrho(t)\,\sqrt{[\dot{\varphi}^2(t) + \dot{\psi}^2(t)]}\,\mathrm{d}t = \int_{t_1}^{t_2} x\varrho\,\sqrt{(\dot{x}^2 + \dot{y}^2)}\,\mathrm{d}t\,, \tag{10'}$$

(c) $$S_x = \int_{\varphi_1}^{\varphi_2} \varrho(\varphi)\,r(\varphi)\sin\varphi\,\sqrt{[r^2(\varphi) + r'^2(\varphi)]}\,\mathrm{d}\varphi\,, \tag{11}$$

$$S_y = \int_{\varphi_1}^{\varphi_2} \varrho(\varphi)\,r(\varphi)\cos\varphi\,\sqrt{[r^2(\varphi) + r'^2(\varphi)]}\,\mathrm{d}\varphi\,. \tag{11'}$$

COORDINATES OF THE CENTRE OF GRAVITY:

$$x_T = \frac{S_y}{M}\,, \quad y_T = \frac{S_x}{M}\,; \tag{12}$$

for example,

$$x_T = \frac{\displaystyle\int_{\varphi_1}^{\varphi_2} \varrho(\varphi)\,r(\varphi)\sin\varphi\,\sqrt{[r^2(\varphi) + r'^2(\varphi)]}\,\mathrm{d}\varphi}{\displaystyle\int_{\varphi_1}^{\varphi_2} \varrho(\varphi)\,\sqrt{[r^2(\varphi) + r'^2(\varphi)]}\,\mathrm{d}\varphi}\,. \tag{13}$$

MOMENT OF INERTIA WITH RESPECT TO THE x- OR y-AXIS:

(a) $$I_x = \int_a^b f^2(x)\,\varrho(x)\,\sqrt{[1 + f'^2(x)]}\,\mathrm{d}x = \int_a^b y^2\varrho\,\sqrt{(1 + y'^2)}\,\mathrm{d}x\,, \tag{14}$$

$$I_y = \int_a^b x^2\varrho(x)\,\sqrt{[1 + f'^2(x)]}\,\mathrm{d}x = \int_a^b x^2\varrho\,\sqrt{(1 + y'^2)}\,\mathrm{d}x\,, \tag{15}$$

(b) $$I_x = \int_{t_1}^{t_2} \psi^2(t)\,\varrho(t)\,\sqrt{[\dot{\varphi}^2(t) + \dot{\psi}^2(t)]}\,\mathrm{d}t = \int_{t_1}^{t_2} y^2\varrho\,\sqrt{(\dot{x}^2 + \dot{y}^2)}\,\mathrm{d}t\,, \tag{16}$$

$$I_y = \int_{t_1}^{t_2} \varphi^2(t)\,\varrho(t)\,\sqrt{[\dot{\varphi}^2(t) + \dot{\psi}^2(t)]}\,\mathrm{d}t = \int_{t_1}^{t_2} x^2\varrho\,\sqrt{(\dot{x}^2 + \dot{y}^2)}\,\mathrm{d}t\,, \tag{17}$$

(c) $$I_x = \int_{\varphi_1}^{\varphi_2} \varrho(\varphi)\,r^2(\varphi)\sin^2\varphi\,\sqrt{[r^2(\varphi) + r'^2(\varphi)]}\,\mathrm{d}\varphi\,, \tag{18}$$

$$I_y = \int_{\varphi_1}^{\varphi_2} \varrho(\varphi)\,r^2(\varphi)\cos^2\varphi\,\sqrt{[r^2(\varphi) + r'^2(\varphi)]}\,\mathrm{d}\varphi\,. \tag{18'}$$

When computing moments of inertia with respect to the origin (i.e. about the z-axis) we must replace x^2 or y^2 by the sum $x^2 + y^2$; particularly, in (18) $r^2 \sin^2 \varphi$ or $r^2 \cos^2 \varphi$ must be replaced by r^2.

(β) *Curves in space.* A curve c is given parametrically by

$$x = \varphi(t)\,, \quad y = \psi(t)\,, \quad z = \chi(t)\,, \quad t_1 \leqq t \leqq t_2 : \quad \frac{\mathrm{d}\varphi}{\mathrm{d}t} = \dot{\varphi}\,, \quad \text{etc.}$$

LENGTH:

$$l = \int_{t_1}^{t_2} \sqrt{[\dot{\varphi}^2(t) + \dot{\psi}^2(t) + \dot{\chi}^2(t)]}\,\mathrm{d}t = \int_{t_1}^{t_2} \sqrt{(\dot{x}^2 + \dot{y}^2 + \dot{z}^2)}\,\mathrm{d}t\,. \tag{19}$$

MASS:

$$M = \int_{t_1}^{t_2} \varrho(t) \sqrt{[\dot{\varphi}^2(t) + \dot{\psi}^2(t) + \dot{\chi}^2(t)]}\,\mathrm{d}t = \int_{t_1}^{t_2} \varrho \sqrt{(\dot{x}^2 + \dot{y}^2 + \dot{z}^2)}\,\mathrm{d}t\,. \tag{20}$$

STATICAL MOMENT WITH RESPECT TO THE xy- OR xz- OR yz-PLANE:

$$S_{xy} = \int_{t_1}^{t_2} \chi(t)\,\varrho(t) \sqrt{[\dot{\varphi}^2(t) + \dot{\psi}^2(t) + \dot{\chi}^2(t)]}\,\mathrm{d}t = \int_{t_1}^{t_2} z\varrho \sqrt{(\dot{x}^2 + \dot{y}^2 + \dot{z}^2)}\,\mathrm{d}t\,, \tag{21}$$

$$S_{xz} = \int_{t_1}^{t_2} \psi(t)\,\varrho(t) \sqrt{[\dot{\varphi}^2(t) + \dot{\psi}^2(t) + \dot{\chi}^2(t)]}\,\mathrm{d}t = \int_{t_1}^{t_2} y\varrho \sqrt{(\dot{x}^2 + \dot{y}^2 + \dot{z}^2)}\,\mathrm{d}t\,, \tag{22}$$

$$S_{yz} = \int_{t_1}^{t_2} \varphi(t)\,\varrho(t) \sqrt{[\dot{\varphi}^2(t) + \dot{\psi}^2(t) + \dot{\chi}^2(t)]}\,\mathrm{d}t = \int_{t_1}^{t_2} x\varrho \sqrt{(\dot{x}^2 + \dot{y}^2 + \dot{z}^2)}\,\mathrm{d}t\,. \tag{23}$$

COORDINATES OF THE CENTRE OF GRAVITY:

$$x_T = \frac{S_{yz}}{M}\,, \quad y_T = \frac{S_{xz}}{M}\,, \quad z_T = \frac{S_{xy}}{M}\,. \tag{24}$$

MOMENT OF INERTIA WITH RESPECT TO THE x- OR y- OR z-AXIS:

$$I_x = \int_{t_1}^{t_2} [\psi^2(t) + \chi^2(t)]\,\varrho(t) \sqrt{[\dot{\varphi}^2(t) + \dot{\psi}^2(t) + \dot{\chi}^2(t)]}\,\mathrm{d}t =$$

$$= \int_{t_1}^{t_2} (y^2 + z^2)\,\varrho \sqrt{(\dot{x}^2 + \dot{y}^2 + \dot{z}^2)}\,\mathrm{d}t\,, \tag{25}$$

$$I_y = \int_{t_1}^{t_2} [\varphi^2(t) + \chi^2(t)]\,\varrho(t) \sqrt{[\dot{\varphi}^2(t) + \dot{\psi}^2(t) + \dot{\chi}^2(t)]}\,\mathrm{d}t =$$

$$= \int_{t_1}^{t_2} (x^2 + z^2)\,\varrho \sqrt{(\dot{x}^2 + \dot{y}^2 + \dot{z}^2)}\,\mathrm{d}t\,, \tag{26}$$

$$I_z = \int_{t_1}^{t_2} [\varphi^2(t) + \psi^2(t)]\, \varrho(t) \sqrt{[\dot\varphi^2(t) + \dot\psi^2(t) + \dot\chi^2(t)]}\, \mathrm{d}t =$$

$$= \int_{t_1}^{t_2} (x^2 + y^2)\, \varrho \sqrt{(\dot x^2 + \dot y^2 + \dot z^2)}\, \mathrm{d}t\,. \tag{27}$$

If a curve is given as the intersection of two surfaces, then for the above computations it is usually better to establish its parametric representation.

(b) Plane Figures.

(a) O is a region bounded by a closed curve k positively oriented with respect to O (Fig. 14.22; see Remark 14.7.1).

If k is given parametrically, i.e. if

$$x = \varphi(t)\,, \quad y = \psi(t)\,, \quad t_1 \leqq t \leqq t_2\,,$$

then as t increases from t_1 to t_2 the point (x, y) moves along the curve k so that O remains on the left-hand side (Fig. 14.22).

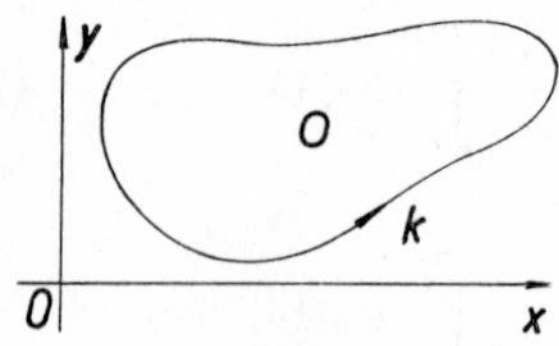

Fig. 14.22.

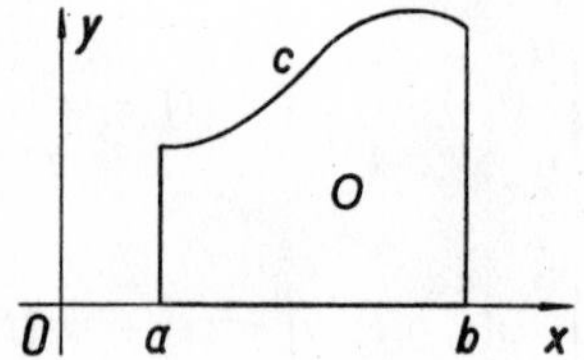

Fig. 14.23.

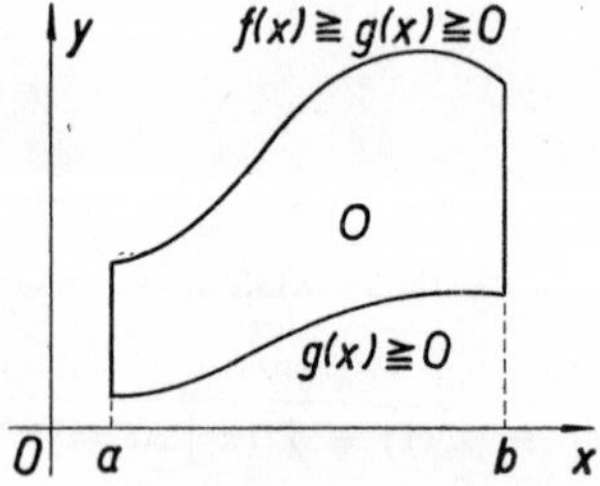

Fig. 14.24.

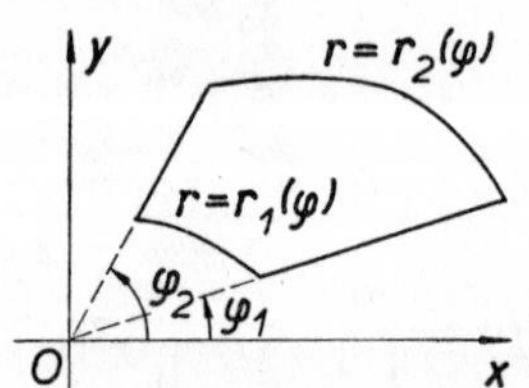

Fig. 14.25.

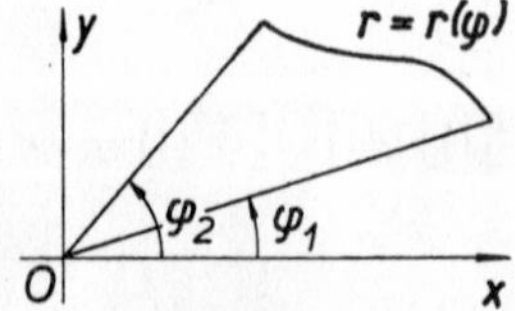

Fig. 14.26.

The specific density (the mass per unit area, g cm^{-2}) is denoted by $\sigma(x, y)$.

(b) The region O is given by a curve c according to Fig. 14.23; the curve c is a graph of a continuous function $y = f(x)$, $f(x) \geqq 0$ in $[a, b]$, or is given parametrically,

$$x = \varphi(t), \quad y = \psi(t), \quad t_1 \leqq t \leqq t_2,$$
$$\psi(t) \geqq 0, \quad \varphi(t_1) = a, \quad \varphi(t_2) = b.$$

The specific density is denoted by $\sigma(x)$ (g cm^{-2}).

(The area of the region O in Fig. 14.24 is obtained, of course, as the difference between areas of regions of the type shown in Fig. 14.23. Similarly for masses, statical moments, coordinates of the centre of gravity, moments of inertia, etc. The parametric representation is also treated in a similar way.)

(c) The region O is given according to Fig. 14.25, $0 \leqq r_1(\varphi) \leqq r_2(\varphi)$ in $[\varphi_1, \varphi_2]$. The specific density is denoted by $\sigma(r, \varphi)$.

(d) If $r_1 \equiv 0$ in (c) we get a sector (Fig. 14.26). (We omit the index in $r_2(\varphi)$ by writing $r(\varphi)$.)

If $\sigma = \sigma(x)$ or $\sigma =$ const., some formulae may be simplified. (If $\sigma(x) =$ const., or $\sigma(x, y)) =$ const., or $\sigma(r, \varphi) =$ const., then, of course, σ can be put in front of the integral sign.)

AREA:

(a)
$$P = \iint_O \mathrm{d}x\,\mathrm{d}y = \tag{28}$$

$$= \int_k x\,\mathrm{d}y = \int_{t_1}^{t_2} \varphi(t)\,\dot{\psi}(t)\,\mathrm{d}t = \tag{29}$$

$$= -\int_k y\,\mathrm{d}x = -\int_{t_1}^{t_2} \psi(t)\,\dot{\varphi}(t)\,\mathrm{d}t = \tag{30}$$

$$= \tfrac{1}{2}\int_k (x\,\mathrm{d}y - y\,\mathrm{d}x) = \tfrac{1}{2}\int_{t_1}^{t_2} [\varphi(t)\,\dot{\psi}(t) - \psi(t)\,\dot{\varphi}(t)]\,\mathrm{d}t\,. \tag{31}$$

In particular, for a sector, see Remark 14.7.14, p. 637.

(b)
$$P = \int_a^b f(x)\,\mathrm{d}x = \int_a^b y\,\mathrm{d}x\,, \tag{32}$$

$$P = \int_{t_1}^{t_2} \psi(t)\,\dot{\varphi}(t)\,\mathrm{d}t = \int_{t_1}^{t_2} y\dot{x}\,\mathrm{d}t\,. \tag{33}$$

(c) $$P = \int_{\varphi_1}^{\varphi_2} \left[\int_{r_1(\varphi)}^{r_2(\varphi)} r \, \mathrm{d}r \right] \mathrm{d}\varphi = \tfrac{1}{2} \int_{\varphi_1}^{\varphi_2} [r_2^2(\varphi) - r_1^2(\varphi)] \, \mathrm{d}\varphi \,. \tag{34}$$

(d) $$P = \tfrac{1}{2} \int_{\varphi_1}^{\varphi_2} r^2(\varphi) \, \mathrm{d}\varphi \,. \tag{34'}$$

MASS:

(a) $$M = \iint_O \sigma(x, y) \, \mathrm{d}x \, \mathrm{d}y \,. \tag{35}$$

(b) $$M = \int_a^b \sigma(x) f(x) \, \mathrm{d}x = \int_a^b \sigma y \, \mathrm{d}x \,. \tag{36}$$

(c) $$M = \int_{\varphi_1}^{\varphi_2} \left[\int_{r_1(\varphi)}^{r_2(\varphi)} r\sigma(r, \varphi) \, \mathrm{d}r \right] \mathrm{d}\varphi \,. \tag{37}$$

(d) For $\sigma = \sigma(\varphi)$ we have

$$M = \tfrac{1}{2} \int_{\varphi_1}^{\varphi_2} \sigma(\varphi) \, r^2(\varphi) \, \mathrm{d}\varphi \,. \tag{37'}$$

STATICAL MOMENT WITH RESPECT TO THE x- OR y-AXIS:

(a) $$S_x = \iint_O y\sigma(x, y) \, \mathrm{d}x \, \mathrm{d}y \,, \quad S_y = \iint_O x\sigma(x, y) \, \mathrm{d}x \, \mathrm{d}y \,. \tag{38}$$

(b) $$S_x = \int_a^b \sigma(x) \frac{f^2(x)}{2} \, \mathrm{d}x = \tfrac{1}{2} \int_a^b \sigma y^2 \, \mathrm{d}x \,, \quad S_y = \int_a^b x\sigma(x) f(x) \, \mathrm{d}x = \int_a^b \sigma x y \, \mathrm{d}x \,. \tag{39}$$

(c) $$S_x = \int_{\varphi_1}^{\varphi_2} \sin \varphi \left[\int_{r_1(\varphi)}^{r_2(\varphi)} r^2 \sigma(r, \varphi) \, \mathrm{d}r \right] \mathrm{d}\varphi \,, \tag{40}$$

$$S_y = \int_{\varphi_1}^{\varphi_2} \cos \varphi \left[\int_{r_1(\varphi)}^{r_2(\varphi)} r^2 \sigma(r, \varphi) \, \mathrm{d}r \right] \mathrm{d}\varphi \,. \tag{40'}$$

(d) For $\sigma = \sigma(\varphi)$ we have

$$S_x = \tfrac{1}{3} \int_{\varphi_1}^{\varphi_2} \sigma(\varphi) \, r^3(\varphi) \sin \varphi \, \mathrm{d}\varphi \,, \quad S_y = \tfrac{1}{3} \int_{\varphi_1}^{\varphi_2} \sigma(\varphi) \, r^3(\varphi) \cos \varphi \, \mathrm{d}\varphi \,. \tag{40''}$$

COORDINATES OF THE CENTRE OF GRAVITY:

$$x_T = \frac{S_y}{M} \,, \quad y_T = \frac{S_x}{M} \,.$$

MOMENT OF INERTIA WITH RESPECT TO THE x- OR y-AXIS, OR TO THE ORIGIN (I. E. ABOUT THE z-AXIS):

(a)
$$I_x = \iint_O y^2\sigma(x, y)\,dx\,dy\,, \quad I_y = \iint_O x^2\sigma(x, y)\,dx\,dy\,, \tag{41}$$

$$I_z = \iint_O (x^2 + y^2)\,\sigma(x, y)\,dx\,dy = I_x + I_y\,. \tag{42}$$

(b)
$$I_x = \int_a^b \sigma(x)\frac{f^3(x)}{3}\,dx = \tfrac{1}{3}\int_a^b \sigma y^3\,dx\,, \tag{43}$$

$$I_y = \int_a^b x^2\sigma(x) f(x)\,dx = \int_a^b \sigma x^2 y\,dx\,, \tag{44}$$

$$I_z = \int_a^b \sigma(x)\left[\frac{f^3(x)}{3} + x^2 f(x)\right]dx =$$
$$= \int_a^b \sigma\left(\frac{y^3}{3} + x^2 y\right)dx = I_x + I_y\,. \tag{45}$$

(c)
$$I_x = \int_{\varphi_1}^{\varphi_2} \sin^2\varphi\left[\int_{r_1(\varphi)}^{r_2(\varphi)} r^3\sigma(r, \varphi)\,dr\right]d\varphi\,,$$

$$I_y = \int_{\varphi_1}^{\varphi_2} \cos^2\varphi\left[\int_{r_1(\varphi)}^{r_2(\varphi)} r^3\sigma(r, \varphi)\,dr\right]d\varphi\,, \tag{46}$$

$$I_z = \int_{\varphi_1}^{\varphi_2}\left[\int_{r_1(\varphi)}^{r_2(\varphi)} r^3\sigma(r, \varphi)\,dr\right]d\varphi = I_x + I_y\,. \tag{47}$$

(d) For $\sigma = \sigma(\varphi)$ we have

$$I_x = \tfrac{1}{4}\int_{\varphi_1}^{\varphi_2} \sigma(\varphi)\,r^4(\varphi)\sin^2\varphi\,d\varphi\,,$$

$$I_y = \tfrac{1}{4}\int_{\varphi_1}^{\varphi_2} \sigma(\varphi)\,r^4(\varphi)\cos^2\varphi\,d\varphi\,, \tag{46'}$$

$$I_z = \tfrac{1}{4}\int_{\varphi_1}^{\varphi_2} \sigma(\varphi)\,r^4(\varphi)\,d\varphi = I_x + I_y\,. \tag{47'}$$

(c) Solids.

(a) The solid is a three-dimensional region T of type A (Definition 14.1.4) (Fig. 14.27). The specific density (mass per unit volume) is denoted by $\varrho(x, y, z)$ (g cm^{-3}).

(b) The solid T is a special region of type A: The lower base O lies in the xy-plane, the upper one is the surface $z = f(x, y)$ and the lateral surface is formed by parallels to the z-axis (Fig. 14.28). The specific mass is denoted by $\varrho = \varrho(x, y)$.

(c) T is a solid with the lower base $O(r_1(\varphi) \leqq r \leqq r_2(\varphi),\ \varphi_1 \leqq \varphi \leqq \varphi_2)$ and the upper base $z = z(r, \varphi)$; the lateral surface is formed by parallels to the z-axis (Fig. 14.29). The specific mass is denoted by $\varrho(r, \varphi)$.

(d) T is a solid with the bases P_1, P_2 perpendicular to the x-axis. P_1 lies in the plane $x = a$, P_2 in the plane $x = b$. The areas p of all cross-sections perpendicular to the x-axis are known, $p = p(x)$ (Fig. 14.30). The specific mass is denoted by $\varrho(x)$.

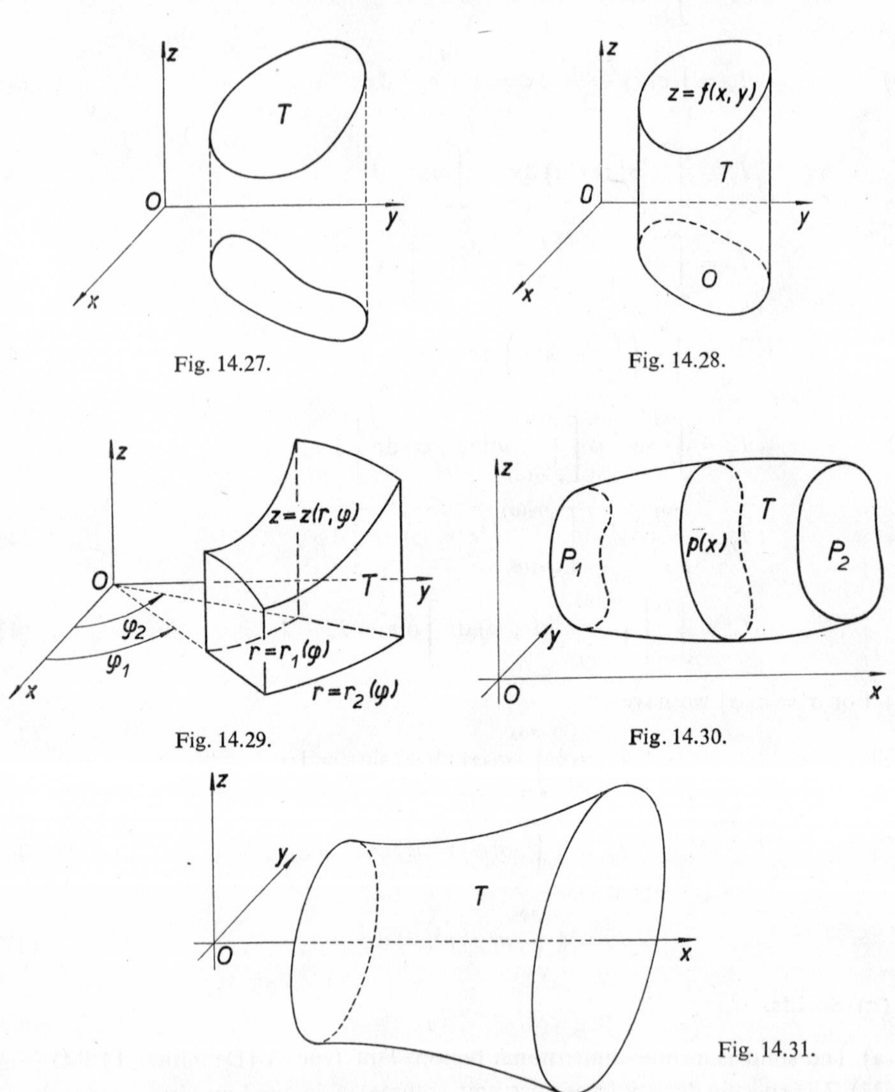

Fig. 14.27.

Fig. 14.28.

Fig. 14.29.

Fig. 14.30.

Fig. 14.31.

(e) T is a solid of revolution, whose lateral surface is formed by rotating the curve $y = f(x)$, $a \leqq x \leqq b$, $f(x) \geqq 0$, around the x-axis (Fig. 14.31). The specific mass is denoted by $\varrho(x)$.

In the cases where ϱ = const., ϱ can be put in front of the integral sign, of course.

(a) $$V = \iiint_T \mathrm{d}x\, \mathrm{d}y\, \mathrm{d}z\,. \tag{48}$$

(b) $$V = \iint_O f(x, y)\, \mathrm{d}x\, \mathrm{d}y\,. \tag{49}$$

(c) $$V = \int_{\varphi_1}^{\varphi_2} \left[\int_{r_1(\varphi)}^{r_2(\varphi)} rz(r, \varphi)\, \mathrm{d}r \right] \mathrm{d}\varphi\,. \tag{50}$$

(d) $$V = \int_a^b p(x)\, \mathrm{d}x\,. \tag{51}$$

If, especially, $p(x)$ is a polynomial of at most the third degree, then

$$V = \frac{b - a}{6} (p_1 + 4p_m + p_2)\,, \tag{52}$$

where p_1, p_2 are areas of the bases P_1, P_2 and p_m is the area of the mean cross-section (i.e. for $x = \frac{1}{2}(a + b)$).

(e) $$V = \pi \int_a^b f^2(x)\, \mathrm{d}x = \pi \int_a^b y^2\, \mathrm{d}x\,. \tag{53}$$

If the lateral surface of a solid of revolution is formed by rotating a curve c around the x-axis and c is given parametrically by $x = \varphi(t)$, $y = \psi(t)$, $t_1 \leqq t \leqq t_2$, $\psi(t) \geqq 0$, $\mathrm{d}\varphi/\mathrm{d}t = \dot{\varphi}(t) > 0$, then

$$V = \pi \int_{t_1}^{t_2} \psi^2(t)\, \dot{\varphi}(t)\, \mathrm{d}t\,. \tag{53'}$$

If $\dot{\varphi} < 0$, then

$$V = \pi \int_{t_2}^{t_1} \psi^2(t) \dot{\varphi}(t)\, \mathrm{d}t\,. \tag{53''}$$

MASS:

(a) $$M = \iiint_T \varrho(x, y, z)\, \mathrm{d}x\, \mathrm{d}y\, \mathrm{d}z\,. \tag{54}$$

(b) $$M = \iint_O \varrho(x, y) f(x, y)\, \mathrm{d}x\, \mathrm{d}y\,. \tag{55}$$

(c) $$M = \int_{\varphi_1}^{\varphi_2} \left[\int_{r_1(\varphi)}^{r_2(\varphi)} r\varrho(r, \varphi)\, z(r, \varphi)\, \mathrm{d}r \right] \mathrm{d}\varphi\,. \tag{56}$$

(d) $$M = \int_a^b \varrho(x)\, p(x)\, dx\,. \tag{57}$$

(e) $$M = \pi \int_a^b \varrho(x) f^2(x)\, dx = \pi \int_a^b \varrho y^2\, dx\,. \tag{58}$$

STATICAL MOMENT WITH RESPECT TO THE xy-, yz- OR zx-PLANE:

(a) $$S_{xy} = \iiint_T z\varrho(x, y, z)\, dx\, dy\, dz\,, \tag{59}$$

$$S_{yz} = \iiint_T x\varrho(x, y, z)\, dx\, dy\, dz\,, \tag{59'}$$

$$S_{zx} = \iiint_T y\varrho(x, y, z)\, dx\, dy\, dz\,. \tag{59''}$$

(b) $$S_{xy} = \tfrac{1}{2} \iint_O \varrho(x, y) f^2(x, y)\, dx\, dy\,, \tag{60}$$

$$S_{yz} = \iint_O x\varrho(x, y) f(x, y)\, dx\, dy\,, \quad S_{zx} = \iint_O y\varrho(x, y) f(x, y)\, dx\, dy\,. \tag{61}$$

(c) $$S_{xy} = \tfrac{1}{2} \int_{\varphi_1}^{\varphi_2} \left[\int_{r_1(\varphi)}^{r_2(\varphi)} r\varrho(r, \varphi)\, z^2(r, \varphi)\, dr\right] d\varphi\,, \tag{62}$$

$$S_{yz} = \int_{\varphi_1}^{\varphi_2} \left[\int_{r_1(\varphi)}^{r_2(\varphi)} r^2\varrho(r, \varphi)\, z(r, \varphi)\, dr\right] \cos\varphi\, d\varphi\,, \tag{63}$$

$$S_{zx} = \int_{\varphi_1}^{\varphi_2} \left[\int_{r_1(\varphi)}^{r_2(\varphi)} r^2\varrho(r, \varphi)\, z(r, \varphi)\, dr\right] \sin\varphi\, d\varphi\,. \tag{64}$$

(d) $$S_{yz} = \int_a^b x\varrho(x)\, p(x)\, dx\,. \tag{65}$$

(e) $$S_{yz} = \pi \int_a^b x\varrho(x) f^2(x)\, dx = \pi \int_a^b x\varrho y^2\, dx\,. \tag{66}$$

COORDINATES OF THE CENTRE OF GRAVITY:

$$x_T = \frac{S_{yz}}{M}\,, \quad y_T = \frac{S_{zx}}{M}\,, \quad z_T = \frac{S_{xy}}{M}\,.$$

MOMENT OF INERTIA WITH RESPECT TO THE x- OR y- OR z-AXIS (see also Steiner's theorem, formula (118)):

(a)
$$I_x = \iiint_T (y^2 + z^2)\,\varrho(x, y, z)\,\mathrm{d}x\,\mathrm{d}y\,\mathrm{d}z\,, \tag{67}$$

$$I_y = \iiint_T (x^2 + z^2)\,\varrho(x, y, z)\,\mathrm{d}x\,\mathrm{d}y\,\mathrm{d}z\,, \tag{68}$$

$$I_z = \iiint_T (x^2 + y^2)\,\varrho(x, y, z)\,\mathrm{d}x\,\mathrm{d}y\,\mathrm{d}z\,. \tag{69}$$

(b)
$$I_z = \iint_O (x^2 + y^2)\,\varrho(x, y,)\,f(x, y)\,\mathrm{d}x\,\mathrm{d}y\,. \tag{70}$$

(c)
$$I_z = \int_{\varphi_1}^{\varphi_2} \left[\int_{r_1(\varphi)}^{r_2(\varphi)} r^3 \varrho(r, \varphi)\, z(r, \varphi)\,\mathrm{d}r \right] \mathrm{d}\varphi\,. \tag{71}$$

(e)
$$I_x = \tfrac{1}{2}\pi \int_a^b \varrho(x)\, f^4(x)\,\mathrm{d}x = \tfrac{1}{2}\pi \int_a^b \varrho y^4\,\mathrm{d}x\,. \tag{72}$$

(d) Surfaces.

(a) The finite simple piecewise smooth surface (Remark 14.1.7) is given parametrically:

$$x = x(u, v)\,, \quad y = y(u, v)\,, \quad z = z(u, v) \tag{73}$$

$(u, v) \in O$. The density (mass per unit area, g cm^{-2}) is denoted by $\sigma = \sigma(u, v)$.

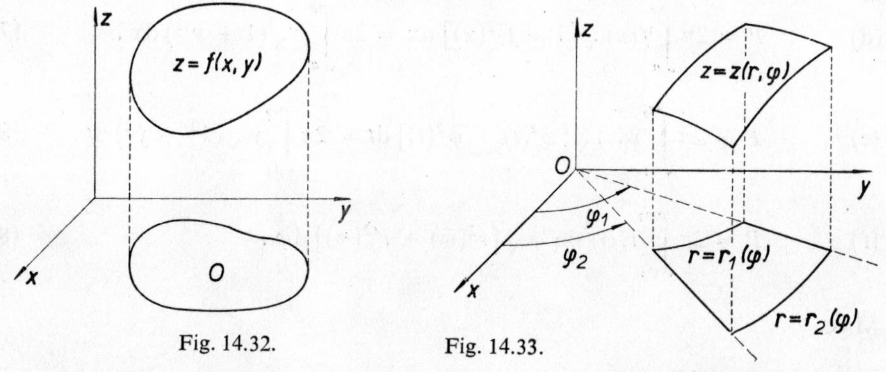

Fig. 14.32. Fig. 14.33.

(b) The surface $z = f(x, y)$ over a region O (Fig. 14.32). The density is denoted by $\sigma = \sigma(x, y)$.

(c) The surface $z = z(r, \varphi)$ over a region $O\,(\varphi_1 \leqq \varphi \leqq \varphi_2, r_1(\varphi) \leqq r \leqq r_2(\varphi))$, Fig. 14.33. The density is denoted by $\sigma(r, \varphi)$.

(d) The surface of revolution obtained by rotating a curve $y = f(x)$, $a \leqq x \leqq b$, $f(x) \geqq 0$ around the x-axis. The density is denoted by $\sigma(x)$.

(e) The surface of revolution obtained by rotating a simple curve $x = \varphi(t)$, $y = \psi(t)$, $t_1 \leqq t \leqq t_2$, $\psi(t) \geqq 0$, $d\varphi/dt = \dot{\varphi}(t) > 0$ around the x-axis. The density is denoted by $\sigma(t)$.

(f) The surface of revolution obtained by rotating a curve $r = r(\varphi)$, $0 \leqq \varphi_1 \leqq \varphi \leqq \leqq \varphi_2 \leqq \pi$ around the x-axis. The density is denoted by $\sigma(\varphi)$.

(If $\sigma =$ const., σ can be put in front of the integral sign.)

AREA:

(a)
$$P = \iint_O \sqrt{(EG - F^2)}\, du\, dv\,, \tag{74}$$

where

$$E = \left(\frac{\partial x}{\partial u}\right)^2 + \left(\frac{\partial y}{\partial u}\right)^2 + \left(\frac{\partial z}{\partial u}\right)^2, \quad G = \left(\frac{\partial x}{\partial v}\right)^2 + \left(\frac{\partial y}{\partial v}\right)^2 + \left(\frac{\partial z}{\partial v}\right)^2, \tag{75}$$

$$F = \frac{\partial x}{\partial u}\frac{\partial x}{\partial v} + \frac{\partial y}{\partial u}\frac{\partial y}{\partial v} + \frac{\partial z}{\partial u}\frac{\partial z}{\partial v}\,. \tag{76}$$

(b)
$$P = \iint_O \sqrt{\left[1 + \left(\frac{\partial f}{\partial x}\right)^2 + \left(\frac{\partial f}{\partial y}\right)^2\right]}\, dx\, dy\,. \tag{77}$$

(c)
$$P = \int_{\varphi_1}^{\varphi_2}\left[\int_{r_1(\varphi)}^{r_2(\varphi)} \sqrt{\left[r^2 + r^2\left(\frac{\partial z}{\partial r}\right)^2 + \left(\frac{\partial z}{\partial \varphi}\right)^2\right]}\, dr\right] d\varphi\,. \tag{78}$$

(d)
$$P = 2\pi \int_a^b f(x) \sqrt{[1 + f'^2(x)]}\, dx = 2\pi \int_a^b y \sqrt{(1 + y'^2)}\, dx\,. \tag{79}$$

(e)
$$P = 2\pi \int_{t_1}^{t_2} \psi(t) \sqrt{[\dot{\varphi}^2(t) + \dot{\psi}^2(t)]}\, dt = 2\pi \int_{t_1}^{t_2} y \sqrt{(\dot{x}^2 + \dot{y}^2)}\, dt\,. \tag{80}$$

(f)
$$P = 2\pi \int_{\varphi_1}^{\varphi_2} r(\varphi) \sin\varphi \sqrt{[r^2(\varphi) + r'^2(\varphi)]}\, d\varphi\,. \tag{81}$$

MASS:

(a)
$$M = \iint_O \sigma(u, v) \sqrt{(EG - F^2)}\, du\, dv \quad \text{(see (75), (76))}\,. \tag{82}$$

(b)
$$M = \iint_O \sigma(x, y) \sqrt{\left[1 + \left(\frac{\partial f}{\partial x}\right)^2 + \left(\frac{\partial f}{\partial y}\right)^2\right]}\, dx\, dy\,. \tag{83}$$

(c) $$M = \int_{\varphi_1}^{\varphi_2} \left[\int_{r_1(\varphi)}^{r_2(\varphi)} \sigma(r, \varphi) \sqrt{\left[r^2 + r^2 \left(\frac{\partial z}{\partial r} \right)^2 + \left(\frac{\partial z}{\partial \varphi} \right)^2 \right]} \, dr \right] d\varphi \, . \tag{84}$$

(d) $$M = 2\pi \int_a^b \sigma(x) f(x) \sqrt{[1 + f'^2(x)]} \, dx = 2\pi \int_a^b \sigma y \sqrt{(1 + y'^2)} \, dx \, . \tag{85}$$

(e) $$M = 2\pi \int_{t_1}^{t_2} \sigma(t) \psi(t) \sqrt{[\dot{\varphi}^2(t) + \dot{\psi}^2(t)]} \, dt = 2\pi \int_{t_1}^{t_2} \sigma y \sqrt{(\dot{x}^2 + \dot{y}^2)} \, dt \, . \tag{86}$$

(f) $$M = 2\pi \int_{\varphi_1}^{\varphi_2} \sigma(\varphi) r(\varphi) \sin \varphi \sqrt{[r^2(\varphi) + r'^2(\varphi)]} \, d\varphi \, . \tag{87}$$

STATICAL MOMENT WITH RESPECT TO THE xy- OR yz- OR zx-PLANE:

(a) $$S_{xy} = \iint_O \sigma(u, v) z(u, v) \sqrt{(EG - F^2)} \, du \, dv \quad \text{(see (75), (76))} \, , \tag{88}$$

$$S_{yz} = \iint_O \sigma(u, v) x(u, v) \sqrt{(EG - F^2)} \, du \, dv \, ,$$

$$S_{zx} = \iint_O \sigma(u, v) y(u, v) \sqrt{(EG - F^2)} \, du \, dv \, . \tag{89}$$

(b) $$S_{xy} = \iint_O \sigma(x, y) f(x, y) \sqrt{\left[1 + \left(\frac{\partial f}{\partial x} \right)^2 + \left(\frac{\partial f}{\partial y} \right)^2 \right]} \, dx \, dy \, , \tag{90}$$

$$S_{yz} = \iint_O x\sigma(x, y) \sqrt{\left[1 + \left(\frac{\partial f}{\partial x} \right)^2 + \left(\frac{\partial f}{\partial y} \right)^2 \right]} \, dx \, dy \, ,$$

$$S_{zx} = \iint_O y\sigma(x, y) \sqrt{\left[1 + \left(\frac{\partial f}{\partial x} \right)^2 + \left(\frac{\partial f}{\partial y} \right)^2 \right]} \, dx \, dy \, . \tag{91}$$

(c) $$S_{xy} = \int_{\varphi_1}^{\varphi_2} \left[\int_{r_1(\varphi)}^{r_2(\varphi)} \sigma(r, \varphi) z(r, \varphi) \sqrt{\left[r^2 + r^2 \left(\frac{\partial z}{\partial r} \right)^2 + \left(\frac{\partial z}{\partial \varphi} \right)^2 \right]} \, dr \right] d\varphi \, , \tag{92}$$

$$S_{yz} = \int_{\varphi_1}^{\varphi_2} \left[\int_{r_1(\varphi)}^{r_2(\varphi)} \sigma(r, \varphi) r \sqrt{\left[r^2 + r^2 \left(\frac{\partial z}{\partial r} \right)^2 + \left(\frac{\partial z}{\partial \varphi} \right)^2 \right]} \, dr \right] \cos \varphi \, d\varphi \, , \tag{93}$$

$$S_{zx} = \int_{\varphi_1}^{\varphi_2} \left[\int_{r_1(\varphi)}^{r_2(\varphi)} \sigma(r, \varphi) r \sqrt{\left[r^2 + r^2 \left(\frac{\partial z}{\partial r} \right)^2 + \left(\frac{\partial z}{\partial \varphi} \right)^2 \right]} \, dr \right] \sin \varphi \, d\varphi \, . \tag{94}$$

(d) $$S_{xy} = 0 \, , \quad S_{zx} = 0 \, , \quad S_{yz} = 2\pi \int_a^b x\sigma(x) f(x) \sqrt{[1 + f'^2(x)]} \, dx =$$

$$= 2\pi \int_a^b x\sigma y \sqrt{(1 + y'^2)} \, dx \, . \tag{95}$$

(e) $S_{xy} = 0\,, \quad S_{zx} = 0\,, \quad S_{yz} = 2\pi \int_{t_1}^{t_2} \sigma(t)\, \varphi(t)\, \psi(t) \sqrt{[\dot{\varphi}^2(t) + \dot{\psi}^2(t)]}\, \mathrm{d}t =$

$$= 2\pi \int_{t_1}^{t_2} \sigma x y \sqrt{(\dot{x}^2 + \dot{y}^2)}\, \mathrm{d}t\,. \tag{96}$$

(f) $S_{xy} = 0\,, \quad S_{zx} = 0\,, \quad S_{yz} = 2\pi \int_{\varphi_1}^{\varphi_2} \sigma(\varphi) r^2(\varphi) \sin\varphi \cos\varphi \sqrt{[r^2(\varphi) + r'^2(\varphi)]}\, \mathrm{d}\varphi\,.$ (97)

COORDINATES OF THE CENTRE OF GRAVITY:

$$x_T = \frac{S_{yz}}{M}\,, \quad y_T = \frac{S_{zx}}{M}\,, \quad z_T = \frac{S_{xy}}{M}\,. \tag{98}$$

MOMENT OF INERTIA WITH RESPECT TO THE x- OR y- OR z-AXIS:

(a) $I_x = \iint_O \sigma(u, v)\, [y^2(u, v) + z^2(u, v)] \sqrt{(EG - F^2)}\, \mathrm{d}u\, \mathrm{d}v$ (see (75), (76)), (99)

$$I_y = \iint_O \sigma(u, v)\, [x^2(u, v) + z^2(u, v)] \sqrt{(EG - F^2)}\, \mathrm{d}u\, \mathrm{d}v\,,$$

$$I_z = \iint_O \sigma(u, v)\, [x^2(u, v) + y^2(u, v)] \sqrt{(EG - F^2)}\, \mathrm{d}u\, \mathrm{d}v\,. \tag{100}$$

(b) $I_x = \iint_O \sigma(x, y)\, [y^2 + f^2(x, y)] \sqrt{\left[1 + \left(\frac{\partial f}{\partial x}\right)^2 + \left(\frac{\partial f}{\partial y}\right)^2\right]}\, \mathrm{d}x\, \mathrm{d}y\,,$ (101)

$$I_y = \iint_O \sigma(x, y)\, [x^2 + f^2(x, y)] \sqrt{\left[1 + \left(\frac{\partial f}{\partial x}\right)^2 + \left(\frac{\partial f}{\partial y}\right)^2\right]}\, \mathrm{d}x\, \mathrm{d}y\,, \tag{102}$$

$$I_z = \iint_O \sigma(x, y)\, (x^2 + y^2) \sqrt{\left[1 + \left(\frac{\partial f}{\partial x}\right)^2 + \left(\frac{\partial f}{\partial y}\right)^2\right]}\, \mathrm{d}x\, \mathrm{d}y\,. \tag{103}$$

(c)

$$I_x = \int_{\varphi_1}^{\varphi_2} \left[\int_{r_1(\varphi)}^{r_2(\varphi)} \sigma(r, \varphi)\, [r^2 \sin^2 \varphi + z^2(r, \varphi)] \sqrt{\left[r^2 + r^2 \left(\frac{\partial z}{\partial r}\right)^2 + \left(\frac{\partial z}{\partial \varphi}\right)^2\right]}\, \mathrm{d}r \right] \mathrm{d}\varphi\,, \tag{104}$$

$$I_y = \int_{\varphi_1}^{\varphi_2} \left[\int_{r_1(\varphi)}^{r_2(\varphi)} \sigma(r, \varphi)\, [r^2 \cos^2 \varphi + z^2(r, \varphi)] \sqrt{\left[r^2 + r^2 \left(\frac{\partial z}{\partial r}\right)^2 + \left(\frac{\partial z}{\partial \varphi}\right)^2\right]}\, \mathrm{d}r \right] \mathrm{d}\varphi\,, \tag{105}$$

$$I_z = \int_{\varphi_1}^{\varphi_2} \left[\int_{r_1(\varphi)}^{r_2(\varphi)} \sigma(r, \varphi)\, r^2 \sqrt{\left[r^2 + r^2 \left(\frac{\partial z}{\partial r} \right)^2 + \left(\frac{\partial z}{\partial \varphi} \right)^2 \right]}\, dr \right] d\varphi\,. \tag{106}$$

(d) $$I_x = 2\pi \int_a^b \sigma(x)\, f^3(x) \sqrt{[1 + f'^2(x)]}\, dx = 2\pi \int_a^b \sigma y^3 \sqrt{(1 + y'^2)}\, dx\,. \tag{107}$$

(e) $$I_x = 2\pi \int_{t_1}^{t_2} \sigma(t)\, \psi^3(t) \sqrt{[\dot\varphi^2(t) + \dot\psi^2(t)]}\, dt = 2\pi \int_{t_1}^{t_2} \sigma y^3 \sqrt{(\dot x^2 + \dot y^2)}\, dt\,. \tag{108}$$

(f) $$I_x = 2\pi \int_{\varphi_1}^{\varphi_2} \sigma(\varphi)\, r^3(\varphi) \sin^3 \varphi \sqrt{[r^2(\varphi) + r'^2(\varphi)]}\, d\varphi\,. \tag{109}$$

(e) The Work Done by a Force Moving Along a Given Curve. The work L done by a force $\mathbf{P}$ in moving along an oriented curve c in a field of force

$$\mathbf{P} = \mathbf{i}P_1(x, y, z) + \mathbf{j}P_2(x, y, z) + \mathbf{k}P_3(x, y, z)$$

is given by the sum of curvilinear integrals

$$L = \int_c P_1(x, y, z)\, dx + \int_c P_2(x, y, z)\, dy + \int_c P_3(x, y, z)\, dz\,. \tag{110}$$

If c is given parametrically and is positively oriented with increasing parameter (Remark 14.7.1, p. 628), we have

$$L = \int_{t_1}^{t_2} P_1(\varphi(t), \psi(t), \chi(t))\, \varphi'(t)\, dt + \int_{t_1}^{t_2} P_2(\varphi(t), \psi(t), \chi(t))\, \psi'(t)\, dt + \int_{t_1}^{t_2} P_3(\varphi(t), \psi(t), \chi(t))\, \chi'(t)\, dt\,. \tag{111}$$

In the plane (where $\mathbf{P} = \mathbf{i}P_1(x, y) + \mathbf{j}P_2(x, y)$):

$$L = \int_c P_1(x, y)\, dx + \int_c P_2(x, y)\, dy = \int_{t_1}^{t_2} P_1(\varphi(t), \psi(t))\, \varphi'(t)\, dt + \int_{t_1}^{t_2} P_2(\varphi(t), \psi(t))\, \psi'(t)\, dt\,. \tag{112}$$

(f) Some Special Formulae. The area of a cylindrical surface $y = f(x)$ cut off by the cylindrical surface $z = g(x)$ and by the planes $z = 0$, $x = a$, $x = b$ ($g(x) \geqq 0$ for $a \leqq x \leqq b$) is given by

$$P = \int_a^b g(x) \sqrt{[1 + f'^2(x)]}\, dx = \int_a^b z \sqrt{(1 + y'^2)}\, dx\,. \tag{113}$$

The area P of a conical surface with the vertex at the origin and whose base is the curve

$$x = x(t), \quad y = y(t), \quad z = z(t) \quad (t_1 \leqq t \leqq t_2)$$

(i.e. the surface composed of lines joining the origin to points on this curve):

$$P = \tfrac{1}{2} \int_{t_1}^{t_2} \sqrt{[(x\dot{y} - y\dot{x})^2 + (x\dot{z} - z\dot{x})^2 + (y\dot{z} - z\dot{y})^2]} \, dt . \tag{114}$$

(g) Guldin's Rules.* Let V be a solid of revolution obtained by rotating a region O (of type A, Definition 14.1.2, p. 603) with boundary H (which does not intersect the x-axis), around the x-axis. Then we have

$$S = 2\pi l y_T, \quad V = 2\pi P Y_T, \tag{115}$$

where S denotes the surface area of the solid, V the volume of the solid, P the area of the region O, l the length of the boundary H, y_T the y-coordinate of the centre of gravity of the boundary H and Y_T the y-coordinate of the centre of gravity of the region O (see Example 2).

Guldin's rules remain true (under obvious assumptions) even in the following more general case: A given profile with the area P and with the length l of boundary is moved so that its plane remains perpendicular to the (spatial) curve described by its centre of gravity. Then

(a) the area of the lateral surface is equal to ld, (116)

(b) the volume of the solid is equal to PD, (117)

where D is the length of the trajectory traversed by the centre of gravity of the profile and d the length of the trajectory traversed by the centre of gravity of the boundary of the profile.

(h) Steiner's Theorem (Parallel Axes Theorem). The moment of inertia I_p of a (not necessarily homogeneous) solid with respect to a given straight line p is equal to the moment of inertia I_r of this solid with respect to the straight line r parallel to p and passing through the centre of gravity of the solid, plus a^2M, where M is the mass of the solid and a is the distance between the straight lines p and r; i.e.

$$I_p = I_r + a^2 M . \tag{118}$$

(i) Examples.

Example 1. Let us calculate the moment of inertia of a homogeneous cone ($\varrho = \varrho_0 = \text{const.}$) with respect to its axis, where v is the height of the cone and r is the radius of the base.

* Otherwise known as Pappus's Rules.

Let the axis of the cone be the x-axis, and let the vertex be at the origin. The lateral surface of the cone is obtained by rotating the segment

$$y = \frac{r}{v} x \quad (0 \leqq x \leqq v)$$

around the x-axis. By (72) we have

$$I_x = \frac{\pi \varrho_0}{2} \int_0^v \frac{r^4}{v^4} x^4 \, \mathrm{d}x = \frac{\pi \varrho_0}{10} r^4 v .$$

Example 2. Let us find the volume and surface area of a *torus* (Fig. 14.34; the torus is obtained by rotating the circle shaded on the figure, around the x-axis).

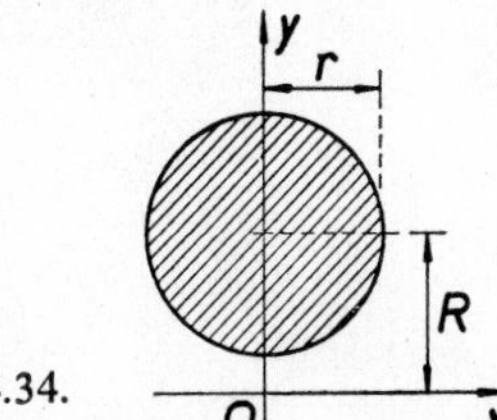

Fig. 14.34.

According to (115) we have

$$S = 2\pi \, . \, 2\pi r \, . \, R = 4\pi^2 r R ,$$

$$V = 2\pi \, . \, \pi r^2 \, . \, R = 2\pi^2 r^2 R .$$

14.10. Survey of Some Important Formulae in Chapter 14

(See also physical and geometrical applications in § 14.9, also Theorem 14.8.9 and Remark 14.8.15.)

1. $$\iint_M [c_1 f_1(x, y) + c_2 f_2(x, y)] \, \mathrm{d}x \, \mathrm{d}y =$$

$$= c_1 \iint_M f_1(x, y) \, \mathrm{d}x \, \mathrm{d}y + c_2 \iint_M f_2(x, y) \, \mathrm{d}x \, \mathrm{d}y$$

and similarly for triple integrals (Theorem 14.2.5).

2. $$\iint_M f(x, y) \, \mathrm{d}x \, \mathrm{d}y = \int_a^b \mathrm{d}x \int_{h_1(x)}^{h_2(x)} f(x, y) \, \mathrm{d}y = \int_c^d \mathrm{d}y \int_{\varphi_1(y)}^{\varphi_2(y)} f(x, y) \, \mathrm{d}x .$$

(Evaluation of a double integral by successive integration, Remark 14.3.5, p. 611. Similar formulae valid for triple integrals are given in Remark 14.5.6, p. 619.)

$$3. \quad \iint_M f(x, y)\,dx\,dy = \iint_N f(x(u, v), y(u, v))\,|D(u, v)|\,du\,dv\,.$$

(Substitution in a double integral, Theorem 14.4.1, p. 615; for polar coordinates $x = r\cos\varphi$, $y = r\sin\varphi$ we have $|D(r, \varphi)| = r$.)

$$4. \quad \iiint_M f(x, y, z)\,dx\,dy\,dz =$$

$$= \iiint_N f(x(u, v, w), y(u, v, w), z(u, v, w))\,|D(u, v, w)|\,du\,dv\,dw$$

(Theorem 14.5.3); for spherical coordinates $x = r\sin\vartheta\cos\varphi$, $y = r\sin\vartheta\sin\varphi$, $z = r\cos\vartheta$ we have $|D(r, \vartheta, \varphi)| = r^2\sin\vartheta$.

$$5. \quad \int_c f(x, y)\,dx = \int_\alpha^\beta f(\varphi(t), \psi(t))\,\dot{\varphi}(t)\,dt\,,$$

$$\int_c f(x, y)\,dy = \int_\alpha^\beta f(\varphi(t), \psi(t))\,\dot{\psi}(t)\,dt\,,$$

$$\int_c f(x, y)\,ds = \int_\alpha^\beta f(\varphi(t), \psi(t))\,\sqrt{[\dot{\varphi}^2(t) + \dot{\psi}^2(t)]}\,dt$$

(Theorem 14.7.2); the curve c is given parametrically by equations

$$x = \varphi(t)\,, \quad y = \psi(t) \quad (\alpha \leqq t \leqq \beta)$$

and is oriented positively for t increasing.

For similar formulae valid for curves in space see Remark 14.7.8, p. 633.

$$6. \quad \int_c f(x, y)\,dx = \pm \int_a^b f(x, g(x))\,dx\,,$$

provided c is given by the equation $y = f(x)$. For more details see Theorem 14.7.3.

$$7. \quad \int_c (P\,dx + Q\,dy) = \iint_O \left(\frac{\partial Q}{\partial x} - \frac{\partial P}{\partial y}\right) dx\,dy$$

(Green's theorem, Theorem 14.7.7, p. 634).

8. $$\iiint_V \left(\frac{\partial P}{\partial x} + \frac{\partial Q}{\partial y} + \frac{\partial R}{\partial z}\right) \mathrm{d}x\,\mathrm{d}y\,\mathrm{d}z = \iint_S (P\,\mathrm{d}y\,\mathrm{d}z + Q\,\mathrm{d}z\,\mathrm{d}x + R\,\mathrm{d}x\,\mathrm{d}y)$$

(Gauss's theorem, Theorem 14.8.5, p. 642).

9. $$\int_c (P\,\mathrm{d}x + Q\,\mathrm{d}y + R\,\mathrm{d}z) = -\iint_S \left[\left(\frac{\partial P}{\partial y} - \frac{\partial Q}{\partial x}\right) \mathrm{d}x\,\mathrm{d}y + \right.$$

$$\left. + \left(\frac{\partial Q}{\partial z} - \frac{\partial R}{\partial y}\right) \mathrm{d}y\,\mathrm{d}z + \left(\frac{\partial R}{\partial x} - \frac{\partial P}{\partial z}\right) \mathrm{d}z\,\mathrm{d}x\right]$$

(Stokes's theorem, Theorem 14.8.6, p. 643).

15. SEQUENCES AND SERIES WITH VARIABLE TERMS (SEQUENCES AND SERIES OF FUNCTIONS)

By KAREL REKTORYS

References: [47], [48], [49], [71], [100], [112], [172], [173], [198], [201], [233], [341], [429].

15.1. Sequences with Variable Terms. Uniform Convergence, Arzelà's Theorem. Interchange of Limiting Processes. Integration and Differentiation of Sequences with Variable Terms. Limiting Process under the Integration and Differentiation Signs

Definition 1. Let a sequence of functions

$$f_1(x), f_2(x), f_3(x), \ldots \tag{1}$$

defined in an interval I be given. The sequence is said to *converge* (or to *tend*) *to the (limiting) function* $f(x)$ *in* I, if for every $x_0 \in I$ a finite limit

$$\lim_{n \to \infty} f_n(x_0) = f(x_0) \tag{2}$$

exists.

REMARK 1. The interval I may be either open or closed, finite or infinite, etc. The definition remains unchanged even if the sequence (1) is given on another set than on an interval.

Definition 2. The sequence (1) will be called *uniformly convergent in* I, if for every $\varepsilon > 0$ a number n_0, *independent of the choice of* $x \in I$, can be found such that for every $n > n_0$ and every $x \in I$ we have

$$|f_n(x) - f(x)| < \varepsilon\,. \tag{3}$$

REMARK 2. Roughly speaking, sequence (1) is uniformly convergent in I, if the functions $f_n(x)$ converge to $f(x)$ "at approximately the same rate" in the whole interval I.

Theorem 1 (*The Bolzano–Cauchy Condition of Convergence*). *The sequence of functions* (1) *is uniformly convergent in I if and only if for every* $\varepsilon > 0$ *a positive* *integer n_0 exists such that*

$$|f_n(x) - f_m(x)| < \varepsilon$$

for every $x \in I$ and for any pair of numbers m, n with $n > n_0$, $m > n_0$.

Theorem 2. *Let the functions (1) be continuous in I and let the sequence (1) be uniformly convergent in I. Then the limiting function $f(x)$ is also continuous in I.*

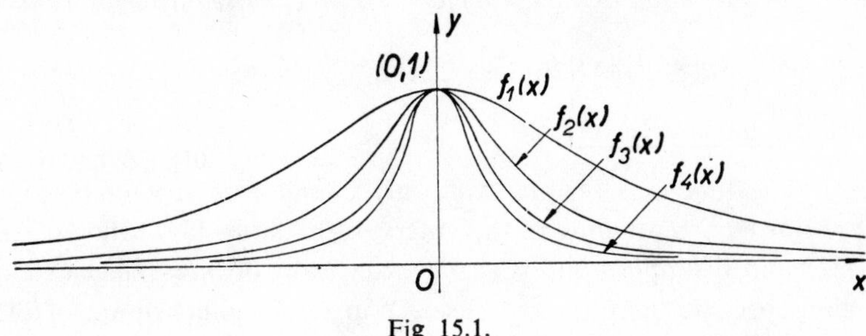

Fig 15.1.

Example 1. Consider the sequence

$$f_n(x) = \frac{1}{1 + n^2x^2} \quad (n = 1, 2, 3, \ldots) \tag{4}$$

in the interval $I = (-\infty, \infty)$ (Fig. 15.1). For every $x \neq 0$, we have

$$\lim_{n\to\infty} f_n(x) = 0$$

since the denominator tends to infinity as $n \to \infty$. For $x = 0$, we have $f_n(0) = 1$ for any n, and consequently,

$$\lim_{n\to\infty} f_n(0) = 1 .$$

Hence, the limiting function $f(x)$ is equal to unity for $x = 0$ and to zero for $x \neq 0$ so that it is not continuous at $x = 0$. Each function (4) is, however, continuous in I. Thus (according to Theorem 2), the sequence (4) cannot be uniformly convergent in I; this fact is also apparent from Fig. 15.1. However, it can be shown that (4) is uniformly convergent in any closed interval which does not contain the point $x = 0$.

Definition 3. The sequence (1) will be called *uniformly bounded in I* if a constant $M > 0$ exists such that

$$|f_n(x)| \leq M \quad \text{for every } n \text{ and every } x \in I . \tag{5}$$

(For example, the terms of the sequence (4) are uniformly bounded by the constant $M = 1$.)

Definition 4. The terms $f_n(x)$ of the sequence (1) are called *equicontinuous in I* if for every $\varepsilon > 0$ a positive constant δ can be found such that

$$|f_n(x_1) - f_n(x_2)| < \varepsilon \tag{6}$$

for every n and every pair of points $x_1, x_2 \in I$ such that $|x_2 - x_1| < \delta$.

Theorem 3 (*Arzelà's Theorem*)*. *Let (1) be a uniformly bounded sequence of equicontinuous functions in I. Then we can choose from (1) a sequence of functions*

$$f_{k_1}(x), f_{k_2}(x), f_{k_3}(x), \ldots \quad (k_1 < k_2 < k_3 < \ldots) \tag{7}$$

which converges uniformly in I.

REMARK 3. The functions (4) are uniformly bounded in the interval $(-\infty, \infty)$, but they are not equicontinuous in this interval; the latter fact follows easily from Theorems 2, 3 and Example 1 and is also geometrically obvious, since as n increases these functions become "steeper and steeper" in the neighbourhood of the origin.

Theorem 4 (*Interchange of Limiting Processes*). *Let the sequence (1) converge uniformly in the interval $(a, a + \delta)$ $(\delta > 0)$ to the function $f(x)$. Let each of the functions $f_n(x)$ tend to a finite limit,*

$$\lim_{x \to a+} f_n(x) = c_n\,. \tag{8}$$

Then there exist finite limits

$$\lim_{n \to \infty} c_n\,, \quad \lim_{x \to a+} f(x)$$

and these are equal, i.e.

$$\lim_{n \to \infty} \lim_{x \to a+} f_n(x) = \lim_{x \to a+} \lim_{n \to \infty} f_n(x)\,. \tag{9}$$

Under similar assumptions, Theorem 4 is also true for $x \to a-$ and for $x \to a$.

Theorem 5 (*Limiting Process under the Integral Sign*). *Let the sequence (1) converge uniformly in $[a, b]$ and let the functions $f(t)$, $f_n(t)$ be integrable in $[a, b]$. Then*

$$\int_a^x f(t)\,dt = \lim_{n \to \infty} \int_a^x f_n(t)\,dt \quad \textit{for every} \quad x \in [a, b]\,, \tag{10}$$

i.e.

$$\int_a^x \lim_{n \to \infty} f_n(t)\,dt = \lim_{n \to \infty} \int_a^x f_n(t)\,dt\,. \tag{11}$$

* Otherwise known as Ascoli's Theorem.

Moreover, the sequence of functions

$$F_n(x) = \int_a^x f_n(t)\,dt$$

is also uniformly convergent in $[a, b]$. (Cf. also Theorem 7.)

Theorem 6 (*Limiting Process under the Differentiation Sign*). *Let* (1) *be convergent at least at one point of the interval* $[a, b]$. *Further, let the functions* (1) *have (finite) derivatives in* $[a, b]$. (*At the points a and b, the derivative from the right or from the left are understood respectively*). *Let the sequence*

$$f_1'(x), f_2'(x), f_3'(x), \ldots \tag{12}$$

be uniformly convergent in $[a, b]$. *Then the sequence* (1) *is also uniformly convergent in* $[a, b]$ *and its limiting function* $f(x)$ (Definition 1) *has a derivative in* $[a, b]$; *moreover, the following relation holds:*

$$f'(x) = \lim_{n\to\infty} f_n'(x)\,, \tag{13}$$

i.e.

$$\frac{d}{dx}\left[\lim_{n\to\infty} f_n(x)\right] = \lim_{n\to\infty} \frac{df_n(x)}{dx}\,. \tag{14}$$

REMARK 4. Note that the uniform convergence assumed here is that of sequence (12) and not of sequence (1), i.e. it is the *derived* sequence that is assumed to be uniformly convergent.

REMARK 5. If the assumption of uniform convergence is not satisfied in Theorems 5 and 6, then equations (10), (11) and (13), (14) need not hold.

Example 2. For the sequence

$$f_n(x) = n^2 x e^{-nx} \tag{15}$$

we have, in the interval $[0, 1]$,

$$f(x) = \lim_{n\to\infty} f_n(x) = 0\,. \tag{16}$$

Furthermore

$$\int_0^1 f_n(x)\,dx = \int_0^1 n^2 x e^{-nx} = [-nxe^{-nx}]_0^1 + n\int_0^1 e^{-nx}\,dx = -ne^{-n} - e^{-n} + 1\,.$$

Hence

$$\lim_{n\to\infty} \int_0^1 f_n(x)\,dx = 1\,. \tag{17}$$

However, $f(x) \equiv 0$ in $[0, 1]$ so that

$$\int_0^1 f(x)\,dx = 0 \tag{18}$$

and (10) is not true. (By Theorem 5, the sequence (15) does not converge uniformly in $[0, 1]$; this fact may, of course, be demonstrated by other methods.)

REMARK 6. The condition of uniform convergence may be replaced by other conditions; for example, we have:

Theorem 7. *Let the functions* (1) *together with the limiting function* $f(x)$ *be integrable in* $[a, b]$ *and let* $|f_n(x)| < M$ *in* $[a, b]$ (i.e. the *functions* $f_n(x)$ *are uniformly bounded in* $[a, b]$, see Definition 3). *Then* (10) *and* (11) *hold.* (Cf. also Theorem 13.14.4, p. 598.)

REMARK 7. Theorem 7 cannot be applied to the sequence (15), since (15) is not uniformly bounded; for example, for $x = 1/n$ we have

$$f_n\left(\frac{1}{n}\right) = n^2 \frac{1}{n} e^{-n/n} = ne^{-1}.$$

15.2. Series with Variable Terms. Uniform Convergence. Integration and Differentiation of Series with Variable Terms

Definition 1. Let a sequence of functions

$$f_1(x), f_2(x), f_3(x), \ldots \tag{1}$$

in an interval I be given, and for each $x_0 \in I$ let the series of numbers

$$f_1(x_0) + f_2(x_0) + f_3(x_0) + \ldots = S(x_0) \tag{2}$$

be convergent. Then the series

$$f_1(x) + f_2(x) + f_3(x) + \ldots \tag{3}$$

is said to be *convergent in* I and to *have the sum* $S(x)$ *in* I.

REMARK 1. For a fixed $x_0 \in I$ (2) is a series of numbers (since $f_1(x_0), f_2(x_0), \ldots$ are numbers). As usual, its *partial sum* $s_n(x_0)$ is defined by the equation

$$s_n(x_0) = f_1(x_0) + f_2(x_0) + \ldots + f_n(x_0).$$

Thus, for any $x_0 \in I$ the relation

$$S(x_0) = \lim_{n\to\infty} s_n(x_0) \tag{4}$$

holds.

The convergence of the series (3) may also be defined as follows: We construct the sequence of partial sums

$$s_1(x) = f_1(x),\ \ s_2(x) = f_1(x) + f_2(x),\ \ \dots,\ \ s_n(x) = f_1(x) + f_2(x) + \dots + f_n(x),\ \ \dots .$$

If the sequence of functions $s_n(x)$ is convergent in I (see § 15.1) and has the limit $S(x)$, then the series (3) is said to be *convergent in I* and to *have the sum* $S(x)$. Obviously, the two definitions are equivalent.

Example 1. Let I be the open interval $(-1, 1)$. For every $x \in I$ we have

$$1 + x + x^2 + x^3 + \dots = \frac{1}{1-x}. \tag{5}$$

Hence,

$$S(x) = \frac{1}{1-x}.$$

Theorem 1. *If the series*

$$|f_1(x)| + |f_2(x)| + |f_3(x)| + \dots \tag{6}$$

is convergent in I, then the series (3) *is also convergent in I.* (If (6) converges, then the series (3) is said to be *absolutely convergent in I.*)

Definition 2. The series (3) is called *uniformly convergent in I*, if the sequence of partial sums $s_n(x)$,

$$s_n(x) = f_1(x) + f_2(x) + \dots + f_n(x), \tag{7}$$

is uniformly convergent in I (cf. Definition 15.1.2).

Theorem 2 (*The Bolzano–Cauchy Condition of Convergence*). *The series* (3) *converges uniformly in I, if and only if corresponding to every $\varepsilon > 0$ a positive integer n_0 (independent of x) can be found such that*

$$|f_n(x) + f_{n+1}(x) + \dots + f_{n+p}(x)| < \varepsilon$$

for every $n > n_0$ and every $x \in I$, p being an arbitrary positive integer.

Theorem 3 (*Weierstrass's M-Test*). *If*

$$|f_1(x)| \leqq A_1,\ |f_2(x)| \leqq A_2,\ |f_3(x)| \leqq A_3,\ \dots$$

for every $x \in I$ and if the series of numbers

$$A_1 + A_2 + A_3 + \ldots \tag{8}$$

is convergent, then the series

$$f_1(x) + f_2(x) + f_3(x) + \ldots$$

is uniformly convergent in I. The series (8) is called a *majorant of the series* (3).

Example 2. The geometrical series (5) converges uniformly, e.g. in the interval $[-0{\cdot}9; 0{\cdot}9]$. As the majorant series (8), we may obviously take the series

$$1 + 0{\cdot}9 + 0{\cdot}9^2 + \ldots .$$

On the other hand, it can be shown that the series (5) does *not* converge uniformly in $(-1, 1)$.

Example 3. The series

$$\zeta(x) = \frac{1}{1^x} + \frac{1}{2^x} + \ldots + \frac{1}{n^x} + \ldots$$

(the so-called Riemann *zeta-function*, frequently used in the theory of numbers) converges uniformly in the interval $[a, \infty)$, where a is any number greater than 1. In fact, for every $x \geqq a$ we have $1/n^x \leqq 1/n^a$, so that the series

$$\frac{1}{1^a} + \frac{1}{2^a} + \ldots + \frac{1}{n^a} + \ldots$$

(convergent according to Example 10.2.7, p. 386) is a majorant of the given series. (The given series, however, does not converge uniformly in the interval $(1, \infty)$.)

Theorem 4. *If the series (3) converges uniformly in I and the functions $f_1(x)$, $f_2(x)$, ... are continuous in I, then $S(x)$ is also a continuous function in I.*

Theorem 5. *Let the functions (1) be continuous in I and let $f_n(x) \geqq 0$ for every n and $x \in I$; then $S(x)$ is continuous in I if and only if the series (3) converges uniformly in I. (Thus, if $f_n(x) \geqq 0$, then the continuity of $S(x)$ implies the uniform convergence of the series (3).)*

Theorem 6 (*Theorem on Integration of Series with Variable Terms*). *Let*

$$f_1(x) + f_2(x) + f_3(x) + \ldots = S(x) \tag{9}$$

be a series of integrable functions in $[a, b]$, which converges uniformly in $[a, b]$; then the series

$$F_1(x) + F_2(x) + F_3(x) + \ldots \tag{10}$$

with

$$F_n(x) = \int_a^x f_n(t)\,\mathrm{d}t \quad (x \in [a, b])$$

also converges uniformly in $[a, b]$ *and has the sum*

$$\int_a^x S(t)\,\mathrm{d}t\,,$$

i.e.

$$\sum_{n=1}^{\infty} \int_a^x f_n(t)\,\mathrm{d}t = \int_a^x \left[\sum_{n=1}^{\infty} f_n(t)\right] \mathrm{d}t\,. \tag{11}$$

Theorem 7 (*Theorem on Differentiation of Series with Variable Terms*). *Let the series*

$$f_1'(x) + f_2'(x) + f_3'(x) + \ldots = f(x) \tag{12}$$

converge uniformly in $[a, b]$ *and let the series*

$$f_1(x) + f_2(x) + f_3(x) + \ldots \tag{13}$$

converge for at least one $x_0 \in [a, b]$; *then* (13) *converges uniformly in* $[a, b]$. *Denoting the sum of the series* (13) *by* $S(x)$, *we have*

$$S'(x) = f(x)\,,$$

i.e.

$$\frac{\mathrm{d}}{\mathrm{d}x} \sum_{n=1}^{\infty} f_n(x) = \sum_{n=1}^{\infty} \frac{\mathrm{d}}{\mathrm{d}x} f_n(x) \tag{14}$$

(cf. also Theorem 8).

REMARK 2. Observe that the uniform convergence of the *derived* series (12), not that of the series (13), is required.

REMARK 3. In Theorems 6 and 7, the condition of uniform convergence, which is rather strong, may often be replaced by a weaker condition, as for example:

Theorem 8. *Let the functions* (1) *together with the sum* $S(x)$ *of the series* (3) *be integrable functions in* $[a, b]$. *Moreover, let the sequence of partial sums* $s_n(x)$,

$$s_n(x) = f_1(x) + f_2(x) + \ldots + f_n(x)$$

be uniformly bounded in $[a, b]$, *i.e. let*

$$|s_n(x)| < M$$

for every n and every $x \in [a, b]$ (*M being a constant*). *Then* (11) *holds, i.e. the given series can be integrated term by term.*

REMARK 4. Sometimes it is convenient to define the sum of a series in another way than as a limit of partial sums. Such series are then called *summable*, in a certain sense (see Remark 10.2.16, p. 392, where summability in the sense of Cesàro was considered). Some series may be summable while they are divergent in the usual sense. This concept of summability is of considerable importance, particularly in the theory of Fourier series.

Example 4. The series

$$a_1 + a_2 + a_3 + \dots \tag{15}$$

is called *summable in the sense* of Euler (or of Abel), if a finite limit

$$\lim_{x \to 1-} \sum_{n=1}^{\infty} a_n x^n \tag{16}$$

exists. This limit is called the *sum of the series* (15) *in the sense of Euler.*

The series

$$1 - 1 + 1 - 1 + 1 - \dots \tag{17}$$

has the sum $1/2$ in the sense of Euler, because

$$x - x^2 + x^3 - x^4 + x^5 - \dots = \frac{x}{1+x} \tag{18}$$

and

$$\lim_{x \to 1-} \frac{x}{1+x} = \frac{1}{2}.$$

In the usual sense, however, the series (17) is divergent. According to Abel's theorem (Theorem 15.3.4), any series which is convergent in the usual sense is summable in the sense of Euler (and has the same sum).

15.3. Power Series

REMARK 1. A *power series* is a particular case of a series with variable terms; it is a series of the form

$$a_0 + a_1(x - x_0) + a_2(x - x_0)^2 + \dots, \tag{1}$$

where the a_n are constants (generally complex).

For $x_0 = 0$ the power series assumes the form

$$a_0 + a_1 x + a_2 x^2 + \dots. \tag{2}$$

Theorem 1. *For any series* (1) *a number* $r \geqq 0$ (*the so-called radius of convergence; the possibility* $r = +\infty$ *is not excluded*) *exists such that* (1) *converges absolutely for each* $x \in (x_0 - r, x_0 + r)$, *and diverges for all* x *lying outside the interval* $[x_0 - r, x_0 + r]$. (*By the notation* $r = +\infty$ *we mean that the series converges for all finite* x.) *Moreover, if* q *is an arbitrary number such that* $0 < q < r$, *then* (1) *converges uniformly in the interval* $[x_0 - q, x_0 + q]$.

REMARK 2. In geometrical terms, (1) converges in an interval which is symmetric about the point x_0. The series (1) is therefore sometimes called a *series with the centre at the point* x_0. For the same reason, (2) is called a *series with the centre at the origin.*

REMARK 3. At the end points of the interval of convergence, i.e. at the points $x_0 - r$, $x_0 + r$, the series may be either convergent or divergent. For example, the series

$$1 + x + \frac{x^2}{2} + \frac{x^3}{3} + \dots \quad (r = 1)$$

is divergent at $x = 1$, but convergent at $x = -1$ (Example 10.2.4, p. 383).

REMARK 4. The radius of convergence may also be zero. For example, it can be shown that the series

$$1 + 1!\,x + 2!\,x^2 + 3!\,x^3 + \dots$$

converges only for $x = 0$ (see also Theorem 2).

Theorem 2. *Let either the limit*

$$\lim_{n\to\infty} \frac{|a_{n+1}|}{|a_n|} = l \quad \textit{or the limit} \quad \lim_{n\to\infty} \sqrt[n]{|a|} = l \tag{3}$$

exist. Then

$$r = \frac{1}{l}. \tag{4}$$

(If $l = 0$, then $r = +\infty$; if $l = +\infty$, then $r = 0$.)

Example 1. For the series

$$1 + \frac{x}{1!} + \frac{x^2}{2!} + \frac{x^3}{3!} + \dots$$

we have

$$\lim_{n\to\infty} \frac{|a_{n+1}|}{|a_n|} = \lim_{n\to\infty} \frac{\dfrac{1}{(n+1)!}}{\dfrac{1}{n!}} = \lim_{n\to\infty} \frac{1}{n+1} = 0\,.$$

Hence, $r = +\infty$. The series in question converges for all finite x.

REMARK 5. If the limits (3) do not exist, Theorem 2 remains true provided the limits (3) are replaced by the limits (see Theorem 10.1.8, p. 378)

$$\overline{\lim_{n\to\infty}}\left|\frac{a_{n+1}}{a_n}\right|, \quad \overline{\lim_{n\to\infty}}\sqrt[n]{|a_n|}\,.$$

Theorem 3. *The series*

$$a_0 + a_1(x - x_0) + a_2(x - x_0)^2 + \ldots, \tag{5}$$

$$|a_0| + |a_1|\,|x - x_0| + |a_2|\,|x - x_0|^2 + \ldots \tag{6}$$

have the same radius of convergence. (In more detail: A single r exists such that both the series (5), (6) are convergent for $|x - x_0| < r$ and divergent for $|x - x_0| > r$.)

REMARK 6. Theorem 3 is very significant from the practical point of view, since the series (6) has positive (or non-negative) terms and its convergence may be tested using the criteria of § 10.2, which are valid for series with positive (or non-negative) terms. Theorem 2 is a consequence (resulting from Theorem 10.2.13, p. 385) of Theorem 3.

Theorem 4 (*Abel's Theorem*). *If the series* (1) *converges for* $x = x_0 + r$ (*i.e. at the right-hand end point of the convergence interval, assuming that* $0 < r < +\infty$), *then its sum* $S(x)$ *possesses a limit from the left at the point* $x_0 + r$, *and this limit coincides with the sum of the series* (1) *for* $x = x_0 + r$. *A similar assertion is true for* $x = x_0 - r$. (Cf. Examples 15.4.1, 15.4.2.)

Theorem 5 (*Arithmetic Operations with Power Series*). *Let the series*

$$S_1(x) = a_0 + a_1(x - x_0) + a_2(x - x_0)^2 + \ldots, \tag{7}$$

$$S_2(x) = b_0 + b_1(x - x_0) + b_2(x - x_0)^2 + \ldots \tag{8}$$

have radii of convergence r_1 *and* r_2, *respectively. Let us use the notation* $r = \min(r_1, r_2)$. *Then in the interval* $(x_0 - r, x_0 + r)$ *we have*

$$S_1(x) \pm S_2(x) =$$
$$= (a_0 \pm b_0) + (a_1 \pm b_1)(x - x_0) + (a_2 \pm b_2)(x - x_0)^2 + \ldots, \tag{9}$$

$$S_1(x)\,S_2(x) =$$
$$= a_0b_0 + (a_0b_1 + a_1b_0)(x - x_0) + (a_0b_2 + a_1b_1 + a_2b_0)(x - x_0)^2 + \ldots, \tag{10}$$

i.e. series (7) *and* (8) *may be added, subtracted and multiplied within their common region of convergence* (cf. Theorems 10.2.3 and 10.2.22). See also formula 27, § 15.6. [The series (10) is known as the *Cauchy product* of the series (7), (8).]

Theorem 6 (*Inversion of a Series*). *Let the series*

$$S(x) = a_0 + a_1(x - x_0) + a_2(x - x_0)^2 + \ldots \tag{11}$$

have a non-zero radius of convergence and let $a_0 \neq 0$. *Then in a certain neighbourhood of the point* x_0 *the function* $1/S(x)$ *can be expanded in a power series*

$$\frac{1}{S(x)} = c_0 + c_1(x - x_0) + c_2(x - x_0)^2 + \ldots . \tag{12}$$

REMARK 7. The coefficients c_n of series (12) may be found, for example, by the method of undetermined coefficients. Multiplying (12) by (11), we get

$$1 = [a_0 + a_1(x - x_0) + \ldots] [c_0 + c_1(x - x_0) + \ldots] =$$

$$= a_0c_0 + (a_0c_1 + a_1c_0)(x - x_0) + \ldots , \tag{13}$$

whence we have

$$1 = a_0c_0 , \quad 0 = a_0c_1 + a_1c_0 \quad \text{etc.} \tag{14}$$

Since $a_0 \neq 0$, we can successively determine $c_0, c_1, \ldots$ from these equations. Cf. also formula 28, § 15.6.

Example 2. We have (see formula 13, § 15.6)

$$\cos x = 1 - \frac{x^2}{2!} + \frac{x^4}{4!} - \frac{x^6}{6!} + \ldots \quad (r = +\infty) . \tag{15}$$

Thus, in a neighbourhood of the point $x_0 = 0$,

$$\frac{1}{\cos x} = c_0 + c_1x + c_2x^2 + c_3x^3 + \ldots . \tag{16}$$

By (13) we have

$$1 = \left(1 - \frac{x^2}{2} + \frac{x^4}{24} - \ldots\right)(c_0 + c_1x + c_2x^2 + c_3x^3 + \ldots) =$$

$$= c_0 + c_1x + \left(c_2 - \frac{c_0}{2}\right)x^2 + \left(c_3 - \frac{c_1}{2}\right)x^3 + \ldots ,$$

so that

$$c_0 = 1, c_1 = 0, c_2 = \tfrac{1}{2}, c_3 = 0, \ldots .$$

(Cf. also formula 28, § 15.6).

Theorem 7. *Let* (7) *and* (8) *be power series with* $r_1 > 0$, $r_2 > 0$, $b_0 \neq 0$. *Then in a certain neighbourhood of the point* x_0 *the function* $S_1(x)/S_2(x)$ *can be expanded in a power series*

$$\frac{S_1(x)}{S_2(x)} = d_0 + d_1(x - x_0) + d_2(x - x_0)^2 + \ldots . \tag{17}$$

REMARK 8. The series (17) may be found, for example, in such a way that the power series for $1/S_2(x)$ is established first by Theorem 6 and Remark 7, and this series is then multiplied by the series for $S_1(x)$. Alternatively, we multiply equation (17) by the series $S_2(x)$ and then use the method of undetermined coefficients in the same way as in Remark 7.

Theorem 8 (*Substituting a Power Series into a Power Series*). *Let the series*

$$S(x) = a_0 + a_1(x - x_0) + a_2(x - x_0)^2 + \ldots, \tag{18}$$

and

$$x - x_0 = g(t) = b_1(t - t_0) + b_2(t - t_0)^2 + \ldots \tag{19}$$

have non-zero radii of convergence. Substituting formally the power series (19) for $x - x_0$ into (18) and arranging the result by powers of $t - t_0$, we get a power series in $t - t_0$. This series converges in a certain neighbourhood of the point t_0 and, in this neighbourhood, its sum is equal to the function $S(x_0 + g(t))$.

15.4. Theorems on Differentiation and Integration of Power Series. Power Series in Two or More Variables

Theorem 1. *Let the series*

$$a_0 + a_1(x - x_0) + a_2(x - x_0)^2 + \ldots = S(x) \tag{1}$$

have radius of convergence r. Then the series

$$a_1 + 2a_2(x - x_0) + 3a_3(x - x_0)^2 + \ldots, \tag{2}$$

obtained from (1) by term-by-term differentiation, also has the same convergence radius r, and its sum in $(x_0 - r, x_0 + r)$ is equal to $S'(x)$ (i.e. to the derivative of the function $S(x)$).

REMARK 1. Since the series (2) is again a power series with radius of convergence r, the series obtained from (2) by term-by-term differentiation again has radius of convergence r and defines the function $S''(x)$ in $(x_0 - r, x_0 + r)$, etc. Hence, the function $S(x)$ defined by the series (1) with radius of convergence r possesses derivatives of all orders in the interval $(x_0 - r, x_0 + r)$.

A function which may be expanded in a power series (1) in an interval $(x_0 - r, x_0 + r)$, is said to be *analytic* in this interval. Thus, every such function possesses derivatives of all orders in $(x_0 - r, x_0 + r)$.

Theorem 2. *Let the series*

$$a_0 + a_1(x - x_0) + a_2(x - x_0)^2 + \ldots = S(x) \tag{3}$$

have radius of convergence r. Then the series

$$a_0(x - x_0) + \frac{a_1}{2}(x - x_0)^2 + \frac{a_2}{3}(x - x_0)^3 + \ldots, \tag{4}$$

obtained from (3) *by term-by-term integration, also has the same radius of convergence r and defines, in* $(x_0 - r, x_0 + r)$, *the function*

$$\int_{x_0}^{x} S(t)\,dt, \tag{5}$$

which is a primitive of $S(x)$.

REMARK 2. In (5) we choose x_0 as the lower limit of integration, since the series (4) has the sum zero for $x = x_0$, and consequently the function defined by this series must vanish for $x = x_0$. If we choose the primitive arbitrarily (i.e. not necessarily equal to zero at the point $x = x_0$), we have to add a constant of integration to the series (4), in general.

Example 1. Let us express the function

$$S(x) = \arctan x$$

by a power series in a neighbourhood of the origin (i.e. by a power series with the centre at the origin).

For the derivative of this function we have

$$\frac{1}{1 + x^2} = 1 - x^2 + x^4 - x^6 + \ldots \quad (r = 1) \tag{6}$$

since the right-hand side is a geometric series with common ratio $-x^2$. By Theorem 2 (note that $\arctan 0 = 0$)

$$\arctan x = x - \frac{x^3}{3} + \frac{x^5}{5} - \frac{x^7}{7} + \ldots \quad (r = 1). \tag{7}$$

The series (7) converges even for $x = \pm 1$ (it is an alternating series, Theorem 10.2.16, p. 388) so that by Abel's theorem (Theorem 15.3.4) the equality (7) is valid also for those values of x. In particular, putting $x = 1$ we get the relationship

$$\frac{\pi}{4} = 1 - \tfrac{1}{3} + \tfrac{1}{5} - \tfrac{1}{7} + \ldots.$$

Example 2. Let us find the sum of the series

$$1 - \tfrac{1}{2} + \tfrac{1}{3} - \tfrac{1}{4} + \ldots. \tag{8}$$

We have

$$1 - x + x^2 - x^3 + \ldots = \frac{1}{1 + x} \quad (r = 1).$$

According to Theorem 2,

$$x - \frac{x^2}{2} + \frac{x^3}{3} - \frac{x^4}{4} + \ldots = \ln(1 + x) \quad (r = 1). \tag{9}$$

The series (9) converges for $x = 1$ (it is an alternating series) so that by Abel's theorem (Theorem 15.3.4) we have

$$1 - \tfrac{1}{2} + \tfrac{1}{3} - \tfrac{1}{4} + \ldots = \ln 2.$$

These methods are often used for finding sums of infinite series.

Theorem 3 (*Power Series in Two Variables*). *Let the double series*

$$S(x, y) = \sum_{m,n=0}^{\infty} a_{mn} x^m y^n \tag{10}$$

be absolutely convergent at a point (x_0, y_0) (cf. Remark 10.2.11, p. 389). *Then* (10) *converges absolutely at any point* (x, y), *whose coordinates satisfy the inequalities*

$$|x| \leqq |x_0|, \quad |y| \leqq |y_0|$$

(*i.e. in a rectangle* $\bar{O}$: $-|x_0| \leqq x \leqq |x_0|$, $-|y_0| \leqq y \leqq |y_0|$). *Moreover,* $S(x, y)$ *is continuous in* $\bar{O}$ *and possesses partial derivatives of all orders inside the rectangle in question. These derivatives may be obtained by term-by-term differentiation of the series* (10). For example,

$$\frac{\partial S}{\partial x} = \sum_{m,n=0}^{\infty} m a_{mn} x^{m-1} y^n = \sum_{m \geqq 1, n \geqq 0} m a_{mn} x^{m-1} y^n. \tag{11}$$

REMARK 3. An analogous theorem is true for power series

$$\sum_{m,n=0} a_{mn}(x - a)^m (y - b)^n, \tag{12}$$

which may, of course, be reduced to the form (10) by the substitution $x - a = \bar{x}$, $y - b = \bar{y}$.

REMARK 4. An analogous theorem is valid for power series in several variables.

REMARK 5. In contrast to power series in a single complex variable, the regions of *absolute* convergence of double series may be of different shapes.

REMARK 6. For double power series (or for power series in several variables) a theorem on addition, subtraction and multiplication of series, similar to Theorem 15.3.5, is true in the region of absolute convergence.

REMARK 7. In Theorem 3, the *absolute* convergence of the series (10) in the sense of Remark 10.2.11, p. 389, is under consideration. If the series (10) converges only in the usual sense (see (14)), then the assertion of Theorem 3 need not hold in general.

We draw the reader's attention to the fact that in technical literature, unless otherwise stated, we always understand by the convergence of the series

$$\sum_{m,n=0}^{\infty} a_{mn} x^m y^n \tag{13}$$

ordinary convergence in the sense of Remark 10.2.12, p. 390, i.e.: the series (13) converges at a point (x_0, y_0) and has the sum $S(x_0, y_0)$, if to each $\varepsilon > 0$ a positive integer P exists such that for any pair of integers M, N with $M > P, N > P$, we have

$$\left| \sum_{m=0}^{M} \sum_{n=0}^{N} a_{mn} x_0^m y_0^n - S(x_0, y_0) \right| < \varepsilon . \tag{14}$$

We interpret similarly the convergence of other double series than power series, e.g. of a Fourier series

$$\sum_{m,n=1}^{\infty} a_{mn} \sin mx \sin ny . \tag{15}$$

(For details as to (15) see Theorem 16.3.5, p. 712.) The same is true for series in several variables.

15.5. Taylor's Series. The Binomial Series

REMARK 1. According to Taylor's theorem we have

$$f(x) = f(a) + \frac{f'(a)}{1!}(x - a) + \frac{f''(a)}{2!}(x - a)^2 + \ldots + \frac{f^{(n)}(a)}{n!}(x - a)^n + R_{n+1}(x) ;$$

the most frequently used forms of the remainder R_{n+1} are given in Theorem 11.10.1, p. 434.

Theorem 1. *If $f(x)$ has a derivative of any order in the interval $[a, x]$ (or $[x, a]$ when $x < a$), then a necessary and sufficient condition for the series*

$$f(a) + \frac{f'(a)}{1!}(x - a) + \frac{f''(a)}{2!}(x - a)^2 + \ldots \tag{1}$$

to converge at given point x and to have the sum $f(x)$ is

$$\lim_{n\to\infty} R_{n+1}(x) = 0 . \tag{2}$$

REMARK 2. For $a = 0$, the series (1) is called a *Maclaurin series:*

$$f(x) = f(0) + \frac{f'(0)}{1!}x + \frac{f''(0)}{2!}x^2 + \ldots .$$

REMARK 3. It can happen that the series (1) converges on a certain set, but that

$$\lim_{n\to\infty} R_{n+1}(x) \neq 0\,;$$

then, of course, its sum is not equal to $f(x)$.

Example 1. By Taylor's formula, we have

$$e^x = 1 + \frac{x}{1!} + \frac{x^2}{2!} + \ldots + \frac{x^n}{n!} + R_{n+1}(x)\,,$$

where

$$R_{n+1}(x) = \frac{e^{\vartheta x}}{(n+1)!}\,x^{n+1} \quad (0 < \vartheta < 1)\,.$$

For any fixed x we have $|e^{\vartheta x}| < M$, where $M = \max(1, e^x)$ and

$$\lim_{n\to\infty} \frac{x^{n+1}}{(n+1)!} = 0\,.$$

Consequently, by Theorem 1 we have, for every x,

$$e^x = 1 + \frac{x}{1!} + \frac{x^2}{2!} + \ldots \quad (r = +\infty)\,.$$

Theorem 2. *If $f(x)$ has derivatives of all orders in the interval $[a, c]$ (or $[c, a]$ when $c < a$) and if these derivatives are uniformly bounded (i.e. a constant M exists such that*

$$|f'(x)| \leqq M\,, \quad |f''(x)| \leqq M\,, \quad |f'''(x)| \leqq M\,, \quad \ldots \text{ in } [a, c] \text{ or } [c, a])\,,$$

then

$$f(x) = f(a) + \frac{f'(a)}{1!}(x-a) + \frac{f''(a)}{2!}(x-a)^2 + \ldots$$

in $[a, c]$ (or $[c, a]$).

Theorem 3 (*The Binomial Series*). *For any real n we have*

$$(1+x)^n = 1 + \binom{n}{1}x + \binom{n}{2}x^2 + \binom{n}{3}x^3 + \ldots \quad (r = 1)\,, \tag{3}$$

where

$$\binom{n}{k} = \frac{n(n-1)\dots(n-k+1)}{k!}.$$

For example

$$\frac{1}{\sqrt{(1-x^2)}} = (1-x^2)^{-1/2} = 1 + \frac{x^2}{2} + \frac{3x^4}{8} + \dots \quad (r = 1), \tag{4}$$

since

$$\binom{-\frac{1}{2}}{1} = -\frac{1}{2}, \quad \binom{-\frac{1}{2}}{2} = \frac{(-\frac{1}{2})(-\frac{3}{2})}{2} = \frac{3}{8}, \quad \text{etc.}$$

Integrating the series (4) and making use of Theorem 15.4.2, we can get the power series for arcsin x (see formula 20, § 15.6).

REMARK 4. Formula (3) may also be used in cases where the first term of the binomial on the left-hand side is not necessarily equal to 1. For example, if $a > |b| > 0$, then

$$(a+b)^n = a^n\left(1 + \frac{b}{a}\right)^n = a^n(1+x)^n \quad (x = b/a).$$

15.6. Some Important Series, Particularly Power Series

(See also § 13.12).

1. $\dfrac{1}{1 \pm x} = 1 \mp x + x^2 \mp x^3 + x^4 \mp x^5 + \dots \qquad (-1 < x < 1).$

2. $\dfrac{1}{(1 \pm x)^2} = 1 \mp 2x + 3x^2 \mp 4x^3 + 5x^4 \mp 6x^5 + \dots \qquad (-1 < x < 1).$

3. $\sqrt{(1+x)} = 1 + \dfrac{1}{2}x - \dfrac{1}{2\,.\,4}x^2 + \dfrac{1\,.\,3}{2\,.\,4\,.\,6}x^3 - \dfrac{1\,.\,3\,.\,5}{2\,.\,4\,.\,6\,.\,8}x^4 + \dots$

$(-1 \leqq x \leqq 1).$

4. $\dfrac{1}{\sqrt{(1+x)}} = 1 - \dfrac{1}{2}x + \dfrac{1\,.\,3}{2\,.\,4}x^2 - \dfrac{1\,.\,3\,.\,5}{2\,.\,4\,.\,6}x^3 + \dfrac{1\,.\,3\,.\,5\,.\,7}{2\,.\,4\,.\,6\,.\,8}x^4 - \dots$

$(-1 < x \leqq 1).$

5. $e^x = 1 + \dfrac{x}{1!} + \dfrac{x^2}{2!} + \dfrac{x^3}{3!} + \dots \qquad (-\infty < x < +\infty).$

6. $e^{-x} = 1 - \frac{x}{1!} + \frac{x^2}{2!} - \frac{x^3}{3!} + \dots \qquad (-\infty < x < +\infty)$.

7. $a^x = 1 + \frac{\ln a}{1!} x + \frac{(\ln a)^2}{2!} x^2 + \frac{(\ln a)^3}{3!} x^3 + \dots \qquad (a > 0, -\infty < x < +\infty)$.

8. $\ln (1 + x) = x - \frac{x^2}{2} + \frac{x^3}{3} - \frac{x^4}{4} + \dots \qquad (-1 < x \leqq 1)$.

9. $\ln (1 - x) = -x - \frac{x^2}{2} - \frac{x^3}{3} - \frac{x^4}{4} - \dots \qquad (-1 \leqq x < 1)$.

10. $\ln \frac{1 + x}{1 - x} = 2 \left(x + \frac{x^3}{3} + \frac{x^5}{5} + \dots \right) \qquad (-1 < x < 1)$.

REMARK 1. This formula is used for the computation of logarithms if the preceding two formulae fail to be applicable (i.e. if $|x|$ is not sufficiently small, so that convergence is slow). For example, for ln 2 we have

$$\frac{1 + x}{1 - x} = 2, \quad x = \tfrac{1}{3}, \quad \ln 2 = 2 \left[\tfrac{1}{3} + \tfrac{1}{3} . (\tfrac{1}{3})^3 + \tfrac{1}{5} . (\tfrac{1}{3^5}) + \dots \right].$$

For $|x| > 1$ we then have

11. $\ln \frac{x + 1}{x - 1} = 2 \left(\frac{1}{x} + \frac{1}{3x^3} + \frac{1}{5x^5} + \dots \right) \qquad (|x| > 1)$.

REMARK 2. For $\ln x$, however, a power series with the centre at the origin does not exist, since $\ln x$ is not analytic in the neighbourhood of the origin. But we have

12. $\ln x = 2 \left[\frac{x - 1}{x + 1} + \frac{1}{3} \left(\frac{x - 1}{x + 1} \right)^3 + \frac{1}{5} \left(\frac{x - 1}{x + 1} \right)^5 + \dots \right] \qquad (x > 0)$.

Furthermore:

13. $\sin x = x - \frac{x^3}{3!} + \frac{x^5}{5!} - \frac{x^7}{7!} + \dots, \quad \cos x = 1 - \frac{x^2}{2!} + \frac{x^4}{4!} - \frac{x^6}{6!} + \dots$

$(-\infty < x < +\infty)$.

14. $\sin^2 x = x^2 (1 - \tfrac{1}{3} x^2 + \tfrac{2}{45} x^4 - \tfrac{1}{315} x^6 + \tfrac{2}{14,175} x^8 - \dots) \quad (-\infty < x < +\infty)$.

15. $\cos^2 x = 1 - x^2 + \tfrac{1}{3} x^4 - \tfrac{2}{45} x^6 + \tfrac{1}{315} x^8 - \tfrac{2}{14,175} x^{10} + \dots$

$(-\infty < x < +\infty)$.

16. $\sin^3 x = x^3 (1 - \tfrac{1}{2} x^2 + \tfrac{13}{120} x^4 - \tfrac{41}{3,024} x^6 + \tfrac{671}{604,800} x^8 - \dots)$

$(-\infty < x < +\infty)$.

17. $\cos^3 x = 1 - \frac{3}{2}x^2 + \frac{7}{8}x^4 - \frac{61}{240}x^6 + \frac{547}{13,440}x^8 - \frac{703}{172,800}x^{10} + \dots$

$(-\infty < x < +\infty)$.

18. $\tan x = x + \frac{1}{3}x^3 + \frac{2}{15}x^5 + \frac{17}{315}x^7 + \dots \qquad (-\frac{1}{2}\pi < x < \frac{1}{2}\pi)$.

19. $\cot x = \dfrac{1}{x} - \dfrac{x}{3} - \dfrac{x^3}{45} - \dfrac{2}{945}x^5 - \dots \qquad (0 < |x| < \pi)$.

20. $\arcsin x = x + \dfrac{1}{2}\cdot\dfrac{x^3}{3} + \dfrac{1\,.\,3}{2^2\,.\,2!}\cdot\dfrac{x^5}{5} + \dfrac{1\,.\,3\,.\,5}{2^3\,.\,3!}\cdot\dfrac{x^7}{7} + \dots \qquad (-1 \leqq x \leqq 1)$.

21. $\arctan x = x - \dfrac{x^3}{3} + \dfrac{x^5}{5} - \dfrac{x^7}{7} + \dots \qquad (-1 \leqq x \leqq 1)$.

22. $\sinh x = \dfrac{x}{1!} + \dfrac{x^3}{3!} + \dfrac{x^5}{5!} + \dfrac{x^7}{7!} + \dots, \quad \cosh x = 1 + \dfrac{x^2}{2!} + \dfrac{x^4}{4!} + \dfrac{x^6}{6!} + \dots$

$(-\infty < x < +\infty)$.

23. $\tanh x = x - \frac{1}{3}x^3 + \frac{2}{15}x^5 - \frac{17}{315}x^7 + \dots \qquad (-\frac{1}{2}\pi < x < \frac{1}{2}\pi)$.

24. $\sin x \sin y = xy[1 - \frac{1}{6}(x^2 + y^2) + \frac{1}{360}(3x^4 + 10x^2y^2 + 3y^4) -$
$- \frac{1}{5,040}(x^6 + 7x^4y^2 + 7x^2y^4 + y^6) +$
$+ \frac{1}{1,814,400}(5x^8 + 60x^6y^2 + 126x^4y^4 + 60x^2y^6 + 5y^8) - \dots]$

$(-\infty < x, y < +\infty)$.

25. $\cos x \cos y = 1 - \frac{1}{2}(x^2 + y^2) + \frac{1}{24}(x^4 + 6x^2y^2 + y^4) -$
$- \frac{1}{720}(x^6 + 15x^4y^2 + 15x^2y^4 + y^6) +$
$+ \frac{1}{40,320}(x^8 + 28x^6y^2 + 70x^4y^4 + 28x^2y^6 + y^8) - \dots$

$(-\infty < x,y < +\infty)$.

26. $\sin x \cos y = x[1 - \frac{1}{6}(x^2 + 3y^2) + \frac{1}{120}(x^4 + 10x^2y^2 + 5y^4) -$
$- \frac{1}{5,040}(x^6 + 21x^4x^2 + 35x^2y^4 + 7y^6) +$
$+ \frac{1}{362,880}(x^8 + 36x^6y^2 + 126x^4y^4 + 84x^2y^6 + 9y^8) + \dots]$

$(-\infty < x, y < +\infty)$.

REMARK 3. The series for $\sin^2 x$, $\cos^2 x$, $\sin^3 x$, $\cos^3 x$, $\sin x \sin y$, $\cos x \cos y$, $\sin x \cos y$ given above may be obtained by multiplying the series corresponding to the individual functions (see Theorem 15.3.5).

In general, if

$$S(x) = a + bx + cx^2 + dx^3 + ex^4 + fx^5 + \dots,$$

then:

27. $S^2(x) = a^2 + 2abx + (b^2 + 2ac)\,x^2 + 2(ad + bc)\,x^3 +$
$+ (c^2 + 2ae + 2bd)\,x^4 + 2(af + be + cd)\,x^5 + \dots$.

28. $$\frac{1}{S(x)} = \frac{1}{a}\left[1 - \frac{b}{a}x + \left(\frac{b^2}{a^2} - \frac{c}{a}\right)x^2 + \left(\frac{2bc}{a^2} - \frac{d}{a} - \frac{b^3}{a^3}\right)x^3 + \right.$$
$$\left. + \left(\frac{2bd}{a^2} + \frac{c^2}{a^2} - \frac{e}{a} - 3\frac{b^2c}{a^3} + \frac{b^4}{a^4}\right)x^4 + \dots\right] \qquad (a \neq 0).$$

29. $$\frac{1}{S^2(x)} = \frac{1}{a^2}\left[1 - 2\frac{b}{a}x + \left(3\frac{b^2}{a^2} - 2\frac{c}{a}\right)x^2 + \left(6\frac{bc}{a^2} - 2\frac{d}{a} - 4\frac{b^3}{a^3}\right)x^3 + \right.$$
$$\left. + \left(6\frac{bd}{a^2} + 3\frac{c^2}{a^2} - 2\frac{e}{a} - 12\frac{b^2c}{a^3} + 5\frac{b^4}{a^4}\right)x^4 + \dots\right] \qquad (a \neq 0).$$

30. $$\sqrt{[S(x)]} = \sqrt{(a)}\left[1 + \frac{1}{2}\frac{b}{a}x + \left(\frac{1}{2}\frac{c}{a} - \frac{1}{8}\frac{b^2}{a^2}\right)x^2 + \left(\frac{1}{2}\frac{d}{a} - \frac{1}{4}\frac{bc}{a^2} + \frac{1}{16}\frac{b^3}{a^3}\right)x^3 + \right.$$
$$\left. + \left(\frac{1}{2}\frac{e}{a} - \frac{1}{4}\frac{bd}{a^2} - \frac{1}{8}\frac{c^2}{a^2} + \frac{3}{16}\frac{b^2c}{a^3} - \frac{5}{128}\frac{b^4}{a^4}\right)x^4 + \dots\right] \quad (a > 0).$$

31. $$\frac{1}{\sqrt{[S(x)]}} = \frac{1}{\sqrt{(a)}}\left[1 - \frac{1}{2}\frac{b}{a}x + \left(\frac{3}{8}\frac{b^2}{a^2} - \frac{1}{2}\frac{c}{a}\right)x^2 + \left(\frac{3}{4}\frac{bc}{a^2} - \frac{1}{2}\frac{d}{a} - \frac{5}{16}\frac{b^3}{a^3}\right)x^3 + \right.$$
$$\left. + \left(\frac{3}{4}\frac{bd}{a^2} + \frac{3}{8}\frac{c^2}{a^2} - \frac{1}{2}\frac{e}{a} - \frac{15}{16}\frac{b^2c}{a^3} + \frac{35}{128}\frac{b^4}{a^4}\right)x^4 + \dots\right] \qquad (a > 0).$$

32. $$\frac{x}{e^x - 1} = D_0 + D_1\frac{x}{1!} + D_2\frac{x^2}{2!} + D_4\frac{x^4}{4!} + D_6\frac{x^6}{6!} + \dots \qquad (|x| < 2\pi).$$

33. $$x \cot x = D_0 - D_2\frac{(2x)^2}{2!} + D_4\frac{(2x)^4}{4!} - D_6\frac{(2x)^6}{6!} + \dots \qquad (|x| < \pi).$$

For $x = 0$ the right-hand sides of the last two equations are equal to the limits of the corresponding functions as $x \to 0$. The numbers D_n are defined by the recurrence formula

$$\binom{n+1}{1}D_n + \binom{n+1}{2}D_{n-1} + \dots + \binom{n+1}{n}D_1 + D_0 = 0\,, \quad D_0 = 1\,.$$

From this relation we have

$$D_1 = -\tfrac{1}{2},\ D_2 = \tfrac{1}{6},\ D_4 = -\tfrac{1}{30},\ D_6 = \tfrac{1}{42},\ D_8 = -\tfrac{1}{30},\ D_{10} = \tfrac{5}{66},\ \dots,$$
$$D_3 = D_5 = D_7 = \dots = 0\,.$$

15.7. Application of Series, Particularly of Power Series, to the Evaluation of Integrals. Asymptotic Expansions

On application to the solution of differential equations, see Chap. 25.

REMARK 1. One of the useful applications of series is to the approximate evaluation of integrals. The integrated function (whose indefinite integral cannot be expressed, for example, by elementary functions) is expanded in a series and integrated term by term according to Theorem 15.4.2, p. 678.

Example 1. Let us consider

$$\int_0^x e^{-t^2}\, dt . \tag{1}$$

By formula 6, § 15.6, we have

$$e^{-t^2} = 1 - \frac{t^2}{1!} + \frac{t^4}{2!} - \frac{t^6}{3!} + \ldots \quad (r = +\infty) . \tag{2}$$

According to Theorem 15.4.2, the series (2) can be integrated term by term over any interval (since $r = +\infty$). Thus,

$$\int_0^x e^{-t^2}\, dt = x - \frac{x^3}{3 \, . \, 1!} + \frac{x^5}{5 \, . \, 2!} - \frac{x^7}{7 \, . \, 3!} + \ldots \quad (r = +\infty) . \tag{3}$$

See also Remark 3 below.

Example 2. Let us consider

$$\int_0^{\pi/2} \frac{\sin x}{x}\, dx . \tag{4}$$

Referring to formula 13, § 15.6, we have

$$\frac{\sin x}{x} = 1 - \frac{x^2}{3!} + \frac{x^4}{5!} - \ldots \tag{5}$$

for every $x \neq 0$. If we define the function $\dfrac{\sin x}{x}$ as 1 at the point $x = 0$ (i.e. by its limit as $x \to 0$), then (5) will be true for all x. Since $r = +\infty$, the series (5) can by integrated term by term over any interval (Theorem 15.4.2). Hence,

$$\int_0^a \frac{\sin x}{x}\, dx = a - \frac{a^3}{3 \, . \, 3!} + \frac{a^5}{5 \, . \, 5!} - \ldots . \tag{6}$$

For $a = \pi/2$ it is sufficient to take the first five terms of (6) into account in order to ensure an accuracy of 5 decimal places (note that (6) is an alternating series, see Theorem 10.2.16, p. 388).

REMARK 2. If one of the limits of integration coincides with an end point of the interval of convergence, Abel's theorem can be used (Theorem 15.3.4, p. 676). See also § 13.12 (application of series to the evaluation of elliptic integrals).

REMARK 3 (Application of Divergent Series; Asymptotic Expansions). The series (3) is suitable for evaluating the integral

$$\int_0^x e^{-t^2}\,dt \tag{7}$$

provided $|x|$ is sufficiently small. If $|x|$ is large, the series (3) is also convergent, but it is evidently not suitable for evaluating the integral (7). In such cases *asymptotic* expansions can advantageously be used, as we proceed to show using the integral (7) as an example. Let us assume $x > 0$. We have

$$\int_0^\infty e^{-t^2}\,dt = \frac{\sqrt{\pi}}{2}, \quad \int_0^x e^{-t^2}\,dt = \int_0^\infty e^{-t^2}\,dt - \int_x^\infty e^{-t^2}\,dt\,. \tag{8}$$

Integrating by parts ($e^{-t^2} = 2te^{-t^2} \cdot (\frac{1}{2}/t)$, so $u' = 2te^{-t^2}$, $v = \frac{1}{2}/t$) we get

$$\int_x^\infty e^{-t^2}\,dt = \frac{e^{-x^2}}{2x} - \frac{1}{2}\int_x^\infty \frac{e^{-x^2}}{x^2}\,dx\,. \tag{9}$$

A repeated integration by parts yields, after n steps,

$$\int_x^\infty e^{-t^2}\,dt = \frac{e^{-x^2}}{2x}\left[1 - \frac{1}{2x^2} + \frac{1\,.\,3}{(2x^2)^2} - \ldots + (-1)^{n-1}\,\frac{1\,.\,3\,.\,5\,.\,\ldots\,.\,(2n-3)}{(2x^2)^{n-1}}\right] + r_n, \tag{10}$$

where

$$r_n = (-1)^n\,\frac{1\,.\,3\,.\,5\,.\,\ldots\,.\,(2n-1)}{2^n}\int_x^\infty \frac{e^{-t^2}}{t^{2n}}\,dt\,. \tag{11}$$

The series in brackets in (10) diverges for every x. But obviously

$$\int_x^\infty \frac{e^{-t^2}}{t^{2n}}\,dt < \frac{1}{x^{2n}}\int_x^\infty e^{-t^2}\,dt$$

and

$$\int_x^\infty e^{-t^2}\,dt < \frac{e^{-x^2}}{2x}$$

so that, by (9), we have

$$|r_n| < \frac{1\,.\,3\,.\,5\,.\,\ldots\,.\,(2n-1)}{2^{n+1}x^{2n+1}}\,e^{-x^2}\,. \tag{12}$$

If x is sufficiently large, then the remainder (12) can be made very small by a proper choice of n. (Divergent series with this property are sometimes called *semiconvergent*, but the more usual term is *asymptotic*.) For example, if $x = 5$, then for $n = 13$ we have $|r_n| < 10^{-20}$. Thus, taking $n = 13$ in (10), the integral

$$\int_5^\infty e^{-t^2}\,dt$$

can be evaluated with an accuracy of 20 decimal places. The former integral $\int_0^5 e^{-t^2}\,dt$ can then be evaluated by (8),

$$\int_0^5 e^{-t^2}\,dt = \frac{\sqrt{\pi}}{2} - \int_5^\infty e^{-t^2}\,dt\,.$$

REMARK 4. The logarithmic integral

$$\operatorname{li}(x) = \int_0^x \frac{dt}{\ln t} \quad (0 < x < 1)\,. \tag{13}$$

may be evaluated in a similar way. By the substitution $t = e^{-u}$, the integral (13) reduces to the integral

$$\operatorname{li}(x) = -\int_a^\infty \frac{e^{-u}}{u}\,du\,,$$

where

$$-\ln x = a \quad (a > 0)\,. \tag{14}$$

Integrating successively by parts, we get

$$\operatorname{li}(x) = -x\left[\frac{1}{a} - \frac{1!}{a^2} + \frac{2!}{a^3} - \dots + (-1)^{n-1}\frac{(n-1)!}{a^n}\right] + r_n\,, \tag{15}$$

where

$$r_n = (-1)^{n+1}\,n!\int_a^\infty \frac{e^{-t}}{t^{n+1}}\,dt\,, \tag{16}$$

so that

$$|r_n| < n!\,\frac{e^{-a}}{a^{n+1}}\,. \tag{17}$$

Thus, if x is sufficiently small and, consequently, a sufficiently large, $|r_n|$ may be made small by a suitable choice of n, and the series (15) can be employed for calculating $\operatorname{li}(x)$.

It is important to note that as $n \to \infty$, $|r_n| \to \infty$; generally, the value of $|r_n|$ starts by decreasing (as n increases) until it reaches a minimum value; thereafter, it increases and, indeed, becomes infinite. The best approximation is obtained, of course, by choosing n so that $|r_n|$ has its minimum value.

REMARK 5. A thorough treatment of asymptotic expansions may be found, e.g., in [173].

15.8. Survey of Some Important Formulae from Chapter 15

(See also §§ 15.6 and 15.7.)

1. $\lim_{n\to\infty} \lim_{x\to a+} f_n(x) = \lim_{x\to a+} \lim_{n\to\infty} f_n(x)$

under the assumptions of Theorem 15.1.4,

2. $\int_a^x \lim_{n\to\infty} f_n(t)\,dt = \lim_{n\to\infty} \int_a^x f_n(t)\,dt$ (Theorem 15.1.5),

3. $\frac{d}{dx}\left[\lim_{n\to\infty} f_n(x)\right] = \lim_{n\to\infty} \frac{df_n(x)}{dx}$ (Theorem 15.1.6),

4. $\sum_{n=1}^{\infty} \int_a^x f_n(t)\,dt = \int_a^x \sum_{n=1}^{\infty} f_n(t)\,dt$ (Theorem 15.2.6),

5. $\frac{d}{dx} \sum_{n=1}^{\infty} f_n(x) = \sum_{n=1}^{\infty} \frac{df_n(x)}{dx}$ (Theorem 15.2.7).

6. The radius of convergence, r, of a power series is given by:

$$r = \frac{1}{l}, \quad \text{where} \quad l = \lim_{n\to\infty}\left|\frac{a_{n+1}}{a_n}\right| \quad \text{or} \quad l = \lim_{n\to\infty} \sqrt[n]{|a_n|}$$

(Theorem 15.3.2, Remark 15.3.5).

If

$$S(x) = a_0 + a_1(x - x_0) + a_2(x - x_0)^2 + \dots \quad (|x - x_0| < r),$$

then

$$\int_{x_0}^x S(t)\,dt = a_0(x - x_0) + \frac{a_1}{2}(x - x_0)^2 + \frac{a_2}{3}(x - x_0)^3 + \dots \quad (|x - x_0| < r),$$

$$S'(x) = a_1 + 2a_2(x - x_0) + 3a_3(x - x_0)^2 + \dots \quad (|x - x_0| < r)$$

(Theorems 15.4.1 and 15.4.2).

16. ORTHOGONAL SYSTEMS. FOURIER SERIES. SOME SPECIAL FUNCTIONS (BESSEL FUNCTIONS, ETC.)

By Karel Rektorys

References: [9], [15], [17], [48], [49], [54], [55], [60], [65], [75], [102], [187], [199], [200], [205], [207], [244], [257], [339], [341], [398], [400], [420], [429].

16.1. Square Integrable Functions. Norm. Convergence in the Mean

Definition 1. The function $f(x)$ is called *square integrable in the interval* $[a, b]$ (notation $f \in L_2(a, b)$, or briefly, $f \in L_2$), if the integrals

$$\int_a^b f(x)\,\mathrm{d}x\,, \quad \int_a^b |f(x)|^2\,\mathrm{d}x \tag{1}$$

exist (and consequently have a finite value).

Example 1. The function $f(x) = 1/\sqrt{x}$ is not square integrable in $[0, 1]$ since (cf. Example 13.8.3, p. 561)

$$\int_0^1 |f(x)|^2\,\mathrm{d}x = \int_0^1 \frac{\mathrm{d}x}{x} = +\infty\,.$$

On the other hand, the function $f(x) = 1/\sqrt[3]{x}$ is square integrable in $[0, 1]$, since

$$\int_0^1 f(x)\,\mathrm{d}x = \int_0^1 \frac{\mathrm{d}x}{\sqrt[3]{x}} = \int_0^1 x^{-1/3}\,\mathrm{d}x = \tfrac{3}{2}$$

and

$$\int_0^1 |f(x)|^2\,\mathrm{d}x = \int_0^1 \frac{\mathrm{d}x}{\sqrt[3]{x^2}} = \int_0^1 x^{-2/3}\,\mathrm{d}x = 3\,.$$

Remark 1. Obviously, every continuous or piecewise continuous function on $[a, b]$ is square integrable on $[a, b]$.

Remark 2. In (1) $|f(x)|^2$ and not $f^2(x)$ was written, since in general complex functions of a real variable are also admitted, i.e. functions of the form

$f(x) = f_1(x) + \mathrm{i} f_2(x)$, where f_1, f_2 are real functions. As $|f(x)|^2 = f_1^2(x) + f_2^2(x)$, $f(x)$ is square integrable in $[a, b]$ if and only if both functions $f_1(x)$ and $f_2(x)$ are square integrable in $[a, b]$. If only real functions are considered (and this is the case most frequently met in applications), it suffices to write $f^2(x)$ in (1).

REMARK 3. The notation L_2 is used for functions square integrable in the *Lebesgue sense* (§ 13.14). Some theorems given below are valid only under the assumption that the functions in question are integrable in the Lebesgue sense. Despite this fact no serious error will result if the integrals involved are understood in the usual (Riemann) sense (by § 13.6).

Theorem 1. *If $f \in L_2(a, b)$, then the integral*

$$\int_a^b |f(x)|\,\mathrm{d}x$$

also exists, i.e. $f(x)$ is absolutely integrable in $[a, b]$.

Theorem 2. *If $f \in L_2(a, b)$, $g \in L_2(a, b)$, then*

A. $c_1 f + c_2 g \in L_2(a, b)$ *(c_1, c_2 being arbitrary constants)*

B. *finite integrals*

$$\int_a^b f(x)\,g(x)\,\mathrm{d}x\,, \quad \int_a^b |f(x)\,g(x)|\,\mathrm{d}x$$

exist.

Definition 2. Let $f \in L_2(a, b)$; the number

$$\|f\| = \sqrt{\left(\int_a^b |f(x)|^2\,\mathrm{d}x\right)} \tag{2}$$

is called the *norm* of $f(x)$.

Example 2. Referring to Example 1 we have $f(x) = 1/\sqrt[3]{x} \in L_2(0, 1)$. For the norm of this function we have, by (2),

$$\|f\| = \sqrt{\int_0^1 \frac{\mathrm{d}x}{\sqrt[3]{x^2}}} = \sqrt{3}\,.$$

Definition 3. If $f \in L_2(a, b)$, $g \in L_2(a, b)$, then the norm of the function $f(x) - g(x)$ i.e. the number

$$\|f - g\| = \sqrt{\left(\int_a^b |f(x) - g(x)|^2\,\mathrm{d}x\right)} \tag{3}$$

is called the *distance* between f and g.

REMARK 4. Due to the form of expression (3), the term *mean square deviation* is often used in the literature for denoting the distance $\|f - g\|$.

The fact that the distance $\|f - g\|$ between two functions is small does not imply that the difference $f(x) - g(x)$ is small *everywhere* in $[a, b]$ (see Fig. 16.1).

Definition 4. The set of all functions which are square integrable in $[a, b]$ (generally in the Lebesgue sense, see Remark 3) with the distance of two functions defined by equation (3), is called the *space of square integrable functions in* $[a, b]$. It is denoted by $L_2(a, b)$.

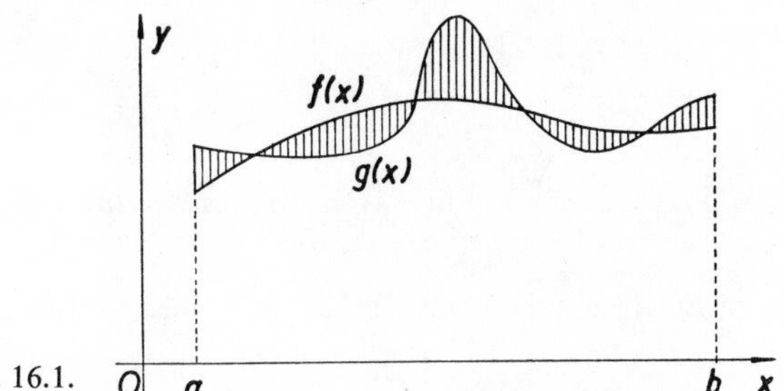

Fig. 16.1.

REMARK 5. Two functions $f \in L_2(a, b)$, $g \in L_2(a, b)$ are considered equal (equivalent) in the space $L_2(a, b)$ not only if $f(x) = g(x)$ for all $x \in [a, b]$, but also if the validity of this equation is violated at points which constitute a set with measure zero [thus, for example, if $f(x) \neq g(x)$ at a finite or at a countably infinite set of points in $[a, b]$, see § 13.14]. This property is often expressed by saying that f and g are "equal almost everywhere" in $[a, b]$.

Definition 5. Let a sequence of functions $f_n \in L_2(a, b)$ be given in $[a, b]$. If a function $f \in L_2(a, b)$ exists such that

$$\lim_{n\to\infty} \|f_n - f\| = 0 \quad \left[\text{i.e.} \lim_{n\to\infty} \int_a^b |f_n(x) - f(x)|^2 \, dx = 0\,, \right. \tag{4}$$

see (3)], the sequence of functions $f_n(x)$ is said to be *convergent in the mean* to the function $f(x)$. (The notation $f_n \to f$ is frequently used.)

Theorem 3. *If $f_n \in L_2(a, b)$ converges in the mean to $f \in L_2(a, b)$, then it cannot converge in the mean to another function.* (In other words, the limit in the sense of convergence in the mean is determined uniquely; however, see Remark 5.)

REMARK 6. If a sequence of functions $f_n(x)$ converges to $f(x)$ *uniformly* in $[a, b]$, then also $f_n \to f$. On the other hand, ordinary convergence (generally non-uniform) does not imply convergence in the mean. Conversely, ordinary convergence does not follow from convergence in the mean.

Example 3. Let

$$f_n(x) = n \quad \text{for} \quad 0 < x < \frac{1}{n},$$
$$= 0 \quad \text{elsewhere in } [0, 1].$$

Obviously, $\lim_{n\to\infty} f_n(x) = 0$ for every $x \in [0, 1]$ so that the sequence of functions $f_n(x)$ converges pointwise (at every point $x \in [0, 1]$) to the function $f(x) \equiv 0$ in $[0, 1]$. However, this sequence does not converge to this function in the mean, i.e. it is not true that

$$\lim_{n\to\infty} \|f_n - f\| = \lim_{n\to\infty} \|f_n\| = 0,$$

since

$$\lim_{n\to\infty} \|f_n\| = \lim_{n\to\infty} \sqrt{\left(\int_0^1 f_n^2(x)\,\mathrm{d}x\right)} = \lim_{n\to\infty} \sqrt{\left(\int_0^{1/n} n^2\,\mathrm{d}x\right)} = \lim_{n\to\infty} \sqrt{n} = +\infty.$$

Example 4. It can be shown (Remark 16.2.13) that the Fourier series corresponding to any function $f \in L_2(-\pi, \pi)$ converges to the function f in the mean. It is known, however, that it need not converge pointwise in $[-\pi, \pi]$.

REMARK 7. The concepts of the space L_2 and of convergence in the mean can be defined in much more general sets than intervals. Here, however, we show only an immediate generalization of the definitions introduced above to a two-dimensional space. Let R be the square $a \leqq x \leqq b$, $a \leqq y \leqq b$. The function $f(x, y)$ will be called square integrable in R (notation $f \in L_2(R)$), if the integrals

$$\iint_R f(x, y)\,\mathrm{d}x\,\mathrm{d}y, \quad \iint_R |f(x, y)|^2\,\mathrm{d}x\,\mathrm{d}y$$

exist (and, consequently, are finite).

The sequence $f_n \in L_2(R)$ will be called convergent in the mean to the function $f \in L_2(R)$, if

$$\lim_{n\to\infty} \|f_n - f\| = \lim_{n\to\infty} \sqrt{\left[\iint_R |f_n(x, y) - f(x, y)|^2\,\mathrm{d}x\,\mathrm{d}y\right]} = 0.$$

16.2. Scalar (Inner) Product. Orthogonal and Orthonormal Systems. Generalized Fourier Series

Definition 1. Let two functions $f \in L_2(a, b)$, $g \in L_2(a, b)$ be given (Definition 16.1.1); then the number

$$(f, g) = \int_a^b f(x)\,\overline{g(x)}\,\mathrm{d}x \tag{1}$$

is called the *scalar product* (*inner product*) of f and g.

REMARK 1. The functions $f(x)$, $g(x)$ are in general complex functions of a real variable, i.e. $f(x) = f_1(x) + \mathrm{i} f_2(x)$, $g(x) = g_1(x) + \mathrm{i}\, g_2(x)$, where f_1, f_2, g_1, g_2 are real functions, and $\overline{g(x)} = g_1(x) - \mathrm{i}\, g_2(x)$. Thus, the scalar product of two functions is in general a complex number and has the value

$$\int_a^b [f_1(x)\, g_1(x) + f_2(x)\, g_2(x)]\, \mathrm{d}x + \mathrm{i} \int_a^b [f_2(x)\, g_1(x) - f_1(x)\, g_2(x)]\, \mathrm{d}x\,.$$

However, nearly always in applications $f(x)$ and $g(x)$ are real functions. Then, of course, the notation of the complex conjugate $\overline{g(x)}$ is superfluous in (1) and we have

$$(f, g) = \int_a^b f(x)\, g(x)\, \mathrm{d}x\,. \tag{1'}$$

Hence the scalar product of two *real* functions is a *real* number.

REMARK 2. From Definition 1 and Definition 16.1.2 it is obvious that

$$\|f\|^2 = (f, f)\,, \quad \text{since} \quad |f(x)|^2 = f(x) \cdot \overline{f(x)}\,. \tag{2}$$

Theorem 1 (*The Schwarz–Cauchy–Buniakowski inequality*). *For any two functions $f \in L_2(a, b)$, $g \in L_2(a, b)$ we have*

$$|(f, g)| \leqq \|f\| \cdot \|g\|\,, \tag{3}$$

i.e.

$$\left|\int_a^b f(x)\, \overline{g(x)}\, \mathrm{d}x\right| \leqq \sqrt{\left(\int_a^b |f(x)|^2\, \mathrm{d}x\right)} \sqrt{\left(\int_a^b |g(x)|^2\, \mathrm{d}x\right)}\,. \tag{4}$$

Definition 2. Functions $f \in L_2(a, b)$, $g \in L_2(a, b)$ are called *orthogonal in* $[a, b]$, if

$$(f, g) = 0\,, \quad \text{i.e.} \quad \int_a^b f(x)\, \overline{g(x)}\, \mathrm{d}x = 0\,. \tag{5}$$

Definition 3. The function f is called *normed* (*normalised*), if its norm equals unity, i.e. if

$$\|f\| = 1\,. \tag{6}$$

Definition 4. Let a system of functions $f_n \in L_2(a, b)$ be given (finite or countably infinite in number, Definition 22.1.12, p. 997). This system is called *orthogonal in* $[a, b]$, if for every two functions from this system we have

$$(f_i, f_k) = 0 \quad \text{provided} \quad i \neq k\,. \tag{7}$$

If, in addition, each of the functions $f_n(x)$ is normed, the system is called *orthonormal*.

REMARK 3. This means that for functions from orthonormal systems we have

$$(f_i, f_k) = \delta_{ik}\,, \tag{8}$$

where δ_{ik} is the *Kronecker delta symbol*, i.e.

$$\delta_{ik} = \begin{cases} 0 & \text{for} \quad i \neq k\,, \\ 1 & \text{for} \quad i = k\,. \end{cases} \tag{9}$$

REMARK 4. From an orthogonal system $f_n(x)$ an orthonormal system $\varphi_n(x)$ is obtained if (for every n) we put

$$\varphi_n(x) = \frac{f_n(x)}{\|f_n(x)\|} = \frac{f_n(x)}{\sqrt{\left(\int_a^b |f_n(x)|^2 \, dx\right)}}\,.$$

Example 1. By integration it can be immediately verified (see also § 13.10) that the system of functions

$$1, \cos x, \sin x, \cos 2x, \sin 2x, \ldots \tag{10}$$

is orthogonal in the interval $[-\pi, \pi]$ (even in every interval $[a, a + 2\pi]$). This system, however, is not orthonormal, since

$$\int_{-\pi}^{\pi} 1 \,.\, dx = 2\pi\,, \quad \int_{-\pi}^{\pi} \sin^2 kx \, dx = \pi\,, \quad \int_{-\pi}^{\pi} \cos^2 kx \, dx = \pi$$

for every k ($k = 1, 2, 3, \ldots$); the corresponding orthonormal system is, according to Remark 4,

$$\frac{1}{\sqrt{(2\pi)}}, \quad \frac{\cos x}{\sqrt{\pi}}, \quad \frac{\sin x}{\sqrt{\pi}}, \quad \frac{\cos 2x}{\sqrt{\pi}}, \quad \frac{\sin 2x}{\sqrt{\pi}}, \quad \ldots\,. \tag{11}$$

In the interval $[-\pi, \pi]$ the systems

$$\sin x, \sin 2x, \sin 3x, \ldots \tag{12}$$

and

$$1, \cos x, \cos 2x, \cos 3x, \ldots \tag{13}$$

are also orthogonal. (This follows from the fact that both systems (12) and (13) are selected from system (10).)

The system (10) is not orthogonal in the interval $[0, \pi]$ (since, for example, we have

$$\int_0^{\pi} 1 \,.\, \sin x \, dx = 2)\,.$$

On the other hand, the systems (12), (13) are each orthogonal in $[0, \pi]$.

Example 2. The system of functions $f_n(x) = e^{inx} = \cos nx + i \sin nx$ (n integral) is orthogonal in $[-\pi, \pi]$, for

$$(f_n, f_m) = \int_{-\pi}^{\pi} e^{inx}\,\overline{e^{imx}}\,dx = \int_{-\pi}^{\pi} e^{inx}\,e^{-imx}\,dx = \int_{-\pi}^{\pi} e^{i(n-m)x}\,dx =$$

$$= \left[\frac{1}{i(n-m)}\,e^{i(n-m)x}\right]_{-\pi}^{\pi} = 0 \quad \text{if} \quad m \neq n\,.$$

In order to make this system a normed one, let us find the norms of the functions $f_n(x)$. According to (2) we have

$$\|f_n\| = (f_n, f_n) = \int_{-\pi}^{\pi} e^{inx}\,\overline{e^{inx}}\,dx = \int_{-\pi}^{\pi} e^{inx}\,e^{-inx}\,dx = \int_{-\pi}^{\pi} 1\,.\,dx = 2\pi\,,$$

so that for the orthonormal system we obtain

$$f_n(x) = \frac{e^{inx}}{\sqrt{(2\pi)}} \quad (n \text{ integral})\,.$$

Definition 5. A system of functions $f_n(x)$ is called *orthogonal in* $[a, b]$ *with a weight function* $\varrho(x)$ ($\varrho(x) \geqq 0$), if for each pair of functions $f_i(x), f_k(x)$ we have

$$\int_a^b \varrho(x)\,f_i(x)\,\overline{f_k(x)}\,dx = 0 \quad \text{provided} \quad i \neq k\,. \tag{14}$$

(As far as the complex conjugate $\overline{f_k(x)}$ is concerned, see Remark 1.) If, in addition, the functions $f_n(x)$ are *normed with the weight function* $\varrho(x)$, i.e. if

$$\int_a^b \varrho(x)\,|f_n(x)|^2\,dx = 1 \tag{15}$$

for every n, then the system of functions $f_n(x)$ will be called *orthonormal with the weight function* $\varrho(x)$. Thus,

$$\int_a^b \varrho(x)\,f_i(x)\,\overline{f_k(x)}\,dx = \delta_{ik} \tag{16}$$

(see Remark 3). (Examples of such systems are given in § 16.6. See also Theorem 16.4.7.)

REMARK 5. From the system of functions $f_n(x)$, which are orthogonal in $[a, b]$ with weight function $\varrho(x)$, a system $\varphi_n(x)$ orthonormal in $[a, b]$ with this weight function $\varrho(x)$ is obtained if we put (for every n)

$$\varphi_n(x) = \frac{f_n(x)}{\sqrt{\left(\int_a^b \varrho(x)\,|f_n(x)|^2\,dx\right)}}\,.$$

REMARK 6. Problems with boundary conditions and eigenvalue problems in differential equations furnish a variety of orthogonal functions (or functions orthogonal with a given weight function); see e.g. Theorem 17.17.2 and Remark 17.17.11, p. 807.

Definition 6. Let an *orthonormal* system of functions

$$\varphi_1(x),\ \varphi_2(x),\ \varphi_3(x),\ \ldots\ (\varphi_n \in L_2(a, b)) \tag{17}$$

be given in $[a, b]$ and let $f \in L_2(a, b)$. The series

$$c_1\varphi_1(x) + c_2\varphi_2(x) + c_3\varphi_3(x) + \ldots, \tag{18}$$

where

$$c_k = (f, \varphi_k) = \int_a^b f(x)\,\overline{\varphi_k(x)}\,dx\,, \tag{19}$$

is called the *(generalized) Fourier series corresponding to the function* $f(x)$. The numbers c_k are called the *(generalised) Fourier coefficients of the function* $f(x)$ *with respect to the system* (17).

Example 3. The Fourier coefficients of the function $f \in L_2(-\pi, \pi)$ with respect to the system (11) are

$$c_1 = \int_{-\pi}^{\pi} f(x)\,\frac{1}{\sqrt{(2\pi)}}\,dx\,, \quad c_2 = \int_{-\pi}^{\pi} f(x)\,\frac{\cos x}{\sqrt{\pi}}\,dx\,, \quad c_3 = \int_{-\pi}^{\pi} f(x)\,\frac{\sin x}{\sqrt{\pi}}\,dx\,, \ldots$$

REMARK 7. In the general case, nothing can be asserted about the pointwise convergence of the series (18) to the function $f(x)$. Convergence *in the mean* will be considered below, particularly in Theorems 4 and 5. See also Theorems 8 and 9.

Theorem 2. *Let a positive integer n and an orthonormal system* (17) *be given. Then from all functions of the form*

$$k_1\varphi_1(x) + k_2\varphi_2(x) + \ldots + k_n\varphi_n(x) \tag{20}$$

the function which has the least distance (i.e. the least square deviation, Definition 16.1.3) *from the function* $f(x)$ *is:*

$$c_1\varphi_1(x) + c_2\varphi_2(x) + \ldots + c_n\varphi_n(x) \tag{21}$$

where the c_k *are given by formula* (19).

REMARK 8. This means that for any set of numbers $k_1, \ldots, k_n$ we have

$$\int_a^b |f(x) - [c_1\varphi_1(x) + c_2\varphi_2(x) + \ldots + c_n\varphi_n(x)]|^2\,dx \leqq$$

$$\leqq \int_a^b |f(x) - [k_1\varphi_1(x) + k_2\varphi_2(x) + \ldots + k_n\varphi_n(x)]|^2\,dx\,.$$

Theorem 3. *For any function $f \in L_2(a, b)$ the following inequality (Bessel's inequality)* is satisfied:

$$\sum_{n=1}^{\infty} |c_n|^2 \leqq \|f\|^2 \tag{22}$$

with the c_n given by (19).

REMARK 9. If both $\varphi_n(x)$ and $f(x)$ are real, then the c_n are also real and the absolute value signs on the left-hand side of inequality (22) are superfluous; hence in this case

$$\sum_{n=1}^{\infty} c_n^2 \leqq \int_a^b f^2(x)\,dx\,. \tag{22'}$$

REMARK 10. Referring to Theorem 10.2.1 (p. 382), the convergence of series (22) or (22′) implies that

$$\lim_{n\to\infty} c_n = 0$$

(for any function $f \in L_2(a, b)$).

REMARK 11. We draw the reader's attention to the fact that Theorems 2 and 3 hold for *orthonormal* systems.

Definition 7. The system (17) is called *complete* (*closed*) *in* $L_2(a, b)$, if for every function $f \in L_2(a, b)$ the corresponding Fourier series (18) converges to f *in the mean*, i.e. if

$$\lim_{n\to\infty} \left\|f - \sum_{k=1}^{n} c_k \varphi_k\right\| = 0\,.$$

Theorem 4. *A necessary and sufficient condition for a system* (17) *to be complete in $L_2(a, b)$ is that for every function $f \in L_2(a, b)$ the corresponding Bessel inequality* (22) *becomes the equality* (*the so-called Parseval relation*)

$$\sum_{n=1}^{\infty} |c_n|^2 = \|f\|^2\,. \tag{23}$$

Definition 8. The system (17) is also called *complete in* $L_2(a, b)$, if there is no non-zero (see Remark 16.1.5) function $g \in L_2$ which is orthogonal to each function from the given system.

Theorem 5. *Definitions* 7 *and* 8 *are equivalent.*

REMARK 12. From Theorem 5 it follows that: If the system (17) is complete according to Definition 7, it is also complete according to Definition 8, and vice versa. Thus, particularly, if (17) is a complete system according to Definition 7, then the relationships

$$(g, \varphi_n) = 0\,, \quad g \in L_2(a, b)\,, \quad n = 1, 2, 3, \ldots$$

imply that $g(x)$ vanishes in $[a, b]$. (In the sense of Remark 16.1.5, i.e. $g(x) = 0$ everywhere in $[a, b]$ except for a set of measure zero.)

Theorem 6 (*Uniqueness Theorem*). *If two functions* $f \in L_2(a, b)$, $g \in L_2(a, b)$ *possess the same Fourier series* (*i.e. the same Fourier coefficients* (19)) *with respect to a complete system* (17), *then they are equal* (in the sense of Remark 16.1.5).

Theorem 7. *Let* (17) *be an orthonormal system* (*not necessarily complete*). *If* $c_1, c_2, \ldots$ *are numbers such that*

$$\sum_{n=1}^{\infty} |c_n|^2 < \infty ,$$

then the series

$$\sum_{n=1}^{\infty} c_n \varphi_n(x)$$

converges in the mean to a function $f \in L_2(a, b)$, *whose Fourier coefficients with respect to the system* (17) *coincide with the numbers* c_n.

Theorem 8. *If the Fourier series* (18) *corresponding to a function* $f \in L_2(a, b)$ *converges in the mean to* $f(x)$ *and, in addition, is uniformly convergent in* $[a, b]$, *then it converges at each point* $x \in [a, b]$ *to the function* $f(x)$ (more accurately, to a function equal to $f(x)$, see Remark 16.1.5). Thus, in particular we have:

Theorem 9. *If the system* (17) *is complete and if the Fourier series corresponding to a function* $f \in L_2(a, b)$ *is uniformly convergent in* $[a, b]$, *then it converges uniformly in* $[a, b]$ *to the function* $f(x)$.

REMARK 13. It is not always easy in practice to prove the completeness of a given orthonormal system of functions. The completeness can often be decided by using the theorems of § 17.17. Typical examples of complete orthonormal systems in L_2 are furnished by the following systems of trigonometric functions:

$$\text{interval } [-\pi, \pi]: \quad \frac{1}{\sqrt{(2\pi)}}, \quad \frac{\cos x}{\sqrt{\pi}}, \quad \frac{\sin x}{\sqrt{\pi}}, \quad \frac{\cos 2x}{\sqrt{\pi}}, \quad \frac{\sin 2x}{\sqrt{\pi}}, \ldots, \tag{24}$$

$$\text{interval } [0, \pi]: \quad \sqrt{\left(\frac{2}{\pi}\right)} \sin x, \quad \sqrt{\left(\frac{2}{\pi}\right)} \sin 2x, \quad \sqrt{\left(\frac{2}{\pi}\right)} \sin 3x, \ldots, \tag{25}$$

$$\text{interval } [0, \pi]: \quad \frac{1}{\sqrt{\pi}}, \quad \sqrt{\left(\frac{2}{\pi}\right)} \cos x, \quad \sqrt{\left(\frac{2}{\pi}\right)} \cos 2x, \quad \sqrt{\left(\frac{2}{\pi}\right)} \cos 3x, \ldots. \tag{26}$$

Thus, the corresponding Fourier series (of a function $f \in L_2(-\pi, \pi)$ or $g \in L_2(0,\pi)$) converges *in the mean* to the function $f(x)$ or $g(x)$, respectively. The same is true for Fourier series transformed on to the interval $[-l, l]$, or $[0, l]$, or $[a, b]$ (§ 16.3). (See also Remark 15, Equation (27) and Remark 17.)

REMARK 14. The concept of completeness was here defined for the space L_2, i.e. with respect to convergence in the mean. In various spaces, however, the character of completeness varies with the sense of convergence introduced in the individual spaces. For example, in the space C (Example 22.2.1, p. 998) the completeness of a given system of functions is understood in the sense of Definition 7 and of the norm in space C, i.e. in the sense of uniform convergence.

REMARK 15. Instead of the expression "complete system" the term "*complete sequence*" is often used. This term is also often used in the following, more general sense:

Let $f_n \in L_2(a, b)$ be a sequence of functions which are linearly independent in $[a, b]$. (This means that every finite number of terms of the sequence constitutes a linearly independent system, Definition 12.8.3, p. 460; see also Remark 16.1.5; the functions $f_n(x)$ need not be pairwise orthogonal.) The sequence $f_n(x)$ is called *complete in* $L_2(a, b)$, if any function $f \in L_2(a, b)$ can be approximated in the mean by linear combinations of functions $f_n(x)$ with any accuracy. In other words: Let a function $f \in L_2(a, b)$ and an $\varepsilon > 0$ be given. Then an integer m and constants $k_1, \ldots, k_m$ can be found such that

$$\|f - (k_1 f_1 + k_2 f_2 + \ldots + k_m f_m)\| < \varepsilon ,$$

i.e.

$$\int_a^b |f(x) - [k_1 f_1(x) + k_2 f_2(x) + \ldots + k_m f_m(x)]|^2 \, dx < \varepsilon^2 .$$

In the same way, the concept of a complete sequence in the sense of uniform convergence may be introduced, etc.

For any interval $[a, b]$, the sequence

$$1, x, x^2, x^3, \ldots \tag{27}$$

is complete in $L_2(a, b)$.

REMARK 16 (*The Schmidt Orthogonalization Process*). Let $f_n \in L_2(a, b)$ be a sequence of functions which are linearly independent in $[a, b]$ (see Remark 15) but not necessarily orthogonal. From this sequence a sequence $g_n \in L_2(a, b)$ orthogonal in $[a, b]$ can be constructed as follows:

First, put $g_1(x) = f_1(x)$.

The function $g_2(x)$ is first sought in the form

$$g_2(x) = f_2(x) + k_1 g_1(x) .$$

We determine the constant k_1 so that $g_2(x)$ should be orthogonal to $g_1(x)$, i.e.

$$(g_2, g_1) = (f_2 + k_1 g_1, g_1) = 0 ;$$

thus, we have

$$(f_2, g_1) + k_1(g_1, g_1) = 0 . \tag{28}$$

As $g_1(x) = f_1(x)$, $(f_1, f_1) > 0$ ($f_n(x)$ are linearly independent), so the constant k_1 is determined uniquely by equation (28). Obviously, $g_2(x)$ is a non-zero function, since $f_1(x)$ and $f_2(x)$ are linearly independent. The function $g_3(x)$ is now sought in the form

$$g_3(x) = f_3(x) + c_2 g_2(x) + c_1 g_1(x)$$

choosing the constants so that $(g_3, g_1) = 0$, $(g_3, g_2) = 0$. Expanding the products we get

$$(f_3, g_1) + c_2(g_2, g_1) + c_1(g_1, g_1) = 0\,,$$
$$(f_3, g_2) + c_2(g_2, g_2) + c_1(g_1, g_2) = 0\,.$$

As $(g_2, g_1) = 0$ and $(g_1, g_1) > 0$, c_1 is determined uniquely by the first equation. For the same reason, c_2 is determined uniquely by the second one. Obviously, $g_3(x)$ is again a non-zero function.

In this manner we obtain a system of non-zero orthogonal functions $g_n \in L_2(a, b)$ from the functions $f_n \in L_2(a, b)$. Finally, by normalizing each of the functions $g_n(x)$ we get an *orthonormal* system.

Example 4. The orthogonalization of the system (27) in $[-1, 1]$ yields the (normalized) Legendre polynomials

$$\left(\tfrac{1}{2}\right)^{1/2},\quad \left(\tfrac{3}{2}\right)^{1/2} x\,,\quad \left(\tfrac{5}{2}\right)^{1/2} . \tfrac{1}{2}(3x^2 - 1)\,,\quad \left(\tfrac{7}{2}\right)^{1/2} . \tfrac{1}{2}(5x^3 - 3x)\,,\quad \ldots$$

(§ 16.5).

REMARK 17. Everything said above about functions of one variable may be extended to functions of two or more variables. Instead of an interval $[a, b]$ a region Ω in a plane or in space, etc., is then considered. The functions dealt with in Theorem 16.4.5 play the role of the system (24) here; see also Chapter 22.

16.3. Trigonometric Fourier Series. Fourier Series in Two and Several Variables. Fourier Integral

REMARK 1. According to Remark 16.2.13, the orthonormal system of trigonometric functions

$$\frac{1}{\sqrt{(2\pi)}},\quad \frac{\cos x}{\sqrt{\pi}},\quad \frac{\sin x}{\sqrt{\pi}},\quad \frac{\cos 2x}{\sqrt{\pi}},\quad \frac{\sin 2x}{\sqrt{\pi}},\quad \ldots$$

is complete in $[-\pi, \pi]$. Thus, taking any function $f \in L_2(-\pi, \pi)$, the corresponding Fourier series (16.2.18), p. 698, converges in the mean to $f(x)$.

An analogous statement is true for the systems (16.2.25), (16.2.26) in the interval $[0, \pi]$.

As far as the *pointwise* convergence of a Fourier series is concerned, we have:

Theorem 1. *Let $f(x)$ be a periodic function with period 2π* (i.e. $f(x + 2\pi) = f(x)$ for every x) *and let $f(x)$ and $f'(x)$ be piecewise continuous in $[-\pi, \pi]$. Define constants a_n, b_n, by*

$$\begin{aligned} a_n &= \frac{1}{\pi}\int_{-\pi}^{\pi} f(x)\cos nx \, dx \quad (n = 0, 1, 2, \ldots), \\ b_n &= \frac{1}{\pi}\int_{-\pi}^{\pi} f(x)\sin nx \, dx \quad (n = 1, 2, 3, \ldots). \end{aligned} \tag{1}$$

Then at each point x, where $f(x)$ is continuous, we have

$$\frac{a_0}{2} + \sum_{n=1}^{\infty}(a_n \cos nx + b_n \sin nx) = f(x); \tag{2}$$

while at each point of discontinuity,

$$\frac{a_0}{2} + \sum_{n=1}^{\infty}(a_n \cos nx + b_n \sin nx) = \tfrac{1}{2}\{f(x + 0) + f(x - 0)\}. \tag{3}$$

(The symbols $f(x + 0)$, $f(x - 0)$ denote the limits from the right and the left, respectively, of the function $f(x)$ at the point x.)

REMARK 2. Equations (2) and (3) hold under far more general assumptions. They are true if: (i) $f(x)$ is integrable in $[-\pi, \pi]$; (ii) the integral

$$\int_{-\pi}^{\pi} |f(x)| \, dx$$

exists ($f(x)$ may be unbounded), and (iii) the point x is interior to an interval, in which $f(x)$ has a bounded variation (Definition 11.3.7, p. 408). Moreover, in the case of existence of the above integral, the series (2) converges uniformly in each interval $[a, b]$, which is interior to an interval on which $f(x)$ is continuous and has bounded variation.

REMARK 3. Brackets in the sum on the left-hand side of equation (2) or (3) cannot in general be omitted.

REMARK 4. For a periodic function with period 2π (Theorem 1) we may take not only the interval $[-\pi, \pi]$ as "basic" interval, but also any interval $[a, a + 2\pi]$ (for example, the interval $[0, 2\pi]$). The lower and upper limits in integrals (1), of course, are then a and $a + 2\pi$, respectively.

Example 1. Let $f(x)$ be a periodic function with period 2π and let $f(x) = x$ on the basic interval $[0, 2\pi)$ (thus at points $x = 2k\pi$, where k is an integer, we have $f(x) = 0$, see Fig. 16.2.)

Integrating by parts we get

$$a_n = \frac{1}{\pi}\int_0^{2\pi} x\cos nx\,dx = \frac{1}{\pi}\left[x\,\frac{\sin nx}{n}\right]_0^{2\pi} - \frac{1}{n\pi}\int_0^{2\pi}\sin nx\,dx = 0 \text{ for } n = 1, 2, 3, \ldots,$$

$$a_0 = \frac{1}{\pi}\int_0^{2\pi} x\,dx = 2\pi\,,$$

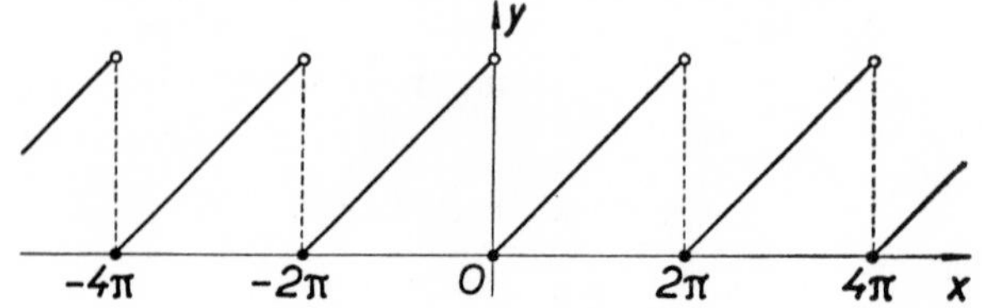

Fig. 16.2.

$$b_n = \frac{1}{\pi}\int_0^{2\pi} x\sin nx\,dx = -\frac{1}{\pi}\left[x\,\frac{\cos nx}{n}\right]_0^{2\pi} + \frac{1}{n\pi}\int_0^{2\pi}\cos nx\,dx =$$

$$= -\frac{2}{n} \quad (n = 1, 2, 3, \ldots).$$

Thus by (2) we have

$$f(x) = \pi - 2\left(\frac{\sin x}{1} + \frac{\sin 2x}{2} + \frac{\sin 3x}{3} + \ldots\right) \tag{4}$$

at any point x different from an integral multiple of 2π. If $x = 2k\pi$ (where k is an integer), $f(x)$ is discontinuous and the sum on the right-hand side of equation (4) equals, in view of (3),

$$\tfrac{1}{2}[f(x + 0) + f(x - 0)] = \tfrac{1}{2}(0 + 2\pi) = \pi\,.$$

This can, of course, readily be verified from (4). In any closed interval which does not contain the point $x = 2k\pi$, the series (4) converges uniformly.

REMARK 5. For a function $f(x)$ periodic with period $2l$ a theorem analogous to Theorem 1 is true. (Remarks similar to Remarks 2, 3 and 4 also hold.) Here, the formula reads:

$$\frac{a_0}{2} + \sum_{n=1}^{\infty}\left(a_n\cos\frac{n\pi x}{l} + b_n\sin\frac{n\pi x}{l}\right) = \begin{cases} f(x) \text{ at a point of continuity of } f(x), \\ \tfrac{1}{2}(f(x + 0) + f(x - 0)) \text{ at a point} \\ \text{of discontinuity of } f(x), \end{cases} \tag{5}$$

where

$$a_n = \frac{1}{l}\int_{-l}^{l} f(x)\cos\frac{n\pi x}{l}\,\mathrm{d}x\,,\quad b_n = \frac{1}{l}\int_{-l}^{l} f(x)\sin\frac{n\pi x}{l}\,\mathrm{d}x\,. \tag{6}$$

((5) and (6) follow from (1), (2) and (3) by the substitution $x = \pi u/l$; in the result x is again written instead of u.)

If $[a, b]$ is a basic interval and $f(x)$ is periodic with period $b - a$, then from (5), (6) and Remark 4 we have (putting $b = a + 2l$)

$$\frac{a_0}{2} + \sum_{n=1}^{\infty}\left(a_n\cos\frac{2n\pi x}{b-a} + b_n\sin\frac{2n\pi x}{b-a}\right) = \begin{cases} f(x)\,, \\ \frac{1}{2}(f(x+0)+f(x-0))\,, \end{cases} \tag{5'}$$

where

$$a_n = \frac{2}{b-a}\int_a^b f(x)\cos\frac{2n\pi x}{b-a}\,\mathrm{d}x\,,\quad b_n = \frac{2}{b-a}\int_a^b f(x)\sin\frac{2n\pi x}{b-a}\,\mathrm{d}x\,. \tag{6'}$$

Formulae (5′) and (6′) are often used for periodic functions with period T and basic interval $[0, T]$ (hence $a = 0$, $b = T$); putting $2\pi/T = \omega$ and writing t instead of x, we have

$$\frac{a_0}{2} + \sum_{n=1}^{\infty}(a_n\cos n\omega t + b_n\sin n\omega t) = \begin{cases} f(t)\,, \\ \frac{1}{2}(f(t+0)+f(t-0))\,, \end{cases}$$

where

$$a_n = \frac{2}{T}\int_0^T f(t)\cos n\omega t\,\mathrm{d}t\,,\quad b_n = \frac{2}{T}\int_0^T f(t)\sin n\omega t\,\mathrm{d}t\,.$$

REMARK 6. If $f(x)$ is an *odd* function of x (i.e. if $f(-x) \equiv -f(x)$), then $a_n = 0$ for $n = 0, 1, \ldots$ and the series involves only sine terms. If $f(x)$ is *even* in x, (i.e. if $f(-x) \equiv f(x)$), then $b_n = 0$ for $n = 1, 2, \ldots$ and the series involves only the constant term $a_0/2$ and cosine terms.

If $f(x)$ is defined in the interval $[0, \pi]$, it can be expressed in this interval by either a sine series or a cosine series. Defining $f(x)$ in $(-\pi, 0)$ by $f(-x) = -f(x)$ (and as a periodic function with period 2π for remaining x), we get $f(x)$ expressed by a sine series. Defining $f(-x) = f(x)$ in $(-\pi, 0)$, we get $f(x)$ expressed by a cosine series. In the first case,

$$\sum_{n=1}^{\infty} b_n \sin nx = \begin{Bmatrix} f(x) \\ \frac{1}{2}(f(x+0)+f(x-0)) \end{Bmatrix} \quad (\text{for } x \in [0, \pi])\,, \tag{7}$$

where

$$b_n = \frac{2}{\pi}\int_0^{\pi} f(x)\sin nx\,\mathrm{d}x\,. \tag{8}$$

In the second case,

$$\frac{a_0}{2} + \sum_{n=1}^{\infty} a_n \cos nx = \begin{Bmatrix} f(x), \\ \frac{1}{2}(f(x+0) + f(x-0)) \end{Bmatrix} \quad (\text{for } x \in [0, \pi]) \tag{9}$$

with

$$a_n = \frac{2}{\pi} \int_0^{\pi} f(x) \cos nx \, dx . \tag{10}$$

If the interval $[0, \pi]$ is replaced by the interval $[0, l]$, formulae (7) and (8) assume the form

$$\sum_{n=1}^{\infty} b_n \sin \frac{n\pi x}{l} = \begin{Bmatrix} f(x), \\ \frac{1}{2}(f(x+0) + f(x-0)) \end{Bmatrix} \quad (\text{for } x \in [0, l]),$$

$$b_n = \frac{2}{l} \int_0^{l} f(x) \sin \frac{n\pi x}{l} \, dx$$

and formulae (9), (10) the form

$$\frac{a_0}{2} + \sum_{n=1}^{\infty} a_n \cos \frac{n\pi x}{l} = \begin{Bmatrix} f(x), \\ \frac{1}{2}(f(x+0) + f(x-0)) \end{Bmatrix} \quad (\text{for } x \in [0, l]),$$

$$a_n = \frac{2}{l} \int_0^{l} f(x) \cos \frac{n\pi x}{l} \, dx .$$

Example 2. For the function $f(x) = x$ in $[0, \pi)$, $f(\pi) = 0$ we get in the first case

$$f(x) = 2\left(\frac{\sin x}{1} - \frac{\sin 2x}{2} + \frac{\sin 3x}{3} - \ldots\right) \tag{11}$$

and the continuation of the function over all real x is plotted in Fig. 16.3.

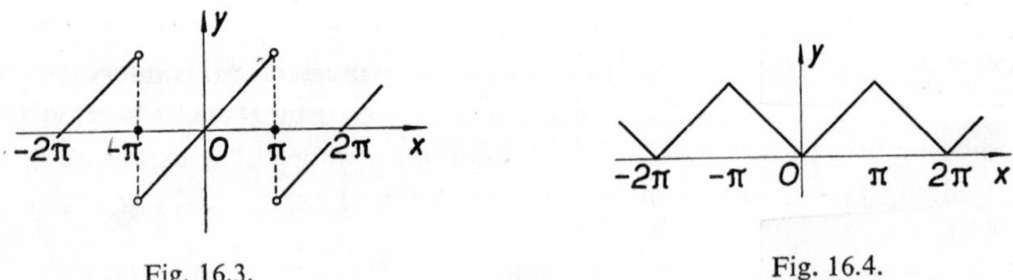

Fig. 16.3.

Fig. 16.4.

In the second case (we define $f(\pi) = \pi$; now the continuation of the function for all x is as plotted in Fig. 16.4),

$$f(x) = \frac{\pi}{2} - \frac{4}{\pi}\left(\frac{\cos x}{1^2} + \frac{\cos 3x}{3^2} + \frac{\cos 5x}{5^2} + \ldots\right). \tag{12}$$

REMARK 7 (*Fourier Expansions of Some Important Functions*).

1. $f(x) = |\sin x|$ (Fig. 16.5):

$$f(x) = \frac{2}{\pi} - \frac{4}{\pi}\left(\frac{\cos 2x}{1.3} + \frac{\cos 4x}{3.5} + \frac{\cos 6x}{5.7} + \ldots\right).$$

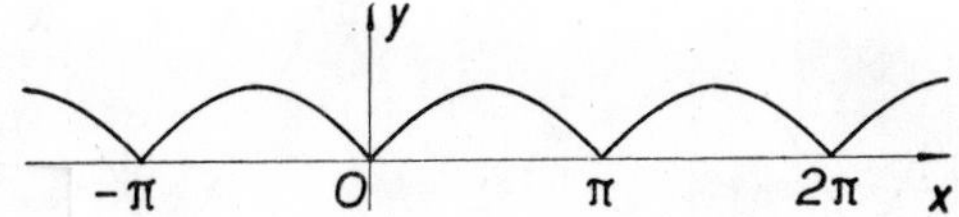

Fig. 16.5.

2. $f(x) = \frac{h}{p} x$ for $0 \leqq x \leqq p$, $f(x) = h$ for $p \leqq x \leqq \pi - p$,

$f(x) = -\frac{h}{p}(x - \pi)$ for $\pi - p \leqq x \leqq \pi$, $f(x + \pi) = -f(x)$ for every x

(Fig. 16.6):

$$f(x) = \frac{4}{\pi}\frac{h}{p}\left(\frac{1}{1^2}\sin p \sin x + \frac{1}{3^2}\sin 3p \sin 3x + \frac{1}{5^2}\sin 5p \sin 5x + \ldots\right).$$

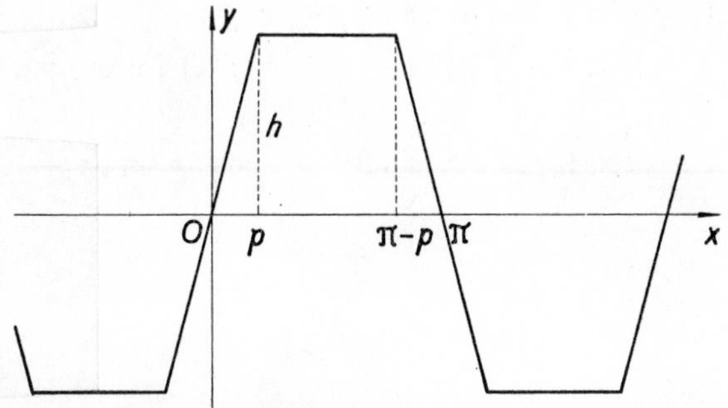

Fig. 16.6.

3. Particularly, for $p = \frac{1}{2}\pi$ (Fig. 16.7):

$$f(x) = \frac{8h}{\pi^2}\left(\frac{\sin x}{1^2} - \frac{\sin 3x}{3^2} + \frac{\sin 5x}{5^2} - \ldots\right).$$

4. $f(x) = h$ for $0 < x < \pi$, $f(0) = 0$, $f(x + \pi) = -f(x)$ for every x (Fig. 16.8):

$$f(x) = \frac{4}{\pi} h\left(\frac{\sin x}{1} + \frac{\sin 3x}{3} + \frac{\sin 5x}{5} + \ldots\right).$$

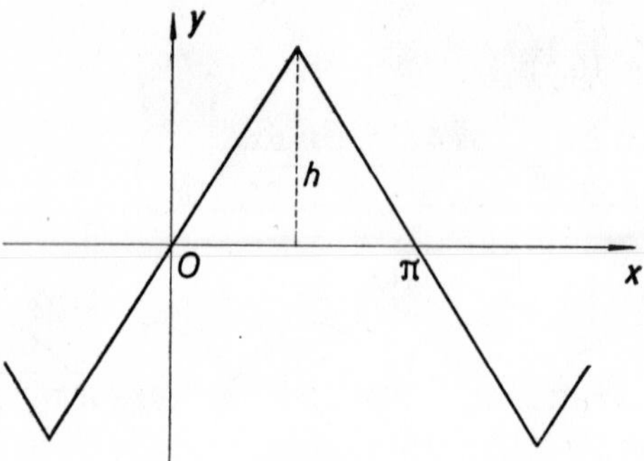

Fig. 16.7.

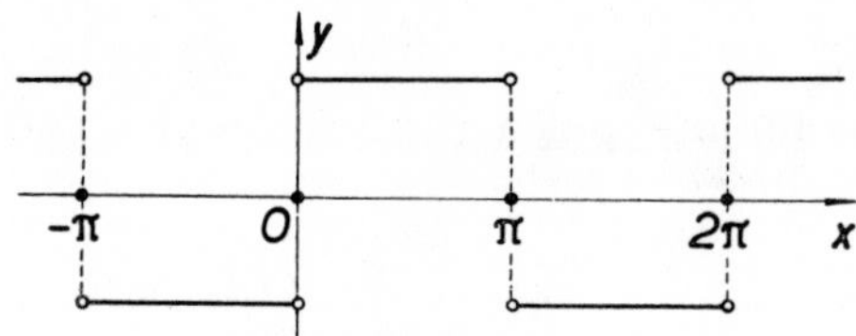

Fig. 16.8.

5. $f(x) = 0$ for $0 \leqq x < p$ and for $\pi - p < x \leqq \pi$,
$f(x) = a$ for $p < x < \pi - p$, $f(x + \pi) = -f(x)$ for every x (Fig. 16.9):

$$f(x) = \frac{4a}{\pi}\left(\cos p \sin x + \tfrac{1}{3}\cos 3p \sin 3x + \tfrac{1}{5}\cos 5p \sin 5x + \ldots\right).$$

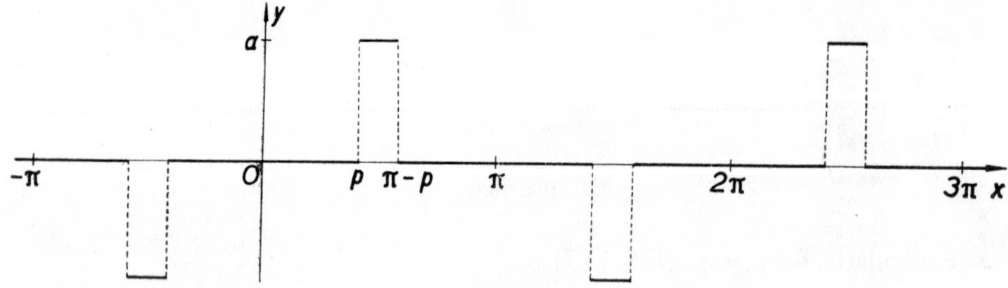

Fig. 16.9.

6. $f(x) = x^2$ for $-\pi \leqq x \leqq \pi$, $f(x + 2\pi) = f(x)$ for every x (Fig. 16.10):

$$f(x) = \frac{\pi^2}{3} - 4\left(\frac{\cos x}{1^2} - \frac{\cos 2x}{2^2} + \frac{\cos 3x}{3^2} - \ldots\right).$$

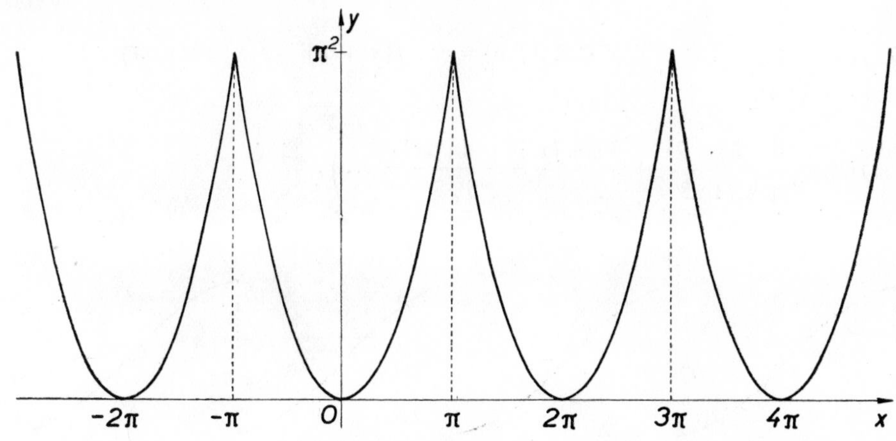

Fig. 16.10.

7. $f(x) = x(\pi - x)$ for $0 \leqq x \leqq \pi$, $f(x + \pi) = f(x)$ for every x (Fig. 16.11):

$$f(x) = \frac{\pi^2}{6} - \left(\frac{\cos 2x}{1^2} + \frac{\cos 4x}{2^2} + \frac{\cos 6x}{3^2} + \ldots\right).$$

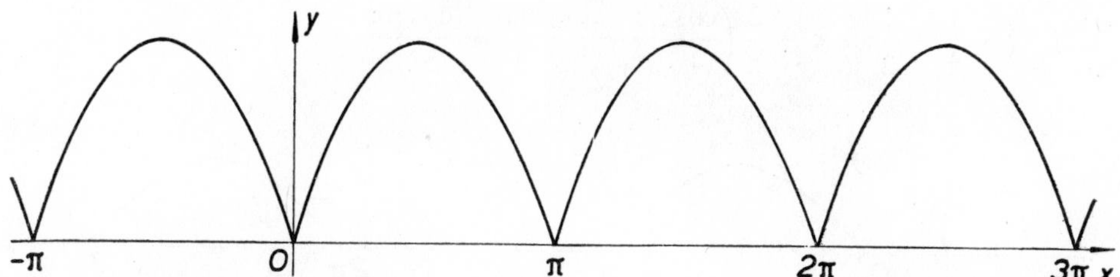

Fig. 16.11.

8. $f(x) = x(\pi - x)$ for $0 \leqq x \leqq \pi$, $f(x + \pi) = -f(x)$ for every x (Fig. 16.12):

$$f(x) = \frac{8}{\pi}\left(\frac{1}{1^3}\sin x + \frac{1}{3^3}\sin 3x + \frac{1}{5^3}\sin 5x + \ldots\right).$$

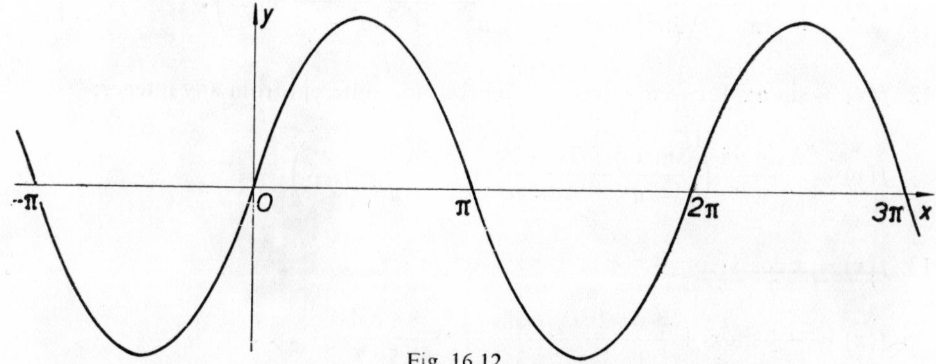

Fig. 16.12.

9. $f(x) = \cos x$ for $0 < x < \pi$, $f(0) = 0$, $f(x + \pi) = f(x)$ for every x (Fig. 16.13):

$$f(x) = \frac{4}{\pi}\left(\frac{2 \sin 2x}{1 . 3} + \frac{4 \sin 4x}{3 . 5} + \frac{6 \sin 6x}{5 . 7} + \dots\right).$$

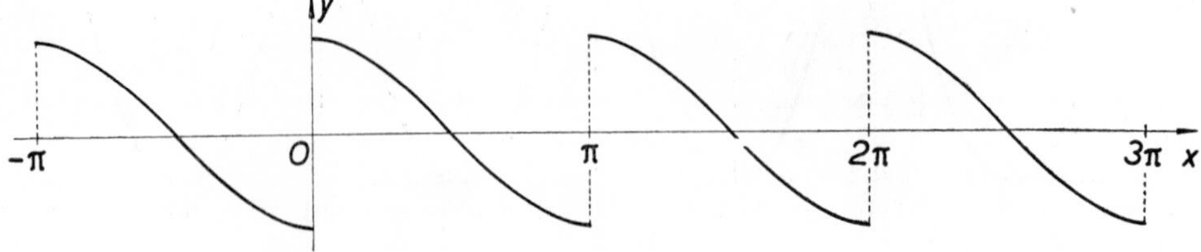

Fig. 16.13.

10. $f(x) = \sin x$ for $0 \leqq x \leqq \pi$, $f(x) = 0$ for $\pi \leqq x \leqq 2\pi$, $f(x + 2\pi) = f(x)$ for every x (Fig. 16.14):

$$f(x) = \frac{1}{\pi} + \frac{1}{2}\sin x - \frac{2}{\pi}\left(\frac{\cos 2x}{1 . 3} + \frac{\cos 4x}{3 . 5} + \frac{\cos 6x}{5 . 7} + \dots\right).$$

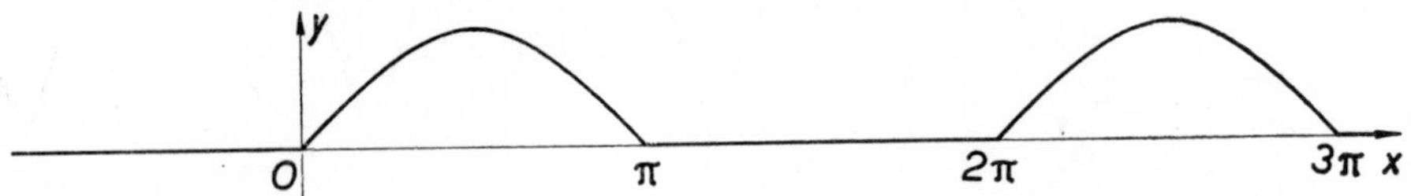

Fig. 16.14.

11. $f(x) = \cos ux$ for $-\pi \leqq x \leqq \pi$, u arbitrary, different from any integer:

$$f(x) = \frac{2u \sin u\pi}{\pi}\left(\frac{1}{2u^2} - \frac{\cos x}{u^2 - 1} + \frac{\cos 2x}{u^2 - 4} - \frac{\cos 3x}{u^2 - 9} + \dots\right).$$

12. $f(x) = \sin ux$ for $-\pi < x < \pi$, u arbitrary, different from any integer:

$$f(x) = \frac{2 \sin u\pi}{\pi}\left(\frac{\sin x}{1 - u^2} - \frac{2 \sin 2x}{4 - u^2} + \frac{3 \sin 3x}{9 - u^2} - \dots\right).$$

13. $f(x) = x \cos x$ for $-\pi < x < \pi$:

$$f(x) = -\tfrac{1}{2}\sin x + \frac{4 \sin 2x}{1 . 3} - \frac{6 \sin 3x}{3 . 5} + \frac{8 \sin 4x}{5 . 7} - \dots .$$

14. $f(x) = -\ln(2\sin\frac{1}{2}x)$ for $0 < x \leqq \pi$:

$$f(x) = \cos x + \tfrac{1}{2}\cos 2x + \tfrac{1}{3}\cos 3x + \dots .$$

15. $f(x) = \ln(2\cos\frac{1}{2}x)$ for $0 \leqq x < \pi$:

$$f(x) = \cos x - \tfrac{1}{2}\cos 2x + \tfrac{1}{3}\cos 3x - \dots .$$

16. $f(x) = \frac{1}{2}\ln\cot\frac{1}{2}x$ for $0 < x < \pi$:

$$f(x) = \cos x + \tfrac{1}{3}\cos 3x + \tfrac{1}{5}\cos 5x + \dots .$$

REMARK 8 (*Fourier Series in Complex Form*). For a periodic function (satisfying the assumptions of Theorem 1 or Remark 2) with basic interval $[0, 2\pi]$ we have

$$\sum_{n=-\infty}^{\infty} c_n e^{inx} = \begin{cases} f(x), \\ \tfrac{1}{2}(f(x+0) + f(x-0)), \end{cases} \quad c_n = \frac{1}{2\pi}\int_0^{2\pi} f(x)\, e^{-inx}\, dx .$$

By a summation from $-\infty$ to ∞ we mean the summation

$$c_0 + \sum_{n=1}^{\infty} (c_n e^{inx} + c_{-n} e^{-inx}) .$$

For a periodic function with basic interval $[0, l]$ we have

$$\sum_{n=-\infty}^{\infty} c_n e^{2\pi inx/l} = \begin{cases} f(x), \\ \tfrac{1}{2}(f(x+0) + f(x-0)), \end{cases} \quad c_n = \frac{1}{l}\int_0^{l} f(x)\, e^{-2\pi inx/l}\, dx .$$

Theorem 2 (*Differentiation and Integration of Fourier Series*). *Let the Fourier series corresponding to a function $f(x)$ be given:*

$$\frac{a_0}{2} + \sum_{n=1}^{\infty} (a_n \cos nx + b_n \sin nx)$$

$$\left(a_n = \frac{1}{\pi}\int_0^{2\pi} f(x)\cos nx\, dx , \quad b_n = \frac{1}{\pi}\int_0^{2\pi} f(x)\sin nx\, dx \right).$$

If $f(x)$ is continuous in $[-\pi, \pi]$, $f(-\pi) = f(\pi)$ and $f'(x)$ is piecewise continuous in $[-\pi, \pi]$, then at each point where $f'(x)$ has a derivative (and consequently is continuous) we have

$$f'(x) = \sum_{n=1}^{\infty} n(-a_n \sin nx + b_n \cos nx)$$

(*term-by-term differentiation*).

If $f(x)$ is piecewise continuous in $[-\pi, \pi]$, then (no matter whether (3) is true or not) we have

$$\int_{-\pi}^{x} f(t)\,dt = \tfrac{1}{2}a_0(x + \pi) + \sum_{n=1}^{\infty} \frac{1}{n}\left[a_n \sin nx - b_n(\cos nx - \cos n\pi)\right] \quad (-\pi \leqq x \leqq \pi)$$

(term-by-term integration).

Similar theorems hold for intervals $[0, l]$ and $[a, b]$.

Theorem 3 (*Riemann–Lebesgue Theorem*). *Let $f(x)$ be a function such that $\int_{-\pi}^{\pi} |f(x)|\,dx$ exists; then $a_n \to 0$, $b_n \to 0$ as $n \to \infty$,* i.e.

$$\lim_{n\to\infty} \int_{-\pi}^{\pi} f(x) \cos nx\,dx = 0\,, \quad \lim_{n\to\infty} \int_{-\pi}^{\pi} f(x) \sin nx\,dx = 0\,.$$

(A similar result is true for the interval $[a, b]$ with functions

$$\cos \frac{2n\pi x}{b - a}, \quad \sin \frac{2n\pi x}{b - a}.)$$

Theorem 4 (a theorem frequently used, for example, in solving partial differential equations by the Fourier method). *If $f(x)$, $f'(x)$, $f''(x)$, $f'''(x)$ are continuous in $[0, l]$ and vanish for $x = 0$ and $x = l$, and if $f^{(4)}(x)$ is piecewise continuous in $[0, l]$, then as $n \to \infty$, the Fourier coefficients*

$$a_n = \frac{2}{l} \int_0^l f(x) \sin \frac{n\pi x}{l}\,dx$$

converge to zero as rapidly as $1/n^4$, i.e. $a_n = O(1/n^4)$ as $n \to \infty$. (In other words, $n^4 a_n$ is bounded for all n.)

The above theorem given for $k = 4$ may be stated for any positive integer k.

REMARK 9. In applications, *Fourier series in two variables* are often encountered. In this case the convergence tests are rather complicated (see e.g. [400]). Let us present a simple criterion:

Theorem 5. *Let the function $f(x, y)$ have continuous derivatives*

$$\frac{\partial f}{\partial x}, \quad \frac{\partial f}{\partial y}, \quad \frac{\partial^2 f}{\partial x\,\partial y}$$

in the rectangle $R(-l \leqq x \leqq l,\ -h \leqq y \leqq h)$. Then at each interior point of the rectangle R we have

$$f(x, y) = \sum_{m,n=0}^{\infty} \lambda_{mn} \left[a_{mn} \cos \frac{m\pi x}{l} \cos \frac{n\pi y}{h} + b_{mn} \sin \frac{m\pi x}{l} \cos \frac{n\pi y}{h} + \right.$$

$$\left. + c_{mn} \cos \frac{m\pi x}{l} \sin \frac{n\pi y}{h} + d_{mn} \sin \frac{m\pi x}{l} \sin \frac{n\pi y}{h} \right], \tag{13}$$

where

$$a_{mn} = \frac{1}{lh} \iint_R f(x, y) \cos \frac{m\pi x}{l} \cos \frac{n\pi y}{h} \, dx \, dy \,, \tag{14}$$

$$b_{mn} = \frac{1}{lh} \iint_R f(x, y) \sin \frac{m\pi x}{l} \cos \frac{n\pi y}{h} \, dx \, dy \,, \tag{15}$$

$$c_{mn} = \frac{1}{lh} \iint_R f(x, y) \cos \frac{m\pi x}{l} \sin \frac{n\pi y}{h} \, dx \, dy \,, \tag{16}$$

$$d_{mn} = \frac{1}{lh} \iint_R f(x, y) \sin \frac{m\pi x}{l} \sin \frac{n\pi y}{h} \, dx \, dy \,, \tag{17}$$

$$\lambda_{mn} = \begin{cases} \frac{1}{4} \text{ if } m = n = 0 \,, \\ \frac{1}{2} \text{ if either } n = 0 \,,\ m > 0 \text{ or } m = 0 \,,\ n > 0 \,, \\ 1 \text{ if } m > 0 \,,\ n > 0 \,. \end{cases} \tag{18}$$

If, in addition, the function $f(x, y)$ has continuous derivatives $\partial^2 f/\partial x^2$, $\partial^3 f/\partial x^2 \, \partial y$ in R and $f(-l, y) = 0$, $f(l, y) = 0$, then the series (13) can be differentiated term-by-term with respect to x and a series with sum $\partial f/\partial x$ is obtained. A similar assertion is true for differentiation with respect to y.

REMARK 10. The exact meaning of the term "convergence of the series (13) to the function $f(x, y)$" is the following: For any chosen $\varepsilon > 0$ an integer N can be found such that for any pair of integers $p > N$, $q > N$ we have

$$|f(x, y) - s_{pq}(x, y)| < \varepsilon \,,$$

where s_{pq} denotes a partial sum of series (13) with m assuming all the values $0, 1, \ldots, p$ and n assuming all the values $0, 1, \ldots, q$ (see Remark 10.2.12, p. 390).

Example 3. For the function $f(x, y) = xy$ on the square $K(-\pi \leqq x \leqq \pi,\ -\pi \leqq \leqq y \leqq \pi)$ we get, by formulae (14) to (17),

$$a_{mn} = b_{mn} = c_{mn} = 0 \,, \quad m, n \text{ non-negative integers} \,,$$

$$d_{00} = d_{01} = d_{10} = 0 \,, \quad d_{mn} = (-1)^{m+n} \frac{4}{mn} \,, \quad m, n \text{ positive integers} \,.$$

Thus, by Theorem 5, at any interior point of K we have

$$xy = 4 \sum_{m,n=1}^{\infty} (-1)^{m+n} \frac{\sin mx \sin ny}{mn} \,.$$

REMARK 11 (*The Fourier Integral*). For $l \to +\infty$ the Fourier series (5) becomes the Fourier integral:

Theorem 6. *If a finite integral*

$$\int_{-\infty}^{\infty} |f(x)| \, dx$$

exists and if $f(x)$ and $f'(x)$ are piecewise continuous functions in any finite interval, then

$$\frac{1}{\pi}\int_0^{\infty} du \int_{-\infty}^{\infty} f(t) \cos\{u(t-x)\} \, dt = \begin{cases} f(x) \text{ at each point of continuity of the} \\ \text{function } f(x), \\ \frac{1}{2}\{f(x+0)+f(x-0)\} \text{ at each point} \\ \text{of discontinuity of the function } f(x). \end{cases} \tag{19}$$

REMARK 12. Expanding the term $\cos\{u(t-x)\}$ in (19), we get

$$\frac{1}{\pi}\int_0^{\infty} \cos ux \, du \int_{-\infty}^{\infty} f(t) \cos ut \, dt + \frac{1}{\pi}\int_0^{\infty} \sin ux \, du \int_{-\infty}^{\infty} f(t) \sin ut \, dt =$$

$$= \begin{cases} f(x), \\ \frac{1}{2}(f(x+0)+f(x-0)). \end{cases}$$

If $f(x)$ is an even function, i.e. if $f(-x) \equiv f(x)$, then

$$\frac{2}{\pi}\int_0^{\infty} \cos ux \, du \int_0^{\infty} f(t) \cos ut \, dt = \begin{cases} f(x), \\ \frac{1}{2}(f(x+0)+f(x-0)); \end{cases}$$

if $f(x)$ is odd, i.e. if $f(-x) \equiv -f(x)$, then

$$\frac{2}{\pi}\int_0^{\infty} \sin ux \, du \int_0^{\infty} f(t) \sin ut \, dt = \begin{cases} f(x), \\ \frac{1}{2}(f(x+0)+f(x-0)). \end{cases}$$

Using complex form, we have:

$$\frac{1}{2\pi}\int_{-\infty}^{\infty} e^{-ixu} \, du \int_{-\infty}^{\infty} f(t) \, e^{iut} \, dt = \begin{cases} f(x), \\ \frac{1}{2}(f(x+0)+f(x-0)), \end{cases}$$

where the first improper integral is to be taken in the sense of the Cauchy principal value, i.e. as the limit $\lim \int_{-a}^{a}$ for $a \to +\infty$.

REMARK 13 (*Harmonic Analysis*). Consider a function $f(x)$ with period b; let us expand it in $[0, b]$ as a Fourier series

$$\frac{a_0}{2} + \sum_{n=1}^{\infty}\left(a_n \cos\frac{2n\pi x}{b} + b_n \sin\frac{2n\pi x}{b}\right), \tag{20}$$

where

$$a_n = \frac{2}{b}\int_0^b f(x)\cos\frac{2n\pi x}{b}\,dx\,,\quad b_n = \frac{2}{b}\int_0^b f(x)\sin\frac{2n\pi x}{b}\,dx \tag{21}$$

(see Remark 5, formulae (5′), (6′), where we have written $a = 0$). The values of the function $f(x)$ are often known only at some discrete points of the interval $[0, b]$ (for example, they may be obtained by measurement). Let these points, for instance, be

$$x_0 = 0\,,\quad x_1 = \frac{b}{12}\,,\quad x_2 = \frac{2b}{12}\,,\quad \ldots,\quad x_k = \frac{kb}{12}\,,\quad \ldots,\quad x_{12} = b \tag{22}$$

and write

$$f(x_k) = f_k\,.$$

If the calculation of a_n, b_n is required, we can proceed by § 13.13 (trapezoidal rule). For example, for a_2 (i.e. $n = 2$) we get, using (21) and the trapezoidal rule,

$$a_2 = \frac{2}{b}\,\frac{b}{12}\left(\frac{y_0}{2} + y_1 + y_2 + \ldots + y_{10} + y_{11} + \frac{y_{12}}{2}\right) =$$

$$= \frac{1}{6}\left(\frac{f_0}{2} + f_1\cos\frac{4\pi\frac{b}{12}}{b} + f_2\cos\frac{4\pi\frac{2b}{12}}{b} + \ldots + f_{10}\cos\frac{4\pi\frac{10b}{12}}{b} + \right.$$

$$\left. + f_{11}\cos\frac{4\pi\frac{11b}{12}}{b} + \frac{f_{12}}{2}\right) =$$

$$= \frac{1}{6}\left(\frac{f_0}{2} + f_1\cos 60° + f_2\cos 120° + \ldots + f_{10}\cos 600° + f_{11}\cos 660° + \frac{f_{12}}{2}\right) =$$

$$= \tfrac{1}{12}\left[(f_0 + f_1 + f_5 + 2f_6 + f_7 + f_{11} + f_{12}) - (f_2 + 2f_3 + f_4 + f_8 + 2f_9 + f_{10})\right]. \tag{23}$$

The procedure just demonstrated is usually called the *Runge method*. There are tables giving computation schemes for respective numbers of points x_k.

Let us present a scheme for the case already given, i.e. for a subdivision of the interval $[0, b]$ into 12 equal parts. In view of periodicity we shall assume that $f_{12} = f_0$. We form sums and differences thus:

	f_0	f_1	f_2	f_3	f_4	f_5	f_6	s_0	s_1	s_2	s_3	d_1	d_2	d_3
		f_{11}	f_{10}	f_9	f_8	f_7		s_6	s_5	s_4		d_5	d_4	
Sums	s_0	s_1	s_2	s_3	s_4	s_5	s_6	σ_0	σ_1	σ_2	σ_3	δ_1	δ_2	δ_3
Differences		d_1	d_2	d_3	d_4	d_5		τ_0	τ_1	τ_2		γ_1	γ_2	

The ensuing calculation follows the pattern given in Table 16.1.

When computing sums I, II, the numbers σ, τ, δ, γ are multiplied by the numbers standing in the respective rows on the left. For example, for $6a_2$ we get

$$\begin{aligned}6a_2 &= \sigma_0 - \sigma_3 + 0{\cdot}5(\sigma_1 - \sigma_2) = s_0 + s_6 - s_3 + 0{\cdot}5(s_1 + s_5 - s_2 - s_4) = \\ &= f_0 + f_6 - f_3 - f_9 + 0{\cdot}5(f_1 + f_{11} + f_5 + f_7 - f_2 - f_{10} - f_4 - f_8) = \\ &= \tfrac{1}{2}(2f_0 + f_1 + f_5 + 2f_6 + f_7 + f_{11} - f_2 - 2f_3 - f_4 - f_8 - 2f_9 - f_{10})\end{aligned}$$

in agreement with (23) (since $f_{12} = f_0$ by assumption).

The coefficients a_n, b_n may also be established by means of graphical methods, method of least squares, etc.

TABLE 16.1

	Coefficients of cosine terms								Coefficients of sine terms					
1	σ_0 σ_2	σ_1 σ_3	τ_0		σ_0	$-\sigma_3$	τ_0	τ_2	δ_3				δ_1	δ_3
0·866				τ_1						δ_2	γ_1	γ_2		
0·5			τ_2		$-\sigma_2$	σ_1			δ_1					
Sums	I	II	I	II	I	II	I	II	I	II	I	II	I	II
Sums I + II	$6a_0$		$6a_1$		$6a_2$		—		$6b_1$		$6b_2$		—	
Differences I — II	$6a_6$		$6a_5$		$6a_4$		$6a_3$		$6b_5$		$6b_4$		$6b_3$	

16.4. Bessel Functions

Definition 1. The *Bessel function of the first kind and order (index)** n is defined by the equation

$$J_n(x) = \left(\frac{x}{2}\right)^n \sum_{k=0}^{\infty} \frac{(-1)^k}{k!\,\Gamma(n+k+1)} \left(\frac{x}{2}\right)^{2k}; \tag{1}$$

* A commonly used convention in English literature on Bessel functions is that in formulae valid for general values of the order, the Greek letter ν is written in place of n; in formulae which are true only for integral values of the order, e.g. (2) below, the Roman letter n is used. But this convention is not adopted here.

the series converges for any real n and all x. (For the interpretation of $(\frac{1}{2}x)^n$ for x complex see Remark 20.6.4, p. 967.)

REMARK 1. $k! = 1$ for $k = 0$. The symbol $\Gamma(u)$ denotes the gamma function defined in § 13.11 (p. 584). At the same time, for $u = 0, -1, -2, \ldots$ we define $1/\Gamma(u) = 0$. In this way formula (1) is meaningful even if n is a negative integer.

Example 1.

$$J_2(it) = \frac{1}{2!}\left(\frac{it}{2}\right)^2 - \frac{1}{1!\,3!}\left(\frac{it}{2}\right)^4 + \frac{1}{2!\,4!}\left(\frac{it}{2}\right)^6 - \ldots = -\left(\frac{t^2}{8} + \frac{t^4}{96} + \frac{t^6}{3{,}072} + \ldots\right)$$

(for every t).

Theorem 1. *For integral n we have*

$$J_{-n}(x) = (-1)^n J_n(x)\,. \tag{2}$$

REMARK 2. From (2) it is clear that for a negative integer n it is not necessary to use formula (1) directly, but to put $n = -m$ (m a positive integer) and compute $J_m(x)$.

Theorem 2. *The function $y = J_n(x)$ satisfies the (Bessel) differential equation*

$$x^2y'' + xy' + (x^2 - n^2)\,y = 0$$

(see § 17.15 and § 17.21, equation 117).

Theorem 3. *Bessel functions with integral orders are coefficients in the expansion of the so-called generating function; that is, for all x and all $t \neq 0$ we have*

$$e^{(x/2)(t-1/t)} = \sum_{n=-\infty}^{\infty} J_n(x)\,t^n = J_0(x) + J_1(x)\,t + J_2(x)\,t^2 + \ldots$$

$$+ J_{-1}(x)\,t^{-1} + J_{-2}(x)\,t^{-2} + \ldots\,. \tag{3}$$

Theorem 4 (*Integral Form*). *For $n = 0, 1, 2, \ldots$ we have*

$$J_n(x) = \frac{1}{\pi}\int_0^{\pi} \cos\,(x \sin \vartheta - n\vartheta)\,d\vartheta\,. \tag{4}$$

REMARK 3. The functions of orders zero and one, i.e.

$$J_0(x) = 1 - \frac{x^2}{2^2} + \frac{x^4}{(2\,.\,4)^2} - \frac{x^6}{(2\,.\,4\,.\,6)^2} + \ldots, \tag{5}$$

$$J_1(x) = \frac{x}{2}\left(1 - \frac{x^2}{2\,.\,4} + \frac{x^4}{2\,.\,4^2\,.\,6} - \frac{x^6}{2\,.\,(4\,.\,6)^2\,.\,8} + \ldots\right) \tag{6}$$

are particularly important in applications; their graphs and some numerical values are shown in Fig. 16.15 and Table 16.2. Further, we have, for functions of half-integral order (sometimes known as *spherical Bessel functions*):

$$J_{1/2}(x) = \sqrt{\left(\frac{2}{\pi x}\right)} \sin x\,, \quad J_{3/2}(x) = \sqrt{\left(\frac{2}{\pi x}\right)}\left(\frac{\sin x}{x} - \cos x\right),$$

$$J_{-1/2}(x) = \sqrt{\left(\frac{2}{\pi x}\right)} \cos x\,, \quad J_{-3/2}(x) = -\sqrt{\left(\frac{2}{\pi x}\right)}\left(\frac{\cos x}{x} + \sin x\right).$$

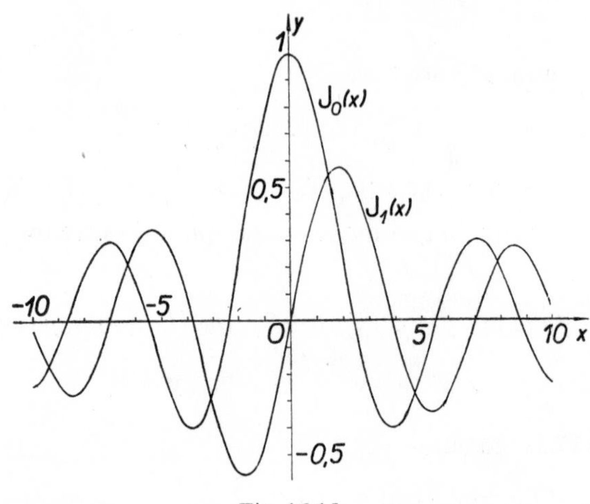

Fig. 16.15.

TABLE 16.2

Bessel functions of orders zero and one

x	$J_0(x)$	$J_1(x)$	x	$J_0(x)$	$J_1(x)$	x	$J_0(x)$	$J_1(x)$
0·0	1·000 0	0·000 0	5·5	−0·006 8	−0·341 4	11·0	−0·171 2	−0·176 8
0·5	0·938 5	0·242 3	6·0	0·150 6	−0·276 7	11 5	−0·067 7	−0·228 4
1·0	0·765 2	0·440 1	6·5	0·260 1	−0·153 8	12·0	0·047 7	−0·223 4
1·5	0·511 8	0·557 9	7·0	0·300 1	−0·004 7	12·5	0·146 9	−0·165 5
2·0	0·223 9	0·576 7	7·5	0·266 3	0·135 2	13·0	0·206 9	−0·070 3
2·5	−0·048 4	0·497 1	8·0	0·171 7	0·234 6	13·5	0·215 0	0·038 0
3·0	−0·260 1	0·339 1	8·5	0·041 9	0·273 1	14·0	0·171 1	0·133 4
3·5	−0·380 1	0·137 4	9·0	−0·090 3	0·245 3	14·5	0·087 5	0·193 4
4·0	−0·397 1	−0·066 0	9·5	−0·193 9	0·161 3	15·0	−0·014 2	0·205 1
4·5	−0·320 5	−0·231 1	10·0	−0·245 9	0·043 5	15·5	−0·109 2	0·167 2
5·0	−0·177 6	−0·327 6	10·5	−0·236 6	−0·078 9	16·0	−0·174 9	0·090 4

Theorem 5 (*some important relationships; n is assumed to be real and* $x \neq 0$, *in general*).

$$2\,J_n'(x) = J_{n-1}(x) - J_{n+1}(x)\,,$$

$$\frac{2n}{x}\,J_n(x) = J_{n-1}(x) + J_{n+1}(x)\,,$$

$$\frac{d}{dx}\left[x^n\,J_n(x)\right] = x^n\,J_{n-1}(x)\,, \quad \frac{d}{dx}\left[\frac{J_n(x)}{x^n}\right] = -\frac{J_{n+1}(x)}{x^n}\,,$$

in particular

$$J_0'(x) = -J_1(x)\,.$$

For $n = 0, 1, 2, \ldots$ *we have*

$$|J_n(x)| \leqq 1\,, \quad |J_n^{(k)}(x)| < 1 \quad \textit{for all } x \quad (k = 1, 2, 3, \ldots)\,,$$

$$\lim_{x \to +\infty} J_n(x) = 0\,.$$

Theorem 6. *For every real n, the function* $J_n(x)$ *possesses an infinite number of positive zeros* $x_1 < x_2 < x_3 < \ldots$, *such that* $x_k \to +\infty$ *as* $k \to \infty$.

Theorem 7. *Let* $\lambda_1 < \lambda_2 < \lambda_3 < \ldots$ *be the positive roots of the equation*

$$J_n(\lambda c) = 0 \quad (n \textit{ fixed, real, non-negative; } c > 0)\,. \tag{7}$$

Then the functions $J_n(\lambda_1 x)$, $J_n(\lambda_2 x)$, $J_n(\lambda_3 x)$, ... *constitute an orthogonal system in the interval* $[0, c]$ *with the weight function x* (Definition 16.2.5), *i.e.*

$$\int_0^c x\,J_n(\lambda_i x)\,J_n(\lambda_k x)\,dx = 0 \quad \textit{for} \quad i \neq k\,. \tag{8}$$

REMARK 4. The same is true, if $\lambda_1 < \lambda_2 < \lambda_3 < \ldots$ are the non-negative roots of the equation

$$\lambda c\,J_n'(\lambda c) = -h\,J_n(\lambda c) \tag{9}$$

(h being a constant not necessarily different from zero).

Theorem 8. *If* $n \geqq 0$, $h \geqq 0$, *then equations* (7), (9) *have only real roots.*

REMARK 5. For the construction of orthogonal systems (Theorem 7 and Remark 4) only non-negative roots λ_k need to be taken into account.

REMARK 6. For problems in applications leading to equations (7) and (9), see § 26.4.

REMARK 7. Orthonormal systems

$$\varphi_{n1}(x)\,, \quad \varphi_{n2}(x)\,, \quad \varphi_{n3}(x)\,, \ldots$$

for the weight function x, i.e.

$$\int_0^c x\, \varphi_{ni}(x)\, \varphi_{nk}(x)\, dx = \begin{cases} 1 \text{ for } i = k, \\ 0 \text{ for } i \neq k, \end{cases}$$

are obtained from the orthogonal systems dealt with in Theorem 7 and Remark 4 in the following manner:

In the case of equation (7),

$$\varphi_{nk}(x) = \frac{J_n(\lambda_k x)}{\sqrt{[\frac{1}{2}c^2\, J_{n+1}^2(\lambda_k c)]}},$$

and in the case of equation (9),

$$\varphi_{nk}(x) = \frac{J_n(\lambda_k x)}{\sqrt{\left[\dfrac{\lambda_k^2 c^2 + h^2 - n^2}{2\lambda_k^2}\, J_n^2(\lambda_k c)\right]}}.$$

(The numbers λ_k are different in the two cases (7) and (9), respectively; consequently the symbol $J_n(\lambda_k x)$ also denotes different functions in the two cases.) For $n = 0$, $h = 0$ and $\lambda = 0$ we have $\varphi_{n1}(x) = \sqrt{(2)}/c$.

Theorem 9 (*Fourier–Bessel Expansion*). *Let*

$$J_n(\lambda_1 x),\quad J_n(\lambda_2 x),\quad J_n(\lambda_3 x),\ \ldots$$

be the functions considered in Theorem 7 and Remark 4, $n \geqq 0$, $h \geqq 0$. (We assume that all the $\lambda_1, \lambda_2, \lambda_3, \ldots$ are positive, or non-negative, roots of equation (7) or equation (9), respectively; n, h being fixed numbers.) Let $f(x)$ and $f'(x)$ be piecewise continuous functions in the interval $[0, c]$. Then

$$\sum_{k=1}^{\infty} a_k\, J_n(\lambda_k x) =$$

$$= \begin{cases} f(x) & \text{at each point of continuity of the function } f(x), \\ \frac{1}{2}\{f(x+0) - f(x-0)\} & \text{at each point of discontinuity of the function } f(x) \end{cases}$$

$$(0 < x < c) \qquad (10)$$

($f(x+0)$, $f(x-0)$ are the limits from the right and the left, respectively, of the function $f(x)$ at the point x). The series (10) converges uniformly in every closed interval interior to an interval in which $f(x)$ is continuous. If the λ_k are the positive roots of equation (7), we have in (10)

$$a_k = \frac{2}{c^2\, J_{n+1}^2(\lambda_k c)} \int_0^c x\, J_n(\lambda_k x)\, f(x)\, dx \quad (k = 1, 2, 3, \ldots); \qquad (11)$$

if the λ_k's are the non-negative roots of equation (9), *we have*

$$a_k = \frac{2\lambda_k^2}{(\lambda_k^2 c^2 + h^2 - n^2)\,J_n^2(\lambda_k c)} \int_0^c x\,J_n(\lambda_k x)\,f(x)\,dx \quad (k = 1, 2, 3, \ldots). \qquad (12)$$

REMARK 8. If $n = 0$, $h = 0$, then $\lambda_1 = 0$ in (12) and consequently

$$a_1 = \frac{2}{c^2} \int_0^c x\,f(x)\,dx\,.$$

REMARK 9. For the validity of equation (10) it is sufficient (provided $f(x)$ is integrable in every interval $[\varepsilon, c]$, $\varepsilon > 0$, arbitrarily small) that the integral

$$\int_0^c \sqrt{(x)}\,|f(x)|\,dx$$

converges and that the point x is interior to an interval in which $f(x)$ has bounded variation. The convergence is then again uniform in every closed interval interior to an interval in which $f(x)$ is continuous.

REMARK 10. Theorem 9 may be generalized for the case $n \geqq -\frac{1}{2}$ (see, e.g., [429], where Bessel functions are thoroughly treated).

REMARK 11. Theorem 9 is of fundamental importance in applications (see § 26.4).

REMARK 12. *Bessel functions of the second kind*, defined by the relationship

$$Y_n(x) = \frac{J_n(x)\cos n\pi - J_{-n}(x)}{\sin n\pi} \quad (n \text{ non-integral}),$$

are of more limited application.

For integral n, $Y_n(x)$ may be defined by the equation

$$Y_n(x) = \lim_{\substack{r \to n \\ r \neq n}} Y_r(x)$$

(i.e. $Y_n(x)$ is approximated by functions $Y_r(x)$ with indices r close to the given integer n). See also § 17.15, p. 797.

Bessel functions of the third kind (*Hankel functions*) *are* defined by the formulae

$$H_n^{(1)}(x) = J_n(x) + i\,Y_n(x)\,,$$

$$H_n^{(2)}(x) = J_n(x) - i\,Y_n(x)\,,$$

where i is the imaginary unit.

For details see [211], [429]. See also § 17·21, equation 117 and further.

16.5. Legendre Polynomials. Spherical Harmonics

Definition 1. The polynomial

$$P_n(x) = \frac{1 . 3 . 5 . \dots . (2n-1)}{n!}\left[x^n - \frac{n(n-1)}{2(2n-1)}x^{n-2} + \right.$$

$$\left. + \frac{n(n-1)(n-2)(n-3)}{2 . 4(2n-1)(2n-3)}x^{n-4} - \dots\right] \tag{1}$$

is called a *Legendre polynomial of degree n.*

Theorem 1. *We have*

$$P_n(x) = \frac{1}{2^n n!}\frac{d^n}{dx^n}(x^2-1)^n \tag{2}$$

(*the Rodrigues formula*).

Example 1. For $n = 2$ we have by (2):

$$P_2(x) = \frac{1}{2^2 . 2!}\frac{d^2}{dx^2}(x^4 - 2x^2 + 1) = \tfrac{3}{2}x^2 - \tfrac{1}{2} .$$

This result, of course, may also be obtained from (1).

Theorem 2. *The function* $y = P_n(x)$ *satisfies the (Legendre) differential equation* (see also § 17·21, equation 129 and further)

$$(1 - x^2)\,y'' - 2xy' + n(n+1)\,y = 0 . \tag{3}$$

REMARK 1. The polynomial $P_n(\cos\vartheta)$ satisfies the equation

$$\frac{1}{\sin\vartheta}\frac{d}{d\vartheta}\left(\sin\vartheta\frac{dy}{d\vartheta}\right) + n(n+1)\,y = 0 , \tag{4}$$

which can be obtained from (3) by the substitution $x = \cos\vartheta$.

Theorem 3. *The first five Legendre polynomials (in the variables x and* ϑ*) are:*

$P_0(x) = 1$,

$P_1(x) = x = \cos\vartheta$,

$P_2(x) = \tfrac{3}{2}x^2 - \tfrac{1}{2} = \tfrac{1}{2}(3\cos^2\vartheta - 1) = \tfrac{1}{4}(3\cos 2\vartheta + 1)$,

$P_3(x) = \tfrac{5}{2}x^3 - \tfrac{3}{2}x = \tfrac{1}{2}(5\cos^3\vartheta - 3\cos\vartheta) = \tfrac{1}{8}(5\cos 3\vartheta + 3\cos\vartheta)$,

$P_4(x) = \tfrac{35}{8}x^4 - \tfrac{15}{4}x^2 + \tfrac{3}{8} = \tfrac{1}{8}(35\cos^4\vartheta - 30\cos^2\vartheta + 3) =$
$= \tfrac{1}{64}(35\cos 4\vartheta + 20\cos 2\vartheta + 9)$.

The graphs and some numerical values of these functions are shown in Figs 16.16, 16.17 and Table 16.3.

Theorem 4 (*Fundamental Properties*).
1. $P_n(-x) = (-1)^n P_n(x)$.
2. $P_n(1) = 1$.

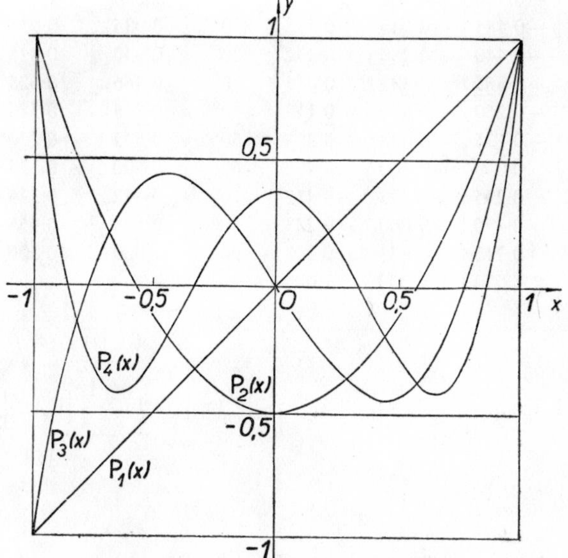

Fig. 16.16.

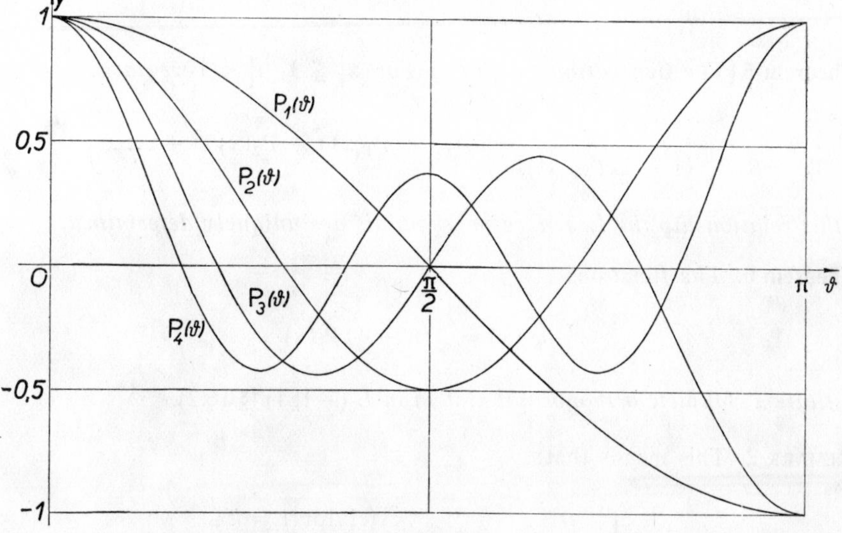

Fig. 16.17.

TABLE 16.3

Legendre polynomials

x	$P_1(x)$	$P_2(x)$	$P_3(x)$	$P_4(x)$	ϑ	$P_1(\vartheta)$	$P_2(\vartheta)$	$P_3(\vartheta)$	$P_4(\vartheta)$
0·0	0·000	−0·500	0·000	0·375	0°	1·000	1·000	1·000	1·000
0·1	0·100	−0·485	−0·148	0·338	10°	0·985	0·955	0·911	0·853
0·2	0·200	−0·440	−0·280	0·232	20°	0·940	0·825	0·665	0·475
0·3	0·300	−0·365	−0·383	0·073	30°	0·866	0·625	0·325	0·023
0·4	0·400	−0·260	−0·440	−0·113	40°	0·766	0·380	−0·025	−0·319
0·5	0·500	−0·125	−0·438	−0·289	50°	0·643	0·120	−0·300	−0·428
0·6	0·600	0·040	−0·360	−0·408	60°	0·500	−0·125	−0·438	−0·289
0·7	0·700	0·235	−0·193	−0·412	70°	0·342	−0·325	−0·413	−0·004
0·8	0·800	0·460	0·080	−0·233	80°	0·174	−0·455	−0·247	0·266
0·9	0·900	0·715	0·473	0·208	90°	0·000	−0·500	0·000	0·375
1·0	1·000	1·000	1·000	1·000					

3. *For* $|x| \leqq 1$ *we have* $|P_n(x)| \leqq 1, \quad \frac{1}{n^2}|P_n'(x)| \leqq 1, \quad \frac{1}{n^4}|P_n''(x)| \leqq 1, \ldots,$

$$\frac{1}{n^{2k}}|P_n^{(k)}(x)| \leqq 1\,.$$

4. *The roots of the equation* $P_n(x) = 0$ $(n = 1, 2, \ldots)$ *lie in the interval* $(-1, 1)$.

5. $P_n(x) = \dfrac{2n-1}{n}\, x\, P_{n-1}(x) - \dfrac{n-1}{n} P_{n-2}(x) \quad (n \geqq 2)\,.$

Theorem 5 (*The Generating Function*). *For* $|x| \leqq 1$, $|t| < 1$ *we have*

$$\frac{1}{\sqrt{(1 - 2xt + t^2)}} = P_0(x) + P_1(x)\,t + P_2(x)\,t^2 + \ldots\,. \tag{5}$$

By this relationship the Legendre polynomials are uniquely determined.

Theorem 6. *The functions*

$$\varphi_n(x) = \sqrt{(n + \tfrac{1}{2})}\, P_n(x)$$

constitute a complete orthonormal system in $L_2(-1, 1)$ (§ 16.2).

REMARK 2. This means that:

1. $$\int_{-1}^{1} \varphi_i(x)\, \varphi_k(x)\, dx = \begin{cases} 1 \text{ for } i = k\,, \\ 0 \text{ for } i \neq k\,. \end{cases}$$

2. The Fourier series (Definition 16.2.6)

$$\sum_{k=0}^{\infty} c_k \varphi_k(x)$$

of any function $f \in L_2(-1, 1)$ converges in the mean to $f(x)$. For pointwise convergence we have:

Theorem 7. *Let $f(x)$ be bounded and integrable in $[-1, 1]$. Let*

$$a_n = \frac{2n+1}{2} \int_{-1}^{1} f(x) \, \mathrm{P}_n(x) \, \mathrm{d}x \quad (n = 0, 1, 2, \ldots).$$

Then at each point x $(-1 < x < 1)$ interior to an interval in which $f(x)$ is of bounded variation, we have

$$\sum_{n=0}^{\infty} a_n \mathrm{P}_n(x) = \begin{cases} f(x), & \text{provided } f(x) \text{ is continuous at } x, \\ \frac{1}{2}(f(x+0) - f(x-0)) & \text{if } f(x) \text{ is discontinuous at } x \end{cases}$$

($f(x+0)$ and $f(x-0)$ denote the limits of the function $f(x)$ from the right and the left, respectively, at the point x).

Remark 3. Thus, in particular, for every function with a continuous derivative in $[-1, 1]$ we have in $(-1, 1)$,

$$f(x) = \sum_{n=0}^{\infty} a_n \mathrm{P}_n(x).$$

Remark 4. The application of Theorem 7 is similar to that of Theorem 16.4.9.

Remark 5 (*Spherical Harmonics*). Denote*

$$\sqrt{[(1-x^2)^m]} \frac{\mathrm{d}^m \mathrm{P}_n(x)}{\mathrm{d}x^m} = \mathrm{P}_{n,m}(x), \tag{6}$$

where $\mathrm{P}_n(x)$ is the Legendre polynomial of the n-th degree. For example, we have

$$\mathrm{P}_{2,1}(x) = \sqrt{(1-x^2)} \frac{\mathrm{d}(\frac{3}{2}x^2 - \frac{1}{2})}{\mathrm{d}x} = 3x\sqrt{(1-x^2)}. \tag{7}$$

$$\mathrm{P}_{2,2}(x) = (1-x^2) \frac{\mathrm{d}^2 \mathrm{P}_2(x)}{\mathrm{d}x^2} = 3(1-x^2). \tag{8}$$

Further, we shall use the notation $\mathrm{P}_n(\cos \vartheta)$, $\mathrm{P}_{n,m}(\cos \vartheta)$ which denotes $\mathrm{P}_n(x)$ and $\mathrm{P}_{n,m}(x)$, respectively, with $\cos \vartheta$ put for x. At the same time, in the case of the func-

* It is more common in English literature to write $P_n^m(x)$ in place of $P_{n,m}(x)$.

tions $P_{n,m}$ we put $\sin \vartheta$ for $\sqrt{(1 - x^2)}$. For example, by (7), (8) we have

$$P_{2,1}(\cos \vartheta) = 3 \sin \vartheta \cos \vartheta , \tag{9}$$

$$P_{2,2}(\cos \vartheta) = 3 \sin^2 \vartheta . \tag{10}$$

The function

$$Y_n(\vartheta, \varphi) = a_{n,0}P_n(\cos \vartheta) + \sum_{m=1}^{n} (a_{n,m} \cos m\varphi + b_{n,m} \sin m\varphi) P_{n,m}(\cos \vartheta) \tag{11}$$

($a_{n,m}$, $b_{n,m}$ being arbitrary numbers) is called a *spherical harmonic of degree n*.

For example (the first index 2 of the constants is omitted),

$$Y_2(\vartheta, \varphi) = a_0(\tfrac{3}{2} \cos^2 \vartheta - \tfrac{1}{2}) + (a_1 \cos \varphi + b_1 \sin \varphi) . 3 \sin \vartheta \cos \vartheta + \\ + (a_2 \cos 2\varphi + b_2 \sin 2\varphi) . 3 \sin^2 \vartheta .$$

REMARK 6. Legendre polynomials (1) are also often called spherical harmonics in the literature.

Theorem 8. *The spherical harmonics are orthogonal on the unit sphere, i.e.,*

$$\iint_K Y_p Y_q \, dS = \int_0^{2\pi} \int_0^{\pi} Y_p(\vartheta, \varphi) \, Y_q(\vartheta, \varphi) \sin \vartheta \, d\vartheta \, d\varphi = 0 \quad (p \neq q) .$$

Theorem 9. *Let $g(\vartheta, \varphi)$ together with its partial derivatives of first and second order be continuous functions of the variables ϑ, φ on the unit sphere K ($r = 1$, $0 \leqq \vartheta \leqq \pi$, $0 \leqq \varphi \leqq 2\pi$). Let us choose, in (11), the constants $a_{n,m}$, $b_{n,m}$ as follows:*

$$a_{n,0} = \frac{2n + 1}{4\pi} \int_0^{2\pi} \int_0^{\pi} g(\vartheta, \varphi) \, P_n(\cos \vartheta) \sin \vartheta \, d\vartheta \, d\varphi ,$$

$$a_{n,m} = \frac{(2n + 1)(n - m)!}{2\pi(n + m)!} \int_0^{2\pi} \int_0^{\pi} g(\vartheta, \varphi) \, P_{n,m}(\cos \vartheta) \cos m\varphi \sin \vartheta \, d\vartheta \, d\varphi ,$$

$$b_{n,m} = \frac{(2n + 1)(n - m)!}{2\pi(n + m)!} \int_0^{2\pi} \int_0^{\pi} g(\vartheta, \varphi) \, P_{n,m}(\cos \vartheta) \sin m\varphi \sin \vartheta \, d\vartheta \, d\varphi .$$

Then

$$g(\vartheta, \varphi) = \sum_{n=0}^{\infty} Y_n(\vartheta, \varphi) , \tag{12}$$

the series (12) *being absolutely and uniformly convergent for* $0 \leqq \varphi \leqq 2\pi$, $0 \leqq \leqq \vartheta \leqq \pi$.

If $g(\vartheta, \varphi)$ is only square integrable on K (Remark 16.1.7, p. 694), *then the series* (12) *converges in the mean to the function g.*

Theorem 10. *For any choice of constants* $a_{n,m}$, $b_{n,m}$ *the function*

$$U_n(x, y, z) = r^n \mathrm{Y}_n(\vartheta, \varphi)$$

is harmonic, i.e. it satisfies the equation

$$\frac{\partial^2 U_n}{\partial x^2} + \frac{\partial^2 U_n}{\partial y^2} + \frac{\partial^2 U_n}{\partial z^2} = 0 .$$

(Here, the relationship between x, y, z *and* r, φ, ϑ *is given by the equations* $x = r \sin \vartheta \cos \varphi$, $y = r \sin \vartheta \sin \varphi$, $z = r \cos \vartheta$,

$$r = \sqrt{(x^2 + y^2 + z^2)} \,.)$$

16.6. Some Further Important Functions (Hypergeometric Functions, Jacobi Polynomials, Chebyshev Polynomials, Laguerre Polynomials, Hermite Polynomials)

Legendre polynomials appear as a particular case of the so-called *hypergeometric function*, namely*

$$F(\alpha, \beta, \gamma, x) = 1 + \frac{\alpha\beta}{1 \,.\, \gamma} x + \frac{\alpha(\alpha + 1)\, \beta(\beta + 1)}{1 \,.\, 2 \,.\, \gamma(\gamma + 1)} x^2 + $$
$$+ \frac{\alpha(\alpha + 1)(\alpha + 2)\, \beta(\beta + 1)(\beta + 2)}{1 \,.\, 2 \,.\, 3 . \gamma(\gamma + 1)(\gamma + 2)} x^3 + \ldots \tag{1}$$

(the series converging for $|x| < 1$ and for γ different from any non-positive integer) which is a solution of the so-called *hypergeometric (Gauss) equation*

$$x(1 - x)\, y'' + [\gamma - (\alpha + \beta + 1)\, x]\, y' - \alpha\beta y = 0 \tag{2}$$

(α, β, γ being constants) (see also § 17.21, equation 140 and further).

If $\alpha = 1$, $\beta = \gamma$, the series (1) reduces to a geometric series

$$1 + x + x^2 + \ldots = \frac{1}{1 - x}$$

(a solution of equation (2) for $\alpha = 1$, $\beta = \gamma$).

For $\alpha = -n$, $\beta = n + p$, $\gamma = q$ ($q > 0$, $p - q > -1$) we get the so-called *Jacobi polynomials*

$$J_n(p, q, x) = F(-n, n + p, q, x) =$$
$$= 1 + \sum_{k=1}^{n} (-1)^k \binom{n}{k} \frac{(n + p)(n + p + 1) \ldots (n + p + k - 1)}{q(q + 1) \ldots (q + k - 1)} x^k .$$

* The slightly different notation $F(\alpha, \beta; \gamma; x)$ is often used.

(It should be noted that the term *Jacobi polynomial* is applied also to a different polynomial in English literature.)

As their particular case we get Legendre polynomials for $p = 1$, $q = 1$ (setting $\frac{1}{2}(1 - x)$ for x),

$$\mathrm{P}_n(x) = J_n\left(1, 1, \frac{1 - x}{2}\right) = F\left(-n, n + 1, 1, \frac{1 - x}{2}\right). \tag{3}$$

Example 1. For $n = 2$ we have from (1) and (3),

$$\mathrm{P}_2(x) = F\left(-2, 3, 1, \frac{1 - x}{2}\right) = 1 + \frac{(-2) \cdot 3}{1 \cdot 1} \cdot \frac{1 - x}{2} + \frac{(-2)(-1) \cdot 3 \cdot 4}{1 \cdot 2 \cdot 1 \cdot 2}\left(\frac{1 - x}{2}\right)^2 + 0$$

(since $\alpha + 2 = 0$ in the fourth term); collecting the powers of x we get

$$\tfrac{3}{2}x^2 - \tfrac{1}{2}$$

in agreement with Theorem 16.5.3.

For $p = 0$, $q = \frac{1}{2}$ the Jacobi polynomials (again setting $\frac{1}{2}(1 - x)$ for x) yield the so-called *Chebyshev polynomials*,

$$\mathrm{T}_n(x) = \frac{1}{2^{n-1}} J_n(0, \tfrac{1}{2}, \tfrac{1}{2}(1 - x)).$$

(It should be noted that in English literature the factor $1/2^{n-1}$ is omitted in the definition of $\mathrm{T}_n(x)$.)

It can be shown that

$$\mathrm{T}_n(x) = \frac{1}{2^{n-1}} \cos(n \arccos x) \quad (|x| \leqq 1)$$

and that $\mathrm{T}_n(x)$ constitute an orthogonal system in the interval $[-1, 1]$ for the weight function $1/\sqrt{(1 - x^2)}$, i.e.

$$\int_{-1}^{1} \frac{1}{\sqrt{(1 - x^2)}} \mathrm{T}_i(x)\, \mathrm{T}_k(x)\, \mathrm{d}x = 0 \quad \text{for} \quad i \neq k.$$

Moreover, $\mathrm{T}_n(x)$ is a solution of the so-called *Chebyshev equation with index n*,

$$(1 - x^2)\, y'' - xy' + n^2 y = 0.$$

Laguerre polynomials are defined by the equation

$$\mathrm{L}_n(x) = \sum_{k=0}^{n} (-1)^k \frac{n^2(n - 1)^2 \ldots (k + 1)^2}{(n - k)!} x^k$$

(where the coefficient of x^n is taken to be $(-1)^n$) or by the equation

$$\mathrm{L}_n(x) = \mathrm{e}^x \frac{\mathrm{d}^n}{\mathrm{d}x^n} (x^n \mathrm{e}^{-x}) .$$

They are solutions of the so-called *Laguerre equation*

$$xy'' + (1 - x)\, y' + ny = 0 ,$$

and are orthogonal in the interval $[0, \infty)$ for the weight function e^{-x}, i.e.

$$\int_0^\infty \mathrm{e}^{-x}\, \mathrm{L}_i(x)\, \mathrm{L}_k(x)\, \mathrm{d}x = 0 \quad \text{for} \quad i \neq k .$$

Hermite polynomials are defined by the equation

$$\mathrm{H}_n(x) = \sum_{k=0}^{h} (-1)^k \frac{n!}{k!(n-2k)!} (2x)^{n-2k} ,$$

where $h = \frac{1}{2}n$ for n even, $h = \frac{1}{2}(n - 1)$ for n odd, or by the equation

$$\mathrm{H}_n(x) = (-1)^n \mathrm{e}^{x^2} \frac{\mathrm{d}^n}{\mathrm{d}x^n} \mathrm{e}^{-x^2} .$$

They satisfy the so-called *Hermite equation*

$$y'' - 2xy' + 2ny = 0 ,$$

and are orthogonal in the interval $(-\infty, \infty)$ for the weight function e^{-x^2}, i.e.

$$\int_{-\infty}^{\infty} \mathrm{e}^{-x^2}\, \mathrm{H}_i(x)\, H_k(x)\, \mathrm{d}x = 0 \quad \text{for} \quad i \neq k .$$

17. ORDINARY DIFFERENTIAL EQUATIONS

By Karel Rektorys

References: [3], [15], [17], [27], [29], [31], [35], [45], [56], [61], [65], [67], [69], [71], [75], [83], [92], [97], [100], [102], [121], [161], [189], [194], [199], [201], [211], [217], [243], [249], [252], [255], [271], [278], [280], [285], [296], [319], [321], [324], [333], [341], [348], [349], [351], [367], [388], [392], [398], [404], [420], [429], [430], [434].

Special mention must be made of the book [211], Vol. I, where the reader will find a great number of solved differential equations (arranged "alphabetically" according to their order, type, etc.). Moreover, that book contains a survey of main results and methods for solving ordinary differential equations, as well as many references.

17.1. Classification of Differential Equations. Ordinary and Partial Differential Equations. Order of a Differential Equation. Systems of Differential Equations

A differential equation is a relation (satisfied in a certain domain) between the unknown function and its derivatives. If the function to be found is a function of one variable only we speak of an *ordinary differential equation*, if it is a function of several variables (so that the equation contains partial derivatives), we speak of a *partial differential equation*.

If m differential equations are given for n unknown functions (not necessarily with $m = n$), we speak of a *system of differential equations*.

The *order* of a given differential equation is that of the derivative of the highest order occurring in it Similarly by the *order of a system* we understand (as usual) that of the derivative of the highest order occurring in the system.

Example 1.

$$y'' + xy'^3 - e^y = 0$$

is an ordinary differential equation of the second order.

Example 2.

$$\frac{\partial^2 u}{\partial x\,\partial y} - \frac{\partial^3 u}{\partial y^3} - u = 0$$

is a partial differential equation of the third order for the unknown function $u(x, y)$.

Example 3.

$$\frac{\partial u_1}{\partial x} - \frac{\partial^2 u_2}{\partial y^2} = 0\,, \quad \frac{\partial^3 u_1}{\partial y^3} - \frac{\partial u_2}{\partial x} = 0$$

is a system of partial differential equations of the third order for two unknown functions $u_1(x, y)$, $u_2(x, y)$.

17.2. Basic Concepts. Integral of a Differential Equation. Theorems Relating to Existence and Uniqueness of Solution. General Integral, Particular Integral, Singular Integral

Definition 1. By an *ordinary differential equation of the first order*, we mean an equation of the form

$$F(x, y, y') = 0 \tag{1}$$

or, in the special case where the equation is solved with respect to y', an equation of the form

$$y' = f(x, y)\,. \tag{2}$$

Definition 2. By an *ordinary differential equation of order n* we mean an equation of the form

$$F(x, y, y', y'', \ldots, y^{(n)}) = 0 \tag{3}$$

or, if the equation is solved with respect to the derivative of highest order, the equation of the form

$$y^{(n)} = f(x, y, y'\ y'', \ldots, y^{(n-1)})\,. \tag{4}$$

Definition 3. By a *solution* or *integral* (or *particular integral* or *integral curve*, if speaking geometrically) *of equation* (3), we mean any function $y = g(x)$ which satisfies equation (3) identically in the domain considered. (See also Remark 3.)

REMARK 1. Most frequently, the domain considered is an interval I. Thus the function $y = g(x)$ is the solution of equation (3) in the interval I, if it has derivatives up to the order n in I and if on substituting $g(x)$ for y, $g'(x)$ for y', etc., in (3), equation (3) is satisfied for all x in I.

Example 1. The function $y = \sin x$ is a (particular) integral of the equation $y'' + y = 0$ in the interval $(-\infty, \infty)$.

REMARK 2. In Definition 3, definitions of the solution of equations (1), (2) and (4) as special cases are included.

REMARK 3. The solution of equation (3) need not be given in the explicit form $y = g(x)$, but possibly in an implicit form by an equation $h(x, y) = 0$. Then the derivatives $y', y'', \ldots$ may be found according to the theorem on implicit functions (Theorem 12.9.1, p. 462, in a region in which the corresponding conditions are fulfilled, of course); $h(x, y) = 0$ is then called a solution of equation (3), if on substituting for $y', y'', \ldots$ in (3), the differential equation is identically satisfied (in the variables x and y) at all the points of the curve $h(x, y) = 0$ (see also Remark 16 and Example 7).

Example 2. The equation $x^2 + y^2 = 4$ constitutes an integral of the equation

$$y' = -\frac{x}{y} \tag{5}$$

in an implicit form, since at every point of the circle $x^2 + y^2 - 4 = 0$, the relation $2x + 2yy' = 0$, holds, i.e. (5) is valid. The points $(-2, 0)$ and $(2, 0)$, where $y = 0$, are exceptions; for this case see Remark 16 and Example 7.

REMARK 4. Geometrical interpretation of equation (2): Let $f(x, y)$ be defined in the region Q. Then by virtue of equation (2), to every point $(x, y) \in Q$ there corresponds a line element (*directional element*) determined by the point (x, y) and by the slope y'. Thus in Q a field of directions is given, the so-called *directional field*. The problem of finding solutions of equation (2) can thus be given the following geometrical interpretation: to find curves in Q, the tangents to which coincide at every point with the corresponding line element [i.e. the slope y' of such a curve at the point considered is prescribed by (2)].

In the case of an equation of the second order

$$y'' = f(x, y, y')$$

there is assigned at every point $(x, y) \in Q$ and for each slope y' the value of the second derivative (and thus the curvature of the integral curve which is to be found). Equations of the higher order have no longer such a simple geometrical interpretation as those of the first and second orders.

Definition 4. By a *solution* (*integral*) *of the system of equations of the first order*,

$$\begin{aligned} y_1' &= f_1(x, y_1, y_2, \ldots, y_n)\,, \\ y_2' &= f_2(x, y_1, y_2, \ldots, y_n)\,, \\ &\ldots\ldots\ldots\ldots\ldots\ldots \\ y_n' &= f_n(x, y_1, y_2, \ldots, y_n) \end{aligned} \tag{6}$$

we mean a system of functions

$$y_1 = g_1(x), y_2 = g_2(x), \ldots, y_n = g_n(x) \tag{7}$$

such that if the functions (7) and their derivatives are substituted into (6), all the equations (6) are identically satisfied (in the domain considered). (On more general systems of equations see also § 17.18).

In this case it is also possible to speak about a solution in implicit form (see Remark 3).

REMARK 5. In the case where $n = 2$, let us denote the functions to be found by y, z instead of y_1, y_2. Then the functions (7),

$$y = g_1(x), \quad z = g_2(x)$$

represent geometrically a *curve* (in three dimensional space). For this reason we often call the system (7) an *integral curve* of the system (6).

REMARK 6. In general, the number of equations and the number of unknown functions in the system (6) need not be equal. The definition of the solution remains unchanged. In this case, however, the theorem on existence and uniqueness of the solution (see below) need no longer be valid in general.

Theorem 1. *Let the system* (6) *and a point*

$$P(a, b_1, b_2, \ldots, b_n). \tag{8}$$

be given. Let the functions $f_1, f_2, \ldots, f_n$ *of system* (6) *be continuous* (*as functions of the* $n + 1$ *variables* $x, y_1, y_2, \ldots, y_n$) *in the neighbourhood* O *of the point* P *and let them have, in* O, *continuous partial derivatives with respect to the variables* $y_1, y_2, \ldots, y_n$. *Then in a certain neighbourhood of the point* a *there exists precisely one system of functions* (7), *which is a solution of the system* (6) *and satisfies the conditions* (the so-called *initial conditions*)

$$g_1(a) = b_1, \quad g_2(a) = b_2, \quad \ldots, \quad g_n(a) = b_n. \tag{9}$$

REMARK 7. In the terminology of Remark 5: If the functions $f_1, f_2, \ldots, f_n$ satisfy the above conditions in O, then there exists precisely one integral curve of the system (6) passing through the point P.

In particular, if the functions

$$f(x, y), \quad \frac{\partial f}{\partial y}(x, y) \tag{10}$$

are continuous in the neighbourhood O of the point $P(a, b)$, then there exists precisely one integral curve of the equation

$$y' = f(x, y)$$

passing through the point P.

REMARK 8. We say that the function $f(x, y)$ satisfies a *Lipschitz condition* with respect to the variable y in the rectangle $R(A \leqq x \leqq B; C \leqq y \leqq D)$, if a constant K can be found such that for every $x \in [A, B]$ and for any numbers y_1, y_2 from $[C, D]$ the inequality

$$|f(x, y_1) - f(x, y_2)| \leqq K|y_1 - y_2| \tag{11}$$

holds. (In particular, condition (11) is satisfied, if $f(x, y)$ has a derivative with respect to y in R and

$$\left|\frac{\partial f}{\partial y}\right| \leqq K \text{ in } R\,.) \tag{12}$$

Example 3. The function $f(x, y) = |y|$ does not possess a partial derivative $\mathrm{d}f/\mathrm{d}y$ at points of the x-axis, but it nevertheless satisfies the Lipschitz condition in the whole plane xy. It is sufficient to choose $K = 1$, since

$$||y_1| - |y_2|| \leqq |y_1 - y_2|\,.$$

Theorem 2. *Let $f(x, y)$ be continuous in the rectangle*

$$Q(a - h \leqq x \leqq a + h, \quad b - k \leqq y \leqq b + k)\,.$$

(*Then* (see Theorem 12.1.5, p. 444) *there exists a constant $M > 0$ such that $|f(x, y)| \leqq$ $\leqq M$ in Q.*) *Let $f(x, y)$ satisfy condition* (11) *in Q. Let us write*

$$d = \min\left(h, \frac{k}{M}\right).$$

Then, in the interval $[a - d, a + d]$, there exists precisely one solution $y = g(x)$ of the equation

$$y' = f(x, y)$$

passing through the point (a, b).

REMARK 9. The Lipschitz condition and a theorem similar to Theorem 2 can be formulated for the system (6); see e.g. [65].

REMARK 10. For the *existence* of the solution, it is sufficient to have only the continuity of the function $f(x, y)$ (or of the functions $f_1, f_2, \ldots, f_n$) in the region concerned. Continuity itself, however, without other supplementary assumptions (continuity of derivatives, or Lipschitz condition, etc.) is not in general sufficient for uniqueness, for instance:

Example 4. The curves

$$y \equiv 0 \quad \text{and} \quad y = \tfrac{1}{27}(x-2)^3$$

are both integral curves of the equation $y' = \sqrt[3]{y^2}$ passing through the point $(2, 0)$.

Theorem 3. *Let the equation*

$$y^{(n)} = f(x, y, y', \ldots, y^{(n-1)}) \tag{13}$$

and the point $P(a, b_1, b_2, \ldots, b_n)$ be given. Let the functions

$$f, \frac{\partial f}{\partial y}, \quad \frac{\partial f}{\partial y'}, \quad \ldots, \quad \frac{\partial f}{\partial y^{(n-1)}}$$

be continuous (as functions of the $n + 1$ variables) in the neighbourhood of the point P. Then, in a certain neighbourhood of the point a, there exists precisely one solution $y = g(x)$ of equation (13), which satisfies the initial conditions

$$g(a) = b_1\,, \; g'(a) = b_2\,, \; \ldots, \; g^{(n-1)}(a) = b_n \tag{14}$$

(i.e. through the point $Q(a, b_1)$ there passess one and only one integral curve of equation (13) satisfying the conditions

$$g'(a) = b_2\,, \; \ldots, \; g^{(n-1)}(a) = b_n)\,.$$

REMARK 11. By employing a Lipschitz condition, it is possible to formulate a theorem relating to equation (13) similar to Theorem 2.

Theorem 3 has a *local* character, i.e. it guarantees existence and uniqueness of the solution in a certain neighbourhood of the point a. In the case of a *linear* equation

$$y^{(n)} + a_{n-1}(x)\, y^{(n-1)} + \ldots + a_1(x)\, y' + a_0(x)\, y = b(x) \tag{15}$$

existence and uniqueness of the solution (satisfying the prescribed initial conditions (14)) is guaranteed in the entire interval I, in which the functions $a_{n-1}(x), \ldots, a_1(x)$, $a_0(x)$, $b(x)$ are continuous and which contains the point a. The interval I may even be infinite. A similar statement holds for the so-called *normal* systems of linear equations (Theorem 17.18.1, p. 819).

REMARK 12. For equation (13) let the initial conditions (14) be prescribed. We say briefly that an $(n + 1)$-dimensional point $P(a, b_1, b_2, \ldots, b_n)$ (cf. Theorem 3) is given. Let Q be an $(n + 1)$-dimensional region composed of such points P, for which equation (13) has precisely one solution according to Theorem 3. We now define the general solution of such an equation.

Definition 5. By the *general integral* (or *general solution*) *of equation* (13) *in the region Q*, we mean a solution of equation (13) containing n arbitrary independent constants, i.e. such that if we choose an arbitrary point $P \in Q$, there is one and only one set of numerical values of these constants for which the solution satisfies conditions (14) at the point P.

REMARK 13. Thus, these constants are independent if it is not possible to replace any of them by others, i.e. if none of them is superfluous.

Example 5. The function

$$y = C_1 e^{2x} + C_2 e^{-x}$$

is the general integral of the equation

$$y'' - y' - 2y = 0 .$$

The function

$$C_1 e^{x+C_2}$$

is not the general integral of the equation

$$y'' - y = 0$$

since evidently

$$C_1 e^{x+C_2} = C_1 e^x . e^{C_2} = K e^x .$$

REMARK 14. In the case of the equation

$$F(x, y, y', \ldots, y^{(n)}) = 0 \tag{16}$$

it is not possible to speak about the general integral exactly in the sense of Remark 12 and Definition 5, since uniqueness of the solution may be in question. For example, given the point (x_0, y_0), the equation

$$F(x, y, y') = 0 \tag{17}$$

can in general be satisfied by several values of y'_0, even when F is a very "reasonable" function. For example, the equation

$$y'^2 - x^4 y^2 = 0 \tag{18}$$

is satisfied by both y' given by equations

$$y' = x^2 y , \quad y' = -x^2 y .$$

Evidently, through every point (x_0, y_0), for which $x_0^2 y_0 \neq 0$, there pass two integral curves of equation (18).

Often we succeed in finding a function (or relation)

$$y = g(x, C_1, C_2, \ldots, C_n) \quad \text{or} \quad h(x, y, C_1, C_2, \ldots, C_n) = 0 \tag{19}$$

such that by a proper choice of the constants $C_1, C_2, \ldots, C_n$ (dependent, for example, on the possible choice of the derivative of the highest order in (16), i.e., for example, on the choice of the sign of the first derivative in (18)), it is possible to satisfy – in a certain region – any initial condition. In such a case, we also speak of a general integral of equation (16).

Definition 6. By a *singular integral* (*singular solution*) of the equation $y' = f(x, y)$ we mean an integral curve of this equation such that at every point of it the uniqueness condition is broken (i.e. through every point of this curve there also passes another integral curve of the equation).

Example 6. The integral curve $y \equiv 0$ is a singular integral of the equation

$$y' = \sqrt[3]{y^2}$$

(see Example 4), since through every point $(a, 0)$ of this curve there passes another integral curve

$$y = \tfrac{1}{27}(x - a)^3$$

of the given equation.

REMARK 15. The general integral of the equation $y' = f(x, y)$ constitutes a one-parameter system of curves. It may be easily proved that the envelope of this system (if it exists) is a singular integral of the given equation.

In the case of the equation $F(x, y, y') = 0$, we define the singular integral in the same way as in Definition 6, with the supplementary condition (see the question of uniqueness of the solution dealt with in Remark 14) that through every point of the singular integral curve there passes also another integral curve with the same tangent.

REMARK 16. For $f(x, y) \neq 0$, the equation

$$\frac{dy}{dx} = \frac{1}{f(x, y)} \tag{20}$$

is equivalent to the equation

$$\frac{dx}{dy} = f(x, y)\,. \tag{21}$$

At points where $f(x, y) = 0$, equation (20) has no meaning, but equation (21) has. Consequently, we often "add" equation (21) to equation (20) and then understand by an integral curve of equation (20) an integral curve satisfying either equation (20) or equation (21).

Example 7. In this sense, the circle

$$x^2 + y^2 - 4 = 0 \tag{22}$$

is an integral curve of the equation

$$y' = -\frac{x}{y}$$

(Example 2) even at the points $(-2, 0)$, $(2, 0)$, since in the neighbourhood of these points it satisfies the equation

$$\frac{\mathrm{d}x}{\mathrm{d}y} = -\frac{y}{x}.$$

17.3. Elementary Methods of Integration of Equations of the First Order. Separation of Variables. Homogeneous Equations. Linear Equations. Bernoulli's Equation. Riccati's Equation

I. The equation

$$y' = f(x)$$

(where the function $f(x)$ is supposed to be continuous in the interval I considered) has the general integral

$$y = \int f(x)\,\mathrm{d}x.$$

(This indefinite integral already contains an arbitrary constant.) The integral, for which $y(x_0) = y_0$ is then

$$y = y_0 + \int_{x_0}^{x} f(t)\,\mathrm{d}t.$$

II. The equation

$$y' = f(y)$$

(where $f(y)$ is continuous in the region considered) can be written, for $f(y) \neq 0$, in the form (Remark 17.2.16)

$$\frac{\mathrm{d}x}{\mathrm{d}y} = \frac{1}{f(y)}.$$

The general integral (in the region where $f(y) \neq 0$) is

$$x = \int \frac{\mathrm{d}y}{f(y)},$$

and the integral curve passing through the point (x_0, y_0) is of the form

$$x - x_0 = \int_{y_0}^{y} \frac{dt}{f(t)}.$$

If $f(y_0) = 0$, then the curve

$$y \equiv y_0$$

satisfies the given equation and the initial condition.

Example 1. Let us consider the equation

$$y' = y^2$$

or (for $y \neq 0$)

$$\frac{dx}{dy} = \frac{1}{y^2}.$$

The general integral is

$$x = -\frac{1}{y} + C$$

or

$$y = \frac{1}{C - x}.$$

The integral curve passing through the point (3, 1) is the curve

$$y = \frac{1}{4 - x}$$

(because, putting $x = 3$, $y = 1$, it follows that $C = 4$). The integral curve passing through the point (3, 0) is

$$y = 0$$

(for in this case $f(y_0) = y_0^2 = 0$). In both cases there is only one solution, in accordance with Theorem 17.2.1 and Remark 17.2.7.

III. (*Separation of Variables.*)

Theorem 1. *Let $f(x)$ be continuous in $[a, b]$, $g(y)$ continuous in $[c, d]$ and let $g(y) \neq 0$ in $[c, d]$. Then the equation*

$$y' = \frac{f(x)}{g(y)}$$

has in $O(a < x < b, c < y < d)$ *the general integral*

$$\int f(x)\,dx = \int g(y)\,dy\,.$$

The integral curve passing through the point $(x_0, y_0) \in O$ *has the equation*

$$\int_{x_0}^{x} f(x)\,dx = \int_{y_0}^{y} g(y)\,dy$$

(and it is the only integral curve of the given equation which passes through this point (x_0, y_0)*).*

REMARK 1. An analogous theorem holds for the equation

$$y' = f(x)\,g(y)\,.$$

The general integral is

$$\int \frac{dy}{g(y)} = \int f(x)\,dx\,.$$

The integral curve passing through the point (x_0, y_0) is

$$\int_{y_0}^{y} \frac{dy}{g(y)} = \int_{x_0}^{x} f(x)\,dx\,.$$

If $g(y_0) = 0$, then the equation considered and the initial condition are satisfied by the integral curve

$$y \equiv y_0$$

(a straight line parallel to the x-axis).

Example 2. Let us solve the equation

$$y' = xy^3 \sin x$$

subject to the condition that $y(0) = 1$.

By the separation of variables we get

$$\int \frac{dy}{y^3} = \int x \sin x\,dx\,.$$

The general integral is then:

$$\frac{1}{2}\frac{1}{y^2} = x \cos x - \sin x + C\,.$$

From the condition $y(0) = 1$ it follows that $C = \frac{1}{2}$, so the solution is

$$y = \frac{1}{\sqrt{(2x\cos x - 2\sin x + 1)}}\,.$$

(The positive sign before the square root follows from the condition $y(0) = +1$.)

The solution of the same equation, but with the condition $y(0) = 0$, is $y \equiv 0$.

IV. An equation of the form

$$y' = f\left(\frac{y}{x}\right) \quad (x \neq 0) \tag{1}$$

is said to be *homogeneous* . (The right-hand side of equation (1) is a homogeneous function of degree zero (Definition 12.6.1, p. 454).)

An example of such an equation is the equation

$$y' = \frac{x^2 + y^2}{xy} \quad \text{or} \quad y' = \frac{1 + \left(\dfrac{y}{x}\right)^2}{\dfrac{y}{x}}\,.$$

Instead of looking for the solution $y(x)$ of equation (1), we try to find a new unknown function $z(x)$, related to $y(x)$ by the equation

$$z(x) = \frac{y(x)}{x} \quad \text{or} \quad y(x) = xz(x)\,.$$

From the second equation it follows that $y' = z + xz'$, so that after substituting into (1) we find that z satisfies the equation

$$z + xz' = f(z) \quad \text{or} \quad z' = \frac{f(z) - z}{x}$$

which is an equation with separable variables. If we find its solution $z(x)$, then the solution of equation (1) is $y(x) = xz(x)$ (see Example 3).

V. Equations of the form

$$y' = \frac{a_1x + b_1y + c_1}{a_2x + b_2y + c_2}, \tag{2}$$

where

$$\begin{vmatrix} a_1 & b_1 \\ a_2 & b_2 \end{vmatrix} \neq 0\,, \tag{3}$$

can be transformed into homogeneous equations by making a substitution of the form

$$x = u + A, \quad y = v + B, \quad \text{so that} \quad \mathrm{d}x = \mathrm{d}u, \quad \mathrm{d}y = \mathrm{d}v, \quad \mathrm{d}y/\mathrm{d}x = \mathrm{d}v/\mathrm{d}u.$$

Equation (2) becomes

$$\frac{\mathrm{d}v}{\mathrm{d}u} = \frac{a_1 u + b_1 v + a_1 A + b_1 B + c_1}{a_2 u + b_2 v + a_2 A + b_2 B + c_2}. \tag{4}$$

Now we choose the constants A, B so that

$$a_1 A + b_1 B + c_1 = 0, \tag{5}$$

$$a_2 A + b_2 B + c_2 = 0,$$

which is possible by virtue of (3). Then equation (4) becomes

$$\frac{\mathrm{d}v}{\mathrm{d}u} = \frac{a_1 u + b_1 v}{a_2 u + b_2 v} = \frac{a_1 + b_1 \dfrac{v}{u}}{a_2 + b_2 \dfrac{v}{u}} \tag{6}$$

and this is a homogeneous equation. Such an equation can be solved by the substitution $v(u) = uz(u)$. After solving this equation we substitute back $z = v/u$ and $v = y - B$, $u = x - A$.

If the determinant (3) is zero, i.e. if $a_2 x + b_2 y = k(a_1 x + b_1 y)$, equation (2) can easily be solved by introducing a new unknown function $z(x)$ given by

$$a_1 x + b_1 y = z, \quad \text{so that} \quad a_2 x + b_2 y = kz,$$

and

$$a_1 + b_1 y' = z', \quad \text{hence} \quad y' = \frac{z' - a_1}{b_1}.$$

We thus obtain an equation for z with separated variables:

$$\frac{z'}{b_1} = \frac{a_1}{b_1} + \frac{z + c_1}{kz + c_2} \quad \text{or} \quad \frac{kz + c_2}{(a_1 k + b_1) z + (a_1 c_2 + b_1 c_1)} \mathrm{d}z = \mathrm{d}x;$$

whose solution is described in III above.

If in (2) $c_1 = c_2 = 0$, then case V becomes case IV.

Example 3.

$$y' = \frac{2x - y + 9}{x - 3y + 2}. \tag{7}$$

By solving equations (5) we get $A = -5$, $B = -1$. Substituting $x = u - 5$, $y = v - 1$ we get (compare (6))

$$\frac{dv}{du} = \frac{2u - v}{u - 3v} = \frac{2 - \dfrac{v}{u}}{1 - 3\dfrac{v}{u}}.$$

By a new substitution

$$z = \frac{v}{u} \quad \text{or} \quad uz = v$$

we have, as shown in IV

$$u\frac{dz}{du} + z = \frac{2 - z}{1 - 3z} \quad \text{or} \quad u\frac{dz}{du} = \frac{2 - 2z + 3z^2}{1 - 3z}.$$

Then, as in III, we get

$$\int \frac{1 - 3z}{2 - 2z + 3z^2}\,dz = \int \frac{du}{u},$$

$$-\tfrac{1}{2}\ln(2 - 2z + 3z^2) = \ln ku \quad (ku > 0),$$

$$\ln(2 - 2z + 3z^2) = -2\ln(ku) = \ln\frac{1}{k^2u^2} = \ln\frac{C}{u^2} \quad (C = 1/k^2 > 0),$$

$$2 - 2z + 3z^2 = \frac{C}{u^2}$$

and after substituting back the original variables $z = v/u$, $u = x + 5$, $v = y + 1$, we obtain

$$2u^2 - 2uv + 3v^2 = C,$$

or

$$2(x + 5)^2 - 2(x + 5)(y + 1) + 3(y + 1)^2 = C.$$

VI. *Linear equations* are equations of the form

$$y' + a(x)\,y = b(x). \tag{8}$$

The equation

$$y' + a(x)\,y = 0 \tag{9}$$

is the so-called *homogeneous linear equation corresponding to equation* (8).

The term "homogenous" here has a completely different meaning from that in IV. However, this terminology is normally used in the literature.

If the functions $a(x)$, $b(x)$ are continuous in the interval I (which can even be infinite), then according to Remark 17.2.11, the existence and uniqueness of the solution of equation (8) or (9) with the prescribed initial condition is guaranteed in the whole interval I.

Equation (9) can be solved by separation of variables:

$$\int \frac{\mathrm{d}y}{y} = - \int a(x)\,\mathrm{d}x\,,$$

$$\ln(ky) = - \int a(x)\,\mathrm{d}x \quad (ky > 0)\,,$$

$$ky = \mathrm{e}^{-\int a(x)\,\mathrm{d}x}\,,$$

$$y = C\mathrm{e}^{-\int a(x)\,\mathrm{d}x}\,(C = 1/k)\,, \tag{10}$$

which is the general integral of equation (9).

The general integral of equation (8) can be obtained from (10) by so-called *method of variation of the parameter* (or variation of the constant). Let us assume that the integral of equation (8) is of the form (10), where C is now a function of the variable x,

$$y = C(x)\,\mathrm{e}^{-\int a(x)\,\mathrm{d}x}\,. \tag{11}$$

Differentiating (11),

$$y' = C'\mathrm{e}^{-\int a\,\mathrm{d}x} - Ca\mathrm{e}^{-\int a\,\mathrm{d}x}\,, \tag{12}$$

and on substituting (11) and (12) into (8), we have

$$C'\mathrm{e}^{-\int a\,\mathrm{d}x} - Ca\mathrm{e}^{-\int a\,\mathrm{d}x} + aC\mathrm{e}^{-\int a\,\mathrm{d}x} = b\,, \tag{13}$$

from which we are able to determine the function $C(x)$:

$$C'\mathrm{e}^{-\int a\,\mathrm{d}x} = b$$

or

$$C(x) = \int b(x)\,\mathrm{e}^{\int a(x)\,\mathrm{d}x}\,\mathrm{d}x\,.$$

Substituting this result into (11), we obtain the general integral of equation (8) in the form

$$y = \mathrm{e}^{-\int a(x)\mathrm{d}x} \cdot \int b(x)\,\mathrm{e}^{\int a(x)\,\mathrm{d}x}\,\mathrm{d}x\,. \tag{14}$$

Example 4.

$$y' + 2xy = x^3\,. \tag{15}$$

First, we solve the corresponding homogeneous equation

$$y' + 2xy = 0:$$

$$\int \frac{dy}{y} = -\int 2x \, dx,$$

$$\ln (ky) = -x^2 \quad (ky > 0),$$

$$ky = e^{-x^2},$$

$$y = Ce^{-x^2} \quad (C = 1/k).$$

The general integral of equation (15) can now be found, according to (11), in the form

$$y = C(x) \, e^{-x^2}. \tag{16}$$

Then

$$y' = C'e^{-x^2} - C \, . \, 2xe^{-x^2}. \tag{17}$$

Substituting (16) and (17) into (15), we get

$$C'e^{-x^2} - 2xCe^{-x^2} + 2xCe^{-x^2} = x^3,$$

$$C' = x^3 e^{x^2}, \quad C = \int x^3 e^{x^2} \, dx = \tfrac{1}{2} e^{x^2}(x^2 - 1) + c.$$

Substituting this result into (16), we obtain the general integral of equation (15),

$$y = \tfrac{1}{2}(x^2 - 1) + ce^{-x^2}. \tag{18}$$

REMARK 2. When solving equation (9) by separation of variables we divide by the unknown function y. Nevertheless, (10) or (14) gives the unique solution, even in the case where the integral curve has to pass through the point (x_0, y_0), where $y_0 = 0$. For example, the solution (and indeed the only solution) of equation (15) passing through the point (0, 0) is according to (18)

$$y = \tfrac{1}{2}(x^2 - 1) + \tfrac{1}{2}e^{-x^2}.$$

REMARK 3. Instead of the method of variation of the parameter it is possible to proceed as follows: The integral of equation (8) is assumed to be of the form

$$y(x) = u(x) \, v(x). \tag{19}$$

Substituting into (8), we get

$$u'v + uv' + auv = b. \tag{20}$$

Let the function $u(x)$ be chosen so as to satisfy the equation

$$u' + au = 0, \tag{21}$$

i.e.

$$u = e^{-\int a(x)\,dx}\,. \tag{22}$$

Since it is possible to write (20) in the form

$$(u' + au)\,v + uv' = b\,,$$

it is sufficient, in view of condition (21), for v to satisfy the equation

$$uv' = b$$

or (see (22))

$$v = \int b e^{\int a\,dx}\,dx\,. \tag{23}$$

Substituting (22) and (23) into (19), we get the previous result (14).

VII. *Bernoulli's equation* is an equation of the form

$$y' + a(x)\,y = b(x)\,y^n \tag{24}$$

($a(x)$, $b(x)$ being continuous functions in an interval I). For $n = 0$, or $n = 1$, we get a linear equation of the form (8) or (9), respectively.

In the following text, let n be an arbitrary real number different from 0 and 1. Dividing by y^n, we get from (24)

$$\frac{y'}{y^n} + \frac{a}{y^{n-1}} = b\,. \tag{25}$$

Substituting

$$z = y^{-n+1}\,, \quad \text{from which} \quad z' = (-n+1)\,y^{-n}y' \quad \text{or} \quad \frac{y'}{y^n} = \frac{z'}{-n+1}\,,$$

we obtain

$$\frac{z'}{-n+1} + az = b\,,$$

i.e. we get a linear equation for the function $z(x)$.

Example 5.

$$y' + xy = xy^3\,. \tag{26}$$

Dividing by y^3 and substituting

$$z = \frac{1}{y^2}\,, \quad \text{so that} \quad z' = -\frac{2y'}{y^3}$$

we have

$$-\frac{z'}{2} + xz = x\,,$$

$$z' - 2xz = -2x\,.$$

As shown in VI the solution of this equation is

$$z = 1 + Ce^{x^2}\,.$$

Hence, the general integral of equation (26) is

$$y^2 = \frac{1}{1 + Ce^{x^2}}\,.$$

VIII. *Riccati's equation* is an equation of the form

$$y' = a(x)\,y^2 + b(x)\,y + c(x)\,. \tag{27}$$

For $a(x) \equiv 0$ we get a linear equation and for $c(x) \equiv 0$ we get Bernoulli's equation.

It is not possible, in general, to solve Riccati's equation by *quadratures*, i.e. it is not possible, in general, to reduce its solution to a mere integration, as was the case in the preceding types. However, if one solution of equation (27) is known, the general integral can be obtained by means of quadratures as follows: Let $y_1(x)$ be a solution of equation (27), i.e.

$$y_1' = ay_1^2 + by_1 + c\,. \tag{28}$$

Introducing a new unknown function $z(x)$ by the relation

$$y = y_1 + z \tag{29}$$

and substituting into (27), we have

$$y_1' + z' = ay_1^2 + 2ay_1z + az^2 + by_1 + bz + c\,;$$

in consequence of (28), the function z then satisfies the equation

$$z' = az^2 + (2ay_1 + b)\,z\,,$$

which is Bernoulli's equation, and which is integrable by quadratures.

Example 6.

$$xy' - 3y + y^2 = 4x^2 - 4x\,. \tag{30}$$

Let us assume the particular solution to be of the form

$$y_1 = Ax + B\,.$$

(In guessing the form of the particular solution, some experience is necessary.) Substituting y_1 in (30) and comparing coefficients of the same powers of x, we get

$$y_1 = 2x .$$

Using the substitution (29), $y = y_1 + z = 2x + z$, we get the equation

$$xz' + (4x - 3)\,z + z^2 = 0\,,$$

which is of Bernoulli's type.

Two following special cases of Riccati's equation are easy to solve:

A. $$y' = ay^2 + by + c\,,$$

where a, b, c are constants; we solve this by the methods of II or III above.

B. $$y' = ay^2 + \frac{b}{x^2}\,,$$

where a, b are constants; this equation may be solved by substituting

$$y = \frac{1}{z}\,,$$

giving

$$y' = -\frac{z'}{z^2}\,,$$

$$-\frac{z'}{z^2} = \frac{a}{z^2} + \frac{b}{x^2}$$

and

$$z' = -a - b\left(\frac{z}{x}\right)^2$$

which is a homogeneous equation and may be solved as shown in IV.

17.4. Exact Differential Equations. The Integrating Factor. Singular Points

REMARK 1. The equation $y' = \varphi(x, y)$ is often of the form

$$y' + \frac{f(x, y)}{g(x, y)} = 0\,, \tag{1}$$

where f and g and their derivatives of the first order are continuous functions in a region O and $g(x, y) \neq 0$ in O. According to Theorem 17.2.1 and Remark 17.2.7, the existence and uniqueness of the solution is guaranteed in O. Written in a differential form, equation (1) becomes

$$f(x, y)\,dx + g(x, y)\,dy = 0\,. \tag{2}$$

Theorem 1. *Let the left-hand side of equation* (2) *be the total differential of a function* $F(x, y)$ *in* O. (In this case equation (2) is said to be *exact*. If O is a simply connected region, then the necessary and sufficient condition for this is that

$$\frac{\partial f}{\partial y} = \frac{\partial g}{\partial x} \text{ in } O\,.) \tag{3}$$

Then the general integral of equation (2) *or* (1) *is*

$$F(x, y) = C\,. \tag{4}$$

REMARK 2. The function $F(x, y)$ may be obtained in the same way as in Example 14.7.2, p. 635.

Example 1.

$$(x^2 - y^2)\,dx + (y^3 - 2xy)\,dy = 0\,.$$

The validity of (3) may be easily verified, because

$$\frac{\partial}{\partial y}(x^2 - y^2) = \frac{\partial}{\partial x}(y^3 - 2xy) = -2y\,. \tag{5}$$

Thus, the given equation is exact. By Example 14.7.2 (p. 635) we find that

$$F(x, y) = \frac{x^3}{3} - xy^2 + \frac{y^4}{4} + k\,,$$

so that the general integral of the given equation is

$$\frac{x^3}{3} - xy^2 + \frac{y^4}{4} = C\,,$$

the constant k being incorporated into the constant C.

Theorem 2. *If the equation*

$$f(x, y)\,dx + g(x, y)\,dy = 0$$

is exact and if the functions f, g are continuous and homogeneous (Definition 12.6.1, p. 454) *of the same degree n ($n \neq -1$), then the general integral of the given equation is*

$$x\,f(x, y) + y\,g(x, y) = C\,.$$

Example 2. The general integral of the equation

$$\frac{2x}{y}\,\mathrm{d}x + \left(1 - \frac{x^2}{y^2}\right)\mathrm{d}y = 0$$

is

$$x\frac{2x}{y} + y\left(1 - \frac{x^2}{y^2}\right) = C\,.$$

REMARK 3. If equation (2) is not exact, we try to find a function $m(x, y) \neq 0$ such that the new equation

$$m(x, y)\,f(x, y)\,\mathrm{d}x + m(x, y)\,g(x, y)\,\mathrm{d}y = 0$$

is an exact equation, i.e. that (under some assumptions on the smoothness of the function m)

$$\frac{\partial(mf)}{\partial y} = \frac{\partial(mg)}{\partial x} \tag{6}$$

holds. The function $m(x, y)$ is called an *integrating factor* of equation (2).

REMARK 4. The existence of an integrating factor can be proved under quite general assumptions.

REMARK 5. From (6) it follows that the integrating factor $m(x, y)$ satisfies a partial differential equation

$$g(x, y)\frac{\partial m}{\partial x} - f(x, y)\frac{\partial m}{\partial y} = \left(\frac{\partial f}{\partial y} - \frac{\partial g}{\partial x}\right)m\,. \tag{7}$$

In general, it is more difficult to integrate this equation than to integrate the given equation (2).

In many cases m can be found as a function of the variable x only. From equation (7) it follows that a necessary and sufficient condition for this is that

$$\frac{\partial f/\partial y - \partial g/\partial x}{g}$$

be a function of the variable x alone; then $m(x)$ can be obtained from the equation (assuming that $m(x) > 0$)

$$\frac{\mathrm{d}\ln m}{\mathrm{d}x} = \frac{\partial f/\partial y - \partial g/\partial x}{g}\,. \tag{8}$$

A similar statement is valid for the case where m is a function of the variable y alone. The result is:

$$\frac{\mathrm{d} \ln m}{\mathrm{d}y} = \frac{\partial f/\partial y - \partial g/\partial x}{-f}. \tag{9}$$

Example 3. The equation

$$(2xy + x^2y + \tfrac{1}{3}y^3)\,\mathrm{d}x + (x^2 + y^2)\,\mathrm{d}y = 0$$

is not exact (the condition (3) is not satisfied). But we note that

$$\frac{\partial f/\partial y - \partial g/\partial x}{g} = \frac{2x + x^2 + y^2 - 2x}{x^2 + y^2} = 1\,.$$

According to (8)

$$\frac{\mathrm{d} \ln m(x)}{\mathrm{d}x} = 1\,, \quad \text{hence} \quad m(x) = \mathrm{e}^x\,.$$

(It is not necessary to write $m(x) = C\mathrm{e}^x$ with a constant C, because we are only trying to find a *particular* solution of equation (7).)

The equation

$$\mathrm{e}^x(2xy + x^2y + \tfrac{1}{3}y^3)\,\mathrm{d}x + \mathrm{e}^x(x^2 + y^2)\,\mathrm{d}y = 0$$

is thus exact, as may also be easily verified by using (3).

REMARK 6 (*Singular Points of Equation* (1)). From Theorem 17.2.1 and Remark 17.2.7 it follows: Let f, g, $\partial f/\partial y$, $\partial g/\partial y$ be continuous in the neighbourhood of the point (x_0, y_0) and let $g(x_0, y_0) \neq 0$. Then exactly one integral curve of equation (1) exists which passes through the point (x_0, y_0). If $g(x_0, y_0) = 0$, but $f(x_0, y_0) \neq 0$, then (see Remark 17.2.16) equation (1) may be written in the form

$$\frac{\mathrm{d}x}{\mathrm{d}y} = -\frac{g(x, y)}{f(x, y)}, \tag{10}$$

and under the supplementary assumption of continuity of $\partial f/\partial x$, $\partial g/\partial x$ again exactly one integral curve passes through the point (x_0, y_0). If the relations

$$f(x_0, y_0) = 0 \quad \text{and} \quad g(x_0, y_0) = 0$$

hold simultaneously, then the point (x_0, y_0) is called a *singular point of equation* (1). In the neighbourhood of a singular point various possibilities can occur as to existence and uniqueness of the solution as well as to the form of the integral curves of the given equation. Let us give some simple examples:

The equation

$$y' = \frac{y}{x} \tag{11}$$

has, one and only one, singular point, namely (0, 0). By separation of variables, the general integral

$$y = Cx \tag{12}$$

can easily be obtained. An infinite number of integral curves of equation (11) (in this case, straight lines) passes through the singular point concerned. A singular point of this type is called a *node.*

The (only) singular point of the equation

$$y' = -\frac{y}{x} \tag{13}$$

is also the origin. The general integral is a system of hyperbolae

$$y = \frac{C}{x} \tag{14}$$

none of which passes through the origin. Their asymptotes,

$$y = 0\,, \quad x = 0\,,$$

pass through the origin and are integral curves of the given equation (13) (the second asymptote is an integral curve in the sense of Remark 17.2.16). This type of singular point is called a *saddle point.*

The equation

$$y' = -\frac{x}{y} \tag{15}$$

has also only one singular point, the origin. The general integral is a system of circles

$$x^2 + y^2 = C\,.$$

None of the integral curves of equation (15) pass through the origin. This type of singular point is called a *centre.*

The reader will find a more detailed analysis in [65] or in [35] where he will also find some results concerning questions on differential equations in a complex domain.

17.5. Equations of the First Order not Solved with Respect to the Derivative. Lagrange's Equation. Clairaut's Equation. Singular Solutions

Equations of the first order not solved with respect to the derivative are equations of the form

$$F(x, y, y') = 0\,, \tag{1}$$

or, if the common notation $y' = p$ is used, of the form

$$F(x, y, p) = 0\,. \tag{2}$$

I. In some cases (see Theorem 12.9.2 on Implicit Functions, p. 465) it is possible to solve (1) with respect to y'.

Example 1.

$$y'^2 = 1 + x^2 + y^2\,. \tag{3}$$

On solving equation (1), we get two equations

$$y' = \sqrt{(1 + x^2 + y^2)}\,, \quad y' = -\sqrt{(1 + x^2 + y^2)}\,.$$

The general integral of each of these equations represents a one-parameter system of curves. The curves of both the first and the second system satisfy equation (3). Thus the general integral of equation (3) consists of two systems of curves.

REMARK 1. Equation (1) is not always so easy to solve. In the domain of real functions, this equation need not have a solution at all, as is to be seen from the simple example of the equation

$$y'^2 + y^2 + 1 = 0\,.$$

In the general case, the problem of solving equation (1), both from a theoretical and practical point of view, is more complicated than in the case of the equation $y' = f(x, y)$. In this paragraph we shall deal only with the most important results and methods of solution.

II. *Method of two parameters.* Under rather simple assumptions about the function F, equation (2) represents a surface. The coordinates of the points (x, y, p) of this surface can be expressed as functions of two parameters u, v:

$$x = f(u, v)\,, \quad y = g(u, v)\,, \quad p = h(u, v)\,. \tag{4}$$

From the relation $dy = p\,dx$ it follows that

$$\frac{\partial g}{\partial u}\,du + \frac{\partial g}{\partial v}\,dv = h(u, v)\left[\frac{\partial f}{\partial u}\,du + \frac{\partial f}{\partial v}\,dv\right]$$

or

$$\frac{dv}{du} = \frac{h\dfrac{\partial f}{\partial u} - \dfrac{\partial g}{\partial u}}{\dfrac{\partial g}{\partial v} - h\dfrac{\partial f}{\partial v}}. \tag{5}$$

This is a differential equation for the unknown function $v(u)$, solved with respect to the derivative dv/du. Finding its general integral $v = t(u, C)$ and substituting into the first two equations (4), we obtain

$$x = f(u, t(u, C)), \quad y = g(u, t(u, C)).$$

Eliminating u from these two equations, we have

$$\varphi(x, y, C) = 0,$$

which, as a rule, gives the general integral of equation (1).

REMARK 2. We have used the phrase "as a rule". The above-mentioned procedure is formal; in particular, the concept of elimination of a variable or parameter is very uncertain. Consequently, the result so obtained must be analysed, especially to verify whether it is really the solution of the given equation. *This remark applies throughout paragraph* 17.5.

It is particularly simple to express the "surface" (2) by means of equations (4), in these two cases:

III. Let equation (2) be of the form

$$y = f(x, p). \tag{6}$$

We choose x and p as parameters. Then, from the equation $dy = p\,dx$, it follows that

$$\frac{\partial f}{\partial x}\,dx + \frac{\partial f}{\partial p}\,dp = p\,dx \quad \text{or} \quad \frac{\partial f}{\partial x} + \frac{\partial f}{\partial p}\frac{dp}{dx} = p. \tag{7}$$

(It is possible to obtain (7) directly by differentiating equation (6) with respect to x, if p is taken as a function of x and p is substituted for y'.) If we find the general integral of (7),

$$p = t(x, C) \tag{8}$$

and substitute for p into (6), we often get (see Remark 2) the general integral of equation (6). (It would not be correct to substitute y' for p into (8) and to try to find the general integral of equation (6) by integrating equation (8) thus obtained. When integrating, a second arbitrary constant would arise. It is now easy to show that by this procedure we would get the general integral of the equation of the second order,

$$\frac{\partial f}{\partial x} + \frac{\partial f}{\partial y'} y'' = y' ,$$

which results from (7) by substituting y' for p.)

IV. Let equation (2) be of the form

$$x = f(y, p) . \tag{9}$$

If we consider p as a function of y (by virtue of a relation between x and y) and differentiate (9) with respect to y, using

$$\frac{dx}{dy} = \frac{1}{dy/dx} = \frac{1}{p} ,$$

we get

$$\frac{1}{p} = \frac{\partial f}{\partial y} + \frac{\partial f}{\partial p}\frac{dp}{dy} . \tag{10}$$

This is an equation for the unknown function $p(y)$. If we find its general integral

$$p = k(y, C) ,$$

and then substitute for p into (9) we get the general integral of equation (9). (See, however, Remark 2.)

REMARK 3. Using the above-mentioned procedure, it is naturally possible to integrate special cases of III and IV,

$$y = f(p) , \quad \text{or} \quad x = f(p) .$$

If, instead of the equation $x = f(p)$, a rather more general equation $g(x, p) = 0$ is given, we can use a method similar to method II. This case becomes simpler, because the equation $g(x, p) = 0$ represents a curve geometrically. Let its parametric equations be of the form

$$x = \psi(t) , \quad p = \chi(t) .$$

Since

$$\frac{dy}{dx} = p = \chi(t) ,$$

then

$$dy = p \, dx = \chi(t) \, \psi'(t) \, dt .$$

Integrating, we get

$$y = \int \chi(t)\, \psi'(t)\, \mathrm{d}t = \mu(t) + C\,,$$

say. Then the equations

$$x = \psi(t)\,, \quad y = \mu(t) + C$$

give the general solution of the given equation.

In a similar way, it is possible to deal with the equation $g(y, p) = 0$ (see Example 2)

Example 2.

$$y = y'^2 + 2\,. \tag{11}$$

Following III, we differentiate with respect to x and write p instead of y'. Then

$$p = 2p\,\frac{\mathrm{d}p}{\mathrm{d}x}\,.$$

If $p \neq 0$ (for $p = 0$ see below), then

$$\frac{\mathrm{d}p}{\mathrm{d}x} = \frac{1}{2}$$

so that

$$p = \tfrac{1}{2}x + C\,.$$

Substituting into (11), we have the general integral of equation (11):

$$y = (\tfrac{1}{2}x + C)^2 + 2\,. \tag{12}$$

It may be easily verified that if we solve (11) to give y', then the general integral of the two equations so obtained is included in (12).

The case $p = 0$ gives the solution $y = 2$ of the given equation.

If we had used the method of Remark 3, we would have got:

$$p = t\,, \quad y = t^2 + 2\,,$$

$$\frac{\mathrm{d}y}{\mathrm{d}x} = p\,, \quad \mathrm{d}x = \frac{\mathrm{d}y}{p} = \frac{2t\,\mathrm{d}t}{t} = 2\,\mathrm{d}t\,, \quad x = 2t + k\,.$$

Thus, the general solution in a parametric form is

$$x = 2t + k\,, \quad y = t^2 + 2$$

and, by eliminating the parameter t,

$$y = \left(\frac{x}{2} - \frac{k}{2}\right)^2 + 2$$

in accordance with (12).

V. *Lagrange's equation*

$$y = \varphi(y')\,x + \psi(y') \tag{13}$$

or

$$y = \varphi(p)\,x + \psi(p) \tag{14}$$

is a special case of type III; it is always possible to integrate it by means of quadratures. Differentiating (14) with respect to x and substituting p for y', we obtain

$$p = \varphi(p) + x\,\varphi'(p)\frac{dp}{dx} + \psi'(p)\frac{dp}{dx}. \tag{15}$$

If we consider x as a function of the variable p, then we obtain from (15) (assuming $\varphi(p) \neq p$; on the case $\varphi(p) \equiv p$ see VI, Clairaut's equation)

$$\frac{dx}{dp} = \frac{\varphi'(p)}{p - \varphi(p)}\,x + \frac{\psi'(p)}{p - \varphi(p)} \quad (\varphi(p) \neq p). \tag{16}$$

Equation (16) is a linear equation for the function $x(p)$. If we substitute its general integral

$$x = t(p, C) \tag{17}$$

into (14), we get

$$y = \varphi(p)\,t(p, C) + \psi(p)\,; \tag{18}$$

(17) and (18) give a parametric form of the general integral of (14). Eliminating the parameter p from (17) and (18), we get the general integral in the form $G(x, y, C) = 0$. (See, however, Remark 2.)

Example 3.

$$y = 2xy' + y'^2 \quad \text{or} \quad y = 2px + p^2. \tag{19}$$

Differentiating with respect to x, we have

$$p = 2p + 2(x + p)\frac{dp}{dx}$$

that is,

$$\frac{dx}{dp} = -\frac{2x}{p} - 2 \quad (p \neq 0). \tag{20}$$

(20) is a linear equation for the function $x(p)$. Its solution is (see Example 17.3.4)

$$x = \frac{C}{p^2} - \frac{2p}{3}. \tag{21}$$

Substituting into the second of equations (19), we get

$$y = \frac{2C}{p} - \frac{p^2}{3} \tag{22}$$

which together with (21) constitute the parametric equations of the general integral of (19).

For $p = 0$ we get, on substituting into (19), the solution $y \equiv 0$.

VI. *Clairaut's equation*

$$y = xy' + \psi(y') \quad \text{or} \quad y = px + \psi(p) \tag{23}$$

(where $\psi(p)$ is a given differentiable function) is a particular case of Lagrange's equation (for the case $\varphi(p) = p$). Differentiating with respect to x, we get

$$p = p + [x + \psi'(p)] \frac{\mathrm{d}p}{\mathrm{d}x} \quad \text{or} \quad \frac{\mathrm{d}p}{\mathrm{d}x} [x + \psi'(p)] = 0 \,. \tag{24}$$

From the equation

$$\frac{\mathrm{d}p}{\mathrm{d}x} = 0$$

we get $p = C$, which substituted into (23) gives the general integral of equation (23):

$$y = Cx + \psi(C) \,. \tag{25}$$

This is a one-parameter system of straight lines.

If the second term on the left-hand side of the second equation (24) is equal to zero,

$$x + \psi'(p) = 0 \,, \tag{26}$$

then (26) determines p as a function of x,

$$p = t(x) \,, \tag{27}$$

(if, for example, $\psi''(p) \neq 0$). Substituting into (23), we get

$$y = xt(x) + \psi(t(x)) \,. \tag{28}$$

This solution may be proved to be the envelope of the system (25). It is a singular integral of equation (24).

REMARK 4. When investigating equations of the first order, it is often important to find the ***singular integral*** of the given equation (Definition 17.2.6). The solution of this problem in its full generality is rather difficult. We introduce here only two special theorems:

Definition 1. By a *discriminant curve* of the equation

$$F(x, y, y') = 0 \tag{29}$$

we mean the curve $G(x, y) = 0$ the points (x, y) of which satisfy (for a certain range of values of α) the equations

$$F(x, y, \alpha) = 0\,, \quad \frac{\partial F}{\partial \alpha}(x, y, \alpha) = 0\,. \tag{30}$$

REMARK 5. It may happen that the curve $G(x, y) = 0$ is not a real curve. For example, for the equation

$$y'^3 + (y^2 + 2)\, y' - xy = 0 \tag{31}$$

we get

$$\frac{\partial F}{\partial \alpha} = 3\alpha^2 + y^2 + 2 \tag{32}$$

(since $F(x, y, \alpha) = \alpha^3 + (y^2 + 2)\,\alpha - xy$), and this cannot be zero for any real values of the variables x, y, α, so that $G(x, y) = 0$ does not represent a real curve.

Theorem 1. *Let a discriminant curve of equation* (29) *exist. Then the singular integral of this equation* (*if it exists*) *is contained in this discriminant curve.*

REMARK 6. From the theorem on implicit functions it follows that the satisfying of the equation

$$\frac{\partial F}{\partial y'}(x, y, y') = 0 \tag{33}$$

at each point of the integral curve concerned is a *necessary* (but not sufficient) condition that the curve be a singular integral of the given equation. If the discriminant curve of equation (29) is an integral curve of this equation, then it need not be the singular integral of this equation.

Theorem 2. *Let*

$$f(x, y, C) = 0 \tag{34}$$

be the general integral of equation (29). *If the curve given by the equation*

$$H(x, y) = 0 \tag{35}$$

which we obtain by eliminating C from the equations

$$f(x, y, C) = 0\,, \quad \frac{\partial f}{\partial C}(x, y, C) = 0\,, \tag{36}$$

is the envelope (§ 9.7) *of the one-parameter system of curves of the general integral* $f(x, y, C) = 0$, *then*

1. *it is an integral curve of equation* (32);
2. *it is the singular integral of this equation.*

Example 4. Let the equation

$$y = x - \tfrac{4}{9}p^2 + \tfrac{8}{27}p^3 \quad (p = y') \tag{37}$$

be solved by the method of differentiation with respect to x (see V). We get (see (16)):

$$\frac{dx}{dp} = \frac{-\frac{8}{9}p + \frac{8}{9}p^2}{p - 1} = \tfrac{8}{9}p\,, \quad \text{so that} \quad x = \tfrac{4}{9}p^2 + C\,. \tag{38}$$

Substituting into (37), we get

$$y = C + \tfrac{8}{27}p^3 \quad \text{or} \quad (y - C)^2 = (\tfrac{8}{27})^2 (\tfrac{9}{4})^3 (x - C)^3$$

(using the second equation (38)), i.e.

$$(y - C)^2 - (x - C)^3 = 0\,. \tag{39}$$

The second of equations (36) gives

$$2(y - C) = 3(x - C)^2\,. \tag{40}$$

If we substitute from here for $y - C$ into the first equation (36), we get

$$\tfrac{9}{4}(x - C)^4 = (x - C)^3\,, \tag{41}$$

so that either

$$x - C = 0 \quad \text{and then also} \quad y - C = 0, \quad \text{consequently} \quad y - x = 0\,, \tag{42}$$

or $(x - C) \neq 0$ and then from (41) we have

$$x - C = \tfrac{4}{9}\,, \quad \text{so that according to (39)} \quad y - C = \tfrac{8}{27} \quad \text{or} \quad y - x = -\tfrac{4}{27}\,. \tag{43}$$

The straight line $y - x = -\frac{4}{27}$ is (see Example 9.7.3, p. 332) the envelope of the system (39) and according to Theorem 2 it is the singular integral of equation (37). The straight line $y - x = 0$ is not the envelope of the system (39) (Example 9.7.3). Substituting into (37), we can easily verify that it is not an integral curve of this equation.

We could reach the same result using Theorem 1 (i.e. without integrating equation (37)): We have

$$\frac{\partial F}{\partial y'} = \frac{8}{9}y' - \frac{8}{9}y'^2\,,$$

so that the second of equations (30) (i.e. equation (33)) is

$$y'(1 - y') = 0. \tag{44}$$

Thus, either $y' = 0$, or $y' = 1$. Substituting into (37), we get

$$y - x = 0, \quad \text{or} \quad y - x = -\tfrac{4}{27},$$

respectively. The first curve does not satisfy equation (37), but the second does.

17.6. Trajectories

Let a one-parameter system of curves be given:

$$F(x, y, C) = 0 \tag{1}$$

and let

$$y' = f(x, y) \tag{2}$$

be the differential equation of this system.

Theorem 1. *The differential equation of a system of such curves $G(x, y, k) = 0$, each of which intersects every curve of the system (1) at right angles (the system of so-called orthogonal trajectories), is*

$$y' = -\frac{1}{f(x, y)}. \tag{3}$$

Example 1. Find orthogonal trajectories of the system of parabolas

$$y = Cx^2. \tag{4}$$

Differentiating (4) with respect to x we get $y' = 2Cx$ and eliminating C from both equations we obtain the differential equation of the system (4):

$$y' = \frac{2y}{x}. \tag{5}$$

According to (3), the differential equation of the orthogonal trajectories is

$$\frac{dy}{dx} = -\frac{1}{2y/x} = -\frac{x}{2y}.$$

Integrating this by separation of variables (§ 17.3, p. 739), we obtain a system of ellipses

$$y^2 + \tfrac{1}{2}x^2 = k. \tag{6}$$

REMARK 1. The differential equation of the system $H(x, y, C) = 0$, every curve of which intersects every curve of the system (1) at an angle $\alpha \neq \pi/2$ (the so-called *isogonal trajectories* or *oblique trajectories*), is given by the equation

$$\frac{dy}{dx} = \frac{k + f(x, y)}{1 - k f(x, y)},$$

where $k = \tan \alpha$ and $y' = f(x, y)$ is the differential equation of the system (1).

17.7. Differential Equations of Order n. Simple Types of Equations of Order n. The Method of a Parameter

On the question of existence and uniqueness of the solution see Theorem 17.2.3.

Throughout this paragraph, the continuity of the functions cousidered and of the respective derivatives is assumed.

I. The equation

$$y'' = f(x) \tag{1}$$

is easy to solve:

$$y' = \int f(x)\,dx\,, \quad y = \int y'(x)\,dx\,. \tag{2}$$

There are two arbitrary constants in the result, because two indefinite integrals are involved. In a similar way we can solve the equation

$$y^{(n)} = f(x)\,.$$

II. The solution of the equation

$$y^{(n)} = f(x)\,, \tag{3}$$

that satisfies the initial conditions

$$y(x_0) = y'(x_0) = \ldots = y^{(n-1)}(x_0) = 0$$

is given by the *Cauchy–Dirichlet Formula*:

$$y = \frac{1}{(n-1)!} \int_{x_0}^{x} (x - z)^{n-1} f(z)\,dz\,. \tag{4}$$

By adding to (4) the expression

$$y_0^{(n-1)} \frac{(x - x_0)^{n-1}}{(n-1)!} + y_0^{(n-2)} \frac{(x - x_0)^{n-2}}{(n-2)!} + \ldots + y_0'(x - x_0) + y_0\,, \tag{5}$$

we get the integral of equation (3) that satisfies the initial conditions

$$y(x_0) = y_0\,,\; y'(x_0) = y'_0\,,\; \ldots,\; y^{(n-1)}(x_0) = y_0^{(n-1)}\,.$$

Example 1. Solve the equation $y''' = \ln x$ under the initial conditions $y(1) = y_0$, $y'(1) = y'_0$, $y''(1) = y''_0$.

We get the solution as a sum of expressions (4) and (5), $y = y_1 + y_2$. Integrating (4) by parts, we get

$$y_1 = \frac{1}{2}\int_1^x (x - z)^2 \ln z\, dz = \tfrac{1}{6}x^3 \ln x - \tfrac{11}{36}x^3 + \tfrac{1}{2}x^2 - \tfrac{1}{4}x + \tfrac{1}{18}\,. \tag{6}$$

According to (5)

$$y_2 = y_0 + y'_0(x - 1) + y''_0\,\frac{(x - 1)^2}{2}\,. \tag{7}$$

The sum of (6) and (7) gives the solution required.

III. The equation of the form

$$y'' = f(x, y')\,, \quad \text{or} \quad F(x, y', y'') = 0\,, \tag{8}$$

in which y does not occur explicitly, can be transformed using the substitution $y'(x) = z(x)$ into an equation of the first order with $z(x)$ as dependent variable. More generally, the equation of the form

$$y^{(n)} = f(x, y^{(n-1)})\,, \quad \text{or} \quad F(x, y^{(n)}, y^{(n-1)}) = 0 \tag{9}$$

can be transformed by the substitution $y^{(n-1)}(x) = z(x)$ into an equation of the first order for $z(x)$. Solving it, we get, by repeated integration (see I and II), the function $y(x)$.

REMARK 1 (*Method of a Parameter*). If a particular case of the second of equations (9),

$$F(y^{(n)}, y^{(n-1)}) = 0\,, \tag{10}$$

is to be solved and if this equation cannot be easily transformed into the form $y^{(n)} = f(y^{(n-1)})$, we can often use the method of a parameter with success. Let us consider, for example, the equation

$$F(y''', y'') = 0\,. \tag{11}$$

First, we express y''' and y'' parametrically,

$$y''' = \varphi(t)\,, \quad y'' = \psi(t)\,,$$

so that equation (11) is identically satisfied in t, i.e. $F(\varphi(t), \psi(t)) \equiv 0$. Further, $dy'' = y''' dx$ holds, so that (under obvious assumptions)

$$dx = \frac{dy''}{y'''} = \frac{\psi'(t)\,dt}{\varphi(t)}, \quad \text{whence} \quad x(t) = \int \frac{\psi'(t)\,dt}{\varphi(t)} = \mu(t) + C_1\,; \tag{12}$$

then

$$dy' = y''\,dx = \frac{\psi(t)\,\psi'(t)}{\varphi(t)}\,dt\,, \quad \text{whence} \quad y'(t) = \int \frac{\psi(t)\,\psi'(t)\,dt}{\varphi(t)} = \chi(t) + C_2\,, \tag{13}$$

$$dy = y'\,dx = [\chi(t) + C_2]\,\frac{\psi'(t)}{\varphi(t)}\,dt\,, \quad \text{whence} \quad y(t) = \int [\chi(t) + C_2]\,\frac{\psi'(t)}{\varphi(t)}\,dt = \varkappa(t, C_2) + C_3\,. \tag{14}$$

Thus we get the general integral of equation (11) in parametric form

$$x = \mu(t) + C_1\,, \quad y = \varkappa(t, C_2) + C_3\,. \tag{15}$$

REMARK 2. In this paragraph, we often use a formal procedure in the same way as in § 17·5, so that even here a remark similar to Remark 17.5.2 is valid.

Example 2.

$$y'''^2 + y''^2 - 4 = 0\,.$$

Let us express y''' and y'' parametrically in the form

$$y''' = 2\cos t\,, \quad y'' = 2\sin t\,.$$

Following (12), (13), (14) we get

$$x = \int \frac{\psi'(t)\,dt}{\varphi(t)} = \int dt = t + C_1\,,$$

$$y' = \int 2\sin t\,dt = -2\cos t + C_2\,,$$

$$y = \int (-2\cos t + C_2)\,dt = -2\sin t + C_2 t + C_3\,.$$

Eliminating t from the equations for x and y, we get the general integral of the given equation:

$$y = -2\sin(x - C_1) + C_2 x + K \quad (K = -C_1 C_2 + C_3)\,.$$

REMARK 3. We can obviously proceed as follows; substituting $y''(x) = z(x)$, we first transform equation (11) into an equation of the form

$$F(z, z') = 0$$

and then integrate this equation using the method of a parameter. From the parametric equations for x and z so obtained, we then eliminate t, in order to get $y(x)$ by a further integration (which is not always a simple matter).

Example 3. Taking the equation of Example 2, we put $y'' = z$ and choose

$$z' = 2\cos u\,, \quad z = 2\sin u\,.$$

From the relation $\mathrm{d}z = z'\,\mathrm{d}x$, or $2\cos u\,\mathrm{d}u = 2\cos u\,\mathrm{d}x$, it follows that $\mathrm{d}x = \mathrm{d}u$, $x = u + C$. Hence

$$z = 2\sin(x - C) = y''\,,$$

and by integrating twice we get the same result as in Example 2.

IV. The equation

$$y'' = f(y) \tag{16}$$

multiplied by $y'(x)$ becomes

$$y'y'' = f(y)\,y'\,, \quad \text{hence} \quad \tfrac{1}{2}y'^2 = \int f(y)\,\mathrm{d}y = F(y) + C\,.$$

In this way, the original equation is transformed into an equation of the first order. A similar procedure may be used when solving an equation of the form

$$y^{(n)} = f(y^{(n-2)}) \tag{17}$$

(after the initial substitution $y^{(n-2)}(x) = z(x)$).

REMARK 4. If equation (16) or (17) is given in implicit form,

$$F(y'', y) = 0\,, \quad \text{or} \quad F(y^{(n)}, y^{(n-2)}) = 0\,, \tag{18}$$

we can use the method of a parameter in a similar way as in Remark 1.

For example, considering the equation

$$F(y''', y') = 0\,, \tag{19}$$

we express, first of all, y''' and y' parametrically:

$$y''' = \varphi(t)\,, \quad y' = \psi(t)$$

(so that $F(\varphi(t), \psi(t)) \equiv 0$). Further, by eliminating $\mathrm{d}x$ from the equations

$$\mathrm{d}y'' = y'''\,\mathrm{d}x\,, \quad \mathrm{d}y' = y''\,\mathrm{d}x\,,$$

we get

$$y''\,\mathrm{d}y'' = y'''\,\mathrm{d}y' = \varphi(t)\,\psi'(t)\,\mathrm{d}t\,, \quad \text{whence} \quad \tfrac{1}{2}[y''(t)]^2 = \int \varphi(t)\,\psi'(t)\,\mathrm{d}t\,.$$

Once $y''(t)$ known, we proceed further according to Remark 1:

$$\left(\mathrm{d}x = \frac{\mathrm{d}y''}{y'''}\,,\quad \mathrm{d}y' = y''\,\mathrm{d}x, \text{ etc.}\right).$$

REMARK 5. A method similar to case IV may also be used for equations of the form

$$y'' = ay'^2 + f(y)\,. \tag{20}$$

Introducing a new unknown function $u(x)$ by the relation $u(x) = y'^2(x)$, we get

$$\mathrm{d}u = 2y'\,\mathrm{d}y' = 2y''\,\mathrm{d}y\,, \tag{21}$$

on using the relations

$$y'\,\mathrm{d}x = \mathrm{d}y\,,\quad \mathrm{d}y' = y''\,\mathrm{d}x\,.$$

Hence, from (20), (21)

$$\mathrm{d}u = 2[au + f(y)]\,\mathrm{d}y\,,$$

which is a *linear* equation for the function $u(y)$. Solving it, we get an equation of the first order

$$y'^2 = u(y)\,.$$

17.8. First Integral of a Differential Equation of the Second Order. Reduction of the Order of a Differential Equation. Equations, the Left-hand Sides of which are Exact Derivatives

REMARK 1. In physics, we often meet the concept of the first integral of a differential equation of the second order. Its meaning is as follows:

Definition 1. We say that the equation

$$g(x, y, y') = C \tag{1}$$

gives the *first integral* of the equation

$$F(x, y, y', y'') = 0\,,$$

if the function $g(x, y, y')$ is constant along each integral curve of the given equation (and, at the same time, is not identically equal to a constant in the variables x, y, y').

Example 1. The equation

$$y'^2 + 4y^3 = C \tag{2}$$

is the first integral of the equation

$$y'' + 6y^2 = 0 \tag{3}$$

because

$$\frac{d}{dx}(y'^2 + 4y^3) = 2y'y'' + 12y^2y' = 2y'(y'' + 6y^2),$$

which, as a result of (3), is zero along each integral curve of equation (3).

REMARK 2. In a similar way, it is possible to define the k-th integral of a given differential equation of order n. Thus we obtain a differential equation of a lower order, which is (at least theoretically) easier to integrate.

Some typical examples of the reduction of the order of an equation follow:

I. The equation

$$F(x, y^{(m)}, y^{(m+1)}, \ldots, y^{(n)}) = 0 \tag{4}$$

is easily transformed by the substitution $y^{(m)}(x) = z(x)$ into the equation

$$F(x, z, z', \ldots, z^{(n-m)}) = 0, \tag{5}$$

the order of which is $n - m$.

II. In the equation

$$F(y, y', y'', \ldots, y^{(n)}) = 0, \tag{6}$$

let us write $y' = p$ and let p be considered as a function of the independent variable y, i.e. $p = p(y)$. Then

$$y'' = \frac{dy'}{dx} = \frac{dp}{dy}\frac{dy}{dx} = p\frac{dp}{dy}, \tag{7}$$

$$y''' = \frac{dy''}{dx} = \frac{d\left(p\dfrac{dp}{dy}\right)}{dy}\frac{dy}{dx} = p^2\frac{d^2p}{dy^2} + p\left(\frac{dp}{dy}\right)^2, \tag{8}$$

. . . .

So we get from (6) an equation of order $(n - 1)$

$$G\left(y, p, \frac{dp}{dy}, \ldots, \frac{d^{n-1}(p)}{dy^{n-1}}\right) = 0.$$

III. Let the left-hand side of the equation $F(x, y, y', \ldots, y^{(n)}) = 0$ be a homogeneous function of the arguments $y, y', y'', \ldots, y^{(n)}$, i.e. let

$$F(x, ty, ty', \ldots, ty^{(n)}) = t^m F(x, y, y', \ldots, y^{(n)}) \tag{9}$$

be satisfied for all t in a certain neighbourhood of the point $t = 1$; m is the degree of homogeneity. Then the order of equation (1) may be reduced by introducing a new unknown function $z(x)$, given by the relation

$$y(x) = e^{\int z(x)dx} . \tag{10}$$

Example 2. The equation

$$x^2yy'' - (y - xy')^2 = 0 \tag{11}$$

is homogeneous (of the second degree) with respect to y, y', y''. Following (10), we get

$$y = e^{\int z\,dx} , \quad y' = ze^{\int z\,dx} , \quad y'' = (z' + z^2)\,e^{\int z\,dx} . \tag{12}$$

Substituting (12) into (11) and dividing by $e^{2\int z\,dx}$, we get the equation

$$x^2(z' + z^2) - (1 - xz)^2 = 0$$

or

$$x^2z' + 2xz - 1 = 0 .$$

The solution of this linear equation (see VI, § 17.3) is

$$z = \frac{1}{x} + \frac{C_1}{x^2} ,$$

so that

$$y = e^{\int z\,dx} = e^{\int(1/x + C_1/x^2)dx} = C_2xe^{-C_1/x} .$$

REMARK 3. The homogeneous linear equation

$$a_n(x)\,y^{(n)} + a_{n-1}(x)\,y^{(n-1)} + \ldots + a_1(x)\,y' + a_0(x)\,y = 0 \tag{13}$$

is also of type III. The substitution (10), however, is not suitable in this case, because it leads, in general, to a non-linear equation. If a (non-zero) particular integral $y = u(x)$ of equation (13) is known, we can reduce the order of the equation by introducing a new unknown function $z(x)$ by the equation

$$y(x) = u(x)\,z(x) . \tag{14}$$

Example 3. One of the solutions of the equation

$$xy''' - y'' + xy' - y = 0 \tag{15}$$

is obviously $u(x) = x$. Using the substitution (14), we get

$$y' = xz' + z , \quad y'' = xz'' + 2z' , \quad y''' = xz''' + 3z'' .$$

Thus, equation (15) is transformed into the equation

$$x^2z''' + 2xz'' + (x^2 - 2)\,z' = 0 \quad \text{or} \quad x^2v'' + 2xv' + (x^2 - 2)\,v = 0 \qquad (16)$$

(putting $z' = v$). This is now an equation of the second order. From (14) and from $v = z'$ it follows that $v = (y/u)'$. Further, if we knew another integral $U(x)$ of equation (15), we could obtain another integral $v = (U/u)'$ of equation (16), so that we could again reduce the order of equation (16). In general: If we know k (*independent*) integrals of equation (14), we can reduce its order by k.

REMARK 4. It is possible to prove, by the method mentioned in Remark 3, that if $y_1(x)$ is a non-zero particular integral of the equation

$$y'' + a_1(x)\,y' + a_0(x)\,y = 0 \qquad (17)$$

then the second function of the fundamental system (Definition 17.11.2) is given by the function

$$y_2 = y_1 \int \frac{1}{y_1^2}\,e^{-\int a_1\,dx}\,dx\,. \qquad (18)$$

Example 4. A particular integral of the equation

$$y'' - y = 0$$

is $y_1 = e^x$. According to (18), the second function of the fundamental system is the function

$$y_2 = e^x \int \frac{1}{e^{2x}}\,e^{-\int 0\,dx}\,dx = ke^{-x}\,.$$

REMARK 5. Another way of reducing the order of a differential equation is to transform the left-hand side of the given equation into a form which is the complete derivative (with respect to x) of an expression of lower order. We shall demonstrate this procedure by an example.

Example 5. Let us consider the equation

$$yy'' - y'^2 = 0\,. \qquad (19)$$

Dividing by y^2, we get

$$\frac{yy'' - y'^2}{y^2} = 0 \quad (y \neq 0)\,. \qquad (20)$$

The left-hand side of equation (20) is obviously a complete derivative:

$$\frac{yy'' - y'^2}{y^2} = \frac{d}{dx}\left(\frac{y'}{y}\right). \qquad (21)$$

Thus the first integral of equation (20) (or (19)) is

$$\frac{y'}{y} = C_1 ,$$

hence

$$y = C_2 e^{C_1 x} .$$

17.9. Dependence of Solutions on Parameters of the Differential Equation and on Initial Conditions

Theorem 1. *Let an equation of the form*

$$y^{(n)} = f(x, y, y', \ldots, y^{(n-1)}) \tag{1}$$

be given. Let us denote by A the point with coordinates $a, b_1, b_2, \ldots, b_n$. Let a function f (as a function of $n + 1$ variables) be continuous in an $(n + 1)$-dimensional region O and let there correspond to each point $A \in O$ precisely one solution of equation (1),

$$y = \varphi(x, y, b_1, b_2, \ldots, b_n) ,$$

satisfying the conditions

$$y(a) = b_1 , \; y'(a) = b_2 , \; \ldots , \; y^{(n-1)}(a) = b_n .$$

Then the function φ is a continuous function of all $n + 2$ variables.

Moreover, if the function f has in O continuous partial derivatives of order r with respect to all the variables, then the function φ has continuous partial derivatives of the order r (and therefore of all lower orders) with respect to all the variables.

REMARK 1. If the function f depends also on parameters $\lambda_1, \lambda_2, \ldots, \lambda_k$ then, of course, the function φ also depends on these parameters. If f is continuously dependent on $\lambda_1, \lambda_2, \ldots, \lambda_k$, then the same holds for φ. A similar statement holds for derivatives with respect to $\lambda_1, \lambda_2, \ldots, \lambda_k$.

REMARK 2. The statement of Theorem 1 (or Remark 1) is very natural: If the function f is "well behaved" then, in the neighbourhood of the point a, its integral curves changes only slightly with small changes in the values of the initial conditions or in the values of the parameters (roughly speaking, the changes are smaller the greater is the number of "well-behaved" derivatives of the function f).

On questions concerning the stability of the solution see § 17.19.

17.10. Asymptotic Behaviour of Integrals of Differential Equations (for $x \to +\infty$). Oscillatory Solutions. Periodic Solutions

REMARK 1. The problem of the asymptotic behaviour of integrals has been most thoroughly studied in the case of linear differential equations

$$y^{(n)} + f_{n-1}(x)\, y^{(n-1)} + \ldots + f_1(x)\, y' + f_0(x)\, y = f(x)\,, \tag{1}$$

where, also, the existence of the solution is guaranteed in the whole interval $(-\infty, \infty)$ if $f_k(x)$ and $f(x)$ are continuous in this interval.

Theorem 1. *Let $f_k(x)$ $(k = 0, 1, \ldots, n-1)$ and $f(x)$ be (real) continuous functions for $x > x_0$ and, for $x \to +\infty$, let the relations*

$$f_k(x) \to a_k \quad (a_0 \neq 0)\,, \quad f(x) \to a \tag{2}$$

hold. Let the equation

$$\varrho^n + a_{n-1}\varrho^{n-1} + \ldots + a_1\varrho + a_0 = 0\,, \tag{3}$$

the coefficients of which are given by the limits (2), have only real, distinct roots. Then equation (1) has at least one solution $y(x)$, which satisfies the conditions

$$y(x) \to \frac{a}{a_0}\,, \quad y^{(m)}(x) \to 0 \quad \text{for} \quad x \to +\infty \quad (m = 1, 2, \ldots, n)\,. \tag{4}$$

If, in addition, all roots of equation (3) are negative, then (4) holds for all solutions of equation (1).

Example 1. Let us consider the equation

$$y'' - y = 3\,.$$

Here, $a_1 = 0$, $a_0 = -1$, $a = 3$. Equation (3),

$$\varrho^2 - 1 = 0\,,$$

has the roots $+1$ and -1. Thus, there exists at least one solution for which

$$y(x) \to -3\,, \quad y'(x) \to 0\,, \quad y''(x) \to 0 \quad \text{if} \quad x \to +\infty\,. \tag{5}$$

(In this example, the solution with this behaviour is obviously the function $y = -3 + e^{-x}$; at the same time, however, we observe that a solution exists that does not have the property (5), namely $y = -3 + e^x$.)

REMARK 2 (*Oscillatory Solutions of Linear Equations of the Second Order*). We shall consider only equations of the form

$$y'' + Q(x)\, y = 0\,, \tag{6}$$

where $Q(x)$ is a continuous function in the interval I considered, since any equation of the form

$$y'' + p(x)\,y' + q(x)\,y = 0$$

can be transformed (see Theorem 17.15.3) by the substitution

$$y = e^{-\frac{1}{2}\int p\,dx}\,z$$

into an equation of the form (6)

REMARK 3. In the following text, the identically zero solution of equation (6) will be excluded from our consideration.

Definition 1. The solution of equation (6), which has, in the given interval I, at most one zero point, is said to be *non-oscillatory* (*non-oscillating*) *in the interval* I; if it has at least two zero-points, it is said to be *oscillatory* (*oscillating*) or to *oscillate* in this interval.

Theorem 2. *If at all points of the interval I the inequality $Q(x) \leqq 0$ holds then every solution of equation* (6) *is non-oscillatory in I.*

REMARK 4. In the following text, we shall consider only the case where $Q(x) > 0$ in I.

Theorem 3 (*Sturm's Theorem*). *Let x_0 and x_1 be two (distinct) consecutive zeros of a non-zero solution $y_1(x)$ of equation* (6). *Then every other solution $y_2(x)$ of this equation, which is not a multiple of the solution $y_1(x)$, has exactly one zero between the points x_0, x_1.*

REMARK 5. Roughly speaking: The zeros of two linear independent solutions of equation (6) are mutually alternating. The solutions $y_1 = \cos x$, $y_2 = \sin x$ of the equation $y'' + y = 0$ may be mentioned as a typical example. (The zeros of such pairs of functions are sometimes said to *interlace*.)

REMARK 6. If one solution of equation (6) has more than two (distinct) zeros in the interval, then all solutions of equations (6) are oscillatory in I.

Theorem 4 (*The Comparison Theorem*). *Let us consider two equations of the form*

$$y'' + Q_1(x)\,y = 0\,, \tag{7}$$

$$z'' + Q_2(x)\,z = 0\,. \tag{8}$$

If the relation $Q_2(x) > Q_1(x) > 0$ holds in I, then between any two (distinct) zeros x_0, x_1 of an arbitrary solution $y(x)$ of equation (7), *there is at least one zero of every solution $z(x)$ of equation* (8). (*The conclusion of the theorem remains valid if in the interval (x_0, x_1) only the inequality $Q_2(x) \geqq Q_1(x) > 0$ holds and if at least in a subinterval of this interval the relation $Q_2(x) > Q_1(x)$ holds.*)

Theorem 5. *Let $x_0, x_1, (x_0 < x_1)$ be two consecutive zeros of the solution $y(x)$ of equation (7). Let the assumptions of Theorem 4 be satisfied (in the interval (x_0, x_1) it is sufficient if the conditions in brackets at the end of Theorem 4 are satisfied). If, at the same time, x_0 is a zero point of the solution $z(x)$ of equation (8), then the next zero x_2 of this solution $(x_2 > x_0)$ lies to the left of the point x_1* (i.e. $x_2 < x_1$).

REMARK 7. Theorems 4 and 5 express – roughly speaking – the fact that, if $Q_2(x) > Q_1(x) > 0$ in I, then the solutions of equation (8) oscillate more rapidly in I than the solutions of equation (7). The simplest example of this statement are the solutions of the equations

$$y'' + ay = 0\,, \quad z'' + bz = 0 \quad (0 < a < b)\,.$$

Theorem 6. *If in the interval I the relation*

$$0 < m \leqq Q(x) \leqq M$$

holds for equation (6), then the distance between two neighbouring zeros of the solution is not greater than $\pi/\sqrt{m}$ and is not smaller than $\pi/\sqrt{M}$.

Theorem 7 (*Kneser's Theorem*). *Let $I = [x_0, \infty)$. If in I the relation*

$$0 < Q(x) \leqq \frac{1}{4x^2}$$

holds, then the solution of equation (6) cannot have an infinite number of zeros in I. If in I the relation

$$Q(x) > \frac{1 + \alpha}{4x^2}\,, \quad \textit{where} \quad \alpha > 0$$

holds, then the solution of equation (6) has an infinite number of zeros in I.

Theorem 8 (*Späthe's Theorem*). *Let the relation*

$$Q(x) = O\left(\frac{1}{x^{k+2}}\right) \quad (k > 0)$$

hold in the interval $I = [x_0, \infty)\,(x_0 > 0)$. (This means (see Definition 11.4.6, p. 414) *that the expression $x^{k+2}|Q(x)|$ is bounded for all $x \in I$.) Then equation (6) has a fundamental system of solutions y_1, y_2* (Definition 17.11.2) *such that*

$$y_1(x) - 1 = O\left(\frac{1}{x^k}\right) \quad \textit{and} \quad y_2(x) - x = O\left(\frac{1}{x^{k-1}}\right) \quad \textit{for} \quad k \neq 1\,,$$
$$= O(\ln x) \quad \textit{for} \quad k = 1$$

(i.e., for large x, y_1 and y_2 differ "by a small amount" from 1 and x, respectively).

Example 2. Let us consider the equation

$$y'' + \frac{1}{x^4}\, y = 0 \, .$$

According to Theorem 8, for large x the functions

$$y \equiv 1 \, , \quad y \equiv x$$

constitute "nearly" the fundamental system of the given equation.

REMARK 8. Remarkable results in the above-mentioned theory of oscillatory solutions have been achieved by applying the so-called *theory of central dispersions.* The work of O. BORŮVKA (Czechoslovak Math. Journal 1953, p. 199) is fundamental in this field. See also e.g. [274].

REMARK 9. For the equation of the n-th order,

$$y^{(n)} + g(x)\, y = 0 \, , \tag{9}$$

the following assertion is true: If the function $g(x) > 0$ is continuous in $I = [x_0, \infty)$ and if

$$\int_{x_0}^{\infty} g(x)\, \mathrm{d}x \quad \text{is divergent} \, ,$$

then 1° for any even n, every solution of equation (9) has an infinite number of zero points; 2° for any odd n, the solution has either an infinite number of zero points or it has zero as its limit as $x \to +\infty$.

Periodic Solutions of Homogeneous Linear Equations

Theorem 9 (*Floquet's Theorem*). *Let us consider the equation*

$$y^{(n)} + f_{n-1}(x)\, y^{(n-1)} + \ldots + f_1(x)\, y' + f_0(x)\, y = 0 \tag{10}$$

in which $f_k(x)$ are holomorphic functions (of the complex variable x (Definition 20.1.9, p. 941)) *on the whole complex plane, and are periodic with a common period ω. Then, there exists at least one non-zero solution $\varphi(x)$ of equation* (10) *such that for a properly chosen constant s (complex, in general) the relation*

$$\varphi(x + \omega) = s\, \varphi(x) \tag{11}$$

holds. (The function $\varphi(x)$ with the property (11) is called *periodic of the second kind* or *pseudo-periodic.*) *The number α, determined by the equation*

$$s = \mathrm{e}^{\alpha\omega} \, ,$$

is called the characteristic exponent. The function

$$\psi(x) = e^{-\alpha x} \varphi(x)$$

is then periodic with period ω.

REMARK 10. Finding the characteristic exponent for the general case is a rather difficult task ([211], volume I, p. 87). When solving an equation of the form

$$y'' = f(x)\, y \tag{12}$$

we can proceed in the following way: Let $y_1(x)$, $y_2(x)$ denote the *normal* fundamental system of equation (12) at the point $x = 0$, i.e. that for which the relations

$$y_1(0) = 1\,, \quad y_1'(0) = 0\,, \quad y_2(0) = 0\,, \quad y_2'(0) = 1 \tag{13}$$

are satisfied. The number s (see (11)) is given by the solution of the equation

$$s^2 - [y_1(\omega) + y_2'(\omega)]\, s + 1 = 0\,.$$

(Since it may be difficult to construct the above-mentioned normal fundamental system even if the function $f(x)$ is relatively simple, the constants $y_1(\omega)$ and $y_2'(\omega)$ are frequently evaluated approximately by numerical integration starting from the known values (13). See also Chap. 25.)

REMARK 11. In the case of linear non-homogeneous equations, the existence of periodic solutions follows in some simple cases immediately from Theorem 17.14.1, p. 787. For the case

$$y'' + a^2 y = g(x)$$

see e.g. [211], volume I, p. 257.

17.11. Linear Equations of the n-th Order

Linear equations of the n-th order are equations of the form

$$y^{(n)} + a_{n-1}(x)\, y^{(n-1)} + \ldots + a_1(x)\, y' + a_0(x)\, y = f(x)\,. \tag{1}$$

The functions $a_0(x)$, ..., $a_{n-1}(x)$, $f(x)$ are assumed to be continuous in an interval I (which can even be infinite). According to Remark 17.2.11 (p. 735), we know that if we choose an arbitrary number $x_0 \in I$ and n arbitrary numbers

$$y_0, y_0', \ldots, y_0^{(n-1)} \tag{2}$$

then there exists exactly one solution of equation (1) satisfying the initial conditions

$$y(x_0) = y_0\,, \quad y'(x_0) = y'_0\,, \quad \ldots, \quad y^{(n-1)}(x_0) = y_0^{(n-1)}\,. \tag{3}$$

This solution is defined in the whole interval I.

REMARK 1. We draw the reader's attention to the fact that this conclusion is not true in general, when the linear equation is of the form

$$a_n(x)\, y^{(n)} + a_{n-1}(x)\, y^{(n-1)} + \ldots + a_1(x)\, y' + a_0(x)\, y = f(x)$$

and $a_n(x_1) = 0$ for some $x_1 \in I$. (As an example of an equation of this form, the equation $x^3 y'' + y = 0$ in the interval $(-1, 1)$ may be mentioned.) However, if $a_n(x) \neq 0$ in the whole interval I, then, dividing through by $a_n(x)$, we get the previous case.

Definition 1. The equation

$$y^{(n)} + a_{n-1}(x)\, y^{(n-1)} + \ldots + a_0(x)\, y = 0 \tag{4}$$

is called the *linear homogeneous equation corresponding to the non-homogeneous equation* (1).

Theorem 1. *If*

$$y_1(x), y_2(x), \ldots, y_k(x)$$

are solutions of equation (4), *then an arbitrary linear combination of them,*

$$y = c_1 y_1 + c_2 y_2 + \ldots + c_k y_k \tag{5}$$

(*where* $c_1, c_2, \ldots, c_k$ *are arbitrary numbers*), *is also a solution of equation* (4). *In particular,* $y_1 + y_2$ *and* $y_1 - y_2$ *are also solutions of the given equation.*

Theorem 2. *Let the functions*

$$f_1(x), f_2(x), \ldots, f_k(x) \tag{6}$$

have $k - 1$ *continuous derivatives in the interval* I *and let they be linearly dependent in* I (Definition 12.8.3, p. 460 and below). *Then the determinant*

$$W(x) = \begin{vmatrix} f_1(x), & f_2(x), & \ldots, & f_k(x) \\ f'_1(x), & f'_2(x), & \ldots, & f'_k(x) \\ \ldots & \ldots & \ldots & \ldots \\ f_1^{(k-1)}(x), & f_2^{(k-1)}(x), & \ldots, & f_k^{(k-1)}(x) \end{vmatrix} \tag{7}$$

(the so-called *Wronskian determinant* (or briefly *Wronskian*) *of the functions* (6)) *is identically zero in* I. Often we write $W(f_1, f_2, \ldots, f_k)$.

Theorem 3. *Let the functions*

$$y_1(x),\ y_2(x),\ \ldots,\ y_n(x) \tag{8}$$

be solutions of equation (4), *linearly independent in the interval* I, *in which the coefficients* $a_0(x), a_1(x), \ldots, a_{n-1}(x)$ *of equation* (4) *are continuous functions of* x. *Then their Wronskian*

$$W(x) = \begin{vmatrix} y_1, & y_2, & \ldots, & y_n \\ y_1', & y_2', & \ldots, & y_n' \\ \ldots & \ldots & \ldots & \ldots \\ y_1^{(n-1)}, & y_2^{(n-1)}, & \ldots, & y_n^{(n-1)} \end{vmatrix} \tag{9}$$

is different from zero in the whole interval I.

REMARK 2. Thus, if we are investigating the linear dependence or independence of functions (8), that are *solutions of equation* (4), it is sufficient to evaluate the determinant (9) at only one point $x_0 \in I$. If $W(x_0) = 0$, the solutions (8) are linearly dependent in I, if $W(x_0) \neq 0$, they are linearly independent.

Example 1. The functions

$$y_1 = e^x, \quad y_2 = e^{-x} \tag{10}$$

are, as can be easily verified, solutions of the equation

$$y'' - y = 0 \tag{11}$$

the coefficients of which are constants. (As the interval I, we may therefore take the interval $(-\infty, \infty)$.) The Wronskian of the functions (10) is

$$W(x) = \begin{vmatrix} e^x, & e^{-x} \\ e^x, & -e^{-x} \end{vmatrix} = -2\,. \tag{12}$$

Thus, the functions (10) are linearly independent in the interval $(-\infty, \infty)$.

Here, the evaluation of the determinant is easy, so that we can easily calculate its value for any x. However, according to Remark 2 it is sufficient to evaluate $W(x)$ at one point $x_0 \in I$, say, at the point $x_0 = 0$ in this case.

REMARK 3. Theorem 3 and Remark 2 follow from the so-called *Liouville's formula* (or *Abel identity*): If the coefficients of equation (4) are continuous in I, $y_1(x)$, $y_2(x), \ldots, y_n(x)$ are arbitrary solutions of equation (4) and x_0 is an arbitrary point of I, then the relation

$$W(x) = W(x_0)\, e^{-\int_{x_0}^{x} a_{n-1}(t)\, dt} \tag{13}$$

holds. Since $e^z \neq 0$ for arbitrary z, $W(x)$ is zero, or is different from zero, in I according to whether $W(x_0) = 0$ or $W(x_0) \neq 0$, respectively.

Example 2. The functions (10) are solutions of equation (11) in the interval $(-\infty, \infty)$. In (11) $a_{n-1}(x) \equiv a_1(x) = 0$, thus

$$W(x) = W(x_0)\, e^{-\int_{x_0}^{x} 0\, dt} = W(x_0) = \text{const.},$$

as may also be seen from (12).

Definition 2. A system $y_1(x), y_2(x), \ldots, y_n(x)$ of n solutions of equation (4), linearly independent (in the interval I), is called a *fundamental system* or a *basis* of this equation.

Theorem 4. *A fundamental system of equation* (4) *having been found, the general integral of* (4) *in I is then of the form*

$$y = c_1\, y_1(x) + c_2\, y_2(x) + \ldots + c_n\, y_n(x) \tag{14}$$

where $c_1, c_2, \ldots, c_n$ are (arbitrary) constants.

REMARK 4. From Theorem 4 it follows that: If we have found n functions $y_1, \ldots, y_n$ such that 1° they are solutions of the homogeneous equation (4) of the n-th order and 2° they are linearly independent in I, then we have found the fundamental system of this equation. Its general integral is, according to (14), a linear combination of these functions.

Example 3. The functions

$$y_1 = \cos x\,, \quad y_2 = \sin x \tag{15}$$

are solutions of the homogeneous equation of the second order

$$y'' + y = 0 \tag{16}$$

in the interval $(-\infty, \infty)$.

We easily verify that (15) is the fundamental system of equation (16), because

$$W(x) = \begin{vmatrix} \cos x, & \sin x \\ -\sin x, & \cos x \end{vmatrix} = 1\,;$$

thus the functions (15) are linearly independent in the interval $(-\infty, +\infty)$. The general integral of equation (16) is then

$$y = c_1 \cos x + c_2 \sin x\,. \tag{17}$$

REMARK 5. In applications we frequently have to solve the following problem: Given a differential equation of the form (4), to find a solution $y(x)$ of this equation satisfying, at a given point $x_0 \in I$, the so-called *initial conditions*

$$y(x_0) = y_0\,, \quad y'(x_0) = y'_0\,, \quad \ldots, \quad y^{(n-1)}(x_0) = y_0^{(n-1)} \tag{18}$$

($y_0, y'_0, \ldots, y_0^{(n-1)}$ being prescribed numbers). If the general integral of equation (4) is known, then the required solution $y(x)$ is of the form (14), where the constants $c_1, c_2, \ldots, c_n$ are uniquely determined by the initial conditions (18).

Example 4. Let us find the solution of the equation

$$y'' + y = 0 \tag{19}$$

satisfying the initial conditions

$$y(0) = 1\,, \quad y'(0) = -2\,. \tag{20}$$

The solution will be of the form (17) (see Example 3). Substituting $x = 0$ into (17) and into the equation $y' = -c_1 \sin x + c_2 \cos x$, which arises by differentiating (17) with respect to x, we obtain from conditions (20)

$$c_1 \,.\, 1 + c_2 \,.\, 0 = 1\,, \quad -c_1 \,.\, 0 + c_2 \,.\, 1 = -2\,,$$

hence $c_1 = 1$, $c_2 = -2$, and the solution is

$$y = \cos x - 2 \sin x\,.$$

(In the general case, the determination of the constants $c_1, c_2, \ldots, c_n$ leads to the solution of a system of n linear (algebraic) equations.)

Theorem 5. *For every equation* (4) *with continuous coefficients in I, there exists at least one fundamental system in I. (In fact there is an infinite number of them.)*

Definition 3. The fundamental system $y_1(x), y_2(x), \ldots, y_n(x)$ for which the relations

$$\begin{array}{lllll}
y_1(x_0) = 1\,, & y'_1(x_0) = 0\,, & y''_1(x_0) = 0\,, & \ldots, & y_1^{(n-1)}(x_0) = 0\,,\\
y_2(x_0) = 0\,, & y'_2(x_0) = 1\,, & y''_2(x_0) = 0\,, & \ldots, & y_2^{(n-1)}(x_0) = 0\,,\\
y_3(x_0) = 0\,, & y'_3(x_0) = 0\,, & y''_3(x_0) = 1\,, & \ldots, & y_3^{(n-1)}(x_0) = 0\,,\\
\multicolumn{5}{c}{\dotfill}\\
y_n(x_0) = 0\,, & y'_n(x_0) = 0\,, & y''_n(x_0) = 0\,, & \ldots, & y_n^{(n-1)}(x_0) = 1
\end{array} \tag{21}$$

hold, is called the *normal fundamental system of equation* (4) *at the point x_0* (or *with respect to the point x_0*).

Theorem 6. *If $y_1, y_2, \ldots, y_n$ is the normal fundamental system of equation* (4) *at the point x_0, then the function*

$$y = y_0\, y_1(x) + y'_0\, y_2(x) + y''_0\, y_3(x) + \ldots + y_0^{(n-1)}\, y_n(x)$$

is the solution of equation (4) *that satisfies the initial conditions*

$$y(x_0) = y_0\,, \quad y'(x_0) = y'_0\,, \ldots, \quad y^{(n-1)}(x_0) = y_0^{(n-1)}\,.$$

REMARK 6. In the general case, it is rather difficult to find a fundamental system of equation (4). However, in the commonly occurring case where $a_0, a_1, \ldots, a_{n-1}$ are constants, this system is easy to find (see § 17.13).

17.12. Non-homogeneous Linear Equations. The Method of Variation of Parameters

Theorem 1. *If the fundamental system* $y_1(x), y_2(x), \ldots, y_n(x)$ *of the equation*

$$y^{(n)} + a_{n-1}(x)\, y^{(n-1)} + \ldots + a_1(x)\, y' + a_0(x)\, y = 0 \tag{1}$$

is known, then the general integral of the corresponding non-homogeneous equation

$$y^{(n)} + a_{n-1}(x)\, y^{(n-1)} + \ldots + a_1(x)\, y' + a_0(x)\, y = g(x) \tag{2}$$

is of the form

$$y = y_p + c_1 y_1 + c_2 y_2 + \ldots + c_n y_n \tag{3}$$

where $c_1, c_2, \ldots, c_n$ *are arbitrary constants and* $y_p(x)$ *is any function satisfying equation* (2).

Theorem 2. *The function* $y_p(x)$ *can be obtained by quadratures (by the so-called method of variation of parameters or variation of constants) when assumed to be of the form*

$$y_p(x) = c_1(x)\, y_1(x) + c_2(x)\, y_2(x) + \ldots + c_n(x)\, y_n(x) \tag{4}$$

where

$$y_1(x), y_2(x), \ldots, y_n(x) \tag{5}$$

is a fundamental system of equation (1).

REMARK 1. Thus, $y_p(x)$ has the form of the general integral of equation (1), where however, instead of constants $c_1, c_2, \ldots, c_n$, we have functions which at the moment are unknown. We have to find these functions so that the function (3) satisfies equation (2).

Theorem 3. *If the functions* $c_1(x), c_2(x), \ldots, c_n(x)$ *satisfy the equations*

$$\begin{array}{l} c_1' y_1 + c_2' y_2 + \ldots + c_n' y_n = 0\,, \\ c_1' y_1' + c_2' y_2' + \ldots + c_n' y_n' = 0\,, \\ \ldots\ldots\ldots\ldots\ldots\ldots\ldots\ldots\ldots \\ c_1' y_1^{(n-2)} + c_2' y_2^{(n-2)} + \ldots + c_n' y_n^{(n-2)} = 0\,, \\ c_1' y_1^{(n-1)} + c_2' y_2^{(n-1)} + \ldots + c_n' y_n^{(n-1)} = g(x)\,, \end{array} \tag{6}$$

then the function y_p, *given by* (4), *satisfies equation* (2).

REMARK 2. The system (6) for the unknown functions $c_1'(x), c_2'(x), \ldots, c_n'(x)$ is uniquely solvable because its determinant is the Wronskian of the fundamental system (5), and is thus different from zero in the interval considered. Integrating $c_1'(x), c_2'(x), \ldots, c_n'(x)$, we get the functions $c_1(x), c_2(x), \ldots, c_n(x)$. (We can write them without constants of integration, because these constants appear in the remaining terms of equation (4).)

Example 1. Let us find the general integral of the equation

$$y'' + y = x^2 . \tag{7}$$

The fundamental system of the equation

$$y'' + y = 0 \tag{8}$$

is

$$y_1 = \cos x , \quad y_2 = \sin x \tag{9}$$

(see Example 17.11.3). The general integral of equation (7) will be of the form (3),

$$y = y_p + c_1 \cos x + c_2 \sin x , \tag{10}$$

where y_p may be found in the form (4),

$$y_p = c_1(x)\, y_1(x) + c_2(x)\, y_2(x) . \tag{11}$$

The system (6) is

$$\begin{aligned} c_1' \cos x + c_2' \sin x &= 0 , \\ -c_1' \sin x + c_2' \cos x &= x^2 , \end{aligned}$$

whence

$$c_1' = -x^2 \sin x , \quad c_2' = x^2 \cos x .$$

Integration by parts yields

$$c_1(x) = (x^2 - 2) \cos x - 2x \sin x , \quad c_2(x) = (x^2 - 2) \sin x + 2x \cos x . \tag{12}$$

Substituting (12) into (11), we get

$$\begin{aligned} y_p(x) = (x^2 - 2) \cos^2 x - 2x \sin x \cos x + (x^2 - 2) \sin^2 x + 2x \cos x \sin x = \\ = x^2 - 2 . \end{aligned}$$

The general integral of equation (7) is then, according to (10),

$$y = x^2 - 2 + c_1 \cos x + c_2 \sin x .$$

REMARK 3. The right-hand side of equation (2) is often of a special form. For example, $g(x)$ may be a polynomial, as it was in the above example. If the left-hand side of equation (2) has constant coefficients, then, for some special forms of the right-hand side of equation (2), y_p can be found in a simpler way than by the method of variation of parameters (see § 17.14).

17.13. Homogeneous Linear Equations with Constant Coefficients. Euler's Equation

Let us consider the equation

$$y^{(n)} + a_{n-1}y^{(n-1)} + \ldots + a_1 y' + a_0 y = 0, \tag{1}$$

where $a_0, a_1, \ldots, a_{n-1}$ are constants. Assuming the solution of equation (1) in the form

$$y = e^{\alpha x},$$

we obtain for the number α (after substituting for $y, y', \ldots, y^{(n)}$ into (1) and dividing the whole equation by $e^{\alpha x}$) the so-called *characteristic* (or *auxiliary*) *equation*

$$\alpha^n + a_{n-1}\alpha^{n-1} + \ldots + a_1\alpha + a_0 = 0. \tag{2}$$

I. If the characteristic equation has distinct roots

$$\alpha_1, \alpha_2, \ldots, \alpha_n,$$

then the fundamental system of equation (1) is given by the functions

$$y_1 = e^{\alpha_1 x}, \quad y_2 = e^{\alpha_2 x}, \quad \ldots, \quad y_n = e^{\alpha_n x}. \tag{3}$$

II. If α_i is an r-fold root, then r functions must correspond to it in the fundamental system (because the fundamental system consists of n functions and n is the number of roots of equation (2), when considering an r-fold root as r roots.) These r functions are

$$y_1 = e^{\alpha_i x}, \quad y_2 = xe^{\alpha_i x}, \quad \ldots, \quad y_r = x^{r-1}e^{\alpha_i x}. \tag{4}$$

Example 1. Let us consider the equation of the form

$$y''' - 3y' + 2y = 0. \tag{5}$$

The characteristic equation

$$\alpha^3 - 3\alpha + 2 = 0 \tag{6}$$

has obviously the root $\alpha = 1$. After dividing by $\alpha - 1$, we get

$$\alpha^2 + \alpha - 2 = 0$$

with the roots 1, -2. Thus, equation (6) has a simple root $\alpha_1 = -2$ and a double root $\alpha_2 = 1$. To α_1 there corresponds the function $y_1 = e^{-2x}$, to α_2 there correspond two functions (because $r = 2$), $y_2 = e^x$, $y_3 = xe^x$. Thus the fundamental system of equation (5) is

$$y_1 = e^{-2x}, \quad y_2 = e^x, \quad y_3 = xe^x .$$

REMARK 1. The roots of the characteristic equation need not always be real. Let the coefficients of equation (1) (and also of equation (2)) be real; then, as is well known from algebra, if the characteristic equation has a complex root $a + ib$ it also has the complex conjugate root $a - ib$, and both these roots are of the same multiplicity. (If, for example, $a + ib$ is a double root, $a - ib$ is also a double root, etc.)

Let the root $a + ib$ (and hence also $a - ib$) be a simple root of equation (2). Then to the two roots

$$a + ib, \quad a - ib$$

there correspond, in the fundamental system, the two functions

$$e^{ax} \cos bx, \quad e^{ax} \sin bx . \tag{7}$$

Let $a + ib$ (and hence also $a - ib$) be a double root of equation (2). Then four functions correspond to these roots in the fundamental system:

$$e^{ax} \cos bx, \quad e^{ax} \sin bx, \quad xe^{ax} \cos bx, \quad xe^{ax} \sin bx ; \tag{8}$$

if $a + ib$ is an r-fold root, then $2r$ functions correspond to the two roots $a + ib$, $a - ib$:

$$\begin{aligned} &e^{ax} \cos bx, \quad xe^{ax} \cos bx, \quad \ldots, \quad x^{r-1}e^{ax} \cos bx, \\ &e^{ax} \sin bx, \quad xe^{ax} \sin bx, \quad \ldots, \quad x^{r-1}e^{ax} \sin bx . \end{aligned} \tag{9}$$

Example 2. Let us consider the equation

$$y^{(V)} - y^{(IV)} + 2y''' - 2y'' + y' - y = 0 .$$

The characteristic equation

$$\alpha^5 - \alpha^4 + 2\alpha^3 - 2\alpha^2 + \alpha - 1 = 0$$

has obviously the root $\alpha_1 = 1$. Dividing by the factor $\alpha - 1$, we get

$$\alpha^4 + 2\alpha^2 + 1 = 0$$

or

$$(\alpha^2 + 1)^2 = 0$$

with double roots $\alpha_{2,3} = \mathrm{i}$, $\alpha_{4,5} = -\mathrm{i}$. The fundamental system, written in the complex form, is (according to (4))

$$y_1 = \mathrm{e}^x, \quad y_2 = \mathrm{e}^{\mathrm{i}x}, \quad y_3 = x\mathrm{e}^{\mathrm{i}x}, \quad y_4 = \mathrm{e}^{-\mathrm{i}x}, \quad y_5 = x\mathrm{e}^{-\mathrm{i}x},$$

written in the real form (according to (8); $a = 0$, $b = 1$), the system is

$$y_1 = \mathrm{e}^x, \quad y_2 = \cos x, \quad y_3 = \sin x, \quad y_4 = x \cos x, \quad y_5 = x \sin x.$$

The general integral (written in the real form) is

$$y = c_1 \mathrm{e}^x + c_2 \cos x + c_3 \sin x + c_4 x \cos x + c_5 x \sin x.$$

REMARK 2. Solving the characteristic equation involves, in general, the solution of an algebraic equation of the n-th degree. For methods of solution of these equations see Chap. 31. We often succeed in finding one root in a simple way (e.g. we look to see if one root of the characteristic equation may perhaps be an integer, as in the case of Example 1 or Example 2, where a negative reciprocal equation was to be solved). Then dividing by the corresponding linear factor, we reduce the order of the characteristic equation.

REMARK 3. By substituting $x = \mathrm{e}^t$, the so-called *Euler equation*

$$y^{(n)} + \frac{a_{n-1}}{x} y^{(n-1)} + \frac{a_{n-2}}{x^2} y^{(n-2)} + \ldots + \frac{a_1}{x^{n-1}} y' + \frac{a_0}{x^n} y = 0 \tag{10}$$

(where $a_0, a_1, \ldots, a_{n-1}$ are constants), can be transformed into an equation with constant coefficients (for $x > 0$; for negative x, we use the substitution $x = -\mathrm{e}^t$). This procedure will be shown in the following example.

Example 3. Let us consider the equation

$$x^2 y'' + 3xy' + y = 0. \tag{11}$$

Then (cf. § 12.11, p. 471)

$$\frac{\mathrm{d}y}{\mathrm{d}x} = \frac{\mathrm{d}y}{\mathrm{d}t}\frac{\mathrm{d}t}{\mathrm{d}x} = \frac{\mathrm{d}y}{\mathrm{d}t}\frac{1}{\mathrm{d}x/\mathrm{d}t} = \frac{\mathrm{d}y}{\mathrm{d}t}\frac{1}{\mathrm{e}^t} = \mathrm{e}^{-t}\frac{\mathrm{d}y}{\mathrm{d}t},$$

$$\frac{\mathrm{d}^2 y}{\mathrm{d}x^2} = \frac{\mathrm{d}}{\mathrm{d}x}\left(\frac{\mathrm{d}y}{\mathrm{d}x}\right) = \frac{\mathrm{d}}{\mathrm{d}t}\left(\frac{\mathrm{d}y}{\mathrm{d}x}\right)\frac{\mathrm{d}t}{\mathrm{d}x} = \frac{\mathrm{d}}{\mathrm{d}t}\left(\mathrm{e}^{-t}\frac{\mathrm{d}y}{\mathrm{d}t}\right)\mathrm{e}^{-t} = \mathrm{e}^{-2t}\left(\frac{\mathrm{d}^2 y}{\mathrm{d}t^2} - \frac{\mathrm{d}y}{\mathrm{d}t}\right).$$

Substituting these results into (11), we get

$$\mathrm{e}^{2t}\mathrm{e}^{-2t}\left(\frac{\mathrm{d}^2 y}{\mathrm{d}t^2} - \frac{\mathrm{d}y}{\mathrm{d}t}\right) + 3\mathrm{e}^t\mathrm{e}^{-t}\frac{\mathrm{d}y}{\mathrm{d}t} + y = 0$$

or

$$\frac{d^2y}{dt^2} + 2\frac{dy}{dt} + y = 0 \tag{12}$$

which is an equation with constant coefficients for the unknown function $y(t)$. The characteristic equation

$$\alpha^2 + 2\alpha + 1 = 0$$

has exactly one double root $\alpha = -1$, so, according to (4), the general integral is

$$y = c_1 e^{-t} + c_2 t e^{-t} = e^{-t}(c_1 + c_2 t).$$

From the substitution $x = e^t$ it follows that $t = \ln x$, thus the general integral of equation (11) is

$$y = \frac{1}{x}(c_1 + c_2 \ln x). \tag{13}$$

In the case of Euler equations of higher order we proceed in a similar way. Instead of transforming the equation into the form (12), we can directly assume the solution of equation (10) in the form

$$y = x^\alpha. \tag{14}$$

The determination of α leads to the solution of a characteristic equation for α. If the roots $\alpha_1, \alpha_2, \ldots, \alpha_n$ are simple, then the fundamental system of equation (10) is

$$y_1 = x^{\alpha_1}, \quad y_2 = x^{\alpha_2}, \quad \ldots, \quad y_n = x^{\alpha_n}.$$

If one of the roots is of multiplicity r, then the corresponding functions of the fundamental system are

$$x^\alpha, x^\alpha \ln x, x^\alpha \ln^2 x, \ldots, x^\alpha \ln^{r-1} x. \tag{15}$$

The complex functions

$$x^{a+ib}, x^{a-ib} \quad \text{or} \quad x^{a+ib} \ln^k x, x^{a-ib} \ln^k x$$

can be replaced by the functions

$$x^a \cos(b \ln x), x^a \sin(b \ln x) \quad \text{or} \quad x^a \cos(b \ln x) \ln^k x, x^a \sin(b \ln x) \ln^k x.$$

Example 4. By substituting (14) into (11), we get

$$x^2 . \alpha(\alpha - 1) x^{\alpha-2} + 3x\alpha x^{\alpha-1} + x^\alpha = 0.$$

Dividing through by $x^\alpha \neq 0$, we get the characteristic equation

$$\alpha^2 + 2\alpha + 1 = 0$$

with a double root $\alpha_{1,2} = -1$. According to (15), the fundamental system is

$$y_1 = x^{-1} = \frac{1}{x}, \quad y_2 = x^{-1} \ln x = \frac{1}{x} \ln x,$$

in agreement with (13).

17.14. Non-homogeneous Linear Equations with Constant Coefficients and a Special Right-hand Side

Let us consider the equation

$$y^{(n)} + a_{n-1}y^{(n-1)} + \ldots + a_1 y' + a_0 y = e^{ax}[P(x) \cos bx + Q(x) \sin bx] \tag{1}$$

where $a_0, a_1, \ldots, a_{n-1}, a, b$ are real constants, $P(x)$ is a polynomial of the p-th degree and $Q(x)$ is a polynomial of the q-th degree, both with real coefficients. (The special cases $P(x) \equiv 0$ or $Q(x) \equiv 0$ are not excluded.)

Equation (1) covers the great majority of differential equations which occur in applications. For example, the equation

$$y'' + y = e^x \sin x$$

is of type (1). Here $a_1 = 0$, $a_0 = 1$, $a = 1$, $b = 1$, $P(x) \equiv 0$, $Q(x) \equiv 1$ (therefore $q = 0$). The equation

$$y'' - 3y' + 2y = x^2$$

is also of the type (1). Here $a_1 = -3$, $a_0 = 2$, $a = 0$, $b = 0$, $P(x) = x^2$ (therefore $p = 2$); $Q(x)$ can be considered as a zero polynomial, $Q(x) \equiv 0$.

The general integral of equation (1) is of the form (cf. Theorem 17.12.1)

$$y = y_p + c_1 y_1 + c_2 y_2 + \ldots + c_n y_n,$$

where $y_1, y_2, \ldots, y_n$ is the fundamental system of the homogeneous equation

$$y^{(n)} + a_{n-1}y^{(n-1)} + \ldots + a_1 y' + a_0 y = 0 \tag{2}$$

(which can be found by the method mentioned in the previous paragraph) and y_p is an arbitrary solution of equation (1). If the right-hand side of the equation is of the form indicated in equation (1) it is possible to assume the function y_p to have a special form.

The characteristic equation corresponding to equation (2) is

$$\alpha^n + a_{n-1}\alpha^{n-1} + \ldots + a_1 \alpha + a_0 = 0. \tag{3}$$

Let us denote by s the greater of the two numbers p, q, where p is the degree of the polynomial $P(x)$ and q is the degree of the polynomial $Q(x)$ in (1). (If $p = q$, then obviously $s = p = q$. If e.g. $Q(x) \equiv 0$, we shall consider $s = p$.)

Theorem 1. *Suppose $a + ib$ is not a root of equation (3) (so that neither is $a - ib$ a root of equation (3)). Then y_p can be assumed to be of the form*

$$y_p = e^{ax}[R(x) \cos bx + S(x) \sin bx] \tag{4}$$

where $R(x)$ and $S(x)$ are polynomials of the s-th degree.

If $a + ib$ is an r-fold root of the characteristic equation, then y_p can be assumed to be of the form

$$y_p = x^r e^{ax}[R(x) \cos bx + S(x) \sin bx] \tag{5}$$

where again $R(x)$ and $S(x)$ are polynomials of the s-th degree.

REMARK 1. The coefficients of these polynomials can be determined by the method of undetermined coefficients, as will be clear from the examples given below.

REMARK 2. If the right-hand side of equation (1) is a sum of terms of the form given in (1), then y_p is a sum of functions of the form (4) or (5). (It is often advantageous to find the integral corresponding to each term of the right-hand side separately; their sum then gives y_p.)

For example, consider the equation

$$y'' + 4y = 2 \sin x + \cos 3x .$$

The right-hand side of this equation cannot be expressed in the form

$$e^{ax}[P(x) \cos bx + Q(x) \sin bx] ,$$

because for each term of the right-hand side of this equation the value of b is different. So we find first the particular integral y_{p_1} of the equation

$$y'' + 4y = 2 \sin x$$

and then the particular integral y_{p_2} of the equation

$$y'' + 4y = \cos 3x .$$

Their sum $y_{p_1} + y_{p_2}$ is the required particular integral y_p.

Example 1. Let us consider the equation

$$y'' + y = x^2 . \tag{6}$$

If we use the notation of (1), then $a = 0$, $b = 0$ (whence $a + ib = 0$), $P(x) = x^2$, $Q(x) \equiv 0$ (therefore $s = 2$); $a + ib$ is not a root of the characteristic equation $\alpha^2 + 1 = 0$ (because this has the roots $\alpha_1 = \mathrm{i}$, $\alpha_2 = -\mathrm{i}$, while $a + ib = 0$), so according to (4) we may assume y_p to be of the form

$$y_p = \mathrm{e}^{0.x}[(Ax^2 + Bx + C)\cos(0 \,.\, x) + (Dx^2 + Ex + F)\sin(0 \,.\, x)] =$$
$$= Ax^2 + Bx + C\,.$$

Hence $y_p'' = 2A$, $y_p = Ax^2 + Bx + C$; substituting into (6), we get

$$2A + Ax^2 + Bx + C = x^2\,.$$

Comparing the coefficients of equal powers of x, we get $A = 1$, $B = 0$, $2A + C = 0$, so that

$$y_p = x^2 - 2$$

in accordance with example 17.12.1; the general integral of equation (6) is therefore

$$y = x^2 - 2 + c_1 \cos x + c_2 \sin x\,.$$

REMARK 3. It may be easily verified that the equation $y'' + y' = x^2$ has no particular solution of the form $Ax^2 + Bx + C$; the function y_p is to be obtained in the form $x(Ax^2 + Bx + C)$, in accordance with (5). The difference between this case and Example 1 lies in the fact that here $a + ib = 0$ is a simple root of the characteristic equation $\alpha^2 + \alpha = 0$.

Example 2. Let us consider the equation

$$y'' - 3y' + 2y = x\mathrm{e}^x\,. \tag{7}$$

Here $a = 1$, $b = 0$ (thus $a + ib = 1$), $P(x) = x$ (therefore $p = 1$), $Q(x) \equiv 0$ (thus $s = 1$). The characteristic equation

$$\alpha^2 - 3\alpha + 2 = 0$$

has the roots $\alpha_1 = 1$, $\alpha_2 = 2$, thus $a + ib = 1$ is a simple root of this equation (i.e. in (5), we have $r = 1$). Following (5), we assume y_p in the form

$$y_p = x\mathrm{e}^x[(Ax + B)\cos(0 \,.\, x) + (Cx + D)\sin(0 \,.\, x)] = x\mathrm{e}^x(Ax + B) =$$
$$= \mathrm{e}^x(Ax^2 + Bx)\,.$$

Substituting for y_p, y_p', y_p'' into (7), we get

$$e^x[x^2(2A - 3A + A) + x(B + 4A - 3B - 6A + 2B) + (2B + 2A - 3B)] = \\ = e^x . x .$$

Dividing by e^x ($\neq 0$) and comparing the coefficients of equal powers of x we get

$$-2A = 1 , \quad 2A - B = 0 \quad \text{whence} \quad A = -\tfrac{1}{2} , \quad B = -1 ,$$

thus

$$y_p = e^x(-\tfrac{1}{2}x^2 - x)$$

and the general integral of equation (7) is

$$y = e^x(-\tfrac{1}{2}x^2 - x) + c_1e^x + c_2e^{-2x} = e^x(-\tfrac{1}{2}x^2 - x + c_1) + c_2e^{-2x} .$$

Example 3. Let us consider the equation

$$y'' + y = 2 \sin x . \tag{8}$$

Here $a = 0$, $b = 1$ (thus $a + ib = i$), $P(x) \equiv 0$, $Q(x) = 2$ (so $q = 0$ and $s = 0$); $a + ib = i$ is a *simple* root of the characteristic equation

$$\alpha^2 + 1 = 0 .$$

According to (5) we have

$$y_p = xe^{0.x}[A \cos x + B \sin x] = Ax \cos x + Bx \sin x .$$

Substituting for y_p, y_p'' into (8), we get

$$-2A \sin x - Ax \cos x + 2B \cos x - Bx \sin x + Ax \cos x + Bx \sin x = 2 \sin x ,$$
$$-2A \sin x + 2B \cos x = 2 \sin x .$$

Comparing coefficients of the linearly independent functions $\sin x$, $\cos x$ we get

$$-2A = 2 , \quad 2B = 0 , \quad \text{i.e.} \quad A = -1 , \quad B = 0 ;$$

consequently

$$y_p = -x \cos x ,$$

and the general integral of equation (8) is

$$y = -x \cos x + c_1 \cos x + c_2 \sin x = (-x + c_1) \cos x + c_2 \sin x .$$

From this example it is clear that even in the case when only a sine term appears on the right-hand side of equation (1) (in this case, of equation (8)), when searching for y_p we must write the expression (4) (or (5)) in the complete form containing both sine and cosine terms.

17.15. Linear Equations of the Second Order with Variable Coefficients. Transformation into Self-adjoint Form, into Normal Form. Invariant. Equations with Regular Singularities (Equations of the Fuchsian Type). Some Special Equations (Bessel's Equation etc.)

In this paragraph equations of the type

$$p_0(x)\, y'' + p_1(x)\, y' + p_2(x)\, y = 0 \tag{1}$$

are considered, where $p_0(x)$, $p_1(x)$, $p_2(x)$ are continuous functions of the variable x on an interval I (which can even be infinite). Unlike § 17.11 we *do not assume* that $p_0(x) \equiv 1$, which leads in some cases to different results from those given in the paragraph mentioned.

Theorem 1. *In any interval in which $p_0(x) \neq 0$, equation (1) can be transformed, by multiplying by the function*

$$\mu(x) = \frac{1}{p_0(x)}\, e^{\int [p_1(x)/p_0(x)] dx}, \tag{2}$$

into the so-called self-adjoint form

$$\frac{d}{dx}\,[p(x)\, y'] + q(x)\, y = 0\,, \tag{3}$$

where

$$p(x) = e^{\int (p_1/p_0) dx}, \quad q(x) = \frac{p_2}{p_0}\, e^{\int (p_1/p_0) dx}. \tag{4}$$

Example 1. For the Bessel equation

$$x^2 y'' + x y' + (x^2 - n^2)\, y = 0\,, \tag{5}$$

$p_0 = x^2$, $p_1 = x$, $p_2 = x^2 - n^2$, thus (carrying through the transformation for $x > 0$)

$$\mu = \frac{1}{x^2}\, e^{\int (x/x^2) dx} = \frac{1}{x^2}\, e^{\ln x} = \frac{1}{x}\,, \quad p = e^{\int (x/x^2) dx} = x\,,$$

$$q = \frac{x^2 - n^2}{x^2}\, e^{\int (x/x^2) dx} = \frac{x^2 - n^2}{x}\,;$$

thus equation (5), multiplied by the function $1/x$, is transformed into the self-adjoint equation

$$(xy')' + \frac{x^2 - n^2}{x}\, y = 0\,. \tag{6}$$

Theorem 2. *Introducing a new variable*

$$u(x) = \int e^{-\int(p_1/p_0)dx}\, dx \tag{7}$$

equation (1) *can be transformed, in any interval in which* $p_0(x) \neq 0$, *into the form*

$$\frac{d^2y}{du^2} + Q(u)\, y = 0 \tag{8}$$

where

$$Q = \frac{p_2}{p_0} e^{2\int(p_1/p_0)dx} \tag{9}$$

and where, on the right-hand side of (9), *u is to be substituted for x according to the relation* (7).

Example 2. In the equation

$$xy'' + \tfrac{1}{2}y' - y = 0\,, \tag{10}$$

$p_0 = x$, $p_1 = \frac{1}{2}$, $p_2 = -1$. Then (for $x > 0$)

$$u = \int e^{-\int(p_1/p_0)dx}\, dx = \int e^{-\frac{1}{2}\ln x}\, dx = \int \frac{1}{\sqrt{x}}\, dx = 2\sqrt{x}\,, \tag{11}$$

$$Q = -\frac{1}{x} e^{2\int(1/2x)dx} = -\frac{1}{x} e^{\ln x} = -1\,.$$

(Here Q is a constant. In the general case, we should substitute $x = \frac{1}{4}u^2$ in accordance with (11).) Consequently, equation (8) becomes

$$\frac{d^2y}{du^2} - y = 0\,. \tag{12}$$

Theorem 3. *By a linear transformation of the form*

$$y = u(x)\, z\,, \tag{13}$$

where

$$u(x) = e^{-\frac{1}{2}\int(p_1/p_0)dx} \tag{14}$$

(*and by dividing by* $p_0(x)$) *we can transform equation* (1) (*in any interval in which* $p_0(x) \neq 0$), *into the form*

$$z'' + I(x)\, z = 0\,, \tag{15}$$

where

$$I(x) = \frac{p_2}{p_0} - \frac{1}{4}\left(\frac{p_1}{p_0}\right)^2 - \frac{1}{2}\left(\frac{p_1}{p_0}\right)'. \tag{16}$$

Example 3. For the equation

$$y'' + \frac{2}{x}y' + y = 0, \tag{17}$$

$p_0 = 1$, $p_1 = 2/x$, $p_2 = 1$. Further

$$u(x) = e^{-\frac{1}{2}\int(2/x)dx} = e^{-\ln x} = \frac{1}{x}, \tag{18}$$

giving

$$I(x) = 1 - \frac{1}{4}\left(\frac{2}{x}\right)^2 - \frac{1}{2}\left(\frac{2}{x}\right)' = 1 - \frac{1}{x^2} + \frac{1}{x^2} = 1. \tag{19}$$

Consequently, equation (15) becomes

$$z'' + z = 0. \tag{20}$$

REMARK 1. Forms (8) and (15) are often called *normal* (or *normalized*) forms of the differential equation of the second order. The expression $I(x)$ is the so-called *invariant* of the given equation. By transforming an equation into the normal form, we often obtain the solution in a simple way, as may be seen from Examples 2 and 3: Equation (12) has the general integral

$$y = c_1 e^u + c_2 e^{-u}$$

whence, applying (11) to equation (10), we get its general integral

$$y = c_1 e^{2\sqrt{x}} + c_2 e^{-2\sqrt{x}}.$$

The general integral of equation (20) is

$$z = c_1 \cos x + c_2 \sin x,$$

whence, applying (13) and (18), we get the general integral of equation (17):

$$y = c_1 \frac{\cos x}{x} + c_2 \frac{\sin x}{x}.$$

Definition 1. Equations of the form

$$(x - a)^2 y'' + (x - a) P_1(x) y' + P_2(x) y = 0, \tag{21}$$

where $P_1(x)$, $P_2(x)$ are functions which can be expanded into power series in a neighbourhood of the point a, are called *equations with a regular singularity at the point a.*

REMARK 2. The name "*equations of Fuchsian type*", is also used although this name is more often used in a rather different sense (see e.g. [35]).

REMARK 3. Equation (21) is often studied in the complex domain (y being a function of the complex variable x). In this case we require that $P_1(x)$, $P_2(x)$ be holomorphic in the neighbourhood of the point a (Definition 20.1.9, p. 941).

REMARK 4. In applications $P_1(x)$ and $P_2(x)$ are very often polynomials (if $P_1(x)$ and $P_2(x)$ are constants, we have (by the substitution $x - a = u$) Euler's equation (Remark 17.13.3, p. 784)). In the general case, we have

$$P_1(x) = \sum_{k=0}^{\infty} \alpha_k (x - a)^k , \quad P_2(x) = \sum_{k=0}^{\infty} \beta_k (x - a)^k . \tag{22}$$

Theorem 4. *Equation* (21) *has always at least one solution of the form*

$$y = (x - a)^\varrho \sum_{k=0}^{\infty} c_k (x - a)^k . \tag{23}$$

REMARK 5. ϱ need not be an integer (it need not even be real). Substituting (23) into (21) and comparing the coefficients of equal powers of $x - a$, we get equations for evaluating ϱ and the unknown constants $c_0, c_1, c_2, \ldots$:

$$\begin{aligned} &c_0 f_0(\varrho) = 0 , \\ &c_1 f_0(\varrho + 1) + c_0 f_1(\varrho + 1) = 0 , \\ &c_2 f_0(\varrho + 2) + c_1 f_1(\varrho + 2) + c_0 f_2(\varrho + 2) = 0 , \\ &\ldots\ldots\ldots\ldots\ldots\ldots\ldots\ldots\ldots\ldots \end{aligned} \tag{24}$$

Here we have used the shortened notation

$$\begin{aligned} f_0(u) &= u(u - 1) + u\alpha_0 + \beta_0 , \\ f_k(u) &= \phantom{u(u - 1) +{}} u\alpha_k + \beta_k , \quad k = 1, 2, 3, \ldots ; \end{aligned} \tag{25}$$

where α_k, β_k are coefficients of the series (22).

Let us determine ϱ so that

$$f_0(\varrho) \equiv \varrho(\varrho - 1) + \varrho\alpha_0 + \beta_0 = 0 . \tag{26}$$

Equation (26) (the so-called *indicial equation* or *fundamental equation* of the given equation (21) at the singularity $x = a$) gives, in general, two roots (not necessarily real), ϱ_1, ϱ_2. In the following text, let us assume that

$$\operatorname{Re} \varrho_1 \geqq \operatorname{Re} \varrho_2 , \tag{27}$$

i.e. that the real part of the root ϱ_1 is greater than (or equal to) the real part of the root ϱ_2. Let us consider first the root ϱ_1 and let us choose $c_0 \neq 0$ arbitrarily (e.g. $c_0 = 1$). In consequence of (27) we get $f_0(\varrho_1 + 1) \neq 0$, $f_0(\varrho_1 + 2) \neq 0$, etc.; thus, the values of $c_1, c_2, c_3, \ldots$ can be uniquely determined from (24).

Theorem 5. *The series*

$$\sum_{k=0}^{\infty} c_k(x - a)^k \tag{28}$$

is convergent in a neighbourhood of the point a. Its radius of convergence is equal at least to the smaller of the radii of convergence of the series (22). (*In particular, if P_1 and P_2 are polynomials, the radius of convergence is infinite.*)

REMARK 6. By this procedure, we obtain the first solution of equation (21),

$$y_1 = (x - a)^{\varrho_1} \sum_{k=0}^{\infty} c_k(x - a)^k . \tag{29}$$

(ϱ_1 need not, of course, be a natural number. Therefore (29) need not be defined in the whole neighbourhood of the point a. More precisely, we should speak about an analytic element of the solution.)

Example 4. For the Bessel equation of order $\frac{1}{2}$, namely

$$y'' + \frac{1}{x} y' + \frac{x^2 - \frac{1}{4}}{x^2} y = 0 ,$$

we have

$$P_1(x) \equiv 1 , \quad P_2(x) = x^2 - \tfrac{1}{4} ,$$

thus

$$\alpha_0 = 1 , \quad \alpha_1 = \alpha_2 = \alpha_3 = \ldots = 0 ,$$

$$\beta_0 = -\tfrac{1}{4} , \quad \beta_1 = 0 , \quad \beta_2 = 1 , \quad \beta_3 = \beta_4 = \beta_5 = \ldots = 0 ,$$

so that by (25)

$$f_0(\varrho) = \varrho(\varrho - 1) + \varrho - \tfrac{1}{4} = \varrho^2 - \tfrac{1}{4} ,$$

$$f_1(\varrho) \equiv 0, f_2(\varrho) \equiv 1, f_3(\varrho) \equiv 0, f_4(\varrho) \equiv 0, \ldots .$$

The fundamental equation

$$\varrho^2 - \tfrac{1}{4} = 0$$

has the solutions $\varrho_1 = \frac{1}{2}$, $\varrho_2 = -\frac{1}{2}$. For $\varrho_1 = \frac{1}{2}$ we have, according to (24),

$$\begin{aligned} &c_1[(\tfrac{1}{2} + 1)^2 - \tfrac{1}{4}] - 0 = 0 , \\ &c_2[(\tfrac{1}{2} + 2)^2 - \tfrac{1}{4}] + 0 + c_0 = 0 , \\ &c_3[(\tfrac{1}{2} + 3)^2 - \tfrac{1}{4}] + 0 + c_1 + 0 = 0 . \end{aligned} \tag{30}$$

If we choose $c_0 = \sqrt{(2/\pi)}$ (cf. Remark 16.4.3, p. 717), we have

$$y_1 = x^{1/2}\sqrt{\left(\frac{2}{\pi}\right)}\left(1 - \frac{1}{6}x^2 + \frac{1}{120}x^4 - \ldots\right).$$

By constructing a recurrence formula for the coefficients,

$$c_{2k+2} = \frac{-c_{2k}}{(\frac{1}{2} + 2k + 2)^2 - \frac{1}{4}}, \quad c_{2k+1} = 0 \quad (k = 0, 1, 2, \ldots)$$

on the basis of (30), we would obtain the result

$$y_1 = \sqrt{\left(\frac{2}{\pi x}\right)} \sin x .$$

REMARK 7. If $\varrho_1 - \varrho_2$ is not an integer, we try to find a second solution of equation (21) in the form

$$y_2 = (x - a)^{\varrho_2} \sum_{k=0}^{\infty} d_k (x - a)^k . \tag{31}$$

For the coefficients d_k, we get the same system (24) as for the c_k but, naturally, ϱ must be given the value ϱ_2. If we choose d_0 arbitrarily ($d_0 \neq 0$), then all other d_k are uniquely determined (because $f_0(\varrho_2 + 1) \neq 0, f_0(\varrho_2 + 2) \neq 0$, etc.). Theorem 5 holds true for the series thus obtained. Moreover, the following theorem is true:

Theorem 6. *The functions* (29) *and* (31) *constitute a fundamental system of the equation* (21).

REMARK 8. If $\varrho_1 - \varrho_2$ is an integer, we cannot use the above procedure because we shall find that, for some $m, f_0(\varrho_2 + m) = 0$. We can, of course, use the formula

$$y_2 = y_1 \int \frac{1}{y_1^2} e^{[-\int P_1/(x-a)]dx} \, dx \tag{32}$$

(see (17.8.18), p. 769). Since, however, we know the solution y_1 in a form of an infinite series, it is often more convenient to use the following theorem:

Theorem 7. *Let $\varrho_1 - \varrho_2$ be an integer and let y_1 be given by equation* (29). *Then a second solution, completing the fundamental system, is of the form*

$$y_2 = (x - a)^{\varrho_1} . A \ln (z - a) \sum_{k=0}^{\infty} c_k (x - a)^k + (z - a)^{\varrho_2} \sum_{k=0}^{\infty} \gamma_k (x - a)^k . \tag{33}$$

The radius of convergence of both infinite series is at least equal to the smaller of the radii of convergence of the series (22).

REMARK 9. In (33), c_k is known (see (28)). We get the conditions for the unknown constants A, γ_k by substituting (33) into (21). If $\varrho_1 - \varrho_2$ is not an integer, the first term in (33) vanishes ($A = 0$). If $\varrho_1 - \varrho_2$ is an integer, then in general $A \neq 0$, although in exceptional cases we may have $A = 0$. This happens, for instance, if $\alpha_1 = \beta_1 = \beta_2 = 0$, which occurs quite frequently in practice. If $\varrho_1 = \varrho_2$, however, then invariably $A \neq 0$.

SOME SPECIAL EQUATIONS WITH VARIABLE COEFFICIENTS

I. *The Bessel equation* (see also § 17.21, equation 117 and others):

$$x^2 y'' + xy' + (x^2 - n^2)\, y = 0\,. \tag{34}$$

A solution (for any x and any real n) is:

$$\mathrm{J}_n(x) = \sum_{k=0}^{\infty} \frac{(-1)^k}{k!\, \Gamma(n + k + 1)} \left(\frac{x}{2}\right)^{n+2k} \tag{35}$$

(*the Bessel function of the first kind of order n*). If n is a negative integer, (35) still has a meaning if we define

$$\frac{1}{\Gamma(t)} = 0 \quad \text{for} \quad t = 0, -1, -2, \ldots$$

($\Gamma(t)$ is the gamma function, see Definition 13.11.1, p. 584).

REMARK 10. For a more detailed treatment see § 16.4.

REMARK 11. We get the series (35) by using the method explained above (calculating c_k from (24), see also Example 4). In Example 4 we had $\varrho_1 - \varrho_2 = 1$, thus $f_0(\varrho_2 + 1) = 0$. When solving the system (30) for $\varrho_2 = -\frac{1}{2}$, this gives the possibility of a free choice of c_1. If we choose $c_1 = 0$ (then all other c_k will be zero for odd k), we shall get a result identical with (35). Similarly, for $\varrho_2 = -\frac{3}{2}$, $\varrho_2 = -\frac{5}{2}$, etc., we choose all odd coefficients equal to zero.

Theorem 8. *If n is not an integer (even if $\varrho_1 - \varrho_2 = 1$, etc.), then*

$$y_1(x) = \mathrm{J}_n(x) \quad \textit{and} \quad y_2(x) = \mathrm{J}_{-n}(x) \tag{36}$$

form a fundamental system of equation (34).

REMARK 12. If n is an integer, then (36) is not a fundamental system because (cf. (16.4.2) p. 717) $\mathrm{J}_{-n}(x) = (-1)^n\, \mathrm{J}_n(x)$. Let n be the positive square root of n^2, $n > 0$. If we determine $\mathrm{J}_n(x)$, then we get the second solution of the fundamental system by applying either (32) or (33):

$$y_2(x) = \mathrm{Y}_n(x) = \frac{2}{\pi}\left(\mathrm{J}_n(x)\ln(x) - \left\{2^{n-1}(n-1)!\,x^{-n} + \right.\right.$$

$$+ \frac{2^{n-3}(n-2)!\,x^{-n+2}}{1!} + \frac{2^{n-5}(n-3)!\,x^{-n+4}}{2!} +$$

$$\left. + \ldots + \frac{x^{n-2}}{2^{n-1}(n-1)!}\right\} - \frac{x^n}{2^{n+1}n!}\sum_{k=1}^{n}\frac{1}{k} +$$

$$\left. + \sum_{m=1}^{\infty}\frac{(-1)^{m-1}\,x^{n+2m}}{2^{n+2m}\,m!(n+m)!}\left(\sum_{k=1}^{m}\frac{1}{2k} + \sum_{k=1}^{n+m}\frac{1}{2k}\right)\right), \quad n \geqq 1; \tag{37}$$

$\mathrm{Y}_n(x)$ is so-called *Bessel function of the second kind*. For $n = 0$ we have $\mathrm{Y}_0(x) = = \dfrac{2}{\pi}\left[\mathrm{J}_0(x)\ln x - \sum_{m=1}^{\infty}\dfrac{(-1)^m}{(m!)^2}\left(\dfrac{x}{2}\right)^{2m}\left(1 + \dfrac{1}{2} + \ldots + \dfrac{1}{m}\right)\right]$.

Note that $\mathrm{Y}_n(x)$ is different from $Y_n(x)$ defined on p. 721. We have $Y_n(x) = = \mathrm{Y}_n(x) + (2/\pi)(C - \log 2)\,\mathrm{J}_n(x)$, where C is Euler's constant, p. 579.

II. *The Gauss (or hypergeometric) equation* (see also § 17.21, equation 140 and further)

$$x(1-x)\,y'' + [\gamma - (\alpha + \beta + 1)\,x]\,y' - \alpha\beta y = 0 \tag{38}$$

has (if γ is not an integer and if $x \in (0, 1)$) the general integral

$$y = C_1F(\alpha, \beta, \gamma, x) + C_2x^{1-\gamma}F(\alpha + 1 - \gamma, \beta + 1 - \gamma, 2 - \gamma, x)\,, \tag{39}$$

where

$$F(\alpha, \beta, \gamma, x) = 1 + \frac{\alpha\beta}{1\,.\,\gamma}x + \frac{\alpha(\alpha+1)\,\beta(\beta+1)}{1\,.\,2\,.\,\gamma(\gamma+1)}x^2 +$$

$$+ \frac{\alpha(\alpha+1)(\alpha+2)\,\beta(\beta+1)(\beta+2)}{1\,.\,2\,.\,3\,.\,\gamma(\gamma+1)(\gamma+2)}x^3 + \ldots \tag{40}$$

is the so-called *hypergeometric series*, convergent for $|x| < 1$; γ is not equal to zero or a negative integer.

III. *The Legendre differential equation* (see also § 17.21, equation 129 and further)

$$(1 - x^2)\,y'' - 2xy' + n(n+1)\,y = 0 \tag{41}$$

(n is a non-negative integer) arises from the hypergeometric equation ($\alpha = -n$, $\beta = n + 1$, $\gamma = 1$) by the substitution $x = \frac{1}{2}(1 - z)$. The function (40) then becomes

$$F\left(-n, n+1, 1, \frac{1-x}{2}\right) = P_n(x) = \frac{1}{2^n n!}\frac{\mathrm{d}^n}{\mathrm{d}x^n}(x^2 - 1)^n \tag{42}$$

(*which is known as the Legendre polynomial of degree n*).

IV. *The Laguerre differential equation*

$$xy'' + (1 - x)\, y' + ny = 0$$

(n being a non-negative integer) has the solution

$$\mathrm{L}_n(x) = \mathrm{e}^x \frac{\mathrm{d}^n}{\mathrm{d}x^n}(x^n \mathrm{e}^{-x})$$

(*the Laguerre polynomial of degree n*).

V. *The Hermite differential equation*

$$y'' - 2xy' + 2ny = 0$$

(n being a non-negative integer) has the solution

$$\mathrm{H}_n(x) = (-1)^n \mathrm{e}^{x^2} \frac{\mathrm{d}^n}{\mathrm{d}x^n} \mathrm{e}^{-x^2}$$

(*the Hermite polynomial of degree n*).

For more details on Bessel functions, Legendre polynomials, etc., see Chap. 16.

17.16. Discontinuous Solutions of Linear Equations

In applications, we encounter cases where the integral of a given equation or its derivative has a prescribed "jump" at various points. (For example, we come across this situation when finding the elastic deflection of a bar which is supporting concentrated loads, etc.)

Definition 1. We say that $f(x)$ is *piecewise n times smooth in* $[a, b]$ if a finite number of points

$$a = x_0 < x_1 < x_2 < \ldots < x_m < x_{m+1} = b$$

exist so that in any interval (x_k, x_{k+1}) $(k = 0, 1, 2, \ldots, m)$ $f(x)$ and its derivatives up to the n-th order inclusive are continuous and have finite limits from the right and from the left at the points $x_0, x_1, x_2, \ldots, x_{m+1}$ (at the point $x_0 = a$ from the right and at the point $x_{m+1} = b$ from the left).

REMARK 1. The difference of these limits is called the *jump* of the function $f(x)$, or of its derivative $f^{(i)}(x)$ at the point x_k $(k = 1, 2, \ldots, m)$, and is denoted by ${}^kS_0$, or ${}^kS_i$, respectively. Thus

$${}^kS_i = \lim_{x \to x_k+} f^{(i)}(x) - \lim_{x \to x_k-} f^{(i)}(x) . \tag{1}$$

(We consider the function $f(x)$ itself as its zero-th derivative.)

REMARK 2. If, in Definition 1, $n = 0$, then $f(x)$ is called a *piecewise continuous function in* $[a, b]$.

REMARK 3. Let an equation of the form

$$y^{(n)} + a_{n-1}(x)\, y^{(n-1)} + \dots + a_1(x)\, y' + a_0(x)\, y = f(x) \tag{2}$$

be given, where $a_j(x)$, $f(x)$ are continuous functions (in the domain considered); let us denote by

$$^k y_0(x),\ ^k y_1(x), \dots,\ ^k y_{n-1}(x) \tag{3}$$

the normal fundamental system of the corresponding homogeneous equation at the point x_k, i.e. the fundamental system, for which

$$\begin{aligned} &^k y_0(x_k) = 1,\ ^k y_1(x_k) = 0,\ ^k y_2(x_k) = 0,\ \dots,\ ^k y_{n-1}(x_k) = 0\,,\\ &^k y_0'(x_k) = 0,\ ^k y_1'(x_k) = 1,\ ^k y_2'(x_k) = 0,\ \dots,\ ^k y_{n-1}'(x_k) = 0\,,\\ &\dots\dots\dots\dots\dots\dots\dots\dots\dots\dots\dots\dots\dots\dots \end{aligned}$$

(see Definition 17.11.3, p. 779).

Let us define the function $^k Y_i(x)$ such that

$$^k Y_i(x) \begin{cases} = 0 & \text{for } x < x_k\,,\\ = {}^k y_i(x) & \text{for } x \geqq x_k\,. \end{cases} \tag{4}$$

Theorem 1. Let $y(x)$ *be piecewise n-times smooth in the interval* $[a, b]$. *Further, in each of intervals* (x_k, x_{k+1}) $(k = 0, 1, 2, \dots, m$, cf. Definition 1) *let it satisfy equation* (2) *and at the points* x_k $(k = 1, 2, \dots, m)$ *let* $y(x)$ *and its derivatives have prescribed jumps* $^k S_i$ (see (1)). *Then* $y(x)$ *is of the form*

$$\begin{aligned} y = {}& y_p + c_1 y_1 + c_2 y_2 + \dots + c_n y_n +\\ &+ \sum_{k=1}^{m} {}^k S_0\, {}^k Y_0(x) + \sum_{k=1}^{m} {}^k S_1\, {}^k Y_1(x) + \dots + \sum_{k=1}^{m} {}^k S_{n-1}\, {}^k Y_{n-1}(x) \end{aligned} \tag{5}$$

where y_p *is a particular solution of equation* (2), $y_1, y_2, \dots, y_n$ *is the fundamental system of the corresponding homogeneous equation and* $^k S_i$, $^k Y_i$ *are defined by equations* (1), (4).

Example 1. Let us find the solution of the equation

$$y'' + y = 3x \tag{6}$$

with initial conditions

$$y(1) = 1\,,\quad y'(1) = -1 \tag{7}$$

such, that at the point $x_1 = 2$, $y(x)$ has the jump $^1 S_0 = -1$ and $y'(x)$ has the jump $^1 S_1 = 1$.

In this case we thus have $m = 1$.

The function $y_p = 3x$ is obviously a particular solution of equation (6), while the general integral of the corresponding homogeneous equation may with advantage be written, in view of the conditions (7), in the form

$$c_1 \cos (x - 1) + c_2 \sin (x - 1).$$

The normal fundamental system with respect to the point $x_1 = 2$ may be written in the form

$${}^1y_0(x) = \cos (x - 2), \quad {}^1y_1(x) = \sin (x - 2).$$

Thus

$${}^1Y_0(x) \begin{cases} = 0, \\ = \cos (x - 2), \end{cases} \quad {}^1Y_1(x) \begin{cases} = 0 & \text{for } x < 2, \\ = \sin (x - 2) & \text{for } x \geqq 2. \end{cases}$$

Applying (5), the solution will be

$$y = 3x + c_1 \cos (x - 1) + c_2 \sin (x - 1), \quad \text{for} \quad x < 2,$$
$$y = 3x + c_1 \cos (x - 1) + c_2 \sin (x - 1) - \cos (x - 2) + \sin (x - 2) \quad \text{for} \quad x \geqq 2. \tag{8}$$

From (8) and (7) it easily follows that $c_1 = -2$, $c_2 = -4$, thus giving the required solution.

REMARK 4. In applications we often come across cases where discontinuities occur not in the required solution, but in the coefficients of the given equation:

Theorem 2. *In equation* (2) *let the functions $a_j(x)$, $f(x)$ be piecewise continuous in the interval $[a, b]$. Let us denote the points of discontinuity by x_k $(k = 1, 2, \ldots, m)$. (With the exception of these points all the functions $a_j(x)$, $f(x)$ are then continuous in $[a, b]$.) If we prescribe at a point $c \in [a, b]$ $(c \neq x_k)$ initial conditions for $y^{(i)}(x)$ $(i = 0, 1, \ldots, n - 1)$, then only one solution of equation* (2) *exists in $[a, b]$ that has $n - 1$ continuous derivatives and a piecewise continuous n-th derivative in $[a, b]$ with points of discontinuity at x_k and which at the same time satisfies the prescribed initial conditions.*

REMARK 5. The procedure of finding the solution may be seen from an example:

Example 2. Let the equation

$$y'' + ay = 0 \tag{9}$$

be given, where $a = 1$ for $x \leqq 0$, $a = 4$ for $x > 0$. Let us find a solution which satisfies the following initial conditions:

$$y(-\tfrac{1}{2}\pi) = 0, \quad y'(-\tfrac{1}{2}\pi) = 1. \tag{10}$$

For $x \leqq 0$, the general integral of equation (9) is

$$y = c_1 \cos x + c_2 \sin x .$$

From conditions (10) we get

$$y = \cos x \quad (x \leqq 0) . \tag{11}$$

Thus

$$y(0) = 1 , \quad y'(0) = 0 . \tag{12}$$

For $x > 0$ the general integral of equation (9) is

$$y = k_1 \cos 2x + k_2 \sin 2x . \tag{13}$$

Since we are trying to find the solution of equation (9) which is continuous, together with its first derivative, in the interval $(-\infty, \infty)$ (we note that, in the notation of Theorem 2, $n = 2$ and the interval $[a, b]$ may be taken as infinite), the function (13) must satisfy conditions (12) for $x = 0$. We easily find that

$$y = \cos 2x \quad (x > 0) . \tag{14}$$

By (11) and (14), the solution is defined in the whole interval $(-\infty, \infty)$. At the point $x = 0$ the second derivative has a jump equal to -3.

17.17. Boundary Value Problems. Self-adjoint Problems. Eigenvalue Problems. Expansion Theorems. Green's Function

INTRODUCTORY REMARK. It is well known that for the problem

$$y'' + \lambda y = 0 , \quad y(0) = 0 , \quad y(1) = 0$$

there exist certain values of λ ($\lambda_1 = \pi^2$, $\lambda_2 = 4\pi^2$, $\lambda_3 = 9\pi^2$, ...), the so-called *eigenvalues* of this problem, for which the given problem has non-zero solutions ($y_1 = \sin \pi x$, $y_2 = \sin 2\pi x$, $y_3 = \sin 3\pi x$, ...), the so-called *eigenfunctions* of this problem (besides the identically zero solution $y = 0$ which exists for all values of λ). Any function $f(x)$, which has a continuous derivative in the interval $[0, 1]$ and fulfils the conditions $f(0) = f(1) = 0$ can be expanded in an absolutely and uniformly convergent Fourier series with respect to these eigenfunctions.

In this paragraph similar questions are treated for more general problems which are important for applications (stability problems for stressed bars, solution of partial differential equations by Fourier's method, etc.).

The results of this paragraph can easily be extended to problems in higher dimensions (cf. § 18.7).

Notation: We shall write the linear equation

$$f_n(x)\, y^{(n)} + f_{n-1}(x)\, y^{(n-1)} + \ldots + f_1(x)\, y' + f_0(x)\, y = f(x) \tag{1}$$

briefly in the form

$$L(y) = f(x)\,. \tag{2}$$

L is thus a *linear differential operator of the n-th order*. Whenever we shall write $L(u)$, we shall automatically assume that $u(x)$ has (in the domain considered, often in some interval I) n continuous derivatives. The functions $f_k(x)$ are assumed to be (real) continuous functions in the domain considered, and $f_n(x) \neq 0$. If necessary, we shall assume that these functions have continuous derivatives of the corresponding order.

Definition 1. The differential expression

$$K(y) \equiv (-1)^n \, [f_n(x)\, y]^{(n)} + (-1)^{n-1} \, [f_{n-1}(x)\, y]^{(n-1)} + \ldots + \\ + [f_2(x)\, y]'' - [f_1(x)\, y]' + f_0(x)\, y \tag{3}$$

is called the *adjoint expression to the expression* $L(y)$. The equation $K(y) = 0$ is called the *adjoint equation to the equation* $L(y) = 0$, and K is called the *adjoint operator to the operator* L.

Theorem 1. *A necessary and sufficient condition for the expression* $u(x)L(y)$ *to be a complete derivative of a differential expression* $V(y)$ *of the n-th order (i.e. that* $u\, L(y) = \mathrm{d}V(y)/\mathrm{d}x$ *for an arbitrary choice of a sufficiently smooth function* $y(x)$*) is that* $u(x)$ *should be a solution of the adjoint equation* $K(u) = 0$.

Definition 2. If $L(y) = K(y)$ for each n-times differentiable function $y(x)$, the expression $L(y)$ (and the operator L) are called *self-adjoint*.

Example 1.

$$L(y) = f_2(x)\, y'' + f_1(x)\, y' + f_0(x) y\,.$$

Then

$$K(y) = (f_2 y)'' - (f_1 y)' + f_0 y = f_2 y'' + (2f_2' - f_1)\, y' + (f_2'' - f_1' + f_0)\, y\,.$$

If $f_1 = f_2'$, i.e. if

$$L(y) = (f_2 y')' + f_0 y\,,$$

then L is a self-adjoint operator. It may be shown that any self-adjoint expression of the second order can be put into the form

$$L(y) = [p(x)\, y']' + q(x)\, y\,. \tag{4}$$

REMARK 1. In applications, self-adjoint expressions of the $2n$-th order

$$L(y) = \sum_{k=0}^{n} (-1)^k \left[f_k(x)\, y^{(k)}\right]^{(k)} = (-1)^n \left[f_n(x)\, y^{(n)}\right]^{(n)} + \\ + (-1)^{n-1} \left[f_{n-1}(x)\, y^{(n-1)}\right]^{(n-1)} + \ldots + f_0(x)\, y \tag{5}$$

play an important role.

Definition 3. Let an equation of the form

$$M(y) - \lambda N(y) = f(x) \tag{6}$$

be given, where $M(y)$, $N(y)$ are self-adjoint expressions of the $2m$-th and $2n$-th order, respectively,

$$M(y) = \sum_{k=0}^{m} (-1)^k \left[f_k(x)\, y^{(k)}\right]^{(k)}, \quad N(y) = \sum_{k=0}^{n} (-1)^k \left[g_k(x)\, y^{(k)}\right]^{(k)}, \quad m > n\,. \tag{7}$$

In the interval $[a, b]$ let the (real) functions $f_k(x)$, $g_k(x)$ have k continuous derivatives (for $k = 0$ this means the continuity of the functions f_0, g_0 themselves), $f_m(x) \neq 0$, $g_n(x) \neq 0$ in $[a, b]$, and $f(x)$ be continuous in $[a, b]$. In addition, let $2m$ linear homogeneous boundary conditions of the form

$$\alpha_{i0}\, y(a) + \beta_{i0}\, y(b) + \alpha_{i1}\, y'(a) + \beta_{i1}\, y'(b) + \ldots + \alpha_{i,2m-1}\, y^{(2m-1)}(a) + \\ + \beta_{i,2m-1}\, y^{(2m-1)}(b) = 0\,, \quad i = 1, 2, \ldots, 2m\,, \tag{8}$$

be prescribed, where α_{i0}, β_{i0}, ... are real constants, not simultaneously equal to zero in any one of the equations (8). Conditions (8) (of which there are $2m$) are supposed to be linearly independent (roughly speaking, none of them is a consequence of the others).

The problem of finding a solution $y(x)$ of equation (6) (with $2m$ continuous derivatives in $[a, b]$) satisfying conditions (8), is called a *boundary value problem* (for equation (6)). If $f(x) \equiv 0$, the problem is called *homogeneous.* To find such numbers λ, for which a *non-zero* solution of the homogeneous boundary value problem exists is called an *eigenvalue problem.*

REMARK 2. Boundary value problems for linear ordinary differential equations can be defined more generally. We do not, however, do this, because our definition involves practically all boundary value problems occurring in applications.

REMARK 3. According to our assumptions, the operator M is of a higher order than the operator N. Often $N(y) = g_0(x)y$, i.e. $n = 0$.

REMARK 4. The number of boundary conditions (8) is equal to the order of the differential equation (6), i.e. equal to $2m$. Derivatives up to the $(2m - 1)$-th order of the required function can occur in (8).

REMARK 5. Obviously, the function $y \equiv 0$ is always a solution of a *homogeneous* boundary value problem. *This function will be excluded from our considerations.*

Example 2. An example of an eigenvalue problem, according to our definition, is the following:

$$ay'' - \lambda y = 0 \quad (a \text{ a non-zero constant}), \tag{9}$$

$$y(0) = 0, \quad y'(1) - 2y(1) = 0. \tag{10}$$

$M(y) = ay''$ is a self-adjoint expression of the second order, because $M(y) = (ay')'$; $N(y) = y$. Conditions (10) are linear and homogeneous (obviously satisfied for $y \equiv 0$) and contain no derivative of a higher order than of the first.

Now suppose that a is not a constant, but $a = a(x) \neq$ const.; then $M(y) = a(x)\,y''$ is not a self-adjoint expression. If, however, $a(x) \neq 0$ in $[0, 1]$, we can divide equation (9) by this function and change it to the form

$$y'' - \lambda \frac{y}{a(x)} = 0. \tag{11}$$

Here $M(y) = y'' = (y')'$ is a self-adjoint expression, while

$$N(y) = \frac{1}{a(x)} \cdot y = b(x)y$$

is also a self-adjoint expression. From this example it can be seen that a simple rearrangement of the given equation may be all that is needed to put it into the form (6).

Definition 4. A number λ for which a non-zero solution $y(x)$ of the eigenvalue problem (i.e. of the homogeneous boundary value problem, Definition 3) exists is called an *eigenvalue* (or a *characteristic value*, or *proper value*) of this problem. The corresponding (non-zero) function $y(x)$ is called an *eigenfunction* (or *characteristic function* or *proper function*) of the problem.

Example 3. The eigenfunctions of the problem

$$y'' + \lambda y = 0, \quad y(0) = 0, \quad y(\pi) = 0 \tag{12}$$

are the functions $y = \sin px$, while the corresponding eigenvalues are $\lambda = p^2$, where p ranges over all integers.

REMARK 6. If $y(x)$ is an eigenfunction, then obviously $cy(x)$, where $c \neq 0$ is an arbitrary constant, is also an eigenfunction.

Definition 5. The (real) function $y(x)$ is called a *comparison function* of the eigenvalue problem, if it has $2m$ continuous derivatives in $[a, b]$ and satisfies the boundary conditions (8).

REMARK 7. A comparison function *need not* be a solution of the differential equation (6). Evidently, any eigenfunction is also a comparison function. For example, the function $y = x(\pi - x)$ is a comparison function of the problem (12), but obviously it is not an eigenfunction of the problem (12) for any λ.

Definition 6. The eigenvalue problem is called *self-adjoint*,* if for any *comparison* functions $u(x)$, $v(x)$ the relations

$$\int_a^b [u\, M(v) - v\, M(u)]\, dx = 0\,, \quad \int_a^b [u\, N(v) - v\, N(u)]\, dx = 0 \tag{13}$$

hold.

Definition 7. The eigenvalue problem is called *positive definite* if for any non-zero comparison function the relations

$$\int_a^b u\, M(u)\, dx > 0\,, \quad \int_a^b u\, N(u)\, dx > 0 \tag{14}$$

hold.

Example 4. The problem

$$y'' + \lambda\, c(x)\, y = 0\,, \quad (c(x) > 0 \text{ in } [a, b]) \tag{15}$$

$$y(a) = 0\,, \quad y(b) = 0 \tag{16}$$

is (as we shall show) self-adjoint and positive definite. We write

$$M(y) = -y''\,, \quad N(y) = c(x)\, y\,.$$

Then, integrating by parts, and using the fact that any comparison function satisfies conditions (16), we have

$$\int_a^b u\, M(v)\, dx = -\int_a^b uv''\, dx = -[uv']_a^b + \int_a u'v'\, dx = \int_a^b u'v'\, dx\,. \tag{17}$$

If we change the roles of u and v, we get similarly

$$\int_a^b v\, M(u)\, dx = \int_a^b u'v'\, dx\,, \quad \text{thus} \quad \int_a^b [u\, M(v) - v\, M(u)]\, dx = 0\,.$$

Moreover

$$\int_a^b u\,.\, cv\, dx = \int_a^b v\,.\, cu\, dx\,, \quad \text{thus} \quad \int_a^b [u\, N(v) - v\, N(u)]\, dx = 0\,. \tag{18}$$

* In the terminology of functional analysis, we should call this problem "symmetric". In the theory of differential equations, the customary terminology is rather different (cf. also the definition of the self-adjoint expression (Definition 2) etc.).

The problem is thus self-adjoint. Further (see (17))

$$\int_a^b u\, M(u)\, \mathrm{d}x = \int_a^b u'^2\, \mathrm{d}x\,, \quad \int_a^b u\, N(u)\, \mathrm{d}x = \int_a^b cu^2\, \mathrm{d}x$$

thus (since we are given that $c(x) > 0$) for any non-zero comparison function $u(x)$, (14) is satisfied. The problem is thus positive definite.

REMARK 8. Similarly the so-called *Sturm-Liouville problem*

$$(py')' + qy + \lambda ry = 0\,, \quad y(a) = 0\,, \quad y(b) = 0$$

can be shown to be self-adjoint and positive definite provided that, in $[a, b]$, $p(x) > 0$, $r(x) > 0$, $q(x) \leqq 0$.

Example 5. In the same way as in Example 4 the problem

$$y'' + \lambda y = 0\,, \quad y'(a) = 0\,, \quad y'(b) = 0$$

can be shown to be self-adjoint. It is not, however, positive definite because, for example, for every comparison function $u(x) = \text{const.} \neq 0$ we have

$$\int_a^b u\, M(u)\, \mathrm{d}x = 0\,.$$

REMARK 9. From Example 4 it can be seen that by using the method of integration by parts we can often decide easily whether a given problem is self-adjoint and positive definite or not. In other cases, we can often make use of the so-called *Green's formula* (*Dirichlet's formula*)

$$\int_a^b [u\, M(v) - v\, M(u)]\, \mathrm{d}x =$$

$$= \Big[\sum_{k=1}^{m} \sum_{l=0}^{k-1} (-1)^{k+l} \{u^{(l)}[f_k v^{(k)}]^{(k-l-1)} - v^{(l)}[f_k u^{(k)}]^{(k-l-1)}\}\Big]_a^b\,, \tag{19}$$

where as usual $[F]_a^b$ denotes $F(b) - F(a)$. For a given k, it is necessary for l in the sum to range through all the integers from 0 to $k - 1$. If $k = 1$, only one value of l, $l = 0$, is concerned.

A similar formula holds true for the operator N (instead of f_k, or m, we have here g_k, or n, respectively). Obviously, the eigenvalue problem is self-adjoint if and only if the boundary conditions are such that the right-hand side of (19) is equal to zero, for every comparison function, for both the operators M and N.

REMARK 10. We very often meet such problems where the self-adjoint operator N has only one term,

$$N(y) = (-1)^n\, [g_n(x)\, y^{(n)}]^{(n)} \tag{20}$$

and the boundary conditions are such that for every two comparison functions u, v the relation

$$\int_a^b u\, N(v)\, \mathrm{d}x = \int_a^b g_n u^{(n)} v^{(n)}\, \mathrm{d}x \tag{21}$$

holds. Then the eigenvalue problem is called *regular*. An example of this is the case where the operator N is of zero degree, i.e.

$$N(y) = g_0(x)\, y$$

(see Example 4). In this very simple case (21) is obviously satisfied, the boundary conditions being arbitrary.

Theorem 2. *If the eigenvalue problem is self-adjoint, then the eigenfunctions $y_s(x)$, $y_t(x)$ corresponding to different eigenvalues λ_s, λ_t are orthogonal in a generalized sense, i.e.*

$$\int_a^b y_s\, N(y_t)\, \mathrm{d}x = 0 \quad \textit{for} \quad \lambda_s \neq \lambda_t\,. \tag{22}$$

REMARK 11. In the special case where $N(y) = g_0(x)y$, equation (22) gives orthogonality with a weight function $g_0(x)$ (Definition 16.2.5, p. 697), i.e.

$$\int_a^b g_0 y_s y_t\, \mathrm{d}x = 0\,.$$

Theorem 3. *If the eigenvalue problem is positive definite, then it can have only positive eigenvalues.*

Theorem 4. *If the eigenvalue problem is self-adjoint and positive definite, then a countable set of positive mutually different eigenvalues of this problem exists* (see also Theorem 5 below).

REMARK 12. To every eigenvalue λ, there corresponds in this case either one or, in general, a finite (not infinite) number p of *linearly independent* eigenfunctions (Definition 12.8.3, p. 460). We say then that λ is a *simple* or a *p-fold eigenvalue*, respectively. (In almost all technical problems λ is simple, i.e. only one linearly independent eigenfunction corresponding to this λ exists, all others being multiples of it.) Linearly independent eigenfunctions corresponding to a given λ (for $p > 1$) can be orthogonalized in a generalized sense, cf. Remark 16.2.16, p. 701. Because, according to Theorem 2, the eigenfunctions corresponding to different λ are orthogonal in the generalized sense, we can thus associate with the set of all eigenvalues λ a system of linearly independent eigenfunctions mutually orthogonal in the sense (22). Let us number the eigenvalues according to their magnitude,

$$\lambda_1 \leqq \lambda_2 \leqq \lambda_3 \leqq \dots \tag{23}$$

and at the same time in such a way that the correspondence between the orthogonal system of eigenfunctions and the system of corresponding eigenvalues be a one-to-one-correspondence. Thus, every eigenvalue λ will occur in the system (23) the same number of times as the number of functions of the above orthogonal system corresponding to it. If e.g., three functions of the orthogonal system correspond to the smallest eigenvalue, then in (23) three numbers $\lambda_1 = \lambda_2 = \lambda_3$ will correspond to them; if to a further number λ two functions of the orthogonal system correspond, then in (23) two numbers $\lambda_4 = \lambda_5$ will correspond to them, etc. For this reason it was necessary to allow the possibility of equalities in the ordering (23).

A typical example of a self-adjoint positive definite problem is the problem (12). In this case the relations

$$\lambda_1 = 1, \; \lambda_2 = 4, \; \lambda_3 = 9, \; \lambda_4 = 16, \; \ldots .$$

hold true.

Every eigenvalue is simple. The system of eigenfunctions is the system

$$y_1 = \sin x, \; y_2 = \sin 2x, \; y_3 = \sin 3x, \; y_4 = \sin 4x, \; \ldots .$$

In the interval $[0, \pi]$ these functions are orthogonal (in the usual sense, i.e. with the weight function 1).

Theorem 5. *Let the eigenvalue problem be self-adjoint and positive definite. Then*

I. $$\lambda_n \to +\infty \quad \text{for} \quad n \to \infty ;$$

the point $+\infty$ is the only point of accumulation of the sequence λ_n.

II. *Let us define the so-called Rayleigh quotient $R(u)$ by the relation*

$$R(u) = \frac{\int_a^b u \, M(u) \, dx}{\int_a^b u \, N(u) \, dx} . \tag{24}$$

Then

$$\lambda_1 = \min R(u) ,$$

if $u(x)$ runs through all comparison functions of the given problem.

III. *More generally:*

$$\lambda_{k+1} = \min R(u) ,$$

where $u(x)$ runs through those comparison functions which are orthogonal (in the *generalized sense) to the first k eigenfunctions* $\varphi_1(x), \ldots, \varphi_k(x)$, *i.e. for which*

$$\int_a^b u\, N(\varphi_i)\, dx = 0 \quad (i = 1, 2, \ldots, k) \tag{25}$$

is true.

REMARK 13. It can be shown that $R(u)$ assumes the minimal value λ_1 exactly for the eigenfunction $\varphi_1(x)$ corresponding to λ_1. If we consider an arbitrary comparison function $u(x)$, which is not an eigenfunction, then $R(u) > \lambda_1$. Thus, $R(u)$ gives an estimate of λ_1 "from above".

Considering, for example, the problem (12), we have $M(u) = -u''$, $N(u) = u$. Let us consider the comparison function $u = x(\pi - x)$. According to (24) we get

$$R(x(\pi - x)) = \frac{\int_0^\pi 2x(\pi - x)\, dx}{\int_0^\pi x^2(\pi - x)^2\, dx} = \frac{\frac{\pi^3}{3}}{\frac{\pi^5}{30}} = \frac{10}{\pi^2} \doteq 1{\cdot}0132\,,$$

which is a good estimate of the first eigenvalue $\lambda_1 = 1$ "from above".

Part II of Theorem 5 provides the possibility of applying variational methods for finding λ_1 or $\varphi_1(x)$, respectively. This is true for part III, as well. See Chaps. 24 and 25.

REMARK 14. The application of variational methods in part III is more difficult, on account of condition (25). The following theorem is then often useful:

Theorem 6 (*Courant's Maximum – Minimum Principle*). *Let the eigenvalue problem be self-adjoint and positive definite. Let* $w_1(x), \ldots, w_k(x)$ *be an arbitrary system of linearly independent integrable functions. Let us denote by* $m(w_1, \ldots, w_k)$ *the minimum (or infimum) of the Rayleigh quotient, if* $u(x)$ *runs through all comparison functions which are orthogonal to all functions* $w_i(x)$, i. e. *for which*

$$\int_a^b u(x)\, w_i(x)\, dx = 0\,, \quad i = 1, 2, \ldots, k\,.$$

Then λ_{k+1} *is equal to the maximum of* $m(w_1, \ldots, w_k)$ *if all systems of (linearly independent and integrable) functions* $w_i(x)$ *are considered.*

Theorem 7 (*a Comparison Theorem*). *Let us consider two self-adjoint positive definite eigenvalue problems*

$$M_1(y) = \lambda\, N_1(y)\,, \quad M_2(y) = \lambda^*\, N_2(y) \tag{26}$$

with the same boundary conditions. For any comparison function $u(x)$ *let the relations*

$$\int_a^b u\, M_1(u)\, dx \leqq \int_a^b u\, M_2(u)\, dx\,, \quad \int_a^b u\, N_1(u)\, dx \geqq \int_a^b u\, N_2(u)\, dx \tag{27}$$

be satisfied. Then (supposing the eigenvalues of both problems to be arranged according to their magnitude, see Remark 12) *we have*

$$\lambda_k \leqq \lambda_k^* \quad (k = 1, 2, 3, \ldots,) . \tag{28}$$

Example 6. If, for example, in the problems

$$(p_1 y')' + q_1 y + \lambda r_1 y = 0 , \quad (p_2 y')' + q_2 y + \lambda^* r_2 y = 0 ,$$
$$y(a) = 0 , \quad y(b) = 0$$

the relations

$$0 < p_1(x) \leqq p_2(x) , \quad r_1(x) \geqq r_2(x) > 0 , \quad 0 \geqq q_1(x) \geqq q_2(x)$$

are satisfied in $[a, b]$, then

$$\lambda_k \leqq \lambda_k^* .$$

Theorem 8. *Let us consider a self-adjoint positive definite eigenvalue problem. Let a system of orthogonal eigenfunctions $\varphi_i(x)$* (Remark 12) *be normalized in the generalized sense,* i.e. *let*

$$\int_a^b \varphi_i N(\varphi_k) \, \mathrm{d}x = \begin{cases} 0 & \text{for} \quad i \neq k , \\ 1 & \text{for} \quad i = k . \end{cases} \tag{29}$$

Let $u(x)$ be an arbitrary comparison function and let the numbers

$$a_k = \int_a^b u \, N(\varphi_k) \, \mathrm{d}x \tag{30}$$

be its generalized Fourier coefficients. Then the series

$$\sum_{n=1}^{\infty} a_n \, \varphi_n(x) , \tag{31}$$

and also each of the series which arise by differentiating it p times term -by-term [$1 \leqq p \leqq m - 1$; on the number m see (7), Definition 3], *converges absolutely and uniformly in* $[a, b]$.

REMARK 15. The sum of the series (31) need not, in general, be equal to the function $u(x)$. The equality holds, however, when the problem is a so-called *closed problem.* The reader will find more details in the work by E. Kamke in Math. Zeitschr. 1940, pp. 275–280. In the following theorem we mention a special case, which, however, covers many cases occurring in applications.

Theorem 9. *If in Theorem* 8 *$N(y)$ is of the form*

$$N(y) = (-1)^n \left[g_n(x) \, y^{(n)}\right]^{(n)} \quad (g_n(x) > 0 \text{ in } [a, b]) \tag{32}$$

and if among the given boundary conditions the conditions

$$y(a) = y'(a) = \ldots = y^{(n-1)}(a) = 0\,, \quad y(b) = y'(b) = \ldots = y^{(n-1)}(b) = 0$$

are included (to which, in general, further conditions must be added according to the degree of the operator M), then the sum of the series (31) is equal to $u(x)$. The series (31) can be differentiated $(m - 1)$ times term by term and the respective sums are equal to the corresponding derivatives of the function $u(x)$.

REMARK 16. The conditions of Theorems 8 and 9 can often be weakened. For example, when investigating the problem

$$(py')' + qy + \lambda ry = 0\,, \quad p(x) > 0\,, \quad r(x) > 0\,, \quad q(x) \leqq 0\,, \tag{33}$$

$$y(a) = 0\,, \quad y(b) = 0 \tag{34}$$

it is sufficient for the function $u(x)$, which is to be developed into the Fourier series (31), to have only the first derivative continuous in $[a, b]$ (and to fulfil the prescribed boundary conditions). Sometimes, it is possible to omit even the condition of satisfying the boundary conditions (which can be advantageous, because we often have to deal with the case where $u(x)$ has a sufficient number of derivatives but does not satisfy the boundary conditions). If, for example, $u(x)$ is a smooth function in $[a, b]$, satisfying the second boundary condition $u(b) = 0$ but not satisfying the first condition $u(a) = 0$ of the problem (33), (34) then the corresponding series (31) can be shown to converge uniformly to $u(x)$ in every interval $[c, b]$ with $a < c < b$.

REMARK 17. The following non-homogeneous boundary value problem will now be considered: To find a solution of the equation

$$L(y) = f(x)\,, \tag{35}$$

where

$$L(y) = \sum_{i=0}^{k} p_i(x)\, y^{(i)}\,,$$

with $p_i(x)$ continuous, $p_k(x) \neq 0$ in $[a, b]$, satisfying the boundaryconditions

$$\sum_{i=0}^{k-1} [{}^l\alpha_i y^{(i)}(a) + {}^l\beta_i y^{(i)}(b)] = 0 \quad (l = 0, 1, \ldots, k - 1) \tag{36}$$

(cf. (8), Definition 3, where $k = 2m$; in our case k need not be an even number).

Definition 8. The so-called *Green's function* $G(x, \xi)$ of the problem (35), (36) is defined as follows:

1. The function $G(x, \xi)$ is defined in a square $a \leqq x \leqq b$, $a \leqq \xi \leqq b$. With the

exception of points lying on the diagonal $x = \xi$ (Fig. 17.1), the partial derivatives

$$\frac{\partial^i G}{\partial x^i} \quad (i = 0, 1, 2, \ldots, k)$$

are continuous functions of both variables (if $i = 0$ the continuity of the function G itself is to be understood), continuously extensible to the boundaries of both triangles into which the given square is divided by the diagonal $x = \xi$.

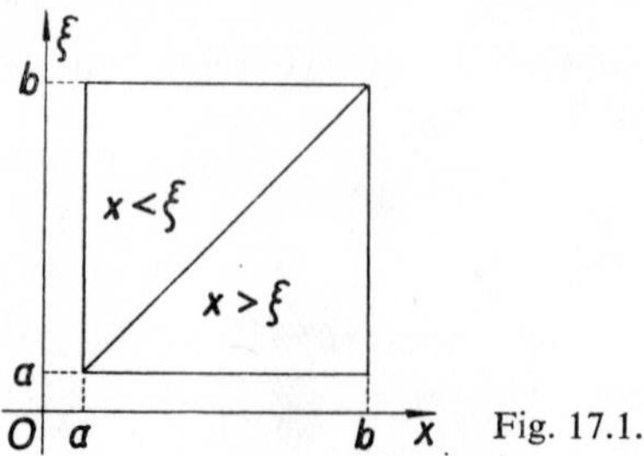

Fig. 17.1.

2. For any fixed $\xi \in (a, b)$, $G(x, \xi)$ satisfies, as a function of the variable x, the equation $L(y) = 0$ everywhere in $[a, b]$ *with the exception of the point* $x = \xi$.

3. $G(x, \xi)$ satisfies, as a function of the variable x, the boundary conditions (36) [for any $\xi \in (a, b)$].

4. $G(x, \xi)$, together with its derivatives with respect to x up to the $(k - 2)$-th order, is a continuous function of the variable x (ξ being fixed). The derivative (with respect to x) of the $(k - 1)$-th order has a jump $1/p_k(\xi)$ at the point $x = \xi$, i.e.

$$\lim_{x \to \xi+} \frac{\partial^{k-1} G}{\partial x^{k-1}}(x, \xi) - \lim_{x \to \xi-} \frac{\partial^{k-1} G}{\partial x^{k-1}}(x, \xi) = \frac{1}{p_k(\xi)}$$

($p_k(x)$ being the coefficient of the highest derivative $y^{(k)}$ in the operator $L(y)$, Remark 17).

REMARK 18. *Construction of the Green's function*: Let

$$y_1(x), y_2(x), \ldots, y_k(x)$$

be the fundamental system of the equation $L(y) = 0$, i.e. of the equation

$$f_k(x)\, y^{(k)} + f_{k-1}(x)\, y^{(k-1)} + \ldots + f_0(x)\, y = 0\,. \tag{37}$$

Let the Green's function be assumed to have the form

$$G(x, \xi) = \sum_{i=1}^{k} (a_i + b_i)\, y_i(x) \quad \text{for} \quad x \leqq \xi\,, \tag{38}$$

$$G(x, \xi) = \sum_{i=1}^{k} (a_i - b_i)\, y_i(x) \quad \text{for} \quad x \geqq \xi\,. \tag{39}$$

For the unknown constants b_i we get a system of equations

$$\begin{aligned} &\sum_{i=1}^{k} b_i y_i^{(l)}(\xi) = 0 \quad \text{for} \quad l = 0, 1, \ldots, k-2\,, \\ &\sum_{i=1}^{k} b_i y_i^{(k-1)}(\xi) = -\frac{1}{2p_k(\xi)} \end{aligned} \tag{40}$$

with a non-zero determinant; this system therefore has always a unique solution, $b_1 = b_1(\xi)$, $b_2 = b_2(\xi)$, …, $b_k = b_k(\xi)$. The unknown constants a_i can then be found with the help of the boundary conditions (36), where we substitute the values of the function G and of its derivatives at the point a by means of (38) (because the form (38) holds true "on the left", i.e. for $x \leqq \xi$) and at the point b according to (39). If the conditions (36) are written briefly in the form

$$U_i(y) = 0\,,$$

then it is possible to show that the determinant of the system of equations for the evaluation of the constants a_i is

$$D = \begin{vmatrix} U_1(y_1), & U_1(y_2), & \ldots, & U_1(y_k) \\ \ldots & \ldots & \ldots & \ldots \\ U_k(y_1), & U_k(y_2), & \ldots, & U_k(y_k) \end{vmatrix}. \tag{41}$$

Example 7. (See also Examples 8 and 9.) Let us consider a non-homogeneous problem

$$-y'' = x^2\,, \tag{42}$$

$$y(0) = 0\,, \quad y(1) = 0\,. \tag{43}$$

Then $L(y) = -y''$. As the fundamental system of the equation $L(y) = 0$ (or $-y'' = 0$) we may take the functions

$$y_1 = x\,, \quad y_2 = 1\,.$$

Since $p_2(x) \equiv -1$ equations (40) are of the form

$$\begin{aligned} b_1\xi + b_2 &= 0\,, \\ b_1 &= \tfrac{1}{2}\,, \end{aligned}$$

whence

$$b_1 = \tfrac{1}{2}\,, \quad b_2 = -\tfrac{1}{2}\xi\,.$$

By virtue of (38), (39) we have

$$G(x, \xi) = (a_1 + \tfrac{1}{2})\,x + (a_2 - \tfrac{1}{2}\xi) \quad \text{for} \quad x \leqq \xi\,, \tag{44}$$

$$G(x, \xi) = (a_1 - \tfrac{1}{2})\,x + (a_2 + \tfrac{1}{2}\xi) \quad \text{for} \quad x \geqq \xi\,. \tag{45}$$

Substituting (44) into the first of equations (43) (we consider (44), because in the first of equations (43) only the "left" point $x = 0$ occurs), we get

$$a_2 = \tfrac{1}{2}\xi\,, \tag{46}$$

then substituting (45) into the second of equations (43) we get (now using (46))

$$a_1 = -\xi + \tfrac{1}{2}\,.$$

Thus

$$G(x, \xi) = -\xi x + x \quad \text{for} \quad x \leqq \xi\,, \quad G(x, \xi) = -\xi x + \xi \quad \text{for} \quad x \geqq \xi\,. \tag{47}$$

Obviously

$$D = \begin{vmatrix} 0, & 1 \\ 1, & 1 \end{vmatrix} \neq 0\,. \tag{48}$$

Theorem 10. *If for the problem (35), (36) the determinant (41) is different from zero, then there exists exactly one Green's function, corresponding to the problem (35), (36). The (unique) solution of the problem is then given, for an arbitrary continuous right-hand side $f(x)$ (it is sufficient if $f(x)$ is piecewise continuous in $[a, b]$) by the equation*

$$y(x) = \int_a^b G(x, \xi)\, f(\xi)\, \mathrm{d}\xi\,. \tag{49}$$

The vanishing of the determinant (41) is a necessary and sufficient condition for the corresponding homogeneous problem (i.e. for the problem (35), (36) with $f(x) \equiv 0$) to have a non-zero solution.

(In this second case the solution of the problem (35), (36) (if it exists) is not unique. For example, a solution of the problem

$$y'' + \pi^2 y = \pi^2(2x - 1)\,, \quad y(0) = 0\,, \quad y(1) = 0$$

is the function

$$y = \cos \pi x + 2x - 1\,,$$

but in addition the function

$$y = \cos \pi x + 2x - 1 + 5 \sin \pi x$$

is also a solution.)

REMARK 19. The Green's function is often called the *influence function*. The reason lies in its technical meaning. Let us consider, for example, a bar supported at both ends a, b and loaded at the point $x = \xi$ by a concentrated load of unit value. Then the Green's function corresponding to this problem is equal to the deflection $y(x)$, caused by this concentrated load. If, at this point, instead of a unit load a load

$f(\xi)\,\Delta\xi$ is acting, the deflection will be equal to $G(x, \xi)\,f(\xi)\,\Delta\xi$. If a continuous load is considered, the deflection will be

$$y(x) = \int_a^b G(x, \xi)\,f(\xi)\,d\xi\,.$$

This is a technical interpretation of equation (49) (in the general sense the meaning is similar).

Example 8. In the case of the problem (42), (43), $D \neq 0$ (see (48)). According to Theorem 10, the solution $y(x)$ is then given by equation (49), and is unique:

$$y(x) = \int_0^1 G(x, \xi)\,.\,\xi^2\,d\xi = \int_0^x (-\xi x + \xi)\,\xi^2\,d\xi + \int_x^1 (-\xi x + x)\,\xi^2\,d\xi =$$

$$= -\frac{x^5}{4} + \frac{x^4}{4} - \frac{x}{4} + \frac{x}{3} + \frac{x^5}{4} - \frac{x^4}{3} = -\frac{x^4}{12} + \frac{x}{12}\,.$$

REMARK 20. Examples 7 and 8 are only illustrative. It is obvious that the problem (42), (43) could easily be solved without the construction of the Green's function.

REMARK 21. Equation (35) often contains a parameter λ (cf. (6), Definition 3). Then the Green's function is a function of λ as well. (Here, the name *Green's resolvent* $G(x, \xi, \lambda)$ or simply *resolvent* is often used.)

Example 9. Let us consider the operator

$$L(y) = -y'' - \lambda y \quad (\lambda > 0)\,, \tag{50}$$

with the boundary conditions

$$y(0) = 0\,, \quad y(1) = 0\,. \tag{51}$$

If we put $\lambda = k^2$, $k > 0$, then the functions $y_1 = \sin kx$, $y_2 = \cos kx$ form a fundamental system of the equation $L(y) = 0$, and using a procedure similar to that used in Example 7 we get

$$\begin{aligned} G(x, \xi) &= \left(\frac{\cos k\xi}{k} - \frac{\sin k\xi}{k}\cot k\right)\sin kx \quad \text{if} \quad x \leqq \xi\,,\\ G(x, \xi) &= \left(\frac{\cos kx}{k} - \frac{\sin kx}{k}\cot k\right)\sin k\xi \quad \text{if} \quad x \geqq \xi\,. \end{aligned} \tag{52}$$

The determinant (41) is equal to $\sin k$ in this case, and is a function of λ. For $\lambda = \pi^2, 4\pi^2, 9\pi^2, \ldots$ we have $D = 0$ and the problem

$$-y'' - \lambda y = 0\,, \quad y(0) = 0\,, \quad y(1) = 0$$

has a non-zero solution (see Theorem 10).

Theorem 11. *If the operator L is self-adjoint* (see Definition 2) *and the boundary conditions are such that*

$$\int_a^b [u\, L(v) - v\, L(u)]\, dx = 0$$

for any two comparison functions $u(x)$, $v(x)$ (i.e. we have a self-adjoint problem), then $G(x, \xi)$ is a symmetric function of the variables x and ξ, i.e.

$$G(x, \xi) = G(\xi, x) . \tag{53}$$

REMARK 22. This implies: If the function $G(x, \xi)$ for $x \geqq \xi$ is known, we obtain $G(x, \xi)$ for $x \leqq \xi$ by writing ξ, x instead of x, ξ.

The self-adjointness of the problems treated in Examples 7 and 9 may be easily verified. The corresponding Green's functions, (47) and (52) respectively, are, in fact, symmetric.

REMARK 23. Let us consider a self-adjoint eigenvalue problem (see Definition 6)

$$M(y) = \lambda\, N(y)$$

with boundary conditions (8). First of all, let

$$N(y) = g_0(x) y\,, \quad g_0(x) > 0 \quad \text{in} \quad [a, b]\,.$$

For the given operator M and the given boundary conditions let us construct the Green's function $G(x, \xi)$. By virtue of Theorem 11, $G(x, \xi) = G(\xi, x)$. Applying (49) (where we write $\lambda\, g_0(x)\, y$ for $f(x)$), we get, for the required function $y(x)$, an integral equation

$$y(x) = \lambda \int_a^b G(x, \xi)\, g_0(\xi)\, y(\xi)\, d\xi\,. \tag{54}$$

If we write now

$$\varphi(x) = \sqrt{[g_0(x)]}\, y(x)\,, \quad K(x, \xi) = \sqrt{[g_0(x)\, g_0(\xi)]}\, G(x, \xi)\,,$$

we get from (54) an integral equation with a *symmetric kernel*

$$\varphi(x) = \lambda \int_a^b K(x, \xi)\, \varphi(\xi)\, d\xi\,. \tag{55}$$

If the problem is not homogeneous, but the equation is of the form

$$M(y) = \lambda\, g_0(x)\, y + g(x)\,,$$

we transform it into the corresponding non-homogeneous equation in the same way by writing $\lambda\, g_0(x)\, y + g(x)$ instead of $f(x)$ in (49). By this procedure it is possible, in this simple case, to transform the investigation of a boundary value problem into the study of an integral equation.

The advantage of this procedure lies in the fact that it is often easy to construct the Green's function corresponding to the operator M itself (with the corresponding boundary conditions). The analysis of the corresponding integral equation is then, as a rule, relatively easy because the theory of integral equations has been extensively developed.

In the case where the problem is regular (Remark 10), it can similarly be transformed into an equation (55), where

$$\varphi(x) = \sqrt{[g_n(x)]}\, y^{(n)}(x)\,, \quad K(x, \xi) = \frac{\partial^{2n} G(x, \xi)}{\partial x^n\, \partial \xi^n} \sqrt{[g_n(x)\, g_n(\xi)]}\,.$$

17.18. Systems of Ordinary Differential Equations

REMARK 1. Let us consider a system of differential equations

$$F_i(x, y_1, y_1', \ldots, y_1^{(m_1)}, y_2, y_2', \ldots, y_2^{(m_2)}, \ldots, y_k, y_k', \ldots, y_k^{(m_k)}) = 0\,,$$
$$i = 1, 2, \ldots, k\,. \tag{1}$$

Equations (1) contain the derivatives of the required function $y_1(x)$ up to the order m_1 (although $y_1^{(m_1)}$ does not necessarily occur in every equation), the derivatives of the function $y_2(x)$ up to the order m_2, etc. The greatest of the numbers m_i is called the *order* of the given system. (In applications we frequently meet the case mentioned above, where the number of equations is equal to the number of unknown functions. In the general case, however, the number of equations need not be equal to the number of unknown functions.)

If the conditions of the theorem on implicit functions are satisfied, the system (1) can be solved with respect to the highest derivatives of the unknown functions and written in the so-called *canonical form*

$$\begin{aligned} y_1^{(m_1)} &= g_1(x, y_1, \ldots, y_1^{(m_1-1)}, y_2, \ldots, y_2^{(m_2-1)}, \ldots, y_k^{(m_k-1)})\,, \\ &\ldots\ldots\ldots\ldots\ldots\ldots\ldots\ldots\ldots\ldots \\ y_k^{(m_k)} &= g_k(x, y_1, \ldots, y_1^{(m_1-1)}, y_2, \ldots, y_2^{(m_2-1)}, \ldots, y_k^{(m_k-1)})\,. \end{aligned} \tag{2}$$

New unknown functions can be introduced by the relations

$$y_{11} = y_1', y_{12} = y_{11}' = y_1'', \ldots, y_{1,\, m_1-1} = y_1^{(m_1-1)}$$

and similarly

$$y_{21} = y_2', y_{22} = y_2'', \ldots, y_{2,m_2-1} = y_2^{(m_2-1)}, \ldots\,,$$

so that we obtain $m_1 + m_2 + \ldots + m_k$ equations of the first order,

$$\begin{aligned} y_1' &= y_{11}\,, \\ y_{11}' &= y_{12}\,, \\ &\ldots\ldots\ldots \\ y_{1,m_1-1}' &= g_1(x, y_1, y_{11}, \ldots, y_{1,m_1-1}, y_2, \ldots, y_{2,m_2-1}, \ldots, y_{k,m_k-1})\,, \\ y_2' &= y_{21}\,, \\ &\ldots\ldots\ldots \end{aligned} \tag{3}$$

The systems (2) and (3) are equivalent, i.e. any solution of the system (2) is also a solution of the system (3) and vice versa. In the following text, we shall consider exclusively systems of the form (3), i.e.

$$\begin{aligned} y_1' &= f_1(x, y_1, y_2, \ldots, y_n)\,, \\ &\ldots\ldots\ldots\ldots\ldots\ldots\ldots \\ y_n' &= f_n(x, y_1, y_2, \ldots, y_n)\,. \end{aligned} \tag{4}$$

The system (4) is called *normal.*

On the solvability of system (4), see Theorem 17.2.1, p. 733. See also Remark 17.2.9, p. 734.

Definition 1. Let the condltion be satisfied that through each point of the $(n + 1)$ dimensional region O there passes exactly one integral curve of the system (4) (we use the terminology of Remark 17.2.5, p. 733). Then by the *general integral* of the system (4), in this region, we mean a system of functions

$$\begin{aligned} y_1 &= \varphi_1(x, C_1, C_2, \ldots, C_n)\,, \\ &\ldots\ldots\ldots\ldots\ldots\ldots\ldots \\ y_n &= \varphi_n(x, C_1, C_2, \ldots, C_n)\,, \end{aligned} \tag{5}$$

such that by a proper choice of constants $C_1, \ldots, C_n$, it is possible to get an arbitrary integral curve of the system (4), which lies in O.

Definition 2. The system

$$\begin{aligned} y_1' &= a_{11}(x)\, y_1 + a_{12}(x)\, y_2 + \ldots + a_{1n}(x)\, y_n + h_1(x)\,, \\ &\ldots\ldots\ldots\ldots\ldots\ldots\ldots\ldots\ldots\ldots\ldots\ldots \\ y_n' &= a_{n1}(x)\, y_1 + a_{n2}(x)\, y_2 + \ldots + a_{nn}(x)\, y_n + h_n(x) \end{aligned} \tag{6}$$

is called *linear.* If all functions $h_i(x)$ are identically zero, the system is called *homogeneous*; if they are not, then it is *non-homogeneous.*

(In modern texts, the system (6) is written briefly in a *vector* (*matrix*) *form* in the following way:

$$\mathbf{y}' = \mathbf{A}\mathbf{y} + \mathbf{h}\,,$$

where $\boldsymbol{A}$ is the square matrix formed by the coefficients $a_{ik}(x)$ of the system (6), $\mathbf{y}, \mathbf{y}', \mathbf{h}$ are matrices formed by the functions $y_i(x), y_i'(x), h_i(x)$:

$$\boldsymbol{A} = \left\| \begin{matrix} a_{11}(x), & a_{12}(x), & \ldots, & a_{1n}(x) \\ a_{21}(x), & a_{22}(x), & \ldots, & a_{2n}(x) \\ \ldots & \ldots & \ldots & \ldots \\ a_{n1}(x), & a_{n2}(x), & \ldots, & a_{nn}(x) \end{matrix} \right\|,$$

$$\mathbf{y} = \left\| \begin{matrix} y_1(x) \\ y_2(x) \\ \vdots \\ y_n(x) \end{matrix} \right\|, \quad \mathbf{y}' = \left\| \begin{matrix} y_1'(x) \\ y_2'(x) \\ \vdots \\ y_n'(x) \end{matrix} \right\|, \quad \mathbf{h} = \left\| \begin{matrix} h_1(x) \\ h_2(x) \\ \vdots \\ h_n(x) \end{matrix} \right\|.$$

The product $\boldsymbol{A}\mathbf{y}$ is a matrix product according to Definition 1.25.3, p. 87. This vector form of writing is more compact and has many other advantages also.)

Theorem 1. *If all functions $a_{ik}(x)$, $h_i(x)$ are continuous in the interval I, then corresponding to any $a \in I$ and to n arbitrary chosen numbers $b_1, b_2, \ldots, b_n$ there exists in I precisely one solution of the system* (6), *defined in the whole interval I and satisfying the given initial conditions, i.e. there exists exactly one system of functions $y_1(x), y_2(x), \ldots, y_n(x)$ defined in the whole interval I, satisfying the system* (6) *in I and also satisfying the conditions*

$$y_1(a) = b_1, \; y_2(a) = b_2, \; \ldots, \; y_n(a) = b_n .$$

Definition 3. Let a *homogeneous* system (6) (i.e. system (6) with $h_1(x) \equiv \ldots \equiv h_n(x) \equiv 0$) be given. Let n solutions of this system be given, i.e. n systems of functions

$$\begin{matrix} {}^1y_1, & {}^1y_2, & \ldots, & {}^1y_n, \\ \ldots & \ldots & \ldots & \ldots \\ {}^ny_1, & {}^ny_2, & \ldots, & {}^ny_n. \end{matrix} \tag{7}$$

(The index i before the function y_k denotes that the function corresponds to a certain i-th solution of the given system). The system of solutions (7) is called a *fundamental system* of the homogeneous system (6) in the interval I, if the determinant

$$D = \left\| \begin{matrix} {}^1y_1, & {}^1y_2, & \ldots, & {}^1y_n \\ \ldots & \ldots & \ldots & \ldots \\ {}^ny_1, & {}^ny_2, & \ldots, & {}^ny_n \end{matrix} \right\| \tag{8}$$

is non-zero in I, i. e. it does not vanish at any point of I.

Theorem 2. *Let the functions $a_{ik}(x)$ be continuous in the interval I. Then $D(x)$ is either non-zero or identically equal to zero in the whole interval I.*

REMARK 2. It is thus sufficient to evaluate the determinant (10) at a single point $x_0 \in I$.

Theorem 3. *Let (7) be a fundamental system. Then the homogeneous system (6) has the general integral*

$$\begin{aligned} y_1 &= C_1\,{}^1y_1 + C_2\,{}^2y_1 + \dots + C_n\,{}^ny_1\,, \\ &\dots\dots\dots\dots\dots\dots\dots\dots \\ y_n &= C_1\,{}^1y_n + C_2\,{}^2y_n + \dots + C_n\,{}^ny_n\,. \end{aligned} \tag{9}$$

REMARK 3. Let us consider a homogeneous system

$$\begin{aligned} y_1' &= a_{11}y_1 + a_{12}y_2 + \dots + a_{1n}y_n\,, \\ &\dots\dots\dots\dots\dots\dots\dots\dots \\ y_n' &= a_{n1}y_1 + a_{n2}y_2 + \dots + a_{nn}y_n\,, \end{aligned} \tag{10}$$

where the coefficients a_{ik} are constants. The assumption

$$y_1 = k_1 e^{\lambda x},\; y_2 = k_2 e^{\lambda x},\; \dots,\; y_n = k_n e^{\lambda x} \tag{11}$$

leads, after substituting (11) into (10), to the system

$$\begin{aligned} (a_{11} - \lambda)\,k_1 + a_{12}k_2 \qquad &+ \dots + a_{1n}k_n &= 0\,, \\ a_{21}k_1 \qquad + (a_{22} - \lambda)\,k_2 &+ \dots + a_{2n}k_n &= 0\,, \\ \dots\dots\dots\dots\dots\dots\dots\dots&\dots\dots\dots\dots\dots\dots \\ a_{n1}k_1 \qquad + a_{n2}k_2 \qquad &+ \dots + (a_{nn} - \lambda)\,k_n &= 0\,, \end{aligned} \tag{12}$$

which has a non-zero solution if and only if (cf. § 1.18, p. 71)

$$\begin{vmatrix} a_{11} - \lambda, & a_{12}, & \dots, & a_{1n} \\ a_{21}, & a_{22} - \lambda, & \dots, & a_{2n} \\ \dots & \dots & \dots & \dots \\ a_{n1}, & a_{n2}, & \dots, & a_{nn} - \lambda \end{vmatrix} = 0\,. \tag{13}$$

I. Let the *characteristic equation* (13) have *n different roots* $\lambda_1, \lambda_2, \dots, \lambda_n$. Substituting λ_1 into (12), we get, on solving the system (12), a system of numbers ${}^1k_1, {}^1k_2, \dots, {}^1k_n$, determined uniquely except for an *arbitrary factor*. (If ${}^1k_1 \neq 0$, then ${}^1k_1, \dots, {}^1k_n$ are uniquely determined if we choose, e.g., ${}^1k_1 = 1$.) The first system of functions of the fundamental system is then

$${}^1y_1 = {}^1k_1 e^{\lambda_1 x},\; {}^1y_2 = {}^1k_2 e^{\lambda_1 x},\; \dots,\; {}^1y_n = {}^1k_n e^{\lambda_1 x}\,. \tag{14}$$

Similarly, on substituting the values $\lambda_2, \dots, \lambda_n$ into (12), we get further systems of functions of the fundamental system required,

$$\begin{aligned} {}^2y_1 &= {}^2k_1 e^{\lambda_2 x},\; {}^2y_2 = {}^2k_2 e^{\lambda_2 x},\; \dots,\; {}^2y_n = {}^2k_n e^{\lambda_2 x}\,, \\ &\dots\dots\dots\dots\dots\dots\dots\dots\dots\dots \\ {}^ny_1 &= {}^nk_1 e^{\lambda_n x},\; {}^ny_2 = {}^nk_2 e^{\lambda_n x},\; \dots,\; {}^ny_n = {}^nk_n e^{\lambda_n x}\,. \end{aligned} \tag{15}$$

The general integral is then, by virtue of Theorem 3,

$$\begin{aligned} y_1 &= C_1\,{}^1k_1 e^{\lambda_1 x} + C_2\,{}^2k_1 e^{\lambda_2 x} + \ldots + C_n\,{}^nk_1 e^{\lambda_n x}\,, \\ &\cdots\cdots\cdots\cdots\cdots\cdots\cdots\cdots\cdots\cdots\cdots \\ y_n &= C_1\,{}^1k_n e^{\lambda_1 x} + C_2\,{}^2k_n e^{\lambda_2 x} + \ldots + C_n\,{}^nk_n e^{\lambda_n x}\,. \end{aligned} \tag{16}$$

Example 1.

$$\begin{aligned} y_1' &= 4y_1 - 2y_2\,, \\ y_2' &= y_1 + y_2\,. \end{aligned} \tag{17}$$

Equation (13),

$$\begin{vmatrix} 4-\lambda, & -2 \\ 1, & 1-\lambda \end{vmatrix} = 0$$

has the roots $\lambda_1 = 2$, $\lambda_2 = 3$. The solution of (12), putting $\lambda = \lambda_1 = 2$,

$$\begin{aligned} 2k_1 - 2k_2 &= 0\,, \\ k_1 - k_2 &= 0\,, \end{aligned}$$

is $k_1 = 1$, $k_2 = 1$, so that

$${}^1y_1 = e^{2x}\,, \quad {}^1y_2 = e^{2x}\,. \tag{18}$$

(It is advisable to verify that the functions (18) satisfy equations (17).) Substituting $\lambda = \lambda_2 = 3$ into (12), we get

$$\begin{aligned} k_1 - 2k_2 &= 0\,, \\ k_1 - 2k_2 &= 0\,, \end{aligned}$$

whence $k_1 = 2$, $k_2 = 1$ and

$${}^2y_1 = 2e^{3x}\,, \quad {}^2y_2 = e^{3x}\,. \tag{19}$$

The general solution is then, according to (16),

$$\begin{aligned} y_1 &= C_1 e^{2x} + 2C_2 e^{3x}\,, \\ y_2 &= C_1 e^{2x} + C_2 e^{3x}\,. \end{aligned} \tag{20}$$

If the initial conditions are, for example, $y_1(0) = 3$, $y_2(0) = -1$, we get from (20)

$$C_1 + 2C_2 = 3\,, \quad C_1 + C_2 = -1$$

so that

$$C_1 = -5\,, \quad C_2 = 4$$

and the required solution is

$$y_1 = -5e^{2x} + 8e^{3x}\,, \quad y_2 = -5e^{2x} + 4e^{3x}\,.$$

REMARK 4. It may happen that the characteristic equation has complex roots. If the coefficients in (10) are real, the solution may nevertheless be found in real form. Let us consider first the case of two simple complex conjugate roots

$$\lambda_1 = \alpha + i\beta\,, \quad \lambda_2 = \alpha - i\beta\,, \quad \alpha, \beta \text{ real.}$$

To these roots, two systems of real functions correspond in the fundamental system,

$$^1y_j = e^{\alpha x}(^1l_j \cos \beta x - {}^2l_j \sin \beta x) \quad (j = 1, 2, \ldots, n)\,, \tag{21}$$

$$^2y_j = e^{\alpha x}(^1l_j \sin \beta x + {}^2l_j \cos \beta x) \quad (j = 1, 2, \ldots, n), \tag{22}$$

where ^1l_j, ^2l_j are numbers given by the equations

$$^1k_j = {}^1l_j + i\,{}^2l_j \quad \text{and} \quad {}^2k_j = {}^1l_j - i\,{}^2l_j$$

and ^1k_j, ^2k_j are the solutions of equations (12) for $\lambda_1 = \alpha + i\beta$, $\lambda_2 = \alpha - i\beta$, respectively.

Similarly, to another pair of simple conjugate values of λ, another two systems of real functions correspond in the fundamental system. For real λ, the corresponding systems of functions can then be obtained according to Remark 3 and Example 1.

Example 2. Let us consider the system

$$\begin{aligned} y_1' &= -7y_1 + \ y_2\,, \\ y_2' &= -2y_1 - 5y_2\,. \end{aligned} \tag{23}$$

Equation (13) is

$$\begin{vmatrix} -7 - \lambda, & 1 \\ -2, & -5 - \lambda \end{vmatrix} = 0$$

or $\lambda^2 + 12\lambda + 37 = 0$, and has roots $\lambda_1 = -6 + i$, $\lambda_2 = -6 - i$. Thus $\alpha = -6$, $\beta = 1$. Substituting λ_1 and λ_2 into (12), we get (apart from a constant factor)

$$^1k_1 = 1\,, \quad ^1k_2 = 1 + i \quad \text{and} \quad ^2k_1 = 1\,, \quad ^2k_2 = 1 - i\,,$$

thus

$$^1l_1 = 1\,, \quad ^2l_1 = 0\,, \quad ^1l_2 = 1\,, \quad ^2l_2 = 1\,.$$

Then the systems of functions that form the fundamental system are, according to (21), (22),

$$^1y_1 = e^{-6x}\cos x\,, \quad ^1y_2 = e^{-6x}(\cos x - \sin x)\,,$$

$$^2y_1 = e^{-6x}\sin x\,, \quad ^2y_2 = e^{-6x}(\sin x + \cos x)\,,$$

and the general integral is

$$y_1 = e^{-6x}(C_1 \cos x + C_2 \sin x)\,,$$

$$y_2 = e^{-6x}[(C_1 + C_2)\cos x + (C_2 - C_1)\sin x]\,.$$

II. If λ_i is an r-fold root of the characteristic equation, then it can be shown that solutions which correspond to this root in the fundamental system are of the form

$$y_j = P_{ij}e^{\lambda_i x} \tag{24}$$

where $P_{ij}(x)$ are polynomials of at most degree $(r - 1)$. The practical procedure will be shown in an example:

Example 3. Let us consider the system

$$\begin{aligned} y_1' &= -y_1 + y_2\,, \\ y_2' &= \qquad\; - y_2 + 4y_3\,, \\ y_3' &= \;\; y_1 \qquad\quad - 4y_3\,. \end{aligned} \tag{25}$$

Equation (13),

$$\begin{vmatrix} -1-\lambda, & 1, & 0 \\ 0, & -1-\lambda, & 4 \\ 1, & 0, & -4-\lambda \end{vmatrix} = -\lambda^3 - 6\lambda^2 - 9\lambda = 0$$

has a double root $\lambda_1 = -3$ and a single root $\lambda_2 = 0$. According to (24), the system of solutions corresponding to the root $\lambda_1 = -3$ will be of the form

$${}^1y_1 = e^{-3x}(a_1 + a_2x)\,, \quad {}^1y_2 = e^{-3x}(b_1 + b_2x)\,, \quad {}^1y_3 = e^{-3x}(c_1 + c_2x)\,. \tag{26}$$

Substituting (26) into (25), we obtain for the unknown constants $a_1, \ldots, c_2$ the following equations

$$\begin{aligned} 2a_1 - a_2 + b_1 &= 0\,, & 2a_2 + b_2 &= 0\,, \\ 2b_1 - b_2 + 4c_1 &= 0\,, & 2b_2 + 4c_2 &= 0\,, \\ -c_1 - c_2 + a_1 &= 0\,, & -c_2 + a_2 &= 0\,. \end{aligned} \tag{27}$$

Choosing $a_2 = C_1$ (C_1 being an arbitrary non-zero constant), the second column of equations (27) gives $b_2 = -2C_1$, $c_2 = C_1$. Further, choosing $a_1 = C_2$, the first column of equations (27) gives $b_1 = C_1 - 2C_2$, $c_1 = C_2 - C_1$. Thus, to the root $\lambda_1 = -3$ there correspond the solutions

$$\begin{gathered} {}^1y_1 = e^{-3x}(C_2 + C_1x)\,, \quad {}^1y_2 = e^{-3x}(C_1 - 2C_2 - 2C_1x)\,, \\ {}^1y_3 = e^{-3x}(C_2 - C_1 + C_1x)\,. \end{gathered} \tag{28}$$

Solutions corresponding to the root $\lambda_2 = 0$ can be obtained either by virtue of Remark 3 and Example 1 or can be supposed, according to (24), to have the form $(r = 1)$

$${}^2y_1 = a\,, \quad {}^2y_2 = b\,, \quad {}^2y_3 = c\,.$$

Substituting into (25), we get

$${}^2y_1 = 4C_3\,, \quad {}^2y_2 = 4C_3\,, \quad {}^2y_3 = C_3\,. \tag{29}$$

The general integral of the system (25) is then, according to (28) and (29),

$$\begin{aligned} y_1 &= e^{-3x}(C_2 + C_1x) + 4C_3 , \\ y_2 &= e^{-3x}(C_1 - 2C_2 - 2C_1x) + 4C_3 , \\ y_3 &= e^{-3x}(C_2 - C_1 + C_1x) + C_3 . \end{aligned}$$

REMARK 5. If the coefficients of the system (10) are real and if the r-fold root λ is complex, i.e. $\lambda = \alpha + i\beta$, then equation (13) has also the r-fold root $\lambda = \alpha - i\beta$ and the solution can be supposed to have the form

$$y_j = e^{\alpha x}(P_{ij} \cos \beta x + Q_{ij} \sin \beta x)$$

where P, Q are polynomials of degree $(r - 1)$ at most (cf. also Remarks 4 and 6).

Theorem 4. *Let a non-homogeneous system of linear equations*

$$\begin{aligned} y_1' &= a_{11}(x)\, y_1 + a_{12}(x)\, y_2 + \ldots + a_{1n}(x)\, y_n + f_1(x) , \\ &\ldots\ldots\ldots\ldots\ldots\ldots\ldots\ldots\ldots\ldots \\ y_n' &= a_{n1}(x)\, y_1 + a_{n2}(x)\, y_2 + \ldots + a_{nn}(x)\, y_n + f_n(x) \end{aligned} \tag{30}$$

be given. Let the functions $a_{ik}(x)$, $f_i(x)$ be continuous in the interval I and let

$$\begin{aligned} &{}^1y_1,\ {}^1y_2,\ \ldots,\ {}^1y_n , \\ &\ldots\ldots\ldots\ldots \\ &{}^ny_1,\ {}^ny_2,\ \ldots,\ {}^ny_n \end{aligned} \tag{31}$$

be the fundamental system (Definition 3) *of the homogeneous system corresponding to system* (30) (*i.e. for $f_i(x) \equiv 0$*). *Then*

I. *the general integral of the system* (30) *is of the form*

$$\begin{aligned} y_1 &= y_{1p} + C_1\, {}^1y_1 + C_2\, {}^2y_1 + \ldots + C_n\, {}^ny_1 , \\ &\ldots\ldots\ldots\ldots\ldots\ldots\ldots\ldots \\ y_n &= y_{np} + C_1\, {}^1y_n + C_2\, {}^2y_n + \ldots + C_n\, {}^ny_n , \end{aligned} \tag{32}$$

where $y_{1p}, y_{2p}, \ldots, y_{np}$ is a particular solution of the system (30) ($C_1, C_2, \ldots, C_n$ *are arbitrary constants*);

II. *the fundamental system* (31) *being known, the particular solution $y_{1p}, y_{2p}, \ldots, y_{np}$ can be obtained by the method of variation of constants:*

$$\begin{aligned} y_{1p} &= C_1(x)\, {}^1y_1 + C_2(x)\, {}^2y_1 + \ldots + C_n(x)\, {}^ny_1 , \\ &\ldots\ldots\ldots\ldots\ldots\ldots\ldots\ldots \\ y_{np} &= C_1(x)\, {}^1y_n + C_2(x)\, {}^2y_n + \ldots + C_n(x)\, {}^ny_n . \end{aligned} \tag{33}$$

It suffices to choose the functions $C_1(x), C_2(x), \ldots, C_n(x)$ so as to satisfy the system of equations

$$\begin{aligned} C_1' \, {}^1y_1 + C_2' \, {}^2y_1 + \ldots + C_n' \, {}^ny_1 &= f_1(x)\,, \\ \ldots\ldots\ldots\ldots\ldots\ldots&\ldots\ldots\ldots \\ C_1' \, {}^1y_n + C_2' \, {}^2y_n + \ldots + C_n' \, {}^ny_n &= f_n(x) \end{aligned} \tag{34}$$

(*which is uniquely solvable*).

Example 4. Let us consider the system

$$\begin{aligned} y_1' &= \quad y_2 + \cos x\,, \\ y_2' &= -y_1 + 1\,. \end{aligned}$$

The fundamental system (obtained in a similar way as in Example 2) is

$$\begin{aligned} {}^1y_1 &= \cos x\,, \quad {}^1y_2 = -\sin x\,, \\ {}^2y_1 &= \sin x\,, \quad {}^2y_2 = \quad \cos x\,. \end{aligned}$$

System (34):

$$\begin{aligned} C_1' \cos x + C_2' \sin x &= \cos x\,, \\ -C_1' \sin x + C_2' \cos x &= 1\,, \end{aligned}$$

whence

$$\begin{aligned} C_1'(x) &= \cos^2 x - \sin x\,, & C_1(x) &= \tfrac{1}{2}x + \tfrac{1}{2}\sin x \cos x + \cos x\,, \\ C_2'(x) &= \sin x \cos x + \cos x\,, & C_2(x) &= -\tfrac{1}{2}\cos^2 x + \sin x\,. \end{aligned}$$

(Constants of integration can be omitted, because we are trying to find only a special particular solution of (30).) Thus, by virtue of (33), we get

$$\begin{aligned} y_{1p} &= \left(\tfrac{1}{2}x + \tfrac{1}{2}\sin x \cos x + \cos x\right)\cos x + \left(-\tfrac{1}{2}\cos^2 x + \sin x\right)\sin x = \\ &= \tfrac{1}{2}x \cos x + 1\,, \\ y_{2p} &= -\left(\tfrac{1}{2}x + \tfrac{1}{2}\sin x \cos x + \cos x\right)\sin x + \left(-\tfrac{1}{2}\cos^2 x + \sin x\right)\cos x = \\ &= -\tfrac{1}{2}x \sin x - \tfrac{1}{2}\cos x\,. \end{aligned}$$

The general integral is

$$\begin{aligned} y_1 &= \tfrac{1}{2}x \cos x + 1 + C_1 \cos x + C_2 \sin x\,, \\ y_2 &= -\tfrac{1}{2}x \sin x - \tfrac{1}{2}\cos x - C_1 \sin x + C_2 \cos x \end{aligned}$$

where C_1, C_2 are arbitrary constants.

REMARK 6. Linear systems of equations can often be solved by transforming them into one linear equation of the n-th order for one unknown function:

Example 5. Let us consider the system (17) (Example 1). From the second equation it follows that

$$y_1 = y_2' - y_2\,, \quad \text{thus} \quad y_1' = y_2'' - y_2'\,. \tag{35}$$

Substituting into the first equation, we get

$$y_2'' - 5y_2' + 6y_2 = 0\,. \tag{36}$$

The characteristic equation $\lambda^2 - 5\lambda + 6 = 0$ is the same as in Example 1. The general solution of equation (36) is

$$y_2 = C_1 e^{2x} + C_2 e^{3x}\,.$$

Substituting into the first equation (35), we then get

$$y_1 = 2C_1 e^{2x} + 3C_2 e^{3x} - C_1 e^{2x} - C_2 e^{3x} = C_1 e^{2x} + 2C_2 e^{3x}\,,$$

which agrees with (20).

This procedure can be used even in the case of non-homogeneous linear equations.

17.19. Dependence of Solutions of Systems of Differential Equations on Initial Conditions and on Parameters of the System. Stability of Solutions

Theorem 1. *Let us consider the system*

$$\begin{aligned} y_1' &= f_1(x, y_1, y_2, \ldots, y_n)\,, \\ &\ldots\ldots\ldots\ldots\ldots\ldots \\ y_n' &= f_n(x, y_1, y_2, \ldots, y_n)\,. \end{aligned} \tag{1}$$

Let O be a region in which existence and uniqueness of the solution is guaranteed, i.e. through any point $A(a, b_1, b_2, \ldots, b_n)$ in O there passes exactly one integral curve of the system (1), *or, in other words, exactly one solution of the system* (1) *exists, satisfying the conditions $y_1(a) = b_1$, $y_2(a) = b_2$, ..., $y_n(a) = b_n$. If the functions $f_1, f_2, \ldots, f_n$ (as functions of $n + 1$ variables) have continuous partial derivatives in O with respect to all variables up to the order k (if $k = 0$, we mean the continuity of the functions themselves), then the solutions*

$$\begin{aligned} y_1 &= \varphi_1(x, a, b_1, b_2, \ldots, b_n)\,, \\ &\ldots\ldots\ldots\ldots\ldots\ldots \\ y_n &= \varphi_n(x, a, b_1, b_2, \ldots, b_n) \end{aligned} \tag{2}$$

of the system (1), *satisfying the given initial conditions, have continuous derivatives with respect to all $n + 2$ variables up to the order k in the region under consideration (i.e. for $(a, b_1, b_2, \ldots, b_n) \in O$, x belonging to a certain neighbourhood of the point a).*

REMARK 1. If, in addition, the functions f_i contain the parameters $p_1, p_2, \ldots, p_s$, then the functions φ_i are also functions of these parameters. If f_i have also continuous derivatives with respect to p_j up to the order k, the same holds true for the functions φ_i.

Definition 1. We say that the solution (2) is *stable* (in the *sense of Liapunov*), if to an arbitrary $\varepsilon > 0$ there exists a $\delta > 0$ such that if the initial conditions $c_1, c_2, \ldots, c_n$ satisfy the inequality

$$\sum_{m=1}^{n} |b_m - c_m| < \delta,$$

at the point a, then for all $x > a$, i.e. in the whole infinite interval $[a, \infty$, the inequality

$$|\varphi_i(x, a, b_1, b_2, \ldots, b_n) - \varphi_i(x, a, c_1, c_2, \ldots, c_n)| < \varepsilon \quad (i = 1, 2, \ldots, n) \tag{3}$$

holds.

Geometrically: If the initial conditions at $x = a$ are altered by a very small amount, then the solutions are altered by very little for all $x > a$.

Theorem 2. *Let us consider the system of equations*

$$\begin{aligned} y_1' &= a_{11}y_1 + a_{12}y_2 + \ldots + a_{1n}y_n + \psi_1(x, y_1, y_2, \ldots, y_n), \\ &\cdots\cdots\cdots\cdots\cdots\cdots\cdots\cdots\cdots\cdots \\ y_n' &= a_{n1}y_1 + a_{n2}y_2 + \ldots + a_{nn}y_n + \psi_n(x, y_1, y_2, \ldots, y_n), \end{aligned} \tag{4}$$

where the a_{ik} are constants (in general, complex). Let the functions ψ_i and the constants a_{ik} be such that:

I. *For* $x \geq 0$, $|y_i| < K$ $(i = 1, 2, \ldots, n;\ K = \text{const.})$, *the functions ψ_i are continuous and*

$$|\psi_i(x, y_1, y_2, \ldots, y_n)| < L(|y_1| + |y_2| + \ldots + |y_n|)$$

$(i = 1, 2, \ldots, n;$ L being a constant). (*In particular* $\psi_i(x, 0, 0, \ldots, 0) = 0$, *and* $y_1 \equiv 0$, $y_2 \equiv 0, \ldots, y_n \equiv 0$ *is a solution of the system* (4).)

II. *Corresponding to an arbitrary* $\varepsilon > 0$ *there exist numbers* δ_ε *and* T_ε *such that for* $|y_1| < \delta_\varepsilon$, $|y_2| < \delta_\varepsilon, \ldots, |y_n| < \delta_\varepsilon$, $x \geq T_\varepsilon$ *the inequality* $|\psi_i(x, y_1, y_2, \ldots, y_n)| \leq \leq \varepsilon(|y_1| + |y_2| + \ldots + |y_n|)$ *holds.*

III. *All roots of the equation*

$$\begin{vmatrix} a_{11} - \lambda, & a_{12}, & \ldots, & a_{1n} \\ a_{21}, & a_{22} - \lambda, & \ldots, & a_{2n} \\ \cdots & \cdots & \cdots & \cdots \\ a_{n1}, & a_{n2}, & \ldots, & a_{nn} - \lambda \end{vmatrix} = 0 \tag{5}$$

have negative real parts.

Then the solution $y_1 \equiv 0,\ y_2 \equiv 0,\ \ldots,\ y_n \equiv 0$ *is stable* (*in the above-mentioned sense of Liapunov*).

Moreover: For any solution $y_1, y_2, \ldots, y_n$ *whose initial values are sufficiently small* (*in absolute value*) *the relations*

$$\lim_{x\to+\infty} y_1 = 0\,, \quad \lim_{x\to+\infty} y_2 = 0\,, \quad \ldots, \quad \lim_{x\to+\infty} y_n = 0 \tag{6}$$

are true.

REMARK 2. Roughly speaking: If the coefficients a_{ik} satisfy the condition III and if the functions ψ_i are sufficiently small for sufficiently small y_j and sufficiently large x, then the solution $y_i \equiv 0$ $(i = 1, 2, \ldots, n)$ is stable, i.e. any other solution with the initial values $y_i(0) = c_i$, where c_i are sufficiently small numbers, differs "very little" from this zero solution in the whole interval $[0, \infty)$. The behaviour for large x is given by (6).

For details see, e.g., Perron's article in Math. Zeitschr. 1929, p. 129, where an application of Theorem 2 is also given.

On the asymptotic behaviour of solutions see, e.g., [211], p. 61.

17.20. First Integrals of a System of Differential Equations

Definition 1. Let us consider the system

$$\begin{aligned} y_1' &= f_1(x, y_1, y_2, \ldots, y_n)\,, \\ &\ldots\ldots\ldots\ldots\ldots\ldots \\ y_n' &= f_n(x, y_1, y_2, \ldots, y_n)\,. \end{aligned} \tag{1}$$

By a *first integral of the system* (1) we mean a differentiable function of the variables $x, y_1, y_2, \ldots, y_n$, which is not identically constant with respect to these variables and which assumes a constant value if we substitute an arbitrary solution of the system (1) for $y_1, y_2, \ldots, y_n$.

REMARK 1. The first integrals of the system (1) are often defined in the following way: Let

$$\begin{aligned} y_1 &= \varphi_1(x, C_1, C_2, \ldots, C_n)\,, \\ &\ldots\ldots\ldots\ldots\ldots\ldots \\ y_n &= \varphi_n(x, C_1, C_2, \ldots, C_n) \end{aligned} \tag{2}$$

be the general integral of the system (1). Let us determine C_i from (2):

$$\begin{aligned} \psi_1(x, y_1, y_2, \ldots, y_n) &= C_1\,, \\ \ldots\ldots\ldots\ldots\ldots\ldots \\ \psi_n(x, y_1, y_2, \ldots, y_n) &= C_n\,. \end{aligned} \tag{3}$$

Then each of the relations (3) is called a *first integral of the system* (1).

These two definitions are not always equivalent in general. In simple cases, however, it can be verified that both of them give the same result.

Example 1. If we calculate C_1 and C_2 from (20) (Example 17.18.1) we get

$$\begin{aligned}(2y_2 - y_1)\,e^{-2x} &= C_1\,,\\ (y_1 - y_2)\,e^{-3x} &= C_2\,.\end{aligned} \tag{4}$$

Each of the relations (4) gives a first integral of the system (17) of the above-mentioned example. The left-hand sides of equations (4) are constants for each solution of (17). For example, for the solution

$$y_1 = -5e^{2x} + 8e^{3x}\,, \quad y_2 = -5e^{2x} + 4e^{3x}$$

of the example under consideration, we have

$$(2y_2 - y_1)\,e^{-2x} = e^{-2x}(-10e^{2x} + 8e^{3x} + 5e^{2x} - 8e^{3x}) = -5\,.$$

Theorem 1. *The necessary and sufficient condition for the equation*

$$\psi(x, y_1, y_2, \ldots, y_n) = C \tag{5}$$

to be a first integral of the system (1) (*in the domain O of the variables $x, y_1, y_2, \ldots, \ldots, y_n$*) *is that*

$$\frac{\partial\psi}{\partial x} + f_1(x, y_1, \ldots, y_n)\frac{\partial\psi}{\partial y_1} + f_2(x, y_1, \ldots, y_n)\frac{\partial\psi}{\partial y_2} + \ldots + f_n(x, y_1, \ldots, y_n)\frac{\partial\psi}{\partial y_n} = 0 \tag{6}$$

be satisfied identically in O. (*It is assumed, naturally, that ψ and f_k are differentiable functions of their respective variables.*)

Example 2. Following Theorem 1, we can easily verify that each of equations (4) gives a first integral of the system (17) in Example 17.18.1. For example, writing down the condition (6) for the second of equations (4), we get

$$-3(y_1 - y_2)\,e^{-3x} + (4y_1 - 2y_2)\,e^{-3x} - (y_1 + y_2)\,e^{-3x} \equiv 0\,.$$

REMARK 2. If only one first integral of the system (1) is known, it is generally possible to determine from its equation one unknown function $y_k(x)$ as a function of x and of the other unknown functions $y_1, \ldots, y_{k-1}, y_{k+1}, \ldots, y_n$. Substituting this into the given system, the order of the system is then reduced by one. Similarly, if j (independent) first integrals are known, the order of the system can be reduced by j. If n (independent) first integrals are known, we can obtain the general integral without any integration; it is sufficient to determine $y_1, \ldots, y_n$ from (3), as functions of $x, C_1, \ldots, C_n$.

It can be shown that corresponding to any system (1) satisfying the conditions of Theorem 17.2.1, p. 733, there exists a system of n independent first integrals. Any system of $n + 1$ first integrals is dependent.

REMARK 3. From Definition 1 it follows that: If (3) are the first integrals of the system (1), then

$$F(\psi_1, \psi_2, \ldots, \psi_n) = C\,,$$

where F is an arbitrary (not identically constant) differentiable function, also constitutes a first integral of the system (1). Since all ψ_k assume constant values for any solution, then F also assumes a constant value for any solution. The same is valid if F is a function of only k of the integrals (3).

REMARK 4 (*construction of first integrals*). The first integrals of the system (1) can often be obtained easily by writing the system (1) in the differential form

$$\frac{dy_1}{f_1(x, y_1, \ldots, y_n)} = \frac{dy_2}{f_2(x, y_1, \ldots, y_n)} = \ldots = \frac{dy_n}{f_n(x, y_1, \ldots, y_n)} = dx \tag{7}$$

and by combining equations (7) in such a way that we obtain complete differentials. For the determination of further first integrals it is possible to use the first integrals already found. In particular, the problem becomes simpler if the functions f_i do not depend explicitly on x. Then, in (7), the last term dx can be omitted and the system be written in the "symmetric form"

$$\frac{dy_1}{f_1(y_1, \ldots, y_n)} = \frac{dy_2}{f_2(y_1, \ldots, y_n)} = \ldots = \frac{dy_n}{f_n(y_1, \ldots, y_n)}\,. \tag{8}$$

The system (8) has $n - 1$ independent first integrals, while, obviously, x does not occur in them. (This fact is important in mechanics, where systems of this type are often met; x then denotes time. The given system has at least $n - 1$ first integrals independent of the time.)

In the case where a system is written in the form (8) we need not assume that all the functions $f_1, f_2, \ldots, f_n$ are different from zero in the region considered, that is to say, (8) is only a brief form of transcription and if, for example, $f_n \neq 0$, the system (8) really represents the system

$$\frac{dy_1}{dy_n} = \frac{f_1}{f_n}\,, \quad \frac{dy_2}{dy_n} = \frac{f_2}{f_n}\,, \quad \ldots, \quad \frac{dy_{n-1}}{dy_n} = \frac{f_{n-1}}{f_n}\,.$$

The points at which all the functions $f_1, f_2, \ldots, f_n$ are simultaneously zero (the so-called *singular points* of the given system) are excluded from our considerations.

Example 3. Let us consider the system

$$y_1' = y_2\,, \quad y_2' = -y_1\,.$$

The system (8) is:

$$\frac{dy_1}{y_2} = -\frac{dy_2}{y_1},$$

whence $y_1\,dy_1 = -y_2\,dy_2$ and the first integral is

$$y_1^2 + y_2^2 = C\,.$$

(A singular point is the point $y_1 = 0$, $y_2 = 0$.)

Example 4. Let us consider the system

$$\frac{dy_1}{dx} = 1 - \frac{1}{y_2}, \quad \frac{dy_2}{dx} = \frac{1}{y_1 - x} \quad (y_1 \neq x,\ y_2 \neq 0)\,. \tag{9}$$

The system can be written in the differential form (7) (after multiplying by dx):

$$dy_1 - dx = -\frac{dx}{y_2}, \quad \frac{dx}{y_1 - x} = dy_2\,. \tag{10}$$

Multiplying together both sides of these equations, we get an integrable combination

$$\frac{dy_1 - dx}{y_1 - x} = -\frac{dy_2}{y_2}$$

whence we obtain the first integral

$$(y_1 - x)\,y_2 = C_1 \quad (C_1 \neq 0)\,. \tag{11}$$

We can use (11) for calculating the second first integral. From (11), it follows that

$$y_2 = \frac{C_1}{y_1 - x}, \tag{12}$$

which, when substituted into the first equation (10), gives

$$dy_1 - dx = -\frac{(y_1 - x)\,dx}{C_1}$$

and

$$(y_1 - x)\,e^{x/C_1} = C_2 \quad (C_2 \neq 0)\,. \tag{13}$$

Now (13) is not a first integral of system (9) (it contains two constants). We therefore, substitute for C_1 from (11) into (13). In this way we get the second first integral of (9),

$$(y_1 - x)\,e^{x/[y_2(y_1 - x)]} = C_2\,. \tag{14}$$

By virtue of Remark 2, the general integral of the system (9) can be determined from (12) and (14). Here, however, it is more advantageous to use (13) and (12). From (13) it follows that

$$y_1 = x + C_2 e^{-x/C_1} \tag{15}$$

and on substituting for $y_1 - x$ into (12), we have

$$y_2 = \frac{C_1}{C_2} e^{x/C_1} . \tag{16}$$

Equations (15) and (16) constitute the general integral of the system (9). (The above procedure can obviously be applied, because (12) and (14) follow from (12) and (13), and conversely.)

REMARK 5. First integrals of ordinary differential equations are frequently used in theoretical mechanics and in the theory of partial differential equations of the first order.

17.21. Table of Solved Differential Equations

See also [211]; in this paragraph m, n denote integers.*

(a) Equations of the First Order.

1. $y' + ay = ce^{bx}$;

$$y = \frac{c}{a+b} e^{bx} + Ce^{-ax} \quad \text{if} \quad a + b \neq 0 ,$$

$$= cxe^{bx} + Ce^{-ax} \quad \text{if} \quad a + b = 0 .$$

---

* Many differential equations which the reader will meet in applications can easily be transformed into equations given in the following table. For example, if we encounter the equation

$$y' + y^2 = 16 ,$$

we can make the substitutions

$$y = 4u , \quad 4x = t ,$$

which give

$$y' = \frac{dy}{dx} = 4 \frac{du}{dx} = 4 \frac{du}{dt} \frac{dt}{dx} = 16 \frac{du}{dt} = 16 \dot{u} .$$

so that the given equation becomes $\dot{u} + u^2 = 1$, which is equation 7 of the table.

2. $\frac{dy}{dt} + \frac{R}{L} y = \frac{E}{L} \sin \omega t$

($y(t)$ denotes the current in a circuit, R the resistance, L the self-inductance, $E \sin \omega t$ the alternating voltage.) If $y = y_0$ for $t = 0$, the solution is

$$y = e^{-(R/L)t}\left(y_0 + \frac{\omega LE}{R^2 + \omega^2 L^2}\right) + \frac{E}{\sqrt{(R^2 + \omega^2 L^2)}} \sin (\omega t - \gamma)$$

where $\tan \gamma = \frac{\omega L}{R}$ and $0 < \gamma < \frac{1}{2}\pi$. For large t, $y \approx \frac{E}{\sqrt{(R^2 + \omega^2 L^2)}} \sin (\omega t - \gamma)$.

3. $y' + 2xy = xe^{-x^2}$;

$y = e^{-x^2}(\frac{1}{2}x^2 + C)$.

4. $y' + y \cos x = e^{-\sin x}$;

$y = (x + C) e^{-\sin x}$.

5. $y' + f'(x) y = f(x) f'(x)$;

$y = f(x) - 1 + Ce^{-f(x)}$.

6. $y' + f(x) y = g(x)$;

$$y = e^{-F(x)} \int g(x) e^{F(x)} dx \quad \text{where} \quad F(x) = \int f(x) dx ;$$

the integral curve passing through the point (x_0, y_0) is

$$y = e^{-F_0(x)}\left(y_0 + \int_{x_0}^{x} g(t) e^{F_0(t)} dt\right), \quad \text{where} \quad F_0(x) = \int_{x_0}^{x} f(t) dt .$$

7. $y' + y^2 = 1$;

$y = \tanh (x + C)$; $y = \coth (x + C)$; $y = \pm 1$.

8. $y' + y^2 - 2x^2 y + x^4 - 2x - 1 = 0$;

substituting $u = y - x^2$, we obtain $u' + u^2 = 1$, see 7.

9. $y' = (y + x)^2$;

$y = -x + \tan (x + C)$.

10. $y' + ay^2 = b$;

the integral curve passing through the point (x_0, y_0) is

$$y = y_0 + b(x - x_0) \qquad \text{if } a = 0,$$

$$= \frac{y_0}{1 + ay_0(x - x_0)} \qquad \text{if } b = 0,$$

$$= \frac{y_0 \sqrt{(ab)} + b \tanh [\sqrt{(ab)}\,(x - x_0)]}{\sqrt{(ab)} + ay_0 \tanh [\sqrt{(ab)}\,(x - x_0)]} \qquad \text{if } ab > 0,$$

$$= \frac{y_0 \sqrt{(-ab)} + b \tan [\sqrt{(-ab)}\,(x - x_0)]}{\sqrt{(-ab)} + ay_0 \tan [\sqrt{(-ab)}\,(x - x_0)]} \qquad \text{if } ab < 0.$$

11. $y' - ax^r(y^2 + 1) = 0$;

$$y = \tan \left(\frac{a}{r + 1} x^{r+1} + C \right) \qquad \text{if } r \neq -1,$$

$$= \tan (a \ln Cx) \qquad \text{if } r = -1.$$

12. $y' + f(x)\, y^2 + g(x)\, y = 0$;

$$\frac{1}{y} = E(x) \int \frac{f(x)}{E(x)}\, dx, \quad \text{where } E(x) = e^{\int g(x) dx}; \quad y = 0.$$

13. $y' = a \cos y + b,\ b > |a| > 0$;

$$\arctan \left[\sqrt{\left(\frac{b + a}{b - a} \right)} \cot \frac{y}{2} \right] + \frac{x}{2} \sqrt{(b^2 - a^2)} = C; \ \cos y = -\frac{b}{a}.$$

14. $y' = \cos (ay + bx),\ a \neq 0$;

substituting $u = ay + bx$, we get $u' = a \cos u + b$, see 13.

15. $xy' + y = x \sin x$;

$$y = \frac{\sin x}{x} - \cos x + \frac{C}{x}.$$

16. $xy' - y = x^2 \sin x$;

$$y = x(C - \cos x).$$

17. $xy' + ay + bx^r = 0, \quad x > 0$;

$$y = Cx^{-a} - \frac{b}{r + a} x^r \quad \text{if } a \neq -r,$$

$$= Cx^{-a} - bx^{-a} \ln x \quad \text{if } a = -r.$$

18. $xy' - y^2 + 1 = 0\,;$

$$y = \frac{1 - Cx^2}{1 + Cx^2}\,; \quad y = \pm 1\,.$$

19. $xy' + xy^2 - y = 0\,;$

$$y = \frac{2x}{x^2 + C}\,; \quad y = 0\,.$$

20. $xy' - y^2 \ln x + y = 0\,;$

$$\frac{1}{y} = 1 + \ln x + Cx\,; \quad y = 0\,.$$

21. $xy' - y(2y \ln x - 1) = 0\,;$

$$\frac{1}{y} - 2(1 + \ln x) = Cx\,; \quad y = 0\,.$$

22. $xy' = x \sin \frac{y}{x} + y\,;$

$$y = 2x \arctan Cx\,.$$

23. $xy' + x \cos \frac{y}{x} - y + x = 0\,;$

$$\cos \frac{y}{x} - (C + \ln x) \sin \frac{y}{x} = 1\,.$$

24. $xy' + x \tan \frac{y}{x} - y = 0\,;$

$$x \sin \frac{y}{x} = C\,; \quad y = 0\,.$$

25. $x^2 y' + y - x = 0\,;$

$$y = e^{1/x}\left(C + \int \frac{1}{x} e^{-1/x}\, dx\right).$$

26. $x^2 y' - y^2 - xy = 0\,;$

$$y = -\frac{x}{\ln Cx}\,; \quad y = 0\,.$$

27. $(x^2 + 1)\, y' + xy - 1 = 0\,;$

$$y = \frac{C + \ln\left[x + \sqrt{(x^2 + 1)}\right]}{\sqrt{(x^2 + 1)}}\,.$$

28. $(x^2 - 1)\, y' + 2xy - \cos x = 0\,;$

$$(x^2 - 1)\, y - \sin x = C\,.$$

29. $(x^2 - 1)\, y' - y(y - x) = 0\,;$

$$y = \frac{1}{x + C\sqrt{|x^2 - 1|}}\,; \quad y = 0\,.$$

30. $(x^2 - 1)\, y' + axy^2 + xy = 0\,;$

$$y = \frac{1}{C\sqrt{|x^2 - 1|} - a}\,; \quad y = 0\,.$$

31. $(x^2 - 1)\, y' = 2xy \ln y\,;$

$$y = e^{C(x^2 - 1)}\,.$$

32. $yy' + xy^2 - 4x = 0\,;$

$$y^2 = 4 + Ce^{-x^2}\,.$$

33. $(y - x^2)\, y' = x\,;$

$$x^2 = y - \tfrac{1}{2} + Ce^{-2y}\,.$$

34. $ayy' + by^2 + f(x) = 0\,;$

substituting $u = y^2$, we obtain the linear equation of the form

$$au' + 2bu + 2f(x) = 0\,.$$

35. $y'^2 + y^2 = a^2\,;$

$$y = a\,\frac{1 - C^2}{1 + C^2}\sin x + a\,\frac{2C}{1 + C^2}\cos x\,; \quad y = \pm a\,.$$

36. $y'^2 = y^3 - y^2\,;$

$$y = \left(\cos\frac{x + C}{2}\right)^{-2}; \; y = 0\,,\; y = 1\,.$$

37. $y'^2 - 4y^3 + ay + b = 0\,;$

$$x = C \pm \int \frac{dy}{\sqrt{(4y^3 - ay - b)}}\,;$$ the integral is an elliptic integral.

38. $y'^2 - 2y' - y^2 = 0$;

$$1 \mp \sqrt{(y^2 + 1)} + y \ln [\sqrt{(y^2 + 1)} \pm y] = y(x + C); \quad y = 0.$$

39. $y'^2 + ay' + bx = 0, \quad b \neq 0$;

the solution (in parametric form) is

$$bx = -t^2 - at, \; by = C - \tfrac{2}{3}t^3 - \tfrac{1}{2}at^2.$$

40. $y'^2 + ay' + by = 0, \quad b \neq 0$;

the solution (in parametric form) is

$$bx = -2t - a \ln t + C, \; by = -t^2 - at.$$

A further solution is $y = 0$.

41. $y'^2 + (x - 2)\, y' - y + 1 = 0$;

$$y = C(x - 2) + C^2 + 1; \; y = x - \frac{x^2}{4}.$$

42. $y'^2 + (x + a)\, y' - y = 0$;

$$y = C(x + a) + C^2, \quad 4y = -(x + a)^2.$$

43. $y'^2 - (x + 1)\, y' + y = 0$;

$$y = Cx + C(1 - C); \quad y = \tfrac{1}{4}(x + 1)^2.$$

44. $y'^2 - 2xy' + y = 0$;

the solution (in parametric form) is

$$x = \tfrac{2}{3}t + \frac{C}{t^2}, \; y = 2xt - t^2.$$

Further solutions are $y = 0$, $y = \tfrac{3}{4}x^2$.

45. $y'^2 + 2xy' - y = 0$;

substituting $u = -y$, the equation is transformed into 44.

46. $y'^2 + axy' = bx^2 + c, \; a^2 + 4b > 0$;

$$y = C - \tfrac{1}{4}ax^2 + \tfrac{1}{4}x \sqrt{[(a^2 + 4b)\, x^2 + 4c]} +$$

$$+ \frac{c}{\sqrt{(4a^2 + b)}} \ln \left[x + \sqrt{\left(x^2 + \frac{4c}{a^2 + 4b} \right)} \right].$$

47. $y'^2 + (ax + b)\, y' - ay + c = 0\,, \quad a \neq 0\,;$

$$y = (ax + b)\, C + aC^2 + \frac{c}{a}\,; \quad 4ay = 4c - (ax + b)^2\,.$$

48. $y'^2 - 2yy' - 2x = 0\,;$

the solution (in parametric form) is

$$x = \frac{t^2}{2} - yt\,, \quad y = \frac{t}{2} + \frac{1}{\sqrt{(t^2 + 1)}}(C - \tfrac{1}{2} \operatorname{arsinh} t)\,.$$

49. $y'^2 - (4y + 1)\, y' + (4y + 1)\, y = 0\,;$

$y = C^2 e^{2x} + Ce^x\,; \quad y = -\tfrac{1}{4}\,.$

50. $y'^2 - xyy' + y^2 \ln ay = 0\,;$

$ay = e^{Cx - C^2}\,; \quad ay = e^{x^2/4}\,.$

51. $y'^2 + 2yy' \cot x - y^2 = 0\,;$

$y(1 \pm \cos x) = C\,.$

52. $ay'^2 + by' - y = 0\,;$

the solution (in parametric form) is

$x = 2at + b \ln |t| + C\,, \quad y = at^2 + bt\,.$

A further solution is $y = 0$.

53. $ay'^2 - yy' - x = 0\,;$

the solution (in parametric form) is

$$x = \frac{t}{\sqrt{(t^2 + 1)}} [\![C + a \ln [t + \sqrt{(t^2 + 1)}]]\!] = \frac{t}{\sqrt{(t^2 + 1)}}(C + a \operatorname{arsinh} t)\,,$$

$$y = at - \frac{x}{t}\,.$$

54. $xy'^2 = y\,;$

$(y - x)^2 - 2C(y + x) + C^2 = 0\,; \quad y = 0\,.$

55. $xy'^2 - 2y + x = 0$;

the solution (in parametric form) is

$$x = \frac{C}{(t-1)^2} e^{2/(t-1)}, \quad y = \frac{x}{2}(t^2 + 1).$$

A further solution is $y = x$.

56. $xy'^2 - 2y' - y = 0$;

the solution (in parametric form) is

$$x = \frac{2t - 2\ln|t| + C}{(t-1)^2}, \quad y = xt^2 - 2t.$$

Further solutions are $y = 0$, $y = x - 2$.

57. $xy'^2 + 4y' - 2y = 0$;

the solution (in parametric form) is

$$x = \frac{2y - 4t}{t^2}, \quad y = \left(\frac{t}{t-2}\right)^2 \left(C + 4\ln|t| + \frac{8}{t}\right).$$

Further solutions are $y = 0$, $y = 2x + 4$.

58. $xy'^2 + xy' - y = 0$;

the solution (in parametric form) is

$$x = Ct^2 e^t, \quad y = C(t+1)e^t.$$

A further solution is $y = 0$.

59. $xy'^2 + yy' + a = 0$;

the solution (in parametric form) is

$$x = \frac{C}{\sqrt{t}} - \frac{a}{3t^2}, \quad y = -C\sqrt{(t)} - \frac{2a}{3t}.$$

60. $xy'^2 + yy' - y^4 = 0$;

$$y(x - C^2) = C.$$

61. $xy'^2 - yy' + a = 0$;

$$y = Cx + \frac{a}{C}; \quad y^2 = 4ax.$$

62. $xy'^2 - yy' + ay = 0$;

the solution (in parametric form) is

$$x = C(t - a)\,e^{-t/a}, \quad y = Ct^2e^{-t/a}.$$

63. $xy'^2 - 2yy' + a = 0, \quad a \neq 0$.

$$16ax^3 - 12x^2y^2 - 12Caxy + 8Cy^3 + C^2a^2 = 0$$

(in parametric form $x = Ct + \dfrac{a}{3t^2}, \quad y = \dfrac{xt}{2} + \dfrac{a}{2t}$);

$$y = \pm \frac{2}{\sqrt{3}}\sqrt{(ax)}.$$

64. $xy'^2 - 2yy' - x = 0$;

$$y = \frac{C}{2}x^2 - \frac{1}{2C}.$$

65. $x^2y'^2 - y^4 + y^2 = 0$;

$$\frac{1}{y} = \sin \ln Cx; \quad y = 0.$$

66. $(x^2 - 1)\,y'^2 = 1$;

$$y = \pm \operatorname{arcosh} x + C.$$

67. $(x^2 - 1)\,y'^2 = y^2 - 1$;

$$x^2 + y^2 - 2Cxy + C^2 = 1; \; y = \pm 1.$$

68. $e^{-2x}y'^2 - (y' - 1)^2 + e^{-2y} = 0$;

$$e^y = Ce^x \pm \sqrt{(1 + C^2)}; \; e^{2y} + e^{2x} = 1.$$

69. $yy'^2 = 1$;

$$4y^3 = 9(x + C)^2.$$

70. $yy'^2 = e^{2x}$;

$$4y^3 = 9(e^x + C)^2.$$

71. $yy'^2 + 2xy' - y = 0$;

$$y^2 = 2Cx + C^2.$$

72. $yy'^2 + 2xy' - 9y = 0$;

the solution (in parametric form) is

$$x = \frac{t}{14} + Ct^{1/8}, \quad y^2 = \frac{t}{9}x + \frac{t^2}{36}.$$

A further solution is $y = 0$.

73. $yy'^2 - 2xy' + y = 0$;

$y^2 = 2Cx - C^2$; $y = \pm x$.

74. $yy'^2 - 4xy' + y = 0$;

$y^6 - 3x^2y^4 + 2Cx(3y^2 - 8x^2) + C^2 = 0$

(in parametric form, $x = y\dfrac{t^2 + 1}{4t}$, $y^3 = \dfrac{C}{t(t^2 - 3)}$).

75. $yy'^2 + x^3y' - x^2y = 0$;

$y^2 + Cx^2 = C^2$.

76. $ayy'^2 + (2x - b)y' - y = 0$;

$y^2 = C(2x - b + aC)$.

Further solutions for $a < 0$ are $\pm 2\sqrt{(-a)}\, y = 2x - b$.

77. $y^2y'^2 + y^2 - a^2 = 0$;

$(x - C)^2 + y^2 = a^2$; $y = \pm a$.

78. $(y^2 - a^2)y'^2 + y^2 = 0$;

$$a \ln \left| \frac{a \pm \sqrt{(a^2 - y^2)}}{y} \right| \mp \sqrt{(a^2 - y^2)} = x + C; \quad y = 0.$$

79. $y'^3 + y' - y = 0$;

the solution (in parametric form) is

$x = C + \frac{3}{2}t^2 + \ln|t|$, $y = t^3 + t$.

A further solution is $y = 0$.

80. $y'^3 + xy' - y = 0$;

$y = Cx + C^3$.

Further solutions for $x < 0$ are $y = \pm 2(-x/3)^{3/2}$.

81. $y'^3 - (x + 5)\, y' + y = 0\,;$

$y = Cx + C(5 - C^2)\,; \quad 27y^2 = 4(x + 5)^3\,.$

82. $y'^3 - axy' + x^3 = 0\,, \quad a \neq 0\,;$

the solution (in parametric form) is

$$x = \frac{at}{t^3 + 1}\,, \quad y = C + \frac{a^2}{6}\,\frac{4t^3 + 1}{(t^3 + 1)^2}\,.$$

83. $y'^3 - 2yy' + y^2 = 0\,;$

the solution (in parametric form) is

$x = C \pm 3\,\sqrt{(1 - t)} + 2 \ln\left[1 \mp \sqrt{(1 - t)}\right], \; y = t\left[1 \pm \sqrt{(1 - t)}\right].$

A further solution is $y = 0$.

84. $y'^3 - axyy' + 2ay^2 = 0\,;$

$$y = \frac{a}{4}\, C(x - C)^2\,; \quad y = \frac{a}{27}\, x^3\,.$$

85. $y'^3 - yy'^2 + y^2 = 0\,;$

the solution (in parametric form) is

$x = t \pm \sqrt[4]{(t^2 - 4t)} \mp \ln\left|\sqrt[4]{(t^2 - 4t)} + t - 2\right| + C\,,$

$y = \frac{1}{2}t^2 \pm \frac{1}{2}t\,\sqrt[4]{(t^2 - 4t)}\,.$

A further solution is $y = 0$.

86. $xy'^3 - yy'^2 + a = 0\,;$

$$y = Cx + \frac{a}{C^2}\,; \quad 4y^3 = 27ax^2\,.$$

87. $ay'^r + by'^s = y \quad (r \neq 1,\; s \neq 1)\,;$

the solution (in parametric form) is

$$x = C + \frac{ar}{r - 1}\, t^{r-1} + \frac{bs}{s - 1}\, t^{s-1}\,, \quad y = at^r + bt^s\,.$$

88. $\sqrt{(y'^2 + 1)} - xy'^2 + y = 0\,;$

the solution (in parametric form) is

$$x = \frac{\sqrt{(t^2 + 1)} - \ln\left[t + \sqrt{(t^2 + 1)}\right] + C}{(t - 1)^2}\,, \quad y = xt^2 - \sqrt{(t^2 + 1)}\,.$$

89. $\ln y' + a(xy' - y) = 0$;

$$y = Cx + \frac{1}{a}\ln C \ (C > 0), \quad ay + 1 + \ln(-ax) = 0 \ (ax < 0).$$

(b) Linear Equations of the Second Order.

90. $y'' + a^2 y = 0 \ (a > 0)$;

$y = C_1 \cos ax + C_2 \sin ax$ (harmonic vibrations).

91. $y'' + a^2 y = b \sin cx \ (a > 0)$;

$$y = \frac{b}{a^2 - c^2}\sin cx + C_1 \cos ax + C_2 \sin ax \quad \text{if} \quad a^2 \neq c^2,$$

$$= -\frac{b}{2c} x \cos cx + C_1 \cos cx + C_2 \sin cx \quad \text{if} \quad a^2 = c^2$$

(undamped forced vibrations).

92. $y'' + a^2 y = b \cos cx$;

$$y = \frac{b}{a^2 - c^2}\cos cx + C_1 \cos ax + C_2 \sin ax \quad \text{if} \quad a^2 \neq c^2,$$

$$= \frac{b}{2c} x \sin cx + C_1 \cos cx + C_2 \sin cx \quad \text{if} \quad a^2 = c^2$$

(undamped forced vibrations).

93. $y'' + y = \sin ax \sin bx$

$$y = \frac{\cos(a - b)x}{2 - 2(a - b)^2} - \frac{\cos(a + b)x}{2 - 2(a + b)^2} + C_1 \cos x + C_2 \sin x$$

$(|a + b| \neq 1, |a - b| \neq 1)$.

94. $y'' - a^2 y = 0$;

$y = C_1 e^{ax} + C_2 e^{-ax} = k_1 \cosh ax + k_2 \sinh ax$.

95. $y'' + \lambda y = 0$;

$$y = C_1 e^{\sqrt{(-\lambda)}x} + C_2 e^{-\sqrt{(-\lambda)}x} = k_1 \cosh \sqrt{(-\lambda)}\, x + k_2 \sinh \sqrt{(-\lambda)}\, x \text{ if } \lambda < 0,$$

$$y = C_1 + C_2 x \quad \text{if} \quad \lambda = 0\,,$$

$$y = C_1 \sin \sqrt{(\lambda)}\, x + C_2 \cos \sqrt{(\lambda)}\, x \quad \text{if} \quad \lambda > 0\,.$$

96. $y'' - (a^2x^2 + a)\, y = 0$;

$$y = e^{ax^2/2}\left(C_1 + C_2 \int e^{-ax^2}\, dx\right).$$

97. $y'' + 2ay' + b^2 y = 0$;

$$y = e^{-ax}(C_1 \cos \alpha x + C_2 \sin \alpha x) \quad \text{if} \quad \alpha^2 = b^2 - a^2 > 0$$

(damped vibrations, see also § 4.13),

$$y = e^{-ax}(C_1 x + C_2) \qquad \text{if} \quad b^2 - a^2 = 0\,,$$

$$y = e^{-ax}(C_1 e^{\beta x} + C_2 e^{-\beta x}) \quad \text{if} \quad \beta^2 = a^2 - b^2 > 0\,.$$

98. $y'' + 2ay' + b^2 y = A \sin \omega x$;

$$y = K \sin(\omega x + \varphi) + y_h\,,$$

where $K = \dfrac{A}{\sqrt{[(b^2 - \omega^2)^2 + 4a^2\omega^2]}}$,

$$\sin \varphi = \frac{-2a\omega}{\sqrt{[(b^2 - \omega^2)^2 + 4a^2\omega^2]}}\,, \qquad \cos \varphi = \frac{b^2 - \omega^2}{\sqrt{[(b^2 - \omega^2)^2 + 4a^2\omega^2]}}$$

and y_h is a solution of the corresponding homogeneous equation (see 97). (Forced damped vibrations; see also § 4.13).

99. $y'' + 2ay' + b^2 y = A \cos \omega x$;

$$y = K \sin(\omega x + \varphi) + y_h\,,$$

where $K = \dfrac{A}{\sqrt{[(b^2 - \omega^2)^2 + 4a^2\omega^2]}}$,

$$\cos \varphi = \frac{2a\omega}{\sqrt{[(b^2 - \omega^2)^2 + 4a^2\omega^2]}}\,, \qquad \sin \varphi = \frac{b^2 - \omega^2}{\sqrt{[(b^2 - \omega^2)^2 + 4a^2\omega^2]}}$$

and y_h is a solution of the corresponding homogeneous equation (see 97). (Forced damped vibrations.)

100. $y'' + xy' + (n+1)\,y = 0$;

$$y = \frac{d^n}{dx^n}\left[e^{-x^2/2}\left(C_1 + C_2 \int e^{x^2/2}\,dx\right)\right].$$

101. $y'' - xy' + 2y = 0$;

$$y = (x^2 - 1)\left(C_1 + C_2 \int \frac{1}{(x^2-1)^2}\, e^{x^2/2}\,dx\right).$$

102. $y'' - xy' + (x-1)\,y = 0$;

$$y = C_1 e^x + C_2 e^x \int e^{(x^2/2) - 2x}\,dx\,.$$

103. $y'' + 4xy' + (4x^2 + 2)\,y = 0$;

$$y = (C_1 + C_2 x)\, e^{-x^2}\,.$$

104. $y'' - 4xy' + (4x^2 - 1)\,y = e^{x^2}$;

$$y = e^{x^2}(1 + C_1 \cos x + C_2 \sin x)\,.$$

105. $y'' - x^2 y' + xy = 0$;

$$y = C_1 x + C_2\left(e^{x^3/3} - x \int x e^{x^3/3}\,dx\right).$$

106. $y'' + \left(\frac{2 - x^2}{4} + n\right) y = 0$;

$$y = C e^{-x^2/4} H_n\left(\frac{x}{\sqrt{2}}\right),$$

where $H_n(x) = (-1)^n e^{x^2} \frac{d^n}{dx^n} e^{-x^2}$ is the Hermite polynomial of degree n.

107. $x(y'' + y) = \cos x$;

$$y = \sin x \int \frac{\cos^2 x}{x}\,dx - \cos x \int \frac{\sin 2x}{2x}\,dx + C_1 \sin x + C_2 \cos x\,.$$

108. $xy'' + y' = 0$;

$$y = C_1 + C_2 \ln|x|\,.$$

109. $xy'' + y' + \lambda y = 0$

with conditions $y(1) = 0$, y bounded for $x \to 0$; the eigenfunctions are $y = C_0 J_0(2\sqrt{(\lambda x)})$, where J_0 is the Bessel function of the first kind (§ 16.4) and the eigenvalues λ are given by the equation $J_0(2\sqrt{\lambda}) = 0$.

110. $xy'' + 2y' - xy = e^x$;

$$y = \tfrac{1}{2}e^x + \frac{1}{x}(C_1 e^x + C_2 e^{-x}).$$

111. $xy'' + 2y' + axy = 0$;

substituting $u = xy$, the equation is transformed into the equation $u'' + au = 0$.

112. $xy'' + (1 - a)\,y' + a^2 x^{2a-1} y = 0$;

$$y = C_1 \cos(x^a + C_2).$$

113. $2xy'' + y' + ay = 0,\ a \neq 0$;

$$y = C_1 \cos\sqrt{(2ax)} + C_2 \sin\sqrt{(2ax)} \quad \text{if} \quad ax > 0,$$

$$= C_1 \cosh\sqrt{(-2ax)} + C_2 \sinh\sqrt{(-2ax)} \quad \text{if} \quad ax < 0.$$

114. $x^2 y'' - 6y = 0$;

$$y = C_1 x^3 + C_2 x^{-2}.$$

115. $x^2 y'' - 12y = 0$;

$$y = C_1 x^4 + C_2 x^{-3}.$$

116. $x^2 y'' + ay = 0$;

$$y = \sqrt{(x)}\,[C_1 \cos(b \ln|x|) + C_2 \sin(b \ln|x|)] \quad \text{if} \quad b^2 = a - \tfrac{1}{4} > 0,$$

$$= \sqrt{(x)}\,(C_1 x^c + C_2 x^{-c}) \quad \text{if} \quad c^2 = \tfrac{1}{4} - a > 0,$$

$$= \sqrt{(x)}\,(C_1 + C_2 \ln|x|) \quad \text{if} \quad \tfrac{1}{4} - a = 0.$$

117. $x^2 y'' + xy' + (x^2 - r^2)\,y = 0$ (*Bessel's equation*; see § 17.15, p. 796);

$y = J_r(x)$ (Bessel function of the first kind, § 16.4, p. 716),

$y = Y_r(x)$ (Bessel function of the second kind, Remark 16.4.12, p. 721, § 17.15, p. 797).

The general integral of Bessel's equation is

$$Z_r(x) = C_1 J_r(x) + C_2 Y_r(x).$$

If r is not an integer, the general integral is

$$Z_r(x) = C_1 J_r(x) + C_2 J_{-r}(x) \text{ (§ 17.15).}$$

By proper transformations, the following equations can be transformed into Bessel's equation:

118. $x^2y'' + xy' - (x^2 + r^2)\,y = 0$ (*modified Bessel's equation*);

$$y = Z_r(\mathrm{i}x) \text{ (see 117)}.$$

119. $x^2y'' + xy' - (x + r^2)\,y = 0$;

$$y = Z_{2r}(2\mathrm{i}\sqrt{x}) \text{ (see 117)}.$$

120. $x^2y'' + xy' + 4(x^4 - r^2)\,y = 0$;

$$y = Z_r(x^2) \text{ (see 117)}.$$

121. $x^2y'' + (1 - 2r)\,xy' + r^2(x^{2r} + 1 - r^2)\,y = 0$;

$$y = x^r Z_r(x^r) \text{ (see 117)}.$$

122. $x^2y'' - [cx^2 + p(p-1)]\,y = 0$;

$$y = \sqrt{(x)}\, Z_{p-\frac{1}{2}}(\mathrm{i}\sqrt{(c)}\,x) \text{ (see 117)}.$$

123. $xy'' + (1 - 2r)\,y' + xy = 0$;

$$y = x^r Z_r(x) \text{ (see 117)}.$$

124. $xy'' - 2py' - cxy = 0$;

$$y = x^{p+\frac{1}{2}} Z_{p+\frac{1}{2}}(\mathrm{i}\sqrt{(c)}\,x) \text{ (see 117)}.$$

125. $y'' - cx^{2p-2}y = 0$;

$$y = \sqrt{(x)}\, Z_{1/2p}\left(\mathrm{i}\sqrt{(c)}\,\frac{x^p}{p}\right) \text{ (see 117)}.$$

126. $y'' + xy = 0$;

$$y = \sqrt{(x)}\, Z_{1/3}(\tfrac{2}{3}x^{3/2}) \text{ (see 117)}.$$

127. $y'' - xy = 0$;

$$y = \sqrt{(x)}\, Z_{1/3}(\tfrac{2}{3}\mathrm{i}x^{3/2}) \text{ (see 117)}.$$

128. $(x^2 - 1)\,y'' + xy' + ay = 0\,;$

$$\left.\begin{aligned} y &= C_1 \cos(\alpha \operatorname{arcosh}|x|) + C_2 \sin(\alpha \operatorname{arcosh}|x|) && \text{if } |x| > 1\,, \\ &= C_1 e^{\alpha \arccos x} + C_2 e^{-\alpha \arccos x} && \text{if } |x| < 1\,, \end{aligned}\right\} a = \alpha^2 > 0\,,$$

$$\left.\begin{aligned} &= C_1 e^{\beta \operatorname{arcosh}|x|} + C_2 e^{-\beta \operatorname{arcosh}|x|} && \text{if } |x| > 1\,, \\ &= C_1 \cos(\beta \arccos x) + C_2 \cos(\beta \arcsin x) && \text{if } |x| < 1\,, \end{aligned}\right\} a = -\beta^2 < 0\,.$$

For $a = -n^2$ (n integral), Chebyshev polynomials

$$y = T_n(x) = 2^{-n+1} \cos(n \arccos x)$$

constitute solutions of 128.

129. $(x^2 - 1)\,y'' + 2xy' - r(r+1)\,y = 0$; *Legendre's equation*, cf. § 17.15, p. 797; by the transformation $x = \cos\xi$, $\eta(\xi) = y(x)$, the equation

130. $\eta'' \sin\xi + \eta' \cos\xi + r(r+1)\,\eta \sin\xi = 0\,,$

also often called *Legendre's equation*, is transformed into 129.

A. If $|x| < 1$ the general integral of equation 129 is

$$y = C_1 F\left(-\frac{r}{2}, \frac{1+r}{2}, \frac{1}{2}, x^2\right) + C_2 x F\left(\frac{1-r}{2}, 1 + \frac{r}{2}, \frac{3}{2}, x^2\right),$$

where $F(\alpha, \beta, \gamma, x)$ is the hypergeometric series (§ 16.6, p. 727).

B. If $|x| > 1$ we define

$$y_r(x) = x^r + \sum_{k=1}^{\infty} (-1)^k \frac{\binom{r}{2k}\binom{r}{k}}{\binom{2r}{2k}} x^{r-2k}\,,$$

$$y_r^*(x) = \lim_{s \to r} \frac{(s-r)\,y_s - \frac{1}{2}\lambda_s(r)\,y_{-s-1}}{s-r}\,,$$

where

$$\lambda_s(r) = (-1)^{r+\frac{1}{2}} \frac{s(s-1)\ldots(s-2r)}{2^{r+\frac{1}{2}}(r+\frac{1}{2})!\,(2s-1)(2s-3)\ldots(2s-2r+2)}\,.$$

The general integral of equation 129 for $|x| > 1$ is then

$y = C_1 y_r + C_2 y_{-r-1}$, if $2r$ is not an odd number,

$ = C_1 y_r^* + C_2 y_{-r-1}$, if $2r = 2p + 1$, $p \geqq 0$ an integer,

$$= C_1 y_r + C_2 y^*_{-r-1}, \text{ if } 2r = -(2n+1), \ n \text{ is a natural number,}$$

$$= C_1 y_{-1/2} + C_2 y^*_{-1/2}, \text{ if } r = -\tfrac{1}{2}.$$

If $r = n$ (n integral), Legendre polynomials (§ 16.5, p. 722).

$$P_n(x) = \frac{1}{2^n n!} \frac{d^n}{dx^n} (x^2 - 1)^n$$

are solutions of equation 129. The following equations may easily be transformed into equation 129. (In equations 131 – 139, $L(x)$ means the solution of Legendre's equation 129.)

131. $(x^2 + 1) y'' + 2xy' - r(r+1) y = 0$;

$y = L(ix)$ (see 129).

132. $(x^2 - 1) y'' + 2(n+1) xy' - (r+n+1)(r-n) y = 0$;

$y = L^{(n)}(x)$ (see 129).

133. $x(x^2 + 1) y'' + (2x^2 + 1) y' - r(r+1) xy = 0$;

$y = L(\sqrt{(x^2 + 1)})$ (see 129).

134. $x(x^2 + 1) y'' + [2(n+1) x^2 + 2n + 1] y' - (r-n)(r+n+1) xy = 0$;

$y = L^{(n)}(\sqrt{(x^2 + 1)}$ (see 129).

135. $x^2(x^2 + 1) y'' + x(2x^2 + 1)y' - [r(r+1) x^2 + n^2] y = 0$;

$y = x^n L^{(n)}(\sqrt{(x^2 + 1)})$ (see 129).

136. $x^2(x^2 - 1) y'' + 2x^3 y' + r(r+1) y = 0$;

$y = L\left(\frac{1}{x}\right)$ (see 129).

137. $x^2(x^2 - 1) y'' + 2x^3 y' - r(r+1)(x^2 - 1) y = 0$;

$y = L\left(\frac{x^2 + 1}{2x}\right)$ (see 129).

138. $x^2(x^2 - 1) y'' + 2x[(1-a) x^2 + a] y' +$

$+ \{[a(a-1) - r(r+1)] x^2 - a(a+1)\} y = 0$;

$y = x^a L(x)$ (see 129).

139. $(x^2 - 1)^2 y'' + 2x(x^2 - 1) y' - [r(r + 1)(x^2 - 1) + n^2] y = 0$

($n \geqq 0$ is an integer);

$y = |x^2 - 1|^{n/2} L^{(n)}(x)$ (see 129).

140. $x(x - 1) y'' + [(\alpha + \beta + 1) x - \gamma] y' + \alpha\beta y = 0$;

the hypergeometric (Gauss's) equation, see § 17.15, p. 797; brief notation:

$H(\alpha, \beta, \gamma, y, x) = 0$.

For $|x| < 1$, the solution is given by the hypergeometric series

$y = F(\alpha, \beta, \gamma, x)$ (see § 16.6, p. 727).

If $0 < x < 1$, the general integral is

$$y = y_1 = C_1 F(\alpha, \beta, \gamma, x) + C_2 x^{1-\gamma} F(\alpha - \gamma + 1, \beta - \gamma + 1, 2 - \gamma, x)$$

if γ is not an integer.

$= C_1 y_1 + C_2 y_2$, if $\gamma = -c$, $c \geqq -1$ being an integer, where

$$y_2 = \lim_{\gamma \to -c} F(\alpha, \beta, \gamma, x) - \frac{\lambda_\gamma}{\gamma + c} x^{1-\gamma} F(\alpha - \gamma + 1, \beta - \gamma + 1, 2 - \gamma, x)$$

if $c \geqq 0$,

$$= \lim_{\gamma \to 1} \frac{1}{\gamma - 1} [F(\alpha, \beta, \gamma, x) - x^{1-\gamma} F(\alpha - \gamma + 1, \beta - \gamma + 1, 2 - \gamma, x)]$$

if $c = -1$, where

$$\lambda_\gamma = \binom{\alpha + c}{c + 1} \frac{\beta(\beta + 1) \ldots (\beta + c)}{\gamma(\gamma + 1) \ldots (\gamma + c - 1)} \quad \text{for} \quad c \geqq 1,$$

$$= \alpha\beta \quad \text{for} \quad c = 0,$$

$$= 1 \quad \text{for} \quad c = -1.$$

The case $\gamma = c$, where $c \geqq 2$ is an integer, can be transformed by the substitution $y(x) = x^{1-\gamma} \eta(x)$ into the above-mentioned case.

In some special cases the solution can be written in a closed form (for notation, see 140):

141. $H(\alpha, \alpha + \frac{1}{2}, 2\alpha + 1, y, x) = 0$, $\alpha \neq 0$;

$y = C_1(1 + \sqrt{(1 - x)})^{-2\alpha} + C_2 x^{-2\alpha}(1 + \sqrt{(1 - x)})^{2\alpha}$ (see 140).

142. $H(\alpha, \alpha - \frac{1}{2}, \frac{1}{2}, y, x) = 0$;

$y = C_1(1 + \sqrt{x})^{1-2\alpha} + C_2(1 - \sqrt{x})^{1-2\alpha}$ (see 140).

143. $H(\alpha, \alpha + \tfrac{1}{2}, \tfrac{3}{2}, y, x) = 0$;

$$y = C_1 \frac{1}{\sqrt{x}} (1 + \sqrt{x})^{1-2\alpha} + C_2 \frac{1}{\sqrt{x}} (1 - \sqrt{x})^{1-2\alpha} \text{ (see 140)}.$$

144. $H(1, \beta, \gamma, y, x) = 0$;

$$y = x^{1-\gamma}(1 - x)^{\gamma-\beta-1}\left[C_1 + C_2 \int x^{\gamma-2}(1 - x)^{\beta-\gamma}\, dx\right] \text{ (see 140)}.$$

145. $H(\alpha, \beta, \alpha, y, x) = 0$;

$$y = (1 - x)^{-\beta}\left[C_1 + C_2 \int x^{-\alpha}(1 - x)^{\beta-1}\, dx\right] \text{ (see 140)}.$$

146. $H(\alpha, \beta, \alpha + 1, y, x,) = 0$;

$$y = x^{-\alpha}\left[C_1 + C_2 \int x^{\alpha-1}(1 - x)^{-\beta}\, dx\right] \text{ (see 140)}.$$

Related equations [notation: $y(\alpha, \beta, \gamma, x)$ is the solution of equation 140]:

147. $x(x - 1)\, y'' + (2x - 1)\, y' - r(r + 1)\, y = 0$;

$$y = y(r + 1, -r, 1, x) \text{ (see 140)}.$$

148. $x(x - 1)\, y'' + [(a + b + 1)\, x + (\alpha + \beta - 1)]\, xy' + (abx - \alpha\beta)\, y = 0$;

$$y = x^{\alpha} y(a + \alpha, b + \alpha, \alpha - \beta + 1, x) \text{ (see 140)}.$$

149. $x(x^2 - 1)\, y'' + (ax^2 + b)\, y' + cxy = 0$;

$$y = y\left(\frac{a - 1}{2} + \sqrt{[\tfrac{1}{4}(a - 1)^2 - c]}, \frac{a - 1}{2} - \sqrt{[\tfrac{1}{4}(a - 1)^2 - c]}, \frac{1 - b}{2}, x^2\right)$$

(see 140).

150. $16(x^3 - 1)^2\, y'' + 27xy = 0$;

$$y = (x^3 - 1)^{1/4}\, y(\tfrac{1}{12}, -\tfrac{1}{4}, -\tfrac{1}{3}, x^3) \text{ (see 140)}.$$

151. $x(x - 1)\, y'' + [(\alpha + \beta + 2n + 1)\, x - (\gamma + n)]\, y' + (\alpha + n)\, (\beta + n)\, y = 0$;

$$y = y^{(n)}(\alpha, \beta, \gamma, x) = y(\alpha + n, \beta + n, \gamma + n, x) \text{ (see 140)}.$$

152. $(x^2 \pm a^2)^2\, y'' + b^2 y = 0$ (the bending flexion of a bar of parabolic cross-section);

$$y = \sqrt{(x^2 + a^2)}\,(C_1 \cos u + C_2 \sin u), \text{ where } u = \frac{\sqrt{(a^2 + b^2)}}{a} \arctan \frac{x}{a}$$

for the sign $+$,

$$= \sqrt{(a^2 - x^2)}\,(C_1 \cos v + C_2 \sin v), \text{ where } v = \frac{\sqrt{(b^2 - a^2)}}{2a} \ln \frac{a + x}{a - x}$$

$(|x| < a)$ for the sign $-$.

153. $(e^x + 1)\, y'' = y$;

$$y = C_1(1 + e^{-x}) + C_2[-1 + (1 + e^{-x}) \ln (1 + e^x)] .$$

154. $xy'' \ln x - y' - xy \ln^3 x = 0$;

$$y = C_1 \left(\frac{x}{e}\right)^x + C_2 \left(\frac{e}{x}\right)^x .$$

155. $y'' \sin x - 2y = 0$;

$$y = C_1 \cot x + C_2(1 - x \cot x) .$$

(c) Linear Equations of Higher Orders. Nonlinear Equations. Systems.

156. $y''' + \lambda y = 0$;

$$y = C_1 + C_2 x + C_3 x^2 \quad \text{for} \quad \lambda = 0 ,$$

$$= C_1 e^{-\sqrt[3]{(\lambda)} x} + e^{\frac{1}{2}\sqrt[3]{(\lambda)} x} \left(C_2 \cos \frac{\sqrt{3}}{2} \sqrt[3]{(\lambda)}\, x + C_3 \sin \frac{\sqrt{3}}{2} \sqrt[3]{(\lambda)}\, x \right) \text{ for } \lambda \neq 0 .$$

157. $y''' + 3y' - 4y = 0$;

$$y = C_1 e^x + \left(C_2 \cos \frac{\sqrt{15}}{2} x + C_3 \sin \frac{\sqrt{15}}{2} x \right) e^{-x/2} .$$

158. $y''' - a^2 y' = e^{2ax} \sin^2 x$;

$$y = C_1 + C_2 e^{ax} + C_3 e^{-ax} +$$

$$+ \left(\frac{1}{12a^3} + \frac{(4 - 11a^2) \sin 2x + 3a(4 - a^2) \cos 2x}{4(a^2 + 1)(a^2 + 4)(9a^2 + 4)} \right) e^{2ax} .$$

159. $y''' - 2y'' - a^2y' + 2a^2y = \sinh x$;

$$y = C_1 e^{2x} + C_2 e^{ax} + C_3 e^{-ax} + \frac{2\sinh x + \cosh x}{3(a^2 - 1)} \quad \text{for} \quad a^2 \neq 1,\ a^2 \neq 4,$$

$$= e^{2x}(C_1 x + C_2) + C_3 e^{-2x} + \frac{2\sinh x + \cosh x}{9} \quad \text{for} \quad a^2 = 4,$$

$$= C_1 e^{2x} + C_2 e^{ax} + C_3 e^{-ax} - \frac{x+1}{4} e^{x} - \frac{3x+1}{36} e^{-x} \quad \text{for} \quad a^2 = 1 .$$

160. $y^{(4)} = 0$;

$$y = C_0 + C_1 x + C_2 x^2 + C_3 x^3 .$$

161. $y^{(4)} + 4y = f(x)$;

$$y = C_1 \cos x \cosh x + C_2 \cos x \sinh x + C_3 \sin x \cosh x + C_4 \sin x \sinh x + y_p,$$

where y_p is a particular integral of equation 161.

162. $y^{(4)} - k^4 y = 0$ (transverse vibrations of a rod); solutions are given in (1)–(6) below for various sets of boundary conditions:

(1) $y(a) = y'(a) = y(b) = y'(b) = 0$;

the eigenvalues are to be calculated from the equation

$$\cos k(b-a) \cosh k(b-a) = 1 , \quad k \neq 0 ;$$

the eigenfunctions are:

$$y = [\cosh k(b-a) - \cos k(b-a)]\,[\sinh k(x-a) - \sin k(x-a)] - [\sinh k(b-a) - \sin k(b-a)]\,[\cosh k(x-a) - \cos k(x-a)] .$$

(2) $y(a) = y'(a) = y(b) = y''(b) = 0$;

the eigenvalues are to be calculated from the equation

$$\cos k(b-a) \sinh k(b-a) = \sin k(b-a) \cosh k(b-a) ;$$

for the eigenfunctions see (1) (the preceding case).

(3) $y(a) = y'(a) = y''(b) = y'''(b) = 0$;

the eigenvalues are to be calculated from the equation

$$\cos k(b-a) \cosh k(b-a) = -1 ;$$

the eigenfunctions are:

$$y = [\cosh k(b-a) + \cos k(b-a)]\,[\sinh k(x-a) - \sin k(x-a)] - [\sinh k(b-a) + \sin k(b-a)]\,[\cosh k(x-a) - \cos k(x-a)] .$$

(4) $y(a) = y''(a) = y(b) = y''(b) = 0$;

eigenvalues: $k = \dfrac{n\pi}{b - a}$, $n = 1, 2, 3, \ldots$;

eigenfunctions: $y = \sin k(x - a)$.

(5) $y(a) = y''(a) = y''(b) = y'''(b) = 0$;

the eigenvalues are to be calculated from the equation

$\cos k(b - a) \sinh k(b - a) = \sin k(b - a) \cosh k(b - a)$;

the eigenfunctions are:

$y = \sin k(b - a) \sinh k(x - a) + \sinh k(b - a) \sin k(x - a)$.

(6) $y''(a) = y'''(a) = y''(b) = y'''(b) = 0$;

the eigenvalues are to be calculated from the equation

$\cos k(b - a) \cosh k(b - a) = 1$;

the eigenfunctions are:

$y = C_1 + C_2 x$ if $k = 0$; for $k \neq 0$, they are:

$$y = [\sinh k(b - a) - \sin k(b - a)] [\cosh k(x - a) + \cos k(x - a)] -$$
$$- [\cosh k(b - a) - \cos k(b - a)] [\sinh k(x - a) + \sin k(x - a)].$$

163. $y^{(5)} + 2y''' + y' = ax + b \sin x + c \cos x$;

$$y = \frac{a}{2} x^2 + \frac{b}{8} x^2 \cos x - \frac{c}{8} x^2 \sin x +$$
$$+ C_1 + C_2 \sin x + C_3 \cos x + C_4 x \sin x + C_5 x \cos x .$$

164. $y^{(6)} + y = \sin \frac{3}{2}x \sin \frac{1}{2}x$;

$$y = \frac{x}{12} \sin x + \frac{1}{126} \cos 2x + C_1 \cos (x + C_2) + C_3 e^{x\sqrt{(3)}/2} \cos \left(\frac{x}{2} + C_4\right) +$$
$$+ C_5 e^{-x\sqrt{(3)}/2} \cos \left(\frac{x}{2} + C_6\right).$$

165. $x^{2n} y^{(n)} = ay$, $a \neq 0$;

$y = x^{n-1} e^{-r/x}$, where $r^n = a$; the n different roots of this equation yield n linearly independent solutions of equation 165.

166. $y'' = a(y'^2 + 1)^{3/2}$;

$$y = C_1 - \sqrt{\left[\frac{1}{a^2} - (x - C_2)^2\right]}.$$

167. $y'' = 2ax(y'^2 + 1)^{3/2}$;

$$y = C_1 + \int \frac{ax^2 + C_2}{\sqrt{[1 - (ax^2 + C_2)^2]}}\,\mathrm{d}x.$$

168. $8y'' + 9y'^4 = 0$;

$(y + C_1)^3 = (x + C_2)^2$.

169. $2y'y''' - 3y''^2 = 0$;

$$y = \frac{C_1x + C_2}{C_3x + C_4}.$$

170. $x'(t) = ay,\ y'(t) = bx,\ a \neq 0,\ b \neq 0$;

$x = C_1 a\mathrm{e}^{\sqrt{(ab)}t} + C_2 a\mathrm{e}^{-\sqrt{(ab)}t},\ y = C_1\sqrt{(ab)}\,\mathrm{e}^{\sqrt{(ab)}t} - C_2\sqrt{(ab)}\,\mathrm{e}^{-\sqrt{(ab)}t}$

if $ab > 0$,

$x = C_1 a \cos\sqrt{(-ab)}\,t + C_2 a \sin\sqrt{(-ab)}\,t$,

$y = C_2\sqrt{(-ab)}\cos\sqrt{(-ab)}t - C_1\sqrt{(-ab)}\sin\sqrt{(-ab)}t$ if $ab < 0$.

171. $x'(t) = ax - y,\ y'(t) = x + ay$;

$x = \mathrm{e}^{at}(C_1 \sin t + C_2 \cos t),\ y = \mathrm{e}^{at}(C_2 \sin t - C_1 \cos t)$

(see § 17.18) .

172. $ax'(t) + by'(t) = \alpha x + \beta y,\ bx'(t) - ay'(t) = \beta x - \alpha y$;

$x = \mathrm{e}^{At}(C_1 \cos Bt + C_2 \sin Bt),\ y = \mathrm{e}^{At}(C_2 \cos Bt - C_1 \sin Bt)$

if $a\beta - b\alpha \neq 0,\ A = \dfrac{a\alpha + b\beta}{a^2 + b^2},\quad B = \dfrac{a\beta - b\alpha}{a^2 + b^2}$;

if $a\beta - b\alpha = 0,\ a^2 + b^2 > 0$, we get

$x = C_1\mathrm{e}^{\lambda t},\ y = C_2\mathrm{e}^{\lambda t}$, where $\alpha = \lambda a,\ \beta = \lambda b$.

173. $x'(t) = -y,\ y'(t) = 2x + 2y$;

$x = \mathrm{e}^t(C_1 \sin t + C_2 \cos t),\ y = \mathrm{e}^t[(C_2 - C_1)\sin t - (C_2 + C_1)\cos t]$.

174. $x'(t) + 3x + 4y = 0,\ y'(t) + 2x + 5y = 0$;

$x = 2C_1\mathrm{e}^{-t} + C_2\mathrm{e}^{-7t},\ y = -C_1\mathrm{e}^{-t} + C_2\mathrm{e}^{-7t}$.

175. $x'(t) = -5x - 2y,\ y'(t) = x - 7y;$

$$x = (2C_1 \cos t + 2C_2 \sin t)\,\mathrm{e}^{-6t},\ y = [(C_1 - C_2)\cos t + (C_1 + C_2)\sin t]\,\mathrm{e}^{-6t}.$$

176. $x'(t) + 2y = 3t,\ y'(t) - 2x = 4;$

$$x = -\tfrac{5}{4} + C_1 \cos 2t - C_2 \sin 2t\,,$$

$$y = \tfrac{3}{2}t + C_1 \sin 2t + C_2 \cos 2t\,.$$

177. $x'(t) + y = t^2 + 6t + 1,\ y'(t) - x = -3t^2 + 3t + 1;$

$$x = 3t^2 - t - 1 + C_1 \cos t + C_2 \sin t\,,$$

$$y = t^2 + 2 + C_1 \sin t - C_2 \cos t\,.$$

178. $x'(t) + y'(t) - y = \mathrm{e}^t,\ 2x'(t) + y'(t) + 2y = \cos t;$

$$x = \mathrm{e}^t + \tfrac{5}{17}\sin t - \tfrac{3}{17}\cos t + C_1 + 3C_2\mathrm{e}^{4t}\,,$$

$$y = -\tfrac{2}{3}\mathrm{e}^t - \tfrac{1}{17}\sin t + \tfrac{4}{17}\cos t - 4C_2\mathrm{e}^{4t}\,.$$

179. $x'(t) = x\,f(t) + y\,g(t),\ y'(t) = -x\,g(t) + y\,f(t);$

$$x = (C_1 \cos G + C_2 \sin G)\,F,\ y = (-C_1 \sin G + C_2 \cos G)\,F\,,$$

where $F = \mathrm{e}^{\int f(t)\mathrm{d}t},\quad G = \int g(t)\,\mathrm{d}t\,.$

180. $tx'(t) + y = 0,\ ty'(t) + x = 0;$

$$x = C_1 t + \frac{C_2}{t},\quad y = -C_1 t + \frac{C_2}{t}\,.$$

181. $tx'(t) + 2x = t,\ ty'(t) - (t + 2)\,x - ty = -t;$

$$x = \frac{t}{3} + \frac{C_1}{t^2},\quad y = -x + C_2\mathrm{e}^t\,.$$

182. $tx'(t) + 2(x - y) = t\,,\ ty'(t) + x + 5y = t^2;$

$$x = \frac{3t}{10} + \frac{t^2}{15} + 2\frac{C_1}{t^3} + \frac{C_2}{t^4},\quad y = -\frac{t}{20} + \frac{2t^2}{15} - \frac{C_1}{t^3} - \frac{C_2}{t^4}\,.$$

183. $x''(t) + a^2 y = 0,\ y''(t) - a^2 x = 0,\ a \neq 0;$

$$x = (C_1 \cos \alpha t + C_2 \sin \alpha t)\,\mathrm{e}^{\alpha t} + (C_3 \cos \alpha t + C_4 \sin \alpha t)\,\mathrm{e}^{-\alpha t}\,,$$

$$y = (C_1 \sin \alpha t - C_2 \cos \alpha t)\,\mathrm{e}^{\alpha t} + (-C_3 \sin \alpha t + C_4 \cos \alpha t)\,\mathrm{e}^{-\alpha t}\,,$$

where $2\alpha^2 = a^2$.

184. $x''(t) - ay'(t) + bx = 0,\ y''(t) + ax'(t) + by = 0,\ a^2 + 4b > 0$;

$$x = C_1 \cos \alpha t + C_2 \sin \alpha t + C_3 \cos \beta t + C_4 \sin \beta t\,,$$

$$y = -C_1 \sin \alpha t + C_2 \cos \alpha t - C_3 \sin \beta t + C_4 \cos \beta t\,,$$

where $2\alpha = a + \sqrt{(a^2 + 4b)},\ 2\beta = a - \sqrt{(a^2 + 4b)}$.

185. $x''(t) - x'(t) + y'(t) = 0,\ x''(t) + y''(t) - x = 0$;

$$x = C_1 e^t + C_2 \alpha e^{\alpha t} + C_3 \beta e^{\beta t},\ y = C_4 - C_2 e^{\alpha t} - C_3 e^{\beta t}\,,$$

where $2\alpha = 1 + \sqrt{5},\ 2\beta = 1 - \sqrt{5}$.

186. $x''(t) = kx/r^3,\ y''(t) = ky/r^3$ ($r^2 = x^2 + y^2$; motion of a particle in a central gravitational field);
on substituting $x = r \cos \varphi,\ y = r \sin \varphi$, we get

$$r^2 \varphi' = C_1,\ r'^2 + r^2 \varphi'^2 = -\frac{2k}{r} + C_2 \left(\varphi' = \frac{d\varphi}{dt},\quad r' = \frac{dr}{dt}\right);$$

$$r[C \cos(\varphi - \varphi_0) - k] = C_1^2 \quad (C^2 = C_2 C_1^2 + k^2)\,,$$

which is the equation of a conic.

18. PARTIAL DIFFERENTIAL EQUATIONS

By KAREL REKTORYS

References: [25], [53], [60], [61], [75], [83], [91], [98], [124], [137], [141], [153], [154], [160], [193], [195], [205], [206], [207], [208], [212], [219], [253], [282], [289], [292], [300], [310], [317], [367], [371], [373], [374], [375], [388], [398], [404], [407], [426], [439].

INFORMATIVE REMARK. The solution of problems in partial differential equations, especially boundary value problems for higher order equations, is usually much more complicated than the solution of corresponding problems for ordinary differential equations. Attempts to find the solution in a closed form have been successful only in very simple cases. At the present time special attention is therefore being paid to numerical methods and their rate of convergence, to error estimates and to questions regarding the most convenient method for the solution of a given problem, etc. The numerical solution of differential equations is dealt with in Chapters 24–28. When studying higher order equations in this chapter we shall investigate the characteristic properties of solutions (and hence shall be dealing with rather theoretical topics which, however, are also of great practical importance). For each important type of equation we shall give a survey of the most suitable numerical methods, making reference to the appropriate chapters. Solutions to some simple problems are given in closed form.

With regard to problems involving equations of *first* order (§ 18.2), almost all cases encountered in applications are solved here.

18.1. Partial Differential Equations in General. Basic Concepts. Questions Concerning the Concept of General Solution. Cauchy's Problem, Boundary Value Problems, Mixed Problems. Kovalewski's Theorem. Characteristics. Well-posed Problems. Generalized Solutions

Definition 1. We shall refer to an equation of the form

$$F\left(x_1, x_2, \ldots, x_n, z, \frac{\partial z}{\partial x_1}, \ldots, \frac{\partial z}{\partial x_n}, \frac{\partial^2 z}{\partial x_1^2}, \ldots, \frac{\partial^k z}{\partial x_1^k}, \ldots\right) = 0 \qquad (1)$$

which relates the unknown function $z(x_1, x_2, \ldots, x_n)$ $(n \geqq 2)$ and its derivatives as a *partial differential equation.*

Definition 2. The highest order of the derivatives which appears in the equation is called the *order* of the equation.

REMARK 1. More generally, a *system of partial differential equations* for unknown functions $z_1, z_2, \ldots, z_r$ may be considered. If the number of equations is greater than the number of functions to be found, then, generally speaking, there is no system of functions satisfying all equations; these equations are not compatible. Certain conditions, the so-called *conditions of integrability*, must be satisfied in order that the solution may exist (see e.g. Remark 18.7.2, p. 911).

Example 1. The equation

$$\frac{\partial^2 z}{\partial t^2} = \frac{\partial^2 z}{\partial x^2} + \frac{\partial^2 z}{\partial y^2}$$

is a partial differential equation of the second order for the unknown function $z(x, y, t)$.

Example 2. The system

$$\frac{\partial z_1}{\partial t} = \frac{\partial z_1}{\partial x} - \frac{\partial^2 z_2}{\partial x^2}, \quad \frac{\partial z_2}{\partial t} = \frac{\partial z_2}{\partial x} + \frac{\partial^2 z_1}{\partial x^2}$$

is a system of partial differential equations (of the second order) for two unknown functions $z_1(x, t)$, $z_2(x, t)$.

Definition 3. Any function $z(x_1, x_2, \ldots, x_n)$ is called a *solution* (or *integral*) of equation (1) (in a given domain) if, on substituting in (1) for z and for its derivatives of the required order, equation (1) is satisfied identically (i.e. for all points $(x_1, x_2, \ldots, x_n)$ of the domain in question).

Definition 4. Similarly, any system of r functions having derivatives of the required orders and satisfying identically all equations of the given system is called a *solution of the system of equations for r unknown functions.*

As in the theory of ordinary differential equations, we may speak about a solution in an implicit form (see Remark 17.2.3, p. 732).

REMARK 2. In the same way as in the theory of ordinary differential equations, *to solve a given partial differential equation* means to determine *all* its solutions. In ordinary differential equations it is possible to find a general solution from which (in the region considered) any other solution may be obtained by a suitable choice of these constants. By contrast, this situation does not occur in the theory of partial differential equations. For some equations only, the most general form of the solution can be found; a detailed analysis shows that this general solution depends on one or more arbitrary functions. However, in the general case it is impossible to find a general solution of a given equation such that any solution can be obtained from it by specifying one or more of the functions involved.

Example 3. On integrating the equation

$$\frac{\partial^2 z}{\partial x\, \partial y} = 0\,, \tag{2}$$

with respect to y, we obtain

$$\frac{\partial z}{\partial x} = f(x)$$

and by a further integration, with respect to x,

$$z = \int f(x)\, \mathrm{d}x + g(y) = F(x) + g(y)\,. \tag{3}$$

This is the *general* solution of equation (1) which depends on two "arbitrary" functions F and g. Of course, one cannot conclude from this example that all solutions of a partial differential equation of the second order may be obtained from a certain function involving two arbitrary functions by a suitable choice of these functions. It can be shown that under certain assumptions equations of the *first order* do possess a general solution depending on an arbitrary function (Remark 18.2.8).

REMARK 3. The analogue of the Cauchy problem for ordinary differential equations (i.e. of the initial value problem), or of the boundary value problem, is here again the *Cauchy problem* and *boundary value problem*, respectively. One often meets with so-called *mixed problems* where initial conditions and boundary conditions are prescribed simultaneously.

Definition 5 (the *Special Cauchy Problem*; for important particular cases see Examples 4 and 5). Suppose we are given the equation

$$\frac{\partial^m z}{\partial x_1^m} = f\left(x_1, x_2, \ldots, x_n, z, \ldots, \frac{\partial^k z}{\partial x_1^{k_1}\, \partial x_2^{k_2} \ldots \partial x_n^{k_n}}, \ldots\right), \tag{4}$$

$$k_1 + k_2 + \ldots + k_n = k \leqq m\,, \quad k_1 < m\,. \tag{5}$$

The *special Cauchy problem* is to find a solution of equation (4) (in a certain region of the variables $x_1, \ldots, x_n$) that satisfies the initial conditions

$$\begin{aligned} z(x_1^0, x_2, \ldots, x_n) &= f_0(x_2, \ldots, x_n), \\ \frac{\partial z}{\partial x_1}(x_1^0, x_2, \ldots, x_n) &= f_1(x_2, \ldots, x_n), \\ &\cdots\cdots \\ \frac{\partial^{m-1} z}{\partial x_1^{m-1}}(x_1^0, x_2, \ldots, x_n) &= f_{m-1}(x_2, \ldots, x_n). \end{aligned} \tag{6}$$

REMARK 4. The variable x_1 (which in applications often denotes time and is then usually denoted by t) plays a special rôle in equation (4) and initial conditions (6); particularly, the equation is assumed to be explicitly solved with respect to $\partial z^m/\partial x_1^m$ which is the highest derivative with respect to x_1 occurring in the equation.

Example 4. An example of the Cauchy problem is finding the solution of the equation

$$\frac{\partial^2 z}{\partial x^2} = f\left(x, y, z, \frac{\partial z}{\partial x}, \frac{\partial z}{\partial y}, \frac{\partial^2 z}{\partial x\, \partial y}, \frac{\partial^2 z}{\partial y^2}\right) \tag{7}$$

subject to the conditions

$$z(x_0, y) = f_0(y), \quad \frac{\partial z}{\partial x}(x_0, y) = f_1(y). \tag{8}$$

Example 5. An other example is the equation of the first order

$$\frac{\partial z}{\partial x} = f\left(x, y, z, \frac{\partial z}{\partial y}\right), \tag{9}$$

with the initial condition

$$z(x_0, y) = f_0(y). \tag{10}$$

REMARK 5. In the case of the problem (9), (10) for equations of the first order we say that we are looking for a solution (i.e. for a surface $z = h(x, y)$) that passes through the curve

$$x = x_0, \quad z = f_0(y). \tag{11}$$

For an equation of the second order (Example 4) a further condition is attached, namely that the first derivative of the function z with respect to x is a prescribed function of y.

REMARK 6. If we choose a point $(x_1^0, x_2^0, \ldots, x_n^0)$, then all the derivatives occurring on the right-hand side of equation (4) and containing a derivative with respect to x_1, are determined, at this point, by conditions (6). For example (see Example 4; the interchangeability of mixed derivatives is assumed)

$$\frac{\partial^2 z}{\partial y\, \partial x}(x_0, y_0) = \left[\frac{\partial}{\partial y}\frac{\partial z}{\partial x}\right]_{\substack{x=x_0\\y=y_0}} = \left[\frac{\mathrm{d}}{\mathrm{d}y}\frac{\partial z}{\partial x}(x_0, y)\right]_{y=y_0} = f_1'(y_0)\,. \tag{12}$$

Let us write

$$\left(\frac{\partial^{k-k_1} f_{k_1}}{\partial x_2^{k_2}\, \partial x_3^{k_3} \ldots \partial x_n^{k_n}}\right)_{x_i = x_i{}^0} = f^0_{k_1,k_2,\ldots,k_n} \quad (i = 2, \ldots, n;\ k_1 + k_2 + \ldots + k_n = k)\,. \tag{13}$$

For instance, in (12) we have $f_1'(y_0) = f^0_{1,1}$ since $k = 2$, $k_1 = 1$, $k_2 = 1$, $f_{k_1} = f_1(y)$.

Theorem 1 (*Kovalewski's Theorem*). *Let the function f appearing in the equation* (4) *be analytic with respect to all variables in the neighbourhood of the point*

$$(x_1^0, x_2^0, \ldots, x_n^0, \ldots, f^0_{k_1,k_2,\ldots,k_n,\ldots}) \tag{14}$$

(i.e. *the function can be developed into a power series in this neighbourhood*) *and let the functions $f_0, f_1, \ldots, f_{m-1}$* (see Definition 5) *be analytic in the neighbourhood of the point $(x_2^0, \ldots, x_n^0)$. Then there exists an analytic solution of the Cauchy problem in a neighbourhood of the point $(x_1^0, x_2^0, \ldots, x_n^0)$ and this solution is unique in the class of analytic functions.*

REMARK 7. Thus, in Example 4, we investigate the function f at the point (see (12) and (13))

$$(x_0, y_0, f^0_{0,0}, f^0_{1,0}, f^0_{0,1}, f^0_{1,1}, f^0_{0,2})\,, \tag{15}$$

where, by (13), we have

$$f^0_{0,0} = f_0(y_0)\,,\ f^0_{1,0} = f_1(y_0)\,,\ f^0_{0,1} = f_0'(y_0)\,,\ f^0_{1,1} = f_1'(y_0)\,,\ f^0_{0,2} = f_0''(y_0)\,.$$

REMARK 8. A typical example of a Cauchy problem is that of solving the equation of a vibrating string

$$\frac{\partial^2 z}{\partial x^2} = a^2 \frac{\partial^2 z}{\partial y^2} \tag{16}$$

(here x denotes time, and y denotes the space variable), subject to the initial conditions

$$z(0, y) = f_0(y) \quad \text{(the initial position of the string)}\,,$$

$$\frac{\partial z}{\partial x}(0, y) = f_1(y) \quad \text{(the initial velocity of the string)}$$

(see § 26.1). Here (as in the majority of cases in practice) the right-hand side of equation (4) is of a very special form; consequently the investigation of the function f at the point (14) can be omitted and the existence of an analytic solution is ensured in in the neighbourhood of the point $(0, y_0)$, provided only that the functions f_0 and f_1 are analytic in the neighbourhood of the point y_0.

REMARK 9. For generalizations of Theorem 1 to systems of equations see, e.g., [317]. The reader will also find there a statement of conditions (which are almost always satisfied in practice) which ensure the uniqueness, but not the existence, of the solution in the class of non-analytic functions.

REMARK 10 (The *Generalized Cauchy Problem*). In this problem, the initial conditions are not prescribed on an $(n-1)$-dimensional hyperplane $x_1 = x_1^0$ (e.g., for $n = 2$, on a straight line) but on an $(n-1)$-dimensional surface (for $n = 2$, on a curve). (For details see [317].) This problem can, in general, be reduced to the previous one by a suitable transformation of coordinates. This reduction, however, is not practicable for some surfaces (curves) typical for the given differential equation; these surfaces (curves) are called *characteristics*. (For a strict definition see [317].) The direction of the normal to such a surface (curve) at a given point is called a *characteristic direction*. In the case of *linear* equations, the direction cosines $\alpha_1, \alpha_2, \ldots, \alpha_n$ of a characteristic direction are determined as follows: Let the equation be of the form

$$\sum_{k_1+k_2+\ldots+k_n=m} A^{(m)}_{k_1,k_2,\ldots,k_n}(x_1, x_2, \ldots, x_n) \frac{\partial^m z}{\partial x_1^{k_1} \partial x_2^{k_2} \ldots \partial x_n^{k_n}} + \ldots = 0 \tag{17}$$

(writing only the highest order terms). Then the direction cosines are given by the equations

$$\sum_{k_1+k_2+\ldots+k_n=m} A^{(m)}_{k_1,k_2,\ldots,k_n}(x_1, x_2, \ldots, x_n)\, \alpha_1^{k_1}\alpha_2^{k_2} \ldots \alpha_n^{k_n} = 0\,, \tag{18}$$

$$\alpha_1^2 + \alpha_2^2 + \ldots + \alpha_n^2 = 1\,. \tag{19}$$

Characteristics of some important equations with constant coefficients.

Example 6.

$$\frac{\partial^2 z}{\partial x^2} - \frac{\partial^2 z}{\partial y^2} = 0\,. \tag{20}$$

Equations (18) and (19) now become $(m = 2,\ A^{(2)}_{2,0} = 1,\ A^{(2)}_{0,2} = -1,\ A^{(2)}_{1,1} = 0)$

$$\alpha_1^2 - \alpha_2^2 = 0\,,\quad \alpha_1^2 + \alpha_2^2 = 1 \Rightarrow \alpha_1 = \pm\frac{\sqrt{2}}{2}\,,\quad \alpha_2 = \pm\frac{\sqrt{2}}{2}\,.$$

The characteristics run perpendicular to these direction; hence they are parallel to the straight lines $y = \pm x$ (Fig. 18.1).

Example 7.

$$\frac{\partial z}{\partial x} - \frac{\partial^2 z}{\partial y^2} = 0\,. \tag{21}$$

Equations (18) and (19) become ($m = 2$, $A^{(2)}_{2,0} = 0$, $A^{(2)}_{0,2} = -1$, $A^{(2)}_{1,1} = 0$):

$$\alpha_2^2 = 0\,, \quad \alpha_1^2 + \alpha_2^2 = 1 \Rightarrow \alpha_2 = 0\,, \quad \alpha_1 = \pm 1\,.$$

The characteristic direction is parallel to the x-axis, so that the characteristics are parallel to the y-axis (Fig. 18.2).

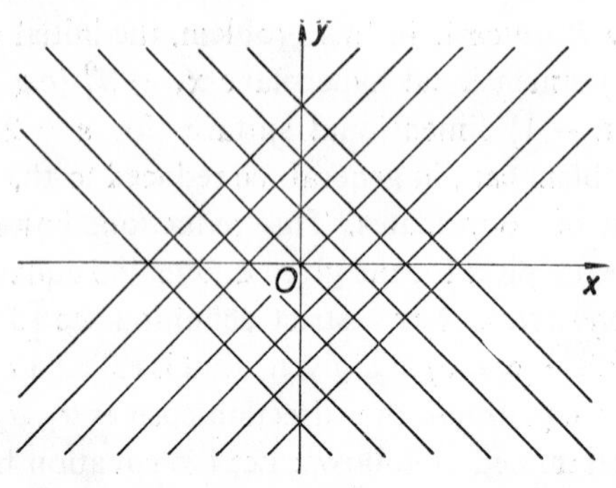

Fig. 18.1.

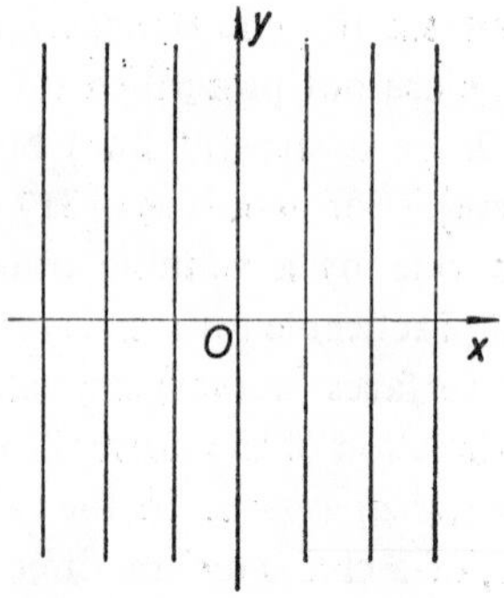

Fig. 18.2.

Example 8.

$$\frac{\partial^2 z}{\partial x^2} + \frac{\partial^2 z}{\partial y^2} = 0\,. \tag{22}$$

$\alpha_1^2 + \alpha_2^2 = 0$, $\alpha_1^2 + \alpha_2^2 = 1 \Rightarrow$ real characteristics do not exist.

Example 9.

$$\frac{\partial z}{\partial x} = 0\,. \tag{23}$$

Equations (18) and (19) now become ($m = 1$, $A^{(1)}_{1,0} = 1$, $A^{(1)}_{0,1} = 0$):

$$\alpha_1 = 0\,, \quad \alpha_1^2 + \alpha_2^2 = 1 \Rightarrow \alpha_1 = 0\,, \quad \alpha_2 = \pm 1\,.$$

The characteristics are parallel to the x-axis.

REMARK 11. For *nonlinear* equations the problem of characteristics is made more complicated by the fact that the characteristics depend not only on the given equation but also on the given solution. For the *quasilinear equation of the first order*

$$a(x, y, z)\,\frac{\partial z}{\partial x} + b(x, y, z)\,\frac{\partial z}{\partial y} = c(x, y, z)$$

the characteristic direction at the point (x, y, z) is given by

$$a(x, y, z)\,\alpha_1 + b(x, y, z)\,\alpha_2 = 0\,, \quad \alpha_1^2 + \alpha_2^2 = 1\,,$$

so that at the given point (x, y) it depends on the particular solution $z(x, y)$.

Theorem 2. *If the initial values are given on a characteristic, then the Cauchy problem has either an infinite number of solutions, or it does not possess any solution.*

Example 10. A solution of the equation

$$\frac{\partial^2 z}{\partial x^2} - \frac{\partial^2 z}{\partial y^2} = 0\,,$$

which takes zero values on the straight line $y = x$ (which is a characteristic; see Example 6) and satisfies also

$$\frac{\partial z}{\partial x} = 0$$

on this line, is

$$z = k(x - y)^2$$

(for any value of k). Thus we have an infinite number of solutions.

Example 11. Find the solution of the equation

$$\frac{\partial z}{\partial x} = 0 \tag{24}$$

such that $z = x$ at points on the x-axis (which is a characteristic; see Example 9). It is readily verified that such a function $z(x, y)$ does not exist, for

$$\frac{\partial z}{\partial x} = 1$$

on the x-axis, while the equation (24) should be satisfied everywhere.

REMARK 12. In applications, the Cauchy problem arises most frequently in the theory of equations of the first order and of hyperbolic and parabolic equations of the second order (§§ 18.5 and 18.6). Boundary value problems arise in connection with many equations, and especially with equations of mathematical physics (where mixed problems are often encountered). An example is the problem of finding a solution of the equation

$$\frac{\partial^2 u}{\partial x^2} + \frac{\partial^2 u}{\partial y^2} = 0\,, \tag{25}$$

taking prescribed values on the boundary Γ of a region O in which equation (25) is to be satisfied (the so-called *Dirichlet problem*). Problems of this kind will be formulated and analysed separately for individual types of equations.

REMARK 13. It is important that the given problem be "well posed". This means – roughly speaking – the following: The problem is *well posed* if a slight change in the boundary and initial conditions produces only a slight change in the solution (we then speak of the *continuous dependence of the solution on the initial and boundary conditions*; physically: if we make only a small error in measuring the boundary conditions or the initial conditions, then the solution will be almost unchanged). What we mean by a "well-posed" problem will be specified for individual types of equations separately. Next we state a definition (one of many possible ones) of a well-posed Cauchy problem for linear equations.

Definition 6. Let the Cauchy problem (4), (6) be given (see Definition 5). Let $\bar{G}$ be a closed region in the space $x_1, x_2, \ldots, x_n$, the boundary of which contains the $(n-1)$-dimensional region O of the hyperplane $x_1 = x_1^0$. For any system of functions $f_0, f_1, \ldots, f_{m-1}$ which are sufficiently smooth in O let there be one and only one solution of the given problem. Suppose that, for every $\varepsilon > 0$, there is a $\delta > 0$ such that the change in values of the solution and of its derivatives up to the order k is less than ε in G if the change in the values of the functions $f_0, f_1, \ldots, f_{m-1}$ and of their derivatives up to the order p is less than δ in O. Then the Cauchy problem for the given equation is said to be *well posed* (more fully: *well posed* (p, k)).

REMARK 14. J. Hadamard proved (see e.g. [317]) that the Cauchy problem for the equation

$$\frac{\partial^2 z}{\partial x^2} + \frac{\partial^2 z}{\partial y^2} = 0$$

is not well posed. It will be shown that almost all the "reasonable" problems arising in physics (the so-called *problems of mathematical physics*) are well posed.

REMARK 15. It will be shown (Remark 18.5.7, p. 904) that: If the boundary conditions or initial conditions are not smooth enough, then, in the case of some types of equations, the classical solution need not exist even if the coefficients are very smooth. That is to say, it is not always possible to find a function having derivatives of appropriate orders and satisfying the boundary conditions and also the given equation in the region considered. Because of this, the so-called *generalized solutions* (see e.g. Definition 18.5.3, p. 904, and also Remark 18.4.24) are introduced. It can be shown that, for some types of equations, the generalized solutions (defined in a similar way as in the definition quoted above) satisfy the equation in the usual sense. For instance the equations

$$\Delta u = 0 \quad \text{(Laplace's equation)}, \quad \Delta u + au = 0 \quad (a = \text{const.}) \tag{26}$$

$$\Delta u - a\frac{\partial u}{\partial t} = 0 \quad (a = \text{const.}), \tag{27}$$

where Δ stands for the Laplace operator (Definition 18.4.1) in n-dimensional space, belong to this type of equations. Solutions of equations (26) and (27) even have derivatives of all orders in the regions considered. In addition, the solutions of equations (26) are *analytic*, i.e. they can be expanded into Taylor series (in n variables) in a neighbourhood of any point of the region in question.

18.2. Partial Differential Equations of the First Order. Homogeneous and Nonhomogeneous Linear Equations, Nonlinear Equations. Complete, General and Singular Integrals. Solution of the Cauchy Problem

Definition 1. The equation of the form

$$a(x, y)\frac{\partial z}{\partial x} + b(x, y)\frac{\partial z}{\partial y} = 0 \qquad (1)$$

is called a *homogeneous linear equation of the first order in two variables.*

Let us assume that $a(x, y)$ and $b(x, y)$ are continuous in the region O in question and do not vanish simultaneously in O.

REMARK 1 (*Solution*). We solve the system of equations

$$\frac{\mathrm{d}x}{a(x, y)} = \frac{\mathrm{d}y}{b(x, y)} \qquad (2)$$

(see Remark 17.20.4, p. 830).

Theorem 1. *Let*

$$\psi(x, y) = C \qquad (3)$$

be the first integral (§ 17.20) *of the system* (2). *Then the function*

$$z = F(\psi(x, y)),$$

where $F(t)$ is an arbitrary differentiable function, is a solution of equation (1). *In particular, if $F(t) \equiv t$, we have $z = \psi(x, y)$.*

REMARK 2 (the *Cauchy Problem*). The solution of equation (1) is to be found such that

$$z(x, y_0) = f(x). \qquad (4)$$

Let the equation

$$\psi(x, y_0) = \varphi, \qquad (5)$$

where ψ stands for the left-hand side of equation (3), be solvable with respect to x giving $x = g(\varphi)$. Then the required solution of the Cauchy problem is given by

$$z = f(g(\psi)). \qquad (6)$$

Example 1. Find the solution of the equation

$$y\frac{\partial z}{\partial x} - x\frac{\partial z}{\partial y} = 0$$

satisfying $z(x, 0) = x^4$ $(1 \leqq x \leqq 4)$.

The first integral of the system

$$\frac{dx}{y} = -\frac{dy}{x}$$

is (see Example 17.20.3, p. 830)

$$\psi(x, y) = x^2 + y^2 = C.$$

According to (5) we have

$$x^2 = \varphi \quad \text{or} \quad x = +\sqrt{\varphi}$$

and by (6)

$$z = (\sqrt{\psi})^4 = [\sqrt{(x^2 + y^2)}]^4 = (x^2 + y^2)^2.$$

REMARK 3. More generally, let the curve

$$x = \varphi(t), \qquad y = \psi(t), \quad z = \chi(t) \tag{7}$$

be given instead of the curve (5). Let us assume that the vector $(\varphi'(t), \psi'(t))$ is not proportional to the vector $(a(\varphi(t), \psi(t)), b(\varphi(t), \psi(t)))$ for any t, i.e. the curve $x = \varphi(t)$, $y = \psi(t)$ is nowhere tangential to the direction field given by the functions a, b. Let

$$\Psi(x, y) = C \tag{8}$$

be a first integral of the system (2). We substitute for x and y from (7) into (8),

$$\Psi(\varphi(t), \psi(t)) = C. \tag{8'}$$

If we obtain t from (8′) as a function of C, i.e. $t = h(C)$, substitute this expression into the equation $z = \chi(t)$ and write the left-hand side of equation (8) instead of C, we obtain the required solution

$$z = \chi(h(\Psi(x, y))).$$

In the case of Example 1 we have: $x = t$, $y = 0$, $z = t^4$. Equation (8): $x^2 + y^2 = C$. Equation (8′): $t^2 = C$. Substituting for t, we obtain $z = C^2$ and using (8), we have

$$z = (x^2 + y^2)^2.$$

REMARK 4. In the general case involving n variables, the investigation of the equation

$$a_1(x_1, x_2, \ldots, x_n)\frac{\partial z}{\partial x_1} + \ldots + a_n(x_1, x_2, \ldots, x_n)\frac{\partial z}{\partial x_n} = 0$$

reduces to that of the system

$$\frac{dx_1}{a_1(x_1, x_2, \ldots, x_n)} = \ldots = \frac{dx_n}{a_n(x_1, x_2, \ldots, x_n)}. \tag{9}$$

Let us assume that the functions $a_1, a_2, \ldots, a_n$ are continuous in the region O considered and that they do not vanish simultaneously anywhere in O. If the equations

$$\begin{aligned} \psi_1(x_1, x_2, \ldots, x_n) &= C_1\,, \\ \psi_2(x_1, x_2, \ldots, x_n) &= C_2\,, \\ &\ldots\ldots\ldots \\ \psi_{n-1}(x_1, x_2, \ldots, x_n) &= C_{n-1} \end{aligned}$$

constitute a system of independent first integrals of the system (9), then

$$z = F(\psi_1(x_1, x_2, \ldots, x_n), \ldots, \psi_{n-1}(x_1, x_2, \ldots, x_n))\,,$$

(where F is an arbitrary differentiable function) is a solution of the given equation. For details the reader is referred to [388] which also shows the construction of the solution of the Cauchy problem.

Definition 2. The equation of the form

$$P(x, y, z)\frac{\partial z}{\partial x} + Q(x, y, z)\frac{\partial z}{\partial y} = R(x, y, z) \tag{10}$$

is called a *nonhomogeneous linear equation of the first order in two variables* x, y. We assume that the functions P, Q, R are continuous in the region O in question, P, Q nowhere simultaneously vanishing, and $R \not\equiv 0$.

(In the theory of first order equations it is common to refer to equation (10) as a *linear equation* though it is, in fact, quasilinear, because the functions P, Q, R involve the unknown function z.)

REMARK 5 (*solution*). To equation (10), there corresponds the system

$$\frac{dx}{P(x, y, z)} = \frac{dy}{Q(x, y, z)} = \frac{dz}{R(x, y, z)}. \tag{11}$$

Theorem 2. *If the equations*

$$\varphi_1(x, y, z) = C_1\,, \quad \varphi_2(x, y, z) = C_2 \tag{12}$$

constitute two independent first integrals (§ 17.20) *of the system* (11), *then the equation*

$$F(\varphi_1(x, y, z), \varphi_2(x, y, z)) = 0, \tag{13}$$

where F stands for an arbitrary differentiable function, is a solution of equation (10).

REMARK 6 (*The Cauchy problem*). Let us find the solution of equation (10) that passes through the curve

$$x = f_1(t), \quad y = f_2(t), \quad z = f_3(t). \tag{14}$$

We assume that these functions do not satisfy the system (11) for any t, i.e. there is no value of t for which the vector $(f_1'(t), f_2'(t), f_3'(t))$ is proportional to the vector

$$(P(f_1(t), f_2(t), f_3(t)), \; Q(f_1(t), f_2(t), f_3(t)), \; R(f_1(t), f_2(t), f_3(t)))$$

(see Remark 3). We substitute from (14) for x, y and z into (12) and obtain C_1 and C_2 as functions of the variable t. If we obtain by eliminating t a relation between C_1 and C_2,

$$g(C_1, C_2) = 0, \tag{15}$$

then substituting (12) into (15), we get the solution:

$$g(\varphi_1(x, y, z), \varphi_2(x, y, z)) = 0.$$

For the linear non-homogeneous equation in n variables

$$a_1(x_1, x_2, \ldots, x_n, z)\frac{\partial z}{\partial x_1} + \ldots + a_n(x_1, x_2, \ldots, x_n, z)\frac{\partial z}{\partial x_n} = b(x_1, x_2, \ldots, x_n, z)$$

the procedure is similar (a_k, b are assumed to be continuous in the region under consideration and the a_k are assumed not to vanish simultaneously anywhere). If the relations

$$\psi_0(x_1, x_2, \ldots, x_n, z) = C_1, \quad \psi_1(x_1, x_2, \ldots, x_n, z) = C_2, \quad \ldots$$
$$\ldots, \quad \psi_{n-1}(x_1, x_2, \ldots, x_n, z) = C_{n-1}$$

constitute first integrals of the system

$$\frac{\mathrm{d}x_1}{a_1} = \frac{\mathrm{d}x_2}{a_2} = \ldots = \frac{\mathrm{d}x_n}{a_n} = \frac{\mathrm{d}z}{b},$$

then

$$F(\psi_0(x_1, x_2, \ldots, x_n, z), \ldots, \psi_{n-1}(x_1, x_2, \ldots, x_n, z)) = 0$$

(F being an arbitrary differentiable function) is an integral of the given equation.

For details see e.g. [388].

Definition 3. A *nonlinear equation of the first order* in two variables x, y is, in general, an equation of the form

$$f(x, y, z, p, q) = 0\,. \tag{16}$$

REMARK 7. We employ the brief notation

$$\frac{\partial z}{\partial x} = p\,, \quad \frac{\partial z}{\partial y} = q\,. \tag{17}$$

The function f is assumed to be differentiable in the region in question.

The previous cases of linear homogeneous and non-homogeneous equations in two variables are special cases of (16).

Definition 4. A function

$$z = \varphi(x, y, a, b), \quad \text{or given by} \quad V(x, y, z, a, b) = 0, \tag{18}$$

satisfying equation (16) and involving two arbitrary (independent) constants is called the *complete integral* (*solution*) of equation (1).

REMARK 8. Either of the equations (18) defines a two-parameter family of surfaces. Equation (18) does not represent the general solution of equation (16); it can be shown that there always exist solutions which cannot be obtained from equation (18) by selecting arbitrary values for the constants a and b. On the other hand, all the solutions may be obtained from (18) by the so-called *method of variation of a parameter*: Let us choose an arbitrary (differentiable) function ω and set

$$b = \omega(a)\,. \tag{19}$$

If we substitute (19) into (18), we obtain (keeping ω fixed) a one-parameter family of surfaces. If the envelope of this family is now obtained from the equations

$$V(x, y, z, a, \omega(a)) = 0\,, \quad \frac{\partial V}{\partial a} + \frac{\partial V}{\partial b}\,\omega'(a) = 0 \tag{20}$$

by eliminating the parameter a, then this surface is an integral surface of equation (16). The system of all integrals obtained by specifying the arbitrary (differentiable) function ω is called the *general integral* of equation (16).

By eliminating both constants a and b from the equations

$$V = 0\,, \quad \frac{\partial V}{\partial a} = 0\,, \quad \frac{\partial V}{\partial b} = 0\,, \tag{21}$$

we obtain an envelope of the two-parameter family (18) (the so-called *singular integral* of equation (16)).

Example 2. One can readily verify by substitution, that the equation

$$(x - a)^2 + (y - b)^2 + z^2 - R^2 = 0 \tag{22}$$

constitutes a complete integral of the equation

$$z^2(1 + p^2 + q^2) = R^2 ;$$

(22) represents a two-parameter family of spheres having their centres at the points $(a, b, 0)$ (lying in the plane xy) and radius R. If we choose the function ω arbitrarily, then (19) defines a curve on which the centres of the spheres of the system (22) lie. If we choose, for instance, $b = \omega(a) = a$, then the one-parameter family obtained is

$$(x - a)^2 + (y - a)^2 + z^2 - R^2 = 0 . \tag{22'}$$

Eliminating a between (22′) and the equation

$$-2(x - a) - 2(y - a) = 0$$

(which is obtained by differentiation of (22′) with respect to a) we get

$$2(x - y)^2 + 4z^2 = 4R^2 .$$

This is a cylindrical surface which envelops the family (22′). The singular integral is obtained from equations (21), namely

$$(x - a)^2 + (y - b)^2 + z^2 - R^2 = 0 , \quad -2(x - a) = 0 , \quad -2(y - b) = 0$$

that is

$$z^2 = R^2 ,$$

and this represents two planes $z = R$, $z = -R$, both enveloping the system (22).

REMARK 9. The complete integral of equation (16) can often be determined in a simple way, if equation (16) is of a particular form. This is done in the following Remark 10 for the cases where equation (16) takes the form (23), (26), or (28), respectively.

REMARK 10. Determination of the complete integral in some special cases.

1.

$$F(p, q) = 0 , \tag{23}$$

We choose $p = a$, solve the equation $F(a, q) = 0$ for q, giving $q = g(a)$, and obtain the complete integral from the relation

$$\mathrm{d}z = p \, \mathrm{d}x + q \, \mathrm{d}y ; \tag{24}$$

in other words

$$z = ax + g(a)\, y + b\,. \tag{25}$$

2.

$$\varphi(x, p) = \psi(y, q)\,. \tag{26}$$

Choosing $\varphi(x, p) = a$, we have $\psi(y, q) = a$; from the first equation, it follows (under obvious assumptions) that $p = g_1(x, a)$ and from the second that $q = g_2(y, a)$. Integrating (24), we have

$$z = \int g_1(x, a)\,\mathrm{d}x + \int g_2(y, a)\,\mathrm{d}y + b\,. \tag{27}$$

3. *The generalized Clairaut's equation,*

$$z = xp + yq + g(p, q)\,, \tag{28}$$

has a complete integral

$$z = ax + by + g(a, b) \tag{29}$$

(see also Example 3).

REMARK 11 (*Solution of the Cauchy Problem using the complete integral*). It is required to find the integral surface of the equation (16) passing through the curve

$$x = f_1(t)\,, \quad y = f_2(t)\,, \quad z = f_3(t) \tag{30}$$

when a complete integral of the equation (16)

$$V(x, y, z, a, b) = 0 \tag{31}$$

is known. Substituting (30) into (31), we obtain an equation relating a, b, t, say

$$G(a, b, t) = 0\,. \tag{32}$$

Then, eliminating t between (32) and the equation

$$\frac{\partial G}{\partial t}(a, b, t) = 0 \tag{33}$$

we obtain a relation between a and b, say

$$H(a, b) = 0 \quad \text{or} \quad b = \omega(a)\,. \tag{34}$$

Substituting (34) into (31) gives us a one-parameter family of surfaces

$$K(x, y, z, a) = 0\,. \tag{35}$$

Finally, eliminating a between (35) and the equation

$$\frac{\partial K}{\partial a}(x, y, z, a) = 0\,, \tag{36}$$

we obtain the equation of the envelope of the system (35), which, as a rule, is the required integral surface.

It must be realized that the procedure employed above to obtain the solution of the given problem is only formal (notice that we have made use of such vague notions as the elimination of variables, etc.) so that the result obtained must be analysed carefully. In particular we must show that it really represents the solution.

The same is true for the Cauchy method (Remark 15).

Example 3. Find the integral surface of the equation

$$z = px + qy + pq \tag{37}$$

passing through the curve

$$x = 0\,, \quad z = y^2\,. \tag{38}$$

By (29) a complete integral is

$$ax + by - z + ab = 0\,. \tag{39}$$

Now parametric equations of the curve (38) are

$$x = 0\,, \quad y = t\,, \quad z = t^2\,, \tag{40}$$

and by substituting these into (39) we get

$$bt - t^2 + ab = 0\,. \tag{41}$$

Equation (33) then becomes

$$b - 2t = 0\,. \tag{42}$$

Next, eliminating t between (41) and (42), we have

$$a = -\tfrac{1}{4}b\,, \tag{43}$$

i.e.

$$b = -4a\,. \tag{44}$$

Substituting into (39), we obtain (as in (35)):

$$ax - 4ay - z - 4a^2 = 0\,. \tag{45}$$

Differentiating this with respect to a,

$$x - 4y - 8a = 0\,, \tag{46}$$

and eliminating a between (45) and (46) (that is, substituting $8a = x - 4y$ from (46) into (45)), we obtain

$$z = \left(y - \frac{x}{4}\right)^2. \tag{47}$$

REMARK 12. For the equation

$$f(x_1, x_2, \ldots, x_n, z, p_1, p_2, \ldots, p_n) = 0, \tag{48}$$

where

$$p_1 = \frac{\partial z}{\partial x_1}, \quad p_2 = \frac{\partial z}{\partial x_2}, \quad \ldots, \quad p_n = \frac{\partial z}{\partial x_n},$$

we define the *complete integral* as follows:

Definition 5. An integral

$$z = \varphi(x_1, x_2, \ldots, x_n, a_1, a_2, \ldots, a_n) \tag{49}$$

or

$$V(x_1, x_2, \ldots, x_n, z, a_1, a_2, \ldots, a_n) = 0 \tag{50}$$

of equation (48) is said to be a *complete integral*, if it is possible to arrive at the equation (48) by elimination of the constants $a_1, a_2, \ldots, a_n$ between equation (49) (or (50)) and the n equations obtained by differentiation with respect to all n variables $x_1, x_2, \ldots, x_n$.

REMARK 13. The complete integral of equation (48) may be defined as an integral involving n independent constants (cf. Definition 4). The independence of these constants has the same meaning as in Definition 5; roughly speaking, these constants are independent if their number cannot be reduced.

REMARK 14. Complete integrals for equations with several variables play an important role in analytical mechanics, e.g. when integrating the so-called *Hamiltonian equations of motion*.

REMARK 15 (the *Cauchy (Lagrange–Charpit) method of solution of the Cauchy problem in two variables*). If we are given the equation

$$f(x, y, z, p, q) = 0 \tag{51}$$

we write down the following system of ordinary equations

$$\frac{dx}{P} = \frac{dy}{Q} = \frac{dz}{Pp + Qq} = -\frac{dp}{X + Zp} - \frac{dq}{Y + Zq} \tag{52}$$

or, which amounts the same thing, the system

$$\frac{dx}{du} = P\,, \quad \frac{dy}{du} = Q\,, \quad \frac{dz}{du} = Pp + Qq\,, \quad \frac{dp}{du} = -X - Zp\,, \quad \frac{dq}{du} = -Y - Zq\,. \tag{53}$$

Here we have employed the abbreviated notation

$$\frac{\partial f}{\partial x} = X\,, \quad \frac{\partial f}{\partial y} = Y, \quad \frac{\partial f}{\partial z} = Z\,, \quad \frac{\partial f}{\partial p} = P\,, \quad \frac{\partial f}{\partial q} = Q\,. \tag{54}$$

Any solution

$$x = x(u)\,, \quad y = y(u)\,, \quad z = z(u)\,, \quad p = p(u)\,, \quad q = q(u) \tag{55}$$

of the system (53) is said to be a *characteristic strip* (or a *characteristic of the first order*) corresponding to equation (51). Its geometric significance is as follows: At every point of the curve

$$x = x(u)\,, \quad y = y(u)\,, \quad z = z(u)$$

the values of $p(u)$ and $q(u)$ are known, thus a "tangent element" of a surface is given. If, for a certain value u_0 of the variable u, the functions (55) of the characteristic strip satisfy the equation (51) (i.e. if for the corresponding values $x_0 = x(u_0), \ldots$, the equation

$$F(x_0, y_0, z_0, p_0, q_0) = 0 \tag{56}$$

is satisfied), then this equation is satisfied at all points of the characteristic strip. In this case, the characteristic strip is called the *integral strip of equation* (51) and the tangent elements which constitute it are called *integral elements of equation* (51). An integral strip is uniquely determined by the initial integral element of the given equation. If two integral surfaces have one common integral element, then they touch each other along the whole integral strip.

Cauchy's method of solving the Cauchy problem consists in the construction of integral surfaces from integral strips. We are given equation (51) and the curve

$$x = f_1(t)\,, \quad y = f_2(t)\,, \quad z = f_3(t)\,. \tag{57}$$

The functions $p(t)$ and $q(t)$ are then determined from the equations

$$\frac{dz}{dt} = p(t)\frac{dx}{dt} + q(t)\frac{dy}{dt}\,, \quad f(x(t), y(t), z(t), p(t), q(t)) = 0\,. \tag{58}$$

(If equation (51) is nonlinear, equations (58) may yield many solutions or may not yield any real solution at all.) Then we find the solution (55) of (53) such that for

$u = u_0$ (and for all t) the conditions

$$x(u_0) = f_1(t), \quad y(u_0) = f_2(t), \quad z(u_0) = f_3(t), \quad p(u_0) = p(t), \quad q(u_0) = q(t)$$

are satisfied (we usually choose $u_0 = 0$). In this way, x, y, z, p, q are obtained as functions of two parameters u and t. Eliminating u and t from the equations

$$x = x(u, t), \quad y = y(u, t), \quad z = z(u, t), \tag{59}$$

we obtain the required relation between x, y, z.

Example 4. Let us solve the problem of Example 3 by Cauchy's method. That is, we have to find a solution of the equation

$$px + qy + pq - z = 0 \tag{60}$$

passing through the curve

$$x = 0, \quad z = y^2 \quad \text{(or, in parametric form, } x = 0,\ y = t,\ z = t^2\text{)}. \tag{61}$$

We have

$$X = p, \quad Y = q, \quad P = x + q, \quad Q = y + p, \quad Z = -1. \tag{62}$$

Equations (53) now take the form:

$$\frac{dx}{du} = x + q, \quad \frac{dy}{du} = y + p, \quad \frac{dz}{du} = (x + q)\,p + (y + p)\,q,$$

$$\frac{dp}{du} = -(p - p) = 0, \quad \frac{dq}{du} = -(q - q) = 0. \tag{63}$$

From the latter equations, it follows that

$$p = C_1 = \text{const.}, \quad q = C_2 = \text{const.}; \tag{64}$$

on substituting these expressions into the first three equations of (63), we get

$$\frac{dx}{du} = x + C_2, \quad \frac{dy}{du} = y + C_1, \quad \frac{dz}{du} = C_1 x + C_2 y + 2C_1 C_2. \tag{65}$$

The first two equations are linear so that

$$x = -C_2 + C_3 e^u, \quad y = -C_1 + C_4 e^u. \tag{66}$$

When we insert (66) into the third equation (65) and integrate we obtain

$$z = (C_1 C_3 + C_2 C_4)\,e^u + C_5. \tag{67}$$

We now choose the constants $C_1, C_2, \ldots, C_5$ depending on t in such a way that the conditions (61) are satisfied. At the same time the conditions (58) should be satisfied. It follows from (61) that

$$\frac{dx}{dt} = 0\,, \quad \frac{dy}{dt} = 1\,, \quad \frac{dz}{dt} = 2t$$

so that the conditions (58) take the form

$$2t = p\,.\,0 + q\,.\,1\,, \quad p\,.\,0 + q\,.\,t + pq - t^2 = 0\,,$$

whence

$$q(t) = 2t\,, \quad p(t) = -\frac{t}{2}\,.$$

For $u = 0$ we then require that

$$x = 0\,, \quad y = t\,, \quad z = t^2\,, \quad p = -\frac{t}{2}\,, \quad q = 2t$$

so that by (64), (66) and (67)

$$C_1 = -\frac{t}{2}\,, \quad C_2 = 2t\,, \quad C_3 = 2t\,, \quad C_4 = \frac{t}{2}\,, \quad C_5 = t^2\,.$$

Hence

$$x = -2t(1 - e^u)\,, \quad y = \frac{t}{2}(1 + e^u)\,, \quad z = t^2\,, \quad p = -\frac{t}{2}\,, \quad q = 2t\,. \tag{68}$$

Eliminating e^u from the first two of equations (68), we have

$$-x + 4y = 4t\,, \quad \text{that is} \quad \left(y - \frac{x}{4}\right)^2 = t^2\,,$$

and by the third equation (68)

$$z = \left(y - \frac{x}{4}\right)^2$$

which agrees with the result of Example 3.

REMARK 16. In the particular case where equation (51) is linear,

$$a(x, y, z)\,p + b(x, y, z)\,q - r(x, y, z) = 0\,, \tag{69}$$

Cauchy's method may be considerably simplified because the first three equations (53),

$$\frac{dx}{du} = a(x, y, z)\,, \quad \frac{dy}{du} = b(x, y, z)\,, \quad \frac{dz}{du} = a(x, y, z)\,p + b(x, y, z)\,q =$$
$$= r(x, y, z)\,, \tag{70}$$

are now sufficient to determine the functions

$$x = x(u, C_1, C_2, C_3), \quad y = y(u, C_1, C_2, C_3), \quad z = z(u, C_1, C_2, C_3). \qquad (71)$$

Equations (70), (71) define the so-called *characteristics* of equation (69). (To be exact, we should, according to Remark 18.1.10, define the characteristics as the projections of the curves (71) on to the xy-plane. However, the above terminology is in common use.) The initial curve

$$x = f_1(t), \quad y = f_2(t), \quad z = f_3(t), \qquad (72)$$

(the projection of which on to the xy-plane is supposed not to touch the projection of any characteristic), together with the characteristics (71), determines the integral surface.

Example 5. Find the solution of the equation

$$xp + yq - z = 0 \qquad (73)$$

passing through the curve

$$x = t, \quad y = t^2, \quad z = t^3 \quad (t > 0). \qquad (74)$$

The first three equations (53) read

$$\frac{dx}{du} = x, \quad \frac{dy}{du} = y, \quad \frac{dz}{du} = xp + yq = z. \qquad (75)$$

The solution satisfying the initial conditions (74) for $u = 0$ is

$$x = te^u, \quad y = t^2e^u, \quad z = t^3e^u; \qquad (76)$$

eliminating t and u we obtain

$$z = \frac{y^2}{x}.$$

REMARK 17. The case of the linear *homogeneous* equation

$$a(x, y)\, p + b(x, y)\, q = 0 \qquad (77)$$

is even simpler, for we need only solve the following two equations:

$$\frac{dx}{du} = a(x, y), \quad \frac{dy}{du} = b(x, y). \qquad (78)$$

The third equation (53) now reads

$$\frac{dz}{du} = 0, \quad \text{or} \quad z = C = \text{const.},$$

so that the characteristics (using the terminology of Remark 16) are parallel to the xy-plane.

REMARK 18. The existence and uniqueness of solutions (in the class of differentiable functions) is ensured for linear equations unless the projection of the curve (72) touches the projection of a characteristic, i.e. unless the numbers dx/dt, dy/dt at a point (x, y, z) are proportional to the numbers dx/du, dy/du given by equations (70) at the same point (x, y, z). On the other hand, if this proportionality holds at all points of (72) and at the same time the relation

$$\frac{dx}{dt} : \frac{dy}{dt} : \frac{dz}{dt} = \frac{dx}{du} : \frac{dy}{du} : \frac{dz}{du} \tag{79}$$

does not hold, then the given problem has no solution. If the relation (79) is satisfied for all points of the curve (72), i.e. the curve (72) itself is a characteristic, then the problem has an infinite number of solutions.

Example 6. Find, first, the solution of equation (73) passing through the straight line

$$x = t, \quad y = t, \quad z = 1. \tag{80}$$

The solution of equations (75) satisfying conditions (80) is

$$x = te^u, \quad y = te^u, \quad z = e^u,$$

and from these equations z cannot be expressed as a function of the variables x, y. (In (80), $x = y$ for every t, so that by (75)

$$\frac{dx}{dt} : \frac{dy}{dt} = \frac{dx}{du} : \frac{dy}{du}$$

for every t; however, the relation (79),

$$\frac{dx}{dt} : \frac{dy}{dt} : \frac{dz}{dt} = \frac{dx}{du} : \frac{dy}{du} : \frac{dz}{du},$$

does not hold in this case.)

If the initial curve is replaced by the curve

$$x = t, \quad y = t, \quad z = t, \tag{81}$$

then the problem has an infinite number of solutions. An integral surface passing through (81) is any surface

$$z = k_1 x + k_2 y\,, \quad \text{where} \quad k_1 + k_2 = 1\,.$$

REMARK 19. In the case of several variables, the characteristic strips (characteristics of the first order) are defined by the equations

$$\frac{dx_1}{P_1} = \ldots = \frac{dx_n}{P_n} = \frac{dz}{\sum_{k=1}^{n} P_k p_k} = -\frac{dp_1}{X_1 + Zp_1} = \ldots = -\frac{dp_n}{X_n + Zp_n}\,, \tag{82}$$

where

$$p_k = \frac{\partial z}{\partial x_k}\,, \quad P_k = \frac{\partial f}{\partial p_k}\,, \quad X_k = \frac{\partial f}{\partial x_k}\,, \quad Z = \frac{\partial f}{\partial z}\,. \tag{83}$$

The Cauchy problem: Given an $(n-1)$-dimensional surface

$$x_k = x_k(v_1, v_2, \ldots, v_{n-1})\,, \quad z = z(v_1, v_2, \ldots, v_{n-1})\,; \tag{84}$$

the problem of finding an integral surface $z = z(x_1, x_2, \ldots, x_n)$ passing through the surface (84), is called the *Cauchy problem*.

When solving this problem, we first determine

$$p_k(v_1, v_2, \ldots, v_{n-1}) \quad (k = 1, 2, \ldots, n) \tag{85}$$

from the equations

$$f(x_1, \ldots, x_n, z, p_1, \ldots, p_n) = 0\,, \quad \frac{\partial z}{\partial v_i} - \sum_{k=1}^{n} p_k \frac{\partial x_k}{\partial v_i} = 0 \quad (i = 1, 2, \ldots, n-1) \tag{86}$$

which correspond to equations (58) and then determine a solution

$$x_k = x_k(u, v_1, \ldots, v_{n-1})\,, \quad z = z(u, v_1, \ldots, v_{n-1})\,,$$

$$p_k = p_k(u, v_1, \ldots, v_{n-1}) \tag{87}$$

of the system (82) such that the functions (87) assume the initial values (84) and (85) for $u = u_0$ (usually we choose $u_0 = 0$). By eliminating the parameters $u, v_1, \ldots, v_{n-1}$ from the first $n + 1$ equations (87), i.e. from the equations for x_k and z, we obtain the desired integral surface, i.e. the relation $g(x_1, x_2, \ldots, x_n, z) = 0$. (See, however, p. 874.)

For details see e.g. [388].

18.3. Linear Equations of the Second Order. Classification

Definition 1. The equation of the form

$$A_{11}(x, y)\frac{\partial^2 z}{\partial x^2} + 2A_{12}(x, y)\frac{\partial^2 z}{\partial x\,\partial y} + A_{22}(x, y)\frac{\partial^2 z}{\partial y^2} + B_1(x, y)\frac{\partial z}{\partial x} +$$

$$+ B_2(x, y)\frac{\partial z}{\partial y} + C(x, y)\,z + D(x, y) = 0 \tag{1}$$

is called a *linear equation of the second order for the function* $z(x, y)$.

REMARK 1. The coefficients $A_{11}, \ldots, D$ are assumed to be continuous (as functions of the variables x, y) in the region O in question.

Definition 2. If everywhere in O

$$\left\{\begin{matrix} A_{11}A_{22} - A_{12}^2 > 0 \\ A_{11}A_{22} - A_{12}^2 < 0 \\ A_{11}A_{22} - A_{12}^2 = 0 \end{matrix}\right\}, \text{ the equation is said to be } \left\{\begin{matrix} \textit{elliptic} \\ \textit{hyperbolic} \\ \textit{parabolic} \end{matrix}\right\} \text{ in } O\,.$$

Theorem 1. *By a suitable transformation of variables, every elliptic, hyperbolic, or parabolic equation in O can be reduced in a neighbourhood of any point $(x_0, y_0) \in O$, to the so-called canonical form* (2), (3) *or* (4), *respectively:*

$$\frac{\partial^2 z}{\partial x^2} + \frac{\partial^2 z}{\partial y^2} + a_1(x, y)\frac{\partial z}{\partial x} + b_1(x, y)\frac{\partial z}{\partial y} + c_1(x, y)\,z + d_1(x, y) = 0\,, \tag{2}$$

$$\frac{\partial^2 z}{\partial x^2} - \frac{\partial^2 z}{\partial y^2} + a_2(x, y)\frac{\partial z}{\partial x} + b_2(x, y)\frac{\partial z}{\partial y} + c_2(x, y)\,z + d_2(x, y) = 0\,, \tag{3}$$

$$\frac{\partial^2 z}{\partial y^2} + a_3(x, y)\frac{\partial z}{\partial x} + b_3(x, y)\frac{\partial z}{\partial y} + c_3(x, y)\,z + d_3(x, y) = 0\,, \quad a_3(x, y) \neq 0\,. \tag{4}$$

REMARK 2. Linear equations of the second order for a function $z(x_1, x_2, \ldots, x_n)$ of more than two variables can be transformed by a suitable change of variables to canonical forms similar to those of (2), (3), (4). However, in the general case it is not possible to find a transformation which reduces the given equation to its canonical form in a whole neighbourhood of the given point but only such that it does so at the point itself. Consequently, an equation for several variables is called elliptic, hyperbolic or parabolic at the point $(x_1^0, x_2^0, \ldots, x_n^0)$ if by a suitable transformation it can be reduced at this point to the form (5), (6) or (7), respectively:

$$\sum_{k=1}^{n} \frac{\partial^2 z}{\partial x_k^2} + \ldots = 0 \tag{5}$$

$$\sum_{k=1}^{n-1} \frac{\partial^2 z}{\partial x_k^2} - \frac{\partial^2 z}{\partial x_n^2} + \ldots = 0\,, \tag{6}$$

$$\sum_{k=2}^{n} \frac{\partial^2 z}{\partial x_k^2} + a_1 \frac{\partial z}{\partial x_1} + \ldots = 0\,. \tag{7}$$

Only the terms involving second order derivatives are written in equations (5) and (6); in equation (7) the coefficient a_1 of the derivative $\partial z/\partial x_1$ is required to be different from zero. If an equation satisfies (5), (6) or (7) at every point of the region in question, it is called elliptic, hyperbolic, or parabolic in this region, respectively.

REMARK 3. If, in equation (6), the minus sign as well as the plus sign stands before more than one of the second order derivatives (while at the same time second derivatives with respect to all n variables are present), the equation is said to be *ultrahyperbolic.* If, in equation (7), more than one of the second derivatives is absent, the equation is said to be *parabolic in a wider sense.*

Example 1. The equation

$$(1 + y^2)\frac{\partial^2 z}{\partial x^2} + (1 + y^2)\frac{\partial^2 z}{\partial y^2} - \frac{\partial z}{\partial x} = 0$$

is an elliptic equation for the function $z(x, y)$ (by Definition 2, because $A_{11}A_{22} - A_{12}^2 = (1 + y^2)^2 > 0$);

the equation

$$\frac{\partial^2 z}{\partial x_1^2} + \frac{\partial^2 z}{\partial x_2^2} - \frac{\partial^2 z}{\partial x_3^2} = 0$$

is a hyperbolic equation for the function $z(x_1, x_2, x_3)$ according to (6);

the equation

$$\frac{\partial^2 z}{\partial x_1^2} + \frac{\partial^2 z}{\partial x_2^2} - \frac{\partial^2 z}{\partial x_3^2} - \frac{\partial^2 z}{\partial x_4^2} = 0$$

is an ultrahyperbolic equation for the function $z(x_1, x_2, x_3, x_4)$ (by Remark 3);

the equation

$$\frac{\partial^2 z}{\partial x_2^2} + \frac{\partial z}{\partial x_1} + \frac{\partial z}{\partial x_3} = 0$$

is a parabolic equation in a wider sense and

the equation

$$\frac{\partial u}{\partial t} = \frac{\partial^2 u}{\partial x^2} + \frac{\partial^2 u}{\partial y^2} + \frac{\partial^2 u}{\partial z^2}$$

is a parabolic equation for the function $u(x, y, z, t)$.

18.4. Elliptic Equations. The Laplace Equation, the Poisson Equation. The Dirichlet and Neumann Problems. Properties of Harmonic Functions. The Fundamental Solution. Green's Function. Potentials of Single- and of Double-layer Distributions

REMARK 1 (*informative*; see also the informative remark at the beginning of this chapter). In this paragraph we deal mainly with the Laplace and Poisson equations. On some generalization of the results and on more general elliptic equations see Remarks 22–24.

Definition 1. The equation

$$\Delta u = 0\,, \tag{1}$$

where

$$\Delta u = \frac{\partial^2 u}{\partial x_1^2} + \frac{\partial^2 u}{\partial x_2^2} + \ldots + \frac{\partial^2 u}{\partial x_n^2}\,, \tag{2}$$

and $n \geqq 2$, is called the *Laplace (differential) equation*. (The symbol ∇^2 is often used in place of Δ in English literature.)

In particular, the Laplace equation for two and three variables reads

$$\frac{\partial^2 u}{\partial x^2} + \frac{\partial^2 u}{\partial y^2} = 0 \tag{3}$$

and

$$\frac{\partial^2 u}{\partial x^2} + \frac{\partial^2 u}{\partial y^2} + \frac{\partial^2 u}{\partial z^2} = 0\,, \tag{4}$$

respectively.

Definition 2. The equation of the form

$$\Delta u = -\frac{2(\sqrt{\pi})^n}{\Gamma\left(\dfrac{n}{2}\right)}\,\varrho(x_1, x_2, \ldots, x_n) \tag{5}$$

is called the *Poisson (differential) equation*. (Γ stands for the gamma function (p. 584).) In particular the Poisson equation for two and three variables reads

$$\frac{\partial^2 u}{\partial x^2} + \frac{\partial^2 u}{\partial y^2} = -2\pi\varrho(x, y) \tag{6}$$

and

$$\frac{\partial^2 u}{\partial x^2} + \frac{\partial^2 u}{\partial y^2} + \frac{\partial^2 u}{\partial z^2} = -4\pi\varrho(x, y, z), \tag{7}$$

respectively.

Definition 3. Let us write

$$r = \sqrt{(x^2 + y^2 + z^2)}.$$

The function $u(x, y, z)$ (defined for all sufficiently large r) is said to *vanish at infinity* if for every $\varepsilon > 0$ there exists R such that $|u(x, y, z)| < \varepsilon$ whenever the point (x, y, z) is such that $r > R$.

For the case of n variables the meaning of the statement "u vanishes at infinity" is similar; r is then defined by

$$r = \sqrt{(x_1^2 + x_2^2 + \ldots + x_n^2)}.$$

Theorem 1. *If the function $\varrho(x, y, z)$ is continuously differentiable in the entire three-dimensional space and if for large r the inequality*

$$|\varrho(x, y, z)| < \frac{A}{r^{2+\alpha}} \tag{8}$$

holds, where A and α are positive constants, then the function

$$u(x, y, z) = \int_{-\infty}^{\infty}\int_{-\infty}^{\infty}\int_{-\infty}^{\infty} \frac{\varrho(\xi, \eta, \zeta)}{\sqrt{[(x-\xi)^2 + (y-\eta)^2 + (z-\zeta)^2]}}\, d\xi\, d\eta\, d\zeta \tag{9}$$

satisfies equation (7) everywhere. Moreover, (9) is the only solution of equation (7) which vanishes at infinity.

REMARK 2. The integral (9) is called *Newton's potential*.

REMARK 3. In contradistinction to the three-dimensional case, the two-dimensional problem of finding a solution of equation (6) that vanishes at infinity is not solvable in general. The integral (similar to that of (9))

$$u(x, y) = \int_{-\infty}^{\infty}\int_{-\infty}^{\infty} \varrho(\xi, \eta) \ln \frac{1}{\sqrt{[(x-\xi)^2 + (y-\eta)^2]}}\, d\xi\, d\eta \tag{10}$$

(the so-called *logarithmic potential*), satisfies equation (6) everywhere but does not, in general, vanish at infinity.

REMARK 4. The integral

$$v(x, y) = \int_{-\infty}^{\infty} \int_{-\infty}^{\infty} \frac{\varrho(\xi, \eta)}{\sqrt{[(x - \xi)^2 + (y - \eta)^2]}} \, d\xi \, d\eta \tag{11}$$

is *not* a solution of equation (6).

Definition 4. The *Dirichlet problem for the Laplace* (or *Poisson*) *equation* is the problem of finding, in a given region, a solution of that equation which assumes prescribed values on the boundary of the region. Precisely: Let a region O be given. Denote its boundary by Γ and $\bar{O} = O + \Gamma$. Let f be a continuous function defined on Γ. We have to find a function u that is continuous in $\bar{O}$, assumes on Γ the prescribed values given by the function f, and satisfies the Laplace, or Poisson, equation, respectively, in O (i.e. u has continuous derivatives of the second order at every interior point P of the region O and satisfies the given equation at P).

Definition 5. The *Neumann problem* is to find a solution of the given equation such that it is continuous together with its first derivatives in $\bar{O}$ and such that the outward normal derivative $\partial u/\partial n$ assumes prescribed values given by a continuous function g.

REMARK 5. The Dirichlet and Neumann problem may also be defined for more general elliptic equations than for the Laplace or Poisson equation. Moreover, other problems may be solved for a given equation; for instance, a relation between the function u and its outward normal derivative may be prescribed (the Newton problem). Mixed problems are also encountered, for instance, when Dirichlet's condition is prescribed on one part of the boundary and Neumann's condition is given on the remaining part.

REMARK 6. We shall be mainly concerned with two and three-dimensional problems that is, with equations (3), (4), (6), (7). If the region O is *bounded* (multiply connected regions are allowed), the Dirichlet or Neumann problems are referred to as *interior problems*; if O is the *exterior* of a simple closed curve (or of a simple closed surface which is the boundary of a simply connected region in three-dimensional space), we refer to an *exterior problem*. The *outward normal* is understood to be oriented (at a given point of the boundary) in a direction out of the given region. (For instance, if O is the exterior of the circle k with centre at the origin then the outward normal points to the origin.) If a_1, a_2, (or a_1, a_2, a_3 in the three-dimensional case) are direction cosines of the outward normal, then

$$\frac{\partial u}{\partial n} = a_1 \frac{\partial u}{\partial x} + a_2 \frac{\partial u}{\partial y} \quad \text{or} \quad \frac{\partial u}{\partial n} = a_1 \frac{\partial u}{\partial x} + a_2 \frac{\partial u}{\partial y} + a_3 \frac{\partial u}{\partial z}. \tag{12}$$

It is naturally possible to define an outward normal also in the n-dimensional case.

Definition 6. A function having continuous derivatives of the second order in a region O and satisfying the Laplace equation in O is said to be *harmonic in* O.

Theorem 2 (*the maximum principle*).
Let the function u be continuous in a bounded closed region $\bar{O}$ and harmonic in O. Write M, m for the maximum and minimum of the function u on the boundary of O, respectively. Then the inequality $m \leqq u \leqq M$ holds everywhere in O, i.e. u attains its maximum and minimum in $\bar{O}$ on the boundary of this region. If u is not constant in $\bar{O}$, then even the strict inequality $m < u < M$ holds everywhere in O. The same is true for the two-dimensional case if O is the exterior of a closed curve provided that u is bounded in O.

REMARK 7 (*uniqueness of the solution of the Dirichlet and Neumann problems*). In this remark, when we are considering an exterior problem (either the Dirichlet or the Neumann problem) we make certain assumptions regarding the behaviour of the function u at infinity. If a plane problem is under consideration, we assume that u is bounded, and in a space problem we assume that u vanishes at infinity (Definition 3).

Uniqueness of the solution of both the interior and the exterior Dirichlet problem for equations (3), (4), (6), (7) is guaranteed as far as the exterior problem is subject to the above conditions.

Under the same conditions, these problems are also *well-posed problems* (that is, the solution depends continuously on the boundary conditions): If the absolute value of the change of the given boundary condition is less than ε, then the same is true for the solution of the given problem in O (in other words, if the boundary conditions have been measured with a small error then the error of the solution is also small).

If the function u is a solution of the Neumann problem, then $u + C$ (where C is an arbitrary constant) is also a solution of the same problem. To be able to guarantee uniqueness of the solution, we have to impose a further condition. Usually it is required that the function u takes a prescribed value at a given point of the region; in the case of an exterior problem, we impose the condition that the function u vanishes at infinity. Then uniqueness of the solution of the Neumann problem (for both the exterior and interior case) for equations (3), (4), (6), (7) is ensured.

REMARK 8 (*existence of solution of the Dirichlet and Neumann problems*). (As in Remark 7, when solving exterior problems we consider only solutions that are, respectively, bounded in O or vanishing at infinity, according to whether a plane or a space problem is in question.) The solution of the *interior* Dirichlet problem for equations (3), (4), (6), (7) always exists, if the boundary of the region O is smooth enough, i.e. if it has a continuously changing tangent everywhere (or a tangent plane, if the three-dimensional problem is considered; see, however, Remark 23) and if the function ϱ has continuous derivatives of the first order in $\bar{O}$ when Poisson's problem is considered.

The solution of the *exterior* Dirichlet problem for the equations considered may be reduced to the solution of the interior problem as follows: Let us consider first the plane problem and let O be the region exterior to the curve k. Let the origin

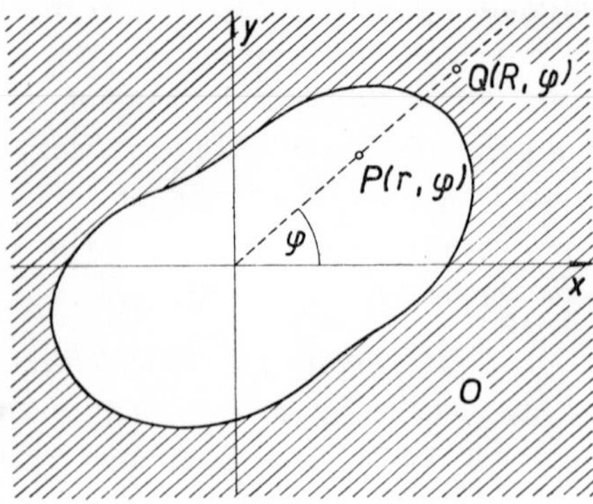

Fig. 18.3.

lie in the interior of that curve. (Otherwise a translation of the coordinate system should be carried out.) Transformation to polar coordinates R, φ gives (see Example 12.11.3, p. 473)

$$\Delta u \equiv \frac{\partial^2 u}{\partial R^2} + \frac{1}{R}\frac{\partial u}{\partial R} + \frac{1}{R^2}\frac{\partial^2 u}{\partial \varphi^2} = \sigma(R, \varphi)\,. \tag{13}$$

The transformation $r = 1/R$ (Fig. 18.3, where the point $P(r, \varphi)$ corresponds to the point $Q(R, \varphi)$) carries the curve k into a curve k' and its exterior O into the interior O' of the curve k'. It may be easily proved that the function

$$u(R, \varphi) = u\left(\frac{1}{r},\ \varphi\right) = v(r, \varphi) \tag{14}$$

satisfies the equation

$$\Delta v \equiv \frac{\partial^2 v}{\partial r^2} + \frac{1}{r}\frac{\partial v}{\partial r} + \frac{1}{r^2}\frac{\partial^2 v}{\partial \varphi^2} = \frac{1}{r^4}\,\sigma\left(\frac{1}{r},\ \varphi\right) \tag{15}$$

in O'. In particular, if $\sigma \equiv 0$ in O, so that u is harmonic in O, then v is harmonic in O'. It follows readily from equation (15) and from what has been said at the beginning of this remark that: If the function $\sigma(R, \varphi)$ tends to zero rapidly enough as $R \to \infty$ so that $r^{-4} \,.\, \sigma(r^{-1}, \varphi)$ is continuous together with its first derivatives in O (more precisely: if $r^{-4} \,.\, \sigma(r^{-1}, \varphi)$ has a removable singularity at $r = 0$), then the existence of the solution of the exterior problem is ensured also for Poisson's equation. The boundary function f prescribed on the boundary k is transformed into a boundary function F on k' in such a way that F assumes on k' the same value as f does at the appropriate point (corresponding to the same value of φ) on k. Hence, having determined the function $v(r, \varphi)$, i.e. the solution of the interior Dirichlet problem for equation (15) under the boundary condition F, equation (14) yields the desired solution $u(R, \varphi)$ directly.

The treatment of the three-dimensional case is similar. We make use of spherical coordinates R, ϑ, φ, carry out the transformation $r = 1/R$ and define the function v by the relation

$$u(R, \vartheta, \varphi) = u\left(\frac{1}{r}, \vartheta, \varphi\right) = rv(r, \vartheta, \varphi). \tag{16}$$

The function v satisfies Poisson's equation

$$\Delta v \equiv \frac{\partial^2 v}{\partial r^2} + \frac{2}{r}\frac{\partial v}{\partial r} + \frac{1}{r^2 \sin^2 \vartheta}\frac{\partial^2 v}{\partial \varphi^2} + \frac{1}{r^2 \sin \vartheta}\frac{\partial}{\partial \vartheta}\left[\frac{\partial v}{\partial \vartheta}\sin \vartheta\right] =$$

$$= \frac{1}{r^5}\sigma\left(\frac{1}{r}, \vartheta, \varphi\right). \tag{17}$$

Again we obtain the appropriate function F on the boundary h' of the region O' easily from the function f by defining $F = Rf$ at corresponding boundary points (i.e. with the same values of ϑ and φ). Having solved the interior Dirichlet problem for Poisson's equation (17), $\Delta v = r^{-5}\,\sigma(r^{-1}, \vartheta, \varphi)$ with boundary condition F, we obtain the desired solution by (16):

$$u(R, \vartheta, \varphi) = \frac{1}{R}\, v\left(\frac{1}{R}, \vartheta, \varphi\right).$$

The existence of the solution of the Neumann problem for the Laplace equation (3) *or* (4): Let the given region have a smooth boundary. (For the exterior problem see the supplementary conditions mentioned at the beginning of this Remark.) A necessary and sufficient condition for the existence of the solution is that the integral of the boundary function g over the boundary h of O be equal to zero, i.e.

$$\int_h g(S)\, dS = 0. \tag{18}$$

(S being a variable point of the boundary.) For plane problems the boundary condition is usually given as a function of the arc length, so that (18) reads

$$\int_0^l g(s)\, ds = 0, \tag{19}$$

where l stands for the length of the boundary; in the three-dimensional case (18) represents a surface integral, see Definition 14.8.2, p. 639. In the case of the exterior problem in three-dimensional space, condition (18) is omitted.

Even if the function ϱ is smooth enough, the solution of the Neumann problem for *Poisson's* equation need not exist. For instance, in the case of the Neumann interior problem in three-dimensional space and for zero boundary condition $g \equiv 0$ a necessary and sufficient condition for the existence of a solution is

$$\iiint_O \varrho(x, y, z)\,dx\,dy\,dz = 0\,. \tag{20}$$

REMARK 9 (*Properties of Harmonic Functions*). The most important property has been formulated in Theorem 2 (the Maximum Principle). Equations (15) and (17) (for $\sigma = 0$) express another property: If $u(x, y)$ is harmonic in a two-dimensional region, then the inversion $r = 1/R$ carries this function again into a harmonic function. In the three-dimensional case the situation is rather different, for, if $u(x, y, z)$ is harmonic, then to obtain a harmonic function v from the harmonic function $u(x, y, z)$ by the inversion $r = 1/R$, the transformed function must be divided by r.

Further properties of harmonic functions are:

1. (*Mean Value Theorem.*) If a function is harmonic in an n-dimensional sphere K and continuous in the closed sphere $\bar{K}$ then its value at the centre of this sphere is equal to its average value over the boundary of the sphere. In particular, for $n = 2$ (when K is a circle of radius R) we have

$$u(x_0, y_0) = \frac{1}{2\pi R} \int_0^{2\pi} u(x_0 + R\cos\varphi, y_0 + R\sin\varphi)\, R\, d\varphi$$

(see also Example 3, p. 898).

2. (*Converse of the Mean Value Theorem.*) Let u be continuous in O and such that its value at the centre of an arbitrary n-dimensional sphere $\bar{K} \in O$ is equal to the mean value over the boundary of the sphere. Then u is harmonic in O.

3. (*The First Harnack Theorem.*) If a sequence of functions u_n, each of which is harmonic inside a bounded region O and continuous in $\bar{O}$, converges uniformly on the boundary of that region, then the sequence u_n converges uniformly in the entire region O and the limiting function is harmonic in O.

4. (*The Second Harnack Theorem.*) If a series $\sum_{n=1}^{\infty} u_n$ of functions u_n, each of which is harmonic and non-negative in a region O, converges at an interior point of that region, then it converges everywhere in O and the limiting function is harmonic in O. The convergence is uniform in any closed bounded part of the region O.

5. (*Theorem on a Removable Singularity.*) If a function u is harmonic and bounded in a neighbourhood of a point P, with the exception of the point P, then the func-

tion u may be defined at the point P in such a way that u will be harmonic in the entire neighbourhood of the point P.

6. A function harmonic and bounded outside an n-dimensional sphere, has a finite limit at infinity.

7. (*Liouville's Theorem.*) A function, harmonic and bounded in the entire n-dimensional space, is a constant. (Hence, if a harmonic function is not constant in the entire n-dimensional space, then it cannot be bounded.)

8. A function harmonic in a region O is *analytic* in that region, i.e. it can be expanded into a power series (in n variables) in the neighbourhood of any point of the region O.

9. A harmonic function bounded in a circle K is angular extensible almost everywhere on the boundary. This means that the following assertion is true for every

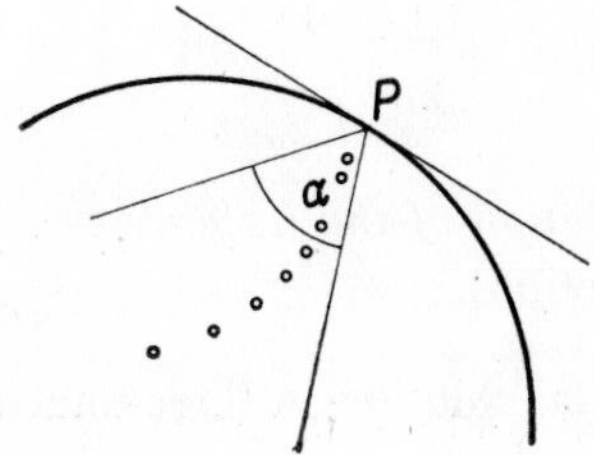

Fig. 18.4.

point P of the boundary with the possible exception of points constituting a set of measure zero: If a sequence of points (x_n, y_n) converges to the point P and if all these points (x_n, y_n) lie in an angle $\alpha < 180°$ the arms of which lie, in a certain neighbourhood of the point P, in the circle under consideration (see Fig. 18.4), then there exists a finite limit $\lim_{n\to\infty} u(x_n, y_n)$ (which is the same for every sequence with the above-mentioned properties).

Theorem 3. *Let two points*

$$P(x_1, x_2, \ldots, x_n), \quad Q(\xi_1, \xi_2, \ldots, \xi_n)$$

be given in n-dimensional space. Let us write

$$r = \sqrt{[(x_1 - \xi_1)^2 + (x_2 - \xi_2)^2 + \ldots + (x_n - \xi_n)^2]}.$$

Then if Q is regarded as fixed, the functions

$$\frac{1}{r^{n-2}} \text{ for } n > 2, \quad \ln\frac{1}{r} \text{ for } n = 2 \tag{21}$$

constitute (as functions of the variables $x_1, x_2, \ldots, x_n$) *a solution of Laplace's equation in the entire space, provided* $P \neq Q$. *The same holds for fixed P, these functions being considered as functions of* $\xi_1, \xi_2, \ldots, \xi_n$, *provided* $Q \neq P$.

Example 1. For $n = 3$ and $P \neq Q$ the function $u = 1/r$ is a solution of the equation

$$\frac{\partial^2 u}{\partial x^2} + \frac{\partial^2 u}{\partial y^2} + \frac{\partial u^2}{\partial z^2} = 0\,.$$

Definition 7. The function

$$\gamma(r) = \frac{\Gamma\left(\frac{n}{2}\right)}{2(n-2)\,(\sqrt{\pi})^n}\,\frac{1}{r^{n-2}} \quad \text{for} \quad n > 2$$

or

$$\gamma(r) = \frac{1}{2\pi} \ln \frac{1}{r} \quad \text{for} \quad n = 2$$

is called the *fundamental solution of the Laplace equation in n-dimensional space* (Γ denoting the gamma function).

Example 2. The fundamental solution in three-dimensional space is thus

$$\gamma(r) = \frac{1}{4\pi}\,\frac{1}{r}\,. \tag{22}$$

Definition 8. Let us consider the interior Dirichlet problem (Definition 4) for Laplace's or Poisson's equation, respectively. The function

$$G(P, Q) = \gamma(r) + v(P, Q)$$

is called the *Green's function* for the appropriate problem provided that $\gamma(r)$ is the fundamental solution of the Laplace equation (see Definition 7) and moreover that $v(P, Q)$ is (for a fixed Q) a harmonic function of the point P in the entire region O in question (including the point Q) and such that $\gamma(r) + v(P, Q)$ vanishes on the boundary of the region O.

REMARK 10. The Green's function for the problem mentioned is therefore a harmonic function of the point P in the entire region O with the exception of the point Q. At this point it has a singularity given by the singularity of the fundamental solution (for example, in three-dimensional space the singularity is given by the singularity of the function (22)). Moreover, the Green's function is zero if P is a point on the boundary.

Theorem 4. The *Green's function is a symmetric function of the points P and Q*, i.e. $G(P, Q) = G(Q, P)$.

REMARK 11. To ensure the existence of the Green's function in a plane or in three-dimensional space it is sufficient to assume that the boundary of the region O is smooth or piecewise smooth (see also Remark 23).

Theorem 5. *Let the Dirichlet problem be given for the Poisson equation with a continuously differentiable right-hand side* $\varrho(Q)$ (*or for the Laplace equation, if* $\varrho = 0$) *and let the region O under consideration have a piecewise smooth boundary S. Let the boundary function* $f(S)$ *be continuous on S. If* $G(P, Q)$ *is the Green's function for this problem, then the solution is given by the formula*

$$u(P) = -\int_S f(S) \frac{\partial G(P, Q)}{\partial n} \, dS - \int_O G(P, Q) \, \varrho(Q) \, dQ \,, \tag{23}$$

where $\partial G/\partial n$ *denotes the outward normal derivative* (Remark 6).

REMARK 12. In the particular case of a plane problem, let the boundary function on the boundary curve k be given as a function of arc length s, i.e. $f = f(s)$. Let us denote by $a_1(s)$, $a_2(s)$ the direction cosines of the outward normal at the point (ξ, η) corresponding to the parametric value s. Then (see equation (12)) formula (23) takes the form

$$u(x, y) = -\int_k f(s) \left[a_1(s) \frac{\partial G(x, y, \xi, \eta)}{\partial \xi} + a_2(s) \frac{\partial G(x, y, \xi, \eta)}{\partial \eta} \right] ds - \\ - \iint_O G(x, y, \xi, \eta) \, \varrho(\xi, \eta) \, d\xi \, d\eta \,. \tag{24}$$

REMARK 13. To find the Green's function for a given region is in general a difficult task. We mention some particular cases:

REMARK 14. Green's function for the interior of a circle O with radius R and centre at the origin of the coordinate system is of the form

$$G(x, y, \xi, \eta) = \frac{1}{2\pi} \ln \frac{1}{r} - \frac{1}{2\pi} \ln \frac{R}{r_1 r_2} \,, \tag{25}$$

where

$$r = \sqrt{[(x - \xi)^2 + (y - \eta)^2]} \,, \quad r_1 = \sqrt{(\xi^2 + \eta^2)} \,,$$

$$r_2 = \sqrt{\left[\left(x - \frac{R^2}{r_1^2} \xi \right)^2 + \left(y - \frac{R^2}{r_1^2} \eta \right)^2 \right]} .$$

REMARK 15. For the interior of a sphere O with radius R and centre at the origin we have

$$G(x, y, z, \xi, \eta, \zeta) = \frac{1}{4\pi r} - \frac{1}{4\pi} \frac{R}{r_1 r_2}, \tag{26}$$

where

$$r = \sqrt{[(x - \xi)^2 + (y - \eta)^2 + (z - \zeta)^2]}, \quad r_1 = \sqrt{(\xi^2 + \eta^2 + \zeta^2)},$$

$$r_2 = \sqrt{\left[\left(x - \frac{R^2}{r_1^2}\xi\right)^2 + \left(y - \frac{R^2}{r_1^2}\eta\right)^2 + \left(z - \frac{R^2}{r_1^2}\zeta\right)^2\right]}.$$

REMARK 16. For $\varrho = 0$ we obtain from (23), (25) and (26) the solution of the Dirichlet problem for the Laplace equation in the interior of a circle or a sphere in the form of the so-called *Poisson's integral*:

$$u(x, y) = \frac{1}{2\pi R} \int_0^{2\pi R} \frac{R^2 - r^2}{R^2 + r^2 - 2Rr \cos \varphi} f(s) \, ds, \tag{27}$$

or

$$u(x, y, z) = \frac{1}{4\pi R^2} \iint_S \frac{R^2 - r^2}{(R^2 + r^2 - 2Rr \cos \varphi)^{3/2}} f(S) \, dS, \tag{28}$$

R standing for the radius of the circle or of the sphere (with centre at the origin), and

$$r = \sqrt{(x^2 + y^2)}, \quad \text{or} \quad r = \sqrt{(x^2 + y^2 + z^2)}, \tag{29}$$

respectively, φ being the angle between the radii drawn to the points (x, y) and (ξ, η), or (x, y, z) and (ξ, η, ζ), respectively. The point (x, y) or (x, y, z) is fixed, the point (ξ, η) (or (ξ, η, ζ)) traces out the circumference of the circle, or the boundary of the sphere, respectively. The function (27) (or (28)) is a solution of the interior Dirichlet problem for an arbitrary continuous f; the point (x, y), or (x, y, z) in (27) denotes an interior point of the circle or of the sphere, respectively. Actual evaluation of the integral may be difficult and is usually carried out approximately.

REMARK 17. We shall now investigate the so-called *potential of a single layer and potential of a double layer* in a plane and in three-dimensional space. We shall assume, without repeating it explicitly, that the curves or surfaces in question are *simple* and *smooth*. Moreover, the curves will be required to have a continuous curvature, and the surfaces will be assumed to be of so-called *Liapunov type*. We do not attempt to give the precise definition of such surfaces (the reader is referred e.g. to [375]); all smooth surfaces we meet in practice are of Liapunov type. The curve k or the surface S is supposed to be closed even though this assumption is not necessary for the definition of potentials. The derivative $\partial u/\partial n$ will always mean the derivative in the direction of the *outward* normal to the closed curve or surface under consideration. The fixed point (x, y) or (x, y, z) at which the potential is evaluated will be denoted by P, the variable point of integration which traces out the curve k or surface S will be denoted by A. $\overrightarrow{AP}$ is the vector with initial point A and end point P, and $\boldsymbol{n}_A$, $\boldsymbol{n}_P$ the vectors of the outward normals at the points A, P, where $P \in k$ (or $P \in S$), respectively. (See Fig. 18.5, p. 898.) The given functions f_1, F_1 or f_2, F_2 (the so-called *densities*) are assumed to be continuous on k or on S, respectively.

Definition 9. The integrals

$$v(P) = \int_k f_1(A) \ln \frac{1}{r} \, \mathrm{d}s \,, \tag{30}$$

and

$$V(P) = \iint_S F_1(A) \frac{1}{r} \, \mathrm{d}S \tag{31}$$

are called the *potentials of a single layer in the plane*, and *in space*, respectively.

The integrals

$$w(P) = \int_k f_2(A) \frac{\cos(\boldsymbol{n}_A, \overrightarrow{AP})}{r} \, \mathrm{d}s \tag{32}$$

and

$$W(P) = -\iint_S F_2(A) \frac{\cos(\boldsymbol{n}_A, \overrightarrow{AP})}{r^2} \, \mathrm{d}s \tag{33}$$

are called the *potentials of a double layer in the plane* and *in space*, respectively (for notation and assumptions see Remark 17); r stands for the distance between the points A, P.

Theorem 6. *The functions* (30)–(33) *are harmonic (as functions of P) in the interior as well as in the exterior of the curve k and the surface S, respectively.*

Theorem 7. *The integrals* (30) *and* (31) *are convergent for* $P \in k$ and $P \in S$, *respectively. The functions v and V are continuous functions of the point P in the entire plane and in the entire space, respectively (including the curve k and the surface S).*

Theorem 8. *The integrals* (32), (33) *converge for $P \in k$ and $P \in S$; however, the functions w, W have a jump on k and on S, respectively. If $P_0 \in k$ or $P_0 \in S$, then the function w, or W, is continuously extensible to the point P_0 from the interior as well as from the exterior of the curve k, or of the surface S, respectively. Let us write w_e or W_e for the value of the continuous extension from the exterior and w_i, or W_i for the continuous extension from the interior of the curve k, or of the surface S,* *respectively. Further, let us denote by* w_0 *or* W_0 *the value at* $P = P_0$ *of the integral* (32) *or* (33), *respectively. Then the following relations hold at* P_0:

$$w_e = w_0 + \pi f_2(P_0), \quad w_i = w_0 - \pi f_2(P_0), \tag{34}$$

$$W_e = W_0 - 2\pi F_2(P_0), \quad W_i = W_0 + 2\pi F_2(P_0). \tag{35}$$

Theorem 9. *Let* $\mathbf{n}_{P_0}$ *be the vector of the outward normal at* $P_0 \in k$, *or* $P_0 \in S$. *The functions* v *and* V, *given by relations* (30) *and* (31), *have derivatives in the direction of* $\mathbf{n}_{P_0}$ *from the exterior as well as from the interior of the curve* k, *or of the surface* S, *respectively. If the derivatives from the interior are denoted by* $(\partial v/\partial n)_i$, $(\partial V/\partial n)_i$ *and from the exterior by* $(\partial v/\partial n)_e$, $(\partial V/\partial n)_e$, *then the following relations hold at* P_0:

$$\left(\frac{\partial v}{\partial n}\right)_e = \left(\frac{\partial v}{\partial n}\right)_0 - \pi f_1(P_0), \quad \left(\frac{\partial v}{\partial n}\right)_i = \left(\frac{\partial v}{\partial n}\right)_0 + \pi f_1(P_0), \tag{36}$$

$$\left(\frac{\partial V}{\partial n}\right)_e = \left(\frac{\partial V}{\partial n}\right)_0 - 2\pi F_1(P_0), \quad \left(\frac{\partial V}{\partial n}\right)_i = \left(\frac{\partial V}{\partial n}\right)_0 + 2\pi F_2(P_0), \tag{37}$$

where

$$\left(\frac{\partial v}{\partial n}\right)_0 = -\int_k f_1(A)\frac{\cos(\mathbf{n}_{P_0}, \overrightarrow{AP_0})}{r}\,\mathrm{d}s, \quad \left(\frac{\partial V}{\partial n}\right)_0 = -\iint_S F_1(A)\,\frac{\cos(\mathbf{n}_{P_0}, \overrightarrow{AP_0})}{r^2}\,\mathrm{d}S. \tag{38}$$

REMARK 18 (*on solving the Dirichlet and the Neumann problems by making use of potentials, leading to a reduction to integral equations*). The interior of the curve k (of the surface S) will be denoted by O, the exterior by H. The Dirichlet problem involves finding a function $u(x, y)$ (or $U(x, y, z)$ in the three-dimensional case) harmonic in O or in H, respectively, which assumes on the boundary the prescribed values f (or F in the three-dimensional case). When solving the Neumann problem, the values g or G of the outward derivative (see Definitions 4 and 5 and Remark 6) are prescribed. In the case of exterior problems the function u should be bounded, while the function U should vanish at infinity.

REMARK 19. Notice that in the above equations as well as in the following equations the normal $\mathbf{n}_A$, or $\mathbf{n}_{P_0}$ points into the exterior of the curve, or of the surface in question, respectively, so that in the case of the exterior Neumann problem it is not an outward normal with respect to H in the sense of Remark 6.

REMARK 20.

A. THE INTERIOR DIRICHLET PROBLEM. The functions u, U are assumed to be of the form (32), (33) ($P \in O$), respectively, where f_2 and F_2 are the functions to be found. Making use of the second equations in (34) and (35), we obtain from the condition

$u(P) \to f(P_0)$, $U(P) \to F(P_0)$ $(P_0 \in k,\ P_0 \in S)$ the following integral equations for f_2, F_2, respectively:

$$f_2(P_0) - \frac{1}{\pi}\int_k \frac{\cos(\mathbf{n}_A, \overrightarrow{AP_0})}{r} f_2(A)\,ds = -\frac{f(P_0)}{\pi} \tag{39}$$

and

$$F_2(P_0) - \frac{1}{2\pi}\iint_S \frac{\cos(\mathbf{n}_A, \overrightarrow{AP_0})}{r^2} F_2(A)\,dS = \frac{F(P_0)}{2\pi}. \tag{40}$$

B. THE EXTERIOR DIRICHLET PROBLEM. The functions u and U are again assumed in the form (32), (33) with unknown functions f_2, F_2, respectively. Now, of course, we have $P \in H$. Making use of the first of equations (34), (35), we obtain for the unknown functions f_2 and F_2 the equations

$$f_2(P_0) + \frac{1}{\pi}\int_k \frac{\cos(\mathbf{n}_A, \overrightarrow{AP_0})}{r} f_2(A)\,ds = \frac{f(P_0)}{\pi}, \tag{41}$$

$$F_2(P_0) + \frac{1}{2\pi}\iint_S \frac{\cos(\mathbf{n}_A, \overrightarrow{AP_0})}{r^2} F_2(A)\,dS = -\frac{F(P_0)}{2\pi}. \tag{42}$$

(The vector $\mathbf{n}_A$ points into the exterior of the curve k or of the surface S.)

C. THE INTERIOR NEUMANN PROBLEM. The functions u, U are assumed in the form (30), (31). By making use of the second of equations (36) and (37), the conditions $\partial u/\partial n = f(P_0)$ and $\partial U/\partial n = G(P_0)$ (for $P \to P_0$ from the interior) yield the following integral equations for the unknown function f_1 or F_1:

$$f_1(P_0) - \frac{1}{\pi}\int_k \frac{\cos(\mathbf{n}_{P_0}, \overrightarrow{AP_0})}{r} f_1(A)\,ds = \frac{g(P_0)}{\pi} \tag{43}$$

or

$$F_1(P_0) - \frac{1}{2\pi}\iint_S \frac{\cos(\mathbf{n}_{P_0}, \overrightarrow{AP_0})}{r^2} F_1(A)\,dS = \frac{G(P_0)}{2\pi}. \tag{44}$$

D. THE EXTERIOR NEUMANN PROBLEM. The functions u, U are assumed in the form (30), (31) (where $P \in H$). The first of equations (36) and (37) imply that

$$f_1(P_0) + \frac{1}{\pi}\int_k \frac{\cos(\mathbf{n}_{P_0}, \overrightarrow{AP_0})}{r} f_1(A)\,ds = \frac{g(P_0)}{\pi}, \tag{45}$$

$$F_1(P_0) + \frac{1}{2\pi}\iint_S \frac{\cos(\mathbf{n}_{P_0}, \overrightarrow{AP_0})}{r^2} F_1(A)\,dS = \frac{G(P_0)}{2\pi}. \tag{46}$$

In (43)–(46), the vector $\mathbf{n}_{P_0}$ points to the exterior of the curve k, or of the surface S, respectively (hence, in the case of the exterior problem it points into the interior of the region H under consideration). In the case of the exterior problem, g or G, respectively is the prescribed derivative of u or U in the direction of the exterior normal to H, that is of the inward normal to the curve h, or to the surface S.

REMARK 21 (*solvability of equations* (39)–(46)). Equations (39), (40), (45), (46) are uniquely solvable for every continuous right-hand side. (In practice, to find the solution may be difficult and numerical methods are usually employed.) Having solved these equations, the functions u, U are given by the integrals (32), (33) and (30), (31), respectively. Of course, we have $P \in O$ for the interior Dirichlet problem and $P \in H$ for the exterior Neumann problem. The integral (30) (for the solution of the exterior Neumann problem) defines a function which is bounded in H if and only if $\int_k g(s)\,\mathrm{d}s = 0$, consequently (30) is a solution in the sense of Remark 7 if and only if this condition is satisfied.

Equations (43), (44) are solvable if and only if

$$\int_k g(s)\,\mathrm{d}s = 0 \quad \text{or} \quad \iint_S G(S)\,\mathrm{d}S = 0\,,$$

respectively, in agreement with (18).

Equations (41), (42) are not, in general, solvable (for an arbitrary continuous right-hand side). By Remark 7, the exterior Dirichlet problem has a solution, this solution, however, need not be of the form (32) or (33), for we do not require that the solution for large r be of the order $1/r$ in the plane or $1/r^2$ in the space. On how to overcome this difficulty, see e.g. [317] (for the plane problem) or [375], (for the three-dimensional problem).

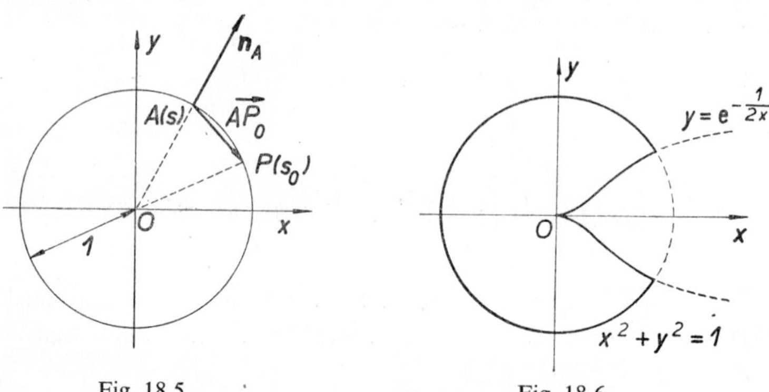

Fig. 18.5. Fig. 18.6.

Example 3. Consider the Dirichlet problem for the unit circle with centre at the origin and with the boundary condition given by a continuous function $f(s)$. Making use of (32) and (39) we shall evaluate the required harmonic function at the origin. It is readily seen from Figure 18.5 that:

$$\cos\left(\boldsymbol{n}_A, \overrightarrow{AP}_0\right) = \cos\left(\frac{\pi}{2} + \frac{s - s_0}{2}\right) = -\sin\frac{s - s_0}{2}\,, \quad r = 2\sin\frac{s - s_0}{2}\,.$$

Equation (39) becomes:

$$f_2(s_0) + \frac{1}{2\pi}\int_0^{2\pi} f_2(s)\,\mathrm{d}s = -\frac{f(s_0)}{\pi}, \tag{47}$$

and its solution is of the form (see § 19.2)

$$f_2(s_0) = -\frac{f(s_0)}{\pi} + k$$

where k is a constant. Substituting this in (47), we get

$$-\frac{f(s_0)}{\pi} + k - \frac{1}{2\pi^2}\int_0^{2\pi} f(s)\,\mathrm{d}s + \frac{1}{2\pi}\int_0^{2\pi} k\,\mathrm{d}s = -\frac{f(s_0)}{\pi}$$

whence

$$k = \frac{1}{4\pi^2}\int_0^{2\pi} f(s)\,\mathrm{d}s\,.$$

If we substitute the expression obtained for $f_2(s)$ into (32), where P denotes the origin, we get (because $r = 1$, $\cos(\boldsymbol{n}_A, \overrightarrow{AP}) = -1$)

$$w(0,0) = \int_0^{2\pi}\left(-\frac{f(s)}{\pi} + k\right).(-1)\,\mathrm{d}s = \frac{1}{\pi}\int_0^{2\pi} f(s)\,\mathrm{d}s - 2\pi k = \frac{1}{2\pi}\int_0^{2\pi} f(s)\,\mathrm{d}s\,.$$

(The above example is, of course, only an illustrative one; see Mean Value Theorem, Remark 9, point 1.)

REMARK 22. Problems involving more general elliptic equations than the Laplace equation may also be reduced to integral equations.

REMARK 23 (*some generalizations of the above-mentioned results*). Existence and uniqueness of the problems mentioned may be proved under very general assumptions. For instance, the Dirichlet problem for the Laplace equation in the plane is (uniquely) solvable (for a given continuous function on the boundary) if the boundary is a Jordan curve (Remark 14.1.3, p. 603).

In three-dimensional space the situation is rather different. Let S be the surface which is formed by revolution about the x-axis of the curve shown in Figure 18.6. The point O is a cusp of this curve. In cases of this type, we encounter the following fact: There exists a function F continuous on S such that the interior Dirichlet problem for the Laplace equation has no solution in the above-mentioned sense (that is, there exists no function U, which is harmonic in the interior $V$of S, continuous in $V + S$ and assumes the values F on S). However, there does exist a bounded function U_1, harmonic in V, continuous in $V + S$ with the exception of the point O and assuming the prescribed values F on S (except at the point O). It can be shown that such a situation cannot occur, however, if the point O may be taken as the vertex of a cone K having a non-zero solid angle at the point O and such that K has only the point O in common with $V + S$. (This was not possible in the foregoing example, since the cross-section passing through the x-axis has a cusp point at O.) In this case, we say that the *exterior cone condition* is satisfied at the point O.

A region V is said to be *regular with respect to the interior Dirichlet problem for the Laplace equation* if the interior Dirichlet problem is solvable for every continuous function prescribed on the boundary of V. In space, every region whose boundary consists entirely of points which satisfy the above-mentioned exterior cone condition is regular with respect to the interior Dirichlet problem; in particular, a region with a smooth boundary is regular. In the plane, a Jordan region is a regular region. In every regular region, there exists a Green's function for the interior Dirichlet problem.

Many results which we have presented for the Laplace and Poisson equation may be generalized to other elliptic equations.

In applications, the prescribed boundary function is frequently discontinuous at some points of the boundary. In this case we must define what is meant by a solution. When solving the Dirichlet problem a solution is, generally, understood in applications to be a bounded function which satisfies the given differential equation and is continuously extensible at all points where the boundary condition is continuous. On existence and uniqueness for special plane regions see, e.g., [332a]; using conformal mapping one can generalize these results for more general regions.

REMARK 24. In recent years there has been a rapid development in so-called *methods of functional analysis* for solving differential equations (not only of elliptic type). The given equation and boundary conditions are then satisfied in a generalized sense (e.g. in an integral sense, in the sense of equality of some functionals, etc.). These *generalized solutions* exist under very general assumptions.

This approach has not yet reached engineering practice (because of its rather high degree of mathematical sophistication) even though it offers greater possibilities for the solution of such problems and often fits the physical nature of the problem better than the classical one. The reader is referred to [282], [374] (see also Chapter 24).

REMARK 25 (*methods of solution*). It is not often possible to find the solution in a closed form. We have presented the solution of Poisson's equation in the entire space (Theorem 1) and of the Dirichlet problem for the Laplace equation in a sphere and in a circle (Remark 16). The solution for a semi-infinite space may be found, for example, in [375]. The Fourier method may often be used with advantage.

In a great majority of cases the solution must be found approximately. It seems that variational methods (Chapter 24) and the finite difference methods (Chapter 27) are most suitable for elliptic equations. Boundary value problems may be solved by reduction to integral equations in the same way as was done for Laplace's and Poisson's equations in Remark 20; these equations are then solved approximately. Solutions of the Laplace equation in the plane may be considered as a real part (or imaginary part) of a holomorphic function and the problem may be solved by the application of integrals of Cauchy's type (see e.g. [281]).

18.5. Hyperbolic Equations. Wave Equation, the Cauchy Problem, the Mixed Problem. Generalized Solutions

REMARK 1 (*informative*). We shall deal only with problems relating to the so-called wave equation, first with the Cauchy problem and then with the mixed problem (involving boundary conditions). When investigating hyperbolic equations, we encounter the following phenomenon which does not occur in the case of elliptic and parabolic equations with sufficiently smooth coefficients: If, as usual, we understand by a solution of the given problem a function which satisfies the given differential equation in the region considered and the prescribed boundary conditions, then a solution need not exist (Remark 7) if the boundary conditions are not sufficiently smooth. Consequently, we introduce so-called *generalized solutions* (Definition 3).

Definition 1. The equation of the form

$$\frac{\partial^2 u}{\partial t^2} = \frac{\partial^2 u}{\partial x_1^2} + \frac{\partial^2 u}{\partial x_2^2} + \ldots + \frac{\partial^2 u}{\partial x_n^2} \tag{1}$$

is called the *wave equation*. In particular, for $n = 1$ we obtain the so-called equation of the vibrating string

$$\frac{\partial^2 u}{\partial t^2} = \frac{\partial^2 u}{\partial x^2}, \tag{2}$$

for $n = 2$ the equation

$$\frac{\partial^2 u}{\partial t^2} = \frac{\partial^2 u}{\partial x^2} + \frac{\partial^2 u}{\partial y^2} \tag{3}$$

and for $n = 3$ the equation

$$\frac{\partial^2 u}{\partial t^2} = \frac{\partial^2 u}{\partial x^2} + \frac{\partial^2 u}{\partial y^2} + \frac{\partial^2 u}{\partial z^2}. \tag{4}$$

Definition 2. The *Cauchy problem* for equation (1) is to find a function $u(t, x_1, \ldots, x_n)$ satisfying equation (1) for $t > 0$, while u and $\partial u/\partial t$ are continuously extensible for $t \to 0$ and the relations

$$u(0, x_1, \ldots, x_n) = \varphi_0(x_1, \ldots, x_n), \tag{5}$$

$$\frac{\partial u}{\partial t}(0, x_1, \ldots, x_n) = \varphi_1(x_1, \ldots, x_n) \tag{6}$$

hold.

REMARK 2. The functions (5), (6) are usually prescribed in the entire hyperplane $x_1, x_2, \ldots, x_n$ (for $n = 1$, for instance, on the entire x-axis). However, they may be prescribed only on a part of that hyperplane.

REMARK 3. The Cauchy problem for the equation

$$\frac{1}{a^2}\frac{\partial^2 u}{\partial t^2} = \frac{\partial^2 u}{\partial x_1^2} + \frac{\partial^2 u}{\partial x_2^2} + \ldots + \frac{\partial^2 u}{\partial x_n^2} \tag{7}$$

(with conditions (5), (6)) may be reduced to the Cauchy problem for equation (1) by the substitution $at = \tau$. We have

$$\frac{\partial^2 u}{\partial \tau^2} = \frac{\partial^2 u}{\partial x_1^2} + \frac{\partial^2 u}{\partial x_2^2} + \ldots + \frac{\partial^2 u}{\partial x_n^2} \tag{8}$$

with initial conditions for $\tau \to 0$:

$$u(0, x_1, \ldots, x_n) = \varphi_0(x_1, \ldots, x_n), \tag{9}$$

$$\frac{\partial u}{\partial \tau}(0, x_1, \ldots, x_n) = \frac{1}{a}\,\varphi_1(x_1, \ldots, x_n). \tag{10}$$

Theorem 1 (*Kirchhoff's formula for $n = 3$*). *If the functions φ_0, φ_1 have continuous partial derivatives of the second order, then the solution of the Cauchy problem* (Definition 2) *for $n = 3$ is*

$$u = \frac{\partial u_{\varphi_0}}{\partial t} + u_{\varphi_1}, \tag{11}$$

where

$$u_\varphi(t, x_1, x_2, x_3) = \frac{1}{4\pi}\iint_{S_t(x_1,x_2,x_3)} \frac{\varphi(\alpha_1, \alpha_2, \alpha_3)}{t}\,\mathrm{d}S_t. \tag{12}$$

Integration is carried out over the surface S_t of the sphere of radius t with centre at the point (x_1, x_2, x_3). To obtain u_{φ_0} and u_{φ_1} and to be able to use (11), *we must, of course, evaluate* (12) *first for $\varphi = \varphi_0$ and then for $\varphi = \varphi_1$.*

REMARK 4. Formula (11) is also applicable for $n = 2$ and $n = 1$. For $n = 2$ (retaining the assumption that φ_0 and φ_1 have continuous derivatives of the second order) the expression

$$u_\varphi(t, x_1, x_2) = \frac{1}{2\pi}\iint_{K_t} \frac{\varphi(\alpha_1, \alpha_2)}{\sqrt{[t^2 - (\alpha_1 - x_1)^2 - (\alpha_2 - x_2)^2]}}\,\mathrm{d}\alpha_1\,\mathrm{d}\alpha_2 \tag{13}$$

is to be substituted in the formula for u (*Poisson's formula*). Here K_t stands for the circle of radius t with centre at the point (x_1, x_2). For $n = 1$ we have

$$u_\varphi(t, x) = \frac{1}{2}\int_{x-t}^{x+t} \varphi(\alpha)\, d\alpha \tag{14}$$

so that

$$u(t, x) = \frac{\varphi_0(x + t) + \varphi_0(x - t)}{2} + \frac{1}{2}\int_{x-t}^{x+t} \varphi_1(\alpha)\, d\alpha \tag{15}$$

(*d'Alembert's formula*).

REMARK 5. For the equation

$$\frac{1}{a^2}\frac{\partial^2 u}{\partial t^2} = \frac{\partial^2 u}{\partial x^2}$$

equation (15) together with (8) and (10) implies that

$$u(t, x) = \frac{\varphi_0(x + at) + \varphi_0(x - at)}{2} + \frac{1}{2a}\int_{x-at}^{x+at} \varphi_1(\alpha)\, d\alpha\,. \tag{16}$$

If it is required to solve the equation

$$\frac{\partial^2 u}{\partial t^2} = \frac{\partial^2 u}{\partial x_1^2} + \frac{\partial^2 u}{\partial x_2^2} + \ldots + \frac{\partial^2 u}{\partial x_n^2} + f(x_1, x_2, \ldots, x_n, t)$$

with initial conditions (5), (6), then the following expressions are to be added to (11):

for $n = 3$:

$$\frac{1}{4\pi}\iiint_{L_t} \frac{f(\alpha_1, \alpha_2, \alpha_3, t - \varrho)}{\varrho}\, d\alpha_1\, d\alpha_2\, d\alpha_3\,,$$

where L_t is the sphere with centre at the point (x_1, x_2, x_3) and radius t,

$$\varrho = \sqrt{[(x_1 - \alpha_1)^2 + (x_2 - \alpha_2)^2 + (x_3 - \alpha_3)^2]}\,;$$

for $n = 2$:

$$\frac{1}{2\pi}\int_0^t d\tau \iint_{L_{t-\tau}} \frac{f(\alpha_1, \alpha_2, \tau)}{\sqrt{[(t - \tau)^2 - \varrho^2]}}\, d\alpha_1\, d\alpha_2\,,$$

where $L_{t-\tau}$ is the circle with centre (x_1, x_2) and radius $t - \tau$,

$$\varrho = \sqrt{[(x_1 - \alpha_1)^2 + (x_2 - \alpha_2)^2]}\,;$$

for $n = 1$:

$$\frac{1}{2}\int_0^t d\tau \int_{x-t+\tau}^{x+t-\tau} f(\alpha, \tau)\, d\alpha .$$

REMARK 6 (*uniqueness and well-posed nature of the Cauchy problem*). Under the assumptions mentioned, the solutions presented in Theorem 1 and Remark 4 are *unique*. Further: the problem formulated in Definition 2 is *well-posed*, i.e. the solution depends continuously on the initial conditions. For $n = 1$ this fact

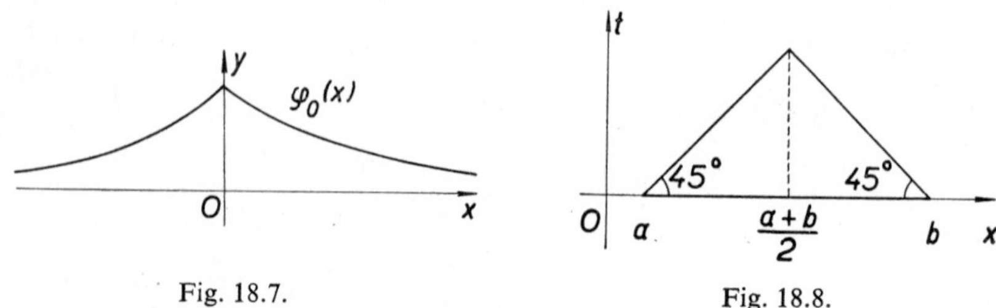

Fig. 18.7. Fig. 18.8.

may be readily seen from (15). In the general case (for n dimensions) the following assertion is valid: Consider the solution in the interval $[0, T]$, T being any finite number. Denote by the symbol $[n/2]$ the greatest integer which is less or equal to $n/2$. (For $n = 1$, $[n/2] = 0$, for $n = 2$ and $n = 3$ we have $[n/2] = 1$, etc.). Then for an arbitrary $\varepsilon > 0$ there exists a $\delta > 0$ such that the change in u (in absolute value) is less than ε whenever the change in φ_0 and φ_1 and their derivatives up to the order $[n/2]$ is less than δ. In particular, for $n = 1$ it suffices to consider only changes in the functions φ_0 and φ_1, as may be seen directly from (15). For $n = 2$, however, it is not sufficient to consider only φ_0 and φ_1; the derivatives must also be taken into account. It can be shown that u may become large even if $|\varphi_0|$ and $|\varphi_1|$ are small since they may, nevertheless, have large derivatives.

REMARK 7 (*generalized solution*). It is readily seen from (15) that if the functions φ_0 and φ_1 are not smooth enough (for example, if φ_0 does not possess a second derivative and φ_1 does not possess a first derivative, see Figure 18.7), then the function u is not a solution (in the ordinary sense), since it does not possess derivatives of the second order. For example, let the functions φ_0 and φ_1 be merely continuous. Since the problem is well posed (Remark 6) the change in the function (15) will be arbitrarily small if the functions φ_0 and φ_1 are replaced (with a sufficient accuracy) by functions ψ_0, ψ_1 having two continuous derivatives. The function (15) corresponding to ψ_0 and ψ_1 will then constitute a solution in the usual sense. In this way we arrive at the concept of a generalized solution:

Definition 3. A function u is said to be a *generalized solution* of the given problem (in the given domain O) if there exists a sequence $u_1, u_2, u_3, \ldots$ of solutions (in the usual sense) of this problem such that as $p \to \infty$, u_p converges to u uniformly in O.

REMARK 8. It is readily seen that if φ_0 and φ_1 are continuous, then (15) represents a generalized solution of the Cauchy problem for equation (2), for instance in every rectangle $-a + T \leqq x \leqq a - T$, $0 \leqq t \leqq T$, while u is continuous in this rectangle. It is sufficient to make use of Remark 10 and in virtue of the Weierstrass Theorem to replace the functions φ_0 and φ_1 by polynomials in the interval $[-a, a]$. (By a suitable choice of T and a, any point of the xt-plane can be included, so that in this sense (15) represents a generalized solution in every bounded part of the plane.)

REMARK 9. Generalized solutions may be introduced in various ways according to the purpose we wish to achieve. (See, for example [375] and [374], where the existence and uniqueness of generalized solutions is proved for a wide class of problems.)

REMARK 10. It follows from (15) that the values of u in the entire triangle T drawn in Fig. 18.8 depend only on the values of φ_0 and φ_1 in the interval (a, b) and are independent of the values of these functions outside that interval. Similarly, it may be shown that the values of the solution (13) in a right circular cone K having its base P in the plane x_1x_2 and formed by generators which make an angle of 45° with the axis of the cone are uniquely determined by the functions φ_0 and φ_1 in P. The sides of the triangle T or the generators of the cone K are obviously characteristics of equation (2), or (3), respectively. A similar result is valid also for $n > 2$.

On the Cauchy problem posed on a characteristic see § 18.1, especially Theorem 2, Examples 10 and 11.

REMARK 11 (*a mixed problem*). In some problems, the required solution u of the wave equation (1) is not only subject to the initial conditions (5), (6), in which $(x_1, x_2, \ldots, x_n)$ denotes a point in the given region O, but must also satisfy certain conditions on the boundary, for all t. Such a problem is said to be a *mixed problem*, and some typical examples follow.

Example 1. Find a solution of the equation

$$\frac{\partial^2 u}{\partial t^2} = \frac{\partial^2 u}{\partial x^2} \tag{17}$$

in the semi-infinite strip $O(0 < x < l, t > 0)$ such that it is continuous for $0 \leqq x \leqq l$, $t \geqq 0$ and satisfies the following initial and boundary conditions:

$$u(0, x) = \varphi_0(x)\,, \tag{18}$$

$$\frac{\partial u}{\partial t}(0, x) = \varphi_1(x)\,, \tag{19}$$

$$u(t, 0) = 0\,, \tag{20}$$

$$u(t, l) = 0\,, \tag{21}$$

where $\varphi_0(0) = \varphi_0(l) = 0$, $\varphi_1(0) = \varphi_1(l) = 0$. (This is the problem of vibration of a string of length l, fixed at both ends; the initial position of the string $u(x, 0)$ being given by the function $\varphi_0(x)$ and the initial velocity $\dfrac{\partial u}{\partial t}(x, 0)$ by $\varphi_1(x)$; for the solution see § 26.1.)

Example 2. In a similar way, the mixed problem can be formulated in the case $n > 1$. For example, if $n = 2$, a solution of equation (3) is to be found which is continuous for $(x_1, x_2) \in \bar{O}$ and $t \geqq 0$ (O being the given region with boundary Γ), satisfies conditions (5), (6), and vanishes for $(x_1, x_2) \in \Gamma$, $t > 0$ (i.e. on the lateral surface of the cylinder in the interior of which the solution is to be found).

REMARK 12. Other conditions may be prescribed on the boundary Γ for $t > 0$; we may, for instance, have $\partial u/\partial n = 0$, or more generally $\partial u/\partial n + \sigma u = 0$, where $\partial u/\partial n$ stands for the outward normal derivative and σ is a non-negative constant (more generally a non-negative continuous function), or the corresponding nonhomogenious conditions. The usual requirement is that not only u but also the derivatives of the first order should be continuous for all $t \geqq 0$ and all points of the closed region $\bar{O}$. These assumptions ensure the uniqueness of the solution of the given problems and also the continuous dependence on initial conditions in the following sense: Two solutions which satisfy the same boundary conditions on Γ for $t > 0$ (i.e. their difference satisfies either $u = 0$ or $\partial u/\partial n = 0$ or $\partial u/\partial n + \sigma u = 0$) are arbitrarily close to each other if the difference of the functions φ_0 and φ_1 and their derivatives up to the order $[\frac{1}{2}n] + 1$ is sufficiently small. Here $[\frac{1}{2}n]$ denotes the greatest integer which is less than or equal to $\frac{1}{2}n$.

REMARK 13 (*methods of solution*). A typical method of solving mixed problems is the Fourier method. Details are given in Chapter 26, especially in § 26.1, where the problem for equation (17) under conditions (18)–(21) is solved. The Fourier method can also be applied to problems concerning more general equations of hyperbolic type and also in multidimensional cases.

Another typical method of finding solutions of mixed problems, especially useful for time-dependent boundary conditions, is the Laplace transformation (see Chap. 28). (Cf. a similar example for the heat-conduction equation presented in Chap. 28.)

The reader will find a variety of practical examples solved by Laplace transformation methods, for example in [61].

Of numerical techniques, the finite difference method (Chapter 27) and the functional analytical methods (Chapter 24) are suitable.

The initial conditions are sometimes prescribed in a different way from that of the Cauchy problem (Definition 2). For instance, we may have to find a solution of equation (2) such that it assumes prescribed values on the curve $y = h(x)$, where $h'(x) < 0$,

$$u\big|_{y=h(x)} = \varphi_0(x)\,, \quad \frac{\partial u}{\partial t}\bigg|_{y=h(x)} = \varphi_1(x)\,.$$

For this case, a suitable method of solution has been proposed by Riemann (see [375], p. 65, for example). A method due to Kirchhoff [375] is suitable for solutions of some multidimensional problems. On some questions concerning hyperbolic systems of equations see, for example, [317].

18.6. Parabolic Equations. The Heat-conduction Equation. The Cauchy Problem. Mixed Boundary Value Problems

REMARK 1. In this paragraph we deal with the heat-conduction equation which for $n = 3$, 2, or 1 is of the form

$$\frac{\partial^2 u}{\partial x^2} + \frac{\partial^2 u}{\partial y^2} + \frac{\partial^2 u}{\partial z^2} = \frac{\partial u}{\partial t} + f(x, y, z, t), \tag{1}$$

or

$$\frac{\partial^2 u}{\partial x^2} + \frac{\partial^2 u}{\partial y^2} = \frac{\partial u}{\partial t} + f(x, y, t), \tag{2}$$

or

$$\frac{\partial^2 u}{\partial x^2} = \frac{\partial u}{\partial t} + f(x, t), \tag{3}$$

respectively.

If $f \equiv 0$ (which means physically that in the region considered no sources of heat are present) the equation is said to be homogeneous.

The equation

$$\frac{\partial^2 u}{\partial x^2} + \frac{\partial^2 u}{\partial y^2} + \frac{\partial^2 u}{\partial z^2} = \frac{1}{a^2} \frac{\partial u}{\partial t} + g(x, y, z, t)$$

is transformed by the substitution $a^2 t = \tau$ to the form (1) in the same way as in Remark 18.5.3.

Definition 1. The following problem is called the *Cauchy problem* for equation (1): To find a function $u(x, y, z, t)$ which is continuous at all points of the three-dimensional space xyz and for $t \geqq 0$ (we shall briefly write for $(x, y, z) \in R_3$, $t \geqq 0$), is bounded in R_3, $t \geqq 0$, satisfies equation (1) for $(x, y, z) \in R_3$, $t > 0$ and assumes values prescribed by a given continuous function $\varphi(x, y, z)$ for $t = 0$, i.e.

$$u(x, y, z, 0) = \varphi(x, y, z). \tag{4}$$

REMARK 2. The Cauchy problem for equation (2) (i.e. in the plane R_2) and for equation (3) (the one-dimensional problem) is defined in a similar way.

Theorem 1. *The function*

$$v(x, y, z, t, \xi, \eta, \zeta, \tau) = \frac{1}{8\pi^{3/2}(t-\tau)^{3/2}} \mathrm{e}^{-r^2/[4(t-\tau)]}, \tag{5}$$

where

$$r = \sqrt{[(x-\xi)^2 + (y-\eta)^2 + (z-\zeta)^2]} \tag{6}$$

when regarded as a function of the variables x, y, z, t *(for fixed* ξ, η, ζ, τ,*) satisfies equation* (1) *whenever* $t > \tau$. *On the other hand, when regarded as a function of* ξ, η, ζ, τ *it satisfies the equation*

$$\frac{\partial^2 u}{\partial \xi^2} + \frac{\partial^2 u}{\partial \eta^2} + \frac{\partial^2 u}{\partial \zeta^2} = -\frac{\partial u}{\partial \tau} \tag{7}$$

for $\tau < t$.

Theorem 2. *For* $t > \tau$ *the relation*

$$\int_{-\infty}^{\infty}\int_{-\infty}^{\infty}\int_{-\infty}^{\infty} v \,\mathrm{d}x\,\mathrm{d}y\,\mathrm{d}z = 1 \tag{8}$$

holds.

REMARK 3. The function (5) is called the *fundamental solution of the homogeneous equation* (1). The physical interpretation of this fundamental solution is, roughly speaking, that it describes the temperature field in space at time t due to a unit heat impulse at the point (ξ, η, ζ) at the instant τ.

REMARK 4. The fundamental solutions of the homogeneous equations (2) and (3), namely

$$v(x, y, t, \xi, \eta, \zeta) = \frac{1}{4\pi(t-\tau)} \mathrm{e}^{-r^2/[4(t-\tau)]} \quad \left(r = \sqrt{[(x-\xi)^2 + (y-\eta)^2]}\right), \tag{9}$$

$$v(x, t, \xi, \tau) = \frac{1}{2\sqrt{(\pi)}\sqrt{(t-\tau)}} \mathrm{e}^{-(x-\xi)^2/[4(t-\tau)]} \tag{10}$$

have quite similar properties.

Theorem 3. *If the functions* $f(x, y, z, t)$, $\varphi(x, y, z)$ *and their derivatives of the first order are bounded and continuous in the entire space* R_3 *and for all* t *under consideration, then the solution of the Cauchy problem* (Definition 1) *is*

$$u(x, y, z, t) = \int_{-\infty}^{\infty}\int_{-\infty}^{\infty}\int_{-\infty}^{\infty} \varphi(\xi, \eta, \zeta)\, v(x, y, z, t, \xi, \eta, \zeta, 0)\,\mathrm{d}\xi\,\mathrm{d}\eta\,\mathrm{d}\zeta -$$

$$- \int_0^t \left[\int_{-\infty}^{\infty}\int_{-\infty}^{\infty}\int_{-\infty}^{\infty} f(\xi, \eta, \zeta, \tau)\, v(x, y, z, t, \xi, \eta, \zeta, \tau)\,\mathrm{d}\xi\,\mathrm{d}\eta\,\mathrm{d}\zeta\right]\mathrm{d}\tau\,. \tag{11}$$

REMARK 5. This solution is *unique* (in Definition 1 we require the function u to be bounded). In addition, Cauchy's problem is well-posed in the sense that small changes of the functions f and φ cause only small changes of the function u.

REMARK 6. It can be shown that if $f \equiv 0$ (i.e. equation (1) is homogeneous), the function u given by formula (11) is continuously differentiable infinitely many times no matter whether φ has derivatives or not. The heat-conduction equation differs essentially in this property from the wave equation (see Remark 18.5.7). On the other hand it should be noted that the Cauchy problem need not possess a solution for $t < 0$.

REMARK 7. The results for equations (2) and (3) are quite similar. The solution of the Cauchy problem is as follows:

$$u(x, y, t) = \frac{1}{4\pi t} \int_{-\infty}^{\infty} \int_{-\infty}^{\infty} \varphi(\xi, \eta) \, e^{-[(x-\xi)^2 + (y-\eta)^2]/4t} \, d\xi \, d\eta -$$

$$- \frac{1}{4\pi} \int_0^t \frac{1}{t - \tau} \left[\int_{-\infty}^{\infty} \int_{-\infty}^{\infty} f(\xi, \eta, \tau) \, e^{-[(x-\xi)^2 + (y-\eta)^2]/4(t-\tau)} \, d\xi \, d\eta \right] d\tau \,, \qquad (12)$$

$$u(x, t) = \frac{1}{2\sqrt{(\pi)}\sqrt{(t)}} \int_{-\infty}^{\infty} \varphi(\xi) \, e^{-(x-\xi)^2/4t} \, d\xi -$$

$$- \frac{1}{2\sqrt{(\pi)}} \int_0^t \frac{1}{\sqrt{(t - \tau)}} \left[\int_{-\infty}^{\infty} f(\xi, \tau) \, e^{-(x-\xi)^2/[4(t-\tau)]} \, d\xi \right] d\tau \,. \qquad (13)$$

REMARK 8. Very frequently *mixed* (boundary value) problems are met with. The following problem is typical: Find a solution of equation (3), continuous and bounded in the region $\bar{O}(a \leqq x \leqq b,\ t \geqq 0)$, such that on the segment $a \leqq x \leqq b$, $t = 0$ it assumes the values of a prescribed function $\varphi(x)$, i.e.

$$u(x, 0) = \varphi(x) \qquad (14)$$

while for $x = a$, $x = b$ it takes the values of other prescribed functions

$$u(a, t) = \psi_1(t) \,, \quad u(b, t) = \psi_2(t) \quad (t > 0) \,. \qquad (15)$$

(This is the problem of heat conduction in an insulated bar of length $b - a$ having initial temperature $u(x, 0) = \varphi(x)$ whose end points are kept at the temperatures $\psi_1(t)$ and $\psi_2(t)$, respectively. The inner sources of heat are characterized by the function $f(x, t)$. See Example 28.2.2 and § 26.3.)

Theorem 4. *Let $f(x, t) \equiv 0$ (i.e. equation (3) is homogeneous). Then the solution of the problem mentioned in Remark 8 has the following property: For any rectangle $a \leqq x \leqq b$, $0 \leqq t \leqq T$ (T being a positive number) the solution takes its maximum and minimum values either on the lower base (for $t = 0$), or on one of its lateral sides ($x = a$, $x = b$).*

REMARK 9. Theorem 4 implies immediately *uniqueness* of the solution. Moreover, the problem is well posed (for the non-homogeneous equation (3) as well) in the following sense: In every rectangle $a \leqq x \leqq b$, $0 \leqq t \leqq T$ the solution u changes only by a small amount if the function f and the functions (14) and (15) change by a small amount. If the function f remains unchanged and if the change of the function (14), (15) is smaller than ε, then the change of u is also smaller than ε.

REMARK 10. In the same way as in Remark 8 the problem may be formulated for equations (1) and (2). In the case of equation (2) a closed region $\bar{O}$ of the plane xy is given instead of the segment $a \leqq x \leqq b$, $t = 0$, in the case of equation (1) a (closed) three-dimensional region is given. Boundary conditions corresponding to those of (15) are not only functions of time but also, in general, functions of position on the boundary. Theorem 4 and Remark 9 are again valid.

REMARK 11. It is possible to pose some other problems different from those formulated in Remark 8. For example, it may be prescribed that for $x = b$, $\partial u/\partial n = 0$ should hold (physical interpretation: the end $x = b$ of the bar $a \leqq x \leqq b$ is insulated) or the condition $\partial u/\partial n + \alpha u = 0$ ($\alpha > 0$; Newton's condition of heat transfer) may be prescribed, or the corresponding non-homogeneous conditions, etc. For a wide class of such problems existence and uniqueness theorems can be proved (see e.g. [317]). As far as the existence and uniqueness of the solution are concerned it does not matter if the boundary conditions are discontinuous at a finite number of points (if, for example, in the problem of Remark 8 the relations $\psi_1(0) = \varphi(a)$ or $\psi_2(0) = \varphi(b)$ do not hold. etc.) provided that we require the solution to be bounded and to satisfy the boundary conditions everywhere except at those points.

REMARK 12 (*methods of solution*). The method of the Laplace transform is widely used. The problem of Remark 8 and problems of a similar nature are typical cases in which this method can be used (see Example 28.2.2). A variety of examples may be found, e.g., in [61]. In the case where several space coordinates are involved, the use of Laplace transforms is complicated by the fact that the transformed equation is again a *partial* differential equation.

Another efficient method is to use Fourier series. If, for instance, we have $\psi_1 \equiv 0$, $\psi_2 \equiv 0$ and also $f \equiv 0$ (the homogeneous equation) in the problem of Remark 8, then using the Fourier method (§ 26.3), we arrive at the result (here $a = 0$, $b = l$)

$$u(x, t) = \sum_{n=1}^{\infty} a_n \sin \frac{n\pi x}{l} \, e^{-n^2\pi^2 t/l^2} \tag{16}$$

where

$$a_n = \frac{2}{l} \int_0^l \varphi(x) \sin \frac{n\pi x}{l} \, dx \,. \tag{17}$$

If $\varphi(x)$ is continuous in $[0, l]$ and $\varphi(0) = 0$, $\varphi(l) = 0$, then (16) is the desired solution. If $\varphi(x)$ is bounded and continuous except at a finite number of points or if

$\varphi(0) \neq 0$ or $\varphi(l) \neq 0$, then the condition (14) is satisfied with the possible exception of those points (see Remark 11).

The finite difference method is practically universal for obtaining solutions of more complicated problems. The choice of space and time differences requires care in the case of parabolic equations (see Remark 27.7.1, p. 1122).

The homogeneous equations (1)–(3) may be solved by transformation into integral equations using so-called *heat potentials* (see e.g. [281]) which are similar to those in Definition 18.4.9. p. 895.

For generalization of some results see [317].

18.7. Some Other Problems of Partial Differential Equations. Systems of Equations. Pfaffian Equation. Self-adjoint Equations. Equations of Higher Order, Biharmonic Equation. Eigenvalue Problems

REMARK 1. The solvability of the so-called *Pfaffian equation* is studied in geometry (Remark 2). The solvability of the system of two equations

$$\frac{\partial z}{\partial x} = A(x, y, z)\,, \tag{1}$$

$$\frac{\partial z}{\partial y} = B(x, y, z) \tag{2}$$

for one unknown function $z(x, y)$ is connected with it. It can be shown that in order that this system should be solvable (i.e. that equations (1) and (2) should be consistent) a necessary condition is:

$$\frac{\partial A}{\partial y} + \frac{\partial A}{\partial z} B = \frac{\partial B}{\partial x} + \frac{\partial B}{\partial z} A \tag{3}$$

(A, B, $\partial A/\partial y$, $\partial A/\partial z$, $\partial B/\partial x$, $\partial B/\partial y$ are supposed to be continuous functions in the region O in question). If (3) is satisfied identically (i.e. for all x, y, z of the region O which is assumed to be simply connected), then there exists a system of surfaces $z(x, y)$ satisfying (1), (2). (One and only one integral surface of the system (1), (2) passes then through each point $(x_0, y_0, z_0) \in O$.)

If (3) is not satisfied identically, it is possible in general to express z as a function of x, y. If the system (1), (2) has a solution it is given by this relation. Whether $z = \varphi(x, y)$ is a solution or not should be verified by inspection.

REMARK 2. The equation of the form

$$P(x, y, z)\,\mathrm{d}x + Q(x, y, z)\,\mathrm{d}y + R(x, y, z)\,\mathrm{d}z = 0 \tag{4}$$

is called the *Pfaffian equation*.

Geometric interpretation: To every point $(x_0, y_0, z_0) \in O$ (O is the simply connected region in question; P, Q, R and their derivatives of the first order are assumed to be continuous in O) there corresponds a vector with components P, Q, R. To solve equation (4) means, geometrically, to find a system of surfaces that are orthogonal to the field of those vectors (in O). The necessary and sufficient condition for such a system to exist is that the condition of integrability

$$P\left(\frac{\partial Q}{\partial z} - \frac{\partial R}{\partial y}\right) + Q\left(\frac{\partial R}{\partial x} - \frac{\partial P}{\partial z}\right) + R\left(\frac{\partial P}{\partial y} - \frac{\partial Q}{\partial x}\right) = 0 \tag{5}$$

should be satisfied identically in O.

If condition (5) is not satisfied, then it can be shown that such a system of surfaces does not exist. However, it is possible to find one-dimensional integral manifolds (curves) of equation (4),

$$y = y(x), \quad z = z(x). \tag{6}$$

One of the functions (6) may be chosen arbitrarily; the other is then determined by solving the ordinary differential equation

$$P(x, y, z) + Q(x, y, z)\frac{\mathrm{d}y}{\mathrm{d}x} + R(x, y, z)\frac{\mathrm{d}z}{\mathrm{d}x} = 0 .$$

REMARK 3. The systems of equations which we meet in physical and engineering problems are usually slightly different in their character. (For some systems see, for example, [317], [75].) The solution of these systems is often reduced to the solution of a single equation of higher order; for example, this is done for the equations of hydrodynamics and for the equations of the electromagnetic field, by introducing scalar and vector potentials, see e.g. [208]. Two-dimensional problems of hydrodynamics are solved fairly simply by using the theory of functions of a complex variable; cf. Example 21.3.3, p. 980).

When solving plane problems of elasticity we meet the system

$$\frac{\partial X_x}{\partial x} + \frac{\partial X_y}{\partial y} = 0, \quad \frac{\partial X_y}{\partial x} + \frac{\partial Y_y}{\partial y} = 0, \quad \Delta(X_x + Y_y) = 0 \tag{7}$$

where Δ stands for the Laplace operator,

$$\Delta u = \frac{\partial^2 u}{\partial x^2} + \frac{\partial^2 u}{\partial y^2}, \tag{8}$$

and X_x, X_y, Y_y are the components of the stress-tensor to be found. It can be shown that the solution of system (7) is equivalent to the solution of the *biharmonic equa-*

tion for Airy's function U, i.e. of the equation (the *biharmonic equation*)

$$\frac{\partial^4 U}{\partial x^4} + 2\,\frac{\partial^4 U}{\partial x^2\,\partial y^2} + \frac{\partial^4 U}{\partial y^4} = 0 \quad \text{or briefly} \quad \Delta\Delta U = 0\,. \tag{9}$$

Every biharmonic function U (i.e. a function satisfying equation (9)) is such that its derivatives

$$X_x = \frac{\partial^2 U}{\partial y^2}\,, \quad X_y = -\,\frac{\partial^2 U}{\partial x \partial y}\,, \quad Y_y = \frac{\partial^2 U}{\partial x^2}$$

satisfy system (7). At the same time we can find the boundary conditions to be satisfied by U from the conditions prescribed for the functions X_x, X_y, Y_y. For example, if the load is prescribed on the boundary of the region in question, the values of the function U and of its normal derivative are prescribed on the region considered.

There are several methods of solving the biharmonic problem. One of the possible methods reduces the search for the function U to that of finding two holomorphic functions of one complex variable. These functions are then found by solving some integral equations, or by variational methods or by conformal mapping. For details see [21], where the existence and uniqueness of solutions is proved for a wide class of problems provided that the boundary of the region considered is sufficiently smooth. For variational methods independent of the use of function theory see Chapter 24; for more details see, e.g. [282].

A widely applied method for solving the biharmonic problem is the finite difference technique (Chapter 27). For the use of infinite series see § 26.5.

For the biharmonic equation there is no maximum principle similar to Theorem 18.4.2 for the Laplace equation.

REMARK 4. Problems in the theory of shells are reduced, essentially by the same method, to the solution of a single equation of higher order [439].

REMARK 5. *Eigenvalue problems* arise also in the theory of partial differential equations. That is to say, we meet the following problem: to find a non-trivial solution of the equation

$$M(u) = \lambda\, N(u) \tag{10}$$

in a region O under homogeneous boundary conditions (which are usually of the form $u = 0$, or $\partial u/\partial n = 0$, or $\partial u/\partial n + cu = 0$, where $\partial u/\partial n$ stands for the outward normal derivative). M and N are self-adjoint linear operators in several variables, N being of lower order than M. Definitions and results are very similar to those quoted in § 17.17.

Definition 1. Let

$$L(u) = \sum_{k=0}^{l} \sum_{j_1,j_2,\ldots,j_k=1}^{m} A_{j_1,j_2,\ldots,j_k}(x_1, x_2, \ldots, x_m)\, \frac{\partial^k u}{\partial x_{j_1}\,\partial x_{j_2} \ldots \partial x_{j_k}} \tag{11}$$

be a linear operator of l-th order in the variables $x_1, x_2, \ldots, x_m$ with

$$A_{j_1,j_2,\ldots,j_k} = A_{i_1,i_2,\ldots,i_k} \tag{12}$$

provided the subscripts $j_1, j_2, \ldots, j_k$ are permutations of the subscripts $i_1, i_2, \ldots, i_k$.

The operator

$$K(u) = \sum_{k=0}^{l} (-1)^k \sum_{j_1,j_2,\ldots,j_k=1}^{m} \frac{\partial^k (A_{j_1,j_2,\ldots,j_k} \cdot u)}{\partial x_{j_1} \partial x_{j_2} \ldots \partial x_{j_k}} \tag{13}$$

is called the *adjoint operator to the operator* $L(u)$.

If $K(u) = L(u)$, the operator L is called *self-adjoint.*

Example 1. It is easy to verify that the operator

$$L(u) = \frac{\partial}{\partial x}\left[A_{11}(x, y)\frac{\partial u}{\partial x}\right] + \frac{\partial}{\partial y}\left[A_{22}(x, y)\frac{\partial u}{\partial y}\right] +$$

$$+ \frac{\partial}{\partial x}\left[A_{12}(x, y)\frac{\partial u}{\partial y}\right] + \frac{\partial}{\partial y}\left[A_{12}(x, y)\frac{\partial u}{\partial x}\right] + A_3(x, y)u \tag{14}$$

is a self-adjoint operator. For, if (14) is written in the form (11), it becomes ($l = 2$, $m = 2$),

$$A_{11}\frac{\partial^2 u}{\partial x^2} + A_{22}\frac{\partial^2 u}{\partial y^2} + 2A_{12}\frac{\partial^2 u}{\partial x\,\partial y} + \left(\frac{\partial A_{11}}{\partial x} + \frac{\partial A_{12}}{\partial y}\right)\frac{\partial u}{\partial x} +$$

$$+ \left(\frac{\partial A_{12}}{\partial x} + \frac{\partial A_{22}}{\partial y}\right)\frac{\partial u}{\partial y} + A_3 u\,. \tag{15}$$

On the other hand, by (13)

$$K(u) = \frac{\partial^2(A_{11}u)}{\partial x^2} + \frac{\partial^2(A_{22}u)}{\partial y^2} + 2\,\frac{\partial^2(A_{12}u)}{\partial x\,\partial y} -$$

$$- \frac{\partial}{\partial x}\left[\left(\frac{\partial A_{11}}{\partial x} + \frac{\partial A_{12}}{\partial y}\right)u\right] - \frac{\partial}{\partial y}\left[\left(\frac{\partial A_{12}}{\partial x} + \frac{\partial A_{22}}{\partial y}\right)u\right] + A_3 u \tag{16}$$

and we readily find that the expressions (15) and (16) are identical.

In particular, the Laplace operator is self-adjoint, for

$$\Delta u = \frac{\partial^2 u}{\partial x^2} + \frac{\partial^2 u}{\partial y^2} = \frac{\partial}{\partial x}\left(\frac{\partial u}{\partial x}\right) + \frac{\partial}{\partial y}\left(\frac{\partial u}{\partial y}\right)$$

so that obviously it is of the form (14).

REMARK 6. More generally: Every second order operator in m variables of the form

$$L(u) = \sum_{i,k=1}^{m} \frac{\partial}{\partial x_i}\left(A_{ik}(x_1, x_2, \ldots, x_m)\frac{\partial u}{\partial x_k}\right), \quad A_{ik} = A_{ki},$$

is self-adjoint.

REMARK 7. In a similar fashion to the case of functions of one variable (§ 17.17) we can define such concepts as a *comparison function* (such a function has a sufficient number of derivatives and satisfies the boundary conditions but, in general, it does not satisfy the given equation); an *eigenfunction* (a non-trivial solution of the eigenvalue problem); a *self-adjoint problem* (the operators in (10) are self-adjoint, the order of N is lower than that of M and for every pair of comparison functions the relations

$$\int_O \{v\, M(u) - u\, M(v)\}\, \mathrm{d}O = 0\,, \quad \int_O \{u\, N(v) - v\, N(u)\}\, \mathrm{d}O = 0 \tag{17}$$

hold), and, finally, a *positive definite problem* (for every comparison function that is not identically equal to zero the inequalities

$$\int_O u\, M(u)\, \mathrm{d}O > 0\,, \quad \int_O u\, N(u)\, \mathrm{d}O > 0 \tag{18}$$

are satisfied).

For example, the problem

$$\Delta u + \lambda c(x, y)\, u = 0 \quad (c(x, y) > 0 \text{ in } O) \tag{19}$$

(where O is a bounded region in the plane xy) with the boundary condition

$$u = 0 \tag{20}$$

prescribed on the boundary Γ of the region O, is self-adjoint and positive definite. In this problem $M(u) = -\Delta u$ and $N(u) = c(x, y)\, u$ are self-adjoint operators (Example 1), the order of M is two and the order of N is zero; furthermore

$$\iint_O \{v(-\Delta u) - u(-\Delta v)\}\, \mathrm{d}x\, \mathrm{d}y =$$

$$= \iint_O \left\{\frac{\partial}{\partial x}\left[u\frac{\partial v}{\partial x} - v\frac{\partial u}{\partial x}\right] + \frac{\partial}{\partial y}\left[u\frac{\partial v}{\partial y} - v\frac{\partial u}{\partial y}\right]\right\} \mathrm{d}x\, \mathrm{d}y =$$

$$= \int_\Gamma \left(u\frac{\partial v}{\partial x} - v\frac{\partial u}{\partial x}\right) \mathrm{d}y - \int_\Gamma \left(u\frac{\partial v}{\partial y} - v\frac{\partial u}{\partial y}\right) \mathrm{d}x$$

(by Green's theorem, Theorem 14.7.7, p. 634) and this expression is zero for each pair of comparison functions u, v (since $u = 0$, $v = 0$ on Γ). For the operator N the condition (17) is obviously satisfied. Finally

$$\iint_O u\,M(u)\,dx\,dy = \iint_O u(-\Delta u)\,dx\,dy =$$

$$= -\iint_O \left\{\frac{\partial}{\partial x}\left(u\frac{\partial u}{\partial x}\right) + \frac{\partial}{\partial y}\left(u\frac{\partial u}{\partial y}\right)\right\} dx\,dy + \iint_O \left\{\left(\frac{\partial u}{\partial x}\right)^2 + \left(\frac{\partial u}{\partial y}\right)^2\right\} dx\,dy$$

and the first integral on the right-hand side is zero for every comparison function by Green's theorem while the second integral is positive for every non-zero comparison function. Hence the first inequality (18) is satisfied, while the second is obvious since $c(x, y) > 0$ in O.

Results similar to those for one variable (§ 17.17) are valid for self-adjoint positive definite problems. See also Chapter 22.

19. INTEGRAL EQUATIONS

By Karel Rektorys

References: [21], [75], [137], [281], [297], [318], [354], [367], [370], [375], [405].

Integral Equations of the Second Kind

19.1. Integral Equations of Fredholm Type. Fredholm's Theorems. Solvability. Systems of Integral Equations

Many problems in mathematical physics lead to the solution of integral equations. (Cf. Remark 17.17.23, p. 816, also [281], etc.)

Definition 1. By the notation $f(x) \in L_2(a, b)$ (or more briefly $f \in L_2$) we mean that $f(x)$ is *square integrable in* $[a, b]$, i.e. that there exist finite integrals

$$\int_a^b f(x)\, dx\,, \quad \int_a^b f^2(x)\, dx\,.$$

Remark 1. Similarly, by the notation $g(x, y) \in L_2(Q)$ (or more briefly $g \in L_2$) we mean that the function $g(x, y)$ is square integrable in the square $Q(a \leqq x \leqq b, a \leqq y \leqq b)$. (For details see § 16.1.)

Theorem 1. *Any function which is continuous in* $[a, b]$ *or in* Q, *is square integrable in* $[a, b]$ *or in* Q, *respectively.*

Remark 2. Two functions $f \in L_2(a, b)$, $g \in L_2(a, b)$ are said to be *equal* in $[a, b]$ if the equation $f(x) = g(x)$ is satisfied at all points of $[a, b]$ with the possible exception of some points constituting a set of measure zero (e.g. with the exception of a finite number of points, see § 13.14). This situation occurs especially if one of the functions $f(x)$, $g(x)$ is not defined at some points of the interval $[a, b]$. In this sense, by the term *zero function* we understand not only a function which is identically equal to zero in $[a, b]$, but also a function which need not be defined or may differ from zero on a set of measure zero (e.g. on a set consisting of a finite number of points). A function taking non-zero values on a set of positive measure (e.g. on the whole interval $[a, b]$ or on some of its subintervals) is called a *non-zero function*.

If the functions $f(x)$, $g(x)$ are *continuous* in $[a, b]$, then obviously $f(x)$ and $g(x)$ are equal if and only if the equation $f(x) = g(x)$ is satisfied for all points of the interval $[a, b]$.

Example 1. The function

$$f(x) \begin{cases} = 1 & \text{for} \quad x = 2 \\ = 0 & \text{for} \quad 2 < x \leqq 4 \end{cases}$$

is a zero function in $[2, 4]$.

Definition 2. The equation

$$f(x) - \int_a^b K(x, s) f(s) \, ds = g(x) \tag{1}$$

is called an *integral equation of the second kind.* The *kernel* $K(x, s)$ is a function defined in the square $Q(a \leqq x \leqq b,\ a \leqq s \leqq b)$, $g(x)$ is defined in $[a, b]$, $f(x)$ is the unknown function, x, s are real variables; $K(x, s)$, $g(x)$ need not be real, i.e. we allow K and g to be complex, so that

$$K(x, s) = K_1(x, s) + iK_2(x, s)\,, \quad g(x) = g_1(x) + ig_2(x)\,,$$

where K_1, K_2, g_1, g_2 are real functions. If $K(x, s) \in L_2(Q)$, then equation (1) is called *Fredholm's equation.*

(The terminology is not uniform. Some authors use the term Fredholm's equation for any equation of the second kind.)

REMARK 3. If $K(x, s) = K_1(x, s) + iK_2(x, s)$ and $g(x) = g_1(x) + ig_2(x)$, then $K \in L_2$ and $g \in L_2$ means the same as $K_1 \in L_2$, $K_2 \in L_2$ and $g_1 \in L_2$, $g_2 \in L_2$, respectively. Thus,

$$\iint_Q |K(x, s)|^2 \, dx \, ds \quad \text{and} \quad \int_a^b |g(x)|^2 \, dx\,,$$

i.e.

$$\iint_Q K_1^2(x, s) \, dx \, ds + \iint_Q K_2^2(x, s) \, dx \, ds \quad \text{and} \quad \int_a^b g_1^2(x) \, dx + \int_a^b g_2^2(x) \, dx$$

are finite numbers.

Definition 3. By a *solution of equation* (1) we understand a function, for which equation (1) is satisfied in $[a, b]$ (in the above-mentioned sense – see Remark 2 – i.e. with the possible exception of points constituting a set of measure zero).

REMARK 4. In the case of two variables, the equation of the second kind is of the form

$$f(x_1, x_2) - \iint_{\Omega} K(x_1, x_2, x_3, x_4) f(x_3, x_4) \, dx_3 \, dx_4 = g(x_1, x_2),$$

where Ω is a given domain; very frequently Ω is the square $a \leqq x_3 \leqq b$, $a \leqq x_4 \leqq b$. The variables x_1, x_2, x_3, x_4 run through the set $Q = \Omega \times \Omega$ (thus $(x_3, x_4) \in \Omega$, $(x_1, x_2) \in \Omega$). In the special case, where Ω is a square, Q is the four-dimensional interval $a \leqq x_i \leqq b$, $i = 1, 2, 3, 4$. If the kernel $K(x_1, x_2, x_3, x_4)$ is square integrable in Q, the equation is called *Fredholm's equation*. The (Fredholm) integral equation for an unknown function of several variables is defined similarly. The theory and computing methods are very much alike for cases both of one and several variables. Therefore, in the following text, we shall deal with the one-dimensional case only. The results obtained may easily be generalized to the case of functions of several variables.

REMARK 5. A parameter λ (generally complex) is often introduced into equation (1),

$$f(x) - \lambda \int_a^b K(x, s) f(s) \, ds = g(x). \tag{2}$$

For $\lambda = 1$ we obtain equation (1) as a special case. If $g(x) \equiv 0$, we obtain the equation

$$f(x) - \lambda \int_a^b K(x, s) f(s) \, ds = 0, \tag{3}$$

the so-called *homogeneous equation corresponding to the equation* (2).

Definition 4. The equation

$$F(x) - \bar{\lambda} \int_a^b \overline{K(s, x)} \, F(s) \, ds = 0 \tag{4}$$

is called the *adjoint equation to equation* (3). The bar above λ and $K(s, x)$ denotes the complex conjugate value (i.e. if $\lambda = \lambda_1 + i\lambda_2$ and $K(x, s) = K_1(x, s) + iK_2(x, s)$, then $\bar{\lambda} = \lambda_1 - i\lambda_2$ and $\overline{K(s, x)} = K_1(s, x) - iK_2(s, x)$, respectively). The kernel $\overline{K(s, x)}$ is called the *adjoint kernel to the kernel* $K(x, s)$.

Example 2. The equation

$$f(x) - \lambda \int_0^2 (x^3 + s) f(s) \, ds = 0$$

is an equation of the Fredholm type (for any value of λ). The equation

$$F(x) - \bar{\lambda} \int_0^2 (s^3 + x) F(s) \, ds = 0$$

is the corresponding adjoint equation. Note the interchange of variables x and s in the kernels of the original and the adjoint equations.

Definition 5. Any number $\lambda = \lambda_0$ for which equation (3) possesses a *non-zero* solution $\varphi(x)$ is called an *eigenvalue* or *characteristic number of equation* (3) (or *of the kernel* $K(x, s)$). The function $\varphi(x)$ is called the *eigenfunction* or the *characteristic function associated with the number* λ_0.

FREDHOLM'S THEOREMS. *Let* (3) *be an equation of the Fredholm type, i.e.* $K(x, s) \in$ $\in L_2(\Omega)$. *Then:*

Theorem 2. *In any bounded part of the complex λ-plane, there exist only a finite number of characteristic numbers; thus the only possible point of accumulation of the eigenvalues is the point $\lambda = \infty$.*

Theorem 3. *At least one eigenfunction is associated with each eigenvalue. The number of linearly independent* (Definition 12.8.2, p. 460) *eigenfunctions associated with a fixed eigenvalue is finite.*

Theorem 4. *If λ_0 is an eigenvalue of equation* (3) (*with kernel* $K(x, s)$), *then the complex conjugate number $\bar{\lambda}_0$ is an eigenvalue of the adjoint equation*

$$F(x) - \bar{\lambda} \int_a^b \overline{K(s, x)} \, F(s) \, ds = 0 . \tag{5}$$

Equations (3) *and* (5) *have the same number of linearly independent solutions* (*corresponding to λ_0 and $\bar{\lambda}_0$, respectively*).

Theorem 5 (*Fredholm's Alternative*). *For the equation*

$$f(x) - \lambda \int_a^b K(x, s) f(s) \, ds = g(x) \tag{6}$$

there are only two possibilities:

either this equation possesses one and only one solution $f(x) \in L_2(a, b)$ for any $g(x) \in L_2(a, b)$ (in particular, $f(x) \equiv 0$ is the only continuous solution for $g(x) \equiv 0$),

or the corresponding homogeneous equation (for $g(x) \equiv 0$) has a non-zero solution.

REMARK 6. In the second case, λ is an eigenvalue for the kernel $K(x, s)$ and hence (Theorem 4) $\bar{\lambda}$ is an eigenvalue for the kernel $\overline{K(s, x)}$. Consequently, the equation

$$\psi(x) - \bar{\lambda} \int_a^b \overline{K(s, x)}\, \psi(s)\, ds = 0 \tag{7}$$

has a finite number (Theorem 3) of linearly independent eigenfunctions. Let us denote them by

$$\psi_1(x), \psi_2(x), \ldots, \psi_k(x) . \tag{8}$$

Then it can be proved that equation (6) is solvable precisely for those functions $g(x) \in \in L_2(a, b)$ which are orthogonal to all functions (8), i.e. for which the relation

$$\int_a^b g(x)\, \overline{\psi_i(x)}\, dx = 0\,, \quad i = 1, 2, \ldots, k \tag{9}$$

holds.

In this case, equation (6) has obviously more than one solution: For if $f(x)$ is a solution of equation (6), corresponding to $g(x)$ ($g(x)$ satisfying conditions (9)) and if

$$\varphi_1(x), \varphi_2(x), \ldots, \varphi_k(x) \tag{10}$$

are linearly independent solutions of the homogeneous equation

$$\varphi(x) - \lambda \int_a^b K(x, s)\, \varphi(s)\, ds = 0\,, \tag{11}$$

then, obviously, the function

$$f(x) + c_1\varphi_1(x) + \ldots + c_k\varphi_k(x)$$

where $c_1, \ldots, c_k$ are arbitrary constants, is again a solution of equation (6).

REMARK 7. The following consequence of Fredholm's alternative is often used: If it is known that equation (11) has (in the domain of square integrable functions) only the zero solution (as can often be expected because of the nature of the technical problem in question), then equation (6) has one and only one solution $f(x) \in L_2(a, b)$ for each function $g(x) \in L_2(a, b)$.

Theorem 6. *If $K(x, s)$ is continuous in Q and $g(x)$ is continuous in $[a, b]$ and if equation (6) is required to be satisfied at all points of the interval $[a, b]$, then its solutions (if they exist) are continuous functions in $[a, b]$.*

REMARK 8. A *system* of two integral equations

$$f_1(x) - \int_a^b K_{11}(x, s) f_1(s)\, ds - \int_a^b K_{12}(x, s) f_2(s)\, ds = g_1(x)\,, \tag{12}$$

$$f_2(x) - \int_a^b K_{21}(x, s) f_1(s)\, ds - \int_a^b K_{22}(x, s) f_2(s)\, ds = g_2(x) \tag{13}$$

may be reduced to a single integral equation as follows:

Instead of the interval $[a, b]$, consider the interval $[a, 2b - a]$ whose length is double that of $[a, b]$ (Fig. 19.1) and define functions $F(x)$ and $G(x)$ by the formulae

$$F(x) \begin{cases} = f_1(x) & \text{for } a \leqq x \leqq b\,, \\ = f_2(x - (b - a)) & \text{for } b < x \leqq 2b - a\,. \end{cases}$$

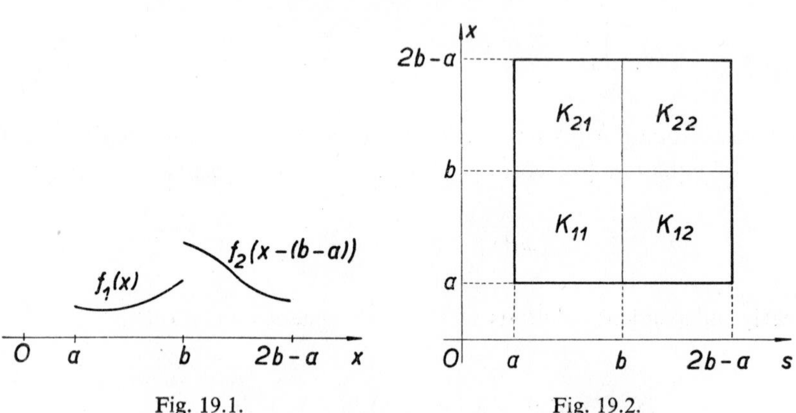

Fig. 19.1. Fig. 19.2.

($f_2(x - (b - a))$ is the function $f_2(x)$ "displaced" by a distance $b - a$ to the right),

$$G(x) \begin{cases} = g_1(x) & \text{for } a \leqq x \leqq b\,, \\ = g_2(x - (b - a)) & \text{for } b < x \leqq 2b - a\,. \end{cases}$$

Similarly (see Fig. 19.2):

$$K(x, s) \begin{cases} = K_{11}(x, s) & \text{for } a \leqq x \leqq b, \quad a \leqq s \leqq b\,, \\ = K_{12}(x, s - (b - a)) & \text{for } a \leqq x \leqq b, \quad b < s \leqq 2b - a\,, \\ = K_{21}(x - (b - a), s) & \text{for } b < x \leqq 2b - a,\ a \leqq s \leqq b\,, \\ = K_{22}(x - (b - a), s - (b - a)) & \text{for } b < x \leqq 2b - a,\ b < s \leqq 2b - a\,. \end{cases}$$

In this notation, we may rewrite equations (12) and (13) in the form

$$F(x) - \int_a^b K(x, s)\, F(s)\, ds - \int_b^{2b-a} K(x, s)\, F(s)\, ds = G(x) \quad (a \leqq x \leqq b)\,,$$

$$F(x) - \int_a^b K(x, s) F(s) \, ds - \int_b^{2b-a} K(x, s) F(s) \, ds = G(x) \quad (b < x \leqq 2b - a)$$

or

$$F(x) - \int_a^{2b-a} K(x, s) F(s) \, ds = G(x) \quad (a \leqq x \leqq 2b - a) . \tag{14}$$

The equation (14) and the system of equations (12) and (13) are equivalent. If $K_{ij}(x, y) \in L_2(Q)$ $(i = 1, 2, j = 1, 2)$, then $K(x, s) \in L_2(R)$ (R is the square $a \leqq x \leqq 2b - a$, $a \leqq s \leqq 2b - a$) and equation (14) is of Fredholm type, so that Theorems 2 – 5 may be applied. Coming back to the system (12), (13) and using Theorem 5 we conclude that:

Theorem 7. *If $K_{ij}(x, s) \in L_2(Q)$ $(i = 1, 2, j = 1, 2)$, then either the system (12), (13) possesses a non-zero solution for zero right-hand sides, or to each pair of functions $g_1(x) \in L_2(a, b)$, $g_2(x) \in L_2(a, b)$ there exists one and only one pair of functions $f_1(x) \in L_2(a, b), f_2(x) \in L_2(a, b)$, which constitute the solution of the system (12), (13).*

19.2. Equations with Degenerate Kernels

The equation of the form

$$f(x) - \int_a^b \left[\sum_{k=1}^n a_k(x) \, b_k(s) \right] f(s) \, ds = g(x) \tag{1}$$

is said to have a *degenerate kernel*. Thus, such a kernel is a finite sum of products, the factors of which are functions of only one variable x or s, respectively. The functions $a_k(x)$ may be assumed to be linearly independent (otherwise the number of terms of the kernel could be reduced).

Theorem 1. *Any solution of equation (1) is of the form*

$$f(x) = g(x) + \sum_{k=1}^n c_k a_k(x) . \tag{2}$$

[For

$$\int_a^b \sum_{k=1}^n a_k(x) \, b_k(s) f(s) \, ds = \sum_{k=1}^n a_k(x) \int_a^b b_k(s) f(s) \, ds = \sum_{k=1}^n c_k a_k(x) .]$$

Substituting (2) into (1) and comparing coefficients of linearly independent functions $a_k(x)$, we obtain a system of n linear algebraic equations for the n unknown coefficients c_k. This system is solvable if and only if the corresponding integral equation is solvable.

Example 1.

$$f(x) - \int_0^1 6(x + s) f(s)\,ds = x^2 . \tag{3}$$

According to (2), the solution is of the form

$$f(x) = x^2 + c_1 x + c_2 \tag{4}$$

(because $a_1(x) = x$, $a_2(x) = 1$). Substituting (4) into (3) we obtain

$$x^2 + c_1 x + c_2 - \int_0^1 6(x + s)(s^2 + c_1 s + c_2)\,ds = x^2 ,$$

$$(c_1 - 2 - 3c_1 - 6c_2)\,x + (c_2 - \tfrac{3}{2} - 2c_1 - 3c_2) = 0 .$$

Comparing the coefficients of corresponding powers of x, we find that

$$-2c_1 - 6c_2 - 2 = 0 , \quad -2c_1 - 2c_2 - \tfrac{3}{2} = 0 ,$$

from which it follows that

$$c_1 = -\tfrac{5}{8} , \quad c_2 = -\tfrac{1}{8} ,$$

hence

$$f(x) = x^2 - \tfrac{5}{8}x - \tfrac{1}{8} .$$

Example 2.

$$f(x) - \lambda \int_0^1 x f(s)\,ds = 0 . \tag{5}$$

Here $a_1(x) = x$, $b_1(s) = \lambda$. So from (2) the solution is of the form

$$f(x) = cx . \tag{6}$$

Substituting (6) into (5), we get

$$cx - \lambda \int_0^1 x \,.\, cs\,ds = 0 ,$$

or

$$cx - \tfrac{1}{2}\lambda cx = 0 . \tag{7}$$

If $\lambda \neq 2$, then $c = 0$ and (5) has as its only solution zero (thus by Fredholm's alternative, Theorem 19.1.5, p. 920, equation (5) has for $\lambda \neq 2$ exactly one solution for arbitrary right-hand side $g(x) \in L_2$).

If $\lambda = 2$, then (7) is satisfied for any c; $\lambda = 2$ is the (only) eigenvalue of equation (5). The corresponding characteristic functions are of the form

$$f(x) = cx \quad (c = \text{const.})$$

(each of them is a multiple of x, so that there is only one linearly independent characteristic function corresponding to the value $\lambda = 2$).

The equation adjoint to equation (5) is of the form (Definition 19.1.4, p. 919)

$$\psi(x) - \lambda \int_0^1 s \,.\, \psi(s)\, ds = 0\,. \tag{8}$$

According to Theorem 19.1.4 (p. 920), $\lambda = 2$ is the (only) eigenvalue of equation (8). The solution of the equation

$$\psi(x) - 2 \int_0^1 s \,.\, \psi(s)\, ds = 0 \tag{9}$$

has the form

$$\psi(x) = a \quad (a = \text{const.}) \tag{10}$$

and, as can easily be verified by substitution, is the solution of (9) for arbitrary a. Solutions of another type do not exist (for, according to Theorem 1, any solution of (9) must be of the form (10)). According to Remark 19.1.6 (p. 921), the equation

$$f(x) - 2 \int_0^1 x f(s)\, ds = g(x) \tag{11}$$

has a solution if, and only if, the function $g(x)$ is orthogonal to the functions (10), i.e. if

$$\int_0^1 a \,.\, g(x)\, dx = 0 \quad \text{i.e.} \quad \int_0^1 g(x)\, dx = 0\,. \tag{12}$$

If we find such a solution, we obtain other solutions by adding an arbitrary multiple of the solution cx of the corresponding homogeneous equation

$$f(x) - 2 \int_0^1 x f(s)\, ds = 0$$

(Remark 19.1.6). Let us choose $g(x) = 1 - 2x$, satisfying condition (12). The form of the solution of equation (11) will then be

$$f(x) = 1 - 2x + c_1 x \tag{13}$$

where c_1 is an arbitrary constant (as may easily be verified, if we substitute (13) into (11)).

19.3. Equations with Symmetric Kernels

The kernel $K(x, s)$ (or the corresponding integral equation) is said to be *symmetric* if

$$K(x, s) = \overline{K(s, x)}, \tag{1}$$

i.e. the kernel equals the function which is obtained by interchanging the variables in $K(x, s)$ and by taking the complex conjugate value. In particular, if the kernel is real (the most frequent case), then the symmetry is expressed by the relation

$$K(x, s) = K(s, x). \tag{2}$$

Example 1. The kernels $x + s$, $\mathrm{i}(x - s)$ are symmetric, the kernels $x^2 + s$, $\mathrm{i}(x + s)$ are not symmetric (in the last case we have

$$K(x, s) = -\overline{K(s, x)}$$

since $\bar{\mathrm{i}} = -\mathrm{i}$).

For the equation of the *Fredholm type* (Definition 19.1.2),

$$f(x) - \lambda \int_a^b K(x, s) f(s)\, \mathrm{d}s = 0, \tag{3}$$

with *symmetric* kernel the following assertions are true:

Theorem 1. *If* $K(x, s)$ *is not a zero function* (see Remark 19.1.2), *then there exists at least one eigenvalue of equation* (3).

Theorem 2. *All eigenvalues of equation* (3) *are real* (*even if* $K(x, s)$ *is not real*).

Theorem 3. *The maximum of the expression*

$$\left| \iint_Q K(x, s)\, \varphi(x)\, \overline{\varphi(s)}\, \mathrm{d}x\, \mathrm{d}s \right|, \tag{4}$$

taken over all functions φ, *which are square integrable and normalized, i.e. for which the relation*

$$\int_a^b |\varphi^2(x)|\, \mathrm{d}x = 1 \tag{5}$$

holds, is equal to $|1/\lambda_1|$, *where* λ_1 *is the eigenvalue with smallest modulus. The function* $\varphi(x)$ *which maximizes expression* (4) *is an eigenfunction corresponding to this eigenvalue* λ_1.

Theorem 4. *The eigenfunctions associated with different eigenvalues are orthogonal*, i.e.

$$\int_a^b \varphi_m(x)\,\varphi_n(x)\,dx = 0 \quad \text{if} \quad \lambda_m \neq \lambda_n\,. \tag{6}$$

REMARK 1. By Theorem 19.1.3 (p. 920), to each eigenvalue there corresponds a finite number of linearly independent eigenfunctions

$$\varphi_1(x),\ \varphi_2(x),\ \ldots,\ \varphi_k(x)\,. \tag{7}$$

We shall assume in the sequel that the functions (7) are *normalized* (i.e. that they satisfy equation (5)) and *orthogonal*, i.e.

$$\int_a^b \varphi_i(x)\,\varphi_j(x)\,dx = 0\,, \quad \text{if} \quad i \neq j\,. \tag{8}$$

The orthogonalization process is described in Remark 16.2.16, p. 701.

It is convenient to number the eigenvalues in such a way that to every eigenfunction of (7) there corresponds exactly one eigenvalue; then to the k linearly independent functions (7) there correspond k characteristic numbers $\lambda^{(1)}, \lambda^{(2)}, \ldots, \lambda^{(k)}$, which of course are all equal, i.e. $\lambda^{(1)} = \lambda^{(2)} = \ldots = \lambda^{(k)}$. Let the eigenvalues corresponding to all normalized and mutually orthogonal characteristic functions of equation (3) be numbered in such a manner that their absolute values form a non-decreasing sequence, i.e.

$$|\lambda_1| \leqq |\lambda_2| \leqq |\lambda_3| \leqq \ldots\,. \tag{9}$$

Theorem 5. *Let $\varphi_1(x), \ldots, \varphi_n(x)$ be eigenfunctions corresponding to eigenvalues $\lambda_1, \ldots, \lambda_n$.* Let *$\varphi(x)$ be any function satisfying the condition* (5) *and also the conditions*

$$\int_a^b \varphi(x)\,\varphi_1(x)\,dx = 0\,, \ \ldots\,, \quad \int_a^b \varphi(x)\,\varphi_n(x)\,dx = 0\,. \tag{10}$$

Then the absolute value of the eigenvalue λ_{n+1} is equal to the reciprocal of the maximum value of the integral (4) *in the class of functions $\varphi(x)$ with the above properties.*

REMARK 2. To determine $|\lambda_1|$ we use (4), and we find also the eigenfunction $\varphi_1(x)$ corresponding to λ_1. To find $|\lambda_2|$ we again look for the maximum of (4), but of all functions φ which are normalized and square integrable in $[a, b]$ we consider only those that are orthogonal to $\varphi_1(x)$. Proceeding in this way, we find successively $|\lambda_3|$, $|\lambda_4|$, etc.

Theorems 3 and 5 are of great importance for the practical evaluation of eigenvalues, because they make possible the application of variational methods (see e.g. § 29.5).

Theorem 6 (*Hilbert–Schmidt Theorem*). *Equation* (3) *has a finite or countably infinite set of normalized and mutually orthogonal eigenfunctions*

$$\varphi_1(x), \varphi_2(x), \ldots, \varphi_n(x), \ldots, \tag{11}$$

which correspond to the eigenvalues $\lambda_1, \lambda_2, \ldots, \lambda_n, \ldots$ $(|\lambda_1| \leqq |\lambda_2| \leqq \ldots \leqq |\lambda_n| \leqq \ldots)$. *For each function* $f(x)$ *which can be expressed in the form*

$$f(x) = \int_a^b K(x, s)\, h(s)\, \mathrm{d}s\,, \quad h(s) \in L_2(a, b)\,, \tag{12}$$

the relation

$$\lim_{k\to\infty} \int_a^b [f(x) - \sum_{n=1}^k a_n\, \varphi_n(x)]^2\, \mathrm{d}x = 0$$

holds, where

$$a_n = \int_a^b f(x)\, \varphi_n(x)\, \mathrm{d}x\,, \tag{13}$$

i.e. the series

$$\sum_{n=1}^\infty a_n\, \varphi_n(x) \tag{14}$$

converges in the mean to the function $f(x)$.

If, in addition, the integral

$$\int_a^b |K(x, s)|^2\, \mathrm{d}s \tag{15}$$

is bounded (*by the same constant for all* $x \in [a, b]$), *then the series* (15) *converges to* $f(x)$ *absolutely and uniformly in* $[a, b]$.

The coefficients a_n *can be expressed in terms of the eigenvalues* λ_n *and of the function* $h(x)$ *as follows:*

$$a_n = \frac{h_n}{\lambda_n} \quad \text{where} \quad h_n = \int_a^b h(x)\, \varphi_n(x)\, \mathrm{d}x\,. \tag{16}$$

19.4. The Resolvent

Definition 1. The number λ is said to be *regular* for the equation

$$f(x) - \lambda \int_a^b K(x, s) f(s)\, \mathrm{d}s = 0 \tag{1}$$

if λ is not an eigenvalue of equation (1).

Definition 2. If for each regular λ and for an arbitrary function $g(x) \in L_2$ the solution of the equation

$$f(x) - \lambda \int_a^b K(x, s) f(s)\, ds = g(x) \tag{2}$$

can be written in the form

$$f(x) = g(x) + \lambda \int_a^b \Gamma(x, s, \lambda)\, g(s)\, ds\,, \tag{3}$$

then the function

$$\Gamma(x, s, \lambda) \tag{4}$$

is called the *resolvent* of equation (2).

Example 1. The solution of the equation with degenerate kernel

$$f(x) - \lambda \int_0^1 (x + s) f(s)\, ds = g(x) \tag{5}$$

is, in view of equation (19.2.2), of the form

$$f(x) = g(x) + c_1 x + c_2\,. \tag{6}$$

Substituting expression (6) into equation (5) and equating corresponding coefficients, we get

$$c_1 = \frac{\lambda(1 - \frac{1}{2}\lambda) \int_0^1 g(s)\, ds + \lambda^2 \int_0^1 s\, g(s)\, ds}{1 - \lambda - \frac{1}{12}\lambda^2}\,,$$

$$c_2 = \frac{\frac{1}{3}\lambda^2 \int_0^1 g(s)\, ds + \lambda(1 - \frac{1}{2}\lambda) \int_0^1 s\, g(s)\, ds}{1 - \lambda - \frac{1}{12}\lambda^2}\,. \tag{7}$$

Substitution of these values of c_1 and c_2 into equation (6) yields

$$f(x) = g(x) + \int_0^1 \lambda \frac{(12 - 6\lambda)\, x + 4\lambda + s(12\lambda x + 12 - 6\lambda)}{12 - 12\lambda - \lambda^2}\, g(s)\, ds\,. \tag{8}$$

Hence

$$\Gamma(x, s, \lambda) = \frac{(12 - 6\lambda)\, x + 4\lambda + s(12\lambda x + 12 - 6\lambda)}{12 - 12\lambda - \lambda^2}\,. \tag{9}$$

(Compare Example 19.2.1, p. 924, where $\lambda = 6$, $g(x) = x^2$.)

For the Fredholm equation (2) the following theorems are valid:

Theorem 1. *For equation* (2) *there exists a resolvent* (*for all regular values of* λ).

Theorem 2. *The resolvent is a meromorphic function* (see Remark 20.4.10, p. 960) *of the complex variable* λ *in the entire* λ*-plane. The eigenvalues are poles of the resolvent.*

Theorem 3. *If* λ *is regular, then*

$$\Gamma(x, s, \lambda) = \frac{D(x, s, \lambda)}{D(\lambda)}, \tag{10}$$

where

$$D(x, s, \lambda) = \sum_{n=0}^{\infty} \frac{(-1)^n}{n!} B_n(x, s) \lambda^n, \tag{11}$$

$$D(\lambda) = \sum_{n=0}^{\infty} \frac{(-1)^n}{n!} c_n \lambda^n \tag{12}$$

and

$$B_0(x, s) = K(x, s), \tag{13}$$

$$B_n(x, s) = \int_a^b \dots \int_a^b \Delta_n \,\mathrm{d}t_1 \,\mathrm{d}t_2 \dots \mathrm{d}t_n, \tag{14}$$

$$\Delta_n(x, s, t_1, t_2, \dots, t_n) = \begin{vmatrix} K(x, s), & K(x, t_1), & K(x, t_2), & \dots, & K(x, t_n) \\ K(t_1, s), & K(t_1, t_1), & K(t_1, t_2), & \dots, & K(t_1, t_n) \\ K(t_2, s), & K(t_2, t_1), & K(t_2, t_2), & \dots, & K(t_2, t_n) \\ \dots & \dots & \dots & \dots & \dots \\ K(t_n, s), & K(t_n, t_1), & K(t_n, t_2), & \dots, & K(t_n, t_n) \end{vmatrix}, \tag{15}$$

$$c_0 = 1, \tag{16}$$

$$c_n = \int_a^b \dots \int_a^b \begin{vmatrix} K(t_1, t_1), & K(t_1, t_2), & \dots, & K(t_1, t_n) \\ K(t_2, t_1), & K(t_2, t_2), & \dots, & K(t_2, t_n) \\ \dots & \dots & \dots & \dots \\ K(t_n, t_1), & K(t_n, t_2), & \dots, & K(t_n, t_n) \end{vmatrix} \mathrm{d}t_1 \,\mathrm{d}t_2 \dots \mathrm{d}t_n. \tag{17}$$

The series (11), (12) *are convergent for all values of* λ.

REMARK 1. If

$$|\lambda| \leq \frac{1}{B} \quad \text{where} \quad B^2 = \int_a^b \int_a^b |K^2(x, s)| \,\mathrm{d}x \,\mathrm{d}s,$$

then there is another expression for the resolvent, namely

$$\Gamma(x, s, \lambda) = \sum_{m=1}^{\infty} K_m(x, s) \lambda^{m-1}, \tag{18}$$

where $K_m(x, s)$ is the *m-th iterated kernel* (corresponding to the kernel $K(x, s)$) given by the recurrence formula

$$K_m(x, s) = \int_a^b K(x, t) K_{m-1}(t, s)\, dt \quad (m \geqq 2), \tag{19}$$

$$K_1(x, s) = K(x, s).$$

REMARK 2. The functions (14) and (17) can be evaluated by applying the following recurrence formulae:

$$c_{n+1} = \int_a^b B_n(s, s)\, ds, \tag{20}$$

$$B_n(x, s) = c_n K(x, s) - n \int_a^b K(x, t)\, B_{n-1}(t, s)\, dt. \tag{21}$$

Example 2. Let us determine the resolvent of equation (5). From (16) and (13) $c_0 = 1$, $B_0(x, s) = x + s$. Using (20) and (21) we find

$$c_1 = \int_0^1 2s\, ds = 1,$$

$$B_1(x, s) = x + s - \int_0^1 (x + t)(t + s)\, dt = \tfrac{1}{2}(x + s) - xs - \tfrac{1}{3},$$

$$c_2 = \int_0^1 (s - s^2 - \tfrac{1}{3})\, ds = -\tfrac{1}{6},$$

$$B_2(x, s) = -\tfrac{1}{6}(x + s) - 2\int_0^1 (x + t)\, [\tfrac{1}{2}(t + s) - ts - \tfrac{1}{3}]\, dt = 0.$$

It follows readily from (20) and (21) that $c_3 = 0$, $B_3(x, s) \equiv 0$, $c_4 = 0, \ldots$. Hence, according to (11), (12) and (10) the resolvent of equation (5) is

$$\Gamma(x, s, \lambda) = \frac{x + s - [\tfrac{1}{2}(x + s) - xs - \tfrac{1}{3}]\, \lambda}{1 - \lambda - \tfrac{1}{12}\lambda^2} \tag{22}$$

in agreement with (9). (This example is used only to illustrate the underlying idea; in fact, the equation has a degenerate kernel and can be solved by the procedure of § 19.2.)

If λ is sufficiently small (more exactly, for $|\lambda| \leqq \sqrt{\frac{6}{7}}$ in view of Remark 1 and of the relation $B^2 = \int_0^1 \int_0^1 (x + s)^2\, dx\, ds = \frac{7}{6}$), then the resolvent of equation (5) can be expressed in the form (18). We have

$$K_1(x, s) = K(x, s) = x + s\,,$$

$$K_2(x, s) = \int_0^1 (x + t)(t + s)\,dt = \tfrac{1}{2}(x + s) + xs + \tfrac{1}{3}\,,$$

$$K_3(x, s) = \int_0^1 (x + t)\left[\tfrac{1}{2}(t + s) + ts + \tfrac{1}{3}\right] dt = xs + \tfrac{7}{12}x + \tfrac{7}{12}s + \tfrac{1}{3}$$

so that in view of (18)

$$\Gamma(x, s, \lambda) = x + s + \left[\tfrac{1}{2}(x + s) + xs + \tfrac{1}{3}\right]\lambda + \left[xs + \tfrac{7}{12}x + \tfrac{7}{12}s + \tfrac{1}{3}\right]\lambda^2 + \ldots\,. \tag{23}$$

We can obtain the same formula from (22) if we write

$$\frac{1}{1 - \lambda - \frac{1}{12}\lambda^2} = 1 + (\lambda + \tfrac{1}{12}\lambda^2) + (\lambda + \tfrac{1}{12}\lambda^2)^2 + \ldots,$$

and consider terms up to and including λ^2.

Having found the resolvent (22), we can easily find the solution of equation (5) for any given (regular) λ and any given $g(x)$; e.g. if $\lambda = 6$, $g(x) = x^2$, then by virtue of (3) we get

$$f(x) = x^2 + 6\int_0^1 \frac{x + s - 6\left[\frac{1}{2}(x + s) - xs - \frac{1}{3}\right]}{1 - 6 - 3}\, s^2\,ds = x^2 - \tfrac{5}{8}x - \tfrac{1}{8}\,,$$

in agreement with example 19.2.1.

Theorem 4. *If $K(x, s)$ possesses continuous partial derivatives of the first order in Q, then for every regular λ the resolvent $\Gamma(x, s, \lambda)$ has continuous partial derivatives of the first order with respect to x and s.*

REMARK 3. From the form of (3) the continuous dependence* of the solution on $g(x)$ clearly follows.

19.5. Equations Involving Weak Singularities. Singular Equations

Definition 1. Equations of the form

$$f(x) - \lambda \int_a^b \frac{H(x, s)}{|x - s|^\alpha} f(s)\,ds = g(x)\,, \tag{1}$$

* The solution $f(x)$ of (2) is "continuously dependent" on $g(x)$ if, roughly speaking, a slight change in $g(x)$ causes a slight change in $f(x)$ for λ fixed and regular.

where $H(x, s)$ is a bounded (integrable) function, $0 < \alpha < 1$, are said to have a *weak singularity*. (If $\alpha < \frac{1}{2}$, we have Fredholm's equation.)

Theorem 1. *All iterated kernels* (Remark 19.4.1) *of equation* (1) *starting from a certain kernel are bounded.*

Theorem 2. *All four of Fredholm's theorems* (Theorems 19.1.2–19.1.5) *hold for equation* (1).

REMARK 1. It can be shown that Fredholm's theorems remain valid not only for equations involving a weak singularity but, more generally, for any equation whose iterated kernels, starting from a certain kernel, are bounded.

REMARK 2. Theorems 1 and 2 hold also for equations in several variables with a weak singularity. For two independent variables such equations are of the form

$$f(x_1, x_2) - \iint_\Omega \frac{H(x_1, x_2, x_3, x_4)}{r^\alpha} f(x_3, x_4)\,dx_3\,dx_4 = g(x_1, x_2), \tag{2}$$

where Ω is a bounded two-dimensional region, $H(x_1, x_2, x_3, x_4)$ is a bounded integrable function, r is the distance between the points (x_1, x_2) and (x_3, x_4), $0 < \alpha < 2$. The equation corresponding to (2) for the case of n variables is of similar form, with $0 < \alpha < n$. For details see e.g. [281].

Definition 2. Integral equations of the form

$$f(x) - \lambda \int_a^b \frac{A(x, s)}{x - s} f(s)\,ds = g(x), \tag{3}$$

where $A(x, s)$ is a differentiable function of the variables x and s, are called *singular integral equations.*

REMARK 3. The integral appearing in (3) may be divergent, in general. If it is considered in the sense of its principal value (Remark 13.8.3, p. 562), an extensive theory for equation (3) can be established. For details see [297]. We shall merely consider two typical cases:

A. The *equation with the so-called Hilbert kernel,*

$$a\,f(x) + \frac{b}{2\pi} \int_0^{2\pi} \cot \tfrac{1}{2}(s - x)\,f(s)\,ds = g(x), \tag{4}$$

possesses for $a \neq 0$, $a^2 + b^2 \neq 0$ (a, b may be complex) the following solution:

$$f(x) = \frac{a}{a^2 + b^2}\,g(x) - \frac{b}{2\pi(a^2 + b^2)} \int_0^{2\pi} g(s) \cot \tfrac{1}{2}(s - x)\,ds + \\ + \frac{b^2}{2\pi a(a^2 + b^2)} \int_0^{2\pi} g(s)\,ds. \tag{5}$$

If $a = 0$, $b \neq 0$, equation (4) becomes an equation of the first kind (see § 19.7) and has a solution if and only if

$$\int_0^{2\pi} g(s)\,\mathrm{d}s = 0\,. \tag{6}$$

The solution is then

$$f(x) = -\frac{1}{2\pi b}\int_0^{2\pi} g(s)\cot\tfrac{1}{2}(s - x)\,\mathrm{d}s + C\,, \tag{7}$$

where C is an arbitrary constant.

When $a^2 + b^2 = 0$ then equation (4) cannot, in general, be solved.

B. The *equation with the so-called Cauchy kernel* is of the form

$$a\,f(z) + \frac{b}{\pi \mathrm{i}}\int_c \frac{f(t)}{t - z}\,\mathrm{d}t = g(z)\,, \tag{8}$$

where c is a simple curve, piecewise smooth, closed and positively oriented with respect to its interior V (Remark 14.7.1, p. 628), while $g(z)$ is a function of the complex variable $z = x + \mathrm{i}y$, and the function $f(z)$ is a function of a point on the curve c. If $a^2 - b^2 \neq 0$, then equation (8) has the solution

$$f(z) = \frac{a}{a^2 - b^2}\,g(z) - \frac{b}{(a^2 - b^2)\,\pi \mathrm{i}}\int_c \frac{g(t)}{t - z}\,\mathrm{d}t\,. \tag{9}$$

REMARK 4. Other kinds of singularities may occur when the given interval is infinite and is transformed to a finite one. For example, the interval $(0, \infty)$ is transformed by the substitution $x = t/(1 - t)$ into the interval $(0, 1)$; using this, simultaneously with the substitution $s = \sigma/(1 - \sigma)$, the new kernel (which is a function of t and σ) may become infinite. In such cases, for existence theorems and also for construction of approximate solutions, it is often convenient to employ the method of successive approximations.

19.6. Equations of Volterra Type

These equations are of the form

$$f(x) - \lambda\int^x K(x, s)\,f(s)\,\mathrm{d}s = g(x) \quad (a \leqq x \leqq b)\,, \tag{1}$$

where the kernel $K(x, s)$ is bounded and integrable. Thus these equations are special cases of equations of the form

$$f(x) - \lambda \int_a^b K(x, s) f(s) \, ds = g(x)$$

in the case where $K(x, s)$ is bounded and is equal to zero for $x < s \leqq b$.

Theorem 1. *If $g(x)$ is absolutely integrable, then equation (1) possesses one and only one solution for every λ. The solution may be obtained as the limit of a uniformly convergent sequence of successive approximations*

$$\begin{aligned} f_0(x) &= g(x), \\ f_1(x) &= g(x) + \lambda \int_a^x K(x, s) f_0(s) \, ds, \\ &\cdots\cdots\cdots\cdots\cdots\cdots \\ f_{n+1}(x) &= g(x) + \lambda \int_a^x K(x, s) f_n(s) \, ds, \\ &\cdots\cdots\cdots\cdots\cdots\cdots \end{aligned} \tag{2}$$

Theorem 2. *Let $|K(x, s)| < M$. The rate of convergence is given by:*

$$|f_{n+1}(x) - f_n(x)| \leqq \frac{|\lambda|^n M^n (b-a)^{n-1}}{(n-1)!} \int_a^b |g(s)| \, ds \,. \tag{3}$$

Remark 1. The inequality (3) is of importance for estimating the error, when using the method of successive approximations.

Remark 2. Volterra's equation can often be reduced to a differential equation (and vice versa).

Example 1. The equation

$$f(x) - \lambda \int_0^x e^{x-s} f(s) \, ds = g(x) \tag{4}$$

differentiated with respect to x ($g(x)$ is assumed to be differentiable) gives

$$f'(x) - \lambda e^{x-x} f(x) - \lambda \int_0^x e^{x-s} f(s) \, ds = g'(x) \,. \tag{5}$$

Eliminating the integral between (4) and (5) we obtain a simple differential equation for $f(x)$

$$f'(x) - (\lambda + 1) f(x) = g'(x) - g(x) \,. \tag{6}$$

If we prescribe for the solution of equation (6) the condition $f(0) = g(0)$, as follows from (4) by setting $x = 0$, then the solution of equation (6) is the solution of equation (4) and conversely.

Example 2. Let us transform the problem

$$y'' + a(x)\,y = f(x)\,, \quad y(0) = y_0\,, \quad y'(0) = y'_0 \tag{7}$$

into an integral equation of the Volterra type.

On integrating the equation (7) twice we get

$$\int_0^x du \int_0^u y''(t)\,dt + \int_0^x du \int_0^u a(t)\,y(t)\,dt = \int_0^x du \int_0^u f(t)\,dt\,. \tag{8}$$

Further

$$\int_0^u y''(t)\,dt = y'(u) - y'(0)\,, \quad \int_0^x [y'(u) - y'(0)]\,du = y(x) - y(0) - xy'(0)\,. \tag{9}$$

According to Cauchy-Dirichlet's formula (§ 17.7.4, p. 762),

$$\int_0^x du \int_0^u a(t)\,y(t)\,dt = \int_0^x (x - u)\,a(u)\,y(u)\,du\,, \tag{10}$$

$$\int_0^x du \int_0^u f(t)\,dt = \int_0^x (x - u)\,f(u)\,du = F(x)\,. \tag{11}$$

Putting (9), (10), (11) into (8), we get

$$y(x) - y(0) - xy'(0) + \int_0^x (x - u)a(u)\,y(u)\,du = F(x)$$

or

$$y(x) + \int_0^x (x - u)\,a(u)\,y(u)\,du = F(x) + y_0 + xy'_0\,. \tag{12}$$

19.7. Integral Equations of the First Kind

Integral equations of the form

$$\int_a^b K(x, s)\,f(s)\,ds = g(x)\,, \tag{1}$$

where $f(x)$ is the unknown function, are called *integral equations of the first kind.*

Theorem 1. *Generally speaking, equation* (1) *has no solution.*

Example 1. The equation

$$\int_0^1 x f(s)\, ds = x^2 \qquad (2)$$

does not possess a solution; for any function $f(s)$, the left-hand side is of the form kx and this is evidently not identically equal to x^2 in the interval $[0, 1]$ for any k. Clearly it is easy to decide whether an equation of the first kind with degenerate kernel is solvable or not.

Equations of the first kind are not encountered in practice as often as equations of the second kind. Equations of the first kind are extensively dealt with in [354].

The following equation, known as *Abel's integral equation,* is of importance:

$$\int_0^x \frac{f(s)}{(x-s)^\alpha}\, ds = g(x) \quad (0 < \alpha < 1)\,.$$

The solution is

$$f(x) = \frac{\sin \alpha\pi}{\pi}\left(\frac{g(0)}{x^{1-\alpha}} + \int_0^x \frac{g'(s)}{(x-s)^{1-\alpha}}\, ds\right).$$

In particular, for the case $\alpha = \frac{1}{2}$ (which frequently occurs), this becomes

$$\int_0^x \frac{f(s)}{\sqrt{(x-s)}}\, ds = g(x)\,, \quad f(x) = \frac{1}{\pi}\left(\frac{g(0)}{\sqrt{x}} + \int_0^x \frac{g'(s)}{\sqrt{(x-s)}}\, ds\right).$$

20. FUNCTIONS OF A COMPLEX VARIABLE

By Karel Rektorys

References: [4], [15], [21], [52], [59], [72], [103], [123], [142], [180], [197], [216], [232], [251], [268], [273], [307], [340], [345], [399], [429].

20.1. Fundamental Concepts. Limit and Continuity. The Derivative. Cauchy – Riemann Equations. Applications of the Theory of Functions of a Complex Variable

REMARK 1. A complex number $z = x + iy$ can be represented by a point with coordinates x, y in the so-called *complex* (or *Gaussian*) *plane* (see p. 48). We usually speak of the *point* z instead of the number z. If we speak of a region M of complex numbers, we mean the corresponding region of points (x, y). A similar meaning is to be given to the statement "*the point z lies on the curve c*".

Operations using complex numbers are discussed in §§ 1.6 and 1.21.

REMARK 2. Complex numbers can also be represented on the so-called *Riemann sphere* touching the Gaussian plane at the origin (which is taken to be the "south pole" of the sphere). If we join an arbitrary point of the Gaussian plane to the "north pole" of the sphere (as centre of projection), then there is a one-to-one correspondence between the points z and z' of the Gaussian plane and of the spherical surface, respectively (the so-called *stereographic projection*, Fig. 20.1). To the "north pole" there corresponds the point $z = \infty$. In this sense, the Gaussian plane has exactly one point at infinity, the point $z = \infty$.

The plane of complex numbers together with the point $z = \infty$ is called the *closed* (or *completed* or *extended*) *plane of complex numbers.*

Definition 1. Let M be a set (of complex numbers) in the Gaussian plane. If a relationship is given, by virtue of which to every point $z \in M$ there corresponds one and only one number w (complex, in general), we say that a *function*

$$w = f(z) \tag{1}$$

is defined on M.

REMARK 3. The function (1) can also be interpreted in the following way: to every point $(x, y) \in M$ (i.e. to every point $z \in M$) there corresponds a complex number $w = u + iv$, i.e.

$$f(z) = u(x, y) + iv(x, y). \tag{2}$$

Thus, the investigation of functions of a complex variable can be reduced to the investigation of two functions u, v of the real variables x, y.

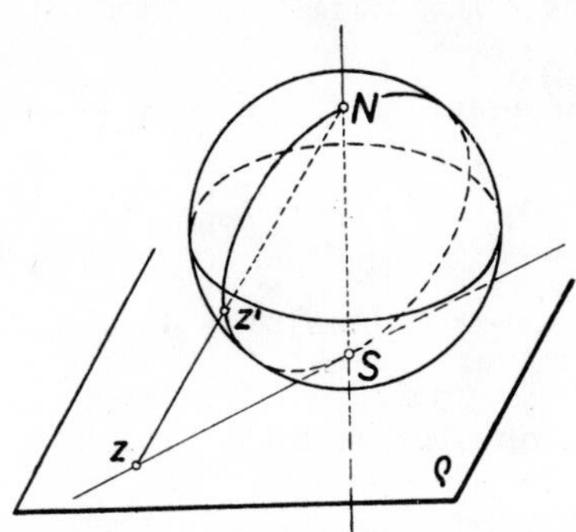

Fig. 20.1.

Example 1.

$$f(z) = z^2 = (x + iy)^2 = x^2 - y^2 + i \,.\, 2xy \,. \tag{3}$$

Hence

$$u(x, y) = x^2 - y^2\,, \quad v(x, y) = 2xy\,. \tag{4}$$

REMARK 4. We write briefly

$$\operatorname{Re} f(z) = u \quad \text{or} \quad \mathscr{R}[f(z)] = u\,;$$

$$\operatorname{Im} f(z) = v \quad \text{or} \quad \mathscr{I}[f(z)] = v$$

(and call these the *real part* and the *imaginary part of the function* $f(z)$). In particular,

$$\operatorname{Re} z = x\,, \quad \operatorname{Im} z = y\,.$$

REMARK 5. A function $w = f(z)$ can be interpreted geometrically as a mapping from one complex plane into another (see "conformal mapping", Chap. 21). We also often illustrate the function $f(z) = u(x, y) + iv(x, y)$ by constructing the surface $t = |f(z)|$, i.e. $t = \sqrt{[u^2(x, y) + v^2(x, y)]}$ over the plane xy.

Definition 2. By a *δ-neighbourhood of a point* z_0 we mean all the points of the complex plane whose distance from z_0 is less than δ, i.e. all the points z satisfying

$$|z - z_0| < \delta\,.$$

(Note that the equation $|z - z_0| = \delta$ characterizes all points whose distance from z_0 is equal to δ, i.e. the circumference of the circle with centre z_0 and radius δ.)

Definition 3. We say that $f(z)$ has a *limit* A (A being in general a complex number) *at the point* z_0 (which need not belong to the domain of definition M), or that $f(z)$ *tends to the limit* A for $z \to z_0$, if for an arbitrary $\varepsilon > 0$ there exists a $\delta > 0$ such that the inequality

$$|f(z) - A| < \varepsilon$$

holds for all $z \neq z_0$ of the δ-neighbourhood of z_0, i.e. for all z satisfying

$$0 < |z - z_0| < \delta .$$

We write

$$\lim_{z \to z_0} f(z) = A .$$

REMARK 6. This definition is equivalent to the following: $f(z)$ has a limit A at z_0 if the relation

$$\lim_{n \to \infty} f(z_n) = A$$

holds for every sequence of points $z_n \neq z_0$ converging to z_0.

Definition 4. If $z_0 \in M$ and if

$$\lim_{z \to z_0} f(z) = f(z_0) , \tag{5}$$

we say that $f(z)$ is *continuous at* z_0.

If $f(z)$ is continuous at every point of a region O, we say that $f(z)$ is *continuous in* O.

Definition 5. If z_0 is not an interior point (Definition 22.1.1, p. 994) of the domain of definition M of the function $f(z)$ and if (5) holds for $z \in M$ (i.e. $|f(z) - f(z_0)| < \varepsilon$ for $|z - z_0| < \delta$ and $z \in M$), we say that $f(z)$ is *continuous at* z_0 *with respect to* M.

REMARK 7. This occurs especially in connection with continuity at boundary points of the domain of definition, or continuity on a given curve, if $f(z)$ is defined only at points on this curve, etc.

Definition 6. If $f(z)$ is continuous in a region O and continuous with respect to the closed region $\bar{O}$ ($\bar{O} = O + c$) at every point of the boundary c, we say that $f(z)$ is *continuous in* $\bar{O}$.

Definition 7. If $f(z)$ is defined only in O (and is continuous in O) and if there exists a function $g(z)$ defined on the boundary c of the region O, such that the function $h(z)$ defined by $h(z) = f(z)$ in O and by $h(z) = g(z)$ on c, is continuous in $\bar{O}$, then we say that $f(z)$ is *continuously extensible on the boundary c to the function* $g(z)$. (If $g(z)$ exists, then it is uniquely determined by $f(z)$.)

Definition 8. If the limit

$$\lim_{z \to z_0} \frac{f(z) - f(z_0)}{z - z_0} \tag{6}$$

exists, we say that $f(z)$ has a *derivative at* z_0 (or that $f(z)$ is *differentiable* or *monogenic at* z_0). We denote the limit (6) by $f'(z_0)$.

Note that, in accordance with definition (3), this means that the limit must be independent of the manner in which $z \to z_0$.

Theorem 1. *In order that a function* $f(z) = u(x, y) + iv(x, y)$ *should have a derivative at* $z_0 = x_0 + iy_0$ *it is necessary and sufficient that* $u(x, y)$ *and* $v(x, y)$ *have total differentials* (see p. 447) *at* (x_0, y_0) *and that their derivatives at this point satisfy the so-called Cauchy–Riemann equations*

$$\frac{\partial u}{\partial x} = \frac{\partial v}{\partial y}, \quad \frac{\partial u}{\partial y} = -\frac{\partial v}{\partial x}. \tag{7}$$

REMARK 8. Hence, if $f(z)$ has a derivative at z_0, then u and v have total differentials at (x_0, y_0) and satisfy (7). Conversely: If u and v have total differentials at (x_0, y_0) and satisfy (7), then the function $u + iv$ has a derivative at $z_0 = x_0 + iy_0$.

Theorem 2. *The derivative of a function*

$$f(z) = u(x, y) + iv(x, y)$$

is given by the formulae

$$f'(z) = \frac{\partial u}{\partial x} + i\frac{\partial v}{\partial x} = \frac{\partial f}{\partial x} \quad or \quad f'(z) = -i\left(\frac{\partial u}{\partial y} + i\frac{\partial v}{\partial y}\right) = -i\frac{\partial f}{\partial y}. \tag{8}$$

REMARK 9. For the evaluation of derivatives of functions of a complex variable, the same rules hold as for functions of a real variable. In particular, $(fg)' = f'g + fg'$, $(z^n)' = nz^{n-1}$ for every natural number n, etc.

Theorem 3. *If* $f(x)$ *has a derivative at* z_0, *then* $f(z)$ *is continuous at* z_0.

Definition 9. If $f(z)$ has a derivative at every point of a region O, it is said to be *holomorphic (regular) in* O.

REMARK 10. A holomorphic function is often called an *analytic function*. However, this name is also used for multi-valued functions (e.g. ln z) which have a derivative (compare § 20.6).*

* In the English literature there are further variations; when consulting other works the reader should carefully examine the definitions used.

Theorem 4. *If the function $f(z) = u(x, y) + iv(x, y)$ is holomorphic in O, then u and v are harmonic in O* (see Definition 18.4.6, p. 887).

Theorem 5. *If $f(z)$ is holomorphic in O, then it has derivatives of all orders in O.*

REMARK 11. We use a bar to denote the conjugate complex number:

$$\bar{z} = x - iy\,, \quad \overline{f(z)} = u(x, y) - iv(x, y)\,.$$

Theorem 6. *If $f(z)$ is holomorphic in O, then $\overline{f(z)}$ is not in general holomorphic in O.*

Example 2. The function

$$f(z) = z^2 = (x + iy)^2 = x^2 - y^2 + i\,.\,2xy \tag{9}$$

(in which $u(x, y) = x^2 - y^2$, $v(x, y) = 2xy$) is holomorphic in the Gaussian plane, for u, v have everywhere continuous partial derivatives and, consequently, the total differential, and they obviously satisfy the Cauchy-Riemann equations (7), since

$$\frac{\partial u}{\partial x} = 2x\,, \quad \frac{\partial v}{\partial y} = 2x\,, \quad \frac{\partial u}{\partial y} = -2y\,, \quad \frac{\partial v}{\partial x} = 2y\,.$$

Further, $f'(z) = 2z$ (see Remark 9). [According to (8), we also have

$$f'(z) = 2x + i\,.\,2y = -i(-2y + i\,.\,2x) = 2(x + iy) = 2z\,.]$$

The functions u and v are obviously harmonic, for

$$\frac{\partial^2 u}{\partial x^2} + \frac{\partial^2 u}{\partial y^2} = 2 - 2 = 0\,, \quad \frac{\partial^2 v}{\partial x^2} + \frac{\partial^2 v}{\partial y^2} = 0 + 0 = 0\,.$$

The conjugate complex function,

$$\overline{f(z)} = \overline{z^2} = x^2 - y^2 - i\,.\,2xy$$

is not holomorphic because it does not satisfy equations (7) (with the exception of the point $z = 0$, where it is monogenic).

REMARK 12. Using equations (7), we can find, corresponding to a given function $u(x, y)$ (or $v(x, y)$), which is harmonic in a simply connected region O, a so-called *conjugate function* $v(x, y)$ (or $u(x, y)$ respectively), i.e. a function such that

$$f(z) = u(x, y) + iv(x, y)$$

is holomorphic in O. This conjugate function is uniquely determined up to an additive constant.

Theorem 5 then implies that every function harmonic in O has derivatives in O of all orders.

REMARK 13. We call a function $f(z)$ *univalent* or *simple** *in a region* O, if for every pair of different points z_1, z_2 of this region the relation $f(z_1) \neq f(z_2)$ holds. Hence a univalent function does not assume the same value at two different points of O. Univalent holomorphic functions are of great importance in conformal mappings where a one-to-one mapping of regions is concerned. The following assertion holds: If $f'(z_0) \neq 0$, then $f(z)$ is univalent in a certain neighbourhood of z_0.

REMARK 14. Functions of a complex variable have wide applications. For example, in the study of two-dimensional potential flow we define a so-called *complex potential of the flow*, i.e. a holomorphic function $f(z) = u(x, y) + iv(x, y)$. Equations $u = \text{const.}$ then represent equipotential curves, while equations $v = \text{const.}$ give the trajectories of the flow (cf. Example 21.3.3, p. 980). In the theory of electricity, functions of a complex variable are used on the one hand in elementary considerations where vectors of basic electrical quantities are expressed by complex numbers, and on the other in solving more complicated problems (e.g. in solving differential equations with the help of the Laplace transform, etc.). We also use functions of a complex variable to solve two-dimensional problems in elasticity, where the so-called Airy stress function is expressed by two holomorphic functions (see e.g. [21]). Properties of series, integral theorems (especially the Residue Theorem, see § 20.5) and conformal mappings are also widely applied.

20.2. Integral of a Function of a Complex Variable. Cauchy's Integral Theorem. Cauchy's Integral Formula

We define the integral of a function of a complex variable in the same way as the line integral of a function of a real variable. Let us consider a simple, finite, piecewise

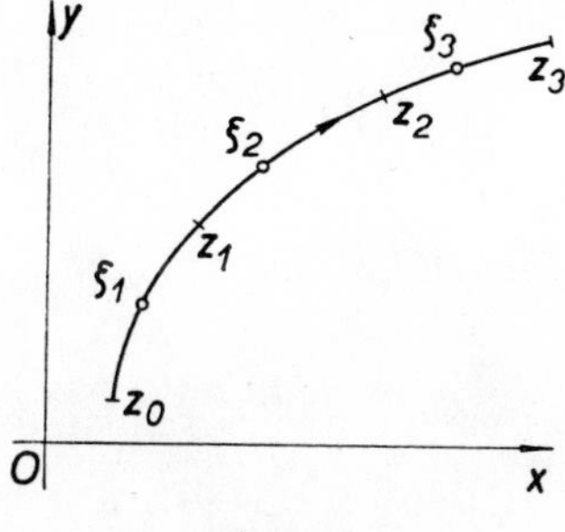

Fig. 20.2.

smooth, oriented curve c (Definition 14.1.1; in the following text, we speak briefly of a curve) with initial point z_0 (Fig. 20.2) and a function

$$w = f(z)$$

* "*schlicht*" in German.

defined on this curve. Let us divide the curve at the points $z_1, z_2, \ldots, z_{n-1}$, numbered in the sense of the positive orientation of the curve, into n arcs $c_1, c_2, \ldots, c_n$. Let us denote by $l_1, l_2, \ldots, l_n$ the lengths of these arcs. The *norm* $\nu(d)$ *of the chosen division* d is defined to be the greatest of the numbers $l_1, l_2, \ldots, l_n$. On each arc c_i let us choose an arbitrary point ζ_i and let us construct the sum

$$\sigma(d) = \sum_{i=1}^{n} f(\zeta_i)(z_i - z_{i-1}) \tag{1}$$

(depending on the division d and on the points ζ_i on c_i).

Definition 1. If there exists a (generally complex) number I such that

$$\lim_{n\to\infty} \sigma(d_n) = I$$

for every sequence of divisions d_n satisfying

$$\lim_{n\to\infty} \nu(d_n) = 0$$

and for every choice of points ζ_i on c_i, then we call this number the *integral of the function* $f(z)$ *along the (oriented) curve* c *and write*

$$\int_c f(z)\,dz = I\,. \tag{2}$$

Theorem 1. *If* $f(z)$ *is continuous on* c, *then the integral* (2) *exists. In particular, if* $f(z)$ *is continuous (or even holomorphic) in a region* O, *then there exists the integral of* $f(z)$ *along every curve* c *(with the properties mentioned above) which lies in* O.

Theorem 2. *The inequality*

$$\left|\int_c f(z)\,dz\right| \leq Ml$$

holds, where l *is the length of the curve* c *and* M *is the maximum* (or. l.u.b., see p. 43) *of* $|f(z)|$ *on* c.

REMARK 1. The integral (2) has properties similar to those of line integrals. In particular,

$$\int_c [k_1 f_1(z) + k_2 f_2(z)]\,dz = k_1 \int_c f_1(z)\,dz + k_2 \int_c f_2(z)\,dz\,,$$

$$\int_c f(z)\,dz = -\int_{c'} f(z)\,dz\,,$$

where c' is the curve c described in the opposite direction.

Definition 2. A function $F(z)$ satisfying the relation

$$F'(z) = f(z)$$

in O is called a *primitive function of $f(z)$ in O.*

Theorem 3. *To every function $f(z)$ holomorphic in a simply connected region O there exists a primitive function.*

REMARK 2. For regions which are not simply connected the assertion of this theorem need not hold.

Theorem 4. *If $F(z)$ is a primitive function of $f(z)$ in O and if c is an arbitrary curve lying in O (with the properties mentioned above) with initial point z_1 and end point z_2, then*

$$\int_c f(z)\,\mathrm{d}z = F(z_2) - F(z_1)\,. \tag{3}$$

REMARK 3. The fact that the value of the integral does not depend on the path of integration in this case but only on the initial and end points of the curve c is often expressed by writing it as

$$\int_{z_1}^{z_2} f(z)\,\mathrm{d}z\,. \tag{4}$$

Example 1. The primitive function of $f(z) = z^2$ is $F(z) = \frac{1}{3}z^3 + C$ in the whole plane. Hence (for every curve)

$$\int_{z_1}^{z_2} z^2\,\mathrm{d}z = \frac{z_2^3}{3} - \frac{z_1^3}{3}\,.$$

REMARK 4. If an integral cannot be evaluated with the aid of a primitive function (i.e. if either the primitive function does not exist, or is difficult to find), we represent the integral by line integrals (see § 14.7); we write $f(z) = u(x, y) + iv(x, y)$, $\mathrm{d}z = \mathrm{d}x + \mathrm{i}\,\mathrm{d}y$, then multiplying formally we obtain

$$\int_c f(z)\,\mathrm{d}z = \int_c [u(x, y)\,\mathrm{d}x - v(x, y)\,\mathrm{d}y] + \mathrm{i}\int_c [v(x, y)\,\mathrm{d}x + u(x, y)\,\mathrm{d}y]\,.$$

We can also often use other methods (see Example 3). In particular, the substitution method and method of integration by parts (under similar assumptions as in § 13.2) can be applied.

Example 2. Let us evaluate the integral

$$\int_c \frac{1}{z}\,\mathrm{d}z\,,$$

where c is the circumference of the circle with centre at the origin and radius a, oriented positively (p. 628) with respect to its interior.

For every point on the circumference we have $z = a(\cos\varphi + i\sin\varphi) = ae^{i\varphi}$. Differentiating, we get $dz = aie^{i\varphi}d\varphi$, so that

$$\int_c \frac{1}{z}\,dz = \int_0^{2\pi} \frac{aie^{i\varphi}\,d\varphi}{ae^{i\varphi}} = i\int_0^{2\pi} d\varphi = 2\pi i\,.$$

The function $\ln z$ could not be used as a primitive function directly, since it is not a *single-valued* function in the region under consideration, and hence is not holomorphic.

This case also shows that the integral of a holomorphic function along a closed curve need not vanish. (Cf. Theorem 5. Here – by contrast with the assumptions of Theorem 5 – the function $f(z) = 1/z$ has a singular point $z = 0$ in the interior of the given circle.)

REMARK 5. The method of integrating by parts leads to the formula

$$\int_{z_1}^{z_2} f_1(z)\,F_2(z)\,dz = [F_1(z)\,F_2(z)]_{z_1}^{z_2} - \int_{z_1}^{z_2} F_1(z)\,f_2(z)\,dz\,, \tag{5}$$

where $F_1(z)$, $F_2(z)$ are primitive functions (Definition 2) of $f_1(z)$, $f_2(z)$, respectively, in O and c is a curve lying in O with initial point z_1 and end-point z_2. (Note that not only F_1 and F_2, but also f_1 and f_2 are holomorphic in O, see Theorem 20. 1. 5.) This formula gives for a closed curve the result

$$\int_c f_1(z)\,F_2(z)\,dz = -\int_c F_1(z)\,f_2(z)\,dz\,, \tag{6}$$

since for a closed curve we have $F_1(z_2)\,F_2(z_2) = F_1(z_1)\,F_2(z_1)$ and the first term of the right-hand side in (5) vanishes. Equation (6) finds frequent application.

Theorem 5 (*Cauchy's Integral Theorem*). *Let c be an oriented closed curve (with the properties mentioned above, i.e. simple, finite and piecewise smooth) lying in a region O and such that its interior lies in O. If $f(z)$ is holomorphic in O, then*

$$\int_c f(z)\,dz = 0\,.$$

(See, however, Example 2). (If O is simply connected, then the curve c need not be simple in order to ensure the assertion of the theorem; it can even intersect itself infinitely many times.)

REMARK 6. Theorem 5 holds under more general assumptions. The curve c need not lie in the interior of the region where $f(z)$ is holomorphic. It is sufficient if $f(z)$ is holomorphic in the interior V of the curve and continuous on $V + c$.

A similar remark holds for Theorems 6 and 7.

Theorem 6. *Let $f(z)$ be holomorphic in a region O* (see the shaded region of Fig. 20.3) *and let c_1, c_2 be two closed curves with the same orientation lying in O, c_2 lying within c_1 (we assume that $f(z)$ is holomorphic in the region the boundary of which is formed by the curves c_1, c_2; $f(z)$ need not be holomorphic in the interior of c_2). Then*

$$\int_{c_1} f(z)\,dz = \int_{c_2} f(z)\,dz\,.$$

(See also Remark 6.)

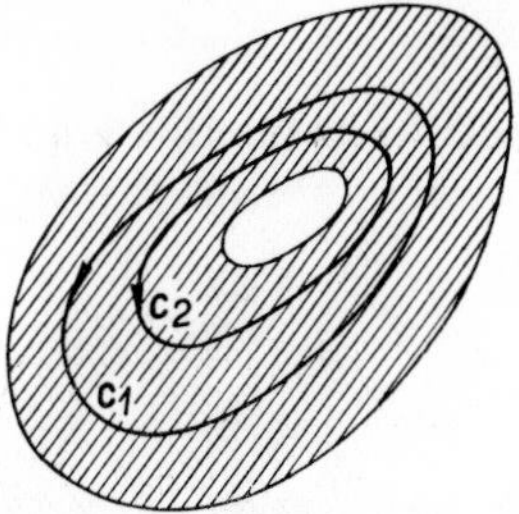

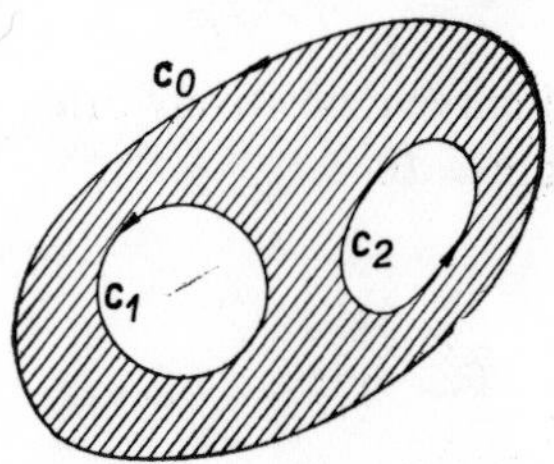

Fig. 20.3.

Fig. 20.4.

REMARK 7. Theorem 6 has important applications. For example, the integral of $f(z) = 1/z$ along any arbitrary closed curve which is positively oriented with respect to its interior, in which the point $z = 0$ is contained, has the value $2\pi i$, since according to 6 this integral is equal to that of Example 2, a being chosen sufficiently small.

REMARK 8. Theorem 6 can be generalized: Let $c_0, c_1, c_2, \ldots, c_n$ be closed curves with the same orientation lying in a region O where $f(z)$ is holomorphic (see Fig. 20.4 where $n = 2$) and let each of the curves $c_1, \ldots, c_n$ lie outside the preceding curves (i.e. $c_2, c_3, \ldots, c_n$ lie outside c_1, etc.) and all of them inside c_0 (we assume that $f(z)$ is holomorphic in the region the boundary of which is formed by the curves $c_0, c_1, \ldots, c_n$; it need not be holomorphic in the interior of $c_1, c_2, \ldots, c_n$). Then

$$\int_{c_0} f(z)\,dz = \sum_{k=1}^{n} \int_{c_k} f(z)\,dz\,.$$

Theorem 7 (*Cauchy's Integral Formula*). *Let a closed curve c, positively oriented with respect to its interior V, lie in a region O in which $f(z)$ is holomorphic* (cf. also Remark 6). *Let $f(z)$ be holomorphic in the interior V of c and let $z_0 \in V$* (Fig. 20.5). *Then*

$$f(z_0) = \frac{1}{2\pi i} \int_c \frac{f(z)\,dz}{z - z_0} \tag{7}$$

and, more generally,

$$f^{(n)}(z_0) = \frac{n!}{2\pi i} \int_c \frac{f(z)\,dz}{(z - z_0)^{n+1}}\,. \tag{8}$$

Here $f^{(n)}(z_0)$ denotes the n-th derivative of $f(z)$ at z_0. (See also Remark 6.)

REMARK 9. If z_0 lies outside the curve c, then $f(z)/(z - z_0)$ is holomorphic in V and according to Cauchy's theorem (Theorem 5) the integrals (7) and (8) are zero.

Example 3. Let us evaluate the integral

$$\int_c \frac{dz}{z^3}$$

along a closed curve c positively oriented with respect to its interior V and let V contain the point $z = 0$.

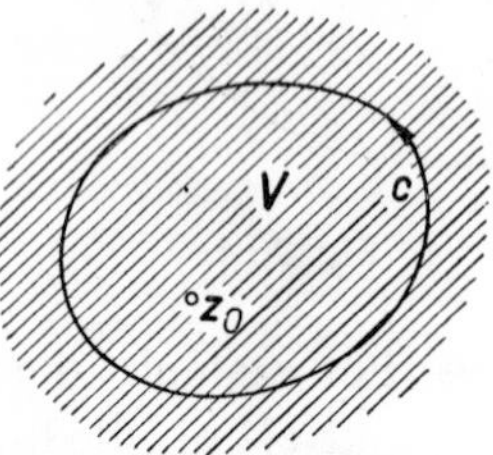

Fig. 20.5.

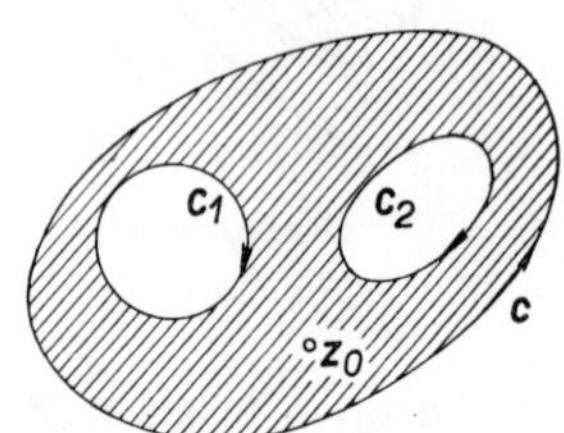

Fig. 20.6.

If we choose $f(z) \equiv 1$ (so that $f(z)$ is holomorphic in the entire plane), and $z_0 = 0$, then use formula (8) for $n = 2$, we have

$$f''(0) = \frac{2}{2\pi i} \int_c \frac{1\, dz}{z^3}$$

and hence, since $f''(0) = 0$,

$$\int_c \frac{dz}{z^3} = 0\,. \tag{9}$$

REMARK 10. Theorem 7 can be modified for the case when $f(z)$ is not holomorphic in the whole interior of the curve c, e.g. if it is not holomorphic in the interior of the closed curves $c_1, \ldots, c_n$ lying in the interior of c and not intersecting each other (see Fig. 20.6 where $n = 2$; note the orientation of these curves). Then, under similar assumptions as in Theorem 7,

$$f(z_0) = \frac{1}{2\pi i} \left(\int_c \frac{f(z)\, dz}{z - z_0} + \int_{c_1} \frac{f(z)\, dz}{z - z_0} + \ldots + \int_{c_n} \frac{f(z)\, dz}{z - z_0} \right). \tag{10}$$

An analogous generalisation holds for equation (8).

20.3. Integrals of Cauchy's Type. Plemelj Formulae

REMARK 1. In this paragraph, by the word *curve* we shall understand a simple, finite, piecewise smooth closed curve, positively oriented with respect to its interior. We always consider these curves in the Gaussian plane and denote their points by t instead of z, i.e. $t = x + iy$. The interior of the curve will be denoted by S^+, the exterior by S^-.

Definition 1. Let a function $f(t)$ be given on a curve c. We say that $f(t)$ has a *derivative with respect to the curve c* (briefly *on c*) at *a point* $t_0 \in c$, if there exists a finite (generally complex) limit

$$\lim \frac{f(t) - f(t_0)}{t - t_0} \tag{1}$$

for $t \to t_0$ along the curve c.

Definition 2. We say that $f(t)$ *satisfies the Hölder condition with the exponent* μ in the neighbourhood of a point $t_0 \in c$ if there exists an arc I on c such that t_0 is an interior point of I and that there exist two constants M and μ ($M > 0$, $0 < \mu \leqq 1$) such that for every two points t_1, t_2 of I the relation

$$|f(t_2) - f(t_1)| \leqq M|t_2 - t_1|^\mu \tag{2}$$

holds. In particular, if there exist constants M and μ ($M > 0$, $0 < \mu \leqq 1$) such that (2) holds for every pair of points t_1, t_2 of the curve c, we say that $f(t)$ *satisfies the Hölder conditions with exponent* μ *on c*.

REMARK 2. We shall briefly say that $f(t)$ *fulfils the condition H* (or, in a more detailed manner, the *condition* $H(\mu)$).

Theorem 1. *If $f(t)$ has a continuous (or, more generally, a bounded) derivative on c, then it satisfies the condition H on c (and hence in the neighbourhood of every point of c) with the exponent* 1.

Definition 3. Let a function $f(t)$ be defined on c. The integral

$$F(z) = \frac{1}{2\pi i} \int_c \frac{f(t)\,dt}{t - z} \tag{3}$$

is called an *integral of Cauchy's type*.

Theorem 2. *If $f(t)$ is continuous on c (it is even sufficient if $f(t)$ is integrable on c), then $F(z)$ is a holomorphic function of z both in S^+ and in S^-* (for the meaning of S^+ and S^- see Remark 1). *The derivatives of $F(z)$ are given by the formulae*

$$F'(z) = \frac{1}{2\pi i} \int_c \frac{f(t)\,dt}{(t - z)^2}, \ldots, F^{(n)}(z) = \frac{n!}{2\pi i} \int_c \frac{f(t)\,dt}{(t - z)^{n+1}} \tag{4}$$

[*which we get by formal differentiation with respect to z under the integral sign in* (3)].

REMARK 3. If we choose a point z on the curve c, then, in general, the integral (3) does not exist (it is divergent). However, the principal value (see Definition 4) of this integral can exist.

Definition 4. Let the point t_0 lie on c. Choose $\varepsilon > 0$ such that the circumference of the circle with centre t_0 and radius ε (and also all circumferences of circles with radius less than ε) intersect the curve c exactly at two points (denote them by

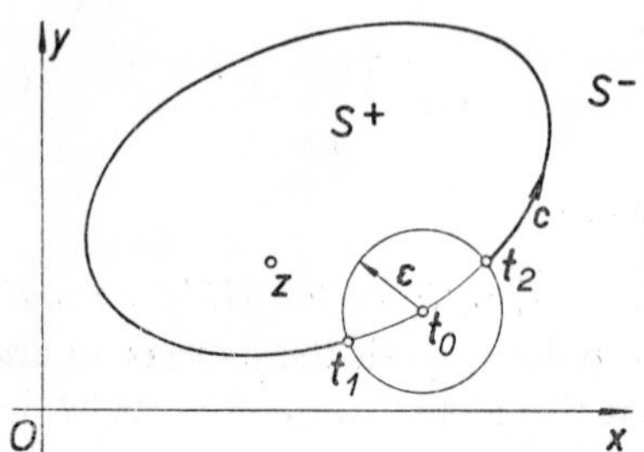

Fig. 20.7.

t_1, t_2; Fig. 20.7). (If the curve c has the properties mentioned in Remark 1, then it can be proved that such an $\varepsilon > 0$ exists for every point $t_0 \in c$). If there exists a finite limit

$$\lim_{\varepsilon \to 0} \int_{c - \widehat{t_1 t_2}} \frac{f(t)\,dt}{t - t_0}, \tag{5}$$

where $\widehat{t_1 t_2}$ is the arc of the curve c lying within the circumference of the circle with centre t_0 and radius ε, we say that the *integral*

$$\int_c \frac{f(t)\,dt}{t - t_0} \tag{6}$$

exists in the sense of its principal value.

REMARK 4. If the point t_0 lies on c, then, in what follows, we shall always interpret the integral (6) to mean its principal value. The function $f(t)$ will always be assumed to be integrable on c.

Theorem 3. *If $f(t)$ has a continuous derivative in the neighbourhood of t_0* (Definition 1) (*it is even sufficient to require only that $f(t)$ fulfils condition H in the neighbourhood of t_0*, cf. Definition 2 and Remark 2), *then the principal value of integral* (6) *exists.*

REMARK 5. By the continuous extensibility (on the curve c) of functions defined in S^+ (or S^-) we mean the extensibility in the sense of definition 20.1.7. Further, we say that a function $g(z)$ defined in S^+ (or S^-) is *continuously extensible on c at the point $t_0 \in c$* if there exists a number A such that

$$\lim_{z \to t_0} g(z) = A, \quad \text{where} \quad z \in S^+ \quad (\text{or } z \in S^-, \text{respectively}).$$

REMARK 6. Hence $g(z)$ is continuously extensible to a value A at $t_0 \in c$ if for every sequence of points $z_n \in S^+$ (or $z_n \in S^-$, respectively) satisfying $\lim_{n \to \infty} z_n = t_0$, we have

$$\lim_{n \to \infty} g(z_n) = A.$$

Theorem. 4 *Let $f(t)$ satisfy the condition H in the neighbourhood of $t_0 \in c$. (A sufficient condition for this is that $f(t)$ have a continuous derivative in a neighbourhood of t_0.) Then the function*

$$F(z) = \frac{1}{2\pi i} \int_c \frac{f(t)\, dt}{t - z} \tag{7}$$

is continuously extensible on the curve c from S^+ as well as from S^- and we have

$$F^+(t_0) = \tfrac{1}{2} f(t_0) + \frac{1}{2\pi i} \int_c \frac{f(t)\, dt}{t - t_0}, \tag{8}$$

$$F^-(t_0) = -\tfrac{1}{2} f(t_0) + \frac{1}{2\pi i} \int_c \frac{f(t)\, dt}{t - t_0}. \tag{9}$$

Here $F^+(t_0)$ and $F^-(t_0)$ stand for the values of the continuous extension of $F(z)$ from S^+ and from S^-, respectively; by integrals (8), (9) we mean their principal values.

Formulae (8) and (9) are called *Plemelj formulae*.

REMARK 7. Formulae (8) and (9) show that if we cross the curve c from S^- to S^+ at the point t_0, then the function (7) changes (in the limit) by a value equal to $f(t_0)$.

REMARK 8. It can be shown that if $f'(t)$ (Definition 1) satisfies the condition H in the neighbourhood of the point t_0, then the function

$$F'(z) = \frac{1}{2\pi i} \int_c \frac{f(t)\, dt}{(t - z)^2}$$

is continuously extensible at t_0 on the curve c from S^+ and from S^-. An analogous assertion is valid for derivatives of higher order.

Theorem 5 (*Cauchy's Theorem*). *Let $\varphi(z)$ be holomorphic in S^+ and continuous in $\bar{S}^+$ (i.e. in $S^+ + c$). Then*

$$\frac{1}{2\pi i}\int_c \frac{\varphi(t)\,dt}{t-z} = \varphi(z) \quad \text{if} \quad z \in S^+\,, \tag{10}$$

$$\frac{1}{2\pi i}\int_c \frac{\varphi(t)\,dt}{t-z} = 0 \qquad \text{if} \quad z \in S^-\,. \tag{11}$$

If $\varphi(z)$ is holomorphic in S^- and at infinity (i.e. if $\varphi(1/z)$ is holomorphic at the origin (cf. Remark 20.4.11)) *and continuous in $\bar{S}^-$ (i.e. in $S^- + c$), then*

$$\frac{1}{2\pi i}\int_c \frac{\varphi(t)\,dt}{t-z} = \varphi(\infty)\,[= \lim_{z\to\infty} \varphi(z)] \quad \text{if} \quad z \in S^+\,, \tag{12}$$

$$\frac{1}{2\pi i}\int_c \frac{\varphi(t)\,dt}{t-z} = -\,\varphi(z) + \varphi(\infty) \qquad \text{if} \quad z \in S^-\,. \tag{13}$$

Example 1. Let the origin belong to S^+. Then

$$\int_c \frac{dt}{t^2} = 0\,, \tag{14}$$

because the function $\varphi(z) = 1/z$ is holomorphic in S^- and at infinity and is continuous in $\bar{S}^-$: According to (12), we have

$$\frac{1}{2\pi i}\int_c \frac{1/t}{t-z}\,dt = 0 \tag{15}$$

for every $z \in S^+$. Putting $z = 0$ in (15) we obtain (14).

Theorem 6. *A function $\varphi(t)$ continuous on c is a continuous extension of some function $\varphi(z)$ holomorphic in S^+ if and only if*

$$\frac{1}{2\pi i}\int_c \frac{\varphi(t)\,dt}{t-z} = 0 \quad \text{for every} \quad z \in S^-\,; \tag{16}$$

$\varphi(t)$ is a continuous extension of some function $\psi(z)$ holomorphic in S^- and at infinity if and only if

$$\frac{1}{2\pi i}\int_c \frac{\varphi(t)\,dt}{t-z} = a \quad \text{for every} \quad z \in S^+\,, \tag{17}$$

where a is a constant. (This constant is equal to the value $\psi(\infty)$.)

REMARK 9. The equation

$$F(z) = \frac{1}{2\pi i} \int_c \frac{f(t)\,dt}{t - z} \tag{18}$$

does *not* imply (even if $f(t)$ has a derivative on c) that $f(t)$ is a continuous extension of $F(z)$ from S^+ (the relation $F^+(t) = f(t)$ need not hold). According to Theorem 6 we can have $f(t) - F^+(t) = g(t)$, where $g(t)$ is a continuous extension of some function holomorphic in S^- and vanishing at infinity.

REMARK 10. The results of this paragraph can be generalized, for instance, to the case of multiply connected regions. Similar results also hold for the case where c is not a closed curve but the x-axis (the axis of real numbers) in the Gaussian plane. S^+ and S^- are then half-planes. The formulation of these results can be found, for example, in [21], §§ 5.7 and 5.11.

20.4. Series. Taylor's Series, Laurent's Series. Singular Points of Holomorphic Functions

In this paragraph we use the notation $a = \alpha + i\beta$, $a_n = \alpha_n + i\beta_n$, where α, β, α_n, β_n are real numbers.

Definition 1. We say that a sequence of complex numbers

$$a_1, a_2, \ldots, a_n, \ldots$$

has the *limit a* (or tends to the limit a) if to every $\varepsilon > 0$ there corresponds a number n_0 such that for every $n > n_0$ we have

$$|a - a_n| < \varepsilon \quad \text{i.e.} \quad \sqrt{[(\alpha - \alpha_n)^2 + (\beta - \beta_n)^2]} < \varepsilon\,.$$

We write

$$\lim_{n\to\infty} a_n = a\,.$$

Definition 2. Let us write

$$s_n = a_1 + a_2 + \ldots + a_n\,.$$

We say that the *series*

$$a_1 + a_2 + \ldots + a_n + \ldots \tag{1}$$

is *convergent* and *has the sum s* if

$$\lim_{n\to\infty} s_n = s\,.$$

Theorem 1. *The series* (1) *is convergent if and only if both series of real numbers*

$$\alpha_1 + \alpha_2 + \ldots + \alpha_n + \ldots, \quad \beta_1 + \beta_2 + \ldots + \beta_n + \ldots$$

are convergent. If the first of them has the sum α *and the second the sum* β, *then*

$$s = \alpha + i\beta .$$

Theorem 2. *The convergence of the series*

$$|a_1| + |a_2| + \ldots + |a_n| + \ldots$$

implies the convergence of the series (1). The series (1) is then said to be *absolutely convergent.*

Definition 3. Suppose we are given a sequence of functions $f_1(z)$, $f_2(z)$, ..., $f_n(z)$, ... The set of those z for which the series

$$f_1(z) + f_2(z) + \ldots + f_n(z) + \ldots \tag{2}$$

converges is called the *domain of convergence* O of this series. The series (2) (more precisely, the sum of the series (2) for every $z \in O$) defines a certain function $s(z)$ in O.

REMARK 1. For series of functions of a complex variable, definitions and theorems can be formulated very similar to those for series of functions of a real variable: Suppose that, to every $\varepsilon > 0$, there corresponds a number n_0 *independent of the choice of* z from the region of convergence of the series (2) (hence depending only on the choice of ε) such that for every $n > n_0$ we have

$$|s_n(z) - s(z)| < \varepsilon$$

(where $s_n(z) = f_1(z) + \ldots + f_n(z)$, $s(z) = \lim s_n(z)$). Then we say that the *series* (2) *converges uniformly in* O. In order that this should happen, it is sufficient that constants $A_1, A_2, \ldots, A_n, \ldots$ exist such that for every $z \in O$ the following relations hold:

$$|f_1(z)| \leqq A_1,\ |f_2(z)| \leqq A_2,\ \ldots,\ |f_n(z)| \leqq A_n,\ \ldots$$

and, at the same time, the series

$$A_1 + A_2 + \ldots + A_n + \ldots$$

is convergent (this is the *Weierstrass test* or the so-called *M-test*).

If the series (2) converges uniformly in every closed region $\bar{B}$ contained in O (corresponding to a given $\varepsilon > 0$ we generally get different numbers n_0 for different regions $\bar{B}$), we say that the series (2) is *almost uniformly convergent in* 0.

Analogous definitions can be given for a sequence $f_1(z), f_2(z), f_3(z), \ldots$.

Theorem 3 (*Weierstrass's Theorem*). *Let the functions* $f_1(z), f_2(z), f_3(z), \ldots$ *be holomorphic in a region* O *and let the series* (2) *be almost uniformly convergent in* O (Remark 1). (*The second assumption is satisfied, for example, if the series* (2) *is uniformly convergent in the whole region* O.) *Then*

1. *The function $s(z)$ defined by the sum of the series (2) is holomorphic in O.*

2. *If we differentiate the series (2) n times, term by term, we obtain a series which is again almost uniformly convergent on O and whose sum in O is equal to the n-th derivative of the function $s(z)$.*

3. *If $z_0 \in O$, $z \in O$, then*

$$\int_{z_0}^{z} s(t)\,dt = \int_{z_0}^{z} f_1(t)\,dt + \int_{z_0}^{z} f_2(t)\,dt + \dots .$$

Definition 4. By a *power series* (in complex variable) we mean the series

$$a_0 + a_1(z - z_0) + a_2(z - z_0)^2 + \dots, \tag{3}$$

where the a_n are constants (generally complex).

REMARK 2. For $z_0 = 0$, a power series has the form

$$a_0 + a_1 z + a_2 z^2 + \dots . \tag{4}$$

Since the series (3) can be transformed into a series of the form (4) by the substitution $z - z_0 = z'$ (translation in the Gaussian plane), it is sufficient to consider the series (4). Theorems valid for the series (4) hold for the series (3) as well, if we write $z - z_0$ instead of z.

Theorem 4. *To every series (4) there corresponds a number $r \geqq 0$ ($r = +\infty$ is also admitted) such that (4) is convergent for all z satisfying $|z| < r$ and divergent for all z satisfying $|z| > r$.* The number r is called the *radius of convergence of the series* (4).

REMARK 3. In order to find the radius of convergence of the series (4), we can use analogous rules to those governing real series. In particular: The series (4) and the series

$$|a_0| + |a_1|\,|z| + |a_2|\,|z|^2 + \dots \tag{5}$$

have the same radius of convergence.

Since (5) is a series with non-negative terms, the criteria of § 10.2 may be applied to determine its radius of convergence. In particular, they imply:

Theorem 5. *If there exists a limit*

$$\lim_{n\to\infty} \frac{|a_{n+1}|}{|a_n|} = l \quad or \quad \lim_{n\to\infty} \sqrt[n]{|a_n|} = l\,,$$

then

$$r = \frac{1}{l}\,.$$

(For $l = 0$ we have $r = +\infty$, for $l = +\infty$ we have $r = 0$.)

Theorem 6 (*Abel's theorem*). *Let a power series*

$$a_0 + a_1(z - z_0) + a_2(z - z_0)^2 + \ldots = s(z)$$

having a radius of convergence r converge at a point t on the circumference of its circle of convergence (i.e. at a point t such that $|t - z_0| = r$). Let S be its sum at this point. Then there exists the angular extension S of s(z) from the interior K of the

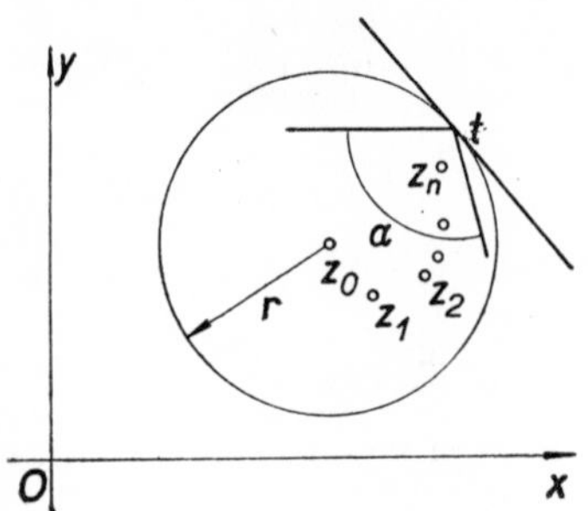

Fig. 20.8.

circle of convergence to the point t. This means: For every sequence of points $z_n \in K$ which converges to t and which lies in the interior of some angle $\alpha < \pi$ with vertex at t whose arms lie, in a neighbourhood of t, inside K (Fig. 20.8), we have

$$\lim_{n \to \infty} s(z_n) = S\,.$$

Theorem 7. *A power series converges absolutely in its circle of convergence K ($|z - z_0| < r$). In addition, it converges uniformly in every closed region contained in K* (i.e. *almost uniformly in K*).

Hence Theorem 3 applies to power series in K. It implies especially: *The sum s(z) of a power series is a holomorphic function in K. Its derivatives can be calculated by differentiating the given series term by term. All these series have the same radius of convergence.*

REMARK 4. Some functions in complex variable may be defined by power series (naturally, in their domain of convergence). The most important are the following:

$$e^z = 1 + \frac{z}{1!} + \frac{z^2}{2!} + \frac{z^3}{3!} + \ldots \quad (r = +\infty)\,,$$

$$\sin z = z - \frac{z^3}{3!} + \frac{z^5}{5!} - \frac{z^7}{7!} + \ldots \quad (r = +\infty)\,,$$

$$\cos z = 1 - \frac{z^2}{2!} + \frac{z^4}{4!} - \frac{z^6}{6!} + \ldots \quad (r = +\infty)\,.$$

For real arguments ($z = x$), these functions coincide with the well-known functions e^x, $\sin x$, $\cos x$. (We say that they are an *extension of these functions into the complex plane* (cf. § 20.6).) For the functions e^z, $\sin z$, $\cos z$, all formulae valid for functions of a real argument still hold. For example,

$$e^{z_1}e^{z_2} = e^{z_1+z_2}, \quad \sin(z_1 + z_2) = \sin z_1 \cos z_2 + \cos z_1 \sin z_2,$$

$$(e^z)' = e^z, \quad (\sin z)' = \cos z, \quad (\cos z)' = -\sin z,$$

etc. From the expansions in series we easily get the so-called *Euler relation*:

$$e^{iz} = \cos z + i \sin z,$$

which is often used for $z = x$ when solving linear differential equations with constant coefficients. Further we have

$$\cosh z = \frac{e^z + e^{-z}}{2} = \cos iz, \quad \sinh z = \frac{e^z - e^{-z}}{2} = -i \sin iz,$$

$$\cos z = \cosh iz, \quad \sin z = -i \sinh iz,$$

$$\sin z = \sin x \cosh y + i \cos x \sinh y, \quad \cos z = \cos x \cosh y - i \sin x \sinh y,$$

$$\cos(z + 2k\pi) = \cos z, \quad \sin(z + 2k\pi) = \sin z, \quad e^{z+2k\pi i} = e^z$$

(k being an integer), etc.

Definition 5. We say that a point z_0 ($z_0 \neq \infty$) is a *regular point* (*ordinary point*) of a function $f(z)$ if there exists a (circular) neighbourhood of z_0 such that $f(z)$ is holomorphic in this neighbourhood. A point which is not regular is called a *singular point* of $f(z)$.

Example 1. The function $f(z) = 1/z$ has only one singular point, namely $z = 0$. Every other point of the Gaussian plane is a regular point of this function.

Theorem 8 (*Taylor's Series*). *Let $z = z_0$ be a regular point of a function $f(z)$. Then in the neighbourhood of z_0 we have*

$$f(z) = a_0 + a_1(z - z_0) + a_2(z - z_0)^2 + \dots, \tag{6}$$

where

$$a_n = \frac{f^{(n)}(z_0)}{n!}. \tag{7}$$

The radius of convergence r of the series (6) *is equal to the distance of the point z_0 from the nearest singular point of $f(z)$.* (*More exactly: The radius r is equal to the* g. l. b. *of the distances of z_0 from all singular points of $f(z)$.*) *The series* (6) *is uniquely determined by the function $f(z)$.*

REMARK 5. It follows from Cauchy's integral formula that the coefficients of the series (6) can also be evaluated as the integrals

$$a_n = \frac{1}{2\pi i} \int_c \frac{f(z)\,dz}{(z - z_0)^{n+1}} \tag{8}$$

along an arbitrary simple closed curve c positively oriented with respect to its interior, lying in a neighbourhood of z_0 in which $f(z)$ is holomorphic and containing the point z_0 in its interior.

Example 2. For $f(z) = 1/z$ and $z_0 = 2$, we have

$$f^{(n)}(2) = (-1)^n \frac{n!}{2^{n+1}}$$

and, according to (6) and (7),

$$\frac{1}{z} = \frac{1}{2} - \frac{1}{2^2}(z - 2) + \frac{1}{2^3}(z - 2)^2 - \frac{1}{2^4}(z - 2)^3 + \dots .$$

The radius of convergence r is equal to the distance of the point $z = 2$ from the singular point $z = 0$, hence $r = 2$.

REMARK 6. Here (and often for other simple rational functions) it is also possible to get the Taylor expansion in such a way that we first rearrange the formula for the functional relation to be of the form of a sum of a geometric series; from the condition for the convergence of this geometric series (the absolute value of the common ratio must be less than 1), we find the radius of convergence r:

$$\frac{1}{z} = \frac{1}{2 + (z - 2)} = \frac{1}{2} \cdot \frac{1}{1 + \dfrac{z - 2}{2}} = \frac{1}{2}\left[1 - \frac{z - 2}{2} + \frac{(z - 2)^2}{2^2} - \frac{(z - 2)^3}{2^3} + \dots\right];$$

$$\left|\frac{z - 2}{2}\right| < 1\,, \quad \text{i.e.} \quad |z - 2| < 2\,, \quad \text{hence} \quad r = 2\,.$$

Theorem 9 (*Laurent's Series*). *Let $f(z)$ be holomorphic in the annulus M with centre z_0, inner radius r_1 and outer radius r_2 (i.e. $r_1 < |z - z_0| < r_2$). Then for $z \in M$ we have*

$$f(z) = \sum_{n=-\infty}^{\infty} a_n(z - z_0)^n \tag{9}$$

where

$$a_n = \frac{1}{2\pi i} \int_c \frac{f(z)\,dz}{(z - z_0)^{n+1}} \quad (n = 0, \pm 1, \pm 2, \dots)\,, \tag{10}$$

c being the circumference of an arbitrary circle with centre z_0, lying in M and positively oriented with respect to its interior.

REMARK 7. By the convergence of the series (9) we mean the convergence of *both* series

$$\sum_{n=0}^{\infty} a_n(z - z_0)^n , \quad \sum_{n=1}^{\infty} \frac{a_{-n}}{(z - z_0)^n} . \tag{11}$$

The first series is called the *regular part*, the second series the *principal part of the Laurent series*. The domain of convergence of series (11) can be a set greater than M. The regular part (as a power series) converges *inside* a certain circle (with centre z_0), the principal part *outside* a certain circle. The series (9) then converges in the common annulus. If $f(z)$ is holomorphic everywhere inside the smaller circle, then the principal part of the Laurent series vanishes (which also follows from Cauchy's integral theorem, since the integrand in (10) is then a regular function for $n = -1, -2, \ldots$) and the Laurent series coincides with the Taylor series.

Theorem 10. *The series* (9) *is uniquely determined by the holomorphic function* $f(z)$.

REMARK 8. It is not always necessary, if we expand a function into its Laurent series, to calculate the coefficients of this the Laurent series according to (10). We can often – especially in the case of simple rational functions – use a similar method as in the case of the Taylor series (Remark 6), i.e. write the given function in a form involving fractions each having the form of the sum of a geometric series.

Example 3. Let us expand the function

$$f(z) = \frac{1}{(z-1)(z-3)}$$

in a Laurent series with centre at $z = 0$ (i.e. in powers of z) and converging in the annulus $1 < |z| < 3$. We have

$$\frac{1}{(z-1)(z-3)} = -\frac{1}{2}\left(\frac{1}{z-1} - \frac{1}{z-3}\right) = -\frac{1}{6}\cdot\frac{1}{1-\dfrac{z}{3}} - \frac{1}{2z}\cdot\frac{1}{1-\dfrac{1}{z}} =$$

$$= -\frac{1}{6}\sum_{n=0}^{\infty}\left(\frac{z}{3}\right)^n - \frac{1}{2z}\sum_{n=0}^{\infty}\left(\frac{1}{z}\right)^n = -\left[\ldots\frac{1}{2z^2} + \frac{1}{2z} + \frac{1}{6} + \frac{z}{18} + \ldots\right].$$

We must be careful to ensure that the common ratio of each of the geometric series be of absolute value less than 1. Therefore we first put the function in the form shown above, for in this example we are considering values of z for which $1 < |z| < 3$. In fact, after this arrangement is made, the common ratios of both geometric series will be of absolute value less than 1, since

$$\left|\frac{z}{3}\right| < 1 , \quad \left|\frac{1}{z}\right| < 1 .$$

Example 4. Let us develop the same function in a Laurent series with centre $z = 1$ (i.e. in powers of $z - 1$), converging for $0 < |z - 1| < 2$. We have

$$\frac{1}{(z-1)(z-3)} = \frac{1}{(z-1)[(z-1)-2]} = -\frac{1}{2(z-1)} \cdot \frac{1}{1 - \frac{1}{2}(z-1)} =$$
$$= -\frac{1}{2(z-1)} \sum_{n=0}^{\infty} \left(\frac{z-1}{2}\right)^n = -\frac{1}{2(z-1)} - \frac{1}{4} - \frac{1}{8}(z-1) - \dots .$$

REMARK 9. Example 4 illustrates an important case of the development of a function $f(z)$ in a Laurent series in the neighbourhood of an *isolated singular point* (i.e. a singular point such that there is no other singular point in a sufficiently small neighbourhood of it). In this case, the inner circle reduces to a point.

REMARK 10. Let (9) be a Laurent expansion of a (holomorphic) function $f(z)$, converging in a neighbourhood O of an *isolated* singular point z_0 of $f(z)$ (we naturally consider the neighbourhood O without the point z_0). Exactly three cases can then arise:

1. If there exists a $k > 0$ such that in (9) we have $a_{-k} \neq 0$ but $a_{-l} = 0$ for all $l > k$ (hence the principal part of the Laurent series has only a finite number of terms), we say that $f(z)$ has *a pole of k-th order* or *a pole of order k at* z_0.

A pole of first order is often called a *simple pole* and a pole of second order a *double pole.*

A (holomorphic) function which has no singular points in the complex plane other than poles is called a *meromorphic function.* It can be shown that if a function $f(z)$ is single-valued in the closed plane (Remark 20.1.2) and has at infinity at most a pole (not an essential singularity, see Remark 11), then it is meromorphic if and only if it is a rational function, i.e. a function which can be expressed in the form

$$f(z) = \frac{a_0 + a_1 z + \dots + a_n z^n}{b_0 + b_1 z + \dots + b_m z^m}.$$

2. If the principal part of the Laurent series has an infinite number of terms, we say that $f(z)$ has an *essential singularity at* z_0.

In this case, in an arbitrary small neighbourhood of z_0 the difference between $f(z)$ and an arbitrarily chosen number A can be made arbitrarily small. More precisely: Given an arbitrary (generally complex) number A we can find a sequence of points $z_1, z_2, z_3, \dots$ converging to the point z_0 such that

$$\lim_{n \to \infty} f(z_n) = A .$$

This is true even for $A = \infty$.

3. If the Laurent series has only a regular part, we say that $f(z)$ has a *removable singularity at* z_0. For example, the function

$$\frac{\sin z}{z} = 1 - \frac{z^2}{3!} + \frac{z^4}{5!} - \ldots$$

(see Remark 4) has a removable singularity at $z = 0$. If we define $f(z)$ at z_0 by the value a_0 (in this example by the value 1), then $f(z)$ is holomorphic in a neighbourhood of z_0. A (holomorphic) function $f(z)$ has a removable singularity at an isolated singular point if and only if f is bounded in the neighbourhood of this point.

These three cases are the only possible types of singularities of holomorphic (single-valued) functions at isolated singular points.

Multi-valued functions (§ 20.6) can have other types of singularities.

REMARK 11. If $f(z)$ is defined for all sufficiently great z (briefly: *in a neighbourhood of infinity*), we speak of a *Laurent series in the neighbourhood of infinity.* By the substitution

$$z - z_0 = \frac{1}{t} \tag{12}$$

we reduce the investigation of a Laurent series in the neighbourhood of infinity to the investigation of another Laurent series at $t = 0$. If this series has a pole, an essential singularity or a removable singularity at the point $t = 0$, we say that the original series (in powers of $z - z_0$) has a *pole*, an *essential singularity* or a *removable singularity at infinity*, respectively. For example, if the original Laurent series has a *regular* part with an infinite number of terms, then the corresponding Laurent series in the variable t has a *principal* part with an infinite number of terms and we say that $f(z)$ has an essential singularity at infinity.

REMARK 12. A holomorphic function given by a power series with an infinite number of terms and radius of convergence $r = +\infty$ (i.e. converging for every finite z) is called an *entire* or *integral transcendental function.* As examples we have the functions $\sin z$, $\cos z$, e^z (Remark 4). An entire transcendental function has its only singular point at $z = \infty$ and this point is an essential singularity. Conversely, every function holomorphic in the entire plane and having an essential singularity at infinity is an entire transcendental function.

Theorem 11 (*Liouville's Theorem*). *If a function $f(z)$ is holomorphic and bounded in the whole plane, then it is merely a constant.*

20.5. The Residue of a Function. The Residue Theorem and its Applications

Definition 1. Let $f(z)$ be a holomorphic function and z_0 an isolated singular point of $f(z)$. Then we can develop $f(z)$ in the neighbourhood of z_0 (for $z \neq z_0$) in its Laurent series (Theorem 20.4.9),

$$f(z) = \sum_{n=-\infty}^{\infty} a_n(z - z_0)^n = \ldots \frac{a_{-2}}{(z - z_0)^2} + \frac{a_{-1}}{z - z_0} + a_0 + a_1(z - z_0) + \ldots . \quad (1)$$

The number a_{-1} is called *the residue of the function* $f(z)$ *at the point* z_0.

Example 1. The residue of the function

$$f(z) = \frac{1}{(z - 1)(z - 3)}$$

at the point $z = 1$ is $-\frac{1}{2}$ (Example 20.4.4).

REMARK 1. According to (20.4.10) we have (since $n = -1$)

$$a_{-1} = \frac{1}{2\pi i} \int_c f(z)\, dz .$$

In some cases we can find the residue of a function more easily:

If $f(z)$ has a pole of the *first* order at z_0, then

$$a_{-1} = \lim_{z \to z_0} (z - z_0) f(z) ,$$

while if $f(z)$ has a pole of the k-th order ($k > 1$) at z_0, then

$$a_{-1} = \frac{1}{(k - 1)!} \lim_{z \to z_0} \frac{d^{k-1}}{dz^{k-1}} [(z - z_0)^k f(z)] .$$

If $f(z)$ has, in the neighbourhood of z_0, the form

$$f(z) = \frac{\varphi(z)}{\psi(z)}$$

[$\varphi(z)$, $\psi(z)$ being holomorphic in the neighbourhood of the point z_0] and if $\varphi(z_0) \neq 0$, $\psi(z_0) = 0$, $\psi'(z_0) \neq 0$, then $f(z)$ has a pole of the first order at z_0. In this case, the residue is equal to

$$a_{-1} = \frac{\varphi(z_0)}{\psi'(z_0)} . \quad (2)$$

Example 2. The function

$$f(z) = \frac{1}{\sin z}$$

has at $z = 0$ a pole of the first order and the residue is $a_{-1} = 1$, for

$$\sin 0 = 0\,, \quad \cos 0 \neq 0\,, \quad a_{-1} = \frac{1}{\cos 0} = \frac{1}{1} = 1\,.$$

Theorem 1 (*Residue Theorem*). *Let c be a simple piecewise smooth closed curve,*

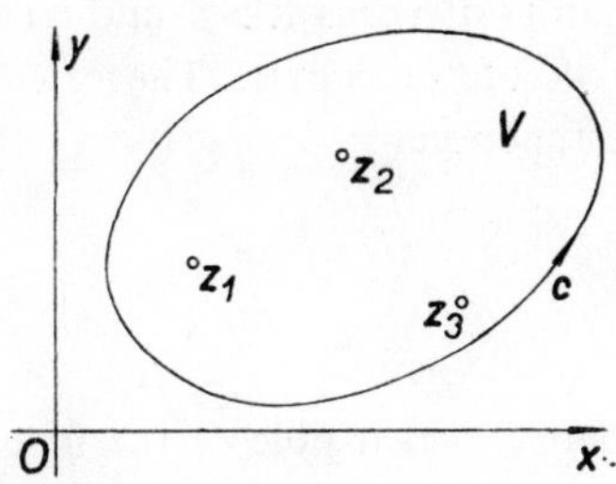

Fig. 20.9.

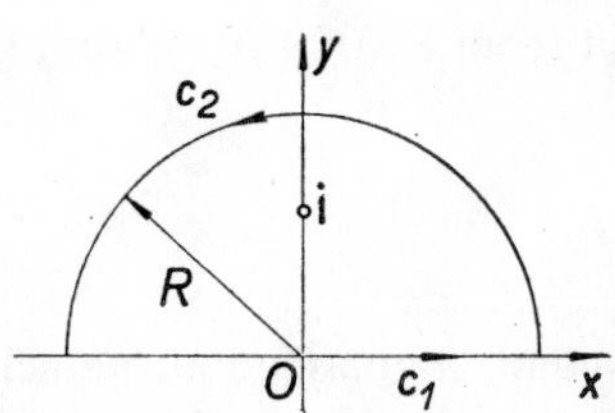

Fig. 20.10.

positively oriented with respect to its interior V. Let $f(z)$ be a function holomorphic in V with the exception of a finite number of singular points $z_1, z_2, \ldots, z_n$ (Fig. 20.9, where $n = 3$) *and continuously extensible on c (i.e. continuous in $\overline{V} = V + c$ with the exception of the points $z_1, z_2, \ldots, z_n$). Then the integral*

$$\frac{1}{2\pi i} \int_c f(z)\, dz$$

is equal to the sum of residues at the points $z_1, z_2, z_3, \ldots, z_n$. In symbols:

$$\frac{1}{2\pi i} \int_c f(z)\, dz = \sum_{k=1}^{n} \operatorname{res}\, [f(z)]_{z=z_k}\,. \tag{3}$$

Remark 2. This theorem has many applications, one of the most important of which is its use in evaluating integrals which cannot be expressed by elementary primitive functions, e.g.

$$\int_0^\infty \frac{\sin x}{x}\, dx\,, \quad \int_0^\infty \frac{x^\alpha}{1 + x}\, dx \quad (-1 < \alpha < 0)\,,$$

etc. Here we shall give an illustrative example showing the fundamental idea of the method:

Example 3. Let us evaluate the integral

$$\int_{-\infty}^{\infty} \frac{dx}{1 + x^2} . \tag{4}$$

(In this case, of course, the integral can be computed by means of the primitive function arctan x; we know that its value is π.)

The integral (4) is convergent and therefore

$$\int_{-\infty}^{\infty} \frac{dx}{1 + x^2} = \lim_{R \to +\infty} \int_{-R}^{R} \frac{dx}{1 + x^2} . \tag{5}$$

Let us draw a semi-circle c_2 with centre at the origin and radius $R > 1$ and write $c = c_1 + c_2$, where c_1 is the segment $-R \leqq x \leqq R$ of the x-axis. The orientation is evident from Fig. 20.10. Within the curve c the function

$$f(z) = \frac{1}{1 + z^2}$$

has only one singular point, namely $z = +\mathrm{i}$, where it has a pole of the first order (Remark 1). The residue can be calculated, for example, by means of (2),

$$a_{-1} = \left[\frac{1}{2z}\right]_{z=\mathrm{i}} = \frac{1}{2\mathrm{i}} .$$

Then from (3), we obtain

$$\int_c f(z)\, dz = 2\pi\mathrm{i} \frac{1}{2\mathrm{i}} = \pi . \tag{6}$$

The relation (6) holds for every $R > 1$, hence

$$\lim_{R \to +\infty} \int_c f(z)\, dz = \pi .$$

Hence (from (5))

$$\int_{-\infty}^{\infty} \frac{dx}{1 + x^2} = \lim_{R \to +\infty} \int_{-R}^{R} \frac{dx}{1 + x^2} = \lim_{R \to +\infty} \int_{c_1} f(z)\, dz =$$

$$= \lim_{R \to +\infty} \int_c f(z)\, dz - \lim_{R \to +\infty} \int_{c_2} f(z)\, dz = \pi - \lim_{R \to +\infty} \int_{c_2} f(z)\, dz . \tag{7}$$

However, for every $R > 1$, we have (according to Theorem 20.2.2)

$$\left|\int_{c_2} f(z)\, dz\right| = \left|\int_{c_2} \frac{dz}{1 + z^2}\right| \leqq \pi R \,.\, \max \left|\frac{1}{1 + z^2}\right|_{c_2} \leqq \pi R \frac{1}{R^2 - 1} , \tag{8}$$

since $|z^2 + 1| \geqq |z^2| - 1$ and on c_2 we have $|z| = R$. Since

$$\lim_{R \to +\infty} \frac{R}{R^2 - 1} = 0$$

we have, using (8),

$$\lim_{R \to +\infty} \left| \int_{c_2} f(z)\, dz \right| = 0$$

and hence also

$$\lim_{R \to \infty} \int_{c_2} f(z)\, dz = 0$$

whence, using (7),

$$\int_{-\infty}^{\infty} \frac{dx}{1 + x^2} = \pi .$$

20.6. Logarithm, Power. Analytic Continuation. Analytic Functions

The so-called *principal branch of the function "logarithm z"* is defined by the relation

$$\ln_0 z = \ln r + i\varphi , \quad 0 \leqq \varphi < 2\pi , \quad r > 0 , \tag{1}$$

where r is the absolute value, φ the argument of the number z, i.e.

$$z = r(\cos \varphi + i \sin \varphi) , \quad 0 \leqq \varphi < 2\pi . \tag{2}$$

The function (1) is holomorphic ($(\ln_0 z)' = 1/z$) in the entire (open) plane with the exception of the point $z = 0$ and all the points on the positive real axis; on this axis it is discontinuous, because its imaginary part has a jump of 2π. In order to eliminate this discontinuity, we define a so-called *second branch of the logarithmic function* by the relation

$$\ln_1 z = \ln_0 z + 2\pi i = \ln r + i(\varphi + 2\pi) .$$

Let Ω be a neighbourhood of an arbitrary point z_0 on the positive real axis which does not contain the point $z = 0$. Let us define, in this neighbourhood,

$$f(z) = \ln_0 z \quad \text{if} \quad \operatorname{Im} z < 0 \quad \text{(i.e. for points of the lower half-plane)} ,$$

$$f(z) = \ln_1 z \quad \text{if} \quad \operatorname{Im} z \geqq 0 \quad \text{(i.e. for points of the upper half-plane)} . \tag{3}$$

Then $f(z)$ is continuous and holomorphic in Ω. We say that $\ln_1 z$ is the *analytic continuation of the function* $\ln_0 z$ *from the lower half-plane into the upper half-plane* through the real axis.

Similarly, we define other branches of the logarithmic function,

$$\ln_n z = \ln_0 z + 2n\pi i = \ln r + i(\varphi + 2n\pi) \quad (n \text{ an integer}).$$

Definition 1. The set of all these branches is called the *multi-valued function* $\ln z$ and the function $\ln_0 z$ the *principal branch* of the function $\ln z$.

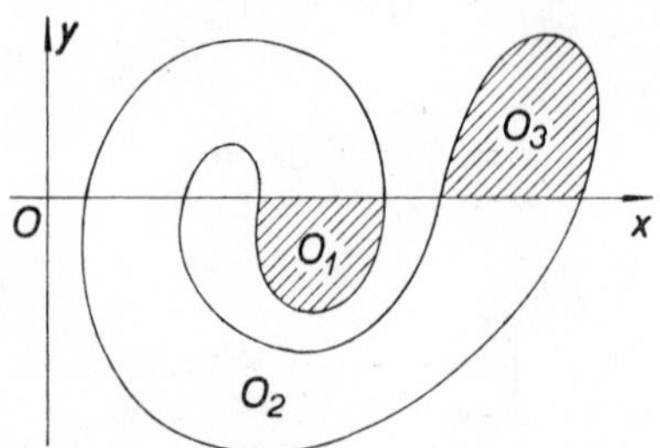

Fig. 20.11.

REMARK 1. If O is an arbitrary *simply connected* region which does not contain the point $z = 0$, then we can assign to each point $z \in O$ a value of the function $\ln z$ (i.e. the value of a certain branch of this function) such that the function $\ln z$ is holomorphic (hence single-valued) in O.

Example 1. In the region O illustrated in Fig. 20.11, we can choose the branches of the function $\ln z$ in the following way:

$$\begin{aligned} \ln z &= \ln_0 z \quad \text{in} \quad O_1 , \\ \ln z &= \ln_1 z \quad \text{in} \quad O_2 , \\ \ln z &= \ln_2 z \quad \text{in} \quad O_3 . \end{aligned}$$

REMARK 2. Each of the branches of the function $\ln z$ is single-valued in the Gaussian plane. Let us consider an infinite number of Gaussian planes $\ldots, R_{-2}, R_{-1}, R_0, R_1, R_2, \ldots$ and let us assign to each branch of $\ln_n z$ the plane R_n. Let us imagine that these planes are made of stiff paper and that they are cut along the positive real axis. We have obtained the function $\ln z$ by "joining" the function $\ln_1 z$ to the function $\ln_0 z$, etc. Similarly, let us "attach" the plane R_1 to the plane R_0 so that we join the right upper half-plane R_1 to the right lower half-plane R_0 along the positive real axis. In the same way, let us "attach" the plane R_2 to R_1, etc., and proceed similarly with the planes $R_{-1}, R_{-2}, \ldots$. Thus we get a surface consisting of an infinity of sheets which is called the *Riemann surface* of the function $\ln z$. To each sheet of this surface, there corresponds a certain branch of the function $\ln z$ (and conversely). If z moves round the point $z = 0$ in the positive sense and does not leave the Riemann surface (i.e. if z, after moving once round the point $z = 0$, passes from R_n to R_{n+1}), then the values of the function $\ln z$ change continuously with z. Since during this movement the function $\ln z$ never returns to the initial value (the imaginary part of $\ln z$ continually increases), the singular point $z = 0$ of $\ln z$ is called a *branch point of infinite order* or a *transcendental branch point* of this function. A singularity of this kind is often called a *logarithmic singularity*.

REMARK 3. The notation $\ln z$, $\ln_n z$ is not uniformly used in the literature. The multi-valued function here denoted by $\ln z$ is often denoted by $\log z$ or $\operatorname{Log} z$. For the argument φ of the function $\ln_0 z$, the interval $(-\pi, \pi)$ is often chosen.* We therefore recommend the reader to study the notation individually in each publication.

REMARK 4. The so-called general power of the function z is a (generally multi-valued) function defined by the relation

$$z^n = e^{n \ln z} = e^{n[\ln r + i(\varphi + 2k\pi)]} \quad (k \text{ an integer}). \tag{4}$$

For real *irrational* n, this function is infinitely multi-valued, the point $z = 0$ being a transcendental branch point, and its Riemann surface has an infinity of sheets. If n is rational, then, since $e^{i.2\pi l} = 1$ if l is an integer ($e^{i.2\pi l} = \cos 2\pi l + i \sin 2\pi l = 1$), we come back after a finite number of rotations of the point z round the origin (i.e. for a certain k in equation (4)) to the same value. The function z^n then has only a finite number of different branches, the point $z = 0$ is a so-called *algebraic branch point* (*branch point of finite order*). The corresponding Riemann surface has then only a finite number of sheets.

For example, the function $\sqrt{z} = z^{1/2}$ has two branches (in (4) we choose $k = 0$ and $k = 1$), each of which is a (-1)-multiple of the other (since $e^{(1/2).2\pi i} = -1$). To each of these, there corresponds one sheet of a two-sheet Riemann surface. Let us denote them by R_0, R_1. We obtain the Riemann surface (cf. Remark 2) by "attaching" the right upper half-plane of the plane R_1 to the right lower half-plane of the plane R_0 and then (after a further rotation of the point z round the origin) the right lower half-plane of the plane R_1 to the right upper half-plane of the plane R_0. (All these operations we naturally perform only mentally.)

Generally, for a natural number m, the function $\sqrt[m]{z} = z^{1/m}$ has m (different) branches, which can be defined (for example) by the following relations:

$$z_1 = e^{(1/m)\ln_0 z}, \quad z_2 = e^{(1/m)\ln_1 z}, \quad \ldots, \quad z_m = e^{(1/m)\ln_{m-1} z}.$$

REMARK 5. Everything which has been said concerning the functions $\ln z$, z^n is also valid for the functions $\ln (z - z_0)$, $(z - z_0)^n$; it is sufficient to substitute $z - z_0 = z'$. Of course, the branch point will now be the point z_0.

Definition 2. Let a region Ω be the intersection of the regions O_1 and O_2 (Fig. 20.12). Let a function $f_1(z)$ be defined in O_1. If there exists a holomorphic function $f_2(z)$

* This is generally the case in English mathematical literature. The cuts in the different Gaussian planes are then made along the *negative* real axis.

in O_2 such that $f_1(z) = f_2(z)$ in Ω, we say that $f_2(z)$ is an *analytic continuation of the function* $f_1(z)$ *from* O_1 *to* O_2 *through* Ω.

REMARK 6. If there exists such a function, then it is the only one, as the following theorem implies:

Theorem 1. *Let M be an infinite set of points lying in a region B and having in B at least one point of accumulation* (Definition 22.1.3). (*For example, a segment*

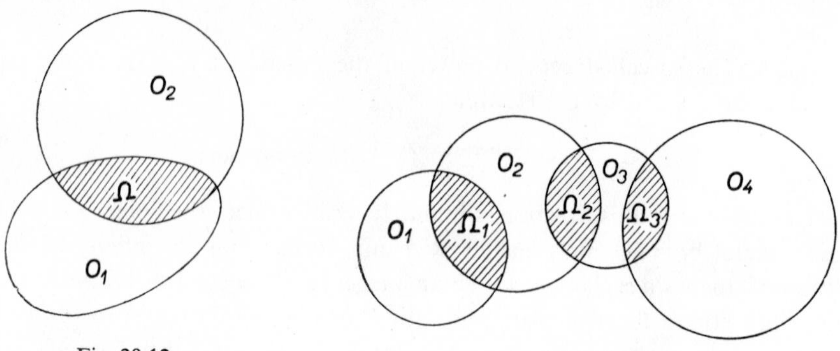

Fig. 20.12. Fig. 20.13.

or a region lying in B satisfies these conditions.) *Let a function g*(z) *be defined on M. If there exists such a holomorphic function G*(z) *on B that* $G(z) = g(z)$ *at the points of the set M, then this function is unique.*

REMARK 7. In definition 2 it is obviously sufficient to take Ω for M. Theorem 1 implies, for example, that the function $\sin z$ defined in Remark 20.4.4 is the only holomorphic function in the Gaussian plane which for real z coincides with the function $\sin x$ defined in the calculus of functions of a real variable.

REMARK 8. Let us have a so-called *chain of regions* $O_1, O_2, \ldots, O_n$, i.e. a system of regions O_k such that each region O_k ($k = 2, 3, \ldots, n - 1$) has non-empty intersections Ω_{k-1}, Ω_k precisely with the regions O_{k-1} and O_{k+1} and that these intersections Ω_{k-1}, Ω_k are simply connected regions, all mutually disjoint (Fig. 20.13).

Let $f_k(z)$ be holomorphic functions defined in O_k, such that $f_k(z) = f_{k+1}(z)$ in Ω_k. Then the function $f_n(z)$ is called an *analytic continuation of the function* f_1 *to the region* O_n *through the given chain of regions.*

REMARK 9. If we take two chains of regions, both leading to the region O_n, it can happen that we reach O_n in each case with a different analytic continuation of the function $f_1(z)$. Or, in other words, if the chain under consideration is closed ($O_n = O_1$), then, after passing through the chain, we come to the region O_n with a different holomorphic function from the one we started with. If, for instance, we start from the circle K_1 (Fig. 20.14) with the value of the function $\ln_0 z$, we return to K_1 after passing through the chain with values of the function $\ln_1 z$.

REMARK 10. We also often denote the function $f_2(z)$ of Definition 2, which is an analytic continuation of the function $f_1(z)$, by $f_1(z)$ and we say that we *have continued the function* $f_1(z)$ (analytically) *from the region* O_1 *to the region* $O_1 + O_2$. We similarly speak of a continuation (extension) in the case of a chain or regions. A function (generally multi-valued) which is an analytic continuation of a holomorphic

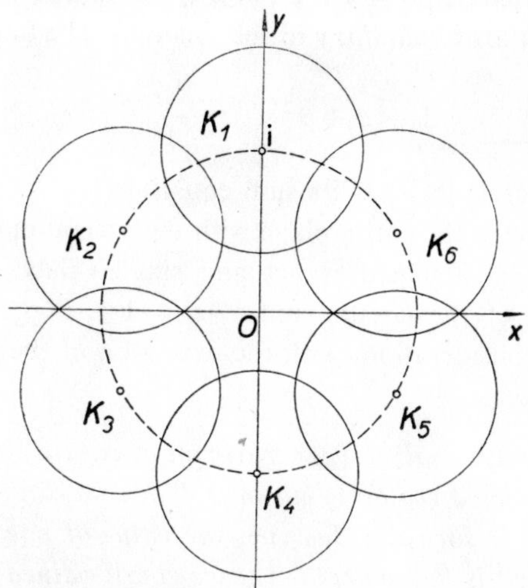

Fig. 20.14.

function to a region D is called an *analytic function in D*. If we carry out all possible analytic continuations of a function $f_1(z)$ from a region O_1 (through all possible chains of regions), we get a so-called *complete analytic function* (generally multi-valued) which cannot be further continued. The domain of existence (domain of definition) of this function is called the *natural domain of the analytic function* (given by the holomorphic function $f_1(z)$ in O_1).

For example, the natural domain of the analytic function $\ln z$ is the whole (open) plane except the point $z = 0$. In the same way (as in the case of the function $\ln z$) we speak also in the general case of *branches* of a (multi-valued) analytic function.

REMARK 11. The idea of a simple method of constructing the analytic continuation of a given function was suggested by Weierstrass: Let $f(z)$ be a holomorphic function in a region O. Let us choose $z_1 \in O$ and expand $f(z)$ in a power series with centre at z_1. This series converges at least in the circle K_1 which lies in O and has a radius equal to the distance of the point z_1 from the boundary of the region O. It can happen that the circle of convergence is greater. Then $f(z)$ is continued to the region $O + K_1$. Further, we choose $z_2 \in O$ and proceed similarly at all the points of O. We then expand the function $f(z)$ thus continued into a power series at the points of the new region, and so on. In this way, we obtain a complete analytic function.

REMARK 12. Let us consider the function $f(z)$ defined by the series

$$f(z) = z + z^2 + z^6 + \ldots + z^{n!} + \ldots. \tag{5}$$

The series (5) converges in the circle $|z| < 1$. It can be shown that the function (5) cannot be continued to any region greater than this circle. We say that no point of the circumference of the circle $|z| = 1$ is a *point of continuability* of the function (5) or that this circle is a *natural boundary* of the function. The function $f(z)$ defined by the series

$$1 - z + z^2 - z^3 + \ldots \tag{6}$$

also converges in the circle $|z| < 1$. Its sum equals $1/(1 + z)$. The function (6) can be analytically continued to the entire plane with the exception of the point $z = -1$. The analytic function so obtained is actually single-valued and is the function $1/(1 + z)$. Every point of the circumference $|z| = 1$ is a point of continuability of the function under consideration, with the exception of the point $z = -1$ which is its only *singular point*.

Theorem 2. *Let $f(z)$ be an analytic (generally multi-valued) function in a region O. Let D be a simply connected region lying in O. Then we can assign to the function $f(z)$ at every point $z \in D$ such a value (i.e. the value of one of its branches) that the function thus defined is holomorphic (hence single-valued) in D.*

REMARK 13. Compare Remark 1 and Example 1. Theorem 2 is a special case of a more general Monodromy Theorem, compare, e.g. [345], p. 256.

21. CONFORMAL MAPPING

By JAROSLAV FUKA

References: [4], [52], [123], [197], [214], [232], [234], [238], [251], [273], [301], [307].

The book [234] which is a dictionary of common conformal mappings is particularly recommended to the reader.

21.1. The Concept of Conformal Mapping

The theory of conformal mapping is of great importance in many branches of technology and physical sciences (for example, in the theories of elasticity, of flow around aerofoils, of two-dimensional stationary vector fields, etc.).

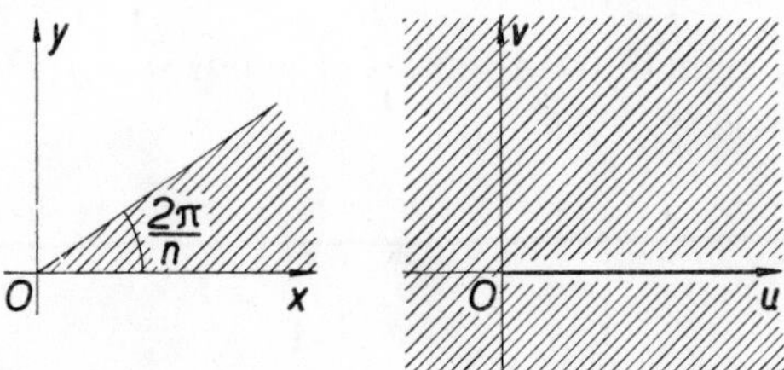

Fig. 21.1. The mapping by the function $w = z^n$.

Definition 1. Let a region G be mapped on a region B with one-to-one correspondence by a function $w = f(z) = u(x, y) + iv(x, y)$. If, moreover, the function $f(z)$ is holomorphic on G (Definition 20.1.9) then the mapping given by the function $f(z)$ is called a *conformal mapping* (or *transformation*) *of the region G on* or *on to the region B*. We also say $f(z)$ *maps the region G conformally on the region B*.

REMARK 1. The function $z = f^{-1}(w)$, inverse to the function $w = f(z)$, maps B conformally on G.

Example 1. The function $w = z^n$ (n a positive integer) maps the sector $0 < \arg z < 2\pi/n$ of the z-plane conformally on the w-plane, from which the positive part of the real axis (including the origin) is excluded (Fig. 21.1).

Example 2. The function $w = e^z$ maps the strip $h_1 < \operatorname{Im} z < h_2, 0 < h_2 - h_1 \leqq 2\pi$, of the z-plane on a sector in the w-plane with vertex at the origin and arms making angles h_1, h_2 with the positive real axis (Fig. 21.2).

REMARK 2. A conformal mapping of a region G on B has the following two basic properties:

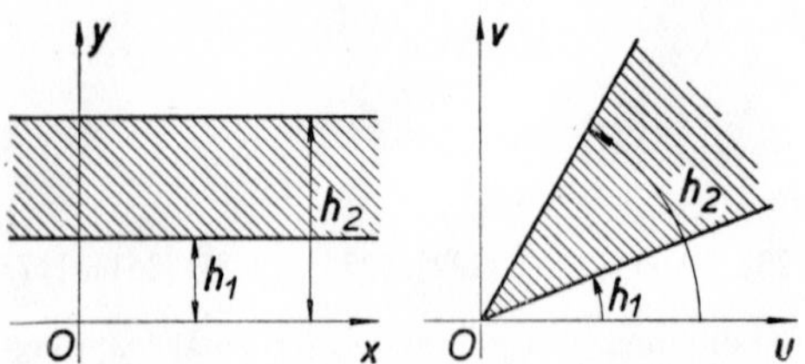

Fig. 21.2. The mapping by the function $w = e^z$.

1. It transforms every circle of infinitely small diameter into a circle of infinitely small diameter. The exact meaning of this assertion is as follows: Let a curve β_r be the image of a circle c_r, the equation of which is $|z - z_0| = r$ (cf. Definition 20.1.2) and which lies in G (Fig. 21.3a). Let us construct a circle k_r in B:

$$|\zeta - f(z_0)| = |f'(z_0)|\, r\,. \tag{1}$$

Let $\varrho(r)$ denote the maximum of the distances of points $w = f(z)$ lying on β_r from the circumference of the circle k_r. Then

$$\frac{\varrho(r)}{r} \to 0 \quad \text{as} \quad r \to 0\,. \tag{2}$$

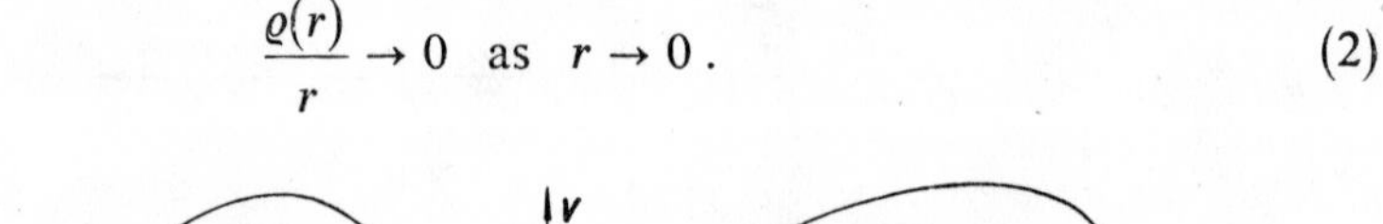

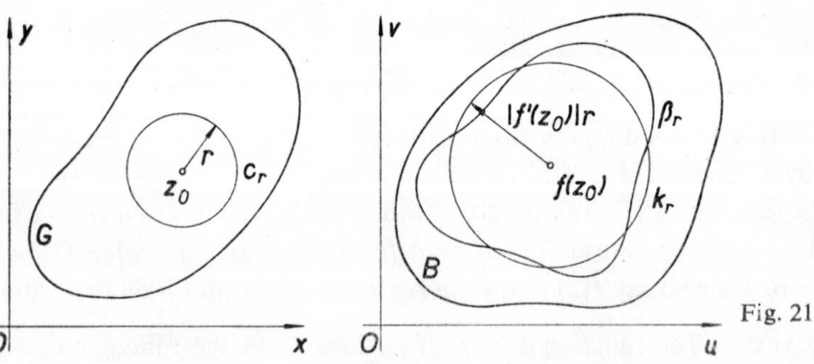

Fig. 21.3a.

2. It preserves angles. More precisely: Let γ_1, γ_2 be smooth curves lying in G and let α be the oriented angle between γ_1 and γ_2 at their point of intersection z_0. Then the images β_1, β_2 of the curves γ_1, γ_2 are also smooth curves and the oriented angle between them (at the point $f(z_0)$) is also α (Fig. 21.3b).

REMARK 3. From (1) and (2), we deduce the geometrical significance of the absolute value of the derivative $f'(z)$: $|f'(z_0)|$ determines the "change of scale" at the point z_0. The number $\arg f'(z_0)$ has the following geometrical significance: Let γ denote an arbitrary smooth curve in the z-plane, passing through the point z_0, and β its image

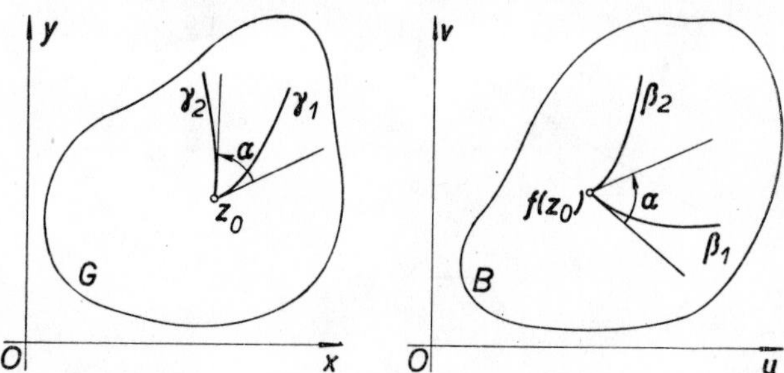

Fig. 21.3b

in the w-plane where $w = f(z)$. If we superimpose the z-and w-planes so that the coordinate axes remain parallel and the points z_0 and $w_0 = f(z_0)$ coincide, then $\arg f'(z_0)$ is the angle between the curves γ and β. We say briefly that $\arg f'(z)$ is the angle of rotation at the point z_0.

REMARK 4. As an important example of conformal mapping, let us mention the mapping of the form $w = (az + b)/(cz + d)$, $ad - bc \neq 0$, which is called *homographic* and has the following properties.

The mapping composed of two homographic mappings is a homographic mapping. The mapping inverse to a homographic mapping is also homographic. The image of a straight line or of the circumference of a circle is again a straight line or the circumference of a circle and the image of a circle or of a half-plane is again a circle or a half-plane. Conversely, every conformal mapping of a circle or a half-plane on a circle or a half-plane is homographic. A homographic mapping maps the closed plane, or the closed plane from which one point is excluded, conformally on the closed plane or on the closed plane excluding one point. Conversely, if $f(z)$ is a conformal mapping of the closed plane, or the closed plane excluding one point, on a region B, then $f(z)$ is a homographic mapping and B is the closed plane or the closed plane excluding one point.

Further: Three mutually different points z_1, z_2, z_3 either determine the circumference k of a circle or lie on a straight line p. Let the circle k (or the straight line p) be oriented curves. Let G denote the interior of k (in this case let the circumference k be oriented in such a way that G lies on the left-hand side of k if we move along k in the positive sense of its orientation). Alternatively, let G denote that one of the two half-planes determined by the straight line p which lies on the left-hand side if we move along p in the positive sense. Let the points z_1, z_2, z_3 follow each other in the order mentioned according to the positive orientation of k or p. Similarly, let the points w_1, w_2, w_3 of the w-plane lie either on a circumference k or on a straight line p', and let us have the same convention concerning the orientation of k' or p' with respect to the corresponding circle B or the half-plane B, respectively, and concerning the ordering of the points w_1, w_2, w_3 on k' or p', respectively (Fig. 21.4). Every conformal mapping of the region G on B is a homographic one. If this mapping

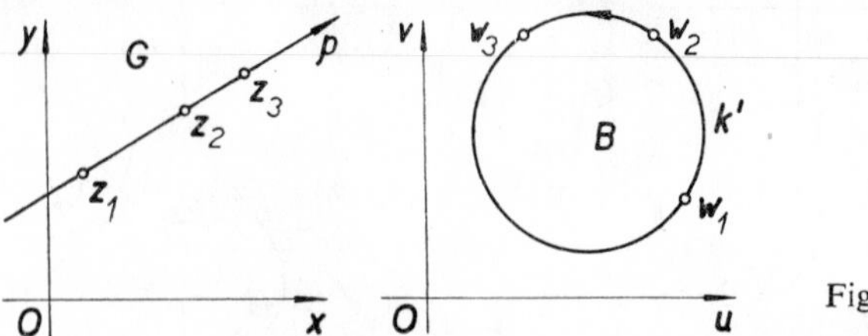

Fig. 21.4.

(let us denote it by $w = f(z)$) maps the points z_1, z_2, z_3 on the points w_1, w_2, w_3 (in this order), then it is uniquely determined (see Remark 21.2.1) by the relation

$$\left(\frac{w - w_1}{w - w_2}\right) \Big/ \left(\frac{w_3 - w_1}{w_3 - w_2}\right) = \left(\frac{z - z_1}{z - z_2}\right) \Big/ \left(\frac{z_3 - z_1}{z_3 - z_2}\right). \tag{3}$$

Example 3. Let us find a conformal mapping of the upper half-plane G of the z-plane on the unit circle B with centre at the origin of the w-plane such that the points $z_1 = -1$, $z_2 = 0$, $z_3 = 1$ of the x-axis are mapped on the points $w_1 = 1$, $w_2 = \mathrm{i}$, $w_3 = -1$ lying on the boundary of the circle B; Fig. 21.5.

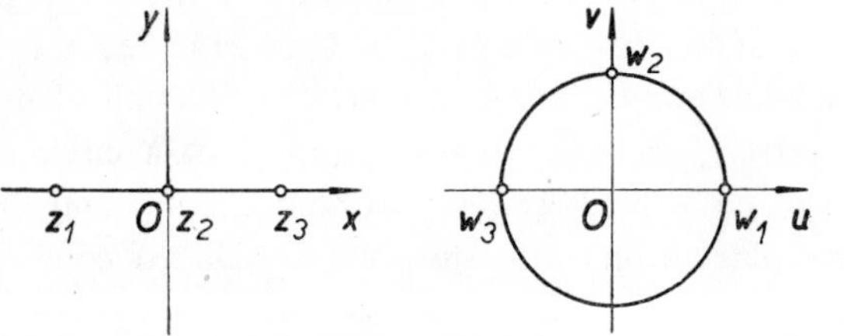

Fig. 21.5.

According to (3), we have

$$\left(\frac{w - 1}{w - \mathrm{i}}\right) \Big/ \left(\frac{-1 - 1}{-1 - \mathrm{i}}\right) = \left(\frac{z - (-1)}{z - 0}\right) \Big/ \left(\frac{1 - (-1)}{1 - 0}\right)$$

and consequently

$$w = \frac{z - \mathrm{i}}{\mathrm{i}(z + \mathrm{i})}. \tag{4}$$

REMARK 5. The formula (3) can be used even if one of the points under consideration is the point at infinity. If e.g. $z_3 = \infty$, then in (3) we put $(z_3 - z_1)/(z_3 - z_2) = 1$, so that the right-hand side of equation (3) is $(z - z_1)/(z - z_2)$. If, e.g., $w_1 = \infty$, then we write the left-hand side of equation (3) in the form

$$\left(\frac{w - w_1}{w_3 - w_1}\right) \Big/ \left(\frac{w - w_2}{w_3 - w_2}\right)$$

and put $(w - w_1)/(w_3 - w_1) = 1$, so that the left-hand side of equation (3) is $(w_3 - w_2)/(w - w_2)$, etc.

In our example it follows from (4) that to the point $z = \infty$ there corresponds the point $w = -\mathrm{i}$. If we now choose the mutually corresponding points

$$z_1 = -1\,, \; z_2 = 1\,, \; z_3 = \infty\,, \quad w_1 = 1\,, \; w_2 = -1\,, \; w_3 = -\mathrm{i}$$

we again obtain, by the method just described, the mapping (4).

21.2. Existence and Uniqueness of Conformal Mapping

If two regions G and B in the plane are given, a fundamental question arises as to whether there exists a conformal mapping of the region G on B. In the case of simply connected regions, the answer is given by the following theorem:

Theorem 1 (*Riemann*'s *Theorem*). *Let G be a simply connected region with a boundary containing at least two points* (see Remark 2) *and let $z_0 \in G$. Then there exists a conformal mapping $w = f(z)$ which maps the region G on the unit circle $|w| < 1$ such that*

$$f(z_0) = 0\,, \quad f'(z_0) > 0\,. \tag{1}$$

The function $f(z)$ is uniquely determined by the conditions (1).

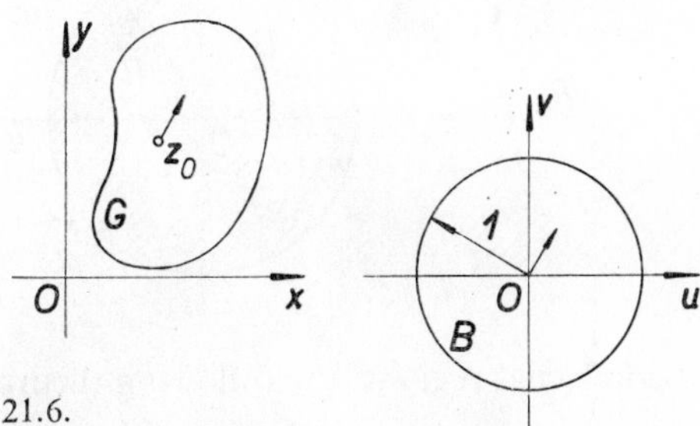

Fig. 21.6.

REMARK 1. Conditions (1) have the following geometrical meaning (Fig. 21.6): two mutually corresponding points $z = z_0$ and $w = 0$ are given and the angle of rotation at z_0 is zero (see Remark 21.1.3). The conditions determining uniquely the function $f(z)$ can also be chosen in other ways; for example, if G and B are Jordan regions (see Remark 14.1.3), then $f(z)$ is uniquely determined if three pairs of mutually corresponding points on the boundaries of the regions G and B are given (Fig. 21.7: see Example 21.1.3). (Cf. also Theorem 21.4.1, p. 985.)

REMARK 2. Theorem 1 gives no information about region, the boundary of which consists of only one point or no point at all, e.g. the closed plane or the closed plane excluding one point (this point can also be the point $z = \infty$). These cases have been

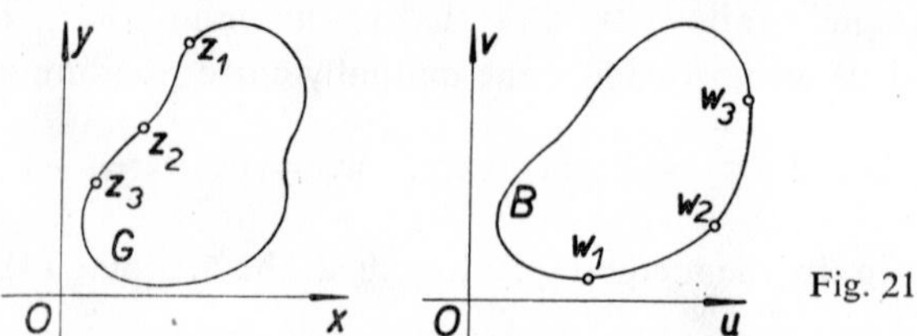

Fig. 21.7.

discussed in Remark 21.1.4. Neither the closed plane, nor the closed plane excluding one point, can be conformally mapped on a circle. The region G of Fig. 21.8 (the closed plane with a circular hole with centre at the origin and radius R) can be mapped, by the so-called "inversion" $w = 1/z$, on the circle with centre at the origin and radius $a = 1/R$. (If the centre is at a point z_0, then the corresponding mapping will be $w = 1/(z - z_0)$: if the hole is not circular and, at the same time, z_0 is an interior point of it, then we can first map the region G by the function $w = 1/(z - z_0)$ on a bounded simply connected region, and then map this region on a circle.) If we exclude one point (e.g. the point $z = \infty$) from the above-mentioned region G (the closed plane with a circular hole and with one point excluded), we can map G in a similar way on a circle excluding one point.

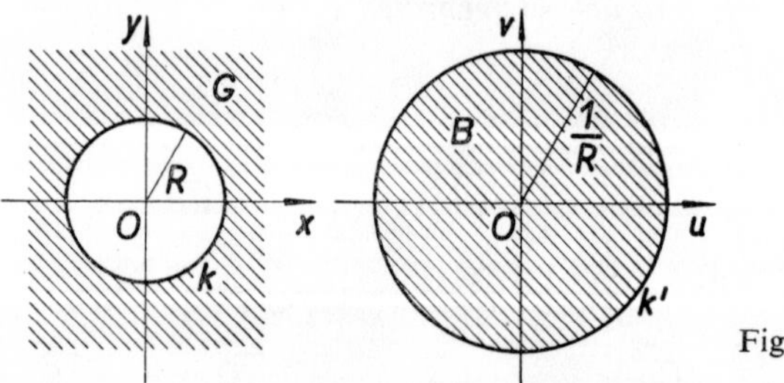

Fig. 21.8.

In the case of multiply connected regions, the following theorem holds:

Theorem 2. *Every n-tuply connected region G $(n \geqq 2)$ can be conformally mapped on one of the following regions B_i $(i = 1, 2, \ldots, 5)$,* where

B_1 *is an annulus with centre at the origin excluding $n - 2$ concentric circular arcs;*
B_2 *is a circle with centre at the origin excluding $n - 1$ concentric circular arcs;*
B_3 *is the plane excluding n concentric circular arcs;*

B_4 *is the plane excluding n segments lying on rays starting from the origin;*
B_5 *is the plane excluding n parallel segments with the same angle* ϑ *between their direction and the x-axis* (Fig. 21.9a, b, c, d, e).

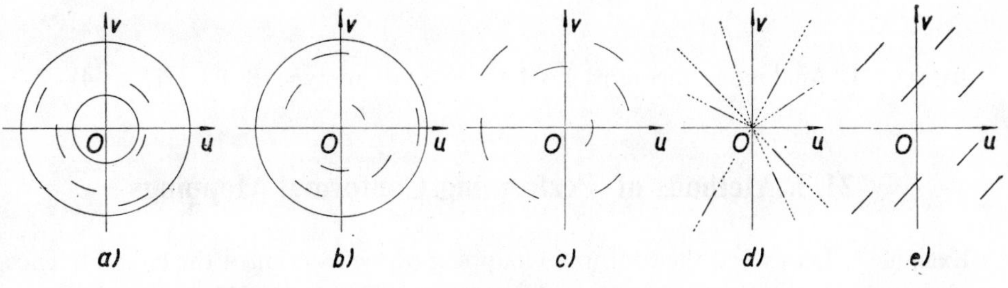

Fig. 21.9.

REMARK 3. The difference between Theorem 2 and Theorem 1 consists in the fact that the shape of the region B_i (e.g. the ratio of the radii of the annulus B_1) cannot be arbitrarily chosen a priori, as distinct from the case of simply connected regions. For example, a conformal mapping of an annulus $r_1 < |z| < r_2$ on an annulus $\varrho_1 < |w| < \varrho_2$ exists if and only if $r_1/r_2 = \varrho_1/\varrho_2$.

The closed plane with two circular holes (Fig. 21.10) can be conformally mapped by the inversion $w = 1/z$ (or $w = 1/(z - z_0)$, see Remark 2) on an eccentric annulus (which can then eventually be mapped on a concentric annulus as in Example 21.3.2). In general, the closed plane with two holes can be mapped conformally on an annulus. The open plane (or the closed plane excluding one point) with two holes cannot be conformally mapped on an annulus. Obviously, an annulus cannot be a conformal image of the closed plane with more than two holes.

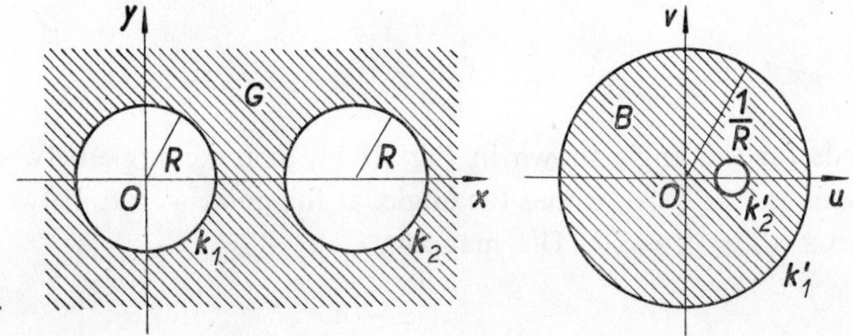

Fig. 21.10.

Theorem 3. *There exists only one function* $w = f(z)$ *which maps G on a region of type* B_5 (cf. Theorem 2) *so that to a given point* $z_0 \in G$ *there corresponds the point* $w = \infty$ *and that the expansion of* $f(z)$ *in the neighbourhood of the point* z_0 *has the form*

$$f(z) = \frac{1}{z - z_0} + \alpha_1(z - z_0) + \alpha_2(z - z_0)^2 + \dots$$

if z_0 is a finite point and

$$f(z) = z + \frac{\beta_1}{z} + \frac{\beta_2}{z^2} + \dots$$

if $z_0 = \infty$.

REMARK 4. Analogous theorems hold for regions of type B_i $(i = 1, \dots, 4)$.

21.3. Methods of Performing Conformal Mappings

Example 1. Let us find the conformal mapping of the exterior of the circumference c of a circle on the plane excluding a circular arc γ (Fig. 21.11). The circumference c has its centre at the point $z = ih$, $h > 0$, and passes through the point a, $a > 0$; the arc γ is given by the points $-a$, ih, a $(-a, a$ being its end-points).

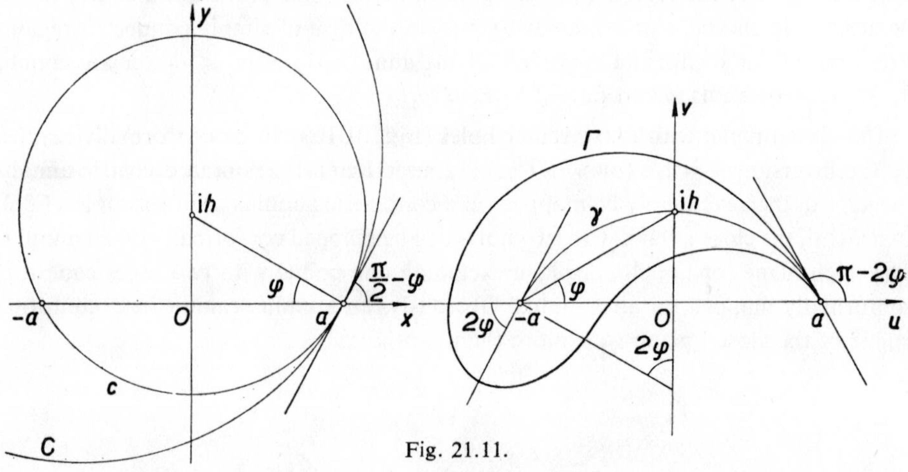

Fig. 21.11.

If φ stands for the angle shown in Fig. 21.11, then the angle between c and the positive x-axis is $\alpha = \frac{1}{2}\pi - \varphi$ and the angle, at the point $w = a$, between γ and the positive u-axis is $\beta = \pi - 2\varphi$. The mapping

$$z_1 = \frac{z - a}{z + a}$$

maps the circumference of the circle c on a straight line passing through the origin of the z_1-plane. (According to Remark 21.1.4, the image of the circumference of a circle is either a straight line or a circle, if a homographic mapping is used. However, for $z = a$ we have $z_1 = 0$, for $z = -a$ we have $z_1 = \infty$, so that a circle is out of the question.) This straight line makes an angle α with the x_1-axis, since the x-axis is mapped on the x_1-axis and conformal mappings preserve angles. Hence, the function

$$z_2 = z_1^2 = \left(\frac{z - a}{z + a}\right)^2$$

maps (see Example 21.1.1) the exterior of the circumference of the circle c on the z_2-plane excluding the ray from the origin making an angle $2\alpha = \beta$ with the positive x_2-axis. By the mapping

$$w_1 = \frac{w - a}{w + a}$$

(see Remark 21.1.4), the complement of the arc γ is mapped on the w_1-plane excluding the ray from the origin making an angle β with the positive u_1-axis, since the u-axis is transformed on to the u_1-axis and conformal mappings preserve angles. Hence, putting $z_2 = w_1$, we get

$$w = \frac{1}{2}\left(z + \frac{a^2}{z}\right). \tag{1}$$

REMARK 1. Let us draw the circumference of a circle (denoted by C) touching c at the point a and lying in the exterior of c (Fig. 21.11). The mapping (1) maps C on a curve Γ containing the arc γ in its interior and having a cusp at a. The function (1) maps the exterior of C on the exterior of Γ. This provides a basis for the study of aerofoils (the so-called *Joukowski aerofoils*).

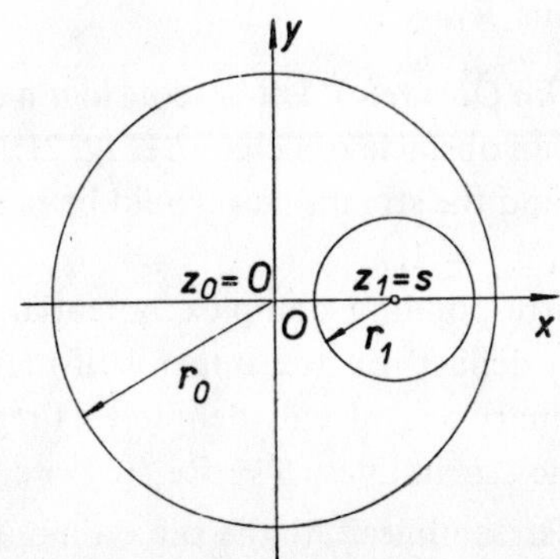

Fig. 21.12.

In Example 1, we have used a simple *method of combining elementary conformal mappings* which is often very effective since simple mappings (such as the homographic mapping) have simple properties which can be well illustrated geometrically.

The following example shows a further simple application of homographic mappings.

Example 2. In electrostatic field theory (when determining the field of an eccentric cylindrical condenser), the conformal mapping of an eccentric annulus on a concentric one is frequently used. If we use the notation of Fig. 21.12 ($r_0 > s + r_1$, $r_1 > 0$, $s \geqq 0$) and write

$$t = \sqrt{\frac{(r_0 + s)^2 - r_1^2}{(r_0 - s)^2 - r_1^2}},$$

$$R_0 = \frac{r_0(t + 1) - (s + r_1)(t - 1)}{(s + r_1)(t + 1) - r_0(t - 1)},$$

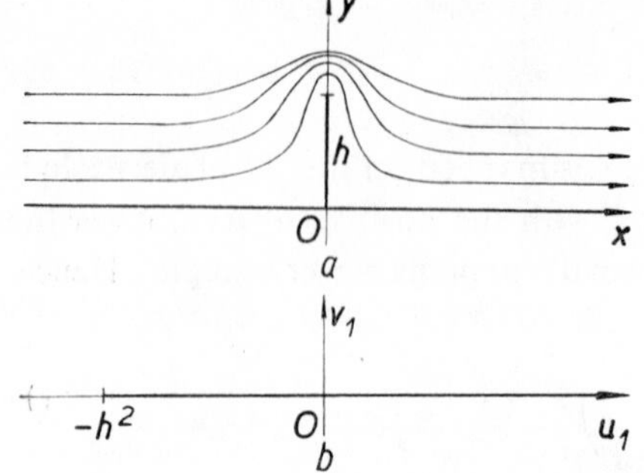

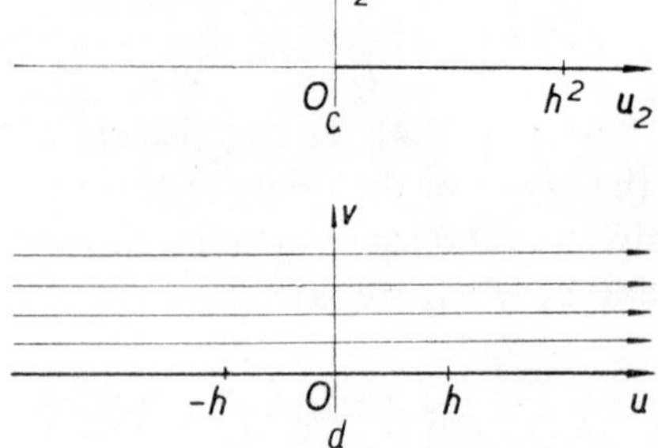

Fig. 21.13.

then the mapping

$$w = R_0 \frac{(t + 1) z - r_0(t - 1)}{-(t - 1) z + r_0(t + 1)}$$

maps the given eccentric annulus on a concentric annulus with centre at $w = 0$ and inner radius R_0, outer radius $R_1 = 1$.

Example 3 (*Flow Round an Obstacle*). Let us consider a steady irrotational flow in the upper u-half-plane with an obstacle of height h (Fig. 21.13a). Without this obstacle the flow would be uniform and the streamlines would be parallel to the x-axis. Let the velocity at infinity be unity, $v_\infty = 1$.

We have to find the corresponding complex potential of the flow, i.e. a holomorphic function $w = f(z)$ defined on the upper half-plane with the necessary slit of length h along the imaginary axis (let us denote this region by O) and such that $v = \operatorname{Im} f(z) = \text{const.}$ are the streamlines, $u = \operatorname{Re} f(z) = \text{const.}$ are the equipotential lines, perpendicular to the streamlines. At the same time, the complex number $\overline{f'(z)}$ at each point z will determine the velocity vector of the flow, not only in direction, but also in absolute value. Obviously, in the limit the boundary of the region O (i.e. the x-axis + the slit) is to be a streamline. Further, the condition $v_\infty = 1$ implies $f'(z) \to 1$ for $z \to \infty$.

We are going to map the region O on the upper half-plane Ω. We shall see that this mapping $w = f(z)$ is the required function. (For more detailed treatment of these problems see, e.g., [142]). We shall apply a combination of elementary mappings.

First, the function $w_1 = z^2$ maps the region O on a region O_1 of the w_1-plane with a slit $[-h^2, +\infty)$ on the real axis, Fig. 21.13b (since the mapping $w_1 = z^2$ doubles angles with vertex at the origin, Example 21.1.1). Next, the function $w_2 = w_1 + h^2$ maps the region O_1 on a region O_2, with a slit $[0, +\infty)$ on the real axis (Fig. 21.13c; this is a translation in the direction of the x-axis). Finally the mapping $w = \sqrt{w_2}$ maps the region O_2 on the upper half-plane Ω of the w-plane (Fig. 21.13d). Hence, the required function is

$$w = \sqrt{(z^2 + h^2)}\,.$$

The flow in the region O corresponds, according to this equation, to the flow in the half-plane Ω. The streamlines $v = \operatorname{Im} \sqrt{(z^2 + h^2)} = \text{const.}$ in O correspond to the streamlines $v = \text{const.}$ in the half-plane Ω. Obviously

$$f'(z) = \frac{z}{\sqrt{(z^2 + h^2)}}\,,$$

so that $f'(z) \to 1$ as $z \to \infty$. For $z \to ih$ we have $f'(z) \to \infty$ (the so-called *defect* on the edge).

In the case where $v_\infty = a$ is specified, the required function is

$$w = a\sqrt{(z^2 + h^2)}\,.$$

Theorem 1 (*Boundary–Correspondence Principle*). *Let $w = f(z)$ be a function continuous on a closed region $\bar{G}$, bounded by a Jordan curve Γ* (Remark 14.1.3, p. 603) *and holomorphic on G.*

If the function $f(z)$ is uniquely invertible on Γ and maps Γ on a Jordan curve β, then $f(z)$ is uniquely invertible on $\bar{G}$ and maps G conformally on the interior of the curve β.

REMARK 2. Theorem 1 holds even if G is the exterior of Γ or if Γ is a generalized Jordan curve (this is a curve which is a stereographic projection (i.e. a projection from the "north pole") of a Jordan curve lying on the Riemann sphere (see Remark 20.1.2) on to the complex plane) and G is one of the regions with the boundary Γ (e.g. if Γ is a straight line, and G is a half-plane). (See e.g. [273], [251].)

Example 4. Let us study the mapping of the upper half-plane, given by the following function (for notation, see Remark 20.2.3):

$$w = f(z) = \int_0^z \frac{dt}{\sqrt{[(1 - t^2)(1 - k^2t^2)]}}\,, \quad 0 < k^2 < 1 \tag{2}$$

(the "elliptic integral of the first kind" in Legendre's normal form).

We take that branch (§ 20.6) of the double-valued function $\sqrt{[(1 - t^2)(1 - k^2t^2)]}$ which, for $t \in (0, 1)$, assumes positive values. (To be precise we denote the positive root of a $(a > 0)$ by the symbol $\underset{+}{\sqrt{}}a$.) Then the function (2) is holomorphic on the upper half-plane and continuous on the closed upper half-plane.

Let us find the image of the whole real axis under the mapping (2) (Fig. 21.14). If $z = x$, $0 < x < 1$, then the value of $f(x)$ lies in the interval $(0, \omega_1)$ where

$$\omega_1 = \int_0^1 \frac{\mathrm{d}t}{\underset{+}{\sqrt{}}[(1 - t^2)(1 - k^2t^2)]}\,.$$

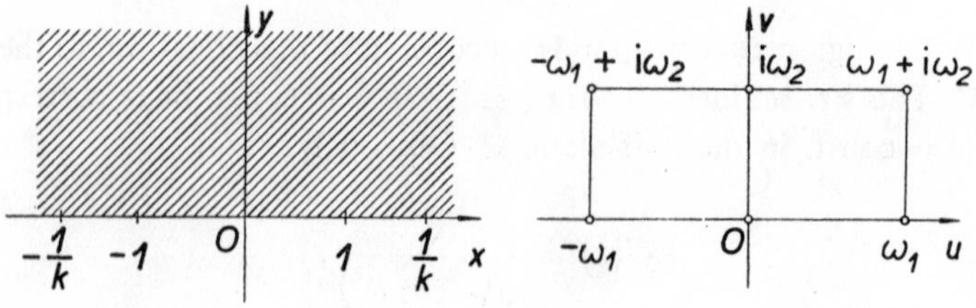

Fig. 21.14.

In the interval $(1, 1/k)$, the integrand is of the form $1/ \pm \mathrm{i} \underset{+}{\sqrt{}}[(t^2 - 1)(1 - k^2t^2)]$. We must choose the sign so that the above-mentioned branch is continuous in the upper half-plane. If we pass from the point $1 - \delta$ to the point $1 + \delta$ along a semi-circle with centre at the point 1 and lying in the upper half-plane (suppose $0 < < 1 - \delta < 1$ and $1 < 1 + \delta < 1/k$), the value of the expression $\varphi(t) = (1 - t^2) . . (1 - k^2t^2)$ changes from + to −, while arg $\varphi(t)$ changes from 0 to $-\pi$, since if t passes along the given semi-circle, the values of $\varphi(t)$ lie in the *lower* half-plane. Thus, arg $\sqrt{}[\varphi(t)]$ becomes equal to $-\pi/2$, and therefore the *minus* sign is to be chosen.

Hence, for $1 < x < 1/k$ we have

$$f(x) = \int_0^x \frac{\mathrm{d}t}{\sqrt{}[(1 - t^2)(1 - k^2t^2)]} = \int_0^1 \frac{\mathrm{d}t}{\underset{+}{\sqrt{}}[(1 - t^2)(1 - k^2t^2)]} +$$

$$+ \mathrm{i} \int_1^x \frac{\mathrm{d}t}{\underset{+}{\sqrt{}}[(t^2 - 1)(1 - k^2t^2)]} = \omega_1 + \mathrm{i} \int_1^x \frac{\mathrm{d}t}{\underset{+}{\sqrt{}}[(t^2 - 1)(1 - k^2t^2)]}\,.$$

The points $f(x)$ therefore lie on the segment parallel to the imaginary axis, with end-points ω_1, $\omega_1 + \mathrm{i}\omega_2$, where

$$\omega_2 = \int_1^{1/k} \frac{\mathrm{d}t}{\underset{+}{\sqrt{}}[(t^2 - 1)(1 - k^2t^2)]}\,.$$

Similarly, it can be seen that $f(x)$ describes the segments from $\omega_1 + \mathrm{i}\omega_2$ to $\mathrm{i}\omega_2$, from $\mathrm{i}\omega_2$ to $-\omega_1 + \mathrm{i}\omega_2$, from $-\omega_1 + \mathrm{i}\omega_2$ to $-\omega_1$ and from $-\omega_1$ to 0, when x describes the segments of the real axis from $1/k$ to $+\infty$, from $-\infty$ to $-1/k$, from $-1/k$ to -1 and from -1 to 0, respectively.

Hence, from Theorem 1 and Remark 2, we may conclude that the function (2) maps the upper half of the z-plane conformally on the rectangle with vertices $-\omega_1$, ω_1, $\omega_1 + i\omega_2$, $-\omega_1 + i\omega_2$ in the w-plane, without being obliged to verify the one-to-one correspondence of the mapping, which would be a rather complicated operation.

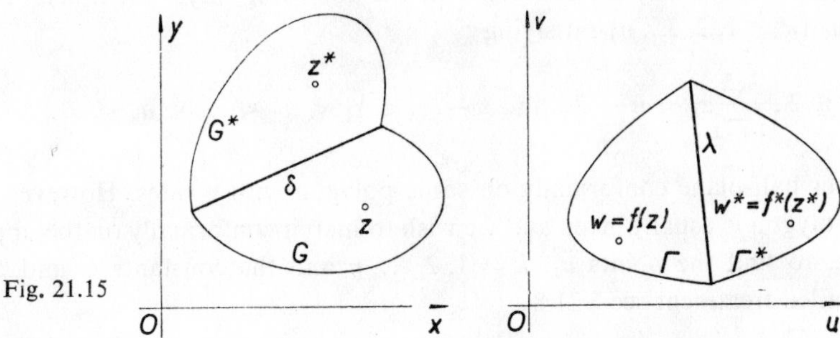

Fig. 21.15

Theorem 2 (*The Riemann–Schwarz Reflection Principle*). *Let G be a region bounded by a Jordan curve Γ and let the curve Γ contain a segment δ. Let $w = f(z)$ be continuous on $G + \delta$ and holomorphic on G and let the segment δ be mapped by the function $f(z)$ on a segment λ* (Fig. 21.15). *Let G^* be the region symmetric to G with respect to the straight line containing δ and let us define the function $f^*(z)$ on G^* in the following way: $f^*(z) = f(z)$ if $z \in \delta$* and *$f^*(z^*)$ is the point in the w-plane symmetric to the point $f(z)$ with respect to the straight line containing λ, if the point z^* is symmetric to the point z in the z-plane with respect to the straight line containing δ. Then the function $f^*(z)$ is the analytic continuation of the function $f(z)$ from $G + \delta$ into the region $G + G^* + \delta$.*

REMARK 3. The segments δ and λ very often lie on the real axis. The values $f(z)$ and $f^*(z^*)$ are then conjugate complex numbers.

The principle of reflection has numerous applications in conformal mapping and can be easily generalized (see, e.g. [273], p. 675; [251], page 148–185). For example, by its help the following (the so-called *Schwarz–Christoffel* theorem) may be proved:

Theorem 3. *If a function $w = f(z)$ maps the upper half-plane $\operatorname{Im} z > 0$ conformally on the interior of a bounded polygon G with angles $\alpha_k\pi$ $(0 < \alpha_k \leqq 2$, $k = 1, 2, \ldots, n$, $\sum_{k=1}^{n} \alpha_k = (n-2))$ so that the vertices of the polygon correspond to the points a_k of the real axis $(-\infty < a_1 < a_2 < \ldots < a_n < +\infty)$, then*

$$f(z) = C \int_{z_0}^{z} (z - a_1)^{\alpha_1 - 1} (z - a_2)^{\alpha_2 - 1} \ldots (z - a_n)^{\alpha_n - 1} \, dz + C_1 \,, \qquad (3)$$

where z_0, C, C_1 are certain constants.

REMARK 4. If e.g. $a_n = \infty$, i.e. if one of the vertices of the polygon corresponds to the point at infinity, then in formula (3), the factor containing a_n is excluded.

REMARK 5. Formula (3) holds even for a polygon with a vertex (or several vertices) lying at the point ∞, if we define the angle between two straight lines at the point ∞ as equal to the angle at their intersection, multiplied by -1.

REMARK 6. Theorem 3 may be converted in the following way: The function (3) with α_k and a_k $(k = 1, 2, \ldots, n)$ satisfying

$$-2 \leqq \alpha_k \leqq 2\,, \quad \sum_{k=1}^{n} \alpha_k = n - 2 \quad \text{and} \quad -\infty < a_1 < a_2 < \ldots < a_n < \infty\,,$$

maps the upper half-plane conformally on some polygon with n sides. However, in practice this polygon is usually given and we wish to map it conformally on the upper half-plane, i.e. to find the points a_k, $k = 1, 2, \ldots, n$ and the constants C and C_1. For more detailed treatment see § 21.8.

An example of the mapping (3) is the mapping (2), where we have $\alpha_1 = \alpha_2 = \alpha_3 = \alpha_4 = \frac{1}{2}$ and the points a_k are ± 1, $\pm 1/k$.

Theorem 4 (*The Green's Function*). *Let a simply connected region G with a boundary Γ and a point $z_0 \in G$ be given. If there exists a function $U(z)$ $(= U(x, y)$, where $z = x + iy)$, harmonic on G and continuous on $G + \Gamma$, which assumes the value $\ln(1/|\xi - z_0|)$ at each point $\xi \in \Gamma$, then the function*

$$f(z) = e^{i\alpha}(z - z_0)\, e^{U(z) + iV(z)}\,,$$

where $V(z)$ is a harmonic function conjugate to $U(z)$ on G (cf. Remark 20.1.12, p. 942) *and α is an arbitrary real number, maps the region G conformally on the unit circle so that $f(z_0) = 0$.*

Conversely, if $f(z)$ is a conformal mapping of the region G on the unit circle such that $f(z_0) = 0$, then $g(z, z_0) = (1/2\pi) . \ln(1/|f(z)|)$ is the Green's function (cf. Definition 18.4.8, p. 892) *of the region G with a pole at z_0.*

Hence, under certain assumptions concerning the smoothness of the boundary of the region G, for every function $u(z)$, harmonic on G and continuous on $G + \Gamma$, the following holds:

$$u(z) = -\int_\Gamma u(\zeta) \frac{\partial g(\zeta, z)}{\partial n}\, ds\,,$$

where Γ is the boundary of the region G and $\partial g/\partial n$ the derivative of the function g with respect to the outward normal.

REMARK 7. Theorem 4 reduces the search for a conformal mapping to the search for the solution of a Dirichlet problem (§ 18.4) together with the problem of finding a harmonically conjugate function, and conversely.

Theorem 5 (*Extremal Properties of Conformal Mappings*). *Let G be a simply connected domain containing the origin and let its boundary contain at least two points. Then from among all the functions $f(z)$, holomorphic on G and such that $f(0) = 0$, $f'(0) = 1$, only the function $f(z)$ mapping the region G conformally on a circle yields*

(a) *the minimum of the value* $M(f) = \sup_{z \in G} |f(z)|$,

(b) *the minimum of the value* $P(f) = \iint_G |f'(z)|^2 \, dx \, dy$ (i.e. the minimum of the area of the image of G),

(c) *the minimum of the value* $D(f) = \int_\Gamma |f'(z)| \, ds$ (i.e. the minimum of the length of the image of the curve Γ; we suppose that Γ is of finite length).

REMARK 8. Also in the case of multiply connected regions, it is possible to seek conformal mappings by means of variational problems.

21.4. Boundary Properties of Conformal Mappings

Theorem 1. *Let $f(z)$ be a conformal mapping of a Jordan region G on a Jordan region B* (Remark 14.1.3, p. 603). *Then the function $f(z)$ is continuously extensible on the boundary of the region G, i.e. a function $g(z)$ exists which is equal to the function $f(z)$ in G and is continuous on the closed region G (i.e. including the boundary). Moreover, g is one-to-one.*

REMARK 1. The theorem holds even for n-tuply connected regions bounded by Jordan curves.

REMARK 2. The study of the correspondence between boundaries of general simply connected regions led C. Carathéodory to the so-called *theory of prime ends* (See, e.g. [273], p. 402–406.)

In the theory of conformal mappings an important question arises, whether a "small" change of the mapped region implies a "small" change of the region into which it is mapped.

Theorem 2. *Let G_i $(i = 1, 2, \ldots)$ be a sequence of regions containing the point $z = 0$ and lying in a certain circle K. Let $G_i \subset G_{i+1}$ for every i (i.e. the G_i form an "increasing" sequence of regions, each of them being contained in all the following ones). Let us write $G = \lim_{n\to\infty} G_n$ (i.e. G is the union of all these regions). Let $\{f_n(z)\}$ be a sequence of functions mapping conformally the regions G_n on the unit circle $|w| < 1$ and normed by the conditions $f(0) = 0$, $f'(0) > 0$* (compare Theorem 21.2.1); *let $\{\varphi_n(w)\}$ be functions inverse to the functions $f_n(z)$ (i.e. mapping $|w| < 1$ on G_n). Let $f(z)$ map the region G conformally (with the conditions mentioned above) on $|w| < 1$ and let $\varphi(z)$ be its inverse function. Then the sequence $f_n(z)$ converges to*

$f(z)$ almost uniformly on G (Remark 20.4.1) *and the sequence $\varphi_n(w)$ converges to $\varphi(w)$ almost uniformly on $|w| < 1$.*

Hence, if the G_n converge to G, then the $f_n(z)$ converge to $f(z)$. Theorem 2 is a special case of a much more general theorem of C. Carathéodory (see [273], p. 380).

NUMERICAL METHODS IN CONFORMAL MAPPINGS

21.5. Variational Methods

REMARK 1. Let G be a bounded simply connected region containing the origin. Let $H_2(G)$ denote the space of all functions $f(z)$ holomorphic on G and such that

$$P(f) = \iint_G f(z)\overline{f(z)}\,dx\,dy < +\infty\,. \tag{1}$$

We define the scalar product of the functions $f(z)$, $g(z)$ in $H_2(G)$ in the following way:

$$(f, g) = \iint_G f(z)\,\overline{g(z)}\,dx\,dy\,.$$

By Theorem 21.3.5, the derivative of the function $\varphi(z)$ which maps G conformally on a circle and fulfils the conditions $\varphi(0) = 0$, $\varphi'(0) = 1$ is the solution of the variational problem

$$P(\varphi') = \iint_G |\varphi'(z)|^2\,dx\,dy = \iint_G \varphi'(z)\,\overline{\varphi'(z)}\,dx\,dy = \min\,. \tag{2}$$

For the solution of this variational problem, we shall use the so-called *Ritz method.*

Let us consider an arbitrary system of linearly independent functions belonging to $H_2(G) : u_0(z), u_1(z), \ldots, u_i(z), \ldots$, with $u_0(0) \neq 0$ and let us seek an approximate solution in the form

$$\varphi_n'(z) = \sum_{i=0}^{n} c_i u_i(z)\,, \tag{3}$$

with the condition

$$\varphi_n'(0) = 1\,, \tag{4}$$

which makes the integral $P(\varphi_n')$ minimal. For this it is necessary and sufficient that the relation

$$\iint_G \varphi_n'(z)\,\overline{\eta(z)}\,dx\,dy = 0 \tag{5}$$

be satisfied for every function of the form (3) satisfying the condition $\eta(0) = 0$. (In fact, if $\psi(z) = \varphi_n'(z) + \eta(z)$ is another function satisfying condition (4), i.e. $\eta(0) = 0$, we have

$$P(\psi) - P(\varphi_n') = \iint_G \varphi_n'\bar{\eta}\, dx\, dy + \iint_G \overline{\varphi_n'\eta}\, dx\, dy + \iint_G \eta\bar{\eta}\, dx\, dy\,,$$

whence it easily follows that $P(\psi) - P(\varphi_n') \geqq 0$ for all "admissible" functions ψ if and only if the condition (5) is satisfied.) If we choose for $\eta(z)$, in particular, $v_i(z) = u_i(z) - [u_i(0)/u_0(0)]\, u_0(z)$, $i = 1, 2, \ldots, n$ (cf. Example 1), we get from (5) a system of equations to determine the coefficients c_i:

$$\sum_{i=0}^{n} \alpha_{ij} c_i = 0\,, \tag{6}$$

where $\alpha_{ij} = \iint_G u_i \bar{v}_j\, dx\, dy$ $(j = 1, 2, \ldots, n)$.

In conjunction with condition (4),

$$\sum_{i=0}^{n} u_i(0)\, c_i = 1 \tag{7}$$

we obtain a system of $n + 1$ equations which has a unique solution.

As usual, it is advantageous to use an orthonormal system $\{u_n(z)\}$. In this case the radius R of the circle on which the region G is mapped is

$$R = \sqrt{\frac{S}{\pi \sum_{n=0}^{\infty} |u_n(0)|^2}}$$

where S is the area of the region G.

In numerical calculations, the evaluation of the integrals α_{ij} and the solution of the above-mentioned system of equations are problems of importance. The calculations can be very laborious even in the case of very simple regions. Calculation of the α_{ij} can be simplified by choosing the $u_i(z)$ so that $u_i(0) = 0$ if $i \neq 0$, $u_0(0) = 1$. (This can evidently interfere with the orthonormality of the system considered.) Then (7) implies $c_0 = 1$ so that the number of unknown quantities is decreased by one; further $v_k(z) = u_k(z)$, and hence

$$\alpha_{ij} = \iint_G u_i(z)\, \overline{u_j(z)}\, dx\, dy\,.$$

Example 1. Let us apply the method of Remark 1 in order to find the conformal mapping of the square $G(-1 \leqq x \leqq 1, -1 \leqq y \leqq 1)$ on a circle. Let us choose $u_n = z^n$, $n = 0, 1, 2, \ldots$ Then

$$\alpha_{ij} = \iint_G z^i \bar{z}^{j}\, dx\, dy\,, \quad i = 0, 1, \ldots, n\,, \quad j = 1, 2, \ldots, n\,.$$

Let us choose $n = 4$. We calculate easily that

$$\alpha_{1,1} = \frac{8}{3}, \quad \alpha_{2,2} = \frac{232}{45}, \quad \alpha_{3,3} = \frac{264}{35}. \quad \alpha_{0,4} = -\frac{16}{15}, \quad \alpha_{4,4} = \frac{8 \, . \, 2131}{25 \, . \, 7 \, . \, 9}.$$

The remaining coefficients vanish. Hence, the system of equations (6), (7), for the unknown quantities c_i, has the form

$$\alpha_{0,j}c_0 + \alpha_{j,j}c_j = 0\,, \quad j = 1, \ldots, 4\,, \quad c_0 = 1$$

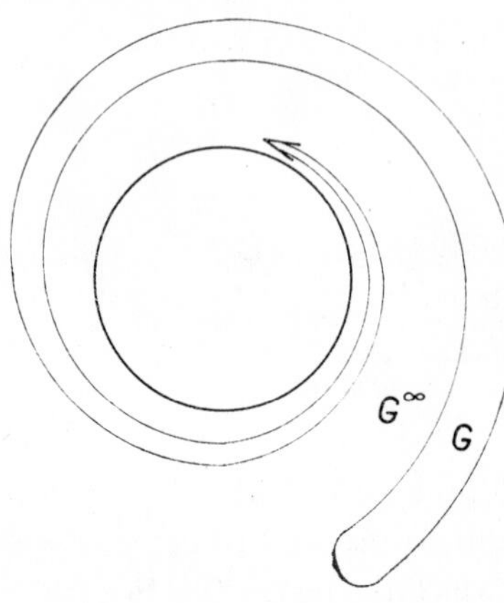

Fig. 21.16.

and therefore: $c_0 = 1$, $c_1 = c_2 = c_3 = 0$, $c_4 = -\alpha_{0,4}/\alpha_{4,4} = 210/2131 = 0{\cdot}098\,55$. Hence, the approximate solution is $\varphi_4'(z) = 1 + c_4 z^4$, $\varphi_4(z) = z + 0{\cdot}01971 z^5$.

Theorem 1. *If the system $u_i(z)$, $i = 0, 1, 2, \ldots$, is complete* (cf. § 16.2 and Remark 22.4.9, p. 1007) *on $H_2(G)$, then*

$$\lim_{n \to \infty} \iint_G \varphi_n'(z)\, \overline{\varphi_n'(z)}\, dx\, dy = \iint_G \varphi'(z)\, \overline{\varphi'(z)}\, dx\, dy\,,$$

$\varphi_n'(z) \to \varphi'(z)$ and $\varphi_n(z) \to \varphi(z)$ almost uniformly on G. Here we have

$$\varphi_n(z) = \int_0^z \varphi_n'(z)\, dz$$

(according to our assumptions, the origin lies in G and $\varphi(0) = 0$).

Definition 1. By a *Carathéodory region*, or briefly by a C-*region*, we mean a simply connected bounded region whose boundary is at the same time the boundary of that region lying in the complement of $\bar{G}$ which contains the point ∞.

REMARK 2. A Jordan region is, therefore, a C-region. The region in Fig. 21.16 (the "interior" of a spiral winding round a circle) is a C-region, but the regions in Figs 21.17 and 21.18 are not C-regions.

Theorem 2. *In a C-region, there exists a complete orthonormal system of polynomials.*

REMARK 3. In the regions of Figs 21.17 and 21.18, neither the polynomials, nor even an arbitrary system of entire functions form a system complete in $H_2(G)$.

Fig. 21.17. Fig. 21.18.

REMARK 4. As we have seen, in a C-region we can seek the function which maps G on a circle by means of the Ritz method; as a complete system we can use, for instance, an orthonormal system of polynomials.

REMARK 5. We can use analogous methods also in case (c) of Theorem 21.3.5.

21.6. The Method of Integral Equations

REMARK 1. In the case of a simply connected region, we can reduce our problem with the aid of Theorem 21.3.4 to a Dirichlet problem which can be solved by the method of integral equations (cf. § 18.4).

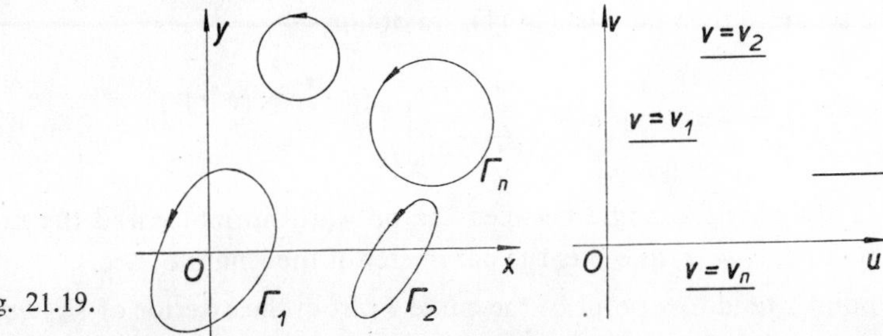

Fig. 21.19.

Example 1. We shall show the procedure for the construction of a system of integral equations for the real part of the function $w = f(z)$ which maps the exterior of a system of oriented curves $\Gamma_1, \Gamma_2, \ldots, \Gamma_n$ with continuous curvature on the plane $w = u + iv$ with slits parallel to the real axis,

$$v = v_k \quad (k = 1, 2, \ldots, n)$$

(Fig. 21.19) so that $f(\infty) = \infty$.

Let us draw a circle C with centre at the origin, sufficiently large in diameter so that all the curves Γ_k lie in the interior of C, and let us apply the Cauchy formula (Remark 20.2.10). We have

$$f(z) = \frac{1}{2\pi i}\left[\int_C \frac{f(\zeta)\,d\zeta}{\zeta - z} - \sum_{k=1}^{n}\int_{\Gamma_k}\frac{f(\zeta)\,d\zeta}{\zeta - z}\right]$$

(the point $z \in G$ lying in the interior of C). Since the expansion of $f(z)$ in the neighbourhood of infinity is

$$f(z) = cz + c_0 + \frac{c_1}{z} + \ldots,$$

we have

$$\frac{1}{2\pi i}\int_C \frac{f(\zeta)\,d\zeta}{\zeta - z} = cz + c_0\,.$$

Hence, the function $f(z)$ will have the form

$$f(z) = cz + c_0 - \sum_{k=1}^{n}\frac{1}{2\pi i}\int_{\Gamma_k}\frac{\mu_k(\zeta)\,d\zeta}{\zeta - z}, \tag{1}$$

where $\mu_k(\zeta) = \xi_k(\zeta) + i\eta_k(\zeta) = f(\zeta)$ for ζ lying on Γ_k. Since $\eta_k(\zeta) = v_k$ on Γ_k and the point z lies in the exterior of Γ_k, we have

$$\int_{\Gamma_k}\frac{\eta_k(\zeta)\,d\zeta}{\zeta - z} = 0\,.$$

Therefore, if we separate the real parts in (1), we obtain

$$\operatorname{Re} f(z) = \operatorname{Re}(cz + c_0) - \sum_{k=1}^{n}\frac{1}{2\pi i}\int_{\Gamma_k}\frac{\xi_k(\sigma)\cos(\boldsymbol{v}, \boldsymbol{r})}{r}\,d\sigma\,,$$

where $z - \zeta = \boldsymbol{r}$, $(\boldsymbol{v}, \boldsymbol{r})$ is the angle between the outward normal $\boldsymbol{v}$ and the radius-vector $\boldsymbol{r}$ at the point ζ, $r = |\boldsymbol{r}|$ and σ is the parameter of the length of arc.

If we let the point z tend to a point of the curve Γ_k from the exterior of Γ_k, we have

$$\int_{\Gamma_k}\frac{\zeta_k(\sigma)\cos(\boldsymbol{v}, \boldsymbol{r})}{r}\,d\sigma \to \int_{\Gamma_k}\frac{\zeta_k(\sigma)\cos(\boldsymbol{v}, \boldsymbol{r})}{r}\,d\sigma - \pi\,\zeta_k(\sigma)$$

(cf. equation (18.4.34), p. 896; in the first integral the point z does not lie on Γ_k, in the second integral it does). Consequently, we get

$$\xi_k(\sigma) = 2\operatorname{Re}(cz + c_0)_{\Gamma_k} - \sum_{j=1}^{n}\frac{1}{\pi}\int_{\Gamma_j}\frac{\xi_j(\sigma)\cos(\boldsymbol{v}, \boldsymbol{r})}{r}\,d\sigma\,. \tag{2}$$

Under the assumption of continuity of the curvature of the curve Γ_k, the kernel $\cos(\mathbf{v}, \mathbf{r})/r$ is continuous on Γ_k. Hence, the Fredholm Alternative (Theorem 19.1.5, p. 920) holds for the system (2). It can be proved that the homogeneous system (2) (i.e. that obtained by setting $c = 0$, $c_0 = 0$ in (2)) has a unique solution $\xi_k = 0$, $k = 1, 2, \ldots, n$. This implies, by the Fredholm Alternative, that the system (2) is solvable for an arbitrary choice of $c \neq 0$, c_0. If we solve the system (2) by any of the approximate methods (cf. Chap. 29), we get $\xi_k(\sigma)$ and then calculate $f(z)$ by using (1), where it is evidently sufficient to write $\xi_k(\zeta)$ instead of $\mu_k(\zeta)$.

21.7. Mapping of "Adjacent" Regions

REMARK 1 (*The Method of a Small Parameter*). Let us have a system of Jordan curves Γ_λ depending on a real parameter λ and containing the origin in their interior. Let the curve Γ_λ be given in the parametric form $z = r(t, \lambda)$. Let the function $w = f(z, \lambda)$ satisfying $f(0, \lambda) = 0$, $f'_z(0, \lambda) = 1$ map the interior G_λ of the curve Γ_λ on a circle $|w| < R$. If $z(t, \lambda)$ is an analytic function of the parameter λ in the neighbourhood of $\lambda = 0$, we can expect, at least in some cases, the function $f(z, \lambda)$ also to be analytic in the neighbourhood of the point $\lambda = 0$ and hence expressible in a Taylor series

$$f(z, \lambda) = f_0(z) + \lambda f_1(z) + \ldots + \lambda^n f_n(z) + \ldots. \tag{1}$$

If we know how to find the functions $f_n(z)$, then we can also find the function $f(z, \lambda)$. Let $\{u_n(z)\}$ be a complete system of functions defined on a region B containing all the G_λ for sufficiently small λ, $u_n(0) = 0$ $(n = 1, 2, \ldots)$, $u'_1(0) = 1$, $u'_n(0) = 0$ for $n > 1$. Then $f(z, \lambda) = \sum_{n=1}^{\infty} \alpha_n(\lambda)\, u_n(z)$. On the boundary, we have, necessarily, $f(z, \lambda)|^2 = R^2$. Hence if we expand the function $|\sum_{n=1}^{\infty} \alpha_n(\lambda)\, u_n(z(t, \lambda))|^2$ into a Fourier series $C_0 + \sum_{n=1}^{\infty} (C_n \cos nt + C'_n \sin nt)$ on the boundary, we obtain, by comparing coefficients, $C_n = 0$, $C'_n = 0$, $C_0 = R^2$. This is an infinite system of quadratic equations which can be solved by successive approximations, taking, as a rule, only a few first coefficients $\alpha_n(\lambda)$. The convergence of the approximation process has been proved only under very particular assumptions (see [214], p. 453–464).

REMARK 2. Let Γ be a curve given in polar coordinates by the equation $r = r(\varphi) = 1 - \delta(\varphi)$ and let $|\delta(\varphi)| < \varepsilon$, $|\delta'(\varphi)| < \varepsilon$, $|\delta''(\varphi)| < \varepsilon$. Then the function

$$w = f^*(z) = z\left(1 + \frac{1}{2\pi}\int_0^{2\pi} \frac{1 + ze^{-it}}{1 - ze^{-it}}\,\delta(t)\,dt\right)$$

differs from the function $w = f(z)$, $f(0) = 0$ (which maps the interior of the curve Γ on $|w| < 1$) by quantities of at least the second order in ε.

REMARK 3. A similar expression for the principal part of the function $f(z)$ holds also for a region adjacent to a half-plane or a region adjacent to another region (see [251], p. 257–360).

21.8. Mapping of the Upper Half-plane on a Polygon

Let K be a polygon with n sides in the w-plane. We wish to find the mapping of the upper half-plane on K, i.e. to find the constants $\alpha_1, \ldots, \alpha_n, a_1, \ldots, a_n, C, C_1$ in formula (21.3.3).

We solve the problem in the following way:

1. We take $\alpha_i = \beta_i$, where $\pi\beta_i$ are the values of the angles of the polygon K.

2. We determine the a_k from the relations

$$\lambda_1 : \lambda_2 : \ldots : \lambda_n = l_1 : l_2 : \ldots : l_n$$

where l_i are the lengths of the sides of K, and

$$\lambda_i = \int_{a_i}^{a_{i+1}} (z - a_1)^{\alpha_1 - 1} \ldots (z - a_n)^{\alpha_n - 1} \, dz \, ;$$

we choose three of the points a_i arbitrarily, e.g. $a_1 = p_1$, $a_2 = p_2$, $a_n = p_n$.

3. The function

$$f(z) = \int_{z_0}^{z} (z - a_1)^{\alpha_1 - 1} \ldots (z - a_n)^{\alpha_n - 1} \, dz$$

maps the upper half-plane on a polygon K^*, similar to K. The constants C and C_1 are then determined by a translation, a rotation and a homothetic transformation so that K^* is transformed into K.

The constants $a_3, \ldots, a_{n-1}$ can be determined by the Newton–Fourier method which we proceed to describe.

The equations for $a_3, \ldots, a_{n-1}$ have the form

$$\lambda_2 = \frac{l_2}{l_1} \lambda_1, \ldots, \lambda_{n-2} = \frac{l_{n-2}}{l_1} \lambda_1 \, . \tag{1}$$

Let us take, for initial values, numbers $a_3^{(0)}, \ldots, a_{n-1}^{(0)}$ which differ little from the numbers $a_3, \ldots, a_{n-1}$, then expand the expressions in (1) into a Taylor series in

$\delta_3^{(1)} = a_3 - a_3^{(0)}, \ldots, \delta_{n-1}^{(1)} = a_{n-1} - a_{n-1}^{(0)}$ taking their first terms. This gives

$$I_k^{(0)} + \delta_3^{(1)} \frac{\partial I_k^{(0)}}{\partial a_3^{(0)}} + \ldots + \delta_{n-1}^{(1)} \frac{\partial I_k^{(0)}}{\partial a_{n-1}^{(0)}} = \frac{l_k}{l_1} \left[I_1^{(0)} + \delta_3^{(1)} \frac{\partial I_1^{(0)}}{\partial a_3^{(0)}} + \ldots + \delta_{n-1}^{(1)} \frac{\partial I_1^{(0)}}{\partial a_{n-1}^{(0)}} \right]$$

$(k = 2, 3, \ldots, n - 2)$, where

$$I_k^{(0)} = \int_{a_k^{(0)}}^{a_{k+1}^{(0)}} (z - p_1)^{\alpha_1 - 1} (z - p_2)^{\alpha_2 - 1} (z - a_3^{(0)})^{\alpha_3 - 1} \ldots$$
$$\ldots (z - a_{n-1}^{(0)})^{\alpha_{n-1} - 1} (z - p_n)^{\alpha_n - 1} \, dz \, .$$

This is a system of linear equations in $\delta_3^{(1)}, \ldots, \delta_{n-1}^{(1)}$ which we solve and then repeat the same process for $a_3^{(1)} = a_3^{(0)} + \delta_3^{(1)}, \ldots, a_{n-1}^{(1)} = a_{n-1}^{(0)} + \delta_{n-1}^{(1)}$, etc. It can be proved that the systems of equations for $\delta_3^{(k)}, \ldots, \delta_{n-1}^{(k)}$ are always solvable and that this process converges for a certain class of initial values. The reader will find a detailed treatment of this method in [214], pp. 540–563.

22. SOME FUNDAMENTAL CONCEPTS FROM THE THEORY OF SETS AND FUNCTIONAL ANALYSIS

By KAREL REKTORYS

References: [10], [68], [70], [75], [124], [168], [170], [213], [236], [247], [254], [267], [282], [298], [299], [337], [365], [367], [374], [395], [397], [413], [442].

22.1. Open and Closed Sets of Points. Regions

In the chapter on functions of two and several variables (Chap. 12), and also in other branches of mathematics, we encounter the concepts of a "region", "interior point of the region under consideration", etc. Let us clarify these concepts first for a plane.

Denote by d the *distance* between two points A and B in the plane. If, in this plane, a cartesian system of coordinates x, y is given (see § 5.1), then

$$d = \sqrt{[(x_2 - x_1)^2 + (y_2 - y_1)^2]}\,, \tag{1}$$

where x_1, y_1 and x_2, y_2 are coordinates of the points A and B, respectively. All points in the plane, which have a distance from a given point P less than a *positive* number δ, constitute the so-called *δ-neighbourhood of the point P*. (The point P belongs to this δ-neighbourhood.)

Definition 1. Let a set M of points in the plane be given. A point $P \in M$ is called an *interior point* of the set M, if a δ-neighbourhood of P can be found which belongs entirely to M (i.e. all its points belong to M.)

Definition 2. A point $P \in M$ is called an *isolated point* of a set M, if a δ-neighbourhood of P can be found such that of all the points of this δ-neighbourhood only the point P belongs to the set M.

Definition 3. P is called a *point of accumulation* (*accumulation point, cluster point, limit point*) of a set M, if *every* δ-neighbourhood of P contains infinitely many points of M.

Definition 4. A point P is called a *boundary point* of a set M, if every δ-neighbourhood of P contains at least one point belonging to M, and at least one point which does not belong to M. All boundary points of a set M constitute the so-called *boundary* of the set M.

Example 1. Let M_1 be the set of all points of a circular disc K_1; let the boundary circle h_1 not belong to M_1 (Fig. 22.1). Each point of the set M_1 is an interior point of M_1. Points of the circle h_1 are boundary points of the set M_1.

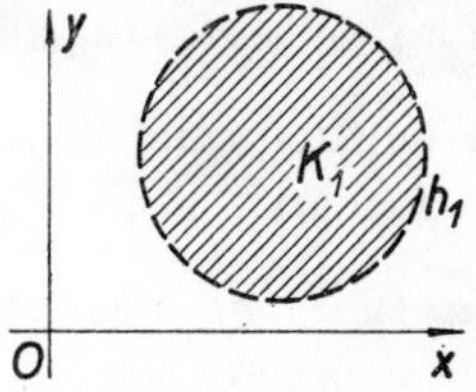

Fig. 22.1.

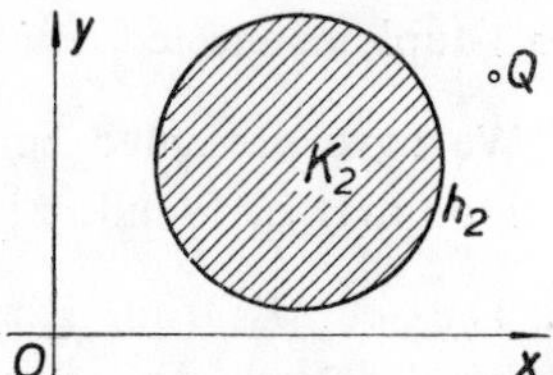

Fig. 22.2.

Example 2. Let the set M_2 consist of the point Q and of all point of a circular disc K_2 including the boundary circle h_2 (Fig. 22.2). The point Q is an isolated point of the set M_2. Points of h_2 are boundary points of the set M_2, while points lying inside the circle h_2 are interior points of the set M_2.

REMARK 1. From Examples 1 and 2 it is clear that a boundary point of a set M may, but need not, belong to M. The same is true for a point of accumulation of a set M.

REMARK 2. A boundary point need not be a point of accumulation, and vice versa. Each interior point is a point of accumulation but not a boundary point. An isolated point is a boundary point but not a point of accumulation.

Definition 5. A set M is called *open* if each point of M is an interior point of the set M (Example 1).

Definition 6. An open set M is called *connected* if every two points of M can be joined by a polygonal line (i.e. by a curve consisting of a finite number of straight line segments) which lies entirely in M (i.e. each point of it belongs to M). An open connected set is called a *region*.

Example 3. Examples of a region are: the set M_1 of Example 1; an annulus with both boundary circles excluded; the plane xy.

REMARK 3. If every two points of a given set M can be joined by a segment lying entirely in M, then the set M is said to be *convex*. The set M_1 from Example 1 furnishes an example of such a set.

Definition 7. A set N obtained by completing a set M by all points of accumulation of M is called the *closure* of M. Notation: $N = \overline{M}$. (The symbol $[M]$ is also used.)

Definition 8. A set M is said to be *closed*, if $\overline{M} = M$ (Example 2).

Definition 9. If M is a region, then $\overline{M}$ will be called a *closed region* (for example, a circular disc with boundary circle included).

Definition 10. A set M is called *bounded*, if a circle c with a finite radius can be found such that M lies in a disc whose boundary is c.

Definition 11. A bounded region is called *k-tuply connected*, if its boundary is formed by k closed curves.

Instead of a 1-tuply connected region we speak of a *simply connected region*.

REMARK 4. We have presented here an intuitive but not quite exact definition. An exact definition may be found in [345]. (See also Remark 8.)

Example 4. The set M_1 from Example 1 (the so-called *open circular disc*) is a simply connected region. An annulus is a doubly connected region. (The inner circle may degenerate into a single point, which, of course, does not belong to the annulus.)

REMARK 5. In three-dimensional (or n-dimensional) space with a Cartesian coordinate system (i.e. the so-called *Euclidean space* E_3 (or E_n)) the distance between two points $A(a_1, a_2, a_3)$, $B(b_1, b_2, b_3)$ (or $A(a_1, a_2, \ldots, a_n)$, $B(b_1, b_2, \ldots, b_n)$) is defined by the formula

$$d = \sqrt{[(b_1 - a_1)^2 + (b_2 - a_2)^2 + (b_3 - a_3)^2]}$$

(or

$$d = \sqrt{[(b_1 - a_1)^2 + (b_2 - a_2)^2 + \ldots + (b_n - a_n)^2]}) . \tag{2}$$

A δ-neighbourhood of a point P is then defined in the same manner as in the two-dimensional case (in the plane E_2). In E_3 a δ-neighbourhood of a point P is an open sphere (the interior of a spherical surface) with centre at P and radius δ, while in E_1 (the x-axis) a δ-neighbourhood of a point P is an open interval $(x_0 - \delta, x_0 + \delta)$. All definitions given above remain completely the same for E_n (see, however, Remark 8). A polygonal line in E_n consists of a finite number of segments; by a segment joining the points $A(a_1, a_2, \ldots, a_n)$, $B(b_1, b_2, \ldots, b_n)$ we mean the set of points with coordinates

$$a_k + (b_k - a_k)\, t\,, \quad 0 \leqq t \leqq 1 \quad (k = 1, 2, \ldots, n)\,. \tag{3}$$

An open sphere (the interior of a spherical surface), a cube without boundary faces, etc., are examples of regions in E_3. In E_1 an open interval corresponds to the concept of a region, a closed interval to the concept of a closed region.

REMARK 6. A complex n-dimensional space K_n, whose elements are complex numbers, is an analogue of the real Euclidean space E_n. The distance between two points $A(a_1 + ic_1, a_2 + ic_2, \ldots, a_n + ic_n)$, $B(b_1 + id_1, b_2 + id_2, \ldots, b_n + id_n)$

(a_i, b_i, c_i, d_i being real numbers) is defined by the formula

$$d = \sqrt{\left[\sum_{i=1}^{n} (b_i - a_i)^2 + \sum_{i=1}^{n} (d_i - c_i)^2\right]}.$$

REMARK 7. The concept of the *complement* of a set M in the space P is often used; the complement is the set $P - M$, i.e. the space P with the points belonging to M removed. For example, the complement of a closed disc in E_2 is an (unbounded) region obtained by removing this closed disc from the xy plane.

REMARK 8. Using the concept of a complement a simply connected region in E_2 (in the xy plane) may be defined as follows: A bounded region $M \subset E_2$ is called *simply connected*, if its complement is a connected set.

For example, the interior of an ellipse is a simply connected region, since the complement is a connected set.

In the space E_3, however, the above-stated definition is not adequate; it fails to express simple connectivity as a property which is to be assumed for ensuring the validity of some important theorems in integral calculus, etc. (see in particular § 14.8; for example, a torus would be a simply connected region in the sense of the above definition). The reader familiar with the fundamentals of topology knows that a region O in E_3 is *simply connected* if it is the so-called homeomorphic image of a sphere. (Very roughly speaking: if O can be obtained from a sphere by a "continuous deformation".)

Definition 12. A set of points is called *countable*, if there is a one-to-one correspondence between all these points and the positive integers 1, 2, 3, ... (i.e. if the points can be ordered in a sequence).

It can be shown, for instance, that the set of points in E_1 with rational coordinates (the set of rational numbers) is countable. The same can be proved for E_n.

22.2. Metric Spaces

In the last paragraph we introduced the Euclidean spaces E_n, the elements of which are points and in which the distance is defined by formula (22.1.2). In a similar way, spaces of a more general nature can be introduced. In Chap. 16 the so-called *space* $L_2(a, b)$ is defined; its elements are functions (in general complex valued, i.e. functions of the form $f(t) = f_1(t) + if_2(t)$ where $f_1(t)$, $f_2(t)$ are real functions) which are square integrable (in the Lebesgue sense) in an interval $[a, b]$. The distance $d(f, g)$ between two elements $f(t)$ and $g(t)$ of $L_2(a, b)$ is defined by the formula

$$d(f, g) = \sqrt{\left[\int_a^b |g(t) - f(t)|^2 \, dt\right]}. \tag{1}$$

The spaces E_n and $L_2(a, b)$ are examples of so-called *metric spaces*.

REMARK 1. In the case of the space $L_2(a, b)$ we often consider the space consisting only of *real* functions with distance defined by the formula

$$d(f, g) = \sqrt{\left[\int_a^b (g(t) - f(t))^2 \, dt\right]}$$

(in contradistinction to equation (1), the notation of the absolute value is clearly superfluous here). This space is referred to as to the *real space* $L_2(a, b)$. The properties of the complex space L_2 given below are preserved also for the real space L_2 so that in the following discussion we refer simply to the space L_2. The same is true for some other spaces introduced later.

Definition 1. A set M is called a *metric space* X if, for each pair of elements x, y belonging to M, the *distance* $d(x, y)$ is defined, and satisfies the following conditions:

$$d(x, y) \geqq 0\,, \quad \text{while} \quad d(x, y) = 0 \quad \text{if and only if} \quad x = y\,, \tag{2}$$

$$d(x, y) = d(y, x)\,, \tag{3}$$

$$d(x, z) \leqq d(x, y) + d(y, z)\,. \tag{4}$$

Example 1. (a) For E_n and $L_2(a, b)$ it can easily be shown that the distance, defined by the equalities (22.1.2) and (1), respectively, satisfies the above three axioms (the so-called *axioms of the metric*).

(b) A more general space than $L_2(a, b)$ is the *metric space* $L_p(a, b)$ *consisting of functions integrable (in the Lebesgue sense) with the p-th power* ($p \geqq 1$); the distance is defined by the formula

$$d(x, y) = \left[\int_a^b |y(t) - x(t)|^p \, dt\right]^{1/p} \tag{5}$$

(and satisfies axioms (2), (3), (4)). For $p = 2$ we get $L_2(a, b)$.

The *space* $C(a, b)$ is a space whose elements are functions continuous in $[a, b]$; the distance between two functions $x(t)$, $y(t)$ in $C(a, b)$ is defined by the formula

$$d(x, y) = \max_{a \leqq t \leqq b} |y(t) - x(t)| \tag{6}$$

(and satisfies axioms (2), (3), (4)).

Definition 2. An element x of a metric space X is called the *limit* of a sequence $x_1, x_2, x_3, \ldots$ of elements from X (or, the sequence $x_1, x_2, x_3, \ldots$ is said to be convergent in the space X to the element x), if

$$\lim_{n \to \infty} d(x, x_n) = 0\,. \tag{7}$$

This fact is denoted by $x_n \to x$.

REMARK 2. Instead of *convergence in* $L_2(a, b)$ we often use the term *convergence in the mean* (see Chap. 16); in this case, equation (7) assumes the form

$$\lim_{n\to\infty} \sqrt{\left[\int_a^b |x(t) - x_n(t)|^2 \, dt\right]} = 0 . \tag{8}$$

Theorem 1. *A sequence $x_1, x_2, x_3, \ldots$ of elements of a metric space X has at most one limit x. If it has a limit, i.e. if it is convergent, then any sub-sequence of it is also convergent and has the same limit.*

Definition 3. The set of all points of a metric space X with distance from a given point $x \in X$ less than δ is called a *δ-neighbourhood of x*. (Instead of δ-neighbourhood we often use the term a *sphere with centre at the point x and radius δ*; in symbols, $S(x, \delta)$.)

Definition 4. Let M be a set of elements of X. The element $x \in X$ is called a *point of accumulation* of the set M, if any δ-neighbourhood of the element x contains infinitely many elements of the set M. (Obviously, a point of accumulation of a set M *need not* belong to M.)

Definition 5. The set $\overline{M}$ obtained by adjoining to M all its points of accumulation is called the *closure* of the set M.

Definition 6. A set M is called *closed*, if $M = \overline{M}$. A set M is called *open*, if the complement $X - M$ (Remark 22.1.7) is a closed set.

Definition 7. A set M is called *dense in X*, if $\overline{M} = X$ (i.e., if each element of the metric space X is either a point of accumulation of the set M or a point of M).

Example 2. It can be shown that the set M of all polynomials (in general, with complex coefficients) is dense in $L_2(a, b)$, i.e. for every function $x(t) \in L_2(a, b)$ and every $\varepsilon > 0$ a polynomial $P(t)$ can be found such that

$$\sqrt{\left[\int_a^b |x(t) - P(t)|^2 \, dt\right]} < \varepsilon .$$

The set of all polynomials is also dense in the space $C(a, b)$, i.e., for every continuous function $y(t)$ and every $\varepsilon > 0$ a polynomial $Q(t)$ can be found such that

$$\max_{a \leqq t \leqq b} |y(t) - Q(t)| < \varepsilon$$

(*Weierstrass's theorem*).

22.3. Complete, Separable, Compact Spaces

Definition 1. A sequence $\{x_n\}$ of elements of a matrix space X is called a *fundamental* (or a *Cauchy*) *sequence*, if for every $\varepsilon > 0$ a number n_0 (depending on the choice of ε) can be found such that

$$d(x_m, x_n) < \varepsilon \tag{1}$$

whenever both numbers m and n exceed n_0.

Theorem 1. *Every convergent sequence is fundamental.*

Definition 2. A metric space X is called *complete*, if any fundamental sequence $\{x_n\}$ (of elements from X) has a limit x *belonging to the space* X.

Example 1. It is known that the number $\sqrt{2}$ is not a rational number. At the same time, a sequence of rational numbers

$$x_1, x_2, x_3, \dots \tag{2}$$

exists in E_1 such that

$$\lim_{n\to\infty} x_n = \sqrt{2}\,. \tag{2'}$$

The sequence (2) is fundamental in E_1 (by Theorem 1). Let X be the set of all rational numbers. Defining the distance between two elements x_1, x_2 in X by the usual formula $d(x_1, x_2) = |x_2 - x_1|$ (i.e. in the same manner as in the space of real numbers), X becomes a metric space. The space X, however, is not complete, as the fundamental sequence (2) has no limit in X ($\sqrt{2}$ does not belong to X).

REMARK 1. It can be shown that if a metric space is not complete, then it can be augmented (completed) by the so-called *ideal elements* so that the space extended in this manner is complete and the original metric is preserved. A typical example is furnished by completing the space of rational numbers just discussed by adding to it irrational numbers. The completed space is then the space E_1.

Theorem 2. *If M is a closed set in a complete space X, then M is a complete metric space.*

Theorem 3. *Spaces E_n, C, L_p (and consequently also L_2) are complete.*

Definition 3. A metric space X is called *separable* if it contains a *countable* set M which is *dense* in X (i.e. if it contains a sequence P such that each element $x \in X$ can be considered as a limit of a subsequence suitably selected from P).

Theorem 4. *Spaces E_n, C, L_p (and consequently also L_2) are separable.*

REMARK 2. In E_n all points with rational coordinates constitute the *countable* set M considered above. So also do all polynomials with rational coefficients in L_2. (If the complex space L_2 (Remark 22.2.1) is under discussion, these coefficients have the form $r + is$, where r and s are rational numbers.)

Definition 4. A set M in a metric space X is said to be *compact in the space* X (or *relatively compact* or *precompact*), if any sequence of elements from M contains a subsequence convergent in X. If, in addition, the limits of all these subsequences belong to M, we say simply that M is *compact* (or, more explicitly, *compact in itself*).

REMARK 3. In particular, a metric space X is called *compact*, if any sequence of elements of X contains a subsequence converging to an element of X.

The terminology introduced in Definition 4 is not consistent in the literature.

Theorem 5. *Every bounded set M in E_n is compact in E_n (i.e. relatively compact); if, in addition, M is closed then M is compact (in itself).*

REMARK 4. Spaces E_n, C, L_p (and also L_2) are *not* compact. To show it, for example, in the space E_1, it is sufficient to consider the sequence of points $x_1 = 1$, $x_2 = 2$, $x_3 = 3, \ldots$.

The interval $[3, 6] \subset E_1$ is compact (in itself).

Theorem 6. *If X is complete, then a necessary and sufficient condition for a set $M \subset X$ to be relatively compact is that to every $\varepsilon > 0$ a finite ε-net N_ε for the set M exists, i.e. a finite set $N_\varepsilon \subset X$ such that to each $x \in M$ an $x_\varepsilon \in N_\varepsilon$ with $d(x, x_\varepsilon) < \varepsilon$ can be found.*

Theorem 7. *A necessary and sufficient condition for a set $M \subset C(a, b)$ to be compact (in $C(a, b)$) is that all functions $x(t) \in M$ be equicontinuous and uniformly bounded.* (Arzelà's theorem, see § 15.1.)

Theorem 8. *A compact metric space is separable.*

Theorem 9. *Let the (real) function $f(x_1, x_2, \ldots, x_n)$ be continuous in a set $M \subset E_n$. Then, if M is compact, f assumes its maximum and minimum in M.*

REMARK 5. Theorem 9 is a generalization of well-known theorems relating to a function continuous in a closed interval or in a closed region. More generally, we have:

Theorem 10. *Let $f(x)$ be a real continuous functional* (see Remark 22.5.2 and Definition 22.5.2) *in $M \subset X$. If M is compact then $f(x)$ assumes its maximum and minimum in M.*

22.4. Linear Space. Normed Space. Hilbert Space. Orthogonal Systems

Definition 1. A set M of elements $x, y, z, \ldots$ is called a *linear space* (*vector space*) if it has the following properties:

If x, y are any two elements of M and a is a number (complex in general), then $x + y$ and ax also belong to M; moreover, these operations obey the usual rules of algebra, i.e.:

$$x + y = y + x\,, \qquad x + (y + z) = (x + y) + z\,,$$

$$a(x + y) = ax + ay\,, \qquad (a + b)\,x = ax + bx\,,$$

$$a(bx) = (ab)\,x\,, \qquad 1\,.\,x = x\,,$$

$$\text{if } x + y = x + z, \text{ then } y = z.$$

REMARK 1. From these rules it follows that for any two elements x, y of M we have $0\,.\,x = 0\,.\,y$. The element $0\,.\,x$ is denoted by 0 and is called the *zero element* of the linear space in question. It can be shown that the familiar rules of algebra are still preserved. For example, if $ax = 0$, $a \neq 0$, then $x = 0$. If $ax = bx$ with $x \neq 0$, hent $a = b$, etc.

REMARK 2. If a metric is introduced into a linear space M (i.e. if the distance between every two elements is defined, see § 22.2) we get a so-called *linear metric space*. Among linear metric spaces, linear normed spaces and Hilbert space are especially important (see Definitions 3 and 5 below).

Example 1. Spaces E_n, $C(a, b)$, $L_p(a, b)$ (and consequently also $L_2(a, b)$) are linear metric spaces. Here, the sum of two functions, and the product of a function and a constant are defined in the usual manner. We know, for example, that if $f(x) \in L_p$, $g(x) \in L_p$ then also $f(x) + g(x) \in L_p$, $cf(x) \in L_p$, where c is an arbitrary number which, of course, is real if real spaces are under discussion. In the space $C(a, b)$ the zero element is the function which vanishes identically in $[a, b]$, while in the space $L_p(a, b)$ it denotes every function equivalent to the zero function, i.e. every function vanishing everywhere in $[a, b]$ except at points which constitute a set of measure zero.

Definition 2. A subset L of a linear metric space X is called a *linear manifold* in the space X, if it has the property that when the elements $x_1, x_2, \ldots, x_n$ belong to L, then any linear combination $c_1x_1 + \ldots + c_nx_n$ also belongs to L. If a linear manifold L is a closed set in X, then L is called a *subspace of the space* X. (In particular, we can have $L = X$.)

Example 2. Constant functions, i.e. functions of the form $f(t) = k$ (k an arbitrary constant) constitute a subspace L of $C(a, b)$. First, they evidently constitute a linear manifold, since any linear combination formed from constant functions is again a constant function. Secondly, it is easy to prove that this manifold L is closed in $C(a, b)$, i.e. that $L = \bar{L}$ (see Definition 22.2.6): Let $f \in \bar{L}$. Then, by the definition of $\bar{L}$, a sequence of elements $f_n \in L$ exists such that $f_n \to f$, i.e. $d(f_n, f) \to 0$. Since $f_n(t) = k_n$ in $[a, b]$ and $d(f_n, f) = \max\limits_{a \leq t \leq b} |k_n - f(t)|$, it follows that $f(t) = k$ in $[a, b]$, where $k = \lim\limits_{n \to \infty} k_n$. Thus $f \in L$; hence L is closed in $C(a, b)$.

Definition 3. A set X is called a *linear normed space*, if

1. X is a linear space (vector space) (see Definition 1);

2. to each element $x \in X$ there is uniquely assigned a real number $\|x\|$ called a *norm* of the element x, which satisfies the axioms

$\|x\| \geq 0$, while $\|x\| = 0$ if and only if $x = 0$ (see Remark 1),

$\|ax\| = |a| \cdot \|x\|$ for every number a,

$\|x + y\| \leq \|x\| + \|y\|$ (triangular inequality).

The distance $d(x, y)$ is defined by the formula

$$d(x, y) = \|x - y\| .$$

Definition 4. A complete linear normed space is called a *Banach space* (*B-space*).

Example 3. Spaces E_n, C, L_p (particularly L_2) are Banach spaces provided the respective norms are defined by the relationships:

$$\text{in } E_n, \quad \|x\| = \sqrt{(x_1^2 + x_2^2 + \ldots + x_n^2)},$$

$$\text{in } C(a, b), \quad \|x\| = \max_{a \leq t \leq b} |x(t)| ,$$

$$\text{in } L_p(a, b), \quad \|x\| = \left(\int_a^b |x(t)|^p \, dt \right)^{1/p} .$$

REMARK 3. By introducing a metric the concept of convergence is defined (see Definition 23.2.2); thus, in our case, the symbol $\lim_{n\to\infty} x_n = x$ (or briefly $x_n \to x$) means that $\lim_{n\to\infty} \|x - x_n\| = 0$, i.e. $\|x - x_n\| \to 0$. This kind of convergence is often called *convergence in the norm*. The notation $x_n \Rightarrow x$ is often used.

Theorem 1. *If* $x_n \to x$, *then* $\|x_n\| \to \|x\|$.

Definition 5. A set H is called a *Hilbert space*, if

1. H is a linear space (vector space) (see Definition 1);

2. to each pair of elements x, y from H there is uniquely assigned a complex number (x, y), called the *inner* (*scalar*) *product* of the elements x and y, such that:

$$(x, y) = \overline{(y, x)}$$

(i.e. $\overline{(y, x)}$ is the complex conjugate of the number (x, y)),

$$(x_1 + x_2, y) = (x_1, y) + (x_2, y),$$

$$(ax, y) = a(x, y),$$

and consequently, $(x, ay) = \bar{a}(x, y)$,

$$(x, x) \geq 0, \text{ while } (x, x) = 0 \text{ if and only if } x = 0;$$

3. the norm of an element x is defined by the formula $\|x\| = \sqrt{(x, x)}$;

4. H is a complete space in the sense of the metric $d(x, y) = \|x - y\|$;

5. H is separable (Definition 22.3.3).

REMARK 4. Hence, a Hilbert space is also a Banach space.

The reader should note that in the definition of a Hilbert space in the literature, separability and completeness are not always required. (Then, of course, the space introduced need not be a B-space).

Example 4. Spaces E_n and $L_2(a, b)$ are Hilbert spaces if the inner product is defined as follows:

$$\text{in } E_n: \quad (x, y) = x_1\bar{y}_1 + x_2\bar{y}_2 + \ldots + x_n\bar{y}_n ;$$

$$\text{in } L_2: \quad (x, y) = \int_a^b x(t)\,\overline{y(t)}\,dt .$$

If *real* spaces E_n, $L_2(a, b)$ are under discussion (Remark 22.2.1), they are referred to as *real* Hilbert spaces.

Theorem 2. *The so-called Schwarz inequality holds:*

$$|(x, y)| \leqq \|x\| \cdot \|y\| .$$

Example 5. In L_2 the latter inequality reads as follows:

$$\left|\int_a^b x(t)\,\overline{y(t)}\,dt\right| \leqq \sqrt{\left[\int_a^b |x(t)|^2\,dt\right]} \sqrt{\left[\int_a^b |y(t)|^2\,dt\right]} .$$

Definition 6. The elements x, y of H are called *orthogonal*, if $(x, y) = 0$. Notation: $x \perp y$.

Theorem 3. *If L is a subspace* (see Definition 2) *of a Hilbert space H, then each element $x \in H$ can be uniquely represented in the form*

$$x = y + z , \tag{1}$$

where $y \in L$, $z \perp L$ (i.e. z is orthogonal to all elements of L). The element y is called a projection of the element x into the subspace L.

REMARK 5. It can be shown that all elements of H, which are orthogonal to a given subspace L, constitute another subspace (the so-called *complementary subspace*), say M. In the sense of equation (1) we write

$$H = L + M \tag{2}$$

(the notation $H = L \oplus M$ is also often used) and read it as: H is a *direct sum* of the subspaces L and M.

REMARK 6. Theorem 2 can be recognized as a generalization of the following well-known relationship in E_3: Construct a plane ϱ passing through the origin of E_3 and a straight line o perpendicular to it. Then each "vector" $x \in E_3$ can be uniquely written in the form $x = x_o + x_\varrho$, i.e. as a sum of two orthogonal vectors, the first of which lies in o, the second one in ϱ. (o and ϱ here play the role of the complementary subspaces L and M.)

Theorem 4. *Let $u_n \in H$ $(n = 1, 2, \ldots)$ and let $u_i \perp u_k$ for $i \neq k$. Then the series*

$$\sum_{n=1}^{\infty} u_n$$

is convergent (in the sense of the metric of the Hilbert space, i.e. in norm, given by Definition 5), if and only if the series

$$\sum_{n=1}^{\infty} \|u_n\|^2$$

is convergent.

Definition 7. We say that the elements

$$x_1, x_2, x_3, \ldots, x_n, \ldots \in H \tag{3}$$

(which may be finite or infinite in number) constitute an *orthonormal system* in H, if

$$(x_i, x_k) = 0 \quad \text{for} \quad i \neq k\,, \quad (x_i, x_k) = 1 \quad \text{for} \quad i = k\,.$$

Definition 8. Let x be an arbitrary element of H. The numbers

$$a_n = (x, x_n) \tag{4}$$

are called the *Fourier coefficients of the element x with respect to the system* (3).

Theorem 5. *A necessary and sufficient condition for a series*

$$\sum_{n=1}^{\infty} b_n x_n \tag{5}$$

(the x_n being elements of the system (3)) to converge is that the series

$$\sum_{n=1}^{\infty} |b_n|^2 \tag{6}$$

converges (see Theorem 4). *If the series (6) converges, then the series (5) converges to a certain element $x \in H$ and the b_n are the Fourier coefficients of x with respect to the system* (3).

Theorem 6. *For any element $x \in H$ we have*

$$\sum_{n=1}^{\infty} |a_n|^2 \leqq \|x\|^2 \tag{7}$$

(the so-called *Bessel inequality*; for a_n see (4)).

REMARK 7. From Theorem 4 it follows that, for any element $x \in H$ the corresponding *Fourier series*

$$\sum_{n=1}^{\infty} a_n x_n \quad (\text{with } a_n \text{ given by (4)}) \tag{8}$$

is convergent (but not necessarily to the element x; see Theorem 7).

Theorem 7. *A necessary and sufficient condition for the series* (8) *to converge to the element* x *is that*

$$\sum_{n=1}^{\infty} |a_n|^2 = \|x\|^2 . \tag{9}$$

Definition 9. If (9) is satisfied for *every* element $x \in H$ (i.e. if for every element $x \in H$ the corresponding Fourier series (8) converges to x), the system (3) is called *complete in H*. Equation (9) is called the *Parseval equation* (or *equation of completeness*).

Definition 10. The system (4) is called *closed* if H does not contain any non-zero element x orthogonal to each element of the system (3).

We note that as far as the terms "complete", "closed" are concerned, terminology is not consistent in the literature (due, perhaps, to the following theorem).

Theorem 8. *A system* (3) *is complete in H if and only if it is closed.*

Theorem 9. *In a Hilbert space H* (which is separable by Definition 5) *there is at least one complete orthonormal system.*

REMARK 8. In $L_2(0, l)$ a typical example of a complete orthonormal system is furnished by the system of functions

$$\varphi_n(t) = \sqrt{\left(\frac{2}{l}\right)} \sin \frac{n\pi t}{l} . \tag{10}$$

For any function $x(t) \in L_2(0, l)$ we have in L_2

$$x(t) = \sum_{n=1}^{\infty} a_n \varphi_n(t) , \quad \text{where} \quad a_n = \sqrt{\left(\frac{2}{l}\right)} \int_0^l x(t) \sin \frac{n\pi t}{l} \, dt , \tag{11}$$

i.e.

$$\left\| x(t) - \sum_{n=1}^{k} a_n \varphi_n(t) \right\| \to 0 \quad \text{as} \quad k \to \infty .$$

(In general, this does not imply *pointwise* convergence in the interval $[0, l]$.)

REMARK 9. Even if the elements

$$x_1, x_2, x_3, \ldots \tag{12}$$

are not orthogonal, the concept of a *complete sequence* may be defined. The elements (12) constitute a so-called *complete sequence* (*fundamental system*) *in* H, if to each element $x \in H$ and to each $\varepsilon > 0$ a positive integer k and constants $b_1, b_2, \ldots, b_k$ can be found such that

$$\left\| x - \sum_{n=1}^{k} b_n x_n \right\| < \varepsilon .$$

Using the familiar orthogonalization process (see § 16.2), a complete orthonormal sequence may be obtained from any complete sequence. For example, the orthonormalization of the sequence

$$1, t, t^2, t^3, \ldots,$$

which is complete in $L_2(-1, 1)$, yields a complete orthonormal system of functions $\sqrt{(n + \frac{1}{2})}\, P_n(t)$, where $P_n(t)$ are the Legendre polynomials.

22.5. Linear and Other Operators in Metric Spaces. Functionals

Definition 1. Let X and X' be two metric spaces. Let there be assigned to each $x \in X$ a uniquely determined element $x' \in X'$. This correspondence is briefly indicated by the relation

$$x' = Ax \tag{1}$$

in which A is called an *operator* (*or mapping*) *defined in the space* X *which maps* X *into* X' (or briefly, an *operator* A *from* X *into* X').

REMARK 1. An operator A need not always be defined in the entire space X but only in a subset $D(A) \subset X$ called the *domain of definition of the operator* A. Two operators A, B are said to be equal (this fact is denoted by $A = B$) if they have the same domain of definition D and if $Ax = Bx$ for every $x \in D$. The set of all $x' \in X'$ obtained from (1) for all $x \in D(A)$ is denoted by $R(A)$ (the so-called *image* of the set $D(A)$, or the *range* of the operator A). If the mapping from $D(A)$ into $R(A)$ is *one-to-one*, then there exists the so-called *inverse operator* A^{-1} with the domain of definition $R(A)$ assigning to each $x' \in R(A)$ exactly that element $x \in D(A)$, for which $x' = Ax$. Then, obviously, we have

$$x = A^{-1}Ax, \quad x' = AA^{-1}x'. \tag{2}$$

Definition 2. An operator A is called *continuous at a point* $x_0 \in D(A)$ if for every sequence $\{x_n\}$ (where $x_n \in D(A)$) with $x_n \to x_0$ in the metric of the space X we have $Ax_n \to Ax_0$, i.e. $x'_n \to x'_0$ in the metric of the space X'.

REMARK 2. If X' is the real space E_1 or the complex space K_1 (see Remark 22.1.6), i.e. if x' is a real or complex number, the operator is called a *functional* (real or complex, respectively). An example of a functional defined in the entire space L_2 (or, more generally, in L_p with $p \geqq 1$) is the operator f given by

$$fx = \int_a^b x(t)\,dt\,, \tag{3}$$

which, therefore, to every function $x \in L_2$ assigns the number (3).

Theorem 1 *Banach's Fixed-Point Theorem (Contraction Mapping). Let the operator A defined in an (entire) complete metric space X map X into itself (i.e. x' also belongs to the space X). Let a number α $(0 < \alpha < 1)$ exist such that*

$$d(Ax, Ay) \leqq \alpha\, d(x, y) \tag{4}$$

for any two elements $x, y \in X$ ($d(a, b)$ being the distance between the elements a, b, see Definition 22.2.1). *Then the equation*

$$x = Ax \tag{5}$$

has exactly one solution x_0 in X. The element x_0 can be obtained by successive approximations as a limit (in the metric of the space X) of the sequence

$$x_2 = Ax_1,\ x_3 = Ax_2,\ x_4 = Ax_3,\ \ldots, \tag{6}$$

where the initial element x_1 may be chosen arbitrarily.

REMARK 3. Using this theorem, the existence (and uniqueness) of solutions of various problems may be proved, such as problems in the field of differential and integral equations, of finite and infinite systems of linear algebraic equations, etc. An application will be demonstrated by the following example:

Example 1. Consider a nonlinear integral equation

$$x(t) = \lambda \int_a^b K(t, s, x(s))\,ds\,, \quad a \leqq t \leqq b\,, \tag{7}$$

where the function $K(t, s, z)$ is continuous and bounded in absolute value by a constant D on a parallelepiped $\bar{Q}(a \leqq t \leqq b,\ a \leqq s \leqq b,\ |z| \leqq k)$ and, in addition, satisfies *Lipschitz condition* with respect to z, i.e. there exists a constant N such that

$$|K(t, s, z_2) - K(t, s, z_1)| \leqq N|z_2 - z_1| \tag{8}$$

for (t, s, z_1), $(t, s, z_2) \in \bar{Q}$. The assertion is that for every λ sufficiently small, or, more accurately, for every λ satisfying both the inequalities

$$|\lambda|\, D(b - a) \leqq k\,, \tag{9}$$

$$|\lambda|\, N(b - a) < 1 \tag{10}$$

there exists one and only one continuous function $x(t)$ satisfying equation (7).

For X let us take the set of all functions continuous in $[a, b]$ such that $|x(t)| \leqq k$, with the metric of the space $C(a, b)$, i.e.

$$d(x, y) = \max_{a \leqq t \leqq b} |y(t) - x(t)| . \tag{11}$$

In view of (11) (uniform convergence), the space X is complete. If condition (9) is satisfied, the operator

$$Ax = \lambda \int_a^b K(t, s, x(s))\, ds \tag{12}$$

maps the space X into itself. Furthermore, in view of (8) we have for every pair of functions x, y from X,

$$d(Ax, Ay) = \max_{a \leqq t \leqq b} \left| \lambda \int_a^b K(t, s, y(s))\, ds - \lambda \int_a^b K(t, s, x(s))\, ds \right| \leqq$$

$$\leqq |\lambda|\, N \int_a^b |y(s) - x(s)|\, ds \leqq |\lambda|\, N(b - a) \max_{a \leqq t \leqq b} |y(t) - x(t)| = |\lambda|\, N(b - a)\, d(x, y) .$$

Thus, if (10) is true, condition (4) is satisfied and equation (7) possesses, by Theorem 1, a unique solution. This solution can be obtained as the limit of a uniformly convergent sequence of successive approximations (6).

Definition 3 (*Operators in Linear Spaces*). Let the domain of definition $D(A)$ of an operator A be a linear manifold in a metric space X (cf. Definition 22.4.2). If for any $x_k \in X$ and any numbers c_k we have

$$A(c_1 x_1 + c_2 x_2 + \ldots + c_m x_m) = c_1 A x_1 + c_2 A x_2 + \ldots + c_m A x_m , \tag{13}$$

the operator A is said to be *additive and homogeneous*. An additive and homogeneous operator, continuous at every point $x_0 \in D(A)$ (see Definition 2), is called a *continuous linear operator* (or, briefly, a *linear operator*).

REMARK 4. The terminology is not consistent in the literature. Some authors call an operator linear if it is additive and homogeneous in the sense of our definition (it need not be continuous); others mean, by the term linear operator, a continuous linear operator. Here, we shall always use the term continuous linear operator.

Theorem 2. *An additive and homogeneous operator continuous at some point $x_1 \in D(A)$ is continuous at every point $x \in D(A)$.*

Definition 4. Let X and X' be normed linear spaces (see Definition 22.4.3), A an additive and homogeneous operator with $D(A) \subset X$, $R(A) \subset X'$. The operator A is called *bounded* if a positive constant C exists such that

$$\|Ax\| \leqq C\|x\| \tag{14}$$

for every $x \in D(A)$. (Here, naturally, $\|x\|$ denotes the norm of the element x in the space X, $\|Ax\|$ the norm of the element $x' = Ax$ in the space X'.)

Example 2. In the space $L_2(a, b)$ with norm

$$\|x\| = \sqrt{\left[\int_a^b |x(t)|^2 \, dt\right]}, \quad x \in L_2, \tag{15}$$

the functional

$$fx = \int_a^b x(t) \overline{y(t)} \, dt, \tag{16}$$

where $y(t)$ is a fixed element from L_2, is a bounded operator. For by the Schwarz inequality (Theorem 22.4.2), we have

$$\|f(x)\| = |fx| = \left|\int_a^b x(t) \overline{y(t)} \, dt\right| \leq \sqrt{\left[\int_a^b |x(t)|^2 \, dt\right]} \sqrt{\left[\int_a^b |y(t)|^2 \, dt\right]} = \|x\| \cdot \|y\| .$$

Thus, it suffices to put

$$C = \|y\| .$$

On the other hand, consider the operator A given by

$$Ax = \frac{dx(t)}{dt}, \tag{17}$$

which is defined in the linear manifold $D(A)$ consisting of those functions from L_2 which are continuous, together with their first derivatives, in $[a, b]$. This operator is *not* bounded, since there exist functions with $\|x\| = 1$, for which $\|Ax\| = \|dx/dt\|$ is as large a number as we please (we can, for instance, put $x(t) = \sqrt{[2/(b-a)]}$. $\sin [n\pi(t-a)/(b-a)]$ with n sufficiently large).

REMARK 5. Among the numbers C for which (14) holds, there exists a unique least number which is called the *norm of the operator* A and is denoted by $\|A\|$ or n_A. We have

$$\|A\| = n_A = \sup_{\|x\|=1,\, x \in D(A)} \|Ax\|, \tag{18}$$

i.e. the norm of the operator A may be found as the least upper bound of the set of numbers $\|Ax\|$ with x ranging over all the unit elements from $D(A)$, or, more concisely, with x ranging over the surface of the unit sphere in $D(A)$. From the definition of the norm of the operator it follows that

$$\|Ax\| \leq \|A\| \cdot \|x\| \tag{19}$$

for every $x \in D(A)$.

Theorem 3. *An additive and homogeneous operator A is continuous in $D(A)$ if and only if it is bounded.*

REMARK 6. In the following text of this paragraph B_1 and B_2 are *Banach spaces* (i.e. *complete* normed linear spaces, see Definition 22.4.4), and A is a *continuous linear operator from B_1 into B_2.*

Theorem 4. *If $D(A)$ is a linear manifold, dense in B_1, the operator A can be (uniquely) extended from $D(A)$ on to the entire space B_1 so that the norm of the operator A is preserved.*

REMARK 7. The operator extended in this manner (which is denoted by A') has the entire space B_1 as its domain of definition, and satisfies the equality $A'x = Ax$ for every element $x \in D(A)$. Moreover, we have $\|A'\| = \|A\|$. The process described in Theorem 4 is referred to as the *continuous extension of the operator.*

Theorem 5. *Let M be the set of continuous linear operators defined in the entire space B_1. For every element $x \in B_1$ let there exist a number $K(x)$ (depending on x) such that $\|Ax\| \leqq K(x) \,.\, \|x\|$ for every operator $A \in M$. Then the system M is uniformly bounded, i.e. a number K (independent of x and A) exists such that for all $x \in B_1$ and all operators $A \in M$ we have*

$$\|Ax\| \leqq K\|x\| \quad (\textit{briefly } \|A\| \leqq K)\,.$$

Theorem 6. *Let the continuous linear operator A map the entire space B_1 in one-to-one correspondence on to (the entire space) B_2. Then the inverse operator A^{-1}* (see Remark 1) *is also a continuous linear (and consequently bounded) operator.*

REMARK 8. A particular class of continuous linear operators are the *linear functionals* (see Remark 2), i.e. additive and homogeneous bounded operators with Ax being a real or complex number (see Example 2). For functionals a more powerful theorem than Theorem 4 is true:

Theorem 7. *If a linear functional f is given in a linear manifold $L \subset B$ (B is a Banach space; L need not be dense in B), then the functional f can be extended on to the entire space B with its norm preserved.*

Definition 5. A sequence $\{x_n\}$ of elements of a Banach space B is said to be *weakly convergent to an element* $x_0 \in B$, if for every linear functional f defined in B we have

$$fx_0 \to fx_n\,.$$

In symbols,

$$x_0 \overset{\mathrm{w}}{\to} x_n\,. \tag{20}$$

Theorem 8. *If a sequence $\{x_n\}$ is convergent to x_0, then it also converges weakly to x_0. The converse statement is not, in general, true.*

REMARK 9. Let us consider all linear functionals f defined in B and define the sum of two functionals and the product of a number and a functional as follows:

$$(f_1 + f_2)\,x = f_1 x + f_2 x\,, \quad (af)\,x = a\,fx\,.$$

Furthermore, define the norm n_f of a functional f by the relation

$$n_f = \|f\|$$

in accordance with Remark 5 (see (18)). Then all these functionals constitute a normed linear space which is complete (and consequently is a Banach space). This space is called the *adjoint space to the space B* and is denoted by B^*.

Definition 6 (*Adjoint Operator*). Let A be a continuous linear operator which maps B_1 into B_2, $x' = Ax$, $x \in B_1$, $x' \in B_2$. Consider a linear functional $f \in B_2^*$ (see Remark 9). Then obviously

$$fx' = fAx = gx \tag{21}$$

is a linear functional. Thus, by means of the operator A, to every linear functional $f \in B_2^*$ there is assigned a linear functional $g \in B_1^*$ by the relation (21), i.e.

$$g = A^*f\,. \tag{22}$$

The operator A^* is called the *adjoint operator to A*.

REMARK 10. The definition of the adjoint operator is simpler and more natural in a Hilbert space, see Remark 22.6.2. See also Example 22.6.2.

Definition 7. A continuous linear operator A is called *absolutely continuous* (or *compact*), if it maps every bounded set $M \subset B_1$ on to a relatively compact set $M' \in B_2$.

Example 3. Let $K(t, s)$ be a function which is square integrable in a square $\bar{Q}(a \leqq t \leqq b,\ a \leqq s \leqq b)$, $K \in L_2(Q)$. It can be shown that the operator

$$Ax = \int_a^b K(t, s)\,x(s)\,\mathrm{d}s\,, \quad x \in L_2(a, b)\,, \tag{23}$$

is absolutely continuous in $L_2(a, b)$, and that its norm is

$$\|A\| = \sqrt{\left[\int_a^b\!\!\int_a^b |K(t, s)|^2\,\mathrm{d}t\,\mathrm{d}s\right]}\,.$$

Theorem 9. *If a sequence of absolutely continuous operators A_n (from B_1 into B_2) is uniformly convergent to a continuous linear operator A (i.e. if $\|A - A_n\| \to 0$ as $n \to \infty$), then A is an absolutely continuous operator.*

REMARK 11. In the theory of integral equations the solvability of Fredholm equations of the form

$$x(t) - \frac{1}{\lambda}\int_a^b K(t, s)\, x(s)\, ds = y(t) \tag{24}$$

is investigated along with problems related to eigenvalues and eigenfunctions (eigenvectors) of these equations. In a similar way, operator equations

$$\lambda x - Ax = y \tag{25}$$

can be analysed. If A is an absolutely continuous operator (for example, the operator (23)), then there hold, for these equations, theorems analogous to Fredholm theorems for integral equations (24) (see Theorem 22.6.6).

22.6. Operators in Hilbert Space

(a) Bounded Operators

REMARK 1. A Hilbert space is (see Remark 22.4.4) a particular case of a Banach space. Consequently, everything stated above in § 22.5 about operators in a Banach space is true also for a Hilbert space. However, in a Hilbert space much more about the operators can often be stated.

In part (a) of this paragraph, we deal with linear *bounded* operators defined in the *entire* space H, a fact which will not be explicitly restated in the following text. (We shall assume that bounded operators defined in a linear manifold L dense in H are already extended into the entire space by Theorem 22.5.4.)

In part (b) we shall consider unbounded operators.

Theorem 1 (*The Riesz–Fischer Theorem*). *Any linear functional f in H can be uniquely represented in the form of an inner product, i.e.*

$$fx = (x, y)\,,$$

where y is an element of H uniquely determined by f, and x is an arbitrary element of H.

Example 1. Thus, in particular, if H is the space $L_2(a, b)$ with the inner product

$$(x, y) = \int_a^b x(t)\, \overline{y(t)}\, dt \tag{1}$$

(Example 22.4.4), then to every linear functional f in H there corresponds exactly one function $y \in L_2$ such that

$$fx = \int_a^b x(t)\, \overline{y(t)}\, dt\,. \tag{2}$$

(Obviously, if y is fixed then (1) is a linear functional.)

REMARK 2. If A is a continuous linear operator in H, then

$$(Ax, y)$$

is a linear functional in H. By Theorem 1 there is exactly one element $y^* \in H$ such that

$$(Ax, y) = (x, y^*) . \tag{3}$$

Thus, by using (3) (with A fixed) we assign to every $y \in H$ a $y^* \in H$ with

$$y^* = A^* y \tag{4}$$

such that

$$(Ax, y) = (x, A^* y) . \tag{5}$$

The operator A^* is called *the adjoint operator to A.*

Example 2. The adjoint operator to the operator A in Example 22.5.3 is given by

$$A^* y = \int_a^b \overline{K(s, t)}\, y(s)\, ds ; \tag{6}$$

thus A^* is an integral operator whose kernel is obtained from the original kernel by interchanging the variables and taking the complex conjugate value. To verify that (5) is satisfied we observe that

$$(Ax, y) = \int_a^b \left(\int_a^b K(t, s)\, x(s)\, ds\right) \overline{y(t)}\, dt =$$

$$= \int_a^b \int_a^b K(t, s)\, x(s)\, \overline{y(t)}\, dt\, ds =$$

$$= \int_a^b \int_a^b x(t)\, K(s, t)\, \overline{y(s)}\, ds\, dt =$$

$$= \int_a^b x(t) \overline{\left(\int_a^b \overline{K(s, t)}\, y(s)\, ds\right)}\, dt = (x, A^* y) .$$

Theorem 2. *The norm of the adjoint operator A^* is equal to the norm of the operator A, i.e.*

$$\|A^*\| = \|A\| .$$

(See also Example 2, where this relation is obvious.)

Definition 1. If $A = A^*$, then A is called a *self-adjoint operator.* Then for every $x \in H$, $y \in H$ we have

$$(Ax, y) = (x, Ay) . \tag{7}$$

Example 3. An integral operator with a real symmetric kernel, for which therefore $K(s, t) = \overline{K(t, s)}$, is a self-adjoint operator (see Example 2).

Theorem 3. *A necessary and sufficient condition for an operator A in a (complex) Hilbert space to be self-adjoint is that (Ax, x) be real for every $x \in H$.*

Definition 2. A self-adjoint operator A is called *positive* or *positive definite,* if for every $x \in H$ we have $(Ax, x) \geqq 0$, while $(Ax, x) = 0$ only for $x = 0$, or $(Ax, x) \geqq m\|x\|^2$ with $m > 0$, respectively.

REMARK 3. Let us consider the operator equation (to be solved in the space H; cf. Remark 22.5.11)

$$Ax - \lambda x = y \quad (\text{or briefly } (A - \lambda E)\, x = y), \tag{8}$$

where E is the identity operator, i.e. $Ex = x$) and also the corresponding homogeneous equation

$$Ax - \lambda x = 0 \quad (\text{or briefly } (A - \lambda E)\, x = 0). \tag{9}$$

A value λ such that the operator $A - \lambda E$ possesses in H a bounded inverse operator (denoted by R_λ) is called a *regular value* (*regular point*) of the operator A; the operator R_λ is called a *resolvent operator* (or *resolvent*). For every regular value λ and every right-hand side $y \in H$ equation (8) has exactly one solution $x \in H$. In particular equation (9) possesses only the trivial solution $x = 0$.

Those values of λ which are not regular constitute the so-called *spectrum* of the operator A.

A value λ such that equation (9) has a non-zero solution is called an *eigenvalue* (*characteristic value*) of the operator A, and the non-zero solutions of equation (9) are called *eigenvectors* (*characteristic vectors, characteristic elements*) corresponding to λ.

Theorem 4. *If A is a self-adjoint operator, then:*

1. *Any complex number λ is a regular value of the operator A. The operator A can thus have only real eigenvalues. If, in addition, A is positive* (Definition 2), *then A can have only positive eigenvalues.*
2. *Eigenvectors corresponding to different eigenvalues are orthogonal.*

Theorem 5. *If A is a self-adjoint absolutely continuous non-zero operator* (Definition 22.5.7), *then A has at least one non-zero eigenvalue λ. Outside every interval $[-\varepsilon, \varepsilon]$, where ε is an arbitrary positive number, there can lie only a finite number of eigenvalues, and to each of them there corresponds only a finite number of linearly independent eigenvectors. An orthonormal system of elements formed by all linearly independent eigenvectors of the operator A including eigenvectors corresponding to the eigenvalue $\lambda = 0$* (Remark 4), *is complete in H.*

REMARK 4. By Theorem 4, the eigenvectors corresponding to different eigenvalues are orthogonal. The linearly independent eigenvectors corresponding to the same number λ can be orthonormalized. Thus, the orthonormal system obtained in this manner is complete by Theorem 5. Because of this fact we often say that a self-adjoint absolutely continuous operator possesses a *pure point spectrum.*

Example 4. If $K(t, s) \in L_2$ is a real symmetric kernel (see Example 3), then the assertions of Theorems 4 and 5 are true for the integral equation

$$\int_a^b K(t, s)\, x(s)\, ds = \lambda\, x(t)\,. \tag{10}$$

There is no uniformity in the terminology and notation used in functional analysis on the one hand and in the theory of integral equations on the other. In the theory of integral equations, λ usually denotes what we have here denoted by $1/\lambda$ (see Chap. 19). The notation used here, however, is more convenient for several reasons (for example, it permits us to formulate easily the assertion on completeness of the orthonormal system of eigenvectors in Theorem 5, etc.).

A convention sometimes adopted is to call the λ of equation (10) a *characteristic value* and its reciprocal an *eigenvalue.*

REMARK 5. Now let A be an *absolutely contiunous* (not necessarily self-adjoint) operator (for example, an integral operator with kernel $K(t, s) \in L_2$, which need not satisfy the equality $\overline{K(x, t)} = K(t, s)$), and let A^* be the adjoint operator to A. Let us consider the equations

$$Ax - x = y\,, \tag{11}$$

$$A^*x - x = y\,, \tag{12}$$

$$Ax - x = 0\,, \tag{13}$$

$$A^*x - x = 0\,. \tag{14}$$

Theorem 6 (*The Fredholm Alternative*). *Let A be an absolutely continuous operator; then equation* (11) *possesses a unique solution for any element* $y \in H$, *if and only if equation* (13) *has only a zero solution. The same statement is true for equations* (12) *and* (14). *If equation* (13) *has a non-zero solution, then equation* (11) *is solvable (but not uniquely) only for those* $y \in H$ *which are orthogonal to all solutions of equation* (14). *A similar statement is true for equations* (14) *and* (12). *Both equations* (13) *and* (14) *have the same number of linearly independent solutions.*

(b) Unbounded Operators

REMARK 6. In Example 22.5.2 an unbounded operator was given. In the following text we shall consider (unbounded) operators defined in linear manifolds $D(A)$ which are *dense* in H. These cases are often encountered in applications of variational methods to the solution of boundary value problems related to ordinary and partial differential equations (Chap. 24).

Example 5. Let us consider the operator

$$Ax = \frac{\mathrm{d}^2 x}{\mathrm{d}t^2} \tag{15}$$

with a domain of definition $D(A)$ consisting of all functions $x(t)$ with continuous second derivatives $x''(t)$ in $[a, b]$, such that $x(a) = 0$, $x(b) = 0$. These functions constitute a linear manifold in H, H being the space L_2 with the inner product (1). It can be shown that this linear manifold is dense in H.

Similarly, the operator

$$Ax = \Delta x = \frac{\partial^2 x}{\partial t_1^2} + \frac{\partial^2 x}{\partial t_2^2}$$

can be defined in a dense linear manifold $D(A)$ consisting of functions $x(t_1, t_2)$ with continuous second derivatives in a closed bounded region $\overline{\Omega}$ with a smooth (or piecewise smooth) boundary, which satisfy the condition $x = 0$ on the boundary of $\overline{\Omega}$. The inner product in H is defined by the relation

$$(x, y) = \iint_{\Omega} x(t_1, t_2)\, \overline{y(t_1, t_2)}\, \mathrm{d}t_1\, \mathrm{d}t_2\,.$$

REMARK 7. For some elements $y \in H$ there exists a corresponding element $y^* \in H$ such that

$$(Ax, y) = (x, y^*) \tag{16}$$

for every element $x \in D(A)$. (The elements $y = 0$, $y^* = 0$, for example, have the property just mentioned.) Since the linear manifold $D(A)$ is dense in H (Remark 6), it can be shown that in such a case the element y^* is uniquely determined by the element y. Thus, on the set (say M) of all these y an operator A^* with

$$y^* = A^* y\,, \quad y \in M \tag{17}$$

is given such that

$$(Ax, y) = (x, A^* y) \tag{18}$$

for every $x \in D(A)$ and every $y \in M = D(A^*)$. The operator A^* (which, clearly, is additive and homogeneous, and unbounded in general) is called the *adjoint operator to A*. If $A = A^*$ (i.e., if $D(A^*) = D(A)$ and $A^* y = Ay$ for every $y \in D(A)$), A is called a *self-adjoint (unbounded) operator*.

We recall the fact that $D(A)$ is assumed to be a linear manifold *dense* in H as stated in Remark 6.

Definition 3. An operator A is called *symmetric*, if the relation

$$(Ax, y) = (x, Ay) \tag{19}$$

holds for every x and y from $D(A)$ (where $D(A)$ is a linear manifold dense in H; see Remark 6).

Definition 4. A symmetric operator A is called *positive*, or *positive definite*, respectively, if $(Ax, x) \geqq 0$ holds for every $x \in D(A)$, while $(Ax, x) = 0$ only for $x = 0$, or if $(Ax, x) \geqq m\|x\|^2$ with $m > 0$ for every $x \in D(A)$.

Example 6. Let A be the operator

$$Ax = -\frac{d^2x}{dt^2}, \tag{20}$$

where $D(A)$ is a linear manifold consisting of all functions $x(t)$ having a continuous second derivative in $[a, b]$ and satisfying the boundary conditions

$$x(a) = 0, \quad x(b) = 0 \tag{21}$$

(see Example 5). Then A is a (symmetric) positive operator. For, on integrating by parts and using conditions (21), we have

$$(Ax, y) = -\int_a^b x''\bar{y}\,dt = -[x'\bar{y}]_a^b + \int_a^b x'\bar{y}'\,dt = \int_a^b x'\bar{y}'\,dt =$$

$$= [x\bar{y}']_a^b - \int_a^b x\bar{y}''\,dt = -\int_a^b x\bar{y}''\,dt = (x, Ay)$$

for every $x \in D(A)$, $y \in D(A)$; further

$$(Ax, x) = -\int_a^b x''\bar{x}\,dt = -[x'\bar{x}]_a^b + \int_a^b x'\overline{x'}\,dt = \int_a^b x'\overline{x'}\,dt = \int_a^b |x'|^2\,dt \geqq 0 \tag{22}$$

for every $x \in D(A)$, while if $(Ax, x) = 0$, then obviously $x = 0$, as follows from (22) and (21).

REMARK 8. It can be shown that the operator just considered is, moreover, positive definite. An analogous assertion is true for the operator $Ax = -\Delta x$ defined in the linear manifold given in Example 5.

For a symmetric operator we have $A \subset A^*$, i.e. $D(A) \subset D(A^*)$ and $A^*x = Ax$ for $x \in D(A)$. (This fact follows immediately from the definition of a symmetric and adjoint operator.)

Remark 9. It can be shown that every symmetric positive definite operator can be extended so that the operator obtained is self-adjoint. This fact is important, for example, in solving problems of differential equations by variational methods (Chap. 24), as self-adjoint unbounded operators possess many valuable properties which permit the use of these methods. (For example, Theorem 4, formerly stated for bounded self-adjoint operators, remains true for self-adjoint unbounded operators; a modified version of Theorem 5 holds for certain suitably extended differential operators.)

23. CALCULUS OF VARIATIONS

By FRANTIŠEK NOŽIČKA

References: [8], [27], [39], [40], [41], [53], [75], [96], [122], [128], [146], [163], [165], [231], [245], [266], [312], [313], [401]

The calculus of variations has numerous applications in physics, especially in mechanics. Usually, the problem is to find – from among functions possessing certain properties – those for which the given integral (functional) depending on the given function and its derivatives, assumes its maximum or minimum values.

It is convenient to divide the problems of the calculus of variations into certain categories which depend upon the type of functional whose extremum is to be found and for which a general theoretical method of solution has been worked out.

A. PROBLEMS OF THE FIRST CATEGORY (ELEMENTARY PROBLEMS OF THE CALCULUS OF VARIATIONS)

23.1. Curves of the r-th Class, Distance of Order r between Two Curves, ε-Neighbourhood of Order r of a Curve

Definition 1. A curve K in the plane described by the equation

$$y = y(x)\,, \quad x \in [a, b]\,, \tag{1}$$

is called a *curve of the r-th class in the interval* $[a, b]$ if the function $y(x)$ has continuous derivatives $y', y'', \ldots, y^{(r)}$ in $[a, b]$.

Definition 2. By the *distance between two curves*

$$y = y_1(x)\,, \quad y = y_2(x)\,, \quad x \in [a, b]\,, \tag{2}$$

we mean the number

$$d(y_1, y_2) = \max_{x \in [a,b]} |y_1(x) - y_2(x)| ,$$

i.e. the maximum of the differences of the y coordinates for all x in $[a, b]$ (Fig. 23.1).

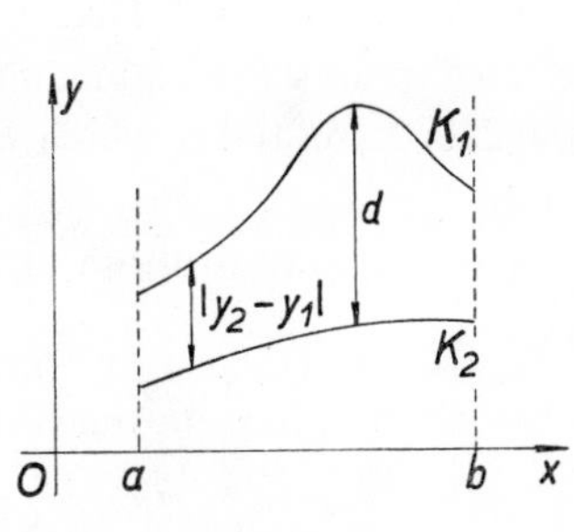

Fig. 23.1.

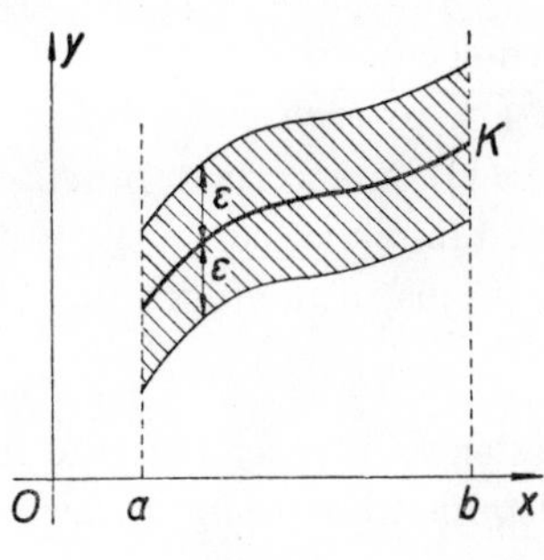

Fig. 23.2.

Definition 3. If the curves (2) are of the r-th class, then the greatest of the numbers

$$\max_{x \in [a,b]} |y_1(x) - y_2(x)| , \quad \max_{x \in [a,b]} |y_1'(x) - y_2'(x)|, \ldots, \quad \max_{x \in [a,b]} |y_1^{(r)}(x) - y_2^{(r)}(x)|$$

is called the *distance of order r* between these curves (we thus take into consideration the distances between not only the functions, but also their derivatives up to order r).

REMARK 1. By an *ε-neighbourhood of a curve* K described by (1) we mean all the curves

$$y = \tilde{y}(x), \quad x \in [a, b] , \tag{3}$$

whose distance (i.e. the distance of order zero in the sense of the previous definition) from the curve K is less than ε ($\varepsilon > 0$). Hence the ε-neighbourhood of the curve K is the family (set) of all the curves (3) which lie in the zone indicated in Fig. 23.2. More generally, by an *ε-neighbourhood of order r of a curve* K we mean all the curves (3) of the r-th class whose distance of order r from the curve K is less than ε. Obviously each curve which lies in an ε-neighbourhood of order r of the curve K lies also in its ε-neighbourhood of order zero.

23.2. Extrema of Functionals of the Form $\int_a^b F(x, y, y')\,\mathrm{d}x$

Let $F(x, u, v)$ be a function continuous for $x \in [a, b]$, $u \in (-\infty, \infty)$, $v \in (-\infty, \infty)$. Let T_1 be the family of all curves (23.1.1) of the first class in $[a, b]$. Then the function

$$\Phi(x) = F(x, y(x), y'(x))$$

is continuous in $[a, b]$, and hence the integral

$$I = \int_a^b F(x, y, y')\,dx \tag{1}$$

defines a certain number I for each curve belonging to T_1. This number I depends on the choice of the curve $y = y(x)$ from the class T_1, i.e. $I = I(y(x))$. Hence, by the relation (1), to each curve belonging to T_1 there corresponds a certain number I. Thus $I(y(x))$ is a certain *functional* (see Remark 22.5.2, p. 1007) which has the considered family of curves as its domain of definition.

The extrema of the functional (1) are defined in a similar way to those of a function:

Definition 1. We say that the functional (1) takes on its *absolute maximum* or *absolute minimum along a curve* K_0,

$$y = y_0(x), \quad x \in [a, b], \tag{2}$$

of a family of curves T_1, if the inequality

$$I(y(x)) \leqq I(y_0(x)) \quad \text{or} \quad I(y(x)) \geqq I(y_0(x)) \tag{3}$$

holds, respectively, for all curves belonging to T_1.

Definition 2. If there exists an ε-neighbourhood (of order zero) of the curve K_0 such that (3) holds for all curves belonging to T_1 which lie in this neighbourhood, we say that the functional (1) takes on a *strong relative maximum* or a *strong relative minimum for* T_1 *along the curve* K_0, respectively.

Definition 3. If there exists an ε-neighbourhood of the first order of the curve K_0 such that (3) holds for all curves belonging to this neighbourhood, we say that the functional (1) takes on a *weak relative maximum* or a *weak relative minimum for* T_1 *along the curve* K_0, respectively.

Obviously the following theorem holds:

Theorem 1. *An absolute extremum is at the same time a strong and a weak relative extremum. A strong relative extremum is at the same time a weak relative extremum. (It is obvious that these assertions do not hold conversely.)*

REMARK 1. The previous fact implies the following result: *In order to find necessary conditions for an extremum of the functional* (1), *it is sufficient to find necessary conditions for its weak relative extremum.*

REMARK 2. Sufficient conditions for the existence of an extremum of a functional are, even in the simplest cases, rather complicated and we are not going to state them in the following paragraphs. In mechanical and physical problems whose problematic is connected with the calculus of variations, this existence of an extremum follows from the formulation of the problem.

23.3. The Variation of a Function and the Variation of the Functional *I*

Let the functional (23.2.1) be given for the family of all the curves belonging to the class T_1 in $[a, b]$ and let the function F – regarded as a function of three variables – have continuous derivatives of the second order in the region $[a, b] \times (-\infty, \infty) \times \times (-\infty, \infty)$.

Definition 1. Let K_0 be a fixed curve belonging to the class T_1, described by the equation $y = y_0(x)$. Let K be an arbitrary curve belonging to T_1, described by the equation $y = y(x)$. Then the difference

$$\delta y_0(x) = y(x) - y_0(x) \tag{1}$$

is called the *variation of the function* $y_0(x)$ *in* $[a, b]$.

Hence the variation of a function is also a function which depends on the choice of the curves K_0, K.

Let us denote

$$I(y(x)) = \int_a^b F(x, y(x), y'(x))\,\mathrm{d}x\,, \tag{2}$$

$$I(y_0(x)) = \int_a^b F(x, y_0(x), y_0'(x))\,\mathrm{d}x\,. \tag{2'}$$

From (1), (2), (2′), we obtain the following expression for the difference $I(y(x)) - - I(y_0(x))$:

$$I(y(x)) - I(y_0(x)) = \int_a^b \{F(x, y_0 + \delta y_0, y_0' + \delta y_0') - F(x, y_0, y_0')\}\,\mathrm{d}x\,, \tag{3}$$

where $\delta y_0' = y'(x) - y_0'(x)$.

If we denote the distance of the first order between the curves K, K_0 in $[a, b]$ by $d^1(y, y_0)$, then – using the mean value theorem for the function $F(x, y, y')$ – we can rewrite the right-hand side of (3) thus:

$$\int_a^b \{F(x, y_0 + \delta y_0, y_0' + \delta y_0') - F(x, y_0, y_0')\}\,\mathrm{d}x =$$

$$= \int_a^b \left\{\frac{\partial F}{\partial y}(x, y_0, y_0')\,\delta y_0 + \frac{\partial F}{\partial y'}(x, y_0, y_0')\,\delta y_0'\right\}\mathrm{d}x + d^1(y, y_0)\,\eta\,, \tag{4}$$

where η is a function depending in general on $y(x)$, $y_0(x)$ which tends to zero as $d^1 \to 0$. Hence the expression

$$\delta I(y_0) = \int_a^b \left\{ \frac{\partial F}{\partial y}(x, y_0, y_0') \, \delta y_0 + \frac{\partial F}{\partial y'}(x, y_0, y_0') \, \delta y_0' \right\} dx \tag{5}$$

is equal to the increment $I(y(x)) - I(y_0(x))$ in the functional I, if we neglect terms of order higher than $d^1(y, y_0)$.

Definition 2. The expression $\delta I(y_0)$ defined in (5) is called the *variation of the functional* (23.2.1) *along the curve* K_0.

Hence $\delta I(y_0)$ is also a functional depending on $y_0(x)$ and $\delta y_0(x)$. It is the principal part of the increment $I(y(x)) - I(y_0(x))$ of the functional I and it is linear in δy_0 and $\delta y_0'$.

REMARK 1. Besides the preceding definition of the variation of the functional I, we can introduce another definition which, in the case considered, is equivalent. Let the curve K_0 described by the equation $y = y_0(x)$ and the curve K described by the equation $y = y(x)$ be chosen from the class T_1. Let δy_0, $\delta y_0'$ preserve their meaning defined above and let I be the functional (23.2.1). Let us define a function of one variable t in the following way:

$$\Phi(t) = I(y_0 + t\delta y_0) = \int_a^b F(x, y_0 + t\delta y_0, y_0' + t\delta y_0') \, dx \, . \tag{6}$$

The increment

$$I(y_0 + t\delta y_0) - I(y_0) = \Phi(t) - \Phi(0)$$

can be expressed – according to (4) – as follows:

$$I(y_0 + t\delta y_0) - I(y_0) =$$

$$= t \int_a^b \left\{ \frac{\partial F}{\partial y}(x, y_0, y_0') \, \delta y_0 + \frac{\partial F}{\partial y'}(x, y_0, y_0') \, \delta y_0' \right\} dx + d^1(y_0, y_0 + t\delta y_0) \, \eta_1 \, ,$$

where d^1/t is a bounded function and $\eta_1 \to 0$ as $t \to 0$. Hence, from (6) and from (5) it follows that

$$\lim_{t\to 0} \frac{\Phi(t) - \Phi(0)}{t} = \Phi'(0) = \lim_{t\to 0} \frac{I(y_0 + t\delta y_0) - I(y_0)}{t} =$$

$$= \int_a^b \left\{ \frac{\partial F}{\partial y}(x, y_0, y_0') \, \delta y_0 + \frac{\partial F}{\partial y'}(x, y_0, y_0') \, \delta y_0' \right\} dx = \delta I(y_0) \, .$$

Hence, the variation δI of the functional I along the curve K_0 is equal to the derivative of the function $\Phi(t)$ with respect to t at the point $t = 0$; consequently, it is possible to *define* the variation of the functional I as the derivative of the function $\Phi(t)$ in (6) when $t = 0$.

REMARK 2. For the variation of the functional (23.2.1) along a certain curve $y = y(x)$ belonging to the class T_1, it is customary to use the notation:

$$\delta I(y) = \int_a^b \{F'_y(x, y, y')\,\delta y + F'_{y'}(x, y, y')\,\delta y'\}\,dx\,. \tag{7}$$

From among all the curves described by the equation $y = \bar{y}(x)$, where $\bar{y}(x) \neq y(x)$ and where $\bar{y}(x)$, $y(x)$ belong to the class T_1 in $[a, b]$, let us consider only

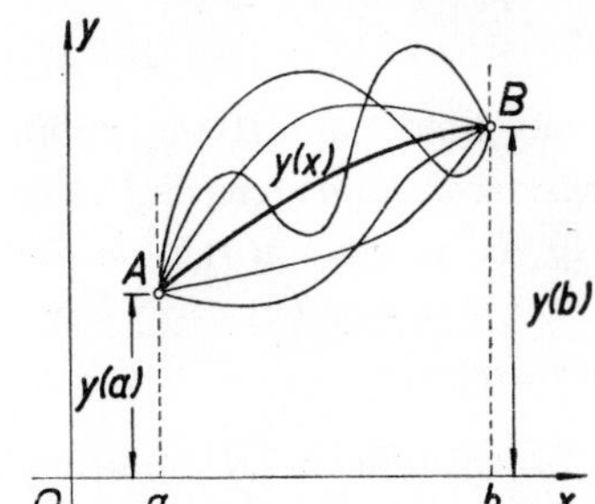

Fig. 23.3.

those curves which satisfy $\bar{y}(a) = y(a)$, $\bar{y}(b) = y(b)$ (Fig. 23.3). Then we have

$$\delta y(a) = \bar{y}(a) - y(a) = 0\,,$$
$$\delta y(b) = \bar{y}(b) - y(b) = 0\,. \tag{8}$$

For those curves K, belonging to the class T_1, which satisfy the boundary conditions (8), the variation of the functional can be written (after integrating (7) by parts) either in the form

$$\delta I(y) = \int_a^b \left\{F'_y(x, y, y') - \frac{d}{dx} F'_{y'}(x, y, y')\right\} \delta y\,dx \tag{9}$$

which is called the *Lagrange form of the variation of the functional* (23.2.1), or in the form

$$\delta I(y) = \int_a^b \{F'_{y'} - N\}\,\delta y'\,dx\,, \quad \text{where} \quad N(x) = \int_a^x F'_y(x, y, y')\,dx \tag{9'}$$

which is called the *Du Bois–Reymond form of the variation.*

23.4. The Necessary Condition for an Extremum of the Functional I

Theorem 1. *Let all the partial derivatives of the second order of the function $F(x, y, y')$ be continuous for $a \leqq x \leqq b$ and for arbitrary y and y'. If a weak* *relative extremum of the functional*

$$I = \int_a^b F(x, y, y')\, \mathrm{d}x$$

occurs along the curve $y = y_0(x)$ *from the set of all curves* $y = y(x)$ *belonging to the class* T_1 *in* $[a, b]$ *which satisfy* $y(a) = y_0(a)$, $y(b) = y_0(b)$, *then the function* $y_0(x)$ *satisfies the differential equation*

$$F'_y - \frac{\mathrm{d}}{\mathrm{d}x} F'_{y'} = 0\,. \tag{1}$$

Further, the following assertions hold: *At all the points of the curve* $y = y_0(x)$ *for which* $F''_{y',y'} \neq 0$, *there exists continuous second derivative* $y''(x)$. *If the curve* $y = y_0(x)$ *gives a minimum of the functional* $I(y)$ *on the set of curves considered, then* $F''_{y',y'} \geqq 0$ *along this curve; if* $y = y_0(x)$ *gives a maximum of the functional* $I(y)$, *then* $F''_{y',y'} \leqq 0$ *along this curve.*

Equation (1) is called the *Euler differential equation* and every solution $y(x)$, $x \in [a, b]$ of this equation is called an *extremal of the given variational problem*; the Euler differential equation (1) is a second-order differential equation; its solution contains two arbitrary constants which may be determined by making use of the boundary conditions.

23.5. Special Cases of the Euler Equation. The Brachistochrone Problem

1. If the given functional has the form

$$I(y) = \int_a^b F(x, y')\, \mathrm{d}x\,,$$

then we have $F'_y = 0$ and the Euler equation (23.4.1) reduces to

$$\frac{\mathrm{d}}{\mathrm{d}x} F'_{y'} = 0\,,$$

whence

$$F'_{y'} = C \quad (C = \text{const.})$$

is the first integral of the Euler equation.

2. If the given functional has the form

$$I(y) = \int_a^b F(y, y')\, \mathrm{d}x$$

(assuming that the function F has continuous second-order derivatives and that $F''_{y',y'} \neq 0$), we have

$$\frac{d}{dx}(F(y, y') - y'F'_{y'}) = F'_y y' + F'_{y'} y'' - y''F'_{y'} - y' \frac{d}{dx} F'_{y'} = y' \left(F'_y - \frac{d}{dx} F'_{y'} \right).$$

If we exclude the case $y' = 0$, i.e. $y = \text{const.}$, then

$$F'_y - \frac{d}{dx} F'_{y'} = 0 \Leftrightarrow \frac{d}{dx}(F(y, y') - y'F'_{y'}) = 0 .$$

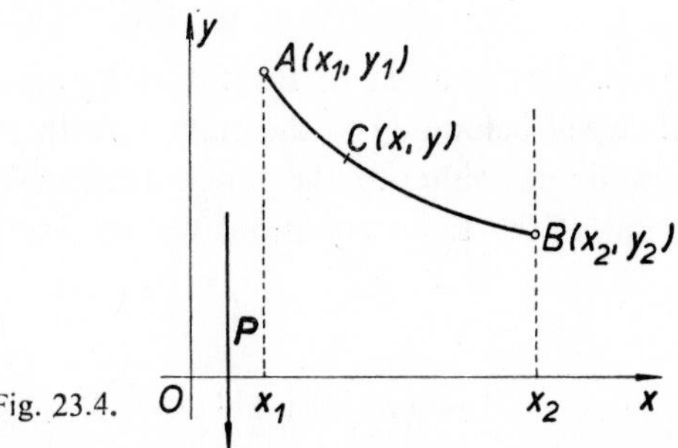

Fig. 23.4.

Whence

$$F(y, y') - y'F'_{y'}(y, y') = C \quad (C = \text{const.}) \tag{1}$$

is the first integral of the Euler equation.

Example 1 (*The Brachistochrone Problem*). From among all the curves joining the points $A(x_1, y_1)$, $B(x_2, y_2)$, $y_1 > y_2$, $x_1 \neq x_2$ (Fig. 23.4), let us find the curve along which a particle moving from rest at the point A to the point B in the gravitation field (described by the vector with components $P_x = 0$, $P_y = -mg$, where g is the gravitational acceleration) reaches the point B in the shortest time. (Practically: we have to find the shape of a groove joining the points A, B such that a particle sliding from A along the groove under the influence of gravitation (friction being neglected) reaches the point B in the shortest possible time.)

Since the force with which the groove acts on the particle is perpendicular to the direction of motion and, consequently, does no work, it follows from the principle of conservation of energy that

$$\tfrac{1}{2}mv^2 = mg(y_1 - y) ,$$

i.e.

$$v^2 = 2g(y_1 - y) . \tag{2}$$

At the same time, we have

$$v = \frac{ds}{dt} = \frac{\sqrt{(1 + y'^2)}\, dx}{dt}$$

along the curve $y = y(x)$ joining the points A, B; here s denotes the arc of this curve, $y' = dy/dx$, t denotes time. Hence, using (2) we obtain

$$dt = \frac{ds}{v} = \frac{\sqrt{(1 + y'^2)}}{\sqrt{[2g(y_1 - y)]}}\, dx\,.$$

This implies that the total time T necessary to traverse the curve is

$$T = \int_{x_1}^{x_2} \sqrt{\frac{1 + y'^2}{2g(y_1 - y)}}\, dx\,. \tag{3}$$

The given physical problem is thus reduced to the following variational problem: To find – from among all curves belonging to the class T_1 with end points A, B – the curve which gives the minimum value to the functional (3). This functional is of type 2. As the first integral of the Euler equation, we obtain, using (1) and (3),

$$\sqrt{\frac{1 + y'^2}{2g(y_1 - y)}} - y' \frac{y'}{\sqrt{[2g(y_1 - y)]}} \frac{1}{\sqrt{(1 + y'^2)}} = C\,,$$

hence, after rearrangement,

$$1 = C \sqrt{[2g(y_1 - y)]}\, \sqrt{(1 + y'^2)}\,,$$

i.e.

$$1 = 2C^2\, g(y_1 - y)\,(1 + y'^2)\,, \quad C \neq 0\,.$$

Let us put $K = 1/2gC^2$. Then we can write

$$\frac{K}{1 + y'^2} = y_1 - y\,. \tag{4}$$

By the parameter method (in the same way as in Remark 17.5.3 and in Example 17.5.2 (pp. 755, 756)), we obtain (putting

$$y' = \tan\varphi\,, \quad \text{so that} \quad y = y_1 - K\cos^2\varphi)$$

the parametric equations of the required extremal:

$$x = \tfrac{1}{2}K(\sin 2\varphi + 2\varphi) + C\,, \quad y = y_1 - \tfrac{1}{2}K(1 + \cos 2\varphi)\,.$$

The constants K, C are determined by the condition that the extremal passes through the points $A(x_1, y_1)$, $B(x_2, y_2)$. An easy analysis leads to the conclusion that the extremal is a cycloid. The physical character of the problem indicates that it makes the functional (3) have a minimum.

B. PROBLEMS OF THE SECOND CATEGORY (EXTREMA OF FUNCTIONALS OF THE FORM $\int_a^b F(x, y_1, \ldots, y_n, y'_1, \ldots, y'_n)\,dx$)

23.6. Fundamental Concepts and Definitions

Definition 1. Let a curve

$$y_1 = y_1(x),\ y_2 = y_2(x),\ \ldots,\ y_n = y_n(x)\,,\quad x \in [a, b]\,, \tag{1}$$

be given in an $(n + 1)$-dimensional Euclidean space E_{n+1}. We say that the *curve* (1) *belongs to the class* T_r $(r \geqq 1)$ if the functions $y_1(x), y_2(x), \ldots, y_n(x)$ have continuous derivatives up to the r-th order in $[a, b]$.

Definition 2. Let K_1, K_2 be two curves lying in E_{n+1} and belonging to T_r, described by the equations

$$K_1\colon\ y_i = y_i(x)\,,\ i = 1, \ldots, n\quad (x \in [a, b])\,,$$

$$K_2\colon\ y_i = \tilde{y}_i(x)\,,\ i = 1, \ldots, n\quad (x \in [a, b])\,.$$

By the *distance of order r between the curves* K_1, K_2 we mean the maximum of the values

$$|\tilde{y}_i(x) - y_i(x)|,\ |\tilde{y}'_i(x) - y'_i(x)|,\ \ldots,\ |\tilde{y}_i^{(r)}(x) - y_i^{(r)}(x)|$$

for $i = 1, \ldots, n$, $x \in [a, b]$.

The *ε-neighbourhood of order r of the curve* (1) *belonging to the class* T_r *in* E_{n+1} is defined in a manner similar to that of a curve in a plane (§ 23.1).

23.7. Formulation of the Variational Problem

Let $F(x, y_1, \ldots, y_n, y'_1, \ldots, y'_n)$ be a function of the variables $x, y_1, \ldots, y_n, y'_1, \ldots, y'_n$ which has continuous partial derivatives of the second order with respect to all its arguments in the region

$$a \leqq x \leqq b\,,\quad y_i \in (-\infty, \infty)\,,\quad y'_i \in (-\infty, \infty)\quad \text{for}\quad i = 1, \ldots, n\,.$$

The problem is as follows:

From among all the curves (23.6.1) belonging to the class T_1 in E_{n+1} with common end points A, B (on each curve the point A corresponds to the value $x = a$, the point B to the value $x = b$), to find the curve which makes the functional

$$I = \int_a^b F(x, y_1, \ldots, y_n, y'_1, \ldots, y'_n)\,dx \tag{1}$$

have an extremum.

REMARK 1. The concepts of *absolute extremum*, *relative strong extremum* and *relative weak extremum of the functional* (1) are defined in a manner similar to §23.2. In the following we shall give only necessary conditions which have to be satisfied along each curve (belonging to the class of curves under consideration) which gives an extremum of the functional (1).

23.8. Necessary Conditions for an Extremum of the Functional I

Theorem 1. *Let K be a fixed curve described by equations* (23.6.1), *belonging to the class T_1 in E_{n+1}, with end points $A(a, y_1(a), \ldots, y_n(a))$, $B(b, y_1(b), \ldots, y_n(b))$. If this curve makes the functional* (23.7.1) *have an extremum on the set of all curves* (23.6.1) *belonging to the class T_1 with end points A, B (and if the function $F(x, y_1, \ldots \ldots, y_n, y'_1, \ldots, y'_n)$ satisfies the assumptions stated in* § 23.7), *then the functions $y_i(x)$ satisfy the following system of differential equations*:

$$\frac{\partial F}{\partial y_i} - \frac{d}{dx}\frac{\partial F}{\partial y'_i} = 0\,, \quad i = 1, \ldots, n\,. \tag{1}$$

REMARK 1. The system of equations (1) is called the *system of Euler equations* belonging to the given variational problem; each curve (23.6.1) satisfying (1) is called an *extremal* of the given variational problem. The general solution of the system of differential equations (1) (if it exists) involves $2n$ arbitrary constants $c_1, c_2, \ldots, c_{2n}$; consequently, it has the form

$$y_i = y_i(x, c_1, c_2, \ldots, c_{2n})\,, \quad i = 1, 2, \ldots, n\,.$$

The constants $c_1, c_2, \ldots, c_{2n}$ are determined by making use of the boundary conditions, i.e. of the condition that the extremal passes through the points A, B.

Theorem 2. *If the curve K lying in E_{n+1} and described by* (23.6.1) *yields a minimum of the functional* (23.7.1) *on a given family of curves, then (provided that the assumptions formulated above are satisfied) the following inequalities hold at* *every point of the extremal K*:

$$F''_{y_1'y_1'} \geqq 0\,, \quad \begin{vmatrix} F''_{y_1'y_1'}, & F''_{y_1'y_2'} \\ F''_{y_2'y_1'}, & F''_{y_2'y_2'} \end{vmatrix} \geqq 0\,, \ \ldots,$$

$$\begin{vmatrix} F''_{y_1'y_1'}, & F''_{y_1'y_2'}, & \ldots, & F''_{y_1'y_n'} \\ F''_{y_2'y_1'}, & F''_{y_2'y_2'}, & \ldots, & F''_{y_2'y_n'} \\ \ldots & \ldots & \ldots & \ldots \\ F''_{y_n'y_1'}, & F''_{y_n'y_2'}, & \ldots, & F''_{y_n'y_n'} \end{vmatrix} \geqq 0\,. \tag{2}$$

If the extremal K gives a maximum of the functional considered, then the inequalities (2) hold at every point of K except that the inequality symbols are alternately $\leqq$, $\geqq$, $\leqq$, $\geqq$ ....

Example 1. Let us find the extremal of the functional

$$\int_0^{\pi/2} (y'^2 + z'^2 + 2yz)\,\mathrm{d}x$$

satisfying the following boundary conditions:

$$y(0) = 0\,, \quad z(0) = 0\,, \quad y(\tfrac{1}{2}\pi) = 1\,, \quad z(\tfrac{1}{2}\pi) = -1\,.$$

In this particular case, the system of Euler equations (1) reduces to two differential equations

$$y'' - z = 0\,, \quad z'' - y = 0\,. \tag{3}$$

If we differentiate the first of these equations twice with respect to x and apply the second one, we get the differential equation

$$y^{(4)} - y = 0$$

the general solution of which is

$$y = c_1 \mathrm{e}^x + c_2 \mathrm{e}^{-x} + c_3 \cos x + c_4 \sin x\,.$$

Hence, using the first of equations (3), it follows that

$$z = c_1 \mathrm{e}^x + c_2 \mathrm{e}^{-x} - c_3 \cos x - c_4 \sin x\,.$$

By making use of the given boundary conditions, we calculate that $c_1 = c_2 = c_3 = 0$, $c_4 = 1$. Hence the required extremal is

$$y = \sin x\,, \quad z = -\sin x \quad (x \in [0, \tfrac{1}{2}\pi])\,.$$

C. PROBLEMS OF THE THIRD CATEGORY (EXTREMA OF FUNCTIONALS OF THE TYPE $\int_a^b F(x, y, y', \ldots, y^{(n)})\,\mathrm{d}x$)

23.9. Formulation of the Problem

Many problems of physics lead to the mathematical problem of finding the extrema of functionals of the type

$$I(y(x)) = \int_a^b F(x, y, y', \ldots, y^{(n)})\,\mathrm{d}x\,, \tag{1}$$

where the integrand depends not only on the first derivative, but also on the derivatives of higher orders, of the function y. In general, the problem is : To find a curve yielding the extremum of the functional (1) in the class of all the curves described by

$$y = y(x)\,, \quad x \in [a, b]\,, \tag{2}$$

in E_2 which belong to the class T_n and satisfy the conditions

$$\begin{aligned} y(a) &= y_0\,, \quad y'(a) = y'_0, \ldots, y^{(n-1)}(a) = y_0^{(n-1)}\,; \\ y(b) &= y_1\,, \quad y'(b) = y'_1, \ldots, y^{(n-1)}(b) = y_1^{(n-1)}\,, \end{aligned} \tag{3}$$

where $y_0, y'_0, \ldots, y_0^{(n-1)}, y_1, y'_1, \ldots, y_1^{(n-1)}$ are given numbers.

23.10. A Necessary Condition for the Extremum of the Functional (23.9.1)

Under the assumption that the function $F(x, y, y', \ldots, y^{(n)})$ in (23.9.1) has continuous partial derivatives of order at least $n + 1$ with respect to all its arguments $x, y, y', \ldots, y^{(n)}$ in the region

$$x \in [a, b]\,, \quad y^{(k)} \in (-\infty, \infty) \quad \text{for} \quad k = 0, 1, \ldots, n\,,$$

the following theorem is true:

Theorem 1. *If the curve K described by (23.9.2), belonging to the class T_n in E_2, makes the functional (23.9.1) have an extremum on the family of all curves belonging to T_n which satisfy (23.9.3), then the equation*

$$F'_y - \frac{\mathrm{d}}{\mathrm{d}x} F'_{y'} + \frac{\mathrm{d}^2}{\mathrm{d}x^2} F'_{y''} - \ldots + (-1)^n \frac{\mathrm{d}^n}{\mathrm{d}x^n} F'_{y^{(n)}} = 0 \tag{1}$$

is satisfied at all points of the curve K.

REMARK 1. The differential equation (1) is called the *Euler–Poisson equation.* In the general case (if $F''_{y^{(n)}y^{(n)}} \neq 0$), it is a differential equation of order $2n$ and its solution involves $2n$ arbitrary constants. Hence the solution of equation (1) has the form

$$y = f(x, \alpha_1, \beta_1, \alpha_2, \beta_2, \ldots, \alpha_n, \beta_n)\,.$$

The arbitrary constants can be determined by making use of the $2n$ conditions (23.9.3) at the end points of the curve. Each solution of equation (1) is called an *extremal of the given variational problem.*

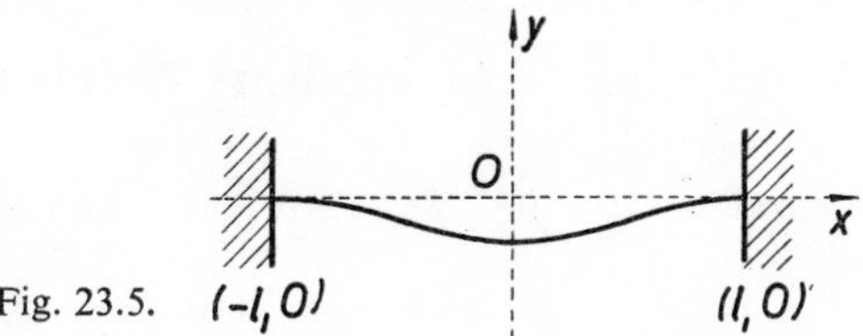

Fig. 23.5.

Example 1. The following problem arises in the theory of elasticity: A cylindrical solid beam, homogeneous and elastic, has its ends fixed at the same height above the ground (Fig. 23.5) and we have to determine the profile of the deflected centre line of the beam.

For the total potential energy E of the beam the formula

$$E = \int_{-l}^{l} \left(\tfrac{1}{2}\mu \frac{y''^2}{1 + y'^2} + \varrho y \sqrt{(1 + y'^2)} \right) \mathrm{d}x$$

can be derived, where μ, ϱ are constants which characterize the given beam. If we assume the deflection of the beam to be very small in comparison with its length, then $1 + y'^2 \approx 1$ and so the formula for the potential energy may be written

$$E = \int_{-l}^{l} (\tfrac{1}{2}\mu y''^2 + \varrho y)\, \mathrm{d}x\,. \tag{2}$$

Now – according to a well-known principle of physics – if a mechanical system is in stable equilibrium, then its total potential energy is minimal. Hence our physical problem leads to the variational problem of finding the minimum of the functional (2) with boundary conditions

$$y(-l) = y(l) = 0\,, \quad y'(-l) = y'(l) = 0\,. \tag{3}$$

In the case of the functional (2) the Euler-Poisson equation reduces to the equation

$$\varrho + \frac{\mathrm{d}^2}{\mathrm{d}x^2}\, \mu y'' = 0\,,$$

i.e.

$$\varrho + \mu y^{(4)} = 0\,. \tag{4}$$

The general solution of the differential equation (4) depends on four arbitrary constants α, β, γ, δ and has the form

$$y = \alpha x^3 + \beta x^2 + \gamma x + \delta - \frac{\varrho}{24\mu} x^4 .$$

Using the boundary conditions (3) we obtain

$$y = \frac{\varrho}{24\mu}(-x^4 + 2l^2x^2 - l^4) .$$

23.11. Generalization to the Case of an Arbitrary Number of Functions

A more general problem similar to the problem formulated in § 23.9 is to find the extremum of the functional

$$I(y_1, \ldots, y_m) = \int_a^b F(x, y_1, y_1', \ldots, y_1^{(n_1)}, y_2, y_2', \ldots, y_2^{(n_2)}, \ldots, y_m, y_m', \ldots, y_m^{(n_m)})\, dx$$

on the family of functions $y_1(x), y_2(x), \ldots, y_m(x)$, where $y_i(x)$ belongs to at least the class T_{n_i} for $x \in [a, b]$, $i = 1, \ldots, m$, satisfying the boundary conditions

$$y_i^{(k)}(a) = y_{i0}^{(k)} , \quad y_i^{(k)}(b) = y_{i1}^{(k)}$$

for $k = 0, 1, \ldots, n_i - 1$ and $i = 1, \ldots, m$. In this general case, we obtain the following *Euler–Poisson system of equations*:

$$\sum_{k=0}^{n_i} (-1)^k \frac{d^k}{dx^k} F'_{y_i^{(k)}} = 0 , \quad i = 1, \ldots, m ,$$

which give necessary conditions for an extremum of the considered functional I.

D. PROBLEMS OF THE FOURTH CATEGORY (FUNCTIONALS DEPENDING ON A FUNCTION OF *n* VARIABLES)

23.12. Fundamental Concepts and Definitions

We shall consider a bounded region O with a boundary H lying in an n-dimensional Euclidean space E_n with Cartesian coordinates $x_1, x_2, \ldots, x_n$.

Definition 1. We say that a *function* $\varphi(x_1, \ldots, x_n)$ *belongs to the class* T_r *in* $\bar{O} =$

$= O + H$ if all the partial derivatives of φ up to and including those of order r are continuous in $\bar{O}$.

Definition 2. Let $\varphi(x_1, \ldots, x_n)$ belong to the class T_r in $\bar{O}$. Let E_{n+1} be the $(n+1)$-dimensional Euclidean space with Cartesian coordinates $x_1, \ldots, x_n, u$. We say that the equation

$$u = \varphi(x_1, \ldots, x_n)\,, \quad (x_1, \ldots, x_n) \in \bar{O} \tag{1}$$

represents a *regular hypersurface in* E_{n+1}.

Definition 3. Let N_1, N_2 be two regular hypersurfaces in E_{n+1} with the following equations:

$$N_1\colon\ u = \varphi(x_1, \ldots, x_n)\,, \quad (x_1, \ldots, x_n) \in \bar{O}\,,$$
$$N_2\colon\ u = \psi(x_1, \ldots, x_n)\,, \quad (x_1, \ldots, x_n) \in \bar{O}\,.$$

By the *distance (of order zero) between the hypersurfaces* N_1, N_2 we mean the number

$$d(N_1, N_2) = \max_{(x_1, \ldots, x_n) \in \bar{O}} |\varphi(x_1, \ldots, x_n) - \psi(x_1, \ldots, x_n)|\,;$$

by the *distance of the first order between the hypersurfaces* N_1, N_2 *in* $\bar{O}$, we mean the number

$$d^1(N_1, N_2) = \max\Big(\max_{(x_1, \ldots, x_n) \in \bar{O}} |\varphi - \psi|,\ \max_{(x_1, \ldots, x_n) \in \bar{O}} |\varphi_1 - \psi_1|, \ldots, \max_{(x_1, \ldots, x_n) \in \bar{O}} |\varphi_n - \psi_n|\Big)\,,$$

where

$$\varphi_i = \frac{\partial \varphi}{\partial x_i}\,, \quad \psi_i = \frac{\partial \psi}{\partial x_i}\,, \quad i = 1, \ldots, n\,.$$

The family of all the regular hypersurfaces

$$\tilde{N}\colon\ u = \tilde{\varphi}(x_1, \ldots, x_n)\,, \quad (x_1, \ldots, x_n) \in \bar{O}\,, \tag{2}$$

which have a distance from the regular hypersurface (1) smaller than ε $(\varepsilon > 0)$ is called the *ε-neighbourhood of the hypersurface N in $\bar{O}$*. In a similar way, by the *ε-neighbourhood of the first order of the regular hypersurface* (1) we mean the family of all the hypersurfaces $\tilde{N}$ in $\bar{O}$ which satisfy $d^1(\tilde{N}, N) < \varepsilon$.

Let the function $F(x_i, \varphi, \varphi_i) = F(x_1, \ldots, x_n, \varphi, \partial\varphi/\partial x_1, \ldots, \partial\varphi/\partial x_n)$ have continuous partial derivatives with respect to all of its $2n + 1$ arguments $x_1, \ldots, x_n, \varphi, \varphi_1, \ldots, \varphi_n$ of at least the second order in the domain $(x_1, \ldots, x_n) \in \bar{O}$, $\varphi \in (-\infty, \infty)$, $\varphi_i \in (-\infty, \infty)$ for $i = 1, \ldots, n$. Let us define the functional

$$I(\varphi) = \int \ldots \int_O F(x_i, \varphi, \varphi_i)\, \mathrm{d}x_1 \ldots \mathrm{d}x_n \tag{3}$$

on the family of all the hypersurfaces belonging to the class T_1 in $\bar{O}$.

Definition 4. We say that the functional (3) takes on its *absolute minimum* (or *absolute maximum*) on the set of regular hypersurfaces (2) belonging to the class T_1 in $\bar{O}$, along the regular hypersurfaces (1), if

$$I(\tilde{\varphi}) \geqq I(\varphi), \quad \text{or} \quad I(\tilde{\varphi}) \leqq I(\varphi), \tag{4}$$

respectively, for each hypersurface of the considered set.

If an ε-neighbourhood of order zero of the regular hypersurface (1) exists such that (4) holds for each regular hypersurface belonging to this ε-neighbourhood, we say that the functional (3) takes on a *strong relative minimum*, or a *strong relative maximum*, respectively, along the hypersurface (1). If (4) holds for an ε-neighbourhood of the first order of the hypersurface (1), we speak of a *weak relative minimum*, or a *weak relative maximum*, respectively, of the functional (3).

Obviously, necessary conditions for a weak relative extremum are at the same time necessary conditions for a strong relative extremum and for an absolute extremum.

23.13. Formulation of the Variational Problem and Necessary Conditions for an Extremum

The variational problem (under the assumptions stated above) is formulated in the following way: From among all the hypersurfaces belonging to the class T_1 in $\bar{O}$ which are described by an equation $u = \varphi(x_1, \ldots, x_n)$, satisfying the boundary condition

$$\varphi(x_1, \ldots, x_n) = f(x_1, \ldots, x_n) \quad \text{for} \quad (x_1, \ldots, x_n) \in H\,, \tag{1}$$

where f is a continuous function given on the boundary H of the region O, we have to find a hypersurface which makes the functional (23.12.3) have an extremum.

Necessary conditions for an extremum of the functional (23.12.3) are given in the following theorem:

Theorem 1. *Let the regular hypersurface* (23.12.1) *make the functional* (23.12.3) *have an extremum on the set of all the hypersurfaces belonging to the class T_1 in O which satisfy the boundary conditions* (1). *Let the function F under the integration sign in* (23.12.3) *have continuous partial derivatives with respect to all its arguments of at least the second order in the domain*

$$(x_1, \ldots, x_n) \in \bar{O}\,, \quad \varphi \in (-\infty, \infty)\,, \quad \varphi_i \in (-\infty, \infty) \quad (i = 1, \ldots, n)\,.$$

Then the equation

$$F'_\varphi - \sum_{i=1}^{n} \frac{\partial}{\partial x_i} F'_{\varphi_i} = 0\,, \tag{2}$$

where

$$F'_\varphi = \frac{\partial F}{\partial \varphi}, \quad F'_{\varphi i} = \frac{\partial F}{\partial \varphi_i} \quad \textit{for} \quad i = 1, \ldots, n,$$

holds at all the points of the hypersurface (23.12.1).

REMARK 1. The partial differential equation (2) is called the *Euler–Ostrogradski equation* and each of its solutions is called an *extremal hypersurface* (an *extremal n-dimensional variety*) of the variational problem considered.

Example 1. The integral

$$I(\varphi) = \int \ldots \int_O \left(\sum_{i=1}^{n} \varphi_i^2\right) \mathrm{d}x_1 \ldots \mathrm{d}x_n$$

is called the *n-dimensional Dirichlet integral.* Let a continuous function $f(x_1, \ldots, x_n)$ be given at the points of the boundary H of the region O. The Euler–Ostrogradski equation (2) for the extremum of this integral, on the set of all the hypersurfaces φ belonging to the class T_1 in $\bar{O}$ which satisfy the boundary condition $\varphi|_H = f|_H$, reduces in this special case to the form

$$\Delta\varphi = \sum_{i=1}^{n} \frac{\partial^2 \varphi}{\partial x_i^2} = 0,$$

i.e. to the Laplace's equation.

E. PROBLEMS OF THE FIFTH CATEGORY (VARIATIONAL PROBLEMS WITH "MOVABLE (FREE) ENDS OF ADMISSIBLE CURVES")

23.14. Formulation of the Simplest Problem

In variational problems belonging to the first three categories, the individual curves of the families considered all have the same prescribed end points. These problems can be generalized in a certain sense. In the following text, we are going to generalize, in a special way, a variational problem of the first category.

Let K_1, K_2 be two curves with the following description:

$$\begin{aligned} K_1&: \quad y = \varphi(x), \quad x \in [a_1, b_1], \\ K_2&: \quad y = \psi(x), \quad x \in [a_2, b_2], \end{aligned} \tag{1}$$

which belong to the class T_1 in their domain of definition. Further, let $F(x, y, y')$ be a function satisfying the assumptions stated in § 23.3 in the region

$$x \in [\min(a_1, a_2), \max(b_1, b_2)],$$
$$y \in (-\infty, \infty), \quad y' \in (-\infty, \infty).$$

The problem is: From among all the curves K belonging to the class T_1 with a description $y = y(x)$, which have one end point $A(x_1, y_1)$ on the curve K_1 and the other

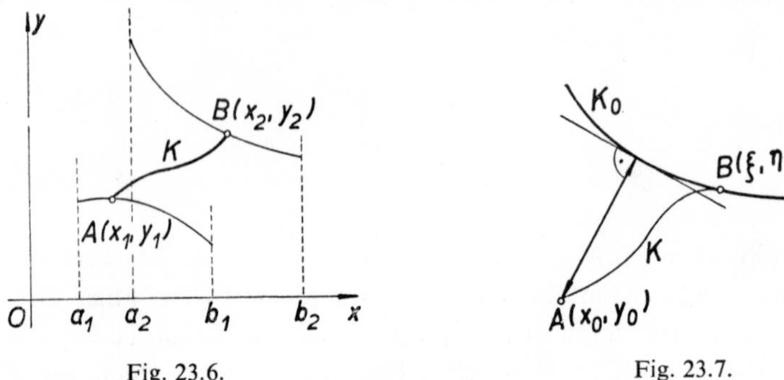

Fig. 23.6. Fig. 23.7.

end point $B(x_2, y_2)$ on the curve K_2 (so that A, B are not fixed chosen points, but A is some point on the curve K_1 and B is some point on the curve K_2), we have to find the curve which makes the functional

$$I(K) = \int_{x_1}^{x_2} F(x, y, y')\,\mathrm{d}x \tag{2}$$

have an extremum (see Fig. 23.6).

23.15. Necessary Conditions for an Extremum

Theorem 1. *If the curve K with the description $y = y(x)$ and with end points $A(x_1, y_1)$, $B(x_2, y_2)$ lying on the curves K_1 and K_2, respectively (K_1 and K_2 are described by (23.14.1), yields an extremum of the functional (23.14.2) with respect to all the curves belonging to the class T_1 with movable end points lying on K_1 and K_2, respectively, then the Euler equation*

$$F'_y - \frac{\mathrm{d}}{\mathrm{d}x} F'_{y'} = 0 \tag{1}$$

is satisfied at all the points of the curve K; moreover, at the end points A, B of this curve we have

$$\{F + (\varphi' - y') F'_{y'}\}_A = 0, \quad \{F + (\psi' - y') F'_{y'}\}_B = 0. \tag{1'}$$

REMARK 1. The conditions (1′) are called the *transversality conditions* of the given variational problem. The symbols $\{\ \}_A$ and $\{\ \}_B$ denote that the coordinates x_1, y_1 of the point A and the coordinates x_2, y_2 of the point B, respectively, are to be substituted into the expression in brackets.

REMARK 2 (*A More General Form of the Transversality Conditions*). Let the curves K_1, K_2 be implicitly described by the equations

$$\Phi(x, y) = 0\,,$$
$$\Psi(x, y) = 0\,,$$

where the functions Φ, Ψ have continuous partial derivatives of at least the first order and the inequalities

$$\left(\frac{\partial \Phi}{\partial x}\right)^2 + \left(\frac{\partial \Phi}{\partial y}\right)^2 > 0\,, \quad \left(\frac{\partial \Psi}{\partial x}\right)^2 + \left(\frac{\partial \Psi}{\partial y}\right)^2 > 0$$

hold at each point (x, y) of the curves K_1, K_2. Then the transversality conditions have the form

$$\begin{aligned} \left\{(F - y'F'_{y'})\frac{\partial \Phi}{\partial y}\right\}_A &= \left\{F'_{y'}\frac{\partial \Phi}{\partial x}\right\}_A\,, \\ \left\{(F - y'F_{y'})\frac{\partial \Psi}{\partial y}\right\}_B &= \left\{F'_{y'}\frac{\partial \Psi}{\partial x}\right\}_B\,. \end{aligned} \tag{2}$$

In particular, if the curves K_1, K_2 are straight lines parallel to the y-axis, then the conditions (2) reduce to

$$\{F'_{y'}\}_A = 0\,, \quad \{F'_{y'}\}_B = 0\,. \tag{3}$$

In the case where the family of curves considered has one common (fixed) end point, while the second end point lies on a given curve, one of the two transversality conditions in (1′) or, alternatively, in (2) or (3) no longer applies.

Example 1. Let K_0 be a curve $y = \varphi(x)$ belonging to the class T_1. We wish to find the smallest distance of a given fixed point $A(x_0, y_0)$ from this curve (where A does not lie on K_0).

We consider all curves K belonging to the class T_1, described by the equation $y = y(x)$, with one end point at the fixed point $A(x_0, y_0)$ and the other end point $B(\xi, \eta)$ moving on the given curve K_0 (Fig. 23.7). The length of a typical curve is

$$I(y) = \int_{x_0}^{\xi} \sqrt{(1 + y'^2)}\, dx\,. \tag{4}$$

Hence we have to find an extremum of the functional (4) with respect to all the curves $y = y(x)$ belonging to the class T_1 which have one end point at the fixed point A

and the other end point on the given curve K_0. In this case, the Euler equation reduces to the equation

$$\frac{d}{dx}\frac{y'}{\sqrt{(1 + y'^2)}} = 0$$

which implies that $y' = c$ ($c = \text{const.}$) and $y = cx + d$ ($d = \text{const.}$). Consequently, the extremal is a straight line. This straight line passes through the point $A(x_0, y_0)$, and consequently the unknown constants c, d satisfy the relation

$$y_0 = cx_0 + d\,. \tag{5}$$

The transversality condition gives in this case

$$\sqrt{(1 + y'^2)} + (\varphi' - y')\frac{y'}{\sqrt{(1 + y'^2)}} = 0\,,$$

i.e.

$$1 + \varphi' y' = 0 \tag{6}$$

at the point B of the extremal lying on the given curve K_0. Hence, it is obvious from (6) that in this particular case the transversality condition is at the same time a condition of orthogonality. At the point B, we have $y' = c$, $\varphi'|_B = \varphi'(\xi)$. Under the assumption $\varphi'(\xi) \neq 0$, (6) implies that

$$c = -\frac{1}{\varphi'(\xi)}\,.$$

The value of the constant d in (5) follows by making use of the condition $\varphi(\xi) = c\xi + d$, i.e.

$$d = \varphi(\xi) - c\xi = \varphi(\xi) + \frac{\xi}{\varphi'(\xi)}\,.$$

With the aid of condition (5) written in the form

$$y_0 = -\frac{1}{\varphi'(\xi)}x_0 + \varphi(\xi) + \frac{\xi}{\varphi'(\xi)} = \frac{1}{\varphi'(\xi)}(\xi - x_0) + \varphi(\xi)$$

we evaluate the first coordinate ξ of the point B; its second coordinate is then given by the relation $\eta = \varphi(\xi)$.

In a similar way, we can determine the smallest distance between two given curves in E_2.

REMARK 3. Variational problems with "movable (free) ends of admissible curves" can be easily generalized to the case of functionals of the form (23.7.1), i.e. functionals defined for curves in a space of arbitrary dimension. Similar problems also arise in the case of functionals depending on derivatives of higher order. The transversality conditions in these two cases are naturally more complicated (see, e.g. [41], [53], [96]).

F. PROBLEMS OF THE SIXTH CATEGORY (THE ISOPERIMETRIC PROBLEM IN THE SIMPLEST CASE)

23.16. Formulation of the Problem

Let $F(x, y, y')$ and $\Phi(x, y, y')$ be two given functions which have continuous second-order partial derivatives with respect to all of their arguments in the region $a \leqq x \leqq b$, $y \in (-\infty, \infty)$, $y' \in (-\infty, \infty)$. The problem is: From among all the curves described by $y = y(x)$, $x \in [a, b]$ and belonging to the class T_1 in $[a, b]$ which keep the functional

$$L(y) = \int_a^b \Phi(x, y, y')\,dx \tag{1}$$

constant, to find the curve on which the functional

$$I(y) = \int_a^b F(x, y, y')\,dx \tag{2}$$

assumes its extremum. This problem can be solved by the procedure shown in the following paragraph, assuming that the curve to be found is not an extremal of the functional $L(y)$ defined in (1), and can be regarded as a basic type of the so-called *category of isoperimetric problems.*

23.17. A Necessary Condition for an Extremum

Theorem 1. *If, making the asumptions of* § 23.16, *the curve K described by* $y = y(x)$ $(x \in [a, b])$ *yields an extremum of the functional*

$$I(y) = \int_a^b F(x, y, y')\,dx$$

with respect to all the curves $y = \tilde{y}(x)$, $x \in [a, b]$, *belonging to the class* T_1 *in* $[a, b]$ *and satisfying the conditions*

$$\text{(i)} \quad \tilde{y}(a) = y(a) = \alpha\,, \quad \tilde{y}(b) = y(b) = \beta\,,$$

$$\text{(ii)} \quad L(y) = \int_a^b \Phi(x, \tilde{y}, \tilde{y}')\,\mathrm{d}x = C\,,$$

where α, β, C *are given constants, and if the curve* $y = y(x)$ *is not an extremal of the functional* $L(y)$, *then a number* λ *exists such that the curve* K *is an extremal of the functional*

$$H(y) = \int_a^b [F(x, y, y') + \lambda\Phi(x, y, y')]\,\mathrm{d}x\,. \tag{1}$$

REMARK 1. Theorem 1 gives a method of solving the isoperimetric problem outlined in § 23.16. The problem reduces to a problem of the first category, i.e. to the problem of finding an extremum of the integral (1) satisfying given conditions. Hence, the extremals of the given variational problem satisfy the Euler differential equation for the extremum of the functional (1), i.e. the equation

$$F'_y - \frac{\mathrm{d}}{\mathrm{d}x} F'_{y'} + \lambda\left(\Phi'_y - \frac{\mathrm{d}}{\mathrm{d}x}\Phi'_{y'}\right) = 0\,. \tag{2}$$

Example 1. From among all the curves described by $y = y(x)$, $x \in [a, b]$, belonging to the class T_1 in $[a, b]$, a curve satisfying the following conditions is to be found:

(a) the curve joins the points $A(a, 0)$, $B(b, 0)$, where $b > a$;
(b) its length is l, where $l > b - a$;
(c) the area enclosed by it and the segment AB is a maximum (Fig. 23.8).

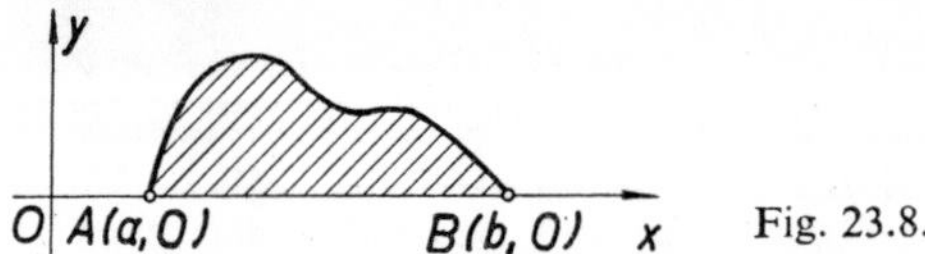

Fig. 23.8.

The area enclosed by the segment AB and the curve $y = y(x)$, $x \in [a, b]$, is given by the integral

$$I(y) = \int_a^b |y|\,\mathrm{d}x\,, \tag{3}$$

the length of the curve by the integral

$$\int_a^b \sqrt{(1 + y'^2)}\,\mathrm{d}x\,.$$

According to our assumptions, we have

$$L(y) = \int_a^b \sqrt{(1 + y'^2)}\, dx = l\,. \tag{4}$$

The boundary conditions for the class of admissible curves are

$$y(a) = y(b) = 0\,. \tag{5}$$

(Notice that the conditions of § 23.16 are not precisely fulfilled; however, a more detailed analysis shows that the given method can be used.) In this particular case, the Euler equation (2) reduces to the form (see § 23.5)

$$|y| + \lambda \sqrt{(1 + y'^2)} - y'\lambda \frac{y'}{\sqrt{(1 + y'^2)}} = \alpha \quad (\alpha = \text{const.})$$

and hence, after rearrangement,

$$|y| \sqrt{(1 + y'^2)} + \lambda = \alpha \sqrt{(1 + y'^2)}\,;$$

this implies that

$$|y| = \alpha - \frac{\lambda}{\sqrt{(1 + y'^2)}}\,. \tag{6}$$

Put $y' = \tan \varphi$, so that

$$|y| = \alpha - \lambda |\cos \varphi|\,. \tag{7}$$

Hence, by differentiating with respect to x, we get

$$y' \operatorname{sign} y = \lambda \sin \varphi \,(\operatorname{sign} \cos \varphi) \frac{d\varphi}{dx}$$

which implies, after substituting $y' = \tan \varphi$, the equation

$$dx = (\operatorname{sign} y)(\operatorname{sign} \cos \varphi)\, \lambda \cos \varphi \, d\varphi$$

and hence

$$x = (\operatorname{sign} y)(\operatorname{sign} \cos \varphi)\, \lambda \sin \varphi + \beta\,. \tag{8}$$

After eliminating the parameter φ between (7) and (8), we obtain the equation

$$(x - \beta)^2 + (|y| - \alpha)^2 = \lambda^2\,,$$

i.e.

$$(x - \beta)^2 + (y - \gamma)^2 = \lambda^2\,, \tag{9}$$

where we have put $\gamma = \alpha \operatorname{sign} y$ for brevity. (9) implies that the required curve is a part of the circumference of a circle. The constants β, γ, λ in (9) are determined by the given conditions (a), (b) of our problem.

REMARK 2. Isoperimetric problems can be generalized to isoperimetric problems with "movable end points of admissible curves", and also to the case of functionals depending upon curves in a space of arbitrary dimension and to functions depending upon higher order derivatives.

FINAL REMARK. Categories I to VI do not cover completely the wide field of problems connected with the calculus of variations. We could mention various other categories, as for instance the so-called *constrained extrema* on more general curves than those belonging to the class T_1, or further variational problems in which the admissible curves or varieties are defined parametrically (see references at the beginning of this chapter, p. 1020).

24. VARIATIONAL (DIRECT) METHODS FOR SOLVING BOUNDARY VALUE PROBLEMS OF DIFFERENTIAL EQUATIONS

By Milan Práger

References: [20], [67], [68], [69], [75], [154], [214], [282], [298], [300], [374], [403].

24.1. Transformation of Boundary Value Problems into Problems of Finding the Minimum of a Quadratic Functional. Eigenvalue Problems

In some cases, boundary value problems of differential equations can be transformed into problems of finding the minimum of a certain functional. If the given differential equation is linear, the corresponding functional is quadratic. Using direct variational methods, we look for the solution of the given boundary value problem as for a function – belonging to a certain class of functions – which makes the corresponding functional minimal. It is necessary to choose this class sufficiently wide in order that the corresponding functional actually attains its minimum in this class; then it is necessary to make clear in what sense the function obtained satisfies the differential equation and the given boundary conditions.

These investigations are easily carried out in the space L_2 of real square integrable functions on the given region Ω (or interval I). In the space $L_2(\Omega)$, the scalar product is given by the well-known formula

$$(u, v) = \int_\Omega uv \, \mathrm{d}\Omega \, .$$

By the space L_2, throughout this chapter, we mean the space L_2 of *real* functions. Let us recall (Chap. 22), that by a linear set (linear manifold) in this space we mean a set M of functions of this space with the property that if it contains the functions u_1, $u_2, \ldots, u_n$, then it also contains any linear combination $u = c_1u_1 + c_2u_2 + \ldots + c_nu_n$ of them. An operator A defined on the linear set M is called *symmetric*, if for each pair of functions u, v of M, the relation

$$(Au, v) = (u, Av)$$

holds. The operator is called *positive*, if for every non-identically vanishing function $u \in M$ the relation

$$(Au, u) > 0$$

is satisfied (see Example 1). In this chapter, we shall assume that the linear set M is *dense* in the space considered (see § 22.2).

Theorem 1. *Let A be a symmetric positive operator on the linear set M of the space L_2 and let f be a given and u an unknown function. If the equation*

$$Au = f \tag{1}$$

has a solution in M (which is then unique, under assumptions of this theorem), then this solution produces on M the minimum of the functional

$$F(u) = (Au, u) - 2(u, f) \tag{2}$$

and conversely, the function producing the minimum of this functional on M, satisfies equation (1).

Example 1. Let us consider the problem

$$-u'' + k^2 u = f\,, \quad k \text{ real}\,, \tag{3}$$

$$u(a) = 0\,, \quad u(b) = 0\,. \tag{4}$$

For the linear set M, we choose all functions u having continuous derivatives of the second order in the interval $[a, b]$ and satisfying conditions (4). It is easy to show that the operator A, given by

$$Au = -u'' + k^2 u\,,$$

is symmetric and positive on M: We have to verify that

$$(Au, v) = (Av, u) \tag{5}$$

for all pairs u, v of M and that

$$(Au, u) > 0 \tag{6}$$

is satisfied for every non-identically vanishing function $u \in M$. In fact, for all $u \in M$,

$v \in M$, we have

$$(Au, v) = \int_a^b (-u'' + k^2 u)\, v \,\mathrm{d}x = -\int_a^b u'' v \,\mathrm{d}x + \int_a^b k^2 uv \,\mathrm{d}x =$$

$$= -\left[u'v\right]_a^b + \int_a^b u'v' \,\mathrm{d}x + \int_a^b k^2 uv \,\mathrm{d}x =$$

$$= \left[uv'\right]_a^b - \int_a^b uv'' \,\mathrm{d}x + \int_a^b k^2\, uv \,\mathrm{d}x =$$

$$= \int_a^b u(-v'' + k^2 v) \,\mathrm{d}x = (u, Av)\,, \tag{7}$$

since $\left[u'v\right]_a^b = 0$ and $\left[uv'\right]_a^b = 0$ as a consequence of (4). Further, putting $v = u$, and making use of (4) we deduce from (7) that

$$(Au, u) = \int_a^b u'^2 \,\mathrm{d}x + k^2 \int_a^b u^2 \,\mathrm{d}x\,. \tag{8}$$

This expression is non-negative and, in view of (4), is zero if and only if $u \equiv 0$. Thus (5) and (6) are satisfied.

The functional corresponding to the problem (3), (4) according to Theorem 1 (using equation (8)), is

$$F(u) = \int_a^b (u'^2 + k^2 u^2) \,\mathrm{d}x - 2 \int_a^b uf \,\mathrm{d}x\,.$$

In Theorem 1, the existence either of the solution of equation (1) or of a function producing the minimum of the functional (2) on the set M is assumed, although it is important first to prove the existence of such a function and only then to try to determine it or an approximation to it.

In some cases, the class M of functions, in which the minimum of the functional (2) is to be found, may be too restrictive so that the problem of finding this minimum can appear to have no solution in M.

Theorem 2. *If the (symmetric) operator A is not only positive, but also positive definite on M (i.e. every function $u \in M$ satisfies $(Au, u) \geqq \gamma^2 \|u\|^2$, where γ is a positive constant* (cf. Definition 22.6.4, p. 1018)), *then it is possible to extend the operator A and the linear set M so that the minimum problem for the functional (2), and consequently equation (1), has a solution.*

This solution is then to be considered as a generalized solution of equation (1) and it must be made clear in what sense equation (1) is satisfied.

REMARK 1. Table 24.1 (pp. 1050, 1051) gives a survey of some positive definite operators frequently occurring in applications and indicates the corresponding functionals and conditions for the class of functions in which the minimum is sought. Among these operators, the operator A of Example 1 is included as a special case of the first problem of Table 24.1 taking $p(x) \equiv 1$, $r(x) \equiv k^2$.

The eigenvalue problems (§ 17.17, p. 801, § 18.7, p. 913) associated with boundary value problems of differential equations can also be solved by using direct methods:

Theorem 3. *Consider the equation*

$$Au - \lambda Bu = 0\,, \tag{9}$$

where A and B are symmetric positive definite operators (*with $D_A \subset D_B$*;* frequently $Bu \equiv u$) *and denote by d the g.l.b. of the functional*

$$R(u) = \frac{(Au, u)}{(Bu, u)}\,.$$

If there exists a function u_0 such that

$$R(u_0) = d\,,$$

then d is the least eigenvalue of equation (9) *and u_0 is the corresponding eigenfunction.* (*Moreover, from the positive definiteness of the operators A, B it follows that $d > 0$.*)

REMARK 2. Thus, the problem of the first eigenvalue of equation (9) is transformed into the minimum problem for the functional $R(u)$. The functional $R(u)$ is called *Rayleigh's quotient.* (See Theorem 17.17.5 and Remark 17.17.13, pp. 808, 809.)

Example 2. Rayleigh's quotient for the eigenvalue problem of the differential equation $-u'' - \lambda u = 0$ with boundary conditions $u(a) = u(b) = 0$ is

$$R(u) = \frac{\int_a^b u'^2 \, \mathrm{d}x}{\int_a^b u^2 \, \mathrm{d}x}\,.$$

The remaining eigenvalues of equation (9) may be found according to the following theorem:

Theorem 4. *With the same assumptions concerning the operators A and B as in Theorem 3 let $\lambda_1 \leqq \lambda_2 \leqq \ldots \leqq \lambda_n$ be the increasing sequence of the first n eigen-*

* D_A and D_B are domains of definition of operators A and B, respectively. In eigenvalue problems associated with differential equations, A and B are differential operators, the operator A being of higher order than the operator B; thus D_A consists of functions "more smooth" than the functions to which the operator B is applied. Consequently, the condition $D_A \subset D_B$ is quite natural.

values of equation (9) *and* $u_1, u_2, \ldots, u_n$ *the corresponding (linearly independent) eigenfunctions.** *Let there exist a function* u_{n+1} *that produces the minimum of the functional* $R(u)$ *under the supplementary conditions*

$$(Bu, u_1) = (Bu, u_2) = \ldots = (Bu, u_n) = 0\,.$$

Then u_{n+1} *is the eigenfunction of equation* (9) *corresponding to the eigenvalue*

$$\lambda_{n+1} = \frac{(Au_{n+1}, u_{n+1})}{(Bu_{n+1}, u_{n+1})}\,.$$

In the increasing sequence of eigenvalues, this eigenvalue is the one following λ_n.

Two types of boundary value problems. The boundary value problem generally means to solve a non-homogeneous differential equation with non-homogeneous boundary conditions. Usually it is possible to find a function v that either satisfies the non-homogeneous boundary conditions or is a particular solution of the given equation. Supposing the solution of the given problem to be of the form $u + v$, we can reduce the problem to one of the following types:

1. non-homogeneous equation with homogeneous boundary conditions,

2. homogeneous equation with non-homogeneous boundary conditions.

In the first case, the minimum problem for the functional (2) is solved in the class of functions with the same homogeneous boundary conditions. Dealing with problems of the second type, let us suppose that a function z exists which satisfies the prescribed non-homogeneous boundary conditions and for which the functional $F(z)$ has a finite value. Then for this fixed z we look for the minimum of the functional $F(z + u) = (Au, u) + 2(Az, u) + (Az, z)$, in the class of all functions u satisfying the corresponding homogeneous boundary conditions. In this way, these problems are transformed into problems of the first type and consequently in the sequel we consider only problems of the first type.

In the following text of this chapter, by "operator" a positive definite operator will always be understood, unless the contrary is stated.

A survey of the most important positive definite operators is given in Table 24.1. We note that functions, among which the minimum of the functional is sought, need not in all cases satisfy *all* the boundary conditions of the given problem. In such cases, the function, minimising the given functional, satisfies these conditions automatically (the so-called *natural boundary conditions*).

---

* It may happen that to the same eigenvalue there correspond more than one linearly independent eigenfunctions. In order to obtain a one-to-one correspondence between eigenvalues and the corresponding (linearly independent) eigenfunctions, the sign = between $\lambda_1, \lambda_2, \ldots$ is admitted.

TABLE 24.1

Differential equation	Boundary conditions	Corresponding functional	Boundary conditions for the class of functions, in which the minimum problem is solved
$-\dfrac{d}{dx}\left[p(x)\dfrac{du}{dx}\right] + r(x)\,u = f(x)$, where $p(x) \geqq p_0 > 0$, $r(x) \geqq 0$	$u(a) = u(b) = 0$	$\int_a^b (pu'^2 + ru^2 - 2fu)\,dx$	$u(a) = u(b) = 0$
the same $p(x) \geqq p_0 > 0$, $r(x) \geqq r_0 > 0$	$u'(a) = u'(b) = 0$	$\int_a^b (pu'^2 + ru^2 - 2fu)\,dx$	none
the same	$u'(a) - \alpha\,u(a) = 0$ $u'(b) + \beta\,u(b) = 0$ $\alpha > 0$, $\beta > 0$	$\alpha\,p(a)\,u^2(a) + \beta\,p(b)\,u^2(b) + {} + \int_a^b (pu'^2 + ru^2 - 2fu)\,dx$	$u'(a) - \alpha\,u(a) = 0$ $u'(b) + \beta\,u(b) = 0$
$\dfrac{d^2}{dx^2}\left[p(x)\dfrac{d^2u}{dx^2}\right] = f(x)$ $p(x) \geqq p_0 > 0$	$u(a) = u(b) = 0$ $u''(a) = u''(b) = 0$	$\int_a^b (pu''^2 - 2fu)\,dx$	$u(a) = u(b) = 0$
the same	$u(a) = u'(a) = 0$ $u''(b) = u'''(b) = 0$	the same	$u(a) = u'(a) = 0$
the same	$u(a) = u'(a) = 0$ $u(b) = u'(b) = 0$	the same	$u(a) = u'(a) = 0$ $u(b) = u'(b) = 0$

the same	$u(a) = u'(a) = 0$ $u(b) = u''(b) = 0$	the same	$u(a) = u'(a) = 0$ $u(b) \;= 0$
$\sum_{k=0}^{m} (-1)^k \frac{\mathrm{d}^k}{\mathrm{d}x^k}\left[p_k(x)\frac{\mathrm{d}^k u(x)}{\mathrm{d}x^k}\right] = f(x)$ $p_m(x) \geqq p_0 > 0$, $p_i(x) \geqq 0$ for $i = 0, \ldots, m-1$	$u(a) = u'(a) = \ldots$ $\ldots = u^{(m-1)}(a) = 0$ $u(b) = u'(b) = \ldots$ $\ldots = u^{(m-1)}(b) = 0$	$\int_a^b \left(\sum_{k=0}^{m} p_k \left(\frac{\mathrm{d}^k u}{\mathrm{d}x^k}\right)^2 - 2fu\right) \mathrm{d}x$	$u(a) = u'(a) = \ldots$ $\ldots = u^{(m-1)}(a) = 0$ $u(b) = u'(b) = \ldots$ $\ldots = u^{(m-1)}(b) = 0$
$-\Delta u = f(P)$	$u\vert_S = 0$	$\int_\Omega \left(\sum_{k=1}^{m}\left(\frac{\partial u}{\partial x_k}\right)^2 - 2fu\right) \mathrm{d}\Omega$	$u\vert_S = 0$
the same; solvable if $\int_\Omega f \,\mathrm{d}\Omega = 0$	$\left.\frac{\partial u}{\partial \nu}\right\vert_S = 0$	the same	none
the same	$\left.\frac{\partial u}{\partial \nu}\right\vert_S + \sigma(P)\, u\vert_S = 0$ $\sigma(P) \geqq 0$, $\sigma(P) \not\equiv 0$	$\int_\Omega \left(\sum_{k=1}^{m}\left(\frac{\partial u}{\partial x_k}\right)^2 - 2fu\right) \mathrm{d}\Omega + \int_S \sigma u^2 \,\mathrm{d}S$	$\left.\frac{\partial u}{\partial \nu}\right\vert_S + \sigma(P)\, u\vert_S = 0$
$\Delta^2 u = f(x, y)$	$u\vert_S = \left.\frac{\partial u}{\partial \nu}\right\vert_S = 0$	$\int_\Omega \left[\left(\frac{\partial^2 u}{\partial x^2}\right)^2 + 2\left(\frac{\partial^2 u}{\partial x\, \partial y}\right)^2 + \left(\frac{\partial^2 u}{\partial y^2}\right)^2 - 2fu\right] \mathrm{d}\Omega$	$u\vert_S = \left.\frac{\partial u}{\partial \nu}\right\vert_S = 0$

24.2. Approximate Methods. Minimizing Sequences

Definition 1. Let us denote by d the minimum of the functional $F(u)$ on the functions of class M (see Theorem 24.1.1). Any sequence $\{u_n\}$ of functions $u_n \in M$ that satisfies $\lim_{n\to\infty} F(u_n) = d$, is called a *minimizing sequence.*

If the operator A is symmetric and positive definite, then the minimizing sequence converges to the desired solution in such a way that $\lim_{n\to\infty} (A(u - u_n), (u - u_n)) = 0$, and hence, as a special case, $\lim_{n\to\infty} (u - u_n, u - u_n) = 0$. (From this, it easily follows that in the case of Poisson's equation $-\Delta u = f$, for example, the functions u_n and their first derivatives converge in the mean (cf. § 16.1), and in the case of the biharmonic equation the functions u_n and their first and second derivatives also converge in the mean, etc.)

Thus, the elements of the minimizing sequence can be taken as approximate solutions of the original boundary value problem, for, if we find u_n so that $F(u_n) - d < \varepsilon$, then $(A(u - u_n), (u - u_n)) < \varepsilon$, i.e. the function u_n is, in this sense, close to the exact solution u.

24.3. The Ritz Method

This method is a basic one for the construction of a minimizing sequence. We choose the sequence $\{\varphi_n\}$ of functions of the linear set M satisfying the given homogeneous boundary conditions and also the following assumptions:

1. φ_n are linearly independent;
2. $\{\varphi_n\}$ is a complete sequence with respect to A; that means, that any function $u \in M$ can be approximated by a linear combination $\sum_{i=1}^{n} a_i\varphi_i$ so that for every $\varepsilon > 0$, $(A(u - \sum_{i=1}^{n} a_i\varphi_i), u - \sum_{i=1}^{n} a_i\varphi_i) < \varepsilon$ holds.

The completeness of a sequence is equivalent to the following condition: If, for some function u, the equality $(Au, \varphi_n) = 0$ holds for all n, then u is necessarily identically zero. (On the choice of complete sequences see § 24.8, p. 1062.)

The elements u_n of the minimizing sequence are assumed to be of the form $u_n = \sum_{i=1}^{n} a_k\varphi_k$, where the constants a_k are to be determined in such a way that, for every fixed n, the functional $F(u)$ attains its minimum. The conditions $\partial F(u_n)/\partial a_j = 0$ ($j = 1, 2, \ldots, n$) for the extreme value yield a system of n linear algebraic equations for the unknown coefficients $a_1, a_2, \ldots, a_n$:

$$\sum_{k=1}^{n} a_k(A\varphi_k, \varphi_j) = (f, \varphi_j), \quad j = 1, 2, \ldots, n . \tag{1}$$

Under the above assumptions, this system always has a unique solution. The sequence constructed in this way is a minimizing one. Denoting by u_0 the exact solution, the estimate

$$\|u_n - u_0\| \leqq \frac{1}{\gamma} \sqrt{[F(u_n) - \delta]}$$

holds, where δ is a number less than d. (This estimate holds even if we give to δ the value d. However, the value of d is usually unknown and then its lower estimate δ must be used.) (Cf. § 24.2; on γ see Theorem 24.1.2.)

Example 1. Let us solve the equation

$$-\Delta u = 1$$

on the rectangle $0 \leqq x \leqq a$, $0 \leqq y \leqq b$, with the boundary condition $u|_S = 0$.

For the system $\{\varphi_n\}$, let us choose the functions

$$\varphi_{kl}(x, y) = \sin \frac{k\pi x}{a} \sin \frac{l\pi y}{b}, \quad k, l = 1, 2, \ldots \tag{2}$$

obviously satisfying the given boundary condition.

For the coefficients a_{kl} we obtain, according to (1), the system of equations

$$\sum_{k,l=1}^{m} a_{kl}(-\Delta\varphi_{kl}, \varphi_{rs}) = (1, \varphi_{rs}), \quad r, s = 1, 2, \ldots, m. \tag{3}$$

According to the well-known properties of orthogonality of the chosen functions φ_{kl}, we obtain on the left-hand side of any of the equations (3) only one term, so that for the given r, s equation (3) becomes

$$a_{rs} \int_0^a \int_0^b \left(\frac{r^2\pi^2}{a^2} + \frac{s^2\pi^2}{b^2}\right) \sin^2 \frac{r\pi x}{a} \sin^2 \frac{s\pi y}{b} \, dx \, dy = \int_0^a \int_0^b \sin \frac{r\pi x}{a} \sin \frac{s\pi y}{b} \, dx \, dy,$$

or

$$a_{rs} = \frac{16a^2b^2}{\pi^4 rs(b^2r^2 + a^2s^2)}$$

if both r and s are odd, and $a_{rs} = 0$ otherwise.

Thus the approximate solution is of the form

$$u_m = \sum_{k,l=1,3,\ldots,m} \frac{16a^2b^2}{\pi^4 kl(b^2k^2 + a^2l^2)} \sin \frac{k\pi x}{a} \sin \frac{l\pi y}{b}$$

from which, letting $m \to \infty$, we obtain the exact solution in the form of an infinite series. (Obviously, in the general case, it is not possible to compute all the coefficients of the series as simply as this. Usually, we have to reduce the computation to a finite number of equations and then solve them numerically.)

Now, let us find the least eigenvalue of the equation $Au - \lambda Bu = 0$, where A and B are symmetric positive definite operators. According to Theorem 24.1.3, this least eigenvalue equals $\min((Au, u)/(Bu, u))$, if the minimum is attained for some function u_0.

Solving this problem approximately by the Ritz method, we again choose a sequence $\{\varphi_n\}$ of functions of the linear set M, satisfying conditions 1 and 2 as above. Approximate solutions are again in the form $u_n = \sum_{k=1}^{n} a_k \varphi_k$ and the coefficients a_k are determined from the condition that (Au_n, u_n) should be minimal for $(Bu_n, u_n) = 1$. This condition leads (after some rearrangement) to the system of equations

$$\sum_{k=1}^{n} a_k[(A\varphi_k, \varphi_m) - \lambda(B\varphi_k, \varphi_m)] = 0\,, \quad m = 1, 2, \ldots, n\,.$$

In order that this system should have a non-zero solution, λ must satisfy the equation

$$\begin{vmatrix} (A\varphi_1, \varphi_1) - \lambda(B\varphi_1, \varphi_1), & (A\varphi_2, \varphi_1) - \lambda(B\varphi_2, \varphi_1), & \ldots \\ (A\varphi_1, \varphi_2) - \lambda(B\varphi_1, \varphi_2), & (A\varphi_2, \varphi_2) - \lambda(B\varphi_2, \varphi_2), & \ldots \\ \ldots & \ldots & \ldots \\ & & \ldots (A\varphi_n, \varphi_n) - \lambda(B\varphi_n, \varphi_n) \end{vmatrix} = 0\,. \tag{4}$$

Under the assumptions made, equation (4) has only real roots. Denoting by λ_n the least root of this equation, the sequence $\{\lambda_n\}$ converges as $n \to \infty$ to $d = \min((Au, u)/(Bu, u))$.

REMARK 1. The Ritz method does not ensure that d is the least eigenvalue of the given problem, i.e. that the minimum of $(Au, u)/(Bu, u)$ is, in fact, attained for some function u_0. This question must be solved in another way, e.g. by a transformation of the problem into an integral equation. For details see [282].

REMARK 2. The other eigenvalues are determined similarly; it is necessary to add further conditions of the type $(Bu_n, u_n^{(i)}) = 0$, where $u_n^{(i)}$ is an approximate value of the i-th eigenfunction. For details see [282].

24.4. The Kantorovitch Method

By this method, the solution of boundary value problems for partial differential equations in several dimensions is reduced to the solution of lower-dimensional (even one-dimensional) boundary value problems. In the latter case, the method leads to the solution of ordinary differential equations. The Kantorovitch method is formally similar to the Ritz method. The approximate solution is assumed to be of the form

$$\sum_{k=1}^{n} \varphi_k(x, y) f_k(x)$$

where $\varphi_k(x, y)$ are known functions satisfying the prescribed boundary conditions. (For simplicity we investigate only the two-dimensional problem.) The unknown functions $f_k(x)$ are determined so that the functional $F(u_n)$ attains its minimum. Instead of a system of linear algebraic equations, as was the case in the Ritz method, we obtain for the unknown functions f_k a boundary value problem for a system of ordinary differential equations in a certain interval $[a, b]$.

REMARK 1. If the given boundary conditions (of the two-dimensional problem) are homogeneous and if the boundary of the region considered includes the segments $x = a$, $x = b$, then it is not necessary to satisfy the boundary conditions on these segments by the proper choice of functions φ_k, since we may satisfy them instead by prescribing them as boundary conditions for the functions f_k. This may be convenient in some problems. For details see [282].

Example 1. We have to find the solution of the following boundary value problem:

$$\Delta u = -1$$

in the rectangle $R(|x| < 2, |y| < 1)$ and

$$u = 0$$

on the boundary.

We look for a solution of the form

$$\bar{u} = (1 - y^2) f(x) .$$

This solution satisfies the boundary condition for $y = \pm 1$. Substituting in the corresponding functional, we obtain

$$F(\bar{u}) = \iint_R \left[\left(\frac{\partial \bar{u}}{\partial x}\right)^2 + \left(\frac{\partial \bar{u}}{\partial y}\right)^2 - 2\bar{u} \right] \mathrm{d}x\, \mathrm{d}y =$$

$$= \int_{-2}^{2} \mathrm{d}x \int_{-1}^{1} [(1 - y^2)^2 f'^2 + 4y^2 f^2 - 2f(1 - y^2)]\, \mathrm{d}y .$$

Integrating with respect to y, we obtain

$$F(\bar{u}) = \int_{-2}^{2} \left(\tfrac{16}{15} f'^2 + \tfrac{8}{3} f^2 - \tfrac{8}{3} f\right) \mathrm{d}x .$$

The corresponding Euler equation (equation (23.4.1), p. 1026) is

$$f'' - \tfrac{5}{2}f = -\tfrac{5}{4},$$

with boundary conditions $f(\pm 2) = 0$.

The solution of this problem is

$$f(x) = \tfrac{1}{2}\left(1 - \frac{\cosh\sqrt{(\frac{5}{2})}\,x}{\cosh\sqrt{10}}\right).$$

Hence, for the approximate solution $\bar{u}$ we obtain

$$\bar{u}(x, y) = \tfrac{1}{2}(1 - y^2)\left(1 - \frac{\cosh\sqrt{(\frac{5}{2})}\,x}{\cosh\sqrt{10}}\right).$$

24.5. The Trefftz Method

Values of the functional $F(u)$ for approximate solutions u_n obtained by the Ritz method are greater than its minimum value. The Trefftz method makes it possible to find a sequence of functions that approach the minimum of the functional $F(u)$ from below. The idea of the Trefftz method consists in looking for approximate solutions, from among the appropriate class of functions that satisfy exactly the differential equation, but do not necessarily satisfy the prescribed boundary conditions (in contrast to the Ritz method, where approximate solutions satisfy the boundary conditions but not necessarily the differential equation). Let us restrict ourselves here to the case of the Dirichlet problem for the Laplace equation, although the Trefftz method can also be applied to other problems.

We have to find, in a region Ω, a harmonic function satisfying the boundary condition

$$u|_S = f(P),$$

where S is the boundary of Ω.

Let us consider a sequence of linearly independent functions φ_n, harmonic in Ω, which is complete in the following sense: For an arbitrary function φ, harmonic in Ω and square integrable together with its first derivatives, and for an arbitrary $\varepsilon > 0$, there exists a linear combination $\sum_{k=1}^{n} \alpha_k \varphi_k$ of functions φ_k, such that

$$\int_\Omega \sum_{i=1}^{m} \left[\frac{\partial(\varphi - \sum_{k=1}^{n} \alpha_k \varphi_k)}{\partial x_i}\right]^2 d\Omega < \varepsilon,$$

where m is the number of independent variables. Suppose an approximate solution in the form

$$u_n = \sum_{k=1}^{n} a_k \varphi_k$$

and let the coefficients a_k be found so that the integral

$$\int_\Omega \sum_{i=1}^{m} \left[\frac{\partial(u - u_n)}{\partial x_i}\right]^2 d\Omega\,,$$

where u is the unknown solution, is minimal. We obtain the system of equations

$$\int_\Omega \left(\sum_{i=1}^{m} \frac{\partial(u - u_n)}{\partial x_i}\,\frac{\partial \varphi_k}{\partial x_i}\right) d\Omega = 0\,, \quad k = 1, 2, \ldots, n\,. \tag{1}$$

This system is independent of the unknown function u. In fact, using Green's Theorem, we obtain from (1)

$$\sum_{j=1}^{n} a_j \int_S \varphi_j \frac{\partial \varphi_k}{\partial \nu}\, dS = \int_S f \frac{\partial \varphi_k}{\partial \nu}\, dS\,, \quad k = 1, 2, \ldots, n\,, \tag{2}$$

where ν is the outward normal to S.

This system always has a unique solution. The sequence of approximate solutions constructed in such a way converges to the exact solution in the following sense:

$$\lim_{n\to\infty} \int_\Omega \sum_{i=1}^{m} \left[\frac{\partial(u - u_n)}{\partial x_i}\right]^2 d\Omega = 0\,,$$

while at the same time

$$\int_\Omega \sum_{i=1}^{m} \left(\frac{\partial u_n}{\partial x_i}\right)^2 d\Omega \leqq \int_\Omega \sum_{i=1}^{m} \left(\frac{\partial u}{\partial x_i}\right)^2 d\Omega\,.$$

One can easily see from system (2) that the Trefftz approximation is determined apart from an additive constant. This constant may be determined in such a way that

$$\int_S (u - u_n)\, dS = 0\,.$$

The difficulty of finding a complete system of harmonic functions in concrete cases is a defect of the Trefftz method. For a two-dimensional simply connected region Ω with a sufficiently smooth boundary, harmonic polynomials form a complete system. In the case of multiply connected regions and in higher dimensions, the problem is difficult and there are still unsolved problems in this field.

Example 1. Let us solve the boundary value problem of Example 24.4.1.

In order to be able to use the Trefftz method, we transform the given problem by the substitution $u = \tilde{u} - \frac{1}{4}(x^2 + y^2)$ to the problem

$$\Delta\tilde{u} = 0$$

with the boundary condition $\tilde{u} = \frac{1}{4}(x^2 + y^2)$.

As coordinate functions, we choose the first two *even* harmonic polynomials (since $\tilde{u}$ is an even function in both arguments). Hence, the approximate solution will be of the form

$$\tilde{u} = a_1(x^2 - y^2) + a_2(x^4 - 6x^2y^2 + y^4).$$

The system (2) yields, in this case,

$$\frac{160}{3}a_1 + 192a_2 = 8$$

$$192a_1 + \frac{11\,904}{7}a_2 = \frac{176}{15}.$$

Its solution is $a_1 = \frac{97}{460} \doteq 0{\cdot}211$, $a_2 = \frac{-7}{414} \doteq -0{\cdot}017$; hence

$$\tilde{u} = 0{\cdot}211(x^2 - y^2) - 0{\cdot}017(x^4 - 6x^2y^2 + y^4)$$

and

$$u = 0{\cdot}211(x^2 - y^2) - 0{\cdot}017(x^4 - 6x^2y^2 + y^4) - \tfrac{1}{4}(x^2 + y^2) + c.$$

The constant c must be chosen in such a way that the mean value of u on the boundary is zero, i.e. $c = -\frac{1}{S}\int_S u \, ds$. We find that $c = \frac{929}{2\,070} \doteq 0{\cdot}449$. Hence,

$$u = 0{\cdot}449 + 0{\cdot}211(x^2 - y^2) - 0{\cdot}25(x^2 + y^2) - 0{\cdot}017(x^4 - 6x^2y^2 + y^4).$$

24.6. The Galerkin Method

This method is closely connected to variational methods and consists of the following procedure:

Let the equation to be solved be $Au - f = 0$, where A *need not be a linear operator*. We choose a sequence of linearly independent functions $\{\varphi_n\}$ (see Remark 1) and look for an approximate solution in the form

$$u_n = \sum_{k=1}^{n} a_k\varphi_k.$$

The coefficients a_k are to be determined from the condition that the expression $Au_n - f$ is orthogonal to each of $\varphi_1, \ldots, \varphi_n$. We thus obtain a system of (generally non-linear) equations

$$(A(\sum_{k=1}^{n} a_k\varphi_k), \varphi_m) = (f, \varphi_m), \quad m = 1, 2, \ldots, n.$$

If the operator A is linear, this system is also linear, i.e.,

$$\sum_{k=1}^{n} a_k(A\varphi_k, \varphi_m) = (f, \varphi_m), \quad m = 1, 2, \ldots, n.$$

If A is a symmetric positive definite operator, the Galerkin method is identical to the Ritz method.

Remark 1. We have explained here only the formal procedure without indicating conditions for convergence of the Galerkin method. For convergence it is necessary to make certain assumptions about the operator A and to require that the sequence $\{\varphi_n\}$ satisfies certain completeness conditions (cf. the Ritz method, § 24.3). For details of these questions we refer the reader to [282].

The Galerkin method applied to the eigenvalue problem

$$Au - \lambda Bu = 0 \tag{1}$$

yields (in the case of linear operators A and B) the following system of equations:

$$\sum_{k=1}^{n} a_k\{(A\varphi_k, \varphi_m) - \lambda(B\varphi_k, \varphi_m)\} = 0, \quad m = 1, 2, \ldots, n,$$

from which the equation

$$\det \|(A\varphi_k, \varphi_m) - \lambda(B\varphi_k, \varphi_m)\| = 0 \tag{2}$$

for the (approximate) eigenvalues follows.

Under certain assumptions about the operators A and B (see e.g. [282]), the sequences of approximate eigenvalues, constructed in this way, converge to the exact eigenvalues of the problem (1), and conversely, every eigenvalue of (1) may be obtained as the limit of some sequence of these approximate eigenvalues.

Example 1. A clamped elliptic plate with semi-axes a and b $(a > b)$ is loaded by a uniform loading $\sigma_x = \sigma_y = -\lambda\, D/h$, acting in the plane of the plate, where h is the thickness and D the rigidity of the plate. Find the least magnitude of the loading which causes loss of stability of the plate.

This problem consists in finding the least eigenvalue of the equation

$$\Delta^2 w + \lambda\, \Delta w = 0 \tag{3}$$

with boundary conditions

$$w = 0\,, \quad \frac{\partial w}{\partial \nu} = 0 \quad (\nu \text{ is the outward normal to the boundary})\,.$$

For the system $\{\varphi_n\}$, we choose the functions

$$x^m y^n \left(1 - \frac{x^2}{a^2} - \frac{y^2}{b^2}\right)^2, \quad m, n = 0, 1, 2, \ldots. \tag{4}$$

If we restrict ourselves to the case $m + n \leqq 2$, we obtain

$$w \approx \left(1 - \frac{x^2}{a^2} - \frac{y^2}{b^2}\right)^2 (a_1 + a_2 x + a_3 y + a_4 x^2 + a_5 xy + a_6 y^2)\,.$$

We substitute this approximate value into the left-hand side of equation (3) and then multiply both sides of the resulting equation successively by those functions of the system (4) for which $m + n \leqq 2$ and, finally, integrate each of these relations over the ellipse in question. In this way, we obtain a system of linear equations for $a_1, \ldots$ $\ldots, a_6$. If we put the determinant of the system equal to zero, we obtain an equation, whose least root approximately determines the critical loading. If we let $a/b = \gamma$, $\lambda a^2 = \varkappa$, this equation is

$$\left| \begin{array}{lll}
\frac{8}{\gamma}(3 + 2\gamma^2 + 3\gamma^4) - \frac{2}{\gamma}(1 + \gamma^2)\varkappa, & 0, & 0, \\
0, & 2\gamma + \frac{5}{\gamma} + \gamma^3 - \frac{1}{5}\left(\frac{1}{\gamma} + \frac{\gamma}{3}\right)\varkappa, & 0, \\
0, & 0, & \frac{2}{\gamma} + 5\gamma + \frac{1}{\gamma^2} - \frac{1}{5}\left(\frac{1}{3\gamma^3} + \frac{1}{\gamma}\right) \\
\gamma\left(\frac{3}{\gamma^2} + 2 + 3\gamma^2\right) - \frac{\gamma}{5}\varkappa, & 0, & 0 \\
0, & 0, & 0, \\
\frac{1}{\gamma}\left(\frac{3}{\gamma^2} + 2 + 3\gamma^2\right) - \frac{1}{5\gamma^3}\varkappa, & 0, & 0,
\end{array}\right.$$

If $\varkappa_1$ is the least root of this equation, the approximate magnitude of the critical loading is $-D\varkappa_1/a^2h$. In particular, when $\gamma = 2$, the value of $\varkappa_1$ is 42,067.

24.7. The Method of Least Squares

We choose a sequence $\{\varphi_n\}$ which satisfies the following two conditions:

1. For every function u (of the linear set M in question) and for every $\varepsilon > 0$, a linear combination of functions φ_n can be found such that

$$\left\| Au - A\sum_{k=1}^{n} \alpha_k \varphi_k \right\| < \varepsilon .$$

2. The functions $\{A\varphi_n\}$ are linearly independent.

An approximate solution of the equation $Au = f$ is required in the form $u_n = \sum_{k=1}^{n} a_k \varphi_k$ where the coefficients a_k are chosen in such a way that the expression $\|Au_n - f\|^2$ is minimal. Thus, we obtain the system of equations

$$\sum_{k=1}^{n} (A\varphi_k, A\varphi_m)\, a_k = (f, A\varphi_m)\,, \quad m = 1, 2, \ldots, n\,,$$

which has a unique solution.

$$\left(\frac{3}{\gamma} + 2\gamma + 3\gamma^3\right) - \frac{\gamma}{15}\varkappa, \qquad 0, \qquad \frac{1}{3}\left(\frac{3}{\gamma} + 2\gamma + 3\gamma^3\right) - \frac{1}{15\gamma}\varkappa$$

$$0, \qquad 0,$$

$$0, \qquad 0,$$

$$\frac{7}{[illegible]\gamma} + \frac{3\gamma^2}{10} + \frac{11\gamma}{15} - \frac{1}{20\gamma}\left(1 + \frac{\gamma^2}{3}\right)\varkappa, \qquad 0, \qquad \frac{1}{10\gamma} + \frac{\gamma}{3} + \frac{\gamma^3}{10} \qquad = 0$$

$$\frac{5}{\gamma^2} + 5\gamma^2 + 6 - \frac{1}{6}\left(\frac{1}{\gamma^2} + 1\right)\varkappa, \qquad 0,$$

$$\frac{1}{[illegible]\gamma^3} + \frac{1}{3\gamma} + \frac{\gamma}{10}, \qquad 0, \qquad \frac{27\gamma}{10} + \frac{3}{10\gamma^3} + \frac{11}{15\gamma} - \frac{1}{20\gamma}\left(1 + \frac{1}{3\gamma^2}\right)\varkappa$$

The method may be applied, even if A is not a symmetric positive definite operator. The approximate solutions converge to the exact solution, if the following conditions are satisfied:

A. The equation $Au = f$ has a solution (in the class of functions concerned).

B. There exists a constant K such that for an arbitrary function $u \in D_A$ the relation

$$\|u\| \leq K\|Au\|$$

holds.

The least squares method ensures in some cases (e.g. for Poisson's equation) the uniform convergence of approximate solutions to the exact solutions (which is not the case in general if we use other methods, e.g. the Ritz method).

24.8. Appendix on the Choice of Complete Systems

It is obvious from the examples in §§ 24.3 and 24.6 that when applying the above-mentioned approximate methods, the basic problem is the choice of sequence of functions $\{\varphi_n\}$ which is complete with respect to the operator A. Here we shall give some simple examples of such complete sequences.

Sequence of trigonometric functions: If the given region is a rectangle $(0, a) \times \times (0, b)$, we choose

$$\varphi_{kl}(x, y) = \sin\frac{k\pi x}{a}\sin\frac{l\pi y}{b}, \quad k, l = 1, 2, \ldots$$

(see e.g. Example 24.3.1). All the functions of this sequence satisfy the boundary conditions

$$\varphi_{kl}(0, y) = \varphi_{kl}(a, y) = \varphi_{kl}(x, 0) = \varphi_{kl}(x, b) = 0$$

and the same holds for all their even derivatives:

$$\frac{\partial^{2n}\varphi_{kl}}{\partial x^{2n}}(0, y) = \frac{\partial^{2n}\varphi_{kl}}{\partial x^{2n}}(a, y) = \frac{\partial^{2n}\varphi_{kl}}{\partial y^{2n}}(x, 0) = \frac{\partial^{2n}\varphi_{kl}}{\partial y^{2n}}(x, b) = 0\,.$$

It can be shown that the sequence chosen in such a manner is complete for differential operators (in two variables) of the second order or of higher even orders as far as the boundary conditions (given by the choice of the linear set M, see § 24.1) are of the indicated form.

If the boundary conditions (or even the given region) are of a different form, it is necessary to modify the choice of functions φ_n, or even to use functions other than trigonometric functions.

Example 1. To the operator of the second order with boundary conditions

$$u(0, y) = u(a, y) = \frac{\partial u}{\partial y}(x, 0) = \frac{\partial u}{\partial y}(x, b) = 0,$$

there corresponds the choice

$$\varphi_{kl}(x, y) = \sin \frac{k\pi x}{a} \cos \frac{l\pi y}{b}, \quad k, l = 1, 2, \ldots.$$

Example 2. To the operator of the second order with boundary conditions

$$u(0, y) = \frac{\partial u}{\partial x}(a, y) = u(x, 0) = u(x, b) = 0$$

there corresponds the choice

$$\varphi_{kl}(x, y) = \sin \frac{(2k - 1)\pi x}{2a} \sin \frac{l\pi y}{b}, \quad k, l = 1, 2, \ldots.$$

Example 3. Let us consider an operator of the fourth order, e.g. the biharmonic operator, with boundary conditions

$$u = 0, \quad \frac{\partial u}{\partial v} = 0$$

on the boundary of the given rectangle, v being the outward normal. These conditions cannot be satisfied by trigonometric functions.

It is easy to apply the above constructions in the case of several dimensions.

Sequences using polynomials: If the boundary of the given region is described by the equation $F(x, y) = 0$ (we again restrict ourselves to two dimensions) and if the prescribed boundary conditions (for an operator of order $2k + 2$) are

$$u = \frac{\partial u}{\partial v} = \frac{\partial^2 u}{\partial v^2} = \ldots = \frac{\partial^k u}{\partial v^k} = 0$$

on the boundary of the region, v being the outward normal, we choose the complete sequence as follows:

$$\varphi_i(x, y) = F^{k+1}(x, y)\, P_i(x, y)$$

where $P_i(x, y)$ is the sequence of all products of powers of the type $x^n y^m$ with non-negative integral exponents (cf. Example 24.6.1). Hence

$$\varphi_1 = F^{k+1}(x, y), \quad \varphi_2 = F^{k+1}(x, y)\, x, \quad \varphi_3 = F^{k+1}(x, y)\, y,$$
$$\varphi_4 = F^{k+1}(x, y)\, x^2, \quad \varphi_5 = F^{k+1}(x, y)\, xy, \quad \varphi_6 = F^{k+1}(x, y)\, y^2, \text{ etc}.$$

The question of whether this sequence is complete or not depends substantially on the structure of the boundary and each case has to be studied separately. However, in practical cases the above-mentioned sequence will be complete, as for as the function $F(x, y)$ has a sufficient number of derivatives.

Example 4. Let us consider the biharmonic operator on a circle with centre at the origin and radius R, with boundary conditions $u = \partial u/\partial v = 0$.

The sequence $\{\varphi_i\}$ will then be of the form

$$\varphi_i(x, y) = (R^2 - x^2 - y^2)^2 P_i(x, y).$$

REMARK 1. It is possible to omit, a priori, some members in the above-mentioned sequences, if the problem possesses symmetry, antisymmetry or rotational symmetry with respect to the variables considered (cf. Example 24.5.1).

25. APPROXIMATE SOLUTION OF ORDINARY DIFFERENTIAL EQUATIONS

By Otto Vejvoda and Karel Rektorys

References: [15], [20], [31], [62], [65], [67], [68], [69], [78], [131], [135], [136], [154], [176], [177], [181], [183], [199], [214], [225], [237], [241], [287], [326], [331], [391], [408], [409], [410], [438].

To *solve approximately* a given differential equation of the n-th order (or a system of n differential equations of the first order), means either to find a function (or a system of n functions) which differs "little" from the exact solution, or to find an approximate value of the solution at a certain number of points of the interval on which the solution is sought. (In the latter case we usually speak of a *numerical solution*.)

The theory of the approximate solution of differential equations is very extensive. It is therefore impossible to state here all known methods and formulae; for more details see [20], [67], [69], [131], [181], [287], etc. In this chapter, three fundamental types of problems are considered; namely

A. initial-value problems,

B. boundary-value problems,

C. eigenvalue problems.

At the end of the chapter, a particular case of boundary-value problems, namely that concerning periodic solutions, is briefly treated.

A. INITIAL-VALUE PROBLEMS

25.1. Notation Used in this Chapter

We deal with the following problems:

(α) The equation

$$y' = f(x, y) \tag{1}$$

with the initial condition

$$y(x_0) = y_0 \,. \tag{2}$$

(β) The equation

$$y^{(n)} = f(x, y, y', \ldots, y^{(n-1)}) \tag{3}$$

with the initial conditions

$$y(x_0) = y_0 \,, \; y'(x_0) = y'_0, \; \ldots, \; y^{(n-1)}(x_0) = y_0^{(n-1)} \,. \tag{4}$$

(γ) The system of equations

$$\begin{aligned} \frac{\mathrm{d}({}^1y)}{\mathrm{d}x} &= {}^1f(x, {}^1y, \ldots, {}^ny)\,, \\ &\cdots\cdots\cdots\cdots \\ \frac{\mathrm{d}({}^ny)}{\mathrm{d}x} &= {}^nf(x, {}^1y, \ldots, {}^ny) \end{aligned} \tag{5}$$

(briefly $\dfrac{\mathrm{d}({}^iy)}{\mathrm{d}x} = {}^if(x, {}^1y, \ldots, {}^ny)\,, \quad i = 1, 2, \ldots, n$) (5′)

with the initial conditions

$${}^1y(x_0) = {}^1y_0, \ldots, {}^ny(x_0) = {}^ny_0 \tag{6}$$

(briefly ${}^iy(x_0) = {}^iy_0\,, \quad i = 1, 2, \ldots, n$). (6′)

It is assumed that the functions f or ${}^1f, {}^2f, \ldots, {}^nf$ satisfy conditions ensuring the existence and uniqueness of the solution in the interval in question (§ 17.2).

Approximate values of the solution y or its derivatives $y', \ldots, y^{(n-1)}$ or ${}^iy$ are denoted by

$$Y, \text{ or } Y', Y'', \ldots, Y^{(n-1)}, \text{ or } {}^iY, \text{ respectively}\,.$$

We write briefly

$$\begin{gathered} y(x_j) = y_j\,, \quad y^{(k)}(x_j) = y_j^{(k)}\,, \quad {}^iy(x_j) = {}^iy_j\,, \\ Y(x_j) = Y_j\,, \quad Y^{(k)}(x_j) = Y_j^{(k)}\,, \quad {}^iY(x_j) = {}^iY_j\,, \\ f(x_j, y_j, y'_j, \ldots, y_j^{(n-1)}) = f_j\,, \quad f(x_j, Y_j, Y'_j, \ldots, Y_j^{(n-1)}) = F_j\,, \\ {}^if(x_j, {}^1y_j, \ldots, {}^ny_j) = {}^if_j\,, \quad {}^if(x_j, {}^1Y_j, \ldots, {}^nY_j) = {}^iF_j\,. \end{gathered}$$

Formulae concerning the equation (1), or (3) or (5), with the corresponding initial conditions are denoted by (α), (β) or (γ), respectively.

Definition 1. The *error of a solution at a point* x_j is defined in case (α) by the expression

$$\varphi_j = \varphi(x_j) = |Y_j - y_j|, \tag{7}$$

in case (β) by the expression

$$\psi_j = \psi(x_j) = |Y_j - y_j|, \tag{8}$$

and in case (γ) by the expression

$$\chi_j = \chi(x_j) = \sum_{i=1}^{n} |{}^iY_j - {}^iy_j|. \tag{9}$$

REMARK 1. In case (β), the definition of the error does not agree with the corresponding definition in case (γ) when equation (3) is replaced in the usual way by the system (5). (See Remark 17.18.1, p. 817.)

REMARK 2. The *error of a solution in the interval* $[a, b]$ is defined by the expressions

(α) $$\varphi(a, b) = \max_{a \leqq x \leqq b} \varphi(x) = \max_{a \leqq x \leqq b} |Y(x) - y(x)|, \tag{10}$$

(β) $$\psi(a, b) = \max_{a \leqq x \leqq b} \psi(x) = \max_{a \leqq x \leqq b} |Y(x) - y(x)|, \tag{11}$$

(γ) $$\chi(a, b) = \max_{a \leqq x \leqq b} \chi(x) = \max_{a \leqq x \leqq b} \sum_{i=1}^{n} |{}^iY(x) - {}^iy(x)|. \tag{12}$$

25.2. The Method of Successive Approximations

Theorem 1 (*case* (α)). *Let the function* $f(x, y)$ *be continuous in the region* $\bar{R}(x_0 \leqq \leqq x \leqq x_0 + a,\ y_0 - b \leqq y \leqq y_0 + b)$, let $|f(x, y)| \leqq M$ *in* $\bar{R}$ *and let* f *satisfy the Lipschitz condition*

$$|f(x, y_2) - f(x, y_1)| \leqq K|y_2 - y_1| \tag{1}$$

in $\bar{R}$. *Let us write* $I = [x_0, x_0 + \alpha]$, *where*

$$\alpha \leqq \min\left(a, \frac{b}{M}\right). \tag{2}$$

Let us form in I the sequence of functions $y_k(x)$ *such that*

$$y_0(x) = y_0, \quad y_k(x) = y_0 + \int_{x_0}^{x} f(t, y_{k-1}(t))\, dt, \quad k = 1, 2, 3, \ldots. \tag{3}$$

Under the stated conditions, as $k \to \infty$, *the sequence* (3) (the so-called *sequence of successive approximations*) *converges uniformly to the solution* $y(x)$ *of the problem* (α),

$$y' = f(x, y), \quad y(x_0) = y_0 . \tag{4}$$

Estimate of error: In the interval I the inequality

$$|y(x) - y_k(x)| \leqq \frac{M\alpha}{k!} \frac{(K\alpha)^k}{1 - K\alpha} \tag{5}$$

holds, provided $K\alpha < 1$.

Theorem 2 (*case* (γ)). *Let the functions*

$${}^if(x, {}^1y, \ldots, {}^ny), \quad i = 1, 2, \ldots, n, \tag{6}$$

be continuous in the region $\bar{R}(x_0 \leqq x \leqq x_0 + a,\ {}^1y_0 - b \leqq {}^1y \leqq {}^1y_0 + b, \ldots, \ldots, {}^ny_0 - b \leqq {}^ny \leqq {}^ny_0 + b)$, let $|{}^if(x, {}^1y, \ldots, {}^ny)| \leqq M$ *in* $\bar{R}$, *and let the functions* ${}^if$ *satisfy in* $\bar{R}$ *the Lipschitz condition*

$$\begin{aligned} |{}^if(x, {}^1y_2, \ldots, {}^ny_2) - {}^if(x, {}^1y_1, \ldots, {}^ny_1)| \leqq K(|{}^1y_2 - {}^1y_1| + \ldots + \\ + |{}^ny_2 - {}^ny_1|), \end{aligned} \tag{7}$$

$i = 1, 2, \ldots, n$. *Let us write* $I = [x_0, x_0 + \alpha]$, *where*

$$\alpha \leqq \min\left(a, \frac{b}{M}\right). \tag{8}$$

Let us form in I the following sequences (the so-called *successive approximations*)

$${}^iy_0(x) = {}^iy_0, \quad {}^iy_k(x) = {}^iy_0 + \int_{x_0}^{x} {}^if(t, {}^1y_{k-1}(t), \ldots, {}^ny_{k-1}(t))\, dt, \tag{9}$$

$i = 1, 2, \ldots, n$, $k = 1, 2, \ldots$. *Under the stated conditions, as* $k \to \infty$ *the sequences of functions* (9) *converge uniformly* (*with respect to* x) *to the solution* ${}^1y(x), \ldots, {}^ny(x)$ *of the problem* (γ), *given by* (25.1.5) *and* (25.1.6) (p. 1066).

Estimate of error: In the interval I the inequality

$$|{}^iy(x) - {}^iy_k(x)| \leqq \frac{M\alpha}{k!} \frac{(Kn\alpha)^k}{1 - Kn\alpha} \quad \text{for} \quad i = 1, 2, \ldots, n \tag{10}$$

holds, provided $Kn\alpha < 1$.

REMARK 1. A similar theorem concerning case (β) may be omitted, since it is completely similar to the above two cases.

Example 1. Let us solve, by the method of successive approximations, the equation

$$y' = y + xy^2 \tag{11}$$

with the initial conditions

$$y(0) = 1 . \tag{12}$$

(This equation is of Bernoulli's type and the exact solution of (11), (12) is given by

$$y(x) = \frac{1}{1 - x} . \tag{13}$$

We have chosen equation (11) so that we can compare the approximate solution with the exact solution.)

According to (3),

$$y_0(x) = 1 ,$$

$$y_1(x) = 1 + \int_0^x (1 + t \,.\, 1^2)\, dt = 1 + x + \frac{x^2}{2} . \tag{14}$$

Using (5) for $k = 1$ with $a = 0{\cdot}5$, $b = 1$, we obtain

$$M = 4 , \quad \frac{b}{M} = \frac{1}{4} , \quad K = \max_{\substack{0 \leqq x \leqq 0\cdot5 \\ 0 \leqq y \leqq 2}} \left| \frac{\partial f}{\partial y} \right| = \max_{\substack{0 \leqq x \leqq 0\cdot5 \\ 0 \leqq y \leqq 2}} (1 + 2xy) = 3 \tag{15}$$

(see Remark 17.2.8, p. 734) and choosing $\alpha = 0{\cdot}2$, we get for $0 \leqq x \leqq 0{\cdot}2$

$$|y(x) - y_1(x)| \leqq \frac{4 \,.\, 0{\cdot}2 \,.\, 0{\cdot}6^1}{1!\,(1 - 0{\cdot}6)} = 1{\cdot}20 .$$

The actual error for $x = \alpha = 0{\cdot}2$ is

$$\left| \frac{1}{1 - x} - \left(1 + x + \frac{x^2}{2}\right) \right|_{x = 0\cdot2} = 1{\cdot}25 - 1{\cdot}22 = 0{\cdot}03 .$$

The estimate of the error given by (5) is obviously very rough.

REMARK 2. The procedure given in Theorems 1 and 2 may also be applied (under the same assumptions) to the left of the point x_0.

25.3. Expansion of the Solution in a Power Series

Theorem 1. *Let $f(x, y)$ be an analytic function in the region $\bar{Q}(x_0 - a \leqq x \leqq$ $\leqq x_0 + a$, $y_0 - b \leqq y \leqq y_0 + b)$. Then, in a certain neighbourhood $O(x_0 - \alpha \leqq$ $\leqq x \leqq x_0 + \alpha)$ (α being sufficiently small) of the point x_0, there exists an analytic solution*

$$y(x) = y_0 + a_1(x - x_0) + a_2(x - x_0)^2 + \ldots \tag{1}$$

of the problem (α),

$$y' = f(x, y), \quad y(x_0) = y_0 . \tag{2}$$

Coefficients of the expansion (1) may be determined either by the successive differentiation of (2) and the application of the Taylor formula or by the substitution of (1) into equation (2) and the comparison of coefficients of the same powers of x. Each partial sum of the series (1) may be regarded as an approximate solution of the problem (α).

The radius of convergence of the series (1) can be determined by known methods of the theory of power series (Chapter 15).

In problems (β) and (γ) (given by (25.1.3), (25.1.4) and (25.1.5), (25.1.6), respectively, we may proceed in an analogous manner.

Example 1. Let us solve again the problem (25.2.11), (25.2.12). We have, using the fact that $y(0) = 1$,

$$y(x) = 1 + a_1x + a_2x^2 + a_3x^3 + \dots, \tag{3}$$

so that

$$y'(x) = a_1 + 2a_2x + 3a_3x^2 + \dots, \tag{4}$$

and substitute (3) and (4) into the given equation $y' = y + xy^2$:

$$a_1 + 2a_2x + 3a_3x^2 + 4a_4x^3 + \dots = 1 + a_1x + a_2x^2 + a_3x^3 + \dots$$

$$\dots + x[1 + 2a_1x + (a_1^2 + 2a_2)\,x^2 + \dots] =$$

$$= 1 + (a_1 + 1)\,x + (a_2 + 2a_1)\,x^2 + (a_3 + a_1^2 + 2a_2)\,x^3 + \dots. \tag{5}$$

Equating coefficients, we get

$$a_1 = 1,\ 2a_2 = a_1 + 1,\ 3a_3 = a_2 + 2a_1,\ 4a_4 = a_3 + a_1^2 + 2a_2,\ \dots,$$

whence

$$a_1 = 1,\ a_2 = 1,\ a_3 = 1,\ a_4 = 1,\ \dots,$$

$$y(x) = 1 + x + x^2 + x^3 + x^4 + \dots.$$

On the other hand, making use of Taylor's formula, we have

$$a_1 = \frac{y'(0)}{1!}, \quad a_2 = \frac{y''(0)}{2!}, \quad a_3 = \frac{y'''(0)}{3!}, \quad \dots.$$

The differentiation of the equation $y' = y + xy^2$ yields

$$y'' = y' + y^2 + 2xyy', \quad y''' = y'' + 4yy' + 2xy'^2 + 2xyy'', \dots.$$

To obtain $y'(0)$, etc., we put $x = 0$, $y = 1$ and finally get

$$y'(0) = 1, \; y''(0) = 2, \; y'''(0) = 6, \; \ldots,$$

giving

$$y(x) = 1 + x + x^2 + x^3 + \ldots .$$

25.4. The Perturbation Method

We often meet the case where the given equation depends on a "small" parameter ε,

$$y' = f(x, y, \varepsilon), \quad y(x_0) = y_0 . \tag{1}$$

If the so-called *unperturbed* (or *limit*) *equation*

$$y' = f(x, y, 0), \quad y(x_0) = y_0 \tag{2}$$

(with $\varepsilon = 0$) can be easily solved, then the problem (1) can be treated in the following way:

Theorem 1. *Let $f(x, y, \varepsilon)$ be continuous in x and analytic in y and ε in the region $\bar{R}(x_0 - a \leqq x \leqq x_0 + a,\; y_0 - b \leqq y \leqq y_0 + b,\; 0 \leqq \varepsilon \leqq \varepsilon_0)$.*

Let the unperturbed problem (2) have for $x \in [x_0 - a, x_0 + a]$ the unique solution $y_0(x)$, where $y_0 - b < y_0(x) < y_0 + b$.

Then there exists a number ε_1, $(0 < \varepsilon_1 \leqq \varepsilon_0)$ such that for all ε, satisfying the condition $0 \leqq \varepsilon \leqq \varepsilon_1$, the problem (1) has a unique solution $y(x)$ in the interval $[x_0 - a, x_0 + a]$. This solution can be written in the form

$$y(x) = y_0(x) + \varepsilon\, y_1(x) + \varepsilon^2\, y_2(x) + \ldots \tag{3}$$

where the functions $y_k(x)$ in the series (3) are given as solutions of the sequence of equations

$$\begin{aligned}
y_0' &= f(x, y_0, 0), \quad y_0(x_0) = y_0, \\
y_1' &= \frac{\partial f}{\partial y}(x, y_0, 0)\, y_1 + \frac{\partial f}{\partial \varepsilon}(x, y_0, 0), \quad y_1(x_0) = 0, \\
&\cdots\cdots\cdots\cdots\cdots\cdots\cdots\cdots\cdots\cdots \\
y_n' &= \frac{\partial f}{\partial y}(x, y_0, 0)\, y_n + \Phi_n(x, y_0, y_1, \ldots, y_{n-1}), \quad y_n(x_0) = 0, \\
&\cdots\cdots\cdots\cdots\cdots\cdots\cdots\cdots\cdots\cdots
\end{aligned} \tag{4}$$

The system (4) may be obtained by substituting (3) into (1), expanding

$$f\big(x, \sum_{k=0}^{\infty} y_k(x)\, \varepsilon^k, \varepsilon\big)$$

into a power series in ε and equating coefficients of the same powers of ε. Note that only the first equation in (4) is non-linear, while all the remaining equations are linear.

If ε is sufficiently small, the solution of k of the equations (4), gives the approximate solution up to terms of order $k - 1$ in ε,

$$Y(x) = y_0(x) + y_1(x)\,\varepsilon + \ldots + y_{k-1}(x)\,\varepsilon^{k-1}\,.$$

In problems (β) and (γ), the method may be used in a similar manner and under analogous assumptions. The same idea may be used in partial differential equations.

Example 1. Let us solve the problem

$$y' = 2y + \varepsilon y^2\,, \quad y(0) = 1\,. \tag{5}$$

Let us substitute (3) into (5):

$$y_0' + \varepsilon y_1' + \varepsilon^2 y_2' + \ldots = 2y_0 + 2\varepsilon y_1 + 2\varepsilon^2 y_2 + \ldots$$
$$\ldots + \varepsilon(y_0^2 + 2\varepsilon y_0 y_1 + \ldots) = 2y_0 + \varepsilon(2y_1 + y_0^2) + \varepsilon^2(2y_2 + 2y_0 y_1) + \ldots\,. \tag{6}$$

By comparison of coefficients of the same powers of ε, we get

$$y_0' = 2y_0,\ y_1' = 2y_1 + y_0^2,\ y_2' = 2y_2 + 2y_0 y_1,\ \ldots \tag{7}$$

with the initial conditions $y_0(0) = 1$, $y_1(0) = y_2(0) = \ldots = 0$. We find that

$$y_0(x) = e^{2x}\,, \quad y_1(x) = 0{\cdot}5e^{4x} - 0{\cdot}5e^{2x}\,,$$

$$y_2(x) = 0{\cdot}25e^{6x} - 0{\cdot}5e^{4x} + 0{\cdot}25e^{2x}, \ldots\,.$$

The function

$$Y(x) = e^{2x}[1 + 0{\cdot}5\varepsilon(e^{2x} - 1) + 0{\cdot}25\varepsilon^2(e^{4x} - 2e^{2x} + 1)] \tag{8}$$

obviously satisfies the initial condition and the given equation up to terms of the third order in ε.

This problem is only an illustrative one since it may be easily solved in a closed form, namely,

$$y = \frac{e^{2x}}{1 + \frac{1}{2}\varepsilon(1 - e^{2x})}\,.$$

REMARK 1. The perturbation method may be used under much less restrictive conditions. Let us formulate a modified procedure for the so-called *weakly nonlinear case*

$$y' = cy + \varepsilon f(x, y, \varepsilon)\,, \quad y(x_0) = y_0\,, \tag{9}$$

where the function $f(x, y, \varepsilon)$, together with its derivative $\dfrac{\partial f}{\partial y}(x, y, \varepsilon)$ is continuous in x and ε, and satisfies the Lipschitz condition with respect to y in the region $\bar{R}$ (cf. Theorem 1).

Then it may be shown that, for sufficiently small ε the successive approximations $y_k(x)$ given by

$$y_0(x) = e^{c(x-x_0)} y_0 ,$$

$$y_{k+1}(x) = e^{c(x-x_0)} y_0 + \varepsilon \int_{x_0}^{x} e^{c(x-\xi)} f(\xi, y_k(\xi), \varepsilon) \, d\xi \tag{10}$$

converge uniformly, in a neighbourhood of the point x_0, to the exact solution of (9). Each function $y_k(x)$ may be taken as an approximate solution of (9).

Formulae (10) may be derived by applying the method of variation of a parameter and the method of successive approximations.

25.5. The Polygon Method

Let us solve the problem (α),

$$y' = f(x, y), \quad y(x_0) = y_0 , \tag{1}$$

in the interval $I = [x_0, x_0 + a]$ in the following way: Divide the interval I by points $x_1 = x_0 + h_1$, $x_2 = x_1 + h_2$, ... (see Fig. 25.1) into k step intervals (generally of unequal length). Of course, $h_i > 0$ and $\sum_{i=1}^{k} h_i = a$. Calculate $y'_0 = f(x_0, y_0)$ and

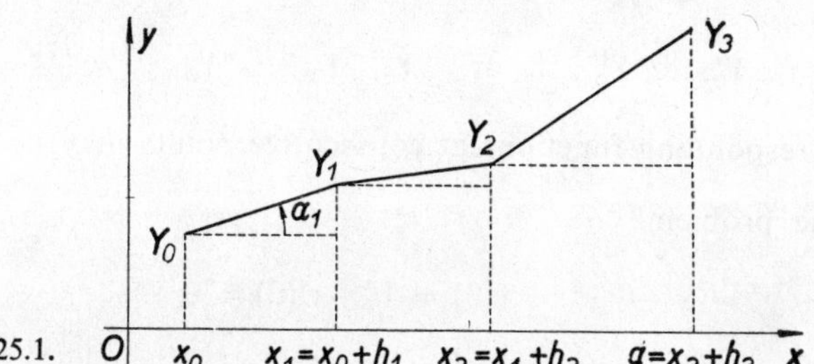

Fig. 25.1.

put $Y_1 = y_0 + h_1 y'_0 = y_0 + h_1 f(x_0, y_0)$. Further, let $Y_2 = Y_1 + h_2 f(x_1, Y_1)$, $Y_3 = Y_2 + h_3 f(x_2, Y_2)$, etc. The integral curve is thus replaced by a polygonal line.

REMARK 1. Very frequently, we choose (for the sake of simplicity of computations) an equidistant division of the interval I; then $h_1 = h_2 = \ldots = h = \dfrac{a}{k}$. In general, we have to choose a finer subdivision in places, where the function $f(x, y)$ changes more rapidly.

In order to estimate the error for an equidistant division let us suppose that for any points (x, y), (x, y_1), (x, y_2) of the region

$$\bar{R}[x_0 \leqq x \leqq x_0 + a,\ y_0 - M(x - x_0) \leqq y \leqq y_0 + M(x - x_0)]$$

we have

$$|f(x, y)| \leqq M\,, \quad |f_x(x, y) + f(x, y) f_y(x, y)| \leqq N\,,$$
$$|f(x, y_1) - f(x, y_2)| \leqq K|y_1 - y_2|\,.$$

Then

$$|Y_j - y_j| \leqq \frac{hN}{2K}[(1 + hK)^j - 1]\,. \tag{2}$$

Example 1. For the problem $y' = y + xy^2$, $y(0) = 1$ (Example 25.2.1, p. 1068), taking $h_1 = h_2 = 0{\cdot}1$, we obtain at the point $a = 0{\cdot}2$ the result:

$$Y_1 = 1 + 0{\cdot}1\,.\,1 = 1{\cdot}1\,, \quad Y_2 = 1{\cdot}1 + 0{\cdot}1\,.\,(1{\cdot}1 + 0{\cdot}1\,.\,1{\cdot}1^2) = 1.2221\,.$$

Hence, for the error φ (see Remark 25.1.2, p. 1067), we get $\varphi(0, 0{\cdot}2) \doteq 0{\cdot}028$, whereas the estimate (2) gives approximately 0·5.

The corresponding formulae for problems (β) and (γ) are given in (3) and (4):

$$Y_{r+1}^{(k)} = \sum_{\varrho=0}^{n-k-1} \frac{h^\varrho}{\varrho!} Y_r^{(k+\varrho)} + \frac{h^{n-k}}{(n-k)!} F_r \quad (k = 0, 1, \ldots, n-1)\,, \tag{3}$$

$${}^iY_{r+1} = {}^iY_r + {}^if(x_r, {}^1Y_r, \ldots, {}^nY_r)(x_{r+1} - x_r) = {}^iY_r + {}^iF_r h_{r+1}\,. \tag{4}$$

(For notation see § 25.1.) Clearly, given the values of

$$Y_0, Y_0', \ldots, Y_0^{(n-1)} \quad \text{or} \quad {}^1Y_0, {}^2Y_0, \ldots, {}^nY_0\,,$$

the values of the corresponding functions at consecutive points may be calculated.

Example 2. Let the problem

$$y'' = x^2 + y^2\,, \quad y(0) = 1\,, \quad y'(0) = 0 \tag{5}$$

be given. Determine approximately $y(0{\cdot}1)$ and $y'(0{\cdot}1)$.

Choose $h = 0{\cdot}1$. Thus $r = 0$, $n = 2$, $F(0) = 0 + 1^2 = 1$. Hence, by (3)

$$y(0{\cdot}1) \approx Y_1 = \frac{0{\cdot}1^0}{0!} Y_0 + \frac{h}{1!} Y_0' + \frac{h^2}{2!} F_0 = 1 + 0 + \frac{h^2}{2!} = 1{\cdot}005\,,$$

$$y'(0{\cdot}1) \approx Y_1' = \frac{0{\cdot}1^0}{1!} Y_0' + \frac{h}{1!} F_0 = 0 + \frac{0{\cdot}1}{1!}\,.\,1 = 0{\cdot}1\,.$$

REMARK 2 (*Improved Polygon Method*). Here, the rule is chosen so that the polygonal line has a point of contact of the second order with the exact solution at the point (x_j, Y_j). For example, in case (γ) the formula corresponding to (4) then becomes

$$^{i}Y_{r+1} = {}^{i}Y_r + {}^{i}f\left(x_r + \tfrac{1}{2}(x_{r+1} - x_r), {}^{1}Y_r + \frac{x_{r+1} - x_r}{2}\,{}^{1}f(x_r, {}^{1}Y_r, \ldots, {}^{n}Y_r),\right.$$

$$\left.\ldots, {}^{n}Y_r + \frac{x_{r+1} - x_r}{2}\,{}^{n}f(x_r, {}^{1}Y_r, \ldots, {}^{n}Y_r)\right)(x_{r+1} - x_r)\,.$$

25.6. The Runge–Kutta Method

Denote: $x_1 - x_0 = h$.

In the case (α) (p. 1065), by the Runge–Kutta formulae we usually mean the following formulae

$$Y_1 = Y_0 + \tfrac{1}{6}h(k_1 + 2k_2 + 2k_3 + k_4)\,, \tag{1}$$

where

$$\begin{aligned} k_1 &= f(x_0, Y_0)\,, \\ k_2 &= f(x_0 + \tfrac{1}{2}h, Y_0 + \tfrac{1}{2}hk_1)\,, \\ k_3 &= f(x_0 + \tfrac{1}{2}h, Y_0 + \tfrac{1}{2}hk_2)\,, \\ k_4 &= f(x_0 + h, Y_0 + hk_3)\,. \end{aligned} \tag{2}$$

Example 1. Let us again solve the problem

$$y' = y + xy^2\,, \quad y(0) = 1$$

and compute the approximate value at the point $x = 0{\cdot}2$. Hence $h = 0{\cdot}2$ and

$$\begin{aligned} k_1 &= 1 + 0\,.\,1^2 = 1\,, \\ k_2 &= 1{\cdot}1 + 0{\cdot}1\,.\,1{\cdot}1^2 = 1{\cdot}221\,, \\ k_3 &= 1{\cdot}1221 + 0{\cdot}1\,.\,1{\cdot}1221^2 \doteq 1{\cdot}248\,, \\ k_4 &= 1{\cdot}2496 + 0{\cdot}2\,.\,1{\cdot}2496^2 \doteq 1{\cdot}562\,, \\ Y_1 &= Y(0{\cdot}2) = 1 + 0{\cdot}250 = 1{\cdot}250\,. \end{aligned}$$

For the accuracy used here (calculating to three decimal places) the result agrees with the exact solution

$$y(0{\cdot}2) = \frac{1}{1 - 0{\cdot}2} = 1{\cdot}250\,.$$

In case (γ), the formulae (1), (2) take the form

$$^iY_1 = {}^iY_0 + \tfrac{1}{6}h({}^ik_1 + 2{}^ik_2 + 2{}^ik_3 + {}^ik_4)\,, \tag{3}$$

where

$$\begin{aligned} {}^ik_1 &= {}^if(x_0, {}^iY_0, \ldots, {}^nY_0)\,, \\ {}^ik_2 &= {}^if(x_0 + \tfrac{1}{2}h,\ {}^1Y_0 + \tfrac{1}{2}h \,.\, {}^1k_1,\ \ldots,\ {}^nY_0 + \tfrac{1}{2}h \,.\, {}^nk_1)\,, \\ {}^ik_3 &= {}^if(x_0 + \tfrac{1}{2}h,\ {}^1Y_0 + \tfrac{1}{2}h \,.\, {}^1k_2,\ \ldots,\ {}^nY_0 + \tfrac{1}{2}h \,.\, {}^nk_2)\,, \\ {}^ik_4 &= {}^if(x_0 + h,\ {}^1Y_0 + h \,.\, {}^1k_3,\ \ldots,\ {}^nY_0 + h \,.\, {}^nk_3)\,. \end{aligned} \tag{4}$$

In case (β), the general formulae are rather complicated (see [69]). We therefore give here the formula for a second-order equation only:

$$\begin{aligned} Y_1 &= Y_0 + hY_0' + \tfrac{1}{6}h^2(k_1 + k_2 + k_3)\,, \\ Y_1' &= Y_0' + \tfrac{1}{6}h(k_1 + 2k_2 + 2k_3 + k_4)\,, \end{aligned} \tag{5}$$

where

$$\begin{aligned} k_1 &= f(x_0, Y_0, Y_0')\,, \\ k_2 &= f(x_0 + \tfrac{1}{2}h,\ Y_0 + \tfrac{1}{2}h \,.\, Y_0' + \tfrac{1}{8}h^2 \,.\, k_1,\ Y_0' + \tfrac{1}{2}h \,.\, k_1)\,, \\ k_3 &= f(x_0 + \tfrac{1}{2}h,\ Y_0 + \tfrac{1}{2}h \,.\, Y_0' + \tfrac{1}{8}h^2 \,.\, k_1,\ Y_0' + \tfrac{1}{2}h \,.\, k_2)\,, \\ k_4 &= f(x_0 + h,\ Y_0 + h \,.\, Y_0' + \tfrac{1}{2}h^2 \,.\, k_3,\ Y_0' + hk_3)\,. \end{aligned} \tag{6}$$

REMARK 1. If the point x_0 is the initial point (so that the values ${}^iY_0$, etc., are not loaded with error by previous calculations), we write, of course, ${}^iy_0$ instead of ${}^iY_0$, etc.

REMARK 2. The formulae (1), (2) or (3), (4), define an approximate solution which has contact of the fourth order at the point x_0 with the exact solution; the formulae (5) and (6) also define an approximate solution which has contact of the fourth order. (For an n-th order equation the contact is of the order $n + 2$ in the corresponding formulae.) For general ideas on the construction of Runge – Kutta formulae see, e.g., [69].

REMARK 3. The estimates of error are complicated and rather unsatisfactory (see [408]).

REMARK 4. In practical calculations, the step length h is chosen so small that the number $|k_2 - k_3|$ constitutes only a small fraction (about 5%) of the number $|k_2|$.

25.7. The Adams Extrapolation Method

Let us divide the interval $[x_0, x_0 + a]$ on which the approximate solution is sought into m equal subintervals by points $x_r = x_0 + rh$, $h = a/m$, $r = 1, 2, \ldots, m$.

In the case (γ), in order to calculate the approximate value ${}^iY_{r+1}$, we replace

${}^{i}f(x, y)$ by the interpolation polynomial ${}^{i}P_s(x; x_r)$ passing through the points $(x_{r-s}, {}^{i}Y_{r-s}), \ldots, (x_r, {}^{i}Y_r)$ and integrate the approximate differential system $\dfrac{\mathrm{d}\,{}^{i}y(x)}{\mathrm{d}x} = {}^{i}P_s(x; x_r)$ with the initial conditions ${}^{i}y(x_r) = {}^{i}Y_r$, over the interval $[x_r, x_{r+1}]$.

TABLE 25.1

$\beta_{n-k,j}$

$n-k$ \ j	0	1	2	3	4	5	6
1	1	$\frac{1}{2}$	$\frac{5}{12}$	$\frac{3}{8}$	$\frac{251}{720}$	$\frac{95}{288}$	$\frac{19{,}087}{60{,}480}$
2	$\frac{1}{2}$	$\frac{1}{6}$	$\frac{1}{8}$	$\frac{19}{180}$	$\frac{3}{32}$	$\frac{863}{10{,}080}$	$\frac{275}{3456}$

Table 25.2

α_{sj}

s \ j	0	1	2	3	4	5
1	$\frac{3}{2}$	$-\frac{1}{2}$				
2	$\frac{23}{12}$	$-\frac{16}{12}$	$\frac{5}{12}$			
3	$\frac{55}{24}$	$-\frac{59}{24}$	$\frac{37}{24}$	$-\frac{9}{24}$		
4	$\frac{1901}{720}$	$-\frac{2774}{720}$	$\frac{2616}{720}$	$-\frac{1274}{720}$	$\frac{251}{720}$	
5	$\frac{4227}{1440}$	$-\frac{7673}{1440}$	$\frac{9482}{1440}$	$-\frac{6798}{1440}$	$\frac{2627}{1440}$	$-\frac{425}{1440}$

(The first approximate values ${}^{i}Y_{r-s}, \ldots, {}^{i}Y_r$ must be calculated by another method as precisely as possible.)

The approximate values ${}^{i}Y_{r+1}$ are then given by

$$ {}^{i}Y_{r+1} = {}^{i}Y_r + h({}^{i}F_r + \beta_1 \nabla_1\, {}^{i}F_r + \ldots + \beta_s \nabla_s\, {}^{i}F_r) \tag{1} $$

or, if the differences are expressed with the aid of the values ${}^iF_r$, by

$$ {}^iY_{r+1} = {}^iY_r + h(\alpha_{s0}\ {}^iF_r + \alpha_{s1}\ {}^iF_{r-1} + \ldots + \alpha_{ss}\ {}^iF_{r-s}) \tag{2} $$

(the so-called ordinate form), where

$$ {}^iF_j = {}^if(x_j, {}^1Y_j, \ldots, {}^nY_j), $$

$\nabla_k\ {}^iF_r$ is the k-th backward difference of ${}^iF_r$,

$$ \nabla_1\ {}^iF_r = \nabla {}^iF_r = {}^iF_r - {}^iF_{r-1}, \quad \nabla_2\ {}^iF_r = \nabla(\nabla {}^iF_r) = {}^iF_r - 2\ {}^iF_{r-1} + {}^iF_{r-2} \tag{3} $$

etc., and

$$ \beta_1 = \tfrac{1}{2}, \quad \beta_2 = \tfrac{5}{12}, \quad \beta_3 = \tfrac{3}{8}, \quad \beta_4 = \tfrac{251}{720}, $$

in general

$$ \beta_j = \int_0^1 \binom{u + j - 1}{j} \mathrm{d}u $$

(see Table 25.1; $\beta_j = \beta_{1j}$). Several of the first values of α_{sj} are listed in Table 25.2. The scheme of calculation for $n = 1$, $s = 3$, is the following:

TABLE 25.3

x	Y	F	$\nabla_1 F$	$\nabla_2 F$	$\nabla_3 F$
x_{-3}	Y_{-3}	F_{-3}			
			$\nabla_1 F_{-2}$		
x_{-2}	Y_{-2}	F_{-2}		$\nabla_2 F_{-1}$	
			$\nabla_1 F_{-1}$		$\nabla_3 F_0$
x_{-1}	Y_{-1}	F_{-1}		$\nabla_2 F_0$	
			$\nabla_1 F_0$		$\nabla_3 F_1$
x_0	Y_0	F_0		$\nabla_2 F_1$	
			$\nabla_1 F_1$		
x_1	Y_1	F_1			

Thus, if for $n = 1$ the values $Y_{-3}, Y_{-2}, Y_{-1}, Y_0$ are known, then by (1) or (2) we have

$$ Y_1 = Y_0 + h(F_0 + \tfrac{1}{2}\nabla_1 F_0 + \tfrac{5}{12}\nabla_2 F_0 + \tfrac{3}{8}\nabla_3 F_0) $$

or

$$ Y_1 = Y_0 + h(\tfrac{55}{24}F_0 - \tfrac{59}{24}F_{-1} + \tfrac{37}{24}F_{-2} - \tfrac{9}{24}F_{-3}), $$

etc.

The *estimate of error* for one equation of the first order ($r \geqq 1$) is obtained from

$$ |Y_r - y_r| \leqq \varepsilon z^r + \frac{\gamma_s}{L} h^{s+1}(z^r - 1) \,.\, \max_{x_0 \leqq x \leqq X} |f^{(s+1)}(x)| $$

where

$$f^{(s)}(x) = \frac{d^s f(x, y(x))}{dx^s},$$

$$|Y_j - y_j| \leqq \varepsilon \quad \text{for} \quad j = -s, -s+1, \ldots, -1, 0,$$

z is the smallest root of the equation

$$z^{s+1} - z^s = Lh \sum_{j=0}^{s} |\alpha_{sj}| \, z^{s-j}$$

larger than 1 (for α_{sj} see Table 25.2), L is the Lipschitz constant of the function f and

$$\gamma_s = \frac{\beta_{s+1}}{\sum_{j=0}^{s} |\alpha_{sj}|} \quad \left(\gamma_1 = \tfrac{5}{24}, \gamma_2 = \tfrac{9}{88}, \gamma_3 = \tfrac{251}{4800}\right).$$

In the case (β), the Adams extrapolation method gives the following formulae:

$$Y_{r+1}^{(k)} = \sum_{j=0}^{n-k-1} \frac{h^j}{j!} Y_r^{(k+j)} + h^{n-k} \sum_{j=0}^{s} \beta_{n-k,j} \nabla_j F_j,$$

where

$$\beta_{n-k,j} = \int_0^1 \int_0^{u_{n-k-1}} \ldots \int_0^{u_1} \binom{u+j-1}{j} du \, du_1 \ldots du_{n-k-1} \quad (j \geqq 1)$$

(see Table 25.1).

REMARK 1. The step length is usually chosen so that the magnitude of the first neglected term $\beta_{s+1} \nabla_{s+1} {}^iF_r$ will not influence the magnitude of ${}^iY_{r+1}$. In the case of less exact calculations we require as a crude rule that $h \left| \sum_{j=1}^{n} \frac{\partial^i f}{\partial y_j} \right| \leqq 1 \; (i = 1, \ldots, n)$.

REMARK 2. Changing the step length causes some difficulties. For instance, if we halve the step length, it is necessary to determine the values ${}^iY_{r-1/2}$, ${}^iY_{r-3/2}$ etc., which requires the aid of one of the interpolation formulae.

REMARK 3. The formulae (1) and (2) are not the only possible forms when using the Adams extrapolation procedure. Making use of other relations, one can get different ordinate formulae (see e.g. [69]).

Example 1 ([69], p. 91). The computation of the current in a coil wound on an iron core with a cubic magnetic characteristic leads to the problem

$$y' = -y - 2y^3 + \sin 2x, \quad y(0) = 0.$$

Expanding $y(x)$ into a power series in the neighbourhood of $x = 0$ (see Example 25.3.1, p. 1070) we find that

$$y = x^2 - \tfrac{1}{3}x^3 - \tfrac{1}{4}x^4 + \tfrac{1}{20}x^5 + \tfrac{13}{360}x^6 - \tfrac{733}{2520}x^7 + \tfrac{1903}{6720}x^8 + \dots \tag{4}$$

Substituting $x = \pm 0{\cdot}1$, $x = \pm 0{\cdot}2$ into (4), we obtain y to an accuracy of six decimal places. Applying formula (1) with $n = 1$, $s = 4$, further values of y are determined from $x = 0{\cdot}3$ to $x = 0{\cdot}6$:

TABLE 25.4

x	Y	$2Y^3$	$\sin 2x$	hF	$h\nabla F$	$h\nabla_2 F$	$h\nabla_3 F$	$h\nabla_4 F$
−0·2	0·042,253	0·000,15	−0·389,42	−0·043,182				
−0·1	0·010,308	0·000,00	−0·198,67	−0·020,898	22,284			
0·0	0	0	0	0	20,898	−1386		
0·1	0·009,643	0·000,00	0·198,67	0·018,903	18,903	−1995	−609	
0·2	0·036,951	0·000,10	0·389,42	0·035,237	16,334	−2569	−574	35
0·3	0·079,082	0·000,99	0·564,64	0·048,457	13,220	−3114	−545	29
0·4	0·132,657	0·004,67	0·717,36	0·058,003	9546	−3674	−560	−15
0·5	0·193,687	0·014,53	0·841,47	0·063,325	5322	−4224	−550	10
0·6	0·257,710	0·034,23	0·932,04	0·064,010				

25.8. The Adams Interpolation Method

In this method, in contradistinction to the preceding one, the interpolation polynomial ${}^iP_s(x; x_{r+1})$ is required to pass also through the point which we are seeking. For $r \geqq 0$, we have

$${}^iY_{r+1} = {}^iY_r + h({}^iF_{r+1} + \beta_1^*\nabla_1\,{}^iF_{r+1} + \dots + \beta_s^*\nabla_s\,{}^iF_{r+1}) = \tag{1}$$

$$= {}^iY_r + h(\alpha_{s0}^*\,{}^iF_{r+1} + \alpha_{s1}^*\,{}^iF_r + \dots + \alpha_{ss}^*\,{}^iF_{r+1-s}) \tag{2}$$

(cf. (25.7.1), (25.7.2), pp. 1077, 1078), where $\beta_j^* = \int_{-1}^{0} \binom{u+j-1}{j}\,du$ $(j \geqq 1)$; β_j^*, α_{sj}^* are listed in Tables 25.5, 25.6, respectively, $\beta_j^* = \beta_{1,j}^*$.

On the right-hand side of equations (1), (2) there appear values ${}^if$ at the (as yet unknown) point $(x_{r+1}, {}^1Y_{r+1}, \dots, {}^nY_{r+1})$. These values are to be determined first. The explicit solution of the system (2) is possible only in the simplest cases, so the following iterative method for computing ${}^iY_{r+1}$ is usually employed: As the zero-approximations to ${}^iY_{r+1}$, we take the numbers ${}_0^iY_{r+1}$ calculated, for example, with the aid of the Adams extrapolation method; the successive approximations are then given by the formulae

$$_{j+1}^{\;\,i}Y_{r+1} = {}_j^iY_r + h({}_j^iF_{r+1} + \beta_1^*\nabla_1\,{}_j^iF_{r+1} + \ldots + \beta_s^*\nabla_s\,{}_j^iF_{r+1})\,.$$

TABLE 25.5

$\beta^*_{n-k,j}$

$n-k$ \ j	0	1	2	3	4	5	6
1	1	$-\frac{1}{2}$	$-\frac{1}{12}$	$-\frac{1}{24}$	$-\frac{19}{720}$	$-\frac{3}{160}$	$-\frac{863}{60{,}480}$
2	$\frac{1}{2}$	$-\frac{1}{3}$	$-\frac{1}{24}$	$-\frac{7}{360}$	$-\frac{17}{1440}$	$-\frac{41}{5040}$	$-\frac{731}{120{,}960}$

TABLE 25.6

α^*_{sj}

s \ j	0	1	2	3	4	5
1	$\frac{1}{2}$	$\frac{1}{2}$				
2	$\frac{5}{12}$	$\frac{8}{12}$	$-\frac{1}{12}$			
3	$\frac{9}{24}$	$\frac{19}{24}$	$-\frac{5}{24}$	$\frac{1}{24}$		
4	$\frac{251}{720}$	$\frac{646}{720}$	$-\frac{264}{720}$	$\frac{106}{720}$	$-\frac{19}{720}$	
5	$\frac{475}{1440}$	$\frac{1427}{1440}$	$-\frac{798}{1440}$	$\frac{482}{1440}$	$-\frac{173}{1440}$	$\frac{27}{1440}$

This process is stopped as soon as two successive approximations are practically identical. We try to choose the degree s of the interpolation polynomial ${}^1P_s(x; x_{r+1})$ and the step length h so that one or two iterations are sufficient to obtain values of ${}^iY_{r+1}$.

In the case (β), the corresponding formulae are valid:

$$Y_{r+1}^{(k)} = \sum_{j=0}^{n-k-1} \frac{h^j}{j!} Y_r^{(k+j)} + h^{n-k} \sum_{j=0}^{s} \beta_{n-k,j}^* \nabla_j F_{r+1},$$

$$\beta_{n-k,j}^* = \int_{-1}^{0} \int_{-1}^{u_{n-k-1}} \dots \int_{-1}^{u_1} \binom{u+j-1}{j} \mathrm{d}u \, \mathrm{d}u_1 \dots \mathrm{d}u_{n-k-1} \quad (j \geqq 1).$$

In this case iterative methods for computing $Y_{r+1}^{(k)}$ are also applied.

The *estimate of error* for one equation of the first order (under the assumption that the iterative process has been stopped when two successive approximations are equal, i.e. when equations (1) are satisfied "exactly") is obtained from the formula

$$|Y_r - y_r| \leqq \varepsilon z^r + \gamma_s^* \frac{h^{s+1}}{L} (z^r - 1) \, . \max_{x_0 \leqq x \leqq X} |f^{(s+1)}(x)| ,$$

where

$$f^{(s)}(x) = \frac{\mathrm{d}^s f(x, y(x))}{\mathrm{d}x^s},$$

$|Y_j - y(x_j)| \leqq \varepsilon$ for $-s \leqq j \leqq 0$, z is the smallest root of the equation

$$z^s - z^{s-1} = Lh \sum_{j=0}^{s} |\alpha_{sj}^*| \, z^{s-j}$$

larger than 1, L is the Lipschitz constant of the function f and

$$\gamma_s^* = \frac{|\beta_{s+1}^*|}{\sum_{j=0}^{s} |\alpha_{sj}^*|} \quad \left(\gamma_1^* = \frac{1}{12}, \gamma_2^* = \frac{1}{28}, \gamma_3^* = \frac{19}{1020}, \gamma_4^* = \frac{27}{2572}\right).$$

25.9. Applicability of Different Methods

The polygonal method is appropriate only if rough knowledge of the behaviour of the solution is required. The method of successive approximations and the power series method are used if a single explicit formula for all x in the interval considered is desired. In practice, both these methods are applicable, with sufficient accuracy, only in the simplest cases.

The Runge–Kutta method and the two Adams methods are used to obtain solutions of a high accuracy. The advantage of the Runge–Kutta method is that all approximate values of the solution are determined by the same rule and that the step length may be easily changed. On the other hand, in case (γ), each step requires the computation of $4n$ values of ${}^if$ (which is usually a very laborious task), whereas when using the Adams extrapolation method or the Adams interpolation method with k iterations, we require n or $(k + 1)n$ (i.e. practically $2n$ or $3n$) values of ${}^i f$, respectively. Another advantage of the Adams methods is that each irregularity in the behaviour of differences $|{}^iY_{r+1} - {}^iY_r|$ automatically reveals a mistake in calculations. The comparatively accurate and applicable estimate of error also constitutes an advantage. However, as far as automatic computers are concerned the advantages of the Runge–Kutta method appear to be of more importance.

B. BOUNDARY-VALUE PROBLEMS

On the theory of boundary-value problems see § 17.17.

Boundary-value problems are, in general, more difficult to solve than initial-value problems since usually no effective criteria exist by means of which we can easily decide on the solvability of the given problem. Hence, a more detailed theoretical discussion is needed before a numerical solution is started. In practice, the solvability and uniqueness of the solution of the given problem is often taken for granted from the physical or technical nature of the problem in question.

25.10. Reduction to Initial-Value Problems

Let the differential equation

$$y^{(n)} + p_1(x)\, y^{(n-1)} + \ldots + p_{n-1}(x)\, y' + p_n(x)\, y = g(x) \tag{1}$$

with boundary conditions

$$h_i(y(a), y'(a), \ldots, y^{(n-1)}(a), y(b), y'(b), \ldots, y^{(n-1)}(b)) = 0\,, \quad i = 1, 2, \ldots, n\,, \tag{2}$$

be given. Let the functions $p_k(x)$, $g(x)$ be continuous in the interval $[a, b]$ and let the h_i be continuous functions of their arguments.

The general solution of equation (1) is

$$y = C_1 y_1 + \ldots + C_n y_n + v\,, \tag{3}$$

where $y_1, y_2, \ldots, y_n$ form a fundamental system of the homogeneous equation

$$y^{(n)} + p_1 y^{(n-1)} + \ldots + p_n y = 0 \tag{4}$$

and v is an arbitrary particular integral of (1).

This circumstance may be used to find the approximate solution of (1) and (2). The procedure will be shown in the case of a second-order equation

$$y'' + p_1(x)\, y' + p_2(x)\, y = g(x) \tag{5}$$

with boundary conditions

$$h_1(y(a), y'(a), y(b), y'(b)) = 0\,, \tag{6}$$

$$h_2(y(a), y'(a), y(b), y'(b)) = 0\,. \tag{7}$$

If y_1, y_2 and v can be found in a closed form, we substitute the expression

$$y = C_1 y_1 + C_2 y_2 + v \tag{8}$$

into (6) and (7). If the given boundary-value problem has a unique solution, the constants C_1 and C_2 are then uniquely determined by equations (6) and (7). These equations (non-linear, in general), may then be solved numerically.

If the functions y_1, y_2 and v cannot be determined in a closed form, we proceed as follows: Choose a system of initial values with a *non-vanishing* determinant

$$D = \begin{vmatrix} y_1(a), & y_2(a) \\ y_1'(a), & y_2'(a) \end{vmatrix} \tag{9}$$

(so that y_1 and y_2 are linearly independent) and calculate the values y_1 and y_2 numerically (by some method given above, e.g. by the Runge–Kutta method). Similarly, for the function v choose arbitrary initial values and determine v numerically. In this way, we find the approximate values of these functions and their derivatives at the point b. The approximate solution is again written in the form (8) and by substituting it into (6) and (7), we obtain equations for C_1 and C_2, which are uniquely solvable if the given problem has a unique solution and if the functions y_1, y_2 and v are computed with sufficient accuracy.

Thus, this method transforms an approximate solution of the boundary-value problem into an approximate solution of an initial-value problem.

REMARK 1. This method may lead to very unsatisfactory results. The reasons for this and methods of overcoming the difficulties are given in [326].

25.11. Finite Difference Method

Let the equation

$$f(x, y, y', \ldots, y^{(n-1)}, y^{(n)}) = 0 \tag{1}$$

with boundary conditions

$$g_i(y(a), y'(a), \ldots, y^{(n-1)}(a), y(b), y'(b), \ldots, y^{(n-1)}(b)) = 0\,, \quad i = 1, 2, \ldots, n\,, \tag{2}$$

be given.

Let the functions f and g_i, together with their first partial derivatives with respect to $y, y', \ldots, y^{(n)}$ or $y(a), \ldots, y^{(n-1)}(a), y(b), \ldots, y^{(n-1)}(b)$, respectively, be continuous (in all their arguments) in the region considered. Divide the interval $[a, b]$ by pivotal points

$$x_r = a + rh\,, \quad h = \frac{b-a}{m} \quad (x_0 = a, x_m = b)$$

into m equal subintervals. The derivatives in (1) and (2) are replaced by

$$\left(\frac{\mathrm{d}y}{\mathrm{d}x}\right)_{x_r} \approx \frac{Y_{r+1} - Y_{r-1}}{2h}\,, \quad \left(\frac{\mathrm{d}^2y}{\mathrm{d}x^2}\right)_{x_r} \approx \frac{Y_{r+1} - 2Y_r + Y_{r-1}}{h^2}\,,$$

$$\left(\frac{\mathrm{d}^3y}{\mathrm{d}x^3}\right)_{x_r} \approx \frac{Y_{r+2} - 2Y_{r+1} + 2Y_{r-1} - Y_{r-2}}{2h^3}\,,$$

$$\left(\frac{\mathrm{d}^4y}{\mathrm{d}x^4}\right)_{x_r} \approx \frac{Y_{r+2} - 4Y_{r+1} + 6Y_r - 4Y_{r-1} + Y_{r-2}}{h^4}\,,$$

$$\left(\frac{\mathrm{d}^{2k}y}{\mathrm{d}x^{2k}}\right)_{x_r} \approx \frac{1}{h^{2k}}\,\Delta^{2k} Y_{r-k}\,,$$

$$\left(\frac{\mathrm{d}^{2k+1}y}{\mathrm{d}x^{2k+1}}\right)_{x_r} \approx \frac{1}{2h^{2k+1}}\left(\Delta^{2k+1} Y_{r-(k+1)} + \Delta^{2k+1} Y_{r-k}\right)$$

and the resulting equations solved for Y_r, $r = 0, 1, \ldots, n$.

For n even, we obtain, in the general case, a system of $m + n + 1$ equations for the same number of unknowns $Y_{-n/2}, \ldots, Y_0, Y_1, \ldots, Y_{m+n/2}$ (of which, of course, only $Y_0, \ldots, Y_m$ are of interest to us). For n odd, the number of equations is less than the number of unknowns. (For more detail about this case see [69].) Often the unsymmetric odd finite differences

$$\left(\frac{\mathrm{d}y}{\mathrm{d}x}\right)_{x_r} \approx \frac{Y_{r+1} - Y_r}{h}\,, \quad \left(\frac{\mathrm{d}^3y}{\mathrm{d}x^3}\right)_{x_r} \approx \frac{Y_{r+2} - 3Y_{r+1} + 3Y_r - Y_{r-1}}{h^3}$$

are used.

REMARK 1. The solvability of the equations for Y_r cannot be guaranteed in general.

REMARK 2. An error estimate is rather complicated. For some special cases see [69].

REMARK 3. In order to increase the accuracy of the finite difference method, we can replace the derivatives by finite-difference expressions of higher accuracy. For another improved method see [69], p. 160.

Example 1 (see [69], p. 145). Let us solve the boundary-value problem

$$y'' = \tfrac{3}{2}y^2\,, \quad y(0) = 4\,, \quad y(1) = 1\,. \tag{3}$$

In order to obtain an approximate solution let us take $h = 0{\cdot}5$. We get

$$Y_0 = 4, \quad Y_2 = 1, \quad \frac{Y_2 - 2Y_1 + Y_0}{0{\cdot}25} = \tfrac{3}{2}Y_1^2. \tag{4}$$

The last equation yields

$$3Y_1^2 + 16Y_1 - 40 = 0,$$

$$Y_1 \doteq \begin{cases} 1{\cdot}855, \\ -7{\cdot}188. \end{cases}$$

The error in Y_1, which approximates to that solution from two existing solutions which may be expressed in terms of elementary functions, is about 4·3%.

25.12. Direct Methods. (Collocation Method, Least-Squares Method.)

Let the equation

$$f(x, y, y', \ldots, y^{(n-1)}, y^{(n)}) = 0 \tag{1}$$

with boundary conditions

$$g_i(y(a), y'(a), \ldots, y^{(n-1)}(a), y(b), y'(b), \ldots, y^{(n-1)}(b)) = 0, \quad i = 1, 2, \ldots, n \tag{2}$$

be given. Let the functions f and g_i, together with their first partial derivatives with respect to $y, y', \ldots, y^{(n)}$, or $y(a), \ldots, y^{(n-1)}(a), y(b), \ldots, y^{(n-1)}(b)$, respectively, be continuous (in all their arguments) in the region considered. Choose a function

$$\varphi(x, a_1, \ldots, a_m) \tag{3}$$

which satisfies the conditions (2) for any choice of parameters $a_1, a_2, \ldots, a_m$. Substituting (3) into the left-hand side of (1) we do not, in general obtain zero but some "error"-function

$$\varepsilon(x, a_1, \ldots, a_m). \tag{4}$$

We determine the parameters $a_1, a_2, \ldots, a_m$ so as to make the error-function as small as possible.

a) *Collocation Method*. We require the function (3) to satisfy equation (1) at m prescribed points. Thus, we get m equations for the m unknowns $a_1, a_2, \ldots, a_m$.

b) *Least-Squares Method*. We require that the relation

$$I = \int_a^b \varepsilon^2 \, dx = \min \tag{5}$$

holds. The necessary conditions to satisfy equation (5) are

$$\frac{\partial I}{\partial a_k} = 0 \quad (k = 1, \dots, m), \tag{6}$$

which again yield a system of m equations for the m unknowns $a_1, a_2, \dots, a_m$. In addition, we must verify that the obtained result really gives the minimum of the integral I.

Example 1. Let us solve the problem

$$L(y) \equiv y'' + (1 + x^2)\, y + 1 = 0\,, \quad y(1) = 0\,, \quad y(-1) = 0\,. \tag{7}$$

Choose

$$\varphi = (1 - x^2)\,(a_1 + a_2 x^2)$$

and require that $L(\varphi) = 0$ at the points $x_1 = \frac{1}{4}$, $x_2 = \frac{3}{4}$ (and hence, because of symmetry also at the points $x_3 = -\frac{1}{4}$, $x_4 = -\frac{3}{4}$). This yields

$$1 - \frac{257}{256}\, a_1 + \frac{5375}{4096}\, a_2 = 0\,,$$

$$1 - \frac{337}{256}\, a_1 - \frac{17{,}881}{4096}\, a_2 = 0$$

whence $a_1 = 0{\cdot}929{,}254$, $a_2 = -0{\cdot}051{,}146$.

Example 2 ([69], p. 184). Let the boundary-value problem

$$-2y'' + 3y^2 = 0\,, \quad y(0) = 4\,, \quad y(1) = 1$$

be given. Choose

$$\varphi = 4 - 3x + a(x - x^2) \tag{8}$$

which satisfies the boundary conditions for an arbitrary a. Then

$$\varepsilon = -2\varphi'' + 3\varphi^2 = 3(16 - 24x + 9x^2) + 2a(2 + 12x - 21x^2 + 9x^3) + \\ + 3a^2(x^2 - 2x^3 + x^4)\,,$$

$$I = \int_0^1 \varepsilon^2 \, \mathrm{d}x\,, \quad \frac{\mathrm{d}I}{\mathrm{d}a} = \tfrac{1}{70}(4a^3 + 303a^2 + 6696a + 19{,}740) = 0\,,$$

which yields the unique real value $a = -3{\cdot}467$. Inserting this value into (8), we get for example when $x = \frac{1}{2}$, $\varphi = 1{\cdot}633\,2$ (cf. Example 25.11.1, p. 1085).

25.13. Method of Successive Approximations

Let the problem

$$f(x, y, \dots, y^{(n)}) = 0\,, \tag{1}$$

$$g_i(y(a), \dots, y^{(n-1)}(a), y(b), \dots, y^{(n-1)}(b)) = 0 \tag{2}$$

be re-written in the form

$$f_1(x, y, \dots, y^{(n)}) = f_2(x, y, \dots, y^{(m)})\,, \tag{1'}$$

where $m < n$, the boundary conditions remaining unchanged, and let the boundary-value problem given by

$$f_1(x, y, \dots, y^{(n)}) = r(x) \tag{3}$$

with boundary conditions (2) be solvable for any continuous function $r(x)$. Let us choose an arbitrary sufficiently smooth function $Y_0(x)$ and construct the sequence of functions $Y_0(x)$, $Y_1(x)$, $Y_2(x)$, ... where $Y_i(x)$ $(i = 1, 2, \dots)$ are solutions of the boundary-value problems

$$f_1(x, Y_{k+1}, \dots, Y_{k+1}^{(n)}) = f_2(x, Y_k, \dots, Y_k^{(m)})\,, \tag{4}$$

$$g_i(Y_{k+1}(a), \dots, Y_{k+1}^{(n-1)}(a), Y_{k+1}(b), \dots, Y_{k+1}^{(n-1)}(b)) = 0\,. \tag{5}$$

The convergence of $Y_i(x)$ to the exact solution $y(x)$ can be proved under rather special assumptions only. The speed of convergence evidently depends on the proper decomposition of the function f into f_1 and f_2. For more details see [69].

25.14. Perturbation Method

The procedure is quite similar to that explained in § 25·4. The problem is assumed to have a unique solution for $\varepsilon = 0$.

Example 1 ([69], p. 188). Let us solve the problem

$$y'' + (1 + \varepsilon x^2)\, y + 1 = 0\,, \quad y(-1) = 0\,, \quad y(1) = 0. \tag{1}$$

The solution is assumed to have the form

$$y(x) = y_0(x) + \varepsilon y_1(x) + \varepsilon^2 y_2(x) + \dots\,, \tag{2}$$

$$y_k(-1) = 0\,, \quad y_k(1) = 0 \quad (k = 0, 1, 2, \dots)\,. \tag{3}$$

Substituting (2) into (1), we get

$$y_0'' + y_0 + 1 + \varepsilon(y_1'' + y_1 + x^2 y_0) + \varepsilon^2(\ldots) + \ldots = 0 . \tag{4}$$

The equating of the coefficients of the powers of ε yields

$$y_0'' + y_0 + 1 \quad = 0 , \quad y_0(-1) = 0 , \quad y_0(1) = 0 , \tag{5}$$

$$y_1'' + y_1 + x^2 y_0 = 0 , \quad y_1(-1) = 0 , \quad y_1(1) = 0 , \tag{6}$$

$$\ldots\ldots\ldots\ldots\ldots\ldots\ldots\ldots\ldots\ldots\ldots\ldots$$

Solving (5) we find that

$$y_0 = \frac{\cos x}{\cos 1} - 1 ,$$

and solving (6) we find that

$$y_1 = x^2 - 2 + \frac{1}{12 \cos 1}\left[(3x - 2x^3) \sin x - 3x^2 \cos x\right] + A \cos x ,$$

where

$$A = \frac{1}{12 \cos^2 1}(15 \cos 1 - \sin 1) ;$$

finally

$$y \approx y_0 + \varepsilon y_1 .$$

25.15. Applicability of Different Methods

It is rather difficult to appraise the individual methods mentioned above. If the problem is linear, use is made of the method of § 25.10. The finite difference method, the collocation method and the least-squares method enable us to obtain, with relatively little labour, a general idea of the behaviour of the solution. The success of a method very often depends on the skilfulness of the solver, e.g. in the methods of § 25.12 it is of high importance to choose a function $\varphi(x, a_1, a_2, \ldots, a_m)$, not only satisfying the boundary conditions but also anticipating the behaviour of the sought solution, whereas in the method of § 25.13 the essential thing is a suitable decomposition of the function f into f_1 and f_2 and an appropriate choice of Y_0, which should be near to the solution, etc.

Besides the methods described here, there exist also many other methods (e.g. the Ritz method, see Chapter 24, etc.). For more details see [69].

C. EIGENVALUE PROBLEMS

25.16. Variational (Direct) Methods

We shall consider here only self-adjoint positive definite problems (see § 17.17), since these occur most frequently in practice. A very simple example of these problems is given by

$$-y'' = \lambda y\,, \tag{1}$$

$$y(0) = 0\,, \quad y(1) = 0 \tag{2}$$

which has the eigenvalues

$$\lambda_1 = \pi^2,\ \lambda_2 = 4\pi^2,\ \lambda_3 = 9\pi^2,\ \dots . \tag{3}$$

In general, the problem given by the equation

$$M(y) = \lambda N(y) \tag{4a}$$

with homogeneous boundary conditions

$$\sum_{k=0}^{2m-1} \left[\alpha_{j,k} y^{(k)}(a) + \beta_{j,k} y^{(k)}(b)\right] = 0 \quad (j = 1, 2, \dots, 2m) \tag{4b}$$

is considered, where M and N are differential operators (for details see Definition 17.17.3, p. 803).

Theorem 1 (*Rayleigh's Quotient*). *Let a self-adjoint positive definite eigenvalue problem be given. To each comparison function* (see Definition 17.17.5, p. 804) *let the number* $R(u)$ (*the so-called Rayleigh's quotient*) *be assigned, according to the formula*

$$R(u) = \frac{\displaystyle\int_a^b u\, M(u)\, dx}{\displaystyle\int_a^b u\, N(u)\, dx}\,. \tag{5}$$

Then, the least eigenvalue λ_1 *of this problem is given by*

$$\lambda_1 = \min R(u)\,, \tag{6}$$

where u *ranges over the set of all comparison functions.*

Example 1. For the problem (1), (2), which is self-adjoint and positive definite, the function

$$u = x(x - 1) \tag{7}$$

is an example of a comparison function. Here $M(y) = -y''$, $N(y) = y$, thus

$$R(u) = \frac{-\int_0^1 x(x-1)\,.\,[x(x-1)]''\,\mathrm{d}x}{\int_0^1 [x(x-1)]^2\,\mathrm{d}x} = \frac{\frac{1}{3}}{\frac{1}{30}} = 10\,. \tag{8}$$

REMARK 1. The Rayleigh quotient yields an upper estimate to the least eigenvalue. In the last example we chose $u = x(x-1)$ as a comparison function and we obtained $R(u) = 10$, whereas the exact value is $\lambda_1 = \pi^2 < 10$.

In the following text, we shall state two theorems (so-called *bracketing theorems*) which yield both upper and lower bounds for λ_1.

Theorem 2 (*The Schwarz Constants*). *Let a self-adjoint positive definite eigenvalue problem be given. Let us consider a sequence of comparison functions* $F_0(x), F_1(x), \ldots$ *such that*

$$M(F_k) = N(F_{k-1})\,, \tag{9}$$

and let the constants a_k *and* μ_{k+1} *(the so-called Schwarz constants and Schwarz quotients, respectively) be defined as follows:*

$$a_k = \int_a^b F_0\,N(F_k)\,\mathrm{d}x\,, \tag{10}$$

$$\mu_{k+1} = \frac{a_k}{a_{k+1}}\,. \tag{11}$$

If the least eigenvalue λ_1 *is simple, then the following relations hold*

$$\mu_1 \geqq \mu_2 \geqq \mu_3 \geqq \ldots \geqq \lambda_1\,, \tag{12}$$

$$\mu_{k+1} - \frac{\mu_k - \mu_{k+1}}{l_2/\mu_{k+1} - 1} \leqq \lambda_1 \leqq \mu_{k+1} \tag{13}$$

where l_2 *is a lower estimate of the second eigenvalue* λ_2 *such that*

$$l_2 > \mu_{k+1}\,. \tag{14}$$

REMARK 2. On the estimate l_2, see Example 2.

REMARK 3. It may be shown that

$$a_k = \int_a^b F_i\,N(F_{k-i})\,\mathrm{d}x\,,$$

where i is an arbitrary natural number between 0 and k. This property may be useful in numerical calculations.

REMARK 4. The inequality (13) enables us to determine both upper and lower estimates of the least eigenvalue, which are not given by the Rayleigh's quotient.

However, the construction of comparison functions satisfying condition (9) is difficult. Hence, for so-called *regular* problems (see Remark 17.17.10, p. 806) the procedure given by the following theorem is more practicable.

Theorem 3 (*The Nečas Theorem*). *Let a self-adjoint positive definite regular eigenvalue problem be given. (In this case, the operator $N(y)$ has the form*

$$N(y) = (-1)^n \left[g(x)\, y^{(n)}\right]^{(n)} .) \tag{15}$$

Let $F_0(x)$ and $F_1(x)$ be sufficiently smooth functions such that

$$M(F_1) = N(F_0), \tag{16}$$

and let $F_1(x)$ be a comparison function ($F_0(x)$ need not be a comparison function). Let us construct the constants

$$a_0 = \int_a^b g(x) \left[F_0^{(n)}(x)\right]^2 \mathrm{d}x , \tag{17}$$

$$a_1 = \int_a^b g(x)\, F_0^{(n)}(x)\, F_1^{(n)}(x) \,\mathrm{d}x , \tag{18}$$

$$a_2 = \int_a^b g(x) \left[F_1^{(n)}(x)\right]^2 \mathrm{d}x , \tag{19}$$

$$\mu_1 = \frac{a_0}{a_1}, \quad \mu_2 = \frac{a_1}{a_2}. \tag{20}$$

Then the inequality

$$\mu_2 - \frac{\mu_1 - \mu_2}{l_2/\mu_2 - 1} \leqq \lambda_1 \leqq \mu_2 \tag{21}$$

holds, where l_2 is a lower bound for the second eigenvalue such that $l_2 > \mu_2$ (see Example 2).

Example 2. Let the eigenvalue problem

$$-y'' = \frac{\lambda y}{E(1 - 0{\cdot}4|x|)^2}, \quad y(-1) = 0, \quad y(1) = 0 \tag{22}$$

be given. (This problem arises from by the consideration of the buckling stress of a bar with a variable cross-section.) Evidently, it is self-adjoint, positive definite and regular. Choose

$$F_1(x) = 1 - x^2 . \tag{23}$$

By (16), we find that

$$-F_1'' = \frac{1}{E(1 - 0{\cdot}4|x|)^2} F_0 , \tag{24}$$

whence

$$F_0(x) = 2E(1 - 0{\cdot}4|x|)^2 . \tag{25}$$

(The ease of determination of F_0 is one of the main advantages of the Nečas theorem.) Further, by (17)–(20), we obtain

$$a_0 = \int_{-1}^{1} \frac{1}{E(1 - 0{\cdot}4|x|)^2} \,.\, 4E^2(1 - 0{\cdot}4|x|)^4 \, dx = 5{\cdot}227E ,$$

$$a_1 = \int_{-1}^{1} \frac{1}{E(1 - 0{\cdot}4|x|)^2} \,.\, 2E(1 - 0{\cdot}4|x|)^2 (1 - x^2) \, dx = 2{\cdot}667 ,$$

$$a_2 = \int_{-1}^{1} \frac{1}{E(1 - 0{\cdot}4|x|)^2} (1 - x^2)^2 \, dx = \frac{1{\cdot}437}{E} ,$$

$$\mu_1 = 1{\cdot}960E , \quad \mu_2 = 1{\cdot}856E . \tag{26}$$

To determine l_2, let us compare the problem (22) with the problem

$$-y'' = \frac{\lambda y}{E \,.\, 0{\cdot}6^2} , \quad y(-1) = 0 , \quad y(1) = 0 , \tag{27}$$

with constant coefficients for which $l_2 = \pi^2 E \,.\, 0{\cdot}6^2 = 3{\cdot}560E$. According to Theorem 17.17.7, p. 809, this is a lower bound for the second eigenvalue of the problem (22). By (21) we find that

$$1{\cdot}856 - 1{\cdot}856 \,.\, \frac{0{\cdot}104}{1{\cdot}704} \leqq \frac{\lambda_1}{E} \leqq 1{\cdot}856 ,$$

$$1{\cdot}743 \leqq \frac{\lambda_1}{E} \leqq 1{\cdot}856 ,$$

which represents quite a satisfatory estimate for practical use. Variational methods are described in more detail in [69] and [282].

25.17. Finite Difference Method

We shall demonstrate the use of the method by an example.

Example 1. Let us again solve the eigenvalue problem of Example 25.16.2,

$$-y'' = \frac{\lambda y}{E(1 - 0{\cdot}4|x|)^2}, \quad y(-1) = 0, \quad y(1) = 0. \tag{1}$$

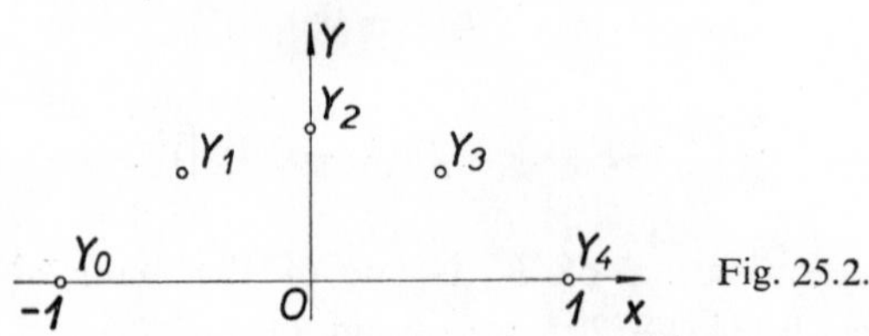

Fig. 25.2.

Divide the interval $[-1, 1]$ into four equal parts, replace the second derivative in (1) by the second difference-quotient and denote by Y_0, Y_1, Y_2, Y_3, Y_4 the values of the approximate solution at the points -1; $-0{\cdot}5$; 0; $0{\cdot}5$; 1 (see Fig. 25.2). (By virtue of the boundary conditions, we have $Y_0 = Y_4 = 0$.) Denoting $\lambda/E = \mu$, we get

$$-\frac{Y_2 - 2Y_1 + Y_0}{0{\cdot}5^2} = \frac{\mu Y_1}{0{\cdot}8^2}$$

$$-\frac{Y_3 - 2Y_2 + Y_1}{0{\cdot}5^2} = \frac{\mu Y_2}{1^2},$$

$$-\frac{Y_4 - 2Y_3 + Y_2}{0{\cdot}5^2} = \frac{\mu Y_3}{0{\cdot}8^2}, \tag{2}$$

or

$$\begin{aligned} (2 - 0{\cdot}39\mu)\, Y_1 \qquad - Y_2 \qquad\qquad &= 0\,, \\ - Y_1 + (2 - 0{\cdot}25\mu)\, Y_2 \qquad - Y_3 &= 0\,, \\ - Y_2 + (2 - 0{\cdot}39\mu)\, Y_3 &= 0\,. \end{aligned} \tag{3}$$

The necessary and sufficient condition for the system (3) to have a non-zero solution is that

$$(2 - 0{\cdot}39\mu)(0{\cdot}0975\mu^2 - 1{\cdot}28\mu + 2) = 0\,, \tag{4}$$

whence

$$\mu_1 \doteq 1{\cdot}810\,, \quad \mu_2 \doteq 5{\cdot}128\,, \quad \mu_3 \doteq 11{\cdot}318\,. \tag{5}$$

These represent approximations to the first three eigenvalues. Inserting μ_1 into (3) and choosing $Y_2 = 1$, we obtain

$$Y_1 = Y_3 \doteq 0{\cdot}773$$

(see Fig. 25.2) and thereby approximately the first eigenfunction. Similarly, substituting μ_2 and μ_3, respectively, into (3), we obtain approximations to the next two eigenfunctions.

The problem of error estimation is rather difficult. For more details on the finite difference and other methods see [69].

D. PERIODIC SOLUTIONS

25.18. Perturbation Method for a Weakly Nonlinear Oscillator

Among boundary-value problems those with periodic boundary conditions play an important role, being equivalent to the problem of finding a periodic solution of an equation. We shall describe here only the so-called *perturbation method* in the simple case of a weakly nonlinear oscillator. The exposition differs here from that of § 25·14 in that we also mention the case when the problem does not possess a unique solution for $\varepsilon = 0$. This case is of particular importance in practice. In systems of the n-th order, the procedure is similar; see [65].

(a) Non-autonomous case. Let the equation

$$\ddot{x} + k^2 x = \varepsilon f(t, x, \dot{x}, \varepsilon) \tag{1}$$

be given, where f is a 2π-periodic function of t (i.e. $f(t + 2\pi, x, \dot{x}, \varepsilon) = f(t, x, \dot{x}, \varepsilon)$), continuous in all its arguments and having continuous partial derivatives, up to the second order, with respect to x and $\dot{x}$ in the region considered.

Theorem 1. *Let $k \neq n$, n an integer. Then for sufficiently small ε* (see Example 1) *equation* (1) *has a 2π-periodic solution, which, when $\varepsilon \to 0$, tends to the trivial solution $x \equiv 0$ of the limit equation $\ddot{x} + k^2 x = 0$.*

REMARK 1. If k is near to an integer n, the theorem is only applicable for very small values of ε and it is more advantageous to put $k^2 = n^2 + \varepsilon a$ and to consider the equation

$$\ddot{x} + n^2 x = \varepsilon[f(t, x, \dot{x}, \varepsilon) - ax] = \varepsilon \tilde{f}(t, x, \dot{x}, \varepsilon)$$

in accordance with Theorem 2.

Theorem 2. *Let $k = n$, n an integer. Let the system of equations*

$$\begin{aligned}
&\int_0^{2\pi} f(s, c_0 \cos ns + d_0 \sin ns, -nc_0 \sin ns + nd_0 \cos ns, 0) \sin ns \, ds = 0\,, \\
&\int_0^{2\pi} f(s, c_0 \cos ns + d_0 \sin ns, -nc_0 \sin ns + nd_0 \cos ns, 0) \cos ns \, ds = 0
\end{aligned} \tag{2}$$

have a real solution $c_0 = c_0^*$, $d_0 = d_0^*$. *Let the Jacobian of the system* (2) *with respect to* c_0 *and* d_0 *be non-zero at the point* $c_0 = c_0^*$, $d_0 = d_0^*$.

Then equation (1) *has a* 2π*-periodic solution which for* $\varepsilon \to 0$ *tends to the solution* $x_0(t) = c_0^* \cos nt + d_0^* \sin nt$ *of the limit equation*

$$\ddot{x} + n^2 x = 0\,.$$

Example 1. Let the equation

$$\ddot{x} + \tfrac{1}{4}x = \varepsilon(x^2 + \sin t) \tag{3}$$

be given. Evidently k is not an integer and the function f satisfies all the assumptions of Theorem 1. Hence, for sufficiently small ε, a 2π-periodic solution of (3) exists and is approximately equal to zero.

In order to determine the solution with higher accuracy, using the fact that the function f is analytic in x, $\dot{x}$ and ε, we assume x to be of the form (cf. § 25.4)

$$x = x_0 + \varepsilon x_1 + \varepsilon^2 x_2 + \ldots\,. \tag{4}$$

Substituting (4) into (3) and comparing the coefficients of powers of ε, we have that

$$\ddot{x}_0 + \varepsilon\ddot{x}_1 + \varepsilon^2\ddot{x}_2 + \ldots + \tfrac{1}{4}(x_0 + \varepsilon x_1 + \varepsilon^2 x_2 + \ldots) =$$
$$= \varepsilon[x_0^2 + 2\varepsilon x_0 x_1 + \varepsilon^2(2x_0x_2 + x_1^2) + \ldots + \sin t]\,,$$
$$\ddot{x}_0 + \tfrac{1}{4}x_0 = 0\,, \tag{5}$$

$$\ddot{x}_1 + \tfrac{1}{4}x_1 = x_0^2 + \sin t\,, \tag{6}$$

$$\ldots\ldots\ldots\ldots\ldots\ldots\ldots\ldots\ldots$$

Solution of (5) gives

$$x_0 = A \cos \tfrac{1}{2}t + B \sin \tfrac{1}{2}t\,,$$

and is 2π-periodic if and only if $A = B = 0$. Equation (6) yields (since $x_0 \equiv 0$)

$$x_1 = -\tfrac{4}{3}\sin t + C\cos\tfrac{1}{2}t + D\sin\tfrac{1}{2}t\,. \tag{7}$$

Putting $C = D = 0$ and further $x_2 = x_3 = \ldots \equiv 0$, we have from (4)

$$x \approx -\tfrac{4}{3}\varepsilon \sin t \tag{8}$$

which represents a more accurate 2π-periodic solution of (3).

(b) Autonomous case.

Theorem 3. *Let the equation*

$$\ddot{x} + k^2 x = \varepsilon f(x, \dot{x}, \varepsilon) \tag{9}$$

be given, where k is a real number and f is continuous in all its arguments and has continuous partial derivatives, up to the second order, with respect to x and $\dot{x}$.

Let the system

$$\begin{aligned} &\int_0^{2\pi/k} f(c_0 \cos ks, -kc_0 \sin ks, 0) \sin ks \, ds = 0 , \\ -v_0 c_0 + &\int_0^{2\pi/k} f(c_0 \cos ks, -kc_0 \sin ks, 0) \cos ks \, ds = 0 \end{aligned} \tag{10}$$

have a real solution $v_0 = v_0^*$, $c_0 = c_0^*$. Let the Jacobian of the system (10) with respect to v_0 and c_0 be non-zero at the point $v_0 = v_0^*$, $c_0 = c_0^*$.

Then equation (9) has, for sufficiently small ε, an ω(ε)-periodic solution, where $\omega(\varepsilon) = \frac{2\pi}{k}(1 + \varepsilon v^*(\varepsilon))$; when $\varepsilon \to 0$, this solution tends to the $\frac{2\pi}{k}$-periodic solution $c_0^* \cos kt$ of the limit equation $\ddot{x} + k^2 x = 0$ and $v^*(0) = v_0^*$.

Example 2. Let us consider *van der Pol's equation*

$$\ddot{x} + x = \varepsilon(1 - x^2)\,\dot{x} \tag{11}$$

(describing oscillations in a certain electronic circuit). Then equations (10) take the form

$$\begin{aligned} &\int_0^{2\pi} (1 - c_0^2 \cos^2 s)\, c_0 \sin^2 s \, ds = \pi c_0(1 - \tfrac{1}{4}c_0^2) = 0 , \\ -c_0 v_0 + &\int_0^{2\pi} (1 - c_0^2 \cos^2 s)\, c_0 \sin s \cos s \, ds = -c_0 v_0 = 0 . \end{aligned} \tag{12}$$

Thus, the system has either the solution $c_0 = \pm 2$, $v_0 = 0$ or the solution $c_0 = 0$, v_0 arbitrary. The Jacobian of (12) has the value

$$\begin{vmatrix} 0, & \pi(1 - \frac{3}{4}c_0^2) \\ -c_0, & -v_0 \end{vmatrix} = \pi c_0(1 - \tfrac{3}{4}c_0^2) .$$

In the first case, it is clearly non-vanishing and for sufficiently small ε there exists a $2\pi(1 + \varepsilon\, v(\varepsilon))$-periodic solution of (11), which for $\varepsilon \to 0$ tends to the solution $2 \cos t$ of the limit equation. On the other hand, in the second case the Jacobian is zero; consequently further investigation is needed in order to determine whether there exists a periodic solution which tends to the corresponding solution of the equation $\ddot{x} + x = 0$ as $\varepsilon \to 0$. (This last solution is now trivial, since $c_0 = 0$.)

26. SOLUTION OF PARTIAL DIFFERENTIAL EQUATIONS BY INFINITE SERIES (BY THE FOURIER METHOD)

By Karel Rektorys

References: [60], [75], [91], [243], [275], [292], [317], [371], [398], [426].

In this chapter we show some typical examples of the so-called *Fourier method* for solving boundary-value problems, otherwise known as the *method of separation of variables* or the *product method.* This method consists – roughly speaking – in assuming the solution of the given boundary-value problem in the form

$$u = \sum_{n=1}^{\infty} u_n\,, \tag{1}$$

where u_n are functions satisfying the given differential equation and some of the given boundary conditions and are assumed to be in the form of a product of functions of one variable only. The method will be thoroughly explained in § 26.1; in the remaining paragraphs of this chapter we shall proceed more briefly.

26.1. Equation of a Vibrating String

A function $u(x, t)$ is to be found, satisfying the differential equation

$$\frac{\partial^2 u}{\partial t^2} = \frac{\partial^2 u}{\partial x^2} \quad (0 < x < l, t > 0)\,, \tag{2}$$

continuous in the domain $\bar{O}(0 \leqq x \leqq l,\ t \geqq 0)$ and satisfying the following initial and boundary conditions:

$$u(x, 0) = \varphi_0(x)\,, \tag{3}$$

$$\frac{\partial u}{\partial t}(x, 0) = 0\,, \tag{4}$$

$$u(0, t) = 0\,, \tag{5}$$

$$u(l, t) = 0\,. \tag{6}$$

The continuity required implies that

$$\varphi_0(0) = 0\,, \quad \varphi_0(l) = 0\,. \tag{7}$$

The physical meaning of the problem is the following: A string of length l with its ends kept fixed (conditions (5), (6)) is put into the (initial) position with amplitude $\varphi_0(x)$ and, after being released at the time $t = 0$, begins to vibrate. (Thus, at this time $t = 0$, its points have zero velocity; condition (4).) The amplitude $u(x, t)$ at each point x $(0 \leqq x \leqq l)$ and for all t $(t \geqq 0)$ is to be determined. (Other problems also lead to the solution of equation (2), for example longitudinal vibration of bars, twisting of bars etc.)

Let us assume the solution of the problem in the form

$$u(x, t) = \sum_{n=1}^{\infty} a_n u_n(x, t)\,, \tag{8}$$

where each of the functions $u_n(x, t)$ has the following properties:

(a) it is of the form $u_n(x, t) = X_n(x)\, T_n(t)$, (9)

(b) it satisfies equation (2),

(c) it satisfies conditions (4), (5), (6), (10)

(d) it is not identically equal to zero.

Thus, each partial sum of the series (8) will satisfy equation (2) and conditions (4), (5), (6). By a suitable choice of the coefficients a_n, we shall satisfy the condition (3).

Since $u_n(x, t)$ satisfies equation (2), on putting (9) into (2) and dividing by the product $X_n T_n$ (supposing $X_n T_n \neq 0$) we find that

$$\frac{X_n''(x)}{X_n(x)} = \frac{T_n''(t)}{T_n(t)}\,. \tag{11}$$

This equation is to be satisfied for all x, t of the region O $(0 < x < l, t > 0)$, since (2) is to be satisfied everywhere in O. The left-hand side of (11) is independent of t, being a function of x only, but it is also independent of x, since it is equal in O to the right-hand side of (11), and this right-hand side is independent of x. Thus, the left-hand and the right-hand sides of (11) are both equal to a (common) constant. Let us denote the constant by $-\lambda_n$. Then, equation (11) yields

$$X_n'' + \lambda_n X_n = 0\,, \tag{12}$$

$$T_n'' + \lambda_n T_n = 0\,. \tag{13}$$

A more detailed analysis shows that equations (12), (13) must be satisfied even at (isolated) zeros of the function $X_n(x)$, or $T_n(t)$, respectively, so that the assumption $X_n T_n \neq 0$ is not essential.

We first find a solution of equation (12), which satisfies (5) and (6) and is not identically zero. Since $T_n(t)$ must not be identically zero (condition (d) to be satisfied by the function $u_n(x, t)$), then from (5) and (6) it follows that

$$X_n(0) = 0\,, \quad X_n(l) = 0\,. \tag{14}$$

It may be easily verified that $\lambda_n < 0$ or $\lambda_n = 0$ yields only the identically zero solution $X_n(x) \equiv 0$. If $\lambda_n > 0$, then the general solution of (12) is

$$X_n = C_n \cos \sqrt{(\lambda_n)}\, x + D_n \sin \sqrt{(\lambda_n)}\, x\,. \tag{15}$$

The first of conditions (14) yields $C_n = 0$; from the second one it follows (since we must have $D_n \neq 0$) that

$$\lambda_n = \frac{n^2\pi^2}{l^2}\,, \quad n \text{ being an integer}\,. \tag{16}$$

It is sufficient to consider

$$n = 1, 2, 3, \ldots$$

only, since for $n = 0$ we get the zero solution and for $n = -1, -2, -3, \ldots$ we do not obtain anything new. If we choose $D_n = 1$, we get

$$X_n(x) = \sin \frac{n\pi}{l}\, x\,. \tag{17}$$

Similarly, using (16), we find that non-zero solutions of equation (13), satisfying condition (4), are

$$T_n(t) = \cos \frac{n\pi}{l}\, t\,.$$

Thus, each of the functions

$$u_n(x, t) = \sin \frac{n\pi}{l}\, x \cos \frac{n\pi}{l}\, t \quad (n = 1, 2, 3, \ldots) \tag{18}$$

(multiplied, eventually, by an arbitrary constant) is a solution of (2), (4), (5) and (6).

Let $u(x, t)$ be of the form (8). Since, for $t = 0$,

$$u_n(x, 0) = \sin \frac{n\pi}{l}\, x\,,$$

it follows from (3), written now in the form

$$\varphi_0(x) = \sum_{n=1}^{\infty} a_n \sin \frac{n\pi}{l}\, x\,, \tag{19}$$

that

$$a_n = \frac{2}{l} \int_0^l \varphi_0(x) \sin \frac{n\pi}{l}\, x\, \mathrm{d}x \quad (n = 1, 2, 3, \ldots) \tag{20}$$

(that is, the a_n are the Fourier coefficients of the function $\varphi_0(x)$ with respect to the functions (17)).

Thus, if the solution of the given problem can be expressed in the form (8), then u_n and a_n are given by (18) and (20) respectively.

If, instead of (4), the condition

$$\frac{\partial u}{\partial t}(x, 0) = \varphi_1(x) \tag{21}$$

is prescribed, then

$$u(x, t) = \sum_{n=1}^{\infty} \sin \frac{n\pi}{l} x \left(a_n \cos \frac{n\pi}{l} t + b_n \sin \frac{n\pi}{l} t \right), \tag{22}$$

where the constants a_n are given by (20) and b_n by the equation

$$b_n = \frac{2}{n\pi} \int_0^l \varphi_1(x) \sin \frac{n\pi}{l} x \, dx . \tag{23}$$

(If $\varphi_1(x) \equiv 0$, we get, of course, the previous result.)

As mentioned above, if (8) is the solution of the boundary-value problem (2) – (6), then the u_n are given by (18), and the a_n by (20). In order that the series (8) (with u_n and a_n determined in this way) may be in fact the solution of the problem, it is sufficient that the function $\varphi_0(x)$ have two continuous derivatives in $[0, l]$ and that $\varphi_0''(0) = \varphi_0''(l) = 0$. If the condition (4) is replaced by (21), then, in addition, the function $\varphi_1(x)$ is supposed to have the continuous derivative in $[0, l]$ and $\varphi_1(0) = \varphi_1(l) = 0$.

In applications, the functions $\varphi_0(x)$ and $\varphi_1(x)$ do not always have all the required properties. For example φ_0 and φ_1 are functions such that φ_0, φ_1, φ_0', φ_1' are continuous in $[0, l]$ and equal to zero for $x = 0$ and $x = l$. In this case the series (22) represents the *generalized solution* of the problem (Definition 18.5.3, p. 904). That is to say, this series is uniformly convergent in the whole domain $\bar{O}(0 \leqq x \leqq l, t \geqq 0)$ and each of its partial sums satisfies equation (1) in O. However, this last assertion need not be true for the sum of this series.

If instead of equation (2) we are considering the equation

$$\frac{\partial^2 u}{\partial t^2} = a^2 \frac{\partial^2 u}{\partial x^2} \quad (a > 0)$$

then (using the substitution $t = \tau/a$) we obtain the solution

$$u(x, t) = \sum_{n=1}^{\infty} \sin \frac{n\pi}{l} x \left(a_n \cos \frac{n\pi a}{l} t + b_n \sin \frac{n\pi a}{l} t \right),$$

where the a_n are given by (20), and the b_n by

$$b_n = \frac{2}{n\pi a}\int_0^l \varphi_1(x) \sin \frac{n\pi}{l} x \, dx .$$

26.2. Potential Equation and Stationary Heat-Conduction Equation

Let us find the solution of the equation

$$\frac{\partial^2 u}{\partial x^2} + \frac{\partial^2 u}{\partial y^2} = 0 \quad (0 < x < a, 0 < y < b) \tag{1}$$

continuous in the rectangle $\bar{O}(0 \leqq x \leqq a, 0 \leqq y \leqq b)$ and satisfying the boundary conditions

$$u(x, 0) = f(x), \quad (f(0) = 0, f(a) = 0), \tag{2}$$

$$u(x, b) \equiv 0, \tag{3}$$

$$u(0, y) \equiv 0, \tag{4}$$

$$u(a, y) \equiv 0. \tag{5}$$

(This is the problem of finding the stationary temperature field in an infinite right prism whose cross-section is a rectangle and whose three faces are kept at zero temperature and the fourth one at the temperature $f(x)$; or a similar problem for the potential.)

As in Example 1, the solution is supposed to be of the form

$$u(x, y) = \sum_{n=1}^{\infty} a_n u_n(x, y),$$

where each of the functions $u_n(x, y)$ is of the form $X_n(x) Y_n(y)$, satisfies equation (1) and the boundary conditions (3), (4), (5), and is not identically zero. Putting $X_n Y_n$ for u_n into (1), we get

$$\frac{X''_n}{X_n} + \frac{Y''_n}{Y_n} = 0$$

so that

$$\frac{X''_n(x)}{X_n(x)} = -k_n, \quad \frac{Y''_n(y)}{Y_n(y)} = k_n, \quad k_n > 0. \tag{6}$$

The general solution of the first of equations (6) is

$$X_n(x) = C_1 \sin \sqrt{(k_n)}\, x + C_2 \cos \sqrt{(k_n)}\, x ;$$

conditions (4), (5) yield $X_n(0) = 0$, $X_n(a) = 0$; thus $C_2 = 0$ and

$$k_n = \frac{n^2\pi^2}{a^2} \quad (n = 1, 2, 3, \ldots)\,. \tag{7}$$

The general integral of the first of equations (6) is (when using (7))

$$Y_n(y) = D_1 \sinh \frac{n\pi}{a} y + D_2 \cosh \frac{n\pi}{a} y\,. \tag{8}$$

It is convenient to choose D_1 and D_2 in (8) in such a way that for $y = 0$ we shall have $Y_n = 1$. Further, for $y = b$ it follows from (3) that $Y_n = 0$. These two conditions are satisfied by the functions

$$Y_n(y) = \frac{\sinh \frac{n\pi}{a}(b - y)}{\sinh \frac{n\pi}{a} b}\,.$$

If, in addition, we put

$$a_n = \frac{2}{a} \int_0^a f(x) \sin \frac{n\pi}{a} x \, dx\,,$$

then, if, for example, $f(x)$ and $f'(x)$ are continuous functions in $[0, a]$, the required solution of the problem (1) – (5) is

$$u(x, y) = \sum_{n=1}^{\infty} a_n \sin \frac{n\pi}{a} x \, \frac{\sinh \frac{n\pi}{a}(b - y)}{\sinh \frac{n\pi}{a} b}\,. \tag{9}$$

If the boundary conditions (3), (4), (5) are not homogeneous (i.e. if temperature is prescribed also on the remaining three faces of the prism), we obtain the solution by superposition of solutions of four problems of this type.

26.3. Heat Conduction in Rectangular Regions

If we solve the equation

$$\frac{\partial u}{\partial t} = a \frac{\partial^2 u}{\partial x^2} \quad (0 < x < l, t > 0)$$

with the boundary conditions

$$u(0, t) \equiv 0\,, \quad u(l, t) \equiv 0\,, \quad u(x, 0) = f(x)$$

by the Fourier method, we get

$$u(x, t) = \sum_{n=1}^{\infty} a_n \sin \frac{n\pi}{l} x \, e^{-a(n^2\pi^2/l^2)t}, \tag{1}$$

where

$$a_n = \frac{2}{l} \int_0^l f(x) \sin \frac{n\pi}{l} x \, dx .$$

Similarly, the solution of the equation

$$\frac{\partial u}{\partial t} = a \left(\frac{\partial^2 u}{\partial x^2} + \frac{\partial^2 u}{\partial y^2} \right) \quad (0 \leqq x \leqq l_1, 0 \leqq y \leqq l_2, t > 0)$$

with the boundary conditions

$$u(x, y, 0) = f(x, y), \quad u(x, 0, t) \equiv 0, \quad u(x, l_2, t) \equiv 0,$$

$$u(0, y, t) \equiv 0, \quad u(l_1, y, t) \equiv 0,$$

is

$$u(x, y, t) = \sum_{m,n=1}^{\infty} A_{mn} \sin \frac{m\pi}{l_1} x \sin \frac{n\pi}{l_2} y \, e^{-a[(\pi^2/l_1^2)m^2 + (\pi^2/l_2^2)n^2]t}, \tag{2}$$

where

$$A_{mn} = \frac{4}{l_1 l_2} \int_0^{l_1} \int_0^{l_2} f(x, y) \sin \frac{m\pi}{l_1} x \sin \frac{n\pi}{l_2} y \, dx \, dy .$$

(The convergence of the series (2) is undertood in the sense of Remark 16.3.10: We say that the series (2) has the sum $u(x_0, y_0, t_0)$ at the point (x_0, y_0, t_0), if corresponding to every $\varepsilon > 0$, there exists an n_0 such that for every pair of numbers M, N for which simultaneously $M > n_0, N > n_0$, the relation

$$\left| \sum_{m=1}^{M} \sum_{n=1}^{N} A_{mn} \sin \frac{m\pi}{l_1} x_0 \sin \frac{n\pi}{l_2} y_0 \, e^{-a[(\pi^2/l_1^2)m^2 + (\pi^2/l_2^2)n^2]t_0} - u(x_0, y_0, t_0) \right| < \varepsilon$$

holds.)

If $f(x)$ has a continuous derivative in $[0, l]$ and $f(0) = f(l) = 0$, or if $f(x, y)$ has continuous first partial derivatives in the rectangle $\bar{O}(0 \leqq x \leqq l_1, 0 \leqq y \leqq l_2)$ and $f(x, y) = 0$ on the sides of this rectangle, then the series (1), (2), respectively, give the solution of the problems in question.

Similar results hold for the three-dimensional case.

26.4. Heat Conduction in an Infinite Circular Cylinder. Application of Bessel Functions

Let us solve the problem

$$\frac{\partial u}{\partial t} = k\left(\frac{\partial^2 u}{\partial r^2} + \frac{1}{r}\frac{\partial u}{\partial r}\right) \quad (0 \leqq r < c, t > 0), \tag{1}$$

$$u(c, t) \equiv 0, \tag{2}$$

$$u(r, 0) = f(r). \tag{3}$$

(This is the problem of heat conduction in an infinite circular cylinder of radius c, the surface of which is kept at zero temperature and the initial temperature of which is independent of φ.)

The solution $u(r, t)$ is also supposed to be axially symmetric (i.e. independent of φ) and to be expressible as an infinite series, the terms of which are of the form

$$R(r)\,T(t) \tag{4}$$

and satisfy the condition (2). Putting (4) into (1), we get

$$\frac{T'}{kT} = \frac{1}{R}\left(R'' + \frac{R'}{r}\right). \tag{5}$$

As in Example 1, we see that the left-hand and the right-hand side of (5) are both equal to a (common) negative constant, say $-\lambda^2$. Thus

$$rR'' + R' + \lambda^2 rR = 0, \tag{6}$$

$$T' + k\lambda^2 T = 0. \tag{7}$$

Using the substitution

$$\lambda r = z \quad \text{or} \quad \frac{dR}{dr} = \lambda\frac{dR}{dz}, \quad \frac{d^2R}{dr^2} = \lambda^2\frac{d^2R}{dz^2},$$

we get from (6)

$$z\frac{d^2R}{dz^2} + \frac{dR}{dz} + zR = 0, \tag{8}$$

which is the Bessel equation of order $n = 0$ (§ 17.15); its solution is the function

$$R(z) = J_0(z)$$

(§ 16.4), so that the solution of (6) is (for a fixed λ)

$$R(r) = J_0(\lambda r). \tag{9}$$

Since the functions of the form (4) must satisfy the boundary condition (2) (for all t), it is necessary for λ to be such that

$$J_0(\lambda c) = 0 . \tag{10}$$

According to Theorem 16.4.8 (p. 719), equation (10) has real roots only, and of these only the positive roots λ_n ($n = 1, 2, \ldots$) need be considered. For a fixed λ_n the function (9) becomes

$$R_n(r) = J_0(\lambda_n r)$$

while

$$T_n(t) = e^{-k\lambda_n^2 t}$$

according to (7).

Let us introduce the constants a_n by

$$a_n = \frac{2}{c^2 J_1^2(\lambda_n c)} \int_0^c r\, J_0(\lambda_n r) f(r)\, dr$$

(see Theorem 16.4.9, p. 720), where J_1 is the Bessel function of order 1. Then if, for example, $f(r)$ is continuous and has a piecewise continuous derivative in $[0, c]$, and $f(c) = 0$, the solution of the problem (1), (2), (3) is

$$u(r, t) = \sum_{n=1}^{\infty} a_n\, J_0(\lambda_n r)\, e^{-k\lambda_n^2 t} .$$

26.5. Deflection of a Rectangular Simply Supported Plate

Let us solve the equation

$$\Delta^2 w \equiv \frac{\partial^4 w}{\partial x^4} + 2\frac{\partial^4 w}{\partial x^2\, \partial y^2} + \frac{\partial^4 w}{\partial y^4} = \frac{q_0}{D} \sin\frac{\pi x}{a} \sin\frac{\pi y}{b} \tag{1}$$

in the rectangle $O(0 \leqq x \leqq a, 0 \leqq y \leqq b)$ with the boundary conditions

$$w = 0\,, \quad \frac{\partial^2 w}{\partial x^2} = 0 \quad \text{for} \quad x = 0\,, \quad x = a\,, \tag{2}$$

$$w = 0\,, \quad \frac{\partial^2 w}{\partial y^2} = 0 \quad \text{for} \quad y = 0, \quad y = b\,. \tag{3}$$

(This is the problem of deflection of a rectangular simply supported plate with loading

$$q = q_0 \sin\frac{\pi x}{a} \sin\frac{\pi y}{b} .)$$

Let us assume the solution in the form

$$w = C \sin \frac{\pi x}{a} \sin \frac{\pi y}{b}, \tag{4}$$

The boundary conditions (2) and (3) are then satisfied. If we put (4) into (1), we get

$$\pi^4 \left(\frac{1}{a^2} + \frac{1}{b^2} \right)^2 C = \frac{q_0}{D};$$

hence

$$w = \frac{q_0}{\pi^4 D \left(\frac{1}{a^2} + \frac{1}{b^2} \right)^2} \sin \frac{\pi x}{a} \sin \frac{\pi y}{b}. \tag{5}$$

If the loading is

$$q = q_0 \sin \frac{m\pi x}{a} \sin \frac{n\pi y}{b} \tag{6}$$

(m, n being positive integers), we obtain similarly

$$w = \frac{q_0}{\pi^4 D \left(\frac{m^2}{a^2} + \frac{n^2}{b^2} \right)^2} \sin \frac{m\pi x}{a} \sin \frac{n\pi y}{b}. \tag{7}$$

Now, let

$$q = f(x, y) = \sum_{m,n=1}^{\infty} a_{mn} \sin \frac{m\pi x}{a} \sin \frac{n\pi y}{b},$$

where

$$a_{mn} = \frac{4}{ab} \int_0^a \int_0^b f(x, y) \sin \frac{m\pi x}{a} \sin \frac{n\pi y}{b} \, dx \, dy$$

(see Theorem 16.3.5, p. 712). Then

$$w = \frac{1}{\pi^4 D} \sum_{m,n=1}^{\infty} \frac{a_{mn}}{\left(\frac{m^2}{a^2} + \frac{n^2}{b^2} \right)^2} \sin \frac{m\pi x}{a} \sin \frac{n\pi y}{b}.$$

In particular, if $q = q_0 = \text{const.}$, we get

$$w = \frac{16 q_0}{\pi^6 D} \sum_{m,n=1,3,5,\ldots} \frac{\sin \frac{m\pi x}{a} \sin \frac{n\pi y}{b}}{mn \left(\frac{m^2}{a^2} + \frac{n^2}{b^2} \right)^2}.$$

REMARK 1. In § 26.5, a non-homogeneous equation with homogeneous (zero) boundary conditions has been solved. Problems with nonhomogeneous boundary conditions are often encountered in applications. In many such cases it is preferable to look for the solution in the form $w = w_1 + w_2$, where w_1 is a special function satisfying the given boundary conditions (but not, in general, the equation in question), while the function w_2 satisfies homogeneous boundary conditions and the equation with a non-zero right-hand side. In this simple way, the problem may often be modified to a form suitable for application of the Fourier method.

27. SOLUTION OF PARTIAL DIFFERENTIAL EQUATIONS BY THE FINITE-DIFFERENCE METHOD

By EMIL VITÁSEK

References: [12], [20], [26], [66], [69], [74], [125], [126], [130], [131], [133], [134], [147], [148], [176], [214], [225], [241], [286], [335], [336], [343], [381], [382], [383], [411], [418], [419].

The method of finite differences is one of most popular methods for the numerical solution of partial differential equations. The majority of technical problems formulated by means of partial differential equations are now solved by this method. The existence of high speed digital computers has brought the method of finite differences into even greater significance.

27.1. The Basic Concept of the Finite-Difference Method

The basic concept of the method of finite differences is very simple. The domain in which the solution of the given differential equation is sought is subdivided by a net with a finite number of mesh points and the derivatives at each mesh point

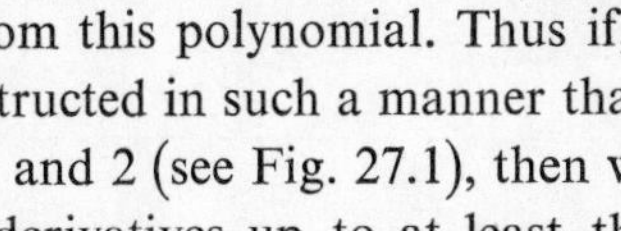

Fig. 27.1. 4 2 0 1 3 (spacing h)

replaced by finite-difference approximations. In fact the function sought is replaced, in the neighbourhood of the given point, by an interpolating polynomial and the derivatives are computed from this polynomial. Thus if, for example, a polynomial of the second degree is constructed in such a manner that it coincides with the given function u at the points 0, 1 and 2 (see Fig. 27.1), then with the assumption that the function u has continuous derivatives up to at least the fourth order, the second derivative of the polynomial at the point 0 is represented by the formula

$$\frac{\partial^2 u}{\partial x^2} = \frac{u_2 - 2u_0 + u_1}{h^2} - R_0 ,$$

TABLE 27.1

Derivative	Scheme	Approximate formulae
$\frac{\partial u}{\partial x}$		$\frac{\partial u_{i,k}}{\partial x} = \frac{u_{i+1,k} - u_{i,k}}{h} + O(h)$
		$\frac{\partial u_{i,k}}{\partial x} = \frac{u_{i+1,k} - u_{i-1,k}}{2h} + O(h^2)$
		$\frac{\partial u_{i,k}}{\partial x} = \frac{u_{i+1,k+1} - u_{i-1,k+1} + u_{i+1,k-1} - u_{i-1,k-1}}{4h} + O(h^2)$
$\frac{\partial^2 u}{\partial x^2}$		$\frac{\partial^2 u_{i,k}}{\partial x^2} = \frac{u_{i+1,k} - 2u_{i,k} + u_{i-1,k}}{h^2} + O(h^2)$
		$\frac{\partial^2 u_{i,k}}{\partial x^2} = \frac{-u_{i+2,k} + 16u_{i+1,k} - 30u_{i,k} + 16u_{i-1,k} - u_{i-2,k}}{12h^2} + O(h^4)$
		$\frac{\partial^2 u_{i,k}}{\partial x^2} = \frac{1}{3h^2}(u_{i+1,k+1} - 2u_{i,k+1} + u_{i-1,k+1} + u_{i+1,k} - 2u_{i,k} + u_{i-1,k} + u_{i+1,k-1} - 2u_{i,k-1} + u_{i-1,k-1}) + O(h^2)$
$\frac{\partial^2 u}{\partial x\, \partial y}$		$\frac{\partial^2 u_{i,k}}{\partial x\, \partial y} = \frac{1}{4h^2}(u_{i+1,k+1} - u_{i+1,k-1} - u_{i-1,k+1} + u_{i-1,k-1}) + O(h^2)$
$\frac{\partial^4 u}{\partial x^4}$		$\frac{\partial^4 u_{i,k}}{\partial x^4} = \frac{1}{h^4}(u_{i+2,k} - 4u_{i+1,k} + 6u_{i,k} - 4u_{i-1,k} + u_{i-2,k}) + O(h^2)$
$\frac{\partial^4 u}{\partial x^2 \partial y^2}$		$\frac{\partial^4 u_{i,k}}{\partial x^2\, \partial y^2} = \frac{1}{h^4}(u_{i+1,k+1} + u_{i-1,k+1} + u_{i+1,k-1} + u_{i-1,k-1} - 2u_{i+1,k} - 2u_{i-1,k} - 2u_{i,k+1} - 2u_{i,k-1} + 4u_{i,k}) + O(h^2)$

where

$$R_0 = \frac{2h^2}{4!}\frac{\partial^4 u(\Theta h)}{\partial x^4}, \quad -1 < \Theta < 1,$$

and the finite-difference approximation to the second derivative of the function at the point 0 is $(u_2 - 2u_0 + u_1)/h^2$. If the given function is interpolated from five points, one obtains (under similar assumptions concerning the smoothness of the function u)

$$\frac{\partial^2 u}{\partial x^2} = \frac{-u_4 + 16u_2 - 30u_0 + 16u_1 - u_3}{12h^2} + R_1,$$

where

$$R_1 = \frac{h^4}{90}\frac{\partial^6 u(\Theta h)}{\partial x^6}, \quad -1 < \Theta < 1,$$

i.e. a formula of higher accuracy. Similarly, the first derivative of the given function may be represented by a finite-difference formula and a remainder which depends on the higher derivatives of the given function. This remainder is then neglected. The finite-difference formulae for the most frequent partial derivatives of a function of two variables are tabulated in Tab. 27.1, where $u(ih, kh) = u_{i,k}$ and, e.g., the symbol $O(h^2)$ denotes that the error is of the order h^2. This means that the error is in absolute value smaller than Ah^2 (where A is some constant) for sufficiently small h. The formulae of Tab. 27.1 are valid if the given function u is sufficiently smooth.

If the derivatives are replaced by finite-difference expressions as indicated, one obtains a system of n (in general non-linear) equations for determining the approximate values of the unknown function in n different points of the net. This system of equations is then solved by appropriate numerical methods.

From this description of the basic concept of the finite-difference method one sees that this method can be applied to the solution of very different types of differential equations. (When solving some special types of differential equations (e.g. partial differential equations of parabolic type) it is often necessary to use special kinds of nets in which the time mesh-size depends on the space mesh-size (see e.g. Tab. 27.3 and Remark 27.7.1).) In practice, the finite-difference method is now used mainly for linear equations, since in this case the corresponding system of finite-difference equations is also linear, and for systems of linear algebraic equations many methods of numerical solution have been elaborated (see Chap. 30). In the non-linear case, there occur theoretical difficulties connected with questions of convergence of the approximate solution to the exact solution and practical difficulties in solving systems of non-linear equations (see § 31.5, p. 1180).

27.2. Principal Types of Nets

The finite-difference method is mainly applied in two-dimensional cases (in multi-dimensional cases, the number of equations is unduly high). We therefore introduce the most frequently used types of plane nets.

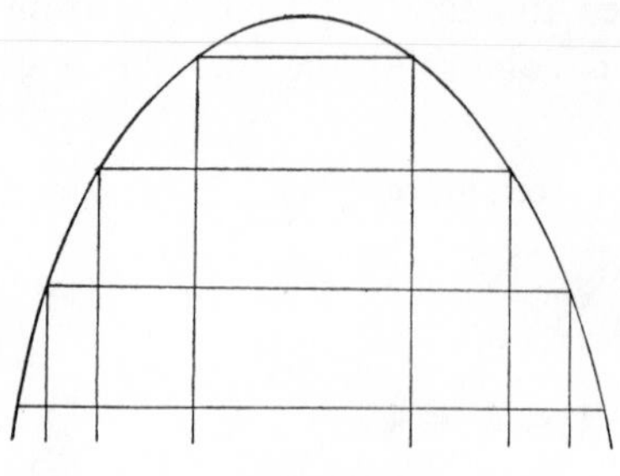

Fig. 27.2.

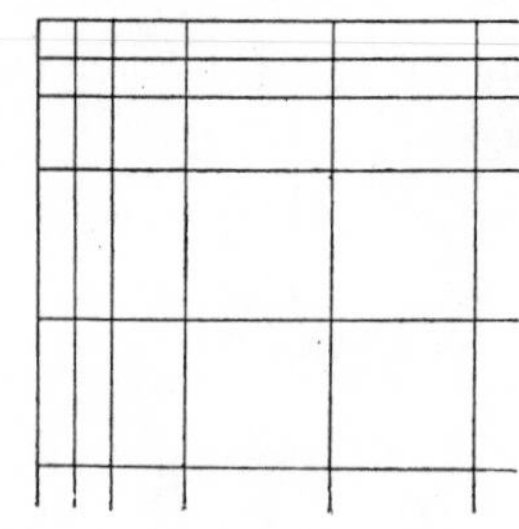

Fig. 27.3.

(a) Rectangular Nets. Nets of this type are now the most widely used. They can be divided into

(α) irregular nets,
(β) regular nets.

(α) *Irregular nets* are formed by different rectangles. They are used in order to simplify the formulation of boundary conditions (with such nets we ensure that the mesh points lie on the boundary of the given domain (see Fig. 27.2 and § 27.5)) and to refine nets (see Fig. 27.3 and § 27.3).

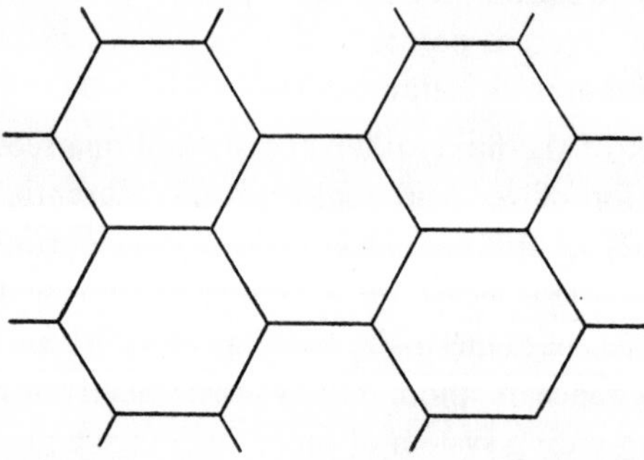

Fig. 27.4.

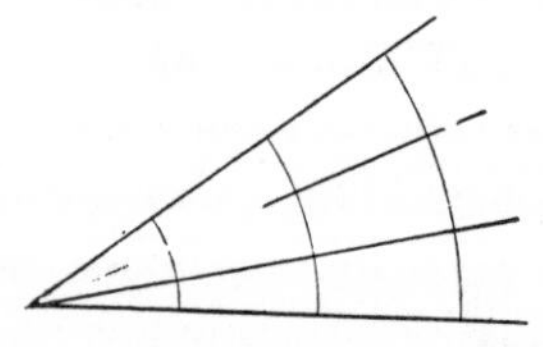

Fig. 27.5.

(β) *Regular nets* are formed by congruent rectangles and squares. Nets of this type are the most frequently used, mainly because the corresponding finite-difference formulae are simple.

(b) Hexagonal and Triangular Nets (Fig. 27.4). These types are seldom used.

(c) Polar Nets (Fig. 27.5). This type is also very seldom used. Such nets are sometimes convenient in special domains as e.g. in sectors of a circle etc. The difference formulae are rather complicated.

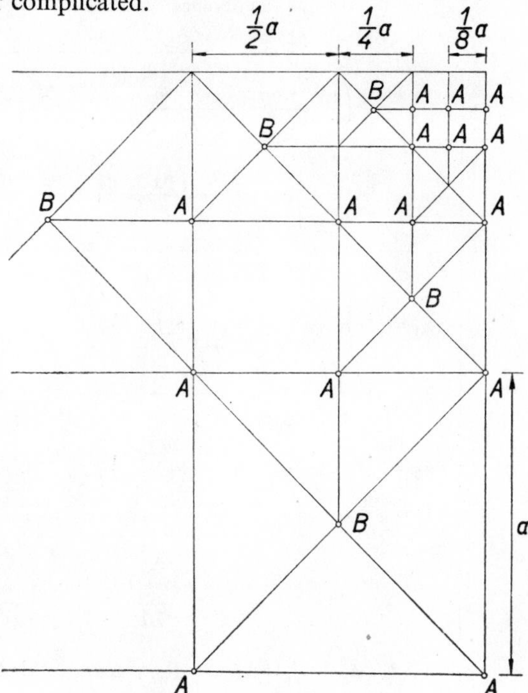

Fig. 27.6.

27.3. Refinement of Nets

As we have seen above, the accuracy of approximation depends on the density of the net. But refinement of the net greatly increases the work of computation. For this reason it is convenient to refine the net only in those regions where we are interested in higher accuracy. The easiest method of refinement of a net consists in using irregular nets (see § 27.2). Moreover, square nets can be refined by means of diagonal nets, as is seen from Fig. 27.6.

27.4. Finite-Difference Formulae for Most Frequently Occurring Operators

1. Laplace's (Poisson's) equation:

$$\Delta u \equiv \frac{\partial^2 u}{\partial x^2} + \frac{\partial^2 u}{\partial y^2} = f(x, y),$$

see Tab. 27.2.

TABLE 27.2

Scheme	Difference equation	Order of accuracy
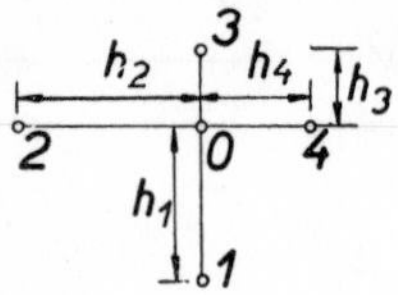	$\frac{2}{h_1+h_3}\left(\frac{u_1-u_0}{h_1}+\frac{u_3-u_0}{h_3}\right)+\frac{2}{h_2+h_4}\left(\frac{u_2-u_0}{h_2}+\frac{u_4-u_0}{h_4}\right)=f_0$	h ($h_i \leqq h$)
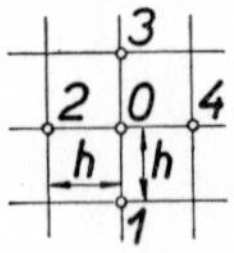	$u_0=\frac{1}{4}(u_1+u_2+u_3+u_4)-\frac{1}{4}h^2f_0$	h^2
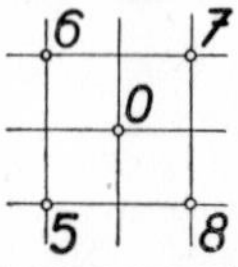	$u_0=\frac{1}{4}(u_5+u_6+u_7+u_8)-\frac{1}{2}h^2f_0$	h^2
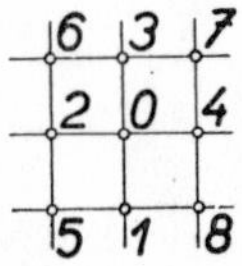	$u_0=\frac{4(u_1+u_2+u_3+u_4)+(u_5+u_6+u_7+u_8)}{20}-\frac{3}{10}h^2f_0-\frac{1}{40}h^4\Delta f_0-\frac{1}{1200}h^6\Delta\Delta f_0-\frac{1}{600}h^6\frac{\partial^4 f_0}{\partial x^2\,\partial y^2}$	h^6
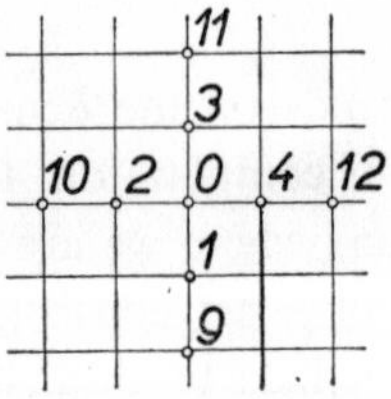	$u_0=\frac{16(u_1+u_2+u_3+u_4)-(u_9+u_{10}+u_{11}+u_{12})}{60}-\frac{1}{5}h^2f_0$	h^4
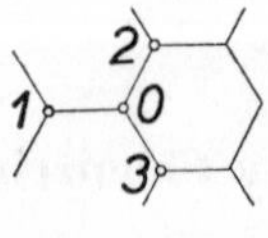	$u_0=\frac{1}{3}(u_1+u_2+u_3)-\frac{1}{4}h^2f_0$	h
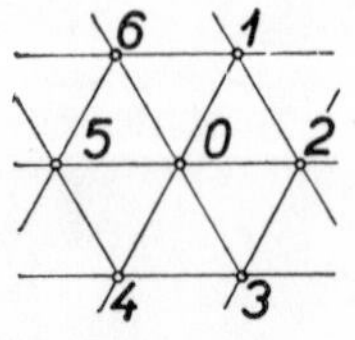	$u_0=\frac{1}{6}(u_1+u_2+u_3+u_4+u_5+u_6)-\frac{1}{4}h^2f_0-\frac{1}{64}h^4\Delta f_0$	h^4

2. The heat conduction equation:

$$\frac{\partial^2 u}{\partial x^2} = a^2 \frac{\partial u}{\partial t}, \quad \Delta u = a^2 \frac{\partial u}{\partial t},$$

see Tab. 27.3 and 27.4, respectively.

TABLE 27.3

Scheme	Relation between τ and h	Difference equation	Order of accuracy
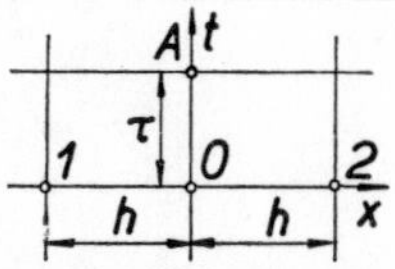	$\tau \leqq \frac{1}{2}a^2h^2$	$u_A = \frac{\tau}{a^2h^2} u_1 + \left(1 - \frac{2\tau}{a^2h^2}\right) u_0 + \frac{\tau}{a^2h^2} u_2$	h^2
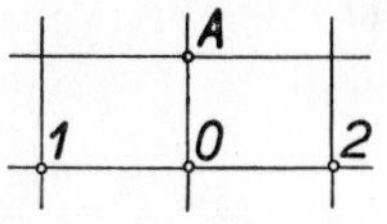	$\tau = \frac{1}{6}a^2h^2$	$u_A = \frac{1}{6}(u_1 + 4u_0 + u_2)$	h^4

TABLE 27.4

Scheme	Relation between τ and h	Difference equation	Order of accuracy
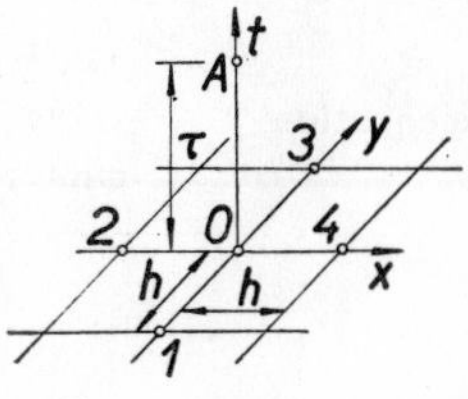	$\tau \leqq \frac{1}{4}a^2h^2$	$u_A = \frac{\tau}{a^2h^2}(u_1 + u_2 + u_3 + u_4) + \left(1 - \frac{4\tau}{a^2h^2}\right) u_0$	h^2
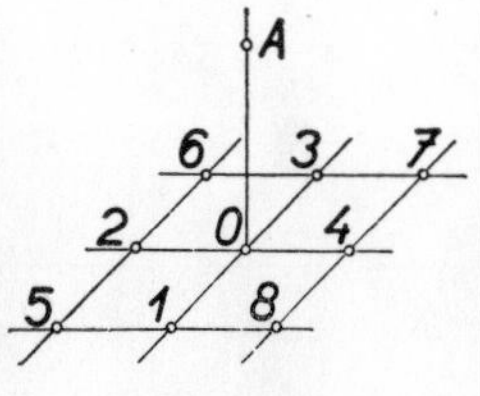	$\tau = \frac{1}{6}a^2h^2$	$u_A = \frac{1}{36}[16u_0 + 4(u_1 + u_2 + u_3 + u_4) + u_5 + u_6 + u_7 + u_8]$	h^4
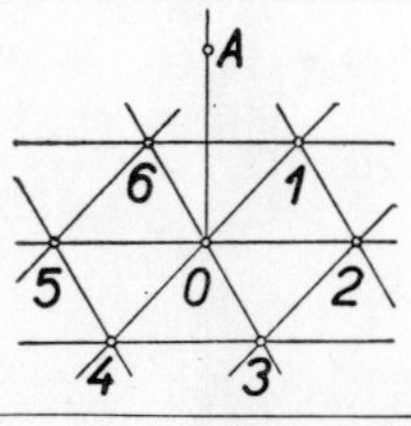	$\tau = \frac{1}{8}a^2h^2$	$u_A = \frac{1}{2}u_0 + \frac{1}{12}(u_1 + u_2 + u_3 + u_4 + u_5 + u_6)$	h^4

TABLE 27.5

Scheme	Difference equation	Order of accuracy
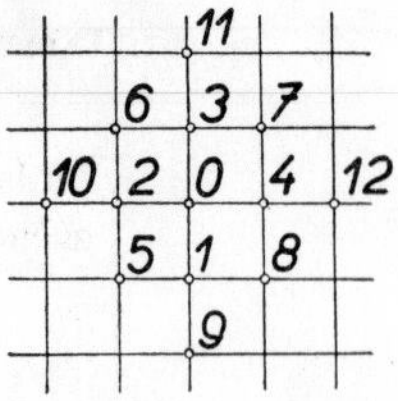	$u_0 = \frac{1}{20}[8(u_1 + u_2 + u_3 + u_4) -$ $- 2(u_5 + u_6 + u_7 + u_8) -$ $- (u_9 + u_{10} + u_{11} + u_{12})] + \frac{1}{20} h^4 f_0$	h^2
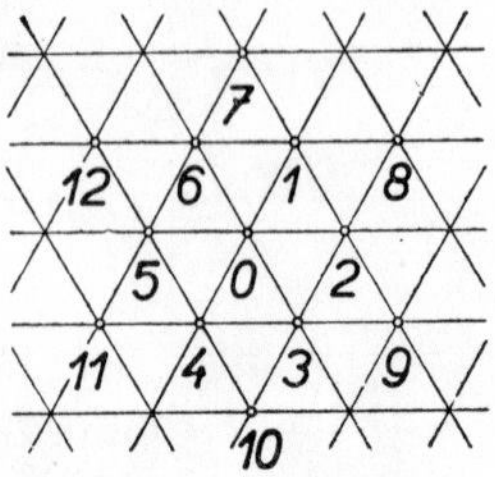	$u_0 = \frac{1}{12}[3(u_1 + u_2 + u_3 + u_4 + u_5 + u_6) -$ $- (u_7 + u_8 + u_9 + u_{10} + u_{11} + u_{12})] +$ $+ \frac{3}{64} f_0 h^4$	h^2

TABLE 27.6

Equation	Scheme	Relation between τ and h	Difference equation	Order of accuracy
$a^2 \frac{\partial^2 u}{\partial x^2} = \frac{\partial^2 u}{\partial t^2}$	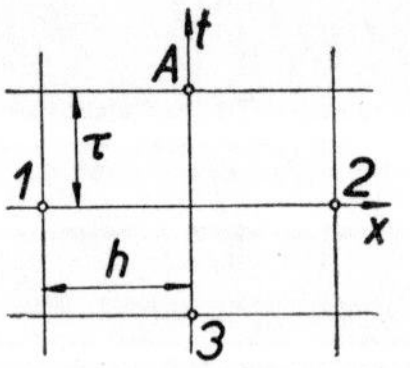	$\tau = \frac{h}{a}$	$u_A = u_1 + u_2 - u_3$	h^2
$a^2 \Delta u = \frac{\partial^2 u}{\partial t^2}$	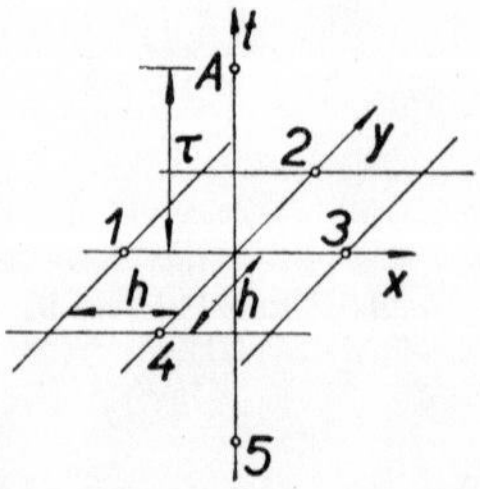	$\tau = \frac{h}{a\sqrt{2}}$	$u_A = \frac{1}{2}(u_1 + u_2 + u_3 +$ $+ u_4 - 2u_5)$	h^2

3. The biharmonic equation:

$$\Delta\Delta u \equiv \frac{\partial^4 u}{\partial x^4} + 2\frac{\partial^4 u}{\partial x^2\,\partial y^2} + \frac{\partial^4 u}{\partial y^4} = f(x, y)\,,$$

see Tab. 27.5.

4. The wave equation:

$$a^2\frac{\partial^2 u}{\partial x^2} = \frac{\partial^2 u}{\partial t^2}\,, \quad a^2\,\Delta u = \frac{\partial^2 u}{\partial t^2}\,,$$

see Tab. 27.6.

In Tab. 27.2–27.6, the order of accuracy indicates that the error when applying the corresponding finite-difference operator to a sufficiently smooth function is of order h^n.

27.5. Formulation of Boundary Conditions

(a) Boundary Conditions Which Do Not Contain Derivatives (the values of the function to be found are given on the boundary of the domain considered). Essentially, two methods are used:

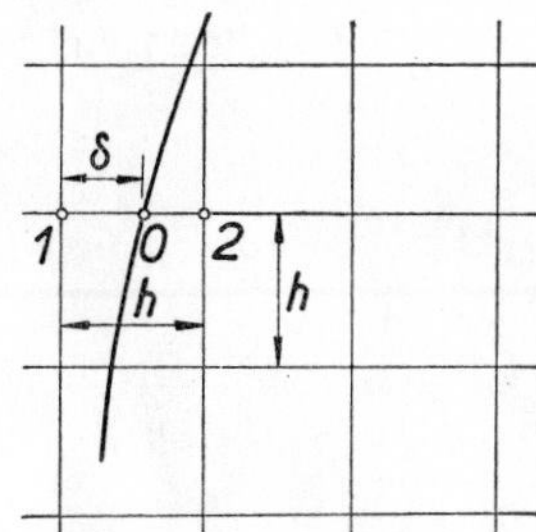

Fig. 27.7.

(α) Collatz's method of linear interpolation ([66]–[69]). Only regular mesh points are used and the boundary condition is transferred to the mesh point nearest to the boundary by linear interpolation or extrapolation (see Fig. 27.7):

$$u_1 = \frac{h\varphi_0 - \delta u_2}{h - \delta}\,,$$

where φ_0 is the value of the given function at the point 0.

(β) The use of irregular nets in such a manner that the mesh points lie on the boundary of the given domain (see Example 27.7.1).

(b) Boundary Conditions Containing Derivatives. This case is treated in a similar manner. The derivatives in the boundary conditions are usually linearly interpolated and replaced by finite differences.

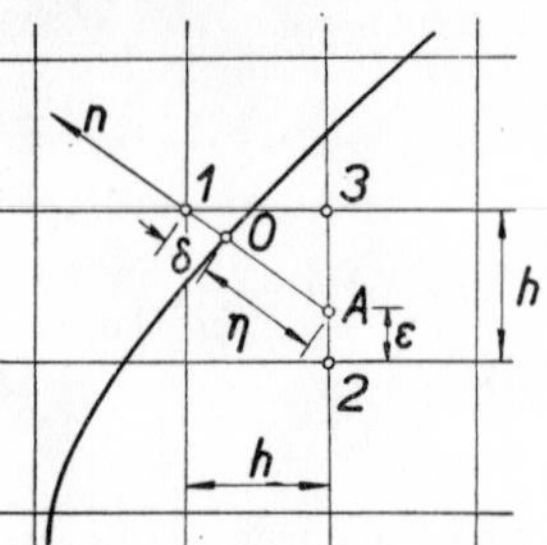

Fig. 27.8.

Let us demonstrate this procedure in the case where a linear combination of the value and the normal derivative of the unknown function are given on the boundary (see [26]):

$$\frac{\partial u}{\partial n} = -k(u - \varphi),$$

where n is the outward normal, φ is the given function, and k is a positive constant.

Let us express the normal derivative by means of finite differences in two ways (see Fig. 27.8):

$$\frac{u_1 - u_0}{\delta} \approx \frac{\partial u}{\partial n}, \quad \frac{u_0 - u_A}{\eta} \approx \frac{\partial u}{\partial n},$$

and substitute in the boundary condition:

$$\frac{u_1 - u_0}{\delta} = -k(u_0 - \varphi_0), \quad \frac{u_0 - u_A}{\eta} = -k(u_0 - \varphi_0).$$

The value u_A at the irregular point of the net is computed by linear interpolation from the values u_2 and u_3:

$$u_A = \frac{(h - \varepsilon)\, u_2 + \varepsilon u_3}{h}.$$

Eliminating u_0 and u_A from the last three equations, we obtain at the point 1 the equation

$$u_1 = \frac{(1 - k\delta)(h - \varepsilon)}{(1 + k\eta)\, h}\, u_2 + \frac{(1 - k\delta)\, \varepsilon}{(1 + k\eta)\, h}\, u_3 + k\, \frac{\delta + \eta}{1 + k\eta}\, \varphi_0 .$$

The treatment in other cases is similar. Moreover, in some cases special formulae which guarantee a higher accuracy are used ([126], [130], [133], [381]).

27.6. Error Estimates

The problem of error estimates when solving partial differential equations by finite-difference methods is very complicated. The majority of estimates which can be found in the literature (see e.g. [147], [418], [419]) are very pessimistic (i.e. they are many times greater than the actual error) and moreover they are very complicated. For this reason, the assessment of error by the so-called deferred correction method is most popular in practice. The basic idea of this method is as follows.

Let us suppose that the order of the error is n (n is usually the order of accuracy to which the derivatives are approximated by corresponding finite-difference formulae, see tables in § 27.4), i.e. let us suppose that a function $\alpha(x, y)$ (independent of h) exists, such that

$$\varepsilon_h(x, y) = u(x, y) - u_h(x, y) = \alpha(x, y)\, h^n + o(h^n),$$

holds where u is the exact solution and u_h the approximate solution computed when using the mesh size h. Then one obtains for the error ε_h the simple formula:

$$\varepsilon_h(x, y) = \frac{u_h(x, y) - u_{2h}(x, y)}{2^n - 1} + o(h^n),$$

where u_{2h} denotes the approximate solution gained by the net with double mesh size $2h$. This formula can often be used even if we only know that $\varepsilon_h = O(h^n)$.

27.7. The Dirichlet Problem for Laplace's Equation. First Boundary Value Problem for the Heat-Conduction Equation. Biharmonic Equation with Prescribed Values of the Function and its First Derivatives on the Boundary

Example 1. Let us solve the problem of a stationary temperature field in a plane plate of the form illustrated in Fig. 27.9 if heat is transferred on the linear part of the boundary ($y = 0$, $-6 < x < 6$) into a medium of known temperature (given by a function $f(x)$) and on the remainder of the boundary a constant temperature $u = 0$ is given.

This problem yields the Laplace differential equation

$$\Delta u = 0$$

in the domain O and the boundary conditions

$$\frac{\partial u}{\partial n}(x, 0) = -\frac{k}{\lambda}\left(u(x, 0) - f(x)\right),$$

where n is the outward normal, λ is the coefficient of heat conductivity and k is the coefficient of heat transfer on the linear part of the boundary, and

$$u = 0$$

on the remaining part of the boundary.

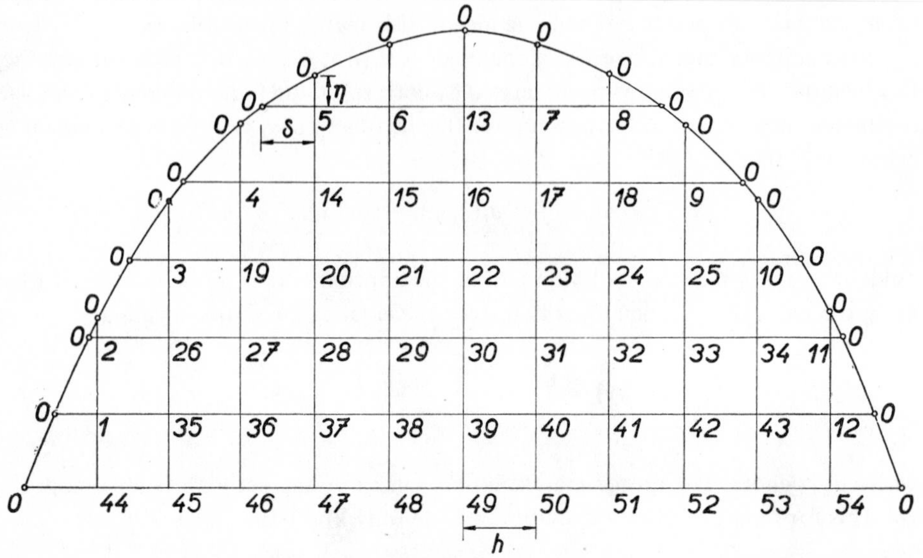

Fig. 27.9.

Let us cover the domain O by a net consisting of the system of lines $x = \pm k$, $y = l$, $k, l = 0, 1, \ldots, 6$ (in our special case $h = 1$). At all mesh points denoted by zeros in Fig. 27.9 the condition $u = 0$ holds.

At mesh points 13–43 we use the second formula of Tab. 27.2 and obtain, for instance, at the point 30 the equation

$$u_{30} = \tfrac{1}{4}(u_{22} + u_{29} + u_{31} + u_{39}),$$

and similarly at other mesh points of this type.

At mesh points 1–12 (i.e. at mesh points adjacent to irregular points of the net) we use the first formula of Tab. 27.2 and obtain, for instance, for the point 5

$$u_5 = \frac{\eta\delta}{(\eta + \delta)(1 + \eta)}\, u_{14} + \frac{\eta\delta}{(\eta + \delta)(1 + \delta)}\, u_6 .$$

At mesh points 44–54 we formulate the boundary condition by the procedure described in § 27·5 (here $\delta = \varepsilon = 0, \eta = h$) which yields, for example, at the point 49

$$u_{49} = \frac{hk\,f(x)/\lambda + u_{39}}{1 + hk/\lambda}\,.$$

Proceeding in this way we obtain a system of 54 equations for 54 unknowns and this system is then solved by some known numerical method for solving systems of linear algebraic equations (cf. Chap. 30). Direct methods as well as iteration methods can be used. Very often the Gauss–Seidel iteration method (also called, in the case of finite-difference method, the Liebmann iteration method) is applied as follows: First, the order of the mesh points in which the computation will proceed must be determined. (In our case, for instance, from left to right beginning with the first row of the net.) Then we write arbitrary numbers at those mesh points for which we are to determine the solution and these numbers are taken to be the zero approximation. The first approximation is then obtained as follows: We begin at the first mesh point (in our case at the mesh point 5) and we replace the value at this mesh point by a value computed from the corresponding equation (here

$$u_5 = \frac{\eta\delta}{(\eta + \delta)(1 + \eta)}\,u_{14} + \frac{\eta\delta}{(\eta + \delta)(1 + \delta)}\,u_6)\,,$$

and proceed to the following mesh point. (When computing the value at the point 6 we use the new value from point 5.) Proceeding in this way, we obtain new values at all mesh points and the procedure is repeated until we obtain the desired accuracy.

Example 2. Let us compute the temperature $u(x, t)$ of a rod of length L, if its temperature at $t = 0$ is $u(x, 0) = p(x)$, where $p(x)$ is a given function and if the temperature at its ends is zero: $u(0, t) = u(L, t) = 0$.

This problem is described by the equation

$$\frac{\partial^2 u}{\partial x^2} = a^2\,\frac{\partial u}{\partial t}$$

in the region $0 < x < L$, $t > 0$ (a^2 being the coefficient of heat diffusivity of the rod) with the initial condition

$$u(x, 0) = p(x)$$

and boundary conditions

$$u(0, t) = u(L, t) = 0\,.$$

Let a positive integer M be chosen and let $h = L/M$. Further let $\tau = \beta a^2 h^2$, where β is an arbitrary number for which $0 < \beta \leqq \frac{1}{2}$ and let us construct, in the domain concerned, a rectangular net formed by a system of lines $x = lh$, $l = 1, 2, \ldots, M - 1$, $t = k\tau$, $k = 1, 2, \ldots$ Let our differential equation be replaced by a finite-difference equation according to the first formula of Tab. 27.3 (see Fig. 27.10):

$$u_A = \beta u_1 + (1 - 2\beta)\,u_0 + \beta u_2\,.$$

The values in the zero time row (i.e. for $t = 0$) are known. From the above formula, the values of the unknown function at inner mesh points of the first time row are computed; at the boundary mesh points the values are known (from the prescribed boundary conditions). Similarly we obtain the approximation to the unknown function for any required time row.

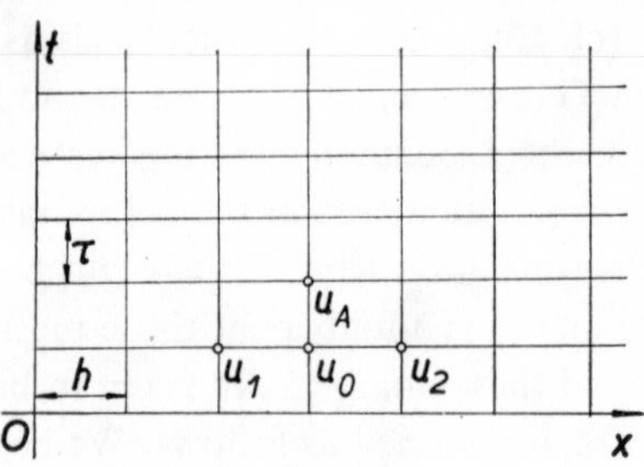

Fig. 27.10.

REMARK 1. The restricting condition $\beta = \tau/a^2h^2 \leqq \frac{1}{2}$ is essential. It can be easily shown that the values of the approximate solution in the k-th time row are linear combinations of terms of the form

$$\left(1 - 4\beta \sin^2 \frac{l\pi h}{2L}\right)^k \sin \frac{l\pi x}{L}, \quad l = 1, 2, \ldots, M - 1 .$$

If $\beta > \frac{1}{2}$, then the expression

$$\left|1 - 4\beta \sin^2 \frac{l\pi h}{2L}\right|$$

is greater than unity for those values of l for which $\sin (l\pi h/2L)$ differs sufficiently little from unity so that the above terms tend to infinity with increasing k (i.e. if the net is refined). Thus the solution to be found cannot be approximated by finite differences of this form.

Example 3. Let us determine the deflection of a square plate of side $2a$ loaded by a uniform loading and clamped on the boundary.

This problem yields the differential equation

$$\Delta\Delta u = f, \quad f = \frac{P}{N},$$

where P is the load on unit area and N is a constant depending on the elastic properties of the plate; the boundary conditions are

$$u = 0 \quad \text{for} \quad x = \pm a, \quad y = \pm a,$$

$$\frac{\partial u}{\partial x} = 0 \quad \text{for} \quad x = \pm a, \quad \frac{\partial u}{\partial y} = 0 \quad \text{for} \quad y = \pm a .$$

In order to write the finite-difference equation at a mesh point (x, y) in the form given in Tab. 27.5 the values at mesh points $(x + h, y)$, $(x, y + h)$, $(x - h, y)$, $(x, y - h)$, $(x + h, y + h)$, $(x - h, y + h)$, $(x - h, y - h)$, $(x + h, y - h)$, $(x + 2h, y)$, $(x, y + 2h)$, $(x - 2h, y)$, $(x, y - 2h)$ are necessary. At boundary mesh points, the value of the solution is known (it is zero). But it is necessary to know the

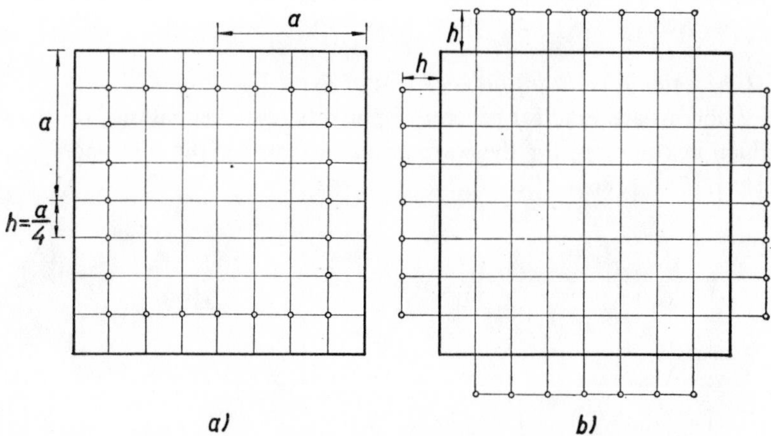

Fig. 27.11.

values of the function u at the mesh points, adjacent to the boundary mesh points, denoted in Fig. 27.11a by small circles. These values can be obtained by linear interpolation from the known boundary values of $\partial u/\partial x$ and $\partial u/\partial y$. A more advantageous procedure is shown in Fig. 27.11b. A second row of mesh points, at which the values are determined by linear interpolation, is here placed outside the domain in which the solution is sought.

27.8. Some Basic Theorems

Theorem 1. *Let u be the solution of the Dirichlet problem, in a domain Ω, for the Laplace equation $\Delta u = 0$ with a continuous boundary condition and where the boundary of Ω is a simple piecewise smooth curve. Let u_h be an approximate solution of this problem computed by the finite-difference method using the finite-difference operators from* Tab. 27.2 *and boundary conditions according to* § 27·5. *Then*

$$u_h \to u \quad \text{as} \quad h \to 0 .$$

Theorem 2. *Let Ω be a domain such that the solution u of the mixed boundary-value problem for the Laplace equation, with four continuous derivatives in $\bar{\Omega}$, exists. (The mixed boundary-value problem is the problem of finding a harmonic* *function in Ω if, on part of the boundary, values of the function u are given, on another part values of its normal derivative are given and on the remaining part linear combinations of values of the function and of its normal derivatives are given.) Let u_h be the approximate solution computed by the finite difference method, i.e. by the formulae of* Tab. 27.2, *treating the boundary conditions as in* § 27.5. *Then the estimate*

$$|u_h - u| \leqq Mh$$

holds in Ω, M being a constant independent of h.

Similar theorems are valid for general elliptic differential equations, for the general heat conduction equation, for the wave equation and for the biharmonic equation (see e.g. [26], [74], [286]).

28. INTEGRAL TRANSFORMS (OPERATIONAL CALCULUS)

By Jindřich Nečas

References: [15], [29], [55], [61], [81], [85], [88], [89], [101], [120], [179], [203], [277], [283], [323], [339], [372], [402], [412], [430].

For solving certain types of ordinary differential equations, particularly those with constant coefficients, and certain types of partial differential equations (e.g. equation of heat conduction, of string and diaphragm vibrations), transform methods may be advantageously used. From among them, the Laplace – Carson transform is formally identical with the operational calculus, and the finite Fourier transform leads to the expansion of a function in a Fourier series.

28.1. One-Dimensional Infinite Transforms (the Laplace, Fourier, Mellin, Hankel Transforms)

Table 28.1

Transform	Image		Inversion formula	
Laplace	$F(p) = \int_0^\infty f(t)\,\mathrm{e}^{-pt}\,\mathrm{d}t$	(1)	$f(t) = \dfrac{1}{2\pi\mathrm{i}}\int_{x-\mathrm{i}\infty}^{x+\mathrm{i}\infty} F(p)\,\mathrm{e}^{pt}\,\mathrm{d}p$	(1′)
Laplace-Carson	$F(p) = p\int_0^\infty f(t)\,\mathrm{e}^{-pt}\,\mathrm{d}t$	(2)	$f(t) = \dfrac{1}{2\pi\mathrm{i}}\int_{x-\mathrm{i}\infty}^{x+\mathrm{i}\infty} \dfrac{F(p)}{p}\,\mathrm{e}^{pt}\,\mathrm{d}p$	(2′)
Bilateral Laplace	$F(p) = \int_{-\infty}^\infty f(t)\,\mathrm{e}^{-pt}\,\mathrm{d}t$	(3)	$f(t) = \dfrac{1}{2\pi\mathrm{i}}\int_{x-\mathrm{i}\infty}^{x+\mathrm{i}\infty} F(p)\,\mathrm{e}^{pt}\,\mathrm{d}p$	(3′)
Fourier	$F(p) = \int_{-\infty}^\infty f(t)\,\mathrm{e}^{-\mathrm{i}pt}\,\mathrm{d}t$	(4)	$f(t) = \dfrac{1}{2\pi}\int_{-\infty}^{\infty} F(p)\,\mathrm{e}^{\mathrm{i}pt}\,\mathrm{d}p$	(4′)

TABLE 28.1

Transform	Image		Inversion formula	
Fourier Cosine	$F(p) = \sqrt{\left(\frac{2}{\pi}\right)} \int_0^\infty f(t) \cos pt \, dt$	(5)	$f(t) = \sqrt{\left(\frac{2}{\pi}\right)} \int_0^\infty F(p) \cos pt \, dp$	(5′)
Fourier Sine	$F(p) = \sqrt{\left(\frac{2}{\pi}\right)} \int_0^\infty f(t) \sin pt \, dt$	(6)	$f(t) = \sqrt{\left(\frac{2}{\pi}\right)} \int_0^\infty F(p) \sin pt \, dp$	(6′)
Mellin	$F(p) = \int_0^\infty f(t)\, t^{p-1} \, dt$	(7)	$f(t) = \frac{1}{2\pi i} \int_{x-i\infty}^{x+i\infty} F(p)\, t^{-pt} \, dp$	(7′)
Hankel	$F(p) = \int_0^\infty J_\nu(2\sqrt{(pt)})\, f(t) \, dt$	(8)	$f(t) = \int_0^\infty J_\nu(2\sqrt{(pt)}\, F(p) \, dp$	(8′)
	(J_ν is the Bessel function of the first kind, $\nu > -1$)			

By each of given integral transforms, given in Table 28.1, to every function $f(t)$ (the so-called *original*) from some class of functions, there is assigned a certain function $F(p)$ (the so-called *image of the function* $f(t)$). For example, the so-called *Laplace image* (i.e. the image by the Laplace transform (1)) of the function $f(t) = e^{3t}$is

$$F(p) = \int_0^\infty e^{3t} \,.\, e^{-pt} \, dt = \int_0^\infty e^{(3-p)t} \, dt = \left[\frac{1}{3-p} e^{(3-p)t}\right]_0^\infty = \frac{1}{p-3} \quad (\operatorname{Re} p > 3) \,. \tag{9}$$

(The Laplace image of a function $f(t)$ is frequently denoted by $\mathfrak{L}\{f(t)\}$. Thus, in our example we have $F(p) = \mathfrak{L}\{e^{3t}\} = 1/(p - 3)$.)

In transforms (5), (6), (8), we have $t \in [0, +\infty)$, $p \geqq 0$; in transform (4), $t \in (-\infty, \infty)$ and p is real; in transforms (1), (2), (7), and (3), $t \in [0, +\infty)$ and $t \in (-\infty, +\infty)$, respectively, p being a complex number (not an arbitrary one; its choice depends on the function $f(t)$).

The improper integrals are understood in the usual sense, for example,

$$\int_0^\infty f(t)\, e^{-pt} \, dt = \lim_{b \to +\infty} \int_0^b f(t)\, e^{-pt} \, dt \,,$$

etc. In order to guarantee the convergence of these integrals, the functions $f(t)$ and the numbers p must have certain properties. For example, integral (9) is convergent for those complex numbers p which satisfy the inequality $\operatorname{Re} p > 3$. Thus, the Laplace image of the function e^{3t} is a function of a complex variable p, defined in the half-plane. Re $p > 3$ of the Gaussian plane (i.e. in the half-plane $x > 3$, if we set $p = x + iy$; for Re $p \leqq 3$ the function $1/(p-3)$ is *not* the Laplace image of the function e^{3t}). If, in addition, the function $f(t)$ has certain particular properties, then also the corresponding image $F(p)$ has certain particular properties (see § 28.3).

In the integral transforms of Tab. 28.1, the following conditions are imposed on the function $f(t)$: $f(t)$ is absolutely integrable

(i) in every finite interval $0 \leqq a \leqq t \leqq b < +\infty$ in case of transforms (1) and (2),
(ii) in every finite interval $-\infty < a \leqq t \leqq b < +\infty$ in case (3),
(iii) in the interval $(-\infty, +\infty)$ in case (4),
(iv) in the interval $[0, +\infty)$ in cases (5) and (6),
(v) in every finite interval $0 < a \leqq t \leqq b < +\infty$ in case (7).

In the case of the transform (8) we assume the function $f(t)\, t^{\nu/2}$ to be absolutely integrable in every finite interval $0 \leqq a \leqq t \leqq b < +\infty$.

By an absolutely integrable function $g(t)$ in an interval (a, b) (or $[a, b]$) we understand a function for which both the integrals $\int_a^b g(t)\, dt$ and $\int_a^b |g(t)|\, dt$ are convergent.

Furthermore, we assume that the functions $f(t)$ are such that

(i) in the case of transforms (1) and (2) a constant $\sigma \geqq -\infty$ exists such that the integral (1) is convergent for Re $p > \sigma$ (in the case (9) we had $\sigma = 3$; not every function possesses the property just mentioned; for example, for $f(t) = e^{t^2}$, the integral (1) is divergent for *every* p);

(ii) in cases (3) and (7) numbers $\sigma_1 \geqq -\infty$, $\sigma_2 \leqq +\infty$ exist such that for $\sigma_1 < \text{Re}\, p < \sigma_2$ the integral (3) or (7) is convergent, respectively.

The inversion formulae (1′) to (8′) assign the original to the image provided certain assumptions are satisfied (§ 28.3). The integral (1′) means

$$\lim_{\omega \to +\infty} \frac{1}{2\pi i} \int_{x-\omega i}^{x+\omega i} F(p)\, e^{pt}\, dp$$

and may also be written in the form

$$\frac{1}{2\pi} \int_{-\infty}^{\infty} F(x + iy)\, e^{(x+iy)t}\, dy\,,$$

and similarly in the other cases.

By the Hankel transform, the transform

$$F(v) = \int_0^\infty u\, J_\nu(vu)\, \varphi(u)\, du$$

(which follows from (8) by the substitution $\sqrt{(2p)} = v$, $\sqrt{(2t)} = u$, $\varphi(u) = f(\tfrac{1}{2}u^2)$) is sometimes understood.

Instead of the Laplace and Fourier transform, one often speaks of the *Laplace* and *Fourier integral*, respectively.

When solving differential equations, we use transforms for reducing the number of independent variables in the differential equation under consideration. If an ordinary differential equation is solved by means of an integral transform, an algebraic equation is obtained; using transforms for solving partial differential equations, the number of independent variables is reduced by one. The type of transform used depends on the equation under consideration and on the corresponding domain of definition. (A thorough treatment of these problems may be found, e.g. in [372], [402]).

The Laplace transform is the most frequently employed owing to the simple relationship which exists between the Laplace image of a function $f(t)$ and of its derivative $f'(t)$, i.e.

$$\int_0^\infty f'(t)\,\mathrm{e}^{-pt}\,\mathrm{d}t = p\,F(p) - f(0)\,. \tag{10}$$

This relationship follows from the theorem on integration by parts provided certain obvious assumptions are satisfied. If we assume that $f(0) = 0$, then to the operation of differentiation of the original there corresponds the algebraic operation of multiplying the image $F(p)$ by p. Upon this fact the Heaviside operational calculus may be theoretically based. To the operation of multiplying by p there corresponds the differentiation of the original, to the operation $1/p$ the integration of the original; hence, p is the inverse operator of $1/p$. This idea has led J. Mikusiński to the definition of the differentiation operator independently of the integral transform methods; this has been well developed and adapted for applications in [283].

28.2. Applications of the Laplace Transform to the Solution of Differential Equations. Examples

Example 1. Let us find the current $i(t)$ in a circuit consisting of an inductance L and resistance R, provided $i(t) = 0$ for $t = 0$ and an electromotive force E is applied at time $t = 0$.

The circuit is governed by the differential equation

$$L\frac{\mathrm{d}i}{\mathrm{d}t} + Ri = E\,. \tag{1}$$

We multiply equation (1) by the factor e^{-pt} and integrate it between limits 0 and $+\infty$. Assuming the function $i(t)$ bounded and $\mathrm{Re}\, p > 0$, and writing $J(p)$ for $\int_0^\infty i(t)\,\mathrm{e}^{-pt}\,\mathrm{d}t$, we have

$$L\int_0^\infty \mathrm{e}^{-pt}\frac{\mathrm{d}i}{\mathrm{d}t}\,\mathrm{d}t = L[i\mathrm{e}^{-pt}]_0^\infty + Lp\int_0^\infty i\mathrm{e}^{-pt}\,\mathrm{d}t = Lp\,J(p)\,,$$

because $i(0) = 0$ and $\lim_{t\to+\infty} i\mathrm{e}^{-pt} = 0$. (Formulae 28.1.10) could also be applied directly, of course.) Further, $\int_0^\infty E e^{-pt} \, dt = E/p$. Thus, for the image $J(p)$ we get from (1) an algebraic equation,

$$Lp \, J(p) + R \, J(p) = \frac{E}{p};$$

from this,

$$J(p) = \frac{E}{p} \frac{1}{R + Lp} = \frac{E}{R} \left(\frac{1}{p} - \frac{1}{p + R/L} \right).$$

Using tables of transform pairs (e.g. the first pair in Tab. 28.2 for $n = 0$, and then the second pair) we find that $1/p$ is the image of the function $f(t) = 1$, and $1/(p + R/L)$ is the image of the function $e^{-(R/L)t}$; consequently, for the desired solution we get

$$i(t) = \frac{E}{R} \left(1 - e^{-(R/L)t}\right).$$

Example 2. Let us find the temperature distribution $u(t, x)$ in a semi-infinite rod with insulated surface, the end of which is kept in a basin with constant temperature, i.e. $u(t, 0) = U_0 = \text{const.}$ Let the initial temperature be zero, i.e. $u(0, x) = 0$, and assume that $\lim_{x \to \infty} u(x, t) = 0$.

The function u satisfies the differential equation

$$\frac{\partial^2 u}{\partial x^2} = k \frac{\partial u}{\partial t}, \tag{2}$$

where k is a positive constant determined by conductivity, specific heat and specific mass of the rod. We multiply equation (2) by the factor e^{-pt} with $\operatorname{Re} p > 0$ and integrate it between the limits 0 and $+\infty$. We thus obtain

$$\int_0^\infty \frac{\partial u}{\partial t}(t, x) \, e^{-pt} \, dt = \left[u(t, x) \, e^{-pt}\right]_0^\infty + p \int_0^\infty u(t, x) \, e^{-pt} \, dt = p \, U(p, x),$$

where $U(p, x) = \int_0^\infty u(t, x) \, e^{-pt} \, dt$ is the image of $u(t, x)$. Further, assuming that differentiation under the integral sign is permitted (Theorem 13.9.9, p. 577; see also [61], p. 167), we obtain

$$\int_0^\infty \frac{\partial^2 u}{\partial x^2}(t, x) \, e^{-pt} \, dt = \frac{d^2 U}{dx^2}(p, x).$$

Thus, for the image $U(p, x)$ we get from (2) the ordinary differential equation

$$\frac{d^2 U}{dx^2}(p, x) - kp \, U(p, x) = 0, \tag{3}$$

with boundary conditions

$$U(p, 0) = \int_0^\infty U_0 e^{-pt}\, dt = \frac{U_0}{p}, \quad \lim_{x\to\infty} U(p, x) = 0 .$$

For a general solution of the differential equation (3) we have

$$U(p, x) = A(p)\, e^{-\sqrt{(pk)}x} + B(p)\, e^{\sqrt{(pk)}x} .$$

The boundary conditions yield

$$U(p, x) = \frac{U_0}{p} e^{-\sqrt{(pk)}x} .$$

In tables of transforms we find that the function $e^{-\sqrt{(pk)}x}/p$ is the image of the function erfc $(x \sqrt{(k)}/2 \sqrt{(t)})$, so that for the desired solution we get

$$u(t, x) = U_0 \operatorname{erfc}\left(\frac{x \sqrt{k}}{2 \sqrt{t}}\right).$$

(The function erfc x is given by the relation $\operatorname{erfc} x = 1 - \operatorname{erf} x = 1 - \frac{2}{\sqrt{\pi}} \int_0^x e^{-u^2}\, du =$ $= \frac{2}{\sqrt{\pi}} \int_x^\infty e^{-u^2}\, du$.)

28.3. Some Results of Fundamental Importance

From the point of view of the applications of transform methods in practice and even from the theoretical point of view the following two problems are of basic importance:

1. to decide whether or not a given function $F(p)$ is the image of some function $f(t)$ (under the transform considered);
2. to find the original for a given image.

Let us present some typical properties of images:

Laplace transform: $F(p)$ is a holomorphic function in the half-plane $\operatorname{Re} p > \sigma$. (If $\sigma = -\infty$, $F(p)$ is holomorphic in the entire plane.)

The same is true for the Laplace–Carson transform.

In the case of the bilateral Laplace transform the image $F(p)$ is holomorphic in the strip $\sigma_1 < \operatorname{Re} p < \sigma_2$.

In the case of the Fourier transform the image $F(p)$ is defined only for real p, and $F(p)$ is a continuous function. (In the case of the cosine and sine transforms p is restricted to have only non-negative values.)

The images of Mellin transforms have the same properties as those given above for the bilateral Laplace transform.

In the case of the Hankel transform the image $F(p)$ is defined only for non-negative p.

The problems 1 and 2 just stated were unsolved until recently. Because the Laplace transform is the most important of the transforms discussed, some of its principal properties will be given.

Theorem 1. *The condition*

$$\sup_{\delta > \gamma} \int_{-\infty}^{\infty} |F(\delta + \mathrm{i}\tau)|^2 \, \mathrm{d}\tau < \infty$$

is necessary and sufficient for a function $F(p)$, holomorphic in the half-plane $\operatorname{Re} p > \gamma$, *to be the Laplace image of a function $f(t)$ satisfying the inequality*

$$\int_0^{\infty} |f(t)|^2 \, \mathrm{e}^{-2\gamma t} \, \mathrm{d}t < \infty \, .$$

Theorem 2. *If $F(p)$ is the Laplace image of an original $f(t)$, then*

$$f(t) = \frac{\mathrm{d}}{\mathrm{d}t} \lim_{\omega \to +\infty} \frac{1}{2\pi\mathrm{i}} \int_{\gamma - \mathrm{i}\omega}^{\gamma + \mathrm{i}\omega} F(p) \frac{\mathrm{e}^{pt}}{p} \, \mathrm{d}p \, . \tag{1}$$

(More accurately, equation (1) holds for almost every t in the interval $[0, \infty)$, i.e. with the possible exception of points which constitute a set of measure zero.) *The integration is performed along a straight line* $\operatorname{Re} p = \gamma$ *with* $\gamma > 0$, *which lies in the domain of definition of the function $F(p)$.* (Thus, it suffices to take γ sufficiently large.)

In (1) the differentiation may sometimes be performed under the integral sign. Sufficient conditions:

Theorem 3. *Let* $\int_0^\infty |f(t)| \, \mathrm{e}^{-\alpha t} \, \mathrm{d}t < \infty$ *for* $\alpha \geqq \gamma$. *Let $f(t)$ be a function of bounded variation in a neighbourhood of a point t ($t \geqq 0$). Then we have*

$$\lim_{\omega \to +\infty} \frac{1}{2\pi\mathrm{i}} \int_{\alpha - \mathrm{i}\omega}^{\alpha + \mathrm{i}\omega} \mathrm{e}^{pt} F(p) \, \mathrm{d}p = \begin{cases} \dfrac{f(t+0) + f(t-0)}{2} & \textit{for} \quad t > 0 \, , \\[2ex] \dfrac{f(+0)}{2} & \textit{for} \quad t = 0 \, , \\[2ex] 0 & \textit{for} \quad t < 0 \, . \end{cases}$$

Here, $f(t \pm 0)$ denotes the limit from the right and the left, respectively, of the *function $f(t)$ at the point t. If $f(t)$ is continuous at the point t, then*

$$\lim_{\omega \to +\infty} \frac{1}{2\pi i} \int_{\alpha - i\omega}^{\alpha + i\omega} e^{pt} F(p) \, dp = f(t) .$$

The originals are in practice calculated by Theorem 3 from the integral

$$\frac{1}{2\pi i} \int_{\alpha - i\omega}^{\alpha + i\omega} e^{pt} F(p) \, dp .$$

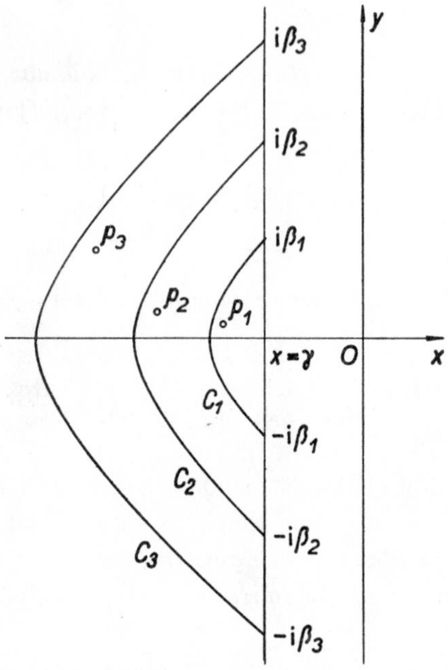

Fig. 28.1.

A formal application of the inverse transform may lead to wrong results (see e.g. [89], p. 193).

For calculation of the original from a given image the residue theorem (Theorem 20.5.1, p. 963) is often used:

Theorem 4. *Let $F(p)$ be the Laplace image of a function $f(t)$ and let*

$$f(t) = \lim_{\omega \to +\infty} \frac{1}{2\pi i} \int_{\gamma - i\omega}^{\gamma + i\omega} e^{pt} F(p) \, dp .$$

Let the function $F(p)$ be holomorphic in the complex plane except for poles $p_1, p_2, \ldots$ (e.g., let $F(p)$ be a rational function) *interior to the half-plane* $\operatorname{Re} p < \gamma$. *Let a sequence of arcs C_n (which do not pass through the poles) be such that each C_n*

meets the straight line Re $p = \gamma$ *at two points* $\gamma + i\beta_n$ *and* $\gamma - i\beta_n$ *and lies in the half-plane* Re $p \leqq \gamma$ *while each arc* C_n *together with the line segment with end points* $\gamma - i\beta_n$ *and* $\gamma + i\beta_n$ *form the boundary of a region* O_n *which contains exactly the poles* $p_1, p_2, \ldots, p_n$ (see Fig. 28.1). *Further, let* $\beta_n \to +\infty$ *as* $n \to \infty$ *and*

$$\lim_{n\to\infty} \int_{C_n} F(p)\, e^{pt}\, dp = 0\,.$$

TABLE 28.2

Image $F(p)$	Original $f(t)$
p^{-n-1}	$\dfrac{t^n}{\Gamma(n+1)}, \quad n > -1$ (If $n \geqq 0$ is an integer, $\Gamma(n+1) = n!$)
$(p+\alpha)^{-1}$	$e^{-\alpha t}$
$\dfrac{\omega}{p^2+\omega^2}$	$\sin \omega t$
$\dfrac{p}{p^2+\omega^2}$	$\cos \omega t$
$\dfrac{e^{-a\sqrt{p}}}{p}$	$\operatorname{erfc}\left(\dfrac{a}{2\sqrt{t}}\right) = \dfrac{2}{\sqrt{\pi}} \displaystyle\int_{a/2\sqrt{t}}^{\infty} e^{-u^2}\, du\,; \quad a \geqq 0$
$\dfrac{e^{-a\sqrt{p}}}{p^{\frac{3}{2}}}$	$2\sqrt{(t)}\, \mathrm{I}\operatorname{erfc}\left(\dfrac{a}{2\sqrt{t}}\right),$ where $\mathrm{I}\operatorname{erfc} x = \displaystyle\int_x^{\infty} \operatorname{erfc} u\, du\,, \quad a > 0$

Then

$$f(t) = \sum_{n=1}^{\infty} \operatorname*{res}_{p=p_n} \left[F(p)\, e^{pt}\right], \tag{2}$$

where $\operatorname*{res}_{p=p_n}\left[F(p)\, e^{pt}\right]$ *denotes the residue of the function* $F(p)\, e^{pt}$ *at the pole* $p = p_n$.

With the aid of this and similar theorems the originals corresponding to various images can be established. In particular, the calculation of the original of every rational function can be reduced to the application of formula (2) (with a finite number of terms on its right-hand side).

Example 1. Let us find the original of the function $\omega/(p^2 + \omega^2)$.

Using (2) and formulae for the calculation of residues on p. 962, we get

$$f(t) = \operatorname*{res}_{p=\mathrm{i}\omega}\left[\frac{\omega e^{pt}}{p^2 + \omega^2}\right] + \operatorname*{res}_{p=-\mathrm{i}\omega}\left[\frac{\omega e^{pt}}{p^2 + \omega^2}\right] = \frac{e^{\mathrm{i}\omega t}}{2\mathrm{i}} - \frac{e^{-\mathrm{i}\omega t}}{2\mathrm{i}} = \sin \omega t\,,$$

in accordance with Tab. 28.2.

Extensive tables of transform pairs are given, e.g., in [89]. In such tables the image $F(p)$ is always given first (see Tab. 28.2).

To tables there is usually attached the so-called *grammar*, which summarizes the basic rules governing relationship between originals and images. In Tab. 28.3 a sample of grammar for Laplace transforms is given.

TABLE 28.3

Image $F(p)$	Original $f(t)$
$\frac{1}{\alpha} F\left(\frac{p}{\alpha}\right)$	$f(\alpha t)$
$p\,F(p) - f(0)$	$f'(t)$
$p^2\,F(p) - p f(0) - f'(0)$	$f''(t)$
$\frac{F(p)}{p}$	$\int_0^t f(t)\,\mathrm{d}t$
$\int_p^\infty F(p)\,\mathrm{d}p$	$\frac{f(t)}{t}$
$F(p - p_0)$	$e^{p_0 t} f(t)$
$F(p)\,G(p)$	$\int_0^t f(\tau)\,g(t - \tau)\,\mathrm{d}\tau$
$F(q(p))\,G(p)$	$\int_0^\infty f(\tau)\,g(t, \tau)\,\mathrm{d}\tau$

where $g(t, \tau)$ is the original for the function $G(p)\,e^{-\tau q(p)}$ with $G(p)$, $q(p)$ being holomorphic functions in a half-plane $\operatorname{Re} p > \sigma$.

Sufficient conditions for validity of the inversion formula for the Fourier transform are stated in the following assertion:

Theorem 5. *Let $f(t)$ be the original of a Fourier image $F(p)$. If $f(t)$ has bounded variation in a neighbourhood of a point t, then we have*

$$\frac{f(t+0)+f(t-0)}{2} = \lim_{\omega\to+\infty} \frac{1}{2\pi} \int_{-\omega}^{\omega} F(p)\,\mathrm{e}^{\mathrm{i}pt}\,\mathrm{d}p\,.$$

If, in addition, $f(t)$ is continuous at the point t, then

$$\lim_{\omega\to+\infty} \frac{1}{2\pi} \int_{-\omega}^{\omega} F(p)\,\mathrm{e}^{\mathrm{i}pt}\,\mathrm{d}p = f(t)\,.$$

If $f(t)$ is the original for the bilateral Laplace transform with image denoted by $F(p)$ and if

$$\int_{-\infty}^{\infty} \left|f(t)\,\mathrm{e}^{-p_0 t}\right| \mathrm{d}t < \infty$$

for some $p_0 = x_0 + \mathrm{i}y_0$, then the function $f(t)\,\mathrm{e}^{-x_0 t}$ is the original for the Fourier transform whose Fourier image is $F(x_0 + \mathrm{i}y)$.

If $f(t)$ is an original for the Mellin transform with the corresponding image denoted by $F(p)$ and if further

$$\int_{0}^{\infty} \left|f(t)\,t^{p_0-1}\right| \mathrm{d}t < \infty$$

for some $p_0 = x_0 + \mathrm{i}y_0$, then $f(\mathrm{e}^{-u})\,\mathrm{e}^{-x_0 u}$ is the original for the Fourier transform with the Fourier image $F(x_0 + \mathrm{i}y)$. (The substitution $t = \mathrm{e}^{-u}$ has been used.)

28.4. Two-Dimensional and Multidimensional Transforms

Definition 1. The *two-dimensional Laplace transform* assigns to an original $f(x, y)$, which is absolutely integrable in every rectangle $0 \leqq x \leqq a < \infty$, $0 \leqq y \leqq b < \infty$, the image

$$F(u, v) = \int_0^{\infty}\int_0^{\infty} f(x, y)\,\mathrm{e}^{-ux-vy}\,\mathrm{d}x\,\mathrm{d}y\,,$$

where u, v are complex numbers.

A more detailed treatment of two-dimensional Laplace transform may be found in [412].

Definition 2. The *n-dimensional Fourier transform* assigns to an original $f(x_1, x_2, \ldots, x_n)$ satisfying

$$\underbrace{\int_{-\infty}^{\infty}\int_{-\infty}^{\infty}\cdots\int_{-\infty}^{\infty}}_{n\text{-tuple}} |f(t_1, \ldots, t_n)| \, dt_1 \ldots dt_n < +\infty$$

the image

$$F(p_1, p_2, \ldots, p_n) = \underbrace{\int_{-\infty}^{\infty}\int_{-\infty}^{\infty}\cdots\int_{-\infty}^{\infty}}_{n\text{-tuple}} f(t_1, \ldots, t_n) \, e^{-i(p_1t_1 + p_2t_2 + \ldots + p_nt_n)} \, dt_1 \ldots dt_n ,$$

where $p_1, \ldots, p_n$ are arbitrary real numbers.

28.5. One-Dimensional Finite Transforms

A one-dimensional finite transform of a function $f(t)$ in one variable assigns the Fourier coefficients of $f(t)$ to this function. The application of a one-dimensional finite transform leads to the expansion of a function in Fourier series, the application in partial differential equations to the Fourier method. More about these transforms may be found in [402].

29. APPROXIMATE SOLUTION OF FREDHOLM'S INTEGRAL EQUATIONS

By Karel Rektorys

References: [21], [214], [281], [354].

Equations with degenerate kernels are solved according to § 19.2. For the theory of integral equations see Chapter 19.

29.1. Successive Approximations (Iterations)

Suppose we have an equation of the Fredholm type

$$f(x) - \lambda \int_a^b K(x, s) f(s) \, ds = g(x) . \tag{1}$$

Write

$$\int_a^b \int_a^b |K(x, s)|^2 \, dx \, ds = B^2 \quad (B > 0) .$$

Let us construct the sequence of functions

$$\begin{aligned} f_0(x) &= g(x) , \\ f_1(x) &= g(x) + \lambda \int_a^b K(x, s) f_0(s) \, ds , \\ f_2(x) &= g(x) + \lambda \int_a^b K(x, s) f_1(s) \, ds , \\ &\ldots\ldots\ldots\ldots\ldots\ldots\ldots\ldots\ldots\ldots \\ f_{n+1}(x) &= g(x) + \lambda \int_a^b K(x, s) f_n(s) \, ds , \\ &\ldots\ldots\ldots\ldots\ldots\ldots\ldots\ldots\ldots\ldots \end{aligned} \tag{2}$$

Theorem 1. *If*

$$\int_a^b |K(x, s)|^2 \, ds \leqq C, \text{ where the constant } C \text{ is the same for all } x, \text{ and } |\lambda| < \frac{1}{B}, \tag{3}$$

then the sequence $f_0(x), f_1(x), f_2(x), \ldots$ converges uniformly in $[a, b]$ to the (unique) solution $f(x)$ of equation (1). The absolute value of the difference between $f_n(x)$ and $f(x)$ does not exceed

$$D \sqrt{(C)} \frac{B^n |\lambda|^{n+1}}{1 - B|\lambda|},$$

where (4)

$$D = \sqrt{\int_a^b |g^2(x)| \, dx}.$$

Example 1 ([281]).

$$f(x) - 0{\cdot}1 \int_0^1 K(x, s) f(s) \, ds = 1, \quad K(x, s) = \begin{cases} x & \text{for } 0 \leqq x \leqq s, \\ s & \text{for } s \leqq x \leqq 1. \end{cases} \tag{5}$$

We easily find

$$B = \frac{1}{\sqrt{6}}, \quad C = \tfrac{1}{3}, \quad D = 1; \quad \lambda = 0{\cdot}1.$$

By (3), the successive approximations are convergent. If we take $n = 2$, then by (4) the error attains at most the value

$$1 \, . \sqrt{(\tfrac{1}{3})} \, . \frac{\tfrac{1}{6} \, . \, 0{\cdot}1^3}{1 - 1/\sqrt{(6)} \, . \, 0{\cdot}1} \doteq 0{\cdot}0001. \tag{6}$$

We have (by (2))

$$f_0(x) = 1, \; f_1(x) = 1 + \tfrac{1}{10}x - \tfrac{1}{20}x^2, \; f_2(x) = 1 + \tfrac{31}{300}x - \tfrac{1}{20}x^2 - \tfrac{1}{600}x^3 + \tfrac{1}{2400}x^4.$$

If we take $f(x) \approx f_2(x)$, then by (6) the error in the entire interval $[0, 1]$ is less than $0{\cdot}0001$.

REMARK 1. For the *resolvent* of an integral equation see § 19.4. Cf. also Example 19.4.2, p. 931.

29.2. Approximate Solution of Integral Equations by Reduction to a System of Linear Algebraic Equations

Using numerical integration (§ 13.13) for the evaluation of an integral we replace the integral by the sum:

$$\int_a^b h(x) \, dx \approx (b - a) \sum_{k=1}^{n} C_k \, h(x_k),$$

where x_k and C_k are points and constants, respectively, defined by the given numerical integration. For example, if we apply to the evaluation of the integral

$$\int_0^1 h(x)\,\mathrm{d}x$$

the trapezoidal rule for three points ($n = 2$), then

$$\begin{aligned} x_1 &= 0\,, & x_2 &= \tfrac{1}{2}\,, & x_3 &= 1\,, \\ C_1 &= \tfrac{1}{4}\,, & C_2 &= \tfrac{1}{2}\,, & C_3 &= \tfrac{1}{4}\,. \end{aligned} \tag{1}$$

In the equation

$$f(x) - \int_a^b K(x, s) f(s)\,\mathrm{d}s = g(x) \tag{2}$$

let us replace the integral by a sum, choosing a certain formula for the numerical integration with dividing points $s_1, s_2, \ldots, s_n$. Let the same partition be chosen in the interval $[a, b]$ for the variable x. Denote these dividing points by $x_1, x_2, \ldots, x_n$. Let us rewrite equation (2), for each x_k, replacing the integral by the corresponding sum. We obtain

$$\begin{gathered} f(x_1) \approx g(x_1) + (b - a) \sum_{k=1}^{n} C_k K(x_1, s_k) f(s_k)\,, \\ \cdots\cdots\cdots\cdots\cdots\cdots\cdots\cdots\cdots\cdots \\ f(x_n) \approx g(x_n) + (b - a) \sum_{k=1}^{n} C_k K(x_n, s_k) f(s_k)\,. \end{gathered} \tag{3}$$

The partition of the interval $[a, b]$ is the same for s as for x, hence $f(x_1) = f(s_1)$, etc. If we write in (3) the equality sign instead of $\approx$, we obtain a system of n equations for n unknown values $f(x_1), f(x_2), \ldots, f(x_n)$ (see Example 1).

REMARK 1. Since the sign $\approx$ in system (3) has been replaced by the sign $=$ the resulting values $f(x_1), f(x_2), \ldots, f(x_n)$ are not the exact values of the unknown function $f(x)$ at the points $x_1, x_2, \ldots, x_n$. Thus let us denote them by $f_n(x_1), f_n(x_2), \ldots, f_n(x_n)$ (specifying thus the dependence on the number n of dividing points of the chosen numerical integration). Now, it can be shown that the values $f_n(x_k)$ tend with increasing n to the exact values of the solution $f(x)$ at the points x_k (under certain assumptions on the smoothness of the kernel and of the function $g(x)$) provided there exists one and only one solution of equation (2); for more detail see for instance [214], where an estimate of the error may also be found.

Example 1. Let us consider the equation

$$f(x) - \int_0^1 (x - s) f(s)\,\mathrm{d}s = x^2\,. \tag{4}$$

Theorem 29.1.1 implies that equation (4) is (uniquely) solvable, for the kernel is bounded and

$$\lambda = 1\,, \quad B^2 = \int_0^1 \int_0^1 (x - s)^2 \,dx\,ds < 1\,, \quad \text{hence} \quad \lambda < \frac{1}{B}\,.$$

Let us choose the above-mentioned numerical integration (1). For $x_1 = 0$, $x_2 = \frac{1}{2}$, $x_3 = 1$ we write down equation (4) and the approximate equations which arise if we replace the integral by a sum:

$$f(0) - \int_0^1 (-s) f(s)\,ds = 0\,, \quad f(\tfrac{1}{2}) - \int_0^1 (\tfrac{1}{2} - s) f(s)\,ds = \tfrac{1}{4}\,,$$

$$f(1) - \int_0^1 (1 - s) f(s)\,ds = 1\,.$$

Hence, for the values of the approximate solution at the points 0, $\frac{1}{2}$, 1, the relations

$$\begin{aligned} f(0) - [\tfrac{1}{4}\,.\,0\,.\,f(0) + \tfrac{1}{2}\,.\,(-\tfrac{1}{2})\,.\,f(\tfrac{1}{2}) + \tfrac{1}{4}\,.\,(-1)\,.\,f(1)] &= 0\,,\\ f(\tfrac{1}{2}) - [\tfrac{1}{4}\,.\,\tfrac{1}{2}\,.\,f(0) + \tfrac{1}{2}\,.\,0\,.\,f(\tfrac{1}{2}) + \tfrac{1}{4}\,.\,(-\tfrac{1}{2})\,.\,f(1)] &= \tfrac{1}{4}\,,\\ f(1) - [\tfrac{1}{4}\,.\,1\,.\,f(0) + \tfrac{1}{2}\,.\,\tfrac{1}{2}\,.\,f(\tfrac{1}{2}) + \tfrac{1}{4}\,.\,0\,.\,f(1)] &= 1 \end{aligned} \tag{5}$$

hold. This is the system (3). After simplification we obtain

$$\begin{gathered} f(0) + \tfrac{1}{4}f(\tfrac{1}{2}) + \tfrac{1}{4}f(1) = 0\,, \quad -\tfrac{1}{8}f(0) + f(\tfrac{1}{2}) + \tfrac{1}{8}f(1) = \tfrac{1}{4}\,,\\ -\tfrac{1}{4}f(0) - \tfrac{1}{4}f(\tfrac{1}{2}) + f(1) = 1\,. \end{gathered} \tag{6}$$

The solution is

$$f(0) = -\tfrac{19}{72}\,, \quad f(\tfrac{1}{2}) = \tfrac{7}{72}\,, \quad f(1) = \tfrac{69}{72}\,. \tag{7}$$

This example is an illustrative one only, for equation (4) may obviously be solved as an equation with a degenerate kernel (§ 19·2).

The exact solution is $f(x) = x^2 + \frac{3}{13}x - \frac{17}{78}$. The values at the points $x = 0$, $x = \frac{1}{2}$, $x = 1$ are

$$f(0) = -\frac{17}{78}\,, \quad f\left(\frac{1}{2}\right) = \frac{11{\cdot}5}{78}\,, \quad f(1) = \frac{79}{78}\,. \tag{8}$$

The difference between (7) and (8) is rather high, for we have made use of a very rough numerical integration. The result may be considerably improved by choosing a more precise formula for numerical integration (e.g. Simpson's Rule or Gauss's formula; see § 13.13) and by choosing a greater number of dividing points.

The above method may also be applied to the approximate evaluation of the eigenvalues.

29.3. Replacement of the Kernel by a Degenerate Kernel

We replace the given kernel $K(x, s)$ by a "close" degenerate kernel $k(x, s)$ and solve the equation with this kernel. For example, we have

$$\sin xs = xs - \frac{x^3 s^3}{3!} + \frac{x^5 s^5}{5!} - \dots .$$

Hence we replace the equation

$$f(x) - \int_0^{\pi/2} \sin xs\, f(s)\, ds = g(x)$$

by the following equation with degenerate kernel:

$$\varphi(x) - \int_0^{\pi/2} \left(xs - \frac{x^3 s^3}{3!} + \frac{x^5 s^5}{5!} \right) \varphi(s)\, ds = g(x) .$$

If the given equation is uniquely solvable and if $k(x, s)$ is sufficiently close to $K(x, s)$, then the corresponding equation with the degenerate kernel $k(x, s)$ is also uniquely solvable (and conversely) and the solution $f(x)$ of the given equation is sufficiently close to the solution $\varphi(x)$ of the new equation. For an exact statement of these facts and an estimate of error see, for example, [214].

29.4. Galerkin's Method (Method of Moments) and Ritz's Method

Let $\varphi_1(x), \varphi_2(x), \dots$ constitute a complete (not necessarily orthogonal) system of (linearly independent) functions in $[a, b]$ (see Remarks 16.2.14, 16.2.15, p. 701). Let us look for the approximate solution of the equation

$$f(x) - \int_a^b K(x, s) f(s)\, ds = g(x) \tag{1}$$

in the form

$$f_n(x) = g(x) + \sum_{i=1}^{n} c_i \varphi_i(x) , \tag{2}$$

where the coefficients c_i are defined by the conditions

$$\begin{aligned} &\int_a^b \left[f_n(x) - \int_a^b K(x, s) f_n(s)\, ds - g(x) \right] \varphi_1(x)\, dx = 0 , \\ &\dots\dots\dots\dots\dots\dots\dots\dots\dots\dots\dots\dots \\ &\int_a^b \left[f_n(x) - \int_a^b K(x, s) f_n(s)\, ds - g(x) \right] \varphi_n(x)\, dx = 0 . \end{aligned} \tag{3}$$

Thus the expression in square brackets is orthogonal to the functions $\varphi_1(x), \varphi_2(x), \ldots$ $\ldots, \varphi_n(x)$. System (3) is a system of n linear equations for n unknown constants $c_1, c_2, \ldots, c_n$. If we solve this system, we obtain the approximate solution. If the system of functions $\varphi_1(x), \varphi_2(x), \ldots$ is orthonormal (Definition 16.2.4, p. 695), then it can be proved (see [214]), that $f_n(x)$ coincides with the approximate solution obtained by the method of § 29.3, the kernel $K(x, s)$ being replaced by the kernel

$$K_n(x, s) = \sum_{i=1}^{n} u_i(s)\, \varphi_i(x)\,, \quad \text{where} \quad u_i(s) = \int_a^b K(t, s)\, \varphi_i(t)\, dt\,.$$

According to § 29.3 this fact may be used to prove the convergence of the approximate solution in question. In particular, if equation (1) is uniquely solvable and if $K_n(x,s)$ tends uniformly in the square $a \leqq x \leqq b$, $a \leqq s \leqq b$ to $K(x, s)$, then $f_n(x)$ tends uniformly in $[a, b]$ to $f(x)$.

System (3) is obtained also if the Ritz method is applied to the approximate solution of integral equations with symmetric kernels. Let $K(x, s) = K(s, x)$ in (1) (we suppose that the kernel is real) and let the approximate solution again be assumed in the form (2), where, now, the constants c_i are such that the function (2) gives a minimal value to the functional

$$I(\varphi) \equiv \int_a^b \varphi^2(x)\, dx - \int_a^b \int_a^b K(x, s)\, \varphi(x)\, \varphi(s)\, dx\, ds - 2 \int_a^b \varphi(x)\, g(x)\, dx$$

(for it is known that the function minimizing the functional $I(\varphi)$ is the solution of the given integral equations). If we substitute $f_n(x)$ for $\varphi(x)$ (and $f_n(s)$ for $\varphi(s)$), then the functional I becomes a function of $c_1, c_2, \ldots, c_n$.

If we set the derivatives

$$\frac{\partial I}{\partial c_1}, \quad \frac{\partial I}{\partial c_2}, \quad \ldots, \quad \frac{\partial I}{\partial c_n}$$

equal to zero (the condition for a minimum), we obtain equations for the unknown constants $c_1, c_2, \ldots, c_n$. These equations are identical with the equations given by system (3). For more details and for examples see [214].

29.5. Applications of the Ritz Method to the Approximate Determination of the First Eigenvalue of an Equation with a Symmetric Kernel

Let $K(x, s)$ be a real symmetric kernel. Then, in accordance with Theorem 19.3.3, p. 926, the equation

$$\left|\frac{1}{\lambda_1}\right| = \max_{\varphi(x) \in L_2(a,b)} \left| \int_a^b \int_a^b K(x, s)\, \varphi(x)\, \varphi(s)\, dx\, ds \right| \tag{1}$$

holds, where

$$\int_a^b \varphi^2(x)\,dx = 1\,. \tag{2}$$

Let $\psi_1(x), \psi_2(x), \ldots$ be a complete sequence of functions defined in $[a, b]$. Let us put in (1)

$$\varphi = \sum_{i=1}^{n} a_i \psi_i(x)\,. \tag{3}$$

If we write

$$\int_a^b \psi_i(x)\,\psi_k(x)\,dx = (\psi_i, \psi_k)\,,$$

$$\int_a^b \int_a^b K(x, s)\,\psi_i(x)\,\psi_k(s)\,dx\,ds = A_{ik} \quad (A_{ik} = A_{ki})\,,$$

then the problem of finding an approximate value $|1/\lambda_1|$ leads to the determination of the maximum of the expression

$$\left|\sum_{i,k=1}^{n} A_{ik} a_i a_k\right| \tag{4}$$

under the condition

$$\sum_{i,k=1}^{n} a_i a_k (\psi_i, \psi_k) = 1\,. \tag{5}$$

We solve this problem conveniently by using the method of Lagrange multipliers. (See Theorem 12.12.3, p. 480. For details see for example [281].)

If we determine $a_1, a_2, \ldots, a_n$ and substitute them into (4), we obtain approximately the value $|1/\lambda_1|$; the process converges as $n \to \infty$ to the exact value $|1/\lambda_1|$. If, moreover, there exists only one eigenfunction corresponding to λ_1 (up to a multiplicative constant), then (3) with the above-mentioned constants $a_1, a_2, \ldots, a_n$ is an approximation to this function. For details see [281].

We frequently meet the following cases:

1. The functions $\psi_k(x)$ constitute an orthonormal system. Then the condition (5) reads

$$\sum_{k=1}^{n} a_k^2 = 1\,.$$

2. The quadratic form

$$\sum_{i,k=1}^{n} A_{ik} a_i a_k \quad (A_{ik} = A_{ki}) \tag{6}$$

is positive definite, hence (6) is everywhere positive (with the exception of the case $a_1 = a_2 = \ldots = a_n = 0$) so that the sign of the absolute value in (4) is superfluous.

It can be shown that in this case the maximum of the form (4) is given by the greatest of the roots of the equation

$$\begin{vmatrix} A_{11} - \sigma, & A_{12}, & \ldots, & A_{1n} \\ A_{21}, & A_{22} - \sigma, & \ldots, & A_{2n} \\ \cdots & \cdots & \cdots & \cdots \\ A_{n1}, & A_{n2}, & \ldots, & A_{nn} - \sigma \end{vmatrix} = 0 . \tag{7}$$

The form (6) is always (i.e. for each n) positive definite if the kernel $K(x, s)$ is positive i.e. if for each function $\varphi(x) \not\equiv 0$ the integral

$$\int_a^b \int_a^b K(x, s)\, \varphi(x)\, \varphi(s)\, dx\, ds$$

is positive. Then all the eigenvalues are positive as well and the evaluated maximum gives directly an approximate value for $1/\lambda_1$.

Example 1. Suppose we have the equation

$$f(x) - \lambda \int_0^1 K(x, s) f(s)\, ds = 0$$

with the kernel

$$K(x, s) = \begin{cases} \frac{1}{2}x(2 - s) & (x \leqq s), \\ \frac{1}{2}s(2 - x) & (x \geqq s). \end{cases} \tag{8}$$

For the sequence $\psi_k(x)$ we choose the orthonormal sequence

$$\psi_k(x) = \sqrt{(2)} \sin k\pi x , \quad k = 1, 2, \ldots .$$

If we take $n = 2$, we obtain

$$A_{11} = \frac{2}{\pi^2}, \quad A_{12} = A_{21} = -\frac{1}{2\pi^2}, \quad A_{22} = \frac{1}{2\pi^2}$$

so that the problem (4), (5) is transformed into the problem of finding the maximum of the function (of two variables)

$$\frac{1}{\pi^2}(2a_1^2 - a_1 a_2 + \frac{1}{4}a_2^2) \tag{9}$$

under the condition

$$a_1^2 + a_2^2 = 1 .$$

(We need not write the sign of the absolute value in (9), since the expression (9) is always positive unless $a_1 = 0$ and $a_2 = 0$ simultaneously. The kernel (8) can be shown to be positive.)

Equation (7) becomes

$$\begin{vmatrix} \dfrac{2}{\pi^2} - \sigma, & -\dfrac{1}{2\pi^2} \\ -\dfrac{1}{2\pi^2}, & \dfrac{1}{2\pi^2} - \sigma \end{vmatrix} = 0\,.$$

The greater of the roots has the value $\sigma \doteq 0{\cdot}218$. Hence

$$\lambda_1 \approx \frac{1}{0{\cdot}218} \doteq 4{\cdot}59\,.$$

In [281] a more precise value for λ_1, $\lambda_1 \doteq 4{\cdot}115$, is determined in a slightly complicated manner. In the same book some other methods of approximate evaluation of eigenvalues are discussed.

For the approximate solution of integral equations of the first kind see e.g. [354].

30. NUMERICAL METHODS IN LINEAR ALGEBRA

References: [12], [23], [51], [67], [104], [105], [129], [132], [140], [143], [149], [176], [183], [286].

A. SOLUTION OF SYSTEMS OF LINEAR ALGEBRAIC EQUATIONS

By Olga Pokorná

Throughout Section A it is assumed that the matrix of the system solved is a square matrix and that a unique solution of the system exists.

30.1. Elimination Method

The given system

$$\sum_{j=1}^{n} a_{ij}x_j = a_{i,n+1}\,, \quad i = 1, 2, \ldots, n\,, \tag{1}$$

is transformed by successive elimination of the unknowns to an equivalent system with a triangular matrix of coefficients (where all the coefficients under the main diagonal are equal to zero). Such a system can then be easily solved by so-called *back substitution.*

(a) *First Modification of the Method.* We subtract the first equation multiplied by a_{i1} from the i-th equation multiplied by a_{11} for $i = 2, \ldots, n$, successively. (If the given system has a unique solution, it is always possible to permute the indices so that $a_{11} \neq 0$.) The first equation remains unchanged, the other $n - 1$ equations form a system that no longer contains the unknown x_1:

$$\sum_{j=2}^{n} a_{ij}^{(1)}x_j = a_{i,n+1}^{(1)}\,, \quad i = 2, 3, \ldots, n\,,$$

where $a_{ij}^{(1)} = a_{11}a_{ij} - a_{i1}a_{1j}$, $j = 2, \ldots, n + 1$. Now, we repeat the process for this system and proceed in this manner until we have one equation for one unknown x_n. The equations, by the help of which we have eliminated the particular unknowns (together with the last equation for the unknown x_n), form the desired system with the triangular matrix:

$$\begin{aligned} a_{11}x_1 + a_{12}x_2 + \ldots + a_{1,n-1}x_{n-1} + a_{1n}x_n &= a_{1,n+1}\,, \\ a_{22}^{(1)}x_2 + \ldots + a_{2,n-1}^{(1)}x_{n-1} + a_{2n}^{(1)}x_n &= a_{2,n+1}^{(1)}\,, \\ \ldots\ldots\ldots\ldots\ldots\ldots\ldots\ldots\ldots\ldots\ldots\ldots\ldots & \\ a_{n-1,n-1}^{(n-2)}x_{n-1} + a_{n-1,n}^{(n-2)}x_n &= a_{n-1,n+1}^{(n-2)}\,, \\ a_{nn}^{(n-1)}x_n &= a_{n,n+1}^{(n-1)}\,. \end{aligned} \tag{2}$$

From the last equation we compute x_n. Substitution of this value for x_n into the last but one equation of (2) enables us to compute x_{n-1}, and so on, until we compute from the first equation the value of x_1 after having substituted for $x_n, x_{n-1}, \ldots, x_2$.

REMARK 1. During the whole computation we can use the following sum-check: To each equation we attach the check element s_i equal to the sum of all its coefficients (including the absolute member). Then during the elimination we apply to every check element s_i the same operations as are applied to other elements of the corresponding equation. After completing each transformation of the equation we determine whether the element obtained by the elimination process from the check element s_i equals the sum of the coefficients of the transformed equation (with an eventual difference owing to rounding-off errors).

Example 1. Let us solve the system

$$\begin{aligned} 8x_1 - x_2 - 2x_3 &= 0\,, \\ -x_1 + 7x_2 - x_3 &= 10\,, \\ -2x_1 - x_2 + 9x_3 &= 23\,. \end{aligned} \tag{3}$$

In Table 30.1 the coefficients (with their signs) are given, the symbols for the unknowns being omitted (Tab. 30.1). The equations I, II and III make up the desired system of the form (2).

REMARK 2. For a symmetric system it is not necessary to compute the elements under the main diagonal during the elimination process, because the systems obtained successively during elimination are also symmetric. This fact can also be used as a check: we compute all the elements and then compare the elements symmetrically situated with respect to the diagonal with one another.

(b) *Second Modification of the Method.* We divide the first equation of the given system by the coefficient a_{11} (if necessary, after a permutation so that $a_{11} \neq 0$), transforming it to the form

$$x_1 + b_{12}x_2 + \dots + b_{1n}x_n = b_{1,n+1} \tag{4}$$

TABLE 30.1

	x_1	x_2	x_3	b	s	
I	8	−1	−2	0	5	The given system with sum-checks
	−1	7	−1	10	15	
	−2	−1	9	23	29	
II		55	−10	80	125	Having eliminated x_1
		−10	68	184	242	
III			3 640	10 920	14 560	Having eliminated x_2
	$x_1 = 1$	$x_2 = 2$	$x_3 = 3$			Solution

where $b_{1j} = a_{1j}/a_{11}, j = 1, 2, \dots, n + 1$. We subtract this equation, multiplied by a_{i1}, from the i-th equation for $i = 2, \dots, n$. A system of $n - 1$ equations for $n - 1$ unknowns is thus formed,

$$\sum_{j=2}^{n} b_{ij}^{(1)} x_j = b_{i,n+1}^{(1)}, \quad i = 2, 3, \dots, n \tag{5}$$

where $b_{ij}^{(1)} = a_{ij} - a_{i1}b_{1j}$. For this system we repeat the process just described and proceed in this way until we reach one equation for one unknown and hence have obtained a system of the desired form

$$\begin{aligned} x_1 + b_{12}x_2 + \dots + b_{1,n-1}x_{n-1} + b_{1n}x_n &= b_{1,n+1}, \\ x_2 + \dots + b_{2,n-1}^{(1)}x_{n-1} + b_{2n}^{(1)}x_n &= b_{2,n+1}^{(1)}, \\ \dots\dots\dots\dots\dots\dots\dots\dots\dots\dots\dots\dots \\ x_{n-1} + b_{n-1,n}^{(n-2)}x_n &= b_{n-1,n+1}^{(n-2)}, \\ x_n &= b_{n,n+1}^{(n-1)}, \end{aligned} \tag{6}$$

from which we compute successively $x_n, x_{n-1}, \dots, x_2, x_1$.

It is possible to use a similar sum-check as described in modification (a). (See Remark 1.)

Example 2. (Tab. 30.2). Let us again solve the system (3). Equations I, II and III make up a system of the form (6).

TABLE 30.2

	x_1	x_2	x_3	b	s	
	8	-1	-2	0	5	The given system with sum-checks
	-1	7	-1	10	15	
	-2	-1	9	23	29	
I	1	$-0{\cdot}125$	$-0{\cdot}250$	0	0·625	Coefficients b_{1j}
		6·875	$-1{\cdot}250$	10	15·625	Having eliminated x_1
		$-1{\cdot}250$	8·500	23	30·250	
II		1	$-0{\cdot}182$	1·455	2·273	Coefficients $b_{2j}^{(1)}$
			8·272	24·819	33·091	Having eliminated x_2
III			1	3·000	4·000	Coefficients $b_{3j}^{(2)}$
	1·000	2·001	3·000			Solution

REMARK 3. Computations of the form of modification (b) may be performed advantageously on a desk-calculator using the scheme described here for the solution of Example 2 (Tab. 30.3). For details see [286]. The computations are so arranged that a minimum number of intermediate results need to be recorded. The scheme contains three tables: A, B and C. The given system (with check sums s_i) is entered into Table A with Tables B and C empty at the beginning of the computation.

The elimination process is reduced to the computation of the values of Table B, which are obtained successively in the order given by indices 1–15 in Table 30.3. At the end of the computations, Table C contains the values of x_i. In the following text we use the symbol $\langle h \rangle$ for the number (with its sign) from the field of index h. The value a_{ij} and s_i are taken from Table A.

TABLE 30.3

A	8	−1	−2	0	5
	−1	7	−1	10	15
	−2	−1	9	23	29
B	1 8	4 −0·125	5 −0·250	6 0	7 0·625
	2 −1	8 6·875	10 −0·182	11 1·455	12 2·273
	3 −2	9 −1·250	13 8·272	14 3·000	15 4·000
C	1·000	2·001	3·000		

The computation procedure:

$$\langle 1\rangle = a_{11};\quad \langle 2\rangle = a_{21};\quad \langle 3\rangle = a_{31};\quad \langle 4\rangle = a_{12} : \langle 1\rangle;$$
$$\langle 5\rangle = a_{13} : \langle 1\rangle;\quad \langle 6\rangle = a_{14} : \langle 1\rangle;\quad \langle 7\rangle = s_1 : \langle 1\rangle$$

(check: $1 + \langle 4\rangle + \langle 5\rangle + \langle 6\rangle = \langle 7\rangle$);

$$\langle 8\rangle = a_{22} - \langle 2\rangle . \langle 4\rangle;$$
$$\langle 9\rangle = a_{32} - \langle 3\rangle . \langle 4\rangle;$$
$$\langle 10\rangle = (a_{23} - \langle 2\rangle . \langle 5\rangle) : \langle 8\rangle;$$
$$\langle 11\rangle = (a_{24} - \langle 2\rangle . \langle 6\rangle) : \langle 8\rangle;$$
$$\langle 12\rangle = (s_2 - \langle 2\rangle . \langle 7\rangle) : \langle 8\rangle$$

(check: $1 + \langle 10\rangle + \langle 11\rangle = \langle 12\rangle$);

$$\langle 13\rangle = a_{33} - \langle 3\rangle . \langle 5\rangle - \langle 9\rangle . \langle 10\rangle;$$
$$\langle 14\rangle = (a_{34} - \langle 3\rangle . \langle 6\rangle - \langle 9\rangle . \langle 11\rangle) : \langle 13\rangle;$$
$$\langle 15\rangle = (s_3 - \langle 3\rangle . \langle 7\rangle - \langle 9\rangle . \langle 12\rangle) : \langle 13\rangle$$

(check: $1 + \langle 14\rangle = \langle 15\rangle$).

The computation of the unknowns (table C):

$$x_3 = \langle 14\rangle,\ x_2 = \langle 11\rangle - \langle 10\rangle \,.\, x_3,$$
$$x_1 = \langle 6\rangle - \langle 5\rangle \,.\, x_3 - \langle 4\rangle \,.\, x_2.$$

REMARK 4. Modification (b) is identical with the so-called *Choleski's (or Banachiewicz's) method* which uses the decomposition of the matrix into the product of two triangular matrices (e.g. [23], [286]).

REMARK 5. The values $\tilde{x}_j$ obtained after solving the given system of equations by one of the described procedures may differ from the exact values x_i owing to rounding-off errors. A more precise solution is obtained by substituting the values $\tilde{x}_j$ into the given system and computing the values $r_i = \sum_{j=1}^{n} a_{ij}\tilde{x}_j - a_{i,n+1}$ $(i = 1, 2, \ldots, n)$. Then we solve the system $\sum_{j=1}^{n} a_{ij}\xi_j = r_i\,(i = 1, 2, \ldots, n)$ which differs from the original one only in the right-hand side members. Such a system may be easily solved since the elimination process for the left-hand sides of the equations need not be repeated. Thus it is performed only for the right-hand side column. If the values ξ_j are solutions of this new system then the values $x_j = \tilde{x}_j - \xi_j$ constitute a more accurate solution of the given system.

As an example let us improve the accuracy of the solution $\tilde{x}_1 = 1{\cdot}0$, $\tilde{x}_2 = 2{\cdot}001$, $\tilde{x}_3 = 3{\cdot}0$ of system (3) obtained by the use of the scheme in Remark 3. We compute $r_1 = -0{\cdot}001$, $r_2 = 0{\cdot}007$, $r_3 = -0{\cdot}001$. To table A we add the column of the new right-hand sides r_i. To table B we add an empty column (the fields 6′, 11′, 14′) into which we enter successively:

$$\langle 6'\rangle = r_1 : \langle 1\rangle,$$
$$\langle 11'\rangle = (r_2 - \langle 2\rangle \,.\, \langle 6'\rangle) : \langle 8\rangle,$$
$$\langle 14'\rangle = (r_3 - \langle 3\rangle \,.\, \langle 6'\rangle - \langle 9\rangle \,.\, \langle 11'\rangle) : \langle 13\rangle.$$

TABLE 30·4

A		−0·001
		0·007
		−0·001
B	6′	0
	11′	0·001
	14′	0

We then compute:

$$\xi_3 = \langle 14' \rangle, \; \xi_2 = \langle 11' \rangle - \langle 10 \rangle \,.\, \xi_3,$$
$$\xi_1 = \langle 6' \rangle - \langle 5 \rangle \,.\, \xi_3 - \langle 4 \rangle \,.\, \xi_2,$$

i.e. $\xi_3 = 0$, $\xi_2 = 0{\cdot}001$, $\xi_1 = 0$, so that a more accurate solution of the given system is $x_1 = 1{\cdot}0$, $x_2 = 2{\cdot}001 - 0{\cdot}001 = 2{\cdot}0$, $x_3 = 3{\cdot}0$.

REMARK 6. It is possible to eliminate the individual unknowns in such an order that the unknown with the coefficient of highest absolute value is always chosen for elimination. In this manner the extent of the rounding-off errors may sometimes be reduced.

(c) *Third Modification of the Method.* (*Modification without Back Substitution.*) We proceed in the same way as in modification (a) or (b) but apply the elimination process at each step not only to all the equations *following* the equation used to eliminate the corresponding unknown but also to all the equations *preceding* this equation. At the end of this elimination process we have determined a system with a diagonal matrix (even with a unit matrix when using process (b)). Thus we obtain the values of the unknowns directly without back substitution.

This modification is practical for the programming of the elimination method for a computer.

REMARK 7. The elimination method may be advantageously used for solving simultaneously several systems with the same matrix and with different right-hand side columns. In such a case the elimination part of the process is performed simultaneously for all the systems to be solved, so that the matrix is treated only once. In this way we can compute, in particular, the inverse to a given matrix $\boldsymbol{A}$ solving n systems with the same matrix $\boldsymbol{A}$ and with the columns of the unit matrix as with the right-hand side columns. The solutions of this systems are the columns of the desired inverse matrix.

30.2. The Ritz Iteration Method

We re-write the given system (30.1.1) in the form

$$x_i = a_{i,n+1} - a_{i1}x_1 - \ldots - (a_{ii} - 1)\, x_i - \ldots - a_{in}x_n\,, \quad i = 1, 2, \ldots, n\,. \tag{1}$$

The initial (zero-approximation) values of the solution, $x_1^{(0)}, x_2^{(0)}, \ldots, x_n^{(0)}$, are chosen arbitrarily and are substituted for $x_1, x_2, \ldots, x_n$ in the right-hand sides of equations (1). The results of this substitution give a better approximate solution $x_1^{(1)}, x_2^{(1)}, \ldots, x_n^{(1)}$:

$$\begin{aligned}
x_1^{(1)} &= a_{1,n+1} - (a_{11} - 1)\, x_1^{(0)} - a_{12}x_2^{(0)} - \ldots - a_{1n}x_n^{(0)}\,,\\
x_2^{(1)} &= a_{2,n+1} - a_{21}x_1^{(0)} - (a_{22} - 1)\, x_2^{(0)} - \ldots - a_{2n}x_n^{(0)}\,,\\
&\ldots\ldots\ldots\ldots\ldots\ldots\ldots\ldots\ldots\ldots\ldots\ldots\\
x_n^{(1)} &= a_{n,n+1} - a_{n1}x_1^{(0)} - a_{n2}x_2^{(0)} - \ldots - (a_{nn} - 1)\, x_n^{(0)}\,.
\end{aligned}$$

In the same way we obtain the values $x_i^{(2)}$ from the values $x_i^{(1)}$, and so on. Such a sequence of approximations $x_i^{(k)}$ converges to the solution of the given system if some of the following conditions are fulfilled (where m_{ij} denote the coefficients of the unknowns on the right-hand side of equations (1) and $\boldsymbol{M}$ denotes the matrix of these coefficients):

(a) $\sqrt{\left(\sum_{i,j=1}^{n} m_{ij}^2\right)} = \alpha < 1$.

Estimate of the accuracy:

$$\sqrt{\left(\sum_{j=1}^{n}(x_j - x_j^{(k)})^2\right)} \leqq \frac{\alpha}{1-\alpha}\sqrt{\left(\sum_{j=1}^{n}(x_j^{(k)} - x_j^{(k-1)})^2\right)} .$$

(b) $\max_i \sum_{j=1}^{n} |m_{ij}| = \beta < 1$.

Estimate of the accuracy:

$$|x_j - x_j^{(k)}| \leqq \beta \max_i |x_i - x_i^{(k-1)}| .$$

(c) $\max_j \sum_{i=1}^{n} |m_{ij}| = \gamma < 1$.

Estimate of the accuracy:

$$\sum_{j=1}^{n} |x_j - x_j^{(k)}| \leqq \gamma \sum_{j=1}^{n} |x_j - x_j^{(k-1)}| .$$

(d) All eigenvalues of the matrix $\boldsymbol{M}$ have absolute values less than 1.

REMARK 1. The convergence of the iteration process may sometimes be accelerated in the following way: when k is large, and we reach an approximation $x_1^{(k+1)}, x_2^{(k+1)}, \ldots$ $\ldots, x_n^{(k+1)}$, we compute the next approximation according to the formula

$$x_j^{(k+2)} = x_j^{(k)} + \frac{x_j^{(k+1)} - x_j^{(k)}}{1 - \lambda} , \tag{2}$$

where

$$\lambda = \frac{x_i^{(k+1)} - x_i^{(k)}}{x_i^{(k)} - x_i^{(k-1)}}$$

for arbitrary i (λ is an approximate value of the largest eigenvalue of the matrix $\boldsymbol{M}$, different from 1). Then we continue according to the original iteration process, starting with the values $x_j^{(k+2)}$. After some further steps we can again use the acceleration process.

Example 1. Let us again solve the system of Example 30.1.1 (in which, this time, all equations are divided by ten). We re-write it in the form (1):

$$\begin{aligned} x_1 &= \phantom{1{\cdot}0 + {}} 0{\cdot}2x_1 + 0{\cdot}1x_2 + 0{\cdot}2x_3\,, \\ x_2 &= 1{\cdot}0 + 0{\cdot}1x_1 + 0{\cdot}3x_2 + 0{\cdot}1x_3\,, \\ x_3 &= 2{\cdot}3 + 0{\cdot}2x_1 + 0{\cdot}1x_2 + 0{\cdot}1x_3\,. \end{aligned}$$

The iteration process converges because, for instance, condition (b) is fulfilled:

$$\max_i \sum_{j=1}^{n} |m_{ij}| = 0{\cdot}5 < 1\,.$$

Estimate of the accuracy: $|x_j - x_j^{(k)}| \leqq 0{\cdot}5 \max_i |x_i - x_i^{(k-1)}|$.

We start with the approximation (Tab. 30.5)

TABLE 30.5

$$x_1^{(0)} = x_2^{(0)} = x_3^{(0)} = 1{\cdot}0\,.$$

k	$x_1^{(k)}$	$x_2^{(k)}$	$x_3^{(k)}$
0	1·0	1·0	1·0
1	0·5	1·5	2·7
2	0·79	1·77	2·82
3	0·899	1·892	2·917
4	0·952	1·949	2·961
5	0·978	1·976	2·981
6	1·001	2·000	2·999
7	1·000	2·000	3·000

After the fifth step the convergence is accelerated by using

$$\lambda = \frac{x_2^{(5)} - x_2^{(4)}}{x_2^{(4)} - x_2^{(3)}} = 0{\cdot}474\,.$$

Without this acceleration the solution after the seventh step would be (Tab. 30.6):

TABLE 30.6

6′	0·989	1·989	2·991
7′	0·995	1·995	2·996

30.3. The Gauss–Seidel Iteration Method

We re-write the given system (30.1.1) in the form

$$a_{ii}x_i = a_{i,n+1} - a_{i1}x_1 - \ldots - a_{i,i-1}x_{i-1} - a_{i,i+1}x_{i+1} - \ldots - a_{in}x_n\,, \quad i = 1, 2, \ldots, n\,. \tag{1}$$

We choose the initial zero-approximation $x_1^{(0)}, x_2^{(0)}, \ldots, x_n^{(0)}$ of the solution arbitrarily. These values are substituted into the right-hand side of the first equation (1) and from here we determine $x_1^{(1)}$, i.e. the first component of the next approximation. Then we substitute this new value $x_1^{(1)}$ as x_1 and the original values $x_2^{(0)}, \ldots, x_n^{(0)}$ for the other components $x_2, \ldots, x_n$ into the right-hand side of the second equation. From here we compute $x_2^{(1)}$, i.e. the second component of the new approximation. Substitute the new values $x_1^{(1)}$ and $x_2^{(1)}$ for x_1 and x_2 and the original values $x_3^{(0)}, \ldots, x_n^{(0)}$ as the other components into the right-hand side of the third equation, and continue the process for the whole system, so that:

$$a_{11}x_1^{(1)} = a_{1,n+1} - a_{12}x_2^{(0)} - a_{13}x_3^{(0)} - \ldots - a_{1,n-1}x_{n-1}^{(0)} - a_{1n}x_n^{(0)}\,,$$

$$a_{22}x_2^{(1)} = a_{2,n+1} - a_{21}x_1^{(1)} - a_{23}x_3^{(0)} - \ldots - a_{2,n-1}x_{n-1}^{(0)} - a_{2n}x_n^{(0)}\,,$$

$$\cdots\cdots\cdots\cdots\cdots\cdots\cdots\cdots\cdots\cdots\cdots\cdots\cdots\cdots$$

$$a_{nn}x_n^{(1)} = a_{n,n+1} + a_{n1}x_1^{(1)} - a_{n2}x_2^{(1)} - a_{n3}x_3^{(1)} - \ldots - a_{n,n-1}x_{n-1}^{(1)}\,.$$

In the same way we obtain the values $x_i^{(2)}$ from the values $x_i^{(1)}$, and so on. Such a sequence of approximations $x_i^{(k)}$ converges to the solution of the given system if the system fulfils any of the following conditions:

(a) $\max\limits_i \sum\limits_{\substack{j=1\\ j\neq i}}^{n} \left|\dfrac{a_{ij}}{a_{ii}}\right| < 1\,.$

Estimate of the accuracy:

$$\max_j \left|x_j - x_j^{(k)}\right| \leq \frac{\delta}{1-\delta} \max_i \left|x_i^{(k)} - x_i^{(k-1)}\right|,$$

where $\delta = \max\limits_i \dfrac{\sum\limits_{j=i+1}^{n} |a_{ij}|}{|a_{ii}| - \sum\limits_{j=1}^{i-1} |a_{ij}|}\,.$

(b) $\max\limits_j \sum\limits_{\substack{i=1\\ i\neq j}}^{n} \left|\dfrac{a_{ij}}{a_{ii}}\right| < 1\,.$

Estimate of the accuracy:

$$\sum_{j=1}^{n} \left|x_j - x_j^{(k)}\right| \leqq \frac{\alpha}{1-\beta} \sum_{j=1}^{n} \left|x_j - x_j^{(k-1)}\right| ,$$

where $\alpha = \max\limits_{j} \sum\limits_{i=1}^{j-1} \left|\dfrac{a_{ij}}{a_{ii}}\right|$, $\beta = \max\limits_{j} \sum\limits_{i=j+1}^{n} \left|\dfrac{a_{ij}}{a_{ii}}\right|$ $(\alpha + \beta < 1)$.

(c) The matrix of the given system is symmetric and definite (positive or negative, see Def. 1.25.6, p. 89, and Def. 1.29.6, p. 105).

REMARK 1. It is cumbersome to determine whether or not a matrix is definite, but for many matrices arising in solving technical problems the definiteness of the matrix follows from its physical meaning. For example, all the systems of linear equations for computations of a framework have positive definite matrices.

REMARK 2. It is possible to accelerate the convergence by using the same acceleration process as used in the Ritz iteration method (see Remark 30.2.1).

Example 1. Let us solve again the system (30.1.3) re-written in the form (1):

$$\begin{aligned} 8x_1 &= \phantom{10 + x_1 +{}} x_2 + 2x_3 , \\ 7x_2 &= 10 + x_1 \phantom{{}+x_2} + x_3 , \\ 9x_3 &= 23 + 2x_1 + x_2 . \end{aligned}$$

The iteration process will converge because, for instance, the relation (a) holds

$$\max_i \sum_{\substack{j=1 \\ j \neq i}}^{n} \left|\frac{a_{ij}}{a_{ii}}\right| = \frac{3}{8} < 1 .$$

Estimate of the accuracy:

$$\max_j \left|x_j - x_j^{(k)}\right| \leqq 0{\cdot}6 \max_i \left|x_i^{(k)} - x_i^{(k-1)}\right| .$$

Let the initial zero-approximation again be $x_1^{(0)} = x_2^{(0)} = x_3^{(0)} = 1{\cdot}0$ (Tab. 30·7).

TABLE 30.7

k	$x_1^{(k)}$	$x_2^{(k)}$	$x_3^{(k)}$
0	1·0	1·0	1·0
1	0·375	1·625	2·819
2	0·908	1·961	2·975
3	0·989	1·995	2·997
4	0·999	1·999	3·000
5	1·000	2·000	3·000

30.4. The Relaxation Method

We write the system (30.1.1) in the form

$$\sum_{j=1}^{n} a_{ij}x_j - a_{i,n+1} = 0\,, \quad i = 1, 2, \ldots, n\,, \tag{1}$$

and arbitrarily choose $x_1^{(0)}$, $x_2^{(0)}$, ..., $x_n^{(0)}$ as the initial zero-approximation of the olution. We substitute these values into equations (1) and obtain

$$\sum_{j=1}^{n} a_{ij}x_j^{(0)} - a_{i,n+1} = r_i^{(0)}\,, \quad i = 1, 2, \ldots, n\,,$$

where the value $r_i^{(0)}$ is called the *residual of the i-th equation.* We select the equation whose residual has the highest absolute value and compute the next approximation n such a way that this residual is reduced to zero: let us denote $\left|r_p^{(0)}\right| = \max_i \left|r_i^{(0)}\right|$. Ve put

$$x_i^{(1)} = x_i^{(0)} \quad \text{for} \quad i \neq p\,, \quad x_p^{(1)} = x_p^{(0)} - \frac{r_p^{(0)}}{a_{pp}}\,.$$

f the residuals for this new approximation are denoted by the symbols $r_i^{(1)}$, the relaion

$$r_i^{(1)} = r_i^{(0)} - a_{ip}\frac{r_p^{(0)}}{a_{pp}}\,, \quad i = 1, 2, \ldots, n\,, \tag{2}$$

ill hold (assuming $a_{pp} \neq 0$), so that now the residual of the p-th equation equals ero.

Among the residuals $r_i^{(1)}$ we select the residual with the highest absolute value and btain a new approximation $x_i^{(2)}$ by changing the corresponding component of the pproximation $x_i^{(1)}$ and so on. Such a sequence of approximations converges to the xact solution for systems with symmetric definite matrices (see Remark 30.3.1).

It is always advantageous, after some steps, to substitute the latest approximations nto the given equations and to let the relaxation process continue with the residuals us obtained. Owing to rounding-off errors, they may differ from the residuals omputed according to formula (2).

Example 1. Let us solve again the system (30.1.3) (Tab. 30.8). At each step we cord only the corresponding values $\Delta_p^{(k)} = -r_p^{(k)}/a_{pp}$ and the new values of the siduals $r_i^{(k+1)}$ (in the row for $k = 0$ we write the approximation values $x_j^{(0)}$ place of $\Delta_j^{(k)}$). If, after some steps, we wish to know the values $x_j^{(k)}$ of the last pproximation, we sum for each j the corresponding values Δ_j and the value $x^{(0)}$ of e initial approximation. The relaxation process converges for the given system ecause the matrix of this system is positive definite. The initial approximation is

TABLE 30.8

k	$\Delta_1^{(k)}$	$\Delta_2^{(k)}$	$\Delta_3^{(k)}$	$r_1^{(k)}$	$r_2^{(k)}$	$r_3^{(k)}$
0	1·0	1·0	1·0	5·0	−5·0	**−17·0**
1			1·889	1·222	**−6·889**	—
2		0·984		0·238	—	**−0·984**
3			0·108	0·022	**−0·108**	—
4		0·015		0·007	—	**−0·015**
5			0·002	(0·003	−0·002	—)
$x^{(5)}$	1·0	1·999	2·999	0·003	−0·006	**−0·008**
6			0·001	0·001	**−0·007**	
7				0	—	−0·001
$x^{(7)}$	1·0	2·0	3·0	0	0	0

again $x_1^{(0)} = x_2^{(0)} = x_3^{(0)} = 1{\cdot}0$. After the fifth step the residuals are replaced (in the table they are written in brackets) by the values of the residuals obtained by putting the approximation $x_j^{(5)}$ into the given equations. The values $x_j^{(5)}$ are written in the columns of $\Delta_j^{(k)}$ in the row denoted by $x^{(5)}$.

REMARK 1. It is possible to modify the relaxation process in various ways so that the equation, the residual of which is to be nullified, is chosen according to a different criterion from that mentioned here. For example, it is possible at each step to nullify the residual which requires, for its becoming zero, the maximal change of the corresponding component, i.e. the residual for which the value $|r_i^{(0)}/a_{ii}|$ is maximal. Moreover it is possible to carry out so-called *group relaxation* where several components of the approximation are changed simultaneously at each step. Experience in using the relaxation method will yield particular procedures to speed up the convergence.

30.5. The Method of Conjugate Gradients

This method starts with an arbitrary approximation of the solution and leads to the theoretically exact solution of the given system in a finite number of steps.

We choose the initial approximation $x_1^{(0)}, x_2^{(0)}, \ldots, x_n^{(0)}$ of the system arbitrarily. We compute its residuals $r_i^{(0)} = a_{i,n+1} - \sum_{j=1}^{n} a_{ij}x_j^{(0)}$ for $i = 1, 2, \ldots, n$. The we compute successively the values $s_i^{(k)}$, $a^{(k)}$, $r_i^{(k)}$, $b^{(k)}$ according to the following recurrence relations (starting with $k = 1$):

$$
\begin{aligned}
s_i^{(1)} &= r_i^{(0)}\,,\\
a^{(k)} &= c^{(k-1)}\Big/\sum_{i=1}^{n} s_i^{(k)} p_i^{(k)}\,, \quad \text{where} \quad c^{(k)} = \sum_{i=1}^{n} \left(r_i^{(k)}\right)^2\\
& \qquad\qquad\qquad\qquad\quad \text{and} \quad p_i^{(k)} = \sum_{j=1}^{n} a_{ij} s_j^{(k)}\,,\\
r_i^{(k)} &= r_i^{(k-1)} - a^{(k)} p_i^{(k)}\,,\\
b^{(k)} &= c^{(k)}/c^{(k-1)}\,,\\
s^{(k+1)} &= r_i^{(k)} + b^{(k)} s_i^{(k)}\,.
\end{aligned}
$$

We stop the process as soon as all the values $r_i^{(m)}$ are equal to zero for certain m. This usually occurs for $m = n$ (if a degeneration does not occur during the process).

TABLE 30.9

$r_i^{(0)}$	$s_i^{(1)}$	$p_i^{(1)}$	$r_i^{(1)}$	$s_i^{(2)}$	$p_i^{(2)}$
−5	−5	−79	3·374	3·109	19·739
5	5	23	2·562	2·827	15·527
17	17	158	0·252	1·153	1·332

$c^{(0)}$	$a^{(1)}$	$b^{(1)}$	$c^{(1)}$	$a^{(2)}$	$b^{(2)}$
339	0·106	0·053	18·011	0·169	0·000

$r_i^{(2)}$	$s_i^{(3)}$	$p_i^{(3)}$	$r_i^{(3)}$	x_i
0·038	0·038	0·312	0·000	1·000
−0·062	−0·062	−0·499	−0·001	2·000
0·027	0·027	0·229	−0·001	3·000

$c^{(2)}$	$a^{(3)}$	$b^{(3)}$	$c^{(3)}$
0·006	0·123	0·000	0·000

Then we find the solution of the given system in the form

$$x_i = x_i^{(0)} = \sum_{k=1}^{m} a^{(k)} s_i^{(k)}, \quad i = 1, \ldots, n.$$

A detailed explanation of this method is given in [104].

Example 1. Let us solve again the system (30.1.3). The initial-approximation is again $x_1^{(0)} = x_2^{(0)} = x_3^{(0)} = 1{\cdot}0$ and the solution is shown in Tab. 30.9.

REMARK 1. The method of conjugate gradients is a special case of the so-called method of conjugate directions (see [104]).

B. NUMERICAL CALCULATION OF EIGENVALUES OF A MATRIX

By KVĚTA KORVASOVÁ

The *eigenvalues* (*characteristic numbers*) $\lambda_1, \lambda_2, \ldots, \lambda_n$ of a matrix $\mathbf{A} = \|a_{ik}\|$ of order n are the roots of the equation

$$\begin{vmatrix} a_{11} - \lambda, & a_{12}, & \ldots, & a_{1n} \\ a_{21}, & a_{22} - \lambda, & \ldots, & a_{2n} \\ \ldots & \ldots & \ldots & \ldots \\ a_{n1}, & a_{n2}, & \ldots, & a_{nn} - \lambda \end{vmatrix} = 0$$

(see § 1.28). For estimates of eigenvalues see Theorem 31.3.1, p. 1172.

Methods for the numerical calculation of eigenvalues can be divided into two groups: *iterative methods* (based on successive approximations) and *direct methods* (based on numerical calculation of the roots of the characteristic polynomial, see Definition 1.28.3, p. 97).

One of the iterative methods and one of the direct method (the so-called *Danilevski method*) will be discussed in the following text.

REMARK 1. The discussion is restricted to real matrices. If the matrix $\mathbf{A}$ is complex, i.e. $\mathbf{A} = \mathbf{U} + \mathrm{i}\mathbf{V}$, we can construct the (square) matrix $\mathbf{K}$ (of order $2n$) which is real,

$$\mathbf{K} = \begin{Vmatrix} \mathbf{U}, & -\mathbf{V} \\ \mathbf{V}, & \mathbf{U} \end{Vmatrix}.$$

If its eigenvalues are denoted by $\eta_1, \eta_2, \ldots, \eta_{2n}$, then, if η_j is a real number, we have $\eta_j = \eta_k$ for some $k \neq j$ and if η_j is complex, then $\eta_j = \bar{\eta}_k$ for some $k \neq j$ (η_j and η_k are complex conjugate numbers). The eigenvalues λ_j (for $j = 1, 2, \ldots, n$) of the matrix

$\mathbf{A}$ are then either real numbers η_j (i.e. $\lambda_j = \eta_j$), or complex numbers $\lambda_j = \eta_j = u + iv$ or $\lambda_j = \bar{\eta}_j = u - iv$, according to whether the determinant

$$D = \begin{vmatrix} \mathbf{U} - u\mathbf{E}, & -\mathbf{V} + \gamma\mathbf{E} \\ \mathbf{V} - \gamma\mathbf{E}, & \mathbf{U} - u\mathbf{E} \end{vmatrix}$$

is equal to zero for $\gamma = +v$, or $\gamma = -v$, respectively.

30.6. Iterative Methods

For the iterative methods, treated in the following text, we make the assumption, that the eigenvalues of the matrix $\mathbf{A}$ *are ordered in a decreasing sequence with regard to their absolute values,*

$$|\lambda_1| \geqq |\lambda_2| \geqq |\lambda_3| \geqq \ldots \geqq |\lambda_n| . \tag{1}$$

Let us construct a sequence of vectors $\mathbf{x}_1, \mathbf{x}_2, \mathbf{x}_3, \ldots$ according to the relation

$$\mathbf{x}_{k+1} = \mathbf{A}\mathbf{x}_k , \quad k = 0, 1, 2, \ldots , \tag{2}$$

where $\mathbf{x}_0 = (x_{0,1}, x_{0,2}, \ldots, x_{0,n})$ is an arbitrary initial vector, at least one coordinate of which is non-zero. Using this sequence, the first eigenvalue (i.e. the maximal one in absolute value) can be found as shown in the following text.

Remark 1. The initial vector used must not be orthogonal to the characteristic vector of the maximal eigenvalue. (However, this property cannot be easily established in advance).

In practice, the following most important cases occur:

1. $|\lambda_1| > |\lambda_2|$. Then, for sufficiently large k, λ_1 is approximately determined by the quotient of the corresponding coordinates of the vectors $\mathbf{x}_{k+1}$ and $\mathbf{x}_k$,

$$\frac{x_{k+1,j}}{x_{k,j}} \approx \lambda_1 \tag{3}$$

(valid for any pair of coordinates of the vectors $\mathbf{x}_k$, $\mathbf{x}_{k+1}$). The vector $\mathbf{x}_k$ is then an approximation to the eigenvector ξ_1 belonging to λ_1. The estimate of the error of the approximation (cf. [104]) is

$$\frac{x_{k+1,j}}{x_{k,j}} = \lambda_1 + O\left(\left(\frac{\lambda_2}{\lambda_1}\right)^k\right). \tag{4}$$

2. $\lambda_1 = -\lambda_2$ and simultaneously $|\lambda_1| > |\lambda_3|$; then for sufficiently large k the approximation λ_1^2 is determined by the quotient of successive coordinates of even

(or odd) vectors of the sequence (2), i.e.

$$\lambda_1^2 \approx \frac{x_{2k+2,j}}{x_{2k,j}}, \quad \lambda_1^2 \approx \frac{x_{2k+1,j}}{x_{2k-1,j}}. \tag{5}$$

3. λ_1 and λ_2 are complex conjugate numbers, $\lambda_1 = u + iv$, $\lambda_2 = u - iv$, where u and v are real numbers. For sufficiently large k, the relations

$$\lambda_1 + \lambda_2 = 2u \approx \frac{x_{k,j}x_{k+3,j} - x_{k+1,j}x_{k+2,j}}{x_{k,j}x_{k+2,j} - x_{k+1,j}^2}, \tag{6}$$

$$\lambda_1\lambda_2 = u^2 + v^2 \approx \frac{x_{k+1,j}x_{k+3,j} - x_{k+2,j}^2}{x_{k,j}x_{k+2,j} - x_{k+1,j}^2} \tag{7}$$

hold. The eigenvalues are, in this case, the roots of the quadratic equation

$$p(\lambda) \equiv \lambda^2 + 2u\lambda + (u^2 + v^2) = 0. \tag{8}$$

The above-mentioned three cases can be distinguished during the calculation of the vectors $\mathbf{x}_k$ and of the quotients (3). If the quotient (3) approximately equals λ_1 for all corresponding coordinates of the vectors x_{k+1} and x_k, then case 1 occurs. If the quotients (3) for $j = 1, 2, 3, \ldots, n$ differ significantly, we examine the quotients (5) (for corresponding coordinates of the vector concerned). If they are equal, then case 2 occurs. If they are not equal, then case 3 may occur. The practice is the following: We calculate the expression $x_{k,j}x_{k+2,j} - x_{k+1,j}^2$ for some of the last vectors $\mathbf{x}_k$. If the quotients (7) differ only slightly for all corresponding coordinates of these vectors, then case 3 occurs.

Other cases can occur besides the three described above, for example the matrix $\mathbf{A}$ may have more eigenvalues of maximal absolute value or eigenvalues which are in absolute value near to the maximal. In these cases, the iterative process is not convenient.

The iterative method is suitable for matrices of higher orders, especially if only a small number of eigenvalues is to be calculated. This process is described e.g. in [67], [104]. These books also include the procedure for calculating further eigenvalues.

REMARK 2. Improvement in the accuracy of the approximation and in the rate of convergence can be obtained by using the following relation

$$\lambda_1 \approx P(u_k) = \frac{\begin{vmatrix} u_k, & u_{k+1} \\ u_{k+1}, & u_{k+2} \end{vmatrix}}{u_k - 2u_{k+1} + u_{k+2}},$$

where

$$u_k = \frac{x_{k+1,j}}{x_{k,j}}, \tag{9}$$

if λ_1, λ_2 and λ_3 are real and $|\lambda_1| > |\lambda_2|$. The estimate of the error for (9) is

$$P(u_k) = \lambda_1 + O\left(\left(\frac{\lambda_2}{\lambda_1}\right)^{2k}\right) + O\left(\left(\frac{\lambda_3}{\lambda_1}\right)^{k}\right) \tag{10}$$

(see for example [104]).

Example 1. Let the matrix

$$\mathbf{A} = \begin{Vmatrix} 2, & 1, & 0 \\ 3, & -1, & 1 \\ 0, & 1, & 2 \end{Vmatrix}, \tag{11}$$

and the initial vector

$$\mathbf{x}_0 = \begin{Vmatrix} 1 \\ 1 \\ 0 \end{Vmatrix}$$

be given. The first three iterative steps and quotients x_{k+1}/x_k are shown in Tab. 30.10. In the fourteenth step we get $\lambda_1 \approx 3$ with error of order 10^{-3} (see Tab. 30.11).

TABLE 30.10

x_0	x_1	x_1/x_0	x_2	x_2/x_1	x_3	x_3/x_2
1	3	3	8	2·6	24	3
1	2	2	8	4	20	2·5
0	1		4	4	16	4

TABLE 30.11

x_{13}	x_{14}	x_{14}/x_{13}
$425 \,.\, 10^3$	$1\,277 \,.\, 10^3$	3·005
$425 \,.\, 10^3$	$1\,277 \,.\, 10^3$	2·998
$422 \,.\, 10^3$	$1\,269 \,.\, 10^3$	3·007

The determination of the characteristic equation by an iterative method. We define the sequence of n iterative vectors $\mathbf{x}_k$ $(k = 1, 2, \ldots, n)$, $\mathbf{x}_0$ being a given initial vector:

$$\mathbf{x}_1 = \mathbf{A}\mathbf{x}_0,\ \mathbf{x}_2 = \mathbf{A}\mathbf{x}_1 = \mathbf{A}^2\mathbf{x}_0,\ \ldots,\ \mathbf{x}_n = \mathbf{A}\mathbf{x}_{n-1} = \mathbf{A}^n\mathbf{x}_0\,. \tag{12}$$

The eigenvalues of the matrix $\boldsymbol{A}$ are the roots of the algebraic equation

$$\lambda^n + a_{n-1}\lambda^{n-1} + \ldots + a_0 = 0\,, \tag{13}$$

where the coefficients a_i are given by the solution of the following system of n unknown linear equations:

$$\begin{aligned} x_{n,1} + a_{n-1}x_{n-1,1} + a_{n-2}x_{n-2,1} + \ldots + a_0x_{0,1} &= 0\,,\\ x_{n,2} + a_{n-1}x_{n-1,2} + a_{n-2}x_{n-2,2} + \ldots + a_0x_{0,2} &= 0\,,\\ \ldots\ldots\ldots\ldots\ldots\ldots\ldots\ldots\ldots\ldots\ldots\ldots\ldots\ldots\ldots\ldots\ldots&\\ x_{n,n} + a_{n-1}x_{n-1,n} + a_{n-2}x_{n-2,n} + \ldots + a_0x_{0,n} &= 0\,. \end{aligned} \tag{14}$$

The solution of system (14) can be obtained by means of one of the methods introduced in the first part of this chapter.

Example 2. For the matrix

$$\boldsymbol{A} = \begin{Vmatrix} 2, & 1, & 0 \\ 3, & -1, & 1 \\ 0, & 1, & 2 \end{Vmatrix}$$

and the initial vector $\boldsymbol{x}_0 = (1, 0, 0)$ the first n iterative steps are as follows:

$$\boldsymbol{x}_1 = \boldsymbol{A}\boldsymbol{x}_0 = \begin{Vmatrix} 2 \\ 3 \\ 0 \end{Vmatrix}, \quad \boldsymbol{x}_2 = \begin{Vmatrix} 7 \\ 3 \\ 3 \end{Vmatrix}, \quad \boldsymbol{x}_3 = \begin{Vmatrix} 17 \\ 21 \\ 9 \end{Vmatrix}.$$

The solution of the system (14),

$$\begin{aligned} a_0 + 2a_1 + 7a_2 &= -17\,,\\ 3a_1 + 3a_2 &= -21\,,\\ 3a_2 &= -\ 9\,, \end{aligned}$$

is $a_0 = 12$, $a_1 = -4$, $a_2 = -3$.

The roots of the characteristic equation $p(\lambda) \equiv \lambda^3 - 3\lambda^2 - 4\lambda + 12 = 0$ are $\lambda_1 = 3$, $\lambda_2 = 2$, $\lambda_3 = -2$.

30.7. The Danilevski Method

Of all known direct methods this method has the smallest possible number of arithmetic operations (see [104]).

Beginning with the matrix $\boldsymbol{A}_0 = \boldsymbol{A}$, similar matrices $\boldsymbol{A}_k$ (§ 1.28) are constructed by means of relations

$$\mathbf{A}_k = \mathbf{B}_k^{-1}\mathbf{A}_{k-1}\mathbf{B}_k\,, \quad k = 1, 2, 3, \ldots, n-1\,, \quad \mathbf{A}_0 = \mathbf{A} \tag{1}$$

until we finally obtain the matrix

$$\mathbf{P} = \left\| \begin{matrix} p_1, & p_2, & \ldots, & p_{n-1}, & p_n \\ 1, & 0, & \ldots, & 0, & 0 \\ 0, & 1, & \ldots, & 0, & 0 \\ \ldots & \ldots & \ldots & \ldots & \ldots \\ 0, & 0, & \ldots, & 1, & 0 \end{matrix} \right\| \tag{2}$$

(the matrix in the so-called *Frobenius normal form*). Its elements p_i $(i = 1, 2, \ldots, n)$ are coefficients of the characteristic equation

$$D(\lambda) \equiv (-1)^n (\lambda^n - p_1\lambda^{n-1} - p_2\lambda^{n-2} - \ldots - p_n) = 0\,. \tag{3}$$

For the first step, the matrices $\mathbf{B}_1^{-1}$ and $\mathbf{B}_1$ have the following form:

$$\mathbf{B}_1^{-1} = \left\| \begin{matrix} 1, & 0, & 0, & \ldots, & 0 \\ 0, & 1, & 0, & \ldots, & 0 \\ \ldots & \ldots & \ldots & \ldots & \ldots \\ a_{n,1}, & a_{n,2}, & a_{n,3}, & \ldots, & a_{n,n} \\ 0, & 0, & 0, & \ldots, & 1 \end{matrix} \right\|,$$

$$\mathbf{B}_1 = \left\| \begin{matrix} 1, & 0, & 0, & \ldots, & 0, & 0 \\ 0, & 1, & 0, & \ldots, & 0, & 0 \\ \ldots & \ldots & \ldots & \ldots & \ldots & \ldots \\ \dfrac{-a_{n,1}}{a_{n,n-1}}, & \dfrac{-a_{n,2}}{a_{n,n-1}}, & \dfrac{-a_{n,3}}{a_{n,n-1}}, & \ldots, & \dfrac{1}{a_{n,n-1}}, & \dfrac{-a_{n,n}}{a_{n,n-1}} \\ 0, & 0, & 0, & \ldots, & 0, & 1 \end{matrix} \right\|$$

for $a_{n,n-1} \neq 0$. The n-th row of the matrix $\mathbf{A}_1$ equals the n-th row of the matrix $\mathbf{P}$. The matrix $\mathbf{A}_2$ is the product

$$\mathbf{A}_2 = \mathbf{B}_2^{-1}\mathbf{A}_1\mathbf{B}_2\,,$$

where

$$\mathbf{B}_2^{-1} = \left\| \begin{matrix} 1, & 0, & \ldots, 0, & 0 \\ 0, & 1, & \ldots, 0, & 0 \\ \ldots & \ldots & \ldots & \ldots \\ a_{n-1,1}^{(1)}, & a_{n-1,2}^{(1)}, & \ldots, a_{n-1,n-1}^{(1)}, & a_{n-1,n}^{(1)} \\ 0, & 0, & \ldots, 1, & 0 \\ 0, & 0, & \ldots, 0, & 1 \end{matrix} \right\|,$$

$$\mathbf{B}_2 = \left\| \begin{array}{cccccc} 1, & 0, & \ldots, & 0, & 0, & 0 \\ 0, & 1, & \ldots, & 0, & 0, & 0 \\ \multicolumn{6}{c}{\dotfill} \\ \dfrac{-a_{n-1,1}^{(1)}}{a_{n-1,n-2}^{(1)}}, & \dfrac{-a_{n-1,2}^{(1)}}{a_{n-1,n-2}^{(1)}}, & \ldots, & \dfrac{1}{a_{n-1,n-2}^{(1)}}, & \dfrac{-a_{n-1,n-1}^{(1)}}{a_{n-1,n-2}^{(1)}}, & \dfrac{-a_{n-1,n}^{(1)}}{a_{n-1,n-2}^{(1)}} \\ 0, & 0, & \ldots, & 0, & 1, & 0 \\ 0, & 0, & \ldots, & 0, & 0, & 1 \end{array} \right\|$$

for $a_{n-1,n-2}^{(1)} \neq 0$. The two last rows of the matrix $\mathbf{A}_2$ are equal to the two last rows of the matrix $\mathbf{P}$. The matrix $\mathbf{A}$ is thus transformed to the form $\mathbf{P}$ by $n-1$ steps. The elements $a_{i,j}^{(k)}$ of the matrix $\mathbf{A}_k$ can be calculated directly according to the following formulae which, for simplicity, are written only for $k = 1$:

$$a_{i,j}^{(1)} = a_{i,j} - \frac{a_{i,n-1}a_{n,j}}{a_{n,n-1}} \quad \text{for all} \quad i, j \neq n-1\,,$$

$$a_{i,n-1}^{(1)} = \frac{a_{i,n-1}}{a_{n,n-1}} \quad \text{for} \quad i \leqq n-1\,,$$

$$a_{n-1,j}^{(1)} = \sum_{k=1}^{n} a_{n,k}a_{k,j} - \frac{a_{n,j}}{a_{n,n-1}} \sum_{k=1}^{n} a_{n,k}a_{k,n-1} \quad \text{for} \quad j \neq n-1\,,$$

$$a_{n-1,n-1}^{(1)} = \frac{1}{a_{n,n-1}} \sum_{k=1}^{n} a_{n,k}a_{k,n-1}\,.$$

The eigenvectors $\mathbf{x}_i$ belonging to the eigenvalues λ_i can then be obtained by the solution of the system of linear equations $\mathbf{A}\mathbf{x}_i - \lambda_i\mathbf{x}_i = \mathbf{0}$.

Some difficulties can arise in using this method. How to remove these difficulties see e.g. [104].

Example 1. For the matrix

$$\mathbf{A} = \left\| \begin{array}{rrr} 2, & 1, & 0 \\ 3, & -1, & 1 \\ 0, & 1, & 2 \end{array} \right\|$$

we obtain

$$\mathbf{B}_1^{-1} = \left\| \begin{array}{ccc} 1, & 0, & 0 \\ 0, & 1, & 2 \\ 0, & 0, & 1 \end{array} \right\|, \quad \mathbf{B}_1 = \left\| \begin{array}{rrr} 1, & 0, & 0 \\ 0, & 1, & -2 \\ 0, & 0, & 1 \end{array} \right\|$$

and

$$\mathbf{A}_1 = \left\| \begin{array}{rrr} 2, & 1, & -2 \\ 3, & 1, & 3 \\ 0, & 1, & 0 \end{array} \right\|.$$

In the second step, we obtain the matrix $\boldsymbol{P}$:

$$\boldsymbol{P} = \boldsymbol{A}_2 = \begin{Vmatrix} 3, & 4, & -12 \\ 1, & 0, & 0 \\ 0, & 1, & 0 \end{Vmatrix}$$

while

$$\boldsymbol{B}_2 = \begin{Vmatrix} \frac{1}{3}, & -\frac{1}{3}, & -1 \\ 0, & 1, & 0 \\ 0, & 0, & 1 \end{Vmatrix}, \quad \boldsymbol{B}_2^{-1} = \begin{Vmatrix} 3 & 1, & 3 \\ 0, & 1, & 0 \\ 0, & 0, & 1 \end{Vmatrix}.$$

The characteristic equation is

$$f(\lambda) \equiv \lambda^3 - 3\lambda^2 - 4\lambda + 12 = 0\,.$$

The coordinates x_1, x_2, x_3 of the characteristic vector ξ_1 corresponding to $\lambda_1 = 3$ are given by the solution of the system of equations

$$\begin{aligned} -x_1 + x_2 \phantom{{}+x_3} &= 0\,, \\ 3x_1 - 4x_2 + x_3 &= 0\,, \\ x_2 - x_3 &= 0\,. \end{aligned}$$

We get $\xi_1 = (1, 1, 1)$. Similarly we determine $\xi_2 = (1, 0, -3)$, $\xi_3 = (1, -4, 1)$.

31. NUMERICAL SOLUTION OF ALGEBRAIC AND TRANSCENDENTAL EQUATIONS

By Miroslav Fiedler

References: [6], [51], [176], [178], [183], [196], [228], [306], [225], [331], [417].

31.1. Basic Properties of Algebraic Equations

An algebraic equation of degree n,

$$f(x) \equiv a_0 x^n + a_1 x^{n-1} + \ldots + a_n = 0 \quad (a_0 \neq 0)\,, \tag{1}$$

with real or complex coefficients $a_0, \ldots, a_n$, has exactly n roots (in general complex) if each root is considered with its appropriate multiplicity (cf. § 1.14, p. 58).

In the following text we consider only the case where all the coefficients in (1) are real. If some coefficients are not real, then we construct the polynomial

$$g(x) \equiv (a_0 x^n + a_1 x^{n-1} + \ldots + a_n)(\bar{a}_0 x^n + \bar{a}_1 x^{n-1} + \ldots + \bar{a}_n)\,,$$

where the $\bar{a}_i$ denote the complex conjugate numbers to a_i; this has only real coefficients; the equation $g(x) = 0$ contains all the roots of the equation $f(x) = 0$.

Remark 1. In some considerations, it is required that all the roots of equation (1) should be simple. Theoretically it is not difficult to find an equation with the same roots as those of (1) but all *simple*. If $d(x)$ is the greatest common divisor of $f(x)$ and its derivative $f'(x)$ (this can be found by using the Euclidean algorithm, cf. Theorem 1.14.7, p. 59), then the quotient $g(x) = f(x)/d(x)$ is a polynomial with the above-mentioned property.

Theorem 1. *Let $\alpha_1, \alpha_2, \ldots, \alpha_n$ be the roots of* (1). *Then,*

(a) $1/\alpha_1, 1/\alpha_2, \ldots, 1/\alpha_n$ *are the roots of the equation* $g(x) \equiv x^n f(1/x) \equiv a_n x^n + a_{n-1} x^{n-1} + \ldots + a_0 = 0$ (*for* $a_n \neq 0$);

(b) $\alpha_1/c, \alpha_2/c, \ldots, \alpha_n/c$ *are the roots of the equation* $g(x) \equiv f(cx) \equiv a_0 c^n x^n + a_1 c^{n-1} x^{n-1} + \ldots + a_n = 0$ (*for* $c \neq 0$);

(c) $-\alpha_1, -\alpha_2, \ldots, -\alpha_n$ *are the roots of the equation* $g(x) \equiv (-1)^n f(-x) \equiv$ $\equiv a_0 x^n - a_1 x^{n-1} + a_2 x^{n-2} + \ldots + (-1)^n a_n = 0$;

(d) $\alpha_1 - a, \alpha_2 - a, \ldots, \alpha_n - a$ *are the roots of the equation*

$$g(x) \equiv f(x + a) \equiv f(a) + \frac{x}{1!} f'(a) + \frac{x^2}{2!} f''(a) + \ldots + \frac{x^n}{n!} f^{(n)}(a) = 0 .$$

In (d), the coefficients $f(a), f'(a), \frac{1}{2!} f''(a)$ etc. can be found, for example, by using Horner's scheme (cf. Remark 1.14.1, p. 60).

31.2. Estimates for the Roots of Algebraic Equations

Theorem 1. *Let in the equation*

$$f(x) \equiv a_0 x^n + a_1 x^{n-1} + \ldots + a_n = 0 \tag{1}$$

a_0 *be positive. Then every real root* α *of* (1) *satisfies the following inequalities:*

(a) $$\alpha < 1 + \frac{|a_i|}{a_0} \quad (\textit{Maclaurin}),$$

where a_i *is the negative coefficient in* (1) *with greatest modulus (if none of the coefficients are negative then* $\alpha \leqq 0$);

(b) $$\alpha < 1 + \left(\frac{|a_i|}{a_0}\right)^{1/r} \quad (\textit{Lagrange}),$$

where a_i *is defined in* (a) *and* a_r *is the first negative coefficient in* (1);

(c) $$\alpha < 1 + \left(\frac{|a_i|}{a_s}\right)^{1/(r-s)} \quad (\textit{Tillot}),$$

where a_i *and* r *are the same as in* (a) *and* (b) *and* a_s *is the greatest of the first* r *positive coefficients in* (1).

Theorem 2. *Every real or complex root* α *of equation* (1) *(with real or complex coefficients) satisfies the inequality*

$$|\alpha| < 1 + \left|\frac{a_i}{a_0}\right|$$

where a_i *is the coefficient of* (1) *with the greatest modulus.*

REMARK 1. Similar estimates of the real roots of (1) from below can be obtained by the application of Theorem 1 to (c) of Theorem 31.1.1; an analogous estimate for the moduli of complex roots of (1) from below can be obtained by application of Theorem 2 to (a) of Theorem 31.1.1.

Example 1. Theorem 1 yields the following estimates for the real roots α_k of the equation $x^3 + 4x^2 + x - 6 = 0$ (here, $i = 3$, $r = 3$, $s = 1$):

(a) $\alpha_k < 7$,
(b) $\alpha_k < 1 + 6^{1/3} \doteq 2{\cdot}8171$,
(c) $\alpha_k < 1 + (\frac{6}{4})^{1/2} \doteq 2{\cdot}2247$.

According to Theorem 2, all roots (real or complex) satisfy the inequality $|\alpha_k| < 7$; if we apply Theorem 2 to the equation $6x^3 - x^2 - 4x - 1 = 0$ with roots $1/\alpha_k$, we obtain $|1/\alpha_k| < 1 + |\frac{4}{6}| = \frac{5}{3}$ so that $|\alpha_k| > \frac{3}{5}$. (The roots of the equation $x^3 + 4x^2 + x - 6 = 0$ are 1, -2 and -3.)

Theorem 3. *Let x_1 be a (complex) number for which $f'(x_1) \neq 0$. Then the circle $|x - \xi| \leqq \varrho$, where*

$$\xi = x_1 - \frac{n}{2}\frac{f(x_1)}{f'(x_1)}, \quad \varrho = \left|\frac{n}{2}\frac{f(x_1)}{f'(x_1)}\right|,$$

contains at least one root of the equation (1).

Example 2. Let us apply this Theorem to the equation $x^3 + 4x^2 + x - 6 = 0$ from Example 1. Choose $x_1 = 0{\cdot}9$, then $f(x_1) = -1{\cdot}141$, $f'(x_1) = 10{\cdot}63$. It follows that in the circle with centre $\xi = 0{\cdot}9 - \frac{3}{2}\,.\,(-1{\cdot}141)/10{\cdot}63 \doteq 1{\cdot}061$ and radius $\varrho = 0{\cdot}161$ there lies at least one root of the equation.

Theorem 4 (*Descartes' Theorem*). *The number of positive roots of equation* (1) *is either equal to the number of changes of sign in the sequence*

$$a_0, a_1, \ldots, a_n,$$

or it is smaller by an even number.

REMARK 2. The number of changes in sign is obtained by ignoring all the zero entries and by determining the number of pairs of consecutive numbers with different signs in the sequence.

Example 3. The corresponding sequence to the equation

$$x^4 + 3x^2 - 1 = 0 \tag{2}$$

is

$$1, 0, 3, 0, -1$$

with a single change of sign $(3, -1)$. It follows that equation (2) possesses exactly one positive root α (since the number of its roots cannot be smaller than one by an even number). According to the estimate (a) in Theorem 1 $\alpha < 2$, according to (b) we also have $\alpha < 2$, according to (c) $\alpha < 1 + \sqrt{\frac{1}{3}} \doteq 1{\cdot}58$.

Theorem 5 (*Budan–Fourier Theorem*). *Let $f(x)$ be a real polynomial, $a_0 > 0$. Let $\alpha < \beta$, $f(\alpha) . f(\beta) \neq 0$ and let $\omega(x)$ denote the number of changes of sign in the sequence*

$$f(x), f'(x), f''(x), \ldots, f^{(n)}(x) .$$

Then, the number of real roots of (1) *in the interval* $[\alpha, \beta]$ *is either equal to* $\omega(\alpha) - \omega(\beta)$, *or it is smaller by an even number.*

REMARK 3. In Theorems 4 and 5 each root is considered with its corresponding multiplicity.

Theorem 6 (*Sturm Theorem*). *Let all the roots of* (1) *be simple* (cf. Remark 31.1.1). *If neither of the real numbers α and β ($\alpha < \beta$) is a root of* (1), *then there are exactly $V(\alpha) - V(\beta)$ real roots of* (1) *in the interval* $[\alpha, \beta]$. *Here, $V(x)$ denotes the number of changes of sign in the so-called Sturm sequence*

$$f(x), f_1(x), f_2(x), \ldots, f_m(x) .$$

For this sequence the following special sequence can be chosen:

$$f(x), f'(x), r_1(x), \ldots, r_s(x) ,$$

where $-r_1(x)$ is the remainder obtained after the division $f(x)/f'(x)$, $-r_2(x)$ the remainder after the division $f'(x)/r_1(x)$ etc., and $-r_s(x)$ is the remainder after the division $r_{s-2}(x)/r_{s-1}(x)$, where $r_s(x)$ is a non-zero constant.

Example 4. Let us apply the Sturm theorem to the problem of finding the number of roots of the equation $x^3 + 4x^2 + x - 6 = 0$ in the interval $[0, 2]$. Here, $f(x) = x^3 + 4x^2 + x - 6$, $f_1(x) = f'(x) = 3x^2 + 8x + 1$, $f_2(x) = \frac{2}{9}(13x + 29)$ ($-f_2(x)$ is the remainder after the division $f(x)/f_1(x)$), $f_3(x) = \frac{324}{169}$. Hence $f(0) = -6$, $f_1(0) = 1$, $f_2(0) = \frac{58}{9}$, $f_3(0) = \frac{324}{169}$, so that $V(0) = 1$; further, $f(2) = 20$, $f_1(2) = 29$, $f_2(2) = \frac{110}{9}$, $f_3(2) = \frac{324}{169}$ so that $V(2) = 0$. Since $V(0) - V(2) = 1$, there is exactly one root of the given equation in the interval $[0, 2]$.

31.3. Connection of Roots with Eigenvalues of Matrices

It is easily seen that if $a_0 = 1$, equation (31.2.1) can be written (with λ instead of x) in the following determinant form:

$$\begin{vmatrix} \lambda, & -1, & 0, & \ldots, & 0, & 0 \\ 0, & \lambda, & -1, & \ldots, & 0, & 0 \\ \cdots & \cdots & \cdots & \cdots & \cdots & \cdots \\ 0, & 0, & 0, & \ldots, & \lambda, & -1 \\ a_n, & a_{n-1}, & a_{n-2}, & \ldots, & a_2, & \lambda + a_1 \end{vmatrix} = 0 . \qquad (1)$$

However, this is the characteristic equation of the matrix

$$\mathbf{A} = \begin{pmatrix} 0, & 1, & 0, & \ldots, & 0, & 0 \\ 0, & 0, & 1, & \ldots, & 0, & 0 \\ \ldots & \ldots & \ldots & \ldots & \ldots & \ldots \\ 0, & 0, & 0, & \ldots, & 0, & 1 \\ -a_n, & -a_{n-1}, & -a_{n-2}, & \ldots, & -a_2, & -a_1 \end{pmatrix}. \tag{2}$$

Hence estimates for eigenvalues of matrices yield immediate estimates for the roots of equation (31.2.1). This is of particular use in numerical solution of algebraic equations.

A well-known estimate for eigenvalues of square matrices has been given by *Gershgorin*:

Theorem 1. *Let $\mathbf{A} = (a_{ij})$ be a square matrix of order n. Then all eigenvalues of $\mathbf{A}$ are contained in the union of the following n circles in the complex plane:*

$$|a_{ii} - z| \leqq \sum_{j \neq i} |a_{ij}|, \quad i = 1, \ldots, n.$$

Moreover, if $|a_{kk} - a_{11}| > \sum_{j \neq k} |a_{kj}| + \sum_{j \neq 1} |a_{1j}|$ for $k = 2, \ldots, n$, the first circle contains exactly one eigenvalue of $\mathbf{A}$.

31.4. Some Methods for Solving Algebraic and Transcendental Equations

Algebraic equations (with one unknown) of degree four at most, binomial equations and some other special types of equations can be solved directly (see §§ 1·20, 1.21, 1.22). To solve algebraic equations of higher degree or transcendental equations, numerical methods are used. Some of these methods will now be described.

(a) Method of Bernoulli and Whittaker. The given algebraic equation of degree n is written in the form

$$x^n = a_1 x^{n-1} + a_2 x^{n-2} + \ldots + a_n. \tag{1}$$

Choose $u_0 = 1, u_{-1} = u_{-2} = \ldots = u_{-(n-1)} = 0$ and compute the numbers $u_1, u_2, \ldots$ according to the recurrence formula ($m = 1, 2, \ldots$)

$$u_m = a_1 u_{m-1} + a_2 u_{m-2} + \ldots + a_n u_{m-n}.$$

Let $\alpha_1, \alpha_2, \ldots, \alpha_n$ be the roots (in general complex and not necessarily distinct) of equation (1), such that

$$|\alpha_1| = |\alpha_2| = \ldots = |\alpha_k| > |\alpha_{k+1}| \geqq \ldots \geqq |\alpha_n|$$

(where thus exactly k roots have the greatest modulus). Then, for integral r sufficiently large, $\alpha_1, \alpha_2, \ldots, \alpha_k$ are approximately equal to the roots of the equation

$$\begin{vmatrix} x^k, & x^{k-1}, & \ldots, & 1 \\ u_{r+k}, & u_{r+k-1}, & \ldots, & u_r \\ u_{r+k+1}, & u_{r+k}, & \ldots, & u_{r+1} \\ \cdots & \cdots & \cdots & \cdots \\ u_{r+2k-1}, & u_{r+2k-2}, & \ldots, & u_{r+k-1} \end{vmatrix} = 0\,. \tag{2}$$

In general, the larger the value of r is, the more accurate is the approximation to the roots. When all the coefficients of (1) are real, the most frequent cases are $k = 1$ (for α_1 real), and $k = 2$ (for α_1, α_2 complex conjugate).

A convenient procedure for practical computation is the following: Compute the first, say, 20 (or more, according to the required accuracy and the magnitude of n) numbers $u_1, \ldots, u_{20}$ and then the ratios $u_{16}/u_{15}, u_{17}/u_{16}, \ldots, u_{20}/u_{19}$. If none of these ratios differs in sign or by more than 5–10% in magnitude, u_{20}/u_{19} can be considered as an approximation to the root α_1 of equation (1), i.e. the case $k = 1$ in (2) has occurred. If the computed ratios differ by more than 10% (or even in sign), the determinants $\Delta_r = u_r^2 - u_{r-1}u_{r+1}$ for $r = 15, \ldots, 19$ and their ratios $Q_r = \Delta_r/\Delta_{r-1}$ for $r = 16, \ldots, 19$ should be computed. If these ratios Q_r are approximately equal and of the same sign, then the roots α_1 and α_2 of equation (1) are approximately equal to the roots of equation (2) for $k = 2$ and $r = 17$. If neither the first, nor the second case occurs, then the first three or more roots of (1) have almost equal moduli. It is then possible to proceed in a similar way for $k = 3$ or to solve another equation in y which has been obtained from the given equation by the substitution $x = y + u$. Here, u can be chosen for example as an approximation to $\sqrt[n]{|a_n|}$.

REMARK 1. It is easily seen (see Theorem 31.3.1) that the eigenvalues of the matrix

$$\mathbf{A} = \begin{pmatrix} 0, & 1, & 0, & \ldots, & 0, & 0 \\ 0, & 0, & 1, & \ldots, & 0, & 0 \\ \cdots & \cdots & \cdots & \cdots & \cdots & \cdots \\ 0, & 0, & 0, & \ldots, & 0, & 1 \\ a_n, & a_{n-1}, & a_{n-2}, & \ldots, & a_2, & a_1 \end{pmatrix}$$

coincide with the roots of equation (1) and that the procedure of § 30.6 applied to $\mathbf{A}$ and to the initial vector

$$\mathbf{x}_0 = \begin{pmatrix} 0 \\ 0 \\ 0 \\ 0 \\ 1 \end{pmatrix}$$

is identical with the method of Bernoulli and Whittaker applied to equation (1).

(b) The Graeffe Method and its Modifications. We have to solve the equation

$$a_n x^n + a_{n-1} x^{n-1} + \ldots + a_0 = 0\,, \quad a_n \neq 0\,. \tag{3}$$

Let us compute the numbers

$$\begin{aligned} a_0^{(1)} &= a_0^2\,, \\ a_1^{(1)} &= -a_1^2 + 2a_0 a_2\,, \\ a_2^{(1)} &= a_2^2 - 2a_1 a_3 + 2a_0 a_4\,, \\ &\ldots\ldots\ldots\ldots\ldots\ldots\ldots\ldots \\ a_{n-1}^{(1)} &= (-1)^{n-1}\, a_{n-1}^2 + (-1)^n \,.\, 2a_{n-2} a_n\,, \\ a_n^{(1)} &= (-1)^n\, a_n^2 \end{aligned} \tag{4}$$

and repeat the computation to obtain

$$a_0^{(2)} = (a_0^{(1)})^2\,, \quad a_1^{(2)} = -(a_1^{(1)})^2 + 2a_0^{(1)} a_2^{(1)}\,, \quad \text{etc.,}$$

and $a_0^{(3)} = (a_0^{(2)})^2$ etc., up to $a_0^{(m)}$, $a_1^{(m)}$, ..., $a_n^{(m)}$, where m is a chosen integer (say $m = 10$).

The polynomial

$$a_n^{(m)} x^n + a_{n-1}^{(m)} x^{n-1} + \ldots + a_0^{(m)}$$

has for its roots the 2^m-th powers of the roots of equation (3). If m is sufficiently large (say $m = 10$), some of the coefficients (in any case $a_n^{(m)}$ and $a_0^{(m)}$) satisfy the approximate equalities

$$a_k^{(m)} \approx (-1)^k\, (a_k^{(m-1)})^2$$

and, moreover, the remaining summands on the right-hand sides of (4) are, for these coefficients, sufficiently small. Suppose that these "well-behaved" coefficients are

$$a_n^{(m)}, a_{k_1}^{(m)}, a_{k_2}^{(m)}, \ldots, a_{k_s}^{(m)}, a_0^{(m)}\,, \quad \text{where} \quad n > k_1 > k_2 > \ldots > k_s > 0\,.$$

Then, the roots of equations

$$\begin{aligned} L_1(x) &\equiv a_n^{(m)} x^{n-k_1} + a_{n-1}^{(m)} x^{n-k_1-1} + \ldots + a_{k_1}^{(m)} = 0\,, \\ L_2(x) &\equiv a_{k_1}^{(m)} x^{k_1-k_2} + a_{k_1-1}^{(m)} x^{k_1-k_2-1} + \ldots + a_{k_2}^{(m)} = 0\,, \\ &\ldots\ldots\ldots\ldots\ldots\ldots\ldots\ldots \\ L_{s+1}(x) &\equiv a_{k_s}^{(m)} x^{k_s} + a_{k_s-1}^{(m)} x^{k_s-1} + \ldots + a_0^{(m)} = 0 \end{aligned} \tag{5}$$

are approximately equal to the 2^m-th powers of the roots of equation (3) in such a sense that one of equations (5) corresponds to a group of roots of (3) with (approximately) equal moduli: the first with moduli r_1, the second with r_2, etc., the last with r_{s+1}. Here, $r_1 > r_2 > \ldots > r_{s+1}$.

In this manner, we obtain the moduli of the roots of equation (3) since for $i = 1, \ldots, s + 1$ (if we put $k_0 = n$, $k_{s+1} = 0$)

$$r_i^{k_{i-1}-k_i} \approx \sqrt[2^m]{\left|\frac{a_{k_i}^{(m)}}{a_{k_{i-1}}^{(m)}}\right|}. \tag{6}$$

To compute the roots themselves, it is necessary to determine which of the 2^m-th roots of the corresponding root of equation (5) satisfies (3). This can easily be done when all the coefficients of (3) are real (we shall assume this from now on) and when the corresponding equation (5) is of degree one. Then, the corresponding root of (3) is real and it is sufficient to find out by substitution into (3) which of the two real 2^m-th roots

$$\sqrt[2m]{\gamma} \quad \text{and} \quad -\sqrt[2m]{\gamma}$$

of the root γ of (5) satisfies equation (3).

If all the equations (5) are linear, we obtain in this manner all roots. If one of equations (5) is quadratic, all the remaining being linear, we can compute the roots which correspond to the linear equations, then use relations for the sum of all roots (equal to $-a_{n-1}/a_n$) and for the product of all roots (equal to $(-1)^n a_0/a_n$), to compute, if $a_0 \neq 0$, the sum and product of the remaining two roots. Other cases lead to complications which can be avoided by the following *modified Lehmer's process*:

We compute, besides the sequence $a_n^{(k)}, a_{n-1}^{(k)}, \ldots, a_0^{(k)}$, $k = 1, \ldots, m$, another set of sequences $b_n^{(k)}, b_{n-1}^{(k)}, \ldots, b_0^{(k)}$, $k = 0, \ldots, m$, as follows:

$$b_n^{(0)} = a_{n-1}, \quad b_{n-1}^{(0)} = 2a_{n-2}, \; b_{n-2}^{(0)} = 3a_{n-3}, \; \ldots, \; b_1^{(0)} = na_0, \; b_0^{(0)} = 0\,;$$

where ($a_i^{(0)} = a_i$, $i = 0, \ldots, n$) and

$$\begin{aligned}
b_0^{(k+1)} &= a_0^{(k)} b_0^{(k)} \; (= 0)\,,\\
b_1^{(k+1)} &= a_0^{(k)} b_2^{(k)} - a_1^{(k)} b_1^{(k)} + a_2^{(k)} b_0^{(k)}\,,\\
b_2^{(k+1)} &= a_0^{(k)} b_4^{(k)} - a_1^{(k)} b_3^{(k)} + a_2^{(k)} b_2^{(k)} - a_3^{(k)} b_1^{(k)} + a_4^{(k)} b_0^{(k)}\,,\\
&\cdots\cdots\cdots\cdots\cdots\cdots\cdots\cdots\cdots\cdots\cdots\cdots\\
b_{n-1}^{(k+1)} &= (-1)^{n-2} a_{n-2}^{(k)} b_n^{(k)} + (-1)^{n-1} a_{n-1}^{(k)} b_{n-1}^{(k)} + (-1)^n a_n^{(k)} b_{n-2}^{(k)}\,,\\
b_n^{(k+1)} &= (-1)^n a_n^{(k)} b_n^{(k)}
\end{aligned} \tag{7}$$

for $k = 0, \ldots, m - 1$. If we denote the following polynomials (with k_i defined as in (5)) by $M_i(x)$, $i = 1, \ldots, s + 1$, where

$$M_i(x) \equiv b_{k_{i-1}}^{(m)} x^{k_{i-1}-k_i} + b_{k_{i-1}-1}^{(m)} x^{k_{i-1}-k_i-1} + \ldots + b_{k_i}^{(m)}\,,$$

then the following assertion holds:

If β is a simple root of $L_i(x) = 0$, then the corresponding root α of equation (3) satisfies the approximate equality

$$\alpha \approx - \frac{M_i(\beta)}{\beta L_i'(\beta)} . \tag{8}$$

If all the polynomials $L_j(x)$ in (5) are of degree at most 2, it is sufficient to use the fact that the sum of those roots of (3) which correspond to the roots of $L_j(x) = 0$ is approximately equal to

$$\frac{b_{k_j}^{(m)}}{a_{k_j}^{(m)}} - \frac{b_{k_{j-1}}^{(m)}}{a_{k_{j-1}}^{(m)}} . \tag{9}$$

If the degree of $L_j(x) = 0$ is equal to one, the number (9) is approximately equal to the corresponding root of (3). If the degree is two, the number (9) is an approximation of the sum, and the right-hand side of (6) an approximation of the product of the two roots of (3). A slight complication can occur in the case where the number (9) is very close to zero. Then two cases are possible: either both roots of (3) are real and approximately equal to r_j and $-r_j$, or they are complex conjugates and approximately equal to $r_j i$ and $-r_j i$. Substitution of r_j into (3) determines the roots.

Example 1. Let us use the Graeffe method to solve the equation

$$4{\cdot}08x^4 - 6{\cdot}03x^3 + 6{\cdot}99x^2 - 9{\cdot}81x + 9{\cdot}72 = 0 .$$

The numbers $a_4^{(k)}$, $a_3^{(k)}$, $a_2^{(k)}$, $a_1^{(k)}$, $a_0^{(k)}$ computed according to formula (4) for $k = 1, \ldots, 10$ ($m = 10$) are listed in Tab. 31.1.

TABLE 31.1

k	$a_4^{(k)}$	$a_3^{(k)}$	$a_2^{(k)}$	$a_1^{(k)}$	$a_0^{(k)}$
—	$4{\cdot}08$	$-6{\cdot}03$	$6{\cdot}99$	$-9{\cdot}81$	$9{\cdot}72$
1	$1{\cdot}665 \,.\, 10^{1}$	$2{\cdot}068 \,.\, 10^{1}$	$9{\cdot}867$	$3{\cdot}965 \,.\, 10^{1}$	$9{\cdot}448 \,.\, 10^{1}$
2	$2{\cdot}772 \,.\, 10^{2}$	$-9{\cdot}909 \,.\, 10^{1}$	$1{\cdot}604 \,.\, 10^{3}$	$2{\cdot}923 \,.\, 10^{2}$	$8{\cdot}926 \,.\, 10^{3}$
3	$7{\cdot}684 \,.\, 10^{4}$	$8{\cdot}794 \,.\, 10^{5}$	$7{\cdot}579 \,.\, 10^{6}$	$2{\cdot}855 \,.\, 10^{7}$	$7{\cdot}967 \,.\, 10^{7}$
4	$5{\cdot}904 \,.\, 10^{9}$	$3{\cdot}914 \,.\, 10^{11}$	$1{\cdot}947 \,.\, 10^{13}$	$3{\cdot}925 \,.\, 10^{14}$	$6{\cdot}347 \,.\, 10^{15}$
5	$3{\cdot}486 \,.\, 10^{19}$	$7{\cdot}671 \,.\, 10^{22}$	$1{\cdot}468 \,.\, 10^{26}$	$9{\cdot}310 \,.\, 10^{28}$	$4{\cdot}028 \,.\, 10^{31}$
6	$1{\cdot}215 \,.\, 10^{39}$	$4{\cdot}350 \,.\, 10^{45}$	$1{\cdot}008 \,.\, 10^{52}$	$3{\cdot}159 \,.\, 10^{57}$	$1{\cdot}622 \,.\, 10^{63}$
7	$1{\cdot}476 \,.\, 10^{78}$	$5{\cdot}572 \,.\, 10^{90}$	$7{\cdot}806 \,.\, 10^{103}$	$2{\cdot}272 \,.\, 10^{115}$	$2{\cdot}631 \,.\, 10^{126}$
8	$2{\cdot}179 \,.\, 10^{156}$	$1{\cdot}994 \,.\, 10^{182}$	$5{\cdot}848 \,.\, 10^{207}$	$-1{\cdot}054 \,.\, 10^{230}$	$6{\cdot}922 \,.\, 10^{252}$
9	$4{\cdot}748 \,.\, 10^{312}$	$-1{\cdot}427 \,.\, 10^{364}$	$3{\cdot}424 \,.\, 10^{415}$	$6{\cdot}985 \,.\, 10^{460}$	$4{\cdot}791 \,.\, 10^{505}$
10	$2{\cdot}254 \,.\, 10^{625}$	$1{\cdot}215 \,.\, 10^{728}$	$1{\cdot}172 \,.\, 10^{831}$	$-1{\cdot}598 \,.\, 10^{921}$	$2{\cdot}295 \,.\, 10^{1011}$

We see that the coefficients a_4, a_2 and a_0 behave regularly. Thus $s = 1$, $k_0 = 4$, $k_1 = 2$, $k_2 = 0$ and (5) consists of two equations of degree 2. We thus compute the coefficients $b_4^{(k)}, b_3^{(k)}, b_2^{(k)}, b_1^{(k)}, b_0^{(k)}$ for $k = 0, 1, \ldots, 10$ using formula (7) (see Tab. 31.2).

TABLE 31.2

k	$b_4^{(k)}$	$b_3^{(k)}$	$b_2^{(k)}$	$b_1^{(k)}$	$b_0^{(k)}$
0	$-6{\cdot}03$	$1{\cdot}398 \,.\, 10^1$	$-2{\cdot}943 \,.\, 10^1$	$3{\cdot}888 \,.\, 10^1$	0
1	$-2{\cdot}460 \,.\, 10^1$	$-7{\cdot}792 \,.\, 10^1$	$1{\cdot}073 \,.\, 10^2$	$9{\cdot}535 \,.\, 10^1$	0
2	$-4{\cdot}096 \,.\, 10^2$	$3{\cdot}155 \,.\, 10^3$	$-1{\cdot}478 \,.\, 10^2$	$6{\cdot}357 \,.\, 10^3$	0
3	$-1{\cdot}135 \,.\, 10^5$	$-3{\cdot}853 \,.\, 10^5$	$-4{\cdot}185 \,.\, 10^6$	$-3{\cdot}177 \,.\, 10^6$	0
4	$-8{\cdot}721 \,.\, 10^9$	$-8{\cdot}430 \,.\, 10^{11}$	$-2{\cdot}697 \,.\, 10^{13}$	$-2{\cdot}427 \,.\, 10^{14}$	0
5	$-5{\cdot}149 \,.\, 10^{19}$	$9{\cdot}214 \,.\, 10^{20}$	$-1{\cdot}546 \,.\, 10^{26}$	$-7{\cdot}592 \,.\, 10^{28}$	0
6	$-1{\cdot}795 \,.\, 10^{39}$	$-1{\cdot}302 \,.\, 10^{46}$	$-1{\cdot}903 \,.\, 10^{52}$	$8{\cdot}409 \,.\, 10^{56}$	0
7	$-2{\cdot}181 \,.\, 10^{78}$	$1{\cdot}542 \,.\, 10^{91}$	$-1{\cdot}573 \,.\, 10^{104}$	$-3{\cdot}352 \,.\, 10^{115}$	0
8	$-3{\cdot}219 \,.\, 10^{156}$	$-4{\cdot}883 \,.\, 10^{182}$	$-1{\cdot}245 \,.\, 10^{208}$	$3{\cdot}477 \,.\, 10^{230}$	0
9	$-7{\cdot}014 \,.\, 10^{312}$	$5{\cdot}141 \,.\, 10^{364}$	$-7{\cdot}293 \,.\, 10^{415}$	$-4{\cdot}953 \,.\, 10^{460}$	0
10	$-3{\cdot}330 \,.\, 10^{625}$	$1{\cdot}472 \,.\, 10^{728}$	$-2{\cdot}497 \,.\, 10^{831}$	$-3{\cdot}441 \,.\, 10^{919}$	0

According to (6) and (9), two roots α_1 and α_2 of the given equation have modulus r_1, where

$$r_1^2 \approx \sqrt[1024]{\frac{1{\cdot}172 \,.\, 10^{831}}{2{\cdot}254 \,.\, 10^{625}}} \doteq 1{\cdot}588$$

and the sum

$$\alpha_1 + \alpha_2 \approx \frac{-2{\cdot}497 \,.\, 10^{831}}{1{\cdot}172 \,.\, 10^{831}} - \frac{-3{\cdot}330 \,.\, 10^{625}}{2{\cdot}254 \,.\, 10^{625}} \doteq -0{\cdot}653 \,.$$

The remaining two roots α_3 and α_4 have modulus r_2, where

$$r_2^2 \approx \sqrt[1024]{\frac{2{\cdot}295 \,.\, 10^{1011}}{1{\cdot}172 \,.\, 10^{831}}} \doteq 1{\cdot}500 \,,^*$$

and the sum

$$\alpha_3 + \alpha_4 \approx \frac{2{\cdot}497 \,.\, 10^{831}}{1{\cdot}172 \,.\, 10^{831}} \doteq 2{\cdot}131 \,.$$

* The absolute values of r_1 and r_2 are very close one to another, which is not convenient for computation. The Bernoulli–Whittaker method — without using the transformation $x = y + u$ — would fail in this case.

It follows that the given equation has two pairs of complex conjugate roots satisfying the quadratic equations

$$x^2 + 0{\cdot}653x + 1{\cdot}588 = 0\,,$$

$$x^2 - 2{\cdot}131x + 1{\cdot}500 = 0\,.$$

Hence

$$\alpha_{1,2} \doteq -0{\cdot}327 \pm 1{\cdot}217\mathrm{i}\,,$$

$$\alpha_{3,4} \doteq \quad 1{\cdot}065 \pm 0{\cdot}605\mathrm{i}\,.$$

(c) Newton's Method. This is a method of obtaining the roots of the equation $f(x) = 0$ where $f(x)$ is a polynomial or, more generally, a function of one variable which has a first derivative in the whole interval (or the whole complex region) containing the roots to be determined. Choose x_0 as an approximation to a particular root and construct a sequence $x_1, x_2, \ldots$ according to the formula

$$x_{k+1} = x_k - \frac{f(x_k)}{f'(x_k)}\,. \tag{10}$$

If the approximation x_0 is close enough to the value ξ of a simple root of $f(x) = 0$, the sequence x_k converges quadratically to ξ. This means that for large k

$$|x_{k+1} - \xi| \leqq C|x_k - \xi|^2$$

for some positive constant C. The numbers x_k can of course be complex (if $f(x)$ is a real function, no real sequence can converge to a non-real root of $f(x) = 0$).

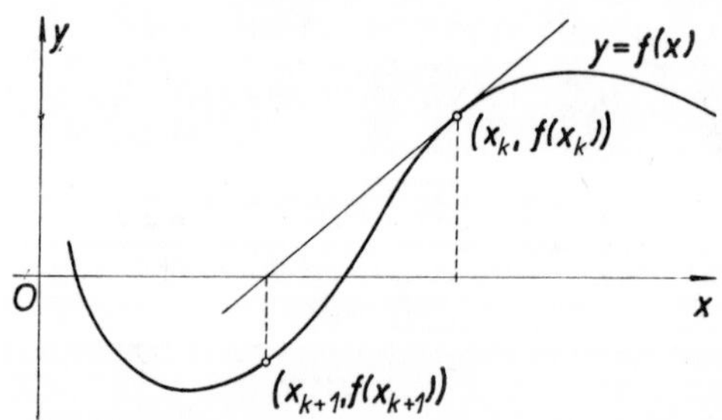

Fig. 31.1.

In the real case, Newton's method has a simple geometric meaning: x_{k+1} is the abscissa of the point at which the tangent to the curve $y = f(x)$, at the point $(x_k, f(x_k))$, intersects the x-axis (Fig. 31.1). If in the whole interval $[a, b]$ the two derivatives $f'(x)$ and $f''(x)$ have unchanged signs and $f(a)f(b) < 0$, then Newton's process converges to some root of $f(x) = 0$ in $[a, b]$ if we choose $x_0 = a$ or $x_0 = b$ according to whether $f(a)$ or $f(b)$ has the same sign as $f''(x)$.

(d) The Regula Falsi Method. This method enables us to solve the equation $f(x) = 0$ ($f(x)$ being a real continuous function in an interval I), if two numbers x_0 and x_1 in the interval I are known such that $f(x_0)$ and $f(x_1)$ have opposite signs, i.e.

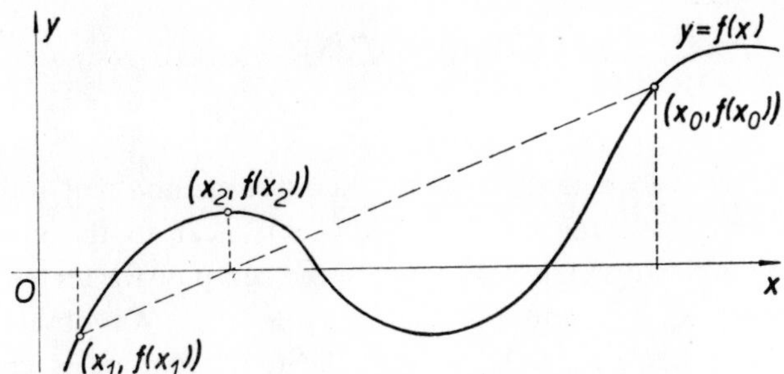

Fig. 31.2.

$f(x_0) f(x_1) < 0$. In this case we construct a sequence $x_1, x_2, \ldots$ in the following manner:

We put

$$x_2 = \frac{x_0 f(x_1) - x_1 f(x_0)}{f(x_1) - f(x_0)} \,. \tag{11}$$

If $f(x_2) = 0$, we have found a root. If $f(x_2) \neq 0$, then either $f(x_0) f(x_2) < 0$ or $f(x_1) f(x_2) < 0$. In the first case x_0 and x_2, or in the second case x_1 and x_2, satisfy the preceding condition and we compute x_3 from (11). Then, from x_3 and either x_2 or one of the previously chosen x_0 or x_1 we compute the number x_4, and repeat the computation to obtain $x_5, x_6, \ldots, x_k$. The sequence $\{x_k\}$ converges, but the convergence is usually slow.

The geometric meaning of this method is based upon the fact that x_2 in (11) is the abscissa of the point where the straight line connecting points $(x_0, f(x_0))$ and $(x_1, f(x_1))$ of the curve $y = f(x)$ meets the x-axis.

A similar process can be obtained if, under the same assumptions, the formula

$$x_2 = \tfrac{1}{2}(x_0 + x_1) \tag{12}$$

is used instead of formula (11).

(e) The General Iterative Method. Let us write the given equation in an equivalent form

$$f_1(x) = f_2(x) \,; \tag{13}$$

here, we choose the function $f_1(x)$ so that the equation $f_1(x) = c$ can easily be solved (for example $f_1(x)$ linear, of the form x^m etc.). Then we take an initial approximation x_0 and construct the recurrent sequence $x_0, x_1, x_2, \ldots$ in such a manner that x_{k+1} is computed from the equation

$$f_1(x_{k+1}) = f_2(x_k) \,. \tag{14}$$

If the sequence $x_0, x_1, x_2, \ldots$ tends to a limit z and if the two functions $f_1(x)$ and $f_2(x)$ are continuous at the point z, then z is a root of the given equation.

If the two functions $f_1(x)$ and $f_2(x)$ have first derivatives in some neighbourhood of the root z for which

$$|f_1'(x)| > |f_2'(x)|,$$

then the sequence $x_0, x_1, x_2, \ldots$ is convergent whenever x_0 is close enough to z.

Example 2. We wish to determine by the iterative method that root of the transcendental equation $x^2 - x \tan x + 1 = 0$ which is near to the number 1. Write the given equation in the form $\tan x = x + 1/x$ and compute the first four terms of the sequence x_k, where $x_0 = 1$ and $\tan x_{k+1} = x_k + 1/x_k$. We obtain $x_1 \doteq 1{\cdot}1071$, $x_2 \doteq 1{\cdot}1092$, $x_3 \doteq 1{\cdot}1093$, $x_4 \doteq x_3$ so that the required root is equal to $1{\cdot}1093$ with error less than 10^{-4}.

31.5. Numerical Solution of (Nonlinear) Systems

We shall consider only the most important case of n equations for n unknowns $x_1, x_2, \ldots, x_n$ $(n \geqq 2)$:

$$\begin{aligned} f_1(x_1, x_2, \ldots, x_n) &= 0\,, \\ f_2(x_1, x_2, \ldots, x_n) &= 0\,, \\ \ldots\ldots&\ldots\ldots \\ f_n(x_1, x_2, \ldots, x_n) &= 0\,. \end{aligned} \tag{1}$$

Here we assume that all the functions f_i have continuous first partial derivatives with respect to $x_1, \ldots, x_n$ in some region containing the roots to be found and that the determinant of the *Jacobi-matrix*

$$J(x_1, x_2, \ldots, x_n) = \begin{Vmatrix} \dfrac{\partial f_1}{\partial x_1}, & \dfrac{\partial f_1}{\partial x_2}, & \ldots, & \dfrac{\partial f_1}{\partial x_n} \\ \dfrac{\partial f_2}{\partial x_1}, & \dfrac{\partial f_2}{\partial x_2}, & \ldots, & \dfrac{\partial f_2}{\partial x_n} \\ \ldots & \ldots & \ldots & \ldots \\ \dfrac{\partial f_n}{\partial x_1}, & \dfrac{\partial f_n}{\partial x_2}, & \ldots, & \dfrac{\partial f_n}{\partial x_n} \end{Vmatrix} \tag{2}$$

(which we shall briefly denote by $\boldsymbol{J}(\boldsymbol{x})$) is not identically zero in this region.

The solution will be obtained either by the (generalized) iterative method, or by the (generalized) Newton method. Both these methods are local, i.e. they yield a solution as a limit of a convergent sequence of n-tuples $({}^kx_1, {}^kx_2, \ldots, {}^kx_n)$, $k = 0, 1, 2, \ldots$, under the assumption that the initial n-tuple ($k = 0$) is already close enough to the solution and that for this solution the determinant of the matrix (2) is non-zero. Usually, the initial n-tuple is chosen either from a knowledge of the problem (physical, technical etc.) from which (1) has arisen, by trial, or, for small n, graphically.

To solve (1) by the iterative method, write the system (1) in some equivalent form

$$\begin{aligned} g_1(x_1, \ldots, x_n) &= h_1(x_1, \ldots, x_n)\,, \\ g_2(x_1, \ldots, x_n) &= h_2(x_1, \ldots, x_n)\,, \\ &\ldots\ldots\ldots\ldots \\ g_n(x_1, \ldots, x_n) &= h_n(x_1, \ldots, x_n) \end{aligned} \tag{3}$$

satisfying two conditions: (a) the functions g_i should also have continuous first partial derivatives in the region mentioned, (b) the system

$$g_i(x_1, \ldots, x_n) = c_i \quad (i = 1, 2, \ldots, n)$$

should be easily solvable for any choice of c_i (for example, the g_i are linear functions or linear in x_k^m, etc.). We choose an initial approximation ${}^0\mathbf{x} = ({}^0x_1, {}^0x_2, \ldots, {}^0x_n)$ and construct a sequence ${}^1\mathbf{x}, {}^2\mathbf{x}, \ldots$ of n-tuples from the recurrence formula

$$g_i({}^{k+1}\mathbf{x}) = h_i({}^k\mathbf{x}) \quad (i = 1, 2, \ldots, n)\,. \tag{4}$$

If the sequence converges (i.e. if all coordinates converge separately), the limit $\mathbf{z} = (z_1, \ldots, z_n)$ is a solution of the system (1).

The following theorem holds: Let a solution $\mathbf{z} = (z_1, \ldots, z_n)$ of (3) have the following properties: (i) $\det \mathbf{J}(\mathbf{z}) \neq 0$ and (ii) all (real or complex) roots λ of the equation

$$\det \left[\lambda \left(\frac{\partial g_i}{\partial x_j} \right)_{(\mathbf{z})} - \left(\frac{\partial h_i}{\partial x_j} \right)_{(\mathbf{z})} \right] = 0$$

are smaller than 1 in modulus. Then there exists an n-dimensional region Ω such that whenever ${}^0\mathbf{x}$ is chosen in Ω, the sequence ${}^k\mathbf{x}$ converges to $\mathbf{z}$.

A special case of this method is Newton's method; here, equations (4) are of the form

$${}^{k+1}\mathbf{x} = {}^k\mathbf{x} - [\mathbf{J}({}^k\mathbf{x})]^{-1}\, \mathbf{f}({}^k\mathbf{x})\,, \tag{5}$$

where

$${}^k\mathbf{x} = \left\| \begin{matrix} {}^kx_1 \\ {}^kx_2 \\ \vdots \\ {}^kx_n \end{matrix} \right\|, \quad \mathbf{f}(\mathbf{x}) = \left\| \begin{matrix} f_1(\mathbf{x}) \\ f_2(\mathbf{x}) \\ \vdots \\ f_n(\mathbf{x}) \end{matrix} \right\|$$

and $\mathbf{J}(\mathbf{x})$ is the matrix in (2). This method converges rapidly in the neighbourhood of the solution.

Example 1. We have to find the minimum of the function

$$f(x, y) = 3x^3 + 2y^2 + xy^2 - 10x - 5y - 1 = 0$$

which lies near the point (1, 1). According to Theorem 12.12.1, this means solving the system $\partial f/\partial x = 0$, $\partial f/\partial y = 0$, i.e. the system

$$9x^2 + y^2 - 10 = 0\,, \quad 4y + 2xy - 5 = 0$$

near the point (1, 1). Let us write this last system in the form

$$x = \tfrac{1}{3}\sqrt{(10 - y^2)}\,,$$
$$y = \tfrac{1}{4}(5 - 2xy)\,.$$

Let us choose $x_0 = 1$, $y_0 = 1$ and find the sequence of pairs (x_k, y_k) for which $x_{k+1} = \tfrac{1}{3}\sqrt{(10 - y_k^2)}$, $y_{k+1} = \tfrac{1}{4}(5 - 2x_k y_k)$ (see Tab. 31.3).

TABLE 31.3

k	x_k	y_k	$10 - y_k^2$	$2x_k y_k$
0	1·0000	1·0000	9·000,00	2·0000
1	1·0000	0·7500	9·437,50	1·5000
2	1·0240	0·8750	9·234,37	1·7920
3	1·0129	0·8020	9·356,80	1·6355
4	1·0196	0·8411	9·292,51	1·7152
5	1·0161	0·8212	9·325,63	1·6688
6	1·0179	0·8328	9·306,44	1·6954
7	1·0169	0·8261	9·317,48	1·6801
8	1·0171	0·8300	9·311,10	1·6884
9	1·0171	0·8279	9·314,58	1·6841
10	1·0173	0·8290	9·312,76	1·6867
11	1·0172	0·8283	9·313,87	1·6852
12	1·0173	0·8287	9·313,24	1·6861
	1·0173	0·8285		

We see that the solution of the system is $x \doteq 1{\cdot}0173$, $y \doteq 0{\cdot}8285$. Since

$$\frac{\partial^2 f}{\partial x^2}\frac{\partial^2 f}{\partial y^2} - \left(\frac{\partial^2 f}{\partial x\,\partial y}\right)^2 > 0 \quad\text{and}\quad \frac{\partial^2 f}{\partial x^2} > 0\,,$$

this is a local minimum. The corresponding value of the given function is $-10{\cdot}086$.

32. NOMOGRAPHY AND GRAPHICAL ANALYSIS. INTERPOLATION. DIFFERENCES

By VÁCLAV PLESKOT

References: [11], [33], [46], [99], [176], [183], [209], [225], [240], [272], [288], [304], [309], [316], [331], [335], [357], [380], [433].

A. NOMOGRAPHY AND GRAPHICAL ANALYSIS

Nomography deals with

(a) the construction of *nomograms*, i.e. of charts and graphs serving to replace cumbersome numerical calculations;

(b) theoretical problems in the analysis of methods of constructing nomograms.

A computing graph considered as a nomographical drawing is composed of geometrical elements denoted (labelled) by number-values of variables or parameters, i.e. of points or lines (curves) to which, according to some rules, the number-values of the variables (labels) are affixed.

Throughout this chapter, $\log x$ stands for $\log_{10} x$.

32.1. Scales

Basic uniform scale. Let us divide an oriented straight line p by a point O into two half-lines (rays). Let us call one half-line *positive* and the other *negative* and let us denote them by $+p$ and $-p$ respectively. If we choose as the unit of length the cm, then to every real number ξ there corresponds a point A_ξ on the straight line p such that

$$OA_\xi = |\xi| \text{ cm}$$

and that A_ξ lies on $+p$, or $-p$ according as $\xi > 0$, or $\xi < 0$ respectively (Fig. 32.1).

The number ξ will be called the *label* of the point A_ξ, or the point A_ξ will be called the *point labelled by* ξ (briefly the *labelled point*). We also say that the point A_ξ is the *image* of the number ξ or that the point A_ξ corresponds to the number ξ.

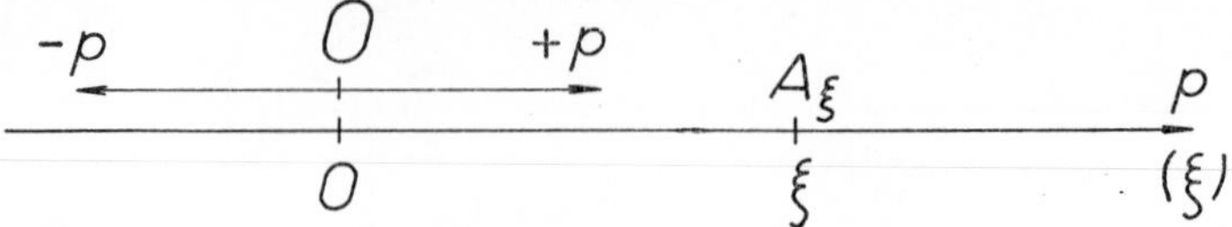

Fig. 32.1. The mapping of a number ξ by the point A_ξ; $\overline{OA} = |\xi|$ cm.

It is evident that, to every real number ξ there corresponds exactly one point of the straight line p, and conversely. The label of the point O is 0 (zero).

Definition 1. The graphical form obtained by the foregoing procedure of labelling points on a straight line with real numbers is called the *number axis* and in nomography the *basic uniform* (or *regular*) *scale*. The straight line p is called the *scale-holder* and is denoted by (ξ).

REMARK 1. For a unit segment we may also choose a segment AB of any length other than one cm. If we perform the labelling of points of a straight line by means of such a segment AB (according to the foregoing procedure) we obtain a so-called *uniform* (or *regular*) *scale* (which is no longer called basic).

The relation between the segment AB and the centimetre is expressed by the equation

$$AB = \alpha \text{ cm} \tag{1}$$

where the number α is called the *scale modulus*.

REMARK 2. The basic uniform scale is used for the drawing of other scales. We have it at our disposal on one edge of a ruler, namely the *centimetre scale*, but to be more exact: on a ruler we have only a part of the scale.

The numbers ξ on the basic uniform scale are called the *coordinates*. The coordinate ξ of a point A_ξ states the number of centimetres that we have to mark off from the point O – the origin of the coordinate system – to the point A_ξ.

THE SCALE. If a strictly monotone function $f(x)$, $x \in [a, b]$, is given, we may plot its functional scale on the scaleholder (ξ) according to the equation

$$\xi = \alpha f(x) \tag{2}$$

(α being the scale modulus) in the following way: corresponding to a given x we compute by (2) the coordinate ξ, then we plot the point A_ξ and mark this point with the number (label) x. Equation (2) is called the *mapping equation of the scale* $f(x)$ or *of the scale* x. The numbers x are the *labels of the scale*.

In practice we choose an oriented scaleholder (ξ) with the coordinate origin O and to a given number x_k we compute from (2) the coordinate $\xi_k = \alpha f(x_k)$. We line up the centimetre scale with the scaleholder (ξ) so that the point $\xi = 0$ coincides with O and then at the point ξ_k of the centimetre scale we mark on the scaleholder (ξ) a point and label it with the number x_k; (if $\xi_k > 0$ we apply the centimetre scale on a ruler to the positive half-line of the scaleholder (ξ) and if $\xi_k < 0$ we apply the ruler to the negative half-line of the scaleholder (ξ)).

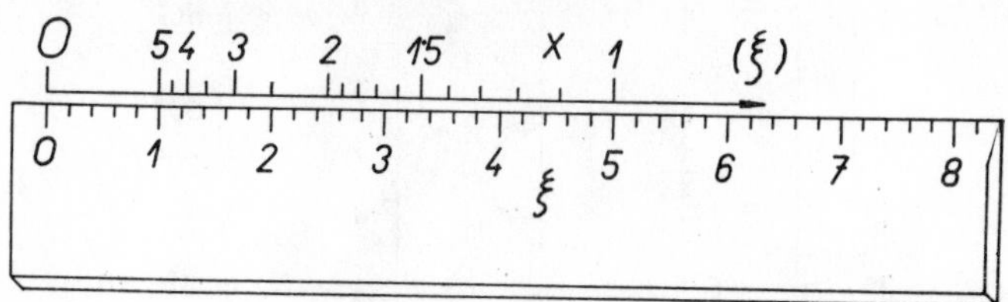

Fig. 32.2. The plotting of the scale $1/x$ for $x \in [1, 5]$ according to the mapping equation $\xi = 5 \,.\, 1/x$ by means of a centimetre scale.

Example 1. In Fig. 32.2 the functional scale of the function $1/x$ for $x \in [1, 5]$ is plotted on the scaleholder (ξ). The mapping equation for the scale modulus 5 is $\xi = 5 \,.\, (1/x)$. The process of plotting the scale by means of the centimetre scale can be seen on Fig. 32.2.

The scaleholder may be a curve; if the scaleholder is a straight line the scale is called *straight*.

Definition 2. A system of labelled points (of points marked by number-values) on a curve is called a *scale* and the curve is the *scaleholder*. The mapping equations of a scale are generally $\xi = \alpha f(x)$, $\eta = \beta g(x)$, where $f(x)$ and $g(x)$ are as a rule differentiable functions, and α and β are the scale moduli.

All points on a scale may be labelled, but the labels are marked only at so-called *points of division of the scale*. The position of a point of division on a scale is marked by means of a short stroke perpendicular to the scaleholder; these strokes are called the *strokes of division of the scale*.

The *scale division* is the part of the scale between two neighbouring scale division points. The *length of a scale division* is given by the length of arc of the scaleholder between two neighbouring scale division points. The *step of a scale division* is the numerical value of the difference between the labels of two neighbouring scale division points.

The length L_x of the scale of a function $f(x)$ which is plotted on a straight scaleholder is expressed by the formula

$$L_x = \alpha(f_{\max} - f_{\min}),$$

where f_{max} and f_{min} are the greatest and least values, respectively, of the function $f(x)$ in the interval $[a, b]$. See Fig. 32.3a.

Example 2. In Example 1 (Fig. 32.2) we have $f(x) = 1/x$, $f_{max} = \frac{1}{1}$, $f_{min} = \frac{1}{5}$ and the length of the scale $L_x = 5(1 - \frac{1}{5}) = 4$.

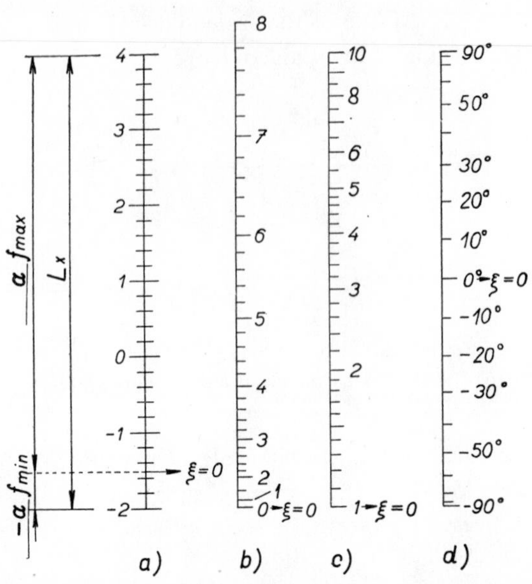

Fig. 32.3. (a) The uniform scale (with mapping equation $\xi = x + 1{\cdot}5$); (b) the quadratic scale (with mapping equation $\xi = 0{\cdot}1x^2$); (c) the logarithmic scale (with mapping equation $\xi = 6 \log x$); (d) the sine scale (with mapping equation $\xi = 3 \sin x$).

The fundamental straight functional scales are:

$f(x) = ax + b, \quad a \neq 0$ (the *uniform scale*, see Fig. 32.3a),

$f(x) = ax^2 + bx + c, \; a \neq 0$ (the *quadratic scale*, only such values of x are considered for which $f(x)$ is strictly monotone; see Fig. 32·3b),

$f(x) = \log x, \; x > 0$ (the *logarithmic scale*; see Fig. 32.3c),

$f(x) = \sin x, \; x \in [-90°; 90°]$ (the *sine scale*; see Fig. 32.4d).

Besides these, the scales of functions $f(x) = (ax + b)/(cx + d)$, $ad \neq bc$, or $f(x) = \log \sin x$ or the scales of other trigonometric functions are particularly important for the construction of nomograms and for different applications.

Theorem 1. *The absolute accuracy (absolute error) of a reading* $|\Delta x|$ *on a scale of a function* $f(x)$ *is expressed by the formula*

$$|\Delta x| \approx \frac{\varepsilon}{\alpha |f'(x)|},$$

where ε *is the length of a segment which the human eye is able to recognize* (*as a rule*, $\varepsilon \approx 0{\cdot}01$ cm) and α *is the scale modulus.* (The formula follows from the relation $\varepsilon = \alpha \,.\, |f(x + \Delta x) - f(x)| \approx \alpha |f'(x)| \,.\, |\Delta x|$.)

Theorem 2. *The relative accuracy of a reading* $\varrho = |\Delta x / x|$ *on a straight scale is expressed by the formula*

$$\varrho \approx \frac{\varepsilon}{\alpha |x\, f'(x)|},$$

the percentage accuracy by the formula

$$\varrho_{\%} \approx \frac{100\varepsilon}{\alpha |x\, f'(x)|}. \tag{3}$$

For a uniform scale we have $f'(x) = \text{const.}$, for a logarithmic scale $f'(x) = \text{const.}/x$. From these relations and from formulae for accuracy the following theorem may be deduced:

Theorem 3. *The absolute accuracy of reading on a uniform scale is constant; the relative accuracy of reading on a logarithmic scale is constant.*

Example 3. On a logarithmic scale of scale modulus 25 (which is in prevailing use on logarithmic slide rules) we read all values with the same relative percentage accuracy, $0{\cdot}1\%$ approximately (for $\varepsilon = 0{\cdot}01$), since $x\, f'(x) = x \,.\, 0{\cdot}434 \ldots / x = 0{\cdot}434 \ldots = 1/2{\cdot}30 \ldots$. For $\varepsilon = 0{\cdot}01$ we obtain by (3) the result $2{\cdot}30 \ldots / 25$, that is $0{\cdot}1$ approximately.

REMARK 3. Scales are useful in the construction of slide rules, graph paper (network charts) and nomograms.

32.2. Graph Paper

Graph paper is the name given to a type of drawing on which a network of scales is plotted.

Graph paper contains two systems of parallel lines which are drawn through the point of division of two scales plotted on the coordinate axis.

Definition 1. *Graph paper* is defined by the mapping equations

$$\xi = \alpha f(x), \quad \eta = \beta\, g(y) \tag{1}$$

(the equations of straight lines that are parallel to the coordinate axis).

According to the choice of functions (1) there are various types of graph paper:

Rectangular graph paper (millimetre paper): $\xi = \alpha x$, $\eta = \beta y$. On such paper the representation of a linear function $y = ax + b$ is the straight line the equation of which in rectangular cartesian coordinates (ξ, η) is $\eta = (\beta/\alpha)\, a\xi + \beta b$ (we obtain this equation by the elimination of the variables x, y from the mapping equations $\xi = \alpha x$, $\eta = \beta y$ and the function $y = ax + b$).

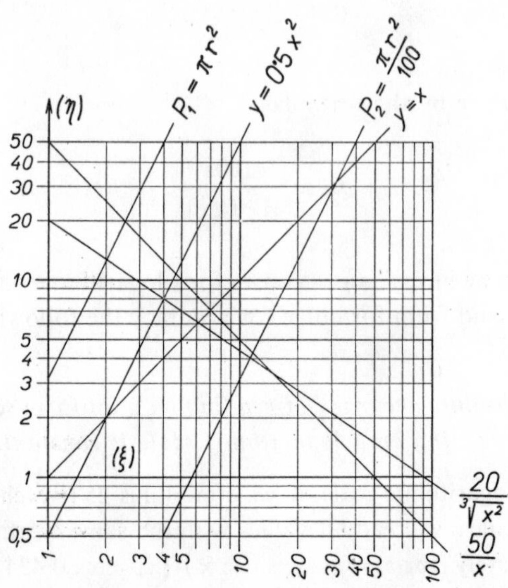

Fig. 32.4. Logarithmic paper (with mapping equation $\xi = 2{\cdot}5 \log x$, $\eta = 2{\cdot}5 \log y$).

Logarithmic paper: $\xi = \alpha \log x$, $\eta = \beta \log y$. On such paper, the graph of the (power) function $y = ax^n$ ($a > 0$, n being a real number) is the straight line $\eta = (\beta n/\alpha)\, \xi + \beta \log a$. (See Fig. 32.4.)

Logarithmic paper is especially suitable for mapping relations with a large range of variability of the variables x and y. The relative accuracy is the same for all parts of such paper.

Semilogarithmic paper: $\xi = \alpha x$, $\eta = \beta \log y$. On such paper the graph of the exponential function $y = ab^x$ ($a \neq 0$, $b > 0$) is the straight line $\eta = [(\beta \log b)/\alpha]\, \xi + \beta \log a$. Because the constant relative change of a function y is mapped in a straight line, semilogarithmic paper is suitable for mapping statistical data, of various characteristics for which the values are changing over a large range, etc.

Sine paper: $\xi = \alpha \sin x$, $\eta = \beta y$. On such paper the representation of a function $y = a \sin x + b$ is the straight line $\eta = (\beta a/\alpha)\, \xi + \beta b$.

There is a great variety of special types of graph paper (e.g. polar paper, probability paper, etc.).

Graph paper is suitable for the construction of empirical formulae because a certain type of functional relationship may be mapped in a straight line on properly chosen graph paper.

REMARK 1. If we plot the graph of a relation $y = f(x)$ on a given type of graph paper by means of the equation that expresses the relation between the coordinates (ξ, η) of a point on the graph, then we say we are *plotting the graph by means of the coordinate method* (see Example 1).

If we plot the graph of a relation $y = f(x)$ on a given type of graph paper by means of the variables x, y from the given relation $y = f(x)$ and by means of the labels on the graph paper, then we say we are *plotting the graph by means of the label method* (see Example 1).

Example 1. In Fig. 32.4 we have logarithmic paper with the scale moduli $\alpha = \beta = 2{\cdot}5$; its mapping equations are $\xi = 2{\cdot}5 \,.\, \log x$, $\eta = 2{\cdot}5 \,.\, \log y$. On this paper the graphs of the following functions are plotted: $y = 0{\cdot}5x^2$, $y = 20/\sqrt[3]{x^2}$, $y = x$, $y = 50/x$, $P_1 = \pi r^2$ and $P_2 = \pi r^2/100$. All graphs are straight lines with slopes 2, $-\frac{2}{3}$, 1, -1, 2, 2, respectively. To draw the graph of the function $y = 0{\cdot}5x^2$ we plot (by the label method) the points $(x = 1,\ y = 0{\cdot}5)$, $(x = 10,\ y = 50)$ or (by the coordinate method) the points $(\xi = 0,\ \eta = 2{\cdot}5 \log 0{\cdot}5 \doteq -0{\cdot}75)$, $(\xi = 2{\cdot}5,\ \eta = 5 + 2{\cdot}5 \log 0{\cdot}5 \doteq 4{\cdot}25)$; in both cases the straight line passing through the plotted points is the graph of $y = 0{\cdot}5x^2$.

32.3. Intersection Charts or Lattice Nomograms

Definition 1. A *nomogram* of a relation between three variables x, y, z

$$F(x, y, z) = 0 \tag{1}$$

is a graphical drawing composed of three families of labelled geometric elements (points or curves) and possesses the following properties:

1. Each variable of the relation (1) is mapped into a family of geometric elements which are labelled by the values of this variable, and to different values of the variable there correspond different elements on the drawing.

2. Between the labelled elements of the drawing there exist relations of position such that, corresponding to given values x_k, y_k of two variables, we may find the value z_k of the third variable, which constitutes, with x_k, y_k, the triple (x_k, y_k, z_k) satisfying equation (1), i.e. $F(x_k, y_k, z_k) = 0$.

3. The reading of the value of any variable must be possible without any further graphical construction on the drawing.

REMARK 1. The definition of a nomogram for relations between more than three variables is an easy generalization of Definition 1.

REMARK 2. We may classify nomograms according to the structure and connection between the labelled elements. The most important classes of nomograms are:

(a) intersection charts or lattice nomograms,
(b) alignment charts or collineation nomograms,
(c) nomograms with a transparency.

The first two classes are only special cases of the third class. By means of nomograms with a transparency, systems of equations may be solved.

Definition 2. A *lattice nomogram* of a relation (1) is a nomogram composed of three families of labelled curves.

Theorem 1. *A triple* (x_k, y_k, z_k) *which satisfies the relation* (1) *is mapped on a lattice nomogram into three curves which intersect at one point* (see Example 1).

Definition 3. The equations which express the families of labelled curves in a lattice nomogram of relation (1),

$$X(\xi, \eta; x) = 0\,,$$
$$Y(\xi, \eta; y) = 0\,,$$
$$Z(\xi, \eta; z) = 0\,, \tag{2}$$

are called the *mapping equations of the lattice nomogram.*

Theorem 2. *Any two equations* (2) *may be suitably chosen and the remaining equation obtained by elimination of the parameter which the chosen equations and equation* (1) *have in common. In other words, select graph paper labelled by the two chosen variables and then plot the graphs of the remaining variables; as a rule, we select the first two equations* (2) *in the form* $\xi = \alpha f(x)$, $\eta = \beta g(y)$.

Example 1. A drawing of a lattice nomogram for the relation

$$V = \frac{\pi}{4} d^2 h\,, \tag{3}$$

where the intervals and variables are

$V \in [15, 800]\ \text{cm}^3$ (the volume of a cylinder),

$d \in [\ 2,\ 10]\ \text{cm}$ (the diameter of the cylinder),

$h \in [\ 5,\ 10]\ \text{cm}$ (the height of the cylinder),

is shown in Fig. 32.5.

1. *Variant.* The mapping equations of the nomogram (we have chosen millimetre graph paper and in Fig. 32.5 the scale moduli are $\alpha = 1$, $\beta = 0{\cdot}01$) are

$$\text{(a) } \xi = \alpha h\,, \qquad \text{(a}'\text{) } \xi = h\,,$$

$$\text{(b) } \eta = \beta V\,, \qquad \text{(b}'\text{) } \eta = 0{\cdot}01\, V\,,$$

$$\text{(c) } \eta = \frac{\beta}{\alpha}\,\frac{\pi d^2}{4}\,\xi\,; \qquad \text{(c}'\text{) } \eta = \frac{\pi}{400}\, d^2 \xi\,.$$

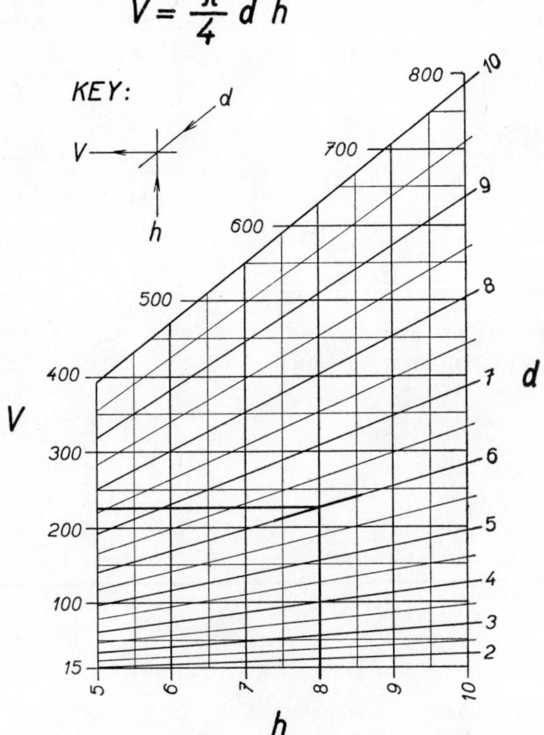

Fig. 32.5. A lattice nomogram for computation of the volume V of a circular cylinder of height h and diameter d. Example: $h = 8$ (cm), $d = 6$ (cm) $\Rightarrow V \doteq 225$ (cm^3).

2. *Variant.* The mapping equations of the nomogram (we have chosen logarithmic paper) are

$$\begin{aligned} &\text{(a) } \xi = 15 \log h\,, \\ &\text{(b) } \eta = 7{\cdot}5 \log V\,, \\ &\text{(c) } \eta = 0{\cdot}5\xi + 7{\cdot}5 \log \frac{\pi}{4}\, d^2\,. \end{aligned} \tag{4}$$

A sketch showing the so-called "working field" of the nomogram in the chosen coordinate system is plotted in Fig. 32.6; the nomogram corresponding to equations (4) is shown in Fig. 32.7.

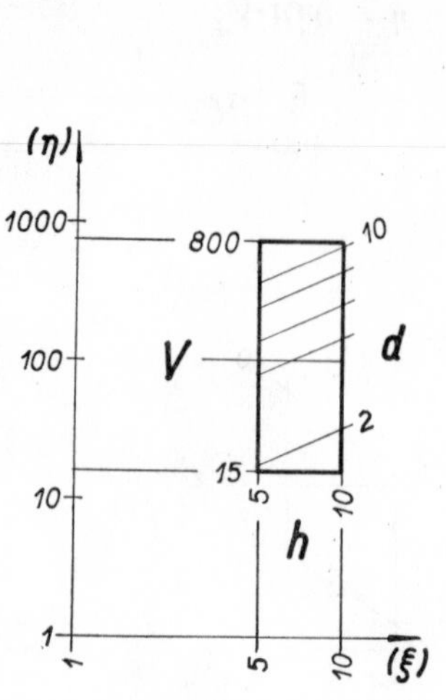

Fig. 32.6. The location of the "working field" of the nomogram of equation (3) with respect to the coordinate axes (ξ), (η).

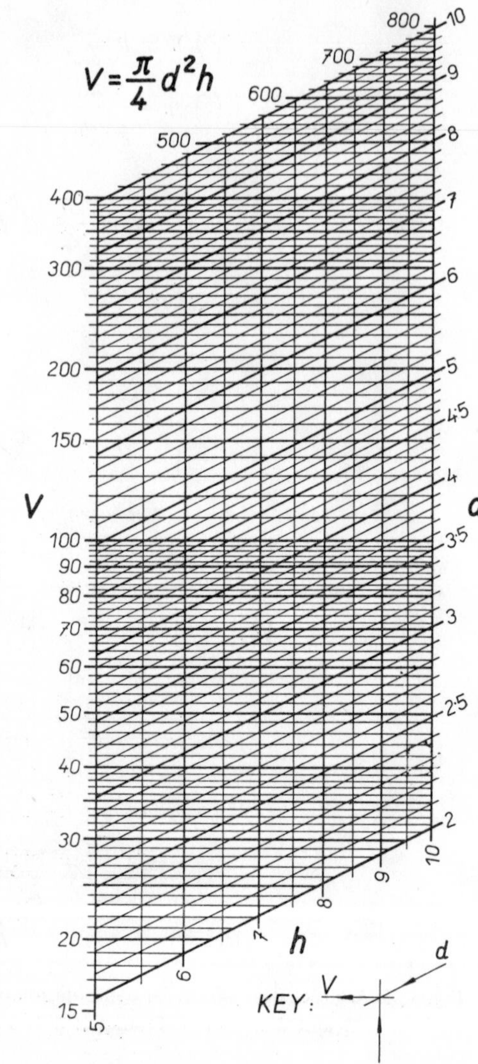

Fig. 32.7. The lattice nomogram for computation of the volume V of a circular cylinder of height h and diameter d. Example: $h = 8$ (cm), $d = 6$ (cm) $\Rightarrow V \doteq 225$ (cm^3).

REMARK 3. Further variants of the lattice nomogram of the relation (3), plotted on millimetre graph paper, may be obtained by a different choice of the first two mapping equations. For $\xi = \alpha d$, $\eta = \beta V$ the variable h is mapped into a family of parabolas, for the choice $\xi = \alpha h$, $\eta = \beta d$ the variable V is mapped into a family of algebraic curves of the third order.

Definition 4. If a relation of the form (1) can be arranged in the form

$$h_1(x) f_3(z) + h_2(y) g_3(z) + h_3(z) = 0\,, \tag{5}$$

we say that it is of *Cauchy's canonical form.*

Theorem 3. *Cauchy's canonical form can be mapped by a lattice nomogram consisting of three families of straight lines (a so-called rectilinear nomogram). It suffices to chooose graph paper with equations*

$$\xi = \alpha\, h_1(x)\,, \quad \eta = \beta\, h_2(y)\,,$$

so that the z-lines consist of the family of straight lines

$$\frac{\xi}{\alpha} f_3(z) + \frac{\eta}{\beta} g_3(z) + h_3(z) = 0\,.$$

REMARK 4. It is conventional to mark the variables only by a suffix added to the letter denoting the function, remembering that the same letter may denote different functions when the suffixes are different. Following this convention we may write Cauchy's canonical form (5) in the form

$$h_1 f_3 + h_2 g_3 + h_3 = 0\,.$$

Theorem 4. *If a relation* (1), *denoted briefly by* $F_{1,2,3} = 0$, *can be put in the form of the so-called Massau's equation*

$$\begin{vmatrix} f_1, & g_1, & h_1 \\ f_2, & g_2, & h_2 \\ f_3, & g_3, & h_3 \end{vmatrix} = 0\,, \tag{6}$$

then it can be mapped into a rectilinear nomogram; the mapping equations of the three families of straight lines are

$$\xi f_1 + \eta g_1 + h_1 = 0\,,$$
$$\xi f_2 + \eta g_2 + h_2 = 0\,,$$
$$\xi f_3 + \eta g_3 + h_3 = 0\,.$$

MAPPING OF RELATIONS WITH SEVERAL (MORE THAN 3) VARIABLES (GROUPING OF NOMOGRAMS). We obtain a relation between more than three variables by means of auxiliary parameters in relations between three variables – so-called *partial relations.* We obtain the initial relation from the elimination of these auxiliary parameters from the partial relations.

Example 2. The product relation (each factor depends on only one variable)

$$f_5 = f_1 f_2 f_3 f_4$$

may be resolved into three partial relations:

$$\text{I. } f_1 f_2 = u\,, \qquad \text{II. } u f_3 = v\,, \qquad \text{III. } v f_4 = f_5\,.$$

The nomograms of two partial relations may be drawn so that the families of curves labelled by the common variable will be one and the same; in this case we say that we *group* the two nomograms.

Definition 5. A nomogram obtained by grouping nomograms of partial relations is said to be a *grouped nomogram.*

Example 3. Let us construct a grouped lattice nomogram for the relation (used in condenser theory)

$$C = \frac{S\varepsilon}{4\pi d}\,.$$

The intervals involved and the meaning of the variables are:

$S \in [5,\ 5\,.\,10^2]$	cm^2	(the surface area of the condenser),
$d \in [5\,.\,10^{-3}, 5\,.\,10^{-1}]$	cm	(the distance between layers),
$\varepsilon \in [1, 10]$		(dielectric constant),
$C \in [10,\ 8\,.\,10^3]$	cm	(capacity of the condenser).

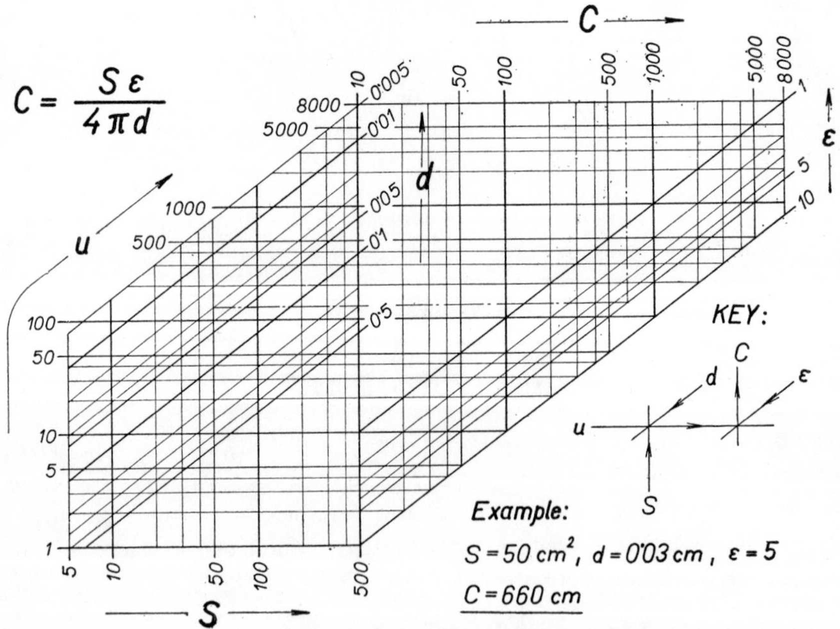

Fig. 32.8. The lattice nomogram (the grouped nomogram) for computation of the capacity C of a condenser.

Since the end-values of the given variables are of different orders, it will be an advantage to construct the mapping on logarithmic paper. We resolve the given relation (using the fact that $\log C = \log S + \log \varepsilon - \log 4\pi d$) into the relations

$$\text{I. } \log S - \log 4\pi d = \log u\,, \qquad \text{II. } \log C - \log \varepsilon = \log u\,.$$

We group the lattice nomograms of both relations by means of their common family of lines labelled with the auxiliary parameter u and plot the nomograms by the mapping equations

$$\text{I. } \xi = 2 \log S\,, \qquad \text{II. } \xi = 2 \log C\,,$$

$$\eta = 1{\cdot}5 \log u\,, \qquad \eta = 1{\cdot}5 \log u\,,$$

$$\eta = \tfrac{3}{4}\xi - 1{\cdot}5 \log 4\pi d\,; \qquad \eta = \tfrac{3}{4}\xi - 1{\cdot}5 \log \varepsilon$$

(see Fig. 32.8).

32.4. Alignment Nomograms or Collineation Nomograms

Definition 1. A *collineation nomogram* of a relation between three variables x, y, z,

$$F(x, y, z) = 0\,, \tag{1}$$

is a drawing composed of three families of labelled points (three scales). To a triple (x_k, y_k, z_k), which satisfies the given relation, there correspond three points in the drawing and these points lie on one straight line (are collinear) (see Fig. 32.9).

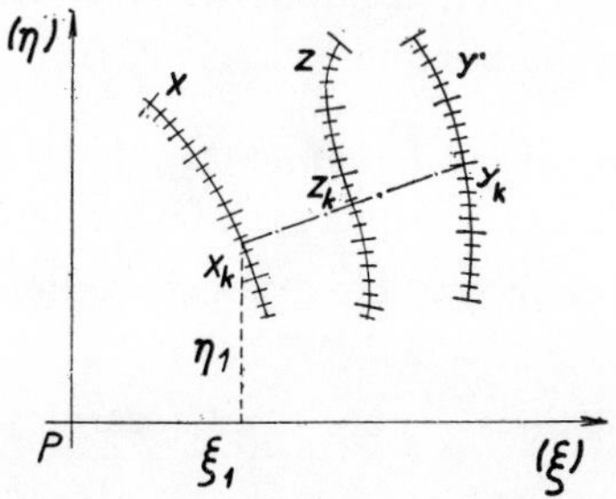

Fig. 32.9. The principle of a collineation nomogram.

Theorem 1. *If a relation* (1) *can be written in the form of Soreau's equation*

$$\begin{vmatrix} F_1(x), & G_1(x), & 1 \\ F_2(y), & G_2(y), & 1 \\ F_3(z), & G_3(z), & 1 \end{vmatrix} = 0\,, \tag{2}$$

then it can be mapped into a collineation nomogram by the mapping equations

$$\begin{aligned} \xi_1 &= F_1(x), \quad \eta_1 = G_1(x), \\ \xi_2 &= F_2(y), \quad \eta_2 = G_2(y), \\ \xi_3 &= F_3(z), \quad \eta_3 = G_3(z), \end{aligned} \tag{3}$$

where ξ_i and η_i $(i = 1, 2, 3)$ are the cartesian coordinates of points on the scales of the nomogram.

This is evident, because for every triple (x_k, y_k, z_k) that satisfies the relation (1) and hence also the relation (2) we obtain three pairs of coordinates from (3), which, because

$$\begin{vmatrix} \xi_1, & \eta_1, & 1 \\ \xi_2, & \eta_2, & 1 \\ \xi_3, & \eta_3, & 1 \end{vmatrix} = 0, \tag{4}$$

lie on a straight line.

The points labelled with the same variable and plotted according to the equations of the corresponding line in (3) lie on a scale of the nomogram (see the sketch in Fig. 32.9).

The conditions for transforming the relation (1) into a Soreau's equation (2) are complicated and are not suitable for application. These conditions are called the *conditions of anamorphosis of equation* (1).

The conditions of anamorphosis are simple for the nomographic rational relations (see Definition 2) and for such relations we can design a very simple method of constructing collineation nomograms.

Definition 2. A relation is said to be *nomographic rational* if it can be arranged in the form of a sum of products in which each factor depends on only one variable.

We have an example of a nomographic rational relation if we expand the determinant in relation (2).

Nomographic rational relations are classified according to so-called nomographic order.

Definition 3. The least possible number of different functions of the same variable in a nomographic rational relation is called the *nomographic order of the relation with respect to that variable.*

Example 1. The relation

$$h_1(x) f_3(z) + h_2(y) g_3(z) + h_3(z) = 0, \tag{5}$$

where $h_3(z) \neq 0$, is of nomographic order 2 with respect to the variable z because it can be written in the form

$$h_1(x)\,\varphi_3(z) + h_2(y)\,\psi_3(z) + 1 = 0\,,$$

where $\varphi_3(z) = f_3(z)/h_3(z)$ and $\psi_3(z) = g_3(z)/h_3(z)$. With respect to the variable x or y, relation (5) is of nomographic order 1.

Definition 4. The *nomographic order of a relation* is the sum of the nomographic orders with respect to each variable.

Example 2. The relation (5) is of nomographic order 4.

Theorem 2. *Nomographic rational relations which can be mapped in collineation nomograms are of nomographic order at least* 3 *and at most* 6.

REMARK 1. To simplify the following text, we shall denote the variables, according to Remark 32.3.4, by adding numeral suffixes to the letters denoting the functions; we write h_1 instead of $h_1(x)$, etc.

Theorem 3. *A relation of nomographic order* 3,

$$Af_1f_2f_3 + B_1f_2f_3 + B_2f_3f_1 + B_3f_1f_2 + C_1f_1 + C_2f_2 + C_3f_3 + D = 0\,,$$

where A_i, B_i, C_i *and* D $(i = 1, 2, 3)$ *are constants, can be transformed by the substitutions* $f_i = (a_i\varphi_i + b_i)/(c_i\varphi_i + d_i)$ $(a_id_i \neq b_ic_i)$, *into one of the following so-called canonical forms:*

$$\varphi_3 = \varphi_1\varphi_2 \qquad \text{(the } \textit{product form}\text{)}\,, \tag{6}$$

$$\varphi_3 = \varphi_1 + \varphi_2 \qquad \text{(the } \textit{sum form}\text{)}\,, \tag{7}$$

$$\varphi_1\varphi_2\varphi_3 = \varphi_1 + \varphi_2 + \varphi_3 \quad \text{(the } \textit{mixed form}\text{)}\,. \tag{8}$$

By a suitable modification, each of the forms (6), (7) and (8) may be transformed into one of the other forms (for example from $\varphi_3 = \varphi_1\varphi_2$ it follows that $\log|\varphi_3| = \log|\varphi_1| + \log|\varphi_2|$).

A relation of nomographic order 4 can be transformed by suitable substitutions into one and only one of the following canonical forms:

$$h_1f_3 + h_2g_3 + h_3 = 0 \quad (\textit{Cauchy's canonical form})\,, \tag{5}$$

$$g_1g_2f_3 + (g_1 + g_2)\,g_3 + h_3 = 0 \quad (\textit{Clark's canonical form})\,. \tag{9}$$

The general equation of nomographic order 5 or 6 is not always amenable to anamorphosis. For these relations the following canonical forms are mentioned:

for a relation of nomographic order 5

$$f_3 = \frac{f_1 + f_2}{g_1 + g_2} \quad (\textit{Soreau's canonical form}), \tag{10}$$

and for a relation of nomographic order 6

$$\frac{f_1 + f_2}{g_1 + g_2} = \frac{f_1 + f_3}{g_1 + g_3}. \tag{11}$$

Every canonical form may easily be transformed into Soreau's form and then by means of this form it is possible to write the mapping equations of the nomogram. First, we obtain the Massau form of the relation

$$\begin{vmatrix} f_1, & g_1, & h_1 \\ f_2, & g_2, & h_2 \\ f_3, & g_3, & h_3 \end{vmatrix} = 0$$

and then modify the determinant so that one column has only unit elements; the transformed relation is the required Soreau's relation (2).

Theorem 4. *Cauchy's canonical form* (5) *may be written in the following Massau form:*

$$\begin{vmatrix} 1, & 0, & -h_1 \\ 0, & 1, & -h_2 \\ f_3, & g_3, & h_3 \end{vmatrix} = 0.$$

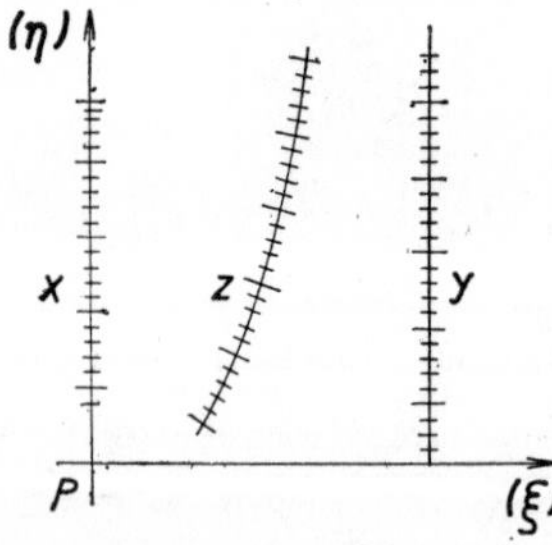

Fig. 32.10. The sketch of a collineation nomogram for Cauchy's canonical form $h_1 f_3 + h_2 g_3 + h_3 = 0$ corresponding to equations (12). (To the suffixes 1, 2, 3, there correspond, in the sketch, the variables x, y, z, respectively.)

The commonly used mapping equations of Cauchy's canonical form which follow from Massau's relation are:

$$\begin{aligned} \xi_1 &= 0, & \eta_1 &= \alpha h_1, \\ \xi_2 &= \delta, & \eta_2 &= \beta h_2, \\ \xi_3 &= \frac{\alpha \delta g_3}{\beta f_3 + \alpha g_3}, & \eta_3 &= -\frac{\alpha \beta h_3}{\beta f_3 + \alpha g_3}. \end{aligned} \tag{12}$$

To the mapping equations (12) there corresponds the sketch in Fig. 32.10, where the variables corresponding to suffixes 1, 2, 3 are denoted by x, y, and z respectively.

Example 3. To construct a collineation nomogram for the relation

$$\sigma_r = \tfrac{1}{2}(\sigma + \sqrt{(\sigma^2 + 4\tau^2)}) , \tag{13}$$

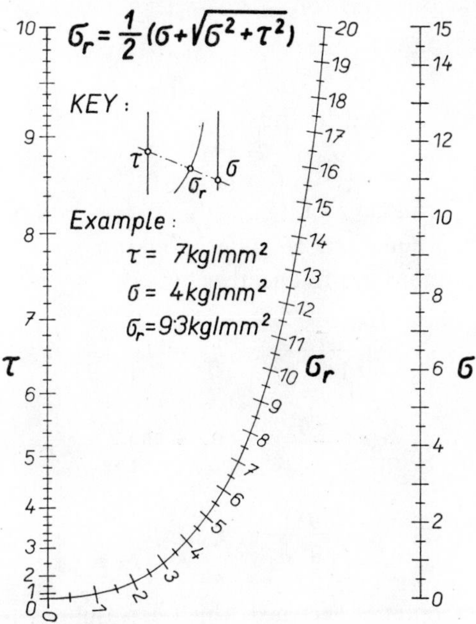

Fig. 32.11. The collineation nomogram for determination of the resultant stress σ_r of butt welds if they are simultaneously stressed by the tension (the compression) σ and the shear τ.

we first transform (13) into

$$\tau^2 + \sigma\sigma_r - \sigma_r^2 = 0$$

and then compare with the Cauchy canonical form (5), so obtaining

$$h_1 = \tau^2 , \quad h_2 = \sigma ,$$

$$f_3 = 1 , \quad g_3 = \sigma_r , \quad h_3 = -\sigma_r^2 . \tag{14}$$

We substitute the right-hand members of (14) into (12) and select the scale moduli $\alpha = 0{\cdot}075$, $\beta = 0{\cdot}5$ and $\delta = 5$, hence obtaining the mapping equations of a collineation nomogram for the relation (13):

$$\xi_1 = 0\,, \qquad \eta_1 = 0{\cdot}075\tau^2\,,$$
$$\xi_2 = 5\,, \qquad \eta_2 = 0{\cdot}5\sigma\,,$$
$$\xi_3 = \frac{3\sigma_r}{4 + 0{\cdot}6\sigma_r}\,, \qquad \eta_3 = \frac{0{\cdot}3\sigma_r^2}{4 + 0{\cdot}6\sigma_r}$$

(see Fig. 32.11).

Theorem 5. *The Clark canonical form* (9) *may be written in the following Massau form:*

$$\begin{vmatrix} 1, & -g_1, & g_1^2 \\ 1, & -g_2, & g_2^2 \\ f_3, & g_3, & h_3 \end{vmatrix} = 0\,.$$

By admissible transformations of Massau's relation into Soreau's form we can obtain various mapping equations of collineation nomograms for relation (9). The following mapping equations are frequently used:

$$\begin{aligned} \xi_1 &= \alpha\frac{-g_1}{1 + g_1^2}\,, \qquad \eta_1 = \beta\frac{g_1^2}{1 + g_1^2}\,, \\ \xi_2 &= \alpha\frac{-g_2}{1 + g_2^2}\,, \qquad \eta_2 = \beta\frac{g_2^2}{1 + g_2^2}\,, \\ \xi_3 &= \alpha\frac{g_3}{f_3 + h_3}\,, \qquad \eta_3 = \beta\frac{h_3}{f_3 + h_3}\,. \end{aligned} \tag{15}$$

In Fig. 32.12 the nomogram is sketched which corresponds to mapping equations (15). The scales x and y have the same scaleholder (in Fig. 32.12 the scaleholder is an ellipse).

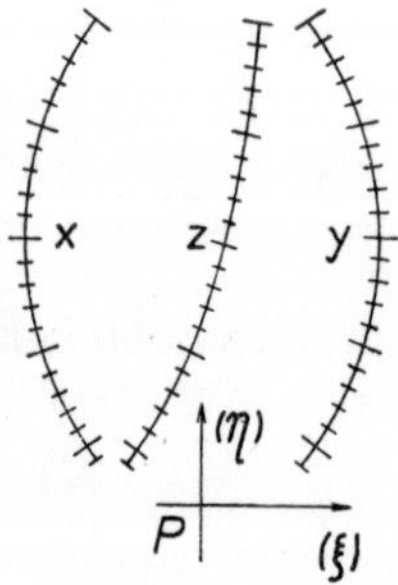

Fig. 32.12. A sketch of a collineation nomogram for Clark's canonical form (corresponding to mapping equations (15)).

Theorem 6. *A relation of nomographic order 3 can be considered as a special case of Cauchy's or Clark's canonical form. The corresponding drawing is called Cauchy's or Clark's nomogram, respectively.*

Theorem 7. *If we compare a relation of nomographic order* 3 *with Cauchy's canonical form, then we can map it into collineation nomograms with three rectilinear scales (e.g. so-called* N-*or* Z-*nomograms and nomograms with three parallel scales). If we compare it with Clark's canonical form, then it can be mapped into collineation nomograms the scales of which lie on a cubic, or into nomograms two scales of which lie on a conic and one scale of which is rectilinear.*

Collineation nomograms constructed for relations of nomographic order 3 may be grouped into eight classes. Nomograms which have scaleholders of the same geometrical

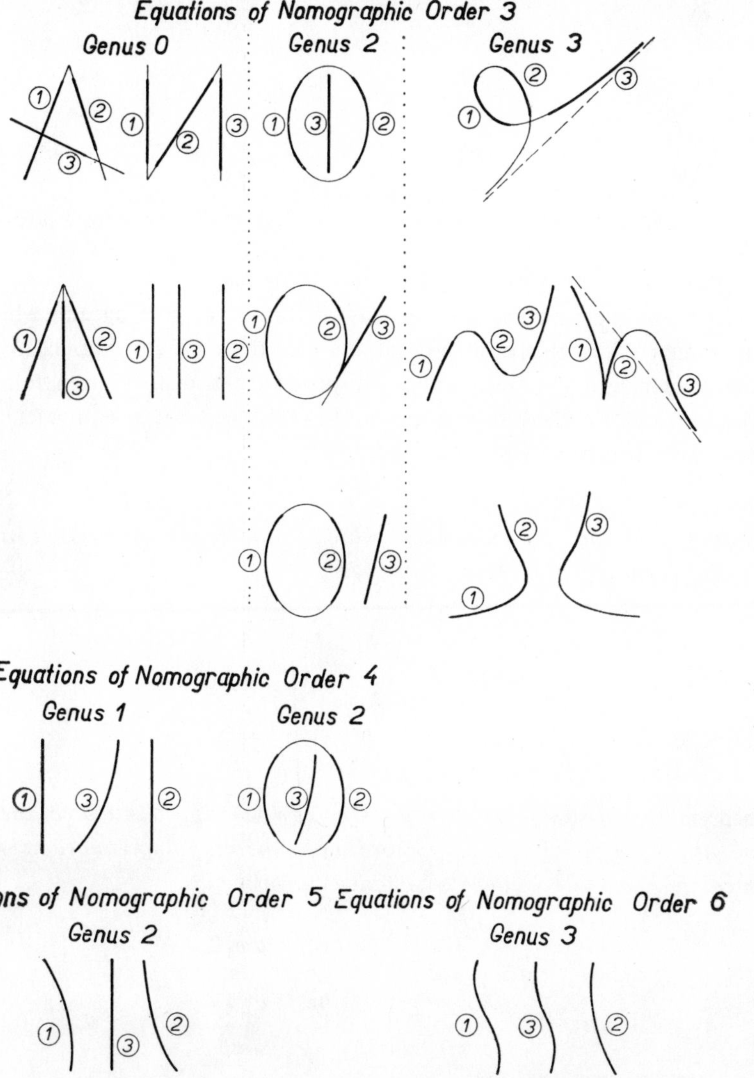

Fig. 32.13. A survey of collineation nomograms of different nomographic orders and genera. The genus of a nomogram = the number of curved scales in the collineation nomogram.

structure belong to the same class; e.g. one class is formed by nomograms the scales of which lie along the sides of a triangle, another class is formed by nomograms two scales of which lie on a conic and the third scale lies on a straight line that does not

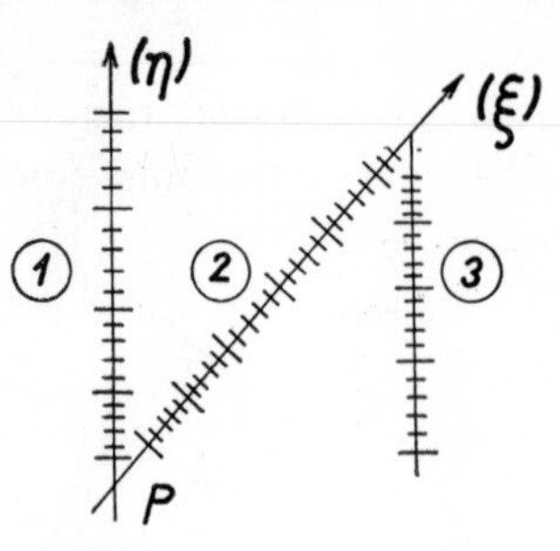

Fig. 32.14. Sketch of an N-nomogram.

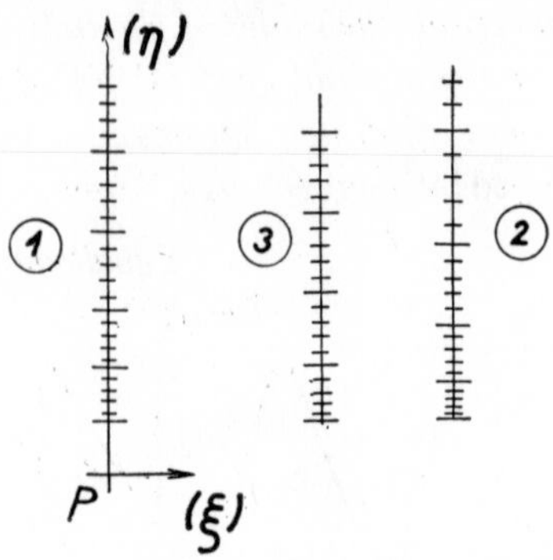

Fig. 32.15. Sketch of a nomogram of three parallel scales.

intersect the conic, and yet another class is formed by nomograms whose scales lie on the same cubic with a double point, etc. (see the survey in Fig. 32.13). A mapping in one class cannot be transformed by projective transformation into a mapping in another class. Finally, nomograms in one class need not be equivalent with respect to projective transformations.

Theorem 8. *The product form* $\varphi_3 = \varphi_1\varphi_2$ *can be mapped into an* N-*nomogram according to the mapping equations which follow from comparison with Cauchy's canonical form* (see Fig. 32.14):

$$\xi_1 = 0\,, \qquad \eta_1 = \alpha\varphi_1\,,$$
$$\xi_2 = \delta\,, \qquad \eta_2 = -\beta\varphi_3\,,$$
$$\xi_3 = \frac{\alpha\delta}{\beta\varphi_2 + \alpha}\,, \qquad \eta_3 = 0\,.$$

Theorem 9. *The sum form* $\varphi_3 = \varphi_1 + \varphi_2$ *can be mapped into a collineation nomogram with three parallel scales according to the mapping equations* (*which follow from comparison with Cauchy's canonical form*):

$$\xi_1 = 0\,, \qquad \eta_1 = \alpha\varphi_1\,,$$
$$\xi_2 = \delta\,, \qquad \eta_2 = \beta\varphi_2\,,$$
$$\xi_3 = \frac{\alpha\delta}{\alpha + \beta}\,, \qquad \eta_3 = \frac{\alpha\beta}{\alpha + \beta}\,\varphi_3$$

(see Fig. 32.15).

Theorem 10. *Soreau's canonical form* (10) *can be reduced to Massau's form*

$$\begin{vmatrix} f_1, & g_1, & 1 \\ f_2, & g_2, & -1 \\ f_3, & 1, & 0 \end{vmatrix} = 0\,,$$

from which the mapping equations of the nomogram can be derived. The scheme of the nomogram is illustrated in Fig. 32.13 (lower left-hand corner).

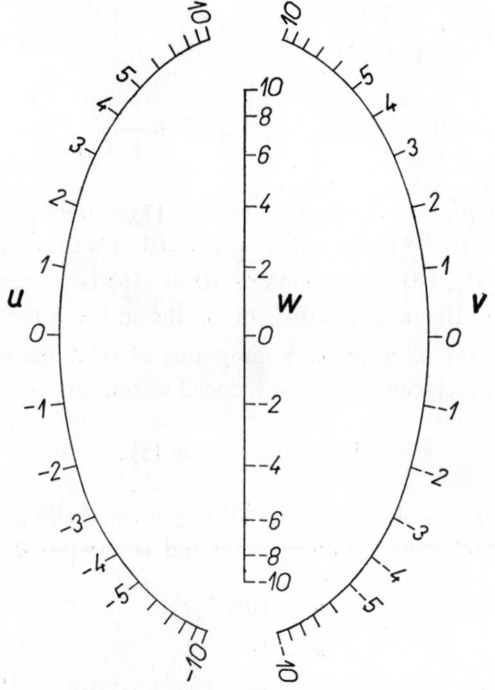

Fig. 32.16. Skeleton for the plotting of the relation $u/10 + v/10 = w/10$.

Theorem 11. *The canonical form* (11) *can be reduced to Massau's form*

$$\begin{vmatrix} f_1, & g_1, & 1 \\ f_2, & g_2, & -1 \\ f_3, & g_3, & -1 \end{vmatrix} = 0$$

from which follow the mapping equations. The sketch of the nomogram is illustrated in Fig. 32.13 (lower right-hand corner).

Definition 5. A collinear nomogram mapping the relation

$$\frac{u}{10} + \frac{v}{10} = \frac{w}{10} \tag{16}$$

is called a *skeleton*.

Fig. 32.16 shows a skeleton plotted according to the mapping equations

$$\begin{aligned}
\xi_1 &= -\alpha \frac{10^{u/10}}{1 + 10^{2u/10}}, & \eta_1 &= \beta \frac{10^{2u/10}}{1 + 10^{2u/10}}, \\
\xi_2 &= \alpha \frac{10^{v/10}}{1 + 10^{2v/10}}, & \eta_2 &= \beta \frac{10^{2v/10}}{1 + 10^{2v/10}}, \\
\xi_3 &= 0, & \eta_3 &= \beta \frac{10^{w/10}}{1 + 10^{w/10}}.
\end{aligned} \tag{17}$$

REMARK 2. We obtain the mapping equations (17), after transforming (16) to the form $10^{u/10} \,.\, 10^{v/10} = 10^{w/10}$ and comparing with Clark's canonical form (9) and the mapping equations (15). (The denominator 10 in (16) is chosen to obtain a more convenient labelling of the points of division on the scales of the skeleton.)

We use a skeleton for plotting the nomograms of relations which are of nomographic order 3 and which can always be reduced to the sum form

$$\varphi_1(x) + \varphi_2(y) = \varphi_3(z)\,. \tag{18}$$

We obtain a nomogram of the relation (18) by a re-labelling of the scales on the skeleton, using the skeleton coordinates u, v and w computed from the equations

$$\begin{aligned}
u &= 10\varphi_1(x)\,, \\
v &= 10\varphi_2(y)\,, \\
w &= 10\varphi_3(z)\,;
\end{aligned} \tag{19}$$

these equations being obtained from (16) and (18). For example, we obtain a point x_k on the scale x in such a way that we compute the skeleton coordinate u_k from $u_k = 10\varphi_1(x_k)$, find the point u_k on the skeleton, and label it with x_k.

Example 4. Let us map the relation $f_{\%} = 100r/a$ into a Clark's nomogram by means of a skeleton. First, we arrange the given relation into the sum form

$$(0{\cdot}5 - \log a) + (\log r - 0{\cdot}5) - \log (f_{\%}/100) = 0 \tag{20}$$

(the constants 0·5 and −0·5 are introduced to obtain a better location of the points of division on the scales a and r). Comparing (16) with (20) we obtain the equations for computation of the skeleton coordinates u, v and w, where

$$u = 10(0{\cdot}5 - \log a),$$
$$v = 10(\log r - 0{\cdot}5),$$
$$w = 10 \log (f_{\%}/100).$$

See Fig. 32.17.

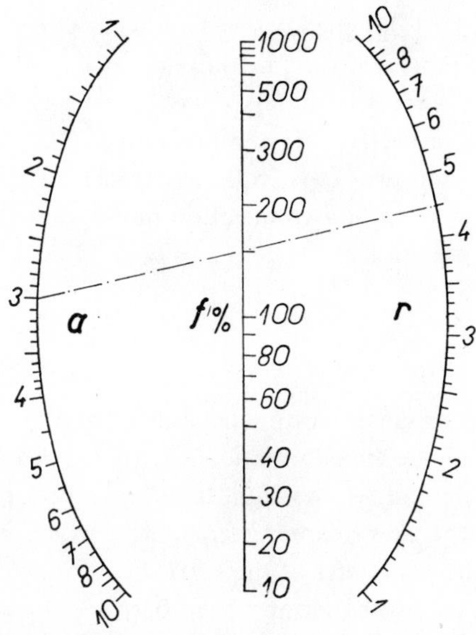

Fig. 32.17. Clark's nomogram for the relation $f_{\%} = 100r/a$ (a — the anticipated result, r — the actual result, $f_{\%}$ — the fulfilment of the plan). Example: $a = 3$, $r = 4{\cdot}5 \Rightarrow f_{\%} = 150$.

The table of skeleton coordinates for Fig. 32.17.

a	u	$f\%$	w
1	5·00	10	−10·00
2	1·99	20	− 6·99
3	0·23	30	− 5·23
4	−1·02	40	− 3·98
5	−1·99	50	− 3·01
6	−2·78	:	:
7	−3·45	100	0·00
8	−4·03	:	:
9	−4·54	900	9·54
10	−5·00	1000	10·00
r	$-v$		

REMARK 3. A translation, extension or compression of the points of division along a scale may be obtained by introducing proper constants into the mapping equation (18), writing

$$c[\varphi_1(x) + a] + c[\varphi_2(y) + b] - c[\varphi_3(z) + (a + b)] = 0;$$

by this arrangement we simultaneously change the skeleton coordinates and induce a change of scales.

M. V. Pentkovskij [316] designed a series of skeletons, each of which is suitable for a different range of values of the functions $\varphi_1(x)$, $\varphi_2(y)$ and $\varphi_3(z)$.

We may improve the reading accuracy of a collineation nomogram by proper collinear transformations, projective or non-projective. Only the relations of nomographic order 3 admit non-projective collinear transformations.

A projective transformation of a collineation nomogram in cartesian coordinates may be defined by the equations

$$\xi_i^* = \frac{a_1\xi_i + b_1\eta_i + c_1}{a_3\xi_i + b_3\eta_i + c_3}, \quad \eta_i^* = \frac{a_2\xi_i + b_2\eta_i + c_2}{a_3\xi_i + b_3\eta_i + c_3} \quad (i = 1, 2, 3,), \tag{21}$$

where ξ_i and η_i are the original coordinates and ξ_i^* and η_i^* are the new coordinates (with the determinant of the transformation $[a_1b_2c_3] \neq 0$).

If we substitute for ξ_i and η_i the right-hand members of the original mapping equations, we obtain the new mapping equations which express the coordinates ξ_i^* and η_i^* in terms of the parameter of the scale. For example, by the transformation of the nomogram (12), the two equations $\xi_1 = 0$, $\eta_1 = \alpha h_1$ will be transformed into the equations

$$\xi_1^* = \frac{b_1\alpha h_1 + c_1}{b_3\alpha h_1 + c_3}, \quad \eta_1^* = \frac{b_2\alpha h_1 + c_2}{b_3\alpha h_1 + c_3}.$$

The choice of so-called *parameters of transformation* a_i, b_i, c_i $(i = 1, 2, 3)$ is very difficult when we want to obtain a nomogram with satisfactory reading accuracy. There are eight parameters to be chosen in equations (21) and it is not easy, as a rule, to estimate their influence on the resulting drawing. It has been found that not all parameters of a transformation significantly affect the character of the scales (i.e. changes in the length of the arc of the scale between neighbouring points of division) in the final drawing. M. V. Pentkovskij obtained excellent results when investigating the transformation of nomograms by introducing projective coordinates, so reducing the number of significant parameters to two (without losing the possibility of making the transformation serve a given purpose). He designed for applications so-called projective nets which facilitate the choice of suitable parameters. See [316].

MAPPING OF RELATIONS BETWEEN SEVERAL (MORE THAN THREE) VARIABLES. The method of grouping the nomograms of partial relations may also be used for collineation nomograms.

Example 5. The construction of a grouped collineation nomogram for the relation

$$C = S\varepsilon/4\pi d\,.$$

(Regarding the intervals, the meaning of variables and the auxiliary parameter u, see Example 32.3.3, p. 1194.)

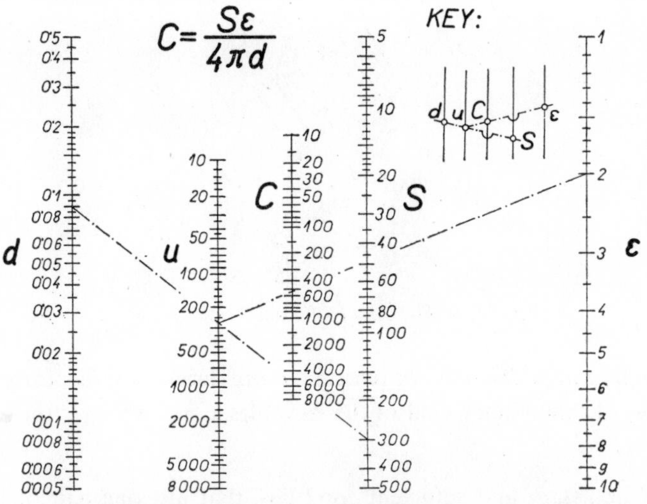

Fig. 32.18. The grouped collineation nomogram.

Resolution of the given relation into partial relations (in logarithmic form):

$$\text{I. } \log d - \log S + \log 4\pi u = 0\,, \qquad \text{II. } \log u + \log \varepsilon - \log C = 0\,.$$

Mapping equations for the construction of the nomogram in Fig. 32.18:

$$\begin{array}{ll} \text{I} & \text{II} \\ \xi_1 = 0\,,\ \eta_1 = 3\log d\,, & \xi_1' = 0\,,\ \eta_1 = -1{\cdot}5\log u \\ \xi_2 = 4\,,\ \eta_2 = -3\log S\,, & \xi_2' = 5\,,\ \eta_2 = -6\log \varepsilon\,, \\ \xi_3 = 2\,,\ \eta_3 = -1{\cdot}5\log 4\pi u\,; & \xi_3' = 1\,,\ \eta_3 = -1{\cdot}2\log C\,. \end{array}$$

For a nomographic mapping of relations of three or more variables we may use a further basic nomographic configuration, the so-called binary field:

Definition 6. A *binary field* of variables x and y is a system of points the coordinates of which are expressed by the parametric equations

$$\xi = f(x, y)\,, \quad \eta = g(x, y)\,, \tag{22}$$

and to every point of the system there corresponds a pair of its parameters (two labels).

In a nomographic drawing of a binary field two families of parametric curves (x_i) and (y_k) are plotted; the values x_i and y_k constituting two sequences of suitably chosen rounded-off numbers. See the sketch in Fig. 32.19.

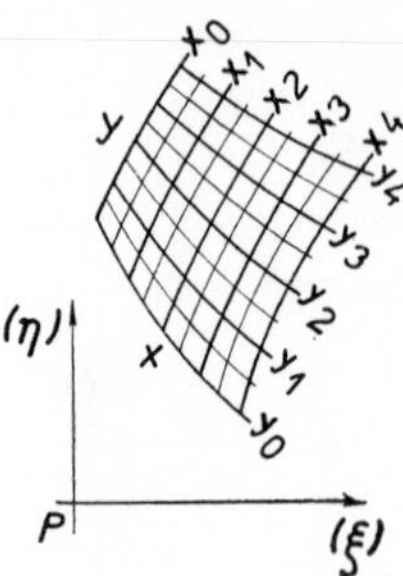

Fig. 32.19. The sketch of a binary field.

REMARK 4. Equations (22) may be written more briefly in the form $\xi = f_{1,2}$, $\eta = g_{1,2}$. Instead of "the binary field of the variables x, y," we say, "the binary field (x, y)".

REMARK 5. A necessary and sufficient condition that just one pair of parametric curves x, y passes through every point of a binary field is that

$$\begin{vmatrix} \dfrac{\partial f}{\partial x}, & \dfrac{\partial g}{\partial x} \\ \dfrac{\partial f}{\partial y}, & \dfrac{\partial g}{\partial y} \end{vmatrix} \neq 0 .$$

Theorem 12. *The "generalized" Cauchy canonical form which can be mapped into a collineation nomogram with a binary field is*

$$h_1 f_{3,4} + h_2 g_{3,4} + h_{3,4} = 0 . \tag{23}$$

The mapping equations of a generalized Cauchy's canonical form are:

$$\xi_1 = 0 , \qquad \eta_1 = \alpha h_1 ,$$

$$\xi_2 = \delta , \qquad \eta_2 = \beta h_2 ,$$

$$\xi_3 = \frac{\alpha \delta g_{3,4}}{\beta f_{3,4} + \alpha g_{3,4}} , \qquad \eta_3 = - \frac{\alpha \beta h_{3,4}}{\beta f_{3,4} + \alpha g_{3,4}} .$$

Example 6. The drawing of a collineation nomogram with a binary field for the relation

$$C = \frac{0{\cdot}0242\varepsilon}{\log(R/r)},$$

where $C \in [0{\cdot}02, 1]$ μF/km (the capacity of the cable),

$\varepsilon \in [1, 5]$ (the dielectric constant),

$R \in [10, 100]$ mm (the outer diameter of the cable),

$r \in [3, 30]$ mm (the inner diameter of the cable),

is shown in Fig. 32.20.

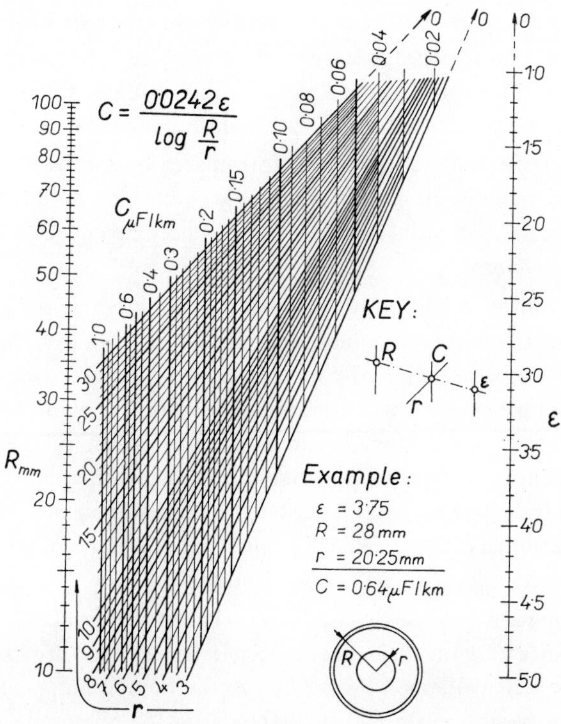

Fig. 32.20. A collineation nomogram with a binary field.

Comparison with the canonical form (23) gives

$$h_1 = \log R, \quad h_2 = -\varepsilon, \quad f_{3,4} = 1, \quad g_{3,4} = \frac{0{\cdot}0242}{C}, \quad h_{3,4} = -\log r.$$

Mapping equations with the scale moduli for Fig. 32.20:

$$\xi_1 = 0\,, \qquad \eta_1 = 7{\cdot}5 \log R\,,$$

$$\xi_2' = 6\,, \qquad \eta_2 = -2\varepsilon\,,$$

$$\xi_3' = \frac{7{\cdot}5\,.\,6\,.\,0{\cdot}0242}{2C + 7{\cdot}5\,.\,0{\cdot}0242} = \frac{10{,}890}{20{,}000C + 1815}\,,$$

$$\eta_3 = \frac{7{\cdot}5\,.\,2\,.\,C \log r}{2C + 7{\cdot}5\,.\,0{\cdot}0242} = \frac{150{,}000C \log r}{20{,}000C + 1815}\,.$$

32.5. Nomograms Using a Transparency

Definition 1. *A nomogram with a transparency (transparent sheet)* is a nomogram which is plotted on two superimposed planes: one plane is fixed (called the *basis* π) and the other plane is movable (called the *transparent plane* π').

We denote all elements plotted on π' with a dash.

A general theory of nomograms may be formulated by means of nomograms with transparencies and includes all classes of nomograms which are plotted on only one plane; systems of equations with several unknowns may be solved by means of nomograms using a transparency.

To solve the required relations by means of a nomogram using a transparency we bring the geometric elements which are plotted on π and π' into contact. The mutual position of two superimposed planes π and π' is characterized by three degrees of freedom. By a simple contact, when a definite point on the plane π falls on a definite curve on the plane π' (or vice versa), one degree of freedom of the planes π and π' is used up; by a double contact when a definite point on the plane π coincides with a definite point on the plane π', two degrees of freedom of the planes π and π' are used up. Three simple contacts define a fixed position of the plane π' with respect to the plane π (such contacts are called *positional contacts*).

The values of the variables which satisfy the relation represented by the nomogram are marked in the drawing on those geometric elements which are in contact and the choice of these contacts is defined by the key to the nomogram.

A scheme of a nomogram using a transparency for the relation

$$F(x_1, x_2, \ldots, x_{12}) = 0 \tag{1}$$

between twelve variables $x_1, \ldots, x_{12}$ may be described in the following way:

Let us consider four binary fields (x_1, x_2), (x_4, x_5), (x_7, x_8) and (x_{10}, x_{11}) on the basis π and four families of curves labelled with the values of variables x_3, x_6, x_9 and x_{12} on the transparent plane π'; see Fig. 32.21.

Given the values $x_1^0, \ldots, x_{11}^0$ of the first eleven variables, let us try to find the value x_{12}^0 of the twelfth variable from the relation (1). We superimpose the transparency π' on the basis π such that the curve $k'(x_3^0) \in \pi'$ passes through the point $P(x_1^0, x_2^0) \in \pi$, the curve $k'(x_6^0)$ passes through the point $P(x_4^0, x_5^0)$ and the curve $k'(x_9^0)$ passes through the point $P(x_7^0, x_8^0)$. These three positional contacts fix the position of the plane π'

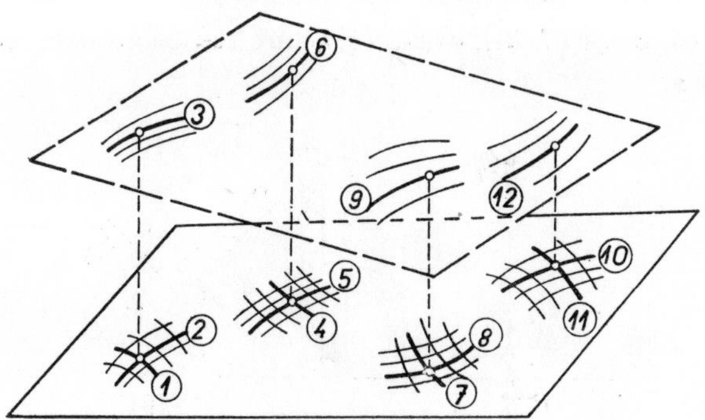

Fig. 32.21. Sketch of a nomogram with a transparency for the solution of a relation of twelve variables.

on the plane π. In this way we can obtain the "solving" contact of the point $P(x_{10}^0, x_{11}^0)$ with a curve $k'(x_{12}^0)$ whose label is the required value of the variable x_{12}.

A simple contact of a point $P(x_i, x_j) \in \pi$ with a curve $k'(x_k) \in \pi'$ is denoted by the symbol

$$P(x_i, x_j) \mapsto\!\!\!\!\dashv k'(x_k) \tag{2}$$

(the positional contact) or by the symbol

$$P(x_i, x_j) \to k'(x_k) \tag{2'}$$

(the solving contact).

The solution of a relation (1) by means of a nomogram using a transparency (with simple contacts) can be expressed, using the symbol, (2) and (2'), by the so-called *symbolic key* (or *structural formula*):

$$P(x_1, x_2) \mapsto\!\!\!\!\dashv k'(x_3)\,, \quad P(x_4, x_5) \mapsto\!\!\!\!\dashv k'(x_6)\,, \quad P(x_7, x_8) \mapsto\!\!\!\!\dashv k'(x_9)\,,$$
$$P(x_{10}, x_{11}) \to k'(x_{12})\,. \tag{3}$$

REMARK 1. If the points $P(x_i, x_j) \in \pi$, $P'(x_k, x_l) \in \pi'$ are coincident we have a double contact which is denoted by the symbol

$$P(x_i, x_j) \models\!\!\!\dashv P'(x_k, x_l)\,.$$

From a structural formula we derive functional relations which can be solved nomographically. The following are typical examples:

1. We express the position of a point of the binary field (x_i, x_j) in the plane π by parametric equations

$$\xi = f(x_i, x_j), \quad \eta = g(x_i, x_j) \tag{4}$$

($(i, j) = (1, 2)$ or $(4, 5)$ or $(7, 8)$), where (ξ, η) are the coordinates in the cartesian system $(P; \xi, \eta) \in \pi$.

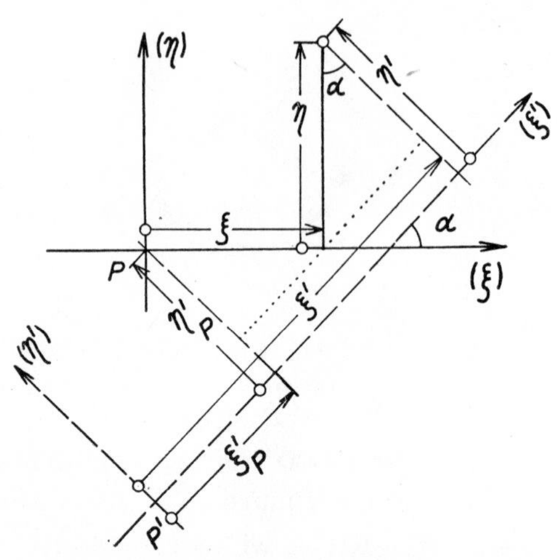

Fig. 32.22. The sketch for the transformation formulae (7).

2. We express the family of curves x_k in the plane π' by an equation with the parameter x_k:

$$\varphi_k(\xi', \eta', x_k) = 0, \tag{5}$$

where (ξ', η') are the coordinates in the cartesian system $(P'; \xi', \eta') \in \pi'$.

3. The positional contact of the point $P(x_i, x_j) \in \pi$ with the curve $k'(x_k) \in \pi'$ implies the validity of the equation

$$\varphi_k(\xi'_P + f(x_i, x_j) \cos \alpha + g(x_i, x_j) \sin \alpha, \eta'_P - f(x_i, x_j) \sin \alpha + g(x_i, x_j) \cos \alpha, x_k) = 0 \tag{6}$$

which we obtain from equation (5), the transformation formulae

$$\xi' = \xi'_P + \xi \cos \alpha + \eta \sin \alpha,$$

$$\eta' = \eta'_P - \xi \sin \alpha + \eta \cos \alpha \tag{7}$$

(see Fig. 32.22) and equations (3).

The triple of indices (i, j, k) in equation (6) is successively equal to (1, 2, 3), (4, 5, 6) and (7, 8, 9) so that equation (6) may be considered as a system of three equations with unknowns ξ'_P, η'_P, α. Let us denote the solution of system (6) by:

$$\xi'_P(x_1, \ldots, x_9)\,, \quad \eta'_P(x_1, \ldots, x_9)\,, \quad \alpha(x_1, \ldots, x_9) \tag{8}$$

(the values ξ'_P, η'_P, α depend on the values of the variables $x_1, \ldots, x_9$, which correspond to the elements that are just now in positional contact).

4. The solving contact of the point $P(x_{10}, x_{11}) \in \pi$ with the curve $k'(x_{12}) \in \pi'$ belonging to the family of curves (labelled with x_{12})

$$F(\xi', \eta', x_{12}) = 0$$

links the variables ξ'_P, η'_P, α, x_{10}, x_{11}, x_{12} by the following equation:

$$F(\xi'_P + f(x_{10}, x_{11}) \cos \alpha + g(x_{10}, x_{11}) \sin \alpha, \eta'_P - f(x_{10}, x_{11}) \sin \alpha + \\ + g(x_{10}, x_{11}) \cos \alpha, x_{12}) = 0\,. \tag{9}$$

If we substitute expressions (8) into (9) we obtain the type of relation between twelve variables which can be solved by a nomogram with a transparency using a symbolical key (3).

32.6. Principles of Graphical Analysis. Graphical Differentiation, Graphical Integration. Graphical Solution of Differential Equations

(a) Graphical Differentiation. Let a cartesian coordinate system $(O; \xi, \eta)$ be given and let the graph of a function $y = f(x)$ be constructed, in this system, according to mapping equations $\xi = \alpha x$, $\eta = \beta y$, respectively, α and β being scale moduli on the coordinate axes. In the system $(O; \xi, \eta)$, the equation of the graph becomes

$$\eta = \beta f\left(\frac{\xi}{\alpha}\right). \tag{1}$$

In Fig. 32.23, this graph is denoted by f. The graph of the derivative $y' = f'(x)$ of the given function can easily be constructed from the graph of f by a simple point construction which will be shown for the argument x, to which there corresponds on f the point $M(\alpha x, \beta f(x))$. Let us construct, at the point M, the tangent to the graph f and draw a parallel tangent through the point $P(-\delta, 0)$, the so-called *pole* with *polar distance* δ. The intersection of this parallel line with the η-axis is denoted by M'_1. Then, let us draw through M'_1 a line parallel with the ξ-axis and denote by M_1 its point of intersection with the parallel line to the η-axis through the point M. It can be easily seen from the triangle OPM'_1 that the η-coordinate η_1 of the point M_1 is equal to

$\delta\eta'$, where $\eta' = \tan\tau$ and τ is the angle between the above mentioned tangent and the ξ-axis. It follows from equation (1) that $\eta' = d\eta/d\xi = (\beta/\alpha)f'(\xi/\alpha)$. If we substitute the right-hand member of this equation into $\eta_1 = \delta\eta'$, we obtain

$$\eta_1 = \frac{\beta\delta}{\alpha} f'(x)\,. \tag{2}$$

It follows from (2) that the η-coordinate of the point M_1 corresponds to $f'(x)$ when the scale modulus is $\beta\delta/\alpha$. The value of the derivative at the point x is thus

$$f'(x) = \frac{\alpha\eta_1}{\beta\delta}\,. \tag{3}$$

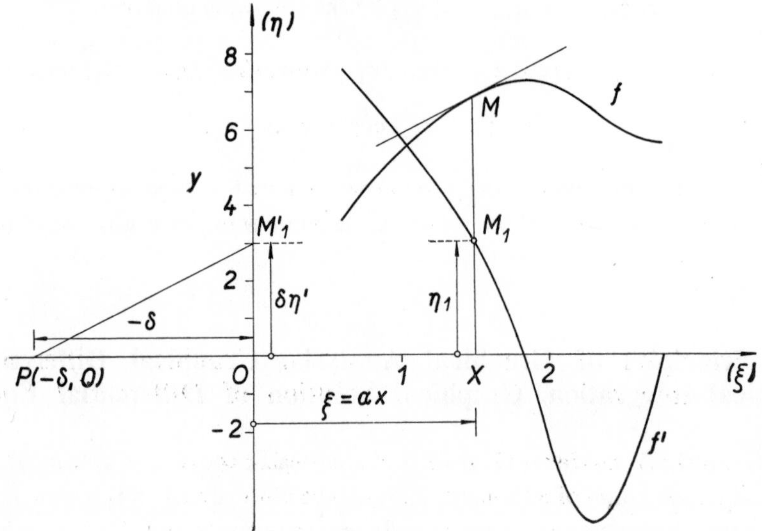

Fig. 32.23. The plotting of the graph of the derivative. The point $M_1(\alpha x, (\beta\delta/\alpha)f'(x))$ on the graph of the derivative $y = f'(x)$ is constructed from the point $M(\alpha x, \beta f(x))$ on the graph of the given function $y = f(x)$ (for the polar distance δ).

Example 1. In Fig. 32.23, the graph of a function $y = f(x)$ is given for scale moduli $\alpha = 2$, $\beta = 0{\cdot}5$. (Its equation in the coordinates ξ, η is, according to (1), $\eta = 0{\cdot}5 f(\xi/2)$.) The graph of the derivative $y' = f'(x)$, constructed for the polar distance $\delta = 3$, has, in the coordinates (ξ, η), the equation $\eta = 0{\cdot}5\,.\,\frac{3}{2}\,.\,f'(\xi/2)$ (see equation (2)), or $\eta = 0{\cdot}75 f'(\xi/2)$. For $x = 1{\cdot}5$, the corresponding point M_1 has coordinates $\xi = 3$, $\eta = 1{\cdot}5$. Thus, according to (3),

$$f'(1{\cdot}5) = 2\,.\,1{\cdot}5/(0{\cdot}5\,.\,3) = 2\,.$$

(b) Graphical Integration. As in the preceding case, let $\eta = \beta f(\xi/\alpha)$ be the equation of the graph of the function $y = f(x)$ in the cartesian coordinate system

$(O; \xi, \eta)$ when the mapping equations $\xi = \alpha x$, $\eta = \beta y$ are used (see equation (1)). Using the given graph (1) we have to construct approximately the graph of the integral

$$I(t) = \int_a^t f(x)\,dx \tag{4}$$

with variable upper limit t. We shall use the formula

$$\int_a^t f(x)\,dx \approx 2h_1 f_1 + 2h_2 f_3 + \ldots + 2h_n f_{2n-1} \tag{5}$$

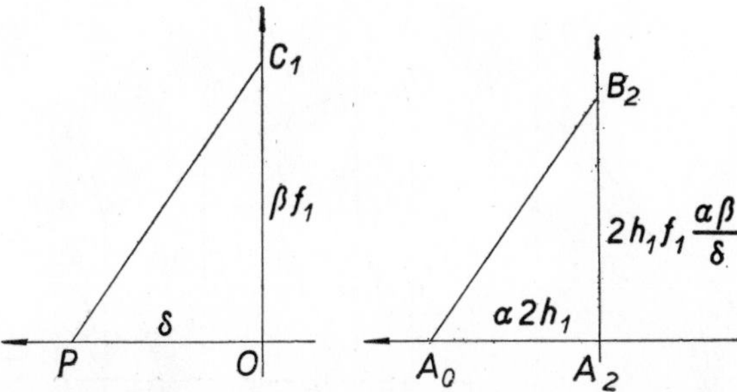

Fig. 32.24. The multiplication of two numbers (βf_1) and $(2h_1)$ constructed for the polar distance δ.

which gives the approximate value of (4). Formula (5) can be interpreted (for non-negative $f(x)$) as follows: The area of the figure whose boundary consists of the ξ-axis, the graph of the function $f(x)$ and ordinates through the points a, t, is approximately equal to the sum of areas of trapezoids with sides (the bases) on lines parallel to the η-axis, with altitudes $2h_1, 2h_2, \ldots, 2h_n$, the arithmetic means of the lengths of bases being $f_1, f_3, \ldots, f_{2n-1}$, respectively; thus, the whole area is divided into n strips with altitudes $2h_1, 2h_2, \ldots, 2h_n$. When constructing the integral graphically according to (5), we multiply successively the numbers $2h_1$ and f_1, $2h_2$ and f_3, ... $2h_n$ and f_{2n-1}, and sum the products so obtained.

The product of two numbers can be obtained by means of two angles with parallel arms as shown in Fig. 32.24, where $A_0A_2 = \alpha \,.\, 2h_1$, $OC_1 = \beta \,.\, f_1$, $OP = \delta$ (δ is the polar distance to the pole P). The similarity of triangles $\triangle POC_1$ and $\triangle A_0A_2B_2$ implies that

$$A_2B_2 = \frac{A_2A_0 \,.\, OC_1}{OP} \quad \text{or} \quad A_2B_2 = \frac{\alpha\beta}{\delta}\, 2h_1 f_1 \,.$$

The segment A_2B_2 represents the number $2h_1f_1$ for the scale modulus $\gamma = \alpha\beta/\delta$. The application of this construction may be seen from Fig. 32.25. The sum of products is performed simultaneously; for example, the product $2h_2f_3$ is added to the former product $2h_1f_1$ in such a manner that the parallel line, corresponding to the product $2h_2f_3$, is plotted directly through the point B_2 (Fig. 32.25) so that the segment A_4B_4 represents $\gamma(2h_1f_1 + 2h_2f_3)$, where $\gamma = \alpha\beta/\delta$. The result of the integration is given by the last segment $A_{2n}B_{2n}$. If we put $A_{2n}B_{2n} = u$, then

$$u = (\alpha\beta/\delta)\int_a^t f(x)\,dx\,,$$

hence

$$\int_a^t f(x)\,dx = \frac{u\delta}{\alpha\beta}\,. \tag{6}$$

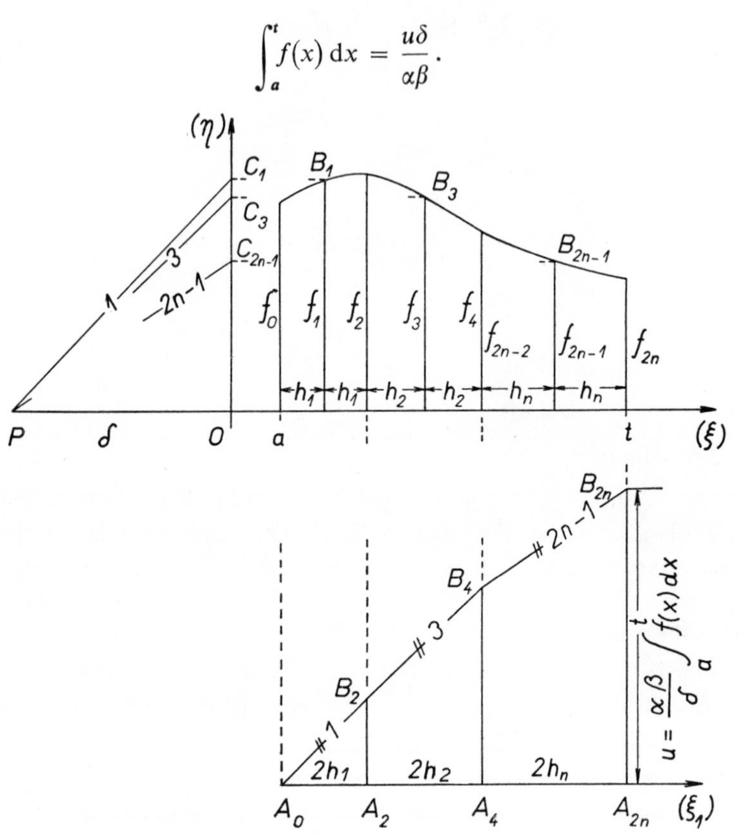

Fig. 32.25. Graphical integration.

Example 2. In Fig. 32.25, the graph of the function $y = f(x)$ is given for the scale moduli $\alpha = 0{\cdot}4$, $\beta = 1{\cdot}5$. (The equation of the graph in the coordinates (ξ, η) is $\eta = 1{\cdot}5\,f(\xi/0{\cdot}4)$.) The integral curve $\int_a^t f(x)\,dx$, where t is the variable upper limit, is given approximately by the polygonal line $A_0B_2B_4B_{2n}$. The second coordinate at the end point t of the domain of integration corresponds to an approximate value of the integral $\int_a^t f(x)\,dx$; let us denote this coordinate by u. In Fig. 32.25 $u = 3{\cdot}9$ and thus for the given $\alpha = 0{\cdot}4$, $\beta = 1{\cdot}5$ and $\delta = 3$, we have according to (6)

$$\int_a^t f(x)\,dx \approx \frac{3{\cdot}9\,.\,3}{0{\cdot}4\,.\,1{\cdot}5} = 19{\cdot}5 \quad (\mathrm{cm}^2)\,.$$

Further methods of graphical differentiation and integration, error estimates, mechanical integration and differentiation, etc., can be found in the literature (see References on p. 1183).

(c) Graphical Solution of Differential Equations. In the following text we mention here two very simple methods of solution of equations

(a) $$y' = f(x, y)\,, \tag{7}$$

(b) $$y'' = g(x, y, y')\,. \tag{8}$$

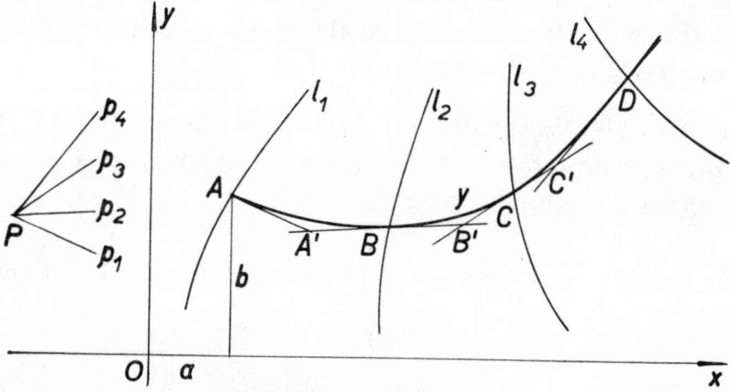

Fig. 32.26. The plotting of the integral curve of the equation $y' = f(x, y)$ with the condition $y(a) = b$ by the method of isoclines.

(a) *The method of isoclines.* Let us solve equation (7) with the initial condition

$$y(a) = b\,. \tag{9}$$

Equation (7) represents, as we know (§ 17·1), a field of line elements, i.e. to every point (x, y) of the domain of definition of the function $f(x, y)$ there corresponds a line element passing through the point considered and having the slope $y' = f(x, y)$ of the tangent to the integral curve going through this point. By drawing line elements at chosen points of the field, we can get a scheme of this field, from which, however, it is difficult to obtain a graphical solution with adequate accuracy. A simple way of solving is given by so-called *method of isoclines.*

Using this method, we let y' be successively equal to constants $h_1, h_2, \ldots, h_n$ and construct the curves

$$f(x, y) = h_i \quad (i = 1, 2, \ldots, n)\,. \tag{10}$$

Curves (10) are called the *isoclines*, for they join line elements of the same direction. Integral curves of (7) passing through the points of an isocline I_i have at every point the same slope h_i. One of the values h_i (say, h_k) is chosen so that on the corresponding isocline the condition $f(a, b) = h_k$ is fulfilled.

To the diagram of isoclines, we add a pencil of rays with the centre P; the rays have the slopes $\tan \alpha_1 = h_1, \ldots, \tan \alpha_n = h_n$. In Fig. 32.26 these rays are denoted by $p_1, \ldots, p_4$. At each point of an isocline I_i, the field of line elements has the same direction, determined by the ray p_i.

To solve equation (7) with the initial condition (9) we proceed as follows: Let the point $A(a, b)$ lie on the isocline I_1. Let us plot the line parallel to p_1 through the point A up to the point A' which lies "in the middle" of isoclines I_1 and I_2. Through the point A', we draw the line parallel to p_2 up to the point B' which also lies "in the middle" of isoclines I_2 and I_3, etc. These parallel lines intersect the isoclines $I_2, I_3, \ldots$ at the points $B, C, \ldots$. If the isoclines are plotted sufficiently densely, by joining the points $A, B, C, \ldots$ we obtain approximately the integral curve of equation (7) satisfying the initial condition (9). At the points $A, B, C, \ldots$, the parallel lines constructed are the tangents to this curve.

REMARK 1. If we plot the isoclines for the scale moduli α, β (cf. (1)), it is necessary to draw the rays p_i with the slopes $(\beta/\alpha)\, h_i$, since it follows from (1) that $\eta' = (\beta/\alpha) \,.\, f'(x)$ or $\eta' = (\beta/\alpha) \,.\, y'$ whence, using (7) and (10), $\eta' = (\beta/\alpha) \,.\, h_i$.

(b) *The Kelvin method.* Let us solve equation (8) with initial conditions

$$y(a) = b\,, \tag{11}$$

$$y'(a) = c\,. \tag{12}$$

We shall explain the method for scale moduli α, β, i.e. for scale equations

$$\xi = \alpha x\,, \quad \eta = \beta y\,. \tag{13}$$

Then at every point of the required integral curve we have

$$\frac{d\eta}{d\xi} = \frac{\beta}{\alpha}\, y' = \tan \tau \quad \left(-\tfrac{1}{2}\pi < \tau < \tfrac{1}{2}\pi\right), \tag{14}$$

$$\frac{1}{\varrho} = \frac{d^2\eta}{d\xi^2}\, \frac{1}{\left[1 + \left(\dfrac{d\eta}{d\xi}\right)^2\right]^{3/2}} = \frac{d^2\eta}{d\xi^2} \cos^3 \tau =$$

$$= \frac{\beta}{\alpha^2}\, y'' \cos^3 \tau = \frac{\beta}{\alpha^2}\, g\left(\frac{\xi}{\alpha}, \frac{\eta}{\beta}, \frac{\alpha}{\beta} \tan \tau\right) \cos^3 \tau\,, \tag{15}$$

where ϱ is the radius of curvature of the integral curve at the point considered.

From (15) it is obvious that if the point of the integral curve and the corresponding tangent (i.e. tan τ) at this point are known, then the radius of curvature at this point is also known. We therefore replace, in a neighbourhood of this point, the integral curve by an arc of the corresponding osculating circle at this point (Fig. 32.27).

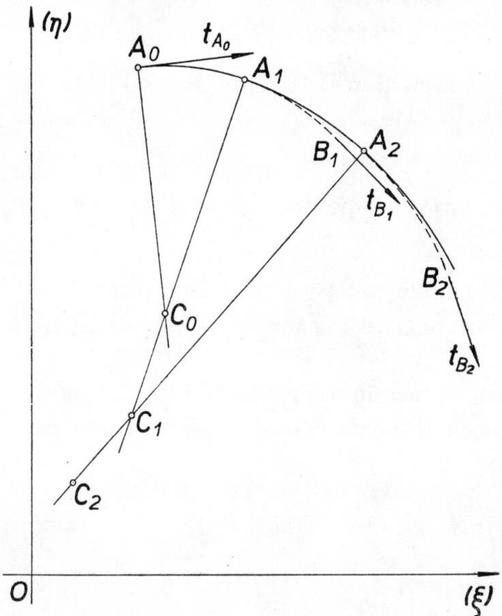

Fig. 32.27. The plotting of the solution of the equation $y'' = g(x, y, y')$ by the Kelvin method.

Condition (11) determines the initial point A_0 of the integral curve, and condition (12) determines the tangent at this point (i.e. τ_0 according to (14)). From (15), we determine ϱ_0 and put this ϱ_0 on the normal constructed at A_0 (i.e. on the perpendicular to t_{A_0}). We get a point C_0. (It is easy to resolve in what sense the ϱ_0 has to be put on the normal, since, depending on y'' being positive or negative, we know whether the curve is convex or concave at the point considered.) We describe a circular arc $\overset{\frown}{A_0A_1}$ round the point C_0 and extend it to the point B_1 (by about a half of the length which is intended to give to the next arc). At the point B_1 we plot the tangent t_{B_1}, by which τ_1 is known. By means of (15), we determine ϱ_1 and put it on the normal at the point A_1, so getting a point C_1. Then we describe a circular arc $\overset{\frown}{A_1A_2}$ round the point C_1, prolong it up to the point B_2, at which we plot the tangent t_{B_2}, then determine ϱ_2 etc. The curve consisting of the arcs $\overset{\frown}{A_0A_1}$, $\overset{\frown}{A_1A_2}$, ... is the approximation to the integral curve to be found (and is, of course, plotted for scale moduli α, β).

REMARK 2. It also is possible to solve graphically equations of higher orders, equations given implicitly (i.e. not solved with respect to the highest derivative), partial differential equations, etc.

B. INTERPOLATION. DIFFERENCES

32.7. The Statement of the Problem

The problem of interpolation consists of

(a) finding, for a given function $f(x)$, a polynomial of the n-th degree $P_n(x)$, which attains for $n + 1$ different values x_k ($k = 0, 1, \ldots, n$), of the independent variable x, the same values as the given function $f(x)$ (thus $P_n(x_k) = f(x_k)$ for $k = 0, 1, \ldots, n$);

(b) estimating the error at the points $x \neq x_k$ if we replace $f(x)$ by the polynomial $P_n(x)$;

(c) evaluating approximate values of the function $f(x)$ for $x \neq x_k$ using the polynomial $P_n(x)$, if $f(x)$ is tabulated for $x = x_k$ (for $k=0,1, \ldots, n$).

Definition 1. The above-mentioned polynomial $P_n(x)$ (put, eventually, into a form suitable for the evaluation of functional values) is called the *interpolation polynomial.*

Theorem 1. *There exists just one interpolation polynomial $P_n(x)$ which attains, at $n + 1$ different points, the same values as the given function $f(x)$.*

Definition 2. The equation

$$f(x) = P_n(x) + R_n(x), \tag{1}$$

where $P_n(x)$ is the interpolation polynomial and $R_n(x)$ is the *remainder term* (briefly the *remainder*), is called the *interpolation formula.* We write

$$f(x) \approx P_n(x). \tag{2}$$

If the arguments x_k are not equidistant, i.e. if the difference $x_{k+1} - x_k$ is not a constant number for all k (see Definition 32.10.1), the equation (1) is called the *general interpolation formula.*

The problem of interpolation generally occurs when

(a) the function $f(x)$ is tabulated and the interpolation formula (2) is more suitable for the evaluation of the values of the function $f(x)$ (or of its derivative) than the definition rule of the function (for example in the case of exponential, logarithmic and trigonometric functions);

(b) the definition rule (functional dependence) is of empirical nature; the interpolation polynomial is then used as an empirical formula by which the required dependence is replaced (if, of course, it may be replaced in this way).

32.8. Divided Differences

Definition 1. The *first divided difference* (the *divided difference of the first order*) is defined (for different x_μ, x_ν) by the relation

$$f(x_\mu, x_\nu) = \frac{f(x_\mu) - f(x_\nu)}{x_\mu - x_\nu} = \frac{f(x_\mu)}{x_\mu - x_\nu} + \frac{f(x_\nu)}{x_\nu - x_\mu}\,, \quad \mu \neq \nu\,;$$

the *second divided difference* (the *divided difference of the second order*) by

$$f(x_\mu, x_\nu, x_\lambda) = \frac{f(x_\mu, x_\nu) - f(x_\nu, x_\lambda)}{x_\mu - x_\lambda} =$$

$$= \frac{f(x_\mu)}{(x_\mu - x_\nu)(x_\mu - x_\lambda)} + \frac{f(x_\nu)}{(x_\nu - x_\mu)(x_\nu - x_\lambda)} + \frac{f(x_\lambda)}{(x_\lambda - x_\mu)(x_\lambda - x_\nu)}\,;$$

and similarly the *n-th divided difference* (the *divided difference of the n-th order*) is defined by

$$f(x_0, x_1, x_2, \ldots, x_n) = \frac{f(x_0, x_1, \ldots, x_{n-1}) - f(x_1, x_2, \ldots, x_n)}{x_0 - x_n} =$$

$$= \sum_{\nu=0}^{n} \frac{f(x_\nu)}{\prod\limits_{\substack{k=0\\ k\neq\nu}}^{n} (x_\nu - x_k)}\,.$$

Theorem 1. *Divided differences are symmetric functions of their arguments.* Thus, for example: $f(x_1, x_2, x_3) = f(x_1, x_3, x_2) = f(x_2, x_1, x_3) = f(x_2, x_3, x_1) = f(x_3, x_1, x_2) = f(x_3, x_2, x_1)$.

Theorem 2. *We introduce constants a, b by the definitions*

$$a = \min\,(x_0, x_1, \ldots, x_n)\,,$$
$$b = \max\,(x_0, x_1, \ldots, x_n)\,.$$

If the function $f(x)$ possesses an i-th derivative in the interval $[a, b]$, then the i-th divided difference can be expressed in terms of the i-th derivative at a point $\xi \in (a, b)$ as follows:

$$f(x_0, x_1, \ldots, x_i) = \frac{f^{(i)}(\xi)}{i!}\,, \quad a < \xi < b\,.$$

REMARK 1. The arguments $x_0, x_1, \ldots, x_n$ need not be ordered according to their value.

Definition 2. If $x_\mu = x_\nu$, we define

$$f(x_\nu, x_\nu) = \lim_{x_\mu \to x_\nu} \frac{f(x_\mu) - f(x_\nu)}{x_\mu - x_\nu} = f'(x_\nu)\,.$$

If $x_0 = x_1 = \ldots = x_\nu$, we define

$$\underbrace{f(x_\nu, x_\nu, \ldots, x_\nu)}_{(i+1)\text{-times}} = \frac{1}{i!} f^{(i)}(x_\nu) .$$

REMARK 2. Divided differences are often tabulated as shown in Tab. 32.1.

TABLE 32.1

x_k	$f(x_k)$	Divided difference		
		of the 1st order	of the 2nd order	of the 3rd order
x_0	$f(x_0)$			
		$f(x_0, x_1)$		
x_1	$f(x_1)$		$f(x_0, x_1, x_2)$	
		$f(x_1, x_2)$		$f(x_0, x_1, x_2, x_3)$
x_2	$f(x_2)$		$f(x_1, x_2, x_3)$	
		$f(x_2, x_3)$		
x_3	$f(x_3)$			

32.9. General Interpolation Formulae

(a) The Lagrange Interpolation Formula. From the form of the divided difference of the $(n + 1)$-th order,

$$f(x, x_0, x_1, \ldots, x_n) = \frac{f(x)}{(x - x_0)(x - x_1)\ldots(x - x_n)} +$$
$$+ \frac{f(x_0)}{(x_0 - x)(x_0 - x_1)\ldots(x_0 - x_n)} + \ldots + \frac{f(x_n)}{(x_n - x)(x_n - x_0)\ldots(x_n - x_{n-1})}$$

there follows the result

$$f(x) = \frac{(x - x_1)(x - x_2)\ldots(x - x_n)}{(x_0 - x_1)\ldots(x_0 - x_n)} f(x_0) +$$
$$+ \frac{(x - x_0)(x - x_2)\ldots(x - x_n)}{(x_1 - x_0)(x_1 - x_2)\ldots(x_1 - x_n)} f(x_1) + \ldots + \tag{1}$$
$$+ \frac{(x - x_0)\ldots(x - x_{n-1})}{(x_n - x_0)\ldots(x_n - x_{n-1})} f(x_n) + R_n ,$$

where

$$R_n = (x - x_0)(x - x_1)\dots(x - x_n) f(x, x_0, x_1, \dots, x_n)\,. \tag{2}$$

Definition 1. Equation (1) is called the *Lagrange interpolation formula* and the polynomial on its right-hand side the *Lagrange interpolation polynomial.*

If $f(x)$ possesses a derivative of the $(n + 1)$-th order in the interval $[a, b]$ (for a, b see Theorem 32.8.2), then

$$|R_n| \leq \frac{1}{(n+1)!} |(x - x_0)(x - x_1)\dots(x - x_n)| \max |f^{n+1}(\xi)|\,, \quad \xi \in [a, b]\,. \tag{3}$$

If we define

$$\varphi_i(x) = (x - x_0)\dots(x - x_{i-1})(x - x_{i+1})\dots(x - x_n) \quad (i = 0, 1, \dots, n)$$

(where the term $(x - x_{-1})$ or $(x - x_{n+1})$ is omitted when $i = 0$ or $i = n$, respectively) then equation (1) may be put in the form

$$f(x) = \sum_{i=0}^{n} \frac{\varphi_i(x)}{\varphi_i(x_i)} f(x_i) + R_n\,.$$

(b) The Newton General Interpolation Formula. The Lagrange interpolation formula, although very important in theoretical considerations, is not, however, suitable for numerical evaluation. The following form of the interpolation polynomial is more convenient:

$$P_n(x) = f(x_0) + (x - x_0) f(x_0, x_1) + (x - x_0)(x - x_1) f(x_0, x_1, x_2) + \\ + \dots + (x - x_0)\dots(x - x_{n-1}) f(x_0, \dots, x_n)\,. \tag{4}$$

Definition 2. The polynomial (4) is called the *Newton general interpolation polynomial.*

The Newton general interpolation formula is then of the form

$$f(x) = P_n(x) + R_n\,,$$

where the interpolation polynomial is of the form (4) and the remainder R_n is given by equation (2). The estimate for the remainder, when $f(x)$ possesses an $(n + 1)$-th derivative in an interval $[a, b]$ containing all the points $x_0, x_1, \dots, x_n$, is expressed by the inequality (3).

Example 1. Using Tab. 32.2, let us evaluate sin 35° and estimate the remainder.

TABLE 32.2

$\alpha°$	$\sin \alpha°$	The first divided difference	The second divided difference
0	0·000,00		
		1667	
30	0·500,00		− 6
		1381	
45	0·707,11		−11
		1059	
60	0·866,03		−13
		447	
90	1·000,00		

Estimation of the remainder. According to (3),

$$|R_2(x)| \leqq \left|\frac{(x - x_0)(x - x_1)(x - x_2)}{3!}\left[\sin\left(\frac{\pi x}{180}\right)'''\right]\right|$$

or

$$|R_2(35°)| \leqq \left|\frac{35 \,.\, 5 \,.\, 10}{6}\right| . \, 1 \, . \left(\frac{\pi}{180}\right)^3 \leqq 2 \,.\, 10^{-3} \,.$$

According to (4) we obtain

$$f(35°) \approx 0{\cdot}000{,}00 + 35[1667 + 5 \,.\, (-6)] \,.\, 10^{-5} = 35 \,.\, 1637 \,.\, 10^{-5} \,.$$

Hence $\sin 35° \approx 0{\cdot}572$, which agrees – to 2 decimal places – with the tabulated value. The example also shows how to evaluate the derivatives of the sine if the argument is expressed in degrees.

REMARK 1. The relation between the Newton interpolation formula and the Taylor formula at the point x_0 is as follows:

If we apply Theorem 32.8.2 to the Newton interpolation formula, we obtain the Taylor formula

$$f(x) = f(x_0) + \frac{f'(x_0)}{1!}(x - x_0) + \frac{f''(x_0)}{2!}(x - x_0)^2 + \ldots + \frac{f^{(n)}(x_0)}{n!}(x - x_0)^n + R_n(x),$$

where

$$R_n(x) = \frac{(x - x_0)^{n+1}}{(n + 1)!} f^{(n+1)}(\xi) \,, \quad \xi = x_0 + \vartheta(x - x_0), \quad 0 < \vartheta < 1 \,.$$

The interpolation polynomial, using the values $f(x_0), \ldots, f(x_n)$, approximates the function $f(x)$ in the interval $[a, b]$, containing the points $x_0, \ldots, x_n$, while the Taylor's polynomial using the values of the function $f(x)$ and of its n first derivatives at the point x_0, approximates the function $f(x)$ in the neighbourhood of the point x_0.

32.10. Interpolation Formulae for Equidistant (Equal) Arguments. Ordinary Differences. The Newton Interpolation Formula for Equidistant Arguments

Definition 1. The arguments x_k are called *equidistant* (or *equal*) if

$$x_k = x_0 + kh \quad (k \text{ integral}) ;$$

the number $h \neq 0$ is called the *step of the argument* (the *step of the table* with respect to a table); the argument x_0 is called *basic.*

For equidistant arguments, the following notation is used:

$$f(x_k) = f(x_0 + kh) = f_k .$$

Definition 2. The difference

$$\Delta f(x_k) = f(x_k + h) - f(x_k)$$

is called the *ordinary* (or *tabular*) *difference of the first order* (or the *first ordinary difference* or the *first forward difference*) *of the function* $f(x)$ *at the point* x_k. We write briefly

$$\Delta_k = f_{k+1} - f_k .$$

REMARK 1. The symbol Δ may be taken as an operator which, when applied to the function $f(x)$, defines a new function $\Delta f(x) = f(x + h) - f(x)$.

Definition 3. The *ordinary difference of the n-th order* (or the *n-th ordinary difference* or the *n-th forward difference*) *of the function* $f(x)$ *at the point* x_k is defined by the relation

$$\Delta^n f(x_k) = \Delta^{n-1} f(x_k + h) - \Delta^{n-1} f(x_k) \tag{1}$$

(or, briefly written

$$\Delta_k^n = \Delta_{k+1}^{n-1} - \Delta_k^{n-1} ,$$

where n is a non-negative integer; $\Delta_k^0 = f_k$ and $\Delta_k^1 = \Delta_k$).

Often only the term "difference" instead of "ordinary difference" or "forward difference" is used. (See also Remark 2.)

Example 1.

$$\Delta_0^2 = \Delta(\Delta_0) = \Delta(f_1 - f_0) = \Delta f_1 - \Delta f_0 = (f_2 - f_1) - (f_1 - f_0) =$$
$$= f_2 - 2f_1 + f_0 .$$

Theorem 1. *The following relations hold for operations with differences:*

$$\Delta(f_k \pm g_k) = \Delta f_k \pm \Delta g_k$$

(*and similarly for differences of higher orders*),

$$\Delta(f_k \cdot g_k) = \Delta f_k \cdot g_k + f_{k+1} \cdot \Delta g_k$$

(*since* $\Delta(f_k \cdot g_k) = f_{k+1}g_{k+1} - f_k g_k = (f_{k+1} - f_k)\, g_k + (g_{k+1} - g_k) f_{k+1}$). *Similarly*

$$\Delta\left(\frac{f_k}{g_k}\right) = \frac{\Delta f_k \cdot g_k - f_k \cdot \Delta g_k}{g_k g_{k+1}} .$$

Theorem 2. *The ordinary difference of the n-th order of the function $f(x)$ at the point x_0 can be expressed, using functional values only, in the form*

$$\Delta_0^n = f_n - \binom{n}{1} f_{n-1} + \binom{n}{2} f_{n-2} + \ldots + (-1)^n f_0 .$$

Theorem 3. *The following relation holds between ordinary differences and divided differences:*

$$\Delta_k^n = n!\, h^n f(x_k, x_k + h, \ldots, x_k + nh) .$$

TABLE 32.3

	$x_0 + kh$	f_k	Δ_k	Δ_k^2	Δ_k^3
	$x_0 - 3h$	f_{-3}			
			Δ_{-3}		
	$x_0 - 2h$	f_{-2}		Δ^2_{-3}	(so called
			Δ_{-2}		Δ^3_{-3} backward
	$x_0 - h$	f_{-1}		Δ^2_{-2}	differences)
			Δ_{-1}		Δ^3_{-2}
	x_0	f_0		Δ^2_{-1}	
$x \rightarrow$			Δ_0		Δ^3_{-1}
	$x_0 + h$	f_1		Δ^2_0	(so-called
			Δ_1		Δ^3_0 forward
	$x_0 + 2h$	f_2		Δ^2_1	differences)
			Δ_2		
	$x_0 + 3h$	f_3			

Theorem 4. *The following relation holds between derivatives and ordinary differences:*

$$\Delta_k^n = h^n f^{(n)}(\xi), \quad x_k < \xi < x_k + nh .$$

In Tab. 32.3 a scheme for computing ordinary differences up to the third order is shown.

THE NEWTON INTERPOLATION FORMULA FOR EQUIDISTANT ARGUMENTS. The interpolation formula for equidistant arguments may easily be derived from (32.9.4). Introducing a parameter t, we express x in the form

$$x = x_0 + th , \quad -1 \leqq t \leqq 1 .$$

The remaining expressions in the polynomial (32.9.4) are then replaced according to the relations

$$x_k = x_0 + kh \text{ (k an integer) and } f(x_0, x_0 + h, \ldots, x_0 + kh) = \frac{\Delta_0^k}{k!\, h^k} .$$

Definition 4. The *first Newton interpolation formula for equidistant arguments* (the so-called *Newton forward interpolation formula*, $t \geqq 0$), is defined as:

$$f_t = f_0 + \frac{t}{1!}\Delta_0 + \frac{t(t-1)}{2!}\Delta_0^2 + \ldots + \frac{t(t-1)\ldots(t-n+1)}{n!}\Delta_0^n + R_n , \quad (2)$$

or

$$f_t = f_0 + \binom{t}{1}\Delta_0 + \binom{t}{2}\Delta_0^2 + \ldots + \binom{t}{n}\Delta_0^n + R_n ,$$

where

$$R_n = \binom{t}{n+1} h^{n+1} f^{(n+1)}(\xi), \quad \xi \in (x_0, x_0 + nh) .$$

In briefer (symbolic) notation (using the operator Δ): $f_t = (1 + \Delta)^t f_0 + R_n$.

For calculation on machines, it is suitable to write formula (2) in the form of a Horner's scheme:

$$f_t = f_0 + t\left\{\Delta_0 + \frac{t-1}{2}\left[\Delta_0^2 + \frac{t-2}{3}\left(\Delta_0^3 + \ldots + \frac{t-n+1}{n}\Delta_0^n\right)\right]\right\} + R_n . \quad (3)$$

REMARK 2. Differences Δ_0^k used in (2) for forward interpolation are known as *forward differences* and lie on the downward sloping diagonal passing through f_0, as in Tab. 32.3.

Estimate of the remainder in (2):

$$|R_n(t)| \leqq \mu_n \,.\, h^{n+1} \,.\, M\,,$$

where μ_n (see Table 32.4) denotes the absolute value of the extremum of $\binom{t}{n+1}$ for $t \in (0, 1)$ and is tabulated for different n, and

$$M \geqq \max \left|f^{(n+1)}(\xi)\right|, \quad \xi \in (x_0, x_0 + nh)\,.$$

TABLE 32.4

n	1	2	3	4	5
μ_n	0·1250	0·0642	0·0417	0·0303	0·0235

If the oscillation of $f^{(n+1)}(x)$ is small, it is possible to write

$$h^{n+1} f^{(n+1)}(\xi) \approx \Delta_0^{n+1}\,, \quad \xi \in (x_0, x_0 + nh)\,.$$

The inaccuracy in evaluation of the interpolated value is influenced by the so-called *tabular inaccuracy* and by the rounding-off of the coefficients in the interpolation formula.

Example 2. The evaluation of log 106 on the basis of values from Tab. 32.5. (Here, log 106 stands for $\log_{10}$ 106.)

TABLE 32.5

x	$\log x$	Δ	Δ^2	Δ^3	Δ^4
100	2·000,000				
		21,189			
105	2·021,189		−985		
		20,204		86	
110	2·041,393		−899		− 9
		19,305		77	
115	2·060,698		−822		− 9
		18,483		68	
120	2·079,181		−754		−10
		17,729		58	
125	2·096,910		−696		
		17,033			
130	2·113,943				

This evaluation proceeds as follows:

$$x = x_0 + th\,, \quad x = 106\,, \quad x_0 = 105\,, \quad h = 5\,,$$

$$t = \frac{106 - 105}{5} = 0{\cdot}2\,,$$

$$f(x_0) = f_0 = \log 105 = 2{\cdot}021,\,189\,.$$

Substituting these results into (3), we get

$$\log 106 = 2{\cdot}021,\,189 + 0{\cdot}2\left\{20,\,204 - 0{\cdot}4\left[-\,899 - \frac{1{\cdot}8}{3}\left(77 - \frac{2{\cdot}8}{4}\,.\,(-9)\right)\right]\right\}.$$

$$.\,10^{-6} + R_n\,.$$

If a calculating machine is used, we proceed from the right to the left as shown in Tab. 32.6.

TABLE 32.6

$-0{\cdot}7\,.\,(-9)$	$= .\cdot \ldots\ldots 6$		$\lfloor 3$
$+$ 77	77		
( )	$=$	83	
$-0{\cdot}6\,.\,(\ \ )$	$=$	$-\ 50$	$\lfloor 8$
	$-899 =$	-899	
	[ ] $=$	-949	
$-0{\cdot}4\,.$	[ ] $=$	$+380$	$\lfloor 6$
$+$	$20{,}204 =$	20,204	
	{ }	20,584	
$0{\cdot}2\,.$	{ } $=$	4,117	$\lfloor 8$
$+2{\cdot}021{,}189$	$=$	$2{\cdot}021{,}189$	
	$\log 106 =$	$2{\cdot}025{,}306$	

The result is: $\log 106 \approx 2{\cdot}025{,}306$.

Estimate of the remainder:

$$|R_4| \leqq \mu_4 h^5 |\max f^{(5)}(\xi)| \leqq 0{\cdot}03\,.\,5^5\,.\,\frac{2\,.\,3\,.\,4\,.\,0{\cdot}434}{105^5} \approx 10^{-7}\,.$$

This result is influenced by the remainder in the sixth decimal place. Thus, it is determined with accuracy up to 6 decimals with the last digit rounded off.

REMARK 3. If the tabulation ends with f_0, the *second Newton interpolation formula* (the so-called *Newton backward interpolation formula*) is used ($x = x_0 - th$):

$$f_{-t} = f_0 - \binom{t}{1} \Delta_{-1} + \binom{t}{2} \Delta^2_{-2} - \binom{t}{3} \Delta^3_{-3} + \ldots + (-1)^n \binom{t}{n} \Delta^n_{-n} + R_n(t),$$

where

$$R_n(t) = \binom{t}{n+1} h^{n+1} f^{(n+1)}(\xi), \quad \xi \in (x_0 - nh, x_0),$$

or

$$f_{-t} = f_0 - t\left\{\Delta_{-1} - \frac{t-1}{2}\left[\Delta^2_{-2} - \frac{t-2}{3}\left(\Delta^3_{-3} - \ldots - \frac{t-n+1}{n}\Delta^n_{-n}\right)\right]\right\} + R_n.$$

REMARK 4. The differences Δ^k_{-k} (so-called *backward differences*) lie on the upward sloping diagonal passing through f_0, as shown in Tab. 32.3.

REMARK 5. Using the operator ∇ (*nabla*), defined by the equation

$$\nabla f(x) = f(x - h) - f(x),$$

the second Newton interpolation formula can be written in the form

$$f_{-t} = f_0 + \binom{t}{1} \nabla_0 + \binom{t}{2} \nabla_0^2 + \ldots + \binom{t}{n} \nabla_0^n + R_n(t)$$

or, more briefly,

$$f_{-t} = (1 + \nabla)^t f_0 + R_n.$$

The reader's attention is drawn to the fact that the definition of the operator ∇ is not unique in the literature.

32.11. Central-Difference Interpolation Formulae (Formulae of Gauss, Stirling, Bessel and Everett)

In different interpolation formulae, the so-called *central differences* are often used, defined as follows:

$$\delta f(x) = f\left(x + \frac{h}{2}\right) - f\left(x - \frac{h}{2}\right),$$

where h is a positive number. We put $\delta^n f(x) = \delta(\delta^{n-1} f(x))$ (n being a positive integer, $n > 1$).

Theorem 1. *Central differences are equal to ordinary differences taken with a proper index.* (*The subscript of the n-th central difference is the arithmetic mean of subscripts of those* $(n-1)$*-th differences from which the n-th difference is formed.*)

Example 1.

$$\delta_0 = f_{1/2} - f_{-1/2}\,,$$

$$\delta_{1/2} = f\left(x_0 + \frac{h}{2} + \frac{h}{2}\right) - f\left(x_0 + \frac{h}{2} - \frac{h}{2}\right) = f_1 - f_0 \Rightarrow \delta_{1/2} = \Delta_0\,;$$

similarly

$$\delta_{k+(1/2)} = f_{k+1} - f_k = \Delta_k\,;$$

in general

$$\delta^n_{k+(n/2)} = \Delta^n_k\,.$$

TABLE 32.7

$x_0 + kh$	f_k	δ	δ^2	δ^3	δ^4
$x_0 - 2h$	f_{-2}				
		$\delta_{-3/2}$			
$x_0 - h$	f_{-1}		δ^2_{-1}		
		$\delta_{-1/2}$		$\delta^3_{-1/2}$	
x_0	f_0		δ^2_0		δ^4_0
		$\delta_{1/2}$		$\delta^3_{1/2}$	
$x_0 + h$	f_1		δ^2_1		
		$\delta_{3/2}$			
$x_0 + 2h$	f_2				

(a) The Gauss Interpolation Formula.*

Definition 1. The interpolation formula

$$f_t = f_0 + \frac{t}{1!}\,\delta_{1/2} + \frac{t(t-1)}{2!}\,\delta^2_0 + \frac{t(t^2-1)}{3!}\,\delta^3_{1/2} + $$

$$+ \frac{t(t^2-1)(t-2)}{4!}\,\delta^4_0 + \frac{t(t^2-1)(t^2-4)}{5!}\,\delta^5_{1/2} + \ldots + R_n(t)\,,$$

where

$$R_n(t) = \frac{t(t^2-1)(t^2-4)\ldots}{(n+1)!}\,h^{n+1}f^{(n+1)}(\xi)\,,$$

* This is normally known as the Gauss Forward Interpolation Formula; there is a similar formula, the "Backward Interpolation Formula", involving the differences δ^0_{n+1} and $\delta^n_{-1/2}$

where $\xi \in \left(x_0 - \left[\frac{n}{2}\right] h,\ x_0 + \left[\frac{n+1}{2}\right] h\right)$ and $\left[\frac{n}{2}\right]$ denotes the integral part of the number $\frac{n}{2}$ (i.e. the symbol $\left[\frac{n}{2}\right]$ denotes the greatest integer that does not exceed $\frac{n}{2}$), is called the *Gauss interpolation formula* (*with central differences*).

The estimate of the remainder is:

$$|R_n| \leqq \mu_n h^{n+1} M$$

or

$$|R_n| \leqq \mu_n \delta_k^{n+1}$$

($k = 0$ if $n + 1$ is even and $k = \frac{1}{2}$ if $n + 1$ is odd).

The μ_n (given in Tab. 32.8) and M have the same meaning as in the case of the Newton formula (p. 1228).

TABLE 32.8

n	1	2	3	4	5	6	7
μ_n	0·125	0·0642	0·0234	0·0119	0·004,89	0·002,46	0·001,07

(b) The Stirling, Bessel and Everett Interpolation Formulae. The Stirling interpolation formula can easily be derived* from the Gauss formula by replacing odd differences by arithmetic means of these differences with the subscript zero according to the equation

$$(\delta_0^{2k+1}) = \frac{\delta_{1/2}^{2k+1} + \delta_{-1/2}^{2k+1}}{2} \quad (k = 0, 1, 2, \ldots)\,.$$

Definition 2. The formula

$$f_t = f_0 + t(\delta_0) + \frac{t^2}{2!}\,\delta_0^2 + \frac{t(t^2-1)}{3!}\,(\delta_0^3) + \frac{t^2(t^2-1)}{4!}\,\delta_0^4 + $$
$$+ \frac{t(t^2-1)(t^2-4)}{5!}\,(\delta_0^5) + \frac{t^2(t^2-1)(t^2-4)}{6!}\,\delta_0^6 + \ldots + R_n\,, \qquad (1)$$

is called the *Stirling interpolation formula*.

* It can also be very easily derived by combining the Gauss forward and backward formulae (see footnote p. 1231).

The remainder is the same as in the case of the Gauss formula.

In a similar way, the Bessel interpolation formula is derived by substituting arithmetic means of even differences

$$(\delta_{1/2}^{2k}) = \frac{\delta_0^{2k} + \delta_1^{2k}}{2} \quad (k = 1, 2, \ldots)$$

into the Gauss formula (and rearranging):

Definition 3. The formula

$$f_t = f_0 + t\,.\,\delta_{1/2} + \frac{t(t-1)}{2!}(\delta_{1/2}^2) + \frac{t(t-1)(t-\frac{1}{2})}{3!}\delta_{1/2}^3 +$$

$$+ \frac{t(t^2-1)(t-2)}{4!}(\delta_{1/2}^4) + \frac{t(t^2-1)(t-2)(t-\frac{1}{2})}{5!}\delta_{1/2}^5 + \ldots + R_n\,,$$

is called the *Bessel interpolation formula*. The coefficients of this formula are tabulated.

The estimate of the remainder can be obtained in the same way as in the case of the Stirling formula.

The Bessel formula will generally be preferable for $t = 0{\cdot}5$ (or for t very near to that value).

Example 2. Using the Stirling formula, let us determine the distance of the point P, with geographic latitude $\varphi = 10°30'$, from the equator, using Tab. 32.9, where S_M (measured in metres) denotes the distance between the equator and the corresponding geographic latitude on the international ellipsoid.

TABLE 32.9

$\varphi°$	S_M	δ	δ^2	δ^3	δ^4
8	884,661·57		5·37		
		110,599·87		0·65	
9	995,261·44		6·02		−0·02
		110,605·89		0·63	
10	1,105,867·33		6·65		0·02
		110,612·54		0·65	
11	1,216,479·87		7·30		

We have

$$x = x_0 + th\,, \quad x = 10°30'\,, \quad x_0 = 10°\,, \quad h = 1°\,,$$

$$t = \frac{x - x_0}{h} = \frac{10{\cdot}5° - 10°}{1°} = 0{\cdot}5\,,$$

$$f_0 = 1{,}105{,}867{\cdot}33\,,$$

$$(\delta_0) = \tfrac{1}{2}(110{,}612{\cdot}54 + 110{,}605{\cdot}89) = 110{,}609{\cdot}22\,,$$

$$(\delta_0^3) = \tfrac{1}{2}(0{\cdot}65 + 0{\cdot}63) = 0{\cdot}64\,.$$

Substituting these results into (1), we have

$$f_t = 1{,}105{,}867{\cdot}33 + 0{\cdot}5\,.\,110{,}609{\cdot}22 + 0{\cdot}125\,.\,6{\cdot}65 + \frac{0{\cdot}5(-0{\cdot}75)}{6}\,.\,0{\cdot}64 + R_3\,,$$

$$f_t \doteq 1{,}161{,}172{\cdot}73\,.$$

The estimate of the remainder is:

$$|R_3| \leqq \mu_3 |\delta_0^4|\,,$$

$$|R_3| \leqq 0{\cdot}0234\,.\,0{\cdot}02\,.$$

Thus, the result, evaluated up to two decimal places, has an adequate accuracy.

Definition 4. The formula

$$f(x_0 + th) = sf_0 + \binom{s+1}{3}\delta_0^2 + \binom{s+2}{5}\delta_0^4 + \binom{s+3}{7}\delta_0^6 + \ldots$$

$$+ tf_1 + \binom{t+1}{3}\delta_1^2 + \binom{t+2}{5}\delta_1^4 + \binom{t+3}{7}\delta_1^6 + \ldots + R_{n+1}(t)\,,$$

where $s = 1 - t$, is called the *Everett interpolation formula*.

The coefficients of the Everett formula are tabulated.

This formula may be derived from the Gauss interpolation formula by eliminating the odd differences. The remainder is estimated as in the case of the Gauss formula.

The scheme of a table for the Everett interpolation is shown in Tab. 32.10.

Remark 1. For the direct evaluation of differences, the formula

$$\delta_i^{2(k+1)} = \delta_{i+1}^{2k} - 2\delta_i^{2k} + \delta_{i-1}^{2k} \quad (k = 0, 1, \ldots)$$

may be used (where the differences of order zero are directly the functional values).

TABLE 32.10

$x_0 + kh$	f_k	δ_k^2	δ_k^4	δ_k^6
x_0	f_0	δ_0^2	δ_0^4	δ_0^6
$x_0 + h$	f_1	δ_1^2	δ_1^4	δ_1^6

For calculation on machines, the Everett formula may be suitably written in the form of a Horner scheme:

$$f(x_0 + th) = sf_0 + tf_1 - \frac{st}{6}\Bigg\{(s + 1)\,\delta_0^2 + (t + 1)\,\delta_1^2 -$$

$$- \frac{(s + 1)\,.\,(t + 1)}{20}\Bigg[(s + 2)\,\delta_0^4 + (t + 2)\,\delta_1^4 - \qquad (2)$$

$$- \frac{(s + 2)\,.\,(t + 2)}{42}\left((s + 3)\,\delta_0^6 + (t + 3)\,\delta_1^6 - \ldots\right)\Bigg]\Bigg\} + R_{n+1}\,.$$

This formula is often used for $n = 2$, so that

$$f(x_0 + th) = sf_0 + tf_1 - \frac{st}{6}\left[(s + 1)\,\delta_0^2 + (t + 1)\,\delta_1^2\right] + R_3\,.$$

TABLE 32.11

N	$\log N$	δ^2	δ^4
538·00	2·73078 22756 66389 17530	1 50044 38924	31
538·01	2·73079 03479 79507 37597	1 50038 81153	31

Example 3. Let us evaluate $\log N$ for $N = 538{\cdot}001$, using Tab. 32.11. (Here, $\log N$ stands for $\log_{10} N$.)

We have

$$x = 538{\cdot}001\,, \quad x_0 = 538{\cdot}00\,, \quad h = 0{\cdot}01\,,$$

$$t = \frac{x - x_0}{h} = \frac{538{\cdot}001 - 538{\cdot}00}{0{\cdot}01} = 0{\cdot}1\,, \quad s = 1 - 0{\cdot}1 = 0{\cdot}9\,.$$

Substituting these results into (2), we have that

$$f_t = 0{\cdot}9 \,.\, 2{\cdot}73078\ 22756\ 66389\ 17530 + 0{\cdot}1 \,.\, 2{\cdot}73079\ 03479\ 79507\ 37597 -$$
$$- \tfrac{1}{6} \,.\, 0{\cdot}1 \,.\, 0{\cdot}9\{1{\cdot}9 \,.\, 1\ 50044\ 38924 + 1{\cdot}1 \,.\, 1\ 50038\ 81153 -$$
$$- \tfrac{1}{20} \,.\, 1{\cdot}9 \,.\, 1{\cdot}1[2{\cdot}9 \,.\, 31 + 2{\cdot}1 \,.\, 31]\} \,.\, 10^{-20} + R_3\,;$$

$$\log 538{\cdot}001 \approx 2{\cdot}73078\ 30828\ 90949\ 08989\,.$$

By estimating the remainder, it may be verified that the procedure of evaluation to twenty decimal places has been carried out with a sufficient accuracy.

32.12. Linear Interpolation

Definition 1. By an *interpolation formula for linear interpolation* we mean a formula containing an interpolation polynomial of the first degree; thus

$$f(x_0 + th) = f_0 + t\Delta_0 + R_1 \quad (\Delta_0 = f(x_0 + h) - f(x_0))\,, \tag{1}$$

where

$$|R_1| \leq \left|\frac{t(t-1)}{2}\right| h^2 \max f''(\xi)\,, \quad \xi \in (x_0, x_0 + h)\,.$$

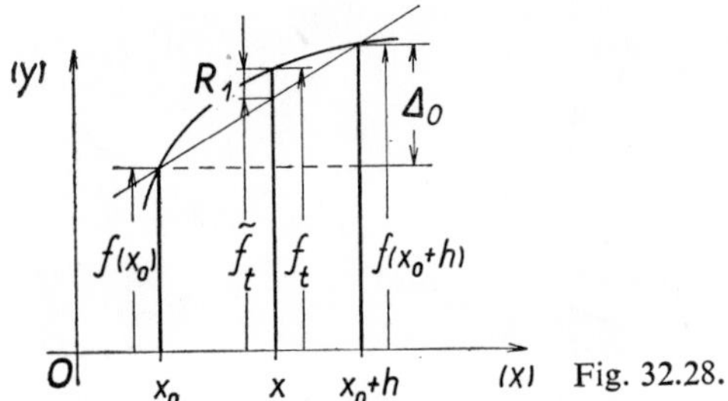

Fig. 32.28.

The remainder is usually estimated as follows:

$$|R_1| \leq \tfrac{1}{8}h^2 \max |f''(\xi)| \quad (\mu_1 = \tfrac{1}{8})\,. \tag{2}$$

Geometrically interpreted, linear interpolation means replacing the graph of the function $f(x)$ with a chord passing through the points $(x_0, f_0), (x_0 + h, f(x_0 + h))$ (Fig. 32.28).

REMARK 1. If a function is to be tabulated so that linear interpolation is permissible we require that the inequality

$$|R_1| < 0{\cdot}5 \,.\, 10^{-k}$$

be satisfied where k is the order of the last digit of tabulated functional values. In practice, however, in order to guarantee sufficient accuracy, it is required that

$$|R_1| < 0{\cdot}1 \,.\, 10^{-k}$$

and from this condition (using (2)), the length of the step h is determined.

32.13. Applications of Interpolation to Approximate Solution of Equations. The Regula Falsi Method, the Iterative Method

Let us solve the following problem: To find x if the value of $f(x)$ is given.

(a) Determination of x by the "regula falsi method"*. Using formula (32.12.1), $f_t \approx f_0 + t\Delta_0$, we determine first the approximate value $t = {}^1t = (f_t - f_0):\Delta_0$ and hence the approximate value ${}^1x = x_0 + {}^1th$ (Fig. 32.29a).

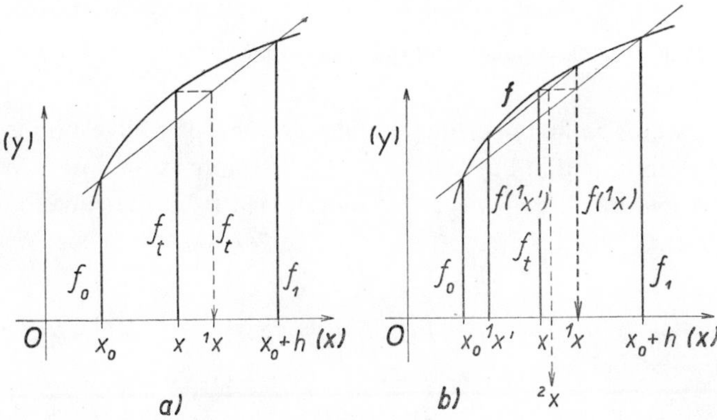

Fig. 32.29. The determination of the argument x by the "regula falsi" method.

We then evaluate (using the interpolation) the value $f(^1x)$. If, for example, $f(^1x) > f_t$, we evaluate $f(^1x')$ with $^1x'$ such that $f(^1x') < f_t$ (Fig. 32.29b), so that

$${}^2t = \frac{f_t - f(^1x')}{f(^1x) - f(^1x')}, \quad {}^2x = {}^1x' + {}^2t(^1x - {}^1x')\,.$$

The calculation is repeated until a further calculation leaves the required number of decimals unchanged.

(b) The iterative method. Applying some interpolation formula, we try to find the value of t corresponding to the given value f_t, using the well-known iterative method for the solution of the equation $F(x) = 0$. The procedure is as follows:

---

* Also known as the method of "false position".

The equation $F(x) = 0$ is put into the form $F_1(x) - F_2(x) = 0$, where $F_1(x)$ and $F_2(x)$ are chosen in such a way that $|F_1'(x)| > |F_2'(x)|$ and the equation $F_1(x) = a$ is easily solvable.

At the point $x = \alpha$, which is the root of the equation $F(x) = 0$, the relation $F_1(\alpha) = F_2(\alpha)$ holds, showing that, at this point the graphs of functions $y = F_1(x)$ and

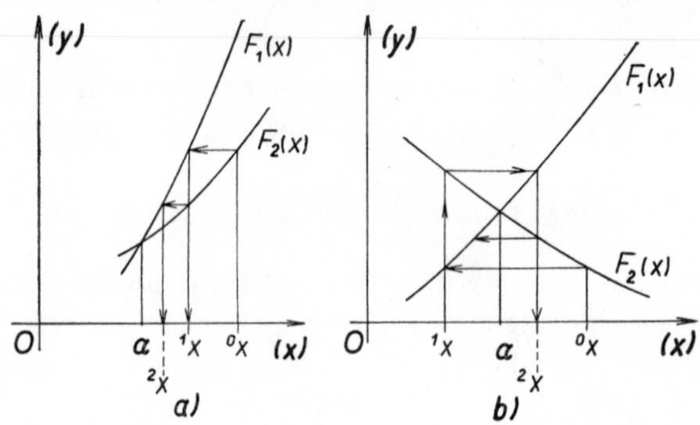

Fig. 32.30. The determination of the argument x by the iterative method.

$y = F_2(x)$ intersect. The first coordinate α of this point of intersection is successively approximated as indicated in Fig. 32.30a,b: the first approximation ^0x being chosen, and the approximations ^kx $(k = 1, 2, \ldots)$ determined from the equations

$$F_1(^1x) = F_2(^0x) \quad \Rightarrow {^1x} = \ldots,$$
$$F_1(^2x) = F_2(^1x) \quad \Rightarrow {^2x} = \ldots,$$
$$\cdots\cdots\cdots\cdots\cdots\cdots\cdots\cdots$$
$$F_1(^nx) = F_2(^{n-1}x) \Rightarrow {^nx} = \ldots$$

(Fig. 32.30a, b). In this way we obtain a sequence

$$^0x,\ ^1x,\ ^2x,\ \ldots,\ ^nx$$

the limit of which (if it exists) gives the required root of the equation $F(x) = 0$.

Example 1. Let us apply the iterative method using the Gauss interpolation polynomial of the third degree.

Rearranging the polynomial, we obtain

$$t\delta_{1/2} = f_t - f_0 - \left[\frac{t(t-1)}{2!}\,\delta_0^2 + \frac{t(t^2-1)}{3!}\,\delta_{1/2}^3\right] \tag{1}$$

which corresponds to the equation $F_1(x) = F_2(x)$.

Initially ${}^0t = 0$ is substituted in the right-hand member, giving

$$ {}^1t = \frac{f_t - f_0}{\delta_{1/2}}, \quad {}^1t = \dots . $$

This newly determined value is then substituted into the right-hand member of (1) giving ${}^2t$ in the same way, and the process is repeated.

The k-th step then gives the approximation ${}^kx = x_0 + {}^kth$.

TABLE 32.12

x	$\log \Gamma(x)$	δ	δ^2	δ^3	δ^4
1·200	9·962,922,503,814		550,415		−2
		−125,253,332		−640	
1·201	9·962,797,250,482		549,775		−2
		−124,703,557		−642	
1·202	9·962,672,546,925		549,133		

Example 2. Using Tab. 32.12 and the Gauss interpolation polynomial, let us determine the argument x of the value $\log \Gamma(x) = 9{\cdot}962{,}811{,}110{,}000$ ($\log \Gamma(x)$ stands here for $\log_{10} \Gamma(x)$; the characteristic -10 need not be considered in carrying out the evaluation).

We have

$$ \begin{aligned} f_0 &= 9{\cdot}962{,}922{,}503{,}814\,, & \delta_{1/2} &= -125{,}253{,}332\,.\,10^{-12}\,, \\ f_t &= 9{\cdot}962{,}811{,}110{,}000\,, & \delta_0^2 &= 550{,}415\,.\,10^{-12}\,, \\ f_t - f_0 &= -0{\cdot}000{,}111{,}393{,}814\,; & \delta_{1/2}^3 &= -640\,.\,10^{-12}\,, \\ & & \delta_0^4 &= -\,2\,.\,10^{-12}\,, \end{aligned} $$

so that

$$ {}^1t = \frac{111{,}393{,}814}{125{,}253{,}332} \Rightarrow {}^1t = 0{\cdot}889{,}3\,, $$

$$ {}^2t = 0{\cdot}889{,}3 + \frac{0{\cdot}889{,}3\,.\,0{\cdot}110{,}7}{2(-125{,}253{,}332)}\,.\,550{,}415 \Rightarrow {}^2t = 0{\cdot}889{,}09\,, $$

$$ {}^2x = x_0 + {}^2th \Rightarrow {}^2x = 1{\cdot}200{,}889\,. $$

32.14. Tabulation of Functions of Two Variables

If a function $f(x, y)$ of two arguments x, y is to be tabulated, we write the value of x in the first column and the values of y in the first row of the table. The values $f(x, y)$, corresponding to the arguments x_i, y_i, are then written in the place where the i-th row (with the argument x_i) and the j-th column (with argument y_j) intersect. The tabulation is usually carried out for equidistant arguments. Denoting the basic arguments by (x_0, y_0) with h and k as the steps of the arguments and $f_{i,j} = f(x_0 + ih, y_0 + jk)$, where i, j are integers, we can present the following table:

TABLE 32.13

x \ y	$y_0 - 2k$	$y_0 - k$	y_0	$y_0 + k$	$y_0 + 2k$
$x_0 - 2h$	$f_{-2,-2}$	$f_{-2,-1}$	$f_{-2,0}$	$f_{-2,1}$	$f_{-2,2}$
$x_0 - h$	$f_{-1,-2}$	$f_{-1,-1}$	$f_{-1,0}$	$f_{-1,1}$	$f_{-1,2}$
x_0	$f_{0,-2}$	$f_{0,-1}$	$f_{0,0}$	$f_{0,1}$	$f_{0,2}$
$x_0 + h$	$f_{1,-2}$	$f_{1,-1}$	$f_{1,0}$	$f_{1,1}$	$f_{1,2}$
$x_0 + 2h$	$f_{2,-2}$	$f_{2,-1}$	$f_{2,0}$	$f_{2,1}$	$f_{2,2}$

DOUBLE LINEAR INTERPOLATION. Let us assume that, in the table of two arguments, the values of the function $f(x, y)$ for the arguments x_0, $x_0 + h$ and y_0, $y_0 + k$, are given and that we have to evaluate the value of the function at the point $(x = x_0 + th, y = y_0 + pk)$, where h and k are the corresponding steps in the table and $t, p \in (0,1)$. The value $f(x_0 + th, y_0 + pk)$ is denoted by $f_{t,p}$.

TABLE 32.14

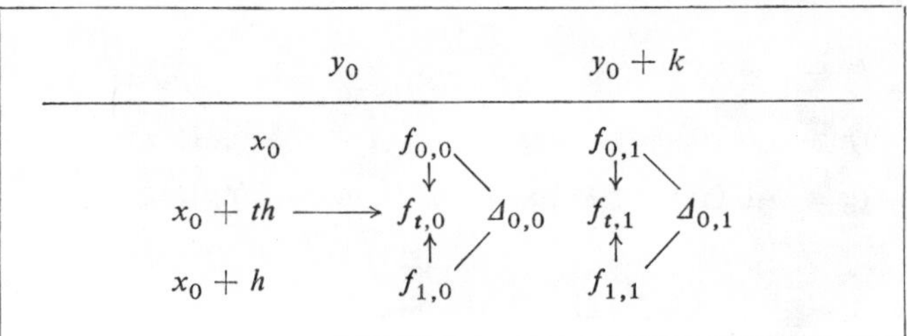

	y_0		$y_0 + k$	
x_0	$f_{0,0}$		$f_{0,1}$	
$x_0 + th$ →	$f_{t,0}$	$\Delta_{0,0}$	$f_{t,1}$	$\Delta_{0,1}$
$x_0 + h$	$f_{1,0}$		$f_{1,1}$	

TABLE 32.15

	y_0	$y_0 + pk$	$y_0 + k$
x_0	$f_{0,0}$		$f_{0,1}$
	$f_{t,0}$ →	$f_{t,p}$	← $f_{t,1}$
$x_0 + h$	$f_{1,0}$	$\Delta'_{t,0}$	$f_{1,1}$

The procedure of evaluation. In the zero column (for $j = 0$), we evaluate, using linear interpolation, the value $f_{t,0}$, and in the same way the value $f_{t,1}$ in the first column ($j = 1$). Then, we apply linear interpolation in the row, where we have already found $f_{t,0}$ and $f_{t,1}$ in order to obtain the value of $f_{t,p}$. Both steps are easily seen from the table for the general case as well as from Tab. 32.14 and 32.15, and from the formulae for interpolation on columns

$$f_{t,0} = f_{0,0} + t\Delta_{0,0} + {}^0R_1\,, \quad \text{where} \quad \Delta_{0,0} = f_{1,0} - f_{0,0}\,,$$

$$f_{t,1} = f_{0,1} + t\Delta_{0,1} + {}^1R_1\,, \quad \text{where} \quad \Delta_{0,1} = f_{1,1} - f_{0,1} \tag{1}$$

and for interpolation in rows

$$f_{t,p} = f_{t,0} + p\Delta'_{t,0} + {}^tR_1\,, \quad \text{where} \quad \Delta'_{t,0} = f_{t,1} - f_{t,0}\,. \tag{2}$$

The remainders ${}^0R_1$, ${}^1R_1$ and ${}^tR_1$ are estimated by a similar method to that used in the case of functions of one argument.

Example 1. The geodetic quantity R, called *geometric rigidity* Δ, is given as a function of α, β by Tab. 32.16:

TABLE 32.16

α \ β	40°	45°	50°
50°	13·8	11·3	9·3
55°	12·2	9·8	7·9
60°	10·9	8·5	6·7
65°	9·7	7·5	5·8
70°	8·8	6·7	5·0

Let us evaluate, using double linear interpolation, the value of R for $\alpha = 62°$, $\beta = 44°$.

We evaluate R, using linear interpolation, for $\alpha = 62°$ in the column for $\beta = 40°$ (i.e. $R(62°, 40°)$), and in the column for $\beta = 45°$ (i.e. $R(62°, 45°)$). Then, again using linear interpolation, we evaluate $R(62°, 44°)$ on the basis of the values $R(62°, 40°)$ and $R(62°, 45°)$. We have that

$$x = x_0 + th \Rightarrow t = \frac{x - x_0}{h} = \frac{62° - 60°}{5°} = 0{\cdot}4\,,$$

in the column for $\beta = 40°$ — in the column for $\beta = 45°$

according to (1)

$$R(62°, 40°) \approx 10{\cdot}9 + 0{\cdot}4(-1{\cdot}2) \doteq \doteq 10{\cdot}9 - 0{\cdot}5 = 10{\cdot}4\,,$$

$$R(62°, 45°) \approx 8{\cdot}5 + 0{\cdot}4(-1) = = 8{\cdot}5 - 0{\cdot}4 = 8{\cdot}1\,.$$

The linear interpolation in the 62° row according to (2):

$$y = y_0 + p\,.\,k \Rightarrow p = \frac{y - y_0}{k} = \frac{44° - 40°}{5°} = 0{\cdot}8\,,$$

$$R(62°, 44°) \approx 10{\cdot}4 + 0{\cdot}8(-2{\cdot}3) = 8{\cdot}56\,.$$

32.15. Arithmetic Operations with Approximate Numbers

In numerical calculations we often operate with numbers the exact values of which are not known: we know only that these numbers lie in a certain interval. This is caused partly by our need to work from inaccurate numbers (e.g. empirical or measured values), and partly by the limited digit capacity of calculating machines (so that we are compelled to round off the given numbers to a smaller number of digits).

The fact that a number x is *approximate* is expressed in the following way:

$$x = X \pm A\,, \tag{1}$$

where X is the *mean approximation of the number* x and A is the so-called *estimate of (the absolute value of) the error of the number* X. The notation (1) states that x lies in the interval $[X - A, X + A]$, that is

$$X - A \leqq x \leqq X + A\,. \tag{2}$$

The estimate of the error is a positive number generally calculated to one or two significant figures. The number X may be calculated such that the last figures in X and A are of the same order. In the case where the number X is rounded off, we have $A = 0{\cdot}5\,.\,10^{-m}$, where m is the number of decimal places of the number X. If there is no danger of misunderstanding we do not write $\pm A$ behind X.

The *estimate of the relative error* is given by the ratio $R = A/|x|$. It does not depend on the order of the number X but it depends substantially on the number of figures. It is one hundreth of the estimation of so-called *percentage error*.

Example 1. Let the distance x between two points be obtained by measurement where $x = 526{\cdot}437 \pm 0{\cdot}016$ m. The estimate of the relative error is $0{\cdot}016/526{\cdot}437 \approx \approx 0{\cdot}016/530 \doteq 3\,.\,10^{-5}$, that is 0·003%.

THE ESTIMATE OF THE ABSOLUTE ERROR OF A SUM OR A DIFFERENCE. In addition and subtraction, the estimates of the absolute errors are summed; thus

$$A(x + y) = A(x) + A(y)\,, \quad A(x - y) = A(x) + A(y)\,. \tag{3}$$

Corollary: In addition or subtraction, the sum, or the difference, has at most as many significant decimal places as the summand with the least number of significant decimal places.

THE ESTIMATE OF THE RELATIVE ERROR OF A PRODUCT OR A QUOTIENT. In multiplication and division, the estimates of the relative errors are added; thus

$$R(xy) = R(x) + R(y)\,, \quad R(x/y) = R(x) + R(y)\,. \tag{4}$$

Corollary: In multiplication or division, the product or the quotient, has at most as many significant figures as the factor with the least number of significant figures.

Example 2. In § 32.14, an exposition of how to tabulate the functional values has been given. Let the tabular values have m decimal places. Thus, the estimate of the error is $0{\cdot}5 \,.\, 10^{-m}$ (so-called *tabular inaccuracy*). The inaccuracy of the first differences lies between the limits $\pm 10^{-m}$, the inaccuracy of the second differences between the limits $\pm 2 \,.\, 10^{-m}$, etc. Therefore, in the case of two-place numbers the round-off errors may be very substantial.

Example 3. When calculating the interpolation coefficient t (by the formula $t = (x - x_0)/h$, p. 1235), we may round off this coefficient to as many figures as the first difference (or one place more). Every term in the interpolation formula may only be calculated to as many places as the corresponding difference (or one place more).

THE ESTIMATE OF THE ABSOLUTE AND RELATIVE ERRORS OF A FUNCTIONAL VALUE. If the argument x is an approximate number, then the accuracy $A(f)$ of the corresponding functional value is given by the formula

$$A(f) = |f'(x)|\, A(x) \tag{5}$$

(as follows from the expression for the differential of a function).

The estimate of the relative error of the functional value is

$$R(f) = \frac{A(f)}{|f|} = \frac{|f'(x)|}{|f(x)|}\, A(x)\,. \tag{6}$$

The following useful results follow from formulae (5) and (6):

1. The absolute accuracy of the common logarithm is approximately half of the relative accuracy of the argument, since $A(\log x) \approx (0{\cdot}434/x) \,.\, A(x) \approx 0{\cdot}5R(x)$. There-

fore, we have to take at least k-place logarithms for a calculation with numbers which have k figures.

2. The relative error of the n-th power is n times greater than the relative error of the argument. That is,

$$R(x^n) = \left|\frac{nx^{n-1}}{x^n}\right| A(x) = nR(x)\,.$$

The argument (and also the estimate of the accuracy) of trigonometric functions is often given in sexagesimal measure of angles (e.g. in seconds). Then the following formulae hold for the estimate of the absolute error of trigonometric functions and its logarithms ($\varrho'' \doteq 206{,}265''$):

$$A(\sin x) = \frac{|\cos x|\, A''(x)}{\varrho''}, \quad A(\cos x) = \frac{|\sin x|\, A''(x)}{\varrho''}, \quad A(\tan x) = \frac{A''(x)}{\varrho'' \cos^2 x}, \tag{7}$$

$$A(\log \sin x) = 0{\cdot}434|\cot x|\, \frac{A''(x)}{\varrho''}, \quad A(\log \cos x) = 0{\cdot}434|\tan x|\, \frac{A''(x)}{\varrho''},$$

$$A(\log \tan x) = \frac{0{\cdot}868 A''(x)}{|\sin 2x|\, \varrho''}\,. \tag{8}$$

(However, in the neighbourhood of a zero of the denominator these formulae do not offer a good estimate.)

Remark 1. If we have to calculate the functional value for the approximate argument $X \pm A$ by an interpolation, then we may find the approximate functional value for the argument X and the error by (5). Instead of $f'(x)$ we may write Δ_0/h.

Example 4. Let us calculate $y = \sin(36°14'15{\cdot}42'' \pm 0{\cdot}13'')$.
We find $\sin 36°14'15{\cdot}45'' = 0{\cdot}591{,}135{,}1$ and then estimate the error by (7),

$$A(\sin 36°14'15{\cdot}42'') \approx \frac{0{\cdot}81\,.\,0{\cdot}13}{2\,.\,10^5} \doteq 5\,.\,10^{-7}\,.$$

The result is

$$y = 0{\cdot}591{,}135{,}1 \pm 5\,.\,10^{-7}\,.$$

33. PROBABILITY THEORY

By Jaroslav Hájek

References: [15], [37], [107], [117], [150], [151], [204], [235], [250], [259], [262], [314], [315], [334], [364].

33.1. Events and Probabilities

Basic assumption: To every *event* observed in a given trial (experiment) there corresponds a *probability* p such that $0 \leqq p \leqq 1$. A *certain event* has probability 1, and an *impossible event* has probability 0.

The event which occurs if and only if the event A does not occur, will be called the *complement* of A.

Theorem 1 (*The Probability of the Complement of an Event*). *If the probability of an event is p, then the probability of its complement is $1 - p$.*

Theorem 2 (*The Addition Rule for Probabilities*). *The probability P of the occurrence of at least one of a finite or countable family of mutually exclusive events with probabilities $p_1, p_2, \ldots$ equals the sum of these probabilities, i.e.*

$$P = p_1 + p_2 + \ldots . \tag{1}$$

If there are m mutually exclusive events, each having the same probability p, then

$$P = mp . \tag{2}$$

Remark 1. The rule for adding probabilities is a basis for the so-called *classical definition of probability.* In this definition, we start with n mutually exclusive elementary events, or cases, each having the same probability $p = 1/n$. Then the probability P of any (composite) event is $P = m/n$, where m denotes the number of cases favourable to this event, i.e. those cases in which the event occurs, and n denotes the number of all the possible cases. The classical definition fails to describe situations in which either the elementary events have different probabilities or the number of the elementary events is infinite.

Theorem 3 (*The Multiplication Rule for Probabilities*). *The probability P of the simultaneous occurrence of n independent events with probabilities $p_1, p_2, \ldots, p_n$* is

$$P = p_1 p_2 \cdots p_n .$$

If all the events have the same probability p, then

$$P = p^n . \tag{3}$$

Example 1. The rule for multiplying probabilities holds only if the events concerned are actually independent. For example, the event composed of the simultaneous occurrence of an event and of its complement is obviously impossible, and hence has probability 0, whereas the product of the corresponding probabilities $p(1 - p)$ is generally different from 0. This discrepancy lies in the fact that each event is mutually exclusive with respect to its complement, and no two mutually exclusive events can be independent.

REMARK 2. The independence of two events is usually inferred from the independence of partial trials. Two partial trials are called mutually independent, if each outcome (elementary event) of the first partial trial may occur simultaneously with each outcome of the other partial trial, and if the probability of the simultaneous occurrence of the two outcomes equals the product of their probabilities. If the occurrence of the event A or of its complement is completely determined by the outcome of the first partial trial, and of the occurrence of the event B or of its complement is completely determined by the outcome of the other partial trial, and if the partial trials are independent, then the events A and B are also independent.

The *Bernoulli experiment* consists of n independent repetitions of a basic trial in which the occurrence of an event A in each partial trial is observed. The occurrence of A will be called a "*success*" for brevity. A success (or its complement) occurs in each partial trial with probability p (or $1 - p$), independently of the outcomes of the other partial trials. (In the English literature the complement of a success is generally called a "failure".) Thus the event, consisting of some x fixed partial trials yielding a success and the remaining $n - x$ partial trials yielding its complement, has the following probability:

$$p^x(1 - p)^{n-x} .$$

The x trials in which there are successes may be chosen in $\binom{n}{x}$ ways, and hence the probability that a Bernoulli experiment yields x successes, no matter in which partial trials they occur, equals

$$\binom{n}{x} p^x(1 - p)^{n-x} \tag{4}$$

since we are dealing with the union of $\binom{n}{x}$ mutually exclusive events, each having the probability $p^x(1 - p)^{n-x}$.

Example 2. The Addition Rule for Probabilities may not hold if the respective events are not mutually exclusive. For example, the occurrence of a success in a partial trial of a Bernoulli trial does not exclude the occurrence of a success in some other partial trial, and, consequently, the probability of the event, that a success occurs in at least one partial trial, does not equal np, but $1 - (1 - p)^n$. Actually, the complement of the event under consideration occurs if and only if a success fails to occur in all partial trials which has a probability of $(1 - p)^n$. The latter probability must add up, with the required probability, to 1.

Example 3 (*Random Sampling without Replacement from a Finite Population*). Let us consider a batch of size N of manufactured articles, in which pN articles are defective and $N - pN$ articles are satisfactory. Let us take a sample of size n from this batch in such a manner that each of the $\binom{N}{n}$ possible samples may occur with the same probability $1/\binom{N}{n}$. Then the probability of the event, that the sample includes x defective articles and $n - x$ satisfactory articles, equals

$$\frac{\binom{pN}{x}\binom{N - pN}{n - x}}{\binom{N}{n}} \tag{5}$$

since x defective articles from the total number pN may be selected in $\binom{pN}{x}$ ways and each such selection may be combined with $\binom{N - pN}{n - x}$ ways of selecting $n - x$ satisfactory articles from the total number $N - pN$.

Example 4. The probability of winning the fourth prize in the Czech lottery called Sportka equals

$$\frac{\binom{6}{3}\binom{43}{3}}{\binom{49}{6}} = 0{\cdot}0176\,.$$

(In this lottery one marks 6 numbers out of 49, and the fourth prize is obtained if exactly 3 numbers coincide with those actually drawn.) We have here used formula (5) with x denoting the number of numbers simultaneously marked and drawn.

Theorem 4 (*Conditional Probability*). *Let p_{AB} denote the probability of the simultaneous occurrence of A and B, let p_A (p_B) denote the probability of A (B) and let $p_{A/B}$ ($p_{B/A}$) denote the conditional probability of A (B) given the evidence that the event B (A) has occurred. Then*

$$p_{AB} = p_B p_{A/B} = p_A p_{B/A} \tag{6}$$

holds.

From (6) we may determine $p_{A/B}$ by means of p_{AB} and p_B, further, p_{AB} by means of p_B and $p_{A/B}$, and, finally, $p_{A/B}$ by means of $p_{B/A}$, p_A and p_B (the *Bayes formula*). If the events A and B are independent, then (6) and the relation $p_{AB} = p_A p_B$ imply that the conditional and nonconditional probabilities are equal, $p_{A/B} = p_A$ and $p_{B/A} = p_B$, i.e. the occurrence of A is not dependent on the event B, and conversely. If the event A is a sub-event of B, then we obtain

$$p_{A/B} = \frac{p_A}{p_B} \quad \text{and} \quad p_{B/A} = \frac{p_{AB}}{p_A} = \frac{p_A}{p_A} = 1\,. \tag{7}$$

33.2. Random Variables

Definition 1. The *distribution law* of a random variable x is a law determining the probabilities of events which may be described by the random variable, e.g. events such as $x = x_a$ or $x_a \leqq x \leqq x_b$, etc.

Definition 2. Random variables will be called *discrete*, if they assume only finitely or countably many different values, e.g. the values 0, 1, 2, ... Their distribution law will also be called *discrete*; it may be given by the probabilities of the individual values $p_a = \mathbf{P}\{x = x_a\}$ (see Fig. 33.1; cf. also Table 33.1, p. 1255).

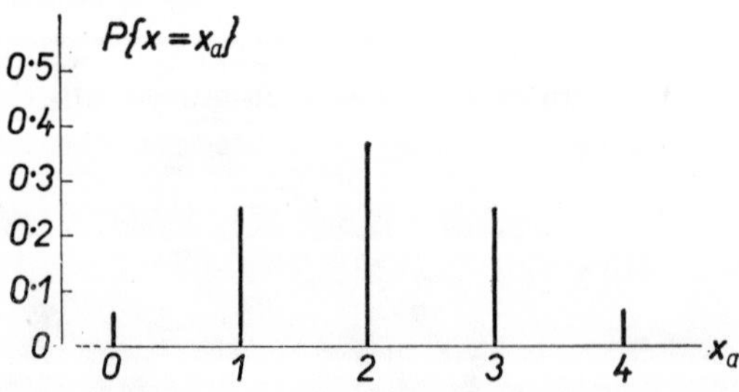

Fig. 33.1. Diagram of the probability distribution of a discrete random variable assuming the values $x = 0, 1, 2, 3, 4$ (the binomial distribution with $n = 4$ and $p = \frac{1}{2}$; see Table 33.1).

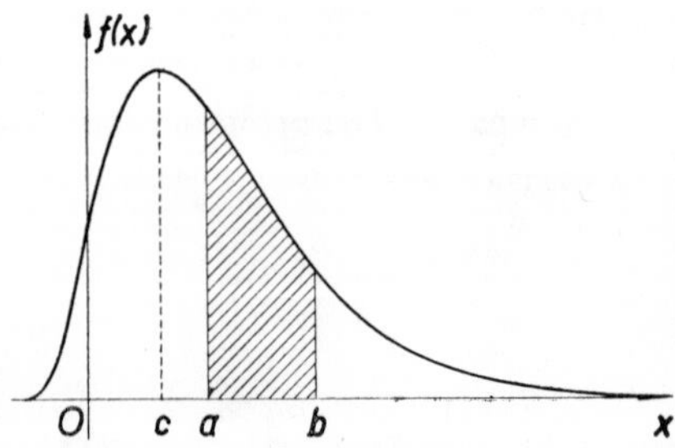

Fig. 33.2. Probability density of a continuous random variable. The shaded area corresponds to the probability $\mathbf{P}(a \leqq x \leqq b)$.

Definition 3. Random variables will be called *continuous*, if their distribution law is given by a *probability density* $f(x) \geqq 0$, the integral of which over an interval $[x_a, x_b]$ yields the probability of the event $x_a \leqq x \leqq x_b$;

$$\mathbf{P}\{x_a \leqq x \leqq x_b\} = \int_{x_a}^{x_b} f(x)\,\mathrm{d}x\,. \tag{1}$$

Such a distribution law will also be called *continuous.*

If $f(x)$ is continuous, then $f(x)\,\mathrm{d}x$ represents the probability that the random variable lies within the interval $[x, x + \mathrm{d}x]$. (See Fig. 33.2; cf. also § 33.3.)

Definition 4. The function $F(x_b)$ determining the probability of the event $x \leqq x_b$ will be called the *distribution function.*

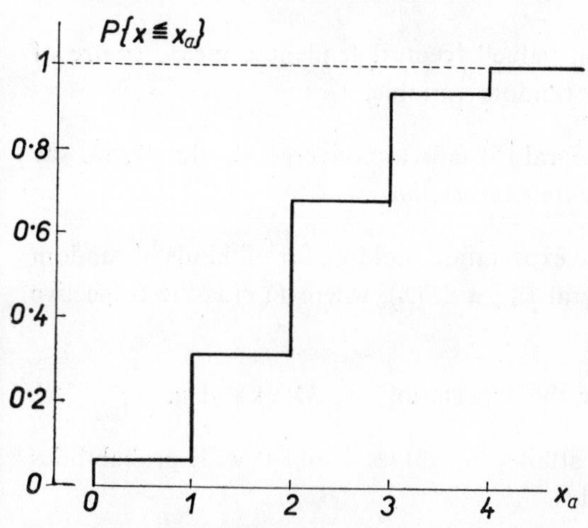

Fig. 33.3. Distribution function of the discrete random variable assuming the values $x = 0,1,2,3,4$ with probabilities shown in Fig. 33.1.

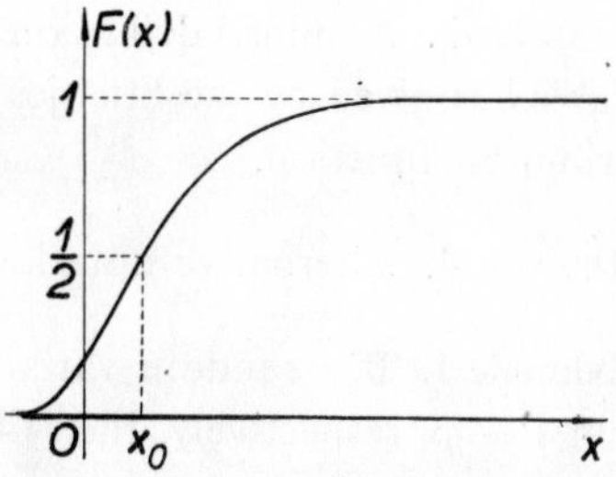

Fig. 33.4. Distribution function of the continuous random variable having the probability density shown in Fig. 33.2.

For discrete and continuous random variables we have

$$F(x_b) = \sum_{x_a \leqq x_b} p_a \quad (p_a = \mathbf{P}\{x = x_a\},\ -\infty < x_b < +\infty) \tag{2}$$

and

$$F(x_b) = \int_{-\infty}^{x_b} f(x)\,\mathrm{d}x\,, \tag{3}$$

respectively. Consequently, for continuous random variables, the relation

$$\frac{\mathrm{d}}{\mathrm{d}x} F(x) = f(x)$$

holds. (See Figs. 33.3 and 33.4; see also § 33.3 and below.)

REMARK 1. The distribution law of every random variable, not only those which are discrete or continuous, is uniquely determined by its distribution function.

Definition 5. The *expectation* of a discrete or continuous random variable is defined by

$$E(x) = \sum_a x_a p_a \tag{4}$$

or

$$E(x) = \int_{-\infty}^{\infty} x\, f(x)\, dx\,, \tag{5}$$

respectively. In formulae (4) and (5), we assume the sum (the integral) to converge absolutely.

The expectation indicates the "mean value" (central tendency, mean, centre of gravity, location) of the corresponding random variable.

REMARK 2. If the sum (4) or the integral (5) fails to converge absolutely, we say that the *random variable does not have an expectation.*

REMARK 3. A unified definition of the expectation, holding for all kinds of random variables, is given by the Stieltjes integral $\int_{-\infty}^{\infty} x\, dF(x)$, where $F(x)$ is the respective distribution function.

REMARK 4. Alternative notations for the expectation are $M(x)$ and μ.

Example 1. If a random variable x attains the values 1 and 0 with probabilities p and $1 - p$, respectively, then (4) yields

$$E(x) = 0\,.\,(1 - p) + 1\,.\,p = p\,.$$

Example 2. To the density

$$\frac{1}{\sigma\sqrt{(2\pi)}}\, e^{-(x-\mu)^2/2\sigma^2}$$

there corresponds the expectation

$$E(x) = \int_{-\infty}^{\infty} x\, \frac{1}{\sigma\sqrt{(2\pi)}}\, e^{-(x-\mu)^2/2\sigma^2}\, dx = \int_{-\infty}^{\infty} (y\sigma + \mu)\, \frac{1}{\sqrt{(2\pi)}}\, e^{-y^2/2}\, dy = \mu\,.$$

Definition 6. The *variance* of a discrete or continuous random variable is defined by

$$D(x) = \sum_a (x_a - \mu)^2\, p_a = \sum_a x_a^2 p_a - \mu^2 \quad [\mu = E(x)] \tag{6}$$

and

$$D(x) = \int_{-\infty}^{\infty} (x - \mu)^2 f(x)\, dx = \int_{-\infty}^{\infty} x^2 f(x)\, dx - \mu^2 , \tag{7}$$

respectively.

REMARK 5. Alternative notations for the variance are $D^2(x)$, $V(x)$ or simply σ_x^2.

Example 3. The random variable considered in Example 1 has the following variance:

$$D(x) = (0 - p)^2 (1 - p) + (1 - p)^2 p = p(1 - p) .$$

Example 4. To the density of Example 2 there corresponds the following variance:

$$D(x) = \int_{-\infty}^{\infty} (x - \mu)^2 \frac{1}{\sigma \sqrt{(2\pi)}} e^{-(x-\mu)^2/2\sigma^2}\, dx = \sigma^2 \int_{-\infty}^{\infty} y^2 \frac{1}{\sqrt{(2\pi)}} e^{y^2/2}\, dy = \sigma^2 .$$

Definition 7.

The *standard deviation*: $\sigma_x = \sqrt{[D(x)]}$.

The *coefficient of variation*: $C(x) = \sqrt{[D(x)]}/E(x)$.

The *median* is a value $\tilde{x}$ such that $\mathbf{P}\{x < \tilde{x}\} = \frac{1}{2}$.

The *P-quantile* is a value x_p such that $\mathbf{P}\{x < x_p\} = P$.

The *mode* of a discrete (continuous) random variable is the value having the maximum probability (probability density).

Theorem 1. *A linear function $ax + b$ of a random variable has the following expectation and variance:*

$$E(ax + b) = aE(x) + b ,$$

$$D(ax + b) = a^2 D(x) .$$

A general function $g(x)$ of a random variable x usually has the following approximate expectation and variance:

$$E(g(x)) \approx g(E(x)) , \tag{8}$$

$$D(g(x)) \approx [g'(E(x))]^2 D(x) , \tag{9}$$

where $g' = dg/dx$. Condition: $g'(x)$ is approximately constant in the neighbourhood of $E(x)$, say for $|x - E(x)| < 3\sqrt{[D(x)]}$.

Theorem 2 (*The Chebyshev Inequality*). *For every random variable with a finite variance and for every* $\varepsilon > 0$ *the inequality*

$$\mathbf{P}\{|x - E(x)| < \varepsilon\} \geqq 1 - \frac{D(x)}{\varepsilon^2} \tag{10}$$

holds.

Theorem 3 (*The Camp-Meidell Inequality*). *If a continuous random variable possesses a unimodal probability density such that* $|\text{Mode} - E(x)| \leqq \sqrt{[D(x)]}$, *then*

$$\mathbf{P}\{|x - E(x)| < \varepsilon\} \geqq 1 - \frac{D(x)}{2{\cdot}25\varepsilon^2}. \tag{11}$$

Theorem 4 (*The Jensen Inequality*). *If a function* $g(x)$ *is convex (or concave) over the region of the x-values, then*

$$\begin{aligned} &E(g(x)) \geqq g(E(x)) \quad (\textit{for convex functions}),\\ &E(g(x)) \leqq g(E(x)) \quad (\textit{for concave functions}). \end{aligned} \tag{12}$$

33.3. The Normal Distribution

Definition 1. The *standard normal distribution* is determined by the following density $\varphi(x)$ and distribution function $\Phi(x)$:

$$\varphi(x) = \frac{1}{\sqrt{(2\pi)}}\,\mathrm{e}^{-x^2/2}, \quad \Phi(x) = \frac{1}{\sqrt{(2\pi)}}\int_{-\infty}^{x} \mathrm{e}^{-t^2/2}\,\mathrm{d}t. \tag{1}$$

It has zero expection and unit variance, and

$$\varphi(-x) = \varphi(x) \quad \text{and} \quad \Phi(-x) = 1 - \Phi(x) \tag{2}$$

holds (see Figs. 33.5 and 33.6).

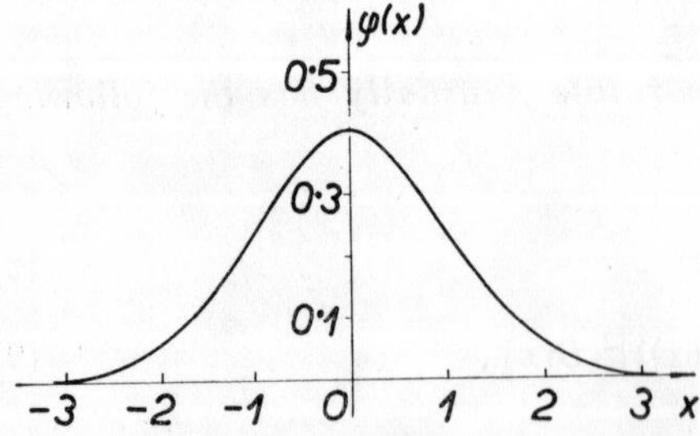

Fig. 33.5. Probability density of the standard normal distribution.

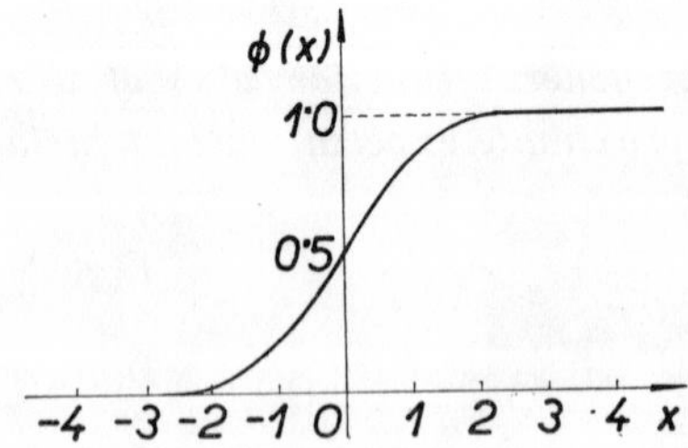

Fig. 33.6. Distribution function of the standard normal distribution.

Definition 2. The *general normal distribution* with expectation μ and variance σ^2, briefly $N(\mu, \sigma^2)$, is determined by the following density:

$$f(x) = \frac{1}{\sigma\sqrt{(2\pi)}}\,e^{-(x-\mu)^2/2\sigma^2} = \frac{1}{\sigma}\,\varphi\left(\frac{x-\mu}{\sigma}\right) \tag{3}$$

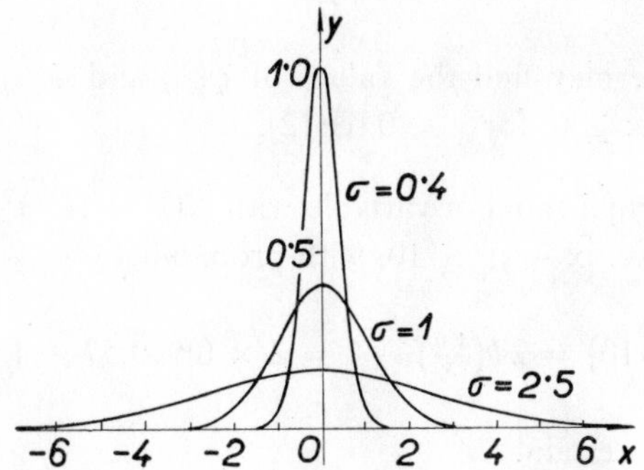

Fig. 33.7. Densities of normal distributions with $\sigma = 0{\cdot}4$; 1; 2·5 and $\mu = 0$.

and by the following distribution function:

$$F(x) = \frac{1}{\sigma\sqrt{(2\pi)}}\int_{-\infty}^{x} e^{-(t-\mu)^2/2\sigma^2}\,dt = \Phi\left(\frac{x-\mu}{\sigma}\right) \tag{4}$$

(see Fig. 33.7).

Theorem 1 (*Probabilities of Various Events Determined by a Random Variable x Having the Normal Distribution* $N(\mu, \sigma^2)$):

$$\mathbf{P}\{x_a \leqq x \leqq x_b\} = \frac{1}{\sigma\sqrt{(2\pi)}}\int_{x_a}^{x_b} e^{-(x-\mu)^2/2\sigma^2}\,dx = \Phi\left(\frac{x_b-\mu}{\sigma}\right) - \Phi\left(\frac{x_a-\mu}{\sigma}\right), \tag{5}$$

$$\mathbf{P}\{|x-\mu| \leqq a\} = 2\Phi\left(\frac{a}{\sigma}\right) - 1\,, \tag{6}$$

$$\mathbf{P}\{|x-\mu| > a\} = 2\left[1 - \Phi\left(\frac{a}{\sigma}\right)\right]. \tag{7}$$

REMARK 1. The right-hand sides remain unchanged, if we replace $\leqq$ or $>$ by $<$ or $\geqq$ on the corresponding left-hand sides.

REMARK 2. It is useful to keep in mind the following crude rules concerning a random variable x with a normal distribution $N(\mu, \sigma^2)$:

(a) $|x - \mu| \leqq \sigma$ holds with probability $\frac{2}{3}$, approximately;

(b) $|x - \mu| \leqq 2\sigma$ holds with probability 0·95, approximately;

(x) $|x - \mu| \leqq 3\sigma$ holds with probability 0·997, approximately.

In various tables we may find the values of $\varphi(x)$ and $\Phi(x)$, or of $2\Phi(x) - 1$ or $\Phi(x) - \frac{1}{2}$, namely for $x \geqq 0$; for $x < 0$ see (2).

Example 1. The normal random variable with $N(\mu = 18, \sigma^2 = 9)$ falls within the interval $8 \leqq x \leqq 28$, i.e. $|x - \mu| \leqq 10$, with probability

$$\mathbf{P}\{|x - \mu| \leqq 10\} = 2\Phi(\tfrac{10}{3}) - 1 = 2 \times 0{\cdot}999{,}57 - 1 = 0{\cdot}999{,}14\,,$$

i.e. the event is almost certain.

Theorem 2. *A linear function $ax + b$ of a random variable with normal distribution $N(\mu, \sigma^2)$ has the normal distribution $N(a\mu + b, a^2\sigma^2)$. In particular, the standard random variable*

$$u = \frac{x - \mu}{\sigma}$$

has the standard normal distribution $N(0, 1)$.

REMARK 3. We say that y_P is a *P-probit*, if $\Phi(y_P - 5) = P$. If $P = P_x$ denotes the value of the normal distribution function $N(\mu, \sigma^2)$ at the point x, i.e. if $P_x = \Phi((x - \mu)/\sigma)$, then the probit is a linear function of x:

$$y_{P_x} = \frac{x - \mu}{\sigma} + 5\,.$$

REMARK 4. The importance of the normal distribution consists in the possibility of using it as an approximation for many distributions, including discrete distributions, and in the fact that an approximately normal distribution is possessed by many observable random variables, the value of which is a result of many independent causes, each having a small effect by itself (see the Liapunov theorem in § 33·9).

In almost all statistical procedures we either assume that the underlying distribution is normal, or we assume that the distribution is not known at all.

33.4. Integral Valued Random Variables

BASIC INTEGRAL VALUED DISTRIBUTIONS. Among discrete random variables the most frequent are those assuming only the values $x = 0, 1, 2, \ldots$, either any of them or only some of them (e.g. zero-one variables assume the values 0 or 1 only). The most common integral valued random variable is the number of occurrences of some event in a series of partial trials, into which the total trial is decomposed. A survey of the basic integral valued distributions together with the respective expectations and variances is given in Table 33.1.

TABLE 33.1

Name of distribution	Probabilities of individual values $x = 0, 1, 2, \ldots$	Expectation $E(x)$	Variance $D(x)$
Alternative (zero-one)	$\mathbf{P}\{x=1\} = p,\ \mathbf{P}\{x=0\} = 1-p$	p	$p(1-p)$
Binomial	$\binom{n}{x} p^x(1-p)^{n-x}$	np	$np(1-p)$
Hypergeometric	$\dfrac{\binom{pN}{x}\binom{N-pN}{n-x}}{\binom{N}{n}}$	np	$np(1-p)\dfrac{N-n}{N-1}$
Poisson	$e^{-\mu}\dfrac{\mu^x}{x!}$	μ	μ
Negative binomial	$\binom{\alpha+x-1}{x} p^x(1-p)^\alpha$	$\dfrac{\alpha p}{1-p}$	$\dfrac{\alpha p}{(1-p)^2}$
Range of parameters	$0 < p < 1,\ \mu > 0,\ \alpha > 0$; $n, N = 1, 2, \ldots,\quad n < N$		

The *binomial distribution* governs, for example, the number of successes in the Bernoulli trial, if the appropriate conditions are satisfied, namely independence of the repetitions of the basic trial and the constancy of the probability of success. For example, the number of "sixes" in throwing six dice is governed by the binomial distribution with $p = 1/6$ and $n = 6$.

The *alternative distribution* is a particular case of the binomial distribution for $n = 1$. Although very simple, it is very important, since the majority of integral valued random variables can be represented as a sum of alternative random variables. For example, the number of occurrences of an event A in a series of trials may be expressed as a sum of numbers of its occurrences in the individual trials, and the last-named numbers may equal either 1 or 0 only.

The *hypergeometric distribution* governs the number of individuals with property A, found in a sample of size n, selected by random sampling without replacement from a population of size N. The number of individuals with the property A in the population equals pN (see formula (33.1.5)).

The *Poisson distribution* is tabulated in great detail, since it possesses only one parameter. It may be utilized to approximate some more complicated distributions, namely the binomial distribution with $p \leqq 0{\cdot}05$ and the hypergeometric distribution for small p and n/N, putting $\mu = np$. The Poisson distribution is sometimes called the *distribution of rare events*, indicating that this distribution governs the number of occurrences of events having small probability, so that they occur in a small number of cases (say smaller than 30) even in a large sequence of observations.

Example 1. The Poisson distribution governs the following random variables: the number of impulses registered by a Geiger-Müller counter; the number of blood corpuscules appearing on a plate when observing a drop of diluted blood by a microscope; the number of defective articles in a batch of products, etc.

REMARK 1. A characteristic property of the Poisson distribution is the equality of the expectation and the variance, $E(x) = D(x) = \mu$. As may be seen from Table 33.1, in the binomial and hypergeometric distributions $E(x) \approx D(x)$ occurs only if p and n/N are small.

The *negative binomial distribution*, as defined in Table 33.1, is appropriate for the "waiting time problems" in which the random variable is the number of independent trials that have to be performed to achieve a preassigned number of successes. In the formula as given the probability of a "success" in each trial is $(1 - p)$ and x denotes the number of further trials that have to be performed, after α trials have taken place, to achieve α successes.

(In the English literature, it is more common to define the negative binomial distribution as

$$\binom{x-1}{x-\alpha} p^{\alpha}(1-p)^{x-\alpha},$$

where the probability of a success in each independent trial is p and x denotes the *total* number of trials needed to achieve α successes $(x = \alpha, \alpha + 1, \alpha + 2, \ldots)$. Then

$$E(x) = \alpha/p\,, \quad D(x) = \alpha(1-p)/p^2)\,.$$

REMARK 2 (*Normal Approximation*). If the variance of the binomial, or hypergeometric, or Poisson, distribution is sufficiently large [say $D(x) \geqq 9$], it is possible to approximate its distribution law within the limits $\mu \pm 2\sigma$ by a normal distribution. We have

$$\mathbf{P}\{a \leqq x \leqq b\} \approx \Phi\left(\frac{b + \frac{1}{2} - \mu}{\sigma}\right) - \Phi\left(\frac{a - \frac{1}{2} - \mu}{\sigma}\right), \tag{1}$$

where Φ denotes the standard normal distribution function, $\mu = E(x)$, $\sigma^2 = D(x)$ and a, b are integers.

REMARK 3. If n_A denotes the number of occurrences of the event A in n trials, then the ratio n_A/n is called the *relative frequency of A*. If the distribution of n_A is binomial with the probability of success $p_A = \mathbf{P}\{A\}$, then

$$E\left(\frac{n_A}{n}\right) = p_A, \quad D\left(\frac{n_A}{n}\right) = \frac{p_A(1 - p_A)}{n}. \tag{2}$$

33.5. Families of Random Variables

The distribution law of a family of random variables $x_1, x_2, \ldots, x_n$ gives the probability of events, the occurrence of which may depend on all the variables $x_1, \ldots, x_n$, as, for instance, in the case of the events $x_1 + \ldots + x_n < a$ or $\max\limits_{1 \leqq i \leqq n} x_i < b$, etc. In order to determine this joint distribution law, it is not sufficient to know the distribution laws of the individual random variables x_i, since these laws give the probabilities only of those events which depend on one random variable from the family, e.g. of events such as $x_1 < x_{1a}$ or $x_2 < x_{2a}$, etc., and not for the simultaneous occurrence of several events. If we know, however, that $x_1, x_2, \ldots, x_n$ are mutually independent, then the distribution of the family is uniquely determined by the probability distributions of the individual random variables, and, according to the multiplication rule for probabilities, we have

$$\mathbf{P}\{x_1 < x_{1a}, \ldots, x_n < x_{na}\} = \mathbf{P}\{x_1 < x_{1a}\} \ldots \mathbf{P}\{x_n < x_{na}\}.$$

The distribution of discrete families is defined by the probabilities of individual values, $p(x_{1a}, \ldots, x_{na}) = \mathbf{P}\{x_1 = x_{1a}, \ldots, x_n = x_{na}\}$; the distribution of continuous families is defined by a probability density $f(x_1, \ldots, x_n)$ such that its integral over a set of points $(x_1, \ldots, x_n)$ yielding the event A gives the probability of A:

$$\mathbf{P}\{A\} = \int \ldots \int_A f(x_1, \ldots, x_n)\, \mathrm{d}x_1 \ldots \mathrm{d}x_n. \tag{1}$$

REMARK 1. The family of n random variables is also called an n-dimensional random variable or a random vector.

THE COVARIANCE AND THE CORRELATION COEFFICIENT OF A PAIR OF RANDOM VARIABLES. The smallest family is a pair of random variables x, y. The covariance

$$\sigma_{xy} = \operatorname{Cov}(x, y) = E((x - E(x))(y - E(y))) = E(xy) - E(x)E(y) \tag{2}$$

(see Definition 33.2.5) and the correlation coefficient

$$\varrho_{xy} = \frac{\sigma_{xy}}{\sqrt{[D(x)\,D(y)]}} \tag{3}$$

(see Definition 33.2.6) are the most usual measures of the dependence (contingency) of x and y. We always have

$$\sigma_{xy}^2 \leqq D(x)\,D(y), \quad \text{i.e.} \quad \varrho_{xy}^2 \leqq 1\,; \tag{4}$$

the case of equality occurs if and only if $y = ax + b$ holds with probability 1. If x and y are independent, then $\sigma_{xy} = 0$. The converse is generally not true.

Definition 1. We say that the distribution of the family $x_1, \ldots, x_k$ of integral valued random variables is *multinomial*, if for $x_1 + x_2 + \ldots + x_k = n$

$$p(x_1, x_2, \ldots, x_k; n) = \frac{n!}{x_1!\,x_2!\ldots x_k!}\,p_1^{x_1}p_2^{x_2}\ldots p_k^{x_k}\,, \tag{5}$$

where the parameters $p_1, \ldots, p_k$ and the integer n are arbitrary numbers such that $0 \leqq p_i \leqq 1, \sum_{i=1}^{n} p_i = 1, n \geqq 1$.

The multinomial distribution is a generalisation of the binomial distribution, and we meet it when observing the frequencies of k mutually exclusive events in a series of n independent repetitions of a basic trial, e.g. when observing the numbers of cases where a random variable falls within one of k disjoint class intervals. The individual random variables x_i of the multinomial family have the binomial distribution with parameters n and p_i $(i = 1, \ldots, k)$.

Example 1. The expectations, variances and covariances are

$$E(x_i) = np_i\,, \quad D(x_i) = np_i(1 - p_i) \quad (i, j = 1, \ldots, k)\,, \tag{6}$$

$$\operatorname{Cov}(x_i, x_j) = \sum_{x_1+\ldots+x_k=n} \cdots \sum (x_i - np_i)(x_j - np_j)\,p(x_1, \ldots, x_k) = -np_ip_j\,,$$

respectively.

Example 2. Suppose we are tossing six dice, and let x_i $(i = 1, \ldots, 6)$ denote the number of dice, on which the number i has appeared. Then

$$E(x_i) = 6 \cdot \tfrac{1}{6} = 1\,, \quad D(x_i) = 6 \cdot \tfrac{1}{6} \cdot \tfrac{5}{6} = \tfrac{5}{6} \text{ and } \operatorname{Cov}(x_i, x_j) = -6 \cdot \tfrac{1}{6} \cdot \tfrac{1}{6} = -\tfrac{1}{6}.$$

Definition 2. The *normal distribution for a pair of random variables* x and y, briefly the *distribution* $N(\mu_1, \mu_2, \sigma_1^2, \sigma_2^2, \sigma_{12})$, is defined by the following probability density:

$$f(x, y) = \frac{1}{2\pi\sigma_1\sigma_2\sqrt{(1-\varrho^2)}} \exp\left(-\frac{1}{2(1-\varrho^2)}\left[\frac{(x-\mu_1)^2}{\sigma_1^2} - \right.\right.$$

$$\left.\left. - 2\varrho\frac{(x-\mu_1)(y-\mu_2)}{\sigma_1\sigma_2} + \frac{(y-\mu_2)^2}{\sigma_2^2}\right]\right). \tag{7}$$

The parameters have the following simple meaning:

$$\mu_1 = E(x), \quad \sigma_1^2 = D(x),$$

$$\sigma_{12} = \operatorname{Cov}(x, y) = \int_{-\infty}^{\infty}\int_{-\infty}^{\infty}(x-\mu_1)(y-\mu_2)f(x, y)\,\mathrm{d}x\,\mathrm{d}y, \tag{8}$$

$$\mu_2 = E(y), \quad \sigma_2^2 = D(y), \quad \varrho = \varrho_{xy} = \frac{\sigma_{12}}{\sigma_1\sigma_2}.$$

33.6. The Expectation of a Sum, a Product and a Ratio of Random Variables

Theorem 1 (*Addition Theorem for Expectations*). *The expectation of a sum of random variables equals the sum of their expectations,*

$$E(x_1 + \ldots + x_n) = E(x_1) + \ldots + E(x_n) \tag{1}$$

or somewhat more generally,

$$E(a_1x_1 + \ldots + a_nx_n + b) = a_1E(x_1) + \ldots + a_nE(x_n) + b. \tag{2}$$

In particular, the expectation of the average of random variables x_i with the same expectation μ is also given by μ:

$$E\left(\frac{x_1 + \ldots + x_n}{n}\right) = \mu. \tag{3}$$

Thus the expectation of a sum of random variables is completely determined by the distribution law of the individual random variables; in particular, it does not depend on whether they are dependent or not.

The addition theorem for expectations enables us to find the expectation of some random variables by decomposing them into a sum of variables, the expectations of which are easy to find. Thus it is not always necessary to compute the expectation directly from its definition (formulae (33.2.4), (33.2.5)); on the contrary, the expectation can generally be ascertained more easily than the corresponding distribution, without knowledge of which the definitions cannot be used. This is illustrated by the following example.

Example 1 (*The Expectation of the Binomial Distribution*). The binomial distribution governs the number of successes x in the Bernoulli trial. Let us denote the number of successes in the i-th repetition of the basic trial by x_i, i.e. $x_i = 1$ if there is a success and $x_i = 0$ otherwise. Then $x = x_1 + \ldots + x_n$. However, the x_i have the alternative distribution with $\mathbf{P}\{x_i = 1\} = p$ so that, according to Table 33.1, $E(x_i) = p$, and, consequently the binomial random variable x possesses the expectation

$$E(x) = E(x_1) + \ldots + E(x_n) = p + \ldots + p = np$$

which is in agreement with the same Table 33.1.

Theorem 2 (*The Expectation of a Product of Independent Random Variables*). *If the random variables $x_1, \ldots, x_n$ are independent, then*

$$E(x_1 x_2 \ldots x_n) = E(x_1)\, E(x_2) \ldots E(x_n) \quad (x_i \textit{ are independent}) . \tag{4}$$

Example 2. If the random variables x and y are independent, then

$$\sigma_{xy} = E(xy) - E(x)\, E(y) = E(x)\, E(y) - E(x)\, E(y) = 0 .$$

The Expectation of the Ratio of Two Random Variables x and y. If the random variable x assumes positive values only, and if its coefficient of variation $C(x) \leqq 0{\cdot}2$, then

$$E\left(\frac{y}{x}\right) \approx \frac{E(y)}{E(x)} .$$

33.7. The Variance of a Sum of Random Variables

Theorem 1 (*The General Formula*). *The variance of a sum of an arbitrary family of random variables $x_1, \ldots, x_n$ with variances $\sigma_i^2 = D(x_i)$ and covariances $\sigma_{ij} = \operatorname{Cov}(x_i, x_j)$ equals*

$$D(x_1 + \ldots + x_n) = \sum_{i=1}^{n} \sigma_i^2 + \sum_{\substack{i=1 \\ i \neq j}}^{n} \sum_{j=1}^{n} \sigma_{ij} , \tag{1}$$

$$D(a_1 x_1 + \ldots + a_n x_n + b) = \sum_{i=1}^{n} a_i^2 \sigma_i^2 + \sum_{\substack{i=1 \\ i \neq j}}^{n} \sum_{j=1}^{n} a_i a_j \sigma_{ij} . \tag{2}$$

Thus the variance of a sum depends on the distribution of the pairs (x_i, x_j) only, and is independent of other relations among the random variables $x_1, x_2, \ldots, x_n$. It is not sufficient, however, to know the distribution of individual random variables, as in the case of expectation.

Theorem 2 (*Addition Theorem for Variances*). *If the random variables are independent, then their covariances satisfy* $\sigma_{ij} = 0$, *and, consequently, the variance of their sum equals the sum of their variances,*

$$D(x_1 + \ldots + x_n) = \sigma_1^2 + \ldots + \sigma_n^2 , \tag{3}$$

$$D(a_1 x_1 + \ldots + a_n x_n + b) = a_1^2 \sigma_1^2 + \ldots + a_n \sigma_n^2 .$$

In particular, if the variances of all the random variables are the same, $\sigma^2 = D(x_i)$, $i = 1, \ldots, n$, *then the variance of their average equals*

$$D\left(\frac{x_1 + \ldots + x_n}{n}\right) = \frac{\sigma^2}{n} . \tag{4}$$

The above formulae enable us to find the variance of some random variables by decomposing them into a sum of variables, the variance of which is easy to find.

Example 1 (*The Binomial Distribution*). Let us decompose a binomial random variable x into the sum of random variables considered in Example 33.6.1. The random variables x_i are alternative and independent, consequently, according to Table 33.1 and (3),

$$D(x) = D(x_1) + \ldots + D(x_n) = p(1 - p) + \ldots + p(1 - p) = np(1 - p)$$

which accords with the value given for the variance of the binomial distribution in Table 33.1.

33.8. The Law of Large Numbers

THE ESSENCE OF THE LAW OF LARGE NUMBERS. The law of large numbers depends upon the fact that the variability of a large number of independent (or weakly dependent) random variables mutually cancels out or compensates itself to such an extent that their sum is relatively (with respect to the number of variables or to its expectation) almost constant.

Example 1. The pressure of a gas on the wall of a container is almost constant, despite being the result of shocks of individual molecules at random instants of time. The course of chemical reactions, though the result of the random behaviour of individual molecules, may be predicted by solving the respective differential equations, because of the large number of molecules and of their small interdependence.

THE ESTIMATION OF AN EXPECTATION BY THE AVERAGE OBSERVATION AND THE ESTIMATION OF A PROBABILITY BY RELATIVE FREQUENCY. The expectation μ of a random variable with a finite variance may be approximated by the average observation of the variable

$$\bar{x} = \frac{x_1 + x_2 \ldots + x_n}{n}.$$

The error $|\bar{x} - \mu|$ of this approximation may be made arbitrarily small with a probability arbitrarily close to 1, if the number of independent observations is sufficiently large. In particular, the probability p_A of an arbitrary event A may be approximated by the relative frequency n_A/n (= number of occurrences/number of trials). To be more exact, we formulate the following:

Theorem 1 (*The Bernoulli Theorem*). *For every* $\varepsilon > 0$

$$\lim_{n\to\infty} \mathbf{P}\left\{\left|\frac{n_A}{n} - p_A\right| < \varepsilon\right\} = 1 \tag{1}$$

and

$$\lim_{n\to\infty} \mathbf{P}\{|\bar{x} - \mu| < \varepsilon\} = 1. \tag{2}$$

REMARK 1. Frequently the law of large numbers does not work well even if the number of random variables is very large. The explanation for this always lies in the absence of the basic assumption of the law of large numbers i.e. of the independence of the individual random variables. For example, the average crop of a certain product varies substantially from year to year, despite being an average of crops on a large number of plots; the dependence of average crops is here due to weather, to the overall state of the agriculture etc. Similarly, the average output may be non-uniform even in large factories, if some important factors affect simultaneously all parts of the factory. Such interdependences are very usual in socio-economic characteristics, which, though related to large populations of persons, may not be governed by the law of large numbers, or are governed by it in a complicated way.

On the other hand a weak (restricted) dependence among the random variables may not render the law of large numbers invalid, though it may make it difficult or even impossible to establish the level of exactness with which the averages (or frequencies) approximate the unknown expectation (or probability).

33.9. The Ljapunov Theorem

The main significance of the normal distribution in probability theory consists in the fact that the sum $x_1 + \ldots + x_n$ of independent random variables, regardless of their individual distributions, has approximately the normal distribution under very general conditions. In Fig. 33.8 there is drawn the frequency distribution of the sum $x_1 + \dots + x_{10}$, the components of which are independent and assume two values only (either $x_i = 0$ or $x_i = i$, both with the same probability $\frac{1}{2}$).

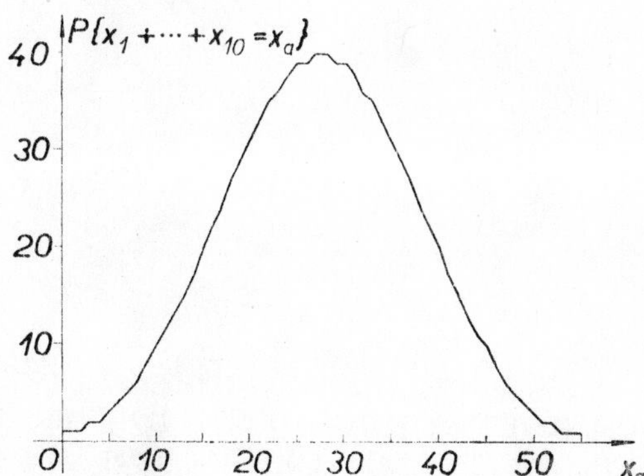

Fig. 33.8. Expected frequency polygon of values taken by the sum $x_1 + \dots + x_{10}$, where the random variables are independent and $\mathbf{P}\{x_i = 0\} = \mathbf{P}\{x_i = i\} = \frac{1}{2}$, $i = 1, \dots, 10$, in a large number of trials

The corresponding exact mathematical assertions have the form of limiting theorems. Let us present the theorem due to A. M. Ljapunov:

Theorem 1 (*The Ljapunov Theorem*). *If the random variables* $x_1, \dots, x_n$ *are mutually independent, then*

$$\lim_{n\to\infty} \mathbf{P}\{x_1 + \dots + x_n - E(x_1 + \dots + x_n) < t\sqrt{[D(x_1 + \dots + x_n)]}\} = \Phi(t) \quad (1)$$

holds provided that

$$\lim_{n\to\infty} \frac{\sum_{i=1}^{n} E(|x_i - E(x_i)|^k)}{[\sum_{i=1}^{n} D(x_i)]^{k/2}} = 0 \quad (2)$$

is satisfied for $k = 3$ *or* $k = 4$ (*or for arbitrary* $k > 2$). *Here* $\Phi(x)$ *denotes the standard normal distribution function.*

REMARK 1. Relation (1) also holds, if the random variables $x_1, x_2, \dots$ have the same distribution with finite variance.

Example 1. Since the binomial random variable x is a sum of n independent random variables x_i, each assuming only the values 1 and 0 with respective probabilities p and $1 - p$, the binomial distribution with fixed p and $n \to \infty$ approaches the normal distribution:

$$\mathbf{P}\{x - np < t\sqrt{[np(1 - p)]}\} \approx \Phi(t) . \quad (3)$$

The exactness of this approximation may be substantially improved if we take into account the fact that the binomial distribution is discrete whereas the normal distribution is continuous (see formula (33.4.1)).

34. MATHEMATICAL STATISTICS

By Jaroslav Hájek

References: [7], [37], [64], [77], [80], [82a], [86], [87], [106], [113], [114], [115], [116], [117], [139], [162], [166], [167], [171], [174], [191], [192], [218], [221], [222], [223], [227], [256], [259], [291], [309a], [329], [330], [350], [352], [358], [361], [364], [414], [415], [416], [425], [427], [428], [435], [432], [443], [444].

34.1. Basic Concepts

Statistics, mathematical statistics and probability theory. Statistics is a science concerned with ascertaining, processing and analysing (evaluating, interpreting) quantitative data, collected either in order to describe large populations or in order to reduce the deviations caused by random factors. Mathematical statistics elaborates methods based on the assumption that the ascertained values are realisations of random variables and that the aim of collecting the data is to establish some unknown features (parameters) in the distribution of these random variables. Thus mathematical statistics has its foundations in probability theory (Chap. 33) and utilizes its concepts (probability, random variable, probability distribution, independence, expectation, variance, etc.).

Observations, statistics and estimates. Actually ascertained (collected, measured, observed) values are usually called *observations*. In a particular situation, they are certain numbers $x_1, \ldots, x_n$, in theoretical considerations they are regarded as random variables $x_1, \ldots, x_n$, which can assume various values according to a certain probability distribution. Values computed from the given observations for purposes of their statistical analysis are called *statistics* (*characteristics*); every statistic is a certain function $s = s(x_1, \ldots, x_n)$ of the observations $x_1, \ldots, x_n$. Statistics computed for the approximate evaluation of parameters are called *estimates* (*estimators*).

Models, parameters, random samples. The working premises (hypotheses) concerning an unknown probability distribution of the given observations are called a *model*. The *parameters* are unknown constants characterizing the source (the conditions of origin, causes) of the ascertained observations; in the model they appear as unknown constants (parameters) in the expressions for probabilities, or probability densities. A specification of these parameters is the main purpose of the analysis. Besides them, the model may contain some further parameters (*nuisance parameters*), the specification of which is irrelevant but which characterise the remaining unknown aspects of the investigated distribution law. The most important example of a model is a random sample, which is defined as a family of n independent observations $x_1, \ldots, x_n$, all with the same distribution, or as a family of pairs $(x_1, y_1), \ldots, (x_n, y_n)$ or of still larger groups of observations, according to the character of the problem. In other words, a random sample is formed by observations on one or more variables under n independent repetitions of the same conditions.

34.2. Calculations, Tables and Charts

FREQUENCIES. Statistical data (observations) are usually classified according to one or more quantitative properties (i.e. variables) or qualitative properties (i.e. attributes). The number of observations falling within a given class is called the *frequency* of this class. If the classification (grouping) is carried out according to ascending values x_a of a certain variable x, then the sum of the frequencies of all values x_b not exceeding x_a is called the *cumulative frequency*. The frequencies divided by the number of all observations are called *relative* (or *proportional*) *frequencies*, or simply *frequencies*, if no ambiguity can arise.

DEFINITIONS OF THE MOST COMMON STATISTICS:

The sample average (mean):

$$\bar{x} = \frac{1}{n} \sum_{i=1}^{n} x_i . \tag{1}$$

The sample variance:

$$s_x^2 = \frac{1}{n-1} \sum_{i=1}^{n} (x_i - \bar{x})^2 = \frac{1}{n-1} \left[\sum_{i=1}^{n} x_i^2 - \frac{1}{n} \left(\sum_{i=1}^{n} x_i \right)^2 \right] . \tag{2}$$

The sample covariance:

$$s_{xy} = \frac{1}{n-1} \sum_{i=1}^{n} (x_i - \bar{x})(y_i - \bar{y}) = \frac{1}{n-1} \left[\sum_{i=1}^{n} x_i y_i - \frac{1}{n} \sum_{i=1}^{n} x_i \sum_{i=1}^{n} y_i \right] . \tag{3}$$

The sample standard deviation:

$$s_x = \sqrt{s_x^2} . \tag{4}$$

The sample coefficient of variation:

$$c_x = \frac{s_x}{\bar{x}} . \tag{5}$$

The sample correlation coefficient:

$$r_{xy} = \frac{s_{xy}}{s_x s_y} . \tag{6}$$

The sample regression coefficient of y on x:

$$b = \frac{s_{xy}}{s_x^2} . \tag{7}$$

For the computation of s_x^2 and s_{xy} the second variant of the formulae (2), (3) is more appropriate.

TABLE 34.1

x	y	$x-y$	x^2	y^2	xy	$(x-y)^2$
0	1	−1	0	1	0	1
13	5	8	169	25	65	64
5	3	2	25	9	15	4
24	8	16	576	64	192	256
20	6	14	400	36	120	196
34	10	24	1156	100	340	576
96	33	63	2326	235	732	1097
$\bar{x} = 16$			$s_x^2 = 158$		$s_x = 12{\cdot}57$	
$\bar{y} = 5{\cdot}5$			$s_y^2 = 10{\cdot}7$		$s_y = 3{\cdot}27$	
$\overline{x-y} = 10{\cdot}5$			$s_{xy} = 40{\cdot}8$		$r_{xy} = 0{\cdot}992$	
			$s_{x-y}^2 = 87{\cdot}1$		$b_{yx} = 0{\cdot}258$	

Example 1. In Table 34.1 (see also Fig. 34.1) a sample of $n = 6$ observations of a pair of random variables x and y is treated. The check is based on the difference $x - y$: $\overline{x - y} = \bar{x} - \bar{y}$, $s_{x-y}^2 = s_x^2 + s_y^2 - 2s_{xy}$.

FREQUENCY TABLES. Samples in which some values occur many times, are reduced (condensed) into frequency tables. The first column of a frequency table contains values which have appeared in the sample at least once (let us denote them by x_a)

and the second column contains the corresponding frequencies (let us denote them by n_a). The sum of the frequencies n_a naturally equals the size of the sample, $n = \sum n_a$, in a frequency table. The statistics $\bar{x}$ and s_x^2 are computed as follows:

$$\bar{x} = \frac{1}{n} \sum x_a n_a ,$$

$$s_x^2 = \frac{1}{n-1} \left[\sum x_a^2 n_a - \frac{1}{n} \left(\sum x_a n_a \right)^2 \right]. \tag{8}$$

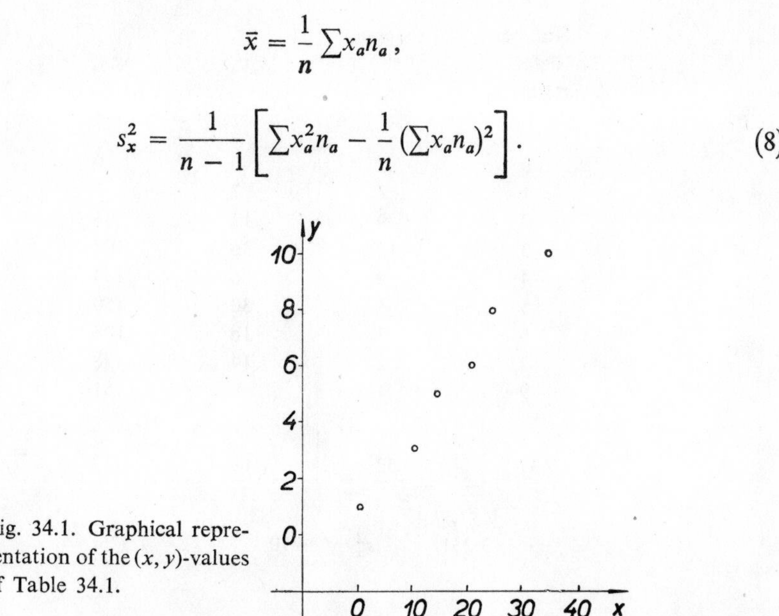

Fig. 34.1. Graphical representation of the (x, y)-values of Table 34.1.

We say that $\bar{x}$ and s_x^2 are the *weighted average* and the *weighted variance* of the values x_a with the weights n_a, respectively. The most frequent value in a frequency table is called the *sample mode.*

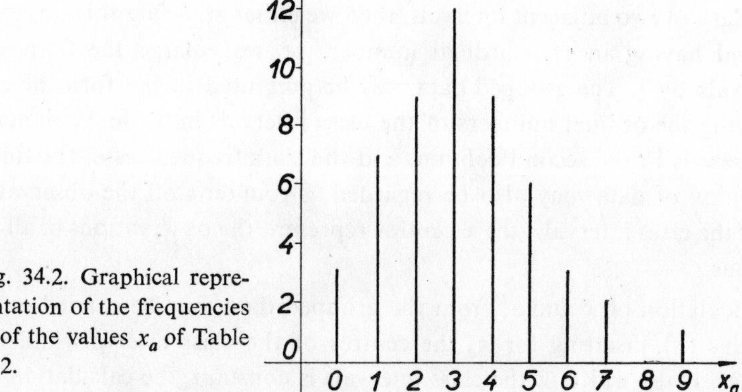

Fig. 34.2. Graphical representation of the frequencies n_a of the values x_a of Table 34.2.

Example 2. In each of a set of 52 shifts the number of defective articles has been recorded. The result is given in Table 34.2 (see also Fig. 34.2), where also $\bar{x}$ and s_x^2 have been calculated. The check may be carried out by repeating calculations for the variable $x + 1$ and by checking the relations $\overline{x+1} = \bar{x} + 1$ and $s_x^2 = s_{x+1}^2$. The sample mode in Table 34.2 equals 3.

TABLE 34.2

Number of defective articles x_a	Frequency of shifts n_a	$x_a n_a$	$x_a^2 n_a$
0	3	0	0
1	7	7	7
2	9	18	36
3	12	36	108
4	9	36	144
5	6	30	150
6	3	18	108
7	2	14	98
9	1	9	81
37	52	168	732
$\bar{x} = 3{\cdot}231$	$s_x^2 = 3{\cdot}710$	$s_x = 1{\cdot}926$	

CLASSIFYING (ROUNDING OFF) THE DATA. If both the sample size and the number of distinct values in it are large, we may simplify the calculation of the statistics by consciously neglecting small differences among the observations. We divide the range of the observed random variable into about 8–20 class intervals and regard all observations falling into the same interval as equivalent. If some observation falls on the boundary of two adjacent intervals, then we either associate this observation with the interval having an even ordinal number, or we enlarge the frequency of the two intervals by $\frac{1}{2}$. The grouped data may be presented in the form of a frequency table, having the ordinal numbers of the class intervals in the first column, the limits of the intervals in the second column, and the class frequencies in the third column. The grouping of data may also be regarded as rounding off the observations to the centres of the class intervals; these centres represent the observations in all subsequent calculations.

The calculation of $\bar{x}$ and s_x^2 from the grouped data may be carried out according to formulae (8), inserting for x_a the centres of the class intervals and for n_a their frequencies. If the width of the class intervals is constant, the calculation of $\bar{x}$ and s_x^2 may be facilitated by introducing an auxiliary variable u attaining the values $u = \ldots, -2, -1, 0, 1, 2, \ldots$ in such a manner that $u = 0$ for some centrally placed interval. Then

$$\bar{x} = x_0 + \Delta\bar{u}\,, \quad s_x^2 = \Delta^2 s_u^2\,, \tag{9}$$

where Δ denotes the width of the intervals and x_0 denotes the centre of the interval for which $u = 0$.

Example 3. The diameters of 200 rollers have been measured in microns and the observations have been grouped into 10 class intervals, proceeding by 5 microns;

TABLE 34.3

No.	Class intervals	Centres of intervals	Frequencies			Relative frequencies	Cumulated frequencies	Relative cumulated frequencies
		x_a	n_a	u_a	u_a^2	n_a / n	h_a	h_a / n
1	30—35	32·5	7	−4	16	0·035	7	0·035
2	35—40	37·5	11	−3	9	0·055	18	0·090
3	40—45	42·5	15	−2	4	0·075	33	0·165
4	45—50	47·5	24	−1	1	0·120	57	0·285
5	50—55	52·5	49	0	0	0·245	106	0·530
6	55—60	57·5	41	1	1	0·205	147	0·735
7	60—65	62·5	26	2	4	0·130	173	0·865
8	65—70	67·5	17	3	9	0·085	190	0·950
9	70—75	72·5	7	4	16	0·035	197	0·985
10	75—80	77·5	3	5	25	0·015	200	1·000
			200					

$\sum n_a u_a = 72$ $\quad \bar{u} = 0{\cdot}36$ $\quad s_u^2 = 3{\cdot}7893$ $\quad s_u = 1{\cdot}947$

$\sum n_a u_a^2 = 780$ $\quad \bar{x} = 54{\cdot}30$ $\quad s_x^2 = 94{\cdot}7337$ $\quad s_x = 9{\cdot}735$

the observations falling on the boundary of two adjacent intervals have been included in the interval having an even ordinal number. The corresponding result is presented in Table 34.3 (see also Fig. 34.3), where also the process of calculating $\bar{x}$ and s_x^2, suitable for a computing machine, is shown. In this case, we carry out the calculation of the sums $\sum u_a n_a$ and $\sum u_a^2 n_a$ directly in the machine without putting down the individual summands. On substituting the sums $\sum u_a n_a$ and $\sum u_a^2 n_a$ into formulas (8) with $x \equiv u$, we obtain $\bar{u}$ and s_u^2. These substituted into (9) ($x_0 = 52{\cdot}5$ and $\Delta = 5$) yield $\bar{x}$ and s_x^2.

Check: We repeat the computation with the only difference that the value $u = 0$ is assigned to another class interval, e.g. to the interval no 6.

The *correlation table* is an analogue of the frequency table for a pair of random variables x and y. In the extreme left column and the uppermost row of the table the values y_b and x_a of the variables x and y are given, respectively, and on the opposite sides their respective (marginal) frequencies n_a and n_b. For grouped data, the values x_a and y_a denote the centres of the class intervals. In the interior of the table

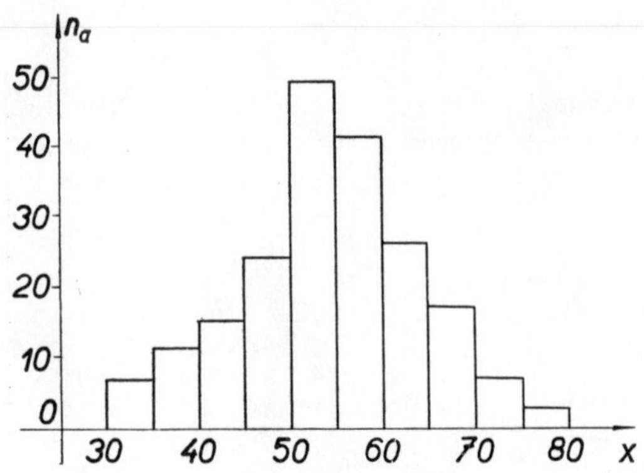

Fig. 34.3. Histogram of the frequencies n_a of the values x_a of Table 34.3.

(within the cells representing the intersection of the column $x = x_a$ and the row $y = y_b$) the frequencies n_{ab} of the observations (x_i, y_i) such that $x_i = x_a$ and $y_i = y_b$ are given. The variance s_x^2 is then computed according to formula (8), and similarly for s_y^2. The covariance s_{xy} may be computed as follows:

$$s_{xy} = \frac{1}{n-1}\left(\sum_b (y_b \sum_a x_a n_{ab}) - \frac{1}{n}\sum_a x_a n_a \sum_b y_b n_b\right). \tag{10}$$

If the possible values of the two variables are equidistant (or if the class intervals are of constant length), the calculation may be facilitated by introducing auxiliary variables u and v, the values of which are integers, and by assigning $u = 0$ and $v = 0$ to some central cell of the correlation table. Then

$$s_{xy} = \Delta_1 \Delta_2 s_{uv} \tag{11}$$

where Δ_1 and Δ_2 denote the successive difference of the x-values and y-values (or the widths of the class intervals), respectively, and s_{uv} is computed by formula (10) with $x = u$, $y = v$.

Example 4. Table 34.4 contains the calculation on a computing machine of averages, variances and of the covariance from 500 observations. The sums $\sum_a u_a n_{ab}$ in the last column have been obtained by adding the products of the values u_a and the respective frequencies n_{ab} from the b-th row. The calculation of $\bar{u}$, $\bar{v}$, s_u^2, s_v^2 and

TABLE 34.4

y_b \ x_a	30	35	40	45	50	55	60	65	70	75	n_b	v_b	v_b^2	$\sum n_{ab}u_a$
0	3	7	7								17	−3	9	−30
10		11	28	4							43	−2	4	−50
20		10	32	38	6	1					87	−1	1	−44
30		2	28	50	31	15	2	1			129	0	0	39
40		2	7	32	43	23	7	2	0	1	117	1	1	113
50				10	15	26	19	14	2		86	2	4	190
60					3	2	6	3	4	1	19	3	9	63
70								2			2	4	16	8
n_a	3	32	102	134	98	67	34	22	6	2	500			
u_a	−3	−2	−1	0	1	2	3	4	5	6				
u_a^2	9	4	1	0	1	4	9	16	25	36				

$\sum n_a u_a = 289$ $\bar{u} = 0{\cdot}578$ $s_u^2 = 2{\cdot}677$ $\bar{v} = 0{\cdot}260$ $s_v^2 = 2{\cdot}089$ $s_{uv} = 1{\cdot}749$

$\sum n_b v_b = 130$ $\bar{x} = 47{\cdot}890$ $s_x^2 = 66{\cdot}925$ $\bar{y} = 32{\cdot}600$ $s_y^2 = 208{\cdot}900$ $s_{xy} = 87{\cdot}450$

$\sum n_a u_a^2 = 1503$ $s_x = 8{\cdot}181$ $s_y = 14{\cdot}453$ $r = 0{\cdot}740$

$\sum n_b v_b^2 = 1076$

$\sum v_b \sum n_{ab} u_a = 948$

s_{uv} has been carried out by formulae (8) and (10), and $\bar{x}$, $\bar{y}$, s_x^2, s_y^2, s_{xy} have then been obtained from (9) and (11) with $\Delta_1 = 5$, $\Delta_2 = 10$, $x_0 = 45$, $y_0 = 30$.

Check: Repeated calculation with a different choice of the cell for which $u = 0$, $v = 0$.

The *sample quantile* is a value chosen so that the proportion of the observations not exceeding it equals a prescribed number, say 5%, 25%, 50%, 75%, 95% etc. In particular, the 50%-quantile is called the *sample median*, the 25%-quantile is called the *lower sample quartile*, and the quantiles corresponding to integral multiples of 10% are called the *sample deciles*. Fig. 34.4 shows a graphical method of establishing the sample quantiles from the polygon of cumulative relative frequencies.

TRANSFORMATIONS OF OBSERVATIONS AND FREQUENCIES. Sometimes it is appropriate to transform the observations $x_1, \ldots, x_n$ by means of a function $f(x)$ to the values $f(x_1), \ldots, f(x_n)$. An example of such a useful transformation is the rounding off performed in grouping of data, or the introduction of the auxiliary variable u in the calculation of $\bar{x}$ and s_x^2. Besides this, the following transformations are often utilized: $\log_{10} x$, $1/x$, $\sqrt{x}$, $\arcsin x$. The average of the values of $\log_{10} x_i$ (or $1/x_i$) corresponds to the geometric average $\bar{x}_G$ (or the harmonic average $\bar{x}_H$) as follows:

$$\log_{10} \bar{x}_G = \frac{1}{n} \sum_{i=1}^{n} \log_{10} x_i\,, \quad \frac{1}{\bar{x}_H} = \frac{1}{n} \sum \frac{1}{x_i}\,. \tag{12}$$

If a sample comes from a normal distribution, or if we wish to verify this fact, then we may represent the cumulative relative frequencies on the probit scale with great

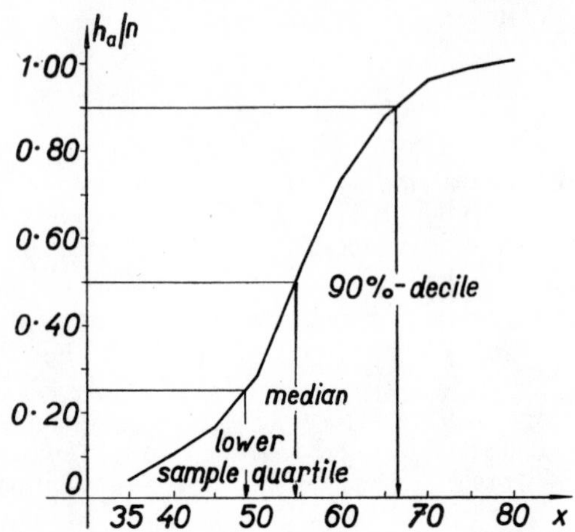

Fig. 34.4. Polygon of cumulated relative frequencies of Table 34.3.

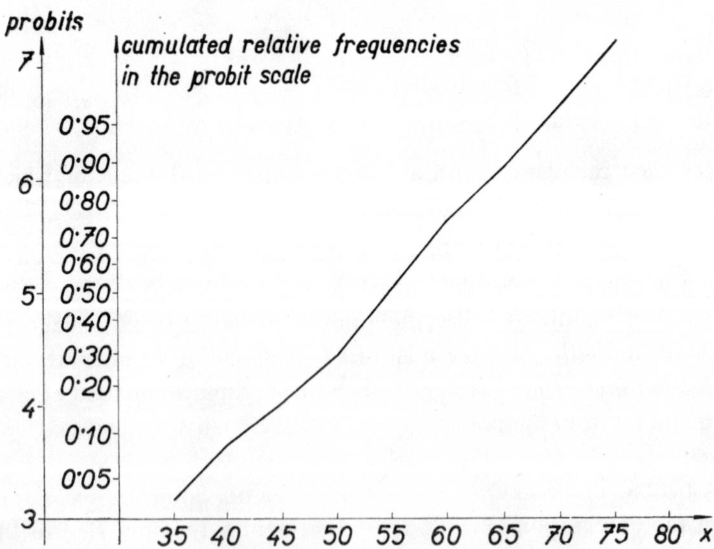

Fig. 34.5. Polygon of cumulated relative frequencies of Table 34.3 in the probit scale.

advantage. This representation may be carried out by plotting the cumulative relative frequencies on probability paper. (For the definition of probits see § 33.3, p. 1254; the graphical correspondence of the probits and cumulative relative frequencies is apparent from Fig. 34.5.)

The main methods of graphical representation of samples are shown in Figures 34.1 to 34.4.

34.3. Tests of Significance

CRITICAL VALUES AND SIGNIFICANCE LEVEL. The test of a hypothesis H_0 on the significance level α is performed by means of a statistic t and of the critical values t'_α and t''_α chosen so that

$$1 - \alpha = \mathbf{P}_0\{t'_\alpha \leqq t \leqq t''_\alpha\} \quad (\text{if } H_0 \text{ holds}) . \tag{1}$$

The hypothesis H_0 is regarded as rejected if either $t < t'_\alpha$ or $t > t''_\alpha$; relation (1) guarantees that H_0 will be erroneously rejected with probability α (in $100\alpha\%$ of cases) only.

REMARK 1. The hypothesis which asserts that the observations have the same distribution, although they were obtained under different conditions, is called the *null hypothesis.* The rejection of the null hypothesis is a necessary prerequisite to the task of estimating the differences (of the expectations, for example) caused by the different conditions.

STUDENT'S t-TEST. This test is appropriate for judging

(a) the difference $\bar{x} - \mu$ between the sample average and the assumed expectation;

(b) the difference $\bar{x} - \bar{y}$ of the averages of two paired samples $(x_1, y_1), \ldots, (x_n, y_n)$;

(c) the difference $\bar{x} - \bar{y}$ of the averages of two independent samples $x_1, \ldots, x_n$ and $y_1, \ldots, y_n$;

(d) the difference $b - \beta$ of the sample regression coefficient (see (34.2.7)) and the assumed value of the theoretical regression coefficient (often called the population regression coefficient). For the hypothesis $\beta = 0$, the test is identical with the test that the correlation coefficient $\varrho = 0$.

The formulas for computing the statistic t and the corresponding degrees of freedom are given in Table 34.5. The value t is regarded as significant at the level α, if $|t| > t_\alpha$, where t_α is the critical value of the Student distribution with the respective number of degrees of freedom for the same significance level α (see tables [116],[309a]). If we expect *a priori* that t will be significantly either positive or negative, we may judge as significant the values for which $t > t_{2\alpha}$ or $t < -t_{2\alpha}$, respectively (one-sided t-tests). The t-test is applicable to samples from approximately normal distributions.

TABLE 34.5

Case	Statistic t for test of significance	Confidence interval	Estimated variance	Number of degrees of freedom
(a)	$t = \dfrac{\bar{x} - \mu}{s_a} \sqrt{n}$	$\bar{x} \pm t_\alpha \dfrac{s_a}{\sqrt{n}}$	$s_a^2 = \dfrac{1}{n-1} \sum_{i=1}^{n} (x_i - \bar{x})^2$	$n - 1$
(b)	$t = \dfrac{\bar{x} - \bar{y}}{s_b} \sqrt{n}$	$\bar{x} - \bar{y} \pm t_\alpha \dfrac{s_b}{\sqrt{n}}$	$s_b^2 = \dfrac{1}{n-1} \sum_{i=1}^{n} (x_i - y_i - \bar{x} + \bar{y})^2$	$n - 1$
(c)	$t = \dfrac{\bar{x} - \bar{y}}{s_c} \cdot \sqrt{\dfrac{m \cdot n}{m + n}}$	$\bar{x} - \bar{y} \pm t_\alpha s_c \cdot \sqrt{\dfrac{m + n}{mn}}$	$s_c^2 = \dfrac{1}{m + n - 2} \left[\sum_{i=1}^{m} (x_i - \bar{x})^2 + \sum_{i=1}^{n} (y_i - \bar{y})^2 \right]$	$m + n - 2$
(d)	$t = \dfrac{b - \beta}{s_d} \cdot \sqrt{\left(\sum_{i=1}^{n} (x_i - \bar{x})^2 \right)}$	$b \pm t_\alpha \cdot \dfrac{s_d}{\sqrt{\left(\sum_{i=1}^{n} (x_i - \bar{x})^2 \right)}}$	$s_d = \dfrac{1}{n-2} \sum_{i=1}^{n} [y_i - \bar{y} - b(x_i - \bar{x})]^2$	$n - 2$

Example 1. In nine successive days two analysts have ascertained the content of ammonia in a certain gas. We ask whether their average results differ significantly or not.

TABLE 34.6

Day	Analyst A	Analyst B
1	4	18
2	37	37
3	35	38
4	43	36
5	34	47
6	36	48
7	48	57
8	33	28
9	33	42

The observations obtained are presented in Table 34.6. The samples are obviously paired since they depend not only on the work of the analysts but also on the actual content of ammonia, which could differ day by day. Thus, we have the case denoted by (b). On carrying out the computations according to Table 34.5, we obtain

$$t = 2{\cdot}03 \quad \text{(the number of degrees of freedom is 8)}\,.$$

For 8 degrees of freedom and the level $\alpha = 0{\cdot}05$ the critical value is $t_{0.05} = 2{\cdot}31$, and, consequently, the obtained value of t is not significant at this level. On the other hand, for $\alpha = 0{\cdot}1$, we have $t_{0.1} = 1{\cdot}86$, so that the difference is significant at this level.

THE F-TEST. This test is suitable, for example, for judging the differences in sample averages $\bar{x}_j = (1/n_j)\sum_{i=1}^{n_j} x_{ji}$ associated with $k\ (> 2)$ independent samples $x_{j1}, \ldots, x_{jn_j}$, $j = 1, \ldots, k$. We compute the statistics

$$F = \frac{(N-k)\sum_{j=1}^{k}(\bar{x}_j - \bar{x})^2 n_j}{(k-1)\sum_{j=1}^{k}\sum_{i=1}^{n_j}(x_{ji} - \bar{x}_j)^2}\,, \quad N = \sum_{j=1}^{k} n_j\,, \quad \bar{x} = \frac{1}{N}\sum_{j=1}^{k}\sum_{i=1}^{n_j} x_{ji}\,; \tag{2}$$

the value so obtained is regarded as significant at the level α, if $F > F_\alpha$, where F_α is the critical value of the F-distribution with $(k - 1, N - k)$ degrees of freedom on the level α (see tables [309a], [116]).

TEST OF SIGNIFICANCE FOR A DIFFERENCE IN RELATIVE FREQUENCIES. Let us assume that an event has occured a times in m repetitions of the conditions A, and b times in n repetitions of the conditions B; let us assume that all $m + n$ repetitions are mutually independent. Then the difference of the relative frequencies, i.e. $a/m - b/n$, is regarded as significant, if

$$\left|\frac{a}{m} - \frac{b}{n}\right| > \frac{1}{2}\left(\frac{1}{m} + \frac{1}{n}\right) + \sqrt{\left(\frac{(a+b)(m+n-a-b)}{mn(m+n)}\right)}\, u_\alpha \tag{3}$$

where u_α is the critical value of the standard normal distribution for the significance level α (see tables [309a], [116]).

Example 2. *Test of significance for the* 2×2 *Contingency Table.* One production method A yielded 18 defective articles among the 180 articles, and another production method B yielded 4 defective articles among 100 articles. We ask whether the difference in the relative frequencies associated with the two production methods, i.e. $\frac{18}{180} - \frac{4}{100} = 0{\cdot}06$, provides sufficient evidence for the conclusion that the production method B is better than A.

If we choose $\alpha = 0{\cdot}05$, we have $u_{0.05} = 1{\cdot}96$, so that the right-hand side of inequality (3) equals

$$\frac{1}{2}\left(\frac{1}{m} + \frac{1}{n}\right) + \sqrt{\left(\frac{(a+b)(m+n-a-b)}{mn(m+n)}\right)}\, u_{0.05} = \frac{1}{2}\left(\frac{1}{180} + \frac{1}{100}\right) +$$

$$+ 1{\cdot}96 \sqrt{\frac{22 \,.\, 258}{180 \,.\, 100 \,.\, 280}} \doteq 0{\cdot}0737 \,.$$

Since this value is not less than the established difference, we cannot regard the difference as significant on this level. The data of this example are presented in Table 34.7, which is an example of a 2 × 2 contingency table. In such a table we usually write $c = m - a$ and $d = n - b$. The test of the significance of the differences in relative frequencies may also be interpreted as the test of the hypothesis that the occurrence of the defective articles does not depend on the production method.

TABLE 34.7

	Method A	Method B	Total
Defective articles	18	4	22
Satisfactory articles	162	96	258
Total	180	100	280

APPROXIMATE TESTS BASED ON THE NORMAL DISTRIBUTION. Testing the hypothesis that the parameter Θ has the value Θ_0, may be approximately carried out by means of an estimate T of Θ and of an estimate $\sqrt{[d(T)]}$ of the standard deviation of this estimate as follows: We compute

$$u = \frac{T - \Theta_0}{\sqrt{d(T)}}$$

and the obtained value is regarded as significant if $|u| > u_\alpha$, where u_α is the critical value of the standard normal distribution for the level α (see tables [309a], [116]). For small samples this test represents a crude approximation, for large samples (more then 30 to 300 observations, according to circumstances), however, it may be quite satisfactory. On the basis of Table 34.8 we may thus test the coefficient of variation, the correlation coefficient, etc.

34.4. Estimation Theory

POINT AND INTERVAL ESTIMATES. Let us take a sample $x_1, \ldots, x_n$ from a distribution determined by probabilities, or probability densities $p(x, \Theta)$ with Θ being an unknown parameter. The parameter may be estimated either by one statistic $T = T(x_1, \ldots, x_n)$ or by a pair of statistics $T' = T'(x_1, \ldots, x_n)$ and $T'' = T''(x_1, \ldots, x_n)$ chosen so that $T' \leqq \Theta$ and $T'' \geqq \Theta$ both occur simultaneously with a high probability. In the former case, we speak about a *point estimate* T, and in the latter case we speak about an *interval estimate* (T', T'').

Let us denote the probabilities corresponding to a value Θ by $\mathbf{P}_\Theta\{\ \}$. If T' and T'' are chosen so that

$$\mathbf{P}_\Theta\{T' \leqq \Theta \leqq T''\} = 1 - \alpha$$

holds for every Θ under consideration, then the interval (T', T'') is called a *confidence interval* with significance level α. The unknown parameter Θ will fail to lie between T' and T'' with probability α (in $100\alpha\%$ of cases) only.

CLASSIFICATION OF POINT ESTIMATES. An estimate is called *unbiassed*, if its expectation equals Θ for every value of Θ under consideration:

$$E_\Theta(T(x_1, \ldots, x_n)) = \Theta \tag{1}$$

where $E_\Theta(.)$ denotes the expectation corresponding to the value Θ. An estimate is called *efficient*, if it is unbiassed and has minimal variance for every sample size n and for every Θ; if this holds in the limit for $n \to \infty$ only, we say that it is *asymptotically efficient*. An estimate is called *sufficient* if, its value being given, the conditional distribution of every other estimate is independent of Θ. Thus, in this sense, the sufficient estimate contains all the information about the parameter provided by the sample. We say that an estimate is *consistent*, if it converges to Θ as $n \to \infty$, i.e. if for every $\varepsilon > 0$

$$\lim_{n \to \infty} \mathbf{P}\{|T(x_1, \ldots, x_n) - \Theta| < \varepsilon\} = 1$$

holds.

THE METHOD OF MAXIMUM LIKELIHOOD. Under broad conditions, it may be shown that good properties are possessed by the estimates $T = \Theta^*$, where Θ^* is the solution of the equation

$$\sum_{i=1}^{n} \frac{\partial \ln p(x_i, \Theta)}{\partial \Theta} = 0 \tag{2}$$

with respect to Θ. If we are dealing with the joint estimation of several parameters $\Theta_1, \ldots, \Theta_k$, we must solve a system of equations of the form (2) for $\Theta = \Theta_i$, $i = 1, \ldots, k$.

Example 1. Let us take a normal sample, i.e.

$$p(x, \mu, \sigma^2) = (2\pi\sigma^2)^{-1/2}\, e^{-(x-\mu)^2/2\sigma^2}\,.$$

Then on solving the equations

$$\sum_{i=1}^{n} \frac{\partial \ln p(x_i, \mu, \sigma^2)}{\partial \mu} = 0\,,$$

$$\sum_{i=1}^{n} \frac{\partial \ln p(x_i, \mu, \sigma^2)}{\partial \sigma^2} = 0\,,$$

i.e. the equations

$$\sum_{i=1}^{n} \frac{x_i - \mu}{2} = 0\,,$$

$$-\frac{n}{2\sigma^2} + \frac{1}{2\sigma^4} \sum_{i=1}^{n} (x_i - \mu)^2 = 0$$

we obtain the estimates

$$\mu^* = \frac{1}{n} \sum_{i=1}^{n} x_i \quad \text{and} \quad \sigma^{*2} = \frac{1}{n} \sum_{i=1}^{n} (x_i - \bar{x})^2 = \frac{n-1}{n}\, s^2\,.$$

The estimated standard errors (deviations) of some important estimates of parameters are given in Table 34.8. They are based on the assumption that the sample refers to an approximately normal distribution; the only exception is in the last row, where a/n denotes the relative frequency of an event A under n independent repetitions of the same conditions (i.e. the Bernoulli trial). By means of this table we can construct an approximate confidence interval

$$T \pm u_\alpha \sqrt{[d(T)]}\,, \tag{3}$$

where u_α is the critical value of the normal distribution on the level α (see tables [309a], [116]). Intervals (3) are appropriate for large samples only (see end of § 34.3).

REMARK 1. The parameters given in Table 34.8 are called the *population* (or *theoretical*) *characteristics* and their estimates are called the *sample* (or *empirical*) *characteristics*.

Confidence intervals based on the Student distribution are used mainly for the estimation of

(a) the expectation μ,
(b) the difference $(\mu - \nu)$ of two expectations in two paired samples,
(c) the difference $(\mu - \nu)$ of two expectations in two independent samples,
(d) the regression coefficient β.

These confidence intervals are given in Table 34.5, where t_α denotes the critical value for the level α of Student's distribution with the corresponding numbers of degrees of freedom (see tables [309a], [116]).

TABLE 34.8

Parameter (characteristic)		Estimate	Estimated standard deviation
	Θ	T	$\sqrt{[\mathrm{d}(T)]}$
Expectation	μ	$\bar{x}$	$\sqrt{\left(\frac{1}{n}\right)}\, s$
Variance	σ^2	s^2	$\sqrt{\left(\frac{2}{n}\right)}\, s^2$
Standard deviation	σ	s	$\sqrt{\left(\frac{1}{2n}\right)}\, s$
Coefficient of variation	C	c	$\sqrt{\left(\frac{1}{2n}+\frac{c^2}{n}\right)}\, c$
Correlation coefficient	ϱ	r	$(1-r^2)\sqrt{\frac{1}{n}}$
Probability of event	p	$\frac{a}{n}$	$\sqrt{\frac{a(n-a)}{n^2(n-1)}}$

34.5. Statistical Quality Control of Manufactured Products

In industry, mathematical statistics is applied mainly in acceptance inspection and in the control of production processes. According to how the inspection or control is carried out, we may distinguish inspection or control by attributes and inspection or control by variables.

SAMPLING INSPECTION. Single sampling inspection plans are defined by two numbers. The first of these is the sample size n, i.e. the number n of items to be selected from the lot under inspection by simple random sampling. The second is the so-called *acceptance number*; with inspection by attributes this is the maximum allowable number of defectives that may be found in a sample for the lot to be accepted; with inspection by variables it is the maximum value of the corresponding statistic that still leads to the acceptance of the lot.

Operating characteristic function and the risks. Every sampling inspection plan may be characterized by its operating characteristic function $L(p)$; this function represents the probability that a lot containing a proportion p of defective items will be accepted. The probability $\beta = L(p_2)$ that a lot with an unacceptable proportion p_2 of defectives will be accepted is called the *consumer's risk*. The probability $\alpha = 1 - L(p_1)$ of rejecting a lot with an acceptable proportion p_1 is called the *producer's risk* (Fig. 34.6).

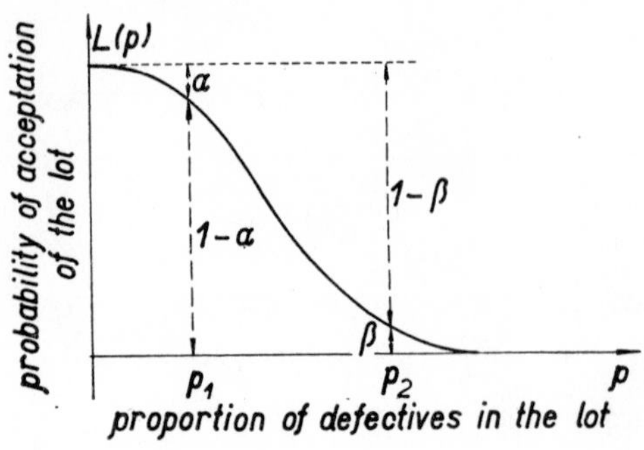

Fig. 34.6.

Sampling inspection by attributes. Sampling inspection of a lot by attributes is carried out as follows:

(a) A sample of n items is drawn at random from the lot of N items submitted for inspection. The items in the sample are then classified as good (i.e. conforming to standard) or defective (i.e. not conforming to standard).

(b) The lot is accepted if the number of detective items z found in the sample does not exceed the acceptance number c.

(c) The lot is returned to the supplier (producer) if $z > c$.

Various tables are available simplifying construction of sampling inspection plans by attributes (see e.g. [87]).

Example 1. Suppose that a producer and consumer have agreed upon the following conditions: Lots containing a proportion of defectives less than $p_1 = 0{\cdot}015$ are to be regarded as completely satisfactory and should be accepted with a probability at least equal to 0·95. Lots containing a proportion of defectives greater than $p_2 = 0{\cdot}08$ are regarded as poor quality, definitely unsatisfactory, and should be accepted with a probability at most equal to 0·10. A sampling inspection plan meeting these requirements is defined by sample size $n = 91$ and acceptance number $c = 3$, as may be found in [82a].

Sometimes other criteria than the values of the operating characteristic curve at two given points are used for the choice of the most suitable sampling inspection plan. Thus, for example, if rejected lots are completely screened and all defective items replaced by good ones, the sample size n and the acceptance number c are determined so as to minimize the average number of inspected items per lot subject to the condition that sufficient protection is offered against accepting bad lots. Sampling inspection plans of this type (so-called *rectifying sampling inspection plans*) have been tabled e.g. by Dodge and Romig ([87]). When the proportion of defectives varies considerably from lot to lot it is sometimes convenient to use double or even multiple sampling. When double sampling is used, two samples are taken in succession; if the first sample indicates definitely high or definitely poor quality, the lot is accepted or rejected, respectively, without taking and inspecting the second sample; the latter is used only if the first sample does not provide sufficient grounds for decision. Multiple sampling permits the taking of even more than two samples before a decision is reached concerning acceptance or rejection of the lot. When a lot of definitely good or definitely bad quality is submitted for inspection, the decision will very probably be reached in the first few samples, thus resulting in reduced inspection costs. (For theory and tables of double and multiple inspection plans see e.g. [87], [350], [358].) For the same purpose so-called *sequential sampling inspection* plans have been developed ([361], [415]).

SAMPLING INSPECTION BY VARIABLES. If the quality of lots is specified in term of the mean value μ of the variable x relevant for the quality of the product, the decision about acceptance or rejection is based on the value of the statistic

$$u = \frac{|\bar{x} - \mu|}{s},$$

where $\bar{x}$ and s are defined by equations (34.2.1) and (34.2.2), respectively. The lot is accepted, if $u \leqq c$ and rejected, if $u > c$. Values of the critical number c corresponding to various sample sizes n, and to various levels of protection against accepting bad lots, are tabled e.g. in [309a], [166]. Since the computation of the standard deviation s is rather time-consuming for routine work, critical values c have been computed also for the modified statistic u with standard deviation s replaced by the range $R = x_{\max} - x_{\min}$ (see e.g. [358]).

If the quality of the lot is expressed in terms of the proportion p of items exceeding an upper tolerance limit T_u or falling below a given lower tolerance limit T_l, the decision is based on the statistics

$$v = \frac{T_u - \bar{x}}{s} \quad \text{or} \quad w = \frac{\bar{x} - T_l}{s},$$

respectively. The lot is accepted if $v \geqq k$ or $w \geqq k$, respectively, otherwise it is rejected. Tables of critical values k corresponding to various sample sizes n are also available (see e.g. [309a] and [166]). As in the case of the statistic u, for the statistics v and w the replacement of the sample standard deviation by the sample range R has been considered and corresponding tables of critical values k computed ([309a], [358]).

CONTROL OF PRODUCTION PROCESSES. Statistical control of production processes is carried out as follows. At regular time intervals samples of n items are taken from the amount produced since the last control, the selected items are inspected and the value of the relevant statistic computed. The result is usually plotted in a special chart called a *control chart* in which so-called *control limits* are drawn. If the value of the statistic falls inside the control limits, no corrective action is required.

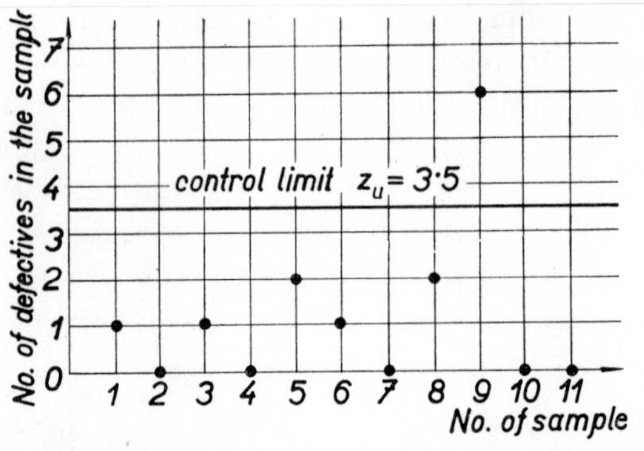

Fig. 34.7.

CONTROL BY ATTRIBUTES. The statistic computed from the sample is the number z of defective items found in the sample. If the number of defectives falls below a prescribed control limit z_u, no corrective action is needed; if it exceeds z_u, the condition of the process should be examined and causes of defectives removed. Fig. 34.7 shows the control chart for the number of defectives.

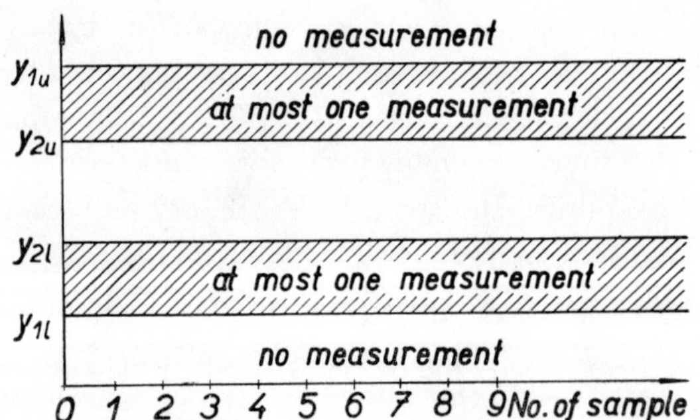

Fig. 34.8.

CONTROL BY VARIABLES. When control by variables is used, values of the relevant characteristic must be ascertained and recorded for every selected item. Suitably chosen statistics indicating the condition of the process are then computed from the data and plotted in the control chart. The most frequently used statistics are:

(a) sample mean $\bar{x}$ and sample range R,

(b) maximal and minimal sample value, $x_{\max}$ and $x_{\min}$,

(c) individual sample values.

Control limits corresponding to these statistics are listed in Table 34.9. If the method based on individual sample values is used, individual measurements are directly plotted in the chart which contains two pairs of control limits, viz. upper and lower outer control limits y_{1u} and y_{1l}, respectively, and upper and lower inner control limits, y_{2u} and y_{2l}, respectively. A scheme of such a chart is indicated in Fig. 34.8. No corrective action is required if the plot satisfies the conditions stated in Fig. 34.8, i.e. if no measurement falls outside the outer control limits and at most one between the outer and inner upper control limit or between the outer and inner lower control limit. Otherwise an investigation of causes of bad quality is called for.

If sufficiently long records of previous results are not available for the construction of control limits (this may be due, for example, to short production series) we may

TABLE 34.9

Method described above under heading	Statistics	Control limits	Modified control limits	Tabled constants
	$\bar{x}$	$\bar{x}_u = \bar{x} + A_2\bar{R}$	$\bar{x}_u = \dfrac{T_u + T_l}{2} + T(0{\cdot}5 - l)$	A_2
(a)		$\bar{x}_l = \bar{x} - A_2\bar{R}$	$\bar{x}_l = \dfrac{T_u + T_l}{2} - T(0{\cdot}5 - l)$	l
	R	$R_u = \bar{R}D_4$	$R_u = D_3 T$	D_4 D_3
	x_{max}	$V_u = \bar{V} + D_5\bar{R}$	$V_u = \dfrac{T_u + T_l}{2} + D_6 T$	
(b)				D_6
	x_{min}	$M_l = \bar{M} - D_5\bar{R}$	$M_l = \dfrac{T_u + T_l}{2} - D_6 T$	
			$y_{1u} = \dfrac{T_u + T_l}{2} + b_1 T$	
			$y_{1l} = \dfrac{T_u + T_l}{2} - b_1 T$	b_1
(c)	individual recorded data		$y_{2u} = \dfrac{T_u + T_l}{2} + b_2 T$	b_2
			$y_{2l} = \dfrac{T_u + T_l}{2} - b_2 T$	

use so-called *modified control limits* (Table 34.9). All formulae listed in Table 34.9 are based on the assumption that the level of the process is located at the centre of the tolerance field and that the random error (deviation from nominal value) is normally distributed.

In this table $\bar{V}$ and $\bar{M}$ denote expectations of maximal and minimal sample value, respectively. $\bar{R} = \bar{V} - \bar{M}$; T_u and T_l are, respectively, the upper and lower tolerance limits; $T_u - T_l = T$ is the tolerance interval. The last column of the table indicates which of the constants appearing in the formulae are available as tables in the literature.

35. METHOD OF LEAST SQUARES. FITTING CURVES TO EMPIRICAL DATA. ELEMENTS OF THE CALCULUS OF OBSERVATIONS

By OTTO FISCHER

References: [7], [77], [82], [114], [145], [158], [167], [221], [329], [352], [363], [366], [428], [436].

The first part of this chapter (§§ 35.1 – 35.4) contains practical rules for fitting curves to empirical data by the method of least squares together with a number of numerical examples for the most common types of curves. The general theory of the least squares method (including properties of estimates by the method of least squares and the distribution theory for these estimates under the assumption that the errors of observations are normally distributed) is discussed in the second part (§§ 35.5 – 35.13). In the third part (§§ 35.14 – 35.16) the results of the first two parts are applied to problems of the calculus of observations.

A. FITTING CURVES TO EMPIRICAL DATA. REGRESSION

35.1. The Principle of Least Squares

Let n points in a plane be given: $(x_1, y_1), (x_2, y_2), \ldots, (x_n, y_n)$, where $x_1, \ldots, x_n$ are some fixed values of the variable x, and $y_1, \ldots, y_n$ are the values obtained by measuring the variable y. The measurement corresponding to the value x_i is denoted by y_i. For example, x may be the temperature and y some variable dependent on the temperature; then y_i denotes the value obtained by measuring the variable y at the given temperature $x = x_i$.

We assume that the variables x and y are related by a functional relationship, $y = f(x)$, where f is a function of known analytic form (e.g. a linear function). If the measurements were free of error affecting the y_i's, we would have $y_i = f(x_i)$ for all $i = 1, 2, \ldots, n$, so that all points (x_i, y_i) would lie on the curve corresponding to the equation $y = f(x)$. However, what is really observed is $y_i = f(x_i) + e_i$ (e_i is the error of measurement) and the points (x_i, y_i) are thus scattered along the curve.

We assume that the function $f(x)$ depends on p ($p < n$) unknown constants called *parameters* which will be denoted by $b_1, b_2, \ldots, b_p$. Thus we may write $f(x) = f(x; b_1, b_2, \ldots, b_p)$. (For example, the linear function $f(x) = b_1 x + b_2$ contains two parameters.) The i-th measurement y_i can then be expressed as

$$y_i = f(x_i; b_1, b_2, \ldots, b_p) + e_i \quad (i = 1, 2, \ldots, n)\,, \tag{1}$$

where $e_1, e_2, \ldots, e_n$ are errors of measurement that are regarded as random variables. It is not required that all the values $x_1, \ldots, x_n$ be different from each other. If, for example, $x_i = x_k$ for some $i \neq k$, then, according to (1), $y_i - y_k = e_i - e_k$, so that two measurements of y corresponding to the same value of x are generally different. Equation (1) is called the *defining equation of the measurement* y_i.

If it is required to fit a curve with the equation $y = f(x; b_1, b_2, \ldots, b_p)$ to the points $(x_1, y_1), (x_2, y_2), \ldots, (x_n, y_n)$, it is necessary to determine, i.e. statistically estimate, the values of the unknown parameters $b_1, b_2, \ldots, b_p$ appearing in the equation of the curve. The estimates should be chosen so that best possible fit to the empirical points (x_i, y_i) is achieved. We shall denote the estimate of b_j by b_j^*. The form of the estimates depends on what criterion of fit is chosen. In the method of least squares the fit is judged according to the sum of squared differences $y_i - f(x_i; b_1^*, b_2^*, \ldots, b_p^*)$ and the estimates $b_1^*, b_2^*, \ldots, b_p^*$ are determined as that set of p-values which minimizes this sum. Thus if we set

$$S = \sum_{i=1}^{n} [y_i - f(x_i, b_1^*, b_2^*, \ldots, b_p^*)]^2\,, \tag{2}$$

the estimates b_j^* ($j = 1, 2, \ldots, p$) will be determined from the condition

$$S = \min\,. \tag{3}$$

Condition (3) expresses the basic principle of the method of least squares. The curve defined by the equation $y = f(x; b_1^*, \ldots, b_p^*)$,the parameters of which are determined so that (3) is satisfied, is said to have been *fitted to the set of points* $(x_1, y_1), (x_2, y_2), \ldots, (x_n, y_n)$ *by the method of least squares.*

Example 1. Suppose that the variables x and y are related by the equation $y = a + bx$, where a and b are unknown parameters. Thus the problem is to fit a straight line to a set of points $(x_1, y_1), (x_2, y_2), \ldots, (x_n, y_n)$ by the method of least squares. According to equation (3), the estimates a^*, b^* will be determined from the condition

$$S = \sum_{i=1}^{n} (y_i - a^* - b^* x_i)^2 = \min.$$

and therefore will be obtained as the solutions of the equations

$$\frac{\partial S}{\partial a^*} = 0 , \quad \frac{\partial S}{\partial b^*} = 0 .$$

We have

$$-\frac{1}{2}\frac{\partial S}{\partial a^*} = \sum_{i=1}^{n} (y_i - a^* - b^* x_i) = \sum_{i=1}^{n} y_i - na^* - \left(\sum_{i=1}^{n} x_i\right) b^* ,$$

$$-\frac{1}{2}\frac{\partial S}{\partial b^*} = \sum_{i=1}^{n} (y_i - a^* - b^* x_i)\, x_i = \sum_{i=1}^{n} x_i y_i - \left(\sum_{i=1}^{n} x_i\right) a^* - \left(\sum_{i=1}^{n} x_i^2\right) b^* .$$

The equations $\partial S/\partial a^* = 0$, $\partial S/\partial b^* = 0$ thus take the form

$$na^* + \left(\sum_{i=1}^{n} x_i\right) b^* = \sum_{i=1}^{n} y_i ,$$

$$\left(\sum_{i=1}^{n} x_i\right) a^* + \left(\sum_{i=1}^{n} x_i^2\right) b^* = \sum_{i=1}^{n} x_i y_i .$$

The solution of these equations (which are called *normal equations*) and corresponding equations in other cases, is described in § 35.3, where a numerical example is also given. The straight line fitted by the least squares method has the equation $y = a^* + b^* x$, where a^*, b^* is the solution of the normal equations.

The function $y = f(x; b_1, b_2, \ldots, b_p)$ is called the *theoretical regression function of y on x*; its graph is called the *theoretical regression curve.* The function obtained by replacing the parameters b_i by their estimates b_i^* is called the *empirical regression function of the variable y on x*. The empirical regression function will be denoted by $f(x; b_1^*, b_2^*, \ldots, b_p^*)$; its graph is called the *empirical regression curve.*

In accordance with the terminology used in mathematical analysis, the variables x and y are sometimes called *independent* and *dependent*, respectively.

For a given value $x = x_i$ to which a measurement y_i corresponds, it is possible to determine from the empirical regression function a value $y_i^* = f(x_i; b_1^*, b_2^*, \ldots, b_p^*)$. The difference $y_i - y_i^*$ which we shall denote by u_i is

$$u_i = y_i - y_i^* = y_i - f(x_i; b_1^*, b_2^*, \ldots, b_p^*) \tag{4}$$

and is called the *residual of the measurement* y_i. Its absolute value, $|u_i|$, represents the distance of the point (x_i, y_i) from the point (x_i, y_i^*) on the fitted curve.

The sum of squares of the residuals which will be denoted by S_0, i.e.

$$S_0 = \sum_{i=1}^{n} u_i^2 = \sum_{i=1}^{n} [y_i - f(x_i; b_1^*, b_2^*, \ldots, b_p^*)]^2 \tag{5}$$

is called the *residual sum of squares* (or the *error sum of squares*).

The residual sum of squares is a very important statistic, since it is used as a basis for statistical estimation of the dispersion of points (x_i, y_i) about the theoretical regression curve.

It should be noted that the quantities e_i in equations (1) need not necessarily be conceived as random variables when applying the least squares method. The method is applicable to any system of points to which a curve with an equation of known form has to be fitted. Thus, for example, the normal equations of Example 1 may be established and solved for a set of points (x_i, y_i), where $y_i = a + bx_i + e_i$ and the e_i's are not random variables. Indeed, the method of least squares is sometimes used in such cases.

However, the assumptions that the errors e_i in the defining equations (1) are of random nature is necessary for applications of the method to physical measurement and to statistical data and we shall use it in the sequel.

35.2. Linear Regression Functions

The regression function is said to be *linear* if it is a linear function of its unknown *parameters.*

The following are examples of linear regression functions (unknown parameters are denoted by $a, b, c, b_0, b_1, b_2, \ldots$):

(a) $y = a + bx$ (*linear regression of y on x*),
(b) $y = a + bx + cx^2$ (*quadratic regression of y on x*),
(c) $y = b_0 + b_1x + b_2x^2 + \ldots + b_rx^r$ (*polynomial regression of y on x*),
(d) $y = a + b/x$ (*hyperbolic regression of y on x*).

The general linear regression function is of the form

$$y = b_1\varphi_1(x) + b_2\varphi_2(x) + \ldots + b_p\varphi_p(x), \tag{1}$$

where $\varphi_1(x), \ldots, \varphi_p(x)$ are functions of the variable x which are of a known form and do not contain unknown parameters.

Example 1. If $p = 3$, $\varphi_1(x) = 1$, $\varphi_2(x) = x$, $\varphi_3(x) = x^2$ and if we put $b_1 = a$, $b_2 = b$, $b_3 = c$, then equation (1) takes the form $y = a + bx + cx^2$.

We suppose that a curve with equation (1) is fitted to a set of points (x_i, y_i) $(i = 1, 2, \ldots, n; n > p)$ so that the following defining equations hold for the y_i's:

$$y_i = b_1\varphi_1(x_i) + b_2\varphi_2(x_i) + \ldots + b_p\varphi_p(x_i) + e_i, \tag{2}$$

where $e_1, e_2, \ldots, e_n$ are random variables.

The estimates $b_1^*, b_2^*, \ldots, b_p^*$ are determined from the condition

$$S = \sum_{i=1}^{n} [y_i - b_1^* \varphi_1(x_i) - \ldots - b_p^* \varphi_p(x_i)]^2 = \min . \tag{3}$$

and are obtained by solving the equations

$$\frac{\partial S}{\partial b_h^*} = 0 \quad (h = 1, 2, \ldots, p) . \tag{4}$$

General properties of estimates b_h^* obtained from condition (3) are stated in § 35.6. If we put

$$a_{hj} = \sum_{i=1}^{n} \varphi_h(x_i)\, \varphi_j(x_i) \quad (h, j = 1, 2, \ldots, p) , \tag{5}$$

$$a_h = \sum_{i=1}^{n} \varphi_h(x_i) y_i , \quad (h = 1, 2, \ldots, p) , \tag{6}$$

we have

$$\frac{\partial S}{\partial b_h^*} = -2(a_h - a_{h1} b_1^* - a_{h2} b_2^* - \ldots - a_{hp} b_p^*) ,$$

so that equations (4) may be written in the form

$$\begin{aligned} a_{11} b_1^* + a_{12} b_2^* + \ldots + a_{1p} b_p^* &= a_1 , \\ a_{21} b_1^* + a_{22} b_2^* + \ldots + a_{2p} b_p^* &= a_2 , \\ \ldots\ldots\ldots\ldots\ldots\ldots\ldots\ldots\ldots\ldots\ldots \\ a_{p1} b_1^* + a_{p2} b_2^* + \ldots + a_{pp} b_p^* &= a_p . \end{aligned} \tag{7}$$

Equations (7) which form a system of linear equations in p unknowns $b_1^*, \ldots, b_p^*$ are called *normal equations*. Some general properties of normal equations are given in § 35.7.

Example 2. In the case of polynomial regression

$$y = b_0 + b_1 x + b_2 x^2 + \ldots + b_r x^r$$

we have $\varphi_j(x) = x^j$ $(j = 0, 1, \ldots, r)$, and hence $a_{hj} = \sum_{i=1}^{n} x_i^{h+j}$, $a_h = \sum_{i=1}^{n} x_i^h y_i$. In this case, the number of unknown parameters is $p = r + 1$. Denoting $\sum_{i=1}^{n} x_i^m$ by S_m, the normal equations will be of the form

$$
\begin{aligned}
nb_0^* + \quad S_1 b_1^* + \ldots + \quad S_r b_r^* &= \sum_{i=1}^{n} y_i\,,\\
S_1 b_0^* + \quad S_2 b_1^* + \ldots + S_{r+1} b_r^* &= \sum_{i=1}^{n} x_i y_i\,,\\
\ldots\ldots\ldots\ldots\ldots\ldots\ldots\ldots\ldots\ldots &\qquad\qquad (8)\\
S_r b_0^* + S_{r+1} b_1^* + \ldots + \quad S_{2r} b_r^* &= \sum_{i=1}^{n} x_i^r y_i\,.
\end{aligned}
$$

By solving the normal equations, we obtain the estimates $b_1^*, b_2^*, \ldots, b_p^*$. With the aid of these we determine the empirical regression function $y^* = \sum_{j=1}^{p} b_j^* \varphi_j(x)$ (i.e. the equation of the curve fitted by the method of least squares) and the residuals $u_i = y_i - \sum_{j=1}^{p} b_j^* \varphi_j(x_i)$. Then we may compute the residual sum of squares $S_0 = \sum_{i=1}^{n} u_i^2$.

However, the expression for the residual sum of squares may be written in the form

$$S_0 = \sum_{i=1}^{n} y_i^2 - b_1^* a_1 - b_2^* a_2 - \ldots - b_p^* a_p\,, \tag{9}$$

where $a_1, a_2, \ldots, a_p$ are the right-hand sides of the normal equations defined by (6) and $b_1^*, b_2^*, \ldots, b_p^*$ constitute the solution of the system (7). Formula (9) for the residual sum of squares is of value especially for the computation of the residual sum of squares from numerically given values (x_i, y_i); using formula (9) we avoid the laborious computation of individual residuals and of their squares.

As already mentioned, the errors $e_1, e_2, \ldots, e_n$ are regarded as random variables. Therefore we must introduce assumptions concerning their probability distribution.

Generally we shall assume that the random variables $e_1, e_2, \ldots, e_n$ all have zero means, that they are uncorrelated and have a common but unknown variance. This variance will be denoted by σ^2. All these assumptions are discussed in more detail in paragraph 35.5; in paragraph 35.10 an additional assumption is made, viz. that the e_i's are normally distributed, and the implications of this assumption are stated.

According to (2), $y_i = \sum_{j=1}^{p} b_j \varphi_j(x_i) + e_i$, where $\sum_{j=1}^{p} b_j \varphi_j(x_i)$ is a constant and e_i is a random variable. It follows from the assumptions concerning the probability distribution of the e_i's that the expectation of y_i is equal to $\sum_{j=1}^{p} b_j \varphi_j(x_i)$, further that $y_1, y_2, \ldots, y_n$ are uncorrelated and that they have the same variance σ^2.

The estimate of the variance σ^2 is given by

$$s^2 = S_0/(n - p)\,, \tag{10}$$

where S_0 is the residual sum of squares. Substituting the expression (9) for S_0 in (10), we obtain

$$s^2 = \frac{\sum_{i=1}^{n} y_i^2 - \sum_{j=1}^{p} b_j^* a_j}{n - p}\,. \tag{11}$$

The number

$$\nu = n - p \tag{12}$$

is called the *number of degrees of freedom for the residual sum of squares.* The quantity

$$s = \sqrt{[S_0/(n - p)]} \tag{13}$$

is an estimate of the standard error (or standard deviation) σ of individual measurements. The quantity $0{\cdot}674{,}5s$ is called the *probable error.*

The estimates b_j^* and the residual sum of squares S_0 are functions of the observations $y_1, \ldots, y_n$. Since the y_i's are random variables, the estimates b^* as well as the residual sum of squares S_0 are random variables possessing probability distributions of their own. For the basic properties of these distributions we refer to paragraphs 35.8 and 35.9, and for their properties under the assumption of normal distribution for the e_i's, to paragraph 35.10.

The variance of the estimate b_j^* which we shall denote by $\sigma_{b_j*}^2$ depends on the unknown variance σ^2 and is also an unknown parameter. Its estimate will be denoted by $s_{b_j*}^2$; the square root s_{b_j*} is then an estimate of the standard error of the estimate b_j^* and $0{\cdot}674{,}5s_{b_j*}$ is the probable error of the estimate.

35.3. Fitting of Some Curves with Linear Regression Equations. Numerical Examples

This paragraph contains a detailed description of the procedures for fitting some types of curves with linear regression equations (unknown parameters are denoted by lower-case letters from the first part of the alphabet). The following types of regression functions are considered:

(a) $y = a$;

(b) $y = bx$ (a straight line passing through the origin);

(c) $y = a + bx$ (a general straight line);

(d) $y = a + bx + cx^2$ (a parabola of the second degree);

(e) $y = a + b/x$ (a hyperbola);

(f) linear regression with several independent variables.

For each of the functions listed above we give the number of unknown parameters, an expression for the sum of squares to be minimized, the normal equations and explicit expressions for their solutions. Further, an expression for the residual sum of squares is given, derived from equation (35.2.9), and formulae for the estimates of standard errors of both individual measurements and estimates of the parameters. It is assumed throughout that the curve is fitted to a set of points $(x_1, y_1), (x_2, y_2), \ldots$ $\ldots, (x_n, y_n)$, where $n > p$. Arithmetic means of x_i and y_i are denoted by $\bar{x}$ and $\bar{y}$, respectively. Thus

$$\bar{x} = \frac{1}{n} \sum_{i=1}^{n} x_i , \quad \bar{y} = \frac{1}{n} \sum_{i=1}^{n} y_i .$$

For all the curves treated in this paragraph we give numerical examples. Empirical values (x_i, y_i) are arranged in tables. In accordance with the assumptions stated above, in any such table $x_1, x_2, \ldots, x_n$ are fixed values of the variable x, and $y_1, y_2, \ldots$ $\ldots, y_n$ are observed values of the random variable y. The tables contain also the values y_i^* calculated from the equation fitted to the data.

(a) $y = a$ $(p = 1)$.

Least squares condition:

$$\sum_{i=1}^{n} (y_i - a^*)^2 = \min .$$

Estimate of the parameter a:

$$a^* = \bar{y} . \tag{1}$$

Residual sum of squares and degrees of freedom:

$$S_0 = \sum_{i=1}^{n} y_i^2 - \frac{\left(\sum_{i=1}^{n} y_i\right)^2}{n} , \quad \nu = n - 1 . \tag{2}$$

Estimates of standard errors:

$$s = \sqrt{\frac{S_0}{n-1}}\,, \quad s_{a*} = \frac{s}{\sqrt{n}}\,. \tag{3}$$

Example 1.

x_i	y_i
0·8	9·3
1·6	4·8
5·4	5·6
3·2	7·9
2·8	3·9
3·6	5·4
6·3	7·8
4·9	7·0
6·0	4·1

Computed from the data in the table:

$n = 9$, $\sum y_i = 55{\cdot}8$, $\sum y_i^2 = 374{\cdot}32$;
hence
$a^* = 6{\cdot}2$, $y^* = 6{\cdot}2$,
$S_0 = 28{\cdot}36$, $\nu = 8$, $s^2 = 3{\cdot}545$,
$s = 1{\cdot}883$, $s_{a*} = 0{\cdot}628$.

(b) $y = bx \quad (p = 1)$.

Least squares condition:

$$\sum_{i=1}^{n} (y_i - b^* x_i)^2 = \min\,.$$

Normal equations:

$$\Bigl(\sum_{i=1}^{n} x_i^2\Bigr)\, b^* = \sum_{i=1}^{n} x_i y_i\,.$$

Estimate of the parameter b:

$$b^* = \frac{\sum_{i=1}^{n} x_i y_i}{\sum_{i=1}^{n} x_i^2}\,. \tag{4}$$

Residual sum of squares and degrees of freedom:

$$S_0 = \sum_{i=1}^{n} y_i^2 - \frac{\Bigl(\sum_{i=1}^{n} x_i y_i\Bigr)^2}{\sum_{i=1}^{n} x_i^2}\,, \quad \nu = n - 1\,. \tag{5}$$

Estimates of standard errors:

$$s = \sqrt{\frac{S_0}{n-1}}\,, \quad s_{b*} = \frac{s}{\sqrt{\Bigl(\sum_{i=1}^{n} x_i^2\Bigr)}}\,. \tag{6}$$

Example 2.

x_i	y_i	y_i^*
2	16·32	14·080
2	13·52	14·080
3	19·32	21·120
3	21·83	21·120
4	31·28	28·160
5	31·54	35·200
6	39·36	42·240
8	57·16	56·320
8	51·85	56·320
12	88·82	84·480

Computed from the data in the table:
$n = 10$, $\sum x_i = 53$, $\sum x_i^2 = 375$,
$\sum y_i = 371$, $\sum y_i^2 = 18{,}666{\cdot}0442$,
$\sum x_i y_i = 2640{\cdot}03$,
hence according to formulae (4) to (6):
$b^* = 7{\cdot}040$, $y^* = 7{\cdot}040x$,
$S_0 = 80{\cdot}0218$, $v = 9$,
$s^2 = 8{\cdot}8913$, $s = 2{\cdot}982$,
$s_{b^*} = 0{\cdot}154$.

(c) $y = a + bx \quad (p = 2)$.

Least squares condition:

$$\sum_{i=1}^{n} (y_i - a^* - b^* x_i)^2 = \min .$$

Normal equations:

$$na^* + \Big(\sum_{i=1}^{n} x_i\Big) b^* = \sum_{i=1}^{n} y_i ,$$

$$\Big(\sum_{i=1}^{n} x_i\Big) a^* + \Big(\sum_{i=1}^{n} x_i^2\Big) b^* = \sum_{i=1}^{n} x_i y_i .$$

Estimates of parameters a and b:

$$b^* = \frac{\sum_{i=1}^{n} x_i y_i - \dfrac{\Big(\sum_{i=1}^{n} x_i\Big)\Big(\sum_{i=1}^{n} y_i\Big)}{n}}{\sum_{i=1}^{n} x_i^2 - \dfrac{\Big(\sum_{i=1}^{n} x_i\Big)^2}{n}} , \tag{7}$$

$$a^* = \bar{y} - b^* \bar{x} .$$

Residual sum of squares and degrees of freedom:

$$S_0 = \sum_{i=1}^{n} y_i^2 - \frac{\Big(\sum_{i=1}^{n} y_i\Big)^2}{n} - b^* \left(\sum_{i=1}^{n} x_i y_i - \frac{\Big(\sum_{i=1}^{n} x_i\Big)\Big(\sum_{i=1}^{n} y_i\Big)}{n} \right) , \tag{8}$$

$$v = n - 2 .$$

Estimates of standard errors:

$$s = \sqrt{\frac{S_0}{n-2}}, \quad s_{a*} = s \sqrt{\left(\frac{\frac{1}{n}\sum_{i=1}^{n} x_i^2}{\sum_{i=1}^{n} x_i^2 - \frac{(\sum_{i=1}^{n} x_i)^2}{n}} \right)},$$

$$s_{b*} = s \Big/ \sqrt{\left(\sum_{i=1}^{n} x_i^2 - \frac{(\sum_{i=1}^{n} x_i)^2}{n} \right)}. \tag{9}$$

Example 3.

x_i	y_i	y_i^*
3	24·82	23·48
4	23·26	21·57
6	14·77	17·75
6	19·06	17·75
6	14·79	17·75
7	17·66	15·83
8	11·83	13·92
9	14·82	12·01
11	5·16	8·18
11	11·12	8·18
12	8·04	6·27
12	2·72	6·27
14	0·74	2·44
16	1·21	−1·39

Computed from the data in the table:
$n = 14$, $\sum x_i = 125$, $\sum x_i^2 = 1309$,
$\sum y_i = 170{\cdot}00$, $\sum y_i^2 = 2853{\cdot}0292$,
$\sum x_i y_i = 1148{\cdot}78$;
hence
$\bar{x} = 8{\cdot}929$, $\bar{y} = 12{\cdot}143$.

Normal equations:

$$14a^* + 125b^* = 170{\cdot}00,$$
$$125a^* + 1309b^* = 1148{\cdot}78.$$

Using formulae (7) to (9) we then compute $b^* = -1{\cdot}913$, $a^* = 29{\cdot}223$, $y^* = 29{\cdot}223 - 1{\cdot}913x$, $S_0 = 82{\cdot}6997$, $v = 12$, $s^2 = 6{\cdot}8916$, $s = 2{\cdot}625$, $s_{a*} = 1{\cdot}827$, $s_{b*} = 0{\cdot}189$.

(d) $y = a + bx + cx^2 \quad (p = 3)$.

Least squares condition:

$$\sum_{i=1}^{n} (y_i - a^* - b^*x_i - c^*x_i^2)^2 = \min .$$

Writing $S_m = \sum_{i=1}^{n} x_i^m$ $(S_0 = n)$, we obtain the system of normal equations in the form:

$$na^* + S_1 b^* + S_2 c^* = \sum_{i=1}^{n} y_i ,$$

$$S_1 a^* + S_2 b^* + S_3 c^* = \sum_{i=1}^{n} x_i y_i ,$$

$$S_2 a^* + S_3 b^* + S_4 c^* = \sum_{i=1}^{n} x_i^2 y_i .$$

Estimates of the parameters a, b, c:

$$a^* = \frac{(S_2S_4 - S_3^2)\sum_{i=1}^{n} y_i + (S_2S_3 - S_1S_4)\sum_{i=1}^{n} x_iy_i + (S_1S_3 - S_2^2)\sum_{i=1}^{n} x_i^2y_i}{\Delta},$$

$$b^* = \frac{(S_2S_3 - S_1S_4)\sum_{i=1}^{n} y_i + (nS_4 - S_2^2)\sum_{i=1}^{n} x_iy_i + (S_1S_2 - nS_3)\sum_{i=1}^{n} x_i^2y_i}{\Delta},$$

$$c^* = \frac{(S_1S_3 - S_2^2)\sum_{i=1}^{n} y_i + (S_1S_2 - nS_3)\sum_{i=1}^{n} x_iy_i + (nS_2 - S_1^2)\sum_{i=1}^{n} x_i^2y_i}{\Delta} \tag{10}$$

where $\Delta = n(S_2S_4 - S_3^2) + S_1(S_2S_3 - S_1S_4) + S_2(S_1S_3 - S_2^2)$.

Residual sum of squares and degrees of freedom:

$$S_0 = \sum_{i=1}^{n} y_i^2 - a^*\sum_{i=1}^{n} y_i - b^*\sum_{i=1}^{n} x_iy_i - c^*\sum_{i=1}^{n} x_i^2y_i, \quad \nu = n - 3. \tag{11}$$

Estimates of standard errors:

$$s = \sqrt{\frac{S_0}{n-3}}, \qquad s_{a^*} = s\sqrt{\left(\frac{S_2S_4 - S_3^2}{\Delta}\right)},$$

$$s_{b^*} = s\sqrt{\left(\frac{nS_4 - S_2^2}{\Delta}\right)}, \qquad s_{c^*} = s\sqrt{\left(\frac{nS_2 - S_1^2}{\Delta}\right)}. \tag{12}$$

Example 4.

x_i	y_i	y_i^*
0	7·1	3·56
3	1·8	6·68
5	12·2	17·92
8	59·2	48·54
8	51·7	48·54
9	59·0	62·42
9	60·9	62·42
10	77·6	78·12
12	109·1	115·03
12	120·2	115·03
16	206·6	210·86
16	214·6	210·86

Computed from the data in the table:

$n = 12,\ S_1 = 108,\ S_2 = 1\,224,$
$S_3 = 15{,}282,\ S_4 = 204{,}564,$
$\sum y_i = 980{\cdot}0,\ \sum y_i^2 = 134{,}679{\cdot}16,$
$\sum x_iy_i = 12{,}299{\cdot}5,$
$\sum x_i^2y_i = 165{,}737{\cdot}1.$

Normal equations:

$$\begin{aligned} 12a^* + 108b^* + 1224c^* &= 980{\cdot}0,\\ 108a^* + 1224b^* + 15{,}282c^* &= 12{,}299{\cdot}5,\\ 1224a^* + 15{,}282b^* + 204{,}564c^* &= 165{,}737{\cdot}1. \end{aligned}$$

Using formulae (10) to (12) we then compute $\Delta = 22{,}676{,}112$, $a^* = 3{\cdot}5616$, $b^* = -1{\cdot}7110$, $c^* = 0{\cdot}9167$, $y^* = 3{\cdot}5616 - 1{\cdot}7110x + 0{\cdot}9167x^2$, $S_0 = 302{\cdot}0369$, $\nu = 9$, $s^2 = 33{\cdot}5597$, $s = 5{\cdot}7931$, $s_{a^*} = 4{\cdot}9932$, $s_{b^*} = 1{\cdot}1898$, $s_{c^*} = 0{\cdot}0669$.

(e) $y = a + b/x \quad (p = 2)$.

Putting $t = 1/x$ we obtain an equation of the form $y = a + bt$. Thus, if a hyperbola with the equation $y = a + b/x$ has to be fitted to a set of points (x_i, y_i), we fit a straight line with the equation $y = a + bt$, to the set of points (t_i, y_i), where $t_i = 1/x_i$.

Example 5. (All values y_i in the same row correspond to the same value of x.)

x	y_i						y_i^*
6	64·79	65·82	68·70	68·22			66·462
8	57·53	61·81	59·45	62·73	60·84	60·77	60·749
10	56·73	56·95	59·23	55·76	57·32	54·90	57·322
14	53·41	54·68	53·08	53·38	54·67	53·93	43·404

The t_i-values are

x	t
6	0·166,667
8	0·125,000
10	0·100,000
14	0·071,429

The line fitted to the points (t_i, y_i) has the equation $y^* = 43{\cdot}6104 + 137{\cdot}1122t$ (the procedure of fitting runs exactly along the same lines as in Example 3), so that the hyperbola fitted to the points (x_i, y_i) has the equation $y^* = 43{\cdot}6104 + 137{\cdot}1122/x$.

(f) *Regression functions with several independent variables.* A general linear regression function with q independent variables $x_1, x_2, \ldots, x_q$ has the form

$$y = \sum_{j=1}^{p} b_j \varphi_j(x_1, x_2, \ldots, x_q), \tag{13}$$

where $\varphi_1, \varphi_2, \ldots, \varphi_p$ are known functions of the variables $x_1, x_2, \ldots, x_q$, which do not involve unknown parameters. If it is required to fit a surface with equation (13) to a set of n points $(x_{1i}, x_{2i}, \ldots, x_{qi}, y_i)$ $(i = 1, 2, \ldots, n)$, where $x_{1i}, x_{2i}, \ldots, x_{qi}$ are fixed values of the variables $x_1, x_2, \ldots, x_q$, we describe the measurements y_i by the following equation:

$$y_i = \sum_{j=1}^{p} b_j \varphi_j(x_{1i}, \ldots, x_{qi}) + e_i\,. \tag{14}$$

The least squares condition is

$$\sum_{i=1}^{n} [y_i - \sum_{j=1}^{p} b_j^* \varphi_j(x_{1i}, \ldots, x_{qi})]^2 = \min.$$

Example 6. Let us fit a plane, having the equation

$$y = a + b_1 x_1 + b_2 x_2 \quad (p = 3)\,, \tag{15}$$

to a set of n points (x_{1i}, x_{2i}, y_i) by the method of least squares.

The least squares condition will be

$$\sum_{i=1}^{n} (y_i - a^* - b_1^* x_{1i} - b_2^* x_{2i})^2 = \min.$$

The normal equations are then as follows:

$$na^* + \left(\sum_{i=1}^{n} x_{1i}\right) b_1^* + \left(\sum_{i=1}^{n} x_{2i}\right) b_2^* = \sum_{i=1}^{n} y_i\,,$$

$$\left(\sum_{i=1}^{n} x_{1i}\right) a^* + \left(\sum_{i=1}^{n} x_{1i}^2\right) b_1^* + \left(\sum_{i=1}^{n} x_{1i}x_{2i}\right) b_2^* = \sum_{i=1}^{n} x_{1i}y_i\,,$$

$$\left(\sum_{i=1}^{n} x_{2i}\right) a^* + \left(\sum_{i=1}^{n} x_{1i}x_{2i}\right) b_1^* + \left(\sum_{i=1}^{n} x_{2i}^2\right) b_2^* = \sum_{i=1}^{n} x_{2i}y_i\,.$$

If we put

$$A_{11} = \sum_{i=1}^{n} x_{1i}^2 - \frac{\left(\sum_{i=1}^{n} x_{1i}\right)^2}{n}\,, \qquad A_{22} = \sum_{i=1}^{n} x_{2i}^2 - \frac{\left(\sum_{i=1}^{n} x_{2i}\right)^2}{n}\,,$$

$$A_{12} = \sum_{i=1}^{n} x_{1i}x_{2i} - \frac{\left(\sum_{i=1}^{n} x_{1i}\right)\left(\sum_{i=1}^{n} x_{2i}\right)}{n}\,, \qquad \Delta = A_{11}A_{22} - A_{12}^2\,,$$

$$C_1 = \sum_{i=1}^{n} x_{1i}y_i - \frac{\left(\sum_{i=1}^{n} x_{1i}\right)\left(\sum_{i=1}^{n} y_i\right)}{n}\,,$$

$$C_2 = \sum_{i=1}^{n} x_{2i}y_i - \frac{\left(\sum_{i=1}^{n} x_{2i}\right)\left(\sum_{i=1}^{n} y_i\right)}{n}\,, \tag{16}$$

we obtain the following expressions for the estimates of a_1, b_1, b_2:

$$b_1^* = \frac{A_{22}C_1 - A_{12}C_2}{\Delta}\,, \quad b_2^* = \frac{A_{11}C_2 - A_{12}C_1}{\Delta}\,, \tag{17}$$

$$a^* = \bar{y} - \bar{x}_1 b_1^* - \bar{x}_2 b_2^*\,,$$

where $\bar{x}_1$ and $\bar{x}_2$ are the means of the values x_{ki}, $k = 1, 2$, respectively, i.e.

$$\bar{x}_1 = \frac{1}{n}\sum_{i=1}^{n} x_{1i}\,, \quad \bar{x}_2 = \frac{1}{n}\sum_{i=1}^{n} x_{2i}\,.$$

Residual sum of squares and degrees of freedom:

$$S_0 = \sum_{i=1}^{n} y_i^2 - \frac{1}{n}\left(\sum_{i=1}^{n} y_i\right)^2 - b_1^* C_1 - b_2^* C_2\,, \quad \nu = n - 3\,. \tag{18}$$

Estimates of standard errors:

$$s = \sqrt{\frac{S_0}{n-3}}\,,$$

$$s_{a*} = s\sqrt{\left(\frac{\left(\sum_{i=1}^{n} x_{1i}^2\right)\left(\sum_{i=1}^{n} x_{2i}^2\right) - \left(\sum_{i=1}^{n} x_{1i}x_{2i}\right)^2}{n\Delta}\right)}\,, \tag{19}$$

$$s_{b_1*} = s\sqrt{\left(\frac{A_{22}}{\Delta}\right)}\,, \quad s_{b_2*} = s\sqrt{\left(\frac{A_{11}}{\Delta}\right)}\,.$$

Example 7.

x_{1i}	x_{2i}	y_i	y_i^*
4	7	12·6	14·603
8	5	38·2	42·126
7	14	20·2	20·303
5	8	20·2	18·847
5	5	27·6	24·120
4	4	20·1	19·876
11	9	49·4	53·100
6	8	26·6	24·848
16	10	83·5	81·352
9	5	51·0	48·128
10	12	38·4	41·824
15	8	81·3	78·866
12	11	56·8	55·586
15	18	62·0	61·287
14	5	75·1	78·137

Computed from the data in the table:
$n = 15$, $\sum x_{1i} = 141$, $\sum x_{1i}^2 = 1579$,
$\sum x_{2i} = 129$, $\sum x_{2i}^2 = 1323$,
$\sum x_{1i}x_{2i} = 1311$, $\sum y_i = 663{\cdot}0$,
$\sum y_i^2 = 37{,}115{\cdot}52$, $\sum x_{1i}y_i = 7581{\cdot}3$,
$\sum x_{2i}y_i = 5916{\cdot}9$,
$\bar{x}_1 = 9{\cdot}4$, $\bar{x}_2 = 8{\cdot}6$, $\bar{y} = 44{\cdot}2$.

Using formulae (16) to (19) we compute $A_{11} = 253{\cdot}6$, $A_{22} = 213{\cdot}6$, $A_{12} = 98{\cdot}4$, $\Delta = 44{,}486{\cdot}40$, $C_1 = 1349{\cdot}1$, $C_2 = 215{\cdot}1$, $b_1^* = 6{\cdot}0019$, $b_2^* = -1{\cdot}7579$, $a^* = 2{\cdot}9002$, $y^* = 2{\cdot}9002 + 6{\cdot}0019x_1 - 1{\cdot}7579x_2$, $s = 2{\cdot}7675$, $s_{a*} = 2{\cdot}0616$, $s_{b_1*} = 0{\cdot}1918$, $s_{b_2*} = 0{\cdot}2090$.

35.4. Nonlinear Regression Functions

If it is required to fit a curve whose equation is not linear in the unknown parameters, we usually encounter difficulties due to the fact that the corresponding normal equations are not in general linear. In such cases we usually try to reduce the regression function to a linear function of the parameters to be estimated.

Sometimes such a reduction may be achieved by a simple transformation of the dependent variable y. Such is the case, for instance, if

$$y = \frac{1}{a + b\,\varphi(x)}, \tag{1}$$

or

$$y = ce^{b\psi(x)}, \tag{2}$$

where a, b, c are unknown parameters and φ, ψ are functions of x of a known form and not involving any unknown parameters. For example, if $\varphi(x) = 1/x^2$ and $\psi(x) = x$, the regression functions (1) and (2) reduce to

$$y = \frac{x^2}{ax^2 + b} \quad \text{and} \quad y = ce^{bx},$$

respectively. If we apply the transformation $z = 1/y$ to (1), we obtain the regression function $z = a + b\varphi(x)$ which is a linear function of a and b; similarly, if in (2) we put $z = \ln y$, we obtain

$$z = a + b\,\psi(x) \quad (a = \ln c).$$

Thus, if we have to fit a curve corresponding, say, to the equation $y = ce^{bx}$ for a set of points (x_i, y_i) $(i = 1, 2, \ldots, n)$, we fit a straight line represented by the equation $z = a + bx$ to the set of points (x_i, z_i), where $z_i = \ln y_i$. The least squares condition

$$\sum_{i=1}^{n} (z_i - a^* - b^* x_i)^2 = \min,$$

then leads to simple normal equations with which we already became familiar in § 35·3, case c). Solving these equations, we determine a^*, b^*, $z^* = a^* + b^*x$; hence $y^* = c^* e^{b^* x}$ $(c^* = e^{a^*})$. By a suitable transformation of the dependent variable it is often possible even to reduce some nonlinear regression functions with more than one independent variable to linear regression functions. Thus, for example, the regression function

$$y = ce^{b_1 \varphi(x_1, x_2)} \cdot (\psi(x_1, x_2))^{b_2},$$

where $\varphi(x_1, x_2)$ and $\psi(x_1, x_2)$ are given functions, is reduced by the transformation $z = \ln y$ to the form

$$z = a + b_1\, \varphi(x_1, x_2) + b_2 \ln \psi(x_1, x_2) \quad (a = \ln c).$$

In cases where the regression function cannot be transformed to a linear one, we may proceed as follows: We first find rough estimates of the parameters b_k in the function $y = f(x, b_1, b_2, \ldots, b_p)$ (e.g. by guessing from the scatter-diagram of the empirical points (x_i, y_i)). Let us denote these first approximations by $b_1^0, b_2^0, \ldots, b_p^0$. Then we expand the function $y = f(x; b_1, b_2, \ldots, b_p)$ (considered as a function of the parameters b_k) in a Taylor series about the point $(b_1^0, b_2^0, \ldots, b_p^0)$. Neglecting terms of the second and higher degrees, we obtain

$$y = f_0 + \sum_{k=1}^{p} f_k(b_k - b_k^0), \tag{3}$$

where

$$f_0 = f(x; b_1^0, b_2^0, \ldots, b_p^0),$$

$$f_k = \frac{\partial f(x; b_1, b_2, \ldots, b_p)}{\partial b_k}\bigg|_{(b_1, \ldots, b_p) = (b_1^0, \ldots, b_p^0)} . \tag{4}$$

We then apply the principle of least squares to the function (3) which is obviously a linear function of $\Delta_k = b_k - b_k^0$. We derive the normal equations for the unknowns Δ_k and, solving them, we obtain corrections $\Delta_k^{(1)}$ to the initial approximations b_k^0. We add these corrections to the values b_k^0 and obtain a second approximation $b_k^1 = \Delta_k^1 + b_k^0$. Then, if necessary, the procedure is repeated with the approximations b_k^1, etc.

B. GENERAL PROBLEMS OF THE LEAST SQUARES METHOD

35.5. Defining Equations

The problem of estimating unknown parameters in a linear regression function together with various problems of the calculus of observations (§§ 35.14–35.16) are, in fact, special cases of more general problems that are frequently encountered in mathematical statistics. These general problems may be described as follows: Random variables $y_1, \ldots, y_n$ are observed, each of which is expressed as a sum of a linear combination of unknown parameters $b_1, b_2, \ldots, b_p$ and a random variable e. Thus the random variable y_i is given by an equation of the form

$$y_i = g_{1i}b_1 + g_{2i}b_2 + \ldots + g_{pi}b_p + e_i \quad (i = 1, 2, \ldots, n), \tag{1}$$

where $g_{1i}, g_{2i}, \ldots, g_{pi}$ are known constants, $b_1, b_2, \ldots, b_p$ unknown constants (parameters), and $e_1, e_2, \ldots, e_n$ are random variables.

Equation (1) is called the *defining equation* of the variable y_i. We shall assume that $p < n$. The matrix $\|g_{ji}\|$ $(j = 1, 2, \ldots, p;\ i = 1, 2, \ldots, n)$ whose i-th column is composed of the coefficients $g_{1i}, g_{2i}, \ldots, g_{pi}$ of the defining equation of y_i will be denoted by

$$\mathbf{G} = \begin{Vmatrix} g_{11}, g_{12}, \ldots, g_{1n} \\ g_{21}, g_{22}, \ldots, g_{2n} \\ \ldots\ldots\ldots\ldots\ldots \\ g_{p1}, g_{p2}, \ldots, g_{pn} \end{Vmatrix}. \tag{2}$$

Sometimes the variables y_i are given by equations

$$y_i = g_{0i} + g_{1i}b_1 + g_{2i}b_2 + \ldots + g_{pi}b_p + e_i \quad (i = 1, 2, \ldots, n),$$

where $g_{0i}, g_{1i}, g_{2i}, \ldots, g_{pi}$ are known constants. In such cases we consider the variables $z_i = y_i - g_{0i}$ for which (1) holds.

Since $p < n$, the rank $h(\mathbf{G})$ of $\mathbf{G}$ does not exceed p, i.e. $h(\mathbf{G}) \leqq p$. We shall assume that

$$h(\mathbf{G}) = p, \tag{3}$$

so that the rows of the matrix $\mathbf{G}$ are linearly independent. (The case $h(\mathbf{G}) < p$ will be mentioned in § 35.13.)

Example 1. (a) If we are going to fit a line with equation $y = a + bx$ to a set of points $(x_1, y_1), \ldots, (x_n, y_n)$, the corresponding defining equation is $y_i = a + bx_i + e_i$ $(i = 1, 2, \ldots, n)$. This is a special case of (1) corresponding to $p = 2$, $b_1 = a$, $b_2 = b$, $g_{1i} = 1$, $g_{2i} = x_i$. Thus we have

$$\mathbf{G} = \begin{Vmatrix} 1, & 1, & \ldots, & 1 \\ x_1, & x_2, & \ldots, & x_n \end{Vmatrix}$$

and $h(\mathbf{G}) = 2$ if at least two of the values x_i are different.

(b) Suppose we measure a quantity taking an unknown value b n times. The measurements are described by the equation $y_i = b + e_i$ $(i = 1, 2, \ldots, n)$, where e_i denotes the error of the i-th measurement. Then

$$\mathbf{G} = \|1, 1, \ldots, 1\|, \quad h(\mathbf{G}) = 1.$$

(c) Suppose we measure an unknown quantity V_1 n_1 times and another unknown quantity V_2 n_2 times. Denote the true value of V_1 by b_1 and the true value of V_2 by b_2. If $y_1, \ldots, y_{n_1}$ denote the results of measurements of V_1, $y_{n_1+1}, \ldots, y_{n_1+n_2}$ the results of measurements of V_2, and $e_1, \ldots, e_{n_1+n_2}$ the corresponding errors of measurement, we have $y_i = g_{1i}b_1 + g_{2i}b_2 + e_i$, where $g_{1i} = 1$, $g_{2i} = 0$ for $i = 1, 2, \ldots, n_1$ and $g_{1i} = 0$, $g_{2i} = 1$ for $i = n_1 + 1, \ldots, n_1 + n_2$. The matrix $\mathbf{G}$ then is

$$\mathbf{G} = \begin{Vmatrix} \underbrace{\begin{matrix}1, & 1, & \ldots, & 1, \\ 0, & 0, & \ldots, & 0,\end{matrix}}_{n_1} & \underbrace{\begin{matrix}0, & 0, & \ldots, & 0 \\ 1, & 1, & \ldots, & 1\end{matrix}}_{n_2} \end{Vmatrix}, \quad h(\mathbf{G}) = 2.$$

It will be assumed throughout this chapter that all the random variables $e_1, e_2, \ldots, e_n$ have zero means,

$$E(e_i) = 0 \quad (i = 1, 2, \ldots, n), \tag{4}$$

that they are uncorrelated,

$$\operatorname{Cov}(e_i, e_j) = 0 \quad (i, j = 1, 2, \ldots, n; i \neq j) \tag{5}$$

and that they all have the same variance

$$D(e_i) = \sigma^2 \quad (i = 1, 2, \ldots, n), \tag{6}$$

where σ^2 is an unknown positive constant. (A more general case with $D(e_i) = \sigma^2/p_i$, where the p_i's are given positive numbers and σ^2 is an unknown constant, will be treated in § 35.12.) These assumptions imply that the random variables $y_1, y_2, \ldots, y_n$ have the following properties:

$$E(y_i) = \sum_{j=1}^{p} g_{ji} b_j \quad (i = 1, 2, \ldots, n), \tag{7}$$

$$\operatorname{Cov}(y_i, y_j) = 0 \quad (i, j = 1, 2, \ldots, n; i \neq j), \tag{8}$$

$$D(y_i) = \sigma^2 \quad (i = 1, 2, \ldots, n). \tag{9}$$

We shall now discuss the problem of finding estimates for the unknown parameters $b_1, b_2, \ldots, b_p$ and for the unknown variance σ^2 under general assumptions about the errors e_i.

35.6. Best Linear Unbiased Estimates

A quantity of the form

$$f = k_1 b_1 + k_2 b_2 + \ldots + k_p b_p \tag{1}$$

where $k_1, k_2, \ldots, k_p$ are given constants and $b_1, b_2, \ldots, b_p$ the unknown parameters of the equations (35.5.1), is called a *parametric function*. Its estimate will be denoted by f^*. It is required to determine the estimate f^* so that

(a) f^* be a linear combination of the random variables

$$y_1, y_2, \ldots, y_n,$$

(b) its expectation $E(f^*)$ be equal to f, whatever the true values of the b_i's may be, i.e.

$$E(f^*) = f \text{ for all vectors } (b_1, \ldots, b_p).$$

(This latter requirement is called the condition of unbiassedness – cf. § 34.4.)

A parametric function f for which an estimate f^* satisfying requirements a), b) exists is called an *estimable parametric function.* Generally, there is an infinity of linear unbiassed estimates for any estimable parametric function; from among them we then choose that particular estimate which has the least variance. This estimate is then called the *best linear unbiassed estimate of the parametric function f.*

The following theorem holds for parametric functions and their best linear unbiassed estimates:

The Gauss-Markov Theorem. *If the matrix* $\mathbf{G}$ *has rank p, then every parametric function* $f = \sum_{j=1}^{p} k_j b_j$ *is estimable. Its best linear unbiassed estimate is unique and is given by* $f^* = \sum_{j=1}^{p} k_j b_j^*$, *where* $b_1^*, b_2^*, \ldots, b_p^*$ *are least squares estimates of the parameters* $b_1, b_2, \ldots, b_p$, i.e. they minimize the sum

$$S = \sum_{i=1}^{n} (y_i - g_{1i}b_1^* - g_{2i}b_2^* - \ldots - g_{pi}b_p^*)^2 . \tag{2}$$

Hence it follows, in particular, that b_j^* is the best linear estimate of the parameter b_j.

Another corollary of the Gauss-Markov Theorem holds for the fitting of curves with linear regression equations. If we are fitting a curve with an equation of the form $y = \sum_{j=1}^{p} b_j\, \varphi_j(x)$ to the set of points $(x_1, y_1), \ldots, (x_n, y_n)$, the defining equation of the observations is $y_i = \sum_{j=1}^{p} b_j\, \varphi_j(x_i) + e_i$ and

$$\mathbf{G} = \left\| \begin{matrix} \varphi_1(x_1), & \varphi_1(x_2), & \ldots, & \varphi_1(x_i), & \ldots, & \varphi_1(x_n) \\ \varphi_2(x_1), & \varphi_2(x_2), & \ldots, & \varphi_2(x_i), & \ldots, & \varphi_2(x_n) \\ \cdots & \cdots & \cdots & \cdots & \cdots & \cdots \\ \varphi_p(x_1), & \varphi_p(x_2), & \ldots, & \varphi_p(x_i), & \ldots, & \varphi_p(x_n) \end{matrix} \right\| ;$$

if $h(\mathbf{G}) = p$, it follows from the Gauss-Markov Theorem that the estimates $b_1^*, \ldots, b_p^*$ determined from the condition

$$\sum_{i=1}^{n} [y_i - \sum_{j=1}^{p} b_j^*\, \varphi_j(x_i)]^2 = \min$$

are the best linear unbiassed estimates of the parameters of the regression equation.

35.7. Normal Equations

The estimates $b_1^*, b_2^*, \ldots, b_p^*$ are obtained by solving the set of equations

$$\frac{\partial S}{\partial b_h^*} = 0 \quad (h = 1, 2, \ldots, p)\,, \tag{1}$$

where S denotes the sum (35.6.2); equations (1) are called the *normal equations.* Introducing the notation

$$a_{hj} = \sum_{i=1}^{n} g_{hi} g_{ji} \quad (h, j = 1, 2, \ldots, p), \tag{2}$$

$$a_h = \sum_{i=1}^{n} g_{hi} y_i \quad (h = 1, 2, \ldots, p), \tag{3}$$

we may write $\partial S / \partial b_h^* = -2(a_h - a_{h1} b_1^* - a_{h2} b_2^* - \ldots - a_{hp} b_p^*)$, so that the normal equations take the form

$$\begin{aligned} a_{11} b_1^* + a_{12} b_2^* + \ldots + a_{1p} b_p^* &= a_1 , \\ a_{21} b_1^* + a_{22} b_2^* + \ldots + a_{2p} b_p^* &= a_2 , \\ \ldots\ldots\ldots\ldots\ldots\ldots\ldots\ldots\ldots\ldots \\ a_{p1} b_1^* + a_{p2} b_2^* + \ldots + a_{pp} b_p^* &= a_p . \end{aligned} \tag{4}$$

Putting $\mathbf{A} = \|a_{hj}\|$, $\mathbf{b^*} = \left\| \begin{matrix} b_1^* \\ \vdots \\ b_p^* \end{matrix} \right\|$, $\mathbf{a} = \left\| \begin{matrix} a_1 \\ \vdots \\ a_p \end{matrix} \right\|$ we can rewrite the system of equations (4) in matrix form as

$$\mathbf{Ab^*} = \mathbf{a} . \tag{5}$$

The following statements concerning the normal equations are true:

(a) It follows from equation (2) that

$$\mathbf{A} = \mathbf{G} . \mathbf{G}' , \tag{6}$$

where $\mathbf{G}'$ is the transpose of $\mathbf{G}$. Thus $\mathbf{A}$ is a symmetric matrix with p rows and columns whose elements depend on the known coefficients g_{ji} of the defining equations. Further, it follows from (6) and (35.5.3) that $\mathbf{A}$ is positive definite.

(b) The estimates b^* are uniquely determined by

$$\mathbf{b^*} = \mathbf{A}^{-1} \mathbf{a} . \tag{7}$$

Equation (3) implies that

$$\mathbf{a} = \mathbf{Gy} , \tag{8}$$

where

$$\mathbf{y} = \left\| \begin{matrix} y_1 \\ \vdots \\ y_n \end{matrix} \right\| .$$

Substituting (8) into (7) we obtain

$$\mathbf{b^*} = \mathbf{A}^{-1} \mathbf{Gy} ; \tag{9}$$

thus the estimates $b_1^*, \ldots, b_p^*$ are expressed as linear functions of the observations $y_1, y_2, \ldots, y_n$.

The estimates b^* as well as the estimates of any parametric functions are themselves random variables. In § 35.5 we have introduced several general assumptions concerning the distribution of the y_i's. From those assumptions it is possible to derive formulae for expectations, variances and covariances of the estimates b^* and of estimates of parametric functions. The results will be summarized in § 35.8.

35.8. Variances and Covariances of Best Linear Unbiased Estimates

Denote by a^{hj} the elements of the inverse of $\mathbf{A}$, $\mathbf{A}^{-1} = \|a^{hj}\|$ $(h, j = 1, 2, \ldots, p)$. Using this notation, we may write

$$E(b_j^*) = b_j\,, \tag{1}$$

$$D(b_j^*) = \sigma^2 a^{jj}\,, \tag{2}$$

$$\mathrm{Cov}\,(b_h^*, b_j^*) = \sigma^2 a^{hj}\,. \tag{3}$$

The expectation and the variance of an estimate $f^* = \sum_{j=1}^{p} k_j b_j^*$ of a parametric function $f = \sum_{j=1}^{p} k_j b_j$ are given by

$$E(f^*) = f\,, \tag{4}$$

$$D(f^*) = \sigma^2 \sum_{h=1}^{p} \sum_{j=1}^{p} k_h k_j a^{hj}\,. \tag{5}$$

The covariance of estimates f_1^*, f_2^* of two parametric functions $f_1 = \sum_{j=1}^{p} k_j b_j$, $f_2 = \sum_{j=1}^{p} l_j b_j$ is given by

$$\mathrm{Cov}\,(f_1^*, f_2^*) = \sigma^2 \sum_{h=1}^{p} \sum_{j=1}^{p} k_h l_j a^{hj}\,.$$

Throughout this paragraph, σ^2 denotes the unknown variance of the variable y_i.

35.9. Residual Sum of Squares. Estimates of Standard Errors

Solving the normal equations, we obtain estimates $b_1^*, b_2^*, \ldots, b_p^*$ with the aid of which we can compute the quantities

$$y_i^* = g_{1i} b_1^* + g_{2i} b_2^* + \ldots + g_{pi} b_p^* \quad (i = 1, 2, \ldots, n)\,. \tag{1}$$

The difference

$$u_i = y_i - y_i^* \quad (i = 1, 2, \ldots, n) \tag{2}$$

is called the *residual of the variable* y_i and the sum

$$S_0 = \sum_{i=1}^{n} u_i^2 = \sum_{i=1}^{n} (y_i - \sum_{j=1}^{p} g_{ji} b_j^*)^2 \tag{3}$$

is called the *residual sum of squares* or the *error sum of squares.*

An estimate s^2 of the unknown variance σ^2 of the errors e_i or of the variables y_i is given by

$$s^2 = \frac{S_0}{n-p}. \tag{4}$$

Its denominator

$$\nu = n - p \tag{5}$$

is the so-called *number of degrees of freedom* for the residual sum of squares.

The estimate s^2 satisfies the condition

$$E(s^2) = \sigma^2, \tag{6}$$

so that it is an unbiassed estimate.

For practical computation of the residual sum of squares it is convenient to rewrite (3) as

$$S_0 = \sum_{i=1}^{n} y_i^2 - \sum_{j=1}^{p} a_j b_j^*; \tag{7}$$

this relation has already been used in §§ 35.2 and 35.3.

Replacing in the formulae of § 35.8 the variance σ^2 by its estimate s^2, we obtain unbiassed estimates of the variances and covariances of the best linear unbiassed estimates. The quantity

$$s = \sqrt{\frac{S_0}{n-p}} \tag{8}$$

is an estimate of the standard error of the variables y_i, and consequently of the errors e_i (0·6745s is called the *probable error* of y_i). Estimates of the standard error of the estimate b_j^* or $f^* = \sum_{j=1}^{p} k_j b_j^*$ are given by the formulae

$$s_{b_j^*} = s\sqrt{(a^{jj})}, \tag{9}$$

$$s_{f^*} = s\sqrt{(\sum_{h=1}^{p} \sum_{j=1}^{p} k_h k_j a^{hj})} \tag{10}$$

The expressions for standard errors given in § 35.3 have been derived from the general formulae (9) and (10).

35.10. Normal Distribution of the Random Variables e_i

The expressions for the expectations, variances and covariances of the estimates of parametric functions, as well as the expression for the expectation of s^2 and the estimates of standard errors, are all derived under the general assumptions about the distribution of the random variables e_i that have been stated in § 35.5. If we make the additional assumption that all the e_i's are normally distributed, we obtain a set of results that is of great importance for applications of the method of least squares to regression theory, to the fitting of curves to empirical data and to the calculus of observations, where the assumption of normality is usually adopted.

For the sake of brevity, we shall denote the normal distribution with mean μ and variance σ^2, the density of which at the point x equals

$$p(x) = \frac{1}{\sigma\sqrt{2\pi}} e^{-(x-\mu)^2/2\sigma^2} \quad (-\infty < x < \infty)$$

by $N(\mu, \sigma^2)$.

For normally distributed random variables, the assumption of zero correlations implies independence. Thus we may state the assumptions concerning the distribution of the random variables e_i (or y_i) in the equations (35.5.1) as follows: $e_1, e_2, \ldots, e_n$ are mutually independent random variables each of which is distributed according to $N(0, \sigma^2)$, where σ^2 is an unknown parameter.

The assumptions just stated imply that $y_1, y_2, \ldots, y_n$ are mutually independent random variables and y_i is distributed according to $N(\sum_{j=1}^{p} g_{ji}b_j, \sigma^2)$.

The following are direct consequences of these assumptions:

(a) the estimate b_j^* $(j = 1, 2, \ldots, p)$ is distributed according to $N(b_j, \sigma^2 a^{jj})$;

(b) the estimate $f^* = \sum_{j=1}^{p} k_j b_j^*$ of the parametric function $f = \sum_{j=1}^{p} k_j b_j$ is distributed according to

$$N(f, \sigma^2 \sum_{h=1}^{p} \sum_{j=1}^{p} k_h k_j a^{hj})\,;$$

(c) the ratio S_0/σ^2, where S_0 is the residual sum of squares, follows a χ^2-distribution with $n - p$ degrees of freedom;

(d) the estimates b_j^* $(j = 1, 2, \ldots, p)$ and the residual sum of squares S_0 are mutually independent random variables;

(e) the estimate $f^* = \sum_{j=1}^{p} k_j b_j^*$ of the parametric function $f = \sum_{j=1}^{p} k_j b_j$ and the residual sum of squares S_0 are mutually independent random variables;

(f) the ratio $r = (b_j^* - b_j)/s_{b_j*}$ $(j = 1, 2, \ldots, p)$, where s_{b_j*} is the estimate of the standard error of b_j^* given by expression (35.9.9), follows Student's t-distribution with $\nu = n - p$ degrees of freedom;

(g) the ratio $t = (f^* - f)/s_{f*}$, where $f^* = \sum_{j=1}^{p} k_j b_j^*$ is the estimate of the parametric function $f = \sum_{j=1}^{p} k_j b_j$ and s_{f*} the estimate of its standard error given by expression (35.9.10), follows Student's t-distribution with $n - p$ degrees of freedom.

The results listed above are of importance in mathematical statistics, where they are used to determine confidence intervals and in testing statistical hypotheses.

The distributions referred to in this paragraph are tabled e.g. in [116].

35.11. Defining Equations with Parameters Restricted by Linear Constraints

In some problems there appear defining equations whose parameters $b_1, b_2, \ldots, b_p$ satisfy s $(s < p)$ linearly independent relations of the form

$$r_{k1}b_1 + r_{k2}b_2 + \ldots + r_{kp}b_p - r_{k0} = 0 \quad (k = 1, 2, \ldots, s), \tag{1}$$

where $r_{k1}, r_{k2}, \ldots, r_{kp}, r_{k0}$ are given constants. The estimates $b_1^*, b_2^*, \ldots, b_p^*$ are then determined so that they satisfy conditions (1). For the determination of the estimates either of the two following procedures can be adopted:

(a) Equations (1) are used to eliminate s parameters from the equations (35.5.1). Thus the modified defining equations contain only $p - s$ unknown parameters that are not restricted by any conditions. The defining equations obtained in this way are then treated with the aid of the methods explained in §§ 35.5–35.10. The residual sum of squares then has $\nu = n - (p - s)$ degrees of freedom.

(b) Methods for determining constrained minima of functions of several variables are applied. The estimates b_j^* are determined as quantities minimizing the sum

$$S = \sum_{i=1}^{n} (y_i - \sum_{j=1}^{p} g_{ji} b_j^*)^2$$

subject to conditions (1), viz.

$$\sum_{j=1}^{p} r_{kj} b_j^* - r_{k0} = 0 \quad (k = 1, 2, \ldots, s).$$

Thus a new function W is introduced,

$$W = S + \sum_{k=1}^{s} \lambda_k (\sum_{j=1}^{p} r_{kj} b_j^* - r_{k0}), \tag{2}$$

where $\lambda_1, \lambda_2, \ldots, \lambda_s$ are Lagrange's multipliers, and the system of equations

$$\frac{\partial W}{\partial b_h^*} = 0 \quad (h = 1, 2, \ldots, p),$$

$$\sum_{j=1}^{p} r_{kj} b_j^* - r_{k0} = 0 \quad (k = 1, 2, \ldots, s) \tag{3}$$

is solved with respect to the unknowns $b_1^*, b_2^*, \ldots, b_p^*, \lambda_1, \lambda_2, \ldots, \lambda_s$.

Example 1. A straight line with equation $y = a + bx$ has to be fitted to a set of points (x_i, y_i) $(i = 1, 2, \ldots, n)$ so as to pass through the point (x_0, y_0). The measurements y_i are described by equations $y_i = a + bx_i + e_i$ whose parameters a, b are constrained by the condition $a + bx_0 - y_0 = 0$. Eliminating the parameter a we obtain $y_i = y_0 + b(x_i - x_0) + e_i$. The estimate of b is

$$b^* = \frac{\sum_{i=1}^{n} (y_i - y_0)(x_i - x_0)}{\sum_{i=1}^{n} (x_i - x_0)^2}$$

and $a^* = y_0 - b^* x_0$. The number of degrees of freedom for the residual sum of squares is $n - 1$.

Denote by S_0 the residual sum of squares corresponding to the defining equations (35.5.1) whose parameters are not restricted by any constraints, and by S_1 the residual sum of squares corresponding to defining equations (35.5.1) if the parameters b are supposed to satisfy the additional conditions (1). Then, obviously, $S_1 \geqq S_0$.

Assuming the random variables e_i to be normally distributed the distributions of residual sums of squares S_0, S_1 have the following properties:

(a) S_0/σ^2 has a χ^2-distribution with $n - p$ degrees of freedom (cf. § 35.10, (c));

(b) S_1/σ^2 has a χ^2-distribution with $n - p + s$ degrees of freedom;

(c) $(S_1 - S_0)/\sigma^2$ has a χ^2-distribution with s degrees of freedom;

(d) S_0/σ^2 and $(S_1 - S_0)/\sigma^2$ are mutually independent random variables.

35.12. Weights

So far we have assumed that the random variables e_i in the equations (35.5.1) have equal variances. We shall now drop this assumption and discuss a more general case, where $e_1, e_2, \ldots, e_n$ are random variables with zero expectations,

$$E(e_i) = 0 \quad (i = 1, 2, \ldots, n), \tag{1}$$

zero covariances

$$\operatorname{Cov}(e_i, e_j) = 0 \quad (i, j = 1, 2, \ldots, n;\ i \neq j)\,, \tag{2}$$

and with variances

$$D(e_i) = \frac{\sigma^2}{p_i} \quad (i = 1, 2, \ldots, n)\,. \tag{3}$$

In equation (3) σ^2 is an unknown parameter and the p_i's are given positive constants. It follows from these assumptions that the random variables $y_1, \ldots, y_n$ have the following expectations, variances and covariances:

$$E(y_i) = \sum_{j=1}^{p} g_{ji} b_j \quad (i = 1, 2, \ldots, n)\,, \tag{4}$$

$$\operatorname{Cov}(y_i, y_j) = 0 \quad (i, j = 1, 2, \ldots, n;\ i \neq j)\,, \tag{5}$$

$$D(y_i) = \frac{\sigma^2}{p_i} \quad (i = 1, 2, \ldots, n)\,. \tag{6}$$

The constant p_i is called the *weight* corresponding to the variable e_i or y_i; it may be noted that to the variable with greater weight there corresponds a smaller variance.

Assuming that $e_1, \ldots, e_n$ are mutually independent and normally distributed according to $N(0, \sigma^2/p_i)$ $(i = 1, 2, \ldots, n)$, then the variables $y_1, y_2, \ldots, y_n$ are also mutually independent, and y_i has the distribution $N(\sum_{j=1}^{p} g_{ji} b_j, \sigma^2/p_i)$.

If the random variable e_i has variance $D(e_i) = \sigma^2/p_i$, then the variable $e_i' = e_i\sqrt{p_i}$ has variance $D(e_i') = p_i D(e_i) = \sigma^2$; thus the transformation

$$z_i = y_i \sqrt{(p_i)} \quad (i = 1, 2, \ldots, n) \tag{7}$$

yields variables $z_1, z_2, \ldots, z_n$ for which the following equation holds:

$$z_i = (g_{1i} b_1 + \ldots + g_{pi} b_p)\sqrt{(p_i)} + e_i \sqrt{p_i}\,, \tag{8}$$

where the variables $e_i' = e_i\sqrt{p_i}$ have equal variances σ^2. We may thus apply all results of §§ 35.5 – 35.11 to the variables z_i.

The estimates $b_1^*, \ldots, b_p^*$ are obtained from the condition

$$S = \sum_{i=1}^{n} \Big(z_i - \sqrt{(p_i)} \sum_{j=1}^{p} g_{ji} b_j^*\Big)^2 = \sum_{i=1}^{n} p_i \Big(y_i - \sum_{j=1}^{p} g_{ji} b_j^*\Big)^2 = \min\,. \tag{9}$$

Putting

$$c_{hj} = \sum_{i=1}^{n} p_i g_{hi} g_{ji} \quad (h, j = 1, 2, \ldots, p)\,, \tag{10}$$

$$c_h = \sum_{i=1}^{n} p_i g_{hi} y_i \quad (h = 1, 2, \ldots, p) \tag{11}$$

we can write the normal equations $\partial S/\partial b_h^* = 0$ $(h = 1, 2, \dots, p)$ in the form

$$\begin{aligned} c_{11}b_1^* + c_{12}b_2^* + \dots + c_{1p}b_p^* &= c_1 , \\ c_{21}b_1^* + c_{22}b_2^* + \dots + c_{2p}b_p^* &= c_2 , \\ \dots\dots\dots\dots\dots\dots\dots\dots \\ c_{p1}b_1^* + c_{p2}b_2^* + \dots + c_{pp}b_p^* &= c_p . \end{aligned} \tag{12}$$

The residual sum of squares S_0 is

$$S_0 = \sum_{i=1}^{n} p_i(y_i - \sum_{j=1}^{p} g_{ji}b_j^*)^2 \tag{13}$$

and it can be expressed as

$$S_0 = \sum_{i=1}^{n} p_i y_i^2 - \sum_{j=1}^{p} c_j b_j^* . \tag{14}$$

An unbiassed estimate of the variance σ^2 is given by

$$s^2 = S_0/(n - p) . \tag{15}$$

An unbiassed estimate of the variance of the variable e_i or y_i is s^2/p_i. An estimate of the standard error of e_i or y_i is equal to

$$s_{e_i} = s_{y_i} = \frac{s}{\sqrt{p_i}} , \tag{16}$$

so that $s = \sqrt{[S_0/(n - p)]}$ is an estimate of the standard error of a variable having weight $p = 1$.

If we denote by $\mathbf{C}$ the matrix of the system of normal equations (12) and by c^{hj} the element in row h and column j of its inverse $\mathbf{C}^{-1}$, we can express the estimate of the standard error of the estimate b_j^* by the formula

$$s_{b_j*} = s\sqrt{(c^{jj})} , \tag{17}$$

and the estimate of the standard error of the estimate $f^* = \sum_{j=1}^{p} k_j b_j^*$ of a parametric function $f = \sum_{j=1}^{p} k_j b_j$ will be given by

$$s_{f*} = s\sqrt{(\sum_{h=1}^{p} \sum_{j=1}^{p} k_h k_j c^{hj})} . \tag{18}$$

Assuming normality of the distribution of the e_i's we can apply all the results of § 35.10 with appropriate modifications.

The generalization of the least squares method to observations with different weights is of great importance especially in the calculus of observations.

Example 1. (a) Suppose that the variables y_i have weights p_i $(i = 1, 2, \ldots, n)$ and are determined by the equation

$$y_i = a + e_i\,,$$

where a is an unknown constant.

The least squares condition is

$$\sum_{i=1}^{n} p_i(y_i - a^*)^2 = \min\,.$$

Hence

$$a^* = \frac{\sum_{i=1}^{n} p_i y_i}{\sum_{i=1}^{n} p_i}\,. \tag{19}$$

Residual sum of squares:

$$S_0 = \sum_{i=1}^{n} p_i y_i^2 - \frac{\left(\sum_{i=1}^{n} p_i y_i\right)^2}{\sum_{i=1}^{n} p_i}\,, \quad \nu = n - 1\,. \tag{20}$$

Estimate of the standard error

$$s_{a^*} = \frac{s}{\sqrt{\left(\sum_{i=1}^{n} p_i\right)}}\,, \quad \text{where} \quad s = \sqrt{\frac{S_0}{n-1}}\,. \tag{21}$$

(b) Suppose that the variables y_i have weights p_i $(i = 1, 2, \ldots, n)$ and are determined by equations $y_i = a + bx_i + e_i$, where a and b are unknown constants and the x_i $(i = 1, 2, \ldots, n)$ have fixed values.

The least squares condition is

$$\sum_{i=1}^{n} p_i(y_i - a^* - b^* x_i)^2 = \min\,.$$

Normal equations:

$$\left(\sum_{i=1}^{n} p_i\right) a^* + \left(\sum_{i=1}^{n} p_i x_i\right) b^* = \sum_{i=1}^{n} p_i y_i\,,$$

$$\left(\sum_{i=1}^{n} p_i x_i\right) a^* + \left(\sum_{i=1}^{n} p_i x_i^2\right) b^* = \sum_{i=1}^{n} p_i x_i y_i\,.$$

Introducing the notation

$$\tilde{x} = \frac{\sum_{i=1}^{n} p_i x_i}{\sum_{i=1}^{n} p_i}, \quad \tilde{y} = \frac{\sum_{i=1}^{n} p_i y_i}{\sum_{i=1}^{n} p_i},$$

$$A = \sum_{i=1}^{n} p_i (x_i - \tilde{x})^2 = \sum_{i=1}^{n} p_i x_i^2 - \frac{\left(\sum_{i=1}^{n} p_i x_i\right)^2}{\sum_{i=1}^{n} p_i},$$

$$B = \sum_{i=1}^{n} p_i (y_i - \tilde{y})^2 = \sum_{i=1}^{n} p_i y_i^2 - \frac{\left(\sum_{i=1}^{n} p_i y_i\right)^2}{\sum_{n=1}^{n} p_i},$$

$$C = \sum_{i=1}^{n} p_i (x_i - \tilde{x})(y_i - \tilde{y}) = \sum_{i=1}^{n} p_i x_i y_i - \frac{\left(\sum_{i=1}^{n} p_i x_i\right)\left(\sum_{i=1}^{n} p_i y_i\right)}{\sum_{i=1}^{n} p_i},$$

we can write the solution of the normal equations as

$$b^* = \frac{C}{A}, \quad a^* = \tilde{y} - \tilde{x} b^* . \tag{22}$$

The residual sum of squares can be written in the form

$$S_0 = B - b^* C, \quad \nu = n - 2 . \tag{23}$$

The estimates of standard errors are

$$s_a^* = s \sqrt{\left(\frac{\sum_{i=1}^{n} p_i x_i}{A \sum_{i=1}^{n} p_i} \right)}, \quad s_{b*} = \frac{s}{\sqrt{A}},$$

where $s = \sqrt{[S_0/(n - 2)]}$.

35.13. Defining Equations, the Matrix of Whose Coefficients is of Rank Smaller than p

In some problems of mathematical statistics we have to deal with random variables $y_1, y_2, \ldots, y_n$, the defining equations (35.5.1) of which have a matrix $\mathbf{G}$ with linearly dependent rows so that its rank r is smaller than p. In such a case not all parametric functions are estimable (for the definition of an estimable parametric function see § 35.6), and the following statements hold true:

If the matrix $\mathbf{G}$ has rank $h(\mathbf{G}) = r < p$, then :

(a) For a parametric function $f = \sum_{j=1}^{p} k_j b_j$ to be estimable, it is necessary and sufficient that the vector

$$\begin{Vmatrix} k_1 \\ k_2 \\ \vdots \\ k_p \end{Vmatrix}$$

be a linear combination of columns of $\mathbf{G}$.

(b) The system of normal equations (35.7.5) (the matrix of which has rank $h = r$) always has a solution; this solution, however, is not unique.

(c) If $f = \sum_{j=1}^{p} k_j b_j$ is an estimable parametric function, then its best linear unbiassed estimate is uniquely determined by the expression $f^* = \sum_{j=1}^{p} k_j b_j^*$, where $b_1^*, \ldots, b_p^*$ is any solution of the system of normal equations (35.7.5) (*The Gauss-Markov Theorem*).

(d) The residual sum of squares $S_0 = \sum_{i=1}^{n} (y_i - \sum_{j=1}^{p} g_{ji} b_j^*)^2$, where $b_1^*, b_2^*, \ldots, b_p^*$ is any solution of the system of normal equations (35.7.5), is determined uniquely as a function of the variables $y_1, y_2, \ldots, y_n$.

(e) An unbiassed estimate of the variance σ^2 is given by

$$s^2 = \frac{S_0}{n - r}.$$

If the variables y_i are normally distributed, then S_0/σ^2 has a χ^2-distribution with $n - r$ degrees of freedom.

C. ELEMENTS OF THE CALCULUS OF OBSERVATIONS

35.14. The Law of Errors

Every measurement is subject to errors. According to their origin and their nature, the errors can be classified into two broad groups, viz. systematic errors and random errors.

Systematic errors have always an assignable and, usually, permanent cause (such as a defect of the equipment, an incorrect setting of the measuring instrument, faulty calibration of scales, etc.). A systematic error distorts all measurements in the same direction and, usually, by the same constant quantity. As a rule, the source of systematic errors can be discovered and removed.

Random errors (or *chance errors*) are due to a large number of unknown causes and they influence the results in quite an irregular way; their size as well as the direction of the corresponding deviation differ from one measurement to another. Unlike systematic errors, they cannot be avoided by any precautions. The calculus of observations is based on the assumption that these errors have the character of random variables and thus that they may be treated by statistical methods.

Let us have a quantity the true value of which is a. If this quantity is measured n times and the individual results are denoted by $y_1, y_2, \ldots, y_n$, we can write

$$y_i = a + e_i \quad (i = 1, 2, \ldots, n),$$

where e_i is the error of the i-th measurement y_i. We assume that the measurements are free of systematic errors and that the e_i's are random variables. The probability distribution of the random variables $e_1, e_2, \ldots, e_n$ is called the *law of error*.

A law of error of fundamental importance is the *Gaussian Law of Error* (*Normal Law of Error*). According to this law the errors $e_1, e_2, \ldots, e_n$ are independent of one another and each has the normal distribution $N(0, \sigma^2)$, the density function of which is given by

$$p(x) = \frac{1}{\sigma\sqrt{(2\pi)}}\, e^{-x^2/(2\sigma^2)} \quad (-\infty < x < \infty). \tag{1}$$

The constant σ^2 (equal to the variance of the distribution (1)) is regarded as an unknown parameter.

Putting

$$h = \frac{1}{\sigma\sqrt{2}} \tag{2}$$

we can rewrite the probability density (1) in the form

$$p(x) = \frac{h}{\sqrt{\pi}}\, e^{-h^2x^2} \quad (-\infty < x < \infty). \tag{3}$$

The constant h is called the *modulus of precision*.

The probability that the error will assume a value in the interval $(-d, +d)$ is equal to

$$\int_{-d}^{+d} p(x)\, dx = \frac{2h}{\sqrt{\pi}} \int_0^d e^{-h^2x^2}\, dx = \frac{2}{\sqrt{\pi}} \int_0^{hd} e^{-u^2}\, du = \Theta(hd)\,.$$

Various tables of the function $\Theta(hd)$ are available.

Referring directly to the measurements $y_1, y_2, \ldots, y_n$ we may reformulate the normal law of error as follows: $y_1, y_2, \ldots, y_n$ are mutually independent random variables each of which is distributed according to $N(a, \sigma^2)$, where a is the true value of the measured quantity.

Generally, all measurements of the same quantity cannot be regarded as equally precise. The quality of individual measurements is then expressed by weights so that greater weight is assigned to more precise measurements. Let us denote the weight of the measurement y_i by p_i $(i = 1, 2, \ldots, n)$; the p_i's are positive numbers. The variance of the measurement y_i is then equal to $\sigma_i^2 = \sigma^2/p_i$ and its precision is $h_i = h\sqrt{(p_i)}$, where $h = 1/\sqrt{(2\sigma)}$. In accordance with the normal law of error, the random variables $e_1, e_2, \ldots, e_n$ are then independent and the error e_i $(i = 1, 2, \ldots, n)$ is normally distributed, its density being

$$p(x) = \frac{h\sqrt{p_i}}{\sqrt{\pi}}\, e^{-p_i h^2 x^2} \quad (-\infty < x < \infty)\,. \tag{4}$$

The weights $p_1, p_2, \ldots, p_n$ are given positive constants and the precision h is an unknown parameter.

REMARK 1. The argument leading to the adoption of the normal distribution as the law of error can be briefly summarized as follows: It is assumed that the total deviation of the measurement from the true value, remaining after the elimination of any systematic errors, is the sum of a large number of independent random variables representing very small "elementary errors" originating from various chance causes. The causes of elementary errors are many uncontrollable factors (such as slight disturbances in the state of the equipment, fluctuations of atmospheric conditions, instantaneous changes in the capability of the observer etc.) which unavoidably influence any measurement. According to the limit theorems of probability theory, the sum of a large number of random variables has, under some assumptions, a distribution very near to the normal distribution.

35.15. Adjustment of Data by the Method of Least Squares

Let a denote the true value of the measured quantity and $y_1, y_2, \ldots, y_n$ n be independent measurements of this quantity. It is not possible to determine the exact value of the measured quantity since the measurements are unavoidably subject to errors. It is, however, possible to use the measurements $y_1, y_2, \ldots, y_n$ to form a statistical estimate a^* of the unknown constant a. The true value of the measured quantity is then inferred from its statistical estimate. In the calculus of observations the estimate a^* is called the *adjusted value of* a. To find the adjusted value and determine its precision, the least squares method is employed. This method has been given a general treatment in §§ 35.10–35.13.

If the measurements y_i have weights p_i $(i = 1, 2, \ldots, n)$, the adjusted value a^* is determined from the condition (see Example 35.12.1a))

$$\sum_{i=1}^{n} p_i(y_i - a^*)^2 = \min ;$$

it is given by equation (35.12.19). Denoting summation by square brackets (this notation is generally adopted in the calculus of observations) we can write the adjusted value in the form

$$a^* = \frac{[py]}{[p]}, \tag{1}$$

where

$$[py] = \sum_{i=1}^{n} p_i y_i \quad \text{and} \quad [p] = \sum_{i=1}^{n} p_i .$$

The precision is generally expressed in terms of the standard error (sometimes called *mean square error*). Of course, a quantity with smaller standard error is considered to be more precise. The standard error of a single measurement y_i is equal to $\sigma/\sqrt{p_i}$, the standard error of the adjusted value a^* is $\sigma/\sqrt{[p]}$. The constant σ represents the standard error of a measurement which has the weight $p = 1$; σ is regarded as an unknown constant whose estimator is the quantity s which is computed from the formula

$$s = \sqrt{\frac{[pu^2]}{n - 1}}, \tag{3}$$

where $u_i = y_i - a^*$ is the residual of the measurements y_i and $[pu^2]$ is the residual sum of squares. (In the calculus of observations the residual $u_i = y_i - a^*$ is often called "the *correction of the measurement* y_i" or "*apparent error of* y_i", the latter term emphasizes the contrast between u_i and the "true error" $e_i = y_i - a$ which is, of course, unknown.)

In terms of the measurements y_i the residual sum of squares is expressed by formula (35.12.20). An estimate of the standard error of a single measurement is

$$s_{y_i} = \frac{s}{\sqrt{p_i}}, \tag{4}$$

and an estimate of the standard error of the adjusted value a^* is

$$s_{a*} = \frac{s}{\sqrt{[p]}} . \tag{5}$$

A generally adopted custom in the calculus of observations is to present the result of the adjustment in the form $a^* \pm s_{a*}$.

The problem that has just been described, viz. that of adjusting a single constant a on the basis of n independent measurements made on it, represents the simplest case of the problem of *adjustment of unconditioned observations.* A more general problem of adjustment of *indirect observations* may be formulated as follows: There are p unknown quantities, the values of which will be denoted by $b_1, b_2, \ldots, b_p$. The quantities cannot be measured directly; it is only possible to measure another quantity whose value y is a given function of the constants $b_1, b_2, \ldots, b_p$, i.e.

$$y = f(b_1, b_2, \ldots, b_p) . \tag{6}$$

The adjustment of unknown constants is carried out on the basis of n $(n > p)$ independent direct measurements of the quantity y.

The least squares method is directly and exactly applicable when y is a linear combination of the constants $b_1, b_2, \ldots, b_p$, thus

$$y = g_1 b_1 + g_2 b_2 + \ldots + g_p b_p , \tag{7}$$

where $g_1, g_2, \ldots, g_p$ are either constants or known functions of one or more variables which do not involve any unknown constants.

Denoting the i-th measurement of the quantity y by y_i, its true error by e_i, and the values of $g_1, g_2, \ldots, g_p$ that correspond to the i-th measurement by $g_{1i}, g_{2i}, \ldots, g_{pi}$, we can write

$$y_i = g_{1i} b_1 + g_{2i} b_2 + \ldots + g_{pi} b_p + e_i \quad (i = 1, 2, \ldots, n) . \tag{8}$$

We assume that $g_{1i}, g_{2i}, \ldots, g_{pi}$ are free of error, so that only the y's are subject to error.

The problem of adjusting the constants $b_1, b_2, \ldots, b_p$ coincides exactly with the problem of estimating unknown parameters in a linear regression equation which has received a detailed treatment in the first three paragraphs of this chapter. When the measurements y_i have different weights, we use the modification of the least squares method dealt with in § 35·12. In that paragraph there is an example showing how to adjust two constants, a and b that are related to the measured quantity by the equation $y = a + bx$, if y_i has weight p_i (Example 35.12.1b)).

When y is not a linear function of the unknown constants $b_1, b_2, \ldots, b_p$, we can have recourse to the approximate method described in § 35.4.

If the unknown constants $b_1, b_2, \ldots, b_p$ in equation (7) are restricted by s $(s < p)$ given linearly independent conditions of the form

$$r_{k1}b_1 + r_{k2}b_2 + \ldots + r_{kp}b_p - r_{k0} = 0 \quad (k = 1, 2, \ldots, s)\,, \tag{9}$$

we say that we are concerned with *conditioned measurements*. The adjustment is then carried out so that the adjusted values $b_1^*, b_2^*, \ldots, b_p^*$ satisfy equations (9).

An example of conditioned measurements is provided by the measurement of angles α, β, γ in a plane triangle. In this case the sum of the angles must equal 180°, i.e. α, β, γ must satisfy the relation $\alpha + \beta + \gamma - 180° = 0°$.

The adjustment must be made so that the adjusted values $\alpha^*, \beta^*, \gamma^*$ satisfy the same relation, $\alpha^* + \beta^* + \gamma^* - 180° = 0°$.

A general method of adjustment of conditioned measurement is contained in § 35.11. In the case of measurements of unequal weights the modified least squares method of § 35.12 is to be used.

Conclusion: According to the Gauss-Markov Theorem, the values adjusted by the method of least squares are best linear unbiassed estimates of the constants involved in the equations (7). An adjusted value is thus a linear combination of n independent observations $y_1, y_2, \ldots, y_n$, the coefficients of which are chosen so as to make its expectation equal to the true value of the measured constant and to minimize its mean square error.

Under the normal law of error, the results reported in § 35.10 hold for adjusted values.

35.16. Functions of Random Variables. The Law of Propagation of Error

(a) Linear functions of random variables. The expectation of a function

$$u = k_0 + \sum_{i=1}^{n} k_i y_i\,, \tag{1}$$

where $y_1, y_2, \ldots, y_n$ are random variables and $k_0, k_1, \ldots, k_n$ constants, is given by

$$E(u) = k_0 + \sum_{i=1}^{n} k_i a_i\,, \tag{2}$$

where a_i denotes the expectation of y_i. The standard error σ_u of u is given by

$$\sigma_u = \sqrt{\left(\sum_{i=1}^{n} k_i^2\sigma_i^2 + 2\sum_{i=1}^{n-1}\sum_{j=i+1}^{n} k_i k_j \sigma_{ij}\right)}\,, \tag{3}$$

where σ_i denotes the standard error of y_i and σ_{ij} the covariance of y_i, y_j.

If the variables $y_1, y_2, \ldots, y_n$ are independent, all the covariances σ_{ij} are equal to zero and thus

$$\sigma_n = \sqrt{\left(\sum_{i=1}^{n} k_i^2 \sigma_i^2\right)}. \tag{4}$$

If all the variables $y_1, y_2, \ldots, y_n$ are normally distributed, then $u = k_0 + \sum_{i=1}^{n} k_i y_i$ itself is normally distributed about the mean given by (2) with a standard error given by (3) or, in the case of independence, by (4).

(b) Expectation and variances of non-linear functions of random variables. If the random variables $y_1, y_2, \ldots, y_n$ are independent and $u = f(y_1, y_2, \ldots, y_n)$, where $f(y_1, y_2, \ldots, y_n)$ is a non-linear function, then the following approximate formulae hold for the expectation and standard error of u:

$$E(u) \approx f(a_1, a_2, \ldots, a_n), \tag{5}$$

$$\sigma_u = \sqrt{\left(\sum_{i=1}^{n} f_i^2 \sigma_i^2\right)}, \tag{6}$$

where a_i and σ_i are the expectation and standard error, respectively, of y_i, and

$$f_i = \left.\frac{\partial f(y_1, y_2, \ldots, y_n)}{\partial y_i}\right|_{(y_1, y_2, \ldots, y_n) = (a_1, a_2, \ldots, a_n)}.$$

(c) The law of propagation of error. Formulae (4) and (6) showing how the errors of the independent random variables $y_1, y_2, \ldots, y_n$ influence the variable $u = f(y_1, y_2, \ldots, y_n)$, are called the *law of propagation of error.*

Given adjusted values $a_1^*, a_2^*, \ldots, a_n^*$ and estimates $s_1, s_2, \ldots, s_n$ of their standard errors, an estimate s_{u*} of the standard error of the quantity $u^* = f(a_1^*, a_2^*, \ldots, a_n^*)$ is obtained by substituting s_i for σ_i and a_i^* for a_i in the formula for σ_u.

REFERENCES

1. Abbott, P., Kerridge, C.: National Certificate Mathematics, Vol. I. London, English Univ. Press 1961.
2. Abbott, P., Marshall, H.: National Certificate Mathematics, Vol. II. London, English Univ. Press 1961.
3. Agnew, R. P.: Differential Equations. New York—Toronto—London, McGraw-Hill 1960.
4. Ahlfors, L. V.: Complex Analysis. New York—Toronto—London, McGraw-Hill 1953.
5. Aitken, A. C.: Determinants and Matrices. Edinburgh, Oliver and Boyd 1946.
6. Aitken, A. C.: On Bernoulli's Numerical Solution of Algebraic Equations — Proc. Roy. Soc. Edinb., 4—6, 1926: 289—305.
7. Aitken, A. C.: Statistical Mathematics, 4th. Ed. Edinburgh, Oliver and Boyd 1945.
8. Akhiezer, N. I.: The Calculus of Variations. New York—London, Blaisdell 1962.
9. Akhiezer, N. I.: Theory of Approximations. New York, Ungar Publishing Co. 1956.
10. Akhiezer, N. I., Glazman, I. M.: Theorie der linearen Operatoren im Hilbert-Raum. Berlin, Akademieverlag 1954.
11. Allcock, H. J., Jones, J. R., Michel, J. G. L.: The Nomogram. London, Pitman 1950.
12. Allen, D. N. de G.: Relaxation Methods. New York—Toronto—London: McGraw-Hill 1954.
13. Andree, R. V.: Selections from Modern Abstract Algebra. New York, Holt, Rinehart and Winston 1962.
14. Andronov, A. A., Chaikin, C. E.: Theory of Oscillations. Princeton, Princeton Univ. Press 1949.
15. Angot, A.: Compléments de Mathématiques à l'usage des ingénieurs de l'électrotechnique et des télécomunications, 3ième ed. revue et augmentée. Paris, Éditions de la Revue d'optique 1957.
16. Apostol, T. M.: Calculus, Vols 1, 2. New York—London, Blaisdell 1961.
17. Arscott, F. M.: Periodic Differential Equations. Oxford—London—New York, Pergamon Press 1964.
18. Artobolevskii, I. I.: Mechanisms for the Generation of Plane Curves. Oxford—London—New York, Pergamon Press 1964.
19. Askwith, E. H.: Analytical Geometry of the Conic Sections. London, Adam and Charles Black 1950.
20. Babuška, I., Práger, M., Vitásek, E.: Numerical Solution of Differential Equations. Prague, SNTL; London—New York, J. Wiley and Sons 1966.
21. Babuška, I., Rektorys, K., Vyčichlo, F.: Mathematische Elastizitätstheorie der ebenen Probleme. Berlin, Akademie-Verlagsgesellschaft 1960.
22. Ball, R. W.: Principles of Abstract Algebra. New York, Holt, Rinehart and Winston 1963.
23. Banachiewicz, T.: Méthode de résolution numérique des équations linéaires, ... — Bull. Int. de l'Ac. Polon., Cl. des Sc. Math. et Natur., Série A, Se. Math., 1938.

24. Barnard, S., Child, J. M.: Advanced Algebra. London, Macmillan 1939.
25. Bateman, H.: Partial Differential Equations of Mathematical Physics. London—New York, Cambridge Univ. Press 1959.
26. Batschelet, E.: Über die numerische Auflösung von Randwertproblemen bei elliptischen partiellen Differentialgleichungen. — Zeitschr. angew. Math. u. Phys., 3, 1952: 165—193.
27. Baule, B.: Die Mathematik des Naturforschers und Ingenieurs, Bd. I—VIII. Leipzig, S. Hirzel Verlag 1955—1959.
28. Beaumont, R. A., Ball, A. W.: Introduction to Modern Algebra and Matrix Theory. New York, Holt, Rinehart and Winston 1954.
29. Beckenbach, E. F.: Modern Mathematics for the Engineer, 1st series 1956, 2nd series 1961. New York—Toronto—London, McGraw-Hill.
30. Bell, R. J. T.: Coordinate Solid Geometry. London, Macmillan, 1954; New York, St. Martin's Press 1959.
31. Bellman, R.: Stability Theory of Differential Equations. New York—Toronto—London, McGraw-Hill 1953.
32. Bermant, A. F.: Course of Mathematical Analysis. Oxford—London—New York, Pergamon Press 1963.
33. Bickley, W. G.: Difference and Associated Operators with some Applications. — J. Math. Phys., 27, 1948: 183—192.
34. Bickley, W. G.: Via Vector to Tensor. London, English Univ. Press 1962.
35. Bieberbach, L.: Theorie der Differentialgleichungen. Berlin, Springer 1933.
36. Birkhoff, G., Maclane, S.: A Survey of Modern Algebra. London, Macmillan 1941.
37. Birnbaum, Z. W.: Introduction to Probability and Mathematical Statistics. New York, Harper 1962.
38. Blaschke, W.: Differentialgeometrie I. Berlin, Springer 1930.
39. Bliss, G. A.: Calculus of Variations. Mathematical Association of America 1944.
40. Bliss, G. A.: Lectures on the Calculus of Variations. Chicago, Univ. of Chicago Press 1946.
41. Bolza, O.: Lectures on the Calculus of Variations. New York, Dover Publications 1961.
42. Bowman, F.: Introduction to Determinants and Matrices. London, English Univ. Press 1962.
43. Brand, L.: Advanced Calculus. New York, J. Wiley and Sons; London, Chapman and Hall 1955.
44. Brand, L.: Vector and Tensor Analysis. New York, J. Wiley and Sons 1947.
45. Brenner, J.: Problems in Differential Equations. San Francisco—London, Freeman and Co. 1963.
46. Brodetzky, S.: A First Course in Nomography. London, Bell and Sons 1938.
47. Bromwich, T. J. I'A.: Theory of Infinite Series. London, Macmillan 1926.
48. Bronstein, I. N., Semendiaev, K. A.: Taschenbuch der Mathematik für Ingenieure und Studenten der technischen Hochschulen, 6. Aufl. Leipzig, Teubner 1963.
49. Burington, R. S.: Handbook of Mathematical Tables and Formulae. New York—Toronto—London, McGraw-Hill 1965.
50. Burkill, J. C.: First Course in Mathematical Analysis. London—New York, Cambridge Univ. Press 1962.
51. Burnside, W. S., Panton, A. W.: Theory of Equations. Dublin, Dublin Univ. Press 1935.
52. Carathéodory, C.: Theory of Functions, Vols I, II, 2nd. Ed. New York, Chelsea Publ. Co. 1958.
53. Carathéodory, C.: Variationsrechnung und partielle Differentialgleichungen erster Ordnung, 2. Aufl. Leipzig, Teubner 1956.
54. Carslaw, H. S.: Fourier Series and Integrals. London, Macmillan 1949.

55. Carslaw, H. S., Jaeger, J. C.: Operational Methods in Applied Mathematics. London, Oxford Univ. Press 1948.
56. Cesari, L.: Asymptotic Behaviour and Stability Problems in Differential Equations. Berlin, Springer 1963.
57. Chaundy, T. W.: The Differential Calculus. London, Oxford Univ. Press 1935.
58. Chevalley, C.: Fundamental Concepts of Algebra. New York, Academic Press 1956.
59. Churchill, R. V.: Complex Variable and Applications. New York—Toronto—London, McGraw-Hill 1960.
60. Churchill, R. V.: Fourier Series and Boundary Value Problems. New York—Toronto—London, McGraw-Hill 1963.
61. Churchill, R. V.: Operational Mathematics in Enginneering, 2nd Ed. New York—Toronto—London, McGraw Hill 1958.
62. Clenshaw, C. W., Olver, F. W. J.: Solution of Differential Equations by Recurrence Relations. — Washington Mathematical Tables and Other Aids to Computation. 5, 1951; 34—39.
63. Coburn, N.: Vector and Tensor Analysis. New York, Macmillan 1955.
64. Cochran, W. G.: Sampling Techniques, 2nd Ed. London—New York, J. Wiley and Sons 1963.
65. Coddington, E. A., Levinson, N.: Theory of Ordinary Differential Equations. New York—Toronto—London, McGraw-Hill 1955.
66. Collatz, L.: Bemerkungen zur Fehlerabschätzung für das Differenzenverfahren bei partiellen Differentialgleichungen. — Zeitschr. angew. Math. u. Mech. 13, 1933: 56—57.
67. Collatz, L.: Eigenwertaufgaben mit technischen Anwendungen. Leipzig, Akademische Verlagsgesellschaft 1949.
68. Collatz, L.: Functional Analysis and Numerical Mathematics. New York, Academic Press 1966.
69. Collatz, L.: The Numerical Treatment of Differential Equations. Berlin, Springer 1960.
70. Cooke, R. G.: Linear Operators. London, Macmillan 1953.
71. Copson, E. T.: Asymptotic Expansions. London-New York, Cambridge Univ. Press 1964.
72. Copson, E. T.: Theory of Functions of a Complex Variable. London, Oxford Univ. Press 1935.
73. Courant, R.: Differential and Integral Calculus, Vols 1, 2. London, Blackie and Sons 1936.
74. Courant, R., Friedrichs, K., Lewy, G.: Über die partiellen Differenzengleichungen der mathematischen Physik. — Math. Ann. 100, 1928: 37—74.
75. Courant, R., Hilbert, D.: Methods of Mathematical Physics. Vols 1, 2. New York, Interscience 1962.
76. Craig, H. V.: Vector and Tensor Analysis. New York—Toronto—London, McGraw-Hill 1943.
77. Cramér, H.: Mathematical Methods of Statistics. Princeton, Princeton Univ. Press 1946.
78. Curtis, J. H.: Numerical Analysis. New York—Toronto—London, McGraw-Hill 1956.
79. Davenport, H.: The Higher Arithmetic. New York, Longmans 1952.
80. Davies, O. L.: Statistical Methods in Research and Production, 3rd Ed. Edinburgh, Oliver and Boyd 1957.
81. Day, W. D.: Tables of Laplace Transforms. London, Iliffe Books 1965.
82. Deming, W. E.: Statistical Adjustment of Data. New York, J. Wiley and Sons 1944; London, Chapman and Hall, 1946.
82a) Deming, W. E.: Some Theory of Sampling. New York, J. Wiley and Sons 1955.
83. Dettman, J. W.: Mathematical Methods in Physics and Engineering. New York—Toronto—London, McGraw-Hill 1962.
84. Dickson, L. E.: Modern Algebraic Theories. Chicago, Sanborn and Co. 1926.
85. Ditkin, V. A., Prudnikov, A. P.: Operational Calculus in Two Variables and its Applications. Oxford—London—New York, Pergamon Press 1963.

86. Dixon, W. J., Massey, F. J.: Introduction to Statistical Analysis. New York—Toronto—London, McGraw-Hill 1957.
87. Dodge, H. F., Romig, H. G.: Sampling Inspection Tables. New York, J. Wiley and Sons; London, Chapman and Hall, 1959.
88. Doetsch, G.: Guide to the Applications of Laplace Transforms. London, van Nostrand 1961.
89. Doetsch, G.: Handbuch der Laplace-Transformation. Basel, Birkhäuser 1950.
90. Dubbels Taschenbuch für den Maschinenbau. Berlin, Springer 1956.
91. Duff, G. F. D.: Partial Differential Equations. Toronto, Univ. Toronto Press; London, Oxford Univ. Press 1956.
92. Dull, R. V.: Mathematics for Engineers. New York—Toronto—London, McGraw-Hill 1941.
93. Durell, C.: A New Trigonometry for Schools. London, Bell and Sons 1963.
94. Ebner, F.: Leitfaden der technisch wichtigen Kurven. Leipzig, Teubner 1906.
95. Eisenhart, L. P.: An Introduction to Differential Geometry. Princeton, Princeton Univ. Press 1940.
96. Elsgol'ts, L. E.: Calculus of Variations. Oxford—London—New York, Pergamon Press 1963.
97. Elsgol'ts, L. E.: Differential Equations. New York, Gordon and Breach (Hindustan Publ. Co.) 1961.
98. Epstein, B.: Partial Differential Equations. New York—Toronto—London, McGraw-Hill 1962.
99. Epstein, L. I.: Nomography. New York, Interscience 1958.
100. Erdélyi, A.: Asymptotic Expansions. New York, Dover Publications 1956.
101. Erdélyi, A., Magnus, W., Oberhettinger, F., Tricomi, F. G.: Tables of Integral Transforms, 2 Vols. New York—Toronto—London, McGraw-Hill 1954.
102. Erdélyi, A., Magnus, W., Oberhettinger, F., Tricomi, F. G.: Higher Transcendental Functions, Vols 1, 2, 3. New York—Toronto—London, McGraw-Hill 1953—5.
103. Estermann, T.: Complex Numbers and Functions. London, Athlone Press. 1962.
104. Fadeev, D. K., Fadeeva, V. N.: Computational Methods of Linear Algebra. San Francisco—London, Freemans 1963.
105. Fadeeva, V. N.: Computational Methods of Linear Algebra. London, Constable 1959.
106. Feinstein, A.: Foundations of Information Theory. New York—Toronto—London, McGraw-Hill 1958.
107. Feller, W.: An Introduction to Probability Theory and its Application, 2nd Ed. London—New York, J. Wiley and Sons 1960.
108. Ferrar, W. L.: Algebra. London, Oxford Univ. Press 1957.
109. Ferrar, W. L.: A Text-book of Convergence. London, Oxford Univ. Press 1938.
110. Ferrar, W. L.: Differential Calculus. London, Oxford Univ. Press 1960.
111. Ferrar, W. L.: Finite Matrices. London, Oxford Univ. Press 1951.
112. Fikhtengol'ts, G. M.: Fundamentals of Mathematical Analysis, Parts 1, 2. Oxford—London—New York, Pergamon Press 1965.
113. Fisher, R. A.: Statistical Methods and Scientific Inference, 2nd Ed. Edinburgh, Oliver and Boyd 1959.
114. Fisher, R. A.: Statistical Methods for Research Workers. New York, Hafner Publishing Co. 1950.
115. Fisher, R. A.: The Design of Experiments, 4th. Ed. Edinburgh, Oliver and Boyd 1947.
116. Fisher, R. A., Yates, F.: Statistical Tables for Biological, Agricultural and Medical Research, 6th Ed. Edinburgh, Oliver and Boyd 1963.
117. Fisz, M.: Probability Theory and Mathematical Statistics, 3rd Ed. London—New York, J. Wiley and Sons 1963.

118. Fleming W. M.: Functions of Several Variables. Cambridge (Mass.), Addison-Wesley 1965.
119. Fletcher, A., Miller, J. C. P., Rosenhead, L., Comrie, L. J.: Index of Mathematical Tables, I, II. Oxford, Blackwell 1962.
120. Fodor, G.: Laplace Transform in Engineering. Budapest, Hungarian Acad. of Science 1965.
121. Forsyth, A. R.: A Treatise on Differential Equations. London, Macmillan 1929.
122. Forsyth, A. R.: Calculus of Variations. New York, Dover Publications 1960.
123. Forsyth, A. R.: Theory of Functions of a Complex Variable. London—New York, Cambridge Univ. Press 1918.
124. Forsythe, G. E., Rosenbloom, P. C.: Numerical Analysis and Partial Differential Equations. New York, J. Wiley and Sons 1958.
125. Forsythe, G. E., Wasov, W. R.: Finite Difference Methods for Partial Differential Equations. London—New York, J. Wiley and Sons 1960.
126. Fowler, C. M.: Analysis of Numerical Solutions of Transient Heat-Flow Problems. — Quart. Appl. Math. 3, 1946.
127. Fowler, R. H.: The Elementary Differential Geometry of Plane Curves. London—New York, Cambridge Univ. Press 1929.
128. Fox, C.: An Introduction to the Calculus of Variations. London, Oxford Univ. Press 1954.
129. Fox, L.: A Short Account of Relaxation Methods. — Quart. J. Mech. and Appl. Math. Vol. I, 1948: 353—380.
130. Fox, L.: Mixed Boundary Conditions in the Relaxational Treatment of Biharmonic Problems (Plane Strain or Stress). — Proc. Roy. Soc. A, 189, 1947: 535—543.
131. Fox, L.: Numerical Solution of Ordinary and Partial Differential Equations. Oxford—London—New York, Pergamon Press 1962.
132. Fox, L.: Practical Solution of Linear Equations and Inversion of Matrices. — National Bureau of Standards Applied Math. Series 39, 1954: 1—54.
133. Fox, L.: Solution by Relaxation Methods of Plane Potential Problems with Mixed Boundary Conditions. — Quart. Appl. Math., 2, 1944: 251—257.
134. Fox, L.: Some Improvements in the Use of Relaxation Methods for the Solution of Ordinary and Partial Differential Equations. — Proc. Roy. Soc. A, 190, 1947: 31—59.
135. Fox, L.: The Numerical Solution of Two-point Boundary Problems in Ordinary Differential Equations. London, Oxford Univ. Press 1957.
136. Fox, L., Goodwin, E. T.: Some New Methods for the Numerical Integration of Ordinary Differential Equations. — Proc. Camb. Phil. Soc., 45, 1949: 373—388.
137. Frank, P., von Mises, R.: Die Differential- und Integralgleichungen der Mechanik und Physik, Vols 1, 2. New York, Rosenberg 1943.
138. Franklin, P.: A Treatise on Advanced Calculus. New York, J. Wiley and Sons 1940.
139. Frazer, D. A. S.: Nonparametric Methods in Statistics. New York, J. Wiley and Sons 1957.
140. Frazer, R. A., Duncan, W. J., Collar, A. R.: Elementary Matrices. London—New York, Cambridge Univ. Press 1946.
141. Friedman, A.: Partial Differential Equations of Parabolic Type. Englewood Cliffs (New Jersey), Prentice-Hall 1964.
142. Fuchs, V. A., Shabat, B. V.: Functions of a Complex Variable. Oxford—London—New York, Pergamon Press 1964.
143. Gantmacher, F. R.: Applications of the Theory of Matrices. New York, Interscience 1959.
144. Gasson, J.: Mathematics for Technical Students, Book II. Cambridge, Cambridge Univ. Press 1958.
145. Geffner, J., Worthing, A. G.: Treatment of Experimental Data. New York, J. Wiley and Sons 1946.
146. Gelfand, I. M., Fomin, S. V.: Calculus of Variations. Englewood Cliffs (New Jersey), Prentice—Hall 1963.

147. Gerschgorin, S. A.: Fehlerabschätzung für das Differenzenverfahren zur Lösung partieller Differentialgleichungen. — Zeitschr. angew. Math. u. Mech. 10, 1930: 373—382.
148. Giese, J. H.: On the Truncation Error in a Numerical Solution of the Neumann Problem for a Rectangle. — J. Math. Phys., 37, 1958: 169—177.
149. Givens, W.: Numerical Computation of the Characteristic Values of a Real Symmetric Matrix. Oak Ridge National Laboratory 1954.
150. Gnedenko, B. V.: The Theory of Probability. New York, Chelsea Publ. Co. 1962.
151. Gnedenko, B. V., Khinchin, A. Ya.: An Elementary Introduction to the Theory of Probability. New York, Dover Publications; London, Constable 1963.
152. Goldberg, S.: Introduction to Difference Equations. New York, J. Wiley and Sons; London, Chapman and Hall 1958.
153. Goldenveizer, A. L.: Theory of Elastic Thin Shells. Oxford—London—New York, Pergamon Press 1961.
154. Gould, S. H.: Variational Methods for Eigenvalue Problems, 2nd Ed. Toronto, Univ. Toronto Press, London, Oxford University Press 1966.
155. Goursat, E.: Course of Mathematical Analysis, Vols 1, 2. Boston, Ginn and Co. 1917.
156. Graves, L. M.: Theory of Functions of Real Variables. New York—Toronto—London, McGraw-Hill 1956.
157. Graustein, W. C.: Differential Geometry. New York, Dover Publications 1947.
158. Graybill, F. A.: An Introduction to Linear Statistical Models, Vol. I. New York—Toronto—London, McGraw-Hill 1961.
159. Green, S. L.: Algebraic Solid Geometry. London—New York, Cambridge Univ. Press 1961.
160. Greenspan, D.: Introduction to Partial Differential Equations. New York—Toronto—London, McGraw-Hill 1961.
161. Greenspan, D.: Theory and Solutions of Ordinary Differential Equations. New York, Macmillan 1960.
162. Grenander, U., Rosenblatt, M.: Statistical Analysis of Stationary Time Series. Stockholm, Almkvist Wiksell 1956.
163. Grüss, G.: Variationsrechnung. Leipzig, Akademische Verlagsges. 1938.
164. Gutsche, O.: Konstruktionen des Krümmungskreises der geraden Strophoide. — Archiv der Mathematik und Physik, 1908, 3. Reihe, Band 13: 197—202.
165. Hadamard, J.: Leçons sur le calcul des variations. Paris, Hermann 1910.
166. Hald, A.: Statistical Tables and Formulas. New York, J. Wiley and Sons; London, Chapman and Hall 1952.
167. Hald, A.: Statistical Theory with Engineering Applications. New York, J. Wiley and Sons; London, Chapman and Hall 1952.
168. Halmos, P. R.: Finite Dimensional Vector Spaces. New York—London, van Nostrand 1958.
169. Halmos, P. R.: Measure Theory. New York—London, van Nostrand 1950.
170. Halmos, P. R.: Naive Set Theory. Princeton, van Nostrand 1960.
171. Hansen, H. H., Hurwitz, W. N., Madow, W. G.: Sample Surveys Methods and Theory, I and II. New York, J. Wiley and Sons; London, Chapman and Hall 1953.
172. Hardy, G. H.: A Course of Pure Mathematics. London—New York, Cambridge Univ. Press 1952.
173. Hardy, G. H.: Divergent Series. London, Oxford University Press 1949.
174. Hartley, M. O., Pearson, E. S.: Biometrical Tables for Statisticians. London—New York, Cambridge Univ. Press 1956.
175. Hartman, S., Mikusiński, J.: Theory of Lebesgue Measure and Integration. Oxford—London—New York, Pergamon Press 1961.
176. Hartree, D. R.: Numerical Analysis. London, Oxford Univ. Press 1955.
177. Hartree, D. R.: Numerical Methods. New York, Academic Press 1958.

178. Head, J. W.: Mathematical Techniques in Electronics and Engineering Analysis. London, Iliffe Books; Princeton, van Nostrand 1964.
179. Head, J. W., Mayo, C. G.: Unified Circuit Theory in Electronics and Engineering Analysis. London, Iliffe Books 1965.
180. Heins, M.: Selected Topics in the Classical Theory of Functions of a Complex Variable. New York, Holt, Rinehart and Winston 1962.
181. Henrici, P.: Discrete Variable Methods in Ordinary Differential Equations. London—New York, J. Wiley and Sons 1962.
182. Henstock, R.: Theory of Integration. London, Butterworths 1963.
183. Hildebrand, F. B.: Introduction to Numerical Analysis. New York—Toronto—London, McGraw-Hill 1956.
184. Hildebrandt, T. H.: Introduction to the Theory of Integration. New York—London, Academic Press 1963.
185. Hlavatý, V.: Differentialgeometrie der Kurven und Flächen und Tensorrechnung. Groningen, Noordhoff 1939.
186. Hlawiczka, P.: Matrix Algebra for Electronics Engineers. London, Iliffe Books; New York, Hayden Book Co. 1965.
187. Hobson, E. W.: Spherical and Ellipsoidal Harmonics. London—New York, Cambridge Univ. Press 1931.
188. Hobson, E. W.: Theory of Functions of a Real Variable. London—New York, Cambridge Univ. Press 1927.
189. Hochstandt, H.: Differential Equations. New York, Holt, Rinehart and Winston 1964.
190. Hodge, W. V., Pedoe, D.: Methods of Algebraic Geometry. London—New York, Cambridge Univ. Press 1954.
191. Hoel, P. G.: Introduction to Mathematical Statistics, 3rd Ed. London—New York, J. Wiley and Sons 1962.
192. Hogg, R. V., Craig, A.T.: Introduction to Mathematical Statistics. New York, Macmillan 1958.
193. Hörmander, L.: Linear Partial Differential Equations. Berlin, Springer 1964.
194. Horn, J.: Gewöhnliche Differentialgleichungen, 6. Aufl. Berlin, De Gruyter 1960.
195. Hort, W., Thoma, A.: Die Differentialgleichungen der Technik und Physik. 5. Aufl. Leipzig, J. A. Barth 1954.
196. Householder, A. S.: Principles of Numerical Analysis. New York—Toronto—London, McGraw-Hill 1953.
197. Hurwitz, A., Courant, R.: Funktionentheorie. Berlin—New York, Springer 1964.
198. Hyslop, J. M.: Infinite Series. Edinbourgh, Oliver and Boyd 1954.
199. Ince, E. L.: Ordinary Differential Equations. New York, Dover Publications; London, Longmans 1944.
200. Jackson, D.: Fourier Series and Orthogonal Polynomials, 6th. Ed. Buffalo—New York, The Mathematical Association of America 1963.
201. Jeffreys, H.: Asymptotic Approximations. London, Oxford Univ. Press 1962.
202. Jeffreys, H.: Cartesian Tensors. London—New York, Cambridge Univ. Press 1931.
203. Jeffreys, H.: Operational Methods in Mathematical Physics. London—New York, Cambridge Univ. Press 1927.
204. Jeffreys, H.: Theory of Probability, 3rd Ed. London, Oxford Univ. Press 1961.
205. Jeffreys, H., Jeffreys, B. S.: Mathematical Physics. London—New York, Cambridge Univ. Press 1956.
206. John, F.: Plane Waves and Spherical Means Applied to Partial Differential Equations. New York, Interscience 1955.
207. Jones, D. S.: Theory of Electromagnetism. Oxford—London—New York, Pergamon Press 1964.

208. Joos, G.: Theoretical Physics, 3rd Ed. London, Blackie and Son 1958.
209. Jordan, C.: Calculus of Finite Differences. New York, Chelsea Publ. Co. 1939.
210. Kamke, E.: Das Lebesgue-Stieltjes Integral. Leipzig, Teubner 1956.
211. Kamke, E.: Differentialgleichungen, Lösungsmethoden und Lösungen, 3., verbesserte Aufl. Leipzig, Akademische Verlagsgesellschaft 1956.
212. Kamke, E.: Differentialgleichungen II, Partielle Differentialgleichungen, 4. Ausg. Leipzig, Geest und Portig 1964.
213. Kamke, E.: Theory of Sets. New York, Dover Publications 1950.
214. Kantorovich, L. V., Krylov, V. I.: Approximate Methods of Higher Analysis. Groningen, Noordhoff, 1958.
215. Kaplan, W.: Advanced Calculus. Cambridge (Mass.), Addison-Wesley 1952.
216. Kaplan, W.: Lectures on Functions of a Complex Variable. Ann Arbor, Univ. of Michigan Press 1955.
217. Kaplan, W.: Ordinary Differential Equations. Cambridge (Mass.), Addison-Wesley 1958.
218. Keeping, E. S.: Introduction to Statistical Inference. Princeton, van Nostrand 1962.
219. Kellog, O. D.: Potential Theory. Berlin, Springer 1929.
220. Kells, L., Kern, W., Bland, J.: Plane and Spherical Trigonometry. New York—Toronto—London, McGraw-Hill 1940.
221. Kendall, M. G.: The Advanced Theory of Statistics. I (1945), II (1946), 3rd Ed. London, C. Griffin.
222. Kendall, D. G., Moran, P. A. P.: Geometrical Probability. London, C. Griffin 1963.
223. Kendall, M. G.: Stuart, A.: The Advanced Theory of Statistics, Vols I (1958), II (1961). London, C. Griffin.
224. Kestelman, H.: Modern Theories of Integration. New York, Dover Publications 1960.
225. Khabaza, I. M.: Numerical Analysis. Oxford—London—New York, Pergamon Press 1965.
226. Khinchin, A. Ya.: A Course of Mathematical Analysis. New York, Gordon and Breach (Hindustan Publ. Co.) 1961.
227. Khinchin, A. Ya.: Analytic Foundations of Physical Statistics. New York, Gordon and Breach (Hindustan Publ. Co.) 1961.
228. Khinchin, A. Ya.: Continued Fractions. Chicago, Univ. Chicago Press 1964.
229. Kiepert, L.: Grundrisse der Differentialrechnung, I, II. Hannover, Helwingsche Verlagsbuchhandlung 1921.
230. Klein, F.: Elementary Mathematics from an Advanced Standpoint, 2 Vols. New York, Macmillan 1932 and 1939.
231. Kneser, A.: Lehrbuch der Variationsrechnung, 2. Aufl. Braunschweig, Vieweg 1925.
232. Knopp, K.: Funktionentheorie I, II, III. Sammlung Göschen. Berlin, Göschen 1944.
233. Knopp, K.: Theory and Applications of Infinite Series. London, Blackie and Son 1951.
234. Kober, H.: Dictionary of Conformal Representations. New York, Dover Publications 1957.
235. Kolmogorov, A. N.: Foundations of the Theory of Probability. New York, Chelsea Publ. Co. 1956.
236. Kolmogorov, A. N., Fomin, S. V.: Elements of the Theory of Functional Analysis, Vols 1, 2. New York, Graylock Press 1957.
237. Kopal, Z.: Numerical Analysis. London, Chapman and Hall 1955.
238. Koppenfels, W., Stallmann, F.: Praxis der konformen Abbildung. Berlin, Springer 1959.
239. Kreyszig, E.: Differential Geometry. London, Oxford Univ. Press 1959.
239a. Kreyszig, E.: Advanced Engineering Mathematics, 2nd Ed. New York, J. Wiley and Sons 1967.
240. Kuntzmann, J.: Méthodes numériques. Paris, Dunod 1959.
241. Kunz, K. S.: Numerical Analysis. New York—Toronto—London, McGraw-Hill 1957.
242. Lamb, H.: Infinitesimal Calculus. London—New York, Cambridge Univ. Press 1919.

243. Lambe, C. G., Tranter, C. J.: Differential Equations for Engineers and Scientists. London, English Univ. Press 1961.
244. Lanczos, C.: Applied Analysis. London, Sir Isaac Pitman & Sons 1957.
245. Lanczos, C.: Variational Principles of Mechanics. Toronto, Univ. Toronto Press 1949.
246. Landau, E.: Differential and Integral Calculus. New York, Chelsea Publ. Co. 1951.
247. Landau, E.: Foundations of Analysis. New York, Chelsea Publ. Co. 1951.
248. Lane, E. P.: Metric Differential Geometry of Curves and Surfaces. Chicago, Univ. Chicago Press 1940.
249. Langer, R. E.: Non-linear Problems. Madison, Univ. Wisconsin Press 1963.
250. Laplace, P. S.: A Philosophical Essay on Probabilities. London, Constable 1951.
251. Lavrentyev, M. A., Shabat, B. V.: Methods of the Theory of Functions of a Complex Variable. Moscow, Gostekhizdat 1958.
252. Lawden, D. F.: A Course in Applied Mathematics. London, English Univ. Press 1960.
253. Lax, P. D.: Partial Differential Equations. New York, New York Univ. Inst. of Mathematical Sciences 1931.
254. Lefschetz, S.: Introduction to Topology. Princeton, Princeton Univ. Press 1949.
255. Lefschetz, S.: Lectures on Differential Equations. Princeton, Princeton Univ. Press 1951.
256. Lehman, E. L.: Testing Statistical Hypotheses. New York, J. Wiley and Sons; London, Chapman and Hall 1959.
257. Lense, J.: Kugelfunktionen. Leipzig, Akademische Verlagsges. 1950.
258. Lichnerowicz, A.: Elements of Tensor Calculus. London, Methuen 1962.
259. Lindley, D. V.: Introduction to Probability and Statistics from a Bayesian Viewpoint. London—New York, Cambridge Univ. Press 1965.
260. Littlewood, D. E.: A University Algebra. London, Heinemann 1950.
261. Lockwood, E. H.: A Book of Curves. London—New York, Cambridge Univ. Press 1961.
262. Loeve, M.: Probability Theory, 2nd Ed. Toronto—New York—London, van Nostrand 1960.
263. Loney, S. L.: Coordinate Geometry, Part I: Cartesian Coordinates. London, Macmillan 1936.
264. Loria, G.: Spezielle algebraische und transcendente ebene Kurven. Teile I (1910) und II (1911). Leipzig—Berlin, Teubner.
265. Lowry, H. V., Hayden, H. A.: Advanced Mathematics for Technical Students, Part II, 2nd London, Longmans 1957.
266. Lusternik, L. A.: Shortest Way; Variational Problems. London, Macmillan 1964.
267. Lusternik, L. A., Sobolev, L. I.: Elements of Functional Analysis. London, Constable 1961.
268. Macrobert, T. M.: Functions of a Complex Variable. London, Macmillan 1933.
269. Mainardi, P., Barkan, H.: Calculus and its Applications. Oxford—London—New York, Pergamon Press 1963.
270. Mangoldt, H. C. F. von, Knopp, K.: Einführung in die höhere Mathematik für Studierende und zum Selbststudium, 9. Aufll. Leipzig, Hirzel 1948.
271. Margenau, H., Murphy, G. M.: The Mathematics of Physics and Chemistry. New York, van Nostrand 1943.
272. Margoulis, W.: Les abaques à transparent orienté ou tournant. Paris, Gauthier-Villars 1931.
273. Markushevich, A. I.: Theory of Analytic Functions. Moscow—Leningrad, Gostekhizdat 1950. (In Russian.)
274. Mařík, J., Raab, M.: Nichtoszillatorische lineare Differentialgleichungen 2. Ordnung. — Czech. Math. J. 13, 1963: 209—225.
275. Massey, H. S. W., Kestelman, H.: Ancillary Mathematics. London, Isaac Pitman & Sons 1958.

276. McCrea, W. H.: Analytical Geometry of Three Dimensions. Edinburgh, Oliver and Boyd 1953.
277. McLachlan, N. W.: Modern Operational Calculus. New York, Dover Publications 1962.
278. McLachlan, N. W.: Ordinary Non-linear Differential Equations. London, Oxford Univ. Press 1950.
279. Michal, A. D.: Matrix and Tensor Calculus. New York, J. Wiley and Sons 1947.
280. Mikeladze, Sh. E.: Numerical Methods of Mathematical Analysis. Moscow, Gostekhizdat 1953. (In Russian.)
281. Mikhlin, S. G.: Integral Equations and Applications. Oxford—London—New York, Pergamon Press 1957.
282. Mikhlin, S. G.: Variational Methods in Mathematical Physics. Oxford—London—New York, Pergamon Press 1963.
283. Mikusiński, J.: Operational Calculus. Oxford—London—New York, Pergamon Press 1959.
284. Miller, K. S.: Elements of Modern Abstract Algebra. New York, Harper and Row 1964.
285. Miller, K. S.: Engineering Mathematics. London, Constable 1956.
286. Milne, W. E.: Numerical Calculus. Princeton, Princeton Univ. Press 1949.
287. Milne, W. E.: Numerical Solution of Differential Equations. New York, J. Wiley and Sons; London, Chapman and Hall 1953.
288. Milne-Thomson, L. M.: Calculus of Finite Differences Lcdnon, Macmillan 1951.
289. Miranda, C. M.: Equazioni alle derivate parziali di tipo ellittico. Berlin, Springer 1955.
290. Modern Computing Methods. National Physical Laboratory, Teddington, England; London, Stationery Office 1961.
291. Mood, A. M., Graybill, F. A.: Introduction to the Theory of Statistics, 2nd. Ed. New York—Toronto—London, MacGraw-Hill 1963.
292. Morse, P. M., Feshbach, H.: Methods of Mathematical Physics Parts I, II. New York—Toronto—London, McGraw-Hill 1953.
293. Mostow, G. D., Sampson, J. H., Meyer, J. P.: Fundamental Structures of Algebra. New York—Toronto—London, McGraw-Hill 1963.
294. Munroe, M. E.: Introduction to Measure and Integration. Cambridge (Mass.), Addison-Wesley 1953.
295. Murnaghan, F. D.: Introduction to Applied Mathematics. New York, Dover Publications 1963.
296. Murray, D. A.: Introductory Course in Differential Equations for Students in Classical and Engineering Colleges. New York, Longmans 1954.
297. Muskhelishvili, N. I.: Singular Integral Equations. Groningen, Noordhoff 1953.
298. Najmark, M. A.: Lineare Differentialoperaroren Berlin, Springer 1960.
299. Natanson, I. P.: Theory of Functions of a Real Variable. London, Constable 1961.
300. Nečas, J.: Les méthodes directes dans la théorie des équations elliptiques. Prague, NČSAV 1966.
301. Nehari, Z.: Conformal Mapping. New York—Toronto—London, McGraw-Hill 1952.
302. Nicholson, M. M.: Fundamentals and Techniques of Mathematics for Scientists. London, Longmans 1961.
303. Nobbs, C.: Trigonometry. London, Oxford Univ. Press 1962.
304. d'Ocagne, M.: Traité de nomografie, 2nd Ed. Paris, Gauthiers-Villars 1921.
305. Olmsted, J. M. H.: Real Variables New York, Appleton-Century-Crofts 1959.
306. Olver, F. W. J.: The Evaluation of Zeros of High-degree Polynomials. — Phil. Trans. Roy. Soc. A, 244, 1952: 385 – 415.
307. Osgood, W. F.: Functions of Real and Complex Variables. New York, Chelsea Publ. Co. 1935.
308. Ostrowski, A.: Vorlesungen über Differential- und Integralrechnung. Basel—Stuttgart, Birkhäuser 1954.

309. Otto, E.: Nomography. Oxford—London—New York, Pergamon Press 1964.
309a. Owen, D. B.: Handbook of Statistical Tables. Palo Alto—London, Addison-Wesley 1962.
310. Page, C. H.: Physical Mathematics. New York, van Nostrand 1956.
311. Parker, W. V.: Eaves, J. C.: Matrices. New York, Ronald Press 1960.
312. Pars, L. A.: An Introduction to the Calculus of Variations. London, Heinemann 1962.
313. Pars, L. A.: Calculus of Variations. London, Heinemann 1962.
314. Parzen, E.: Modern Probability Theory and its Applications. London—New York, J. Wiley and Sons 1960.
315. Parzen, E.: Stochastic Processes. San Francisco, Holden-Day 1962.
316. Pentkovskii, M. V.: Nomographie. Berlin, Akademie-Verlag 1953.
317. Petrovskii, I. G.: Lectures on Partial Differential Equations. London, Interscience 1954.
318. Petrovskii, I. G.: Lectures on the Theory of Integral Equations. Moscow, Gostekhizdat 1948. (In Russian.)
319. Petrovskii, I. G.: Vorlesungen über die Theorie der gewöhnlichen Differentialgleichungen. Leipzig, Teubner 1954.
320. Phillips, E. G.: A Course of Analysis. London—New York, Cambridge Univ. Press 1948.
321. Piaggio, H. T. H.: Differential Equations. London, Bell and Sons 1942.
322. Plumpton, C. and Chirgwin, B. H.: Course of Mathematics for Engineers and Scientists, Vols 1—6. Oxford—London—New York, Pergamon Press 1963.
323. Pol, B. van der, Bremmer, H.: Operational Calculus Based on the Two-Sided Laplace Integral. London—New York, Cambridge Univ. Press 1950.
324. Pontryagin, L. S.: Ordinary Differential Equations. Cambridge (Mass.), Addison-Wesley 1962.
325. Porter, A., Mack, C.: New Methods for the Numerical Solution of Algebraic Equations. = Phil. Mag., 40, 1949: 578—585.
326. Prager, M., Babuška, I.: Numerisch stabile Methoden der Lösung von Randwertaufgaben. Numerische Mathematik 1962.
327. Primrose, E. J. F.: Plane Algebraic Curves. London, Macmillan 1955.
328. Raab, M.: Asymptotische Formeln für die Lösungen der Differentialgleichung $y'' + q(x)\, y - = 0$. — Czech. Math. J. 14, 1964; 203—221.
329. Rao, C. R.: Advanced Statistical Methods in Biometric Research. New York, J. Wiley and Sons 1952.
330. Rao, C. R.: Linear Statistical Inference and its Applications. New York, J. Wiley and Sons 1965.
331. Redish, K. A.: An Introduction to Computational Methods. London, English Univ. Press 1961.
332. Rektorys, K.: Die Lösung des ersten Randwertproblems im Ganzen für eine nichtlineare parabolische Gleichung mit der Netzmethode. — Czech. Math. J., 12, 1962: 69—103.
332a. Rektorys K.: Two Theorems on the Solution of the Equation $\partial u/\partial t = \partial^2 u/\partial x^2 + \partial^2 u/\partial y^2$. Čas. pěst. mat. 79 (Prague 1954), 4: 333—366. (In Czech, German and Russian Summaries.)
333. Relton, F. E.: Applied Differential Equations. London, Blackie 1948.
334. Rényi, A.: Wahrscheinlichkeitstheorie mit einem Anhang über Informationstheorie. Berlin, Wiss. Vlg. 1962.
335. Richardson, C. R.: An Introduction to the Calculus of Finite Differences. Princeton, van Nostrand 1954.
336. Richtmayer, R. D.: Difference Methods for Initial-Value Problems. New York, Interscience Publishers 1957.
337. Riesz, F., Nagy, B. Sz.: Leçons d'analyse fonctionnelle. Budapest, Académie des sciences de Hongrie 1952.
338. Rogosinski, W. W.: Volume and Integral. Edinburgh, Oliver and Boyd 1962.

339. Romanovskii, P. I.: Mathematical Methods for Engineers and Technologists. Oxford—London—New York, Pergamon Press 1961.
340. Rose, M.: Einführung in die Funktionentheorie. Sammlung Göschen. Leipzig, Göschen 1912.
341. Rothe, R.: Höhere Mathematik für Mathematiker, Physiker und Ingenieure, I—V, 13. Aufl. Leipzig, Teubner 1954.
342. Rudin, W.: Principles of Mathematical Analysis. New York—Toronto—London, McGraw-Hill 1964.
343. Runge, C.: Über eine Methode die partielle Differentialgleichung $\Delta u =$ const. numerisch zu integrieren. — Z. Math. Phys., 96, 1908: 225—232.
344. Rutherford, D. E.: Vector Methods. Edinburgh, Oliver and Boyd 1957.
345. Sachs, S., Zygmund, A.: Analytic Functions. Warszawa: Polskie Towarzystwo Matematyczne 1952. (In English.)
346. Saks, S.: Theory of the Integral. New York, Hafner 1937.
347. Salmon, G.: A Treatise on Conic Sections. New York, Chelsea Publ. Co. 1954.
348. Salvadori, M. G., Baron, M. L.: Numerical Methods in Engineering. Englewood Cliffs (New Jersey), Prentice-Hall 1956.
349. Salvadori, M. G., Schwarz, R. J.: Differential Equations in Engineering Problems. Englewood Cliffs (New Jersey), Prentice-Hall 1954.
350. Sampling Inspection (Columbia University). New York—Toronto—London, McGraw-Hill 1948.
351. Sansone, G., Conti, R.: Non-linear Differential Equations. Oxford—London—New York, Pergamon Press 1964.
352. Scheffé, H.: The Analysis of Variance. London—New York, J. Wiley and Sons; London, Chapman and Hall 1959.
353. Scheffers, G.: Einführung in die Theorie der Flächen. Leipzig, Veit 1913.
354. Schmeidler, W.: Integralgleichungen mit Anwendungen in Physik und Technik. Leipzig, Akademie Verlagsgesellschaft 1950.
355. Schouten, J. A.: Tensor Analysis for Physicists. London, Oxford Univ. Press 1951.
356. Schreier, O., Spencer, E.: Modern Algebra and Matrix Theory. New York, Chelsea Publ. Co. 1951.
357. Schwerdt, H.: Lehrbuch der Nomografie. Berlin, Springer 1924.
358. Selected Techniques of Statistical Analysis (Columbia University). New York—Toronto—London, McGraw-Hill 1947.
359. Semple, J. G., Kneebone, G. T.: Algebraic Curves. London, Oxford Univ. Press 1959.
360. Semple, J. G., Kneebone, G. T.: Algebraic Projective Geometry. London, Oxford Univ. Press 1952.
361. Sequential Analysis of Statistical Data: Applications. New York, Columbia Univ. Press 1954.
362. Serret, J. A., Scheffers, G.: Lehrbuch der Differential- und Integralrechnung, 7. Aufl. Leipzig, Teubner 1915.
363. Shchigolev, R. M.: Mathematical Analysis of Observations. London, Iliffe; New York, Elsevier 1965.
364. Shewart, W.: Economic Control of Quality of Manufactured Product. New York, van Nostrand 1931.
365. Shilov, G. Ye.: Mathematical Analysis. Oxford—London—New York, Pergamon Press 1965.
366. Smart, W. M.: Combination of Observations. London—New York, Cambridge Univ. Press 1958.
367. Smirnov, V. I.: Course of Higher Mathematics. 5 Vols. Oxford—London—New York, Pergamon Press 1964.

368. Smith, C.: An Elementary Treatise on Conic Sections by the Methods of Coordinate Geometry. London, Macmillan 1943.
369. Smith, C.: An Elementary Treatise on Solid Geometry. London, Macmillan 1946.
370. Smithies, F.: Integral Equations. London—New York, Cambridge Univ. Press 1962.
371. Sneddon, I. N.: Elements of Partial Differential Equations. New York, McGraw-Hill 1957.
372. Sneddon, I. N.: Fourier Transforms. New York—Toronto—London, McGraw-Hill 1951.
373. Sneddon, I. N.: Introduction to Partial Differential Equations. New York—Toronto—London, McGraw-Hill 1957.
374. Sobolev, S. L.: Applications of Functional Analysis in Mathematical Physics. Providence, R. I., American Math. Soc. 1963.
375. Sobolev, S. L.: Partial Differential Equations of Mathematical Physics. Oxford—London—New York, Pergamon Press 1964.
376. Sokolnikoff, I. S.: Advanced Calculus. New York—Toronto—London, McGraw-Hill 1939.
377. Sokolnikoff, I. S.: Tensor Analysis. New York, J. Wiley and Sons; London, Chapman and Hall 1951.
378. Sommerville, D. M. Y.: Analytical Conics. London, Bell and Sons 1951.
379. Sommerville, D. M. Y.: Analytical Geometry of Three Dimensions. London—New York, Cambridge Univ. Press 1947.
380. Soreau, R.: Nomographie ou traité des abaques. Paris, Ciron 1921.
381. Southwell, R. V.: Plane Potential Problems Involving Specified Normal Gradients. — Proc. Roy. Soc. A, 182, 1943.
382. Southwell, R. V.: Relaxation Methods in Engineering Science. London, Oxford Univ. Press 1940.
383. Southwell, E. V.: Relaxation Methods in Theoretical Physics. London, Oxford Univ. Press 1946.
384. Spain, B.: Analytical Conics. Oxford—London—New York, Pergamon Press 1957.
385. Spain, B.: Tensor Calculus. Edinburgh, Oliver and Boyd 1953.
386. Spain, B.: Vector Analysis. London, van Nostrand 1965.
387. Spiegel, M. R.: Theory and Problems of Advanced Calculus. New York, Schaum Publ. Co. 1963.
388. Stepanov, V. V.: A Course of Differential Equations. Moscow, Gostekhizdat 1949. (In Russian).
389. Stewart, C. A.: Advanced Calculus. London, Methuen 1940.
390. Stigant, S. A.: Elements of Determinants, Matrices and Tensors for Engineers. London, Macmillan 1959.
391. Stoker, J. J.: Non-linear Vibrations in Mechanical and Electrical Systems. New York, Interscience 1950.
392. Struble, R. A.: Non-linear Differential Equations. New York—Toronto—London, McGraw-Hill 1962.
393. Synge, J. A., Schild, A.: Tensor Calculus. Toronto, Univ. Toronto Press 1949.
394. Tarasov, N. P.: A Course of Advanced Mathematics for Technical Schools. Oxford—London—New York, Pergamon Press 1961.
395. Taylor, A. E.: Introduction to Functional Analysis. New York, J. Wiley and Sons 1958.
396. Teixeira, F. G.: Traité des courbes spéciales remarquables planes et gauches. Tome 1 (1908), 2 (1909), 3 (1915). Coimbre, Imprimérie de l'Université.
397. Thrall, R. M., Tornheim, L.: Vector Spaces and Matrices London, J. Wiley and Sons 1957.
398. Titchmarsh, E. C.: Eigenfunction Expansions, Vols 1, 2. London, Oxford Univ. Press 1946, 1958.
399. Titchmarsh, E. C.: Theory of Functions. London, Oxford Univ. Press 1952.
400. Tolstov, G. G.: Fourier Series. Englewood Cliffs (New Jersey), Prentice-Hall 1962.

401. Tonelli, L.: Fondamenti di calcolo delle variazioni, Vol. 2. Bologna, Zanichelli 1921—1923.
402. Tranter, C. J.: Integral Transforms in Mathematical Physics. London, Methuen; New York, J. Wiley and Sons 1954.
403. Trefftz, E.: Konvergenz und Fehlerabschätzung beim Ritz'schen Verfahren. — Mathematische Annalen, 100, 1928: 503—521.
404. Tricomi, F. G.: Differential Equations. London, Blackie and Son 1961.
405. Tricomi, F. G.: Integral Equations. New York, Interscience 1957.
406. Turnbull, H. W.: Theory of Determinants, Matrices and Invariants. New York, Dover Publications 1960.
407. Tychonov, A. N., Samarskii, A. A.: Partial Differential Equations of Mathematical Physics. San Francisco, Holden-Day 1964.
408. Vejvoda, O.: Estimate of Error of the Runge-Kutta Formula (in Czech, German Summary). — Apl. Math. (Prague), 2, 1957, 1: 1—23.
409. Vejvoda, O.: Perturbed Boundary-Value Problems and their Approximate Solution. — Proc. Rome Symp. Num. Treatment etc. (Basel), 1961: 37—41.
410. Vejvoda, O.: On Perturbed Nonlinear Boundary-Value Problems. — Czech. Math. J. (Prague), 1961 11 (86),: 323—364.
411. Viswanathan, R. V.: Solution of Poisson's Equation by Relaxation Methods, Normal Gradient Specified on n Curved Boundaries. — Mathematical Tables and Aids to Computation, 11, 1957: 67—78.
412. Voelker, D., Doetsch, G.: Die zweidimensionale Laplace-Transformation. Basel, Birkhäuser 1950.
413. Vulich, B. Z.: Introduction to Functional Analysis for Scientists and Technologists. Oxford—London—New York, Pergamon Press 1963.
414. Waerden, B. L. van der: Mathematische Statistik. Berlin, Springer 1957.
415. Wald, A.: Sequential Analysis. New York, J. Wiley and Sons; London, Chapman and Hall 1948.
416. Wald, A.: Statistical Decision Functions. New York, J. Wiley and Sons; London, Chapman and Hall 1950.
417. Wall, H. S.: Continued Fractions. London, van Nostrand 1948.
418. Walsh, J. L., Young, D.: On the Accuracy of the Numerical Solution of the Dirichlet Problem by Finite Differences. — J. Res. Nat. Bur. Stand., 51, 1953: 343—363.
419. Wasow, W.: On the Truncation Error in the Solution of Laplace's Equations by Finite Differences. — J. Res. Nat. Bur. Stand., 48, 1952: 345—348.
420. Watson, G. N.: Bessel Functions. London—New York, Cambridge Univ. Press 1966.
421. Weatherburn, C. E.: Advanced Vector Analysis. London, Bell and Sons 1949.
422. Weatherburn, C. E.: Differential Geometry. London—New York, Cambridge Univ. Press 1933.
423. Weatherburn, C. E.: Elementary Vector Analysis. London, Bell and Sons 1943.
424. Weatherburn, C. E.: Riemannian Geometry and Tensor Calculus, London—New York, Cambridge Univ. Press 1938.
425. Weber, E.: Grundriss der biologischen Statistik für Naturwissenschaftler, Landwirte und Mediziner, 2. Aufl. Jena, G. Fisher 1956.
426. Webster, A. G.: Partial Differential Equations of Mechanics and Physics. New York, Stechert 1933.
427. Weiss, L.: Statistical Decision Theory. New York—Toronto—London, McGraw-Hill 1961.
428. Whittaker, E. T., Robinson, G.: The Calculus of Observations. London—Glasgow, Blackie and Son 1948.
429. Whittaker, E. T., Watson, G. N.: A Course of Modern Analysis, 4th Ed. London—New York, Cambridge Univ. Press 1958.

430. Widder, D. V.: The Laplace Transform. Princeton, Princeton Univ. Press 1946.
431. Wieleitner, H.: Spezielle ebene Kurven. Leipzig, Göschen 1908.
432. Wiener, N.: Cybernetics, 2nd Ed. Boston, Massachusetts Inst. of Technology 1961.
433. Wiener, N.: Extrapolation, Interpolation and Smoothing of Stationary Time Series. With Engineering Applications. New York, J. Wiley and Sons 1949.
434. Wilcox, C. H.: Asymptotic Solution of Differential Equations. London—New York, J. Wiley and Sons 1964.
435. Wilks, S. S.: Mathematical Statistics. London—New York, J. Wiley and Sons 1962.
436. Wilks, S. S.: Mathematical Statistics. Princeton, Princeton Univ. Press 1946.
437. Williamson, J. H.: Lebesgue Integration. New York, Holt, Rinehart and Winston 1962.
438. Wilson, E. M.: A Note on the Numerical Integration of Differential Equations. — Quart. J. Mech. and Appl. Math., 2, 1949: 208—211.
439. Wlasov, W. S.: Allgemeine Schalentheorie und ihre Anwendung in der Technik. Berlin, Akademieverlag 1958.
440. Wylie, C. R.: Plane Trigonometry. New York—Toronto—London, McGraw-Hill 1955.
441. Yates, R. C.: Curves and their Properties. Ann Arbor, J. B. Edwards 1959.
442. Yosida, K.: Functional Analysis. Berlin, Springer 1965.
443. Yule, G. U.: An Introduction to the Theory of Statistics. London, C. Griffin 1917.
444. Yule, G. U., Kendall, M. G.: An Introduction to the Theory of Statistics. London, C. Griffin 1947.
445. Zaanen, A. C.: An Introduction to the Theory of Integration. Amsterdam, North Holland 1958.

INDEX

Where an indented subheading begins with a capital letter this indicates (except in the case of a personal name) that a more detailed break-down of the subject will be found under a main heading which begins with that letter